The Wiley

ENCYCLOPEDIA OF Packaging Technology

Second Edition

R

The Wiley

ENCYCLOPEDIA OF Packaging Technology

Second Edition

Edited by

Aaron L. Brody

Kenneth S. Marsh

A Wiley-Interscience Publication

John Wiley & Sons, Inc.

New York / Chichester / Weinheim / Brisbane / Singapore / Toronto

This text is printed on acid-free paper.

Library of Congress Cataloging in Publication Data:

The Wiley Encyclopedia of packaging technology / Aaron L. Brody,
Kenneth S. Marsh. — 2nd ed.
p. cm.
ISBN 0-471-06397-5 (cloth : alk. paper)
1. Packaging—Dictionaries. I. Marsh, Kenneth S.
TS195.A2W55 1997
688.8′03—dc20 96-44725

Printed in the United States of America

10 9 8 7 6 5 4 3 2 1

EDITORIAL STAFF

Editors: **Aaron L. Brody,** Rubbright·Brody, Inc.
Kenneth S. Marsh, Kenneth S. Marsh and Associates, Ltd.

CONTRIBUTORS

Timothy Aidlin, *Aidlin Automation,* Air Conveying
Jim Arch, *Scholle Corporation,* Bag-in-Box, Liquid Product
George Arndt, *Pet, Inc.,* Leak Testing
Robert Bakish, *Bakish Materials Corporation,* Metallizing, Vacuum
M. Henri Barthel, *EAN International,* Code, Bar
R. W. Bassemir, *Sun Chemical Corporation,* Inks
Catia Bastioli, *Novamont S.p.A,* Biodegradable Materials
Allan M. Baylis, *Amber Associates,* Bags, Multiwall
A. J. Bean, *Sun Chemical Corporation,* Inks
Colin Benjamin, *Florida A&M University,* Materials Handling
Dane Bernard, *National Food Processors Association,* HAACP
Barbara Blakistone, *National Food Processors Association,* HAACP
T. J. Boedekker, *IWKA,* Tube Filling
Terrie K. Boguski, *Franklin Associates,* Life-Cycle Assessment
John K. Borchardt, *Shell Development Company,* Recycling
Robert Borchelt, *University of Wisconsin,* Integrated Packaging Design and Development
Raymond Bourque, *Ocean Spray Cranberries Inc.,* Hot Fill Technology

Roger Brandt, *American Can Company,* Tubes, Collapsible
Richard Brasington, *Buckhorn,* Boxes, Rigid, Plastic
Thomas Brighton, *Tenneco Packaging,* Films, Stretch
Aaron Brody, *Rubbright·Brody, Inc.,* Packaging, Food
Robyn Brody, *Roy F. Weston, Inc.,* ISO-14000
W. E. Brown, *The Dow Chemical Company,* Vinylidene Chloride Copolymers
Richard Cabori, *Angelus Sanitary Can Machinery,* Can Seamers
Richard Carter, *Husky Injection Molding Systems, Ltd.,* Injection Molding
Stephen J. Carter, *Solvay Polymers,* Polyethylene, High Density
Joseph Cavanagh, *Joe Cavanagh Associates,* Glass Container Manufacturing
Dennis Cocco, *The Geon Co.,* Polyvinyl Chloride
John P. Colletti, *John P. Colletti and Associates,* Marine Environment and Export Packaging
Al Corning, *A. Corning Consulting,* Managing the Packaging Function
Bruce Cuthbertson, *Flexible Intermediate Bulk Container Association,* Flexible Intermediate Bulk Containers
Ray E. Davis, Jr., *Chesebrough Pond's USA,* Robots
Brian Day, *Campden and Chorleywood Food Research,* Modified Atmosphere Packaging, Markets, Europe
Phillip DeLassus, *The Dow Chemical Company,* Barrier Polymers; Vinylidene Chloride Copolymers
Robert Demorest, *MOCON,* Testing, Permeation and Leakage
Mark Dickens, *Cryovac,* Cook/Chill
David Dixon, *Packaging Research Laboratory,* Boxes, Wirebound
LeRoy H. Doar, Jr., *Clemson University,* Bags, Heavy-Duty, Plastic
Phil Dodge, *Quantum Chemical Company,* Rotational Molding
Anil G. Doshi, *Borden Inc.,* Film, Flexible PVC
T. W. Downes, *School of Packaging, Michigan State University,* Economics of Packaging
Thomas J. Dunn, *Printpack,* Multilayer Flexible Packaging
Joseph O. Eastlack, Jr., *Saint Joseph's University,* Consumer Research
Mark Eubanks, *MBE Associates,* Total Quality Management
F. B. Fairbanks, *Horix Manufacturing Co.,* Filling Machinery, Liquid, Still
Robert Fiedler, *Robert Fiedler and Associates,* Shock
Eric Finson, *BOC Coating Technology,* Surface Treatment
Vaseem Firdaus, *Mobil Chemical Company,* Polyethylene, Linear and Very Low Density
Ed Fisher, *CMS Gilbreth Packaging Systems,* Bands, Shrink
Jacques Fonteyne, *European Recovery and Recycling Association,* Recycling, Europe
Gene A. Foster, *Middlesex Container Company, Inc.,* Boxes, Corrugated
Ronald H. Foster, *EVAL Corp. of America,* Ethylene-Vinyl Alcohol Copolymers
Steve Fowler, *Fowler Associates, Inc.,* Electrostatic Discharge Protective Packaging
William E. Franklin, *Franklin Associates,* Life Cycle Assessment
H. Sean Fremon, *Meech Static Eliminators,* Static Control
Paul Fry, *Franklin Associates, Inc.,* Life-Cycle Assessment
Alicia Garrison, *Perry Johnson, Inc.,* ISO 9000
Chester Gaynes, *Gaynes Testing Laboratories,* Testing, Packaging Materials
Jack Giacin, *School of Packaging, Michigan State University,* Permeability and Diffusion Aromas, Solvents
Joseph Giacone, *Kraft General Foods,* Microwave Pasteurization and Sterilization
R. Goddard, *Pira International,* European Packaging
Gerald A. Gordon, *Sonoco Products Company,* Drums, Fiber; Intermediate Bulk Containers
Michael J. Greely, *Amplas Inc.,* Stand-Up Flexible Pouches
Jennifer Griffin, *Polaroid Corporation,* Point-of-Purchase Packaging
Joseph Grygny, *Midwest Packaging Consultant,* Fiber, Molded
Karen Hare, *Nutrition Services, Inc.,* Nutrition Labeling
David L. Hartsock, *Phillips 66 Company,* Styrene-Butadiene Copolymers
Eric Hatfield, *James River Corporation,* Coextrusions for Flexible Packaging
Marvin Havens, *Cryovac,* Electrostatic Discharge Protective Packaging
Otto L. Heck, *Angelus Sanitary Can Machinery,* Can Seamers
Peter Henningsen, Bulk Packaging
Ruben Hernandez, *School of Packaging, Michigan State University,* Metrication and Unit Systems for Packaging; Permeability and Diffusion Aromas, Solvents; Polymer Properties
Russell J. Hill, *BOC Coating Technologies,* Film, Transparent Glass on Plastic Food Packaging Materials
George L. K. Hoh, *DuPont P&IP,* Ionomers
Larry Horvath, *James River Corporation,* Coextrusions for Flexible Packaging
B. A. Howell, *Central Michigan University,* Vinylidene Chloride Copolymers
John Hull, *Hull Corporation,* Thermosetting Plastics, Processing Systems for
George Huss, *Rubbright·Brody, Inc.,* Microwaveable Packaging
Jim Iciek, *Institute of Packaging Professionals,* Contract Packaging
Ronald C. Idol, *Multiform Desiccants,* Oxygen Scavengers
Christopher Irwin, *Johnson Controls, Inc.,* Blow Molding
Bill Jenkins, *Flexel, Inc.,* Cellophane
Montfort A. Johnsen, *Montfort A. Johnsen & Associates, Ltd.,* Aerosol Containers; Pressure Containers; Propellants, Aerosol
Bert Johnson, *MOCON,* Testing, Permeation and Leakage
Charles R. Jolley, *Cryovac,* Films, Shrink
Gunilla Jonson, *Lund University,* Edge-Crush Concept
Harry Kannry, *Dussek-Campbell,* Waxes
Stephen L. Kaplan, *4th Stage,* Surface Treatment
Irving Kaye, *National Starch and Chemical Company,* Adhesives
Huston Keith, *Keymark Associates,* Trays, Barrier Foam; Trays, Foam
Allen G. Kirk, *DuPont,* Films, Plastic
Ted Klein, *Montebello Packaging,* Tubes, Collapsible
Dan Kong, *Mobil,* Film Nonoriented Polypropylene
David Kopsick, *CMS Gilbreth Packaging Systems,* Bands, Shrink
Fred Kraus, *Crown Cork & Seal,* Cans, Steel
John M. Krochta, *University of California, Davis,* Film, Edible
Curt Kuhr, *Kliklok Corporation,* Cartoning Machinery, Top-Load
Garry Latto, *Dussek-Campbell,* Waxes
Brad Leonard, *CAPE Systems,* Computer Applications, Pallet Patterns
Robert Luciano, *Luciano Packaging Technologies, Inc.,* Consulting
Robert Luise, *Hytem Consultants, Inc.,* Liquid Crystal Polymers
Burton Lundquist, Career Development, Packaging
Paul R. Lund, *BP Chemicals,* Nitrile Polymers
Michael S. Mabee, Aseptic Packaging
Luis Madi, *CETEA,* Brazil Packaging
Donald W. Mallik, *Virtual Image (a Printpak co.),* Holographic Packaging
Andrea S. Mandel, *Andrea S. Mandel Associates,* Bottle Design, Plastic
Raymond Mansur, *US Army Natick, Research Development and Engineering Center,* Military Packaging
Norma Maraschin, *Union Carbide Corporation,* Polyethylene, Low Density
Jorge Marcondes, *San Jose State University,* Vibration
Carl Marotta, *Tolas Health Care Packaging,* Medical Packaging
Kenneth S. Marsh, *Kenneth S. Marsh & Associates, Ltd.,* Shelf Life
Joseph P. McCaul, *The BP Chemical Company,* Nitrile Polymers
Alfred H. McKinlay, *Consultant,* Distribution Packaging; Testing, Shipping Containers
Richard C. Miller, *Montell USA,* Polypropylene
Eldridge M. Mount, III, *Mobil Chemical Company,* Film, Oriented Polypropylene; Film, Nonoriented Polypropylene
P. V. Narayanan, *Indian Institute of Packaging,* India Packaging
Benjamin Nelson, *Nelson Associates,* Code Marking and Imprinting in the Computer Age
John Newton, *ICI,* Film, Oriented Polyester
N. F. Niedler, *Anheuser-Busch, Inc.,* Cans, Aluminum
Richard Norment, *Steel Shipping Container Institute,* Drums/Pails, Steel
Robert F. Nunes, *International Packaging Machines,* Wrapping Machinery, Stretch Film
Paul Obolewicz, *Rand Whitney Packaging,* Cartons, Folding
Jim Ohlinger, *Tipper Tie, Inc.,* Netting, Plastic
Lorna Opatow, *Opatow Associates, Inc.,* Testing Consumer Packages for Marketing Effectiveness
Mary Alice Opfer, *Fibre Box Association,* Transportation Codes
Octavio Orta, *Riverwood International,* Carriers, Beverage
Richard Perdue, Vacuum Packaging
Alexander M. Perritt, *Perritt Laboratories, Inc.,* Child-Resistant Packaging

William Pflaum, *Packaging Education Forum, Institute of Packaging Professionals,* Education, Packaging; Exhibitions
Richard C. Randall, *Randall's Practical Resources,* International Standards
Stephen A. Raper, *University of Missouri, Rolla,* Integrated Packaging Design and Development; Materials Handling
Michele Raymond, *Raymond Communications,* Environmental Regulations, International; Environmental Regulations, North America
Terre Reingardt, *Ball Corporation,* Cans, Aluminum
David Reznick, *Raztek Corporation,* Cans, Corrosion
Michael Rooney, *CSIRO Division of Food Science and Technology,* Active Packaging
Dominick V. Rosato, *Plastics FALLO,* Additives, Plastic; Thermosetting Polymers
Jack L. Rosette, *Forensic Packaging Concepts, Inc.,* Forensic Packaging; Tamper-Evident Packaging
Richard G. Ryder, *Klöckner-Pentaplast of America,* Film, Rigid PVC
Claire Koelsch Sand, *Performance Development Systems,* Specifications and Quality Assurance
D. Satas, *Satas and Associates,* Coating Equipment
H. H. Schueneman, *Westpak,* Testing, Product Fragility
Ronald B. Schultz, *RBS Technologies, Inc.,* Decorating, In-Mold Labeling
Darrin S. Seeley, *MAS Technologies,* Aroma Barrier Testing
Susan E. Selke, *School of Packaging, Michigan State University,* Environment
R. L. Sheehan, *3M Corporation,* Tape, Pressure-Sensitive
Jean Silbereis, *SLCC,* Metal Cans Fabrication
Ralph A. Simmons, *Keller and Heckman,* Laws and Regulations, U.S.
R. Paul Singh, *University of California, Davis,* Time-Temperature Indicators
Paul Sowa, *Signode Corporation,* Strapping
Burt Spottiswode, *DuPont P&IP,* Skin Packaging
Douglas Stewart, *Montebello Packaging,* Tubes, Collapsible
Jean Storlie, *Nutrition Labeling Solutions,* Nutrition Labeling
Andrew Streeter, *CPS International,* Nippon Packaging
John Sugden, *The Dow Chemical Co.,* Polystyrene
K. W. Suh, *The Dow Chemical Co.,* Foam Plastics
J. T. Sullivan, *Rohm & Haas,* Acrylic Plastics
Arthur J. Taggi, *DuPont Printing and Publishing,* Printing, Gravure and Flexographic
Tamerica Products, Paper, Synthetic
Stephen R. Tanny, *DuPont Polymers,* Adhesives, Extrudable
George Tarulis, *Crown Cork & Seal,* Cans, Steel
P. A. Tice, *PIRA International,* Laws and Regulations, Europe
Paul Tong, *Mobil Chemical Company,* Polyethylene, Linear and Very Low Density
Art Trefry, *R. A. Jones & Co.,* Carton Terminology Brief
M. H. Tusim, *The Dow Chemical Co.,* Foam Plastics
Diana Twede, *School of Packaging, Michigan State University,* Bags, Bulk, Flexible Intermediate Containers; Economics of Packaging; Logistical Distribution Packaging
J. Keith Unger, *Chesebrough-Pond's USA,* Robots
H. J. G. Van Beek, *Klöckner-Pentaplast of America, Inc.,* Film, Rigid PVC
John R. Wagner, Jr., *Mobil Chemical Company,* Film, Oriented Polypropylene
J. Robert Wagner, *Philadelphia College of Textiles and Science,* Nonwovens
Phillip A. Wagner, *The Dow Chemical Co.,* Foam, Extruded Polystyrene; Polystyrene
Peter Walker, Printing: Gravure and Flexographic
Graham Wallis, *Datamark,* Brazilian Packaging Industry
Walter Warren, *Salwasser Manufacturing Company, Inc.,* Case Loading
Ray White, *PIRA International,* European Packaging; Laws and Regulations, Europe
John Wininger, *Eastman Chemical Company,* Sheet, PET-G
George D. Wofford, *Cryovac,* Film, Shrink
Fritz Yambrach, *Rochester Institute of Technology,* Package-Integrity Issues for Sterile Disposable Healthcare Produc
Dennis E. Young, *Dennis Young & Associates,* Distribution Hazard Measurement
Devon Zagory, *Devon Zagory and Associates,* Modified Atmosphere Packaging
Paul Zepf, *Zarpac, Inc.,* Conveying; Glossary

PREFACE

Welcome to the second edition of the *Wiley Encyclopedia of Packaging Technology*.

With the rapid changes in packaging technology that have occurred and are happening globally, many readers are asking why this second edition is ten years after the acclaimed first edition. During these 10 years, packaging technology has changed profoundly. We have been apprised by environmentalists of the impact of packaging in the environment; influenced by economics which dictate low, low and lower; driven by dramatic alterations in the distribution structure for consumer and industrial goods which alter the manner of protection and communication to be offered by packaging; hit by the new computer-controlled and managed engineering; surprised by too many different innovations from around the world; and repeatedly disrupted by change. We operate on a small planet where a development in Japan is known in Germany in the morning and is being reengineered in North America by the afternoon.

Ten years ago, the Institute of Packaging Professionals had not yet been born, polyester was a carbonated beverage bottle material, servo drives were for airplanes, sous vide was French for vacuum and Crown Cork and Seal was an also ran in the world of can makers.

Packaging technology in all of its manifestations has more than shifted; it has radically changed. Function and performance are critical, and economics are less crucial to satisfying consumer and customer needs. Technology does not hold all the answers to product protection through packaging, but few better places to begin are found. With the exponential increases in developments, packaging newsletters and periodicals have multiplied, but how does any rational person keep track of all of the news items? The few specialized books track specific topics, but none captures the totality of packaging technology as comprehensively and concisely as the *Wiley Encyclopedia of Packaging Technology*. We have created what all of us believe is a classic reference for our times.

We have struggled to embrace the world, enlisting professionals from everywhere not only as authors but as members of our hard-working editorial advisory board. We have endeavored to incorporate all the relevant and contemporary topics, but it is inevitable that some have been omitted or deemphasized. The A to Z format of traditional encyclopedias is employed and is enhanced by a mammoth cross index to permit the user to find everything the book has on any topic, regardless of where it is hidden.

Each of the more than 250 experts and authors who invested his or her time to prepare and refine articles for this undertaking is owed a debt of gratitude by the people of packaging.

The editorial advisory board of this second edition of *Wiley Encyclopedia of Packaging Technology* warrants the deep appreciation not only of the publisher and the two editors, but the entire packaging community for their contributions to making this the best packaging technology book ever, by far. Bill Armstrong, Lejo Brana, Rob Clarke, Sophia Dilberakis, Dr. Richard Eidman, Bob Esse, Ron Foster, Dr. Bruce Harte, Doug Howe, Melissa Larson, Burt Lundquist, Bob Luciano, Dick Perdue, Judy Rice, Dr. Gordon Robertson, Neil Robson, Alan Skeddle, Doug Stewart and Paul Zepf represent a diversity of professional interests and demographics within packaging technology. Even as many of these great professionals changed their positions and even their countries during the long period between conception and publication, each continued to fulfill his or her accepted responsibilities.

Not to be overlooked are the hundreds of organizations—companies and universities—which encouraged and permitted the authors and editorial advisory board members to work on this project. We are grateful for their tolerance in this effort.

We enjoyed the labor and even more its outcome in orchestrating this second edition of the *Wiley Encyclopedia of Packaging Technology*. We sincerely hope it will be an effective tool for seasoned professionals, novices, students and casual readers. And we thank the publishers and their patient and gracious staff members for really doing the hard work.

Duluth, Georgia — Dr. Aaron L. Brody
Woodstock, Illinois — Dr. Kenneth S. Marsh

INTRODUCTION

Packaging is the science, art and technology of protecting products from the overt and inherent adverse effects of the environment. Packaging is the integration of elements of materials, machinery and people to erect and maintain barriers between the product and those external forces inexorably seeking to revert the contents back to their essential components. The package is the physical entity that functions as the wall between the contents and the exterior.

If we had no products that required movement in space or transition in time, there would be no need for packaging. If products were themselves infinitely resistant to the ravages of moisture, oxygen, stress, impact, biological vectors, and even people inadvertently or otherwise attempting to see or touch, packaging would not be needed.

But the Industrial Revolution and events prior to that period dictated that populations divide their work and geographies, leading to the imperative to move goods to and from them. Most indispensable has been food, grown, harvested, killed or caught in regions remote in time and place from people diligently toiling at producing automobiles, furniture, music, artificial intelligence, and encyclopedias. Drugs to prevent and cure illness and disease, cleaning materials, playthings to relieve the pressures of everyday life, clothing, components for remanufacture and millions of other items are daily transported across factories, cities, countries and oceans for use by others. It is packaging that facilitates, nay, permits the safe and economical movement.

Packaging is a $300 billion worldwide industry with about one third or $100 billion in the United States. The expenditure of this much money has been carefully and frequently reviewed by independent financial professionals and found to be cost-effective. Some might lament that since packaging is not a part of the product, it is not needed. Others have cried, and continue to state, that the package materials are useless after their immediate protective function has been fulfilled and therefore should be restricted or banned. These arguments have been heard for decades from persons who continue to benefit from the performance of packaging. When they eat, they are safeguarded because their food has been protected by cans or bottles or pouches or bags or cartons. When they are sick, their pharmaceuticals are sterile and efficacious because of metal foils or plastic films. The paper, or now even the plastic, upon which they write their messages, their computer discs, their shoes, and virtually everything that constitutes their everyday life is surrounded by package structures prior to use.

Packaging is employed not to fill the coffers of steel, aluminum, glass, plastic or paper manufacturers and converters, but rather to ensure that the products of the world reach the consumers of the world. And, if nearly $60 billion of packaging is invested in our U.S. food supply, consider that this food is the safest and most diverse in world history. Consider that less than 10% of the retail price paid for food is packaging. Consider that we have long recovered and reused large proportions of our spent package materials. Indeed packaging is among United States' best investments.

Definitions of packaging now range into realms that few among us would ever dared to venture a half century ago when the disciplines of packaging were first formalized. The word "packaging" and its derivatives is found in the areas of real estate, financial planning and execution, descriptions of attractive humans, entertainment, legal documents, computers and electronics, alcoholic beverages, and probably places few of us have ventured. Obviously, in a broad diversity of disciplines, there has been a recognition that integration of a multiplicity of elements is best described by the word "packaging," the very same word used by we whose profession is to protect and preserve products.

Technology is a word describing the application of science. Although packaging technology has been employed for many years, no professional society or college curriculum uses the word in its title as of yet. But the concept is clear: the application of all the elements of science, chemistry, physics, biology and mathematics, to the protection of products.

The origins of an encyclopedia of packaging technology were in the original work of the editors and publishers of a now memorable trade magazine called *Modern Packaging*. Lost in history is the genesis of their annual *Modern Packaging Encyclopedia*, but some of us are old enough to remember. Some libraries still shelve dusty, and in some instances, not so dusty, since they continue to be read by practitioners—copies of hard cover volumes edited by Bill Simms containing dozens of articles and endless tabulations that served as our reference from the 1940s until the early 1980s. These references are sometimes referred to even today. The disappearance of that great publication following the passing of its editor created a void which was not filled until publisher John Wiley & Sons recognized the gap and cast Marilyn Bakker into the role of assembling their initial venture into the packaging encyclopedia. This diligent, dedicated, and meticulous professional stamped the first edition with her own very special brand and style.

Almost as soon as the first edition was published to great acclaim in the profession, portions were, of course, rendered obsolete, and so a second edition was essential. Institute of Packaging Professionals was recruited to act as an endorsing agency and to aid the new editors in gathering data and information that would ultimately constitute the second edition of the *Wiley Encyclopedia of Packaging Technology.* Institute of Packaging Professionals (IoPP) is the United States' professional society for packaging technologists and their colleagues. With membership approaching 10,000 from among the estimated 50,000 to 100,000 women and men who function as the engineers, technologists, technicians, managers, and resources in the U.S. packaging community, IoPP is the largest packaging professional organization in the world. Further, it is formally affiliated through the World Packaging Organization and the International Association of Packaging Research Institutes with all of the world's packaging technical groups.

The two editors of the 2nd edition have between them about sixty years of experience in packaging which is not as much as a few but more than most. Both of us have been members of IoPP and its predecessor organizations for decades and we are both Fellows of the Institute. Dr. Marsh was educated at Rutgers, the State University of New Jersey, in food and packaging and worked in packaging for years with such companies as Lipton, Pillsbury, Ball, and Quaker Oats prior to forming his own consulting firm, Kenneth S. Marsh and Associates, Ltd., to specialize in solving food and drug packaging problems. Dr. Brody, who was inducted into the Packaging Hall of Fame, was IoPP Member of the Year and Institute of Food Technologists' first Industrial Scientist and one of the first recipients of their Industrial Achievement Award. He is also a Fellow of Institute of Food Technologists, the only person to achieve the dual distinction of being a fellow of the two preeminent food and packaging technology societies. A founder of Institute of Food Technologists' Food Packaging Division, he received their Riester-Davis Award for lifetime achievement in food packaging. Currently he is Managing Director of Rubbright•Brody, Inc., the world's finest professional food and packaging consultancy, and a visiting professor at the University of Georgia.

The editors take this opportunity to express their appreciation to all who have made this encyclopedia possible. The contributing authors represent the greatest collection of packaging expertise ever assembled into a single volume. Their contributions have made the encyclopedia both useful and possible. In addition, the support of colleagues and organizations surrounding these men and women who supplied time, information, and other support is gratefully acknowledged.

The staff at John Wiley & Sons deserves substantial thanks for coordinating and tracking the efforts of nearly 300 authors throughout this effort.

On a personal level, Dr. Marsh hopes that the efforts represented in this volume will lead to packaging that improves the quality of life for people around the world. Since this is a concern of his wife, Janet and their son, Abe, this will be partial compensation for enduring the birth of this second edition. It could not have been done without their support and patience.

Since this document is a foundation of the future, Dr. Brody's grandchildren will be the greatest beneficiaries of his efforts. They represent our tomorrows. Thank you for making it all worth the investment! Thanks also to their parents, our children and to their Grandma, the wife who has lived these many years with Aaron and grown more beautiful and gracious as a result. Words cannot express the emotion of gratitude and love for all of the Brody bunch.

The Wiley ENCYCLOPEDIA OF Packaging Technology

Second Edition

A

ACRYLIC PLASTICS

Polymers based on acrylic monomers are useful in packaging as a basis for printing inks and adhesives and as modifiers for rigid PVC products.

Acrylic-Based Inks

Paste inks. Acrylic solution resins are used in lithographic inks as dispersing or modifying letdown vehicles (see Inks; Printing). A typical resin (60% in oil) offers excellent dot formation, high color fidelity, exceptional print definition, nonskinning, and good press-open time. Set times are fast (ca 60–90 s), and a minimal level of starch spray (75% of normal) is effective. Coatings on cartons, fabrication stocks, and paper are glossy and exhibit good dry resistance.

Solvent inks. Because of their resistance to heat and discoloration, good adhesion, toughness, and rub resistance, acrylics are widely used in flexographic inks on paper, paperboard, metals, and a variety of plastics (1). These inks also give block resistance, resistance to grease, alcohol, and water, and good heat-sealing performance (see Sealing, heat). With some grades, adding nitrocellulose improves heat sealability, heat resistance, and compatibility with laminating adhesives. This family of methacrylate polymers (methyl to isobutyl) has broad latitude in formulating and performance. Solid grades afford low odor, resist sintering, and dissolve rapidly in alcohol-ester mixtures or in esters alone (gravure inks). Solution grades (40–50% solids) are available, as well as nonaqueous dispersions (40% solids) in solvents such as VMP naphtha, which exhibit fast solvent release and promote superior leveling and hiding. They are excellent vehicles for fluorescent inks.

Water-based inks. The development of waterborne resins has been a major achievement. Their outstanding performance allows them to replace solvent systems in flexographic and gravure inks and overprint varnishes on corrugated and kraft stocks, cartons, and labels. The paramount advantage of aqueous systems is substantial decrease in environmental pollution by volatile organics. Aqueous acrylic colloidal dispersions (30% solids) and a new series of analogous ammonium salts (46–49% solids) are effective dispersants for carbon blacks, titanium dioxide, and organic pigments. Derived inks give crisp, glossy impressions at high pigment loading, good coverage and hiding, and water resistance. The relatively flat pH–viscosity relationship assures formulation stability on presses despite minor loss of volatiles. Adjusting the alcohol–water ratio controls drying rate, and quick-drying inks can be made for high-speed printing. The resins are compatible with styrene–acrylic or maleic dispersants and acrylic or styrene–acrylic letdown vehicles. Blends of self-curing polymer emulsions are excellent overprint varnishes for labels and exhibit a good balance of gloss, holdout, slip, and wet-rub resistance.

Some aqueous acrylic solutions (37% solids) combine the functions of pigment dispersant and letdown resin and serve as ready-to-use vehicles for inks on porous substrates like kraft and corrugated stocks and cartons. They afford excellent color development, excellent heat-aging resistance in formulations, and fast drying. The flat pH–viscosity relationship gives the same benefit as the dispersants.

Acrylic Adhesives

Pressure-sensitive adhesives. Solution copolymers of alkyl acrylates and minor amounts of acrylic acid, acrylonitrile, or acrylamide adhere well to paper, plastics, metals, and glass and have gained wide use in pressure-sensitive tapes (2). Environmental regulations, however, have raised objections to pollution by solvent vapors and are requiring costly recovery systems. This opportunity has encouraged the development of waterborne substitutes, such as emulsion polymers, which eliminate these difficulties and offer excellent adhesion, resistance to wet delamination, aging, and yellowing, and, like the solvent inks, need no tackifier. In packaging applications, the emulsion polymers provide high tack, a good balance of peel adhesion and shear resistance, excellent cling to hard-to-bond substrates, and clearance for food packaging applications under FDA Regulations 21 CFR 175.105, 21 CFR 176.170, and 21 CFR 176.180. Their low viscosity makes formulation easy, and the properties of the adhesives can be adjusted by adding surfactants, acrylic thickeners, and defoamers.

Resins are available that are designed specifically for use on polypropylene carton tapes (3). They are ready-to-use noncorrosive liquids applicable to the corona-treated side of oriented polypropylene film using knife-to-roll, Mayer rod, or reverse-roll coaters (see Coating equipment; Film, oriented polypropylene; Surface modification). A release coating is unnecessary because the adhesive does not stick to the untreated side and parts cleanly from the roll. The tapes are used to seal paperboard cartons with high-speed taping machines or handheld dispensers. Adhesion to the cartons is instantaneous and enduring. The colorless tape is well suited for label protection. The material adheres well to other plastics and metals.

Hot-melt adhesives. These adhesives offer obvious advantages over solvent or waterborne materials if equivalent performance is obtainable. Acrylic prototypes gave better color and oxidative stability than rubber-based products, but exhibit poor adhesion quality. New improved grades are providing an impressive array of adhesive properties and superior cohesive strength at elevated temperatures in addition to stability and low color. The action of the adhesives involves a thermally reversible cross-linking mechanism that gives ready flow at 350°F (177°C), rapid increase in viscosity on cooling, and a stiff cross-linked rubber at ambient temperature. The resins give durable peel adhesion, good shear resistance, resistance to cold flow, and excellent photostability in accelerated weathering. On commercial machinery these resins have displayed excellent coatability on polyester film at high line speeds (see Film, oriented polyester). A wide variety of possible applications, including packaging tapes, is envisaged for these materials.

PVC Modifiers

Acrylics have played a major role in the emergence of clear rigid PVC films and bottles (4,5). Acrylic processing aids provide smooth processing behavior in vinyl compounds when passed through calenders, extruders, blow-molding machinery, and thermoforming equipment (see Additives, plastics). One member of this group is a lubricant–processing aid that prevents sticking to hot metal surfaces and permits reduction in the level of other lubricants, thereby improving clarity. Other benefits of acrylics are low tendency to plateout and a homogenizing effect on melts to give sparkling clarity and improved mechanical properties. The usual level in vinyl compounds for packaging is about 1.5–2.5 phr.

In a second group are the impact modifiers, which are graft polymers of methyl methacrylate–styrene–butadiene used in the production of clear films and bottles. The principal function of impact modifiers is to increase toughness at ambient and low temperature. Levels of 10–15 phr, depending on modifier efficiency, are normal.

Many acrylics are cleared for use in food-contact products under FDA Regulation 21 CFR 178.3790. The regulation stipulates limits in the permissible level of modifiers relative to their composition. Processors should seek advice from suppliers on the makeup of formulations. Many modifiers are fine powders that may produce airborne dust if handled carelessly. Above 0.03 oz/ft^3 (0.03 mg/cm^3), dust is a potential explosion hazard and its accumulation on hot surfaces is a fire hazard. The recommended exposure limit to dust over an 8-h period is 2 mg/m^3. Eliminate ignition sources, ground equipment electrically, and provide local exhaust ventilation where dusting may occur (6). Workers may wear suitable MSHA-NIOSH respiratory devices as protection against dust.

BIBLIOGRAPHY

1. B. V. Burachinsky, H. Dunn, and J. K. Ely in M. Grayson and D. Eckroth, eds., *Encyclopedia of Chemical Technology,* Vol. 13, Wiley, New York, 1981, pp. 374–398.
2. D. Satas, ed., *Handbook of Pressure-Sensitive Adhesives,* Van Nostrand Reinhold, New York, 1982, pp. 298–330, 426–437.
3. W. J. Sparks, *Adhes. Age* **26**(2), 38 (1982).
4. J. T. Lutz, Jr., *Plast. Compd.* **4**(1), 34 (1981).
5. *Bulletin MR-112b,* Rohm and Haas Company, Philadelphia, Jan. 1983.
6. American Conference of Governmental Hygienists, Cincinnati, *A Manual of Recommended Practice,* 1982; American National Standards Institute, New York, N.Y., *Fundamentals Governing the Design and Operation of Local Exhaust Systems,* ANSI Z-9.2, 1979.

J. T. SULLIVAN
Rohm and Haas Company

ACTIVE PACKAGING

Packaging is described as active when it performs some desired role other than to provide an inert barrier between the product and the outside environment. Therefore, active packaging differs from conventional passive packaging in that one or more forms of interaction are planned, usually to offset a deficiency in an otherwise suitable package. The active component may be part of the packaging material or may be an insert or attachment to the inside of the pack. Active packaging is largely an innovation of the 1980s and 1990s, although there are examples that have been in use for over a century. The tinplate can, for instance, provides a sacrificial layer of tin that protects the food from accumulation of catalytically active iron salts. Antioxidant release from waxed-paper packs for breakfast cereals has been used for several years as has been the impregnation of cheese wraps with sorbic acid.

It was only in 1987 that the term "active packaging" was introduced by Labuza (1). Prior to that time terms such as "smart," "freshness preservative," and "functional" were used to describe active-packaging materials. Sachets of iron powder have been described as "deoxidizers," "free oxygen absorbers," and "oxygen scavengers" (see Oxygen scavengers). Active packaging is designed to enhance the matching of the properties of the package to the requirements of the product. Therefore, the forms and applications of active packaging are diverse, addressing specific situations in the protection and presentation of foods and other products.

Problems Addressed by Active Packaging

Active packaging can be used to address problems classified according to the mechanism of deterioration of the packaged product. These mechanisms can be broadly classified as either biological or physicochemical.

Biological deterioration may result from insect attack as occurs, for instance, in foods, furs, fabrics, and museum specimens. Elevated temperatures and humidities enhance the rate of activity at various stages in the life cycles of insects. Chemical fumigation is possible in some cases but is becoming more tightly controlled with foods such as grains and dried fruits. Accordingly, use of modified-atmosphere packaging (MAP) is becoming more popular, especially in Europe. Since low levels of oxygen and/or high carbon dioxide levels are required to suppress growth, packaging systems or adjuncts that assist in achieving such atmospheres can contribute to quality maintenance. Such adjuncts are oxygen scavengers, desiccants, and carbon dioxide emitters.

The other generically common cause of biological deterioration is microbial growth. This is usually enhanced by the same variables, but there is also danger from anaerobic pathogenic bacteria, such as clostridia, that grow at very low oxygen levels or in the absence of oxygen. Hence, the removal of oxygen is not necessarily a solution to all microbial growth problems. Antimicrobial treatments such as the release of carbon dioxide, ethanol, other preservatives, or fungicides can play a role in reducing microbial growth. Similarly, desiccants can assist in providing the "hurdle" of reduced water activity, especially in foods. Where liquid water is formed by condensation on the packages of fresh produce, the use of humidity buffers or condensation control films can be useful. Where tissue fluids from fish or white and red meats is unsightly, the use of drip absorbent pads is becoming commonplace.

Biological deterioration of fresh produce also occurs naturally as part of the process of senescence. Reduction in the rate of senescence can be achieved in many cases by reduction of the respiration rate by reducing equilibrium oxygen concentrations to ~2%. Ethylene synthesis that accelerates ripening and senescence can be suppressed by elevated carbon dioxide concentrations. Existing plastic packaging films seldom allow beneficial equilibrium modified atmospheres to be developed, so some form of active packaging is needed. Transpiration of water by produce leads to condensation when temperatures fluctuate slightly. Furthermore, ethylene release by one or more damaged or ripe fruit can cause rapid ripening of others. This is akin to the "one rotten apple in the barrel" situation. Ethylene removal is therefore a highly desirable property of produce packaging.

Chemical deterioration vectors act on the widest range of packaged products. These include especially foods and beverages (lipid and nutrient loss, off-flavor generation), but also pharmaceuticals. Industrial chemicals such as amines, and particularly some printing inks are oxidized on storage. Microelectronic components, some metals and a variety of unrelated items can be subject to attack by oxygen. Often the rate of loss can be reduced adequately by inert-gas flushing and barrier packaging. However these treatments are not always effective, convenient, or economical, particularly when oxygen levels below 0.5% are desired (2). Nitrogen flushed packs of dry foods often have residual oxygen levels of 0.5–2%. Chemical forms of in-pack oxygen scavenging have been introduced both to reduce these residual levels further and to deoxygenate air headspaces without the use of inert-gas flushing or evacuation.

Fried snacks are particularly susceptible to oxidation depending on their moisture content. This has been demonstrated in the case of potato chips (3). Although sliced, processed meats are packaged commercially under vacuum, improved presentation using MAP can be achieved when an oxygen scavenger is present. The pink nitrosomyoglobin is damaged by even low quantities of oxygen in the package. The flavor of alcoholic beverages such as beer and white wines is particularly sensitive to oxygen, so the relatively high oxygen permeability of polyester [polyethylene terephthalate (PET)] bottles makes them unsuitable for packaging most wines and beers. The presence of oxygen in glass bottles is usually offset by addition of sulfur dioxide to the beverage. However, oxidative loss of this antioxidant still limits the shelf life of beer and white wines and limits their packaging options. A similar sulfur dioxide loss occurs in dried apricots. In these cases the presence of an oxygen scavenger that does not react with this acidic gas is required. Porous adsorbents in current oxygen scavengers may also remove some of the sulfur dioxide.

The flavor of some foods changes on storage because of effects other than oxidation. Tainting is a recurrent problem. Moldy taints can result from long voyages in shipping containers. Methods of odor interception without the use of expensive barrier packaging are needed for the transportation of low-valued primary products. Besides interception of external taints, there is also a need for removal of food breakdown products that can be formed during storage. These include amines or thiols formed rapidly in fish or rancid odors in oil-containing foods. Such compounds can be present in trace amounts that are significant organoleptically but may not constitute a health hazard. The bitter principle in some orange juices, limonin, is formed on standing and needs to be removed if the juice is not to become unacceptable rapidly.

Two physical properties of a product that can potentially be affected by active packaging are heating and cooling. Thus the microwave heating of packaged multicomponent entrees offers a challenge for uniform heating in spite of varying layer thicknesses and water contents (see Microwave pasteurization and sterilization). Canned drinks, such as sake and coffee, supplied via vending machines in Japan are frequently consumed warm. Other drinks may need to be cooled, and so dispensing from the one machine may necessitate building the temperature-changing capacity into the can itself.

Goals of Active Packaging

Active packaging is chosen to enhance the ability of conventional packaging to help deliver the product to the user in a desired state. The decision to use some form of active packaging will often be based on one or more of the following considerations (see also Shelf life).

1. *Extension of shelf life.* This extension may exceed the presently accepted limits as with sea shipment of some fresh produce.
2. *Less expensive packaging materials.* Packaging of limited-shelf-life products may require enhancement of only one property for a fixed period. This can include bakery products, metal components shipped by sea, or chilled meats.
3. *Simpler processing.* Introduction of additional microbiological "hurdles" can allow MAP to be achieved without use of expensive equipment.
4. *Difficult-to-handle products.* Oxygen can be removed from tightly packaged products such as cheeses that are subject to mold growth.
5. *Allowing particular types of package to be used.* This could include retortable plastics for products with multiyear shelf lives or PET wine bottles.
6. *Presentation.* Heating by microwave susceptors and other adjuncts has allowed packaging innovation for convenience foods.

Other goals are developing as the potential is being realized. Indicators of time–temperature and temperature abuse are presently available. The composition of the package headspace can potentially indicate chemical, physiological, or microbiological state or the potency of the packaged product.

Forms of Active Packaging

The active components in packaging can exist either as part of an otherwise unmodified package or as an elaborate adjunct or design modification. The major form in use at present is the insertion of sachets of various scavengers or emitters. These are followed by plastics blends or compounds, and to a lesser extent by composite packages of various forms.

Sachets. Silica gel has been supplied for protection of packaged goods from water for many years. A range of sachets and porous canisters as well as saddles are manufactured in sizes from grams to kilograms by companies such as Multiform Desiccants of Buffalo, NY or United Desiccants-Gates. Silica gel has a capacity when dried for taking up 40% of its own weight of water vapor. An alternative is lime (calcium oxide) that takes up (28%). Both are used largely in the shipment of goods through humid atmospheres to protect against corrosion (steel, aluminum computers), caking (pharmaceuticals), or mold growth (foods). In Japan these are used with some snacks such as rice crackers to give a high level of crunchiness, as well as a sticky, dehydrating sensation on the tongue. Many variants in form have facilitated new uses for these well-known materials. Sachets are marked "Do not eat" and are often between the primary and secondary package. Less severe desiccants can be also used for condensation control in the wholesale distribution of produce, particularly where the carton liner bag is heat-sealed to generate a modified atmosphere. A few products such as tomatoes are packed with large microporous sachets of salts, like sodium chloride, which absorb excess water at the high relative humidities experienced in such closed packages. The relative humidity can be lowered from ~95% to 80%. This is a first-generation approach to humidity buffering, and various carton liners containing humectants are starting to appear on the market in Japan (5).

Oxygen scavenging sachets were introduced in Japan in 1969 initially containing sodium dithionite and lime. This followed early work by Tallgren in Finland in 1938 using iron and other metals (6). Mitsubishi Gas Chemical Co. introduced Ageless sachets in 1977 containing reduced iron powder, salt, and trace ingredients. This technology has developed with a wide variety of formulations being provided by Mitsubishi and 10 other companies in Japan (6). In 1989 it is estimated nearly 7 billion such sachets were manufactured in Japan. These sachets are marketed in the United States by W. R. Grace Ltd. Multiform Desiccants manufactures the Fresh Pax series of iron-based oxygen absorbers and these are also marketed in the United Kingdom (Britain). Standa Industrie of Caen manufacture a range of sachets under the name ATCO in France. The growth rate of sales was around 20% per annum in 1989, so that sales are estimated to be on the order of 7–10 billion in Japan, several hundreds of millions in the United States, and several tens of millions in Europe.

The oxygen scavenging materials can now be bonded to the inside of the package, resulting in even less chance of accidental ingestion or incorporation into food preparations. Mitsubishi Gas Chemical Co. have introduced a hot-melt adhesive system for sachets, and Multiform Desiccants Inc. now market an adhesive label (FreshMax), which is thin, so that it can be applied with conventional labeling machinery. This innovation is shown in Figure 1. The contents of oxygen scavenging sachets differ depending on the relative humidity of the product, usually food. Some are designed to operate at refrigerator or even freezer temperatures. Characteristics of some of the more commonly used sachets are shown in Table 1. The form of triggering is one of the key aspects of oxygen scavengers of any type. It is essential that the scavenging composition can be activated only when required as premature reaction with atmospheric air leads to failure in the sealed package.

Combination sachets are also available from Mitsubishi Gas Chemical Co. and Toppan Printing Co. Some of these release carbon dioxide while taking up oxygen. These are normally based on ascorbic acid and sodium bicarbonate. Ageless E sachets contain lime as well as iron to absorb CO_2 and oxygen and was used in the Maxwell House roasted-coffee packs.

Ethanol emitting sachets. Low concentrations of ethanol, 1–2% in bakery products, have been shown to suppress the growth of a range of common molds. Higher levels are necessary to suppress bacteria and yeasts, and the effectiveness is dependent on the water activity of the product. The Freund Industrial Co. Ltd. (Japan) has developed two forms of ethanol emitting sachets which release ethanol vapor in response to the absorption of water vapor from the food headspace. Ethicap sachets contain food—grade ethanol (55%) adsorbed in silica powder (35%). The sachets consist of films of varying permeabilities to provide some control of the rate of ethanol release. Sachets are available from Freund in sizes of 0.6–6G containing 0.33–3.3 g of ethanol. The size of the sachet required can be calculated from knowledge of the water activity and weight of the product and the shelf life desired (Freund Technical Information).

Figure 1. FreshMax oxygen-absorbing label attached to the inside of processed meat package. Courtesy of Multiform Desiccants, Inc.

Food packages containing ethanol releasing sachets should have an ethanol vapor permeability of <2 g/m^2 per day at 30°C (Freund). Packaging films used with ethanol generators can be as simple as oriented polypropylene/polypropylene, but polyethylenes are too permeable for use. Ethicap has been investigated with pita bread, apple turnovers, strawberry layer cakes, and madeira and cherry cream cake. It is used widely in Japan with semimoist or dry fish products.

The second type of ethanol emitting sachet is a combined oxygen scavenger and ethanol emitter marketed by Freund under the name Negamold. This type of sachet is not widely used. Ethanol emitting sachets have been manufactured by other companies in Japan, including Nippon Kayaku (Oitech) and Ueno Seiyaku (ET Pack). The estimated annual production in 1990 was around 10 million sachets by the latter two companies and 60 million Ethicap sachets, at an average weight of 1.5 g (6). Use of such sachets with foods is not permitted in North America. It has been suggested that regulatory approval for uses other than for brown-and-serve products is unlikely (8).

Ethylene absorbing sachets, sometimes made of steel mesh, are available and follow from the variety of porous slabs and blankets developed for ethylene removal in cool stores and shipping containers. Several minerals are used to

Table 1. Properties of Some Oxygen Scavenging Sachets[a]

Type	Trigger	A_w	Time, Days at 25°C (other)	Substrate Base	Additional Effect
Fresh Pax					
B	Water	>0.65	0.5–2	Fe	
D	Self	>0.7	0.5–4 (2 → −20)	"	
R	Self	All	0.5–1	"	
M	Self	>0.65	0.5–2	"	$+CO_2$
Ageless					
Z	Self	>0.65	1–3	"	
S	Self	>0.65	0.5–2	"	
SS	Self	>0.85	2–3 (0 → −4)	"	
			10 (−25)	"	
FX	Water	>0.85	0.5–1	"	
G	Self	0.3–0.5	—	Ascorbic acid	$+CO_2$
E	Self	<0.3	3–8	Fe/lime	$-CO_2$
Freund					
Negamold	Water	>0.85	—	Fe/ethanol	Ethanol

[a] Data from Technical Information from manufacturers and references 6 and 8.

contain potassium permanganate in the form of purple beads or in other shapes. Typical inert substrates include perlite, alumina, silica gel, and vermiculite containing 4–6% potassium permanganate. The manner in which these might be used should be checked as potassium permanganate is toxic. There are many manufacturers such as Air Repair Products Co. of Stafford, TX; Ethylene Control, Inc. of Salinas, CA; Purafil Co. of Chalamblee, GA; and Nippon Green Co. of Japan. The efficiency of such absorbers will depend on the product, the surface area of the substrate, and possibly any water condensation.

Ethylene absorbing sachets based on other principles are also available. "SendoMate" made by Mitsubishi Chemical Co. of Japan contains a palladium catalyst for destruction of the ethylene. Nonspecific absorbents are also marketed in sachet form in Japan for removal of gases such as ethylene, carbon dioxide, and unwanted odors from food packs. Products based on activated carbon have been marketed by many companies, both large and small, and include Kuraray Co. Ltd. (Fresh Keep), Toppan Printing Co. (Sendo-Hojizai-V) and Mitsubishi Gas Chemical Co. (Ageless-C, which includes slaked lime) (6). The capacity of such absorbents for ethylene at physiological concentrations (eg, 1 ppm, 95%RH) needs to be defined.

Sulfur dioxide releasing pads are available for use in the transportation of cartons of table grapes. Grapes are readily separated from their stalks by the action of fungi in the moist atmosphere of polyethylene-lined cartons. Microporous pads containing sodium metalbisulfite (~7 g) placed on top of the fruit release sulfur dioxide as water vapor is absorbed. If the uptake of water vapor is too rapid, as is often the case, the rapid premature hydrolysis results in excessive levels of sulfur dioxide, resulting further in bleaching of the grapes, commencing at the bottom of the berries. Such pads are largely manufactured in Chile by companies such as Productions Quimicos & Alimenticos Osku SA, of Santiago. Various pads are widely distributed internationally, for instance, by ICI Australia.

Film Composites

Moisture control. Moisture in packages may be in the form of liquid (condensate or drip/weep) or as the vapor. Desiccants remove both forms of water, although they are designed to remove the vapor. The simple form of liquid moisture sorption has been provided by drip-absorbent sheets consisting of two layers of nonwoven polyolefin, divided by heat seals into pouches containing polyacrylate superabsorbent polymers. These sheets are used under chicken or turkey pieces and sometimes under red meats to absorb drip during display. Other uses are to absorb drip from seafood, especially when air-freighted to avoid corrosion of airframes caused by spilling. These sheets are widely available from companies such as Toppan Printing Co. Japan (Toppan Sheet).

Although superabsorbent polymers can absorb up to 500 times their own weight of water, they do not function as such rapid absorbents for water vapor. Condensation can be prevented by use of two-layer plastic sheets containing a humectant between the layers. Several forms of moisture vapor absorbent sheets are available in Japan. Neupalon Sheet is manufactured by Sekisui Chemical Co. of Osaka and consists of a sealed microporous envelope containing a sheet of paper coated with a layer of a ceramic powder and a water absorbent. This is claimed to take up 40 mL of ethylene per square meter of sheet as well as 500–1000 times its own weight of water (5).

A somewhat similar design is used by the Mantsune Co. (Japan) for their corrugated fibreboard "TM Corrugated case," which has a polyethylene liner coated onto the linerboard. This polyethylene contains a ceramic to absorb some ethylene and is laminated to a layer of paper containing an acrylic polymer to absorb water and buffer the humidity (5).

There is at least one water vapor absorbent sheet for domestic use. This consists of an envelope of polyvinylalcohol film sandwiching a glycol and carbohydrate in a strong water vapor absorber (see Fig. 2). It is sold as a perforated role and as packs of single sheets. This product, Pichit, is manufactured by Showa Denko Ltd. for wrapping food portions in domestic refrigerators.

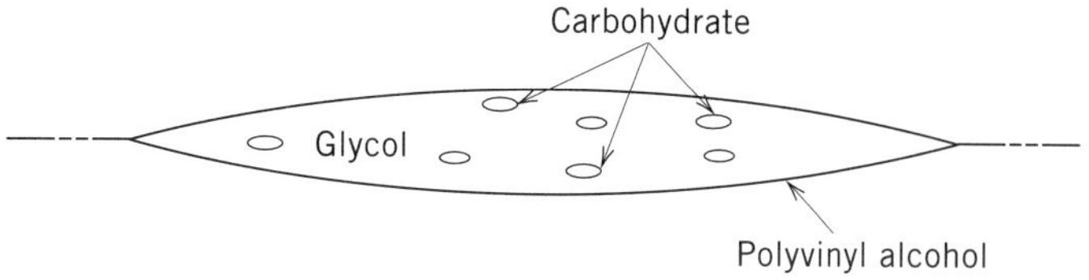

Figure 2. Pichit bilayer sheet for absorbing water from food portions Courtesy of Showa Denko Ltd.

Oxygen scavenger films have been a goal of packaging industry researchers since the work of the American Can Co. in 1976 with the palladium-catalyzed reaction of oxygen with hydrogen. That package, marketed by American Can Co. as Maraflex, was not widely used commercially because of its complexity and its requirement for flushing with a nitrogen/hydrogen moisture. Oxygen scavenging films or other plastic materials offer the opportunity to prevent oxygen ingress to the package by permeation as well as removing that originally present. They also offer the potential for package fabrication, filling and sealing without the need for insertion or attachment of a sachet.

Despite the substantial international R&D effort over the last two decades very few developments have been commercialized (9). The Ox-Bar process of the U.K. CMB Technologies plc for making PET bottles oxygen impermeable while scavenging oxygen from the packaged beverage was announced but not commercialized pending any future demonstration that it met regulatory requirements. Advanced Oxygen Technologies Inc. of Alameda, CA (formerly Aquanautics Inc.) has announced the development of Smart Mix, a master batch for blending with other plastics. This followed their introduction in 1991 of SmartCap, a beverage bottle closure liner containing a blend of ascorbic acid (or its isomer) and a transition-metal catalyst. The W. R. Grace Company has developed a closure liner containing up to 7% sodium sulfite and 4% sodium ascorbate in a polyolefin base (10). The Grace compositions exemplified by Daraform 6490 have been used by Heineken in France in 1989 and more recently in Anheuser-Busch beer produced under license in Britain. The closure seals are under trial in U.S. and European businesses (10). Toyo Seikan Kaisha Ltd. of Yokohama, Japan (Chicago, IL) have taken a different approach using a reduced iron base for reaction with oxygen. The crown closure consists of three layers with the middle, reactive layer separated from the beer by a microporous polymer layer. The scavenging reaction involves water vapor from the beer, especially during pasteurization, and premature reaction is presented by keeping the composition dry prior to use. The closure sealant designs can be compared by reference to Figure 3, which represents the Grace and Advanced Oxygen Technologies approaches (top), and the Toyo Seikan Kaisha approach (bottom).

The first thermoformable oxygen scavenging film has been commercialized in 1994 by Toyo Seikan Kaisha Ltd. for use in retortable plastics trays. The oxygen scavenging layer is between the EVOH (ethylene vinyl alcohol) oxygen-barrier layer and the inner, permeable, polypropylene layer. Figure 4 shows this structure diagramatically. This tray has been adopted by Sato Food Industry Co. Ltd. (Japan) for all their products made in a line planned to operate at 100,000 units per day.

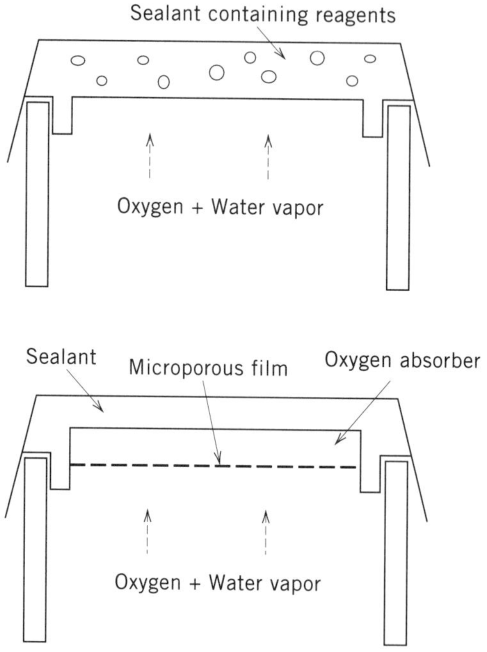

Figure 3. Oxygen absorbing closure liners for bottles: *top*—W. R. Grace or Advanced Oxygen Technologies Inc./Zapata Industries, Inc. type; *bottom*—Toyo Seikan Kaisha Ltd. type.

Antimicrobial films. Antimicrobial agents, fungicides, and natural antagonists are applied to harvested produce in the form of aqueous dips or as waxes or other edible coatings. Their roles and their U.S. regulatory status have been tabulated (11). Besides produce, foods with cut surfaces are subject to largely superficial microbial attack and some cheeses are packaged with wrappings or separating films (sliced cheese) containing sorbic acid. Although many foods are subject to rapid attack at the cut surfaces, potentially useful antimicrobial packaging films are still largely a subject of research.

Mitsubishi Gas Chemical Co., in association with Shinanen New Ceramic Co. in Japan, have produced a synthetic zeolite, Zeomic, which has silver ions bonded into the surface layers of the pores. The zeolite is dispersed in, for instance, a polypropylene or polyethylene 3–6-μm-thick layer and protrude into the package from this layer as indicated in Figure 5 (5). Other layers provide package strength and permeation barrier as required. Liquid in the food is meant to have access to the zeolite, and it appears that the mode of action is uptake of silver ion that dissolves in the aqueous phase (12). The

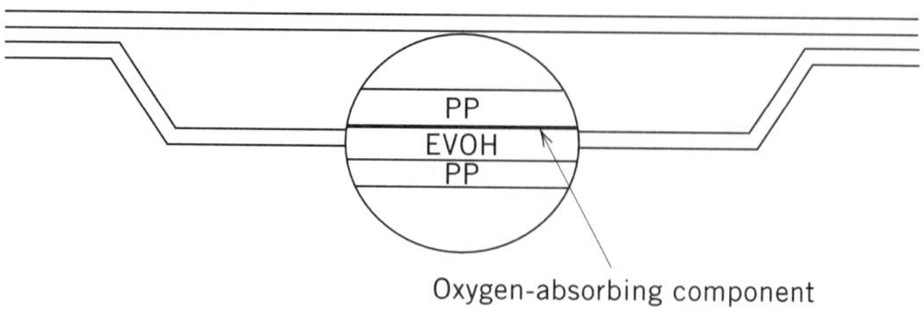

Figure 4. Oxygen-absorbing thermoformed multilayer tray for semiaseptic rice (PP—polypropylene, EVOH—ethylene vinyl alcohol copolymer). Courtesy of Toyo Seikan Kaisha Ltd.

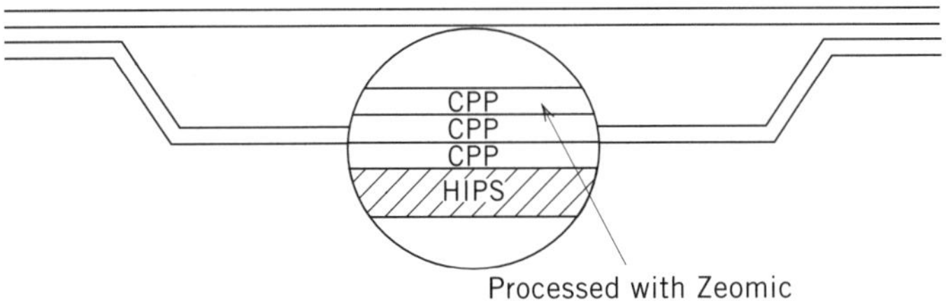

Figure 5. Antimicrobial thermoformed tray containing Zeomic silver zeolite heat-seal layer. (CPP—cast polypropylene, HIPS—high-impact polystyrene). Courtesy of Mitsubishi Gas Chemical Co./Shinanen New Ceramic Co.

zeolite has been trialed and has been found to be highly effective against several vegetative bacteria, especially dispersed in water, saline solution, or oolong tea. The effect of amino acids in food proteins is the subject of research (13). Since the antimicrobial action seems to require the dissolution of silver ion in the food, the regulatory status of such packaging will need to be determined.

Odour absorbent films. Since odors can be sensed at very low levels, there is the opportunity to use packaging materials to reduce the concentrations of these components in otherwise acceptable foods (see Aroma barrier testing). Dupont have released Bynel 1PX101, a masterbatch for coextruded or laminated films, having the ability to bind aldehydes, such as hexanal or heptanal, which are formed on autoxidation of fats and oils (14). The polyethylene layer containing the masterbatch should not be the food-contact layer, and the latter should be permeable to the odor compound if the latter is to be removed readily from the food.

A different approach has been taken in Japan, where amine or sulfur-compounds need to be removed from fish in domestic refrigerators. The Anico Company Ltd. of Japan markets ANICO BAG, which contains ascorbic acid and an iron salt dispersed in the plastic. These bags are marketed for preservation of food "freshness." The components—ferrous sulfate and ascorbic acid or citric acid—have regulatory approval or GRAS (generally regarded as safe) status, but the effectiveness of the necessary reactions in a plastic matrix need demonstration. Numerous other "freshness" preservative plastics have been available in Japan for several years, often on the basis of inclusion of adsorbent powders, but their effectiveness also needs to be demonstrated.

Packages that heat or cool. Microwavable packages containing foods with differing reheating requirements can be made to crisp or brown some components by use of susceptors and reduce the heating of other components by use of foil shields. Susceptors normally consist of a vacuum-deposited layer of aluminum, typically with a light transmission of 50–60%, or a 12-μm-thick film of biaxially oriented PET. The film is laminated to paper or paperboard by means of an adhesive. In a microwave field susceptors have reached a temperature of 316°C in the absence of food, or 223°C in pizza packs (15). These temperatures have caused regulatory authorities to investigate the stability of all components of susceptor films, particularly the adhesives. The microwave field strength can be intensified by specific distributions of foil patches in the domes lids of microwavable packs. Alcan Ltd. have produced packs of this type under the name MicroMatch.

Beverage cans can be made either "self-heating" or "self-cooling" by means of chemical reactions in compartments separated from the beverage (16). Sake is heated by the exothermic reaction of lime with water in aluminum cans. This process is potentially valuable in the vending machine market. Cooling is achieved by the endothermic dissolution of ammonium nitrate and ammonium chloride with water. Both of these thermal effects are brought about by shaking and so are unsuitable for use with carbonated beverages.

Research and Development

Since active packaging is largely a series of innovation of the 1980s and 1990s, there is still a substantial amount of innovation in progress. In oxygen scavenging alone there have been over 600 primary patent applications for both chemical principles and package designs. Very few of these have yet led to commercial products. Conversely, several of the "freshness preservative" films introduced commercially in Japan have not been supported by unambiguous, independent research results demonstrating their effectiveness. A great deal of attention is being given to plastic packaging incorporating in-film chemistry. Methods of activating chemical systems that are stable during thermal processing are particularly interesting. The benefits of using active packaging, such as the suggested involvement of far-infrared radiation from ceramic particles, need to be established clearly.

Summary

The introduction of active packaging requires reappraisal of the normal requirement that the package should not interact with the packaged product. Since plastic packages came into common use, undesired interactions have become common, and so the decision to utilize desired interactions is now easier to make. Active packaging introduced so far represents substantial fine-tuning in the matching of packaging properties to the requirements of the product. Accordingly, it will be seen increasingly in niche markets and in wider applications in which specific problems are inhibiting the marketing of the product. Indeed, the specific examples being introduced are too numerous to describe here, and the reader is referred to the Bibliography.

BIBLIOGRAPHY

1. T. P. Labuza and W. M. Breene, *J. Food Proc. Preservat.* **13,** 1–69 (1989).
2. Y. Abe and Y. Kondoh, "Oxygen Absorbers" in A. L. Brody, ed., *Controlled/Modified Atmosphere/Vacuum Packaging of Foods,* Food and Nutrition Press, Trumbull, CT, 1989, pp. 149–174.
3. P. Waletzko and T. P. Labuza, *J. Food Sci.* **41,** 1338–1344 (1976).
4. B. V. Chandler and R. L. Johnson, *J. Sci. Food Agric.* **30,** 825–832 (1979).
5. P. J. Louis and J. P. de Leiris, *Active Packaging,* International Packaging Club, Paris, 1991, pp. 197–202.
6. Y. Abe, "Active Packaging—a Japanese Perspective" in *Proc. Intl. Conf. Modif. Atmos. Pkg.,* Campden Food and Drink Research Assoc., Chipping Camden, U.K., 1990, Part 1.

7. Anonymous "No-mix-type Mould Inhibitor" in *Freund Technical Information,* Freund Industrial Co. Ltd., Tokyo, 1985, pp. 1–14.
8. J. P. Smith, H. S. Ramaswamy, and B. K. Simpson, *Trends Food Sci Technol.* 111–118 (Nov. 1990).
9. M. L. Rooney, "Overview of Active Packaging" in M. L. Rooney, ed., *Active Food Packaging,* Blackie Academic and Professional, Glasgow (U.K.), 1995, pp. 1–37.
10. F. N. Teumac, "The History of Oxygen Scavenger Bottle Closures" in M. L. Rooney, ed., *Active Food Packaging,* Blackie Academic and Professional, Glasgow, 1995, pp. 193–202.
11. S. L. Cuppett "Edible Coatings as Carriers of Food Additives, Fungicides and Natural Antagonists" in J. M. Krochta, E. A. Baldwin, and M. Nisperos-Carriedo, eds., *Edible Coatings and Films to Improve Food Quality,* Technomic Publishing, Lancaster, PA, 1994, pp. 123–124.
12. J. H. Hotchkiss, "Safety Considerations in Active Packaging" in M. L. Rooney, ed., *Active Food Packaging,* Blackie Academic and Professional, Glasgow, 1995, pp. 238–255.
13. T. Ishitani, "Active Packaging for Foods in Japan" in P. Ackermann, M. Jägerstad & T. Ohlsson, eds., *Food and packaging Materials—Chemical Interactions,* Royal Society of Chemistry, Cambridge (UK), 1995, pp. 177–188..
14. Anonymous, "Bynel IPX101 Interactive Packaging Resin" in *Temporary Product Data Sheet,* Du Pont Polymers, Wilmington, DE, 1994.
15. G. L. Robertson, *Food Packaging: Principles and Practice,* Marcel Dekker, New York, 1991, pp. 419–430.
16. T. Katsura, *Packaging Japan,* 21–26 (Sept. 1989).

General References

T. P. Labuza and W. M. Breene, *J. Food Proc. Preservat.* **13,** 1–69 (1989).

A. L. Brody, *Controlled/Modified Atmosphere/Vacuum Packaging of Foods,* Food & Nutrition Press, Trumbull, CT (1989).

P. J. Louis and J. P. de Leiris, *Active Packaging—Etude Technologique,* International Packaging Club, Paris, 1991.

B. Ooraikul, ed., *Modified Atmosphere Packaging of Food,* Ellis Horwood Publishers, New York, 1993.

M. L. Rooney, ed., *Active Food Packaging,* Blackie Academic and Professional, Glasgow, 1995.

MICHAEL L. ROONEY
CSIRO Division of Food Science and Technology
North Ryde, NSW, Australia

ADDITIVES, PLASTICS

Plastic packaging materials (as well as practically all plastic products) make extensive use of different additives to meet different plastic performance requirements and/or processing capabilities. Basically, additives are physically dispersed in a plastic matrix without significantly affecting the molecular structure of the matrix. They are normally classified according to their specific functions rather than on a chemical basis. It is also convenient to classify them into groups and to subdivide them according to their specific functions. Examples include mechanical properties modifiers—plasticizers, toughening agents, interfacing agents (makes additives and plastics compatible), blowing agents, reinforcements, etc; processing aids—lubricants (internal and/or external), processing stabilizers, melt flow promoters, etc; physical properties modifiers—blowing agents, flame retardants, etc; surface properties modifiers—adhesion promoters, slip agents, antiblock agents, antiwear agents, antistatic agents, etc; optical properties modifiers—nucleating agents, dyes, pigments, etc; antiaging modifiers—fungicides, antioxidants, UV stabilizers, etc; electricial properties modifiers—mineral and/or metallic agents, glass microballoons, etc; reducing formulation and manufacturing costs—particulate fillers, blowing agents, reinforcements, etc; others—biodegradable agents, preset explosive agents, etc (1–9).

Unfortunately, there is no "ideal additive" since each of the infinite number of end uses will call for a particular set of characteristics, including diverging properties. To be realistic, the most important requirement of any additive is that it be effective, for the purpose for which it was designed, at an economical level. One should recognize that, like a seesaw, improvements in one property can lead to deterioration in others. The effectiveness of compounding additives also depends on correct procedure of incorporation into the plastic.

The target is to set up the parameters required in a compound and when required, develop a compromise. Factors such as cost, availability, surface wetting and bonding, oil absorption, chemical resistance, strength, color, shape consistency, density, thermal expansion, thermal properties, and permeability should be considered.

The compatibility and diffusibility of additives in plastic compounds are normally assessed by trial and error with experience when available. Theories on their behaviors may exist so that can be used in the preliminary concepts to meet specific performances. The number of additives available and their possible combinations are enormous, and their compositions are continually changing. The basic theories and knowledge of solution thermodynamics may be used to determine potential compatibility.

Antiblocking agents. These agents are substances that prevent or reduce blocking, such as with film or sheet, when added to the plastic compound or applied to their surface. The blocking action is an adhesion between touching layers of plastic, such as one that may develop under pressure during storage or use. The extent of blocking depends on temperature, pressure, humidity, physical properties of the plastic itself, and processing conditions. If the plastic has a low softening point or if it picks up moisture readily, it will have a greater tendency to block than will a plastic which has a high softening point and does not pick up moisture. The physical properties of the plastic itself on which blocking depends are as follows: (1) smooth surfaces adhere more readily than rough; (2) adhesion will depend on the amorphous or crystalline character of the plastic, with amorphous having a greater tendency to block; (3) if one surface is readily wet by the other, the tendency to block is increased; (4) if the melting point is low, there will be an increased tendency to block; (5) if a surface shows flow under pressure, the tendency to block may be severe; (6) blocking is promoted by the tendency of the film to pick up water vapor; and (7) the film and sheet that develop static electricity readily adhere to each other.

The concentration of the antiblocking agent used varies from 0.1 to 1 wt%. Such chemicals as stearamide, stearoguanamine, copolymers of vinyl acetate (with either maleic anhydride, ethylene, or ethyl acrylate), metal salts of fatty acids (calcium stearate), sodium dioctsulfosuccinate, alkylamines, and alkyl quaternary ammonium compounds are usually incorporated into the plastic before processing. Stearamide is widely used to prevent blocking with polyethylene. Other materials such as colloidal silica, clays, starches, silicones, guanidine stearate, and long-chain alkyl quaternary ammonium compounds may be applied to the surface either during or after processing to reduce blocking.

Antifogging agents. Moisture droplets can obstruct the view through film or sheet. To permit visibility, these droplets can be dispersed by additives such as nonionic ethoxylates or hydrophilic fatty acid esters (glyceryl stearate, etc), with promote the deposition of continuous films of moisture. Antifog action converts dropwise condensation to filmwise condensation through reduced surface tension.

Antimicrobial agents. Growth of microorganisms on the surface of certain packaging materials can occur. As an example, high densities of fungi and bacteria can develop on plasticized PVC shower curtains. These preventive agents include algicides, bactericides, fungicides, copper-8-quinoleate, and *N*-(trichloromethylthiophthalimide.

Antioxidant agents. These agents are of major importance because they extend the plastic's useful temperature range and service life. The variety and the specific uses of antioxidants available are extensive. An *antioxidant* is defined as a substance that opposes oxidation or inhibits reactions promoted by oxygen or peroxide. In the specific case of plastics, they retard atmospheric oxidation or the degradative effects of oxidation when added in small proportions. For this reason, they are also known as *aging retardants.* Oxidative degradation by ozone, however, is controlled by antiozonant agents.

Two major areas of plastic deterioration protection are of concern: during processing and product use. These protecting agents during processing and storage are often called *stabilizers.* In general, agents used to protect and preserve properties of fabricated products are referred to as *antioxidants* in elastomer technology and *stabilizers* in plastic technology. Agents that protect plastics from photochemical or other kinds of deterioration are often also referred to as stabilizers. An example is one that prevents hydrogen chloride elimination from PVC. In many applications stabilizers for the processed plastics also serve as an antioxidant in the finished product.

This undesirable deterioration may be prevented using different additives. Examples are hindered phenols, such as butylated hydroxytoluene (BHT). The hindered phenols act as free-radical scavengers or polymer-chain terminators.

Antiozonant agents. This term, in its broadest sense, denotes any additive that protects elastomeric plastics from ozone deterioration. Most often, the protective effect results from a reaction with ozone, in which case the term used in *chemical antiozonant.* Ozone is generated by electrical discharge, but a more important source is the photolysis of oxygen on sunny days. Plastics can be protected against ozone by the addition of waxes, inert plastics, or chemically inactive antiozonants. Waxes are generally of the paraffin or microcrystalline type. Many of the chemical types contain nitrogen.

Antislip agents. These agents are also called *slip depressants.* It is usually desirable to have a certain degree of blocking between surfaces, for example, where filled plastic bags cannot be stacked because they have a tendency to slide off the piles. Copolymers of ethylene and maleic anhydride, colloidal silica, finely powdered sand, mica, and other minerals effectively reduce slippage when compounded into the plastic or sprayed onto the surface. Colloidal silica solutions are used extensively for spraying surfaces; they are usually diluted with water prior to application. Agents when compounded into plastics usually are of small particle size ($<1\ \mu m$) and the amount used is generally <1 wt%.

Antistatic agents. These additives can be applied to the surface of plastics products or incorporated in the plastic. Its function is to render the surface of the plastic free of or less susceptible to the accumulation of electrostatic charges that attract and hold fine dirt or dust on the surface of the plastic. There are two basic methods to minimize static electricity: (1) metallic devices that come into contact with the plastic and conduct the static to earth and (2) chemical additives that give a degree of protection. The term *static electricity* (SE) is a misnomer, insofar as it seems to indicate a special kind of electricity.

There is only one kind of electricity. SE denotes the group of phenomena associated with the accumulation of electrical charges, in contrast to the phenomena connected with rapid transport of charges, which is the subject of electrodynamics. SE was discovered long before the other aspects of electricity were studied. Its simplest manifestation is the attraction that certain bodies exhibit for each other as well as other bodies after they are brought together and then separated.

Biodegradable–biocide environment. Controlling degradation of plastic packaging materials is rather routine. Biodegradable plastics are available with the ability to break down or decompose rapidly under natural conditions or processes. Plastics containing additives (organic compounds, etc) are available that permit digesting the plastic via microorganisms in the environment.

Certain natural and synthetic plastics are subject to attack by biological agents. At some time during their life there may be a possible place for living organisms, a barrier to be penetrated in the search for sustenance, a substitute subject to enzymatic or chemical byproduct attack, or simply a material for shelter or attachment during the life cycle of certain organisms. This may or may not manifest itself in an attack on the object, depending on the inherent properties, including the susceptibility of the plastic or its component parts to penetration or degradation. The degree of susceptibility, the anticipated end use of the material, cost, and other factors determine whether a chemical or other protective treatment may be applied.

Active chemical compounds used as preservatives control degradation in several ways; those that produce death are termed *biocides.* The more rigorous definition of *biocide* as an agent that kills living organisms precludes the somewhat more common usage of the term, which implies that a biocide is an agent that kills, repels, or inhibits the growth of the biological environment in which the plastic exists. Many chemicals may inhibit the reproduction and growth of organisms without causing death. These agents may be exemplified in the two common classes of control agents known as bacteriostats and fungistats. With higher organisms (some of the mammals), the repellent qualities of a chemical agent may minimize damage adequately. These protective agents render the plastic less susceptible to biological attack and prolong the useful life of the material by preservation of its physical properties and aesthetic appearance.

Plastics such as polyolefins, polyesters, or vinyls are considered to be resistant to biological attack, for the most part particularly in relation to the plastic additives in their compounded mixture. As more general use is made of plastics, they may be subjected to different and exotic biological environments that may affect their resistance to biological attack. Plastics are seldom used alone. Other materials are added in amounts from minor to major proportions. These include the additives mentioned in this section that may change the biological resistance of the compounded plastics.

Blowing and foam agents. For the production of packaging foams or cellular plastics, depending on the basic material and process, different blowing agents (also called *foaming agents*) are used to produce gas and thus generate cells or pockets of gas in the plastics. They can produce rigid to flexible types and may be divided into two broad groups: physical blowing agents (PBAs) and chemical blowing agents (CBAs). PBAs are represented by both compressed gases and volatile liquids. The CBAs most often used are nitrogen and carbon dioxide. These gases are injected into a plastic melt under pressure (higher than the melt processing pressure) and form a cellular structure when the melt is released to atmospheric pressure or low pressure, such as in a mold cavity with a short shot (molding products such as a foamed packaging container). The volatile liquids are usually aliphatic hydrocarbons that may be halogenated and include material such as hexane and methylene chloride. The chlorofluorocarbons previously used have been phased out (reported to cause ozone problems). The liquids act as a gas source by vaporizing during processing. Regardless of their physical form, they rely solely on pressure for controlling gas development.

CBAs, generally solid materials, are of two types: inorganic and organic. Inorganics include sodium bicarbonate, by far the most popular; and carbonates such as zinc or sodium. These materials have low gas yields and do not yield as uniform a cell structure as do organic CBAs. Organics are mainly solid materials designed to evolve gas within a defined temperature range, usually referred to as the *decompostion temperature range.* This is their most important characteristic and allows control over gas development through both pressure and temperature. This increased control results in a finer and more uniform cell structure as well as better surface quality of the foamed plastic. There are over a dozen types available that decompose at temperatures from 100°C (212°F) to at least 371°C (700°F). Many of these CBAs can be made to decompose below their decomposition temperature through the use of activator additives.

Typical activators include compounds such as zinc oxide, various vinyl heat stabilizers and lubricants, acids, bases, and peroxides. The reduction of decomposition temperature achieved is dependent on the particular activator used and its concentration, and the particle size of the blowing agent. (See also Blow Molding.)

Catalyst agents. These "curing agents" change the rate of a chemical reaction without themselves undergoing permanent change in composition or becoming a part of the molecular structure of a product. They markedly speed up the cure of a compound when added in a minor quantity, compared to the primary reactants. There rarely exists a single polymerization process in which certain accelerating regulating and modifying ingredients are not used with great advantage even though the might be present in only very small quantities. In the early years of producing plastics, where there did not yet exist a well-founded understanding of the mechanism of polymerization processes, the action of these ingredients and additives so much resembled the phenomenon of normal catalysis that they were termed *catalyst.*

With the development in the theory of polymerization reactions, it became evident that in most cases the role of these materials during formation of macromolecules does not fall in the domain of the classic definitions of the terms *catalysis* or *catalyst.* However, since the misnomer became well established and such correct expressions as initiator, transfer agent, terminator or telomer, cross-linking agent, accelerator, curing agent, hardener, inhibitor, or promoter are frequently used interchangeably with the general term *catalyst.*

The latest very popular and useful catalysts are the different types of metallocene catalyst. They produce more uniform PE (and other plastics) and there may be property or performance advantages such as producing more efficient lower densities (PE can span at least from <0.91–0.97 g/mL and melt indices of 0–100 g/10 min) improved mechanical strength (tensile, toughness, etc).

Colorant agents. Colorants are generally divided into dyes and pigments. They may be naturally present in, mechanically admixed with, or applied in solution in a material. A valid distinction between dyes and pigments is almost impossible to draw. Some have established it on the basis of solubility or on physical form and method of application. In industry, the terms *black and white dyes* and *pigments* are used. Like all colorants, each has many different shades. Included as colorants are heavy metals (lead, cadmium, mercury, etc) that pose no harm to the consumer when properly processed and used, but present a potential problem in waste disposal. They can constitute a toxic residue following incineration if they are not properly handled and disposed of. Safer alternatives are being used based on environmental requirements.

The principal colorants are carbon black, white titanium dioxide, red iron oxide, yellow cadmium sulfide, molybdate orange, ultramarine blue, blue ferric ammonium, ferrocyanide, chrome green, and blue and green copper phthalocyanines. (See also Colorants, plastic.)

Coupling agents. Certain additives, as well as fillers and reinforcements, do not bond well with plastics and may result in a reduction of product performance. *Coupling agents* are defined primarily as materials that improve the adhesive bond of dissimilar surfaces. This action must involve an increase in true adhesion, but it may also involve better wetting, rheology, and other handling properties. The agent may also modify the interphase region to strengthen the organic and inorganic boundary layers.

Electrically conductive agents. Some plastics are excellent electrical insulators and others are electrically conductive. Plastics are used extensively in electrical applications mainly because they are excellent electrical insulators. The most significant dielectric properties are dielectric strength, dissipation factor, dielectric constant, and resistivity. There are plastics with certain chemical structures that display unusual electrical conductive properties such as low-energy optical transitions, low ionization potentials, and high electron affinities. The result is a class of plastics that can be oxidized or reduced more easily and more reversibly than conventional plastics. Charge-transfer additive agents (dopants) effect this oxidation or reduction and in doing so convert an insulating plastic to a conducting plastic with near metallic conductivity in many cases.

Typically, metallic conductive fillers and/or reinforcements such as carbonaceous powders or aluminum or steel powders are used with plastics to make them conductive. Different constructions have been used since the 1940s such as the use of metallic coated chopped glass fibers. As an example, there are high strength and electrically conductive reinforced plastics using high-aspect-ratio reinforcements (fiber construction). They produce a combination of desired conductivity of metal with the processing ease and economy of plastic.

Flame retardant agents. Plastics compounded with certain chemical additives can reduce or eliminate their tendency to burn. As an example, PE and similar plastics are compounded with antimony trioxide and chlorinated paraffins. However, certain plastics, such as fluorinated types, do not or do not tend to burn. One of the least toxic and most widely used flame retardants is alumina trihydrate (ATH), which acts as a heat sink by producing steam when heated. Rigid PVC is more flame resistant than PE (and other polyolefins) and PS, but the plasticized (flexible) PVC are combustible. Flame retardants must be added to the flexible PVC if flammability is a problem.

Many packaging plastic materials are combustible. Their lack of flame resistance is disregarded in most packaging applications. The flammability of bulky packaging materials, such as rigid foams, should be considered. As an example, foamed PS cups and plates are combustible. An important aspect is that one must balance the risk of fire versus the possible toxicity of flame retardant additives.

Fragrance enhancer agents. These additives are used in plastic compounds to eliminate undesirable odors. Other applications include films such as those used in trash bags that may absorb an undesirable odor from contact. These additives are also used in blow-molded bottles to enhance the odor of the contents.

An important property of many substances is manifested by a physiological sensation due to contact of their molecules with the olfactory nervous system. Odor and flavor are closely related, and both are profoundly affected by submicrogram amounts of volatile compounds. Many compounds have a characteristic order that is an effective means of identification.

Heat stabilizer agents. The term *stabilizer* basically represents materials added to a plastic to impede or retard degradation; see section in this article on stabilizing agents for details on different types of stabilizers. With regard to heat stabilizers, the major types used are barium–cadmium, organotin, and lead compounds. To a lesser amount others used are calcium–zinc and antimony compounds. To meet environmental requirements, action has been taken to use, where required, more environmentally friendly heat stabilizers.

An example in the use of heat stabilizers is with PVC. Films and bottles made from PVC can degrade when heated at moderately high temperatures or subjected to gamma radiation–sterilization or UV irradiation unless proper additives are used. These heat stabilizers retard the decomposition of PVC into hydrogen chloride and dark degraded plastics. Additives in PVC bottles used for cooking oil and other food products require FDA clearance. Octyltin mercaptide, calcium–zinc compounds, and methyltin are examples of additives having FDA clearance. Epoxidized soybean and linseed oil, which have FDA clearance, are used as a secondary additives to supplement the effectiveness of metal-compound stabilizers. Antimony compounds and methyltin are also used in PVC pipes for potable water.

Impact modifier agents. This is a general term for any additive, usually an elastomer or a different type of plastic, incorporated in a plastic compound to improve impact resistance. For example, when 7.5 g of acrylic polymer (ACR) is added to 100 g of PVC, a product with good room-temperature impact resistance results. For film- or bottle-grade PVC with good resistance at 0°C (32°F), the concentration of ACR is raised to 12.5 phr. Excellent impact modifiers for PVC include ABS, chlorinated polyethlene (CPE), polyethylene–*co*-vinyl acetate (EVA), poly(methyl methacrylate–costyrene) (MBS), and poly(methyl methacrylate)–*co*-acrylonitrile–*co*-butadiene–*co*-styrene (MABS). Examples for impact modifiers used in polystyrene include hydroxyl terminated polyethers and styrene–butadiene block copolymers.

Lubricant agents. These compounded additives can be classified into two areas: internal and external. The *internal lubricants* promote plastic melt flow without affecting fusion properties of the compound. *External lubricants* promote metal release facilitating the smooth flow of melt over die surfaces, mold cavities, etc. In mold cavities adhesion is prevented between plastic and metal; these lubricants, used in injection molding (and other processes), is generally called *release agents.* The tendency for plastics, such as PVC and polyolefins, to stick to metal parts during processing is an example in which sticking is reduced. Lubricants used include fatty-acid esters and amides, and paraffin and polyethylene waxes. Metallic stearates, such as zinc stearates, are also used.

There are lubricants, also called *slip promoters* such as silicones, perfluoroalkyl esters, stearate soaps, and clays, that are effective in reducing blocking by imparting high lubricity to plastic films and sheets causing them to slide easily over each other.

Mold release agents. These are also called *lubricating agents,* parting agents, dusting agents, and release agents depending on the type of material and/or process involved. See section on lubricant agents, reviewed above, for details. The external release agents can be lubricating liquids or solids (including dusting powder) substances that are applied to the mold cavity during molding, casting, and other procedures. Examples of products used include silicones, calcium stearates, zinc stearates, sodium dioctylsulfosuccinates, long-chain alkyl quaternary ammonium compounds, talcs (soap-stones), micas, flours, clays, and—very important—waxes. To improve or ensure satisfactory part release, keep mold cavity clean. Use releases that do not interfere with the plastic or part performance. For example, using silicone with electrical connections could cause contamination or corrosion of metal pins; also silicone usually will cause difficulty in bonding or printing on the molded part.

Nucleating agents. Chemical additives when incorporated in certain plastics form nuclei for controlled growth of crystals in the melt. In polypropylene, for example, a higher degree of crystallinity and more uniform crystalline structure is obtained by adding a nucleus such as adipic and benzoic acid, or certain of their metal salts. Colloidal silicas are used in nylon, seeding the material to produce more uniform growth of spherulites. These additives are also used to control cell size in foamed plastic (1).

Plasticizer agents. The primary role of these agents is to reduce the rigidity of certain packaging films and containers, to render them more flexible. PVC, for which plasticizer agents are widely used, can thus be obtained in a wide range of stiffnesses, from rigid and somewhat brittle types to very flexible rubberlike types. Phthalic acid esters, such as dioctyl phthalates (DOPs), are principally used. In addition to providing flexibility, plasticizers also provide workability and distensibility. These additives may reduce the melt viscosity, lower the glass-transition temperature, increase elongation and impact strength, reduce tensile strength and hardness, and lower the elastic modulus of the plastic. In addition to the DOPs, others used include diisononyl phthalate (DINP), diisoheptyl phthalate (DIHP), or di(2-ethylhexyl) terephthalate (DOTP); these are typically the external plasticizer types. Internal plasticizers are also used; they may be brought about by copolymerization of vinyl chloride with monomers, such as vinyl acetate, ethylene, or methyl acrylate.

Low-cost plasticizer, added mainly to reduce the cost of compounds when the level of stiffness is not important, are sometimes called *extenders.* All plasticizers allow the long molecular chains to move more easily relative to each other when a strain is imposed, thus the lower strength; this also allows for easier processing. Plasticizers can age or migrate out of the plastic with time, and the plastic can lose its original toughness; this condition was particularly true common before ~1960. It was rather common to see loss on vinyls in automobile, upholstery and steering wheels. This migration of the plasticizer was usually accompanied with a greasy feeling. Most of these instances have disappeared with improved plastic technology, which allowed internal plasticization through the use of blends and copolymerization. Some thin films of polyethylene or other polyolefins and bottles made from polyethlene terephthalate (PET) are somewhat flexible and do not require plasticizers as additives. A plastic that has adequate flexibility and toughness without the use of plasticizers is always preferred; cost definitely influences choices.

Processing aid agents. During the conversion of plastic materials into desired parts, various melt processing procedures are involved. The inherent viscoelastic properties of each plastic type can lead to certain undesirable processing defects, so additives are used to ease these processing related problems. Heat and light stabilizers, antioxidants, and lubricants are well-defined, well-established additives. Process aids are another important plastic additive product. Processing aids can be made of small molecules, oligoers, or high-molecular-weight plastics.

As an example, after its initial synthesis in Germany, PVC remained an academic curiosity for a long time until the development of plasticizers in the early 1930s. They permitted PVC processing at temperatures below the degradation temperature, and soft, flexible, rubbery products rather than rigid glasslike products. By the 1960s acrylic processing copolymers with very high molecular weights were developed that did not compromise the softening effects on the physical properties of PVC. Today commercial processing aid components are high-molecular-weight copolymers of methyl methacrylate (MMA) and alkyd acrylates, preferably ethyl acrylates (EA) or butyl acrylates (BA) where MMA is the main component. The glass-transition temperature of these plastic compounds is generally greater than that of PVC.

Processing aids, such as ethoxylated fatty acids, are used as viscosity depressants in PVC plastisols, and flow promoters, such as polyisobutylene, are used to improve the processability of polyolefins (PE, PP, etc). Acrylics, which are incompatible with PVC at elevated temperatures, actually improve the processability and clarity of PVC films.

Reinforced plastic low-profile agents. In packing containers special polyester plastic systems for reinforced plastics (RPs) are combinations of thermoset (TS) and thermoplastic (TP) materials, rather than the usual all-TS polyester that in certain applications cannot provide a class A finish (no surface waveness). Although the terms *low-profile* and *low-shrink* are sometimes used interchangeably, there is a difference. Low-shrink plastics contain up to 30 wt% TP, while low-profile plastics contain 30–50 wt%. Low-shrink offers minium surface waviness in the molded part (as low as 25 μm or 1 mil/in. mold shrinkage; low-profile offers no surface waviness (12.7–0 μm or 0.5–0 mil/in. mold shrinkage).

Slip agents. A slip agent is a modifier that acts as an internal lubricant that exudes to the surface of the plastic during or immediately after processing; a nonvisible coating blooms to the surface to provide the necessary lubricity to reduce coefficient of friction and thereby improve slip characteristics (see sections on antiblocking agents and antislip agents).

Stabilizing agents. All plastics (polymers), natural or synthetic, degrade under normal use conditions with progressive loss in aesthetic appearance, mechanical strength, dielectric integrity, etc. Time periods for the different plastics can range from less than an hour to many decades. Degradation, the result of irreversible chemical reactions or physical changes, can ultimately lead to product failure. These agents vary considerably in the rate at which they degrade. For many applications, degradation must be inhibited by stabilizers to assure the required life expectancy.

Stabilization may be required in all stages of a plastic's life cycle. Plastics that are particularly sensitive to oxidative degradation often require protection during processing and fabrication into final products. Plastics are fabricated at elevated temperatures, and oxygen is always present. Stabilizers have been developed to protect the plastic under extreme conditions encountered during processing. To be effective, these short-term stabilizers must be sufficiently mobile within the molten plastics to reach sites of incipient degradation before extensive reaction can occur. Therefore, the short-term stabilizers are low-molecular-weight compounds that are usually lost through evaporation or extraction during long-term exposure under normal use conditions. Long-term stabilizers must be incorporated into the plastic to protect it from degradation during extended exposure. These stabilizers are of higher molecular weight and thus are not as readily lost by evaporation or extraction.

There are two general approaches to the stabilization of plastics: structural modification and the use of additives. In the first case, the chemical structure of the plastic is modified to eliminate or reduce trace impurities or weak molecular links. Bonding of a stabilizer to plastic molecules should also be classified as stabilization through modification of plastic structure, but this approach has some limitations. The second general approach to stabilization is the most commonly used and involves additives blended into a plastic mass. A wide variety of additives has been developed to stabilize plastics against the several modes of degradation. As an example, there are additives to protect against thermal oxidation, UV degradation, burning, attack of ozone, and other conditions.

Ultraviolet stabilizing agents. An important type of stabilizer is the UV stabilizer. Many plastics, such as films, are deteriorated in sunlight. Like human skin, sunscreens are used to prevent this oxidation. Light itself does not directly harm plastics, but the radiations and particularly UVs tend to initiate or catalyze chemical degradation. Result is oxidation, a process often globally referred to as photooxidation. Various plastics are susceptible to photooxidation (PE, PMMA, PP, PS, PVC, etc). Absorbers and/or stabilizers are used to provide long-term outdoor exposure, such as at least half a century for PMMA. Chemists attempting to elucidate the mechanisms of photooxidation, and its reduction by suitable additives often distinguish between UV absorber (UA), which reduces the essential factor of degradation, and light (or photooxidation) stabilizers (LS), which control its progress; although the distinction appears sometimes subtle. There are many classes of UAs and LSs for the different plastics. Amount used ranges from 0.1 to 0.6 wt%. UAs are also called *UV inhibitors.*

UV stabilizers include phenyl salicylate, derivatives of 2-hydrooxybenzophenone, etc. The major use of the stabilizers (about 75 wt%) are used to stabilize polyolefins (PE, PP, etc) against sunlight. Polyarylates, which contain 2-hydroxybenzophenone groups, can be used in PCs. There are additives such as nickel complexes and nickel salts that act as quenchers or energy-transfer agents and convert the excess energy from the sunlight to harmless low heat energy. The efficiency of these quenchers, which are about 400% more effective than 2-hydroxybenzophenones, is independent of product thickness. Pigments, such as titanium dioxide, in combination with zinc oxide are also effective screens. Carbon black has been used extensively to protect plastics against UV induced degradation.

BIBLIOGRAPHY

1. D. V. Rosato, *Rosato's Plastics Encyclopedia and Dictionary,* Hanser, 1993.
2. R. Gachter and H. Muller, *Plastics Additives,* Hanser, 1990.
3. F. Hensen, *Plastics Extrusion Technology,* Hanser, 1988.
4. J. I. Kroschwitz, *Concise Encyclopedia of Polymer Science and Engineering,* Wiley, New York, 1990.
5. D. V. Rosato, *Designing with Plastics and Composites,* C&H, 1991.
6. D. V. Rosato, *Plastics Processing Data Handbook,* C&H, 1990.
7. D. V. Rosato, *Environmental Effects on Polymer Materials,* Vol. 1—*Environment,* Vol. 2—*Materials,* Wiley, New York, 1968.
8. D. V. Rosato, "Role of Additives in Plastics: Function of Processing Aids," *SPE-IMD Newsl.* (Nov. 1987).
9. *Polymer Additives for Injection Molding and Extrusion Applications,* SPE-RETEC, Oct. 18–19, 1995.

DONALD V. ROSATO
Plastics FALLO
Waban, Massachusetts

ADHESIVE APPLICATORS

Adhesive applicating equipment used in packaging applications is available in a vast array of configurations to provide a specific means of sealing containers. The type of adhesive equipment chosen is determined by a number of factors: the class of adhesive (cold waterborne or hot-melt), the adhesive applicating unit and pump style that is most compatible with the adhesive properties, and production line demands. The variables in the packaging operation are matched with the available adhesives and equipment to achieve the desired results.

Packaging Adhesives

Adhesives used in packaging applications today are primarily cold waterborne or hot-melt adhesives (see Adhesives).

Cold waterborne adhesives can be broadly categorized into natural or synthetic. Natural adhesives are derived from protein (animal, casein) and vegetable (starch, flour) sources. Synthetic-based adhesives (primarily resin emulsions) have been gradually replacing natural adhesives in recent years. The liquid "white glue" is generally composed of protective poly(vinyl alcohol) or 2-hydroxyethyl cellulose colloids and compounded with plasticizers, fillers, solvents, or other additives. Also, new copolymers have been developed and used to upgrade performance of cold emulsion adhesives in dispensing characteristics, set time, and stability.

Cold adhesives have good penetration into paper fiber and are energy-efficient, especially when no special speed of set is required. Hot-melt adhesives are thermoplastic polymer-based compounds that are solid at room temperature, liquefy when heated, and return to solid form upon cooling. They are blended from many synthetic materials to provide specific bonding characteristics. Most hot melts consist of a base polymer resin for strength, a viscosity control agent such as paraffin, tackifying resins for greater adhesion, and numerous plasticizers, stabilizers, antioxidants, fillers, dyes, and/or pigments.

Hot melts are 100% solid; they contain no water or solvent carrier. This offers several advantages: rapid bond formation and short set time because heat dissipates faster than water evaporates, shortened compression time, and convenient form for handling and long-term storage. Being a thermoplastic

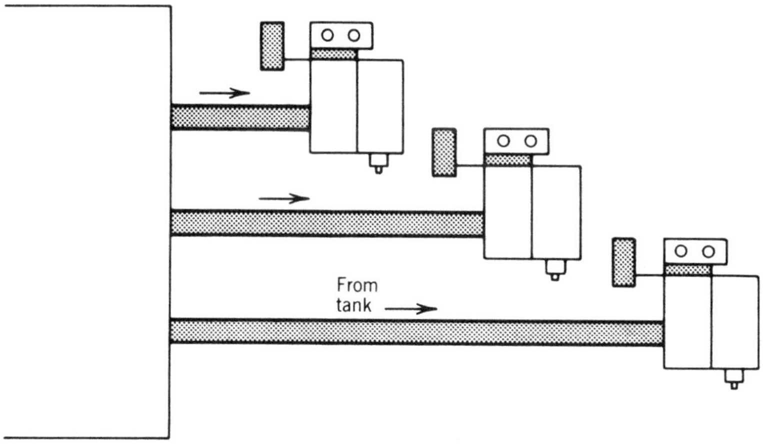

Figure 1. Noncirculating gun installation (parallel) system.

material, hot melts have limited heat resistance and can lose their cohesiveness at elevated temperatures.

Adhesive Applicating Equipment Classification

Both cold waterborne and hot-melt adhesive application systems are generally classified as noncirculating or circulating. Noncirculating systems are the most common (see Fig. 1). They are identified easily because each gun in the system is supplied by its own hose. The noncirculating system is often referred to as a "dead-end system" because the hose dead ends at the gun. An offshoot of the noncirculating system is the internally circulating hot-melt system (see Fig. 2). There is circulation between the pump and manifold, but from the manifold to the gun it is the same as a dead-end system.

Circulating systems are used to some extent in applications requiring a standby period, as in a random case sealing operation, when some setup time is needed. A circulating system is identified by the series installation of the hoses and guns (see Fig. 3). In the typical circulating installation, a number of automatic extrusion guns are connected in series with the hot-melt hose. Molten material is siphoned out of the applicator tank and pumped into the outlet hose to the first gun in the series. The material then flows from the first gun to the second gun and continues on until it passes through a circulation valve and back into the applicator's tank. The circulation valve permits adjustment of the flow of material.

Cold-Glue Systems

Most cold-glue systems consist of applicator heads to apply adhesive either in bead, spray, or droplet patterns; fluid hoses to carry the adhesive to the applicator head from the tank; a pressure tank of lightweight stainless steel that can include a filter, quick-disconnect couplings for air and glue, a pressure relief valve and an air pressure gauge; and a timing device to control the adhesive deposition.

The applicator heads are controlled by either an automatic pneumatic valve or a manually operated hand valve. The bead, ribbon, or spray patterns can be dispensed using multiple-gun configuration systems with resin or dextrin cold adhesives. Cold adhesive droplet guns dispense cold mastic and plastisols, and come in a wide variety of configurations for spacing requirements. In bead and ribbon cold glue extrusion, the tips either make contact or close contact with the substrate. Spray valves emit a mistlike pattern without touching the substrate's surface.

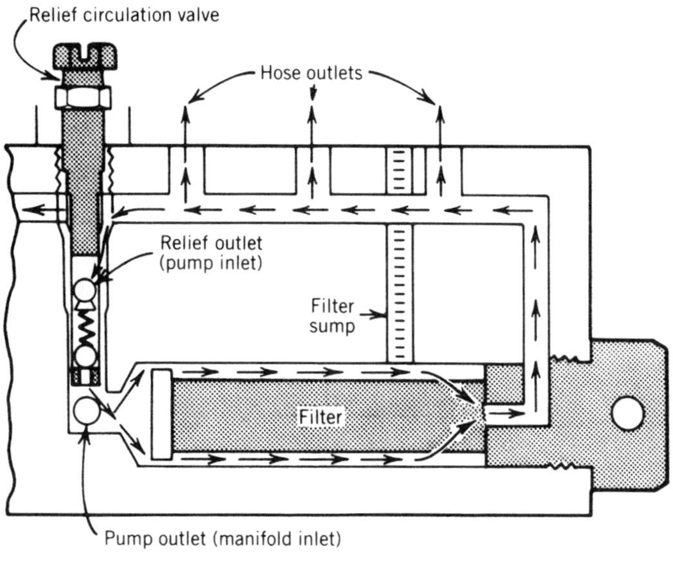

Figure 2. Internally circulating system.

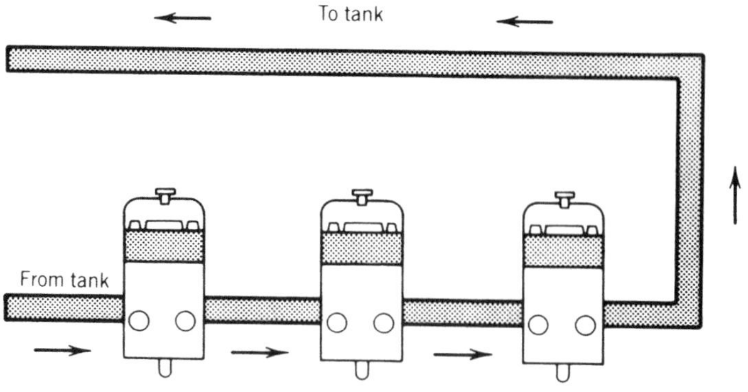

Figure 3. Circulating gun installation (series) system.

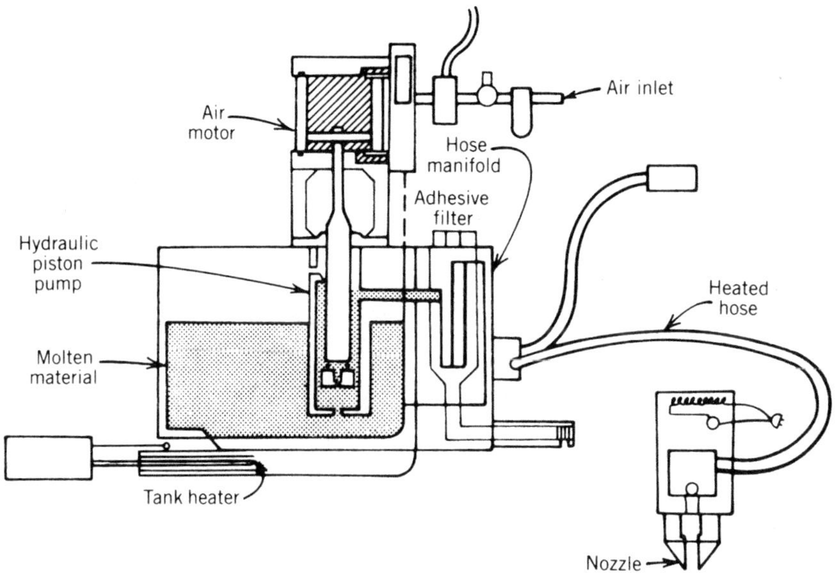

Figure 4. Tank-type hot melt unit.

Hot-Melt Systems

Hot-melt application equipment performs three essential functions: melting the adhesive, pumping the fluid to the point of application, and dispensing the adhesive to the substrate in a desired pattern.

Melting devices. Tank melters are the most commonly used melting unit in packaging applications. Best described as a simple open heated pot with a lid for loading adhesive, a significant feature of tank melters is their ability to accept almost any adhesive form. The tank melter is considered the most versatile device for accepting hot melts with varying physical properties of adhesion and cohesion.

In tank melters, tank size is determined by melt rate. Once melt rate objectives or specifications are determined, tank size is fixed. Larger tanks have greater melt rates than smaller tanks. Holding capacities range from 8 lb (3.9 kg) in the smaller units to more than several hundred pounds (>90 kg) in the larger premelting units (see Fig. 4).

The tanks are made of highly thermal conductive material such as aluminum, and are heated by either a cast-in heating element or a strip or cartridge-type heater. The side walls are usually tapered to provide good heat transfer and to reduce temperature drop.

Adhesive melts first along the wall of the tank as a thin film. Internal circulation currents from the pumping action assist in transferring heat throughout the adhesive held in the tank. Even when the hot melt is entirely liquid, there will be temperature differences within the adhesive. Under operating conditions, adhesive flows along tank surfaces and absorbs heat at a faster rate than it would if allowed to remain in a static condition. Even though adhesive in the center of the tank is cooler, it must flow toward the outer edges and pick up heat as it flows into the pumping mechanism.

Grid melters are designed with dimensional patterns resembling vertical cones, egg crates, honeycomb shapes, and slotted passages arranged in a series of rows. Such an arrangement creates a larger surface area for heat transfer. The grid melter is mounted above a heated reservoir and pump inlet (see Fig. 5).

The grid melting process is exactly the same as the tank melting process; however, the film of adhesive flows along with surfaces and flows through ports in the bottom of the grid. In this way, the solid adhesive will rest above the grid and force the molten liquid through the grid. The grid is designed for deliberate drainage of liquid adhesive to maintain a thin film adjacent to the heated surfaces. This thin film provides for a greater temperature difference than normally found in a tank, and a much larger heat flow is attainable. The grid melter also achieves a much greater melt rate for a given size or area of melter. It is able to heat higher-performance adhesives because it provides relatively uniform temperature within the melter itself, which also minimizes degradation. These features are achieved with some sacrifice of versatility since the adhesives must normally be furnished as pellets or other more restricted geometries.

Between each row of patterned shapes are passageways that open into a reservoir beneath the melter. The passageways ensure a constant and unobstructed flow of molten material to the reservoir below.

The reservoir has a cast-in heating system similar to that of most tank applicators. The temperature control can be separate from the grid melter. The floor of the reservoir is sloped so that molten material is gravity-fed toward the pump inlet.

Grid melters are available with optional hopper configurations. The major difference, other than capacity, is the ability to keep the material "cool" or "warm" before it reaches the grid. The cool hopper merely supplies adhesive and the material becomes molten at the grid. The molten time of the material is shortened before actual application. The cool hopper works well with hot melts of relatively high softening and melting points.

Warm hoppers are insulated and attached to the grid so that heat is radiated from the hopper through the hopper casting. Materials that are better formulated to melt in a zone-heating process, such as hot melts with medium to low melting points, and pressure-sensitive materials work well with the warm hopper design.

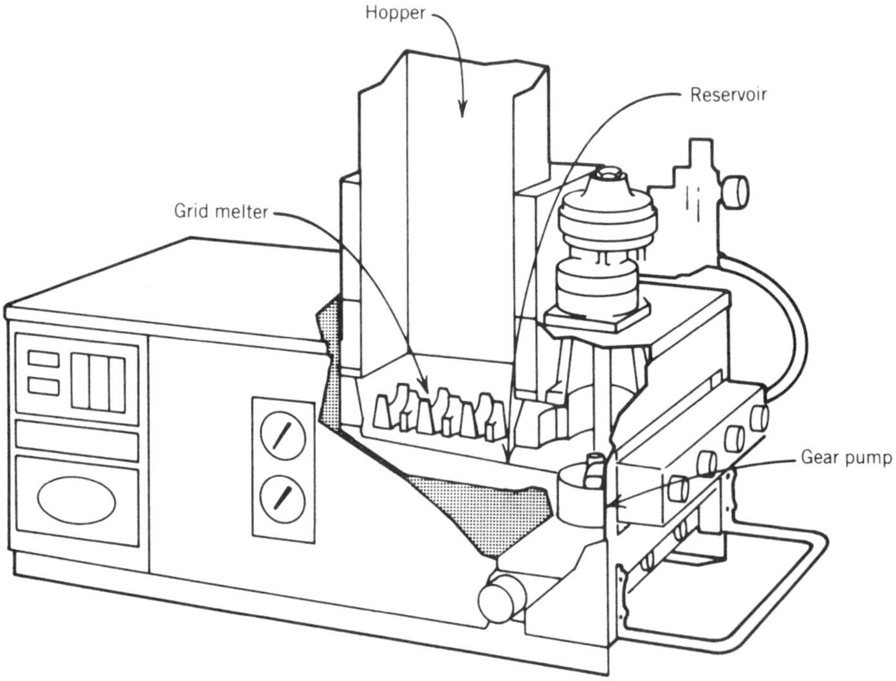

Figure 5. Grid melter hot melt unit.

The tank capacities of both tank and grid melting devices can be extended with premelt tanks. They may be equipped with their own pumping devices or act on a demand signal from a level sensor in the applicator tank; however, all perform like tank units to keep hot melt materials molten at controlled temperatures. One of the newer premelting devices is in the bulk melter unit (see Fig. 6).

Bulk melting systems are designed to dispense hot-melt adhesives and other highly viscous thermoplastic materials in applications requiring a high volume or rapid delivery of material. Units can be used as direct applicators or as premelters as part of a central feed system.

The material is pumped directly from the drum or pail in which it is shipped. This provides ease of handling and lower material costs of bulk containers. In premelting applications, the bulk melter system preheats the material before it is pumped into heated reservoirs. The material is then pumped from the reservoirs to the application head on demand.

The electrically heated platen is supported by vertical pneumatic or hydraulic elevating posts. The platen melts the hot melt material on demand directly from the container and forces it into the pump inlet.

The platen can be a solid one-piece casting, or in the larger units, several grid or fin sections. It is important that the platen size match the ID (inside diameter) of the drum or pail. The platen is protected by one or more seals to help prevent leakage.

Pressurized melters, or screw extruders, are among the earliest designs used in hot-melt applicators. Imitating injection-molding machinery (see Injection molding), early screw-extruder and ram-extrusion handgun systems had limited success because they were designed only for continuous extrusion. The closed-system and screw-extrusion design allows for melting and pumping of high viscosity, highly degradable materials.

Extruder equipment is now adapted to intermittent applications, it consists of a hopper feeder, a high-torque dc-drive system, a heated barrel enclosing a continuous flight screw, and a manifold area (see Extrusion; Extrusion coating). Heating and drive control systems can be independently controlled

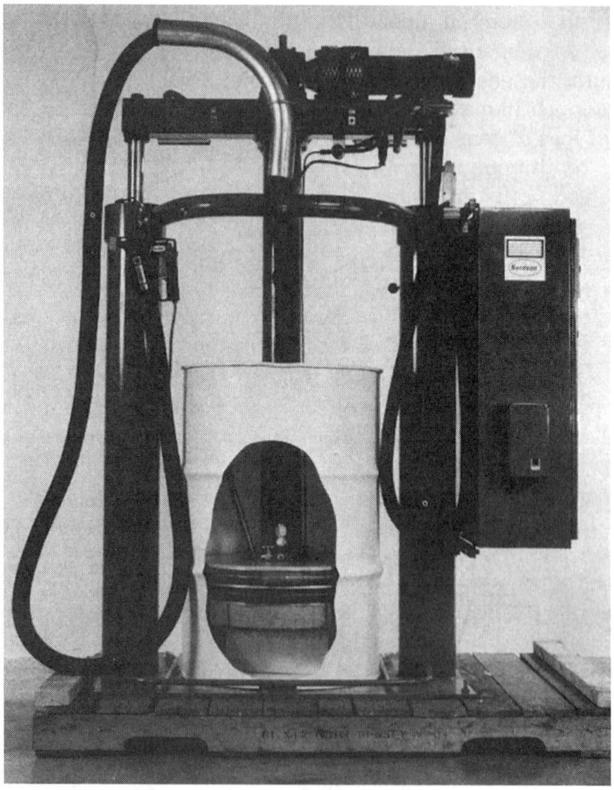

Figure 6. Cutaway of drum showing bulk melter system.

by a microprocessor. Temperatures and pressure are monitored by digital readouts. Adapted for high-temperature, high-viscosity, or degradation-sensitive adhesives, the new technological advances in extruder equipment give greater potential for adhesive applications such as drum-lid gasketing, automotive-interior parts, and self-adhering elastic to diapers (see Fig. 7).

Pumping devices and transfer methods. Once the hot-melt material is molten, it must be transferred from the tank or reservoir to the dispensing unit. Pumping mechanisms are of either piston or gear design.

Piston pumps are air-driven to deliver a uniform pressure throughout the downstroke of the plunger. Double-acting piston pumps maintain a more consistent hydraulic pressure with their ability to siphon and feed simultaneously. Piston pumps do not provide complete pulsation-free output, but they are well suited for fixed-line speed applications.

Gear pumps are available in several configurations: spur gear, gerotor, and two-stage gear pump.

Spur gear pumps have two counterrotating shafts that provide a constant suction and feeding by the meshing action of the gear teeth (see Fig. 8). They are becoming more common because of their versatility in handling a variety of high viscosity materials and their efficient performance in high speed packaging.

Gerotor gear pumps have a different arrangement of gears and larger cavities for the transfer of materials. The meshing action, which occurs on rotation, creates a series of expanding and contracting chambers (see Fig. 9). This makes the gerotor pump an excellent pumping device for high-viscosity hot-melt materials and sealants.

The latest patented two-stage gear pump introduces an inert gas in a metered amount into the hot melt. When the adhesive is dispensed, exposing the fluid to atmospheric pressure, the gas comes out in solution, which foams the adhesive much like a carbonated beverage (see Fig. 10).

All types of gear pumps provide constant pressure because of the continuous rotating elements. They can be driven by air motor, constant speed electric motors, variable-speed drives, or by a direct power takeoff from the parent machine. PTO and SCR drives allow the pump to be keyed to the speed of the parent machine. As the line speed varies, the amount of adhesive extruded onto each segment of substrate remains constant.

Variations and modifications to the pumping devices incorporate improvements to their transfer efficiency and performance. Multiple-pump arrangements are also offered in hot-melt systems to meet specific application requirements.

The transfer action of the pumping device moves the hot-melt material into the manifold area. There, the adhesive is filtered and distributed to the hose or hoses. In the manifold there is a factory-set, unadjustable relief valve, which protects the system from overpressurization. The adhesive then passes through a filter to remove contaminants and is directed through the circulation valve. The circulation valve controls the hydraulic pressure in the system. The material circulates to the hose outlets and out the hose to the dispensing devices.

For hot-melt systems requiring a fluid link between the melting–pump station and the point of application, hot-melt hoses provide a pipeline for transferring the adhesive. Some methods of dispensing (eg, wheel-type applicators) do not use hoses and will be discussed later.

To be able to withstand operating hydraulic pressures up to 1600 psi (lb/in.^{2}) (11.3 MPa), hot-melt hoses are constructed of air-craft-quality materials. They are flexible, electrically heated, and insulated, and come in various lengths to accommodate particular installation requirements.

Hot-melt hoses are generally constructed of a Teflon innertube that is surrounded by a stainless steel wire braid for pressure resistance (see Fig. 11). Noncirculating hoses maintain temperature with a heating element spirally wrapped around the wire braid throughout the length of the hose. Circulating hoses sometimes use only the wire braid to maintain heat. Wrapped layers of materials such as polyester felt, fiberglass, and vinyl tape provide insulation. For abrasion resistance, the entire hose is covered with a nylon braid.

Hose temperature can be independently controlled and is monitored inside the hose by a sensing bulb, thermistor, or resistance temperature detector.

Dispensing devices. There are several methods of depositing adhesive onto a substrate once the material is in a molten state. The applicating devices can be categorized as follows:

1. Extrusion guns or heads (automatic and manual).
2. Web-extrusion guns.

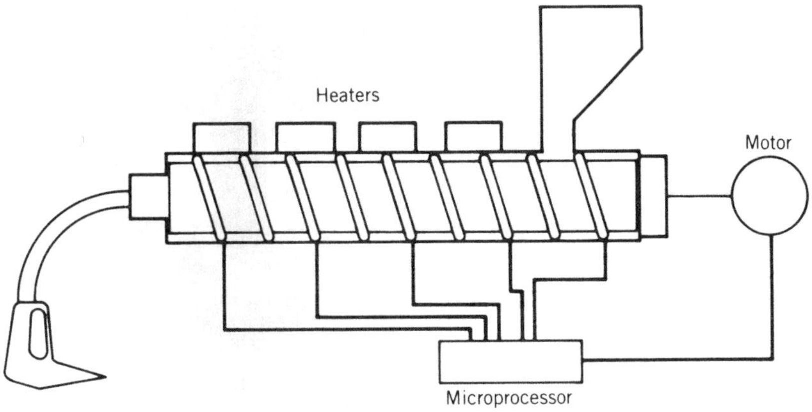

Figure 7. Screw extruder.

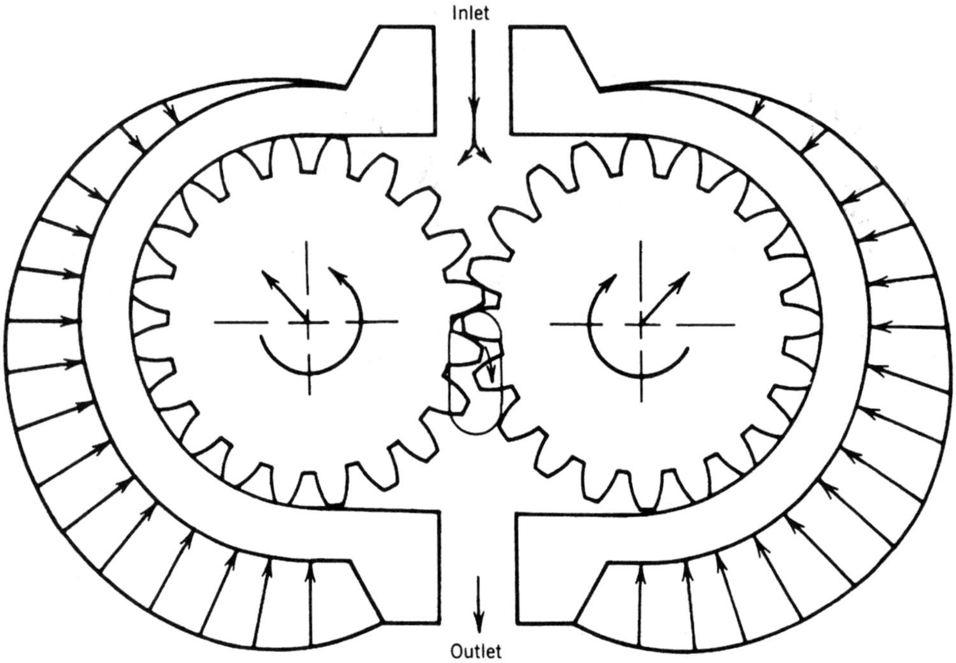

Figure 8. Spur gear pump.

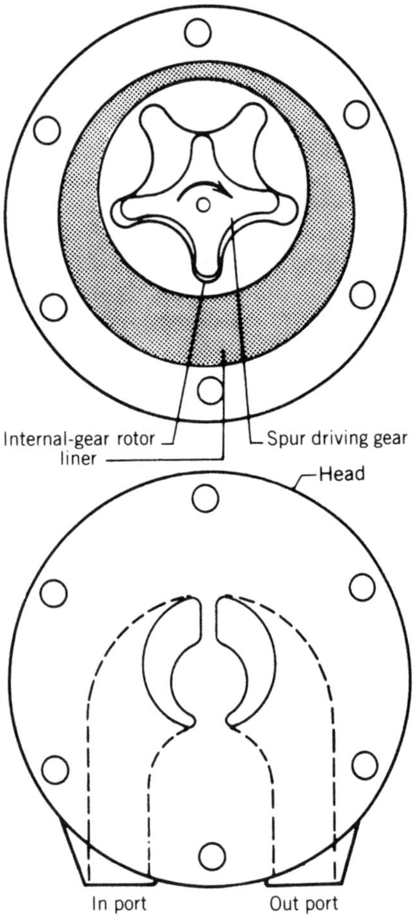

Figure 9. Gerotor gear pump.

3. Wheel and roll dispensers.

Extrusion guns are used on most packaging lines. This extrusion method entails applying beads of hot melt from gun and nozzle. The gun is usually fed from the melting–pumping unit through the hose or directly from the unit itself.

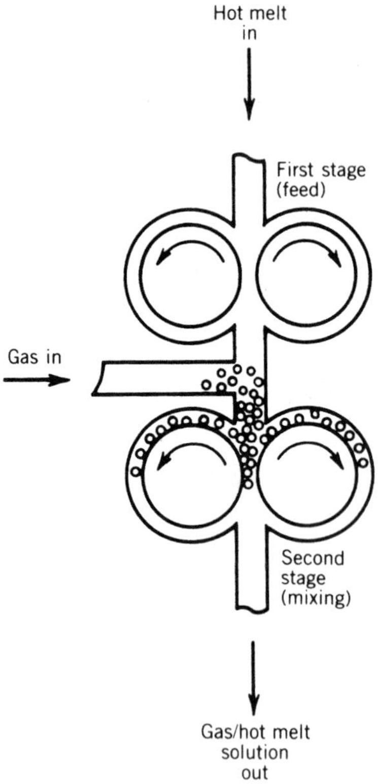

Figure 10. Foaming process of two-stage gear pump.

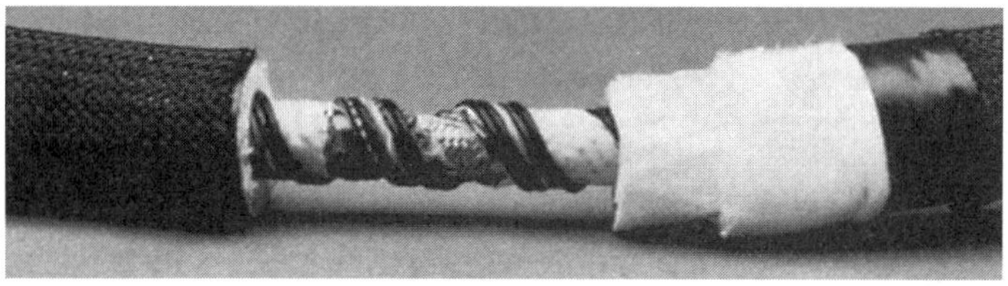

Figure 11. Cutaway of hot-melt hose.

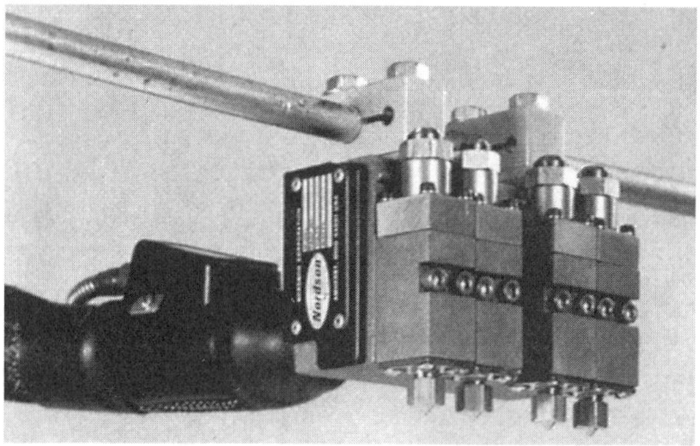

Figure 13. Automatic gun with four modules.

Most high-speed applications use automatic guns that are triggered by timing devices or controllers on line with the parent machinery to place the adhesive on a moving substrate.

Automatic guns are actuated by pressure that forces a piston or plunger upward, lifting the attached ball or needle off the matched seat. Molten adhesive can then pass through the nozzle as long as the ball or needle is lifted off its seat by the applied pressure. The entire assembly can be enclosed in a cartridge insert or extrusion module (see Fig. 12). Either style, when fitted into or on the gun body, allows for multiple extrusion points from one gun head. Modular automatic guns with up to 48 extrusion modules are possible (see Fig. 13).

The pressure to actuate automatic guns is either electropneumatic by means of a solenoid or electromagnetic with a solenoid coil electrically signaled.

Gun temperatures can be thermostatically controlled with cartridge heaters to a maximum of 450°F (230°C). A maximum operating speed of 3500 cycles per minute (58.3 Hz) is possible.

Handgun extrusion is based on the same principles, but with manual rather than automatic triggering. A mechanical linkage operated by the gun trigger pulls the packing cartridge ball from its seated position to allow the adhesive to flow through the nozzle.

The extrusion nozzle used on the head or gun is the final control of the adhesive deposited and is used to regulate the bead size. It is designed for varying flow rates that are determined by the nozzle's orifice diameter and length. Classified as low-pressure–large-orifice or high-pressure–small-orifice, they provide different types of beads. Low-pressure–large-orifice nozzles are specified for continuous bead applications with the large orifice helping to limit nozzle clogging.

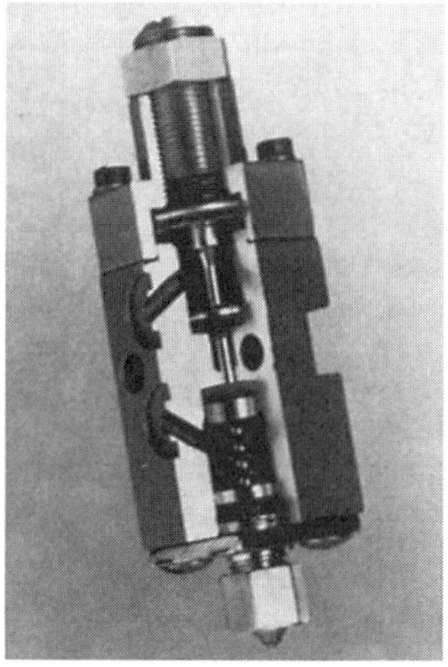

Figure 12. Cutaway of extrusion module.

High-pressure–small-orifice nozzles are better adapted to applications requiring clean cutoff and rapid gun cycling. Drooling and spitting must be controlled for applications such as stitching.

Patterns can be varied further by selection of multiple-orifice designs, right-angle nozzles for differences in positioning, and spray nozzles for a coated coverage.

Heated in-line filters can be installed between the hose and gun to provide final filtering before adhesive deposition. Independent temperature control helps keep the temperature constant. However, further options are available to provide optimum control by combining a heat exchanger and filter integrally with the gun (see Fig. 14). These specialized guns can precisely elevate the temperature of the adhesive material and can hold adhesive temperature within ±2°F (±1.1°C) of the set point. This allows the rest of the system to be run at lower temperatures, thereby minimizing degradation of the material. The filter assembly incorporated into the service block catches contaminants not trapped by the hot melt system filter.

Efforts to prevent nozzle clogging and drool have resulted in a zero-cavity gun that replaces the traditional ball-and-set seat assembly with a tapered needle and precision-matched nozzle seat. In traditional guns, the ball-and-seat assembly interrupts adhesive flow some distance from the nozzle, allowing the adhesive left in the nozzle to drool from the tip and char to lodge in the nozzle orifice. With the zero-cavity gun, no separate nozzle is needed. A microadjust feature adjusts the needle for precise flow control. When the needle closes into the nozzle seat, any char is dislodged. In addition, with no nozzle cavity area the cutoff is clean and precise.

Web-extrusion guns have adapted extrusion dispensing technology to deposit a film of hot melt on a moving substrate. Better known as *slot nozzles* or *coating heads,* they are well suited to continuous or intermittent applications. Mounted on an extrusion gun, a heated or nonheated slot nozzle extrudes an adhesive film of varying widths, patterns, and thickness. Pattern blades can be cut to desired patterns. Film thickness is adjustable by using different thickness blades, by stacking of blades in the slot nozzle, or by varying the adhesive supply pressure (see Fig. 15).

Web extrusion is well suited to coating applications such

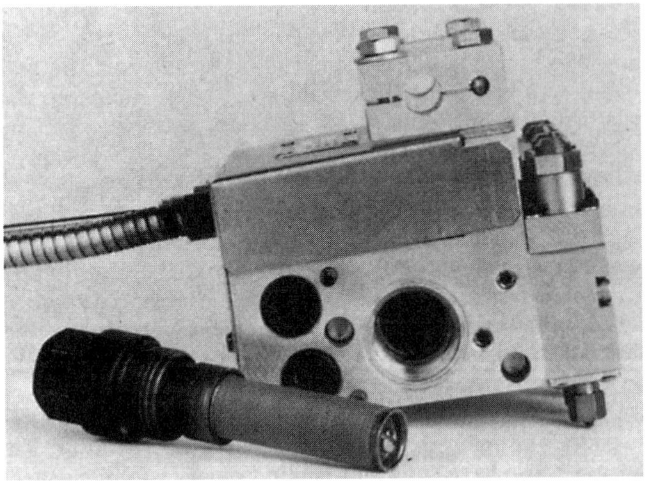

Figure 14. Heat exchanger gun.

as labeling, tape/label, envelopes, business forms, and web lamination as in nonwovens.

For temperature-sensitive adhesives in continuous web extrusion, the slot nozzle is used with a heat-exchanger device to minimize temperature exposure of the material.

Wheel and roll dispensers are the predecessors of present coating extrusion. Wheels or rolls are mounted in a reservoir of molten adhesive. The wheels or rolls are finely machined and may be etched, drilled, or engraved with desired patterns for specific pattern transfer. As the wheel rotates in the reservoir, it picks up the hot melt and transfers it to the moving substrate by direct contact (see Fig. 16). The reservoir may be the primary melting unit or fed by an outside melting device.

Roll coaters involve a series of unwinding and rewinding units for paper coating, converting, and laminating, plus wide-web applications for tape and label applications.

Timing and controlling devices. Automatic applications require installation of one or more devices to control the placement of adhesives on the moving substrate. Such devices normally include a sensor or trigger to detect the presence of the substrate in the gluing station, and a timer or pattern control to measure the predetermined intervals between

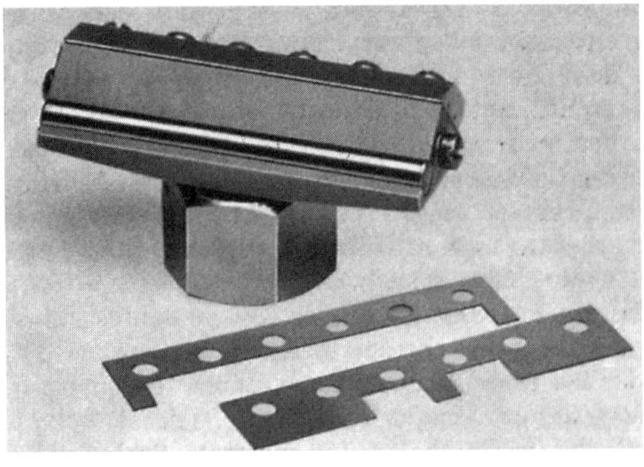

Figure 15. Two-inch slot nozzle with pattern blades.

beads of adhesive or the substrate and to activate extrusion guns at the proper moments (see Fig. 17).

The sensor may be operated mechanically (as with a limit switch activated by the substrate or a cam on the packaging machine) or optically (as with photo eyes or proximity switches).

Timers or pattern controls may be used at constant line speeds to time delay extrusion intervals or produce stitched beads. When line speed varies, pattern controls equipped with line speed encoder or tachometer must be used to compensate for changes in line speed so that bead lengths remain the same. Such devices are highly reliable but are more complex and correspondingly more expensive than duration controls. They can usually control bead placement accurately at line speeds up to 1000 ft/min (5.1 m/s).

Some pattern controls can also be equipped with devices that electronically vary air pressure to the hot-melt applicator to control adhesive output, thus maintaining constant bead volume as well as placement at varying line speeds. Other accessories can be obtained to count the number of packages that have been glued, check for missing beads, and allow the same device to control guns or different lines. Advanced controls are often modular in construction, user oriented, and include self-diagnostic features.

System selection. Choosing the correct system to produce the results desired on the packaging line is not difficult once the variables are identified. Some of the primary variables to consider in specifying equipment for a hot-melt system include rate of consumption, rate of deposition, adhesive registration, and control. Trained factory representatives for adhesive applicating equipment can identify and recommend the best system to fit those variables.

The melting device selected must be capable of handling the pounds-per-hour demand of the packaging operation. The unit must also have sufficient holding capacity to prevent the need for frequent refilling of the adhesive tank or hopper.

Adhesive consumption is affected by line speed, bead size, and pattern. The adhesive consumption rate and maximum instantaneous delivery rate of the pump must be matched to the application requirements (see below).

The pattern to be deposited will determine the dispensing device. Also, the pattern size and registration of the adhesive deposit must be matched with the cycling capabilities of that device.

The system must fit neatly into the entire operation with spacing considerations for mounting unit, gun, and hoses; location to point of application; and accessibility for maintenance.

Calculating Maximum Instantaneous Delivery Rate

The acronym MIDR (maximum instantaneous delivery rate) is used interchangeably with IPDR (instantaneous pump delivery rate). MIDR is the amount of adhesive that a pump would need to supply if its associated guns were fired continuously for a specified period of time. Useful units for measuring MIDR are pounds per hour (lb/h) and grams per minute (g/min).

To calculate the MIDR for a specific application, follow the steps in the example below. Refer to Figure 18 for a visual

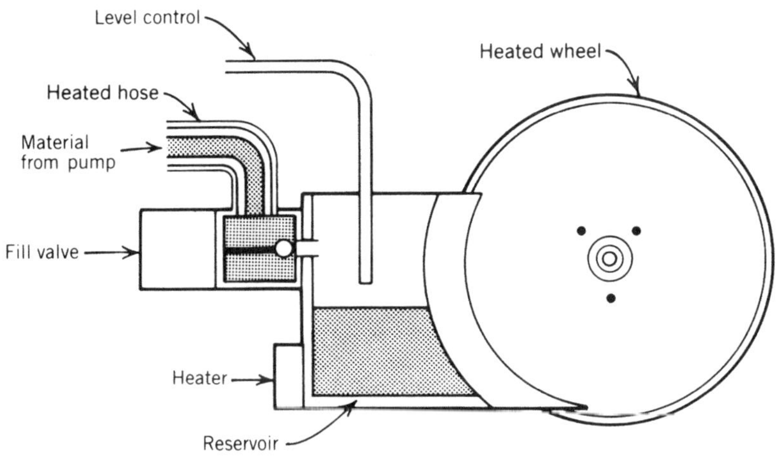

Figure 16. Wheel-type applicator.

description of the physical parameters used in the calculation.

Given:

1. Four beads per flap, top and bottom sealing, for a total of 16 beads per case.
2. Case length of 16 in. (41 cm).
3. Production rate of 20 cases per min at 100% machine efficiency.
4. Bead length of 4 in. (10 cm), 1 in. (2.5 cm) from case end, with a 6-in. (15-cm) gap between beads.
5. Eight-in. (20-cm) gaps between cases.
6. Adhesive "mileage" of 700 lineal ft · lbf (949 J).

"Mileage" is a function of bead size and the specific gravity of the adhesive. The 700 lineal ft · lbf used above is based on a 3/32-in. (2.4 mm) half-round bead (standard-size packaging bead) and melt density of 0.82 g/cm^3 (melt density of standard packaging adhesive). Adhesives of this type yield approximately 30 $in.^3/lb$ (1 cm^3/g).

Calculation:

1. Determine the total bead length in in./h (cm/h) and convert to ft/h (m/h).

$$(20\ \text{cases/min}) \times (4\ \text{in./bead}) \times (16\ \text{beads/case}) \times (60\ \text{min/h}) = 76{,}800\ \text{in./h}\ (195{,}000\ \text{cm/h})$$
$$(76{,}800\ \text{in./h}) \times (1\ \text{ft}/12\ \text{in.}) = 6400\ \text{ft/h}\ (1950\ \text{m/h})$$

2. Determine the adhesive consumption rate at 100% machine efficiency using total bead length per hour and adhesive mileage:

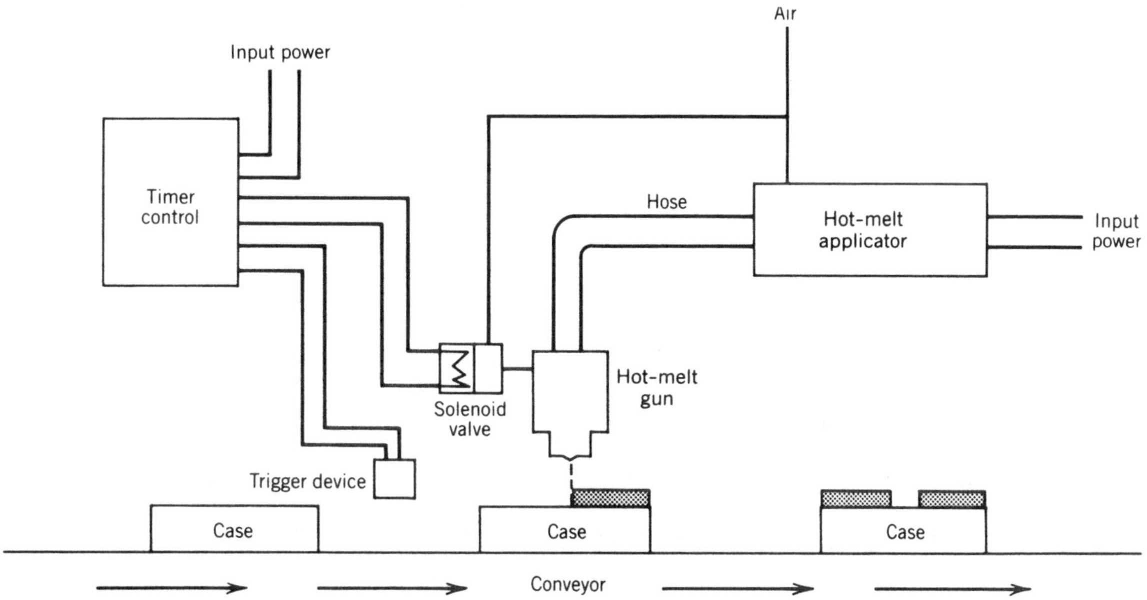

Figure 17. Timer application process. The timer control sequence is activated when the trigger device senses the leading edge of the case. The timing sequence controls preset adjustable delay and duration gun actuations.

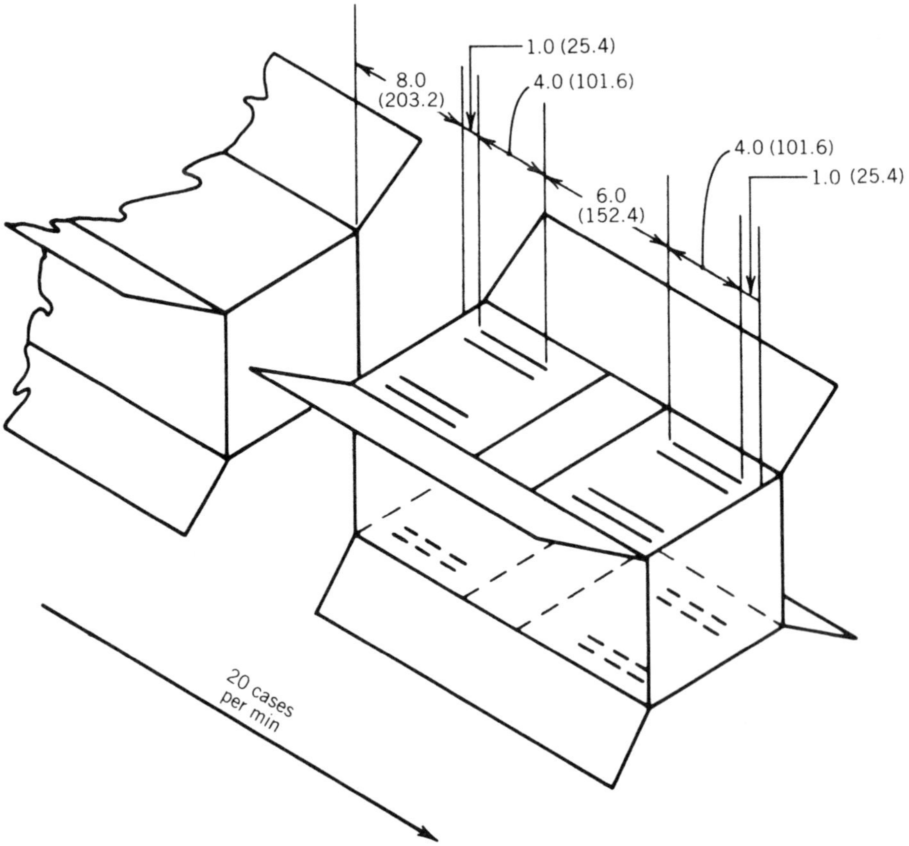

Figure 18. Calculating the MIDR for a specific application. Dimensions are given in inches (millimeters in parentheses).

$$(6400 \text{ ft/h}) \times (1 \text{ lb}/700 \text{ ft})\ (1 \text{ g}/1.54 \text{ m}) = 9.14 \text{ lb/h}\ (4.15 \text{ kg/h})$$

3. Determine duty cycle using length of bead during machine cycle and machine cycle (length between case leading edges).

$$\frac{\text{8-in. bead length}}{\text{24-in. cycle length}} = 0.333$$

4. Determine the MIDR using the inverse of the duty cycle and the adhesive consumption rate from step 2.

$$\frac{1}{0.333} \times (9.14 \text{ lb/h}) = 27.42 \text{ lb/h}\ (207 \text{ g/min})\ \text{MIDR}$$

This example demonstrates the difference between the delivery rate and the adhesive consumption rate. When the guns are firing, the pump is delivering adhesive at a rate of 27.42 lb/h. This is the MIDR. Speed reducers are rated and selected according to the MIDR they are capable of providing. This figure should not be confused with the consumption rate, since the consumption rate is an *average* measure of consumption that includes the time when adhesive demand is zero.

Several equations may prove helpful in some applications:

Bead length, line speed, and duration:

$$\frac{5 \times \text{bead length (in.) (cm)}}{\text{line speed (ft/min) (cm/s)}} = \text{duration (s)}$$

$$\frac{\text{line speed (ft/min) (cm/s)} \times \text{durations (s)}}{5} = \text{bead length (in.) (cm)}$$

$$\frac{\text{bead length (in.) (cm)} \times 5}{\text{duration (s)}} = \text{line speed (ft/min) (cm/s)}$$

BIBLIOGRAPHY

"Adhesive Applicators" in *The Wiley Encyclopedia of Packaging Technology,* 1st ed. Wiley, New York, by D. H. Shumaker and C. H. Scholl P&A Division, Nordson Corporation, 1986, pp. 4–14.

General References

G. L. Schneberger, *Adhesives in Manufacturing,* Marcel Dekker, New York, 1983.

"Definitions of Terms Relating to Adhesives and Sealants." *Adhes. Age* (May 31, 1983).

I. Kaye, "Adhesives, Cold, Water-Borne;" T. Quinn, "Adhesives, Hot Melt;" and C. Scholl, "Adhesives Applicating, Hot Melt" in *The Packaging Encyclopedia 1984,* Vol. 29, No. 4, Cahners Publishing, Boston.

1981 Hot Melt Adhesives and Coatings, Technical Association of the Pulp and Paper Industry, Norcross, GA. (Short course notes presented at the TAPPI Conference, San Diego, July 1–3, 1981.)

"Calculating Maximum Instantaneous Delivery Rate," *Components*

Catalog, P&A Division, Nordson Corporation, Norcross, GA, 1984, p. A2.

ADHESIVES

The total world market for adhesives and sealants in 1989 was 13 billion U.S. dollars, with a volume of 6000 metric tons. Of this, about 35% went into paper and packaging application, the largest single market segment (1). Although a large part of the tonnage was cornstarch used in the manufacture of corrugated board, a significant portion was used in the forming, sealing, or labeling of almost every package in the marketplace. The principal uses include the forming and sealing of corrugated cases and folding cartons; the forming and sealing of bags; the winding of tubes for cores, composite cans, and fiber drums; the labeling of bottles, jars, drums, and cases; the lamination of paper to paper, paperboard, and foil; and the lamination of plastic films for flexible packaging.

There are two significant trends in recent years in packaging adhesives. First, there is a strong movement to replace solvents and solvent-borne adhesives in both the United States and Europe, to minimize volatile organic emissions. This has led to a growth in waterborne and hot-melt adhesives at the expense of solvented types. Second, and more recent, has been a strong interest in recyclable adhesives based on requirements for recycled paper content in both America and Europe. The requirements for recyclability are not completely clear at this time, but it appears likely that standard waterborne or hot-melt adhesives, possibly modified to improve dispersibility and/or filterability in papermaking operations, will be able to meet these requirements.

There are many types of packaging adhesives used, frequently for the same end-use applications, with the choice dictated by cost, productivity factors, the particular substrates involved, special end-use requirements, and environmental considerations (see also adhesives, extrudable). To help clarify this complex picture involving many different chemical types, it is useful to classify packaging adhesives into three physical forms: waterborne, hot-melt, and solvent-borne systems.

Waterborne Systems

This is the oldest, and still, by far, largest-volume class of adhesive used in packaging. These adhesives share the general advantages of ease and safety of handling, energy efficiency, low cost, and high strength. Waterborne adhesives can be further divided into two categories, natural and synthetic.

Natural waterborne adhesives. The earliest packaging adhesives were based on naturally derived materials—indeed, almost all were until the 1940s—and they still constitute a large segment of the market. However, the last four decades have seen their gradual replacement in many applications.

Starch. The largest class of natural adhesives is based on starch, and in the United States, this means cornstarch. Some potato starch is used in Europe primarily because of economics since they are heavily subsidized. Adhesives are produced from raw flour or starch, but more frequently the starch molecule is broken down into smaller chain segments by acid hydrolysis. Depending on the conditions of that reaction, the resulting material can be a fluidity or thin-boiling starch or a dextrin. These can then be further compounded with alkaline tackifiers such as borate salts, sodium silicate, or sodium hydroxide, with added plasticizers or fillers.

The single largest use of starch adhesives is in the manufacture of corrugated board for shipping cases. The standard process involves suspending ungelatinized cornstarch in a thin carrier-starch cook as a vehicle. When the bond line is subjected to heat and pressure, the cornstarch gelatinizes almost instantly, forming a bond between the flutes and the linerboard at very high rates of production. Additives are usually used to improve adhesion, lower gel temperature, increase water resistance, and further increase speed of bond formation.

Other important uses of modified starches and dextrins are in the sealing of cases and cartons, winding of spiral or convolute tubes, seaming and forming of bags, and adhering the seams on can labels. Glass bottles are frequently labeled with a special class of alkaline-treated starch adhesive called a "jelly gum." These have the special tacky, cohesive consistency required on some moderate-speed bottle labeling equipment. Specially modified starches based on genetically bred high-amylose strains are also used as primary ingredients in the remoistening adhesive on gummed tape used for box sealing.

There are many strong points to recommend starch-based adhesives. They are regarded as very easy to handle, clean-machining, easy to clean up, and, above all, inexpensive. Starch has excellent adhesion to paper, and, being nonthermoplastic, it has outstanding heat resistance. Starch also has a green, environmentally friendly image, being "natural" and based on a renewable resource. The negatives are the relatively slow rate of bond formation, the limited adhesion to coatings and plastics, and poor water resistance.

There has been recent interest and activity in developing starch-based hot melts, to take advantage of the perceived environmental benefits, but to date this is semicommercial at best and remains to be proven.

Protein. Another class of natural adhesives is based on animal protein. Once very widely used, these materials now have specific narrow areas of use in packaging where synthetics have not been able to match their performance.

Animal glue. This is one of the earliest types of adhesives. It is derived from collagen extracted from animal skin and bone by alkaline hydrolysis. When used as a heated colloidal suspension in water, animal glues have an unusual level of hot tack and long, gummy tack range. However, because of fluctuating availability and cost, and the development of improved synthetics, there are only two significant uses of animal glue in packaging: (1) as a preferred ingredient in the remoistening adhesive on reinforced gummed tape use for box sealing and (2) as the standard adhesive used in forming rigid setup boxes.

Casein. This is produced by the acidification of skimmed cow's milk. The precipitated curds thus produced form the basis of casein adhesives. Casein has not been manufactured in the United States because the U.S. milk price support program applies only to food products. The main sources for casein are Australia, New Zealand, and Poland, but it is also produced in Argentina and the Scandinavian countries.

There are two packaging applications where casein is used in large volume. One is in adhesives for labelling glass bottles, particularly on newer high-speed labelers where they outperform starch-based adhesives. They are especially favored for beer bottles where casein provides the resistance to cold-water immersion required by brewers, together with removability in alkaline wash when the bottles are returned. The second use is as an ingredient in adhesives used to laminate aluminum foil to paper. Combined with synthetic elastomers such as polychloroprene or styrene-butadiene lattices, casein provides a unique balance of adhesion and heat resistance (2).

Natural rubber latex. This is extracted from the rubber tree, *Hevea brasiliensis,* and is available in several variations of concentration and stabilization. One major use in packaging is as a principal ingredient in adhesives for laminating polyethylene film to paper, as in the construction of multiwall bags. Natural rubber latex also finds use in a variety of self-seal applications, since it is the only adhesive system that will form bonds only to itself with pressure. This property is used in self-seal candy wraps (where it is called *cold-seal*), and press-to-seal cases, as well as on envelopes.

Synthetic waterborne adhesives. Synthetic waterborne adhesives are the most broadly used class of adhesives in general packaging. Almost all are resin emulsions, specifically poly (vinyl acetate) emulsions-stable suspensions of poly(vinyl acetate) particles in water. These systems usually contain water-soluble protective colloids such as poly(vinyl alcohol) or 2-hydroxyethyl cellulose ether, and may be further compounded by the addition of plasticizers, fillers, solvents, defoamers, and preservatives.

These emulsions are supplied in liquid form (the ubiquitous "white glue") in a range of consistencies from thin milky fluids to thick, nonflowing pastes. They are used in a broad range of packaging applications, to form, seal, or label cases, cartons, tubes, bags, and bottles. In most of these uses they have replaced natural adhesives because of their greater versatility. They can be compounded to have a broad range of adhesion not only to paper and glass but also to most plastics and metals. They can be rendered very water-insensitive for immersion resistance, or very water-sensitive to promote ease of cleanup and good machining. They are the fastest-setting class of waterborne adhesives, facilitating increased production speeds. They are low in odor, taste, color, and toxicity and have excellent long term aging stability. They are tough, with an excellent balance of heat and cold resistance. The equipment used to apply them is relatively simple and inexpensive to purchase and to operate. Finally, they are economical and reasonably stable in cost.

The utility of these emulsion systems has broadened in recent years with the greater use of copolymers of vinyl acetate. Copolymerizing vinyl acetate with ethylene or acrylic esters in particular has greatly improved the adhesion capabilities of these emulsions, particularly where adhesion to plastics or high-gloss coatings is required. For example, cross-linking acrylic–vinyl acetate copolymer emulsions have replaced polyurethane solution systems for laminating plastic films for snack packages. The largest areas of use for vinyl emulsions, however, are in case and carton sealing, forming the manufacturers joint on cases and cartons, and the spiral winding of composite cans.

All acrylic emulsion pressure-sensitive adhesives have to a significant degree replaced acrylic or rubber solution products in the manufacture of pressure-sensitive labels.

Polyurethane dispersions are beginning to find acceptance in medium-performance flexible packaging applications laminating plastic films together where some chemical resistance is required.

The use of other synthetic waterborne systems is quite minor and specialized. Some synthetic rubber dispersions are used in film adhesives and, in conjunction with casein, for the lamination of aluminum foil to paper. There is some use of tackified rubber dispersions as pressure-sensitive masses on tapes and labels replacing solvent-borne rubber–resin systems.

Sodium silicate was once widely used in many paper packaging applications, ranging from corrugating to case sealing, but today the primary use of silicate adhesives is in tube winding, especially in the convolute winding of large drums or cores where it produces a high degree of stiffness.

Hot-Melt Adhesives

Hot melts have been the fastest-growing important class of adhesives in packaging for the last 25 years. Most of their volume goes into high-speed large-volume case and carton sealing. Hot melts can be defined as 100% solids adhesives based on thermoplastic polymers, that are applied heated in the molten state and set to form a bond on cooling and solidification. Their chief attraction is the extremely rapid rate of bond formation, which can translate into high production rates on a packaging line.

The backbone of any hot melt is a thermoplastic polymer. Although almost any thermoplastic can be used, and most have been, the most widely used material by far is the copolymer of ethylene and vinyl acetate (EVA). These copolymers have an excellent balance of molten stability, adhesion, and toughness over a broad temperature range, as well as compatibility with many modifiers. The EVA polymers are further compounded with waxes and tackifying resins to convert them into useful adhesives. The function of the wax is to lower viscosity and control set speed. Paraffin, microcrystalline, and synthetic waxes are used depending on the required speed, flexibility, and heat resistance. The tackifying resins also function to control viscosity, as well as wetting and adhesion. These are usually low-molecular-weight polymers based on aliphatic or aromatic hydrocarbons, rosins, rosin esters, terpenes, styrene or phenol derivatives, or any of these in combination. The formulations always include stabilizers and antioxidants to prevent premature viscosity change and char or gel formation that could lead to equipment stoppage.

Two recent variations on traditional EVA hot melts have recently become commercially significant. First, the recent availability of very low-molecular-weight EVA copolymers has made possible EVA hot melt that can be run at much lower temperatures, 250°F (121°C), rather than the traditional 350°F (177°C). This allows for much safer running conditions as well as energy savings. Second, an analog of EVA, ethylene-butyl acrylate, has been introduced as the backbone polymer in some packaging hot melts (3), providing advantages in both adhesion and in heat and cold resistance.

Another class of hot melts used in packaging is based on lower-molecular-weight polyethylene, compounded with natural or synthetic polyterpene tackifiers. These lack the broader adhesion capabilities of the EVA-based hot melts as well as their broader temperature resistance capabilities. However, they are economical and are adequate for many paper bonding constructions, and find application in case sealing and bag seaming and sealing.

A third type of hot melt is based on amorphous-poly-α-olefin (APAO) polymers. Originally these were based on amorphous polypropylene, which was widely available as the byproduct of the polymerization of isotactic polypropylene plastics. As a byproduct, it was inexpensive, but suffered from low strength and was limited to applications such as lamination of paper to paper to produce water resistant wrapping material on two-ply reinforced shipping tape. Improvements in polypropylene polymerization catalysts have almost eliminated this byproduct, but several producers noted the market and began producing on-purpose APAO polymers, albeit at somewhat higher costs.

A more recent class of hot melt is based on thermoplastic elastomers: block copolymers of styrene and butadiene or isoprene. These find primary application in hot-melt pressure-sensitive adhesives for tapes and labels replacing solvent-borne rubber systems. More recently, their broad adhesion and low temperature and impact resistance are finding use such as the attachment of polyethylene-base cups to polyester soft-drink bottles (4), and the sealing of film laminated frozen-food cartons.

Even more specialized applications use hot melts based on polyamides or polyesters when specific chemical or heat resistance requirements have to be met, but their high cost and relatively poor molten stability have precluded their widespread use to date.

The most recent, and highest-performance hot-melt technology, is moisture curing polyurethane hot melts, but these have been limited to higher-requirement product assembly applications and have found few uses in packaging.

All hot melts share the same basic advantages, based on their mechanism of bond formation by simple cooling and solidification. They are the fastest-setting class of adhesives—indeed, preset before both substrates can be wet is the most frequent cause of poor bonds with hot melts. Because of the wide range of polymers and modifiers used, they can be formulated to adhere to almost any surface. With no solvent or vehicle to remove, they are generally safe and environmentally preferred. They are excellent at gap filling, since a relatively large mass of material can "freeze" in place, thus joining poorly mated surfaces with wide dimensional tolerances.

However, all hot melts share the same weakness, which is the rapid falloff in strength at elevated temperatures. Therefore, properly formulated hot melts can be suitable for almost all packaging applications, but they are not appropriate for very hot-fill or bake-and-serve applications.

Solvent-Borne Adhesives

By far the smallest, and most rapidly declining, of the three classes of adhesive used in packaging, solvent-borne adhesives find use in specialized applications where waterborne or hot-melt systems do not meet the technical requirements.

Rubber–resin solutions are still used as pressure-sensitive adhesives for labels and tapes. However, factors of cost, safety, productivity, and, above all, compliance with clean-air laws have led to a strong movement toward waterborne or hot-melt alternatives. Such alternatives are available to meet most requirements and most knowledgeable observers predict an almost total disappearance of rubber–resin solvent-borne pressure sensitives for packaging tapes and labels over the next decade.

Solvented polyurethane adhesives are widely used in flexible packaging for the lamination of plastic films. These multilayer film constructions find application in bags, pouches, wraps for snack foods, meat and cheese packs, and boil-in-bag food pouches. They have the ideal properties of adhesion, toughness, flexibility, clarity, and resistance to heat required in this area. However, here, too, alternative systems are being introduced to eliminate the costs, hazards, and regulatory problems associated with solvent-borne systems. Cross-linking waterborne acrylic polymers have gained acceptance in the large snack food laminating market for constructions such as potato chip bags. Polyurethane dispersions and (100% solids) "warm melt" systems are starting to find use in some of the more demanding food packaging applications.

Solvent-borne ethylene–vinyl acetate systems found use in some heat-seal constructions, such as the thermal strip on form–fill–seal pouches, or on lidding stock for plastic food containers such as creamers or jelly packs.

BIBLIOGRAPHY

1. J. Dahs, "The Adhesive and Sealants Industry—Existing Structures and Global Perspective," *J. Adhes. Sealant Counc.* **19**(2), 1990.
2. U.S. Pat. 2,754,240 (July 10, 1956), W. B. Kinney (to Borden Company).
3. U.S. Pat. 4,816,306 (March 28, 1989), F. Brady and T. Kauffman (to National Starch and Chemical Company).
4. U.S. Pat. 4,212,910 (July 15, 1980), T. Taylor and P. Puletti (to National Starch and Chemical Company).

General References

I. Skeist, ed., *Handbook of Adhesives,* 3d ed., Van Nostrand Reinhold, New York, 1990.

K. Booth, ed., *Industrial Packaging Adhesives,* CRC Press, Boca Raton, FL, 1990.

IRVING KAYE
National Starch and Chemical Company
Murray Hill, New Jersey

ADHESIVES, EXTRUDABLE

In its broadest definition, extrudable adhesives are polymeric resins that can be processed by standard extrusion processes and are useful for bonding together various substrates. In practice, extrudable adhesives are commonly polyolefin materials useful in processes such as blown and cast film, blow molding, and extrusion coating.

Overview

The variety of polymeric materials that are available to both the industrial user and the consumer that are useful as adhesives include a diverse number of materials almost equal to the number of polymer types themselves (see also Adhesives). Adhesives are developed from such polymers as polyvinyl acetate, polyvinyl alcohol, polyamides, polyesters, and many others. Adhesives can be applied as solvent solutions, aqueous dispersions, pastes, spray coatings, tapes, and thermally activated films. One special subset of adhesives, called *extrudable adhesives,* are different in that, in their application, they are applied in an extrusion process where they are melted, conveyed, and inserted between the substrates that are to be bonded together. While there are many adhesive application methods that require the melting of a polymeric adhesive, extrudable adhesives are distinguished from other adhesive types used in processes such as powder coating, flame spraying, and the thermal lamination of adhesive films and webs. As a second distinction, extrudable adhesives are also distinguished from hot-melt adhesives that generally require a resin viscosity not suitable for traditional extrusion processes. Thus, extrudable adhesives are those materials specifically designed to function in processes such as coextrusion blown and cast film, monolayer and coextrusion coating and lamination, coextrusion cast sheet, coextrusion blow molding, and coextrusion tubing.

Types of Extrudable Adhesive

Extrudable adhesives are most often polyolefin based compositions. The most common use of these polyolefin extrudable adhesives is as a specific layer in a multilayer coextrusion. Coextrusion is a technique that allows the creation of a plastic composite, in a single operation, combining the benefits of a number of different materials. The plastic composite may be, for example, a packaging film combining the properties of an oxygen-barrier resin with a heat-sealable layer on one side and an abuse-resistant layer on the other side. The purpose of the extrudable adhesive is to bond together the diverse plastic materials in the construction that would not, under ordinary circumstances, bond to each other. The polyolefin adhesive is designed to bond to similar polyolefins by a diffusion process. That is, during the extrusion process the molten extrudable adhesive resin comes in contact with the molten polyolefin. Their molecules diffuse together creating a strong bond between the materials (1). The polyolefin extrudable adhesive is designed to bond to other polymeric materials, such as oxygen-barrier resins like polyamides and ethylene vinyl alcohol, through a chemical reaction between a functional group on the adhesive and a functional group on the barrier resin. Often the functional group on the extrudable adhesive can be chosen for bonding to specific materials of interest.

There are many types of extrudable adhesive. Polyethylene can be considered to be an extrudable adhesive. Commonly polyethylene is used in extrusion lamination of paper to aluminum foil. The polyethylene is extruded at very high temperatures. The melt is oxidized by contact with air creating polar functionality on the surface of the melt. This provides chemical bonding to the aluminum oxide on the surface of the foil (2). The low viscosity of the melt and its polar nature allows for good wetting on the paper and encapsulation of the individual fibers. Copolymers of ethylene and vinyl acetate are also useful extrudable adhesives and capable of bonding polyethylene to polyvinyl chloride. However, the most sophisticated extrudable adhesives are either polyolefins with either acid or anhydride functionality. Acids and anhydrides are particularly reactive and can create strong bonds to a number of different materials in extrusion processes. Examples of acid modified polyolefins are the copolymers of ethylene with acrylic acid or methacrylic acid. Variations include the partially neutralized acid copolymers with metal ions referred to as ionomers or terpolymers of ethylene, an acid and an acrylate such as methyl acrylate or isobutyl acrylate. Acid-containing extrudable adhesives are widely used to bond to aluminum foil (3,4). Examples of anhydride modified polyolefins include terpolymers of ethylene, maleic anhydride, and acrylates such as ethyl acrylate or butyl acrylate and the anhydride grafted polyolefins.

The anhydride grafted polyolefins are created by combining the polyolefin with an anhydride, most commonly maleic anhydride. The anhydride is added to the polyolefin with a free-radical initiator in a solvent or the melt. This allows the attachment of the highly reactive anhydride to polyolefins such as high-density polyethylene, linear low-density polyethylene, polypropylene, ethylene vinyl acetate copolymers, or ethylene propylene rubbers. While the polar functionality of the anhydride and the polyolefin backbone are the necessary ingredients for the extrudable adhesive to function, almost all commercially available extrudable adhesives are formulated with other polymers. Formulations of anhydride modified extrudable adhesives will generally contain two or three basic components. Two-component extrudable adhesives will contain the anhydride graft blended into a second, or matrix polyolefin (5). The purpose of the second polyolefin is to lower the overall cost of the adhesive and to control the viscosity, modulus, tensile, thermal, and other properties of the adhesive. The three-component formulation will contain the anhydride graft, the matrix polyolefin, and a modifier (6). The purpose of the modifier is to enhance the peel strength characteristics of the bonded composite. In most cases the efficiency of the extrudable adhesive is judged by peeling the bonded composite apart in a "T" peel mode. The modifier affects the peel strength characteristics of the adhesive by dissipating the force at the interface where the composite is being peeled apart. The result of this is that it takes more work to peel apart the bonded composite (7). These extrudable adhesives can also have other additives such as antioxidants, slip agents, or resin tackifiers.

While the formulation of an extrudable adhesive is very important to its utility, many other factors also affect how well an extrudable adhesive will bond together different materials. How the extrudable adhesive is processed, its thickness, and whether the bonded composite is oriented, shaped, or exposed to aggressive environmental conditions can all affect the performance of the adhesive (8,9).

Commercial Offerings

There are many producers of extrudable adhesives around the world. Some of these producers will produce only a few types of extrudable adhesive. Others have a broader product line. Some manufacturers market their adhesives internationally; others, only regionally or only to specific market areas or for use in specific applications. Manufacturers of acid copolymers include Dow Chemical Company as Primacor and E. I. DuPont de Nemours as Nucrel. DuPont also manufactures acid terpolymers. Ethylene terpolymers of anhydride with acrylate are produced by Atochem under the name Lotader. DuPont manufactures extrudable adhesives based upon anhydride graft technology under the name Bynel. Quantum Chemical Company also makes and sells extrudable adhesives under the name Plexar. Other manufacturers include Mitsui Petrochemical Company with Admer, Morton International with Tymor, Atochem with Orevac, and DSM with Yparex.

Extrudable Adhesive Applications

Extrudable adhesives are used primarily in coextrusion processes (see Coextrusion). The major market area is food packaging. Examples include the coextrusion coating of an oxygen-barrier material, extrudable adhesive, and polyolefin onto paperboard to create high-barrier, nonscalping fruit juice cartons. Oxygen barriers, extrudable adhesives, and ionomers are coextruded by either cast-film or blown-film processes to produce packaging films for hot dogs, bacon, and other processed meats. A third example is the coextrusion blow molding of polyolefin, extrudable adhesive, and oxygen barrier to produce high-barrier ketchup bottles. Extrudable adhesives are also used in applications that do not involve the packaging of foods. Extrudable adhesives are used to bond high-density polyethylene to ethylene vinyl alcohol in a coextrusion blow-molding process to produce automotive gas tanks (see Blow molding). Extrudable adhesives are also used to bond cross-linked polyethylene to ethylene vinyl alcohol through a coextrusion crosshead tubing process to make radiant hot-water heating pipes.

The selection of the proper extrudable adhesive for any particular application may be a complex problem. The first consideration is always the materials that need to be bonded together. The adhesive must be able to bond to these materials. When bonding a polyolefin to an oxygen barrier such as ethylene vinyl alcohol or polyamide, extrudable adhesives with anhydride functionality are usually the adhesive of choice. The type of anhydride modified polyolefin will depend on the polyolefin being coextruded with the barrier. For example, if polyamide is being coextruded with polyethylene, an anhydride-modified polyethylene or anhydride modified ethylene vinyl acetate will be the resin best able to perform in the application. If polypropylene is coextruded with polyamide, an anhydride modified polypropylene will be chosen. Second, the adhesive must be processable in the equipment that the converter intends to use. This may mean choosing a resin with a relatively lower viscosity for coating applications and a relatively higher viscosity for blow-molding applications. If the bonded composite is going to see a specific environment, such as oil or grease, it should be resistant to that product. If the composite will be exposed to either very high or very low temperatures, the adhesive must be functional at those temperatures. Finally, some extrudable adhesive resins may have to have a specific regulatory compliance depending on the application for the bonded composite. Most manufacturers of these extrudable adhesives are prepared to help the converter select the best adhesive for their application.

BIBLIOGRAPHY

1. L. Lee, *Fundamentals of Adhesion,* Plenum Press, New York, 1992.
2. A. Stralin and T. Hjertberg, *J. Adhes. Sci. Technol.* **7**(11), 1211–1229 (1993).
3. M. Finlayson and B. Shah, "The Influence of Surface Acidity and Basicity on Adhesion of (Ethylene-*co*-acrylic acid) to Aluminum" in *Acid–Base Interactions: Relevance to Adhesion Science and Technology,* VSP BV, Utrecht, 1991, pp. 303–311.
4. G. Hoh, S. Sadik, and J. Gates, "1989 Polymers" in *Proc. Laminations and Coatings Conf.,* Sept. 1989, Tappi, Atlanta, pp 361–366.
5. U.S. Pat. 4,230,830 (Oct. 28, 1980), S. Tanny and P. Blatz (to DuPont).
6. U.S. Pat. 4,198,327 (April 15, 1980), H. Matsumoto and H. Niimi (to Mitsui Petrochemical Industries, Ltd.).
7. S. Wu, *Polymer Interface and Adhesion,* Marcel Dekker, New York, 1982.
8. B. Morris, *Tappi J.* **75**(8), 107–110 (1992).
9. B. Morris *Eng. Plast.* **6,** 96–107 (1993).

Stephen R. Tanny
DuPont Polymers
Wilmington, Delaware

AEROSOL CONTAINERS

The first aerosol cans were heavy steel "bombs," consisting of two shells about 0.090 in. (2.3 mm) thick, brazed together at the lateral centerline. They were known at least since the early work of Eric A. Rotheim (Oslo, Norway, 1931) and gained fame during 1943–1945 as insecticides for U.S. troops fighting in such places as Guadalcanal and other South Pacific areas. After the war these products were made available to the public, but acceptance was very poor, due to the high initial expense and the aspect of having to return the emptied unit for refilling. It was obvious that a lightweight, disposable can was needed.

In 1946/47 Harry E. Peterson developed such a can in the laboratories of the Continental Can Corporation in Chicago. It consisted of a 2.68-in.-diameter solder side seamed can body, to which were seamed a pair of concave end sections. The top section, assembled to the body by the canmaker, carried a small valve, soldered at the centerline. The can was designed to be filled upside-down with highly refrigerated (−45°F or −43°C) aerosol concentrates and propellants, after which the end was double-seamed to the body. The final unit stood 4.8 in. tall and had a capacity of about 12.2 fluid oz (361 mL). In an almost concurrent but independent development, Earl Graham of Crown Cork & Seal Company developed a higher-strength modification of the "Crowntainer" beer can. This was a two-piece container. The base was double-seamed onto a drawn steel shell that contained a soldered valve. The early valves were manufactured by Bridgeport Brass Company, Continental Can Corporation, and many other firms. Most were outrageously costly and inefficient. Slightly later, pioneers such as Robert Alplanalp (Precision Valve Corp.) and Edward Green, Sr. (Newman-Green, Inc.) patented more efficient types.

A major innovation occurred about 1951, when Crown Cork & Seal Company engineers developed the "1-in." ("25.4-mm") hole, along with a corresponding valve cup. The cup could carry the valve components within the central pedestal (except for actuator and dip tube), and could be crimped (swaged) onto the curl or bead that surrounded the "one-inch" ("1-in.") hole. This secondary plug-type closure rather quickly displaced the soldered valve units—the last of which were filled during October 1953. The Continental Can Company developed a can dome—called a "cone" in the United Kingdom—to provide the needed "1-in." hole, while at the same time enlarging their can somewhat and making it taller.

The budding aerosol industry had to make do with these two nominal 12-oz cans until 1953, when the 2.12-in.-diameter can size was developed as a nominal 6-oz container.

At the insistance of a fast-growing industry, the two can companies introduced some additional can heights. They were joined in 1955 by the American Can Company, who entered the market with their "Regency" line of 2.47-in.-diameter cans in four heights. The National Can Corporation followed soon afterward.

The impact extrusion technology for drawing aluminum beer and beverage cans was well developed by the 1950s. The Peerless Tube Company produced aerosol units as early as 1952 and perhaps even earlier. The American Can Company made a unique two-piece aluminum can (Mira-Spray and Mira-Flo) in just two 6-oz sizes. Other very early entrants were Victor Tube, White Metal, and Hunter-Douglas Corporation. (See also Cans, aluminum; Cans, fabrication.)

Glass aerosols were made, first by the Wheaton Glass Company (Mays Landing, NJ) about 1953, and then by Ball Brothers and several other firms. The valve was incorporated into a ferrule by Risdon Manufacturing Corporation, Emson Research, Inc. and Precision Valve Corporation and the ferrule was sealed to the glass finish by means of clinching—a method then also used to seal metal caps on beer bottles.

During 1954 Wheaton developed a method for encasing their bottles in a heavy skin of PVC. The plastic envelope helped the bottle withstand minor falls to hard surfaces, but if breakage should occur, it contained the glass shards and kept them from flying outward and possibly injuring persons nearby. Later on, a bonded film was developed and made an unofficial industry standard on glass bottles of >1-oz (30-mL) capacity.

Current Canmaking Technology

Tinplate, in a number of thicknesses of steel and tin, is routinely delivered to the canmaking plant in the form of large rolls typically 34 in. (860 mm) wide and 48 in. (1.22 m) in diameter. A roll may weigh 12,000 lb (5450 kg) and contain about 4.5 mi (7.25 km) of tinplate, depending on the thickness. After sending the roll through a straightener, a slitter cuts it into sheets best suited for making can bodies with a minimum of waste. At the same time, a scroll cutter produces strips from which dome and base section circles can be cut—again with a minimum of waste (see also Cans, steel).

The body sheets are first lined with epoxy–phenolics or other materials, baked in huge ovens, and then lithographed. At this point the individual can bodies are cut apart. A bare metal fringe is allocated for the area to be welded. The WIMA or other type of bobymaker acts to roll up the can body and tackweld it every inch or two to hold it in place with just the right lapover. After this the welding process produces either a "standard" or "full" (tin-free) weld line. Weld nuggets ($\sim$31 in.$^{-1}$) form the basic structure. While still extremely hot, the overlap thickness is reduced by heavy compression to only $\sim$1.4 times the average plate thickness. The cylindrical can body is then flanged top and bottom for a "standard" or straight-wall can, or is both necked-in and flanged for the "necked-in" containers. The latter have several advantages and are increasingly popular.

Meanwhile, can domes and bases are being formed in multi-stage presses. They are either used directly, or are "sleeved" into long paper tubes for later fabrication. Quite often one large plant will make sleeve packs of can ends, for shipment to satellite locations where the assembly process is undertaken. The same is true for can bodies, which can be easily shipped in the flat to other facilities.

The final can is assembled using double seamers that typically operate at about 350 cans per minute. As a rule, the finished cans are then tested using a large wheel-like device that pumps a significant air pressure into each one and then checks for pressure leakage, if any. Such cans are automatically shunted aside and scrapped. Finally, the cans are tiered onto pallets of about 40-in. × 44-in. size, strapped, plastic shrink-wrapped, and warehoused for delivery to fillers.

Aluminum cans, which occupy about a 16% of the U.S. market in 1995 (and growing), are made quite differently. Pure (99.70%) aluminum slugs or pucks are lubricated with zinc stearate in a tumbler, then conveyed to an extruder, where they are formed into a "cup." The cup may be drawn and ironed (draw–ironed), in some advanced operations, to obtain a more uniform wall thickness and lighter structure. After trimming and vigorous cleaning, the lining and exterior decoration coatings are applied. The top is then formed in a number of stages (the number increasing with can diameter) and finally convoluted into either an outside or inside curl configuration. The outside curl is more common. Both have different advantages. Curl machining is sometimes done to smooth the curl surface of larger-diameter cans. Finally, they are strapped into typically 96-pack hexagon shapes and loaded onto 40-in. × 44-in. wood pallets. After strapping and shrink-wrapping they are ready for delivery.

Very small numbers of cans are made by other methods. The Sexton Can Company produces one-piece steel cans by an extrusion process. For larger diameters, they extrude a sheel shell, then double-seam a can bottom to it. The company is able to make very strong cans by this process, able to withstand pressures up to 650 psig (lb/in.^{2} gauge) (45 bars). They are used to pack such higher-pressure products as HCFC-22 refrigerant, which generates 302 psig at 130°F (21 bars at 54.4°C). A special permit from the U.S.DOT (Department of Transportation) is required for such high pressures. As part of the development process, Sexton learned how to produce these cans with a bottom indentation, able to open at about 425 psig (30 bars), and thus long before heating could cause bursting. The orifice lets the product come out with a fair degree of control; otherwise, the dispenser might eventually burst with the brissance and concussive effects of a grenade.

A new canmaking company is known as the Dispensing Containers Corporation (DCC) (Glen Gardner, NJ). They also produce tinplate cans, but have aluminum capability as well. Their small, one-piece (35- and 38-mm) cans are extruded, then draw–ironed. Additionally, they produce 52- and 65-mm cans by extruding a shell, which is draw–ironed to a thickness of $\sim$0.0038 in. (0.096 mm) before the dome section is double-seamed into place. The significant source reduction of metal enabled the firm to suggest that they could manufacture finished cans for 15–25% less than the factory cost of same size three-piece ETP counterparts. Rather interestingly,

DCC has no plans to decorate, beyond the application of a white base coat. They have installed an American Fugi-Seal, Inc. machine to allow application of either a shrink-sleeve or a wraparound label with reverse-print rotogravure and high-gloss appearance. As a note, fillers cannot use a vacuum crimping process on this thin-walled can; air must be partly removed by purging techniques. Otherwise the partly evacuated cans will crumple inward.

Tinplate Options

The electrotinplated steel sheet stock from the tin mill is available in a modest variety of plate thicknesses and tin coating weights. The steel for aerosol cans will typically be ~0.007–0.015 in. (0.18–0.38 mm) thick, according to intended use. The thinnest plate is used for bodies of small (45–52-mm)-diameter cans. The bodies of larger cans (57–76 mm in diameter) will typically be made from 75–85-lb ETP—which is a canmaking term for stock of ~0.0083–0.0094 in. (0.21–0.24 mm) thick. Tops and bottoms require still heavier plate, to prevent premature buckling (eversion) and subsequent unwrapping of the top or bottom double seam, leading to a burst event. End sections of the smaller cans typically use 112-lb ETP, or plate that is 0.0123 in. (0.31 mm) thick. The largest-diameter aerosol can is a nominal 3.00-in. (76-mm) size, and requires 135-lb ETP, or 0.015-in. (0.38-mm)-thick plate. Valve cups are almost always made of 95-lb ETP; eg, 0.0105-in. (0.266-mm) plate.

Since aluminum is notably softer and more deformable than steel, these cans are extruded to have thicker metal. The thickness must be increased as diameters are made larger. The thinnest part of a typical 52-mm-diameter aluminum aerosol can will be about 25% up on the body wall, measuring about 0.017 in. (0.43 mm). The base of the largest aluminum can (66 mm) may easily get to 0.080 in. (2.0 mm). Aluminum valve cups average 0.016 in. (0.41 mm) thick.

In the past, tinplate could be ordered with very heavy tin coatings: up to 1.35 lb of total tin weight per basis box area of 31,360 in.^{2} on each side. This is equivalent to a tin coating of 15.1 g/m^2 on each side. But today, with economic considerations forcing lower inventories, plus improvements in the tinplating process, it is rare to see tinplate of greater than 0.50 lb—5.6 g/m^2 per side. Some tinplates are made with the so-called "kiss of tin" (0.05-lb ETP) having a nominal coating weight of only 0.56 g/m^2 per side. The thickness then averages only 0.00000303 in. (0.077 μm) per side. The dark gray color of the steel and $FeSn_2$ alloy layer can be easily seen through this ultrathin coating.

With electrotinplating methods it has been possible to obtain differentially coated tinplates; that is, plate having coatings of different thickness on each side. For example, D50/25-lb ETP (more accurately noted as D0.50/0.25-lb ETP) will carry 0.25 lb of tin on one face and 0.125 lb of tin on the other (5.6 + 2.8 g/m^2). This type of plate is generally used with the heavier tin-coated area turned toward the aerosol product, to provide corrosion protection.

Over half of all tinplate aerosol cans and virtually 100% of all aluminum cans have organic linings. Single linings are the most common, but double linings and (for a few tinplate can bodies) even triple linings can be ordered. A variety of lining materials are used. The most common are the epoxyphenolics, used for about 70% of tinplate cans and around 78% of aluminum cans. Other options include the vinyl organosols, polyamideimide (PAM), and now the polyimideimide (PIM). The PAM and PIM coatings are relatively costly and often more difficult to apply, especially on tinplate. They are extremely resistant to permeation. Finally, there are the pigmented epoxyphenolics and the vinyls. The latter do not adhere well to metal substrates, and are used as a second or top coating, when extra performance is needed. They are unaffected by water, but quickly dissolved by methylene chloride, oxygenated solvents, and certain other solvents.

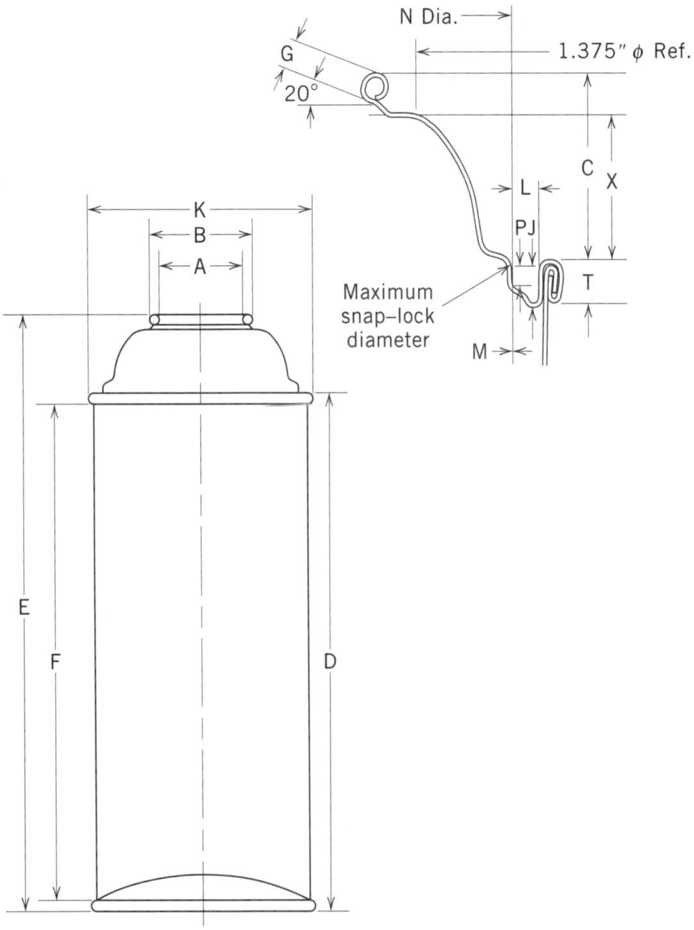

Figure 1. Straight-wall aerosol can (standard tinplate).

The canmaking industry has strived for uniform dimensions of cans, both from different plants of a given supplier and between suppliers. This acts to save fillers from making time-consuming adjustments to crimping and gassing machines when moving from one lot of cans to the next. The CSMA (Aerosol Division) Commercial Standards Committee has now developed about 13 key dimensions and their tolerances for tinplate cans—both standard and necked-in—and about 10 more for aluminum cans (see Figs. 1 and 2). During 1996 dimensions and tolerances will be promulgated for a wide variety of different tinplate and aluminum can sizes, although certain ones are still controversial and will be discussed only between canmakers and their customers.

Well-known examples of can dimensions are "A," which is 1.000 ± 0.004 in. (25.4 ± 0.1 mm) for all can sizes and metals, and "B," which is 1.232 ± 0.010 in. (31.3 ± 0.25 mm) for tin-

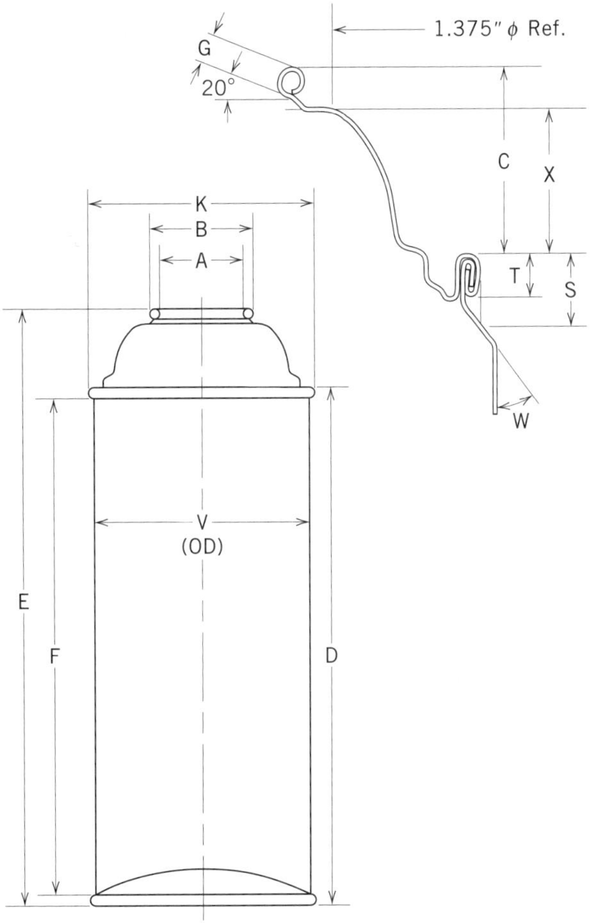

Figure 2. Necked-in aerosol can (tinplate).

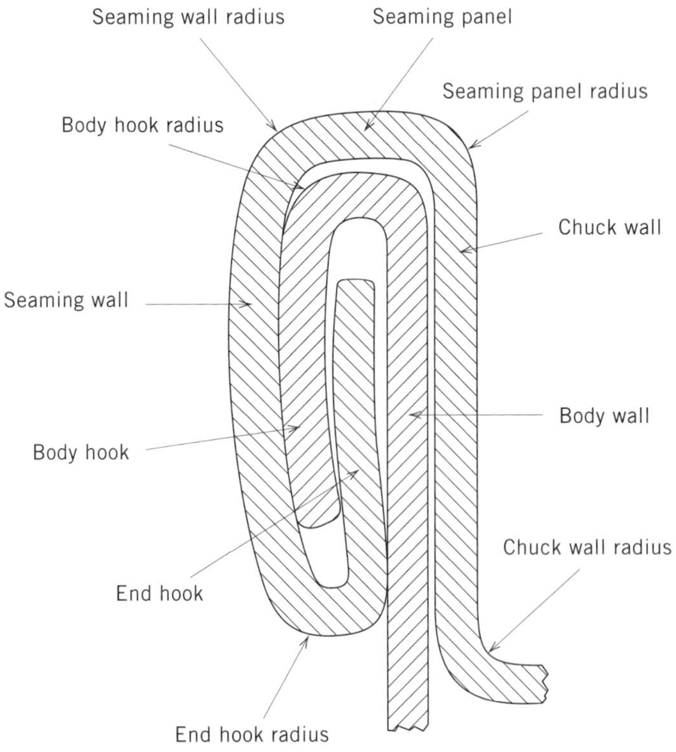

Figure 3. Anatomy of a double seam.

plate cans of all sizes. These particular dimensions are quite critical because the valve cup must fit rather perfectly into the "1-in." can opening to avoid jamming or scraping, and to allow a good hermetic seal when crimped. The outer wall diameter of the valve cup is ~0.992 ± 0.003 in. (25.2 ± 0.08 mm), and this leaves a contingency clearance of only 0.001 in. (0.025 mm) between the largest cup and the smallest hole. This is important not only for fit, but to anticipate traces of out-of-round, metal dimpling and other factors. The "B" dimension is important in making the top rim and shirt of the standard valve cup fit snugly to the can bead, increasing the statistical probability of a good seal.

The tinplate can has three seams, which can occasionally become matters of concern. Aside from aesthetics, if can leakage occurs, it will often be at the side seam, and more rarely at the top or bottom double seams. The side seam is also a favored site for can corrosion, due to exposed iron and the relatively poor coverage of side-seam enamel stripes, if applied. The dimensions of the top and bottom double seams are about the same, for a given can diameter. Small diameter side seamed cans (35, 38, and most importantly 45 mm) have smaller-size double seams than the larger cans. In fact, the 35- and 38-mm cans actually have no top double seam; the metal (side seam and all) is smoothly formed into the "inch-inch" curl. Figure 3 can be used to illustrate the general shape and important elements of a typical side seam. In general, these seams are about 0.125 in. (3.2 mm) high; from sealing wall radius to end hook radius, if the can is 52 mm or larger in diameter.

The United States and Canada use a "sales description" method for indicating can overall dimensions that is based on the English (inch) measurement system. England is changing over to the metric system, to conform to the ISO descriptions used in the rest of Europe, and generally throughout the rest of the world. Australia and Canada have taken steps in the same direction.

The U.S. "can description" system can be best described by illustrations. The 202 can is one with a body diameter of 2²⁄₁₆ in.—or 2.125 in. The ISO diameter (actually the inside diameter of the body) would be 52 mm. The U.S. can height is dimension "D" (see Fig. 1 or 2), measured as total body height over the two seams. A 612 can would then have a body height of 6⁴⁄₁₆ in., or 6.250 mm. The system extends to such descriptions as 211 × 1208 and 207.5 × 605. Seeds of change are being sown in the United States by the United Nations and other international groups, but the system is deeply ingrained and is likely to persist, even into the next millennium.

Future Canmaking Trends

The three-piece aerosol can is highly serviceable and is produced by the daily use of heavy equipment valued in the multi-billion-dollar range. There are no massive changes predicted during the next decade. The aerosol dispenser has been accused of having about the same cylindrical shape it had over 40 years ago. Minor improvements—such as necking-in, welding side seams, plastic labeling (to cover the side seam scar), and the "ecogorge" indentation on some aluminum cans, for the attachment of full-diameter caps—have been relatively unnoticed by consumers. To them the aerosol is a cylindrical package, although sometimes with a variously convoluted top portion.

The cylindrical image is an architectural necessity for a pressure-resistant dispenser (see also Pressure containers). Even though glass aerosols had somewhat wider limits, the underlying shape limitations have been a factor in having nearly all the perfume and cologne business transfer to non-pressurized glass pump sprayers and other containers. However, some advances have been offered recently, in the metal can area.

In the United Kingdom the Wantage Research Center of CarnaudMetalBox, plc (a firm now being purchased by Crown Cork & Seal Co.) engineers have developed a process by which a finished (plain or lithographed) three-piece tinplate can is placed momentarily in a heavy steel mold cell and expanded against the contoured sidewalls of the cavity by pressurization with some 1200 psig (83 bars) of filtered dry air. The volume increase is limited to about 12–18%, depending on relative can length. The emerging cans are necked-in, and may have pleated, quilted, crestlike, ergonomic finger depressions or other debossings in the body wall. In general, these are never more than ~0.15 in. (3.8 mm) deep. Round-the-can lateral depressions must not be too sharply defined, or the can will increase in height when pressure-tested during later canmaking checks or in filler hot tanking. The fact that this is presently an extracost operation has thus far prevented marketer acceptance—but one is mindful that such innovations as welded seams and necked-in profiles were will engineered decades before marketers paid them much attention.

There is a trend toward aluminum cans that has been quite noticable during 1992–1995, when these cans grew at about 16% per year, and thus much more than the more proseic 2–3% increase per year for the total aerosol market. This is thought to relate to the greater aesthetics of aluminum, more than anything else. Marketers who use tinplate cans are responding by increasing the trend toward necked-in types, and by permanently covering, at least the unsightly valve cup, and ideally the entire can dome with a spray cap or foam spout. In Europe and Japan, starch and fabric finish products are offered with "pistol-grip handles" that are integral with a full-diameter, non-removable spray cap. The accoutrement not only provides aesthetics, but reduces hand and finger fatigue.

The activities of Dispensing Containers Corporation (DCC) and similar firms are being monitored by the larger can companies, since these extruded, draw–ironed steel and aluminum cans are economically interesting. Environmentalists like the source reduction fact that they use up to 35% less metal than do present three-piece cans. In fact, further weight reductions are suggested, as these firms look for weight reductions for the dome segment. The base is already lightweight, having the "small dome" contour so successfully used in the beer and beverage industry.

Aerosol formulations weave their effects into the fortunes of the steel and aluminum canmakers. Because of environmental considerations centering on the issue of clean ambient air, and the reduction of emissions that directly or indirectly produce air pollutants, the U.S. aerosol industry has been obliged to reformulate most of their products toward those that have reduced amounts of volatile organic compounds (VOCs). The state governments of California and New York have been very active in limiting the VOC content of aerosols and other products. As a result, many hair sprays and other products now incorporate significant amounts of water, or else use new propellants, such as (non-VOC) HFC-152a (1,1-difluoroethane). Quite often, these new formulations favor aluminum cans, from both a corrosion resistance and a smaller package standpoint. In some other countries the percentage of aluminum cans is in the area of 40–60% of the total, and it is possible that the United States and Canada may slowly approach this high ground, as time goes by.

MONTFORT A. JOHNSEN
Montfort A. Johnsen & Associates, Ltd.
Danville, Illinois

AEROSOLS. See PRESSURE CONTAINERS.

AIR CONVEYING

The fast changes within both lightweight packaging and plastics technology during the 1980s brought forth the development of a new technology: the air conveyor. Because of the backpressure created on chain conveyors, the new lightweight packages and bottles were getting crushed and marred—but with the advent of the air conveyor, those problems were no longer of great concern. The air conveyor is faster, easier, cleaner, and safer than its predecessors, the belt, cable, and chain conveyors. For these reasons, air conveyor has quickly become the conveying method of choice.

Air Conveying

How does it work? All air conveyors share several basic operating principles:

- Use of air as the transport medium
- Containers are moved using a high volume of low-pressure air to transport the product along the conveyor path

The major differences among the leading air conveyor manufacturers are

- Where the air is directed against the product, container, or bottle
- How the airflow is created (design of motors and fans for maximum efficiency)
- Where the airflow is created: one large blower vs multiple small blowers
- Construction details and "user-friendly" features of the air conveyor

See Figures 1–3 for air conveying operations.

How does air conveyor solve the problems found with chain conveyor and why is it the better method of conveying? It eliminates crushing. The air conveyor moves the packages by a directed flow of air against the containers. The flow of air can be controlled throughout the air conveying system. In this way, crushing of packages is eliminated as follows:

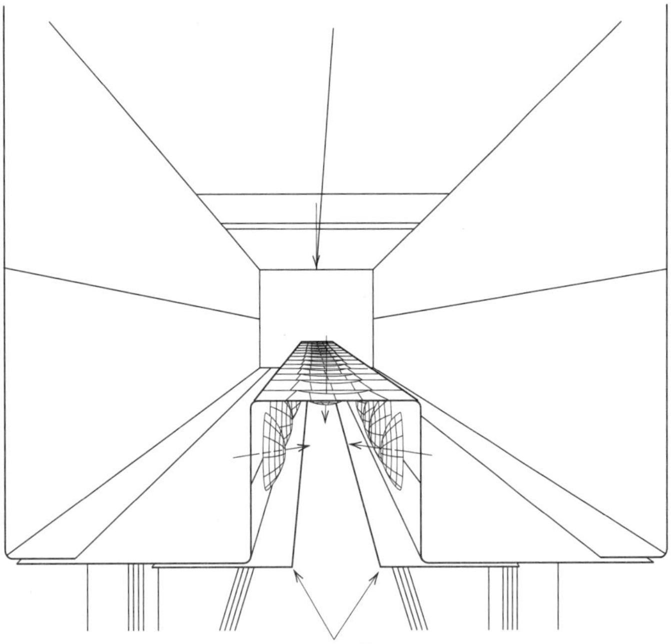

Figure 1. Airflow inside plenum.

- By reducing the backpressure force against the first product in a long line of products in a backfeed condition.
- By controlling the velocity of the product through the use of manual or automatic baffles, one can reduce the impact force of a product arriving from upstream and reduce damage.
- By introducing lift holes in combination with louvers, the product can hover while using the conveyor for accumulation on flat-top air conveyor—there is no marking of crushing from the friction caused by the belt or chain.

In beverage applications, the bottles cannot fall:

- The bottles cannot fall when the bottle is transported hanging by the neck support ring.
- Eliminating bottles falling down on the conveyor and getting caught in the starwheels or timing screws.

As a result, users of air conveyors derive the following benefits:

- *Gapless filling.* At the higher production speeds, even small inefficiencies are costly and unacceptable. Air conveyors will virtually eliminate the possibility of a missed cap or container by assuring that there will be a suffi-

Figure 2. Plenum, cutaway view.

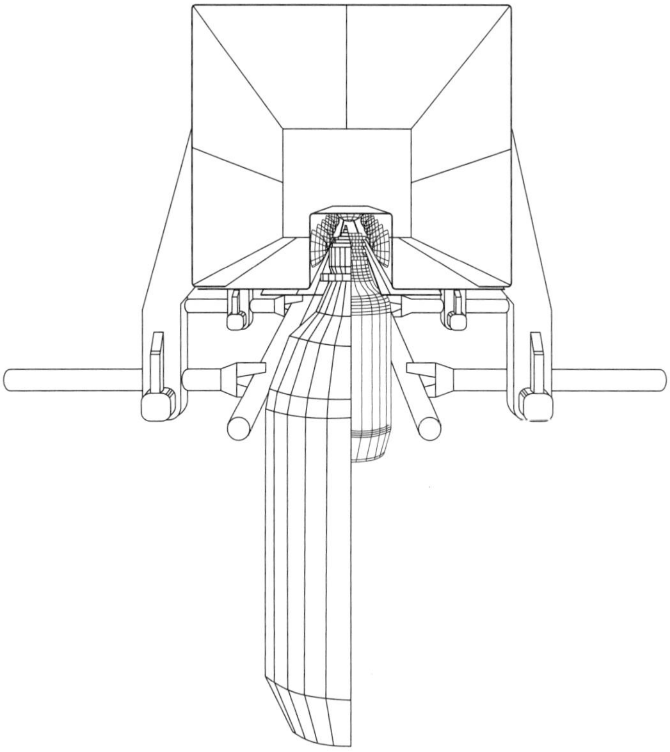

Figure 3. Single-lane air conveyor, adjustable bottle guide.

cient supply to the infeed of the filler, rinser, capper, or packaging equipment. In the event of a "hiccup" of the upstream equipment, bottles, caps, or other products from upstream will be conveyed quickly, allowing the line to recover.

- Shorter surge areas. Faster conveying speeds allows quicker recovery from the upstream supply of a product to the rinsers, fillers, cappers, or packaging equipment. As a result, users of air conveyors will be able to greatly reduce the lengths of conveyor to accommodate surges.
- *Reduction in buffer areas (BIDI tables)*. For the same reasons as described above, bidirectional accumulating tables can be eliminated. There may still be the need for accumulation, as, for example, where the labelers need to be changed over. For beverage applications we recommend side lengths of air conveyor where the bottles can be diverted for additional accumulations as the most effective means of accumulating and preventing bottles from falling over.

The net result of these advantages are less overall length of conveyor in the plant. This also results in both lower cost for the conveyor installation and freeing up of valuable plant space at floor level.

Characteristics of a Properly Designed Air Conveyor

Construction materials. We recommend and use stainless steel construction of the conveyor; this will result in lowered maintenance costs, easier washdown. Most importantly, the use of stainless steel guiderails on neckring a air conveyor avoids the need to replace guiderail wear strips.

Airflow. In cap conveying applications on flat-top air conveyors, center louvers are the most efficient. On heavier parts, the addition of lift holes combined with louvers aids the product flow, especially where accumulation is required. The three methods used most commonly are neck blow, shoulder blow, and sidewall blow, in neckring air conveyor applications. In these applications, we feel that neck blow is the preferred method because most bottles have the same neck dimensions as compared to body diameter dimensions. Using the neck dimension as the criterion, adjustment for different sized containers is reduced.

Flat-top air conveyor applications. The food industry found a solution in flat-top air conveyor and in turn is a major user of it. Because of the flat-top air conveyor's ability to convey with virtually no damage to the product, the manufacturers of candy, for instance, turned away from conventional tabletop conveyors and turned toward air conveyors. Because of the flat-top air conveyor's ability to lift as well as convey, the air conveyor can be used to accumulate products, eliminating the damage caused by friction from a belt or chain. These features make air conveyor the product transportation vehicle of choice.

The beverage industry has changed significantly with the growth of the 16- and 20-oz bottles. Use of 28- and 38-mm plastic beverage caps has grown as well. The beverage industry was now looking for a versatile method of conveying them while maintaining orientation to the capper. Their solution was to use an air conveyor coupled with a cap feeder/orientor. In general, the bottlers were now able to convey their caps quickly, efficiently, and 100% oriented to their cappers at speeds to 1800 Hz from remote locations, such as warehouses and production facilities.

Beverage industry applications

Faster filling speeds. Filling line speeds have and will continue to increase. Our company is now involved with projects where the required line speeds are:

- 2-L container: 800/min
- 20-oz bottle: 1100/min
- 16-oz bottle: 1200/min

Our company's Airtrans Air Conveyor installations have been to provide better control of the containers out of the depalletizers that are running at higher speeds.

One-piece and single-serving bottles. Because of their higher center of gravity and thinner-walled, lighter design, most one-piece bottles are inherently less stable than the base-cupped bottle. As well, in recent years there has been an increase in market share for the 16-oz and now the newly introduced 20-oz PET bottles. These containers are less stable on filling lines than the 2-L bottle and they fall over more easily than the glass bottles they replaced. The higher filling line speeds of the smaller bottles (up to 800 bottles per minute) further aggravates the problem of the bottles falling over and jamming on the conveyors. The neckring air conveyor virtually eliminates these problems by conveying quickly, cleanly, and with little to no jamming or tipping. Because the bottles are suspended by the neckring, the possibility of tipping is eliminated. As well, because the bottles are conveyed with air, there is much less back pressure exerted on the new, thin-walled bottles, significantly reducing damage. (see also Carbonated beverage packaging)

Marketability. A major trend has been the expansion of blow-molding plant outputs, in terms of both total output and numbers of packaging lines. This has resulted in an increased need to eliminate cable or chain conveyors and the increased implementation of air conveyors. Bottlers are splitting the output of their blow molding machined into several streams to downstream packaging equipment and combining the output of multiple blow molders into their packaging equipment. However, as shown in the applications of flat-top air conveyors, the applications of air conveying are not limited to blow molding applications by any means.

Problems of conventional tabletop conveyors needing solution. The following problems of cable and chain conveyors are commonly known by the industry:

- Greater nonproductive costs of mechanical conveyors due to higher maintenance of high-speed chain conveyors—mechanical conveyors have moving parts that are subject to wear and mechanical failure.
- Spare parts required:
  - Chain
  - Wear strips
  - Gears
  - Sprockets
  - Bearings
- Expense for disposable lubricants and wear parts.
- Unsanitary—grease and constant soap on bottles and product.
- Higher unplanned downtime of tabletop conveyors—these mechanical components will fail unexpectedly.
- Greater crushing of lightweight plastic bottles and delicate packages.
- Bottles fall over.
- Bottles jam on tabletop conveyors.

Benefits of Air Conveyors and Elimination of Popular Myths

Expense. An air conveyor system can actually be more cost-effective than chain, belt, or cable conveyor for the same application. While initial costs of an air conveying system may be more, the reduced maintenance, amount of spare parts, and downtime make air conveying a overall less expensive method of transport. Basic capital expenditures of the conveyor are only part of the complete cost picture. The true and total cost of the conveying system also includes:

- *Spare parts.* The air conveyor needs and uses less. There are few moving parts on an air conveyor so, the initial capital for spare parts and yearly additional parts' costs are less than for mechanical conveyors.
- Maintenance labor costs are less than for mechanical conveyors. Because there are few moving parts, there is less likelihood of breakdown and need of repair.
- Air conveyors can be more readily located overhead. This savings in floorspace costs may be applied as cost saving.

Flexibility. Air conveyors offer significantly more flexibility than chain conveyors for

- *Revisions to floorplan.* Both flat-top and neckring air conveyor systems are furnished in sections that bolt to each other in a continuous path. The modules typically are combinations of straight sections, horizontal curves, vertical curves, and gates for merging and diverging. Any of these modules can be reconfigured in a different combination and can be added or deleted. Our company reconfigured a system that had been shipped 4 years previously, and by adding additional sections in the original system and adding other new sections, we provided a totally different conveyor system layout. Very few of the old sections were wasted, but rather they were reused elsewhere in the new conveyor line. The entire system was reconfigured with less than 1 week of installation and dismantling.
- *Multiple sizes of containers.* The flat-top air conveyor can generally carry 5 lb/ft^2. As well, our flat-top air conveyor can be designed to accommodate many differently sized products. Whether it be through a dual-lane, multilane, or single-lane flat-top air conveyor, from unwrapped candy to boxes to caps, a virtually endless variety of products can be transported using this system.
- *Multiple sizes of beverage containers.* Multiple sizes of containers are accommodated using several different techniques. Different heights of containers are accommodated by adjusting the height of the air conveyor through hand wheel adjustment or automatically. The air conveyor transports bottles hanging by the neck support ring. The Aidlin Airtrans has hinged end sections on the infeed and discharge ends of the air conveyor. Similarly, the height would be adjusted when discharging bottles to the infeed screw of the filler. Infeed and exit plenums are hinged to allow adjustment of bottle height in the air conveyor. Also, bottle heights from the same supplier can have height variations for which the conveyor may need to be adjusted. Fixed height neck rails obviously do not have the necessary adjustability for this condition. Different neck diameters are accommodated by Aidlin's Dual-Lane Airtrans. The Dual-Lane Neck Ring Air Conveyors is a double-lane neck rail. One lane is set up for 16-oz bottles, while the other is set for 2-L containers; similarly, one lane could be set for 28-mm neck finishes and the other lane set up for 38-mm neck finishes (as on 3-L containers). A single air plenum is switched over to supply either set of neck rails as required.

Interfacing with other equipment. One significant advantage of both flat-top and neckring air conveyor is their ability to easily interface with other packaging equipment (see also, Blow holding; Labels and labeling machinery; Palletizing).

Mechanical interfacing

- The flat-top air conveyor can transport from and to most equipment: from the orienting orientors to the cappers, liners, or decorators; or to a wrapper, cartoner, or case packer. When needed, the air conveyor can be fit to virtually any line.
- *Blow molders*—bottles are received either through a bot-

tle collector conveyor (as on the Cincinnati, Magplas, Nissei) or directly from the output neck rails (as on the Sidel and Krupp).

- *Palletizers*—an escapement is mounted to the discharge of the air conveyor to stabilize the bottle and match the container's speed to be the same as the infeed conveyor to the palletizer.
- *Depalletizers*—containers are received off the outfeed conveyor directly to the split neck rails of the air conveyor. The bottles are accelerated and conveyed away from the palletizer at a faster line rate than the depalletization. In this way, there is no possibility of bottles falling down.
- *Labelers*—containers can be placed directly into the in feed starwheel or timing screw of the labeler.
- *Fillers*—bottles are placed directly into the infeed timing screw of the filler. By assuring a proper backpressure and constant supply of bottles, maximum filling speed is achieved.

Maintenance. The total maintenance factor of air conveyor is significantly less than that for mechanical conveyors. For example, the normal maintenance in the Airtrans system consists of replacing, in less than one minute, the 5-μm fan filters as needed. In our flat-top air conveyor, both the top guiderail and Lexan covers are hinged for easy cleaning.

Less contamination to products. Based on R&D done at Aidlin Automation in Bradenton FL, the air transporting the bottles or products is filtered, in our case, to 5-μm. The net result: bottles or products such as food or caps remain cleaner than in the typical plant where the neck, cap, or product is open to unfiltered ambient air.

Conclusion. As one can clearly see, air conveying provides the alternative to chain, cable, or belt conveyors. Air conveying provides clean, consistent, and predictable performance. The new generation of conveying technology is here and in great demand. In order to be profitable in the quickly changing industry, one must keep up to date with new technology—and that is the air conveyor.

TIMOTHY AIDLIN
Aidlin Automation
Sarasota, Florida

AMPULS AND VIALS, GLASS

Ampuls and aluminum-seal vials are glass containers used primarily for packaging medication intended for injection. Ampuls are essentially single-dosage containers that are filled and hermetically sealed by flame-sealing the open end. Vials, which contain single or multiple doses, are hermetically sealed by means of a rubber closure held in place with a crimped aluminum ring.

An ampul is opened by breaking it at its smallest diameter, called the *constriction.* A controlled breaking characteristic is introduced by reproducibly scoring the glass in the constriction, or by placing a band of ceramic paint in the constriction. The ceramic paint has a thermal expansion that differs from the glass, thus forming stress in the glass surface after being fired. This stress allows the glass to break in a controlled fashion at the band location when force is applied. Medication is then withdrawn by means of a syringe.

Medication can be withdrawn from a vial by inserting the cannula of a syringe through the rubber closure. Since the rubber reseals after cannula withdrawal, multiple doses can be withdrawn from a vial.

Both ampuls and vials are fabricated from glass tubing produced under exacting conditions. The glass used for these containers must protect the contained product from contamination before use, and in the case of light-sensitive products, from degradation due to excessive exposure to light. In addition, the glass must not introduce contamination by interacting with the product.

Table 1. Compositions of Soda–Lime and Borosilicate Glasses, wt%

Constituent	Soda–lime	Borosilicate
SiO_2	68–72	70–80
B_2O_3	0–2	10–13
Al_2O_3	2–3	2–7
CaO	5	0–1
MgO	4	
Na_2O	15–16	4–6
K_2O	1	0–3
typical forming temperatures	1796–1895°F (980–1035°C)	2066–2264°F (1130–1240°C)

Glasses

The most important property of a glass used to contain a parenteral (injectable) drug is chemical durability; that is, the glass must be essentially inert with respect to the product, contributing negligible amounts of its constituents to the product through long-term contact before use. The family of glasses that best meets chemical durability requirements is the borosilicates. These glasses also require higher temperatures for forming into shapes than other glass types.

When glass–product interactions are far less critical, the soda–lime family of glasses can be used to fabricate vials. These glasses can be formed at lower temperatures than borosilicates, but do not nearly have their chemical durability. Typical compositions are shown in Table 1. Borosilicate and soda–lime glasses contain elements that facilitate refining, but borosilicates generally do not contain arsenic or antimony.

Both borosilicate and soda–lime glasses can be given a dark amber color by adding small amounts of coloring agents, which include iron, titanium, and manganese. The amber borosilicate and soda–lime glasses can then be used to package products that are light-sensitive.

The interior surface of containers formed from soda–lime glass is often subjected to a treatment that enhances chemical durability without affecting the desirable lower melting and forming temperatures typical of soda–lime glass. For very critical applications, borosilicate ampuls and vials can be treated to improve their already excellent chemical durability.

For pharmaceutical packaging applications (see Pharmaceutical packaging), the various types of glass have been codified into groups according to their chemical durabilities, as specified by the *United States Pharmacopeia* (USP) (1). The glasses are classified by the amount of titratable alkali extracted into water from a crushed and sized glass sample during steam autoclaving at 250°F (121°C). Thus borosilicate glasses are typical of a USP Type I glass, and most soda–lime glasses are typical of a USP Type III glass. There are some soda–lime glasses that are less chemically durable than Type III glass, and these are classified as USP Type NP.

USP Type III (soda–lime) containers that have had their interior surface treated to improve durability can be classified as USP Type II if they meet the test requirements. The test in these cases is performed on the treated container instead of a crushed sample, using a similar steam autoclave cycle.

The pharmacopeiae of other nations have also classified glass into groups according to their chemical durability. These classifications are generally similar to those specified by USP.

Forming Processes

Ampuls and vials are formed from glass tubing. The glass tubing is formed by processing in a glass furnace and by a tube-forming operation. The glass furnace operation consists of bulk batch preparation, continuous batch melting, and refining (see Glass-container manufacturing). The tube forming is done to exacting specifications in either a Danner process or a downdraw process. The Danner process involves continuous streaming of molten glass onto an angled rotating sleeve that has an internal port for inflation air. The inflation air controls the tubing outside diameter (OD). The downdraw process is an extrusion process through an annular area. The inner core has an inflation air hole. The inflation air serves the same purpose as in the Danner process. In either process the tubing wall weight is controlled by adjusting the rate of glass withdrawal and supply. Typical ampul and vial tubing dimensions and tolerances are shown in the Figures 1 and 2.

The tubing is formed in a continuous-line process. Various devices are used to support the tubing during pulling. A device, normally consisting of pulling wheels on belts and a cutting mechanism, is situated downstream to pull and cut the tubing. The tubing is cut to prescribed lengths and used in vertical- or horizontal-type machines for converting the tubing into vials or ampuls.

Many machines are rotary and either index or operate with a continuous action. The tubing is placed in the machines and is handled in a set of chucks. Heat is applied in the space between the chucks, and forming of the ampul or vial occurs throughout the machine rotation cycle.

Ampuls are formed on continuous-motion rotary machines. One sequence is shown in Figure 3. The process consists of sequentially heating and pulling (elongating the glass) to form the constriction, bulb, and stem contours of the ampul. The ampul contours are controlled primarily by proper temperature patterns in the tubing and by pulling rate of the tubing. Mechanical tooling of the glass can be used to assist in constriction contour forming. The forming process accurately controls the seal plane diameter, which controls ampul closing after filling. After the basic ampul is formed on the machine, the ampul blank is separated from the tubing and is transferred to a horizontal afterforming machine. On the afterforming machine the ampul is trimmed to length, glazed,

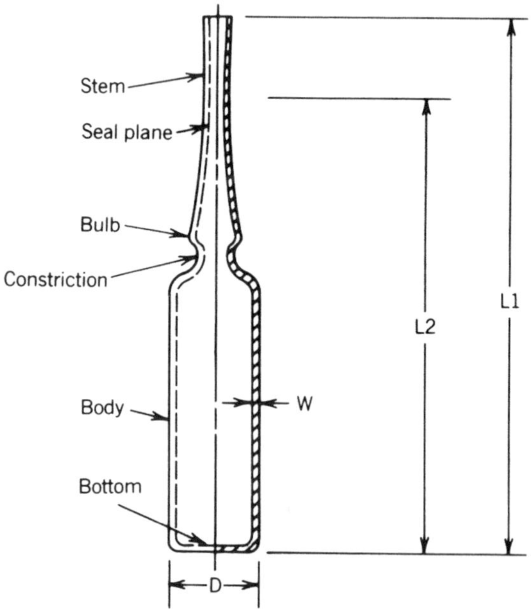

Figure 1. Standard long-stem ampul.

Capacity, mL	Diameter (D) mm	Width (W), mm	Length (L1), mm ± 0.50 mm	Length (L2), mm
1	10.40–10.70	0.56–0.64	67	51
2	11.62–12.00	0.56–0.64	75	59
5	16.10–16.70	0.61–0.69	88	73
10	18.75–19.40	0.66–0.74	107	91
20	22.25–22.95	0.75–0.85	135	120

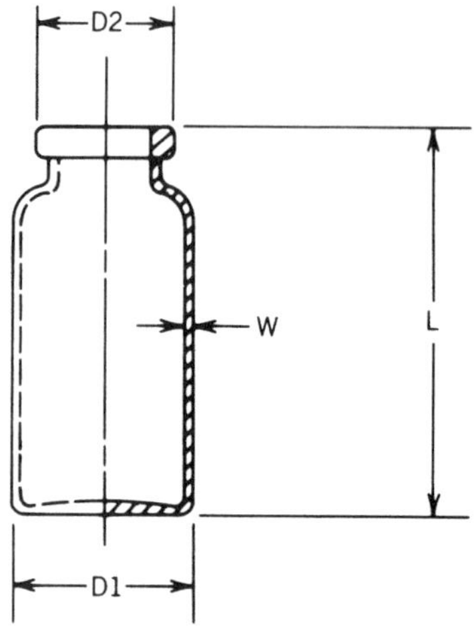

Figure 2. Standard tubular serum vial.

Capacity, mL	Diameter (D1), mm	Width (W), mm	Length (L), mm ± 0.50 mm	Diameter (D2), mm
1	13.50–14.00	0.94–1.06	27	12.95–13.35
2	14.50–15.00	0.94–1.06	32	12.95–13.35
3	16.50–17.00	1.04–1.16	37	12.95–13.35
5	20.50–21.00	1.04–1.16	38	12.95–13.35
10	23.50–24.00	1.13–1.27	50	19.70–20.20
15	26.25–27.00	1.13–1.27	57	19.70–20.20

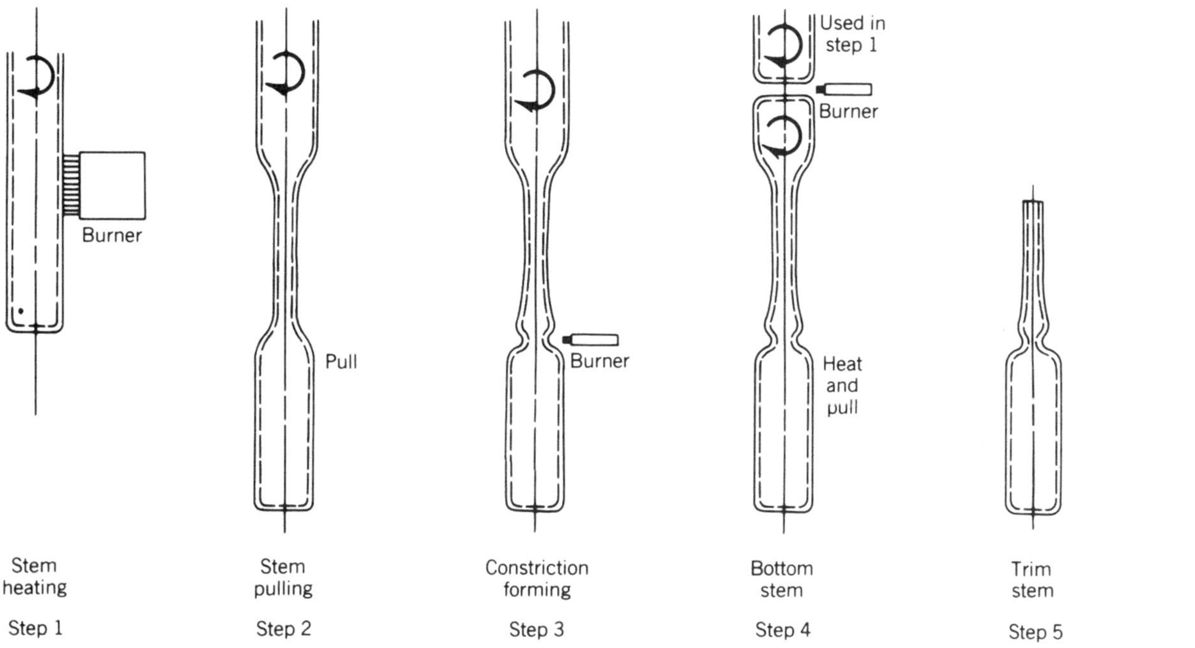

Figure 3. Ampul contour-forming sequence.

and treated if necessary. Also, the ampul constriction can be either color banded with a ceramic-base paint or scored to control opening properties. The ceramic paint and scoring cause stress concentrations in the constriction which assist in obtaining desirable opening force and fracture characteristics. Identification bands are applied and the ampul is annealed to relieve the strains caused by the thermal forming of the ampul. The completed ampuls are then transferred into a packing area where the ampuls are accumulated, inspected manually or automatically, and packed into clean trays for distribution (see Fig. 4).

Vial forming is done on vertical machines that either index or have a continuous motion. A vertical forming sequence (see Fig. 5) consists of a parting (separation operation) wherein a narrow band of glass is heated to a soft condition and the vial blank and the tubing are pulled apart. After parting, the finish-forming operations occur. The finish forming consists of heating and mechanically tooling the glass in sequential steps. Normally, multiple heating and tooling operations are necessary to form the closely held tolerances of aluminum-seal finishes. The tooling is done with an inner plug to control the contour and diameter of the finish bore and with outer contoured round dies that control the contour and diameter of the finish outer surface. The vial bottom contours are formed in the lower chucks while finish forming occurs for another vial in the upper chucks. After tooling, the vial length is set by a mechanical positioner. The process then continually repeats itself until the whole tubing length is consumed. After fabrication, the vial blank is transferred to a horizontal afterforming machine. The operations that are normally performed on an afterformer are dimensional gauging, vial treatment, and annealing. The vials are then transferred to a packing area where they are accumulated, inspected manually or automatically for cosmetic conditions, and packed in clean containers (see Fig. 6).

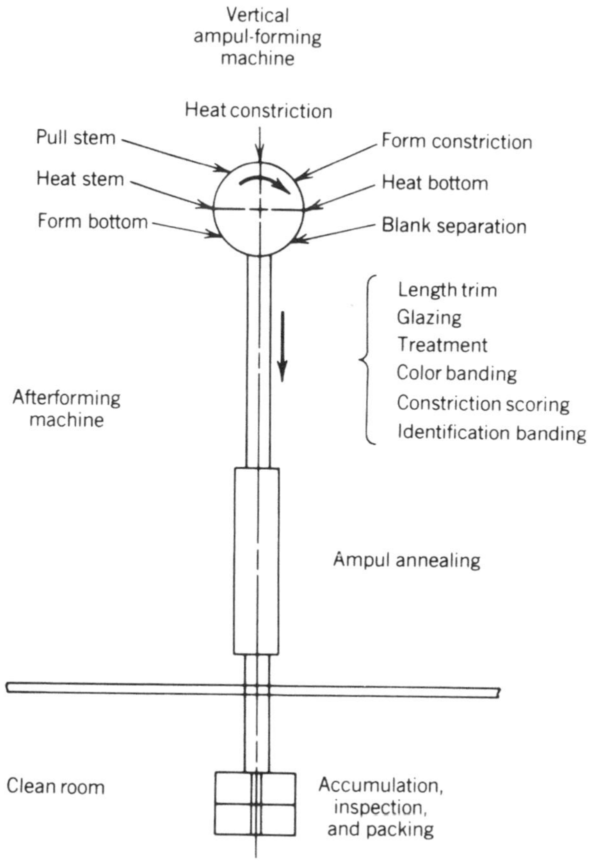

Figure 4. Tube converting for ampul manufacture.

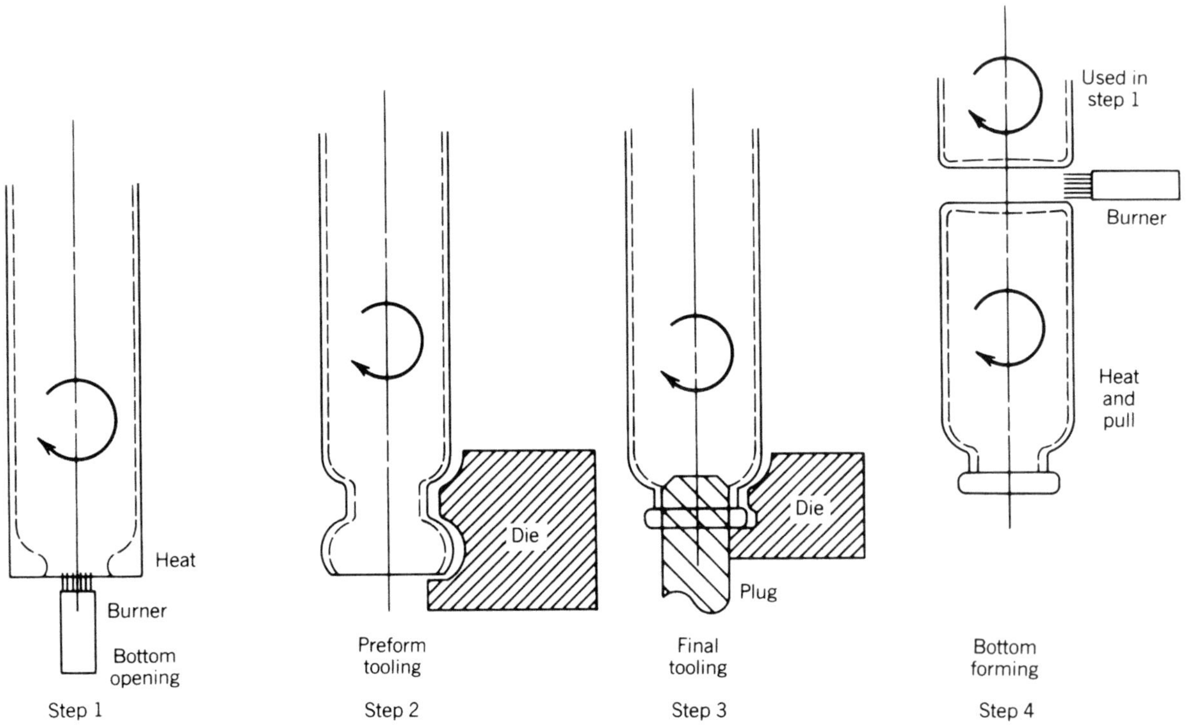

Figure 5. Vial contour-forming sequence.

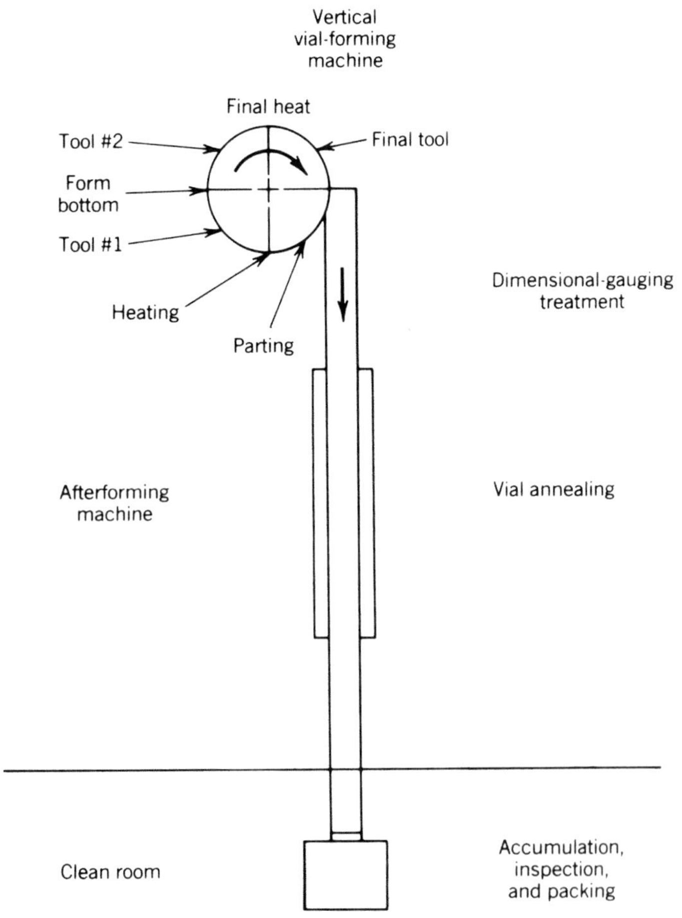

Figure 6. Tube converting for vial manufacture.

BIBLIOGRAPHY

"Ampuls and Vials," in *The Wiley Encyclopedia of Packaging Technology,* 1st ed., Wiley, New York by R. P. Abendroth and J. E. Lisi, Owens-Illinois, Inc, 1986, pp. 16–20.

1. *The United States Pharmacopeia XXII,* and the *National* Formulary. The United States Pharmacopeial Convention, Inc., Rockville, MD, 1990

AROMA BARRIER TESTING

An aroma can be described as an odor, perfume, scent, fragrance, bouquet, or redolence. Characterizing a package's ability to either prevent or promote aroma transfer has been a hot topic for some years now. With the introduction of newer, lightweight, low-cost polymer materials replacing much of the traditional packages, such as glass and aluminum, data about aroma transfer have become increasingly popular and necessary. Loosing aroma from a package or keeping harmful organics from the product is extremely important when trying to extend shelf life, or limit off-odor and liability issues. Until recently, most experimentation has been conducted by the major universities and industry leaders using benchtop apparatus. Differing techniques have led to discussions on proper procedural methodologies, and therefore driven the instrumentation industry to develop a commercial instrument to measure aroma barriers using an accepted technique.

The use of moisture or oxygen permeability experimentation can certainly assist in a material or design decision. Yet, using these results to interpret aroma permeability has been found to be very unreliable and in some cases dangerous (1). The detection of aroma transfer can be a much slower process than moisture or oxygen barrier measurement, often requir-

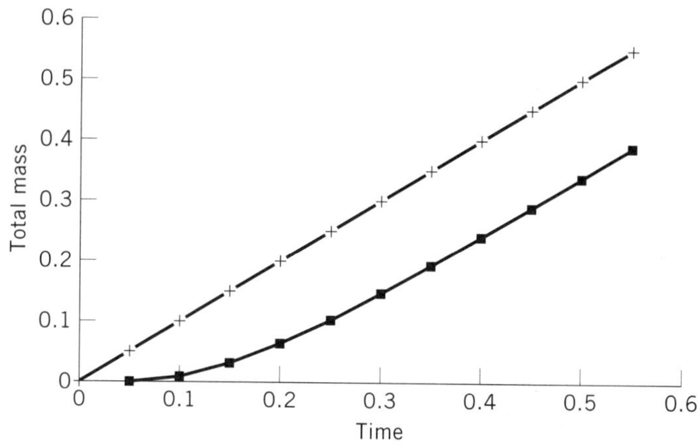

Figure 1. Mass transport across a membrane, with *S* and *D* set to 1. ■, Mass transport; +, permeation.

ing days, weeks, or months until a final result is obtained. Shelf-life studies and taste panel evaluations will always remain as the final word in package design and source guiding, yet these methods are also extremely slow and often very expensive. It stands to reason that prescreening packaging materials to fit a particular need before shelf-life studies and/or their final application is the most cost-effective method.

Theory

Aroma or organic permeation occurs directly through a packaging material by first absorbing into the high-aroma-concentration side of the material, transversing the median and desorbing off the low-aroma-concentration side. Over time, as the aroma diffuses into the packaging material, an equilibrium in the material solubility isotherm occurs. This is defined as steady state, and is the condition at which the amount of permeation is measured.

When dealing with a flat, planar packaging material such as films, sheets, coated papers, and laminates, steady state conditions will simplify the interpretation of permeation data. Although steady state may be acceptable for moisture or oxy-

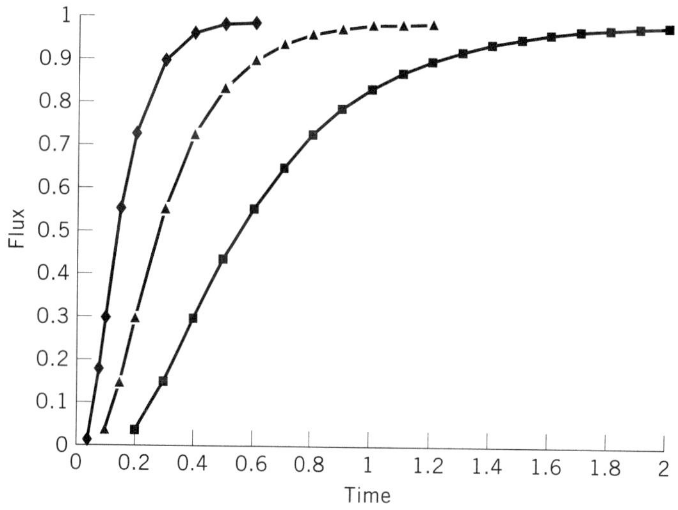

Figure 3. Permeated flux vs diffusion. ◆, D = 1; ▲, D = 0.5; ■ D = 0.25.

gen permeation, aroma permeation typically will not reach steady state. Therefore, data should be obtained about the transient portion of the experiment, as steady state is being reached, namely, diffusion and solubility. Diffusion is the rate of molecular transfer and can be thought of as the length of time to reach steady state, which pertains to shelf life and solvent release rate estimates. The basic theory of the diffusion process, proposed by Fick in 1855, states that molecular travel occurs from a high concentration of molecules to a low concentration, and that the rate of transfer will be proportional to the concentration difference (2). Solubility is a density per unit volume value, representing the amount of permeant absorbed by the material. Flavor loss is therefore a combination of diffusion and solubility. (See Figs. 1 and 2.)

Most measurements today solve for a permeation coefficient, yet to really understand what is going on, diffusion and solubility must also be defined (see Permeability and diffusion aromas, solvents). It does not matter if aromas are lost be-

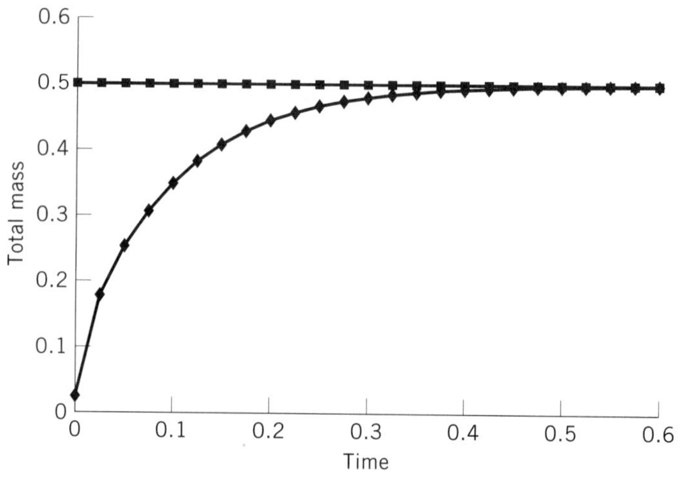

Figure 2. Mass absorbed by a membrane *D* and *S* equal to 1. ◆, Mass transport, ■ permeation.

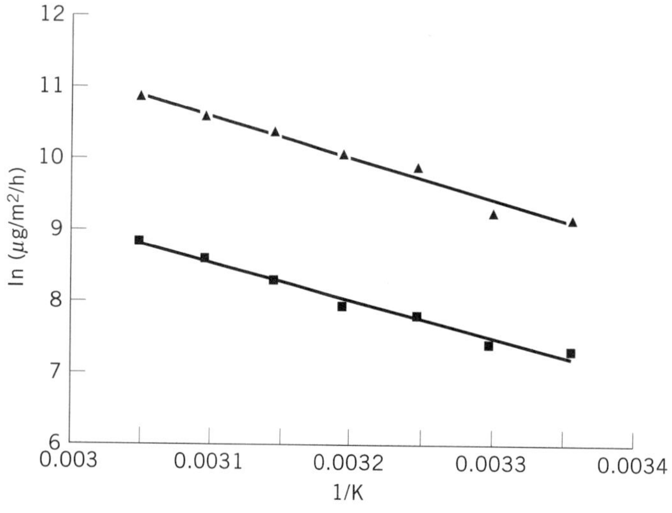

Figure 4. Permeation coefficients. ■, met OPP; −met OPP; ▲, OPP; −OPP.

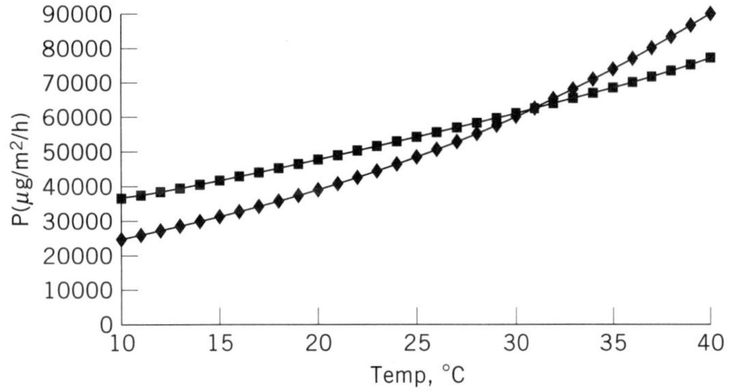

Figure 5. Permeation vs temperature D-limonene vapor. ■, Film 1; ◆ Film 2.

cause of permeation or solubility, any migration is considered a loss. For example, situations can occur where permeation is zero, yet flavor loss exists due to high solubility. All three values are related by $P = D * S$. Therefore, it is possible for three different packaging materials to have the same permeation rate, yet different diffusion and solubility rates. (See Fig. 3.)

Effects of Temperature

Controlling temperatures is critical to measurements of this nature. In fact, permeation and diffusion data are exponentially related to the temperature of the material. Failure to control temperature during testing can result in low sensitivity, often too low to measure the small values associated with testwork of this nature. On the other hand, accurately controlling the material's temperature can yield significant results. By testing at accelerated temperatures, test time can be dramatically reduced without sacrificing precision. Careful notation of the polymer's glass transition temperature is critical and testing should be conducted below this point. Since temperature is exponentially related to permeation and diffusion, the Arrhenius law states that the log value of these coefficients are linearly related to the inverse of temperature in degrees Kelvin (3). (See Fig. 4.)

Testing at accelerated temperatures provide results across a wide range of end-use conditions and are necessary, since permeation, diffusion, and solubility values between materials may be identical at specific environmental conditions, yet completely different at others. (See Fig. 5.)

By testing at multiple temperatures, the user is able to extrapolate the activation energy values, the true value by which barrier structures should be compared.

Apparatus

The issue is how to obtain results in a reasonable timeframe. Numerous apparatus have been designed, to different degrees of success, to obtain this information. Typical systems were usually quasi-isostatic, gravimetric, or isostatic. Quasi-isostatic techniques employ an accumulation process in which a material sample was mounted in a permeability cell above either a static sample of liquid permeant or continuous flow of gaseous permeant. As the permeant diffuses across the membrane, headspace samples are collected from above the

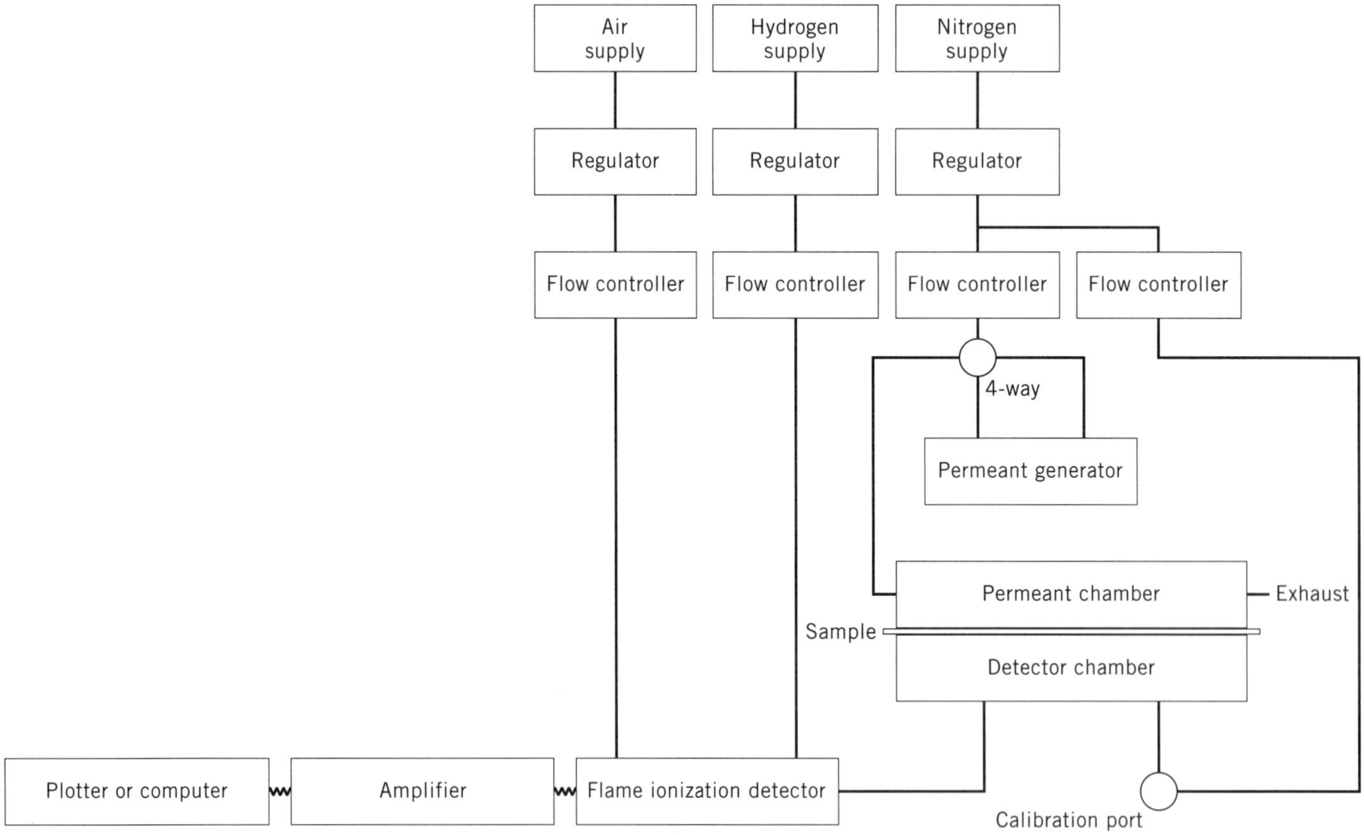

Figure 6. Isostatic apparatus.

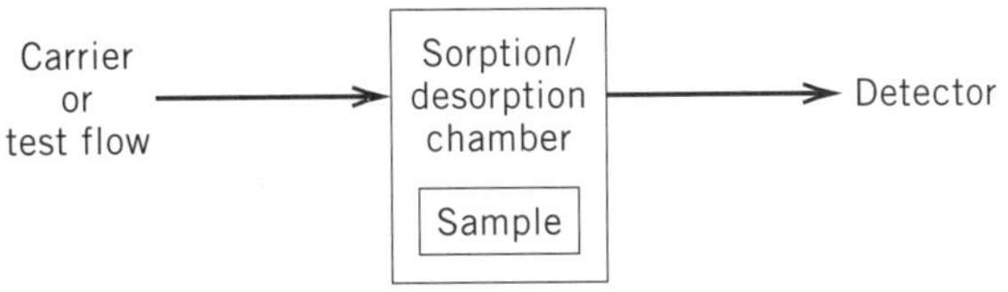

Figure 7. Desorption technique.

material and quantified using gas chromatography. This indirect process is often slow and tedious. Gravimetric methods are similar to quasi-isostatic methods, except instead of quantifying by means of chromatography, a microbalance is employed to actually weigh the material over time as it is exposed to a permeant. Isostatic methods employ a continuous stream of carrier or test gas to both sides of the cell. (See Fig. 6.)

The measurement value is therefore a flux expressed in terms of the number of moles per unit area per unit time rather than total mass. The flux value is continuously time-stamped and direct, real-time measurements of permeation are typical. With the advent of the microprocessor and by employing mass-transport theory, data about flavor permeation, diffusion, and solubility can all be obtained as they actually happen.

Test Vapor Generation

One primary concern when dealing with aroma barrier testing is generating an accurate test vapor. Since diffusion and solubility are specific to a given material and a single compound, accurate data require precise control of the test permeant concentration. Since aromas are a complex mixture, the use of an aroma as a test permeant is typically difficult.

Although aromas consist of many compounds, only a handful will typically interact with the packaging material. The first step in aroma–package interaction is therefore to identify which specific compounds solubilize within the package material. This can be accomplished by exposing the material to the aroma of interest in an enclosed chamber via a constant flow rate at a constant temperature and then purging the sample, observing which compounds desorb back off the material. Identification is accomplished with mass spectrophotometry or gas chromatography. (See Fig. 7.)

Once the specific problem compounds have been identified, each can be accurately analyzed for their affinity to permeate, diffuse, and solubilize in the material since controlling the concentration of a single or a group of two or three compounds is very easily accomplished.

The MAS 2000 Organic Permeation System (designed by MAS Technologies, Inc., Zumbrota, MN) is a commercially available high-speed instrument that employs full mass-transport theory combined with full analysis of the effects of temperature. Results regarding permeation, diffusion, and solubility are standard. MAS Technologies, Inc. is also in the process if finalizing an ASTM method for testing of this nature.

BIBLIOGRAPHY

1. P. T. DeLassus, "Permeation of Flavors and Aromas through Glassy Polymers," *Tappi J.* **77**(1), 109 (1994).
2. J. Crank, *The Mathematics of Diffusion,* 2nd ed., Oxford University Press, New York, 1975, pp. 32, 52.
3. R. M. Felder and G. S. Huvard, *Methods of Experimental Physics,* Vol. 16c, Academic Press, New York, 1980, pp. 324, 344.

General References

J. Ylvisaker, "The Application of Mass Transport Theory to Flavor and Aroma Barrier Measurement," *Tappi Polymers, Laminations & Coatings Conf. Proc.,* 1995, Book 2, p. 533.

W. E. Brown, *Plastics in Food Packaging: Properties, Design and Fabrication,* Marcel Dekker, New York, 1992.

H. Maarse, *Volatile Compounds in Food and Beverages,* Marcel Dekker, New York, 1991.

Darrin Seeley
MAS Technologies
Zumbrota, Minnesota

ASEPTIC PACKAGING

Aseptic packaging is an alternative to conventional canning in the production of shelf-stable packaged food products. The term *aseptic,* derived from the Greek word *septicos,* means the absence of putrefactive microorganisms. Aseptic packaging technology is fundamentally different from that of traditional food processing systems. Canning processes commercially sterilize filled and sealed containers, while aseptic systems commercially sterilize the product separately from the package; which is sterilized during filling and sealing operations. In aseptic packaging, presterilized product is filled into sterilized containers that are hermetically sealed in a commercially sterile environment. Aseptic processing and packaging systems are separate but integrated operations where the packaging step relies on the processor to provide a quality,

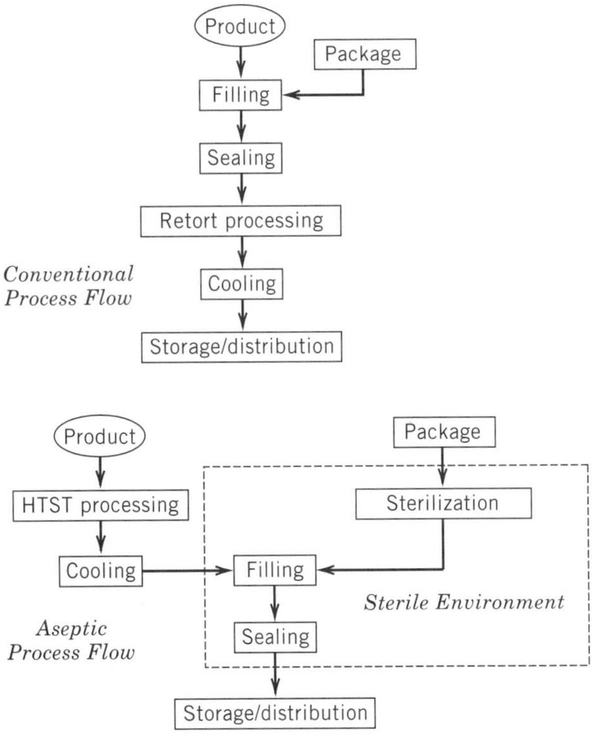

Figure 1. Simplified comparison of conventional and aseptic processes.

sterile product. Figure 1 is a simple illustration comparing the basic differences between conventional canning and aseptic packaging processes for the production of shelf-stable food products. (See also Shelf-life).

The Thermal Process

Aseptic systems typically utilize a continuous-flow heating process referred to as *high temperature short time* (HTST). HTST processes raise the product to very high temperatures for a few seconds and rapidly cool the product prior to packaging. This form of heat treatment enables the commercial sterilization of the product and results in minimal loss of product quality and nutrition. In contrast, conventional retort processing systems thermally process at relatively lower temperatures for much longer periods of time (minutes) to achieve the same degree of commercial sterility. Unlike the HTST systems, process times can vary with retort processes depending on the size of the container. Thus, a smaller can of tomato sauce would receive a shorter thermal process than the same product packaged in a larger can. These differences in time are associated with the slower transfer of heat required to adequately process larger masses of product. There are, therefore, practical limits to the size of containers that can be treated by conventional process systems without severely cooking the product. Conversely, the aseptic HTST thermal process is constant for each product treated and is independent of package size. Larger containers such as drums and even tankers can be aseptically filled with HTST processed product without the problems of thermal degradation and loss of quality.

The Package

An important distinction of aseptic technology is that the container does not have to withstand the extreme conditions of heat and pressure encountered during retort processing. The package material and structural requirements are therefore reduced to that needed for the maintenance of structural integrity, hermeticity, and barrier protection of the product. Pioneering work in aseptics utilized metal cans for packaging. In these early applications, container sterilization was achieved through the use of steam. More recently, the acceptance of chemical sterilants has enabled the use of a wide range of packaging materials for aseptic applications. Aseptic packaging has taken advantage of the use of plastics and composite paperboard materials for the production of lightweight containers in a variety of sizes and shapes.

Additional information on the technical aspects of aseptic processing and packaging is available in two excellent reference books (1,2).

The History of Aseptic Packaging

The first practical application of thermal processing for the canning of packaged foods is credited to Nicholas Appert, a French cook and confectioner, who in the 1800s developed shelf stable foods packed in glass jars for Napoleon's navy (3). The significance of his work was discounted by the scientific community until Pasteur finally disproved the doctrine of spontaneous generation in 1861 (4). It was the understanding of thermal treatment for the preservation of food in sealed containers that led to the development of the canning industry.

Recognition that product sterilization of corn and other vegetables prior to packaging was effective in improving the quality of the finished product was reported by Plummer and Stare in 1902 (5). Dunkley, in 1918, patented a process that was one of the forerunners of modern aseptic canning. In this process, product was sterilized in tanks at elevated pressure and temperature using live steam. After the product temperature was reduced it was transferred to an aseptic storage vessel prior to filling. Can and lid sterilization were achieved with live steam. Filling and closing were accomplished in a steam chest under atmospheric conditions (6). William Martin developed an improved aseptic canning process in the 1940s (7) that became known as the Martin–Dole process. This is generally recognized as the first commercialized version of aseptic canning to gain a wide acceptance in the industry. Martin utilized an HTST process for the product, and superheated steam (350–500°F) for sterilizing the container and lid, and for sterility maintenance in the sterile zone during filling and sealing. Numerous improvements to the process have been made and aseptic canning is widely used in the United States today.

In 1981 the U.S. Food and Drug Administration accepted the use of hydrogen peroxide as a chemical sterilant for aseptic packages comprising a low-density polyethylene product contact surface (8). Subsequent acceptance of other plastic materials and other chemical sterilants has led to dramatic growth and diversity of the domestic aseptic industry.

Aseptic Process Systems

Aseptic processing equipment consists of a heating section to bring the product to the required temperature, a holding section to maintain the product temperature for a minimum period, and a cooling section to reduce the product to the desired final temperature prior to sterile filling and sealing. Since aseptic process systems are continuous, they require product flow control to assure that thermal process requirements are met. Depending on product requirements, deaeration and aseptic homogenization may also be incorporated into the process.

Aseptic processes fall into two broad categories; direct and indirect. The heating section in direct systems incorporates the sterilizing medium (steam) directly into the product, while the heating section of indirect systems separates the product and sterilizing medium by a stainless steel heating surface.

Direct systems. There are two types of direct steam processing systems: *steam injection,* which heats product by injecting steam into the product flow in order to achieve a very rapid temperature increase; and *steam infusion,* a process involving the direct introduction of product into a steam atmosphere resulting in a similar rapid increase in product temperature. Both processes utilize a flash (vacuum) vessel to rapidly cool the product and remove condensed water, and an indirect cooling section to reduce the final temperature to ambient or lower. Direct systems have the advantage of very rapid heating and cooling, and a minimum problem with product scorching. These processes are well suited to liquid products including milk, where flavors are affected by high process temperatures.

Indirect systems. Plate, tubular, and scraped-surface heat exchangers represent the commonly used indirect systems for aseptic HTST processing. The heating medium used for indirect systems may be hot water or steam and the cooling medium is typically chilled water. The choice of indirect system is largely determined by the physical properties of the product to be processed. Indirect process systems encounter some degree of product fouling on the heat-transfer surfaces, which can impact product flavor, and process efficiency.

Plate systems comprise a number of parallel, closely spaced, corrugated stainless steel plates compressed together in a steel frame. Product flows through the heat exchangers as a thin film that passes between alternate plates. The porting design of the plates directs the flow of the product to one side and the heating or cooling medium to the other. Plate systems are efficient and often utilize regenerative heating sections in the process. These systems are typically used to process liquid products, and are limited by product viscosity and system pressure.

Tubular systems are available in a variety of design configurations. These systems incorporate stainless steel tubes where product flow is directed through the inside while that of the heating or cooling medium is applied on the outside. Some designs incorporate product regeneration. Tubular heat exchangers are designed to handle liquid and viscous food products as well as those containig particulate pieces. These systems are limited by product particulate size based upon tube diameter.

Scraped-surface heat-exchange (SSHE) systems consist of large jacketed cylinders equipped with blades that rotate and scrape the interior cylinder walls. Product flows through the cylinder while the heating or cooling medium is applied to the outside jacketed area around the cylinder. Rotation of the scraper blades minimizes product fouling on the cylinder walls and agitates the product as it advances through the cylinder. SSHE systems are capable of processing very viscous products as well as products containing large particulate pieces (9,10).

Other systems. Ohmic, or electrical resistance, heating has been shown to be an effective means for aseptically processing foods containing particulate material (10). Sterilization results from the passage of electrical current, usually from a series of electrodes, into the product as it flows through the ohmic heater section. High temperatures are achieved by high system pressures. Product cooling is accomplished with indirect heat exchangers such as tubular or SSHE systems. The advantage of the ohmic process is that the particle and the liquid phase of the product heat almost simultaneously as compared to other heating methods that rely on thermal convection and conduction for the liquid and solid phases. The disadvantage of this process is the dependence on electrical conductivity and resistance in the product. Consideration must be given to product composition and formulations to assure the correct heat treatment from ohmic processes from product to product, or batch to batch.

Pulsed electric field (PEF) processing is an emerging food processing technology that utilizes high voltage for product sterilization. In contrast to Ohmic heating, PEF processes generate pulsed high-voltage electrical energy that incites damage directly to the cell membrane of vegetative bacterial cells. Sporicidal capabilities are being studied. There is little or no product temperature rise in PEF processes. Research is currently under way at The Ohio State University in evaluating the PEF technology for use by the U.S. Army (11).

Another relatively new process for treating foods utilizes isostatic high pressure under ambient or warm conditions to sterilize products package in flexible containers (12). This process has been utilized in Japan for the production of packaged fruit products, and is under evaluation in the United States. Isostatic pressures of 65,000 to several hundred thousand pounds have been demonstrated as effective in preserving high acid products. Temperature increases to 100–150°F improve the lethality of the process without cooking the product. Product enzymes are also denatured by this process.

Aseptic Filling Systems

There is considerable variety in aseptic filling machines as well as packaging in use in the United States. Packages cover the spectrum of sizes and materials ranging from 1-oz form–fill–seal plastic cups such as dairy creamers, to 300-gal plastic bags in containers used for bulk production of food products (13). As of 1995, there were 22 aseptic machine manufacturers whose filling systems were in commercial use in the United States (14). Those aseptic filling systems with multiple installations that are currently in production are summarized in Table 1. The data presented include the type of filling system, material supply, finished package form, package sterilization mode, and the U.S. location of the manufacturer if additional information desired.

Regardless of package size, all aseptic filling machines have certain requirements in common. They must be capable of being sterilized and of sterilizing the package (if not presterilized). Machine and package sterilization can be accomplished through the use of steam or chemicals such as hydrogen peroxide. These modes of sterilization are proprietary to the filling machine manufacturer. In addition, aseptic filling machines must maintain the package sterility during filling and sealing for extended periods of production. Maintenance of sterility is achieved through the use of sterile air overpressure in the sterile zone. These requirements are particularly important for products with pH >4.6 (low-acid foods). Not all aseptic filling machines in use in the United States are capable of producing shelf-stable low-acid products. Capability is determined by microbiological challenge testing to validate the sterilization and sterility maintenance requirements. Filling machines that meet the compliance requirements of low-acid products can be used for high-acid products as well. Those filling systems listed in Table 1 have been or are currently seeking acceptance by the U.S. Food and Drug Administration for low-acid use. (see Form/fill/seal, vertical and horizontal.)

Aseptic Packaging Materials

Requirements. The versatility of aseptic technology has given rise to the use of a variety of plastic and polyolefin materials for packaging as illustrated in Table 1. Since aseptic applications require both product preservation, and utility from the package, there are several basic requirements that these relatively new packaging materials must meet for suc-

Table 1. Summary of Aseptic Filling Machines in the United States

Filler Type and Manufacturer	Packaging Material Supply	Filled Package	Package Sterilization
Bag-in-container			
Alfa (Tetra) Laval StarAsept, Tetra Pak Plastic, Linconshire, IL	Presterilized–preformed bags	Bag-in-box	Radiation presterilized bag Spout–steam
Fran Rica, FR Manufacturing, Stockton, CA	Presterilized–preformed bags	Bag-in-box/drum	Radiation presterilized bag Spout–steam
Liqui-Box, Liqui Box Corp., Worthington, OH	Presterilized–preformed bags	Bag-in-box	Radiation presterilized bag Spout–peroxide fog
Scholle, Scholle Corp., Irvine, CA	Presterilized–preformed bags	Bag-in-box/bin	Radiation presterilized bag Spout–steam
Preformed containers			
Bosch, Robert Bosch Corp., South Plainfield, NJ	Glass/plastic bottles	Glass/plastic bottles	Hot peroxide spray/Hot air Cap–steam
Dole Aseptic Can, Graham Engineering, York, PA	Metal/composite cans	Metal/composite cans	Superheated steam
Preformed cups			
Metal Box, FMC Corp., Madera CA (Service only)	Plastic cups	Plastic cups	Hot peroxide/Hot air Lid—presterilized
Thermoform–fill–seal cups			
Bosch, Robert Bosch Corp., South Plainfield, NJ	Roll stock	Plastic cups	Hot peroxide/Hot air
Hassis, Hassia USA Inc., Somerville, NJ	Roll stock	Plastic cups	Saturated steam
Form–fill–seal bags			
Bosch Robert Bosch Corp., South Plainfield, NJ	Roll stock	Plastic bags	Hot peroxide/Hot air
Inpaco, Inpaco Inc., Nazareth, PA	Roll stock	Plastic bags	Hot peroxide/Hot air
Form–fill–seal cartons			
Combibloc, Combibloc Inc. Columbus, OH	Composite paperboard sleeves	Rectangular cartons	Hot peroxide fog/Hot air
Fuji, International Paper Co., Raleigh, NC	Composite paperboard roll stock	Rectangular cartons	Hot peroxide bath/Hot air
Tetra Brik, Tetra Pak Inc., Chicago, IL	Composite paperboard roll stock	Rectangular cartons	Hot peroxide bath/Hot air

cessful application in the marketplace, and most are product- or usage-dependent:

1. The packaging material must be acceptable for use in contact with the intended product, and must comply with applicable material migration requirements.
2. Physical integrity of the package is necessary to assure containment of the product and maintenance of sterility. The term *integrity* applies to the structural integrity of the container itself as well as that of the closures and seals to assure package soundness and hermeticity during handling and distribution.
3. The package material must be able to be sterilized and be compatible with the method of sterilization used (heat, chemical, or radiation).
4. The package must provide the barrier protection necessary to maintain product quality until it is used. Barrier protection means control over the transmission of oxygen, moisture, light, and aroma through the package as required by the product.

Package structure and composition. Compared to metal and glass containers, the structure and composition of aseptic packaging are more complex and vary depending on product application, package size, and package type. Factors such as seal strength and integrity, package shape, stiffness, and durability, as well as barrier properties determine the choice and/or combination of materials required. In most applications, aseptic packages incorporate more than one material in the structure that is assembled by lamination or coextrusion processes. Examples of some materials commonly used in aseptic packaging are presented in Table 2. The functional attributes of the materials selected are presented in the table to show similarities and differences among those examples. It should be noted that cost is a major factor in the material selection for a given attribute when more than one material available exhibits the same functional properties. (See also Coextrusion; Barrier polymers and permeability.)

Applying the information provided in Table 2 for illustration, the bag, in bag-in-container packaging, might require a generic laminate comprising nylon, metalized film, and low-density polyethylene in order to meet the functional requirements of the package and product. The bag would likely have to be tough, puncture-resistant, and durable, with barrier protection that is resistant to cracking, and be able to be hermetically sealed using a relatively inexpensive product contact surface. In order to determine the most appropriate material selection for a particular package and application, the reader is referred to the following suppliers for a more detailed description of resinous and polymeric materials used

Table 2. Functional Attributes of Some Aseptic Packaging Materials

Material	Barrier Property			Seal Quality and Adhesion	Durability		
	Oxygen	Moisture	Light		Stiffness	Tear	Puncture
Paperboard			X		X		
Aluminum foil	X	X	X				
Metallized film	X		X				
Ethylene acrylic acid				X			X
Low-density polyethylene		X		X			
Linear low-density polyethylene		X		X		X	X
Nylon						X	X
Polypropylene		X		X		X	
Polystyrene					X		
Polyvinylidine chloride	X	X					
Ethylene vinyl alcohol	X						

in aseptic packaging applications: Chevron Chemical, Orange, TX; Dow Chemical, Freeport, TX; E. I. Du Pont, Wilmington, DE; EVAL Company, Lisle, IL; and Quantum, USI Division, Cincinnati, OH. See also the various articles in the *Encyclopedia* on the resins listed in Table 2.

BIBLIOGRAPHY

1. H. Reuter, ed., *Aseptic Packaging of Food,* Technomic Publishing, Lancaster, PA, 1989.
2. S. D. Holdsworth, *Aseptic Processing and Packaging of Food Products,* Elsevier Science Publishing, New York, 1992.
3. *Nicholas Appert,* Grolier Electronic Publishing, Danbury, CT, 1993.
4. T. Brock, ed., "On the Organized Bodies Which Exist in the Atmosphere; Examination of the Doctrine of Spontaneous Generation, by L. Pasteur" English translation, in *Milestones In Microbiology,* Prentice-Hall, Englewood Cliffs, NJ, 1961.
5. U.S. Pat. 699,765 (May 13, 1902), C. H. Plummer and F. T. Stare.
6. U.S. Pat. 1,270,797 (July 2, 1918), M. H. Dunkley.
7. U.S. Pat. 2,549,216 (April 17, 1951), W. McK. Martin (to James Dole Engineering Co.).
8. Anonymous, *Fed. Reg.* **46,** 2342 (Jan. 9, 1981).
9. V. R. Carlson, *Aseptic Processing,* Technical Digest CB-201, a publication of Waukesha Cherry-Burrell Corp. Louisville, KY, 1977.
10. *Aseptic/Extended Shelf Life Processing Handbook,* a publication of APV Crepaco, Inc., Rosemont, IL, 1992.
11. Howard Zhang, Dept. Food Science and Technology, The Ohio State University, Columbus, OH, Personal Communication.
12. Richard Muzyka, ABB Autoclave Systems Inc., Columbus, OH, Personal Communication.
13. J. Mans, "Showcase: Aseptic Packaging," *Prepared Foods,* 106–109 (March 1988).
14. T. E. Szemplenski, *Aseptic Packaging in the United States,* a publication by Aseptic Resources Inc. for Packaging Strategies Inc., West Chester, PA, Jan. 1, 1995.

MICHAEL S. MABEE
Mason, Ohio

ASIAN PACKAGING. See NIPPON PACKAGING

B

BAG CLOSURES.
See CLOSURES, BAG.

BAG-IN-BOX, DRY PRODUCT

When the bag-in-box concept is applied to dry products, it generally involves a bag inside a folding carton (see Cartons, folding). In order to appreciate the impact of the bag-in-box (BIB) concept for dry products, one must understand the history of the folding carton. The turn of the century marked the first use of the folding carton as a package when National Biscuit Company introduced the "Uneeda Biscuit" (soda cracker). Instead of opting for the conventional bulk method of selling crackers, Nabisco decided to prepackage in smaller boxes, using a system that would prolong freshness. The paperboard carton shell with creased score line flaps had recently been developed, along with a method for bottom and top gluing on automatic machinery (see Cartoning machinery). Waxed paper (see Paper; Waxes) was to be added manually to the inside of the carton. So was born the "lined carton."

The evolutionary process eventually culminated in two basic methods of producing lined cartons. The first was a machine to automatically open a magazine-fed side-seamed carton and elevate it around vertically indexed mandrels where glue is applied to the bottom flaps and folded up against the end of the mandrel with great pressure. The result is a squarely formed open carton with a very flat bottom surface capable of being conveyed upright to a lining machine which plunges a precut waxed-paper sheet into it by a system of reciprocating vertical mandrels. The lining is overlapped at the edges and sealed together to form an inner barrier to outside environmental factors. It protrudes above the carton top score line by a sufficient amount of paper to be later folded and sealed at a top-closure machine. Straight-line multiple-head filling machines are used to fill premeasured product into the carton. Initial fill levels are often above the carton-top score line and contained within the upper portion of the lining, which eventually settles with vibration before top sealing takes place. This fact has relevance with respect to the bag-in-box concept.

The second method involves the use of a double-package maker (DPM), which combines the carton forming/gluing operation with a lining feed mechanism which wraps the lining paper around a solid mandrel prior to the carton feed station. In this instance the carton blank is flat and is side-seamed on the DPM. The lined carton is then discharged upright onto a conveyor leading to the filler and top closing machine.

These packaging lines are typically run at up to 80 packages per minute (in some special cases 120/min). They are considered to be very complex machines requiring skilled operating personnel and are usually restricted up to a single size.

Although the time reference is rather vague, it would appear that the bag-in-box concept began with the refinement of vertical form/fill/seal (VFFS) machinery in the 1950s (see Form/fill/seal, vertical). Packaging machinery manufacturers and users saw an alternative to the DPM in the horizontal cartoner coupled with VFFS equipment. The idea of automatically end-loading a sealed bag of product into a carton offers the following important advantages compared to lined cartons: simplicity (ie, fewer and less-complicated motions); flexibility (ie, size changes more easily and quickly accomplished); higher speed (ie, up to 200 packages per minute is theoretically possible with multiple VFFS machines in combination with a continuous-motion cartoner); lower packaged cost (ie, higher speeds and lower priced machinery); improved package integrity (ie, bags are hermetically sealed using heat-seal jaws); reduced personnel (ie, possible for one operator to run line); requires less floorspace (ie, more compact integrated design); and wider choice of packaging materials (ie, unsupported as well as supported films can be handled).

Although bags are sometimes inserted manually, the high-speed methodology (see Fig. 1) employs multiple VFFS machines stationed at right angles to the cartoner infeed, dropping filled and sealed bags of product onto an inclined conveyor that carries them to a sweep-arm transfer device for placement into a continuously moving bucket conveyor. The VFFS machines are electrically synchronized with the cartoner, and the transfer device is mechanically driven by the bucket conveyor. When the bag reaches the carton-loading area it is gradually pushed out of the bucket by cammed push rods through guides and into the open mouth of the box which is contained within chain flights traveling at the same speed adjacent to the bucket. A guide plate drops into the bucket from overhead to confine the bag during insertion into the box. Once the bag is in the box the end flaps are glued and rail closed before entering compression belts for discharge to the case packer.

When difficulties occur on this system, it is usually at the insertion station, caused by misshaped or rounded bags with a girth larger than the carton opening. An attempt is made to condition the bag on the inclined conveyor by redistributing the product within the bag more evenly and, once in the bucket, by vibrating tampers to flatten it. However, if there is too much air entrapment in the bag, or a high product fill level, or an improperly shaped bag, these devices become futile. Increasing the carton size would be a simple solution, but the packager is often not free to do this. Marketing departments are generally reluctant to change the size of a carton that has been running satisfactorily on DPM equipment. The main problem is that the BIB manufacturer must allow more clearance of bag to box than is required on a close-fitting lining. The usual BIB bag-sizing rule of thumb is a gusseted bag having a width 3/8 in. (9.53 mm) less than the carton face panel and 1/4 in. (6.4 mm) less than carton thickness. This can vary somewhat, but the bag must be small enough to transfer positively into the bucket and subsequently into the box without interference.

One manufacturer has attempted to deal with the problem by wrapping the carton blank around the bucket after the bag has been top loaded into it (see Fig. 2). In wraparound cartoning, flat blanks are used and glue is applied to the manufacturer's joint and side seamed against the mandrel by rotating-compression bars. The mandrel/bucket is withdrawn leaving the bag in the box ready for flap gluing and closing.

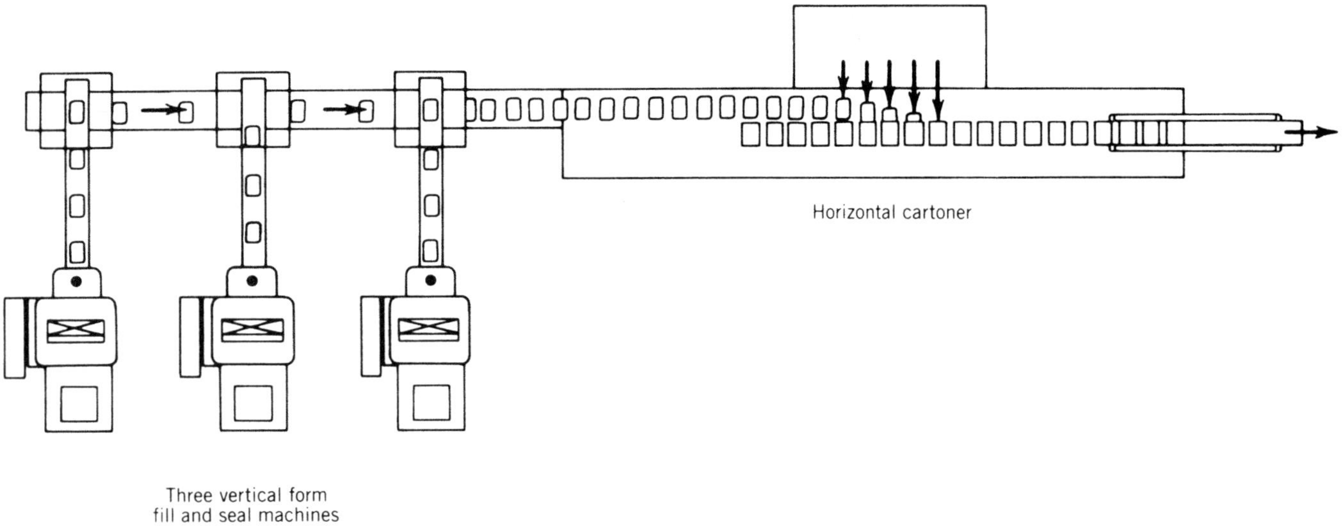

Figure 1. Bag-in-box packaging system with horizontal cartoner.

This approach is more forgiving than end-loading, but is still susceptible to bag sizing problems because additional clearance must be allowed to compensate for the gauge of the three-sided bucket walls. Speeds of ≤140 per minute are possible.

The "vertical load concept has gone a long way toward overcoming bag insertion problems (see Fig. 3). It relies on simple gravity and special bag-shaping techniques to drop the bag directly into the box from a film transport-belt-driven type of VFFS machine. The carton shell is formed from a flat blank and bottom glued on a rotary four-station mandrel carton former. The upright carton is conveyed to a starwheel-timed flighted chain indexing device to position it squarely under the VFFS rectangular forming tube. Two VFFS machines with electronically synchronized motor drives operate independently of the carton former. A prime line of empty cartons initiates the operation of each VFFS machine.

The bag-forming parts consist of a rectangular forming shoulder and tube which has been manufactured to produce a bag with cross-sectional dimensions ¼ in. (6.4 mm) less than carton face panel and ⅛ in. (3.2 mm) less than the side panel. This very close fit is made possible by a combination of several mechanisms. First a gusseting device creates a true flat-bottom bag using fingers that fold in the film from the sides as contoured cross-seal jaws close on the bag. The bottom of the rectangular forming tube is within ¼ in. (6.4 mm) of the seal jaws and acts as a mandrel around which the bag bottom is formed. The bag is thereby given a sharply defined rectangular shape, which is maintained as it is filled with product and lowered through a shape-retaining chamber. This chamber is vibrating so as to present a moving surface to the bag to reduce frictional contact and, more important, to settle the product before the top seal is made. Since the bag is confined and not allowed to round out, headspace between prod-

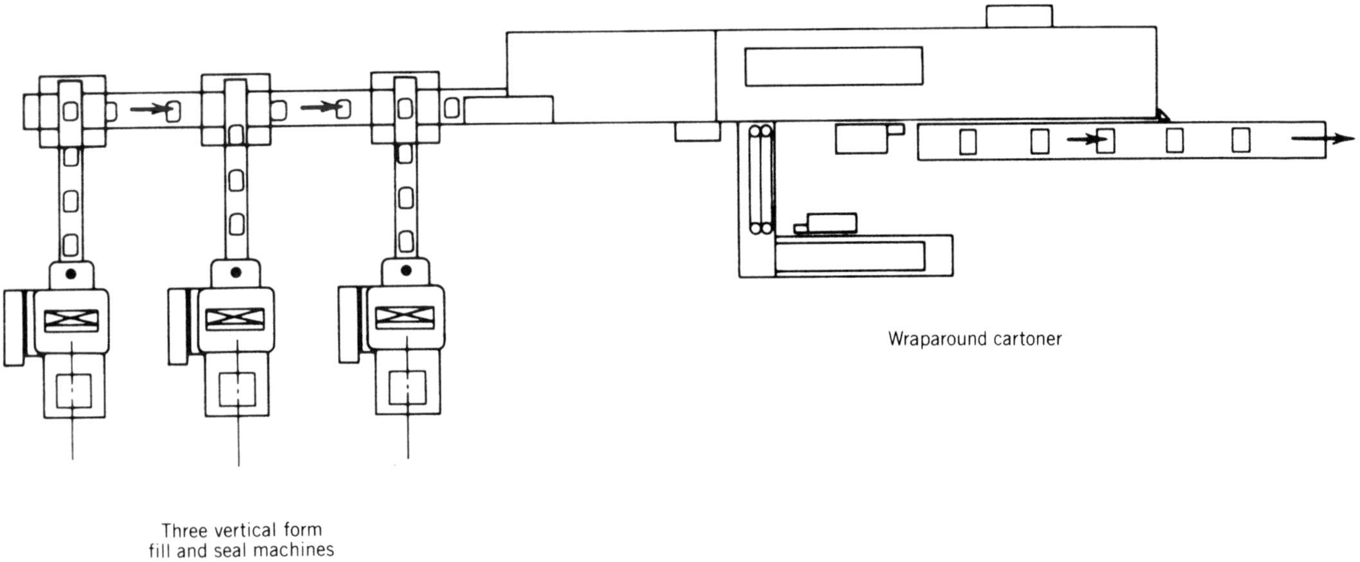

Figure 2. Bag-in-box packaging system with wraparound cartoner.

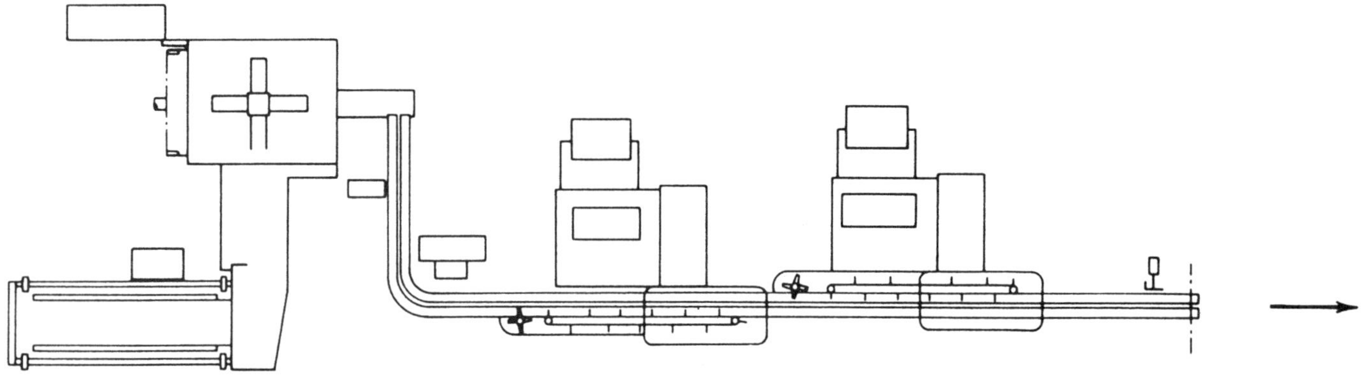

Figure 3. Bag-in-box packaging system with mandrel carton former.

uct and top seal is kept to a minimum, thereby reducing air entrapment to manageable levels. Flap spreaders ensure that there is unobstructed access into the box.

The shaped filled bag slips freely into the box, allowing the four corners of the bag to settle snugly into the bottom, making maximum use of available volume. The bag cutoff is determined by product fill level and, if necessary, can be made so that top seal protrudes over the score line by several in. (cm) when fully seated in the box, which is neatly pressed down at the next station. While the carton is contained in the flighted chain it is indexed through a top sealer where hot-melt glue is applied to the flaps, which are railed over before passing under compression rollers. The top sealer is integrated mechanically and electrically with each VFFS machine, and because the operation is performed immediately after bag insertion, the carton never has a chance to bulge, which is eventually important for efficient case packing.

This packaging system is rated at up to 100 cartons per minute and two lines can be mirror-imaged for higher speeds. A recent adaptation of the vertical-load system integrates a carton erector for side-seamed blanks with one VFFS and a top-and-bottom gluer into a single compact module rated at up to 50 boxes per minute.

Another BIB system close couples a carton former (preglued or flat blanks) with a pocket conveyor which indexes the box to a series of bag insertion, filling, sealing, and carton top/bottom gluing stations. The bag is formed using vertical form/fill techniques utilizing a rectangular forming tube. A unique gripping mechanism engages the bottom seal of the bag and pulls it into the box from underneath. The bag top is left open to be filled with product at succeeding stations. This system offers the advantages of multiple-stage filling for optimum accuracy, checkweighing, and settling before the bag top is sealed. This is a very effective way of dealing with the problem of high fill levels. Outputs of 65 to 80 packages per minute are claimed for this type of machinery.

Related concepts. A field that has grown is institutional bulk packaging of large quantities of product in 10–20 lb (4.5–9.1 kg) sizes which usually require corrugated-box materials (see Boxes, corrugated). The VFFS unit is an expanded machine capable of producing large deeply gusseted bags and is usually set up for vertical bag loading. Speeds are in the range of 15–30 cartons per minute. In a related process, not technically "bag-in-box," horizontal pouch (three-side-seal) machines are close coupled to a cartoner to automatically insert one or multiple pouches into a carton at high speeds.

BIBLIOGRAPHY

"Bag-In-Box, Dry Product," In *The Wiley Encyclopedia of Packaging Technology,* 1st ed., Wiley, New York, by Norman H. Martin, Jr. Pneumatic Scale Corporation, 1986, pp. 24–26

General Reference

S. Sacharow, *A Guide to Packaging Machinery,* Magazines for Industry, Inc., 1980.

BAG-IN-BOX, LIQUID PRODUCT

Bag-in-box is a form of commercial packaging for food and nonfood, liquid and semiliquid products consisting of three main components: (*1*) a flexible, collapsible, fully sealed bag made from one or more plies of synthetic films; (*2*) a closure and a tubular spout through which contents are filled and dispensed; and (*3*) a rigid outer box or container, usually holding one, but sometimes more than one, bag (see Fig. 1).

The bag-in-box concept appeared in the United States in the late 1950s. As early as 1957 (*1*), the package was introduced into the dairy industry in the form of a disposable, single-ply bag for bulk milk. By 1962, it had gained acceptance as a replacement for the returnable 5-gal (19-L) can used in institutional bulk-milk dispensers (*2*). One of the first nonfood items to be offered in a bag-in-box package was corrosive sulfuric acid used to activate dry-charge batteries.

During this early period, bags were manufactured from tubular stock film by labor-intensive methods. Initially, the physical properties of monolayer films (chiefly low-density polyethylene homopolymers) limited applications, and filling equipment was slow and often imprecise. This situation changed significantly with the introduction in the 1960s of ethylene–vinyl acetate (EVA) copolymer films that provided

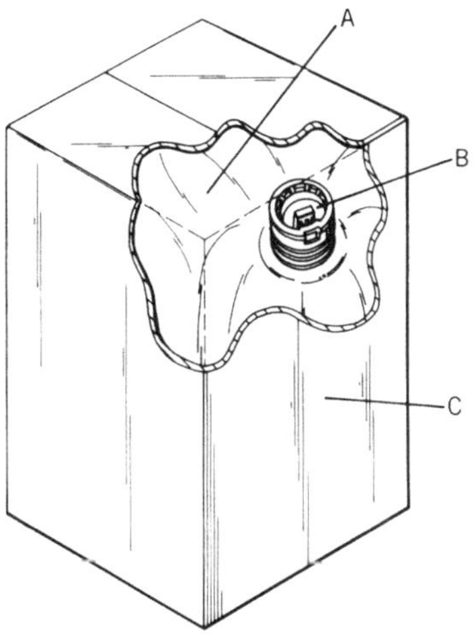

Figure 1. A bag-in-box package for liquid products consists of a bag (A), a closure and tubular spout (B), and a rigid outer container (C).

added sealability and resistance to stress and flex cracking. By 1965 faster, dual-head fillers became available featuring semiautomatic capping capabilities. Developments in automated box forming and closing kept pace. Also by 1965, proprietary bag-manufacturing equipment capable of making bags from single-wound sheeting sealed on all four sides had been developed (see Sealing, heat). In the mid-1970s, filling machines that automatically loaded filled bags into boxes came on-line. This was followed by totally automatic filling equipment that accepted a continuous feed of bags in strips (*3*), separated, filled, and capped them, then placed them in outer boxes. Meanwhile, barrier films (*4*) (see Barrier polymers) with improved handling and storage characteristics were being developed. Multilayer films of polyethylene coextruded with poly(vinylidene chloride) (PVDC) to provide an oxygen barrier had become commercially available as Saranex (Dow Chemical Co.) in the early 1970s (see Vinylidene chloride copolymers). The application of this barrier film permitted packaging of oxygen-sensitive wines and highly acidic foods such as pineapple and tomato products. Beginning in 1979, multiple-ply laminates (*5*) (see Multilayer flexible packaging) combining foil or metallized polyester substrates were introduced and in wide use by 1982 (see Film, polyester). Such laminations are thermally or adhesively bonded and in some instances by hot melt extruding the adhesive layer (see Adhesives; Extrusion coating). The most commonly used barrier film in the United States today is a three-ply laminate consisting of 2-mil (51-μm) EVA/48-gauge (325-μm) metallized polyester/2-mil (51-μm) EVA. The barrier properties of metallized polyester are directly proportional to the optical density of the metal deposit. Multilayer coextruded films combining the barrier properties of ethylene–vinyl alcohol copolymer (EVOH), the strength of nylon, and sealability of linear low-density polyethylene are being successfully used for some bag-in-box applications. It is the sensitivity to moisture by EVOH that is limiting the films' wider use (see Ethylene–vinyl alcohol; Nylon; Polyethylene, low density).

Fully automatic filling machines with as many as six heads (*6*) handling up to 40 two-liter (~2 qt) bags per minute are in operation. Films have been refined and specialized to meet tight packaging specifications for such procedures as hot fill at 200°F (93.3°C) temperatures. Bags are also presterilized by irradiation for filling with a growing number of aseptically processed products for ambient storage without preservatives (see Aseptic packaging). Outer boxes have become not only stronger, but more attractive and appealing as consumer sales units.

Manufacturing Process

In general, large producers of bag-in-box packaging design, develop, and manufacture packages to specifications meeting customer requirements. Containers vary in capacity from small, consumer and institutional sizes (or 2–20 L), to large, process and transportation packs of 55–300 gal (or 200–1000 L).

Bag. The principal considerations in choosing a film or laminate for the bag construction are strength and flexibility, with permeability and heat resistance added critical factors in an increasing number of applications. In the case of laminates, the bond between the layers of dissimilar materials must be maintained at a high level. Minimum requirements of over 1.1 lbf (500 gf) per inch (2.54 cm) (ie, ~193 N/m) is not uncommon. Films and laminates must be abuse resistant and must also, of course, be compatible with the product being packaged.

Once the appropriate film compositions have been determined for a specific application, the typical manufacturing procedure is as follows: Referring to Figure 2, two or three pairs of rolls of single-wound sheeting (A) are unwound on a machine where the webs advance intermittently, holes are punched (B) and spouts are sealed (C) into one of the duplex or triplex film sheets at predetermined points depending on finished size, and the bags are formed by sealing the two films together along the sides (D) then at the ends (E). Therefore, the bags for liquids are of a "flat" nature, not gusseted. Because precise time and temperature must be maintained to generate the seal between the thin films, the uneven thickness resulting from wrinkles or folds must be avoided because the resulting "darts" will not be fully sealed. A removable closure is applied to the spout (F), and after passing by the draw rolls (G) the bags are either cut apart as the final operation, or are perforated (H) for subsequent machine separation at time of filling.

Bag size. The box size must be measured first, and then the exact sizing of the inner bag can be determined. The bag must occupy virtually all of the interior space of the box without unfilled corners or potentially damaging excess headspace resulting from an oversize box.

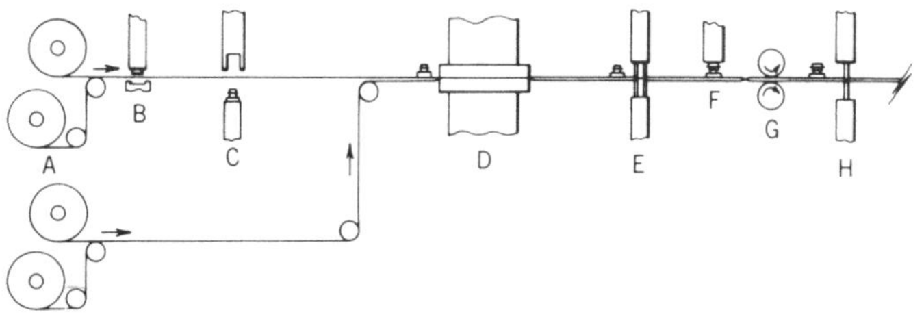

Figure 2. Typical bag-in-box manufacturing procedure.

Spout and closure. The spout is the filling port of the bag. Together with the closure, they are designed to mate with filling heads and must be able to withstand the mechanical shock of the closure being removed and replaced during the filling operation without damage to the spout or closure. Spouts are generally molded of polyethylene with a thin flexible flange to which the bag film is sealed. The spout has handling rings for holding the spout during the filling sequence, and the closure likewise has rings for the same purpose. Figure 3 shows the filling head in the fill position. The closure (A) has been removed and lifted up and away at an angle permitting the filling nozzle (B) to come downward and enter the spout (C) which is being held firmly in position.

Because the design of bag-in-box packages provides for the contents always to be in contact with the spout, the spout–closure fit must be leakproof. When high oxygen-barrier properties are essential, the spout and closure appear as the weak area in the bag because the materials used are highly permeable. This can be minimized by certain design considerations and by the application of barrier coatings.

There are many designs for spouts and closures, depending on function. In packages destined for consumer use, the closure typically is a simple, one-piece, flexible valve which opens and closes as a lever is activated. Such a combination is shown in Figure 4. Two layers (A,B) of film are sealed to the spout sealing flange (C). Around the tubular wall (D) are one or more spout handling rings (E). The closure also has a handling ring (F) that facilitates its removal by the filler capping head. The closure is retained on the spout by a mating groove and head (G), and the liquid seal is achieved by a plug-like fit (H) between the two components. The one-piece closure is molded from a resilient material and becomes a dispensing valve by flexing the toggle (I), creating an opening to an orifice (J), and providing a path for the contents to exit.

For food-service use, such as in restaurants, the closure may have a dispensing tube attached or be compatible with a quick-connect–disconnect coupler leading to a pump. For other uses, just a cap may need to be removed prior to emptying of contents.

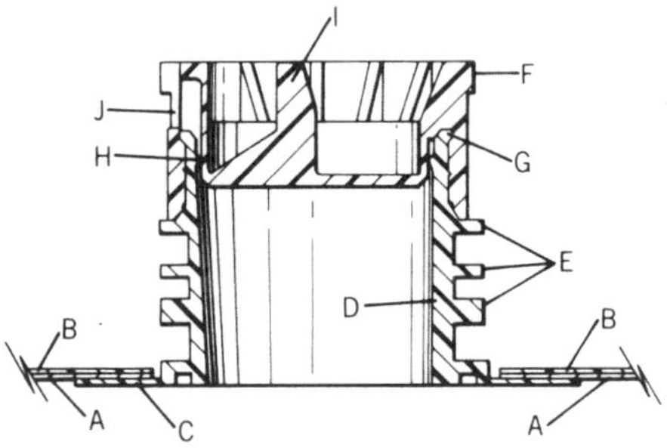

Figure 4. A flexible-valve closure. See text.

Box. In some dairy applications, the bags are transported in returnable plastic crates. Most typically a wide variety of materials in many forms may be used for the nonreturnable box of bag-in-box packages. For smaller sizes [1–6 gal (4–23 L)], the outer box is usually made of corrugated board in a conventional cubic configuration. For larger sizes [30–55 gal (114–208 L)], rigid plastic and metal containers, or even cylindrical drums, may be employed. The required strength must be designed into the box as dictated by the specific application.

Outer boxes may be manufactured with built-in handholds or locked-in-place handles, and some have special wax or plastic coatings for moisture protection. Most boxes are built with punch-out openings for easy access to the spout and closure.

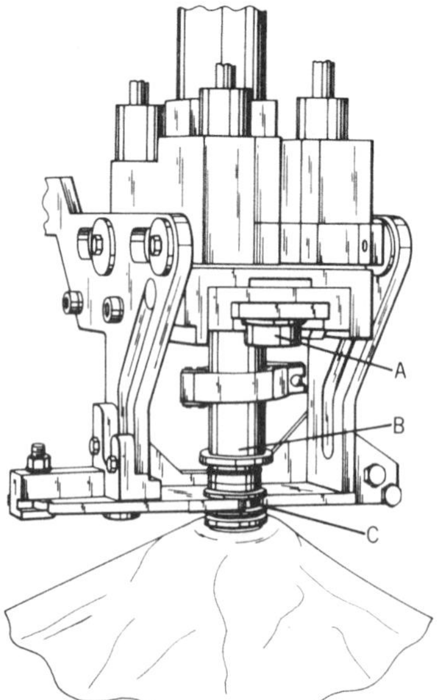

Figure 3. Filling head in fill position. The closure (A) has been removed to permit the filling nozzle (B) to come down and enter the spout (C).

Filling

Filling bag-in-box packages may be a manual or semi- to fully automatic operation and is adaptable to a wide range of standard industrial processing procedures, including cold, ambient, high-temperature, and aseptic filling. The basic design of a typical filling machine incorporates a flow meter, filling head or heads, an uncap–draw vacuum–fill–recap sequence, and filled bag discharge (see Filling machinery, still liquid). Bags may be manually loaded into the film head or automatically fed in strip form into more sophisticated models. Advanced filling equipment can also be provided with such devices as a cooling tunnel where hot-filled bags are agitated and cooled by jets of chilled water; a specialized valve to allow passage of liquids with large particulates; steam sterilizing and sterile air chambers for aseptic filling; or other modifications as determined by application.

Five-gallon (19 L) bags can be filled at speeds that range from four per minute (1200 gal/h or 76 L/min) to 20 per minute (6000 gal/h or 303 L/min). Low speed filling is done on single-head, worker-attended fillers; high-speed filling on multihead equipment, comparing favorably with line speeds of conventional rigid-container operations. Complete systems including box former, conveyors, automatic bag loading, and top sealers are available to support the automated large-capacity fillers.

Shipping and Storage

Bag-in-box packaging offers significant weight- and space-saving economies. Before filling, components are shipped flat;

after filling, the basic cubic shape of most bag-in-box outer boxes occupies less space and tare weight than cylindrical metal containers of comparable volume. Limitations include restrictions on palletizing and stacking height, where content weight may exceed outer box ratings, especially those constructed of corrugated board; vulnerability of uncoated boxes to humidity and moisture; possibility of flex cracking of the bag from the effects of long, transcontinental shipments; and damage potential to surrounding packages from a leaking unit.

Dispensing

Dispensing may be accomplished in one of three basic ways: uncapping and discharging contents; attaching one or more packages to a pumping system; or activating a small volume, user-demand closure often referred to as a dispensing valve.

In single-bag packages, the spout–closure is contained within the outer box for protection and withdrawn prior to use through a perforated keyhold opening in the box. During dispensing, the bag collapses from atmospheric pressure as contents are expelled without the need for air to be admitted. When completely empty, bag-in-box package components, except those outer boxes or containers specifically designed for reuse, are fully disposable. Corrugated board and polyethylene are easily incinerated, and metallized and foil inner bags compact readily in landfills.

Applications

With advancements constantly being made in bag film capabilities, along with filling and dispensing techniques, practically every commercial product is either being considered for or is now available in bag-in-box packages. Major users include the dairy industry, especially in the packaging of fluid milk for restaurant/institutional use, and of soft ice cream mixes; the wine producers, who offer 4-L (3.8-qt) bag-in-box packages to consumers and 12- and 18-L (11.4- and 17-qt) sizes to restaurant/institutional outlets; and the fast-food markets, where condiments are quickly dispensed onto menu items from bag-in-box packages and soft drinks are prepared from fountain syrup pumped from a bag-in-box arrangement (7) which eliminates the need for recycling and accounting for metal transfer containers.

BIBLIOGRAPHY

1. U.S. Pat. 2,831,610 (Apr. 28, 1958), H. E. Dennis (to Chase Bag Company).
2. U.S. Pat. 3,090,526 (May 21, 1963), R. S. Hamilton and coworkers (to The Corrugated Container Company).
3. U.S. Pat. 4,120,134 (Oct. 17, 1978), W. R. Scholle (to Scholle Corporation).
4. "Films, Properties Chart" in *Packaging Encyclopedia 1984,* Vol. 29, No. 4, Cahners Publication, pp. 90–93.
5. J. P. Butler, "Laminations and Coextrusions," in *Packaging Encyclopedia 1984,* Vol. 29, No. 4, Cahners Publication, pp. 96–101.
6. U.S. Pat. 4,283,901 (Aug. 18, 1981), W. J. Schieser (to Liqui-Box Corporation).
7. U.S. Pat. 4,286,636 (Sept. 1, 1981), W. S. Credle (to Coca-Cola Company).

General References

Glossary of Packaging Terms (Standard Definition of Trade Terms Commonly used in Packaging). Compiled and published by The Packaging Institute, U.S.A., New York.

JIM ARCH
Scholle Corporation

BAGS, BULK, FLEXIBLE INTERMEDIATE BULK CONTAINERS

Bulk bags, also known as *flexible intermediate bulk containers* (FIBCs), contain about 1 ton of dry bulk product such as grains, powders, and pellets. *Intermediate bulk containers* (IBCs) are *intermediate* between smaller packages, such as shipping sacks and drums, and larger carload bulk quantities. There are also rigid IBCs. Filled bulk bags occupy about the same space as a typical palletized unit load. They are usually filled from the top, lifted from the top, and dispensed through the bottom (see Fig. 1). Many are reused.

Bulk bags were first adopted in Japan and Europe, as an economical way to ship intermediate quantities in logistical systems where shippers and their industrial customers cannot handle a whole tank-car of product, and prefer not to manually handle multiple shipping sacks or drums. Bulk bags have become increasingly popular in the United States, and they are very popular in Europe and Japan, primarily replacing 50–100-lb (25–50 kg) paper and plastic bags. Growth is expected to continue, especially given the present trends of legislation to reduce the maximum weight that a person may lift, and the lower cost (and natural resource conservation) compared to smaller sacks or drums.

The advantages of bulk bags, compared to buying smaller packages, include reduced costs for package purchase (from about $20 per bulk bag), filling, and handling. They also offer a customer service in handling savings because of the ease of dispensing the product.

Typical Uses

Bulk bags are used for a wide variety of dry products, primarily for ingredients intended for further processing. Typical products include chemicals, minerals, dyes, resins, feed, seed,

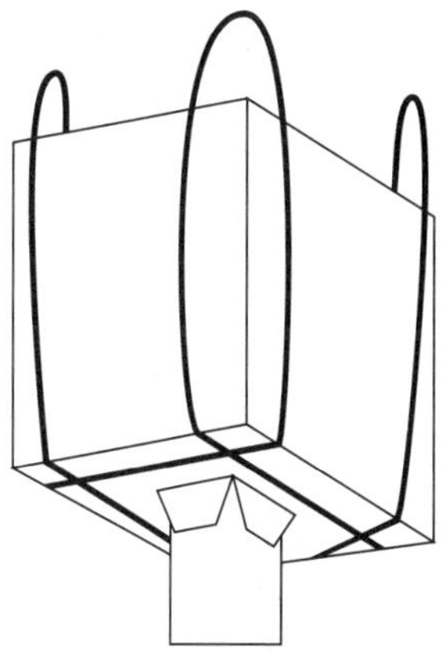

Figure 1. The most popular style of bulk bag: filled from the top, lifted from the top, and dispensed through the bottom.

grain, flour, sugar, salt, nuts, detergents, sand, clay, cement, hazardous materials, building materials, pharmaceuticals, fertilizers, and other commodities in powder, pellet, flake, or granular form. Liquid products need more support than a flexible bag alone can provide. Intermediate quantities (200 gal) of liquids can be packaged in a "composite" IBC, with the bag supported by a metal frame, or in rigid IBCs, such as pallet boxes or small tanks that may be lined with a bag. However, some "flexible tanks" for over 5000 gal of liquid, without rigid supports but with transport restraint attachments, have been introduced in Europe (1).

The filled bag weight depends on the size of the bag and the density of the product. The heavier the product, the more strength is needed. The standard FIBC holds 1100–2200 lb, and bags from heavier fabric can contain up to 5 metric tons. FIBC manufacturers offer bags with a volume of 10–100 ft^3. Footprint sizes range from 29 in. × 29 in. to 41 in. × 41 in., and heights of $\leq$ 88 in. empty size. When bags are filled, they have a tendency to settle into a more circular shape. Examples of bulk bag use can be found in many industries. Sugar and flour are shipped to food processing plants in FIBCs. Sandblasters can receive sand in 5500-lb FIBCs. Plastic resin is shipped to converters in 1-ton FIBCs. About 80% of fertilizer in the United Kingdom currently goes into FIBCs. One of the largest markets in the United States is for exported commodities.

Materials

Most bulk bags are made of plastic fabric or films with a very high tensile strength. Most are made from densely woven polypropylene (PP) flat fabric and have side seams, but they can also be made from circular-woven fabric (see Film, nonoriented PP; Film, oriented PP). Other materials include high-density polyethylene (HDPE) and polyester. Some European FIBCs are made from fiber-reinforced paper or polyvinyl chloride (PVC) (see Film, rigid PVC). The most heavy-duty FIBCs are usually made from PVC-coated PE fabric. Bulk bags can be printed to identify the contents. Filled bulk bags are usually cube-shaped with a square footprint, but some have a circular footprint. Some desjgns have antibulging reinforcement features.

Selection of the appropriate materials and structure depends on the properties of the product and the distribution system. Waterproof FIBCs are coated with PVC or latex and have heat-sealed (rather than sewn) seams. If the bags will be stored outside, the fabric is formulated to resist ultraviolet degradation. Some products generate static electricity when they flow, and antistatic (or static conductive/dissipative) fabric or liners are available. If bulk bags are intended to be reused, they are usually stronger than those intended for a single trip. Often the liner is disposable (for sanitation reasons) and the outer bag is reused.

Many bulk bags have plastic liners. The choice of liner material depends on the barrier properties needed or the tendency of the products to dust or leak. Products that attract moisture, or are sensitive to water, require moisture vapor transmission resistance. Polyethylene (2–4 mils) is common, but it is not a complete moisture or gas barrier. Saran and aluminum foil laminate liners are better barriers, and can be vacuum-packed. Liners are available in single or multiple plies. The liner construction matches the outer bag construction with respect to dimensions and placement of filling and dispensing ports. Liners come in tube shape or are custom-made to fit, with fill and discharge spouts. The liner can be designed to be manually or mechanically inserted.

There are various lifting and dispensing features available. For example, the most standard European FIBCs are for fertilizer, with single-point lifting and no discharge spout. The most common U.S. FIBCs have four lifting points and a spout. The choice of features depends on the application.

Filling and Dispensing

Bulk bags are usually filled from the top and discharged from the bottom. Most bulk bags have a spout at the top and one on the bottom. The spouts are closed with a tie, clamp, tape, or stitching. Clamps or ties can also be used to tie off a partially discharged bag or to regulate the rate of discharge. Intermittent flow controls are also available. However, some bags have no spouts and are simply filled or emptied through an open "duffle" that is gathered and tied to close. Some single-use FIBCs have no discharge spout, and are simply slit to empty.

Bulk bags normally require special filling and dispensing fixtures and equipment. Gravity directs the flow. Filling and discharge procedures and efficiency are influenced by the product's angle of repose and flow characteristics. During filling, the bag needs to be suspended from the top so that the product will completely fill the corners. Some filling systems incorporate a vibrating or settling device, deaeration and dust control measures, or a bag squaring method (although they rarely stay square once they are handled). Most are filled to weight using either a batch weigh hopper above the bag or a load cell to monitor the bag's weight as it is filled. Bulk bags are quick to fill; a two-operator filling station can fill a 2200-lb-capacity bag in about 30 s with a dense free-flowing product. But a light powdery product, which may require extra handling or vibration to compact, may take up to 30 min to pack.

A typical discharge fixture positions the bulk bag above the awaiting hopper, conveyor, pump, or tank receptacle where the bottom is opened so that the product can flow out. Special discharge equipment has been developed to reduce dust, improve flow, meter and reclose, and improve sanitation (especially important for food products).

Safe Handling, Transport, and Storage

Bulk bag handling and transport require special systems to ensure safety and efficiency. FIBCs are very heavy and can be unstable during handling, storage, and transport. This instability can result in danger to materials-handling workers and damage to the bag or its contents.

Bulk bags are lifted from the top, although they may also be handled on pallets. The most common lifting design incorporates four loops at the cube-shaped bag's four top corners. These loops may extend from the top to cradle the bottom like a sling, sewn onto the fabric. There are also center-lift designs that incorporate a single sleeve or loop on the top center, and two-sleeve designs with sleeves along two top edges. One-loop bags are more popular in the United Kingdom and France, especially for fertilizer. Some bags have a combination design with center-lift features to aid in discharge and corner lifting to aid in filling. The most heavy-duty FIBCs have steel lifting devices.

The loops or sleeves are lifted by the two forks of a forklift (or a single bar, in the case of the center-lift bags) that may be inverted to shorten the height of the lift. For most bags, it is necessary for one worker to position the loops onto the forks and a second worker to maneuver the forklift. Several can be lifted simultaneously by a ship's tackle and the appropriate stevedore fixtures. It is important for the loops or sleeves to be strong enough to support the weight during handling, including jerks caused by the lift equipment. Correct handling procedures are often printed on the bag or label.

Bulk bags can be transported by flatbed trucks, enclosed trailers, boxcars, flatcars, ship, or barge. Some restraint may be necessary to prevent shifting in transit. Typical restraints are made from straps and fabric, attached to the trailer and boxcar walls. Bulk bags must be strongly restrained if stacked in transit (2). Bulk bags can be stored and stacked on pallets, but shelving is recommended, because stacks can be unstable. FIBCs can also be used for temporary storage within a plant, and some can be used for outdoor storage.

Disposal and Reuse

Empty bulk bags can be discarded in landfills or incinerated if they have no hazardous residue, or they can be recycled (shredded and re-extruded) with compatible materials. But many bulk bags are returned and reused.

FIBCs have excellent return and reuse properties because they are strong when filled and yet lightweight (usually < 10 lb) and can be folded small when empty. The life of a reusable FIBC depends on its construction, the nature of the contents, and the handling and transportation method used, but is typically 5–10 trips. Many shippers reuse the outer bag but discard the liner. Reusable FIBCs are usually stronger than one-way bags. Returnable packages add costs for return shipment, cleaning, tracking, and inspection. Most are reused in a closed-loop system. They require a close relationship with customers to ensure their timely return in order to minimize the packaging investment.

In some cases, there is a market for used bulk bags, similar to the market for reconditioned drums. But an FIBC is usually designed for a specific type of product and use, and is best reused for the same type of product. Buyers and sellers of used shipping containers should certify the identity and compatibility of previous contents and document that the package was cleaned, inspected, and certified as to its safe reuse capacity. For example, lime for steelmaking can be packed in used bags because the residue is burned up and vaporized (along with the whole bag, which is not emptied but is added to the furnace intact).

Testing and Standards

Safety and performance standards have always been a concern of the FIBC industry, evidenced by the creation of the industry's self-policing associations such as the Flexible Intermediate Bulk Container Association (FIBC) and the European Flexible Intermediate Bulk Container Association (EFIBCA). These associations work with other standards groups like the British Standards Organization (BSO), the International Standards Organization (ISO), the American Society for Testing and Materials (ASTM), the American National Standards Institute (ANSI), the United Nations (UN), and the European Technical Committee for Packaging Standardization (CEN/TC261/SC3/WG7 covers IBCs) to better control FIBC safety and quality. Performance standards for FIBCs vary by country and regulatory agency. The first performance standards were developed in the United Kingdom by the EFIBCA, and were later incorporated into the British Standard Institute's BS 6382 (BSI/PKM 117). These form the basis for later standards adopted by other countries, including the United States, Australia, Japan, and throughout the European Economic Community (EEC). It is also the basis for the International Standards Organization's ISO TC 122/SC 2 N 238. The EFIBCA has also standardized the information that appears on each bag to include name, date and address of manufacture, construction identification, standard to which the bag is produced, test certification, class of bag (eg, single-trip), safe working load, safety factors, handling pictograms, and the contents' identification.

FIBCs used for hazardous materials are the most highly regulated. In 1990/91, the UN Chapter 16 for the carriage of dangerous goods, accepted bulk bags for UN Class 4.1 flammable solids, Class 5.1 oxidizing substances, Class 6.1 toxic substances and Class 8 corrosives, providing that they conform to the particular modal requirement and have passed drop, topple, righting, top-lift, stacking, and tear tests. In the United States, FIBCs have been accepted by the U.S. Department of Transportation (DOT) for Groups II and III hazardous materials, and performance specifications are given in HM 181-E (3). (See also Transportation codes.)

FIBC materials vary, so most standards are based on performance. Weight-bearing performance tests usually specify testing they bag's safe rated capacity with a much heavier weight—6 times more heavy for standard FIBCs, 8 times for more heavy-duty FIBCs, and 5 times for single-use FIBCs—than the expected contents. Filled bag performance should also be judged with respect to tear resistance, stacking, toppling, dropping (2–4 ft), dragging, righting (by one or two loops), and vibration. The most prevalent forms of damage are split side seams, broken loops, torn fabric, abrasion, and sifting. Damage can also be dangerous, if hazardous materials are spilled or if a stack topples onto a worker. In addition to performance tests, there are relevant material tests for the bag fabric and liner, including tensile strength and moisture vapor transmission. Material tests are used primarily for quality control.

Conclusion

Flexible intermediate bulk containers (FIBCs) provide a safe and cost-effective system for handling and transporting a wide range of bulk materials. They are "intermediate" between bulk handling and using shipping sacks for dry flowable commodities. FIBCs offer advantages compared to bulk handling and require a low initial investment compared to the special transport and handling equipment used for bulk handling. Some can be stored outdoors without the need for a warehouse or silo. They reduce product waste and the contents can be easily and accurately metered. An FIBC can perform the function of a mobile hopper. Compared to conventional 50-lb bags, FIBCs are less labor-intensive and do not involve manual handling that can cause back injuries. They are less expensive and require less space to pack. They are quicker to fill, handle, and discharge. And they reduce the risk of product loss, contamination, and pilferage.

BIBLIOGRAPHY

1. "A Heavy Commitment," *Packaging Week* (U.K.) **8**(41), 6 (April 29, 1993).
2. "Damage Claims Drop as Loads are Secured," *Packaging* **72** July, 1991).
3. United States Government, "Intermediate Bulk Containers for Hazardous Materials, Final Rule, 49 CFR Part 171 (HM 181-E)," *Fed. Reg.* (1994).

General References

"As Easy as FIBC," *Hazardous Cargo Bull.,* 62–63 (Sept. 1993).

British Standards Institute, "Flexible Intermediate Bulk Containers: Specification for Flexible Intermediate Bulk Containers Designed to be Lifted from Above by Integral or Detachable Devices," BS 6382, Part 1.

"Bulk Delivery," *Pack. Rev.* (U.K.), 97–101 (May 1980).

J. Clifton, "FIBC Quality and Safety—Mandatory European Standards Debate Begins," *Internat'l. Bulk J.* **11**(3), 36–50 (1991).

Flexible Intermediate Bulk Containers / Big Bags, Loadstar Publications, London, 1988.

A Guide to Selecting Bulk Bags and the Equipment to Utilize Them, Taylor Products, Parsons, Kansas, 1986.

EFIBCA Agree Standards, *Hazardous Cargo Bull.* (U.K.), **6**(6), 32 (June, 1985).

"Flexible Intermediate Bulk Container Association" in *Safe Handling Instructions and Glossary of Terms for Flexible Bulk Containers,* P.O. Box 2206, Macon, GA, 31203-2206, tel. (912)757–1006.

"Guidelines for Selecting Bulk Handling Containers," *Modern Materials Handling,* **40**(9) 62–64 (1985).

International Standards Organization, "Packaging. Flexible Intermediate Bulk Containers," ISO/TC 122/SC 2 N 238.

Japanese Industrial Standards, "Flexible Intermediate Bulk Containers," 1988.

H. Mostyn, "Principal Types of IBCs and Their Standardization" in *IBCs in a Competitive Environment,* PIRA, Leatherhead, U.K., 1993.

"Present State of Flexible Containers in Japan," *Pack. Jpn.* **4**(13), 50–53 (Jan. 1983).

D. Reid, "Flexible IBCs" in *IBCs in a Competitive Environment,* PIRA, Leatherhead, U.K., 1993.

United Nations, "United Nations Recommendations on the Transport of Dangerous Goods" in *Intermediate Bulk Containers,* 1983.

DIANA TWEDE
Michigan State University School of Packaging
East Lansing, Michigan

BAGMAKING MACHINERY

Heavy-duty bags, ie, shipping sacks, of multiwall paper or single-wall mono- or coextruded plastic are used to package such dry and free-flowing products as cement, plastic resin, chemicals, fertilizer, garden and lawn-care products, and pet foods. These bags typically range in capacity from 25 to 100 lb (11.3–45.4 kg), although large plastic bulk shipping bags may hold as much as a metric ton (see Bags, paper; Bags, heavy-duty plastic; Intermediate bulk containers).

Although there are dozens of variations in heavy-duty bag constructions, there are only two basic styles: the open-mouth bag and the valve bag. The former is open at one end and requires a field-closing operation after filling. Valve bags are made with both ends closed, and filling is accomplished through an opening called a valve. After filling, the valve is held shut by the pressure of the bag's contents.

Multiwall-bag Machinery

Traditionally, multiwall bags are manufactured in two operations on separate equipment lines. Formation of tubes takes place on the *tuber.* Closing of one or both ends of the tubes to make the bags is done on the *bottomer.* Multiwall bags have two to six plies to paper. Typical constructions are three and four plies. Polyethylene (PE) film is often used as an in-between or innermost ply to provide a moisture barrier.

Tube forming. The tuber (Fig. 1) starts with multiple giant rolls of kraft paper of a width that will finish into the specific bag width. At the cross-pasting station, spots of adhesive are applied between the plies to hold them together. The material is then formed into a tube that is pasted together along the seam. The tube may be formed with or without gussets. During seam-pasting, the edges of the various plies form a shingle pattern. When they are brought together to form a seam, these edges interweave so that each ply glues to itself. This provides optimal seam strength.

Flush-cut tubes are cut to the appropriate sections by a rotating upper and lower knife assembly. With stepped-end tubes, perforating knives are used to cut stepping patterns on both ends of the tube. The tube sections are then snapped apart along perforations that were made prior to cross-pasting. This snapping action is accomplished by sending the tubes through two sets of rollers, with the second set moving slightly faster than the first. Once the tubes have been flush-cut or separated, they proceed to the delivery section of the line.

About one-third of the multiwall-bag market is accounted for by bags with an inner, or intermediate, ply of plastic film. Flat film, used as an inner or intermediate layer, is formed into a tube along with the paper plies and pasted or, if necessary, hot-melt laminated in place. Another possibility is the insertion on the tuber of open-mouth film liners, the open end of which can project beyond the mouth of the paper sack. Stepped-end and flush-cut tubes are usually made on differently equipped tubers, but there is also a universal model which can be adapted to produce either type.

Flush-cut vs stepped end. Flush cutting is the most inexpensive tubing method in terms of both original equipment investment and tubing productivity, but these gains are lost in the subsequent bagmaking operations. The bottom of a flush-cut tube is normally sewn, and sewing is also the traditional method of field closure for many products such as seed and animal feeds. There was some use of flush-cut tubes as valve bags, particularly in Europe, but they are not widely used today because a pasted flush-cut bottom is structurally weak. The gluing of the bottom takes place only on one ply. To compensate for this weakness, a patch would normally be added to the bottom of the bag. Sewing is a widely used bottoming method in the United States because there is a great deal of flush-cut tubing and sewing equipment in place, and replac-

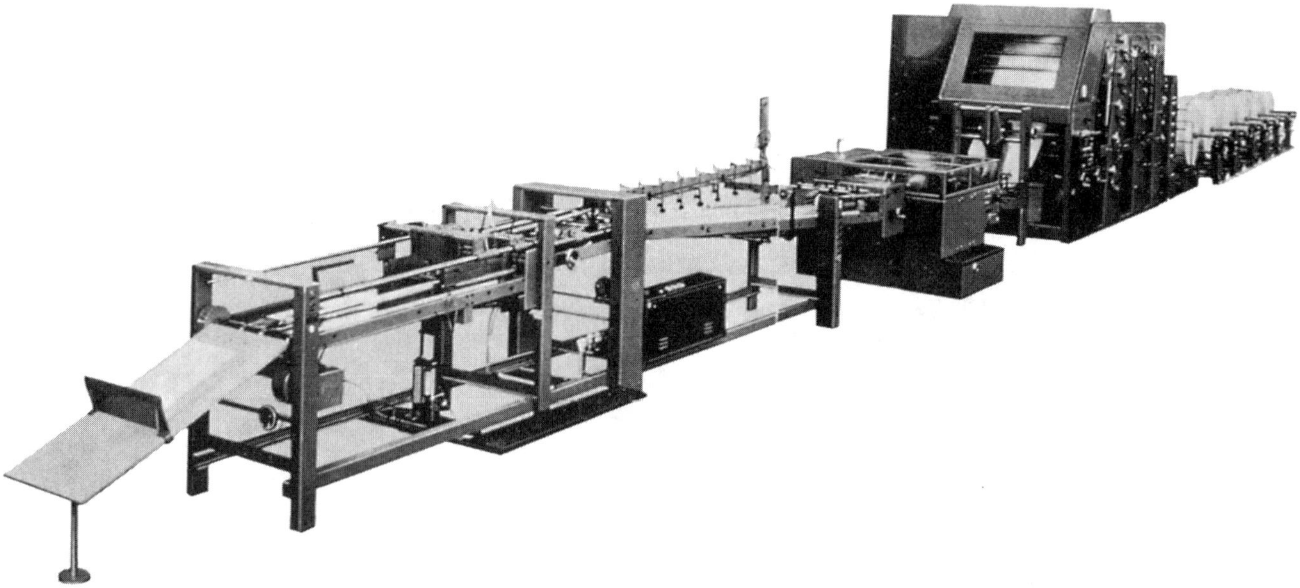

Figure 1. A universal tubing machine. Figure insert shows components: 1, flexoprinter; 2, unwind stations with reel-change arrangements; 3, automatic web brake; 4, web-guider path rollers; 5, web guider; 6, vertical auxiliary draw; 7, perforation; 8, cross pasting; 9, longitudinal register rollers; 10, seam pasting and auxiliary draw; 11, tube forming; 12, cut/register regulator; 13, cutting and tearoff unit; 14, variable-length drive-unit; 15, packet delivery unit; 16, takeoff table.

ing it in many instances would result in only a marginal return on investment. Unfortunately, sewing has many drawbacks. It is labor-intensive and, because the equipment has a large number of delicate moving parts, maintenance and repair costs are high. Also, the needle holes created by sewing weaken the bag, allow sifting, and make the bag more accessible to rodents and other pests.

The stepped-end tube makes the strongest bag. The ends of the bag have shinglelike stepping patterns that intermesh at the gluing points. In the bottoming process, ply one is glued to ply one, ply two to ply two, etc. Generally speaking, all bag manufacturers have their own stepping-pattern designs.

Bottoming equipment. The finished tube sections are converted into bags by closing one or both of the tube ends in any of the following three ways:

(*1*) One end of the tube is shut, forming a sewn open-mouth (SOM) bag. After filling, the top of the bag is closed by means of a portable field sewing unit.

(*2*) A satchel bottom is formed on each end of the tube, with one of the bottoms provided with an opening or valve through which the bag is filled by insertion of the spout of an automatic filling machine or packer. The valve is closed by the pressure of the bags contents. Additional means are available to make the bag more siftproof. In a valve bottomer (Fig. 2), tubes advance from a feeder to a tube aligner and a diverting unit for removing incorrectly fed tubes. The tubes pass through a series of creasing stations, and needle holes may be added under the valve for proper venting of the bag during filling. At the opening section, the tube is opened and triangular pockets are formed. Valves are inserted at a valving station. Valves are automatically formed by a special machine unit and then inserted into the bottom. They may be inserted and folded simultaneously along with the bottom or performed and automatically inserted. Preformed valves permit the use of a smaller valve size in proportion to the bottom of the bag. In Europe, reinforcing patches are customarily applied to both ends of the bag for added strength. The bags are discharged to a press section where they are conveyed in a continuous shingled stream. Powerful contact pressure of belts (top and bottom) ensures efficient adhesion. In most instances, the final station is an automatic counting and packeting unit.

(*3*) Stepped-end tubes with gussets and a special step pattern can be converted into pinch-bottom bags on which beads of hot-melt or cold adhesives (see Adhesives) are applied to the steps in the bottom (see Fig. 3**a**). These, in turn, are folded over and pressed closed to make an absolutely siftproof bottom (Fig. 3**b**). Beads of hot melt applied to the steps at the top of the bag are allowed to cool and solidify. After the bag is filled, a field-closure unit reactivates the hot-melt adhesive, folds over the top of the bag, and presses it closed.

In most instances, bags are collected in packets or bundles palletized for shipment to the end user. However, it is also possible to collect the bags on reels for efficient loading of automatic bag-feeding equipment in the field. The reeled bags form a shingled pattern held in place by the pressure of two plastic bands that are wound continuously around the reel along with the bags.

Other developments. With conventional equipment, it usually takes two bottomers to keep pace with one tuber. This

Figure 2. A valve bottomer. Figure insert indicates components: 1, rotary or double feeder; 2, tube aligner; 3, diverter for removing incorrectly fed tubes; 4, diagonal creasing and needle vent hole arrangements; 5, bottom center creasing stations with slitting arrangement for bottom flaps; 6, bottom opening station; 7, bottom creasing station; 8, unwind for valve patch; 9, bottom turning station; 10, valve unit; 11, bottom pasting; 12, bottom closing station; 13, bottom capping unit or intermediate pressing station; 14, flexo printers for bottom caps; 15, unwinds for bottom caps; 16, delivery, optionally with incorporated counting and packeting station.

fact has generally discouraged the development of in-line multiwall bagmaking systems in the United States. For example, tubers for cement bags typically operate at speeds from 270 to 320 tubes/min, whereas old-style bottomers run at 120–150 bags/min. Newer bottoming equipment can achieve speeds up to 250 bags/min, enabling one-to-one operation of tuber and bottomer on an in-line system. The tuber operates at less than maximum output, but the in-line system still produces more finished bags because of the increased efficiency resulting from the bottomer being continually fed with fresh tubes. Tubes where the paste has dried become stiff and difficult to handle. As paper is unwound from a roll, it quickly loses its moisture content and becomes less workable. These types of problems are alleviated with in-line bottoming.

In-line tube forming and tube bottoming also lend themselves to significant improvements in manpower utilization. The U-shaped in-line pinch-bottoming system shown in Figure 4 is capable of reducing the personnel requirements from nine to four. The key to the system is a unique turning station which rotates the axis of the tube by 90° for proper alignment with the bottomer. The "factory end" of the bottomer may be heat-sealed in-line (see Sealing, heat). On the "customer" end of the bag, hot melt or cold glue can be applied, or this end of the bag can be flush-cut for sewing in the field. A sewn top with a pinch bottom offers strength and siftproofness in this bottom style while allowing the customer to retain existing closing equipment. For a consumer product such as pet food, the pinch bottom allows the bag to be stacked horizontally on the shelf, still presenting a large graphics display area for the shopper.

An out-of-line double feeder-equipped pinch bottomer produces pinch-bottom bags for the manufacturers not anticipating having the volume to fully utilize the more productive in-line system. The trend for bag users to reduce inventories and place more small orders is expected to continue indefinitely. For the converter, this has meant decreased productivity because of a disproportionate amount of time being spent in changeovers. New CNC (computer numerical control) bottoming equipment promises to reduce changeover time from an average of about 3 h to about 30 min. All gross adjustments of machinery for a particular set-up are stored in the microprocessor and made on the machine by way of stepping motors. Although minor fine-tuning is still required, the starting adjustment point of each operator is the same, and settings are optimized according to a logical sequence designed into the control (see Instrumentation). In addition to faster setups, standardization of tuning procedures should result in more consistent and improved product quality.

Plastic Bag Machinery

The procedure for making all-plastic heavy-duty bags is similar to the procedure for multiwall bags, ie, various bottoming

Figure 3. (**a**) adhesives being applied; (**b**) pressing station (bags are folded over and pressed closed).

techniques are used to transform a tube into a finished bag, generally either an open-mouth or valve bag, with or without gussets. The three basic differences are described below:

(*1*) Plastic bagmaking almost always uses a single ply of material, either mono-extruded film, coextruded film, or woven fiber instead of the multiple plies used in paper shipping sacks.

(*2*) All bagmaking operations are performed on a single converting line. If the bag is made from flat sheet, the tubing and bottoming operations are integrated into a single bagmaking line. Bags are often made from tubes of blown film or circular woven fibers, and no tubing step is necessary.

(*3*) The plastic-bagmaking line may incorporate in-line printing, although the outer ply of kraft paper used in a multiwall bag is typically preprinted off-line.

Woven-bag machinery. Economy of raw materials and toughness are two features that make the woven plastic bag an attractive packaging medium for goods mainly intended for export.

A typical line for converting woven high-density polyethylene (HDPE) or polypropylene (PP) material into heavy-duty shipping sacks includes the following: unwind units for sheet or tubular webs, jumbo or normal size; a flexographic printing machine (see Printing) designed for in-line operation; a wax-application unit (see Waxes) to apply a hot-melt strip across uncoated material at the region of subsequent cutting to pre-

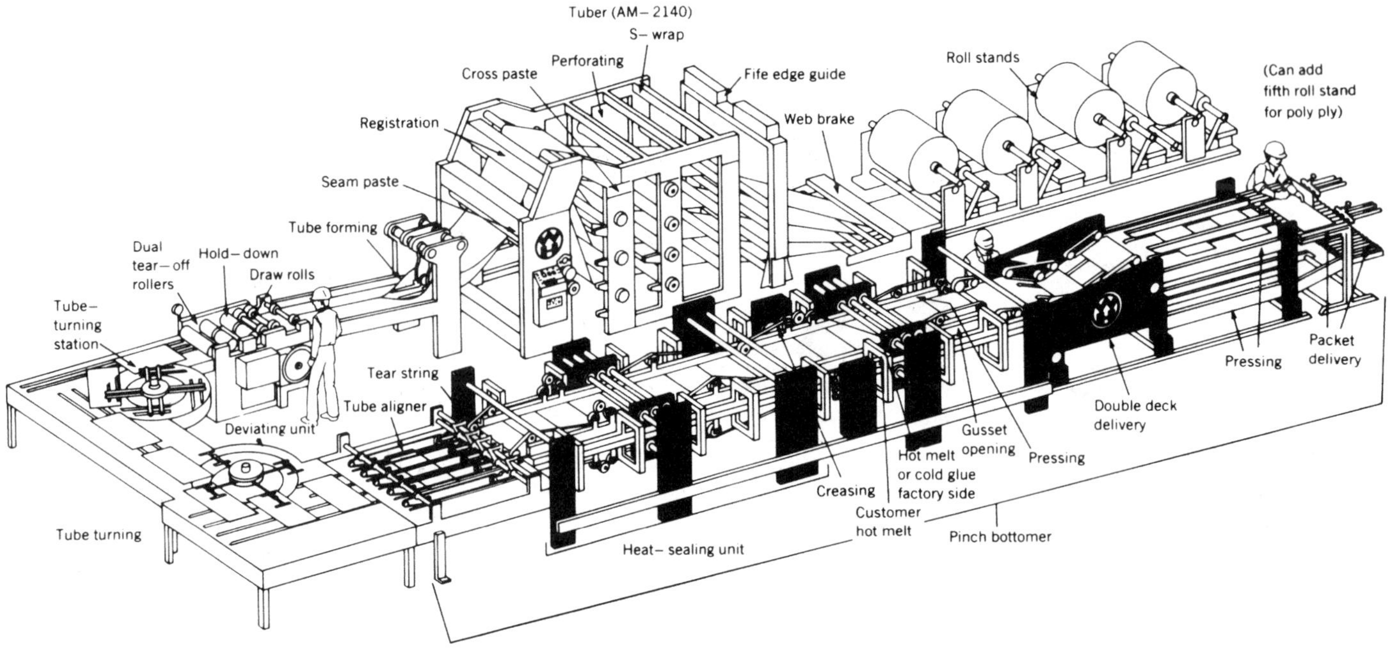

Figure 4. In-line tube forming and pinch bottoming.

vent fraying; and a flat and gusseted tube-forming unit. The flat sheet of coated or uncoated material is longitudinally folded into tubular form. Some machines have the ability to do this without traditional tube-forming parts. A longitudinal seam is sealed by an extruded bead of plastic. Output of the extruder is matched to the web speed by a tachogenerator.

A PE liner unit can be arranged above the tube-forming section to apply a PE liner to the flat web automatically. The principal element of this unit is a welding drum with rotating welding segments that provide the reel-fed PE with a bottom weld at the correct intervals. A Z-folding device enables a fold to be made in the crosswise direction for the provision of a liner which is longer than the sack. In addition, there is a crosscutting unit that cuts the outer web and the PE insert, usually by means of heated rotating knives. In the bottoming unit, cut lengths are transferred to the bottoming equipment by conveyor. Bottoming is accomplished either by sewing or the application of a tape strip. Instead of folding the tape over the open end of the sack, the sack end can be folded once or twice and the tape applied in flat form over the folds. The delivery unit collects finished sacks into piles for manual or automatic unloading.

Plastic valve sack machinery. Plastic valve bags operate by the same principle as multiwall valve bags. On filling, the pressure of the product closes a valve that has been inserted in either the bottom or the side of the bag. If the material is granular (not pelletized), channels along the bottom of the valve sack would allow some of the product to sift out. These channels can be made siftproof by closing them off with two beads of hot wax during the bottoming operation. Only 5–10% of the plastic valve sacks made in the United States require this feature. Therefore, most plastic valve sacks are produced on high-speed lines that produce sacks at about twice the speed of the siftproof machinery.

A typical system for the production of pasted PE bags from either flat film or blow tubes (Fig. 5) consists of the following equipment:

1. An unwind unit for flat film or tubing incorporating automatic tension and edge-guide controls.
2. A tube former in which folding plates form flat film into a tube. A longitudinal seam is bonded by an extruded PE bead.
3. A rotary cross cutter in which the formed tube is separated into individual lengths by the perforated knife of the rotary cross cutter. Fraying of woven materials can be eliminated with a heated knife that bonds the tapes together.
4. A turning unit in which, after the cross cutter, the tubes are turned 90° to bring the cut ends into position for the following processes.
5. A tube aligner and ejector gate in which exact alignment of tubes in longitudinal and cross direction is achieved by means of stops affixed to circulating chains and obliquely arranged accelerating conveyor bands. Photocells monitor the position of the tube lengths. In the event of misalignment, leading to malformed bottoms and, therefore, unusable sacks, the photocell triggers an electropneumatic gate, which, in turn, ejects the tube length from the line.
6. Pasting stations in which each tube end is simultaneously pasted by a pair of paste units using a special adhesive.
7. An enclosed drying system evaporates and draws off solvent from the adhesive.

8–10. Creasing, bottom-opening, and fixing of opened bottom in which a rotating pair of bars hold the tube length ends by suction and the rotary movement

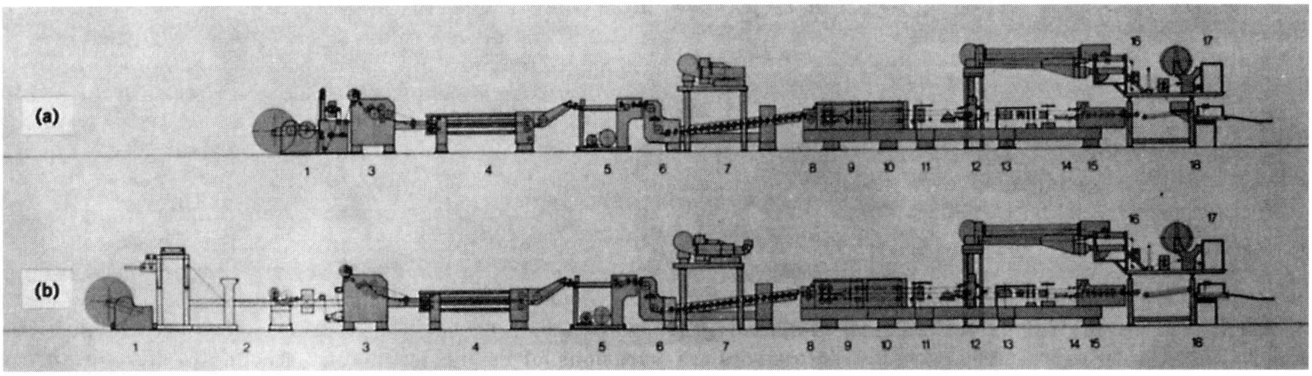

Figure 5. Systems for making all-plastic heavy duty bags from (**a**) tubular film or (**b**) tubular or flat film. Components are described by number in the text.

pulls the tube ends open sufficiently to enable rotating spreaders to enter and complete the bottom-opening process. The diagonal folds of the pockets are fixed by press rolls to avoid subsequent opening of the pockets.

11. A valve-patch unit forms the valve from rolls of flat film and places it in the leading or trailing pocket, as required.
12. The bottom-closing station, where, after the valve is positioned, the pasted bottom flaps are folded over, one to the other, and the sack bottom is firmly closed.

13–17. The bottom-patch unit, bottom-turning station, flexo-printing units for bottom patches, pasting stations with drying, and unwind for bottom-patch film, in which patches are formed from two separate rolls of film, flexo-printed (if required), and pasted to both sack bottoms. The bottom geometry is checked by photocells and faulty sacks are ejected through a gate. Just before they reach the delivery section, the bottoms are turned from a vertical to horizontal processing plane.

18. Delivery with counter and packeting station, in which good adhesion of the cover patch to the sack bottom is assured by applying pressure to the shingled sacks with staggered spring-loaded disks. Having reached a predetermined count, the conveyor accelerates the shingled sacks to the packing station, where the counted sacks are collected into packets and discharged. To accommodate a user's automated filling line, equipment is also available to wind the plastic valve sacks onto reels.

Continuous bagforming and bagfilling. Plastic valve bags have been used extensively, especially in Europe, for products such as plastic resin. However, continuous systems for forming, filling, and closing flat and gusseted plastic bags are becoming increasingly popular in the resin market. Such systems typically use prefabricated tubing for high strength. The tubular material is usually preprinted with random printing. Since resin weight varies from day to day, depending on ambient conditions and other factors, random printing allows the bag length to be adjusted according to the prevailing resin volume–weight relationship. In this manner, a tight and graphically appealing package is formed.

An integrated system for forming, filling, and closing of shipping bags would contain the following stations: unwind unit; compensator roller; hot-emboss marking unit; sealing station for bottom seam; bag shingling; separation of bags; introduction of bag-holding tongs; bag filling; supply of the filling product; sealing station for closing seam; bag outfeed conveyor; and control panel. Such a system can produce up to 1350 filled sacks per hour.

One-way flexible containers. One-way bulk shipping containers are becoming very popular in Europe. These oversized bags are designed for handling by forklift trucks equipped with one of several specially designed transport devices. Called intermediate bulk containers, they range in capacity from 1100 lb to about a metric ton (0.5–1 t), and are constructed of woven PP or HDPE. Tubes are generally woven on a circular loom because elimination of the longitudinal seam gives the bag exceptional strength. The advantage of this bag is that it represents an exceptionally economical and efficient method of handling bulk quantities. Acceptance of the concept has been relatively slow in the United States because it requires bag producers, product manufacturers, and product customers all to invest in special equipment for bag making, product filling, or handling.

Electronic Controls

Today's bagmaking equipment is following the overall industrial trend toward the use of programmable microprocessor control systems of increasing complexity. Ancillary equipment such as printing presses and extruders already have a high level of control, and other units on the bagmaking line are quickly being adapted to the computer. The first objective in conversion to programmable control is replacement of cumbersome mechanical logic. The next is storage of set-up and processing parameters for subsequent reuse. Microprocessors are being used for controlling temperatures, web tension, surface-tension treatment, adhesive application, ink, and registration. The most recent stage of automation has been provision of multiple outputs so that lines may be monitored or controlled by hierarchal computers.

BIBLIOGRAPHY

"Bagmaking Machinery," in *The Wiley Encyclopedia of Packaging Technology,* 1st ed., Wiley, New York, by Richard H. John, Windmoeller & Hoelscher Corporation, 1986, pp. 29–34.

BAGS, HEAVY DUTY, PLASTIC

In the spring of 1959, a load of ammonium nitrate fertilizer left the Spencer Chemical Company plant in Pittsburg, Kansas packed in printed polyethylene (PE) bags, each holding 50 lb of product. The bags were open-mouth style, heat-sealed at both ends, and 10 mils (0.001 in.) thick. These "10-mil bags," as they were called, had been fabricated by Chippawa Plastics Company, Chippawa Falls, WI using a special low-density PE resin manufactured by Spencer at its Orange, TX plant.

The cost of the bags used in 1959 exceeded that of the multiwall bags they replaced, but Spencer management had decided that the quickest path the market development of this new package, and the subsequent sale of PE resin to produce them, was to "marry" the two products, fertilizer and polyethylene, that were both produced by Spencer.

Early Applications

If free-flowing granules of fertilizer could be filled in the 10-mil bags, surely pellets of PE resin could also be packed. Shipments of PE resin in the new bags followed soon after the fertilizer bags. Domestic users were curious and accepted the bags provided that no additional costs were incurred by them. When Spencer's international sales agents, Omni, found out about the availability of the new package, they immediately recognized the added protection afforded overseas resin shipments and requested that they be used. It is interesting to note that one of the largest markets for heavy-duty plastic bags in 1996, almost 40 years after their introduction, are still overseas shipments of plastic resins (see Fig. 1) (Bona Packaging, Tyler, Texas).

The obvious advantages of using heavy-duty PE bags for outdoor applications dictated that early marketing development efforts would concentrate in the areas of (*1*) fertilizer, (*2*) potting soil, (*3*) mulch, and (*4*) lawn and garden products.

The shift of populations to the suburbs in the decades after World War II coincided with the growing of manicured lawns, decorative shrubbery, small flower and vegetable gardens, and the entire do-it-yourself trend. Covered storage of numerous large bags of "outside"-type products was neither available nor welcome in fastidious sheds or garages. A package that could remain in the yard and yet still protect the product was an obvious target. As shown in Figure 1, it still is. In fact, in terms of number of bags, the lawn and garden market for PE bags exceeds that for paper bags by a factor of 3 (The Omega International Group, Clemson, SC).

Technical Problems and Solutions

A bag 10 mils thick was certainly tough enough to withstand handling and shipping, but it presented problems in the areas of extrusion, fabrication, sealing, and economics. When film is blown, the bubble must extend some distance from the extruder die to permit effective cooling and limited orientation. The shear weight of a film bubble 10 mils thick-required new takeoff equipment that handled the film in a specialized manner. When an extruder is called on to pump so much material, the unit output volume in terms of numbers of bags or area is reduced significantly even though the number of pounds per hour may be at the limit of the machine. When the bag is fabricated from an extruded tube, the output is further limited because the width of the bubble can be no greater than the flat width of the bag.

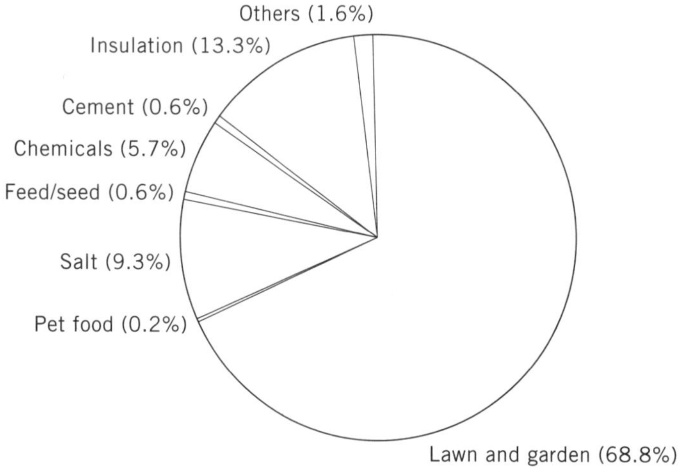

Figure 1. Plastic shipping sack markets (total usage = 880 million sacks):

Markets	Number (Sacks)
Lawn and garden	605
Pet food	2
Salt	82
Feed/seed	5
Chemicals	50
Cement	5
Insulation	117
Others	14

(The Omega International Group.)

If a tube is to be printed on front and back, either two passes through the press or special printing techniques are required. Generating sufficient heat and pressure to seal through 20 mils of film presents a bottleneck at both producing and filling ends. The use of gussets and the incorporation of valves are severely restricted.

The most obvious target for change was the bag thickness. The development of new low-density PE resins, the subsequent emergence of linear low-density (LLDPE) resins and most recently, the new metallocene catalyst systems are producing films whose strength and toughness is several times that of the original. The average thickness of a heavy-duty plastic bag in 1996 is in the four to six mil (0.004–0.006-in.) range.

To expand into areas dominated by MW paper sacks, the plastic bags had to compete in terms of appearance and style. This meant full-color process printing of all sides, the incorporation of valves for filling, and most recently, the forming, filling, and sealing of printed flat film all in one operation at continuously increasing speeds.

Technical Innovations

Co-extrusion was first introduced by USI Film Products in 1961 for use in bread overwrapping. Since USI (the operation is now owned by Bonar Packaging) was an early producer of heavy-duty bags, it was natural for this technology to find its way into the bags. The result was even further reduction in film thickness and differential slip characteristics between the inside and the outside of the bags. (Often films of different colors were used to highlight their differences.)

The inherent toughness of LLDPE film was a welcome innovation in the heavy-duty bag field. With the subsequent downgauging of the film and the fewer pounds of resin per bag, costs began to fall dramatically. Industries that had been out of the reach of PE bags suddenly found bag costs more nearly approaching those of MW paper sacks. (See Polyethylene, linear and very low density.)

Metallocene catalysts, introduced by Exxon and others, have further increased the toughness of heavy-duty PE film so that bag thickness again can be reduced. No only is less resin needed per bag, the film is easier to handle, form, and seal.

Other technical innovations range from the use of fiber filaments embedded in the plastic film to cross-laminated high-density film of the VALERON (trademark of Val Lier Flexibles, Inc., Houston) type. Although appreciably higher in price, the toughness of these bags places them not only beyond the reach of MW paper but well above the so-called conventional heavy-duty plastic bags as well.

Automatic Packaging

Coupled with the advent of thinner films and increased performance has been the development of high-speed form/fill/seal equipment designed to meet the demands of large plastic bags and product loads of $\geq$50 lb. Just as these types of machines signaled the end of waxed paper in snack-food packaging in the early 1930s, so may they expand the use of heavy-duty plastic bags as the year 2000 approaches.

The current leader among the machinery makers is Windmoeller & Hoelscher Corporation of Lincoln, Rhode Island. With three different machines in commercial use, this equipment is capable of producing as many as 1200 bags per hour. Weight control is reported to be $\pm$5 g. Other companies in this field are Hanner & Boecker, Bag Line, and Expomatic Bemis, a supplier of both MW paper as well as heavy-duty plastic bags, is reported to be offering an attachment to MW filling machines to convert them to plastic bags.

The Future

As newer materials are developed and combinations of materials are tailored to meet specific product demands, the market for heavy-duty plastic shipping sacks can only expand. Work is under way to control the surface slip characteristics, to permit "breathing" or the expulsion of internal gases and air, and to provide better protection against penetration of grease. Coinciding with these projects are efforts to improve toughness that will permit reductions in thickness (ie, cost) and to increase both bag production rates and filling rates.

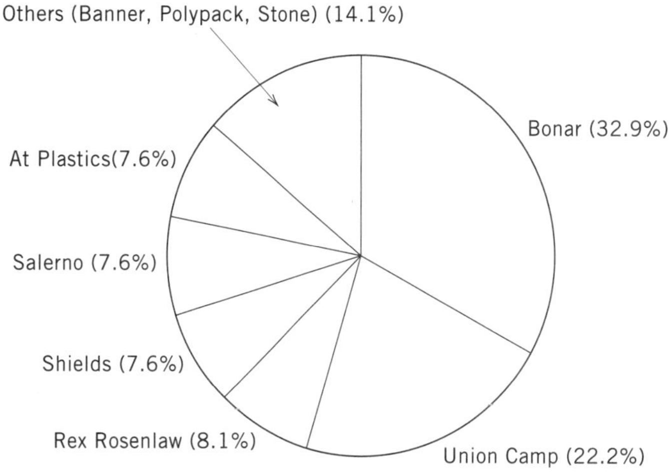

Figure 2. Plastic shipping sacks—producers (total usage = 880 million sacks):

Producers	% (Sacks)
Bonar	33.00
Union Camp	22.20
Rex Rosenlaw	8.10
Shields	7.60
Salerno	7.60
At Plastics	7.60
Others (Banner, Polypack, Stone)	14.10

Note: Growth rate = 3.4% per year.

(The Omega International Group.)

Major converters of plastic films have committed themselves to this market, a step that all but insures increased usage and sales. For market share of current suppliers, see Figure 2.

L. H. Doar, Jr.
Clemson University
Clemson, South Carolina

BAGS, MULTIWALL

Multiwall bags occupy an important niche for the packaging and transportation of loose and free-flowing products. Like many other systems of packaging, they satisfy a need that had been waiting for a long time.

Prior to the United States Civil War, two industrious merchants in upstate New York, Arkell and Smiths, had been looking at cotton sacks to ship flour and grains to the larger southern cities, and from this vantage point, they were also investigating the use of paper sacks for marketing flour. Arkell received the first U.S. patent for paper sack making machines in the 1860s; this was just in time to permit paper sacks to replace the difficult to obtain cotton bag fabric during the War.

At the same time salt was being packaged and marketed in cotton bags, until the 1890s, when Adelmer Bates developed a concept wherein a "valve" was incorporated into one corner of the cotton bags. Its advantages provided a faster method of filling cotton bags with salt, and it was patented in 1899. In 1901 John Cornell worked with Bates on the development of a valve sack filling machine, and in 1902 they were awarded a patent on a valve for paper sacks, but almost two decades passed before the paper shipping sack was widely used.

In 1919 R. M. Bates, Adelmer's brother, was visiting in Norway, where he saw a paper sack of a truly different construction. Having five walls made from long strips of kraft paper with each edge of the strip pasted to its adjacent edge, it formed tubes of paper. These were cut to length and were open at both ends. At the place where they were to be filled, one end was gathered together and tied off and when filled, the top end was also tied with twine or wire. These were flat paper tubes.

While Bates was working on his ideas of filling these tubes through the valve corner, his partner Cornell conceived the idea of having valve corners inserted into a gusseted tube, or "in-fold" feature. These were first made in the mid-1920s, and their advantages were promptly recognized. During the first year of production millions were manufactured and sold, and by 1927 about 130 million were made annually.

While the cotton bags with their paper-filling sleeves had solved the bag-filling problem, the other problem of their return and reuse to reduce bag costs still existed. Into this breach came the multiwall paper sacks (either with or without filling sleeves), and thus the paper shipping sack industry was born.

Portland cement and masonry cement plants, grain products and flour factories, and saltworks were the earlier multiwall sack users, because these new bags solved the dusty conditions that had been so dangerous, healthwise. Now the workers were freed from breathing undesirable dust and airborne mineral deposits, and working conditions were greatly improved.

The terms *multiwall bags* and *paper shipping sacks* are often used synonymously. Other industries followed suit as soon as more producers of multiwall bags were established, and the shipping sack industry prospered accordingly. World War II caused severe shortages in burlap and manila fibers due to the extremely long distance from the primary sources in Southeast Asia, and concurrently cotton fiber for bags was in short supply, due to U.S. military and naval needs (as well as for civilian clothing). These shortages resulted in a large expansion of the use of paper shipping sacks, and the cotton bag never recovered its earlier dominance, just as cotton and burlap had replaced the rigid containers such as wood boxes, crates, and barrels.

During World War II the multiwall bag really came into common useage. With a wide range of special papers to control moisture and insect infestation plus the ability to tailor-make whatever package strength was required to fit the rough military handling, multiwalls came to be the approved specification packages for government purchasing.

From these early beginnings the sack industry has grown into a multi-billion-dollar business with shipments of over 3 billion sacks annually for over 2000 different products. These end uses fall into four major categories: agricultural products and supplies, food products, chemical products, and rock and mineral products.

More than 50 bag plants in 25 states can provide a steady and reliable source of supply of the containers they produce. These plants ship about 900,000 tons of product annually, and when converted into finished paper bags, they would encircle the earth about 55 times.

The primary advanges of paper shipping sacks are low tare weight, flexibility, ease of filling and handling, low cost, minimum storage space, biodegradability, and good graphics. Other advantages include the basic material, which is a renewable resource; protection of contents from moisture absorption; control of contents from moisture loss; protection of contents from chemical action; control of seepage or penetration of hot-packed products; provision of loss of gas or vapor; prevention of product sifting or contamination; good stacking and utilization of warehouse space; FDA approval for human food products; excellent graphics; and ease of use in merchandising displays.

Multiwall bags exclude single-ply bags and bags of duplex construction, but do include those with three or more plies of paper or other barrier protectors. They are produced by combing several layers of paper (or other substrate) over a metal former that nests the walls into a long continuous tube (flat or gusseted). These walls are bonded together with adhesives so that each tube is independent of the other tubes, and therein lies the bag's strength. The tubes may be cut straight (all at once) or in stepped-end formation where each wall is first perforated, then simultaneously snapped apart.

Next, in order for the tubes to be filled and used, the tubes are delivered to bottoming machines in the bag plant where they are either sewn closed at one or both ends. Or, the tubes may go to a pasting bottomer to form either pasted open-mouth, stepped-end pasted valve, or pinch-bottom open-mouth sacks (see Fig. 1).

Types of Multiwall Sack

As we have described in the manufacturing process, there are tow basic multiwall types: the open-mouth type and the valve type bags. Multiwalls are custom-made to order according to the customer's requirments and for adequate strength to ensure the safe arrival of their contents to the destination locations.

Whether the product is packaged with high-speed open-mouth weighing and filling machines with automatic closing and sealing machinery, or with even faster, valve packers and automatic bag placers, a proper and successfully designed and printed multiwall bag may be readily developed.

Most multiwall bag suppliers have very well trained sales personnel, and experienced factory technical representatives with a wide range of user experience who are well qualified to recommend the best bag for any product requirement.

Open-mouth bag type

Sewn open mouth (SOM). This style, with any number of walls, has a factory sewn bottom and open top, and may be either flat or with side gussets. These bags are used primarily to package granular or large-particle products. The product is delivered into the open top, and the top may be closed by sewing or other means. If a polyethylene liner ply is specified, this may also be heat-sealed. These types of packaging lines may be fully automated.

Pasted open mouth (POM). This style has a pasted bottom and open-mouth top, and is sealed after filling by folding and pasting, or in some cases, sewing the top. This latter style is not preferred for packaging and shipping free-flowing prod-

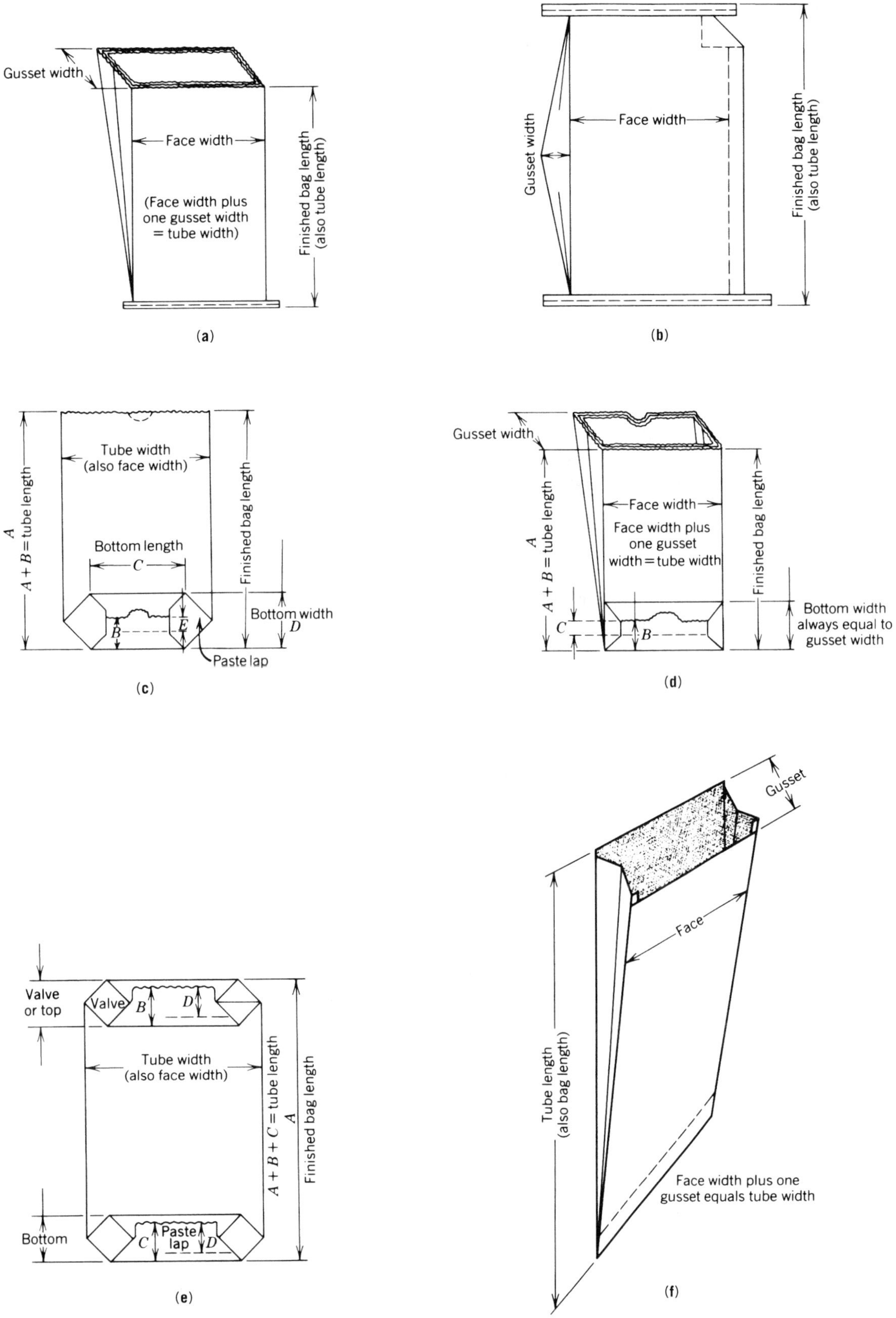

Figure 1. Common bag styles in current use: (**a**) sewn open-mouth; (**b**) sewn valve, (**c**,**d**) pasted open-mouth; (**e**) pasted-valve stepped-end; (**f**) pinch-bottom open-mouth.

ucts because a finlike (nonflat) package is obtained. But it is ideal for containing a number of previously filled and sealed unit bags, eg, 12.5-lb sugar, or 5–10-lb bags of potatoes. These are called *master container* or *baler bags.*

Pinch-bottom open-mouth (PBOM). This style has the plies cut in a stepped fashion with the bottom sealed by gluing or heat sealing at the bag factory. After filling at the user's plant, the open mouth top is folded over and sealed by reactivating preapplied hot-melt adhesive. This style of bag provides completely moistureproof and siftproof packages. PBOM bags are the fastest-growing segment of the total multiwall bag industry because of the secure and strong closure ends, and full control of sifting or infestation. These were developed during the 1960s and have come into great demand in the 1970s and 1990s.

Valve bag type

Sewn valve (SV). These are made with the plies at the top and bottom ends sewn at the bag plant usually with sewn-through crepe paper tape. The bags are filled with a valve packing machine by forcing the product through a built-in valve corner. For small particles and large powders, the contents may be retained without any leakage, by hand-folding in the paper valve extension.

This old style with sewn tape at both ends has fallen out of favor in recent years, and the modern pasted valve style has supplanted it.

Pasted valve stepped end (PVSE). As in the pinch-bottom open-mouth bag, the plies are cut off in a stepped configuration; then the bag factory bottoming machine folds and, with adhesive, seals both top and bottom ends of the tubes. An added valve sleeve is built into a corner, permitting rapid bag placement on a valve packer; and when the filled package falls to a take-away conveyor, the valve automatically is closed by the internal pressure of the product.

PVSE-type bags, introduced in 1956, are the present standard package for most automatic or high-speed operations such as: cements, powders, feeds, fertilizers, or other chemical or building products. When stacked or palletized, they provide stable and attractive three-dimensional containers.

Valve bags may also be purchased with paper tuck-in sleeves that, when folded in, will control all possible leakage.

Sizing Multiwall Bags

First, the weight to be packaged and the product density must be determined. This latter may be done by using a box of known dimensions (1 ft × 1 ft × 1 ft) or 1 ft^3 to determine the weight per cubic foot. With the known weight and volume requirements and with the bag type selected (by product characteristics) an educated estimated set of dimensions can be offered. Your supplier can fine-tune these numbers by offering several handmade factory samples. When test-filled, these may be adjusted to supply a larger number of machine-made test packages to finalize all specifications such as dimensions, filling, handling and shipping, and palletizing conditions.

Constructions

The paper most commonly used to fabricate multiwall bags is natural-colored virgin pulp brown kraft. Made primarily in the Southern U.S. states using a modified sulfate pulping process, it affords longer fibers that enhance the paper's cross- and machine-direction strength. Typically brown, it may be lightened through bleaching, or semibleaching sulfate pulping.

Kraft is usually designated in basis weights ranging from 40 to 80 lb per ream of 500 sheets of 24 in. × 36 in. in area. These grades are known as *multiwall kraft* and have higher strength specifications than grocery-bag kraft (NK) used for light-duty brown paper bags.

Other popular grades used in multiwall bags are extensible (XTK) and free-dry (FDK) process papers. These developments during the 1960s and 1980s resulted in improved cross- and machine-direction stretch specifications and greater overall bag strength. Other grades, such as high-finish, rough, calendered, machine-finish, and wet-strength, are available for specific end-use purposes (see Paper).

Subsequent to the mid-1940s with the development of tailor-made plastics, polyethylene (PE) films have been integrated into many bag constructions. Prior to this, asphalt laminated kraft (ALK) provided moisture barrier properties to multiwall bags, but it was difficult to fabricate on the tuber and bottoming equipment, and it is no longer in use.

Multiwall paper bags may be lightly coated with high-shear, low-stress adhesives to overcome bag slippage when stacked on pallets or in storage. This inexpensive treatment is invisible yet very effective.

Heavy-duty single-wall plastic film bags are also available, however, then compared in cost with multiple-ply kraft or a kraft paper ply plus a lightweight poly ply or polyethylene-coated kraft, the single-wall plastic bags are not cost-effective. They also are slippery, and palletized filled bags cause shipping difficulties (see Bags, plastic).

Depending on the physical characteristics of the product being packaged in multiwall bags, they can be custom-designed to provide all the protection that is necessary. Low- or high-density polyethylene, polyvinyl chloride, polyvinyl dichloride, nylon, saran, foil, and polypropylene are some of the available substrates.

Most multiwall bag suppliers have well-trained sales personnel and specially trained factory technical representatives who are well qualified to recommend the best construction for any particular requirement.

Specifications

With the complete knowledge of a product's physical and flow characteristics, a bag style may be selected. This is only a minor portion of the information required to design a new multiwall bag that will deliver the product in acceptable condition. The product density and the desired weight to be bagged must be known; from this, the required capacity (in cubic inches) is calculated.

General rules are usually followed to select the total kraft paper basis weight and the number of kraft plies, For instance, 50 lb of product might require three walls of 50-lb natural kraft to total 150 lb, and 80 lb would need three or four walls of combined 40- and 50-lb basis weight multiwall kraft, totaling ~180 lb.

The requirements for product protection must be carefully evaluated. This might include moisture protection and control, gas or odor control, grease or oil control, siftproofness, chemical resistance, mold protection, and toxicity or hazardous chemical protection.

Each of these factors can be controlled by the use of one or more of the following available sheets: low-density PE (LDPE), high-density PE (HDPE), polyethylene-coated kraft paper (PEK), saran, Tyvek, saran-coated polypropylene (SCPP), polyfoil–polykraft laminate (PE-AL-PE-K) biaxially oriented polypropylene (BOPP), or even polycoated crepe kraft tape, or heat-sealed waxed kraft tape. About 25% of the multiwalls now being produced specify some kind of special liner ply.

Special barrier plies are commonly used in multiwall bags to protect the product from gaining or losing its intended moisture content. This is stated in terms of its water vapor transmission rate (WVTR). A wall of material may be laboratory-tested, to determine its WVTR rating in terms of g/m^2 per 100°F at 95% RH (relative humidity) in 24 h. The more effective barriers usually are the heavy 2.5–4.0-mil HDPE and LDPE, or 0.75-mil BOPP film, while those most frequently needed are the less expensive 0.5-mil HDPE and 0.8–1.0-mil. LDPE films.

Some products have specific needs, such as controlled porosity to permit proper air or moisture transfer, grease or oil penetration, wet-strength protection with urea formaldehyde, or melamine formaldehyde, which enable the paper to retain its original strength when saturated by water. All these and more—toxicity, hazardous nature, odor control, acidity or alkalinity—can be controlled.

Or, for a really attractive consumer package, the manufacturer might select a supercalendered high-finish full-bleached white sheet with gravure-quality printing.

Multiwall prices are developed by applying cost factors for bag style, bag size (sheet width and tube length), basis-weight factors for kraft, and the plies of other substrates. To these the supplier adds the cost for printing by area and number of colors required. Bags are usually ordered by the carload or trailerload to obtain the lowest unit price.

In cases where a product is highly aerated and moisture protection is only nominal, one or more kraft plies may require allover pinhole perforations (AOPHP) of different diameter and spacing. These allow the internal pressure to be expended during bag filling and they prevent bag breakage.

Discussing the particular needs that are required for your product with a multiwall bag supplier will direct you to the correct and complete bag specifications from the weighing and filling stages to the customer's ultimate place of use.

Packaging Equipment

For each style and size of bag used, there is a wide choice of equipment available for weighing, filling, and closing the packages. There are two main styles of bagging equipment: open-mouth and valve.

Open-mouth packers. The older, original method of open-mouth bagging required a scale, a vertical delivery spout, and a closing function. With today's modern high-speed open-mouth equipment, there are automatic and accurate weighing scales, automatic bag placers, and finally, automatic bag closures.

These systems usually run up to 20 or more preweighed charges of up to 100 lb of material per minute with only token supervisory requirements. There are two preferred methods of closing open-mouth multiwall bags: sewn closure or heat sealing.

Most common of the sewn types is the stitching of the bag mouth with cotton or polypropylene sewing thread with an industrial sewing head. With more secure or special product requirements, either sewn-through tape, adhesive, or heat-sealed tape over sewing may be provided.

Valve bag packers

Valve. With the invention of the valve packer in the mid 1920s, the valve-style bag may be filled by forcing the product through a spout (usually horizontal) into the valve corner and then down into the bag. The product is either preweighed and then forced into the bag, or it is weighed in the package (gross-weighed).

Choices are available when weighing and filling valve bags, and they depend on the product's physical characteristics, price, and production speed requirements.

Most valve packers use either the impeller-wheel method or the belt-feed method. With the former, rotating high-speed paddles force the material into the package. This causes a great amount of aeration to occur, and the PV or PVSE bags must be capable of withstanding great pressure. One operator with a four-spout impeller valve packer may produce up to thirty 94-lb (206-kg) filled bags per minute.

With the belt valve packer, the product is thrust between a moving belt and a rotating grooved pulley and then into the valve opening. This is gentler on the material, but not as fast as the impeller method. Fertilizers and small-pelleted materials are usually packaged on belt-style valve packers.

Three other less popular valve packer styles are available auger, gravity, and airflow. The latter is more modern than the other two and has the advantages of handling non-free-flowing finely powdered materials, without degradation with much better accuracy. The auger packer utilizes a horizontal screw within a hollow tube. These work well for non-free-flowing products, and since the screw's rate of rotation can be adjusted, better weight accuracies are obtained. The last and least popular valve packing method is the vertical-gravity style. But, since it is used for very free-flowing or inexpensive materials, accurate bag weighments are sometimes difficult to obtain.

Graphics

Previously, the fabrics of multiwall paper bags, predecessors were printed with simple yet effective designs to identify the manufacturer, the product, the net weight, and some methods for using the contents.

Cotton and burlap bags had a second life, since they must either be returned, cleaned and filled for reshipment, or in the case of the cotton bags, be washed, and then sewn into towels or items of apparel at home.

With the conversion from fabric to paper bags came the important advantage of superior printing capabilities. With smooth surfaces and all-around six-sided printing of up to four colors, a company's marketing message could be extolled, thereby differentiating itself from all its competitors.

If brown kraft is not attractive enough, a supplier can readily furnish semibleached (SBW) or full-bleached (FBW) kraft. Smoother or brighter papers and coatings in conjunction with modern printing capabilities provide appealing and forceful advertising that make the bags effective billboards for marketing their contents. Whether they were small-unit consumer bags, or large industrial-size bags, they can utilize flexographic printing to carry the intended messages.

If more definition is required, prepress gravure printing using up to six colors will result in photographic quality printing, with the following advantages:

Excellent product and protection
Unlimited supplier capabilities
Biodegradeable materials
Proven shipping and storage environments

Pet-food and charcoal briquet bags are excellent examples of effectively printed packages found in all supermarket shelves today.

Multiwall bags are readily accepted and desirable as shipping containers in all domestic and export markets, including hundreds of uses for agricultural products, building materials, chemicals, food products, and minerals.

Transportation

In the past 40 years the shipment of filled multiwall bags to the market where they are used has swung almost completely from railroad boxcars to over-the-road trailers or straight trucks. Also pallets have replaced the individual handling of filling bags. Coincidentally, plastic film has in many cases been incorporated into the bag's construction. These changes have resulted in lighter basis-weight requirements and lower costs.

Helpful Hints

1. In the initial planning stage of a new use for multiwall shipping sacks, obtain the guidance of bag suppliers.
2. Develop accurate product physical and chemical properties.
3. Finalize decision on desired package weights.
4. Determine annual production requirements.
5. Select proper bag style and features.
6. Consider styles of packaging equipment available.
7. Decide on the transportation and distribution modes.
8. Determine probable bag sizes and constructions.
9. Plan bag transporting methods to provide minimum individual handlings.
10. Select an ISO 9000–grade bag supplier.

Conclusion

With today's concerns relative to bag shipments, when changing from specification-oriented to performance-oriented standards, there is no reason why paper shipping sacks cannot continue to be designed and produced to meet these new standards. The only true test of the effectiveness of any package is its ability to deliver the product to the market in a safe and acceptable manner.

Paper shipping sacks have a proven history of ability to deliver products at low cost to many markets, and are expected to continue to provide these requirements for many years to come. And, with the increased use of special sheets in multiwall bag constructions, there is no limit to the end uses and products that may be packaged in paper shipping sacks in the future.

BIBLIOGRAPHY

1. Friedman and Kipness, *Distribution Packaging,* Krieger Publishing, Huntington, NY, 1977.
2. J. F. Hanlon, *Handbook of Package Engineering,* McGraw-Hill, New York, 1971.
3. Paper Shipping Sack Manufacturers Assoc., Inc., *Reference Guide for the PSSMA,* Tarrytown, NY, 1991.

Allan M. Baylis
Amber Associates
Jamesburg, New Jersey

BAGS, PAPER.

See Bags, Multiwall.

BAGS, PLASTIC

Plastic bags, available in virtually all shapes, sizes, colors, and configurations, have replaced paper in most light-duty packaging applications. Paper has been more difficult to replace in heavy-duty applications (see Bags, heavy-duty plastic). Light-duty plastic bags are generally described in one of two ways: by the sealing method or by application. This article explains the methods used to produce plastic bags and defines the various types of plastic bags in terms of their intended use. A plastic bag is defined here as a bag manufactured from extensible film (see Films, plastic) by heat-sealing one or more edges and produced in quantity for use in some type of packaging application.

Methods of Manufacture

By definition, all plastic bags are produced by sealing one or more edges of the extensible film together. The procedure by which this heat-sealing occurs (see Sealing, heat) is typically used to identify and categorize types of plastic bags. There are three basic sealing methods in use today: sideweld, bottom-seal, and twin-seal.

Sideweld seal. A sideweld seal is made with a heated round-edged sealing knife or blade that cuts, severs, and seals two layers of film when the knife is depressed through the film material and into a soft rubber backup roller. The materials are fused by a combination of pressure and heat (see Fig. 1).

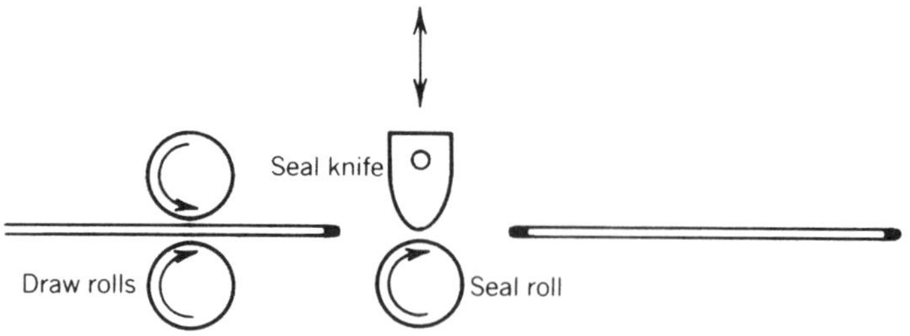

Figure 1. Sideweld-seal mechanism.

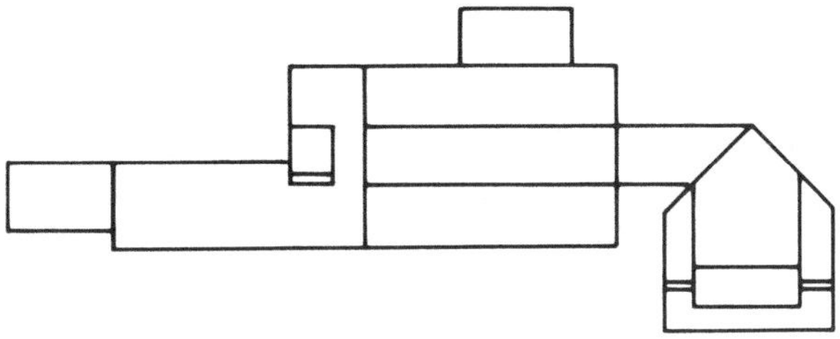

Figure 2. Sideweld process (top view).

The sideweld seal is the most common bag-sealing method. Typical high usage bags, eg, bread bags and sandwich bags, are produced using the sideweld technique. The term sideweld is derived from the fact that many of the bags produced in this fashion pass through the bag machine with the length (or depth) of the finished bag perpendicular to the machine floor. The film fed into the machine is either prefolded, ie, J stock, or folded during the in-feed process (see Fig. 2).

Bottom seal. The bottom-seal technique seals the bag at the bottom only. Tube stock is fed into a bag machine, the single seal is produced at the bottom of the bag, and the bag is cut off with a knife action that is separate from the sealing action. A bottom seal is generally made by a flat, heated sealing bar which presses the layers of film to be sealed against a Teflon (Du Pont)-covered rubber pad, ie, seal pad (see Fig. 3), or another hot-seal bar (see Fig. 4). A separate cutoff knife is used to separate the bag from the feedstock while the seal is made or immediately thereafter. Both bottom-seal mechanism designs produce a bag with only one seal, unless the tube has been manufactured by slit-sealing (see below). The small amount of unusable, wasted film between the edge of the seal and the cutoff point, called the "skirt" of the bag, is an important factor in the total cost of the bag. Any disadvantages caused by the presence of the unwanted skirt are usually offset by greater control of the sealing process. The sideweld method actually melts through the plastic film, and overheating of the film resins can change the physical structure of the plastic molecules when the seal cools. In contrast,

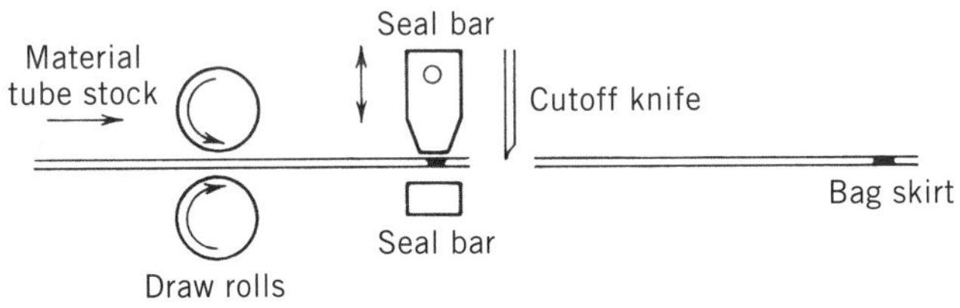

Figure 3. Bottom-seal mechanism (top heat only).

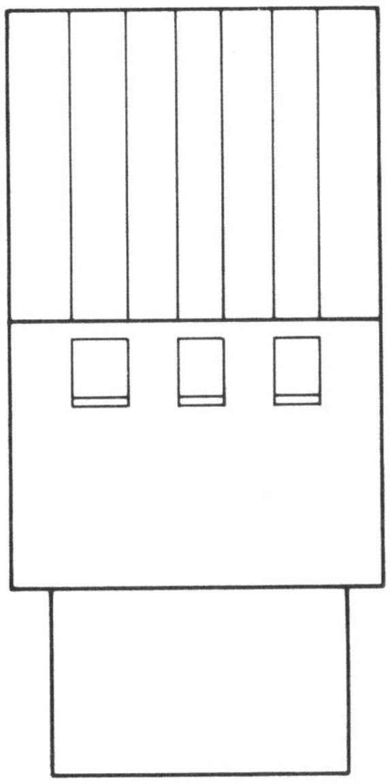

Figure 5. Bottom-deal bag machine, three-level.

the bottom-seal method controls the amount of heat and the dwell time, ie, the time that the heat is applied to the film, to produce a seal that does not destroy the film or change its physical properties. In addition, the total amount of film sealed together is usually larger since the seal bar has a fixed width and none of the film material is melted or burned away.

The bottom-seal method is commonly used to produce HDPE merchandise bags and LDPE industrial liners, trash bags, vegetable and fruit bags, and many other types of bag supplied on a roll. In contrast to the sideweld method, designed primarily for high-speed production of bags made from relatively light-gauge films, ie, 0.5–2.0 mil (13–51 μm), bottom-seal methods are often used to produce bags from film from 0.5 to $\geq$6 mil (13–$\geq$152 μm) at slower production speeds. Bags manufactured by bottom-seal methods are delivered through and out of the bag machine with the length (or depth) of the bag parallel to the machine direction. Since all of the bags are produced from tube sock, multiple-lane production of bags is limited only by the widths of the machine and the bags and the film-handling capability of the bag machine (see Fig. 5).

Twin seal. The twin-seal method employs a dual bottom-seal mechanism with a heated or unheated cutoff knife located between the two seal heads (see Fig. 6). The unique feature of the twin-seal mechanism is that it supplies heat to both the top and bottom of the film material and makes two completely separate and independent seals each time the seal

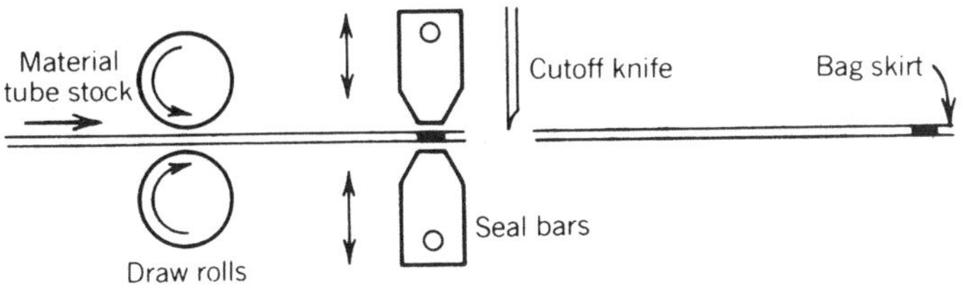

Figure 4. Bottom-seal mechanism (top and bottom heat).

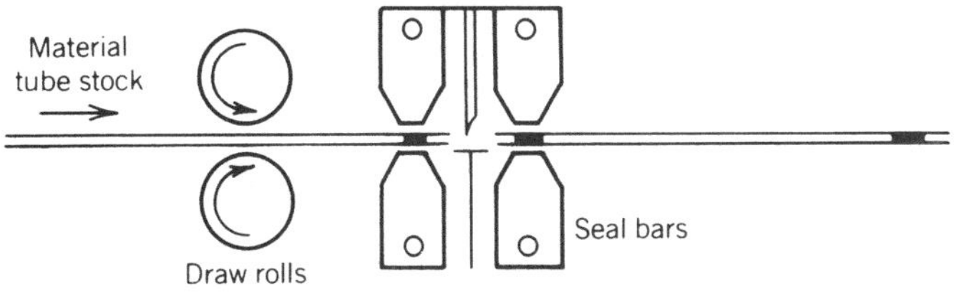

Figure 6. Twin-seal mechanism.

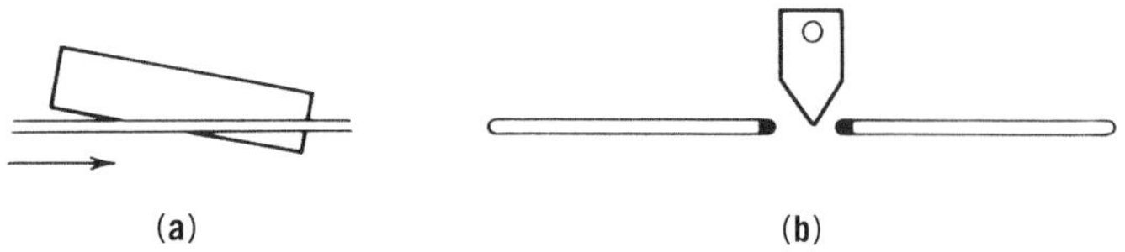

Figure 7. Slit sealer (hot-knife type). (**a**) side view; (**b**) top view.

head cycles. Like the bottom-seal method, the twin-seal technique can supply a large amount of controlled heat for a given duration. This makes the twin seal useful in sealing heavier-gauge films as well as coextrusions and laminates. Since two seals are made with each machine cycle, the twin-seal method can be used to make bags with the seals on the sides of the finished bag, ie, like sideweld bags, or on the bottom of the bag in some special applications, such as retail bags with handles. Many special applications call for the use of a twin-seal-type sealing method, but it is most often used in production of the plastic "T-shirt" grocery sack.

Slit seal. Another type of sealing method, the slit seal, involves sealing two or more layers (usually only two) of film together in the machine direction through the use of a heated knife, hot air, laser beam, or a combination of methods. The slit-seal technique is usually used to convert a single large tube of film into smaller tubes. In the production of grocery sacks, for example, a single extruded 60-in. (152-cm) lay-flat tube of film (see Extrusion) can be run through two slit sealers in line with the bag machine. This results in three tubes of 20-in. (51-cm) lay-flat material being fed to the bag-making system (see Fig. 7).

Applications

Most plastic bags are characterized in terms of their intended use, eg, sandwich bags, primal-meat bags, grocery sacks, handle bags, or bread bags. To provide some order to this user-based classification and definition system, it is convenient to separate the bags into commercial bags and consumer bags.

Commercial bags. A commercial plastic bag is used as a packaging medium for another product, eg, bread. Typical commercial bags and seal methods are listed in Table 1.

Table 1. Commercial Plastic Bags

Sealing Method	Types of Bag
Sideweld	Bread bags, shirt and millinery bags, ice bags, potato and apple bags, hardware bags
Bottom seal	Vegetable bags on a roll, dry-cleaning bags, coleslaw bags, merchandise bags
Twin seal	Primal-meat bags, grocery sacks[a]

[a] With dual bottom seal.

Table 2. Consumer Plastic Bags

Typical Sealing Method	Types of Bag
Sideweld	Sandwich bags, storage bags
Sideweld or bottom seal	Trash bags, freezer bags, can liners
Bottom seal	Industrial liners

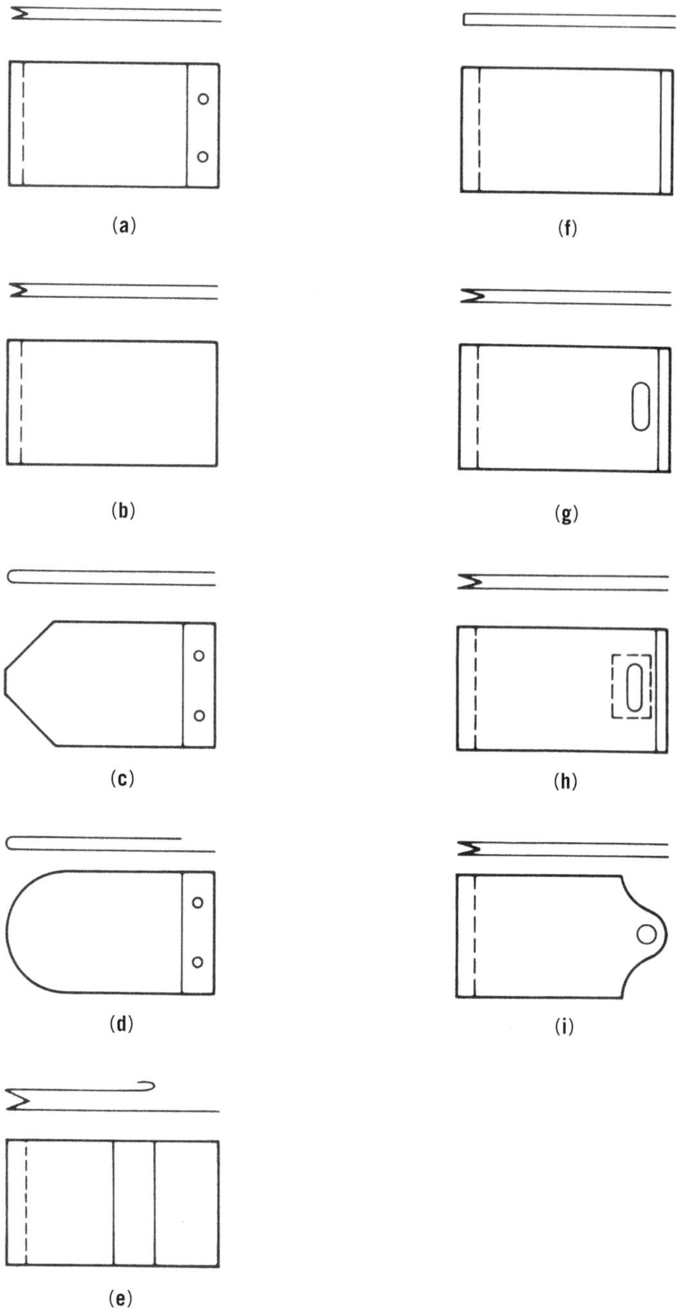

Figure 8. Bags, plastic: (**a**) conventional wicket bag; (**b**) conventional bag; (**c**) square-bottom bag; (**d**) round-bottom bag; (**e**) sandwich bag; (**f**) trash bag; (**g**) handle bag; (**h**) patch-handle bag; and (**i**) sine-wave bag.

Consumer bags. Consumer bags are purchased and used by the consumer, eg, sandwich bags, trash bags. The plastic bag is the product. Typical consumer bags and seal methods are listed in Table 2.

In addition to the conventional bags, there is an enormous variety of special bags that have been created by bag producers to meet commercial and consumer needs. These bags are usually described as specialty bags, but they are all variations of the standard bags described above. Specialty bags include rigid-handle shopping bags, sine-wave handle bags, pull-string bags, patch-handle bags, double-rolled bags, square-bottom bags, round-bottom bags, deli bags, etc. Examples of both common and specialty bags are shown in Figure 8.

BIBLIOGRAPHY

"Bags, Plastic," in *The Wiley Encyclopedia of Packaging Technology,* 1st ed., Wiley, New York, by L. L. Claton, FMC Corporation, 1986, pp. 39–42.

General References

"Bag Making: Inline or Off-Line," *Plast. Technol.* **30**(2), 55 (Feb. 1984).

L. R. Whittington, *Whittington's Directory of Plastics,* 2nd ed., Technomic Publishing, Lancaster, PA, 1978.

Polyethylene Film Extrusion, U.S. Industrial Chemicals Co., New York, 1960.

BANDS, SHRINK

Shrink labels and bands provide today's packaging companies with a unique and versatile method for labeling their products. Developed in Japan during the 1960s, shrink bands became popular in the United States in the early 1980s, mainly as unprinted or one-color tamper-evident neckbands. Today the shrink band is used as a primary label, decoration, for multipack applications, as well as a traditional tamper-evident feature. The high-quality printing and gloss inherent in shrink labels assure packagers that their product will generate tremendous visual shelf appeal.

Manufacturing

Shrink bands are generally produced using special poly(vinyl chloride) (PVC) films, although PETG films, made from a glyco-modified polyester, make up a small but growing portion of the market. Film thicknesses generally run from 1.25 to 3.0 mils, with 1.50 and 2.0 mils the most prevalent. Material is available in two varieties: seamed film, and tubing. *Seamed film* is converted by bonding a flat sheet of printed or unprinted material into a sleeve configuration. This is achieved through a process of solvent seaming into the desired dimension or flat width. *Tubing* is extruded into the desired flat width before printing. Seamed film can be reverse printed with 360° graphics and generally provides more consistent dimensions in both flat width and gauge profile. Tubing is generally less expensive, and is used for unprinted or one-color work.

Important dimensional measurements associated with shrink bands are flat width, which represents the dimension relating to the diameter or circumference of the container; and cut height or impression height, which represents the length of the container or the length of the section of the container to which the shrink band will be applied. These dimensions are usually expressed in millimeters.

The inherent shrink is imparted into the material in a heated stretching process. With tubing, this is achieved during the initial extrusion process. With seamed film, a process known as *tentering* is used. A tenter frame is a modified oven that uses a combination of airflow and temperature zones to orient the material. For example, a 30-in.-wide roll of 5.0-mil material will be converted into a 60-in. roll of 2.0- or 2.5-mil film. Shrink ratios of 50–70% can be achieved by adjusting the process. Various ratios are used depending on the contour of the container being labeled.

Shrink bands are generally processed using the gravure printing method. Reverse rotogravure allows the image or copy to be printed on the back side of the film, which results in a glossy look to the package as well as protection of the image from scratching or scuffing that may occur in final packaging or distribution. Gravure printing provides excellent color reproduction consistency, high speed, and productivity, and is an excellent method for printing smooth film surfaces. Flexo printing is beginning to make inroads into this market as the technology improves, but gravure is still dominant.

After printing, several other features can be added to the shrink band. Vertical and horizontal perforations, as well as tear tape strips, can be added to make the band removable for the consumer. Some packagers use a horizontal perforation to make part of the label removable, the tamper-evident feature is removed, and the primary label stays on the container. Some examples of this are toothpaste pumps, lip balms, syrups, and salad dressings. Bands can be provided in continuous roll-fed form, or in individual cut pieces for manual application.

Application

Bands can be applied manually or by automatic machinery. Machinery on the market today is capable of applying bands at speeds of 500 per minute or more depending on container size, container contour, and label dimensions. For example, a lip-balm label of approximately 27 mm flat width and 60-mm length can be applied faster than a 190-mm-flat width, 40-mm length band for a large dairy container, due to both the label size and the container size. Machinery is dominant in the dairy, packaged-food, and pharmaceutical markets. Most machinery applies shrink bands in a horizontal method, but recent developments have allowed for vertical application of pen barrels, lip balms, and other small cylindrical objects. Most machinery requires material of 2.0 mils or more for processing, and the use of a wedge or other device to open the material before application is common. After application, the container is sent through a heat tunnel, which uses a combination of airflow and heat to shrink the band securely onto the container. A dwell time of 3 s at roughly 300°F is used for most applications, although some products require an elaborate system of varying temperatures and product rotation to achieve the desired effect.

A recent development in machinery is the ability to pack two containers together with one band. This process is used in lieu of an overall bundling method, which then may require the application of another printed label for bar coding or to convey the promotional message. Manual applications are used when product volume is small, or in market trial introductions.

Since shrink bands adhere using shrink energy, no glue or glue applicating systems are required to apply shrink bands.

Marketing and Use

Shrink bands are generally used in three ways: as tamper-evident neck bands, as primary labels and decorations, or for

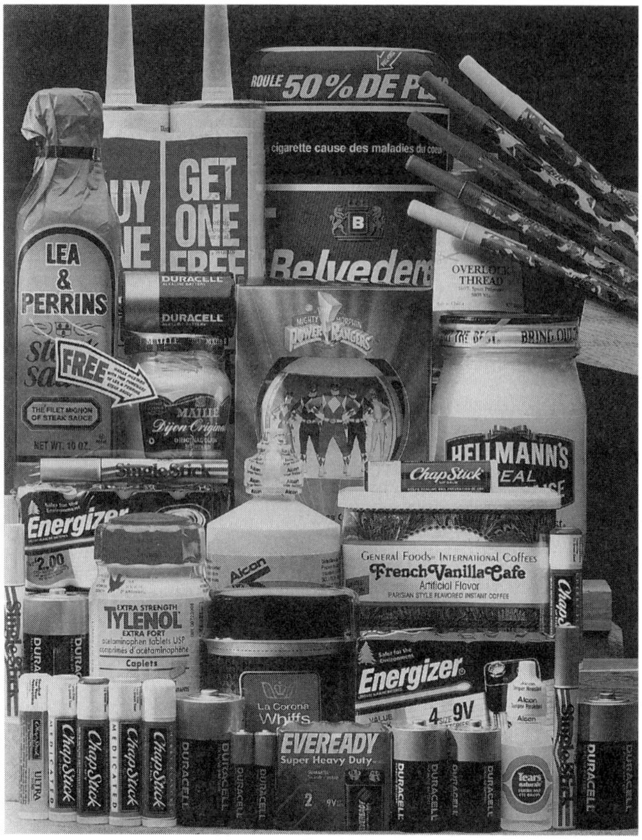

Figure 1. Examples of commercial uses of shrink bands.

promotional multipacking (see Fig. 1 for examples). Food and pharmaceutical packagers have been using neckbands as a tamper-evident feature ever since the Tylenol incident in Chicago in 1982. They are often used in conjunction with vacuum packaging, inner seals, and breakaway closures for product integrity. One advantage for consumers is that neckbands provide evidence of tampering before the consumer brings the product home, as any attempt at removal is evident on the store shelf. In the food-packaging market, yogurt, sour cream, salad dressing, mayonnaise, syrup, mustard, and jelly are just some examples of products that use neckbands. In the pharmaceutical and health–beauty arena, use of neckbands is even more prevalent. Eye care, mouthwash, cough syrup, pain relief, and vitamins as well as many ethical medicines are among the products that utilize this tamper-evident feature (see Tamper-evident packaging).

In the primary label and decorating market, shrink bands provide a versatile and unique way to package various products. The reverse printing capability allows graphics to stand out on the store shelf as well as provide durability. Batteries have been labeled using shrink film for years, as the fine copy and metallic look required on these products are achievable with shrink labels. Lip-balm labeling combines a primary and tamper-evident label using a horizontal perforation for easy removal of the tamper-evident feature. Since this product is often sold loose at drug and convenience-store checkout counters, the lip-balm package can be easily displayed without any further packaging. In recent years, writing instruments decorated with multicolor graphics have become popular, especially those with designs targeted to the teenage market. One advantage of shrink labels on this product is that the reverse printing protects the image from the oils and dirt that occur naturally on people's hands, which can erode an image directly printed on a writing instrument. Other products using shrink bands as primary labels or decorations are Christmas ornaments, Easter egg decorating kits, children's shampoos and soaps, deodorant sticks, plastic baseball bats, golf clubs, yarn and thread spools, craft paints, and tobacco containers.

A third use of shrink labels is for multipacking purposes. Two or more products are bundled together using a shrink band. Printing the bands eliminates the need to apply a paper label later in the distribution process. Two products are placed side by side, and the label is applied vertically over the top. It is not uncommon to put two different products together. Examples of this would be 16 oz of shampoo and a trial size of conditioner, or mouthwash and a toothbrush. When three containers are banded, placing the products in a triangle configuration provides for a more secure package, as this prevents the third product from falling out of the middle. Just about any product can be packaged this way; some common uses are for packaging of hair-care products, vitamins, car-care products, lubricants, peanut butter, caulking compounds, and cooking sauces. An innovative use that has just recently come to the multipack market is where three tubs of baby wipes are stacked on top of each other and vertically sleeved using a shrink band. Most multipacks are done manually, and almost any product or container allows for this banding method.

Shrink bands will continue to provide packaging companies with a unique method of labeling their products. The double benefit of increased tamper evidence and integrity along with attractive graphic impact is beneficial for almost any product. Shrink bands will continue to grow as a labeling method in the beverage, industrial, and toy markets as well as maintaining an important position in the traditional food, health and beauty, and pharmaceutical markets.

David R. Kopsilk and Ed Fisher
CMS Gilbreth Packaging Systems
Trevose, Pennsylvania

BAR CODE. See Code, Bar.

BARRELS

A barrel, or cask, is a cylindrical vessel of wood that is flat at the bottom and top, with a slightly bulging middle. The three primary parts of a barrel are heads (bottom and top), staves (sides), and hoops (rings that bind the heads and staves together) (see Fig. 1). Specifications are contained in the Department of Transportation regulations (1).

In architecture and physics, the arch is probably the stongest possible structure. The more pressure or weight exerted on the top (keystone) of the arch, the stronger the arch becomes.

The wooden barrel is designed according to the double-arch principle of strength. Like an egg shell, it is doubly arched, both in length and girth. The bend in the stave's length is the first arch and the bilge circumference of the stave's width is the second arch. These arches impart great strength.

There are three basic cooperage operations: logging the timber, milling the logs, and assembling the barrel staves and heading material. Saws reduce the logs to length, and produce edge-grained pieces of cylindrically shaped and jointed wood for the staves, and flat pieces of wood for the heading. In recent times, staves have been quarter-sawn as opposed to earlier cylindrical-sawn staves. The quarter-sawn straight staves are planed interiorly and exteriorly throughout their thickness to achieve a stave of cylindrical width. After the wooden material has been air- and/or kiln-dried to approximately 12% moisture content, the staves and heads are assembled into steel-hoop-bound barrels. Assembly operations include: setting up staves; steaming and winching staves to achieve the belly, bilge, or circumference arch; heating to make wood pliable and give one last drying after being steam bent; tapping out for uniform thickness; trussing to tighten stave joints; crozing interior grooves in each end of the staves where the heads will be inserted; heading up by inserting heads in the croze at each end of the staves; hooping up by driving riveted-steel hoops onto exterior of staves; boring for testing, lining, and future filling; bunging up the bored hole; and rolling the barrel out to the marketplace.

Dozens of species of timber from all over the world, have been used to make tight (for liquid) and slack (for nonliquid) cooperage. Hardwood barrels for spirits and wine include oak timber from Limousin and Nevers in France; Alastian and Italian oak; and fork-leafed American white oak, found principally in the slow growing forest regions of the United States of Missouri, Indiana, Tennessee, Kentucky, and Arkansas. Virtually all of the wooden barrels made in the United States today [1–2 million (10^6)/yr] are 50-gal (189-L) capacity barrels used by the bourbon whisky trade. Barrels for bourbon are charred interiorly about 1⁄16 in. (1.6 mm) of their 1 in. (25.4 mm) thickness to bring out the tannin in the wood. Tannin aids in the coloring and flavoring of spirits and wine.

Wooden barrels have had numerous names, depending on their size and use. A small sampling of these names include: pickled-pigs-feet kit; fish pail; one-quarter; one-half, and full-beer ponies; hogshead; salmon tierce; tallow cask; rum puncheon; and port wine pipe. A list of common international cask sizes is presented in Table 1 (2).

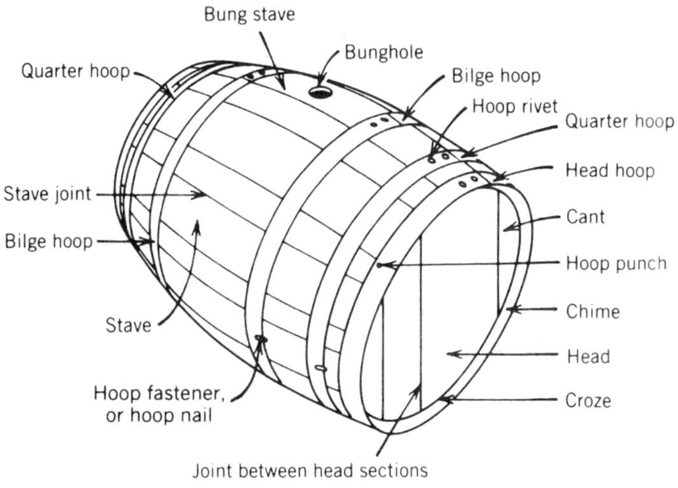

Figure 1. The wooden barrel.

Table 1. Wooden Casks for International Shipments of Alcoholic Beverages

Beverage	Name	Gallons[a]	Liters[a]
		Wine	
Sherry	Butt	137.5/140	500
	Hogshead	67.5/69	250
Port	Pipe	145/147.5	53
	Hogshead	72.5/74	265
Vermouth	Hogshead	67.5/70	250
Burgundy	Hogshead	57.5/60	215
		Spirits	
Rum	Puncheon	137.5/144	520
	Hogshead	67.5/72.5	255
	Barrel	50	180
Brandy	Hogshead	77.5/79	280
	Quarter case	39/40	140
Beer	Butt	135	500
	Hogshead	67.5	245
	Barrel	45	165
	Kilderkin	22.5	82
	Firkin	11	41
	Pin	6	20

[a] Gallon (U.S.) and liter sizes are not equivalent.

Just as the wooden barrel replaced the crude basketry used centuries ago, many other types of container have replaced the wooden barrel: steel and fiber drums, plastic pails, aluminum and steel cans, fiberglass and cement tanks, etc; aluminum and stainless steel replaced wood for beer barrels (3).

To date, no industrial engineer has come up with a blueprint to replace the strength of a wooden barrel. In tests involving high stacking, they can perform better than steel drums.

BIBLIOGRAPHY

"Barrels," in *The Wiley Encyclopedia of Packaging Technology,* 1st ed., Wiley, New York, by Frank J. Sweeney, Sweeney Cooperage Ltd., 1986, pp. 47–48.

1. *Code of Federal Regulations,* Title 49, Sect. 178.155–178.161.
2. F. A. Paine, *The Packaging Media,* Blackie, Glasgow and London; Wiley, New York, 1977.
3. H. M. Broderick, *Beer Packaging,* Master Brewers Association of the Americas, Madison, WI, 1982.

BARRIER POLYMERS

Modern synthetic polymers have been used for over fifty years as barriers to mass transport. The overwhelming majority of applications have been for food storage. Recently, emphases on longer term storage and microwavability have become important. Research remains heavy in industry and academia to offer improved polymers and fabrication methods and to address the issues of conservation such as recycle, source reduction, and degradability. Barrier polymers remain attractive packaging alternatives, because they offer the advantages of light weight, formability into useful and attractive shapes, shatter resistance, transparency, and low cost.

Traditionally, the definition of a barrier polymer has been strongly attached to the oxygen permeability. Barrier polymers had oxygen permeabilities less than about 2.0 nmol/m · s · GPa (= 1.0 cc(stp) · mil/100in^2 · day · atm). (See Table 1 for unit conversions.) This definition unnecessarily limits the range of barrier polymers. Some polymers with higher oxygen permeabilities are useful barriers for other molecules.

The Permeation Process

A basic understanding of the permeation process can help clarify the barrier characteristics of polymers. A permeant molecule moves through a barrier in a multistep process. First, the molecule collides with the polymer surface. Then, it must adsorb and dissolve into the polymer mass. In the polymer, the permeant "hops" or diffuses randomly as its own thermal kinetic energy keeps it moving from vacancy to vacancy as the polymer chains move. The random diffusion yields a net movement from the side of the barrier polymer that is in contact with a high concentration or partial pressure of permeant to the side that is in contact with a low concentration of permeant. After crossing the barrier polymer, the permeant moves to the polymer surface, desorbs, and moves away.

In virtually every case, the permeation is controlled by the solution and diffusion steps. The diffusion coefficient, D, is a measure of the speed of molecules moving in the polymer. The solubility coefficient, S, is an indication of the number of permeant molecules that are diffusing. Together, the diffusion coefficient and the solubility coefficient describe the permeability coefficient, commonly called the permeability, P.

$$P = D \times S \tag{1}$$

A low permeability may result from a low diffusion coefficient or a low solubility coefficient or both.

The permeability for a given polymer–permeant combination can be used to describe the steady-state transport. Equation 2 is Fick's First Law adapted for packaging

$$\frac{\Delta M_x}{\Delta t} = \frac{PA\Delta p_x}{L} \tag{2}$$

where $\Delta M_x/\Delta t$ is the steady-state rate of permeation of permeant x through a polymer film with area A and thickness L. P is the permeability, and Δp_x is the difference in pressure of the permeant on the two sides of the film.

Equation 2 shows why reliable tables of permeabilities are important. The packaging engineer has no control over the Δp_x since the conditions of the environment and the contents are fixed. Mechanical, economic, and containment requirements limit the allowable ranges of area and thickness. Only P has a wide range of possibilities.

One caveat must be considered before applying equation 2. The permeation must be at steady state. With small molecules such as oxygen, steady state is usually attained in a few hours or less depending the polymer and the thickness. However, with larger molecules in barrier polymers, especially glassy polymers, the time to reach steady state can be very long, possibly exceeding the anticipated storage time. The time to reach steady state, t_{ss}, can be estimated with equation 3.

$$t_{ss} \cong L^2/4D \tag{3}$$

Equation 2 should only be used when t_{ss} is small compared to the storage time.

Water vapor transmission is treated differently. The industry has arrived at a standard condition for reporting and comparing performance, 37.8°C (100°F) and 90%rh difference. Equation 4 shows how the rate of water vapor transmission can be calculated using the value of the water vapor transmission rate (WVTR) and the package geometry.

$$\frac{\Delta M_{H2O}}{\Delta t} = \frac{WVTR \cdot A}{L} \tag{4}$$

Table 1. Permeability Units[1] with Conversion Factors

Multiply → to obtain ↓	$\frac{\text{nmol}}{\text{m·s·GPA}}$	$\frac{\text{cc·mil}}{\text{100 in.}^2\text{·d·atm}}$	$\frac{\text{cc·mil}}{\text{m}^2\text{·d·atm}}$	$\frac{\text{cc·cm}}{\text{cm}^2\text{·s·atm}}$	$\frac{\text{cc·cm}}{\text{cm}^2\text{·s·cm Hg}}$	$\frac{\text{cc·20 }\mu\text{m}}{\text{m}^2\text{·d·atm}}$
$\frac{\text{nmol}}{\text{m·s·GPa}}$	1	2	0.129	4.390×10^{10}	3.336×10^{12}	0.1016
$\frac{\text{cc·mil}}{\text{100 in.}^2\text{·d·atm}}$	0.50	1	6.452×10^{-2}	2.195×10^{10}	1.668×10^{12}	5.08×10^{-2}
$\frac{\text{cc·mil}}{\text{m}^2\text{·d·atm}}$	7.75	15.50	1	3.402×10^{11}	2.585×10^{13}	0.787
$\frac{\text{cc·cm}}{\text{cm}^2\text{·s·atm}}$	2.278×10^{-11}	4.557×10^{-11}	2.939×10^{-12}	1	76.00	2.315×10^{-12}
$\frac{\text{cc·cm}}{\text{cm}^2\text{·s·cm Hg}}$	2.998×10^{-13}	5.996×10^{-13}	3.860×10^{-14}	1.316×10^{-2}	1	3.046×10^{-14}
$\frac{\text{cc·20 }\mu\text{m}}{\text{m}^2\text{·d·atm}}$	9.84	19.68	1.27	4.32×10^{11}	3.283×10^{13}	1

[a] Throughout the *Encyclopedia* cm^3 (or mL) is used in preference to cc. However, the advantage of using cc here is an obvious visual aid in the complex units and there are further comments regarding cc vs cm^3 in the text.

Table 2. Water Vapor Transmission Rate Units with Conversion Factors

Multiply → to obtain ↓	$\frac{\text{nmol}}{\text{m·s}}$	$\frac{\text{g·mil}}{100\ \text{in.}^2\text{·d}}$	$\frac{\text{g·cm}}{\text{m}^2\text{·d}}$
$\frac{\text{nmol}}{\text{m·s}}$	1	0.253	6.43
$\frac{\text{g·mil}}{100\ \text{in.}^2\text{·d}}$	3.95	1	25.40
$\frac{\text{g·cm}}{\text{m}^2\text{·d}}$	0.155	3.94×10^{-2}	1

When the actual conditions differ from the standard, the WVTR can be adjusted with great care. Ideally, data have been reported at the actual conditions; otherwise adjustments for both the relative humidity difference and temperature must be made. If the polymer is known to be insensitive to humidity, such as a polyolefin, the humidity adjustment is merely multiplication by the actual relative humidity difference on the two sides of the film divided by 90%rh. The temperature effect will be discussed later.

Units

The units for permeability are complex, and many correct combinations are used in the literature. Lamentably, many incorrect combinations are used too. Table 1 contains conversion factors for several common units for the permeability. In these units the quantity of permeant is a molar unit, typically a cc(stp). For the permeability of flavors and aromas, a unit using the mass of the permeant is useful. A modified unit, the MZU (= 10^{-20} kg·m/m^2 s Pa), can be converted to a molar unit according to equation 5 where *MW* is the molecular weight of the permeant in daltons (g/mol).

$$P \text{ in MZU} \times (10/MW) = P \text{ in nmol/m} \cdot \text{s} \cdot \text{Gpa} \quad (5)$$

Table 2 contains conversion factors for several common units for WVTR.

Permeability Data

Table 3 contains the oxygen permeabilities for several polymers at 20°C and 75% rh. The polymers are ranked roughly in order of increasing permeability. This list contains polymers that meet the traditional criterion for barrier polymers and several that do not. The range of permeabilities here is more than 4 orders of magnitude. Table 3 also contains the permeabilities of nitrogen and carbon dioxide. Generally, the permeabilities of N_2, O_2, and CO_2 are in the ratio 1:4:14.

Table 4 contains diffusion coefficients and solubility coefficients for oxygen and carbon dioxide in many of the same polymers. These will be useful for comparison with flavor, aroma, and solvent permeation.

Table 5 contains WVTR data for a similar list of polymers. The polymers are ranked roughly in order of increasing WVTR. The rankings in Table 5 have some notable differences from the rankings in Table 3. The polyolefins, polyethylene and polypropylene, are good water vapor barriers in contrast to their oxygen status. The vinylidene chloride copolymers remain as water vapor barriers while some other polymers have moved down the barrier list.

Table 6 contains data for the transmission of flavor, aroma, and solvent (F/A/S) molecules in a few polymers. These data are for 25°C at 0% rh and very low activity (or partial pressure) of the permeant. This list represents only a small fraction of the virtually limitless combinations of F/A/S compounds and polymer films. Reliable data at low activities, as typically encountered in foods, are difficult to find. However the data in Table 6 are consistent with some general rules. First, polyolefins are not good barriers for F/A/S compounds. Second, vinylidene chloride copolymers and ethylene–vinyl alcohol copolymers are good barriers. A third rule is not apparent from Table 6. A polymer below its glass-transition temperature (T_g), ie, a glassy polymer, is an excellent barrier for F/A/S compounds. Data are extremely rare; hence, only a few data are given for an ethylene–vinyl alcohol copolymer. The problem is that the diffusion coefficients are so low that the experiments take too long to do with accuracy. The author and associates have been unsuccessful in many attempts to test polystyrene, PET, and nylons.

In these tables of data, different polymers occupy the top barrier positions. For oxygen, vinylidene chloride copolymers and ethylene–vinyl alcohol copolymers are the best barriers. For water vapor, vinylidene chloride copolymers and the polyolefins are the best barriers. For F/A/S compounds, while vinylidene chloride copolymers remain good. Some glassy polymers that are not given are the best barriers. Yet of all these polymers, only a few meet the traditional definition of a barrier polymer.

Factors Affecting Permeability

The permeability increases with increasing temperature for all known cases. A plot of logarithm *P* versus 1/*T* in Kelvin yields a straight line with a slope proportional to the activation energy for permeation. Usually the slope is steeper above T_g than below T_g. Hence, knowledge of the permeability at two temperatures allows calculation of the permeability at a third temperature provided T_g is not in the range. For many polymers the oxygen permeability increases about 9% per °C above T_g and about 5% per °C below T_g. The temperature sensitivity is greater for larger-permeant molecules. The temperature sensitivity for the WVTR is theoretically a little more complicated; however, it is about the same as for the oxygen permeability.

Humidity can affect the permeability of some polymers. When a polymer equilibrates with a humid environment, it absorbs water. The water concentration in the polymer might be very low as in polyolefins or it might be several weight percent as in ethylene–vinyl alcohol copolymers. Absorbed water does not affect the permeabilities of some polymers including vinylidene chloride copolymers, acrylonitrile copolymers, and polyolefins. Absorbed water increases the permeabilities in some polymers including ethylene–vinyl alcohol copolymers and most polyamides. A few polymers show a slight decrease in the oxygen permeability with increasing humidity. These include PET and amorphous nylon. Since humidity is inescapable in many packaging situations, this effect cannot be overlooked. The humidity in the environment is often above 50%rh, and the humidity inside a food package can be nearly 100%rh.

Table 3. Permeabilities of Selected Polymers[a]

Polymer	Gas permeability nmol/m·s·GPA		
	Oxygen	Nitrogen	Carbon dioxide
Vinylidene chloride copolymers	0.02–0.30	0.005–0.07	0.1–1.5
Ethylene-vinyl alcohol copolymers, dry	0.014–0.095		
at 100% rh	2.2–1.1		
Nylon-MXD6[b]	0.30		
Nitrile barrier polymers	1.8–2.0		6–8
Nylon-6	4–6		20–24
Amorphous nylon (Selar[c] PA 3426)	5–6		
Poly(ethylene terephthalate)	6–8	1.4–1.9	30–50
Poly(vinyl chloride)	10–40		40–100
High density polyethylene	200–400	80–120	1200–1400
Polypropylene	300–500	60–100	1000–1400
Low density polyethylene	500–700	200–400	2000–4000
Polystyrene	500–800	80–120	1400–3000

[a] Ref. 1; see Table 1 for unit conversion.
[b] Trademark of Mitsubishi Gas Chemical Co.
[c] Trademark of E. I. du Pont de Nemours & Co., Inc.

Table 4. Diffusion and Solubility Coefficients for Oxygen and Carbon Dioxide in Selected Polymers at 23°C, Dry[a]

Polymer	Oxygen		Carbon dioxide	
	D, m^2/s	S, nmol/(m^3·GPa)[b]	D, m^2/s	S, nmol/(m^3·GPa)[b]
Vinylidene chloride copolymer	1.2×10^{-14}	1.01×10^{13}	1.3×10^{-14}	3.2×10^{13}
Ethylene-vinyl alcohol copolymer[c]	7.2×10^{-14}	2.4×10^{12}		
Acrylonitrile barrier polymer	1.0×10^{-13}	1.0×10^{13}	9.0×10^{-14}	4.4×10^{13}
Poly(ethylene terephthalate)	2.7×10^{-13}	2.8×10^{13}	6.2×10^{-14}	8.1×10^{14}
Poly(vinyl chloride)	1.2×10^{-12}	1.2×10^{13}	8.0×10^{-13}	9.7×10^{13}
Polypropylene	2.9×10^{-12}	1.1×10^{14}	3.2×10^{-12}	3.4×10^{14}
High density polyethylene	1.6×10^{-11}	7.2×10^{12}	1.1×10^{-11}	4.3×10^{13}
Low density polyethylene	4.5×10^{-11}	2.0×10^{13}	3.2×10^{-11}	1.2×10^{14}

[a] Refs. 1
[b] For unit conversion, see equation 5.
[c] 42 mol % ethylene.

Table 5. Water-Vapor Transmission Rates of Selected Polymers[a]

Polymer	WVTR, nmol/m·s)[b]
Vinylidene chloride copolymers	0.005–0.05
High density polyethylene (HDPE)	0.095
Polypropylene	0.16
Low density polyethylene (LDPE)	0.35
Ethylene–vinyl alcohol, 44 mol % ethylene[c]	0.35
Poly(ethylene terephthalate) (PET)	0.45
Poly(vinyl chloride) (PVC)	0.55
Ethylene–vinyl alcohol, 32 mol % ethylene[c]	0.95
Nylon-6,6, nylon-11	0.95
Nitrile barrier resins	1.5
Polystyrene	1.8
Nylon-6	2.7
Polycarbonate	2.8
Nylon-12	15.9

[a] At 38°C and 90% rh unless otherwise noted.
[b] See Table 2 for unit conversions.
[c] Measured at 40°C.

Additives are blended into polymers to improve mechanical or chemical properties such as flexibility, cling, and thermal stability. When the additive is a small molecule and it is soluble in the polymer, the polymer is likely to be plasticized. This effect increases the diffusion coefficient for permeant molecules. The solubility coefficient is unaffected. The effect can be small if the additive is included at less than about 1 wt%. However, at larger concentrations the effect can be large. The oxygen permeability of poly(vinyl chloride) increases by about ten times when enough plasticizer is added to make the resulting film flexible. The phenomenon of antiplasticization is under investigation. The potential exists that, for some small molecule additives, the diffusion coefficient can be decreased.

If the additive is not soluble in the polymer, the result is more complicated. For an inorganic filler, the permeability might increase if the polymer does not wet the filler. The permeability will decrease if the polymer wets the filler. However, the effect is likely to be small unless the loading of filler is greater than about 20 wt%. Loadings this high are avoided since these composites are typically difficult to handle.

Filler in the form of platelets can lower the permeability

Table 6. Examples of Permeation of Flavor and Aroma Compounds in Selected Polymers at 25°C[a], Dry[b]

Flavor/aroma compound	Permeant formula	P, MZU[c]	D, m^2/s	S, $kg/(m^3 \cdot Pa)$
		Vinylidene chloride copolymer		
Ethyl hexanoate	$C_8H_{16}O_2$	570	8.0×10^{-18}	0.71
Ethyl 2-methylbutyrate	$C_7H_{14}O_2$	3.2	1.9×10^{-17}	1.7×10^{-3}
Hexanol	$C_6H_{14}O$	40	5.2×10^{-17}	7.7×10^{-3}
trans-2-Hexenal	$C_6H_{10}O$	240	1.8×10^{-17}	0.14
d-Limonene	$C_{16}H_{16}O$	32	3.3×10^{-17}	9.7×10^{-3}
3-Octanone	$C_8H_{16}O$	52	1.3×10^{-18}	0.40
Propyl butyrate	$C_7H_{14}O_2$	42	4.4×10^{-18}	9.4×10^{-2}
Dipropyl disulfide	$C_6H_{14}S_2$	270	2.6×10^{-18}	1.0
		Ethylene–vinyl alcohol copolymer		
Ethyl hexanoate		0.41	3.2×10^{-18}	1.3×10^{-3}
Ethyl 2-methylbutyrate		0.30	6.7×10^{-18}	4.7×10^{-4}
Hexanol		1.2	2.6×10^{-17}	4.6×10^{-4}
trans-2-Hexenal		110	6.4×10^{-17}	1.8×10^{-2}
d-Limonene		0.5	1.1×10^{-17}	4.5×10^{-4}
3-Octanone		0.2	1.0×10^{-18}	2.0×10^{-3}
Propyl butyrate		1.2	2.7×10^{-17}	4.5×10^{-4}
		Low density polyethylene		
Ethyl hexanoate		4.1×10^{6}	5.2×10^{-13}	7.8×10^{-2}
Ethyl 2-methylbutyrate		4.9×10^{5}	2.4×10^{-13}	2.3×10^{-2}
Hexanol		9.7×10^{5}	4.6×10^{-13}	2.3×10^{-2}
trans-2-Hexenal		8.1×10^{5}		
d-Limonene		4.3×10^{6}		
3-Octanone		6.8×10^{6}	5.6×10^{-13}	1.2×10^{-1}
Propyl butyrate		1.5×10^{6}	5.0×10^{-13}	3.0×10^{-2}
Dipropyl disulfide		6.8×10^{6}	7.3×10^{14}	9.3×10^{-1}
		High density polyethylene		
d-Limonene		3.5×10^{6}	1.7×10^{-13}	2.5×10^{-1}
Menthone	$C_{10}H_{18}O$	5.2×10^{6}	9.1×10^{-13}	4.7×10^{-1}
Methyl salicylate	$C_8H_8O_3$	1.1×10^{7}	8.7×10^{-14}	1.6
		Polypropylene		
2-Butanone	C_4H_8O	8.5×10^{3}	2.1×10^{-15}	4.0×10^{-2}
Ethyl butyrate	$C_6H_{12}O_2$	9.5×10^{3}	1.8×10^{-15}	5.3×10^{-2}
Ethyl hexanoate		8.7×10^{4}	3.1×10^{-15}	2.8×10^{-1}
d-Limonene		1.6×10^{4}	7.4×10^{-16}	2.1×10^{-1}

[a] Values for vinylidene chloride copolymer and ethylene–vinyl alcohol are extrapolated from higher temperatures.
[b] Permeation in the vinylidene chloride copolymer and the polyolefins is not affected by humidity; the permeability and diffusion coefficient in the ethylene–vinyl alcohol copolymer can be as much as 1000 times greater with high humidity (1).
[c] MZU = $(10^{-20} kg \cdot m)/m^2 \cdot s \cdot Pa)$; see equation 5 for unit conversions.

Table 7. Effect of Orientation on Oxygen Permeability for Certain Polymers[a]

Polymer	Degree of orientation, %	Oxygen permeability,[b] $nmol/m \cdot s \cdot GPa)$[c]
Polypropylene	0	300
	300	160
Polystyrene	0	840
	300	600
Polyester	0	20
	500	10
Copolymer of 70% acrylonitrile and 30% styrene	0	2.0
	300	1.8

[a] Ref. 2.
[b] At 23°C.
[c] See Table 1 for unit conversions.

more than filler with a more compact shape. If the platelets tend to lie in the plane of the film, permeant molecules must make wide detours (tortuous path) while traversing the film. This gives a greatly reduced effective diffusion coefficient.

Crystallinity is an overrated contributor to barrier in polymers. First, if a polymer has crystallinity, the level of crystallinity typically exists within a narrow range with only modest variation allowed from fabrication variables. Hence, crystallinity is not a strong design parameter. Second, the same properties that lead to crystallinity also lead to efficient packing in the amorphous phase. Efficient packing in the amorphous phase gives a low diffusion coefficient. Polyolefins are glaring exceptions because they have considerable crystallinity and high diffusion coefficients.

Orientation is frequently cited as a contributor to barrier in polymer films. If the polymer molecules truly are oriented in the plane of the film, either uniaxially or biaxially, the permeability is probably lower than in an unoriented film. However, sometimes the word "oriented" merely means that the film has been stretched. If the polymer molecules relax during stretching, little orientation results and little effect is expected. A few cases have been noted where stretching has created micro-fissures in the polymer and the permeability has increased. Table 7 contains permeability data for elongated films. The results vary. A practitioner is wise to test permeability before concluding that a fabrication involving elongation will lower the permeability. When elongation does lower the permeability, the causes are a combination of better packing among parallel molecules, difficulty in moving perpendicular to the alignment of the polymer molecules, and, when crystallinity is present, the tendency for crystallites to act as platelets aligned in the plane of the film.

Multilayer barrier structures are commonly used for both flexible and rigid applications. These structures can be the result of coextrusion, lamination, or coating. Typically, one of the layers provides most of the barrier while other layers provide inexpensive mechanical integrity, printability, opacity, sealability, formability, adhesion, or merely a place to locate reground scrap. The total barrier performance of a multilayer structure can be estimated with equation 6

$$\frac{L_t}{P_t} = \frac{L_1}{P_1} + \frac{L_2}{P_2} + \frac{L_3}{P_3} + \ . \ . \ . \tag{6}$$

where L_1, L_2, and L_3 are the thicknesses of layers and P_1, P_2, and P_3 are the respective permeabilities of the layers. L_t is the total thickness, and P_t is the effective permeability of the multilayer structure. The quantity L_t/P_t is the "permeance." L_t and P_t may be used in equation 2 to calculate the expected performance.

Polymer Composition

Although a case has been made for a situational definition for barrier polymers, common practice still focuses on a rather small set of polymers. Figure 1 contains schematic chemical structures of polymers with low permeabilities to permanent gases, especially oxygen.

Finding common traits among this diverse group is difficult. However, each has some polarity which leads to chain-to-chain interactions that give good packing in the amorphous phase. Also, frequently, sufficient symmetry exists to allow crystallinity to develop. Again, this can lead to good packing in the amorphous phase.

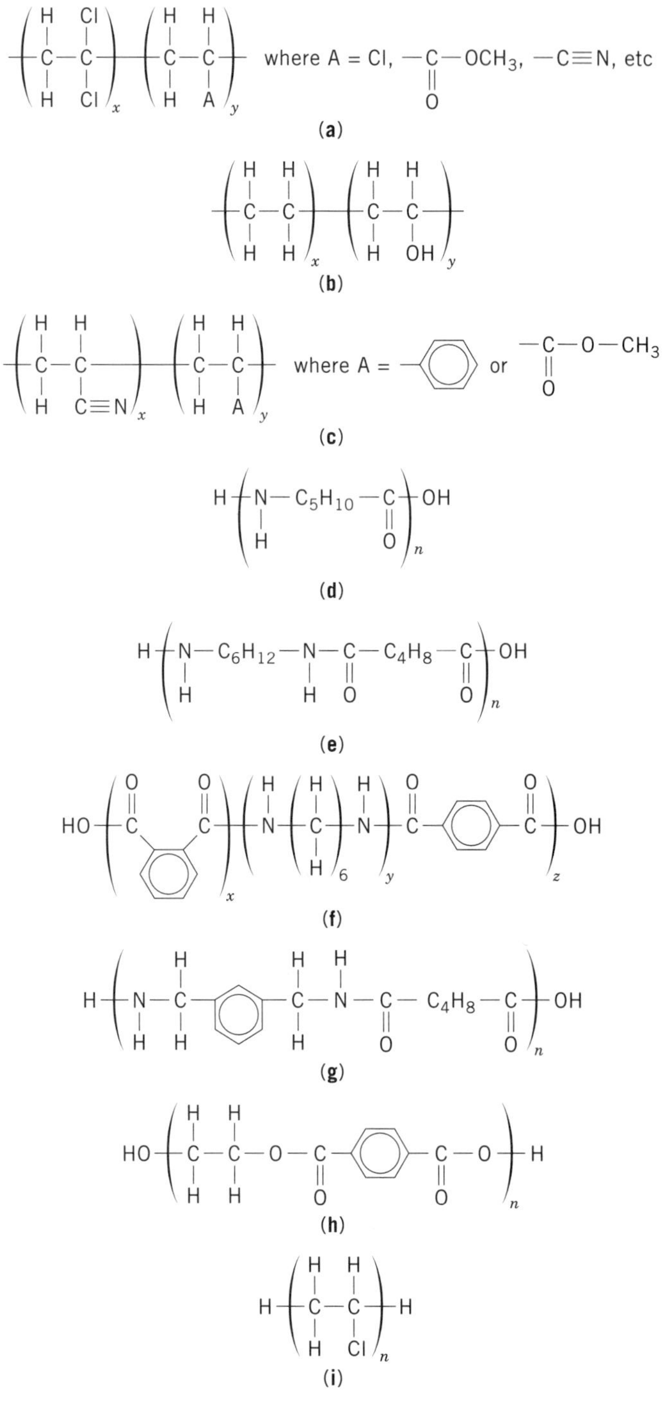

Figure 1. Chemial structures of barrier polymers. (**a**) Vinylidene chloride copolymers; (**b**) hydrolyzed ethylene–vinyl acetate (EVOH); (**c**) acrylonitrile barrier polymers; (**d**) nylon-6; (**e**) nylon-6,6; (**f**) amorphous nylon (Selar PA 3426), y = x + z; (**g**) nylon-MXD6; (**h**) poly(ethylene terephthalate); and (**i**) poly(vinyl chloride).

Availability

Barrier polymers are available as resins for extrusion, resins for dissolution and coating, and latices for coating. Each form has its own advantages and disadvantages. Resins for extrusion can be made into monolayers or multilayers with thicknesses to give adequate barrier for demanding applications. However, extrusion can give a severe thermal stress to the polymer which could lead to degradation. For semicrystalline polymers the extrusion temperature must exceed the melting temperature. When more modest total barrier is needed, a coating may be adequate. Typically coatings more than 3 μm (= 0.1 mil) are difficult to achieve. However, semi crystalline polymers can be used in solvents well below the melting temperature. Hence, polymers with marginal thermal stability may be used. Solvent recovery and management must be considered. For latices, the particles are too small to allow crystallinity to develop; hence, semicrystalline polymers remain amorphous until after coating and drying. Latices cannot be stored indefinitely and must be used before coagulation occurs.

BIBLIOGRAPHY

1. P. T. DeLassus, "Barrier Polymers," in J. I. Kroschwitz, ed., *Kirk-Othmer Encyclopedia of Chemical Technology,* 4th ed. Vol. 3, John Wiley Sons, New York, 1992, pp. 931–962.
2. M. Salame and S. Steingiser, *Polymer-Plastics Technology and Engineering,* **8**(2), 155–175 (1977).

PHILLIP DELASSUS
The Dow Chemical Company
Freeport, Texas

BIODEGRADABLE MATERIALS

Biodegradable polymers constitute a loosely defined family of polymers that are designed to degrade through the action of living organisms. They offer a possible alternative to traditional nonbiodegradable polymers when recycling is impractical or not economical. The main driving forces behind this technology are the solid-waste problem, particularly with regard to the decreasing availability of landfills; the litter problem; and pollution of the marine environment by nonbiodegradable plastics. Technologies such as composting used for the disposal of food and yard waste, accounting for 25–30% of total municipal solid waste, are the most suitable for the disposal of biodegradable materials together with soiled or food-contaminated paper.

International organizations such as the American Society for Testing and Materials in connection with the Institute for Standards Research (ASTM/ISR), the European Standardisation Committee (CEN), the International Standardisation Organisation (ISO), the German Institute for Standardisation (DIN), the Italian Standardization Agency (UNI), and the Organic Reclamation and Composting Association (ORCA), are all actively involved in developing definitions and tests for biodegradability in different environments and compostability (1,2).

Even if actively debated, a standard worldwide definition for biodegradable plastics has not been yet established. The definitions used by the ASTM for "degradable" and "biodegradable plastics" are reported as examples.

Biodegradable plastic: "Plastic which degrades as a result of the activity of naturally-occurring microorganisms such as bacteria, fungi and algae."

Degradable plastic: "A plastic that undergoes a change in its chemical structure under specific environmental conditions that results in a loss of properties that may vary as measured by standard test methods appropriate to the plastic and the application in a period of time that determines its classification."

CEN, ORCA, and DIN have already defined, at draft level, the basic requirements for a product to be declared compostable based on

- Complete biodegradability of the product, measured through respirometric tests such as ASTM D5338-92, ISO/CD14855, and CEN proposal xx1 or the modified Sturm test ASTM 5209, in a time period of some months, compatible with the selected disposal technology.
- Disintegration of the material during the fermentation phase
- No negative effects on compost quality, and in particular no toxic effects of the compost and leachates to the terrestrial and aquatic organisms
- Control of laboratory-scale results on pilot/full-scale composting plants.

These requirements set forth a common base for a universal marking system to readily identify products to be composted. An important driving force for the development of biodegradable materials and the related compostability label is the recently adopted European Directive 94/62/EC on packaging waste where composting of packaging waste is considered a form of material recycling.

The future role of biodegradable polymers in the world's solid-waste management strategy is still being developed. The rate of market growth for biodegradable polymers depends primarily on

- The speed at which legislature requiring degradables is enacted at international, national and regional level
- How quickly governments and industry can develop infractructures for collection and composting
- Consumer attitudes toward protecting the environment
- The availability of truly biodegradable materials able to degrade in timeframes compatible with specific disposal infrastructures

Cost is a major obstacle to wide-scale use of biodegradable polymers.

It follows a short description of the main classes of biodegradable materials at industrial or development stage.

Traditional Plastics

Some of the most cited studies on the biodegradation of synthetic polymers such as polyethylene, PVC, and PS (3) showed that, among high-molecular-weight polymers, only the aliphatic polyesters were biodegradable.

In general, molecular weight is a critical factor in the biodegradation process, since below critical numbers (approximately 500), the oligomers of polyethylene are biodegradable while those of polystyrene are not. Albertsson has reported the very slow mineralization of 14C-polyethylene with initial rates of 0.005–0.1% per month depending on culture cultivation (4). Attempts to improve the biodegradation rate of polyethylene have been made introducing small amounts of starch as a filler and/or adding prooxidants.

A biodegradation rate of 14C-polyethylene, after irradiation, when in presence of pro-oxidants was found of the order of magnitude of 4.5% in 10 years (4).

In all the studies performed up to now, polyethylene cannot significantly degrade in timeframes compatible with viable disposal options such as composting and sewage disposal. Under no one of the different ASTM procedures already approved (1), simulating different biodegradation environments, has polyethylene been proved as biodegradable, even in presence of prooxidants.

Photolabile chemical groups (eg, carbonyl moieties, benzophenone) have been introduced into polyolefins to accelerate their ultraviolet, light-catalyzed depolymerization (Norrish type II) mechanism in the environment via a free-radical process. For some applications and disposal routes this may be a viable option; however, concerns remain with this approach since complete biodegradation has been not yet proved.

Starch-Based Materials

Starch is an inexpensive product available annually from corn and other crops, produced in excess of current market needs in the United States and Europe. It is totally biodegradable in a wide variety of environments and can permit the development of totally degradable products for specific market demands. Degradation or incineration of starch products recycles atmospheric CO_2 trapped by starch producing plants and does not increase potential global warming.

All these reasons excited a renewed interest for starch-based plastics in the last years. Starch graft copolymers, starch plastic composites, starch itself, and starch derivatives have been proposed as plastic materials.

Starch is constituted by two major components: amylose, a mostly linear α-D-(1,4)-glucan; and amylopectine, an α-D-(1,4) glucan that has α-D-(1,6) linkages at the branch point. The linear amylose molecules of starch have a molecular weight of 0.2–2 million, while the branched amylopectine molecules have molecular weights as high as 100–400 million.

In nature starch is found as crystalline beads of about 15–100 μm in diameter, in three crystalline design modifications: A (cereal), B (tuber), and C (smooth pea and various beans), all characterized by double helices—almost perfect left-handed, six-fold structures, as elucidated by X-ray-diffraction experiments.

Starch as a filler. Crystalline starch beads can be used as a natural filler in traditional plastics (5); they have been used particularly in polyolefines. When blended with starch beads, polyethylene films biodeteriorate for exposure to a soil environment. The microbial consumption of the starch component, in fact, leads to increased porosity, void formation, and loss of integrity of the plastic matrix. Generally, starch is added at fairly low concentrations (6–15%); the overall disintegration of these materials is obtained, however, by transition metal compounds, soluble in the thermoplastic matrix, used as prooxidant additives to catalyze the photo and thermooxidative process (6).

Starch-filled polyethylenes containing prooxidants are commonly used in agricultural mulch film, in bags, and in six-pack yoke package. According to St. Lawrence Starch technology, regular cornstarch is treated with a silane coupling agent to make it compatible with hydrophobic polymers, and dried to less than 1% of water content. It is then mixed with the other additives such as an unsaturated fat or fatty-acid autoxidant to form a masterbatch that is added to a commodity polymer.

The polymer can then be processed by convenient methods, including film blowing, injection molding, and blow molding.

Thermoplastic starch. Starch can be made thermoplastic according to extrusion cooking technology. Extrusion cooking and forming are characterized by sufficient work and heat being applied to a cereal-based product to cook or gelatinize completely all the ingredients. Equipments used for high-pressure extrusion heat materials being processed, and continually compress them. Gelatinized materials with different starch viscosity, water solubility, and water absorption have been prepared by altering the moisture content of the raw product and the temperature or the pressure in the extruder. An extrusion cooked starch can be solubilized without any formation of maltodextrins, and the extent of solubilization depends on extrusion temperature, moisture content of starch before extrusion, and the amylose/amylopectine ratio (7).

There are two different conditions for loss of crystallinity of starch: at high water volume fractions (>0.9) described as gelatinization; and at low water volume, fractions (<0.45) with a real melting of starch (8,9) (destructurized starch or thermoplastic starch (10,11).

Thermoplastic starch alone can be processed as a traditional plastic; its sensitivity to humidity, however, makes it unsuitable for most of the applications.

Thermoplastic starch composites. Starch can be destructurized in combination with different synthetic polymers to satisfy a broad spectrum of market needs. Thermoplastic starch composites can reach starch contents higher than 50%.

EAA (ethylene-acrylic acid copolymer)/thermoplastic starch composites. EAA/thermoplastic starch composites have been studied since 1977 (12). The addition of ammonium hydroxide to EAA makes it compatible with starch. The sensitivity to environmental changes and mainly the susceptibility to tear propagation precluded their use in most of the packaging applications; moreover, EAA is not biodegradable at all.

Starch/vinyl alcohol copolymers. Starch/vinyl alcohol copolymer systems, depending on the processing conditions, starch type, and copolymer composition, can generate a wide variety of morphologies and properties. Different microstructures were observed: from a dropletlike to a layered one (13), as a function of different hydrophylicity of the synthetic copolymer. Furthermore, for this type of composite, materials containing starch with an amylose/amylopectine ratio of >20/80 w/w do not dissolve even under stirring in boiling water. Under these conditions a microdispersion, constituted by microphere aggregates, is produced, whose individual particle diameter is <1 μm.

The morphology of materials in film form, containing starch with an amylose/amylopectine ratio of <20/80 w/w, gradually looses the dropletlike form, generating layered structures. In this case the starch component becomes partially soluble.

Fourier transform infrared (FTIR) second-derivative spectra of starch/vinyl alcohol copolymer systems with dropletlike structure, in the range of starch ring vibrations between 960 and 920 cm^{-1}, provides for an absorption peak at about 947 cm^{-1} as observed for amylose when complexed (V-type complex) by low-molecular-weight molecules such as butanol and fatty acids. Vinyl alcohol copolymers, as well as butanol, leave the amylopectine conformation unchanged. A model has been proposed considering large individual amylopectine molecules interconnected at several points per molecule as a result of hydrogen bonds and entanglements by chains of amylose/vinyl alcohol copolymer V complexes. This structure has been defined in the literature as "interpenetrated."

The biodegradation rate of starch in these materials is inversely proportional to the content of amylose/vinyl alcohol complex.

The products based on starch/EVOH show mechanical properties good enough to meet the needs of specific industrial applications. Their moldability in film blowing, injection molding, blow-molding thermoforming, foaming, etc is comparable with that of traditional plastics such as PS, ABS, and LDPE (14). The main limits of these materials are in the high sensitivity to low humidities, with consequent enbrittlement. The biodegradation of these composites has been demonstrated in different environments (15). A substantially different biodegradation mechanism for the two components has been observed:

- The natural component, even if significantly shielded by the interpenetrated structure, seems, first, hydrolyzed by extracellular enzymes.
- The synthetic component seems biodegraded through a superficial adsorption of microorganisms, made easier by the increase of available surface that occurred during the hydrolysis of the natural component.

The degradation rate of 2–3 years in watery environments remains too slow for considering these materials compostable.

Aliphatic polyesters/thermoplastic starch. Starch can also be destructurized in presence of more hydrophobic polymers such as aliphatic polyesters (16).

It is known that aliphatic polyesters having low melting point are difficult to be processed by conventional techniques for thermoplastic materials, such as film blowing and blow molding. It has been found that the blending of starch with aliphatic polyesters allows improvement of their processability properties and biodegradability properties thereof. Particularly suitable polyesters are poly-ε-caprolactone and its copolymers, or polymers at higher melting point formed by the reaction of glycols as 1,4-butandiol with succinic acid or with sebacic acid, adipic acid, azelaic acid, decanoic acid, or brassilic acid. The presence of compatibilizers between starch and aliphatic polyesters such as amylose/EVOH V-type complexes (13), starch grafted polyesters, and chain extenders such as diisocyanates, and epoxydes is preferred. Such materials are characterized by excellent compostability, excellent mechanical properties, and reduced sensitivity to water.

Thermoplastic starch can also be blended with polyolefines possibly in presence of a compatibilizer. Starch/cellulose derivative systems are also reported in the literature (15). The combination of starch with a soluble polymer such as polyvinyl alcohol (PVOH) and/or polyalkylene glycols was widely considered since 1970. In the last years the system thermoplastic starch/PVOH has been studied mainly for producing starch-based loose fillers as a replacement for expanded polystyrene.

The results obtained in the field of thermoplastic starch in combination with polymers or copolymers of vinyl alcohol and with aliphatic polyesters and copolyesters in terms of biodegradation kinetics, mechanical properties and reduced sensitivity to humidity make these materials ready for a real industrial development starting from film and foam applications.

The main producers are Novamont with the Mater-Bi trademark; Novon International, which acquired Warner-Lambert technology; and Melitta, with the Natura trademark.

Cellulose

Cellulose, formed by plants and bacteria, is the most abundant naturally occurring polymer in the biosphere. It is not soluble in water or in most organic solvents. Cellulose is regularly mineralized by microorganisms, due to the activity of the cellulase enzyme complex, which results in the formation of cellobiose and glucose leading to mineralization in cellular biochemical cycles. Cellulose cannot be thermally processed, undergoing thermal decomposition before melting, due to the extensive hydrogen bonding.

Cellophane is one of the common forms of cellulose used in packaging and is used in laminated wrappers for a variety of foods because of its barrier properties to oils and its transparency. Cellophane is usually coated with nitrocellulose or acrylates to enhance barrier properties, although these coatings are not biodegradable.

Chemically substituted celluloses are not always biodegradable, and usually the degree of substitution impacts the degradation kinetics (17). Examples of chemical modifications include acetylation with acetic anhydride or other anhydrides, cyanoethylation with acrylonitrile, phosphorylation, nitration, and grafting.

Cross-linking also decreases its mineralization rate. Generally, either steric hindrance or a decrease in hydrophilicity is responsible for the decrease in rate of degradation with increasing degree of substitution. Nitrocellulose (with di- or trinitro substitutions per monomer) is generally regarded as resistant to microbial attack, as are cellulose triacetate, cellulose tripropionate, and cellulose tributyrate. Recent studies with cellulose triacetate indicate evidence for mineralization by microorganisms (18), particularly in presence of specific plasticizers such as ε-caprolactone.

As reported in article on starch composites, cellulose-acetate has been also used in combination with starch in order to maximize the biodegradation rate.

Cellulose has been also combined with chitosan, giving films with good gas barrier and water resistance.

Chitin and Chitosan

Chitin is a linear (1,4)-aminoacetyl glucan synthesized mainly by insects and crustacea as hardened composite structures for exoskeletons, and by filamentous fungi as part of their cell walls (19). The chemical treatment of chitin for deacetylation produces chitosan (β-1,4-linked 2-amino-2-deoxy-D-glucose). Chitin and chitosan are potentially useful polymers because of their excellent mechanical properties in film and fiber forms, as well as their low oxygen permeability. Chitosan, derived from the base-catalyzed chemical deacetylation of chitin from crustacean waste materials, has been commercialized by a number of companies. Applications for these two polymers include flocculants, functional food ingredients, and coatings or films. Polymer complexes with chitosan and poly(acrylic acid), alginates, and other polyanions have been reported, primarily for food coagulants and in wastewater treatments.

A variety of chemically modified chitins and chitosans has been synthesized. One reaction involves the chemical cross-linking of chitosan with epichlorohydrin; the reaction improves water stability for the polymer while maintaining biodegradability at a slower rate. The rate of enzymatic hydrolysis of partially *N*-acetylated chitosans is more rapid than fully substituted chitosan (chitin) according to studies performed with a series of chitinases and lysozyme.

Chitin and chitosan are readily converted to their xanthates with CS_2, suitable for fibers and films.

Pullulan

Pullulan is constituted by maltotriose α-1,4-linked D-glucose units linked α-1,6 produced by *Aureobasidium pullulans* (20).

Pullulan is depolymerized by a number of enzymes, and the glucose monomers are subsequently mineralized. Pullulan is used primarily as a food coating because of its excellent oxygen-barrier properties. Pullulan can be produced in high yields from a variety of carbon sources; product recovery and purification is relatively simple, since it is an extracellular polymer. Many chemical derivatives of pullulan have been synthesized, including esters and ethers, to reduce water solubility depending on the degree of substitution.

Proteins

Many protein-based polymers have been considered as possible materials, particularly for their inherent biodegradability due to ubiquitous protease. Many of these polymers have been used particularly in microspheres for encapsulation or slow release of pharmaceuticals. Some of these proteins offer future potential as fibers or fiber composites because their excellent mechanical properties (21), although in many cases costs of production will have to be reduced for wider markets to become a reality (eg, silks, elastins, fibrinogen). Options for genetic manipulation of these and other proteins are becoming a reality and may impact costs.

Collagen. This is a major structural protein in animal tissue characterized in part by the repeating trimer glycine-X-Y, where X and Y are often proline or hydroxylproline (22).

Collagen films, formed by water or solvent processes, are in some cases cross-linked, and are used primarily for medical applications.

Gelatin. This is a biodegradable animal byproduct, consisting primarily of proteins, which can be readily cross-linked (23). Gelatin is derived from the partial hydrolysis of collagen followed by hot-water extraction. The molecular-weight distribution is generally between 15,000 and 250,000. Gelatin is used extensively in the food and pharmaceutical industries. Graft copolymers of gelatin and poly(ethyl acrylate) in film form are reported to be biodegradable (24).

Polyesters

Polyhydroxyalkanoates. The bacterial polyhydroxyalkanonates are aliphatic polyesters, homo- or copolymers of (*R*)-β-hydroxyalkanoic acids produced by microorganisms (25).

Prokaryotic organisms, as bacteria and cyanobacteria, accumulate poly(3-hydroxy butyrate), as inclusions in the cytoplasmic fluid, amounting to 30–80% of the cellular dry weight, when their growth is limited by the depletion of an essential nutrient such as nitrogen, oxygen, phosphorus, sulfur, or magnesium. These microbial polyesters can be defined as intracellular storage products, and yield, composition, and molecular weight are influenced by carbon source and nutrients. Long-side, novel chain polymers have also been produced by the addition of the appropriate substrates in the culture medium (26).

The different polyesters obtained with this technology can be thermoplastic or elastomeric, and those containing valeric and butyric monomers are mineralized.

The physical properties and biodegradability of microbial polyesters may be regulated by blending with synthetic or natural polymers. Poly(3-hydroxybutyrate–valerate) is miscible to some extent with nylon, polyethyleneoxide, polyvinyl acetate, and polyvinyl chloride. Its rate of hydrolytic degradation is dramatically affected by the presence of polysaccharides. The mechanism of biodegradation of poly(3-hydroxybutyrate) has been studied and involves the enzyme poly (3-hydroxybutyrate depolymerase), which depolymerizes the polymer to the corresponding acid (27). Recent efforts to express the copolymer in *Escherichia coli* and in plants (28) is drawing interest because of the possibility of reducing production costs and simplyfying purification procedures.

High-molecular-weight poly(3-hydroxybutyrate) and copolymers with poly(3hydroxyvalerate) can also be produced synthetically from racemic β-butyrolactone and β-valerolactone, respectively, with an oligomeric alumoxane catalyst.

The synthetic polyesters are less susceptible to enzymatic degradation than the bacterial polyester, due to the difference in structure: a isotactic one for the bacterial copolyesters with random stereosequences, and a block one for the synthetic polyesters, only partially stereoregular.

In this last case the (*S*)-stereoblock hinders the enzymatic degradation, making it difficult for depolymerase to penetrate into the surface and access the available (*R*)-stereoblocks. Polyhydroxybutyrate-valerate is produced under the BIOPOL trademark by Zeneca, with a production capacity of about 300 tons/yr.

Poly(lactic acid)/poly(glycolic acid). The linear aliphatic polyester, poly(lactic acid), is a thermoplastic polymer, chemically synthesized by polycondensation of the free acid or by catalytic ring-opening polymerization of the lactide (dilactone of lactic acid) (29). The enantiomeric monomers for the synthesis of L-lactic acid (naturally occurring) and D-lactic acid can be produced via biological or chemical methods. The ester linkages in the polymer are sensitive to both enzymatic and chemical hydrolysis. Poly(lactic acid) is hydrolyzed by many enzymes including pronase, proteinase K, bromelain, ficin, esterase, and trypsin (30). Rates of biodegradation of poly(lactic acid), along with the bacterial polyesters, can be accelerated with the addition of cinnarizine and clonidine to increase hydrolysis. Copolymers of glycine and D,L-lactic acid have been synthesized and are biodegradable.

Poly(lactic acid) in the past has been considered primarily for medical implants and drug delivery, but broader applications in packaging and consumer goods are also targeted (29). An attractive feature of this material is the potential relatively low cost of the monomer, lactic acid, which can be derived from biomass (fermentation), coal, petroleum, or natural gas. Poly(ether–ester) block copolymers with polyethylene oxide (600–6000 molecular weight range) and poly(lactic acid) have also been synthesized and shown to be highly hydrophilic.

Copolymers of glycolide L(−)lactide have been commercialized for biomedical applications and are high-strength biodegradable thermoplastic materials. Poly(glycolic acid) and copolymers with D,L-lactides are presumed to be biodegradable, although the role of chemical hydrolysis vs enzymatic depolymerization in this process remains open to debate.

Other materials with a possible future commercial interest are the copolymers of lactic acid and ε-caprolactone.

The market for high-molecular-weight polymers from lactic acid is still in its early stages. A large growth for the polymer market requires additional sources of lactic acid as raw material.

Cargill started up a 10-million-lb/yr lactic acid/polylactide plant in 1994 and announced a large plant with a capacity as high as 250 million lb/yr in the next years. Other producers of poly(lactic acid) are Mitsui Toatsu Chemical, which has developed a polycondensation process of lactic acid; Shimatzu; Dainippon Ink; and Neste.

The mechanical properties of polylactic acid are similar to those of polyethylene terephthalate, the limits to be solved stay in the too-low glass-transition temperature, the limited stability to hydrolysis, and the slow crystallization rate.

The main applications at the development stage are in the sectors of bottles and thermoformed containers for food, films, and fibers.

Polycaprolactone. Polycaprolactone is a biodegradable, aliphatic polyester that is made by ring-opening polymerization of ε-caprolactone.

This polymer is used with a molecular weight of up to several thousand, in the form of a waxy solid or viscous liquid as polyurethane intermediate, a reactive diluent for high solid coatings or plasticizer for vinyl resins; with a molecular weight of >20,000, it is used as a thermoplastic polymer with mechanical properties similar to those of polyethylene. A number of studies have clearly stated the biodegradability of poly(ε-caprolactone) by fungi (31). The rate of hydrolysis depends on the molecular weight and degree of crystallinity. ε-Hydroxy caproic acid was detected as an intermediate during the degradation process by *Penicillum* sp. (32). Molecular weight and degree of crystallinity of polycaprolactone influence its biodegradation rates. Chemical hydrolysis rate is very slow for the homopolymer, particularly when compared with poly(glycolic acid) and poly(glycolic acid)–*co*–(lactic acid); the rate can be increased by copolymerization. Cross-linked polycaprolactone films with a lower degree of crystallinity than the un-cross-linked material were also found to be biodegradable.

Union Carbide, Daicel, and Interox are the three main producers of poly-ε-caprolactone worldwide.

High-molecular-weight poly-ε-caprolactone may be processed by a variety of techniques, including film blowing and slot casting. The main applications in the field of biodegradable plastics are in combination with thermoplastic starch in films, sheets, and injection-molded parts.

Polyamides

Examples of polyamides are synthetic copolymers consisting of an α-amino acid, such as glycine; and ε-aminocaproic acid (nylon 2/nylon 6). This copolymer is biodegradable (33). Similar results were found for serine and ε-aminocaproic acid. Of particular interest is the finding that for the nylon 2/nylon 6 copolymer, the two respective homopolymers (polyglycine and nylon 6) were inert under the same test conditions.

Various modifications of these polymers have been studied in an attempt to enhance rates of biodegradation, including substituting different amino acids, or the incorporation of methyl, hydroxyl, or benzyl groups (34). Polyamide esters and polyamide urethanes synthesized from amino alcohols are generally depolymerized by protease enzymes, with the lower-melting esters more rapidly depolymerized than the urethanes. Recently, the bulk low-cost synthesis of poly(amino acids) such as polyaspartic acid in 95% yield was demonstrated (35). The polymers are biodegradable and could replace polyacrylic acid as dispersants in paints and detergents, scaling in piping, and absorbants in diapers and medical products.

Other approaches to the synthesis of poly(amino acids) are possible, some based on the use of trifunctional aminoacids as starting materials. New biodegradable functional polymers will be forthcoming that may prove useful as fibers for rigid composites for packaging or compatibilizers. The costs associated with some of these processes will also have to be addressed.

Polyurethanes

Polyurethane foams or sponges are used in a variety of applications in the sector of packaging. Polyurethanes are synthesized from diisocyanates (aromatic and aliphatic) and diols. Polyester polyurethanes, synthesized with straight- and branched-chain diols such as poly-ε-caprolactone and polyethylene glycol and polyester diols constitued by adipic acid, ethylene glycol, 1,3-propanediol, and 1,4-butanediol are generally considered biodegradable (36). Some of the polyether urethanes are biodegradable, depending on whether there is a sufficiently long carbon chain between urethane links to impart chain flexibility to the polyurethane. Polyurethanes derived from poly(alkylene tartrate)s and cellulose hydrolyzates are biodegradable, as are those synthesized from 1,6-hexane diisocyanate and poly(caprolactone diol)s (34).

Poly(vinyl alcohol) and Its Copolymers

Poly(vinyl alcohol) (PVOH) is a water-soluble synthetic product that has widely been used worldwide in a variety of end uses. PVOH is used mainly in textile, warp sizes; as binder or thickening agent for emulsions and suspensions; and as soluble films for packaging or fibers.

It is prepared by polymerization of vinyl acetate to poly(vinyl acetate) followed by alcoholysis. Compared to other vinyl polymers [eg, polystyrene, polyethylene, polypropylene, poly(vinyl chloride)], poly(vinyl alcohol) is the most readily biodegraded. Respirometric assays with mixed culture activated sludges have shown poly(vinyl alcohol) is mineralized (37).

The literature contains extensive reports of studies on the purification, characterization, and mechanism of degradation of poly(vinyl alcohol) by enzymes isolated from *Pseudomonas* sp. The role of crystallinity on the biodegradation behavior of PVOH has been also investigated (15).

It has been shown that isolated cultures cannot grow on liquids containing PVOH as the sole carbon source, but that mixed cultures of *Pseudomonas* sp. and *Alcaligenes* sp. and *Pseudomonas putida* can easily grow.

For most applications poly(vinyl alcohol) must be dissolved in water prior to use, but the addition of additives can allow PVA copolymers to be processed as thermoplastics.

Potential uses include packaging application, especially for fast-food restaurants, hazardous material, contaminants, and removable protective coatings. A particularly strong market growth for these films is in hospital laundry bags, where the handling of contaminated liners may increase exposure to infection.

Japan is the largest producer of PVA worldwide with approximately 45% of the world capacity.

Poly(ethylene-*co*-vinyl alcohol)

This is a thermoplastic polymer used extensively in laminates for food containers because of excellent film-forming and oxygen-barrier properties. In the literature there is evidence for the biodegradability of polyethylene–vinyl alcohol. The biodegradation rate is slow and very sensitive to the environmental conditions (38).

It is assumed that the size and distribution of the ethylene blocks and the extent of crystallinity will determine whether the copolymer is biodegradable.

Poly(vinyl acetate) and poly(ethylene-*co*-vinyl acetate) are slowly biodegraded (34), particularly when there is a high percentage of vinyl acetate in the copolymer. Growth of fungi on ethylene vinyl acetate copolymers correlates directly with vinyl acetate content.

Polyethylene Oxide

Polyethylene oxide is a high-molecular-weight, water-soluble, thermoplastic biodegradable polyester. It is synthesized by the polymerization of ethylene oxide and is used in adhesives and as a thickener in many applications. Polyethylene oxide can be used as a component of polyesters, polyamides, and polyureas, increasing their flexibility and biodegradability.

BIBLIOGRAPHY

1. *ASTM Standards on Environmentally Degradable Plastics,* ASTM Publication Code Number (DCN) 03-420093-19 (1993).
2. CEN TC 261 SC4 W62 draft "Requirements for Packaging Recoverable in the Form of Composting and Biodegradation. Test Scheme for the Final Acceptance of Packaging," March 11, 1996.
3. J. E. Potts, "Aspects of Degradation and Stabilization of Polymers" in H. H. G. Jellinek, ed., Elsevier, New York, 1978, pp. 617–658.
4. A. C. Albertsson and B. Ranby, *J. Appl. Polym. Sci.,* **35,** 423–430 (1979).
5. G. J. L. Griffin, U.S. Pat. 4016117 (1977).
6. G. Scott, U.K. Pat. 1,356,107 (1971).
7. C. Mercier and P. Feillet, *Cereal Chem.* **52**(3), 283–297 (1975).
8. J. W. Donovan, *Biopolymers* **18,** 263 (1979).
9. P. Colonna and C. Mercier, *Phytochemistry* **24**(8), 1667–1674 (1985).
10. J. Silbiger, J. P. Sacchetto, and D. J. Lentz, Eur. Pat. Appl. 0 404 728 (1990).
11. C. Bastioli, V. Bellotti, and G. F. Del Tredici, Eur. Pat. Appl. WO 90/EP1286, (1990).
12. F. H. Otey, U.S. Pat. 4133784 (1979).
13. C. Bastioli, V. Bellotti, M. Camia, L. Del Giudice, and A. Rallis "Biodegradable Plastics and Polymers" in Y. Doi, K. Fukuda, Ed., Elsevier, 1994, pp. 200–213.
14. C. Bastioli, V. Bellotti, and A. Rallis, "Microstructure and Melt Flow Behaviour of a Starch-based Polymer," *Rheologica Acta* **33,** 307–316 (1994).
15. C. Bastioli, V. Bellotti, L. Del Giudice, and G. Gilli, *J. Environ. Polym. Degradation* **1**(3), 181–191 (1993).
16. C. Bastioli, V. Bellotti, G. F. Del Tredici, R. Lombi, A. Montino, and R. Ponti, Internatl. Pat. Appl. WO 92/19680, (1992).
17. B. Rossall, *Internatl. Biodeterioration Bull.* **10,** 95–103 (1974)
18. C. M. Buchanan, R. M. Gardner, and R. J. Komarek, *J. Appl. Polym. Sci.,* **47,** 1709–1719 (1993).
19. S. Arcidiacono, S. I. Lombardi, and D. L. Kaplan, "Chitin and Chitosan" in G. Skjak-Brack, T. Anthosen, and P. Sandford, eds., London Elsevier, 1988, pp. 319–332.
20. J. M. Mayer, M. Greenberger, D. H. Ball, and D. L. Kaplan, *Proc. Am. Chem. Soc. Div. Polym. Materials: Sci. Eng.* **63,** 732–735 (1990).
21. K. Grohmann and M. Himmel, "Assessment of Biobased Materials" in H. L. Chum, ed., Report SERI/TR-234-3610, Solar Energy Research Institute, Colorado, 1989, pp. 11.1–11.10.
22. S. Gorham, in *Biomaterials: Novel Materials from Biological Sources,* Stockton Press, New York, 1991, pp. 55–122.
23. S. J. Huang, J. P. Bell, J. R. Knox, H. Atwood, D. Bansleben, M. Bitritto, W. Borghard, T. Chapun, K. W. Leong, K. Natarjan, J. Nepumuceno, M. Roby, J. Soboslai, and N. Shoemaker, in *Proc. 3rd Internatl. Biodegradable Symp.,* J. M. Sharpley and A. M. Kaplan, eds., Applied Science, London, 1976, pp. 731–741.
24. G. S. Kumar, V. Kalpagam, and U. S. Nandi, *J. Appl. Polym. Sci.,* **29,** 3075–3085 (1984).
25. A. Steinbuchel, "Polyhydroxalkanoic Acids" in *Biomaterials: Novel Materials from Biological Sources,* Stockton Press, New York, 1991, pp. 123–124.
26. B. A. Ramsay, I. Saracova, J. A. Ramsay, and R. H. Marchessault, *Appl. Environ. Microbiol.* **57,** 625–629 (1991).
27. Y. Doi, Y. Kensawa, M. Kunioka, and T. Saito, *Macromolecules* **23,** 26–31 (1990).
28. Y. Poirier, D. E. Dennis, K. Klomparens, and C. Somerville, *Science* **256,** 520–522 (1992).
29. E. S. Lipinsky and R. G. Sinclair, *Chem. Eng. Progress* **82,** 26–32 (1986).
30. D. F. Williams, *Eng. Med.* **10,** 5–7 (1981).
31. P. Jarrett, C. Benedict, J. P. Bell, J. A. Cameron, and S. J. Huang, *Polym. Preprints, Am. Chem. Soc. Div. Polym. Chem.* **24,** 32–33 (1983).

32. Y. Tokiwa, T. Ando, and T. Suzuki, *J. Fermentation Technol.* **54,** 603–608 (1976).
33. W. J. Bailey, Y. Okamoto, W. C. Kuo, and T. Nanta, *Proc. Internatl. Biodegradation Sympo.,* 1976, pp. 765–773.
34. S. J. Huang, *Encycl. Polym. Sci. Eng.* **2,** 220–243 (1985).
35. L. Koskan, *Industrial Bioprocessing,* 1–2 (May 1992).
36. R. T. Darby and A. M. Kaplan, *Appl. Microbiol.* **16,** 900–905 (1968).
37. J. P. Casey and D. G. Manly, *Proc. Internatl. Biodegradation Sympo.* 1976, pp. 819–833.
38. J. Roemesser, presentation at Plastics Waste Management Workshop, American Chemical Society, Polymer Chemistry Division, December, New Orleans, LA 1991.

CATIA BASTIOLI
Novamont S.p.A.
Novara, Italy

BLISTER PACKAGING. See CARDED PACKAGING.

BLOW MOLDING

Blow molding is a process used to produce hollow articles and bottles, such as those shown in Figure 1, from thermoplastic materials. The basic process involves the manufacture, by either extrusion molding or injection molding, of a "preshape" usually called a *preform* or *parison.* While still warm, the parison is inflated, that is, "reshaped" with air pressure inside a female mold cavity of the bottle.

The most common materials used for bottle packaging applications are high-density polyethylene (HDPE) and polyethylene terephthalate (PET). Other materials often used are polypropylene (PP), low-density polyethylene (LDPE), polystyrene (PS), and poly(vinyl chloride) (PVC).

Figure 1. An array of extrusion blow-molded bottles. Courtesy of Johnson Controls, Inc.

History

The first attempt to blow-mold hollow plastic articles, about 120 years ago, was with two sheets of cellulose nitrate clamped between two mold halves. Steam injected between the sheets softened the material, sealed the edges, and expanded it against the mold cavity (1). The highly flammable nature of cellulose nitrate limited the usefulness of the technique.

In the early 1930s the availability of more suitable materials, polystyrene and cellulose acetate, led to the development of automated equipment, based on glass-blowing techniques, by the PLAX Corp. and Owens-Illinois (1,2). Unfortunately, the plastic bottles offered no advantage over glass bottles; however, the availability of low-density polyethylene in the early 1940s provided the needed advantage. The "squeezability" of this material gave the plastic bottle a feature glass could not match.

Nonetheless, the real beginning of blow molding came in the late 1950s with the development of high-density polyethylene and the availability of commerical blow-molding equipment (1). High-density polyethylene provided the stiffness needed for many bottle applications, and commerical equipment provided the opportunity for many firms to start blow molding. Until that time, all blow molding was done by a select few, using proprietary technology.

Basic Process

Figure 2 illustrates the basic blow-molding process. Although the extrusion method for creating the parison or preform is shown, that detail does not alter the fundamental blow-molding principle of a process based on an injection method. Nonetheless, the extrusion method is the most common.

In view **a** the hot tube-shaped parison is extruded downward between the two halves of a bottle blow-mold cavity. The blow pin in this example is shown inside the parison.

In view **b** the blow-mold cavity has closed pinching the parison flat. The inside is sealed at the bottom and around the top and blow pin. The pinching of the parison creates flash material that must later be trimmed away and recycled. This flash is often 20–50% of the bottle weight.

In view **c** the pinched parison is inflated into the bottle shape by air flowing through the blow pin. The air pressure is held inside until the newly molded bottle is cooled by the mold. Water, circulating through channels, is used to cool and maintain mold temperature.

In view **d** the newly molded bottle is ejected from the mold cavity. Removal is either downward or to the side. Once outside the molding area the flash is trimmed away.

Extrusion Blow Molding

Extrusion blow molding is divided into two broad categories: continuous extrusion and intermittent extrusion. These, in turn, are divided into other subcategories. Bottles, containers, or drums of virtually any size can be made by extrusion blow molding. Compared to injection blow molding, extrusion blow-mold tooling is relatively inexpensive. Finally, extrusion blow molding has relatively few shape restrictions, and it is the only low-cost method for manufacturing a bottle with a handle.

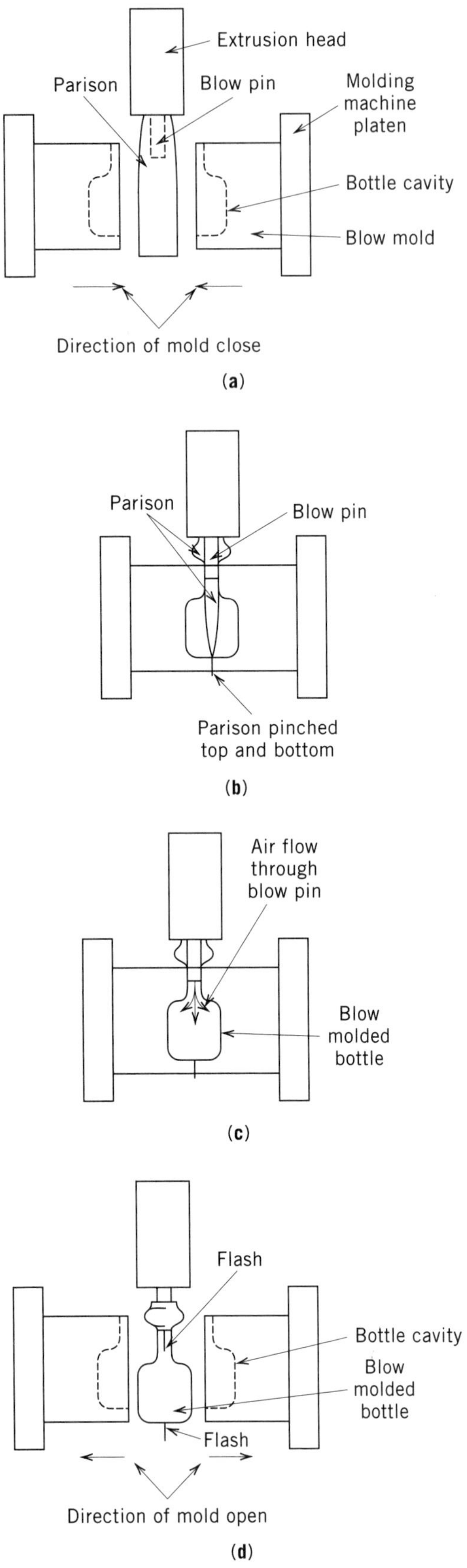

Figure 2. Basic blow-molding process (**a**) parison extrusion; (**b**) mold close and parison pinch; (**c**) parison inflation and bottle cool; (**d**) mold open and bottle eject.

Continuous extrusion. The parison in this process is continuously formed at a rate equal to the rate of bottle molding, cooling, and removal. To avoid interference with parison formation, the mold-clamping mechanism must move quickly to capture the parison and return to the blowing station where the blow pin enters. While a few very large containers, for example, a 35-gal trash container, have been made utilizing the process, generally equipment design limits container size to 10 L (2.5 gal) or less. In fact, most continuously extruded blow-molded bottles are ≤ 4 L.

All the commonly used plastic bottle materials are molded with the continuous extrusion process. The process is particularly well suited for molding heat-sensitive PVC materials. Thermal degradation of the PVC is minimized by the relatively slow uninterrupted low-pressure flow of the polymer material through the extruder and parison extrusion head.

Several machinery configurations exist, but two approaches generally predominate: shuttle continuous and vertical rotary continuous. Less common approaches include rising-mold continuous and horizontal rotary continuous.

The shuttle continuous extrusion approach is illustrated in Figure 3. With this method, a blowing station is located on one or both sides of the parison extrusion head. As the parison reaches the proper length, the blow mold and clamp quickly shuttle to a point under the extrusion head, capture and cut the parison, and return to the blowing station where the blow pin enters an open end of the parison to form the bottle finish area. With dual-sided machines, as shown in the illustration, two clamp systems shuttle on an alternating basis. For increased bottle production output, multiple parison extrusion heads with multiple mold cavities are used.

The shuttle process equipment layout offers a number of advantages. The process allows critical dimensions of the bottle finish to be "prefinished" in the mold with a water-cooled blow pin. This eliminates the need for postmold secondary sizing or machining of the finish area. Machinery operators also value the easy mold tooling and process setup. Job changes on shuttle continuous equipment can be relatively quick. On the other hand, the relatively slow parison extrusion rate requires the plastic material to have exceptional melt strength. That is, the parison must not change shape from material flow caused by its own weight as it is extruded from the head. The shuttle process requires the parison to hang for a relatively long time.

Recently a number of shuttle continuous-extrusion machinery suppliers have begun offering "long-stroke" machines. These are single-sided shuttle machines with relatively large multicavity molds. These machines have 6–12 cavities and, of course, require an extrusion head with the same number of parisons. In comparison a more conventional shuttle machine can be up to 6 cavities, but rarely is more than 4.

Single-sided "long-stroke" machines can simplify the integration of in-mold labeling equipment and downstream bottle-handling equipment. On the other hand, the distance that the mold and platen must travel between the parison extrusion head and the blowing station will take more time to traverse, and the added mass of the mold, machine platens, and clamping system is more difficult to accelerate and decelerate. All things being equal, an alternating shuttle with half as many parisons and tooling with half as many cavities per side with half the shuttle transfer distance will provide a faster per cavity production rate than will a long-stroke single-sided shuttle

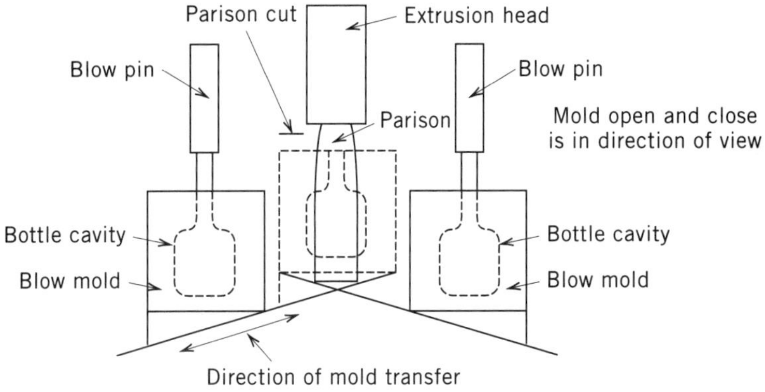

Figure 3. Basic alternating shuttle continuous extrusion blow molding process.

machine. Figure 4 shows a typical alternating shuttle continuous-extrusion blow-molding machine. It features a four-parison extrusion head and eight (four per side) blow-mold cavities.

The vertical rotary continuous extrusion approach is illustrated in Figure 5. In this process, simultaneously, a parison is captured, bottles are molded and cooled, and a cooled bottle is removed as the wheel rotates past the parison extrusion head. Usually only one bottle cavity is provided in each mold-clamping station; however, two cavities, in-line and face-to-face with a single parison or side-by-side with two parisons in parallel, have been used. Typically machines are fitted with up to 24 mold-clamping stations. These stations can open and close either axially, radially, or as a book with an edge of each mold half-hinged to each other. Figure 6 shows a vertical rotary machine with a "book"-style mold-closing system.

The rotary approach has many advantages, primarily for bottle applications requiring high production output. The single parison extrusion head helps to maintain product consistency between all the mold cavities, particularly for applications requiring a complicated multilayer or coextruded structure. The machine shown in Figure 6 is fitted with a multilayer parison extrusion head connected to five individual extruders and manifold systems. Another advantage is that in-mold labeling systems, when used, are less complicated because the same unit is used for all cavities. Finally, the vertical rotary approach has the unique capability of holding the parison fast at the top and the bottom, allowing the opportunity to manipulate parison material distribution in ways not as easily done with processes where the bottom end of the parison is free.

On the other hand, a needle blow pin is used to puncture the parison in a dome area above the finish. The molded dome is later cut off with the other flash trimmings. To establish final dimensions, the finish must now be "faced" or machined. This may allow plastic chips to fall inside the container and potentially contaminate the product. Another disadvantage is that tooling job changes are complicated and time consuming.

Intermittent extrusion. The parison in this process is quickly extruded after the bottle is removed from the mold. The mold-clamping mechanism does not transfer to a blowing

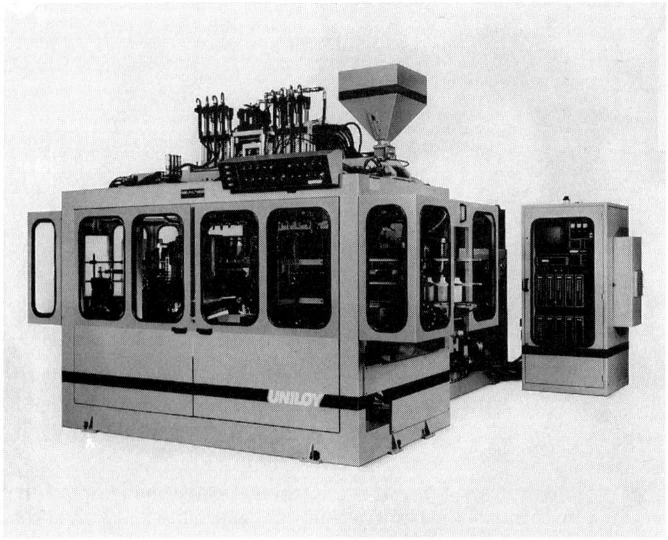

Figure 4. An alternating shuttle continuous-extrusion blow-molding machine featuring four blow-mold cavities on each side. Courtesy of Johnson Controls, Inc.

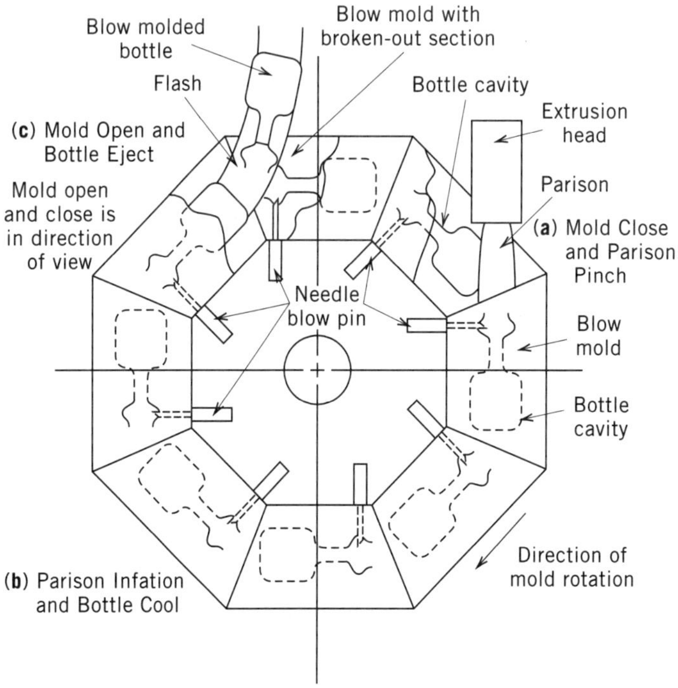

Figure 5. Vertical rotary continuous extrusion.

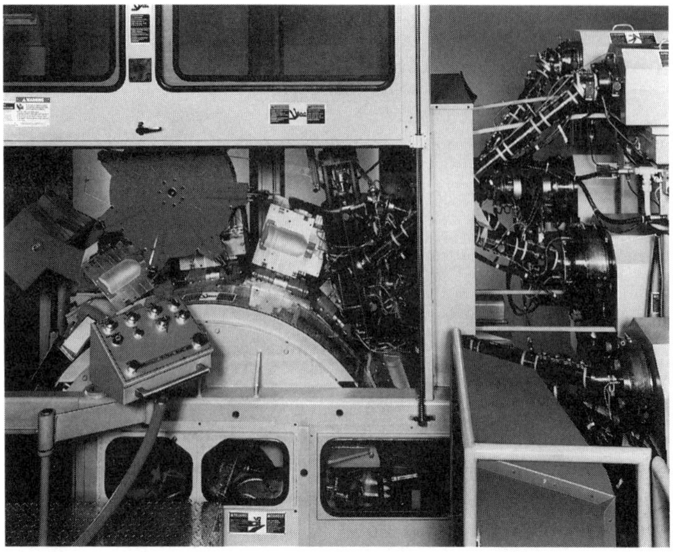

Figure 6. A vertical rotary continuous-extrusion blow-molding machine featuring a "book"-style mold clamping system and a multilayer parison extrusion head connected to five individual extruder systems. Courtesy of Johnson Controls, Inc.

station. Bottle molding, cooling, and removal all take place under the parison extrusion head. This arrangement allows the clamping system to be more simple and rugged. Bottles and containers of every size from 100 mL (4 fl oz) to 210 L (55 gal) are commonly made utilizing the process.

Polyolefin plastic materials, particularly HDPE, are well suited for the intermittent process, but the stop/start aspect and extremely rapid rate of the parison extrusion can make the molding of heat-sensitive materials difficult.

Several machinery configurations exist, but today two approaches predominate: reciprocating screw and accumulator head. A less common approach is the side ram accumulator.

The reciprocating-screw intermittent-extrusion approach is typically used for bottles of ≤10 L (2.5 gal). After the parison is quickly extruded and the mold closes, the extruder screw moves backward, accumulating plastic melt in front of its tip. When the molded bottle has cooled, the mold opens, the bottle is removed, and the screw quickly moves forward pushing plastic melt through the parison extrusion head forming the next parison.

The machinery configuration allows critical bottle dimensions to be "prefinished" during the molding process. Unique to the process is the ability to position the blow pin inside the parison prior to mold close. This permits a choice of two prefinishing approaches. The first, called a "ram-down" or "calibrated" prefinishing system, provides a result identical to the system used in the shuttle continuous extrusion process described earlier. The second, called a "pullup" prefinishing system, is unique to intermittent extrusion. The pullup system blow pin shears plastic material in a critical finish inside diameter area. Virtually all plastic milk bottles made in the United States utilize the technique. Figure 7 shows a reciprocating-screw blow molder with six parison extrusion heads. In the foreground is a bottle-trimming system used to remove tail and head flash from the molded bottle. The trimming unit also checks the bottle for pinhole leaks. Machines have been made that simultaneously extrude up to 12 parisons.

The accumulator-head intermittent-extrusion approach is used for heavy-weight containers and drums of >10 L in size and for other noncontainer articles such as parts for toy tricycles, automotive ductwork, and highway traffice cones. A melt reservoir, tubular in shape, is a part of the parison extrusion head itself. Plastic melt that enters the head first is first to leave. A tubular plunger quickly extrudes the parison through a head-tooling annulus.

Parison programming and head-tooling ovalization. Extrusion blow-molding process equipment usually features a capability to program the parison and often utilizes ovalized or shaped head tooling. Parison programming and head-tooling ovalization allow the blow-molding machine operator to maximize bottle performance while minimizing bottle weight and cost. Parison expansion in the blow-mold cavity is not uniform. Some areas will require more expansion than others, thus causing the parison to thin more. With parison programming plastic material distribution in the wall of the tubular parison is shifted so as to even material distribution in the final blow-molded bottle.

Parison programming alters the gap between the die and mandrel tooling. Following a predetermined profile during parison extrusion, a hydraulic cylinder moves either the die or the mandrel to change position relative to the other. The parison wall becomes "ringed" with areas of thicker or thinner

Figure 7. A reciprocating-screw intermittent-extrusion blow-molding machine featuring six parison extrusion heads. This machine is configured for molding 1 gal (3.79 L) milk bottles. In the foreground is a bottle-trimming system for removing head and tail flash. The trimming unit features a leak tester to examine bottles for pinhole leaks. Courtesy of Johnson Controls, Inc.

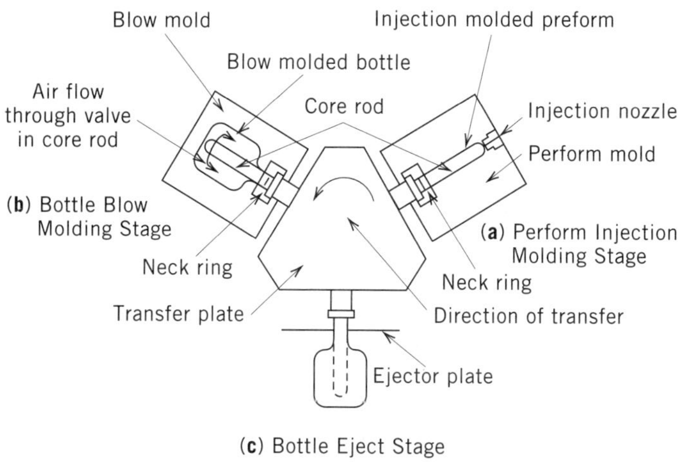

Figure 8. Basic injection blow molding process.

material located to correspond with areas of the bottle that require more or less material.

Head-tooling ovalization also changes parison wall thickness. Here a round mandrel or die is shaped or "ovalized" slightly. This causes the parison during extrusion to become "striped" with axial sections of thicker or thinner material. As before, these stripes are located to correspond with areas of the bottle that require more or less material.

Injection Blow Molding

In the injection blow molding process, melted plastic is injected into a parison cavity around a core rod. The resulting injection-molded "test tube" shaped parison, while still hot, is transferred on the core rod to a bottle blow-mold cavity. Air passing through a valve in the core rod inflates the parison. When cooled, the bottle is removed.

Early injection blow molding techniques, all two position methods, were adaptations of standard injection molding machines fitted with special tooling. A shortcoming of the approach was that the injection and blow-molding stations stood idle while the bottle was removed from the machine. This led to the 1961 Gussoni invention, in Italy, of the three-position method which uses a third station for bottle removal. Figure 8 illustrates the basic three-position injection blow-molding process. Three core rods are required for each injection/blow-mold cavity set.

Today virtually all injection blow molding is done on three-position machines; however, two- and four-position machines are in use. For the four-position machine, the extra position can be used for a variety of purposes, such as detection of a bottle not ejected, core rod temperature conditioning, bottle decoration, and parison preblow. Core rod temperature conditioning is the most common.

The injection blow-molding process has a number of advantages. The process is flash or scrap-free. Molded bottles do not require secondary trimming. Critical neck finish dimensions are injection molded with high accuracy, an important consideration for complex child-resistant and snap-on closure systems. Bottle weight control is extremely precise, accurate within a range of 0.2 g. On the other hand, machinery configurations limit bottle size to 4 L (1 gal) or less. Because of the relative high cost of mold tooling, the injection blow-molding process is usually not economically justifiable for typical bottle applications of >500 mL in volume, but for typical bottle applications of <250 mL, the injection blow-molding process is almost always less expensive. For vials (for example, 1, 5, 10 mL) and other very small bottles, injection blow molding is the only practical, cost-effective process choice. Finally, the process limits bottle shape and proportion. Although this is not a problem for most applications, the process is not well suited for extremely flat or handleware bottle shapes. Table 1 contrasts the key advantages of injection blow molding vs extrusion blow molding.

The injection blow-molding process is used mostly for pharmaceutical and cosmetics bottles because the bottles are frequently small, precise neck finishes are often important, and the process is usually more efficient than the extrusion blow-molding alternative. Furthermore, the high value of these products usually demands an equally high-value, high-quality package possible by injection blow molding. High- and low-density polyethylene, polypropylene, and polystyrene are the plastic materials most often used. In the case of polystyrene, a degree of molecular orientation is accomplished that enhances impact resistance. Unfortunately, other materials rarely achieve any degree of orientation. Polyethylene terephthalate and polyvinyl chloride materials are also injection blow-molded.

Extrusion–Injection-Molded Neck Process

A hybrid proprietary process initially developed by Owens-Illinois 40 years ago and still in use today. It combines the many advantages of injection blow molding and extrusion blow molding. It is a unique approach, in which the bottle neck finish is injection-molded, providing excellent dimensional detail; and the bottle body is extrusion-molded, allowing some freedom with bottle shape and proportion. The method, in some cases, will allow a handled bottle to be molded with an injection-molded finish. The method is also ideally suited for the integration of in-mold labeling equipment.

Stretch Blow Molding

The molecules of polymer materials prefer to be coiled together and arranged randomly in all directions. The careful, proper processing of some plastic materials, such as polyethylene terephthalate (PET) and polypropylene (PP), will permit the orientation or partial uniaxial or biaxial alignment of the material's molecules to be established and retained. Stretch blow molding, sometimes known as *biaxial-orientation blow molding,* is a process for achieving a biaxial molecular alignment within the sidewall of a blow-molded bottle.

The biaxial orientation of a polymer's molecular structure can significantly improve bottle impact strength, transparency, surface gloss, stiffness, and gas-barrier performance. However, these performance advantages are generally not significant in <250 mL bottles.

Fundamentally the process requires a parison, also known as a preform, to be carefully conditioned to a temperature warm enough to allow the parison to be inflated and its molecular structure aligned but cool enough to retard rerandomization of its molecular structure once aligned. The parison can be either extrusion-molded or injection-molded; however,

Table 1. Injection versus Extrusion Blow Molding

Injection Blow Molding	Extrusion Blow Molding
Used for small (typically <500 mL) heavier bottles	Used for larger bottles, typically >250 mL
Best process for polystyrene; most plastic resins can be used	Best process for polyvinyl chloride; many plastic resins can be used provided adequate melt strength is available
Scrap-free: no flash to recycle, no pinchoff scar on bottle, no post-mold trimming	Fewer bottle shape limitations permitting extreme dimensional ratios; long and narrow, flat and wide, doubled-walled, offset necks, molded-in handles, and other odd shapes
Injection-mold neck provides more accurate neck finish dimensions and permits special shapes for complicated safety and tamper-evident closures	Low-cost tooling often made of aluminum; ideal for short or long production run situations
Accurate and repeatable bottle weight control	Adjustable weight control ideal for prototyping
Excellent bottle surface finish or texture	

the precision provided by injection molding has allowed that approach to dominate in bottle manufacturing today.

Many plastic materials are capable of being biaxially oriented, but two, namely PET and PP, are commonly used. Others include PVC, acrylonitrile-based copolymers, and recently polyethylene naphthalate (PEN), a high-performance polyester similar to PET.

Figure 9 illustrates the relationship of the parison to the bottle. Note the parison (or preform) is significantly shorter than the bottle. In comparison, extrusion and injection blow-molding parisons are the same length as the bottle. Air pressure is used to stretch the parison in the "hoop" or horizontal direction. Air pressure, and usually a stretch rod, together are used to stretch the parison in the "axial" or vertical direction. The air pressure needed can be extreme; PET applications often require 600 psi (4100 kPa). In comparison, most extrusion blow-molding applications require about 90 psi (600 kPa).

Machinery suppliers have developed three basic methods for producing stretch blow-molded containers. With rare exception most of the equipment is intended for production of <4 L bottles.

The one-step method. In this method the production stages of parison injection molding, temperature conditioning, and blow molding take place in the same machine. The one-step method is considered best for wide-mouth jar applications and is ideal for bottles with extreme oval or other unusual cross-sectional shapes. The method minimizes blemishes to provide a more pristine bottle appearance. Furthermore, not needed is a preform handling ring, which can also distract from bottle aesthetic appearance. Tooling and machinery setup is relatively easy, making the method ideal for short-production-run applications. Finally, the method potentially saves energy in that the preform is not reheated (see Fig. 10).

The two-step method. In this method, also known as "reheat and blow," the production stage of parison injection molding is carried out in an operation separate from parison temperature conditioning and blow molding. The two-step method allows the production of the preform injection molding and bottle blow molding to be independently optimized. Furthermore, bottle performance, for example, impact resistance, weight, gas barrier, is not compromised. Extremely high-

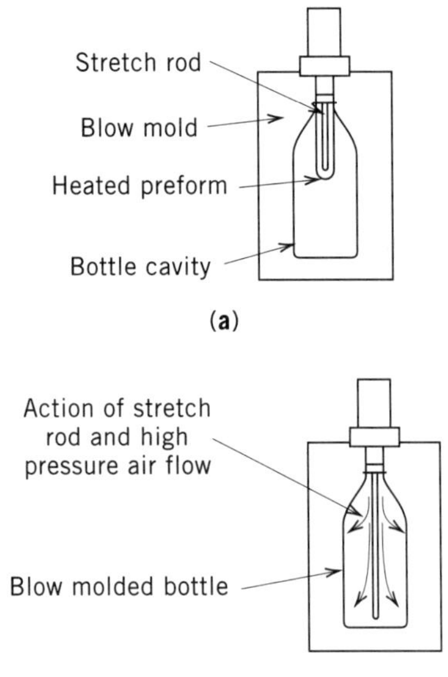

Figure 9. Basic stretch blow molding process (**a**) Mold close on preform; (**b**) stretch blow molding. Note: For many PET material applications, preform is typically half the length of the bottle.

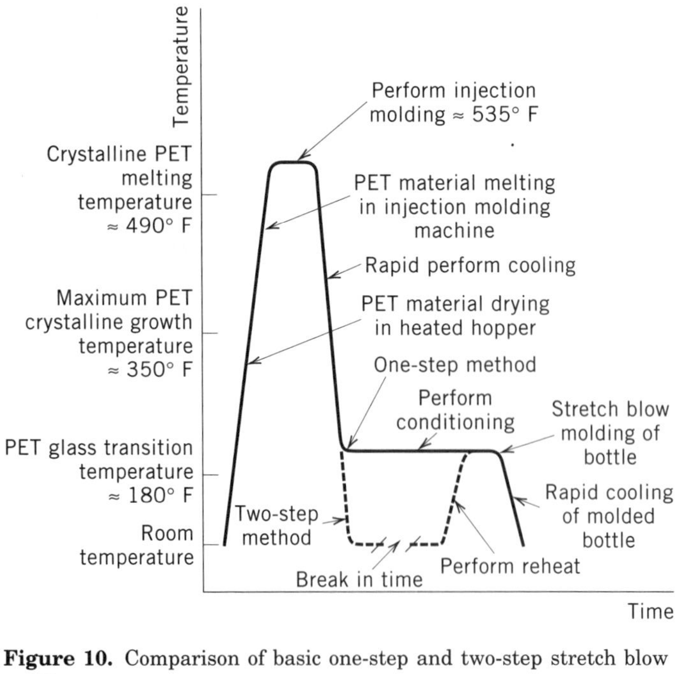

Figure 10. Comparison of basic one-step and two-step stretch blow molding processes.

Figure 11. An injection-molding machine and tooling for producing PET bottle preforms. The machine features a side-entry robot providing part removal and postmolding cooling. This is a high-output system utilizing an injection-mold tool with 96 cavities. Courtesy of Husky Injection Molding Systems.

output equipment is available. Figure 11 shows a high-output PET preform injection-molding machine and tooling with a robotic removal and postmold preform cooling system. Figure 12 shows a high-output horizontal rotary stretch blow-molding machine with a preform temperature-conditioning oven. Often production of preforms and bottles are physically separated in different facilities, which, in turn, can be across town, in a separate town, or a separate state.

In comparison with proper preform design and temperature conditioning, the one-step method is also capable of optimizing bottle performance. Unfortunately, machinery configurations feature the same number of preform and bottle cavities when production capabilities of these cavities are rarely in balance. Overall productivity is usually determined by the injection-molding stage and occasionally by the temperature-conditioning stage. As a result, preform design is often altered to achieve a faster production rate but compromising bottle performance.

The hybrid method. In this method the production stages of parison injection molding are carried out in an operation separate from parison temperature conditioning and blow molding, but directly linked with a conveyor. For some bottle manufacturing applications the hybrid method promises to better balance the productivity of the preform injection molding stage with the blow-molding stage, but the main advantage is the convenience of preform handing between the stages.

Multilayer and Coextrusion

All materials—whether metal, glass, paper, or plastic—have certain strengths and weaknesses. Often two or more materials can be layered to economically overcome a shortcoming. Multilayer blow molding is a process where the strengths of two or more plastic polymers are combined to economically package a product far better than any of the polymers could individually.

Ideal characteristics of many bottles are low cost, strength, clarity, product compatibility, and gas barrier. Polypropylene (PP), for example, is a relatively low-cost plastic suitable for food contact with excellent water-vapor barrier and good heat and impact resistance. The polymer is also a poor oxygen barrier, making it unsuitable for packaging oxygen-sensitive foods requiring a long shelf life. Polyethylene vinyl alcohol (EVOH), on the other hand, is a relatively high-cost polymer with excellent oxygen barrier but sensitive to water, which can deteriorate its properties. A thin layer of EVOH, with two layers of adhesive sandwiched between two layers of PP, provides a very effective package solution.

Figure 12. A horizontal rotary two-step stretch blow-molding machine with preform oven. This is a high-output system featuring 40 blow-mold cavities capable of 40,000 bottles per hour of output. Courtesy of Groupe Sidel.

A multilayer parison can be produced by either extrusion or injection-molding techniques. Both depend on the laminar flow characteristics of polymers, that is, the tendency of polymers to flow in layers. Figure 13 shows the interior polymer flow channels of a typical six-layer multilayer extrusion-head arrangement. Figure 6 shows an extrusion blow-molding machine fitted with a multilayer head, and further shows the manifold pipes connecting the end of each material plasticizer to the parison extrusion head.

Economically improving the gas-barrier performance of polyolefin polymers was the initial market opportunity stimulating the development of the multilayer processes. Structure **a** and **c** of Figure 14 show a typical PP-based multilayer structure and a typical PET-based multilayer structure for food product applications requiring high oxygen gas barrier. Without the adhesive layers, the EVOH-barrier layer can easily delaminate from the PP. The PET structure, while still easy to delaminate from the EVOH, provides adequate integrity.

Recently multilayer structures have become an important method for recycling postconsumer plastic packaging materials. Commingled HDPE, for example, is a mixture of homopolymer and copolymer materials in a rainbow of colors that typically become an unpleasing yellow-greenish-bluish-gray color when molded. Nonetheless, this material can often be reused in new bottles for household chemical products, but the commingled material provides poor chemical stress-crack resistance, a critical product requirement. Placing the recycled HDPE between two layers of virgin HDPE provides the solution (see structure **b** of Fig. 15). In this example, the inner layer (of a proper copolymer HDPE) creates a chemical resistant barrier while the outer layer hides the unpleasing color. The middle layer of recycled material provides structure. In some areas postconsumer PET materials are being recycled in much the same way, but, in this case, the recycled PET is clean and clear, and the new bottle is for a food packaging application. The outer and inner layers provide an extra barrier between the product and the clean recycled PET to provide an extra measure of confidence.

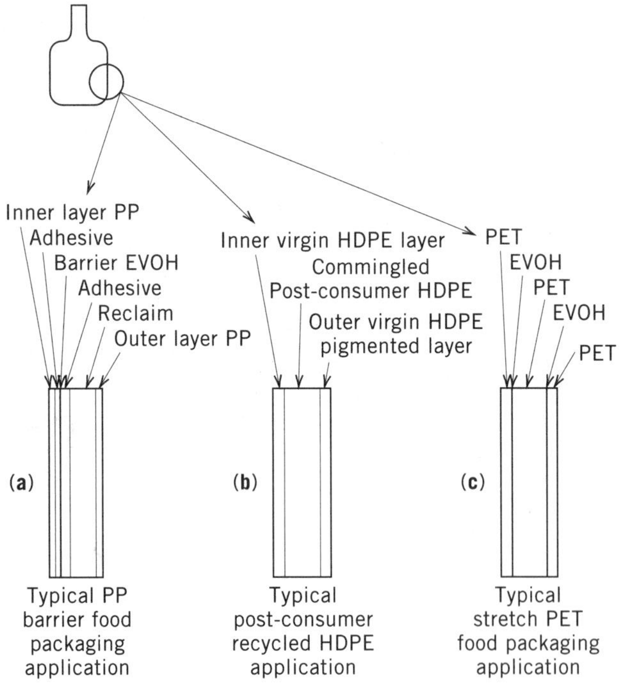

Figure 14. Common multilayer wall structures.

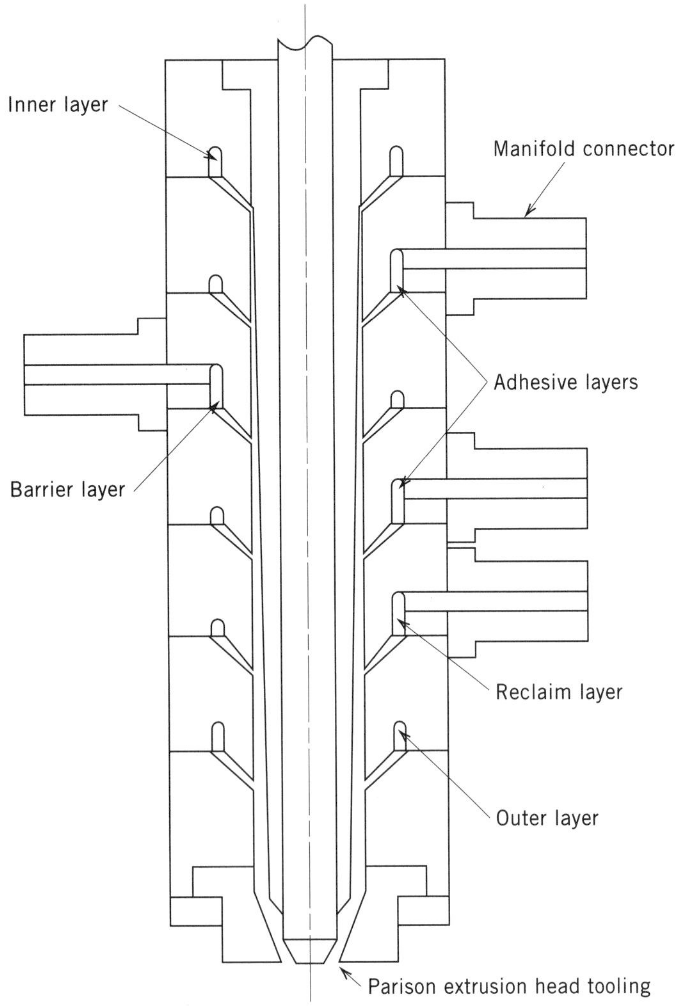

Figure 13. Typical multilayer parison extrusion head arrangement.

Many pigmented HDPE bottles have a coextruded view stripe of unpigmented material. The tubular cross-sectional form of the parison is established inside the parison head from a pigmented plastic material as the material flows through. Prior to the actual parison exit from the head, this tube-shaped flow is split at one side with a flow of unpigmented material, thus creating a pigmented parison with an unpigmented stripe. This stripe is usually positioned so that it is located along the side of the blow-molded bottle. Graduation marks are often provided to allow the consumer to easily judge content volume.

Heat-Resistant Polyester Bottles

Heat-resistant bottles, also known as heat-set or heat-stabilized bottles, are usually made from PET or other similar polyester materials. Product applications include high acid foods, such as, fruit drinks and sports or isotonic drinks that are processed, filled, and sealed at elevated temperatures, typically 175–185°F (80–85°C). A normal PET stretch blow-molded bottle cannot be filled much above 150°F (65°C) without causing bottle shrinkage from molded in strains created during the initial stretch blow molding and biaxial molecular orientation of the PET material.

Fundamentally the technique used to manufacture a heat-resistant PET bottle involves increasing the crystallinity of the plastic material to about 28% or more. Material crystallinity can be increased in two ways. First is strain-induced crystallization, that is, by stretch blow molding. Second is heat-induced crystallization (3).

The amount achievable with heat is far more significant. Strain-induced crystallization alone (~20–28%) is not adequate for most heat-resistant bottle applications. In a stretch blow-molded PET bottle the crystals formed by either strain or heat tend to be "rodlike" in shape. These "rodlike" crystals are small enough to allow the material to retain its clarity. On the other hand, heat-induced crystallization without molecular orientation tends to be spherulitic in shape (3). These are larger crystals that causes the clear PET to become a white opaque color.

Bottle producers and machinery manufacturers have developed two basic approaches for improving bottle hot-fill performance with heat. The first, a single-mold method, is to stretch blow-mold the preform into a hot mold and hold for a sufficient period of time to allow stresses to relieve and the necessary crystals to form. The second, a dual-mold method, is to stretch blow-mold the preform into a first mold with a first shape. This is followed by a reheating this shape

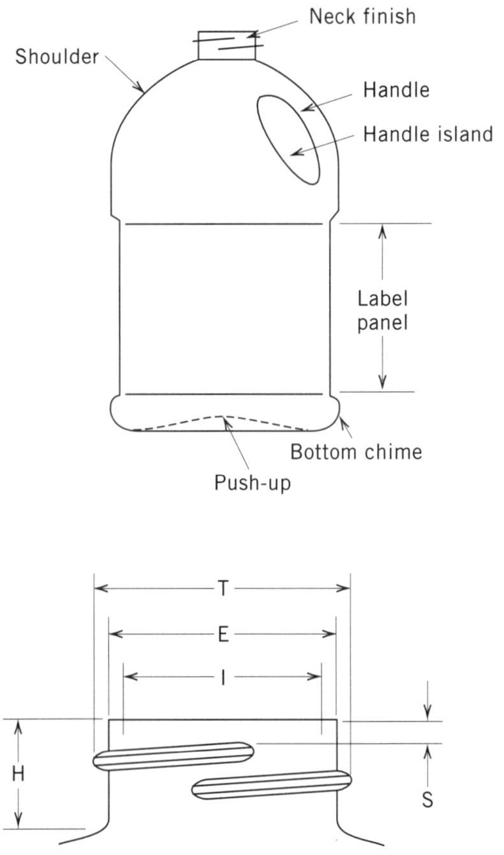

Figure 15. Bottle and finish nomenclature. T = thread diameter; E = external diameter; I = inside diameter; H = finish height; 5 = thread start.

in an oven to relieve stresses and to shrivel the first shape into a second shape. This second shriveled shape is then re-blow-molded in a second mold to produce a final bottle shape.

A crucial area is the performance of the neck finish. The neck finish must not be allowed to deform. The PET material within the neck finish is amorphous (noncrystalline) in nature and will often provide heat resistance unacceptable for many product applications. The best way to provide heat stability is to deliberately crystallize the finish with heat to create a white spherulitic crystalline structure. Another approach is to increase the wall thickness and material weight of the finish. This second approach works because in a filling line environment, utilizing a postfill cooling tunnel, and coupled with the generally poor thermal conductivity of plastic materials, the heavy finish area does not get thoroughly hot enough, long enough to allow a deformation to occur.

Beside heat resistance, another consideration for hot-fillable PET bottles is vacuum collapse. For the most part, materials, when heated, tend to expand in volume; fruit and other juices are no different. Hot-fill processing can be considered a modified "canning" process. That is, the bottle is filled hot and sealed with little or no product headspace. Following seal, the packaged product is slowly cooled to about room temperature. The thermal contraction of the product creates vacuum-induced forces against the container, which a typical PET bottle cannot withstand without distorting. To eliminate an unsightly bottle appearance, hot-fillable bottles are molded with "vacuum panels" that manage these distortions in a uniform fashion.

Aseptic Blow Molding

Aseptic blow molding is usually based on the extrusion process where the bottle is blow-molded in a commercially sterile environment with highly modified equipment. In many cases the product filler is combined with the blow molder.

Aseptic blow molders are generally divided into two subgroups.

The blow-and-hold method. The bottle is molded and sealed in the blow-molding machine. Moving plates in the mold seal the blow pin opening after the bottle is molded. Next the bottle is stored for a period of time, which could be from a few minutes to a few days. While in storage, the bottle is not in a sterile environment. At the filler, the outside of the bottle is resterilized and then the top-seal area is cut off. Next the bottle is filled with sterilized product and resealed.

The blow-fill-seal method. This, in turn, can be accomplished in two ways. In one approach the bottle is blown and filled in the mold cavity. One to two seconds after initial bottle blow molding, a measured volume of product is placed inside. A few seconds later the blow pin and product filler tube withdraw and another small mold cavity, just above the main bottle cavity, closes to seal the parison and bottle. Bottles made by this approach often have a molded-in twistoff seal that can be removed without a knife or scissors. In another approach, the bottle is molded by conventional means and then immediately transferred to a filling station. Bottle blow molding, filling, and sealing steps take place in the same controlled commercially sterilized environment.

Equipment. Aseptic blow-molding equipment modifications generally include the use of special stainless steel and plated materials throughout. The molding/filling area is enclosed in a cabinet. Positive air pressure with laminar flow characteristics is maintained inside. All internal surfaces, passageways, hoses, blow pins, valves, and so forth are sterilized with special "clean-in-place" fixturing. Once the molding process begins, nothing inside can be touched with human hands.

Other Blow-Molding Process Additions

Several secondary processes are combined with the basic blow-molding process: in-mold labeling, fluorination treatment, and internal and postmold cooling.

In-mold labeling (IML). In this process a paper- or plastic-based printed label is placed into the blow-mold cavity just prior to mold closing and parison inflation. The label is held in place by a vacuum system in the mold until the expanding hot parison can contact and establish a 100% bond with the label's heat-activated adhesive.

Fluorination treatment. This is a process for improving the barrier of polyethylene to nonpolar solvents (4) found in certain paints and agricultural chemicals. The barrier is created by a chemical a reaction of the fluorine and the polyethylene, forming a thin 200–400-Å fluorocarbon layer on the bottle surface. Today a few blow-mold processors still use the in-mold system where a small percentage of fluorine is mixed with nitrogen and used to expand the parison, thus creating a barrier layer on the inside of the bottle. While effective, use of the treatment process is limited. Fluorine is a dangerous gas requiring special handling by trained individuals.

Internal-cooling systems. These systems are used to shorten the blow-molding process cycle time, thus increasing equipment productivity. Normally a blow-molded bottle is cooled externally by the mold cavity requiring heat energy to travel through the entire bottle wall thickness. Usually this is not a major problem for most bottle applications, but for heavy-wall bottles (or heavy sections within a bottle or container or drum), blow-molding cycle time can be quite long. Internal-cooling systems attempt to remove some of the heat energy from the inside.

Several approaches for internal-cooling have been developed with varying levels of effectiveness and cost, namely: liquefied gas, supercold air with water vapor, and air-exchange methods. Each application requires a separate analysis. Often system expense is greater than the benefit.

Postmold cooling systems. These systems also represent an attempt to accelerate the blow molding process. Here the bottle is removed from the blow-mold cavity before it is completely cooled and then placed into a secondary nest that continues the cooling process. Often this is combined with the trim tooling used to remove flash and utilizes several air jets directed toward a localized thick area.

Product Design Guidelines

Good product design begins with a clear understanding of process and tooling limitations. In injection blow molding, for example, available core rod strength to minimize deflection will limit its length to about 10 times its diameter. The means that the bottle can be no more that 10 times its finish diameter in height. Limits in parison wall thickness and the ability to properly temperature condition will restrict the blowup ratio to 3 times its diameter. That is, bottle diameter or the largest dimension must be 3 times or less the finish diameter. Bottle cross-sectional ovality must also be restricted. Problems in maintaining a consistent wall distribution in an oval bottle limit the major diameter to no more than 2 times the minor diameter.

Extrusion blow molding, by comparison, is not so limited. Nonetheless, the blowup ratio is often held to 4 times the parison diameter or less. This rule of thumb applies not only to overall bottle shape but also to isolated sections. For example, bottle handle designs that are deeper than they are wide across the parting face are difficult to mold and often thin and weak.

In general, most blow-molded bottles perform better with a uniform or nearly uniform wall thickness. The designer can help achieve this by remembering to radius, slant, and taper all surfaces. Square, flat surfaces with sharp corners do not work very well. Corners become thin and weak; heavy side panels become thick and often distorted. Because of different material shrinkages, flat panels are never flat, and flat shoulders offer little top-load strength. Likewise, highlight accent lines should be "dull." Because of trapped air between the mold cavity and the parison, the parison will not penetrate into a sharp accent. This not only leaves a "trapped air" mark on the bottle but also can weaken the side wall limiting drop impact performance.

Ribs do not always stiffen. Often, blow-molded ribs create more surface area to be covered by the parison, which, in turn, thins the bottle wall thickness. A bellows or accordion effect is created, which flexes easier. The designer must analyze where flexing is likely to occur and where the "hinge" action is likely to be. The bottle design can then be altered to interrupt this action.

Blow-molding process conditions can influence bottle dimensions and bottle volume. Processors of HDPE bottles have identified, several process conditions that significantly change bottle volume. HDPE bottles shrink over time, with about 80% occurring in the first 24 hs. Lighterweight bottles are not only bigger on the inside from less plastic; they also bulge more. A 5 g weight reduction of a typical 1 gal (3.79 L) milk bottle results in a volume increase of about 12 mL (5 mL for plastic, 7 mL for bulge). Faster cycle times, lower parison inflation pressure, and higher melt and mold temperatures also reduce bottle volume. Finally, bottle storage temperature is important. After 10 days, very significant volume changes can occur in bottles stored at 140°F (60°C) (5). There are usually differences between bottles made and shipped in the winter and those made and shipped in the hot summer. *Figure 15* illustrates basic bottle and finish nomenclature.

BIBLIOGRAPHY

1. G. P. Kovach, *Forming of Hollow Articles* in E. C. Bernhardt, ed., *Processing of Thermoplastic Materials,* R. Krieger Publishing, Huntington, NY, 1959, (reprinted 1974), pp. 511–522.
2. R. Holzmann, *Kunststoffe* **69**(10), 704 (1979).

3. M. Bakker, "Heat-Stabilized PET" in *Bottlemaking Technology and Marketing News* **1**(2), 7–17 (March 1993).
4. "Surface Treatment Improves Polyethylene Barrier, Properties," *Pack. Eng.* 64 (Nov. 1981).
5. *Operators Guide—Controlling Shrinkage of HDPE Bottles,* Dow Chemical Co., Midland, MI, 1979.

General References

C. Irwin, "Blow Molding" in M. Bakker, ed., *Encyclopedia of Packaging Technology,* 1st ed., Wiley, & New York, 1986, pp. 54–66.

N. C. Lee, ed., *Plastic Blow Molding Handbook,* Van Nostrand Reinhold, New York, 1990.

D. V. Rosato and D. V. Rosato, eds., *Blow Molding Handbook,* Hanser Publishers, Munich, 1989.

Christopher Irwin
Johnson Controls, Inc.
Manchester, Michigan

BOTTLE DESIGN, PLASTIC

Like many of today's popular packaging techniques, blow molding of plastic bottles became popularized soon after World War II (see Blow molding). The original applications took advantage of the flexibility of plastic material to create squeeze bottles for dispensing of deodorants or medicines. The availability of reasonably priced higher-density polyethylene for rigid containers in the 1950s led to the widespread use of plastic bottles for detergents (1).

As both molding and plastic material technology developed, conversion to plastic bottles expanded beyond household products into health and beauty aids, foods and beverages, and general goods. Two major conversions in more recent years were glass to plastic for soda bottles and metal to plastic for oil and other automotive products.

The advantages of plastic over glass bottles are abundant. They are safer, lighter, easier to handle for consumer and manufacturer, easier to manufacture and offer improved versatility in design. However, many of the earlier conversions suffered because of the marketer's attempt to imitate the original glass bottle as closely as possible. It was feared that the consumer (especially in beauty aids) would perceive the change to plastic as strictly a cost savings measure resulting in reduced quality. Today, the majority of plastic bottles are designed to take advantage of the material's unique properties, and the consumer has come to expect all the inherent advantages in the items they purchase.

In the 1970s and 1980s, plastic bottle designers and manufacturers were faced with widely fluctuating plastic material and additive costs and availability. The later 1980s and 1990s have added the additional challenge of environmental concerns. The package must not only perform until the end of the life of the product but also take into account its eventual disposal.

Basic Steps in the Plastic Bottle Design and Development

There are several basic steps for the proper design and development of plastic bottles. They can be summarized as follows:

1. Define bottle requirements—product to be contained, use, distribution, aesthetics, and environmental issues.
2. Define manufacturing and filling requirements—types of molding available and filling and packaging systems.
3. Select materials.
4. Rough drawings.
5. Part drawing.
6. Model.
7. Mold drawing.
8. Unit cavity.
9. Unit cavity sampling and testing.
10. Finalize drawings.
11. Production mold.
12. Production mold sampling and testing.
13. Production startup.

This is a basic list of steps for bottle production. Some items may be done in a different order, and some can be performed concurrently to save time. Many items will be performed more than once as the design is refined. In addition, mating fitments and secondary packaging are frequently being developed in the same timeframe and become part of the critical path. Figure 1 shows a typical timeline for a plastic bottle project.

Defining Bottle Requirements

What will the bottle hold? Information about the product to be contained is of primary importance for bottle design, particularly in the selection of material and neck finish. Will the product be liquid, powder, or solid? What is its viscosity? Is it homogenous, or does it tend to separate out? Will the product have to be shaken to be used? What plastic materials and additives is it compatible with? How sensitive is it to moisture or oxygen gain or loss? Does the product have components (such as those found in detergents) that could make the bottle prone to stress cracking? Will the product outgas into the headspace and cause a pressure buildup? Will the product absorb oxygen from the headspace and tend to collapse the bottle? Will the product be filled into the bottle hot or cold? How much product should the container hold (in weight or volume)?

What are the special requirements for bottle performance?

Product end use. How will the product be used by the ultimate end user for dispensing and storage? Examples of this are squirt bottles, rollon bottles, trigger-spray bottles, drainback closure/measuring cup systems, bottles with integral funnels, and bottles that are stored on their caps. What are the storage conditions (temperature, humidity, pressure) of opened and unopened containers?

Secondary packaging. Will the primary package be sold in an intermediate secondary package? Examples of this are cartons, blisters, and trays.

Distribution requirements. Will the unitized product be sold in shipping case or partial-height tray? What type of inner partitions or shrink wrap can be used? Do intermediate customers require smaller unitized packs and if so, of what type? Will the bottle itself

	Total	Week number																																																
	weeks	1	2	3	4	5	6	7	8	9	10	11	12	13	14	15	16	17	18	19	20	21	22	23	24	25	26	27	28	29	30	31	32	33	34	35	36	37	38	39	40	41	42	43	44	45	46	47	48	49
1. Define bottle requirements	2	X	X																																															
2. Define manufacturing requirements	2			X	X																																													
3. Select materials	2					X	X																																											
4. Rough drawings	2							X	X																																									
5. Part drawing	3									X	X	X																																						
6. Model	2												X	X																																				
7. Mold drawing	3														X	X	X																																	
8. Unit cavity	8																	X	X	X	X	X	X	X	X																									
9. Unit cavity sampling and testing	4																									X	X	X	X																					
10. Finalize drawings	2																													X	X																			
11. Production mold	14																															X	X	X	X	X	X	X	X	X	X	X	X	X	X					
12. Production mold sampling/testing	4																																													X	X	X	X	
13. Production start-up	1																																																	X
TOTAL	49																																																	

Figure 1. Typical blow-molded bottle development timetable. Assumes sequential activities, no consumer testing, and no need to resample and retest.

contribute to stacking strength? Does the final outer package have to fit on a standard GMA 40-in × 48-in. pallet (consider load optimization and stack heights)? Will cases be shipped in mixed loads at any time? Will shipment generally be by full truckload, partial truckload (LTL), or individual shipment by ground or air (such as United Parcel Service or Federal Express)? Does the product travel in a rack system or bulk package for any part of its distribution? Are there shelf size requirements for the retail market or for storage?

Aesthetic requirements. The plastic bottle is frequently required to act as a salesperson at point of purchase and after the sale. Common concerns include

1. Shelf facing size—maximization of width and/or height vs depth while maintaining package stability.
2. Clarity, opacity, and color
3. Label area and label qualities
4. Requirement for recycled and/or recyclable material

What are the special requirements for bottle manufacturing and filling?

Options for bottle manufacture. Ideally, the designer should be able to select the best manufacturing technique for the particular bottle. However, because of cost, available capital equipment, and other factors, it may be necessary to compromise if possible. The major choices are

1. Molding process—injection blow, extrusion blow, stretch blow (with variations)
2. New or modified molds (or parts of molds and mold bases)

The various advantages and disadvantages of the different plastic bottle manufacturing methods and molding processes are summarized as follows:

1. *Process:* extrusion blow molding.

a. Description: A tube (parison) is extruded through an annular die. Two halves of a bottle mold are clamped over the parison, sealing the top and bottom except for a hole for air injection. Air is injected, expanding the parison to match the mold. Clamped material (necks and tails) is removed and commonly put back into the system as regrind.

b. Advantages

(*1*) Relatively inexpensive mold, mold modification and equipment costs for a basic system.
(*2*) Multilayer bottles and extruded side stripes possible.
(*3*) Larger size containers are economical.
(*4*) In-mold labeling available.
(*5*) Good for handleware or other designs requiring a molded-in "hole."

c. Disadvantages

(*1*) High built-in waste due to necks and tails.
(*2*) Requires in-mold or postmold trimming of necks and tails.
(*3*) Finish dimensions and quality are not as consistent.

d. Variations

(*1*) Extrusion dies can be ovalized to improve material distribution in oval or rectangular bottles.
(*2*) Extrusion dies can be programmed to open and close while extruding, providing a top to bottom variation in wall thickness.
(*3*) For high-volume large-container production, continuous wheel machines can be used instead of molds that shuttle in and out.
(*4*) Instead of direct blow into a bottle, a dome may be created that must be trimmed off the finish. A calibrated neck, on the other hand, uses the blow pin to help mold the finish. The top of the finish may be further improved by posttrim reaming.
(*5*) High blow ratio (finish to maximum width) or off-center neck bottles can sometimes be improved by blowing "outside" the neck; a wide parison is used and waste is trimmed off the sides of the finish on the parting line.

2. *Process:* injection blow molding

a. Description: The neck finish and "cigar"-shaped body are injection blow-molded in the first phase. This parison mold is then moved to a blow station containing a mold shaped like the bottle body. Air is injected through the finish, and the completed bottle is removed from the pin.

b. Advantages

(*1*) Good quality and control of finish dimensions due to injection process.
(*2*) Capability of control and design inside the finish area (as long as there are no undercuts).
(*3*) Economical for large volumes of smaller containers.
(*4*) Waste not built into molding process (injection mold is hot runner).
(*5*) Can use PET as material.
(*6*) Finished container does not have a "pinchoff" area at the tail that can provide a weak point for drop test or stress-crack failure.

c. Disadvantages

(*1*) Additional need for injection molds increases cost and lead times.
(*2*) In-mold labeling and handles (with "holes") not generally available.
(*3*) Not good for larger containers or high blow ratios.

d. Variations

(*1*) Existing parison molds can sometimes be utilized to save cost and time.
(*2*) Injected parison mold can be ovalized for better plastic distribution in oval or rectangular bottles.

3. *Process:* stretch blow molding

a. Description: This is a variation on injection blow molding wherein the hot parison mold, or preform, is stretched in length by a push rod placed in the bottom prior to blow molding.

b. Advantages

(*1*) Improved bottle impact and cold strength.
(*2*) Improved transparency, surface gloss, stiffness, and gas barrier (2).
(*3*) Good for pressure containers such as soda bottles.
(*4*) Materials such as PEN (polyethylene naphthalate) promise improved hot fill and oxygen barrier over PET (3).

c. Disadvantages

(*1*) Commonly used PET requires special handling and drying process.
(*2*) Extra step required in process.
(*3*) High capital equipment costs.
(*4*) Additional need for injection molds increases cost and lead times.
(*5*) In-mold labeling and handles (with "holes") not generally available.

d. Variations

(*1*) Preforms can be molded as part of one process or made and stored for later use.
(*2*) Existing parison molds can sometimes be utilized to save costs and time.

Bottle filling and packing. Bottle design needs to take into account the filling and packing operations for the bottle. Important criteria include

(*1*) *Current equipment and change parts.* There are frequently size limitations (height, width, and depth on equipment). Likewise, change parts and changeover times can be minimized by determining common critical dimensions or easier adjustments (frequently height) with other bottles that might run frequently on the same line.

(*2*) Forces and conditions imparted by the filling and packing operation on the bottle. Common conditions include downward compression due to insertion of items in the neck (such as plugs and balls) or on the finish (such as snap-on closures); bottle body torsion and neck distortion due to high-torque cap application; side-to-side compression due to labeling and starwheel pinch points; multidirectional compression and abrasion due to bulk handling such as in a bottle unscrambler; and bottom drop impact due to automatic drop packing into cases or trays (Fig. 2). In addition, aesthetic damage can be done at various points to a labeled or unlabeled bottle due to abrasion at transfer points and handling throughout the line.

Bottle Design and Specification

Once the special requirements for the bottle have been determined, sketches and models of various bottle designs meeting

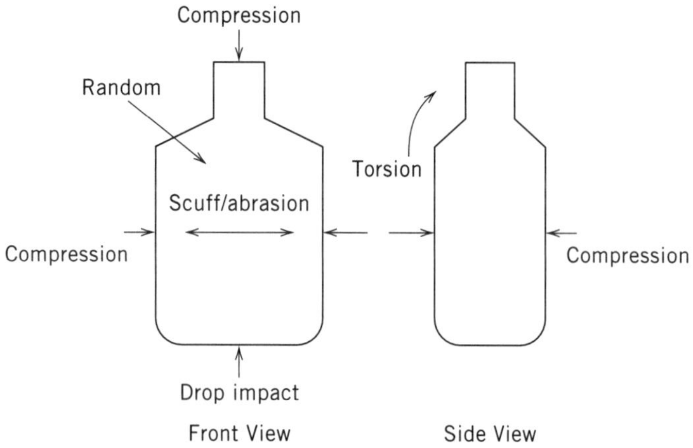

Figure 2. Common stresses on a bottle on a filling and packing line.

these criteria can be produced. As the process continues, the finalists need to become more detailed (dimensions, angles, finish) and begin to incorporate good plastic bottle design principles.

General design considerations

(*1*) *Sharp edges and changes in direction should be avoided.* Surfaces should be blended and generous radii used as much as possible at bends and corners (4). Sharp edges can lead to thin and high-stress areas, resulting in stress cracking or failure on drop testing.

(*2*) *The bottle must have minimal or no undercuts.* Thought must be given to how the bottle is to be removed from the mold. The inside of the finish must be stripped from the core rod or blow pin, and the front and back halves of the mold need to open up from the outside of the bottle. Any plastic that would hook onto the tooling or molds can cause the bottle to distort or even break. Very small and rounded undercuts, such as small retaining rings on the inside of the neck, can sometimes be successfully molded; trial and error may be necessary for a particular design and process to see how much is possible.

(*3*) *Wall thickness minimum and distribution should be considered.* Thin spots can lead to stress cracking, failure on drop test, and even pinholing; thick sections can act as heatsinks after the bottle is molded, and the uneven cooling and shrinkage can lead to bottle distortion. Unfortunately, wall thickness is not easily controlled, and specification is frequently limited to minimum wall thickness and general guidelines or ranges.

(*4*) *Bottle stability is critical.* This is especially true for many of the tall, thin (depth) bottles frequently used in consumer products to maximize shelf impression. In injection blow molding, the height of the pushup (the center of the base of the bottle) can be adjusted. The center of the bottle base must be high enough to not "belly out" for a full or empty bottle. Various aids to stability include molding bottom rings (radiused or with a flat land area) or feet (three or four). It is a good idea to put a slight depression in the vicinity of the parting line so that a slightly raised parting line will not contribute to instability. The best solution depends on the bottle's general shape and center of gravity and on the controls available in the molding process.

(*5*) *Embossed or debossed decorations.* These include logos, which can be a free or inexpensive way to add information to the bottle. However, care should be taken to avoid thin spots, undercuts, or sharp edges.

(*6*) *Information that is frequently molded into the bottom of the bottle.* This includes material identification symbol (recycling logo, legally required to be on the bottle in several states), mold and cavity number (for quality checks and troubleshooting), and a molder or company logo. Placement should be planned ahead of time, especially as this may affect bottle stability. These should generally not touch or cross the parting line.

(*7*) *The bottle sealing area should be closely specified.* For standard continuous thread closures, lined or unlined, the land seal area at the top of the container is critical. It needs to be flat, horizontal, and free of dips or nicks. If the flat surface is going to be angled, care should be taken to specify a continuous surface that will create a seal. A minimum land width (flat neck thickness at the top) can be specified. If a valve seal closure is to be used, the circumference inside the finish where it meets the valve is critical.

(*8*) *Labeling and decoration.* These are a major consideration in bottle design. A label panel must be flat or have a surface that curves in only one direction. Many label areas are recessed or provide top and bottom "bumpers" to protect the label against scuffing during normal handling and distribution. The maximum label area is determined by a combination of the tolerances of the label placement equipment, the label dimensions, and the usable label panel dimensions. The additive tolerances need to be subtracted from the specified label panel size. Major labeling or decoration methods include (*a*) *in-line labeling* (*filling line*)—labels (generally pressure-sensitive or plain-paper and glue) are applied on the filling line; (*b*) *postmold labeling*—labels (generally pressure-sensitive or heat-transfer) are applied soon after the bottle is molded (further shrinkage in the bottle, particularly over the next 24 h, must be taken into account); (*c*) *in-mold labeling*—labels (either plastic or paper) are applied inside the mold during the molding process (bottle vs label shrinkage must be designed into the bottle tooling, especially for paper labels); and (*d*) *direct application to bottle*—instead of applying a label, the container is decorated by printing or silk screening directly onto the bottle.

HDPE or PP bottles usually require flame treatment prior to decoration or labeling. Round bottles may need a rotational guide in the bottom to help control them during either process. (See also Labels and labeling machinery.)

Industry standards. As in any purchased item, specifications and tolerances should be set up between the customer and the supplier. However, the designer is greatly aided by sets of industry standards published by The Plastic Bottle Institute, Division of The Society of the Plastics Industry, Inc. (5). The Institute publishes a bulletin (AU-124) with standard neck finish dimensions, threads, and tolerances, and tolerances for various ranges of bottle capacity and body dimensions. The location and methodology of measurements is also described. The existence of standard finishes greatly facilitates the interchangability of various stock and custom closures that might be available. The standard tolerances also provide a good starting point for specifications. The designer might wish to tighten some of these tolerances as required, but should be prepared to work closely with any vendors ahead of time to ensure manufacturing capability.

A typical dimensioned bottle is shown in Figure 3. Table 1 describes a typical SP-400 bottle with a 33-mm diameter and the importance of various dimensions.

Materials and colorants. Critical criteria for selection of bottle materials include the following:

1. Cost and availability—should be calculated by the bottle, not just the pound.
2. Environmental—recyclability and availability of recycled material, and toxicity.
3. Clarity or opacity.
4. Stiffness or squeezability.
5. Water or moisture vapor transmission rate (WVTR or MVTR).
6. Gas permeability.
7. Chemical resistance to type of chemicals required.
8. Temperature range.
9. Impact strength.
10. Environmental stress–crack resistance (ESCR)

Materials generally used for bottles in the United States include high-density polyethylene (HDPE), polypropylene (PP), low-density polyethylene (LDPE), linear low-density polyethylene (LLDPE), polyvinyl chloride (PVC), polyethylene terephthalate (PET), and polystyrene (PS). Other materials, including engineering plastics such as polycarbonate, are used for special applications. Individual material properties can be found in detail under their separate articles. Many materials can have their properties significantly modified by additives such as plasticizers, fillers, impact modifiers, antistatic agents, and UV (ultraviolet radiation) inhibitors. Copolymers and multilayer coextrusions can further enhance a material.

Colorants are added to provide color and/or opacity to the material. Pigments, an opacity agent (such as titanium dioxide) and a compatible base material are compounded and added to the regular plastic during the molding process. The colorant loading, wall thickness of the plastic, and color and opacity of the product all affect the final aesthetics of the filled bottle (see Colorants, plastic).

Different materials and to some extent different colorants all affect the shrinkage and final performance of a bottle. This must be taken into consideration for testing and qualification.

Computer utilization in the design process. The now ubiquitous computer is an integral part of the bottle design process, and new ways of using it are being added continuously. Uses include:

Specifications. Bottle specifications are computerized at both the customer and supplier level.

Computer-aided design/manufacturing (CAD/CAM). Bottle drawings are performed on CAD systems, with improved opportunity for analysis and exploration of options and variables. Mold makers frequently use a CAM system for production of the mold. Some forms of rapid prototyping can also be performed.

Analysis and data recording. Test data can be both directly recorded and analyzed on the computer. Sophisticated engineering programs, such as finite-element analysis, can be used to compute and visualize stresses at various points.

Pallet loads. Several programs, such as those produced by CAPE (6) or TOPS (7), allow quick calculations of finished pallet options based on bottle sizes. This allows for pallet load optimization at the initial design stage.

In addition, there is various software to do everything from evaluating heat transfer in mold design to visualizing product placement on the supermarket shelf.

Prototyping and Testing

Once the bottle is designed and unit-cavity samples have been produced in the proper materials, a testing program must be initiated. A testing protocol should be established that checks bottles against both their specifications and the requirement

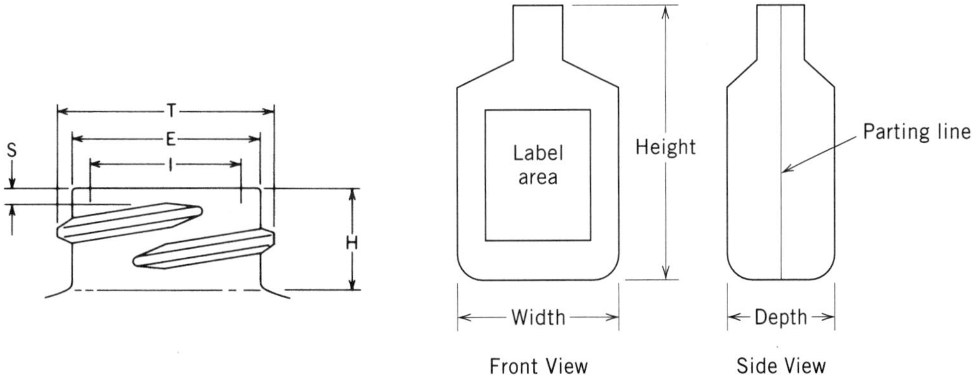

Figure 3. Typical dimensions of a plastic bottle. Bottle and finish nomenclature. T, diameter of thread; E, diameter of root of thread; I, diameter of inside; S, distance of thread start to top of finish; and H, height of finish.

Table 1. Typical Bottle Dimensions

Dimension	Description	Sample Dimension (in.)
T	Thread outer diameter	1.241–1.265
Purpose: Cap fit and torque control.		
E	Diameter between threads	1.147–1.171
Purpose: Cap fit and torque control.		
H	Height from shoulder to top of finish	0.388–0.418
Purpose: Bottom of closure must clear shoulder at maximum torque for proper sealing. The maximum minus minimum H (ΔH) should also be specified to limit neck "cocking", which will also adversely affect the seal.		
S	Height from top of thread to top of finish	0.031–0.061
Purpose: Thread engagement.		
I	Inside diameter of finish	Minimum 0.791
Purpose: Minimum is for filling tube clearance only. Custom detail should be added for placement of special fitments.		
Ovality	$T_{max} - T_{min}$ or other area	Maximum TBD[a]
Purpose: Ovality in the finish can affect closure fit, torque control, or fitment integrity and fit. Specification should be based on requirements and manufacturing capabilities.		
Land width	Sealing width on top of finish	Minimum TBD
Purpose: Cap seal and torque control. Specification should be based on requirements and manufacturing capabilities.		
Overflow capacity	Volumetric capacity of bottle to top	Varies with size
Purpose: Consistent filling of bottle with proper amount of product and correct aesthetic fill height.		
Overall dimensions	Height, width, and depth at maximum point	Varies with size
Purpose: Proper machinability on filling line and fit in secondary packaging. Additional dimensions are specified in article drawing.		
Weight	Weight of empty bottle	Varies with size
Purpose: Cost control, to ensure sufficient material, and fill weight control.		
Wall thickness	Thickness of wall at minimum and specified points	TBD
Purpose: Thin spots can lead to drop failure or stress crack. Wall thickness should be specified at the absolute minimum found on the bottle (this can vary as to location) and at other critical locations if necessary.		

[a] To be determined.

list that was generated. The following literature is published by the Plastic Bottle Institute (8):

Technical Report	Catalog No.
Visual Evaluation of Reflected Color	AU-123
Finish Dimensions with Nine Drawings	AU-124
Vertical Compression Test	AU-134
Drop Impact Resistance Test	AU-135
Weight Change and Compatibility of Packaged Goods Test	AU-136
Evaluating Flame Surface Treatment of PE and PP Bottles	AU-137
Closure Torque Test	AU-138
Capacity Determination—Fill Point and Overflow	AU-139
Soot Accumulation Test	AU-140
Decoration Adhesion Test	AU-141
Top Load Stress Crack Resistance Test	AU-142
Dead Load Compression Test—Water Filled and Capped	AU-143
Compression Test for Corrugated Boxes and Inserts	AU-144
Stacked Box Compression Test	AU-145
Recommended Practice for Measuring Specular Gloss of Transparent Plastic Bottles	AU-147
Recommended Practice for Measuring Haze of Transparent Plastic Bottles	AU-148
Recommended Practice for Measuring Color of Transparent Plastic Bottles	AU-149
Visual Evaluation of Optical Clarity of Transparent Bottles	AU-150
Optical Clarity Reference Chart Only (use with AU-150)	AU-151
Closure/Bottle Finish Leakage Test	AU-152
Recommended Procedures for Measuring Plastic Bottle Dimensions	AU-153
How to Measure Plastic Bottle Dimensions (Live Action)	AU-162
Method for Determining the 24-Hour Gas (Air) Space Acetaldehyde Content of Freshly Blown PET Bottles	AU-155
Inherent Viscosity of Polyethylene Terephthalate	AU-156
Voluntary Guidelines, Plastic Bottle Material Code System-Mold Modification Drawings	AU-158
Dimensional Stability of Plastic Carbonated Beverage Bottles	AU-167
Good Practices Manual for Packaging in Plastic Bottles with Glossary of Plastic Bottle Terminology	AU-103
Recommended Industry Specifications for Recycled HDPE	AU-175

In addition, specific tests for various industries and materials are included under those topics in this encyclopedia.

Specialty Bottle Requirements

A number of specialty bottles have distinctive requirements. Some of these include

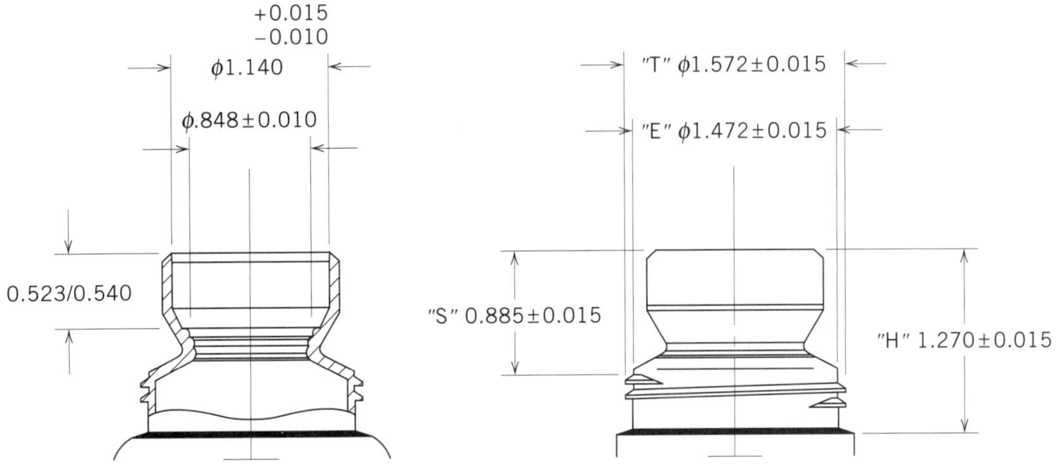

Figure 4. Typical special rollon finish.

Rollon. A round ball is inserted in the finish; rolling the ball against a surface (clothes, parts of the body) dispenses a liquid. The ball is held in place between two rings in the finish. The most common method of sealing is to torque the cap so that it pushes the ball down tightly against the lower ring (Fig. 4).

Plug. A small orifice plug is placed into the finish so that product dispensing can be controlled (eye or nose drops, creams). The plug must be inserted without crushing the bottle body or neck finish and yet be difficult to remove. Controlled dimensions or retention rings are some of the solutions.

Drain-back closures. These fitments provide pour spouts and matching caps for detergents and fabric softeners. The drain-back section is sometimes placed inside the finish similar to a plug, but frequently before filling.

Child-resistant closure (CRC). There are a number of different types of closures on the market, generally requiring special finishes. The popular "push and turn" instruction requires sufficient *H* dimension to allow the cap to be pushed down for thread engagement. The "line up the arrows" instruction requires a ring with an interruption under the arrow. Protocol testing includes both closures and containers (see Child-resistant packaging.)

Nonremovable closure. Some dispensing closures are designed to prevent removal. One design has ratchets that can easily be overridden for tightening but prevent loosening (see Closures, bottle and jar).

Tamper band. Many products (drugs and foods) have tamper bands to show tamper evidence. The band must be able to anchor itself to something on the bottle—a ring is sometimes added to the outside under the finish area (see Bands, shrink; Tamper-evident packaging.)

Environmental Concerns

Whether driven by legislation, consumer preference, or the desire to be a good corporate citizen, today's packager has to take environmental concerns into account when producing a package. A new plastic bottle presents a special challenge and opportunity. The following is a brief look at plastic bottles and the ways of handling solid waste other than the landfill:

(*1*) *Reduce.* The preferred way of handling the problem—creating less waste in the first place—has another major advantage. The less plastic we use, the more money we save. This can be accomplished through optimizing structural design, eliminating oversizing, and taking advantage of the best materials for the job.

(*2*) *Reuse.* Although returnable bottles may seem like an obvious answer, this is not necessarily the case; "to withstand the process of return and refill, a refillable container must be about twice as heavy as a one-way container of the same material." This adds impact to energy usage and eventual disposal also (9). However, a number of consumer products, notably detergents and cleaning products, are sold in bottles that the consumer refills without having to undergo this process. The refill packages are designed with less plastic and frills, or are made of lighterweight or more easily crushable materials.

(*3*) *Recycle* (*recyclable*). Everything can be recycled by somebody, somehow, somewhere. However, unless a substantial percentage of the bottles can realistically be recycled in the near future, this doesn't mean too much. Although the percentages of bottles that are recycled or could be from programs in place changes constantly, the bulk of recycling is with PET (mostly soda bottles, although some areas have added capabilities to accept other bottles) or HDPE (mostly homopolymer milk bottles, although some areas now accept other containers including copolymer detergent bottles).

(*4*) *Recycle* (*recycled*). The relatively high cost of collecting, sorting, cleaning, and processing good-quality recycled HDPE has made it more expensive pound for pound than virgin resin. As the process becomes more efficient and if virgin continues to climb (up about $0.09/lb. in 1994) (10), this may change. Because of the unpredictability of the color and properties of much regrind, it is frequently used as a middle layer in a multilayer coextruded bottle or as a small percentage in a dark bottle holding nonaggressive product. Food and drug primary packages may not be permitted to use reground plastic directly, although repolymerization or special considerations can be investigated.

(*5*) *Incineration and conversion to energy.* Plastics produce heat when burned, which, in turn, can be converted to other forms of energy. The presence of heavy metals (such as lead or cadmium used in some colorants) in a bottle will leave toxic waste residue. Several states have outlawed the use of these materials in the last few years, and it is critical to avoid them. Older incinerators may also have a problem with chloride containing materials such as PVC (11).

BIBLIOGRAPHY

1. R. J. Kelsey, *Packaging in Today's Society,* St. Regis Paper Company, 1978, p. 69
2. C. Irwin, "Blow Molding," in M. Bakker, ed., *The Wiley Encyclopedia of Packaging Technology,* Wiley, New York, 1986.
3. J. Meyers, "Process Technologies Expand Markets for Stretch Blow Molded Bottles," *Modern Plastics* **70**(13) (Dec. 1993).
4. S. Levy and J. H. DuBois, *Plastic Product Design Engineering Handbook,* Van Nostrand Reinhold, New York, 1977, pp. 169–170.
5. The Plastic Bottle Institute, a Division of the Society of Plastic Engineers, Inc., 1275 K Street Northwest, Suite 400, Washington, DC 20005.
6. CAPE Systems, Inc., 2000 N. Central Expressway, Plano, TX 75074.
7. TOPS Engineering Corporation, 1721 W. Plano Parkway, Plano, TX 75075.
8. The Plastic Bottle Institute, a Division of the Society of Plastic Engineers, Inc., 1275 K Street Northwest, Suite 400, Washington, DC 20005.
9. L. Erwin and L. H. Healy, Jr., *Packaging and Solid Waste Management Strategies,* American Management Association, New York, 1990, p. 26.
10. "Pricing Update," *Plastics World* **52**(10), 68 (Oct. 1994).
11. L. Erwin and L. H. Healy, Jr., *Packaging and Solid Waste Management Strategies,* American Management Association, New York, 1990, p. 37.

General References

R. J. Kelsey, *Packaging in Today's Society,* St. Regis Paper Company, 1978.

C. Irwin, "Blow Molding," in M. Bakker, ed., *The Wiley Encyclopedia of Packaging Technology,* Wiley, New York, 1986.

J. Meyers, "Process Technologies Expand Markets for Stretch Blow Molded Bottles," *Modern Plastics,* **70**(13) (Dec. 1993).

S. Levy and J. H. DuBois, *Plastic Product Design Engineering Handbook,* Van Nostrand Reinhold, New York, 1977.

The Plastic Bottle Institute, a Division of the Society of Plastic Engineers, Inc., 1275 K Street Northwest, Suite 400, Washington, DC 20005.

CAPE Systems, Inc., 2000 N. Central Expressway, Plano, TX 75074, publications available.

TOPS Engineering Corporation, 1721 W. Plano Parkway, Plano, TX 75075, publications available.

L. Erwin and L. H. Healy, Jr., *Packaging and Solid Waste Management Strategies,* American Management Association, New York, 1990.

"Pricing Update," *Plastics World* **52**(10), (Oct. 1994).

S. Selke, Ph.D., *Packaging and the Environment,* Technomic Publishing Inc., Lancaster, PA, 1990.

G. Graff, ed., *Modern Plastics Encyclopedia '94,* McGraw-Hill, New York, 1993.

"Special Report on Labeling," *Packaging* **38**(12) (Nov. 1993).

B. Hepke, "The In-Mold Labeling Decision," *J. Pack. Technol.* **3**(4), (Oct./Nov. 1989).

J. Hanlon, *Handbook of Package Engineering.* McGraw-Hill, New York, 1971

J. Bergerhouse, "New Resin Development Improves LLDPE Blow Molding Properties," *Pack. Technol. Eng.* **2**(4) (Nov. 1993).

R. Walker, "Clarifying Advances in PP Lead to Third-Generation Product," *Pack. Technol. Eng.* **3**(4) (May 1994).

ANDREA S. MANDEL
Andrea S. Mandel Associates,
Packaging Consulting Services
West Windsor, New Jersey

BOXES, CORRUGATED

Corrugated containers (erroneously termed *cardboard boxes*) became the shipping package of preference in the early 1900s as a replacement for wooden crates. This strong, lightweight, durable, and economical product has expanded its use to a 1995 U.S. domestic market that will consume 350 billion ft^2 (31.5 billion m^2), equaling 3.5 million truckloads of corrugated board. There are 2129 box plants producing this volume, ranging from huge integrated companies to very small sheet plants.

The corrugated shipping container and related inner packaging is multifunctional. Its varied uses include wrapping, enclosing, protecting, cushioning, indexing, stacking, and displaying approximately 90% of goods shipped. U.S. citizenry consumes approximately one truckload of corrugated per 100 people.

Raw Materials

Kraft Linerboard derives its name from the Swedish word *kraft,* meaning strength. It is produced for its facial stiffness (stackability), and pucture resistance. These characteristics are achieved through use of soft-wood fibers that are long and resilient. Linerboard becomes the facings of corrugated board.

The basic raw material for liner is cellulose fiber extracted from wood chips. The chips are digested in a cooking liquor to remove lignin and resin, leaving the fiber that is fibrilated to enhance the tendancy for the fibers to mat together. The fibers are pulped with water and spread onto the screen of a fourdrinier paper machine. Water is drained, leaving the mass of fiber to form a sheet of paper that is dried and wound into rolls. Many fourdrinier machines are over 20 ft (6 m) wide. Linerboard is graded by *basic weight,* which will be explained in the section on liner grades.

The natural color of pulp is tan or kraft. A small percentage of the pulp is bleached white in a chlorine bath and then converted into fully bleached white paper. Bleached white pulp is also furnished as a topping for kraft liner to produce mottled white linerboard. Paper mills sell linerboard to corrugated converting plants in units of tons.

The medium is the paper rollstock that is converted into the fluted portion of corrugated board. The medium is also known as "nine point" because it calipers 0.009 in. It is also termed "semichem" because the pulp is cooked with chemicals and then mechanically ground into fibers. The pulp, like linerboard, is converted into paper on the fourdrinier machine;

however, pulp for the medium is selected from hard-wood trees that have short fibers that hold a set better when they are steamed and fluted on the corrugator. Technical advances have made possible the use of roots and recycled containers as percentage additives to medium stock.

The universal weight for the medium is 26 lb/1000 ft^2 (11.7 kg/90 m^2); however, minimum quantities of 33- and 40-lb medium are used to improve flat crush, stacking, and moisture resistance in finished board.

Adhesives. Contemporary corrugator adhesives are starch-based with additives and are selected for their flow, tack, absorbtion, evaporation, and set qualities. Specialty additives may be blended with the base adhesive to improve wet strength and moisture resistance. Modern corrugators use as little as 2 lb (0.9 kg) of adhesive per 1000 ft^2 (90 m^2) to combine single-wall board. Fingerless single-facers have eliminated the old problem of adhesive buildup lines (finger lines) in the finished product. The manufacture of corrugator adhesive is highly technical and is controlled through a series of extreme specifications.

Joint adhesives can be plastic based hot melts or starch-based cold melts. Both are used successfully to produce high-speed manufacturer's joint closure (see Adhesives).

Inks. Are a product of earth elements and chemical formulations. The base carrier for modern flexographic printing inks is water laced with additives designed to enhance drying speed, produce image clarity, inhibit smearing, and eliminate spotting or blotching. Inks must be carefully scrutinized for their pH factor because acidic inks differ in performance from base inks. For example, basic inks tend to smear. Ready-to-run inks are supplied to the corrugated printers in drums of 50-lb (22.5-kg) pails. Many press rooms are now developing in-house ink kitchens, which permits them the flexibility of mixing raw materials delivered from the ink manufacturers into custom batches in which the corrugated printers can control tones, viscosities, additive selections, and many other custom characteristics that will enhance the final printed image (see Inks).

The industry has maintained a minimum range of colors through a *Glass Container Manufacturers Index* (GCMI); however, modern ink kitchens using Pantone Matching System (PMS) Service have opened an endless range of available ink colors.

Printing plates. Modern printing plates are supplied to the corrugator mounted on poly backing, ready to be locked into the press. The industry has moved from a rather slow, methodical art form to a highly technical and efficient producer. The engraved rubber plate (die) has been replaced by a photopolymer plate that is rapidly produced through a series of computer imaging, and photoprocessing. The photopolymer plate is presently subject to ultraviolet (UV) deterioration; however, chemical suppliers of the base poly material will solve this problem soon.

Cutting dies. Corrugated products with scores or slots with angles other than 90° must be die-cut. Cutting dies are produced for the purpose by imposing an image onto a plywood sheet, jigsawing or laser-cutting the imaged lines, and filling the eradicated space with a cutting knife or scoring rule as desired. The dies are made on curved or flat plywood depending on the style of diecutter to be used. Stripping forms accompany cutting dies if they are to be used on automatic equipment.

Labels. High-graphics products incorporate the use of labels in conjunction with or in lieu of printing. Labels are supplied to the corrugated converters by label manufacturers. Typical specifications requested for labels are number of printed colors, paper basis weight, stock color, finish, and grade. Labels are adhered to the corrugated stock manually, semiautomatically, or fully automatically.

Additives. Finished corrugated blanks may be treated by many additives or coatings to improve water repellency, scuff resistance, and petrochemical resistance. Waxes and polymers are favored additives.

Board Construction

Corrugated board is a sandwich of one or more linerboard sheets adhered to a fluted medium. Single-face construction incorporates one linerboard adhered to the medium. Double-face, better known as *single-wall,* has a linerboard adhered to both sides of the medium. Additional media and linerboards yield double-wall and triple-wall. (See Fig. 1.)

Combinations of linerboard grades and flute configurations are used to generate the many variations of corrugated board. Liner weights are increased to improve board bursting and stacking strength, and flutes are modified to accommodate various compression, stack, and printing features. Two important requirements of corrugated board are flat crush and stacking strength. *Flat crush* measures the pounds per square inch (psi) resistance to pressure applied at a 90° angle to a horizontal sheet, thus establishing the rigidity of the flute structure. This property is changed by varying the flute outline and linear density. It can also be revised by changing the basis weight of the fluted medium. Flat crush supplies the internal resistance to squeezing forces such as feed rollers in presses and gluers. It also supplies resistance to gravitational forces imposed upon bottom sheets in stacks and bottom cartons in units.

Stacking strength measures the ability of a vertical panel to resist bowing, buckling, or collapsing from pressure exerted in line with that panel. This is regulated by varying the linerboard weights, changing flute heights, or both. Stacking strength is probably the most sought after characteristic in corrugated cartons.

Flutes. Flutes are most essential to the characteristic of corrugated board. They supply the rigidity to the board that imparts strength with minimal weight and density. Fluting makes the product economical. Flutes are designed by height (thickness) and density (number per linear foot). Higher flutes produces a physically stronger columnar stack in line with the flutes. Denser flutes, ie, flutes that present more images per linear foot, produce more resistance across the flutes. Three most used sizes are B, C, and E flute. A was the flute of choice into the 1940s, when C flute was selected as a compromise between A and B. Other, minimally used, flutes include J (jumbo), which is larger than the others, as the name suggests; S flute; and F flute, which is the lowest flute height.

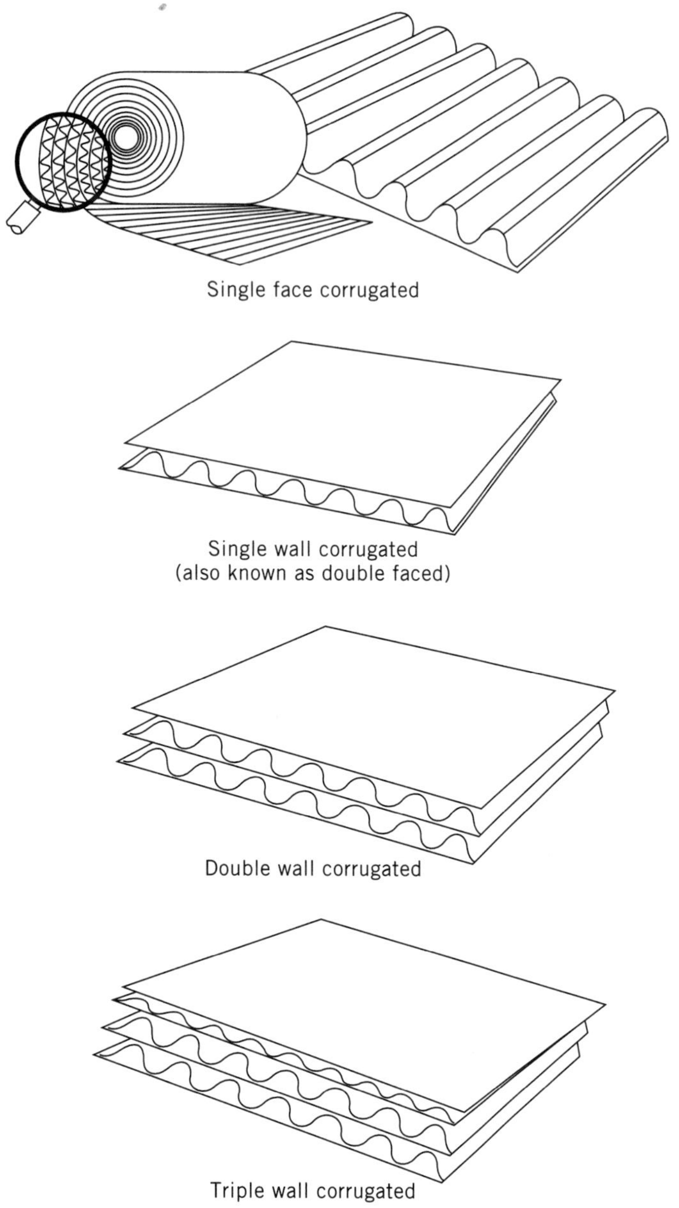

Figure 1. Various examples of corrugation.

Table 1. Standard U.S. Corrugated Flutes[a]

Flute Designation	Flutes per Linear Foot	Flutes per Linear Meter	Flute Thickness, in. (mm)	Flutes per Cross Section
A flute	33 ± 3	108 ± 10	3/16 (4.8)	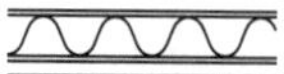
B flute	47 ± 3	154 ± 10	1/8 (3.2)	
C flute	39 ± 3	128 ± 10	5/32 (4.0)	
E flute	90 ± 4	295 ± 13	1/16 (1.6)	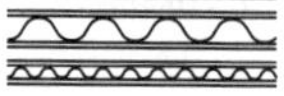

[a] Corrugator equipment manufacturer's single-face flute roll dimension may vary slightly to accommodate user preference.

Higher (less dense) flutes in addition to stacking strength provide softer cushion characteristics, while lower (more dense) flutes provide greater flat crush resistance, smoother print surfaces, and crisper score lines. Flutes are often used in combinations such as BC double wall, which provides an overlap of required characteristics. B and C flutes are the most commonly produced flutes, and E is considered sparingly. Flute selection is determined by shipper needs. A fragile-decorations shipper would choose C flute for its stacking strength and cushioning qualities. A canned beverage filler would choose B flute for its end-to-end compression attributes; and a point-of-purchase display designer would select E flute for its superior printing surface. (See Table 1.)

Mullen vs Edge Crush

Carrier regulating agencies require a board upgrade to accommodate increased loads and increased box sizes analogous to the idea that a ¾-in. (19-mm) plywood sheet is stronger than a ¼-in. (6.35-mm) sheet. Paper mills fulfill this requirements by producing various-weight liner grades, corrugated converters combine board using proper heavier grades to accommodate the need for stronger containers.

Liner grades. The universally accepted test for corrugated board has been the *Mullen test* (TAPPI test method T-810), which measures the resistance of the board to withstand a puncturing pressure measured in pounds per square inch (psi) or metric kilopascals (kPa). Linerboard grades have been manufactured to meet specific Mullen requirements. Increased mullen demands increased linerboard strength, which is accomplished by increasing the mass of the linerboard measured in pounds per thousand square feet (lb/1000 ft^2) (kg/90 m^2).

A major modification to containerboard was initiated in the 1980s, when paper mills developed a process that enhanced linerboard material, giving it an ability to withstand crush or compression examinations at lower basis weights. The following list compares STFI crush performance of standard to high-performance liners. Note the improved STFI in high-performance liner.

Liner Basis Weight		STFI	
lb/1000 ft^2	(kg/90 m^2)	lb/in. width	(kg/cm width)
26	11.8	12	13.9
33	14.9	15	17.4
38	17.2	17	19.7
42	19.0	19	22.0
69	31.2	29	33.6
90	40.7	38	44.0
	Medium		
26	11.8	11	12.8
33	14.9	15	17.4
40	18.0	19	22.0
	High-Performance Liner		
35	15.8	20.3	23.6
45	20.3	27.5	31.9
57	25.8	33.7	39.1
72	32.6	45.0	52.2

The product *high-performance linerboard* provides a serious potential for user satisfaction and improved economics. New linerboard concepts prompted the corrugated industry trade associations to sponsor proposals to truck- and rail-carrier committees to consider classifications recognizing edge crush as a test criteria. Proponents for the new test methods argued that edge crush is more customer-oriented and that it permits more latitude for manufacturers to design and supply boxes that meet customer performance criteria. Rules committees of truck and rail carriers accepted the proposals and in 1991 approved the edge-crush test (ECT) as an alternative test method for containerboard.

Edge-crush test. This measures the resistance of the *containerboard* to edgewise compression. Ring-crush and STFI tests measure the resistance of the *linerboard* and medium. These tests are performed to give the linerboard and corrugated manufacturers relative material comparisons. The edge-crush test includes the value of the medium's resistance to compression as well as the linerboard. It should be noted that these tests can, and are, performed on standard weight liners as well as high-performance liners (see Edge-crush concept).

Medium grades. General industrial principle requires use of the same medium weight (26 lb/1000 ft^2) (11.7 kg/90 m^2) for all board grades; ie, liner weights change to generate increasing board tests, but the medium remains the same. There are some approved exceptions that call for upgrading medium to improve flat crush and moisture resistance. Two upgraded medium weights are 33 and 40 lb.

Regulations

Freight carriers and government agencies control the regulations for the container industry. The major regulation is encompassed in Rule 41 (*Uniform Freight Classification*) established by the National Railroad Freight Committee. This regulation is closely paralled by Item 222 (National Motor Freight Carriers). Table 2 outlines minimum standards. Triple-wall and solid-fiber standards have not been included in this chart.

Rules prescribed by these agencies outline requirements that determine proper liner weights and manufacturing specifications. All corrugated boxes made to conform to the regulations carry a printed certificate identifying the boxmaker and appropriate board grade as shown in Figure 2. Note the apparent difference between the burst-test and the edge-crush certificates.

Several other regulatory agencies are included depending on shipper requirements, including

Department of Transportation—Code of Federal Regulations No. 49 (Transportation)-Pointing specifically at Hazardous materials.

International Air Transport Authority (IATA)—concerned with air transport

International Maritime Authority—water cargo

United Parcel Service (UPS)—ground deliveries of <150 lb

U.S. Postal Service—Domestic Mail Manual

Converting Operations

Integrated operations are large, publicly owned, multiplant corporations. These companies own forests, farm the forests, convert trees into paper at their mills, and convert the paper into corrugated board for ultimate delivery to the consumer. *Independent manufacturers* purchase paper roll stock and begin their operations at the corrugator. Sheet plants purchase flat sheets (containerboard) from corrugator operators and convert the board into finished boxes and interior parts. *Sheet feeders* purchase roll stock and produce flat stock for delivery to sheet plants.

Contemporary box plants vary widely in size and capability. A small sheet plant may ship one truck load of cartons per day while the megacontainer plant will ship 50 truckloads per day. Corrugated requirements are so diverse that a wide variety of company sizes is a desirable asset to the industry.

Equipment

Corrugator. This is the major piece of equipment in a box plant. The machine converts mill supplied roll stock (linerboard and medium) into flat sheets. It varies from 50 to 100 in. plus (12.7–25.4 m) in width, and most are over 300 ft (91.5 m) long. Wide computer-managed corrugators operating at speeds of up to 1000 lineal feet per minute are capable of producing four truckloads of corrugated board per hour.

The single-facer (wet end) of the corrugator converts the medium into fluted paper and adheres it to the inside liner of the corrugated sandwich. Next, a double backer roll is applied as the outside liner, and finally the sandwich is run over dryer plates and delivered to the dry end. Cross corrugation scores may be applied and the board is trimmed into two-dimensional blanks in preparation for delivery to the plant for further conversion into a box.

Scheduling (trimming) a corrugator requires skill and training because variations of liner combinations, quantity requests, and blank dimensions are endless. Modern computer-trimmed corrugators are capable of producing several hundred setups per day.

Printer-slotters. These machines print, slot, and score flat banks in preparation for folding and joining into finished boxes. Sized blanks are delivered from the corrugator prescored for flap and depth dimension. The operator, generally one of a two- or three-worker crew, sets scoring and slotting heads at proper dimensions to cut required length, width, and joint dimensions into the carton. The operator then hangs the premounted printing plates on the print cylinders, fills the ink wells with desired colors, and begins production.

These machines are referred to by the number of colors they are capable of printing in one pass plus cylinder diameter and machine width. Small-dimension machines are obviously better suited for producing small boxes, etc. A majority of presses today are two-color; however, presses in up to six colors are available.

Corrugated printer-slotters are letterpress printers; however, most modern equipment is a flexographic variation that incorporates the use of an analox roller with water-base inks vs the old doctor and impression rollers with oil-based ink. Flexographic printing offers many advantages over letterpress, including operating speed, clarity, registration, trapping, drying, and cleanup. Printer-slotters employing flexographic printing can produce an average size-box (eg, a 24/12-oz beverage box) at approximately 15,000 per hour.

Flexofolder-Gluer. Most printer-slotter equipment has been

Table 2. Minimum Standards for Construction of Corrugated Boxes[a]

Maximum Weight of Box and Contents, lb (kg)	Maximum Outside Dimensions, Length, Width, and Depth Added Inches (cm)	Table A: Minimum Bursting Test, Single-wall, Double-wall, psi (See Note 1) (MPa)	Table A: Minimum Combined Weight of Facings, Including Center Facing(s) of Double-wall, lb/1000 F^{t2} (gm/m^2)	Table B: Minimum ECT, lb/in., width (See Note 2) (kg/cm width)
		Single-wall Corrugated Fibreboard Boxes		
20 (9)	40 (102)	125 (0.862)	52 (254)	23 (26.3)
35 (16)	50 (127)	150 (1.034)	66 (322)	26 (29.7)
50 (22)	60 (152)	175 (1.206)	75 (366)	29 (33.1)
65 (29)	75 (190)	200 (1.379)	84 (410)	32 (36.5)
80 (36)	85 (216)	250 (1.723)	111 (542)	40 (45.7)
95 (43)	95 (241)	275 (1.896)	138 (674)	44 (50.3)
120 (54)	105 (267)	350 (2.412)	180 (879)	55 (62.8)
		Double-wall Corrugated Fibreboard Boxes		
80 (36)	85 (216)	200 (1.379)	92 (449)	42 (48)
100 (45)	95 (241)	275 (1.896)	110 (537)	48 (54.9)
120 (54)	105 (267)	350 (2.412)	126 (615)	51 (58.2)
140 (63)	110 (279)	400 (7.757)	180 (878)	61 (69.6)
160 (72)	115 (292)	500 (3.445)	222 (1084)	71 (81.1)
180 (81)	120 (305)	600 (4.135)	270 (1318)	82 (93.6)

[a] *Note 1*—Burst test: (*a*) Tests to determine compliance with the bursting test requirements of Table A must be conducted in accordance with Technical Association of Pulp and Paper Industry (TAPPI), Official Test Method T-810; (*b*) a minimum of six bursts must be made three from each side of the board, and only one burst test will be permitted to fall below the specified minimum value. Board failing to pass the foregoing test will be accepted if in a retest consisting of 24 bursts, 12 from each side of the board, not more than four burst tests fall below the specified minimum value. *Note 2*—edge-crush test: (*a*) tests to determine compliance with the edge crush requirements of Table B must be conducted in accordance with Technical Association of Pulp and Paper Industry (TAPPI), Official Test Method T-811; (*b*) a minimum of six tests must be made and only one test is permitted to fall below the specified minimum value, and that one test cannot fall below the specified minimum value by more than 10%. Board failing to pass the foregoing will be accepted if in a retest consisting of 24 tests, not more than four tests fall below the specified minimum value, and none of those tests fall below the specified minimum value by more than 10%.

updated to flexofolder-gluers. This equipment incorporates the addition of an automatic folder-gluer system with the printing unit, which permits the completion of most cartons on one machine. The equipment has been available since the 1960s, when flexographic printing with high-speed drying made it possible for cartons to be folded immediately after printing without smearing. These machines may also be equipped with automatic bundeling and unitizing systems.

Folder-gluers. These are designed to finish boxes in a straight line or right angle. Both are extremely efficient at folding panels and applying adhesive to produce finished cartons that require minimum or no final sealing. Predecorated, slotted, and or die-cut blanks are belt fed through the folding sections; glue is applied at required spots; and finished cartons are stacked at the finish end. Most equipment is designed to accept a wide variety of sizes both in the machine direction and across the machine.

Die cutters. Die cutters are required for all parts that are scored or slotted other than 90° or in line with press direction. Intricate die cut interior parts and box designs are a very common part of the industry.

Flatbed die cutters are designed to produce accurate products. They vary from small hand-fed machines that are limited in speed to the operators performance (about 500 blanks per hour) to high-performance automatic equipment that include multicolor printing selections. Flatbed cutting dies are reasonably priced. The process for flatbed die cutting closely resembles cookie cutting. The die board is locked into a chase that holds it in a firm position. Corrugated board is registered under the die, and the machine is closed striking the die impression into the board.

Rotary die cutters are best used to produce parts at high speeds. They are not as accurate as flatbed machines; however, they can be designed to accept large blanks. Rotary cutting rule is inserted into curved plywood and then locked onto the die-cutting cylinder. An impression is made into the board when a blank is passed between the cut die cylinder and an opposing thick polyblanketed impression cylinder.

Slitters. Slitters have a series of wide rotating shafts to which scoring and slitting collars (heads) are attached. These heads can be moved by the operator to change dimensions. Most slitters are manually fed; however, some are equipped with automatic feed sections. Slitters are used to reduce large sheets to smaller blank dimensions, add scores to existing blanks, and prepare small runs for further processing. Slitters are very versatile machines. Small sheet plants may be totally equipped with a slitter a printer–slotter and a taper.

Partition slotters. These machines slot corrugated pads in preparation for assembly into partitions.

Partition assemblers. These machines automatically assemble slotted racks into cell divided partitions for use in separating delicate parts such as glass bottles.

Tapers. These machines apply tape fed from rolls onto pre folded cartons to form a manufacturer's joint. Standard tape widths are 2 in. (5.08 cm) or 3 in. (7.62 cm). Tapers are hand-fed semiautomatic or hopper-fed automatic. The machine is

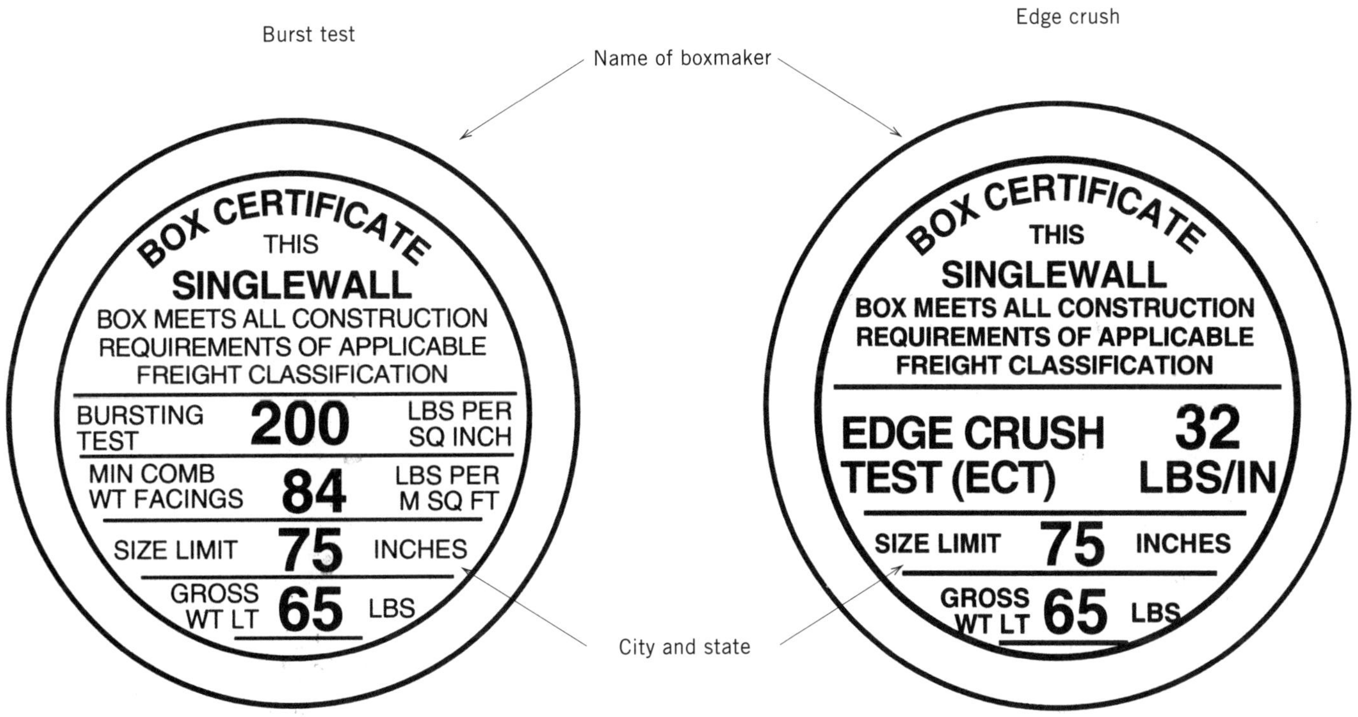

Figure 2. Boxmaker certificate.

an economical piece of equipment that can be adjusted and set very quickly. Taped joints represent a minimal amount of todays production because of their relatively slow production speeds and higher costs compared with glued joints. Glued joints also perform better in regulatory transit tests.

Stitchers. These devices apply staples cut from a roll and are driven into the joints of prefolded boxes. Stitched joints have limited use in today's market; they are used mostly with government-grade boards and specialty wet-strength containers.

Coaters. These machines are operated at box plants or, in many cases, at specialty plants that process coatings on finished containers. These specialty plants function as a separate entity to the corrugated industry, ie, they seek customers for their process, mostly from the fish, poulty, and produce industries; purchase finished containers from corrugated converters; and add the required coatings (waxes and plastics) to the containers. The industry specializes in wet-strength and moisture-barrier containers.

Curtain coaters feed box blanks under a curtain of liquid coating (molten poly) that coats one side of the blank and passes it to a drying section. Some coaters are capable of flipping the blank to deposit a coating on both sides. *Cascaders* pour wax coatings onto finished containers. *Wax dippers* immerse finished containers into tanks of molten wax and then hang them to dry. The process looks very much like your local laundry.

Laminators. *Litholaminators* have become an important adjunct to the high-graphics corrugated producer. This equipment rolls adhesive onto a printed litholabel, registers the label under the substrate, drops the substrate onto the label and rolls the combined stock to achieve 100% adhesion and eliminate air bubbles. Litholaminators are designed to feed flat banks, including joined (knocked-down) boxes. *Stock-laminators* are designed to laminate varied substrates together to achieve added thickness or uncommon stock variations. They simply pass a flat blank over a surface roll that is revolving in a pan filled with adhesive, register the now adhesive coated substrate with a second substrate, and pass the combined sheet through compression rollers. The finished product is usually stacked immediately, flipping every other handful to reduce warp.

Miscellaneous equipment. A well-equipped corrugated plant will require additional materials handling and finishing equipment, including lift trucks, conveyors, bundelers, unitizers, bailers, load turners, eccentric slotters, quick sets, strippers, and band saws, plus many custom machines such as riviters.

Manufacturer's Joints

Joints applied by the corrugated manufacturer provide the most practical way to convert a two-dimensional product into a third-dimensional one. *Manufacturer's joints,* as the name implies, are created by the box manufacturer. Box users torque containers open, fill them, and seal them closed by gluing, stapeling, or taping. Edges that meet during closure are called *seams.*

Joints are glued, taped, or stitched. Glued and stitched containers employ a 1¼-in. lap that is a contiguous part of a

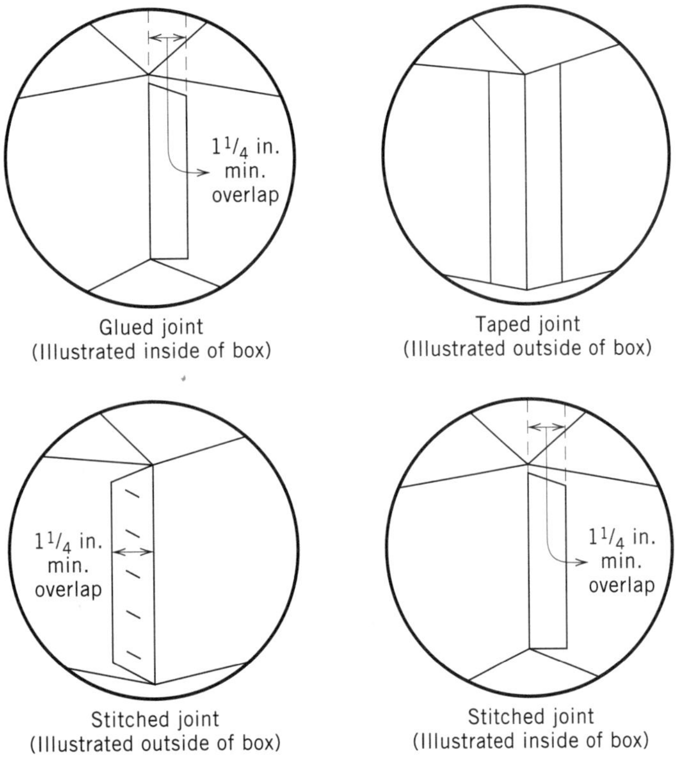

Figure 3. Examples of joints.

length or width panel that will be glued or stitched to the opposing panel when the panels are folded to meet. Glued joints became the joint of preference in the 1960s. Almost all high-speed flexo presses and folders are equipped to run glued joints. Some volume box users are equipped with wraparound machines that fold flat blanks around a product and seal joints and seams in place. This process eliminates the need for a manufacturer's joint from the box manufacturer, which in some cases saves cost and in all cases produces a more tightly wrapped package.

Glued and stitched joints are produced with the lap on either the inside or the outside of the container. Both have their advantages. Inside laps present a finished outside edge that presents a better appearance and allows more print area. Outside laps leave a smooth surface on all four inside panels of a carton. (See Fig. 3.)

Dimensioning

Three-dimensional cartons are designated by length × width × depth. Length and width are the longer and shorter dimensions of the opening of a box, and depth is the third dimension. Two-dimensional parts are designated by supplying the flute direction as the first dimension.

Corrugated board has a definite caliper, which was noted in Figure 1; consequently, scoring allowances must be considered when dimensioning boxes and interiors. One-half thickness of board is lost for each 90° bend. See an example in Figure 4.

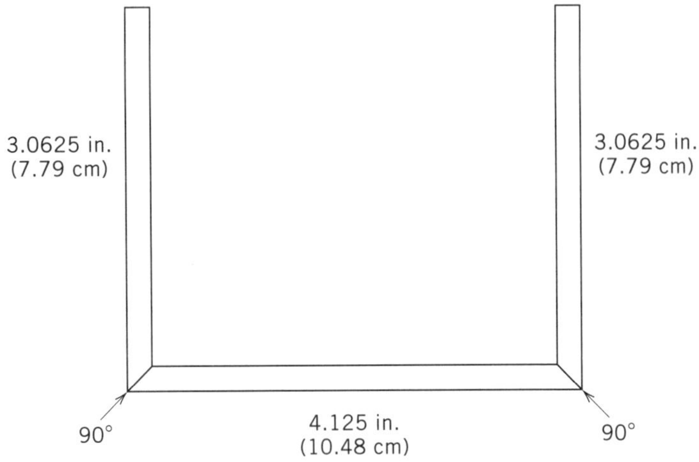

Figure 4. Score and fluting.

Score B flute ⅛ in. (3.175 mm) thick at 3 1/16 in. × 4⅛ in. (7.79 cm × 10.48 cm × 7.79 cm) to achieve inside dimensions of 3 in. × 4 in. × 3 in. (7.62 cm × 10.16 cm × 7.62 cm).

Corrugated Economics

Carton economics demands that solid geometric figures (pertaining to three dimensions, $L \times W \times D$) must be reduced to plain geometric figures (pertaining to two dimensions, $L \times W$) for manufacturing specifications and cost analysis (see Fig. 5).

Having established the single-plane dimensions of a container ($L \times W$), the area per piece can be established. This

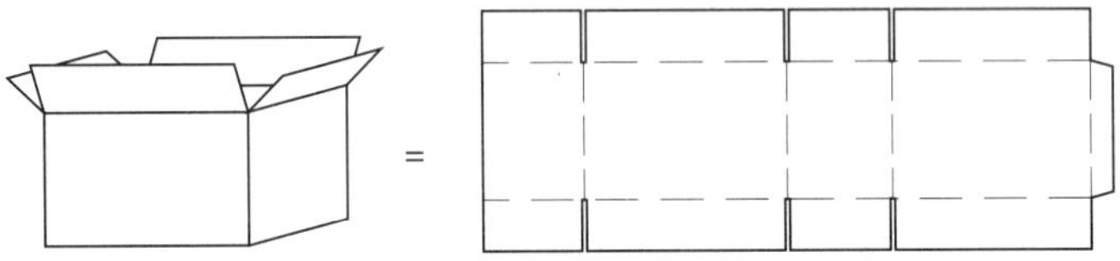

Figure 5. Example of reduction to plain geometric figure.

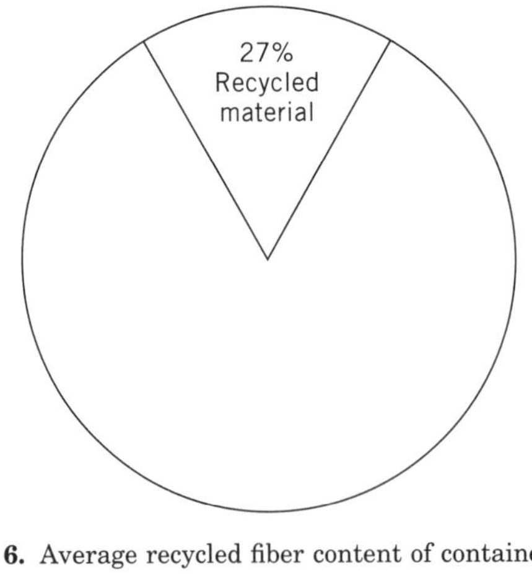

Figure 6. Average recycled fiber content of container board.

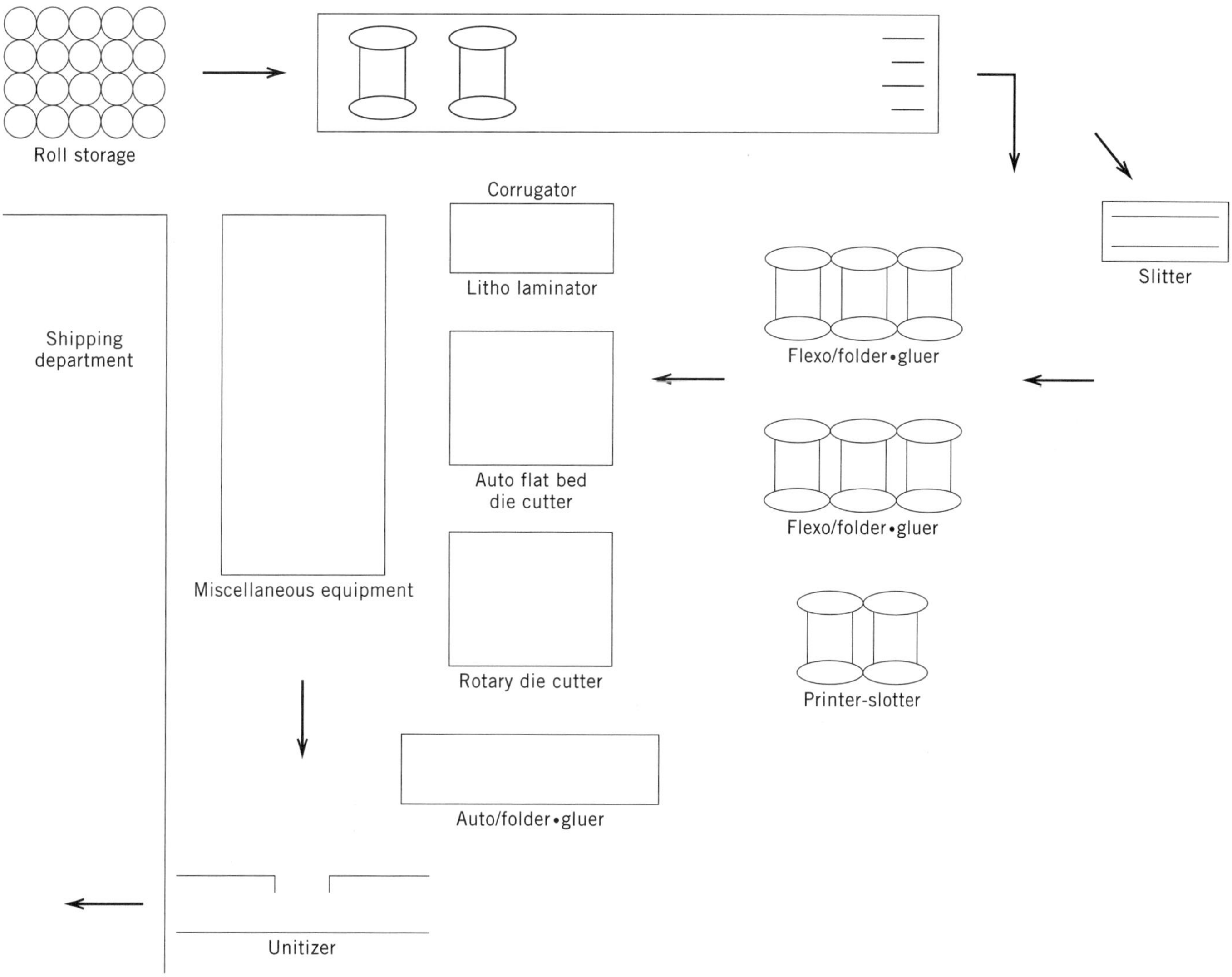

Figure 7. Typical plant floorplan.

area is calculated in square inches per piece, which is generally referred to in square feet per 1000 blanks, ie, L in. × W in. divided by 144 × 1000 = square feet (F^{t2})/1000 pieces.

Corrugated box costs are functions of *raw material,* plus manufacturing, administrating, and delivering costs; consequently, a major objective in box design is to maximize fill area and minimize the blank dimension required to do so. Several principles to consider that will accomplish this maxim follow:

1. The most economical RSC (regular slotted container) formula is $L = 2W = D$.
2. Rectangular RSCs are more economical than square RSCs.
3. Square tubes are more economical then rectangular tubes.
4. Width is the least economical dimension of an RSC.
5. Depth is the most economical dimension of an RSC.

Box Styles

Box styles and interiors are too numerous to discuss in this article, and can best be referred to in the *Fibre Box Handbook* (1) listed in the Bibliography. Most designs are designated by their closure feature. The most used box is the *regular slotted container* (RSC), which features the most economical use of board from the corrugator. All flaps are equal in depth, and the width of the outer flaps equals one-half that of the containers, so they meet to form a seam at the center of the box when folded.

Carton designers using CAD/CAM computer programs generate intricate new designs that enhance packing programs and support marketing schemes. Today's world of discount stores requires containers that function as sales tools that fully describe the product enclosed as well as protecting the product for delivery to the consumer.

Recycling

Used corrugated containers are very adaptable to recycling; in fact, they represent the major source of recycled paper today. Over 50% of recycled containers are reclaimed for paperboard used to make more corrugated. Recycled corrugated is also used in the manufacture of boxboard and other paper products. Twenty-six percent of all corrugated is made from recycled fibers. Recycling along with new high-performance linerboard, which requires less basis weight to accomplish comparable packaging functions, will work together to reduce the drain on our forest products. (See Fig. 6.) Figure 7 illustrates a rough floorplan of a typical box plant.

Industry Future

The corrugated industry is mature, meaning that it has developed a full share of the possible market and that most volume advances come through volume increase in existing markets. Newly opened world markets combined with the continuous increase in population will escalate growth in the industry. Higher profit products are becoming more available, including exotic flutes (E & F), and process printing, and labeling will upgrade considered use.

Undoubtedly the most beneficial adjustment the consumer will receive is the move to high-performance linerboard and the corresponding change from burst test to edge-crush test. This very significant change will ultimately result in not only lower direct costs but also lower freight and handling costs. The combination of stronger paperboard at lower basis weights will also result in a reduction of waste tonnage and pulp consumption. High-performance linerboard will improve machine efficiencies and printing capabilities. Given time, this new product will promote a major industry improvement.

High-graphics containers are claiming a larger share of the market. Packagers are accepting the corrugated container as a sales tool in addition to its use as a shipping container. Exotic labeling equipment and preprinted liner laminates are becoming more common to the industry, and the trend continues toward faster in-line converting machines that incorporate printing, slotting, die cutting, folding, and gluing into one piece of equipment.

The rapid change to today's knowledgeable society has occasioned many industrial collapses; however, the outlook for the corrugated industry appears to be one projecting a steady and continuous climb.

BIBLIOGRAPHY

1. *Fibre Box Handbook,* Fibre Box Association, 2850 Golf Road, Rolling Meadows, IL.
2. *Corrugated Board Edge Crush Test Application & Reference Guide,* Fibre Box Association, 2850 Golf Road, Rolling Meadows, IL.
3. *Uniform Freight Classification 6000-F,* National Railroad Freight Committee, Chicago.
4. *National Motor Freight Classification,* American Trucking Association, Washington, DC.
5. *Code of Federal Regulations,* Dept. Transportation, Washington, DC.

General References

Official Board Markets (weekly publication), Advanstar Communications Inc., 7500 Old Oak Blvd., Cleveland, OH.

Paperboard Packaging (monthly publication), Advanstar Communications Inc., 7500 Old Oak Blvd., Cleveland, OH.

The Packaging Encyclopedia, Cahners Publishing, Des Plains, IL.

Performance and Evaluation of Shipping Containers, Jelmar Publishing, Plainview, New York.

Packaging Data Sources, National Railroad Freight Committee, Burlingame, CA.

Gene A. Foster
Middlesex Container Company, Inc.
Milltown, New Jersey

BOXES, RIGID, PAPERBOARD

Rigid-paperboard boxes are also called "setup" boxes. Unlike folding cartons (see Cartons, folding) they are delivered to the packager setup and ready to use. The rigid box was originally used by the Chinese, who were among the earliest to discover a process for making strong and flexible paper from rice fiber. The first known use for a paper box was tea. The word *box* generally means a receptacle with stiff sides as distinguished from a basket, and so-called because it was first made from a tree called *Box* or *Boxwood.*

Boxes for gift-giving became popular over 2000 years ago when the Roman priests encouraged people to send presents during the seasons of rejoicing. The paperboard box of current use originated in the 16th century, with the invention of pasteboard. In Europe, one of the earliest types of paper boxes was commonly known as a *band box.* It was a box highly decorated by hand and was used to carry bands and ruffles worn by the Cavaliers and Ladies of the Court. It was not until 1844 that setup boxes were manufactured in the United States. Starting with a machine that cut the corners of the box, Colonel Andrew Dennison soon found that manufacturing boxes by hand was tedious work and developed the Dennison Machine which led to the creation of the Dennison Manufacturing Company. The Colonel's invention was revolutionary, but until the Civil War, most consumer products were packaged in paper bags or wrapped in paper. There were only about 40 boxmakers in the country and most boxes were made by hand. For these 40 craftsmen the box business was merely an adjunct to other lines of business, which varied from printing to the manufacture of the consumer items they would eventually pack.

In 1875, John T. Robison, who had worked with Colonel Dennison and others, developed the first modern scoring machine, corner cutter, and shears. These three machines still form the machinery basis for most box shops, but it was not until the end of World War II that significant progress was made to improve the production of machinery for the industry. Today's machinery takes a scored piece of blank boxboard through to the finished covered box.

Manufacturing Process

The process starts when sheets of paperboard are sent through a machine known as a *scorer.* The scorer has circular knives that either cut through or partially cut through the paperboard and form the box blanks from the full sheet. The scorers must be set twice, once for the box and once for the lid, since there is usually a variation of ⅛–⅜ in. (3.2–9.5 mm) between the box and the lid. After the individual box blanks are broken from the full sheet they are stacked and prepared for corner cutting. Once the corners are cut the basic box blank is ready. Diecutting, usually performed on a platen press and an alternative method of cutting blanks, is economically justifiable for large orders.

The blanks are now ready for staying. For small quantities the box blanks can be sent through a single stayer where an operator must first bend all four sides of the blank to prebreak the scores. The single-staying machine will then glue a strip of ⅞-in. (22-mm)-width kraft paper of the required length to each corner of the box. For greater quantities the boxmaker uses a quad stayer. The quad stayer feeds the box blanks automatically under a plunger that has the same block size as the box or cover to be formed. With each stroke of the machine, the box blank sides are turned up and stay paper is applied to all four corners at one time.

After staying, the box is in acceptable form, requiring only an attractive outer wrap. In most operations, the paper box wrap is placed onto the conveyor gluer, which applies hot glue to the back of the wrap and then places it on a traveling belt. The wrap is held in place by suction under the belt. As the wrap travels on the belt, it is either removed from the belt by a machine operator, who manually spots the paperboard box on the wrap, or it is automatically spotted by machine. After spotting, the box and wrap move on to another plunger mechanism where the wrap is forced around the box. Simultaneously, nylon brushes smooth the paper to the four sides of the box. Just before the plunger reaches the bottom of the stroke, the wooden block splits allowing metal "fingers" to push the paper in the box. The wooden block then closes together and completes its downstroke where felt-lined blocks press the sides and assure the gluing of the paper to the inside as well as the outside of the box.

The manufacturing process described above is for the simplest box, but the setup box can accommodate unusual requirements with regard to windows, domes, embossing, platforms, hinges, lids, compartments, and other variations. Standard variations of the rigid box include the telescope box, the ended box, the padded-cover box, special shapes (eg, oval, heart-shaped), slide tray, neck or shoulder style, hinged cover, slanted side, full telescope, box-in-box, specialty box, interior partition, extension bottom, three-piece, slotted partition, and interior platform. This versatility is extremely valuable in meeting the merchandiser's demands for quality, quantity and convenience. Figures 1 and 2 show a familiar candy box and an unusual configuration for cosmetics.

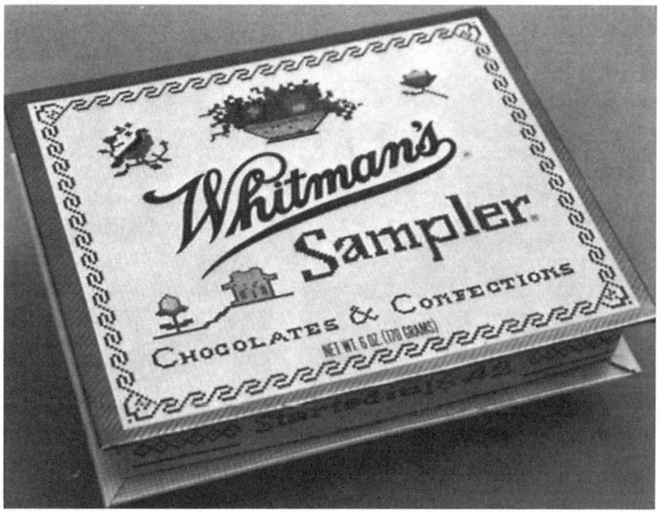

Figure 1. A familiar example of a rigid paperboard box.

Materials

Four primary materials are needed for manufacturing the rigid box: chipboard (for the rough box), stay paper (to hold the sides of the box together), glue (to hold the outer wrap to the box), and outer wraps (for the decorative appearance).

There are four common types of chipboard used in boxmaking: plain, vat-lined book-lined, and "solid news." Plain chipboard, is made entirely of waste paper, with the minimum of selection or de-inking of waste. Vat-lined chipboard provides a cleaner appearance. It is primarily chipboard with a liner (made of low-grade white waste paper) that is applied on the board machine at the mill. Book-line chipboard is chipboard with a liner of book or litho paper pasted to one or two sides as a separate operation. "Solid news" differs from plain chipboard in that the waste used in this sheet is a little more selected and is made mostly from newspaper waste which has been de-inked. Other boards are available, including glassine-lined, foil-lined, and folding grades of boxboard, but the four listed above are those that are primarily used.

To properly serve the customer, a boxmaker must stock various sizes and weights of board. Setup boxboard is measured in basis weight: a 50-lb (22.7-kg) bundle of 50-basis-weight boxboard contains 50 standard 26 × 40-in. (66–102-cm) sheets; a 50 lb (22.7 kg) bundle of 40 basis weight boxboard contains 40 sheets. The larger the basis weight, the thinner the sheet.

Stay paper is almost always ⅞-in. (22-cm) wide kraft or white paper. Glues are either animal or starch-based, formulated to dry fast or slow. They are available in dry form for mix-it-yourself, or flexible form to melt as used. Most glue is hot-melted to give a faster drying and lay-flat quality to the paper with minimal warping (see Adhesives).

The final ingredient in box manufacturing is the outer

Figure 2. The high-fashion look of a lacquered wooden box is copied with style in the glossy finish and fine detail work of this rigid box designed for Lancôme cosmetics. Especially noteworthy is the registration of the gold-stamped border on the wrap of each separate drawer. Brass drawer-pulls complete the illusion. Shelves are sturdy and wrapped, allowing the drawer to slide in and out smoothly.

wrap. There are thousands of stock papers, most of them available in 26-in. (66-cm) or 36-in. (91.4-cm) rolls. The wraps are generally paper, but foil and cloth wraps are also used. Through the use of artwork, photography, and good printing, a boxmaker can also produce a distinctive custom-made wrap.

Applications

The rigid (setup) box has stood the test of time and competes well within its selected markets. It protects, it builds image, it displays, and it sells. The setup box possesses unique qualities that satisfy the specific needs of all four segments of the marketing chain: the consumer, the retailer, the marketer, and the product packager. The boxes are delivered setup and ready to load. Small "market-test" or emergency quantities can be prepared quickly at low cost, and individual custom designs are available without expensive investment in special tools, jigs or dies. The rigid "feel" creates consumer confidence and the manufacturing process permits utilization of varied overwraps to reflect product quality. The boxes are easy to open, reclose, and reuse without destroying the package. Rigid boxes are recyclable and made from recycled fibers. Production runs can be small, medium, or large, and volume can fluctuate without excessive economic penalities. In addition, these boxes provide lasting reuse features for repeat advertising for the seller. They are customized to provide product identity. Rugged strength protects the product from plant to consumer. Reinforced corners and dual sides provide superior product protection and minimize damage losses.

The markets for rigid boxes are as follows: textiles apparel and hosiery; department stores and specialty shops; cosmetics and soaps; confections; stationery and office supplies; jewelry and silverware; photographic products and supplies; shoes and leather; drugs, chemicals, and pharmaceuticals; toys and games; hardware and household supplies; food and beverages; sporting goods; computer software; educational material; and electronic supplies.

BIBLIOGRAPHY

"Boxes Rigid-Paperboard" in *The Wiley Encyclopedia of Packaging Technology,* 1st ed., Wiley, New York, by Larry Lynch and Julia Anderson, National Paperbox and Packaging Association, 1986, pp. 76–78.

BOXES, RIGID, PLASTIC

Rigid plastic containers have been used for many years in a wide variety of consumer and industrial applications. Some of the more familiar uses have been as the bread trays, milk crates, and beverage cases one might see in a supermarket; as cosmetic and shampoo bottles; or as medicine bottles or blister packs used to display consumer goods on a store rack. Some of these and less-familiar uses are touched on in this section, and the processes used to produce this wide range of containers are covered as well. Finally, container configurations and applications in the emerging category of reusable shipping containers are described in some depth.

There are five major *processes* used to manufacture rigid plastic containers: injection molding, compression molding, blow molding, rotational molding, and thermoforming. Each process is suited to the production of a range of geometries with a variety of materials at different costs. A complete description of these processes, materials, and applications is beyond the scope of this article, but an overview is possible and will be helpful for later discussions.

Types of Molding

Injection molding. In this process a heated, softened plastic material is forced from a cylinder into a cool mold cavity, which results in a plastic product matching the geometry of the cavity of the mold. Generally molds are expensive, but they produce the widest variety of part shapes at very high rates of production (see Injection Molding).

Compression molding. This process uses preheated thermosets and high mechanical pressure to shape the material between male and female portions of a die. The mold cavities are preheated themselves, and the parts are removed after they have been cured under pressure. Many fiberglass and phenolic parts are made with this method (see Compression Molding).

Blow molding. This is most often used to shape hollow items like bottles. A tube of hot melt, called a *parison,* is extruded between the open halves of a chilled mold. The mold halves are clamped together, pinching off the tube at the ends. Air or an air and water mist is injected into the cavity, inflating the soft tube against the mold surfaces. Cooling time is quick, and production rates are extremely high (see Blow Molding).

Rotational molding. This is also a process for producing hollow, seamless products. In this method, a powder or liquid plastic is placed in a mold; the mold is heated and rotated about two perpendicular axes simultaneously and then cooled. After the material solidifies, the mold is opened and the part is removed. Tooling and equipment costs are low, but production rates are slow, and the geometry of the parts produced is limited.

Thermoforming. This involves elevating the temperature of a thermoplastic sheet material to a workable level and forming it to shape. The forming process draws the sheet by a vacuum into an open, chilled mold. Like rotational molding, the tool costs are low, but production rates are slow, and there are limitations to the complexity of the part geometry.

All of these processes can be used to manufacture reusable shipping containers, but by far the most often used method is injection molding. This process is best suited for reusable shipping containers because it allows intricate shapes to be molded at high rates of production. The molds last a long time and relatively low unit costs can be achieved.

The reusable shipping container (RSC) concept is simple: a container is loaded with product and shipped to its destination, where it is emptied. The empty container is sent back to the supplier, refilled with product, then shipped again to its end-user destination. The cycle is then repeated over and over again.

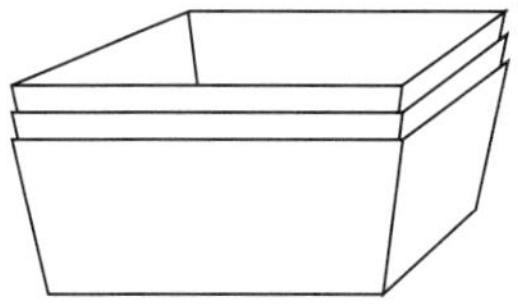

Figure 1. Nest only.

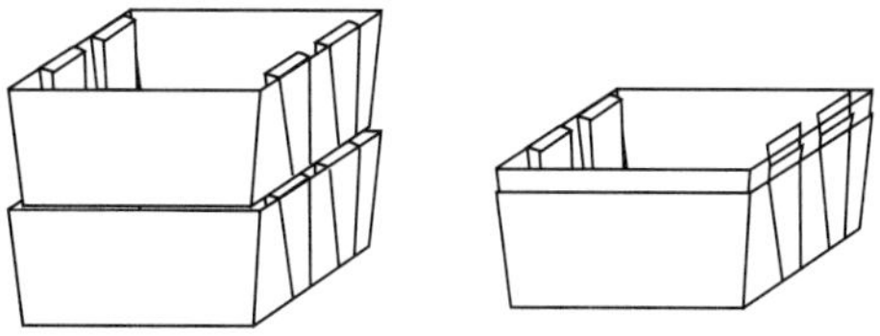

Figure 3. 180° stack and nest.

While the concept is not new, its application in a wide variety of industries has spurred increased usage. Replacing expendable corrugated packaging with RSCs saves money, reduces worker injuries, and helps the environment. Money is saved by purchasing plastic containers once and reusing them instead of buying corrugated cartons each time a shipment is made. Worker safety and health are improved by providing easy-to-lift boxes with ergonomic handles. And the environment is helped by using recyclable plastic in the containers and eliminating the dumping of corrugated cartons in the landfills.

Plastic reusable shipping containers have been used for years by wholesale bakeries shipping bread, dairies shipping milk, and by soft-drink bottlers shipping 2-L bottles into supermarkets. More recently, drug chains, hardware stores, chain restaurants, and durable goods manufacturers have realized the benefits of reusables. Much of the recent growth in this category can be attributed to the automotive industry and their suppliers that ship component parts into assembly plants.

Types of Container

Various container types can be used in these applications, and most are injection molded from high-density polyethylene (HDPE). Generally, there are four categories of RSCs: nest only, stack only, stack and nest, and collapsible. Each type has its advantages and disadvantages as outlined in the remainder of this article.

Nest-only containers. Nest-only containers (see Fig. 1) have tapered walls that allow them to fit inside each other or "nest" when they are empty. These are inexpensive containers to make, but they require racks or shelves to sit on when they are full of product.

Stack-only containers. Stack only containers (see Fig. 2) have straight walls and good interior room for optimal utilization of interior space. They have excellent stacking stability and make good over-the-road containers. Because they stack only, they take up a lot of space and are inefficient when stored empty. Lids are not necessary for stacking, but could be used to keep dirt out of the containers.

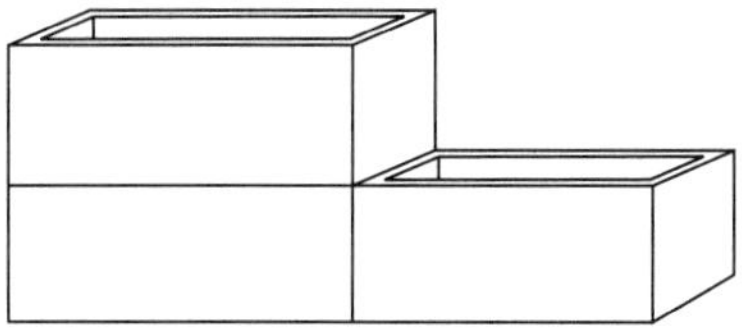

Figure 2. Stack only.

Stack-and-nest containers. There are five types of stack-and-nest containers:

(*1*) *90° stack-and-nest.* By turing a container 90° containers can either stack on top of each other or nest inside one another. They do not need a lid and are generally inexpensive. Depending on the configuration, they can be good for over-the-road transport.

(*2*) *180° stack-and-nest containers* (see Fig. 3. Similar to the 90° version, with a 180° turn a container can be either nested when empty or stacked when full. Stacking posts on the inside ends of the container allow this versatility, but they take up interior room and take up potential product space.

(*3*) *Attached-lid containers* (see Fig. 4). These are nest-only containers with lids that are attached, usually by a metal hinge wire. They are ideal over-the-road shipping containers and provide an excellent unitized load. They stack when the lids are closed and nest when the lids are open and are one of the most popular RSC.

(*4*) *Detached-lid containers.* These are basically nest-only containers with lids that make excellent over-the-road shipping containers. They stack when the lids are snapped on and nest when the lids are removed. These are the easiest lidded containers to clean, but do require keeping track of both boxes and separate lids.

(*5*) *Bail containers* (see Fig. 5). These are nest-only containers with "bails" that flip in to allow stacking and flip outside the container to allow nesting. They make good transportation containers, are easy to clean, and do not require the tracking of separate lids and boxes.

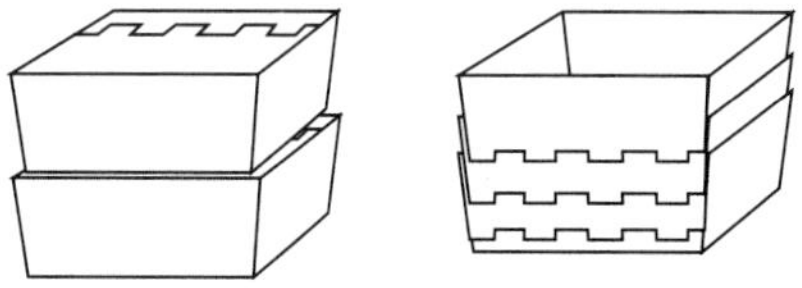

Figure 4. Attached lid.

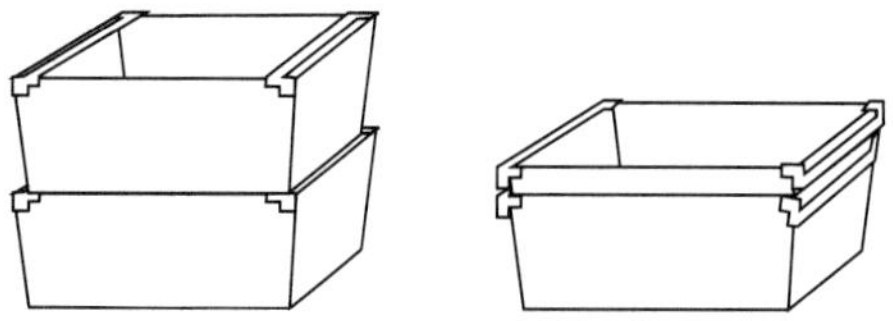

Figure 5. Bail.

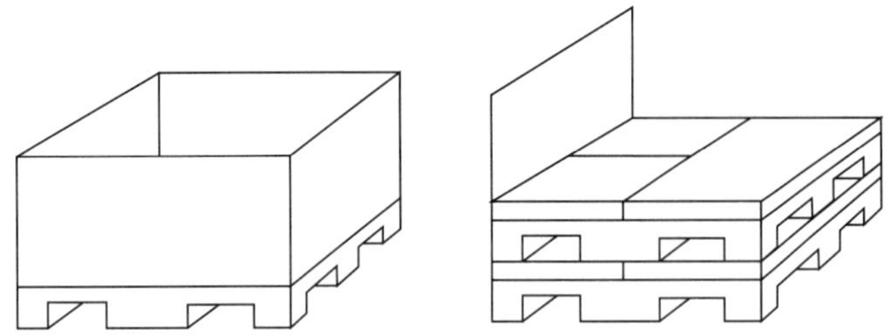

Figure 6. Collapsible bulk box.

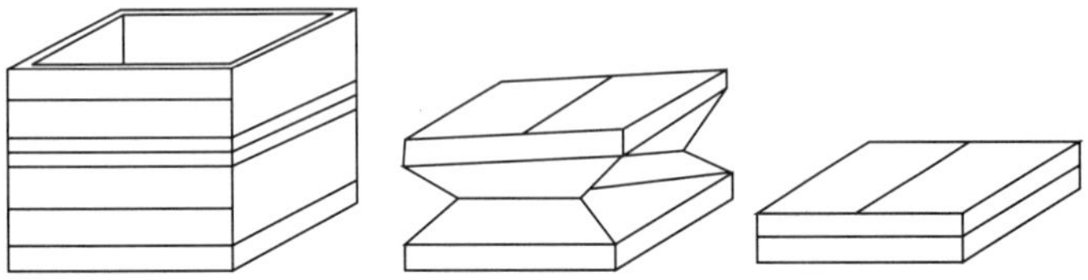

Figure 7. Collapsible tote.

Collapsible containers. There are several types of collapsible containers. For bulk boxes (see Fig. 6), which are generally built on a pallet base and 30 in. high, the most efficient design has side and end walls that are hinged at the base and fold inward flat on themselves. With straight walls, these offer excellent interior space utilization, and when they are empty and collapsed they offer efficient storage. They generally come with optional detached lids.

Smaller collapsible totes (see Fig. 7) offer the same interior space efficiencies and collapsible storage benefits as do bulk boxes. They usually are designed with an accordion collapsing design and can be open-topped or have a detached or attached lid.

BIBLIOGRAPHY

General References

E. A. Leonard, *Packaging Economics,* Books for Industry, New York, 1980.

S. Sacharow and R. C. Griffin, Jr., *Plastics in Packaging,* Cahners Publishing, Boston, 1973.

J. F. Hanlon, *Handbook of Package Engineering,* 2nd ed., McGraw-Hill, New York, 1984.

K. Auguston, "A Selection Guide to Returnable Containers," *Modern Materials Handling,* 42–43 (Nov. 1995).

Richard M. Brasington
Buckhorn
Milford, Ohio

BOXES, SOLID-FIBER

Solid-fiber containers are used almost exclusively for applications in which container return and reuse are desirable and where return can be controlled by the distributor. Without such control, the impetus to use the multitrip shipping containers, which are more costly than corrugated boxes, would not exist.

As a rule of thumb, the solid-fiber box costs two to three times as much as a comparable-size general-purpose corrugated shipper. However, the solid-fiber container can be used an average of 10–15 times before retirement. The economics are obvious, but only in a "closed-loop" distribution system.

In the United States, annual shipments of solid fiber total about 900 million (10^6) ft^2 (83.6 km^2). That is less than 1% of those recorded by corrugated board for the same general use category (1). Nevertheless, solid fiberboard possesses intrinsic performance values that have assured a continuity of acceptance dating back to the late 19th century, when solid fiber and corrugated board began to supplant wooden cases in world commerce.

Solid fiberboard differs from corrugated board in several significant respects (see Boxes, corrugated). As its name implies, the former is a solid (nonfluted) structure consisting of two or more plies of containerboard. Four plies generally are used to manufacture board from which solid-fiber boxes are to be produced.

Solid-fiber sheets are constructed by gluing roll-fed containerboard plies together on a machine called a *laminator.* The plies are bonded under controlled pressure to form a sheet that comes off the line as a continuous strip that is subsequently cut to predetermined lengths.

Caliper of the finished board is the result of the number and thickness of individual plies. It varies according to the needs of the customer market. For most applications, finished sheet thickness ranges between 0.035 and 0.135 in. (0.089 and 0.343 cm). (In the industry lexicon, 0.035 in. is called 35 points). For special heavy-duty applications, solid fiberboard of 250-point thickness can be produced. This, however, is the exception. For the largest market (shipping containers), board thickness averages 70–80 points (0.18–0.20 cm).

When the continuous web of solid fiberboard exits the laminator, it is cut to length. Converting equipment prints, diecuts, and, if needed, coats the material with polyethylene or other protective finish. The converted board is then set up or assembled into containers, usually at the customer's plant.

Brewery use of solid-fiber shipping containers encompasses a variety of constructions, both closed-top and opentop. A common feature is a die-cut handhold at each end to facilitate carrying.

Breweries are the largest users of solid-fiber shipping containers, largely because of that industry's relatively manageable distribution controls. Overall, shipping containers take about 40% of the total annual volume of solid fiberboard, and 80% of all solid-fiber shippers are used to carry bottled beer. The rest of the solid-fiber box business is shared mainly among producers of meat, poultry, fish, and fresh produce (for so-called wet shipments). The U.S. government also uses solid fibre containers for "C" rations.

Soft-drink bottlers once were heavy users of returnable solid-fiber shippers. This market, however, has diminished for two reasons. One is competition from molded plastic crates. The other is the decline of returnable softdrink bottles.

Most industry authorities agree that local or state deposit bills that mandate return of empty bottles or cans will have little long-term effect on solid-fiber board use trends. Consensus is that only a nationally enforced litter law will bring about a meaningful increase in sales to the beverage industry.

A principal difference in construction between solid-fiber and corrugated boxes is that rail and highway shipping-authority rules limit solid-fiber boxes to only two styles of manufacturer's joints; stitched and extended-glued. Standard methods for testing solid-fiber boxes have been developed and published by the American Society for Testing and Materials (ASTM) and the Technical Association of the Pulp and Paper Industry (TAPPI). Common-carrier requirements for solid-fiber box performance (burst strength, size, and weight limits) have been established and are fully detailed in Rule 41 of the Uniform Freight Classification (rail shipment) and Item 22 of the National Motor Freight Classification (truck shipment) (see Laws and regulations).

After containers, the largest use of solid fiberboard is for slip sheets (see Slip sheets). They account for about 45% of annual solid-fiber sheet production. Slip sheets are gaining wide acceptance in materials-handling applications, chiefly as replacements for bulkier and more expensive wood pallets. Designed for forklift handling and requiring minimal warehouse space, solid-fiber slip sheets are easy to use and store.

The remaining 15% of annual solid-fiberboard volume goes into a variety of applications. Point-of-purchase displays display is one. Furniture is another (sheets are affixed to the wooden framework to provide a firm backing for upholstery). Mirror backing is another. Automative applications include non-load-bearing interior bulkheads. Drums, railcar dunnage, multitier partitions for glass-container shipments, and wire-and-cordage reels are other miscellaneous uses in which solid fiberboard's strength, damage resistance, and machinability are cost-effective.

A major source of information on solid fiberboard is the Fibre Box Association, with headquarters in Chicago, Ill. Available information includes annual statistical reports on shipments, inventory, etc, and the *Fibre Box Handbook* (1). This comprehensive reference includes industry definitions, details on box constructions (see Fig. 1) and testing procedures, and a full discussion of all regulations affecting solid-fiber boxes.

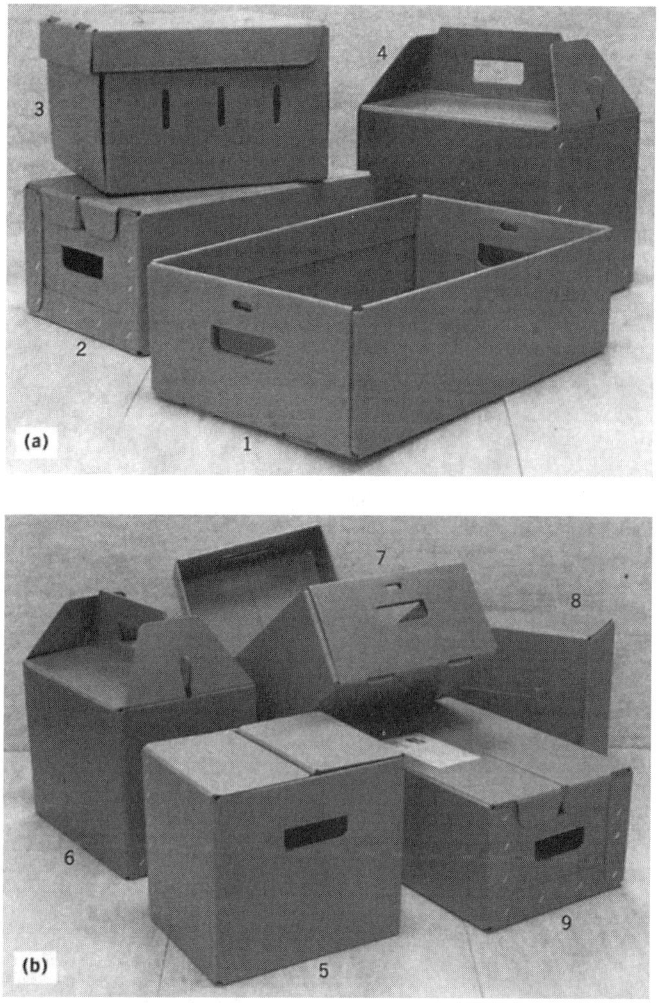

Figure 1. Solid-fiber beverages cases and tote boxes. (**a**) 1. One-piece bin box with locking feature. 2. Three-piece 24/12-oz (355-mL) beverage case. 3. One-piece vegetable box. 4. Attaché-style tote box. (**b**) 5. One-piece 4/1-gal (3.785-L) beverage case. 6. Attaché-style tote box. 7. One-piece lidded tote box. 8. One-piece self-locking tote box. 9. Three-piece 24/12-oz (355-mL) beverage case.

BIBLIOGRAPHY

"Boxes, Solid Fibre" in *The Wiley Encyclopedia of Packaging Technology* 1st ed., Wiley, New York by Robert Quinn, Union Camp Corporation, 1986, pp. 80–81.

1. *Fibre Box Handbook,* Fibre Box Association, Chicago.

General References

A. R. Lott, "Solid and Corrugated Fibreboard Cases" in F. A. Paine, ed., *The Packaging Media,* Wiley, New York, 1977.

BOXES, WIREBOUND

Wirebound boxes and crates are high strength-to-weight ratio, resilient containers designed to support heavy stacking loads even under adverse humidity and moisture conditions. They are available worldwide and are used for a variety of agricultural and industrial products. Wirebounds are fabricated and delivered in flat form to conserve shipping and warehouse storage space and may be easily assembled, as required, by the user with simple hand tools or automated equipment when volume warrants.

The basic material used in wirebound construction is wood integrally combined with steel binding wires fastened to the wood elements by staples. In a relatively small number of design applications corrugated fiberboard is substituted for some wood components, and plastic strapping replaces steel binding wire. The many design options available by combining the various components of the composite container present an opportunity to custom design wirebounds for specific products, weight, sizes, and distribution hazards. The high-speed automated machinery used to fabricate wirebounds is custom built for specific size ranges and styles of wirebounds.

Wirebound Styles

The most frequently used style of wirebound is the *all-bound* (Fig. 1) used in crate form for shipment and storage of fresh fruits and vegetables, and in box form (Fig. 2) for industrial and military applications. The openings between faceboards in the Figure 1 style all-bound provide needed ventilation for such produce items as sweet corn, celery, beans, cabbage, and citrus (1). Outside the United States this style wirebound is often called the *bruce* box. When the all-bound is intended for industrial or military products, it is normally specified with heavier face material, extra binding wires, and/or interior cleats (compare Figs. 1 and 2). The all-bound is characterized by binding wires on all six faces of the made-up container, and a cleat framework in the vertical plane. Both assembly and closing is accomplished by engaging the wire loops at the ends of each binding wire. These loops are called *rock fasteners.* Two simple handtools are required to engage the rock fasteners: the bon ender for assembly of the all-bound and the sallee closer for final closure. If high volumes of all-bounds are to be assembled or closed at one location, automated equipment is available. A variation of the all-bound, the rock fastener box, is less frequently used. The rock fastener box ends do not have binding wires and may be of solid plywood, linered or battened construction.

A completely different style of wirebound, the wirebound pallet box (Fig. 3) is used for bulk storage and/or shipment of industrial products including auto and plane parts, chemicals, castings and forgings, etc (2). The bottom horizontal cleats are either fastened to or lock onto a two- or four-way pallet base for fork trunk handling and stacking. Cleats may be located inside the pallet container as shown or outside if the nature of the load requires a smooth interior.

Although all-bounds and wirebound pallet boxes are often manufactured, inventoried, and sold as stock containers, many wirebounds are custom designed and manufactured on order for specific applications. These applications include cast-iron bathtubs, insulators, garden tractors, machine tools,

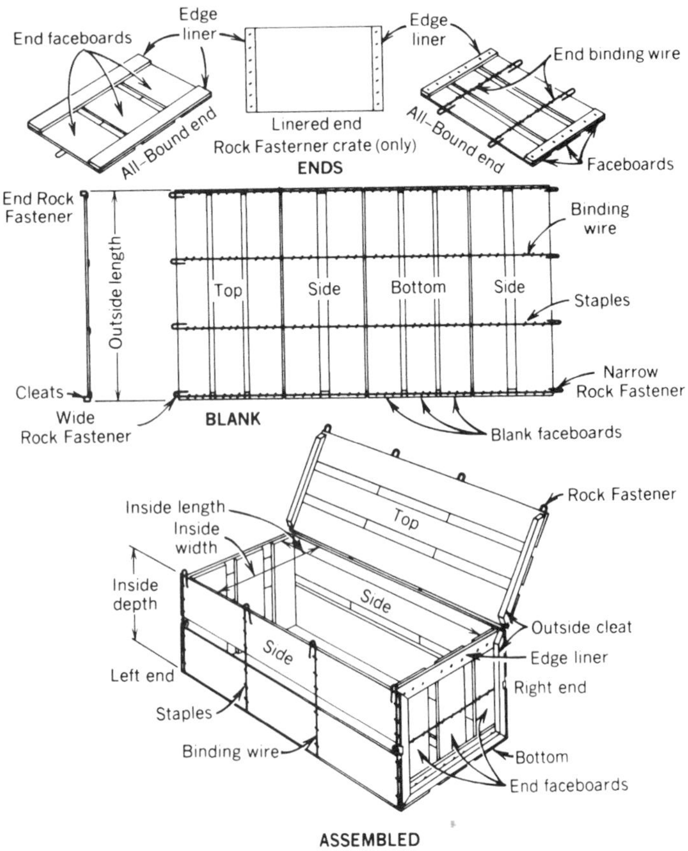

Figure 1. Standard reference for all-bound wirebound crates.

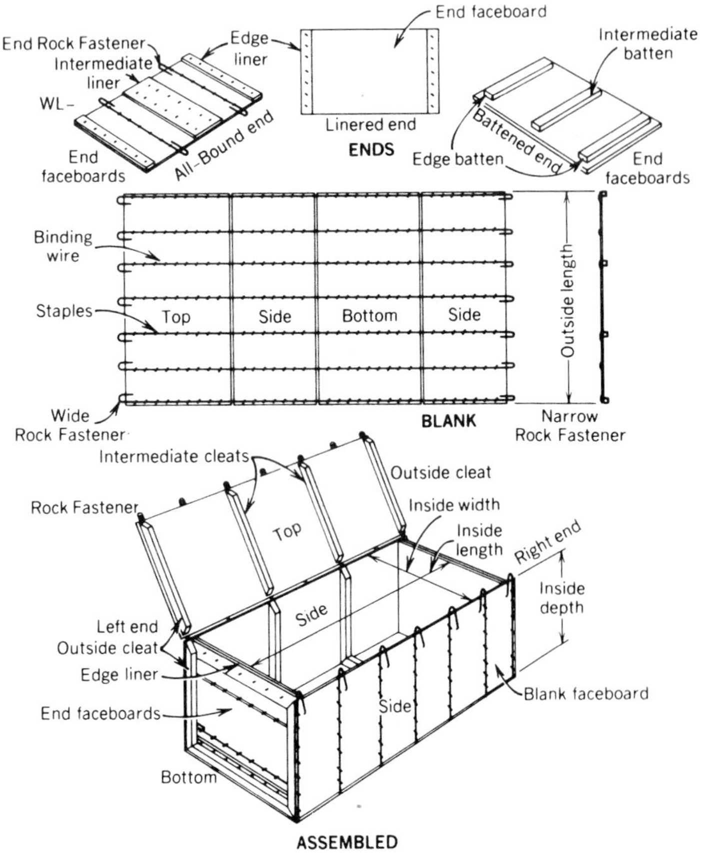

Figure 2. Standard reference for all-bound wirebound boxes. When linered end and battened end are used, the wirebound style is rock fastener box.

etc (3). Net weights carried range between 100 and 6000 lb (45 and 2722 kg). There is no size limit since blank sections may be joined together if required. Wirebounds for military use are generally covered under Federal specifications PPP-B-585-C, *Boxes, Wood, Wirebound* and Military specification MIL-B-0046506-C, *Boxes, Ammunition Packing, Wood, Wirebound.*

Wirebound Material

Any of the deciduous wood species in Group IV or III (4) may be used for wirebound containers. Selected Group II and I

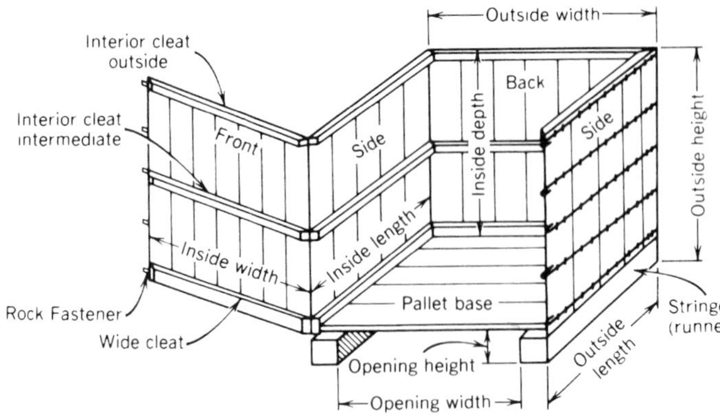

Figure 3. Wirebound pallet box.

woods, from higher density coniferous trees, are permissible in specific design situations. The faceboard members may be generated directly from the log on high-speed veneer lathes equipped with a backroll and spur knives, or from resawing sawmill-supplied rough cut lumber using band saws. Thicknesses from the first method generally range from 1/9 in. to 3/16 in. (2.8 to 4.8 mm) and the resawn material, from 1/4 to 3/8 in. (6.4 to 9.5 mm). In general, the thinner veneer faceboards are used in wirebounds for produce and the heavier resawn faceboards for industrial and military applications. Both forms of faceboards are dried to 15–20% moisture content before fabrication to prevent subsequent mold growth. Wirebounds for long-term storage or export may be dipped in a wood preservative. The cleat structure is sawn from air-dried planed lumber on specialized saw equipment.

The steel wire used for both binding the wirebound and to form staples has unique characteristics that permit machine runability and functional performance (5). Depending on its use in the wirebound, the low-carbon steel wire has a tensile strength between 45,000 and 125,000 psi (310 and 861 MPa) and an elongation of 0.5–10%. A simple testing device, the Rockaway wire tester, is used by wire suppliers and wirebound manufacturers to ensure wire performance.

BIBLIOGRAPHY

1. *Wirebound Boxes and Crates for Fresh Fruits and Vegetables,* Package Research Laboratory, Rockaway, NJ, 1980.
2. *Engineered Wirebound Pallet Containers,* Package Research Laboratory, Rockaway, NJ, 1972.
3. *Versatile Wirebounds,* Package Research Laboratory, Package Research Laboratory, Rockaway, NJ, 1983.
4. *Wood Handbook,* U.S. Department of Agriculture, Forest Products Laboratory, Madison, WI., revised 1974.
5. Wire for Wirebound Boxes and Crates, Package Research Laboratory, Rockaway, NJ, 1979.

DAVID DIXON
Packaging Research Laboratory
Rockaway, New Jersey

BOXES, WOOD

The use of wooden boxes and crates dates back to the Industrial Revolution, when the building of roads and railways led to their development as the first "modern" shipping contain-

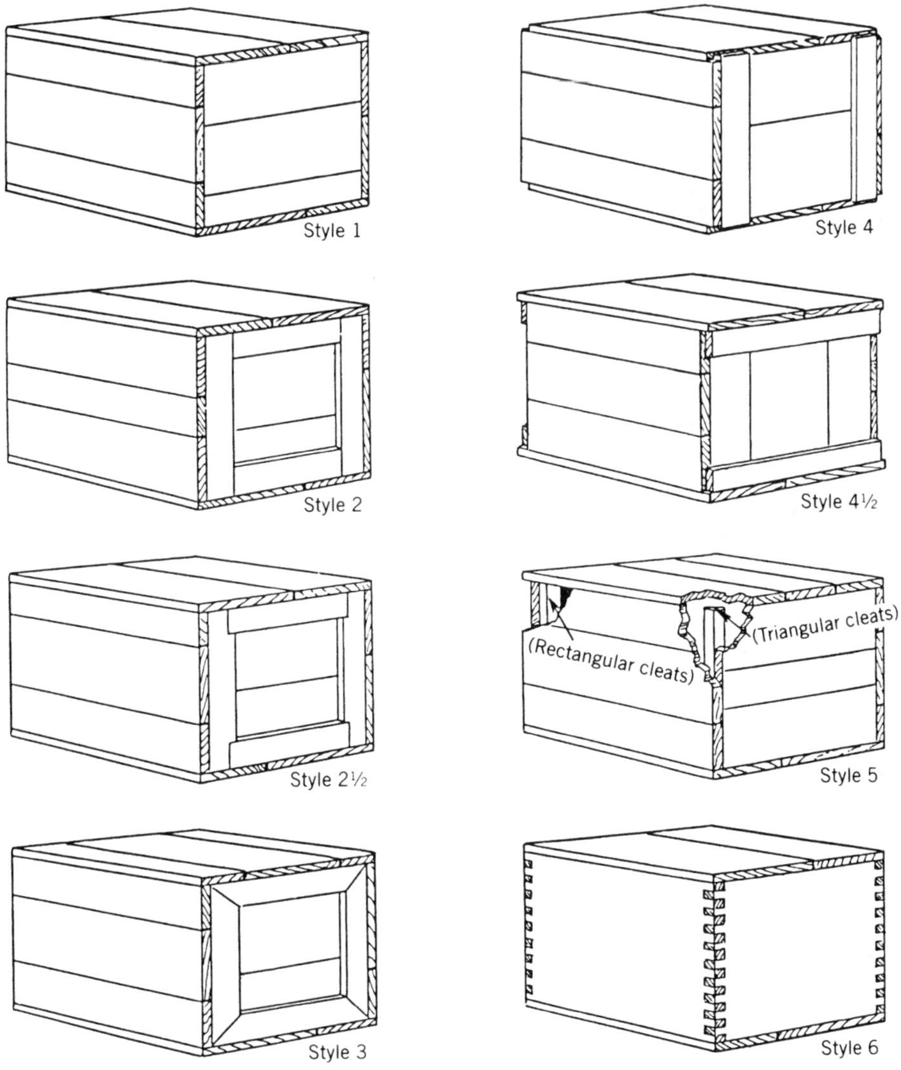

Figure 1. Styles of wooden boxes (3, 5). Style 1: uncleated ends. Style 2: full-cleated ends, butt joints. Style 2½: full-cleated ends, notched cleats. Style 3: full-cleated ends, mitered joints. Style 4: two exterior end cleats. Style 4½: two exterior end cleats. Style 5: two interior end cleats. Style 6: lock corner.

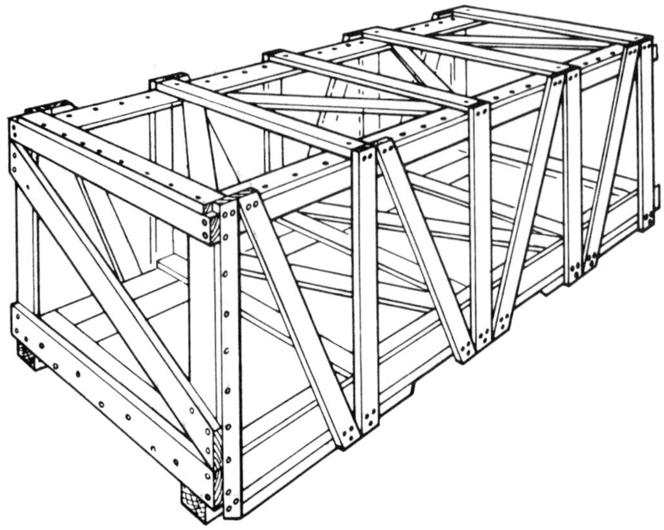

Figure 2. Simple wooden crate.

ers (1). (See also Boxes, wirebound; Pallets, wood.) They are still used today for products that require the strength and protection that only wood can provide. The difference between a box (or case) and a crate is that a box is a rigid container with closed faces that completely enclose the contents. A crate is a rigid container of framed construction. The framework may or may not be enclosed (sheathed) (2).

Boxes

Box styles. Wooden boxes are of either "nailed construction" or "lock-corner construction." Of the eight basic box styles shown in Figure 1 (3), Style 6 is the only lock-corner construction. The others are nailed (cleated).

Wood. Specifications for wooden boxes refer to the categories developed by the United States Forest Products Laboratory, which relate to strength and nail-holding power (see Table 1). Groups 1 and 2 are relatively soft; Groups 3 and 4 are relatively hard. For a given box of a given style, the thickness of the wood and cleats depends on the type of wood.

Fastenings. The strength and rigidity of crates and boxes are highly dependent on the fastenings: nails, staples, lag screws, and bolts. Nails are the most common fastenings in the construction of boxes. The size and spacing of the nails depends on the type of wood (5).

Loads. The type of load is determined by the weight and size of the contents and its fragility, shape, and capacity for support of, or damage to, the box. Load types are classified as

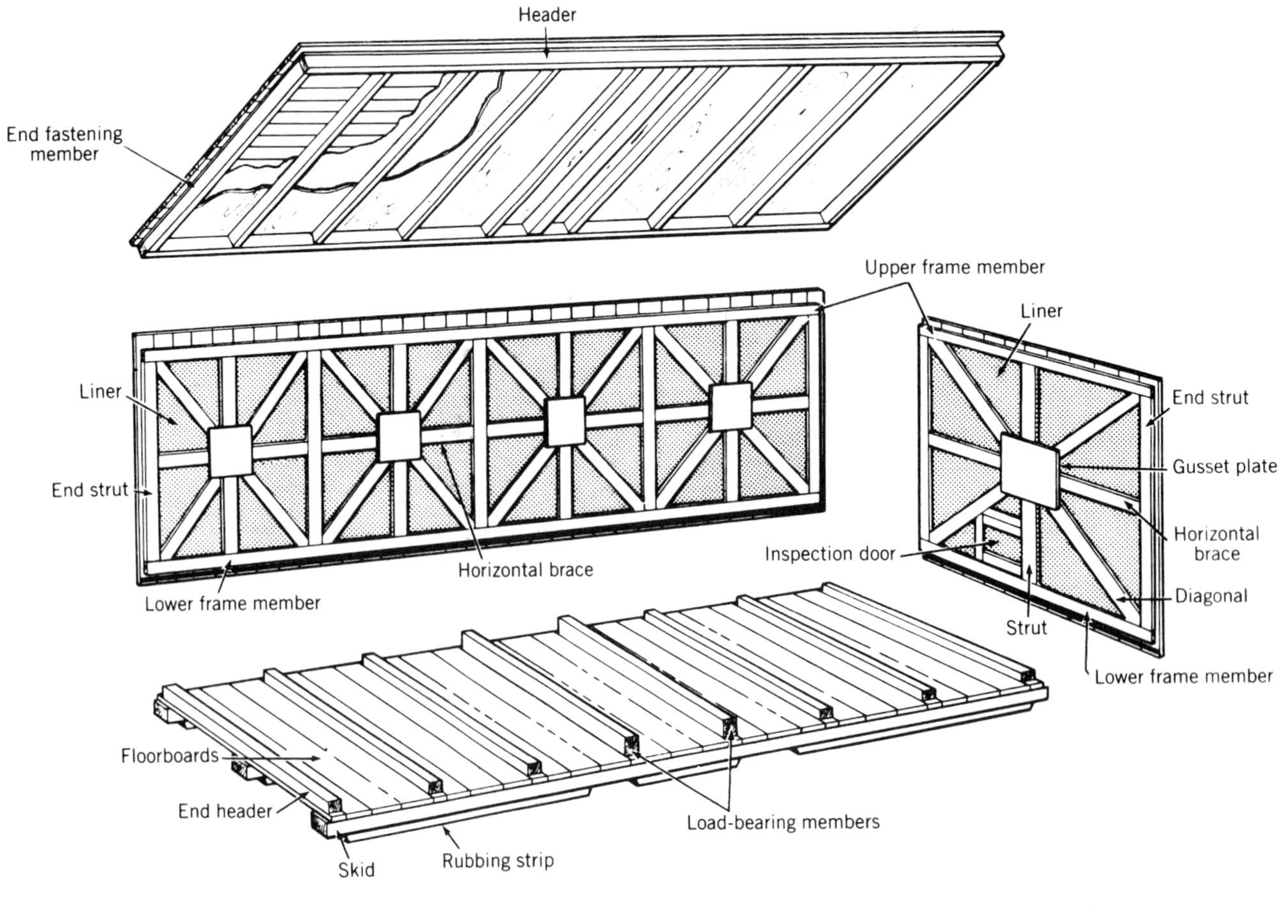

Figure 3. Plywood-sheathed crate.

Table 1. Commercial Box Woods[a]

Group 1		Group 2	Group 3	Group 4
Alpine fir	Lodgepole pine	Douglas fir	Black ash	Beech
Aspen	Magnolia	Hemlock	Black gum	Birch
Balsam fir	Nobel fir	Larch	Maple (soft or silver)	Hackberry
Basswood	Norway pine	North Carolina pine	Pumpkin ash	Hickory
Buckeye	Redwood	Southern yellow pine	Red gum	Maple (hard)
Butternut	Spruce	Tamarack	Sap gum	Oak
Cedar	Sugar pine		Sycamore	Rock elm
Chestnut	Western yellow pine		Tupelo	White ash
Cottonwood	White fir		White elm	
Cucumber	White pine			
Cypress	Willow			
Jack pine	Yellow poplar			

[a] Ref 4.

Type 1 (easy); Type 2 (average); or Type 3 (difficult). Descriptions and examples of each load type are contained in Ref. 5.

Crates

A wood crate is a structural framework of members fastened together to form a rigid enclosure which will protect the contents during shipping and storage. The enclosure is usually rectangular, and may or may not be sheathed (4). A crate differs from a nailed wood box in that the framework of members in sides and ends must provide the basic strength. A box relies for its strength on the boards of the sides, ends, top, and bottom. A crate generally contains just a single item, and its dimensions are not subject to standardization. The function of a crate is to protect a product during handling and shipping at the lowest possible cost. A simple enclosing framework is shown in Figure 2 (4).

Some products and shipping conditions require greater protection. The value of the contents or the likelihood of top loading may dictate the use of a sheathed crate (6). The sheathing can be lumber or plywood. A plywood-sheathed crate is shown in Figure 3 (4).

BIBLIOGRAPHY

1. F. A. Paine, *The Packaging Media, John Wiley & Sons, Inc., New York, 1977.*
2. *ASTM D-996: Standard Definitions of Terms Relating to Packaging and Distribution Environments,* ASTM, Philadelphia.
3. National Wooden Box Association, *Specifications for Nailed Wooden and Lock Corner Boxes for Industrial Use,* Washington, DC, 1958.
4. *Wood Crate Design Manual, Agriculture Handbook No. 252,* U.S. Department of Agriculture, Forest Service, Feb. 1964.
5. *Federal Specification PPP-B-621C,* Oct. 5, 1973.
6. American Plywood Association, *Plywood Design Manual: Crating,* Tacoma, WA, 1969.

BRAZILIAN PACKAGING INDUSTRY, THE

A new area that is expected to improve the socioeconomic development in Brazil is described in this article.

Introduction

In developed, industrialized countries, almost all goods are packaged, and most national and international distribution systems could not operate without the use of packaging. This also means that packaging costs are an integrated element in everyone's daily life.

It is currently estimated that 5–50% of retail product prices is represented by packaging (in the food sector the average is 16%).

The socioeconomic value of packaging technology is only today getting the recognition of society in developed, industrialized countries, but in developing countries, this recognition is just beginning.

Packaging, however, usually represents about 2–3% of any country's GNP (gross national product). The aggregate value of packaging production and services puts the packaging industry in fourth or fifth place in the rank list of all industries (at least in industrialized countries).

Corresponding figures from developing countries are not readily available, but, in general, it can be said that packaging is a much more important factor in the national economies of all countries than is commonly believed.

Brazil is one of the few countries trying to identify this importance and demonstrate to the society the "packaging factor." DATAMARK Ltda., every year updates the *Brazilian Packaging Industry Study,* an independent multiclient study on consumer products and packaging in Brazil, with the 9th edition available in 1994.

CETEA, the packaging technology center through the membership system established in 1988 and today with 150 members (Associados), is motivating the packaging related industry to understand and benefit from the packaging factor.

The Lawson/Mardon Group (LMG) estimate, on the basis of OECD/IMF data, that the packaging industry worldwide spent U.S.$ 400,000 million in 1992 in packaging materials and packaging machinery (see Table 1).

In most of the industrialized countries the packaging materials and machinery industry represents 1.7–2.0 of the GNP. This number can also be visualized in the per capita comparison of packaging materials consumption in the world (Table 2).

In this introduction we would like to emphasize the state-

Table 1. Packaging Materials and Machinery 1992 Sales—U.S.$000 Million

Countries	U.S.$000 Million	%
Europe	150	37.5
North America	100	25.0
Asia	100	25.0
Latin America	30	7.5
Africa	20	5.0
Total	400	100.00

Table 2. International Consumption of Packaging Materials

	Brazil			
	Economic Activity	Total	USA	Europe
Plastics	6.2	2.5	23.4	15.1
Paper	25.0	10.2	104.7	43.9
Steel	13.3	5.4	15.4	9.1
Aluminium	0.5	0.2	5.1	1.0
Glass	14.1	5.7	59.8	37.5

Source: DATAMARK Ltda, Brazil.

ment of IAPRI (International Association of Packaging Research Institutes):

> In modern society packaging is playing an important role. Thanks to packaging technology it is possible to distribute products over a wide geographical area as well as over a long range of time without losing major product qualities. Packaging technology also allows this process to be conducted within an acceptable financial profit. The process itself is generating a subtle cooperation between product and packaging with the aim to fulfill the needs of the product end use as well as the needs of manufactures distributers.

The packaging chain (raw material–end user/waste management) is an area where availability and demand is continuously subject to changes in consumer habits, logistical developments, new technological developments, new materials, and increase in the international distribution volumes, and—last but not least—increasing environmental constraints.

Consequently, the packaging chain faces problems from an increasing complexity on a strategic as well as technological level. It requires an integrated approach based on a multidisciplinary knowledge to be able to bring these problems to an adequate solution. This, in turn, requires a high level of understanding of the total packaging area in combination with a deeper disciplinary knowledge in the areas of materials, product sensitivity/fragility, production technology, logistics, ecology, marketing, etc.

Brazilian Socioeconomic Data

Brazil, with an area of 8.5 million km^2, is the fifth largest country in the world. The nation comprises 24 states, 2 territories, and the Federal District. The country extends from the Equator in the north to latitude 30′ in the south.

The most economically important region is the southeast, where 43% of the total population lives and most of the industry is concentrated.

In 1995 the population was estimated at 156 million, 76% of which lived in towns and cities. The total population is expected to grow by 19% up to the year 2000; and the urban population, by 27%.

Another aspect of Brazil's population, which bears on the

Table 3. Balance of Payments in Brazil

U.S.$ Billion	1986	1987	1988	1989	1990	1991	1992	1993	1994	1995
Exports	22.4	26.3	33.8	34.4	31.4	31.6	36.2	38.8	43.6	46.5
Imports	12.9	15.1	14.7	18.3	20.4	21.0	20.5	25.7	33.2	49.6
Trade balance	9.5	11.2	19.1	16.1	11.0	10.6	15.7	13.1	10.4	3.1

Source: DATAMARK Ltda., Brazil.

Table 4. Principal Economic Indicators in Brazil

U.S.$ Billion	1986	1987	1988	1989	1990	1991	1992	1993	1994	1995
Population (10^6)	133	136	138	141	144	146	149	152	154	156
Economically active population (10^9)	57	60	61	62	64	62	62	62	64	72
GDP (U.S.$ 10^9)	335	358	371	400	398	418	425	456	578	563
GDP growth rates (%)	7.6	3.6	(0.1)	3.3	(4.4)	1.1	(0.9)	5.0	5.8	4.2
Real sectorial change										
Agriculture (%)	(8.0)	15.0	0.8	2.9	(3.7)	2.8	5.3	(1.9)	9.1	5.9
Industry (%)	11.8	1.1	(2.6)	2.9	(8.0)	(0.5)	(3.6)	9.0	6.9	2.0
Services (%)	8.2	3.3	2.3	3.8	(0.8)	2.2	(0.1)	3.5	4.2	5.7
Per capita income, U.S.$	2.466	2.581	2.638	2.787	2.722	2.808	2.732	3.001	3.429	3.609
Minimum wage, U.S.$/month	56	48	58	76	74	68	64	69	80	98
Inflation (%), IGDPI	65	416	1.038	1.783	1.477	480	1.158	2.708	1.094	15

Source: DATAMARK Ltda., Brazil.

Table 5. Industrial Production in Brazil

	1986	1987	1988	1989	1990	1991	1992	1993	1994	1995
Industrial electrical power (tWh)	97.1	97.4	100.5	103.0	99.3	102.7	103.5	107.4	107.1	113.0
Crude oil ($m^3/10^6$)	33.2	32.8	32.2	34.5	36.4	36.1	36.4	37.0	38.3	40.0
Steel (tons 10^6)	21.2	22.0	24.5	25.0	20.5	22.6	23.1	23.9	25.7	25.0
Vehicles (units 10^3)	1.054	918	1.070	1.019	918	962	1.074	1.391	1.583	1.640
Refrigerators (units 10^3)	1.963	1.907	1.651	1.931	1.910	2.117	1.382	1.656	2.400	3.031
TVs (units 10^3)	2.905	2.641	2.697	2.700	2.851	2.990	2.624	3.824	5.450	6.204

Source: DATAMARK Ltda., Brazil.

purchase of value-added products, is the number of economically active people.

The 1995 estimates showed that of the economically active population of 62 million, 65% were men.

In order to better visualize the Brazilian packaging industry, Tables 3, 4, and 5, present the following data: balance of payments, principal economic indicators, and the industrial production.

The Brazilian Packaging Industry

The Brazilian packaging industry was estimated at U.S.$5597 million (3.8 million tons in total) in 1993, equivalent to 1.2% often gross national product. The breakdown of the various packaging materials is presented in Table 6.

Values complementing the figures given in Table 6 are presented in Tables 6–9 (data of the Brazilian packaging market).

It is important to emphasize that all these figures presented by the Brazilian packaging market were extracted from the work *Brasil Pack '94* (The Brazilian packaging industry), a multiclient work updated done annually by DATAMARK Consultores S/C Ltda.

At U.S.$5.5 billion annual sales in 1993, the Brazilian Packaging Industry compares well in size with European packaging markets with the added advantage that the Brazilian market still has an enormous potential for growth. The forecast for the year 2000 is of the order of U.S.$7 billion based on projected growth of the current situation, excluding new types of packaging or substitutions.

The use of packaging is well developed with large companies such as Nestle, Unilever, Coca Cola, and Brahma (a major brewer) purchasing well over U.S.$100 million annually. In all, over 60 companies purchase over U.S.$10 million of packaging annually, 30 of which represents over U.S.$20 million. Over 200 companies purchase more than U.S.$1 million per year.

On the supply side, although there are relatively few companies (perhaps 10) with sales in excess of U.S.$100 million, over 100 companies have sales of over U.S.$1 million.

The range of packaging supplied is also considerable. Brazil is essentially self-sufficient in most conventional types of packaging, although there are opportunities for the introduction of more recent packaging innovations. As far as quality is concerned, standards are generally high, with print quality approaching European levels. Several packaging suppliers have either achieved ISO 9000 classification or are in the process of implementation. In general, Brazil has more ISO 9000 certificates than the rest of Latin America put together.

Until the earlier 1990s, the most practical way to sell machinery in Brazil was to manufacture locally. With the open-

Table 6. Packaging Material Consumption in Brazil in 1995[a]

	Volume, %	Value, %
Corrugated cases	28.7	21.9
Duplex/triplex	6.5	9.3
Flexibles	4.8	18.2
Metals	18.9	13.1
Paper	5.3	3.8
Plastics	11.1	22.7
Glass	24.8	11.0

[a] Excluding crates and supermarket bags
Source: DATAMARK Ltda., Brazil.

Table 7. Packaging Market by Materials in Brazil

	Food			Nonfood		
Material	Tons	%	US$$10^6$	Tons	%	US$ 10^6
Flexibles	183.071	81.0	1.303	42.862	19.0	326
Metals						
Aluminum	67.912	89.5	346	7.969	10.5	87
Tinplate/blackplate	579.473	83.6	527	113.837	16.4	103
Steel	23.633	19.1	21	99.813	80.9	91
Paper						
Kraft/MG paper	118.510	47.3	158	132.008	52.7	179
Duplex/triplex	61.696	20.2	162	243.268	79.8	675
Corrugated cases	532.181	39.3	772	821.624	60.7	1.191
Plastics	452.067	62.5	1.536	270.901	37.5	938
Glass	913.134	94.1	473	57.229	5.9	74
Total	2.931.677	62.1	5.298	1.789.512	37.9	3.665

Source: DATAMARK Ltda., Brazil.

Table 8. Packaging Market by End-User Market in Brazil (by Volume, in tons)

Products	Flexibles	Metals	Paper	Plastics	Glass
Food					
Drinks	29.529	148.937	12.611	193.968	735.535
Meat and vegetables	13.711	140.851	7.018	52.207	63.163
Cereals and flour-based products	26.600	13.022	65.463	85.331	6.708
Sugar and chocolate	30.750	34.374	48.953	48.578	40.556
Dairy products and fats	82.481	333.834	46.161	71.958	67.173
Nonfood					
Electrical	51	8.439	10.482	3.428	—
Health and beauty	15.925	8.549	40.321	58.341	56.426
Personal and leisure	25.496	—	113.885	4.637	53
Cleaning and household	788	25.652	90.446	82.537	—
Chemicals and agriculture	602	178.979	120.142	121.984	751
Total	225.933	892.637	555.482	722.969	970.364

Table 9. Packaging Market by End-User Market in Brazil [by Value, in millions (10^6)]

Products	Flexibles	Metals	Paper	Plastics	Glass
Food					
Drinks	242	411	30	780	309
Meat and vegetables	109	136	18	136	85
Cereals and flour-based products	284	12	128	222	18
Sugar and chocolate	267	31	85	129	21
Dairy products and fats	402	305	60	269	39
Nonfood					
Electrical	—	8	26	10	—
Health and beauty	154	83	104	229	74
Personal and leisure	161	—	394	14	—
Cleaning and household	7	23	177	300	—
Chemicals and agriculture	4	168	153	386	—
Total	1.628	1.176	1.175	2.474	547

ing up of the economy, this has changed and today there is a better mix of imported and locally manufactured equipment. Local manufacture is principally concentrated in plastics conversion: extrusion, injection molding, and blow molding with several international companies present in the market. There are also several well-known international manufacturers of filling machines, both liquid and solid manufacturing equipment, in Brazil.

Brazil, where perhaps only 30 million people out of a population of 150 million actually consume packaged products, has an enormous potential for growth, particularly in liquid packaging, where the combination of high annual average growth rates and the substitution of returnable containers by single-trip packaging has produced a boom in PET and aluminum can production.

The Future of the Brazilian Packaging Industry

In the past 10 years the Brazilian packaging industry suffered the consequences of a lack of consistency in the Brazilian economy, a high rate of inflation, and a closed market.

As a result, the packaging industry was restructured, the sectorial associations and unions were organized, and a new era began in 1995.

Following this growth, CETEA, considered the "model" of packaging research center in developing countries, changed its name to "Packaging Technology Center" in order to work in the pharmaceutical, cosmetics, electronic, dangerous goods, and other important areas that will demand research and technical assistance in the future.

As mentioned before, the future of the Brazilian packaging industry is very promising. The market is expected to grow, and the country is already prepared for the worldwide integration and joint ventures.

Luis Madi
CETEA
Grahan Wallis
DATAMARK Ltda. Campenas SP, Brazil

BULK PACKAGING

Bulk packaging is an economical solution for the shipment of many different types of product. Continuing advancements in the materials applicable to constructing bulk containers contribute to increasing acceptance in their use. These factors, together with lower materials-handling and freight costs, combine to create an effective shipping container system meeting the needs of the product, the shipper, and the customers.

Bulk Packaging Considerations

Almost all types of product present possibilities for use of bulk containers. Successful applications include chemicals,

Table 1. Comparative Attributes of Container Materials

General Material Category	Reuse	Disposability	Tare Weight	Durability	Resistance to Shipping Environment	Container Cost Advantage	Freight Cost Advantage
Corrugated	Good	Best	Best	Good	Good	Best	Best
Bags	Better	Good	Best	Better	Better	Better	Best
Rigid plastic	Best	Good	Better	Better	Best	Good	Better
Metal	Best	Best	Good	Best	Best	Good	Good
Wood	Better	Better	Good	Better	Better	Better	Better

liquids, pastes, metallic powders, resins, food items, automotive parts, and grains. Although metal, plastic, and fiber drums are a form of bulk containers, they are not included here as they are specifically covered in other articles of this encyclopedia. Planning for use of bulk containers should include the following analysis.

Product needs. These include physical and chemical properties, such as sensitivity to moisture gain or loss, flow characteristics, loading and unloading, compatibility with packaging materials, vapor emissions, protection from oxygen, retention of various gases, physical protection, and normal production volumes.

Product weights. Depending on the product and the container used, shipping weights of 2000 lb (907 kg) and more are common, and some dry, flowable products weighing up to 8000 lb (3629 kg) are being sucessfully shipped.

Container costs. Weight and size of product, production volumes, loading and unloading requirements, and methods of storage and transportation are all key elements in the design process. It is essential to design for optimum cube in storage and in the transportation vehicle to realize the most benefits.

Lightweight products may reach vehicle weight limits before cubing out. For heavy products, stacking strength to utilize warehouse space effectively is a major point. Container base dimensions must relate efficiently to both storage and transport vehicle size and incorporate a solid base for handling. Typical sizes are 48 in. × 40 in. (1219 mm × 1016 mm) and 44 in. × 35 in. (1118 mm × 889 mm). Many other sizes are used, depending on distribution requirements.

Container types. Markets served and distribution needs determine selection of the most cost-effective container. Woven plastic films, corrugated containers, rigid plastics, metal, and wood and wood products are materials commonly used. Reusability reduces total costs, and ease of disposability at the end of container life is important. The automotive industry makes great use of returnable bulk containers for parts and assemblies feeding their production lines. Ingredient packaging for food products is another major use.

Customer requirements for use of the product are important. Special features can be designed into containers to facilitate filling and emptying. All containers must be equipped to accept applicable materials-handling devices—forklifts, cranes, hoists, slings, pallet jacks, etc—used to load, store, move, and unload the containers at all points in the distribution cycle.

Materials Used for Bulk Containers

The most commonly used materials are described below. Table 1 shows a comparison of some attributes of various container materials.

Corrugated. Widely used for bulk containers, corrugated containers can be designed to compensate for bulge resistance, puncture resistance, and stacking strength as needed. Variations in board construction to provide necessary properties include one-piece liners, two-piece inners, and three-ply and/or eight-sided (octagonal) shapes. Board used can be single, double, or triple wall or combinations thereof. Within supplier equipment limitations [up to ~180 in. (4572 mm) in perimeter and 55 in. (1397 mm) in depth] many different sizes and shapes are possible. Box liners are used as needed to ensure product protection. Containers are easily disposed of and can be designed for reuse. Container surfaces readily accept graphics for product and company identification.

Bags. Many uses for bags are possible because of the multitude of films and laminations available for fabrication. Most common are woven polypropylene fabrics. Use of rubberized plastic films increases bag strength greatly. If needed, special extrusion coatings or additional protective liners can be used for special product needs. Most uses are for dry products. Bag handling is usually accomplished through four corner lifting loop straps sewn into bags at the top. Common bag sizes are usually up to 66 ft^3 (7.8 m^3). However, recent designs using rubberized nylon and DuPont Kevlar provides the ultimate with bags up to 9 ft (2.74 m) in diameter and 10 ft (3.05 m) high holding up to 25 tons (2268 kg). Bags are collapsible when empty, light in weight, and can be designed for reusability (see Bags, plastic).

Rigid plastics. Designs for rigid plastic containers can be made to meet all expected requirements for bulk packaging. They can be collapsible, lightweight, and usually up to 34 in. (864 mm) in depth. They are adapted to product needs by selection of compatible plastic resins for molding and/or addition of protective liners. These containers are generally more costly than other types, but can be justified in a total system concept through recycling and reuse. Plastic containers are resistant to normal environmental hazards, such as moisture and water. Temperature extremes—low or high—are addressed by specific material selection. Some plastics may be adversely affected by exposure to sunlight. Plastics are often used in captive systems for handling and moving product between multiple plants of a single parent company.

Metal. Metal containers in the form of drums were among the first types of bulk container. The possibilities for metals to meet many different product needs—high or low temperatures, corrosion resistance, longevity, chemical compatibility, shipping weight and cube, etc—are significant. They can be closed-wall, cage-wall, collapsible, and even expandable by adding tiers to a base container. Metal returnables are commonly used for handling and shipping automotive parts and assemblies from suppliers to assembly plants. Quantities involved make disposing of packaging materials very costly. Efficient container design, including internal components, provides for reuse and physical protection. Metal containers are usually heavier than other container types.

Wood. Wood is adaptable to the design of bulk containers to meet many different product demands. Styles include open crates, cleated plywood, and wooden boxes. Containers can be designed for knocked-down shipment when empty to reduce space and freight costs. Reusable fastening devices are available to close many types of box construction. Box liners in various types are used to satisfy special product needs. Wirebound boxes represent a specially designed box or crate to reduce weight and increase strength and are successfully used in bulk product shipments.

Container Testing

ASTMD-4169 provides for definition of test elements and the applicable test procedures for shipping units to meet specific distribution patterns. Performance can be measured against any one of three three assurance levels as determined by the shipper. Preshipment testing is essential for bulk container validation.

BIBLIOGRAPHY

General References

Fibre Box Handbook, Fibre Box Association, Rolling Meadows, IL, 1992.

G. G. Maltenfort, Corrugated Shipping Containers: An Engineering Approach, Jelmar Publishing Co. Plainview, NY, 1988.

Wood Crate Design Manual, Agricultural Handbook No. 252, U.S. Department of Agriculture Forest Services, Feb. 1964.

Annual Book of Standards, Vol. 15.09, American Society for Testing and Materials, Conshohocken, PA, 1996.

R. J. Kelsey, *Packaging in Today's Society,* 3rd ed., St. Regis Paper Company, New York, 1989.

Peter Henningsen
Minnetonka, Minnesota

C

CANNING, FOOD

Canning may be defined as the packaging of perishable foods in hermetically sealed containers that are to be stored at ambient temperatures for extended times (months or years). The objective is to produce a "commercially sterile" food product. *Commercially sterile* does not mean that the food is free of microorganisms, but rather that the food does not contain viable organisms that might be a public health risk or might multiply under normal storage conditions and lead to spoilage. Canning processes do not necessarily kill all microorganisms present in a food, and it may be possible to isolate viable organisms from canned foods. The food product may be made commercially sterile either prior to or after filling and sealing. Three conditions must be met for canning safe and wholesome food:

1. Sufficient heat must be applied to the food to render it commercially sterile.
2. The container must prevent recontamination of the product.
3. The filled and sealed container must be handled in a manner which prevents loss of integrity.

Canning was invented as a means of food preservation in 1810 in response to a prize offered by Napoleon. The original containers were corked glass; handmade tinplate "canisters" (shortened to "cans") were introduced shortly afterward. Cans used prior to 1990 were manually produced from a cylindrical body, an end unit or disk, and a top ring. All seams were formed by dipping in hot solder. The food was filled through the hole in the top ring and a plate containing a small hole was soldered over the opening. Cans were heated to exhaust the headspace so that a partial vacuum would be created after sealing and cooling. A drop of solder was used to seal the small hole and the can was then thermally processed.

Around the turn of the century, the process for manufacturing the three-piece open-top can became widely available. This container used the same double-seamed ends that are in use today. The second end was not put on until the can was filled, which meant that food no longer had to be forced through the hole in the ring.

Today the tin-plated steel double-seamed can is still the predominant food canning package (see Can seamers; Cans, steel). Glass is also used for some products (see Glass container design). Very recently, flexible pouches (see Retortable flexible and semi rigid packages), rigid plastics (see Cans, plastic), and thin aluminum (see Cans, aluminum) cans have been used to can foods. Processes have also been developed in which the food and container are commercially sterilized separately, often by different methods, and the container filled and sealed without recontamination (see Aseptic packaging).

Food canning accounts for just over 30% of U.S. metal can shipments (1) and just under 30% of glass container shipments. Approximately 1700 canning plants process about 36 billion (10^9) pounds (16.33×10^6 metric tons) of food per year (2). The importance of canning in marketing food products varies widely. Virtually all tuna is canned, as is 90% of the tomato crop (2). Other foods rely less on canning.

Process Description

The processing of canned food must produce a commercially sterile product and minimize degradation of the food. The container must also withstand the process and prevent recontamination of the product after processing and up to the time of use, often months or years after processing. The most common sequence of events in canning is that the food product is prepared for canning, the container is filled and hermetically sealed, and the sealed container is thermally processed to achieve commercial sterility. The thermal process necessary to commercially sterilize a canned product depends on the acidity of the food.

Role of pH. High acid foods such as fruits and fruit juices, pickled products, and products to which acid is added in sufficient amounts to give a pH of 4.6 or lower require considerably less heat treatment than low acid foods (pH >4.6). Low-acid foods include most vegatables, meats, fish, poultry, dairy, and egg products. High-acid foods may be processed at boiling water temperature (212°F or 100°C) after sealing. High-acid liquid foods, such as fruit juices, may also be sufficiently processed by "hot-filling" the container with product near the boiling point and allowing slow cooling after sealing. Low-acid foods (pH >4.6) must be processed at temperatures above the boiling point of water. Most often this is accomplished in a pressurized vessel called a retort or autoclave, which contains water or steam at 250°F at 15 psi (121°C at 103 kPa).

Time and temperature requirements. The process or scheduled process refers to the specific combination of temperature and time used to render the food commercially sterile. Several factors affect this process including the nature of the product, shape and dimensions of the container, temperature of the retort, the heat-transfer coefficient of the heating medium, the number and type of microorganisms present, and the thermal-death resistance of these microorganisms.

The relationship between the heat destruction of specific organisms, heating time, and temperature has been intensely studied since the 1920s and equations have been derived for several organisms, the most important of which is *Clostridium botulinum* (3). This spore-forming organism is found in soil, is ubiquitous, and grows in anaerobic environments such as canned foods and produces a deadly toxin. Its spores are also highly heat-resistant. When commercially sterilizing a canned product, it is the temperature profile of the coldest spot in the container, which must be known before the correct process can be calculated. This is accomplished by placing a thermocouple inside the can, usually at the geometric center for products heated by conduction or in the lower portion of the can for products heated by convection. The temperature is recorded during heat processing and used to calculate the proper process time under the given conditions of product, container size and geometry, and retort temperature. This information coupled with the thermal death characterstics of *Cl. botulinum* or a more heat-resistant organism is used to determine the correct process time. Often these processes are described in terms of *F* values.

***F* value.** An *F* value is the time in minutes to heat-inactivate a given number of certain microorganism at a fixed temperature (4). If the temperature is 250°F (121°C) and the organism is *Cl. botulinum,* the *F* value is called F_0. This value is the number of minutes required to kill a given population of *Cl. botulinum* spores at 250°F (121°C). Combinations of times and temperatures other than 250°F and F_0 minutes can inactivate the same number of spores; temperatures lower than 250°F for longer time periods or temperatures greater than 250°F for shorter times have an equal ability to inactivate the spores. An F_0 = 2.45 min reduces the population of *Cl. botulinum* spores by a factor of 10^{12} (5). In practice, F_0 values of greater than 3 are used as a safety measure. In order to prevent overprocessing (overcooking) of the food, the spore inactivation (called lethality) is summed up during the time the coldest spot in the can is coming up to the retort temperature. Some viscous conduction-heated foods may never completely reach the common retort temperature of 250°F (121°C) yet still receive the proper F_0 treatment. The Food and Drug Administration regulations require that these tests and calculations be carried out only by recognized authorities (6).

Interest in thermal processes which resulted in techniques that commercially sterilize fluid foods in continuous-flow heat-exchange systems before packaging (7). The thermal death calculations described above still apply to these processes, and proper *F* values must be achieved (8). The continuous-flow commercial sterilization procedures have the advantage that products can be heated and cooled more rapidly for shorter times with equal lethality. This can give a higher quality product. These products must be filled into presterilized containers (see Aseptic packaging).

Canning Operations

The canning process requires several unit operations that normally take place in a set sequence (9).

Product preparation. As soon as the raw agricultural product is received at the canning plant, it is washed, inspected, sorted to remove defective product, and graded. Often, the edible portion is separated from nonedible as in the case of peas or corn. Fruits and vegetables are subjected to a blanching operation by exposing them to either live steam or hot water at 190–210°F (88–99°C). Blanching serves to inactivate enzymes which would otherwise cause discoloration or deterioration in the product. It also softens, cleans, and degases the product. Peeling, coring, dicing, and/or mixing operations may be carried out next. These operations prepare the product for filling into the can.

Container preparation. Containers must be thoroughly washed immediately prior to filling. Cans are washed inverted so that any foreign objects and the excess water can drain out. The container is now ready for filling. Accurate and precise filling is necessary to meet minimum labeled fill requirements yet leave sufficient headspace for development of the proper vacuum after closure. Too large a headspace results in an underweight container, while overfilling can result in bulging or domed ends after processing. Excessive headspace may also suggest that large amounts of oxygen remain in the can which accelerates product deterioration and can corrosion. Liquid or semiliquid products including small pieces are filled by automated equipment. Larger, more fragile products such as asparagus, are packed by hand or by semiautomated equipment. In most products, brine, broth, or oil is added along with the product. This liquid excludes much of the air between the particles and provides for more efficient heat transfer during thermal processing.

Vacuum. Proper application of the closure after filling is one of the most critical steps in the canning operation. The two-step seaming operation must not only produce a sound, well-formed double seam at speeds of several hundred cans per minute, but must also produce an interior vacuum of 10–20 in. Hg (34–68 kPa) (10). This vacuum reduces the oxygen content and retards corrosion and spoilage, leaves the can end in a concave shape during storage, and prevents permanent distortion during retorting. Proper internal vacuum can be achieved by several methods. Containers that are sealed while the food is at or near the boiling point develop a vacuum when the product cools. This preheating or hot fill also serves to sterilize the container when high-acid foods are packaged. Products which are cool when filled can be heated in the container prior to sealing with the same result as the hot fill. This is often termed "thermal exhaust." Internal vacuum may also be achieved by mechanical means. The filled, unsealed container is fed into a vacuum chamber by means of an air lock and the closure sealed while under vacuum. This system has the disadvantage that flashing of the liquid may occur if air is entrapped in the food or high levels of dissolved air are found in the liquid. The most common method of producing internal vacuum is by displacing the air in the headspace with live steam prior to and during double seaming the cover. The steam in the headspace condenses and forms a vacuum as the container cools.

Retorting. In conventional canning operations of low-acid foods, the sealed containers are next thermally processed at 250°F (121°C) in retorts. Recent regulatory agency rule changes allow specific flexible containers to be processed at 275°F (134°C).

There are several distinct types of commercially manufactured retorts for thermally processing canned food (11). Although all, by necessity, operate at pressures above 15 psi (103 kPa), the design characteristics of each type are considerably different. At least six design variables exist: (*1*) discontinuous (batch) types versus continuous container processing, (*2*) the heating medium used to transfer heat to the container, (*3*) the agitation or nonagitation of containers during processing, (*4*) the layout of the pressure vessel (vertical vs horizontal), (*5*) the method used to load and unload the containers from the retort, and (*6*) the cooling procedures used after thermal processing.

The simplest retorts are batch (discontinous) retorts that use pure steam as the heating medium and do not have provisions for mixing (agitation) of the container contents during

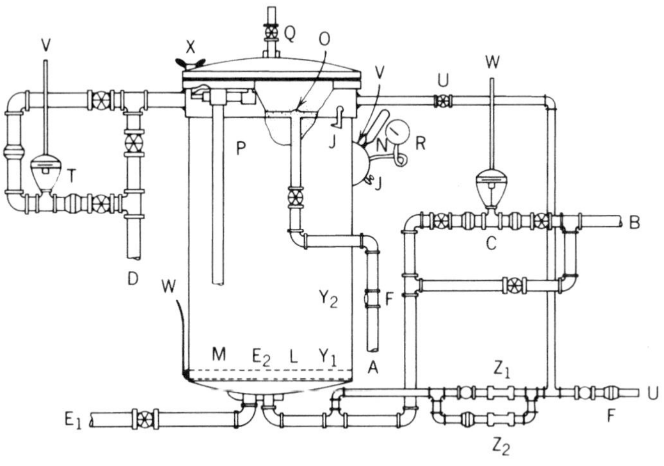

Figure 1. Vertical still retort (21 CFR, Part 113).

A—Water line.
B—Steam line.
C—Temperature control.
D—Overflow line.
E_1—Drain line.
E_2—Screens.
F—Check valves.
G—Line from hot water storage.
H—Suction line and manifold.
I—Circulating pump.
J—Petcocks.
K—Recirculating line.
L—Steam distributor.
M—Temperature-controller bulb.
N—Thermometer.
O—Water spreader.
P—Safety valve.
Q—Vent valve for steam processing.
R—Pressure gauge.
S—Inlet air control.
T—Pressure control.
U—Air line.
V—To pressure control instrument.
W—To temperature control instrument.
X—Wing nuts.
Y_1—Crate support.
Y_2—Crate guides.
Z—Constant flow orifice valve.
Z_1—Constant flow orifice valve used during come-up.
Z_2—Constant flow orifice valve used during cook.

processing. These retorts are termed still retorts (Fig. 1). Temperature inside still retorts is maintained by automatic control of the steam pressure.

Loading and unloading the containers from discontinuous still retorts are accomplished by preloading containers into crates, baskets, cars, or trays. "Crateless" systems randomly drop containers into the retort vessel which is filled with water to act as a cushion and prevent container drainage (Fig. 2). The water is drained prior to processing. The orientation of the retort depends on the type of container handling system. Systems using crates or baskets and the crateless systems, by necessity, use vertical vessels whereas car handling necessitates a horizontal orientation.

Glass, semirigid, and flexible containers must be processed in still retorts which have been designed to accommodate the fragility of these containers at retort temperatures and pressures. These retorts operate at pressures greater than the 15 psi (103 kPa) of steam required to reach 250°F (121°C) in order to counterbalance the internal pressure developed in the container. This is termed "processing with overpressure." The pressure buildup inside individual containers during processing would result in the loss of seal integrity in heat-sealed containers (see Sealing, heat) and could loosen the covers of glass containers or permanently distort semirigid plastic containers. Four design changes in still retorts must be made to process with overpressure:

1. Either steam or air overpressure must be automatically controlled. Pressures of 25–35 psi (172–241 kPa) are typical.
2. Control of the retort temperature must be independent of retort pressure.
3. Mixed heating media of either steam–air, water–air, or water–steam are used in place of pure steam. Heat transfer is less efficient in these mixed heating media. For this reason some means of circulating or mixing the heating medium is necessary. For steam–air mixtures, fans may be provided. Water–air and water–steam systems use circulating pumps.
4. Provisions to prevent stress on the containers due to motion during processing are made.

These designs may be incorporated into either vertical or horizontal retorts depending on how the containers are handled. Glass containers are typically loaded into crates or baskets for processing in vertical water–air or water–steam retorts. Flexible retortable pouches are often loaded into trays and cars (which also serve to maintain the proper shape of the pouch) and moved into horizontal retorts for water–steam or steam–air processing (see Fig. 3).

Still retorts, whether designed for metal cans or other containers, are batch (discontinous) systems. The hydrostatic retort is technically a still retort (product is not agitated) that continuously processes containers. The retort operates at a constant temperature (and pressure) as the containers are carried through the retort by a continuously moving chain (Fig. 4). The required 15 psi (103 kPa) of steam pressure inside the retort (or steam dome) is maintained by two columns of water which also serve as pressure locks for incoming and outgoing containers. These columns of water (called feed and discharge legs) must be greater than 37 ft (11.3 m) high in order to maintain at least a minimum 15-psi (103-kPa) steam pressure. Hydrostatic retorts have the highest throughputs, are efficient in their use of floorspace, steam, and water, can process a variety of container sizes and types (including flexible), and are highly automated. They have the disadvantage of high capital costs and are therefore applicable only to high volume operations.

The heating time necessary to ensure that the coldest spot in the container receives the proper lethality depends somewhat on the consistency of the product. For viscous products such as canned pumpkin and baked beans, the primary heat-transfer mechanism is conduction. Products that have a thin consistency or are packed with brine (canned peas) are heated by convection. The transfer of heat to the center of the container can be greatly facilitated if internal mixing occurs in the can during retorting. This results in a shorter processing time and higher-quality product.

Retorts that are designed to increase convection heating by container motion during processing are termed agitating retorts. Both end-over-end and axial rotation are used but the latter predominates. Glass, semirigid, and flexible containers are not agitated because of fragility. Agitating retorts may be either batch or continuous types. Continuous retorts predomi-

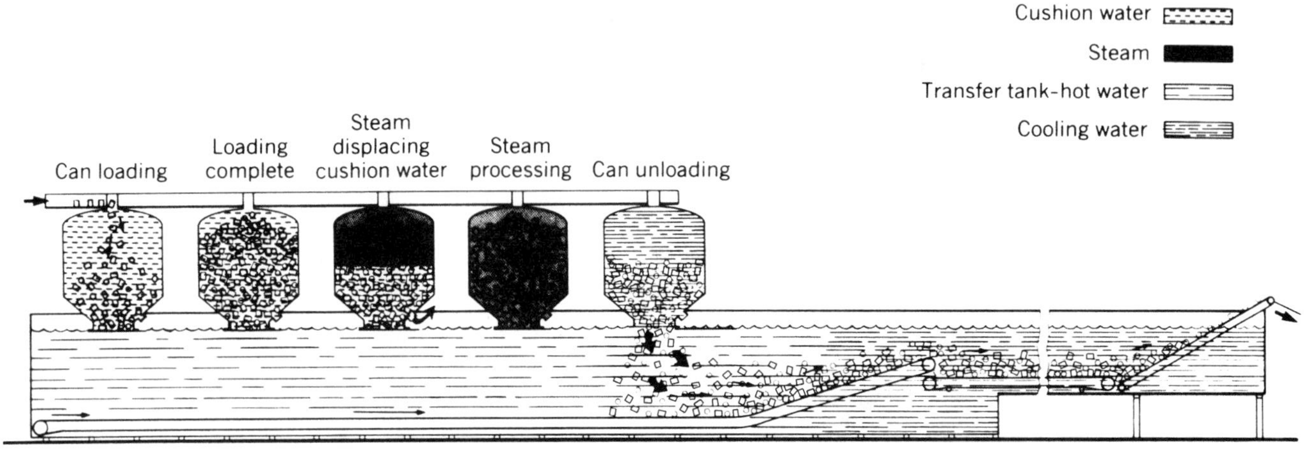

Figure 2. FMC crateless retort system. (Courtesy of FMC Corporation, Food Processing Machinery Division.)

nate because they have greater efficiency and throughout but are less easily adapted to changes in container size.

The continuous rotary cooker–cooler (see Fig. 5) has become widely used for large volume operations in which convection-heated products such as vegetables in brine are packed in metal cans. This system feeds individual cans into and out of the pressurized vessels by means of rotary pressure lock valves. Cans are rotated around the inside of the vessel's shell by means of an inner rotating reel and a series of spiral channel guides attached to the shell (Fig. 5). This system provides for intermittent agitation of the cans by providing rotation about the can axis during a portion of the reel's rotation inside the vessel. This system has the disadvantage that container size cannot be easily changed.

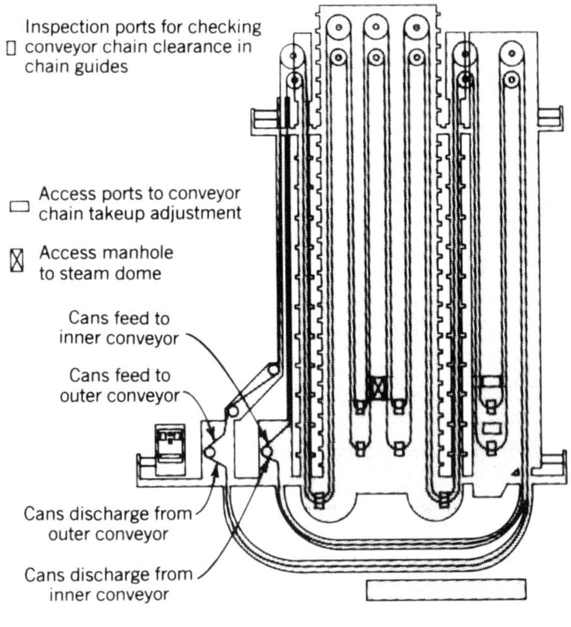

Figure 4. Flow diagram of a hydrostatic sterilizer for canned foods. (Courtesy of FMC Coporation, Food Processing Machinery Division.)

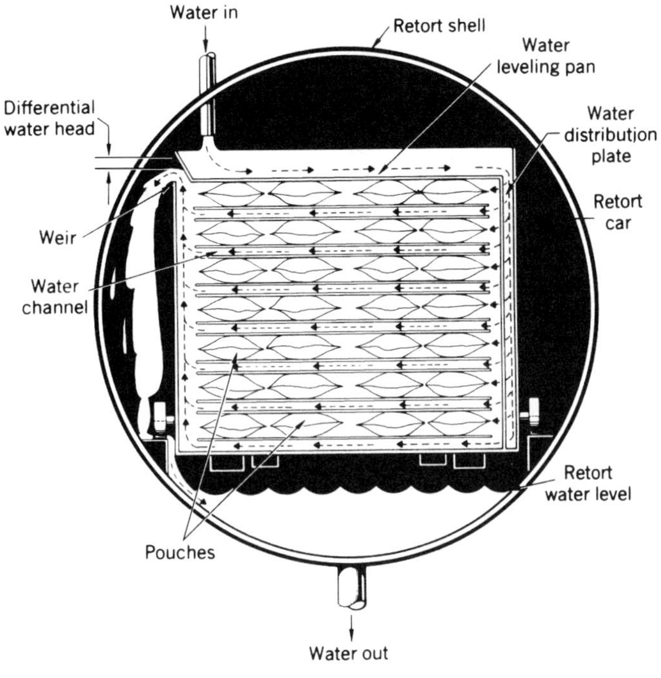

Figure 3. FMC convenience foods sterilizer, showing water flow and pouch restraints. (Courtesy of FMC Corporation, Food Processing Machinery Division.)

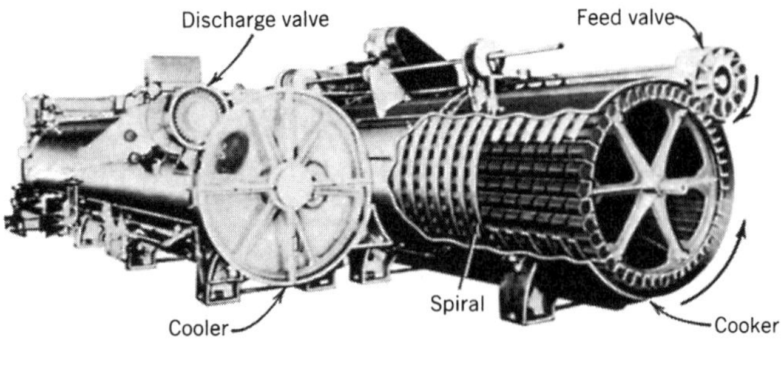

Figure 5. Cutaway view of a continuous rotary cooker–cooler's turning wheel and interlock. (Courtesy of FMC Corporation, Food Processing Machinery Division.)

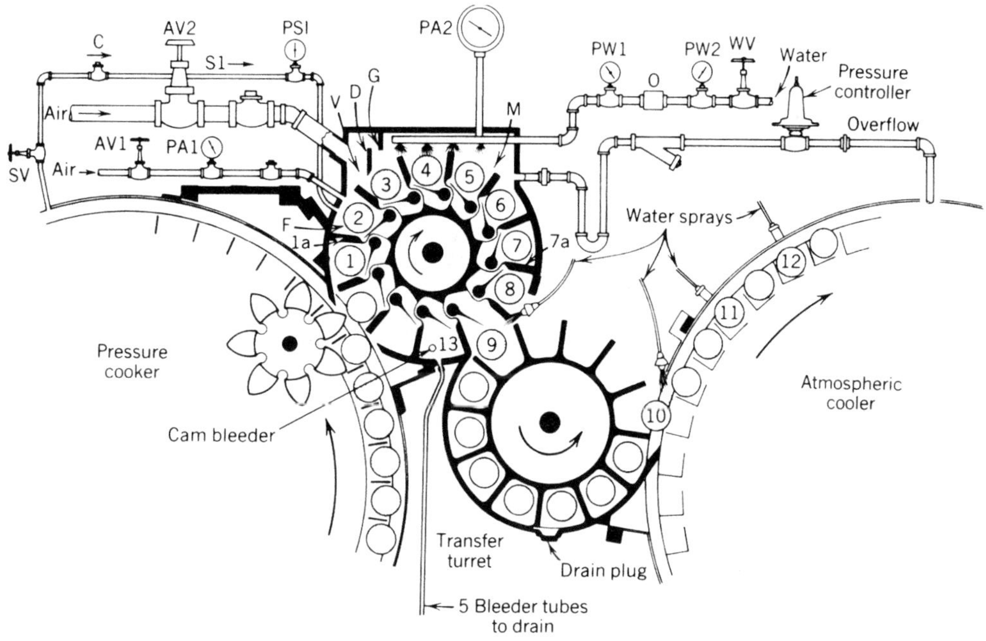

Figure 6. Transfer valve between cooking and cooling vessels of a continuous cooker–cooler.

R	Rotor containing can pockets and ejector paddles.	PW1	Downstream water pressure gauge.
SV	Forespace steam valve.	O	Orifice for flow control.
M	Microcooling chamber.	Overflow	Release line for excess air or water.
F	Forespace for pressure equilization.	S1	Steam line.
D	Water dam.	C	Check valve.
G	Splash guard.	PSI	Pressure vacuum gauge.
Air	Pressure regulated air supply.	1.	Hot can.
V	Vestibule.	1a.	Valve-leading edge.
AV1	Forespace regulated air inlet valve.	2.	The can receive some water splash.
AV2	Microcooling space air inlet valve.	3.–7.	The cans are fully exposed to water flow.
PAI	Forespace air pressure gauge.	7a.	Valve-trailing edge.
PA2	Microcooling space air pressure gauge.	8.	Can is about to leave microcooler valve.
Water	Water inlet and perforated distributor.	9.–12.	Cans are subject to sprays of water.
WV	Water inlet regulator valve.	13.	Drain pocket.
PW2	Upstream water pressure gauge.		

(Courtesy of FMC Corporation, Food Processing Machinery Division.)

Regardless of retort design, consideration must be given to cooling containers after processing. For glass, flexible, and semirigid containers, cooling with overpressure is necessary. These containers would fail due to the internal pressure developed during heating should the external pressure drop. Even metal cans may buckle and panel if brought to atmospheric pressure while the contents are at 250°F (121°C). In batch-type still retorts, overriding air pressure with water cooling is used. Hydrostatic retorts (continuous still retorts) cool containers by removing heat from the water in the discharge leg. If further cooling is necessary, an additional cooling section is added and cool water is cascaded over the containers. Continuous agitating (rotary) systems cool under pressure by transferring containers to pressurized cooling vessels by means of rotating transfer valves (Fig. 6). The second-stage vessels are maintained at elevated air pressures while the containers are cooled with water. A third-stage atmospheric cooler may also be incorporated.

In addition to the pressurized heating and cooling vessels, all retort systems require a set of precise instruments and controls. Regulations require a direct reading mercury-in-glass thermometer as well as temperature recording devices. A continuous temperature controller must be installed. For retorts using pure steam, this may be a pressure controller; processing with overpressure requires a direct temperature controller. Retorts require reliable sources of steam, air, and water. A pressure reading device is required as well as an accurate recording timing device so that the scheduled process can be insured and the proper records maintained. All instruments must undergo periodic calibration.

Regulation

The canning of foods is carefully regulated by the FDA, or in the case of canned meats and poultry, the USDA. These agencies recognize the serious public health implications of improperly processed foods. The FDA has developed a complete set of regulations commonly referred to as the Good Manufacturing Practices (GMPs) for canning foods. These regulations govern the type of equipment used to can foods and the procedures, the frequency of inspection of containers and equipment, and the records which must be kept, and they provide for the filing of individual processes prior to production. The regulations pertaining to food canning are contained in Title 21 of the Code of Federal Regulations (CFR) under the following sections: (*1*) 21 CFR Part 108, "Emergency Permit Control"; (*2*) 21 CFR Part 113, "Thermally Processed Low-acid Foods Packaged in Hermetically Sealed Containers"; and (*3*) 21 CFR Part 114, "Acidified Foods."

Part 108 stipulates that food canning plants must register their establishments and specific processes with the FDA. This section also contains provisions for issuing emergency permits to firms that the FDA believes do not fully meet the regulations. Part 113 is the most extensive section and details the equipment, procedures, process controls, establishment of correct process, critical factors, and necessary records for canning low acid foods. This section also details the procedures to be used in evaluating the integrity of the double seams. Part 114 describes the GMP requirements for packaging high acid foods (pH $\leq$4.6). This section includes general provisions as well as specific requirements for production and process control.

The USDA's Food Safety and Inspection Service (FSIS) has regulatory authority over canning poultry and meat products and has promulgated a series of regulations under Title 9 of the CFR. Current FSIS regulations are considerably more general than FDA regulations and have not kept pace with changes in canning technology. Meat and poultry canning operations are subject to continuous inspection in a manner similar to other FSIS regulated plants. These regulations are contained in sections 318.11 and 381.49, which deal with the cleaning of empty containers, inspection of filled containers, coding, use of heat-sensitive indicators (see Indicating devices), and incubation of processed products. In 1984, FSIS proposed a more detailed set of regulations similar to those promulgated by the FDA for canning low acid foods. The sections of the CFR which will deal, in part, with meat and poultry canning, and related requirements will be 9 CFR 308, 318, 320, and 381.

Trends

During the last 15 years, the food canning industry has undergone substantial changes, most notably in the area of containers (12). A major trend is to move away from tinned and soldered metal cans into tin-free steel with welded side seams or two-piece drawn–redrawn cans (13) (see Cans, steel). A majority of the cans produced in the United States today are lead-free because solder is not used. These trends are due to the unfavorable cost and availability of tin and the desire to make cans without lead. Cans made today are also significantly lighter in weight than cans made just a few years ago.

Metal as well as glass cans will have increasing competition from plastics and composite materials (see Retortable flexible and semirigid containers) (14). Although the retort pouch, in its present form, has not been a large success, second- and third-generation pouches may increase the use of these multilaminate flexible, retortable containers. The same thermal processing technology used for pouches is being used to process large one-half and one-fourth steam-table trays (15) (see Trays, steam-table). These large flat containers hold foods used in institutional feeding that would be difficult to place in conventional cans. The containers are reheated before opening by submerging in boiling water and are kept warm on steam tables.

The most dramatic trend is the further development of thermal processes in which the food is commercially sterilized before packaging. This allows food to be thermally processed in continuous-flow heat-exchange systems which can result in higher quality products and allow the use of less expensive containers based on paperboard or thin plastics. This technology is currently used for juices, drinks, and milk. However, a great deal of active research is underway to apply the same principles to other products such as soups, stews, and vegetables (8).

BIBLIOGRAPHY

"Canning Food," in *The Wiley Encyclopedia of Packaging Technology,* 1st ed., Wiley, New York, by J. H. Hotchkiss, Cornell University, 1986, pp. 86–91.

1. S. R. Friedman in W. C. Simms, ed., *The Packaging Encyclopedia—1984,* Cahners Publishing, Boston, **29**(4), 334 (1985).
2. A. Lopez, *A Complete Course in Canning,* 11th ed., Book 1, *The Canning Trade, Inc.,* Baltimore, 1981, p. 9.
3. I. J. Pflug and W. B. Esselen in J. M. Jackson and B. M. Shinn, eds., *Fundamentals of Food Canning Technology,* AVI Publishing, Westport, CT., 1979, pp. 10–94.
4. N. N. Potter, *Food Science,* 3rd ed., AVI Publishing, Westport, CT., 1978, pp. 177–193.
5. Ref. 2, p. 330.
6. *Establishing Scheduled Processes,* Code of Federal Regulations, Title 21, Part 113.83, U.S. Government Printing Office, 1983, Washington, DC, p. 112.
7. D. Wernimont, *Food Eng.* **55**(7), 87 (July 1983).
8. A. A. Teixeira and J. E. Manson, *Food Technol.* **37**(4), 128 (Apr. 1983).
9. Ref. 4, pp. 550–557.
10. Ref. 2, pp. 217–219.
11. *Canned Foods: Principles of Thermal Process Control, Acidification and Container Closure Evaluation,* 4th ed., The Food Processors Institute, Washington, DC, 1982, p. 162.
12. B. J. McKernan, *Food Technol.* **37**(4), 134 (Apr. 1983).
13. "Welded Can Expected to Capture 3-Piece Can Market," *Food Prod. Manage.* **104**(12), 12 (June 1982).
14. J. Haggin, *Chem. Eng. News.* **62**(9), 20 (Feb. 27, 1984).
15. "The Optimum Container for Mass Feeding," *Food Eng.* **54**(8), 59 (Aug. 1982).

General References

A. Lopez, *A Complete Course in Canning, 11th ed., Books 1 and 2,* The Canning Trade, Inc., Baltimore, 1981, pp. 556.

J. M. Jackson and B. M. Shin, *Fundamentals of Food Canning Technology,* AVI Publishing, Westport, CT, 1979, pp. 406.

Canned Foods: Principles of Thermal Process Control, Acidification and Container Closure Evaluation, 4th ed., The Food Processors Institute, Washington, DC, p. 246.

CAN SEAMERS

Can seamers are machines that mechanically attach component ends to can bodies in a reliable manner. Around 1900, the sanitary can made its appearance in Europe, where both top and bottom ends were double-seamed to the can body. The term "sanitary" indicated that solder was not used in the ends being double-seamed, but only on the outside of the can body side seam (1). In 1910 Henry Louis Guenther, inventor and manufacturer, introduced can seamers for double seaming that met the requirements of modern food and beverage processing. His products, which are sold under the trade name of "Angelus," were so well introduced that they are now used by the largest can manufacturing and packing companies in the United States as well as abroad (2). Basically, there are two categories of can seamers; in can manufacturing they are called *can shop machines* which attach the first end on a three-piece can, and in product filled cans they are called *closing machines,* which attach the last end on either a two-piece or three-piece can.

A double seam requires two operations to produce a quality seam and is formed by seaming rolls as the can body and can end are held together on a seaming chuck by a vertical load applied to the lower lifter or baseplate table as the can parts move through the machine. During this seaming cycle the can end and can body meet, and the first-operation seaming roll contacts the can end and begins curling the can end around the can body flange. A second-operation seaming roll follows, which tightens and irons out the seam between the can body and can end forming an airtight hermetic seal between the two parts (see Fig. 1).

The double seam is a critical can component. Every angle, radius, and dimension must be correct to ensure a hermetic seal (3). The double seam is defined as "The curl on the can end containing sealing compound and the flange on the can body are indexed and rolled flat, forming five folds of metal. Sealing compound between folds gives an air-tight seal (4)."

Method of Seaming

There are two basic seamer designs: can spin and can stand still. Practically all closing machines designed in the early years were of the can stand still type incorporating up to four seaming heads and operating at speeds ranging from 25 to 275 cans per minute. The can stand still design is still used extensively and in many cases is a necessity due to the products being closed, such as shortening. In this design the nonrotating can body and can end are assembled between the knockout rod pad and the lower lifter or base plate table. The knockout rod pad keeps the can end firmly in place as the lower lifter, which is synchronized with the knockout pad, raises the can body and can end into the seaming position on the seaming chuck. First operation seaming rolls, which are diametrically opposite each other in a seaming head, revolve around the stationary seaming chuck, and pressure is applied through cam action to form the first operation. After the first operation is completed to the proper thickness, the second-operation seaming rolls, which are diametrically opposite each other, iron out the double seam to the proper thickness. After the seaming operations are completed and the second-operation seaming rolls have been released, the knockout rod follows the seamed can away from the seaming chuck as it is being lowered to the discharge position by the lower lifter (5).

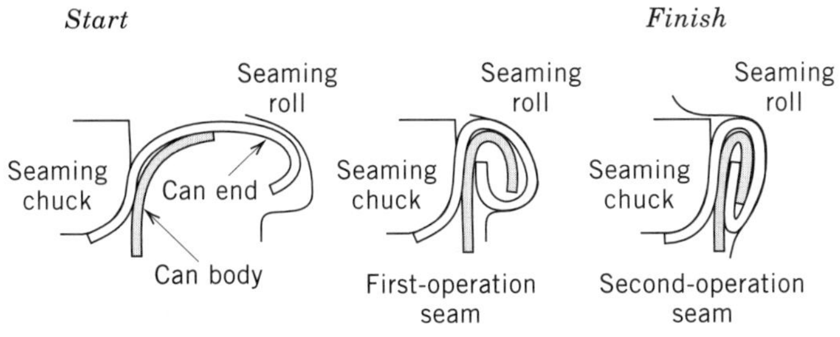

Figure 1. Double-seam process. Courtesy of Angelus Sanitary Can Machine Company.

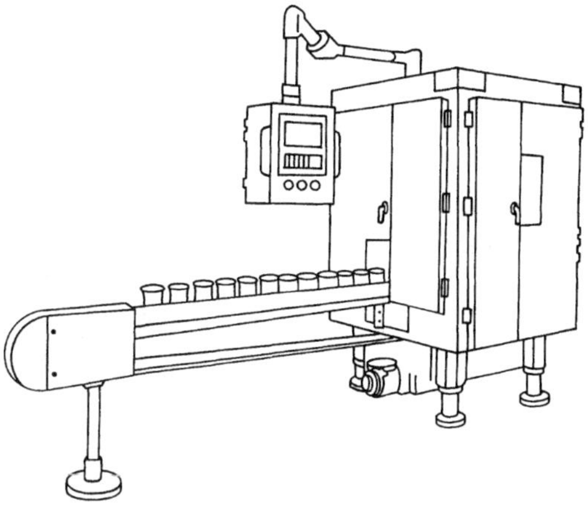

Figure 2. Can seamer. Courtesy of Angelus Sanitary Can Machine Company.

As canning speed requirements increased, can spin or rotating-can-type seamers were developed using a multistation design. Machines with 4, 6, 8, 10, 12, 14, and 18 seaming stations are in use providing production speeds of 100–2300 cans per minute (see Fig. 2). Can ends are automatically separated and mated with the can bodies in the seaming position. The knockout rod pad contacts the can end while the lower lifter, synchronized with the knockout rod pad, lifts the can body and can end into position on the seaming chuck. As contact is made with the seaming chuck, the can starts to revolve with the rotating seaming chuck. On thin-walled lightweight cans the machine design incorporates driven seaming chucks and driven lower lifters to prevent *can skid* or *can buckling*. One first operation seaming roll actuated by a cam forms the first operation seam. After completion of the first operation seam, the second operation seaming roll actuated by a similar cam action irons out the seam to the proper thickness.

Machine Type by Application

Can seamers are typically manufactured in five machine types for seaming can ends to can bodies: can shop, atmospheric, mechanical vacuum, steam vacuum, and under cover gassing.

1. *Can shop.* For can manufacturing, attaches the first end on three-piece cans.
2. *Atmospheric.* For can closing products not requiring removal of oxygen for preservation, ie, soaps, petroleum products, frozen products, hot products >180°F.
3. *Mechanical vacuum.* Evacuates oxygen from headspace

of can for preservation of product at slow speeds, ie, vegetables, specialized powdered products.

4. *Steam vacuum.* Evacuates oxygen from headspace of can with steam for preservation of product at high speeds, ie, fruits, vegetables, soups, fish, meat products, juices.
5. *Under cover gassing.* Displaces oxygen in headspace of can with a gas, such as carbon dioxide or nitrogen to extend shelf life and/or increase internal can pressure for thin-walled aluminum cans, ie, beer and soft drinks.

Generally, during the process of modern automated canning operation cans are filled with a measured amount of product, then transferred from the filler to the can feed table of the seamer. This transfer from the filler to the seamer is critical, and must be timed in such a manner to avoid can damage, product spill, or extreme product agitation rendering a smooth flow of cans and product into the seamer. As the cans move into the seamer, they are sensed by a mechanical, electrical, or optical device that triggers a signal to separate one can end from a stack and feed it in a synchronized rotary manner to an incoming can. Generally, the timing of machine operations is mechanically controlled by cams and there is a dwell period for steam vacuum and under cover gassing applications as steam or gas is injected between the top of the open can and the can end prior to their contact. Filling the headspace of cans with steam or gas displaces the air, preserving the quality of the canned product. After the can parts meet, they move through the seaming cycle of the first and second seaming operations and then the cans are discharged from the machine.

Seamer Setup

Can seamers are designed to double-seam a given range of can diameters, can heights, and speeds. Can diameters and can heights are expressed in both inches and millimeters, but generally use an industry nominal diameter, such as 202, 207.5, 208, 211, 300, 301, 303, 307, 401, 404, 502, and 603. Industry nominal diameters are defined as follows: The first digit equals inches. The second and third digits equal 1⁄16 fraction of an inch. For example, a can with a nominal 211 diameter would be the equivalent of 2 11⁄16 in., and a can with a nominal 307 diameter would be the equivalent of 3 7⁄16 in.

It is very important that seamers be set up to the correct specification for the type of can end and can body, diameter of the can end and can body, and material thickness of end and body. Important setup procedures include.

1. Checking the fit of seaming chuck to can end.
2. Initially installing only the first operation seaming rolls. Remove second-operation seaming rolls if in place.
3. Installing seaming chucks and checking that the first operation seaming rolls do not interfere with the seaming chucks when in the seaming position.
4. Setting lower lifter assemblies to correct height relationship with can feed table.
5. Setting lower lifter spring pressure to proper load with an appropriately calibrated instrument, such as a Dillon force gauge or force cell gauge.
6. Setting pin height with an appropriately calibrated instrument, such as a pin height gauge or planer gauge.
7. Setting the first operation seaming rolls on each station to a specified seam thickness using a wire gauge of the proper diameter. Run samples of first-operation seamed cans to verify quality of seam. Visually inspect the seam measuring the seam thickness, seam width, and countersink depth.
8. Installing and properly adjusting the second operation seaming rolls to the seaming chucks on all stations to the specified seam thickness. Run samples of the finished second operation seam. Visually inspect the seam measuring the seam thickness of the finished seam to given specifications using a properly calibrated seam micrometer. Finally, tear down the second operation seam for further inspection.

Key Features and Attachments

Depending on the application and production requirements, can seamers are equipped with features and various attachments to meet the demands of industry processors.

Automatic stops. For safety reasons, machines are equipped with mechanical, electrical, or optical sensors that cut power to the motor, actuate the clutch release unit, and apply the brake to stop the machine rapidly. These safety devices are located at critical areas on the machine.

Filler drive seamer safety clutch. A safety overload clutch is used to protect the seamer in case of a severe can jam or mechanical failure in the filler.

In-motion timer. This is a timing attachment located between the seamer and filler that synchronizes the transfer of cans from the filler pockets to the seamer feed chain fingers during machine operation.

Can coding markers. Basically there are two types of markers—mechanical and ink-jet—used to place the processor's identification code on can ends. Mechanical markers use type dies that are capable of debossing and embossing identification characters on can ends. Debossing is where the characters are indented into the top of the can end, and embossing is where the characters are raised on the top of the can ends. Mechanical markers are driven by the seamer and have speed limitations up to 1000 cans per minute. They are used primarily on sanitary food cans. Ink-jet markers are not driven by the seamer and use a nozzle assembly device to print droplets of ink to make up characters forming alphanumerics or bar codes. Ink-jet markers apply clear codes to virtually any surface at nearly any production speed, using a programmable controller and software to monitor the ink quality, size, font, and lines of print.

Automatic lubrication. Metered amounts of grease and/or oil are automatically delivered by pumps to designated machine areas requiring lubrication while the machine is operating.

Automatic oil lubrication recirculating and filtration system. This system continually filters water and particles from the recirculating oil. The lubricating oil is pumped and recirculated through the machine, reducing the amount used and the environmental concerns of discarding cycled oil.

Programmable controller. This provides electrical monitoring of seamer functions and operates auxillary equipment.

Driven lower lifters. Also referred to as *driven lower chucks* or *driven baseplate tables.* They accept incoming can bodies or product-filled cans from can infeed devices before being raised by a cam action to meet can ends to be seamed at the makeup area. They are gear-driven and rotating in a synchronized design with the seaming chucks to provide stability and enhance can control during the seaming cycle process. Driven lower lifters use a preset spring pressure, which is a vital component in the formation of a double seam.

Seam Tightness Evaluation

Generally, in order to stand the rigors of processing, handling, damage by abuse, distribution, and to ensure product shelf-life, the tightness of the seam is critical and should be evaluated carefully. During the formation of a double seam the proper tightness assures that the sealing compound will fill all the spaces not occupied by metal.

Seam tightness is normally evaluated by the degree of waviness or wrinkle found in the cover hook. This wrinkle is formed by compression of the curled outer edge of the can end as it is folded back under the body flange. By increasing the pressure of the seaming rolls the wrinkle can be ironed out to a smooth strip, and by loosening the rolls, the wrinkle is increaesd (6). Wrinkles may be classified by a tightness (wrinkle) rating as shown in Figure 3. A 70% wrinkle rating equates to a 30% wrinkle in the cover hook or 70% of the cover hook is wrinkle free. There are other numerical cover wrinkle rating systems used; the Dewey and Almy wrinkle rating uses a 0 to 10 scale in which absence of wrinkle rates as zero and a full-width wrinkle rates as ten, another uses a 0 to 3 rating system. The rating for each end component is based on the worst or deepest wrinkle, for it is at this point or area that the seam is most vulnerable to abuse, leakage, and penetration by bacteria.

In hemming a straight edge of metal, no wrinkles are formed. On curved edges, wrinkling increases as the radius of curvature decreases. For this reason, different wrinkle ratings are specified for small diameter cans as compared to large diameter cans (7). A 100% wrinkle rating of a 211 diameter can indicates that the seam may be too tight and should be watched for possible defects such as cut-overs, droops, and unhooking. A 100% wrinkle on a 603 diameter can would not necessarily be too tight. With most seamers using standard seaming roll profiles, the ideal seam almost invariably shows a slight wrinkle, except on 603 diameter and other large diameter cans.

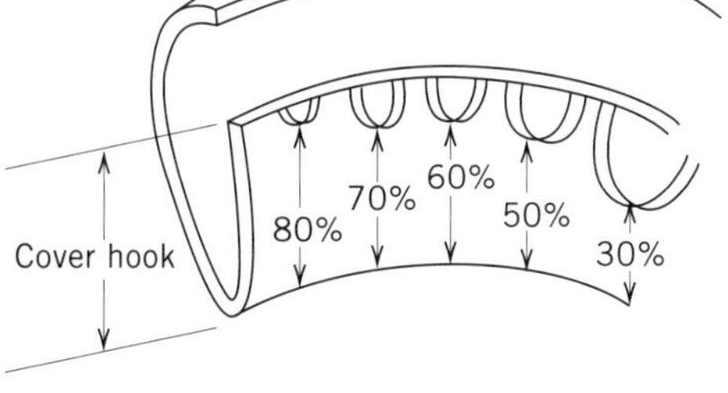

Figure 3. Wrinkles are rated by the percentage of tightness. Courtesy of Angelus Sanitary Can Machine Company.

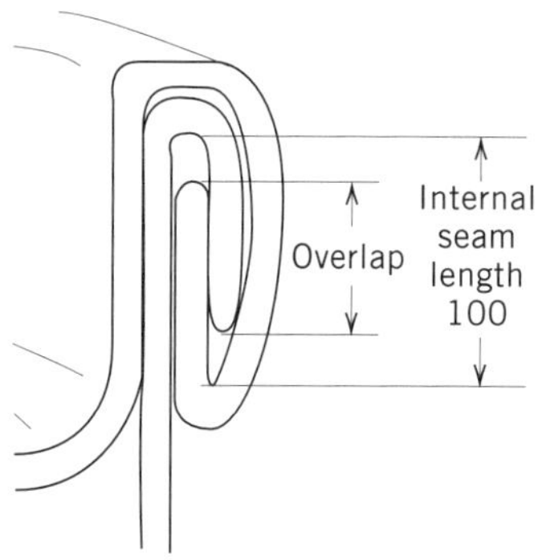

Figure 4. Overlap measurement. Courtesy of Angelus Sanitary Can Machine Company.

Measurement of Percentage Overlap

The overlap of a double seam is expressed as a percentage of the maximum possible overlap. The percentage overlap of the seam is established by first measuring the internal seam length, a measurement between the inside of the cover hook and the inside of the body hook and rate this length as 100. The measured length of the actual overlap is then a portion or percentage of that length (see Figure 4).

Profiles of Seams

Properly formed first and second operation double seam requires the correct adjustments of pressure of the seaming rolls and lower lifter or base plate table. The shape and conformity of the finished seam is determined by the taper and fit of the seaming chuck to the top of the end component and the contoured profile of the seaming rolls. The seaming roll profile is a groove around the circumference of the roll which varies with the diameter of the can to accommodate variations in material, material thickness, cover curl, body flanges, and seam specifications of the user. For any given can diameter there may be a number of roll profiles which will properly form the double seam. First and second operation roll profiles are uniquely different. As the first operation seaming roll profile contacts the cover curl, the flange of the can bends over to form the body hook and the edge of the cover tucks underneath to form the cover hook. The second operation seaming roll profile completes the seam formation by compressing the seam so that the hooks interlock tightly and any metal voids or spaces are filled with the sealing compound.

BIBLIOGRAPHY

1. F. L. Church, *Modern Metals,* 28 (Feb., 1991).
2. *The National Cyclopedia of American Biography,* Vol. XXXIII, James T. White & Co., New York, 1947.
3. W. Soroka, *Fundamentals of Packaging Technology,* Institute of Packaging Professionals, Herndon, Virginia, 1995.

4. CMI Can Standards Manual, Can Manufacturers Institute, Washington, DC, 1992.
5. *Canning Industry History, Terminology & Reference Manual,* Angelus Sanitary Can Machine Company, Los Angeles, CA, 1968.
6. *Evaluating A Double Seam,* W. R. Grace & Co., Cambridge, Mass.
7. *Top Double Seam Inspection & Evaluation,* American Can Company, Geneva, New York, 1966

Otto L. Heck
and Richard Cabori
Angelus Sanitary Can Machine Co.
Los Angeles, California

CANS, ALUMINUM

Over 98% of all aluminum cans are drawn-and-ironed (D & I) cans used for beer and soft drinks (see Cans, fabrication; Carbonated-beverage packaging). The other 1–2% is accounted for by small shallow-draw food cans. In the United States, production of aluminum beverage cans has reached about 100 billion/yr. Virtually all beer cans are aluminum, and 100% of all soft-drink cans. Until 1965, the three-piece soldered can was the only can used for beer and beverages (see Cans, fabrication; Cans, steel). It was generally made of 75-lb per base box (16.8 kg/m^2) tinplate (see Tin-mill products). Aluminum does not have the strength of steel per unit weight and cannot be soldered. Therefore, early in the development of the aluminum can, attention was focused on improving the properties of aluminum and perfecting two-piece D & I technology. More recently, the emphasis has been on saving metal, because the cost of metal is the single largest component of final product cost. One of the chief reasons for the success of aluminum cans has been their scrap value for recycling (see Energy utilization; Recycling). This article pertains to the technical developments that have led to today's aluminum beverage can.

The significant developments took place in the late 1960s. Earlier, in 1958, Kaiser Aluminum attempted to make a 7-oz (207-mL) aluminum can using a 3003 soft-temper aluminum of the type used for aluminum-foil production, but the effort was not a commercial success. In the universe of aluminum alloys, the 3-series alloys contain a small percentage of manganese as the principal alloying element. The success of the aluminum can depended on the development of the 3004 alloy for the can body, which contains manganese along with a slightly lower amount of magnesium. (The softer 5182 alloy for can ends contains a higher amount of magnesium.) Reynolds Metals Co. began making production quantities of 12-oz (355-mL) seamless D & I cans in 1964 using a 3004 alloy. The walls of these first cans were straight, with a top diameter of "211": $2^{11}/_{16}$ in. (68.26 mm). The starting gauge was 0.0195 in. (0.495 mm); can weight was 41.5 lb/1000 (18.8 kg/1000).

Reynolds introduced the first necked-in cans (see Fig. 1) in 1966, reducing the top diameter from 211 to 209 ie. $2^{9}/_{16}$ in. (65.09 mm). This represented a break-through in technology and container performance, particularly as it is related to cracked flanges. Cracked flanges were a serious problem in both double-reduced tinplate and straight-walled aluminum cans. The introduction of carbide knives at the slitter essentially eliminated cracked flanges on tinplate cans. Eliminating them on aluminum cans required a change from die flanging to spin flanging, and necking-in before flanging, which does not stretch the metal beyond its elastic limits.

Figure 1. A necked-in can.

In 1968, a new harder-temper (H19) 3004 alloy was introduced for aluminum cans. Although the 3004–H19 combination had been available since the 1950s for other purposes, it was not until 1968, when Alcoa and Reynolds were in commercial production with full-hard-temper can sheet, that it could be used to effect significant weight reductions. Since then, weight/1000 cans has decreased from 41.5 lb (18.8 kg) in the mid-1960s to 34 lb (15.4 kg) in the mid-1970s and to less than 30 lb (13.6 kg) in the mid-1980s.

Gauge reductions have increased the point where design techniques have become critical to sustaining the can's ability to hold the product. With few exceptions, U.S. brewers pasteurize beer in the can. This generates high internal pressures, and most cans used for beer must be designed so that they have a minimum bottom buckle strength of 85–90 psi (586–620 kPa), depending on the carbonation level. Brewers are also asking for cans with minimum column strength, ie, vertical crush, of 300 lbf (1330 N).

A revolution has taken place in bottom profiles (see Fig. 2). The original D & I bottom had a rather generous bottom-heel radius. In order to meet the 90-psi (620 kPa) minimum bottom buckle-strength requirement, a 211 can with this configuration would have to be made with a starting gauge of 0.016–0.0165 in. (406–419 μm). The next profile development, basically the Alcoa B-53 design, is widely used

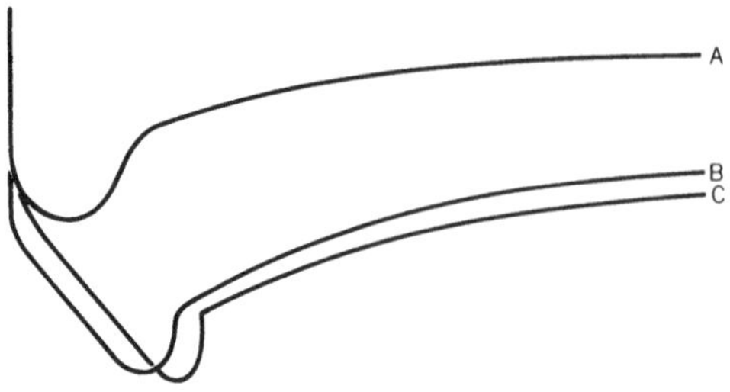

Figure 2. Can-bottom profiles: A, original bottom profile; B, Alcoa B-53, V-bottom profile; C, Alcoa B-80. (Courtesy of Alcoa.)

throughout the beer industry today because it allows the starting gauge to go as low as 0.013 in. (330 μm), in most cases without sacrificing the 90-psi (620-kPa) minimum bottom buckle strength. The newest entry is the Alcoa B-80 profile, which allows starting-gauge reduction to ca 0.0126 in. (320 μm).

An "expandable bottom" design, still on the drawing boards at the aluminum and can companies might permit use of a 0.010-in. (254 μm) starting gauge for pasteurized product. In contrast to the current dome profiles, an expandable bottom is essentially a flat bottom with small pods located near the perimeter to provide stability to the pressurized can, reduce drag, and increase mobility. A totally flat bottom would drag too much on the filling line. The "expandable" bottom is designed to flex outward during pasteurization, relieving some of the generated pressure.

Sidewalls are also being redesigned (see Fig. 3). In the 1970s, the so-called nominal thinwall, ie, the area of the can that has been thinned most, generally ranged from 0.0052 to 0.0053 in. (132–135 μm). More recent versions of the D & I can have reduced the nominal sidewall to 0.0045 in. (114 μm). A reduction of this magnitude represents substantial cost reduction. It also means a corresponding reduction in the overall column strength (vertical crush) of the can, not below the minimum 300 lbf (1330 N), but in terms of overall operating average. Column strength is very critical to the brewers, who ship long distances by truck and rail. Extensive testing by the can companies and the beer and beverage industry in general has shown that the 300-lbf (1330-N) minimum is satisfactory. Because of the reduction in body-wall thickness, dents that were acceptable before have now become critical owing to their influence on reducing the can's column strength. Can makers and brewers are monitoring their handling systems for empty and full cans to minimize denting wherever possible.

In the 1980s, can suppliers have reduced costs further by double "necking-in" (see Fig. 4). These configurations reduce costs primarily because of the diameter reduction of the lid. In 1984, further activity with respect to necking-in began to occur. Cans with three or four die necks are now being run commercially (see Fig. 4). In Japan, a can with eight necks is being tested. Metal Box (UK) has introduced a spin-neck can (see Fig. 4) which essentially produces the same 206 top diameter as the triple- or quadruple-neck can, ie, 2$\frac{6}{16}$ in. (60.33 mm).

Another advantage of aluminum is the "split gauge." The industry used to sell coils in 0.0005-in. (12.7-μm) increments, eg, 0.0130 in. (330 μm), 0.0135 in. (343 μm), etc. A new pricing structure introduced in 1983 allows can-stock buyers to order gauge stock in 0.0001-in (2.54-μm) increments. A can manufacturer can reduce costs by taking advantage of these slight gauge reductions. In addition, there has been a change in the gauge tolerance as rolled by the aluminum mills. In the 1970s, the order gauge was subject to a $\pm$0.0005-in. (12.7 μm) tolerance; today, it has been reduced to 0.0002 in. (5.1 μm). This permits further gauge reduction because it allows the can manufacturer to reduce the order gauge without changing the minimum bottom buckle strength.

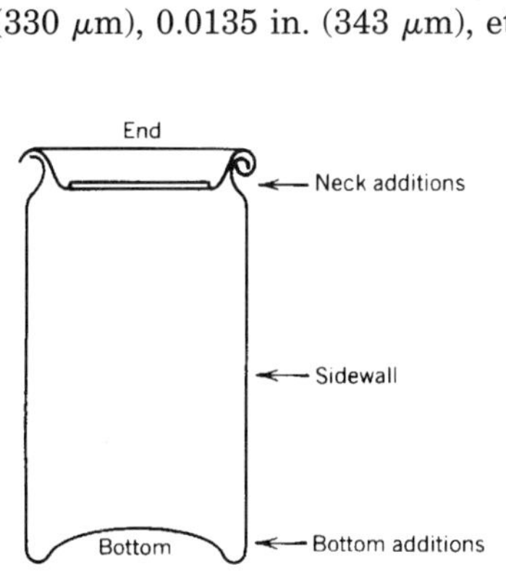

Figure 3. Design elements.

Figure 4. Double-, triple-, quadruple-, and spin-neck can.

In the years since this article was written, a number of changes or improvements have been introduced. Today 100% of all U.S. beer and soft-drink cans are fabricated from aluminum. The industry has moved from 206-diameter necks on 211-diameter bodies to 204- and 202-diameter necks. These diameter changes have reduced the metal gauge and net weight from 0.0108 in. and 7.3-lb/1000 ends (for 206) to 0.0088 in. and 5.3 lb/1000 for 202 ends.

Net can weights have also been reduced from approximately 30 lb/M at 0.0125 in. to 23 lb/M at 0.0108 in. This has been accomplished primarily by creating new stronger dome geometry, along with dome postreforming technology to improve performance criteria, such as dome buckle, drop resistance, and dome growth. These new dome designs are more difficult to manufacture, but provide improved performance and stacking at reduced gauges.

Can sidewalls have been further reduced from 0.0045 in. down to 0.0039 in. nominal, and the number of die necks utilized to produce finished necks has increased to as many as fourteen for the 202 diameter. Multidie necks combined with spin necking are still popular for all three neck diameters, and several canmakers are using multidie necking with Spin Flow necking.

BIBLIOGRAPHY

General References

J. F. Hanlon, *Handbook of Package Engineering,* 2nd ed., McGraw Hill, New York, 1984.

Drawn & Ironed Aluminum Cans, Aluminum Company of America, Pittsburgh, PA, 1975.

Tin Mill Products Manual, U.S. Steel Corp., Pittsburgh, PA, 1970.
Brewing Industry Recommended Can Specifications Manual, United States Brewers Association, Inc., Washington, DC, 1981.

TERRE REINGARDT
Ball Corporation
Westminster, Colorado
N. F. NIEDER
Anheuser-Busch, Inc.

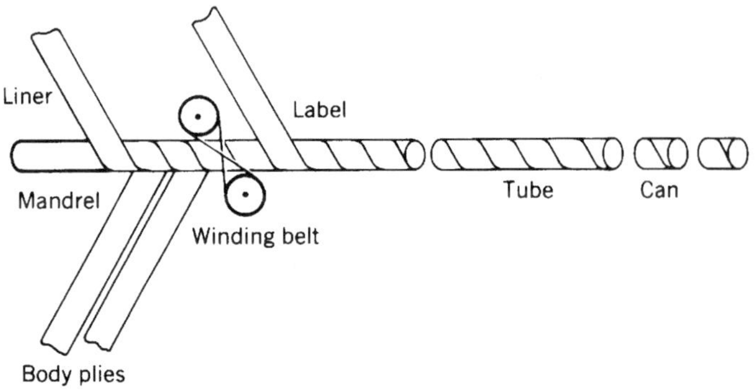

Figure 2. Spiral-wound composite-can fabrication.

CANS, COMPOSITE

The "composite can" is broadly defined as a can or container with body and ends made of dissimilar materials. In commercial practice the composite can has several more focused descriptions: cylindrical or rectangular shape; rigid paperboard (or plastic) body construction; steel, aluminum, or plastic end closures; generally employing inside liners and outside labels; and generally delivered with one end attached and one end shipped separately to be attached by the user. Today the most common form is the cylindrical paperboard can with a liner, a label, and two metal ends. There are other packaging forms similar to traditional composite cans that are sometimes called *composites,* but which are more closely related to folding cartons. These are single-wrap fiber cans made from blanks and mainly designed for the users' in-plant production (see Cans, composite, self-manufactured).

The composite can is not a new package. Early applications included refrigerated dough and cleansers. As technology improved, motor oil and frozen juice concentrates were converted to composites. In the last decade, snack foods, tobacco, edible oils, shortening, powdered beverages, pet foods, and many other items have been added to the list. Figure 1 illustrates some of the products currently available in composite cans.

Since their introduction, composite cans have generally been marketed and used as a lower-cost packaging form relative to metal, plastic, and glass. This emphasis has overshadowed other positive attributes such as the frequent use of recycled materials, weight advantages, noise reduction, improved graphics, and design flexibility. Because of its early applications in cleansers and oatmeal, the composite can once suffered from a low performance image associated with the term "cardboard can." This term cannot begin to describe the current and potential properties of the paperboard and other materials that go into today's composite cans.

Composite cans are normally available in diameters of 1–7 in. (3–18 cm) and heights of 1–13 in. (3–33 cm). Dimensional nomenclature for composites has been adapted from metal cans, and nominal dimensions are expressed in inches and sixteenths of an inch (see Cans, steel). Hence, a 404 diameter can has a nominal diameter of $4\frac{4}{16}$ in. (10.8 cm). Likewise, a height of $6\frac{10}{16}$ in. (16.8 cm) is expressed as 610, etc.

Figure 1. Examples of composite cans in commercial use.

Manufacturing Methods

Composite-can bodies are produced by two basic methods: spiral winding and convolute winding. Figure 2 shows a schematic drawing of the spiral process. Multiple webs including a liner, body plies, and label are treated with adhesive and wound continuously on a reciprocating mandrel. The resulting tube is trimmed and the can bodies are passed on to flanging and seaming stations. Figure 3 depicts the convolute process, wherein a pattern is coated with adhesive and entered onto a turning mandrel in a discontinuous process. Trimming and finishing operations for the convolute and spiral systems are virtually the same. Most composite-can manufacturers favor the spiral process in situations where long production runs and few line changeovers are involved.

Body Construction

Paperboard. The primary strength of the composite can is derived from its body construction, which is usually paperboard (see Paperboard). Body strength in composite cans is

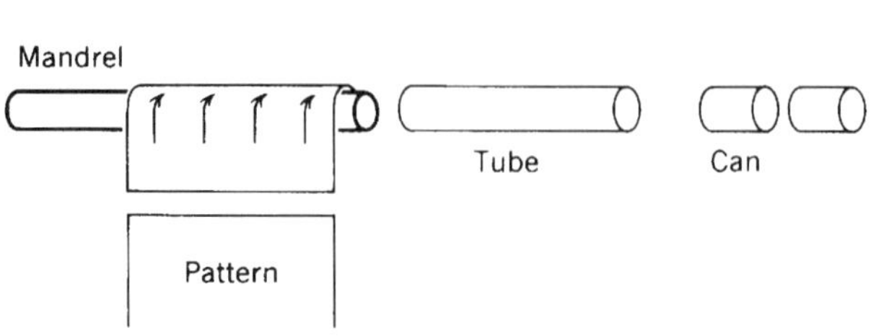

Figure 3. Convolute composite-can fabrication.

Table 1. Adhesives Used in Composite-Can Manufacture

Adhesives	Properties
Poly(vinyl alcohol–acetate) blends	Good initial tack, good runnability, moderate to good water resistance
Dextrin	Fast tack, poor water resistance
Animal glue	Good tack, vulnerable to insect attack
Polyethylene	Requires heat, good dry bond, moderate water resistance
Hot melts	Require heat, difficult to handle, good water resistance, good water bond

an attribute that has improved over the years and can be varied to meet many application demands.

In the early years of composite can development, it was common for can manufacturers to start with a readily available body stock such as kraft linerboard or tube-grade chip. These boards are adequate for most applications, but new boards with special qualities have also been developed for more demanding end-use requirements.

Research and development efforts in the combination paperboard field encompass a number of areas of expertise. Examples are engineering mechanics concepts used to develop structural criteria and to develop tests to assure that the paperboards possess the necessary resistance to bending buckling, and creep; surface chemistry used to predict the resistance of the board to penetration of adhesives, coatings, and inks and, similarly, protection from environmental conditions such as rain, high humidity, and freezing temperatures; and process engineering used to optimize paperboard manufacturing and converting and to assist in quality assurance programs (see Specifications and quality assurance).

The scientific and engineering efforts have, in many instances, supplemented the artisanlike judgment of yesterday's papermaker. However, in many cases the new technological approaches have been blended with the papermaker's art to achieve the best of both worlds. As a result, the following advances have taken place: super-high-strength board that can be converted into composite cans with reasonable wall thicknesses that can resist implosion when subjected to near-perfect vacuum; resin-treated paperboards that

Table 3. Physical Characteristics of Composite Can Liners

Composite[a]	WVTR[b] at 100°F (37.8°C) and 90% rh: Flat	Creased	O_2 permeability[c], $cm^3/(m^2 \cdot d)$
100 ga PP/adh/100 ga AF/LDPE/25# MGNN kraft	< 0.001	<0.001	<0.001
100 ga PP/adh/35 ga AF/LDPE/25# MGNN kraft	0.06		<0.02
12# ionomer/35 ga AF/LDPE/30# XKL kraft	0.01		<0.02
1# PET slipcoat/35 ga AF/casein/25# MGNN kraft	0.09	0.93	<0.001
1# PET slipcoat/35 ga AF/LDPE/25# MGNN kraft	0.06	0.09	<0.001
12# HDPE/20# MGNN kraft	15.35		153
14.4# HDPE/20# MGNN kraft	12.09		126

[a] key: # = lb/ream = 454 g/ream.
100 ga = 0.001 in. = mil = 25.4 μm.
MGNN = Machine-grade natural Northern.
XKL = extensible kraft linerboard.
[b] ASTM Test Method E 96-80.
[c] ASTM Test Method D 3985.

Table 2. Physical Characteristics of Commonly Used Liner Films[a]

Physical characteristics	Polypropylene: Oriented	Polypropylene: Oriented PVDC coat	Polypropylene: Nonoriented	Polyester	LDPE	Ionomer
Properties						
Tear strength, gf (N)	3–10[b] (0.03–0.10)	3–10 (0.03–0.10)	50–MD[b] (0.5) 300–XD[b] (3)	12–27[b] (0.12–0.27)	50–150[b] (0.5–1.5)	50–150[b] (0.5–1.5)
Burst strength, psi (kPa)				55–80 (379–551)	10–12 (69–83)	10–12 (69–83)
WVTR, g·mil/(100 in.²·24 h) [g·mm/(m²·d)]	0.75 [29.5]	0.3–0.5 [11.8–19.7]	1.5 [59]	1.5 [59]	2.0–3.0 [78.7–118.1]	2.0–3.0 [78.7–118.1]
O_2 rate, cm³·mil/(100 in.²·24 h) [cm·m/m²·d)]	160 [630]	1–3 [3.9–11.8]	160 [630]	3.0–4.0 [11.8–15.7]	500 [1970]	250–300 [985–1180]
elongation, %	35–475	35–475	550–1000	60–165	100–700	400–800
Product resistance[c,d]						
Strong acids	G	G	G	G	G	G
Strong alkalines	G	G	G	G	G	G
Grease and oil	G	G	G	G	G	E

[a] Source: Sonoco Products Company, Hartsville, SC.
[b] gf/mil (0.386 N/mm) thickness.
[c] G = good.
[d] E = excellent.

retain their structural integrity when thoroughly wet; chemically treated paper that resists penetration by water over long time intervals; and paperboards that can be distorted, rolled, and formed in high-speed converting equipment without "creeping" back to their original shape.

Adhesives. The adhesives (see Adhesives) and coatings used in the manufacture of composite cans have also been improved to provide better heat and water resistance plus increased operating efficiency. Most product applications require precision gluing equipment to control the amount and position of the adhesive on each web. The most commonly used adhesives in today's composite can production are listed in Table 1.

Liners. Like all successful packages, the composite can must contain and protect the products. For that reason, continuous improvement is sought in liner materials. By combining materials such as LDPE (see Polyethylene, low-density), HDPE (see Film, high-density polyethylene; Polyethylene, high-density), PP (see Film, nonoriented polypropylene; Film, oriented polypropylene; Polypropylene), ionomer (see Ionomers), PVDC (see Vinylidene chloride copolymers), and PE (see Film, oriented polyester; Polyesters, thermoplastic) with aluminum foil (see Foil, aluminum) or kraft (Paper; Paperboard), the barrier properties of composites can be matched with a broad range of product requirements (see Laminating, Multilayer flexible packaging). The polymers may be included as film or coatings, or both. Tables 2 and 3 illustrate the physical properties of some of the more common liner films and complete liner constructions.

The laminates shown in Table 3 vary significantly in cost. Depending on can size, the difference in cost per thousand units can be substantial. The foil-based laminates, which provide virtually 100% water and gas barrier, are becoming quite expensive. If less than total impermeability is acceptable, it pays to investigate coated- or plain-film alternatives. For example, packers of frozen juice concentrates have gradually moved away from foil liners to laminations of PE or ionomer/PE and kraft.

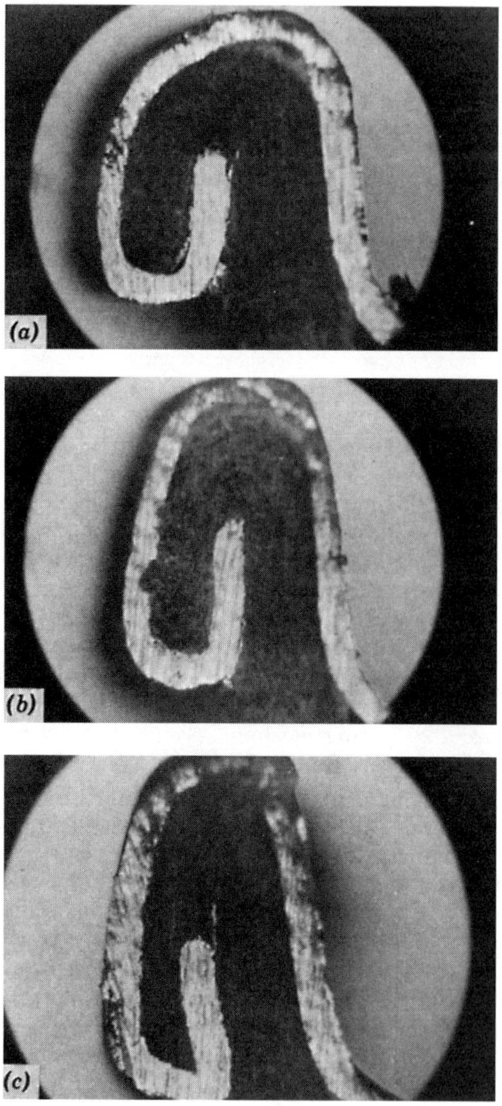

Figure 4. Composite-can double seams. (**a**) Loose; (**b**) correct; and (**c**) tight.

End closures. A critical process in the manufacture of a composite can is double seaming the metal end (see Can seamers). Since a composite can body is typically thicker than a metal can for any given package size, metal-can specifications for finished seam dimensions cannot be followed. Figure 4 shows composite double-seam profiles that are correctly and incorrectly made. Careful attention should be given to compound placement, selection of first- and second-stage seamer rolls, seamer setup, chuck fit, and base plate pressure if a satisfactory double seam is to be achieved.

In the future, end-seam configurations will probably be substantially different. Present double seams tend to creep under a continuous pressure, and minimum abuse will cause most composites to leak at pressures exceeding 15 psi (103.4 kPa). The present trend in end development is to combine mechanical and chemical systems to improve double-seam integrity.

A variety of steel and aluminum ends with solid panel, removable tape, or other easy-opening features are available, as well as plastic-end closures with easy-opening and sifter tops. In composites, the most expensive components are usually the metal ends. With this in mind, gastight, puncture-resistant membrane closures for composite have been test marketed and evaluated by several companies (see Fig. 5). They are considerably less expensive than aluminum full-panel removable ends, and they eliminate the cut-finger hazard posed by both the center panel and score residual on rigid ends. In addition, the membrane end eliminates the metal fines that can be produced by can openers.

Labels. The outer label on a composite can supplies additional package protection and, more importantly, enhances the can's aesthetic appeal and provides required consumer educational-instructional information. Composite labels include coated papers, foil/kraft laminates, and film constructions based on polyethylene or polypropylene. Flexographic, rotogravure, or offset printing (see Printing) are used, depending on cost and quality requirements.

Figure 5. Composite with peelable membrane closure.

Other Developments

The composite can for shortening, introduced in 1979, illustrated the improved containment capacity of fiber cans. Traditionally in metal, shortening can now be packaged in composite cans with a PP/foil kraft liner. The can body consists of two plies of paperboard with a foil/kraft outer ply. The label can be supplied convolutely after filling, or it can be wound as the outer ply. It has been shown that powdered milk and other oxygen-sensitive food products can remain stable for at least one year. It is expected that more powdered and granular foods will be packed in composites in the future through the use of nitrogen-purge systems or similar processes.

Certain products, such as tobacco and snacks, are already packed with nitrogen, and the concept was considered for coffee as well. It would eliminate the vacuum (see Vacuum coffee packaging) that has held back the use of composite cans thus far. Coffee and other vacuum-packed products, eg, powdered milk and nuts, are now being tested in composite cans. These cans must be capable of holding a 29.5-in. Hg (100-kPa) vacuum for periods of up to one year and performing not only under static conditions but under actual abuse conditions of packing and distribution.

The packaging of single-strength juice and juice drinks represents an exciting opportunity for composite cans in both hot-fill and aseptic processes (see Aseptic packaging). Extensive technical work begun in the 1970s has resulted in can constructions and filling techniques that have proven successful in commercial canning and distribution systems.

BIBLIOGRAPHY

"Cans, Composite" *The Wiley Encyclopedia of Packaging Technology*, 1st ed., Wiley, New York, by M. B. Eubanks, Sonoco Products Company, 1986, pp. 94–98.

CANS, COMPOSITE, SELF-MANUFACTURED

There are two basic types of self-manufactured composite cans produced in-house by packagers. One is the traditional spiral-wound or convolute composite can, the familiar form used for motor oil, orange juice, and cocoa-based products (see Cans, composite). A more recent development is the utilization of folding-carton material for the production of self-manufactured paperboard cans using composite materials in the folding-carton base stock. An even newer innovation is a paperboard can that is hermetically sealed and capable of holding a gas environment.

Composite cans. A typical example of self-manufacturing of a traditional composite can is a system known as Sirpack. This system, from Sireix (France), installed in the packager's manufacturing facility, produces composite cans that can be round, square, rectangular, or oval. The cans are produced from a continuous form-and-seal process from four reels of material in a horizontal fashion. The materials are shaped around a forming mandrel and sealed, generally by a hot-melt adhesive (see Adhesives). The inside liner is heat-sealed for excellent moisture and liquid tightness (see Sealing, heat). The reels of material are slightly staggered to each other so that the sealing lines on the four materials do not superimpose. The inside materials that provide moisture and liquid tightness are generally made from a plastic film or aluminum foil (or both), according to the protection needed for the product. The outside laminate can be printed by web rotogravure, or web offset (see Printing).

The bottom is generally a metal end, but it can also be made of a composite material matching that of the sidewall. It can also be one of the many plastic closures available. Generally, the top is a heavy injection-molded plastic that is heat-sealed to the upper edges of the container. In most cases the cover incorporates an easy-open device allowing a separation or opening of the lid and can be reclosed after removing a portion of the product.

The main advantages of the in-plant Sirpack system are that it permits the option of various sizes (material options) and the ability to save the conversion cost normally paid to the converter of the composite cans. Added to this are other savings, such as the savings in floor space and warehousing of empty composite cans prior to utilization in the filling and closing process. The machinery is relatively compact and easy to maintain. It has been designed for in-plant production with the average mechanic in mind.

Paperboard cans. Several companies provide systems for in-plant manufacture of paperboard cans. They include folding-carton companies such as Westvaco (Printkan), International Paper (CanShield), and Sealright (Ultrakan). All of their packages are low-to-moderate barrier paperboard cans that can be manufactured in the packager's facilities. Another paperboard can, Cekacan, offered in the United States by Container Corporation of America, provides a high barrier, gastight supplement to the paperboard-can list. All the paperboard cans are formed from a flat blank and assembled into either a straight-wall can or tapered-wall can. In some cases, eg, CanShield, the use of paper-cup technology has been employed to develop the vertical-wall paperboard can.

A paperboard can is described as a semirigid container

with the body sidewall fabricated from a single sheet of folding-carton-based material, wound once and sealed to itself, with either or both ends closed by a rigid or semirigid closure. The typical paperboard can has three pieces: a single-sheet single-wound body, a single sheet of base material fixed to one end, and a closure. The system is almost always made in line with the packager's filling and closing operations, but it is also possible to manufacture to storage. The can body can be cylindrical, rectangular, or combinations thereof, but most are cylindrical. Commercial or prototype paperboard cans range in size from 2 to 10 in. (0.8–3.9 cm) high, and up to 6 in. (2.4 cm) in dia, but mostly are confined to a maximum 5.25-in. (2-cm) dia and 3–10-in. (1.2–3.9-cm) heights.

In the CanShield construction, the paperboard sidewall is rolled into a cylindrical shape and the two edges are overlapped and sealed. Continuous thermofusion along the seal is effected by bringing the coating on one or two edges of the blank to a molten state, ie, by direct contact with a heated plate or more recently by blowing hot air onto the edge, bringing the two edges together, and applying pressure. Usually a polyolefin, eg, polyethylene, is used on the surface of the board to provide the adhesive factor. A disk of paperboard with a diameter approximating that of the cylinder is crimp-folded around its perimeter to form an inverted shallow cup. This base piece is positioned in one end of the hollow cylinder so that the bottom edge is about ¼ in. (0.64 cm) below the edge of the cylinder. The edge of the inner periphery is heated and crimpfolded over to lock and seal the disk in place. In this manner, the outer peripheral of the base is sealed to the inner perimeter of the cylinder wall. The segment of the outer wall extended beyond the inner disk is then heated and folded over to come in contact with the inner side of the disk. A spinning mandrel applies pressure to the base of the cylinder to effect the final seal.

The result is a primary seal between the bottom disk and the sidewall and a secondary seal, wherein the sidewall is folded over, capturing the disk with an additional seal. For the rectangular version, the base piece is forced against the body wall under pressure using an expanding mandrel to seal the base to the body. Here too, the material for sealing is usually polyethylene, and hot air is used to bring the material to a molten state to act as a sealant. On the round containers, the top is usually rolled out and the closure, eg, a foil membrane, is adhered to the top rolled edge after the product has been placed into the paperboard can. On the rectangular version, the top rolled edge is generally closed by a rim closure that clamps onto the periphery of the opening and is sealed into place by induction or by glue. The rim then acts as a holder for a full panel closure.

The Ultrakan concept is similar to the CanShield in that the body wall is wrapped around a mandrel and the two edges are overlapped and heat-sealed to each other. The bottom disk is inserted in the container, and the body wall and bottom disk are heated and crimped or rolled together. Thermoplastic hot-melt adhesive may be used for added security and seal strength. The interior edge of the sideseam can be skived to enhance WVTR or greaseproofness of the container, or both. The top of the Ultrakan container can be finished in a variety of ways: rolled outward (to accept a membrane seal); flared (for a variety of seamed metal ends); rolled inward (for special thermoformed or injection-molded plastic closures); or gently flared (for insert rotor/dispenser style closures). The Ultrakan system also provides the option of customizing by special bottom techniques which offer dispensing features for granulated products, powders and paste, or semiliquid sauces or condiments.

A high barrier paperboard can has been introduced from Sweden (Cekacan). By incorporating the use of polyolefin laminates (see Laminating) along with foil and a special means of sealing the package, a hermetic seal has been demonstrated, making the package virtually impermeable to gas, liquid, fat, etc (see Table 1). The Cekacan system involves both a can-forming operation and a can-closing operation. In the forming operation, the sidewall is wrapped around a mandrel and butt-seamed (not overlapped). Just prior to the wrapping operation, a foil-laminated tape is induction-sealed to one edge of the blank. With the seam butted, and induction sealer affixes the tape to the interior of the can in such a way as to provide a continuous hermetic seal along the longitudinal seam.

The package is transfered to an end-closing device wherein a top or bottom closure is affixed. In this case, the closure is inserted into the can with the closure sidewalls flanged to the vertical position. Through the use of pressure and induction sealing, the disk is hermetically sealed into place. The package is then discharged for filling and brought back to the second piece of equipment, which inserts and hermetically seals the final closure. Closures are available that provide easy opening without compromising the gastight integrity of the package. Through the use of a butt seam, held together with sealable tape, there are no discontinuous joints to bridge. Ends are inserted into and fused by induction sealing to the smooth interior wall. During the induction-sealing process, the fluid flow of the internal coating, usually polyolefin, fills any short gaps that may occur. The equipment is simple to operate and does not require special expertise.

A special attribute of all the in-plant paperboard-can packaging systems is the reduction of materials storage and handling. Since the body walls are shipped flat along with the bottom disk and top closures, a minimal amount of storage

Table 1. Comparative Water-vapor Transmission[a]

Product	Conventional Package	Rate of Moisture Pickup at 75°F (24°C) and 100% rh, wt % per week: Conventional Package	Cekacan
Dehydrated sweetened beverage powder	26.5-oz (751-g) composite can	0.03	0.0075
Powdered soft-drink mix	34.0-oz (964-g) composite can	0.02	−0.0025
Sweetened cereal product	12.0-oz (340-g) bag-in-box	0.2	0.04
Snack	7.5-oz (213-g) composite can	0.01	−0.01

[a] Courtesy of the Center for Packaging Engineering, Rutgers, The State University of New Jersey.

space is necessary. The average space needed to contain the paperboard can in its flat form represents approximately 97% savings over a similar number of composite or metal cans or glass jars. Additional savings are realized by the reduced cost of shipping the container components to the plant and also in the weight of the final product.

Materials. Basic paperboard-can-body structures are made of laminations of paperboard, aluminum foil, and polyolefins. End structures are analogs that might omit the paperboard for some applications. The generalized structure is paperboard/bonding agents/aluminum foil/polyolefin (outside-to-inside). Engineering the components to each other and the structure to the package has been a significant advance. The material components must be functional, economic, structurally sound, and compatible with the contained product.

The paperboard component of the lamination is not critical to the hermetic function of the Cekacan, but it is essential to the commercial value of the system. The exterior surface must be smooth and printable and the interior surface must be sufficiently tied to ensure adhesion to the adjacent layer. Since the paperboard-can body is composed of a single-ply material (as opposed to the multiple plies in composite cans), ranging from 0.016- to 0.032-in. (406–813-μm) finished caliper, the appropriate finished caliper must be chosen to meet the physical stress. At the same time, the economics of the additional caliper board must be weighed against the cost.

Structures can be engineered for each product's specific requirements. Interior polyolefins may be polyethylenes or polypropylenes. They can be applied by extrusion, coextrusion, extrusion–lamination, or adhesion–lamination (see Extrusion coating; Laminating). In the case of a gastight container, a crucial variable is the bonding of the interior polyolefin to the aluminum foil or paper substrates. This adhesion must be maintained above preestablished minimums in converting, body erecting, sealing, and operation. This is a very demanding requirement.

BIBLIOGRAPHY

"Cans, Composite Self-Manufactured" *The Wiley Encyclopedia of Packaging,* 1st ed., Wiley, New York, J. M. Lavin, Container Corporation of America, 1986, pp. 98–100.

CANS, CORROSION

Corrosion in food cans, manifested in failure phenomena such as detinning, hydrogen swell, and enamel delamination, has troubled the food industry ever since canning was practiced. Much research work has been invested in this subject, but problems still exist with only little relief. The major reason for this fact is that the mechanisms involved in the corrosion phenomena are not well known. Molecular hydrogen is well known as a product of the corrosion process, but its origins and its role along the corrosion process in atomic form is not well established, or even considered (1,2).

There are various corrosion phenomena that the existing theory cannot explain, and therefore cannot provide solutions to corrosion problems. There are many indications that all corrosion phenomena involve hydrogen activity. The source of the hydrogen is from water HOH, in which the radicals H and OH are interdependent, similar to the ions H^+ and OH^-. All the four species are involved in the oxidation and reduction reactions in nature. The state of the art corrosion theories consider mainly the activity of the ions H^+ and OH^-, but ignore the major role of the radicals H and OH. The glass electrode that measures the pH defining the H^+ and OH^- activity is very common, but the redox potential, defining the ratio of H and H^+ activities, is rarely measured and considered. The pH indicates the degree of acidity or alkalinity of the liquid. The redox indicates to what degree the liquid is oxidizing or reducing.

Both H and H^+, and perhaps OH and OH^- have some qualitative dimension, as expressed for example, by the NMR (nuclear magnetic resonance) of the hydrogen. This energetic property differentiates hydrogen atoms based on its source. Hydrogen atoms and ions from different sources may have a different effect on corrosion and many other phenomena. This may be why different acids have a different corrosive effect, and the pH alone is far from providing an explanation to corrosion failures.

The source of corrosion is water, which contains the four species participating in the corrosion process. These are H and OH radicals as well as the well-known H^+ and OH^- ions. Corrosive materials affect some qualitative dimension of the hydrogen atoms of the water. This is why extremely small quantities of some material added to the water may have a dramatic effect on corrosion, as well as on many biological reactions. A simple known example for such an additive is sulfur dioxide, that in trace concentration may lead to detinning of fruit cans in a matter of days or even hours. The effect of such traces on pH are negligible.

These observations lead to the conclusion that water is not just a solvent or carrier. It is the corrosive agent itself, affected however, by materials dissolved in the water, and even by materials that the water was in contact with.

The major tool used for corrosion studies in the food packaging industry is still the test pack, by which a product or medium is packed, stored, and evaluated. This method should have been used to verify a mechanism, rather than as the research tool. In fact, very little was learned from test packs to ameliorate the theories of corrosion mechanisms. In many cases the conclusions from test packs were inconclusive and misleading.

Many analytical methods in corrosion studies (3) are actually based on oxidation reduction reactions where specific hydrogen is involved (4). Hydrogen in molecular form was detected on the external wall of cans during storage (1,5), but its origin and permeability (6,7), as well as its reactions in the bulk of the metal (8,9) have not been considered in can corrosion studies.

Enamel adhesion is based on hydrogen bonds (10), and reactions affecting the hydrogen bonding (11) affect the performance and the shelf life of enameled cans.

The corrosion mechanism involves oxidation reduction reactions (12). In the case of detinning in plain cans (13), where the food is in contact with the tin coated steel, the tin and some iron are oxidized. This is the tangible bottom-line result, but it can be shown that the active element leading to this result is the atomic hydrogen from water. The oxidation of the food product in enameled cans, manifested in color and flavor changes, can be explained by hydrogen loss (12).

Oxidation of organic matter is caused by loss of hydrogen atoms. Organic matter that gains hydrogen atoms is reduced. Oxidation of metals is usually defined as loss of electrons. Metallic ions can gain electrons and deposit as metals. It can be shown that the hydrogen atom, in water or in the food compounds, is the element that is transferred or gives and takes electrons and in its absence there will be no changes in organic and inorganic matter.

The Corrosion Mechanism

The basic reaction of oxidation reduction is

$$H \rightleftharpoons H^+ + e^-$$

The reaction to the right expresses the oxidation of the hydrogen atom to H^+ (15). The reversible reaction to the left is the reduction of the hydrogen ion H^+ to H atom. The Nernst equation derived from this equation is

$$E = E^\circ + \frac{RT}{F}\ln\frac{[H^+]}{[H]}$$

where E is the electrochemical potential of the hydrogen electrode, E° is the standard hydrogen potential, R and F are constants, and T is the temperature. The expression $[H^+]$ is the activity of H^+, and [H] is the activity of atomic hydrogen. The means to measure this potential is basically a platinum surface, to which the atomic hydrogen is absorbed, and a reference electrode, such as a calomel half-cell.

Replacing platinum with tin, the reading will be different, because of the very different activity of hydrogen on tin. In conclusion, the redox potential measurement indicates the potential of hydrogen on the electrode at the conditions of the measurement. The galvanic cell, therefore, expresses the difference of hydrogen activity on the two metals.

The reaction on the cathode in a galvanic cell is known to be $H^+ + e^- \rightarrow H$, but it is not common consideration, that the opposite reaction takes place on the anode, where H, from the solution, gives its electron to the anode and turns into H^+. The latter may take an electron from the metal and turn back into H, and the metal will dissolve as positively charged ions. If the anode cannot lose electrons, the solution will be oxidized, because it lost hydrogen that turned into H^+. The net result is hydrogen formation on the cathode and H^+ or metal ions formation at the anode. According to the above, the hydrogen in its different forms is the major player in transferring electrons, or in oxidation reduction reactions.

In a corrosion process in plain tinplate cans, the metal is usually oxidized, and the product is reduced. In the case where hydrogen is formed on the cathode and there is no metal dissolution from the anode, the product itself is oxidized, because it lost atomic hydrogen.

$$\text{On the cathode } 2H^+ + 2e \rightarrow H_2$$
$$\text{On the anode } 2H \rightarrow 2H^+ + 2e$$
$$5n + 2H^+ \rightarrow Sn^{2+}$$

The oxidation, therefore, can affect the solution as well as the metal. Loss of H from water creates excess OH radicals that may oxidize the product. Pears packed in plain cans will be slightly bleached, and some tin will dissolve. The product is reduced, and the tin is oxidized. In enameled cans pears will darken because there is no metal dissolution.

Water is the source of H, which is in equilibrium with H^+ and the respective counter groups OH and OH^-. H is therefore the major parameter in the corrosion mechanism. Any additive to water will somehow affect the activity of the above four species, thus indirectly contributing to the corrosion process, through the effect of the additives on the water.

Radical and ionic reactions. Electrons can also be given to the anode by negative ions, such as chloride and hydroxyl ions. The chloride at the anode gives an electron to the anode and turns into chlorine radical, that attacks the water and forms HCl and OH radicals. The OH radicals may attack the metal, take electrons, and turn into hydroxyl ions OH^-.

$$Cl^- \rightarrow Cl + e-$$
$$Cl + HOH \rightarrow H^+Cl^- + OH$$
$$Me + OH \rightarrow Me^+ OH^-$$
$$\text{or } H + OH \rightarrow HOH$$

The reaction of chlorine with water is not an ionic reaction; it is a radical reaction, similar to the attack of OH on the metal. Indeed, dry chlorine can attack metals directly in absence of water, but such reactions are out of the scope of the common corrosion in food cans.

The common approach in corrosion research is that the mechanism is ionic only and that the water is just the solvent and the carrier. This approach is incorrect and misleading. In fact, corrosion of metals in distilled water may be more severe than in tap water. The biochemical and pharmaceutical industries realize severe corrosion of stainless steel piping in distilled water.

The electrochemical potential. The galvanic corrosion has been explained on the basis of potential difference between two dissimilar metals. This is the driving force of the electrical current and therefore of the corrosion process. The electrochemical potential of a metal has been defined in textbooks as the tendency of the metal to dissolve, or to give electrons. Actually, the potential is not of the metal, but of the hydrogen on the metal. The potential difference of the galvanic cell or the bi-metal is therefore, the difference of the hydrogen potential on the two metals. The potential of hydrogen depends on the solid to which it is adsorbed and the solution in which the solid is immersed. Also the pressure above the solution has an effect on the potential (15). This is another indication that the potential of a metal has to do with hydrogen activity. The reactivity of the hydrogen with the metal will determine the type and extent of the corrosion. If the two metals are noble the solution will be oxidized near the anode, where hydrogen gives its electron.

Potential difference does not require two different metals. Nonuniform mechanical and chemical composition leads to nonuniform absorption of hydrogen atoms to the metal and thus to potential differences. The potential of hydrogen on a stressed site on the metal is different than on a non stressed area. The stressed metal site will usually serve as the anode, by accumulating hydrogen atoms that turn into ions by donating their electrons to the cathodic sites. The next step, as explained above, is the loss of electrons from the metal to the hydrogen ions. As a result, the metal dissolves as ions and the hydrogen atoms are reformed. Trapped hydrogen atoms in the bulk of the metal may lead to corrosion at the traps. It is well known that stresses in steel lead to trapping of hydrogen, as well as the fact that stressed areas tend to corrode. The type and rate of the corrosion will also depend on the metal composition and the distribution of the impurities on the surface and bulk on the metal.

A good indication that the electrochemical potential of metals is actually the potential of hydrogen on the metal is that enameled metal surface has a potential very close in value to that of the uncoated metal. Coated tinplate has a different potential than coated TFS (tin-free steel). This means that hydrogen and protons, which can migrate through organic coatings, may reach the metal, where also electrons may flow. The fact that metal does not dissolve into the food, does not mean that oxidation reduction reactions do not occur. A possible result of such potential difference is hydrogen swell and food discoloration.

Polarization. The activity of the hydrogen species in the solution is changing in the course of the corrosion process. Near the anode there is a depletion of H atoms, and near the cathode there is a depletion of H^+ and formation of H atoms. This leads to a decrease of the anode potential and increase of the cathode potential, which means a decrease in the potential difference and the driving force for electrons flow. This phenomenon is termed *polarization*. Oxidizing materials, which tend to combine with hydrogen atoms, will counteract polarization on the cathode. Reducing materials, which provide H atoms, will counteract the anodic polarization. Materials that counteract polarization are called *depolarizers*.

The role of oxygen. The term *oxidation* is perhaps the most misleading term in science. It has misled everyone, including scientists to relate oxidation to oxygen. While the combination of matter with oxygen is an oxidation process, most oxidation processes do not involve molecular oxygen. Oxygen is perhaps the mildest oxidizer in nature; however, its combination with hydrogen is forming the OH radical, which is the oxidizer in nature.

It can be shown that rust formation requires water or moisture and not oxygen, as it is commonly thought. Steel will rust under vacuum and high moisture, but it will not rust in dry air. Oxygen can enhance rust formation, by combining with hydrogen atoms adsorbed to the metal surface, and form OH radicals. Hot steel surface, charged with hydrogen under vacuum, will form immediately a layer of rust, on exposure to air.

Oxygen is a mild cathode depolarizer, but in presence of atomic hydrogen and a suitable catalyst oxygen can form OH radicals that are strong oxidizers, as these tend to combine with hydrogen atoms to form water. Four hydrogen atoms are oxidized to water by one molecule of oxygen. Such activity, however, depends on the ability of the metal to catalyze the combination of molecular oxygen with atomic hydrogen. A good illustration of the mild oxidizing nature of oxygen can be demonstrated by the fact, that bubbling air into orange juice for a few hours does not lead to discoloration. On the other hand, orange juice discolors in enameled cans sealed under vacuum. Food product that lost hydrogen is oxidized. The hydrogen can be lost to solids, such as coated tinplate, that absorb hydrogen atoms. The oxygen may combine with these absorbed hydrogen atoms to form OH radicals that will directly oxidize the product. Oxidation of metals is defined as loss of electrons. It is quite well established that the electrons are lost in the corrosion process to H^+ or to OH radicals, rather than oxygen.

On the basis of these observations, the oxygen is a secondary player, which enters into the game at a later phase in the corrosion mechanism, when the depolarization is the governing mechanism. The combination of molecular oxygen and hydrogen requires a suitable catalyst. Steel seems to be a good catalyst for such a reaction, but it also is a strong absorber of hydrogen atoms.

Hydrogen activity in metals. Hydrogen embrittlement and stress corrosion are formed by absorption of hydrogen into the metal. Hydrogen trapped in steel can form extremely high pressures in the metal, leading to cracking, blistering, and pitting. The hydrogen trapped in the steel can react with many of the noniron elements, such as carbon, sulfur, and phosphor. The result is pitting corrosion, decarbonization, and loss of strength.

Redox potential of water and aqueous solutions. The quality parameters of water in contact with the metal should include its redox potential that takes into consideration not only the pH, but also the hydrogen activity. Water is composed from H and OH radicals. The multiplication of the activities of the H and OH radicals is constant, similar to the relationship of H^+ and OH^- from which the pH values are derived. When the H activity is higher than the OH activity, the water is reducing, and when the activity of OH is higher, the water is oxidizing. In neutral water the activity of the H and the OH is the same. Neutral water in respect of redox is not the same as neutral water at pH 7, and vice versa. Water at pH 7 can be very oxidizing and can be made very reducing by bubbling hydrogen in the water. Orange juice, for example, is acidic and reducing. Tap water may be at pH 7 and very oxidizing. The redox potential, as defined by the Nernst equation given above, includes the effect of the pH, but the pH does not include the redox. Many materials added to the water may affect the redox but not the pH. A good example is the addition of ozone and hydrogen peroxide to water, which will affect the redox, but not the pH. The corrosivity of the water may sharply change by such additives. It is very important to consider the redox potential as well as the pH of the water used for the product and the process, in order to minimize internal corrosion, and that of the cooling water to avoid external corrosion.

The Specificity of Hydrogen

The empirical corrosion research clearly indicates, that the pH alone cannot explain nor predict corrosion phenomena. The corrosivity of various food products having the same pH may be very different. Acetic acid, for example, is much more corrosive than citric acid in contact with steel, but the opposite is true for tin that is not attacked by acetic acid. In lack of a scientific explanation to these facts, this phenomenon is explained by the term “affinity.” Even the consideration of the redox potential does not enable prediction of corrosion. Two solutions at the same pH and the same redox may exhibit different corrosion effects on the same type of can.

Haggman (2) showed that steel with high sulfur content is attacked by foods containing sulfates. Similarly, foods containing phosphates are corrosive to steel containing high phosphor levels. All of these confusing facts, which cannot be explained by the common theories of corrosion mechanisms, are indicative that some parameter, having a major effect on corrosion, is not known and therefore not considered.

Researching the mechanism of biological oxidation reduction provides a clue that might point out a possible direction for explaining the puzzling phenomena described above.

The specificity of enzymes is well known. Enzymes are responsible for the oxidation reduction reactions in biological systems. Each biological reaction requires a specific enzyme. The catalyst for the reduction of carbon dioxide into sugar in the photosynthesis reaction is the chlorophyll. No other catalyst in nature will do it. This reaction also requires very specific light energy, at a very specific narrow range of wavelength or frequency. The physicist relates to energy also through its qualitative properties, while others usually consider energy quantitatively only. Energy has a dimension of intensity, but also a qualitative dimension expressed in wavelength or frequency.

Antioxidants donate hydrogen atoms, serving therefore, as reducers. However, such reducing activity is quite specific. Vitamins C and E are antioxidants, but each one is responsible for the reduction of specific systems. This means that the hydrogen donated by vitamin E is qualitatively different from that of vitamin C, or other antioxidants. In other words, if antioxidants had not have this specificity, there would have been only one antioxidant in nature.

The summary of all these facts leads to the conclusion that hydrogen atoms and ions differ in their qualitative energetic properties. This is the source of the multiple forms of corrosion, and this must be the explanation for the lack of understanding and control of some corrosion phenomena.

The NMR spectroscopy is based on the theory that the active hydrogen atom in an organic molecule has a specific energetic property. This property is specific to every material and therefore can be used for the purpose of identification of materials. This knowledge is applied in medicine to identify, for example, cancerous cells in biological tissues. Applying NMR spectroscopy to corrosion research may lead to explanations as to why atomic and ionic hydrogen from different sources, has a different corrosion effect on a certain metal.

Common Corrosion Problems

The mechanisms explained above can be used to enlighten some of the common corrosion failures.

Pitting corrosion. Microscopic examination of the metal in cans exhibiting "hydrogen swell" and sulfide black reveals pits in the steel. This type of corrosion is common in cans with foods containing sulfur. It has been realized that the pits occur more on stressed areas, in steels containing sulfur.

Hydrogen atoms that had been in contact with sulfur bearing compound permeate into steels and are trapped at preferred specific points. Such points contain extra amounts of sulfur, that is irregularly distributed in the steel. The permeation and trapping of hydrogen in such sites is enhanced, if the sites are mechanically stressed. The hydrogen trapped in the steel develops very high local pressures in the steel. The trapped hydrogen reacts with the nonmetallic components of the steel, and also recombines to form molecular hydrogen gas. Such activity in the steel occurs mainly in enameled cans. This process requires permeation of hydrogen into the steel, and it is therefore more common in enameled tin-free steel and low tin-coated tinplate. The permeating hydrogen may leave to the atmosphere, unless trapped by nonuniform distribution of impurities and mechanical stresses.

Sulfide black. The trapping of hydrogen in the steel depends on the steel and the food composition. The gas emanating from the pits may be only hydrogen, but usually, in case of pit formation, it is composed of a mixture of compounds containing iron, sulfur, and other elements from the steel. These erupting compounds appear as black spongy lumps, termed by canners as "sulfide black." This phenomenon occurs mainly in enameled cans, and the sulfide black appears mainly on the coated side seam and along beads and scratches, where irregular stresses are formed. The food product triggers the formation of the sulfide black, but the steel composition and mechanical stresses are the controlling factors in the mechanism of sulfide-black formation.

Enamel peeling and under film corrosion. The bonding between the enamel and the metal is based on hydrogen bonds. These bonds are formed by hydrogen atoms shared by the metal and the coating. The H atom, coming usually from the organic coating, shares its electron with the metal and the coating. For example, the hydrogen atom of an OH group in the coating may share its electron with an oxygen atom on the metal surface. The hydrogen bond can be described as

$$\text{Metal}-\text{O}\cdots\cdots\text{H}\cdots\cdots\text{O}-\text{polymer}$$

Such bonds can be destroyed by addition of a hydrogen atom to the bond. By that, two OH groups that repel each other will be formed, and the metal will repel the polymer. This means that the delamination is caused by a reduction mechanism. A complete delamination, with or without any corrosive attack on the metal, can occur when the food product is very reducing, the polymer is very permeable to hydrogen atoms, and the metal surface is extremely impermeable to the hydrogen. By this excess of atomic hydrogen will be formed between the tin and the polymer, leading to reduction of the metal and the loss of the hydrogen bond. Such complete delamination was frequent when heavy hot-dipped tin coatings were coated with phenolic and other types of polymer. Partial delamination, followed by corrosion under the enamel in tomato cans is still a common phenomenon. It appears usually near scratches and on stressed areas, such as beads and side seams.

The hydrogen bond can also be destroyed by oxidation, which means, by removing the hydrogen atom that is forming the bond. Pulling the hydrogen out chemically or electrochemically may lead to loss of the bonding hydrogen from the polymer. The loss of adhesion in such cases is expressed in small blisters to the enamel, pit formation, some metal dissolution, and frequently by formation of hydrogen gas in the container. The later leads to vacuum loss and even hydrogen swell.

Hydrogen swell. The formation of hydrogen gas and severe tin dissolution in plain cans is well known. This is usually explained as the attack of H^+ on the tin. The mechanism is much more complicated, and it involves the redox potential of the food and the specificity of the hydrogen in relationship to the tin and steel composition. The pH alone is far from explaining why a few parts per million (ppm) of sulfite, which have no significant effect on the pH, will lead to very rapid detinning. The tin is attached by H^+, which, as mentioned above, has also a qualitative dimension besides its concentration in the food. The H^+ could also be formed from atomic hydrogen that entered the bulk of the tin. The fact that large quantities of hydrogen gas are formed may indicate that the atomic hydrogen in the tin delivers the electrons to cathodic sites in the steel. Analysis of cases of such failure supports the conclusion that the interaction of the tin with specific hydrogen is the controlling factor in rapid detinning. The redox potential of the product and the specificity of the hydrogen of the product, as well as depolarization agents in the product, are the important parameters to be considered in such failure studies.

Hydrogen swell occurs also in enameled cans. The mechanism is commonly attributed to the porosity of the enamel. The enamel coverage and its thickness affect the resistance to hydrogen permeation into and out of the enamel. The resistance is only one of the factors affecting hydrogen permeability. At zero driving force, the resistance is meaningless. The formation of the hydrogen pressure and its permeability depend mostly on the characteristics of the metal and the food product. There is a certain optimal enamel coating that should be applied. This optimum depends on a few factors related mainly to the metal and the product than the enamel. Hydrogen atoms can diffuse through the enamel and dissolve and accumulate in the metal, but they can also migrate to the external wall. They can then react with the impurities in the steel, form pits, recombine to form molecular hydrogen, and diffuse through the enamel. It is possible to have hydrogen swell without significant or equivalent metal dissolution.

Darkening of light-colored fruit. *Corrosion* in cans refers to the oxidation of the metal. The role of hydrogen in its atomic and ionic form in this corrosion process was explained above. Oxidation of the food product may be regarded as the corrosion of the product, and it is explained as loss of atomic hydrogen. In many cases the hydrogen that was lost from the food is the hydrogen that has led to the corrosion. In other words, the metal can, enameled or plain, can lead to oxidation of the food product by absorbing atomic hydrogen from the product.

Darkening of the food is the result of oxidation of some compounds in the food. The oxidation is due not to addition of oxygen, but rather to hydrogen loss. The composition and the physiochemical properties of the food, the organic coating and the metal, in respect of hydrogen activity, are the factors involved in the mechanism of the oxidation of both the metal and the food.

Conclusion

Hydrogen in its atomic, ionic, and molecular form participates in any corrosion process. There is no corrosion without hydrogen. The electrochemical potential of the metal is the potential of hydrogen. The redox potential of the food defines the hydrogen activity of the food and its tendency to lose or gain hydrogen atoms.

The specific energetic properties of the active hydrogen will determine the rate of corrosion. The hydrogen activity and its effects have to do with the composition of the materials involved. The interaction of the materials with hydrogen determines the type and rate of corrosion. The properties of materials, such as hydrogen permeability through the polymers and the metals, should be studied and correlated with performance.

The source of hydrogen is water that acts through its H and OH radicals. The redox potential of water, together with the pH and the specificity of the hydrogen, have to be considered in order to understand and control corrosoin. The water used for processing and the product water should be considered along the above parameters.

The food product composition and any additive will have an effect on the corrosivity of the product. The packaging materials should be studied in view of their interactions with hydrogen, taking into account the specific properties of the hydrogen atoms in the product and the affinity of the specific hydrogen with the specific packaging materials.

This article is a summary of the author's research through a new approach to corrosion studies. The research is based on the quite well-verified assumption that hydrogen activity is the major parameter controlling corrosion. The research is far from being complete and may not offer immediate solutions to all corrosion phenomena and failures. It may serve as a new direction for researchers in this field who seek better understanding of the corrosion mechanisms and means to avoid failures.

Corrosion failure analysis through the presented theory can spread light on many unexplained corrosion phenomena, such as detinning, enamel peeling, and underfilm corrosion, as well as sulfide black and pitting corrosion. The key to understanding and solving corrosion problems is *thinking hydrogen.*

BIBLIOGRAPHY

1. J. Haggman and Oy G. W. Sohlberg, *Proc. 2nd Internatl. Tinplate Conf.,* International Tin Research Institute, London, 1980, pp. 400–405.
2. J. Haggman, Oy G. W. Sohlberg, *Proc. 4th Internatl. Tinplate Conf.,* 1988, pp. 294–304.
3. S. C. Britton, *Tin versus Corrosion,* ITRI Publication 510, 1975.
4. D. Reznik, "Porosity of Coatings," *The Canmaker* (Feb. 1990).
5. G. Serra and G. A. Perfetti, *Food Technol.* 57–61 (March 1963).
6. N. V. Parthasaradhy, *Plating,* 57–61 (Jan. 1974).
7. D. Reznik, *Proc. 4th Internatl. Tinplate Conf.,* 1988, pp 286–294.
8. I. M. Bernstein and A. W. Thompson, *Hydrogen Effects in Metals,* 1981.
9. T. Zakroczymski, *Hydrogen Degradation of Ferrous Alloys,* Noyes, New Jersey, pp. 215–250.
10. G. Pimental and A. L. McClellan, *The Hydrogen Bond,* 1960.
11. W. C. Hamilton and J. A. Ibers, *Hydrogen Bonding in Solids,* 1968.
12. D. Reznik, "Oxidation of Foods in Enameled Cans," *The Canmaker,* 32–33 (March 1993).
13. D. Reznik, "Corrosion of Tinplate," *The Canmaker* (July 1991).

14. D. Reznik, "Recent Research on Side Seam Striping," *The Canmaker,* 47–48 (Sept. 1990).
15. A. White, P. Handler, et al., *Principles of Biochemistry,* McGraw-Hill, New York, 1959, pp. 35–37.

DAVID REZNICK
Raztek Corporation
Sunnyvale, California

CANS, FABRICATION.

See METAL CAN FABRICATION.

CANS, PLASTIC

Cans are defined here as open-mouthed cylindrical containers, usually made of aluminum (see Cans, aluminum) or tinplated steel (see Cans, steel). This form can also be made of plastics, by injection molding (see Injection molding), blow molding (see Blow Molding) or forming from sheet (see Thermoforming). Such cans have been made for many years, and have taken small shares of some markets, but are not dominant in any. Plastics are much less rigid than metals, thicker walls are needed for equivalent performance, and container weight and cost can become excessive. For carbonated beverages (see Carbonated beverage packaging), cans are rigidified by internal pressure, but other properties such as burst strength, creep, and gas barrier become more critical.

Tensile strength is needed, especially in beverage cans, and can be achieved through selection of appropriate plastics plus orientation (stretching). Orientation in the circumferential (hoop) direction is hardest to do, and this explains why blow molding from a narrow parison may give better performance than thermoforming or direct injection to size. Heat resistance may be needed: either around 122°F (50°C) to resist extreme warehousing conditions; or 140°F (60°C) for 30 min (pasteurization); or 185–212°F (85–100°C) for a few seconds (hot filling); or 257°F (125°C) for 20–40 min (retort sterilization cycle). A can for processed foods must withstand the internal pressure generated when a closed can is heated. In a sterilizing retort, compensating overpressure may be needed outside the can to prevent failure. Not all retorts are capable of such overpressure (see Canning, food). Gas and/or moisture barrier and chemical resistance are needed for most packages. The can shape is advantageous for barrier, as it offers a low surface-to-content ratio, but it also introduces the possibility of failure from chemical attack in some places, particularly the stressed flange area and the bottom edge.

Product-design centers on these two features: the flange and the base. A precise flange is needed to guarantee a perfect seal, which is absolutely critical for sterilized cans, and certainly desirable for contained liquids, especially under pressure. If a plastic end is used, heat-sealing is possible, but the plastic end will be less rigid and may require a heavier flange. Common metal ends may be used, but flanges must be very flat. Also, no matter what end is used, stresses and later stress cracking in that region must be anticipated and avoided.

Vertical sidewalls are desired for easy transport in filling lines. Tapered walls do allow nesting, which has storage advantages, but for mass applications the filling speed is more important. Another design concern is the necked-in end now customary for beverage cans, which allows tighter six-packing and cheaper ends. This can be done with plastics, but mold design is more complicated to permit the undercut needed.

Potential applications. The beer and soft-drink markets beckon like gold to the alchemists, but mass use is still well in the future. There are thermofolded 250-mL PET cans used in Britain (Plastona), but they are nonvertical (they nest), and they are not coated to enhance barrier. In the United States, Coca-Cola has been working with "Petainers," which are cans drawn in solid phase from molded PET cups developed in Europe by Metal Box and PLM A.B. These can be coated with PVDC (see Vinylidene chloride copolymers) for improved barrier, but such use would require special attention to recycling, both in-plant and at consumer level. In Italy, some beverage cans are made by thermoforming, and some by blowing bottles and cutting/flanging the tops. The latter should give the best properties for given weight because of more orientation.

For heat-processed foods, polypropylene is the preferred material because it has the highest heat resistance of the commodity thermoplastics. Polycarbonate (see Polycarbonate) and other engineering plastics have been suggested, but are much more costly. Both PP and PC are poor oxygen barriers, however, and would need a barrier layer for many foods. American Can's "Omni" is a coinjection-blow PP/EVOH/PP can, first used by Hormel for single-service heat-and-eat meals (see Multilayer plastic bottles) and many companies have produced and offered such cans thermoformed from multilayer sheet (see Coextrusion, flat; Coextrusion for semirigid packaging). Metal Box has also worked with extruded tubes with top and bottom flanged ends, and resin suppliers have made cans on pilot lines, but there is still no large-scale commercial use. One of the problems is that other forms of plastic (eg, trays, bottles, pouches and bowl shapes) are competing for the same markets.

In motor oil, the long-neck HDPE bottle is taking over and the can shape is somewhat old-fashioned now. Wide-mouth blow-molded HDPE oil cans have been in the market for many years, but with very limited success. All-plastic paint cans appear from time to time, and the consumers are said to welcome an easy closure, but market penetration is still low (around 1% in 1984), due to intense competition in metal, plus considerable captive metal production. They are generally injection-molded, but they can also be blow-molded. For frozen orange juice, composite cans have most of the market (see Cans, composite) although Tropicana injection molds its own polystyrene cans, and some injection-molded HDPE cans are in the market as well. Injection-molded HDPE cans are used for cake frostings, and dry beverages are sometimes seen in plastic cans, notably the heavy injected PP cans used by Star in Italy for coffee.

BIBLIOGRAPHY

General Reference

"Serving up a Better Package for Foods," *Chemical Week,* 100–104 (Oct. 16, 1985).

CANS, STEEL

In 1975, France was not only involved in a revolution but was also at war with several hostile European nations. The

French people as well as the armed forces were suffering from hunger and dietary diseases. Consequently, a prize of 12,000 francs was offered by the government to any person who developed a new means for the successful preservation of foods. Napoleon awarded this prize to Nicholas Appert in 1809. Mr. Appert's discovery was particularly noteworthy because the true cause of food spoilage was not discovered until some 50 years later by Louis Pasteur. Appert had nevertheless recognized the need for utter cleanliness and sanitation in his operations. He also knew or learned the part that heat played in preserving the food, and finally, he understood the need for sealed containers to prevent the food from spoiling. The containers he used were wide mouthed glass bottles that were carefully cooked in boiling water (1,2).

A year after the recognition of Appert's "canning process," in 1810, Peter Durand, an Englishman, conceived and patented the idea of using "vessels of glass, pottery, tin (tinplate), or other metals as fit materials." Thus, the forerunners of modern food packages were created. The original steel cans had a hole in the top end through which the food was packed. A disk was then soldered onto the top end. The disk had a small hole in it to act as a vent while the can was cooked. The vent hole was soldered and closed immediately after cooking (2).

Durand's tinplate containers were put together and sealed by soldering all the seams. The techniques were crude but nevertheless, with good workmanship, afforded a hermetic seal. A *hermetically sealed container* is defined as a container that is designed to be secure against the entry of microorganisms and to maintain the commercial sterility of its contents after processing. Commercial sterility is the inactivation of all pathogenic organisms and those spoilage organisms that grow in normal ambient distribution and home-storage temperatures. No technological advance has exerted greater influence on the food habits of the civilized world than the development of heat treatment and the use of hermetically sealed containers for the preservation of foods (1,3).

Foods canned commercially by modern methods retain nearly all the nutrients characteristic of the original raw foods. Several investigations showed that good canning practices and proper storage and consumer preparation improve the retention of the nutritive value of canned foods (4–6).

The steel container provided a reliable lightweight package that could sustain the levels of abuse that were common in packing, distribution, and sale of products. It has also been necessary to improve and develop new concepts to keep up with the advances in packing procedures, materials handling, and economic pressures.

Evolution of the Can

A four-track evolutionary road developed. One uses solder to seal all the can seams. This evolved from the hole-in cap container to one that holds evaporated or condensed milk. These "snap-end cans" retain the vent hole in the top that permits filling the can with a liquid. They are then sealed with a drop of solder or solder tipping. The second track combines the soldering of the side seam with the mechanical roll crimping of the ends onto the body. The attachment procedure, known as *double seaming,* was patented in 1904 by the Sanitary Can Company. This invention significantly improved opportunities to increase the speeds of can manufacture and packing operations. Today, double seamers or closing machines seal cans in excess of 2000 cans per minute with filling equipment capable of matching this task. The third track utilizes a press operation that stamps out cuplike structures with an integrated body and bottom, with the lid or top attached by double seaming. These began as shallow containers like sardine cans. These drawn or two-piece cans evolved into taller cylindrical cans that are popular for many food products and carbonated beverages. The primitive shallow cans are fabricated on simple presses. The taller two-piece cans go through multiple press operations like the draw–redraw technique or through a drawn press and wall iron operation. The fourth track is the incorporation of a welded side seam for three-piece cans, which provided greater body strength and more double seaming latitude in comparison to the soldered side seam.

The growth of the steel can caused the manufacturing function to change. At first, the packers manufactured their own cans. This was understandable, since the same craftspersons and equipment were necessary to seal the can as well as to make them. When the double-seamed can, or "sanitary can," was accepted, the can manufacturing function coalesced into large manufacturing organizations. This came about because the double-seamed can lent itself to mechanized production whereas the all-soldered can remained a manual operation. Hence, it was economically attractive to invest large sums of money for improving the sanitary can. A few large packers could afford to be self-manufacturers, and they invested large sums to keep up with the developing technology. However, their attempts to produce cans was often short-lived because of economic changes which forced them to become dependent on the can companies for their container needs.

There still remains today some packaging companies that make their own containers.

Shapes and Sizes

A wide variety of styles and sizes have grown out of the tremendous usage of steel containers (see Tables 1 and 2). The first figures of the container dimension represent the diameter of the container measured overall across the double seam; the second figure, the height, which is the vertical overall height measured perpendicular to the ends of the can. The first digit in each number represents inches; the second two digits, sixteenths of an inch. Thus, a 307 × 113 can is $3\frac{7}{16}$ in. in diameter and $1\frac{13}{16}$ in. high. For rectangular cans, the first two sets of digits refer to base dimensions; the third set, to can height.

Some can sizes are given various "names" by which the can is known. Some of these names that are identified in Table 2 are very old, dating back to the early history of canmaking. Today most cans are identified by their dimensions and not by name. Some of those names and can sizes listed are no longer applicable.

International standards for can sizes have been developed under auspices of the International Standards Organization, and cans are named in a number of countries on the basis of ISO standards. The specifications, measurements, and nomenclatures in ISO standards are used to describe cans in a number of countries. As can sizes listed in the ISO standards are the least likely to be restricted in international trade, it would be advisable, when developing a new product line or

Table 1. Steel Can Styles and Sizes

Style	Dimensions in. (cm)	Capacity	Some Uses	Convenience Features
Aerosol cans (*1*)	202 × 214–300 × 709 (5.4 × 7.3–7.6 × 19.2)	3–24 oz (85–680 g)	Foods, nonfoods	Designed for fit of standard valve cap
Beer-beverage cans (*2*)	209.211 × 413 (6.5/6.8 × 12.2)	12 or 16 fl oz (355 or 473 cm^3)	Soft drinks, beer	Easy-open tab top, unit-of-use capacities
	209.211 × 604 (6.5/6.8 × 15.9)			
	207.5/209 × 504 (6.3/6.5 × 13.3)			
Crown-cap, cone-top can	200 × 214–309 × 605 (5 × 7.3–9 × 16)	4–32 oz (113–907 g)	Chemical additives	Tamperproof closure, easy pouring
Easy-open oblong can	405 × 301 (11 × 7.8) #¼ oblong	4 oz (113 g)	Sardines	Full-paneled easy-open top
Flat, hinged-lid tins (*3*)	112 × 104.5 × 004–212 × 205 × 003.75 (4.4 × 104.5 × 0.64–7 × 5.9 × 0.6)	12–30 tablets	Aspirin	Easy opening and reclosure
Flat, round cans	213 × 013 (7.1 × 2.1)	1½ oz (43 g)	Shoe polish	Friction closure
Flat-top cylinders (*4*)	401 × 509–610 × 908 (10.3 × 14.1–16.8 × 24)	1–5 qt (946–4730 cm^3)	Oil, antifreeze	Unit-of-use capacities; tamperproof since it cannot be reclosed
	211 × 306 (6.8 × 8.6)	8 fl oz (237 cm^3)	Malt liquor	
	211 × 300 (6.8 × 7.6)	8 oz (227 g)	Cat food	
	300 × 407 (7.6 × 11.3)	15 oz (425 g)	Dog food	
Hinged-lid, pocket-type can			Tobacco, strip bandages	Firm reclosure
Key-opening, nonreclosure can (*5*)			Sardines, large hams, poultry, processed meats	Contents can be removed without marring product
Key-opening, reclosure cans (*6*)	307 × 302–502 × 608 (8.7 × 7.9–13 × 16.5)	½–2 lb (0.2–0.9 kg)	Nuts, candy, coffee	Lugged cover reclosure
	401 × 307.5–603 × 712.5 (10.3 × 8.8–15.7 × 19.8)	1–6 lb (0.45–2.7 kg)	Shortening	Lid is hinged
	211 × 301–603 × 812 (6.8 × 7.8–15.7 × 22.2)	¼–5 lb (0.11–2.2 kg)	Dried milk	Good reclosure
Oblong F-style cans (*7*)	214 × 107 × 406–610 × 402 × 907 (7.3 × 3.7 × 11.1–16.8 × 10.5 × 24)	1⁄16–1 gal (237–3785 cm^3)	Varnish, waxes, insecticides, antifreeze	Pour spout and screw-cap closure
Oblong key opening can (*8*)	314 × 202 × 201–610 × 402 × 2400 (9.8 × 5.4 × 5.2–16.8 × 10.5 × 61)	7 oz–23.5 lb (0.2–10.7 kg)	Hams, luncheon meat	Wide range of sizes, meat-release coating available
Oval and oblong with long spout (*9*)	203 × 014 × 112–203 × 014 × 503 (5.6 × 2.2 × 4.4–5.6 × 2.2 × 13.2)	1–4 fl oz (30–118 cm^3)	Household oil, lighter fluid	Small opening for easy flow control
Pear-shaped key-opening can (*10*)	512 × 400 × 115–1011 × 709 × 604 (14.6 × 10.2 × 4.9–27.1 × 19.2 × 15.9)	1–13 lb (0.45–5.9 kg)	Hams	Easy access through key-opening feature, meat-release coating available
Round truncated			Waxes	Screw-cap simple reclosure
Round, multiple-friction cans (*11*)	208 × 203–610 × 711 (6.4 × 5.6–16.8 × 19.5)	1⁄32–1 gal (118–3785 cm^3)	Paint and related products	Large opening, firm reclosure, ears and bails for easy carrying of large sizes
Round, single-friction cans	213 × 300–702 × 814 (7.1 × 7.6–18.1 × 22.5)	≤ 10 lb (4.54 kg)	Paste wax, powders, grease	Good reclosure
Sanitary or open-top can (three-piece) (*12*)	202 × 214–603 × 812 (5.4 × 7.3–15.7 × 22.2)	4 fl oz–1 gal (118–3785 cm^3)	Fruits, vegetables, meat products, coffee, shortening	Tamperproof, ease of handling, large opening
Slip-cover cans			Lard, frozen fruit, eggs	Simple reclosure
Spice can, oblong (*13*)	Wide range	1–16 oz (28.4–454 g)	Seasonings	Dredge top, various dispenser openings
Square, oval, and round-breasted containers (*14*)			Powders	Perforations for dispensing, reclosure feature
Two-piece drawn redrawn sanitary can	208 × 207/108 (6.4 × 6.2/3.8)	3–5½ oz (85–156 g)	Food	Improved can integrity, stackability
	307 × 111 (8.7 × 4.3)	6¾ oz (191 g)		
	211 × 214 (6.8 × 7.3)	7½ oz (213 g)		
	404 × 307 (10.8 × 8.7)	1½ lb (680 g)		

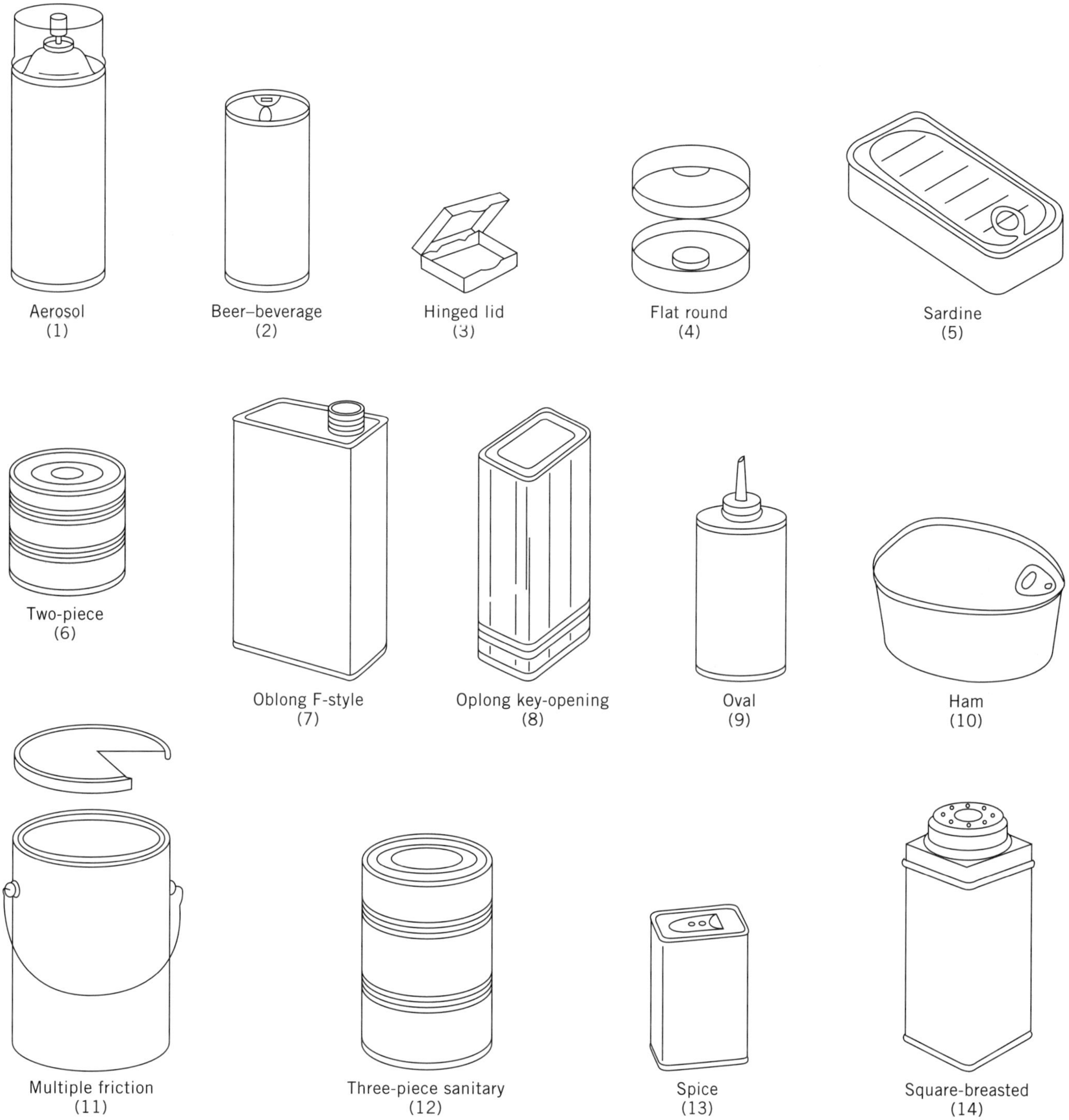

Table 1. ***(Continued)***

changing package sizes to consider can sizes from the ISO standards. In order to comply with the ISO standards, the Can Manufacturers Institute has now recommended that U.S. can sizes be identified in metric measurements according to their body plug diameter rounded to the nearest whole number of millimeters, and their height rounded to the nearest whole number of millimeters as shipped from the can factory. For example, the 307 × 113 is identified as 83 × 46 mm.

Normally during the selection of a container size for a specific item the concerns are not with providing a can with minimum surface area, but rather, to minimize package metal weight (i.e., cost) within practical limitations. As an example, the filling, processing, and end attachment operations demand that the circular ends be made of thicker metal than the cylindrical body. The cost impact tends to reduce end diameter while increasing body height compared to the theoretical dimensions for minimum surface area. Further, the ends are attached to the body with mechanically overlapped metal formed into the body and ends. Again, the requirement for thicker metal for the ends would favor smaller diameter.

Similarly, the body beads (for strength against can collapse) and end beads (for rigidity) increase the effective surface area. This again favors a smaller diameter to minimize metal weight. In addition, the ends are punched out of a rectangular sheet of metal in the manufacturing process. Not only must the weight of metal used in the ends be minimized, but

Table 2. Popular Can Sizes

Dimensions	Popular name	Inches	(mm)	Capacity[a], (oz)	(cm³)
202 × 214		2.13 × 2.88	(54 × 73)	4.60	(137)
211 × 413	12 oz	2.69 × 4.81	(68 × 122)	12.85	(380)
300 × 407	#300	3.00 × 4.44	(76 × 113)	14.60	(432)
303 × 407	#307	3.19 × 4.38	(81 × 111)	16.20	(479)
307 × 409	#2	3.44 × 4.56	(87 × 116)	19.70	(583)
401 × 411	#2½	4.06 × 4.69	(103 × 119)	28.60	(846)
404 × 700	40 oz	4.25 × 7.00	(108 × 178)	49.55	(1465)
603 × 700	#10	6.19 × 7.00	(157 × 178)	105.10	(3108)

[a] Completely filled.

the scrap produced from manufacturing round ends from square sheets has to be held to a minimal level.

Other limitations must also be considered. For example, the can manufacturer provides to the customer a can with one end attached (factory-finished can). The customer, in turn, fills the empty can and conveys it to another machine, where the top end (customer/packer's end) is attached. To avoid spilling product until the top end can be attached, the can is not fully filled but a small "headspace" remains. Besides preventing product loss this void provides for product expansion during processing and allows the consumer to open the can without splashing. This need for a headspace requires the can to be slightly taller than the theoretical height. In conveying, the laws of physics (inertia) state that the liquid juice will try to remain in place while the can begins moving. Without adequate headspace, juice would flow out of the open can. At a given conveying speed, less headspace is required for a smaller-diameter can than for a larger one. Therefore, it would be required to either further increase can height for a larger-diameter can, or customers would have to slow their lines to avoid spilling product. Increasing the can height would make air removal more difficult. Increased air (actually the oxygen in the air) content would have a somewhat negative effect on product quality.

Can Performance

No other container has all the attributes of the can: economy, strength, durability, absence of flavor or odor, ease in mass handling without breakage, compactness, lightness, and no light damage to product; the can also provides a hermetic seal, and can be produced at exceptionally high speeds and be lithographed.

Food cans are expected to have tightly drawn-in ends. A can with swelled ends or with metal feeling springy or loose is not merchantable. Consumers associate such appearances with spoiled contents. Gases produced by microbial action can, indeed, cause such appearances. It is also important to remove air from food cans to retard adverse internal chemical reactions. Therefore, end units must be properly engineered to cope with their many environments. They must act as diaphragms that expand during thermal processing and return to their tight drawn-in appearance when vacuum develops upon cooling. The necessary strength is built into the end units by the use of adequate plate materials and design of end profile (see Fig. 1).

Food ends are either profiled structures requiring a mechanical means for opening or designed with scores for easy

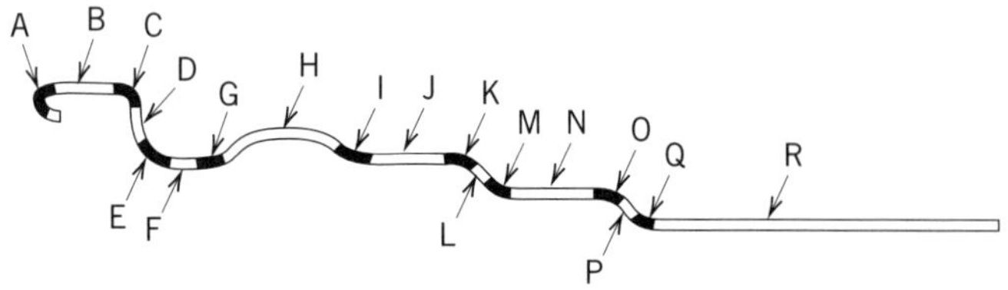

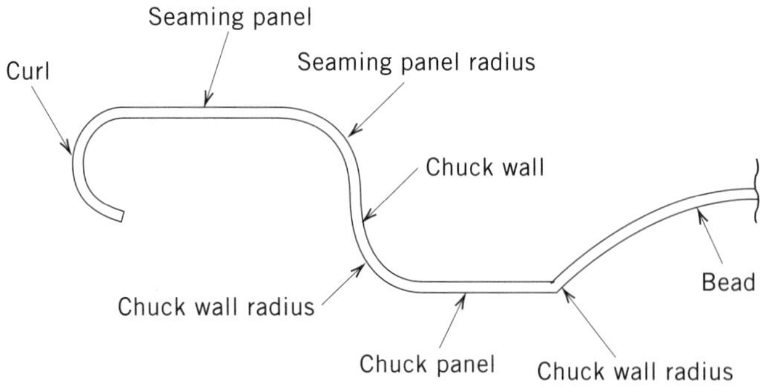

Figure 1. End profiles (A—curl, B—seaming panel, C—steaming panel radius, D—chuck wall, E—chuck wall radius, F—chuck panel, G—chuck panel radius, H—bead, I—bead-edge radius, J—first expansion panel, K—first expansion panel radius, L—first expansion panel step, M—first expansion panel step radius, N—second expansion panel, O—second expansion panel radius, P—second expansion panel step, Q—second expansion panel step radius, R—center expansion panel).

opening. Since the introduction of the easy-opening/convenience end in the mid-1960s, a broad variety of products have been packaged in cans having this feature. These easily removed ends are available in various diameters, metals, coatings, and configurations to meet product needs. Units can be removed as a whole with separation occurring around the perimeter of the end or only a portion of the top can be removed. For those products where consumer protection is desired, easy opening ends can be provided with safety features.

Ends used for carbonated beverages and aerosol products and those used for noncarbonated beverages packed using liquid-nitrogen pressurization, need little or no vacuum accommodation but rather resistance to high internal pressure. An end unit will permanently distort or buckle when the internal pressure exceeds its capability. This also renders the can unmerchantable because the distortion could affect the double seam integrity.

Cans with inadequate body strength will panel because of internal vacuum or collapse under axial load conditions. A panel is the inward collapse of the body wall due to pressure differential. The condition may appear as a single segment, many flat segments, or panels that develop around the circumference of the cylindrical can body. Axial load is the vertical pressure which a packed container will be subject to during warehousing and shipping. The axial load capacity of a can is greatest when the cylinder wall is in no way deformed. A casual dent or designed structure that breaks the integrity of the straight cylinder will, in most cases, reduce the axial load capability of the container. Panel resistance and axial load capability are direct functions of the metal specification.

Panel resistance can be enhanced by the fabrication of beads in the body wall. This, in effect, produces shorter can segments that are more resistant to paneling. However, such beads predispose the container to early axial load failure. The deeper the bead is, the greater the paneling resistance, but the greater the reduction in axial load capability. Many shallow beads can provide additional paneling resistance with less reduction in axial load capability, but labeling problems are often associated with cans having such bead configurations. Consequently, there are many bead designs and arrangements, all of which are attempts to meet certain performance criteria (7).

The steel container with proper material and structural specifications possesses, within limits, good abuse resistance. Excess abuse causes obvious damage and severely dented cans are unacceptable in the marketing of canned products. There are also insidious events that run parallel with excessive abuse. The double seams may flex momentarily, permitting an equally short-term interruption of the hermetic seal. Under some unsatisfactory conditions, this lapse can permit the entrance of microorganisms that cause spoilage. This leakage can also admit air that accelerates adverse chemical reactions within the container. Many cans showing some level of damage are in the food distribution chain, and the safety of using such products is frequently questioned. To prohibit the sale of all containers that have insignificant amounts of damage would be a waste of large amounts of very acceptable food (8).

Can Corrosion

The steel container is not chemically inert and, therefore, can react with its environment and its contents. Steel's major ingredient, iron, is a chemically active metal that readily takes a part in reactions involving water, oxygen, acids, and a host of other elements and compounds that can participate in oxidation reduction reactions (see also Cans, corrosion).

The application of tin to the surface of sheet significantly increases the corrosion resistance of steel. Nevertheless, the potential for corrosion attack persists. Although numerous modifying factors produce varied patterns of attack, the chemical fundamentals for oxidative corrosion are the same. The basic requirements are always (*1*) differences in potential between adjacent areas on an exposed metallic surface to provide anodes and cathodes, (*2*) moisture to provide an electrolyte, (*3*) a corroding agent to be reduced at the cathode, and (*4*) an electrical path in the metal for electron flow from anodes to cathodes.

Under normal conditions tin forms the anode of the couple, going into solution at an extremely slow rate and thus providing protection to the canned food. Under some conditions, iron forms the anode with resultant failure due to perforations or the development of hydrogen and subsequent swelled cans (hydrogen springers). Under still other conditions, as when depolarizing or oxidizing agents (ie, nitrates) are present, the removal of tin will be greatly accelerated with a consequent significant reduction in shelf life. Hydrogen is formed by two distinct processes: (*1*) at exposed steel areas that are protected cathodically by the tin–steel couple current and (*2*) as the steel corrodes, either because it is not completely protected by the tin or after the tin has been consumed. When perforations occur, they are usually the result of the same process in which hydrogen is developed except the steel is consumed in a localized area (9–11).

Product Compatibility

Products can be loosely categorized in terms of their susceptibility to chemical reaction with the can. Oils and fatty products seldom react with the metal surface of can interiors. However, when small quantities of moisture are present, either by design or accident, adverse reactions can develop. Highly alkaline products, usually nonfoods, will rapidly strip off tin or organic coatings but will not corrode the base steel. Acid products are corrosive, and highly colored acid foods and beverages are also susceptible to color reduction or bleaching, due to the reaction of the product with tin.

In this regard, when tin is exposed to some food products a bleaching action occurs. Although this is very objectionable with many products, such as the red fruits, and is avoided by the use of a suitable can lining, this bleaching action is desirable in certain instances. This is particularly true with the lighter-colored products such as grapefruit juice and grapefruit segments and sauerkraut. A slight bleaching action keeps the color light and compensates for the normal darkening effect, which may result from the processing or sterilization. Peaches and pears packed in cans completely enameled inside will be somewhat darker in color and slightly different in flavor than if packed in nonenameled plain tin cans. Although some individuals may prefer peaches packed in all enameled cans, it is doubtful that such cans will be produced under present conditions because the presence of an appreciable area of plain tin greatly increases the shelf life of this canned product.

Foods with $pH > 4.6$ often have sulfur-bearing constituents (eg, protein) and react with both tin and iron. This has been recognized as a problem for as long as such products have been canned, and the cause has been sought by many investigators. These low-acid foods form a dark staining of the tin surface and react with the iron to form a black deposit that adheres to both the can interior and the food product or can cause a general graying of the food product and liquor. This is often referred to as *black sulfide discoloration* or *sulfide black.* It can be very unsightly but harmless. Although sometimes exclusively a can headspace phenomenon, any interior container surface can be affected. Also, the black sulfide discoloration condition generally occurs during or immediately following heat processing, but it occasionally develops during storage. It is believed to be an interaction of the volatile constituents from the food product and/or oxidation–reduction agents in the food product with an oxidized form of iron from the tinplate. The staining of the tinplate is not part of this reaction (13,14).

Although tin sulfide staining and black sulfide discoloration is harmless from a product standpoint and has no detrimental effect on container or product shelf life, it has been found to be objectionable from an aesthetic standpoint.

In addition to container and product appearances, some food products and beverages are highly sensitive to off-flavors caused by exposure to can metals (12).

Can Metal

The term "tin can" is somewhat of a misnomer, because tin cans are made of steel sheets that have either no tin or coated with a very thin film of tin. The steel products commonly produced for container components have a theoretical thickness ranging from 0.0050 to 0.0149 in. (0.127 to 0.378 mm) expressed as weights per base box* of 45–135 lb. Depending on end use, the plate (blackplate) can be processed by the electrolytic deposition of metallic chromium and chromium oxides and coated with a lubricating film (tin-free steel) or coated with tin by electrodeposition. Improvements in electrolytic tinplate, since its commercial introduction in 1937, has increased both its versatility and uniformity of performance. Electrolytic plate has entirely replaced hot-dipped plate in present day cans. The result of these changes/improvements has allowed for reductions in tin coating weights. Tin plate carries a coating of tin that may very from 0.000015 to 0.000100 in. in thickness, depending on the grade. Differential coated electrolytic tinplate, having different coating weights of tin on each side of the steel baseplate, has been used commercially since 1951. Containers made of differential coated tinplate have an inside tin coating of sufficient thickness to withstand the corrosive attack of processed foods and an outside coating adequate to withstand the rigors of processing and atmospheric conditions. The thickness of tin coating is designated by the total amount of tin used on coating one base box of plate. For example, #25 plate is electrolytic tinplate on which 0.25 lb of tin was electroplated on one base box, covering both sides of the sheet. Thus, the coating on one side of the sheet is one half that amount of tin, or 0.125 lb.

For structural integrity, base weight or gauge thickness is a prime consideration, but any increment of base weight is an increment in container cost. Very often temper or design of can components can provide added strength without the need to add base weight. The chemistry and general metallurgy of steel plate has considerable influence on the performance of steel containers. Before the advent of continuously annealed and cold-rolled double-reduced steel plate, which influence plate stiffness or temper, it was necessary to add ingredients to the steel to produce high tempers. Temper as applied to tin-mill products is the summation of inter-related mechanical properties such as elasticity, stiffness, springiness, and fluting tendency. During the development of the use of cold-reduced steel for mill products the fabricators determined that various steel compositions and degrees of hardness in the base metal would permit the production of containers for a wider variety of products. Although there are certain limitations to the Rockwell hardness test,† it has been adopted as an industry standard because of its simplicity and overall good correlation with fabrication requirements. As a result, ranges of Rockwell hardness were developed to guide the steel manufacturers and these Rockwell hardness ranges have become accepted as temper ranges.

Rephosphorized steel was necessary to fabricate beer can ends with sufficient buckling resistance. When the market opened for canned carbonated soft drinks, rephosphorized steel was not acceptable because of the corrosive nature of soft drinks and the very high susceptibility of this plate to corrosion. Test pack experience indicated that some could perforate the endplate in 4–6 weeks. The availability of continuously annealed plate without the corrosion sensitivity properties of rephosphorized steel permitted the canning of soft drinks. It was still necessary to use heavier base weight material for the high carbonation drinks. The added stiffness afforded by cold-rolled double reduced plate permitted the canning of all soft drinks with carbonation ranging from 1.0 to 4.5 volumes in cans having the same basis-weight ends. These same economies have been applied to other products where container fabrication could be adapted to the degree of stiffness characteristic of this plate (14).

Can Fabrication

One who is not associated with the container industry seldom realizes that the manufacture of cans is classed as one of the most mechanized industries, and, in addition to the large amount of automatic equipment used, the speed at which cans are made no doubt rates among the faster automatic operations (see also Metals Cans, Fabrication).

The most common type of steel can produced today are three-piece, consisting of a body having a welded or soldered side seam in conjunction with two end components (see Fig. 2). Another popular structure is the two-piece can, which is produced by either a single-draw, draw–redraw (DRD), or drawn-and-iron (D & I) process. Some limited quantities of cemented side-seam cans are produced for dry and nonthermal processed type products.

The majority of three-piece tinplate cans currently made have wire-welded side seams. The traditional soldered can is now in the minority and only used for irregular-type meat cans and specialty cans. The solder used in canmaking is now generally composed of tin and silver. In July 1995 the U.S. Food and Drug Administration (FDA) has issued a final rule prohibiting the use of lead solder in the manufacturing of food cans, including imported products. The tin/silver solder is categorized as a soft solder and has a relatively low melting point, usually below 450°F, (232°C). Tinplate cans are easily soldered because the tin solder alloy readily fuses with the tin on the surface of the steel. In addition to providing a hermetic seal, solder also contributes to the mechanical strength of the seam by forming a metallurgical bond with the tinplate. Solder also has a certain amount of ductility and can be plastically deformed within certain limits. This characteristic permits the soldered laps of the body to be flanged and then incorporated into the double-seaming process. The speed of the can manufacturing line, the temperature of the molten solder, and the length of cooling all play an important role in good soldering operations. The soldered side seam can has been replaced for many can structures with an electrical resistance–welded seam. Benefits include a clean compact process, a much narrower weld area, allowing virtually wraparound decoration and greater integrity, particularly for

* *Base weight*—approximate thickness in pounds per base box: (base weight × 0.00011 = theoretical thickness). *Base box*—unit of area: 112 sheets, 14 in. × 20 in. = 31,360 in.^{2} (217.78 ft^2); dimensions in increments of 1/16 in.

Package: 112 sheets of any dimensions. *Number of base boxes in a package:* use ratio tables (ASTM A623) or L(in.) × W(in.) × 112 = 31,360

Number of base boxes in a coil: L(in.) × W(in.) = 31,360.

† A device used to measure the surface hardness of canmaking plate. A 1/16-in.-diameter ball penetrator is impressed into the plate surface with a 30-kg load. The measurement expressed as a Rockwell 30T reading, is inversely proportional to the penetration. An arbitrary scale is used to convert Rockwell readings into plate temper values. Stiffness increases from T-1 to T-6.

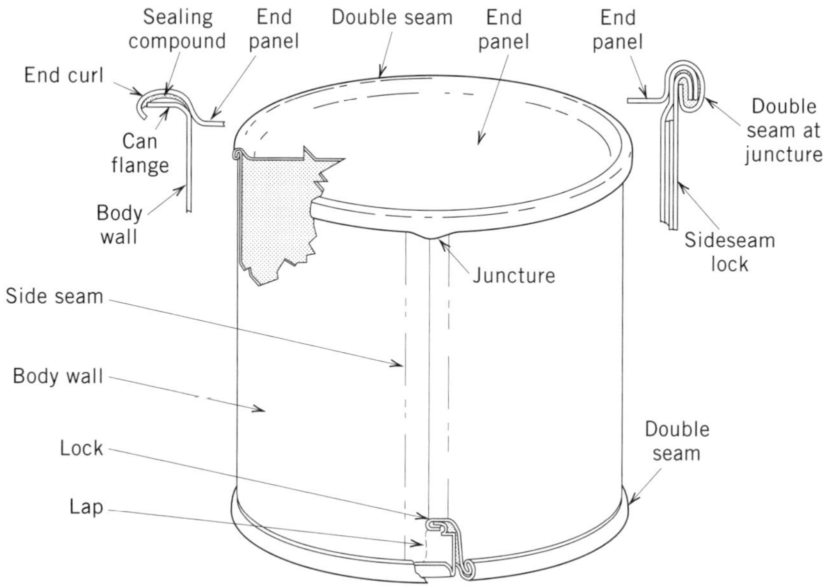

Material	Application	Method
Metal plate	Can units (bodies and ends)	Bodymaker; end press
Solder	Side seam sealling and bonding	Side seamer
Flux	Soldering aid	Bodymaker and/or side seamer
Sealing compound	To can ends for double-seam sealing	Comound liner
Protective coating	Metal plate protection	Roll coater or spray
Body and ends	Affix ends to body	Double-seamer

Figure 2. Double-seam three-piece can.

seaming-on of ends, since there is much less metal overlap thickness at the weld, compared with the soldered lock and lap seam employed on soldered cans. The welding unit is easily integrated and ancillary equipment used on solder bodymaker lines.

Two reliable welding systems were developed and used commercially: The forge–wheel welding system developed by Continental Can Company, and the wire weld system was developed by Soudronic AG (15,16). Both are resistance-welding systems and variations of a forge weld (see Fig. 3). The materials to be welded together are heated by an electric current passing through the materials. The resistance of the overlapped side seam produces the required welding heat. To obtain a weld, the heated material must be pressed together or clamped. This clamping capability is provided by wheels in the welding process. Because welded seams are basically lap seams, a raw edge exists, which generally requires a side-seam stripe coating application to minimize metal exposure.

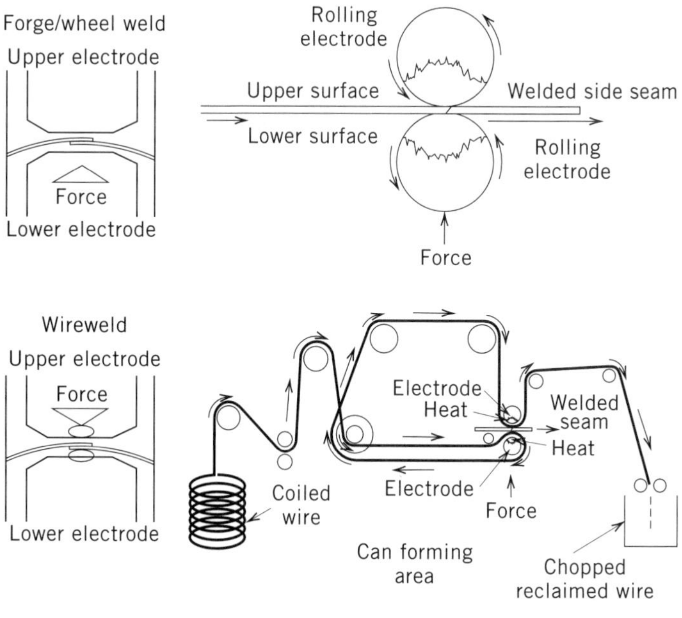

Figure 3. Welding process.

For economic reasons, the forge–wheel welding process is no longer a viable commercial operation. Such forge-welded cans utilized tin-free steel (TFS) plate and started with the grinding of the body blank edges to remove oxide and chromium to ensure acceptable welds. After blanks are formed into cylinders, the edges are tack welded together to produce an overlap, and the seam is made by rolling electrodes that weld the interior and exterior of the lap simultaneously.

In the wire welded operation that is used extensively today, the welding process consists of passing the seam with a small overlap between two electrode wheels over which runs a copper wire. The use of copper wire as an intermediate electrode is necessary to remove the small amount of tin picked up from the tinplate during the welding process, which would otherwise reduce the welding efficiency. A constantly renewable copper surface is presented to the weld area. To obtain a weld, the heated material is pressed together or clamped by wheels in the welding process (see Fig. 3).

Another innovation has been the introduction in the early 1960s of the seamless two-piece can (integral base and body, with one customer end). Metal forming technology has been used to produce these cans from flat sheet. Drawn and ironed cans employ tinplate and are manufactured by first drawing a shallow cup and then extending the sidewalls by thinning the metal between two concentric annular dies in an "ironing" process. This results in a can with a normal base thickness and thinner walls, and is economical in materials usage. The capital cost of D & I canmaking plants is high, but unit costs are the lowest of all container types if the throughput is sufficiently great. Because of the thin walls, body beading is essential to maintain rigidity for vacuum-packed food cans. The other two-piece can achieves final can dimensions through a series of consecutive drawing operations. In these draw–redraw (DRD) cans the thickness of the bottom end and the side walls is largely the same and there is not the same oppor-

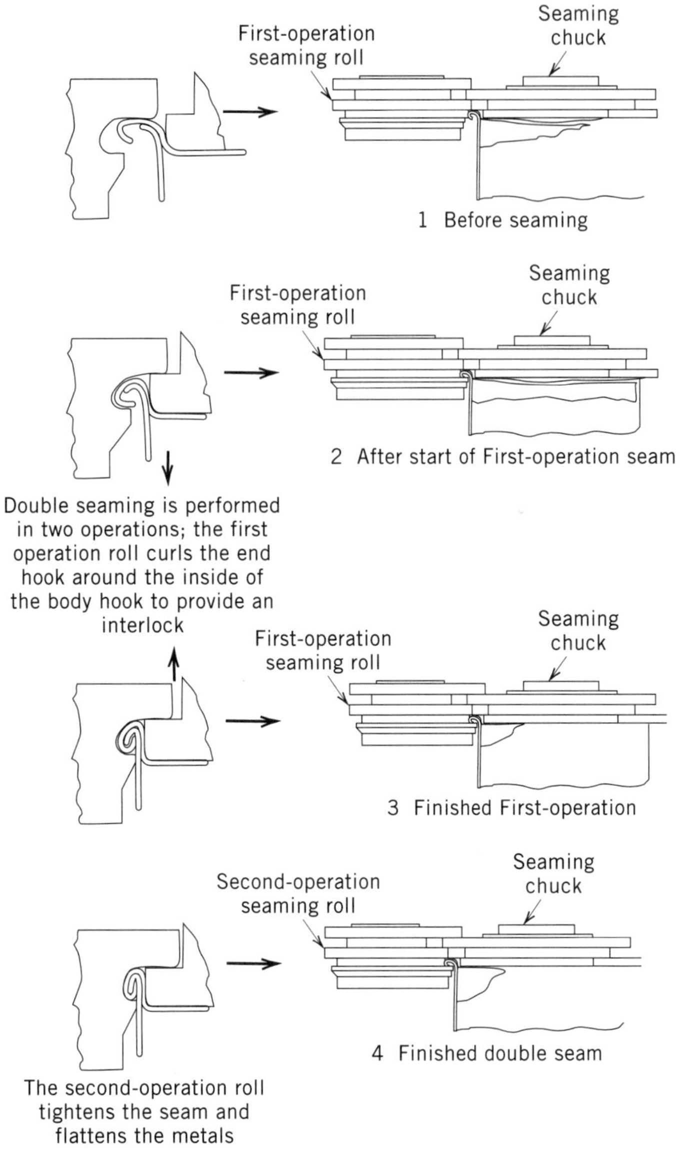

Figure 4. Sequence of operations in seaming a can end onto a can body.

tunity for saving on metal. Unlike the D & I can, which uses tinplate and inside spray coatings, the DRD can uses TFS-precoated sheet or coil plate.

Organic sealing materials have been used as side-seam sealants mostly before the advent of the welded can. They have been used in situations where soldered structures were not compatible with the product, when special wraparound lithography is desired, and when tin-free steel is used. For the most part, polyamides and organosols are used as the basic resin of organic sealing materials. They are normally utilized for a variety of can sizes, shapes, and styles that are utilized for dry and nonfood products. However, a polyamide hot-melt adhesive has been successfully used in oblong meat cans that are subjected to a high-temperature thermal process.

Much consideration has been given to cost reductions of the metal can. In conjunction with thinner-gauge and higher-temper plates, reduction of the can diameter of the two-piece

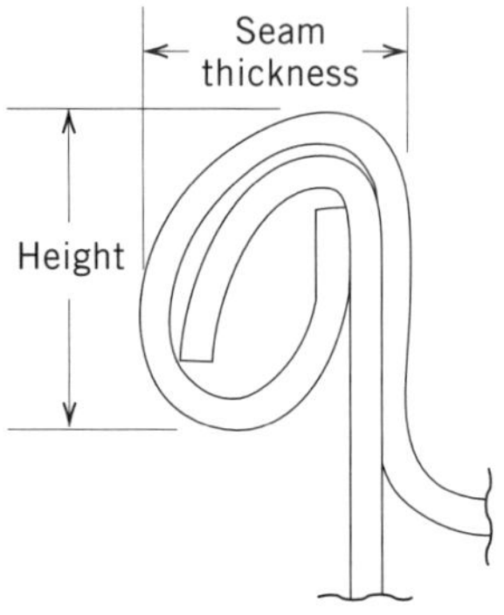

Figure 5. First operation.

beverage cans in the area of the end unit has reduced cost of the metal cans significantly. This body diameter reduction is usually accomplished by a die necking-in process, which is performed prior to flanging. The necking-in process tends to alter the can height, so careful body cut edge and/or can trim diameter must be calculated to afford these accommodations. With three-piece cans, this necked-in structure can be fabricated on both ends; however, it is done on only one end of two-piece cans used for liquid-type products.

The popularity of steel cans has been due in large measure to the double seam's ease of fabrication and robust resistance to physical abuse. This seam structure is the portion of a can that is formed by rolling the curled edge of the end and the body together to produce a strong leakproof structure. This seam is normally formed by a set of first and second operation seaming rolls while the end and the can body are held together by a seaming chuck. The first operation seaming roll forms the seam by interlocking the curled edge of the end with the flange of the can body. The second-operation seaming roll compresses the formed seam to make a hermetic seal (Figs. 4–6).

Since metal-to-metal contact does not produce good sealing capability, a rubberlike material, known as *sealing compound,* is applied to the loose end-unit seaming panel that becomes engaged to the body flange to form the double seam. This material acts like a sealing gasket. Compound placement and compatibility with the product are important in providing for a hermetic seal. End-sealing compounds are generally one of two types: (*1*) water base—rubber/water dispersions and (*2*) solvent base—rubber/solvent dispersions.

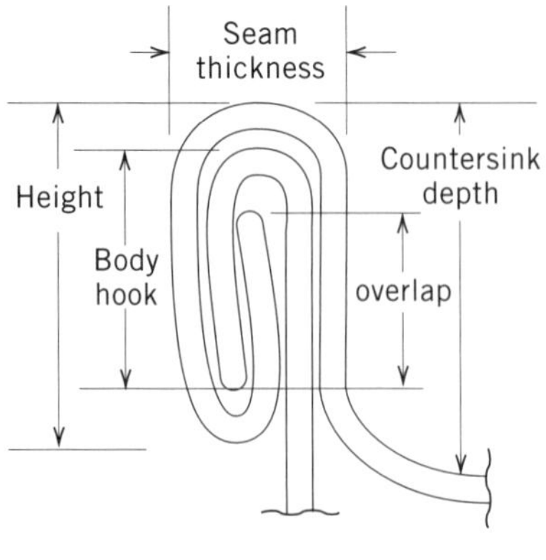

Figure 6. Second operation.

Coatings

Organic protective coatings are used as can linings to provide the additional protection above that provided by the metal substrate. These organic materials are applied to flat plate or coil coated and cured by means of continuous ovens. Spraying after complete or partial fabrication is also done when low-metal-exposure cans are required for aggressive-type products. It is desirable to use can linings that possess a reasonable degree of latitude in film weight and bake. The roller coating machines and the spraying machines used to apply linings provide means for close control of film weight. The films, of course, are relatively thin, about 0.0001–0.0005 in. thick or having a weight of 2–6 mg/in.^{2}. Control limits generally allow for variation of ±0.5 mg/in.^{2}.

Can linings must be nontoxic, meet government standards, be free from odors and flavors, and be readily applied and cured. They also must withstand the mechanical operation of canmaking, must provide the required barrier between canned product and metal, and finally, must be economical. Accelerated tests are used as screening or control tests and evaluations under commercial conditions are normally required before final approval of a can lining is standardized. When required, product test packs will be made and monitored over a period of time to confirm that the coating material is compatible with the intended product and/or container components.

Coatings accomplish several purposes. They make possible the use of less costly steels that permit the use of different types of tin-mill products in the can body and ends. They protect the steel against corrosion and the product from the steel and on the can exterior, minimize rusting, and serve as a background to or improve adhesion for lithography. Inorganic protective coatings that often contain phosphates, chromates, or both are usually applied to tin-free steel at the tin mill to prevent rust and provide a good surface for organic materials. If tin or iron is detrimental to a product, a suitable protective coating can be applied. These coatings can prevent bleaching discoloration by tin to dark pigments of foods such as strawberries, blueberries, blackberries, and cherries. Also, protective coatings can reduce the metal exposure for metal-sensitive products such as beer and certain soft drinks. Other products would react with the unprotected can interior to cause corrosion or cause discoloration of both container and product. Finally, a protective coating can mask unsightly discoloration of the can interior.

Currently the major trend in coating technology is toward waterborne coatings to replace conventional organic coatings. Typical organic coatings contain as much as 70% solvents, and are subject to stringent environmental restrictions. One answer is the aqueous coating, which contains only a small amount of organic solvents and can use the same coating and curing equipment. These aqueous coatings, which were commercialized for food cans in the early 1990s, exhibit the same good flow and leveling characteristics as found with solvent-base coatings.

Metal exposure in the side seam area is a common occurrence with all three-piece can structures and a special protective coating material (a "side stripe") is sometimes applied, to the inside and outside side seam area by spray or roller, online, followed by curing. The majority of stripes used are organic liquid materials, which provide only minimal coverage. Polyester powder (thermoplastic) electrostatic spray stripes give superior performance. These types of stripe coating provide 100% side-seam coverage, but they are more costly than liquid stripes. However, powder stripes have poor heat resistance and are not recommended for use in cans that will be subjected to aseptic process conditions that employ superheated steam.

Organic protective coatings were first applied to can interiors to protect the red fruit colors. Coatings used to combat this problem were often called "regular" or "R-enamels." Now they are generally identified by a number, letters, or a combination of numbers and letters.

If ordinary oleoresinous can linings are used for foods containing appreciable quantities of protein such as corn, peas, and fish, a black tin sulfide stain will form under the enamel; and under certain conditions, black sulfide discoloration will form at iron-exposed small scratches and fractures. To minimize black sulfide discoloration, oleoresinous enamels used for such products contain about 6% of a fine-particle-size zinc oxide, added for its chemical reactivity and not as a pigment. The sulfides that form during the processing or heat sterilization of the protein-containing foods react with the zinc oxide, which "de-zincs" and allows the formation of zinc sulfide. Zinc sulfide is white and generally goes unnoticed. Can linings containing zinc oxide were first developed for corn cans and were called "corn enamels"; their use was soon extended to other products, and they were then called "C-enamels."

Sulfide staining appears as a brown or brownish blue stain on the surface of the tin coating and cannot be removed by rubbing. Sulfide staining occurs not only on exposed tin surfaces but also in tin surfaces protected by many of the enamels since the volatile sulfur compounds involved can permeate certain types of enamel coating. To prevent underfilm sulfide staining, enamels impervious to volatile sulfur compounds, are used where possible. However, such enamels are not always compatible with the product being packed, in which case it is necessary to use a pigmented enamel that will either inhibit or mask the sulfide staining. These methods of inhibiting or masking sulfide staining are limited to their application, because not all enamels combine satisfactorily with a given pigment or additive.

There are many coatings now in use and being developed for steel cans used for a variety of food and nonfood products. Some of those materials, which are supplied by coating suppliers to the application facility and are now being utilized, are acrylic, alkyd, epoxy amine, epoxyphenolic, oleoresinous, oleoresinous w/zinc oxide, phenolic, polybutadiene, vinyl, and vinyl organosol. Each material has its own unique performance characteristics with regard to its application, fabrication, product compatibility, flavor and odor, process resistance, etc.

It has been apparent since the early days of canning green vegetables that the bright green darkens as a result of the required thermal process, that is, the bright green chlorophyll undergoes chemical change to an olive green pheophytin. There have been several studies conducted to inhibit this chemical change. They all demonstrate an inhibition capability by using additives containing magnesium ions and careful control of pH by the addition of hydroxyl ions. All of these procedures cause some toughening of the product, which probably discourage any commercial pursuits. The Crown Cork and Seal Company, Inc. has a patented process, "Veri-green," which incorporates additives to the product as well as the protective coating. The materials added are specific for particular products, and pretesting is necessary (17–21).

Decoration

Decorative lithographic designs afford an external protective coating. The process of application is known as *offset lithography.* Lithography is a printing process based on the fact that oil and water do not mix. The decorative design to be printed on the tinplate is etched onto a plate, known as a *master plate,* in such a manner that the image area to be printed is ink-receptive whereas the portion to be blank is water-receptive. A lithographic design is a system or a series of coatings and inks printed on plate in a particular sequence. The order of laydown is determined by the purpose of the coat, the kind of coating or ink, and the baking schedule for each coat or print. The print process is usually carried out with sheet stock prior to slitting into can body blanks or scroll shearing into end stock. The sequence and type of equipment used depends upon the design and the ultimate use of the can (see article on Decorating).

With the advent of two-piece cans, less elaborate designs are used on beer and soft-drink cans because it was necessary to use presses that print completely fabricated cans. The finishing varnishes are usually the external protective coatings. They not only protect against corrosion but must also be rugged enough to resist scuffing and abrasion.

For drawn-and-iron beaded food cans, the outside is coated with a water-base colorless wash coat material. After packing the customer glues a paper label to the body wall, which identifies the contents.

Technological Developments

The steel can has grown despite competitive pressures. During war years, tin and iron become controlled materials and other more available packaging materials usually make inroads into the steel-can market. These same constricting conditions often give rise to the development of methods that reduce the consumption of the controlled materials. During World War II, electrolytic deposition of tinplate was commercialized, which resulted in substantial reductions in tin consumption. Competition from aluminum in the beverage-can business gave impetus to pursue the use of tin-free steel to afford more favorable economics. Successful cost reductions with no reduction in performance, however, has been the cause for the continued growth of the steel can. The advent of cold-rolled double-reduced plate permitted significant reductions in base weight. To accommodate the fabricating characteristics of this material, new approaches in end manufacture and double seaming had to be developed. Another direction to cost reduction was to reduce materials usage. This was accomplished by using smaller-diameter ends and necking-in the can bodies to accommodate the reduced-diameter ends. This saving can be directed to one or both ends on three-piece cans.

The use of convenience features has been an important stimulus to the growth of steel cans. Key-opening lids and rip strips have been part of fish and meat cans for generations. This style was also standard for coffee and shortening cans, but economic pressures caused a change to less expensive open-top food cans. Development of the integral rivet and scored end solved many can opening problems, particularly in the beverage and snack-food businesses.

There have been dramatic developments in the overall operation of can-making plants with increasing use of automated procedures and computerized control systems and techniques to optimize production processes. There has been a reduction in the number of operatives but the working conditions of those remaining have been greatly improved. These new developments are ensuring that the can remains cost-effective and reliable.

With thinner-gauge plates and faster production speeds, canmaking equipment has become increasingly precise in its operations. Reliability is all important and quality-control measures have been put in place at every critical step to ensure container integrity. These include gauge measurements, weld monitors, and leak testing. Also there is more use of sophisticated statistical control charts, statistical analysis, and design of experiments to solve complex problems, particularly in manufacturing and technical areas.

There have been innovations in coatings technology such as introduction of waterborne coatings, coil coating, powder coating in electrostatic spray systems, and organic resins that are deposited electrophoretically on the can components from an aqueous solution. Increasing use is being made of the new ultraviolet systems, which dry sheet coatings and inks in seconds at ambient temperatures. Also, laminate polymer films have been introduced to compete with liquid organic coatings to metal substrates that are used for can components.

Early in the 1970s health authorities and the FDA became concerned about the increment, if any, of lead and other heavy metals that are picked up by foods packed in soldered cans. The manufacturers of baby foods and baby-formula foods or ingredients, such as evaporated milk, were the first to be asked to reduce the level of lead in their canned products. Better care during soldering operations and ventilation resulted in major reductions but did not eliminate all lead in the respective foods. Currently all soldered cans being produced are utilizing lead-free solders. Also the introduction of the three-piece welded and two-piece cans have virtually eliminated the concern of lead contamination (22).

BIBLIOGRAPHY

1. *Canned Foods, Principles of Thermal Process Control, Acidification and Container Closure Evaluation,* 3rd ed., Food Processors Institute, Washington, DC, 1980, pp. 7, 141.
2. American Can Company, *Canned Food Reference Manual,* McGraw-Hill, New York, 1947, pp. 25–29.
3. S. C. Prescott and B. E. Proctor, *Food Technol.* **4,** 387 (1937).
4. E. J. Cameron and J. R. Esty, *Canned Foods in Human Nutrition,* National Canners Association (NFPA), Washington, DC, 1950.
5. G. A. Hadaby, R. W. Lewis, and C. R. Ray, *J. Food Sci.* **47,** 263–266 (1982).
6. B. K. Watt and A. L. Merrill, "Composition of Foods—Raw, Processed, Prepared" in *U.S. Department of Agriculture Handbook,* 8th ed., Washington, DC, 1963.
7. Crown Cork and Seal Company, Inc., unpublished information.
8. *Safety of Damaged Can Food Containers,* Bulletin 38-L, National Canners Association (NFPA), Washington, DC, 1975.
9. R. R. Hartwell, *Adv. Food* Rev. 3, 328 (1951).
10. R. P. Farrow, J. E. Charboneau, and N. T. Loe, *Research Program*

on Internal Can Corrosion, National Canners Association (NFPA), Washington, DC, 1969.

11. N. H. Strodtz and R. E. Henry, *Food Technol.* **8,** 93 (1954).
12. J. S. Blair and W. N. Jensen, "Mechanism of the Formation of Sulfide Black in Non Acid Canned Products," paper presented on June 12, 1962 at the 22nd Annual Meeting of Food Technologists, Miami Beach, FL.
13. J. E. Chabonneau, The Cause and Prevention of "Sulfide Black" in *Canned Foods,* National Food Processors Research Foundation, Washington, DC, 1978.
14. "Tin Mill Products" in *Steel Products Manual,* American Iron and Steel Institute, New York, 1968.
15. U.S. Pat. 3,834,010 (Sept. 10, 1974), R. W. Wolfe and R. E. Carlson (to Continental Can Co.).
16. Soudronic AG, CH-8962, Bergdietikon, Switzerland.
17. F. A. Lee, *Basic Food Chemistry,* Avi Publishing, Westport, CT, 1975, pp. 163–165.
18. U.S. Pat. 2,189,774 (Feb. 13, 1940), J. S. Blair (to American Can Co.).
19. U.S. Pat. 2,305,643 (Jan. 5, 1942), A. E. Stevenson and K. Y. Swartz (to Continental Can Co.).
20. U.S. Pat. 2,875,071 (May 18, 1955), Malecki and co-workers (to Patent Protection Corp.).
21. U.S. Pat. 4,473,591 (Sept. 25, 1984), W. P. Segner and co-workers (to Continental Can Co.).
22. Bakker, "The Competitive Position of the Steel Can" in *Technology Forecast,* Avi Publishing, Westport, Conn., 1984.

FRED J. KRAUS
GEORGE J. TARULIS
Crown Cork & Seal Company, Inc.
Alsep, Illinois

CANS, TIN.

See CANS, STEEL.

CAPPING MACHINERY

In categorizing capping machinery that applies closures to bottles and jars, the best place to start is with the closure itself (see Closures). The different types of machinery for applying these closures have features in common (eg, straight-line vs rotary). This article provides a basic description of machinery used for continuous-thread (CT) closures, vacuum closures, roll-on closures, and presson closures.

Cappers for CT Closures

There are four basic types of capper for CT closures: hand cappers and cap tighteners; single-spindle (single-head) intermittent-motion cappers; straight-line continuous-motion cappers; and rotary continuous-motion cappers.

Torque control. All of the automatic cappers apply the closures mechanically, but they differ in their approaches to torque control. In general, torque control is achieved with chucks or spinning wheels (rollers). Straight-line continuous-motion cappers generally control torque mechanically, but all of the other types use either pneumatic or mechanical means, or combinations thereof.

A pneumatic chuck contains a round flexible ring with a hole in the center (1). The cap enters the hole when the ring is in its relaxed state. When air pressure is applied, the ring compresses and grips and holds the cap while it is moved to the bottles and screwed on. The air pressure may be applied by the downward movement of an air piston, or it may be directed into the space between the ring and the wall of the chuck (see Fig. 1).

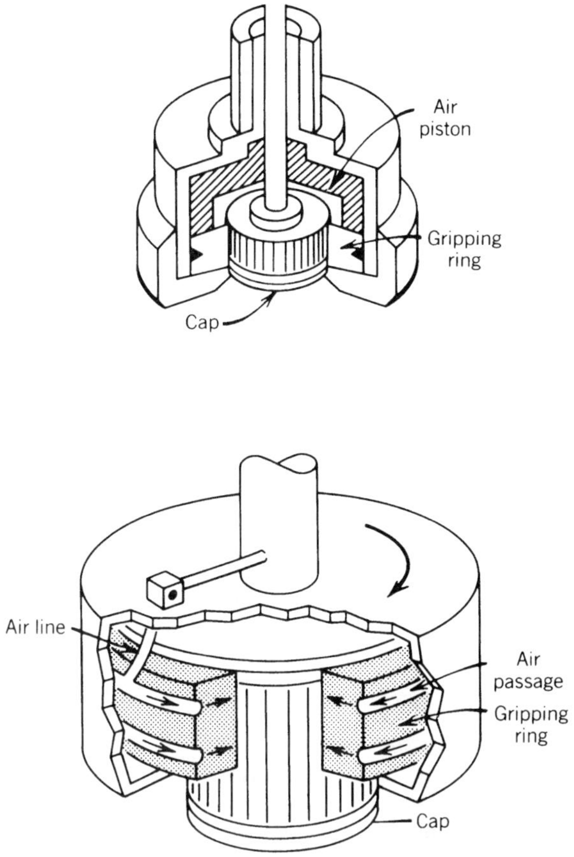

Figure 1. Pneumatic chuck (1).

The amount of torque is controlled by a pneumatic clutch operated by pressure from low level pneumatic lines (see Fig. 2). It contains two or more sets of disks that are pressed

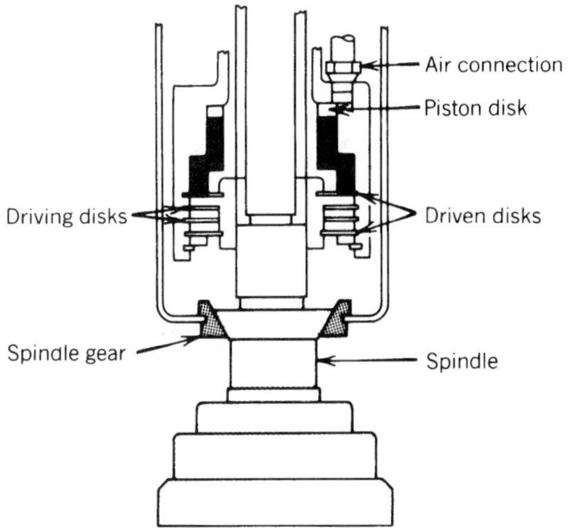

Figure 2. Pneumatic clutch (1).

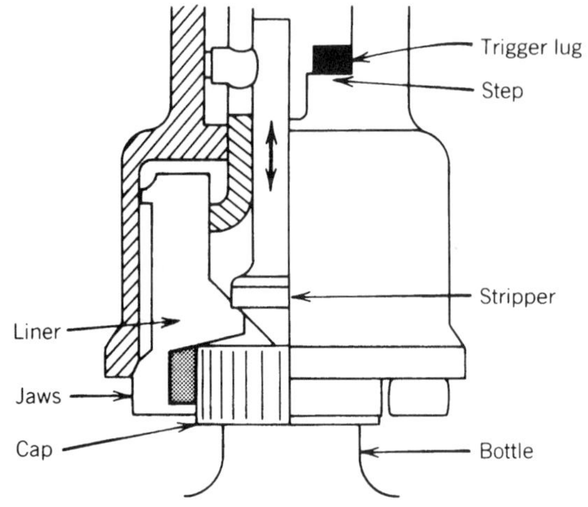

Figure 3. Mechanical chuck with jaws (1).

against each other when air is applied, connecting the chuck and the drive shaft. When the cap is screwed on to the point that the torque being applied equals the force being applied to the disks by air pressure, the disks start to slip and the drive shaft is disconnected.

A typical mechanical chuck has jaws that close around the skirt of the cap to maintain a grip until the closure application is complete (see Fig. 3). A mechanical chuck can be controlled by a pneumatic clutch, but it can also be controlled by a spring-loaded clutch, or a barrel cam. The pneumatic clutch is similar to the clutch used for a pneumatic chuck. A spring-loaded clutch (see Fig. 4) uses a spring to disconnect the chuck from the power source when the preset amount of torsion has been applied to the cap (1). Torque can be increased by compressing the spring by screwing down the collar on top of the spring; it can be decreased by moving the collar to loosen the spring. The chuck opens to release the bottle when the torsion on the cap matches the torsion on the spring.

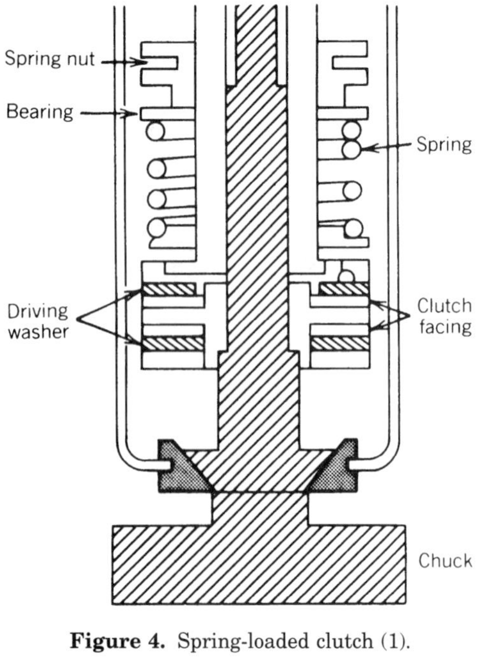

Figure 4. Spring-loaded clutch (1).

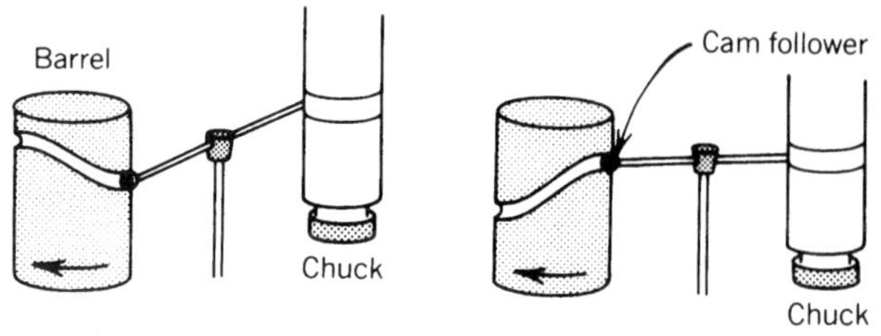

Figure 5. Barrel cam arrangement (1).

On some capping machines, the heads are raised and lowered by a follower riding on a barrel cam (see Fig. 5). Manufacturers of capping machinery have different approaches to torque control based on the principles described above, or combinations thereof.

Hand cappers/cap tighteners. With operating speeds of up to 20 caps/min, these are used for very low production volumes or unusual caps. Hand cappers work with some kind of handheld chuck. Cap tighteners, a step up in automation, are useful for pump and trigger-cap applications, retorqueing after induction sealing, and short production runs in general.

Single-spindle intermittent-motion cappers. Using either pneumatic or mechanical torque control, these cappers can theoretically apply up to 60 caps/min, and they are useful for relatively short production runs. Their versatility is limited by the intermittent motion that would tend to cause spillage from widemouth containers.

Straight-line continuous-motion cappers. On these machines, torque-control clutches are built into multiple spindles that turn the rollers (disks) that apply the caps (see Fig. 6). In contrast to the single-spindle cappers, the bottles never stop and the speed of capping is limited only by the speed of the conveyor. The capability of the machines has traditionally been quoted as 60–300 caps/min (in contrast to rotary machines which can be more than twice as fast), but by using multiple spindles speeds can be increased greatly; for example, with eight spindles at four stations, production can be as high as 600/min or more, depending on the size of the cap and the container. The major advantage of straight-line (vs rotary) cappers is that they generally do not require change parts, and cap and container sizes can be changed with very little downtime.

Rotary continuous-motion cappers. Rotary cappers use many heads in combination in a rotary arrangement that permits very high speeds: 40–700/min (see Fig. 7). The speed of application of one cap by one head is limited, but rotary cappers can achieve 700/min through the use of 24 heads. (Production speeds are always related to the size of container and cap. Rotary machines are available with more than 24 heads, but 700/min is a rough upper limit).

Extra features. Most of the automatic machines can be supplied with optional equipment for flushing with inert gas. There was a time when application torque could be measured only by testing removal torque, but today's capping machin-

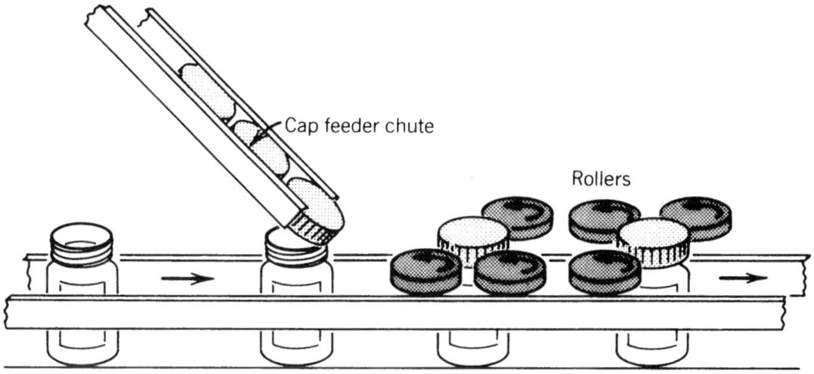

Figure 6. Roller screw capping (1).

ery is equipped with sensors that provide continuous measurement and constant readouts.

Cappers for Vacuum Closures

There are three basic types of vacuum closure: pryoff side-seal; lug; and presson twistoff. The pryoff side-seal closure, the earliest type of vacuum closure, has a rubber gasket to maintain the vacuum in the container. This type of closure has been displaced in the United States by the other two types. The vacuum is generally achieved by a steam flush (2), which also softens the plastisol to facilitate sealing.

Lug closures. Lug closures have 4–6 lugs that grip onto special threads in the bottle finish as well as a flowed-in plastisol liner (gasket). The lugs mate with the threads with a half turn. This is almost always done by straight-line machinery that incorporates two belts moving at different speeds. Maximum production speed with this type of closure and machinery is about 700/min.

Press-on twist-off closures. These closures are used primarily on baby-food jars, but they are being used now on other products as well. One of the reasons for their increasing popularity is the very high production speeds attainable. Like lug closures, they have a plastisol liner, but the liner is molded for high precision. Unlike lug closures, the skirt of the closure is straight. The seal is achieved when the plastisol softens and conforms to the bottle threads. Unlike threaded closures, which require multiple turns, and lug closures, which require a half-turn, presson twistoff closure require no turn at all. They are applied by a single belt that presses the closure onto the bottle at very high speed (eg, over 1000/min on baby food jars). These closures are held on by vacuum, which is achieved by sweeping the headspace with steam. The use of these closures was limited at one time by a requirement of reduced pressure of at least 22 in. of mercury (74.5 kPa) to hold them on. Not all products benefit from that degree of vacuum. A recent development is the ability to use these closures with a much lesser degree of vacuum.

Cappers for Rollon Closures

A rollon closure has no threads before it is applied. An unthreaded (smooth) cap shell (sleeve) is placed over the top of the bottle, and rollers in a chuck (see Fig. 8) form the threads to conform with the threads on the bottle finish. The chucks are raised and lowered by capper heads. These closures are available with and without a pilferproof band that is a perforated extension of the skirt. If there is such a band, it is rolled on by a special roller in the chuck. Operating speeds of the machines are in roughly the same range as rotary cappers for CT closures: up to 700/min.

Cappers for Presson Closures

The familiar crown closure for carbonated beverage bottles is one type of presson closure. They are applied by rotary machines that can operate with production rates of over 1000/min (see Fig. 9). Other types of presson closures are snap-fit caps, with or without a tamper-evident band; dispensing caps, which are often secondary closures; and overcaps.

Like screw caps, press-on caps can be applied by chucks or rollers; but the chucks and rollers do not need twisting action, nor clutches for torque control. There are several approaches to controlling the amount of pressure applied by chuck-type cappers. The capper head can be spring operated, with tension controlled by a collar adjustment; or a pneumatic clutch can

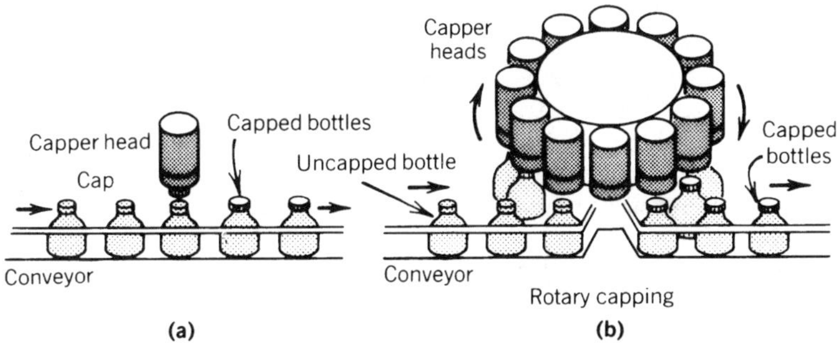

Figure 7. (a) Straight-line and (b) rotary capping (1).

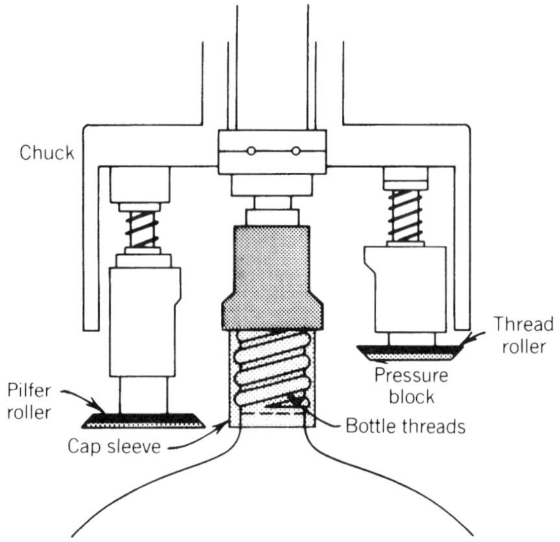

Figure 8. Applications of roll-on closures (1).

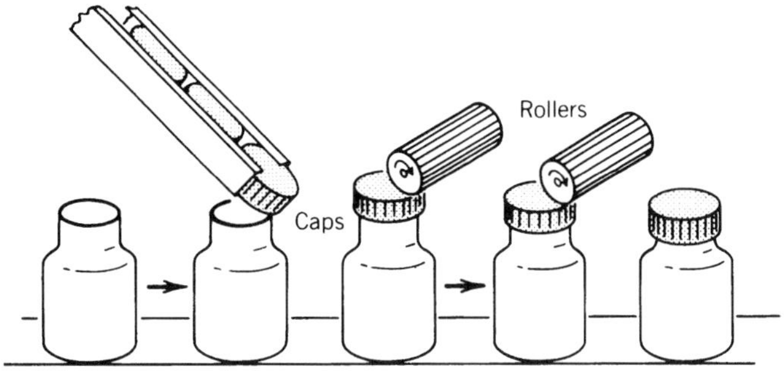

Figure 9. Roller application of presson closures (1).

control it with air pressure; or a barrel cam can be used. On a roller-type presson capper, the cap passes under one (shallow cap) or more (deeper cap) rollers that press the cap on tight.

BIBLIOGRAPHY

1. C. Glenn Davis, "Bottle Closing" (Packaging Machinery Operations course book), Packaging Machinery Manufacturers Institute, Washington, DC, 1981.
2. *Canned Foods: Principles of Thermal Process Control, Acidification, and Container Closure Evaluation,* The Food Processors Institute, Washington, DC.

CAPS. See CLOSURES.

CARBONATED BEVERAGE PACKAGING

Carbonated beverages represent the biggest single packaging market, with over 100 billion (10^9) beer and soft-drink containers filled each year in the United States. The packages cost substantially more than the contents, and are actually the beverage producer's largest expense item. Performance requirements are quite severe, as the contents are under pressure, they must hold their carbonation, and must withstand summer storage and (for beer) pasteurization temperatures that may reach 55–60°C. Furthermore, the highly competitive nature of this business has brought container design and manufacturing to a highly refined level in order to keep cost to a minimum. Even $0.001 per container becomes big business when multiplied by the billions (10^9) of containers made and used.

Package Types

There are four major categories of carbonated beverage container: plastic bottles, nonreturnable glass, refillable glass, and metal cans.

Plastic bottles. These are mainly 2-L soft-drink bottles, which first appeared in the mid-1970s and in just a few years took over around one-third of the soft-drink sales volume. The bottles are stretch-blowmolded (see Blow molding) from PET polyester (see Polyesters, thermoplastic). The stretching (orientation) is needed to get maximum tensile strength and gas barrier, which in turn enables bottle weight to be low enough to be economical. Because material cost is about two-thirds of total manufacturing cost, such weight savings are essential. Typical 2-L bottle weights are around 60 g PET, with some as low as 50 g, plus a high density polyethylene base cup (18–24 g), a label (3 g), and a closure. One design uses no base cup; instead, the bottle itself forms five petal-like feet at the base. It uses slightly more PET (65–70 g) but total weight is less because of the absence of the base cup.

A 3-L PET bottle has been used, with weights around 80–85 g PET plus 22–28 g for the base cup, or around 90 g without the base cup. Although the 3-L may not be as popular as the 2-L bottle in terms of size (too large for most refrigerator-door shelves), it is more economical.

Half-liter PET has been offered since 1979 but has never caught on because it does not provide enough economic incentive to change, and in small sizes, the advantages of safety and light weight are not as significant. Instead of 0.5-L, 16-oz (473-mL) PET is now offered, to compete more directly with 16-oz (473-mL) glass, and to save a gram or two of bottle weight because of the smaller volume (6% less than the half-liter). Typical bottle weights for these small PET containers are 28–30 g plus a 5-g base cup. Some cupless designs have been developed but have only minor commercial use thus far.

In the deposit-law states, small plastic bottles are more successful because come consumers do not like the handling and return of glass and are unwilling to pay more for metal cans. In these nine states, however, which represent over 20% of the population, the 2-L PET bottle has been the real winner, as many people prefer to pay one deposit instead of six (see Recycling).

It is believed that an improved plastic barrier to CO_2 loss (see Barrier polymers) would allow lighter (hence cheaper) bottles and thus spark the large-scale use of single-service sizes for soft drinks, and also open up the beer market to plastic. Such containers are the object of much development, both in can and bottle forms, but there has been no commercial activity beyond a few small test markets (see Cans, plastic).

Plastic bottles of all sizes usually carry full wraparound labels, either a shrunk polyolefin sleeve or a glued polypropylene/paper laminate (see Labels and labeling). Some plastic bottles have foamed plastic labels, which have a desirable nonslip feel and offer a little thermal insulation. These used to be very large, extending down over the base cup for maximum advertising effect (Owens-Illinois Label-Lite) but lately have been in conventional wraparound form and size (Owens-Illinois Plasti-Grip).

Nonreturnable glass. These are mainly 10- or 16-oz (296- or 473-mL) size, with a Plastishield (Owens-Illinois and licensees) foam-plastic protective label, or a paper/polyolefin roll-fed or an all-plastic shrink sleeve (see Bands, shrink). Typical bottle weights are around 130–140 g and 180–185 g, plus 3–4 g for the label. The glass industry has strongly promoted these containers in competition with metal cans, both in retail markets and in vending machines. Glass has an apparent cost advantage, which carries through to retail level and has managed to hold on to its market share in this size. However, the advantage in container cost is eroded by labels, closures, secondary packaging, filling speeds, and occupied space, plus the possibility of breakage and costs of clean-up. In all of these aspects, cans have an edge.

Prelabeling has become standard. The labels are preprinted and come to the bottler with no need for a further printing operation. Patch labels are cheaper and better for short runs, but are disliked by retailers as the bottles can be turned to obscure the labels.

The old crown closure that had to be pried off has all but disappeared for nonreturnables and is seen only on refillable glass to allow use of bottles with corresponding threadless necks (see Closures). The nonreturnables and even some refillables use rollon aluminum screw caps on threaded necks, with a tamper-evident ring (see Tamper-evident packaging). The threaded screwoff crowns that are crimped over a fine-threaded neck remain, but in diminishing numbers despite low cost. They are harder to open and are thus used mainly for beer (stronger hands), and their lack of tamper protection has discouraged their use.

The use of plastic closures for both glass and plastic bottles increased in 1984 after a decade of trial use and design refinement. The two most popular designs use a separate liner, but there are linerless designs as well. These closures weigh around 3 g in standard 28-mm size and are slightly more expensive than competitive aluminum rollons. All unscrew and fit over the same threads as the rollons and have some visible indication of tampering. Initial successes have been with soft drinks; a few have been tried for beer, but they have not been accepted due to their inability to withstand pasteurization.

Refillable glass. The old standby, the refillable glass bottle, still accounts for about a third of the soft-drink business, and about a fifth of the beer market. The main sizes are 16-oz (473-mL) for soft drinks, and 12-oz (355-mL) for beer, with some soft drinks also in sizes around a quart. It is usually the cheapest way to buy soft drinks on a unit price basis (except where 2-L bottles are being discounted), but most consumers prefer to pay a little more for convenience. In deposit states, refillables do better but remain a minor factor, after 2-L PET and metal cans. Typical bottle weight is around 300 g for the 16-oz (473-mL) soft-drink bottle, and 250 g for a 12-oz (355-mL) beer bottle that is used in bars and hotels where economy is foremost and delivery/return is not a problem.

Labels on major-brand soft drink refillables may be permanently silk-screened. Paper/plastic patch labels are used on most refillable beer bottles. Full wraparound labels are seldom seen because the containers are usually sold in six-packs or cases which cover most of the label area.

The economics of refillables are very different from other containers, as the bottler must support a "float" of containers and cases (which may cost as much as the containers themselves), as well as a larger fleet of delivery trucks (less compact, must stop to pick up as well as deliver), and the appropriate cleaning and washing machinery. For this reason, soft-drink franchises have been a classic home for local investor/entrepreneurs and this spirit persists in the soft-drink industry today despite the predominance of nonreturnable packaging.

Refillables are often promoted for their environmental benefits. The concept of refill/reuse may be laudable as it supports a resource conservation ethic, but the actual use of refillables brings its own environmental problems: water pollution from washing, air pollution from less efficient truck usage, and sanitation problems in both shops and homes.

Proponents of nonreturnables also point to the success of aluminum recycling. Opinions and emotions are very strong in these areas, but an impartial examination of the facts leads to no firm conclusion. In fact, much depends on locality; that is, the nature of the water supply, the extent of recycling possible, the degree of urbanization, and the number of trips made by each refillable. There is no easy answer to this question and neither refillables nor nonreturnables offer a clear advantage on environmental grounds.

Metal cans. The 12-oz (355-mL) aluminum can with easy-open end has become the primary small carbonated-beverage package, despite apparently higher container costs and retail price compared to either glass or plastic (see Cans, aluminum). Container weight and design have been the subject of much development work, with the weight of a modern can now down to around 18 g including the end. There is at least one necked-in ridge at the seam to allow wall-to-wall contact, which helps on the filling lines, makes firm six-packs, and uses smaller, cheaper ends. Double- and triple-necked cans are sold to get even smaller ends, and a new spin-neck design has a conical top section which achieves the same effect.

Nondetachable ends are now quite common, with a ring-like tab to pull, but without the ring coming off to create litter and a safety hazard. They are mandatory in certain states and are often used in other areas to avoid manufacture and stocking of both types, and to present an environmentally supportive image. A reclosable end for a flat-topped can has been announced but is not yet on the open market. Another reclosable metal can had a special polypropylene bottle-sized closure. It was introduced in 1983, but later withdrawn.

Steel cans were much more common than aluminum before the early 1960s, but their use has declined, especially for beer, and the old three-piece can with soldered side seam and separate top and bottom ends is seldom seen any more in this market (see Cans, steel). There are some three-piece cans with welded side seams and even more two-piece steel cans, drawn from steel much as aluminum cans are made (see Cans, fabrication). These may weigh as little as 37 g (with 5 1/2 g aluminum end). Although the steel can is heavier, the material is cheaper, and the steel industry had expected to keep more of the beverage market than it did with the two-piece cans. But fabrication is more expensive, coatings are more critical (to prevent rust), and aluminum did a fine job of selling its recyclability on a consumer level. (Steel can be recycled, too, of course, and is easy to separate from solid waste streams by magnets. However the economics are not favorable, and its does not always pay to get the cans back into the new-material stream, which is what really counts.) Formerly, the need for short runs of preprinted cans for house brands and other low-volume products kept three-piece steel in the running, but the growth of cooperative large-volume canning, improved machinery to coat and print finished cans, and the disappearance of many smaller brands have all contributed to steel's decline.

If steel is doing so poorly in beverages, why is it still the main can material for foods, juices, coffee and many other products? The answer is that the internal pressure of carbonated beverages stiffens the filled container, and makes steel's great advantage in rigidity of little importance. For the other products, this advantage counts, and makes aluminum more costly on an equal-performance basis.

Beer vs Soft Drinks

The beer market is approximately the same size as the soft-drink market and uses similar containers, but there are also some important differences. The market for large-size containers such as the 2-L PET bottle is very small for beer, and will remain so because beer quickly goes stale once the container is opened and exposed to the oxygen in the air. Thus, it must be consumed quickly and cannot be reclosed and finished later, and the large-size market is limited to parties where rapid consumption can be expected. But even at these occasions, there is some preference for cans and small bottles (what one does not drink now, one can drink later), and on the other end, some competition from kegs and even a plastic sphere that is set in a waterproof carton surrounded with ice. These bulk packages are also used in bars, of course, and hold a fairly steady 13% of the annual United States beer market of around 58 billion (10^9) 12-oz (355-mL) equivalents (estimated 1984 sales). The remainder is divided among cans (mostly aluminum, 34.5×10^9), and nonreturnable and refillable glass (12.4 and 3.6×10^9, respectively).

Similar figures for soft drinks are as follows:

	12-oz equivalents, 10^9
Unpackaged soft drinks (fountain sales)	26.0
PET bottles	19.5
Metal cans	31.3
Nonreturnable glass	12.5
Refillable glass	18.7
Total	*108.0*

Almost all soft-drink cans are actually 12 ounces (355 mL) in capacity, as are most beer cans; but some beer is sold in 8-, 10-, and 16-oz (237-, 296-, and 473-mL) sizes, and even a 32-oz (946-mL) size has been used. The nonreturnable glass includes 12-oz (355 mL) and 32-oz (946-mL) for beer, and 16-oz (473-mL) and 28–32-oz (829–946-mL) for soft drinks. Refillables are 12-oz (355-mL) for beer, and 16- and 32-oz (473- and 946-mL) for soft drinks.

Another addition to both beer and soft-drink marketing was the 12-pack of cans (and bottles), boxed in carry-home secondary packaging, and offering an economical compromise between the six-pack and the 24-unit case (see Carriers, beverage). Secondary packaging is an important aspect of beverage packaging, with shrink-film, plastic can-holding rings, paperboard carriers, molded-plastic bottle carriers, and corrugated and molded-plastic cases all vying for their share, and making their contribution to the total system price of the primary package.

Plastic containers have not yet entered the huge single-service beer market because no container has been offered that is cheap enough and still able to withstand pasteurization conditions, with a good-enough oxygen barrier to assure desired shelf-life under the most unfavorable storage conditions. The shelf-life problem is made worse by the presence of some oxygen in the beer as brewed and in the headspace of the container. In effect, any oxygen permeability at all puts pressure on the brewing operation to tighten their oxygen-excluding procedures even more.

Despite these problems, many companies are working to develop a plastic beer container, as the sheer size of this market gives these efforts a huge potential. Plastic beer containers do exist in Britain, where PVDC-coated (see Vinylidene chloride copolymers) 2-L PET bottles for products that are not pasteurized in the bottle are sold. Plastic beer cans have also been tested in Britain. Elsewhere in Europe, PVC bottles have been used at soccer games and institutions where breakage and cleanup are serious problems. In Japan, 2-L and 3-L elaborate beer bottles are made from PET, but they are virtually gift items and far too expensive for any mass market. The Japanese also have small (11½-oz 340-mL) PET beer bottles on the market on a limited scale. In Sweden, the "Rigello" container was used for beer for 15 years, but was later discontinued. This was a unique can/bottle made by welding two thermoformed halves together, surrounding with a paperboard cylinder, and topping with an injection-molded polyethylene closure.

Beer and soft drinks differ in carbonation content, which affects pressure requirements and rate of CO_2 loss. Pressure is typically expressed in volumes or in g/L of carbon dioxide. One volume equals approximately 2 g/L. At room temperature, each volume produces about one atmosphere (0.1 MPa) of internal pressure, but this changes with temperature, so that a 4-volume beverage such as a cola rises to 7 atmospheres (0.7 MPa) pressure at 100°F (38°C) and to 10 atmospheres (1 MPa) at maximum storage/pasteurization temperatures. The carbonation levels of some common beverages are

	Volumes of CO_2	g/L
Club soda and ginger ale	5	10
Common cola drinks	4	8
Beer	3	6
Citrus and fruit soft drinks	1½	3

The beer industry differ from the soft-drink industry in still another, very important way: there is no franchise system. In the franchise system, a parent company licenses a large number of local bottlers, who run independent businesses under the supervision of the parent, and buy their flavor concentrate from that parent. In the beer industry, on the other hand, there is great concentration, with 90% of the beer made by the top 10 companies. Most of these have some captive container capacity, so that introducing a new container may mean idling of existing capacity, and not just a simple cost comparison. The distribution systems also differ greatly. Soft drinks are much more local and even the big franchises and cooperative canning plants are still regional. Brewers, however, distribute over wider areas, and some ship to more than half the country from a single location. Such distances make container compactness important, discourage breakable glass and less-than-perfect closures, and make refilling (but not recycling) less economical.

Deposit Laws

No discussion of carbonated beverage packaging would be complete without comment on the deposit laws which are in effect in some states at the present time. In these states, all beverage containers carry a deposit, typically 5 or 10¢. Despite many complaints, the industry and the public have learned to live with such laws, and argue over the relative merits and troubles that they bring. It is fairly well agreed, however, that the cash value for discarded containers does keep most of them off the highways and gets more of them into the recycle streams. The aluminum-can industry, which has a widely-publicized recycle system in operation in both deposit and nondeposit states, now claims that more than 53% of all aluminum beverage cans are recovered in this way! There were some attempts to repeal the deposit laws in a few states, all unsuccessful, and attention then turned to the financial side: compensation for handling of the containers, who is obliged to refund money to whom, the scale of deposits, exempt containers, the status of refillables, and the like.

On a national scale, a deposit law was initially proposed in the early 1960s, with no short-term likelihood of passage, but always a possibility. Later, pro-deposit supporters planned campaigns to change the laws in other states. If a few more large states do, indeed, change their laws, the pressure for a nationwide law to avoid fragmented and often conflicting regulations among states will certainly increase.

BIBLIOGRAPHY

Carbonated Beverage Packaging in *The Wiley Encyclopedia of Packaging Technology* 1st ed., Wiley, New York, by A. L. Griff, Edison Technical Services, 1986, pp. 519–522.

General References

Beer Packaging, Master Brewers' Association of the Americas, Madison, WI, 1982.

Plastic Containers for Beer, Soft Drinks, and Liquor, Edison Technical Services, Bethesda, MD, 1980.

CARDED PACKAGING

Marrying plastic materials with paperboard to produce visual, self-vending packages is one of the most important and fastest-growing methods of merchandising products today.

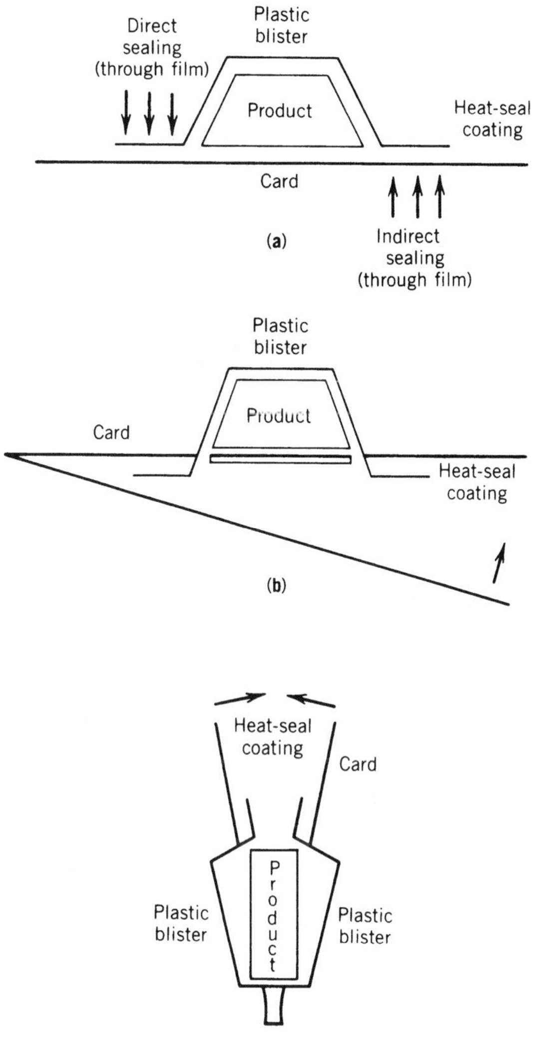

Figure 1. Blister packaging: (**a**) conventional surface seal; (**b**) foldover-card variation; (**c**) hinged blister/foldover-card variation.

The rapid growth of self-service retailing created a demand for innovative packaging that protects the product and also provides sales appeal in terms of product visibility and instructions for use. The styles and forms of visual carded packaging are too numerous to discuss in detail, but they can be grouped in two categories: blister packaging and skin packaging.

Blister Packaging: Components and Assembly

Components. Three of the many styles of blister packages are shown in Figure 1. The basic components of a blister pack are the preformed plastic blister, heat-seal coating, printing ink, and paperboard card.

Preformed plastic blister. An important key to package success is the selection of the right plastic film for the blister: property type, grade, and thickness. Consideration must be given to the height and weight of the product, sharp or pointed edges, impact resistance, aging, migration, and cost. The plastic must also be compatible with the product. Heat-sealing properties, ease of cutting and trimming the formed blisters, and other factors influencing production and speed of assembly must be taken into account.

Heat-seal coatings. Heat-seal coatings (see Sealing, heat) provide a bond between the plastic blister and the printed paperboard card. These solvent- or water-based coatings can be applied to rolls or sheets of printed paperboard using roll coaters, gravure or flexographic methods, knives, silk-screening or gang sprays (see Coating equipment). Whatever the system, it is essential that the proper coating weight be applied to the paperboard card for optimum heat-sealing results. In addition, the heat-seal coating must be compatible with the paperboard card and the plastic blister. For example, the type of coating used with an acetate blister is not the same as one used with polystyrene (PS). With proper coatings, strong fiber-tearing bonds can be obtained. The heat-seal coating also protects the printed areas of the card and provides a glossy finish.

Printing inks. Printing inks provide graphics and aesthetic appeal, and must also provide a bond between the paperboard card and the heat-seal coating. Inks may be applied to the paperboard by letterpress-, gravure-, offset-, flexography-, or silk-screen printing processes (see Inks; Printing). The inks must be compatible with the heat-seal coating and paperboard card, must resist high heat-sealing temperatures, abrasion, bending, and fading, and must be safe for use with the intended product. Inks should not contain excessive amounts of hydrocarbon lubricants, greases, oils, or release agents. Qualification tests should always precede production runs.

Paperboard. Paperboard (see Paperboard) for blister packaging must be selected according to the size, shape, and weight of the product as well as the style of package to be produced. It provides the base or main structural component of the package. Paperboard for blister packaging ranges in caliper, ie, thickness, from 0.014 to 0.030 in. (0.36–0.76 mm), but 0.018–0.024 in. (0.46–0.61 mm) is the most popular range. The surface must be suitable for printing by the required process and inks and also compatible with the heat-sealing-coating process. Paperboard must be able to meet the stresses imposed by the printing processes, ie, primarily delamination of offset-printing presses, and still provide good fiber-tearing bonds when heat-sealed to a plastic blister. Clay coatings are added to paperboard to enhance printing results and heat-seal-coating holdout. Heat-sealing and printability are both important considerations in blister packaging, and the paperboard must offer the best workable compromise.

Assembly. The normal sequence of assembly involves loading the blister with product, placing a paperboard card over the blister, and heat-sealing the package. This can be a simple manual operation, semiautomatic, or fully automated. All heat-sealing methods mate the blister and card under constant pressure for a specified time during which heat is applied. The mating surfaces fuse and bond, setting almost instantaneously when heat input stops. Heat-sealing machines and methods should provide the necessary production rate and should also be able to be adjusted to maintain constant conditions of dwell time, temperature, and pressure. Heat-sealing machines usually employ heated dies, produced when an electric current is passed through a resistance-type heating element in the die (see Sealing, heat). Some heat-sealing machines utilize impulse, electronic, or high-frequency heating. Both direct (heat applied through the blister) and indirect (heat applied through the paperboard card) heating techniques are successful. Direct sealing is faster than the indirect method and minimizes scorching or warping of the paperboard card.

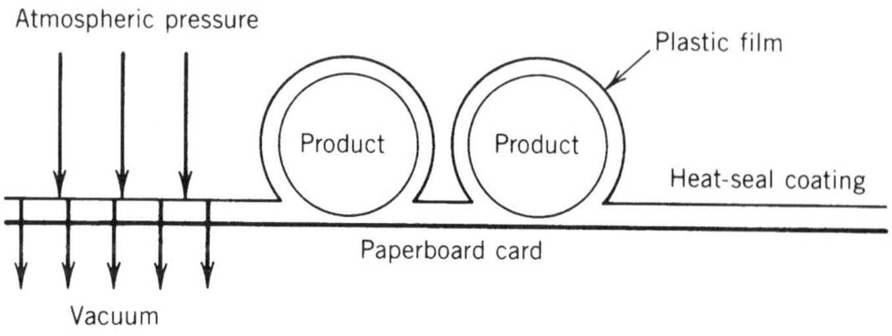

Figure 2. Skin packaging.

Skin Packaging: Components and Assembly

Skin packaging is a special form of visual carded packaging. It differs from blister packaging in that the product itself becomes the mold over which the heated plastic film or "skin" is drawn by vacuum and heat-sealed to a paperboard card (see Fig. 2). As in blister packaging, there are four principal components in a skin package: the plastic film, the heat-seal coating, the printing ink, and the paperboard card.

Plastic film. There are three types of flexible plastic film used in skin packaging: LDPE (see Polyethylene, low density), PVC (see Film, flexible PVC), and ionomer (see Ionomers). Which film is used depends on the application requirements. The film must be compatible with the heat-seal coating on the paperboard card to ensure a fiber-tearing bond or an otherwise acceptable heat seal. Normally, skin-packaging films are heated, draped and formed, and bonded to the paperboard card in one operation.

Heat-seal coatings. Heat-seal coatings for skin packaging must be compatible with the inks, paperboard, and plastic film used in the package. They may be solvent- or water-based. Methods of application are the same as those for blister packaging.

Printing inks. Inks for skin packaging must provide a bond between the paperboard surface and the heat-seal coating. Inks should not contain any compounds that may inhibit heat-sealing, such as waxes, oils, and release agents, and must be resistant to heat and fading and compatible with the product.

Paperboard. Paperboard for skin packaging must be selected according to the caliper, stiffness, and other strength characteristics necessary to support the product. High-porosity paperboard should be used to allow proper drawdown of the film and good contact with the card. Paperboard for skin packaging is generally not clay-coated. Clay coatings provide good printing surfaces and holdout properties for heat-seal coatings, but they also decrease porosity and interfere with vacuum drawdown during heat sealing on skin-packaging equipment. Certain inks and heat-seal coatings may also decrease porosity, and in that case, the paperboard must be perforated to increase air flow through the sheet during vacuum draw-down and heat-sealing.

Assembly. After proper selection and processing of the skin-packaging components, assembly of the package can take place. In effect, the product to be packaged is laminated between the paperboard card and the plastic film. The product is usually positioned on the printed heat-seal-coated paperboard card, which then moves onto the base or platen of the skin-packaging machine which contains air passages connected to a vacuum system. Plastic film is held in a frame above the product/paperboard card. The film is heated to a softening temperature, and at the proper time in the heating cycle the frame carrying the film is dropped, allowing the film to drape over the product and paperboard card. Vacuum is then applied through the platen and card, bringing the film in contact with the card. The residual heat in the plastic film activates the heat-seal coating, which fuses and forms a fiber-tearing bond. Skin-packaging machinery is available in many designs ranging from manual to fully automated. Control of preheat, timing cycle, and postheat are all critical variables that must be controlled.

Plastic Films

Blister films. There are three principal types of rigid plastic films used in blister packaging: cellulosics, styrenics, and vinyls. A copolyester is being used now as well (see Polyesters, thermoplastic). The most popular cellulosic films are acetate, butyrate, and propionate. All three films have excellent clarity and thermoforming characteristics and heat-seal well to properly coated cards. Sealing temperatures are generally higher for the cellulosics than they are for the other films. Cellulosics do not have exceptional cold strength, but they are reasonably shock and craze resistant.

Oriented polystyrene (OPS) has low resistance to impact and shatters easily. Low temperature performance is also poor. Impact polystyrene has good impact resistance and cold-temperature properties. The styrenics seal well to coated paperboard under the proper conditions. Clarity of OPS is excellent (see also Styrene–butadiene copolymers).

Both plasticized and unplasticized vinyl films (PVC) are made (see Film, flexible PVC; Film, rigid PVC). The amount of plasticizer affects cold-temperature resistance and impact strength. PVC heat-seals well to properly coated paperboard. Vinyls vary in clarity from excellent to good, and in color from slightly yellow to slightly blue. Thermoforming characteristics are excellent to good.

Skin-packaging films. Flexible films that conform to product shape are used in skin packaging. The flexible films are LDPE, PVC, or Surlyn (DuPont) ionomer. LDPE is the least expensive of the group. It is generally not as clear as the others, requires more heat, and because it shrinks more upon cooling, board "curl" can result. LDPE is strong in both impact and tensile properties, and it adheres well to heat-seal-coated paperboard cards. PVC film was used in the past for skin packaging more than it is today, having been replaced to a great extent by ionomer. PVC clarity is excellent, and the slight yellow or blue tint is not objectionable. It heat-seals well to coated paperboard and conforms well to intricately shaped products. Compared with polyethylene (PE), it requires less heat and shrinks less. Ionomer, a relatively recent entry in the skin-packaging field, offers excellent clarity and color, fast heating, and good adhesion to a properly coated paperboard. Preheating time for ionomer is the same as for PVC and much faster than for PE, and it has exceptional strength. It is more expensive, but because of its strength, relatively thin ionomer films can replace heavier-gauge alternatives and be cost-effective on an "applied" basis.

Heat-seal Coatings

Heat-sealable coatings for blister- and skin-packaging cards are perhaps the most critical component in the entire system. The appearance and physical integrity of the package depends upon the quality of the heat-seal coating.

Blister-card coatings. A successful blister-board coating must have good gloss, clarity, abrasion resistance, and hot tack and must seal to the various blister films. Hot tack is particularly important because the product is usually loaded into the blister and the board heat-sealed in place face down onto the blister. The package is ejected from the heat-seal jig and the entire weight of the package must be supported by the still-warm bond line. A relatively low heat-seal temperature is desirable for rapid sealing and to prevent heat distortion of the blister film. Blister-board coatings are still predominantly solvent-based vinyls because of their superior gloss. Some in-roads are being made by water-based products, but these must be carefully evaluated for hot tack, gloss retention, adhesion to specific inks, and sealability to selected blister films. The rheology and flow properties of the coating must be appropriate for roll-coater application and holdout on clay-coated solid-bleached-sulfate (SBS) board. The final test of a blister-board coating is a destruct bond to the printed board.

Skin-board coatings. Heat-seal coatings for skin board must have special properties unique to this application. In skin packaging, the board is placed face up on a vacuum plate and the product positioned on the board. The film is heated to the proper temperature and dropped down on the product, and a vacuum is pulled through the board. The hot film conforms to the product and adheres to the heat-seal coating on the board. Here again, the skin film must form a fiber-tearing, destruct bond to the printed board. Unlike the clay-coated SBS board used for blister cards, skin board is either uncoated SBS or combination, ie, recycled board. Because these substrates are porous, they have poor holdout for coatings. Heat-seal coatings for skin board must have good holdout nevertheless, and must be heat-sealable to the films at low temperatures. Gloss and clarity of the coating must be sufficient to permit accurate identification of graphics after printing. The appearance of the final package is more a function of the skin film than of the coating, but the coating must not detract from the clarity of the film. Conventional solvent-based coatings tend to soak into the board and interfere with porosity. Mechanical perforation is usually necessary when this type of coating is used. Water-based coatings are now available with excellent hold-out and porosity, thus eliminating the need for mechanical perforation.

Another package that presents a coating problem is the foldover blister card. The heat-seal coating is applied on the back of the board, which is die-cut and folded around the blister. The coated surfaces of the board are heat-sealed to themselves. This type of card is usually combination board which is extremely porous and has poor holdout for coatings. When conventional solvent-based products are used, several passes through the roller coater are often necessary before adequate coating buildup is obtained. Fortunately, there are water-based coatings available now which have good holdout on this type of porous board. These products offer economies through lower material costs and faster production rates.

Several different types of heat-sealable coatings are used for carded packaging. Nitrocellulose (NC) lacquers have long been noted for their gloss and abrasion resistance. Their use is limited to the cellulosic blisters, however. NC lacquers have good coatability and leveling by most application methods, including roller coating.

Gel lacquers are used by some board converters for both skin and blister board. These coatings, based on ethylene–vinyl acetate (EVA) copolymers, have good gloss and adhesion to most films used in carded packaging. Skin-board coated with a gel lacquer must be mechanically perforated for adequate porosity. The chief difficulties in using gel lacquers are that they must be heated for application and that most require the use of aromatic solvents for thinning and cleanup.

Some of the most widely used coatings for blister cards are solvent-based vinyls. These lacquers have excellent gloss and abrasion resistance and are especially favored for use with PVC blisters. They are also used for skin packaging with PVC film, but the board must be perforated after coating. Adhesion to olefinic skin films, eg, LDPE and ionomer, is marginal at best. Most vinyls lend themselves to roller coating.

Among the newer coatings for blister board are water-based acrylics (see Acrylics). These approach the solvent-based vinyls in gloss and abrasion resistance, but they require more care in coating to achieve optimum appearance. Water-based EVA dispersions have become quite popular for coating skin board. They have good sealability to most skin-packaging films, and their high porosity eliminates the need to perforate the coated board. Most of these products are solvent-free, thus eliminating the odor and fire hazards associated with solvent-based products.

Paperboard

Solid bleached sulfate (SBS). The basic raw material for SBS paperboard is bleached wood pulp (see Paperboard). On a fourdrinier machine, a single layer of pulp is laid down on the wire and becomes the base for building the sheet. Internal sizing is added at the wet end of the machine. Surface sizing is added at the size press (both sides) and wet calenders (one side or both sides). This may be followed by clay coatings on one side or on both sides. SBS paperboard is produced in this manner for various packaging end uses, including folding cartons, and as a base stock for laminations.

Recycled paperboard. Recycled paperboard is generally made with multiple layers. This may be accomplished on machines such as multicylinders and vats, multifourdrinier and headboxes, multicylinders and headboxes, fourdriniers with multiple headboxes, etc. Fiber furnish is normally recycled fiber from waste collection. The multi-ply concept allows numerous variations in sheet construction, eg, a sheet may contain a recycled center ply or plies with a white liner on both sides. Clay coatings can also be added for printability and holdout. Recycled paperboard is suitable for setup boxes (see Boxes, rigid paperboard), some folding cartons, visual carded packaging, and other packaging applications.

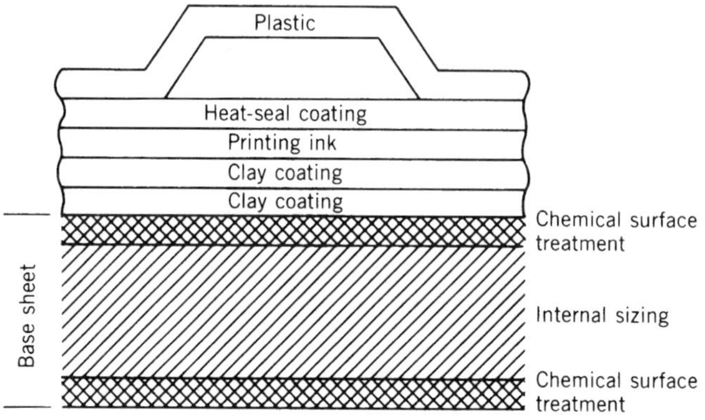

Figure 3. Cross section of a carded package.

Paperboard variations. Visual carded packaging places many demands on the paperboards, which is always the base on which the final package is built. The finished package is comprised of a variety of materials (see Fig. 3).

First, there is a wood-pulp base sheet with internal and surface sizing on both sides. The next layer (or layers) is clay coating, followed by printing ink, heat-seal coating, and plastic film or blister. The main objective is to produce a package in which the product is visible, securely contained, with the base sheet as the weakest link.

Depending on the package style, the selection of paperboard can be crucial. For a blister pack (see Fig. 4**a**) with intricate process printing on one side and line printing on the other (see Fig. 4**b**), a double-clay-coated one-side sheet would be selected. This sheet should accept multicolor printing, usually by the offset process, and withstand heat-seal coating, die cutting, and heat sealing to a plastic blister, adhering with such force as to tear fiber from the paperboard base sheet when the blister is removed.

Another variation is a blister sealed to a die-cut card (see Fig. 4**c**). Heat sealing of the blister flange to the back of the card may require clay coating or special heat-seal coating for the uncoated side of the paperboard. For a hinged or foldover card, which is die-cut and folded over with the blister protruding through the die-cut opening (see Fig. 4**d**) which is to be heat-sealed, a coated two-sided sheet would probably be selected. This would permit printing on one side and application of a heat-seal coating on the other side, which would be held out by the clay coating on the paperboard. Fiber tear is not so important. If the foldover card is not heat-sealed, it can be glued or stapled. The sandwich or double card (see Fig. 4**e**) is a variation of Figure 4**d** without the hinge.

Machinery

Blister-packaging machinery. Blisters are formed of 5–15-mil (127–381-μm) film in epoxy or water-cooled aluminum molds. After vacuum-forming (see Thermoforming), the blisters are die-cut and placed in a heat-sealing tool. After the product is loaded into the blister, a solvent- or water-based coated card is placed face down on the tool. Locating pins ensure accurate register of card to the blister. The heat-seal tool is then moved into the heat-seal area where pressure and temperatures of 250–400°F (121–204° C) are applied for 1.5–3 s and the film is bonded to the card.

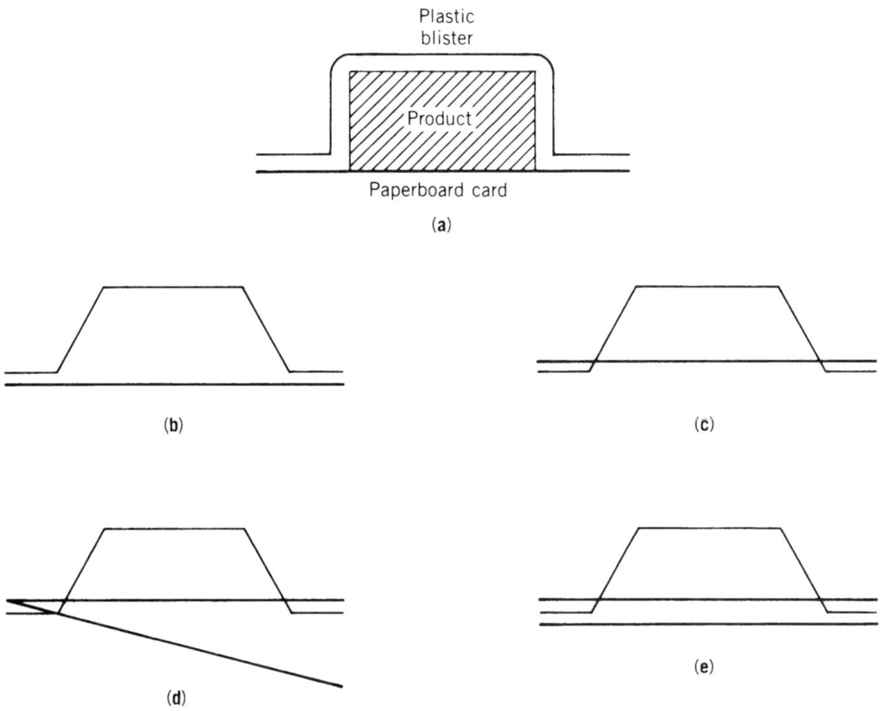

Figure 4. Blister packs. (**a**) Basic configuration. (**b**) Blister adhered to single card. Paperboard requirements: clay-coated on one side for multicolor printing, line printing on uncoated side; rigidity; flatness; heat seal on face down through board to fiber-tear endpoint. If multi-color printing is required on both sides, clay-coated two side may be used. (**c**) Blister sealed to die-cut card. Paperboard requirements: same as (**b**). Heat seal on backside of card may require clay coating or special heat-seal coating for uncoated side of paperboard. (**d**) Foldover card. Paperboard requirements: clay-coated on two sides for multicolor offset printing on one side and heat-seal-coating holdout on opposite side; rigidity; flatness; good heat seal. Foldover style may be hot-melt- or resin-glued or stapled, in which case clay-coated one side board may be used. (**e**) Sandwich or double card. Paperboard requirements: clay-coated on two sides for multicolor offset printing on one side and heat-seal coating on opposite side; rigidity; flatness; good heat seal. For resin glue, hot-melt, or staples, clay-coated on one side may be used.

When heat is applied through the card (indirect method), it is advisable to use a milled upper sealing die to apply pressure and heat only in the blister flange area on the back of the card. This speeds sealing time and reduces card warping. Manual and semiautomatic sealers are available as multistation rotary machines, shuttle-type machines, and indexing belt-type sealers. The indirect sealing tools are constructed of wood or aluminum with cork or rubber mounted to them in the configuration of the bottom blister flange.

Direct (through the flange) sealing tools use Teflon (DuPont) to prevent the film from sticking to the heated tool. Fully automatic machines are multistation rotary or belt type. They can be equipped with in-line forming/cutting, blister feeding, and card feeding. Speeds vary from 10 packages/min on smaller sealers to 150 packages/min on large automatic sealers. Blisters are formed by small users on small-bed, multipurpose vacuum-forming machines, and on large-bed or in-line vacuum-forming machines by large users. Multipurpose vacuum-forming machines can be used for both blister forming and skin packaging. Depending on speeds and blister configurations, options on machines may include blister blowoff (air used to lift blister off molds), water-cooled molds (higher speed operations), film assists (on higher blisters to eliminate film webbing), and plug assists (for deep-draw applications).

Roller-type and ram-type die-cutters are used to die-cut sheets of blisters apart after forming. For accurate cutting, care must be taken in rule-die construction to allow for variations in film gauge and shrinkage rates of different films after removal from the vacuum former. In the blister-packaging process, forming molds, blister-cutting dies, and heat-sealing tooling are required. Platform blister molds are sometimes used to reduce the number of cutting dies and blister sealing tools required for a multiple-item product line. Platform molds are designed to have a common blister flange and reduce setup times.

Skin-packaging machinery. Skin packaging is a vacuum-forming process where the product being packaged acts as a forming mold. Sheets of multiple products are processed and die-cut or slit apart into individual packages. Skin packaging is used for visual retail display and industrial product protection. The principal objective of industrial product protection is to immobilize the product for handling and shipping. Skin packaging costs much less than dunnage materials used in industrial packaging and it requires much less storage space. Since industrial products are generally relatively large and heavy, 7.5–15-mil (191–381-μm) PE and treated corrugated are commonly used. Display skin packaging typically uses ionomer film and coated printed boxboard, which is somewhat denser than corrugated board.

In the skin-packaging process, products are positioned on a boxboard or corrugated sheet and automatically conveyed or placed manually on the machine's vacuum platen. A heater bank is energized, and when the film is softened to a pliable, thermoformable state, the film frame descends, stretching film over the product. The vacuum system initiates the exhaust of air below the film and between the products. Vacuum draws the film in tight conformity to the product, encapsulating each in a tough, transparent plastic skin. Film is simultaneously heat-sealed to the surface of the board. When the film frame opens, the packaged products are discharged from the machine, which automatically rethreads the film for the next cycle. When the film frame is closed, the film is cut, and the frame elevates to its raised position, permitting entry of the preloaded sheet for the next cycle.

For industrial applications using heavy-gauge PE film, turbine pumps are used. They are capable of removing large volumes of air with a softer vacuum, which is generally required on odd-shaped or undercut items. Display packaging using ionomer film generally uses positive-displacement pumping systems, which consist of a pump and a storage tank which, when activated, can evacuate air between the film and substrate rapidly with a much stronger force. If certain items in a product line have deep recesses or undercuts, a combination system may be needed. The positive-displacement system is also used in blister forming.

Heating ovens are constructed with either metal-sheathed elements or nichrome ribbon. A film frame is required to hold the film during the heating cycle. These two-piece clamping frames or vacuum frames can be operated manually or automatically with automatic film cutoff for higher volume applications.

Since skin packaging is a sheet-processing system, the board converter and the machinery manufacturer should be involved at the early stages of a new program. The board converter and the machinery manufacturer can often lay out sheets that maximize printing-press and skin-packaging-machine efficiency and substantially reduce material costs.

Skin-packaging equipment is available in a wide variety of sizes and configurations, ie, from 18 in. × 24 in. (45.7 cm × 61 cm) to 30 in. × 96 in. (76.2 cm × 244 cm). Manual and semiautomatic systems are capable of running 2–3 sheets/min on smaller sizes and 1½–2 sheets/min on larger machines. Fully automatic machines can run 3–8 cycles/min. These speeds are for 3–5-mil (76–127-μm) ionomer display packages. Speeds of 1–2/min are generally achieved with 10-mil (254-μm) PE for industrial-packaging applications. Selection of size, infeed systems, and options is dictated by application and volume. Roller-type die cutters are used to die-cut sheets apart and die-cut hanger holes.

Blister Packaging vs Skin Packaging

The choice between blister and skin packaging is not always clear-cut. When the size of the product and blister is small, eg, a ballpoint pen, in relation to the card size, total blister-pack cost is generally less. If the product and blister are large in relation to the card, total skin-packaging cost is generally less. For this reason, large items such as antennae, hand tools, kitchen implements, and plumbing accessories are generally skin-packaged. Totally automatic blister-sealing lines have been available for 20 years and are used for high volume items such as razor blades, batteries, and pens.

Until recently, round items such as batteries, screwdrivers, and fasteners were blister-packaged since they moved on the card during the skin-packaging process. These now can be skin-packed through use of magnets and recessed moving platens. Fully automatic skin-packaging systems are a relatively recent development.

BIBLIOGRAPHY

"Carded Packaging" in *The Wiley Encyclopedia of Packaging Technology,* 1st ed., Wiley, New York, by J. M. Gresher, John D. Clarke Company, 1986, pp. 124–129.

CAREER DEVELOPMENT, PACKAGING

Career planning and development should be a lifetime occupation. Without a plan or specific development activities, one's career will tend to drift. There must be a conscious effort to set and achieve goals. Planning and implementation of your plan are probably the most important things you can do to maximize your capability and realize the success you desire. It should be noted that plans should be flexible and should be changed and refined to fit new goals.

One of the most important steps in career development is a career book to document the plan and keep a record of accomplishments. This book should be started as early in the career as possible in order to maintain a complete record of all important accomplishments. Most packaging professionals that have been active for 10 or 20 years have forgotten many of their most important accomplishments. They have no records to show how or what was accomplished. Even recent graduates cannot remember all their important activities. I personally believe that every packaging student should have a career book to keep copies of the results of important projects. There is also an opportunity for packaging directors to give each new employee a career book. I personally required each of my packaging engineers to maintain their book and keep it current. With each annual review they had to present their up-to-date book. There is great satisfaction for everyone in reviewing all that has been accomplished during the year. It should be noted that the career book is a great tool to present your credentials when interviewing for a new job or even a new position with your current employer.

Your career will span many years and possibly many different jobs and companies. Therefore, you can expect to have more than one volume. Be sure to buy a high-quality book because it must last a lifetime. Have your name embossed in gold, and use plastic inserts to protect your documents and specimens. Some of the most important things to include in your book are your college degree; certificates from continuing education; Institute of Packaging Professionals Certification; copies of patents; awards, letters of recommendation, and congratulations; samples of packages that you developed or were involved with; ads and promotional material for new products with new packages; copies of papers that you presented, publications, proposals, and prototypes of new packages; photographs of projects, co-workers' employees, family members, etc. Your book is a personal record, so the types of record put in the book may vary with different individuals. The important part is to maintain as complete a record as possible.

Building Relationships

The author has worked as a packaging professional for over 40 years and has learned that it is impossible to make any meaningful accomplishments by working alone. Real accomplishments are made by people working together, and working together builds relationships. Relationships should be built with all the people with whom you interact: your manager, employees, co-workers, suppliers, members of professional organizations, universities and educators, and packaging professionals in other companies, including your competitors. Use these relationships to make things happen. The success of most projects is dependent on using the right people. Use this resource to explore new ideas, help solve problems, identify new contacts and new relationships, organize new projects, implement new systems, and bring projects to a successful conclusion. Relationships are of tremendous value when you network to make a job change. The same networking can be used to identify candidates for a position you may wish to fill. It is the best way to identify speakers for professional seminars, authors to contribute to publications, and as candidates for professional organizations. Relationships will bring richness to a career.

Active participation in professional organizations is one of the most important places to build relationships. Professional organizations provide an important opportunity for association with many people who share common interests and goals. Membership is not enough. The maximum benefit is obtained only with active participation. This may be at the national level or at the local chapter level. The Institute of Packaging Professionals (IOPP) is the primary organization for packaging professionals. There are 41 local chapters across the country and 21 technical councils and task-force groups. IOPP "Who's Who in Packaging" is published annually. It lists 94 packaging-related trade and professional organizations. These organizations may be of interest to people who want to be active in their own industry. Other organizations, such as the Institute of Food Technologists, have a packaging committee. IOPP also lists 49 international associations for professionals outside the United States. All of these professional organizations provide the opportunity for the packaging professional to interact and build relationships with other professionals and establish lifelong friendships.

Continuing Education

This is probably the most important part of career development. It is essential for continued growth and expansion of your knowledge base and overall capability. We all start with a base to build on, and how we build and expand on that base determines the extent of growth of our overall capability. Recognizing that packaging technology utilizes many different disciplines, there are many different resources we can use to expand our knowledge, and also learn how to develop new disciplines. Many universities provide continuing education. These may be in the form of night school courses or short courses. Other sources are seminars at Packaging Expositions, Institute of Food Technology and other professional meetings, regularly scheduled seminars by the Institute of Packaging Professionals and other professional organizations, and seminars by suppliers. The professional journals are a source of information on current developments.

Some of other types of training that will support career development are public speaking, assertive training, report writing, writing patent disclosures, and learning management skills. An important skill to consider is to learn a second language. Next to English, Spanish is the most commonly used language in the United States. NAFTA will expand the need for people who can speak Spanish.

A development area few professionals have recognized is the need to learn to use the right side of the brain. Most of us primarily use the dominant left side of the brain. Dr. Betty Edwards, in her book *Drawing on the Right Side of the Brain,* states "The left side of the brain analyzes, abstracts, counts, marks time, plans step by step procedures, verbalizes and makes rational statements based on logic. It is said to be more closely linked to thinking, reasoning and the higher mental functions." It is recognized that most of our training and education has been directed at the dominant left side. The right side is the creative side. "It uses imagination, the dreamer, the artificer, the artist. Using the right side we dream and create new combinations of ideas." Most of the training on the use of the right side of the brain appears to be directed at learning to draw. This does provide a means for self-expression. It is said that learning to use the right side of the brain will release you from stereotypic expression. This release in turn will open the way for you to express your individuality in all aspects of your career. Seek out art training that teaches learning to draw using the right side of your brain. If you are lucky, you may find courses in imagination, visualization, in perceptual or spatial skills, or inventiveness. The right and left sides of the brain do communicate. You can develop your capability to the maximum when both sides of your brain are working together.

Sharing Knowledge

Sharing knowledge with others is very important to career development. It provides a means for dialog with your peers about what you have learned, new research and development, unique ideas and patent disclosures, and areas you want to explore. Sharing leads to better organization and understanding of ideas. Sharing is also an important part of learning. If we don't share with others, we do nothing to expand the information base. Take every opportunity to share through publications and presentations at professional meetings, by participating in IOPP and other groups such as the American Society of Testing and Methods (ASTM) and Technical Association of Pulp and Paper. You can be one of the continuing education resources for new professionals starting their career. This sharing will provide another opportunity to build relationships.

Manager's Responsibility

Managers and supervisors have a special responsibility to assist their staff in developing their career to the maximum of their capability. They need to be involved with career books, helping their staffs build professional relationships, and providing opportunities for continuing education and sharing. They should support programs that will help all aspects of career development. In many cases the supervisor is the key to the entire career development process.

BIBLIOGRAPHY

General References

T. Buzan, *Use Both Sides of Your Brain,* Dutton, New York, 1989.

B. Edwards, *Drawing on the Right Side of the Brain,* Putnam Publishing Group, New York, 1989.

D. Mac Crimmon Mac Kay, *Behind the Eye,* Basil Blackwell, Cambridge, MA, 1991.

Institute of Packaging Professionals, 481 Carlisle, Herndon, VA 22070.

Institute of Food Technology, 221 N. LaSalle Street, Chicago, 60601.

American Society of Testing and Methods, 1916 Race Street, Philadelphia, PA 19103.

Technical Association of Pulp and Paper, P.O. 105113, Atlanta, GA 30348.

BURTON R. LUNDQUIST
Director of Packaging
Armour Food Company, Conagra (retired)
Cave Creek, Arizona

Figure 1. Plastic ring carriers.

CARRIERS, BEVERAGE

Beverage carriers have traditionally been designed to group primary beverage containers into retail units of sale, as well as to protect the primary containers and merchandise or market the liquid contained.

The grouping function is important, as it must securely provide easy portability of a plurality of single cans, bottles, cups, tubs, or paperboard containers of beverages in units of sale that meet the needs of consumers. Typically, multiples of 4, 6, 8, 10, or 12 are commonly found; but larger multiples of 15, 18, 20, 24, 30, and even 36 are growing in popularity.

The protection function has developed to mean protection to the primary container against breakage, leakage, denting or disfigurement; protection to the beverage as in the case of providing a barrier against ultraviolet-ray degradation of beer; and protection to the consumer against personal injury from accidents while carrying or using the package.

The merchandising or marketing function has also evolved because of the development of brand proliferation and flavor segmentation in all beverages. The introduction of "new-age beverages," teas, micro brewed beers, etc, has established a need for the secondary package to establish brand identity, have on the shelf point-of-purchase appeal, establish differentiation among brands and flavors, and evoke media advertising.

Finally, the beverage carrier has evolved to provide a means of displaying retail unit Universal Product Codes (UPCs) reading through modern laser scan checkout machines while masking UPCs from the individual primary containers.

As beverages have been formulated to appeal to different tastes, primary containers have changed, and so have beverage carriers. Metal cans with thinner wall thicknesses; glass bottles with thinner wall thicknesses; PET (polyethylene terephthalate), and other plastic bottles in various sizes and shapes; plastic tubs, cups, and containers of various configurations; and paperboard "boxes" of various sizes and shapes have resulted in an evolution of the various types of carriers used as beverage secondary packages.

Can Multipacks

Several different raw materials are used in the multiple packaging of cans. Since over 55% of beverages in the United States are sold in cans, these various multipacks are the most commonly purchased today.

Plastic Ring Carriers

The Hi-Cone (Illinois Tool Works) plastic ring carrier is still widely used to package beverage and food cans into 4s, 6s, 8s, and other multiples. This carrier consists of a series of plastic rings that carry cans by the rim and grips them throughout the distribution and retail cycles. The rings are equipped with cutout "finger holes" for gripping and have tearout strips in order to access the cans themselves. These rings are applied by a machine furnished by Illinois Tool Works and applied in line at the canning line between the filler and the palletizers. (See Fig. 1.)

Paperboard Carriers

Paperboard carriers for cans are used in two different varieties, the wraps and the fully enclosed sleeves. In the United States, fully enclosed sleeves account for 62% of the can multipacks. Sold in 12, 15, 18, 24, 30, and even 36 multipacks, the popularity of these has grown significantly. Originally sold in 3×4 up to 4×6 can configurations, these carriers have recently developed to be double layered stacks of cans with an interstitial pad between the cans. Thus, multiples of $2 \times 3 \times 4$, $2 \times 3 \times 5$, and $2 \times 3 \times 6$ have been growing as a percent of the total retailed cans. (See Fig. 2.)

As an example in the United States soft-drink industry, 6-packs have been declined from over 60% of the total volume of packaged cans in 1989 down to 43% in 1994, while 12- and

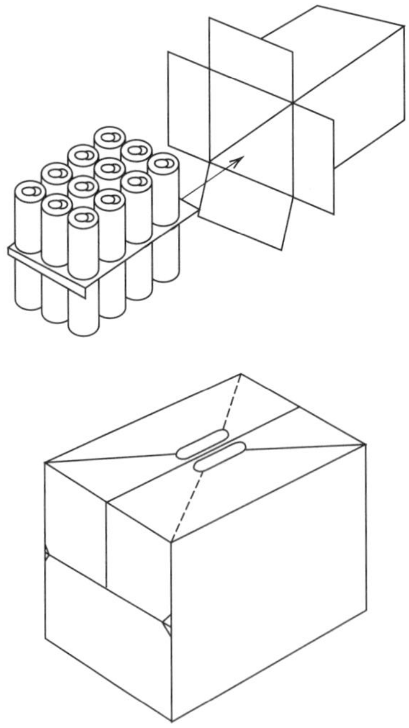

Figure 2. Paperboard carriers.

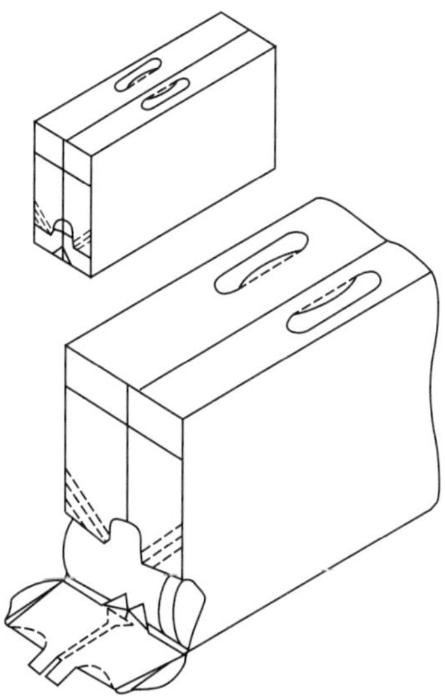

Figure 3. Fully enclosed sleeve packs.

24-packs increased from 30% of the total in 1989 to over 54% in 1994.

Typically, these fully enclosed sleeve packs are formed from a preglued sleeve of paperboard that is then removed from a magazine in a packaging machine, and collated single or double layers of cans are packed inside these sleeves in a parallel and continuous motion at speeds of ≤ 3000 cans per minute. The ends of the sleeves are then folded and glued or locked in place. (See Fig. 3.)

While the fully enclosed sleeves are the most popular form of secondary packaging for beverage cans in the United States, can wraps continue to be popular in Europe, Asia, and other parts of the world. Paperboard can wraps are printed and die-cut blanks of paperboard, which are removed from a magazine in a packaging machine, brought over the cans in line and in continuous motion; then the side wing panels are folded down, the bottom flaps folded in, and the cartons are locked or glued in line. (See Fig. 4.)

Both the fully enclosed sleeves and the can wraps are formed principally from paperboard formulated with 70–80% virgin fiber and 20–30% recycled fiber and containing high resistance against wet tear.

The most common machinery used for forming fully enclosed sleeves is supplied by Riverwood International, The Mead Corporation, and R. A. Jones Company, while the majority of the wet-strength carrierboard used in manufacturing these packages is supplied by Riverwood International and/or The Mead Corporation.

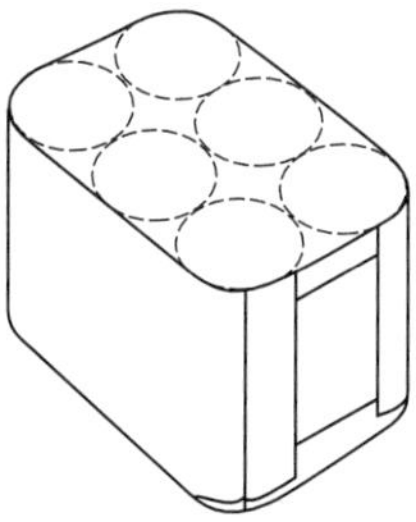

Figure 4. Can wrap.

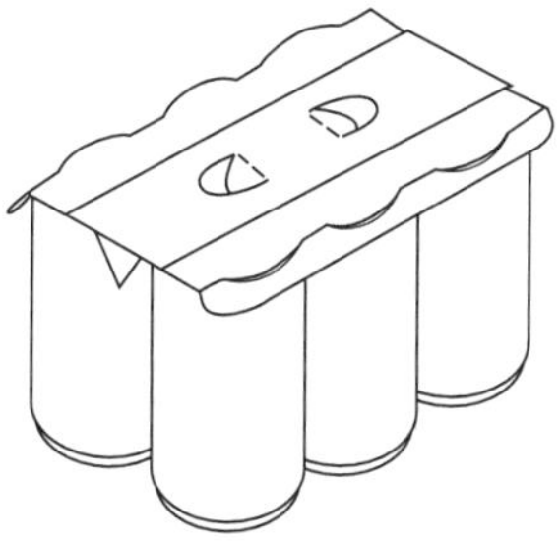

Figure 5. Paperboard clips.

Other forms of can multipackaging are paperboard "clips," which grip the rims of cans, holding them by this rim through distribution and retail; and shrink film, utilizing thermoplastic films that shrink around the cans with the application of heat. Both of these forms of can multipackaging have been tried in applications in the United States, but have not been able to sustain measurable shares of the total can packaging market segment. (See Fig. 5.) Riverwood International's, International Paper's "Triton" and "Eclipse" are primary examples of these forms of can multipackaging.

Bottle Multiple Packaging

Returnable glass bottles have decreased significantly as a portion of the beverage multiple packaging in the United States, dropping to less than 6% of brewery volumes and less than 2% in soft drink in 1994. The most common form of beverage carrier for returnable bottles is the performed basket-style carrier fabricated from paperboard.

Basket-style carriers are also a popular form of multipackaging nonreturnable or one-way glass bottles, especially in smaller multiples such as 4, 6, or 8 packs, and particularly in premium brands. Basket-style carriers have also been chosen by the Coca-Cola Company to package their proprietary PET contour-shaped bottle with special die-cut designs registered to make their design proprietary to the Coca Cola Company. Thus, the popularity of basket-style carriers continue, whether to package returnable or nonreturnable glass bottles, as well as for packaging plastic bottles (See Fig. 6.)

Basket-style carriers traditionally are die-cut and preglued from virgin or recycled paperboard, and all provide cells for each bottle, sides and ends that are attractively printed for merchandising and brand identification, and a central handle panel that rises above the necks of the bottles and provides for convenient and easy portability. Riverwood International,

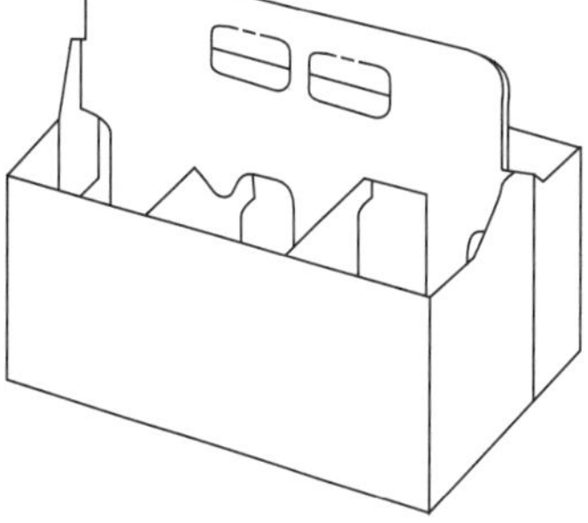

Figure 6. Bottle basket.

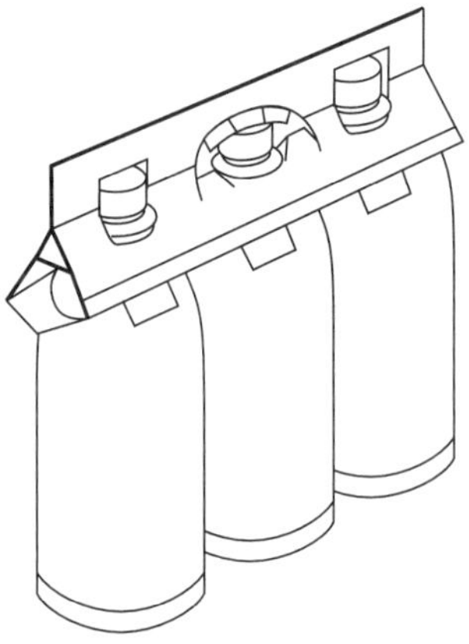

Figure 7. Bottle grips.

Mead Packaging, International Paper, Jefferson Smurfit Corporation, Zumbiel, and Standard Paper Box are the primary suppliers of these carriers.

Outside the United States, nonreturnable bottles are more commonly multipacked in paperboard wraparound carriers, which, similar to can wraps, are formed, wrapped, and glued or locked in place by in-line wrap machines. These wraps, most commonly used to form 4-, 6-, and 8-packs, are applied by in-line machines supplied by various companies such as Riverwood International, The Mead Corporation, and Certipak, a division of Kliklok.

Plastic bottles are also multipacked in plastic rings supplied by Illinois Tool Works. In the case of bottles, the rings fit snugly around the middle portion of the bottles and provide a handle protruding from one side in order to carry the bottles. These rings commonly form 6 or 8 packs of bottles.

Other forms of multipackaging nonreturnable bottles appear in the form of clips that grip the bottles by the neck or under the crowns and provide a handle with which to carry the bottles. (See Fig. 7.)

These clips are commonly used in countries of the European Economic Community (EEC), where the concern for environmental wastes has driven retailers and consumers alike to seek minimal packaging while offering portability and some merchandising possibilities. Clips can be formed from virgin paperboard and are also available in plastic.

Finally, for larger multiples of nonreturnable glass bottles, the fully enclosed paperboard sleeve is still the most widely utilized multipack, particularly in the brewing industry. In this package, the preglued sleeve, die-cut from wet-tear-strength board, is applied in-line and in a continuous motion by a packaging machine in an operation similar to the can fully enclosed package machines previously described. Operating at speeds of up to 1800 bottles per minute, these machines can operate with or without inserts that provide separations between the bottles.

These fully enclosed packs are commonly used for premium brand product and serve as an ultraviolet-ray barrier to the product as well as a merchandiser on the retail shelf.

BIBLIOGRAPHY

General References

B. Bynum, "New Twist on The Old Paper or Plastic," *Mississippi Business J.* Sec. 1, p. 1 (May 9, 1994).

L. Jabbonsky, "Points of Difference," *Beverage World,* Interbev 1994 ed., Keller International Publishing, p. 66.

L. Jabbonsky, "Image is Everything," in *Beverage World,* June 1992 ed., Keller International Publishing, p. 52.

E. Selane, J. Selane, and G. S. Kolligian, *Packaging Power-Corporate Identity and Product Recognition,* AMACOM (a division of American Management Associations), New York.

M. Wahl, *In Store Marketing,* Sawyer Publishing Worldwide, New York, 1992.

Octavio Orta
Coated Board Systems Group
Riverwood International Corporation
Atlanta, Georgia

CARTONING MACHINERY, END-LOAD.

See Machinery, Cartoning, End-load.

CARTONING MACHINERY, TOP-LOAD

Top-load cartoning employs flat paperboard blanks that have been die-cut by a carton manufacturer (see Fig. **1a,b**) to produce specific shapes and sizes when formed or folded into finished trays or hinge-cover packages. This style generally includes no pregluing by the carton converter, so flat blanks can be stacked directly on a pallet and shipped without secondary shipping cases. Hence, top-load cartons are very economical. They also allow product placement through an opening on the largest panel of the carton, which can greatly simplify the loading operation (see Machinery, cartoning, end-load; Cartons, folding).

Carton Forming

The heart of any top-load packaging operation is the carton-forming machine (see Figs. 2**a,b**). Although various configurations exist; the most common is a vertical system that provides overhanging delivery to outfeed conveyors or packing conveyors. Generally, these forming systems incorporate an inclined, gravity-advance magazine or powered horizontal hopper from which individual die-cut blanks are fed. Carton blanks are retained by small projections or tabs that extend slightly from the sides of a gate frame at the front of the magazine. Vacuum cups, mounted on a reciprocating feed bar, pull the individual carton blanks from the magazine and transport them in a downward arc. As vacuum is released, they are deposited in a registered position on top of a forming cavity. The carton blank is rotated from vertical to a horizontal plane during this feed cycle, with the vacuum cups contacting its inside surface. As the feed bar moves upward to feed the next blank, a plunger or mandrel moves downward to force the blank through the forming cavity (see Fig. 3). The plunger is designed so that the carton body conforms to its shape. The carton is folded, guided, and manipulated by a series of metal or composite plastic fingers and plows installed within the forming cavity. As the plunger completes each forming stroke and begins moving upward, spring-loaded mechanical traps retain the carton and strip it or remove it from the plunger. In some cases, special carton coat-

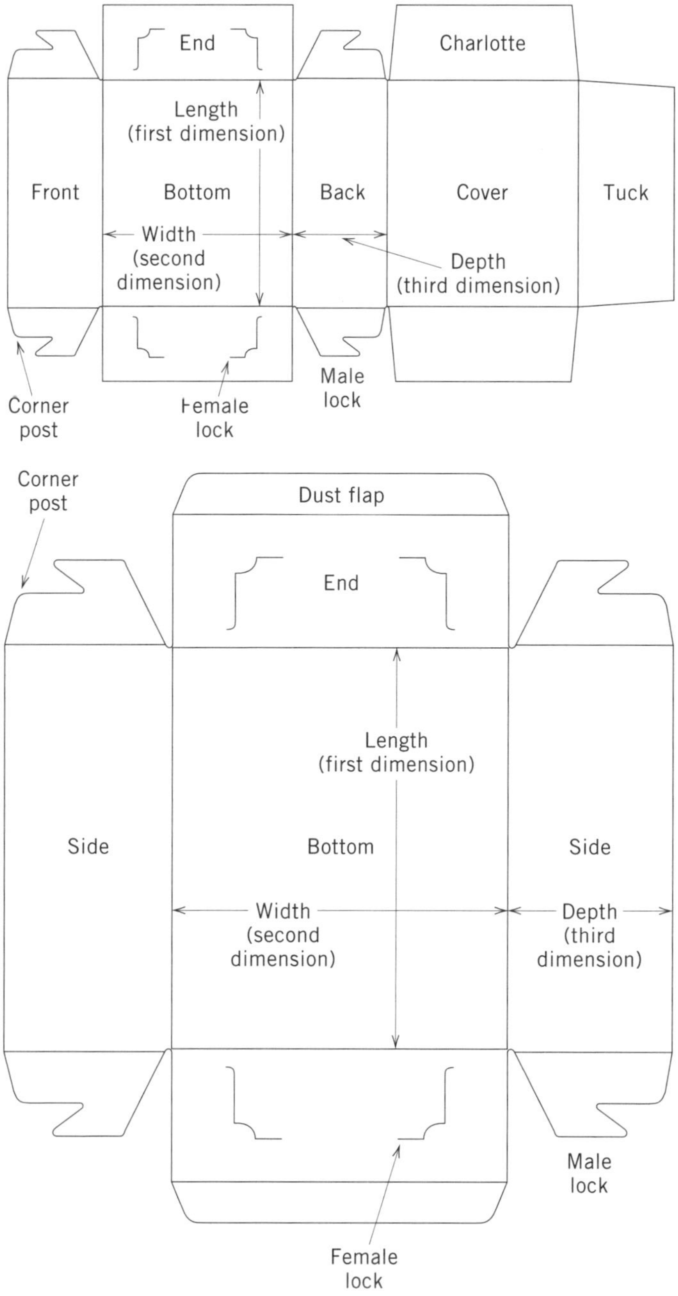

Figure 1. Diagrams of typical **(a)** hinge-cover carton and components and **(b)** dust-flap-style paperboard tray.

ings, shapes, or speeds dictate the use of a timed air blast system to positively eject the carton onto takeaway conveyors.

A wide range of carton sizes can be accommodated in styles ranging from simple rectangular corner-lock trays to a myriad of special shapes such as triangles, hexagons, octagons, and complex hollow-wall or shadow-box structures. The key to top-load carton forming lies in the special tooling that is designed and manufactured for each carton size and style. This removable tooling is commonly referred to as *a forming head.* It can be a relatively simple device or a complicated unit requiring various cams and actuation for flap folding and manipulation of the paperboard panels. The forming head consists of (1) the forming cavity, (2) a plunger or mandrel, and (3) various components for the hopper or magazine. By interchanging this tooling on any given forming machine, it can erect different carton styles, shapes, and sizes within a specified size range. For standard designs such as simple

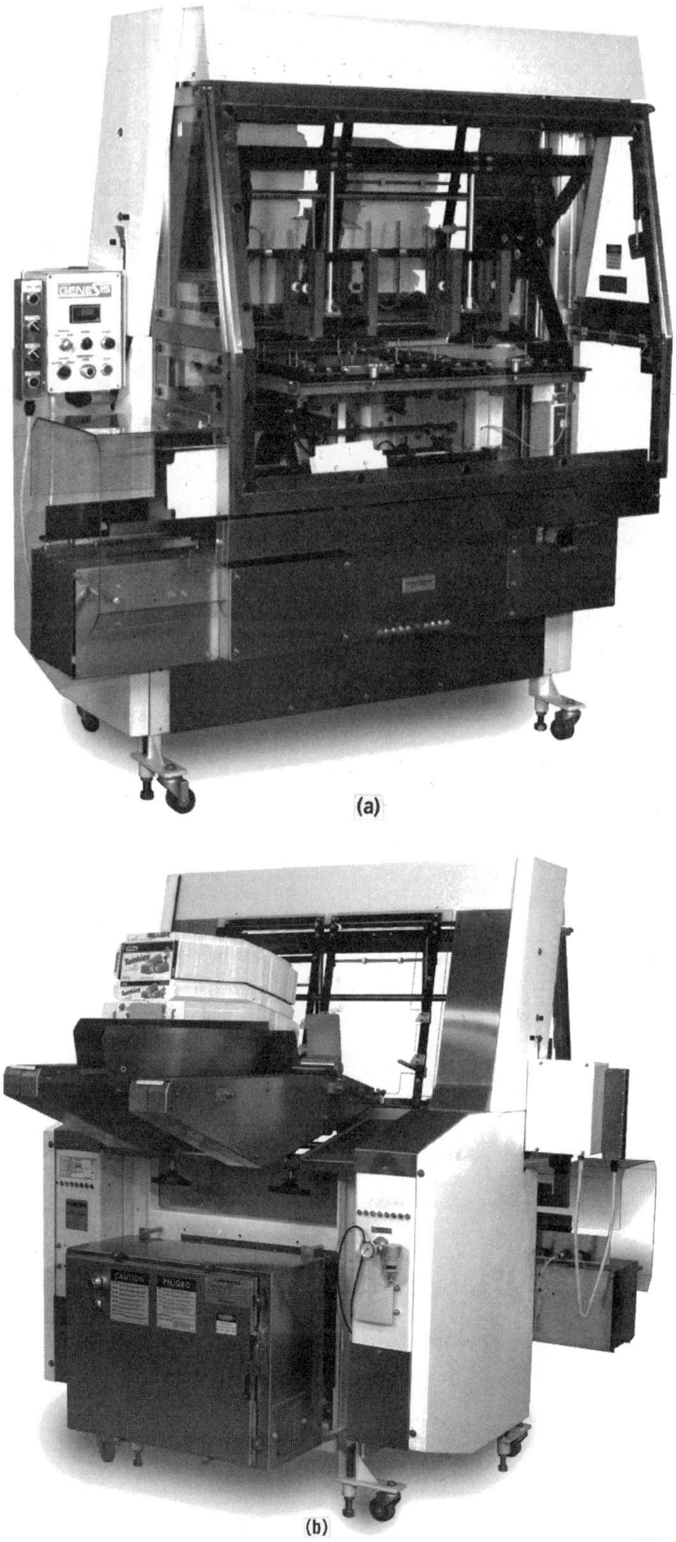

Figure 2. **(a)** Top-load carton forming machine with double-forming head; **(b)** rear of carton forming machine with powered carton hopper.

Figure 3. Lock-style carton forming head.

trays or hinge-cover cartons, the tooling can generally be changed in 10–15 min. In many cases, a multiple-head configuration is employed to erect several cartons with each forming stroke and increase the output of cartons per minute. Depending on the size of the package to be formed, machines may accommodate double, triple, or even quadruple forming heads.

The carton body can be formed utilizing locks, adhesive, or heat sealing.

Lock Forming

Dozens of different lock designs are available for forming trays and cartons to meet various packaging requirements. Most often, the lock design consists of a vertical and/or horizontal slit in the upright panels or walls of the carton body through which a specially shaped corner post or tab is inserted during the forming operation (see Fig. 4). The opening of the die-cut slits in the vertical walls and folding and insertion of the locking tabs are accomplished using specially designed and fabricated fingers or guides that are an integral part of the forming cavity. Normally, a mechanical actuation is also incorporated in the plunger to pull in the locking limb and ensure positive engagement.

Glue Forming

Glue forming of flat die-cut paperboard blanks can be accomplished using either hot-melt or cold-vinyl adhesives (see Adhesives). Hot-melt adhesive is generally applied in one of two ways. The simplest method employs open heated reservoirs, mounted in the machine directly below the forming cavity, to melt and contain the adhesive. This is commonly referred to as an "open pot" system. Applicator blades are mounted on a shaft running across the top of each reservoir. As the carton blank is placed on top of the forming cavity, the shaft is mechanically actuated and the blades rise upward from within the adhesive. They apply a series of dots or lines of adhesive to the underside of a flap or corner post located in each corner of the carton blank. As the carton is plunged through the forming cavity, these flaps are folded inside the vertical walls of the carton. Spring-loaded rollers in the cavity provide compression using the plunger inside the carton for backup. This compresses the adhesive and dissipates heat to allow quick bonding for reasonably high forming speeds.

The other commonly used method of hot-melt glue forming employs an enclosed, pressurized system where adhesive is supplied through heated hoses to guns or nozzles that spray a pattern on the inside of the carton's vertical body panels. Depending on the size of the carton and the pattern desired, this may be performed by first feeding cartons onto a short section of horizontal conveyor and applying adhesive as they are shuttled into position on top of the forming cavity. This avoids the need to mount necessary feed hoses and nozzles within the constraints of the forming cavity. It also allows greater pattern flexibility, and fewer nozzles can be used to apply glue to opposing corners of the carton as it travels beneath them. In other cases, where only simple glue patterns are required, applicator nozzles may be permanently fixed at each corner of the carton above the cavity. After adhesive is applied, the blanks are plunged vertically through the cavity and compressed in the manner described above.

Cold-vinyl carton forming was virtually eliminated with the advent of hot-melt adhesives because of the extended compression times required for setup. Carton styles were generally limited to specific compatible designs, such as outward tapered trays, which could be nested for extended compres-

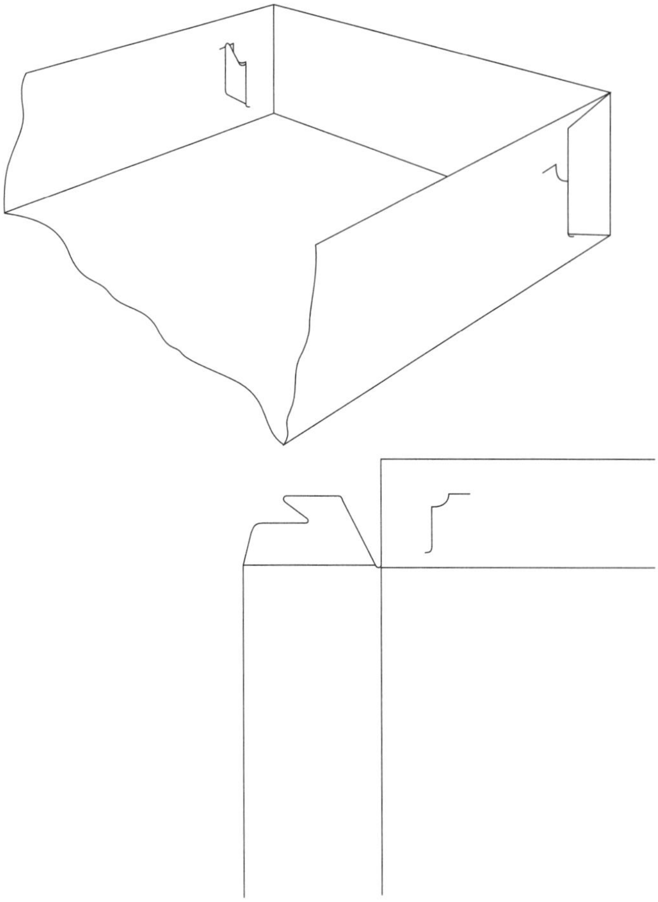

Figure 4. Common stripper-lock design: **(a)** assembled; **(b)** unfolded.

sion times after the package was discharged from the forming machine. However, recent developments in applicator technology and new end-use markets have created a renewed interest in this forming method. Today, it is frequently used to produce paperboard scoops, boats, and clamshells for the fast-food and food-service industries, as it suits their requirement for economical, preformed packages with a bonding medium that is not sensitive to heat. This heat resistance has also led to applications in connection with "ovenable" cartons. In the forming operation, a carton blank is fed from the magazine and deposited on top of a forming cavity. Specially designed adhesive application nozzles, mounted on the forming cavity, are then mechanically actuated to deposit small dots of cold-vinyl adhesive on the inside of the carton blank. Adhesive is supplied under pressure from a single remote tank. A plunger forces the blank through the forming cavity and, in this case, deposits the formed trays into a nest or stack in order to keep them under pressure until the adhesive has set. This method allows forming at speeds of up to ~80 strokes per minute or more than 300 cartons per minute with multiple heads, in order to satisfy the substantial production volumes required by the fast-food industry.

Heat-Seal Forming

Heat-seal forming utilizes special coatings or preapplied hot-melt adhesives on the paperboard as a bonding medium in the carton forming operation. Overall board coatings that can be heat-seal-formed include single-side polyethylene, double-side polyethylene, polyester, and polypropylene. Special pattern-applied hot-melt adhesive coatings, placed only in the area to be bonded, can also be used. The heat-seal system is designed to force air through electrically heated quartz elements into specially designed nozzles on the forming cavity. Depending on the carton coating, air temperatures ranging from 400 to 800°F (200–425°C) are directed over specific areas of the carton blank, where the coating on the paperboard stock is melted or activated. The carton blank is plunged through the forming cavity, and the board coating serves to bond the appropriate carton flaps. Very high speeds—up to a maximum of ~90 strokes per minute—can be achieved, depending on variables such as coating thickness, carton size, and carton style. Heat-seal forming operations require some degree of cooling to prevent heat buildup in various parts of the forming head and to accelerate carton compression times. Water or special refrigerants are plumbed to various components within the forming cavity and plunger. The extent of cooling generally depends on the bonding medium used, but becomes most extensive when double-polyethylene coatings are involved. Heat-seal forming is employed most often in the frozen-food industry, where thermoplastic coatings are otherwise included for moisture-barrier protection, graphic enhancement, or a degree of leak resistance necessitated by very wet products.

Top-Load Forming Capabilities

Various carton forming machine models are produced, each designed for a given size range and speed rating. The same basic machine chassis can usually be modified with special attachments and equipped as a dedicated lock, glue, or heat-seal system. As indicated, machines can be equipped with double, triple, or quadruple heads to feed and form multiple carton blanks simultaneously. Forming speeds generally range from 20 to 120 strokes per minute, and, with multiple forming heads, a single machine can produce more than 300 cartons per minute.

Figure 5. Flighted carton packing conveyor, hand packing.

Carton Conveying

After forming, the top-load carton is typically carried on a conveyor for loading either manually or automatically. For slow- to moderate-speed hand packing, simple flat-belt or plastic tabletop chain conveyors are frequently employed and offer the most economical approach. One end of the conveyor is generally placed below the forming cavity and is independently driven with no electrical or mechanical connection between the conveyor and the forming machine. After forming, trays or cartons drop onto the conveyor and are carried downstream for product loading. The alternative to this method is a conveyor with chain flights or lugs (see Fig. 5). Generally, flighted conveyors are either attached to and mechanically driven by the forming machine or electronically synchronized through the use of intelligently controlled independent drive motors. Flighted conveyors offer the advantage of pacing the operators, since they're not able to individually retard cartons for loading as they can on a flat belt. Flighted systems are required to achieve adequate carton control in any high-speed operation. They also allow the carton cover to be controlled during the packing operation by either maintaining a vertical position or folding it back almost 180° to permit loading from either side of the conveyor. Packing conveyors should be designed so that the bottom of the carton is approximately 30–34 in. (76–86 cm) from the floor. This helps optimize the efficiency of operators who are placing product into the cartons by hand.

Manual Product Loading

For hand-pack operations, product can be presented to the operator in many different ways. These range from tote bins to product conveyors running up to the carton conveyor. The most efficient method involves bringing the product in on a flat belt or tabletop chain that runs parallel and adjacent to carton flow. The bottom of the product should be elevated just slightly above the top of the carton to allow for simple sweep loading into the largest opening of the carton. This is the most efficient and reliable method of hand packing.

Automatic Product Loading

Automatic loading of products into top-load cartons can be accomplished using many different standard and or highly customized systems. Free-flowing products, such as individually quick-frozen vegetables, are often filled automatically using a volumetric system that is integrated mechanically or electronically with the carton conveying system. Where net-weight filling is desirable, a variety of different systems can be employed. These are normally interlocked electrically with the carton forming and conveying system in order to sense the presence of a formed carton and to signal the scales to dump. For net weighing or automatic loading of some products into top-load cartons, the carton conveyor must sometimes operate on an intermittent-motion basis. This allows the carton to stop or dwell momentarily beneath a filling device, providing sufficient time to completely load the product charge. Alternatively, special traveling or reciprocating-funnel systems can be incorporated at the point of product loading to allow continuous motion. The top or throat of the funnel is designed to provide a continuous open target beneath the filler discharge while the bottom aligns and travels in synchronization with the carton on the conveyor.

Many products are loaded into top-load cartons by count, in a specific pattern. These range from things such as spark plugs or hardware items to bare frozen hamburger or potato patties. Many other products such as wrapped candy, pouched mixes, boil-in-bag items, overwrapped baked goods, health-care and pharmaceutical items, office supplies, tobacco products, and all types of frozen foods have been automatically loaded into top-load cartons. The loading technique depends on the product and can vary from a simple mechanical shuttle to advanced units that automatically align, accumulate, group, and transfer product into the carton. Characteristics and consistency in the product's weight, size, shape, texture, temperature, and surface traits all have significant influence on the method of automatic loading. The transfer of such products is most frequently accomplished using a high-volume, low-pressure vacuum pickup system (see Fig. 6). A reciprocating-vacuum manifold transports items into the carton after they have been previously organized into required patterns. In some instances, advanced robotics, incorporating sophisticated vision systems, are used to locate product from a randomly positioned supply and automatically transfer it into cartons.

Figure 6. Automatic product loading, vacuum transfer.

Carton Closing

When hinge-cover or self-cover designs are used in any top-load cartoning application, the method of carton closing becomes a final major consideration. General categories of carton closers include flighted or lugless models. On flighted systems, actuation and sealing functions are timed according to predefined spacing between conveyor flights. On lugless systems, intelligently controlled motors and other functions react to the presence and position of a carton as it is photosensed on conveyor belts. In broad terms, flighted systems generally provide a higher degree of carton control, while lugless systems offer greater speed, flexibility, and ease of maintenance. For slow to moderately high-speed flighted operations, the closing machine is usually an independent unit with its own drive motor. It is equipped with a special infeed assembly that accepts cartons at random from an upstream packing conveyor and automatically times them into the flights of the closing machine. When flighted systems are required to operate at high speeds ($>$ 200 cartons per minute), it is desirable to eliminate the infeed section and drive the entire packaging line from the carton forming machine. This "line-driven" method requires the use of flighted packing conveyor and ensures that positive carton control is maintained throughout. The need to retime cartons into the closer at high speeds is eliminated.

Lugless systems are, inherently, independent as opposed to line-driven. Cartons are received at random and conveyed with sequential belts that have special surface traction characteristics. Both overhead and underlying conveyor belts may be employed to maximize control. Each sequence of the operation is powered by its own individual, intelligently controlled drive motor, so speeds and relationships between various machine functions can be readily adjusted through program controls. This allows a much greater degree of flexibility and fine-tuning without the need for physical changeover and tool adjustments.

Dust-Flap-Style Closure

Dust-flap-style cartons require closure of only a single tuck panel. As such, they can be closed on a straight-line operation where it is unnecessary to turn or rotate the carton. A static plow folds down the leading dust flap while a rotating paddle or wheel assembly, timed to the carton-conveying chain, serves to "kick" the trailing dust flap forward. After this has taken place, the cover is plowed down using guide rods, belts, or rollers. As the cover is plowed down, the front tuck score is prebroken in preparation for final closure. Dust-flap-style cartons can be closed either by inserting and locking the front tuck inside the body of the carton, or, by applying hot-melt adhesive to the inner surface of the tuck and adhering it to the outside of the carton's front panel.

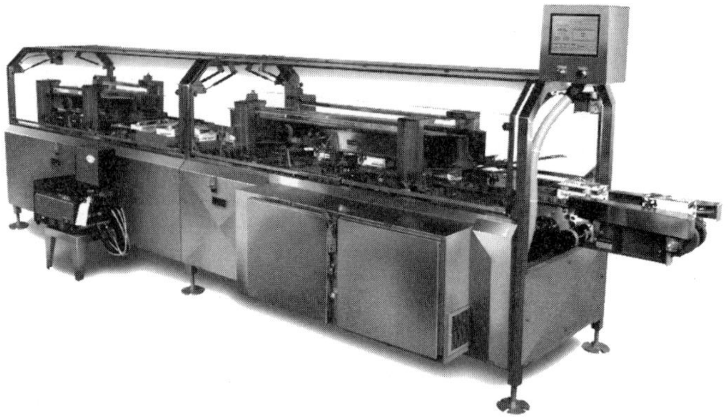

Figure 8. Straight-line lugless top-load carton closer.

Triple-Seal Style Closure

The hinge-cover carton design most frequently used is the triple-seal ("tri-seal") style, also called the "three-flap" or "charlotte" style. For this carton style, three primary closer designs are available: vertical, right-angle, and straight-line. For slow-to moderate-speed operations, where space is a limiting factor, a very compact machine design consists of an intermittent motion unit with vertical carton compression. Cartons are indexed by shuttling them at a right angle into a single hot-melt adhesive application station. Adhesive is applied to all three carton cover flaps simultaneously.

This type of closer generally uses an open-reservoir style of application system, with blades rising from within the adhesive to apply solid lines or dots of adhesive as required. It can also be equipped with gun- or nozzle-style applicators that spray adhesive onto the carton flaps. After adhesive application, the entire carton is elevated vertically through a compression tunnel that folds down all three flaps and discharges finished cartons at the top of the machine.

For most applications, tri-seal cartons are closed using either right-angle or straight-line machine configurations. In the right-angle operation (see Fig. 7), the carton is indexed into the closer, the front tuck is sealed to the carton body, and the carton is shuttled through a 90° change of direction for sealing the charlottes or end flaps. With the straight-line closer design (see Fig. 8), the carton body is rotated or turned 90° after closure of the front tuck, as it continues to travel in a straight line. This positions the charlottes parallel to the line of travel to facilitate application of adhesive or hot-air sealing and subsequent compression. To achieve proper registration between the cover and the carton body, it is best to fold down and trap the trailing charlotte before front-tuck closure. This is generally executed with an overhead, rotating tucker-paddle. This paddle is positioned above a chain transfer area, where a new set of lugs then traps the trailing charlotte against the rear of the carton and holds it in position through the front-tuck closing section. On lugless systems, the speed of overhead belts can simply be adjusted in relation to the carton-carrying belts in order to precisely position the cover.

Figure 7. Right-angle top-load carton closer.

Lock Closure

Closure of tri-seal cartons can be accomplished using locks, adhesive, or heat sealing. Lock closure is effected by utilizing specially designed plows, guides, and tucking fingers and is employed most often in the frozen-vegetable industry for plain-paperboard shells that are later overwrapped with printed paper.

Adhesive Closure

Adhesive closure can be accomplished with either hot-melt or cold-vinyl adhesives. Hot-melt is most frequently employed, and the adhesive is commonly applied using either open-pot or enclosed, pressurized nozzle systems. Depending on the glue pattern required, a thin wheel or cluster of wheels can be used to apply adhesive from an open reservoir to the carton flaps. When special patterns are desired, intaglio-wheels or nozzle-type applicators supplied by a remote tank, generally offer greater flexibility.

Historically, cold-vinyl adhesive found little application in on-line closing operations because of extended compression times required for setup. Recently, however, interest has been prompted by the requirement for ovenable paperboard packaging, where most hot-melts are unsuitable because they tend to soften and release during cooking. As a result, systems have been developed for polyester-coated ovenable board stocks. These systems apply a finely atomized spray of cold-vinyl adhesive, which is followed by the application of electrically generated hot air in a manner similar to that used for heat sealing. The hot air accelerates the water evaporation process and allows use of carton compression sections that are comparable in length to those of hot-melt systems. Sufficiently high production speeds (~150 cartons per minute) are achieved. This system has been used with polyester-coated paperboards for dual-oven applications, where the product is intended for preparation in either microwave or conventional ovens. Among the various thermoplastic polymers used for carton coatings, polyester is generally the only one considered suitable for dual-oven use, because of its high-temperature compatibility and limited heat solubility. By the same token, these characteristics necessitate high temperatures and extreme compression when the coating is used as a heat-sealing medium to bond the carton. Although the coating can be utilized for carton forming, where a plunger inside the carton provides necessary backup for compression, it is not practical for conventional heat-seal closure since only limited compression can be applied against the hollow package.

Heat-Seal Closure

Heat-seal closure can be used for cartons that include thermoplastic coatings, heat-sealable wax coatings (see Waxes), or preapplied hot-melt adhesives. Here again, the board coating or sealing medium is activated using electrically generated hot air at very high temperatures. The package then proceeds through compression to bond the flaps as the coating solidifies. For wax-coated cartons, the compression section must consist of refrigerated bars to prevent smearing of the wax. This method also results in quick bonding of the paperboard flaps, within a relatively short compression section.

Optional Functions

Top-load cartoning systems can be equipped with a wide variety of options, including many types of coding devices, leaflet feeders, labelers, sensors, and computers to facilitate effective control and management of the entire packaging system.

C. R. Kuhr
Kliklok Corporation
Decatur, Georgia

CARTON TERMINOLOGY BRIEF: A LOOK AT CARTONS ON JONES LARGE CENTER CARTONERS

General Information

Machine characteristics. The Jones cartoning machine will feed and open the carton, close the side flaps opposite the loading side and carry the carton along in a horizontal transport for the automatic insertion of the product. The major carton flaps are then closed, after which the carton is discharged from the machine.

Dimension standards. Carton dimension standards have been established by the Folding Box Association of America. Dimensions are score-to-score and do not include flaps (see Fig. 1).

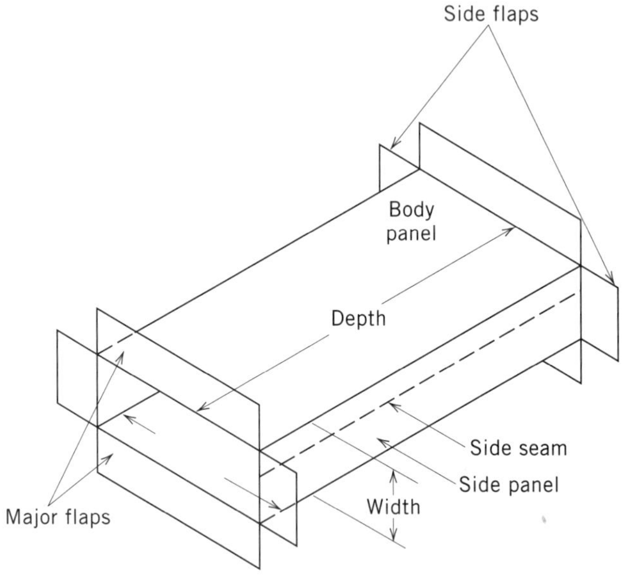

Dimension sequence:
length × width × depth

Length: Measurement between side flaps or along the top or bottom flap (dimension between the lugs)

Width: Measurement along the side with the side flap (dimension between inner and outer guiderails)

Depth: Measurement between top and bottom (of the carton) (dimension does not include side flap length)

Figure 1. Dimension standards.

Constant carton line (CCL). The score lines at the end of the carton on the loading side of the machine always stay in the same lateral position. This lateral position is the constant carton line (CCL). When the carton depth changes, the nonloading side of the machine is adjusted laterally to compensate (see Fig. 2).

Constant-opening line (COL). The constant-opening line is the position for all cartons to be opened. The carton's common point is the lower trailing score line.

Definition for left or right-hand machines. The Jones Criterion® cartoner may be designed to load the cartons from the left or the right. You will find this listed on the specification section of your machine under "machine hand." To check your machine, stand at the discharge end of the machine. When looking "upstream" at the discharge of the cartoner, if the barrel loader is on the *left,* your machine is a *left*-handed machine. Right-handed machines (Fig. 3) will have the barrel loader on the right.

Elements from R. A. Jones (RAJ) carton development drawing. Carton development drawings are useful tools. They are the result of many years of experience and should be passed on to carton suppliers as a guide. None of these suggestions are

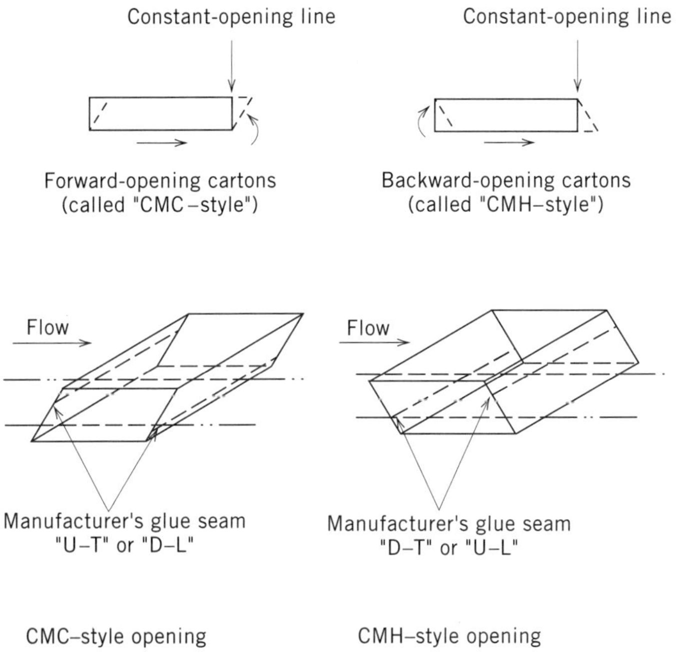

Figure 2. Opening lines (D-T = downward and trailing; U-L = upward and leading).

anything more than generally accepted standards in the cartoning industry.

Board thickness. Caliper is the thickness of board expressed in $\frac{1}{1000}$ in. and written as a decimal of an inch. Board with a caliper of 0.025, for example, is generally referred to as "25-point"; "points" are $\frac{1}{1000}$ of an inch.

Your cartoning machines will run in board thicknesses of 0.015–0.026 in. (or between 16 and 26 (16–26-point). Thicknesses down to 0.015 in. and up to 0.028 in. are also machinable if emphasis is given to prebreaking, scoring techniques (cut scores, skip cuts, etc), and an acceptable board stock.

Scoring. Good deep scoring is very important. All the carton flaps must break crisply at the score lines for efficient operation. If good scoring is not present, then "bowing" will take place at the score line and tucking inefficiencies become apparent.

Skip-cut or cut scores are often required on heavier boards or on cartons with very small width dimensions.

Grain direction. Carton board grain direction should always be perpendicular to the natural scorelines of the carton. It should always be shown on carton development drawings.

Carton finish. Generally, a lacquered finish is desirable as it is impervious to marking and seems to give the cartons "crispness." Some cartons are not lacquered, and these cartons are vulnerable to marking, especially when the cartons are clay-coated.

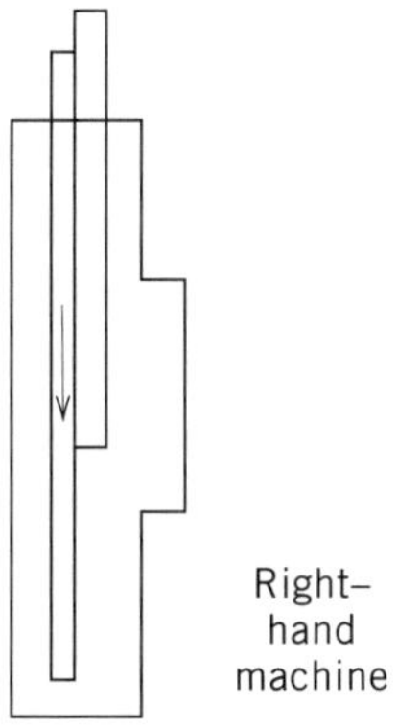

Figure 3. Right-hand machine.

How to Pack Cartons

(1) *Preglued Cartons in Shipping Containers.* Cartons must be furnished on edge in their corrugated shipping container. They should also be "loosely packed" in these shippers to retain carton prebreak. (Can you stick your finger in the shipping container? If not, the cartons are packed too tightly.) Cartons should be stored in low-humidity environments and with short shelf life to ensure maximum machinability. (See the discussion of "Fresh" cartons later in this article.)

(2) *Flat Blank Cartons on Skids.* See guidelines for handling blanks and pallet packaging for the Jones side seam gluer later in this article.

Industry standards for carton storage. The following are generally accepted industry standards that we offer for your consideration. Our customers who adhere to these general storage principles report the highest cartoning efficiencies:

1. Control conditions in your plant (in finishing area): 65°/75°F with 35–45% relative humidity.
2. Preferred conditions for storing your cartons: cartons in corrugated cases stored at 60–80°F with 60–70% humidity.
3. Carton optimum (industry standards established by the paperboard industry): best results occur when carton board moisture content is held at 4.5–6.0%.

Additional Carton Terminology

Barrier Coatings. Coatings applied to packaging boards to provide resistance to moisture penetration. Typically wax or polyethylene.

Basis Weight. The amount of fiber deposited per given unit of packaging board area. Typically, pounds per 3000 ft^2 of area or pounds per 1000 ft^2 of area.

Bleached Board. Packaging board produced at 100% bleached kraft or sulfate pulped fibers. Generally used when carton appearance is of prime importance.

Clay-Coated Board. A coating applied to packaging board to provide a smooth surface necessary for printing.

Dust Flaps. Inner carton flaps located on panels opposite outer flaps. Closed initially before application of glue and glue flap closure.

Glue Flap. A narrow longitudinal flap attached to one of the carton panels used to overlap and construct the manufacturer's seam.

Glue-Style Carton. Cartons using adhesive to secure the end flaps.

Grammage. Basis weight expressed in grams per square meter.

Manufacturer's Seam or Side Seam. The longitudinal carton seam used to create the "tube shape" of the opened carton. Typically applied by the carton manufacturers before shipment to customers, although R. A. Jones can provide equipment (side-seam gluer) to complete this seam at the cartoner.

Outer Flaps. Outer flaps that are closed after glue has been applied. The outside portion of these flaps shows on finished carton and therefore is decorated or printed.

Point. The thickness of the board in thousands of an inch (0.001). Also known as a *caliper* (eg, 0.20 in. = 20-point).

Prebreak. Prebending the carton at joints prior to closing the manufacturers seam to assist in opening the carton at a later date.

Recycled Board. Packaging board made primarily from fiber that was once a carton or piece of writing paper that has been returned to the papermaking cycle.

Scoring. Mechanically creasing the carton packaging board to create bending lines where the carton will open.

Tuck-Style Carton. Cartons using a mechanical flap interlock on end closure.

Guidelines and Standards for the Folding Paper Box Industry

Objective. To present a set of voluntary industry standards and specifications recommended for the guidance of both users and suppliers of folding paper cartons.

Guidelines and Standards by Folding Paper Box Association of America.

Applicability of standards. Because the entire field of folding cartons cannot be adequately covered by one set of standards, the ones set forth below are necessarily limited and intended to apply principally to cartons from board up to 0.024 caliper; of a brightness not less than 70 points on the GE Reflectance Meter, and excluding boards coated with barrier coatings or those with laminated structure. While these standards have general application to all cartons, they are not applicable in all respects and in every instance outside of the ranges indicated above.

Cartons outside the limitations described above are not intended to be excluded from the application of these standards, but rather are intended to be reviewed individually by user and supplier to determine the special specifications and wider tolerances necessary to meet the various end-use requirements.

Applicable conditions

1. Sampling techniques are recognized, and such samples taken shall be representative of the lot.
2. Measurements are to be made on cartons conditioned according to TAPPI (Technical Association of the Pulp and Paper Industry) standards T-402.
3. It is recognized that because boxboard changes with time and exposure, all measurements shall be made within 30 days from the time of carton manufacture, and cartons shall be stored during the interim under proper conditions.

Industry Standards

1. *Physical Dimensions.* Shall be within ± 1/64 in. any single dimensions, and ± 1/32 in. for any combination of dimensions in a straight line, or 1%, whichever is greater.
2. *Caliper.* ±5% or ±0.0001, whichever is greater.
3. *Brightness.* ±2 points on GE Reflectance Meter.
4. *Appearance of Scores.* Allowance of one surface check or crack not over 1/8 in. long or an accumulation of not over 2% of the length of any one score of a carton, with such occurrence being limited to appearance on not over 2½ in. of the total quantity of cartons involved.
5. *Stiffness.* Where required, ±5% of agreed standard using Gurly or taber test.
6. *Strength and Integrity of Glue Bond.* Carton must be strong enough to withstand setting up and filling operations that the carton normally can be expected to encounter.
7. *Straightness of Gluing.* On a simple one-compartment carton, an allowance of the lesser of 1/16 or 2% of the glue flap length for misalignment. On cartons involving compound folds, tolerances must be increased.
8. *Machine-Filled Cartons.* Cartons shall conform to the requirements of customer's filling machines and carton blueprints for same (within the limitations of these industry standards). Machine trails of full die forms are strongly recommended.
9. *Color Matching.* After initial agreement on an acceptable range of colors within a given production run of cartons, then not more than 5% of the carton shall fall outside that range.
10. *Defects in Printing.* When the specific characteristics of critical major and minor defects have been acceptably defined by reference by visual samples, then the tolerances (or allowance range) shall be determined by agreement between user and supplier.
11. *Rub Resistance.* User and supplier should agree on acceptable test results.
12. *Standard of Carton Weights.* The weight per thousand cartons shall be controlled −10% of the agreed standard.
13. *Handling and Storage*
    a. Storage should not be extended beyond 90 days of date of manufacture, due to the technical characteristics of folding cartons, including machinability of scores.
    b. Curl limitations: It must be recognized that paper board, because of its hygroscopic nature and the fact that its two surfaces are generally dissimilar in porosity (printed vs unprinted), will curl when moved from one set of temperatures and humidity conditions to another. Generally, a period of aging at the new conditions will reduce the curl.
    c. It will be recognized that an enlightened policy of product handling and storage can result in positive benefit to users of folding cartons. The following should receive particular attention:
       (1) Reasonable control of temperature and humidity
       (2) Reasonable height of stacking of cartons
       (3) Use of weight-distributing style of pallets
       (4) Attention to position of master cases on pallets
       (5) Proper proportioning of shipping containers

Other Cartoning Aspects: Guidelines for Blanks, Handling and Pallet Packaging for the Jones Side-Seam Gluer (SSG)

Finished cartons (as they are discharged from a SSG) are a blending of three factors:

1. The SSG
2. Hot-melt glue
3. Carton blanks

Of these factors, the most variable (and critical) is the carton blanks.

The carton blanks should arrive at the SSG in a flat condition. Warp or bow in cartons is normally the biggest cause of inefficient machine performance. Carton warp or bow generally has two main causes:

1. In-plant handling of blanks
2. Pallet packing of blanks (by the carton company)

Pallet packing. There are a number of ways to pack pallets of cartons successfully. It may be prudent for you to allow your carton blank supplier to choose their own method—as long as it is successful. Success means that the cartons arrive at the SSG with the pallet intact, the blanks flat, and the blanks protected from moisture and physical harm.

With the preceding recommendations in mind, successful flat blank pallet packaging could include (see Fig. 4 for an exploded view of a typical pallet) the following:

1. A wooden pallet on the bottom.
2. Slip sheets placed approximately every 12 in. within the stack of flat blanks. These slip sheets provide stability for the load.

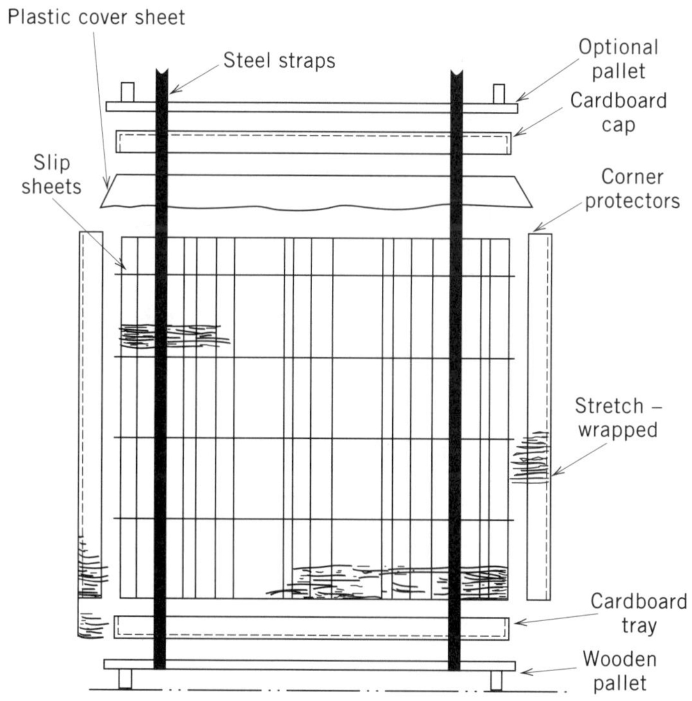

Figure 4. Typical pallet.

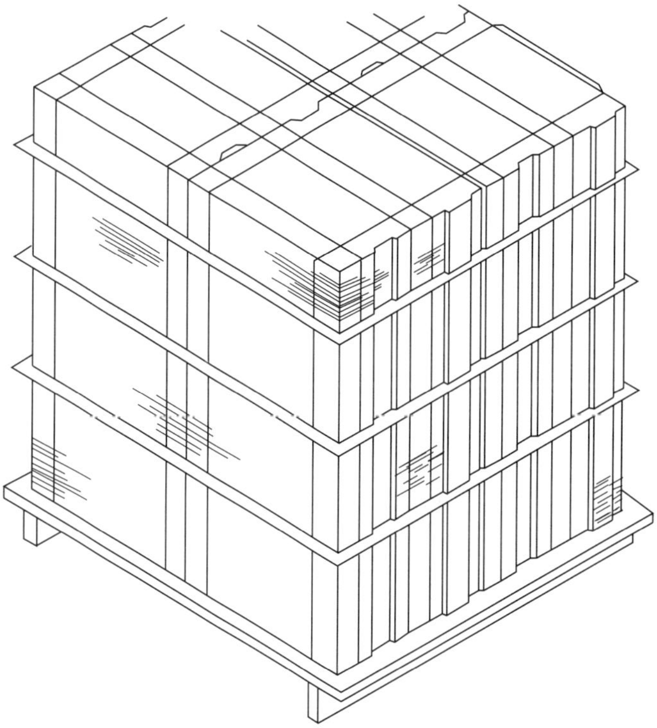

Figure 5. Cartons stacked in columns.

3. A corrugated cardboard bottom tray—directly on the pallet.
4. Pressboard corner protectors at each corner of the pallet.
5. Pallet that is stretch-wrapped.
6. A plastic cover sheet on top of the carton to give moisture protection from the top.
7. A corrugated cardboard cap over the loose plastic cover sheet.
8. Sufficient steel straps to hold everthing together.
9. An optional pallet on top of the load—this will help prevent physical damage during shipping and also permit you to stack carton pallets two or three high in your warehouse.

In addition, we suggest that cartons on a pallet be packaged in straight columns (see Fig. 5) and not staggered periodically. This method provides solid support for the carton flaps—minimizing warp and bow.

In-plant handling. There are only three suggestions:

1. Open only one pallet at a time. There is no need to let the blanks "breathe"—they will only start warping.
2. Drape a thin plastic sheet over any opened pallet to be left overnight. The plastic sheet restricts airflow (airflow can add or take away moisture from the blanks on top, causing bow and warp).
3. Remove any carton blanks from the SSG magazine if the SSG is to be left overnight. Restack these blanks on a pallet or other flat surface.

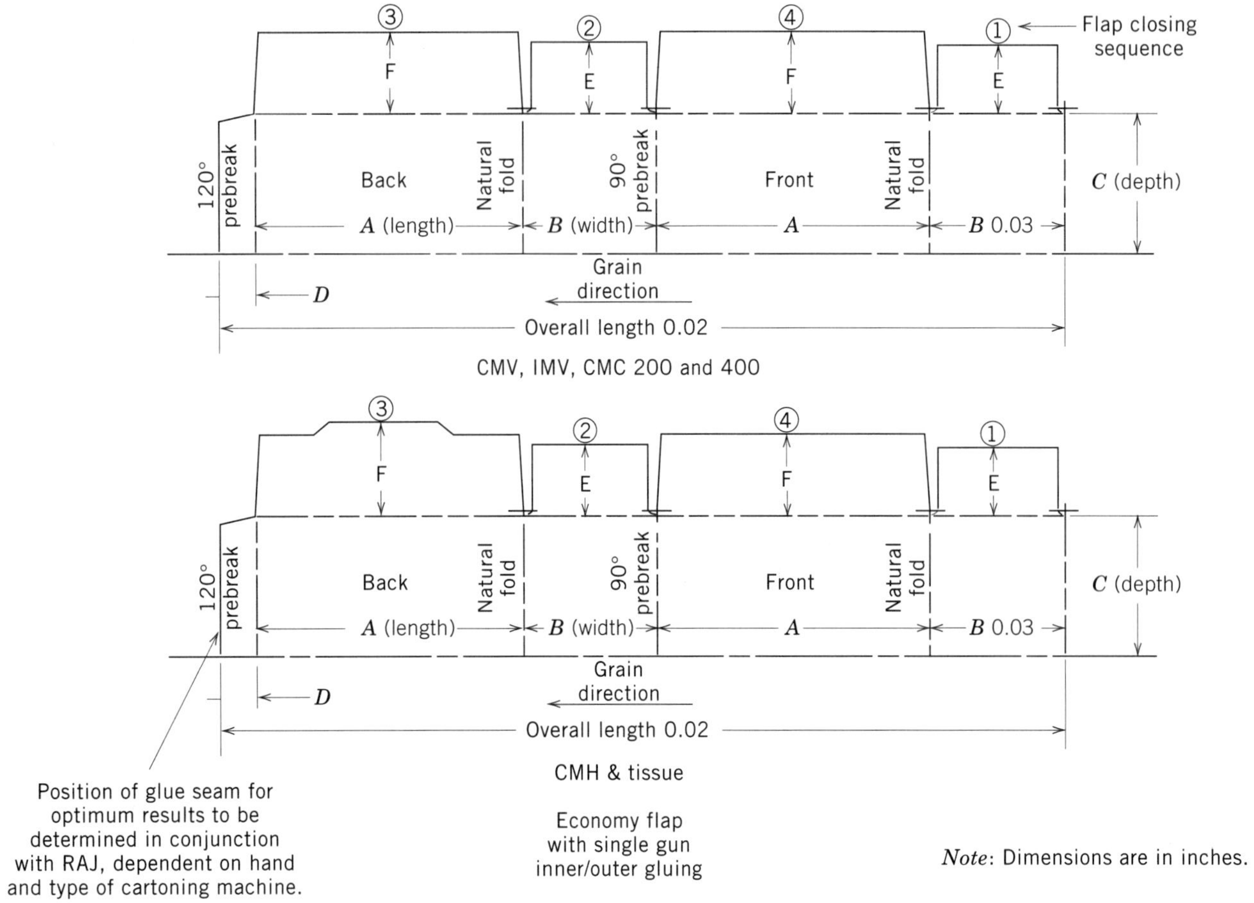

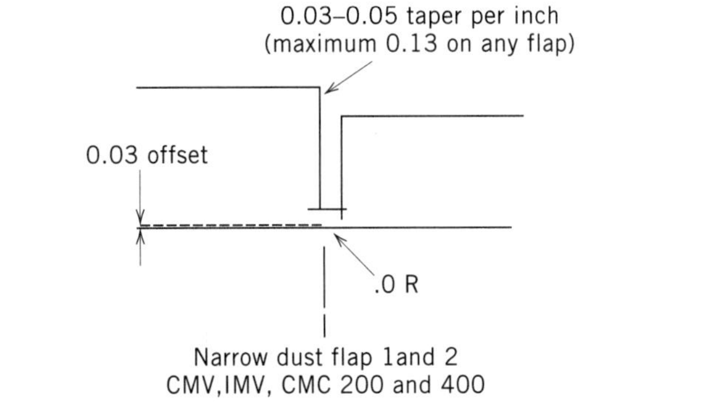

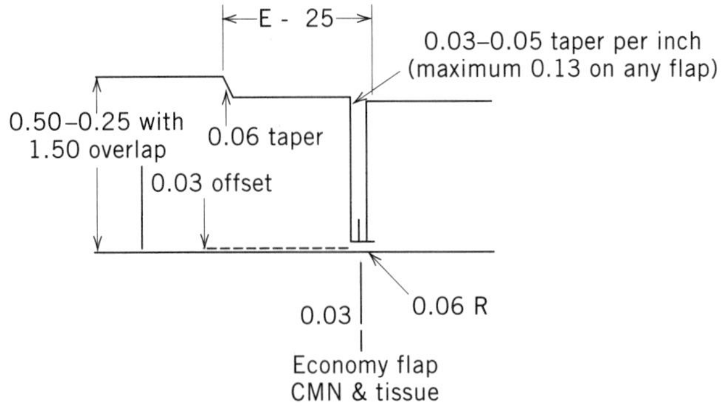

Note: For development of tuck flaps, see drawing no CMV- 10132-A

Development key:

A Length of carton
B Width of carton
C Depth of carton
D 50 Minimum glue flap
E F- 0.13 minimum for CMV and IMV dust flap
F- 0.38 minimum for CMH
F 1/2 of *B* dimension + 0.25 minimum

Note: The Length of *E* should be adjusted to ensure
-That 2× *E* = *A* 0.13 minimum or sap
-That 2× *E* = *A* 0.25 minimum or overlap

These are general guidelines to improve machine performance; these dimensions and recommendations are for boards up to 24-point; for other points, contact R.A. jones.

Offset score lines improve carton appearance and are more necessary on heavy boards.

General notes:

1. Score lines must be deep and accurate.
2. Cartons must be prebroken if preglued.
3. Glue seams must be consistently straight.
4. Carton to be furnished on edge and loosely packed to retain prebreaks.
5. Flat blanks should be protected on a skid with a flat board on top to prevent warping.
6. Skids of blanks should be brought to ambient temperature 24 hours before use.

Figure 6. Carton development.

Carton blank tolerances. Overall carton blank length (in the direction of travel through the SSG) is important. It should be uniform. Carton companies tell us that a carton blank should have a length tolerance of ±0.002 in. (0.05 mm) at die cutting. Allowing for moisture absorption and the like, an overall length of ±1⁄32 in. (0.8 mm) is a reasonable variation as the blanks are ready to run through the SSG.

In many cases, you are purchasing your carton blanks from more than one source. In order to minimize machine adjustments between pallets of cartons supplied from various sources, the preceding carton blank tolerances should be held to among your suppliers—ie, *all* blanks should have an overall length variation of ±1⁄32 in. (0.8 mm) *regardless* of supplier source.

Overall Carton Development

Refer to Figure 6.

Art Trefry
R. A. Jones & Company
Cincinnati, Ohio

CARTONS, FOLDING

Folding cartons are containers made from sheets of paperboard (see Paperboard) that have been cut and creased for forming into a designed shape. There is evidence that paper was used by Egyptian merchants to wrap goods for their customers as early as 1035 A.D. The modern-day folding-carton industry began in 1839 when Colonel Andrew Dennison began producing commercial folding cartons to complement and protect jewelry sold in his retail store. By the mid-1890s, automatic machines were in widespread use for the production of cartons (1). From these beginnings in the 19th century, the folding-carton industry has grown into an economic segment with 1994 sales of \$4.9 billion (10^9) that utilized 3.2 million (10^6) tons of paperboard, excluding milk packaging (2).

Merchandise displays at supermarkets and drug, hardware, automotive, and department stores demonstrate the extent to which folding cartons are used today. Cereal, crackers, facial tissue, detergent, dry mixes, frozen food, ice cream, butter, bacon, bar soap, candy, cosmetics, toys, cigarettes, canned beverages, carryout foods, and pharmaceuticals represent the broad range of products for which folding cartons are commonly utilized. The use of folding cartons is widespread because of the ability of this packaging format to satisfy the functions of protection, utility, and motivation. Protection from crushing, bending, contamination, sifting, grease, moisture, and tampering can all be built into folding cartons. For the producer, utility is achieved through high speed automatic packing (see Cartoning machinery). For the end user or consumer, utility is provided by opening, reclosing, and dispensing features. In some cases, the carton even serves as the cooking utensil. High-quality graphic reproduction, excellent billboard presentation of the graphics design, and the ability to take on unique and varied shapes provide the carton user with the means to motivate the consumer to purchase products packaged in folding cartons.

Paperboard Selection

Successfully meeting the needs of a folding-carton user begins with choosing the paperboard best suited for the job. In general, this means selecting the grade with the lowest cost per unit area that is capable of satisfying the performance requirements of the specific application. Economics and performance dictate careful selection of paperboard grades for each use.

Selection criteria. A variety of criteria are commonly used in the selection of paperboard grades. The Technical Association of the Pulp and Paper Industry (TAPPI) has published standardized test methods for many of these criteria (3) (see Testing, packaging materials). TAPPI Standard Methods are widely used and accepted by the industry. The most important and widely used criteria are shown below.

FDA/USDA compliance. This is a nondiscretionary criterion for food products and is dependent on the type of food and the type of contact anticipated between the food and the paperboard or coatings on the paperboard.

Color. Color is typically chosen for marketing reasons. The side of the paperboard that becomes the outside of the carton is generally white, but the degree of whiteness varies among grades. Depending on the materials-selection and processing strategies of suppliers, outside board color can be blue-white or cream-white. These shades are noticeably different and can limit substitution of grades. Board color on the inside of cartons varies from white to gray to brown.

Physical Characteristics. It is possible to establish minimum levels for each carton application that allow the package to satisfactorily withstand the rigors of packaging machinery, shipping, distribution, and use by the consumer. Physical properties commonly used to predict suitability of board for a given use include stiffness, tear strength, compressive strength, plybond strength, burst strength, tensile strength, elongation, and tensile energy absorption. Physical criteria normally define the basis weight and thickness of paperboard that is used to produce a carton.

Printing characteristics. Following the selection of a specific graphic design and printing method for the carton, a paperboard is selected based on these criteria: smoothness; coating strength; ink and varnish gloss; mottle resistance; and ink receptivity. Not all criteria are important for every printing technique.

Barrier. The most common barrier requirements are for cartons to provide protection against moisture and grease. The choice of a barrier material and application method influences board choice. For example, if polyethylene is to be applied to the carton, a board with a treatment that holds the P.E. on the board surface can have economic and processing advantages over an untreated board. Materials and application methods are described below.

Paperboard types. In the United States, the three most widely used types of paperboard are identified as follows:

Coated Solid Bleached Sulfate (*SBS*). 100% virgin, bleached, chemical furnish, clay-coated for printability.

Coated Solid Unbleached Sulfate (*SUS*). 100% virgin, unbleached, chemical furnish, clay coated for printability.

Coated recycled. Multiple layers of recycled fibers from a variety of sources, clay coated for printability.

Coated recycled boards are the most widely used, with 42% market share followed by SBS with a 25% share. Coated SUS has a 18% share with the remainder split among other recycled grades (2).

Overall treatments or coatings are applied to webs of paperboard to provide specific functions. Clay-based coatings to provide high-quality printing surfaces are the most common treatment applied on the paperboard machine. Grease-resistant fluorochemicals are applied on board machines as well, either as furnish additives, surface treatments, additives to clay coatings, or in combination. Mold-inhibiting chemicals are also applied to boards designed for bar-soap packaging, to prevent moisture in the product from initiating mold growth. Surface treatments applied on other-than-board production equipment are discussed below under Carton Manufacturing Processes.

Paperboard is the overwhelming choice as the substrate for folding cartons. However, a segment has developed that utilizes plastic sheet as a substrate. These cartons are normally produced from clear, impact grades of PVC sheet (see Film, rigid PVC) using specialized heated-scoring techniques to achieve acceptable folding characteristics. Unique product visibility is the primary reason for the use of this more costly substrate for specialty folding-carton applications, such as cosmetics and soft goods (see Boxes, rigid plastic).

Another segment making an impact on the folding carton market is E-flute, and the newer F-flute is corrugated. (See F-flute, corrugated.) E-flute is a transitional medium between corrugated and folding cartons. While E-flute height, depending on the profile of the manufacturing ranges within 43/1000–5/1000 in. F-flute is typically ~30/1000 in. The fluted material has positive attributes for its stacking strength, low weight, stiffness, as well as insulating and shock-absorbing properties. It is used typically in cartons for products such as perfumes, glassware, household products, candies, fast foods, and countertop displays.

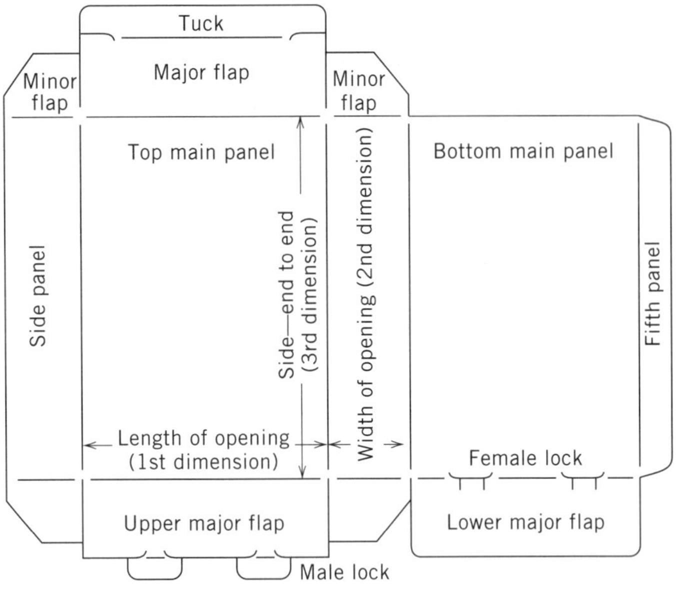

(a)

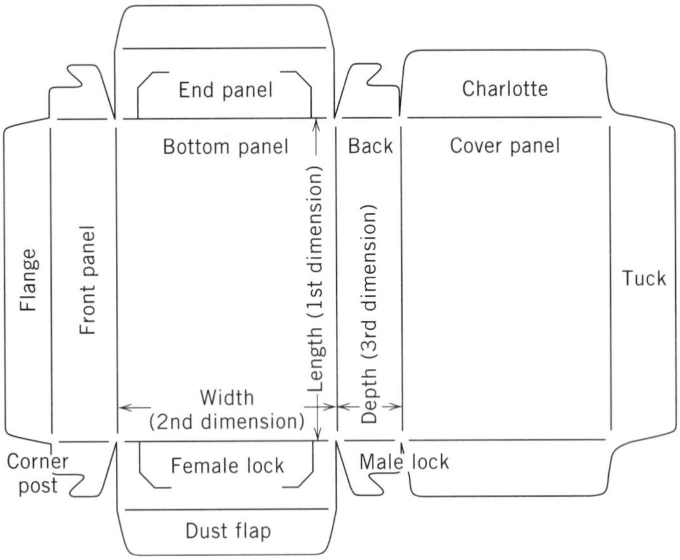

(b)

Figure 1. Terminology.

Carton Styles

As the demand for cartons grew, so did demands for additional features. These demands catalyzed the development of new and unique ways to cut and fold sheets of paperboard to produce cartons. The records of the U.S. Patent and Trademark Office contain many thousands of patents granted to protect folding-carton structures.

Three broad classifications are commonly used to categorize folding-carton styles: tube (end load); tray (top load); and special construction. Figure 1 describes accepted terminology for the various parts of tube (**a**) and tray (**b**) cartons (4), as well as the order in which dimensions are listed in carton specifications. Compliance with these standards prevents confusion.

Tube style. Tube (shell) constructions are the most common style in use today. Figure 2 shows a typical sealed-end carton in various stages of production and filling. These cartons are characterized by a fifth panel glue seam in the depth direction, yielding a side-seamed shell that folds flat for transportation. The cross section of the carton opening is normally rectangular, and product may be loaded either horizontally or vertically (see Cartoning machinery). Tube-style cartons are well suited for very high speed automated filling lines, but they are also used for manual filling applications.

Figure 3 shows the treatment of end flaps on a tuck-end carton. Other end treatments are in common use, including zippers and similar opening features for sealed-end cartons. Internal shelves and panels are often included to secure and protect the product. This is particularly done when the product is irregular in shape or the carton is much larger than the product for improved graphic presentation or shoplifting deterence. When heavy granular products are packaged, bulging of main panels can be a problem. The use of bridges connecting the two main panels increases the carton's integrity

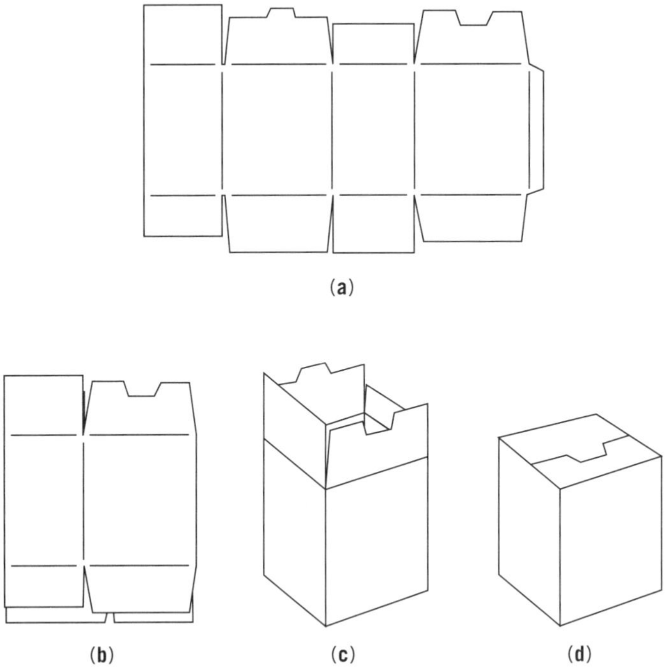

Figure 2. Sealed-end carton: (**a**) blank; (**b**) side-seamed shell; (**c**) carton erected for loading; (**d**) filled and sealed carton.

required to resist bulbing. These bridges can be made from paperboard and attached during the gluing operation.

Tube-style cartons are commonly used for granular or pourable solid products such as detergent, cereal, and dry mixes. Dispensing features are often designed by special cuts and creases in the paperboard. End opening is preferred when inner bags are employed (see Bag-in-box, dry product). Large products packed one to a carton, such as pizzas, frozen dinners, pot pies, cosmetics, and pharmaceuticals are packed in end-opening tube-style cartons. End-loaded cartons are also designed for opening and product removal through the main panel; a cream cheese carton is a good example of this approach. Gabletop milk cartons also fall into the tube or shell category (see Cartons, gabletop). They incorporate liquid-tight sealing and a reclosable pour spout. From ice cream to lipstick, tube-style cartons satisfy many diverse packaging needs.

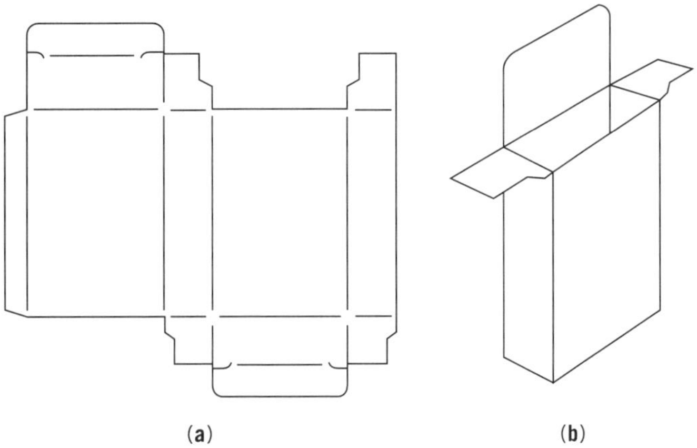

Figure 3. Tuck-end carton: (**a**) blank; (**b**) erected carton for loading.

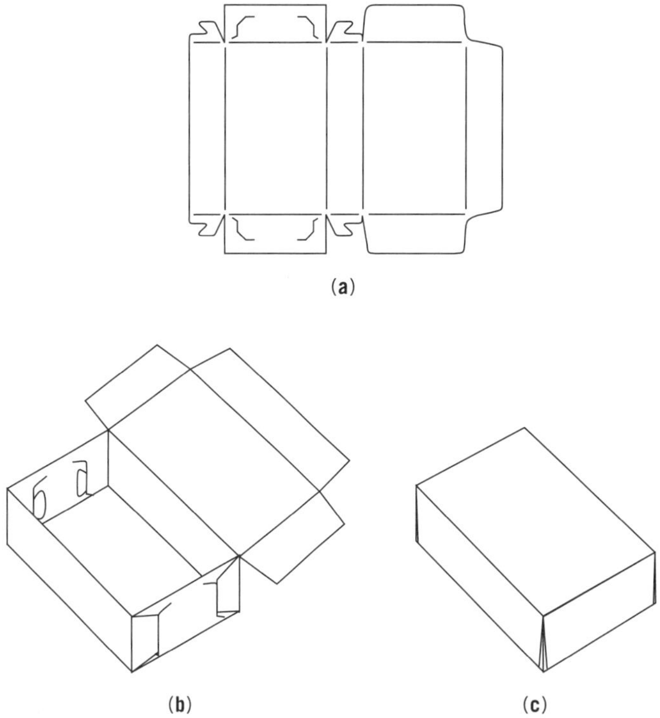

Figure 4. Locked corner hinge cover carton: (**a**) blank; (**b**) carton; (**c**) tuck-end carton.

Tray style. Tray or top-load cartons are characterized by a solid bottom panel opposite the product-loading opening. As shown in Figure 1(**b**), panels are connected to each edge of that bottom panel. Tray cartons are especially useful for manual or automatic loading of multiple products. Figure 4 contains schematic drawings of a tray carton blank, the carton setup for loading, and the completely closed and sealed carton. In this example, the front and back panels are connected to the end panels using mechanical locks. Panels are also commonly connected using adhesives or heat sealing (see Adhesives; Sealing, heat).

Where additional resistance to leakage is desired, web corners are employed. Figure 5 shows a web-corner tray with folded double sidewalls that provide finished sidewall edges. A similar, slightly larger tray could be used to cover the tray following product loading, yielding an extremely crush-resistant package. Figure 6 shows a six-corner carton that can easily be set up by hand. Diagonal scores permit the tray to be glued and delivered in a collapsed form. Web corners could be incorporated with additional modifications.

Tray cartons that require no gluing at the point of use (eg, the six-corner carton of Fig. 6) are used extensively for manual loading. The cake, pie, or pastry carton employed by the local bakery is the best example of the utility of these designs. Garments and other dry goods are often packed, especially when purchased as gifts, using two-piece cartons that comprise two collapsible glued trays. Tray cartons are also widely used for products that can be automatically packaged. Doughnuts, and fish sticks are examples of products commonly packaged in tray-style cartons with attached covers. Cartons similar to that shown in Figure 4 are used for products of this type. Most display cartons for smaller candy packages also fall into the tray category.

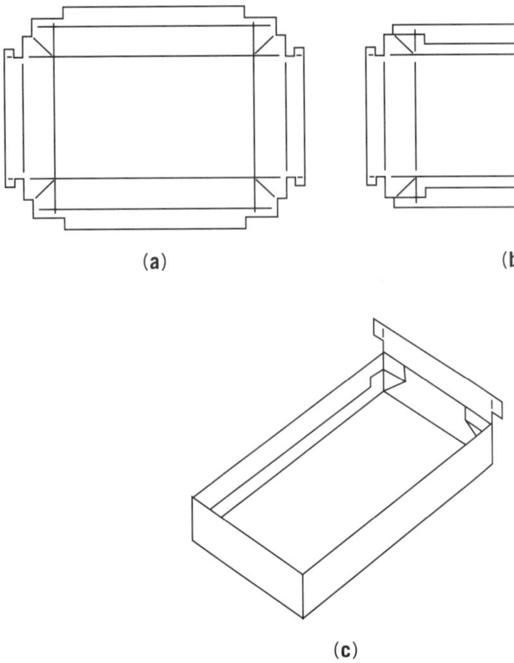

Figure 5. Web-corner tray: (**a**) blank; (**b**) sidewalls glued; (**c**) final panel folding.

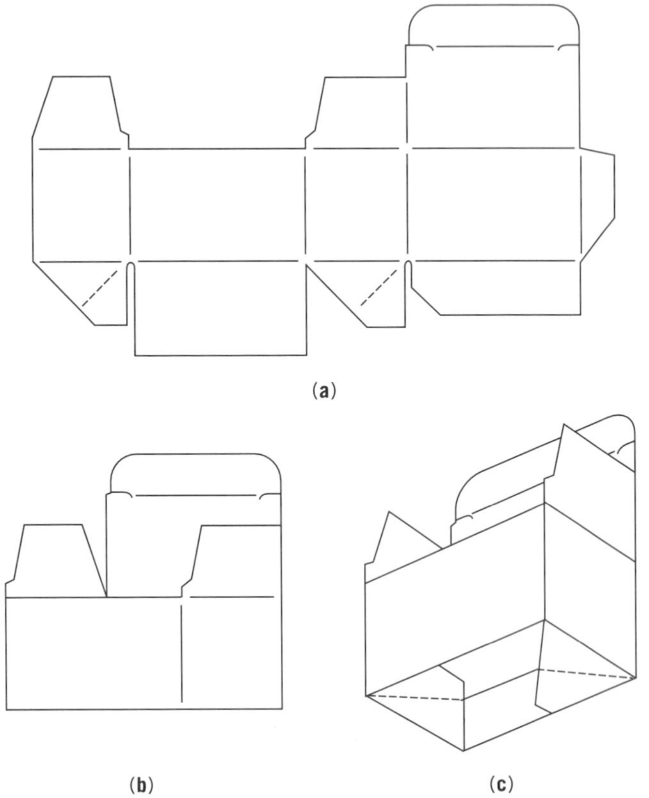

Figure 7. Automatic-bottom carton: (**a**) blank; (**b**) carton glued for shipping; (**c**) bottom view of erected carton.

Special construction. Special construction is a classification employed for cartons that do not fit tray or tube descriptions, or that represent sufficient departures from normal tray or tube practice. A blister package that employs a combination of heat-seal-coated paperboard and a clear thermoformed plastic blister is a good example of a special construction (see Carded packaging). The automatic-bottom carton shown in Figure 7 combines elements of a side-seamed tube carton with those of a top-load carton and requires no manipulation to form the carton bottom. When the collapsed shell is opened, the bottom panels lock into place. This carton is used extensively for fast-food carryout packaging. It is also popular for hardware items.

The bottle-wrap carton in Figure 8 is an example of the many wraparound carrier cartons used for multipacks of bottles, cans, or plastic tubs, (see Carriers, beverage). These cartons are either locked or glued after being wrapped around the primary packages.

The tube, tray, and special carton styles depicted and described here are broadly representative of the great variety in shapes and sizes produced by the folding-carton industry. Customization of design for function or appearance is a significant advantage of folding cartons.

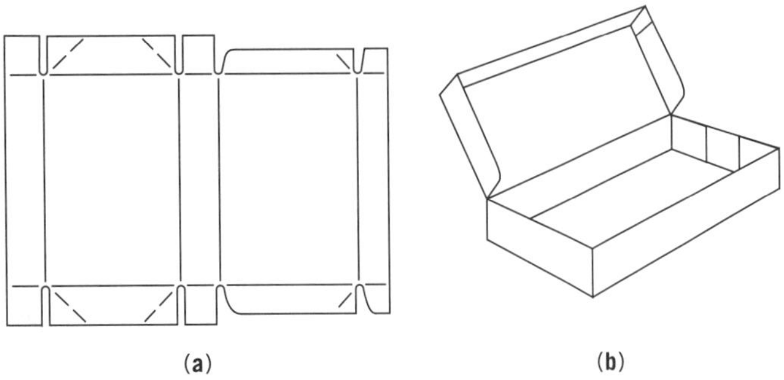

Figure 6. Six-corner carton: (**a**) blank; (**b**) carton erected for loading.

Carton Manufacturing Processes

After a paperboard grade has been selected for a specific carton style and use, a variety of manufacturing options are available for converting that board into cartons. Although it is a highly unusual carton that requires each one of the steps or stages described below, all are commonly employed to produce folding cartons in today's market.

Extrusion coating. This technique involves the coating of one or both sides of the paperboard web with a relatively thin (generally less than 0.001-in. (25.4-μm)) layer of a thermoplastic polymer (see Extrusion coating). Low-density polyethylene (LDPE) is the most commonly used extrusion coating for folding cartons and provides a cost-effective means of obtaining excellent protection against water resistance as well as a fairly good water-vapor barrier. LDPE is also used as a heat sealant (see Sealing, heat) particularly when two-side coatings are employed. When the use temperature of the package exceeds 150°F (65°C), HDPE or PP can be used to raise the acceptable use temperature to 250°F (121°C). These two polymers also provide improved grease resistance. Coating board with PET can raise the use temperature to over 400°F (204°C), suitable for most "dual-ovenable" applications. Coextrusion, in which back-to-back layers of two plastics are laid onto paperboard, makes it possible to take advantage of

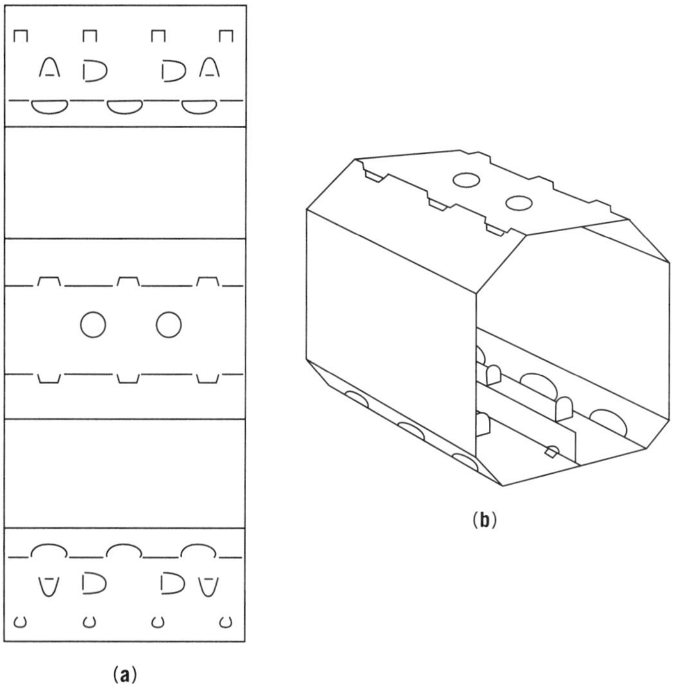

Figure 8. Bottle wrap carton: (**a**) blank; (**b**) erected carton. (Bottles omitted for clarity.)

the special properties of some exotic plastics, such as nylon, that by themselves will not adhere to paperboard.

Laminating. The earliest means of significantly enhancing the properties of paperboard was the combination with other materials through lamination (see Laminating). The most commonly used laminating adhesives are water-based glues (see Adhesives), or thermoplastic polymers. Materials laminated to paperboard include high-quality printing paper for enhanced graphics capabilities (see Paper), grease- or water-resistant paper for improved barrier, aluminum foil for barrier or aesthetics (see Foil, aluminum), and film (sometimes metallized) for barrier or aesthetics (see Metallizing).

Printing. Prior to the printing operation, paperboard is handled in web form. A decision must be made to continue in web form or convert the web to sheets before printing and die cutting. This choice is primarily dictated by the printing technique chosen (see Printing). Sheeting is most often done at the carton-producer's facility. A small segment of the industry purchases board sheeted at the paperboard mill.

Sheet-fed offset *lithography* is the most popular printing technique used to produce folding cartons. Its primary advantage is its ability to accommodate wide variations in the size of the sheet to be printed and thus, the sizes and shapes of the cartons on those sheets. The process is economical for producing short- to medium-length runs of high-quality processprinted cartons of varying sizes. Printing plates can easily be remade for every order, permitting graphics changes to be accomplished at relatively low cost. This is especially useful for printing sheets of mixed graphic designs with common carton shape and size. A different distribution of graphic designs can be printed for each order.

Web-fed *rotogravure* printing technique is well suited for large orders and repeat orders not requiring many copy changes. Excellent print quality and color consistency may be obtained, but a greater sensitivity to board smoothness exists than with lithography. The web-fed nature of the process yields higher output than a sheet-fed process. Cylinders cost more than offset plates, but can often be used for several million (10^6) impressions before reengraving is required. Web-fed offset lithography provides excellent print quality and has U.S. application for fixed-sized cartons requiring frequent copy changes. It is typically used on tobacco, cereal, beverage and the recording industries. As a process, flexography has made great technological advances over the past decade.

The machinery, printing plates, and inks manufactured have all contributed to close the quality gap between flexo and offset.

From a general standpoint, flexo has a great deal in common with the gravure process in terms of basic method of ink distribution. In addition to a somewhat similar doctor blade application, both processes use similar premixed fast-drying inks and can be operated at equivalent press speed ranges. Flexo is generally considered to be less expensive than gravure in terms of:

1. *Capital outlay.* Gravure capital costs are running approximately 2:1 ratio when compared to flexo equipment.
2. *Total carton preparation.* The cost differences vary depending on each individual job and its complexities however. Flexo is generally considered less expensive.
3. *Pollution compliance and regulating demand.* Although both flexo and gravure converters must be sized by these guidelines, the flexo connector is in a better position to satisfy regulatory codes without additional capital expenditures, for absterment equip.

Likewise in comparing and analyzing the flexo and litho printing processes, it is clear the litho plate costs are currently lower than those of flexo, but flexo does not require sophisticated inking and dampening systems. In addition, with the advancing technology of platenmaking, mounting, and remounting, the amortized cost comparison is greatly narrowed. Generally, however, flexo is less capital-intensive, labor-intensive and waste-intensive. *Varnishes* or overcoats providing varying levels of smoothness, gloss, rub resistance, coefficient of friction, and water resistance are used in rotogravure, offset, and flexo operations for folding cartons.

Cutting and Creasing. Following the printing operation, individual cartons are cut from webs or large sheets and creased or scored along desired folding lines. Reciprocating flat-bed or platen cutting is almost invariably used to cut and crease sheets printed by offset lithography. In this technique, an accurately positioned array of steel cutting knives and scoring rules (see Fig. 9) is pressed against a printed sheet of paperboard. The knives penetrate through the paperboard to cut out the pattern of the carton. Rules force the board to deform into channels in the counter plate producing controlled lines of weakness (scores) along which the board will later predictably bend or fold. Alternatively, scores can be produced by cutting partially through the paperboard or by

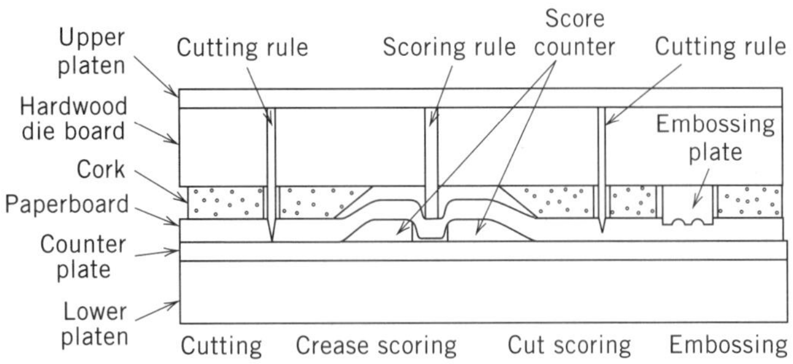

Figure 9. Flat-bed cutting-die schematic.

alternating uncut segments with completely cut-through segments.

In years past, knives and rules were separated and held in place by hand-cut blocks of dimensionally stable hardwood plywood. Hundreds of individual blocks were required for dies incorporating 10 or more carton positions. Greater accuracy and consistency as well as substantially reduced die preparation time is being used today through the use of computer-controlled laser die cutters. The laser beam is used to cut slots in large sheets of the same special plywood. Knives and rules are cut and bent automatically or manually and placed by hand into the slots. Crease or score quality has been improved through the use of computer-controlled counter-plate machining and accurate pin-registration systems.

Transport of printed sheets in a sheet-fed platen cutter is accomplished with mechanical grippers that hold the leading edge of the sheet. Small nicks in the cutting knives result in uncut areas that keep the full sheets intact and permit transport from the cutting and creasing station to the stacking station. Large stacks are then removed from the machine and unwanted pieces of board, called broke, are stripped from the cut edges of the cartons, yielding stacks of printed carton blanks. Newer sheet-fed platen cutters incorporate automatic stripping of broke between the cutting and stacking stations. Platen cutters are also employed to cut and crease paperboard printed in web form. In the past, these cutters were often placed out of line from normally faster-running presses. Speed increases permit economical in-line placement, which is common today.

Rotary cutting and creasing offers the advantage of higher speeds than reciprocating platen cutting, but at greater cutting die cost. Matched machined cylinders used for fixed carton sizes are most often placed in-line with printing operations. Both electrical discharge and mechanical machining are used to produce the knives, rules, and score channels in these matched cylinders.

Rotary cutting dies have also been developed for sheet and web cutting and creasing that are produced by pattern chemical etching of thin metal plates. Cutting and creasing patterns are coated with chemical-resistant materials and chemicals are used to reduce the thickness of the plate in the unprotected areas, resulting in raised rules. Creases are formed by pushing the paperboard with the rule of one plate into a channel formed between two rules of the second plate, a configuration quite analogous to that of platen creasing. Cutting, however, is quite different; cutting rules on opposing plates are offset slightly from each other. As these rules rotate, approaching each other closely but not touching, they compress the board. Compressive forces within the board cause it to rupture, yielding cut edges. For sheet-fed cutting, these etched flexible dies are mounted on large cylinders which, like sheet-fed offset plate cylinders, leave a gap between leading and trailing edges. This gap accommodates sheet feedup and variable repeats. For web cutting, however, leading and trailing edges must be butted to correspond to the continuously printed webs. Die mounting techniques as well as carton layout on the web are the keys to the successful operation of this approach.

Two additional specialty converting steps are accomplished on cutting and creasing equipment: foil stamping and embossing. Foil stamping involves the use of heat and pressure to transfer a thin metallic or pigmented coating from a carrier film to the carton surface to obtain patterned decorative effects. When this is done in combination with embossing, reflectance and gloss are combined with raised image effects for enhanced graphic presentation. Embossing alone can generally be accomplished on standard die-cutting equipment. Foil stamping and detailed, deep embossing requires the ability to heat the stamping and embossing plates. This is most commonly accomplished on a second pass through specialized equipment.

Hot-Melt Application

Hot melts (see Adhesives) can be preapplied at this or a later production point using knurled wheels or timed guns. The hot melt is later heat-activated on the packaging machinery to effect sealing. Although hot melt application on the packaging machine is common, some carton users find it advantageous for the carton manufacturer to preapply the hot-melt adhesive.

Windowing or couponing. When product visibility is desired, a hole is cut out of the carton blank. To protect the product or prevent it from spilling out of the carton, pieces of an adhesive pattern are applied around the edge of the opening, a rectangular piece is cut from a roll of film, and then pressed in place. Registered application of printed films or printed coupons in roll form to interior or exterior surfaces adds value and function to the carton. Devices are also available that adhere coupons supplied in sheet form.

Gluing. Although more and more packaging machinery is designed to accept flat carton blanks (see Cartoning Machinery), gluing still represents a major and important converting operation. The simplest operation converts a flat blank into a side-seamed tube or glued shell (see Fig. 2**b**). Carton blanks are removed one at a time from a stack and carried by sets of endless belts. Stationary curved plows move one or more panels of the blank out of the original plane to either prebreak scores or form the glue seam. Prebreaking of scores assists packaging-machine operation, since the force required to bend a previously bent score is greatly reduced. Sealing is accomplished with cold glues, hot melts, or heat sealing of polymers. Side-seamed cartons are discharged into a shingled delivery that provides compression and time to set the bond; case or bulk packing for shipment follows.

Gluers in which the cartons move in a continuous straight line, transported by belts, are known as straight-line gluers. Although straight-line gluers are most commonly used to produce glued shell-type cartons, attachments provide the ability to produce automatic-bottom as well as certain collapsible-tray styles. Paper or paperboard bridges can be attached to main panels during straight-line gluing. For simple styles, the feeding of carton blanks into the gluer does not need to be timed into specific folding actions. Complicated folding devices may dictate that blank feeding be timed, which generally reduces speeds. Compound folds in both directions on the blank cannot be handled by straight-line machines.

For more complicated carton and collapsible-tray styles, right-angle gluers are employed. As the name implies, midway through the machine the travel direction of the blank is changed by 90°. All parallel folds can be made in the direction of blank travel, resulting in simplified machine setup and more positive and accurate folding. Generally speed is limited by the transfer section, which changes blank travel direction. Right-angle gluers combine flexibility and precision in the manufacture of complex folding cartons.

Setup and nested tapered trays are also produced in folding-carton manufacturing plants for shipment to customers. These trays are produced on plunger-type gluing equipment that is designed to accept either blank or roll feeds. Blank-fed machines first apply adhesive, then form the tray as a moving plunger forces the previously printed and creased blank through a stationary folding and forming device. Roll-fed machines incorporate printing as well as cutting and creasing units in line prior to gluing. Nested trays are not as space efficient as unglued blanks; they do, however, have application in uses for which it would be uneconomical or impractical to operate a forming device at the location of use. Paperboard french-fry scoops and sandwich containers used by fast-food outlets are good examples of these trays.

BIBLIOGRAPHY

1. D. Hunter, *Papermaking—The History and Technique of an Ancient Craft,* Dover Publications, New York, 1978, pp. 471, 552, 577.
2. *Paperboard Packaging Council Marketing Guide.*
3. *1984–1985 Catalog—TAPPI Official Test Methods, Provisional Test Methods, Historical Test Methods,* TAPPI Press, Atlanta, 1984.
4. *Kliklok Packaging Manual,* Kliklok Corporation, Greenwich, CT, 1983 revision, p. I.

General References

Paperboard Packaging, Magazines for Industry, Cleveland, OH. A monthly trade magazine for solid and corrugated paperboard containers.

Boxboard Containers, Maclean Hunter Publishing, Chicago. A monthly trade magazine for solid and corrugated paperboard containers.

Package Printing, North American Publishing, Philadelphia. A monthly trade magazine including printing and die-cutting articles.

The Folding Carton, Paperboard Packaging Council, Washington, DC, 1982. Short booklet.

J. Byrne and J. Weiner, "Paperboard II. Boxboard," *IPC Bibliographic Series Number 236,* The Institute of Paper Chemistry, Appleton, WI, 1967. The most complete historical bibliography of paperboard and its use in folding cartons.

J. Weiner and V. Pollock, "Paperboard. II. Boxboard," *IPC Bibliographic Series Number 236 Supplement I,* The Institute of Paper Chemistry, Appleton, Wisc., 1973. Update of previous bibliography.

Paul Obolewicz
Rand Whitney Packaging
West Chester, Pennsylvania

CARTONS, GABLETOP

The gabletop folding format is one of the oldest and most basic end closures possible for a paperboard package. The first patent dates back to 1915 (1), but 20 years passed before the first commercial installation began to operate at a Borden Company plant, after the patent was acquired by Ex-Cell-O Corp. Today a number of manufacturers supply machinery to make gabletop cartons for milk and other still liquids.

Early gabletop milk packages were precision-cut folding boxes with an adhesively sealed side-seam and bottom closure and a stapled top closure (see Fig. 1). Semiformed cartons were dipped in hot paraffin for sanitization and moistureproofing prior to filling. Tops were stapled. The first packages had no convenient opening device. Subsequent designs had convenience openings based on secondary patch seals adhesively secured to either the inside or outside of a side panel. The secondary patch was eventually eliminated in favor of an integral pouring spout.

The modern gabletop carton retains a simple basic geometry but includes design refinements acquired over 50 years of developement and commercial use. The transition from the wax-coated carton to precoated paperboard came in 1961, necessitating several new developments in package and materials technology (see Extrusion coating; Paperboard). The use of precoated board eliminated paraffin, wire, and adhesives from a filling plant's inventory. Also eliminated were the asso-

Figure 1. Early gabletop container. It is paraffin coated with a patch-type opening device.

Figure 2. Current gabletop container. It is polyethylene coated with a pitcher-spout opening device.

ciated mechanical systems, including carton-coating chiller units, wax melters, adhesive applicators, and related instrumentation. Precoated blanks simplified the form/fill/seal process and permitted the design of faster, more-efficient equipment with filling rates up to 300 cartons per minute. To retain the essential pouring-spout feature, an antisealant or *Abhesive* was developed. This allowed the carton top seal to be tightly sealed, yet easily opened (see Fig. 2).

A typical blank for a quart (946.25 mL) gabletop container is shown in Figure 3. With a panel width of 2.764 in. (7.02 cm) and a body height of 7.375 in. (18.73 cm), the apparent contained volume of 56.34 in.^{3} (923.25 cm^3) falls short of the quart volume of 57.75 in.^{3} (946.35 cm^3). The needed extra volume is found in the bulge of the side panels after filling, which leaves the filled product line below the top horizontal score, providing "headspace" necessary to compensate for foam generated during filling and to allow for a certain amount of splash as the filled container is conveyed through the top heat-sealing machine function.

The standard square cross section of the quart (946.25 mL) carton is used for a full range of containers from 6 fl. oz (177.4 mL) through the Imperial quart (1182.8 mL). Other cross sections in the same carton format have panel widths of 2.240 in. (5.69 cm), 3.3764 in. (8.576 cm), and 5.531 in. (14.05 cm), with container volumes from 4 fl. oz (118.3 mL) through one gallon (3.785 L). For quart-series containers, the typical paperboard structure consists of 195–210 lb (88.5–95.3 kg) per ream paperboard with a coating of 0.0005 in. (12.7 μm) polyethylene on the outside surface and 0.001 in. (25.4 μm) polyethylene on the inside surface. Other structures that include aluminum foil, ionomer (see Ionomers), and other barrier materials are also possible (Fig. 4).

Coated-paper containers for liquids provide relatively short shelf life. It is possible, however, to tailor the container to its contained product. Liquids with high solids content, such as milk or fountain syrups, are relatively easy to contain, since there is little product penetration of cut edges or random flaws in the coating. Other products require near hermetic seals, dictating a continuous high barrier such as aluminum foil, and the elimination of cut edges in the finished containers.

This can be accomplished by a number of mechanical techniques including skiving and hemming of the cut edge of the side seam, and refining the folding of the bottom closure to protect the cut edge from liquid contact. These techniques permit the successful packaging of oils and alcohol-bearing liquids with little problem. Special treatment of the paperboard may also be necessary to assure package stability. To-

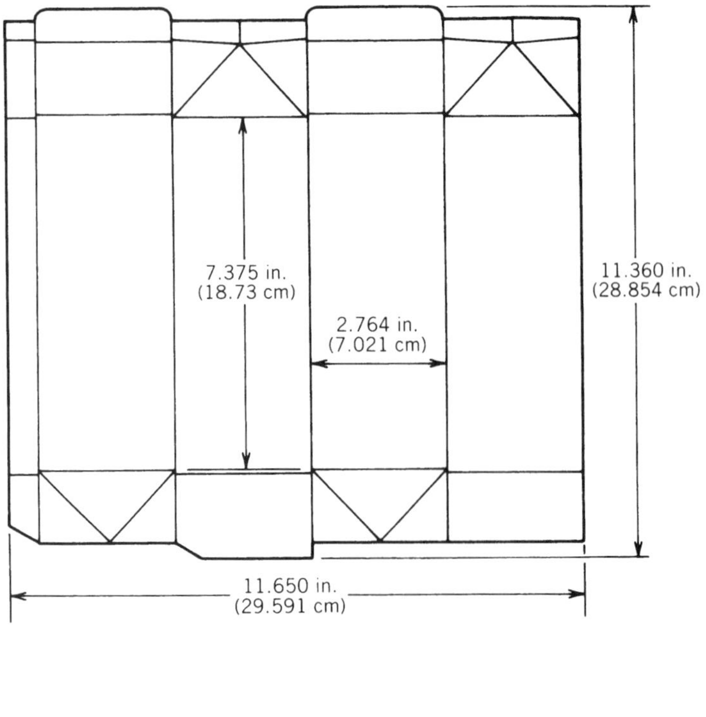

Figure 3. Typical profile for the quart (946-mL) series.

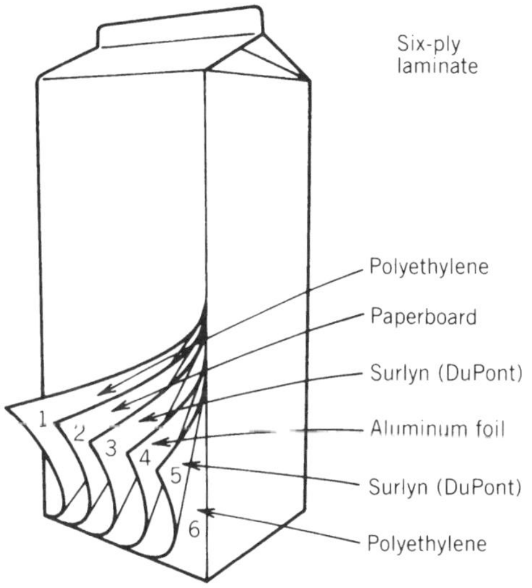

Figure 4. A typical six-ply laminate.

day's form/fill/seal equipment (see Fig. 5) contributes to shelf life by enclosing the processing line. Some machines feature air filtration and carton sanitization systems.

At its inception, the gabletop packaging system bordered on the brink of failure because the boxmaker's inability to deliver precision-cut blanks necessitated the development of high-speed rotary converting equipment. Now an industry standard, this type of equipment delivers thousands of blanks per minute, cut and creased to machine-tool accuracy. Rigid quality control is essential to the system. Blanks are routinely checked for profile, cut, and crease alignment, and other critical factors affecting container durability and machinability.

In the near term, manual procedures will be augmented by an automatic inspection system. Currently available is a new device, controlled by a small computer, with the capability of making several hundred evaluations of cut and crease accuracy, symmetry, and formation. Data are displayed instantly, recorded and printed out on demand.

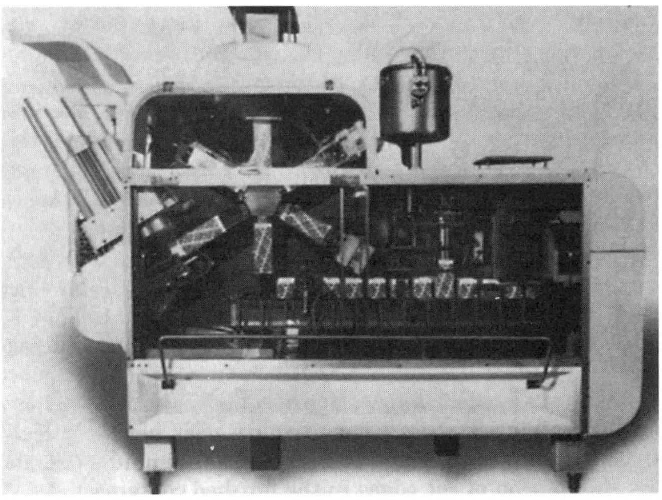

Figure 5. Automatic form/fill/seal machine for gabletop cartons. Courtesy of Ex-Cell-O Corp.

At its peak, the gabletop carton was the premier package for milk and other fluid products in the United States. Process refinements permitted its use for other likely products, such as fruit juice and fountain syrups, and a few unlikely products, such as candy and epsom salts. In the United States, the plastic blow-molded bottle (see Blow molding) has become the container of choice for gallons (128 fl oz or 3.785 L) of milk but the paperboard gabletop carton is still favored for smaller sizes. Worldwide, the carton is the popular package for still-liquid products, with over 30 billion (10^9) sold each year.

BIBLIOGRAPHY

"Cartons, Gabletop" in the *Wiley Encyclopedia of Packaging Technology,* 1st ed., by R. E. Lisiecki, Ex-Cell-O Corporation, pp. 152–154.

1. U.S. Pat. 1,157,462 (Oct. 19, 1915) J. Van Wormer.

CASE LOADING

The corrugated case is still the most universally accepted method of packaging and shipping products from one destination to another (see Boxes, corrugated). It offers excellent product protection in both storage and shipping with the additional benefit of full-panel graphic identification or advertisement exposure. The terms *case packing, case loading,* and *casing* all refer to the method of placing product into corrugated shipping containers. This can be accomplished by the very fundamental method of hand loading or by semiautomatic or automatic case-loading machinery. This article is structured to demonstrate a normal sequence of conversions from hand loading to fully automatic loading. Cases, whether manually or automatically loaded, are classified as top-, side- or end-load. The top-load case has flaps in the largest panel and is the most expensive because of the large flap area. An end-load case, with flaps on the smallest panel, is the least expensive. Proper machinery selection depends on many variables such as type of package, style of case, production rates, automatic versus semiautomatic machinery, floor space, and rate of investment return. The following basic case-loading methods are discussed below: hand loading; horizontal semiautomatic case loading; horizontal fully automatic case loading; vertical drop-load and gripper-style case loading; wraparound case loading; and tray former/loader.

Hand Loading

This is the simplest version of case loading. It requires limited machinery involvement, but it is highly labor-intensive. As packages are delivered to the packing area, personnel manually open, load, and seal the corrugated cases utilizing a variety of closing methods including cold glue, hot-melt glue (see Adhesives), tape (see Tape, pressure sensitive), or metal staples (see Staples). Cold and hot-melt adhesive applicators are most common (see Adhesive applicators). Replacing the case-closing operation with an automatic top-and-bottom case-sealing machine is the first step in automation. This unit can always be utilized later as the packaging line converts to fully automatic. To decrease labor and increase production,

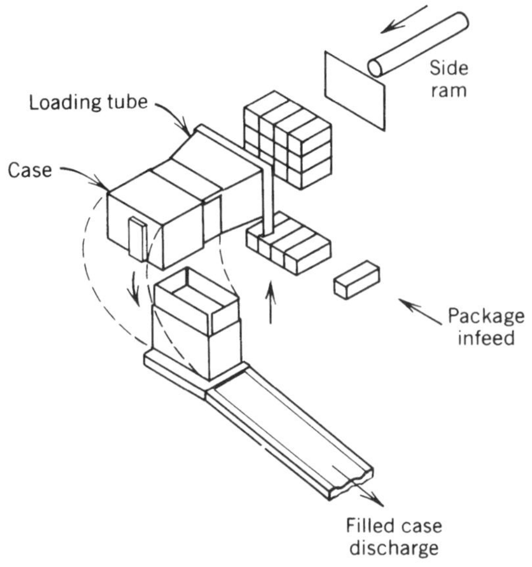

Figure 1. Horizontal semiautomatic case loader.

the manual loading operation is replaced with a semiautomatic case-loading machine integrated with the existing top-and-bottom case sealer.

Horizontal Semiautomatic Case Loader

With this type of equipment, product is loaded from the side (see Fig. 1). This method offers considerable flexibility in obtaining the desired case pattern, which is defined as the position or orientation of cartons grouped inside the case. It is ideally suited for handling cartons, cannisters, cans, or any product that takes a rigid or semirigid shape. The most common product handled is consumer-type cartons, which are loaded in a variety of configurations and counts. The semiautomatic case loader requires an operator to manually open a flat premade corrugated case, fold the bottom flaps, and place it on the machine-loading tube or funnel ready for package insertion. All other machine functions are performed automatically. Typically, an operator can open approximately 10–12 cases/min. Higher case rates would require fully automatic case-loading machinery, described below. Package speeds in excess of 500/min are obtainable, but the number of cartons per case and the dexterity of the operator dictate if it can be done semiautomatically.

The basic unit includes an infeed conveyor to receive packages and deliver to the machine accumulator section, where the product is grouped or stacked to the prescribed case pattern and loaded into the already opened and formed corrugated case. Single and multitier case patterns are easily accommodated. Multitier applications require the cartons to be stacked prior to loading. Functionally, the infeed conveyor delivers cartons to a lifter plate, where the prescribed number is accumulated. Through a pneumatically operated cylinder, the cartons are lifted and deposited on stacker bars. The lifter returns and continues this cycle until the correct number of cartons have been grouped in front of the loading tube. A pneumatically operated side-ram cylinder pushes the final load into the case. Because of the slower case rates of semiautomatic machines, air cylinders are normally used for the product lifting and side-ram load motions. Mechanical cam/crank and/or servo drives usually perform these same motions on automatic high-speed machines. Many machine configurations and accessories are available including multilane units for higher package production rates. Standard upstream filling and packing machinery usually discharge product in a single lane; the use of multilane casers would require some type of package lane dividing systems. Many packages must be repositioned to coincide with the case pattern and this is accomplished through rail twisters, upenders, turn-pegs, etc. Converging equipment is also available if the output from several upstream packaging machines must be converged into the caser single-lane infeed. After final loading, the filled case is lowered onto a short discharge conveyor ready for final case sealing. The existing top-and-bottom case sealer can be utilized, or any other type of sealer. This approach to case loading is ideal for lowercase rate applications. It requires minimal capital investment, but is a major step in automation. At this point, the horizontal semiautomatic case loader has eliminated the hand-load operation and reduced the personnel to one operator. The next sequence in automation replaces the operator with an automatic corrugated-case erector, forming a fully automatic case-loading system.

Steps Toward Fully Automatic Case Loading

There are normally two approaches to automatic case loading: a fully automatic integrated system including a case erector, loader, and sealer or a case erector loader utilizing an existing case sealer. Most case-loading machinery is manufactured in modular design to allow the proper equipment selection for the application. Consideration should always be given to how existing equipment can be used in conjunction with new equipment. A decision to automate, and how to do it, is based on many considerations: case rates in excess of 10–12 min; high package-production rates; packages more easily loaded automatically; large-size cases more easily handled automatically; labor reduction; floor space reduction; and increased line production and efficiency.

Case erector/loader. This machine is equipped with a flat-corrugated storage magazine that will, on demand, extract a case from the magazine, open it, fold in bottom flaps, and automatically place it on the caser loading tube for final package insertion (see Fig. 2). After loading, the filled case is lowered onto a discharge conveyor and transferred into a new or existing case-sealing machine. Case extraction and opening are the most critical functions of an automatic erector. Corrugated cases have a built in memory (resistance) and proper blank scoring will increase opening efficiency. Although more expensive, experience has proven that equipment offering mechanisms to prebreak or restrict the case back panel during opening are well worth the investment. Generally, vacuum/pneumatic mechanisms appear to function well up to approximately 20 cases/min. Higher case rates normally require a mechanical/vacuum/pneumatic combination with several stations for case extracting, opening, and loading. Both horizontal and vertical case magazines are available. The automatic case erector can be added to an existing semiautomatic case loader with its separate sealer, or it can be offered as part of a new integrated system interfaced with the existing case sealer. The case erector eliminates the operator

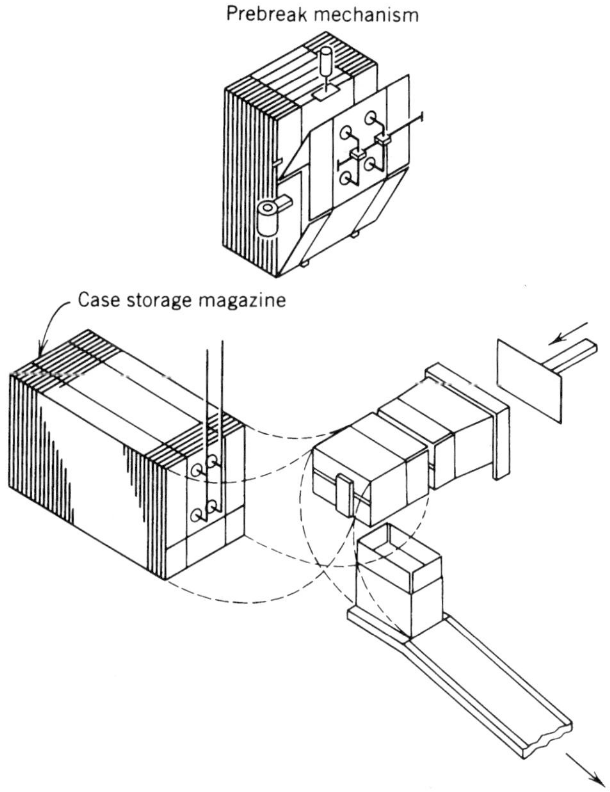

Figure 2. Case erector loader.

who would otherwise open cases manually; therefore, the erector flat-case storage magazine must have sufficient capacity for at least 30–60 minutes supply. If floor space permits, additional magazine capacity is encouraged. Vertical-style magazines are offered with bulk-storage feed systems where several stacks of cases are loaded on a floor-level conveyor and on demand, feed automatically to the magazine (see Fig. 3). A variety of case erectors are offered for various case rates of ≤30 cases/min.

Horizontal Automatic Caser Erector/Loader/Sealer

The final approach in automation incorporates a completely integrated system. A new automatic case extractor/sealer can be integrated with an existing semiautomatic caser to form a fully automatic line, but the case erector/loader/sealer is normally purchased new as a part of the complete packaging line. Case extracting/loading functions are the same as those discussed above, but after loading, the filled case is transferred horizontally through the glue-application section and into the compression unit using heavy-duty continuous-motion cleated chains. Minor case flaps are folded closed and major flaps are opened ready for adhesive application. After gluing, stationary plow rods fold in major flaps as the case is deposited into the intermittently driven side-sealing compression unit (see Fig. 4). A secondary set of top chains may be employed to ensure that the case is presented squarely to the compression unit. Vertical compression units for use in overhead filled-case conveying systems reduce initial floor-space requirements. The compression-section length is a function of the type of adhesive used and its corresponding drying time. Both hot-melt and cold-glue adhesives are commonly used (see Adhesives). Hot-melt adhesive has a faster setup time and requires a relatively short compression section, usually 4–5 ft (1.2–1.5 m). Cold-glue adhesive takes longer to set and requires more compression-section length. The hot-melt adhesive unit takes less space, but it is somewhat more expensive than the cold-glue system. PVC sealing tapes are becoming an attractive sealing method for various reasons, and most automatic machines can be equipped with tape heads in place of glue heads (see Tape, pressure-sensitive). Compression-section length can usually be reduced, because no drying time is required. As with all automatic machines, the flat-corrugated case magazine storage capacity should be large enough to ensure that an operator is not constantly replenishing the supply hopper. A complete automatic system offers many advantages including higher case rates, increased line efficiencies, labor reduction, and the operational technology of programmable logic controllers (PLC). The machine functions are now computerized and programmed accordingly. This information can be coordinated into the main control center, providing valuable information to the production department. This new electronic technology offers many specialized options, such as operator interface panels, data highway information systems, and troubleshooting diagnostic displays. In summary, the horizontal fully automatic case opener, loader, and sealer is capable of receiving product from upstream packaging equipment and delivering that product to the shipping department in a sealed corrugated case. This is all accomplished in a relatively small area at speeds of ≤30 cases/min. Continuous-motion machines are available for case rates in excess of 50/min but require considerably more floor space.

Vertical Case Loaders

This method of case loading is used primarily in the beverage, glass, can, and plastic container industries, where fragile or irregular-shaped containers require some special packing considerations. As with the horizontal case packers, the product is delivered to the machine infeed conveyor from upstream filling equipment to the accumulator section. Tabletop chain is commonly used in delivering the product to reduce back-

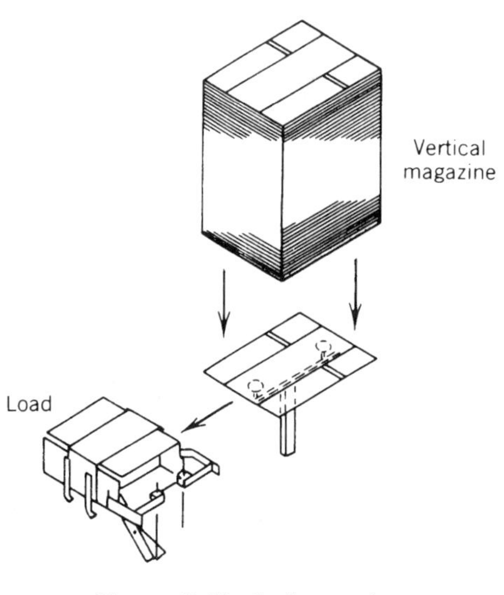

Figure 3. Vertical magazine.

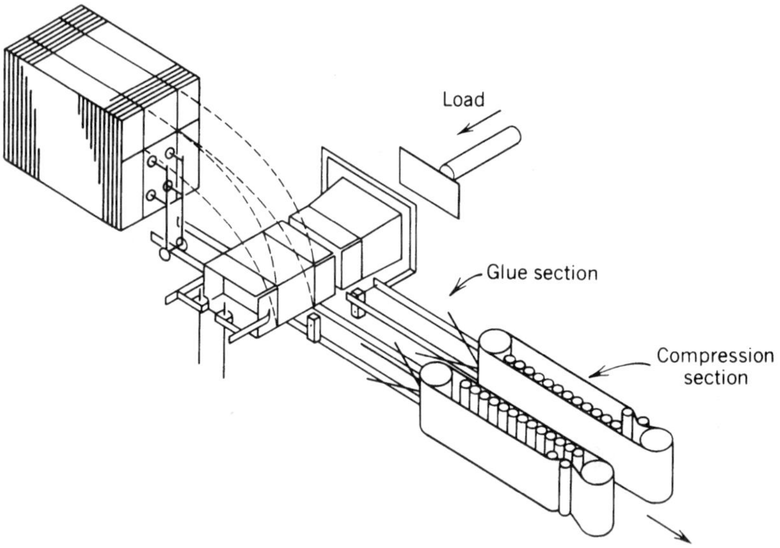

Figure 4. Horizontal fully automatic case loader.

pressure during the load cycle and for infeed washdown applications. Cylindrical-type products are divided automatically into several lanes using oscillating or vibrating dividers to form the accumulated load pattern. Irregular-shaped containers such as blow-molded plastic bottles must be divided by special equipment. When all lanes are filled in the accumulator area, a formed corrugated case is positioned underneath the loading mechanism ready for depositing. At that point, retractable shifter plates in the accumulator area move out, allowing the containers to drop vertically through fingers into the cells of the case (see Fig. 5). Special fingers guide and reduce side shock to the containers during the load cycle. The use of cells or corrugated partitions inside the case to eliminate container contact is based on product-protection requirements. Usually glass containers have partitions, and plastic containers do not. Case rates of up to 25/min are achieved for intermittent motion machines while newly developed continuous motion equipment approaches 40–50/min. The vertical case loader can be interfaced with many different kinds of corrugated-case erecting equipment. Manual case erecting and placement under the load area tied into a case sealer is one alternative. Another, used by the glass and plastic bottle industries, is to ship empty bottles in cases to the filling plant where they are emptied, filled, and loaded back into the reshipper cases utilizing a top case sealer. A third method is to incorporate an automatic case erector, vertical loader, and sealer. The machinery selection is based on floor space, capital investment, type product to be handled, and most important, case-rate requirement.

Another vertical load method for handling fragile or heavy containers and flexible pouches is one that utilizes vacuum or mechanical grippers to lower the containers into the case (see Fig. 6). This approach is ideal for containers that can be gripped at the top, such as glass or plastic bottles. The con-

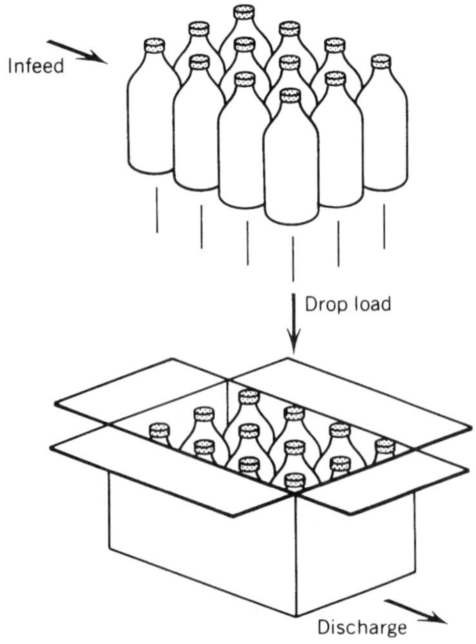

Figure 5. Vertical drop-load case packer.

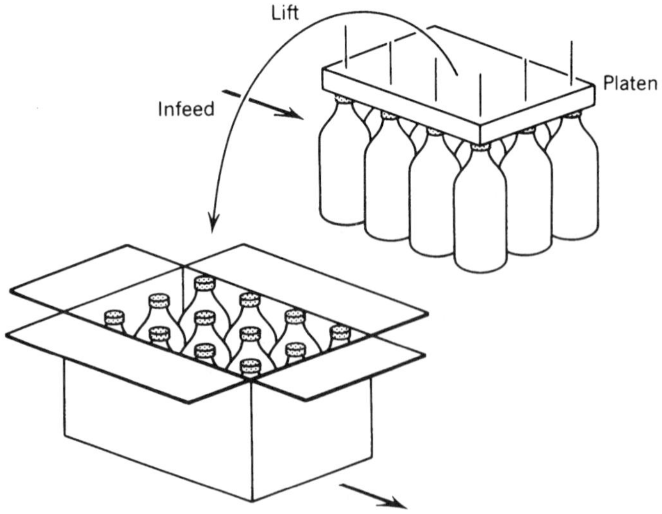

Figure 6. Vertical gripper-style case packer.

tainers are delivered and accumulated in the same manner as with the vertical drop loader, but a special plate or load head incorporating vacuum or mechanical grippers moves down and picks up approximately 12 containers at a time and places them in the opened corrugated case. Generally, both the drop-load and vacuum or mechanical gripper-style machines are offered in multiload station modules to obtain speeds of 40–50 cases/min with automatic case erectors. Although slower, flexible pouches, cartons and irregular shaped containers are currently being loaded in the same manner using robotic multiaxis motions with a variety of vacuum and mechanical grippers. Robotic case loading is emerging as a viable alternative to conventional vertical loading due to its product handling versatility, less floor space and quick changeover.

Equipment is also available that loads cartons, cans, tapered cannisters, etc, vertically up through the bottom of a case. This method is usually limited to top-load cases, but it does offer the advantage of eliminating possible package repositioning. A case is extracted and opened in the conventional method over the accumulated product. On demand, the product is lifted up into the case, which is then transferred horizontally into the sealing section. For the most part, both horizontal and vertical case-loading equipment are of the intermittent-motion design, which is somewhat speed limited. Requirements for higher speed have led to the development of faster filling machinery and continuous-motion casing equipment that runs in excess of 2400 cpm and 1200 bpm. This special machinery is an integrated system handling multiple cases. Continuous-motion horizontal case-loading equipment using both premade RSC and wraparound blank cases (see Boxes, corrugated) have exceeded the 50/min range. These rates apply to some special tray forming/loading applications as well. Because upstream filling equipment for cartons and flexible packages has not achieved the high rates of the can and bottling fillers, the requirement for continuous-motion carton-type casers has been limited.

Automatic Wraparound Case Loading

An entirely different approach to case loading utilizes a five-panel corrugated blank instead of a flat premade corrugated case with the manufacturer's joint already glued. Vertical or horizontal corrugated blank-storage magazines are employed that extract the blank and position it between chain lugs by either vacuum or mechanical mechanisms. During this motion, both side panels are folded into a vertical position forming half the case. The blank is then positioned in front of the loading machine where the product is either pushed onto the blank or dropped vertically (see Fig. 7). After loading, the top panel is folded down over the product and final flap folding

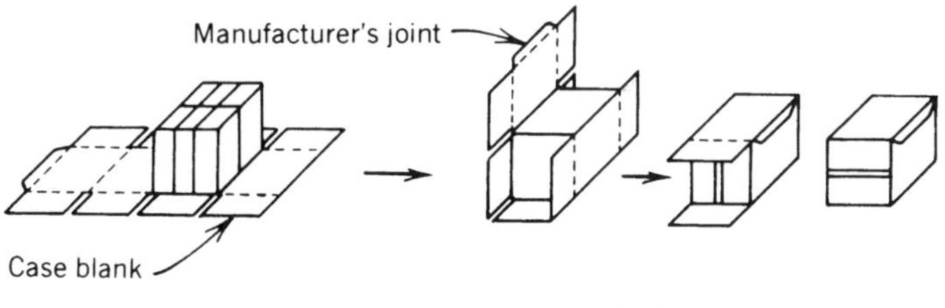

Figure 7. Wraparound case loader.

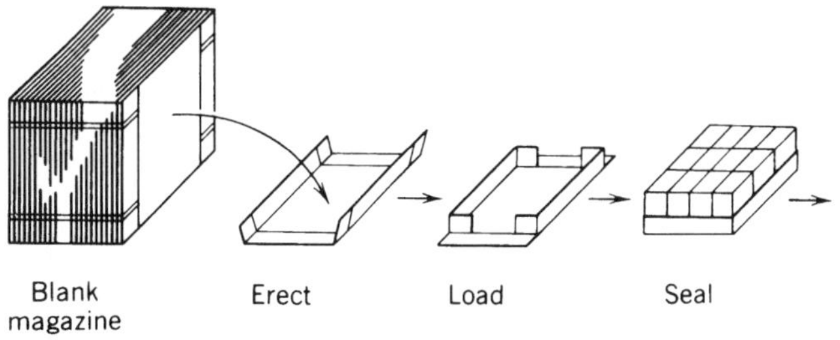

Figure 8. Tray former/loader.

and gluing is completed. Depending on the type equipment, the manufacturer's joint is then glued and folded down to one of the vertical panels for final sealing. Wraparound casers are usually larger and more complex than loaders for premade cases owing to the additional functions that must be performed, but depending on the size of the case, there may be some economical board-cost advantages using a five-panel blank. Pneumatically operated machines can achieve speeds up to 20 cases/min; higher rates of approximately 30 cases/min would require a more mechanical/pneumatic design. Continuous-motion wraparounds are available for special applications with speed requirements in the 40–50/min range. Both horizontal and vertical blank storage magazines are available as well as vertical compression sections suited to overhead case-conveyor systems (see Conveyors). Selection of a wraparound versus a premade case machine requires close scrutiny of the particular application requirements including type of case, potential board savings, machine cost, floor space, case rate, size case, etc.

Tray Former/Loader

The concept of replacing a corrugated case with a shrink-wrapped tray was developed as part of an ongoing effort to reduce packaging material costs. The tray former/loader (see Fig. 8) with a shrink wrap is ideally suited for those products eg, cans and bottles, that do not require the complete product protection provided by a full case. Corrugated trays that are 1–2 in. (2.5–5.1 cm) high are preferred, but there are a host of different tray designs suited to the product, distribution, display, etc. Many machine alternatives are available, including equipment to make a tray and present it to a vertical or drop-load caser. There are also integrated systems that extract a blank from the magazine, fold up two or three sides of the blank, index to the load area (where the product is deposited horizontally or vertically), and transfer, folding remaining panels and gluing to form the final tray. At this point the filled and sealed tray is indexed into a shrink-wrapping machine (see Wrapping, shrink) where film (see Films, shrink) is completely wrapped and shrunk around the complete load. The film unitizes and holds the product in the tray and provides protection from the external environment. The cost for the corrugated tray and shrink film is considerably less than a full corrugated case, but consideration must be given to product protection and warehouse stacking strengths because now the product must bear the full vertical load, warehouse identification, etc. Tray rates for intermittent-motion machines are approximately 30/min, while continuous-motion machines can run in excess of 70/min.

Other Concepts

The case-loading field is extremely broad, and many other different types of machines are available for special markets and applications too numerous to describe in detail. For example, there is automatic case-opening/bottom-gluing machinery that presents an opened case with the bottom flaps glued for the manual loading of large or irregular-shaped items. This same equipment can also present an opened/glued case to a vertical or drop-load caser, increasing the operating speeds of hand-packed lines. Automatic case openers are available that bring a case down vertically over the accumulated product and then fold and seal the flaps or that lift the acculated product vertically up through the bottom of a corrugated case, folding and sealing flaps accordingly. There are case loaders that role the product into the case and manual machines that are used for handling irregular-shaped products in a wrap-around blank. All these variations and concepts are necessary in providing the unique machinery to meet the changing markets and products of today.

BIBLIOGRAPHY

General References

The PMMI 1984–1985 Packaging Machinery Directory, Packaging Machinery Manufacturers Institute, Washington, DC.

S. Sacharow. *A Guide to Packaging Machinery,* Harcourt-Brace Jovanovich, New York, 1980.

Walter Warren
Salwasser Manufacturing Company, Inc.,
Decatur, Georgia

CASKS. See Barrels.

CELLOPHANE

Background

The word *cellophane* was derived from the first syllable of cellulose and the final syllable of diaphane, meaning transparent. It was invented in the early 1900s in France and introduced in this country in 1924 by E. I. DuPont de Nemours & Company, Inc.

In the early stages, cellophane was somewhat of a curiosity that was very expensive, and its use was limited to the packaging of luxury items.

The growth of cellophane paralleled the growth and development of the entire flexible-packaging industry, from printing presses and inks to automatic packaging machinery. For 30 or more years, the dominant flexible packaging material was cellophane because it was so well suited to offer the marketplace a wide variety of characteristics adaptable to product needs at reasonable costs. The large markets were in the areas of baked goods, candies, and tobacco products.

The advent of plastic materials such as polyethylene and polypropylene started the corrosion of cellophane consumption that resulted in the closing of seven cellophane plants, with only one remaining in operation today in the United States, owned by Flexel, Inc. of Atlanta, Georgia (see Films, plastic). Worldwide consumption for 1995 is estimated to be 320×10^6 lb, with large producers located in Mexico, Europe, Russia, China, and Japan.

Cellophane is a thin, flexible, transparent material used worldwide mainly in packaging applications. The primary raw material used in manufacturing cellophane is "dissolving" wood pulp purchased from wood pulp suppliers. Cellophane has two unique characteristics for a flexible, transparent packaging film. It is both compostable and made from a renewable resource. Since its introduction in 1924, additional types of cellophane have been introduced to meet changing packaging needs. The primary markets for cellophanes are the food, pharmaceutical, and healthcare product markets.

Features

Cellophane has various attributes that have accounted for its continued acceptance and ongoing utilization in flexible-packaging applications. These include

- *Dead-Fold.* Once shaped in certain packaging applications, cellophane, unlike plastic films, can maintain its shape. This is especially true in twist-wrap applications such as packaging for hard candies.
- *Ease in Tearing.* Differentiated tensile strength within cellophane allows for ease in tearing and opening products that utilize this material for packaging and for tape.
- *Machinability.* Cellophane can be cut and sealed easily and economically. Many competing flexible-packaging materials require more sophisticated and expensive packaging equipment to process them.
- *Appearance.* Cellophane has a high level of gloss and haze versus certain competing flexible films. These factors are important to customers desiring a premium appearance for their packaging.
- *Resistance to High Temperatures.* Cellophane can be used in temperature ranges above those of many common plastic films, which is critical for hot-fill applications and for use in shrink tunnels.
- *Barrier to Air and Moisture.* Cellophane, when coated with saran or other barrier resins, has increased strength, seal, and barrier properties. In addition, some coated plastic films can be more expensive than coated cellophane products.

Film Types

Cellophane film is produced in various types, which vary with respect to (1) film thickness, (2) film width, (3) type and degree of coating, and (4) combination with other materials.

1. *Film Thickness.* Cellophane can be supplied in varying thickness depending on customers desire for strength, flexibility, and resistance to air and moisture.

2. *Film Width.* The films can also be produced in numerous widths, which are custom cut to customers' specific packaging needs.

3. *Type and Degree of Coating.* A large percentage of cellophane products are coated, generally on both sides of the film. These coatings increase the cellophane's durability and seal qualities. Two of the major coating materials are polyvinylidene chloride copolymer (PVdC) and nitrocellulose (NC).

PVdC offers more durability and higher moisture and gas barriers than NC. NC-coated films are typically used to package cookies, snacks, cheese, gum, and cough drops.

4. *Combination with Other Materials.* Cellophane film can be laminated to other films such as biaxially oriented polypropylene (BOPP) or metallized polyester film. These products offer enhanced qualities of both cellophane and the particular reinforcement material that is used. The end result is a flexible film that has excellent resistance to breakage and a high-quality appearance. Reinforced cellophane is used primarily to package pretzels, popcorn, chips, nuts, meats and cheeses.

Physical Properties

The physical properties of cellophane are closely related for all the film types. They are differentiated according to coatings, reinforcing structures, and thicknesses. All of them are clear with exception of two of the reinforced structures: metallized polyester and white opaque polypropylene core.

The nitrocellulose-coated films range from 16,000 to 18,000 psi (lb/in.^{2}) tensile strength in machine direction and 8000–9000 psi in transverse direction. Elongation ranges 15–25% in machine direction and 30–45% in transverse direction. The heat-sealable coatings have a wide sealing range, usually requiring temperatures of 225–350°F depending on machine speed and pressure.

Water-vapor transmission rate (WVTR) for the moistureproof film types averages about 0.5 gm/100 in. per h, and the breathable types range between 30 and 50. Oxygen permeability averages about 2 mL/(100 in.) (24 h) (atm) for most two-side-coated nitrocullulose film types.

The PVdC- or saran-coated films utilize very similar and in some cases identical cellulose base sheets, but as previously described, the PVdC coating results in improved keeping characteristics.

Machine direction (MD) and transverse direction (TD) tensile strengths and elongations are roughly equivalent to the nitrocellulose-coated films. PVdC coated films exhibit superior gas barrier properties to nitrocellulose coated films. Where the average oxygen barrier for nitrocellulose is about 2 mL/100 in. per h, the PVdC-coated films are about 0.5. Coefficient of friction is about 0.30–0.35 for both PVdC- and NC-coated films.

The third major type of cellophane film is plain, uncoated film. This film is used most widely in producing pressure-sensitive tapes, and also for applications where high-quality printing is desired. This uncoated film prints most easily with almost any flexographic printing ink. The reason for this wide range of ink receptivity is that the surface being uncoated readily absorbs liquids.

Cellophane Production

Cellophane is a regenerated cellulose film derived from chemically purified wood pulp known in the trade as "dissolving pulp." Cellulose, the primary repeating molecule of regenerated cellulose film, is shown below:

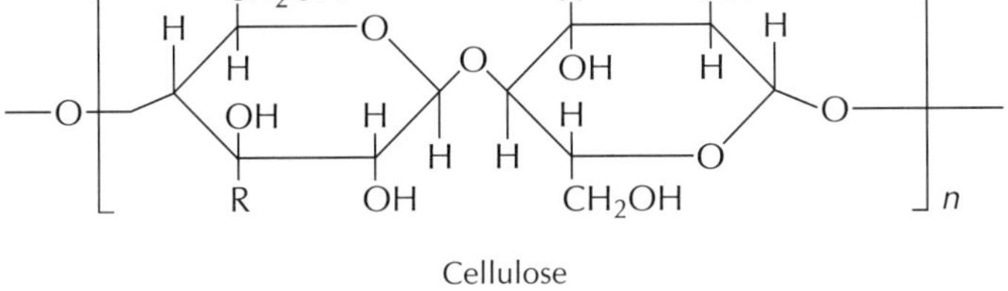

Cellulose

The cellophane production process involves five major steps: (1) soaking of the wood pulp in sodium hydroxide solution for several hours to form alkali cellulose; (2) polymerization of the alkali cellulose, under controlled conditions; (3) reaction of the alkali cellulose with carbon disulfide to form an alkali-soluble sodium xanthate; (4) formation of viscose by dissolving sodium xanthate in a solution of sodium hydroxide; and (5) extrusion of the viscose through a slit die into a water/sulfuric acid/sodium sulfate coagulating bath. After the regeneration of the cellulose, the process is completed by neutralizing, washing, and drying the film.

BILL JENKINS
Flexel, Inc.
Atlanta, Georgia

CELLULOSE, REGENERATED.

See CELLOPHANE.

CHECKWEIGHERS

Automatic, in-line checkweighers have long been used to perform basic weight-inspection functions. These units, as distinguished from static, off-line check scales, perform 100% weight inspection of products in a production process or packaging line. Furthermore, these machines perform their functions without interrupting product flow and normally require no operator attention during production.

These charactertistics allow the machines to be used in several different ways. The applications can be broadly categorized as weight-regulation compliance, process control, and production reporting. This article describes a number of checkweigher applications in these categories. The emphasis is on problem solutions and benefits, although some description of the machine hardware and features employed is necessary.

Weight-Regulation Compliance

The earliest use of in-line checkweighers was to help producers guard against shipment of underweight products, and today various regulatory agencies set standards for weight compliance of packaged goods. Most of these standards are adopted from recommendations made by the National Bureau of Standards (NBS) in their various handbooks. In general, these standards recommend that the average product weight shipped by the producer be equal to or greater than the declared weight and that no unreasonable underweights be shipped. The in-line checkweigher provides assurance to the producer that these general requirements are met. With proper adjustment of checkweigher and process, an additional benefit of reduced product giveaway can be enjoyed.

The equipment required to carry out the basic weight-compliance requirement consists of a weighing element (weighcell), associated setup controls and indicators, a product transport mechanism for continuous- or intermittent-motion product movement, and a reject device for diverting

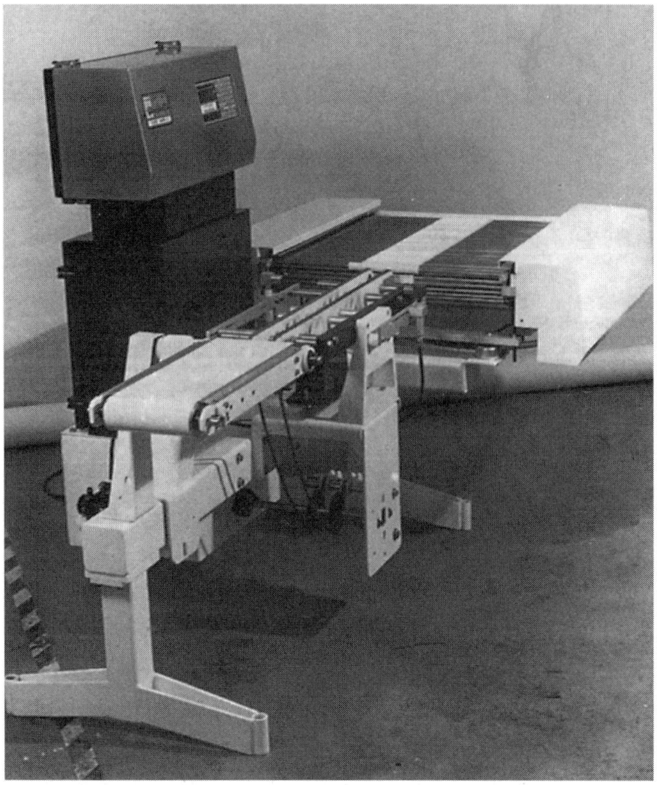

Figure 1. A typical checkweigher.

off-weight items out of the production stream. A typical machine comprising these elements is shown in Figure 1.

The central portion of this machine shows a weighcell consisting of a strain-gauge loadcell that is connected to the integrally mounted electronics enclosure. Associated with the weighcell is a product transport, in this case consisting of a pair of stainless-steel roller chains. Shown on the left (upstream) side of the unit is an infeed belt conveyor, which serves as a product-spacing (speedup) device and assures that only one product at a time passes over the weighcell. On the right (downstream) side of the unit is a channelizer product-reject device. This unit receives signals from the checkweigher and provides a gentle lateral displacement of rejected product by carrying it to the side on sliding carrier plaques. Other commonly used reject devices include air blasts, air pushers, swing gates, and drop-through mechanisms.

Rejection of underweight products (and in some cases overweight products such as critical pharmaceutical packages or expensive products) provides the required consumer protection. The added benefit to the producer of reduced product giveaway frequently provides additional incentive to install the checkweighing machine. The information from the typical modern checkweigher allows the producer to control the target weight of the process and assures minimum average product weight, and it minimizes rejected underweight products. The following application example illustrates these points.

New-contents weight. Company A prepares a variety of expensive frozen-food products. These products are clearly labeled showing net-contents weight and come under close scrutiny from regulatory agencies. Some ingredients, such as chunky pieces, are of somewhat nonuniform piece weight. Until an accurate in-line checkweigher was installed, the producer was forced to overfill most packages to prevent underweight shipments. The company installed a number of checkweighers that provide the required controls. Now, by means of digital displays and easily adjustable reject cutpoints on the checkweigher, the processor is able to closely monitor the filling process and establish the best compromise between product giveaway and excessive rejects.

Product-weight distribution. Figure 2 illustrates a typical product-weight distribution curve before and after a suitable checkweigher is installed.

Before checkweighing, the mean product weight was maintained at approximately 293 g to ensure against underweight shipments. After checkweigher installation, the mean product weight is reduced to approximately 289 g. This results in less than 1% underweight rejects but saves an average of 4 g per package. The economics of this checkweigher installation is illustrated below:

Package labeled weight	283 g
Average overweight per package	10 g
Possible reduction in average overweight	4 g
Annual production	900,000 kg
Value of product	$0.75/kg
Cost of checkweigher (including freight and installation)	$10,000

RESULTS

Product savings	1.4%
Product saved per year	12,720 kg
Savings per year	$9540
Payback period, approx.	1 yr

Process Control

Many products that are sold on the basis of a declared (labeled) weight are packaged by a volumetric filler or count or other nongravimetric processes. These processes frequently yield products of varying weight because of product density changes. Density changes typically occur with hygroscopic products under conditions of changing humidity. Also, material-handling methods can cause bulk-density changes in

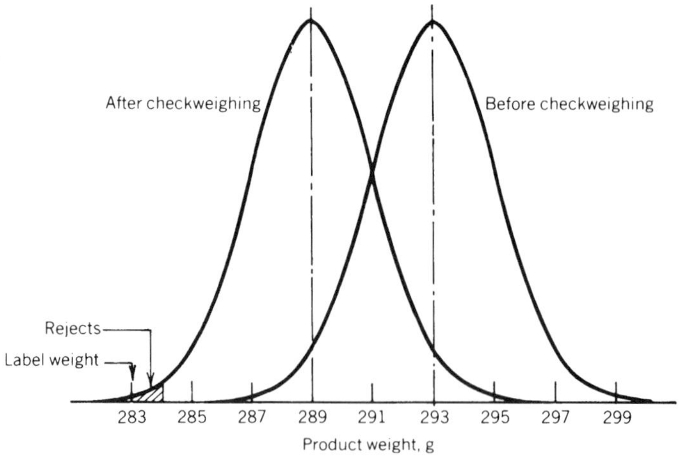

Figure 2. Weight distributions before and after checkweighing.

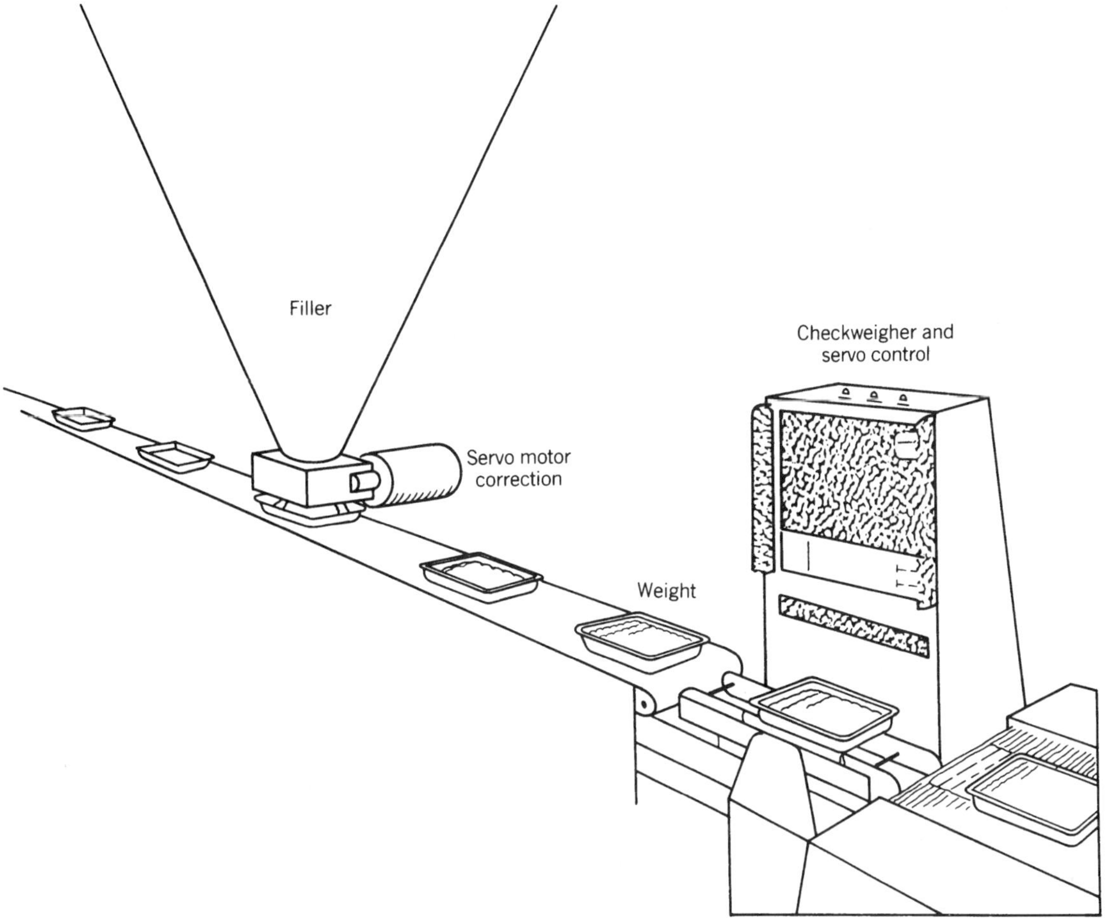

Figure 3. Checkweigher feedback-control to filler.

other products. Frequently, the density changes occur in the form of a drift or periodic change. These changes lend themselves well to automatic-feedback control from a weight sensor (checkweigher) back upstream to the process controller. Figure 3 diagrams a checkweigher/filler feedback-control system.

This feedback loop continually adjusts the process setpoint and maintains delivered product weight within acceptable limits, even under conditions of density variations. This in turn reduces the long-term weight standard deviation of the packaged products. This allows the process setpoint to be set at a lower average value, while minimizing the number of underweight (rejected) packages. The result is product savings. Two applications are described to illustrate the use of checkweighers with feedback for process control.

Feedback control. Company B packages a low density, hygroscopic product using a volumetric filler. Spot checks revealed that moisture variations from batch to batch and humidity changes throughout the day resulted in substantial density changes. These changes often went unnoticed until a large number of off-weight packages were produced, causing either excessive product giveaway or an excessive number of underweights being rejected by the old checkweigher. In the former case, the giveaway represented considerable revenue loss, and in the latter case resulted in product waste due to the impracticability of recycling and repackaging the rejected product.

The company purchased a replacement checkweigher equipped with feedback-control features for signaling their existing filler. The checkweigher is located as close as possible to the filler to achieve maximum responsiveness of the feedback loop. In this case, the feedback signals are electronically adjusting the setpoint of the filler. However, other control arrangements are possible including signaling a servo motor attached directly to the control shaft of a filler. This company was able to justify purchase of the checkweighing system solely on the basis of product savings. Figure 4 illustrates how long-term standard deviation reduction allows lowering of average setpoint with resultant product savings. Short-term standard deviation resulting from package-to-package filler errors is inherently beyond the control capability of the checkweigher.

Company C packages sliced luncheon meats. The "stick" or "load" of luncheon meat is fed through a slicer that counts a preset number of slices per draft to be packaged. Because of density changes of the meat, the correct slice count did not always yield the declared weight of the draft within allowable limits. With no on-line way to detect these density changes, the company had to supply an extra number of slices to assure labeled weight or take a chance on packaging underweights with correct slice count. It installed an in-line

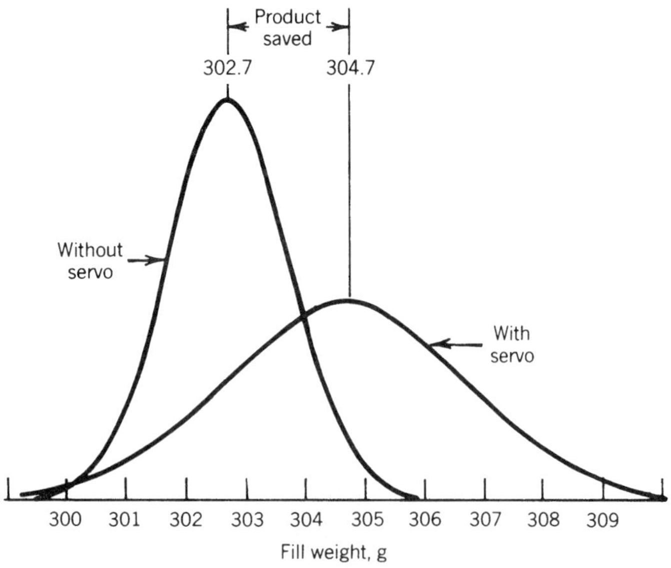

Figure 4. Filler setpoint reduction with servo system.

checkweighing machine that very accurately weighs each draft of sliced meat just after it leaves the slicer. Error signals, representing departures of delivered weight from target weight, are fed back to the slicer controls. The controls in turn cause slice thickness to be adjusted so that the preset slice count results in correct draft weight within very narrow limits. The result was control over the slicing process that economically justified the checkweigher quickly.

Production Reporting

Automatic production reporting by a checkweigher can provide running analysis of packaging-line performance. Checkweigher-generated production data can be transmitted to printers, computers, terminals, and other data receivers. Whatever the reporting medium, the result is virtually instantaneous indications of production-line performance or automatically kept records of production data. Productivity gains and material savings result from such uses. Here are two applications of the checkweigher for production recording:

Company D requires that a statistical sampling procedure be carried out on each of its production lines. Historically, quality assurance personnel periodically removed a number of product samples from the production line, weighed them on manual scales, recorded the weights, and returned the samples to the production line. Using the raw data they had collected, quality-assurance personnel manually calculated statistical information about the samples. From this information, they prepared a product sampling report. Because of the frequency of the sampling procedure and the number of lines involved, a full-time quality-assurance (QA) person was employed to perform this task. Furthermore, the manual nature of the weighing and recording resulted in reports of questionable accuracy.

Microprocessor-controlled checkweigher. The company purchased a modern microprocessor-controlled checkweigher that is capable of accumulating production-weight data in real time and performing the required sampling calculations for transmission to a printer in the QA supervisor's office. Data reported to the supervisor include date and time of report, product code, operator identification number, average weight, standard deviation, and total weight for both accepted and rejected product over the selected sample size. Additionally, verification of checkweigher setup parameters such as target weight and reject cutpoint settings are recorded. Installation of the checkweighing system results in cost savings in personnel for acquiring the sample data and virtually instantaneous feedback of production data to QA and production personnel.

Company E ships large quantities of product to customers for further processing. To corroborate material-amount reports between supplier and customer, accurate total production records were required. In addition to actual weight of shipped product, an accounting of underweight (rejected) packages would yield important production-efficiency data.

The company installed an automatic checkweighing system and appropriate data-collection capabilities along with a printer for location in the accounting office. On command, accumulated data are transmitted from the checkweigher to the printer, which provide subtotals and totals of average weight, total weight of product shipped and rejected, package counts, and a figure for "yield" (shipped product divided by total production) on the basis of both weight and package count. The result of this installation is a true and accurate summary report for the accounting department of product shipped to each of their customers, thereby minimizing protests. The additional benefit is a more accurate report of actual production efficiency.

The application examples presented here describe a few of the many kinds of packaging lines that can and should benefit from the use of in-line automatic checkweighers. A few additional applications are suggested below:

Piece-count verification in packages and cases.

Product-safety assurance by detecting undesirable underfills or overfills.

Multizone classification of natural products such as chicken parts, fish, etc.

Detection of missing components in "recipe" food packages.

A survey of product weights in any processing/packaging line may reveal opportunities for substantial cost savings through use of an automatic checkweigher.

BIBLIOGRAPHY

"Checkweighers" in *The Wiley Encyclopedia of Packaging Technology,* 1st ed., by J. F. Jacobs, Ramsey Engineering Company, Icore Products, pp. 163–166.

General References

C. S. Brickenkamp, S. Hasko, and M. G. Natrilla, *Checking the Net Contents of Packaged Goods,* National Bureau of Standards, Washington, DC, 1979.

C. Andres, "Microprocessor-Equipped Checkweighers," *Food Process.* **40,** 152 (Jan. 1979).

C. Andres. "Special Report: Expanding Capabilities of Checkweighers," *Food Process.* **42,** 58 (Mar. 1981).

N. W. Rhea, "Are You Ready For Checkweighing?," *Mater. Handl. Eng.* **5,** 74 (May 1983).

Automatic Checkweigher Test Procedure, Scale Manufacturers Association, Inc., Washington, DC, Jan. 1977.

"Checkweighing Scales," *Mod. Mater. Handl.* **37**(3), 66 (Feb. 1982).

U.S. Department of Commerce/National Bureau of Standards, *NBS Handbook 130, 1979 (Draft),* U.S. Government Printing Office, Washington, DC, 1979.

U.S. Department of Commerce/National Bureau of Standards, *NBS Handbook 133, Checking the Net Contents of Packaged Goods,* U.S. Government Printing Office, Washington, DC, 1981.

U.S. Department of Commerce/National Bureau of Standards, *NBS Handbook 44, Specifications, Tolerances, and Other Technical Requirements for Commercial Weighing and Measuring Devices, (Fourth Edition),* U.S. Government Printing Office, Washington, DC, Dec. 1976.

CHILD-RESISTANT PACKAGING

Child-resistant (CR) packaging, a term synonymous with poison-prevention packaging (PPP) and special packaging (SP), can be defined as a container that precludes entry by children under the age of 5 years but not adults to hazardous substances such as drugs, household cleaning agents, and pesticides. Packaging of this nature may take a number of forms, including bottles, drums, pouches, and blister packs. Many children have been saved from bodily harm and even death as a result of CR packaging.

Historical Aspects

Designs for CR packaging can be traced back to 1880, when the first U.S. patent was issued for a CR package. The U.S. Congress began to take direct interest in 1966, in response to public concern about the large number of children gaining access to harmful substances in the home. As a result of Congressional hearings that year, the commissioner of the FDA appointed Dr. Edward Press to be chairman of a committee to review the "state of the art of safety packaging." The Press committee, as it came to be known, comprised members from U.S. industry and government. The committee reviewed the 63 patents on CR packaging that had been awarded between 1880 and 1966 and decided that the most realistic and practical approach to this problem was to establish a performance standard, using children to test the units. A series of closure studies was conducted, involving more than 1000 panelists: adults from 18 to over 65 years of age and children between the ages of 18 and 52 months. From the data obtained in these studies a Protocol to evaluate child-resistant packaging was derived. This protocol, submitted to the FDA in 1970, is reflected in the Protocl cited in the Code of Federal Regulations (CFR) (1), with the exception that the Press committee protocol contained two 3-min test periods for adults, one before a demonstration and one after a demonstration. The FDA revised Protocol, as it appeared in the CFR, cited one 5-min test period for adults without a demonstration. This was later changed in 1995 to a 5-min, 1-min test. (2)

The U.S. Poison-Prevention Packaging Act (PPPA), signed into law December 30, 1970, was under the jurisdiction of the FDA. The protocol for the evaluation of poison-prevention packaging appeared in the Federal Register in July 1971 (3). FDA standards began to appear in the Federal Register in 1972, first for aspirin, and then for controlled drugs, methyl salicylate, and furniture polish. In May 1973, jurisdiction was transferred from the FDA to the newly formed U.S. Consumer Product Safety Commission (CPSC), and additional products came under regulation. The CPSC was given the responsibility for medicines and household substances, and in 1979 the EPA was delegated to administer the compliance of pesticides.

Household pesticides of a hazardous nature were regulated in March 1981 (4). The initial EPA regulation was more stringent than the CPSC regulation, but much of the initial draft regulation was deleted to a point where the EPA regulation (5) approximates the CPSC regulation regarding testing requirements and standards. The major difference between the two regulations, at this point, has been that the CPSC regulation makes provisions for a noncompliance package that can be used by older adults and/or households without children, and the EPA regulation does not. In addition, the 1995 PPPA regulation allows the EPA to merely require the previous test procedure employing adults, 18–45 years of age, with 5-min and 1-min test periods, if it is appropriate.

Because of the difficulty older adults were having trying to open CR packaging, in the summer of 1995 the CPSC revised the CR packaging test procedure to include older adults between the ages of 50 and 70 years [Fed. Reg., **60**(140), 37710–37744 (July 21, 1995)]. This will be discussed in detail later.

Although there is worldwide awareness of child-poisoning problems in the home, only three other countries in the mid to late 1970s had followed the lead of the United States in regulating the need for certain hazardous substances to require CR packaging. These countries are Canada (6), Germany (7), and the United Kingdom (8). Regulations for these countries were not as extensive as those for the United States, in that the U.S. regulations covered drugs, household-type cleaning agents, pesticides, and other products, whereas the regulations of the other countries were aimed primarily at pharmaceutical products. Their CR packaging standards and test procedures are comparable, however.

Members of the International Standards Organization (ISO) had been working on a proposed child-resistant packaging standard since 1981, and a standard was enacted in 1989 (9). There are some differences in the ISO standard and the U.S. Protocol, which are discussed below.

Effect of Regulation

There are 24 substances regulated or about to be regulated by CPSC. They are listed in Table 1. Regulations were placed on these substances because of their harmful nature to small children and the large number of ingestions noted. The substances appear in this table by virtue of the ingestion patterns they have demonstrated from the data generated by 73 poison-control centers located in the United States and compiled by the CPSC. Because of the increased sophistication of data collection since the government began collecting this information, in some cases it is difficult to measure actual reductions in ingestions by children. However, it was possible to present comparative ingestion data up until 1982 showing considerable ingestion reductions in many regulated products with reductions of up to 86% as shown in Table 2. Currently, tables of mortality rates can be used to demonstrate the reduction of deaths because of CR packaging. These figures can be noted in Table 3, which cites deaths of children under age

Table 1. Regulated Substances—CPSC

Aspirin	>45 grains
Furniture polish	≥10% petroleum distillates
Methyl salicylate	>5% by weight
Controlled drugs	all oral forms
Na or K hydroxide	≥10% by weight
Turpentine	≥10% by weight
Kindling and illuminating preparations	≥10% petroleum distillates
Methyl alcohol	≥4 by weight
Sulfuric acid	≥10% by weight
Prescription drugs	all oral forms
Ethylene glycol	≥10% by weight
Iron-containing drugs	≥250 mg elemental iron
Dietary iron supplements	≥250 mg elemental iron
Solvents	≥10% by weight
Acetaminophen	>1000 mg acetaminophen
Diphenhydramine	>66 mg diphenhydramine
Glue removers	>500 mg acetonitrile
Permanent-wave neutralizers	>600 mg Na bromate >50 mg K bromate
Ibuprofin	>1000 mg ibuprofin
Loperamide	>0.45 mg loperamide
Lidocaine (April 10, 1996)	≥5 mg
Dibucaine (April 10, 1996)	≥0.5 mg
Mouthwash (ethanol)	≥3 g
Naproxen sodium (February 6, 1996)	≥20 mg

5 involving household products and aspirins. This table indicates that there has been a considerable reduction in child deaths as a result of the ingestion of either household products or aspirins.

Testing Procedures

The first U.S. test procedures protocol for the evaluation of child-resistant packaging were published in the *Federal Register* in 1971. As a result, the packaging field has had over twenty years of experience in generating data utilizing this protocol. As stated earlier, poisoning statistics demonstrated the use of CR packaging has been shown to reduce child ingestions and deaths from harmful materials. Nevertheless, as data gathering methodology has become more sophisticated, it appears that there is an area still to be addressed when it comes to child-resistant packaging. It is the use of CR packaging by older adults. This point is demonstrated by the fact that the ingestion statistics for prescription drugs shows the highest level of the regulated products. Older adults are the major users of these products and were assumed to be the ones for problems in the child ingestion. Further, studies conducted under contract to CPSC found this to be true (10). In interviews with these older adults it has been found that they misuse CR packaging. They either do not secure the packaging properly, leave the caps off completely, or transfer their drugs to an easily opening unsafe container. Recognizing that older adults have problems opening CR packaging, the U.S. CPSC has tried to devise new test procedures to encourage the design and manufacture of easy to open so-called "senior friendly" CR packaging.

Initially, the CPSC let contracts to three universities to design new CR packaging concepts that would satisfy "senior friendly" criteria, and then a mockup and a unit cavity production package was prepared from the best concept. From this point over a period of time changes were made to the test protocol to ensure that senior adults could access newly designed packaging of the universities design as well as packaging designed by individual inventors and designs developed within the packaging industry. Concepts that were devised included ones that required cognitive skills to activate. That is, they relied on the mental ability of the adults to determine how to open units versus high forces as push and turn or pull as have been required with the ongoing type CR packaging systems. In addition to the new cognitive skill concepts, package manufacturers refined their current CR packaging design to make them easier to open for senior adults yet still difficult for children.

In a selected group of 100 seniors 50–70 years of age, 70% female, effectiveness rates were

25%	50–54
25%	55–59
50%	60–70

Table 2. A Comparison of the Number of Regulated Substances by Children Under 5 Years Old

Regulated Substances	Effective Year	Ingestions During the Effective Year	Ingestions During 1982	Decrease Since Effective Year, %
Aspirin	1972	8146	1753	78
Controlled drugs	1972	1810	541	70
Methyl salicylate	1972	161	49	70
Furniture polishes	1972	697	229	67
Illuminating and kindling preparations	1973	1736	452	74
Turpentine	1973	777	110	86
Lye preparations	1973	508	69	86
Sulfuric acid[a]	1973	7	1	86
Methanol[a]	1973	37	21	44
Prescription drugs (oral)	1974	4180	2251	46
Ethylene glycol	1974	138	59	57
Iron preparations	1977	359	163	55
Paint solvents	1977	641	180	72
Acetaminophen	1980	1511	1118	26

[a] Since these figures are based on a few cases, the percent decrease may not be reliable.

Table 3. Deaths of Children Under Age 5 Involving Household Products[a]

Deaths Due to All Household Chemicals			Deaths Due to Aspirin Products		
Year	No. deaths	Decline (%) since 1972	Year	No. deaths	Decline (%) since 1972
1972	216	—	1972	46	—
1973	149	31	1973	26	43
1974	135	38	1974	24	48
1975	114	47	1975	17	63
1976	105	51	1976	25	46
1977	94	56	1977	11	76
1978	81	63	1978	13	72
1979	78	64	1979	8	83
1980	73	66	1980	12	74
1981	55	75	1981	6	87
1982	67	69	1982	5	89
1983	55	75	1983	7	85
1984	64	70	1984	7	85
1985	56	74	1985	0	100
1986	59	73	1986	2	96
1987	31	86	1987	3	93
1988	42	81	1988	3	93
1989	55	75	1989	2	96
1990	49	77	1990	1	98
1991	62	71	1991	2	96

[a] Data obtained from the National Center for Health Statistics in Washington, DC.

and

1. 5-mins opening: screening units used if not opened
2. 1-min opening and closing:
   a. 90% performance standard
   b. Minimum of three testers
   c. Minimum of five test sites

As stated earlier, the first U.S. CR packaging test procedure (protocol) appeared in the *Federal Register* in the spring of 1971, and this protocol was adhered to up until the summer of 1995, at which time a new and modified CR protocol appeared in the *Federal Register.* The original 1971 protocol cited the need to employ 200 children between the ages of 42 and 51 months of age equally divided according to age and sex with an allowance of 10% variance, and also 100 adults between the ages of 18 and 45 years representing 70 women and 30 men. The children were to be tested in pairs in familiar surroundings and would have two 5-min test periods, one without a demonstration and one with a demonstration and they were told they could use their teeth if they had not used them in the first 5-min test period. The adults were allowed a 5-min test period and told they should open and resecure (if appropriate) the packaging according to the directions.

Standards for the children were set as 85% unsuccessful before demonstration and 80% after demonstration. Ninety percent of the adults had to open and close (if appropriate) the package in a 5-min test period.

The 1995 CR protocol takes into consideration a number of features that have been observed over the years that would help standardize the testing procedure from one test agency to the next. In addition, it includes major changes in the ages of the adult test group. These features will be discussed later. The ages of the child panelists has remained 42–51 months, but instead of immediately utilizing a group of 200 children, 42–51 months of age, the test is conducted in groups of 50 children on a sequential basis up to 200 if it is required according to an established pass–continue–fail pattern based upon package openings (see Table 4). In order to make it easier to obtain child panelists, the number of chronological groupings has been changed from 10 to 3. The three selected groups are to include

30% 42–44 months of age
40% 45–47 months of age
30% 48–51 months of age

The standard for the child test has remained the same 85% unsuccessful before demonstration and 80% unsuccessful after demonstration.

Included in the U.S. testing procedure for CR packaging from its inception in 1971 is a provision for testing nonreclosable packaging (unit packaging, as blister packs, strip packs, and pouches). The test procedure follows the same pattern as the reclosable packaging, but instead of just access to a container, an individual package failure with the nonreclosable package for the children is the case whereby the child gains access to greater than eight units of the package (blisters, packs, or pouches) or to the concentration of product that would constitute a toxic dose for a 25-lb child, whichever is less.

The ISO standard for CR packaging merely covers reclosable packaging. The inclusion of nonreclosable packaging in the ISO standard has been a very controversial topic. An initial draft of the ISO standard did include a provision for nonreclosable CR packaging, but was deleted when the standard was brought to a vote in 1987. The Europeans take a different approach as to how they view nonreclosable CR packaging. It is their attitude that blister and strip packages are CR if they are opaque and are therefore inherently CR. The European concept has been proved wrong on a number of occasions

Table 4. Number of Openings: Acceptance (Pass), Continue Testing, and Rejection (Fail) Criteria for the First 5 min and the Full 10 min of the Children's Protocol Test

		Package Openings					
		First 5 min			Full 10 min		
Test Panel	Cumulative Number of Children	Pass	Continue	Fail	Pass	Continue	Fail
1	50	0–3	4–10	11+	0–5	6–14	15+
2	100	4–10	11–18	19+	6–15	16–24	25+
3	150	11–18	19–25	26+	16–25	26–34	35+
4	200	19–30	—	31+	26–40	—	41+

when this type of packaging has been tested employing the U.S. methodology. Nevertheless, the controversy continues, perhaps due to the economics of converting to true CR packaging from what is currently manufactured in Europe.

The truly major changes in the new CR protocol involves the changes in the ages of the 100-member adult panel and the time frame in which they are tested. The new age groups are as follows:

25% 50–54 years of age
25% 55–59 years of age
50% 60–70 years of age

Of these, 70% of the 50–59 year-olds are to be female and 70% of the 60–70 year-olds are to be female. The standard for the adult test is a 90% successful pattern.

As for the time of the test, in order to be become familiar with new CR packaging concepts, they are tested as follows. The adults are tested individually with a 5-min test period to learn how to open and close the test unit. They are then given a 1-min test period to open and close the test unit. A panelist who is unsuccessful in opening the unit in the first 5-min test period is given a non-CR screw cap and a non-CR snap cap to open to attempt to open and close. In this case an opening and closing of the two non-CR units would constitute a failure in the case of the test unit. On the other hand, a panelist who fails to open and resecure the non-CR units is not included in the test unit pattern.

The U.S. Environmental Protection Agency (EPA) will follow the lead of the CPSC in the regulation of CR packaging, but the EPA may also rely on an adult panel of 18–45 years of age to evaluate CR packaging with a 5-min test period and 1-min test period.

Up until 1992 only four other countries besides the United States had CR packaging regulations: Germany, United Kingdom, Netherlands, and Canada. Then in 1992 the European Economic Community adopted the ISO standard 8317 (12) as the CR packaging regulation for Europe designating the standard as EN28317.

ISO 8317 titled "Child-resistant packaging—Requirements and Testing Procedures for Reclosable Packages" was first made official in 1989 and was reviewed and voted on again by the ISO member nations in 1994. The ISO CR packaging standard as cited covers merely reclosable packaging, not nonreclosable packaging as unit packaging that includes blister pack, strip packs, and pouches. The ISO standard as presented is similar to the U.S. standard in the manner in which the procedures are carried out, but differs considerably in the sequential test pattern for the children and the age grouping for the adults. Efforts have been made to either include nonreclosable CR packaging in this standard or have a separate nonreclosable CR packaging ISO standard, but member ISO countries could not agree on this topic.

Basically, the ISO standard include children 42–51 months that are tested in pairs using two 5-min test periods before and after demonstrations as cited in the U.S. regulation. However, the children are not informed they can use their teeth as they are in the U.S. standard. Testing procedures are conducted on a sequential basis utilizing a special sequential statistically devised charts to fill in the test results and determine the pass–fail pattern. Using the ISO test chart, it is possible to complete the child portion of a CR study by employing only 30 children to test the CR unit.

The adult portion of the ISO CR test employs up to 100 adults between the ages of 18 and 65 years. Twenty of the adults are to be 61–65 years of age, and 70% of the test population are to be female.

In the course of the test the adults are allowed 5 min to open and resecure the test package. A 90% success pattern is required. If the sequential test pattern is utilized, as few as 36 adult panelists could be employed to complete the test.

The first Canadian standard for CR packaging (CSA Standard Z76) appeared in 1976 and was comparable to the U.S. standard as cited in 16 CFR 1700 employing 200 child panelists, 42–51 months and 100 adults, 18–45 years of age. Then in 1990 Canadian standard CAN/CSA Z76.10M90 was published. This standard was similar to the above standard, but included more details of the test procedures as well as an option to utilize a sequential test scheme like the one utilized in the ISO 8317 standard. It also suggests the use of some American Society for Testing and Materials mechanical CR packaging tests to evaluate certain test units.

Another feature that was included in the U.S. standard, which does not appear in any other standard, is an "adult-resecuring effectiveness test." This test is utilized to confirm that adults properly resecure the test closure in the safe position after they have opened the unit. For the most part, this test applies primarily to continuous thread CR closures that depend on-torque to make them CR, but it may apply to others as well.

The adult resecuring effectiveness test is conducted by taking the test units that the 100 adults have opened and resecured according to the directions and further testing the units that were opened and apparently resecured by the adults with the appropriate number of children. A 90% success pattern has to be achieved in this test. The success pattern is calculated with a special formula that takes into consideration an allowance for a 80% unsuccessful rate for the children and a 90% success pattern with the adults.

Enforcement

CPSC and EPA are the two federal agencies in the United States that are responsible for enforcing the CR packaging regulations: the CPSC for household products and medications and the EPA for pesticides utilized in and around the home. Both agencies have their own legal staffs, which can act with the Justice Department against companies whose products are not in compliance with packaging regulations. The U.S. federal government relies on product samplings from retail stores and warehouses, complaints from the public, and the number of child ingestions to determine which products should be evaluated for compliance purposes. Producers of products found not to be in compliance are approached by the agency are informed of their problem and advised to improve their packages. If the package is still found not to be in compliance, the federal agency involved can go to the extent of halting further production of the package form and ordering a product recall. This, of course, can result in loss of revenues and undesirable publicity for package and product manufacturers. Voluntary recalls of products by manufacturers have occurred in the past because of packages not satisfying the regulation.

The Canadian, British, and German regulations are based on registration and certification of CR packaging used for specific products. Once this is established, it is up to the companies manufacturing the products to make certain that the packaged product is CR. Compliance efforts to the degree utilized in the United States are not employed. The ISO and Comité Européen de Normalisation standards present a similar approach to adhering to CR characteristics for compliance purposes.

The United States has recognized that CR characteristics of packaging can change because of dimensional problems encountered in the manufacture of components of the packaging to render the packaging non-CR. These usually have been the result of a poor quality-control program by the package manufacturer and/or the package user. Rigid compliance sampling and testing has resulted from problems observed with faulty CR packaging.

Features Affecting CR Effectiveness

In the course of evaluating many CR packages, it has become apparent that certain features can affect the CR characteristics of a package:

- Closures generally perform better on plastic or metal containers than on glass containers.
- Smaller sizes (18–24 mm) of closures are generally easier for children to open than larger sizes (≥33 mm).
- Different-size bottles may perform differently with exactly the same closure size.
- Different closure liners (see Closure liners) perform differently in the same closure.

In recognition of these factors, as well as other bottle/closure arrangement characteristics, the EPA regulations published in 1981 (4) include a testing scheme to be carried out if modifications or changes were made after original testing in the following respects: packaging shape; packaging material; volume of package; closure material; and/or cap liners. These requirements, although realistic, were later deleted (10). Nevertheless, these features should be considered in the design and modification of CR packaging.

Classification

Some of the designs of CR packaging in the late 1960s and early 1970s were based on the need for strength to open them. Because these units prevented child entry, but adult entry as well, they were not of much value. Other designs requiring two dissimilar simultaneous motions in order to activate or open a unit, such as push-and-turn, squeeze-and-turn, pull-and-turn, and turn-and-push, were developed. These designs have been quite successful since small children appear to lack the ability to readily perform the two motions at the same time.

In addition, there are units that require implements, such as screwdrivers or scissors, to open. This concept has been effective since the protocol disallows giving child panelists implements not included with the unit in the course of the test pattern. These and other types of units appear in the ASTM classification standard.

The ASTM D10.31 subcommittee has established what can be considered the official classification system for child-resistant packaging. This ASTM standard, *D3475-76 Classification of Child-Resistant Packages,* is based on the forces required to open the packages (10). Nine major types of packages are included in this classification system, each with a number of subgroups within the types. Examples of the different types of packages, along with their producers are presented in the standard. A total of 54 different kinds of packaging appear in this document with as many or more manufacturers of packaging. The types of packaging presented in the classification system are *continuous thread closure, lug finish closure, snap closure, unit packaging—flexible, unit packaging—rigid, unit reclosable packaging, aerosol packaging, nonreclosable packaging—semirigid* (*blister*), and *mechanical dispensers.*

The ASTM CR packaging classification standard was originally published in 1976 and has been reviewed and updated, in this classification system, each with a number of subgroups within the types. Examples of the different types of packages along with their producers are presented in the standard. A total of 54 different kinds of packaging appear in this document with as many or more manufacturers of packaging. With the revised U.S. CR packaging protocol in 1995 leaning toward "senior-friendly" CR packaging, it was anticipated that there should be an addition of new packaging concepts and perhaps some deletions of older designs from this CR packaging list.

With the major changes that are taking place in CR packaging, because of the new regulation, it may take some time to update this classification standard.

BIBLIOGRAPHY

1. *Code of Federal Regulations, Title 16, part 1700,* CPSC Regulation, U.S. Consumer Product Safety Commission, Washington, D.C.
2. *Fed. Reg.* **60**(140), 37710 (July 21, 1995).
3. *Fed. Reg.* (July 20, 1971).
4. *Fed. Reg.* **46**(41), 15106 (Mar. 3, 1981).

5. *Code of Federal Regulations, Title 40, Part 162,* EPA Regulation, U.S. Environmental Protection Agency, Washington, DC.
6. *CSA Standard Z76-1979, Child-Resistant Packaging,* Canadian Standards Association, 1979
7. DIN 55 559, *Child-Proof Packaging/Requirements Tests,* FRG Standard, 1980.
8. BS 5321, *Reclosable Pharmaceutical Containers Resistant to Opening by Children,* British Standard, 1975.
9. "Closures, Child-Resistant 15.09," in *Annual Book of ASTM Standards,* American Society for Testing and Materials, Philadelphia, 1983.
10. *Fed. Reg.* **47**(179), 40659 (Sept. 15, 1982).
11. *Child-Resistant Packaging,* ASTM Technical Publication 609, American Society for Testing and Materials, Philadelphia, 1976.
12. ISO 8317, *Child-Resistant Packaging—Requirements and Testing Procedures for Reclosable Packaging.*

General References

R. L. Gross and H. E. White, *Identification of Selected Child-Resistant Closures,* U.S. Consumer Product Safety Commission, 1978, U.S. Government Printing Office, Stock No. 052-011-00194-5.

Safety Packaging in the 70s, proceedings of a conference sponsored by the Scientific Development Committee, The Proprietary Association, New York, 1970.

P. Van Gieson, "ASTM History in Child-Resistant Packaging," *ASTM Stand. News* 26 (Apr. 1983).

ALEXANDER M. PERRITT
Perritt Laboratories, Inc.
Hightstown, New Jersey

CHUB PACKAGING

The term "chub" as related to packaging originated in the processed meat industry as a term used to describe large "chubby" sausages similar to bratwurst or kielbasa. The term "chub package" is used to describe sausage-shaped packages used for a wide range of semiviscous products (see Fig. 1).

The chub package is made by forming a web of flexible packaging film around a mandrel, sealing the overlapped edges to form a tube. A clip or closure is applied to the bottom of the tube as product is introduced through the interior of the mandrel. Finally, a clip or closure is applied to the top of the tube, forming a closed finished package (see Fig. 2). Alternative methods of chub package formation involve the use of preformed tubes of flexible packaging film that are filled and closed at the ends but retain the cylindrical sausage shape of the classic chub package.

Automatic form/fill/seal equipment such as the Kartridg Pak Chub Machine (see Fig. 3) provide a means to automatically and continuously produce chub packages. Package size ranges from about 0.6 in. (15 mm) to 6 in. (150 mm) in diameter and up to about 48 in. (1220 mm) in length. Depending on package size, production rates of up to 100 packages per minute are common. Semiautomatic means of producing chub packages are also available.

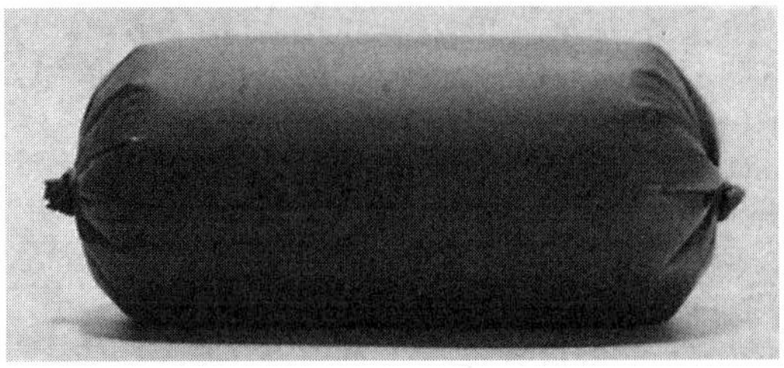

Figure 1. A chub package.

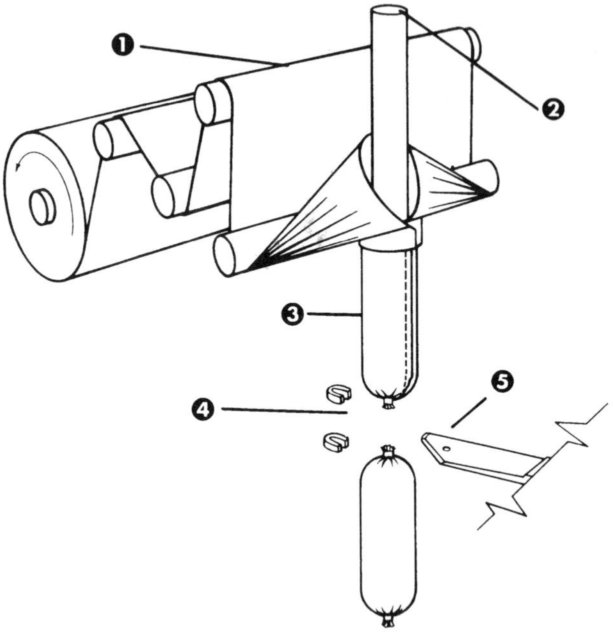

Figure 2. Formation of chub package: (**1**) film is formed around mandrel; (**2**) the overlapped edges are sealed (**3**) to form a tube and the product is introduced through the mandrel interior; (**4**) closures or clips are applied to the ends of the package; (**5**) the finished package is discharged.

Chub-packaged products have found wide acceptance in the consumer and industrial marketplace. Almost any pumpable viscous product can be chub packaged. A listing of chub-packaged products appears in Table 1.

Many types of flexible-packaging film are used to produce

Table 1. Chub-Packaged Products

Barbecue beef	Epoxy resin (2-part)	Pizza topping
Barbecue sauce	Explosive slurries	Pork role
Beef roll	Frosting	Pork sausage
Bologna	Frozen bread dough	Poultry rolls
Butter	Frozen juice	Precooked pork sausage
Caulking compound	Fruit preserver	Processed cheese
Chili	Ground beef	Refried beans
Chocolate fudge	Ham roll	Sandwich spread
Citrus pulp	Ham salad	Sauce bases
Cooked salami	Ice cream	Scrapple
Cookie dough	Lard	Snuff
Cornmeal mush	Mashed potatoes	Soup
Cream cheese	Mushroom gravy	Tamales
Dental impression material	Pet food	Vegetable shortening
Drywall compound	Pie dough	
Elongated hard-boiled eggs	Pizza Sauce	

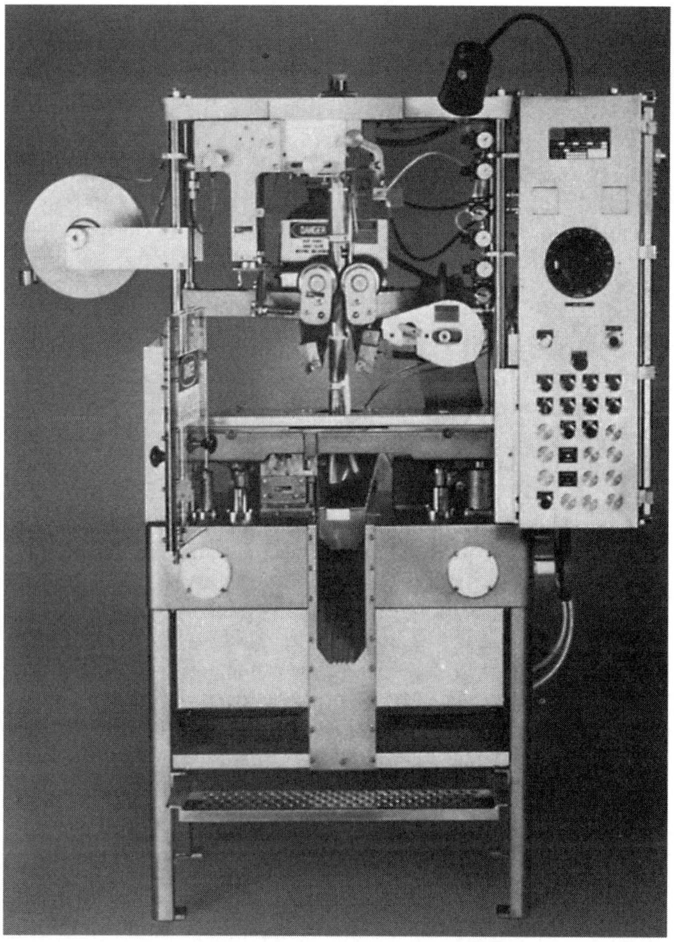

Figure 3. Automatic chub-packaging machine. (Courtesy of The Kartridg Pak Co.)

chub packages. Films are generally selected on performance and cost considerations. Aesthetic considerations such as finished package appearance and printability may also influence packaging film selection.

Chub-packaged products such as fresh pork sausage, ground beef, cold pack cheese, or slurry explosives are packaged in essentially the same form in which they are delivered to the user. Some chub-packaged products such as soups, thermoplastic poultry rolls, and imitation cheese are packaged at elevated temperatures, usually not higher than 194°F (90°C), in a viscous state and set or become solid when chilled. Products such as liver sausage, thermoset poultry rolls, precooked pork sausage, and cooked sausage are heat-processed after packaging. Chub packages have not been aseptically processed commercially.

BIBLIOGRAPHY

"Chub Packaging" in *The Wiley Encyclopedia of Packaging Technology,* 1st ed., by H. J. Sievers, The Kartridg Pak Co, pp. 170–171.

General References

The Wonderful World of Chub Packaging, sales brochure. The Kartridg Pak Co., Davenport, Iowa, 1983.

CLOSURE LINERS

The proper choice of a closure liner (see Closures) can often make the difference between the success and failure of a product. Most closure liners contain two basic components: a backing and a facing. The backing provides compressibility, resiliency, and resealability, and the facing provides barrier protection. In some liners, one component serves both functions, and in some two-component liners, each component contains more than one material.

Selection of the proper lining material generally involves one or more of the following considerations: adequate chemical resistance to the product; sealing ability against volatile loss; sealing ability against moisture loss or gain; sealing ability against air or oxygen; compliance with FDA regulations; freedom from odor and taste contamination; ability to withstand sterilization by various methods; avoidance of "overpackaging" by selection of the most economical material for the required performance; sealing ability for products prone to leakage; sealing for products that develop pressure; sealing for hazardous and highly corrosive products; sealing for products which are vacuum packed; compatability with the container and the closure; and filled product stability. Selection is typically based on field experience with a similar product, and knowledge of the properties of specific materials and their reaction to various products. Performance requirements are often appraised differently by individual packagers. The differences are based on marketing approach, merchandising practice, and the rate of product turnover and expected shelf-life. Responsibility for the selection always rests with the packager.

Liner Materials

The state of the art in liner technology can best be described as "crowded." Available choices of liners currently number in the hundreds and new structures continue to appear. Although there are more than 40 basic structures, they generally contain one or more of the following: polyethylene, PVC, PVDC, polyester, aluminum foil. Along with pulp, chipboard or newsprint backings, these five substrates, in combination with various waxes and adhesives, constitute the basic "standard" materials list. The many options for facings fall into about six major categories: coated paper; paper laminations; unsupported foil; coated foil; film/foil; and plastic film.

Each of these categories contains subgroups: paper laminations include paper/foil, paper/OPET, paper/foil/PVC, paper/PVDC, paper/foil/PE/, paper/foil/PVDC, etc. The coated-foil category includes foil-PE, foil-PVC, foil-PVDC, foil-PP, foil-ionomer, and foil-EAA. Manufacturers of lining materials generally use their own trade names and grade designations to describe structures: for example. PET-coated aluminum foil is available in the United States under at least four separate brand designations.

Extruded lining materials have gained wide acceptance in the past 10 years. These include solid monolayer (eg, LDPE, EVA) and coextruded plastics; monolayer foam plastics; and coextruded solid/foam structures. The polymers are generally either polyolefins or PVC, which provide outstanding chemical resistance and relatively low water-vapor transmission rates (WVTR). They can be tailored to density, thickness, width, color, and surface texture. The solid/foam coextrusions typically consist of solid polymeric skins on the top and bottom, with a foamed core in the middle. Most are combinations of homopolymer and copolymer polyolefins. All of the synthetic materials, when used as backings, can be combined with all of the common facings used with pulp backings.

Innerseals

Innerseals had been used for many years for barrier and leakage protection before they attracted attention as tamper-evident packaging materials (see Tamper-evident packaging). There are three types of innerseals: induction, pressure-sensitive, and glued-glassine. Heat-induction seals are particularly useful as tamper-evident devices. Potential users can select from over 50 structures. Typically, heat-induction foils are specified in one of three ways: wax-adhered (temporarily) to a separating base material; permanently adhered to chipboard; or coated unsupported foil for use in dispensing closures. The heat-induction (see Sealing, heat) foils vary with respect to foil gauge, thickness of coating/film, type and thickness of backing material, foil alloy, level of adhesion to various types of container materials, melting and sealing temperature, strength, appearance, destruction properties, and cost. Because seals applied by heat induction cannot be removed without breaking the seal, they are particularly useful as tamper-evident features.

Pressure-sensitive innerseals provide functional sealing and a measure of tamper evidence to glass and plastic containers without the need for heat induction or a secondary adhesive application. Typically, these structures consist of a foamed polystyrene (see Foam, extruded polystyrene) with a surface adhsive that seals under torque-activated pressure. Although these structures are not recommended for oils, hydrocarbons, and solvents, they do provide an inexpensive contribution to seal integrity. Glassine (see Glassine, greaseproof, parchment) the oldest of the three innerseal types, is generally wax-bonded (see Waxes) to clay-coated pulp board and adhered to the container with adhesive. Unlike heat-induction structures, the bond between glassine and pulp is broken when the consumer removes the closure.

Before the regulations for tamper-resistant packaging went into effect, glassine was commonly used in the pharmaceutical industry. With the general trend toward heat-induction foils, the largest remaining market for glassine is in instant coffee and a variety of powdered drink mixes and nondairy coffee creamers.

The performance characteristics of all closure lining materials are based on the following criteria: WVTR, gas transmission rate and blocking properties; chemical resistance; stability and compatability with the container, the closure, and the product being filled; and containment of the product.

When evaluating available structures for a specific application, the selection process does not necessarily result in an obvious solution. The primary objective should be to achieve an equitable balance between meeting the performance criteria and remaining cost-effective.

BIBLIOGRAPHY

"Closure Liners" in *The Wiley Encyclopedia of Packaging Technology*, 1st ed., by R. F. Radek, Selig Sealing Products, Inc., pp. 171–172.

General References

Technical data books with structural breakdown, FDA compliance, and suggested uses and limitations of closure liners can be obtained from suppliers such as: Cap Seal Division, 3M Co., St. Paul, Minnesota; Selig Sealing Products, Oakbrook Terrace, Illinois; Tekni-Plex Inc., Brooklyn, New York; and Sancap Inc, Alliance, Ohio.

Information about extruded and coextruded closure liners may be obtained from suppliers such as: J. P. Plastics, Naperville, Illinois and Tri-Seal International, Blauvelt, New York.

CLOSURES, BOTTLE AND JAR

The cork stopper and the continuous-thread cap represent two epochs in closure evolution. The cork stopper began its slow ascendency 25 centuries ago, attaining its broadest use by the middle of the 19th century. With the arrival of the standardized continuous-thread cap and the introduction of plastic closures, both in the 1920s, the modern closure era was underway.

Cork provided an incomparable friction-hold seal. A material of high cellular density, cork is compressible, elastic, highly impervious to air and water penetration, and low in thermal conductivity; a natural panacea for the elementary problems of closure. Historical antecedents of the cork stopper are found within the Roman Empire. The art of glass-blowing matured commercially there, resulting in a vast commerce of vases, jars, bottles, and vials. Bottles and jars were more common during the Roman Empire than at any period before the 19th century (1). The use of cork floats and buoys by the Romans, with subsequent applications as bungs (large stoppers) for casks, suggests the likelihood of fabricated cork bottle-stoppers. Yet with the fall of the Roman Empire, glass-blowing and the use of the cork stopper declined until after the Renaissance. Other sealing methods used at the same time as the rise of the cork stopper in the 16th century include a Near Eastern method of covering the container with interlaced strands of grass, or strips of linen, and applying a secondary seal of pitch. Western Europeans used glass stopples and various lids of glass and clay before the common use of cork. Wax was a very common closure, inserted into the neck and covered with leather or parchment. Raw cotton or wool, sometimes dipped in wax or rosin, was also employed, frequently covered with parchment or sized cloth, which was then bound to the neck (2).

The aftermath of the Industrial Revolution was characterized by a heightened quest for technical sophistication. Steam and, later, electric power provided quicker realization of more complex goods. The ethic of the economy of scale, "the more you produce, the less costly it is to produce it," became the momentum for the sudden explosion in manufacturing technology. During this time the world was colonized on vast scales, and population doubled in 150 years. These forces of urbanization and industrialization created an unprecedented demand for bottled goods by the 19th century. Closure evolution of this period reflects the search for a practicable seal through a variety of mechanical devices. By mid-19th century cork was the predominant closure, providing a friction seal for foods, beverages, and patent medicines.

Many attempts were made then to attune the concept of threaded closure to the demands of a new industry. A major contributor was John Mason. His 1858 patent of the Mason

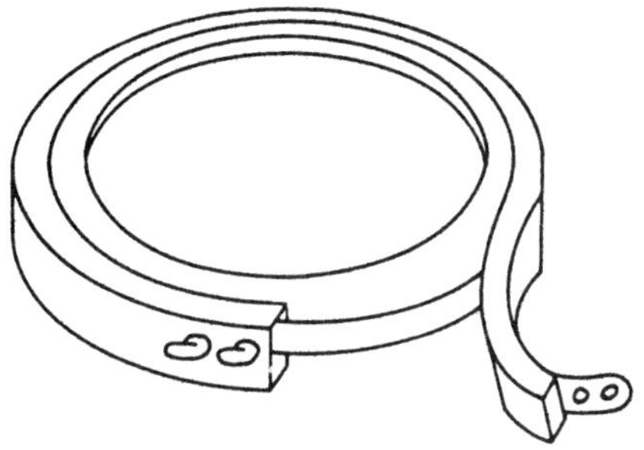

Figure 1. The Phoenix band closure.

Jar redesigned glass threads to accomplish a tighter and more dependable seal. Developing closure and container industries were to remain a chaos of varying pitches, lengths, and thicknesses throughout the century as manufacturers tried to perfect some particular feature which would require their exclusive manufacture. Intermediate to the development of the standardized continuous thread cap was the Phoenix band cap (see Fig. 1), a popular closure among food packagers because it could provide a hermetic seal in a world of imperfect finishes. Invented in France by Achille Weissenthanner in 1892, the closure provided adjustability to finishes by means of a slit-and-tongue neckband, an improvement over the original neckband closure patent of 1879 by Charles Maré of France (3). This cap also introduced the custom which later became standard in the industry: measuring closures in millimeters (mm).

By the end of the century some form of external mechanical fastener, such as Henry W. Putnam's bailed clamping device known as the "Lightning Fastener" was the leading closure for beer and ale. Internal stoppers provided the seal for most carbonated soft drinks, most notably Hiram Codd's glass-ball stopper and Charles Hutchinson's spring stopper (4).

In the early 1900s Michael J. Owens successfully automated the production of jars and bottles, which in turn created latent market demand for inexpensive, easy to use, standardized closures. The crown cap, devised by William Painter aroun 1890, provided a solution for the beverage industry. Shortly after WWI glass and cap manufacturers, through the Glass Container Association, designed and standardized the shallow, continuous-thread cap. Subsequent forces shaping today's closure were the emphasis on package styling created in the 1930s and the development of thermoplastic-molding technologies. During the 1970s, plastic closures showed a 60% market increase (5). In the 1980s the forces of consumer demand for convenience, society's need for access control, and industry's need for cost-efficient innovation continue to redefine the closure.

Closure Functions

A closure is an access-and-seal device which attaches to glass, plastic, and metal containers. These include tubes, vials, bottles, cans, jars, tumblers, jugs, pails, and drums. About 80 billion (10^9) closures are produced annually in the United States (6). The closure works in conjunction with the container to fulfill three primary functions: to provide protective containment through a positive seal; to provide access and resealability according to varying requirements of convenience and control; and to provide a vehicle for visual, audible, and tactile communications.

Protective containment. Protective containment and seal are achieved when closure and container are integrated to form a unified protection system for the product during its cycle of use. Protective containment has a two-fold meaning: containing the product so that neither the contents nor its essential ingredients escape; and providing a barrier against the intrusion of gases, moisture vapor, and other contaminants.

A positive seal. A packaged product is vulnerable to many forms of natural deterioration, including migration of water or water vapor, contamination by oxygen or carbon dioxide, and assaults by microbiological life. The packaged product is further challenged by extremes of heat and cold, dryness and humidity, and by physical stresses imposed upon it during the distribution cycle (7).

A positive seal is attained when the contact points of the closure and the top of the container (its "land" surface) are pressed together to form a seal. Frequently a resilient lining material, compressed between the closure and the container, provides a tighter, more secure seal. A liner may be made of paper, plastic, or metal foil, and is often a composite of many materials (8). A seal may also be formed by caps containing flow-in compounds where a gasket is devised by pouring a liquid sealing compound into the closure. A variety of linerless thermoplastic closures utilize molded-in sealing devices. These embossed or debossed features press against the land surface and provide a seal when the closure is applied and tightened. Sealing specifications may range from mere containment to the preservation of highly sensitive food, pharmaceutical, household, and industrial products. Three common types of seal applications are sterilized, vacuum, and pressurized.

Two closure methods provide containment and seal: friction-fitting closures, including snap-ons, stoppers, crowns, and press-ons; and thread-engagement closures, including continuous-thread and lug caps. A positive seal depends upon such factors as the type of product, closure, container, and seal desired, the resiliency of the liner, the flatness of the sealing surface, and the tightness or torque with which the closure is applied (9).

Access. Contemporary closure design is shaped by the demands of a pluralistic marketplace where strong consumer preferences for convenient access exist alongside legal mandates for access control. Many packages today are ergonomically designed systems capable of easy opening and dispensing, and also affording critical access control. Closure technology has always sought to provide "a tight seal with easy access", but today's simultaneous demands for easy access and access control are the most polarized in the industry's history. Access to a product can be said to exist on a continuum of convenience. This may range from the knurls (vertical ribbing) on the side of a continuous-thread cap, designed to provide assistance in cap removal, to what can be called convenience closures, which have a variety of spouts, flip-tops, pumps, and sprayers to facilitate easy removal or dispensing of the product.

Control. Concurrent to greater demand for convenient, often one-handed, access to a product, legal mandates and consumer preferences press for more access controls. These access controls are of two major types: tamper-evident (see Tamper-evident packaging) and child-resistant (see Child-resistant packaging). Regulated tamper-evident (TE) closures may be breakable caps of metal, plastic, or metal/plastic composites. In one variety the closure itself is removable but a TE band remains with the neck of the bottle. In another, the TE band is torn off and discarded. Another system not specifically addressed in the regulations incorporates a vacuum-detection button on the closure. Other TE systems include paper, metal foil, or plastic innerseals affixed to the mouth of the container. The FDA has stated 11 options for making a package tamper-evident, two of which apply to closures (10). Child-resistant closures (CRCs) are designed to inhibit access by children under the age of five. This is frequently accomplished through access mechanics involving a combination of coordinated steps which are beyond a child's level of conceptual or motor skills development. Of these closures, 95% are made of plastic; the remaining 5% combine metal with plastic (11).

Verbal and visual communications. The closure is a focal point of the container. As such, it provides a highly visible position for communications, an integral aspect of today's packaging. Three communication forms include styling aesthetics, typography, and graphic symbols. Since the closure is handled and seen by the consumer every time the product is used, the audible, visual, and tactile message (often subconscious) becomes very important to the packager.

Styling aesthetics. Aesthetics are an important consideration because package design has the same basic goals as advertising: to promote brand awareness leading to brand preference. The closure and the container provide a visual symbol of the product, creating imagery through aspects of styling. Three important aspects are form, surface texture, and color. The form of a closure can be utilitarian to suggest value, or it can assume elaborate and elegant forms to suggest luxury. The surfaces of glass, metal, and plastic can provide a variety of surface textures unique to the materials used. Metal caps and decorative overshells of steel, aluminum, copper, or brass, can be burnished, painted, screened, or embossed. Plastics can be molded in vivid colors, anodized to assume metallic sheens, or printed, hot-stamped, screened, or embossed. Glass can provide the kind of design statement exclusive to the glass arts, creating imagery of luxury or elegance. Many closures today are styled simply, with brand-identification or functional embossments (eg stacking rings) appearing on the closure top.

Color is the most pervasive form of closure decoration. A closure may be purely functional in form yet, with color, it can take on dramatic significance. With the advent of color-matching systems in industry, such as the Pantone Matching System (Pantone, Inc.), the closure, container, label, and point-of-purchase display can be coordinated to produce a strong emotional reaction. The emotion may be one of action and excitement, as the hot primary colors used in soap and detergent packaging, or cool and subdued colors that characterize many cosmetics and fragrances. In addition to its decorative aspects, color can provide functional assistance. Color contrast directs the eye to areas of emphasis, and this direction can be important in teaching the mechanics of container access and use. In a crowded environment of dispensing options found at point-of-purchase, color can help identify a closure as one that the consumer already knows how to use. a great many dispensing closures today, for example, differentiate the spout from its surrounding fitment by strong color contrast. Closure color can also be used to identify the flavors of a food product or beverage and help to differentiate these flavors quickly within a product line.

Typography. Common forms of written communications found on a closure may include brand identification, a listing of ingredients, nutritional information, access instructions, or consumer advisories. These can be printed, screened, hot stamped, or molded onto a closure. For purposes of impact at point of purchase, a brand name frequently appears on the closure top.

Graphic symbols. A graphic symbol frequently found on the closure is a company or product logo. Another common graphic is the arrow, a symbol which has gained importance with the advent of safety, convenience, and control mechanisms of modern closures. Arrows direct the consumer to proper disengagement of the closure, indicate engagement points where access is possible, or signify the direction of dispension (eg in the control tips of spray-type mechanisms). The scannable bar code is a more recent functional graphic to appear on closures (see Code, Bar) (12).

Methods of Closure

Removable closures attach to containers by two principal methods: thread engagement and friction engagement. Threaded closures include continuous-thread caps, lug caps, and metal roll-on caps. Friction-fit closures include crowns, snap-fit, and press-on types. Thread engagement is the most widely used method of attaching a closure to a container (13).

Thread-engagement types. Three closure types provide a seal through thread engagement: continuous thread (CT), lug, and roll-on caps. CT designs attain a seal through the attachment of a continuously threaded closure to a compatibly threaded container neck. The lug cap uses an abbreviated thread design, with access and reclosure accomplished in one-quarter turn. The roll-on is supplied as a blank unthreaded metal shell which then becomes a closure on the capping line when it is compressed to conform to the finish of a bottle. The press-twist closure also has its threads formed after it is applied to the container.

The CT Closure. Threaded closures were standardized in the 1920s and continue to prosper due to the basic soundness of their principle, which offers a mechanically simple means of generating enough force for effective sealing, access, and resealing (14). Today the CT design is manufactured in plastic, tin-free steel, tin plate, and aluminum. Some CT control-closures combine metal and plastic by using a regular CT metal cap and a plastic overshell. The CT closure provides a seal for the container by engagement of its threads with the corresponding threads of the container (see Fig. 2). As the thread structure, (or "finish"), is designed on an inclined plane, the engagement and application torque cause the threads to act like the jaws of vise, forcing closure and container into contact to form a positive seal. Critical sealing applications typically include a liner placed between the closure

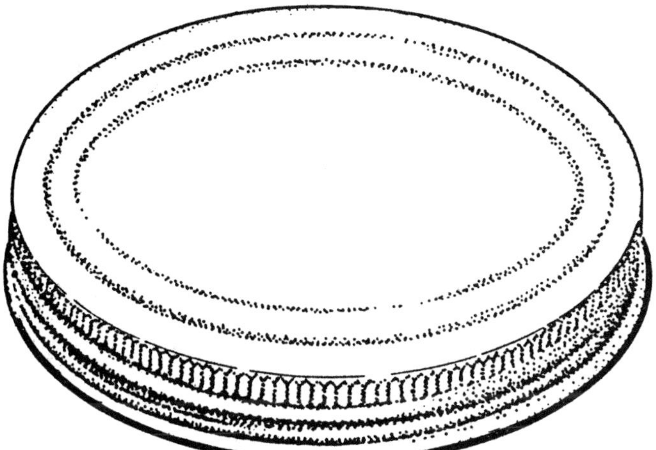

Figure 2. A metal CT closure.

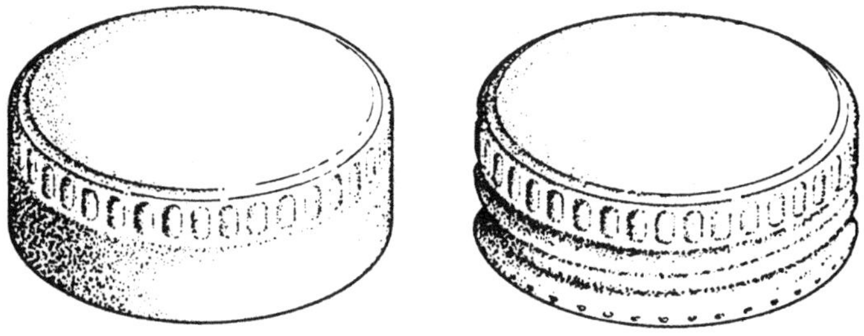

Figure 4. Roll-on closure.

and the container. When tightened, the material is compressed between the sealing surfaces to form a seal. The thermoplastic "linerless closure" employs a variety of molded devices which can also provide a positive seal. All CT closures are designated by diameter as measured in mm followed by the finish series number. A closure with the designation number "22-400" refers to a closure with an inside diameter of 22 mm designed in the 400 finish of a shallow continuous thread.

Lug cap. The lug cap operates on the principle of thread engagement as does the CT cap. The thread structure of the lug is not continuous, but consists of a series of threads which may be inclined or horizontal depending upon finish variations. The lugs of the cap are positioned under corresponding threads on the container finish (see Fig. 3). When tightened, the engagement of two, four, or six lugs pulls the closure and lining material onto the container. The lug cap is the most popular steel vacuum-closure today (15). Frequently a flow-in plastisol inner gasket is used as the liner. The lug cap is used extensively for vacuum packs in the food industry, and is suitable for use on many products packaged in glass containers. The lug cap's design allows application and removal with a one-quarter turn. This not only means consumer convenience, but also quick capping. The common finish designations for lug caps are 120, 140, and 160 referring to 2, 4, and 6 lug finishes respectively (16).

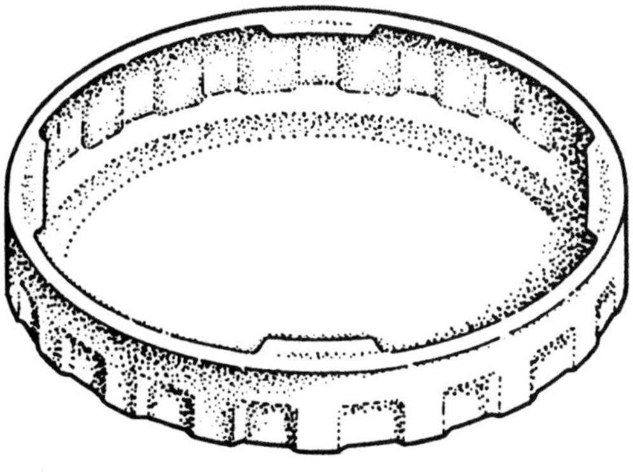

Figure 3. Lug closure.

Roll-on. The aluminum roll-on cap was an innovative method of closure that won immediate acceptance within the packaging industry in the 1920s (17). Although frequently categorized separately from threaded closures, the roll-on nonetheless utilizes thread engagement to accomplish seal and reseal. What makes this closure unique is the capping process. A lined, unthreaded shell (or "blank") is furnished to the packer. During capping the blank is placed on the neck of a container and the capping head exerts downward pressure, which creates a positive seal as the liner is pressed against the container finish. Next, rollers in the capping head shape the malleable aluminum shell to conform to the contour of the container thread (see Fig. 4).

The roll-on closure is used in the food, carbonated beverage, and pharmaceutical markets where pressure sealing is required. It is considered one of the most versatile sealing devices for normal, high and low pressure seals. A widely used version today is the tamper-evident roll-on. A tear band perforated along the bottom of the closure skirt is tucked under a locking ring during capping by a special roller. When opened by the consumer, the band separates from the closure to provide visible evidence of tampering. The roll-on cap can be applied at high speeds, approaching 1200 bottles per minute. In the standard bottle finishes, the 1600 series designates roll-on.

Press-twist. The "press-twist" is another closure which attains its threads on the capping line (see Fig. 5). Primarily used for baby foods it also provides closure for sauces, gravies, and juices. Applied in a steam atmosphere, the plastisol side gasket in the heated cap forms thread impressions when pressed against the glass finish. As it cools, permanent impressions are formed in the compound so the cap can be twisted to open and reseal similar to a CT cap.

Friction-fit closures. Many bottles are sealed with simple metal or plastic closures that are pressed onto the top and held in place by friction. The four basic types of friction-fit closures are crowns, snap-fit caps, press-on caps, and stoppers.

Crowns. The crown beverage cap was a major innovation in friction-fitting closure. It has been widely used since the turn of the century for carbonated beverages and beer. Crowns are made of tin-free steel and tin plate. Matte-finish tin plate is used for soft drinks and a brighter finish is used for beer. The crown has a short skirt with 21 flutes which are crimped into locking position on the bottle head. The flutes

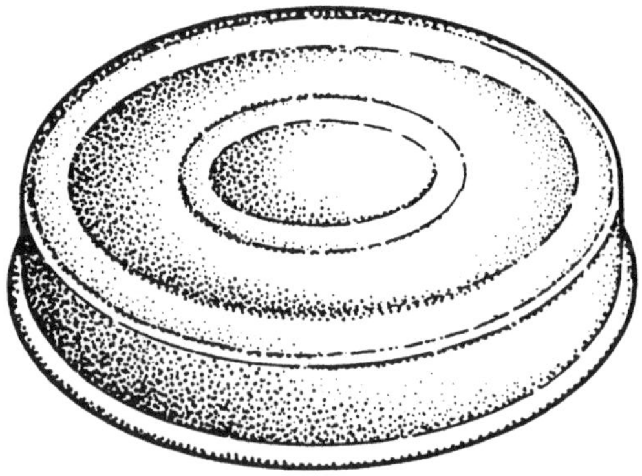

Figure 5. Press-on/twist-off vacuum closure.

are angled at 15° in order to maintain an efficient seal (see Fig. 6) (18).

The crown contains a compressible lining material, which over the years has included solid cork, composition cork or plastic liners, and foil and vinyl spots. Though simple in concept, the crown provides a friction-fit sufficient to seal pressurized beverages. The flared cap skirt in conjunction with the smoothness of the bottle neck provides easy access through the prying motion of the bottle opener. As convenient access came to be demanded in the marketplace, easy-open crowns were developed. The twist-off beverage crown first appeared on beer bottles in 1966 and gained in popularity (19). Designed for convenient access, it could be twisted off by hand or removed with a bottle opener. Bottlers switched to this cap because no special capping equipment was required for the new closure. In 1982 many companies moved away from this cap because of potential tampering. The crown is currently capable of application speeds exceeding 1000 units a minute.

Snap-fit caps. The snap-fit caps are simple lids that can be pressed onto the tops of bottles. They may be held in place by the friction of a tight fit or by supplementary flanges, ridges, or grooves that grasp the bottle finish (20). The skirts of some metal snap-fit caps are rolled under to form a spring action as the cap is pushed against the lip of the bottle. Some types of snap-fit caps have ridges on the inside of the caps that match the grooves in the ridges on the bottle. The simplest form is designed to fit so snugly that the friction between the surfaces of the cap and bottle is sufficient to hold the cap in place. Snapfit caps may be made of metal or plastic. They are used for such food products as jellies and for over-the-counter medicines such as headache remedies. An important variation is the press-on TE closure, frequently used on milk and juice products. When these caps slide over a ridge near the bottom edge, they fit so tightly that they cannot be pulled off. The bottle is opened by pulling a tear tab located above the ridge. This separates the top of the cap from the bottom portion that was locking it in place.

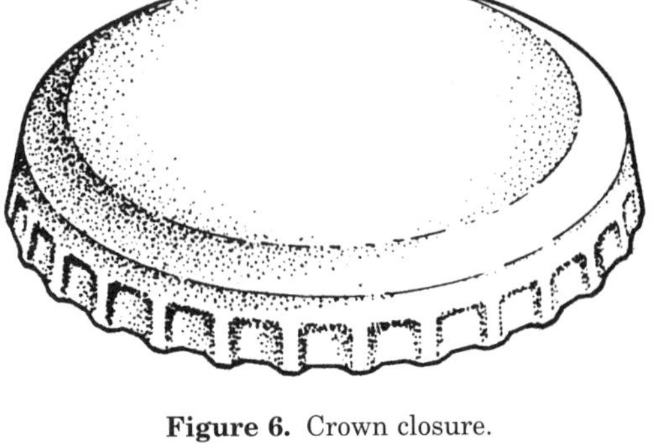

Figure 6. Crown closure.

Press-on vacuum caps. Sealing for the press-on vacuum closure is obtained from atmospheric pressure when air is withdrawn from the headspace of the container by steam or mechanical sealing methods. Sometimes further security for this type of seal is provided when the edge of the cap is forced under a projection of the glass finish or held under a similar projection by snap lugs. In many vacuum caps, it is the pressure of a gasket against the top of the container finish which provides a seal. In others, atmospheric pressure alone is adequate. A recent press-on variation is the composite cap, a gasketed metal disk and plastic collar used in sealing dry-roasted nuts and seafood products. Its plastic collar is used as a TE device.

Types of Closure

It is impossible to place all closures into clean-cut categories where there is no overlap of functions. Yet despite these limitations a classification can provide focus for understanding contemporary closure trends. As defined by their utility, the four classes of contemporary closures are: containment, convenience, control, and special purpose.

Containment closure. Though all closures provide containment, a containment closure is here defined as a one-piece cap whose primary function is to provide containment and access on vast production scales. CT caps (for general-purpose sealing), crowns and roll-ons (for sealing of pressurized beverages), lug and press-on caps (for vacuum sealing of foods) are within this class of containment closure.

Convenience closure. Closure development in recent years has been in response to consumer preferences for convenient access to the product. Convenience closures provide ready access to liquids, powders, flakes, and granules for products that are poured, squeezed, sprinkled, sprayed, or pumped from their containers. There are five types of convenience closures: spout, plug-orifice, applicator, dispensing-fitment, and spray and pump types.

Fixed-spout closures. A spout is a tubular projection used to dispense liquid and solid materials. It may be fixed or movable, and capable of dispensing a product in a wide ribbon or a fine bead depending on size and configuration of the orifice. Fixed-spout caps incorporate a cylindrical or conical projection into the center of a threaded or friction-fitting closure. Spouts on reusable containers are often sealed by a small sealer tip on the end of the spout. On some sealed spouts, dispensing control can be attained by cutting the spout at various heights, thereby providing different orifice sizes. A more contemporary form of fixed-spout closure is molded with a smaller sealed spout on the top of the cap. Called "snip-tops", they are one of the most inexpensive forms of dispensing clo-

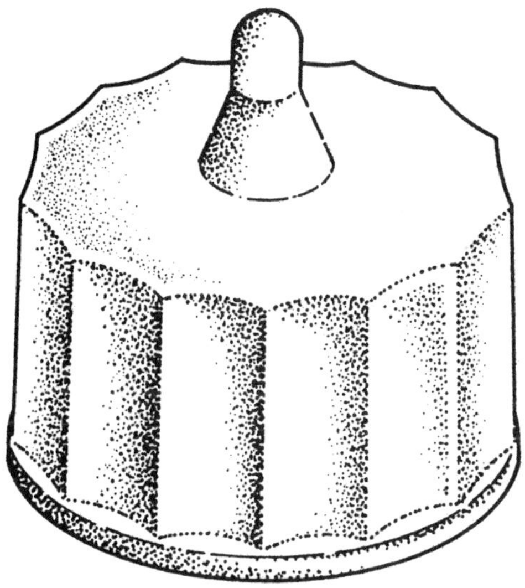

Figure 7. Snip-top closure.

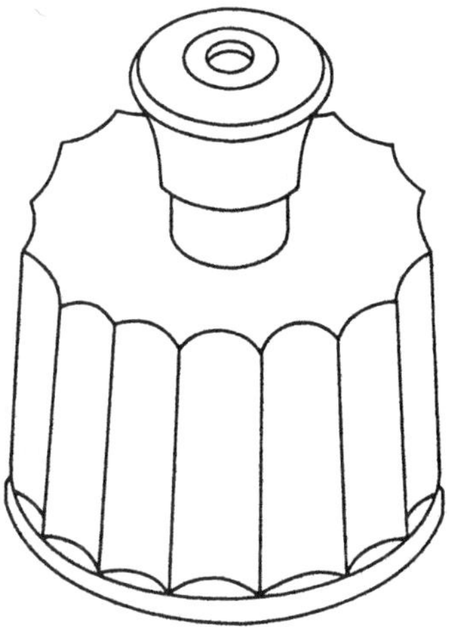

Figure 9. Push-pull closure.

sure (see Fig. 7). Contemporary "dripless pour spouts" have recently been introducted to provide "No-Mess" dispensing of viscous, sticky products packaged in large containers, such as liquid detergents and fabric softeners. Large pour spouts are protected by screw-on overcaps which double as measuring caps. Upon reclosure, the measuring overcap is designed to drain residual product directly into the container through the spout.

Movable-spout closures. Also referred to as turret, swivel, or toggle types, the movable-spout concept features a hinged spout which can be flipped into operating position and reclosed with the thumb alone to provide one-handed access and reseal (see Fig. 8). Most movable spouts are two-piece constructions, though a one-piece swivel spout design requiring one manufacturing operation has recently been introduced. Newer refinements of this type include the incorporation of a tear band across the spout to provide tamper evidence. Valve-spout closures, such as the "push-pull" closure, are opened and closed in a straight-line, vertical fashion (see Fig. 9). "Twist spouts" employ a tapered flange design and open and close by a twisting motion.

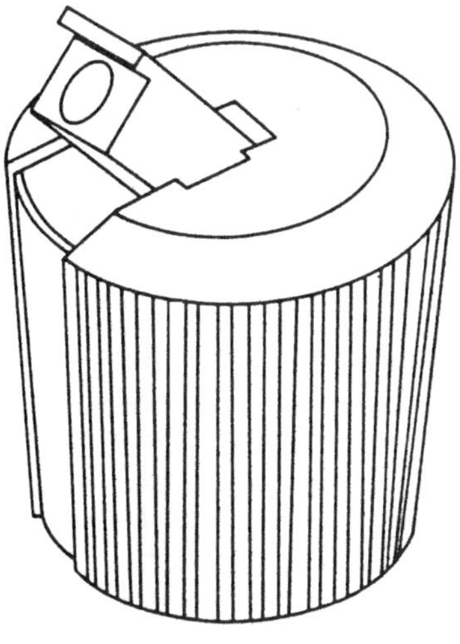

Figure 8. Flip-spout closure.

Plug-orifice closures. These closures first aided in the dispensing of personal-care and cosmetic products, and are now used in conjunction with multilayer high-barrier plastic bottles for convenient dispensing of food products (see Multilayer plastic bottles). To some, the hinged-top designs represent the wave of the future in food packaging (21). The closure consists of a dispensing orifice incorporated into a screw-on base closure, and a plug, or "spud", hinged within the top of the closure or molded into a flip-up hinged cap (see Fig. 10). In the polypropylene plastic versions, the plug and orifice provide a friction-fitting seal which produces an audible "snap" when engaged, an instance where a closure can communicate its sealed state by sound. The top of a "snap-top hinged closure" swings open on two or three external hinges. The "disk closure" is another plug-orifice type, a two-piece design consisting of an orifice closure base and a plug fitment hinged to a round disk which is set into the closure top. By pressing upon the access point, the disk fitment swings up upon its hinges, deactivating the plug seal from the orifice for one-handed dispension. Some of these designs also produce audible "snap" upon engagement and disengagement.

Applicators. There are many different kinds of convenience applicators, many specialized for particular product applications. Four major types are brushes, daubers, rods, and droppers. Brush caps range from small cosmetic brushes to large applicators used for applying adhesives. Sponge, cotton, felt, or wool pads affixed to applicator rods are used to apply a wide variety of household and cosmetic products and are known as dauber caps. Glass and plastic rods are used in the drug and cosmetic industries, such as the balled-end rod used to apply medicines. Glass and plastic droppers, with straight, bent, and calibrated points, are frequently used to provide precise dosages for medicinal products. The three components of the dropper are an elastomeric bulb, the cap, and the pipette.

Fitment closures. Fitments and fitment closures are designed to regulate the flow of liquids, powders, flakes, and granules. Fitments are inserted into the neck of the container

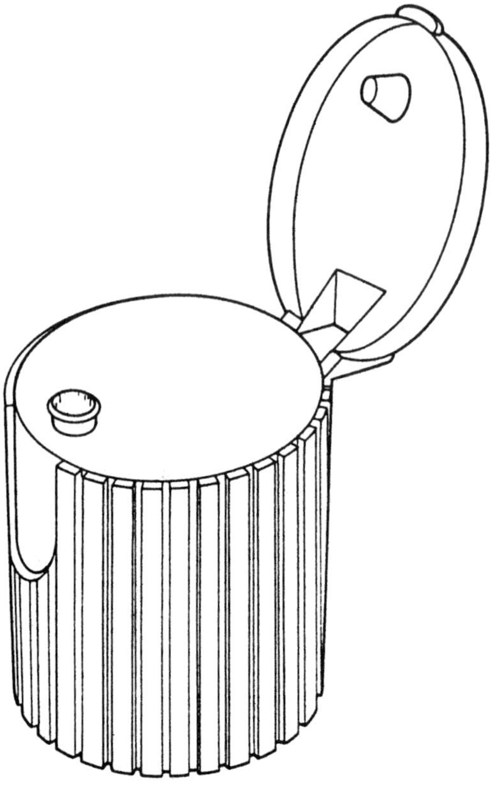

Figure 10. Hinged plug-orifice closure.

or are permanently attached. Fitment closures incorporate regulating devices into screw-on or press-on caps. Those which plug inside the neck finish include dropper and flow-regulating fitments. "Dropper tips", used with squeeze-type bottles, dispense liquid in increments of one drop and are usually covered with a protective overcap. "Pour-out fitments" control the splashing of liquids by retarding their flow, a frequent problem when precise, small-volume pouring is required from cumbersome containers. Those fitments or closures which regulate solid materials include sifter and shaker designs. Most contain a number of sifter holes in which powdered material can be dispensed evenly. Others incorporate options which permit the material to be shaken-out, poured-out, or spooned. "Shaker caps" and "powder sifter caps" dispense powdered or granular products, sometimes incorporating revolving fitments which provide containment. A wide variety of sliding panels or hinged covers provide the consumer with sprinkling, pouring, or spooning options. Many variant designs combine spouts, sifters, dial disks and sliding covers.

Spray and pump dispensers. "Regular sprayers" operate on a basic pump-principle and function with a piston, accumulator, or cylinder. Dispensing a heavier particle size, these sprayers are often used for household and personal-care products. Generally, the smaller the orifice size, the finer the spray pattern. "Fine mist sprayers" dispense in finer particle size as required by some personal-care products. "Trigger sprayers" are larger and more complex in design, offering convenient dispensing for large volumes of liquids. The bulb-and-piston-drive trigger-sprayer units emit spray patterns ranging from a fine mist to a stream. The amount of product delivered by piston-driven "pump dispensers" depends upon its viscosity; the more viscous, the more strokes to prime and the lower the output per stroke (22). Regular dispensing pumps dispense in volumes from less than 0.5 cm^3 to slightly more than 1 cm^3 per stroke in water. Large-volume dispensers range in capacity from ⅛ oz (3.7 mm) to 1 oz (29.6 mm) per stroke.

Control closure. The first "clerkless" food store appears to have been opened in 1916 by Clarence Saunders. By 1930 there were 3000 of them, soon to be known as "supermarkets" (23). These stores raised new problems of hygiene as more and more products were available in unit packs that had to be capable of withstanding repeated handling, attempts at sampling, and occasional malicious intrusion. The need for consumer product safety grew along with this new concept in food retailing, a need which would become a matter of increasing concern for the U.S. Congress as the number and variety of products increased. Among the legal mandates developed to protect the public against harmful substances are those which specify access controls for containers.

Tamper-evident closures. Tamper-evident caps have been in use for years, though earlier they were referred to as "pilferproof caps". Today these metal and plastic caps provide visible evidence of seal disruption and are used for over-the-counter (OTC) drugs, beverages, and food products. The two kinds of TE closures are "breakaway" or "tear band" closures used for pressurized and general sealing applications, and TE vacuum designs for vacuum-sealing applications. The closure user can also fulfill tamper-evident requirements through the use of innerseals which cover the container mouth (see Tamper-evident packaging).

Breakaway caps. In 1982 the FDA established requirements for tamper-evident packaging for over-the-counter (OTC) drug products. The agency defined such packaging as "having an indicator or barrier to entry which, if breached or missing, can reasonably be expected to provide visible evidence to consumers that tampering has occurred" (24). The FDA did not issue rigid standards of compliance but instead listed 11 suggested and approved methods from which a packer may choose. Regulation #6 describes paper or foil bottle seals covering the container mouth under the cap as a means to provide tamper-evidence. Regulation #8 describes breakable caps as another option. The FDA defined these caps as being plastic or metal that either break away completely when removed from the container or leave part of the cap on the container. The breakaway cap is the most common form of tamper-evident closure. With other options, such as shrink bands and strip stamps, a packager needs additional operations and equipment to achieve tamper evidence. Two forms of tamper-evident caps are mechanical breakaway and tear bands.

Mechanical breakaway. These are threaded caps with perforations along the lower part of the skirt which form a "break line" in the closure (see Fig. 11). When the closure is twisted for removal, the band, which is locked to the finish by crimping or rachets, separates from the closure along the break line. The cap is removed and the lower part of the skirt remains on the container neck. The breakaway cap can be efficiently applied, is highly visible, familiar to consumers, and is durable enough to maintain its integrity throughout distri-

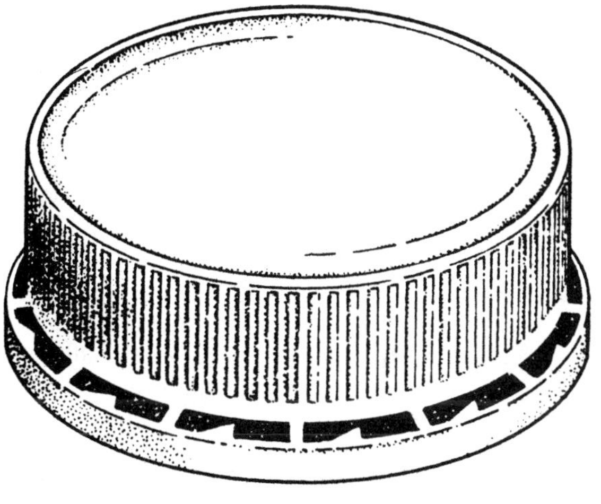

Figure 11. Mechanical breakaway closure.

bution. Metal closures of this type frequently crimp the band to the container neck for a friction hold. Variations in this type of TE closure include different band designs and methods of off-torque resistance; for example, ratchets on the band that lock to mated protrusions on the finish. TE bands on some plastic closures are shrunk by heat to form a tight fit around the container neck.

Tear bands. These types, frequently called tear tabs, employ a locked band to prevent cap removal (see Fig. 12). Access is accomplished by completely removing the band from the container. Frequently a protruding tab is evident for the consumer to grasp and commence tearing. Many non threaded TE closures utilize this type, such as the press-on friction fit closures found on milk containers. The closure is removed by tearing off the lower skirt, which overrides a bead on the container finish. Most of the removable-band types are made of plastic, usually polyethylene.

TE vacuum caps. Marketing leverage, rather than legal mandate, accounts for the expansion of TE into food packaging (25). These measures are not referred to as "tamper-evident" in label or closure communications, but are placed in a more positive light, such as "Freshness Sealed" or "Safety Sealed". The two major types of TE vacuum closures are vacuum button and vacuum tear-band caps.

A popular TE option for food products packaged in glass containers under vacuum is the "button-top closure". These include lug versions used for jellies, sauces, and juices, and the threaded-seal version popular with the baby-food industry. A safety button, or coin-sized embossment on the top of the cap, pops-up as the jar is opened and its vacuum is lost. Accompanying this is the "pop" which serves as audible evidence of an undisrupted seal. When capped the embossed button is held down by vacuum pressure, providing the consumer with visual evidence that the container has not been opened. Another type, the "vacuum tear-band closure", is a two-component closure used for the packaging of nuts and condiments. It consists of a metal vacuum lid inserted into a plastic tear-band closure skirt. Protrusions molded into the plastic collar provide friction-fitting resealability for the container.

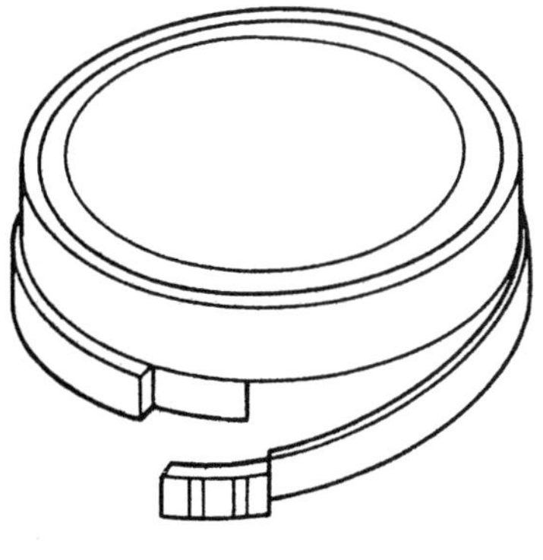

Figure 12. Tear band.

Child-resistant closures. Alarmed by the increasing number of children being harmed through accidental poisoning, Congress acted in 1970 to pass the Poison Prevention Packaging Act. This act, Public Law 91-601, established mandatory child-resistant closures for rigid, semirigid, and flexible containers of such compounds as the Consumer Product Safety Commission (CPSC) deemed dangerous to children (see Child-resistant packaging). Studies revealed that the accidental-poisonings curve in children peaked at 18–24 mo, but the manual dexterity curve increased with age, peaking in 4–5 yr olds. Therefore, protocol testing requires testing of children aged 42–51 mo. Four major premises were considered in the development of child-resistant closures (26): children 42–51 mo of age could not perform two deliberate and different motions at the same time; children of that age could not read, nor could they determine alignments, but they can learn quickly by watching; children are not as strong as adults, but through ingenious use of teeth, table edges, or other tools around them, their persistence would give them leverage to make up for their strength; and although their hands and fingers are smaller than adults, childrens teeth and fingernails are thin and sharp and can slide under and into gaps. Childhood deaths involving all household chemicals have declined 75% since the first regulation under the act was passed. Packages designed to protect children, however, have come under fire for restricting access by the handicapped and the elderly. Most OTC drug packagers agree that CR devices are generally more burdensome than TE devices. Efforts are underway to develop new CR closures.

Child-resistant measures are defined by the CPSC as packaging that is designed to be significantly difficult for children under five years of age to open within a reasonable time, yet not difficult for adults without overt physical handicaps to use properly. Under current regulations, a package fails to be child-resistant if more than 20% of a test panel of 200 children are able to gain access, or if more than 10% of a test panel of 100 adults are unable to open and properly resecure the test package (27). The three most frequently used child-resistant closure types are press-turn, squeeze-turn, and combination-lock. The "press-turn" cap is removed by applying downward force while simultaneously turning the cap. "Squeeze-and-turn" caps employ a free-rotating soft-plastic overcap which engages an inner threaded cap or disengages a locking mechanism when sidewall pressure is applied (see Fig. 13).

The "combination-lock" caps use interrelated components formed into the cap which must be oriented before the cap can be removed. A common low cost variety of this closure is the one-piece "line-up, snap-off cap" (see Fig. 14). A slight

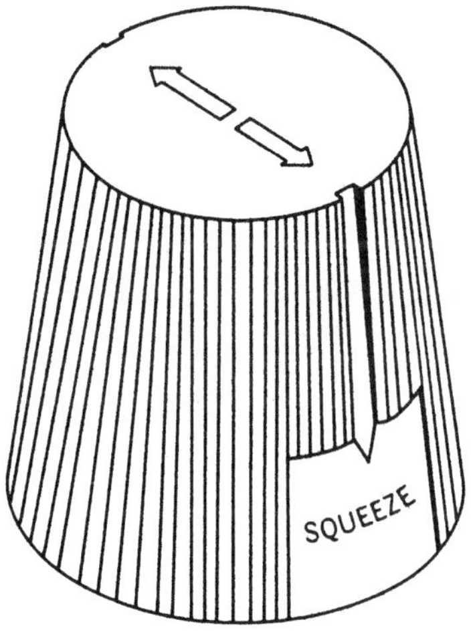

Figure 13. Child-resistant "squeeze-and-turn" closure.

interruption of the thread on the container serves as an engagement point for these caps. A protrusion on the cap fits under the single thread. When the cap is turned it becomes "locked" onto the container. As with TE closures, many packagers add CR closures to packages not required to have them, even though the cost may generally add 1–5% to the cost of the package.

Special-purpose closure. Special-purpose closures are those which are of specialized application or premium design. These include aesthetic closures, special-function closures, stoppers, and overcaps.

Aesthetic closure. The aesthetic closure is an important sales-promotion aspect of the package. It is designed to communicate clearly and powerfully by imagery. Original private-mold glass stoppers used in fragrance bottles, some with lavish sculptural representations, are an example of this type. These are frequently the most expensive forms of closure.

Special-function closures. These closures serve a specialized function in the marketplace. There are, for example, closures that vent containers which sustain a pressure build-up. Such pressures can cause the container to rupture, or violently expel the product when the cap is removed. Since venting closures can leak when the container is not in an upright position, they must be used in controlled circumstances. Many manufacturers require a "hold-harmless" agreement as a condition of sale for such closures. Another special-function closure is the twist-off closure for injectables, a two-piece aluminum cap used on parenteral vials.

Stoppers. The wine and champagne industry is the largest user of stoppers. Cork stoppers are standardized by size and grades, the latter according to the degrees of product vintage (28). Stoppers of natural rubber, synthetic silicone rubbers, and thermoplastic materials provide closure in some chemical and biological applications. Rubber plug closures are crimped onto ampules with metal bands and allow for the insertion of a hypodermic needle in medical uses.

Overcaps. The overcap is a secondary cap designed to protect the primary closure, dispenser, or fitment of a container. Metal or plastic overcap designs attach to the container by friction-fit or thread engagement, and are used to protect aerosol and dispensing fitments. Overcaps frequently double as measuring caps for mouthwash, liquid detergents, and fabric softeners.

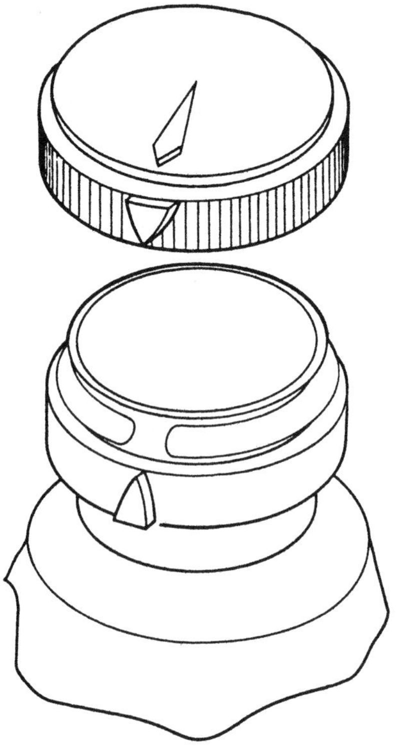

Figure 14. Child-resistant "snap-off closure.

Sealing Systems

Though often the smallest aspect of a package, the seal is responsible for keeping the entire concept intact. If the seal is not maintained by the closure, liner, and container working together, the success of the product is at stake.

Liners. Today's lining material is either a single substance (usually paperboard or thermoplastic) or a composite material. Synthetic thermoplastic liners include foamed and solid plastics of varying densities. A composite lining material consists of a backing and a facing. The backing, usually made of cellulose or thermoplastic, is designed to provide the proper compressibility to affect the seal and proper resiliency for resealing. Facing materials, representing the side of a composite liner that comes into direct contact with the product, are numerous, as are the variables of product chemistry with which they must contend. Generally, facing materials are thermoplastic-resin-coated papers, laminated papers of foil or film, or multilayer types devised for special applications (see Closure liners).

Innerseals. The innerseal affords TE protection by sealing the mouth of the container. Three common types are inserted by the closure manufacturer into the cap (29). A waxed-pulp backing and glassine innerseal is common within the food industry. After the filling operation the container runs under a roller system which applies an adhesive to the lip of the container, and then the cap is applied. Upon removal, the glassine adheres to the container while the pulp backing remains in the closure. Pressure-sensitive innerseals, generally a foamed polystyrene, adhere to the lip upon application and require several hours to set. Heat-induction innerseals are plastic-coated aluminum foils, often adhered to a waxed pulp base liner. After the cap is applied, the container passes under an electromagnetic field which causes the aluminum to generate heat. The plastic facing on the aluminum subsequently melts and adheres to the container.

Linerless closures. Plastic linerless closures provide a positive seal in certain circumstances, foregoing the need for intermediary materials and secondary liner-insertion operations. To many packagers, the cost savings provided by the linerless closure can be considerable. The seal of a linerless closure is achieved by molded embossments forming diaphragms, plugs, beads, valve seats, deflecting seal membranes, or rings which press upon, grasp, or buttress the sealing surfaces of the container. Over a dozen types of linerless closures are in common use, each designed to provide a seal at one or more critical sealing surfaces of the container, which include the land surface, the inside edge of the land surface, or the outside edge of the land surface. Some form of land seal in conjunction with a valve or flange represents one type of effective linerless closure design. The land is typically the most consistent sealing surface. A land-seal ring can bite into plastic container finishes or deflect on glass finishes. An inner buttress can correct ovality problems in plastic containers by forcing such off-round finishes back into proper shape.

Closure Materials

Closures are made of plastics, metal, or glass.

Plastic closures. Molded plastic closures are divided into two groups: thermoplastics (eg, polyethylene, and polypropylene, and polystyrene), and thermosets (eg, phenolic resins and urea components) (see Polymer properties). Thermoplastic materials can be softened or recycled by heat; thermoset materials cannot be recycled once they are molded.

Thermoplastics. In general, thermoplastic closures offer the packager light weight, versatility of design, good chemical resistance to a wide range of products, and economical resins and manufacturing processes. Their relative flexibility is essential to contemporary closure design with its emphasis on convenience and control devices. Thermoplastics provide good application and removal torque. They maintain a good seal and tend to resist back-off. Unlike thermosets, thermoplastics can be pigmented in the full-color spectrum in strong, fade-resistant intensities.

Most thermoplastic closures are produced by injection molding (see Injection molding), although some are made by thermoforming (see Thermoforming). Polypropylene and polyethylene account for about 90% of all thermoplastic closures.

Polypropylene. Polypropylene (see Polypropylene) has unusual resistance to stress-cracking, an essential characteristic of hinged closures. In thin hinged sections, it has the quite remarkable property of strengthening with use. The homopolymer has limited impact resistance, but it can be modified for better performance. It has excellent resistance to acids, alkalies, oils and greases, and most solvents at normal temperatures. It has the best heat resistance of the polyolefins, with a high melting point suitable for sterilized products, but it becomes embrittled at low temperatures. Polypropylene has better printability than polyethylene but both are inferior to polystyrene or thermosetting plastics in that respect. As a relatively rigid molded material it has outstanding emboss and deboss potential for closure communications.

Low density polyethylene. LDPE (see Polyethylene, low density) is resilient and flexible. It is relatively tasteless and odorless, although some organoleptic problems are more prevalent with LDPE than with polypropylene. It provides outstanding moisture protection, but it is not a good gas barrier. LDPE's economy as a closure material is provided by low-cost resins and relatively short injection-molding cycle times. Though it is considered to have good resistance to stress cracking, problems may occur in the presence of certain chemicals such as detergents. Communications embossments or debossments are good but limited by the softness of the material.

High density polyethylene. Compared to LDPE, HDPE is stiffer, harder, and more impermeable (see Polyethylene, high density). It is tasteless, odorless, and impact-resistant, but it will stress-crack in the presence of some products such as detergents unless it is specially formulated. Its heat resistance and barrier properties are superior to LDPE. HDPE resin is more expensive than LDPE, but it is still considered a relatively low-cost material. A particular drawback to HDPE closures is a potential for warpage and loss of torque.

Polystyrene. Polystyrene (see Polystyrene) is used for about 10% of the closures produced today. Polystyrene homopolymer is attacked by many chemicals, is very brittle, has relatively low heat resistance, and does not provide a good barrier against moisture or gases. Many of the disadvantages of polystyrene are overcome by rubber modification and/or copolymerization.

Thermosets. Phenolic and urea compounds have a wide range of chemical compatibility and temperature tolerances. Some thermosets can sustain sub-zero temperature without embrittlement, and survive at temperatures higher than 300°F (149°C) (30). The density and rigidity of thermosetting plastics give the material its heavy weight and guard against slippage over threads, a problem with softer thermoplastics such as LDPE. Thermosets cannot provide the color range or intensity of thermoplastics, but they accept vacuum metallizing decoration in silver and gold with superior adhesion qualities. During the molding process, thermosets undergo a permanent chemical change and cannot be reprocessed as thermoplastics can. Thermoset closures are manufactured by compression molding (see Compression molding). Cycle time for thermosets is generally longer than thermoplastics, (30–120 s), depending upon thickness of the product and additives.

Phenolics. Phenol–formaldehyde closures are hard and dense. They are the stiffest of all plastics, but are relatively brittle and low in impact strength. The properties of phenolics depend to a large extent upon the filler material used. Wood flour improves impact resistance and reduces shrinkage. Cotton and rag fiber additives increase the impact strength; asbestos and clay additives improve chemical resistance. Phenolics are resistant to some dilute acids and alkalies and attacked by others, especially oxidizing acids. Strong alkalies will decompose phenolics, but they have excellent solvent resistance. Their heat resistance is outstanding. Phenolics cost less than ureas, and are easier to fabricate, but they are limited in color to black and brown unless decorated.

Urea. Urea–formaldehyde is one of the oldest plastic packaging materials, first used in the early 1900s. The resin produces extremely hard, rigid closures with excellent dimensional stability. It has the highest mar resistance of plastics discussed, but is the most brittle. Urea compounds are odorless and tasteless, with good chemical resistance. They are not affected by organic solvents but are affected by alkalies and strong acids. They show good resistance to all types of oils and greases. They will withstand high temperatures without softening. Urea compounds are available in white and a wide range of colors, but with muted intensities compared to thermoplastics. Urea compounds, like phenolic resins, do not build up static electricity which leaves them free of dust. They are the most expensive of the plastic closure materials.

Metal. Metal caps, the strongest of closures, are used today for general, vacuum, and pressurized applications. Tin-plate and tin-free steel (see Tin mill products) are used in the production of continuous thread, and vacuum press-on closures, lugs, overcaps, and crown caps. The largest market for steel closures is vacuum packaging. Aluminum closures are primarily continuous thread caps and roll-on designs.

Steel closures. Steel closures are of two materials: tinplate and tin-free steel. Tinplate closures are plated steel with a thin coating of tin on both sides that helps protect the base steel from rust and corrosion. There are limitations to tin's protective abilities, however, for tinplate is susceptible to rust when exposed to high humidity. Additional coating operations offer increased protection. Tinplate is graded according to temper. Temper T-1 is soft, and T-6 is quite hard. Closures can be fabricated in any temper, but are predominantly T-2 to T-5. The more common base weights used include single-reduced 80- and 90-lb (36.3- and 40.8-kg) with a cost-reducing trend toward double-reduced 55- and 65-lb (25- and 29.5-kg) plate (15).

Tin-free steel shows promise of becoming the dominant steel closure material. Crown closures are now made primarily of tin-free steel produced as single-reduced stock in a 90-lb (40.8-kg) plate weight for conventional crowns and a lighter, 80-lb (36.3-kg) plate for twist-off crowns.

Aluminum closures. Light weight, malleability, and resistance to atmospheric corrosion characterize aluminum closures. Some products more corrosive to aluminum than tinplate require special coatings for optimum protection (31). The composition of aluminum alloys varies according to intended use with up to 5% magnesium and lesser elements such as manganese, iron, silicon, zinc, chromium, copper and titanium (32).

Metal-overshell closures. Steel, aluminum, copper, or brass shells slip over plastic closures to form composite "overshell" caps. Freed from finish contours, the smooth, often-tall sidewalls of polished and burnished metals provide aesthetically-pleasing characteristics much in demand by the cosmetic and fragrance industries. There is a greater willingness to pay a premium for appearance in these industries because the closure assumes a greater role in sales promotion.

Glass. Glass stoppers are used in commercial glassware and in cosmetic and fragrance packaging. Frequently a polyethylene base cap assists in friction-fitting the stopper into the bottle. Stoppers for the premium fragrance industries represent superlative designs in molded glass.

Closure Selection

Selection. Who selects and specifies the closure depends upon the size, nature, and organization of a company. It may be a president or general manager, a brand manager, package engineer, purchasing agent, or a packaging committee. General guidelines for closure selection are provided below by "The 5 Cs of Closure" (33).

Containment. The essential requirements of containment are product compatibility and the ability to provide functional protection. This objective is reached by evaluated choices in closure method, type, material, and sealing system. Determining the sealing system, for example, may involve decisions as to whether lining materials or a linerless closure will resist permeation of the product to standards. Other important variables arise in the interaction of closure and container and how they affect the efficacy of engagement and seal. Torque considerations include seal pressure (the amount of pressure exerted on sealing surfaces), and strip torque (the torque at which a closure slips over the container threads). As torque is affected by different coefficients of friction between the liner surface and the container, as well as by materials used in closure manufacture, each closure system should be individually evaluated to ensure it meets applicable performance criteria.

Convenience. Opportunities for convenient dispensing may begin with a containment closure that provides a reduced number of turns, or broader "knurls" on the wall of the closure skirt to provide surer opening and closing. Convenience closures provide many options including simple spouts, plug-orifice snap caps, and, at the mechanically complex end of the continuum, variable-dispersion sprayers and pumps. The method of closure engagement, the requirements of containment, the type of sealing system required, and the premium placed upon convenience will determine options in dispensing closures.

Control. A variety of substances are mandated by law to be packaged as tamper-evident or child-resistant. Cost and sealing needs will determine options in control closures, as well as whether secondary sealing systems are required. More consumer complaints result from inadequate opening and closing of product containers than any other package function. Careful review and testing of control-closure and lining system can prevent potential access problems with elderly or handicapped consumers.

Communications. The shelf appearance of a closure is perceived as a reflection of product quality. The closure communicates this by style and brand signature. In addition it often gives a detailed list of ingredients, and sometimes instructs on disengagement. The larger the closure the more it may augment the label in communicating ingredients or nutritional information. Today's closure not only communicates visually, but audibly as well. Steel vacuum button closures "pop" to confirm the freshness of a product, and polypropylene plug-orifice types "snap" when the seal is engaged. Determining the kinds of communications required, and selecting the graphic options which maximize readibility and impact, specify the final appearance, or point-of-purchase impact, of the closure.

Cost. Some cost considerations depend on production requirements. Thermoplastic-mold costs, for example, are generally more expensive than those for thermosets, but faster cycles and resin economy may prove more economical in larger volumes. A cost savings can be realized through the use of linerless closures if they can maintain seal integrity. Lightweighting the closure by selection of an appropriate material has helped to reduce transportation costs for packagers. Another cost consideration is whether a "stock" or a "privately-tooled" closure is required. A privately tooled closure is far more expensive to produce, but, again, the packaging concept or production volumes may "economize" it. There is a trend toward specifying stock closures with market-tested designs. Many manufacturers and distributors offer extensive stock-closure lines to the packer.

Closure Specification

A closure is designated by a series of numbers and/or letters. An example is the designation 48–400. The first number refers to the inside diameter of the closure as measured in millimeters. Common closure diameters range from 22–120 mm. The second set of numbers, the 400, is the finish designation.

The "finish" of a closure is its thread design, and the size, pitch, profile, length, and thickness of the engagement threads on plastic and metal closures and containers. Today there are over 100 glass-finish designations for a great variety of glass containers (see Table 1). Series designations for the most popular CT closures are 400 and 425 for shallow continuous thread designs, 410 for medium CTs, and 415 for tall CTs. For all glass finishes, tolerances have been established by the Glass Packaging Institute, whose closure committee became the Closure Manufacturers Association (CMA) in 1980. Voluntary standards for closures have been issues by CMA which include closures for both glass and plastic container finishes (see Table 2) (34).

Closures developed for glass containers were used for plastic containers when the latter were introduced, but it was soon realized that the contour of a glass bottle thread is not an optimum profile for the plastic bottle. It does not provide accurate closure centering on the finish nor does it permit higher capping torques required to provide a positive seal on plastic containers (35). The Dimensional Subcommittee of the Society of the Plastics Industry developed specific finish dimensions, tolerance, and thred contours for blown plastic bottles. The two basic contours are the M-style and the L-style. Where a typical glass thread is rounded in contour, the M-style thread engaging surfaces are angled at 10° and the L-style is angled at 30°. Both contours increase sealing abilities for closures on plastic bottles.

Table 1. Common Finishes and Descriptions

Finish Designation	Description
120	2-lug Amerseal quarter-turn finish
140	4-lug Amerseal quarter-turn finish
160	6-lug Amerseal quarter-turn finish
326	pour-out snap cap CT combination
327	snap cap CT combination
400	shallow CT finish
401	wide sealing surface on 400 finish
405	depressed threads of 400 finish at mold seam
410	medium CT concealed-bead finish
415	tall CT concealed-bead finish
425	8 to 15 mm shallow CT
430	pour-out CT
445	deep S CT finish
450	deep CT Mason finish
460	home-canning jar finishes
600	beverage crown finish
870	vacuum side seal pry-off
1240	vacuum lug-style finish
1337	roll-on pilferproof finish
1600	roll-on finish
1620	roll-on pilferproof finish
1751	twist-off vacuum seal

Four critical closure dimensions are represented by the four letters, T, E, H, and S (see Fig. 15). T is the dimension of the root of the thread inside the closure. E is the inside dimension of the thread in the closure. H is the measurement from the inside top of the closure to the bottom of the closure skirt. S is the vertical dimension from the inside top of the closure to the starting point of the thread. These critical closure dimensions and tolerances for metal and plastic closures designed for glass and plastic containers are represented in the voluntary standards for closures as issued by CMA.

Closure Trends

Closure concepts seem to have changed more in the last 40 yr than in the last 4000 yr. Yet changes in state-of-the-art concepts, materials and manufacturing processes do not represent the real driving forces behind today's closure developments. The industry is consumer-driven. More and more, a premium is readily paid for convenience. The industry has also been awakened, sometimes with great shock and alarm, to the powers of human foible, which demands a redefinition of access control.

Functional trends. Today's consumer has been characterized as oriented toward health, diet, appearance, longevity, and convenience (36). Households with two working spouses, increased single households, and retiree households all command a market for convenience packaging (37). The dispensing closure is no longer a functional appendage, but is seen as an integral part of the total package (38). Today's convenience closure is time and labor-saving. It prevents spills, leaks, and drips. It provides measured-dose dispensing, and can visually signal tampering (39). As plastic containers continue to penetrate the food market, the closure will broaden squeeze-dispensing. Other functional trends include the expansion of TE

Table 2. CT Closure Finishes and Pitch

Finish	Description	Sizes, mm	Threads per in., Pitch, (per cm)
400	shallow continuous thread	18, 20, 22, 24	8 (3.2)
		28, 30, 33, 35, 38, 40	6 (2.4)
		43, 45, 48, 51, 53, 58	6
		60, 63, 66, 70, 75, 77	6
		83, 89, 100, 110, 120	5 (2.0)
410	medium continuous thread	18, 20, 22, 24	8
		28	6
415	tall continuous thread	13, 15	12 (4.7)
		18, 20, 22, 24	8
		28	6
425	shallow continuous thread	8, 10	14 (5.5)
		13, 15	12
430	pour-out continuous thread	18, 20, 22, 24	8
		28, 30, 33, 38	6
445	deep "S" continuous thread	45, 56, 58, 63, 73, 75	6
		77, 83	5
450	deep CT Mason finish	70, 86, 96, 132	4 (1.6)
455	CT for 455 glass finish	28, 33, 38	8
460	Home canning jars	70, 86	4
470	CT for GPI 470 glass finish	70, 86	4
480	CT for GPI 480 glass finish	24, 28, 33, 38	6
485	deep "S" fitment cap	28, 33, 35, 38, 40	6
		43, 48, 53, 63	
490	deep "S" larger "H" fitment cap	18, 20, 22, 24, 28, 30	8
		33, 35, 38, 43, 48, 63	
495	CT for GPI 495 glass finish	28, 33, 38	8
SP 100	CT for plastic SP-100 finish	22, 24, 26, 28, 38	8
SP 103	CT for plastic SP-103 finish	26	8
SP 200	CT for plastic SP 200 finish	24, 28	6
SP 444	CT for plastic SP 444 finish	24, 28, 33, 38, 43, 45, 48	6
		53, 56, 58, 63, 73, 75	6
		70	4
		83	5

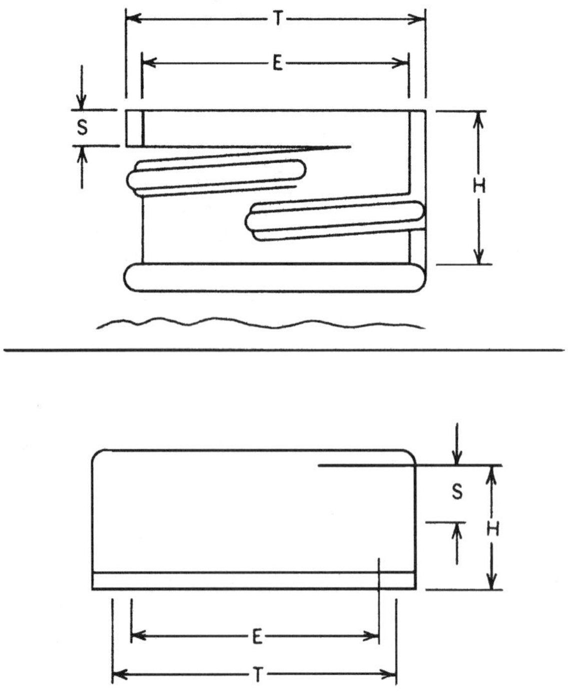

Figure 15. T, E, H, and S dimensions.

food packaging, larger closure sizes, increased use of stock caps to avoid private-mold costs, and new concepts in linerless closure design. Since innovative packaging can increase market share, special emphasis is being placed on improved tamper-evidence, child-resistance, and convenience designs. These functions will no doubt become more and more integrated into one cap. Closures are now being marketed which provide for both TE and CR.

Conclusion. The ductility of plastics accounts for the fast progress of plastic closures, which will undoubtedly take a still-larger share of the market in years ahead (40,41). Polypropylene represents the largest volume and highest growth plastic material with 200 million (10^6) lb (90,700 t) used in the production of plastic closures in 1984, an increase of 33% within four years (42). Polystyrene has shown slight growth in recent years, with 72 million lb (32,700 t) used in closure production in 1984. Closures accounted for 66 million lb (29,900 t) of HDPE consumption, an amount that has remained relatively stable over recent years. LDPE (37×10^6 lb or 16,800 t) and PVC (35×10^6 lb or 15,900 t) have both remained stable in recent years. As for thermosets, 15×10^6 lb (6800 t) of phenolic resins and 11×10^6 lb (5000 t) of urea compounds were used in closure production in 1984 reflecting little growth in recent years.

Metal thread-engagement closures continue to assume a position in the food and pharmaceutical industires due to new fabricating, plating, and light-weighting technologies which will keep steel and aluminum closures competitive.

BIBLIOGRAPHY

"Closures, Bottle and Jar" in the *Wiley Encyclopedia of Packaging Technology,* 1st ed., by J. F. Nairn and T. M. Norpell, Phoenix Closures, Inc., pp. 172–185.

1. H. McKearin and K. M. Wilson, *American Bottles and Flasks and their Ancestry,* Crown Publishers, New York, 1978, p. 17.
2. *Ibid.,* pp. 210–212.
3. H. Higdon, ed. *The Phoenix Flame,* (a house magazine of the Phoenix Cap Company, Chicago, now Phoenix Closures, Naperville, Ill., **XV,** 32, (Feb. 1940).
4. A. Lief, *A Close-Up of Closures,* Glass Container Manufacturers Institute, New York, 1965, p. 16.
5. Closure Manufacturers Association, "Closures for Bottles, Cans, Jars," *Packaging Encyclopedia 1984,* **29,** 153 (March 1984), Cahners Publ. Co., Boston, Mass.
6. "Closures for Containers," *Current Industrial Reports, M34H (85)-1,* U.S. Department of Commerce, Washington, D.C., April, 1985, p. 1.
7. R. J. Kelsey, *Packaging in Today's Society,* St. Regis Paper Co., Ridgewood, New Jersey, 1978, pp. 20–38.
8. T. Tang, "Closures, Liners and Seals," *Packaging Encyclopedia 1984,* **29,** 158–160 (March 1984).
9. J. F. Hanlon, *Handbook of Package Engineering,* McGraw-Hill, New York, 1971. Sect 9, pp. 5–6.
10. "Tamper-Evident Packaging Requirements," *Fed. Regist.* **47,** (215), 50442 (Nov. 5, 1982).
11. J. B. Carroll, *Memorandum from the Closure Manufacturers Association,* McLean, Va. April 19, 1985, p. 1.
12. B. Knill, *Food and Drug Packaging,* **48,** 5 (Dec., 1984).
13. "Quantity and Value of Metal and Plastic Closures," *Current Industrial Reports, M34H(83)-13* U.S. Department of Commerce, Washington, D.C., Oct. 1984, p. 1.
14. Ref. 5, p. 152.
15. *Steel Cans,* Report by the Committee of Tin Mill Products Producers, American Iron and Steel Institute, Washington, D.C., 1984, p. 8.
16. Ref. 9, Sect 6, p. 20.
17. Ref. 4, p. 29.
18. P. Zwirn, "The Crown is Still King," *Canadian Packaging,* **37**(6), 24 (June, 1984).
19. "The Best Ideas in Packaging," *Food and Drug Packaging,* **39,** (Nov., 1984).
20. C. G. Davis, *Packaging Technology,* **27,** 27 (Oct./Nov. 1982).
21. R. Heuer, *Packaging,* **29,** 34 (April, 1984).
22. R. L. Harris, "Closures, Dispensing Systems," *Packaging Encyclopedia 1984,* Cahners Publishing Co., Boston, Mass, **29** 157 (March 1984).
23. T. T. Williams, *A History of Technology,* Clarendon Press, Oxford, 1978, p. 1411.
24. "Tamper-Resistant Packaging Requirements," *Fed. Regist.* **47,** (215), 50444 (Nov. 5, 1982).
25. H. Forcinio, *Food and Drug Packaging,* **29,** 35 (Sept. 1984).
26. *Child Resistant Packaging,* American Society for Testing and Materials, Philadelphia, Pa., pp. 20–21.
27. "CPSC Replies to Queries on CR Rules Compliance," *Food and Drug Packaging,* **40,** 20 (March 1985).
28. Ref. 9, Sect. 9, p. 21.
29. *Options for Successful Medical/Pharmaceutical Packaging, Bulletin PB-484,* Phoenix Closures, Inc., Naperville, Ill., April 1984, p. 2.
30. L. Roth, *An Introduction to the Art of Packaging,* Prentice-Hall Inc., Englewood Cliffs, New Jersey, 1981, p. 153.
31. Ref. 9, Sect 7, p. 16.
32. Ref. 9, Sect. 7, p. 31.
33. "The 5 C's of Closure," *Marketing Communications Bulletin,* Phoenix Closures, Inc., Naperville, Ill., Dec. 1, 1984.
34. *CMA Voluntary Standards,* Closure Manufacturers Assoc., McLean, Va. 1984.
35. J. Szajna, *Food and Drug Packaging,* **29,** 12 (May 1984).
36. A. J. F. O'Reilly, *Food and Drug Packaging,* **29,** 72 (Sept. 1984).
37. H. K. Foster, *Food and Drug Packaging,* **29,** 46 (Nov. 1984).
38. Ref. 21, p. 36.
39. B. Miyares, *Food and Drug Packaging,* **29,** 72 (Nov. 1984).
40. R. Graham, *Food and Drug Packaging,* **29,** 42 (Sept. 1984).
41. H. Peter Aleff, "Comparison of Thermoplastic with Thermoset Closures, *Packaging Technology* **12**(2), 18 (April 1982).
42. "Materials 1985," *Mod. Plast.,* **63**(1), 69 (Jan. 1985).

General References

H. L. Allison, "High Barrier Packaging: What are the Options?," *Packaging,* **30,** 25 (March, 1985).

J. Agranoff, ed., *Modern Plastics Encyclopedia,* **60,** McGraw-Hill, Inc., New York, 1984. Note: The Engineering Data Bank section contains extensive data on properties of plastics, design data, chemicals and additives information, and data on molding machinery.

L. Barail, *Packaging Engineering,* Reinhold Publishing Corporation, New York, 1954.

The Closure Industry Report, Closure Manufacturers Association, McLean, Va., Fall, 1983 and Summer, 1984.

E. F. Dorsch, "Closures and Systems," *Packaging Encyclopedia,* **30,** 126–128 (1985) Cahners Publishing Company, Boston, Mass., 1985.

R. C. Griffin, Jr., and S. Sacharow, *Principles of Packaging Development,* The Avi Publishing Co., Westport, Conn., 1972.

K. Hannigan, "Baby Food Closure Grows Up," Chilton's *Food Engineering,* Chilton Company, Radnor, Pa., Jan., 1984.

G. Jones, *Packaging: A Guide to Information Sources,* Gale Research Company, Detroit, Mich., 1967.

Kline's Guide To The Packaging Industry, Charles H. Kline & Co., Inc., Fairfield, N.J., 1979.

E. A. Leonard, *Packaging Economics,* Books for Industry, New York, 1980.

"Mack-Wayne Introduces New One-Piece Dispensing Closure and New Breakaway Closure," *Packaging Technology,* **14**(1), 34 (July/Aug. 1984).

"Packaging: Big Changes Ahead," *Prepared Foods,* **153,** 60–70 (June, 1984).

"Plastics Hit the Market with Tamper Evident Closures," Chilton's *Food Engineering,* **56,** 60 (April, 1984).

N. C. Robson, "The State of the Anti-Tampering Art," *Packaging Technology,* **13.** 26 (June, 1983).

S. Sacharow, *A Guide to Packaging Machinery,* Books for Industry, New York, 1980.

S. Sacharow, *A Packaging Primer,* Books for Industry, New York, 1978.

B. Simms, ed., "Closures, Dispensers and Applicators," *Packaging En-*

cyclopedia, Vol. 29, Cahners Publishing Company, Boston, Mass. 1984.

Drawings for figures 1–14 courtesy of S. Kiefer and B. Zemlo, Phoenix Closures, Inc.

CLOSURES, BREAD BAG

Packaging bread in polyethylene bags is an almost universal method of bread packaging. Customer preference for easy opening and reclosing the package influenced the development of bag closures at the very start of bread bagging. Closing by tape, which had been used in connection with early bagging methods, was quickly discarded when the bulk of the bakery production lines changed to the new polyethylene bagging and closing equipment. Tape was neither convenient for the consumer nor fast and dependable for the bakers. Two types of closures and automatic equipment systems have emerged as the standards of the baking industry. Developed simultaneously, these are the wire tie and the plastic-clip closure.

Wire ties. The predominant automatic wire-tie equipment now being used by the United States baking industry is from the Burford Corporation, Maysville, Oklahoma. Different models are required to attach to the different types of installed bread and bun baggers. This type of automatic bag-closing equipment is limited to about 60 packages per minute. The wire ties used in automatic application are 4-in. (10.2-cm) long. The wire comes on reels of 6,000 ft (1,829 m), 18,000 closures per reel, and is cut to length as it is applied to the package. Colors are available for color-coding purposes (see Fig. 1).

The wire tie has a left- or right-hand twist, depending on the production equipment. Wire tires are available with all-plastic, laminated plastic/paper, and all-paper covers over the wire core. The quality of the wire tie must be carefully maintained or separations of the laminations occur leaving the bare wire exposed. Source of the wire ties is Bedford Industries, Worthington, Minn.

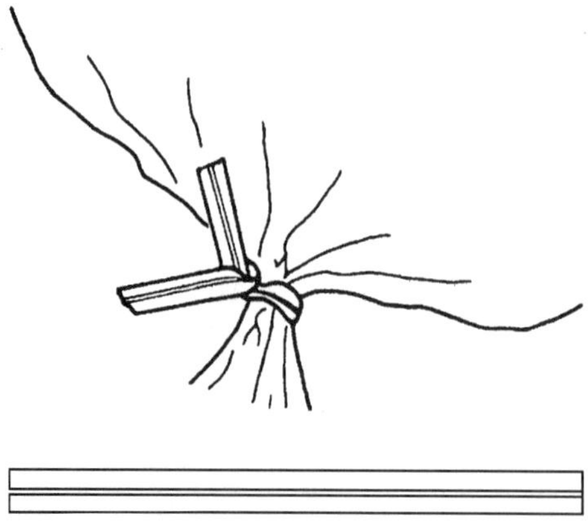

Figure 1. A wire tie for a bread bag.

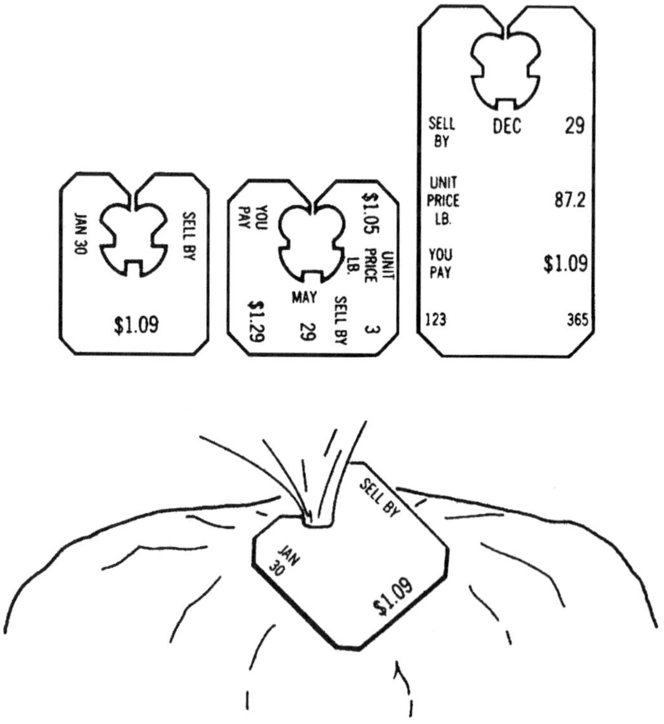

Figure 2. A plastic-clip closure for a bread bag.

Plastic-clip closure. Plastic-clip closures are produced from medium-impact polystyrene (see Polystyrene). The material has a resiliency that allows it to return to its original shape after many reuses by the consumer (see Fig. 2). They are furnished with various aperture sizes to accommodate the different polyethylene bag widths and thicknesses. The proper selection of aperture size can result in an almost air-tight package. Colors are available for color-coding purposes. The plastic-clip closures are provided in reels of 4000 or 5000 closures per reel.

The Striplok closures and closure-applying equipment are manufactured by the Kwik Lok Corporation, Yakima, Wash. The equipment is simple and dependable, with production speeds up to 120 bakery bags per minute. The Kwik Lok Corporation manufactures two basic machines that attach to any of the common bagging machines from Formost Packaging Machines, Inc., Woodinville, Wash.; United Bakery Equipment, Compton, Calif. and AMF, Inc., Union Machinery Division, Richmond, Va. By mounting an imprinter on the closing machine, imprinting can be done on the plastic clip. Many bakeries are required to print on the package the price per pound, the unit price of the package, and the "sell by" date (see Code marking and imprinting). Printing on the polyethylene bag lacks legibility and is often obscured by the package graphics. Three sizes of Striplok closures are available to facilitate compliance with regulatory coding and dating requirements.

BIBLIOGRAPHY

"Closures, Bread, Bag," in *The Wiley Encyclopedia of Packaging Technology,* 1st ed., by G. E. Good, Kwik Lok Corporation, pp. 185–186.

COATING EQUIPMENT

Coating equipment is used to apply a surface coating that may serve as a barrier, decorative coating, or other purpose. It may be used to apply adhesive for laminating one web to another or for manufacturing of pressure-sensitive tapes and labels. It also may be used for saturation of a porous web, such as paper, in order to improve its resistance to moisture or grease penetration, or to improve its strength.

Coating equipment consists of three main components: (*1*) a coating head, (*2*) a dryer or other coating solidification station, and (*3*) web handling equipment and accessories (drives, winders, edge guides, coating thickness controls, etc).

Coating when applied must be sufficiently fluid to be spread into a uniformly thin layer across the web. Therefore, coatings are applied as solutions in organic solvents, as aqueous solutions or emulsions, as a hot melt (solid molten or softened by heat), or as a reactive liquid that solidifies by a polymerization reaction induced either thermally or by radiation. Extrusion coating, which is similar to hot-melt coating, is discussed separately (see Extrusion coating).

Coating may be applied directly to the substrate, or it may be cast to another surface, dried, and later transferred to the substrate of choice. This transfer coating process is used for manufacturing of pressure-sensitive label stock: the adhesive is first applied to a silicone-coated release liner, dried, and then laminated to the label face stock. Coating may be applied to the web material wound in rolls. This process requires unwinding the web, applying coating, drying, and then rewinding the web again. Less frequently it is applied to precut and sometimes printed sheets (such as in manufacturing of blister packaging cards). Handling and coating of sheeted material requires different coating and handling equipment.

Coating Heads

The coating head accomplishes two functions: applying the coating to the substrate and distributing a metered amount uniformly over the surface. Metering may be combined with the coating application, or it may be carried out separately immediately following the coating deposition. Coating equipment and coating processes have been reviewed in several books (1–4). There are many designs of coating heads, but they fall into three major categories: (*1*) roll coaters; (*2*) knife, blade, and bar coaters; and (*3*) extrusion and slot-orifice coaters. General characteristics of most important coating heads are shown in Table 1.

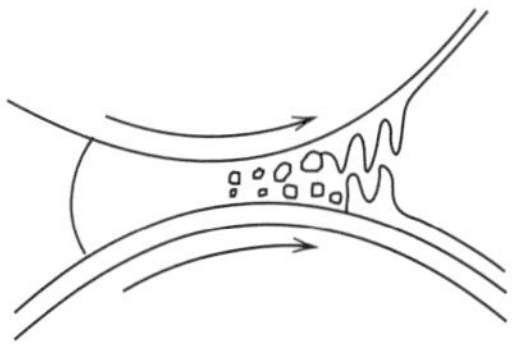

Figure 1. Coating splitting.

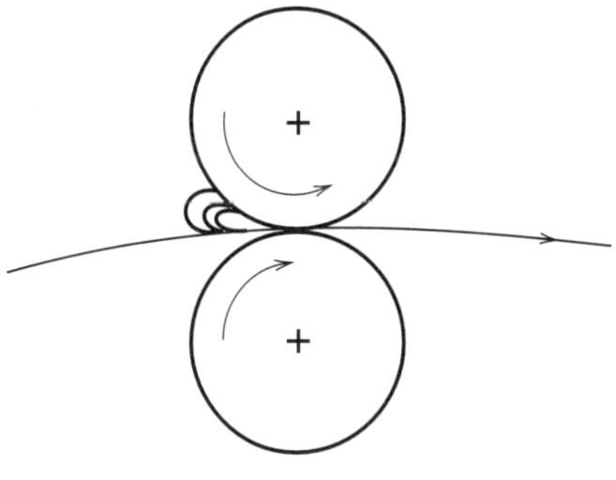

Figure 2. Squeeze coater.

Roll coaters. Roll coaters are most widely used and can be subdivided into several major types: direct, reverse, gravure and calender coaters.

Direct roll coaters. The web carrying roll and the coating applicator roll rotate in the same direction and at the same speed in the direct roll-coating process. The coating is split about evenly between the roll and the substrate as shown in Figure 1. Coating splitting produces uneven surface which sometimes can be a problem. Many different coating heads utilize the direct roll-coating principle. A squeeze-type roll coater is shown in Figure 2. Direct roll coaters can be used to coat sheeted material as shown in Figure 3. The coating is metered in a gap between accurately machined metering and applicator rolls. Sheets are fed through a nip formed by applicator and carrier rolls. Direct roll coaters are used widely for packaging materials: paper sizing, paper color coating, overcoating of blister packaging board, and heat seal coating.

Kiss roll coaters apply the coating to the web from a pan (Fig. 4). The amount of coating deposited is not controlled by the coater; therefore, kiss roll coaters usually employ a metering device, such as wirewound rod, to remove the excess coating. Such a rod is installed immediately after the coater and is a part of the coating head. Kiss roll coaters can be also run

Table 1. Characteristics of Various Coaters

Coating System	Deposit Weight Range (g/m²)	Coating Speed Range (m/min)	Viscosity Range (cP)
Knife-over-roll	10–100	150	1000–30,000
Floating knife	2–30	150	500–15,000
Air knife	3–30	500	100–1000
Blade	5–20	100–1200	2,000–10,000
Wirewound rod	5–20	150	100–2,500
Reverse roll	25–300	150	300–20,000
Kiss/squeeze	10–30	300	50–2,000
Transfer roll	0.3–50	300	300–150,000
Gravure	2–25	10–400	100–10,000
Slot orifice	20–600	300	400–200,000

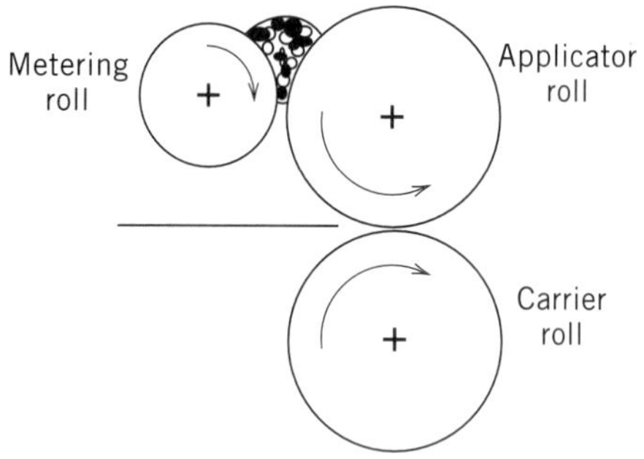

Figure 3. Direct roll sheet coater.

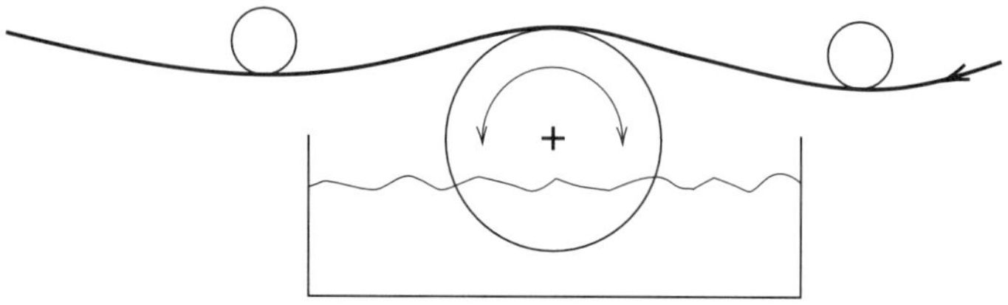

Figure 4. Kiss-roll coater.

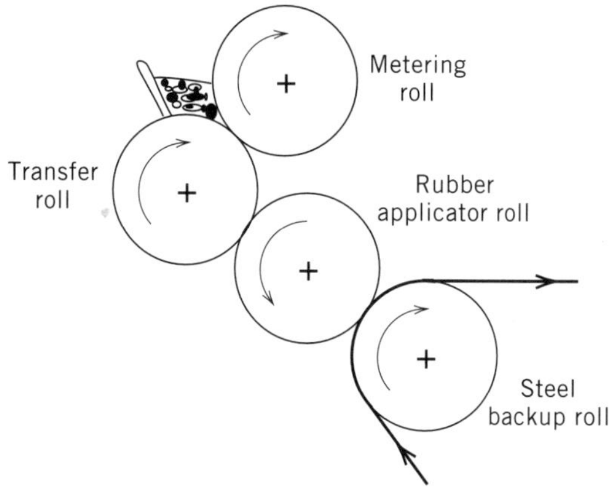

Figure 5. Transfer coater.

in the direction opposite to the web travel. In such cases the web wipes the roll clean.

Transfer roll coaters of modern design are used for deposition of very lightweight coatings accurately (Fig. 5). One of the important uses is the application of 100% solids radiation curable silicone release coatings, which are used for pressure-sensitive adhesive labels and other products. The transfer roll coater is a direct roll coater. The rubber-covered applicator roll is run 5–25 times faster than the transfer roll, and this decreases the amount of coating transferred to the substrate by that factor.

Reverse roll coaters. These coaters are versatile machines that can handle a wide range of coating thickness and viscosities. Solvent solutions, aqueous coatings, and less frequently hot melts are coated on these machines. Reverse roll coaters are expensive; rolls must be accurately machined. These coaters are available in several designs: feed location may be varied from nip to pan, three or four rolls may be used. In the case of a nip fed coater as shown in Figure 6, the coating is delivered to the nip between metering and applicator rolls, and dams are used to contain the coating. A gap set between these two accurately machined rolls determine the amount of coating carried by the applicator roll. The main feature of the reverse roll coater is that the applicator roll rotates in the direction opposite to the web travel (therefore reverse roll) and transfers the coating by wiping. The web usually runs faster than the applicator roll, and the amount of coating deposited depends on the ratio between the speed of these two rolls (wipe ratio) as well as on the gap between metering and applicator rolls.

Gravure coaters. These machines are inexpensive and highly reliable, but limited to the coating thickness and viscosity that can be handled. An engraved chrome-plated (sometimes ceramic coated) copper roll is wetted with the coating, excess is removed by a doctor knife, and the coating remaining in the engraved cells below the roll surface is transferred to the web at the gravure roll/backup roll nip (Fig. 7). Three engraving patterns are commonly used for coating purposes: pyramidal, quadrangular (truncated pyramid), and trihelical. The amount of coating depends on the liquid volume retained in the cells, ie, on the cell depth and their density, referred to as *screen* or *ruling* (the number of cells per unit

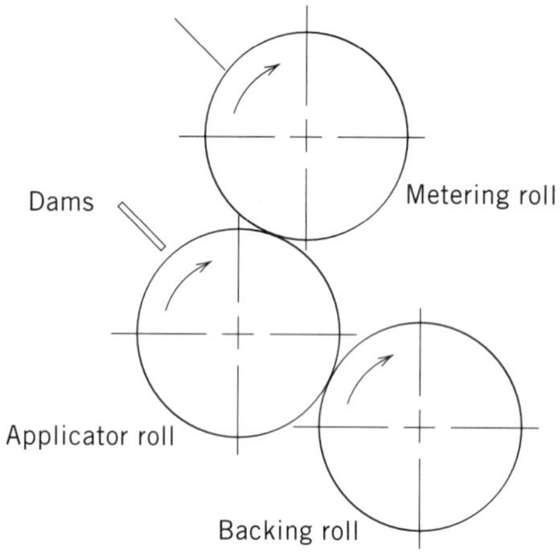

Figure 6. Nip fed reverse roll coater.

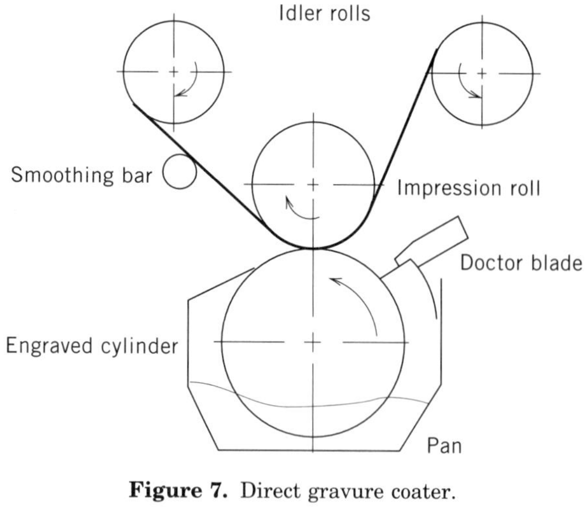

Figure 7. Direct gravure coater.

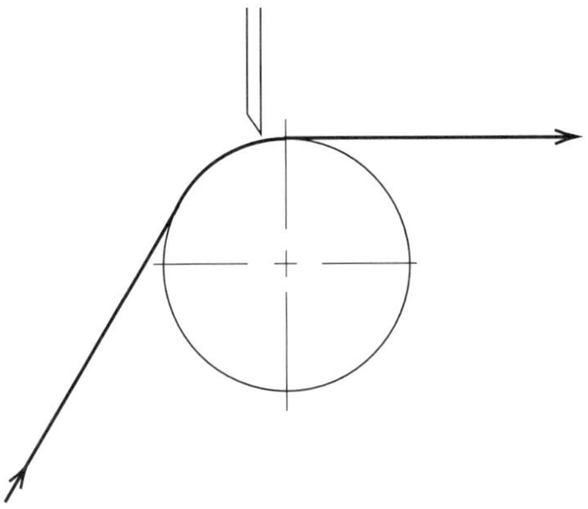

Figure 8. Knife-over-roll coater.

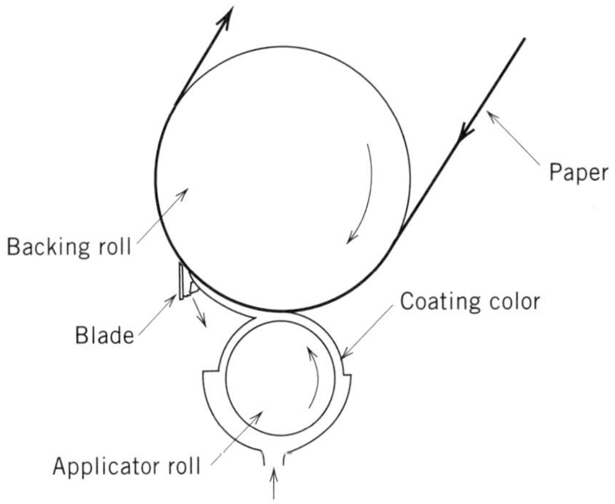

Figure 9. Blade coater.

length). The coarser the screen, the larger are the diameter and depth of cells and the higher the deposit is. The fluid viscosity must be sufficiently low to allow the transfer of the coating from the cells to the web at nip pressure.

In the direct gravure arrangement (Fig. 7), the engraved roll contacts the web directly. In offset gravure the coating is first transferred to a rubber-covered roll and then transferred to the web. Offest gravure allows the use of higher nip pressure and is therefore more suitable for coating rough-surface substrates. It also allows the coating to level on the roll surface before its transfer to the web.

Gravure coating is widely used for various decorative and functional lightweight coatings on plastic films, paper, and other packaging substrates. The same technique is suitable for printing, ie, depositing noncontinuous coating.

Calender coaters. The calendering process involves squeezing a polymeric material between steel rolls into a thin sheet. The formed sheet may be then laminated to a substrate. Rubber and PVC are most often used for calender processing. Calendering is rarely used for manufacturing packaging materials.

Knife and bar coaters. Knife and bar coaters are metering devices that remove excess coating and allow only a predetermined amount to pass through.

Knife-over-roll coaters. A knife-over-roll coater consists of a straight-edged knife placed against a roll. The coating weight is adjusted by setting a gap between the roll and the knife. An excess coating is delivered to the bank before the knife, and the desired amount is metered by the gap. An accurately machined steel roll and knife are used (Fig. 8). In another version of a knife-over-roll coater, a rubber-covered roll is used and the coating weight is determined by the gap, and rubber hardness (its capability to deform because of hydraulic pressure). The knife-over-roll coater is a widely used inexpensive machine most suitable for thicker coatings of a higher viscosity.

Other knife coaters. Knife coaters are used in many other configurations: floating knife, knife-over-blanket, knife-over-channel, and inverted knife. These methods are used more frequently in the textile industry.

Blade coating. Flexible-blade coating is the dominant process for applying pigment coating over paper and board. It has replaced the air knife for paper clay coating. Clay coatings give a smooth and printable surface. Blade coating processes are suitable for high-speed application such as required in paper converting. The coating head consists of a coating applicator, such as roll or fountain, which applies an excess of coating to the paper. The excess coating is removed by a blade (Fig. 9). Several modifications of blade coaters are available employing different blade designs, different methods of applying and regulating blade pressure, and different ways of applying the coating to the paper. Machines capable of applying coating to both sides of the web are available.

Air-knife coaters are being replaced by more recently developed blade coaters for paper-coating applications. In air-knife coating an excess of material is applied to the web surface, usually by a kiss-roll applicator, and the excess is removed by the air knife, which consits of a head with a narrow slot. Pressurized air is forced into the head and is accelerated at the slotted nozzle. The escaping airstream impinges on the coated web and removes excess coating (Fig. 10).

Wirewound-rod coater. A wirewound-rod coater (Meyer rod) is a simple metering device widely used in applying lightweight coatings over film and paper packaging materials. The coating is applied by an applicator, usually a kiss roll, and the excess of coating is removed by a rod wound spirally with stainless-steel wire. The rod wipes the surface clean, except what escapes through the spaces between the wires. The larger the wire diameter, the heavier is the coating. The rod is rotated to help to dislodge any particles that might become trapped between the wire and the web and to insure uniform wear of the wire.

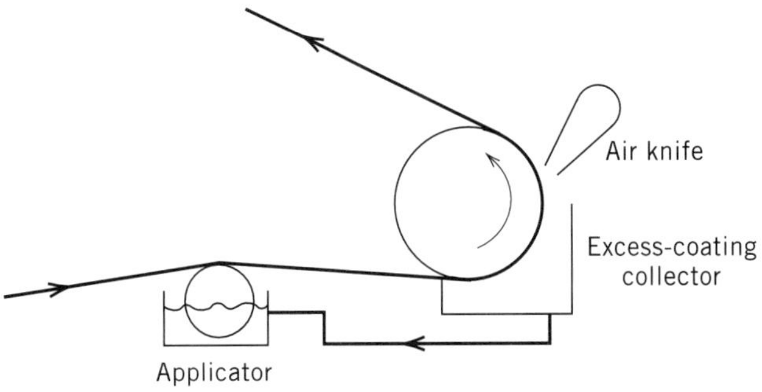

Figure 10. Air-knife coater.

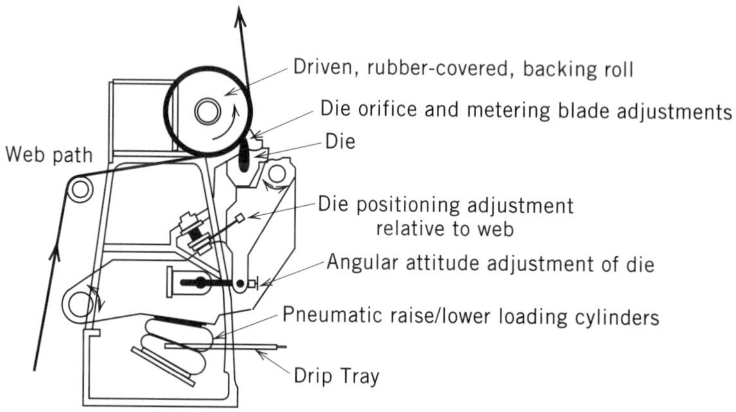

Figure 11. Hot-melt slot-orifice coater. (Courtesy of Black Clawson Co.)

Slot-orifice and curtain coaters. Slot-orifice coating is a variation of extrusion coating, except the viscosities are much lower. The coating is forced through a narrow slot extending the full width of the web. The slot die is designed to give a uniform coating thickness across the width. This technique is preferred for hot-melt coating. A slot-orifice coater is widely used for manufacturing hot-melt pressure-sensitive adhesive packaging tapes (5) and for applying waterproof coatings over paper. It also found applications in aqueous emulsion coating. Slot orifice coater is shown in Figure 11.

Curtain coater. Curtain coating is similar to slot-orifice coating, except that the liquid curtain coming from the die is allowed to drop down some distance to the material to be coated. The slot width is adjustable and the flow is regulated by the slot width, by the liquid level (in weir-type coaters), or by pressure and pump speed in enclosed-head coaters, and by the distance between the slot and the coated material, which determines the acceleration of the curtain. This technique is used for heavier coatings. In the packaging area curtain coating may be employed to deposit a hot wax coating over corrugated board.

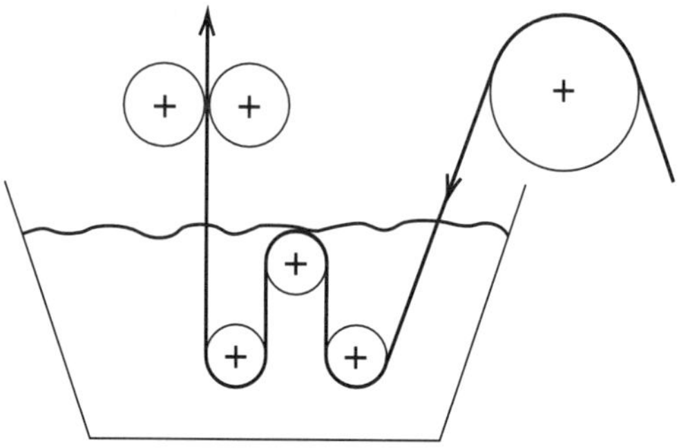

Figure 12. Sawtooth saturator.

Saturators

The saturation or impregnation process is used to treat paper and paperboard with a polymeric binder to improve the web's strength, barrier properties as packaging material, and resistance to water and grease. The process consists of immersion of the web into a coating bath, or applying an excess of coating on both sides and then squeezing or scraping to remove the excess. The coating may penetrate the web or most of it may remain on the surface depending on the product needs.

Saturating machines consist of a web-immersion section and a metering section. Several saturating arrangements are used. Figure 12 shows a conventional sawtooth saturator followed by squeeze-roll metering. Other types of metering arrangements are inflateable bars, bar scrapers, doctor blades, and similar devices.

Drying

Coatings applied as solutions or emulsions must be dried in order to remove the liquid vehicle. Heat and mass transfer take place simultaneously during the drying process, and the

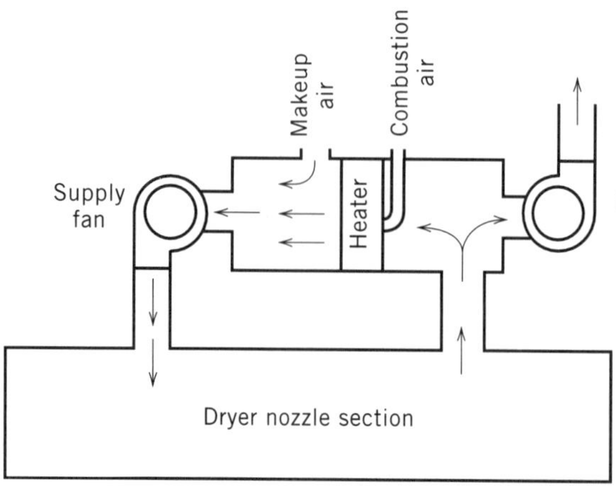

Figure 13. Airflow in a convection oven.

Table 2. Drying Equipment

Heat Transfer	Web Handling
Convection dryers	
Parallel air flow	Idler-supported dryers
Impingement air	Conveyer dryers
Through dryers	U-type dryers
Infrared radiation dryers	Arch dryers[a]
Near infrared (electric)	Tenter frame dryers
Far infrared (electric or gas)	Floater dryers
Conduction dryers	
Hot roll dryers	

[a] An arch dryer including winding, coating, and laminating stations is shown in Figure 13.

Figure 14. Coating–laminating line with an arch dryer. (Courtesy of Faustel, Inc.)

Figure 15. Carton blank coater. (Courtesy of International Paper Box Machine Co.)

heat is transferred by convection in air dryers, radiation in infrared radiation dryers, or conduction in in contact drum dryers. The drying equipment also has a means of vapor removal and recirculating and heat-exchange equipment to conserve energy. Figure 13 schematically shows the airflow in a convection dryer. If a coating from a solution in an organic solvent is used, the solvent vapor must be removed from the exhaust in order to satisfy the environmental laws and to decrease solvent costs. Solvent adsorption on activated carbon, incineration, or condensation in an inert-gas dryer are used. Drying equipment (see also Fig. 14) may be subdivided according to the heat-transfer mechanism or according to web handling as listed in Table 2.

Extruded, hot-melt, and wax coatings do not require drying and are solidified by chilling. Such coating machines require considerably less space than do the coating lines with drying ovens.

Some coatings are applied as reactive monomers or polymer/oligomer/monomer blends and may be cured by either ultraviolet or electron-beam irradiation. Such irradiation units are incorporated into the coating line.

Web Handling

Coating machines may apply the coating or the adhesive to the packaging material supplied as a continuous web on a roll, and the finished product is rewound after completion of the operation (see Roll handling). Unwind and rewind stands, web-carrying equipment, and web controls and other accessories are used (6). Controls are needed to track the web properly on the machine and may consist of tension-sensing devices and means of controlling the tension and of edge-sensing devices and means of keeping the web centered on the coating machine.

Many packaging materials are coated as sheet, requiring sheet-feeding and sheet-handling devices. There is less of a choice between various coating heads for sheet coating. A carton blank coater is shown in Figure 15.

BIBLIOGRAPHY

1. D. Satas, ed., *Web Processing and Converting Technology and Equipment,* Van Nostrand Reinhold, New York, 1984.
2. H. L. Weiss, *Coating and Laminating Machines,* Converting Technology Co., Milwaukee, WI, 1977.
3. G. L. Booth. *Coating Equipment and Processes.* Lockwood Publishing, New York, 1970.
4. E. D. Cohen and E. B. Gutoff, eds., *Modern Coating and Drying Technology,* VCH Publishers, New York, 1992.
5. D. Satas, ed., *Handbook of Pressure Sensitive Adhesive Technology,* Van Nostrand Reinhold, New York, 1982.
6. D. R. Roisum, *The Mechanics of Winding,* TAPPI Press, Atlanta, 1994.

D. Satas
Satas and Associates
Warwick, Rhode Island

CODE, BAR

A *bar code* may be defined as a series of bars and spaces arranged according to the encodation rules of a particular specification in order to represent data. Its purpose is to represent information in a form that is machine-readable. Bar codes are read by scanning devices that are programmed to analyze the structure of the bars and spaces and transmit the encoded data in electronic format. These data can then be stored on a file or transmitted to a computer for processing.

Other techniques than bar codes achieve the same objective: capturing automatically data encoded using a particular technology. These include optical character recognition, magnetic stripe, and radiofrequency identification. The concept of encoding and reading data automatically is called *automatic data capture* (ADC).

Benefits of Bar Codes

The main benefits of bar codes are *speed* and *accuracy.* Capturing data automatically by reading a bar code can be done in fraction of a second, much faster than manual key entry. It is commonly agreed that an operator doing key entry makes one error for every 300 characters typed. Reading bar codes makes data capture almost error-free. The error rate depends on the type of bar code and equipment being used but usually it is lower than one error per 1,000,000 readings.

Bar-Code Symbologies

A *bar-code symbology* is a set of rules describing the way bar and spaces have to be organized to encode data characters.

Since the invention of the bar code concept in the United States in the late 1940s, hundreds of bar-code symbologies have been developed, but only few of them are actually being used on a large scale.

Typically, a symbology is qualified as being discrete or continuous. In a discrete symbology, the spaces between symbol characters do not contain information as each character begins and ends with a bar. In a continuous symbology, there is no intercharacter gap, ie, the final element of one symbol character abuts the first element of the next symbol character, and all the elements carry data contiguously. The most popular bar-code symbologies are briefly described below.

Code 39 was launched in 1975. It is widely used for industrial applications. Code 39 is a discrete, variable-length symbology encoding the 36 numeric and uppercase alpha characters (A–Z, 0–9) and seven special characters: space, dollar sign ($), percent (%), plus (+), minus (−), dot (.), and slash (/). A symbol character is composed of nine elements, five bars and four spaces. An element is either wide or narrow. There are three wide elements and six narrow elements in a symbol character. A Code 39 symbol begins with a start character and ends with a stop character. It can be read from the right to the left and from the left to the right.

0 5 4 1 2 3 4 5 6 7 8 9 0 8

Code 39.

Interleaved two of five (abbreviated ITF) has been found to be well adapted to the materials and printing conditions frequently used on fiberboard cases. It is a continuous symbology encoding only numeric digits. A pair of digits is represented by five bars and five spaces. One of the pair is represented by the dark bars and the other by the light bars, and the dark and light bars are interleaved. As the digits are represented in pairs, the symbol can only encode even number of digits. In addition to the digit characters there are two auxiliary characters used as guard bars at the beginning and at the end of the digit representation. The symbol is designed to be read bi-directionally by fixed or portable scanners.

05412345678908

Interleaved two of five.

Code 128 was introduced in 1981 in response to the need for a compact alphanumeric code symbol that could be used to encode complex data. The fundamental requirement called for a symbology capable of being printed by existing data-processing printers. Code 128 uniquely addresses this need with the most compact, complete, alphanumeric linear symbology available. In addition, Code 128 has been designed with geometric features to improve scanner reading performance and to be self-checking.

05412345678908

Code 128.

EAN/UPC was developed in the late sixties when researches were conducted in the United States to improve the efficiency of checkout operations in retail stores. EAN/UPC is a continuous symbology encoding fixed-length numeric digits. Several variants exist, known as EAN-13, UPC-A, EAN-8, and UPC-E. In addition, the symbology enables to encode two small symbols encoding two and five digits. These are called add-ons as the information they contain supplement the main symbols. A symbol character is composed of seven modules, two bars, and two spaces. A bar or a space is composed of one to four modules. An EAN/UPC symbol begins and ends with a guard pattern. In the EAN-13, UPC-A and EAN-8 version, a centre pattern separates the symbol into segments that can be read separately by a decoding equipment, thus making the symbol omnidirectionally readable. The EAN/UPC symbology is widely used to encode the identification number of consumer products.

5 412345 678908

EAN/UPC.

The symbologies described above are all linear symbologies. The symbol is always formed of a single row of symbol characters. Since the early 1990s, two-dimensional symbologies have been developed. Some are qualified as multirow symbologies consisting of two or more vertically adjacent rows of symbol characters. Others are known as *matrix symbologies,* which take the form of a two-dimensional graphic that is decoded in its entirety and not row by row.

Two-dimensional symbologies can encode a large amount of data in a small amount of space. As the development of these symbologies is relatively recent and because the specifications are not fully stable yet, they are not used on a large scale.

PDF417 is a two-dimensional, stacked bar-code symbology. In PDF417, the basic data unit or minimum segment containing interpretable data is called a codeword. Every codeword in the symbol is the exact same physical length, and each codeword can be divided into 17 equal modules. Within every codeword, there are four bars and four spaces. The minimum number of modules of any bar or space is one; the maximum is six. The PDF417 symbology defines 929 distinct codewords and supports 12 modes. Each mode specifies the meaning of the codewords. The standard modes are Extended Alpha Numeric Compaction Mode, Binary/ASCII Plus Mode, and Numeric Mode. The number of data characters that can be encoded in a PDF417 symbol depends on the mode being used. In the extended alphanumeric compaction mode, the maximum number of ASCII characters per symbol is 1850. In numeric mode, a symbol can encode a maximum of 2725 digits.

PDF417.

Data Content

The purpose of bar-code applications is to capture data automatically and to process these data in computer applications. Rules are therefore required to specify the way data should be encoded in a symbol. Similarly, when reading a bar code, there must be a way to know accurately what data has been captured. Three methods exist to specify the rules of encoding and decoding data in bar-code symbols:

1. The first method is to establish a one-to-one relationship between the symbology and the data content. In this case, a particular symbology is exclusively reserved to carry certain types of data. The EAN/UPC symbology is an example of this method. EAN/UPC bar codes always carry a number that is a unique identifier of the item on which the bar code is affixed.

2. The second method is based on conventions defined either by a party for its own applications or established on the basis of mutual agreements between two or more parties. The conventions describe the symbology to be used and specific rules indicating the way data elements are to be encoded. This method is appropriate only in internal or closed applications, because the rules are relevant only to the parties agreeing to follow the convention.

3. The third method is based on the concept of *data identifiers,* which are prefixes used to define data fields. Each prefix uniquely identifies the meaning and the format of the data field following it. Two sets of identifiers have been standardized and are used in many bar-code applications.

In the late 1980s, a body called FACT (Federation of Automated Coding Technologies) was formed in the United States to examine existing standards and to produce materials that would permit coexistence of standards from all industries. A dictionary of FACT data identifiers has been put together and was formally approved as an American National Standard by the American National Standards Institute (ANSI) in 1991. FACT identifiers are composed of one alpha character preceded by up to three digits. The FACT identifiers have usually to be complemented by industry guidelines providing additional clarifications regarding the definition and format of the data fields.

Following the requirements of their membership, EAN and UCC jointly developed a system of *application identifiers,* permitting the encodation of a wide range of information. The EAN/UCC system is characterized by the management of a unique identification scheme for products, services, and locations, by a clear and unambiguous definition of data elements and by the strict recommendation of protected symbologies, offering a high level of security.

When considering the use of data identifiers for their barcoding applications, users should consider the following guidelines:

- *Made-to-Order Products.* When the products are manufactured for and shipped to individual customers operating in a specific industry sector, the FACT data identifier set can be suitable.
- *Products for General Distribution.* When products are traded with more than one customer and possibly with more than one industry sector, the EAN/UCC application identifier set should be used.

Applications

The bar-coding technology has gained wide acceptance in numerous applications. Today, virtually all packages, from the smallest units intended for sale to a consumer to the biggest transport units, bear one or several bar codes carrying their identification number and other data relevant to the parties shipping, carrying, or receiving goods.

Scanning at retail point of sale is a major application using bar-code technology. At the end of 1994, the statistics published by EAN International showed that close to 650,000 stores around the world had implemented scanning systems relying on the EAN/UPC identification number and the associated EAN/UPC bar-code symbol. Scanning at point of sale enables to automatically register the sales through price-lookup files. It also opens up the opportunity to implement a wide range of applications such as inventory management, automatic reordering, and sales analysis.

A rapidly growing field of applications using bar-coding technologies lies within the supply chains. Goods ready for shipment by a supplier are packed, and each package is numbered and bar-coded with a unique number. Before the physical delivery of the merchandise, the supplier sends an electronic message to the delivery point, advising about the arrival of the goods. This electronic message contains the unique identification number of each package and the description of its contents. When processed by the receiver, the electronic message is matched against the original purchase order and stored in a computer database. When goods arrive, a bar-code reading device scans the unique number identifying the goods, and the computer makes the link with the information previously stored. The system is then able to show what has to be delivered, and the actual delivery can be checked. Inventories can then be updated automatically, as the information is already available in electronic form.

Printing Bar Codes

Virtually any printing technology can be used to print bar codes, provided it is accurate enough to achieve the right level of required quality. The printing processes fall into two categories: commercial and on-site. The choice between these two approaches is determined by the nature of the information to be encoded and the number of codes to be printed. Typically, if the information is static (eg., the identification number of a product to be placed on a package), and if the number of codes to be printed is large, the traditional commercial method using film masters is appropriate. If the information is variable (ie, different for each item or short series of items or if the quantity required is small), then on-site printing processes should be used.

The commercial printing techniques use a master image of the bar code on the printing plate. Film masters are very accurate and are available on the market from commercial companies. The printing process itself generally generates a print gain due to the ink viscosity, the pressure of the printing plate, or the type of substrate being printed. This print gain can be evaluated and compensated in the film master itself by reducing the size of the bars. The party ordering the film, the film master supplier, and the printer must work closely together to achieve a quality bar code.

The on-site printing techniques may use a piece of hardware acting as graphics controller between the computer and the printer. They may also use "intelligent" printers incorporating the controller equipment. Finally, software is commercially available to generate the picture of the bar code and send it to a printer. The printing systems may be based on character-by-character impact, on serial dot-matrix, or on linear-array technologies. The choice of the appropriate technology depends on many parameters linked to the application requirements.

Reading Bar Codes

Many types of devices are commercially available to read bar codes. They all illuminate the symbol and analyze the resulting reflectance. Areas of high reflectance are interpreted as spaces, while areas of low reflectance do represent bars. The reflected pattern of bars and space is converted to an electric signal that is then digitized. The decoder assigns a binary value to the signal and forms a complete message. The message is checked by the decoder's software and transformed into data, according to the appropriate decoding algorithm relevant to the symbology being read.

Fixed-beam readers depend on external motion to read the symbol. This can be provided by an operator moving the reader across the symbol or by providing movement to the symbol in front of the reader. A low-cost popular reading device is the handheld contact scanner. It requires an operator to move the reader smoothly over the symbol.

Moving-beam readers use a mirrored moving surface to provide the illumination. The light source appears as a continuous line of light. The most common moving-beam readers in use are generally referred to as *laser scanners.*

Imaging devices are also used to read bar codes. They operate similarly to a camera. The reflected image of the bar code is projected onto photodiodes composed of many photodetectors. The photodetectors are sampled by a microprocessor and produce a video signal that is then decoded.

The choice of a reading device is dictated by many parameters linked to the application and to economic criteria.

BIBLIOGRAPHY

1. EAN General Specifications, available from EAN International, rue Royale 145, B-1000 Brussels, Belgium.
2. UCC manuals, available from the Uniform Code Council, 8163 Old Yankee Road, Suite J, Dayton, OH 45459 (U.S.A.)
3. C. K. Harmon, *Lines of Communication, Bar Code and Data Collection Technology for the 90s,* Helmers Publishing, NH, 1994.
4. R. C. Palmer, *The Bar Code Book, Reading, Printing and Specification of Bar Code Symbols,* Helmers Publishing, NH, 1991.
5. W. H. Erdei, *Bar Codes, Design, Printing & Quality Control,* McGraw-Hill, Interamericana de Mexico, 1993.

HENRI BARTHEL
EAN International
Brussels, Belgium

CODE MARKING AND IMPRINTING IN THE COMPUTER AGE

Since the last edition of this encyclopedia, many changes have taken place in the packaging industry. Nutritional information on packages is now a law. Noncontact printing has come of age. Line speeds are higher. New and better packaging materials extend the life of products. But perhaps the most significant has been the adaptation of the computer to the on-line printing and coding process. Date codes, lot numbers, line time, sell by data, shift information, bar codes, selling prices, and other unique customer and producer information are routinely downloaded automatically from a remote, or local computer or a hand controller directly into the printer or coder. This change from requiring the operator to monitor equipment, change type, set dials, or key in new data has done much to eliminate error, increase productivity, and, of course, reduce cost. One maker of on-line coders recently advertised, "Set up, start running, and walk away from." Certainly a welcome change from the labor-intensive equipment of yesterday. All plants, large and small, are installing automatic equipment to enable them to remain competitive. There is little doubt that good packaging with attractive graphics and easily found product information increases sales.

Today, most products are required to have some kind of machine readable code, either the U.P.C Code for over-the-counter sales, a code like PDF-417 for electronic data interchange, or an industrial bar code (Code 128, for example; see Code, bar) used for inventory control. Accompanying most of these codes will be the human-readable equivalent. These codes will be read by electronic scanners placed beside conveyors, or even in some remote location, and they have to be correct. Both format and content must meet rigid requirements of both the accepted standards for the code, and the required data of the producer and/or customer. A means must be provided to verify the codes. It makes no sense to print variable bits of information on a product if the printed information is incorrect or the format is wrong. Many of the major chains, and large producers, are penalizing suppliers who place incorrect data on their products with fines that can range up to $100,000. Computer-controlled printing has helped immeasurably to make sure all is correct before the product is shipped. When one realizes that the manufacturer, the wholesaler, and the end user all must read the same codes, and each may extract different information using a variety of different makes and models of scanners and computers, the need for accuracy becomes apparent.

Another problem that the printer must address is maintaining the integrity of the packaging material. There are many new materials being developed to help keep products fresher longer, prevent tampering, and facilitate opening and resealing. The printing method chosen must be able to place the required information on the package, make sure that it will remain readable as long as the consumer has the product, and additionally, assure that the printing process does not fracture or damage the material in any way.

You may have to consider another problem if you print directly on the product. For example, inks used to identify a candy or pharmaceutical item may have to be approved for human consumption. It is very important that the inks, cleaners, thinners, or other chemicals that you use in the printing process do not migrate through the packaging material. No contamination of item or package can be allowed to take place either during the printing process or later during the shelf life of the item. A good example is the plastic bag used by the Red Cross for blood drawings and transfusions. There is printing and labeling directly on the bag and the associated tubes. The adhesive used on the labels must be approved, and the printing and all chemicals involved must remain on the surface of the labels and plastic parts. Nothing is allowed to migrate through the label or bag. The printing must be resistant to a minor rub with alcohol, and the label or adhesive must not separate, allowing the label to fall off. Many other products have similar requirements.

Liability is a continuing concern. A growing list of regulations, and regulatory agencies, make it even more important to have correct information on the package in the event a recall is necessary. For example, a recent mandate by the Food and Drug Administration (FDA) requires 100% on-line code verification for some products. You should to be able to identify only those products and lots that may be affected, and not have to recall everything you ever produced.

In addition to open dating, nutritional labeling in now a requirement. Are you going to preprint this information directly on the package, add the data on-line, or will you apply a label? The same printer used for coding and dating may not be able to print a nutrition label.

If you have not made a recent visit a trade show covering packaging, be prepared for some surprises. You will find computers and programs designed to control some part of your operation in virtually every booth. Companies specializing in software have developed programs especially for the packaging industry to enable everyone to meet accepted standards and constantly changing requirements. Product-specific labels or data unique to one customer are no longer a problem. An order going to the automotive industry requires Code 39. The same product going to a wholesaler may require Code 128 for warehouse inventory, and when it is sold over the counter, you may need U.P.C. You may need to preprint, then label, and finally print variable data on-line. These new software programs allow you to do this, plus design special labels or add unique data often while the equipment is running. It is no longer necessary to design a label, send it out for printing plates, and then wait several days before you can start printing. A central computer can send current or custom label information to local PCs on an "as needed" basis. The local computer then downloads to each printer or print/apply stations as required and not only decides which label to print but also keeps track of the number of labels printed for real-time inventory information.

If you print your products and then palletize and shrink-wrap, you may not be able to scan the labels through the plastic. You need a pallet label. How do you get the new label? One way is to scan a product label into your printer controller, and a new label containing all the information about the pallet load will be printed ready for you to apply either automatically or by hand.

Now that some of the requirements of the modern packager are apparent, it's time to look at a few of the ways you can get these necessary marks on your products.

A quick scan of the more popular trade magazines that cover the packaging field shows that four package identification techniques predominate: ink jet, hot stamp, wet ink, and labels. There are dozens of manufacturers, vendors, oem's, var's all vying for your business. How do you choose? What should you look out for? What system is best for you? Following is a brief look at each of these identification methods. For details of any of the systems you should refer to the information sources following this article or contact manufacturers directly.

Ink-jet printing is growing in popularity. Major advances in this noncontact printing method have taken place in recent months. For years continuous ink jets that emitted a constant stream of ink droplets were all that were available. The area around the print station tended to be rather messy, and the print quality was often marginal. The system required a great deal of operator attention and used inks, cleaners, and thinners that contained solvents. The newer ink-jet printers are *drop-on-demand,* ejecting ink only as required, and are much more efficient; print finer, more clearly defined characters; and use water-based inks and chemicals that are environmentally safe. Multiple drop-on-demand jets in one printing head are common, and those jets using hot-melt inks eliminate the need for all cleaners and solvents. The inks available for most ink-jet systems are now water-based and environmentally safe.

A typical application is a food-packaging company with 200 different products packaging under several different brand names. They print 1.2 million legends per day 220 days per year. The company has installed nearly 100 drop-on-demand multiple-jet hot-melt printers, all computer-controlled. This type of ink jet can print up to 750 characters per second. The printers are mounted on label applicators that traverse back and forth over a three-lane Multivac packaging machine. Preprinted labels are used, and the ink jets add the best before date, weight, price, and packaging date. As the printers are computer-controlled, the day's production is downloaded from a computer in production control and all required changes to legends are made automatically. If you use wet-ink printers, some drying time must be allowed to prevent smudging the prints; with hot melt the prints dry instantly. It should be noted that it is very difficult to print bar codes, on-line, with a narrow bar width (below ~0.40) as the vibration of the package passing by the ink-jet head tends to distort the bars and render the codes unscannable.

Hot stamping or *foil printing* is another popular method. The process uses a dry ink bonded to a carrier film. A heated metal plate carrying the legend presses the film against the packaging material, and the image is transferred. Although a separate printing plate is required for each legend change, the printing plates often have a slot to allow the operator to insert metal type to change dates or lot numbers. The hot-stamp prints are instantly dry and will not rub off or smudge. The system is clean, and no chemicals are required for either printing or clean up. Printed results are very impressive with bright golds and metallics creating exceptionally high-quality graphics. Hot stamping is normally used for medium to long runs, and print speeds vary depending on application. When printing detailed ingredient lists, about 120 per minute is possible. For date codes only, 300 or more is common. Very high-quality printing is possible, and bar codes with a narrow (0.0075) bar are routinely printed.

A company specializing in private-brand gourmet coffees has installed hot-stamping units on their form/fill/seal machines running at 60 bags per minute. They have 75 different coffee flavors in package sizes ranging from 1-oz pot packs to 1-lb bags. Both vacuum and valve bags are used with nitrogen flushing to extend shelf life. The imprinters are used to vary the product names, U.P.C number, and ingredient statements, and to add the date codes.

Prior to the installation of the hot-stamp printers, three different labels had to be applied to each bag. All variables are now added to the preprinted roll stock before the bag is formed and all labels eliminated at substantial savings.

Wet-ink printers fall into three broad categories: rotary, reciprocating, and platen printers (see also Printing). A variety of inking systems are used, including ink fountains, hot melt, preinked cartridges, and flexographic systems (see also Inks). Printing bar codes with wet ink is marginal unless the bars are very wide ($\geq$0.040) and the printers are kept very clean and in perfect adjustment.

Roller coders are a common type of rotary printer and often used to print larger bar codes, human-readable information, and rather crude graphics on corrugated cartons on conveyor lines. These printers use rubber printing plates and are driven by a drive wheel or by the printing plate that runs on the carton being printed. Print quality is acceptable if the printers are kept clean and the wear on the plate monitored. Print speed is controlled by conveyor line speed. Other versions of these coders are smaller and can be mounted under a conveyor to place date codes on the bottom of cases, cartons, or cans. The ink supply is often a preinked cartridge. Still a third type is mounted on form/fill/seal machines to mark the web prior to forming the package. These also use preinked cartridges, may have electronic drives, and cycle at up to 700 prints per minute. Typical use is to place date codes, sell-by dates, or prices on candy-bar wrappers. Time must be allowed after the print station for the ink to dry.

Reciprocating printers rarely have a print area that exceeds 2 in. $\times$ 4 in., and are designed for intermittent packaging lines. As the package breaks a photoelectric beam the printhead moves forward over the ink roll and places an imprint on the product. The head retracts and waits for the next package. Printing plates can be rubber, photopolymer, or metal. The inking system can be either preinked rolls, a felt roll inked by hand, a fountain, or an open plate. As the print is made with wet ink, there must be time for the ink to dry before the product is handled to prevent smudged prints. Some of the newer printers use a hot-melt inking system that eliminates the need for thinners, dries instantly without the danger of smudging, and greatly reduces the need for operator attention. Print speeds are in the 60–90-min^{-1} range.

Platen printers are very similar to reciprocating printers, except they may have a much larger print area. The same inking systems may be used, but the difference is that an ink roller moves back and forth over the type rather than the type moving over a fixed ink roller. A typical use is to place data on each package on a sheet of unit dose pharmaceuticals. Printing plates can be either photopolymer or rubber. On any printer using wet ink, print quality is dependent largely on the attention paid by the operator to ink viscosity, print pressure, cleaning the printing plates, and keeping everything in correct adjustment.

Laser printers are becoming more popular to "burn" date codes into plastics or onto labels. A caution is that some types of labels or products tend to be *laser-transparent;* that is, the laser mark does not show up, so sample printing should be considered before purchasing equipment. Laser printers are fast, silent, use no type, no inks, thinners, or cleanup chemicals. Almost without exception, lasers are computer-controlled, and safety features are built in so there is little or no danger to the operators from stray laser light.

Other types of printers seen less frequently are *pad printers,* which use wet ink and are often used for uneven surfaces. Typical uses are printing logos and fixed data on the beavertail paintbrush handle with compound curved surfaces, and on golf balls. Pad printers have the ability to place multicolored logos on very rough surfaces without distortion. *Flexographic* equipment is used for long runs of labels, or to preprint rolls of packaging film, and is used more frequently by commercial printers than for in-house applications. Numbering boxes for sequential numbers or bar codes are available for some printers that either emboss the data into the package surface, or print directly on the product. Other accessories are available for all printing lines to dry the prints, provide consecutive numbering, and in some cases print consecutive bar codes. Combinations of equipment to allow you to perform more than one operation in one pass through the system are also available, for example, print–laminate–die-cut in one pass. Special handling and feeding systems such as vibratory bowls and robotic handling devices are common.

You may have to use labels, as some products do not lend themselves to printing. For example, a very fragile item that must be placed in a bag or box may have to be hand-labeled, or a product may require special graphics and multicolored printing and logos. If you do use labels, will you make them yourself as required, buy them all printed from a label house, or buy them partially preprinted and add variables as you apply them? Will you apply them by hand or with an applicator? What kind of variable information will you want to add—bar codes, date codes, part numbers, or all three? Compliance labeling is becoming much more important. Shipping labels need to be designed so that everyone can make use of the same data to take advantage of electronic data transfer for reorders, bills of lading, and invoices. Fortunately there are many good products and suppliers available to solve all these problems.

Thermal–thermal transfer printers are the most common types of printers for producing labels in house or on-line. As this is written (August 1996), it is estimated that over 50% of all printers sold worldwide are thermal transfer. Most print at 300 dots per inch (dpi), but there is a move to 600 dpi to improve graphic quality, allow for more compact bar codes, and to be able to print the new matrix codes in micro sizes. Print quality is excellent, and typical speeds are 3–5 in./s. Pre-printed labels with logos and colors can be purchased and the bar codes and variables added on line. Thermal printers use a heat-sensitive label stock that tends to turn black on prolonged exposure to bright sun or heat. Labels should be fully life-tested where there maybe a question about the environment. Thermal transfer uses most any label stock and a one-time ribbon, labels will not fade or turn black in bright sunlight, and is a much better choice where long life or difficult conditions may be encountered.

All printing, and especially bar codes, should be printed to the industry standards. As the printed information is now used by second, third, or even fourth parties, it is mandatory that all required data be present in the correct format. Those printing bar codes should have some means of verifying the correctness of their printing. There are numerous handheld devices available to check bar codes for accuracy that sell in the $2000–$4000 range. These devices are programmed to the accepted industry standards for the bar codes and assure that everyone can read the symbols that you print. If you purchase bar-code labels, you should at least spot-check them for accuracy before applying them to your products. A general rule is that if you don't inspect your bar codes, your customers will.

Although there are many ways to place data on your products, there is no one best way. You should evaluate your needs carefully, investigate several different systems and suppliers, get references, then go over your needs again. There is much help available to assist you in choosing the right system. The same printer or system that is correct for receiving may not be the same type as needed for production, inventory control, or shipping. Take your time, go slow, do it right, and do it once.

SOURCES OF ADDITIONAL INFORMATION

Packaging Machinery Manufacturers Institute
38 North Fairfax Drive, Suite 600
Arlington, VA 22203 (703) 243-8555
(Packaging machinery and printers.

Automatic Identification Manufacturers Association
42 Alpha Drive
Pittsburgh, PA 15238-2802 (412) 963-8588
(Bar-code and scanner standards, printers, consultants.)

Distribution Codes Incorporated
46 Old Yankee Road, Suite J
Dayton, OH 45459 (513) 435-3870
(U.P.C. information and standards.)

Benjamin Nelson
Nelson Associates
East Swanzey, New Hampshire

COEXTRUDED BOTTLES.

See Blow Molding.

COEXTRUSION MACHINERY, FLAT

Multilayer coextruded flat film and sheets are produced on single-slot T dies. The overall process is similar to that used for single-layer products of the same dimensions (see Extrusion). The specialized design considerations for coextrusion are discussed below.

Machinery

Extruders. Each product component requires a separate extruder. Several layers may be produced by the same extruder using suitable feed block or die connections. Systems range from two extruders for a simple BA or ABA structure to five or six extruders for high barrier sheet (see Coextrusions for semi-rigid packaging).

Since all of these extruders feed one die, the area behind the die can become a crowded place. Extruders are therefore built as narrow as possible. Vertical gearboxes permit tuck-under motors that reduce space requirements and provide good service access to the motor and other components.

Smaller extruders can be mounted overhead at various angles. Larger machines present an access and height problem when located overhead. The most effective and accessible arrangement is usually a fan layout of larger extruders with some small machines overhead (Fig. 1).

Thermal expansion requires that all but one and sometimes all machines be mounted on wheels with expansion capability both axially and laterally. Height adjustments must also be provided to permit accurate alignment to the interconnecting piping.

PVDC requires special corrosion-resistant extruder construction (see Vinylidene chloride copolymers). High-nickel cylinder lining and Z nickel screws are essential to avoid corrosion and polymer degradation. The optimum ratio is 24 *L/D* length/diameter for this heat-sensitive material. Everything associated with the PVDC extruder is critically streamlined, and all flow surfaces normally contacting or possibly contacting PVDC must be nickel. No screens or breaker plates are used. A long conical tip with matching adapter assures streamlined flow. A PVDC extruder can run over barrier materials equally well with different suitable screw designs.

All other extruders can be of conventional materials of construction. A ratio of 30 *L/D* is desirable for best performance in most cases. These can also be vented for devolatizing when necessary to remove entrapped air or moisture. Venting cannot be used with high backpressures or low screw speeds. Screen changers are used on most extruders to avoid laborious disassembly when screens are plugged. Good screening is essential in barrier sheet extrusion to avoid plugging critical flow passages and pinholes in some layers. Remote-control hopper shutoffs help quick startup and shutdown.

Ethylene–vinyl alcohol (EVOH) requires predrying. Otherwise, it degrades during extrusion with a reduction of melt viscosity. An increase in melt index will disturb layer distribution (see Ethylene–vinyl alcohol).

Scrap or recycled material is often used as a 100% constituent of one layer. This may require special feed handling such as a grooved feed section or the addition of a crammer feeder. If the scrap contains PVDC, the extruder must also include

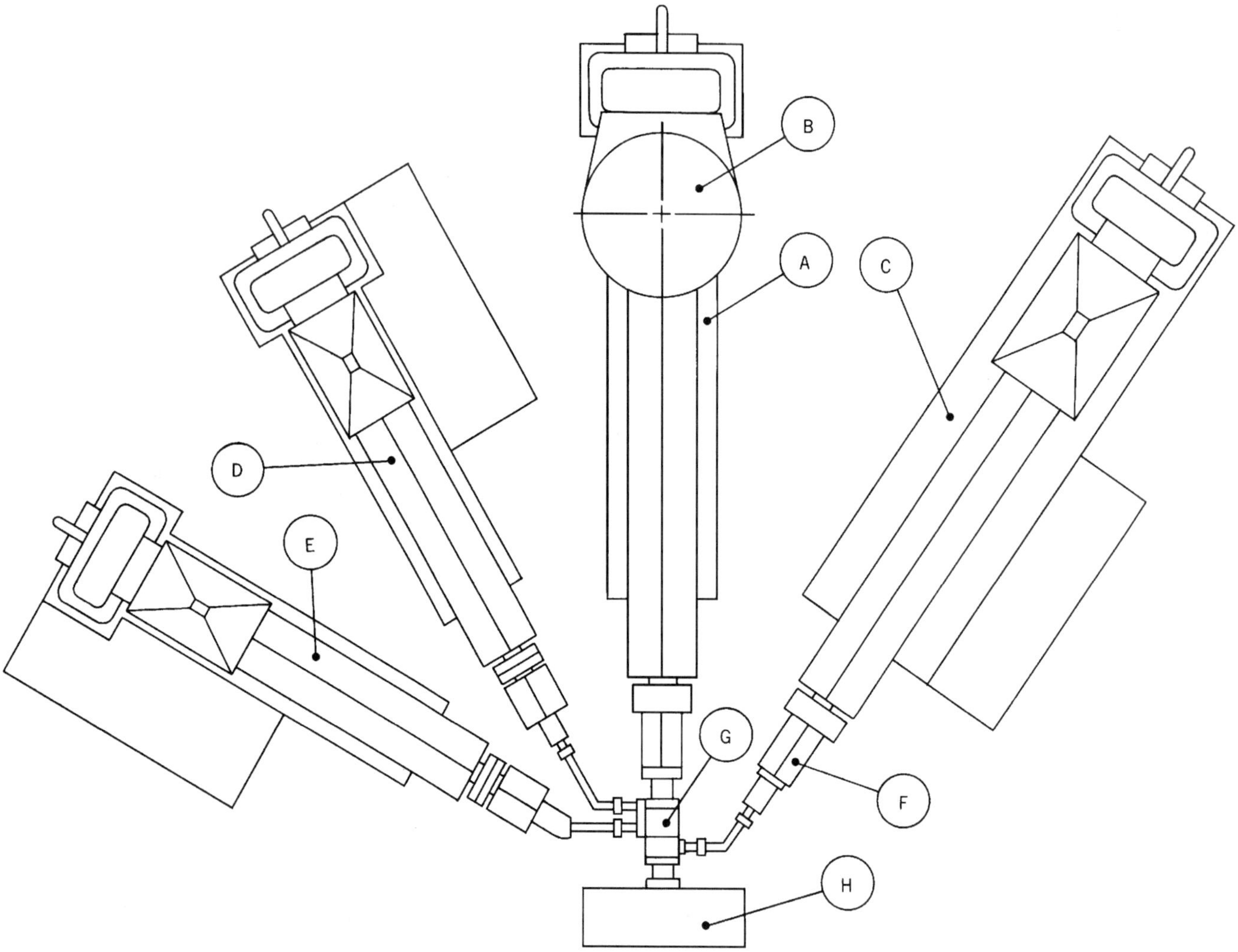

Figure 1. A typical layout for coextrusion (1): A, recycled layer extruder; B, crammer feeder; C, virgin layer extruder; D, glue layer extruder; E, barrier layer extruder; F, static mixer; G, feed block; H, sheet die.

the appropriate materials of construction and streamlined design.

Good mixing with special screws is necessary to homogenize the components.

Melt quality and uniformity are absolutely critical for good multilayer coextrusion. Small variations that are invisible in single-layer sheet can cause severe disturbances in coextrusion resulting from layer interactions. It therefore cannot be assumed that an extruder that works well in single-layer service is suitable for critical coextrusion work.

Cylinder cooling system. Melt viscosity is the major factor controlling layer distribution. The ability to control melt temperature level upward and downward to some extent is essential to permit layer distribution control. This requires conservative speed extruder operation. Complex sheet extruders should therefore always be larger for a given capacity than those used for single-layer extrusion. Closed-loop liquid cooling rather than air cooling is desirable on larger extruders to achieve desired melt temperatures.

PVDC extruders of any size should be liquid-cooled for fast cooling in case of problems. Automatic fail-safe liquid cooling is often used to cool the extruder in case of a power failure.

Layer uniformity and stability can be improved by two devices that are relatively new to extrusion.

Gear pump. Gear pumps provide positive output delivery systems for extruders. They permit accurate control of the content of each layer and ensure that all layers are present in the preselected proportions. The pump is run at an accurately controlled speed. The extruder speed is automatically regulated to maintain a constant feed pressure into the gear pump. All variations in extruder output are therefore automatically compensated. The regrind extruder, which is subject to the largest variability, should be fitted with a gear pump.

Static mixer. Static mixers play an important part in stabilizing melt uniformity. This is particularly important with viscous polymers such as PP and HDPE, which have long stress–relaxation times.

The mixer also provides an extended residence time at low

Figure 2. Feed block and piping.

shear rate for stress relaxation after the high shear in the extruder and gear pump. The static mixer is best installed as the last element prior to the feed block.

Piping. The extruder output is conveyed to the feed block or die through a feed pipe. There are a number of important considerations regarding this technical plumbing (Fig. 2):

It should be as short as possible with a minimum of bends. All bends should be smooth to avoid material hangup.

The internal diameter should be large enough to avoid large pressure drops, which cause a rise in melt temperature, but not so large as to create stagnation.

The wall thickness should not only take operating pressure into consideration but also act as a good heat sink and distributor. Polymer-filled pipes can be subjected to enormous thermal expansion pressures during heating.

Heating must be very uniform to avoid hot and cold spots.

Low voltage density heaters with almost complete pipe coverage are desirable. Heater tapes are dangerous owing to the possibility of poor uniformity of heat distribution.

Control thermocouples must be carefully located to sense the actual pipe temperature.

The construction material must suit the polymer to be conveyed.

Pipes should be easily and quickly disconnected for cleaning and access. Longer pipes should be sectionally assembled to help with this. C clamps are ideal for coupling feed pipes. These also permit rotational motion and compensation for minor misalignment without leakage.

Methods

Two different methods are used to coextrude flat film and sheet: multimanifold die and feed-block coextrusion.

Multimanifold die. The molten polymer streams are fed to separate full-width manifolds in a T die. They are merged prior to exiting from a common slot. These dies are complex and expensive but provide for accurate adjustment of individual layer profiles. The number of layers is limited by the die design, and five appears to be the practical upper limit. The layer capability can be increased by using a feed block on one of the manifolds. Layer adjustment is tedious owing to the great number of adjustment points, and these dies are usually limited in use to single-purpose applications.

Feed-block coextrusion. The product is coextruded on a conventional single manifold T die preceded by a feed block in which the layers are formed. This is the most frequently used process for complex structures. It has been the object of many patents and much litigation.

Feed blocks combine the polymer layers in the structure arrangement desired for the finished sheet in a narrow width and a relatively thick cross section. This makes the layer assembly fairly easy. Thereafter, laminar and nonturbulent flow in the die is necessary to maintain the desired structure.

Viscosity matching of components is essential, that is, the viscosities of the separate components must be alike. Higher-viscosity material displaces lower viscosity material at the edges of the die. Even materials having apparently identical viscosity may not flow evenly because of interfacial slip or die surface drag. In spite of this, viscosity matching works very well for many complex structures. Viscosity differences can often be compensated by temperature adjustments. Feed blocks also incorporate mechanical compensating devices. Consequently, layer distribution uniformity in the 1% range is attainable across the sheet.

There are three major feedblock systems in use commerically. Each has its advantages and disadvantages. All use the principle of nonturbulent laminar flow through the die to achieve good results. The difference between the systems is in how the layers are assembled before the die.

The *Dow system* (Dow Chemical Co.) uses a square die entrance with the height of the die manifold. The layers are assembled in one plane through a series of streamlined flow channels. Details are covered by secrecy agreements and therefore cannot be disclosed. This coextrusion technology has been developed around Dow's saran PVDC and excels in this field. The Dow feed block incorporates a number of flow adjustments. These feed blocks are available from Dow machinery licensees who include most sheet extrusion system builders (1).

The *Welex system* uses a circular flow passage, which is the usual die entrance configuration for single-layer extrusion (see Fig. 3). The layers are assembled sequentially in and around the cylindrical flow. The system is modular so that

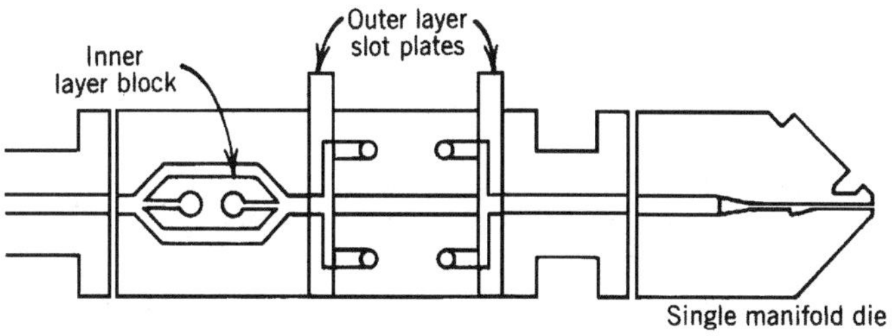

Figure 3. Welex feed block (2).

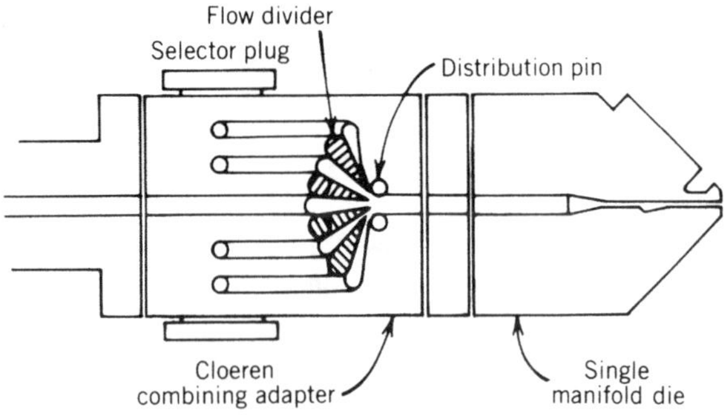

Figure 4. Cloeren feed block (3).

further layers can be added to a given feed block at any time. Inner- and outer-layer feed components are fixed but are removable for correction or adjustment (2).

The *Cloeren system* uses a rectangular die entrance with the height of the die manifold and a width of about 4 in. (100 mm) (see Fig. 4). The block is essentially a miniature multilayer sheet die with movable separating vanes. These permit adjustment of the relative flow gaps during operation. The number of layers is predetermined by the number of flow channels. The feed to each channel can be selected by interchangeable inlet plugs. They are available from Cloeren Company or through builders of sheet extrusion systems. Cloeren also builds multimanifold dies similar in structure to the feed block. These can be fed by a feed block into the central manifold. Although this may seem complex and expensive, it offers a solution to structures with widely differing melt viscosities (3).

Encapsulation and Lateral Adjustment

It is generally desirable to limit the width of the barrier layer to less than the full sheet width. This is essential with PVDC to avoid contact with the die surfaces and much of the feed block. This not only permits the use of normal materials of construction but totally eliminates degradation of PVDC by stagnation on a metal surface. This problem is solved by encapsulating or totally surrounding the barrier layer with other polymers so that it is floated through the die.

Accurately adjustable width of the barrier layer permits major savings in barrier-material cost and reduces recycling problems. Preferably, this layer should extend only to the tooling width in the thermoformer. Edge trim from the extrusion operation and from the thermoforming should be single layer.

All three feed-block systems incorporate adjustments for this purpose. This is usually achieved by a flow mechanism in the feed block which adjusts the flow width of the barrier layer. Other layers such as the glue layers can be similarly controlled.

Other Equipment

Controls. The most important aspect of multilayer sheet and film extrusion is control. A system includes a tremendous number of variables and adjustments that affect the layer thickness and distribution. Since 100% inspection is impossible, reliance is placed on the consistency and stability of operation.

A microprocessor control system is ideal to help maintain good control and to alarm process deviations. It can also be easily programmed for rapid product changes and for critical startup and shutdown procedures.

Gauging. Single-layer gauging and control have reached a high state of perfection. Tolerances better than 1% are readily achieved. Measurement of individual layers is possible in certain cases. In thin transparent structures, selective infrared absorption bands permit the separate measurement of widely differing polymers. This method does not work on thick and opaque products, and it does not give the layer location within a structure. Continuous nondestructive-layer measurement remains under intensive development.

Downstream equipment. Multilayer sheet and film use conventional sheet and film takeoff equipment. Nickel-plated rolls are preferably used when processing product containing PVDC to avoid damage to chrome plating in the event of a breakdown. Multiple-edge trimming is sometimes used to separate single-layer from multiple-layer material to reduce scrap recycling problems.

BIBLIOGRAPHY

"Coextrusion Machinery, Flat" in *The Wiley Encyclopedia of Packaging Technology* 1st ed., by F. R. Nissel Welex, Inc., pp. 193–196.

1. U.S. Pat. 3,557,265 (Jan. 19, 1971), D. F. Chisholm et al. U.S. Pat. 3,479,425 (Nov. 18, 1969), L. E. Lefevre et al. (to Dow Chemical U.S.A.).
2. U.S. Pat. 3,833,704 (Sept. 3, 1974), U.S. Pat. 3,918,865 (Nov. 11, 1975), and U.S. Pat. 3,959,431, May 25, 1976, F. R. Nissel (to Welex Incorporated).
3. U.S. Pat. 4,152,387 (May 1, 1979), and U.S. Pat. 4,197,069, (Apr. 8, 1980), P. Cloeren.

General References

L. B. Ryder, "SPPF Multilayer High Barrier Containers," *Proceedings of the Eighth International Conference on Oriented Plastic Containers,* Cherry Hill, N.J. 1984, pp. 247–281.

F. Nissel, "High Tech Extrusion Equipment for High Barrier Sheeting," *Proceedings of the Second International Ryder Conference on Packaging Innovations,* Atlanta, Ga., Dec. 3–5, 1984, pp. 249–279.

J. A. Wachtel, B. C. Tsai, and C. J. Farrell, "Retorted EVOH Multilayer Cans with Excellent Barrier Properties," *Proceedings of the Second International Ryder Conference on Packaging Innovations,* Atlanta, Ga., Dec. 3–5, 1984, pp. 5–33.

COEXTRUSION MACHINERY, TUBULAR

Tubular coextrusion for packaging applications is generally referred to as *blown-film coextrusion,* distinguishing it from other similar tubular processes that produce products such as pipe and heavy-wall tubing. Blown-film coextrusion therefore refers to the process of forcing more than one molten polymer stream through a multimanifold annular die to yield a film consisting of two or more concentric plastic layers (see Extrusion; Films, plastic). The laminar characteristic of polymer

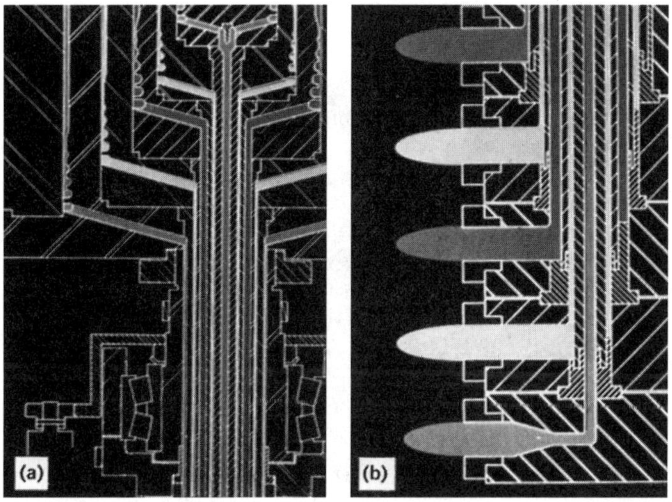

Figure 1. Typical rotating tubular five-layer configuration: (**a**) die structure; (**b**) extruder inlet arrangement.

flow permits the maintenance of discrete layer integrity such that each polymer in the film structure can fulfill a specific and individual purpose (see Coextrusions for flexible packaging).

Coextruded film structures are designed to incorporate one or more of the following objectives: heat sealability; barrier against gas or moisture transmission; high strength (ie, tensile, impact, and tear); color differential; surface frictional properties; adhesive between layers; stiffness (modulus of elasticity); optical quality (clarity and gloss); and reclaim carrier. Combinations of these properties can be achieved by the arrangement of polymer layers in which each polymer exhibits the specific desired property.

Process Equipment

A tubular coextrusion process fundamentally consists of the extruders, die, air ring, collapsing mechanism, haul-off, and winder. These elements are similar to those of single-layer film extrusion except for the die, which must contain more than one flow manifold, ie, layer channel, for extrusion (see Fig. 1).

The added complexity of multilayer die components, coupled with the inherently superior quality requirements for coextruded films, make the die the highest design priority of the extrusion system. The most critical die-design considerations for multilayer applications are (1) structural integrity, ie, the hardware's ability to withstand typical internal pressures of 3000–6000 psi (21–41 MPa); (2) dimensional integrity, the interlocking of and machining precision related to mating parts defining flow-stream concentricity; (3) polymer-flow distributive quality, in order to utilize a range of diverse materials; and (4) reduction of design-flow restriction, permitting extrusion of high-viscosity polymers.

Closely related to the die issue is the frequent need to rotate (or, preferably, oscillate) the die assembly for the purpose of randomizing film-thickness variations across the entire windup width. In the seal section, where there is an interface between fixed and oscillating members, polymer pressure is large, ie, typically 5000 psi (34.5 MPa). Because the seal must act against this force, the seal design must be well qualified, and thrust-bearing and seal-maintenance costs are likely to be high.

Alternative methods sometimes used for thickness randomizations are oscillating haul-off assemblies, rotating winders, or rotating extrusion systems. Each method poses some significant technical difficulty worthy of extensive selection and design consideration.

Specialized Process Design

Because coextruded films often employ plymers uncommon to those used in single-layer extrusion, some unusual process design criteria, discussed below, are added for multilayer systems.

Degradable polymers. Most of the gas-barrier resins are vulnerable to temperature degradation. This imposes a need for specialized die streamlining and extruder-feedscrew design. The feedscrew configuration is critical in minimizing melt temperature and ensuring uniformity of temperature and viscosity across the melt-flow stream.

High modulus of elasticity (film). Many barrier and high-strength polymers exhibit modulus, ie, stiffness, characteristics that cause unique web-handling and winding difficulties. The elimination of web wrinkles and flatness distortions becomes a critical design objective related especially to collapsing geometry, idler and nip-roll size, and line-drive quality. The handling of stiff webs usually entails relatively high equipment costs because higher tension levels are required, along with more precise tension control. Use of highly accurate regenerative dc drive equipment is usually advisable for coextruded films. The high-modulus webs also dictate greater hardware rigidity and tighter roll-alignment tolerances.

Four-side treatment capability. Because of layer-thickness structure considerations in coextrusion, the surface to be printed may be extruded as the inner layer of the bubble. This shifts corona-treatment requirements (for subsequent ink adhesion) from the outer to the inner layer. Two treater stations are sometimes installed on coextrusion systems; one is for the treatment of the inner layer downstream of the web separation.

High film surface coefficient of friction (COF). For many high speed sealing applications, as well as such products as stretch film, multilayer film surfaces are abnormally tacky. One of the principal advantages of the coextrusion process is its ability to create such properties with relatively low additive content concentrated in individual layers. However, very high resultant COF values can cause unusual web-handling difficulty, especially in relation to the bubble-collapsing function. This aggravated geometric problem, ie, flattening a cylinder (bubble) into a single plane, can be alleviated with the use of very low friction collapsing means. In contrast to the more conventional wood-slat configuration, low friction systems employ rollers on ball bearings or air-cushion surfaces to minimize film-surfacing drag forces during collapsing.

Quality-Control Requirements

Multilayer films, because of their enhanced physical properties, frequently command premium selling prices; however, these films also necessitate several added cost factors and engineering complications related to process design. The cost differential is due to the film's added value, exemplified by more stringent thickness-uniformity and winding-quality standards. These are described below.

Temperature control. In addition to plant-space problems associated with multiple extruders, tighter film-quality specifications dictate improved temperature control in a smaller control console. Most recent coextrusion-system plans employ microprocessors to save space and take advantage of digital-control logic. Achieving process-temperature stability is often essential for coextruded films, in contrast to the fluctuations and errors normally tolerated in single-layer processes.

In-line blending. Because of the many types of raw material used in coextrusion, the purchase and storage of specialized-resin blends is impractical. There is an advantage to in-line blending of additives, and investment plans for a complex coextrusion system generally include a high priority for blending equipment. This is also logical in view of the relatively high per pound (kg) cost of the special resins and additives involved and the particularly high quality demanded of multilayer products.

Layer-thickness control. Individual layer thickness must be carefully monitored either by tedious off-line measurement or in-line by gravimetric (weigh-feeding) extruder loading. A difficult technical objective unique to coextrusion, layer-thickness control is a key process-control priority that provides the opportunity to achieve cost savings or the liability to waste raw material and produce defective film. Although layer-thickness measurement can be achieved with spectrophotometers, weigh feeding seems the most practical and reliable means of controlling layer percentages.

Roll-winding quality. Roll-conformation requirements associated with multilayer films are usually severe, representing a more costly and complicated winder configuration. In-line slitting is a common cost-saving requirement, encouraging the use of advanced web-handling technology in the categories of alignment precision, web spreading concepts, and tension and speed control (see Slitting and rewinding machinery). Typical high multilayer line speeds, often 200–500 ft/min (61–152 m/min), dictate the incorporation of automatic cut and transfer mechanisms. Manual roll transfers are not practical at these speeds and with multiple slits. Additionally, the broad range of film elasticity, stiffness, thickness, and surface tack encountered in coextruded applications demands extraordinary winder versatility and performance quality.

Economic Factors

Most blown-film coextrusion systems operate in an output range of 200–1000 lb/h (91–454 kg/h). A typical average rate is 300 lb/h (136 kg/h). Although some two-, four-, and five-layer systems exist, a common installation utilizers three extruders, even when producing 2-layer products. Usual extruder combinations include 2.5-in (6.4-cm) dia and 3.5-in. (8.9-cm) sizes, although many 4.5-in. (11.4-cm) extruders are also used. Some lines operate at 2000 lb/h (907 kg/h) with 6-in. (15.2-cm) extruders.

In the United States, the coextrusion industry consists of a large population of in-line multilayer bag operations in addition to those requiring film winding. Investment levels and process-quality requirements are usually not as high for the bag operations. A three-layer 300-lb/h (136-kg/h) in-line bag extrusion system, for example, costs approximately $300,000; a film-winding version with the same output specification would probably cost at least $400,000.

Operating costs for coextrusion systems are similar to those of single-layer extrusion except for the higher initial investment, ie, typically 50% higher for coextrusion of the same output category. Energy costs are equivalent ($0.03–0.05/lb or $0.07–0.11/kg) to those of single-layer extrusion, and manpower requirements vary only slightly. Labor costs per unit weight are often higher for coextrusion, not as a function of manpower requirements but because of more elaborate processes require greater skills.

A frequent important economic incentive for the manufacture of coextruded films is that premium film pricing reduces the cost percentage of raw materials, eg, resin. Therefore, multilayer products are generally reputed to offer higher profit margins than their single-layer counterparts.

Scrap reclaim can often be an economic disadvantage with coextrusion. It may be limited or prohibited by incompatibilities between the polymers of corresponding layers, a complication particularly prevalent among specialty food-packaging films that contain gas-barrier resins. In these cases, reclaimed scrap may only be eligible for insertion into a thin adhesive layer, thus severely limiting reclaim percentages. Conversely, some coextruded films are designed specifically to exploit high scrap-input potential. In these cases, high loadings of scrap or reprocessed resin are sandwiched between skin layers of virgin polymer.

BIBLIOGRAPHY

"Coextrusion Machinery, Tubular" in *The Wiley Encyclopedia of Packaging Technology,* 1st ed., by W. D. Wright, Western Polymer Technology, Inc., pp. 197–199.

General References

W. J. Shrenk and R. C. Finch, "Coextrusion for Barrier Packaging"; R. C. Finch, "Coextrusion Economics"; *Papers Presented at the SPE Regional Technical Conference (RETEC)*, Chicago, June 1981, The Society of Plastics Engineers, Inc., Brookfield Center, CT, pp. 205–224 and pp. 103–128.

R. Hessenbruch, "Recent Developments in Coextruded Blown and Cast Film Manufacture," *Papers Presented at COEX '83,* Düsseldorf, FRG, Schotland Business Research, Princeton, NJ, 1982, pp. 255–273.

N. S. Rao, *Designing Machines and Dies for Polymer Processing with Computer Programs,* Macmillan, New York, 1981

R. L. Crandell, "CXA—Coextrudable Adhesive Resins for Coextruded Film"; G. Burk, "On Line Measurement of Coextruded Coated Products by Infrared Absorption", *Papers Presented at TAPPI Coextrusion Seminar,* May 1983, TAPPI, Atlanta, pp. 89–90.

Properties of Coextruded Films, TSL #71–3, E. I. duPont de Nemours & Co., Inc., Wilmington, DE.

C. D. Han and R. Shetty, "Studies of Multi-layer Film Coextrusion," *Polym. Eng. Sci.* **16**(10), 697–705 (Oct. 1976).

D. Dumbleton, "Market Potential for Coextrudable Adhesives," *Pa-*

pers Presented at COEX '82, Düsseldorf, FRG, 1982, Schotland Business Research, Princeton, NJ, 1982, pp. 55–74.

G. Howes, "Improvements in the Control of Plastics Extruders Facilitated by the Use of Microprocessors, *Papers Presented at the TAPPI Paper Synthetics Conference '81,* TAPPI, Atlanta, pp. 21–31.

"Coextrusion Coating and Film Fabrication," TAPPI Press Report 112, Atlanta, 1983.

COEXTRUSIONS FOR FLEXIBLE PACKAGING

Coextrusion technology continues to be one of the leading growth areas for new flexible packaging materials. Coextruded films continue to replace many coated paper products such as wax-coated glassine. They are expected to replace many laminations and solvent coatings because of the elimination of processing steps, thus reducing product costs. Coextrusion coating and lamination have expanded in use because of cost and additional flexibility in product attrbutes when compared to monolayer coating and lamination.

A coextruded film is best defined by distinct layers of different polymers formed by a simultaneous extrusion of the polymers through a single die. The resultant film is called a multilayer structure. Coextrusion coating–lamination is defined by simultaneous extrusion of distinct layers of different polymers through a single die, whereas the extrudate forms a coating on a substrate or laminates two substrates together.

The principal reason for development and utilization of coextrusion technology in films is cost reduction. A coextrusion allows one to prepare multifunctional packaging materials in one manufacturing step as opposed to the traditional multistep processes, thereby reducing cost while retaining functional properties. For example, a single-step coextrusion of high-density polyethylene (HDPE) and ethylene–vinyl acetate (EVA) can replace an lamination of oriented polypropylene film to low-density polyethylene (LDPE) film for some packaging applications. Such a lamination requires four manufacturing processes: polypropylene film manufacture, polypropylene film orientation, polyethylene film manufacture, and the lamination process. The potential for cost reduction is apparent (see also Multilayer flexible packaging).

Coextrusion coating and lamination development is also driven by cost reduction and to some extent improved product performance. For example, replacing a monolayer extrusion of polyethylene with a coextrusion of polyethylene and ethylene acrylic acid copolymer will significantly improve adhesion to aluminum foil containing structures. Coextrusion, either in producing multilayer film, coatings, or laminations, can be used to produce optimum flexible-packaging products while reducing costs in materials. The potentials of cost reduction and improved product performance will continue to drive the usage of coextrusions in flexible-packaging products.

Advantages of Coextrusions Over Blends

The layers of a coextruded film are generally composed of different plastic resins, blends of resins, or plastic additives. The difference between a coextruded film and a resin blend lies in the existence of distinct layers in the coextruded film as opposed to the blend. Figure 1 illustrates the differences.

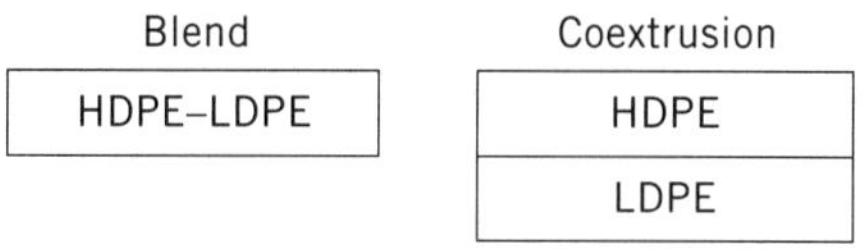

Figure 1. Cross section of film composed of resin blends and coextrusion.

Some structures would not function as blends, but perform very well as coextrusions. For example, a film requiring aroma barrier and easy sealability would be very difficult to make as a blend. As a coextrusion the product could look something like the structure shown in Figure 2.

If the structure shown in Figure 2 were run as a blend, the EVA would degrade at the temperature required to melt the nylon, and the nylon would lose much of its aroma barrier because it would be contaminated by the EVA.

Principal Manufacturing Processes

Five principal manufacturing processes utilize coextrusion technology in producing flexible-packaging material: cast-film coextrusion, blown-film coextrusion (tubular), coextrusion coating, coextrusion lamination, and oriented coextruded films.

The coextrusion processes for cast film, extrusion coating, and laminations are similar in that the coextrusions pass through a flat die. However, there are differences in the remainder of each process. The cast-film process requires extruding the molten extrudate on to a chill roll, quenching it into a multilayer film and eventually winding into a roll.

In extrusion coating, the molten extrudate is extruded onto a substrate such as paper or foil, cooled, and wound into a roll. A coextrusion lamination occurs when the molten polymer is extruded between two substrates gluing them together, quenching the polymer and then winding the lamination into roll form. A typical example of a coextrusion lamination process is illustrated in Figure 3.

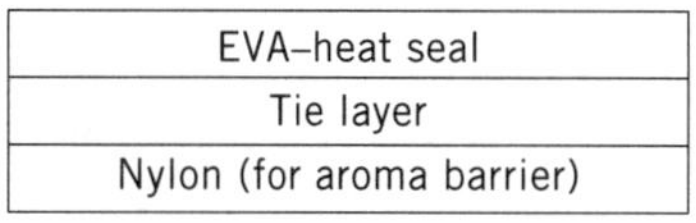

Figure 2. Cross section of functional coextruded film.

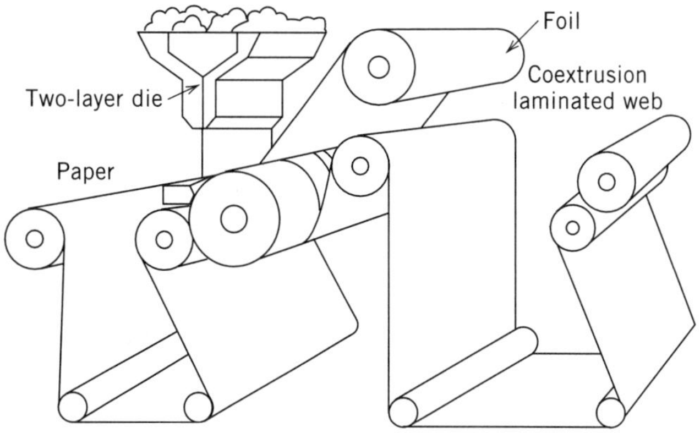

Figure 3. A typical coextrusion lamination process.

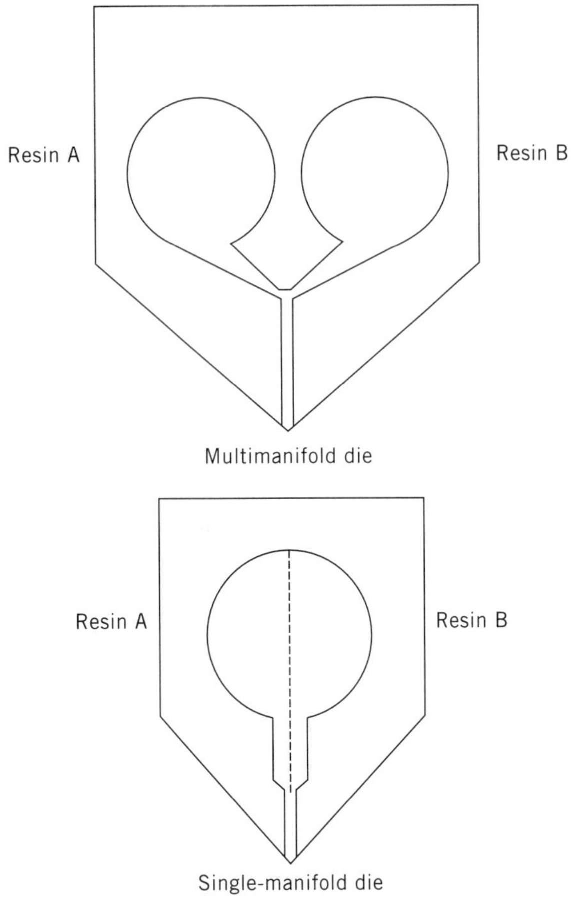

Figure 4. Schematic of a single-manifold die and multimanifold die.

Oriented coextruded films can be made by either the cast or the blown process but require additional processing before being wound into a finished roll.

Cast-film process. The main focal points of technology in the cast processes are the designs of the dies and melt-flow properties of the resins.

Two types of design are used: a multimanifold die and a single-manifold die with an external combining adapter. A schematic of each die design is shown in Figure 4.

In the single-manifold die, separate resin melt streams are brought together in a common manifold. The resin streams are combined in a combining adapter (or feedblock) prior to the die where distinct layers are maintained. In a multimanifold die, resins streams flow in separate channels and are combined inside the die after attaining full width. The important advantages of the single manifold are capital costs, flexibility of operation, thinner layers, and a larger number of possible layers. Coextrusions of over 2000 layers have been reported (1). The design of feedblocks and control of laminar flow of the various components are critical to successful operation of this process.

A potential disadvantage of the single-manifold die is the need to carefully select materials with relatively similar melt-flow properties. The multimanifold system allows for easier processing of dissimilar materials. In addition to a broader range of melt properties, a greater differential of temperatures between layers is possible. In practice, almost all coextrusion is done with the combining adapter and the single-manifold die.

Blown-film process. The blown-film coextrusion process is illustrated in Figure 5. In this process, separate resins are extruded into a circular die. The molten-resin streams are blown into a bubble, cooled by air rings, and collapsed in the primary nip. The tubular film is generally slit for specific packaging applications. The die design for blown film, in addition to being circular, is different from the cast process in that separate melt streams are combined near the die exit or external to the die.

Compared to cast films, blown films generally have more balanced physical strength properties, higher moisture barrier, and greater stiffness. Optical properties such as clarity and gloss, however, are generally inferior to those of the cast process because of the slower rate of crystallization in the blown process. (Blown films are more crystalline than the corresponding cast films.)

Coextrusion coating and laminating processes. Die designs for coextrusion coatings and laminations are similar to those used in the manufacture of cast film. The polymers used are usually lower viscosity and are extruded at higher temperatures, resulting in easier flow properties. Therefore the dies are usually smaller in bulk and rigidity when compared to cast-film dies. The higher temperatures are required to achieve good adhesion to the various substrates.

Oriented coextruded processes. Coextruded films are often oriented to enhance their physical or barrier properties. Orientation of the film is the result of stretching the film after it is quenched. Coextrusions can be oriented in three ways: (*1*)

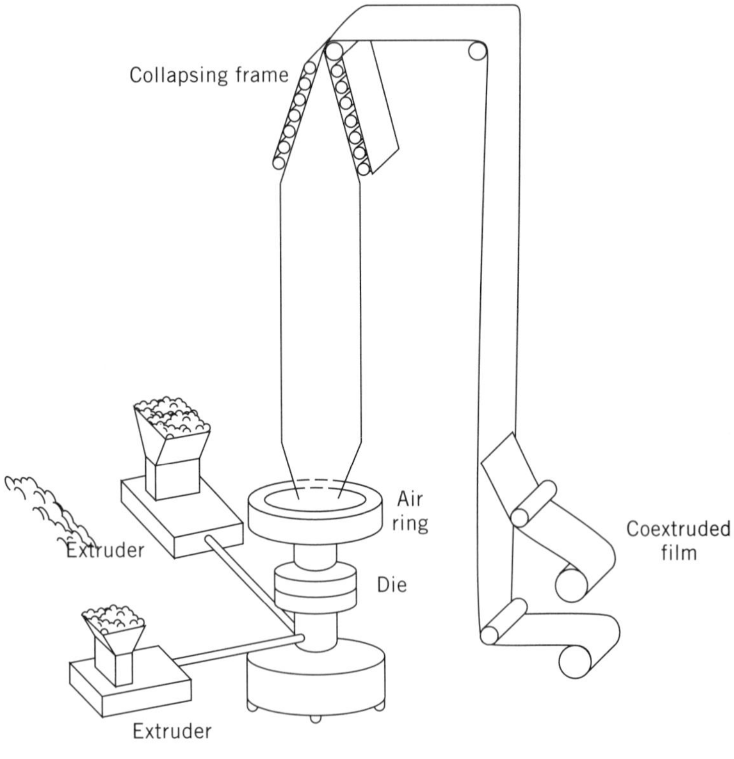

Figure 5. Blown-film coextrusion process.

Table 1. Typical Coextruded Film Structures

Outside Layer	Core Layer	Inside Layer	Remarks
LDPE	White LDPE + recycle	LDPE	Virgin skin layers control surface quality
HDPE	HDPE + recycle	EVA	EVA provides rapid fin seal machinability
EVA	LLLDPE + recycle	EVA	EVA increases lap seal cycle time
LDPE	LLDPE + recycle	LDPE	LDPE limits film's elongation under load
EMA	OPP	EMA	Oriented polypropylene sealability poor without coextruded or coated skin layers

monodirectional—either MD (machine direction) or TD (transverse direction-cross machine), (*2*) sequentially MD and TD, or (*3*) simultaneously MD and TD. The method of orientation chosen is dependent on the materials used and the desired final properties.

Principal Raw Materials

The majority of coextruded structures are made up of polyolefins (polyethylene and polypropylene). This class of material is preeminent because of low cost, versatility, and easy processability. LDPE–LLDPE resins (see Polyethylene, low-density) are used extensively in coextruded structures for their toughness and sealability. HDPE resins are selected for their moisture barrier and machinability characteristics (see Polyethylene, high-density). Polypropylene is chosen for its ability through orientation to provide machineable films with high impact and stiffness properties.

Although the polyolefins are the workhorse grades for coextruded packaging, they are almost always combined with other resins to achieve multilayer functionality. Copolymers of ethylene–vinyl acetate (EVA), ethylene–acrylic acid (EAA), and ethylene–methacrylic acid (EMA) are regularly used as skin layers for their low-temperature sealing characteristics. When oxygen, aroma, or flavor protection is necessary, polymers such as polyvinylidene chloride (PVDC), nylon, and ethylene–vinyl alcohol (EVOH) for clarity are used (see Ethylene–vinyl alcohol copolymers) are used (see Nylon; vinylidene chloride copolymers). Nylon and EVON do not readily adhere to polyolefins, so an adhesive or tie layer is necessary to hold the coextruded structure together. Other polymers such as polycarbonate (see Polycarbonate) or polyester (see Polyesters, thermoplastic) many be used as skin layers to provide unusual thermal integrity for packaging machine performance.

Structures

Coextruded flexible-packaging applications include coextruded films (see Table 1 for a list of structures), laminations, and coatings. In general, coextruded films are preferred to coextruded laminations and coatings because of their cost-effectiveness in use. Because lamination and coating require an extra value-added stage, they tend to cost more.

Coextruded multilayer structures can be divided into three categories: single-resin, unbalanced, and balanced. Many structures that are based on the performance properties for a single resin are coextruded for performance or cost reasons. Unbalanced structures typically combine a functional layer with a heat-seal resin. Balanced structures generally have the same heat-sealable resin on both sides of the film.

Single-resin structures. Single-resin films are coextruded for a variety of reasons. Many commodity film applications may not appear to be multilayer films, yet they actually have three or more distinct layers. Bakery, produce, and trash-bag films, for example, are often three-layer structures. The core material may contain pigment or recycled material, while virgin skin layers control surface quality and machinability. Single-resin coextrusions can also provide a differential coefficient of friction on the two surfaces.

Unbalanced structures. Typical of the unbalanced structures are films designed for vertical form/fill applications with a fin seal. A base resin such as HDPE is augmented by an EVA skin layer for sealability. For horizontal wrappers a polypropylene skin layer is sometimes selected for its higher thermal resistance. In another important unbalanced application, cast polypropylene, which has a limited sealing range, is combined with more sealable polyethylene for single-slice cheese wrappers.

There are multilayer films using only one polymer (AAA), unbalanced coextruded films with two or more polymers (ABC), and balanced multilayer structures with two or more polymers (A/B/C/B/A).

Balanced structures. Balanced coextruded structures typically have a core resin selected for its functionality plus two skin layers that are heat-sealable. Oriented polypropylene films, for example, are increasingly coextruded instead of coated to attain machinable surfaces (see Film, oriented polypropylene). Frozen-food films are typically constructed with an EVA skin layer for enhanced sealability. Heavy-wall bags are regularly coextruded with LLDPE (linear LDPE) cores for impact strength and LDPE skins to limit the film's elongation under load. Primal meats are packaged in PVDC shrink film with EVA skins for seal integrity.

Two main applications that appear to be shifting from monolayer films to coextrusions are overwrap and stretch wrap (see Wrapping machinery, stretch film). Horizontal overwrap machines typically use an MDPE film or an LDPE–HDPE blend.

Coextrusions can provide comparable overwrap machinability at lower gauge. Stretch wrap is difficult to produce as a single-layer structure without blocking. By splitting stretch wrap into a multilayer structure, its LLDPE core can be provided with controlled tackiness on the surface layer.

BIBLIOGRAPHY

1. J. A. Wheatley and W. J. Schrenk, "Polymetic Reflective Materials (PRM)," *J. Plast. Film Sheeting* **10** (Jan. 1994).

General References
Coextrusion patents

U.S. Pat. 3,222,721 (Dec. 14, 1965), M. Reynolds, Jr. (to Anaconda).

U.S. Pat. 3,223,761 (Dec. 14, 1965), G. E. Raley (to Union Carbide Corp.).
U.S. Pat. 3,321,803 (May 30, 1967), H. O. Corbett (to USI).
U.S. Pat. 3,308,508 (Mar. 14, 1967), W. J. Schrenk (to The Dow Chemical Co.).
U.S. Pat. 3,398,431 (Aug. 27, 1968), H. O. Corbett (to USI).
U.S. Pat. 3,320,636 (May 23, 1967), H. O. Corbett (to USI).
U.S. Pat. 3,400,190 (Dec. 3, 1968), H. J. Donald (to The Dow Chemical Co.).
U.S. Pat. 3,477,099 (Nov. 11, 1969), R. E. Lee and H. J. Donald (to The Dow Chemical Co.).
U.S.Pat. 3,479,425 (Dec. 18, 1969), L. E. Lefevre and P. Dreidt (to The Dow Chemical Co.).
U.S. Pat. 3,476,627 (Nov. 4, 1969), P. H. Squires (to E.I. du Pont de Nemours & Co., Inc.).
U.S. Pat. 3,448,183 (June 3, 1969), D. C. Chisholm (to The Dow Chemical Co.).
U.S. Pat. 3,440,686 (Apr. 29, 1969), H. O. Corbett (to USI).
U.S. Pat. 3,524,795 (Aug. 17, 1970), N. R. Peterson (to The Dow Chemical Co.).
U.S. Pat. 3,557,265 (Jan. 19, 1971), D. S. Chisholm and W. J. Shrenk (to The Dow Chemical Co.).
U.S. Pat. 3,583,032 (June 8, 1971), L. O. Stafford (to Beloit).
U.S. Pat. 3,365,750 (Jan. 30, 1968), H. J. Donald (to The Dow Chemical Co.).
U.S. Pat. 3,611,492 (Oct. 12, 1971), R. Scheibling (to Siamp-Cedap).

Die design

R. J. Brown and J. W. Summers, *Plast. Eng.* **37**(9), 25 (Sept. 1981).

Melt rheology

J. H. Southern and R. L. Bullman, *J. Appl. Polym. Sci.* **20,** 175 (1973).
J. D. Han, *J. Appl. Polym. Sci.* **17,** 1289 (1973).
A. E. Everage, *Trans. Soc. Rheol.* **17,** 629 (1973).

General extrusion processes

G. R. Moore and D. E. Kline, *Projection and Processing of Polymers for Engineering,* Prentice-Hall, Englewood Cliffs, NJ, 1984

Eric Hatfield
Larry Horvath
James River Corporation
Milford, Ohio

COEXTRUSIONS FOR SEMIRIGID PACKAGING

This article pertains to flat semirigid coextruded sheet which is a minimum of 0.010 in. (0.25 mm) thick (see Coextrusion machinery, flat). These coextruded sheet structures are thermoformed to produce high-barrier plastic packages (see Barrier polymers; Thermoforming). A similar concept is used to produce high-barrier plastic bottles, except that the bottles are formed from coextruded multilayer tubes instead of flat sheet (see Blow molding).

The production of coextrusions for semirigid packaging was made possible by technology developed in the late 1960s and early 1970s (1,2). Utilization of this technology was initially limited to "simple" structures such as two-layer systems (a general-purpose polystyrene cap layer on a high-impact polystyrene base layer) for drink cups. Commercialization of high barrier coextrusions occurred in the 1970s in Europe and Japan. Large-scale commercial barrier coextrusion applications did not surface in the United States until the 1980s. For purposes of this discussion, barrier materials are defined as those that exhibit an oxygen transmission rate of less than 0.2 cm^3 · mil/(100 in^2 · day · atm) [0.777 c^3 · μm/(m^2 · d · kPa)] (see Barrier polymers). Other techniques that can be used to produce multilayer barrier structures are coating and lamination (see Coating equipment; Laminating). Some advantages coextrusion offers versus these other two methods are thicker barrier layer capability, single-pass production, barrier layer sandwiched between cap layers, and generally lower cost. The potential markets for packages formed from these high-barrier coextrusions include both low- and high-acid food products sterilized by aseptic, hot-fill, or retort methods. These markets obviously represent a significant opportunity for barrier coextrusions.

Table 1. Barrier Materials

Resin	O_2 Transmission Rate[a]	Water-vapor[b] Transmission Rate	Mid-1985 Price, $/lb ($kg)
EVOH (Eval F, Kuraray)	0.035 [0.136]	3.8 [1.50]	2.41 [5.31]
PVDC (Saran 5253, Dow Chemical)	0.15 [0.583]	0.10 [0.04]	1.02 [2.25]

[a] cm^3 · mil/(100 in.^{2} · d · atm) [cm^3 · μm/(m^2 · d · kPa)] at 73°F (23°C), 75% rh.
[b] g · mil/(100 in.^{2} · d) [g · mm/(m^2 · d)] at 100°F (38°C), 90% rh.

Barrier Materials

On the basis of the barrier definition above, only two commercially available thermoplastic resins can be considered as barrier resin candidates for these extrusions: ethylene–vinyl alcohol (EVOH) (see Ethylene–vinyl alcohol) and poly(vinylidene chloride) (PVDC) (see Vinylidene chloride copolymers). The barrier properties of specific grades of these two materials are listed in Table 1. The resins identified in the table are currently the highest barrier commercially available coextrudable resins of their respective polymer classes. Other formulations of both resin types are available offering certain property and processing improvements at the sacrifice of barrier properties.

The most significant technical issue concerning the use of EVOH as a barrier material is its moisture sensitivity. The material is hygroscopic, and its barrier properties are reduced as it absorbs moisture. The importance of this property to the food packager is dependent on the sterilization process, food type packaged, and the package storage conditions. The most severe conditions are encountered during retort processing (see Canning, food). Special consideration to coextrusion structure design and postretorting conditions may be required to achieve the desired oxygen barrier for packages produced from EVOH coextrusions (3).

PVDC is not moisture sensitive and does not exhibit the deterioration of barrier properties shown by EVOH. The challenges associated with using heat-sensitive PVDC are faced by the coextruded sheet producer. Equipment and process design are critical to the production of coextrusions containing PVDC. Concern relating to the reuse of scrap generated in the production of coextrusions based on PVDC is a real economic issue. Development of new material forms and recycle-containing structures is underway with commercialization

Table 2. Structural Materials

Resin	Maximum Process Temperature, °F (°C)	Mid-1985 Price \$/lb (\$kg)
Polystyrene	195 (90.6)	0.49–0.51 (1.08–1.12)
Polypropylene	260 (127)	0.43–0.47 (0.95–1.04)
High-density polyethylene	230 (110)	0.44–0.50 (0.97–1.10)
Low-density polyethylene	170 (77)	0.40–0.44 (0.88–0.97)
Polyester, thermoplastic (heat-set)	>260 (>127)	0.63–0.67 (1.39–1.48)
Polycarbonate	>260 (>127)	1.69–1.81 (3.73–3.99)

targeted for 1985 (4). In the meantime, resin manufacturers are working on the development of other types of barrier materials for coextrusion applications (5).

Structural Materials

The materials generally used to support the barrier resins in coextrusions are listed in Table 2. The maximum process temperature listed is the highest sterilization temperature that packages based on these resins should experience. Polystyrene, polypropylene, and the polyethylenes are the predominant structural materials used in coextrusions for semirigid packaging applications. Structural resin selection is dependent on use requirements, coextrusion processability, and container-forming considerations.

Polystyrene (see Polystyrene) exhibits excellent coextrudability and thermoformability. It can be used in applications requiring low-temperature processing and in some hot-fill applications. Polypropylene (see Polypropylene) is also excellent from a coextrusion-processing standpoint, but it requires special forming considerations. Deep-draw containers from polypropylene-based sheet are most commonly formed using solid-phase forming techniques. Polypropylene can be retorted; but some grades exhibit poor low-temperature impact characteristics which limit their use in applications requiring resistance to refrigerated or freezing temperatures.

High-density polyethylene (see Polyethylene, high-density) offers a significant improvement in low-temperature properties compared to polypropylene, but its suitability in applications requiring retort processing is marginal. Low-density polyethylene would be incorporated in coextrusions requiring good heat sealability (see Sealing, heat) for applications involving low-temperature-fill conditions.

Although coextrusions based on crystallizable polyester (see Polyesters, thermoplastic) and polycarbonate (see Polycarbonate) are not commercially available at this time, these materials are included as structural materials because of their future potential in retort applications. The success of these relatively expensive materials will be dependent on the cost and performance achieved. Considerable developments of coextrusion and forming techniques need to be completed prior to commercialization of coextrusions based on polyester and/or polycarbonate.

Applications

Three representative commercially coextruded structures are shown in Table 3. The transition layers in these structures are materials used to ensure the integrity of the coextrusion. The technology of transition layers is complex and maintained as proprietary by coextrusion manufacturers. The first structure, which uses polystyrene as both cap layers, finds use in form/fill/seal applications because of the particularly good thermoformability of polystyrene (6) (see Thermoform/fill/seal). The second structure has one polystyrene cap layer to maintain thermoformability and one polyolefin cap layer. The polyolefin layer in this case would be the food-contact layer. This structure would comply with the current FDA regulations for aseptic H_2O_2 package sterilization (see Aseptic packaging). The resins that comply with current FDA regulations for H_2O_2 sterilization are polyethylenes, polypropylenes, polyesters, ionomers (see Inonomers), and ethylene vinyl acetates (EVA). Petitions have been submitted for FDA clearance of polystyrene and ethyl methyl acrylate (EMA) as food-contact layers as well. Containers formed from this structure, with polypropylene as the food-contact surface, can also be hot-filled (7).

The last structure shown in Table 3 has the most potential of those listed because it can be used in applications including retort processing. The primary market target for coextrusions with polypropylene as the cap layers is processed foods currently in metal cans (8,9).

In addition to the food-packaging markets, barrier coextrusions can be utilized in the medical (see Healthcare packaging), pharmaceutical (see Pharmaceutical packaging), and industrial packaging markets where barriers to oxygen, moisture, and hydrocarbons are required.

Economics

Simply utilizing resin prices to calculate a material cost for a coextruded sheet structure can be unreliable in determining the economics of barrier plastic package. Using material

Table 3. Commercial Coextrusions

Structure	Application
Polystyrene Transition Barrier Transition Polystyrene	Form/fill/seal Preformed containers Hot fill
Polystyrene Transition Barrier Transition Polyolefin	Form/fill/seal Preformed containers H_2O_2 aseptic Hot fill
Polypropylene Transition Barrier Transition Polypropylene	Preformed containers H_2O_2 aseptic Hot fill Retort

prices only to compare the economics of several coextruded sheet structures on the basis of different resins can result in erroneous conclusions. Items such as required equipment costs, coextrusion output rates, package-forming method and rates, amount of scrap generated, amount of scrap reutilized, container design, and container performance are some of the cost considerations that can be dissimilar for different coextruded sheet structures. Economic comparison of various coextruded barrier packages with alternative packaging materials should be based on a total packaging systems analysis. The current commercial applications and market tests underway show that packages from coextruded sheet offer economic and/or performance advantages versus other packaging materials.

BIBLIOGRAPHY

"Coextrusions for Semirigid Packaging" in *The Wiley Encyclopedia of Packaging,* 1st ed., by R. J. Dembrowski, Ball Corporation, pp. 201–204.

1. U.S. Pat. 3,479,425 (Nov. 18, 1969), L. E. Lefeure and P. Braidt (to Dow Chemical Company).
2. U.S. Pat. 3,557,265 (Jan. 19, 1971), D. Chisholm and W. J. Schrenk (to Dow Chemical Company).
3. K. Ikari, "Oxygen Barrier Properties and Applications of Kuraray EVAL Resins," presented at Coex 1982, sponsored by Schotland Business Research, Inc., Princeton, NJ.
4. W. J. Schrenk and S. A. Marcus, "New Developments in Coextruded High Barrier Plastic Food Packaging," presented at SPE–RETEC, Cleveland, Ohio, April 4–5, 1984.
5. R. McFall, "New High Barrier Polyester Resins for Coextrusion Applications," presented at Coex 1984, sponsored by Schotland Business Research, Inc., Princeton, NJ, Sept. 19–21, 1984.
6. "Cheese Invades Europe," *Packag. Dig.,* 45 (Feb. 1981).
7. "Industry/Newsfocus," *Plast. Technol.,* p 114 (Sept. 1984).
8. "Campbell's Plans for Plastics: 'mm, mm, good'," *Plast. World,* 6 (June 1984).
9. S. A. Marcus, *Food Drug Packag.,* 22 (Aug. 1982).

General References

S. E. Farnham, *A Guide to Thermoformed Plastic Packaging,* Cahners Publishing Company, Boston, 1972.

S. Sacharow and R. C. Griffin, *Basic Guide to Plastics in Packaging,* Cahners Publishing Company, Boston, 1973.

J. A. Cairns, C. R. Oswin, and F. A. Paine, *Packaging for Climatic Protection,* Newnes–Butterworths, London, 1974.

Proceedings of Coex '81, '82, '83, and '84, Schotland Business Research, Inc., Princeton, NJ.

R. J. Kelsey, *Packaging in Today's Society,* St. Regis Paper Company, New York, 1978.

Proceedings of the Seventh International Conference on Oriented Plastic Containers, Ryder Associates, Inc., Whippany, NJ, 1983.

L. B. Ryder, *Plast. Eng.,* May 1983.

S. Hirata and N. Hisazumi, *Packag. Japan,* 25 (Jan. 1984).

A. Brockschmidt, *Plast. Technol.,* 67 (Sept. 1984).

COLLAPSIBLE TUBES.

See TUBES, COLLAPSIBLE.

COLORANTS

Colorants for packaging materials fall into two broad categories: pigments and dyes. Used for both decorative and utilitarian purposes, their diversity is at least as broad as the diversity of packaging materials. This article focuses on their use in inks, plastics, and paperboard. The emphasis is on pigments, which are far more prevalent than dyes in packaging applications.

Pigments are black, white, colored, metallic, or fluorescent organic or inorganic solids that are insoluble and remain essentially unaffected by the medium into which they are dispersed or incorporated. They are small in particle size, generally in the range of 0.01–1.0 μm diameter. Pigments produce color by selective absorption of light, but because they are solids, they also scatter light. Light scattering is undesirable in a transparent material, but desirable if opacity is the goal. Some organic pigments that are extremely small in particle size scatter very little light and therefore act like dyes; for example, Benzimidazolone Carmine HF3C, with a particle size of 0.05–0.07 μm. Some colorless pigments, relatively large in particle size (up to 100 μm), are used as fillers or extenders.

Organic pigments are characterized by high color strength, brightness, low density, high oil absorption, transparent and translucent properties, bleeding in some solvents, and heat and light sensitivity. In the world at large, the major user of organic pigments is the printing-ink industry. In packaging, they are useful for numerous applications such as printing on cartons, labels, and flexible bags. Naphthol reds, for example, are used for soap- and detergent-carton printing because of bleed resistance. Barium lithol is a most important red for packaging flexo and gravure inks, and alkali blue is used in glycol-type inks (see Inks; Printing).

Compared to organic pigments, inorganic pigments are more opaque, less bright, and weaker in tint; but they are more resistant to heat, light, chemical attack, bleed, migration, and weathering. They have higher density, lower cost, and less antioxidant effect. The major use of inorganic pigments is the paint industry. In packaging, they are useful for printing on cartons, bags, and glass bottles. Examples are molybdate orange for gift wrap and vinyl film, titanium dioxide for glass beverage bottles, and cadmium reds for plastics. Metallic pigments such as gold, platinum, and silver help vivify colorants for glass bottles.

Dyes are intensely colored solubilized organic substances that are retained by the medium which they color by chemical bonding, absorption, or mechanical retention. Dyes produce color by absorption of light, without affecting transparency and high optical purity. The major user of dyes is the textile industry. In packaging, dyes are used to some extent in inks for special effects, for coloring paperboard, and to produce tinted transparent plastic containers or films.

Pigments in Packaging

Properties of pigments are a function of the chemical composition as well as other physical and chemical parameters such as particle size, particle shape, particle-size distribution, and the nature of the pigment's surface. Particle size affects a number of pigment properties. Lightfastness improves with increasing particle size, and oil absorption and strength de-

Table 1. Some Factors Involved in Selection of a Pigment

For Printing Ink	For Plastics	For Paper and Paperboard
Color	Color	Whiteness
Masstone	Nature of resin	Brightness
Tintone	End use	Opacity
Printone	Toxicity	Rheological properties
Density	Heat resistance	Bulk
Rheological behavior	Resistance to migration	Specific gravity
Opacity	Bleeding	Transparency
Oil absorption	Crocking	Refractive index
Texture	Bronzing	Use cost
Chemical resistance	Plateout	Color
Acid	Lightfastness	Color migration
Alkali	Weatherability	Flocculation
Solvent resistance	Dispersibility	Gloss
Heat resistance	Electrical properties	Mechanical properties
Oil, fat, grease, soap resistance	Allowance for additives such as antioxidants, uv absorbers	Ink absorbency
Lightfastness	Morphological properties	Sheet strength
Resistance to sterilizing	Filtration characteristics	Abrasion
Bake stability	Effect on mechanical properties	
Pearlescence	Use cost	
Iridescence	Tensile strength	
Viscosity		
Bulk		
Transparency		
Use cost		
Particle size		

crease. Hue is also affected by particle size; for example, an orange pigment usually appears yellower as the size decreases. Narrower particle-size distribution leads to cleaner hue, higher gloss, and lower oil absorption and viscosity. A pigment's light absorption, light-scattering power, and particle size contribute to determining the hiding power of the pigment. Opacity is also affected by refractive index differences between the pigment and the dispersing medium. Selection of a pigment for a specific application depends on a great many physical properties and characteristics. Some of the factors involved in pigment selection are listed in Table 1. Table 2 lists most of the pigments used in packaging materials.

Special-effect pigments. Nacreous pigments, like basic lead carbonate and titanium-coated mica, are used for luster effects for cosmetic containers. Fluorescent pigments and dyes are used in gravure inks for carton printing and special effects in gift-wrap printing. They are also used as colorants for blow-molded bottles, closures, tubs, cartons, and pails. Other special-effect pigments include luminescent and phosphorescent pigments, as well as metallics (eg, aluminum flake and various copper bronzes which vary in shade depending on their chemical composition).

Dyes in Packaging

When dyes are used for coloration, as in plastics, they must be checked for migration, heat stability, lightfastness, and sublimation.

For plastic materials, dyes that are widely used are azo dyes, anthraquinone dyes, xanthene dyes, and azine dyes that include induline and nigrosines. Azo dyes such as Solvent Red 1, 24, and 26, Solvent Yellow 14 and 72, and Solvent Orange 7 are colorants for polystyrene, phenolics, and rigid PVC. Better heat stability and better weatherability are obtained from anthraquinone dyes such as Solvent Red 111, Blue 56, Green 3, and Disperse Violet 1 in the coloration of acrylics, polystyrene, and cellulosics. Basic Violet 10 is a xanthene dye used in phenolics. Solvent Green 4, Acid Red 52, Basic Red 1, and Solvent Orange 63 are used for polystyrene, and rigid PVC. Azine dyes produce exceptionally jet blacks and are used in ABS, polypropylene, and phenolics. A perinone dye, Solvent Orange 60, has good light and heat stability for ABS, cellulosics, polystyrene acrylics, and rigid PVC. ABS, polycarbonate, polystyrene, nylon, and acrylics may be colored with quinoline dye. Methyl Violet and Victoria Blue B, two basic triphenylamine dyes find limited use in phenolics.

For printing inks, five dye families are of particular interest: azo, triphenylmethane, anthraquinone, vat, and phthalocyanine. *Certified food colorants* are used in packaging applications where the printed surface is in direct contact with food. Dyes for paper include acid, basic, (including resorcine and alizarine), and direct dyes. Basic directs are used to color containers, boxboards, wrapping paper, and multiwall bags.

Colorants for Printing Inks

Printing inks are used for a broad range of packaging items such as cartons, bags, labels, metal cans, rigid containers, and decorative foils. Table 2 highlights the pigments used for printing (see Inks). Important qualities of pigments for printing inks are contribution to printing properties (ie, rheology and bleeding), print appearance (ie, the color has sales appeal), and useful service life (ie, resistance to fading and chemicals). In general, most ink pigments can be utilized in most types of inks; but there are minor differences in a pigment's performance that must occasionally be considered. For example, pigments for lithographic inks must not bleed in water or very mild inorganic acidic (phosphoric acid) solutions. Glossy finishes require small-particle-size pigments. Most printing-ink pigments are organic.

Extender pigments are essentially transparent or translucent and contribute neither to hiding power nor color. They are formulated into inks to extend the covering power of strong pigments and enhance the working properties (eg, to increase viscosity without affecting color).

Colorants for Plastics

In selecting a colorant for plastics, in addition to the colorant characteristics listed in Table 1, the resins, its properties and compatibility, and the method by which it is processed are critical factors. Because PVC and its copolymers liberate acid at high temperature, acid-sensitive colorants like cadmium reds and ultramarine blue must be used with care. Hansa yellow tends to crock (smudge) in polyethylene. In polyethylene, impurities in pigments such as cobalt and manganese should be avoided. Polyacetals and thermoplastic polyesters are sensitive to moisture. Molten nylon acts as a strong acidic reducing agent and can decolorize certain dyes and pigments.

Colorants must have enough heat stability to withstand processing temperatures, which are sometimes very high, eg,

Table 2. Listing of Pigments for Utilization in Packaging Materials

Common Name	*Colour Index* Name Number	CAS Registry Number	Application[a]	Color Permanency,[b] Indoor Fadeometer, Max h		Plastic Applicability[c]		Some Other Data
				Masstone	Tint	Wide Use	Limited Use	
				White Pigments				
Zinc oxide[d]	White 4 77947	[*1314-13-2*]	2			A,B,C,D, E,F,G, H,I,J, K,L, M,N,O	P	Refractive index 2.01; embrittles oleoresinous film
Lithopone	White 5 77115	[*1345-05-7*]	1			A,B,C,E, G,H,I, J,K,L, M,N, O,P	D	Refractive index 1.84
Titanium dioxide, (anatase, rutile)[d]	White 6 77891	[*13463-67-7*]	1,2,3	250	250	A,B,C,D, E,G,H, I,J,K, L,M, N,O, P,Q,	F	Refractive index 2.76 (rutile) Refractive index 2.55 (anatase)
Zinc sulfide	White 7 77975	[*1314-98-3*]	1,2			A,B,C,D G,H,I, J,K, L,M	N,O,P	Refractive index 2.37
Calcium carbonate[d]	White 18 77220	[*471-34-1*] [*1317-65-3*]	1,3					Refractive index 1.48–1.65; brightness 85–95% (nat.) 92–98% (syn.)
Kaolin clay, bentonite[d]	White 19 77004	[*1332-58-7*] [*8047-76-5*]	1,3					Refractive index 1.56; low brightness
Blanc fixe, process white[d]	White 21 77120	[*7727-43-7*]	1					Refractive index 1.64
Aluminum hydrate[d]	White 24 77002	[*1332-73-6*]	1					Refractive index 1.57; rheology modifier for inks
Talc, French chalk[d]	White 26 77718	[*8005-37-6*] [*14807-96-6*]	1					Refractive index 1.54–1.59
Silica[d]	White 27 77811	[*7631-86-9*] [*14808-60-7*] [*61790-53-2*] [*63231-67-4*]	1					Refractive index 1.45–1.55; brightness 91–96% (syn.), < 90% (nat.)
				Black Pigments				
Aniline black	Black 1 50440	[*13007-86-8*]	1	120	20–30			Gives deep matt black or velvety finish
Lamp and vegetable black	Black 6 77266	[*1333-86-4*]	1					
Carbon black[d] Furnace black Channel black	Black 7 77266	[*1333-86-4*]	1,2	>240	>240	A,B,C,D, E,F,G, H,I,K, L,M, N,O, P,Q	J	Excellent stability to light, chemicals and heat; good uv absorption
Iron titanate Brown spinel	Black 12 77543	[*68187-02-0*]	2			A,B,C,D, F,G,H, J,K,L, M,N,O	E,I,P,Q,	Excellent heat, light, and chemical resistance
Iron copper Chromite black Spinel	Black 23 77429		2			A,B,C,D, F,G,H, J,K,L, M,N,O	E,I,P,Q	Certain plastics embrittled by iron

Table 2. (*Continued*)

Common Name	*Colour Index* Name Number	CAS Registry Number	Ap-plica-tion[a]	Color Permanency,[b] Indoor Fadeometer, Max h		Plastic Applicability[c]		Some Other Data
				Masstone	Tint	Wide Use	Limited Use	
Manganese ferrite Black spinel	Black 26 77494	*[68186-94-7]*	2			A,B,C,D, F,G,H, J,K,L, M,N,O	E,I,P,Q	Certain plastics embrittled by Mn and Fe
Copper chromite Black spinel	Black 28 77428	*[68186-91-4]*	2			A,B,C,D, F,G,H, J,K,L, M,N,O	E,I,P,Q	Excellent chemical and heat resistance
				Red Pigments				
Naphthol Red FRR	Red 2 12310	*[6041-94-7]*	1,3	20–80	15–20			Printing inks for packaging; excellent chemical resistance
Toluidine Red	Red 3 12120	*[2425-85-6]*	3	40–140[D]	5–20[F]		D,H,J	
Chlorinated Para Red[d]	Red 4 12085	*[2814-77-9]*	1,3	40–120[DL]	5–30[F]	N,O	D,H,J	Bleeds in organic solvents and overstripes
Naphthol Carmine FB	Red 5 12490	*[6410-41-9]*	1	60–120	20–40			Excellent chemical resistance; gravure inks for packaging
Naphthol Red F4RH	Red 7 12420	*[6471-51-8]*	1,3	80–160[F]	60[F]		C,E,J,K, L,M, N,O,Q	Packaging printing inks; excellent chemical resistance
Naphthol Red FRLL	Red 9 12460	*[6410-38-4]*	1,3	80–120	40–60			Excellent chemical resistance
Naphthol Red FRL	Red 10 12440	*[6410-35-1]*	1,3	60–80[F]	25–35			
Naphthol Red (medium shade)	Red 17 12390	*[6655-84-1]*	1,3	30–80[F]	10–30[FL]		A,D,E,H, J,K,N, O	Excellent chemical resistance; packaging printing ink
Naphthol Red (light yellow shade)	Red 22 12315	*[6448-95-9]*	1,3	40–80[DFL]	15–30[FL]		A,D,E,H J,K,N, O	Printing ink for packaging, superior chemical resistance
Naphthol Red (dark blue shade)	Red 23 12355	*[6471-49-4]*	1,3	60[DFL]	30[FL]		A,D,E,H, J,K,N, O	Printing inks for packaging
Pyrazolone Red	Red 38 21120	*[6358-87-8]*	1	50–75[DF]	15–50[F]	A,P	C,D,E,H, J,K,N, O	Metal decorating and packaging printing inks
Dianisidine Red	Red 41 21200	*[6505-29-9]*	1				K,P,Q	Packaging printing inks
Permanent Red 2B (barium)	Red 48:1 15865:1	*[7585-41-3]*	1,2,3	10–30[DFL]	10–20[FL]	K,L,M, N,O,P	A,D,E,J, Q	Carton and label printing inks; excellent brightness
Permanent Red 2B (calcium)	Red 48:2 15865:2	*[7023-61-2]*	1,2	20–100[DFL]	10–50[FL]	E,K,P,Q	D,J,L,M, N	Printing inks for labels and cartons; bright and good tint strength
Permanent Red 2B (strontium)	Red 48:3 15865:3	*[15782-05-5]*	1,2	10–30[D]	5–30[F]			Excellent brightness; solvent-based printing inks
Permanent Red 2B (manganese)	Red 48:4 15865:4	*[5280-66-0]*	1	80–120[DFL]	30–40[DFL]			Poor alkali and soap resistance
Lithol Red (sodium)	Red 49 15630	*[1248-18-6]*	1	5–10	2–5			Excellent tint strength
Lithol Red (barium)	Red 49:1 15630:1	*[1103-38-4]*	1	5–40[DL]	2–20[FL]		P	Resination increases transparency; poor alkali and soap resistance

Table 2. ***(Continued)***

Common Name	*Colour Index* Name Number	CAS Registry Number	Ap-plica-tion[a]	Color Permanency,[b] Indoor Fadeometer, Max h		Plastic Applicability[c]		Some Other Data
				Masstone	Tint	Wide Use	Limited Use	
Lithol Red (calcium)	Red 49:2 15630:2	*[1103-39-5]*	1	2–5[DL]	2–5[FL]			Excellent brightness and tint strength
Red 2G (calcium)	Red 52:1 15860:1	*[17852-99-2]*	1,2	10–15[D]	5–10[FL]	N	B,C,D,E, H,J,K, O,P,Q	Process magenta for printing inks
Red Lake C[d] (barium)	Red 53:1 15585:1	*[5160-02-1]*	1,2	5–50[DFY]	1–25[F]	A	C,D,E,H, J,K,L, N,O	Standard warm red; foil coatings
Lithol Rubine[d] (calcium)	Red 57:1 15850:1	*[5281-04-9]*	1,2	15–50[DFL]	5–25[FL]	E,K	C,D,H,J, L,M, N,O,P, Q	Standard process magenta; foil coatings
Pigment Scarlet (barium)	Red 60:1 16105:1	*[15782-06-6]*	1,2	25–50[D]	20–30[B]	A,C,E,K, L,N,O, P	D,H,J,M, Q	Printing ink for gloss labels, waxed papers, metal decorating; foil coatings
Anthosine Red 3B (Ba, Na)	Red 66 18000:1	*[68929-13-5]*	1					Metal decorating printing inks
Anthosine Red 5B (Ba, Na)	Red 67 18025:1	*[68929-14-6]*	1					Metal decorating printing inks; transparency
Rhodamine Y (PTMA)	Red 81:1 45160:1	*[12224-98-5]*	1	15–30[D]	5–10[F]			Brillant, color purity, good tint strength
Rhodamine Y (SMA)	Red 81:3 45160:3	*[63022-06-0]*	1	15–30[D]	5–10[F]			Process magenta printing inks
Rhodamine Y (PMA)	Red 81:x 45160:x	*[63022-07-1]*	1	15–30[D]	5–10[F]			Poor alkali and soap resistance
Alizarine Red B	Red 83 58000:1	*[72-48-0]*	1	120	30		A,D,E,H, J,N,O, P,Q	Metal decorating inks, butter and soap packages
Thioindigold Red	Red 88 73312	*[14295-43-3]*	2	120–160	80–120	P,Q	D,H,J,K, L,M, N,O	Clean color with excellent fastness
Phloxine Red (Lead)	Red 90 45380:1	*[1326-05-2]*	1	<20[DF]	<20[F]			Poor chemical, light, solvent and heat resistance
Synthetic Red[d] iron oxide	Red 101 77491	*[1309-37-1]*	1,2			A,E,H,J, K,L,N, P,Q	C,F,G,M	Foil coatings, Fe embrittles certain plastics
Molybdate Orange	Red 104 77605	*[12656-85-8]*	1,2	20–160[D]	20–160	D,H,J,P, Q	A,B,K,L, M,N,O	Poor alkali and acid resistance
Cadmium Sulfoselenide Red	Red 108 77202	*[58339-34-7]*	2	500	200[GF]	A,B,C,D E,G,H, I,J,K, L,M, N,O,P, Q	F	Bright, clean, intense colors
Cadmium Sulfoselenide Lithopone Red	Red 108:1 77202:1	*[58339-34-7]* and *[7727-43-7]*	2	500	150[GF]	A,B,C,D, E,G,H, I,J,K, L,M, N,O,P, Q	F	Sensitive to mineral acids
Naphthol Red FGR	Red 112 12370	*[6535-46-2]*	1,2,3	60–160	40–160			Excellent brightness, very good lightfastness; paper coatings

Table 2. (*Continued*)

Common Name	*Colour Index* Name Number	CAS Registry Number	Application[a]	Color Permanency,[b] Indoor Fadeometer, Max h		Plastic Applicability[c]		Some Other Data
				Masstone	Tint	Wide Use	Limited Use	
Mercadium Red	Red 113 77201	*[1345-09-1]*	2			A,B,C,D, E,F,G, H,J,K, L,M, N,O,P, Q		Poor lightfastness when light and moisture present
Cadmium mercury Lithopone Red	Red 113:1 77201:1	*[1345-09-1]* and *[7727-43-7]*	2			A,B,C,D, E,F,G, H,J,K, L,M, N,O,P, Q		Low tint strength
Quinacridone Magenta Y	Red 122 73915	*[980-26-7]*	1,2	140–160[D]	80–120[F]	K,L,M,P, Q	A,B,C,D, H,J,O	Soluble in nylon and certain plastics
Perylene Vermilion	Red 123 71145	*[24108-89-2]*	1,2			A,B,C,E, J,K,P, Q	L,M,N	Transparent; good fastness properties
Disazo Red	Red 144	*[5280-78-4]*	1,2	160	100–140	C,E,H,J, K,L, M,N, O,P,Q	D	High-performance pigment
Naphthol Carmine FBB	Red 146 12485	*[5280-68-2]*	1,2,3	60[D]	30[D]	C,E,H,J, K,L, M,N, O,P,Q	D	Packaging and metal decorating inks, paper coatings
Perylene Red BL	Red 149 71137	*[4948-15-6]*	1,2	40–80[D]	20–80	A,B,C,E, H,I,J, K,L, M,N, O,P,Q	G	Metal decorating
Disazo Scarlet	Red 166	*[12225-04-6]*	1,2			C,E,H,J, K,L, M,N, O,P,Q	D	High-performance pigment
Brominated Anthanthrone Red	Red 168 59300	*[4378-61-4]*	1			J,K,L,M, P,Q		Metal decorating printing ink
Rhodamine 6G	Red 169 45160:2	*[12224-98-5]*	1	30[D]	10[F]			Gravure printing inks, excellent tint strength
Naphthol Red F5RK	Red 170 12475	*[2786-76-7]*	1	80–120[D]	60[D]	C,L,M, N,O,Q	A,E,F,I, J,K,P	Brilliant excellent chemical resistance
Benzimidazolone Maroon HFM	Red 171 12512	*[6985-95-1]*	1	120	120	C,K,L,N, P,Q	A,B,E,G, J	Foil coatings; very transparent
Benzimidazolone Red HFT	Red 175 12513	*[6985-92-8]*	1,2	160[F]	120[F]	C,K,L,N, P,Q	A,B,E,G, J	Highly transparent; inks for packaging and metal decorating
Benzimidazolone Carmine HF3C	Red 176 12515	*[12225-06-8]*	1,2	80[D]	40[F]	C,K,L,N, O,P,Q	A,E,F,I, J,M	Transparent, bright, chemical resistant; packaging and metal decorating inks
Anthraquinoid Red	Red 177 65300	*[4051-63-2]*	2	120	70–100	C,D,G,H, J,K,L, M,P,Q	B,E,I	Transparent
Perylene Maroon	Red 179 71130	*[5521-31-3]*	2			A,B,C,E, J,K,P, Q	L,M	Excellent fastness properties
Naphthol Rubine F6B	Red 184		1	60[F]	15[F]			Printing ink for packaging

Table 2. (*Continued*)

Common Name	*Colour Index* Name Number	CAS Registry Number	Ap-plica-tion[a]	Color Permanency,[b] Indoor Fadeometer, Max h		Plastic Applicability[c]		Some Other Data
				Masstone	Tint	Wide Use	Limited Use	
Benzimidazolone Carmine HF4C	Red 185 12516	[*61951-98-2*]	1,2	60[D]	30[F]	C,K,L,N, O,P,Q	A,E,F,I, J,M	Process magenta for metal decorating
Naphthol Red HF4B	Red 187 12486	[*59487-23-9*]	1,2	80[F]	40[F]	K,L,N,P, Q	C,E,I,J, M	Bright, transparent; packaging and metal decorating inks
Naphthol Red HF3S	Red 188 12467	[*61847-48-1*]	1,2	80[D]	60[D]	K,L,N,P, Q	C,E,I,J, M	Superior chemical resistance; packaging inks
Perylene Scarlet	Red 190 71140	[*6424-77-7*]	2					Transparent, dull tints
Rubine Red (calcium)	Red 200 15867	[*58067-05-3*]	1,2	20–70[FL]	10–15[FL]			Oil-based printing inks, poor soap, solvent and alkali resistance
Quinacridone Scarlet	Red 207	[*1047-16-1*] and [*3089-16-5*]	1,2	120–320[D]	80–120[F]			High-performance pigment
Benzimidazolone Red HF2B	Red 208 12514	[*31778-10-6*]	1,2	80[F]	40[F]			Bright medium red; packaging and metal decorating inks
Quinacridone Red Y	Red 209 73905	[*3089-17-6*]	1	120[D]	120[F]			Soluble in nylon
Naphthol Red F6RK	Red 210		1	60[D]	*30*[F]	C,J,K,L, M,N, O,Q	A,E,F,I, J,K,P	Packaging printing inks
Perylene Red Y	Red 224 71127	[*128-69-8*]	2	>500	>500			Transparent; very good strength and fastness properties
				Orange Pigments				
Dinitraniline Orange[d]	Orange 5 12075	[*3468-63-1*]	1,3	40–80[DFL]	5–10[DFL]			Good chemical resistance
Pyrazolone Orange	Orange 13 21110	[*3520-72-7*]	1	10–60[DFL]	5–10[FL]	K,L,N,O, P,Q	A,C,D,H, J,M	
Dianisidine Orange	Orange 16 21160	[*6505-28-8*]	1,2	25–75[DF]	5–50[F]	J,P,Q	K,L,M	
Persian Orange Lake (A1)	Orange 17:1 15510:2	[*15876-51-4*]	1					Inks for waxed bread wrappers
Pure Cadmium Orange; Cadmium Sulfoselenide Orange	Orange 20 77202	[*12556-57-2*]	2	500	100[GF]	A,B,C,D, E,F,G, H,J,K, L,M, N,O,P, Q		Lightfastness needs protection from moisture
Cadmium Sulfoselenide Orange Lithopone	Orange 20:1 77202:1	[*12556-57-4*] and [*7727-43-7*]	2	500	100[GF]	A,B,C,D, E,F,G, H,J,K, L,M, N,O,P, Q		
Mercadium Orange	Orange 23 77201	[*1345-09-1*]	2			A,B,C,D, E,F,G, H,J,K, L,M, N,O,P, Q		Fades in presence of light and moisture

Table 2. (*Continued*)

Common Name	*Colour Index* Name Number	CAS Registry Number	Application[a]	Color Permanency,[b] Indoor Fadeometer, Max h		Plastic Applicability[c]		Some Other Data
				Masstone	Tint	Wide Use	Limited Use	
Mercadium Lithopone Orange	Orange 23:1 77201:1	[*1345-09-1*] and [*7727-43-7*]	2			A,B,C,D, E,F,G, H,J,K, L,M, N,O,P, Q		
Disazo Orange	Orange 31	[*5280-74-0*]	2					
Diarylide Orange	Orange 34 21115	[*15793-73-4*]	1,2,3	20–60^{D}	15–30^{F}	K,L,N,O, P,Q	A,C,D,H, J,M	Bleeds in some overstripes
Benzimidazolone Orange HL	Orange 36 11780	[*12236-62-3*]	1	120^{D}	80^{F}	A,C,E,J, K,L, M,N, O,P,Q	B,G,I	
Naphthol Orange	Orange 38 12367	[*12236-64-5*]	1,2	80–120^{D}	10–40	K,L,N,O P,Q	C,E,I,J, M	Printing inks for metal decorating and packaging
Perionone Orange	Orange 43 71105	[*4424-06-0*]	1	160^{D}	120^{D}	J,K,L,M, P,Q		
Ethyl Red Lake C (barium)	Orange 46 15602	[*67801-01-8*]	1,2	5–30^{FL}	2–20^{FL}			Metal decorating inks
Quinacridone Gold	Orange 48	[*1047-16-1*] and [*1503-48-6*]	2	120–320^{D}	80–120^{F}			
Quinacridone Deep Gold	Orange 49	[*1047-16-1*] and [*1503-48-6*]	2	120–320^{D}	80–120^{F}			
Benzimidazolone Orange HGL	Orange 60		1,2			E,J,K,P, Q	A,B,C,F, H,L, M,N,O	Transparent
Tetrachloro-isoindolinone Orange	Orange 61		2	200–500	200–500			High-performance pigment
Benzimidazolone Orange H5G	Orange 62		1	>160	60–160			Oil-based printing inks
Orange GP	Orange 64		2					
				Yellow Pigments				
Arylide Yellow G	Yellow 1 11680	[*2512-29-0*]	1,3	60–200^{DFG}	20–40^{F}			Printing inks requiring alkali resistance, aqueous dispersions for paper
Arylide Yellow 10G	Yellow 3 11710	[*6486-23-3*]	1,3	120–200^{DF}	20–60^{F}			Printing inks requiring alkali resistance, aqueous dispersions for paper
Diarylide Yellow AAA	Yellow 12 21090	[*6358-85-6*]	1	10–60^{DFL}	2–30^{FL}		A,C,D,E, H,J,K, M,N, O,P	
Dairylide Yellow AAMX	Yellow 13 21100	[*5102-8-30*]	1,2	20–60^{F}	10–40*F*	A,P	D,E,G,H, I,J,K, L,N,O	
Diarylide Yellow AAOT	Yellow 14 21095	[*5468-75-7*]	1,2	10–60^{DFL}	5–40^{FL}	G,K	A,C,D,E, H,J,L, M,N, O,P,Q	
Permanent Yellow NCG	Yellow 16 20040	[*5979-28-2*]	1,2	80–120^{DF}	30^{F}	C,E,K,L, M,N, O,Q	A,I,P	Printing inks for packaging

Table 2. (*Continued*)

Common Name	*Colour Index* Name Number	CAS Registry Number	Application[a]	Color Permanency,[b] Indoor Fadeometer, Max h		Plastic Applicability[c]		Some Other Data
				Masstone	Tint	Wide Use	Limited Use	
Diarylide Yellow AAOA	Yellow 17 21105	*[4531-49-1]*	1,2	20–80[DF]	10–40[F]	G,K,L, M,P,Q	A,C,D,H, I,J,N, O	
Chrome Yellow (primrose, light, medium)	Yellow 34 77600 77603	*[1344-37-2]*	1	10–160[D]	10–160[FG]	D,H,J,P, Q	A,B,I,K, L,M, N,O	
Cadmium Zinc Yellow (primrose, lemon, golden)	Yellow 35 77205	*[12442-27-2]*	2	400	100[GF]			
Cadmium Zinc Yellow Lithopone	Yellow 35 : 1 77205 : 1	*[12442-27-2]* and *[7727-43-7]*	2	400	80[GF]			
Cadmium Yellow	Yellow 37 77199	*[1306-23-6]*	2	400	70–100[GF]	A,B,C,D, E,F,G, H,I,J, K,L, M,N, O,P,Q		
Cadmium Lithopone yellow	Yellow 37 : 1 77199 : 1	*[1306-23-6]* and *[7727-43-7]*	2	400	80[GF]	A,B,C,D, E,F,G, H,I,J, K,L, M,N, O,P,Q		
Synthetic Yellow[d] Iron oxide	Yellow 42 77492	*[12259-21-1]* *[51274-00-1]*	2					
Nickel antimony Titanium Yellow rutile	Yellow 53 77788	*[8007-18-9]* *[71077-18-4]*	2	1000[F]	1000[F]	A,D,E,F, G,H,J, K,L, M,N, O,P,Q	B,C,I	
Diarylide Yellow AAPT	Yellow 55 21096	*[6358-37-8]*	1	35–60	25–40			Printing inks for waxed food wrappers
Arylide Yellow 4R	Yellow 60 12705	*[6407-74-5]*	1	60–100	35–50[F]			Alkali resistance and lightfast printing inks
Arylide Yellow RN	Yellow 65 11740	*[6528-34-3]*	1,3	80–150[DF]	60–100			Lightfast and alkali resistance printing inks; aqueous dispersions for paper
Arylide Yellow GX	Yellow 73 11738	*[13515-40-7]*	1,3	70–120[DF]	30–40[F]			Lightfast and alkali resistance printing inks; aqueous dispersions for paper
Arylide Yellow GY	Yellow 74 11741	*[6358-31-2]*	1,3	70–120[DF]	20–60[F]			Lightfast and alkali resistance printing inks; aqueous dispersions for paper
Diarylide Yellow H10G	Yellow 81 21127	*[22094-93-5]*	1	60–160	30–60[F]	C,E,K,L, N,O,P, Q	A,M	Printing inks for packaging
Diarylide Yellow HR	Yellow 83 21108	*[5567-15-7]*	1,2	70–240[DF]	20–60[DF]	P,Q	C,D,E,H, J,K,L, M,N	
Disazo Yellow G	Yellow 93	*[5580-57-4]*	1,2	120	80	A,K,L, M,N, O,P		
Disazo Yellow R	Yellow 95	*[5280-80-8]*	1,2	120	80	A,K,L, M,N, O,P		Metal-free printing inks

Table 2. (*Continued*)

Common Name	*Colour Index* Name Number	CAS Registry Number	Ap-plica-tion[a]	Color Permanency,[b] Indoor Fadeometer, Max h: Masstone	Tint	Plastic Applicability[c]: Wide Use	Limited Use	Some Other Data
Permanent Yellow FGL	Yellow 97 11767	[*12225-18-2*]	1	120[D]	120[F]	A,C,E,J, K,L, M,N, O,Q	B,G,N,I	Heat, lightfast, and alkali-resistant printing inks
Arylide Yellow 10GX	Yellow 98 11727	[*12225-19-3*]	1	80–120[F]	40–60[F]			
FD&C Yellow No. 5[d] aluminum lake	Yellow 100 19140:1	[*12225-21-7*]	1			C,F,G,K, L,O,P	E	Colorant for food, drugs, cosmetics, and food-contact surfaces; metal decorating
Fluorescent Yellow	Yellow 101 48052	[*2387-03-3*]	2					
Diarylide Yellow GGR	Yellow 106	[*12225-23-9*]	1	30–50[F]	20–30[F]			Printing inks for packaging
Tetrachloro-isoindolinone Yellow G	Yellow 109		2	200	100–200	H,J,K,L, M,P,Q	A,D,G, N,O	
Tetrachloro-isoindolinone Yellow R	Yellow 110		2	200–500	200–500	E,H,J,K, L,M,P, Q	A,D,G, N,O	
Diarylide Yellow H10GL	Yellow 113 21126	[*14359-20-7*]	1	60[F]	30[F]	C,E,K,L, M,N, O,P,Q	A,G,I	Printing inks for packaging
Diarylide Yellow G3R	Yellow 114		1	40[F]	15[F]			Oil-based ink for packaging
Azomethine Yellow	Yellow 117	[*21405-81-2*]	1					Food packaging
Zinc ferrite brown spinel	Yellow 119 77496	[*12063-19-3*] [68187-51-9] [*61815-08-5*]	2					
Diarylide Yellow DGR	Yellow 126		1	25–35[F]	10–20[F]			
Diarylide Yellow GRL	Yellow 127		1	30–50[F]	20–30[F]	A,K,L,N, O,Q	C,D,E,H, J,M,P	Inks for packaging and metal decorating
Disazo Yellow GG	Yellow 128		2					
Quinophthalone Yellow	Yellow 138		2					
Isoindoline Yellow	Yellow 139		2			C,D,E,J, K,L, M,N, P,Q		
Nickel Yellow 4G	Yellow 150	[*68511-62-6*]	2					
Benzimidazolone Yellow H4G	Yellow 151		1,2	160[F]	120[F]	A,C,E,J, K,L, M,N, O,P,Q	F,G,H,I	
Diarylide Yellow YR	Yellow 152	[*20139-66-6*]	1	20–40[F]	10–20[F]			Lead chromate replacement
Benzimidazolone Yellow H3G	Yellow 154		1,2	160[F]	120[F]	C,E,J,K, P,Q	A,F,H,L, M,N,O	Inks for metal decorating
Nickel niobium Titanium Yellow Rutile	Yellow 161 77895	[*68611-43-8*]	2					
Chrome niobium titanium buff rutile	Yellow 162 77896	[*68611-42-7*]	2					

Table 2. (*Continued*)

Common Name	*Colour Index* Name Number	CAS Registry Number	Application[a]	Color Permanency,[b] Indoor Fadeometer, Max h		Plastic Applicability[c]		Some Other Data
				Masstone	Tint	Wide Use	Limited Use	
Manganese antimony titanium buff rutile	Yellow 165 77899	*[68412-38-4]*	2					
				Green Pigments				
Brilliant Green (PTMA)	Green 1 42040:1	*[1325-75-3]*	1,3	10–15[D]	5–10[F]			Poor alkali and soap resistance
Brilliant Green (PMA)	Green 1:x 42040:x	*[68814-00-6]*	1,3	10–15[D]	2–10[F]			Excellent brilliance, color purity
Permanent Green (PTMA)	Green 2 42040:1 49005:1	*[1328-75-3]* and *[1326-11-0]*	1,3	10–15[D]	5–10[F]			Poor alkali and soap resistance
Permanent Green (PMA)	Green 2:x 42040:x 49005:1	*[68814-00-6]* and *[1326-11-0]*	1,3	10–15[D]	5–10[F]			For lustrous appearance in printing
Malachite Green (PTMA)	Green 4 42000:2	*[61725-50-6]*	1,3					Poor alkali and soap resistance
Phthalocyanine Green	Green 7 74260	*[1328-53-6]*	1,2	120–320[DFL]	120–160[FL]	A,B,D,E, G,H,I, J,K,L, M,N, O,P,Q	C	Standard green for printing inks
Nickel Azo Yellow (green gold)	Green 10	*[51931-46-5]*	2	>70	>70	E,N,O,P	C,D,F,H, I,J,K, L,M,Q	Very lightfast pigment
Chrome Green	Green 15 77510 77603	*[1344-37-2)* and *[25869-00-5]*	2,3	60–80[DG]	20–40	J	C,D,H,I, N,O, P,Q	
Cobalt chromite green spinel	Green 26 77344	*[68187-49-5]*	2	>160	>160			Outstanding chemical and light stability
Phthalocyanine Green (Cl, Br)	Green 36 74265	*[14302-13-7]*	1,2	160–320[FL]	120–160[FL]	A,B,D,E, H,I,J, K,L, M,N, O,P,Q	C,G	High-performance pigment
Cobalt titanate green spinel	Green 50 77377	*[68186-85-6]*	2					Outstanding chemical, light, and heat stability
				Blue Pigments				
Victoria Blue (PTMA)	Blue 1 42595:2	*[1325-87-7]*	1,3	20–40[D]	5–10[F]			Outstanding brilliance
Victoria Blue (SMA)	Blue 1:2 42595:3	*[68413-81-0]*	1,3	30–40[D]	5–10[F]			SMA salt is stronger than PMS/PTMA but not as clean
Victoria Blue (PMA)	Blue 1:x 42595:x	*[68409-66-5]*	1,3	15–40	5–10[F]			Printing inks with lustrous appearance
Permanent Blue (PMA) Peacock	Blue 9:x 42025:x	*[68814-07-3]*	1,3					Excellent color purity and strength
Phthalocyanine Blue Alpha (red crystallizing)[d]	Blue 15 74160	*[147-14-8]*	1,2	120–320[FLZ]	120–160[FL]	A,B,C,D, E,G,H, I,J,K, L,M, N,O,P, Q	F	Excellent transparency, chemical resistance, and lightfastness
Phthalocyanine Blue Alpha (R, NC)	Blue 15:1 74160 74250	*[147-14-8]* *[12239-87-1]*	1,2	120–320[FLZ]	120–160[FL]	A,B,C,D, E,G,H, I,J,K, L,M, N,O,P, Q		

Table 2. (*Continued*)

Common Name	*Colour Index* Name Number	CAS Registry Number	Ap-plica-tion[a]	Color Permanency,[b] Indoor Fadeometer, Max h: Masstone	Tint	Plastic Applicability[c]: Wide Use	Limited Use	Some Other Data
Phthalocyanine Blue Alpha (R, NCNF)	Blue 15:2 74160 74250	*[147-14-8]* *[12239-87-1]*	1,2,3	120–320[FLZ]	80–160[FL]	A,B,C,D, E,G,H, I,J,K, L,M, N,O,P, Q		
Phthalocyanine Blue Beta (G, NC)	Blue 15:3 74160	*[147-14-8]*	1,2	120–320[FLZ]	120–160[FL]	A,B,C,D, E,G,H, I,J,K, L,M, N,O,P, Q		Standard process cyan.
Phthalocyanine Blue Beta (G, NCNF)	Blue 15:4 74160	*[147-14-8]*	1,2	120–320[FLZ]	80–160[FL]	A,B,C,D, E,G,H, I,J,K, L,M, N,O,P, Q		
Phthalocyanine Blue (metal free)	Blue 16 74100	*[574-93-6]*				A,G,K,L, M,N, O,P		
Fugitive Peacock Blue (Ba)	Blue 24 42090:1	*[6548-12-5]*	1					
Iron Blue, Milori Blue[d]	Blue 27 77510	*[25869-00-5]*	1	>160[G]	20–80[C]	K	B,C,D,H, J,L,N, O,P,Q	
Cobalt Blue[d]	Blue 28 77346	*[1345-16-0]* *[68186-86-7]*	2	>100[F]	>100[G]	B,C,D,E, F,G,H, J,K,L, M,N, O,P,Q	I	Outstanding lightfastness, chemical resistance
Ultramarine Blue[d]	Blue 29 77007	*[57455-37-5]*	1,2	>240	>240	A,C,E,G, H,K,L, M,N, O,P,Q	F,I,J	
Cobalt chromite blue-green spinel	Blue 36 77343	*[68187-11-1]*	1	>100[F]	>100[G]	B,C,D,E, F,G,H, I,J,K, L,M, N,O,P, Q		
Alkali Blue G	Blue 56 42800	*[6417-46-5]*	1					For printing inks needing alkali and soap resistance
Indanthrone Blue	Blue 60 69800	*[81-77-6]*	2	120–160[F]	120–160[F]	O,P,Q	B,C,D,E, H,J,K, L,M,N	
Alkali Blue, Reflex Blue	Blue 61 42765:1	*[1324-76-1]*	1	2–10[FL]	2–10[FL]			Poor alkali and soap resistance
Victoria Blue (CFA)	Blue 62 42595:x		1	20[D]	10[F]			
				Violet Pigments				
Rhodamine B (PMA)	Violet 1:x 45170:x	*[63022-09-3]*	1	<20[D]	<20[F]			Brilliant, color purity and tint strength
Methyl Violet (PTMA)	Violet 3 42535:2	*[1325-82-2]*	1,3	15–30[DFL]	5–10[FL]		A,D,H,J, N	Printing inks with lustrous appearance
Fugitive Methyl Violet	Violet 3:3 42535:5	*[68308-41-8]*	1	2–5[FL]	2–5[FL]			
Methyl Violet (PMA)	Violet 3:x 42535:x	*[67989-22-4]*	1,3	15–30[DG]	5–10[F]			Standard purple for printing inks

Table 2. (*Continued*)

Common Name	Colour Index Name Number	CAS Registry Number	Application[a]	Color Permanency,[b] Indoor Fadeometer, Max h		Plastic Applicability[c]		Some Other Data
				Masstone	Tint	Wide Use	Limited Use	
Cobalt violet phosphate	Violet 14 77360	[13455-36-2]	2					Decolorizer for clear and white plastics
Ultramarine Violet and Pink[d]	Violet 15 77007	[12769-96-9]	2			A,D,E,G, H,I,J, K,L, M,N,O	B,C,P,Q	Counteracts yellowing in plastics on heating
Manganese Violet[d]	Violet 16 77742	[10101-66-3]	2					
Quinacridone Violet	Violet 19 73900	[1047-16-1]	1,2	120–320[D]	80–120[F]	I,J,K,L, M,N, O,P,Q	A,B,C,D, E,H	High-performance pigment
Crystal Violet (CFA)	Violet 27 42535:3	[12237-62-6]	1	20[d]	10[F]			
Benzimidazolone Bordeaux HF3R	Violet 32 12517	[12225-08-0]	1,2	60	40	K,L,N,O, P,Q	C,E,I,J, M	Inks for packaging and metal decorating
Cobalt lithium violet phosphate	Violet 47 77363	[68610-13-9]	2					Decolorizer for clear and white plastics
Cobalt magnesium red-blue borate	Violet 48 77352	[68608-93-5]						Decolorizer for clear and white plastics
				Brown Pigments				
Monoazo Brown (copper)	Brown 5 15800:2	[16521-34-9]	1,2	>160	>100			Amber effects in plastics
Magnesium ferrite	Brown 11 77495	[12068-86-9]	2					
Diazo Brown	Brown 23		1,2			K,L,P,Q	A,M,	High-performance pigment
Chrome antimony titanium buff rutile	Brown 24 77310	[68186-90-3]	2	>1000	>1000	A,C,D,E, F,H,J, K,L,N, O,P,Q	G,I,M	Outstanding light, heat, and chemical stability
Benzimidazolone Brown HFR	Brown 25 12510	[6992-11-6]	1,2	80[F]	80[F]	K,L,N,O, P,Q	C,M	Printing inks for metal decorating; highly transparent
Zinc iron chromite brown spinel	Brown 33 77503	[68186-88-9]	2					Excellent heat and light resistance
Iron chromite brown spinel	Brown 35 77501	[68187-09-7]	2					Excellent heat and light resistance
				Metal Pigments				
Aluminum Flake[d]	Metal 1 77000	[7429-90-5]	1,3			B,C,D,E G,J,K, L,M, N,O,P, Q		Nonleafing for decorative inks, paper coatings
Copper powder, bronze powder[d]	Metal 2 77400	[7440-50-6] [7440-50-8] [7440-66-6]	1,2			B,C,D,G, J,K,L, N,O, P,Q	E,I,M	Decorative printing inks for packaging, labels

[a] Key to application: 1 = printing inks; 2 = plastics; 3 = paper and paper coatings.
[b] Key to permanency failures: F = fades; D = darkens; L = loses gloss; B = turns bluer; G = turns gray or greener; Y = turns yellower; Z = bronzes.
[c] Key to plastic: A = ABS; B = acetal; C = acrylic; D = amino resins; E = cellulosics; F = fluoroplastics; G = nylons; H = phenol–formaldehyde; I = polycarbonate; J = polyester or alkyd; K = polyethylene, low density; L = polyethylene, high density; M = polypropylene; N = polystyrene, general-purpose; O = polystyrene, impact-resistant; P = flexible vinyl; Q = rigid vinyl.
[d] Pigments having an FDA status.

Table 3. Some Pigments for Use in Paperboard and Paper

Filler	Composition	Refractive Index	Brightness	Use
Titanium dioxide (rutile, anatase)	TiO_2	2.7 for rutile 2.55 for anatase	98–99 98–99	Board, waxing stock, board liner, paper, specialties
Clay (kaolin)	Aluminum silicate	1.87–1.98	80–85	Board, papers, specialties
Calcium carbonate (natural and precipitated)	$CaCO_3$	1.56	95–97	Printing, cigarette papers
Talc	Magnesium silicate	1.57	70–90	Board, printing papers
Gypsum	$CaSO_4$	1.57–1.61	80–90	Boards, specialties
Diatomite (natural)	Diatomaceous earth	1.40–1.46	65–75	Improves bulk and drainage of board stocks

~600°F (316°C) for injection-molded polycarbonate. Some organic pigments processed at 350°F (177°C) for 15 min begin to show signs of darkening, but some cadmium pigments can withstand 1500°F (816°C) without noticeable color change. When processing temperatures are particularly high, the choices for colorants are few. Colorants may also interact with other additives used in the plastic such as heat and light stabilizers; for example, sulfur-containing cadmium pigments react with the nickel-bearing stabilizers used in some film-grade polypropylenes.

Dispersion. Dispersing the pigment to develop color strength and maximum optical properties is an extremely important issue that is continually being addressed. Dispersion of dyes is not as great a problem because of their solubilizing nature. Pigments need to be properly wetted by vehicles or the plastic medium. Poor dispersion can lead to processing difficulties for thin films. Agglomeration and aggregation in pigments are due to many reasons, such as different particle sizes and shapes, presence of soluble salts or impurities, and improper grinding. Improvements in pigment dispersion have been obtained through surface treatments of pigments and better grinding techniques.

Pigments for Paper and Paperboard

Two reasons for the use of pigments, fillers, and extender pigments for paper goods are to load or fill the sheet during manufacture to increase bulk and improve such properties as opacity, printability, and brightness; and to coat the paper to provide opacity, black out defects, or color and provide a receptive surface for printing. Table 3 lists some pigments and extender pigments used in the manufacture of paper goods for packaging.

Supply Options

Colorants are supplied in a number of different forms.

Dry powder. Pigments and dyes are sold in a dry powder form. Pigments are ground to suitable working particle sizes for ink manufacture or for dispersion into a plastic. The maximum working particle size of most dry pigments is about 44 μm. These pigments should pass through a 325-mesh (44-μm) screen with less than 1% retention.

Presscake form. This is an undried form in which the presscake may typically contain 25–50% solids. Presscake is used for the preparation of water-based inks.

Flushed colors. Pigments are dispersed in a varnish or mineral oil forming a paste which has a pigment content of ≥30%. The flushing operation involves exhanging water in a presscake for the organic vehicle by a kneading action in a Sigma blade mixer.

Chip dispersion. Pigments dispersed in resin with little or no solvent content. Usually prepared by milling on a two-roll mill.

Resin-bonded pigment. This is a dry flush in which the vehicle phase is a resin.

Easy-dispersing pigments and stir-in pigments. Pigments treated with surfactants or polymeric materials to make them readily dispersible in various ink vehicles, particularly gravure ink types.

Color concentrates. Colorant dispersed in resin which is let down with virgin resin to make final product. Color concentrate is supplied in chip or pellet form. Liquid- and paste-color concentrates are used for vinyls.

Slurry. Titanium dioxide and other white pigments or extenders are supplied to the paper industry in this manner.

Regulatory Requirements

In the United States, the FDA and USDA administer the laws of interest to colorists. The Federal Food, Drug and Cosmetic Act, the Federal Hazardous Substances Act, and the Poison Prevention Packaging Act of 1970 are relevant FDA statutes. The Meat Inspection Act and the Poultry Inspection Act of the USDA relate to colored plastics where food contact is a concern. The Food, Drug, and Cosmetic Act (1938) was modified by the Food Additives Amendment (1958) and the Color Additives Amendment (1960). Pigments that have an FDA status are noted in Table 2 (see Food, drug, and cosmetics regulations).

The regulations do not deal with colorants in printing inks. Four groups of colorants are presently permitted in the coloring of plastics for food, drugs, and cosmetics: (*1*) certified colorants are those in the list of FD&C certified dyes and alumina lakes; (*2*) purified nonaniline colors include iron oxides, carbon black, and titanium dioxide for use in and on plastics; (*3*) use of a noncertified colorant may be petitioned–responsibilities for compliance with regulations rests with user; and (*4*) a colorant is not subject to the color-additive amendment if there is an impermeable barrier between colorant and food, drug, or cosmetics, and no chance of contact.

Lead pigments and other toxic pigments have been eliminated from inks for food packaging. For nonfood packaging, the three major lead pigments for inks are chrome yellow, molybdate orange, and phloxine red. No other heavy metal pigments are used in significant quantities.

BIBLIOGRAPHY

"Colorants" in *The Wiley Encyclopedia of Packaging Technology,* 1st ed., by R. C. Schiek, CIBA-GEIGY Corporation, pp. 204–217.

General References

T. C. Patton, ed., *Pigment Handbook,* Wiley-Interscience, New York, 1973

T. B. Webber, ed., *Coloring of Plastics,* Wiley-Interscience, New York, 1979.

M. Ahmed, *Coloring of Plastics, Theory and Practice,* Von Nostrand Reinhold Co., New York, 1979.

Modern Plastics Encyclopedia, Vol. 56, McGraw-Hill, New York, Oct. 1980.

R. P. Long, *Package Printing,* Graphic Magazines, Inc., Garden City, NY, 1964.

Pigments, Vol. 4 of *Raw Materials Data Handbook,* NPIRI, 1983.

COMPOSITE CANS. See Cans, Composite.

COMPRESSION MOLDING

Compression molding is used to produce parts from thermosetting plastics that cannot be processed by thermoplastic processing methods (see Blow molding; Extrusion; Injection molding; Thermoforming). In compression molding, polymerization takes place in a closed mold. Under heat and pressure, the materials fills the mold cavity, and with continued heat and pressure, the part hardens as crosslinking takes place (see Polymer properties). The irreversibility of the process rules out thermoplastic-processing methods, but it also imparts excellent heat resistance and dimensional stability to the part. These properties are essential for many nonpackaging applications (eg, electrical components, automotive and aircraft body panels). The use of compression molding in packaging is very limited. Thermosets rarely provide the best balance of properties and compression molding is not economical for high-volume production.

Compression molding has some advantages compared to injection molding, including lower tooling costs and fewer stresses in the part; but design flexibility and production rates are limited (1). For many years, the only significant packaging application for thermosets was closures (see Closures). At one time phenolic and urea–formaldehyde molding compounds were the only plastics available for the purpose. Thermoplastics have replaced them to a great extent, but in 1984, United States consumption of these compounds for closures still totaled about 24×10^6 lb (10,900 metric tons) (2).

There has been a resurgence of interest in thermosets and compression molding in recent years because of the heat required for "ovenable" packages. Compression-molded glass-filled thermoset polyesters, which had been used for some time for heat-and serve airline trays, moved into the consumer market with the advent of frozen dinners that could go from the freezer to the convection or microwave oven, and then to the table (3).

BIBLIOGRAPHY

"Compression Molding" in *The Wiley Encyclopedia of Packaging Technology,* 1st ed., pp. 217–218.

1. G. A. Tanner, "Compression Molding," *Modern Plastics Encyclopedia,* Vol. 61, no. 10a, 1984–1985.
2. *Mod. Plast.* **62**(1), 69 (Jan. 1985).
3. "Dish Sales Sizzle for Ovenable Frozen Dinners," *Plast. World* **42**(9), 59 (1984).

COMPUTER APPLICATIONS: PALLET PATTERNS

The Pallet Pattern

A *pallet pattern* is a grouping of finished products in a single layer that is organized onto a pallet base or slip sheet. The most common form of finished product container is the corrugated shipping case. Multiple layers of product on a pallet or slipsheet are used to create a unitized pallet load. The efficiency of the pallet pattern, and the resulting pallet load, is determined by the quantity and size of the product in the corrugated shipping case. For example, a 16-in. × 12-in. case gets 10 cases per layer in a pallet pattern on a 48-in. × 40-in. pallet surface. However, a $16\frac{1}{8}$-in. × 12-in. case can only get 8 cases per layer, a 20% decrease. Therefore, optimizing each case size, the product inside, the pallet pattern and the number of layers can have a dramatic effect on the utilization of space within the entire distribution system. Typically, companies average 80–85% utilization on the pallet. This under utilization of space offers many industries the opportunity to realize tremendous savings.

The Use of the Pallet

Pallets were first used for handling finished goods in the distribution environment in the 1930s. Modern wooden pallets were introduced on a larger scale during World War II by the military. Then, in 1946, the food-processing industry along with transportation companies, terminal warehouse companies, and the pallet manufacturing industry recommended the adoption of the 40-in. × 32-in. and the 48-in. × 40-in. pallet sizes.

Today, hundreds of million of pallets are sold each year, to support the handling and movement of all types of manufactured products. Pallets are produced in many shapes and sizes; however, the most common size in North America is 48-in. × 40-in. The most common materials are wood, plastic, corrugated board, and metal.

The Pallet Pattern Process: Manual Method

Traditionally, companies used laborious manual methods for calculating pallet patterns or they used "pallet pattern charts" and lookup tables to create very basic pattern styles. This manual method of calculation was slow and inaccurate. Manual methods also proved to be inconsistent. They considered only a few alternatives and were narrow in scope, and communicating the results made preplanning difficult. With increasing mass-production techniques and the growth of product distribution on a nationwide basis, manual methods were simply inadequate.

Computers in Packaging

The earliest use of computers, in the traditional packaging function, dates back to the early 1970s. At that time, the mainframe computer was a large room-sized machine used to crunch numbers for activities related to accounting, sales, and inventory control. Its use by the packaging department was limited primarily to providing simple pallet pattern configurations and a small amount of secondary package design work. Both of these tasks were performed without the use of scaled diagrams or graphics. However, mainframe programs were expensive, and not readily available. Therefore, the computer did not play a significant role in packaging until much later.

The introduction of the first PCs in the mid-1980s brought about a very different opportunity as the use of these personal computers became more prominent in the business world. The PC, with the advantage of its small size, ease of use, and computing power, brought about many changes. One of the most important changes was the quick acceptance of its emerging role in the area of packaging, where an ideal use was the creation of pallet patterns and packaging design.

The Pallet Pattern Process Using the Computer

In the packaging process, the pallet base is usually considered to be fixed in size and cannot be easily changed. However, there are many variables relating to the loading of the pallet that can be evaluated to produce the optimum pallet pattern. These parameters include the maximum load height and weight, underhang and overhang requirements, minimum area and cube utilization, the size of the corrugated case, and the dimension loaded vertical on the pallet. The final compression strength of the corrugated case and the various pattern styles that can be considered for each pattern type are also important factors to be considered.

To add to these requirements, in 1988 a Joint Industry Shipping Container Committee (drawn from the membership of the Food Marketing Institute, the Grocery Manufacturers of America, and the National American Wholesale Grocers' Association) published the *Voluntary Industry Guidelines for Dry Grocery Shipping Containers.* These guidelines were aimed at reducing the damage from poor palletization standards. The two major components of these guidelines were to standardize on the 48-in. × 40-in. pallet size and request that all pallet loads be designed toward the 48-in × 40-in. pallet surface without the use of any overhang. This move also prompted other industries to adopt similar guidelines.

What Benefits the Computer Can Provide

It is virtually impossible for the human mind or a simple lookup chart to consider the combinations of all these factors and individual user requirements. This is why today's computers and pallet loading programs are ideally suited to the task.

In a PC-based pallet pattern/loading program, the user simply types in the basic information required for the specific analysis. This information usually consists of the corrugated case size, the dimension vertical for stacking on the pallet, the weight of the filled case, the pallet size and type, allowable underhang and overhang, the maximum finished load height and weight, and the pattern types to be considered. This information is then saved in a file format that can be retrieved for later use, and from which the various solutions are generated.

From the information provided by the computer, the user can then determine which pallet pattern arrangement will best meet their needs. Hundreds, even thousands, of potential solutions can be considered in the process. The computer's calculations are consistently accurate and enable the rapid evaluation of all patterns and load plans that fit within the restrictions applied by the user.

The major advantages of using a computer to calculate pallet patterns and pallet loading efficiency are speed, consistency, ease of use, the ability to replicate actual situations, more alternatives can be considered, and the results can be easily reproduced and communicated to other functions and departments.

From the list of available solutions, the user can select one or more options for viewing and printing (both text and graphical reports) showing the exact location of each case, the layout of the cases within a layer, and the arrangement of each layer on the pallet. After viewing a satisfactory solution, the user can then use the software to model the compression strength performance of the corrugated case.

Using modern computers to calculate pallet patterns is highly effective. Studies have shown that a 10% improvement in pallet load utilization is common. Often improvements are much higher than 10%. This is especially important as these improvements directly affect the profitability of an organization. In larger organizations such savings can be many hundreds of thousands of dollars annually.

The Future of Computerized Pallet Patterns

Today's users of computerized pallet pattern/loading programs are looking for even more sophistication and reality in their ability to create reports and share this information with others.

Current pallet loading programs are far more sophisticated. They are used throughout the world to evaluate different pallet pattern options, the best pallet size to use, and which packing medium will do the best job. All of these tasks are now calculated in seconds. Each solution created can be viewed as three-dimensional color diagrams. Graphics technology now allows users to select the rotation and stacking of individual pattern layers. Even the layout of individual cases can be edited to meet any special pattern layout requirements. High-quality, very detailed, and customized reports can be produced in seconds.

Modern programs can also be used to export palletizing information directly to other programs. Examples are word-processing applications, spreadsheets, and specification systems. Information from pallet loading programs can even provide the necessary information to drive robotic and mechanical palletizing equipment. Such information can then be shared with other departments to create a finished-product specification which can be easily communicated, controlled, and monitored for future updates and modifications.

Manufacturing companies are now beginning to understand the importance of multiple product pallet patterns, for retail store shelf replenishment and as end-aisle displays for promotional use. Programs to deal with these "multiproduct loading" situations work with many of the same inputs as standard pallet loading software. The one exception is that many different products are calculated on the same pallet at the same time, adding an extra level of complexity to the final solution.

For many multinational companies, the key to future growth is exporting their products to other countries. This means considering how to palletize their products on at least two different pallet sizes: one for domestic use and one for export. This type of application is also ideally suited to modern computers and pallet loading programs.

Technology is now available that allows pallet loading programs to merge artwork and graphic images onto the surfaces of both cases and pallets. This brings a new and extremely powerful "visual reality" to the world of computerized pallet patterns and allows the most comprehensive pallet pattern reports and specifications to created "at the touch of a button."

Summary

The use of computers for creating pallet patterns has come a very long way in the last 20 years. Because of the increasing pressures for modern industries to continue to change and become even more competitive, there are still many opportunities that lie ahead. As technology continues to advance, and computers become more powerful, the computerization of pallet patterns will grow in many directions.

Brad Leonard
Cape Systems, Inc.
Plano, Texas

CONSULTING

In the packaging field it is often necessary to hire a consultant. Frequently this need arises because there is an emergency, or crisis situation in one aspect or another of a packager's packaging program. The consultant is necessary to provide a timely answer to the problem at hand. However, it is not always necessary for an emergency to exist to establish a need for consulting assistance.

There are many legitimate reasons for retaining the assistance of a qualified consultant. Packaging is a very diverse field, and the need for expertise can arise from many different sectors. Consultants are available in many disciplines within packaging. The Institute of Packaging Professionals (IoPP) Council of Packaging Consultants publishes an annual directory. This directory alone lists areas of expertise in some 28 different industries, 36 different materials and package forms, as well as 18 different types of processes. There are experts listed in such diverse areas as aerospace, medical devices, glass bottles, metal cans, foil trays, packaging line design, machine design, robotics, and many more areas. Literally, there are consulting experts available for even the most diverse among packaging problems.

Given this diversity, there are three major problems to be resolved:

- When should a consultant be used?
- What are the attributes of a qualified consultant?
- What is the proper procedure for hiring a consultant?

The answer to these questions are many faceted but there are logical sequences to go through to attain the correct answers.

When to Hire a Professional Consultant

Packaging consultants are, in the main, technical experts in one field or another. They are also, by the nature of their business, accustomed to working on a temporary basis. Hence, the most common situation for hiring a consultant is when there is a specific, short-term business problem—one that will be best solved by someone with specific packaging expertise, pertinent to the problem at hand. This is especially true in situations for which there is no in-house talent available.

Another major use for an expert's services is when independent advice and opinions are needed. The ideal situation for using a consultant can exist in this regard because true consultants are ethically answerable to their own ideals and not to a boss or department head of the packager. They can, and should, "tell it like it is." This aspect is particularly useful in identifying new markets and strategies or when the need arises to reinvigorate operations with new ideas or perspectives. The creation of new operations or the reorganizing of old ones are areas that can certainly benefit from independent input. Another controversial area that can be effectively resolved by independent expertise is the performance of a search for cost-savings opportunities in packaging operations or materials. This is a classic use for packaging consulting services. Independence of thought is also a valuable asset in the resolution of internal technical or managerial conflicts.

By definition, another excellent reason for using a qualified consultant is as an expert witness. Should the need arise, a large variety of packaging consultants are available to serve as part of a legal team in situations requiring expert testimony. And, finally, packaging consultants can serve simply as additional help when business conditions create a heavy, but short-term workload—a workload that requires experienced help to supplement the efforts of in-house staff.

The Attributes of a Qualified Consultant

Most packaging consultant firms are small; in fact, most are, principally, one-person enterprises or, at most, there might be an additional single-person office or clerical staff with, possibly, one or two technical aides. Many individual consultants will form a loose network with other individual consultants to form a team expressly tailored to meet the needs of a larger particular job. For instance, a machinery consultant might team up with a materials and a process consultant to form a team for a new packaged product introduction. On the other hand, some of the bigger firms might have a full staff ready to carry out the complete program with their own in-house team. For either case, the prospective client should be expressly concerned with the qualifications of the exact individual major players on the team. Hence, when looking at either a smaller or larger firm, the judgmental process is basically the same. First, determine who the individuals are who will play the leading roles in the project. Then, perform a detailed analysis of those individual's qualifications.

There are many attributes to look for in judging these qualifications, and all the attributes are important; however, four attributes stand out as probably the most important:

- Experience
- Strong ethics
- Complete objectivity
- Strong references

There are more attributes than these, and the remaining ones are discussed later; however, these four deserve the most attention. If a consultant fails on any of them, the prospective client should be wary.

In the area of experience, remember that the product being paid for in any consultation is expert advice. Advice that is usually concentrated in the very narrow field being consulted on. Make sure that the prospective consultant's credentials in that field are sufficient in scope and specific in nature. If a consultant is being hired to advise on cartoning machinery, for instance, make sure that there is a heavy dose of cartoner experience in that individual's past. A consulting assignment is neither the time nor the place for on-the-job training, not to mention the higher costs involved for consulting fees. This brings up the next of the primary attributes of a consultant qualified to perform in a specific area—strong ethics. Truly ethical consultants will never take on projects that they determine themselves to be unqualified for through either experience or training. The good consultant will refer such a project to another, properly qualified consultant rather than take on a job where the learning curve will be too steep. Strong ethics form one cornerstone of a consultant's business life and are a necessity for long-term success in the field. In one sense hiring a consultant is like hiring any other employee where honesty and integrity are necessary ingredients for proper employment.

The third of the primary attributes that a consultant must bring to a job is the quality of complete objectivity. To properly perform on a project, it is essential that a consultant not be compromised by preconceived ideas or become emotionally involved in the decision making processes. A large part of what a client is paying for are decisions based on impartial judgment. For instance, the consultant must be able to advise that aluminum is a better packaging material of choice than plated steel, or vice versa, for a particular application without regard to emotional bias or feelings based on any preexisting client–vendor relationships. In its best sense, objective judgment should be synonymous with the word *consultant*.

Given that these three attributes form the cornerstones of the qualities that make a successful consultant, the trick is how to determine whether a prospective candidate possesses them. A good way to find out, possibly the only way, is to ask for strong references from people that the consultant has performed for before. It is proper to request such references from a prospective candidate, and it is just as proper to contact the references and question them thoroughly. The most important questions to ask would concern the three important attributes that are pointed out above.

Along with the attributes of experience, ethics, and objectivity, there are several additional qualities that should be looked for at the same time, for their possession rounds out the profile of a truly qualified consultant:

- *Strong Analytical Skills.* Analysis is at the heart of any consultation. When interviewing a prospective candidate, this is one of the important qualities to look for. Before a meaningful search for a solution can begin, it is first necessary that the real problem can be articulated, fully and correctly. Make sure the candidate is capable of developing a full and accurate picture of the problem, as well as, the solutions and results.
- *Creativity.* A good consultant should always mirror back to the client, the client's assessment of the problem to ensure that a mutual understanding exists. However, this is only a starting point. Boilerplate suggestions are not enough. Look for signs of innovation, uniqueness, and a different perspective.
- *Personality.* The odds are that the client will be working very closely with the consultant. A good match of personalities helps, immeasurably to ensure a successful project.
- *Full-Time Attention.* This can be a key factor in a client–consultant relationship. Part-time or "moonlighting" consultants usually have a poor track record. In the same manner, too large a consulting firm handling too small an assignment can also be dangerous. For best results, the job at hand must be important to the consultant—important enough to demand full-time attention.
- *Ability to Transmit Knowledge.* Excellent results will serve very little purpose if they are not fully communicated to all those concerned with the project. For a successful consultation to occur, communications is a key ingredient. Look for good communication skills in a prospective consultant.
- *Good Listener.* While it is very important that a good consultant can talk and communicate, it is also just as essential that the consultant exactly understand the problem in the first place. This means that, especially in the beginning phase, the consultant must listen intently and understand. Beware of candidates who do not take the time or who are not smart enough to listen.
- *Cost.* Finally, cost must be a consideration. Just remember that with consultant services, one usually gets what one pays for. The lowest cost is seldom the least expensive investment. As a general rule, the better consultants receive the best pay. Any consultants who are not smart enough to know their own self-worth perhaps do not know enough in that particular field in the first place. Be prepared to pay for good service.

After measuring the attributes of what constitutes good consultancy, the next task is to determine how to actually hire the consultant.

How to Hire the Right Consultant

The first step in hiring a packaging consultant is for the prospective client to be able to define the actual project to be performed. The client must be able to articulate the job to be done in as precise terms as possible. If possible, a written document outlining the requirements of the proposed project is the preferred vehicle of communication. Once the scope of the project is known, the next step is to locate a select group of consultants who are qualified to do the job. The best source is, of course, to know someone in the packaging field who can recommend a good consultant. Preferably, the recommender would have used those sources previously. If this is not possible, because of inability to locate any previous users, the next step would be to consult the directories in the field for possible sources of consultants. The best directory in this regard is published by the Institute of Packaging Professionals (IoPP), who publish an annual *Directory of Packaging Consultants* that is very representative of who is available and in what area. ASTM also publishes a general *Directory of Scientific & Technical Consultants and Expert Witnesses,* and the *Packaging Sourcebook,* published by North American Publishing Company, also lists packaging consultants. Once a selection of likely candidates is chosen, contact each candidate. Question them about their professional backgrounds and request two important items:

- Their professional resumes or curriculum vitae.
- A choice of references. Preferably these references should be packaging specialists who have had experience that is close to the project under consideration.

Questions to ask concern prior experience, length of time in business, the size of the consultant's firm, what size jobs have been performed in the past, how fees are typically determined, what are the fees, how many accounts are repeat business and—possibly most important—your personal assessment on how the candidate's personality will fit with your organization.

After the interviewing process, the next step is to obtain proposals for the actual work. If the project is an involved one, especially, one that requires significant creative input, it is not unusual to expect to pay a fee and expenses for the proposal. This would almost certainly be true if it required the consultant to travel to your location. Be careful in this step. It is always tempting to go with a consultant who does not charge for the initial visit. The most qualified consultants usually do. This initial savings could end up costing a considerably larger amount of money if it results in the best-qualified candidate not being selected.

Make sure that the final proposal contains the following items in a well-defined manner:

- Scope of work (this should contain a reiteration of the problem along with an abstract of the proposed program)
- Services to be provided
- Description of fees and fee schedule
- Terms of payment

Resist the temptation to base the final decision on price alone. The best consultants frequently cost more.

After selecting the best candidate, fine-tune the details. Make sure that the proposed "scope of work" meets the project's requirements. Then sign a contract or issue a purchase order. Frequently an initial down payment is involved.

Once hired, the consultant needs to be worked into your organizational fabric in the best way. The important point to consider is that the staff should not be intimidated or feel threatened by the consultant's presence. The best way to accomplish this is to be completely open about the job the consultant is supposed to do. Introduce the consultant to all concerned personnel on the project.

During the consultant's tenure, be sure to provide ongoing support and interest in the progress of the project. One other caveat. The higher the level within the organization that the consultant reports to, the better. It should be someone with sufficient authority to ensure the full cooperation of everyone concerned with the project.

Robert A. Luciano
Luciano Packaging Technologies, Inc.,
Somerville, New Jersey

CONSUMER RESEARCH

It goes without saying that strong and effective packaging has to be an integral part of the marketing mix. A strong package is equity. A strong package sells a product and a strong package justifies premium pricing.

But who defines a strong package? The shopper, because the shopper is the individual who ultimately picks up and buys the product or quickly bypasses or totally ignores it. From both the shoppers' and the marketers' point of view, the shopping experience is a battle. The shopper has to battle through competing items, through heavy shopper traffic, and through the checkout lines. The marketer, on the other hand, has to break through the clutter to get a few precious seconds of consideration. A number of startling facts about shopping in the United States are now available. For example

- The average supermarket contains over 25,000 items, with 17,000 new products being introduced each year.
- The average shopper is spending approximately 24 min in his/her normal shopping trip. This translates to 1440 s to consider 25,000 items. Not surprisingly, approximately one-third of the packages on the shelf were being completely ignored. Even with more than 8200 products bypassed, the U.S. supermarket shopper is still faced with 16,000+ items being considered in 1440 s; not much time for a marketer to sell his/her product.

In the United States, it is now generally acknowledged that 80% of the decisions made in today's supermarkets are in-store decisions and, of equal importance, 60% are impulse purchases. This means that the packaging and its effectiveness in breaking through the clutter and conveying the right imagery can determine, and does determine, a product's success or failure.

How do marketers research packaging to ensure that our products will receive a fair share of nonplanned or impulse purchases? Initially, we must define an effective package. An effective package

- Is simple.
- Quickly communicates what the product is.

- Makes use of focal points.
- Stands out from the competition.
- Makes selection within a product line easy.
- Has the right quality impression.
- Reflects the image of the product.
- Retains a visual connection with the past when redesigned.

Package research must address each of these criteria. Effective packaging research must also be conducted with the appropriate target audience—the same target audience that is used when developing marketing strategy, when planning TV commercials, and when producing print ads.

Good Packaging Research

What is good packaging research? Let's begin with the premise that the packaging must be compatible with the long-term strategy and positioning of the brand. The packaging must close a sale in the store.

If we can define what the package should be, then certainly we should be able to design research to uncover if the packaging is delivering in all the key areas.

Research should uncover shoppers' attitudes and feelings toward packaging, if the shopper takes the time to pick the package up, to consider it, and, hopefully, to buy it. We recognize that a designer can create a gimmick package that gets attention but does not close a sale. Conversely, a marketer can have a unique product with unique benefits, yet its packaging may be lost or buried on the cluttered store shelf. A strong package must be strong in many areas. It must have stopping power, it must generate readership and involvement with the labeling, and it must convey the imagery that helps close a sale.

Packaging for a well-established brand represents equity. Changing that package represents a risk. Packaging research must profile the equity the marketer has in his/her existing packaging and the risk he/she may encounter by making a change. Thus, packaging research must be evaluative. This means that it must profile to the marketer the strengths as well as the limitations of his/her package and, at the same time, it must be diagnostic. It must generate information or fine-tune packaging designs, if fine tuning is necessary.

Conducting Research

There are basically three methods of conducting packaging research:

- Focus group sessions
- Mall intercept interviewing
- Test market auditing

Focus groups have a number of advantages as well as serious limitations. They alow you to see and to hear consumer reaction. They can also provide valuable diagnostic input. However, we're all familiar with one dominant respondent totally controlling the focus group and biasing the responses of the other attendees. Many of you have also learned from attending focus groups that we often hear what we choose to hear. So often, four or five observers walk out of a focus group session with four or five different interpretations.

We noted earlier the importance of shelf visibility. It is impossible to measure the shelf prominence of a package in a focus group session. To ask a shopper if he/she would or would not take notice of a package on the shelf is naive at best. What you gain from focus group sessions is an overall insight of "acceptable or unacceptable." If you're looking for serious negatives, the focus group will begin to uncover problems. If you're looking for quantitative decisionmaking information, the focus group session is not appropriate.

Mall Interviewing

Mall intercept interviewing is widely used. In the shopping malls, we're generally able to reach target shoppers and interview them individually. The mall intercept interview is conducted on a one-to-one basis (shopper and interviewer). Accordingly, one strongly opinionated shopper cannot make or break a package, as so often occurs in the focus group session.

Areas that should be covered in the questionnaire include

- Aesthetic appeal of the packaging
- Perceived product imagery conveyed by the packaging
- Believability of claims
- Effectiveness of the package in stimulating interest in trial
- Functionality of the packaging (easy to store, easy to pour, easy to hold)
- Confusion (if any) with labeling claims and instructions

Test market auditing, on face value, appears to be an effective way of documenting the impact of packaging in the store. In actuality, it is probably the least efficient method, since it is subject to so many uncontrollable variables: competitive pricing, positioning on the shelf, number of facings and in-store sales. In addition, in-store auditing requires the packaging to be produced in finished form and in sufficient quantities to stock the shelves. Unfortunately, if your packaging is deficient, the marketer will not find out until he/she has wasted a lot of time and money.

Measuring Shelf Impact

Researchers have tried a variety of tools to measure shelf impact. Some have strengths and some have serious limitations. Let's consider:

- Tachistoscopic research
- Findability tests
- Recall questioning
- Eye tracking

Tachistoscopic research. This is simple to administer and provides a measure of quick recognition. The shopper is exposed to a series of scenes at brief time intervals (0.2, 0.5, 1 s) and asked to identify what he/she saw. The package that is identified most quickly is generally considered to be the best.

Unfortunately, there are many serious limitations to T-scope research. The most fundamental is the arbitrary time the researcher chooses to show each package. Does it really matter that package A communicates faster (in 0.2 s) than package B? Who's to determine whether 0.2 s ($\frac{1}{5}$ of a second) is a relevant timeframe?

An even greater drawback is the aspect of familiarity. As you might imagine, the familiar brands are generally those that are identified faster. Thus, when a marketer is considering a packaging modification, the T-scope may put his/her new design at a distinct disadvantage.

Findability tests. Findability tests are simply those that ask a shopper to look at a cluttered in-store shelf scene and find specific products. Again, it is assumed that the package that can be found fastest is the most effective. We mentioned earlier that 80% of the purchases made in the supermarket are non-planned. Thus, the findability tests that ask the shopper to locate a specific brand he/she is looking for is relevant only to those 20% who are going into the store looking for that brand. In actuality, the marketer's concern is with the other 80% who might make a nonplanned purchase. The 20% who plan to buy your product will find it, regardless of where it's located on the shelf.

Recall questioning. A third commonly used measure of shelf impact is recall questioning. Recall, like T-scope research, is often influenced by familiarity. The well-known brands receive the higher recall scores. The new products or low-awareness brands suffer from a lack of previous exposure.

Eye-tracking research. Eye-tracking research to document shelf prominence overcomes many of the limitations of the T-scope, findability tests, and recall. The eye tracker enables you to observe the shopper behavior, to see what they see, to see what they consider and, most importantly, to observe what they ignore. Eye tracking is not hampered by the need to select an arbitrary viewing time. You are able to observe what people do and how they shop the category, be it one second, one minute, or one hour. The eye tracking is not biased by familiarity.

One might even argue that the uniqueness of new brands would have an advantage, rather than a disadvantage, in drawing shopper attention. Many major marketers in the United States are using the eye tracking to develop planograms and to pinpoint the pros and cons of additional shelf facings and horizontal and vertical layouts, and even to uncover the competitive products that are strong on the shelf.

Some may argue that the eye-tracking viewing situation is artificial, and they are correct. However, the reality is that no one can duplicate the "real world," for it differs from store to store. How many times have we seen beautiful planograms never implemented in the supermarkets?

Often we refer to eye-level shelf placement as being optimal. But does eye level assume the shopper is 6 ft 2 in., 5 ft 10 in., or 4 ft 9 in.? With eye tracking we can watch the shopper discriminate and, most importantly, see what he/she ignores. We know from eye tracking the packaging that is breaking through the visual shelf confusion.

Label Readership

Label readership is another vital area that comprehensive packaging research must address. The designer has positioned key elements on the label to be seen and to be read. The three research tools commonly used to evaluate label readership are T-scope, recall, and eye tracking. The T-scope and recall measures suffer from the limitations described earlier: a fixed viewing time, contamination by familiarity and, in many instances, shopper guessing. The eye tracking shows how shoppers read the package labeling. It quantifies the advantages of top of the package versus bottom. It shows the shoppers' discriminating process and, most importantly, it allows the shopper to ponder and thoroughly examine a package if he/she chooses.

Researching packaging attributes can be accomplished only through a comprehensive and well-thought-out questionnaire. Unfortunately, many researchers attempt to answer these questions through paired comparisons, ie, showing two potential packaging alternatives side by side. This is a simple approach for forcing a winner because one package will test better than the other. However, in actuality, neither may meet marketing objectives, and the end result may be the best of the worst.

Designing Research

With all the above in mind, let us consider a few of my do's and don'ts when designing packaging research:

1. Don't show different packaging alternatives side by side to a shopper. The shopper will never see two executions for the same package side by side in the store. His/her frame of reference is competition. Thus, each packaging execution should be tested against competition.
2. Never control the amount of time you let a shopper look at a package. Remember, a designer is trying to develop a package that is a stopper, a package that is involving, a package that a shopper will want to take a second or third look at. The instant a researcher controls the amount of time he/she lets the shopper look at a package is the instant you can no longer measure involvement. Speed of communication is not the key. Effective communication is far more relevant.
3. Don't live by hard-and-fast rules. Packaging designers have demonstrated time and time again that being different can pay great dividends at the cash register. Who would have guessed years ago that an orange-juice company could use the color black as their primary packaging identifier? Yet Minute Maid has done it and done it with enormous success.
4. Don't rely on traditional advertising research recall scores. Many researchers have a tendency to rely on that magical thing called recall. Yet recall per se is irrelevant when it comes to packaging research. A package is not on the shelf to be recalled. A package is there to be seen, to be considered and to sell.
5. Don't tie the package designer's hands. Respect the creativity and excellence of the packaging design industry. They're creative, they're innovative, they're insightful, and most of all, they approach their task with a marketing frame of reference. To advise a package designer that certain colors or shapes or designs will not work within a category is shortsighted. Utilize their talents, allow them free rein, and explore all creative opportunities before accepting or rejecting an innovation in packaging.
6. Don't forget to look at competition. The shopper consid-

ers competition prior to making his/her purchase decision. The marketer should do the same. All too often, there is a tendency to "follow the leader." If he/she is using red, we should use red. If his/her packaging is horizontal, ours should be the same. Keep in mind, breaking through clutter and getting attention is the first step to a sale.

Effective packaging is an integral part of the marketing mix. An effective package catches the consumer's eye and entices the shopper to give the product a try. Successful packaging leads to successful businesses. Don't underestimate the influence of packaging. Your package represents your product. Your package is equity. A major change in your package is a risk. Research the risk thoroughly and logically. Remember that your package is your product to the consumer at the point of sale.

Joseph O. Eastlack, Jr.
Saint Joseph's University
Philadelphia, Pennsylvania

CONTRACT PACKAGING

Packagers Outsource Resource Responds to Expanding Market

Herndon, Virginia, March 1, 1996: In the crunch of the 1990s, with tighter and tighter operating restrictions, new technologies, equipment, and partners are providing an important boost in flexibility and productivity for the nation's manufacturers.

One element of this is contract packaging, a vibrant, growing industry at a time when businesses find themselves downsizing or, worse yet, closing altogether.

A small specialized segment of the industry 10 years ago, contract packagers now report an annual growth rate of more than 10% a year, according to a recent survey completed by the Contract Packagers Association (CPA). CPA says most industry experts expect that growth rate to continue through the balance of the decade.

Product manufacturers are increasingly turning to the contract packaging option because

- Their packaging volume dictates a need for short-term capacity relief for their own packaging lines.
- They have a short-term business problem that is best solved by someone with packaging experience or equipment.
- Manufacturers are promoting a product with trial sizes or promotional insertions requiring special machinery or labor-intensive work.
- The pressure of new business or a new product creates a heavy, short-term workload.
- Manufacturers are entering a foreign market and need specialized packaging.
- They need expert help to train their staff in new areas of packaging.
- They are marketing or distribution companies with no production or packaging facilities. Contract packages are also contract manufacturers able to produce and package your product.

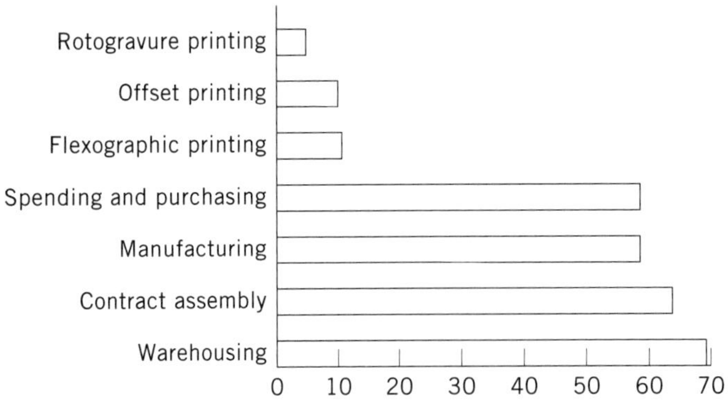

Figure 1. Additional services.

The profile of the average contract packaging company is anything but average. Most contract packaging companies operate one plant and run one shift. These companies on average maintain a staff of 50 full-time employees and 12 part-time employees. The staff comprises, on average, seven senior executives and 43 line and support staff. Contract packaging companies also employ an average of 12 part-time employees during the course of a year.

Better than 70% of the contract packagers have a warehousing operation (see Fig. 1). Fifty percent offer services such as contract assembly, manufacturing, spending, and purchasing.

Eleven percent of the industry offers flexographic printing, 10% offer offset printing, and 5% have the capability of handling rotogravure printing.

While these companies may offer customers additional services, 90% of all their annual sales volume comes from their primary business: contract packaging.

New contract packaging companies start up every month, but the average experience of the companies in the industry is 16.5 years, notes CPA.

Contract packagers package products in a multitude of end-use markets. Figure 2 presents an abbreviated list of the contract packaging markets and the percentage of companies that serve these markets.

The CPA survey revealed that 67%, or nearly 7 out of 10

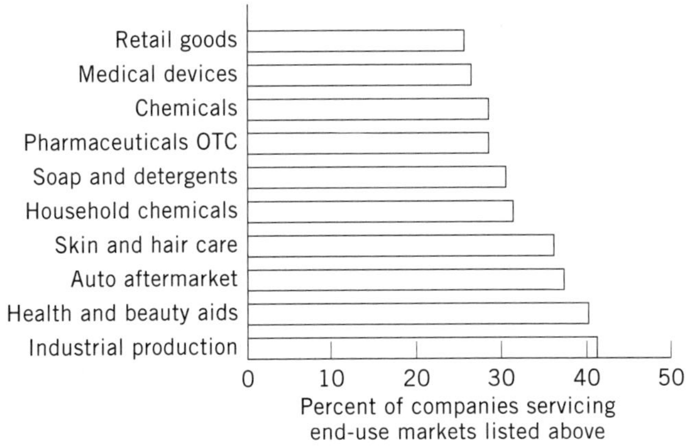

Figure 2. Products packaged.

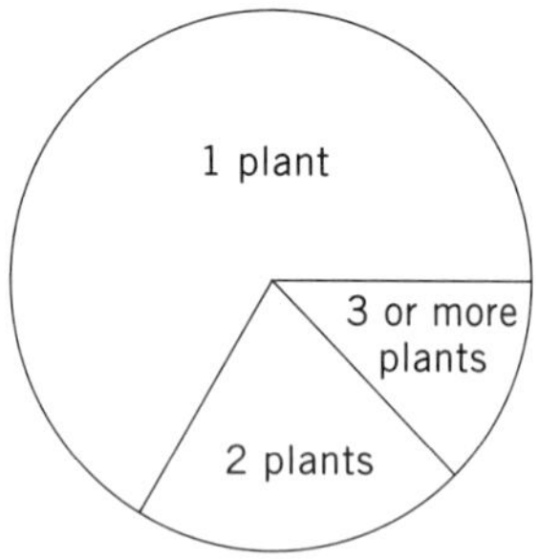

Figure 3. Number of plant locations.

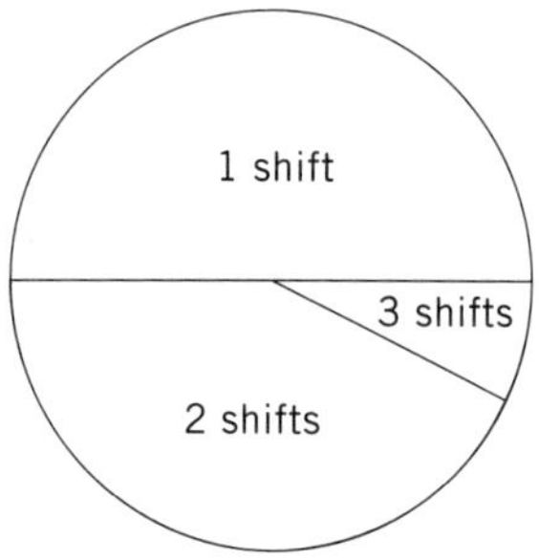

Figure 4. Number of shifts per day.

companies, operate one plant (Fig. 3). The survey also revealed that nearly 20% operate two facilities. The remaining 13% have three or more plants in operation.

Of all contract packagers, 50% run one shift a day (Fig. 4). Roughly 4 out of 10 companies (42%) operate two shifts a day. Just under 10% have their facility operating 24 h per day.

Average gross sales volume of contract packaging companies is $2.1 million per year, ranging from less than $2 million to over $25 million.

The outlook for the coming years is equally impressive, with companies forecasting sales volume to grow an annual average basis of 12% per year. Some companies were even more optimistic and predicted sales to improve by 25% per year on an average annual basis!

BIBLIOGRAPHY

General Reference

A *Directory of Contract Packagers and Their Facilities* is published by and available from the Institute of Packaging Professionals.* This 150-page directory lists the firms, and their special expertise and management contacts, and describes their facilities. The directory sells for $35 plus shipping and handling and can be ordered by calling The Packaging Matters! Customer Service Center at 1-(800) 432-4085.

Jim Iciek
Contract Packagers Association
Herndon, Virginia

* For more information on the contract packaging industry or membership in the Contract Packagers Association, or to purchase a copy of the *Directory of Contract Packagers,* contact the Contract Packagers Association at 481 Carlisle Drive, Herndon, VA 22070, tel. (703) 318-8969, fax (703) 318-0310.

CONVEYING

Definition

The Conveyor Equipment Manufacturers Association (CEMA) of the United States defines a conveyor as a horizontal, inclined or vertical device for moving or transporting bulk materials (cereals, aggregate, powders, etc.) or objects (such as bottles, cases, cartons, products, etc) in a path predetermined by the design of the device and having points of loading and discharge fixed, or selective; included are skip hoists, vertical tractors and trailers, tiering machines (truck type), cranes, monorail cranes, hoists, power and hand shovels, power scoops, bucket drag lines, any conveyor designed to carry people, and highway or rail vehicles.

Basically a conveyor is made up of one or more sections that transfers (via an intermittent or continuous media) such items as inputs (bottles, bags, caps, cartons, product, etc), packages, cases, or pallets from one location or position to another. Each section may be driven independently at varying speeds or employ gravity as a method of transfer. The *conveyor speed* is the surface speed of the conveyor in feet per minute (fpm) or meters per minute (mpm), while "line speed" or more properly actual run speed is the transfer rate, given in pounds or kilograms per minute or parts per minute (ppm) or containers per hour (cph).

Two or more rows of containers carried side by side on the same conveyor are collectively referred to as *mass flow,* while carrying only a single row is referred to as *single-lane* or *inline* flow. For mass flow, the number of adjacent rows is referred to as the *line width* and is given as an integer, while *conveyor width* refers to the width of a section of moving conveyor element expressed in inches or meters. Pitch is defined as the center-to-center spacing of containers, given in inches or meters. The pitch in the direction of motion is referred to as the longitudinal pitch or travel distance, while the spacing across the width of the conveyor is the transverse pitch or sideways distance.

The integration of all packaging machinery with their conveyors is critical to produce a seamless transfer and assembly system that results in the consistent regular flowing movement of inputs, packages, and cases that enter into the packaging line at specific points and are assembled, identified, verified, and shipped to fulfill customer needs. Integration of conveyors to conveyors or conveyors to packaging machinery should follow the principles as stated below in order for the packaging line to be successful.

Integration of conveyors with machinery should consider the following factors:

1. Selection and design of proper conveyor system to fulfill needs
2. Design and installation of *guiderails* and *handling control components.*
3. Speeds, feeds, dynamics, and loads to achieve the needed results.
4. Compatibility of *interconnecting machinery* and other conveyors.
5. Installation of conveyors and machinery.

There are many other components that can be attached to or

are part of a conveyor system that perform special functions such as

1. Quality inspection (Vision, Dud, Fill Height, No Cap, etc)
2. Coolers, freezers, conditioning ovens, cooker ovens
3. All types of coding units
4. Staging or allowances for manual operations
5. Hard and soft reject stations
6. In-process storage (buffers) areas
7. Combiners, dividers, turners, orientors, etc

All of these devices require some form of conveyor or transfer equipment as the main critical element to manipulate or identify or test the products and/or inputs. Conveyors are therefore an integral part of the packaging process and are elements or machines which have their own reliability issues.

Design and Installation of Conveyor Systems

Traditionally, conveying systems were designed without paying much attention to drive dynamics, the forces imposed on the conveyor and product, or the effects of these forces on container stability. Design criteria were based on the average running-speed requirements and mechanical longevity. In fact, typical design formulas considered only rough estimates of the loading (forces) combined with factors that represent the number of startups and the amount of product slippage. Most present chain manufacturers are improving the nature and extent of engineering for their products. Chain manufacturers such as Rexnord and MCC have technical engineering data and software programs available to assist in designing effective conveying systems related to their products. Also, engineering houses such as Zarpac Inc. have developed specialized software programs for conveyor systems.

In recent years, increasing demands, economics, and environmental concerns have placed enormous pressure on production facilities to increase operating speeds, reduce wastage, reduce rework, and cut manpower requirements. These changes have pushed traditional design methods to the limit of their effectiveness. In addition, the aesthetic appeal of complex input shapes (bottles, products, cartons, cases, etc), the use of lightweight materials, and input thin walling to reduce material weight has resulted in a new generation of stability and control problems associated with the transport and manipulation of containers by conveyor.

With current conveyor technology and the demand for high-productivity packaging lines, simple conveyors must be designed, built, and installed to run at a run utilization of over 99.5%. Conveyors with attachments must be designed, built, and installed to run at a run utilization of over 99.0% (buffers, combiners, dividers, etc).

For most cases, conveyors should be designed and run with minimum backpressures on the inputs or products in order to minimize part damage, maximize control, and maintain product conformity for optimum quality.

This article only *summarizes* the concerns and methods of container or input transport by conveyors.

Conveying Systems: Overview

There are many types of conveying systems in common use today. Most conveyor types and styles fall into 1 of 10 designs. These conveyor designs are listed and described in the following paragraphs.

Tabletop chain design. This consists of

1. Magnetic hinged-joint chain conveyors or magnetic tabletop chain conveyors
2. Steel or stainless-steel hinged-joint chain conveyors or steel or stainless-steel tabletop chain conveyors.
3. Plastic hinged-joint chain conveyors or plastic tabletop chain conveyors
4. Magnetic tabletop roller chain conveyors
5. Steel or stainless tabletop roller chain conveyors
6. Plastic-top roller chain conveyors
7. Stainless-steel multiflex chain conveyors
8. Plastic multiplex chain conveyors for small-curve-radius requirements
9. Side-flex hinged chain conveyors or side-flex tabletop chain conveyors (Fig. 1)
10. Flex link hinged conveyors or flex link tabletop chain conveyors. (See Figs. 2–4).

These conveyor types are composed of individually hinge connected flat slats that form a smooth surface (*1–3, 9*) or plastic flat slats that are connected to standard roller chain usually number 60 chain (*4–6*). They function at any level, including substantial elevations, although the most common usage height is 36–42 in. (910–1070 mm) and running horizontal.

Tabletop conveyors, sometimes called *flat-top conveyors,* can run at very high speeds and are widely used, for example, in the filling and packaging of bottles and cans, where production requirements constantly push the chain to their design limits. These chains are also used for many industrial conveying requirements, such as the manufacture and packaging of bearings, small mechanical components, blow-molded parts, and injection-molded parts. They can even interface effectively with robotic systems used in manufacturing and packaging. Tabletop conveyors use plastic or metal or a combination of the two materials in their construction (excluding the link pins, which are stainless-steel or ferritic steel).

Plastic chain is usually made of predominantly three materials:

1. Acetal or delrin chain for frictional requirements needing an approximate coefficient of friction 0.25–0.3 dry and 0.2–0.25 lubricated.
2. Low friction (LF) (about 18% Teflon) chain for frictional requirements needing an approximate coefficient of friction 0.15–0.2 dry and 0.1–0.15 lubricated.
3. High-performance (HP) (about 26% Teflon) chain for frictional requirements needing an approximate coefficient of friction 0.1–0.15 dry and 0.05–0.1 lubricated.

The proper plastic tabletop chain, which depends on friction, temperature, and speed requirements, is normally used for conveying aluminum cans or plastic containers and other nonabrasive products.

Lubrication takes the form of a spray mist on the idle side of the chain. The spray mist consists of soft water mixed with minute amounts of special liquid soaps.

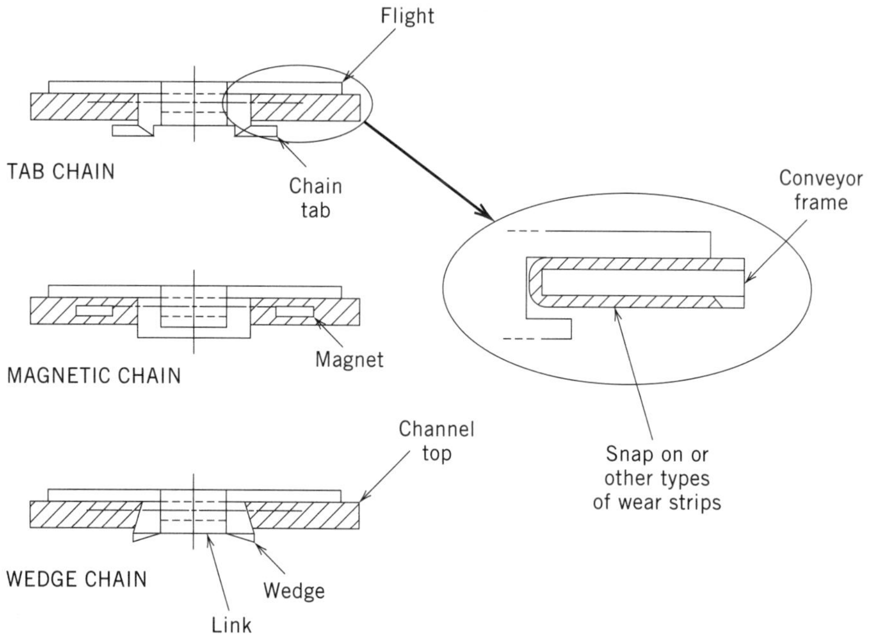

Figure 1. Side-flexing conveyor chain types.

Glass containers, steel cans, and abrasive-type products are generally conveyed on carbon or stainless-steel tabletop chains. Sometimes plastic tabletop chains can be used as an economy measure.

Hinged tabletop chain or roller-driven tabletop chain comes in the following standard widths to fit the needs of the packaging process. All widths are narrow as compared to mattop or mesh chain. Standard widths are $3\frac{1}{4}$, $4\frac{1}{2}$, 6 (not common) 7.5, 10, or 12 in. (89, 108, 152, 190, 254, and 305 mm) wide, and containers are carried at medium to high speed in single lane or mass flow. Speeds over 300 ft (90 m) per minute are possible for specific chain materials and under certain conditions.

The chain or hinge runs in a "channel," which is specially designed to support the chain or hinge on both sides and can carry both the loaded and return sections. The channel is fitted with wear strips to reduce friction and prevent long-term damage to the channel and tabletop chain. The channel is designed to allow the hinged section to pull through on top, feed through the drive and idle sprockets, and return below the channel smoothly without hesitation. There is always a caternary or return chain slack at the drive end, and its profile is very critical to torque loading.

Tabletop chain may be either straight-running or side-flexing. Straight-running chain is simple in design but requires special devices such as turntables or dead plates in order to move product through any type of turn. A side-flexing conveyor is more expensive but provides smooth, continuous turns and transfer of products. Side flexing gives more control and flexibility than straight-running tabletop chain.

There is a fundamental problem with the use of side-flexing chain in that side forces are imposed on the chain, which can cause it to twist and lift out of the channel. There are three common designs of side-flexing chain: tab, wedge, and magnetic.

Tab. Tab chain uses short lips on the underside of the links to hold the chain in the channel and keep it from lifting. Unfortunately, the tabs cause increased side loading and a longer maintenance time for chain repair or replacement.

Wedge. The body of the links are tapered, decreasing in width near the top surface of the chain. The channel has a mating taper that prevents the chain from twisting or lifting. Unfortunately, the wedge causes increased side loading but less than tabs and a longer maintenance time for chain repair or replacement.

Magnetic. Magnetic chain has permanent magnets located in either the chain or the channel. The magnetic attraction between the channel and the chain holds the chain into the channel. Plastic chain can be used in plastic channels with magnetic impregnated materials only if the link pins are ferritic.

Belt design. This consists of

- Troughed belt conveyor for bulk granular transfers
- Spiral belt conveyors
- Flat straight belt conveyors
- Magnetic flat belt conveyors
- Retracking belt conveyors
- Magnetic spiral belt conveyors
- Flighted belt conveyors

(See Figs. 5 and 6.)

A *belt conveyor* is defined as an endless fabric, rubber, plastic, leather, or metal belt operating over a suitable drive, tail end, and bend terminals. In the case of handling bulk materials, packages, or objects placed directly on the belt, the belt operates over belt idlers or a slider bed. Belt conveyors use a belt as a carrying medium for the controlled movement of a great variety of regular- and irregular-shaped commodities, from light and fragile, to heavy and rugged, to granular and to

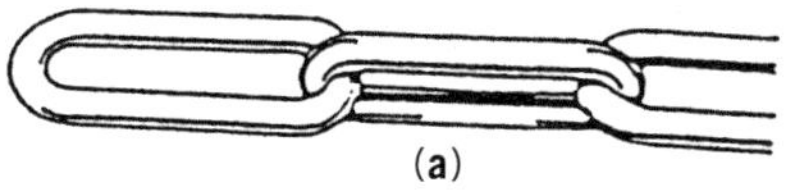

(a)

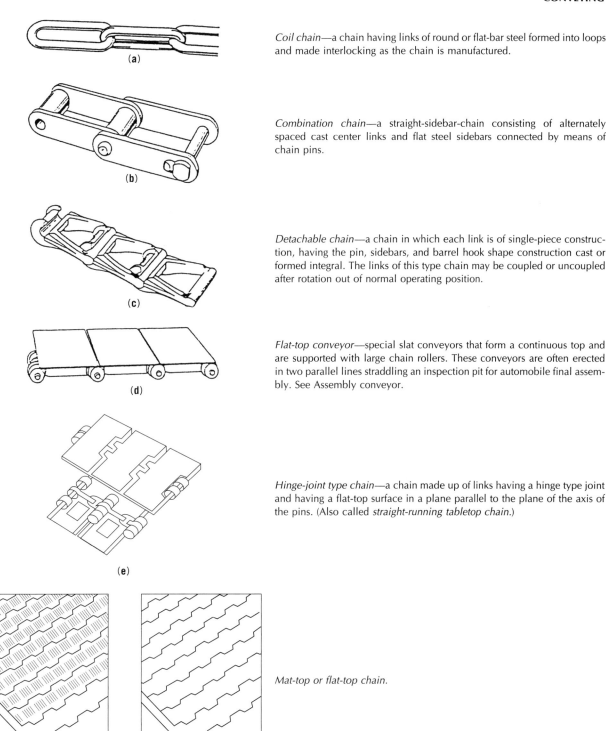

Coil chain—a chain having links of round or flat-bar steel formed into loops and made interlocking as the chain is manufactured.

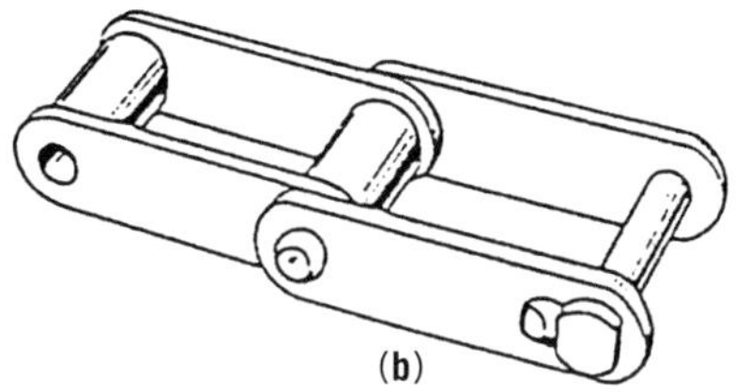

(b)

Combination chain—a straight-sidebar-chain consisting of alternately spaced cast center links and flat steel sidebars connected by means of chain pins.

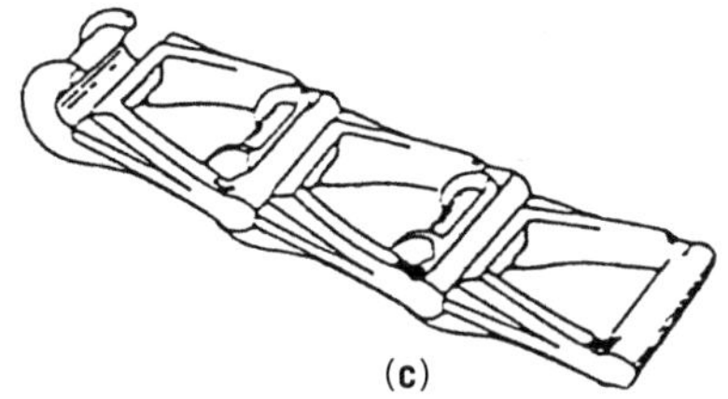

(c)

Detachable chain—a chain in which each link is of single-piece construction, having the pin, sidebars, and barrel hook shape construction cast or formed integral. The links of this type chain may be coupled or uncoupled after rotation out of normal operating position.

(d)

Flat-top conveyor—special slat conveyors that form a continuous top and are supported with large chain rollers. These conveyors are often erected in two parallel lines straddling an inspection pit for automobile final assembly. See Assembly conveyor.

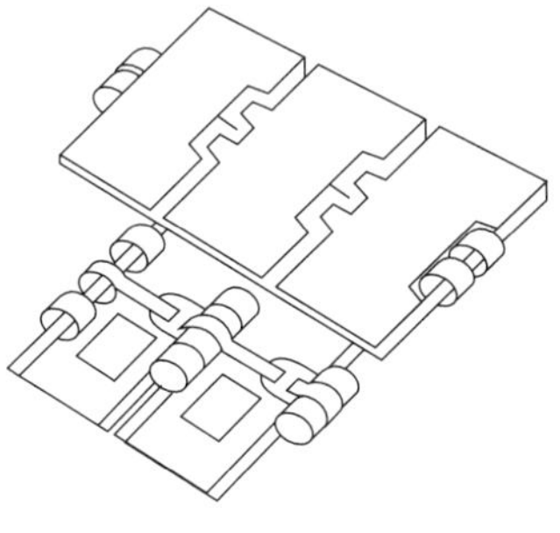

(e)

Hinge-joint type chain—a chain made up of links having a hinge type joint and having a flat-top surface in a plane parallel to the plane of the axis of the pins. (Also called *straight-running tabletop chain*.)

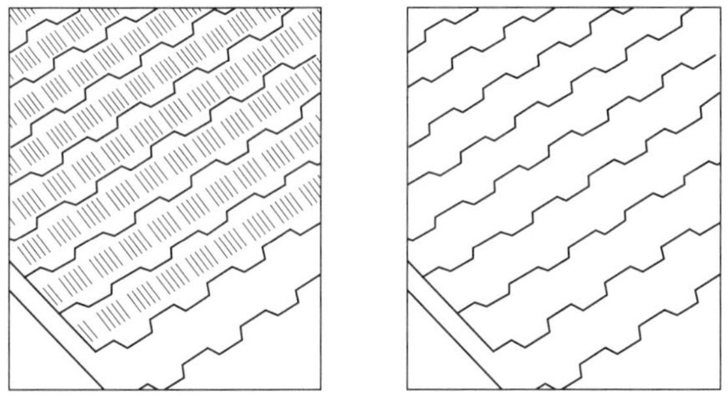

(f)

Mat-top or flat-top chain.

Figure 2. Chains.

solid unit loads. The belt conveyor can be level or angled up or down. The angle of transfer is limited mainly by the stability of the commodity and strength of the product being moved. Belt conveyors are connected together usually via a butt transfer or overlap in-line transfer.

A belt conveyor consists of a wide flat belt wrapped around two end rollers (one powered, one idle) and supported along its length on both the loaded and return sections. The "loaded section" is the portion of the belt that carries product (from idler roller to driver roller), while the return section runs from the driver back to the idler underneath the top section to form a closed loop. Belt conveyors can be driven at the drive or discharge end or at most locations on the return leg, generally near the discharge end.

Mesh-top or open-top chain.

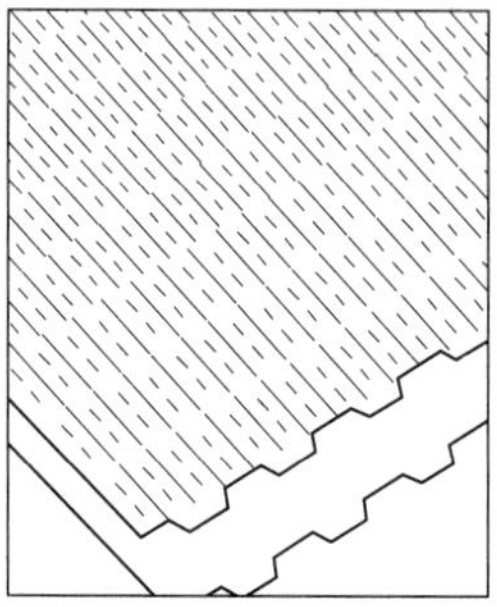

(a)

Pintle chain—a type of offset chain in which the barrel is cast integral at one end between a pair of offset sidebars.

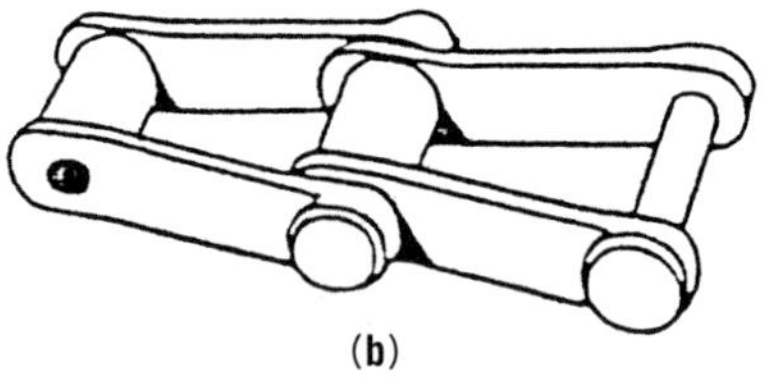

(b)

Rivetless chain—a series of pins, side links, and center links that can be assembled or disassembled without the use of tools.

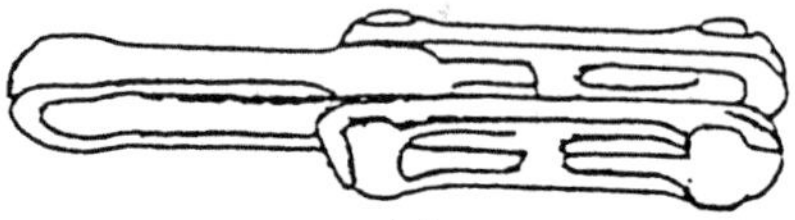

(c)

Figure 3. Chains.

Support for the top section may take the form of a fixed sheet or metal on which the belt slides or a series of equally spaced free-turning rollers or slider bars. Support for the return section takes the form of free-turning rollers or a stationary metal strip or horn.

Belt conveyors run at a variety of speeds and conditions. Belt conveyors can be used

1. To transport granular bulk, boxes, cartons, confectionery products, or randomly distributed containers.

Steel sidebar bushed chain—a fabricated all-steel chain made up of either successive offset links with bushing barrels or alternate center links with bushing barrels and outside links connected by means of chain pins.

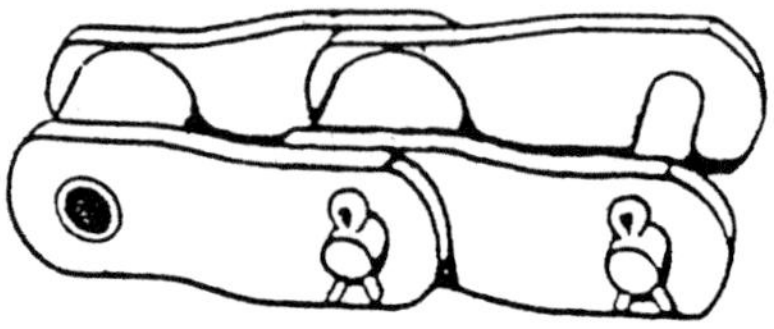

(a)

Transmission roller chain—a type of steel roller chain manufactured to relatively close clearances and tolerances and with highly finished surfaces.

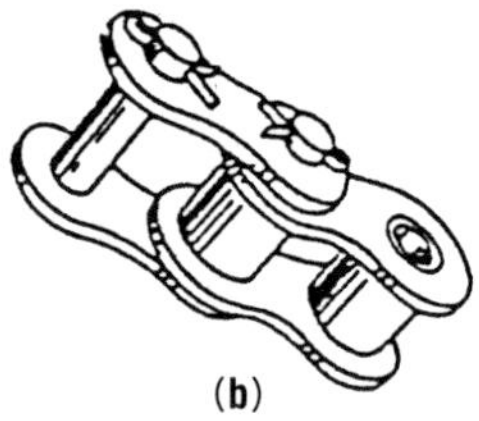

(b)

Troughing idler—a belt idler consisting of two or more rollers arranged to turn up the edges of the belt to form it into a moving trough.

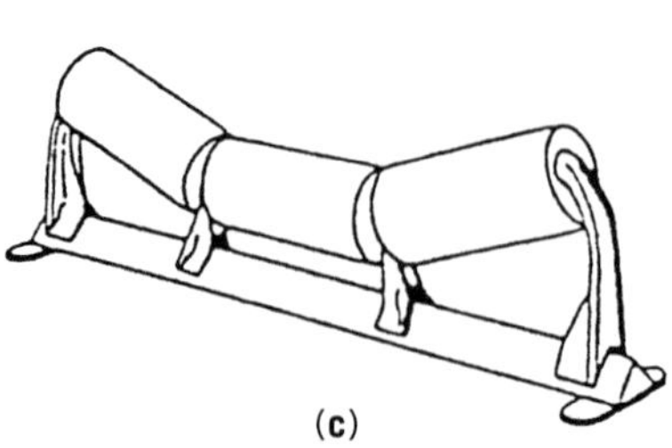

(c)

Figure 4. Chains.

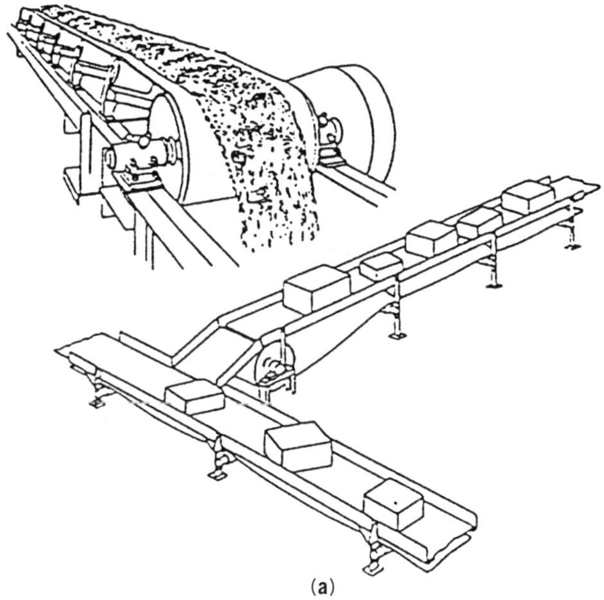

(a)

Belt conveyor—an endless fabric, rubber, plastic, leather, or metal belt operating over suitable drive, tail-end, and bend terminals and over belt idlers or slider bed for handling bulk materials, packages, or objects placed directly on the belt.

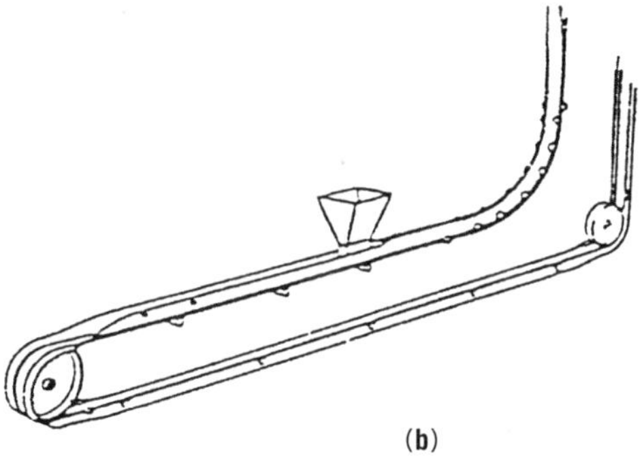

(b)

Closed-belt conveyor—moving, endless, flexible belt, or belts, which may be formed into a tubular shape by joining of edges, and which are opened while in motion to receive load, closed to convey or elevate, and opened to discharge. See also *Hugger-belt conveyor.*

Figure 5. Belt conveyors.

2. To set the pace of assembly operations.
3. As a timing medium for integrated handling systems or subsequent machine operation.
4. For controlling and/or spacing products through the use of top and bottom or side-to-side grip belts.

Mesh-top or open-top chain design. This consists of

- Metal (steel or stainless steel) straight mesh conveyors
- Metal (steel or stainless steel) curved mesh conveyors
- Plastic straight mesh conveyors

(See Fig. 3.)

Mesh chains, like mat-top chains, form wide-running conveyors. They are constructed from steel or stainless-steel wire that is formed into interlocking loops or injection-molded plastic wide links connected together by stainless-steel or plastic long pins (pins are as wide as the chain). Mesh chains are manufactured in various mesh patterns. They are designed for slow mass transport of products such as bakery items, solid food products, small metal products, and some confectionary items. However, they are usually designed for the mass transport of stable inputs needing no orientation requirements. The metal mesh chains are excellent to use in hot and some corrosive environments. Mesh conveyors are connected together usually via a butt transfer or overlap inline transfer. Curved or side-flexing configurations are possible.

New plastics and designs have now resulted in plastic mesh-type chains, giving low friction, reduced weight and torque requirements, wet capability, excellent drainage or fall through, better economy, and effectiveness for products where steel or stainless mesh or mats either cannot be used or are not economical.

Mat-top or flat-top chain design. This consists of

- Mat metal (steel or stainless steel) conveyors
- Mat-type plastic conveyors

(See Fig. 2.)

Mat-top chain, like belts, are wide-running conveyors. However, they are usually designed for the mass transport of stable inputs needing no orientation requirements. The chain is constructed using one wide link per longitudinal pitch as opposed to many adjacent narrow links. The wide links are connected by stainless-steel or plastic long pins so that two adjacent links resemble a "panel hinge" or "piano hinge." The

Portable conveyor—any type of transportable conveyor, usually having supports that provide mobility.

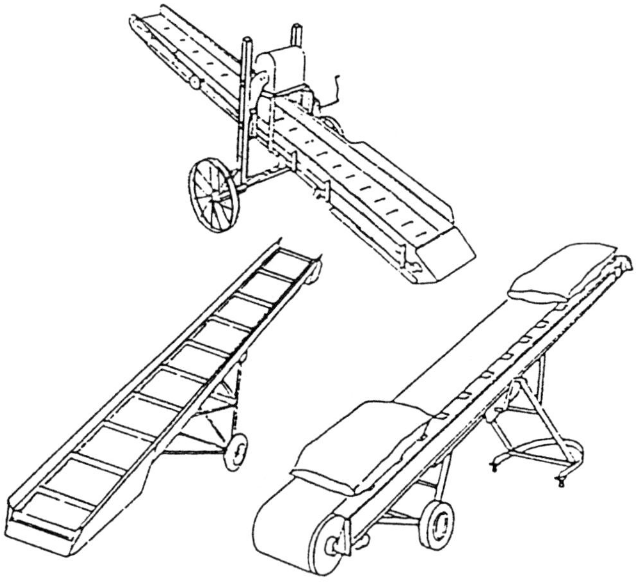

(a)

Trimmer conveyor—a self-contained, lightweight, portable conveyor, usually of the belt type, for use in unloading and delivering bulk materials from trucks to domestic storage, and for trimming bulk materials in bins or piles.

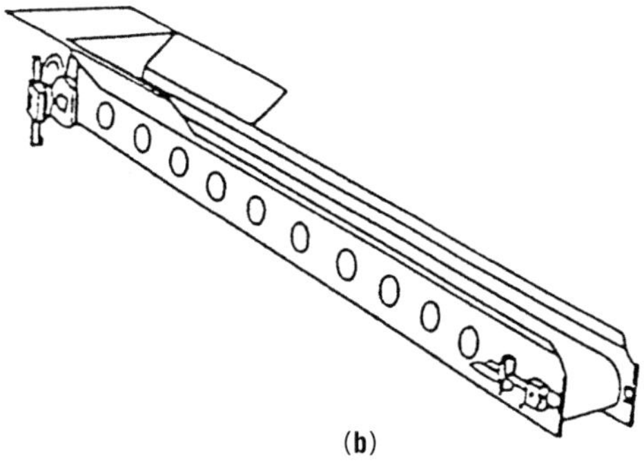

(b)

Figure 6. Belt conveyors.

links may have slots or holes for drainage. Mat conveyors are connected together usually via a butt transfer or overlap inline transfer. Curved or side-flexing configurations are possible.

New plastics and designs have now resulted in plastic mat-type chains, giving lower friction, reduced weight and torque requirements, wet capability, better economy and effectiveness for products where steel or stainless-steel mats cannot be used or are not economical.

Lug or bar chain design. This consists of

- Car-type conveyor
- Crossbar conveyor
- Drag chain conveyor
- Flight or bucket conveyor
- Floor chain conveyor
- Overhead trolley conveyor
- Pan or apron conveyors
- Pallet-type conveyor, indexing or continuous
- Pin-type slat conveyor
- Pocket conveyor
- Pusher bar conveyor
- Pusher chain conveyor
- Reciprocating-beam conveyor
- Rope and button conveyor
- Shuttle conveyor
- Sliding-chain conveyors
- Walking-beam conveyor
- Vertical elevator chain conveyor

(See Figs. 7–12.)

Lug or bar chain conveyors transport items via hooks, brackets, or bars attached to the top or bottom sections of most types of chains at a predetermined pitch, which will pull or push the input or product with no slippage.

Roller design. This consists of

- Nonpowered roller or wheel conveyors or gravity roller or wheel conveyors
- Chain-driven live roller conveyors
- Roller spiral gravity conveyor

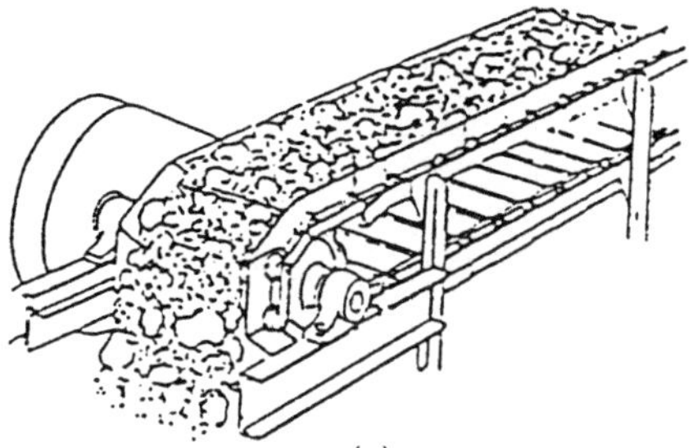

(a)

Apron conveyor—a conveyor in which a series of apron pans form the moving bed.

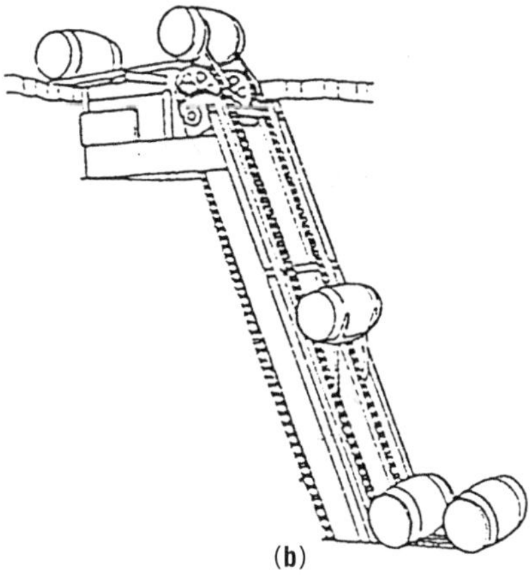

(b)

Arm conveyor—a conveyor consisting of an endless belt, one or more chains, to which are attached projecting arms, or shelves, for handling packages or objects in a vertical or inclined path.

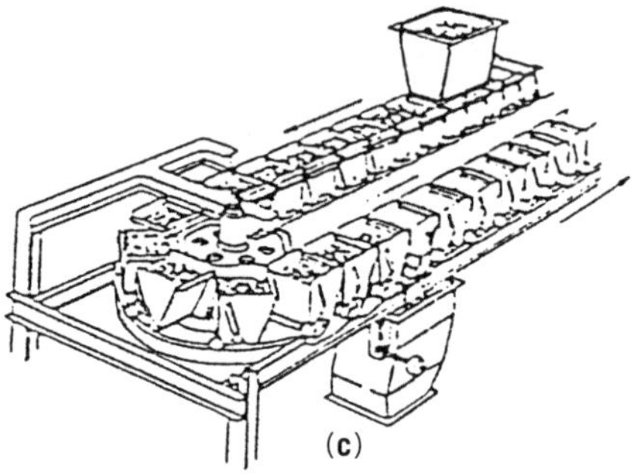

(c)

Bottom-discharge bucket conveyor—a conveyor for carrying bulk materials in a horizontal path consisting of an endless chain to which roller-supported, cam-operated conveyor buckets are attached continuously.

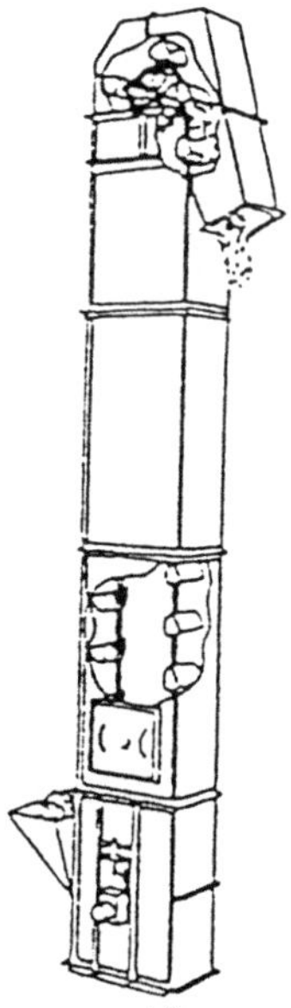

(d)

Bucket elevator—a conveyor for carrying bulk material in a vertical or inclined path, consisting of endless belt, chain or chains, to which elevator buckets are attached, the necessary head and boot terminal machinery and supporting frame or casing. See types: *Centrifugal-discharge; Continuous; Double-leg; Internal-discharge; Positive-discharge; Single-leg; Spaced; Supercapacity.*

Figure 7. Lug or bar chain conveyors.

Car-type conveyor—a series of cars attached to and propelled by an endless chain or other linkage running on the horizontal or a slight incline.

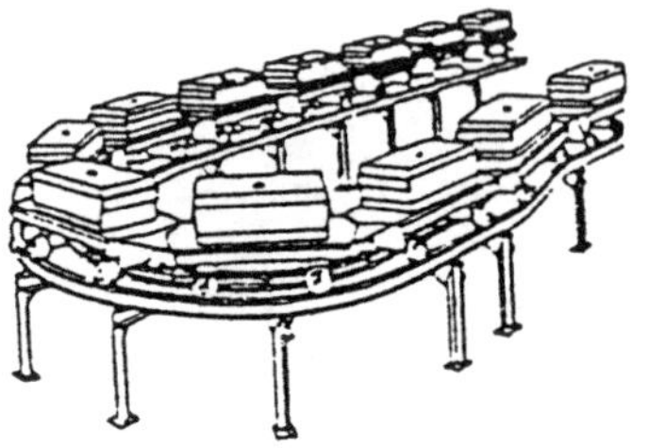

(a)

Crossbar conveyor—a type of conveyor having endless chains supporting spaced cross-members from which materials are hung or festooned while being processed.

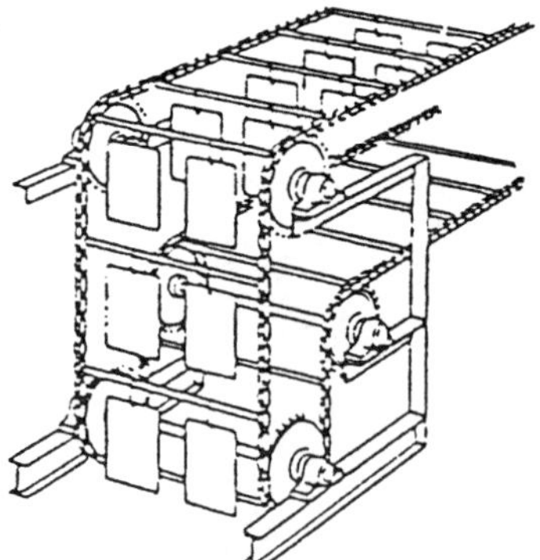

(b)

Drag chain conveyor—a type of conveyor having one or more endless chains that drag bulk materials through a trough. See *Sliding-chain conveyor.*

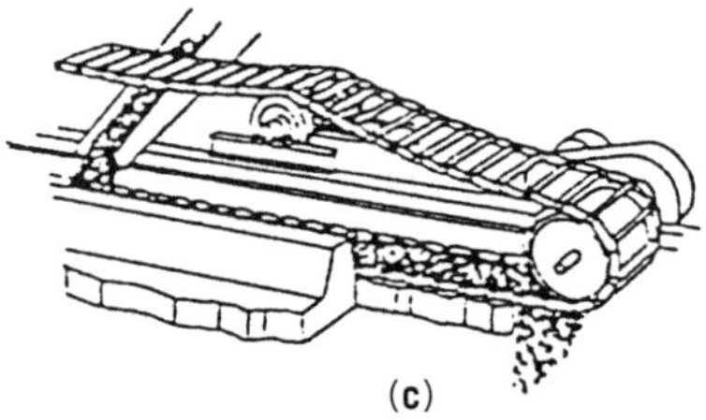

(c)

Flight conveyor—a type of conveyor comprised of one or more endless propelling media, such as chain, to which flights are attached, and a trough through which metal is pushed by the flights.

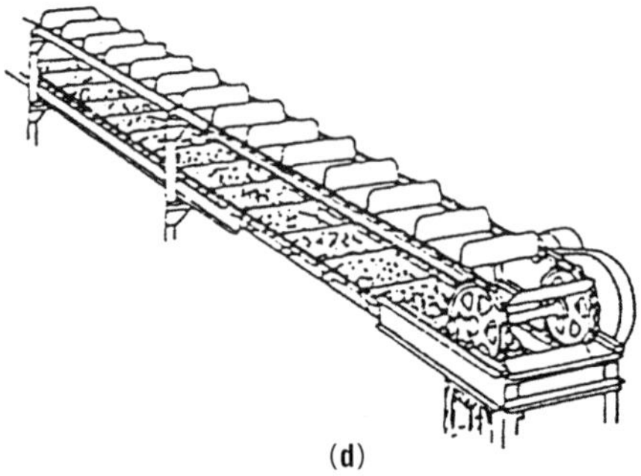

(d)

Gravity-discharge bucket conveyor—a type of conveyor using gravity-discharge buckets attached between two endless chains and that operate in suitable troughs and casings in horizontal, inclined, and vertical paths over suitable drive, corner, and takeup terminals. Buckets are designed to contain material on the vertical lifts and to scrape material along a trough on horizontal runs. Discharge is effected by gravity.

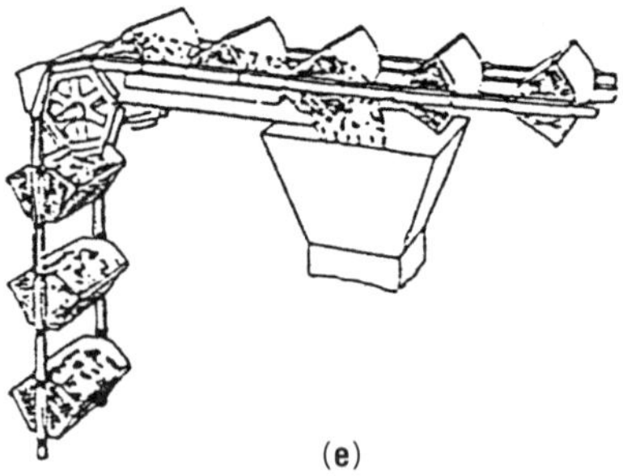

(e)

Figure 8. Lug or bar chain conveyors.

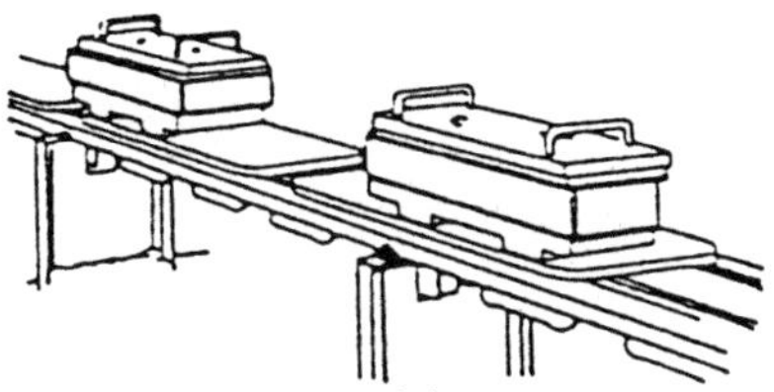

(a)

Pallet-type conveyor—a series of flat or shaped wheelless carriers propelled by and attached to one or more endless chains or other linkage.

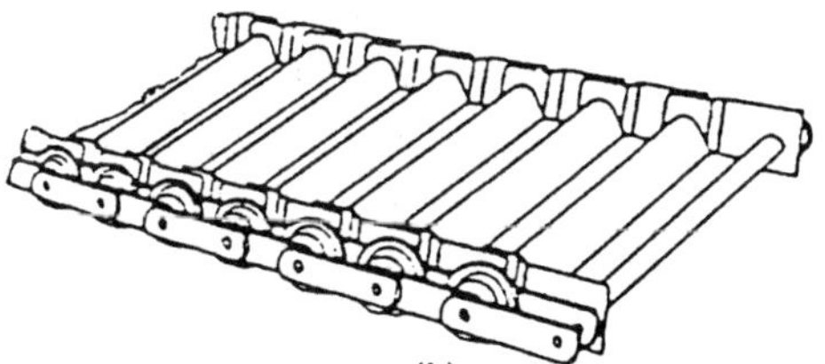

(b)

Pan conveyor—a conveyor comprised of one or more endless chains or other linkage to which overlapping or interlocking pans are attached to form a series of open-topped containers. (Some are called *apron conveyors.*)

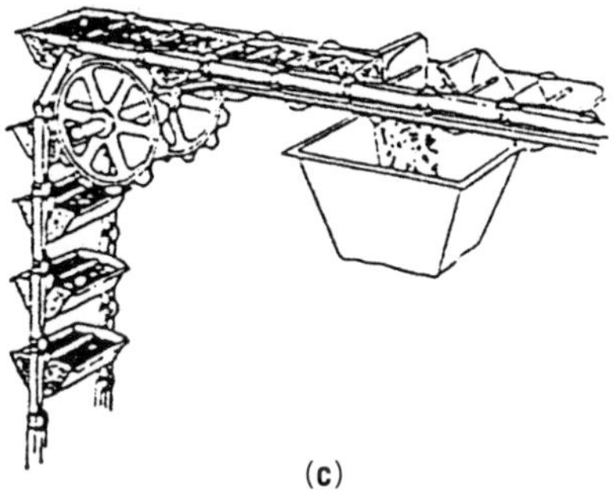

(c)

Pivoted-bucket conveyor—a type of conveyor using pivoted buckets attached between two endless chains that operate in suitable guides or casing in horizontal, vertical, or inclined, or a combination of these paths over drive, corner, and takeup terminals. Buckets are carried on and free to pivot about trunions on cross-rods carried on the chains. The buckets remain in the carrying position until they are tipped or inverted to discharge.

(d)

Pocket conveyor—a continuous series of pockets formed of a flexible material festooned between cross-rods carried by two endless chains or other linkage that operate in horizontal, vertical, or inclined paths.

Figure 9. Lug or bar chain conveyors.

- Belt-driven live roller conveyor
- Line-shaft roller conveyor—straight, curved, and spiral
- Skate-wheel spiral conveyor—straight, curved, and spiral
- Skate-wheel conveyor

(See Figs. 13–15.)

This type of conveyor system is used mainly for transporting large and/or heavy inputs or components such as large cartons (empty or full), cases, pallets, or unit loads.

Screw design. This consists of

- Screw conveyors
- Ribbon flight screw conveyor
- Gravity chutes or spiral cage

(See Fig. 16.)

Mostly granular or dough-type products are transported and/or mixed via screws that are fully or partly contained in a tube or trough.

Cable design. This consists of

- Floor cable conveyors
- Steel cable conveyors
- Steel-sheathed cable conveyors
- Urethane cable conveyors

(See Fig. 17.)

Another basic conveyor design employs a cable as the transporting vehicle. The cable conveyor is used as an inexpensive method of moving lightweight containers, primarily empty cans and plastic bottles, in single-file mode only, at speeds of ≤800–1000 containers per minute. The method is

Power-and-free conveyor—a conveying system wherein the load is carried on a trolley or trolleys that are conveyor-propelled through part of the system and may be gravity- or manually propelled through another part. This arrangement provides a means of switching the free trolleys into and out of adjacent lines. The spur or subsidiary lines may or may not be powered.

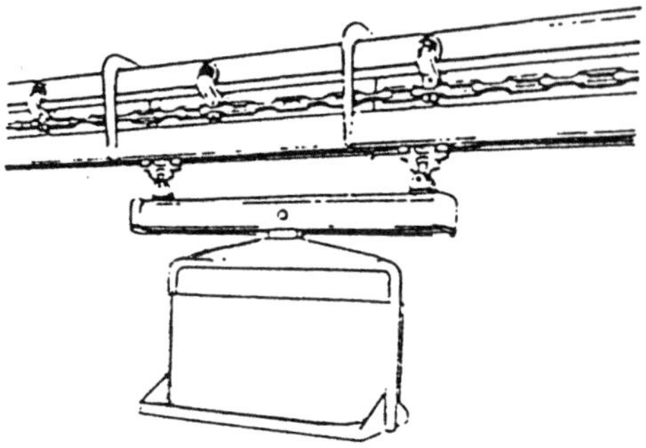

(a)

Pusher bar conveyor—two endless chains cross-connected at intervals by bars or rotatable pushers that propel the load along the bed or trough of the conveyor.

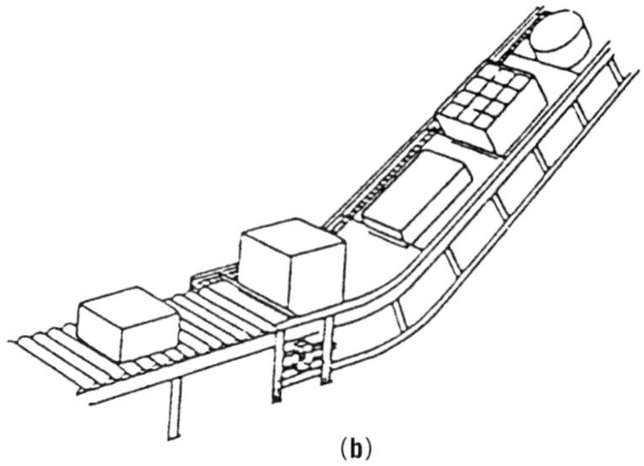

(b)

Reciprocating-flight conveyor—a reciprocating beam or beams with hinged flights arranged to advance bulk material along a trough.

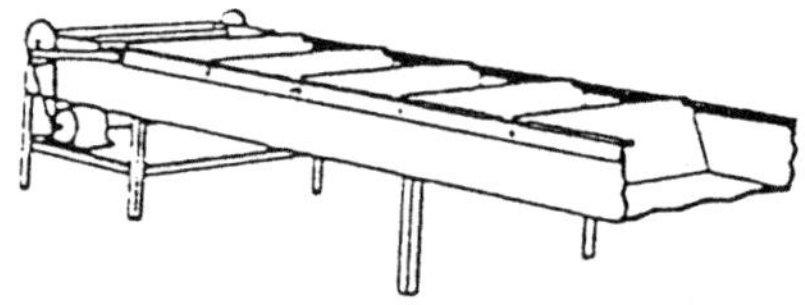

(c)

Rolling chain conveyor—a conveyor consisting of one or more endless roller chains on which packages or objects are carried on the chain rollers. The speed of transportation is double that of the chain speed.

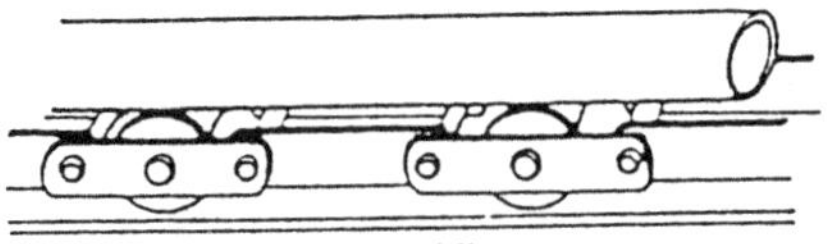

(d)

Slat conveyor—a conveyor employing one or more endless chains to which nonoverlapping, noninterlocking spaced slats are attached.

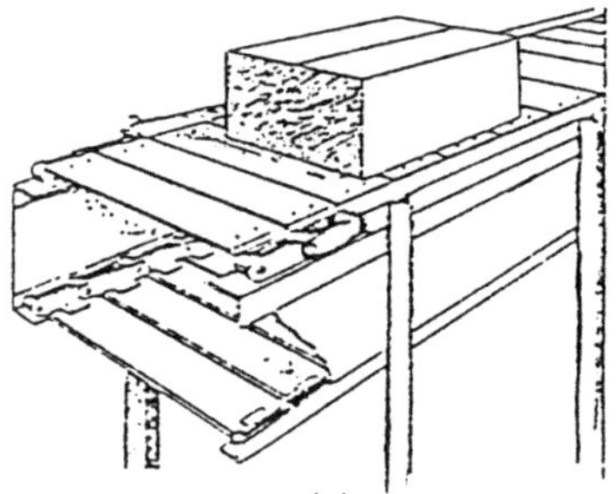

(e)

Figure 10. Lug or bar chain conveyors.

also used in the food industry for handling full containers (usually cans) generally at speeds under 500 containers per minute. In operation, a ⅜-in. (9.5-mm)-diameter, nylon-coated steel cable driven by a motor is supported on small-diameter sheaves along its length. The containers are seated on top of the cable and held in position by adjustable side guiderails made of stainless steel, aluminum, or plastic. The cable provides the drive and friction to move the container.

This method provides relatively low-cost conveying, notably for conditions where lengthy distances must be economically traversed in a plant environment. A capacity for long pulls can minimize transfer points. Although about one-third the cost of flat-topped conveyors, the system cannot be used effectively for accumulating containers.

Also, urethane has made it possible to make short multiple cable transfer and positioning conveyors for a wide variety of products and inputs, especially cartons, pouches, and bags.

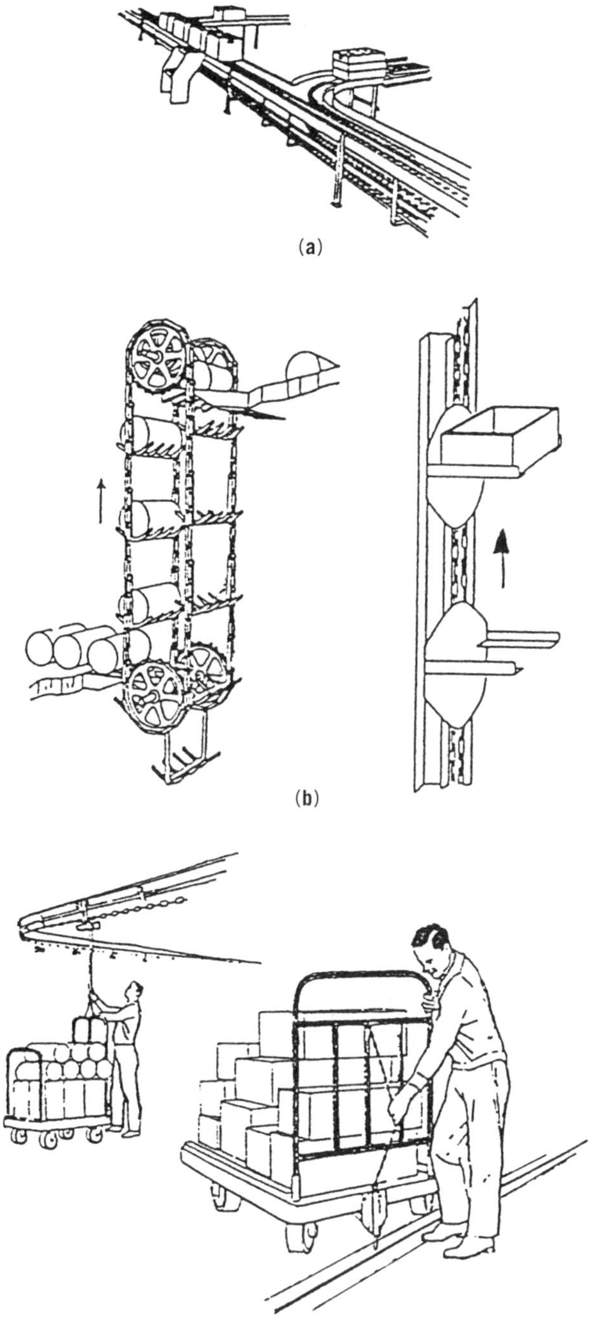

(a)

Sliding-chain conveyor—one or more endless chains sliding on tracks on which packages or objects are carried.

(b)

Suspended-tray conveyor—a vertical conveyor having one or more endless chains with suitable pendant trays, cars, or carriers that receive objects at one elevation(s) and deliver them to another elevation(s).

(c)

Tow conveyor—an endless chain supported by trolleys from an overhead track or running in a track at the floor with means for towing floor-supported trucks, dollies, or carts.

Figure 11. Lug or bar chain conveyors.

Air of vacuum design. This consists of

- Airveyor (air-driven conveyor)
- Vertical or angled air chutes
- Vacuum bulk-transfer tube conveyor
- Vacuum holddown conveyor
- Air-transfer tube conveyor for parts
- Pneumatic conveying

(See Fig. 17.)

In the last few decades packaging technology has developed new methods of stabilizing lightweight and/or some oddly shaped containers by drawing the product down onto flat-topped chain or belts through the use of a vacuum plenum. This permits conveying the containers at much higher speeds. The same concept can be applied to the operation of elevators and lowerators, but is not common. (See also Air conveying)

Inputs or products can be very effectively moved in mass or single file on a thin film of air. Pressurized air is blown into a plenum, and small holes are drilled at an angle into a stationary plate. The containers are then supported by air pressure, with their motion over the plate induced by air com-

Trolley—an assembly of wheels, bearings, and brackets used for supporting and moving suspended loads or for carrying load connecting and conveying elements such as chain, cable, or other linkage.

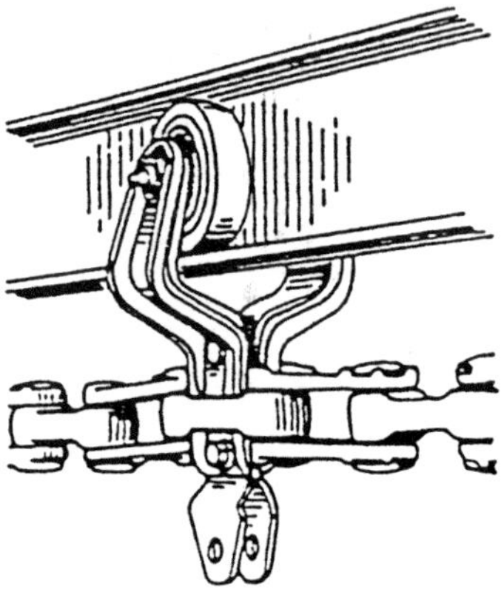

(a)

Vertical chain conveyor, opposed-shelf type—two or more vertical elevating–conveying units opposed to each other. Each unit consists of one or more endless chains whose adjacent facing runs operate in parallel paths. Thus each pair of opposing shelves or brackets receive objects (usually dish trays) and deliver them to any number of stations.

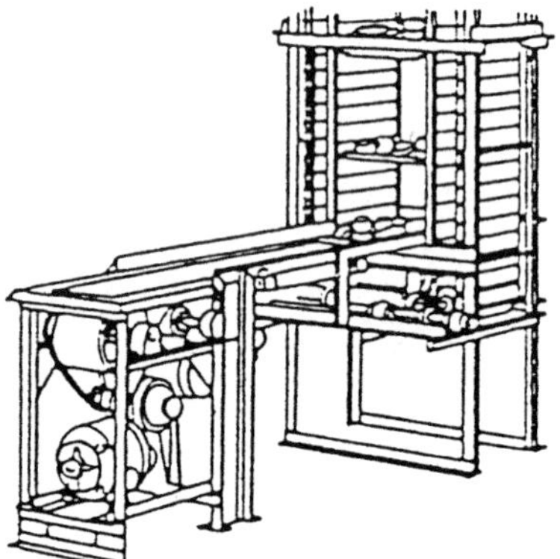

(b)

Figure 12. Lug or bar chain conveyors.

ing out of the holes' raked angle. A major advantage of this low-friction system is the reduction in the number of moving parts and reduced maintenance. Disadvantages are the large blower requirements, which cause noise and power consumption issues as well as plant air contamination and pressures. Air transfer is effective for inputs or products that have light weight, low center of gravity, and a difficult shape, such as cartons, toilet tissue, and bags.

Pneumatic tubes are very common for the transfer of granular or powder products, and some solid objects can also be transferred effectively such as caps.

Vibratory design. This consists of

- Vibrating spiral conveyors
- Vibrating straight transfer conveyors

(See Fig. 17.)

Most vibratory designs are used to convey granular-type products short distances. Cereal companies are common users of this type of conveyor.

Power Transmission Components

The following analyses are applicable to almost all types of conveyor designs. Each conveying system must have a "drive." The term "drive" refers to the apparatus that supplies motive power to the conveyor or is the energy source for the motive power. It usually consists of an electric motor (AC or DC), a gearbox, and possibly a mechanical drive train (shaft, chain, or belt). The most common configuration is still an AC electric motor, gearbox, and chain drive. Note also that the electric motor may be supplied by a VFO or variable frequency drive, which is becoming more of the standard method of control.

Gearboxes. The purpose of a gearbox is to change the rotational speed and torque of a rotating shaft. Most gearboxes are *fixed-ratio* designs, where the ratio of input speed to output speed is constant. Fixed-ratio gearboxes are best with an AC inverter, and mechanically variable gearboxes should normally not be used with and AC inverter.

Belt-pulley power transmission. A flexible connector (belt) is wrapped around two or more pulleys mounted on shafts. Power is transmitted from one pulley to another by a difference in tension in different sections of the belt.

There are two types of belt drives: positive and nonpositive. In non-positive belt arrangements, friction is relied on to transmit the peripheral force from the driving pulley to the belt and then to the driven pulley. This design permits a small amount of slippage between the belt and the pulleys. In positive belt drives, the peripheral force is transmitted by positive locking of "teeth" on the belt and pulleys. There is no slippage in this design. Examples of positive drives are gearbelts (standard or HTD). Gearbelts provide minimum noise levels and smooth operation. Care must be taken when high impact loading is required.

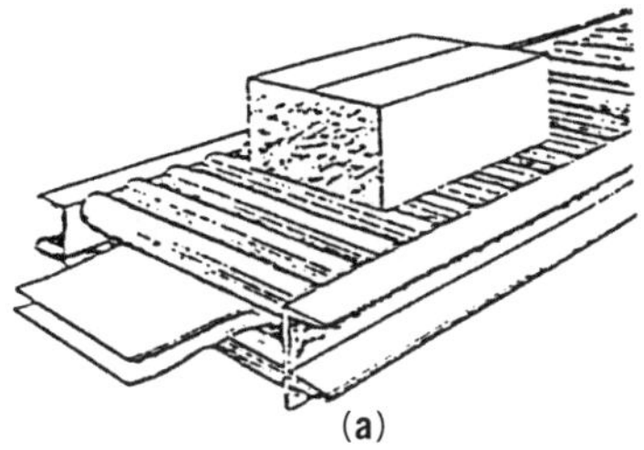

(a)

Belt-driven live roller conveyor.

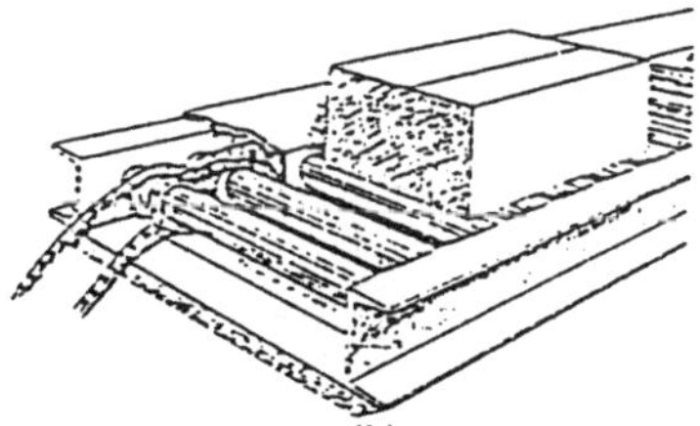

(b)

Chain-driven live roller conveyor.

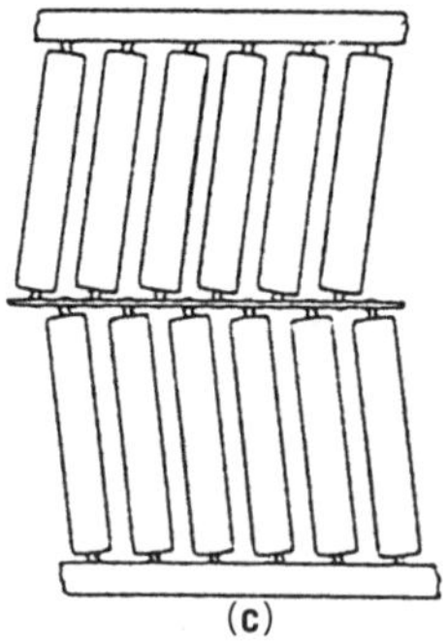

(c)

Herringbone roller conveyor—a roller conveyor consisting of two parallel series of rollers having one or both series skewed. See *Roller conveyor.*

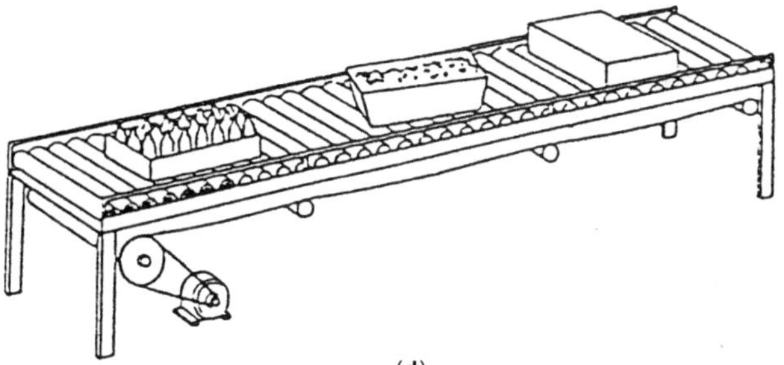

(d)

Live roller conveyor—a series of rollers over which objects are moved by the application of power to all or some of the rollers. The power-transmitting medium is usually belting or chain.

Figure 13. Roller conveyors.

Chain–sprocket power transmission. Chain drives are similar in concept to positive belt drives except that the connector is made from a number of rigid links that are connected by pins. The links rotate about the pins providing the flexibility necessary to wrap the chain around the sprocket. The pins engage the spaces between the teeth of the sprocket. The most common type of chain used is SAE roller number 40 and 50 steel chain.

Transfers between Conveyors Using Tabletop Chains

The most common types of transfers as used in tabletop hinged or *roller* chain conveyors and some special lug, mesh, and mat-top conveyors are

1. Butt end
2. Side transfer
3. S-in-line transfer.

(See Fig. 18.)

In a *butt-end transfer,* two conveyors are literally put together inline or end to end in such a manner that one chain is short of touching the other chain by at least $\frac{1}{2}$ in. A dead plate or small rollers are placed between the gap to allow the product to run or be pushed across. Also, the upstream chain can overlap and be higher than the downstream chain. Therefore the inputs or product will fall or cascade onto the downstream conveyor. Belt, chain, tabletop, mat, and mesh-type conveyors can all be butt-ended.

It is not advisable to use only a dead plate transfer for the following applications:

- Unstable packages with a difficult shape and/or high center of gravity

Minimum-pressure-accumulation conveyor—a type of conveyor designed to minimize buildup of pressure between adjacent packages or cartons. See *Accumulation conveyor.*

Nonpowered roller conveyor.

Reciprocating beam conveyor—one or more parallel reciprocating beams with tilting dogs or pushers arranged to progressively advance objects.

Roller conveyor—a series of rollers supported in a frame over which objects are advanced manually, by gravity, or by power. See types: *Controlled velocity; Herringbone; Hydrostatic; Live; Skewed; Spring-mounted; Troughed.*

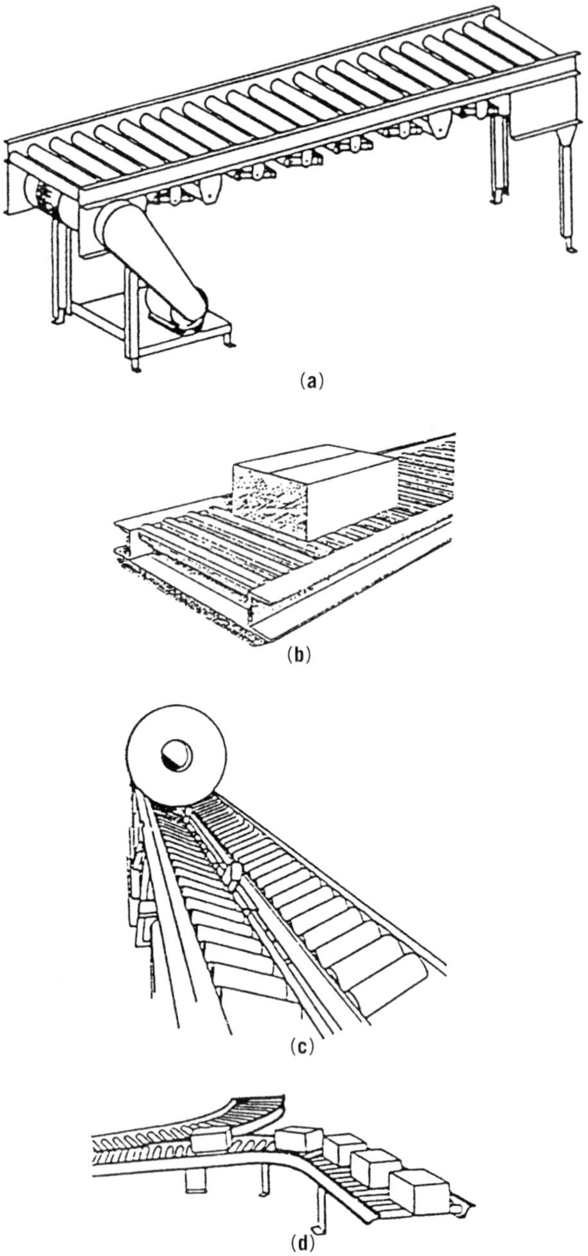

Figure 14. Roller conveyors.

- Conveyor speeds above 150 ft per minute
- Package is shorter than the dead plate or transfer roller such that the package will hang up and wait for the next package to push it off
- Packages whose base has sharp edges

Twin side-driven belt conveyors can be used to grip most types of containers (especially with vertical side faces) and laterally carry them over the dead spot on the butt-end transfer. High-speed transfers can be accomplished under specific conditions.

In a *side transfer,* two conveyors are placed side by side so that the package is guided via angled guiderails from one conveyor onto the other. This ensures that the package will be controlled and powered from one conveyor onto the next. The items one should consider when designing side transfers are

- The gap between chains should never exceed ¼ in. (6 mm); ⅛ in. is optimum.
- The downstream conveyor should be about 1/32–1/16 in. (1 mm) lower than the upstream conveyor for good transfer.
- The angle of cross transfer should not be greater than 15°. For difficult shapes or higher centers of gravity, less than 10° is advisable.
- Guiderail shape can be critical for difficult shapes.
- Conveyor speed differentials can be very critical.
- Because the transfer guiderails are angled to the con-

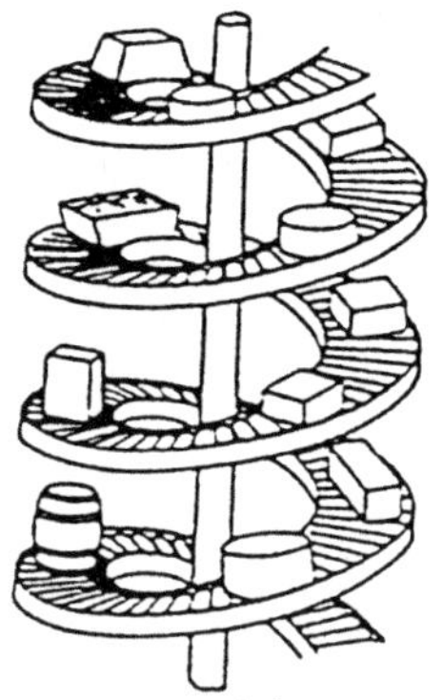

(a)

Roller spiral—an assembly of curved sections of roller conveyor arranged helically and over which objects are lowered by gravity.

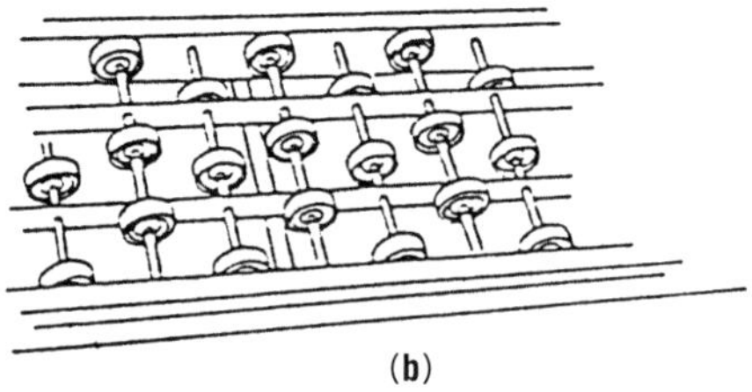

(b)

Skate-wheel conveyor—a type of wheel conveyor making use of series of skate wheels mounted on common shafts or axles, or mounted on parallel spaced bars on individual axles.

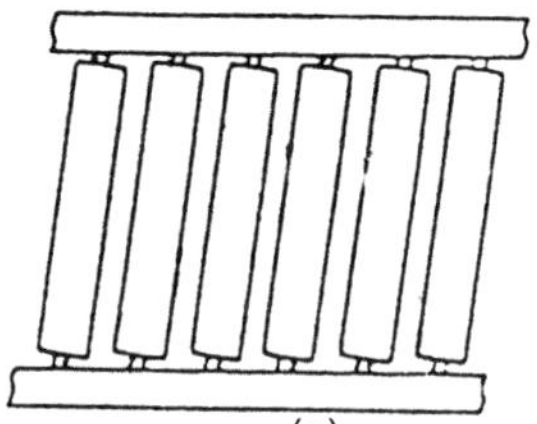

(c)

Skewed roller conveyor—a roller conveyor having a series of rollers skewed to direct objects laterally while being conveyed.

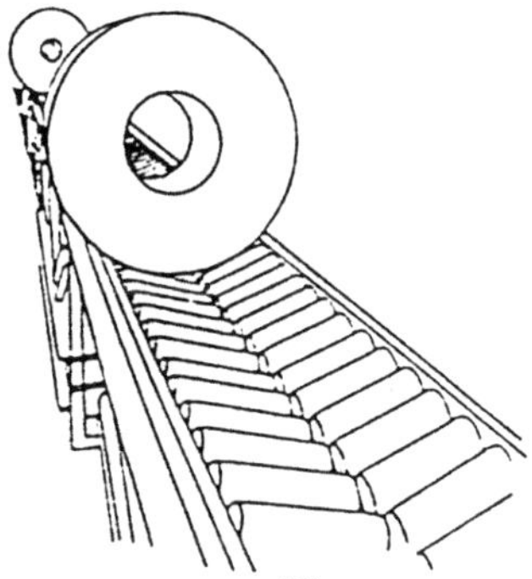

(d)

Troughed roller conveyor—a roller conveyor having two rows of rollers set at an angle to form a trough over which objects are conveyed.

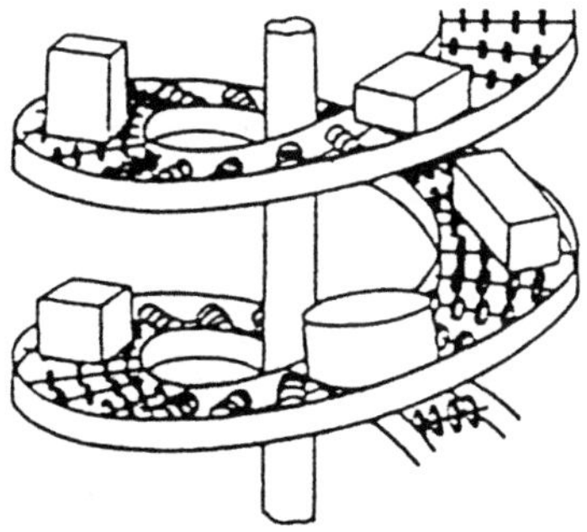

(e)

Wheel spiral—helically wound, curved sections of wheel conveyor that lower objects by gravity.

Figure 15. Roller conveyors.

Ribbon flight screw conveyor—a screw conveyor having a ribbon flight conveyor screw. See *Screw conveyor.*

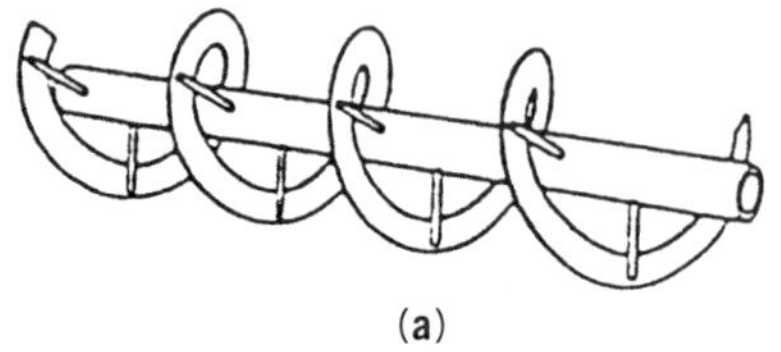

(a)

Screw conveyor—a conveyor screw revolving in a suitably shaped stationary trough or casing that may be fitted with hangers, through ends or other auxiliary accessories. See types: *Internal ribbon; Rotating casing; Vertical.*

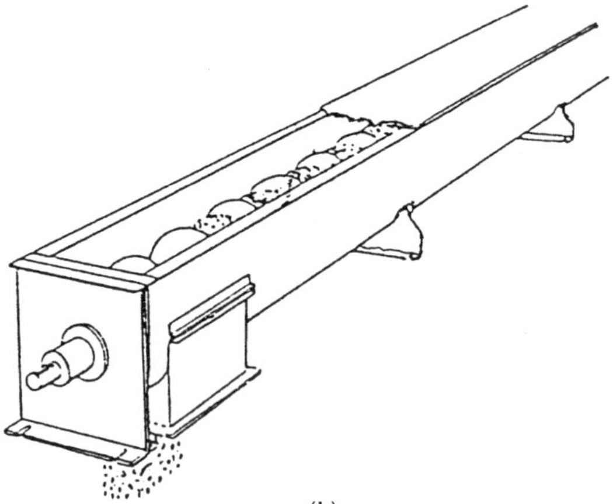

(b)

Screw-type mixing conveyor—a type of screw conveyor consisting of one or more conveyor screws, ribbon flight, cut-flight, or cut-and-folded flight conveyor screws with or without auxiliary paddles. See *Screw conveyor.*

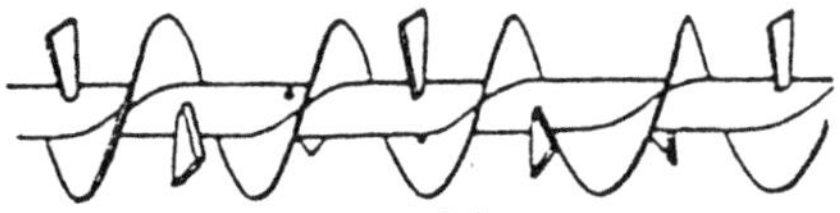

(c)

Figure 16. Screw conveyors.

Vibrating conveyor—a machine that transports material using an oscillating or vibrating motion. The machines may be designed with a wide range of frequencies and strokes. See types: *Balanced; Natural-frequency.*

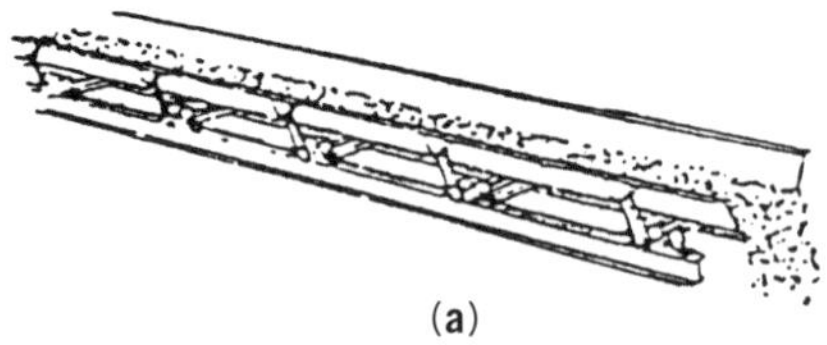

(a)

Figure 17. Miscellaneous conveyors.

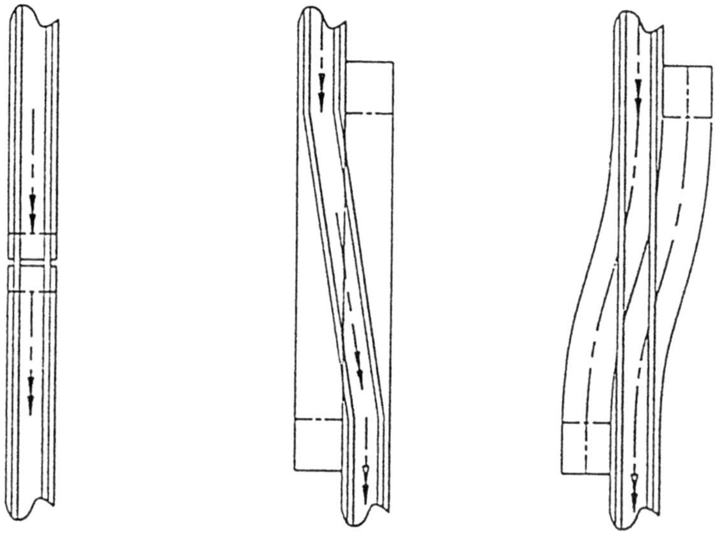

Figure 18. Transfers.

veyor pull direction, rotational forces on the container or input can be substantial.

In an *S-curve transfer,* two conveyors operate parallel together in an elongated S form so that the guide rails are straight and the conveyors flex at the transfer. This ensures that the package will be controlled and powered from one conveyor onto the next in line with minimal side forces and maximum control. Although this type of conveyor is more expensive than the traditional side transfer, the control for difficult shapes is superior and guiderail changeover to other sizes is quicker and more accurate. The items one should consider when designing S-transfers are

1. The gap between chains should never exceed ¼ in. (6 mm).
2. The downstream conveyor should be about 1/32–1/16 in. (1 mm) lower than the upstream conveyor for good transfer.
3. Conveyor speed differentials are a minor consideration.
4. The side flex radii should exceed minimum chain specifications. Only side-flex conveyor chains can be used.
5. The idle and drive ends should be straight sections for about 12 in.

In general, conveyors are critical elements in any packaging process and are grossly misunderstood and poorly manufactured. This is due mainly to nonpackaging people and some packaging people thinking that conveyors are non-value-added items that are

- Not a major item or consideration.
- The last thing considered in packaging line design and the last item purchased.
- Cheap units for moving inputs from machine to machine.
- One conveyor type is as good as another.

This thinking can be disastrous for many inputs and packages that have difficult or irregular shapes, sizes, and weights.

Design and Installation of Guiderails and Handling Control Components

In general, guiderails are critical control parts attached to every conveyor system. They are very important parts in any packaging process and are grossly misunderstood, poorly manufactured, poorly set up, and usually not robust enough. This is mainly because most people think that guiderails are like the conveyors themselves, non-value-added items that are

- Not a major item or consideration.
- The last thing considered in line design and are poorly specified.
- Cheap units for just keeping inputs from falling over the conveyor.
- One guiderail is as good as another.

This thinking can be disastrous for many package shapes and sizes. Today, many packaging engineers require quick positive position guiderails that are robust in adjustment features. This is imperative on modern packaging lines that have multiple changeovers.

Speeds, Feeds, Dynamics, and Loads for Interconnecting Machinery

The common thread throughout all the operations described above is the use of conveyors to transport inputs from machine to machine, and finally to storage facilities or shipping positions. In general, each machine or staging point will have different requirements in terms of container spacing, linear infeed and outfeed speeds, method of infeed and input, or package or case orientation. For the packaging line to function smoothly, the conveying system must provide the required changes in pitch and speed, and must also serve at times as a reservoir of inputs to help level out fluctuations in machine operation. In addition, the conveying system may also have to convert from single to multilane or mass flow and visa versa. All of these requirements must be met without tipping, spilling, jamming, or damaging the inputs, packages, cases, or pallet loads.

In the past, packages or case transfer rates were relatively low (eg, ≤300 ppm or <20 cases per minute) and the packages were geometrically simple with low centers of gravity (eg, short, round bottles). Recently, however, higher speed and performance requirements and more sophisticated machines have resulted in increased line rates (≤2000 bottles per minute in the case of breweries). Also, the aesthetic appeal of complex shapes and the use of lightweight materials and thin walling have resulted in a new generation of stability and quality problems associated with conveyance systems.

Traditionally, line speeds have been dictated by the operating capacity of the critical machine in the line such as the filling station. As a result of these changes, however, the performance of today's lines are limited not by the capabilities of the machinery employed but by the conveying systems used for package transport and manipulation.

Conveyor Design

A packaging line should try to balance its machinery and conveyors with compatible function type. *Function type* means the method of operation, which can be either continuous or intermittent. Most low-speed lines generally have intermittent-motion machinery. High-speed lines usually have all continuous motion machinery. Medium speed lines could be a mix of both. There usually are tremendous problems when a intermittent machine is coupled to a continuous-motion machine. The best way to solve the mix is to place a buffer between the two elements. These two motions are almost impossible to line-shaft mechanically but could be electronically line-shafted, but with great difficulty. It is possible to effectively run an intermittent motion on top of or within a continuous-motion machine or vice versa. These types of new high-technology machines require the correct selection of conveyor system for optimal results.

All packaging machines and equipment require some form of conveyance system to transport product or inputs from one machine to the next. Conveyors are therefore an integral part of the packaging process and are critical to the function of the packaging process.

The primary objective in the design or selection of the conveyor is that it function as efficiently and with as little maintenance as possible. Factors to consider are materials, load capacity, type of drive, speed range, and capability of the conveyor, not only with the objects being transported but also with the other equipment in the production line.

Many industries have developed specialized conveyors to suit particular needs. Often they travel at very slow speeds. Overhead trolley conveyors, for example, are used in heavy industrial applications, such as in automative plants and in other heavy manufacturing facilities. Drag-type chain conveyors connect under an automobile's chassis and intermittently and very slowly pull the vehicle along its manufacturing and assembly cycle.

Interfacing with feedscrews (feedscrews are sometimes called *scolls, timing screws, worms*) can be very critical. Normally, conveyors should run about 10% faster than the discharge feedscrew pitch.

Conveyor Technology Related to Speed

The beer–beverage industry is a prime example of reliance on advanced conveying and packaging technology to meet steadily increasing demands for greater speeds, production efficiencies, and marketing innovations. The developments in current high-speed conveyor systems are a response to this need. Today's state-of-the-art high-production conveyor systems can achieve smooth, damage-free travel of containers from delivery to final packaging. The design challenge is to sustain high production rates through the proper integration of the various conveyor types despite any temporary interruptions in container flow that may occur at any points.

High-speed can–bottle conveyor technology is a combination of up-to-date mechanical, electromechanical, and electrical/electronic techniques, resulting in production capabilities of up to 2000 cpm and plastic bottle rates of 200–1000 bottles per minute (bpm). The high production derives from container transport and accumulation techniques combined with microcircuit-controlled, start–stop, and speed modulation that compensates for intermittent container-flow variations.

Most high-speed container packaging lines have one or more critical machines in the line, such as the filling machine, which dictates the flow parameters for the remainder of the system. For smooth continuous production, all functions upstream and downstream of the critical machine(s) must be designed to assure an uninterrupted supply of inputs in and out of the critical machine(s). The conveyor system then must isolate discontinuities in input flow so that the critical machine(s) will be neither short of inputs from the upstream side nor slowed or shut down because of inputs that are backed up downstream. Continuous movement of the inputs in and out of the critical machine(s) is the best indication that the conveyor system is functioning properly, barring unreliability in upstream and downstream machines.

Backpressure is a multiple of the product weight times the coefficient of friction between the chain and product. Factors affecting chain pulls include the type of chain, container weight, and whether the process is dry or wet depending on the product being handled. Smooth side transfers to and from the buffer areas and coordination of conveyor speeds with the number of lanes also are critical to the maintenance of the high production rates.

Handling Lightweight Containers

Conveyors for delicate inputs or packages such as lightweight aluminum cans and thin-walled plastic bottles, confectionery goods, bakery goods, and medical and high-cost quality products must handle the inputs gently, with as little contact and pressure as possible. This is accomplished by combining the buffer techniques and container traffic patterns with sensing and interlock devices that prevent jamming of containers and assure smooth, shock-free flow and minimal surging (which is critical for some types of plastic containers). Conveyor surfaces that reduce friction, techniques that maintain container-separation gaps, dimensional precision in fabrication of conveyor components, and interlocked motion controls are also a requirement to handling lightweight containers.

The container material, type, and shape are equally important factors in the design of a conveyor. Flow paths and buffer areas must be sensitive to the increased jamming potential of nonround shapes. Lightweight plastic bottles, particularly without base cups, are more likely to tip over. Usually nonround lightweight plastic container designs might function best with vacuum conveyors that stabilize the packages by pulling them and holding them onto conveyor surface for optimum control. Another is to use airveyors and convey plastic bottles via their necks.

Stages in Conveying

Today most production packaging lines receive a continuous supply of containers from single or multiple feed lines. The latest designs use a programmable controller that contains the electronic logic for operation of the solenoids, motors, clutches, brakes, and other control components to monitor the supply lines and determine the routing of inputs into and through the packaging line and the routing of finished goods to the truck or warehouse.

Sometimes, to prevent impact shocks downstream, notably at the critical machine, a comparatively slow, usually double or triple 7½-in. (19-cm)-wide mass conveyor is used to spread the large volume of inputs, such as round stable containers, over a wider surface area. The high production rate can then be maintained, and the conveyor velocity is limited to up to 50 ft/min (15 m/min.). Experience has shown this rate to be the maximum manageable at this stage of the production line to minimize package damage.

Single Filing from Mass Flow

Conventional methods of single filing containers have been limited to round units, with the volume of containers from multiple lanes or mass flow being directed into a single lane with the aid of converging guiderails (which have a shaker bar or use small plastic roller chain in the vertical position to facilitate merging and reducing friction or binding) mounted over multiple-speed combining conveyors.

With new computer-controlled, low-pressure techniques, however, nonround shapes can be conveyed at high speed by photoelectric monitoring of the gaps between the containers at the output end of the station. Chain speeds are adjusted automatically to maintain the desired gap between the containers and the optimum single-file exit rate. In addition to being much more tolerant of nonround container shapes, this method is significantly quieter and assures gentler handling of less-sturdy containers. Care must be taken in using this technique, because it will work only for a specific range of nonround containers.

System Design

The following factors affect the type and design of a conveyor system:

1. Input or package material, shape, size, weight, and center of gravity
2. The conveyor system's speed of operation
3. The space available and the production line layout and flow
4. The best location for the major pieces of equipment and their interrelationships
5. The location points of supply for the system
6. The routing of finished products to warehouse and/or marketplace
7. The production volume needed by the marketing department (present and future needs)
8. The cost justification for the system
9. The probable return on the investment
10. Locating points of labor needs and the most efficient use of the labor supply
11. Access routes for primary and secondary packaging supplies, labeling, and maintenance
12. The power, air, and/or hydraulic services that are available for systems operation
13. The logic requirements of the electrical power and control system
14. The systems basic pacing factors—whether the design is based on maximum process speed or on the projected sales volume

Profits of successful consumer products companies are heavily influenced by the proper design and installation of conveyors in their packaging lines.

BIBLIOGRAPHY

General References

Available from Conveyor Equipment Manufacturer's Association (CEMA) [9323 Hungerford Drive, #36, Rockville, MD, 20850, tel. (301) 738-2448]:

- *Conveyor Performance Terminology*—CEMA Standard No. 705.
- *Pocket Glossary of Pneumatic Conveying Terms*—CEMA Standard No. 805.
- *Conveyor Terms and Definitions*—CEMA Standard No. 102-1988.
- *Belt Conveyors*—CEMA Standard No. 402-1992.
- *Roller Conveyors—Non Powered*—CEMA Standard No. 401-1985.
- *Slat Conveyors*—CEMA Standard No. 405-1985.
- *Chain Driven Live Roller Conveyors*—CEMA Standard No. 404-1985.
- *Belt Driven Live Roller Conveyors*—CEMA Standard No. 403-1985.
- *Safety Standard for Case & Bag Palletizers, Unitizers, Depalletizers*—CEMA Standard No. 605-1992.

ANSI B155.1-1994: *American National Standard for Packaging Machinery and Packaging Related Converting Machinery—Safety Requirements for Construction, Care, and Use.* ANSI, American National Standards Institute, Inc., 1430 Broadway, New York, NY, 10018.

ANSI B20.1-1990: *American National Standard for Safety Standard for Conveyors and Related Equipment.* ANSI, American National Standards Institute, Inc., 1430 Broadway, New York, NY 10018.

F. Nippard, *Dynamic Analysis of Container Conveyance Systems,* Zarpac Inc. (388 Speers Road, Oakville, Ontario, L6K 2G2), 1991.

F. Nippard, *Kinematic Analysis of Container Conveyance Systems,* Zarpac Inc. (388 Speers Road, Oakville, Ontario, L6K 2G2), 1990.

P. J. Zepf, *How to Analyze Packaging Line Performance,* Institute of Packaging Professionals (IOPP) (481 Carlisle Drive, Herndon, VA 22070), 1993.

PAUL J. ZEPF
Zarpac Inc.
Oakville, Ontario, Canada

COOK/CHILL FOOD PRODUCTION

Cook/chill is a shorthand description of three closely related fresh-food preparation methods pioneered by Cryovac in the early 1980s. Each has been used extensively to produce prepackaged refrigerated and frozen foods for both the retail and institutional markets. Consumer demand for fresh, minimally processed foods that are ready to eat with little preparation has led to the rapid commercial development of these new processes and packaging technologies.

The dollar volume of these products sold at retail has been estimated at over $80 billion. Food-service and institutional volumes are often pegged at six to seven times that amount.

The basic process is simply cooking food in large quantities and rapidly chilling it for extended storage in flexible packaging. It is increasingly the production method of choice for high-volume feeding, replacing traditional, small-batch, cook-and-serve food service. Cook/chill is a major resource of prepared foods in supermarkets, hospitals, prisons, and school systems. Recently, multiunit restaurant operators have begun developing cook/chill facilities, either as large central commissaries that serve regional locations or smaller, in-store operations that may serve nearby satellite locations.

According to the *Technology Assessment of Cook/Chill Foodservice in Schools, Hospitals and Correctional Facilities,* a study commissioned by Southern California Edison and performed by the Architectural Energy Corporation, "There seems to be little doubt that cook/chill will be the primary mode of food production at some point in the future, at all scales of institutional and commercial operation."

Reasons for the growth of cook/chill include consistency of product, improved sanitation, better control over product yield, and a reduction of food and labor costs at point of service. The result is better quality control, an advantage that helps retain consumers who demand a predictable eating experience.

Backroom advantages include virtually 100% yield from the containers and improved safety. The flexible packaging materials have no jagged edges like those found with opened

No. 10 cans, for instance. Pouches and bags are also more compact and easier to ship. Storage costs can be reduced by as much as 40%, and additional savings can be realized in shipping and handling. The solid waste stream created by discarded containers can be dramatically reduced, too.

The three basic cook/chill methods are (*1*) fill/cook/chill, (*2*) cook/fill/chill, and (*3*) cook/chill/fill. Each has unique packaging and processing requirements, and a basic knowledge of the preparation techniques will help bring an understanding of their appropriate uses.

Equipment

All three methods require specialized kitchen layouts to achieve optimum output. The first two (F/C/C and C/F/C) use a variety of high-volume production equipment. Steam-jacketed kettles with vertical sweep-and-fold tilting mixers are used for soups, sauces, taco meats, and puddings. The same kinds of kettles equipped with horizontal stationary mixers are used for thicker foods such as mashed potatoes and batters.

Tilting skillets are used for searing, browning, frying, poaching, and boiling. Ground and stew meats, fried foods, and poached fish are generally prepared with this equipment. After cubed meats are seared or browned, they are often transferred to kettles for further preparation as ingredients in soups or stews.

Metered filling stations (pump/fill equipment) are used to pump premeasured amounts of cooked product into clipped casings or premade pouches. After filling, the containers are closed with a metal clip. Moderate volume production of low- and high-viscosity foods with particulates of <1 in. diameter are packaged with this kind of equipment.

For larger volumes of pumpable products, vertical form/fill/seal (VF/F/S) equipment is used, creating pouches from roll stock in a continuous operation. These packages are heat-sealed as part of the filling process.

Blast chillers or tumble chillers are used to quickly reduce the temperature of these products. The choice is based on product, the package style, and cooling temperatures. Blast chillers, for instance, are most often used to reduce already chilled product to a frozen state. Tumble chillers are the most effective method of reducing cooked food temperatures from pasteurization levels to 40°F. They can be aggressive, however, and are generally used with foods that have little or no particulates that may be damaged in the process and packaged in abuse-resistant casings.

Continuous-belt chillers operate by transferring the packaged product through a cold-water bath, chilled brine, or spray. A preferred cooling method for VF/F/S pouches, they are less abusive and can be custom-designed to specific lengths and with multiple passes to ensure proper dwell time to reach the desired temperature. For applications such as meat prepared in cook-in casings, cook/chill tanks are used. This specialized piece of equipment cooks the product in a hot-water bath, then drains the water and replaces it with cold water for quicker and more efficient cooling.

Production with the third system (C/C/F) can be accomplished with standard cooking equipment found in most commercial kitchens. A method to quickly cool the just-cooked products may need to be added, however, if cold storage space or ice baths are not able to handle the volume. Product can be placed in holding pans for chilling in cold storage or immediate use. These products are generally packaged in laminate materials or shrinkable vacuumized bags. A variety of packaging equipment from many sources is available.

Fill/Cook/Chill

F/C/C is basically a cook-and-ship technology in which raw items such as ham, beef, ribs, or poultry products are placed in flexible cook-in materials that allow the product to be cooked, chilled, and distributed in the same package. This technology has been most often employed for high-volume HRI uses and supermarket delicatessen-style meats. However, growing demands for center-of-the-plate items at retail have expanded its use into that arena.

The process requires that raw products be portioned and seasoned, then packaged and placed in a hot-water bath or a high-humidity oven. Cooking time and temperature depend on the product and are federally mandated. After cooking, the products must be quickly chilled to <40°F before distribution.

Specialized flexible-packaging materials have been developed to meet the handling and production requirements of this method. For some meat products, slight adhesion of film to product to retain moisture and reduce purge is desirable. A high oxygen barrier to maintain eating quality over the longest possible shelf life is necessary. Good abrasion resistance to withstand the rigors of distribution is also important. Depending on the packaging equipment, bags, casings, and roll stock are available.

F/C/C prevents postprocessing contamination and retains product quality attributes. The process can deliver a refrigerated shelf life of 21–150 days, depending on the product. Products commonly produced are

Delicatessen ham	Bologna
Luncheon meat	Turkey and chicken roll
Cooked salami	Sausage
Corned beef	Roast beef
Pizza toppings	Whole-muscle meats

Cook/Fill/Chill

This process is most often used for pumpable foods such as soups, sauces, and toppings. The term "hot fill" is often used to describe this process. C/F/C products are usually prepared in large, steam-jacketed kettles and thermalized to a minimum temperature of 185°F. Products are then transferred via pump/fill stations into preformed clipped casings, or through a vertical form/fill/seal (VF/F/S) machine that uses roll stock to create a flexible pouch. Heating and filling while maintaining that temperature ensures a pasteurized product, extending shelf life.

For maximum shelf life, the casings or pouches must be voided of air during the packaging process then quickly chilled. For food safety, FSIS directives require a temperature reduction to 80°F in less than 1.5 h and 40°F in less than 5 h. Quality considerations, however, dictate a much shorter cooling cycle. Stopping the cooking process as quickly as possible ensures a better-tasting product when it is rethermalized and reduces the chance for undesirable microbial growth. Tumble chillers are generally most effective in cooling products placed in casings; blast chillers are most often used on VF/F/S packages.

Kettles, other cooking equipment and pump/fill stations are available from companies such as Chester-Jensen, Cleveland Range, and Groen. VF/F/S machinery and flexible-packaging materials are supplied by Cryovac, DuPont, and a few other companies. The proper choice of cooking and packaging equipment is based on the product and the volume requirements. Pump/fill stations and clipped casings work best with low to moderate volumes of production. Sustained, high-volume production of a single product is done most efficiently with VF/F/S equipment and roll stock film.

Pouch packaging can be sized to fit the commercial need. Pouches can be produced containing as little as an ounce of product to more than two gallons (>2 gal). Custom size to fit specific requirements (a package containing the exact amount of filling for one pie, for instance) can be easily made, improving sanitation and inventory management, and reducing storage requirements.

Soups and similar products are usually rethermalized in a hot-water bath at the point of consumption. They are then placed in another container for serving. Sauces, condiments, and toppings are often dispensed straight from the pouch, which can be formed for that purpose with special fitments available from Cryovac.

Because of the temperature extremes of this system (from fill temperatures of >200°F to freezer storage <0°) and the shock of tumble chilling, abuse resistance is very important for C/F/C packaging materials. Excellent barrier properties for pouches and roll stock, as well as quick sealability for high-speed VF/F/S materials, are also required characteristics. In addition, they must be able to withstand the rigors of distribution as well as the stress of rethermalization.

The quality and safety of the foods are retained by reaching pasteurization temperatures when the product is cooked and maintaining them during the filling/packaging process. Rapid chilling then preserves the flavor and nutritional characteristics of the food. For maximum shelf life of fresh products and food safety, storage temperatures between 28 and 32°F are recommended.

Products commonly produced are

Soups	Sauces
Toppings	Condiments
Pie filings	Stews
Syrups	Fruit-based beverages
Salsas	Chili
Taco meat	Pizza sauce

Cook/Chill/Fill

This process, commonly used for baked goods, fresh pastas, fresh vegetables, and multicomponent meal items, relies on a modified atmosphere for extended shelf life. It has experienced rapid growth recently with the rise of specialty and convenience items at retail.

C/C/F products are cooked and chilled in near-cleanroom conditions to reduce the chance of postprocess contamination. After preparation, foods are immediately packaged in a gas-flushed environment using a high barrier material. The gas mixture is selected to fit the particular requirements of the product. Good manufacturing practices can help deliver up to 28 days of shelf life (as much as 400% improvement) on a properly prepared and packaged item.

This preparation method requires increased vigilance because postprocess contamination is a greater risk. Microorganisms present in the environment, packaging materials, and cooking and packaging equipment can transfer to the product after it has been cooked (pasteurized) and cause spoilage. Problems with sanitation during production or inadequate refrigeration afterward can render the food unsafe for consumption.

Gas mixtures are usually some combination of carbon dioxide and nitrogen. Carbon dioxide retards microbial growth. Nitrogen acts as a filler, replacing oxygen to reduce oxidative rancidity and minimize the growth of molds and aerobic bacteria. For products such as red meat that must "bloom" to be visually acceptable to consumers, oxygen is included. Studies also show that 2–5% oxygen as part of the atmosphere surrounding cooked meats can help prevent the growth of anaerobic bacteria. For other products, oxygen absorber sachets can be included in the package to reduce O_2 to an absolute minimum.

Products commonly produced are

Cooked chicken breasts	Precooked vegetables
Breads	Pastries
Individually portioned meats	"Ready meal" components
Fresh pasta	Pizza toppings

BIBLIOGRAPHY

1. A. C. Beiler, and M. A. Howe, (Cryovac Division, W. R. Grace & Co.), V. S. Pat. 4,218,486 "Process for Packaging, Cooling and Storing Food Items" (1980).
2. R. J. Kelsey, "Microbiology and the New Food Packaging," Proc. Pack Alimentaire, 1989.
3. J. Rice, "Modified Atmosphere Packaging," *Food Process.* (March 1995).
4. L. L. Young (Cryovac Division, W. R. Grace & Co.), "Cooking Chilling and Package Engineering," *Proc. Internat. Food Technologist's Short Course on Minimally Processed Refrigerated Foods.*
5. "Foodservice Equipment 2000, The Cook/Chill Connection," *Restaurant Business,* (July 1995).
6. "Packaging of Refrigerated Foods," *Activities Report and Minutes of Work Groups & Sub-Work Groups of the Research & Development Associates,* Vol. 45, No. 1, 1993 (report by K. R. Deily, Director of Applications, Development & Support, Cryovac Division, W. R. Grace & Co.).

Mark Dickens
Cryovac Division, W. R. Grace & Co.
Duncan, South Carolina

CORRUGATED BOXES. See Boxes, Corrugated.

CORRUGATED PLASTIC

Traditional corrugated packaging board is paper-based and in many instances, treatment of these materials with waterproofing or chemical-resistant compounds is necessary to suit specific applications (see Boxes, corrugated). Methods have been devised to form a corrugated plastic profile, which suit the markets where traditional materials were inadequate.

Table 1. Mechanical Properties Extruded Profile of Polypropylene Copolymer[a]

Property		Value		
Thickness	in. (mm)	0.157 (4.0)	0.157 (4.0)	0.196 (5.0)
Weight	lb/ft^2 (g/m^2)	0.143 (700)	0.159 (775)	0.205 (1000)
Impact strength[b]				
73.4°F (23°C)	lbf/in. (N/cm)	90.7 (159)	97.9 (171)	126.5 (222)
32°F (0°C)	lbf/in. (N/cm)	88.5 (155.0)	100.3 (175.7)	129.4 (226.6)
−4°F (−20°C)	lbf/in. (N/cm)	62.7 (109.8)	69.4 (121.5)	89.6 (156.9)
Tensile strength				
Load	lbf (N)	62	68.6	88.6
Yield point	lbf/in. (N/cm)	661.5	732.4	945.9
Point of failure	lbf/in.^{2} (N/cm^2)	3417	3783.2	4886.3
Elongation	%	166.3	165.8	165.3
Compression strength[d] flat				
Load	lbf (N)	36 (160.1)	39.5 (175.7)	51.4 (228.6)
Compression	lbf/in.^{2} (N/cm^2)	9.2 (6.3)	10.2 (7.0)	13.2 (9.1)
Strain	%	1.04	1.06	1.08
Vertical flute				
Load	lbf (N)	87.7 (390.1)	97.1 (431.9)	125.4 (557.8)
Compression	lbf/in.^{2} (N/cm^2)	280.8 (193.3)	310.2 (213.9)	401.1 (276.5)
Strain	%	2.37	2.52	2.71
Horizontal flute				
Load	lbf (N)	6.4 (28.5)	7.1 (31.6)	9 (40.0)
Compression	lbf/in.^{2} (N/cm^2)	21.3 (14.7)	23.6 (16.3)	30.5 (21.0)
Strain	%	1.7	1.7	1.4

[a] Tests conducted by Tokan Kogyo Co., Ltd., Japan, on extruded profile.
[b] DuPont Impact Tester to ASTM D781-59T. Test specimen 1.9685 × 1.9685 in. (50 mm × 50 mm).
[c] Instron Material Tester to ASTM D828-60. Test specimen 9.8425 × 0.5905 in. (250 mm × 15 mm).
[d] Tensilon Material Tester to ASTM D695-69. Test specimen 1.9685 × 1.9685 in. (50 mm × 50 mm).

Theoretically, almost any plastic material can be formed into a corrugated profile, but costs can be prohibitive. In the United States market, polypropylene copolymer (see Polypropylene) and high-density polyethylene (see Polyethylene, high-density) are the materials in common use, although there is a small quantity of polycarbonate (see Polycarbonate) material imported for specilized outdoor applications. The board is marketed under the following trademarks: Cor-X (I.C.C. Primex Plastics Corp., Garfield, NJ), Coroplast (Coroplast, Inc., Granley, Quebec), Corrulite (Southbay Growers, Inc., Southbay, FL).

Cor-X and Coroplast are extruded profiles. Corrulite is laminated from three separate sheets and has the characteristic S-shaped flute of standard fiberboard. In all cases, the formability and mechanical properties are very similar, although the printability of the extruded sheet is superior. The

Table 2. Comparative Tests Between Plastic and Paper[a]

Test #1 Box size	in. (mm)	12.4 × 9.4 × 11.8 (315 × 240 × 300)		
Material		*Polypropylene*		*Paper*
Board thickness	in. (mm)	0.157 (4)	0.197 (5)	0.205 (5.2)
Board weight	lb/ft^2	0.150 (730)	0.191 (930)	A Flute[b]
Compression load	lbf (N)	485 (2157)	1455 (6472)	661 (2940)
Test #2 Box size	in. (mm)	23.6 × 19.7 × 16.1 (600 × 500 × 410)		
Material		*Polypropylene*		*Paper*
Board thickness	in. (mm)	0.157 (4)	0.197 (5)	0.299 (7.6)
Board weight	lb/ft^2 (g/m^2)	0.150 (730)	0.191 (930)	double wall[c]
Compression load	lbf (N)	717 (3189)	1482 (6592)	1753 (7798)
Distortion	in. (mm)	0.630 (16)	0.787 (20)	0.512 (13)
Test #3 Box size	in. (mm)	15.7 × 9.4 × 8.8 (400 × 240 × 225)		
Material		*Polyethylene*		*Paper*
Board thickness	in. (mm)	0.150 (3.8)		0.191 (4.85)
Compression load	lbf (N)	794 (3532)		708 (3149)
Distortion	in. (mm)	0.472 (12)		0.630 (16)

[a] Testron #2000, compression speed 0.472 in./min (12 mm/min). Ten samples cases at 68°F (20°C).
[b] A Flute (B-240)(B-240)(SCP-135)(B-240).
[c] Double-wall corrugated (B-240)(SCP-135)(SCP-135)(SCP-135)(B-240).

plastic corrugated board has advantages over standard fiberboard, but certainly cannot be used as a substitute for all applications. The packaging designer should consider the following opposing criteria:

Advantages	Disadvantages
Long life	Cost
Chemical resistance	Formability
Insulation	Temperature resistance
Multiple color choice	Ultraviolet degradation
Strength: weight ratio	
Waterproofness	

Test data. Plastic corrugated board utilizes test data from both the fiber and plastics industries, and this can lead to confusion or, at least, lack of pertinent data. One example of this can be illustrated by extruded board, which is categorized by thickness, and not by test or flute specifications, yet laminated board is identified by lb/1000 ft^2 (g/m^2). It is also possible to vary the weight: thickness ratios on the plastic corrugated board. Mechanical properties of extruded PP copolymer board are shown in Table 1.

Table 2 compares the performance of plastic and paper boxes.

Printing techniques. Plastic corrugated board is supplied in various base colors by blending pigment into the plastic resin. It can be printed by screen printing or flexography (see Decorating; Printing) if the extruded board is flat enough. The polyolefins are nonabsorbent and have poor adhesion surfaces unless they are corona treated or flame treated prior to printing (see Surface modification). Minimum film thickness of ink is essential to expedite ink drying (see Inks). Therefore, ink viscosity should be approximately 10% higher than normal. Squeegee pressure is normal with the squeegee medium sharp to sharp. Halftones and transparencies are possible using direct emulsion screens or indirect photo-films on a fine monofilament fabric (245–305 mesh). When force-drying, care should be taken to keep the oven temperature below 110°F (43°C), and to prevent sharp variations in air temperature. Corrugated plastic tends to be relatively rigid and, therefore, is best printed on flat bed types of automatic and semiautomatic equipment.

Forming methods. Standard boxmaking techniques can be used to fabricated corrugated plastic board. Generally, flatbed presses using cam action or single stroke are used to diecut, score, crease, or fold the material. Three-point or four-point, single-side bevel-edge rule is used for cutting. Six-point creasing is used for creasing parallel with the flutes, three-point for creasing across the flutes, to obtain a 90° bend. The packaging designer must bear in mind that the polyolefins have a "memory" and, unlike paperboard, will generally attempt to return to their previous shape. This characteristic calls for modified bending and creasing techniques, but difficulties can be overcome.

High-frequency welding has been the most successful method of joining the material. Because of the nature of the polymer, glues are not generally successful; but lap joints have been accomplished using corona-treated board with silicone-type or hot-melt adhesives (see Adhesives). Metal stitching can be used, but this creates a weak spot immediately surrounding the staple (see Staples).

Conductive containers. Changes in the electronic industry over the past decade have resulted in a requirement for different packaging materials. Plastic resins have been formulated to prevent, or dissipate, a static-electricity charge that would normally build up in the material. In the past, a carbon-loaded film was printed onto the surface of corrugated paperboard, but this had a tendency to slough off easily. The sloughing rate can be reduced by dipping the entire material, but this is another area in which plastics have an advantage. If the carbon is introduced into a polymer before extrusion, the wear factor is sharply reduced and the board can be used in near-"clean-room" conditions. The electronic industry has requirements for conductive containers for dip tubes, kitting trays, stackable tote boxes, multitrip shipping containers, dividers, covers, and lids.

BIBLIOGRAPHY

"Corrugated Plastic" in *The Wiley Encyclopedia of Packaging Technology*, 1st ed., by Neil Ferguson, I.C.C. Primex Plastics Corporation, pp. 226–228.

CRATES, WOOD.

See BOXES, WOOD.

CUSHIONING, DESIGN

Some products do not require cushioning to protect them from environmental shock and vibration hazards during distribution. This is best determined through fragility assessment (see Testing, product fragility). This article deals with those items that require some degree of cushioning to be able to survive the handling and transportation environment (see Distribution hazards). Many design decisions are based on the results of testing the cushioning materials themselves. A cushion in a package mitigates the force of impact by providing space through which a body in motion is brought to rest and dissipates some of the energy of impact. Proof of the cushion design's ability to protect the product must be through laboratory testing that subjects the packaged product to anticipated "worse-case" handling and transportation hazards, such as drops and impacts. In addition to providing protection from mechanical shocks the cushion must be capable of attenuating damaging vibration inputs. This is most commonly accomplished by a cushion system with a resonant frequency different from that of critical components of the product (see Testing, product fragility). The effects of vibration must also be verified by testing, because many variables determine the actual resonant frequency of a cushion system within a shipping container. Additional testing may also be needed to prove the ability of the package to survive static and dynamic compression, heat, humidity, and other hazards (see Testing, shipping containers).

Cushion Characteristics

Although many factors are used to assess the success of cushion design, certain characteristics of the cushion are particu-

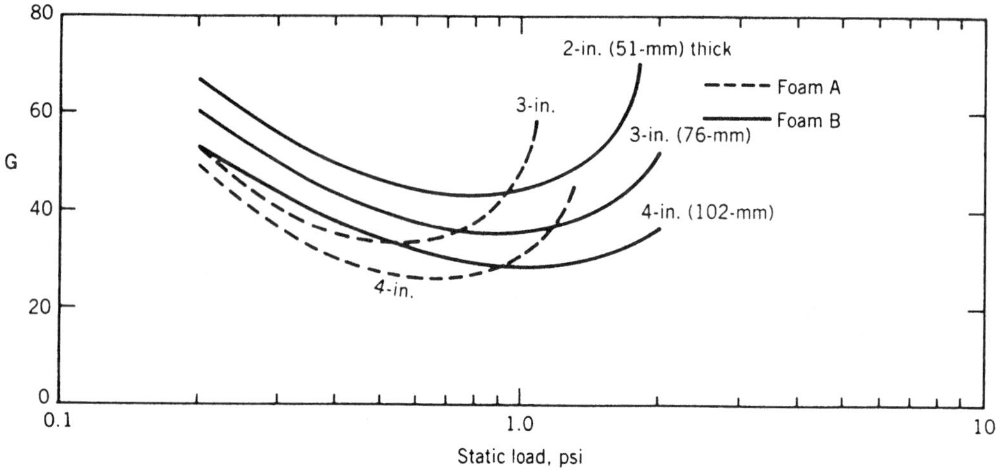

Figure 1. Dynamic cushion curve, 30-in. (76-cm) drop. G is a dimensionless ratio of specific deceleration: acceleration due to gravity. To convert psi to kPa, multiply by 6.893.

larly important at the materials-selection stage; the material's ability to mitigate shock pulses from shock levels that will cause product damage to lower levels that will not, attenuation of damaging vibration inputs from the transportation environment, resistance to compressive creep during storage, resistance to buckling under dynamic loading, and resistance to deformation under the load of the product. Design data are available from multiple sources. The most commonly used are U.S. Department of Defense MIL-HDBK-304B (1) and test data made available by manufacturers of cushioning materials (2–4). The designer must be able to apply the test data available to the problem at hand.

Shock transmission. The ability of a material or device to function as a cushion is usually presented as a dynamic cushion curve. These curves are developed through such test methods as ASTM D 1596 (5). They provide the designer with the level of shock transmitted at various static loadings as a function of cushion thickness for a particular drop height (Figure 1). Typically the designer chooses a set of curves for various materials at a specific drop height, based on the anticipated environmental hazards. Combining the shock transmission values with the product's fragility, and the product's available bearing surface area, the designer is able to decide which cushioning materials are able to function effectively for the design contemplated.

Vibration attenuation. This may be done by choosing a cushion system having a different natural frequency than the critical components of the product, or by shifting the natural frequency of the cushion by various means, such as changing static loading or the shape of the cushions. Vibration data are developed through test methods outlined in MIL-HDBK-304B (6). The data are typically presented as transmissibility-versus-frequency plots (Figure 2). They give the designer information regarding the range of frequencies through which possibly damaging amplification of the input vibration may be expected. Due to the manner in which these data are generated a considerable shift in resonant frequency may be experienced in an actual shipping package. Because of this shift, the completed package should always be subjected to laboratory tests if vibration damage due to a critical component resonance is a concern.

Creep resistance. Excessive creep would increase shock transmission by reducing the thickness of the cushion. It also leads to looseness in the package, which can alter the overall ability to resist damage through loss of overall compressive strength of the package (through reduction in support of the shipping container), altered vibration response, and secondary impacts during drops. Creep data are developed to allow the designer to anticipate and compensate for changes in cushion thickness due to long-term static loading. The data generally follow a format of percent loss in thickness versus time. They may also be presented as strain: in./in. (cm/cm) versus time. In either case the designer can quickly deter-

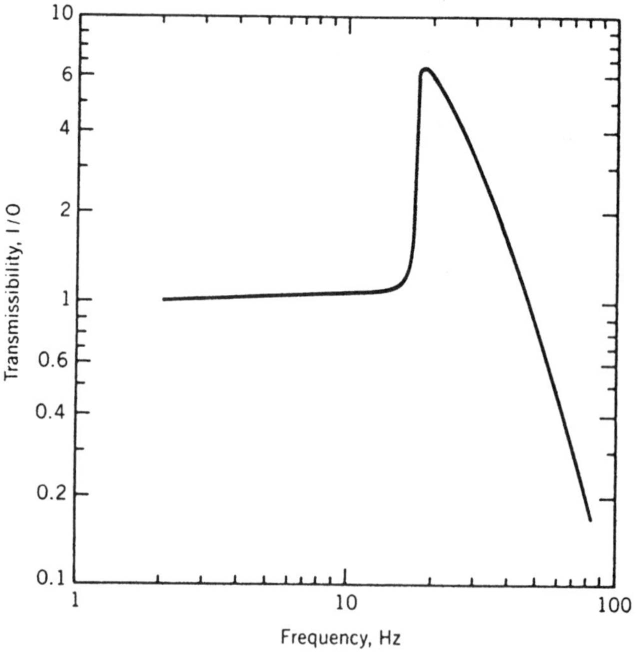

Figure 2. Vibration transmissibility plot of Foam B.

mine if the loss in thickness at a given static load over a period of time presents a problem. In most cases a 5–10% loss may be expected at normal loadings. In cases where the originally selected cushion thickness is minimal, this loss in thickness may be unacceptable. In these cases the designer can opt to increase the original cushion thickness by the anticipated loss.

Buckling resistance. A ratio of cushion thickness to bearing area determines the ability of the cushion to resist buckling. When buckling occurs, the product may not return to its original position in the package prior to the next impact. When using cushions with small bearing areas the designer should be aware of the possibility of buckling. In this mode the cushions have a tendency to buckle rather than compress throughout their thickness giving unpredictable results. A widely used formula for determining if there is adequate area to avoid buckling is Minimum bearing area = $(1.33 \times \text{Design thickness of cushion})^2$ (7). When the bearing area exceeds this minimum the possibility of buckling is generally eliminated.

Resistance to deformation under load. This is assessed by compressive stress-strain information, which may also be combined with shipping-container compressive-strength data to help determine stacking limits for the finished packages (see Testing, shipping containers). Although compressive stress-strain curves are available from the same sources mentioned above, their use has been limited by the availability of dynamic data. In cases where the cushion is to contribute to the overall compressive strength of the package the compressive strength of the cushion must be determined as a part of the completed package rather than relying on test data for the material itself.

Design Constraints

The design of the cushion system is determined by many factors. Some of these are described below.

Product size. The overall linear measurements of the product, such as length, width, and height, or diameter and height, affect the design of the cushion system and may help define the shipping environment.

Product weight. The design must take into account the gross weight of the product in lb (kg) and the location of the center of gravity of the product. If much of the product weight is substantially off center, the cushioning medium will have to be distributed accordingly.

Product fragility. The designer must know what level of mechanical shock causes damage. Damage may be cosmetic, functional, or both. In any case, it represents an unacceptable change in the product. Mechanical shock fragility is sometimes expressed as the G-factor of a product, where G is the dimensionless ratio between a specific deceleration and the acceleration due to gravity. Table 1 contains values that are often used as indicators. They should not replace actual product data as determined through testing (8) (see Testing, product fragility).

Table 1. Approximate Fragility of Typical Packaged Articles, G

Extremely fragile	
Missle-guidance systems, precision-test instruments	15–25
Very delicate	
Mechanically shock-mounted instruments and some electronic equipment	25–40
Delicate	
Aircaft accessories, electric typewriters, cash registers, and office equipment	40–60
Moderately delicate	
Television receivers, aircraft accessories	60–85
Moderately rugged	
Major appliances	85–115
Rugged	
Machinery	>115

Distribution environment hazards. To use the fragility information the designer must know what types of hazards are present in the shipping environment. Several sources of this information are available. Among the most useful are MIL-HDBK-304B (9) and Forest Products Laboratories Reports (10). The data presented in these sources can help the designer determine the realistic levels of shock and vibration to be expected. These data are often presented in formats such as those shown in Figures 3 and 4.

In some cases the data provided for environmental hazards may be superseded by the need to pass specific test criteria. For example, the designer may find that shipments are made via a parcel delivery service that requires test procedures set forth by the National Safe Transit Association. Other test criteria determining design drop are performance-oriented test specifications such as ASTM D-4169 "Standard Practice for Performance Testing of Shipping Containers and Systems" (11). This recently issued performance test provides good correlation between the type of shipping mode, degree of hazard presented, and tests needed to determine the ability to successfully ship through that environment.

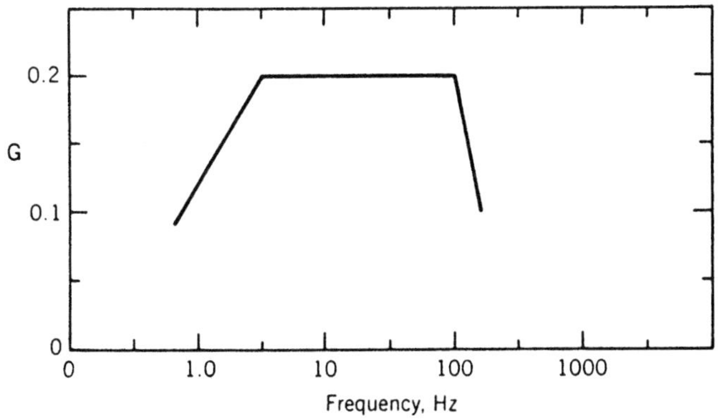

Figure 3. Typical environmental vibration envelope.

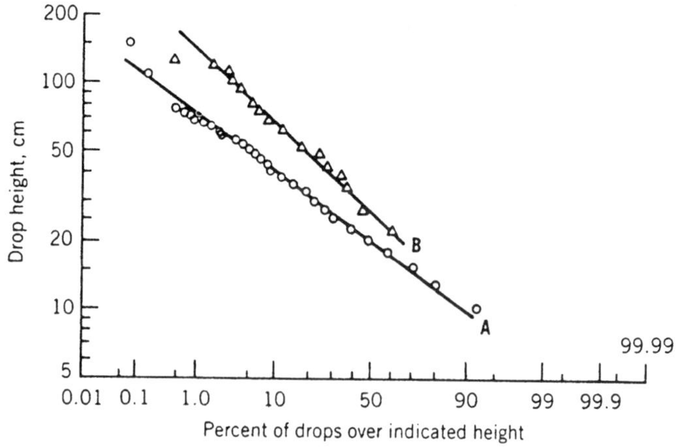

Figure 4. Drop height vs percent over indicated height. A, 43 lb (20 kg) plywood box 19 × 19 × 19 in. (48 × 48 × 48 cm), 862 drops above 3 in. (7.6 cm) (8). B, 25 lb (11.3 kg) fiberboard box 16 × 12 × 20 in. (41 × 30 × 51 cm), 312 drops above 6 in. (15.2 cm) (6).

Design Procedure

Cushion design proceeds according to the following general procedure.

1. Obtain product information: size; gross weight, including center of gravity; and product fragility, based on both mechanical shock and vibration.
2. Obtain information about environmental hazards/test criteria needed for "proof testing" the design.
3. Obtain other data that influence the final design, but are secondary in importance to the cushion design. Typically these might include cost constraints, storage/warehousing/handling constraints, and marketing/sales constraints.
4. From available cushion-curve data, determine which materials satisfy the shock-mitigation needs of the product.
5. From vibration-transmissibility curves, determine which of the materials selected in step 4 satisfy the vibration-attenuation needs of the product.
6. From the selections made in steps 4 and 5 calculate the size (bearing area) and thickness (amount of cushioning) dictated by the product.
7. Determine if the chosen static stress causes excessive compressive strain or undesirable compressive creep.
8. Determine the approximate cost of the cushion system. Compare this with guidelines if appliable.
9. Determine the effect on storage, handling, and transportation. See if minor modifications may be made to improve the design in these areas without affecting the performance of the cushion system.
10. Prepare a sample package for testing.
11. Subject the packaged product to performance tests.
12. Redesign or approve, and document as applicable.

Example problem. The foregoing procedure is illustrated by the following example. The product is a 19 × 5 × 15 in. (48 × 13 × 38 cm) rectangular prism with no protrusions (see Figure 5). It has a gross weight of 30 lb (14 kg).

Fragility assessment has revealed that the product is subject to damage when incurring mechanical shocks higher than 40 G in magnitude (see Figure 6).

Critical resonant frequency is 90 Hz. The company plans to produce 1000 of these per quarter. The shipping environment is primarily common carrier. One of the carriers requires that a package of this weight must withstand a series of ten drops (six faces, one corner, three edges) from a height of 30 in. (76 cm) if freight claims are to be honored.

Design Data
Product size: 9 × 5 × 15 in. (48 × 13 × 38 cm)
Product weight: 30 lb (14 kg)
Design drop height: 30 in. (76 cm)
Mechanical shock fragility: 40 G
Critical resonant frequency: 90 Hz

The next step is choice of the right foam (see Foam cushioning). Consulting cushion-curve data (see Fig. 1), the designer sees that a 3-in. (7.6-cm) thickness of Foam B transmits the required ≤40 G at static load values of 0.5–1.5 psi (3.4–10.3 kPa). With respect to Foam B, the curves show that a 2-in. (5.1-cm) thickness does not transmit a value low enough to be considered further; a 4-in. (10.2-cm) thickness transmits the required 40 G or less over a wider range, but its higher cost is not warranted. The most cost-effective cushion of Foam B is obtained by choosing the point at which the 40 g line intercepts the 3-in. (7.6-cm) cushion curve at the highest static loading: in this case, 1.5 psi (10.3 kPa). Similar information is available for Foam A. In practice, many materials may be examined using curves supplied by vendors or sources such as MIL-HDBK-304B.

The next step is the calculation of the required foam-bearing area. This is the area of the foam that distributes the static load of the product to the shipping container. The bearing area on any face is calculated by dividing the weight of the product by the static load selected from the cushion curve: in this case, 1.5 psi (10.3 kPa).

Required bearing area = 30 lb/1.5 psi = 20 in.^{2} (129 cm^2) (In this calculation allow lb to be the same as lbf).

As noted above, the static-load value could have been anywhere between 0.5 and 1.5 psi; but the use of values lower than 1.5 psi would increase the cost of the cushioning unnecessarily. For example, a choice of 0.5 psi would indicate a need for 60 in.^{2} (387 cm^2). the 20 in.^{2} bearing area derived from the calculation above must be large enough to prevent buckling. Its adequacy is shown by the following calculation (see discussion above).

Minimum bearing area = $(1.33 \times 3 \text{ in.})^2 = 3.99^2 = 15.92$ in.^{2} (102.7 cm^2). Figure 2 shows that the resonant frequency of the foam and its attenuation range complement the critical resonance of the product. In other words, the foam will effectively attenuate in the range of frequencies that may cause product damage.

It is good practice to anticipate how the unit will be packed and unpacked in placing the cushion pads. This gives the designer insight into how much room may be needed to reach into the box and remove the cushioned product. Because of this and the fact that the product is symmetric, with its mass centered, the cushion pads are placed toward the outer cor-

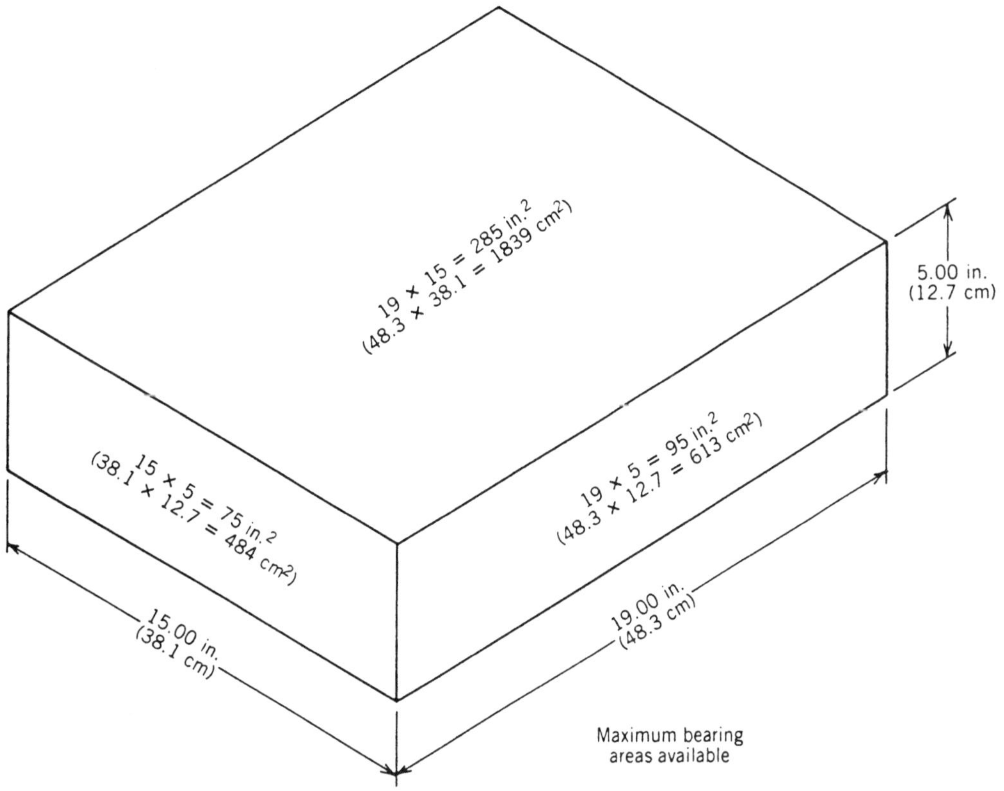

Figure 5. "Product."

ners of the product. To provide even support for the unit two cushion pads are used for each of the vertical faces of the product (front, back, left and right sides) and four pads for the top and bottom. Four acceptable solutions are described below.

Solution 1. Using die-cut slab material made from Foam B, each of the end cushions must have 10 in.^{2} (64.5 cm^2) of bearing area per pad for the front and back, left and right sides,

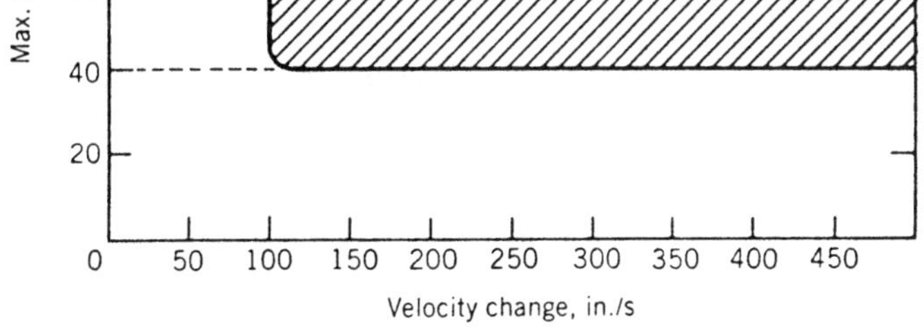

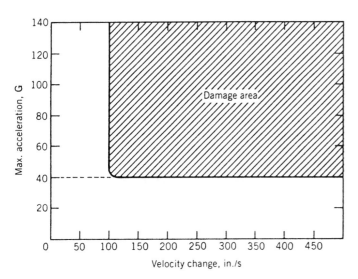

Figure 6. Damage boundary plot. To convert in./s to cm/s, multiply by 2.54.

top and bottom. The shape of the trapezoidal cushion itself is determined by noting the available material slab thicknesses of 2, 3, and 4 in. The most economical thickness is 2 in. (5.1 cm). A 2 × 5 in. or two 2 × 2.5 in. pads result. In order to place this near the outside of the product when viewed from the end or sides, the corners may be cut at 60° angles. This angle is not critical, but does allow the cushions to compress more easily at the beginning of impact, and compress more slowly as the duration of the pulse increases. At least 1 in. of material is used to form the framework holding the cushion pads together. The effective bearing area should be within the outline of the product in each axis. To make the best use of material possible and hold cost to a minimum, an attempt should be made to obtain the left and right end cushions that are heat sealed to the basic frame from the material that would otherwise be scrap. End cushions are generally available in the area cut out for the end profile of the product (see Figure 7).

Solution 2. Many rectangular products may be cushioned using simple corner pads manufactured from many materials. These may include molded EPS, die-cut foams etc. In the case of the above choice of materials we can use two types of corner pads: hinged one-piece or heat-sealed filled-corner type. The concept of bearing area applies in that each corner pad should have 5–15 in.^{2} (32–97 cm^2) available for each axis (see Figure 8).

Solution 3. One of the major drawbacks to the use of corner pads is the difficulty in packing the product without disturbing the placement of the corner pads. One of the draw-

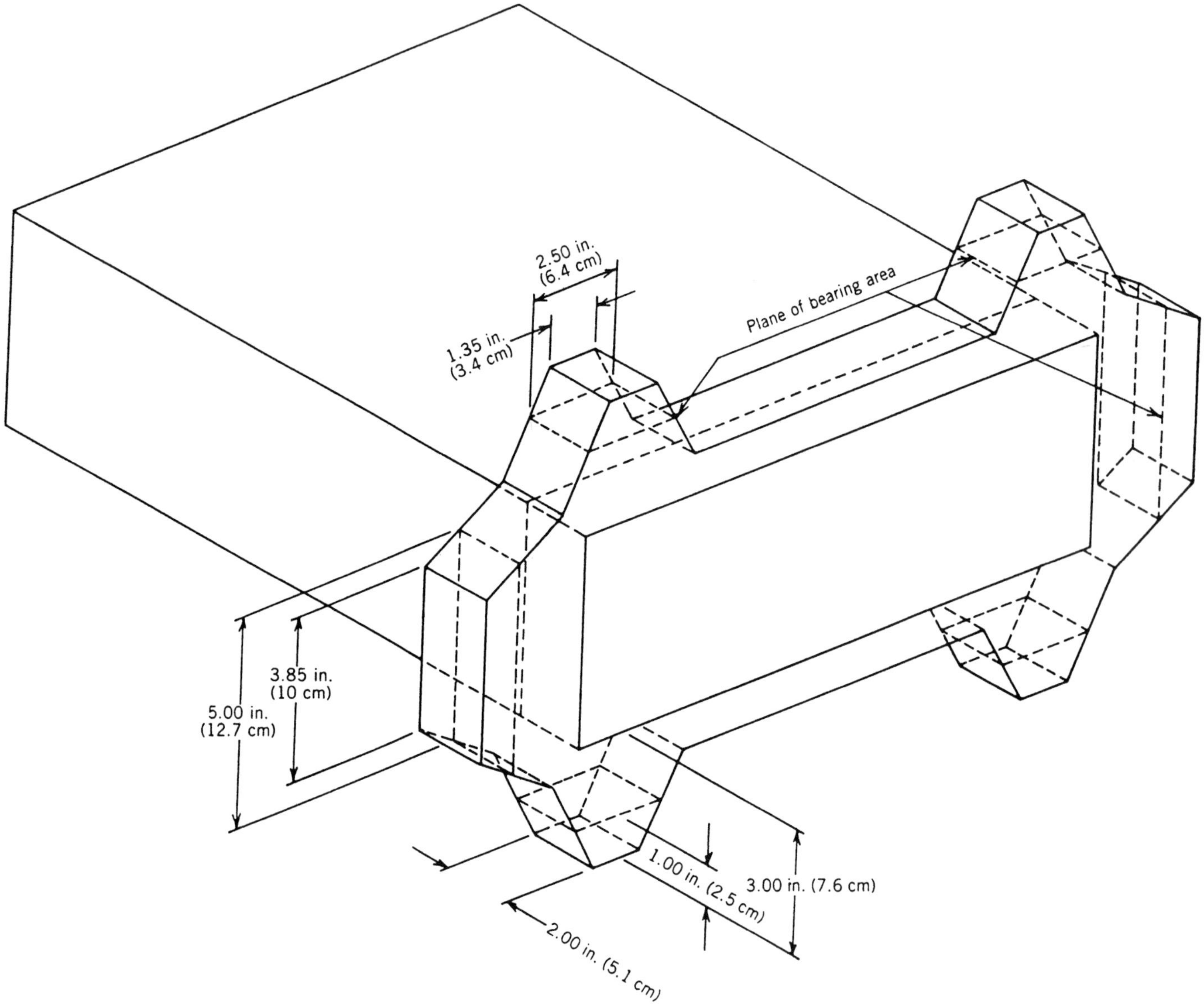

Figure 7. Die-cut cushion set [end pads not shown].

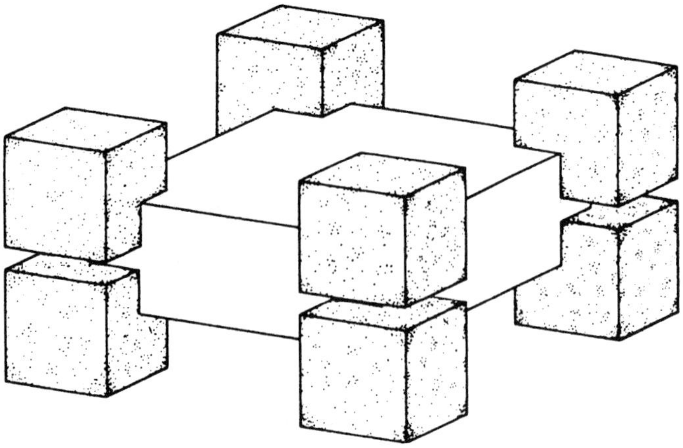

Figure 8. Typical corner pad solution.

backs to die cut cushions is their indulgent use of materials. Both of these problems may be overcome by forming a die cut sheet of corrugated (fiberboard) into a folder around the product and bonding minimum sized cushions to the surfaces of the corrugated (see Figure 9).

Solution 4. When volume warrants the investment in tooling (sometimes as low as 50 units per week for high value products), many items can be cushioned using molded foams. These may be of several types; expandable polystyrene (EPS), polyurethanes (both self-skinned and supported by a polyethylene outer layer), expandable polyethylene, and a few new types of proprietary moldable copolymers. The last category represents the response of plastics suppliers to produce foams with the appearance qualities of EPS with a higher degree of resilience. One of the major advantages of moldable materials is their ability to reduce the amount of material needed as a framework holding the cushion pads together to a minimum.

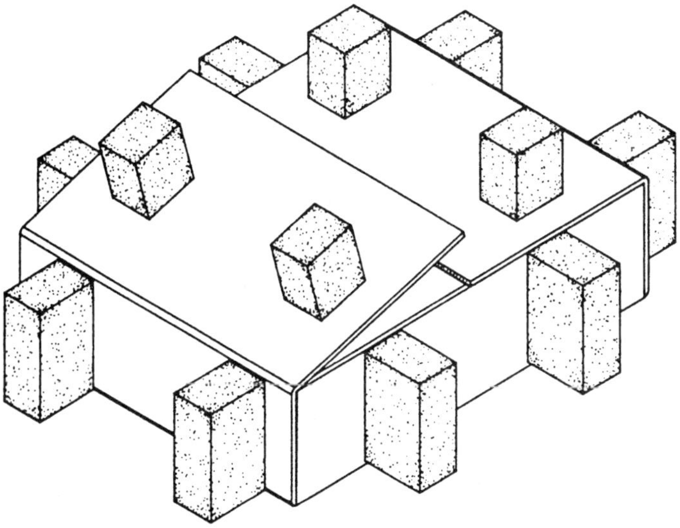

Figure 9. Foam pads bonded to corrugated fiberboard folder.

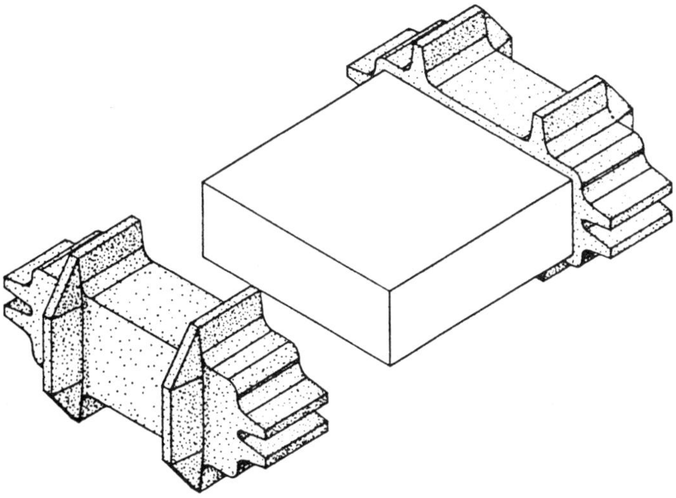

Figure 10. Molded end caps.

It also allows such options as full encapsulation at reasonable cost. The same rules for designing these cushions pads apply (see Figure 10).

Trends

With the availability of moldable resilent foams, some of the high volume applications of resilient slab stock and bun foams will be assumed by these materials, in spite of their higher cost. EPS has been underused as a cushioning material, in part due to the need to meet test criteria that call for repeated exposure to high drop levels. As more knowledge of the shipping environment is developed, newer specifications require that the package be tested in a manner that more closely replicates the environment; that is, more drops at lower heights combined with specifically oriented design height drops.

Microcomputer programs that simplify the repetitive mathematics of the design process are available. Some of these programs also have mathematical models of cushion curves and recommend material selection and cost. This type of modeling can also be programmed into some of the commercially available formula-handling software for many microcomputers. Computer-aided design (CAD) systems with software developed for three-dimensional modeling will allow the cushion designer more freedom to experiment with cushion designs before they must be committed to expensive handmade prototypes. When this is combined with finite element modeling (FEM) software, the designer will be able to perform a portion of the vibration and shock testing with the computer model, thus discarding unworkable designs early in the process.

BIBLIOGRAPHY

"Cushioning, Design," in *Wiley Encyclopedia of Packaging Technology,* 1st ed., by Robert Peache, Wang Laboratories, Inc., pp. 228–230.

1. U.S. Department of Defense, *Military Standardization Handbook, Package Cushioning Design,* MIL-HDBK-304B, 1978. Contains an exhaustive bibliography and works very well as a guidebook.
2. Arco Chemical Company, *Expanded Polystyrene Package Design,* Philadelphia, Pa., 1984.
3. BASF Wyandotte Corporation, *Styropor Protective Packaging, Properties and Design Fundamentals,* Technical Center, Jamesburg, N.J., June, 1985.
4. Dow Chemical U.S.A., *Product and Design Data for ETHAFORAM Brand Polyethylene Foam,* Functional Systems and Products, Midland, Mich., Feb., 1986.
5. *ASTM D 1596, Standard Text Method for Shock Absorbing Characteristics of Package Cushion Materials,* American Society for Testing and Materials, Philadelphia, Pa., 1984.
6. Ref. 1, chap. 4.
7. Ref. 1, p. 45.
8. *ASTM d 3332, Standard Test Methods for Mechanical-Shock Fragility of Products, Using Shock Machines,* American Society for Testing and Materials, Philadelphia, Pa., 1984.
9. Ref. 2, Chap. 2.
10. F. E. Ostrem and W. D. Godshall, *An Assessment of the Common Carrier Environment,* General Technical Report U.S. Forest Products Laboratory, FPL 22, Madison, Wis., 1979. The complete series of Forest Products Laboratories Reports cover virtually all area of cushioning testing and as well as a historical perspective.
11. *ASTM D 4169, Standard Practice for Performance Testing Shipping of Containers and Systems,* American Society for Testing and Materials, Philadelphia, Pa., 1984.
12. C. M. Harris and C. E. Crede, *Shock and Vibration Handbook,* McGraw-Hill, Inc., New York, 1976. An excellent reference for the mathematics of cushion design.

D

DATING EQUIPMENT. See CODE MARKING AND IMPRINTING.

DATING REGULATIONS. See SHELF LIFE.

DECALS. See DECORATING.

DECORATING

Heat-transfer labeling, 235
Hot stamping, 237
In-mold labeling, 238
Offset container printing, 239
Pad printing, 241
Screen printing, 243

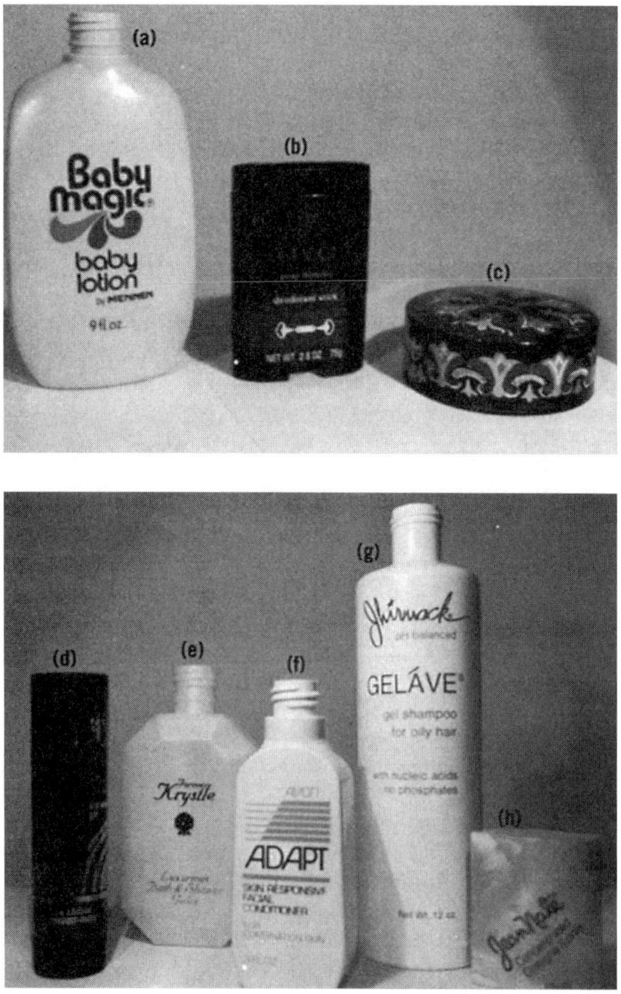

Figure 1. (**a**) Therimage heat transfer; (**b,c**) Nissha heat transfer; (**d,e**) hot stamping; (**f**) hot stamping and silk screening; (**g**) silk screening; and (**h**) had printing.

The use of plastic containers for all types of consumer products is increasing dramatically from year to year. One of the reasons for their popularity is design flexibility. A number of resins can be processed by a variety of techniques to produce unusual shapes. Pigments (see Colorants) can provide a broad range of visual effects. In addition, plastic containers provide the package designer with many decorating options. Package graphics convey information, but they can do much more. They catch the eye of the consumer and project product quality. Metals are lithographed (see Cans, aluminum, Cans, fabrication; Cans, steel; Printing), glass bottles and jars are screen printed (see Glass container design; Glass container manufacturing), and separate labels can be used on all types of containers (see Labels and labeling). Plastic containers are decorated by heat transfer labels or labeling, hot stamping, in-mold labeling, offset printing, pad printing, and screen printing. The method depends on factors that include package shape and productivity requirements (see Fig. 1).

HEAT-TRANSFER LABELING

Heat-transfer labeling is a decorating process that permits the single-pass application of single- and multi-colored copy and designs which have been preprinted on a paper or polyester (see Film, oriented polyester) carrier by a combination of heat, dwell time, and pressure. Preprinting is accomplished by gravure printing, silk screening, flexography, and more recently, selective metallization (see Metallizing; Printing).

The more significant advantages of this process, when compared to other decorating processes, include tighter color-to-color registration, fewer holding fixtures, fewer decorating machines, lack of drying or curing ovens, fewer operators, and lower scrap rates. These advantages are primarily a result of the fact that this is a dry and indirect process of application as opposed to direct methods such as silk screening. As a result it is not subject to ink smearing and part spillage resulting from multiple handling of the individually applied passes. The tighter registration is accomplished because the side to be decorated must be fixtured only once, and dimensional variation from fixture to fixture is not a factor in color-to-color registration. Disadvantages include limited flexibility with copy or design changes owing to label lead times, and costs involved with maintaining label inventories (1) (see Labels and labeling).

Two types of labels in common use today are wax-release and sizing-adhesive. Wax-release labels are applied at linear speeds as great as 750 in./min (19 m/min). Application is accomplished through a combination of heat and pressure, with the greater emphasis on heat. This type of label system depends upon the inks and/or lacquer to form the bond to the part to be decorated. Sizing-adhesive labels are applied at linear speeds of 50–150 in./min (1.3–3.8 m/min) with application more a function of dwell time and pressure. Sizing-adhesive labels depend on the sizing, which is heat reactive, to form the bond to the part to be decorated. The bond with sizing-adhesive labels is usually much stronger than the bond that can be achieved with wax-release labels.

All labels consist of a series of coating layers that are deposited or printed on a paper or polyester carrier web. Paper carriers are generally used with wax-release labels and polyester carriers are normally used with sizing-adhesive labels. Paper costs less than polyester, but polyester offers a significantly finer print surface as well as resistance to tearing during application. The first coating layer is commonly called the release coat, which allows the subsequent layers to break cleanly from the carrier. The second layer may include a protective lacquer coating as well as the actual decoration. The final layer normally includes a bonding or adhesive coating that is formulated specifically for the substrate to be decorated (2). Since wax-release systems depend on the inks and/or lacquer to form the bond, a separate adhesive coat is not required (3).

As an adjunct to developing a bond to polyethylene and polypropylene substrates, most label systems require an oxidized decorating surface. This can be accomplished by flame treatment or the corona-discharge treatment methods (see Surface modification).

The selection of appropriate coatings is done by the label manufacturer. It is important, therefore, to provide the manufacturer with as much information as possible, such as durability testing requirements and samples of the exact substrate to be decorated.

Transfer is achieved when heat and pressure are applied to the unprinted side of the carrier, causing the inks and/or adhesive coatings to become fluid and mechanically bond to the part. As the pad or roller is separated from the label and item being decorated, the carrier cools and the inks and/or adhesive coatings bond to the substrate.

The choice of the equipment used for application depends on the release coating that is used. Hot-stamping machinery is widely used as the means of applying sizing-adhesive labels. Equipment exists for roll-on as well as direct-hit application.

Specialized machines and handling equipment have been developed specifically for use with wax-release label systems. These specialized machines feature modular tooling systems. When combined with the relatively high decorating rates, this becomes the most flexible and widespread system in use today.

Registration or positioning of the copy on the carrier in relationship to the part to be decorated is accomplished by use of a metering-hole system that is prepunched in paper carriers during printing or by a reference mark that is normally printed outside the copy area for detection by a photoelectric label advance system during decoration. Registration can vary from $\pm$ 1/64 to $\pm$ 1/16 in. ($\pm 0.4 - \pm 1.6$ mm) depending on the registration system used, the accuracy of the tooling and machinery, and the dimensional consistency of the parts to be decorated.

Recent developments. Several new developments in the area of heat-transfer technology offer a much wider range of options to packaging designers. One is a sizing-adhesive-type label manufactured by Nissha Printing Co. Ltd. of Japan. This label, which offers a high-gloss, scuff-resistant finish, not only includes the effects achieved through gravure printing but also the highly desired mirror-metallic effect of hot stamping. The mirror-metallic effect is made possible by selectively metallizing the carrier with the desired copy or design. Other features are also available such as tortoise-shell, woodgrain, or marbleized finishes. An additional advantage of this process is that surface oxidation of polyethylene and polypropylene substrates is not a requirement for developing satisfactory label adhesion. The combination of these special effects along with the exceptionally tight color-to-color registration that this label offers can change the entire image of a product (see Fig. 1).

Another development is offered by Dennison Mfg. Co., a pioneer and leader in the field of heat-transfer technology. This innovation combines the pad-printing process with gravure-printed transfers. This process, which uses a flexible heated silicone pad, can be used to decorate sculptured or recessed surfaces. The preprinted carrier is routed past a preheater to a heated platen. The silicone pad presses against the carrier allowing the heat-softened label to adhere to the

Figure 1. Nissha heat-transfer labels for decorated plastic tubes.

pad. It is then pressed against the part to be decorated where it bonds upon removal from the pad (4).

A third breakthrough in transfer technology is a universal decorating machine offered by Permanent Label Machinery Corp. for printing items of noncircular cross section.

The Acrobot as modified for heat-transfer application is a patented, computerized machine that allows 360° decoration on noncylindrical parts. Shape tables which are stored on cassette, disk, or ROM are used to control the various axes of motion during application. This machine has opened up a new dimension to package designers by removing the traditional copy design and layout restrictions intrinsic to noncylindrical shapes. By accurately controlling label application velocity and maintaining tangential contact of the surface to be labeled to the print roller or pad, multifaceted or elliptical shapes can be continuously labeled in a single pass as if they were cylindrical. Prior to the advent of this machine, these shapes could only be considered by the use of exotic and expensive tooling and machinery that was designed specifically for the shape to be labeled (5).

BIBLIOGRAPHY

"Decorating: Heat Transfer Labeling" in *The Wiley Encyclopedia of Packaging Technology* 1st ed., by Douglas H. Contreras, Permanent Label Machinery Corp., pp. 235–236.

1. W. S. Anderson, Jr., *The 1983 Plastics Design and Processing Manual,* Lake Publishing Corp., Libertyville, IL, 1982.
2. S. Glazer, *Modern Plastics Encyclopedia,* McGraw Hill, New York, 1983.
3. W. La Voncher, *Decorating Plastics RETEC,* Society of Plastic Engineers, Cherry Hill, NJ, 1983
4. "Toiletries Packagers Warm up to Heat Transfer," *Packaging,* 60–61 (Feb. 1984).
5. U.S. Pat 4,469,022 (Sept. 4, 1984), N. E. Meador (to Permanent Label Corporation).

HOT STAMPING

The term *hot stamping,* as applied to decorating, is a very broad category. One definition of hot stamping is the application of characters and/or designs to a surface through the use of heat and pressure to press a web (foil) onto the surface for a finite period of time. The web is removed, leaving the desired characters and/or designs permanently attached to the surface.

The process known as heat transfer would appear to fall under the above definition, but there is a difference. In hot stamping, the decoration is on the hot die; in heat transfer, the web is preprinted by various processes and then registered to the surface to be decorated. Heat transfer and hot stamping have many common characteristics and there is some overlapping of techniques.

Materials. The web (foil) has several layers of material which are built up by successive application of ink and/or coatings in one or more passes (see Fig. **1a**) (1). Each layer has a special function (2). The manufacture of hot-stamp foil is a highly technical and difficult process, and successful hot stamping depends on selecting the proper foil for the particular job. The material to be printed (substrate) and the kind of artwork (fine or broad line) are the primary factors in selecting the correct foil.

Polyester is the most common carrier material in thicknesses of 0.0005–0.001 in. (13–25 μm) (see Film, oriented polyester). The thickness, which affects the printing characteristics, is chosen based on the end process. Other carrier materials, such as cellophane, are also used. The release coat, which can be clear or a translucent color, also controls the speed and printing characteristics of the foil. The terms *tight* and *loose* refer to how the print medium transfers from the carrier to the item to be decorated: a tight foil, used for fine line copy, breaks cleanly at the edges of the die; a loose foil, used for broad areas, tends to bridge over small surface imperfections in the die or the part. Foils are available over the complete range from loose to tight. The protective lacquer is the most important factor in both product resistance and abrasion resistance since it is the outermost part of the decoration when applied to the part. For a mirror-metallic effect, a tint can be added to the protective lacquer to get the exact metallic color desired.

The print medium can be either a continuous color or a reflective metallic layer. If a continuous color is desired, it is generally applied in one or more gravure-type coating steps. The thickness of all the various coated layers is critical to proper printing performance. In vacuum metallizing for a reflective metallic layer, a thin coating of aluminum is deposited on the transparent (or translucent) protective lacquer (see Metallizing). This results in a mirrorlike surface that is still somewhat flexible. The sizing, a heat activated adhesive, is the layer which sticks to the item to be decorated. It is very important to consider the substrate to be printed before a foil sizing is selected. The foil manufacturer is the best source for suggesting what foil will adhere to a particular substrate. Technicians generally acknowledge that the only sure way to determine if a particular foil works on a particular substrate is to try it.

Methods. There are two general methods of hot stamping: direct stamping and rollon stamping (see Fig. **1b,c**).

Direct stamping is the process of making contact with the entire area to be printed at one time. The time of contact, called dwell time, is generally controlled by either a timer or a pressure switch. The die contacts the entire part and after proper dwell, leaves the entire part as one motion. With this method, the die is contoured to match the part to be printed (including flat).

Rollon stamping is the process of making line contact between the part and the die by rolling the part across the die surface. The dwell time is controlled by the deflection of the die (or part) and the speed of traverse from the beginning to the end of the print. The sketch shows a flat die; it can also be curved or even a continuous round. In most cases, foils that are made for direct hit stamping are not good for roll-on stamping and vice versa.

Before making the selection of a foil, it is necessary to know what material is to be used for the die. The two most common die materials are metal and silicone rubber. Any ma-

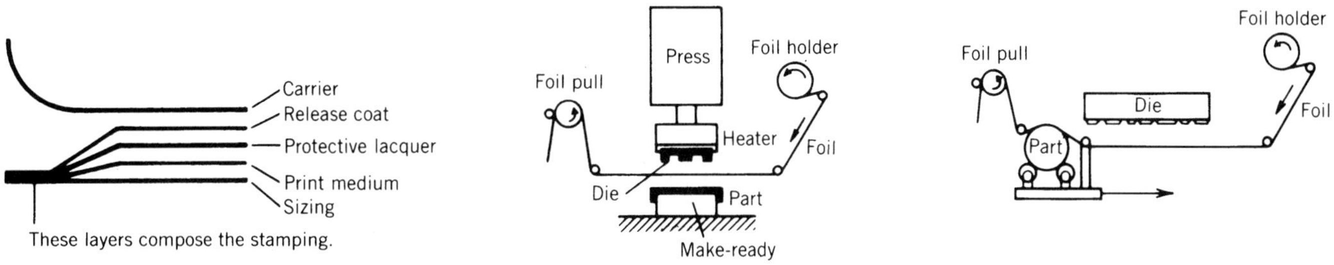

Figure 1. Hot stamping: (**a**) hot-stamp foil contains many layers; (**b**) direct stamping; (**c**) rollon stamping.

terial can be used if it can be molded, etched or engraved, and then heated repeatedly without distorting. Many factors must be considered including the length of run, detail of the printed copy, hardness and smoothness of the substrate, and the skill of the technician who sets the machine. Different printed effects can be obtained by choosing a harder or softer die. The surface finish (degree of polish) and heat conductivity of the die are also significant factors.

The hot-stamp machine is the most important factor in the hot-stamp process. The important factors in selecting a machine are rigid construction and accurate controls. There are many good hot-stamp machine manufacturers, and most make a variety of sizes. The maximum clamping force needed and the size of part that can be handled govern size selection. As the size increases, the power required to run the machine increases, and the maximum speed of operation decreases. The clamping force required is directly proportional to the surface area being printed. Metal dies require more force per area transferred than rubber dies. Roll-on stamping can be done with low clamping forces because the area being transferred at any particular instant is small.

The hot-stamp process is becoming increasingly popular for the decoration of plastic parts because of the advantages that are unique to this process. Hot stamping is a dry process; that is, no wet ink or post-treatment (drying) is required. The printed part is complete as it leaves the print station. Hot stamping can provide a mirror-metallic print; and the addition of simulated gold or silver adds marketing appeal. Recent developments in the foils, dies, and the machines to do hot stamping are making it possible to hot-stamp items that could not be decorated this way just a few years ago. Hot stamping is a developing technology that can be used to advantage on high-speed production lines. A recently patented, computer driven hot-stamp machine (3) now permits 360° decoration on noncircular parts without the need for costly gears or cams.

BIBLIOGRAPHY

"Decorating: Hot Stamping" in *The Wiley Encyclopedia of Packaging Technology,* 1st ed., by Arthur C. Peck, Permanent Label Machinery Corp., pp. 237–238.

1. L. Kurz, *Application Techniques, Hot Stamping Process, D–8510,* Furth, FRG, Aug. 1981.
2. A. Panzullo, "Hot Stamp Foils Update," *Papers Presented at the Regional Technical Conference of the Society of Plastics Engineers,* Cherry Hill, NJ, Oct 5, 1983, pp. 39–44.
3. U.S. Pat. 4,469,022 (Sept 4, 1984), N. A. Meador (to Permanent Label Corp.).

IN-MOLD LABELING

In-mold labeling (IML) is a process for prelabeling plastic containers during the blow-molding operation. IML differs from other prelabeling methods, such as heat-transfer and pressure-sensitive, methods in that the labeling mechanism is built into the blow-molding machine. No additional labeling equipment or labor is required because the containers are ejected from the blow-molding machine with the labels already in place. The most important advantage that IML offers the packager is cost reduction. Use of containers that come from the blow molder already labeled allows packagers to eliminate their labeling machines and flame treaters along with the labor and maintenance costs for that equipment. Other advantages include reduced in-house container inventories, a plastic gram weight savings over postmold-decorated containers, higher packaging line speeds, and improved appearance. Procter and Gamble recognized these advantages and was the first consumer products manufacturer to use IML on HDPE bottles in the late 1970s.

The majority of IML containers are made on continuous extrusion blow-molding equipment. In-mold labeling requires some modifications to be conventional blow-molding procedure (see Blow molding): (*1*) in-mold labels have a heat-activated adhesive on the backside; (*2*) the molds are fitted with vacuum ports to hold the labels in place during the blow-molding operation; (*3*) label magazines are added to hold stacks of die-cut labels, usually one for the face and one for the backside of the container; and (*4*) finally, a robotic pick-and-place device is added to transfer labels from the magazines into the mold cavity.

In a continuous-extrusion blow-molding IML process, the pick-and-place device takes a label from the magazine and places the printed side against the wall of the open mold where it is held in place by the vacuum ports. As the mold closes, it pinches off a section of a molten plastic tube (the parison) as it comes out of the extruder. The closed mold then moves away from the extruder and high-pressure air is injected into the hot parison, inflating it like a balloon. When the hot plastic contacts the back of the label, it activates the

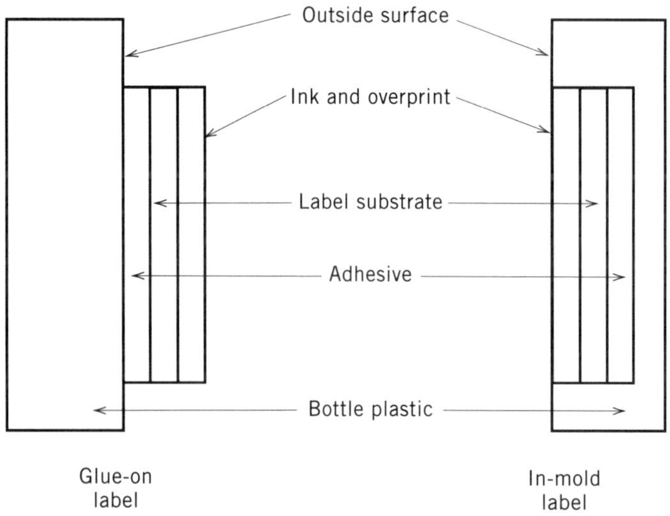

Figure 1. Diagram of in-mold versus glue-applied label.

adhesive and flows up and around the edges of the label. The label is flush with the surface of the container and becomes an integral part of it. A volume of plastic is displaced that is equal to the volume of the label (Fig. 1).

The largest number of IML containers produced are HDPE bottles made on rotary or "wheel" blow-molding machines. These machines are used primarily for long runs of containers ranging in size from 8 oz to large, economy-size laundry detergent jugs. Volume of production is a function of container size and rotational speed of the wheel. Shorter runs of IML containers are made on shuttle blow-molding machines. These can be single- or double-sided with multiple mold cavities on each side. Recent developments in label-insertion device technology have permitted retrofitting of older, non-IML shuttle machines to render them IML-capable. Lesser quantities of in-mold labeled PET and PP bottles are made by injection-stretch blow molding.

The greatest percentage of in-mold labels are still produced by integrated converters on large gravure presses using 60-lb, coated two-side (C2S) litho stock. What makes these in-mold labels unique is the ethylene–vinyl acetate (EVA)-based heat seal adhesive applied to the backside of the web on the last station of the press. Proper choice of adhesive is critical to the successful functioning of the label during the blow-molding operation. Label defects such as blistering can result if the adhesive does not activate completely during blow molding. The printed rolls are cut into sheets and stacks of individual labels are die cut from them using "high-die" equipment.

Gravure printed in-mold labels can have up to 10 colors and offer the highest-quality graphics. Advances in print quality now make if feasible to print in-mold labels by other methods such as sheet-fed offset, web offset, and a variety of narrow-web combination presses. These narrow-web presses can combine flexo, uv flexo, rotary letterpress, and hot-foil stamping to create the type of highly unusual effects desired for labeling of personal care products. These printers use paper or film label stock that has been precoated with heat-seal adhesive by another converter. Labels printed on narrow-web equipment are usually rotary die-cut in-line rather than high die-cut off-line.

The evolution of IML technology has included a transition to film label substrates. The most successful film in-mold labels are polyolefin-based materials supplied in both white opaque and clear versions. Many of these are coextrusions of modified HDPE that incorporate a sealable layer and thus do not require application of a liquid adhesive during the converting process. Film in-mold labels offer many advantages over paper IML. These include improved moisture and product resistance, fewer label defects, easier recycling, and the ability to create a "no label" look. A film in-mold labeled container also has little or no label panel bulge typical of a paper IML application. However, film labels still cost at least 50% more than paper labels and are more difficult to print and handle because of static and other problems unique to unsupported plastic substrates. Other disadvantages of the IML process are higher tooling and setup costs, cycle time penalty for label placement and bottle cooldown, and the need to remove paper labels for recycling.

Film in-mold labels have captured a large share of the market and are now used extensively on laundry-product, household-cleaner, premium orange-juice, and motor-oil containers. They have replaced pressure-sensitive and heat-transfer labels in a number of health and beauty aid applications where the "no label" look is desired.

In-mold labeling of blow-molded containers continues to grow in both new and existing applications. Investigational work is currently in progress on injection in-mold labeling of polypropylene tubs and other open top containers for soft dairy products. Although this technology is very strong in Europe, it has not yet attained a significant market position in the United States.

BIBLIOGRAPHY

General References

R. B. Schultz, *The ABC's of IML,* seminar guide presented by RBS Technologies, Inc., March 1995.

M. Knights, "Is In-Mold Labeling in Your Cards?" *Plas. Technol.* **41**(5), 48–53 (May 1995).

Ronald B. Schultz
RBS Technologies, Inc.
Skokie, Illinois

OFFSET CONTAINER PRINTING

The dry offset process is the most satisfactory method for the high-speed, large-volume printing of multicolored line copy, halftones, and full process art on preformed containers (see Table 1). Dry offset is used primarily to print products such as tapered cups, tubs, and buckets as well as tubes, jars, aerosol and beverage cans, bottles, and closures (lids, can ends, caps).

Dry offset printing is similar to offset lithography in that a rubber blanket is used to carry the image from the printing plate to the container surface (see Printing). As in letterpress, the image area is raised above the surface of the plate. Ink is distributed through a series of rollers onto the raised surface of the plate (see Inks). The plate transfers the image to the

Table 1. Selected Vocabulary Used in Connection with Dry Offset Printing

Halftone	The printing plate gives the illusion of tones of one color by means of a dot pattern or screen.
Makeready	Adjustment of printing pressure of plate to blanket to accomplish better contact, or more controlled, and therefore higher-quality printing.
Mandrel	Workholding tooling for material handling of container (called *spindle* in tube decorating).
Offset	Printing process where image is lithographed or letterpressed onto rubber printing blanket, from which it is transferred to paper or container surface.
Photoengraving	Printing plate produced by photochemistry for relief printing.
Process color	Duplication of full-color original copy by means of optically mixing two or more primary colors to create another. Black is used for highlighting.
Reverse plate	Printing plate which gives the effect of white letters, which are really the white of the container, with the plate "printing" the background color. This is the way to compensate for dry offset's inability to print white ink on a colored surface.

blanket, which then prints the entire multicolor copy taken from one to as many as six plate cylinders on the container in one operation. The "dry" denotation of this offset system serves to differentiate it from the offset system that uses the incompatibility of water and inks to dampen the surface of the plate or substrate to prevent ink transfer.

Every day offset-printing system contains two distinct sections: the print section, where the high-quality image is reproduced on the printing blanket; and the material handling section, where the container to be decorated is prepared and then positioned for contact with this blanket for the act of printing.

Print section. One to six colors can be printed in a single pass over the container, with all colors applied simultaneously by the same blanket (see Fig. 1). The inking station on each color head contains the fountain and the individual set of rollers (A–D). The number of rollers and their arrangement guarantees the finest, most uniform distribution of ink to the individual plate cylinder (E). Ink laydown on the container is so minimal that, on average, one pound of ink gives coverage on approximately 150,000 in^2. (or 213 m^2/kg ink).

The offset blanket is inked in turn by each plate cylinder, each plate cylinder having all of the copy for one color. The plate cylinder holds the printing plate, which is secured to the cylinder by clamps or through magnetism. Each plate cylinder has very fine micrometer-adjustment: 0.001 in. (25.4 μm) in any direction. The inked printing plates (E) deposit their image in sequence and in registration on a common printing blanket (F). The printing plate is prepared by a photochemical process, providing extremely fine reproduction of delicate artwork. The paste type ink used in this process, ultraviolet or conventional, allows wide latitude in choice of colors that are resistant to scuffing and moisture.

The blanket platen holds the rubber printing blanket, which is secured through the use of either "sticky back" material or ratchet clamps. The blanket transfers all of the images and copy to the container in one pass. Various blanket materi-

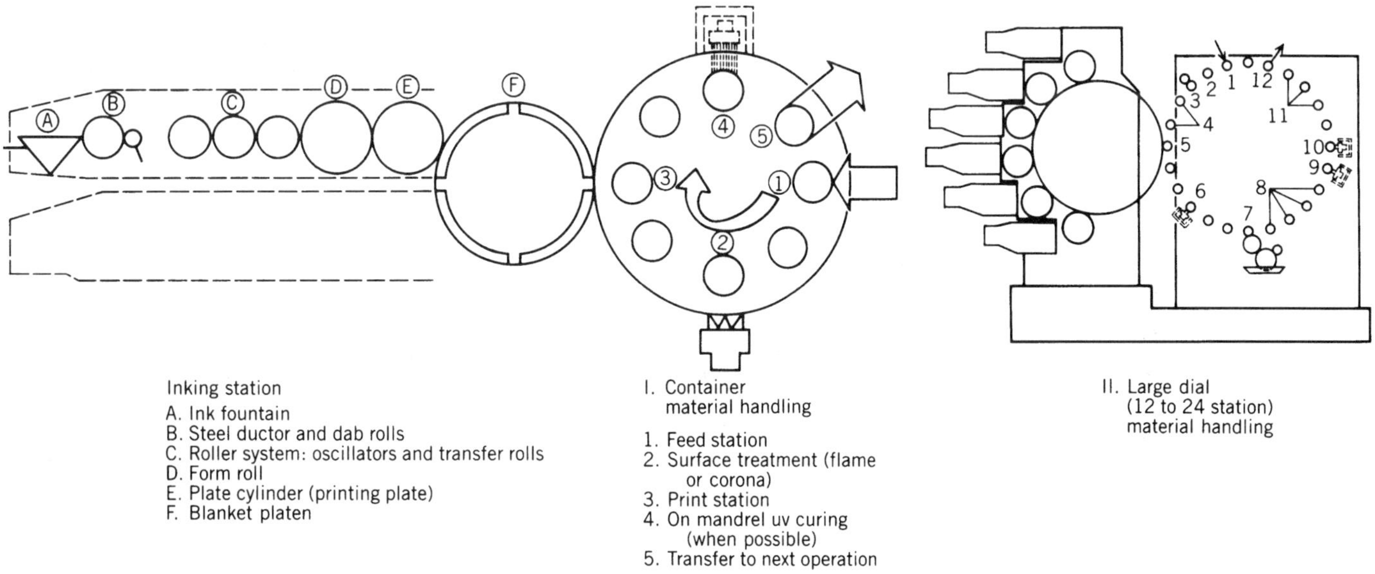

Figure 1. Dry offset container printing.

als and thicknesses are available for varying printing requirements on different containers.

Material-handling section. The container shape and its tolerances are important, not only for the mechanical nature of its handling, but also for the quality of print that can be transferred to it.

The material-handling section of the decorating line is customized for the individual style container. The indexing or constant motion turrent holds the specific container tooling, commonly called *mandrels* (or *spindles*), during all operations including printing. The material-handling section or the individual container tool can be swiveled for container taper, and positioned for printing pressure. Because of the pressure required for printing, containers must be supported by a mandrel or inflated with air pressure.

Containers can be cleaned by rotating them in front of a vacuum destaticizer which uses ionized air, or with surface treatment, which is accomplished by a flame burner head system or an electric corona discharge (see Surface modification). When required, surface treatment promotes the adhesion of ink to the substrate. The use of flame treatment has the added advantage of producing a warm container surface, beneficial in both transfer of ink from the blanket to the container and in cure reaction time in ultraviolet drying. When required, stations for orientation are provided which use a container lug, side seam, or bottom notch to register the container prior to the print station. After printing, the parts pass through the print curing unit (for "on mandrel" drying) or are transferred to a conveyor for transport through an appropriate dryer oven (ultraviolet or infrared) (see Curing).

Printer production speeds are high: tapered plastic containers such as yogurt cups are handled by indexing equipment up to 400 parts per minute; constant motion two-piece beverage decorating equipment operates at up to 1200 cans per minute.

Auxiliary equipment for automatic operation such as feeders, takeoff devices, and dryer ovens are all important adjuncts to an offset-printing system for containers. No system is universal. Most are designed for size changeover among product lines or families of containers, but few permit the modification required to handle various types of containers (eg, jars, tubes, and bottles).

The state of the art. Ultraviolet-curing technology has been a leading factor in the progress made by the dry offset process over the last 15 years. The first ultraviolet dryers were conveyor type, giving extra curing time with existing material handling systems. The improvement of the uv-drying systems led to the realization of the true potential of ultraviolet: the "on mandrel" curing of the decoration while the container is still on its respective mandrel, prior to transfer. With the reduction in container handling and the elimination of many of the transfers required by conveyor systems, production rates and overall efficiency rose dramatically. For example, an 8 oz (237 mL) yogurt container that was printed and dried at 300 containers per minute using a uv-conveyor system can now be "on mandrel" cured and handled at speeds up to 400 per minute.

Another important improvement is in the quality and detail of the copy. Most of it is multicolor line copy, but the use of intricate halftones and full four-color process work is increasing. That customers are demanding this quality of work is certainly the impetus; that it is being accomplished is the result of improvements in printing presses, color separation and platework, ink technology, and printing-room personnel. Printing presses use advanced technologies, not only in their electronics (eg, programmable controllers) but also in printing adjustment capabilities and operator controls. Platemaking companies have improved the color-separation techniques used to accommodate the dry offset process, and they have added photographic makeready systems to help print more intricate artwork. Most importantly, for without it all of the above improvements amount to nothing, the pressmen on the job floor have learned to implement this technology into the everyday workplace.

For the dry offset process, the near future seems to hold a continuation of forces already in progress. Increased electronics, used in controlling and monitoring the mechanical printing process, involve larger, more powerful programmable controllers. Production speeds have continued to increase as a result of this redesign of the printing presses.

Although new container designs and their continuing weight reduction sometimes slow the progress to higher production speeds, at some point the limitation of mechanical indexing equipment will be met. Whether plastic containers, such as the metal beverage cans, will embrace the continuous-motion material-handling systems that have propelled can decorating to speeds up to 1200 containers per minute remains to be seen.

BIBLIOGRAPHY

"Decorating: Offset Container Printing" in *The Wiley Encyclopedia of Packaging Technology,* 1st ed., by Charles Simpson, American Production Machine Co., pp. 239–241.

General References

Decorating Polyolefin Molded Items, U.S.I. Chemicals Corporation, Division of National Distillers and Chemical Corporation, New York, 1980, pp. 12,13,20,21,27,28.

R. J. Kelsey, "Choosing the Right Decoration Process," *Food and Drug Packaging,* 10 (Sept. 1983).

"Processing Handbook," *Plast. Technol.* 247 (mid-Oct, 1968).

Wilfag/Polytype Information Review, Wifag Research Department, Berne, Switzerland, May 1974, pp. 34,40.

E. C. Arnold, *Ink on Paper 2,* Harper and Row, New York, 1972 and G. A. Stevenson, *Graphic Arts Encyclopedia,* McGraw-Hill, New York, 1968. These are two fine books on printing in general. No specific work on offset container decorating exists.

Technical papers on finishing process, Association for Finishing Process of SME (Society of Manufacturing Engineers), One SME Drive, PO Box 930, Dearborn, Mich., 48128.

PAD PRINTING

Pad printing, the unique process for decorating three-dimensional objects was invented hundreds of years ago in Europe to print watch-face plates.

The printing plates were made of steel and engraved by hand or by the use of a pantograph.

The transfer pads were made of gelatin. They were very fragile and sensitive to humidity, and therefore limited in their durability. A good operator was able to produce 25–30 good parts per hour, and from a production stand point, that was a ridiculously small amount. All these factors limited the pad-printing process and restricted its further commercial and industrial use.

The first step toward large-scale use of the process was made when Tampoprint/Germany developed the silicone rubber pad as it is used today. Then related items were eventually developed: printing plates, inks, and special machines.

The process. Pad printing is a deceptively simple process, especially if one discounts the highly developed parts-handling system that is often part of the equipment. The basic machine consists of a parts holding device, a soft, oddly shaped silicone pad that picks up the ink image from an engraved, flat steel plate called a *cliche.* The cliche is engraved with the design to a depth of approximately 0.001 in. (25.4 μm). All the processing steps (except the placement of the substrate) are automatic and generally operator independent. The pad-printing printers usually operate with a reciprocal motion of the pad, which alternately picks up and delivers the image to the substrate, and an ink spatula keeps the cliche filled with ink.

Rotary-pad printers, used for decorating round objects, which can turn around a given axis, operate with a gravure-like small cylinder, and the design is transferred by a rotating pad.

The events that take place during pad printing in their sequence of operation are as follows:

In pad printing, which is often called an *indirect gravure process,* ink is deposited on an etched metal plate (cliche), where it fills the etched portions of the plate with ink and is cleared from the nonetched portions by a "doctor blade". The etched portions represent the imprint.

The ink is picked up in total by a silicone pad, which positions itself over the item being printed, descends and deposits ink on the substrate. The pad is so soft and flexible that it conforms to almost any shape and is so resistant to ink that it deposits every bit on the substrate, leaving it clean and ready for the next printing cycle.

Multicolor work is done in the usual sequence, one color at a time, but with the great potential advantage of printing wet on wet.

Tools of the trade. Pad printing utilizes the following components:

Cliché. The most conventional element in pad printing is the cliché, ie, the engraved printing plate. Like any gravure cylinder, it is produced by photographic etching. If the plate material is steel, as it should be for long runs and extremely fine details, it is coated with a photosensitive emulsion, exposed and etched with fairly conventional equipment.

There are also photosensitive polymers (plastic cliché, express cliché) available that can be easily processed by the users of pad-printing equipment. These "plastic materials," similar to the ones used in flexography and letterpress printing, provide nearly the same quality as steel, with the exception of reduced production life cycle.

Inks. The inks used in pad printing are closely related to the inks used in screen printing. Pad printing also borrowed a wealth of inks from screen printing which have been reformulated for the various substrates, with slight modifications in pigmentation and tack. Higher pigmentation is necessary to compensate for the limited thickness of the ink deposit common in pad printing. The added tack assures the unique transferring capabilities of the process.

Pad. The silicone pad, equivalent to a printing plate, is made in several durometer sizes and geometric shapes. The durometer of the pad is directly proportional to its life and inversely proportional to its wraparound capabilities on complicated three dimensional forms. In addition the durometer also determines the amount of contact force necessary to transfer the design. Higher durometers require higher pressures [ranges are available from 10 ozf to 500 lbf/in.^{2} (4.3–3400 kPa)]. The size of the pad is determined by the size of the image to be transferred and its shape is dictated by the shape of the substrate.

There are more than 400 shapes and sizes available to accomplish the necessary rolloff effect to release the image.

Physics. The softness of the pad may explain the contour-following capability of the process, but it does not reveal the reasons for its flawless transferring capability. Anyone who has tried to use a rubber stamp to transfer a design manually would attest to the fact that the transfer is usually severely distorted, with extremely poor ink distribution. In addition, there is a curious paradox in the ink transfer: the ink is tacky enough to adhere to the pad in the first place, although it separates completely from the pad and adheres to the substrate.

As the cliché is wiped with the doctor blade during pad printing, the ink surface within the engraved areas is exposed to the atmosphere and changes rapidly, due to the loss of solvents. The surface becomes more viscous and tacky than the rest of the ink layer. The smooth surface of the silicone pad has a unique property: a high critical surface tension relative to most of the inks used for decorating. Because of this, initial wetting of the pad by the ink is easily accomplished when the pad is pressed against the cliché. Although the ink has very weak lateral adhesion to the pad (it can easily be wiped off at any time) it has sufficient vertical strength to stay with the pad as it lifts off the cliché.

Depending on the cohesiveness of the ink, only a portion (approximately 75%) of the ink leaves the cliché. The cliché retains the ink due to adhesion and the vacuum created by the upward motion of the pad. In pad printing just as in gravure printing, the ink deposit is much more dependent on the cohesiveness of the ink than on the depth of the engraved design. The ink on the surface of the pad rapidly changes its properties, and it loses solvents from the exposed surface becoming tacky and viscous; but it rewets the interface between the pad and the ink, reducing the adhesion between them. When the pad is moved over to and pressed down on the substrate to be decorated, the adhesion is greater between the substrate and the ink than between the pad and the ink.

In pad-printing inks, there is a delicate balance among the solvents, the drying, and the process.

The most frequently occurring problems in pad printing are either that the ink is not picked up by the pad or not released from it. To correct these problems, air is sometimes blown across the surface of the cliché (increasing the tackiness) to promote the ink's adhesion to the pad, and sometimes air is blown on the pad to promote ink adhesion to the product. In either of these cases, too much air causes the ink to dry so much that it does not transfer to either the pad or to the product. The fast drying characteristics of pad-printing inks make it possible to print wet on wet.

The thin deposits of ink, coupled with fast drying, allow for four-color process printing, usually without the need for any drying system between various colors.

Even with the appropriate "switching" in adhesion, the ink transfer occurs because of the unique properties and geometric design of the pad. As in any printing process, the key to successful ink transfer is the point-by-point or line-contact separation between the substrate and the printing plate. In screen printing, for example, it is accomplished by off contact; in gravure, through the tangential contact of the substrate and the printing cylinder. In pad printing, the shape of the pad is designed so that it rolls away from the surface by means of a geometrical deformation. Ideally, no portion of the pad presents a flat surface to the substrate that could separate as an area rather than a line. In other words, the contact angle between the pad and the substrate is never zero within the image area.

The zero angle is excluded because the pad is a solid surface; any entrapped air between it and the substrate will prevent ink transfer. Since pad printing is used almost exclusively for three-dimensional decoration, changes in the angle of the printed surface could result in air entrapment, unless careful consideration is given to the design of the pad's shape. There are hundreds of differently shaped pads available, there is no "standard" pad in the industry. Beyond using a pad already tried for a given shape, there are no guarantees inherent in any design. To date, the manufacturers of pad-printing equipment must also design the pads since they are the most crucial element for the successful use of pad-printing equipment.

Summary. Compared to other methods of product decoration, pad printing is clearly the most suitable for three-dimensional objects. Its capability to wrap around and print close to edges and corners makes it in some cases the only logical alternative to manual decoration. Its basic limitations are that of image size (most useful in the range of 0.01–50 $in.^2$ or 0.06–323 cm^2) and ink deposit thicknesses (0.0005–0.001 in. or 13–25 μm). Large, solid print areas can be printed only with the help of a screened cliché. Owing to the few elements required by pad printing (cliché and pad), nearly all the processing parameters can be built into the equipment, making its operation extremely easy and efficient.

Pad-printing machines lend themselves well to automation not only for handling the products, but for incorporating other necessary auxiliary functions such as corona treatment, drying, sorting, and packaging.

BIBLIOGRAPHY

"Decorating: Pad Printing" in *The Wiley Encyclopedia of Packaging Technology,* 1st ed., by John F. Legat, Tampoprint American, Inc., pp. 241–243.

General References

P. Wasserman, "Looking at Pad Transfer Printing," *Screen Printing,* 122 (Oct. 1979).

"Rotary Printing for Mennen Speed Sticks," *Packaging Digest,* 83 (June 1983).

J. F. Legat, "Pad Printing—The Magic is Gone," talk presented to the Society of Plastics Engineers, Decorating Division, RETEC, Itasca, Ill., Oct. 17, 1984.

J. F. Legat, "Pad Printing," talk presented to the Society of Plastics Engineers, Decorating Division, Elmhurst, Ill., Sept. 10, 1984.

SCREEN PRINTING

The ancient Chinese and Egyptians employed open screens and inked brushes for applying ornamental decorations to fabrics, wallpaper, and walls, although no one place of origin is credited with the discovery of screen printing. Their screens were like stencils made of impervious materials cut to form an open design. The early Japanese screens consisted of oil-treated heavy paper sheets through which openings were cut in rather intricate detail. To keep the isolated parts of the cut paper from dropping out, the Japanese painstakingly glued a spiderlike network of human hair across the openings, keeping the integral parts of the screen intact.

The modern screen-printing process, no matter how rudimentary or automated, consists of a screen stretched over a wood or metal frame, a very viscous ink that is flooded on the screen and pressed through the screen mesh to the ware with a rubber squeegee (see Inks). A fixture is normally used to hold the ware to be printed and a method for curing the ink once printed is required.

Screens are made of a woven fabric of silk, polyester, nylon, or wire cloth stretched tightly over a metal or wood frame. The mesh count of the screen fabric can range from 80–520 per linear inch (30–200/cm). The screen is then completely coated on both sides with a light-sensitive emulsion. A positive image of the graphics to be screen-printed is placed upon the outside of the emulsion-coated screen. The unit is then exposed to intense light, curing (hardening), the emulsion not shielded from the light by the positive. The positive is then removed and the noncured emulsion is washed away. The nonblocked mesh area of the screen allows ink to pass through the fine mesh of the screen while the cured area remains closed to the passage of ink. A fairly viscous ink is used to flood the entire surface of the screen and is contained by the screen frame. Only when pressure from a rubber blade or squeegee is applied and passed over the screen surface is the ink forced through the unblocked print area. Just as an artist selects a brush for size, shape, and firmness to create varying effects, the screen printer has a number of variables to consider in the selection of the ideal screen and squeegee for a particular job.

The selection of the appropriate ink is just as important as screen and squeegee construction. The screen printer can select from a wide array of epoxy, vinyl, acrylic, enamel, and ultraviolet inks. The decision must take into consideration the chemical formulation of the substrate to be printed since various ink formulations are required to obtain a good bond between the substrate and the ink. In addition, it is necessary to pretreat polyolefin containers by flaming or electrostatic oxidation to obtain a good resin-to-ink bond (see Surface modification). The effect of the product contained within the package when it comes in contact with the screening is a further consideration in ink selection. No other printing process offers the depth, richness, and spectrum of color obtainable with screen printing.

The proper design of fixtures and the selection of equipment enable the screen printer to print effectively a wide variety of oval, round, flat, and tapered parts. Due to the relatively low cost of fixtures and screens and the speed with which they can be made, screen printing is the overwhelming choice for small runs; yet screen printing can remain highly competitive with alternative printing methods on larger production runs that use fully automated equipment capable of producing at speeds in excess of 100 pieces per minute.

The curing systems used for silk-screen inks range from air drying, electric and gas-fired ovens, to ultraviolet chambers (see Curing). The ware to be cured can be placed on a metallic or fabric belt, into specially designed baskets or on pins. Generally speaking, the conventional inks are cured at temperatures of 135–220°F (57–104°C) for 10–20 min. Ceramic inks used to print glass need to be cured in high-temperature ovens at 1100°F (593°C). Organic inks for glass are cured at 250°F (121°C) for two-part inks with a catalyst and 500°F (260°C) for one-part inks.

Recent developments that are providing screen printers with improved process capabilities attest to the longevity of the screen-printing process. They include the introduction and perfection of ultraviolet inks with improved product-resistant capabilities and superior scuff-resistance. The ultraviolet inks are more energy efficient and save valuable floor space needed to dry conventional inks. In addition, the compact ultraviolet-curing ovens can be placed between stations on automatic equipment enabling screen printing to be more competitive on multicolor work. Automated screen printing equipment designed to print more than 100 parts per minute on rounds and ovals is being marketed by several suppliers. In addition, a patented computer controlled screen printer designed to print nonround cross sections as if they were round, is being marketed. This machine concept further expands the shape printing capabilities of screen-printing technology.

BIBLIOGRAPHY

"Decorating: Screen Printing" in *The Wiley Encyclopedia of Packaging Technology,* 1st ed., by Eugene E. Engel, Permanent Label Machinery Corp., p. 243.

General References

J. I. Biegeleisen, *The Complete Book of Screen Printing Production,* Dover Publications, New York, 1963.

F. Carr, *Guide to Screen Process Printing,* Pitman Publishing, New York, 1962.

R. Mastropolo, personal interview, Sept. 28, 1984.

J. Agranoff, ed., *Modern Plastics Encyclopedia,* McGraw-Hill, New York, 1983–1984.

DISTRIBUTION HAZARD MEASUREMENT

To be successful in the marketplace, products must be available to consumers. Most manufacturers produce products in one or several central locations, gaining the efficiency of larger-scale operations. Markets and consumers, however, are located in widely separated areas, even thousands of miles away from the point of manufacture. This characteristic of most producers and consumers generates the need for physical distribution, a key part of the logistics system of any manufacturer. As global markets open and distribution channels grow in length and complexity, logistics takes on increasing importance.

Physical distribution in its simplest form combines three related logistic activities. Transportation moves products from point to point, utilizing a plethora of modes and equipment. Warehousing buffers the uncertainties of production and demand and helps insure availability to satisfy demand. Handling connects the elements of the system; production to warehouse, warehouse to truck, truck to retail, etc. Each of these elements includes hazards to the safe passage of products and packages. In each case, the hazards are characteristic of the operations performed within that distribution element.

Transportation requires vehicles, and vehicles produce vibration as a consequence of their motion. The discontinuities of transit media; road roughness, rail irregularities, water waves and air currents, are among the original sources of vibration. The vehicle reacts to these irregularities, amplifying some types and reducing other types. Vehicle suspension and structure play a role in vibration modification. The result is a complex mix of vibration frequencies and intensities, changing in response to immediate conditions. Figure 1 shows an example of actual vehicle vibration, demonstrating the change of vibration intensity over time. These transit vibrations may produce fatigue, abrasions, crumbling, separation and other types of damage in products and packages.

Warehouses gain efficiency by the careful utilization of space. This usually involves stacking one product and package on another to fill the available cube of the building. Unit loads of products may be stacked several high in floor loading warehouses. Even with storage that uses racks to contain packages and unit loads, stacking often occurs, albeit with lower heights. The loads imposed on packages by the warehousing system can exist for weeks, or even months. These warehouse loads may produce collapse, denting, buckling, bending, and other types of damage in products and packages.

Handling operations include the manual movement and placement of packaged products in sorting, transferring, and vehicle loading and unloading. Persons are often assisted by mechanical tools in this handling, including lift trucks, dollies, conveyors, and similar devices. Some handling operations are fully automated and utilize robotics or purpose-built devices to lift and place packages. Vehicles are also handled, and vehicle impacts such as railcar switching and truck docking may be hazardous. Inherent in handling is the possibility of accidental or purposeful drops, including severe mishandling. The shocks experienced by the product and package when dropped in handling may produce breakage, shatter, bending, misalignment, chipping, and other types of damage in products and packages.

Along with packaging material performance and product fragility, knowledge of the hazards of distribution is critical to successful protective packaging development. Data on package drops, vehicle vibration, compressive loads, and atmospheric conditions are central to determining the target performance of packages. If the general probability of occurrence of these hazards is known, then intelligent decisions on

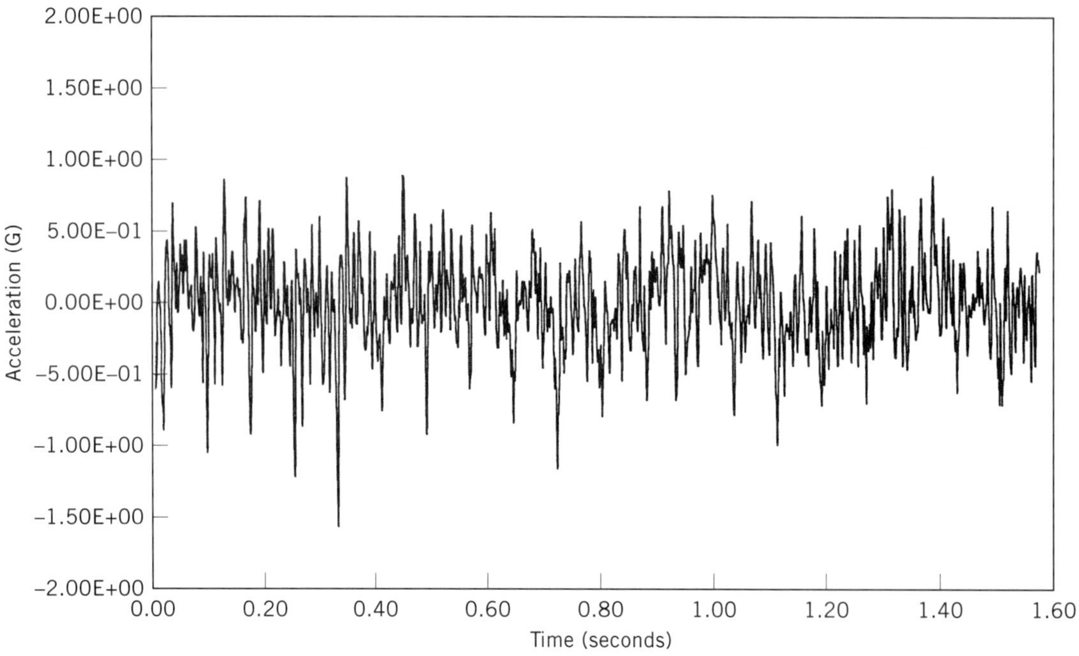

Figure 1. Example of truck vibration.

level of protection are possible. With developments in measurement technology, these benefits are within the reach of package developers.

Small, battery-powered instruments, capable of accurate measurements and recording of transit hazards, are available from several manufacturers. Instruments capable of measuring shock, drop, impact, and vibration usually employ accelerometers, or acceleration transducers. Likewise, instruments for detecting compressive loads would use some version of a load cell or weighing transducer. Temperature and humidity measurement requires appropriate sensors. In all cases, power from batteries and signal conditioning matched to the sensors are needed. Data is converted from the continuous or analog form, into small, discrete steps by internal electronics. This analog-to-digital conversion (ADC) allows the data to be stored and understood by digital devices, including computers. Once detected, hazard data are stored in computer-compatible memory within the instrument, available for eventual transfer to a personal computer for analysis.

The process of measuring the hazards of distribution requires careful planning. The following steps are recommended:

- Observation
- Measurement
- Analysis
- Specification
- Validation

The first step to a quantitative understanding of logistic hazards is qualitative. Observation of the distribution system details the elements of distribution; handling, transit, warehousing, and assists in understanding how these pieces fit together into a system. Attention should be paid to possible sources of handling drops or impacts. A block diagram of how a package travels from manufacture to consumption is a useful tool to develop. With these observations, targets for the measurement step may be selected. A certain handling operation, or a transportation mode or route, or an entire trip or system, may be selected. Independent variables, such as position in the vehicle, or weight of package may be targeted at this stage.

The measurement step concentrates on data collection. The location of the measurement system and transducers is critical at this point. Measurement needs to be taken at a point where eventual tests will be controlled. For example, if the goal is a vibration test, then the test system will be taking the place of the vehicle bed. The test system will be programmed to simulate the motion of this vehicle. Accordingly, the measurement system must monitor the vehicle bed. Measurements taken of the response vibration, inside the package, will be interesting, but not useful for development of test specifications.

Drop height data are collected by an instrument inside of the package, as are data on compressive load. Shock and impact data collection points depend on how the eventual test will be controlled. Temperature and humidity data may be taken at any point, but with due consideration of the mitigating effects of the package. Data should be time- and date-stamped when collected if possible. This allows the user to evaluate the location of each data event: truck terminal, warehouse, road or rail section, flight number, or loading dock.

The analysis step focuses on converting raw data to information. Data from an electronic recorder are organized and stored by events: individual readings or packets of time. Slowly changing data, such as temperature, atmospheric pressure, or relative humidity, may be sampled by individual

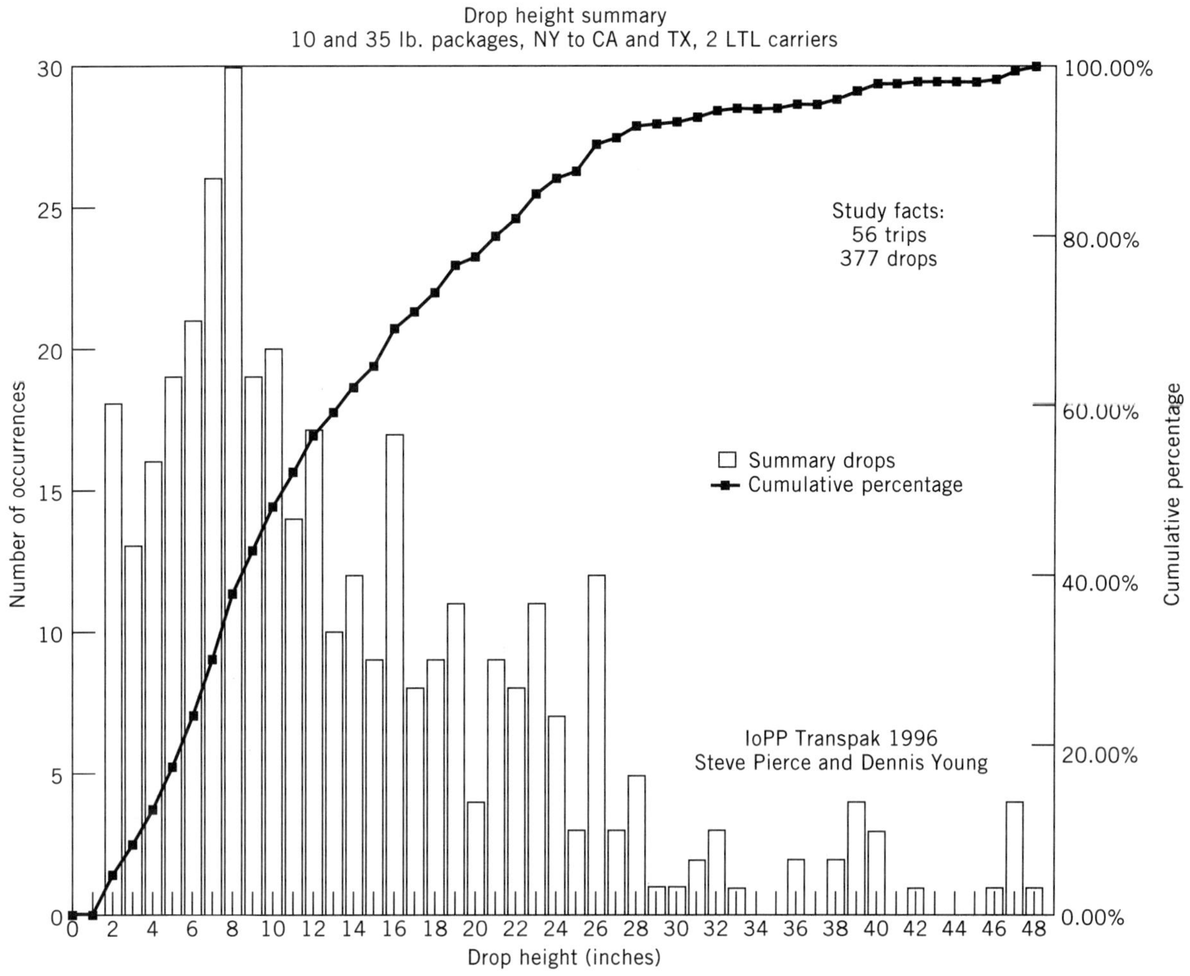

Figure 2. Example of drop-height data analysis.

readings. Dynamic data, ie, shock and vibration, need to be captured in time packets so that the analog nature of the original event may be observed. Drop heights may be estimated by the instrument software, and vibration data may be sorted for frequency content.

Data analysis should begin with careful consideration of the goal of the measurement program. If the goal is the development of a random vibration test, then spectral analysis is the typical form. If data taken by a recorder inside a package are to be used to set up a drop test specification, then data analysis should include number of drops, orientation of drops, and drop-height distribution. A useful form of data analysis for drop-height distribution is a drop height histogram with cumulative percentages. Figure 2 shows an example from data collected in shipments by less-than-truckload (LTL) mode (1). Figure 3 shows a summary analysis of truck vibration in a truckload, Interstate environment. This information should serve as examples of analysis form and content, not as definitions of these shipment conditions.

One key advantage of laboratory testing is the ability to compress time in the evaluation of package performance. A shipment that might take several days of elapsed time can be simulated in the laboratory in a matter of hours. The specific technique for time compression depends on the test being performed.

Time compression for drops is clear. The time between drops in actual shipment may range from minutes to days. In the lab simulation, drops are conducted with minimal waiting time, so the elapsed time is compressed. Time compression for top-load compression tests, to simulate the effects of long term storage in a short test time, is being developed by some users.

Time compression for vibration is more complex. In general, increased test amplitude intensity (G level) may be traded for reduced time. There are undoubtedly practical limits to such compression. A shipment of 20-h transit time probably cannot be effectively simulated in 10 min. Random vibration test durations of one to ≥6 h have proved successful in replicating field damage. Shorter or longer vibration test durations may be required to achieve desired results. The default vibration time of 3 h used by ASTM D4169 is suggested as an initial target. Within limits, trip time may be compressed to lab time using the following relationship (2):

$$L_2 = L_1 \left(\frac{T_1}{T_2}\right)^a$$

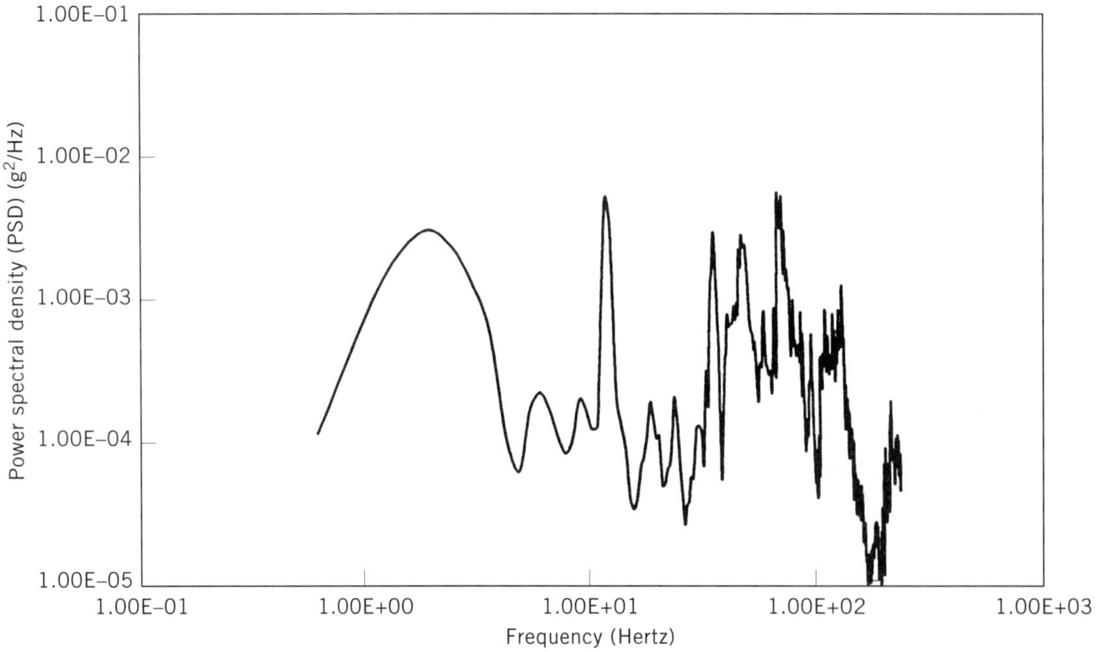

Figure 3. Example of vibration analysis.

where L_2 = level of lab test
T_2 = time duration of lab test
L_1 = level of field measurement
T_2 = time duration of field measurement
a = exponent, typically 0.3–0.5

Test specifications should also include consideration of the diversity of distribution environment hazard levels. For example, a series of drop heights may be constructed to reflect the actual occurrences measured in the field. Rather than a series of drops all at the same drop height, a series should be constructed with different drop heights. Such a series might include several low heights, a few midlevel drops and one high drop. This is modeled on the actual hazard, where high drop heights occur seldom, and lower drops occur more often. Table 1 demonstrates the concept based on the data shown in Figure 2.

This level diversity technique is also valuable for vibration testing. Data may be sorted by event amplitude and grouped. The estimated actual time that vibration occurred at each group level is then calculated. One background level, reflecting the majority of the events recorded may be combined with one of more higher level tests of shorter durations. Time compression may be applied to some or all of the test segments that result. This multilevel test specification should provide a more robust simulation of the logistic environment hazard than a single-level test.

Table 1. Demonstration of Varied Drop Heights in Test Specification

Drop Height	Number of Drops
11 in. (28 cm)	8
19 in. (48 cm)	7
46 in. (117 cm)	1

The validation step is a potentially time-consuming, but necessary, effort. During validation the user compares actual field results to laboratory results and makes any required adjustments. This may include selected testing of products with known shipping histories. Of particular interest would be products that exhibit repeatable, definable damage during transit. A laboratory test that produces very similar results is the goal. If the test specification under development achieves this end, then it has demonstrated a degree of validity. A number of such test cases will increase the validity and confidence in this specification. In general, exercise caution when using new test specifications that have not undergone validation, especially in the evaluation of new products and packages where no history is available.

All stages of the process, from observation through validation, should be carefully documented. In this way, future users may extend the development process to new logistical operations, or assess the need for additional measurement and analysis efforts.

BIBLIOGRAPHY

1. S. R. Pierce and D. E. Young, "Package Handling in Less-Than-Truckload Shipments: Focused Simulation Measurement and Test Development," *Proc. IoPP Educational Symp. Transport Packaging,* IoPP, Herndon, VA, 1996.
2. Tining

General References

ASTM, *1995 Annual Book of ASTM Standards,* Vol. 15.09, The American Society for Testing and Materials, Philadelphia, 1995 (ref.: ASTM D4169).

Procedure 1A, For Testing Packaged Products Weighing Under 100 Pounds or 45.4 kgs., International Safe Transit Association, East Lansing, MI.

A. J. Curtis, N. G. Tinling, and H. T. Abstein, *Selection and Performance of Vibration Tests,* SVM8, The Shock and Vibration Information Center, USDOD, Washington, DC, 1971.

S. R. Pierce, *Development of a Shock Measuring Telemetry System,* Multi-Sponsor Research Project 1, Michigan State University School of Packaging, 1969.

S. R. Pierce, *Modular Approach to Package Performance Testing,* Proceedings of IoPP Educational Symposium on Transport Packaging, IoPP, Herndon, VA, 1993.

S. R. Pierce and D. E. Young, *Development of a Product Protection System,* Annual Conference Proceedings, Shock and Vibration Information Center, 1972.

S. P. Singh, A. Cheena, and H. El Khateeb, *A Study of the Packaging Dynamics in the Overnight Parcel Delivery System of Federal Express, United Parcel Service and the United States Post Office,* The Consortium for Distribution Packaging, Michigan State University, E. Lansing, MI, 1991/92.

D. E. Young, "Field-to-Lab: Applying Available Technology to Package Performance Testing," *1st Internatl. Transit Packaging Conf. Proc.,* Pira International, Leatherhead, Surrey, UK, 1993.

D. E. Young, "Strategic Transport Packaging Performance," *Proc. IoPP Educational Symp. Transport Packaging,* IoPP, Herndon, VA, 1995.

D. E. Young and C. D. Pierce, *Developing Package Vibration Tests: The Field-to-Lab Technique,* TEST Engineering and Management, Oct./Nov. 1993.

DENNIS E. YOUNG
Dennis Young & Associates, Inc.
Charlotte, Michigan

DISTRIBUTION PACKAGING

The objective of distribution packaging is preservation of the product in its delivery from point of manufacture to the customer. Without packaging, most products would have a difficult and expensive trip through handling and transportation, many of them delivered to customers in a damaged condition. Distribution packaging actually adds to the value of the product, by lowering the cost for customers to obtain possession of the product from its origination.

Distribution packaging is known as *transport packaging* outside North America and includes the shipping container, interior protective packaging, and any unitizing materials for shipping. It does not include packaging for consumer products such as the primary packaging of food, beverages, pharmaceuticals, and cosmetics. In the United States distribution packaging represents about one-third of the total purchases of packaging; the balance is attributable to primary packaging for consumer products.

The goal in distribution packaging is to provide the correct design for packaging such that its contents will arrive safely at the destination, without using too much or too little packaging material. In other words, the package designer must assure that this equation is maintained:

$$\text{Product} + \text{package} = \text{distribution environment}$$

The following text will show how to determine the right amount of each of the three variables so that the equation will always balance, with no excessive overpackaging cost or loss from damage.

Functions and Goals of Distribution Packaging

The functions of distribution packaging can be summarized as follows:

Containment. The basic purpose of packaging is to contain the product. Packaging permits products to move from their source to the customer, supplying use value to products that are otherwise useless to the customer, who is usually remote from the source.

Protection. Most products require some degree of protection from the hazards in distribution. Packaging furnishes the degree of protection needed to safely transport products from source to customer.

Performance. Packaging aids in transportation, handling, storing, selling, and use of the product. This function includes such things as orientation of the product, ease of identification, appropriate quantity, ease of disposal, and handling features.

Communication. The package must identify its contents and inform about package features and handling requirements. It generally provides space for shipping information as well.

To design a distribution package, one must have goals in mind. These will vary with products, customers, distribution systems, manufacturing facilities, etc, but most distribution packaging should address the following objectives:

Product Protection. The primary purpose of any distribution package is to ensure the integrity and safety of its contents through the entire distribution system.

Ease of Handling and Storage. All parts of the distribution system should be able to economically move and store the packaged product.

Shipping Effectiveness. Packaging and unitizing should enable the full utilization of carrier vehicles and must meet carrier rules and regulations.

Manufacturing Efficiency. The packing and unitizing of goods should utilize labor and facilities effectively.

Ease of Identification. Package contents and routing should be easy to see, along with any special handling requirements.

Customer Needs. The package must provide ease of opening, dispensing, and disposal, as well as meet any special handling or storage requirements the customer may have.

Environmental Responsibility. In addition to meeting regulatory requirements, the design of packaging and unitizing should minimize solid waste by any of the following: reduction–return–reuse–recycle.

Since distribution packaging must always be economical, all the above goals should be balanced to achieve the lowest overall cost.

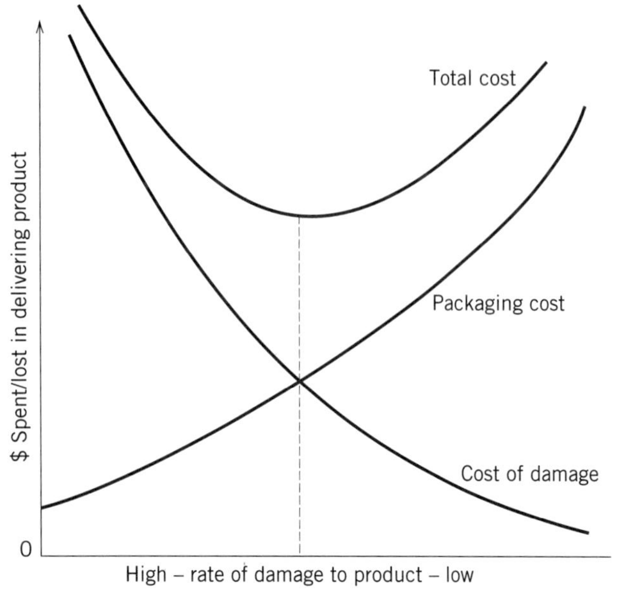

Figure 1. Cost of product protection is a tradeoff between packaging and damage.

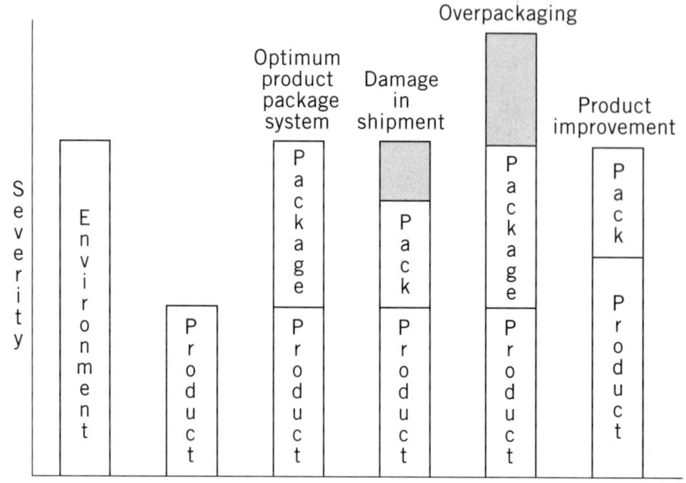

Figure 2. Concept of a protective package.

Taking a Total System Approach to Package Design

The scope of design in distribution packaging must consider all aspects of the distribution system, including customers, carriers, and distributors as well as the manufacturing plant, packaging line, warehousing, and shipping. Successful package design is a total-system approach.

Once created, a package does not magically form around the product, float through shipping, travel hundreds of miles in isolation, arrive at the customer's site, and then disappear. It has an influence on and is influenced by everyone and everything it encounters. Many of these encounters will affect manufacturing and distribution costs, or product integrity with indirect impact on sales. Therefore such events should be considered in the design process.

Unfortunately, there is often too much focus on the cost of packaging materials to the exclusion of other factors, including cost-related ones in handling, storage, and transportation. If the package is slightly larger and/or heavier than it really has to be, costs in all three areas will be higher than necessary, perhaps producing an even greater effect on profits than would higher packaging costs for a smaller and/or lighter package. For instance, although each industry and company is different, a general rule of thumb states that transportation will cost between 3 and 10 times as much as packaging on average for all shipments. A small reduction in package size or weight could mean much more savings in transportation plus handling and storage. For example, a small kitchen appliance in bulky wraparound die-cut may involve less material cost, but molded packaging may pack faster and require less cube, permitting more pieces per unit load, fewer handling trips, more units per storage cube, and more units per truckload, for an overall savings.

An inverse relationship exists between packaging cost and maintaining product integrity with low damage rates, as shown in Figure 1. Other factors being equal, an increase in packaging will provide more protection to the contents and therefore lower the potential for damage. Or, conversely, cutting packaging costs without other improvements generally means less protection and higher damage rates.

The real cost of getting the product safely to market is the *sum* of packaging and damage. Optimizing the total is the true goal of packaging design. As damages rise too high on the left of the graph, both replacement–repair costs and potential for loss of customer goodwill and cancellation of orders increase. For companies where loss of sales and customer satisfaction are more important than costs, there is not much room for movement to the left of the package–damage intersection.

No matter where packaging design takes place—in engineering, manufacturing, shipping, or at the supplier—all factors mentioned above must be considered, costed, and involved in a *total-system approach.*

The Protective-Package Concept

The equation presented earlier is used here to explain the concept of protective packaging for distribution:

$$\text{Product} + \text{package} = \text{distribution environment}$$

Figure 2 depicts this equation and the consequences of an imbalance in the equation, what happens when a product plus its package are not exactly what is needed to survive in the distribution process.

Reading from left to right, here is an explanation of the bar chart. *Severity* is the quantitative measure of the *environment,* which can be any one or combination of hazards in distribution. Here are examples of hazards and their severity: the rough-handling hazard to a 20-lb package is determined to be 30 in. of drop on any of six package surfaces; or the compression (storage) hazard is determined to be 10 packages high in warehousing; or the high-temperature hazard is 140°F.

Product represents the measured level of resistance to damage of the product. In the rough-handling example, the product has been tested and has shown the ability to survive six drops from 15 in. with no damage, while higher drops will cause product impairment.

The third bar depicts the equation whereby the *product's* measured level of damage resistance plus the *package's* measured abilities to protect the product are exactly equal to the expected environmental hazard(s), an optimum solution. For the example, a product with 15-in. drop resistance is packaged in 15-in. drop protection which total the 30 in. of drop height the packaged product is expected to encounter in the distribution environment.

When the package provides less protective capacity than needed for the environment, as shown in the shaded area of the fourth bar (Fig. 2), this underpackaging will result in *damage in shipment*. For example, the package supplies only 9 in. of drop-height protection instead of the required 15 in.

The next bar defines *overpackaging*. The package protection level is higher than the environment requires, and the shaded area shows amount wasted. Instead of designing for 15 in. of drop protection in our example, packaging supplies 6 in. more than needed.

At times, *product improvement* is an alternative to more packaging. Instead of accepting a certain level of product ruggedness of a product, product engineers may be able to raise the level as shown in the bar chart to the far right. The result is a reduction in packaging needed. Perhaps the product had a component that malfunctioned above 15-in. drop heights, but a redesign improved ruggedness to the 20-in. level. Now packaging need only supply 10 in. of protection, a 33% reduction from the original.

The most elusive part of the equation is the distribution environment. The most difficult part of defining the environment is not in identification of the types of hazards, but what each hazard's expected level is and its probability of occurrence. Dedicated package designers continually search for better methods of defining the environment on the right side of the equation so they can solve for the parameters of product and package on the left side of the equation.

The 10-Step Process of Distribution Packaging Design

Here is a proven 10-step procedure that will assure that a distribution package design will provide maximum performance at least overall cost.

1. Identify the Physical Characteristics of the Product. Product knowledge means more than knowing simply its dimensions and weight. The package designer must be aware of surface characteristics and susceptibility to abrasion or corrosion, the ability to hold a load in compression, internal characteristics affected by vibration, and particularly the product's fragility. Guessing about any of these factors will surely lead to potential problems.

2. Determine Marketing and Distribution Requirements. Package design must incorporate marketing and distribution requisites in addition to product characteristics. One must know the number of units that will ship in a container, the composition and attributes of the primary package, the identity of customers and their handling and storage requirements, the package disposal criteria, total volume expected per shift per day per year, expected life cycle, the planned modes of transport, types of distribution channels, etc.

3. Learn about the Environmental Hazards Your Products Encounter. It was emphasized earlier that knowledge of the distribution environment is key to designing an optimum package. Major hazards to be expected in the environment are rough handling, in-transit vibration and shock, compression in storage or in transit, high humidity and water, temperature extremes, and puncturing forces. Learning about them may include observation, reading other's research, or conducting measurements.

4. Consider Alternatives Available in Packaging and Unitizing. Many alternatives are available for shipping containers, interior packaging, and unit loads. All should be considered and reviewed before selecting the final types for further development. Tradeoff analysis techniques such as make vs buy often help. Rather than considering only materials that one has experience with, comparing paper vs plastic vs wood vs metal is a good exercise at times to assure the best for the particular project. Once the basic materials have been selected, detailed work on design can begin.

5. Design the Shipping Unit (Container, Interior Packaging, and Unit Load). With basic materials and information established in steps 1–4, the designer can now scientifically engineer a distribution package, and unit-load where appropriate. Each component of the shipping unit (container–interior–unit load) is analyzed for strength and other required properties and compared to technical data available from suppliers. Some packaging materials have good design data available, but most do not. The designer frequently must rely on experience to reach a successful solution; the novice may find that lack of information makes it difficult to arrive at an optimum solution. Trial and error can be shortened for the novice, as well as the experienced designer, by conducting engineering tests in package development. Impact, vibration, and compression testing in the lab not only identifies shortcomings but helps to fine-tune to the optimum solution.

6. Determine Quality of Protection through Performance Testing. After the shipping unit is designed, perhaps with the aid of engineering development tests, it should then be performance-tested. This consists of subjecting the unit to a sequence of anticipated hazards and/or tests in the laboratory for the purpose of a pass/fail decision. Will the shipping unit protect its contents all the way through distribution? The performance test methods should be based on industry standards. Such standards have considerable experience and history behind their development and use, and a successful completion of the test sequence almost guarantees damage-free shipments. The most widely used standard is the International Safe Transit Association's Project 1 and 1A, in use since 1948 (1). The 1982 approved ASTM D4169 (2) provides a more complete array of distribution possibilities and identified hazards with corresponding test sequences, and permits the user some flexibility in selecting test intensities. For users who can clearly define their distribution cycles, but find them different than the standard cycles in D4169, the ASTM standard also provides a means of developing a unique sequence of tests, resulting in performance tests that can more precisely simulate the actual conditions.

7. Redesign the Shipping Unit until It Successfully Passes All Tests. There is an old saying which goes—One Test Is Worth 1000 Expert Opinions. Often performance test results fool even the most experienced engineers, and it is necessary to repeat an entire cycle of redesign and retesting as many times as required to reach a "pass" decision.

8. Redesign the Product Where Indicated and Feasible. Occasionally testing reveals a product weakness that can be compensated for by protective packaging, but at excessive cost. If at all feasible, the product should be redesigned to correct the weakness rather than redesigning the package. This is particularly important when the cost of the redesigned product is less than that of the increased packaging. It is usually difficult for package designers to bring about product redesign when they are located organizationally in other than the product engineering group. If this is the case, the packaging designer should attempt to establish a continuing line of communication with the product engineers. Sometimes this means educating product engineers in the hazards of distribution and showing them how to correct product weaknesses.

9. Develop the Methods of Packing. An important part of package design is packing of the product in the shipping container and unitizing of containers. Although this may be the responsibility of someone else, ie, industrial engineering, the designer must be aware of cost factors and the appropriateness of mechanizing or automating all or part of the operations. Sometimes a tradeoff in package design must be implemented to achieve overall system economics.

10. Document All Work. One step repeatedly overlooked in the design process is documentation. This includes documenting test results, specifications, drawings, and methods of packing. Drawings should be in company-standard formats with appropriate designations for reference in the corporate specification system. Relying on supplier sketches or drawings as the reference document is not a wise idea. They should be transferred to company format so that purchasing, manufacturing, and engineering can reference them.

On any package design project, follow these 10 steps and check your work against the checklist below. Doing so will significantly reduce potential of an unpleasant surprise when shipments begin.

Checklist for each package design project.

1. Considered the solid-waste aspects of the package and unitload, and their alternatives, to minimize impact on the environment?
2. Pondered the use of returnable or reusable containers and dunnage?
3. Contemplated all cost factors in the distribution cycle: handling, storage, and transportation?
4. Checked cost of this package versus company or plant average for similar products?
5. Considered all possible alternatives in materials and methods?
6. Used industry standards for materials and design criteria where possible?
7. Performance-tested the design against accepted industry standards or regulatory requirements?
8. Documented the design in company's spec system?
9. Checked damage and customer complaints on this product line?
10. Satisfied all rules and regulations applying to this product for all distribution modes it is expected to encounter?

The Components of Distribution Packaging

The components of distribution packaging may be broadly classified in the following categories:

Shipping containers
Interior protective packaging
Closure and securement materials
Unitized load shipping bases

Together these represent about one-third of the total U.S. packaging costs.

By far the largest component of distribution packaging is shipping containers, representing an estimated 75% of the total purchased. Corrugated boxes represent about 80% of the total spent on shipping containers and probably close to 90% of total units of shipping containers used. Other types of container include paper and plastic shipping sacks, wooden boxes and crates; metal and plastic drums and pails, and composites for textiles.

Interior packaging takes many forms, but the most widely used for cushioning, positioning, or simply to fill space are corrugated pads, die-cuts, partitions, foldups, molded and fabricated expanded polystyrene forms, molded and fabricated resilient foams, foam-in-place, loose-fill, air-bubble and foam wraps, variations of paper waddings, and a variety of specialty devices and forms. Barriers against water, water vapor, and various gases and to prevent electrostatic damage are another class of interior materials. Corrosion preventatives are also in this category, such as preservatives, volatile corrosion inhibitors, and desiccants.

Closure and securement includes adhesives, tape, and staples for corrugated box closure, and strapping, stretch wrapping, and shrink bags for unitizing purposes.

Shipping bases include wood, corrugated, or plastic pallets; corrugated, solid fiber, or plastic slip sheets; and wood skids.

BIBLIOGRAPHY

1. *Pre-shipment Test Procedures.* International Safe Transit Association, Chicago, 1990.
2. *Annual Book of ASTM Standards,* Vol. 15.09, D4169, American Society for Testing and Materials, Philadelphia, 1995.

Alfred H. McKinlay
Consultant
Pattersonville, New York

DRUMS, FIBER

A *fiber drum,* by definition of the Hazardous Materials Regulations (HMR) of the U.S. Department of Transportation (DOT) (1) and of the nongovernmental rail (2) and motor freight classification agencies (3), is a shipping container made by the convolute (not spiral) winding of multiple plies of paperboard into a tubular body, to which are attached headings that may be made from solid fiberboard, metal, plastic, natural wood, or plywood. No single ply of the sidewall may be less than 0.012 in. (0.30 mm) thick. The sidewall plies must be firmly glued together and may include protective layer(s) of metal foil, plastic, or other appropriate materials. The outer ply may be waterproofed, or, at the least, must be sized to resist the effect of casual water. Fiber drums provide high strength and low tare weight for the packaging of industrial commodities. A fiber drum intended for heavy-duty use may weigh as little as half that of a comparable steel drum.

Construction

Sizes. Fiber drums are manufactured in a wide range of diameters and capacities: in diameter from about 8-in. (200-mm) to 23-in. (584-mm) in $1\frac{1}{2}$-in. (38-mm) increments, and in capacity from 0.75 gal (2.8 L) to 75 gal (285 L), with almost infinite variability, since capacity is controlled by the slit width of the tube, which can be adjusted in $\frac{1}{8}$-in. (3-mm) increments.

Stacking. The stacking strength of a fiber drum is determined by the number of plies of paperboard in the sidewall. This commonly varies from 4 plies for lightweight drums to 11 plies for large drums intended to carry heavy loads (these ply counts are for 0.012-in. board; the increasingly common use of heavier board requires fewer plies). The stacking strength is also a function of the moisture content of the paperboard; the higher the humidity, the lower the stacking strength. Taking these and other construction factors into account, the bottom drum in a stack of large, heavy-duty fiber drums might safely support as much as 2000 lb (900 kg).

Linings and barriers. The basic structure of a drum may be modified by the insertion of a laminated barrier board (a sandwich made from aluminum foil and/or polyethylene between two plies of lightweight paperboard) into the sidewall to reduce the moisture vapor transmission rate (MVTR). Alternatively (or sometimes additionally) a laminated integral lining made of paperboard, plastic film, and sometimes aluminum foil, may be used as the interior surface of the tube, to impart MVT resistance, oxygen transport resistance, and/or chemical resistance to the drum. The plastic part of such a laminated lining may include polyethylene (as barrier layer or as adhesive for a different barrier), polyester, polyvinyl alcohol, silicone release lining, or a laminate of a special barrier polymer with polyethylene. A lightweight polyethylene lining or a loose polyethylene bag is also frequently used in the packaging of dry products to reduce MVTR. A "skin" (pigmented polyethylene–coated board) or label (decorated stock) may be wound as the outside ply of the drum, with the colored plastic or decorated stock to the outside, to provide protection from moisture or physical abuse, or to project information, using the drum surface as a "billboard."

Table 1 (4) shows that the MVTR of fiber drums may vary over a factor in excess of 300, depending on lining or barrier construction.

Table 1. Water-Vapor Transmission Rate of Chime-Style Fiber Drums[a]

Construction	Weight increase
Silicate adhesive, no lining	10,000+ (est.)
Silicate adhesive, polyethylene barrier (0.7 mil)	600
Silicate adhesive, polyethylene lining (2 mils)	350
Flame activated polyethylene adhesive, no lining	175
Silicate adhesive, aluminum-foil barrier	125
Silicate adhesive, polyethylene/aluminum foil lining	30

[a] Total weight increase of a desiccant inside a 55-gal fiber drum with rubber gasketed steel cover, in grams of water per 30 days, with the drum exposed to 100°F and 90% rh. The values listed above are for comparative purposes only and should not be understood as absolute values that would be picked up by a specific product.

Adhesive. The adhesive most commonly used in the manufacture of fiber drums is sodium silicate, or water glass. Silicate adhesive has the advantages of low cost and general effectiveness, but the disadvantage of being water-soluble, and therefore producing drums with poor water resistance. Thus, a silicate drum subjected to a 5-day exposure in a water-spray test will lose about 90% of its stacking strength. Other adhesives (eg, polyvinyl alcohol and polyvinyl acetate), available at higher cost, will produce drums with modestly improved water resistance, but are not widely used. Such drums should not be used for stacked storage in a high-humidity environment or for outdoor storage (see also Adhesives).

Weatherproof adhesive. Use of polyethylene as the sidewall adhesive, either as a roll-coated hot melt, or preferably, extrusion-coated onto the kraft linerboard and flame-activated during the winding process, produces far superior weather resistance in the drums because the polyethylene essentially encapsulates the paper fibers in a hydrophobic matrix. A 10-ply drum wound with polyethylene as the adhesive would lose only 10% of its stacking strength in the water-spray test described above, because only the outermost ply would become saturated and lose its strength. Such drums are eminently suitable for outdoor or high-humidity stacked storage.

Drums with intermediate levels of water resistance may be produced by winding several plies of polyethylene barrier board as the outside layers of the tube, using silicate or other (nonpolyethylene) adhesive, with or without a skin on the outside. While this produces fair-to-good water resistance when the drum is new, should the outer ply or plies become abraded, punctured, or torn with use, the weather resistance would be markedly reduced. Therefore major applications for such drum constructions involve only short-term moisture contact, such as for the packaging of products intended to be refrigerated (for protection from water damage from condensation on removal from the refrigerator), and for products where the drum is washed down after filling.

Applications

Fiber drums are commonly used for the packaging of a broad range of products, from apple juice to zinc powder. Major markets include dry and solid chemicals, liquid and hot-melt adhesives, paints and coatings, liquid textile chemicals, such foods as tomato paste and dried onions, and rolls of plastic

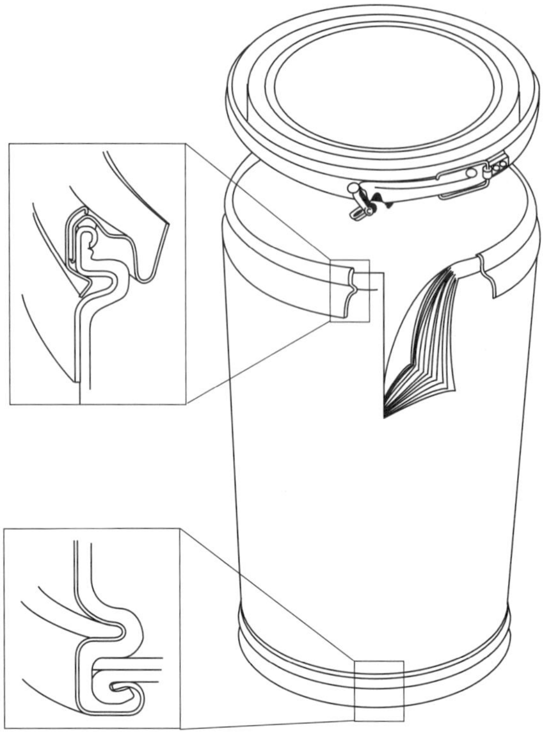

Figure 1. Drum with metal chimes.

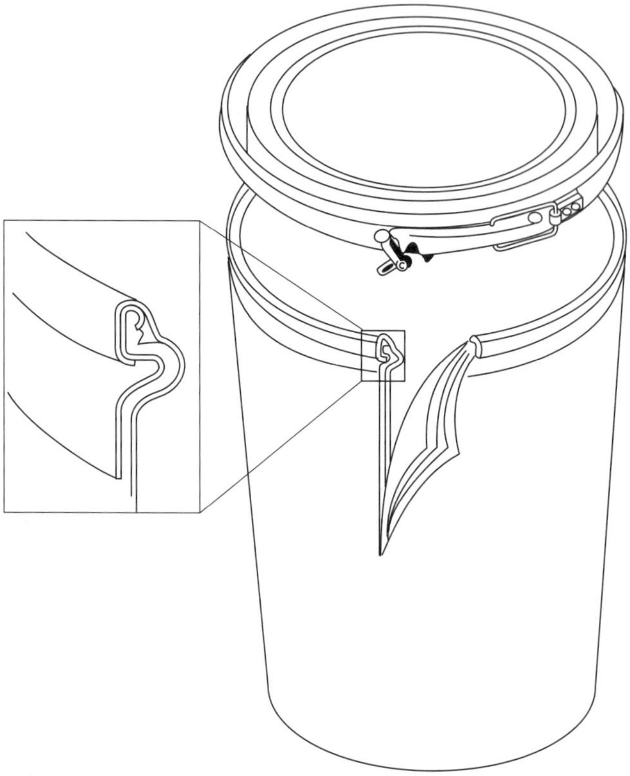

Figure 2. Open-head composite fiber drum.

film and carpeting. Fiber drums are also used for the transport and dispensing of electrical wire and for the transport of hazardous wastes to incinerators and landfills.

Drum Styles

Drums with metal chimes. Figure 1 shows the most common style of fiber drum, with metal chimes on the top and bottom of the drum. These chimes are mechanically formed during drum assembly to attach the bottom heading to the sidewall (bottom inset) and to provide an attachment site for a toggle-action locking band to hold the cover onto the drum (top inset). This drum style can provide the highest level of abuse resistance and product protection by a fiber drum. Many variations are commercially available for this versatile drum style.

Drums for packaging liquids. If made with an interior lining, caulking compound in the bottom juncture, and a gasketed steel or, preferably, plastic cover, a metal-chime drum can be used for the packaging of liquids, ranging from self-sealing latex adhesives to water-based chemical solutions, to solvent-based cleaners and fiberglass gelcoats. Such drums can also be provided with a spun-on locking band to produce a tight-head drum that provides an even better top seal. The use of fiber drums to package liquids has grown significantly with advancements in lining materials and manufacturing processes that have expanded the holding capabilities of the drums.

Composite fiber drums. Figures 2 and 3 respectively show open-head and tight-head composite fiber drums intended for the packaging of liquids. Some drums have a self-supporting plastic insert, or inner receptacle, inside each drum to contain the liquid. The inner receptacle is an integral part of the packaging and is filled, shipped, and emptied as such.

Straight-sided drums. Figure 4 shows the straight-sided chime of a drum intended for the packaging of hot-melt adhesives and other products that are dispensed by use of a platen pump. Since a close-fitting platen must slide into the drum to dispense the product, the inverted groove design typical of other drums with metal chimes cannot be used. Instead of reducing the diameter at the groove to provide purchase for a lock band as in the standard design, the chime and fiber tube of a straight-sided drum must be expanded outward by about 8%. Since paper typically can be elongated only about 3% before failure, the development of this drum style was not a simple task. This drum style may also be made with two tops, ie, two removable headings, so that when the platen pump has removed as much product as it can reach, the drum may be inverted, the bottom removed, and the residual heel removed and placed on top of the next drum for dispensing. At several dollars per pound, this can provide a significant saving for the adhesive user.

Drums for packaging wire. Figure 5 shows a drawing of a drum designed for the packaging and dispensing of wire. The drum is made with a central core, around which the wire is coiled and from which it is dispensed.

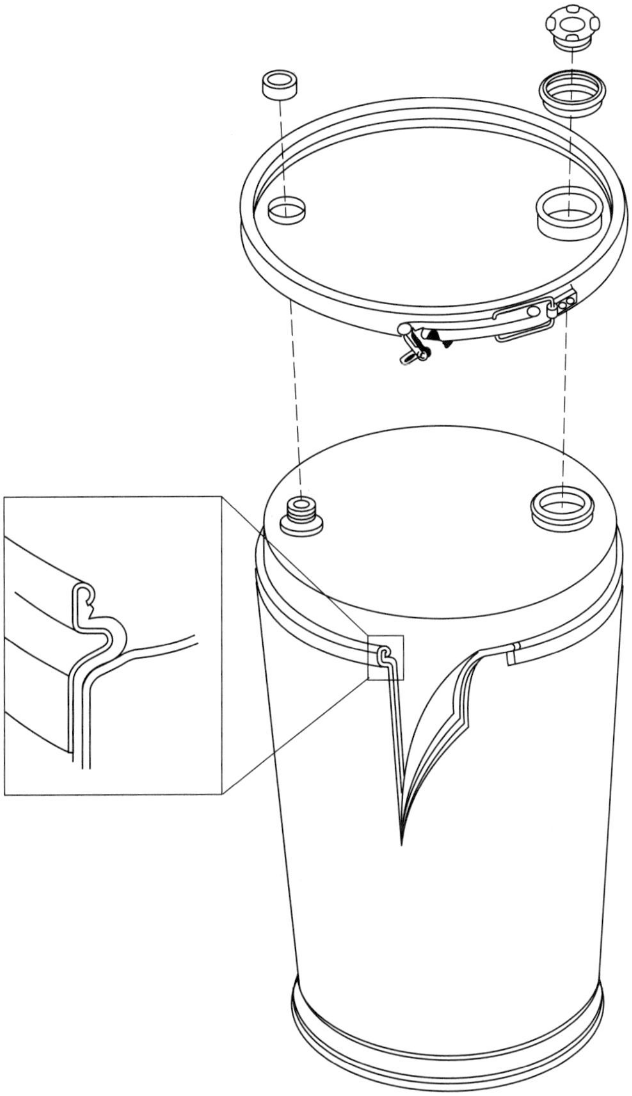

Figure 3. Tight-head composite fiber drum.

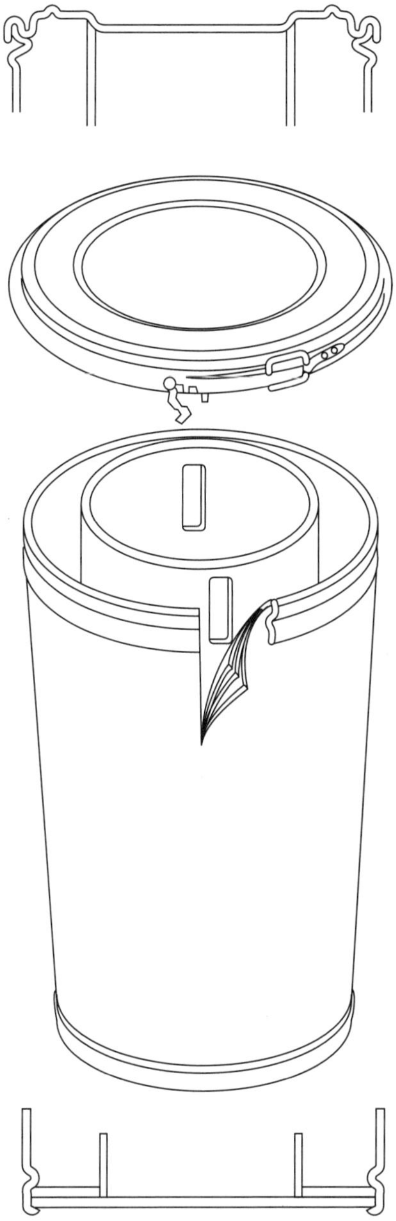

Figure 5. Drum for packaging wire.

Drums for the aseptic packaging of foods. The aseptic packaging of foods at ambient temperatures requires a containment system that provides a good barrier to both oxygen and biological contaminants that might cause food spoilage. In this system the food is packaged in a high-strength polyester/polyethylene/aluminum-foil bag, which is then placed inside the drum. The drum itself is of heavy ply construction, frequently using weatherproof adhesive, to allow outdoor storage. The drum functions to provide protection from mechanical abuse and moisture. The bag provides the barrier to chemical and/or biological degradation of the food (see also Aseptic Packaging).

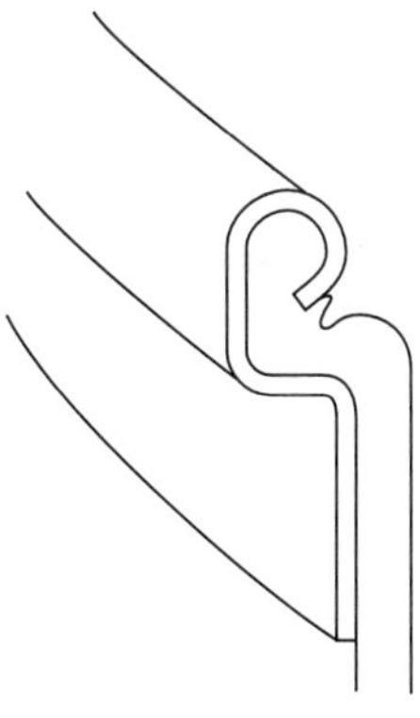

Figure 4. Crimp of straight-sided drum.

Drums without chimes. Figure 6 shows the simplest of drum designs: a fiber tube with steel or plastic headings attached by stitching, taping, adhesively bonding, or the use of mechanical clips. The top and bottom headings are designed to interlock to facilitate stacking. This inexpensive drum design, generally with an inner polyethylene bag, is commonly used

Figure 6. Drum without chimes.

Figure 7. All-fiber drum.

for the packaging of such low-cost products as detergent and ice-melt compound in relatively small sizes.

Nestable fiber drums. A striking variation of this drum style is a nestable fiber drum. A frustroconical tube is made by the winding of precut sheets of paperboard, one for each ply. An integral polyethylene lining and plastic headings (adhesively bonded on the bottom, taped on the top) complete the structure. The advantage of this construction lies in the reduction of shipping cost and storage space for empty drums, but at the sacrifice of some stacking strength. This drum design was developed for the storage and transport of hazardous medical waste for incineration.

All-fiber constructions. Drums made only from fiberboard have somewhat greater structural integrity than do the chimeless drums described above. As shown in Figure 7, in the manufacture of such drums the bottom-end portion of a tube is notched and/or folded in, and then covered inside and out with fiber disks that are stitched and/or glued to produce a strong and secure heading. Covers, designed for slip fit, are made in the same way, but from a short piece of tube.

Drums of this style are available in several design variations. The simplest has a drum bottom with a cover of somewhat larger diameter, which slides over the drum, leaving a step or ridge at the juncture between cover and drum. The cover is often taped in place, but this may leave an unsightly ridge. This drum design can also be made with a tube and two covers, often used for the shipment of rolls of plastic or other materials that need protection to prevent damage to the roll edge.

In a somewhat stronger all-fiber design, the drum body is composed of two components (Fig. 7): a tube that is glued into a somewhat shorter shell of somewhat larger diameter. The cover, having the same diameter as the shell, slides on over the exposed portion of the tube, leaving no ridge at its juncture with the shell. In a fourth variation of this style, the drum body is squared off to provide for more efficient use of storage space. The tradeoff for this gain is some reduction in stacking strength as compared to the right circular cylindrical form of other drums. The primary application for all-fiber drums is for the packaging of dry powders, frequently with an inner polyethylene bag.

Regulations

The manufacture and use of fiber drums are regulated by a wide variety of governmental and nongovernmental national and international agencies.

NMFC and UFC. The National Motor Freight Classification (NMFC) and the Uniform Freight Classification (UFC) are nongovernmental agencies concerned with the shipment of all commodities, hazardous or not, by common carrier by highway and rail, respectively, in the United States. As part of their function of setting freight rates for such shipments, these agencies also set construction specifications for the packaging to be used.

DOT, IMO (IMDG), ICAO, IATA, RID/ADR, and the UN RECOMMENDATIONS. These agencies [the U.S. Department of Transportation (DOT), the International Maritime Organization (IMO), which publishes the International Maritime Dangerous Goods Code (IMDG), the International Civil Aviation Organization (ICAO), the International Air Transport Association (IATA), and RID/ADR, the organizations regulating international shipments of dangerous goods by road and rail in Europe] have several things in common. First, they are all concerned only with the shipment of hazardous materials (the transport of dangerous goods). As far as these agencies are concerned, nonhazardous materials may be shipped in any packaging that the shipper desires to use. Second, they have all based their regulations, to a large extent, on the UN *Recommendations on the Transport of Dangerous Goods.* These are performance-based regulations. Basically this means that a packaging must be able to pass certain performance tests before it may be used for the shipment of hazardous materials. Hazardous materials are categorized into three Packing Groups—I, II, and III—in decreasing order of hazard level. Packagings are then rated by Performance Standard—X, Y, or Z—where X-marked packagings must pass the most rigorous performance tests and are allowed for use in the shipment of Packing Group I, II, or III commodities; Y-marked packagings must pass performance tests of intermediate severity and may be used in the shipment of Packing Group II and III commodities; and Z-marked packagings must pass the mildest tests and may be used only with Packing Group III commodities. Fiber drums intended for the packaging of dry and solid products have been constructed strongly enough that they met the requirements of Performance Standard X when tested with a gross mass of as much as 200 kg (X200), or a net weight of about 425 lb. Such drums have also successfully passed performance tests which allow them to be rated at as much as Y260 (555 lb net) and Z280 (605 lb net). Of fiber drums intended for the packaging of liquids, only tight-head composite fiber drums have been shown to be capable of passing the required performance tests, which are different from, and more rigorous than, those required for dry products.

BIBLIOGRAPHY

1. Hazardous Materials Regulations of the U.S. Department of Transportation, Code of Federal Regulations, 49 *CFR* 178.508, 1991 or later.
2. Rule 51, *Uniform Freight Classification,* 6000-K, National Railroad Freight Committee, Chicago, 1994, pp. 351–357.
3. Item 291, *National Motor Freight Classification,* NMF 100-U, American Trucking Association, Alexandria, VA, 1994, pp. 234–236.
4. W Swihart, private communication, Oct. 1987.

GERALD A. GORDON
Sonoco Products Company
Industrial Container Division
Lombard, IL

DRUMS, PLASTIC

Chemicals and other industrial products are shipped mainly in pails and drums. In most parts of the world, 1–6-gal (4–23-L) open-top plastic containers, generally injection-molded (see Injection molding), are called pails (see Pails, plastic). In North America, 1–6-gal (4–23-L) closed-head, ie, bung-type, blow-molded containers (see Blow molding) are called pails as well, or jerrycans. The term jerrycan is often used in other countries to describe bung-type containers in the 1–16-gal (4–61-L) size range. In general, however, the word drum applies to open- and closed-head containers larger than 6 gal (23 L). In North America, the standard drum sizes are 15, 20, 30, 35, 55, and 57 gal (57, 76, 114, 132, 208, 216 L). In western Europe and most other parts of the world, standard drum sizes are 30, 60, 120, and 216 L (roughly 8, 16, 32, 55, and 57 gal).

Polyethylene liners for steel and fiber drums were blow-molded and rotationally molded in the United States and western Europe in the 1950s, but there were no self-supporting plastic alternatives until the 1960s. In 1963, U.S. production of all-plastic 5-gal (19-L) pails and 16-gal (61-L) drums began. In western Europe, production of 16-gal open-top and bung-type drums started at about the same time, and a 32-gal (121-L) open-top drum soon followed. The introduction of all-plastic 55-gal (208-L) drums did not come until the early 1970s. The development of large all plastic drums took many years because they required special resins and processing equipment. Market acceptance has also required special designs.

Plastic containers have excellent performance characteristics. They are strong, lightweight, durable, corrosion-free, and weather-resistant, and, in accordance with international transport regulations, are authorized to carry a great number of hazardous materials. Most plastic drums are used for chemicals, but they are also used in the food-processing industry for the shipment and storage of products that include concentrated fruit juice, vegetable pulps, and condiments.

Resins

Self-supporting plastic drums are made of extra-high-molecular-weight (EHMW) high-density polyethylene (see Polyethylene, high-density). The molecular weight of these resins is so high that their flow rates cannot be expressed in terms of melt index (MI) as measured according to ASTM 1238, Condition E (44 psi or 303 kPa). The MI of relatively low molecular weight HDPE injection-molding resins ranges from 1 to 20 g/10 min; the MI of higher molecular weight blow-molding bottle resins is less than 1 g/10 min. Measuring the flow of EHMV resins requires higher pressure (Condition F, 440 psi or 3 MPa), and the values obtained are expressed in terms of high-load-melt index (HLMI). The HLMI of resins used for self-supporting drums range from 1.5 to 12 g/10 min. Design trends in plastic drums are related to the availability of EHMW resins. In the United States, 10 HLMI became the standard resin for drums with separate plastic- or metal-handling rings. The development of drums with integral handling rings required higher-molecular-weight resins (HLMI 1.5–3) that were available in Western Europe before they were produced in the United States.

All properties of polyethylene depend on three important factors: molecular weight (length of molecule chains), density (degree of crystallinity), and molecular weight distribution (distribution of longer and shorter chains). As molecular weight increases (and MI or HLMI decreases), toughness and resistance to stress cracking increases. The tradeoff for these improved properties is difficulty in processing. HDPE has crystallinity of 60–80% at density of 0.942–0.965 g/cm³. (In contrast, LDPE has crystallinity of 40–50% at density of 0.918–0.930 g/cm³.) As density increases, toughness and stress crack resistance decrease, but stiffness, hardness, and resistance to oils and chemicals improves. Molecular-weight distribution is related primarily to processing.

Chemical resistance. Chemical resistance is particularly important in drum design. All 55-gal (208-L) drums use EHMW high-density resins, but there are variations within that category. Drums used for chemicals that are compatible with HDPE generally use relatively high-molecular-weight, eg, 1–3-HMLI, and relatively high-density, ie, >0.95-g/cm³, resins. Where stress cracking is a potential problem, relatively low-molecular-weight, eg, 6–10-HLMI, and relatively low-density, ie, <0.95-g/cm³, resins give better performance. Although no chemical dissolves polyethylene, particularly high-molecular-weight polyethylene, the effects of certain chemicals include strong swelling action by penetration into the container walls, stress cracking, oxidation, degradation by destruction of the macromolecules, or permeation through the container wall. Resistance tests should be made based on laboratory samples and on containers in use.

The resistance to stress cracking must be examined before using self-supporting plastic drums. The ESCR test prescribed in ASTM D1693 can be used for material selection, but tests must be performed on finished containers. This can be done by storing a drum filled with 5% surfactant for more than 3 months at temperatures higher than 40°C. Stress cracking may occur if there are stresses in the wall. Tensile and compressive stresses can be avoided by using an optimum wall-thickness distribution in production or in the design of the blow-molding tool. They may also occur when the products contain surface-active substances such as wetting agents at concentrations up to 20%. HDPE's resistance to stress cracking depends on the density and the molecular weight of the raw materials. As density increases, ESCR decreases. As molecular weight increases, ESCR increases. High-molecular-weight blow-molding resins with densities of 0.947–0.954 g/cm³ have a high degree of stress-crack resistance and are ideal for drums (1).

HPDE drums can safely package a wide variety of chemicals. Uses within the chemical industry include dairy, agricultural, electronic, specialty, photographic, and oil-well applications, as well as those for organic and inorganic chemicals and natural flavorings.

Permeability. HDPE is susceptible to permeation by certain chemicals. Special attention should be paid to inorganic chlorinated hydrocarbons (eg, per- and trichloroethylene), and aromatic hydrocarbons, (eg, benzene, toluene, and xylene). Normally, inorganic chlorinated or aromatic hydrocarbons cannot be packed in HDPE drums owing to permeation, but the permeation rate can be reduced by several methods of surface modification (see Surface modification). Within the HDPE family, high-molecular-weight grades have relatively high resistance to permeation. The risk is also reduced by using relatively thick walls.

Uv resistance. The service life of HDPE containers cannot be accurately predicted because it largely depends on climatic conditions. Different colorings (see Colorants), especially

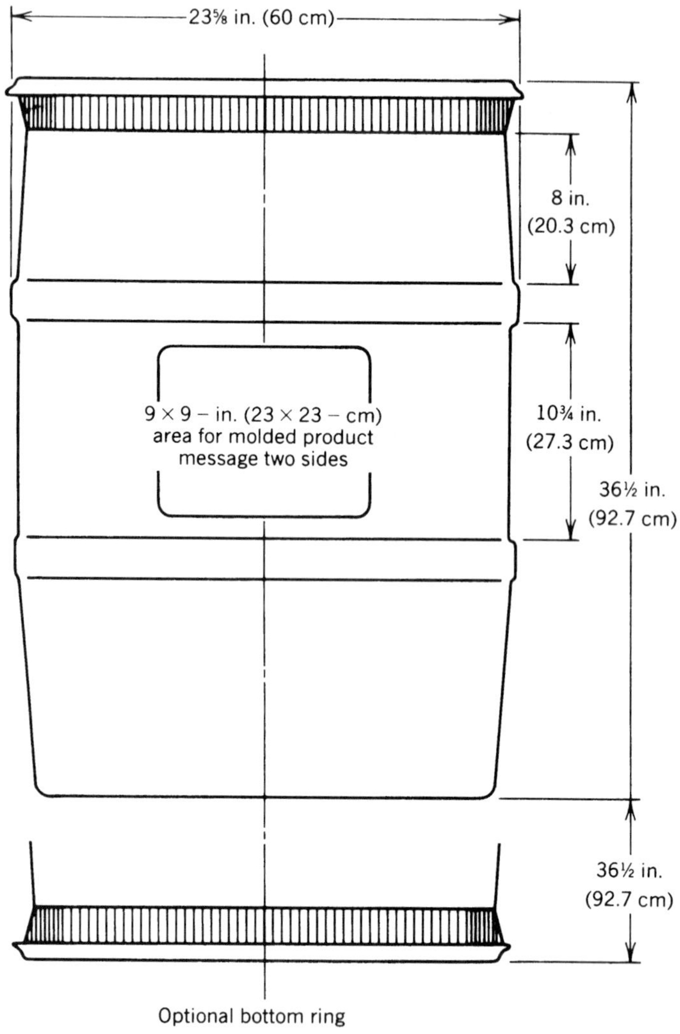

Figure 1. Dimensions of a 55-gal (208-L) drum. (Courtesy of Container Corporation of America.)

black (with carbon black), blue, green, white, and gray, increase resistance to weathering and protect the product from light. Depending on climatic conditions, additional uv stabilizers must be added.

Design

Closed-head drums. Closed-head drums, also called tight-head drums, are available in different designs, with 2-in. (51-mm) and ¾-in. (19-mm) bungs, in 15-, 30-, and 55-gal 57-, 114-, and 208-L) sizes (see Fig. 1). They are used where handling equipment is available to accommodate them. This is important because large self-supporting drums are designed to replace composite steel drums (steel with plastic liners or coatings), for which mechanical handling equipment is available worldwide. In the United States, plastic drums were initially designed with metal handling rings so that the drums could package heavy liquids, eg, up to 1.8 g/cm³ or 825 lb/55 gal, and still be handled by traditional steel-drum handling equipment, particularly the "parrot's beak."

A different approach was taken in Europe and other parts of the world, where the L-ring drum was 30- and 55-gal (114-

Figure 2. Handling closed-head L-ring drums with parrot beak equipment. (Courtesy of Mauser Werke GmbH.)

and 208-L) capacities has become the standard. L-ring drums are produced by blow-molding the drum and integral rings in one step. The use of this configuration required the development of a wider parrot's beak, and at first, it was approved only for liquids with densities less than 1.2 g/cm^3 or 550 lb/55 gal. The L-ring is now approved for heavier liquids (see Fig. 2).

Figure 3. An open-top 55-gal (208-L) HDPE drum. (Courtesy of Mauser Werke GmBH.)

Open-top drums. For general applications, open-top drums are provided in 8-, 15-, 30-, and 55-gal (30-, 51-, 114-, and 208-L) sizes (see Fig. 3). The advantages of the standard open-top drum are easy handling, absolute tightness, good stacking properties, and high radial rigidity. They are used most often for water-based products such as glues, softeners, and liquid soaps, as well as for foodstuffs. The standard open-top drum can be cleaned easily and reused.

Shipment of Hazardous Materials

International regulations for transport of dangerous goods (hazardous materials in the United States) are one of the central issues regarding the use of plastic drums because of heightened worldwide concern for environmental protection and the increasing number and volume of dangerous goods being shipped. HDPE is a safe and durable material for shipping dangerous goods.

United Nations regulations. International regulations concerning packaging tests for hazardous materials are contained in the *United Nations Transport of Dangerous Goods* (2) and in the *International Maritime Dangerous Goods Code* (IMDG Code), which regulates substances shipped overseas. The recommendations do not specify how a package is to be made. They stipulate package-performance tests for each dangerous substance and modification of the test procedures based on the degree of hazard and some physical properties of the substance. They require a certification mark to show that the package has passed the tests. Self-supporting plastic drums can be used for a wide variety of dangerous substances. A typical drum marking under these regulations would be

Ⓤ

1H1/Y.1.8/150/84/D/Mauser 824

Ⓝ

where UN = United Nations; 1H1 = plastic drum, small opening; Y.1.8 = Group II products up to 1.8 g/cm^3; 150 = test pressure, kPa; 84 = year of production; D = FRG; Mauser = producer; and 824 = registration no.

U.S. regulations. The Materials Transportation Bureau (MTB) of the Department of Transportation (DOT) has expanded Specification 34 (CFR, Title 49, Part 178.19) to include 55-gal (208-L) plastic drums (3). Previously, only drums of up to 30 gal were included, and 55-gal (208-L) drums needed special exemptions. The revision eliminates the need for those exemptions. They are now marked DOT-34-55. Some familiar commodities have been written into the regulations with no restrictions, but many of the chemical listings in Part 173.24(d) prescribe channel compatibility and permeation tests. Reference to specific types of polyethylene has been deleted from the most recent revision. Minimum wall thickness is specified, but this does not create significant differences between drum construction in the United States and other countries.

European regulations. European regulations for rail and road transport of hazardous materials in plastic containers up to 250 L (66 gal) include testing procedures based on test data obtained with model liquids such as acetic acid, nitric acid, water, white spirit, surface-active agent, and *n*-butyl acetate. On the basis of the dangerous properties of these materials, all other dangerous substances can be approved and assimilated as long as their individual requirements are taken into account in testing. Plastic containers are included in the packaging regulations as follows: up to 15 gal (57 L) for Class-1 dangerous substances (very dangerous); up to 55 gal (208-L) for Classes 2 (dangerous) and 3 (less dangerous). Apart from the requirements concerning approval of the containers, quality assurance in production plays an important role in the safe transportation of dangerous goods.

Plastic drums will continue to gain importance, particularly if worldwide standardization occurs.

BIBLIOGRAPHY

"Drums, Plastic" in *The Wiley Packaging Encyclopedia Technology,* 1st ed., by Bruno Poetz and Ernest Wurzer, Mauser Werke GmbH, pp. 247–249.

1. D. L. Peters, P. E. Campbell, and B. T. Morgan, "High Molecular Weight High Density Polyethylene Powder for Extrusion Blow Molding of Drums and Other Large Parts," *Proc. 42nd SPE Annual Technical Conference and Exhibition,* 1984, Society of Plastics Engineers, Inc., Brookfield Center, CN, 1984, p. 939.
2. "General Recommendations on Packing," *United Nations Transport of Dangerous Goods—Recommendations of the Committee of Experts on the Transport of Dangerous Goods,* 3rd rev. ed., United Nations, New York, 1984, Chapter 9.
3. *Fed. Regist.* **49**(116), 24684 (June 14, 1984) and **49**(199), 40033 (Oct. 12, 1984).

DRUMS/PAILS, STEEL

Despite centuries of innovation, no one has devised a more useful and adaptable medium-sized container for liquids and semisolids than a cylindrical container. Its shape, based on the circle, provides maximum strength; when fully laden, it can be tipped over and rolled. Early ocean shippers employed heavy, easily breakable clay and ceramics. They knew smelting and metalworking, but evidently could not produce metal containers much larger than pots and pans. From the Middle Ages until recent times, the standard container material was wood, usually oak, formed into metal-bound, stave-constructed barrels and kegs (see Barrels). The wooden barrel had no weight advantage but was far stronger and could be manufactured anywhere of materials widely available at low cost. Design differed little from the early jars, featuring the same bilged sides for maximum strength. In almost every trading nation, the wooden barrel reigned as the universal shipping container for liquids and semisolids until the late 19th century, when the first steel barrels appeared in Europe.

Impetus for the invention and development of the modern steel drum had begun years earlier with the Great Oil Rush of 1859. Oil-drilling technology improved so rapidly that by 1869 U.S. wells were producing 4,800,000 barrels (7.6×10^5 m^3) a year. That introduced a much tougher problem: how to store and ship the oil. The only container available was the wooden barrel. Demand soon outstripped the production capacity of the cooperage firms, and oak became scarce and very expensive. The immediate solution was the development of pipelines, railroad tank cars, and the tank truck. But for smaller packages, oil-industry shippers still had to struggle along for another 40 yr with the time-honored wooden barrel. Kerosene was the most important barreled product. The wooden barrel's chief problem was leakage. Designed to hold 50 gal (0.189 m^3), it commonly lost enough per trip to arrive at its destination containing only about 42 gal (0.159 m^3), a figure that has remained the standard "barrel" measurement of the oil industry.

A few years after their development in Europe, the first steel barrels were produced commercially in the United States by Standard Oil at Bayonne, NJ, in 1902. Constructed from 12–14 gauge terne steel, they were heavy, clumsy, bilged affairs with riveted or soldered side seams, and were anything but leakproof. Extremely rugged, many of them lasted 20–30 years. They were also expensive compared with wooden barrels. Despite the need, steel barrels were slow to catch on; yet technical developments came rapidly. In about 1907, the welded side seam was introduced, which curtailed the leakage problems of riveted barrels and reduced costs. Shortly afterward, the first true 55-gal (208-L) drum was introduced, its characteristic straight sides contrasting markedly with the bilged barrel. Rolling hoops were introduced soon thereafter, both expanded and attached, the latter utilizing an I-bar section. The mechanical flange was invented after 1910. These improvements were far-reaching and gave the new container added advantages. Among them was lighter weight, which reduced shipping costs and reduced the amount of steel required. The new drum design permitted use of 16- and 18-gauge steels instead of the far heavier gauges used previously. These drum developments were followed in 1914 by the first true steel 5-gal (18.9-L) pail featuring the first lug cover.

The use of steel containers grew slowly before 1914, despite their cost, weight, and safety advantages over wooden containers. The advent of World War I marked the beginning of the end for the wooden barrel, and the eventual dominance of the steel drum and pail. Wartime demand also spurred many improvements in manufacturing techniques and equipment. After the war, many innovations appeared, including pouring pails, agitator drums for paints, and new, colorful decorating techniques. Manufacturers began to use steel containers for products other than petroleum products and chemicals. Toward the end of the 1930s, the steel-container industry started gearing up for a second wartime effort. This time, however, the demands were far more stringent, requiring fuel containers for a highly mechanized war on more than one front. Innovation took a back seat to production considerations as war machines on the ground and in the air consumed vast amounts of fuel and chemicals. The 55-gal (208-L) 18- and 16-gauge drums were indispensable to the fuel supply of island bases and assaults in the Pacific, frontline mechanized operations in Europe, and to air and ground operations in East Asia. Apart from its ruggedness, the fact that a cylindrical drum could be rolled by one man was an important feature.

Although a downturn in steel drum and pail production occurred at the end of World War II, it was of short duration. Resumption of business created a demand from industry, agriculture, and consumers. The acceleration in chemical and pharmaceutical product development and output provided new markets for steel drums and pails. Demand for paints, lacquers and varnishes, adhesives, inks, foodstuffs, and other products made the steel shipping container industry one of the largest users of cold-rolled sheet steel (approximately 1 million tons in 1993). Despite the introduction of competitive containers made of other materials and of intermediate-bulk containers, U.S. production increased from 2.4 million drums in 1922 to 32.35 million drums in 1993.

Drums range in size from 13 to 110 gal (49–416 L), but the 55-gal drums account for 80% of annual production. Over 75% of all new drums are used for liquids and the rest, for viscous and dry products. About 70% of the market is accounted for by five broad product categories: chemicals (35%); petroleum products, including lubricants (15%); paints, coatings, and solvents (10%); food and pharmaceutical products (5%); and janitorial supplies, cleaning compounds, and soaps (5%).

As a result of increased environmental awareness, drum manufacturers, drum users, reconditioners, pail and drum recyclers, and steelmills have developed programs to collect, recondition, and/or recycle used steel drums and pails. In addition the recycled content of steel has surpassed an average of 25% per container. In fact, more steel is recycled than all other packaging materials combined (53 million tons of steel scrap in 1993) (1). Thus, choosing steel packaging conserves energy and natural resources and reduces waste. Each year over 40 million drums are reconditioned, thus prolonging the useful life of steel drums.

The industry's growing involvement in drum and pail reclamation is an important factor in purchasing agents' or packaging engineers' decisions to select the appropriate container for their company's products.

Drum and Pail Construction

Steel drums [13–110 gal (49.2–416 L)] and pails [1–12 gal (3.8–45.4 L)] are generically fabricated from cold-rolled sheet steel in a range of thicknesses from 0.0946 in. (2.4 mm) (formerly 12-gauge) to 0.0115 in. (0.292 mm) (formerly 29-gauge). They consist of a cylindrical body with a welded side seam and top and bottom heads. The thickness of steel used in pails and small drums usually range from 0.0115 in. (0.3 mm) to 0.0269 in. (0.7 mm), while thicker steel, ie, from 0.030 in. (0.8 mm) to 0.0533 in. (1.4 mm), are used for larger, reconditionable drums (see Table 1).

Most drums are made of commercial-grade cold-rolled sheet steel, but stainless steel, nickel, and other alloys are used for special applications. Only about 45% of all new drums are lined with interior protective coatings, but the percentage is much higher, ie, 80%, for drums used for chemicals. Over the years, the cost and weight of steel drums have been reduced owing to technological advances, such as the introduction of the triple-seam chime in the early 1980s. Improvements in cold-rolled steel chemistry, surface quality, and gauge control have also contributed to a reduction in cost and weight. Until the early 1960s, most tight-head drums were made of steel 0.0428 in. (1.1 mm) thick (formerly 18-gauge). There has since been a shift to a lighter-gauge drum with a steel thickness of 0.043 inch (1.1 mm) in the top and bottom heads and 0.030 in. (0.8 mm) in the body (formerly known as the 20/18 drum, now marked as 1.1/.8/1.1). Currently, 55-gal (208 L) drums of 0.0378-in. (1.0-mm) steel thickness are being manufactured to transport hazardous materials, and 55-gal drums of 0.030-in. (0.8-mm) thickness or thinner are used for nonhazardous materials. Other popular sizes are 30 and 16 gals. In addition, the 85-gal drum, knwon as the "salvage drum," is used to transport leaking or damaged packagings and debris from hazardous materials accidents. Each size can fit the non-bulk packaging needs of the drum user. [For a complete list of standard drum sizes and dimensions, consult *ANSI NH2-1991* (2).]

Table 1. Sheet Steel Thicknesses vs Gauge No.[a]

Gauge no.	Minimum thickness, in.	Minimum thickness, mm	Nominal thickness marking[b], mm
12	0.0946	2.40	2.4
16	0.0533	1.35	1.4
18	0.0428	1.09	1.1
19	0.0378	0.960	1.0
20	0.0324	0.823	0.8
22	0.0269	0.683	0.7
24	0.0209	0.531	0.5
26	0.0159	0.404	0.4
28	0.0129	0.328	0.3
29	0.0115	0.292	0.3

[a] Sheet steel thickness is measured at any point no less than ⅜ in. (9.53 mm) from the edge. New DOT regulations that went into effect for new packaging manufacturers on Oct. 1, 1994 no longer refer to steel thicknesses in gauges but rather in millimeters.

[b] Nominal thickness markings are those applied as part of the durable and permanent UN marks on the drum and pail (4). Consult ISO Standard 3574 for nominal thickness tolerances.

Styles

Two basic styles of drums exist: the tighthead (or nonremovable head), with permanently attached top and bottom heads, and the open head (or removable head), in which the removable top head or cover is secured by using a separate closing ring with either a bolted or lever-locking closure (see Fig. 1).

Expanded rolling hoops, ie, swedges, in the drum body stiffen the cylinder and provide a low friction surface for rolling filled containers.

Tight-head drums (and pails) have their top and bottom heads mechanically rolled (seamed) in multiple layers to the body using a nonhardening seaming compound to form a joint (chime). Two openings, one 2 in. (51 mm) and the other ¾ in. (19 mm), for filling and venting are usually provided in the top head, although side openings and other opening combinations and sizes are sometimes used. The openings are fitted with mechanically inserted threaded flanges conforming with American National Pipe thread standards. Threaded plugs for insertion in the flanges are made of steel or plastic and have resilient gaskets where appropriate. On full-removable-head drums, the top of the body sidewall is rolled outward to form a follow curl (false wire) to which the top head or cover is attached using a gasket of resilient material and a separate closing ring.

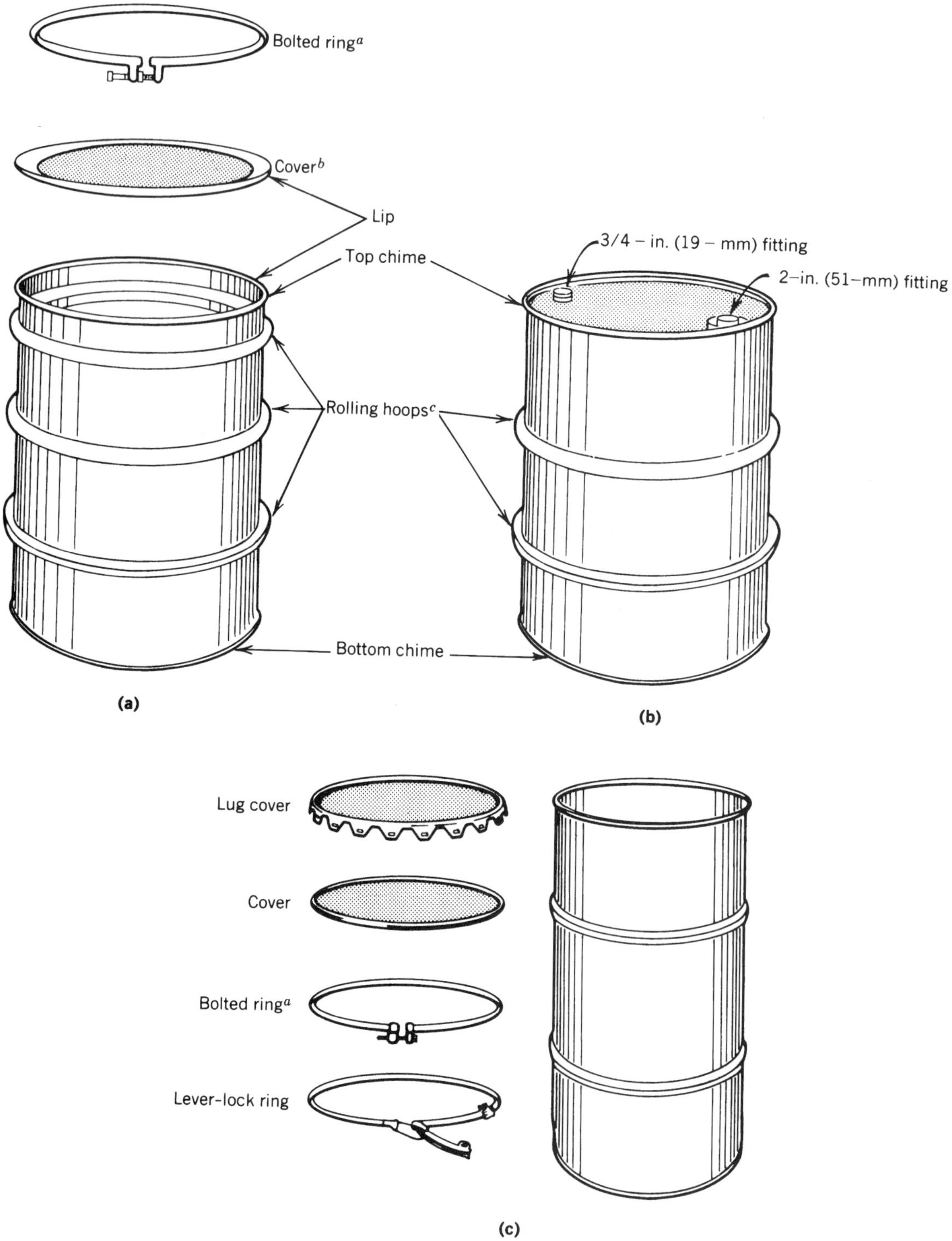

Figure 1. Steel drum designs: (a) open-head 55-gal (208-L) drum; (b) tight-head 55-gal (208-L) drum; (c) open-head 16-gal (606-L) drum. [a] Lever-lock ring may be used; [b]lip of cover is turned down to fit over lip of drum; [c]two or three rolling hoops may be used depending on size and material packaged. Hoops are equally spaced horizontally. Actual drum dimensions may vary.

Steel pails are generally of the same configuration and style as the large-capacity steel drums, but are usually of thinner metal and may have only one expanded body hoop. A bail handle or carrying grip is often provided for handling purposes. A common closure for open-head pails is a lug cover that is crimped in place around the top curl and is removed by lifting the lugs (see Fig. 1c).

Protection and Linings

Most steel pails and drums are fabricated from steel treated to resist rusting owing to moisture in the air. Steel is a nonpermeable, biodegradable material that is compatible with most chemicals and petroleum-based products

Coatings are applied to the inside of drums (linings) and the outside (paint) to provide additional protection and decoration. State and federal environmental regulations control the amount of volatile organic compounds (VOCs) emitted during the application of the linings and the paint. Because conventional linings and paint contain some degree of organic solvent or heavy-metal pigments, these are being replaced by water-based or high-solids linings and nontoxic paint. In many instances steel drum and pail manufacturers use afterburners to incinerate all vapors emitted in the paint booths, thereby reducing VOCs. Companies also recapture spray paint for remixing and reapplication on containers to conserve paint and protect the environment.

Interior. Linings are used for protection against acids, alkalies, and some organic chemicals. Phenolics provide protection against certain acids, and epoxies offer protection against alkalies. Linings consisting of varying percentages of epoxy and phenolic materials are most commonly used today. In some instances, the needed protection is supplied by a flexible or semirigid polyethylene liner insert.

Exterior. New steel containers can be painted, lithographed, or silk-screened to provide an attractively decorated and durable finish. Enamels are sprayed or roller-coated, baked, and oven-cured to give a scuff-resistant exterior coating. Black is generally the standard color, but other colors are available as well. Product and manufacturer information for merchandising or to satisfy transportation needs is applied by lithography, silk-screening or stenciling.

Standardization

National standards for steel pails and drums have been developed in the United States within the American National Standards Institute (ANSI) by a Committee on Steel Pails and Drums sponsored by the Steel Shipping Container Institute.

These dimensional standards have received international acceptance and have provided many advantages in the areas of filling, handling, storage, and shipping. There are presently standards for the 55-, 30-, and 16-gal open-head and tight-head drums, as well as for the 20-L and 5-gal tight-head, nesting-lug cover, and straight-sided lug cover pails. The ANSI standard also includes a thorough glossary of terms related to packaging. The ANSI Committee is presently in the process of revising the current edition of the standard, *ANSI MH2-1991* (2), to incorporate the construction requirements set forth under the Department of Transportation's *Performance-Oriented Packaging Standards,* published in 1990 (see under "Regulations" below).

Regulations

All-steel pails and drums used in the United States for the transport of hazardous materials must comply with the Department of Transportation's Hazardous Materials Regulations (DOT) (4). For nonhazardous products, these containers usually comply with the minimum requirements of the specifications set forth in the railroads' Uniform Classification Committee (UCC) (5) and the highway carriers' National Classification Committee (NCC) (6). Noncompliance with these latter two organizations' specifications, known respectively as the *Uniform Freight Classification* (UFC) (5) and the *National Motor Freight Classification* (NMFC) (6), may lead to higher insurance costs for the packager and/or shipper.

Significant regulatory changes have taken place at the DOT. Hazardous materials, governed by DOT, include flammable liquids, gases, and solids; oxidizing agents and organic peroxides; poisons; explosives; radioactive materials; corrosive materials; and certain marine pollutants and hazardous wastes. Decades-old DOT design specifications (such as the DOT-17E, -17H, and -17C containers) have been replaced by new *Performance-Oriented Packaging Standards* (also known as HM-181 for its DOT docket number), based on the United Nations' *Recommendations on the Transport of Dangerous Goods.* The reasons for this shift were harmonization of packaging requirements with international regulations and development of package safety criteria based on the performance of the container rather than on its design. This has entailed a complete restructuring of the way packagings are specified. The SSCI's manual *Understanding HM-181 for Steel Drums* (3) summarizes the DOT regulations found in Title 49 of the *Code of Federal Regulations* (4) as they pertain to steel drums and pails.

Packagers must now provide their drum and pail suppliers with the following information: Packing Group, product vapor pressure (if liquid), net mass (if solid), and specific gravity (if liquid). The SSCI *Buyer's Guide* (7) provides a checklist for the packager. The steel drum and pail manufacturer marks the container, after having performed the following tests: drop, leakproofness, stacking, hydrostatic pressure (if liquid), and vibration. These tests are meant to minimize the risk of leakage that might result from normal handling, shipping, storage, and accidents.

A sample mark for an open-head steel drum of 1 millimeter thickness manufactured in 1994 by manufacturer M1234 and authorized to carry a Packing Group II or III solid with a gross mass of 300 kg (or less) is

UN 1A2/Y300/S/94/USA/M1234 1.0

where UN = United Nations, 1 = drum, A = steel, 2 = open head, Y = Packing Group II or III, 300 = maximum gross mass in kg (net mass of solid plus mass of drum), S = solid, 94 = year of manufacture, USA = country of manufacture, M1234 = manufacturer's number of symbol, and 1.0 = thickness of millimeters.

A tighthead drum with a nominal 1.1-mm-thick head and

bottom and 0.8-mm body manufactured in 1994 and authorized to carry a Packing Group II or III liquid with a specific gravity of 1.8 or less and with a product vapor pressure of 230 kPa (or less) at 55°C is marked

UN 1A1/Y1.8/230/94/USA/M1234 1.1/.8/1.1

where UN = United Nations, 1 = drum, A = steel, 1 = tight-head, Y = Packing Group II or III, 1.8 = specific gravity (relative density of material to water), 230 = maximum hydrostatic pressure tested in kPa, 94 = year of manufacture, USA = country of manufacture, M1234 = manufacturer's number of symbol, and 1.1/.8/1.1 = thickness of top, body and bottom in millimeters.

Steel drums of 1.0 mm thickness or more (also 1.1/.8/1.1) are permitted to be reconditioned and reused to transport hazardous materials, thereby extending the life of the container. Title 49 of the *Code of Federal Regulations* (CRF), Parts 100–199, give the full details of DOT hazardous materials regulations. These regulations govern shipment by land, sea, and air. It is the responsibility of the shipper to ensure that they are using containers tested and marked in accordance with the minimum requirements for the material to be transported.

International. As stated above, the DOT's new POP standards are based on the *Recommendations of the United Nations Committee of Experts on the Transport of Dangerous Goods,* acting under the direction of the United Nations Economic and Social Council (8). Chapter 9 contains "General Recommendations on Packing," which details packaging requirements, types of packagings, and marking and testing requirements.

Members of the UN Committee are committed to adopt these recommendations into their respective nation's transportation regulations as closely as possible to the original, although differences are permitted. The UN recommendations are not regulations per se, but guidelines for regulations. Yet they are usually part of the law in the country of export or import if not both. For example, Canada has adopted the UN recommendations into its Transportation of Dangerous Goods Act and Regulations and Related Performance Packaging Standards, while Mexico has nearly completed the process of writing the POP standards required under its hazardous materials transportation law of 1993.

Two international codes do have the force of law for member states: the International Maritime Organization's (IMO) *International Maritime Dangerous Goods Code* (IMDG Code) governing hazardous materials transportation by water and the International Civil Aviation Organization's (ICAO) *Technical Instructions for the Safe Transport of Dangerous Goods by Air.* These two codes have adopted the UN recommendations.

As of January 1, 1995, European road and rail regulations (ADR and RID) conform to the most recent revisions of the UN recommendations.

Steel Pails

About 73 million new steel pails are currently produced in the United States each year; sizes range from 1 to 12 gal (3.8–45.4 L).

About 80% is accounted for by the popular 5-gal (18.9-L) pail. They are made in four basic configurations: full open-head, straight side, lug cover; full open-head, nesting, lug cover; tight-head, straight side; and tight-head dome top (see Fig. 2). They are constructed of 0.0115-in. (0.3-mm) or thicker steel. Pail heights vary by volume, but diameters are 6 and 8 in. (15.2 and 20.3 cm, respectively) for 1–2½-gal (3.8–9.5-L) sizes, 11.4 in. (28.6 cm) for 3–7 gal (11.4–26.5 L), and 13¹⁵⁄₁₆ in. (35.4 cm) for larger capacities. The two open-head designs account for about 75% of the total pail production.

Both types of open-head pails have a liquid-tight, welded side seam on the pail body and a bottom affixed by double seaming (see Can seamers). Two side "ears" are welded or riveted to the body, and a galvanized wire bail handle is attached. Handles are furnished with or without a grip, which can be wood or contoured plastic. The straight-sided pail normally has one strengthening body head, ie, swedge, to add rigidity to the top of the cylinder. The tapered type usually has a second bead that, in nesting, rests on the top curl of the pail below and limits nesting depth.

The tight-head pail, accounting for some 25% of sales, is often used for the shipment of low viscosity or free-flowing liquids. It embodies a welded side seam, double-seamed top and bottom ends, and a carry handle, usually a D-ring, of galvanized wire spot-welded to the head. This container can be fitted with a variety of pouring and venting apparatus. An offshoot of the tight-head pail is a domed-top or utility pail, especially popular in 2½- and 5-gal (9.5- and 18.9-L, respectively) sizes for petroleum products.

Pails are used for liquids, viscous products, powders, and solids. Pail markets include paint and printing inks; chemicals; adhesives, cements, and roofing materials; petroleum products; janitorial supplies, eg, cleaners, and waxes; abrasives; cosmetics; fasteners and stamping; foods; insecticides; marine supplies; pharmaceuticals; powdered metals; and scores of other products and materials.

Because of their ability to withstand high temperatures, pails are the container of choice for the transportation and indoor storage of flammable and combustible liquids. The range of classes of flammable and combustible liquids allowed for warehousing and storage is greater than the range for like-sized containers made of plastic. See, for example, NFPA Code 30 on Flammable and Combustible Liquids (9).

In addition to varying capacities, steel thicknesses, and container construction a host of options, fittings, and accessories are available to design a pail to a buyer's exact requirements.

Open-head pails can take two types of covers. The lug cover, usually incorporating 16 wide lugs around its circumference, can be applied at production-line speeds by automatic crimping equipment, although hand-operated and semiautomatic crimping tools are also available. The lug cover is opened with standard hand tools. The second cover, a ring seal, is best for resealing purposes. It consists of a formed disk that sits on the top curl of the container and is clamped to it either by a separate ring band or with rings that lock by lever action or bolt-tightening.

Another option is a combination of steel thicknesses. Cover and ends can be made of steel of different thickness than the body for different requirements of strength and economy. Lids and bottom ends can be strengthened by using embossed circumferential beads to provide increased rigidity.

On both tight-head and open-head pails, a wide range of

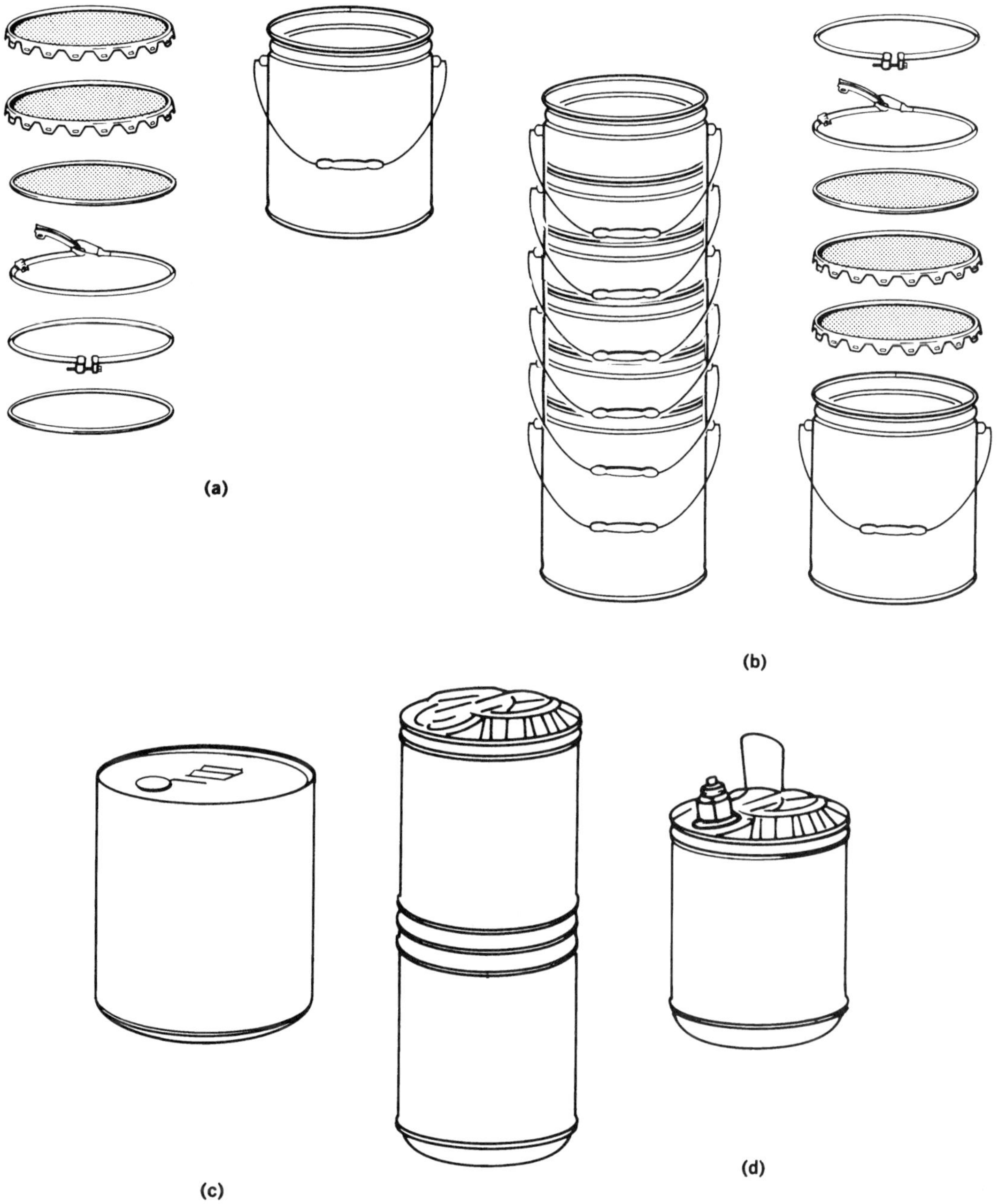

Figure 2. Designs for 5-gal (18.9-L) steel pails: (**a**) open-head, straight side, lug cover; (**b**) open-head, nesting, lug cover; (**c**) tight-head, straight side; (**d**) tight-head dome top. (Courtesy of SSCI.)

opening sizes, pouring spouts, and cap closures is available. To cut costs, covers can have a simple, threaded pouring nozzle topped by a screw cap. Even simpler are pails furnished with just a dust cover over the pour opening to keep the interior clean, with the user clinching on or pressing in a pouring fitting after filling. Various metal and plastic pouring devices are offered, mostly of the pull-up style; these are covered by a cap during shipment. Some pour fittings incorporate vent openings that eliminate the need for a separate vent opening on the cover. Tamperproof seals, consisting of a steel cap clinched directly onto the pail fitting, are often used.

BIBLIOGRAPHY

1. Steel Shipping Container Institute, *Steel Shipping Containers: The Choice for the Environment,* Washington, DC, 1994.
2. *ANSI MH2-1991,* American National Standards Institute, New York, 1993.

3. Steel Shipping Container Institute, *Understanding HM-181 for Steel Drums,* Washington, DC, 1993.
4. U.S. Department of Transportation, *Code of Federal Regulations,* Title 49, Parts 100–199 (Oct. 1993).
5. Uniform Classification Committee, *Uniform Freight Classification,* National Railroad Freight Committee, Atlanta.
6. National Classification Committee, *National Motor Freight Classification,* National Motor Freight Traffic Association, Alexandria, VA.
7. Steel Shipping Container Institute, *A Buyer's Guide to New Steel Drums & Pails,* Washington, DC, 1994.
8. United Nations Committee of Experts on the Transport of Dangerous Goods, *Recommendations on the Transport of Dangerous Goods,* 8th ed., United Nations, New York, 1993.
9. *ANSI/NFPA 30,* National Fire Protection Association, Quincy, MA, 1993.

RICHARD B. NORMENT
Steel Shipping Container Institute
Washington, DC

DUAL OVENABLE PACKAGING.

See MICROWAVEABLE AND DUAL OVENABLE PACKAGING.

E

ECONOMICS OF PACKAGING

Packaging affects every part of our market economy. The most obvious component is the packaging supply industry. It may be less obvious that packaging affects the cost and demand of every product and factor of production, from food to building materials, to auto parts. All products are packaged in some way and may have been repacked several times before use. Every manufacturing operation purchases packaging. Disposal costs accrue to all customers (households and industrial customers). Most service businesses buy or use packages and incur costs related to them. Retailing and distribution industry economics are also affected by packaging.

This article outlines macroeconomic aspects of packaging in the United States, the behavior of U.S. supply industries, and the microeconomic effect of this behavior on the supply and demand for products in a market economy. It concludes with an exploration of costs from the standpoint of a firm that buys, fills, and distributes packaged products, including the costs associated with disposal.

Packaging Macroeconomics

Packaging represents a large segment of the U.S. economy. Included in the packaging industry are the manufacturers of packaging materials, the manufacturers of packaging machines, and the packaging operations in every manufacturing firm.

Industry shipments of packaging materials in 1989, for example, were $67.4 billion, and sales of packaging machines were $2.498 billion. Combined, these represent 2.5% of all U.S. Manufacturing Shipments (see Table 1) (2). This estimate does not include the nation's packing operations costs, since the data are not available. The packaging industry performance parallels the national economy's performance, since almost all products are packaged.

The amount of packaging materials used per product varies enormously, and depends on the product characteristics and marketing requirements. Consumer products use 78% of the packaging produced and industrial products use 22%. Of consumer products, the largest segment is the food and beverage industry, purchasing over $36 billion/year (1). For the top 100 packaged brands, producers spent over $12 billion for packaging in 1993, and 20% of the nation's total packaging material expenditures (3).

In a developed market economy, packaging plays a major role in differentiating products. Per capita packaging expense is relatively high, about $280/year, in order to provide good protection and communication in long and complex marketing channels.

Packaging plays a significant role in a nation's economic development. Generally, the expenditure on packaging is much lower in less developed countries. Most materials and graphics are less sophisticated, the process to make and fill packages is more labor-intensive, and machinery is slower and less efficient than in industrialized countries. Food is not shipped long distances and is more likely to be purchased fresh rather than processed and packaged. There is less variety and much greater losses due to spoilage. But the packaging that exists is generally economical, material recovery systems are efficient, and there are many creative packaging applications using indigenous materials and appropriate technology. Packaging varies enormously throughout the developing world.

Improvements in packaging can facilitate economic growth by increasing the efficiency of marketing food and other products, and by adding value to exports. Packaging can reduce the cost of food and increase its supply by preventing losses, which can be as much as 50% for some foods in developing countries. The best way to improve the value of exports is to package products rather than shipping in bulk. There is a growing demand for more sophisticated packaging materials and methods to be used for export packaging. There is increasing realization by government authorities that investments in packaging technology can yield economic benefits, and the United Nations has implemented programs to develop human resources in packaging and make information more easily accessible (4).

Packaging Supply Industries

Packing is produced by a diverse group of manufacturing industries using different raw materials and manufacturing processes. In the United States, the largest segment (by sales), 39%, is paperboard-based. Metal represents 24%; plastics, 19%; and paper and glass, each 7%. Table 2 shows sales, by packaging material type, for a ten-year span, from 1984 to 1994 estimates.

Packaging supply is a mature industry in the United States. Its growth rate is generally equal to all manufactur-

Table 1. Gross Domestic Product 1984–1989 versus Packaging Expenditures (in millions of current U.S. dollars)

	1984	1985	1987	1989
U.S. value of all manufacturing shipments[a]	$2,253,429	$2,280,184	$2,475,901	$2,793,000
U.S. value of packaging shipments[b]	$53,798	$54,285	$60,569	$67,448
Percentage: packaging for all shipments	2.39%	2.38%	2.44%	2.41%
U.S. packaging machinery shipments[c]	$1,712	$1,835	$2,040	$2,498
Percentage (packaging material + machinery) for all shipments	2.46%	2.46%	2.53%	2.5%

[a] U.S. Department of Commerce Bureau of the Census, *1987 Census of Manufactures,* Table 1; and *Statistical Abstract of the United States 1992,* No. 1243.
[b] Rauch Associates, *The Rauch Guide to the U.S. Packaging Industry,* Rauch Associates, Bridgewater, NJ, 1986, 1990.
[c] Packaging Machinery Manufacturers Institute, Arlington, VA, 1994.

ing. However, unlike other mature industries, it is not concentrated, and packaging producers are highly competitive. The largest 50 packaging suppliers account for only 58% of total packaging shipments, with many companies supplying only one type of packaging material. The packaging industry's profitability varies, but is generally below the average for other manufacturing industries. Commodity markets for metals, paperboard, and plastic resins used in packaging are characterized by low profit margins and high capital investments. Packaging conversion is, likewise, a low-profit sector. In 1989 the average margin on sales was 5.6%, below the 6.4% average for all manufacturing (1). Intermaterial compe-

Table 2. Shipments of the U.S. Packaging Industry 1984, 1985, 1987, 1989, Estimated 1994 (in millions of current U.S. dollars)

	1984	1985	1987	1989	Est. 1994
		Paperboard and Molded Pulp			
Corrugated containers	11,643	12,660	15,570	16,534	21,000
Folding cartons	3,850	4,452	4,794	6,879	8,000
Sanitary food containers					
Cartons and trays	865	590	559	635	740
Milk and beverage cartons	545	117	149	222	255
Liquidtight containers	57	60	61	70	85
Fiber and composite packaging					
Cans	835	587	513	565	535
Drums	320	333	312	350	365
Rigid boxes	567	481	518	530	585
Molded pulp products	160	269	294	339	395
	18,842	19,540	22,770	26,124	31,960
		Metal			
Cans	11,200	11,644	10,318	11,243	13,030
Shipping containers					
Drums	588	600	1,004	1,085	1,045
Pails	176	182	310	133	665
Miscellaneous	215	240	775	890	1,150
Aerosols	670	710	707	758	830
Crown and closures	831	801	350	330	355
Strapping	360	360	196	220	235
Foil containers	145	155	1,117	1,155	1,300
Flexible packaging					
Converted flexible packaging	65	70			
Wrappers	40	50	82	89	75
Collapsible tubes	95	84	20	20	20
Pallets	22	22			
	14,638	15,002	14,879	15,923	18,705
		Plastics			
Containers					
Blow-molded bottles	2,390	2,683	3,040	3,687	4,950
Miscellaneous	1,120	1,807	1,867	2,042	2,485
Shipping containers	375	415	620	725	950
Squeeze tubes	40	75	125	250	305
Flexible packaging					
Specialty bags	1,680	755	883	970	1,180
Converted wraps	2,250	1,930	2,045	2,212	2,750
Wrappers	395	275	300	350	420
Shipping sacs	90	120	150	210	255
Closures	522	585	1,108	1,306	1,480
Cushioning	375	395	997	1,033	1,140
Other components	200	230	—	—	350
Strapping	195	210	230	260	50
Pallets	40	50	50	50	
	9,672	9,320	11,415	13,095	16,315

Table 2. ***(Continued)***

	1984	1985	1987	1989	Est. 1994
		Paper			
Flexible packaging					
Converted wraps					
All-paper	485	500	562	620	775
Paper and foil	530	275	308	340	410
Wrappers	205	240	300	324	375
Specialty bags	500	375	346	374	440
Labels and tags	1,375	1,375	1,350	1,360	1,450
Heavy-duty bags	900	885	1,187	1,498	1,820
Tapes	395	170	225	284	310
Wadding	42	45	55	60	75
	4,432	3,865	4,333	4,860	5,655
		Glass			
Containers	3,850	4,100	4,520	4,592	4,950
		Wood			
Pallets and skids	1,150	1,210	1,434	1,567	1,770
Containers					
Nailed boxes and crates	340	330	323	343	375
Wire bound boxes and crates	125	127	110	118	130
Veneer and plywood containers	67	68	60	61	60
Cooperage	90	93	46	52	50
Excelsior	18	20	20	20	20
	1,790	1,848	1,993	2,161	2,405
		Textile			
Bags	450	486	539	580	700
Twine	95	100	90	80	80
Flock	29	30	30	33	30
	574	610	659	693	810
	53,798	54,285	60,569	67,448	80,800

Source: Rauch Associates, *The Rauch Guide to the U.S. Packaging Industry,* Rauch Associates, Bridgewater, NJ, 1986 and 1990. Used by permission.

tition is often as significant as competition between the suppliers of a single material, since more than one packaging form can often be used for the same product.

Paperboard packaging. Paperboard-based packaging is the largest segment of the packaging supply industry. It is also the most vertically integrated; companies producing 50% or more of their own paperboard account for 80% of the shipments of corrugated containers and folding boxboard cartons. Most of the large companies making paper-based flexible packaging also make their own paper (1) (see also Paperboard).

Corrugated fiberboard accounts for almost 75% of the paperboard tonnage used in packaging. Since most products (in the United States) are shipped in corrugated fiberboard boxes, the price is sensitive to the overall economy's performance. Periods of growth directly increase the demand for boxes by every industry, causing suppliers to reach full production capacity, which leads to higher prices.

Paperboard packaging prices are also affected by the cost of wood (or recycled raw materials) and energy. Wood has always been relatively plentiful in the United States, and corrugated fiberboard boxes are one of the most highly recycled packaging materials. Energy is a large component of the production cost, and the 1974 and 1979 "energy crisis" resulted in large price increases.

Metal packaging. Metal containers account for about 24% of total packaging shipments. Most (70%) are cans (tinplated steel and aluminum). For many uses, steel and aluminum cans are interchangeable, and usage depends on relative prices. Aluminum is more expensive, per pound, than steel. But aluminum cans are lighterweight, and most innovations have emphasized further weight reductions. To reduce the cost of raw materials, the aluminum industry has facilitated recycling. Since 1981, aluminum cans have outnumbered steel cans; aluminum is more popular for beverages, but steel still predominates for food (see Cans, aluminum; Cans, steel).

The largest use for metal cans (71%) is for carbonated beverages. The metal can is such a vital input to beer production that several breweries (and food canners) have their own captive can producing facilities. Food manufacturers use 26% of the cans produced. Other metal packaging includes steel drums and pails, aluminum foil in flexible packaging, bottle caps, and collapsible tubes.

Packaging is a small portion (only 5%) of total steel production, but is over 25% of total aluminum production. The metal packaging industry is one of the most concentrated of packaging industries, with the top 10 can producers accounting for most of the volume (1). Metal can producers generally do not own sources of metals, except for Reynolds Metals, which produces aluminum.

Plastics packaging. Plastics are the fastest-growing segment in packaging. Plastics have won market shares from all other packaging materials, converting glass bottle users to plastic bottles, paper bag users to plastic bags, fiberboard boxes to plastic wraps, and steel drums to plastic drums. Plastic packages generally use less material, are less costly to fabricate, and weigh less, thus reducing transport costs.

The largest-volume plastic used for packaging is low-density and linear low-density polyethylene used for film (LDPE and LLDPE = 38% of plastics packaging), followed by high-density polyethylene (HDPE = 31%), polypropylene (PP = 11%), polyethylene terephthalate (PET = 7%), polyvinyl chloride (PVC = 6%), and polystyrene (PS = 5%). Packaging is an important usage for plastics comprising half of all shipments of LDPE, LLDPE, HDPE, and PET, but less than half of all PP, PS and PVC. Bottles are the highest-volume plastic containers, with 35.6 billion units of bottles, including 4.5 billion in captive production by manufacturers of detergents, bleach, and beverages (1).

As the plastics industry has matured, it has developed more specialty applications. Lamination, coextrusion, and barrier coatings have improved barrier and strength properties, and thereby increased the market for plastic packaging. Furthermore, most composite packages (and even packages not normally thought of as composites, like coated cans and cartons) rely on an essential layer of plastic that adds strength, sealability, or barrier to other materials in the structure.

The production of plastic resins is concentrated in oil producing firms like Dow Chemical, ARCO, and Dupont. But the plastics converting industry is much more diverse, with few barriers to entry. Many oil companies have sold their container business to independent convertors. Economical production scale is much smaller than that for converting paper, metal or glass. Market leadership for a container type is often governed by proprietary technology that provides a special value. Since plastic forming is relatively easier than forming other packaging materials, there is more captive production of standard containers by filling companies, especially for plastic bottles, thermoforms, and flexible packages that are produced in a form/fill/seal operation. (see Blow Molding and the articles on the individual plastics).

Glass packaging. Glass packaging is the smallest material segment in the U.S. packaging industry. Glass has suffered from declining sales as glass bottles and jars have been replaced by plastic, metal, and composite packages for many products. The 1980s was a period of consolidation and mergers, in an effort to make the industry more competitive with other materials. Only the lowest-cost producers have survived.

The raw materials for glass are relatively inexpensive, but glass production requires a high amount of energy and has high labor costs. Very few bottle producers own the sources of raw materials. The increase in recycling glass has reduced energy and material costs. Innovations in glassmaking have been focused on weight reduction, including improving the uniformity of glass distribution and plastic coatings, in order to reduce material and shipping costs. In addition, the efficiency of glassmaking has improved considerably (see also Glass container manufacturing).

The largest uses for glass containers are for food, beer, and soft drinks, which account for 86% of total shipments; 4% are captive production by beer companies (1). However, aluminum cans have displaced glass bottles for beer, and PET bottles have displaced glass for soft drinks and liquor. Glass applications are growing only for food and drinks where its "prestige" image and high clarity are desired.

Packaging Affects Consumer Demand

Packaging affects the demand for products. It can increase the quantity demanded of a product by reducing its cost. For example, as developments in packaging technology have reduced the cost of protecting processed food, the market has grown to include more low-income consumers.

Packaging can also increase the absolute demand for a product by providing features that attract a new category of consumers. For example, the market for paint was increased by the introduction of "spray paint" to include consumers who had never painted before.

Packaging is used as a tool to differentiate products for market segmentation strategies. Buyers have unique needs and can be segmented into broad classes. For example, consumers with physical disabilities need easy-to-open packages. Some of the primary packaging benefits that are used to differentiate products are:

- The amount in the package, to match various consumption rates
- Low-cost minimal packaging for frugal shoppers
- Convenient package opening, reclosing, and dispensing features for a variety of consumer use behaviors
- Longer shelf life for consumers who want to store the product for future use, including packaging to reduce oxidation of fats and decay of fresh fruits and vegetables
- Special packaging for special occasions (gifts, holidays, etc.)
- Appeal to a consumer's psychographic image
- Fit with a lifestyle
- Package recyclability for environmentally conscious consumers

Of these benefits, the most universal demand is for low-cost improvements in package opening, reclosing, and dispensing features (5).

Packaging affects all stages of the buyer decision process—from problem recognition to postpurchase behavior (6)—especially for routine purchases of low-involvement goods. Seeing a package can stimulate recognition of a problem that could be solved by the product. Package graphics can facilitate information search, evaluation of alternatives, and the purchase decision, by showing the attributes that differentiate the product inside. After the product is purchased, packaging shows how to use the product, and encourages a repeat purchase. Packaging can also play a role in purchases that require more extensive problem solving without the help of a salesperson.

Packaging Microeconomics

The cost of packaging per product, expressed as a percentage of selling price, varies widely, from 1 to 40%. For some products, like bottled drinks, perfumes, and aerosols, the package may cost more than the products' ingredients. These products depend on the package for their very existence. For other products such as durable goods, where the package is simply a means to facilitate distribution, the relative cost of the package is low.

Packaging costs depend on the materials and production methods employed. The choice of materials generally depends on the protection required and the marketing requirements. Protection and preservation needs depend on the nature of the product and its logistical system; fragile products and packages that will be roughly handled during distribution require more protection than do rugged products; and for perishable products to be stored, more preservation is required.

There is a growing recognition that package system development should begin early in the product development process. Packaging-related cost tradeoffs are much easier to explore before the product is fully developed. For example, product modification to reduce fragility may be more economical than improving package protection.

Material costs. Packaging materials include the primary package and its closure, the shipping container, and unitization materials. The cost of packaging materials depends on the cost of the raw materials (plastic, paper, wood, glass, and/or metal) plus the cost of conversion into packages to be filled. Most of the raw materials are competitively priced commodities. Prices for raw materials can be found in publications such as *Plastics News* and *Official Board Markets*.

The conversion cost varies, depending on the material. The percentage of the sales price represented by raw-material cost fluctuates, depending on raw-material pricing dynamics. In 1992 U.S. manufacturers of paperboard packaging spent 59% of the sales price on raw materials; metal containers, 64%; paper and plastic bags, 57%; wooden containers, 61%; and glass, 39% (7). More complex conversion processes, such as molding, coating, or laminating, add a greater percentage to the cost of the finished package. The conversion setup cost can also vary by container type. Conversion processes that require tooling, dies, or molds add fixed costs that are generally amortized over an initial production period.

The prices of packaging products are also affected by competition, vertical integration, and opportunities for intermaterial substitution. Prices are affected by general economic conditions such as recessions that result in oversupply and growth cycles that strain production capacity. The export demand for goods and for packaging materials also affects prices.

Packages can be purchased directly from the convertors or from independent distributors. While a convertor may offer a lower price for high-volume orders, it is rare to be able to purchase all packaging components from a single supplier. Bottles may be purchased from the bottle manufacturer, but caps, labels, shipping containers, pallets, and stretch-wrap will be purchased from other sources (although in the United States, glass bottles are often sold in corrugated fiberboard "reshipper" boxes). Independent distributors offer various components as well as entire packaging systems.

Most packaging is purchased competitively, especially when it is more of a commodity. For example, corrugated fiberboard boxes have standardized properties and are very similar when purchased from different suppliers. In order to ensure low cost, a purchaser may encourage competing firms to bid against one another. Packaging innovations, like lightweighting or material substitutions, are often introduced by one supplier competing against another. But there is a countervailing trend to closer partnerships between packaging suppliers and purchasers; in exchange for a single-source contract, suppliers provide services such as design, quality guarantees, inventory reduction, just-in-time delivery, and customized logistics. Likewise, packages with special designs are more likely to be purchased from a single source. The supplier may work in partnership to design the package to the user's needs. For example, custom-molded bottles, trays, or plastic foam cushioning are available only from the supplier with the custom mold.

Packaging machinery costs. Packaging machines can be either purchased or leased. The purchase decision is evaluated like any other capital investment, judged on its net present value, by subtracting the initial investment from all cash flows from the machine's expected life of production, discounted for the time value of money. Expected income cash flows are forecast by a firm's marketing department and the finance department crunches the numbers.

Selection of packaging machinery capacity and capability should match the expected production volume and lot sizes for the product's expected life cycle. Extremely high-volume filling operations have many dedicated single-purpose filling lines controlled by a single computer. But production plans can dramatically change in response to competitive conditions or packaging material substitutions. In order to reduce the risk of obsolescence, packaging machinery increasingly has the flexibility to run different materials and to change over quickly for different products with various lot sizes.

Package filling economics. Package filling operations are accounted for like any other factory operation, including inputs of direct labor, materials, energy, overhead, and shrinkage due to defects. Traditional approaches to cost reduction have focused on productivity (output/input) improvements, especially by reducing labor and energy inputs by increasing automation, and by increasing machine efficiency and speed. Since successive packing machines are often linked together (eg, filling, capping, labelling, cartoning and palletizing), packaging line efficiency depends on a smooth flow of materials, often including master controls and a plan for accumulation to cover variation between adjacent machines. New activity-based costing methods allow for the cost of specific operations to be more closely monitored, in order to compare alternative methods.

Many organizations also now pursue strategies to reduce the time needed to accomplish tasks and improve package and product quality. Planning economical production quantities, quick changeovers, and time phasing the delivery of materials can dramatically reduce costs. Improved quality control, using real-time statistical evaluation of data gathered by automated monitoring equipment on the packaging line, can reduce scrap and improve overall profitability.

In some cases it is less expensive to hire a contract packaging firm to fill packages. If production need is irregular, does not fit with current production, or is an uncertain test market, contract packaging may be more cost-efficient than filling packages in house (see also Contract packaging).

Package distribution economics. Packaging can dramatically affect the cost of distribution. Transportation costs are directly related to packaging cube and weight efficiency. Methods to reduce the size and weight of packaging include concentrating or nesting products, shipping products unassembled, lightweighting containers, improving the efficiency of stacking patterns in unit loads and vehicles, substituting slipsheets for pallets, and reducing the volume of cushioning materials by decreasing the fragility of products. Sometimes postponing packaging to a later time and place, for example packing to order, can reduce packaging cube and investment dramatically. Material-handling cost is also related to packaging; the productivity of operations like vehicle unloading and order picking depends on packaging configuration. If packages are manually handled, the cost will be much higher than if the packages are unitized in order quantities. The cost of the unitization materials is often offset by more economical handling.

Returnable packaging can often reduce distribution packaging costs if the shipping cycle is short in time and distance, and if the shipper and consignee can work out the partnership details of ownership and cost sharing. Most returnable packages are initially more costly than single-trip packages, but the per trip cost can be lower. Potential returnable packaging investments should be evaluated on the basis of net present value to judge the investment's profitability. Factors to be evaluated when comparing returnable to expendable packaging include purchase cost, expendable packaging disposal cost, number of containers required for the logistical cycle, packaging management costs, and return sorting, cleaning, and transport costs.

The cost of distribution damage is directly related to packaging protection. But damage is not necessarily related to packaging cost. Often the cost of damage and packaging can be reduced simultaneously with a redesign that uses less materials or by substituting less expensive, yet more protective, materials. Sometimes it is more cost-effective to strengthen the product. Measurement and control of distribution damage costs is an important step in reducing packaging-related costs (see Logistical/distribution packaging).

Disposal costs. It has been estimated that 64.4 million tons of packaging materials went into the U.S. waste stream in 1990, accounting for one-third of all municipal solid waste (MSW). MSW, generated from residences and commerical establishments, accounts for 2.2% of the total solid waste (before recycling) generated; the other 97.8% includes waste from agricultural production, mining, demolition, hazardous materials, etc. In the United States in 1990, most MSW (66.6% by weight) was disposed of in sanitary landfills; 16.3% was incinerated, and 14.9% was recycled (8).

Disposal of municipal solid waste incurs costs to local governments, consumers, and businesses. Businesses clearly pay the direct cost for disposal, and as a result have a higher rate of recycling and more reusable packaging than do consumers. There is a trend to making consumer disposal costs more explicit, rather than hiding them in the general tax base, in an effort to encourage trash reduction and recycling. There are also "external" social costs, especially for landfilling and incineration, including possible pollution and the fact that many communities will not permit landfills or incinerators to be sited nearby. The siting problem resulted in skyrocketing disposal costs; landfill tipping fees doubled in the period from 1987 to 1991, especially in highly populated areas in the Northeast. As a result, recycling of packaging materials has become more attractive. Recycling can be economically viable, but only when the reuse value of recycled material and the disposal avoidance cost exceeds the cost for collection, sorting, transporting, and reprocessing.

Packaging development economics. The decision whether to develop a new product or package is a marketing responsibility. In 1990, over $142 billion was spent on research and development in the United States (9). Most businesses get a high percentage of their sales and profits from new products.

Most packaging development activities are performed by personnel in the firm that makes a product and buys and fills the packages. The costs are associated with the time spent by the packaging professional and the team responsible for decisions and implementation. Tasks include establishing criteria, identifying concepts, designing prototypes, selecting and evaluating samples, issuing specifications, and setting up and starting production. Setting up and starting production are the highest-cost steps, often requiring the purchase of new equipment.

Vendors of packaging systems provide a great deal of development assistance, and the services of a design firm may be required. Vendors who are established suppliers often provide design service at "no charge." But when a prospective package will be competitively supplied by several vendors, it may be necessary to contract for design and sample preparation.

Conclusion

Packaging affects every part of our market economy. There is a growing awareness at the financial level in major processing–packaging corporations that packaging represents a significant proportion of the cost of doing business. Even at 5–10% of the retail price, packaging costs can amount to hundreds of millions of dollars for a multi-billion-dollar food or beverage firm. It should be recognized that packaging costs must be analyzed in a systems approach, where the basic component purchase price is only one factor among many.

Other costs include labor, machinery, distribution, development, impact of one component's cost on others', and research.

But it is increasingly recognized that packaging is more than a cost center. Packaging also directly impacts sales and profitability. A packaging innovation may have a higher purchase price than the container it replaces but can increase profits by adding value, increasing product quality, reducing damage, and improving efficiency of production and distribution.

BIBLIOGRAPHY

1. Rauch Associates, *The Rauch Guide to the U.S. Packaging Industry,* Bridgewater, NJ, 1990.
2. Packaging Machinery Manufacturers Institute, Arlington, VA, 1994.
3. "Top 100 Packaged Brands Buy $12 Billion," *Packaging* 37 (Jan. 1994).
4. N. C. Robson, "Packaging as a Factor in Progress: The Needs of Developing Countries," and W. C. Pflaum, "International Consultation Produces Consensus Recommendations," *IOPP Technical J.* 4–12 (summer 1991).
5. M. Spaulding and G. Erickson, "In the Mind of the Consumer; Is Low Price Everything?" *Packaging,* 54–56 (June 1993).
6. J. A. Howard and J. N. Sheth, *The Theory of Buyer Behavior,* Wiley, New York, 1969.
7. Rauch Associates, *The Revised Rauch Guide to the U.S. Packaging Industry,* Bridgewater, NJ, in press.
8. U.S. Environmental Protection Agency, *Characterization of Municipal Solid Waste in the United States: 1992 Updated,* EPA/530-R-92-109, 1992.
9. U.S. Department of the Census, *Statistical Abstract of the United States,* 1992.

General References

L. M. Guss, *Packaging is Marketing,* American Management Association, New York, 1967.

E. A. Leonard, *Packaging Economics,* Books for Industry, New York, 1980.

F. A. Paine and H. Y. Paine, "The Economics of Primary Packaging," in *A Handbook of Food Packaging,* 2nd ed., Kapitan Szabo, Washington, DC, 1992.

S. Sacharow and A. L. Brody, *Packaging: AN Introduction,* Harcourt Brace Jovanovich, Duluth, MN, 1987.

S. E. M. Selke, *Packaging and the Environment,* Technomic, Lancaster, PA, 1994.

DIANA TWEDE
T. W. DOWNES
Michigan State University
School of Packaging
East Lansing, Michigan

EDGE-CRUSH CONCEPT

Background

Corrugated board is a widely used packaging material because of its flexibility in box design to meet different needs, its possibility to adopt to different performance requirements as well as its high strength:price ratio, and its high strength:weight ratio.

Corrugated board is constructed as a sandwich that is characterized as a material with two facings, called *liners,* that provide bending stiffness and a lightweight corrugated core, called *fluting,* that separates the facings and provides shear stiffness. (1).

The corrugated-board properties are derived from the types of paper used and the formation of the fluting. The manufacture of corrugated board as well as the conversion of the corrugated board into different box designs will influence the performance of each box.

Compression strength has become the test method best found to express the performance of individual boxes (see Fig. 1).

Much interest has been focused on the relationship between separate components—liners and fluting—of corrugated board and the corrugated board itself as well as the relationship between the corrugated board and the compression strength of the box.

Corrugated Board as an Engineering Material

Paper and corrugated board are not regarded as construction materials, as the raw material has a greater influence on the performance than the manufacturing process of the board. However, this does not influence the possibility of applying standard engineering principles to box design work.

The four engineering characteristics that influence the structural performance of corrugated board are

- Papers and their fiber orientations
- Flute height
- Number of flutes and the spacing between the flute tops
- Integrity of the glue lines

In corrugated board the liners provide the bending stiffness while the fluting provides the shear stiffness (1). The flute height, the number of flutes, and the integrity of the glue lines influence the bending stiffness of the corrugated board to greater or lesser extent (2).

A major obstacle to the application of engineering principles to corrugated-board has been the difficulty of accurately measuring the mechanical and physical properties of paper and corrugated board. The true caliper of paper is, for example, difficult to accurately measure because of the surface roughness of the paper. Edgewise compression strength of paper is also difficult to measure because of edge effects of the

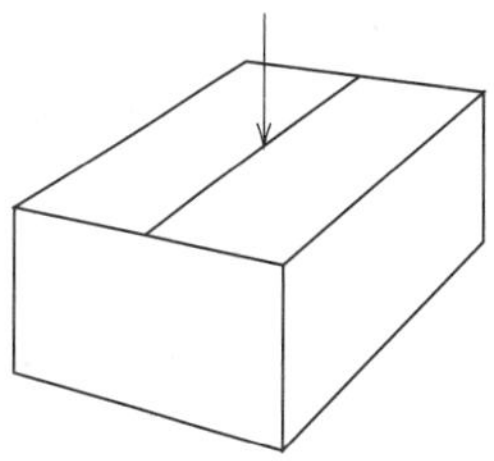

Figure 1. Compression strength test method.

paper as well as the influence of the clamps used in test fixture needed to hold the material during the testing.

A reluctance to apply engineering principles probably also relates to the fact that the structural characteristics of paper are affected much more rapidly by factors such as atmospheric conditions and rate of loading than are other more commonly used engineering materials, for example, steel. The result is that few accurate methods, that measure and describe corrugated board and box performance, have been made available.

When corrugated board is manufactured, twist and warp may occur during and after the manufacturing and conversion of corrugated board, due to differences and variations in the hygroscopic properties of the papers used (1). Washboarding (3) occurs also. It gives an uneven board surface due to too high pressure on the flute tops (3), and this causes deformations around the flute tops. These phenomena influence the measurements of corrugated-board properties and box performance and make measurement even more difficult. Efforts have nevertheless continued to provide the tools to ensure proper use of corrugated-board materials.

Corrugated board was initially considered as an engineering material in 1939 (4), at which time a correlation between elasticity and the strength properties of the separate paper components and the corrugated board was established.

Box Compression Strength

In corrugated-board design it is the stacking strength of the box that is considered as a measurement of the box performance as the load-bearing ability of the box has great importance in modern distribution. The stacking strength of a box is the basis for good distribution system efficiency.

The stacking strength of a corrugated-board box is measured as the compression strength according to a standard test method. This test gives a box compression test (BCT) value. The BCT test measures the pure top to bottom load of an empty corrugated-board box.

The box is placed centrally between flat parallel plates in a compression tester, where the plates are compressed with a constant rate, usually 10–13 mm/min until failure. The force and the strain are recorded continuously until compression failure occurs (5). The maximum force attained at collapse of the box is reported as the compression strength of the corrugated-board box.

A number of standard test methods describe in detail how to test, what conditions to use, and how to report the results. Examples of standard methods are TAPPI T804 om81, ASTMD642-76(83), FEFCO No. 50, and DIN 55440 (6).

Models For Predicting BCT

Many scientists and researchers have worked on models to predict the box compression strength from material properties and box geometry. The McKee formula (7) is the best-known and most commonly applied model.

McKee original formula. The McKee formula predicts the box compression strength from the edge-crush strength and bending stiffness of the corrugated board and the perimeter of the box.

This McKee formula is based on the semiempirical equation

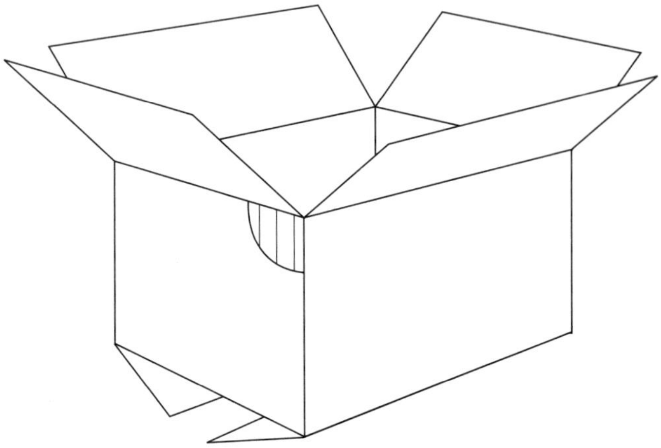

Figure 2. A regular slotted container.

$$\frac{P}{P_{cr}} = c\left(\frac{P_m}{P_{cr}}\right)^d \qquad (1)$$

where P is the collapse load for a simply supported panel and P_{cr} and P_m are the critical loads and the compression strength in the loading direction for a panel, respectively.

The box compression strength is the sum of the collapse loads of its panels. The parameter c is experimentally determined. The McKee formula is calculating the compressive strength of a regular slotted container (RSC) (see Fig. 2). The formula states (7):

$$\mathrm{BCT} = k_1\,\mathrm{ECT}^b\,S^{1-b}\,Z^{2b-1} \qquad (2)$$

and is in general adapted for corrugatedboard boxes as

$$\mathrm{BCT} = k_1\,(\mathrm{ECT}^{0.75})\,S^{0.25}\,Z^{0.5} \qquad (3)$$

where S is the geometric mean stiffness given by

$$S = (S_{\mathrm{MD}}\,S_{\mathrm{CD}})^{1/2} \qquad (4)$$

where the bending stiffness has been tested in MD (machine direction) and CD (cross direction) of the sample and Z is the perimeter of the box. The term k_1 is a constant chosen in the McKee formula to give the BCT in N alternatively lbf (1bf = 4.44822 N).

The properties in the McKee original formula. The *edge-crush compression test* (*ECT*) is in the McKee equation defined as the maximum compression force a test sample will sustain without failure. As for BCT the load is applied on the sample with a given loading speed, usually 10–15 mm/min. The test sample is standing with the flutes perpendicular to the plates (8) (see Fig. 3).

The maximum edgewise crush resistance is most often given as the force per length unit, kN/m or alternatively lbf/in (1 lbf/in = 0.1751268 N/m). Examples of standard methods are TAPPI T811 om83 and T823 pm84, ASTMD2808-69(84), BS 6036, FEFCO No. 8, and DIN 53149 (6).

To measure ECT correctly, it is essential to prepare the samples properly. The cutting of the samples for the ECT is

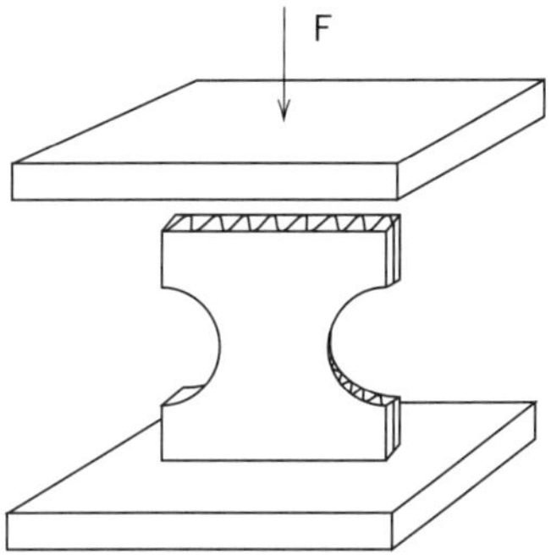

Figure 3. FPL method.

critical. The most important thing is to avoid force concentration on the end surfaces of the sample.

It is also important to recognize that different standards use different principles to ensure that the true edgewise compression strength is reported. This means also that the standards may give different values. Efforts are made to ensure that the differences are eliminated. The FPL (Forest Product Laboratory, Madison, WI) method using a sample as shown in Figure 3 has been found to provide the most accurate calculations of the BCT value according to the McKee original formula (9). FPL is also best related to the paper testing methods commonly used today. ECT is actually built from the compression strength (CS) of all papers in the corrugated board, which means that (10).

$$\mathrm{ECT} = k\,(\text{sum of liners CS} + \alpha \times \text{fluting CS}) \qquad (5)$$

where α flute A is 1.54; for flute B, 1.34; and for flute C, 1.45. The k factor depends on the test methods used.

The *Bending-stiffness test* in the McKee formula is determined by the thickness of the corrugated board and the ability of the outer and inner liners to resist tensile and compression forces. The bending stiffness is nearly independent of the fluting (10).

Several test methods are also used to measure the bending stiffness. Examples of test methods are TAPPI T820 cm85, BS 3748, and DIN 53121 (6).

The simplified McKee formula. To avoid the problems encountered in measuring the bending stiffness, McKee developed a simplified version of his formula (7), where the bending stiffness is replaced by the caliper of the corrugated board, t:

$$\mathrm{BCT} = k_2\,\mathrm{ECT}\,t^{0.5}\,Z^{0.5} \qquad (6)$$

This is natural as (10)

$$S \sim 0.5\,\mathrm{TS}\,t^2 \qquad (7)$$

where TS is the tensile stiffness of the liners and

$$\mathrm{TS} = E\,t \qquad (8)$$

where E = modulus of elasticity and t = caliper of the liner. In the McKee simplified formula k_2 is a constant chosen to give the BCT in N alternative lbf (1 lbf = 4.44822 N).

The use of the McKee formula. The behavior of a box that is subject to a compressive top–bottom load until it collapses, has been extensively studied and tested.

It has then been found that the panels in a box usually buckle before they collapse. At buckling, the middle part of the panels bends, decreasing its ability to carry in-plane load. The corners do not bend because of the scorelines and can therefore carry further load (1).

McKee and Gander (11) have verified experimentally that a regular slotted container will fail first at a vertical edge and then propagate into the middle of the panel.

The general rule is that if the structure is intact, the failure of a package is initiated by a compression failure of the liner on the concave side of the panel, which can be either the inside or the outside of the box (1).

The maximum load bearing of a corrugatedboard box will therefore depend not only on the corrugated board itself but also on the ability to resist buckling. The higher bending stiffness of the corrugated board, the better bearing of the load and the better resistance to buckling. Therefore McKee warned against using the simplied formula uncritically (7). The main reason was that the simplification might imply that it is needed only to evaluate ECT when BCT is determined.

At this point it is also important to remind the corrugatedboard user that although BCT is a corrugatedboard performance, other corrugatedboard properties are no less important. There has always to be a balance between material properties to provide best performance—either stackability, containment, rigidity, or all in combination—of the corrugated board (12).

Defects in corrugatedboard manufacturing and conversion must also get attention. BCT as calculated by McKee gives the potential compression strength possible to attain. Defects in the corrugated board—often the flute formation and height—may reduce the actual performance. But as machine and process technologies improve, the possibility of achieving the calculated BCT will increase, and the ECT value will also become a more accurate measurement of actual corrugatedboard performance.

Conclusions

The objective of all package testing is to ensure fulfilment of market demands. The test methods must be simple to carry out. The underlying principles behind must also be clear and easy to understand. The test procedures should be internationally accepted to facilitate communication between companies and countries. And the selection of test methods and tests should be such that they relate to the package/product performance.

ECT is a very important property for estimating the corrugated box performance. It relates to both stacking strength and overall performance of a corrugatedboard box. It can also be used to follow the manufacturing process and ensure that the final corrugatedboard box is performing well.

However, great attention must be paid to the sample production, as this is the basis for correct results.

BIBLIOGRAPHY

1. K. Hahn, *A Study of the Compressive Behavior of Corrugated Board Panels,* Lund Institute of Technology, Lund University, Lund, Sweden, 1991, p. 4.

2. G. Jönson, *Corrugated Board Packaging,* Pira International, Leatherhead, Surrey, UK, 1993, pp. 127–128.
3. T. Nordstrand, and A. Andersson, *Measuring the Washboarding,* SCA Research, Sundsvall, Sweden, 1991.
4. T. A. Carlson, "A Study of Corrugated Fibreboard and Its Component Parts as Engineering Materials," *Fibre Containers* **24**(7), 22–35 (1939).
5. H. Markström, *Test Methods and Instruments for Corrugated Board,* L&W, Stockholm, Sweden, 1988, p. 9.
6. G. Jönson, *Corrugated Board Packaging,* Pira International, Leatherhead, Surrey, UK, 1993, p. 108.
7. R. C. McKee, J. W. Gander, and J. R. Wachuta, "Compression Strength Formula for Corrugated Boxes, *Paperboard Packag.* **48**(8), 149–159 (1963).
8. H. Markström, *Test Methods and Instruments for Corrugated Board,* L&W, Stockholm, Sweden, 1988, p. 17.
9. H. Markström, *Test Methods and Instruments for Corrugated Board,* L&W, Stockholm, Sweden, 1988, p. 24.
10. G. Jönson and S. Ponton, "Utilization of liner/fluting weight in Corrugated Board for Best Box Performance," TAPPI Corrugated Containers Conference, Atlanta, Oct. 1984.
11. R. C. McKee, and J. W. Gander, "Top-Load Compression," *TAPPI* **40**(1), 57–64 (1957).
12. G. Jönson, *Economy and Product Protection of Corrugated Board Containers,* Ph.D. thesis, Chalmers University of Technology, Göteborg, Sweden, 1974.

GUNILLA JÖNSON
Centre of Packaging Logistics
Lund University Sweden

EDUCATION, PACKAGING

Packaging technology is not nearly as demanding as it should be. That is a promising premise on which to start an article on packaging education. Nonetheless, it is true. The basic questions are

1. Is technical innovation in packaging keeping pace with innovation in other fields?
2. If not, what role can education play in furthering the innovation in packaging?

The answer to the first question is "It depends." That is an answer clearly reflective of the diversity of packaging itself. While generalizations are dangerous, let us assume that packaging does, in fact, lag in its response to today's demanding requirements for technical innovation in materials science and machinery design as well as in the design, production, and distribution functions.

Let us also assume that education can, in fact, play a significant role in enabling that improvement—but *only if* the practicing packaging professional demands it. Fortunately, the case for an increased emphasis can be easily made—packaging altogether is a huge business that protects and preserves a product while often adding value to it as well. That is the positive side. The negative side that has to be considered is that it gobbles resources, both natural and financial. That last factor alone should alert informed corporate management to the importance of packaging education.

Turning to the practical side of what is available and where, there are three basic methods of education applicable today: university, industry, and self. We will look at each separately.

University

Packaging is a discipline staffed, on the technical side, primarily by scientists and engineers whose principal degree is in one of the primary sciences, usually chemistry, or one of the old-line engineering disciplines, especially chemical, industrial, or mechanical.

But beginning in the 1950s, a new discipline, packaging technology, was formulated at Michigan State University and has since spread in fits and spurts to a number of other universities. Today, there are nine schools offering degrees—or, at least, majors—in packaging:

Larry Gay, Ph.D.
Packaging Program Coordinator
Industrial Technology Department
California Polytechnic State University
San Luis Obispo, CA 93407
Tel. (805) 756-2058;
Fax (805) 756-6111

Robert F. Testin, Ph.D.
Chairman, Department of Packaging Science
Clemson University
223 Poole Agricultural Center
Clemson, SC 29634-0371
Tel. (864) 656-3397;
Fax (864) 656-0331

J. Paik, Ph.D.
Coordinator of Packaging, Department of Industrial Technology
Indiana State University School of Technology*
6th & Cherry Streets, Room 118
Terre Haute, IN 47809
Tel. (812) 237-3371;
Fax (812) 237-7607

Daniel L. Goodwin, Ph.D.
Chairman, Packaging Science
Packaging Science Department
Rochester Institute of Technology*
1 Lomb Memorial Drive, P.O. Box 9987
Rochester, NY 14623-0887
Tel. (716) 475-2278;
Fax (716) 475-5555

James D. Idol, Ph.D.
Director, Rutgers University Center for Packaging Science
Busch Campus, Bldg. 3529
Piscataway, NJ 08855
Tel. (908) 445-3224;
Fax (908) 445-0777

Donald J. Betando
San Jose State University School of Applied Arts & Sciences Department of Nutrition & Food Science
One Washington Square
San Jose, CA 95192-0058
Tel. (408) 924-3100;
Fax (408) 924-3114

Stephen A. Raper, Ph.D.
Program Coordinator, University of Missouri—Rolla*
223 Engineering Management Department
Rolla, MO 65401-0209
Tel. (573) 341-6569;
Fax (573) 341-6567

* For information on courses, conferences, trade shows, publications, and books, contact the Institute of Packaging Professionals, 481 Carlisle Drive, Herndon, VA 22070, tel. (703) 318-8970, fax (703) 318-0310.

Bruce R. Harte, Ph.D.
Director/Professor
Michigan State University*
School of Packaging
135 Packaging Building
East Lansing, MI 48823-1223
Tel. (517) 355-9580;
Fax (517) 353-8999

Ken E. Neuburg
Lecturer, Packaging Program
University of Wisconsin—Stout
Department of Technology
School of Industry & Technology
Menomonie, WI 54751
Tel. (715) 232-1246;
Fax (715) 232-1624

Of those, seven are oriented to packaging technology and two are incorporated into engineering departments.

The debate continues as to which approach is better, although it has always seemed nonproductive to have such a debate when reality is shouting both. That is, packaging needs people equipped to apply technology originated by others also aware of packaging's needs and applications, even if educated in any of the other technical fields. It needs qualified scientists and engineers to do the heavy-duty work of basic research and, ultimately, product innovation.

These nine universities offer differing approaches to coursework, but each remains open to positive reinforcement from industry. Each has a more-or-less active industry advisory council, and that council in the best of the programs is very instrumental, especially in providing sound advice on tomorrow's packaging needs and tuning senior administrators into the educational needs and opportunities our field provides.

Several of the universities offer postgraduate work. One, the Rochester Institute of Technology, several years ago formulated an "Executive Masters" program for hard-working industry professionals who desired more education (and recognition), yet could not afford to leave work to achieve it.

The Michigan State School of Packaging is providing a doctorate in packaging, a move that will help elevate the field with the technical disciplines as well as provide qualified university teaching talent for the future.

Industry

Industry itself is a major provider of educational opportunities. Starting with the "Fundamentals of Packaging Technology," a multisession course spread over several months, the Institute of Packaging Professionals has assumed a leading position in this type of educational offering.

Its reputation has been established by a series of "Fundamentals of . . ." courses formulated in response to its perception that so many qualified individuals came to packaging needing specific technical information in a relatively small field. It also provides several annual conferences attended by leaders in their respective areas, including Transpack for distribution, and the IoPP Advanced Packaging Technology Conference for latest developments.

There are many other societies, associations, and even private providers of seminars and conferences in packaging. Organizations such as TAPPI and the Institute of Food Technologies have defined specific packaging-related areas, but none has the broad mandate to cover the field "from A to Z" as IoPP does.

Of course, trade shows are a form of industry education as is publishing. Again, in recent years, it has been IoPP that has stepped forward to, first, consolidate the packaging publications into its "IoPP Bookstore" so that interested parties could even find where to purchase the texts; then, next, to enter the publishing business aggressively itself. Its offerings now span from the general text, *Fundamentals of Packaging Technology,* to books on specific topics such as packaging sales, the environment, and food packaging; to conference proceedings.

Self-Instruction

To date, there is no rush to publish in packaging in any of the new technologies—CD-ROM, for example. Even books printed in the "old" style using paper and ink are not formatted for self-help, although there are a number of commercial "publications" of this sort provided by suppliers to the field.

Whether there is a real market here and whether any of the current providers will step in is questionable. The problem seems to be one that generally plagues all packaging publishing—the field is so broad and the specializations so narrow that there seldom are enough people in any one of them to merit the development of full-text presentations.

Where to Go

If you are a packaging manager, try visiting one or more of the universities. Take time to understand what is going on there and how it may apply to and benefit your business. If, after reflection, you conclude it will not be of benefit, tell someone! Tell the professor. Write or call the Packaging Education Forum!

Consider how you can help. Supply materials or a machine? How about a summer job or internship? Best of all—how about a job for a graduate?

If you are looking for an advanced degree for yourself, note the schools with an asterisk.

If you are a practicing professional developing a need for information in an area in which you have not personally worked, try contacting IoPP for help. If it does not have a course to help, it may be able to suggest others that do.

And, of course, read the industry press, both for the education in the pages and for the activities calenders that may list a course in which you have interest.

William C. Pflaum
The Institute of Packaging Professionals and Packaging Education Forum
Herndon, Virginia

ELECTROSTATIC DISCHARGE PROTECTIVE PACKAGING

Introduction

Friction generates electrical charges that, if separated quickly enough or isolated by sufficient resistance, result in what is known as *static electricity*. Indeed, the root word for electricity is *elektron,* Greek for "amber," because the Greeks found static electricity after rubbing amber with fur or cloth. This experiment is still repeated in school science demonstrations. Thus, static electricity can be considered to be an old and reliable scientific observation. However, the technical details were little explored until static electricity was found to be a hazard to modern electronic components.

Historically, static electricity has been a serious fire and explosion hazard for many industries from milling grain to manufacturing explosives. Dust control, choice of materials, and simple procedures were adequate solutions for these industries. Early electronics were large and robust enough to be immune to static electricity damage. However, with smaller and more dense solid-state devices, even subtle static electricity problems could destroy an entire device. Military reliability and economic losses estimated as high as $15 billion (1–3) have driven the technical understanding and development in the area now known as *electrostatic discharge* (ESD) protective packaging.

History

People have seen static electricity for millennia. Lightning, the most spectacular example, results when charges are generated by friction between air and water droplets and where water carries one charge away. When sufficient field strength accumulates for the separated charges to neutralize each other, lightning strikes. This is the same basic effect observed by walking across a carpet in the winter with rubber-soled shoes and reaching for the doorknob. In the dry air, a spark will often be observed and felt at about 10,000 V. In humid air, the voltages are less and not often felt.

Early electrical components, like resistors, capacitors, inductors, and vacuum tubes, were simple and robust enough to be virtually immune to physical damage from static electricity. The first transistors were also relatively insensitive to static because of their physical size. However, as solid-state components became smaller and faster, static sensitivity became a minor but increasing problem. Early static-control methods such as good grounding, choices of clothing materials, relative humidity control, and air ionizers, were sufficient for a while. As solid-state technology progressed to ever-smaller, more complex circuits with metal oxide semiconductor technology (like CMOS and TMOS) or field-effect devices such as field-effect transistors (FETs), static became recognized as a serious problem in both manufacturing and operation. Some MOSFETs are known to be susceptible to electrostatic discharge damage at voltage thresholds as low as 150–250 V, values well below human sensitivity to static (4). While it is possible to design some static electricity protection into a chip, the best static-safe solutions for manufacturing, shipping, and storage of electrical components include electrostatic discharge protective packaging.

The military and NASA were the first to technically define and quantify static electricity problems. Not only was static electricity a problem with electronics reliability; there were some unfortunate accidents believed to be caused by static, such as several missile explosions and, perhaps, the 1967 Apollo command module fire. The Department of Defense issued the original MIL-Specifications that set the basis for further investigation and understanding of electrostatic discharge protective packaging (5).

Classification of Types

Based on the type of protection required and the materials available, the original MIL-Spec divided electrostatic protective packaging into two types. In later recognition of performance differences between foils and metallized films, Type I was subdivided for a total of three types. Although MIL-Spec nomenclature and tests for electrostatic protective packaging are being phased out in favor of industry standards, they are still widely recognized.

Type I—uses a metal or foil as a moisture barrier layer within a multilayer plastic pouch. Their electrical function is to form a metallic Faraday cage around the components to provide protection against high-voltage fields, eg, static, and attenuate at least 25 dB or essentially all RF (radiofrequency) interference. Although the metal forms a Faraday cage, commercial Type I materials are multilayer composites of metal foils, usually aluminum, sandwiched between films with Type II surface resistivity and heat-sealing properties. This structure provides good Faraday cage protection from the metal with an antistatic isolation between the electronic components and the foil from the Type II film. Type I materials are commonly referred to today as *electrostatic barrier materials.*

Type II—uses a plastic film with a strong antistatic agent either compounded within or coated onto it. The antistatic agent must have an appropriate hydroscopic nature so that it will absorb and hold a trace of atmospheric moisture that actually provides the electrostatic discharge protective layer. These materials are now called *static-dissipative materials.*

Type III—uses a metallized layer, often vapor deposited nickel or aluminum on a polyester film, to provide a Faraday cage effect. However, because the vapor-deposited metal is thin enough to have limited transparency, the RF attenuation is limited to >10 dB. This difference in RF attenuation is the major difference between Type I and III. As with Type I packaging, commercial Type III materials are multilayer sandwiches with surface properties of Type II films. In recognition of their lesser attenuation than Type I or electrostatic barrier films, Type III films are called *static shielding materials.*

Terms and Test Methods

Triboelectricity. The Greek word *tribein,* meaning "to rub," combined with *elektron,* Greek for "amber," defines the phenomenon we know as triboelectricity. Although the phenomena is ancient, it has been difficult to quantify and is better explained in relative terms. Table 1 is a triboelectric series of materials arranged from positive to negative, based on their charge polarities after being rubbed together. In general, when two materials on the list are rubbed together, the upper one will take the positive charge; the lower, the negative charge. Except for illustration, the triboelectric series is of little practical value since materials near each other can

Table 1. Triboelectric Series (29)

Positive	Human hands
+	Rabbit fur
	Glass
	Mica
	Human hair
	Nylon
	Wool
	Fur
	Lead
	Silk
	Aluminum
	Paper
	Cotton
	Steel
	Wood
	Amber
	Sealing wax
	Hard rubber
	Nickel, copper
	Brass, silver
	Gold, platinum
	Sulfur
	Rayon
	Polyester
	Celluloid
	Polyurethane
	Polyethylene
	Polypropylene
	PVC
	KEL F
−	Silicon
Negative	Teflon

switch relative charge position and since the amount of charge between the two materials can vary widely (6).

However, the major point of the list is to show that common semiconductor materials and packaging materials are very low in the triboelectric series with metals higher and humans being much higher. Thus it is easy to see how a printed wiring circuit board with metal conductors, components, and connectors, all loosely held in a plastic bag and handled by human hands, can produce static electricity with possible electrostatic damage to the electronics.

Triboelectricity and its measurement has been discussed at length by the ESD Association (7). Most of the measurement techniques use a rolling or sliding test specimen that drops into a Faraday cup where the accumulated charge measured as nanocoulombs. One method uses large (1-in × 1-in.) cylinders of either PTFE or quartz, another uses smaller (½ × 1-in.) cylinders of PTFE, quartz, or polished brass that are rolled down an inclined plane into a Faraday cup. Another method uses chips in rail magazines that are slid into the Faraday cup. Other tests pull cards from bags or rub materials in a programmed way and measure the field strength developed as volts. As expected, all these tests are done with controlled humidity and temperature conditions.

Surface resistivity. Surface resistivity is the resistance in ohms measured between two parallel electrodes on opposite sides of a square. The geometry in Figure 1 illustrates why the common units of surface resistivity are ohm/square for a

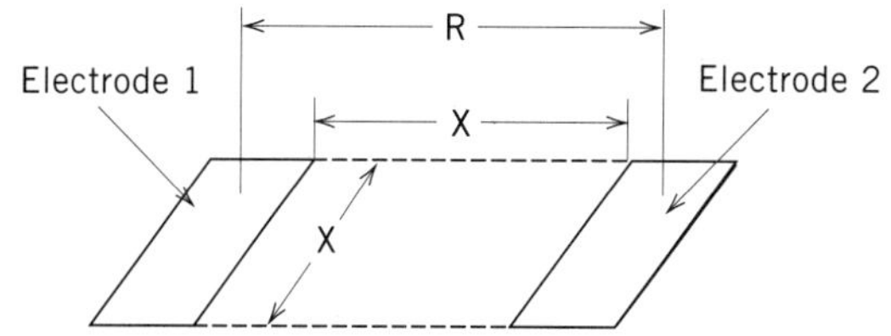

Figure 1. Surface resistivity geometry with parallel electrodes.

measurement that is otherwise physically dimensionless. Some commercial surface resistivity measuring devices use concentric circles with a calculated "area factor" as seen in Figure 2. In either case, a given applied voltage allows a current to be measured that calculates the surface resistivity. Although test voltages are not spelled out in the specifications, relatively low voltages (50–100 V) are used, which, for the lower limits of the measurement correspond to picoampere current measurements (8).

Surface resistivity can be used to broadly classify materials as conductors such as metals, insulators such as most ordinary plastics, or semiconductors as illustrated in Figure 3. Materials with surface resistivity values of $<10^5$ Ω/square are classified as conductors and above 10^{12} Ω/square to be insulators. Static dissipative materials are defined to have surface resistivity values between 10^5 and 10^{12} Ω/square.

There has been some discussion that surface resistivity is

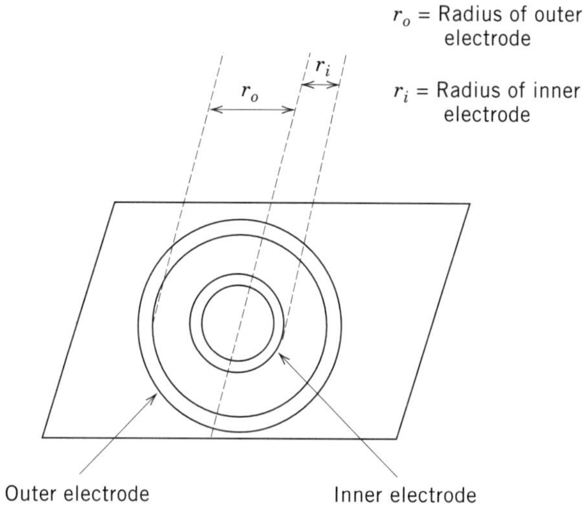

Figure 2. Alternative geometry for surface resistivity using concentric ring electrodes.

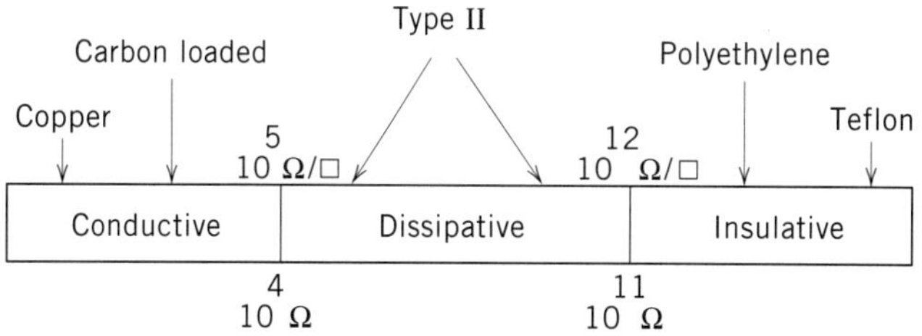

Figure 3. Surface resistivity and surface resistance ranking of common materials.

too simple a measurement for several factors such as the effect of film thickness and the use of multiple layers of differing resistivity values. For example, a more conductive layer below a less conductive layer can suppress the measured charges. This has resulted in a recent test method for surface resistance.

Triboelectricity and surface resistivity. While triboelectricity and surface resistivity have some relationship for the generation and separation of electrical charges, no good overall experimental correlation between the two has been found. One paper took an extensive variety of materials and test conditions and found no direct correlation (6). On these data, it may be best to associate triboelectricity with the generation of static electricity and surface resistivity with its dissipation.

Static decay time. Static decay time is the time for a charge impressed on a film sample to decay to a nominal lower value. This is most often measured as the time for 5000 V to decay to 1% or 50 V. Commercial instruments often use the fixture geometry illustration in Figure 4. They usually operate where the voltage cuts off the timer. One static decay-time measurement variant uses a digital oscilloscope to take voltage versus time values, which are then fed through an exponential decay curve-fit program that calculates the static decay time with a minimum error (9).

Surface resistivity versus static decay time. While surface resistivity and static decay-time measurements are simple and straightforward lab procedures, there have been comparisons and discussions about their relative merit. One relatively extensive work showed data that, for homogeneous bulk additive loaded Type II films, indicated a reasonable correlation between the surface resistivity and static decay time as illustrated in Figure 5. However, this correlation does not hold for any films with more than a single layer. Such films include the Types I and III structures with metal foils or metallized layers for RF attenuation or multilayer and coated Type II film materials (10).

Surface resistance. Because of concerns that surface resistivity was oversimplified, the ESD association prepared a new specification for surface resistance. ESD Association S11.1.1 uses a fixed annular geometry and a fixed 100 V to measure resistance across the surface of the material to calculate resistance in ohms. This method separates the test result from any inherent resistive properties of the film. In special cases, it is possible to convert this test fixture geometry to a loose estimation of surface resistivity by

Surface resistance (Ω) × 10 = surface resistivity (Ω/square)

However, surface resistance is increasingly the preferred term over surface resistivity.

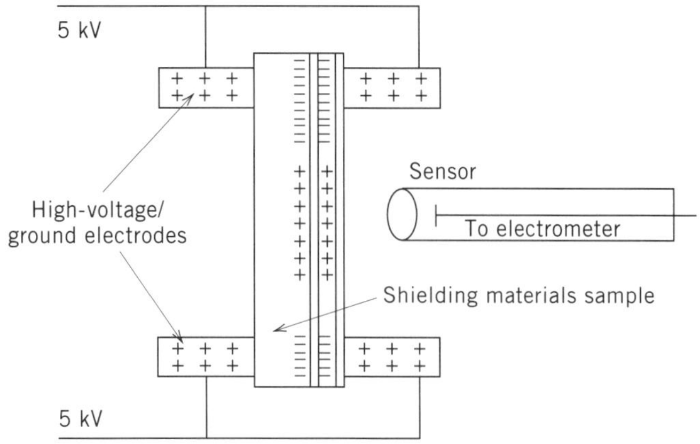

Figure 4. Static decay time fixture (5).

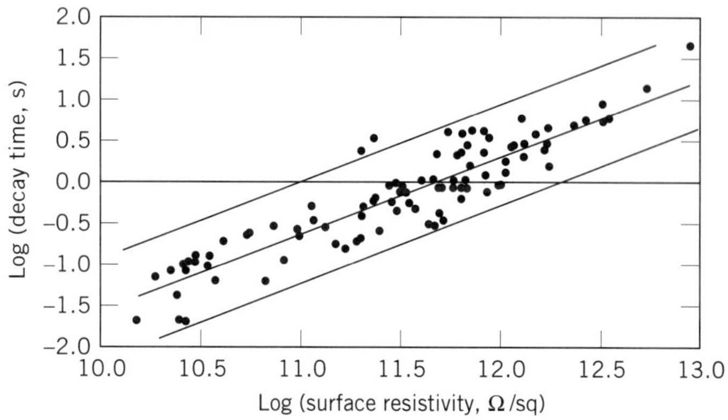

Figure 5. Correlation of surface resistivity and static decay for monolayer materials Only (10).

Volume resistivity. Volume resistivity is the resistance of current passing through a section of film. However, even for simple homogeneous composition monolayer materials, like early Type II films, the static dissipative properties are based on atmospheric moisture absorbed on the surface, not a bulk material property. Also, because most advanced electrostatic discharge protection materials are of multilayer construction, all Types I and III plus many Type II films, *volume resistivity* is now an essentially meaningless term.

Static shielding. Static shielding measures the energy that penetrates the test bag relative to the source energy. With an impressed 1000 V onto the outside of a bag containing a capacitive sensor, the current and time are plotted with a fast digital oscilloscope. The current and time are then used to calculate the energy transferred inside as nanojoules (11).

Effect of relative humidity. Most additives achieve either low surface resistivity or static decay-time function by attracting low levels of moisture to the film surface. For this reason, it was quickly apparent that the electrical properties would be related to relative humidity (12). In addition, moisture is quickly absorbed and slowly desorbed by strongly hydrophilic antistatic agents. Both of these were addressed by the test methods that specify the use of 10 ± 3% rh (relative humidity) at 72°F and holding the samples in a controlled environment chamber for 48 h before measuring surface resistivity or surface resistance and static decay time.

Effect of accelerated aging. Early Type II materials based on bulk loading of an antistatic additive into a monolayer film were found to have diminished electrostatic protection over time. This was due to the low but finite volatility of the additives used. Because fresh films would perform well, but older films would fail, an accelerated aging test procedure was

Table 2. Electrical Standards for Types I–III Materials

Property	Surface resistivity (Ω/square)	Surface Resistance (Ω)	Static decay time (ss)	RF attenuation, EIA (dB)	Static shielding, ESD-S11.3.1 (joules)
Type I	$<10^{12}$	XXX	<2	>25	XXX
Type II	$<10^{12}$	XXX	<2	—	—
Type III	$<10^{12}$	XXX	<2	>10	XXX

needed to simulate field use. MIL-Specs were developed that used held samples at 160°F for 3 days before the surface resistivity and static decay-time measurements were done. For monolayer bulk additive loaded Type II films, it is estimated that this accelerated aging test predicts performance for films between 6 and 12 months old. While other methods of achieving good surface resistivity films are less susceptible to activity loss with time, the test method has remained.

Corrosion. Because antistatic additives function by absorbing atmospheric moisture and weakly ionizing it, corrosion can be a problem with many metals used with electronics (13,14). The MIL-Spec considered seven metals, low carbon steel, 2024 aluminum alloy, copper, silver-plated copper, SN63 solder-coated copper, 314 stainless steel, and Kovar, to be important. The test methods check for corrosion and/or surface discoloration when test films are in contact with clean coupons of these metals for 72 h at 120°F and 65% rh (5).

Polycarbonate compatibility. Field service people from a major electronics equipment manufacturer found that polycarbonate components, like circuit board connectors, would "craze" and then crack (15). After a costly worldwide recall, the problem was traced to a Type II bag additive that could attack polycarbonate. EIA S-564 standard for polycarbonate compatibility had the effect of nearly eliminating certain antistatic additives from the market films (16).

Current Standards

Standards have changed over the years as additional information accumulates. On the basis of this information and the three different types of materials discussed above, the current accepted standard properties are listed in Table 2. Besides the electrical, corrosion, and polycarbonate standards, there are many more that cover heat seals and optical transmission for reading product codes inside the bag.

Technical Solutions

Early Type II additives and mechanism. Because Type II properties and materials are used either by themselves or as outer layers in other types of electrostatic protective packaging for good antistatic performance, they will be discussed as the base technology.

The original static-dissipative materials were based on then-current additive technologies for antifog and antistatic properties in plastic films. These additives are generally characterized as relatively long molecules with two distinctive molecular ends that can be considered "heads" and "tails" as illustrated in Figure 6. Most of the molecule's length is the tail, which is a chain of roughly 10–18 carbon atoms surrounded by hydrogen. This tail part of the molecule is nonpolar and is considered hydrophobic or polymerphilic because it prefers to "dissolve" in the polymer. The other "head" end is polar and hydrophilic and wants to diffuse out of the nonpolar polymer to the surface or "bloom." On the surface, the polar end is capable of absorbing atmospheric moisture and weakly ionizing it to create the modestly conductive electrical path necessary for static dissipation as in Figure 7. While these films use additives that were homogeneously compounded into the plastic and then extruded as a single layer, even simple "homogeneous" composition films function with distinct layers because of additive blooming to the surface.

Because the first MIL-Specs required a visually distinct identification for these materials, a pink tint was added, which is why these early Type II materials are often called "pink poly."

There were some problems with the early materials. Some of these problems, like limited lifetime and water solubility of the additive, have been noted in the discussion of terms. The limited effective lifetime problem is related to a combination of blooming and the nature of additive "reservoirs" within the bulk film. As long as there is sufficient additive concentration within the film, additive diffusion to the surface will maintain a surface concentration sufficient for good surface resistivity. However, when the interior reservoir becomes depleted, diffusion to the surface cannot maintain the necessary surface concentration and therefore the loss of surface electrical properties as illustrated in Figures 8 and 9. It was found that static dissipation required a higher additive concentration than for an antifog property, which made some pioneer films

Figure 6. Antistatic additive molecules showing dual characteristics.

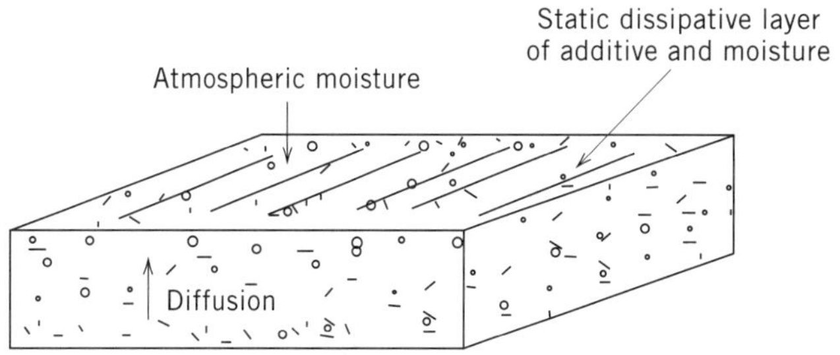

Figure 7. TYPE II film (17).

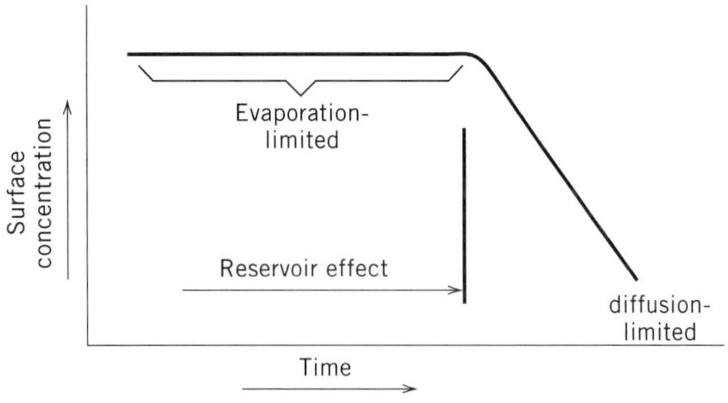

Figure 8. Reservoir effect.

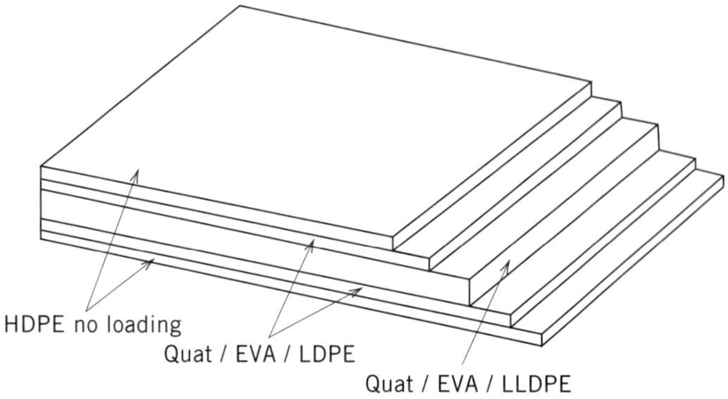

Figure 10. Multiply static dissipative film with clean skins.

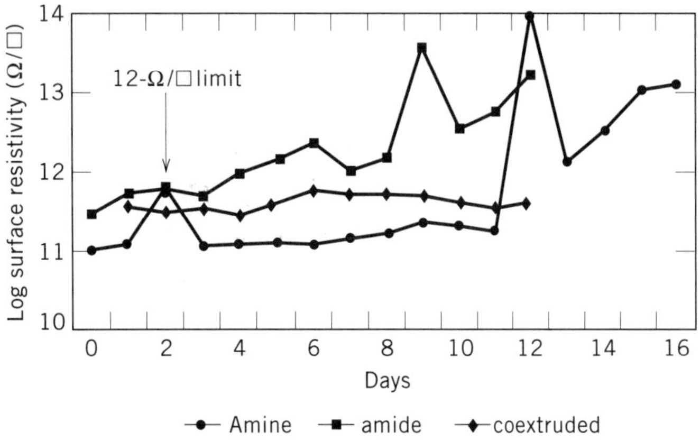

Figure 9. Log of surface resistivity versus time—reservoir effect is seen the loss of surface resistivity when the film interior becomes depleted of additive (17).

feel greasy. Refinement of the additives has improved this situation (17,18).

Multilayer Type II ESD protective films. After recognizing that the static dissipative properties of even simple Type II films were based on distinct layers, it was inevitable that coextrusion and/or coated films would be adapted to ESD protective films to maximize the properties.

To deal with the diffusion-related limitations of the early Type II films, coated or topical antistatic films were developed. The first used quaternary amines (or quats), which are excellent antistatic agents but, because they are nearly incompatible with nonpolar polymers, must be topically applied. These quats were added to a compatible polar acrylate polymer that was electron beam–cured to form a static-dissipative surface matrix (19–21). While this cured coating solved some problems with diffusion and evaporation of the additive, it was found that any contact with water would leach out the antistatic additive and, thus, the static-dissipative electrical property. There are some current materials that have improved compatibility between the additive and the coating matrix that resist leaching with modest water exposure.

The next materials evolution utilized coextrusion of multiple layers in which different static-dissipative properties could be incorporated. One of the first of this type used high-density polyethylene (HDPE) skins over a quat-loaded polar polymer matrix blended into less polar polymers as illustrated in Figure 10. This structure buried the static-dissipative layer behind insulating layers. This material had static decay times that met the then-current MIL-Spec, which required that a Type II film have either static decay time of <2 s *or* surface resistivity between 10^5 and 10^{12} Ω/square. This structure of insulators over static-dissipative layers was a major advantage because the film surface was virtually free of contamination. This same coextrusion technology was used to make low-surface-resistivity film shown in Figure 11 (17).

Additive chemistry and polycarbonate compatibility. As discussed previously, some expensive problems related to additives crazing polycarbonate were encountered. In these, ethoxylated amines were found to be the worst for their effect on polycarbonate. Polymers, like other chemistry, follows the rule "like dissolves like." While amines can begin to dissolve into polycarbonate, the polycarbonate is so rigid that it cannot swell to accept the amine molecule; instead, it forms a surface crack or "craze." These surface cracks can expand until the polycarbonate piece breaks. Fortunately, ethoxylated amides and quaternary amines were found to be more polycarbonate compatible as well as have good antistatic properties. Typical chemical structures are shown in Figure 12. The polycarbonate compatibility is illustrated in Table 3 (17).

Types I and III mechanism via Faraday cage. Types I and III materials have a common MIL-Spec heritage that is differentiated only by the degree to which they attenuate RF interference. They function by creating a Faraday cage effect where electrical charges repel each other to the maximum extent

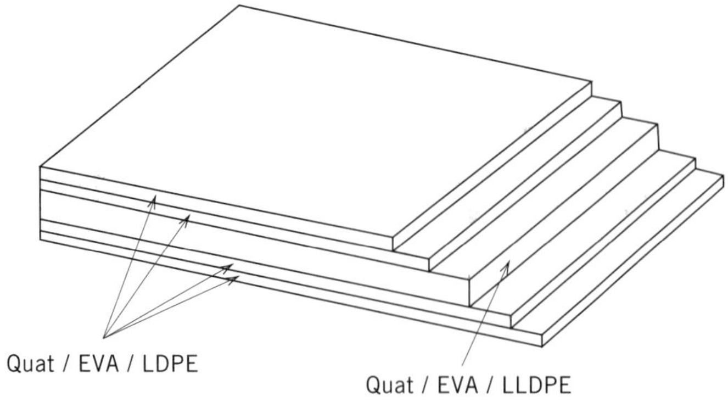

Figure 11. Multiply static dissipative film with low surface resistivity skins.

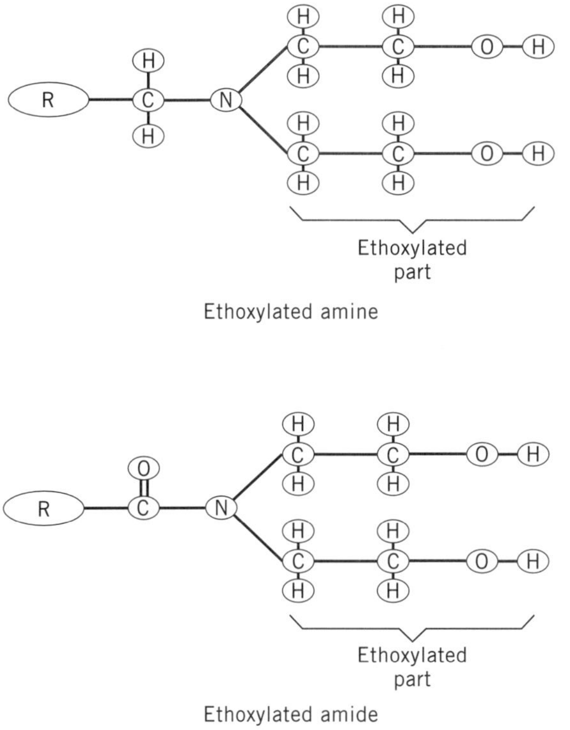

Figure 12. Typical antistatic additive molecules.

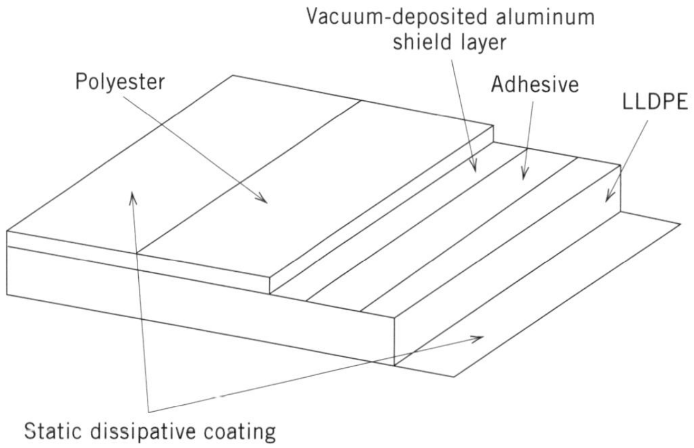

Figure 13. Typical Type III metallized bag laminated construction.

possible, that is, on an exterior surface. The electrical field inside the package is essentially nil. This is visible with metal foil containing Type I films because the metal is opaque. The foil is a barrier to more than just static. With Type III, the vapor deposited metal is controlled to a thickness that is somewhat transparent but achieves static shielding of at least 10 dB. This allows one to read some label or bar-code information through the bag without opening it and still have a high level of static dissipation. The polyester substrate film also often provides more physical protection than a foil in a polyethylene-based film sandwich (22).

Because foils and metallized films are relatively fragile against printed circuit boards with many edges and devices and because their metallic nature can help transmit some charges into the contents, commercial Types I and III materials are laminated with Type II films. This provides physical protection and a modestly conductive static-dissipative layer between the electronics components and the shielding metal layer as illustrated in Figure 13 (23). Some Type I and III bags are laminated with extruded Type II films; others use static-dissipative coatings on regular films.

Table 3. Polycarbonated Compatibility (17): Highest Compatible Stress Level (psi)

	Temperature			
Material	73°F	120°F	158°F	185°F
Ethoxylated amine	2500	1700	<1000	<1000
Ethoxylated amide	3400	2500	1700	1000
Quaternary amine	3400	3400	2500	2000
Control (air)	3400	3400	2500	2000

Heat-shrinkable ESD protective packaging. Most ESD protective packages are loose-fitting bags that function by surface static dissipation but offer no mechanism to minimize static generation by friction within the bag. To address this need, static-dissipative chemistry and structures have been adapted to heat-shrink films that, when heat-shrunk tight onto the electronics, essentially eliminate motion within the package. The exterior is then protected by the static-dissipative surface of the shrink film. Electrostatic discharge protective heat-shrink films have not been possible using metals or metallized materials and are therefore all Type II in their performance.

The first antistatic shrink films were based on multiply (multiple-ply) coextrusion, which encapsulates the additive containing layer within two "clean" skin layers. This trapped the additive and effectively eliminated corrosion and polycarbonate compatibility problems but is not currently on the market (4,24). It has been replaced, in part, by straightforward commercial shrink films that are coated with special antistatic additives.

Inherently static-dissipative polymers for ESD protective packaging. Several classes of inherently antistatic polymer have been utilized experimentally in electrostatic discharge protective packaging (25–27). While the most public attention is given to inherently conductive polymers, like polyacetylene, these polymers are often so rigid or insoluble that their packaging applications are limited. Also, in the case of polyacetylenes, these often need to be doped with some other material like iodine to achieve their conductivity. Most dopants like these are also inherently corrosive and therefore unacceptable.

One class of inherently antistatic materials that was found to have workable properties is poly(amide ethylene oxide) block copolymers. On the molecular level, these polymers incorporate hydrophilic poly(ethylene oxide) (PEO) block segments into a polyamide (nylon) structure. The polyethylene oxide absorbs and holds a sufficient amount of water to achieve static dissipation. The PEO segment length and choice of nylon are important to balance the moisture absorption for surface resistivity versus moisture that blocks heat seals (27).

In all cases discussed so far, the major drawback to the inherently static-dissipative polymers has been cost. Few materials can compete effectively with combinations of additives compounded into or coated onto inexpensive resins like polyethylene and metallized polyesters or foils that are sandwiched with static-dissipative polyethylene.

Polymer systems with conductive fillers. Conductive fillers in conventional polymers were considered a means of achieving high conductivity for static dissipation. These fillers have included aluminum flakes, carbon powders, carbon flakes, nickel-coated carbon fibers, nickel-coated mica flakes and stainless-steel fibers, to name a few. In general, once a conductive filler is added in sufficient quantity to physically connect particle to particle, the resin matrix becomes essentially metallic in electrical conductivity. The practical problems have been numerous and include opacity, reliable extrusions, electrical properties, and relative cost. In general, these systems are applicable to injection or compression molding but have limited use with films (28). A carbon-loaded film was successfully marketed for a while with a modest surface resistivity. However, it fell into disfavor when it was noticed that the black bag would leave pencil-like black streaks that were conductive. Conductive fillers have not been satisfactory solutions.

Organizations Concerned with ESD Protective Packaging and Appropriate Standards

Given the safety and economic issues behind electrostatic-dissipative protective packaging, it was not surprising that a group of organizations have participated in opening communications and formulating standards. The three groups listed below have issued pertinent standards for the industry.

American Society for Testing Materials (ASTM)

D257, *Standard Test Methods for DC Resistance or Conductance of Insulating Materials*

D-991 *Standard Test Method for Rubber Property–Volume Resistivity of Electrically Conductive and Antistatic Products*

Department of Defense

Federal Test Method Standard 101C, *Method 4046.1 Test Procedures for Packaging Materials*

MIL-B-81705 Military Specifications, *Barrier Materials, Flexible, Electrostatic-Free, Heat Sealable*

MIL-HDBK-263 Military Handbook, *Electrostatic Discharge Control Program for Protection of Electrical and Electronic Parts, Assemblies and Equipment*

MIL-HDBK-773 Military Handbook, *Electrostatic Discharge Protective Packaging*

MIL-STD-883, *Test Methods and Procedures for Microelectronics*

MIL-STD-1686, *Electrostatic Discharge Control Program for Protection of Electrical and Electronics Parts, Assemblies and Equipment*

Electronic Industries Association*

EIA-541, *Packaging Material Standards for ESD Sensitive Items*

EIA-564, *EIA*

EIA-583, *Packaging Material Standards for Moisture Sensitive Items*

ESD Association†

ADV 1.0 *For Electrostatic Discharge Technology—Glossary*

S 3.1, *Ionization*

S 8.1, *ESD Awareness*

S 11.1.1, *Surface Resistance Measurement of Static Dissipative Planar Materials*

S 11.2.1, *For Protection of Electrostatic Discharge Susceptible Items—Triboelectric Charge Accumulation Testing*

S 11.3.1, *For Evaluating the Relative Performance of Electrostatic Shielding Bags*

* EIA address: 2500 Wilson Blvd., Arlington, VA 22201-3834.

† ESD (formerly Electrical Overstress/Electrostatic Discharge Association or EOS/ESD Assoc.) address: 7902 Turin Road, Suite 4, Rome, NY 13440.

BIBLIOGRAPHY

1. "The ESD Control Process is a Tool for Managing Quality," *Electron. Packag. Prod.,* 50–53 (April 1990).
2. J. Jesse, "ESD Control," *EOS/ESD Technolo.,* 13 (Dec.–Jan. 1988).
3. R. J. Pierce, "Static Control Pays," *EOS/ESD Technolo.,* 14–19 (April/May 1990).
4. M. R. Havens and D. L. Hines, "Static Protection through Shrink Packaging," *Proce. Technical Program,* Vol. II, NEPCON West, Anaheim, CA, Feb. 23–27, 1992, pp. 821–827.
5. MIL-B-81705, Military Specification, *Barrier Materials, Flexible, Electrostatic Protective, Heat Sealable.*
6. S. L. Fowler, "Triboelectricity and Surface Resistivity Do Not Correlate," *EOS/ESD Symposia Proc.,* in press.
7. ESD Assoc. S 11.2.1, *For Protection of Electrostatic Discharge Susceptible Items—Triboelectric Charge Accumulation Testing,* ESD Assoc., Rome, NY (1995).
8. W. Klein, "On Resistivity," *Threshold* pp. 6–9 (EOS/ESD Association Newsletter (now the ESD Assoc.) (March 1990).
9. S. Fowler, J. Lovin, and B. Carson, "Computer Analysis of Static Decay Curves," *Threshold,* 1–3 (Sept. 1987).
10. S. L. Fowler, "Surface Resistivity and Static Decay Do Not Correlate," *EOS/ESD Symposia Proc.,* New Orleans, Sept. 26–28, 1989, pp. 7–11.
11. ESD Assoc. S 11.3.1, *For Evaluating the Relative Performance of Electrostatic Shielding Bags.*
12. L. E. Walp, T. E. Breuer, and S. E. Golyer, "Modification of the Electrostatic Behavior of Polyolefins," *ANTEC 86 Proc.,* SPE Annual Technical Conference, 1986, pp. 1141–1143.
13. J. M., Kolyer and J. D. Guttenplan, "Corrosion and Contamination by Antistatic Additives in Plastic Films," *EOS/ESD Symposia Proc.,* 1988.
14. B. Borodoli, "Characterization of Corrosivity of Antistatic Packaging Materials," *EOS/ESD Symposia Proc.,* 1988.
15. W. Kropf and S. Royce, "Qualifying ESD-Protective Materials," *EOS/ESD Technol.,* 20–24 (Feb./March 1990).
16. EIA-564.
17. M. R. Havens, "Understanding Pink Poly," *EOS/ESD Symposia Proc.,* New Orleans, Sept. 26–28, 1989, pp. 95–101.

18. M. R. Havens, "The Chemistry of Pink Polyethylene," *EOS/ESD Technol.* pp. 8–18 (Feb./March 1990).
19. A. H. Keogh and G. T. Sydney, "The Versatility of Electron Beam Processing and the Conversion of Medium and High Performance Films for ESD Protection," *EOS/ESD Symposia Proc.,* Sept. 11–13, 1990.
20. A. H. Keogh, "Electron Beam Radiation Cured Coatings for Static Control," *EOS/ESD Symposia Proc.,* in press.
21. A. H. Keogh, "Antistatic Resin Composition," U.S. Pat. 4,623,594 (Nov. 18, 1986).
22. "Transparent Bags Foil Static Electricity's 'Zap'," *Package Eng.* pp. 104–107 (Jan. 1983).
23. C. L. Mott, "Antistatic Sheet Material and Package," U.S. Pat. 4,756,414 (Sept. 15, 1992).
24. D. L. Hines and W. V. Duncan, "Shrink Film Packaging Evaluation," *EOS/ESD Symposia Proc.,* Dallas, Sept. 16–18, 1992, pp. 5A1.1–5A1.3.
25. T. E. Fahey and G. F. Wilson, "Inherently Dissipative Polymer Films," *EOS/EST Symposia Proc.,* Dallas, Sept. 16–18, 1992, pp. 5A3.1–5A3.6.
26. T. R. Maas, "ESD Polymer Alloys: An Alternative Approach for Producing Permanently Antistatic Polyethylene," *EOS/ESD Symposia Proc.,* Sept. 11–13, 1990.
27. M. R. Havens, "Inherently Static Dissipative Packaging Films," *EOS/ESD Symposia Proc.,* Las Vegas, Sept. 24–26, 1991, pp. 204–209.
28. "Conductive Thermoplastics Compounded for Static Control Applications," *Compliance Eng.* (in press).
29. MIL-HDBK-263A, *ESD Control Handbook for Protection of Electrical and Electronic Equipment,* Naval Sea Systems Command, Washington, DC.

Marvin Havens
Cryovac Division, W. R. Grace & Company
Duncan, South Carolina

Steve Fowler
Fowler Associates Inc.
Duncan, South Carolina

ENVIRONMENT

The term *environment* subsumes many concepts related to how packages and packaging systems function in the broader context of our planet, its people, and its ecosystems. In recent years packaging has often been attacked as being a prime contributor to environmental problems, especially in the area of solid waste. While it is certainly true that much of this criticism was unjustified, it is also true that until the last decade packaging designers and manufacturers often paid little or no attention to the environmental impacts of their packages. This has now changed. Driven by consumer and legislative pressure, as well as the desire to be good corporate citizens, many companies have made consideration of environmental impacts a routine part of the package design process. In this article, we discuss the broad topic of environmental impacts of packaging, along with a brief overview of some of the major current environmental concerns impacting packaging, and a more detailed look at solid-waste issues, which is the environmental issue the public has, during the past several years, most commonly associated with packaging. Other environmental issues of importance include resource depletion, pollution, effects on the ozone layer, and global warming.

Resource Depletion, Conservation, and Sustainable Use

Depletion of natural resources may occur when nonrenewable resources are utilized in the production of packaging materials. For example, the manufacture of aluminum cans may consume aluminum ore, which is a finite resource. (Ore is not consumed if the cans are made from 100% recycled aluminum.) Whether this is a serious concern is dependent on the relative scarcity or abundance of the resource in question, and also on the amount that is used for packaging. For example, large amounts of sand and sandstone are used in the production of glass bottles, but the earth's supply of this resource is abundant, so resource depletion is not a serious concern. In contrast, the earth's supply of chromium is rather small, but only relatively tiny amounts of chromium are used in the manufacture of tin-free steel for packaging, so while the overall global depletion of chromium may be a concern, depletion of chromium due to packaging is not a serious problem. In principle, use of renewable resources is less of a concern than use of nonrenewable resources since renewable resources, by definition, can be replenished. The major raw materials used for all the major packaging materials (wood, paper, aluminum, steel, glass, and plastics) are, fortunately, relatively abundant at this time, at least to the extent that serious shortages are not likely to arise in the near future, given current and projected patterns of use. Recycling of materials, of course, can extend their availability.

In addition to material resources, packaging is also a consumer of energy resources. The typical mix of energy resources used is different for different packaging materials. In the production of paper, renewable biomass (wood) is a major energy source, as well as being the raw material. On the other hand, in production of paper from recycled fiber, petroleum products are the primary energy source. For plastics, petroleum and natural gas are raw materials, as well as energy sources. The aluminum industry is heavily dependent on electrical power, and production of aluminum metal from ore is highly energy-intensive. The glass industry relies primarily on natural gas and electricity as energy sources. The manufacture of steel from ore uses coal as the primary energy source. Transporting of packaging materials of all types uses primarily petroleum-based energy. All packaging materials use electricity at various points in their manufacture. Electricity is not a primary energy source, and depends on a mix of energy technologies, including hydroelectric and nuclear power, but the United States depends most heavily on coal. Modification of packaging processes to make them more energy-efficient can result in significant cost savings, as well as environmental benefits. Beginning in the late 1970s and early 1980s, increasing fuel prices and consequent interest in energy efficiency led to a number of studies comparing the energy use of various alternative packaging systems for products (1–3). These are the forerunners of more recent life-cycle assessment studies.

Waste-to-energy incineration can recover valuable embedded energy from combustible packaging materials, and thus diminish dependence on other fuel sources.

Pollution

Emission of air and water pollutants during the manufacture, distribution, and disposal of packaging also impacts the environment. Different packaging materials are associated with different types of pollution concerns. Manufacture of paper, for instance, produces water effluents with a high load of organic materials, producing a significant biological oxygen demand (BOD). In the manufacture of plastics, on the other hand, air emissions of volatile organics are generally a more serious concern than water emissions. For all types of packaging materials, proper control and treatment of effluent streams is important for protection of the environment, and is mandated by government regulations. This is true not only for manufacture of packaging materials but also for the converting processes that change these raw materials into packages. The practice of waste minimization (defined as reduction of the production of toxic materials), commonly practiced in the chemical industry, can also have benefits for the packaging industry in dealing with requirements for proper treatment of effluent streams. Simply put, if a toxic material is not used or produced, it will not be present in effluent streams, and thus will not require treatment for its removal. An example that is commonplace in packaging is the substitution of water-based inks and adhesives for solvent-based systems in a variety of printing and converting operations. This substitution enabled many converters to avoid the necessity of installing expensive air-pollution-control systems to collect volatile organics when new air emission regulations were enacted.

Three particular types of pollution concerns have been given substantial attention in recent years.

Ozone depletion. In the mid-1980s, it was discovered that an "ozone hole" was formed over Antarctica in the spring, and a vast body of scientific evidence has since accumulated linking the destruction of stratospheric ozone with human-made (synthetic) chlorofluorocarbons and similar chemicals. As a result, there is an international agreement to phase out production of these chemicals. After 1996, no fully halogenated CFCs are to be manufactured in industrialized countries (4).

Well before this deadline, the U.S. packaging industry responded to this environmental threat. Earlier fears about ozone depletion had lead to legislation banning them from most uses as aerosol propellants in the United States in 1978. By 1990, CFCs were no longer being used for food service expanded polystyrene packaging, and remaining packaging uses quickly disappeared.

Global warming. In the late 1980s, a number of scientists became increasingly concerned about the possibility that increases in the atmospheric concentration of carbon dioxide and other "greenhouse gases" was acting to increase average global temperatures. Consequences of such a temperature rise could include increases in ocean levels and consequent flooding, changes in rainfall patterns, change in ecosystems with adverse effects on animal and plant populations, and a variety of other effects, including increases in the frequency and severity of tropical storms (5).

Use of fossil fuels is associated with increased atmospheric carbon dioxide, as is slash-and-burn cultivation of agricultural land and destruction of tropical rainforests, among others.

While our current understanding of the determinants of global climate is far from complete, there is a growing body of scientific experts who feel there is real cause for concern.

Remedies proposed to slow down the rise in greenhouse gases often include attempts to stabilize CO_2 emissions by decreasing use of fossil fuels, perhaps by increasing their price, as well as attempts to provide more CO_2 sinks by reforestation efforts. The packaging industry could be affected, as could other industries, by mandates for increased fuel efficiency, and certainly would be affected by increases in energy cost.

Chlorine and chlorinated organics. In recent years, there has been increasing concern about the effects of chlorine, and in particular of chlorinated organic compounds, in our environment. Initial attention focused largely on dioxins. Considerable concern was expressed about incineration of polyvinyl chloride (PVC) because of fear that presence of PVC in incinerators would lead to increased dioxin production. Despite evidence that PVC is at most a minor factor in dioxin production in incinerators, with combustion and emission-control system conditions being much more significant, opponents of incineration still raise this as an issue. More recently, concern has been raised about possible hormonal disruption effects associated with a large variety of chlorinated organics. PVC has again been targeted as a contributor to these problems. Organizations such as Greenpeace have called for elimination of chlorine-based chemical production processes. The International Joint Commission on the Great Lakes has called for a phaseout of chlorine feedstocks, the International Whaling Commission and the Barcelona Convention on the Mediterranean Sea have called for the elimination of all organohalogens, and the American Public Health Association has called for phaseout of chlorine-containing compounds. On the other hand, the U.S. Environmental Protection Agency and Canada's Ministry of Environment both oppose a phaseout of chlorine feedstocks, and the Vinyl Institute, a number of PVC manufacturers, and others are actively trying to defend the environmental image of PVC. Paper manufacturers have also been involved in the controversy, as dioxins and other chlorinated organics are also associated with paper production, especially when chlorine-based bleaching agents are used.

The controversy about the relative risks and benefits of chemical process involving chlorinated organics is likely to remain active for some time (6).

Solid-Waste Issues

Much of the criticism of packaging in recent years has been related to its disposal. During the 1980s, a number of municipalities in the United States, particularly on the East Coast, began to encounter significant problems with lack of availability of disposal facilities for municipal solid waste (MSW). Packaging accounts for approximately one-third of the MSW stream by weight, a fraction that has been relatively constant in recent years (7), so it became a natural target for efforts to decrease the amount of MSW requiring disposal. At the same time, there was a change in the general philosophy about solid-waste management techniques from a reliance on landfilling to a more complex mixture of options.

In the United States, this was expressed, by the EPA and others, as the philosophy of integrated solid-waste management incorporating the solid-waste management hierarchy. This hierarchy placed waste reduction at the top, as the most desirable option, followed by reuse (sometimes incorporated in reduction), recycling, incineration, and landfilling, in that order. As composting began to grow in importance, it was generally identified as a recycling option. The goal then became to handle waste materials, to the extent possible, at the top rather than the bottom of the hierarchy.

Waste reduction. Waste reduction, or source reduction, is regarded as the best option. Avoidance of packaging prevents impacts related not only to waste disposal but also to production, processing, and transportation of materials. Of course, this must be balanced against the function of the package. Elimination of packaging would mean the elimination of many of the products we enjoy, as well. Similarly, underpackaging can lead to significant product damage, thus increasing waste. The packaging industry has a long tradition of source reduction, accomplished for the most part with the goal of saving money by using less material for packages. Glass bottles, plastic bottles, metal cans, and many other package forms have become thinner and lighter over the years.

Reuse. Reuse of packages is an old idea that is seeing new life in some applications. We long had a system of refillable bottles for carbonated beverages, milk, and other products. In the United States, refillable bottles for milk nearly disappeared, only to return on a small scale as refillable polycarbonate bottles in school lunch programs. Refillable beer and soft-drink bottles have been declining slowly but steadily, and have yet to show any significant resurgence. Reuse of corrugated boxes is infrequent but does occur. Some industries, notably automotive parts and office furniture, have switched from corrugated to plastic or metal crates or bins, in large part to permit increased reusability. Plastic pallets have captured a small but significant market share for the same reason. Consumer reuse of laundry detergent and other bottles by refilling them from a source-reduced refill package is a combination of source reduction and reuse.

Recycling. Much of the attention focused on changes in package design has been directed at increasingly the recyclability of packages or increasing their use of recycled materials. There are significant differences between packaging materials in average recycling rates and in the potential for incorporation of recycled content (see Recycling; Recycling, Europe).

Aluminum is currently recycled in the United States at the highest rate of any packaging material. Aluminum beverage cans receive the largest use of aluminum in packaging. The recovery rate for these containers in 1993 was estimated at 63.4%. The overall recovery rate for aluminum packaging was 53.0%, since recycling of other forms of aluminum packaging was much less prevalent (7). A substantial portion of the aluminum beverage cans collected come from the nine states with some form of deposit or refund value for these containers (one state exempts aluminum cans from its deposit). Most of the remainder come from a network of buyback centers, set up largely by aluminum companies, and from curbside collection programs. Aluminum cans are often one of the most valuable materials collected in curbside recycling programs. New aluminum cans are an important end market for recycled aluminum. Using recycled aluminum is economically attractive, due to both the high cost of virgin raw materials (bauxite), and to the large energy savings associated with making aluminum sheet from recycled aluminum rather than from bauxite (95% energy savings).

Steel cans are also recycled at an appreciable rate, 46.3% in 1993 (7). While steel cans are also commonly collected in curbside recycling programs, recovery of steel from commingled garbage associated with resource recovery facilities (waste-to-energy incineration) is another important source. This is made possible by the magnetic nature of steel, which allows it to be fairly easily separated from mixed waste streams. While steel cans generally contain some recycled content, most collected cans are used in nonpackaging applications. Used cans are being increasingly recognized as a source of high-quality steel scrap.

The recycling rate for glass containers is considerably lower than for aluminum and steel, about 24.6% in 1993 (not including refilling) (7). In contrast to aluminum, raw materials for glassmaking are relatively low in cost. While energy savings do accrue from using recycled glass (cullet) in making new containers, they are significantly more modest than with aluminum. The energy required to process the cullet and transport it to the glassmaking facility may consume a significant amount of this savings. An additional problem with glass recycling is the requirement for separation of the glass by color. This is generally accomplished by hand sorting, although automated equipment for this purpose does exist. Sorting can be accomplished fairly readily by either of these methods for whole containers, but broken glass presents a serious problem. For safety reasons, it cannot be sorted by hand. Automated equipment that is capable of performing sorting of broken glass with enough accuracy to yield container-quality cullet is not yet commercially available. Facilities for sorting commingled recyclables typically lose a substantial fraction of their incoming glass stream to breakage. Uses for mixed-color cullet are being found in the construction and related industries, where it can substitute for aggregate in concrete or asphalt, for example. Container manufacturing can accommodate very little color contamination in the glass, but remains the major market for recycled glass. Production of fiberglass has also become an important market, especially in California, where manufacturers are required to incorporate recycled content. California requires all glass containers sold or manufactured in the state to contain 35% recycled content, effective January 1, 1996, increasing in steps to 65% on January 1, 2005 (8). The major sources of collection of glass containers for recycling are deposit/refund-value laws for beverage containers in 10 states, and curbside collection programs. Dropoff programs are an additional source.

Recycling of paper and paperboard packaging materials is heavily dependent on the particular form of packaging involved. Corrugated has the highest recycling rate, 55.5% in 1993. Most of this material originates in the business sector, rather than from individual consumers. Retail stores, factories, etc can often achieve savings in their waste-disposal bills by participating in programs to source-separate corrugated boxes for recycling. In addition, some facilities divert selected loads of waste, known to contain a high percentage of corrugated, to sorting facilities where hand sorting is used to recover this material. An additional source of material is dropoff collection sites. Little corrugated is collected at curbside. Major end-use markets for corrugated containers include manufacture of new corrugated containers, boxboard, and kraft paper for grocery sacks. While in general paper fibers are shortened and weakened by the processing required to repulp and clean the fibers, it is possible to make recycled grades of board with properties near those of virgin materials, especially when the recycled content is blended with virgin fiber. Kraft bags and sacks had a recycling rate of about 15.9% in the United States in 1993, considerably lower than the rate for corrugated. Some of this is collected through dropoff sites or curbside, often along with corrugated or newspapers. Recycling of folding cartons was almost unheard of before 1990, but reached a rate of 14.2% in 1993. Recycling of milk cartons and aseptic "juice boxes" is just beginning to emerge now. There is very little recycling of other types of paper or paperboard packaging. The overall recycling rate for paper and paperboard packaging in 1993 was 44.2%. In addition to domestic uses, export of recovered paper is significant (7).

Recycling of plastic packaging differs considerably by resin and container type. The most-recycled plastic package is PET soft-drink bottles, which were recycled at a 48.6% rate in 1994 (along with the HDPE base cups on many of those bottles) (9). The 10 deposit/refund states and curbside collection programs provide the major source of this material. Custom PET bottles, which now account for over a third of all PET bottles, are recycled to a smaller extent than soft-drink bottles, bringing the overall PET bottle recycling rate down to 34% in 1994 (10). The primary source of custom bottles for recycling is curbside collection programs. The major use for recycled PET is fiber applications such as fiberfill, carpet, and clothing. Other uses include nonfood containers, automotive parts, and paint brushes. Recycled PET is accepted for some types of direct food contact, such as egg cartons and vegetable trays. PET can also be used in food containers if the recycled material is in an inner layer, with the food protected from direct contact by a layer of virgin resin (11,12). Recently Johnson Controls was granted a letter of nonobjection by the FDA for essentially unlimited direct food contact for PET produced by their patented physical recycling progress. PET is also recycled by tertiary recycling processes (methanolysis and glycolysis) that chemically break the polymer down into monomers, purify them, and use them as feedstocks for production of PET that is identical to virgin resin (13). Demand for recycled PET has generally exceeded supply in recent years.

The next most widely recycled plastic packaging material is natural HDPE bottles, such as those for milk, which were recycled at a nearly 26% rate in 1994 (14). These are collected predominantly through curbside programs. The high-quality unpigmented homopolymer is suitable for a variety of uses, including applications in packaging, housewares, toys, pipe, and traffic cones. Significant packaging applications include motor-oil bottles (blended with virgin resin), household chemical bottles (as a buried inner layer), and bins and pallets. Some recycled HDPE is also used in film form, primarily for shopping bags. Recovery of pigmented HDPE is less than for unpigmented material, 10.8% in 1994 (14), and its uses are more limited because of the complications caused by the mixtures of colors. Motor-oil bottles are one major market. The demand for recycled HDPE tends to be strongly influenced by ups and downs in the supply (and cost) of virgin resin.

A significant amount of plastic film is recycled from two major sources, stretch wrap and bags. Pallet stretch wrap is widely collected for recycling, although precise recycling rates are not readily available. This material is collected in the same way as most corrugated materials—through commercial sources. A large number of grocery and other retail stores maintain dropoff bins for collection of plastic merchandise bags, which are recycled at a rate of 10–15%, mostly by this mechanism (15). A major market for recycled plastic bags and film is trash bags. Other markets include housewares, plastic lumber, agricultural film, and merchandise bags. The U.S. EPA estimated the overall recycling rate for plastic bags and sacks as 1.9% and for wraps as 1.6% in 1993 (7).

Polystyrene packaging is being recycled to a limited extent in the United States, primarily by the National Polystyrene Recycling Company, which concentrates on food-service EPS, and by the members of the Association of Foam Packaging Recyclers, which concentrate on cushioning materials. Reliable recycling rates for PS packaging are not readily available. The U.S. EPA listed recovery from packaging sources as "negligible" in 1993, while Leaversuch estimated at least 50 million lb recovered from cushioning materials and insulation alone (7,16).

Little recycling of other plastic packaging materials is occurring in the United States. PVC is widely recycled in Europe, but not much in the United States. Some recycling of commingled plastics occurs; the major products are substitutes for wood or concrete, often for outdoor use. Examples include park benches, fences, landscape timbers, and parking stops.

As can be seen above, the most desirable way to collect materials for recycling is significantly influenced by the type of material to be collected, and where it is most prevalent. For corrugated boxes and stretch wrap, collection efforts focus on the commercial sources that discard most of those materials. For beverage bottles, deposit legislation, passed as a litter-prevention measure, is a very effective method of collection, in terms of recovery rate, averaging ≥90% return of covered containers. For aluminum, buyback centers are able to offer a financial incentive sufficient to accomplish significant returns, although considerably less than those achieved by deposits or by California's refund system. For most consumer packages, curbside collection appears to be the most effective system, achieving recovery rates considerably higher than dropoff centers. (17).

Composting. Composting is the managed biodegradation of waste materials. It is growing rapidly as a disposal option for yard waste, but is still very new, in the United States, as an organized system for disposal of MSW. Only the biodegradable fraction of MSW can be handled by composting. This includes most paper packaging, but almost no plastic packaging, and no metal or glass. In operation of MSW composting facilities, odor problems are a significant concern, so they are often designed for in-vessel composting, in contrast to yard waste facilities, which are most often open-air. Markets for MSW compost are uncertain at this time (18).

Incineration. The next step down the solid-waste management hierarchy is incineration. As practiced in the United States today, it is always coupled with energy recovery, and thus offers not only a reduction in the volume of solid-waste requiring landfilling but also the generation of energy, which can displace some combustion of coal. To avoid unacceptable air pollution, sophisticated and expensive pollution control equipment is required. The ash generated from the unburned residue and from the pollution-control equipment may contain levels of heavy metals high enough to require disposal as a hazardous waste. While incineration has grown as a waste management option since the early 1980s, its growth has slowed considerably, in the face of substantial public opposition to picking a site these facilities.

The growth in incineration was instrumental in the development of U.S. state legislation prohibiting the deliberate introduction of lead, cadmium, mercury, and hexavalent chromium into packaging materials (except in narrowly defined circumstances) and in limiting their inadvertent presence. These laws, which have passed in at least 18 states, were based on model legislation promulgated by CONEG, the Coalition of Northeastern Governors, and are commonly referred to as Toxics in Packaging legislation (8).

Landfill. Landfilling is at the bottom of the solid-waste management hierarchy, seen as the least-desirable option, but nevertheless one that we cannot do without. There are always residual materials that cannot effectively be handled by the options higher in the hierarchy. The major drawbacks of landfilling are the space it uses and its potential for groundwater contamination. On the positive side, landfills are often the least-expensive option for many types of waste, and can serve as a source of methane that can be used as a fuel.

Other Environmental Impacts

In addition to the effects discussed above, packaging can be involved in the production of a variety of other environmental impacts, such as destruction of natural habitats, loss of scenic beauty, and noise pollution, For the most part, packaging plays only a minor role in such impacts.

Life-Cycle Assessment

The goal of environmentally responsible packaging design is to minimize the adverse environmental impacts of packages, their associated systems, and the products they contain. While this can be easily stated, it is not easy to do. The concept of life-cycle assessment is precisely the analysis of the environmental impacts of product systems, including their packaging, with the goal of selecting the system that has the least overall adverse environmental impacts, or of modifying systems to diminish these impacts.

When we apply these concepts to packaging, we must not forget that changes in packaging can lead to changes in rates of product damage, to modification of packaging lines, to changes in transportation systems, etc, which can also have environmental impacts. Similarly, changes in products can permit or result in changes in packaging that modify overall environmental impacts associated with the package system.

Life-cycle assessment, as formulated by the Society for Environmental Toxicology and Chemistry (SETAC), has three components: the *inventory analysis,* which lists inputs and outputs for the product or process life cycle being analyzed; the *impact analysis,* which seeks to evaluate the effects of these inputs and outputs; and the *improvement analysis,* which searches for ways to decrease the environmental effects associated with the product or process (19).

The whole science of life-cycle assessment is still developing. The inventory analysis is the most developed; and the impact analysis, less developed. The improvement analysis can proceed, in at least some cases, without completion of the other two stages, based on the simple concept that a decrease in resource use or emissions, without substantial other changes, is generally a positive step. The U.S. EPA is supporting research on the continued development of this tool. It is also being used to a considerable extent in Europe.

Legislation

No discussion of packaging and the environment would be complete without a mention of legislation, which has been, in many cases, the primary driving force behind the increased attention given to environmental impacts. The mid-1990s have brought a slowing in the pace of legislative demands for packaging changes, but they remain an important factor. In the United States, legislation at the state level remains more significant than federal legislation. In Europe, both EEC (European Economic Community) and individual country legislation are significant (20). (See also Environmental Regulations; International Environmental Regulations, North America).

Green Marketing

The environmental attributes of a package are sometimes used as a marketing tool. In a number of cases, enforcement activities have been undertaken against companies making such claims. It is extremely important for any business contemplating environmental claims on a package in the United States to ensure that the claims are in compliance with state regulations and with the federal guidelines issued by the U.S. Commerce Department (8).

BIBLIOGRAPHY

1. L. L. Gaines, *Energy and Materials Use in the Production and Recycling of Consumer-Goods Packaging,* ANL/CNSV-TM-58, U.S. Dept. Commerce, National Technical Information Service, 1981.
2. Arthur D. Little, Inc., *The Life Cycle Energy Content of Containers: 1991 Update,* report to The American Iron and Steel Institute, The Steel Can Recycling Institute, Ref. 67276, Arthur D. Little, Inc., 1991.
3. I. Boustead and G. F. Hancock, *Energy and Packaging,* Ellis Horwood, Chichester, UK, 1981.
4. P. Zurer, *C&EN,* 8–18 (May 24, 1993).
5. B. Hileman, *C&EN,* 18–23. (Nov. 27, 1995).
6. I. Amato, *Garbage,* 30–39 (Summer 1994).
7. U. S. Environmental Protection Agency, *Characterization of Municipal Solid Waste in the United States, 1994 Update,* EPA 530-R-94-042, Washington, DC, 1994.
8. *Environmental Packaging: U.S. Guide to Green Labeling, Packaging and Recycling,* Thompson Publishing Group, Washington, DC, 1995.
9. R. Woods, *Recycling Times,* 6 (June 13, 1995).
10. S. Apotheker, *Resource Recycling,* 27–34 (Sept. 1995).
11. M. Bakker, *Resource Recycling,* 59–64 (May 1994).
12. T. Ford, *Plast. News,* 6 (Aug. 28, 1995).
13. A. Bisio and M. Xanthos, eds., *How to Manage Plastics Waste: Technology and Market Opportunities,* Hanser, Munich, 1994.
14. L. Rabasca, *Recycling Times,* 11 (Aug. 22, 1995).
15. P. McCreery, *Recycling Times,* 10 (April 4, 1995).
16. Anon., *Packaging Technol. Eng.,* 16 (Nov./Dec. 1994).

17. BioCycle, *The BioCycle Guide to Maximum Recycling,* The JG Press, Emmaus, PA, 1993.
18. BioCycle, *The BioCycle Guide to the Art & Science of Composting,* The JG Press, Emmaus, PA, 1991.
19. B. W. Vigon et al., *Life-Cycle Assessment: Inventory Guidelines and Principles,* EPA/600/R-92/245, NTIS, Washington, DC, 1993.
20. *State Recycling Laws Update,* Raymond Communications, Riverdale, MD, 1995.

General References

F. Lox, *Packaging and Ecology,* Pira International, Leatherhead, Surrey, UK, 1992.

M. Murphy, ed., *Packaging and the Environment,* Pira International, Leatherhead, Surrey, UK, 1992.

E. Rzepecki, *Packaging and Environmental Issues,* St. Thomas Technical Press, St. Paul, MN, 1991.

S. Selke, *Packaging and the Environment: Alternatives, Trends and Solutions,* rev. ed., Technomic, Lancaster, PA, 1994.

E. J. Stilwell, R. C. Canty, P. W. Kopf, A. M. Montrone, *Packaging for the Environment: A Partnership for Progress,* American Management Assoc., 1991.

Tellus Institute, *Tellus Packaging Study,* Tellus Institute, Boston, 1992.

SUSAN E. SELKE
Professor, School of Packaging
Michigan State University
East Lansing, Michigan

ENVIRONMENTAL REGULATIONS, INTERNATIONAL

Introduction

Outside of the United States more than 20 countries have producer responsibility for packaging. This article covers the international regulations. Most of the material has been taken from *Recycling Laws International* (see ref. 1).

Europe

European lawmakers have a different approach to ensure recycling than those in the U.S. Unlike U.S. legislatures that have created hundreds of laws to encourage companies to use recycled material, those in Europe apparently believe that it makes more sense to force manufacturers to take more of a direct responsibility for packaging waste.

The assumption is that the less environmentally sound containers will cost more to manage and consumers will choose the better materials, and that manufacturers will supposedly use the material in their new products.

In Europe, sending the package back to a local manufacturer is common for some beverages, such as beer. Container deposits (2) are common and accepted. In some areas, most beer is sold in refillable containers which are returned through retailers.

The European Union enacted the Liquid Food Containers Directive in 1985; (which was not implemented in most countries). In 1991, the European Union started work on a Directive on Packaging and Packaging Waste, as concern mounted over dwindling landfill space and wise use of resources. In December 1994, the Directive passed the European Parliament.

The Directive requires that by July 2001, countries recover 50% minimum of their used packaging; the material recycling rate must be 25% minimum, with no material recycled at less than 15%. Member states recovering more than 65% or recycling more than 45% must demonstrate they have enough capacity to handle the material.

Exceptions: Greece, Ireland and Portugal have until the end of 2005 to meet these targets and will only be held to 25% recovery by July 2001.

While the Directive's intent was to harmonize national, measures on package recycling, about the only thing constant will be the goals, and eventually the required symbols, on the packages. Each country can set up its own "economic instruments" (i.e., taxes, deposits, industry collection fees, or a combination), and each country can stress different types of packaging from different sources. It is likely there will most likely be separate systems for sales and transport packaging complicating the patchwork.

All packaging affected. The Directive applies to all packaging. There are no specific exemptions for institutional, commercial, or other packaging that may be difficult to recycle. Every company supplying companies within the European Union with raw materials for packaging, finished packaging, packaging components, or packaged goods will be affected by the Directive, as will distributors of packaged goods and companies engaged in the collection, sorting, recovery, or disposal of packaging waste.

The Directive also covers disposable cups, plates, and cutlery used in the food service sector, but does not include the large road, rail, ship and air containers. In principle, the Directive covers all packaging placed on the market within the EU and all packaging waste, whether disposed of at industrial or commercial sites or in private homes.

However, the European Commission and a committee of government experts from the member states (the "Article 21 Committee"), will be deciding what to do about any problems in applying the Directive. In particular, primary packaging for medical devices and pharmaceutical products, small packaging, and luxury packaging will be examined. Meanwhile, medical packaging is *covered* by the Directive until they decide on exemptions. This is not an exclusive list and it is open to industry to propose other exemptions.

The Directive is binding on national governments, not on individual companies. Companies will simply be responsible for complying with whatever legal requirements are laid down at the national level.

National governments must ensure that systems are set up for the return or collection of used packaging, to ensure that it is effectively reused or recovered. For packaging which ends up in the home, recovery organizations are already in full operation in Germany and Austria, while France and Belgium have fully functioning organizations funding and coordinating pilot programs.

The legal basis for these operations varies: they can be "Take back," which obliges a company to take back directly or contract with a third party; "Give back," where companies are legally obliged to ensure that the used packaging material in their possession is either "valorized" (which means either recycled, composted, or sent for incineration with energy recovery), or passed to an organization which will take responsibility for valorizing it; or "Join up," companies are legally obliged to join (and pay a fee) to an approved recovery organization, which is responsible for meeting mandated targets.

Member states must ensure that packaging complies with certain essential requirements. These include minimization of packaging weight and volume to the amount needed for safety, hygiene, and consumer acceptance of the packed product; of noxious or hazardous constituents; and suitability for reuse, material recycling, energy recovery, or composting.

Packaging which complies is guaranteed free access to Community markets from July 1996 and packaging which does not comply will be banned from January 1998. CEN, the European standards organization, has been invited by the Commission to draw up standards relating to the "essential requirements," to fill out these "motherhood" statements into more specific criteria.

The Directive contains certain "optional extras." Member states *may* (but do not have to) encourage environmentally sound reuse systems, provided these conform with the EU Treaty; and where appropriate member states are to encourage the use of materials recovered from recycled packaging waste in the production of new packaging and other products.

Member states have until July 1996 to put any laws or other measures in place to comply with the Directive. The Directive will not apply to packaging manufactured before July 1996. Provided this packaging conforms with existing national law, member states cannot exclude it from the market until January 2000. *There will be 15 different programs by the year 2000 for those exporting into Europe.*

Marking a Quagmire. The European Union is supposed to agree on a marking and identification system by January 1997. While a Commission Decision on Material Identification, was agreed to, it was held up for legal reasons in early 1996.

The Council of Ministers of European Parliament are responsible for a Directive on the Marking of Packaging, which will go through the full law-making process. The draft proposal for a round circle arrows was held up in May 1996 by high level EC officials apparently responding to protests from industry. The draft would have banned all other symbols including the popular "Mobius Loop" or chasing arrows.

Germany's system the most expensive. Germany enacted the world's most stringent, and well-known, packaging Ordinance in 1991. It forced a direct take-back obligation on the entire packaging chain (manufacturers and distributors, including wholesalers and distributors).

The obligation is to take back used packaging and ensure it is reused or recycled as a material. This obligation is absolute (ie, it applies to 100% of the packaging handled). Manufacturers and distributors may appoint third parties to fulfil their obligations.

Quantified targets only apply to sales packaging within the DSD system. Manufacturers and distributors taking part in a "Dual System" which collects, sorts, and passes on used sales packaging for recycling are exempted from the general take-back requirement, provided the Dual System meets the targets laid down. The DSD does *not* cover transport packaging or industrial packaging, where the absolute take-back requirement applies.

Germany's ordinance is the best-known outside Europe, and it has been studied by researchers in Europe and the U.S. (where its strongest critics are). The system is costing about $3 billion per year; the most expensive in the world. For example, the DSD charges nearly 90 cents per pound for every pound of plastic sales packaging, since it covers collection and recycling costs.

In a 1996 conference address, British Consultant David Perchard summed up the problems with the German system. (3) He says many of the problems stemmed from the legal framework of the law itself and are as follows:

"The emphasis is on material recycling for all materials—energy recovery is not allowed, even for paper and flexible plastics packaging in the household waste stream.

High recycling targets: The law currently requires DSD to ensure that *64%* of the used plastics packaging it covers is collected and sorted ready for recycling!

Short time frame: The system had to be up and running throughout Germany within 18 months, which meant that DSD was in a weak bargaining position when negotiating with municipalities and waste haulers.

The emphasis on the household waste stream: Jurisdictions aiming to keep costs down have allowed industry to concentrate on transport packaging, which is cheaper to collect and more readily recyclable, building up systems for consumer packaging much more gradually; DSD fees cover *all* costs of collection and sorting: Other member states (except Austria, Belgium and maybe later Spain) only expect industry to meet the *additional* costs of the new collection and sorting systems.

DSD is a monopoly, so there are no competitive pressures to keep costs down.

Pioneer disadvantage: Germany was unable to learn from experience elsewhere.

Because of the high mandated targets and broad nature of the Ordinance, Germany's DSD had to export millions of tons of mixed plastics and paper to other European countries and the Far East. This wreaked havoc with its European neighbors, as temporary market distortions appeared. In 1994 Germany agreed to stop "dumping" on European countries, though it still exports to the Far East.

No other country has precisely copied Germany's strict manufacturer's responsibility law, though a few have come close (Austria, and perhaps Belgium and later Spain). The Green Dot symbol (Fig. 1), which means a company has paid

Figure 1. Germany's "Green Dot."

its DSD license fees, has become well-known in Europe. DSD has licensed its symbol to Austrian, French, and Belgian organizations, and a number of other countries are interested.

The confusing issue for exporters is that use of the symbol on signifies payment of appropriate fees in *each country* where required. There is no pan-European program. Some countries, such as the Netherlands, do not want to use it. Using the symbol without payment of fees can bring lawsuits for fraud.

Many countries will move in the direction of shared responsibility rather than full manufacturers responsibility as in Germany.

For example, in France, manufacturers did not have to set up their own curbside collection system; government continues to handle the residential collection end. France has a goal to "valorize" 75% of packaging waste, but it is taking it slow, allowing for pilot programs. Manufacturers still have to either join a third-party organization such as ecomballages, take back directly, or set up a deposit system. There is a separate decree for transport packaging that places an obligation on end users.

Germany does not allow incineration with energy recovery to count towards a recovery or recycling goal as France and many other countries do. (However, since the DSD is stuck collecting a wide variety of mixed plastics with little market value, it has set up a deal for a steel company to use much of the mixed plastics as a replacement for coke as a feedstock. Americans may not consider this "recycling" but rather "resource recovery." Apparently this is acceptable to German Ministry officials. Of interest, it appears that Japan may skip mechanical recycling of plastic resins other than PET, and go the steel feedstock route.)

Each European country appears to be moving in its own direction, as of 1996, and those with existing laws seem reluctant to make major changes because of the EU Directive.

In 1991, the Dutch set up a voluntary agreement or "covenant" to nearly eliminate landfilling of packaging waste by 2000, since the Netherlands has a low water table which makes for poor landfills. Most packaging and many other durable and nondurable wastes have been banned from landfill. While there is a draft regulation to conform to the EU Directive, the government wants to allow flexibility for further industry covenants.

Belgium enacted a complex eco-tax law on beverage containers and many disposable items such as razors and cameras in 1993, although the original law was never implemented. The law continues to be amended. Meanwhile, the power lies with regional governments, which are working on a separate co-operation agreement for recycling of all packaging.

Refillables quotas under debate. Refillable containers are popular for beer and other beverages in some European countries. Germany has a refillables "quota" which has been challenged by the EU on trade grounds. Portugal recently enacted a refillables quota in its packaging law. Critics charge these quotas are designed to protect the market share of local beverage makers, not protect the environment. Meanwhile, the patchwork of deposits, voluntary and mandatory, are pushing soft drink makers to use refillable plastic bottles.

Eastern Europe

The countries of Eastern Europe are highly influenced by Germany, thus, several are moving in the manufacturers responsibility direction, despite their early stages of development. Packaging taxes and/or take-backs are on the books or proposed in Hungary, Latvia, Poland, Slovakia, Slovenia, and Ukraine.

Pacific Rim/East Asia

While Latin American countries will be slow to place any mandates on packaging, several East Asian countries have already moved in the direction of Europe.

Japan. Japan has enacted a packaging producers responsibility law in June 1995. The law is applied to PET bottles, glass bottles, aluminum cans and steel cans from fiscal 1997, and paper containers and other plastic containers from fiscal 2000.

There will be a designated third party organization. The new TPO would be in charge of collecting fees from manufacturers to fund recycling of packaging. Local governments will collect residential packaging, and turn those items with no market value over to company organizations to ensure recovery.

It does not appear that there will be much emphasis on mechanical recycling of any plastic packaging other than PET. The Ministry of International Trade and Industry is spending millions on research for plastics-to-fuel technologies, and is pushing Japanese governments to purchase items with recycled content.

Taiwan. Taiwan has laws for 18 different items including packaging on the books since 1989. The laws set up varying recycling quotas for different materials, ie, 65% for PET packaging by 1995. Third party organizations have been set up for most packaging materials except non-PET or PS plastics. However, many materials have been unable to meet the quotas, and observers question the validity of the official recycling rates. There is a voluntary deposit on PET beverage containers. The government plans to amend the law to make it more enforceable.

Hong Kong, Singapore, and Korea. All have active recycling systems, though they do not yet have direct "take-backs" for packaging. Malaysia is considering the idea, as its government has consulted with Germany's Minister of the Environment.

China. China has recently enacted a comprehensive waste law, although there appears to be no direct mandates on packaging yet. There is an ecolabeling law which apparently restricts companies from making green claims without government sanction.

The new law does restrict imports of wastes to those on a list. Baled plastics are not yet on the lists. U.S. and European recyclers have a major interest in export of baled plastics, especially PET for fiber production. If the Chinese government stops such imports there will be dislocations in all three markets.

Australia. Australia has no national recycling mandates, and defers most regulation to its states. It does have national guidelines for environmental labeling.

Most states have voluntary programs, and voluntary

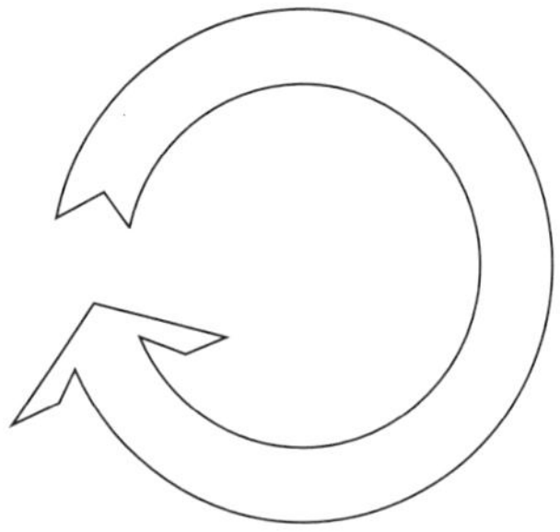

Figure 2. European Union's proposed recycling symbol.

agreements with industry. New South Wales lawmakers grew impatient with voluntary goals in 1996, and passed a law that will push each industry sector to come forward with its own plan to meet stringent recycling and recovery goals.

Ecolabeling an Issue

A dozen countries worldwide have voluntary ecolabeling programs for various products. In most cases, the country sets up a quasi-governmental organization to set up criteria for certain products, and companies meeting the criteria can pay a fee, show proof, and use the ecolabel. Critics claim the criteria are arbitrary and favor local companies.

The European Union passed up a pan-European Ecolabel program in 1991, with the intention of eliminating the need for country seals. However, only ten criteria have gone through the slow, complex approval process and only three ecolabels have been awarded. In 1996, the U.S. and other paper industries strongly protested a criteria for copy paper, claiming U.S. firms do not do the same tests and could not pass European criteria.

While there will be no European Ecolabel for packaging, the issue affects labeling in Europe. Unlike the U.S., most European countries have no laws or guidelines on environmental labeling, so numerous labels proliferate. (See Figs. 2 and 3.)

Standards

There are no internationally agreed-to standards for green labeling terms or symbols, although committees of ISO 14000 are working to get them approved by April 1997.

It should be noted that East Asian countries are jumping on ISO 14000 (eco auditing standards) bandwagon. It appears that many ISO 14000 standards, including those for green labeling and symbols, could find their way into law in many Asian countries.

Figure 3. European Union's ecolabel symbol.

Other Laws

It should be noted that recycling laws are only part of the regulatory picture. The design and labeling of packaging also can be impacted by hazardous waste and hazardous product laws, which are more well-developed internationally; specialized labeling laws; local and international standards; as well as food-contact laws and customs.

Conclusion and Summary

Even though U.S. manufacturers are staunchly against the concept of "producer responsibility," it appears the idea is firmly entrenched in governments outside the U.S. These concepts are being implemented as close as Canada.

More than 20 countries now have packaging take-backs or related taxes on the books, and more become interested every year. This is partly because most countries import most of their packaging and there is a temptation for governments to avoid raising taxes directly to pay for recycling and new waste management systems. North American manufacturers will increasingly be asked to foot the bill and pass on the cost, creating a more costly patchwork of regulation for packaging designers and compliance officers to follow.

On the positive side, Asian countries will have to invest billions to clean up their environment, and this will create opportunities for U.S. and European environmental companies.

BIBLIOGRAPHY

1. D. Perchard, ed., *Recycling Laws International, Country-by-Country Summaries,* Raymond Communications, Inc., Riverdale, MD, Feb. 1996.
2. D. Perchards, *Packaging Legislation in Europe, An Update,* Raymond Communications, Inc., Riverdale, MD, April 1996.
3. "Germany was just the Beginning," Address to the Take it Back! Conference, sponsored by Raymond Communications, Inc., and Kranson Industries, Columbia Hilton, May 2, 1996.

MICHELE RAYMOND
Raymond Communications, Inc.
Riverdale, Maryland

ENVIRONMENTAL REGULATIONS, NORTH AMERICA

Environmentalists started the trend towards recycling mandates on packaging because of the "Stop Styro" campaign, originating in Berkeley, California in 1987. The greens used the McDonald's foam clamshell as a "symbol," something consumers could easily identify. They told consumers to stop using polystyrene foam (often mis-identified as *styrofoam,* a Dow Chemical trade name). Thus, eschewing PS foam became a simple way consumers could do their part to save the earth.

The original rationale for banning foam was that it was

not degradable in landfills, not made of a renewable resource, used toxic chemicals in manufacture, killed sea life when ingested, (allegedly styrene leaches in to food from the containers,) and that foam uses harmful CFCs.

The City of Berkeley, California banned retail PS foam in 1988, and Suffolk County, N.Y. enacted a broader ban later that year. The Suffolk County law banned plastic grocery sacks, all retail PS containers, utensils, and related packaging also made of polyvinyl chloride.

The plastics industry formed a new group (Council for Solid Waste Solutions) to lobby plastics waste issues in general. CSWS (Now American Plastics Council) had to hire lobbyists throughout the country to combat the new wave of PS foam bans.

Industry won round one, but ultimately lost at the appeals court level. Cities had the right to ban certain packaging. Meanwhile, a total of 42 local governments nationwide banned or restricted retail PS food packaging between 1988 and 1991.

The states of North Carolina and Iowa enacted 1989 laws that required the PS industry to meet certain recycling rates on retail PS packaging, or face future ban. These laws were repealed in 1995.

Recycling aseptic packaging was taken up in right earnest ever since Maine banned aseptic juice packs in 1989. The industry continues to support recycling where possible even though Maine repealed the ban in 1994, and other states have showed no inclination toward a ban. This article is adapted from Ref. 1.

Recycled Content Mandates

In 1991, Oregon and California enacted laws that could affect 60,000 product makers by the middle of 1996. Both SB 66 in Oregon and SB 235 in California require that the plastics industry meet an aggregate 25% recycling rate on rigid plastic containers by 1995. If the rate is not met, then every plastic package must meet one of the following criteria: be reusable five times (OK to reuse within the home), be source-reduced by 10% (without a material switch), or contain 25% post-consumer recycled resin.

Oregon allowed an exemption if a company has made a substantial investment; is close to reaching the content goal. Towns only had to pick up plastics if there are stable markets that pay more than 75% of collection.

In December, 1992, the Oregon Environmental Quality Commission surprised industry by recommending that SB 66 be strengthened. Rather than allow a blanket exemption for FDA containers, the board recommended fees for companies not in compliance. Oregon ended up interpreting its law to include even thermoformed cookie trays, and companies were not allowed to figure compliance across product lines—theoretically, each container was supposed to comply if the state did not meet the 25% recycling rate.

In both states the food industry lobbied heavily to repeal or loosen requirements on food containers. They had little credibility with most Democratic lawmakers, who found little logic in their protests that they could not use any recycled material because of FDA requirements or could not source-reduce anymore.

This was partly because Coca Cola and Pepsi announced trials of using 25% post consumer PET in soda bottles. This PET was re-polymerized, so it was like virgin. (Of interest, this was just a test but it gained a lot of publicity, and made the soft drink industry look good. It sent a message to regulators that other food companies could do the same thing, which was not the case. Coke and Pepsi ultimately found they could not continue using the recycled material because consumers were very price-sensitive, and the fancy recycled resin had a 10% up charge that became intolerable.)

California allows an exemption for "hardship," and in feasibility. There is an escape clause if less than 65% of single family households have access to curbside recycling of beverage containers by 1994, but the state met its goal. A material or type of container can be exempt if it reaches a 45% recycling rate in California. PET must achieve a 55% recycling rate to be exempt.

Both laws exempted drug and related containers, but not food and cosmetics. In 1994 California amended law allowing an extra two years for food and cosmetic firms to comply, but required them to report, either individually or through their associations by December 1995. The state regulators also decided to *exempt fast-food-type retail containers* (those which hold food for a short period). This was because there were too many entities to regulate with restaurants. This was a victory for the food service packaging industry.

Outcome of Content Mandates

Oregon. Oregon ultimately exceeded the 25% recycling rate for rigid plastics (it is now at around 33%) partly because of the state's container deposit law. Counting the recycling rate proved to be rather time-consuming, political, and expensive. Meanwhile, Republicans were elected to the Legislature and were ready to amend the rigid plastic law. Even though the point may have been moot, the lawmakers passed exemptions for certain hazardous materials and one for all solid food containers, thus, only liquid foods and non-food containers were included in the law.

Now that manufacturers do not have to take any action in Oregon, state regulators are looking at ways to repeal the law and replace it with another. The plastics industry has no interest in discussions, since the current requirements are moot. It is unclear if there will be another change with the next election back to lawmakers more sympathetic to recycling.

OSPIRG, the environmental group which pushed the rigid plastic law, is circulating a petition to expand the state's deposit law to most liquid beverages. The group is picking up considerable support from labor, consumer groups, and local grocery chains. The major chains and soft drink industry are opposed and will spend heavily. The politics and media coverage of this could impact future Oregon legislation that will affect packaging and it could put packaging back on the agenda.

California

Food/cosmetics extension. Of the potential 60,000 companies affected by Californias rigid plastics mandate, 10–12,000 are in food or cosmetics. No reporting is required for most companies, but records must be kept. About 300 food and cosmetic companies have reported to CIWMB on their progress toward using recycled plastics.

CIWMB staff recommended the board approve compliance

reports from a consortium of food processors, the Cosmetic Toiletry and Fragrance Association (CFTA), the American Health and Beauty Aids Institute (AHBAI), and the American Plastics Council (APC).

Of interest, the CFTA and the food consortium reported spending about $15 million on research for source reduction and use of post-consumer material. The food and cosmetics firms may use the recycled plastics in other items, such as pallets and office items, rather than finicky bottles. However, few have reported on this option, sources said.

Food and cosmetic companies that are in compliance with the law are not required to report to CIWMB. (The author notes that nearly all large consumer product companies seem to be in compliance or working on the issue.)

The food consortium included the California League of Food Processors, California Dairy Institute, Grocery Manufacturers of America, National Food Processors Association, Distilled Spirits Council, and American Frozen Foods Institute.

The associations did not receive reports from all of their members. For example, 93 companies with 98 subsidiaries or affiliates, out of 500 members of NFPA reported. Tim Willard of NFPA said it appears that these 95 represent the majority of rigid container users that are members.

None of the reports specifically identified how members would be in compliance by 1997. The barriers listed included source reduction (no more opportunities); FDA limitations; contaminations issues on refills; problems for small businesses that order small quantities—no options from suppliers for them.

Since there are limits on the use of recycled content in food packaging or reusable containers because of food safety concerns, it will be difficult to fully comply.

On the other hand, the AHBAI, which represents 17 manufacturers of ethnic hair care and cosmetic products, made no excuses on "feasibility" grounds. Members claimed to be making progress towards compliance by 1997.

Six of their members have source-reduced at least 10% since 1990, making them in compliance. One major change is moving from jars to dairy-style tubs and lids which saves 30–50%. Five members reported using containers with 25% PCR, while a similar number are using reusable containers.

APC, which represents 27 plastic resin manufacturers, reported its members spent $1 billion since 1990 on recycling and source reduction activities. The report identified 102 programs in which post-consumer resin is used by resin makers and their customers.

Recycling rate. If the overall rate is met, manufacturers are off the hook. If it is not, CIWMB will do selective enforcement of this law.

The growth in custom PET bottles and other applications may even have a negative effect on recycling rates because recycling cannot keep pace with growth of plastics. Note that if the current rigid plastic law remains in effect, the 25% recycling rate must be met *each year.* Thus, if manufacturers are off the hook in 1995–1996, and many abandon new markets for the recycled bottles, the rate could drop for 1996–1997. This means enforcement of the content/source reduction mandate.

Proposed repeal. But while packagers were awaiting word on whether the 25% recycling rate would be met in California, a bill SB 1155 (Ken Maddy, Republican) was used as a placeholder bill for language to repeal the mandate itself. However, lawmakers may be loathe to kill a pro-recycling bill before it was known if the goals are realistic or not.

Meanwhile, many plastics recyclers support continuation of the law. They claim that without the incentive the law provides, their higher-end markets will crumble. Moreover, major haulers now have a vested interest in recycling, so they may lobby to ensure there is no repeal as well.

Wisconsin. Wisconsin has a law which requires 10% recycled content in rigid plastic containers eight ounces and over. This law would include many thermoformed containers. However, the law allows pre-consumer material to be used and there is no real enforcement of the law. Of interest is Wisconsin's sweeping landfill ban law.

Advance Disposal Fees

From 1993 through October 1995, Florida had an advance disposal fee on what amounted to only rigid plastic containers, five ounces to one gallon. Manufacturers that were able to recover 25% of their containers in the state through some means, or those that contained 25% recycled content were exempt. This did provide an incentive for some companies to set up third party programs; and a few recyclers decided to locate in the state. However, the program sunsetted, as the Legislature became more conservative, while the state's key environmental group decided to pull away from packaging issues for the time being.

The program was not in effect long enough to really determine its exact effect on markets for recycled plastics. It did bring in a large amount of revenue (more than $30 million/yr) which the state did not really need for that purpose, since it already had $22 million allocated for recycling grants. The effect of the ADF was similar to California's law: companies that were trying to comply with the California law and were using recycled content had to try and track what was going into Florida (a cost factor) and then apply for exemptions.

All companies had to report to their retailers on number of containers that fit in the ADF so the retailers could pay the fees. In the case of small retailers, accounting did not allow such tracking and smaller stores had to pay the fee out of pocket.

The soft drink and beer industries started to like the ADF because it kept away efforts to impose a deposit and they found they could comply. It provided major brands using plastic with a one cent advantage over some competitors and made them look good.

Landfill Bans

Several states have rather sweeping bans still on the books, and North Carolina has about 32 local landfill bans. Many communities go beyond state landfill bans. This is catalogued for about 120 local governments in detail in Ref. 2.

Massachusetts is coping with its sweeping bans of packaging (including most rigid plastic containers and clamshells) and paper (passed by regulation in 1991) by requiring municipalities to either offer curbside or drop-off collection of certain items. If the cities meet the requirements, then they are exempt from having to inspect for the landfill bans. State regulators have managed to get about 85% compliance from local town governments (there are no counties or regular cities). However, the other 15% may be liable to inspect at landfills if they do not improve their recycling programs, so DEP is holding meetings on the issue.

Wisconsin's sweeping bans of all types of packaging and paper (from the original solid waste law of 1990) went into effect in 1995. There is a one-year extension on packaging made of #3–#7 plastics (ie, polypropylene, PVC, other), which means that everything else is still in the law. This law includes all corrugated, office papers, PET and HDPE containers, steel, aluminum, paper, and glass containers.

NR 544 requires the "responsible entity" must have a mandatory recycling ordinance, multifamily and commercial collection of a certain number of materials, appliances must be recycled, staff must be hired, and volume-based fees must be put into place (unless the city is recycling 25% by volume of its waste already). It also requires a per-capita minimum recycling of each material. If all these requirements are met, then the community is exempt from enforcing the landfill bans.

It does not look like the Wisconsin Legislature will move on a permanent exemption for #3–#7 plastics from the landfill ban in 1996. Gov. Tommy Thompson signed a bill that will allow state regulators more flexibility on the landfill bans. They can allow contaminated loads to be landfilled and extend the exemption for #3–#7 plastics for another year.

North Carolina remains the benchmark state for local landfill bans. A total of 32 communities now have some sort of local landfill bans in this #2 manufacturing state. Some just ban corrugated containers (OCC), others ban all kinds of recyclables. The bans have forced businesses to recycle and push haulers to either set up routes or pull out OCC at landfills when the price is right. In a few cases, the bans have inspired on-site MRFs.

Deposits

Nine states have traditional bottle deposits while California has a complex redemption system. The issue in existing bottle-bill states has been to expand it to juices, ice teas, and waters. Soft drink interests continue to support curbside collection hoping that lawmakers will lose interest in deposits. Interest in deposits remains high, as surveys show the majority of consumers favor them. The deposit system tends to increase return rates of beverage containers and feedstocks tend to be cleaner than curbside materials. However, industry remains opposed to the concept. Environmentalists in Massachusetts and Oregon are pushing to expand deposits to all types of beverages, since the popularity of bottled waters and teas have become more popular.

Maine has an expanded deposit law, which has proved expensive for industry, because only the soft drink industry has exclusive territories allowed. The law has led to over redemption in border towns, though the state apparently has no intention of backing away from the complex law.

Heavy Metals Bans

The model Toxics in Packaging bill, developed by the Coalition of Northeastern Governors, (CONEG) is law in 18 states. In 1995 six states moved to amend the law in an effort to clean up some loose ends.

The model bill states that no package may be sold if any packaging component (including inks, dyes, pigments, adhesives, stabilizers, or other additives) contains lead, cadmium, mercury or hexavalent chromium intentionally introduced during manufacture or distribution. The law usually gives manufacturers two years to phase down to 600 ppm; then 250 ppm, 100 ppm after four years. The model law allows state regulators to later institute bans on other heavy metals if they are found to be a threat.

Green Labeling

Green claims were popular from 1990–1993, until manufacturers found they did not really help sales. Laws are on the books in 18 states, however, and many actions have been taken against companies that had misleading claims.

In 1992 a proliferation of conflicting state proposals led industry to push the FTC into issuing guidelines for environmental labeling. This in turn has influenced states to back away from their laws, or at least defer and compliance action if a claim complies with FTC's guides.

In 1995, California managed to get court clearance to regulate green claims in a full truthful sentence, just as the California Legislature enacted a law backing away from its stringent definitions of terms such as "recyclable" and "recycled."

However, many companies continue to use unqualified terms such as "recyclable," "recycled," and "non-toxic" on their packages or products despite many complaints brought by the Federal Trade Commission. FTC reviewed its guidelines in 1995 for possible revisions to be issued in 1996.

Resin Coding

The resin identification code developed by the Society of the Plastics Industry in 1989 was intended to assist plastic recyclers in sorting bottles by resin type. Even though there are thousands of different resin grades used by industry, a committee decided on some basic identification numbers and abbreviations for seven types is shown in Figure 1.

Some of the states objected to the codes at first because of a concern that the code would be used as a recycling symbol, and not all resins are welcome in curbside bins. Nevertheless, the SPI model bill was pushed and passed in 39 states. While there still is much confusion, industry and consumer/government interests have been unable to agree on an acceptable change in the code.

Many companies inappropriately use the SPI code as a green claim on bags; on nonretail bottles and on clamshells,

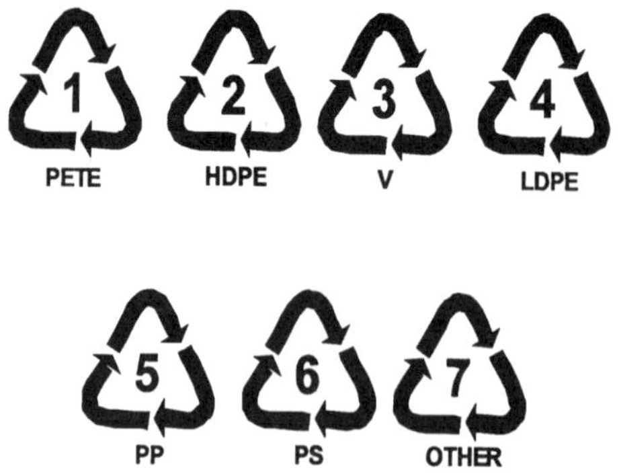

Figure 1. Some identification and abbreviations for seven types of resin grades.

which are unwelcome in curbside bins. Even though two states technically require the code on plastic clamshells, legal experts say there have never been any enforcement actions against companies not using the code.

If companies are using blends of polymers or multilayer bottles that recycle, then experts say it is generally acceptable to put that resin number (eg, PET) on the bottle. Many large companies have in practice.

Conclusions/Summary

In 1988, the famous "garbage barge" from New York brought media attention to the landfill issue, and all 50 states enacted comprehensive recycling laws. In 1996, lawmakers realize there is plenty of landfill space at present. But the courts have ruled that cities and governors can have little control over where the trash goes because of the Interstate Commerce clause. The Fresh Kills landfill in New York must close by 2000, and so far, no other state wants this garbage. Congress has thus far failed to act on this "flow control" issue, so the overall solid waste "crisis" could loom again in some states.

In 1996, recycling laws are not as much of a "threat" to packaging makers in the U.S. as they once were. Lawmakers have learned that banning items that take up little landfill space will do little for the environment. Prices have softened for used plastic containers in 1996, as companies begin to shy away from using the material. But curbside programs now collecting in 6,000 locations cannot be turned off like a spigot. Local governments tend to stockpile materials or take a loss. If prices for certain materials collapse and stay low, lawmakers could turn their attention back to packaging again. Lobbyists warn that if packaging makers let their guard down and stop using recycled material and looking into new technologies for reuse, they could be stuck with more threatening legislation in the late 1990s.

Moreover, a proliferation of "producer responsibility" laws for packaging, enacted outside the U.S., in some form as close as Canada, could have an influence on the thinking of U.S. legislators.

BIBLIOGRAPHY

1. M. Raymond, ed., *State Recycling Laws Update,* Raymond Communications, Inc., Riverdale, MD, year-end edition 1996, April 1996.
2. *Disposal Bans in America: Who's Banning What and Why,* Raymond Communications, Inc., Riverdale, MD.

MICHELE RAYMOND
Raymond Communications, Inc.
Riverdale, Maryland

ETHYLENE–VINYL ALCOHOL COPOLYMERS (EVOH)

Ethylene–vinyl alcohol copolymers are hydrolyzed copolymers of ethylene and vinyl alcohol. The polymer poly(vinyl alcohol) has exceptionally high barrier properties to various gases, solvents, and chemicals, but it is water-soluble and difficult to process. By copolymerizing vinyl alcohol with ethylene, the high barrier properties are retained and improvements are achieved in moisture resistance (EVOH is not water-soluble) and processability.

The resulting EVOH copolymer has the following molecular structure:

$$\text{----}(CH_2 - CH_2)_x \text{----}(CH_2 - \underset{\displaystyle OH}{\underset{|}{CH}})_Y\text{---}$$

Ethylene unit   Vinyl alcohol unit

There are currently three suppliers of EVOH copolymers worldwide. Kuraray Company LTD. [EVAL (registered trademark) resins] and Nippon Gohsei (Soarnol resins) are both Japanese companies with manufacturing facilities in Japan. EVAL Company of America, a division of Kuraray American Inc., has manufacturing facilities in the United States. Manufacturing capacity for EVOH resins worldwide is approximately 60 million lb annually. Table 1 lists the range of EVOH resins available.

Table 1. Range of EVOH Resins

Property	Range
Melt index (g/10 min)	0.7–20
Density (g/mL)	1.12–1.20
Ethylene content (mol %)	27–48
Melting point (°C)	156–191

EVOH copolymer resins first became commercially available in 1972 in Japan. However, widespread use of these resins did not occur until 1983 when U.S. food producers started using the resin for all-plastic, squeezable bottles.

The primary use of these resins is still food packaging, where the excellent gas-barrier properties are utilized to prevent oxygen degradation of food products. However, recent trends are utilizing EVOH's superior resistance to solvents, chemicals, and hydrocarbons to expand uses into nonpackaging, industrial applications such as the agricultural chemical, automotive, appliance, and protective-clothing areas.

EVOH copolymers are highly crystalline in nature, and their properties are highly dependent on the relative concentration of the comonomers. Figure 1 shows the relationship

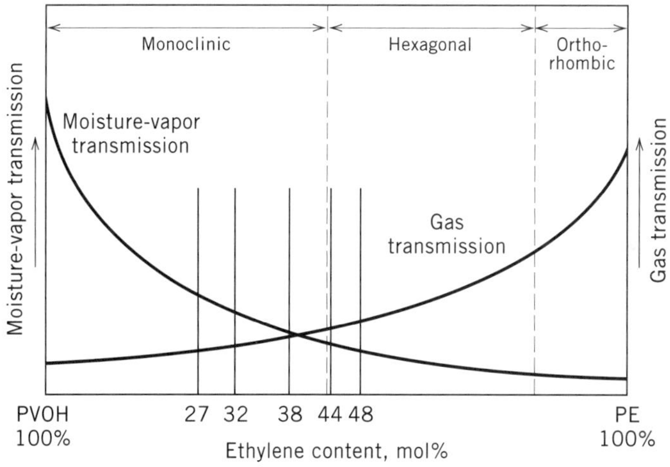

Figure 1. Comonomer concentration vs barrier properties.

Table 2. Gas Transmission Rate[a]

	(mL · mil)/(100 in.^{2} · 24 h · atm)				(mL · 25 μm)/(m^2 · 24 h · atm)			
	O_2	N_2	CO_2	He	O_2	N_2	CO_2	He
32 mol% EVOH	0.02	0.002	0.06	18.8	0.31	0.03	1.0	291.2
38 mol% EVOH	0.03	0.005	0.15	22.0	0.39	0.08	2.3	341.3
44 mol% EVOH	0.08	0.008	0.23	25.3	1.2	0.12	3.5	391.5
PVDC	0.15	0.01	1.1	27.3	2.3	0.19	17.4	423.2
Oriented nylon	3.6	1.4	13.2	232.0	55.2	21.7	205.2	3596
Oriented PET	4.6	0.46	19.6	180.0	71.3	7.1	303.8	2790

[a] At 23°C (73°F), 65% rh.

between comonomer concentration and moisture and gas transmission barrier properties, and Figure 2 shows the relationship between relative humidity and oxygen transmission rate.

It is this combination of exceptional barrier properties that permit the use of thin layers of these materials in multilayer packaging and nonpackaging structures.

EVOH Copolymer Properties

The general characteristics of EVOH copolymer resins are discussed in the following paragraphs.

Gas-barrier properties. EVOH resins are best known for their outstanding barrier to the permeation of gases such as oxygen, carbon dioxide, nitrogen, and helium. The use of EVOH copolymers in a packaging structure enhances flavor and quality retention by preventing oxygen from penetrating the package. For packages utilizing modified-atmosphere packaging (MAP) technology, EVOH resins effectively retain the carbon dioxide, oxygen, or nitrogen used to enhance the product. (See also Barrier polymers.)

Table 2 compares the gas transmission properties of EVOH copolymers with other commonly used packaging materials.

Hydrocarbon and solvent resistance. EVOH resins have a very high resistance to hydrocarbons and organic solvents. The weight-percent (wt%) increase in EVOH resins after immersion for 1 year at 20°C in various solvents in 0% for solvents such as cyclohexane, xylene, petroleum ether, benzene, and acetone; 2.3% for ethanol; and 12.2% for methanol.

Table 3. Protective Glove Performance[a]

Chemical or Solvent	Time	Chemical or Solvent	Time
Acetone	1440	MMA	1440
Acetonitrile	1440	MEK	1440
Aniline	1440	Methylene chloride	1440
Benzene	1440	Nitrobenezene	1440
Carbon disulfide	1440	Sodium hydroxide	1440
Chloroform	1440	Styrene	1440
Diethylamine	60	Sulfuric acid 93%	1440
DMF	1440	Tetrachloroethylene	1440
Ethyl acetate	1440	Toluene	1440
n-Hexane	1440	Trichlorethylene	1440
Methanol	1440	Xylene	1440

[a] Breakthrough time in minutes (1440 min = 24 h).
Source: Evaluated by ASTM F739.

Table 3 shows the performance of protective gloves containing an EVOH layer, and Table 4 reflects the resistance of EVOH to various fuels. These properties make EVOH resins an excellent choice for the packaging of oily foods, edible oils, agriculture pesticides, organic solvents, and other industrial applications.

Flavor, fragrance, and odor protection. Protecting food products is becoming increasingly important, especially with the current trend away from product additives. Protection can mean different things in different situations. One of the areas

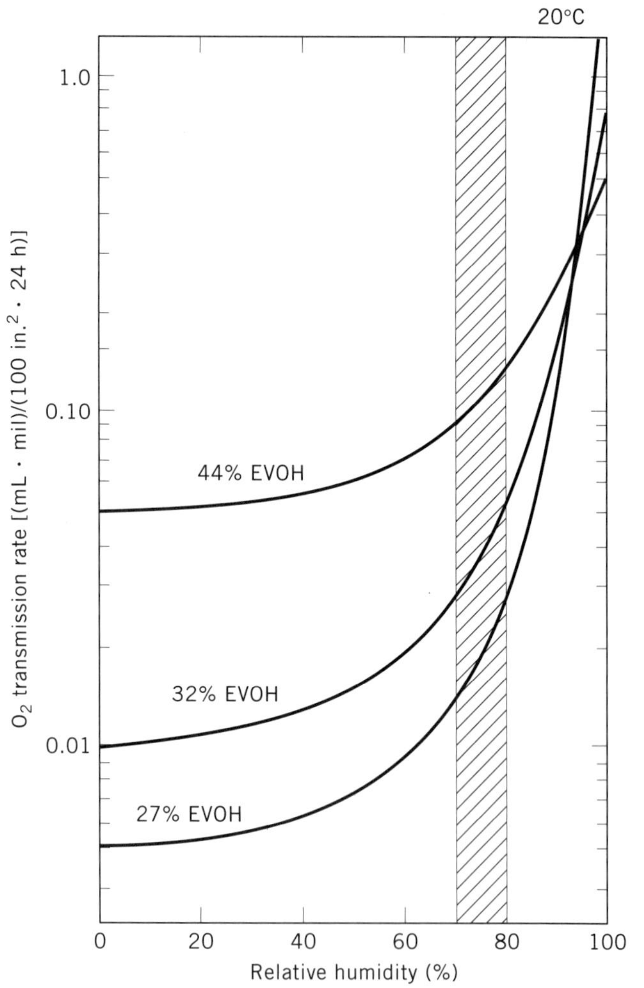

Figure 2. Oxygen transmission rate vs relative humidity.

Table 4. Resistance to Fuels[a]

	[(g · mil)/(100 in.^{2} · 24 h · atm) at 40°C]			
Fuel or Component	32% EVOH	27% EVOH	Nylon	HDPE
REF Fuel C	0.0009	0.0008	0.015	208
CM15	0.61	0.46	5.0	168
MTBE15	0.0007	0.0006	0.012	166
MTBE100	0.0002	—	0.003	26
Toulene	0.001	—	—	301
Isooctane	0.003	—	—	24

[a] *Note:* REF Fuel C—100% hydrocarbon; CM15—REF Fuel C + 15% methanol; MTBE15—REF Fuel C + 15% MTBE.

receiving increased attention in food packaging is the flavor–aroma barrier. (See Aroma Barrier Testing)

The food industry is becoming increasingly aware of the importance of flavor barriers in containing natural flavors and preventing extraneous flavors from contaminating the product. Some food products are notorious for their ability to absorb extraneous flavor or aromas. Another important flavor consideration is commonly called "flavor scalping," in which some plastics, such as polyolefins, selectively absorb certain flavor constituents from the product.

EVOH resins provide excellent protection to the migration of flavors and the absorption of D-Limonine, which is the primary flavor constituent in citrus juices, as can be seen in Table 5 and Figure 3.

Where food products are concerned, aroma barrier is somewhat analogous to flavor barrier. Food products must be protected from outside aromas in the distribution chain, grocery store, and home. Other food and nonfood products such as garlic, agricultural chemicals, pesticides, insecticides, or perfumes can be highly aromatic. In this case, the desire is to retain the aroma in the package and not let it escape into the surrounding environment.

As shown in Table 6, EVOH can provide protection to aroma migration.

Mechanical and optical properties. Although not intended for use as structural resins, EVOH resins have high mechanical strength, elasticity, and surface hardness, and excellent weatherability. These films are highly antistatic in nature, preventing the buildup of dust and/or static charges. EVOH films have a very high gloss and low haze, resulting in outstanding clarity characteristics. Table 7 compares the mechanical and other properties of EVOH resins.

Processability. EVOH resins are thermoplastic polymers easily processed on conventional fabrication equipment without special modifications. Using commercially available equipment, EVOH resins, in conjunction with other resins, are suitable for use in the following processes:

- Blow film
- Cast film
- Sheet extrusion
- Blow molding
- Profile extrusion
- Extrusion coating

Depending on the requirements, EVOH resins can be coextruded with all types of polyolefins, nylon, polystyrene, polyvinyl chloride, and polyester. Downstream processing such as

Table 5. Flavor Permeability

	Permeation [g/(m^2 · 24 h)]			
	73°F, 0% rh		73°F, 75% rh	
Structure	D-Limonene	Methyl salicylate	D-Limonene	Methyl salicylate
A	0.001	0.001	0.01	0.01
B	0.001	0.0003	0.01	0.002
C	0.0003	0.0002	—	—
D	0.040	0.009	0.01	0.002
E	0.032	2.16	0.01	6.79
F	0.040	0.87	0.01	2.15

Note: A—HDPE/tie/EVOH/tie/EVA; B—LDPE/tie/PA/EVOH/PA/tie/LDPE; C—PA/tie/EVOH/tie/LDPE; D—LDPE/tie/EVOH/tie/LDPE; E—PP/LDPE/tie/PETG/tie/LDPE/PP; F—PVDC concentrated OPP.

Table 6. Aroma Protection

		Days to Aroma Leakage			
Construction	Thickness (μm)	Vanillin (vanilla)	Menthol (peppermint)	Piperonal (heliotropin)	Camphor
PET/EVOH/LDPE	12/15/50	15	25	27	<30
OPP/EVOH/LDPE	18/15/50	30	<30	27	<30
PET/EVOH	12/15	<30	<30	30	<30
PET/LDPE	12/50	2	16	5	<30
OPP/LDPE	17/50	6	2	1	13

Table 7. Mechanical Properties of EVOH Resins

Property	Unit	EVOH resin grades				
Ethylene content	mol %	27	32	38	44	48
Melting point	°C	191	181	175	164	156
Crystalization temperature	°C	167	161	151	142	134
Glass-transition temperature	°C	72	69	62	55	48
Melt index at 210°C	g/10 min	3.0	3.8	3.8	13.0	14.7
Density	g/mL	1.20	1.19	1.17	1.14	1.12
Tensile strength, break	psi	10,385	10,385	6,685	7,395	5,405
Tensile strength, yield	psi	13,655	11,235	9,385	8,535	6,260
Elongation, break	%	200	230	280	380	330
Elongation, yield	%	6	8	6	7	3
Young's modulus	psi × 10^4	45.5	38.4	34.1	29.9	29.1
Rockwell hardness	M	104	100	93	88	—
Tabor abrasion	mg	—	1.2	1.2	2.0	2.2

thermoforming, vacuum forming, and printing is easily accomplished with EVOH resin- or film-containing structures.

Regulatory approvals. Depending on the resin grade, EVOH resins and films have FDA approval for direct food contact as specified in the code of Federal Regulations, Title 21, Section 177.1360. EVOH resins and films can also be used in high-temperature laminates (retortable applications) provided a functional barrier layer is used to separate the EVOH layer from the food product.

Various other approvals related to the use of EVOH resins and films in meat and cheese products and in other countries such as Latin America, Canada, and Europe have been granted by the appropriate regulatory agencies.

EVOH resins are considered nontoxic according to the USP Class VI test protocol.

Packaging Structures

Throughout the world today, the primary use for EVOH resins and films is food packaging. All plastic multilayer structures containing EVOH provide a cost-effective alternative to traditional forms of packaging such as glass or metal. In addition, they also offer other advantages such as convenience of use and source reduction of packaging materials.

Rigid and semirigid containers such as bottles, trays, bowls and tubes, flexible films, and paperboard cartons containing EVOH as the functional barrier layer are commercially available today. These structures utilize the multilayer concept to protect the barrier layer (EVOH) from the effects of moisture. This concept is also used to produce an economical structure by using relatively inexpensive materials, such as polyolefins, as the bulk of the structure. Most multilayer structures used today have five or six layers. Using feedblock coextrusion technology, seven- and nine-layer structures are being produced for special applications.

When using EVOH in multilayer structures, it is necessary to use an adhesive layer to gain adequate bonding strength to the other polymers. Commercially available adhesive resin such as Admer (Mitsui Petrochemical Industries Ltd.), Plexar (Quantum Chemical), Tymor (Morton Chemical, or Bynel (DuPont Company) are suitable for use with EVOH resins.

Table 8. EVOH Film Properties

Property		Unit	EVAL EVOH Grades				
			EF-XL	EF-F	EF-E	ON	PVDC/concentrated BOPP
Thickness		mils	0.6	0.6	0.8	0.6	0.9
Tensile strength	MD	psi	29,700	11,600	10,150	23,450	24,100
at break	TD		28,300	7,550	6,400	31,290	31,200
Elongation	MD	%	100	180	260	90	140
at break	TD		100	140	190	90	60
Young's modulus	MD	psi × 10^4	51.0	28.4	28.4	24.2	28.4
	TD		51.0	27.0	27.0	21.3	37.0
Tear strength	MD	g	260	380	460	500	300
	MD		330	300	440	450	200
Dimensional stability	MD	%	−4.0	−2.7	−1.6	−1.5	−10.4
at 14°C, 1 h	TD		−0.5	−0.9	−1.2	−0.9	−12.5
Melting point		°C	181	181	164	220	165
O_2 transmission rate at 20°C		mL/100 in.^{2}					
65%		24 h · atm	0.01	0.02	0.08	1.92	0.55
85%			0.04	0.08	0.17	5.4	0.55
100%			0.23	1.0	0.52	19.0	0.55
H_2O transmission rate		g/100 in.^{2}	2.6	6.5	2.3	16.8	0.3
40°C, 90% rh		24 h · atm					

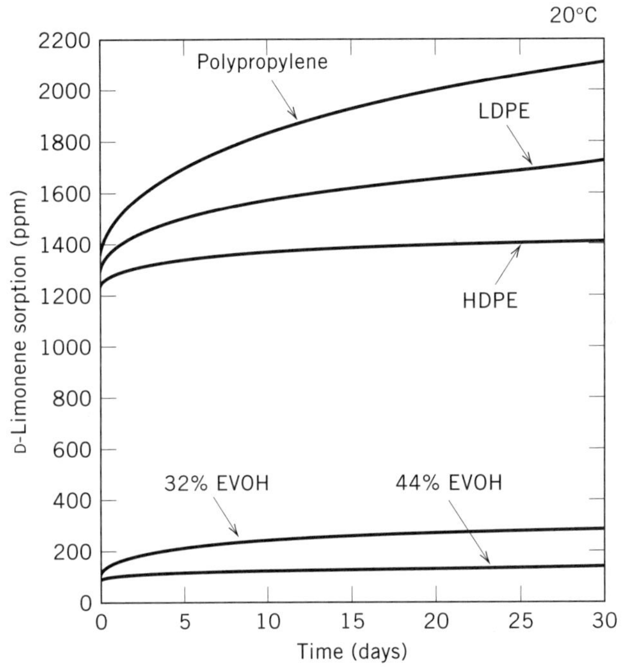

Figure 3. D-Limonene sorption.

Rigid and semirigid containers. The most common methods of producing these containers are thermoforming, blow molding, and profile extrusion. Most thermoformed containers are produced in a two-step process. First a coextruded multilayer sheet is produced. This sheet, containing a barrier layer (EVOH), is then formed into the final container using either melt-phase or solid-phase thermoforming technology. There are two exceptions to the two-step process: the use of rotary thermoforming and the American National Can "omni" process, which is a variation of the injection blow-molding process.

Multilayer bottles are produced using either the extrusion blow-molding or injection blow-molding technology. Likewise, tubes can be produced using either the direct-profile coextrusion or blow-molding coextrusion process.

Figure 4 shows typical multilayer structures used for rigid and semirigid containers.

Flexible films. Flexible films consume the largest volume of EVOH of any package structure. Flexible films are produced by several methods. These include solution coating, laminating coextrusion coating, and coextruding either tubular or flat (cast) films. Converting operations such as vacuum forming, printing, form, fill, seal, and heat sealing are used to fabricate packaging structures from these films. Figure 4 shows typical multilayer flexible-film constructions. Monolayer EVOH films, either biaxially oriented or nonoriented, can be used in conjunction with the laminating or coextrusion coating process to also produce flexible films.

Paperboard containers. In recent years, the use of EVOH to replace aluminum foil as the barrier layer in paperboard cartons has grown rapidly. These containers, used primarily for juice products but also for nonfood products, utilize the EVOH in several ways. Besides serving as an oxygen barrier, the EVOH layer can also be used to prevent "flavor scalping" (fresh juice) or as an aroma barrier (laundry products). Coextrusion coating onto paperboard is the primary process for producing these containers. Typical structures may be seen in Figure 4.

EVOH Films

In addition to EVOH resins, monolayer EVOH films are also available. These films, either biaxially oriented or nonori-

Table 9. EVOH Applications

Fabrication process	Application	Structure
Cast coextrusion	Processed meats, natural cheese, snacks, bakery	PP/nylon/EVOH/nylon/LLDPE Nylon/nylon/EVOH/nylon/surlyn Nylon/EVOH/surlyn PET/LDPE/EVOH/surlyn
Blown coextrusion	Processed meats, bag-in-box, red meat, pouches	Nylon/LLDPE/EVOH/LLDPE LLDPE/EVOH/LLDPE LLDPE/EVOH/LLDPE/surlyn Nylon/EVOH/LLDPE
Lamination	Coffee, condiments, snacks, lidstock	OPET/EVOH/LDPE/LLDPE PP/nylon/EF-XL/nylon/LLDPE OPET/EVOH/OPET/PP
Coextrusion coating	Juice, bakery, laundry products	LDPE/paperboard/LDPE/EVOH/LDPE LDPE/paperboard/LDPE/EVOH/LDPE/EVOH OPP/LDPE/EVOH/LDPE/EVA
Thermoforming	Vegetables, fruit sauce, entrees, pudding	PP/regrind/EVOH/regrind/PP PS/EVOH/LDPE
Coextrusion blow molding	Ketchup, sauces, cooking oil, salad dressing, agricultural chemicals, juice	PP/regrind/EVOH/PP HDPE/regrind/EVOH/HDPE HDPE/regrind/EVOH PET/EVOH/PET/EVOH/PET
Profile coextrusion	Cosmetics, toothpaste, condiments, pharmaceuticals	LDPE/EVOH/LDPE LDPE–LLDPE/EVOH/LDPE-LLDPE

Note: Tie layers are omitted for clarity.

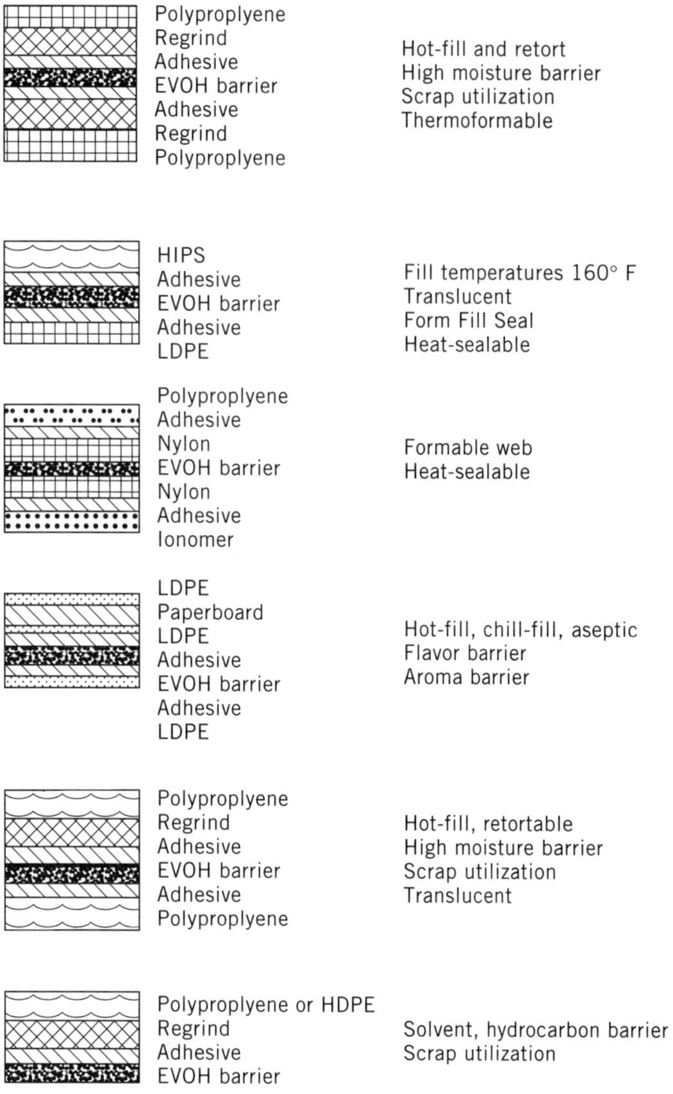

Figure 4. Barrier constructions.

ented, provide a unique balance of attractive appearance of toughness, dimensional stability, barrier properties, and economics. When combined with other types of polymeric films in a laminating or laminating–coating process, the resulting structure provides properties similar to those of a coextruded structure. Table 8 lists the properties of several monolayer EVOH films.

Applications

Since the introduction of EVOH resins and films in Japan during the 1970s and in the United States and Europe in the 1980s, the use of these materials has continued to grow. Table 9 shows a portion of the commercial applications utilizing EVOH as a barrier layer.

Economics. To determine the most economical barrier polymer, the price of obtaining the barrier needed for a given application at a given set of use conditions and a given packaging structure must be compared. Factors such as price per pound of the polymer, polymer yield, oxygen transmission rate, and the use of recycled scrap must be considered.

When the low barrier cost of EVOH resins is combined with their recyclability and the new developments in coextrusion technology, the packaging specialists has available a means to design a packaging structure that will not only compete with glass and metal containers on a performance basis but also provide a definite economic advantage.

BIBLIOGRAPHY

General References

M. Salame, "The Use of Low Permeation Thermoplastics in Food & Beverage Packaging," *Coatings and Plastics Preprints of the 167th Meeting of the American Chemical Society, Division of Organic Coatings and Plastics Chemistry,* **34**(1), 515 (April 1974).

R. Foster, "Plastics Barrier Packaging—The Future is Bright," *Proc. Society of Plastics Engineers, Inc.,* Regional Technical Conference, April 1984.

C. R. Finch, "Coextrusion Economics," in *Coextrusion Comes of Age,* Society of Plastics Engineers, Extrusion Division Regional Technical Conference, Chicago Section, June 1981.

K. Ikari, "Oxygen Barrier Properties and Applications of Kuraray EVAL Resins," *Proceedings of the Second Annual International Conference on Coextrusion Markets and Technology,* Schotland Business Research, Inc., Princeton, NJ, Nov. 1982.

S. Sonsino, *Packaging Design,* Van Nostrand Reinhold, 1990.

RONALD H. FOSTER
EVAL Corp.
Lisle, Illinois

EUROPEAN PACKAGING

As an entity, "Europe" is not easily defined because its boundaries can be taken as geographic, cultural, or political. It occupies about 7% of the world's landmass, and if the whole of Russia is included, has a population of about 800 million (almost one-fifth of the world).

Figures in Table 1 show western Europe's relationship with other industrialized regions of the world in terms of packaging consumption. Dramatic changes have taken place in Europe since 1990 with the fall of communism. The countries of western Europe had strong packaging industries, whereas in the communist bloc industry was organized into large conglomerates, with packaging often being manufactured on site as part of a total operation. With the breakdown of the iron curtain, many multinational food and industrial companies have been setting up plants or buying the newly privatized companies in these former communist countries; this has heralded a change in packaging and the packaging industries. The eastern bloc countries are seen as major growth markets as they strive to improve their overall standard of living to reach that of the West, which means an increased use of packaging materials. At the end of 1994 there were no reliable packaging statistics for packaging production or use in eastern Europe. The old system of 5-year plans and targets meant that companies overstated production statistics to ensure that targets were achieved, but now, in certain countries, they tend to be understated to avoid paying tax.

Table 1. Total Packaging Consumption, 1992

	U.S.$ billion	% of total	Population million	Packaging expenditure per capita (U.S.$)
Western Europe[a]	90	32	340	265
North America	70	25	240	292
Japan	65	23	120	542
Rest of world	55	20	4800	11

[a] Western Europe only has been included because figures for Europe as a whole are not available.
Source: Various.

Other countries still regard economic statistics as a state secret, and the figures are not published.

In western Europe the economic downturn of 1992/93 seriously affected packaging production, although 1994 showed significant improvement, especially in the United Kingdom, Germany, and the Netherlands. Industry sources put European consumer packaging at approximately U.S. $70 billion annum (Table 2).

The materials used are paper and board, plastics, metal, and glass (Fig. 1), with paper and board the clear leader.

Packaging material output in western Europe is valued at approximately U.S. $104 billion. The disparity between production and consumption is accounted for by other materials such as wood, composites, and the methods of data collection. The production data (Table 3) show Germany to be the leading country, with 26% of western Europe's output in value terms, followed by Italy, France, and the United Kingdom, all four accounting for 75% of total output.

The packaging industry has been affected by global recession and factors such as environmental pressures, legislation, and political change in eastern Europe. Companies have taken action that has led to reorganization and restructuring, not least among Europe's major packaging companies, with a concentration on core business. Many announcements were made throughout 1993 and 1994. The manufacture of consumer products is increasingly becoming a global business with large companies combining into larger production units. This has forced packaging manufacturers to consider similar rationalization so that they can use their size to apply pressure to raw materials and machinery suppliers and to meet their customers' needs on quality, price, and service.

Packing Materials Paper

In the paper sector overall growth has been slow, although it is still the largest by volume of the packaging materials used. The industry has worked hard to address what are seen as areas of weakness: use of forest resources, energy consumption and mill effluents, especially those related to chlorine bleaching. In the long term, the biggest issue may be the availability of raw materials for the production of paper as world demand increases.

Recycling is growing at a high rate, but there are economic, performance, and even environmental limits to how far this can continue. In some European countries a utilization rate of >60% is claimed. There is a growing interest in the use of nonwood sources of cellulose fiber from crop residues such as straw and other cereals. These offer extra benefits in resolving difficult disposal problems.

Molded pulp is also being used because of its perceived environmental benefit. It provides another outlet for the increasing quantities of wastepaper collected.

Plastics. In the plastics sector all polymers except PVC have shown growth; the largest growth rates are for PET and polypropylene. Regenerated cellulose has steadily been replaced by the latter, and future growth for polypropylene will come in the label market and for overwrapping as a single material and in flexible packaging.

Metals. Tinplate shows modest potential growth, but aluminum is outstripping this by an order of magnitude, albeit from a much smaller base. Tinplate and a variety of new structures, some based on steel sheet laminated with thin plastic films, may improve its competitive position vis-à-vis aluminum. The introduction into Europe of both iron foil and all-steel easy-open ends for beverage cans may change the materials performance in the long run.

Glass. Glass consumption has risen in recent years after a poor performance in the 1980s. Unless there are major developments in plastics technology or other influencing factors, glass should sustain its present level.

Table 2. Estimated End Use of European Consumer Packaging Market, 1992

	U.S.$ billion
Food	28
Drink	13
Healthcare	12
Household chemicals and other products	17
	70

Source: Goldman Sachs.

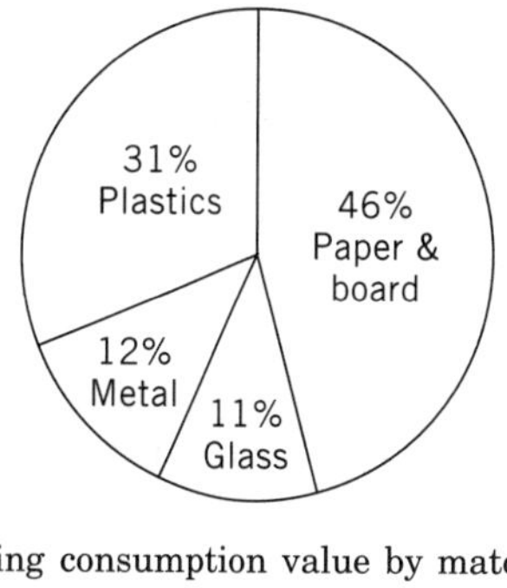

Figure 1. Packaging consumption value by material 1992. (*Source:* Pira International.) European Packaging (White)

Table 3. Estimated Packaging Material Production per Capita, 1992

Country	Production (U.S.$)	%	Production per capita (U.S.$)
Austria[a]	2,126	2.0	269
Belgium	3,110	3.0	311
Denmark	2,227	2.2	438
Eire[a]	350	0.3	100
Finland	1,560	1.5	312
France	17,510	16.9	305
Germany	27,262	26.3	338
Italy	17,890	17.2	310
Netherlands	5,119	4.9	337
Norway	724	0.7	168
Portugal[a]	1,200	1.2	122
Spain	6,744	6.5	172
Sweden	2,065	2.0	237
Switzerland	900	0.9	130
United Kingdom	14,912	14.4	258
	103,749	100.0	281

[a] 1991 figures.
Source: Pira International.

Multilayer materials. Multimaterial forms of packaging are under pressure, especially those containing aluminum foil. Other mixed materials at risk include blister packs and window cartons—window packs are already a diminishing part of the carton market because of environmental demands.

Packaging Reduction

The combined effect of the 1992–1994 recession and environmental concerns has been to concentrate on the reduction of overall packaging consumption. Some of this has led to improved performance materials, allowing down gauging, and in other areas there has been a recognition that certain pack forms may be either significantly reduced by using different materials or designs or even eliminated entirely. A prime example is the launch by Tetra Pak of a new lightweight flexible pack "Tetra Pouch" in Switzerland.

Simplicity was the key in 1994, and a number of complex consumer packs on the market are expected to disappear by 2000. Despite the criticism of packaging, it does not account for a high proportion of total materials usage, and there is no real likelihood of a shortage of resources before 2015. Alternative materials and technologies are already available. (See also Environmental Regulation, International.)

Technology

Packaging technology has a worldwide application, and no technologies are specific to Europe. Some technologies may be more widely used, eg, gravure printing for cartons with Europe taking the lead in developing some technologies such as uv inks for printing, modified-atmosphere packaging, and electron-beam curing for coating systems.

Eastern Europe

The greatest changes have taken place in eastern Europe. The overall consumption of packaging has been much lower than in the West, resulting in higher losses, especially of agricultural products. Paper is the most important material used, and the use of plastics is much less than in the West. For instance, PET bottles were not introduced into Hungary until 1992.

Glass was the main material for packaging beer, soft drinks, and mineral water. With the advent of the multinationals setting up in these countries, canned drinks have been introduced, with the consequent reduction in production of glass bottles. Several major companies are setting up beverage-can plants with a view to their introduction into breweries.

Quality and a lack of marketing has been a major problem in all the eastern bloc countries. This is now being addressed, and with major investments in new equipment, many of the countries now have the nucleus of a thriving and developing packaging industry that will equal that of western Europe.

BIBLIOGRAPHY

General References

Packaging 2000, Pira International European Packaging 1993/94, published by Pira International.

R. White
R. Goddard
Pira International
Leatherhead, Surrey
England

EXHIBITIONS

Packaging trade shows are a regular part of the diet of every packaging professional the world over, but in today's highly competitive economies, there may be more choices than a really healthy diet should include. Thus, the problem is more one of selection than of availability!

There are basically three kinds of shows to consider in addition to the one that just happens to be occurring when you just happen to have a specific problem to solve: large, multinational shows; more restricted regional and local shows; and vertical industry shows with some packaging. This is somewhat tricky definition, admittedly, since a show that sounds local, say, Salon d'Emballage in France, is many times larger than a titled regional show such as Asia Pack.

Then one has to question the packaging component of the vertical industry shows. Usually, whatever packaging form and vendors that already dominate that particular industry predominate. If you are looking for other ways to see the world, whether it's your vision or a vendor's, you're not likely to find it at these shows and, instead, you're obligated to attend at least one major *packaging* show somewhere in the world every year.

To do this, one can look to the truly multinational shows such as our own Pack Expo in Chicago in November of the even years. Internationally, one turns to Dusseldorf every 3 years for Interpack, reputedly the largest packaging show in the world; or to France for Salon d'Emballage; or to Britain for PAKEX; or to Tokyo for Tokyo Pack. These are shows of the first rank—1000 or more exhibitors of machinery, materials, components, and services with real attendance nearing 100,000 people, if not well above. None is an annual show, however, so you have to plan your calendar at least 3 years in advance to see a pattern of shows you may want to attend.

Then consider the host of second tier shows—second tier not because the answers one seeks may not be available, but because the major manufacturers tend to concentrate the announcement of new technologies in the larger shows mentioned above.

However, WestPack here in the United States, for example, known historically as a West Coast show, is nonetheless a really large show and is now broadening its horizons east and west to encompass a more national orientation as well as some Pacific Rim interest. The Packaging Machinery Manufacturers Institute (PMMI) created in 1995, a counterpart of its big show, Pack Expo, for the Western market, and it, too, provides an important machinery forum. PMMI is also involved with the Mexican show and, while still and a small show, it serves the market and brings people together for important business conversations necessary to an exploding market potential. The Canadian show, PACEX, is very decently sized now, too. Several shows are going well in Latin America, especially in Argentina and Brazil; Asia has a number of opportunities as well, although the shows in Japan are by far the largest and most important in that region.

As is most often the case, the decision on which show(s) to attend will be made easier once the information is available. So where to find information on what show is when? Short term, a good bet is to watch the industry magazines. Their calendar of events is usually reliable and provides direct information for each show listed. Alternatively, the World Packaging Organization maintains contact with the major show producers and attempts to keep an international calendar up to date as well. It is a particularly good information source for shows in new and building economies such as those in the former Soviet block, Asia, and Latin America. WPO members are the national packaging institutes, and most not only are involved in production of the national trade show but also provide another easy access to market and technical information.

The key is planning ahead, then, and remaining flexible. As noted above, every packaging professional should attend at least one of the big shows every year. Additional attendance, particularly at shows produced in locations abroad, can add immeasurably to a packaging professional's experience and understanding of global trends, local customs, and emerging technologies.

So our advice is to make your commitment, then begin serious planning to gain maximum benefit from your show attendance. Here's suggested routine to follow to that end:

1. *Make your travel plans well in advance.* There usually are no problems in arranging transportation to the city where the show will be, but getting a hotel during one of the large shows can be challenging! Hotel space should be reserved at least 6 months in advance and done when you advance register for the show itself. *Do* advance-register for the show! It makes no sense to arrive at a show travel-weary, then stand in line for even a minute, let alone the hour or more, at many shows waiting to register.

2. *Do your homework.* Survey your department, your division, and/or corporate headquarters to make sure you are in touch with production and distribution, with marketing and with R&D. Identify the packaging problems in each major area of concern. Try to put numbers on them: How much is it costing the company to have x percentage breakage, for example? Or check to see if annual maintenance requirements indicate a need for new machinery. Has new technology provided sufficient productivity gains to prove cost-efficient for you? And make sure you know about any new product plans or special promotions requiring new packaging ideas.

3. *Don't reinvent the wheel.* Most of the major packaging trade publications and many of the packaging-related magazines publish "preshow" issues. Be sure to take advantage of the good work that the editorial staffs of these magazines have done on your behalf. Some print booth-by-booth previews of the show. Some highlight only the new developments scheduled to be unveiled at the show. All of them offer important information that you can put to practical use before you travel.

4. *Remember that a show is usually bigger than the exhibit hall.* Many institutes and even private providers sponsor seminars and conferences in and around the show. Make sure you get a list of such conferences from the show producers or by watching the trade publications. Again, WPO or the in-country institute can often help you with this information.

5. *When packing your suitcase, think of a packaging show as a long, well-dressed hike.* Focus on comfort rather than style, and you'll be OK. It is not unusual, for instance, to see a vice president or director of packaging for a major *Fortune* 500 company wandering a show floor in a suit and walking shoes.

6. *Assemble your team.* Many departments prefer to send several individuals to a show of the size and importance of Pack Expo, Interpack, and so forth. Make sure you know who's coming, and why. Don't forget your foreign affiliates, suppliers, and other contacts. Let them know you're on the way and schedule some time together to talk shop.

7. *Schedule your priorities.* Using the survey information you've gained, combined with knowledge of the individual strengths (and weaknesses) of any team members, begin to formulate your primary and secondary objectives. You need

to specify as closely as possible the kind of information you need, by material and by machinery function.

8. Divide and conquer. Divide the responsibilities, if you can, among your team members. This will probably best be done along functional lines as opposed to your own product lines. So, if you need code date machines for many product lines, have one person in charge of reviewing all the code daters, rather than having one person review all the machinery requirements for one product.

9. All work and no play . . . . Most events have a social side—or two! Everything from formal welcoming receptions to awards banquets to supplier hospitality suites are part of your very long day at shows. So plan on it. Contact key suppliers in advance so that you know where they are—arrange to meet friends and associates at the receptions—and have in mind the areas of your primary concern so you'll pick up on the conversations going on all around you.

10. Welcome. On arrival at the show, pick up a show directory, sit down, and study it. Check function by function the cross-reference listing of exhibitors. (*Careful:* There may be well over 100 machinery functions, a wide variety of material classifications and so forth listed, especially in the big shows.) Use a colored ink pen or "hi-liter" to indicate on the floorplan map in the directory each booth that you must visit. It may be helpful to use different colored ink for each major machinery and material function so you can complete your research by functional requirements, rather than by happenstance of exhibit floor organization. Consider using a second and even a third color of ink to indicate second- and third-priority booths for you.

11. Follow the road signs. Now, map a "most-direct path" or paths through the show to accomplish your primary objectives. If you are arriving early in the morning, consider starting your show tour in the rear of one exhibit area since most new registrants just start by walking into the front of the hall, which therefore quickly becomes more crowded.

12. Check off. Keep a record of where you've been and where you've got to go. After a couple of hours in and among several thousands machines, it's sometimes difficult to remember where you're going next—or why!

13. Review your work. After you've completed your planned circuit, sit down again and review the data you've collected against the objectives you established before arriving. This is a good time to clear you head. Try a cup of coffee, or even return to your hotel for some peace and quiet. "Do I need to return to any exhibit for more information?" is an important question to ask yourself now.

14. Not again! Force yourself, regardless of the pain, to walk the entire show floor. Keep up a good pace, and keep a pencil and paper handy. As you pass quickly by, scan for new machines, different applications, and bright ideas that you may not have contemplated in your original planning. After you've browsed the entire show, go back for details on the exhibits you jotted on your note pad.

15. Debrief with your team. This is best done periodically throughout your show tour, but don't waste a lot of time waiting for a member who may be involved in a critical conversation aisles away with a very good, and very hard-to-corner technical representative and is thus late for the meeting.

16. Write a report. This is an often neglected chore after a trip because of "the press of other business." But remember, you've just used a fair amount of corporate resources to finance your trip, let alone for all the planning that went into it. Don't compromise now! The report will translate the thoughts of your team into conclusions. It will also provide a record for the future if a brand-new packaging problem pops up before the next show.

Packagers who pay careful attention to preplanning their visits to trade shows will do much better than those who simply arrive to "see what's going on." The suggestions listed above are a place to start the planning process.

Additional Information

The World Packaging Organization and the Institute of Packaging Professionals now share office space, and both are good sources of information about future show schedules.

The WPO's *World Packaging Directory,* published in 1995, lists key contact information in 51 countries. Included in each country's listing, as appropriate, are the addresses of the packaging institute, major trade associations, testing labs, and packaging magazines and directories. It is an essential desk reference for any one working internationally. Available from the IoPP Bookstore for a nominal price. Contact (800) 432-4085.

Also listed below are the U.S. offices of the major packaging show producers. The Reed Exhibition Company is part of the worldwide Reed/Elsevier complex, which has interests in packaging shows in the United States, Europe, and Asia.

World Packaging Organization
481 Carlisle Drive
Herndon, VA 20170
Tel. (703) 318-5512; Fax (703) 318-0310

Institute of Packaging Professionals
481 Carlisle Drive
Herndon, VA 20170
Tel. (703) 318-8970; fax (703) 318-0310

United States: East Pack, South Pack, West Pack, and several vertical industry shows of interest; plus contact for shows in Europe, Asia and Latin America:

Reed Exhibition Company
383 Main Street
Norwalk, CT 06905
Tel. (203) 840-5933; fax (203) 840-9349

United States: Pack Expo, Pack Expo West; also the show in Mexico:

Packaging Machinery Manufacturers Institute
4350 North Fairfax Drive, Suite 400
Arlington, VA 22203
Tel. (703) 243-8555; fax (703) 243-8556

Germany, United States Contact: Interpack, K, others

Dusseldorf Trade Shows
150 North Michigan Avenue, Suite 2920
Chicago, IL 60601
Tel. (312) 781-5180; fax (312) 781-5188

William C. Pflaum
Herndon, Virginia

EXPORT PACKAGING

Export shipments have been a major business for thousands of years. Even today, export cargo is carried by surface ships in the holds and on decks. Since the worldwide acceptance of containerization, a large percentage of exports is moved in cargo containers. A smaller, but growing, quantity of high-value and high-priority items is shipped by air as break bulk or in air containers.

There are no standards, rules or regulations, codes, references, or guidelines for the export packaging of specific products. In a sense, every item, destination, and form of shipment dictates the requirements. When packaging for export, every detail must be carefully considered. The products and packages will be completely out of the control of the shipper, and they may be "somewhere out there" for weeks. Export shipping is the test. The package fails at its weakest point. It is almost impossible to overpack. Export shipments must arrive in good order. They may have taken months to arrange and to produce. They cannot be easily replaced or repaired at destination.

Hazards of Export Shipments

The packaging engineer must understand and meet the requirements and hazards of various forms of shipments (see Distribution hazards).

1. Break bulk cargo in the holds and on the docks of ocean-going surface vessels of many types, sizes, flags, and ages.
2. Unitized loads in the holds of surface ships.
3. Containerized loads on container ships, on the decks and in the holds of cargo vessels.
4. Transport by lighters and barges; on open boats through surf to sandy beaches.
5. Rollon, rolloff (Ro-Ro).
6. Containerized air shipments plus ground handling and transportation.
7. Individual packages and unitized loads shipped by air freighters and in the holds of passenger planes.
8. Miscellaneous combinations of modes of transportation, handling, and intermittent storage in foreign places and unknown conditions.

Export packaging encounters at least two domestic movements that involve the usual hazards of handling, loading, dropping, compression, and moisture, plus the hazards of transportation. One major exposure is during the movement to the seaport or the gateway airport. The second involves the customs procedures and storage of the country of debarkation, multiple handling, unforeseen conditions of transshipment, and eventual delivery. Products must be packed for the toughest part of the journey and survive all of the exposures.

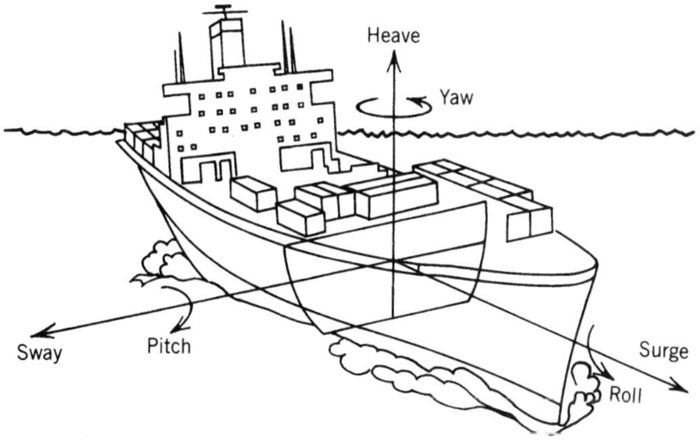

Figure 1. The six directions of motion of a ship at sea.

Only about 30% of the cargo losses can be classed as fortuitous losses resulting from sinking, stranding, fire, collision, seawater, or heavy weather. One ship is lost, somewhere in the world, almost every day. Approximately 70% of the losses may be preventable:

10%	Water damage—freshwater, sweat, condensation, and saltwater.
20%	Theft, pilferage, and nondelivery
40%	Handling and stowage—container damage, breakage, leakage, crushing, contact with oil and other cargo, contamination, and failure of refrigeration or other equipment

During transit, cargo must withstand the conditions of rough seas, turbulent air, and substandard roads.

At sea, a ship may move in six different directions at the same time. A top-loaded container may travel 70 ft (~20 m) in each direction with each 40° roll as often as eight or nine times per minute. Figures 1 and 2 diagram the movements involved. The center point is least subject to these move-

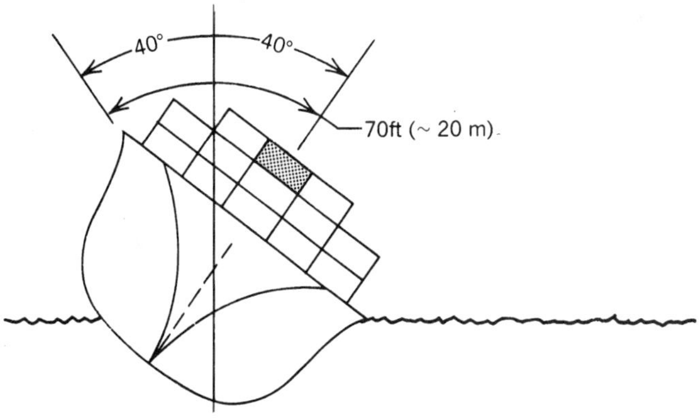

Figure 2. The distance (~20 m) traversed by a top-loaded container in a 40° roll.

ments, but the shipper does not control the location and stowage of the cargo or of individual containers.

Specific Conditions

With the above as a brief introduction to the hazards of international cargo movements, the packaging engineer should further study this subject and the conditions that may exist at the ports of embarkation and debarkation.

Brief descriptions of port conditions and equipment are provided in *Ports of the World* (1,2) and similar publications. Information can be obtained from the cargo carriers regarding their equipment, methods of handling, routes, stops, transshipment, and other details. They can also provide some information about local conditions at the port facilities. Some marine underwriters may have constructive suggestions. The consular staffs and the commercial development departments of each country are well informed. Foreign sales agents can provide details about the customers and their facilities, needs, and preferences. Too often the shipper is uninformed and relies on gossip rather than acquiring a realistic understanding of this complex subject.

Product Analysis

Every facet of the products to be exported must be analyzed, and the potential problems, even marginal ones, must be identified. Export handling and shipping expose the weaknesses if they are overlooked.

Corrosion and mildew. All metals must be protected against water and moisture for surface transportation and even for air shipment from or to humid and tropical areas. All natural materials (eg, leather, cotton, wool, and paper) must be protected against moisture, fungus, and mildew.

Problems can begin because the products are not thoroughly cleaned for fingerprints, cutting or cooling oils, dirt, and foreign matter, or because of unnoticed rust, mildew, or local moisture in the packages. In transit and in storage, the moisture and mildew, condensation, or sweat may find the weak point in the packaging and cause irreversible damage. If the product surface is critical or if the item is very expensive or choice, the damage may be total.

When the problems are identified, corrosion inhibitors (special oils and petrolatums) or properly used vapor-phase inhibitors and waterproof wrapping are applied. For electronic equipment and very critical metal surfaces or hidden areas, desiccants with sealed moisture-vapor barriers are used. If possible, the critical sections or elements are removed from the larger units and packaged separately (inside the larger box or a sheathed crate). A small vapor-phase inhibitor or moisture-vapor barrier package provides better protection more economically than a thorough package for an entire unit. Containerization does not assure protection against water damage or moisture. Air shipment does not offer full protection because of entrapped atmosphere, and the cargo is on the ground more than in the air.

Pilferage and nondelivery. It is essential to recognize that the exported products may be days or weeks in conditions where pilferage is easy. Therefore, the packages should not identify the contents as worth pilfering for private use or resale. Identification of the product on the outside of the package by brand name, manufacturer's name, or the shape and size of the box is an open invitation to steal. A damaged package exposes the contents for pilferage.

Cigarettes, liquor, cameras, stereos, jewelry, furs, small appliances, and many other items are ideal targets for pilferage. Many consignees use code identification because their names on the packages would encourage pilferage. Some even have the items delivered to a "front" to avoid identification.

With the potential problem of pilferage identified, steps should be taken to assure special handling and accountability. Packages should be stored in separate locked areas and on board in lockers or safes.

Nondelivery occurs when the packages are stolen, destroyed, lost, unloaded and left at the wrong destination, or misdelivered. Adequate, permanent, and prominent markings help to avoid some of these problems. Unitizing or using master overpacks is also helpful.

Breakage. The internal weaknesses of electronic products; the fragile construction, design, or materials of many products; any part that extends; and any item that is not in balance are all subject to breakage or physical damage. It may be necessary to protect the package from the product because of a dense weight load or severe imbalance.

Any product that is hard to handle, tends to fall over, or is especially awkward for any reason must be given special consideration. Often, production equipment requires special handling, and provision must be made in the package. Skids and other devices may help to keep the equipment on its base and handled properly.

Cushioning, blocking, and bracing are all essential (see Cushioning design). Built-in lifting eyes or other devices, special skids and pallets, and guide marks for the stevedores are helpful. On large items the balance and the lifting points should be marked. Occasionally, an experienced exporter includes special instructions and photographs with the documentation and on the outside of the packages or crates.

Contamination. Some products can be spoiled or damaged because they can absorb the odors or fumes from the other items that may be in the same container or nearby in the holds or warehouses. Often the people who work with these items ignore or forget these problems.

Precaution can be taken to specifically request the carriers not to expose the items to adverse conditions. Sealed moisture-vapor barriers may give further assurance. Packages should be marked to indicate that they should not be stowed with potentially harmful cargo.

If photographic supplies and other sensitive items are exposed to excess heat, light, or moisture, damage can result. This type of problem must be identified, the product shielded in the packages, and full instructions provided to the carriers.

Hazardous materials. The shipper must identify any hazardous materials. These should be separated and given special documentation and packing under strict rules and specifications. It is essential that each hazardous material is classified and properly identified on the packages, with the correct and legal labels.

It cannot be assumed that compliance with domestic regulations will assure foreign acceptance of the shipments. This

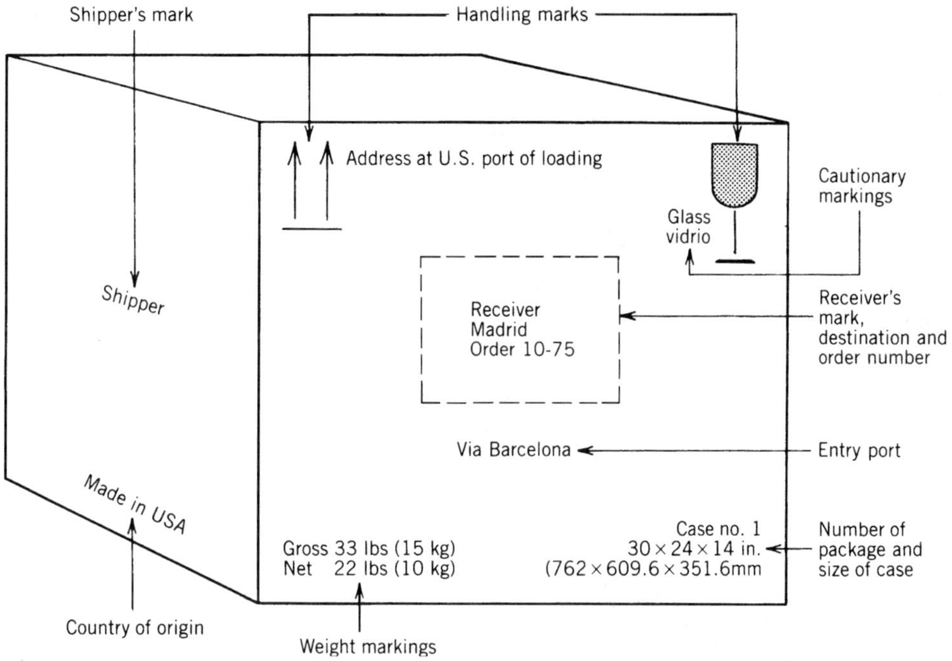

Figure 3. Assignment of codemarks for shipping.

requires special study. Improperly labeled cargo may not be shipped and can cause great confusion (see Standards and practices).

Marks and Symbols

It is essential that proper markings and symbols be used. Even a poor package, adequately closed and marked, has a good chance of being delivered in reasonable order. An excellent package that is not adequately marked may never reach its destination.

Only those markings that are essential and appropriate should be used. Any other markings, or too many, can be confusing and serve no purpose. New, clean packages with no advertising or other printing should be used. The selected markings can be printed by the manufacturer of the packages or stenciled permanently. Crayon, chalk, marking pens, tags, and cards should not be depended upon.

Code marks for the name of the consignee and of the shipper (if used) are best for products that might be pilfered (see Code marking and imprinting). These and port marks should be large, clear, and applied on the side, end, and top of each package (Fig. 3). Required weight and dimension information should be clear.

Handling precautionary instructions should be printed or stenciled on the outside in the language of the destination country (Fig. 4). Many times the cargo handlers cannot read the language of the country in which they are working. Pictorial precautionary markings may be the most helpful (Fig. 5). In general use, they are recognized and have replaced a wide variety of symbols used in the past. Regardless of the markings, it is essential to package adequately and protect the products being shipped because the precautionary marks are frequently ignored.

When a large number of units are being moved in a single shipment, it helps to corner-code mark them with a distinctive symbol or to color-code the opposite top corners with triangles or stripes. If a series of shipments is being sent to the same destination to be collected and staged for further inland shipment, the same method of coding should be used. This helps to reduce the number of stray packages.

When a number of packages are being shipped together, they should be numbered on the packages with the same numbers as the documentation. If there are 18 packages in a shipment, they should be numbered on the packages as 1/18, 2/18, 3/18, . . . , 18/18. Thus, if a package is missing, it can be checked on the documents for size, weight, and contents.

Packaging for Export

Some exporters use their domestic packages for export. This is not good practice, but it may be economical. Preparation of domestic packages for export includes the following:

1. Waterproofing by liners, overwrapping, or overbagging.
2. Master packing small to medium size packages.
3. Unitizing a quantity of packages with suitable strapping, adequate pallets, and shrink or stretch films.

Unless conditions are thoroughly understood, this applies even for air shipments because of ground handling, storage, and transshipment.

Packaging for break bulk. This is the traditional form of shipment for small and large items stowed in holds or on the decks of surface ships. The individual packages are handled by the ship's gear individually or on stevedore pallets. They are pushed or moved manually, or sometimes by roller conveyor or lift truck, out of the square of the hatch to wings. They are stacked with other cargo and used to help brace the

English	French	German	Italian	Spanish	Portuguese	Swedish	Japanese	Chinese	Arabic
Handle With Care	Attention	Vorsicht	Manne-giare con Cura	Manejese Con Cuidado	Tratar Com Cuidado	Varsamt	取扱注意	小心處理	بانتباه
Glass	Verre	Glas	Vetro	Vidrio	Vidro	Glas	ガラス	玻璃製品	زجاج
Use No Hooks	Manier Sans Crampons	Ohne Haken handhaben	Non Usare Ganci	No Se Usen Ganchos	Nao Empregue Ganchos	Begagna inga kroka	手鉤無用	勿用鈎子	عدم استعمال خطاطيف
This Side Up	Cette Face En Haut	Diese Seite oben	Alto	Este Lado Arriba	Este Lado Para Encima	Denna sida upp	天地無用	此面向上	هذه الجهة فوق
Fragile	Fragile	Zerbrech-lich	Fragile	Frágil	Fragil	Omtaligt	壊物注意	易碎貨物	قابل للكسر
Keep in Cool Place	Garder En Lieu Frais	Kuehl auf-bewahren	Conservare in luogo fresco	Mantén-gase En Lugar Fresco	Deve Ser Guardado Em Lugar Fresco	Forvaras kallt	冷暗所蔵	保持陰涼	احفظ بمكان بارده
Keep Dry	Proteger Contre Humidite	Vor Naesse schuetzen	Preservare dall umidita	Mantén-gase Seco	Nao Deve Ser Molhado	Forvaras torrt	水気厳禁	保持乾燥	احفظ بمكان جاف
Open Here	Ouvrir Ici	Hier offnen	Lato da Aprire	Abrase Aqui	Abra Aqui	Oppnas har	取出口	由此開啟	افتح هنا

Figure 4. Precautionary handling instructions.

loads against movement as the vessel rolls and pitches. There is a loading plan specifying that the lightweight items should go on top of the heavier and more rugged cargo and that special classes of products should be given special treatment. The plan must also consider late cargo and the sequence of discharge at ports of call. Packages must withstand a static load of similar material 20 ft (6.1 m) high without distortion or rupture throughout the intended voyage. This applies in the lateral as well as the vertical direction. Problems of shock and vibration may be 50 times those normally experienced in domestic transit. Moisture is usually encountered during a sea voyage at deckside, in customs, and in lighters. Wooden boxes, sheathed crates, and cleated "export" plywood are appropriate for break bulk shipments when properly constructed, packed, and secured.

Unitized loads. In addition to the details already covered, it is necessary to note that the outside boxes in unit loads may be chaffed by other cargo during the ship's movement. The 20-ft (6.1-m) static load rule applies. It may be necessary to provide a wood or heavy-duty cover to distribute the superimposed weight and to protect the individual packages.

Containerized loads. Cargo containers provide physical protection so the 20-ft (6.1-m) guideline does not apply. The moisture problem can sometimes be intensified because of minimum air circulation. Containers should be secured and loaded evenly from end to end. Containers should be thoroughly cleaned and inspected before use. There are special containers for unusual loads and conditions. The packages may be removed at portside and then face another domestic shipment to the consignees. Customs inspectors in some foreign ports do not restuff containers the way they were shipped. They sometimes repack with the larger and heavier items on top of the smaller items, for example.

Lighters, barges, and open boats. Some modern barges may be similar to large containers. Lighters and open boats may provide the most difficult tests of the trips. The handling is doubled and may be crude or rough, partly because of greater exposure to sea water, fog, mist, and spray. Sound packaging is required.

Rollon, rolloff (Ro-Ro). In this method of shipment, highway trailers and other units on wheels can be loaded and secured in the vessel for relatively short voyages. These modern ships offer good conditions, and the handling is minimized; however, the loading, unloading, customs, and storage are portside.

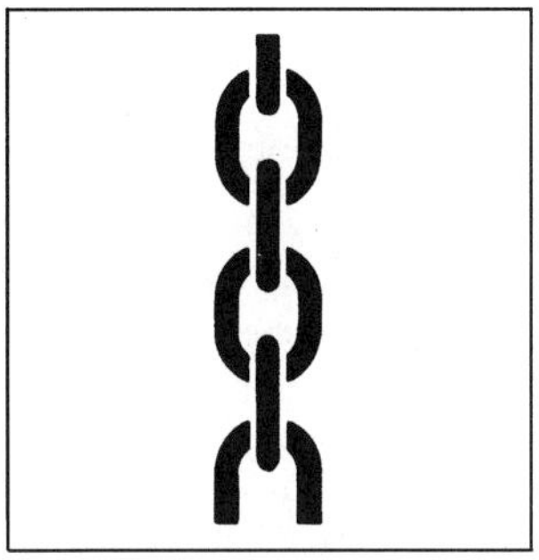

Sling here

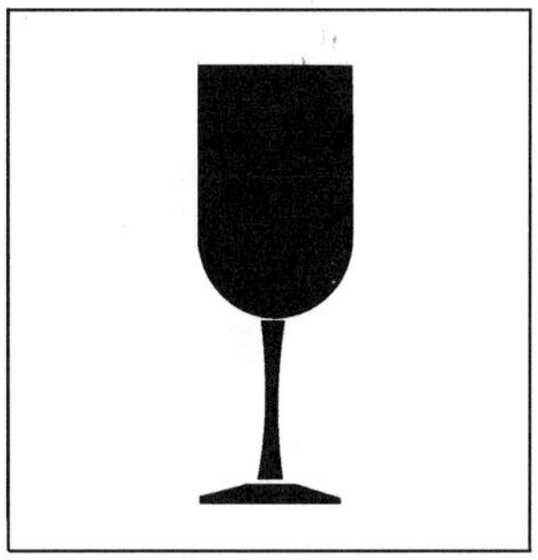

Fragile. Handle with care.

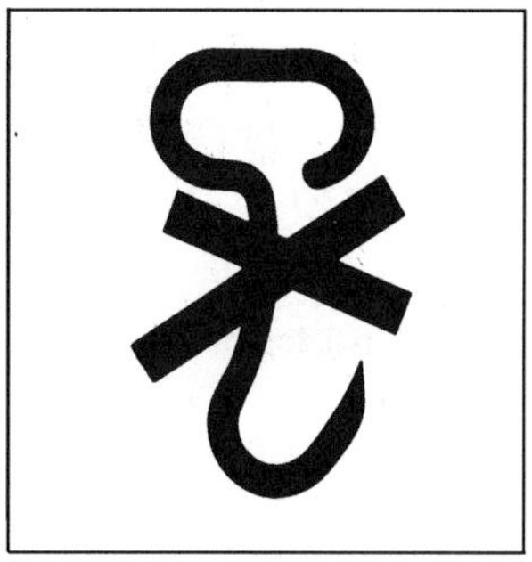

Use no hooks

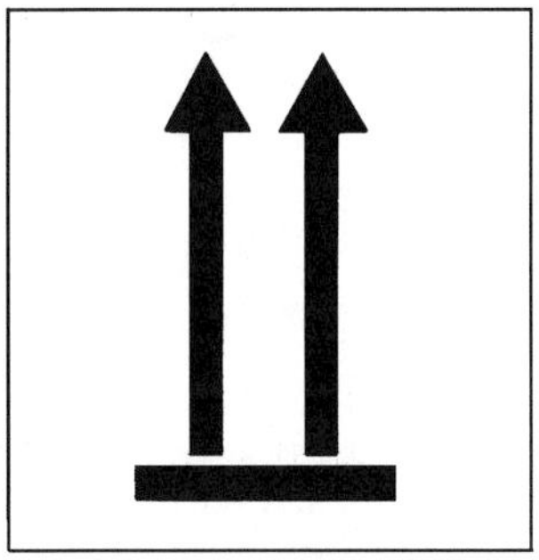

This way up

Keep away from heat.

Keep dry

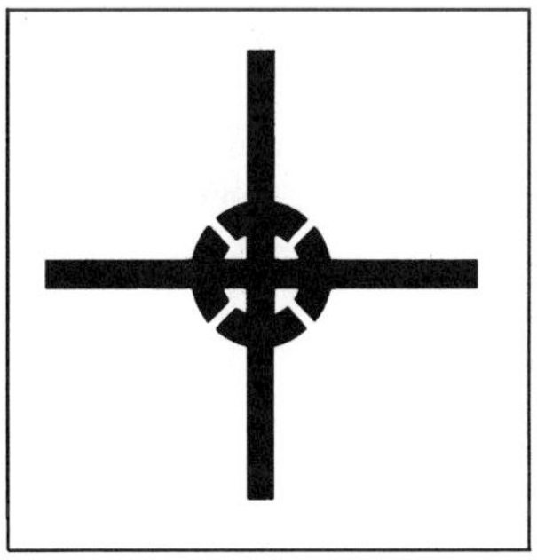

Center of gravity

U.S. STANDARDS

Do not roll

Hand truck here

Keep away from cold

Figure 5. Pictorial representation of precautionary markings.

Containerized air shipments. If the containers are properly loaded and precautions are taken against moisture in the packages and on the ground, this may be the safest method of shipment. Domestic packaging can often be used if the delivery is known to be normal.

Individual units by air. Unit loads by air freighters and individual packages unitized by the airlines for shipment on air freighters may be moved automatically and have little or no handling or moisture problems. Individual packages in the holds of passenger planes, however, may experience many rough handlings. In the cargo space, the handler usually cannot stand up straight and often resorts to crude methods of moving packages. The small items may be rolled or thrown and will not be thoroughly secured.

Miscellaneous modes. Packages shipped to inland points may experience a good deal of rough handling, storage under unexpected conditions, and exposure to pilferage and corrosion.

Guidelines

If shipments can be fully controlled by the shipper and the consignee by containerization, documentation, representation at both ends, and complete understanding by all parties, there can still be serious trouble unless adequate packaging is used and properly implemented. It is almost impossible to do packaging that is too good for export shipment.

BIBLIOGRAPHY

"Export Packaging" in *The Wiley Encyclopedia of Packaging Technology* 1st ed., by Frank W. Green, Point O'View, pp. 276–282.

1. *Ports of the World,* Benn Publications, Ltd., London, annual.
2. *Ports of the World, A Guide to Cargo Loss Control,* Insurance Company of North America, Philadelphia, Pa, 13th ed 1984.

General References

Preservation, Packaging and Packing of Military Supplies and Equipment, 2 volumes, Defense Supply Agency, Department of the Army, the Navy, the Air Force, Washington, DC.

EXTRUSION

Molding and extrusion are the basic techniques of forming polymers into useful shapes. The molding process, which is normally intermittent, can fix three dimensions (height, width, length) of an object. The continuous extrusion process through a die can fix only two (height, width). These two processes are usually complementary rather than competitive and produce a wide variety of products as diverse as pigmented pellets, threaded closures, and refrigerator liners (1).

The extrusion process in which an Archimedean screw rotates within a cylindrical barrel is probably the most important polymer processing technique used today. It is used to manufacture continuous profiles such as fibers, tubing, hose, and pipe; to apply insulation to wire; and to coat or laminate paper or other webs (see Extrusion coating). This article, however, deals primarily with single-screw extruders for compounding polymers and producing pellets, for producing rigid or foam sheet, and for making blown or cast films.

The extrusion principle was first employed about 1795 for the continuous production of lead pipe. The first patents on an Archimedean screw extrusion machine were granted to Gray in England (2) and Royle in the United States (3). During the nineteenth century the machinery became refined for manufacturing rubber, gutta-percha, cellulose nitrate, and casein products. Modern extrusion technology as applied to synthetic thermoplastic polymers began in about 1925 with work on PVC (see Poly(vinyl chloride)). The first screw extruder designed specifically for thermoplastic materials appears to have been made by Paul Troester in Germany in 1935 (4).

Single-Screw Extruders

Modern single-screw extruders designed to process thermoplastic resins normally are <1 in.–12 in. dia., although larger extruders have been built. The most common diameters for production-sized machines are 2–8 in. (51–203 mm). Figure 1 shows an 8-in. vented extruder with a barrel length of 32 dia., and a 600-hp (447.4 kW) drive. The main features of an extruder are shown cross-sectionally in Figure 2.

The solid polymer fed to the extruder may be in the shape of powder, beads, flakes, pellets, or combinations of these forms. The extruder conveys, melts, mixes, and pumps the polymer at high temperature and pressure through a specially shaped die. The die's configuration, and the solidifying or cooling process, determine the shape of the product.

All extruders consist of a barrel, a screw, a drive mechanism, and controls. The heart of the extrusion process is the screw. It is fashioned with a helical thread or threads, and varying channel depth. The function of the screw is to convey material and generate pressure in order to produce pellets or other shapes. In the case of a solids-fed screw the function is expanded to include solids conveying, compression, and melting. Rotation of the screw accomplishes all these functions.

Successful operation of an extruder depends on the design of the screw. The depth and length of each zone of the screw is determined by the product to be run. Barrier flights and/or mixing sections are sometimes built into the screw to improve its efficiency in melting and delivering a homogeneous polymer to the die at the proper temperature and pressure. The profile of the melting process in a mixing screw is shown in Figure 3.

An extruder interior schematic is shown in Figure 4. Here the solid resin is introduced into the hopper, and through the action of the rotating screw is conveyed into the heated barrel. The screw in this section is feeding or conveying the solids and hence is quite deep. The geometry of the screw is such that the depth decreases in the transition zone, and the solids start to melt. Melting results from the shearing action of the screw as motor horsepower is converted into frictional heat. Barrel heaters are used for start-up and to supplement the melting process. The melt continues to be pumped toward the discharge, or die end, of the extruder through the metering section of the screw. (Metering is pumping at a given rate in a uniform manner, within close temperature and pressure tolerances.) The die then forms the polymer in the desired

Figure 1. An 8-in. (203-mm) 32 : 1 L/D vented extruder.

shape. Downstream cooling equipment solidifies and maintains that shape.

Screws are cut from alloy steel. The tops of the conveying flights are hardened or surfaced with special alloys for extended life. The clearance between screw and barrel is close. In operation, the screw floats in the barrel on a layer of melted polymer.

The barrel, or hollow cylinder in which the screw rotates, is manufactured from machined steel and built to withstand pressures of 7500–10,000 psi (51.7–68.9 MPa). Barrels are also lined with special alloys or hard-surfaced to extend life. The length of a barrel is defined as a multiple of its diameter (ie, L:D ratio of 32 : 1 = a barrel : screw 32 dia. long). The length of an extruder barrel is determined by the polymer and process involved.

Extruders must be heated and cooled. Electrical heating or fluid heating can be used on the barrel. In electrical heating, resistance heating elements in various forms surround the barrel. Tubes for cooling fluid, cast in aluminum, also contain heating elements. Some barrels are built with jackets through which heating and cooling fluid can be circulated. Extruder barrels are usually divided into zones of specific lengths, each of which can be set at a desired temperature. The zones are controlled by instruments or by microprocessors. Thermocouples are normally used to sense temperature, and to signal the action of the controller (see Instrumentation).

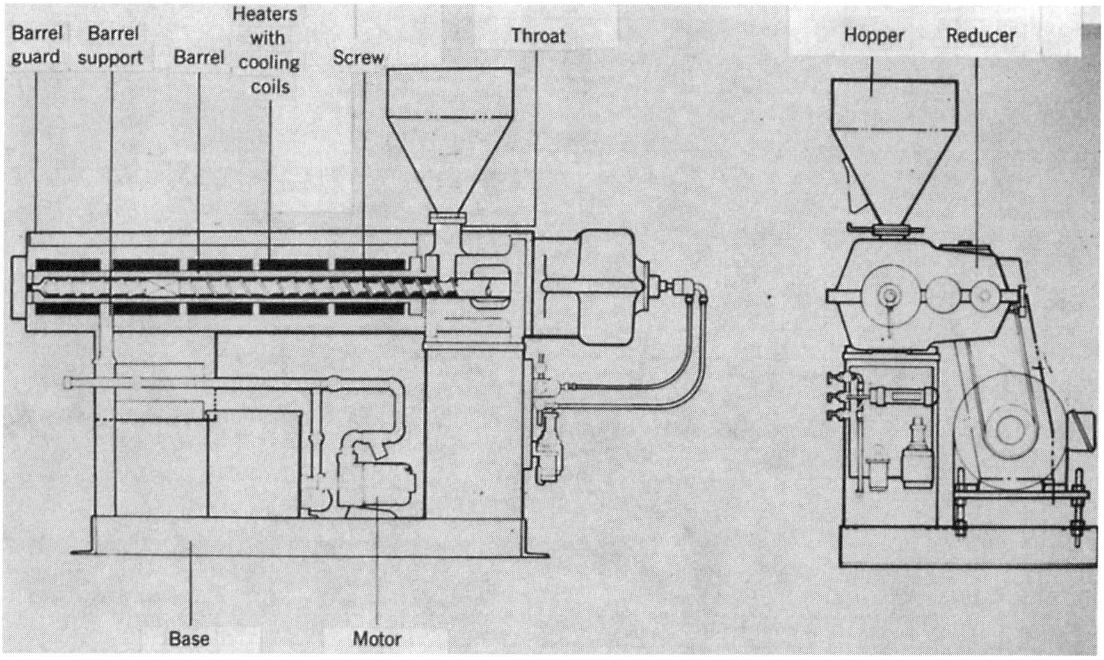

Figure 2. A cross section of an extruder.

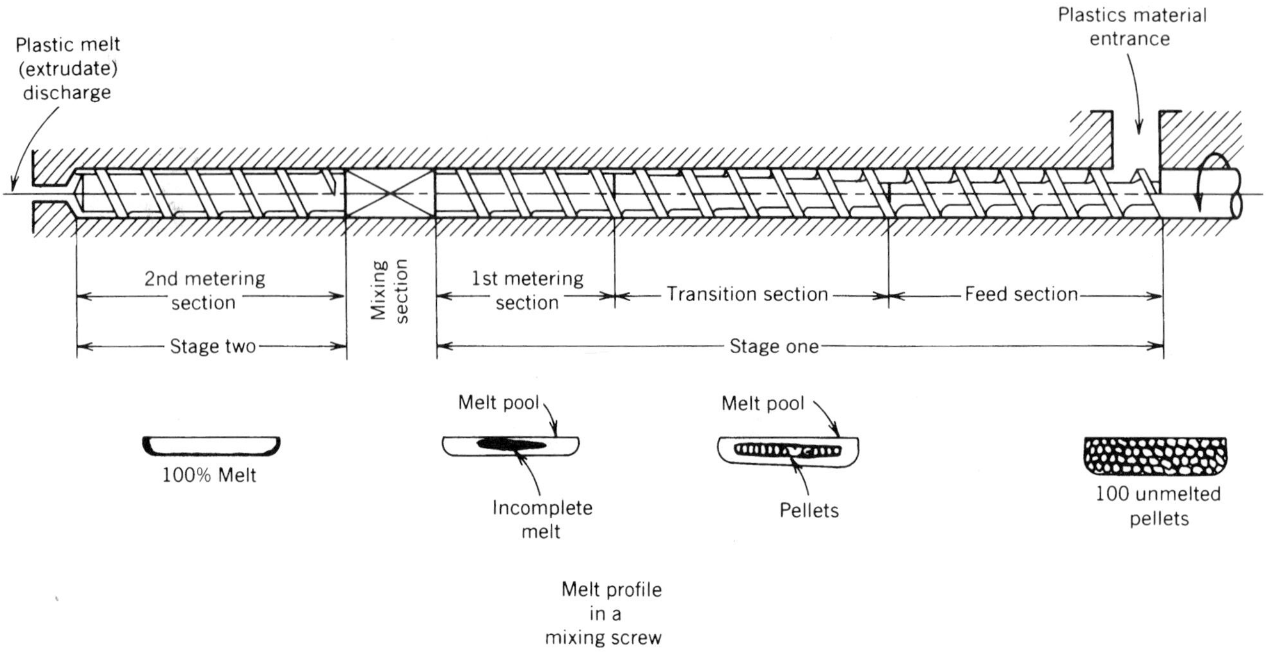

Figure 3. Melt profile in a mixing screw.

The drive mechanism consists of a motor and a gear reducer. The motor is usually a variable-speed dc drive system with the ability to run a speed range of slow to fast in a ratio of approximately 1:20. The gear reducer is used to lower the speed of the motor output shaft (eg, 1750–2000 rpm top speed) to the desired screw rotation speed. Most extruders operate in a variable screw speed range of up to 200 rpm, but speeds considerably lower are used with certain polymers, especially on very large machines. The screw rotation speed depends on the diameter of the extruder, the polymer to be extruded, and the production rate desired. The limiting factor in a given-size extruder is quality of the product, which is dependent on the melt quality. As screw speed increases and more rate is achieved, the melt quality deteriorates because of nonuniform mixing or excessive temperatures, or because of degradation of the polymer from excessive heat generated by high shear.

There are many variations of the basic extruder design, employed to perform specific operations. The extruder and the process are in a constant state of evolution to suit requirements of new polymers being developed for new products.

Compounding

Extrusion compounding as it relates to packaging consists of preparing polymers for use in specific product applications. Mixtures of polymers, filled polymers, pigmented polymers, and a host of other polymer additives constitute a huge market. Compounding with a single-screw extruder consists of mixing and dispersing one or more minor constituents (eg, pigments, stabilizers) into a major constituent, a polymer. The product of a compounding line is pellets. These are used by the converting industry. Converting is the process of melting the pellets and producing extruded sheet, film, injection-molded parts, etc.

Compounding can be separated by functions: resin-plant extruding, blending, reclaim, and devolatization.

Resin plant extruders take the products from a polymerization operation and make pellets. The feed to these machines can be powder, granules, other irregular shapes, or even a melt. A melt extruder has a molten feedstock and only generates pressure, whereas a plasticating extruder has the job of turning solids into a molten mass and then generating pressure. During pelletizing, stabilizers and processing aids are combined with the polymer (see Additives, plastic).

A blending extruder is used to mix feedstocks of compatible polymers, or different viscosities of the same polymer. The blending operation tailors the physical properties to meet a specific end use.

Reclamation of polymers is a rapidly growing segment of the plastics industry. The problems are numerous because the feedstock for the extruder comes in many sizes and shapes. Some examples are polypropylene battery cases, polyester X-ray film, soft-drink bottles, bread wrappers, unusable foam products, fibers, filaments, and off-grade film of all kinds. Specially designed extruders must be used for processing these materials.

Some polymers require extraction of moisture or gases from the melt before a satisfactory product can be produced from the die. To accomplish this, a vent hole with vacuum pump is introduced along the extruder barrel, and the screw is specially designed. Devolatizing extruders are used where residual monomer, water, or other unwanted materials must be removed from the extrudate.

The machinery required for the aforementioned processes must be specifically designed for the application. A wide variety of design features are utilized (5).

Specially designed extruders are used for the addition of short fibers to thermoplastic polymers (6). This is done to im-

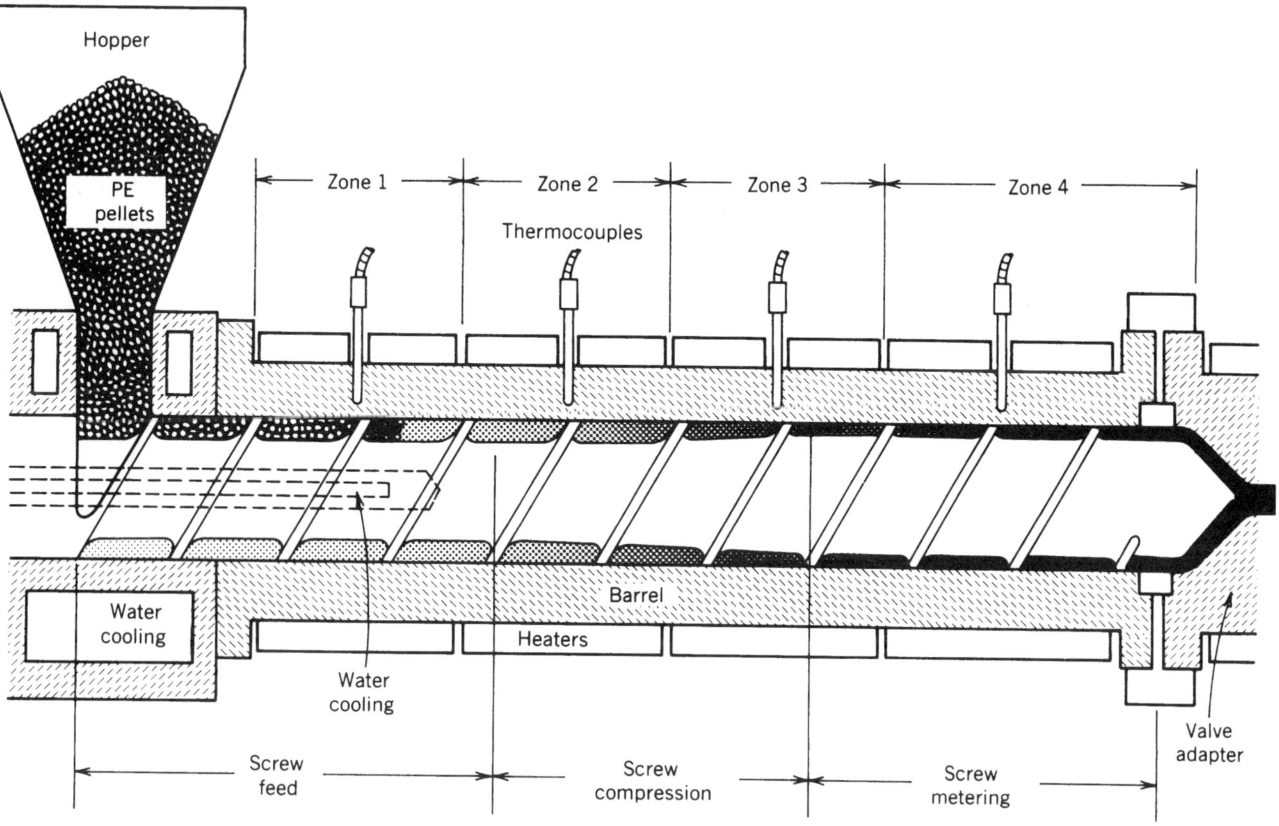

Figure 4. Schematic of the interior of an extruder.

prove the physical, mechanical, and structural properties of the virgin plastic. Cost-per-unit volume can also be improved. The extruder uses a three-stage screw. The polymer is melted in the first stage. The short fiber, usually chopped fiberglass, is screw conveyed into the side of the extruder barrel. The glass and polymer are mixed in the second stage. The third stage allows venting and pressure generation for the die at the exit. A cross-sectional view of a side-fed extrusion system is shown in Figure 5. Figure 6 shows a 3.5-in. (89-mm) extruder with side feeder.

Another unique design is used for reclaiming. Because the bulk density of most of the scrap-plastic items is low, use of a dual-diameter extruder has become prevalent. The feed end of the screw has a larger diameter so there is a greater volume for the entering light fluffy feedstock. This facilitates high production rates.

Vertical extruders meet special needs such as limited floor space, or other plant layout requirements.

Blown and Cast Film

Film is a relatively thin (usually ≤10 mil (≤254 μm) flexible web made from one or more polymers, either blended or coextruded (see Coextrusion, flat; Coextrusion, tubular) but not to be confused with a fabricated extrusion-coated or laminated web (see Laminating; Multilayer flexible packaging).

Films are used in flexible packaging (for overwraps, bags) as industrial wraps (stretch and shrink films), in medical and health care products (disposable diapers, backings, hospital bed liners), in agriculture (mulch films), for sacks (drum liners, garbage bags), and as laminates (aseptic container stock). (Some packaging overwrap films are produced by casting a solvent solution of PVC resin on a stainless steel belt and evaporating the solvent as the belt travels through a heated chamber. This process normally uses a pump to distribute the solution through the die. Solvent casting is beyond the scope of this discussion.)

The same basic extrusion processes are used for producing both blown and cast film. The first two steps, melting and metering, are part of the extrusion process described above.

Forming. In the forming process, the polymer is squeezed through a die as it leaves the extruder, to form a thin uniform web. The cast-film process produces a flat web. In the blown-film process the die shapes the polymer into a tube. This latter process is more versatile because it can produce not only tubular products (bags), but flat film as well, simply by slitting open the tube. Key to the success of both processes is the die, which must distribute the polymer uniformly.

Orientation. In the orienting stage of the blown-film process, the tube is blown up into a bubble that thins out, or "draws down" the relatively thick tube to the required product gauge (thickness). In certain blown-film processes, the polymer is blown downward to produce films with special properties. The ratio of the diameter of the blown bubble to the diameter of the die is called the *blowup ratio.* Most LDPE

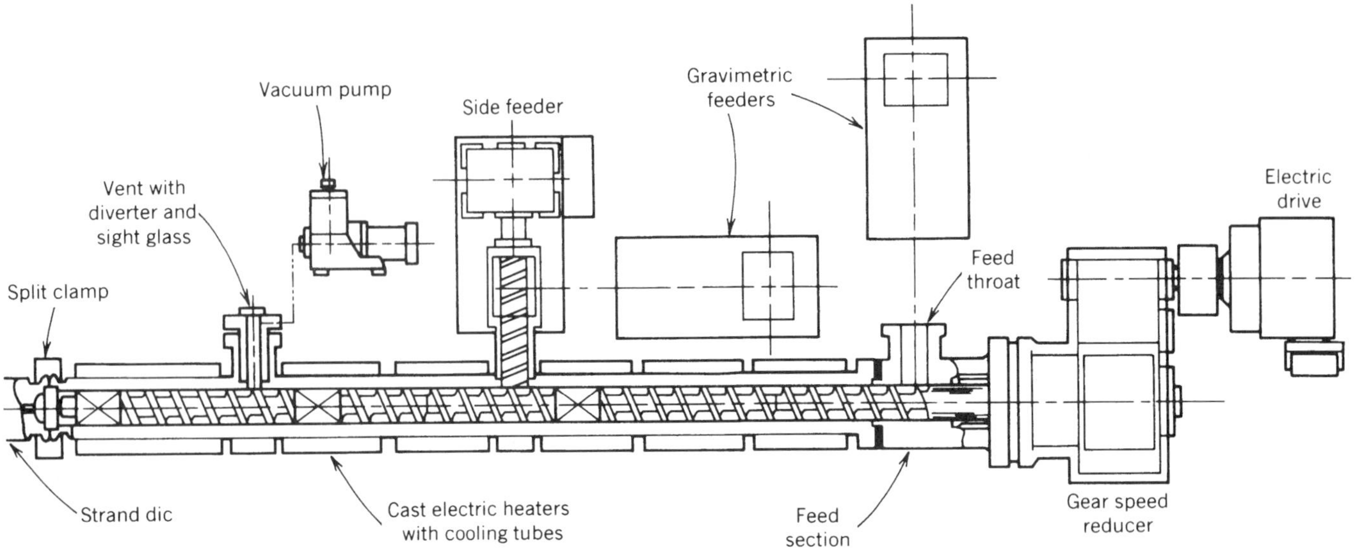

Figure 5. Cross section of a side-fed extrusion system.

Figure 6. A 3.5 in. (89-mm) $L:D$ 40:1 vented extruder with a 3-in. (76-mm) side feeder.

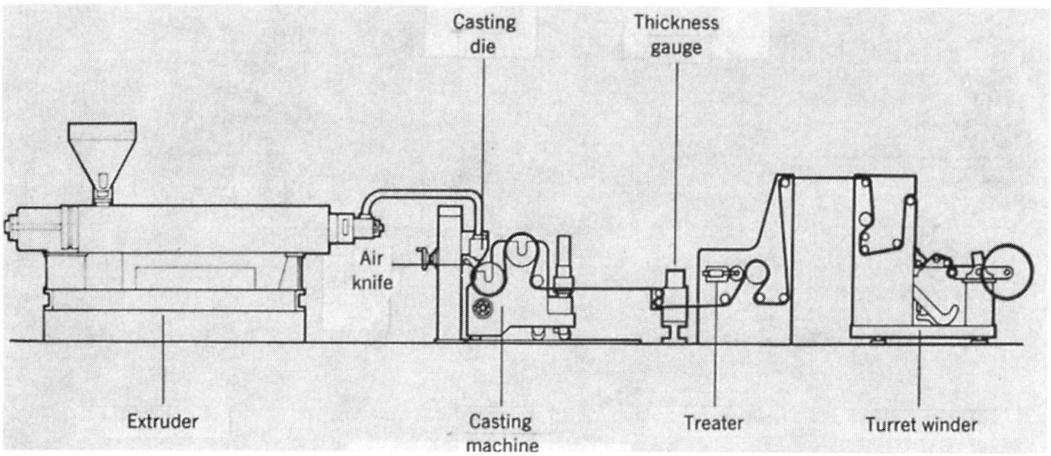

Figure 7. Cast-film line.

(see Polyethylene, low density)-blown films used in packaging are made using blowup ratios of 2.0 : 1–2.5 : 1. The blowup ratio is changed depending on the characteristics of the resin being extruded and the properties desired in the film. In cast film, the molten polymer is also drawn down to the desired finished gauge. Drawdown ratios between 20 : 1 and 40 : 1 are typical (If a polymer exits a die at 40 mils (1 mm) thick and finishes up 1 mil (25 μm) thick, the overall drawdown ratio is 40 : 1.) In the orientation process the long molecules of the polymer line up in the stretching direction, which improves the film's strength in that direction. A key difference between the two filmmaking methods lies in the manner of orientation.

Because both edges are free in cast film, it is drawn down only in the direction the material exits the die (machine direction). Because cast film is drawn in one direction only, it usually exhibits excellent physical properties in the machine direction, and poor properties in the cross-machine direction. The cast-film process is shown in Figure 7. In the blown-film process, the extruded tube is stretched in two directions: as it is blown into a bubble and as it is drawn from the die in the machine direction by the adjustable speed drive system. This results in strength properties that are more uniform and can be balanced depending on the blowup ratio and the takeoff speed. Figure 8 illustrates the blown-film process.

Certain films can be biaxially oriented to enhance properties for specific packaging uses such as shrink films (see Films, shrink) or overwrap. The blown process produces a thick tube that is then reheated and blown out while increasing the take-away speed to maximize orientation in both directions. The cast process extrudes a thick, flat sheet that is chilled, then reheated and stretched in the machine direction and then reheated and stretched across its width by means of a tenter frame. The most common plastic materials to be biaxially oriented are polystyrene (see Polystyrene), homopolymers and copolymers of polypropylene, usually coextruded (see Film, oriented polypropylene); and polyester (see Film, oriented polyester). These films are then coated to enhance heat-seal or barrier properties.

Quenching. After the polymer has been extruded, it must be solidified into finished film. In blown film the quenching (or cooling) process is achieved by convection; by blowing air on the outside and sometimes on the inside of the bubble. Air rings at the die exit direct and distribute air uniformly to the bubble. In cast film the web leaves the die and is deposited on the surface of a driven cooled roll. There are usually several rolls in series (normally called "chill rolls"), arranged to cool the polymer by conduction, or direct contact.

Conduction cooling is quicker than convection cooling and this has an effect on the clarity of the film. Because convection cooling (quenching of blown films) is relatively slow, more and larger crystals form in the film, as compared to those formed in the casting method. Because interfaces between crystals scatter light, blown film tends to be more hazy than cast film. This "haze factor" normally rules out use of blown film where clarity is very important, such as food overwrap applications. Because conduction cooling is more uniform and rapid than convection cooling, cast film has less gauge variation than blown film. This superior flatness means the film can be handled better in subsequent converting operations such as multicolor printing and laminating. These operations are performed at high speeds and cast film is preferred to minimize scrap.

Gauge randomization. In practice, perfectly flat film cannot be made, due to die geometry and machine tool constraints. Blown-film thickness variations of ±7% and cast-film thickness variations of ±3% are typical. Variations in thickness are frequently evidenced by gauge bands. If relatively small variations become significant at the film winder or at a later converting process (printing, laminating), adjustments must be made to distribute them. In the blown method, variations in the film are usually randomized by rotating or oscillating the die to distribute the gauge variations over the finished web. In cast film, because a flat web with two free edges is produced, the downstream winder with edge trim slitters is normally oscillated across the film, winding only a portion of the cast web (see Slitting and rewinding). This generates waste film that must be recycled or scrapped.

It is common now to measure sheet and flat-film thickness automatically after the die, compute and average thickness, and use the signal to control screw speed and thus control thickness in the linear direction. Special casting and sheet

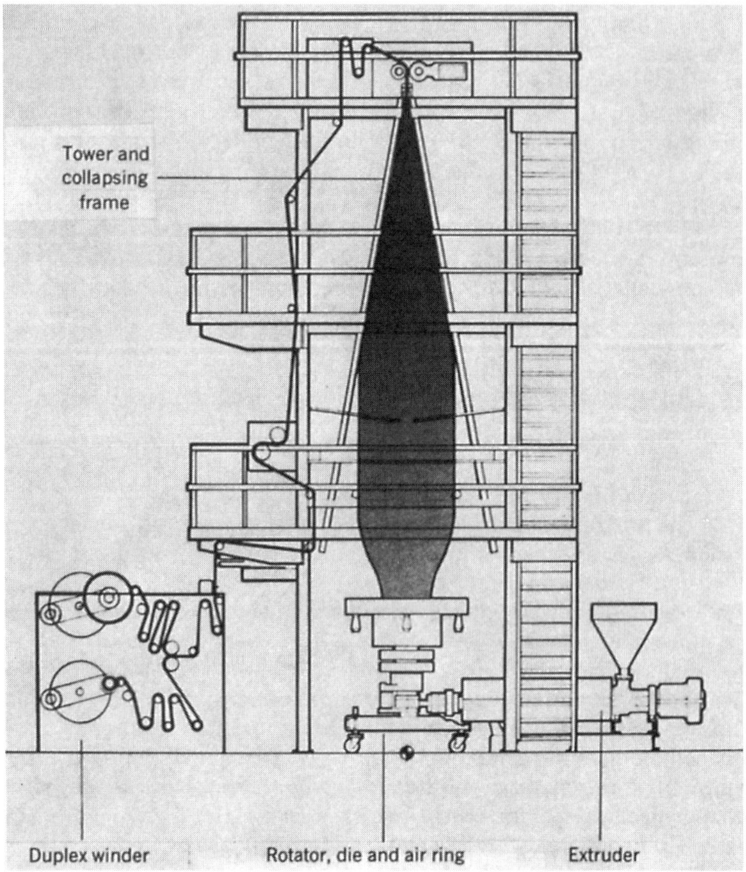

Figure 8. Blown-film line.

dies that operate in conjunction with a computer and a thickness-measuring gauge automatically control the film thickness across the width of the die as well.

Rigid Sheet Extrusion

Film thicker than 0.010 in. (0.25 mm) is normally defined as sheet (see Films, plastic). It is thermoformed (see Thermoforming) into objects that hold their shape, a property that film does not possess. Extruded sheet is thermoformed into cups, lids, containers, packaging blisters, automotive panels, signs, and windows. The machinery required for the manufacture of sheet usually extrudes the polymer horizontally into a nip formed by two hardened cooling rolls that define the final product thickness and surface. Additional rolls and a conveyor for more cooling and pull rolls and a winder or shear complete the sheet extrusion line (see Fig. 9). The extruder is often vented to remove low levels of moisture from polystyrene and ABS. Sheet is traditionally extruded horizontally from a die similar to a flat-film die, but with specially designed interior flow surfaces to suit the particular polymer. Restrictor bars are usually used for added gauge uniformity (see Fig. 10). Sheet dies are often more massive, to minimize distortion.

The takeoff unit for extrusion of sheet usually consists of a cooling and polishing unit (C & P unit) having three driven,

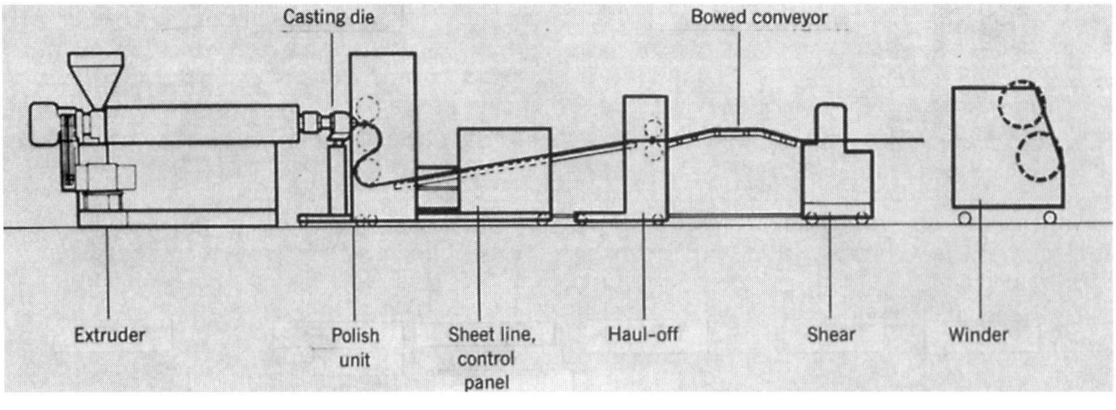

Figure 9. Sheet line.

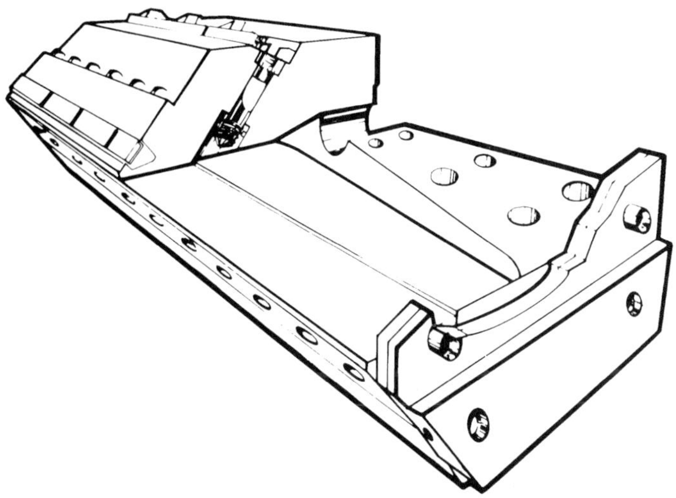

Figure 10. Cutaway of sheet die with restrictor.

highly polished, chrome-plated rolls; a roller conveyor; and a pair of driven rubber-covered pulloff rolls. The C & P unit serves three functions: cooling, polishing, and gauge control. In some cases one or two of the chrome-plated rolls are embossed to yield a sheet with specific surface qualities. Roll diameter is contingent on the output of the extruder, the linear speed of the equipment, and the level of heat transfer required. High-capacity multiple-zone temperature control units are often built into the C & P unit, which must be of rugged construction to eliminate vibration. Sheets up to approximately 0.050-in. (1.3-mm) thick can be wound onto rolls; thicker sheet is cut to desired lengths. In some cases the sheet is pulled directly into a thermoforming machine, providing an in-line, pellet-to-part operation.

Foam Sheet Extrusion

Extruded polystyrene foam (see Foam, extruded polystyrene) sheet material used for making egg cartons, meat and vegetables trays for fast-food packaging, and similar applications continues to find new uses, ranging from decorated picnic ware to coated and laminated sheets. Polyolefin foams are used for packaging materials (see Foam cushioning), insulation, and wire coverings. Most of the world's extruded foam is produced on tandem-extrusion equipment. Although the extrusion process is relatively straightforward, special equipment and controls are needed (see Fig. 11).

The first extruder has a long barrel and is used to mix a nucleating agent uniformly throughout the melt of a base polymer. The nucleator, typically a selected filler, controls foam cell quality. In effect, it creates imperfections in the polymer melt, forming nucleation centers for cells to originate.

About two-thirds down the primary extruder barrel, the gas blowing agent is introduced. At this point the melt is homogeneous and at a pressure of 3500–4000 psi (24.1–27.6 MPa). Fluorocarbons are the usual agents, often blended with hydrocarbons to reduce costs. Carbon dioxide blended up to 35% with either fluorocarbons or hydrocarbons reduces material costs still further.

The product mix is fed through a screen changer for filtering. Then the mix, still under pressure, is fed into a larger extruder that cools and discharges the product under conditions to allow extrusion through the annular die. This is achieved through use of a low-speed screw rotating in a barrel cooled by high flow rates of water. Foaming occurs only outside the die lips.

The foamed tube is expanded to 3.5–4 times its diameter and extruded horizontally over an internal cooling and sizing mandrel that cools and orients the foam and supports the tube as it leaves the die. After the tube passes along the mandrel and a slitting unit, the two webs are pulled through nip rolls. From the nip rolls the webs are wound on either dual-spindle turret type or cantilevered winders. Large-diameter reels are required to handle foam sheet.

Sheet weight per inch (or centimeter) is governed by the amount of blowing agent incorporated in the mix. Sheet thickness is determined by adjustment of the die lips and the take-off speed of the nip rolls. Sheet orientation is controlled by a combination of die gap, blow-up ratio, and line speed.

Accurate metering equipment is needed for a good finished product. The blowing agent system calls for sophisticated controls to safely handle high-pressure gas products on the production line. The difficulties of operating a two-extruder system have been reduced through the use of microprocessors that automatically monitor and control a multitude of functions on the extrusion line.

Figure 11. Tandem-extruder foam-extrusion line.

BIBLIOGRAPHY

"Extrusion" in *The Wiley Encyclopedia of Packaging Technology,* 1st ed., by J. A. Gibbons, Egan Machinery Company, pp. 282–289.

1. J. M. McKelvey, *Polymer Processing,* Wiley, New York, 1962.
2. Brit. Pat. 5056 (1879), M. Gray.
3. U.S. Pat. 325,360 (1885), V. Royle and J. Royle, Jr.
4. Z. Tadmor and I. Klein, *Engineering Principles of Plasticating Extrusion,* Reinhold, New York, 1970.
5. E. C. Bernhardt, *Processing of Thermoplastics Materials,* Reinhold, New York, 1959.
6. U.S. Pat. 4,006,209 (Feb. 1, 1977), J. J. Chiselko and W. H. Hulbert (to Egan Machinery Co.).

General References

J. H. Du Bois and R. W. John, *Plastics,* 5th ed., Van Nostrand Reinhold, New York, 1974.

R. Barr, "Solid Bed Melting Mechanism: The First Principle of Screw Design," *Plast. Eng.* **37**(1), 35 (1981).

C. I. Chung, "A Guide to Better Extruder Screw Design," *Plast. Eng.* **33**(2), 28 (1977).

C. Rauwendaal, "Optimal Screw Design for LLDPE Extrusion," *Plast. Technol.* **30**(9), 61–63 (Aug. 1983).

EXTRUSION COATING

Extrusion coating is a process in which an extruder forces melted thermoplastic through a horizontal slot die onto a moving web of material. The rate of application controls the thickness of the continuous film deposited on the paper, board, film, or foil. The melt stream, extruded in one or several layers, can be used as a coating or as an adhesive to sandwich two webs together.

Equipment for extrusion coating and laminating lines is normally associated with product groups, with some overlapping between groups. Substrates or web handling characteristics distinguish the difference among plastic films, paper, and paperboard combinations.

Three types of lines for extrusion coating and laminating are: thin-film or low tension applications at operating web tension levels of 8–80 lbf (35.6–356 N); paper and its combinations in the middle range of 20–200 lbf (89–887 N); and high tension for paperboard applications at 150–1500 lbf (667–6672 N).

In extrusion laminating, a film of molten polymer is deposited between two moving webs in a nip created by a rubber pressure roll and a chrome-plated steel chill roll. In this continuous operation, rolls of material are unwound, new rolls are automatically spliced on the fly, and the surface of the substrate is prepared by chemical priming or other surface treatment to make it receptive to the extrusion coating, and to help develop adhesion between the two materials (see Fig. 1).

Pressure and temperature on the web and extrudate combine to produce adhesion. The substrate normally provides the mechanical strength to the resultant structure, and the polymer provides a gas, moisture, or grease barrier.

As materials, especially for food packaging, become more complex with ever increasing performance standards, coating lines become more complicated. The requirements for new extrusion coating lines are high productivity, extreme flexibility, and labor-saving computerized and robotized equipment. Modern extrusion coating lines must be able to process the new speciality resins that offer greater adhesion, allowing line speeds to be increased.

Applications

Products from extrusion coating/laminating lines have six main market classifications: liquid packaging; flexible packaging; board packaging; industrial wraps; industrial products; and sacks.

Liquid packaging. Liquid packaging utilizes a single web-coated lightweight board, or a combination of board, plastic, and aluminum foil, for semirigid containers for milk, juices, water, oils, processed foods, sauces, cheese products, and aseptic packaging of liquids. The polyethylene-coated milk carton was largely responsible for the emergence of the extrusion coating industry in the 1960s, and as more commercial uses were found for polyethylene-coated materials, the industry grew rapidly (see Polyethylene; Cartons, gabletop).

In the 1980s, aseptic packaging is making strong inroads in replacing traditional metal and glass containers (see Aseptic packaging). The sterile flexible "paper bottle," which extends the shelf life of dairy products for months without refrigeration, has emerged as a major alternative form of packaging (1,2) (see Shelf life).

Although aseptic packaging systems differ, most of them use paperboard–foil–plastic composite material that is formed to shape, sterilized, and filled with a sterile liquid or semiliquid product under the sterile conditions. Customized extrusion-coating lines, complete with in-line laminating stations, are used to produce an almost unlimited variety of shapes, sizes, and printing options for aseptic and other packaging materials.

Flexible packaging. The flexible-packaging classification covers the combination of plain, printed, or metallized films, papers, polymers, and foil, used for protection, unitizing, dispensing, or holding of commodities. These include medicine and pharmaceutical supplies, foods, chemicals, hardware, liquids, notions, sterile products, and primal meats. Flexible packages also include wrappers for fast food, the bag for "bag-in-box" containers, and the multilaminated web for Glaminate tube packaging (3) (see Bag-in-box, dry; Bag-in-box, liquid; Tubes, collapsible; Multilayer flexible packaging).

Flexible packaging lines are processing progressively thinner substrates of polyester, oriented polypropylene, and metalized materials (see Film, polyester; Film, oriented polypropylene; Metallizing). The light-gauge preprinted substrates used for snack foods require minimum tension to assure that preprinted webs are not distorted. Machines to create these new structures are becoming increasingly complex. Thinner coating layers are more difficult to extrude on the coating machine, and often the coating head itself must be engineered to handle a variety of coating materials (see Fig. 2).

Flotation drying, using air on both sides to float the web, is widely used in flexible package manufacturing because it handles light films well. Improved drying efficiency compared with roll-support dryers, allows higher line speeds.

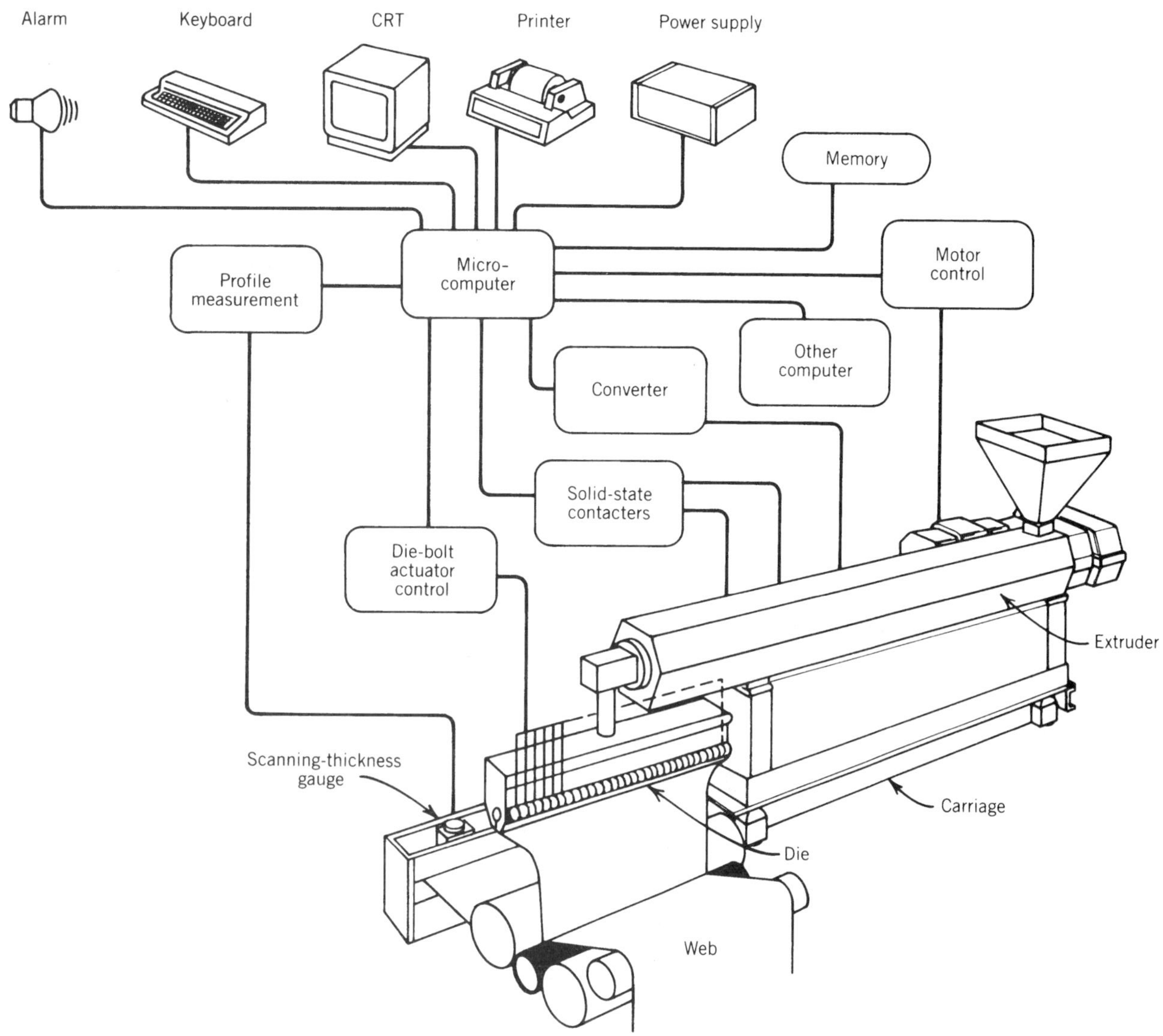

Figure 1. Simple extrusion-coating line.

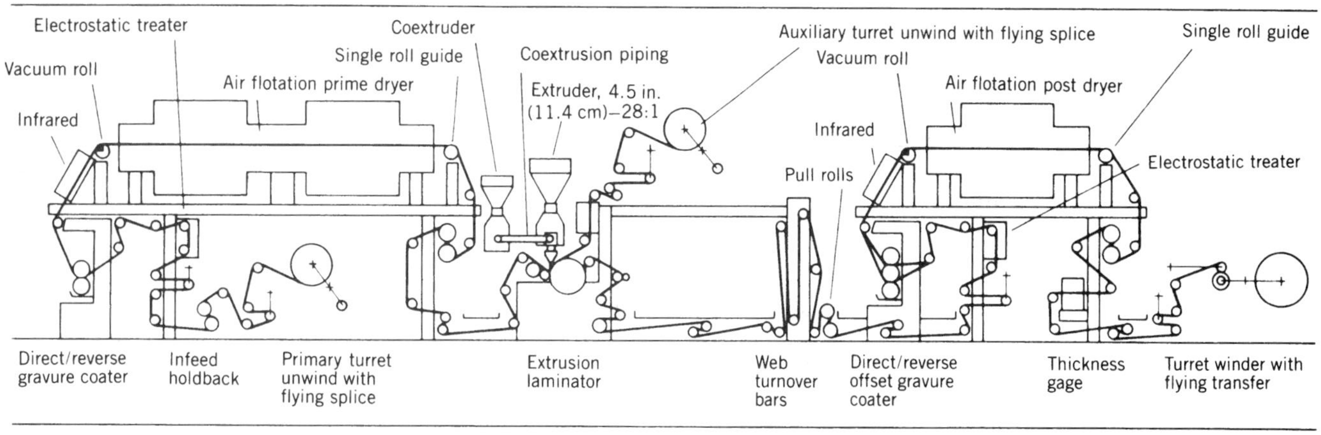

Figure 2. High-speed extrusion-coating line used to produce flexible-packaging-type materials.

Board packaging. In board packaging, heavyweight boards are coated, laminated, and then formed into boxes (folding cartons) for packaging detergents, tobacco, liquor, frozen foods, and bakery products (see Cartons, folding). "Ovenable" trays and ice cream cartons are also in this category. The plastic-coated containers protect against grease, moisture, and gas. Release characteristics can also be provided.

Industrial wraps. Industrial wraps cover the range of heavy or reinforced papers, films, or boards used for products in which the extrusion-coated material may be added to other media used for products such as composite cans, drum liners, soap wrappers, and sheet overwraps for a variety of baled materials (see Cans, composite). The coated product is not necessarily used as a unit container, but as a wrapper or part of a protective structure.

Industrial products. The industrial products classification takes in extrusion coated or laminated material serving various industry requirements. Products include photographic-base papers, substitutes for bitumen coatings, base papers for silicone coatings, insulation backing, automative carpet coating, and metallized film balloons. Also in this category are functional laminates like credit cards and printed circuit boards, decorative laminates such as wallpaper, and disposables like tablecloths and various hospital and surgical supplies. As with industrial wraps, these products may or may not be associated with packaging.

Sacks. Sacks cover materials for multiwall paper bags and plastic-coated raffia. Intermediate tension lines produce coated scrim that is woven from tapes of oriented polyethylene or polypropylene. These materials are used for heavy-duty sacks and tarpaulins, and have applications in building, recreational, and agricultural areas (see Bags, paper; Bags, heavy-duty plastic).

A miscellaneous classification would include coated foam used in the fast-food industry.

Machinery

Obviously, no single coating line can produce all the foregoing products. Today various types of machinery are manufactured to produce coated and laminated products using substrates ranging from 0.4-mil (10-μm) film to 246 lb/3000 ft^2 (400 g/m^2) board, with coating weights 4.3–49 lb/3000 ft^2 (7–80 g/m^2), at widths 76.2–787 mil (300–3100 μm), and at speeds 1.1–33 ft/s (0.33–10 m/s).

The most important consideration in web processing or web handling equipment is the determination of the practical range for the system—the maximum and minimum unwind and winder roll diameters, maximum and minimum web-tension forces, splicing speeds, core diameters, and other process needs.

The unwind basically takes material in roll form and processes it continuously over a series of idler or driven rolls with a suitable amount of tension in order to minimize wrinkling yet not produce deformation. A dancer or transducer roll can be used for tension control; dc regenerative drives and electric or pneumatic brakes are used where applicable. Similar considerations apply to the infeed holdback when levels of tension required differ from those of the in-line operations. These sections tend to isolate tension transients from the unwinding roll.

Electrostatic treatment and flame treatment are available for enhancing surface tension or wettability conditions of the inert substrates prior to applying aqueous solutions. The direct/reverse gravure coater can be used for either priming, coating, or printing. Chemical priming is used mostly in flexible-packaging lines to promote adhesion between the extrudate and substrates such as cellophane, polyester, ionomer, nylon, or polyolefin films (see Film articles). Infrared preheating and vacuum rolls provide the means to dry the PVDC-coated web and effect proper web handling.

Air-flotation, driven-roll, idler-roll, or drum-support dryers are selected depending on the strength, support, or tension required for the substrates. Recirculation of heated air in the dryers is a common energy conservation practice in all these dryers. The single-roll web guide at the dryer exit and chill or pull rolls are needed for special web processing requirements.

The extrusion laminator along with the extruder-and-die system is the heart of the process. The backup chill roll, rubber roll, and large-diameter chill roll form a three-roll system. Two-roll laminators can be used for heavy substrates or paper-board applications. As the moving web enters the nip section, it is coated, laminated, or both. Ozone in close proximity to the entering web is used for oxidation of the molten polymer for improved adhesion in high bond level applications. Most of the heat is removed from the coating or laminate by the chrome-plated chill rolls. Chill rolls normally are steel and are constructed with a double-shell arrangement and spirally baffled. Outer shells of aluminum have been used for high coating weights. High-velocity chilled water is circulated to maintain a temperature rise between inlet and outlet of 2–4°F (1–2°C).

The coated or laminated structure is normally edge-trimmed at the laminator by razor, score, or shear cutting. Trim removal systems are installed just after the laminator. Slitting can also be done just prior to winding at a turret or single-drum winder.

Auxiliary unwinds can be located on, near, or over the extrusion laminator to provide a secondary substrate for laminating at the nip where the extrudate acts as an adhesive. These unwinds can consists of single-position or turret assemblies with flying splices for aluminum foil, oriented polypropylene or polyester film, paper, or paperboard substrates.

Web turnover systems, pull rolls, coaters, infrared heating, dryers, and web processing steps after the extrusion laminator are designed according to product needs. Thickness measuring devices include infrared and scanning of clear webs.

There are two basic winding techniques. The turret winder or center wind system is used for most flexible packaging materials. Tension is controlled by a dancer or transducer roll. The same design criteria for unwinds also applies to winders. The type of web, operating speed, tension range, and roll buildup must be properly controlled to wind up a satisfactory roll. Paper and paperboard products can be wound by surface methods on a single-drum winder.

All-plastic constructions require more advanced web controls. Many converters utilizing traditional wood cellulose substrates are specifying that their new coating lines must be able to handle all-plastic films. Wider tension ranges and air flotation dryers are two principal requirements of these convertible systems.

Other features being incorporated into various lines include dc-regenerative unwinds and infeed holdback drives for prices and low-level tension, direct/reverse gravure coaters for aqueous PVDC coating, infrared preheating, and vacuum rolls for web controls.

In the production of photographic-base papers, exacting specifications and special criteria for pigmented polymers are needed to produce coated materials that constantly provide high quality photographs. The concept of tandem operations or coating two sides of a substrate in one pass can be applied to many flexible-packaging lines that produce combinations of paper, extrusion lamination to aluminum foil, and extrusion coating a polymer for heat sealing. Higher-operating line tensions can be used in producing structures with paper for granulated or powdered mixes and freezer-wrap or sugar-pouch materials. Polyethylene is not the only resin used for lamination or coating. Polypropylene, ionomer, nylon, ethylene–acrylic acid (EAA), ethylene–methacrylic acid (EMA), and ethylene–vinyl acetate (EVA) can also be part of a converter's inventory of resins.

Single-unit pilot coating lines feature an entire coating system preassembled and prewired at the factory and mounted on a structural steel base. These lines can be completely enclosed and have applications for the development of products such as the retort pouch, aseptic packaging, vacuum packaging, and other extended shelf-life products used to replace conventional glass and metal-can packages; they can also be used in the development of many types of medical-grade extrusion coatings (3).

Stainless steel is used when extreme cleanliness is required. The "cleanroom" machines are designed so that any metallic particles generated by machine friction are either contained or swept away by laminar air flow. Stainless steel is also used when lines are frequently washed with solvents that could remove conventional paint.

A typical pilot coating line consists of an unwind, coating heat, air-flotation dryer, dryer exit tension control, cooling station, extrusion coater, and rewinder, all aligned on a one-piece steel frame. Pilot coating lines are designed to handle narrow web widths and can be built so that components are cantilevered instead of being supported by traditional side-frames. The spindles, idler rolls, force transducers, and air-flotation bars are all mounted on a vertical backplate.

Extrusion-coating lines are experiencing increased automation. Raw material and roll stock can now be selected from a controlled inventory, delivered to the line, and handled through robotics (see Robots). The entire operation can be monitored and controlled by computer. (see Instrumentation/controls).

Drives are also coming under computer process control, and there have been advances in digital drives and in energy-efficient ac inverters. While a number of different drive systems have been installed and operated, the multimotor dc system is predominantly used for extrusion coating equipment. These drives can consist of as many as ten motors in one line with a single control to bring the web up to operating speed. The tandem follower is another drive or computer feature whereby the extruder will increase or decrease in rate with line speed in order to maintain a fixed coating weight as the line speed is changed.

BIBLIOGRAPHY

"Extrusion Coating" in *The Wiley Encyclopedia of Packaging Technology,* 1st ed., by Michael G. Alsdorf, Extrusion Group, Egan Machinery, pp. 289–293.

1. W. Schoch, "Aseptic Packaging," *Tappi,* 56 (Sept. 1984).
2. L. J. Bonis, *Correlation of Coextruded Barrier Sheet Properties with Packaged Food Quality,* Composite Container Corp., Medford, Mass. (presented as *Proc. 1984 Polymer, Laminations and Coatings Conf.,* TAPPI, Atlanta, Sept. 24–26, 1978, pp. 319–328).
3. M. Schlack, "Extrusion Coaters Gear for New Packaging Action," *Plast. World,* 42 (July 1984).

General References

H. L. Weiss, *Coating and Laminating Machines,* Converting Technology, Inc., Milwaukee, WI.

H. L. Weiss, *Control Systems for WEB-FEB Machines,* Converting Technology, Inc., Milwaukee, WI.

F

FIBER, MOLDED

Molded fibers and recycled paper have been in commercial use for about 90 years. The idea to use wetted or other secondary fibers in some form probably dates back to the time of the early Egyptians. Over the years, processes and machinery, similar to those used in papermaking, have evolved. Now various grades of wastepaper, including newspaper and corrugated paperboard, are mixed with water to form a slurry to manufacture a variety of packaging products. Prior to the development of the use of expanded polystyrene (EPS) as a packaging material, molded pulp very likely had a bright future. Moreover, EPS was inexpensive, and clean-looking, and its performance properties were well defined. This essentially relegated molded pulp products to uses such as flower, shrub and tree containers, and egg cartons. The generally unknown performance characteristics and somewhat drab appearance kept molded fiber from the mainstream of designed packaging products. Environmental issues have now created a strong interest in molded paper-pulp packaging.

There are now about 35 molded pulp product producers in the United States as well as a number in Europe, Asia, and South America. Most of these are producing items such as trays, end caps, blocking devices, pads, and corner-and-edge protectors as well as egg cartons, drink trays, fluorescent-tube end pads, etc. New technology, demand and applied research are increasing the number of producing operations. The differences in these packaging items are such that we could arbitrarily categorize molded pulp into two types: (*1*) precision-molded and (*2*) slush-molded. The basic manufacturing principles are the same for both, but there are significant operational, marketing, and product differences.

Neither of these commonly used terms adequately defines the finished products. The process for type 1—precision molded product manufacturing—is more technically advanced than the slush method (type 2) and is generally geared to producing very high volumes ($\geq$200,000,000) of products such as the egg cartons, paper-plate-type items, and various other custom packaging products. Some molding machines are producing egg trays and other products at the rate of 500 per minute and are capturing markets for packaging products that had been using polystyrene.

Compared to the slush-molded products, precision-molded items are thin-walled ($\leq \frac{3}{16}$ in.) and have the capability of finer detail, patterns, rib structure, and cavity arrangements. Precision molding also results in a product with a relatively smoother surface more suitable for printing. All of these conditions are largely a function of the design and quality of the precision molds, which are significantly more expensive than slush molds. With the precision process, a transfer molding method is frequently used with a mating takeoff mold. The forming section is the mold producing the product, and the mate is the transfer mold. In operation the transfer mating mold pulls the finished product from the forming mold using vacuum. In the molded pulp process, the top-mating mold moves into a position to "pop" or blow the finished molded product off the transfer mold on to a tray or conveyer for oven curing at temperatures well above 300°F for a number of hours.

This transfer process results in some "pressing" of the molded product as well as removal of water. The precision-molding process is somewhat more sensitive to the pulp slurry formulation and requires fairly expensive mold sets. In some cases production volumes need to be in the millions in order to justify the costs. There are about 10 precision-molded pulp product companies in the United States, each capable of consuming many thousands of tons of wastepaper per year.

Slush-molded products are being manufactured with many types of equipment, mostly custom-made or modified machines using a variety of systems. In many cases there is no transfer mold and the product is formed and "popped off" or blown off the forming mold. Since no mating or transfer mold is used and the slush process is somewhat different, and mold costs can be significantly less.

Packaging products made with the slush-molding process are generally used for heavier items where the thin-walled precision-molded material may be inadequate. The slush products have thicker walls ($\leq \frac{3}{4}$ in.) and have less detail. The surfaces are not as smooth, and one side is frequently very rough. Slush-molded packaging products are in use for interior packing such as corner pads, edge protectors, end caps and blocking devices.

A slush-molded reusable pallet-sized nesting tray shown at Pack-Expo 1994 in Chicago won an Ameristar packaging award. Another slush-molded item winning an award was an interior blocking device to protect truck grills packed in a box. Applications are for furniture, appliances, automobile parts, electrical equipment, and plumbing ware. As with the precision parts, slush-molded products are replacing polystyrene in some applications. Depending on the type of molding machine and mold design requirements, pulp slurry formulations can use a wide variety of grades of wastepaper with the slush process. Little appears to be known about the specific packaging product performance characteristics with respect to the slurry formulations; however, new studies are helping to identify uses and limitations. Althuogh the molding of products from wastepaper pulp is basically a simple process, many of the techniques have been developed empirically and remain "tricks of the trade."

A basic description of the molding process is as follows. A mold, consisting of a variety of materials, frequently metal or plastic, is designed and manufactured according to performance requirements and in the shape and size of the desired packaging part.

The mold, which is porous to water, has a fine filter screen in the shape of the mold attached to the shaped surface. A rigid tube or pipe is attached to the opposite side of the mold, or the mold is mounted into the machine, providing a conduit or manifold for air and water. The mold assembly mounted in the molding machine is connected to a source of switchable air pressure and vacuum. The machine immerses the mold into a tank containing a pulp slurry, and vacuum is applied to the tube. The slurry is drawn into the mold, depositiong pulp fiber on the mold screen while water is extracted. The fibers build up in the shape of the mold on the screen, forming the packaging product. The thickness of the part depends somewhat on the immersion time and the type of slurry. After the mold is removed from the slurry tank, the vacuum is turned off and a controlled blast of air is applied to the mold (or transfer mold), blowing the formed pulp part off the mold.

The finished product at this point is still very wet and must be treated carefully. The part is placed, or conveyed, into a circulating-air oven to drive off the remaining moisture, completing the basic process. Other secondary operations such as pressing, trimming, die-cutting, and printing may be performed.

The molded fiber packaging product industry is growing and now has a solid position as functional packaging products. Continued growth is expected because of consumer demand and new developments, such as the establishment of performance data, new slurry fibers, slurry additives, new machines and processes, and reduced mold costs. New uses for molded fiber packaging are continuously developing. Packaging designs using molded fiber products in conjunction with other types of packaging materials are also gaining in the marketplace.

Joseph Grygny
Midwest Packaging Consultants
Milwaukee, Wisconsin

FILLING MACHINERY, BY COUNT

Accurate measuring is imperative in packaging to avoid costly overage and shortages that are now prohibited by law. As a basis for measurement, most packaging lines use either the weight of the package (see Checkweighing) or the number of pieces in the package. This article pertains to methods used to produce packaging containing a specified number of pieces. Modern automatic counting systems are based on concepts that evolved in ancient times. They are all comprised of three basic functions: parts representation, parts detection, and product handling.

Parts Representation

Counting is used to determine the amount of a specified batch, and the method used to achieve this begins by selecting a basic system of representation. Systems of representation are used in all forms of counting. Units can be represented by fingers and toes, or knots in a rope, or, as in the packaging industry, they can be pulses of electrical current generated from specially designed detection units. Over the years, humans have engineered quick and accurate ways of counting, but none is as accurate as a single-file count of an individual unit of product.

Parts Detection

The next step in counting is detecting the unit of product to be counted. A person detects the product either by sight or by touch. Machines are designed to operate on the same principles, and use either optical systems (sight) or nonoptical systems (touch).

Optical systems. An optical system operates much like a human eye. A photosensitive receiving device is established and the unit of product to be counted is passed within detection distance of it. There are many models of optical systems used in packaging, but most are based on either a simple digital photocell system or a very intricate electronic analog detection unit.

In the photocell system, the breaking of a light beam fed to the photocell receiver indicates that a unit of product has passed through the detection zone. The break is then recorded as the counting of one unit of product. This method of detection is perfect for most product applications that meet specifications for light-blocking systems, but it is limited by the fact that the light source fed to the receiver must be completely blocked out before detection is recognized. This method is not very efficient with transparent, overlapping, or bicircular objects such as clear plastic, two pieces of material riding on top of each other, or objects with holes, which may trigger the photocell more than once.

Another approach to optical detection is the analogue photooptic system. This system is fast, flexible, and accurate since certain parameters must be established and met before detection is recorded. The Photo Optic Shadow Detector (Sigma Systems, Inc.), for example, detects the dimensions of a specified shadow made by the unit of product when fed through the detection zone. Each shadow of a detected unit must meet certain parameters before being recorded. The parameters can be entered into the computer portion of the counting system to compensate for objects such as flat washers and O rings, which would trigger a photocell system twice, or clear plastic, which would not block out a light source but would cast a shadow that could be detected.

Nonoptical systems. The nonoptical methods of parts detection are similar to human touch. The touch methods generally involve escapement devices, electrical contact, or magnetic-field contact. The escapement device is usually fully mechanical and is similar to an analog machine. Each unit of product must be of a specific shape and size. The product is fed into a receiving device that fits those exact parameters and then is discharged and recorded as a specific unit of product. This is equivalent to placing dominoes in a box made just for dominoes. If only ten dominoes fit, the unit of product would be recorded as ten units. This system is very accurate, but it offers little flexibility. It is commonly used for high-speed counting of uniform products such as pills and tablets.

Another nonoptical method is the electrical-contact method, in which an electrical switch is triggered each time a part comes in contact with the switch. The part is then recorded as one unit of product. The response time and the ability of the product to actuate the mechanical switch-triggering device greatly affects the accuracy and flexibility of this type of system. The *magnetic system* is another touch-system method. Each unit of product to be counted must come in contact with or disturb the magnetic flow being transmitted from a magnetic source. As each unit of product is fed through the magnetic field, it is recorded as a counted unit.

Coupled with the detection system is a process known as *discrimination* (eg, a farmer counting cows knows how to exclude sheep). This process has hampered the automated counting system greatly, for once this vital function leaves the dependability of the human senses, accuracy often suffers. Only two of the detection systems mentioned above are able to discriminate: the photo-optic-analogue system and the escapement-device system. In both, certain parameters must be met before a unit of product is recorded; the other systems record any item that is detected. Engineers have been working for years to improve the efficiency of the detection system to ensure an accurate count.

Product Handling

In almost every counting system, the process of getting the product to the detection zone and then moving it away from the zone must be achieved, whether the detection zone goes to the product or the product comes to the zone. This system of product movement is known as product handling. The product is usually brought to the detection zone, and in packaging, all of the methods bring the product to the zone in single file. There are various ways to do this. One is vibratory feeding, which is designed to vibrate a track or bowl filled with product, which, in turn, causes the product to vibrate along a designated path or ramp. The tracks or ramps narrow as the product nears the detection zone in order to create a single file. These are the most flexible of feeding systems since most objects lend themselves to vibration.

Another method is the *belt* or *V-belt system.* The product is placed in a master container and discharged onto a belt system. The width of the belt track is narrowed to allow only one part to pass at a time in order to achieve single-file feeding. Product handling plays a very important role in the counting process, for even the most refined counting system will be inaccurate if the product is not presented to the detection zone in a manner acceptable to the detection device.

There are many approaches to product feeding, but most counting systems allow free entry and exit of product through the detection zone. In packaging, however, it is sometimes necessary to retain all or part of the amount counted for a specific function. The retention is known as accumulation or partial accumulation. The accumulation functions are usually determined or predetermined by a manual function in which an accumulation parameter, ie, the amount of product desired, is assigned to the counting unit. The counting unit usually must meet the assigned parameters before permitting the accumulation functions to discharge the retained product. This function is essential when a manufacturer wants to place a predetermined amount of product into a specific-size container, eg, accumulating 20 tablets, then discharging them into a package or bottle. It also plays an important role in the packaging process since most packaging machines are intermittently cycled by the signal received from the counting system when the accumulation function discharges product. Most containers are aligned under the accumulation chute to receive the allotted amount of product.

As electronic technology advances, the ability to count and discriminate parts will improve. However, the attainable operating speeds will depend on the speed with which the unit of product can be fed into the detection mechanism and the speed of the product-handling system. The performance of any counting system must be objectively evaluated on its ability to count accurately and its adaptation to the intended use.

BIBLIOGRAPHY

"Filling Machinery, by Count" in *The Wiley Encyclopedia of Packaging Technology,* 1st ed., by David Madison, Sigma Systems, Inc., pp. 294–295.

FILLING MACHINERY, DRY-PRODUCT

The objective of dry-product filling machinery is to fill a container with a product to a specified weight or volume, meeting legal or formula requirements. This should be accomplished at an economic balance of fill accuracy, product cost, package cost, production rate, and overall equipment and operating cost.

The accuracy of fill is measured by the standard deviation known as *sigma.* This is a calculated statistical number and indicates the shape of a normal curve (see Fig. 1). The majority of weighing plots approximate a normal curve so sigma is used to indicate the shape of this plot. A commonly used index of performance is plus or minus three sigma, which indicates that 99.73% of all weights measured fall within this range. As the value of sigma is improved or decreased, it indicates that the average weight can be targeted closer to the label or formula weight with less risk of over- or underweights.

Sigma can be determined for the overall system over a period of time or as a machine capability parameter. In statistical weight procedures, target weight calculations are based on a system-performance sigma. Determination of machine performance is made by comparing the machine-capability sigma with the system-performance sigma.

Overfill, or "giveaway," is determined by target weight and is a function of filling equipment accuracy.

The basic elements of a dry-product filling system are product feed, product measurement, and package- or container-forming or handling system.

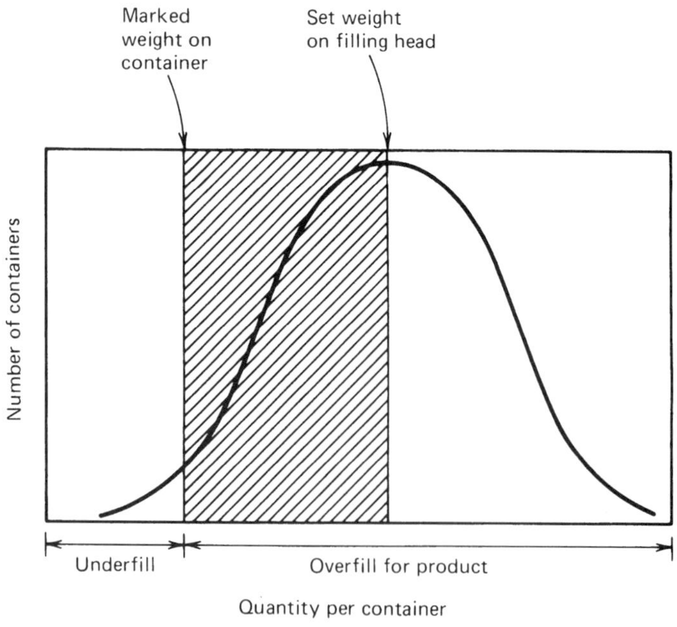

Figure 1. Tolerance weight (shaded area) added to nominal fill weight stated on container.

Product-Feed Systems

Product-feed systems commonly utilized in dry-product filling equipment are

1. *Augers.* For free-flowing and non-free-flowing products.
2. *Gravity Flow.* For free-flowing products.
3. *Vibratory Feeders.* For free-flowing, non-free-flowing, and friable products.
4. *Belts.* For free-flowing, non-free-flowing, friable, and sticky products.
5. *Screw-Type Units.* For free-flowing and non-free-flowing products.
6. *Vibrating-Bin Outlet.* For free-flowing and non-free-flowing products.
7. *Cascade-Filling Systems.* For free-flowing non-free-flowing, and sticky products.
8. *Vacuum-Filling Systems.* For free-flowing and non-free-flowing powders.

Flowability is basically measured by the angle of repose of a product on a flat plate after a gravity fall from a fixed orifice from a given height. There are standard tests to give a quantitative value to flowability.

Product measurement can be determined by either volumetric or gravimetric methods. If density and flow characteristics are consistent, volumetric fillers give excellent results. Normally, volumetric fillers are less expensive, easier to maintain, and operate at higher production rates than weighing units.

Weighing units give improved fill accuracy when density and flow characteristics vary. Problems develop with weighing units if density varies too widely. High densities cause slack container fills, which can exceed legal or quality limits. Low densities can cause container overflow. Variable container vibration is frequently used to package a wider range of densities.

Examples of volumetric measuring methods include varying auger rotation by time or angular measurement; varying time or flow rate of gravity-flow systems, vibrators, screw feeders, and belts; utilizing container as volume measurement and varying fill density by vibration or vacuum; and varying intermediate flask volumes.

Augers. Augers provide a method for uniformly feeding both free-flowing and nonfree-flowing powders, flakes, granules, and other types of products (see Fig. 2). Augers are normally mounted in the vertical position but have also been designed to operate on an angle or in a horizontal position. Augers may be designed with rotating agitators in the feed hopper to give positive product into the auger screw area and prevent "rat holeing." Variable screw pitches are utilized to provide product compaction and minimize product density variations. Augers for free-flowing products are equipped with spinner plates at the discharge of the auger to prevent product runout. The discharge of the auger can be provided with a compartmented divider head to split the main flow into multiple streams. The auger drive can be intermittent or continuous, depending on the application. A timed cycle or controlled rotation through a preset number of revolutions are normally used to control volumetric fills.

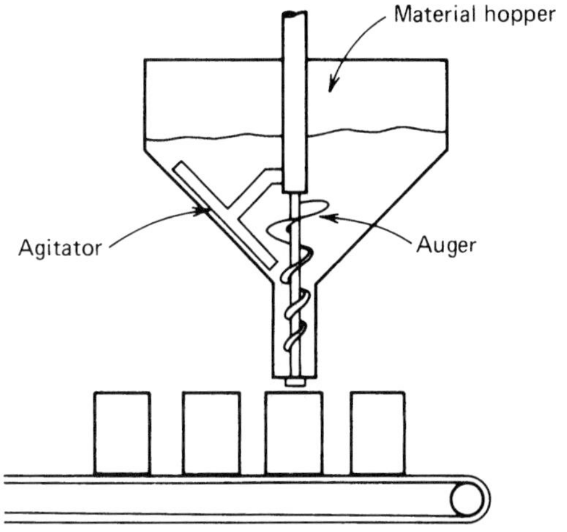

Figure 2. An auger filler.

Continuous-feed augers are used on continuous-motion packaging machinery where pockets pass, at a constant rate, under the auger discharge. These pockets are a part of a rotary table or affixed to a chain drive. They are designed to overlap so there is no product spillage. Fill volume is regulated by changing the speed of the auger or the rate of the pockets passing through the auger flow.

Integrated auger systems are used on multiple lane machines such as horizontal and vertical pouch machines, thermoform machines, and form/fill/seal units (see Form/fill/seal, horizontal; Form/fill/seal, vertical; Thermoform/fill/seal). Each auger is usually controlled by its own clutch brake system, and fill volume is individually set.

Augers can be used as feeders for net-weigh systems. The auger rate can be varied to provide a bulk and dribble feed, or dual augers are utilized to provide this feature. Cutoff points are set by the weight-sensing unit.

Gravity-flow systems. Gravity-flow designs are the simplest of feeding methods and are used on free-flowing products such as coffee, tea, nuts, rice, sugar, and salt. Flow rate is controlled by the angle of flow, product characteristics, pressure or "head" of the feed system, and orifice sizing. Normally a bulk and dribble rate is used to achieve improved weight accuracy. Gravity-flow systems can feed other volumetric units such as pockets, and if these are continuously moving under the flow, the fill is determined by the flow rate and pocket velocity. Gates can be used as product-shutoff devices, and these can operate on a timed or product-level-sensing control system.

Vibratory feeders. Vibrating-feed systems are used on free-flowing and relatively nonfree-flowing products to achieve controlled feed rates (see Fig. 3). These units can handle a wide range of products, such as powders, friable materials like snack foods, large discrete particles such as those in the vegetable industry, and abrasive materials, to name a few.

Feed rates are controlled by varying vibration frequency or amplitude. Vibrators can be driven electrically, mechanically, or hydraulically. In many installations, vibrators are mounted in series to obtain more consistent flow rates. Fill volumes can be varied by changing the flow rate to the vibra-

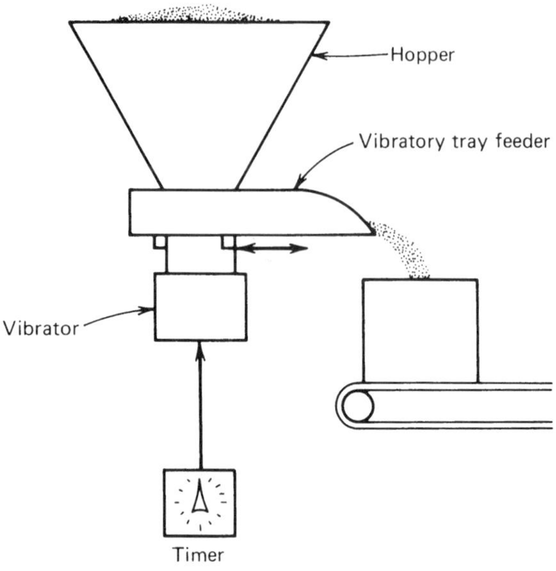

Figure 3. A timed vibrator filler.

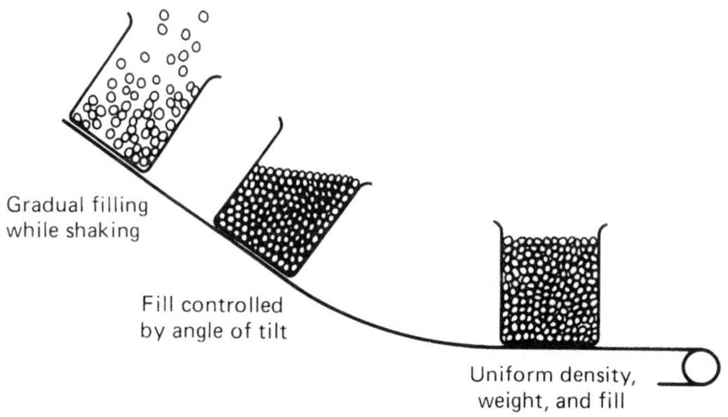

Figure 4. A cascade-filling system.

tor, time of feed, frequency, amplitude of vibration, or pitch angle of the vibrator tray. Vibrators are sometimes equipped with shutoff gates at the discharge of the pan to prevent product from sliding off after the fill weight has been reached.

Vibrator shapes can be a flat pan, tube, serrated, or V-shaped, depending on the product and the rate to be fed. In many instances the pans are coated with a material such as Teflon (DuPont Co.) to prevent product buildup. Some pans are constructed entirely of plastic to minimize and facilitate cleaning. In some designs, where buildup is a factor, the pans are designed to clip apart for easy replacement and immersion in cleaning solutions. Vibrators are commonly used to feed scale systems because of their versatility and ease of control.

Belt feeders. Belts are used to feed free-flowing, non-free-flowing, friable, and sticky products. They can be provided with scrapers and washing systems. There are a wide range of belt materials and sizes to handle an extremely broad range of products. Belt materials range from tough multiple construction for use on items such as minerals and abrasive chemicals to sanitary belting for food products. Lightweight belts of polyester film (see Film, polyester) and fiberglass-coated belts are used on sensitive products similar to those in the pharmaceutical industry. Belts can be supplied with molded pockets and side walls and can operate horizontally or on an incline. They can function on an intermittent or continuous basis and controls are basically similar to those outlined for augers or vibrators.

Weight-belt systems use weighing devices to measure the mass flow across the belt. A wide range of weigh-belt designs is commercially available. The most common design supports the belt on a load cell, and either totalizes the weight across the belt or varies the rate to maintain a constant mass flow.

Screw feeders. There are two basic types of screw feeders: the horizontal type that delivers a consistent flow rate and flexible feeders that are used to move product from a floor hopper to the filling system. Some horizontal units are designed to vibrate to improve the consistency of the product feed. There are designs where the side walls of the feed hopper are made of an elastomeric material and flexed to improve the flow of difficult materials. The screws can be solid with constant or varying pitch similar to an auger. Some screws are made of a coil of round or flat wire, depending on product requirements. Feed rates are varied by changing the rotational speed of the screw.

Vibrating-bin discharge. Vibrating-bin discharge systems are primarily used for filling large containers such as drums or bags. The design normally consists of a relatively shallow angle-dished bottom that is flexibly connected to a bin. A vibrating system is mounted to this section, and when activated, provides a uniform stream of product from the bottom orifice. A gate can be provided to obtain a positive shutoff. Container fill is volumetric or by weight-control systems.

Cascade-filling systems. In this system the product is fed in waterfall fashion to the open-mouth containers (see Fig. 4). The containers fill as they move along a conveyor in the path of the falling product. The containers are tilted at a variable angle and vibrated to settle the product. The final volume is determined by the container speed, the angle of container tilt, and the vibration amplitude and frequency. The product overfill is recycled back to the original product stream. This system provides high filling rates for free-flowing and non-free-flowing products.

Vacuum-filling systems. There are two basic types of vacuum-filling systems. In one type the container is evacuated and the product flows through an orifice into the container (see Fig. 5). The vacuum is pulled through a fine mesh screen inside the gasket that seals the container. The vacuum is pulsed to compact the product within the container. The amount of vacuum and the number of pulses determine the final weight. The base weight is established by the amount the vacuum screen protrudes into the container. This may be varied by changing the gasket thickness. The containers must have a relatively constant volume and be rigid enough to withstand the vacuum without flexing to obtain reasonable weights.

In another form of vacuum-filling system, the product is filled into a rotating pocket that has been evacuated (see Fig. 6). The pocket is mounted on a wheel that moves in a vertical

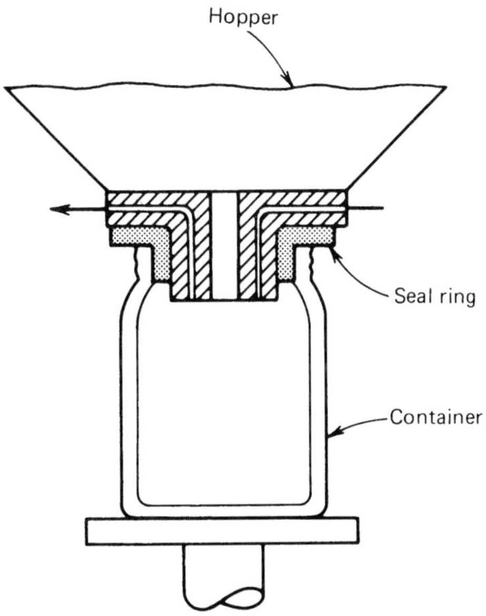

Figure 5. A vacuum-filled container filler.

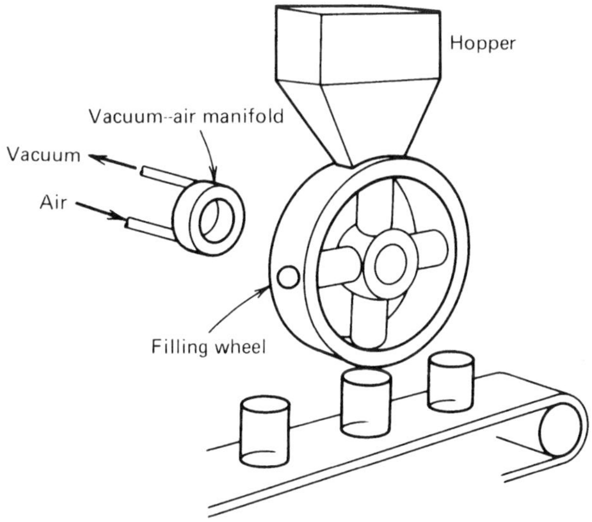

Figure 6. A vacuum-filled pocket filler.

plane. The vacuum holds the product in the pocket and discharges it to a container or funnel that moves below the wheel. The volume is adjusted by varying the telescoping pocket depth.

Weighing Systems

A weighing system basically senses an imbalance due to the addition of product in a weight receptacle. The feed to the weighing system is critical to its performance and could consist of many of the product-feed systems outlined above. The weighing or sensing elements vary considerably in detail. In its simplest form the imbalance reaches a set point that unlatches a mechanical stop to discharge the product. Sensing elements used in advanced weighing systems are

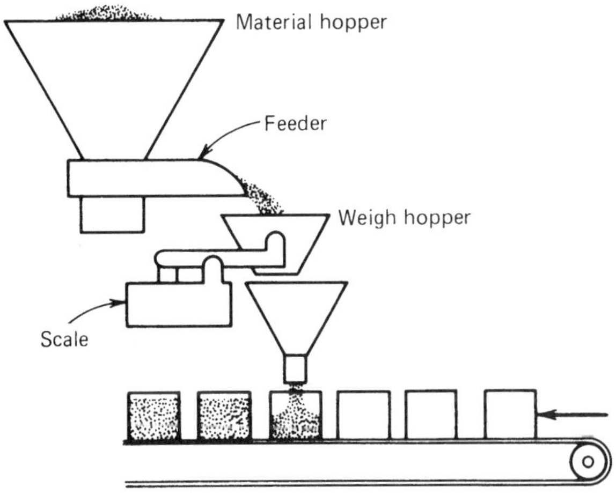

Figure 7. A net-weighing filler.

1. Linear-voltage-differential transformers, commonly known as LVDT
2. Low-capacity strain gauges
3. Load cells
4. Pneumatic-pressure-differential switches
5. Proximity switches
6. Photoelectric sensors
7. Limit switches
8. Reed switches
9. Mercury switches

Most weighing systems dampen the action of the weight-sensing element with an oil-filled mechanical dashpot. Electronic systems require constant voltage systems, which are usually designed into the circuitry. Most units also have temperature compensation circuits to maintain accuracy under fluctuating environmental conditions.

Almost all of the net-weigh systems operate on a bulk- and dribble-feed principle (see Fig. 7). The majority of the product is fed at a high rate until the bulk set point is reached. The bulk feed is halted, and the dribble rate is continued until the final-weight set point is reached and the feed is stopped. With this system there is always some material in suspension when the final set point is reached. For improved accuracy this amount should be kept to a minimum.

Weighing can take place in an intermediate hopper (net weighing) or in the final container itself (gross weighing). Gross weighing is commonly used where the product buildup on a weigh bucket would severely impair the accuracy of the weighing system. In order to obtain accurate gross weights, the container weights must be consistent or the empty containers must be weighed and the final weight compensated for the weight of the empty container (see Fig. 8).

Computer-combination net weighing is a weighing concept that was introduced commercially in 1971. The system utilizes multiple weigh buckets that are filled to a portion of the total net weight. A microprocessor analyzes individual bucket weights and selects the combination of weigh buckets that yields the weight closest to the target weight. Since the product is completely weighed when the selection is made, there is no variation owing to product in suspension or change in

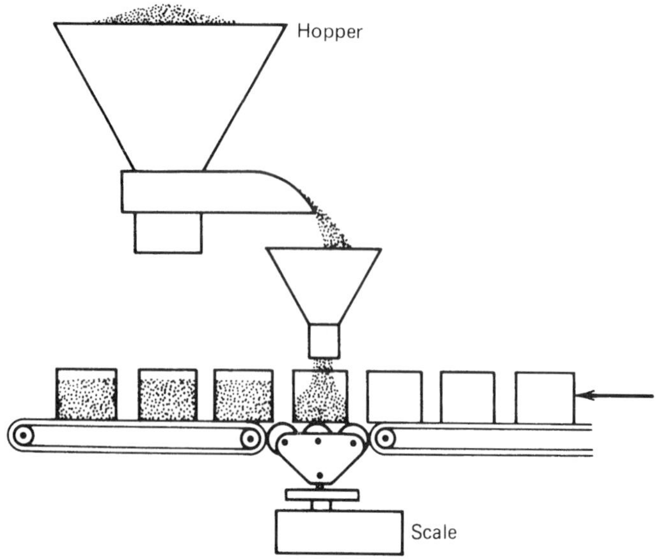

Figure 8. A gross-weighing filler.

product characteristics. This system provides extremely accurate weights, but is particularly effective where discrete pieces of varying size and weight are being packaged. They have shown remarkable weight improvement in the snack, vegetable, and shellfish food-packing areas.

These units are available for automatic or manual product feed. Rates of up to 120 weighings per minute are available with automatic feed, depending on product characteristics. Manually fed systems are currently designed to operate at rates of 25 weighings per minute.

In the 120-per-minute systems the weigh buckets are mounted in a circular position around a central discharge point. Automatically fed straight-line systems are available now as well. The manually fed systems are in-line configurations that discharge to a belt conveyor.

The majority of the sensing elements used in modern electronic weighing systems provide electrical outputs that are directly proportional to the weight applied. Multiple cutoff points can be used to control product feed and improve accuracy. The use of electronic circuitry in conjunction with microprocessor controls and computer interface can provide auxiliary functions such as

Automatic Checkweighing. The final weight can be sensed, recorded or displayed, and refilled to a proper weight if below an underweight reject point (see Checkweighing).

Tare Weighing. The empty weigh container can be checked for tare between weighings, and the fill weight compensated for a change in tare.

Feed Timing. Overall cycle timing can be monitored and feed rates adjusted for optimum performance. Bulk and dribble times can be electronically compared to preset values.

Informational Readouts. Weight performance can be fed to a computer interface to generate statistical and operating data. This data can be indicated utilizing a CRT, LED, or LCD (cathode-ray tube, light-emitting diode, liquid-crystal display) system, or hard copy. Actual weights from each bucket can be displayed along with performance data for a predetermined period of time. Based on actual performance, target weights can be automatically recomputed and the machine sensing unit reset to meet these new targets. When tare-weight changes exceed a given limit, a signal can be given that the buckets need to be cleaned. Total average weight, product giveaway, and machine efficiency can be displayed and recorded. Normal curve distributions can be visually displayed on a CRT with each weighing indicated. These electronic systems can monitor other machine functions not associated with the weighing equipment.

Bulk-Product Feed

The primary feed system that delivers product to the packaging equipment is critical to system performance. For both volumetric and gravimetric filling equipment, changes in density, flowability, and particle size have a detrimental effect on fill performance.

A major cause of changes in product characteristics comes from variation in production processes or natural variations in agricultural products. These variations may be minimized by the use of blending systems. Blending can be done in batch units, belts, or blending screws. Even with these blending systems, segregation can occur in bins feeding the packaging equipment. Static-blending storage systems and desegregation cones are used to minimize product variations. Breakage in mechanical feeders causes changes in product characteristics and the product handling system should be designed to take this into account. Another factor that affects fill accuracy is the level of product in the feed hopper. Adequate level control is required to maintain consistent feed.

Packaging Equipment

The variety of packaging equipment utilizing the systems outlined is extensive. A partial listing of major types is given below:

Vertical form/fill/seal machines
Horizontal form/fill/seal machines
Rotary form/fill/seal machines
Rotary volumetric fillers
Rotary net-weighing fillers
In-line volumetric net-weigh machines
Thermoforming equipment

In many cases there may be multiple feeding systems used on one machine. For instance, there are indexing pouch machines and vertical form/fill/seal equipment that have up to five separate feeding systems to provide a blended product in the final package.

Gravimetric equipment utilizes the volumetric feed systems outlined earlier and controls the feed cutoff by a weight-sensing unit. There is equipment available where a combination of volumetric and gravimetric units are utilized both in series and in parallel. One type of machine fills the container below its label weight with an auger volumetric unit. The container is indexed to a weighing platform and the fill is completed to the target weight. To improve the production rate, the container can be preweighed after the bulk fill and this weight stored in a memory circuit to control the final weight addition. The container tare weight has to be consistent to obtain accurate fills.

An example of a parallel system is a rotary filling unit that fills adjustable volumetric flasks. One flask discharges product to a weigh bucket prior to filling the container. This weight is used to control all flask volumes to meet the required weight. This system combines the simplicity of a volumetric system with accuracies approaching that of a complete scale system.

Within a given category of packaging machines there are significant design differences For instance, there are rotary scale sytems where the scale elements are fixed and the containers move under the scale discharge. In another design the scales rotate with the container. There are advantages and disadvantages to each design. Product differences would dictate the proper selection.

The physical installation of certain types of filling equipment is critical to fill performance and efficiency. Equipment and building vibration has a major effect on accuracy. Location within the packaging layout affects efficiency and also fill variation. Start–stop operation of a filling machine gives poor weight performance.

Selection of filling equipment for a particular package and product is a major study in itself and the subject of many articles. Factors to be considered, among others, are production rate, product cost, package type and cost, machine efficiency, maintenance cost, equipment cost, operating cost, and layout limitations.

Careful analysis of all variables is necessary to determine the optimum system for any given situation.

BIBLIOGRAPHY

"Filling Machinery, Dry-Product" in *The Wiley Encyclopedia of Packaging Technology,* 1st ed., by Robert F. Bardsley, Bard Associates, pp. 295–300.

General References

E. R. Ott, *Process Quality Control,* McGraw-Hill, New York, 1975.

J. M. Juran in J. M. Juran, F. M. Gryna, Jr., and R. S. Bingham, Jr., eds., *Quality Control Handbook,* McGraw-Hill, New York, 1974.

C. S. Brickenkamp, S. Hasko, and M. G. Natella, *Checking the Net Contents of Packaged Goods,* U.S. Department of Commerce, National Bureau of Standards, U.S. Government Printing Office, Washington, DC, 1981.

D. L. Winegar, *Package Engineering Encyclopedia,* Cahners Publishing, 1984, pp. 247–250.

J. Blackwell, "Machinery, Filling, Dry Products" in *Package Engineering Encyclopedia,* Cahners Publishing, 1981.

F. C. Lewis, *Package Engineering Encyclopedia,* Cahners Publishing, 1984, pp. 247–250.

LVDT Transducers for Weight, Dimension, Pressure in *Cataglog A-100 June 1980,* Automatic Timing and Controls, Co., King of Prussia, PA, 1980.

"Weighing and Proportioning" in M. Grayson and D. Eckroth, eds. *Kirk-Othmer Encyclopedia of Chemical Technology,* 3rd ed., Vol. 24, Wiley–Interscience, New York, 1984, pp. 482–501.

FILLING MACHINERY, LIQUID, CARBONATED

The method for filling carbonated liquids, primarily beer and soft drinks, differs from other filling techniques (see Filling machinery, still liquid) in that it is accomplished under pressure and uses the container as part of the control of net contents. Carbonated beverages, which tend to foam, require filling techniques that ensure the retention of the required carbonation levels in different sizes of cans and bottles. It is imperative that the carbonated liquid be processed in a way that prevents excess foaming, which would result in uncontrolled filling levels as well as impaired closing of the vessel.

The filling machine (filler) consists of a rotating bowl with a valve and CO_2 pressurization control, filling valves that attach to the perimeter of the bowl, a tabletop that contains the controlling technique for feeding in and taking away the container, a closing section that applies either a bottle closure or a can end, and a drive system that keeps all components in proper synchronization with each other.

The bowl must control the pressure on the liquid within it, yet be at a pressure lower than that of the system that feeds it. In this way, a continuous supply of product is assured, with the least amount of turbulence. The level within the bowl is maintained by a simple float valve or similar device.

The filling valves embody the applied technology involved in filling carbonated products. Almost all of the machinery manufacturers of note employ the same principles, with some variation in the mechanical interpretation of the concepts. To understand the valve's function, it is necessary to follow it through the various filling stages described in the following paragraphs.

The vessel is presented to the filling valve to assure complete intimacy between both surfaces. This is controlled by a pneumatic pressure system that holds the vessel firmly against the valve, yet tenderly enough to avoid crushing thin metal cans or light plastic bottles.

The container rotates with the filler valve, and a mechanical trip actuates the valve to permit CO_2 from the upper part of the bowl (above the liquid level) to pressurize the container. The pressure in the container is now equal to or somewhat lower than the pressure in the bowl.

The valve is then actuated into the next stage, permitting the product to flow into the prepressurized container gravimetrically. During the filling phase (as fluid enters the vessel) it is essential that the CO_2 in the container be displaced. This is accomplished through a vent tube, which is normally in the center of the valve and protruding downward into the vessel. As the product fills the vessel, it ultimately rises to seal off the vent tube or ball check. This stops the filling process, since pressures in the bowl above and in the container below have reached an equilibrium.

While the container is still in an intimate seal with the valve, another external latch actuates an internal chamber in the valve which closes the connection to the bowl (both the liquid portion and the CO_2 charge above it) and simultaneously vents the container to atmosphere. This step maintains control of the product in the container during the depressurizing step and assures that the product will be virtually foam-free when the container is removed from contact with the valve.

The tabletop, which employs an exit star wheel (not generally used on can fillers), sweeps away the lowered package from the rotating bowl, and transfers it into the closing section at tabletop height. If the container is a can, it is closed in the closing machine by an "end" which is rolled and seamed in place (see Can seamers), integrally with the can body, with forces great enough to withstand high internal pressures. Bottles can be closed with a pryoff or twistoff crown; a rolled-on aluminum closure, which is threaded in place using the threaded portion of the bottle as a mandrel; or a prefabricated threaded closure applied by standard technique (see Capping machinery; Closures).

Carbon dioxide dissolves more readily in cold water than in hot water; thus, to keep foaming at a minimum and filling speeds at a maximum, it has been common to run beer and soft drinks as close to 32°F (0°C) as is practicable. To save the energy consumed by refrigeration, fillers have been introduced that operate at ambient temperatures. Controlling the CO_2 in the carbonating and filling stages means these operations must be accomplished at above-normal pressures, which complicates the internal parts of the filling valve and reduces rates of speed. The production rates of modern fillers have increased to approximately 2000 12-oz (355-mL) cpm and reportedly 1500 16-oz (473-mL) glass bpm. These increased rates of output are made possible by the use of programmable computers and new advanced instrumentation. Precise fill heights in the containers are controlled by the length of a vent tube in a bottle filler and ball check technique in a can filler. However, because of minor variations in the dimensions of glass and plastic bottles, there is more variability in volumetric content than in the exact volumetric measurement technique sometimes employed in "still" liquid filling.

BIBLIOGRAPY

"Filling Machinery, Liquid, Carbonated" in *The Wiley Encyclopedia of Packaging Technology,* 1st ed., by W. R. Evans, Coca-Cola Bottling of New York, p. 300.

FILLING MACHINERY, LIQUID, STILL

This article deals with the filling of noncarbonated liquids intended for packaging and distribution into rigid and semirigid preformed containers, such as glass bottles, sanitary cans, plastic bottles, and preformed paper cartons. Form/fill/seal packaging is not discussed (see Form/fill/seal, horizontal; Form/fill/seal, vertical; Thermform/fill/seal), and the information relates only indirectly to such applications as the filling of paper cups in vending machines.

Liquid-filling machines are classified here in terms of two fundamental characteristics: filling principle employed (see Table 1) and container-positioning method utilized (see Table 2). Except for a few specific restrictions (discussed below), the two characteristics are independent, but certain combinations are not commercially available. Packagers can select from a wide range of fillers, however. Among the members of the U.S. Packaging Machinery Manufacturers Institute (PMMI) alone, 61 companies offer liquid fillers (1), and this figure does not include machines made in other countries.

Table 1. Methods of Container Filling

Type of Container	Filling Method
Sealed container	Balanced pressure
	Gravity
	Gravity–vacuum
	Counterpressure
	Unbalanced pressure
	Vacuum
	Prevacuumizing
	Gravity
	Pressure
Unsealed container	Level sensing
	Piston volumetric
	Cylinder vertical, closed ends
	Cylinder vertical, open-end inlet
	Cylinder horizontal, single-ended
	Cylinder horizontal, double-ended
	Rolling-diaphragm volumetric
	Displacement-ram volumetric
	Volume cup
	Turbine-meter volumetric
	Positive-displacement-pump volumetric
	Peristaltic-pump volumetric
	Weight
	Gross weight
	Net weight
	Time
	Controlled-pressure head
	Constant-volume flow
	Overflow

Table 2. Container Positioning Methods and Configurations

Positioning Method	Configuration
Manual	
Automatic, in-line	Single- or dual-lane
Automatic, rotary	Single- or dual-lane

Methods of Filling

In Table 1, methods of fill are divided into two primary categories: the *sealed-container system,* in which the filling device seals positively against the container; and the *unsealed-container system,* in which the container is left open to the atmosphere during the fill process.

Sealed-container filling system. In sealed-container filling, there are seven distinctly identifiable types of fillers. All sealed-container fillers fill to a controlled level in the container.

Balanced-pressure fillers. The first three of the seven mentioned above are balanced-pressure fillers, in which product flows through a valve into the container from a tank of liquid located above the container, and air from the container is vented back to the headspace in the tank through the same valve. The typical embodiment of such a filling system, *gravity filling*—one of the simplest and most reliable—is illustrated in Figure 1. As the filling takes place through the sleeve-type valve illustrated, liquid flows from the tank through the liquid port into the container, and air within the container flows up the vent tube to the top of the tank. The

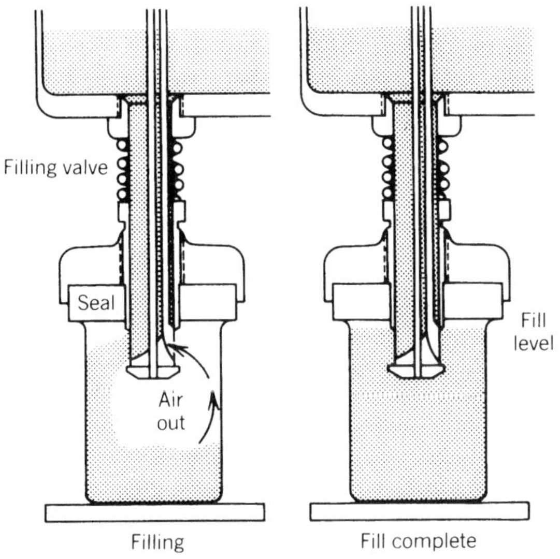

Figure 1. Pure gravity filling. (Courtesy of Horix.)

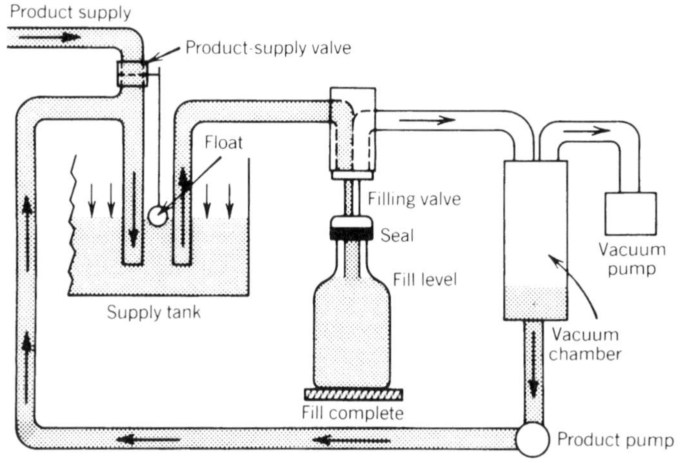

Figure 2. Pure vacuum filling. (Courtesy of PMMI.)

container fills product to an exact level determined by the position of the vent port relative to the bottle. Any liquid in the vent from a previous filling cycle is returned to the product in the tank by the air flowing up the vent tube. If necessary, air or vacuum may be used to clean the vent before the start of fill.

A modification of pure gravity filling is the *gravity–vacuum filler.* In such a system, a low vacuum is maintained in the headspace in a sealed tank. When the container is brought into sealing contact with the filling valve and the valve is opened, the pressure of air in the container helps force any product in the vent back into the tank, accelerating the start of the filling process. The use of gravity–vacuum fillers also prevents the loss of product that would occur if a chipped or slightly broken container were filled. This savings is of particular advantage when filling more expensive products.

To fill thin-walled plastic containers, such as 1-gal (3.8-L) milk bottles, a pulsating vacuum in the tank is sometimes used to cause the container walls to flex in and out, assisting the foam in moving up the vents. The pulsations must be timed so that a container is not flexed inward at the position at which it is ready to break away from the filling valve, because this could cause underfill.

Thousands of different styles of filling nozzles, utilizing single ports, multiple ports, screens, sliding tubes, or check valves, are offered by various manufacturers. All these styles are designed to achieve the maximum production rate with the fewest number of filling valves and provide greater accuracy of fill. The selection of nozzle type is probably best left to the judgment of the machinery manufacturer, on the basis of experience and product testing.

Counterpressure fillers for carbonated beverages are discussed elsewhere (see Filling machinery, carbonated liquid).

Unbalanced-pressure fillers. Unbalanced-pressure fillers utilize a difference in pressure between that on the liquid to be filled and on the vent that permits air in the container to escape during the filling process. The usual combinations are listed in Table 1. The use of unequal pressure permits higher rates of product flow than possible with the balanced-pressure fillers. Unequal pressure is particularly advantageous when filling containers with small openings, viscous products, or large containers. Unbalanced-pressure filling has the disadvantage of requiring an overflow-collection/product-recirculation system, in contrast to the relative simplicity of balanced-pressure fillers. Higher liquid-flow rates do not necessarily result in faster filling because the additional foam generated by rapid entry of product into the container must be drawn off through the overflow system to obtain accurate filling-level control.

A schematic diagram of a typical *vacuum filler* is shown in Figure 2. The supply tank may be located either above or below the container to be filled. After the filling valve seals against the container and the valve opens, the vacuum on the vent draws the liquid into the container up to the filling level. Usually, a substantial quantity of liquid is drawn into the vent, which leads to an overflow tank. Product is recovered in the overflow tank and then recycled.

The *prevacuumizing filler* is a special form of vacuum filling. On such a filler, a vacuum is first drawn in the container, evacuating the air. The valve then permits liquid to enter the container. Because such a system is complex and expensive, it is normally used only when liquid is being added to solids already in the container. Certain solids, such as peach halves, trap air. Such air entrapment may be eliminated by use of a prevacuumizing filler.

In an unbalanced-pressure *gravity filler,* the product-supply tank and the overflow tanks are open to the atmosphere, but the product tank is located above the container and the overflow tank is located below the container, permitting the differential pressure achieved by the difference in elevation to cause product flow. Such a filler is necessarily rather restricted in its ability to adapt to varying products and containers since the pressure difference is established solely by the product-tank and overflow-tank locations. Fillers of this type are not very common.

A *pressure filler* is very similar to a vacuum filler except that pressure is applied to the product. This may be achieved either by pressurizing the headspace over a tank or by direct pumping of the product to the filling valve. In the most common form of pressure filling, the product is pressurized and the overflow tank open to the atmosphere. Such a system

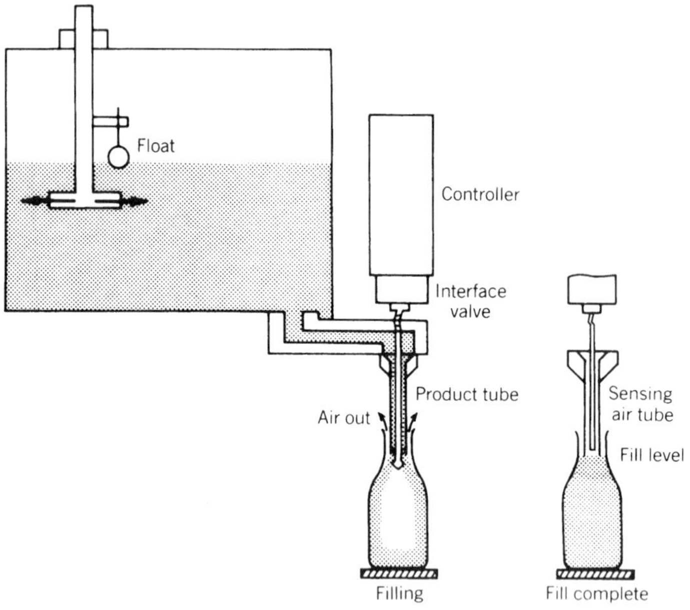

Figure 3. Level-sensing filling. (Courtesy of PMMI.)

allows unbalanced fill without vacuum. This is desirable when vacuum cannot be drawn on the product.

For example, drawing a high vacuum on alcoholic beverages can reduce the alcoholic content of the beverage. Applying a vacuum to a hot product, such as juice at 200°F (93°C) causes the liquid to flash. If desired, both the product and the vent can be maintained above atmospheric pressure, but with a higher pressure on the product. Such a filler is often used for filling lightly carbonated products, such as certain wines, using the pressure to retain the low carbonation in the product.

Unsealed-container Filling Systems

Level-sensing fillers. Level-sensing fillers fill containers to a controlled level without sealing the container, as shown in Figure 3. Such a filling technique eliminates product recirculation and allows filling to a level in plastic containers that would bulge out or flex inward if pressure or vacuum were applied to the sealed container. A level-sensing filler uses some type of sensing means, typically a flow of ultra-low-pressure air. Rising liquid level in the container blocks airflow, triggering a control system that shuts off product flow to the container. Such control mechanisms, which are required at each filling nozzle, are expensive, but high rates of fill may be achieved because there is no product overflow and no foam to be removed. Electronic sensing is also available.

Piston volumetric fillers. At present, unsealed-container fillers are the most common, and volumetric filling is a frequently used method. In view of this popularity, many different kinds of volumetric fillers are available. For volumetric filling, piston fillers are most widely used. Table 1 indicates four subclasses of piston fillers, depending on the orientation of the pistons, specifically vertical or horizontal, and inlet arrangement.

1. *Cylinder Vertical, Closed Ends.* One station of a typical vertical-cylinder rotary filling machine is illustrated in Figure 4. The valve(s) controlling the product flow between the supply tank, measuring chamber, and dispensing nozzle may have either a rotary or a reciprocating motion. The rotary style, which is more common, is illustrated. The product is drawn into the cylinder from the liquid-supply tank when the piston moves upward. The valve then rotates to permit the premeasured volume in the cylindrical chamber to flow into the container. Usually, either a direct mechanical drive from a cam track or an air cylinder is used to stroke the piston. If an air cylinder is used to drive the piston, controls are usually such that the piston does not cycle if there is not container in place. This eliminates moving the liquid back and forth between the measuring chamber and the supply tank, a situation that is usually undersirable and may cause product breakdown with foods such as mayonnaise. It is not easy to uncouple a mechanically driven piston.

2. *Cylinder Vertical, Open-End Inlet.* In an alternative design, product enters vertical volume chambers through cylin-

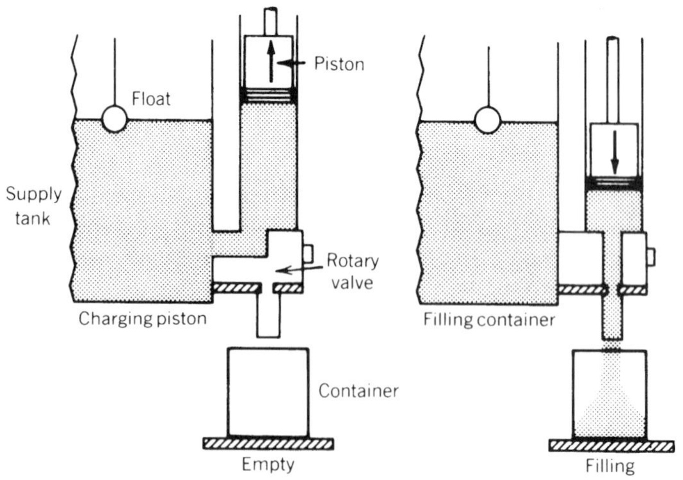

Figure 4. Piston volumetric filling. (Courtesy of PMMI.)

ders open at their upper ends, with a cam drive located below. A nonrotating plate with an orifice allows product to enter the open-ended cylinders at the appropriate position during the rotation of the filler bowl. Such a configuration is mechanically complex and generally considered difficult to clean. Fillers of this type have the advantage, however, of being able to handle products containing sizable solids in suspension.

3. *Cylinder Horizontal.* Measuring cylinders may be mounted horizontally. Usually single-ended, they are very similar to vertical cylinders in operating principle. They are frequently found on in-line, large-volume fillers, in order to avoid excessive height for such machines. A volume cylinder may also be double-ended, with inlets and outlets at both ends. Product under pressure flows into one end, causing a floating piston to move and expel the liquid in the opposite end of the cylinder into the container. In some fillers, the double-acting cylinder may be used for a single fill, and the first half of the fill comes from one end. The flow pattern is then reversed to discharge product from the other end of the cylinder for the second half of the fill while the first end is being filled for the next cycle.

Rolling-diaphragm volumetric fillers. The volumetric fillers described above normally have some type of a seal, such as V- or O-rings, between the pistons and cylinder walls. An alternative method for measuring volume is to use a rolling diaphragm; a typical arrangement can be seen in Figure 5. The diaphragm provides an absolute seal and also eliminates the friction contact of a seal with a cylinder wall. Such sliding causes abrasion and particle generation, which is seldom important, but minimizing particulate generation is very important in the packaging of intravenous solutions and injectable drugs. Rolling-diaphragm volumetric fillers often employ flexible diaphragms or pinch valves to control product flow to and from the measuring chambers.

Displacement-ram volumetric fillers. Another method for dispensing a specific volume of liquid is to use a displacement ram. The ram enters one end of a cylinder through a seal of some type, but the displacement ram does not touch the inside of the cylinder wall. An external air cylinder drives the ram, which displaces a controlled volume from the cylinder into the container. Such fillers are easy to clean.

Volume-cup fillers. Cup-type volumetric fillers operate by first transferring the product from an open tank into measuring cups of precise volume. Depending on the design, each cup may be filled to a level matching that of the tank, or the cup may fill to overflow and then rise above the level of liquid in the tank. Each valve then opens at the bottom, permitting liquid to flow from the cup into a container. Although such fillers are appropriate only for low-viscosity liquids that do not cling to the sidewalls of the cups, for suitable applications the volume-measuring method is accurate, inexpensive, and reliable. Various volume-adjustment systems are provided. These fillers are usually rotary and are often designed to be easily changed, by means of a change of filling valves, to gravity or gravity–vacuum filling.

Turbine-meter volumetric fillers. The amount of liquid dispensed from a nozzle can be measured by placing a turbine flowmeter in the line ahead of the nozzle. Such meters, which include an electronic counting and control system to start and

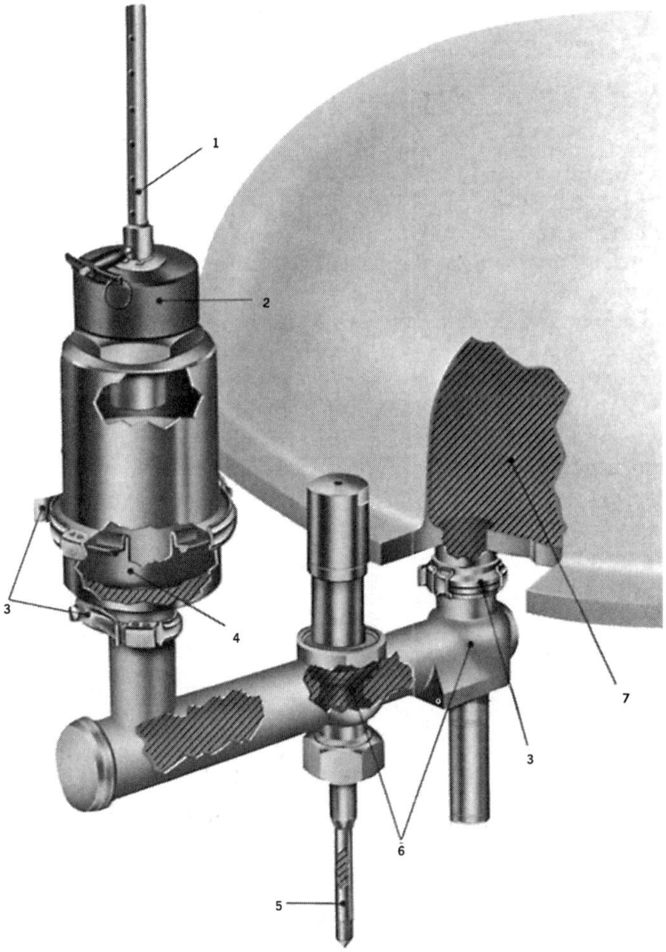

Figure 5. Rolling-diaphragm volumetric filling: (1) volume adjustment rod; (2) precision volume adjustment; (3) quick-disconnect design; (4) volume-chamber diaphragm; (5) product tube; (6) valve seats; (7) product supply manifold. (Courtesy of Horix.)

stop flow, are accurate but expensive, and generally used only for filling large containers, ie, ≥5 gal (≥19 L).

Positive-displacement volumetric fillers. Viscous liquids may be moved directly from a supply system through a positive-displacement pump into a container. Volumetric measurement is achieved by accurately controlling the number of revolutions made by the pump. An auger may be used as the pump mechanism.

Peristaltic-pump volumetric. Peristaltic pumps are often used to fill sterile liquids, as they are easy to clean and have no sliding parts that might contaminate the product being filled.

Weight fillers. On *gross-weight fillers,* each fill station is fitted with a weighing device, typically a beam scale, which acts to shut off product flow when a predetermined weight has been reached. Each scale is set for the maximum weight of container and contents. Adjustment for other containers and content weight may be made by adding special weights to each filling platform.

By using a load cell as the weighing device, the weight is electronically measured continuously, and liquid flow is stopped at a predetermined weight. With microprocessor controls, the change from one weight to another is quite simple. *Net-weight fillers* employ more advanced load-cell weighing devices. The tare weight of each container is measured, and then product fill proceeds until the proper net weight is in the container. Net weight filling is the most accurate method for filling free-flowing products. Such fillers are readily available with simple, easily cleanable, nonsliding contact valves. They are usually relatively expensive, as each filling station requires a load cell and its associated electronic interface. But, such costs are often justified by the reduction of product overfill possible with the increased accuracy.

Time-fill fillers. Time-fill filling consists of delivering liquid under pressure to an orifice that is open for a controlled length of time. Such a filler may be either of the controlled-pressure-head type or the constant-volume-flow type. On multistation *controlled-pressure-head* fillers, which are most frequently in-line, all orifices are opened for approximately the same length of time. Minor adjustments may be made at each station to compensate for individual orifice characteristics. Until quite recently, such time-fill fillers depended for their accuracy on maintaining a precise pressure head on the liquid at the orifice. This was achieved with pressurizing tanks or a liquid level established by flowing the product over a dam. Small pressure fluctuations are now acceptable, because a microprocessor controls the time each orifice is open, on the basis of measured pressure variations. Such fillers are designed for easy cleaning and changeover from one product to another. To date, such fillers have not generally been used for high-speed production runs. They are probably best suited to filling pint (473 mL) or smaller containers with free-flowing liquids.

The *constant-volume-flow* type of time filler is almost always adapted to a rotary filler. In a typical embodiment (see Fig. 6), product is delivered continuously to the filler at a constant flow rate, using constant-displacement pumps or their equivalent. The amount of product entering each container is proportional to the length of time a nozzle is under the liquid ports. The time is determined by the rotational speed of the filler. Constant-volume-flow time fillers are relatively simple and inexpensive. They are capable of reasonable accuracy, particularly with products of medium to high viscosity. Leakage between plates is hard to control with low-viscosity liquids. The product flow must be simultaneously altered any time the filler is stopped or started. It is difficult to do this and maintain consistent filling accuracy. Since a no-container/no-fill mechanism is rarely provided, any missing containers cause product overflow, which must be collected and either returned for reuse or discarded.

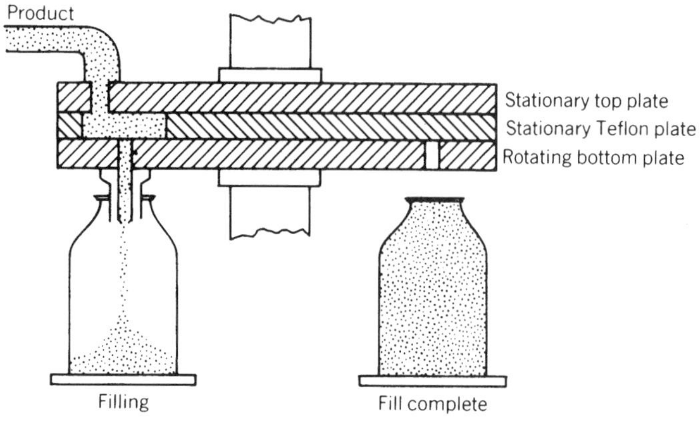

Figure 6. Constant-volume-flow time filling. (Courtesy of PMMI.)

Overflow fillers. Some products can be filled by filling open containers, usually sanitary cans or widemouth glass bottles, to overflow. The liquid may flow from a pipe or over a barrier. In more advanced overflow fillers, the liquid flow into open containers is directed by moving funnels synchronized with the movement of containers. The headspace in the container, typically very small, may be established in various ways. When brine is added to pickles, for example, the headspace is usually created by displacement pads that enter the container and establish the desired headspace. If solids are present, the headspace pads also ensure that the solids are properly down into the container. Another method of establishing headspace is to tilt the containers slightly, permitting liquid to pour out. With some overflow fillers, an upwardly directed curtain of air prevents the overflowing liquid from contacting the outside of the containers. With careful adjustment of liquid flow rate and container speeds, the amount of fluid that is overflowed and recirculated may be limited to a very small proportion. Tilted-container overflow fillers are relatively fast and inexpensive. They are frequently used for filling juice in cans, but they normally cannot be used to fill narrow-neck containers.

Container Positioning

Filling machines may be characterized by the way they deliver containers to the liquid-dispensing mechanisms and remove them after filling (see Table 2).

Manually loaded fillers. The oldest and simplest method of container delivery and removal is by manual means. Filling occurs after one or more containers are in place. The containers may be raised to the filling valves or the valves lowered to the containers, or no relative motion may be required. Because of the amount of labor involved, manual filling is usually limited to small prodution runs.

In-line fillers. The simplest automatic fillers are single-lane in-line machines. In a typical machine of this type, containers, standing on a conveyor, are delivered to the filler. One or more containers back up behind a stop or gate. The barrier then opens, and a controlled number of containers move under the filling heads, where they are positioned by another barrier of some type. Conveyor motion is usually intermittent, with the conveyor stopped during the fill cycle to prevent tipping the containers. After the filling is completed, the positioning barrier opens, and the filled containers leave while unfilled packages enter.

The size and shape of a container are factors that determine how many containers may be filled simultaneously on an in-line intermittent-motion filler. Increasing the number of filling stations increases the total output of the machine, but this approach runs into space limitations, and, at a certain point, an additional valve is not cost-effective. Sixteen stations appear to be the maximum commercially feasible number of valves. If the containers are not straight-sided, there usually is difficulty in backing up any significant number behind a positioning stop; even with straight-sided containers, the process is limited by container dimensional tolerances. The size of the containers determines the position of the last container in a row relative to the first container. If the lead container is properly positioned under a nozzle, the accumulation of container dimensional tolerances may cause the trailing container not to be positioned under a nozzle.

An in-line filler may be used for in-case filling. A multiple array of filling valves is used to fill all the containers in a case at the same time. As with individual containers, two or three cases may be backed up for simultaneous filling. Such filling is possible only if the cases have reasonably consistent dimensional control.

A variation of in-line intermittent-motion filling is occasionally used to raise the containers relative to the filling nozzles. This is done (eg, in filling mayonnaise or peanut butter) to change the position of the container relative to the nozzle during fill. On such a filler, the containers are moved by means of a pusher mechanism from the infeed conveyor onto a platform under the filling heads, and the platform is then raised. After the containers have been filled and the platform lowered, successive unfilled containers may be used to push the filled packages onto a discharge conveyor running behind and parallel to the infeed conveyor, or individual mechanisms may move the containers between the infeed and discharge conveyors. Some manufacturers offer a rising-platform device with a straight-through conveyor arrangement.

Dual-lane straight-line fillers permit more efficient utilization of filling stations. Containers move on two parallel conveyors. The filling nozzles, in a row, fill in one lane while container movement occurs in the other lane. The use of dual-lane fillers is generally considered only if the limit of the number of valves in a single lane has been reached. They are typically used for small containers because fill time is short relative to the time required for container movement.

Valve utilization is the percentage ratio of actual filling time, ie, the time the valve is fully open, to the total cycle time for that valve from the beginning of filling one container to the start of filling the next container. Valve utilization is low on in-line fillers, ie, 25–50% of cycle time, dependent on conveyor speed, container diameter, and actual unit-filling time required for the fill.

Rotary fillers. The most common system for filling containers are moderate to high speeds is rotary filling. Containers arrive continuously on a conveyor and are spaced by some means into a rotating infeed star that delivers them, properly separated, to filling stations on the main rotary assembly. The timing mechanism normally employed to take containers coming at random to the filler is a feed screw; for large containers and low-speed operation, timing fingers, escapement wheels, or like devices may be used. The discharge from the main rotary is usually by means of a second star wheel with the same diameter as the first. However, if the liquid level is high in the container, either a large-diameter star or a tangential conveyor should be used to prevent spill.

On rotary fillers, valve utilization is almost independent of container size but is a function of filler diameter. For example, a 22-in. (56-cm)-pitch-diameter gravity filler for 32-oz (946-mL) containers has a valve utilization of 49%, whereas a similar gravity filler having the valves on a 60-in. (152-cm) pitch diameter has a valve utilization of 73%. The time that the valve is closed is necessary for the transfer of containers into and out of the rotary section and for the relative movement between container and filling valve needed to open and close the valve.

The concept of dual-lane rotary fillers, involving two lanes (inner and outer) on the main rotary, was proposed and patented many years ago. Except, perhaps, for large-diameter carbonated-beverage fillers, the system is not practical. Two infeed stars and two discharge stars (or a tangential system) would be needed, and such an arrangement greatly reduces valve utilization. The fact that valve spacing is limited in the inner row is a disadvantage as well, unless unequal production speeds are desired. In one configuration study, the valve utilization on an 80-valve, dual-lane, 60-in. (152-cm)-outer-lane-pitch-diameter machine is only 66%, compared with the normal 85% utilization of a 48-valve, 60-in. (152-cm)-pitch-diameter filler. Adding 67% in number of valves adds only 29% to the production capability.

Continuous-motion in-line fillers use a timing screw to position the containers under multiple filling heads, which move in synchronization with the containers.

Multihead in-line weight fillers may have a conveyor system that separates and stops a group of containers. Weighing platforms then lift the containers above the conveyor for the filling cycle. Alternatively, they may use a lateral transfer device to place the containers on the scale platforms.

Relative motion. The relative motion between the container and the liquid-dispensing device can significantly affect filler performance. The various relative motions possible are none, raise container, lower valve, and raise container and lower valve.

In sealed-container filling (see Table 1), relative motion between the container and the valve is required to bring the valve into contact with the container. With manual container placement, the container can be raised against the valve, or the valve(s) can be lowered to one or more containers on platforms.

The valve utilization of rotary sealed-container fillers can be increased by lowering the valve (see Fig. 7). With a descending valve, the valve can begin to enter the container at position 1, and it must be clear of the container at position 1′. If the container is to be raised up to the valve, the rise may not begin until position 2 because space must be allowed for the container platform to clear the infeed star. Likewise, the

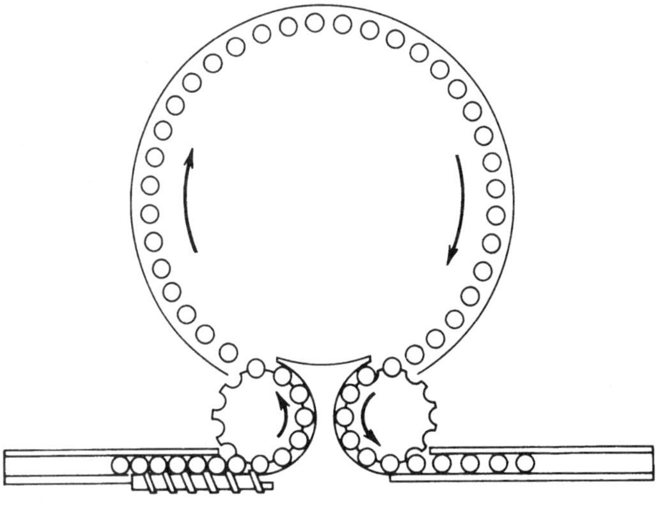

Figure 7. Rotary-filler arrangement.

container platform must be in the full-down position by position 2′. As an example, with 40 valves on a 60-in. (152-mm) pitch diameter, arc 1–1′ is 327, whereas arc 2–2′ is 306. This 6% difference may be advantageous, but movement of each valve imposes other difficulties on filler design, often including the need for a flexible liquid connection to each valve. Such flexible connections often make cleaning more difficult. The combination of raising the container and lowering the filling valve is very rarely used, as it requires mechanisms for moving both elements, with all the disadvantages of each system.

In-line, sealed-container fillers are rather slow and not very common, except for filling containers in a case. Most of those available operate by lowering the valves to the container, which permits the containers to remain on the conveyor. As with sealed-container fillers, level-sensing fillers require relative motion between the filling valves and the container. Rotary level-sensing fillers typically raise the container; in-line fillers bring the nozzle down to the package.

Volumetric, weight, and time-fill fillers do not require relative motion between the dispensing nozzle and the package. Thus, for simplicity and economy, most of these fillers operate with the nozzle located above the container opening. There are, however, two principal exceptions:

1. In filling containers with relatively small openings, such as long-neck plastic bottles, greatly increased rates of flow are possible if the product flow emerges from the side of the tube and is directed against the container wall (see Fig. 1). Filling straight down into that style of container from above may cause air entrapment with consequent possible splashout or overflow at the end of the filling cycle. The lost time, if any, required to raise the container or lower the filling valve may be more than offset by the faster flow rates possible.
2. Viscous products such as mayonnaise and peanut butter are best filled volumetrically using a flow tube open at the bottom. To prevent air entrapment, the flow tube is maintained at a position very close to the rising surface of the liquid by lowering the container during the filling cycle.

Considerations in Design and Selection of Fillers

Many factors must be considered in selecting filling machinery. These include such obvious factors as operating and maintenance costs, efficiency, reliability, size, speed, and materials of construction. Several considerations, discussed below, are very specific to liquid-filling machinery and very important to the satisfactory operation and use of such equipment.

Accuracy. Accuracy of fill is important for two reasons: (1) it is necessary in order to comply with state or other regulations as to labeled content [the regulatory agencies generally refer to the filling tolerances recommended in the *National Bureau of Standards Handbook 44* (2)] and (2) in addition to meeting legal requirements, however, packagers of most products desire to keep overfill to a minimum, thus minimizing product giveaway. Even with relatively inexpensive products, the amount given away with inaccurate overfilling on high-speed production lines can usually justify the higher cost of purchasing more accurate fillers.

Changeover and cleaning. As a part of operating costs, the time and labor required for cleaning and preparing the filler for daily operation, changeover from one product to another or one container to another, and cleanup at the end of each day's production should be carefully evaluated. User needs may range from fillers that operate 24 h per day with no need for product or container-size change, to fillers for 30-min production runs, after which the filler must be completely sterilized and adjusted for a different container size. Comments are made above concerning the influence of filling techniques. The manufacturer's descriptions of container-positioning methods contain comments on cleanability, container-size flexibility, and ease of adjustment. Specific aspects of each filler design must be carefully considered in choosing the proper liquid filler for a particular application.

No-container/no-fill system. In general, it is desirable to have a filler equipped with some type of no-container/no-fill system. In most sealed-container fillers, the filling-valve mechanism is designed to automatically provide a no-container/no-fill system in conjunction with the action of the container and the filling valve coming together. Likewise, with most of the unsealed-container filling techniques, a relatively simple container-detection device can be used to prevent product flow if no container is present. The constant-volume time filler does not permit start or stop of product flow, but means such as drains in the container platform to catch the product flow may be provided. Alternatively, particularly on in-line fillers, a control system may be provided to ensure that a full row of containers is present.

Drive location. The location of the main drive for the filler should be considered in filler selection. Usually, the main motor drive, or, alternatively, a lineshaft synchronization system, is located near floor level. Thus, the heavy drive components are conveniently supported and reasonably accessible for routine maintenance. Because splash of corrosive products may significantly harm drive components even if they are protected against splash, some filler manufacturers locate the main drive above the product. This may not be desirable for filling food or drug products because of the potential for product contamination from drive lubricants.

Equipment standards. Various industry-consensus standards are applicable in the United States:

ANSI B155.1-1994	*Safety Requirements for the Construction, Care, and Use of Packaging and Packaging-related Converting Machinery,* American National Standards Institute
3-A Sanitary Standard 17-06	*3-A Sanitary Standards for Fillers and Sealers of Single Service Containers for Milk and Fluid Mild Products,* International Association of Milk, Food, and Environmental Sanitarians; U.S. Public Health Service; The Dairy Industry Committee

The first standard is general but applicable to fillers. The 3-A Standard is specifically for milk fillers.

The packaging of certain products is regulated by government agencies. The packaging of fluid-milk products in the United States is usually regulated by local (ie, county or state) authorities. They normally adopt the 3-A Standards without modification, but a few jurisdictions impose their own requirements, usually stricter. If fresh milk is packaged in flexible plastic containers, the container is usually the legally specified measuring device. In such cases, volumetric tolerances normally follow the recommendation of the National Bureau of Standards (2).

The packaging of liquids in meat and poultry establishments subject to inspection by the USDA Food Safety and Inspection Service is governed by guidelines issued by the department (3,4). Equipment for packaging food and drug products is regulated in the United States by the FDA (CFR Title 21, Chapter I), and the packaging of distilled spirits by the Bureau of Alcohol, Tobacco, and Firearms (BATF).

BIBLIOGRAPHY

1. *The PMMI 1994-1995 Packaging Machinery Directory,* Packaging Machinery Manufacturers Institute, Washington, DC, 1994.
2. *Specifications, Tolerances, and Other Technical Requirements for Weighing and Measuring Devices, National Bureau of Standards Handbook 44,* NBS, U.S. Department of Commerce, Washington, DC, 1987.
3. *Accepted Meat and Poultry Equipment,* Meat and Poultry Inspection Technical Services, Food Safety and Inspection Service, U.S. Department of Agriculture, Washington, DC, May 1984, pp. 1–8.
4. *Guidelines for Aseptic Processing and Packaging Systems in Meat and Poultry Plants,* Meat and Poultry Inspection Technical Services, Food Safety and Inspection Service, U.S. Department of Agriculture, Washington, DC, June 1984.

General Reference

C. G. Davis, Product Filling, Vol 1. of *Packaging Machinery Operations,* Packaging Machinery Manufacturers Institute, Washington, DC, 1987.

F. B. Fairbanks
Horix Manufacturing Company
Pittsburgh, Pennsylvania

FILM, EDIBLE

Edible films formed as coatings on food products, such as fruits and candies, have been used for centuries to prevent moisture migration and improve appearance. Present uses of edible films include wax coatings on fruits and vegetables to prevent moisture loss and improve appearance; shellac and zein coatings on candies to provide moisture resistance and improve appearance; collagen casings on sausages to inhibit moisture loss and oxygen transport and provide structural integrity; hydroxypropylmethyl cellulose (HPMC) pouches for food ingredients; and HPMC, shellac, and zein coatings and gelatin capsules for pharmaceuticals to improve product appearance, structural integrity, ingestibility and stability. However, edible films are generally seen as having considerable potential for applications far beyond current usage for improving product quality and, in some instances, reducing use of synthetic packaging films.

These examples involve forming an edible film as a coating directly on a food or drug product or filling a food or drug product into a preformed edible-film casing, pouch, or capsule. Such edible films are not intended to eliminate the need for nonedible protective packaging. Rather, they are intended to work with conventional packaging to improve product quality and shelf life. To the degree that the edible-film coating, casing, pouch, or capsule functions to protect the product from the environment, the amount and complexity of nonedible protective packaging can be reduced, leading to source reduction and improved recyclability. In addition, after packaging is opened, the edible film can continue to protect the product. In addition to controlling moisture and/or oxygen transport and protecting products from mechanical forces, edible films can reduce aroma loss or gain and lipid migration, as well as act as a carrier for edible antioxidants, antimicrobials, and other additives. Furthermore, besides acting as coatings, casings, capsules, and pouches, edible films can be positioned within food products (eg, a pizza) to reduce migration of moisture, lipids, and solutes from one food component (eg, tomato sauce) to another (eg, pizza crust). Use of materials that are either GRAS (generally recognized as safe) or are sanctioned by the *U.S. FDA Code of Federal Regulations* and/or the *U.S. Pharmacopoeia / National Formulary* (or equivalent) is essential. The objective of this article is to review the materials, manufacture, properties, applications, and trends of edible films. The area of microencapsulation of food and drug ingredients is not covered.

Film Composition

Film formers. Film-forming materials available for edible films fall generally into the categories of polysaccharides, proteins, and lipids derived from plants and animals. Polysaccharide film-forming materials include cellulose derivatives, starch and starch derivatives, carrageenan, alginate, pectinate, and chitosan. Protein film formers include collagen, gelatin, casein, whey protein, corn zein, wheat gluten, and soy protein. The hydrogen-bonding character of polysaccharides and proteins produces films that have (1) high moisture permeability, (2) low oxygen and lipid permeabilities at lower relative humidities, and (3) compromised barrier and mechanical properties at higher relative humidities. Edible lipids include carnauba wax, candelilla wax, beeswax, shellac, triglycerides (eg, milkfat fractions), acetylated monoglycerides, fatty acids, fatty alcohols, and sucrose fatty-acid esters. Lipid materials are not polymers and, therefore, do not generally form coherent stand alone films. However, they can provide gloss and a moisture barrier on a food or drug surface, or constitute the moisture-barrier component of a composite film. Composite films can consist of a lipid layer supported by a polysaccharide or protein layer, or lipid material dispersed in a polysaccharide or protein matrix.

Plasticizers. Like synthetic polymer films, edible films often require incorporation of low-molecular-weight plasticizers to improve film flexibility and durability. Plasticizers generally interrupt polymer chain-to-chain interactions and lower the glass-transition temperature (T_g), resulting in greater flexibility but also, unfortunately, increased film permeability.

Plasticizing materials acceptable for edible films include sucrose, glycerol, sorbitol, propylene glycol, polyethylene glycol, fatty acids, and monoglycerides. Water is also a plasticizer for polysaccharide and protein edible films. Thus, film moisture content as affected by content of other plasticizers and the surrounding relative humidity can have a dramatic effect on edible-film properties.

Other additives. Edible films hold promise for being effective carriers and providers of antioxidants, antimicrobials, nutrients, flavors and colors to enhance food safety, nutrition, and quality. For some food-coating applications, addition of surfactant to a film formulation may be necessary to ensure good surface wetting, spreading, and adhesion.

Film Manufacture

Solvent casting. This approach to forming edible films from aqueous or water–ethanol solutions or dispersions is similar to production of certain synthetic films (eg, polyvinyl chloride) by solvent casting. The solution or dispersion is spread on a smooth surface. After the solvent evaporates, a film can be stripped from the surface. With the exception of collagen, corn zein, and wheat gluten, most of the polysaccharide and protein film-forming materials are soluble in water. This can be an advantage in terms of film manufacture and use, but makes films whose properties are especially vulnerable to moisture. Corn zein and wheat gluten films must be formed from water–ethanol solution or from an aqueous dispersion, but provide a somewhat improved moisture barrier compared to other protein-based films formed from aqueous solution. Edible-film production that requires ethanol necessitates appropriate safety measures and solvent recovery for commercialization. Insoluble whey protein films result from heat-treated aqueous whey protein solutions due to denaturation and subsequent intermolecular crosslinking of individual whey protein molecules.

Edible films from methyl cellulose (MC) and hydroxypropylmethyl cellulose (HPMC) are manufactured commercially by solvent casting on a continuous belt, with subsequent drying, removal, and winding of the resulting film. Such films can be formed into edible, water-soluble pouches for food ingredients. Gelatin capsules cast from aqueous solution find broad use in the pharmaceutical industry. Hard capsule halves are formed on steel pins by dipping into gelatin solution, followed by drying and removal from the pins. After a drug is filled into a hard capsule half, two interlocking halves are joined to form the full capsule. Soft capsules containing a drug or vitamin are formed from two previously formed sheets of plasticized gelatin by injection of the drug at the moment when the two sheets are brought together between the rotating halves of roller dies.

Polysaccharide–lipid bilayer films with excellent moisture-barrier properties have been produced from water–ethanol solutions of cellulose derivatives with lipids. As the solvent evaporates, phase separation results in bilayer formation. Lipid layers or coatings can be formed directly from ethanol solution or aqueous emulsion. Stable emulsions of lipids in proteins or polysaccharides produce well-dispersed composite films with improved moisture barrier compared to protein or polysaccharide films alone, but inferior moisture barrier compared to bilayer films. Research efforts continue to develop composite edible films with good moisture-barrier properties.

Molten casting. Lipid films, layers, and coatings can be produced by cooling of a melt to produce a solid structure. Challenges include the high temperatures required for some melts, film thickness control, adhesion, and brittleness of some materials. Forming a lipid layer from a melt on a precast polysaccharide or protein film is an alternative approach for manufacturing an edible composite film with good moisture-barrier properties.

Extrusion. Water-insoluble, edible collagen film sausage casings and meat wraps are made by regenerating collagen that is extracted from animal hides. Collagen casings and wraps are produced by extruding a viscous, 4–10% solids aqueous suspension of purified acidified collagen into a neutralizing coagulation bath, followed by washing, plasticizing, and drying. Some polysaccharides and proteins display thermoplastic behavior, but this property has generally not been explored or exploited for edible-film production. Successful, efficient production of edible films using conventional extrusion equipment would improve commercialization opportunities.

Food and Drug Coating

Formation of edible films as coatings on foods and drugs generally consists of solvent casting directly on the food or drug surface. Application of film-forming solution, suspension, or emulsion can be by dipping, spraying, or dripping.

Fruits and vegetables. Edible coatings on fruits and vegetables generally serve to reduce moisture loss, improve glossy appearance, reduce abrasion, and, in some cases, control oxygen and carbon dioxide exchange or carry a fungicide. Materials commonly used include beeswax, carnauba wax, candelilla wax, shellac, and mineral oil. These materials cannot form stand alone films, but form continuous coatings on the fruit or vegetable surface. Currently, film coatings based on polymeric materials such as cellulose derivatives, chitosan, and proteins are being developed to reduce oxygen and carbon dioxide exchange for respiration and ripening rate reduction. Coating solution or emulsion is most typically sprayed or dripped onto the fruit or vegetable and/or onto a bed of brushes rotating either below or above the produce, followed by drying. Thorough distribution of the coating over the entire fruit or vegetable is dependent on a clean, dry produce surface and suitable surface tension and viscosity of the coating formulation.

Candies, nuts, and drug tablets. Edible coatings on candies, nuts, and tablets serve to resist moisture change, reduce oxygen uptake, improve structural integrity, and make the product more palatable by adding gloss and/or flavor, or, in the case of drugs, by masking flavor and improving ingestibility. Materials commonly used include shellac, corn zein, cellulose derivatives, waxes, and acetylated monoglycerides. The most common method of coating involves controlled spraying and drying of coating formulations on the product surface in a rotating, horizontal drum-like chamber called a "pan." Formation of a uniform coating depends on good spray coverage, thorough product tumbling action to help distribute the coat-

Table 1. Water-Vapor Permeabilities of Edible Films Compared to Low-Density Polyethylene

Film	Test Conditions[a]	Permeability [(g · mm)/(m^2 · day · kPa)]	Reference
HPMC : PEG (9 : 1)	25°C, 85/0% rh	6.5	1
SA : PA : HPMC : PEG	25°C, 85/0% rh	0.048	1
BW/SA : PA : MC : HPMC : PEG	25°C, 97/0% rh	0.058	2
HPMC	27°C, 0/85% rh	9.1	3
HPMC : SA (1.25 : 1)	27°C, 0/85% rh	0.026	3
Amylose	25°C, 100/0% rh	31.6	4
Zein : Gly (4.9 : 1)	21°C, 85/0% rh	9.6	5
Gluten : Gly (3.1 : 1)	21°C, 85/0% rh	53	5
WPI : Gly (4 : 1)	25°C, 0/77% rh[b]	70	6
WPI : BW : Sor (3.5 : 1.8 : 1)	25°C, 0/98% rh[b]	5.3	7
Na caseinate	25°C, 0/81% rh[b]	37	8
Ca caseinate : BW (1.7 : 1)	25°C, 0/97% rh[b]	3.6	8
Shellac	30°C, 0/84% rh	0.72	9
BW	25°C, 0/100% rh	0.021	10
LDPE	38°C, 90/0% rh	0.079	11

[a] Relative humidities are those on top and bottom sides of film (top/bottom).
[b] Corrected rh shown; HPMC = hydroxypropylmethyl cellulose, MC = methyl cellulose, SA = stearic acid, PA = palmitic acid, BW = beeswax, PEG = polyethylene glycol, WPI = whey protein isolate, Gly = glycerol, Sor = sorbitol, LDPE = low-density polyethylene.

ing formulation, and rapid drying to allow gradual buildup of the coating. Successful coating again depends on a clean, dry product surface and suitable coating-formulation surface tension and viscosity.

Film Physical Properties

Depending on which food or drug application is considered, different barrier and mechanical properties are of most interest. Tables 1–3 compare the permeability and mechanical properties of edible films made from representative materials presently used or proposed for use. It is essential to understand that plasticizer content and test conditions (temperature and relative humidity) have important effects on film properties. Increasing relative humidity increases the moisture content of polysaccharide- and protein-based films, resulting in increased permeability and decreased strength. Thus, comparison of edible-film properties must be approached with caution.

Polysaccharide and protein films are poor moisture barriers compared to the low-density polyethylene (LDPE) packaging film commonly used to protect food and drugs from moisture (Table 1). However, when polysaccharides or proteins are combined with edible waxes or fatty acids in composite films where the polysaccharide or protein provides film structural integrity, films with good moisture-barrier properties result. HPMC- and MC-based composite films have been developed which are bilayer in nature, with resulting excellent moisture barrier properties. This occurs because the water-vapor permeabilities of waxes and fatty acids rival that of LDPE. In the case of wax coatings on fruit and vegetable surfaces, it is not necessary to use the film-forming and lipid-supporting nature of an edible polymer.

Limited data exist on oxygen barrier properties of edible films, especially at relative humidities other than 0% (Table 2). Edible-film materials provide better oxygen barriers at 50% rh than does LDPE, which is a poor oxygen barrier. Protein edible films appear to be better oxygen barriers than polysaccharide or lipid films, with permeabilities approaching that of the excellent oxygen barrier provided by ethylene–vinyl alcohol copolymer (EVOH) at low to intermediate relative humidity (see Ethylene–vinyl alcohol copolymer). In

Table 2. Oxygen Permeabilities of Edible Films Compared to Synthetic Polymer Films

Film[a]	Test Conditions	Permeability [(cm^3 · μm)/(m^2 · day · kPa)]	Reference
HPMC	24°C, 50% rh	272	12
MC	24°C, 50% rh	97	12
Collagen	RT, 63% rh	23.3	13
WPI : Gly (2.3 : 1)	23°C, 50% rh	76.1	14
WPI : Sor (2.3 : 1)	23°C, 50% rh	4.3	14
Shellac	29°C, 55% rh	212	9
BW	25°C, 0% rh	931	10
LDPE	23°C, 50% rh	1870	15
EVOH (70% VOH)	23°C, 0% rh	0.1	15
EVOH (70% VOH)	23°C, 95% rh	12	15

[a] HPMC = hydroxypropylmethyl cellulose, MC = methyl cellulose, WPI = whey protein isolate, Gly = glycerol, Sor = sorbitol, BW = beeswax, LDPE = low-density polyethylene, EVOH = ethylene–vinyl alcohol copolymer, VOH = vinyl alcohol.

Table 3. Mechanical Properties of Edible Films Compared to Synthetic Polymer Films

Film[a]	Tensile Strength (MPa)	Elongation (%)	Reference
HPMC	69	10	12
MC	62	10	12
Starch	49	7	4
Amylose	70	23	4
Collagen : Cell : Gly (3.4 : 0.8 : 1)	3–11	25–50	16
Zein : PEG + Gly (2.6–5.9 : 1)	3–28	6–213	17
Gluten : Gly (2.5 : 1)	3	276	18
WPI : Gly (2.3 : 1)	14	31	14
LDPE	9–17	500	19
Polystyrene	35–55	1	20, 21

[a] HPMC = hydroxypropylmethyl cellulose, MC = methyl cellulose, cell = cellulose, Gly = glycerol, PEG = polyethylene glycol, WPI = whey protein isolate, LDPE = low-density polyethylene.

addition, the molecular composition and limited literature data for polysaccharide and protein edible films suggest that they are good lipid and aroma barriers.

Edible films appear to have tensile strengths and elongations between those of polyethylene and polystyrene (Table 3). Plasticizer level has a dramatic effect on properties, with tensile strength decreasing and elongation increasing with increased plasticizer content. Pouches are being produced successfully from HPMC; and casing and wraps are being produced from collagen. Edible films formed as coatings on food or drug products do not require the mechanical properties of stand alone films used as wraps, pouches, or casings.

Material and Application Trends

Much research in recent years has been devoted to quantification of the barrier and mechanical properties of edible films produced from both old and new materials. Although most effort has been devoted to moisture and oxygen permeabilities and to mechanical properties, increasing attention is being given to flavor and aroma and to lipid-barrier properties. Availability of such data should lead to more rational selection of edible-film systems for food and drug design. Besides the various waxes, acetylated monoglycerides, shellac, zein, collagen, gelatin, MC and HPMC in current use in food and drug products, wheat gluten, soy protein, casein, whey protein, triglycerides, fatty acids, alginate, carrageenan, and chitosan are among materials that have received increased attention as potentially useful materials in edible films. Of particular interest have been multicomponent film-forming formulations that combine the advantages of the individual components. For example, multicomponent composites have potential for combining the film-forming and oxygen-barrier properties of a polysaccharide or protein with the moisture-barrier properties of a lipid.

Considerable current research is also being devoted to improving the efficiencies of forming edible films as coatings on foods and to quantification of the effects of coatings on food moisture gain or loss, food oxidation, and fruit and vegetable respiration. Advantages could be derived from ability to extrude edible films as an alternative to solvent casting for some applications. In addition to the current uses mentioned earlier, edible films have been studied or proposed for applications ranging from prevention of moisture loss, browning and microbial growth with fresh-cut fruits and vegetables, to prevention of moisture migration and oxidation in frozen foods.

Continuing interest in increased food quality and decreased packaging, along with improved edible-film properties and economics, will likely result in increased use of edible films in the future.

BIBLIOGRAPHY

1. S. L. Kamper and O. Fennema, *J. Food Sci.* **49,** 1478–1481, 1485 (1984).
2. J. J. Kester and O. Fennema, *J. Food Sci.* **54**(6), 1383–1389 (1989).
3. R. D. Hagenmaier and P. E. Shaw, *J. Agric. Food Chem.* **38**(9), 1799–1803 (1990).
4. I. A. Wolff, H. A. Davis, J. E. Cluskey, L. J. Gundrum, and C. E. Rist, *I&EC* **43**(4), 915–919 (1951).
5. H. J. Park and M. S. Chinnan, *J. Food Eng.* **25,** 497–507 (1995).
6. T. H. McHugh, J. F. Aujard, and J. M. Krochta, *J. Food Sci.* **59**(2), 416–419, 423 (1994).
7. T. H. McHugh and J. M. Krochta, *J. Food Process. Preserv.* **18,** 173–188 (1994).
8. R. J. Avena-Bustillos and J. M. Krochta, *J. Food Sci.* **58**(4), 904–907 (1993).
9. R. D. Hagenmaier and P. E. Shaw, *J. Agric. Food Chem.* **39**(5), 825–29 (1991).
10. I. K. Greener, *Physical Properties of Edible Films and Their Components,* Ph.D. thesis, University of Wisconsin, Madison, WI, 1992.
11. S. A. Smith, "Polyethylene, Low Density" in M. Bakker, ed., *The Wiley Encyclopedia of Packaging Technology,* Wiley, New York, 1986.
12. *A Food Technologist's Guide to Methocel Premium Food Gums,* The Dow Chemical Co., Midland, MI, 1990.
13. E. R. Lieberman and S. G. Gilbert, *J. Polym. Sci.* **Symp. No. 41,** 33–43 (1973).
14. T. H. McHugh and J. M. Krochta, *J. Agric. Food Chem.* **42**(4), 841–845 (1994).
15. M. Salame, "Barrier Polymers" in M. Bakker, ed., *The Wiley Encyclopedia of Packaging Technology,* Wiley, New York, 1986.
16. L. L. Hood, "Collagen in Sausage Casings" in A. M. Pearson, T. R. Dutson, and A. J. Bailey, eds., *Advances in Meat Research,* Van Nostrand Reinhold, New York, 1987.
17. B. L. Butler and P. J. Vergano, *Degradation of Edible Film in Storage,* ASAE Paper No. 946551, ASAE, St. Joseph, MI, 1994.

18. A. Gennadios, C. L. Weller, and R. F. Testin, *Modification of Properties of Edible Wheat Gluten Films,* ASAE Paper No. 90-6504, ASAE, St. Joseph, MI, 1990.
19. J. H. Briston, "Films, Plastic" in M. Bakker, ed., *The Wiley Encyclopedia of Packaging Technology,* Wiley, New York, 1986.
20. J. S. Houston, "Polystyrene" in M. Bakker, ed., *The Wiley Encyclopedia of Packaging Technology,* Wiley, New York, 1986.
21. J. F. Hanlon, *Plastics* in *Handbook of Package Engineering,* Technomic Publishing, Lancaster, PA, 1992.

General References

K. R. Conca and T. C. S. Yang, "Edible Food Barrier Coatings" in *Activities Report and Minutes of Work Groups & Sub-Work Groups of the R&D Associates,* Research and Development Associates for Military Food and Packaging Systems, Inc., 1993.

R. Cook and M. Shulman, "Aqueous, Ultrapure Zein Lattices as Functional Ingredients and Coatings" in *Corn Utilization Conference V,* 1994. Bedford, MA.

R. Daniels, *Edible Coatings and Soluble Packaging,* Noyes Data Corp., Park Ridge, NJ, 1973.

F. Debeaufort and A. Voilley, *J. Agric. Food Chem.* **42,** 2871–2875 (1994).

A. Gennadios and C. L. Weller, *Food Technol.* **44**(10), 63–69 (1990).

A. Gennadios and C. L. Weller, *Cereal Foods World* **36**(12), 1004–1009 (1991).

S. Guilbert, "Technology and Application of Edible Protective Films" in M. Mathlouthi, ed., *Food Packaging and Preservation—Theory and Practice,* Elsevier Applied Science Publishers, New York, 1986.

Y. C. Hong, C. M. Koelsch, and T. P. Labuza, *J Food Process. Preserv.* **15,** 45–62 (1991).

J. J. Kester and O. R. Fennema, *J. Food Sci.* **40**(12), 47–59 (1986).

C. Koelsch, *Trends Food Sci. Technol.* **5,** 76–82 (1994).

J. M. Krochta, "Control of Mass Transfer in Foods with Edible Coatings and Films" in P. R. Singh and M. A. Wirakartakusumah, eds., *Advances in Food Engineering,* CRC Press, Boca Raton, FL, 1992.

J. M. Krochta, E. A. Baldwin, and M. Nisperos-Carriedo, eds., *Edible Coatings and Films to Improve Food Quality,* Technomic Publishing, Lancaster, PA, 1994.

J. M. Krochta, "Edible Protein Films and Coatings" in S. Damodaran and A. Paaraf, eds., *Food Proteins and their Applications in Foods,* Marcel Dekker, New York, in press.

T. R. Lindstrom, K. Morimoto, and C. J. Cante, "Edible Films and Coatings" in Y. H. Hui, ed., *Encyclopedia of Food Science and Technology,* Wiley, New York, 1992.

T. H. McHugh and J. M. Krochta, *Food Technol.* **48**(1), 97–103 (1994).

T. H. Shellhammer and J. M. Krochta, "Edible Coating and Film Barriers" in F. D. Gunstone and F. B. Padley, eds., *Lipids—Industrial Applications and Technology,* Marcel Dekker, New York, in press.

J. A. Torres, "Edible Films and Coatings from Proteins" in N. S. Hettiarachchy and G. R. Ziegler, eds., *Protein Functionality in Food Systems,* Marcel Dekker, New York, 1994.

P. D. Wood, "Technical and Pharmaceutical Uses of Gelatin" in A. G. Ward and A. Courts, eds., *The Science and Technology of Gelatin,* Academic Press, New York, 1977.

John M. Krochta
University of California
Davis, California

FILM, FLEXIBLE PVC

Flexible poly(vinyl chloride) (PVC) film is well established as a versatile, cost-effective packaging material. The ability of PVC to accept and respond to a range of additives makes it a unique and popular packaging material. A variety of production processes are used to manufacture PVC film for a variety of applications. Film can be produced by blown or cast extrusion, calendering or solution casting in thicknesses ranging from 0.0004 in. to >0.004 in. It can be produced as a highly flexible, permeable stretch film or an oriented shrink film.

PVC resins, the principal component of PVC films, are made by polymerizing vinyl chloride monomer using suspension, emulsion, or bulk polymerization. PVC resin is a free-flowing white powder. Of the approximately 11 billion lb of PVC resin produced in the United States in 1994, about half a billion pounds went into packaging applications (1). Flexible PVC packaging film applications are estimated to have consumed about 200×10^6 lb of PVC resin.

Composition

The ingredients used in manufacturing of flexible PVC films depend on the intended applications. The selection of the proper additive type and level are key parameters that influence the characteristics of the PVC packaging film. Food packaging, the single largest application for flexible PVC films, requires the use of ingredients sanctioned by the Food & Drug Administration (FDA). Non-food-packaging applications offer a broader range of additive selection.

Resin

Suspension polymerized PVC resin, having a K value of about 66–68, is typically used to produce flexible PVC packaging films. The resin is usually classified as film-grade, indicating good absorbtivity, a very low level of gels and contamination, and having very good heat stability.

Plasticizer

Plasticizers are the major additives, and their prime purpose is to impart flexibility (2). Essentially, all PVC packaging films contain one or more plasticizers. Food-packaging applications require the use of only FDA-sanctioned plasticizer. Because of the balance of plasticizing efficiency, cost, and availability, adipate plasticizers are most commonly used. For non-food-packaging applications, the plasticizer choice is vast, but the majority that are used are nontoxic and often FDA-sanctioned. Phthalate and adipate esters are dominant in such applications. Most flexible PVC film formulations also use an oil epoxide, typically epoxidized soybean oil. This additive, frequently called a *secondary plasticizer,* imparts flexibility as well as considerable heat stability during processing.

Heat Stabilizers

The main function of a heat stabilizer is to prevent discoloration during processing (3). Heat-stabilizer selection for flexible PVC film is a complex matter, dictated by the application, manufacturing process and interaction with other additives. For food-contact applications, stearates of calcium, magnesium, and zinc are most often used. They are relatively inefficient and are often boosted by oil epoxies and phosphites. For nonfood applications, organometallic salts of barium and zinc are popular and offer excellent heat stability.

Lubricants

Lubricants are added to PVC compounds to facilitate processing and control processing rate (4). The effectiveness of lubricants depends on their marginal or complete insolubility in PVC compounds. External lubricants reduce sticking of the melt to hot-metal surfaces; montan ester waxes, paraffins, and low-molecular-weight polyethylenes are commonly used. Internal lubricants affect the frictional properties of the resin particle surface during processing and thus control the fusion rate of the resin. Fatty acids, esters, and metallic soaps are frequently used.

Other Components

Most other components of a flexible PVC film are added to impart a specific physical property. Esters of multifunctional alcohols, with the proper hydrophobic–hydrophilic balance, are added at low levels to impart antifog and antistatic properties. Inorganic additives such as clay or talc, and organic additives such as amides can be added for slip and antiblock properties. Pigments may be added for tinting and coloring.

Methods of Manufacture

Essentially all flexible PVC packaging films are produced from externally plasticized resin, containing other additives, that is extruded through a film-forming die. Two basic methods are used: blown-film extrusion and slot-die-cast extrusion. The blown film process dominates PVC packaging film production.

PVC resin is blended with selected additives using a high-intensity mixer. Mixing time varies with the actual formulation used as well as the equipment. Mixing is complete when the compound (blend of resin and additives) is once again dry and free-flowing and the additives are completely dispersed. The compound is typically heated, through shear energy, to 230–260°F and then discharged to a ribbon blender to be further mixed and cooled to the desired temperature. This compound is often referred to as *dry blend.*

Dry blend may be pelletized for later use or fed directly to a blown-film extruder. Equipment needs to be expressly designed for the heat-sensitive PVC compound, requiring streamlined flow with no "dead spots." The film is wound using surface or center winders and sold in roll form, in a variety of widths. Flexible PVC film can also be preferentially or biaxially oriented to provide shrink characteristics. This is usually accomplished by using a two-roll orienting unit or bubble inflation under controlled temperatures and strain rates. Tentering equipment can also be used, usually with slot-die-cast extrusion equipment.

Markets

Food packaging. Most flexible PVC film is used for packaging food products, particularly fresh red meat. Properly formulated flexible PVC stretch films features high oxygen permeability, toughness, resilience and clarity, making it ideal for in-store wrapping of fresh meat. Packaging of fresh fruits and vegetables in flexible PVC film is primarily performed in-store. Films are formulated for high gas transmis-

Table 1. Typical Properties of 1-mil (25.4 μm) Flexible PVC Film

Property	ASTM Test Method	Flexible PVC Meat Package Stretch Film	Flexible PVC Dispenser Film	Flexible PVC Shrink Bundle Film
Specific gravity	D1505	1.23	1.27	1.3
Yield, in.^{2}/(lb·mil) [m^2/(kg·mm)]		22,400 [1254]	21,600 [1210]	21,400 [1198]
Haze, %	D1003	1.2	1.0	2.5
Tensile strength, psi (MPa)	D882 MD[a]	5000 (34.5)	5500 (37.9)	15,000 (103)
	TD[b]	4500 (31)	5500 (37.9)	5500 (37.9)
Elongation, %	D882 MD	275	300	90
	TD	375	325	275
Tear strength, gf/mil (N/mm)	D1922 MD	300 (116)	325 (125)	335 (129)
	TD	450 (175)	500 (193)	575 (222)
Change in linear dimensions at 212°F (100°C) for 30 min, %	D1204			
	MD	NA	NA	45
	TD			10
Service temperature °F (°C), range		−20–150 (−29–66)	0–150 (−18–66)	10–150 (−12–66)
Heat-seal temperature °F (°C), range		290–320 (143–160)	290–340 (143–171)	280–330 (138–166)
Oxygen permeability, (cm^3 · mil)/(100 in.^{2} · day · atm) [(cm^3 · μm)/(m^2 · day · kPa)] 73°F (23°C), 50% rh	D1434	860 [3342]	340 [1321]	NA
Water-vapor transmission rate, (g · mil)/100 in.^{2} · day) [(g · mm)/(m^2 · day)], 100°F (38°C), 90% rh		16 [6.3]	10 [3.9]	NA

[a] MD = machine direction.
[b] TD = transverse direction.

sion rates and provide excellent shelf life to the respiring produce, including sensitive mushrooms.

Packaging of chilled, tray-packed poultry parts is often done with flexible PVC film. The toughness and resilience of PVC film, plus its good heat sealability in a wet environment make it a material of choice in this centralized packaging application.

The ability to be readily printed with solvent- and water-based inks, without the need for any special surface treatment, have figured prominently in both meat and poultry packaging, for brand identification and merchandising. A recent development has been the imprinting of a safe handling label, mandated by USDA, on the film itself. This ensures 100% compliance at the store level on all packages of raw meat.

Institutional packaging is the second-largest food-wrap application for flexible PVC film. These films, in a thickness range of 0.4–0.6 mil (10–15 μm), are wound on rolls and placed in dispenser boxes with a cutter blade. They are used in institutional kitchens, cafeterias, restaurants and caterers to overwrap food trays, salads, glassware, and utensils. The film feature excellent cling, clarity, stretch, and dispensability.

Properly formulated flexible PVC films are also used for frozen-food storage, as well as in boxed beef operations and in-store bakeries and delicatessens.

Nonfood packaging. Oriented PVC films are widely used to overwrap boxed goods such as games and toys. One of the newer applications for oriented, flexible PVC is in the bundling of multiple aseptic packs. Ranging in thickness from 0.6 to 1.2 mils (15–30 μm), such films provide high shrink at low temperatures, resulting in a firm unitized pack, without any distortions, offering merchandising appeal. Heavy-gauge flexible PVC, 2–6 mils (51–152 μm), has been used as a component of windowed cards for many years. Toughness, formability, good dielectric sealability, and merchandisability are properties that lend themselves to the packaging of hardware, flashlights, and automobile parts.

Table 1 lists typical physical properties of the main types of flexible PVC packaging films.

BIBLIOGRAPHY

1. *Modern Plast.* p. 64 (Jan. 1994).
2. J. K. Sears and J. R. Darby, *The Technology of Plasticizers,* Wiley, New York 1982, pp. 35,305.
3. H. A. Sarvetnick, *Polyvinyl Chloride,* Van Nostrand Reinhold, New York, 1969, p. 88.
4. *Ibid.,* p 1214.

ANIL G. DOSHI
Borden, Inc., Resinite Division
North Andover, Massachusetts

FILM, FLUOROPOLYMER

Fluoropolymers are a family of materials that have a general paraffinic structure with some or all of the hydrogen atoms replaced by fluorine. All members of the fluoropolymer family are available in film form, but only Aclar film is used extensively in specialty packaging applications. Allied Corporation is a supplier of Aclar film. Some of its useful properties are extremely low transmission of moisture vapor and relatively low transmission of other gases, inertness to most chemicals, outstanding resistance to forces that cause weathering (uv radiation and ozone), transparency, and useful mechanical properties from cryogenic temperatures to temperatures as high as 300°F (150°C). The initial interest in Aclar film was for packaging military hardware, but its water-vapor-transmission properties led to interest by pharmaceutical companies for packaging moisture sensitive products (see Pharmaceutical packaging). Available in several grades, Aclar is more expensive than most thermoplastic films.

Composition and extrusion. Aclar is an Allied Corporation trademark for film made from Aclon, a modified PCTFE (polychlorotrifluoroethylene) fluoropolymer, that contains greater than 95 wt% chlorotrifluoroethylene (1,2). It is converted to film by melt extrusion (see Extrusion). It is a difficult thermoplastic to extrude because it has a very high melt viscosity and a low, critical, shear rate for melt fracture. The commercially available films have different machine-direction and cross-direction properties because some orientation is induced during fabrication. Aclar 22 films have good tear strength and can be used unsupported as well as in thermoformable laminates (see Laminating; Thermoforming). Aclar 33, made from a different copolymer, offers superior dimensional stability and resistance to chemicals and water vapor transmission.

Fabrication. Aclar film can be heat sealed (see Sealing, heat), laminated, printed, thermoformed, metallized, (see Metallizing), and sterilized (see Health-care packaging). The unsupported and laminated varieties can be handled and processed on most common converting and packaging machines. Unsupported and laminated films can be heat-sealed by machines that employ constant heat, thermal impulse, radiofrequency, or ultrasonic energy. In some cases, special precautions must be observed.

Most Aclar film for packaging is converted to some type of laminate. It can be laminated to paper, polyethylene (low and medium density), and to preprimed PVC, aluminum foil, polyester, nylon 6, and cellophane. A typical extrusion lamination (see Extrusion coating; Multilayer flexible packaging) uses molten LDPE as an adhesive between the fluoropolymer film and one of the substrates listed above. An MDPE tie layer produces a laminate with better properties at elevated temperatures. Laminates can also be produced by adhesive lamination. In that case the Aclar film should be preprimed. Preprimed film can be bonded to polymeric substrates with two-component urethane adhesives. The adhesive system is applied in water or organic solvent, and the liquid is evaporated in an oven. Because of its low surface energy, Aclar film does not have acceptable bond strength to most substrates unless the film is corona-treated (see Surface modification) to increase its surface energy. With corona treatment, the film can be laminated to aluminum and steel foils by using an epoxy-polyamide curing adhesive.

All grades and gauges can be thermoformed. Typical thermoforming temperatures are 350–400°F (175–205°C). It is

Table 1. Approximate Comparative Transmission Rates

Films	Moisture[a] Vapor, g · mil/(100 in.^{2} · d) [g · mm/(m^2 · d)]	O_2[b]	N_2[b]	CO_2[b]
		cm^3 · mil/(100 in.^{2} · d · atm) [cm^3 · μm/(m^2 · d · kPa)]		
Aclar 33C	0.025 [0.01]	7 [27.2]		16 [62.2]
Aclar 22C	0.045 [0.018]	15 [58.3]	2.5 [9.7]	40 [155]
Aclar 22A	0.041 [0.016]	12 [46.6]	2.5 [9.7]	30 [117]
PVDC	0.20 [0.78]	0.8–1.0 [3.1–3.9]	0.12–0.16 [0.47–0.62]	3.0–4.6 [11.7–17.9]
Polyethylene				
Low-density	1.0–1.5 [0.4–0.6]	300–700 [1166–2720]	130–260 [505–1010]	1400–2800 [5440–10,879]
Medium-density	0.4–1.0 [0.16–0.4]	170–500 [661–1943]	100–120 [389–466]	500–1500 [1943–5828]
High-density	0.3–0.7 [0.12–0.28]	34–250 [132–971]	40–55 [155–214]	250–720 [971–2798]
Nylon-6	19–20 [7.5–7.9]	2.6 [10.1]	0.9 [3.5]	9.7 [37.7]
Fluorinated ethylene polypropylene (FEP)	0.4–0.5 [0.16–0.20]	750–1000 [2914–3886]	300–400 [1166–1554]	1600–2000 [6217–7771]
Poly(vinyl fluoride) (PVF)	2.0–3.2 [0.79–1.3]	3.0–3.3 [11.7–12.8]	0.25–0.70 [1.0–2.7]	11–15 [42.7–58.3]
Polyester	1.0–3.0 [0.4–1.2]	4.0–8.0 [15.5–31]	0.7–1.0 [2.7–3.9]	12–25 [46.6–97.1]

[a] ASTME 96 at 100°F (38°C) and 90% rh.
[b] Dry gas at room temperature.

done close to, but not above, the melting point of the film. One of the important uses of Aclar film is the use of thermoformed laminated Aclar 22A/PVC in blister packs for ethical drugs. Heavier gauges (5–10 mil or 127–254 μm) of unsupported Aclar film can be thermoformed using ceramic or quartz heating units. Thermoformed film and laminates should be quick-quenched to maintain low crystallinity and prevent brittleness.

Most applications require transparency, but Aclar film can be metallized with aluminum. Corona-treated film can be printed with polyamide-based inks (see Inks). Best results are obtained if the film is corona-treated in-line prior to printing.

Properties. Aclar film can be sterilized with steam and with ethylene oxide (ETO) systems. It is being tested to determine if radiation sterilization of Aclar film will deteriorate its properties in the cobalt-60 dosage range of 2.5–5 × 10^6 rad (2.5–5 × 10^4 Gy). Tests at 1 × 10^6 rad (1 × 10^4 Gy) (3) indicate that little or no property loss occurs at that dosage (see Radiation, effects of). Aclar films have outstanding barrier properties to water vapor and to other gases (see Barrier polymers). Properties at a wide variety of temperatures can be determined by a method that is described in Ref. 4. Table 1 compares transmission rates of Aclar films with a number of other films.

Aclar film is inert to acids, bases, strong oxidizing agents, and most organic chemicals. It exhibits excellent dimensional stability in inorganics, including water, salt solutions, strong acids, and bases. Some polar organics, especially hot polar solvents, diffuse into Aclar film and act as a plasticizer. These solvents cause it to become more flexible and sometimes hazy. There are no known solvents that dissolve the film at temperatures up to 250° (120°C). Mechanical properties are usually not an important factor in packaging applications because the important mechanical properties are those of the laminates. Aclar films are not particularly strong or tough, but they have outstanding abrasion resistance, important in clean-room packaging of military hardware.

Applications. Aclar film is used in military, pharmaceutical, electrical/electronic, and aircraft/aerospace component applications. It serves as a key component of a transparent laminated barrier construction that meets the requirements of the MIL-F-22191, Type I specification (see Military packaging). This laminate is used for packaging moisture-sensitive military hardware. Aclar also passes the liquid oxygen (LOx) compatibility impact test under NASA specification MSFC-106A. It is used for packaging components designed for service in liquid oxygen and other oxidizers used in spacecraft applications.

The major commercial applications are in packaging moisture-sensitive drugs. Laminates are used for rigid and semirigid blister packs, unit-dose packages, and aseptic peel-packs. They are also used in medical applications; for example, as an overwrap for plastic containers for biomedical specimens and/or pathology specimens. This application requires a film that is sterilizable by heat and ETO and that does not absorb or denature biological fluids (5).

Safe handling considerations. Aclar film is inert and nontoxic and safe to handle at ordinary temperatures. It is thermally stable for up to one hour at temperatures as high as 446°F (230°C). It can be processed for a few seconds at temperatures as high as 554°F (290°C) provided the machine is ventilated with an exhaust fan. Vacuum forming is typically done at 374°F (190°C) with exhaust-fan ventilation. If Aclar film is continuously processed at temperatures above 446°F (230°C), the processing machinery should be equipped with ventilating equipment or the work should be performed in an exhaust hood. The film should not be disposed of by burning. Exposed to flame, it degrades to fluorochlorocarbon gases, some of which are toxic. In the presence of oxygen and olefins or polyolefins, it may form HF and/or HCl when exposed to flame. These acids are toxic if inhaled and corrosive to metals.

BIBLIOGRAPHY

"Film, Fluoropolymer" in *The Wiley Encyclopedia of Packaging Technology,* 1st ed., by A. B. Robertson and K. R. Habermann, Allied Fibers and Plastics, pp. 311–313.

1. J. P. Sibilia and A. R. Paterson, *J. Polym. Sci. C* **8**, 41–57 (1965).
2. A. C. West, "Polychlorotrifluoroethylene" in M. Grayson and D. Eckroth, eds., *Kirk-Othmer Encyclopedia of Chemical Technology*, 3rd ed., Vol. 11, Wiley, New York, 1980, pp. 49–54.
3. King, Broadway, and Pallinchak, *The Effect of Nuclear Radiation on Elastomeric and Plastic Components and Materials,"* Report 21, Addendum AD 264890, Radiation Effects Information Center (now defunct), Battelle Memorial Institute, National Technical Information Series of the Department of Commerce, Aug. 31, 1964.
4. N. Vanderkooi and M. Ridell, *Mater. Eng.* p. 58 (March 1977).

FILM, HIGH-DENSITY POLYETHYLENE

HDPE film has been used for a variety of specialty applications since the mid-1950s, but it was not until the 1970s that significant growth began. United States consumption in 1984 was about 1.8×10^5 t (1). More than half of that amount was for grocery sacks and merchandise bags and a few other nonpackaging applications. The other half was accounted for by a variety of packaging uses that require the unique property profile that HDPE films provide.

In 1990 the estimated sales of HDPE films for packaging applications were 3.45×10^5 t (2).

In packaging applications, HDPE films compete with LDPE and LDPE/LLDPE blends, cast PP (see also Film, nonoriented polypropylene), and oriented PP (see Film, oriented polypropylene). These polyolefin films have many properties in common, but with some significant differences in degree. Typical HDPE film properties are shown in Table 1.

HDPE, by virtue of its higher density, exhibits higher tensile strength, lower WVTR, and greater stiffness than LDPE or LLDPE/LDPE blends. PP films can offer higher tensile strength because of a higher degree of crystallinity and the ability to be biaxially oriented. PP films also have excellent optical properties due to fine crystal structure and biaxial orientation behavior. LDPE and LLDPE/LDPE blends exhibit good clarity due to their lower degree of crystallinity. HDPE is the most opaque of the three polyolefin types. This is an advantage if opacity is desired, because a relatively low amount of pigment is required to achieve that opacity (see Colorants). LLDPE and LLDPE/LDPE blends exhibit higher tendency to elongation under load because of their lower crystallinity and hence, lower tensile modulus. This property is useful in stretch-wrap applications (see Films, stretch) but it is a disadvantage if low elongation is required (as in handled grocery sacks). It is important to recognize, however, that the properties of all polyolefin films can be tailored during polymerization or through the use of blends or coextrusions.

Packaging applications. Two of the most important applications for HDPE film are retail grocery sacks and merchandise bags, which are not considered packaging applications in this *Encyclopedia.* Other important nonpackaging applications are trash-can liners and typewriter ribbons.

In packaging, HDPE films have replaced large quantities of glassine for packaging cereals, crackers, and snack foods. In these applications, clarity is not desirable, and the translucency or opacity of HDPE films is an advantage. They are used in conjunction with bag-in-box vertical form/fill/seal equipment (see Bag-in-box, dry product; Form/fill/seal, vertical). Rubber-modified HDPE film is used for medical overwraps. HDPE films are also used to produce shipping sacks. Heavy-gauge coextrusions of HDPE with low density polyethylene are used for all-plastic shipping sacks (see Bags, heavy-duty plastic), and thin-gauge HDPE films are used as moisture-barrier plies in multiwall bags (see Bags, multiwall). Very thin embossed HDPE film is also as a replacement for tissue paper, and without embossing it is used as an inner wrap for delicatessen products.

Film polymers. Three principal polymer properties can be adjusted to tailor the performance of HDPE films (see Polyethylene, high-density). Density plays the most significant role as a measure of the degree of polymer crystallinity. Crystallinity has a direct influence on tensile strength, WVTR, hardness, opacity, and coefficient of friction. HDPE film resins are 0.941–0.965 g/cm^3, depending on comonomer type and concentration. The most common comonomers are butene and hexene. As density increases, tensile modulus and tensile

Table 1. Typical Properties of High Density Polyethylene Film, 1-mil (25.4 μm)[a]

Property	ASTM Test Method		HDPE-HMW (blown)	HDPE-MMW (blown)
Specific gravity	D1505		0.950	0.950
Yield, in.^{2}/(lb·mil) [m^2/(kg·m)]			29,200 [1635]	29,200 [1635]
Haze, %	D1003		78	78
Tensile strength, psi (MPa)	D882	MD/XD, Yield	5200 (35.9)/5000 (34.5)	4600 (31.7)/4400 (30.3)
		MD/XD, Failure	7500 (51.7)/7000 (48.3)	6500 (44.8)/6100 (43.1)
Elongation, %	D882	MD	450	350
		XD	500	550
Tensile modulus, 1% secant, psi		MD	125,000 (862)	120,000 (828)
(MPa)		XD	130,000 (897)	125,000 (862)
Tear strength, gf/mil (N/mm)	D1922	propagating, MD/XD	20 (7.7)/150 (57.9)	15 (5.8)/100 (38.6)
Service temperature °F(°C), range			200–250 (93–121)	200–250 (93–121)
WVTR, g · mil/(100 in^2·day) [cm^3·mm/(m^2·day)]; 38°C, 98% rh			0.8 [0.3]	0.8 [0.3]
COF, face-to-face	D1894		0.3	0.3

[a] COF = coefficient of friction; MD, XD = machine, cross-direction.

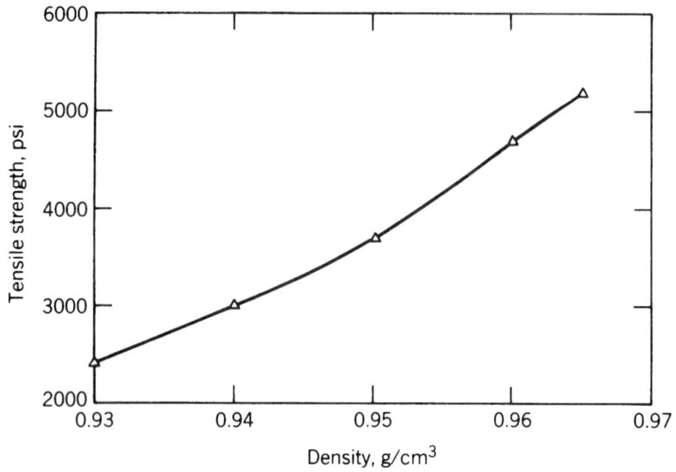

Figure 1. Influence of density on tensile strength. To convert psi to MPa, divide by 145.

strength increase as well (see Fig. 1). The same is true for film hardness, and to some extent for water-vapor transmission rate (see Fig. 2).

The molecular weight (MW) of the polymer has a significant influence on toughness and impact strength. For many years, HDPE films were made from medium-molecular-weight (MMW) or low-molecular-weight (LMW) polymers, for applications that required only medium-to-low toughness. The development of high-molecular-weight (HMW) polymers changed the range of applications available to HDPE films (3). The influence of MW on film impact strength is significant (impact resistance increases with molecular weight), but the processing of the resin is critical as well (4).

Molecular-weight distribution (MWD) is the third parameter influencing resin design, because it influences processing. Broad-MWD resins are easier to process than narrow-MWD resins. The ability to process HDPE at economic speeds and at controlled melt temperatures requires careful resin design. High flow LMW film resins can be narrow or medium in MWD and still be extruded at high output rates without excessive shear that would degrade the polymer. MMW resins require some careful control for high-speed extrusion (5). The use of HMW-HDPE film resins requires extreme MWD control to ensure both acceptable output rates and control of melt temperatures. Thermal degradation must be avoided to maintain the long molecular chains necessary to impart high impact strengths. This MWD has been achieved in some cases by bimodal polymerization technology. The polymer chemist has certain secondary design criteria available to fine-tune the HDPE film polymers. These include special process stabilizers, high-heat stabilizers, antistatic agents (see Additives, plastics), different comonomer types and the physical form in which the resins are supplied (ie, pellets or coarse powders).

Film manufacturing. HDPE films are produced using two principal extrusion techniques: blown and cast (see Extrusion). The most common is blown-film extrusion, in which $1\frac{1}{2}$–6-in. (3.8–15.2-cm) extruders extrude the resin in tubular form. The tube is cooled, collapsed to a flat film, and wound either in tubular form or slit and folded. The properties of blown HDPE film are highly dependent on the type of resins used and the design of the extrusion equipment employed (6). MMW-HDPE films, monolayer or coextruded, are generally made with extruders in the $2\frac{1}{2}$–6-in. (6.4–15.2-cm) diameter range operating at relatively low screw speeds yet achieving high output at moderate melt temperatures. The screw $L:D$ ratio is normally 25–30 and the barrel design is normally smooth. These conventional extruders do not allow full development of the resin capability in terms of optimum toughness, but they serve a useful role in providing high outputs at economic rates for applications which do not require the full strength potential of HDPE.

In contrast, newer designs for HMW–HDPE film extrusion are based on forced-feed grooved sections. They operate at high screw speed, using relatively short (18–21) $L:D$ ratios (7). These extruders achieve high output with the minimum residence time required to produce a homogeneous melt for the subsequent blowing and orientation processes necessary to develop the optimum toughness and tensile strength. For HMW film it is imperative to use a relatively long neck length (frost height) (6–10 × die diameter) and high blowup ratio

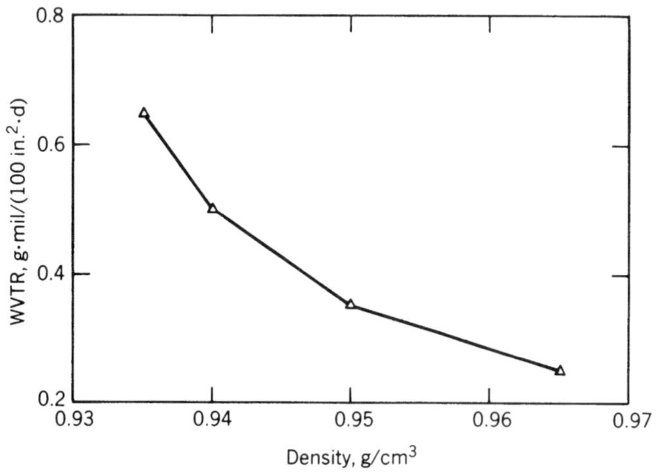

Figure 2. Influence of density on WVTR. To convert g·mil/(100 in.^{2}·d) to g·mm/(m^2·d), multiply by 0.3937.

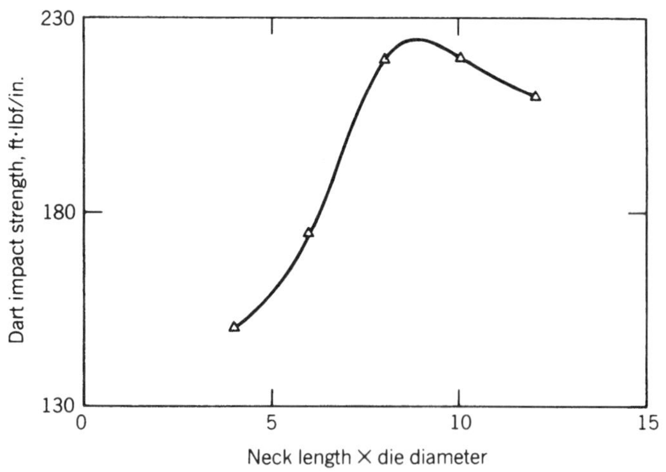

Figure 3. Effect of neck length on film toughness. To convert ft·lbf/in. to J/m, multiply by 53.38 (see ASTM D256).

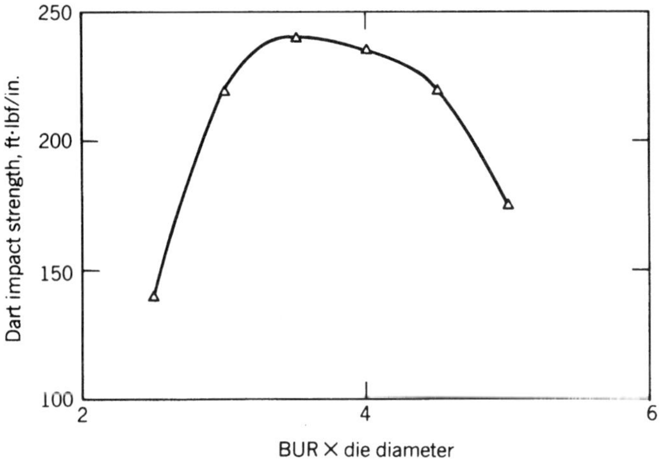

Figure 4. Effect of blowup ratio (BUR) on film toughness. To convert ft·lbf/in. to J/M, multiply by 53.38.

(3–5 × die diameter) to achieve balanced properties. The influence of these processing parameters on the optimum toughness of the film is illustrated in Figures 3 and 4.

LMW-HDPE films are generally made with cast-film technology, in which free-flowing polymer melts are extruded from flat dies on to substrates or chill rolls. These polymers do not have the melt strength necessary for the blown film process and their applications are limited.

BIBLIOGRAPHY

"Film, High Density Polyethylene," in *The Wiley Encyclopedia of Packaging Technology,* 1st ed., by R. H. Nurse and J. S. Siebenaller, American Hoechst Corporation, pp. 313–315.

1. J. S. Siebenaller and R. H. Nurse, "Packaging Developments Utilizing HMW-HDPE in the USA Market," presented at SPI/SPE, June 1984 Conference, Philadelphia.
2. K. J. MacKenzie, "Film and Sheeting Materials," in J. I. Kroschwitz, ed. *Kirk Othmer Encyclopedia of Chemical Technology,* 4th ed., Vol. 10. Wiley, New York, 1993, p. 784.
3. A. Brockschmidt, *Plast. Technol.* **30,** 64 (Feb. 1984).
4. H. R. Seintsch and H. E. Braselmann, "Further Evaluations of HMW-HDPE Film Processing and Effects on Film Properties," TAPPI Paper Synthetics Conference, Washington, DC, 1979.
5. H. E. Braselmann and R. E. Barry, "HMW-MMW HDPE Extrusion and Fabrication Techniques for Grocery Sacks and Merchandise Bags," TAPPI 1984 Polymers, Laminations and Coatings Conference, Boston.
6. H. R. Seintsch and H. E. Braselmann, *Paper, Film Foil Convertor,* **55,** 41 (March 1981).

General References

E. F. Giltenan, "The International Flap over Packaging," *Chem. Business,* **6**(10), 36–40 (Oct. 1984).

D. Sommer, "It's Plastics vs. Paper at Grocery Checkout Counters," *Sales Market. Manage.* **133**(5), 59 (Oct. 8, 1984).

R. R. MacBride, "HMW-HDPE Extruder Runs Ultra-Thin Film," *Modern Plast.* **57,** 12 (Sept. 1980).

FILM, NONORIENTED POLYPROPYLENE

Nonoriented polypropylene film is made primarily by the chill-roll cast process using multilayer coextrusion systems (Fig. 1). Other methods can also be used to generate nonoriented film such as tubular processes which are commercially available. Because nonoriented polypropylene film does not go through the orientation process (see Film, oriented polypropylene), the barrier properties and mechanical strength are not as well developed as in mono- or biaxially oriented polypropylene film. But the dollar investment to install a cast film line is much less than that for an orientation film line. Also, some applications, which do not require the improvement in barrier, low temperature durability, and mechanical strength, may be suitable for nonoriented polypropylene film. In the United States, nonoriented polypropylene films, which use homopolymer and copolymer resins, have a 20% market share in the polypropylene film market.

Nonoriented polypropylene films are being equally used in packaging and nonpackaging applications. In the packaging area, two-thirds of applications are food related. In the food packaging, cast polypropylene film accounts for more than 50% share of the candy twist wrap application, because cast polypropylene film is particularly suitable for newer high-speed twist wrap machines. In cheese wrapping applications, polypropylene/low-density polyethylene (PP/LDPE) coextruded film is used because of improved moisture barrier and film stiffness supplied by the cast polypropylene layer over single-layer LDPE. But cast polypropylene film in this market has ceased to grow and lost market share to LDPE because of the low cost of single-layer LDPE film. Nonoriented polypropylene film laminations with outer web for snack-food packaging represents a large potential market in North America.

Other applications include tapes, labels, diaper components, photograph holder, page protectors, medical packaging, and textile packaging. The advantages for using nonoriented polypropylene film in textile packaging are faster bag production rates and better contact clarity for PP over LDPE. The use of nonoriented polypropylene film for medical packaging is a potential growth area, due to PP's higher temperature resistance in steam sterilization. Nonoriented polypropylene film laminated with high-temperature-resistant outer web is particularly suitable for retort pouch applications due to the high-temperature resistance for nonoriented polypropylene film in retort process.

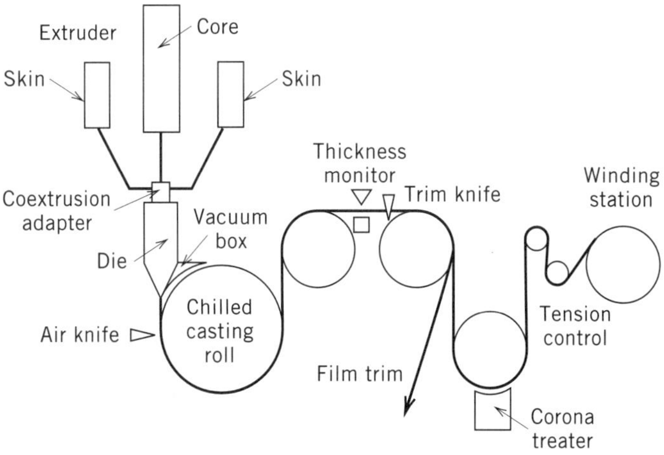

Figure 1. Three-layer cast-film line.

Table 1. Typical Properties for Nonoriented Polypropylene Film (1 mil)

Properties	Copolymer	Homopolymer	ASTM Test Method
Density, g/mL	0.89	0.90	D792
Haze, %	2	2	D1003
Gloss, %	86	86	D2457
Ultimate tensile strength, psi			
MD	8600	9200	D882
TD	5200	6300	
Elongation, %			
MD	500	600	D882
TD	650	650	
1% secant modulus, psi			
MD	70,000	100,000	D882
TD	70,000	100,000	
Heat-seal temperature range, °F	270–330	330–370	
Water-vapor transmission rate, g · mil/100 in^2 · 24 h 90% rh/100°F	0.8	0.7	Mocon
Oxygen permeability, C(mL · mil)/(100 in.^{2} · 24 h/atm 23°C, 0% rh	240–300	240–300	Mocon

Film Fabrication

Nonoriented polypropylene film is commonly made by chill-roll casting drum process. Typical extrusion melt temperatures are from 480 to 520°F with a screw $L:D$ ratio of 32:1 and compression ratio 2–4:1. Today many new products are made by the coextrusion process using melt adapters and/or multicavity flat-sheet dies. The extrudate emerging out of the flat die is immediately quenched on the chill roll with an air knife or vacuum box, forcing the extrudate into intimate contact with the chill roll (see Fig. 1). Increasing the chill-roll temperature, and/or increasing chill-roll takeoff speed, normally improves film stiffness but decreases optical clarity. After the quenching station, there is an edge trimming station to slit off the thickened edges. Then, a noncontacting, film-thickness monitoring device, with β-radiation source, is repetitively moved across the film on a track to scan and report the transverse-direction (TD) gauge profile. Automatic control systems then adjust the die opening to control film thickness to specified thickness tolerances. Between the thickness monitoring station and the film winding station, a corona treatment station is located to enhance film surface energy of the film. This is done to improve the wetability of the film surface for better ink printability or lamination adhesion. Tension control for nonoriented polypropylene film winding is very important due to low modulus of the film. With very high tension, the film tends to stretch as it is wound. When such a roll is unwound for use, it will not lay flat and is unsuitable for subsequent slitting, printing or lamination. Also, winder oscillation in the transverse direction is added to randomize the gauge profile to avoid the buildup of the gauge band in the same location. After winding up, the mill rolls normally need post-treatment (aging) to stabilize the film properties. A general aging condition is 40°C for 72 h. Film surface coefficient of friction (COF) can, and often must, be modified with antiblock particles and migratory slip additives, such as fatty-acid amides, to improve the film-handling characteristics on packaging machines. However, those additives reduce film clarity by increasing film internal and surface haze. The balance of film clarity and surface COF must be optimized to insure a balance between appearance and film machinability. Coextrusion technology is used to minimize the additives to the surface layers, maximizing the surface effects while minimizing the additive effect on clarity.

Film Properties

Nonoriented polypropylene films can be a monolayer or multilayer coextruded film to balance barrier, printing, heat seal, COF, and other properties. Typical properties for nonoriented polypropylene film are shown in Table 1. Oriented polypropylene film is superior to nonoriented polypropylene film in barrier properties and mechanical strength. However, nonoriented polypropylene film has excellent tear resistance and is cost effective. Each type of film has benefits and disadvantages, which must be weighed in determining which film is most appropriate ("fit for use") for a particular application.

BIBLIOGRAPHY

General References

K. R. Osborn and W. A. Jenkins, *Plastic Films,* Technomic Publishing, Lancaster, PA, 1992.

J. H. Briston and L. L. Katan, *Plastics Films,* Longman Scientific & Technical, London, 1989.

K. M. Finlayson, *Plastic Film Technology,* Technomic Publishing, Lancaster, PA, 1989.

Dan Kong
Eldridge M. Mount III
Mobil Chemical Company,
Macedon, New York

FILM, ORIENTED POLYESTER

Biaxially oriented polyester film is considered a high performance film. It is used in numerous applications, including graphic arts, information storage and display, labels, solar control, membrane touch switches, pressure-sensitive tapes, hot-stamping foils, photoresists, fiberglass-reinforced panels, postlamination, ID and smart cards, and last but not least, packaging.

World supply of PET film continues to expand. However,

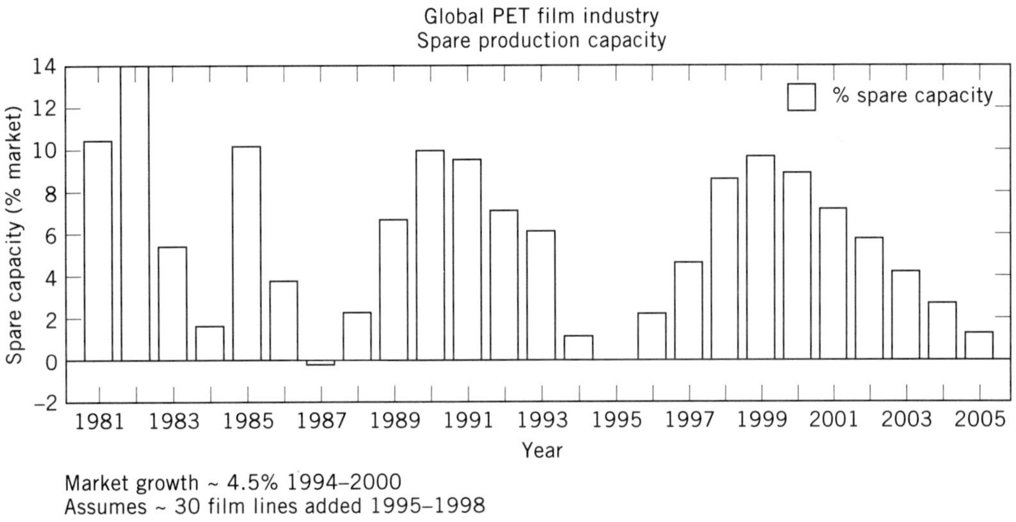

Figure 1. Global PET film supply–demand balance.

in 1995, the demand outstripped the capacity available (see Fig. 1). New capacity is scheduled to come on stream in 1995 (5 lines), 1996 (10 lines), 1997 (9 lines), and 1998 (9 lines). Most of those lines are scheduled to produce thin films, typically used for packaging. The consumption of PET film for packaging is expected to grow globally at 7% (see Fig. 2).

The domestic manufacturers of PET film for the merchant market are ICI Films (Melinex), DuPont Films (Mylar), Hoechst Diafoil (Hostaphan), Toray (Lumirror), Rhone-Poulenc (Terphane), Bemis Converting Films (Esterfane), and 3M (Scotchpar). Kodak manufactures PET film for internal use only.

Film Manufacturing Process

Polyester is produced by combining ethylene glycol and either terephthalic acid or dimethyl terephthalate in a condensation reaction. PET film can be either cast or oriented. The oriented variety can either be blown or tentered (stentered). Tenter-oriented PET film is the most versatile because the process imparts an improved combination of physical properties.

In the cast-tentered process, the molten resin is extruded onto a cooled casting drum (see Fig. 3). The film is oriented in the machine direction by heating, then stretching the film to three to four times its original length. The film is then heated again in a tenter (a large oven consisting of several heating zones) and drawn three to four times in the transverse direction (see Fig. 4). The film proceeds to the hottest zone of the oven, where the now biaxially oriented film is heat-set, or annealed.

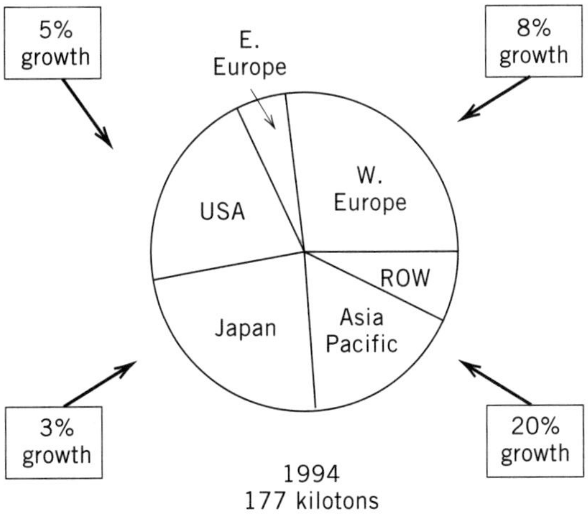

Figure 2. Global PETF packaging market.

Film processing conditions affect shrinkage, gloss, tensile strength, elongation, and other characteristics. The film can be treated or coated, within the manufacturing process or afterward, to change or enhance its properties.

Film Properties

PET has an outstanding combination of properties that make it valuable to the converting and packaging industries. Base PET film offers mechanical strength, dimensional stability, moisture resistance, chemical resistance, clarity, stiffness, and barrier properties. It handles well and can be printed or laminated (see Fig. 5). In response to customer needs, polyester film manufacturers have developed many technologies that change film properties to achieve useful effects (see Fig. 6).

Thermal. Polyester film gains excellent thermal properties through the heat-set process, allowing it to be processed and used in a wide range of temperatures. It tolerates temperatures ranging from −70 to 150°C for several hours or more, and can withstand even higher temperatures for shorter time periods, making it suitable for extended drying ovens, hot stamping, and packaging processes that form, fill, and seal.

Barrier. Flavor preservation is a major concern in the packaging industry. Barrier-type films constitute 50 percent of current U.S. PET film consumption. Polyester's barrier properties are second only to aluminum foil. Polyester film has demonstrated an excellent ability to retain flavor and exclude odor from outside the packaging (see Figs. 7 and 8). Metallizing or coating with PVC can provide greatly increased barrier properties.

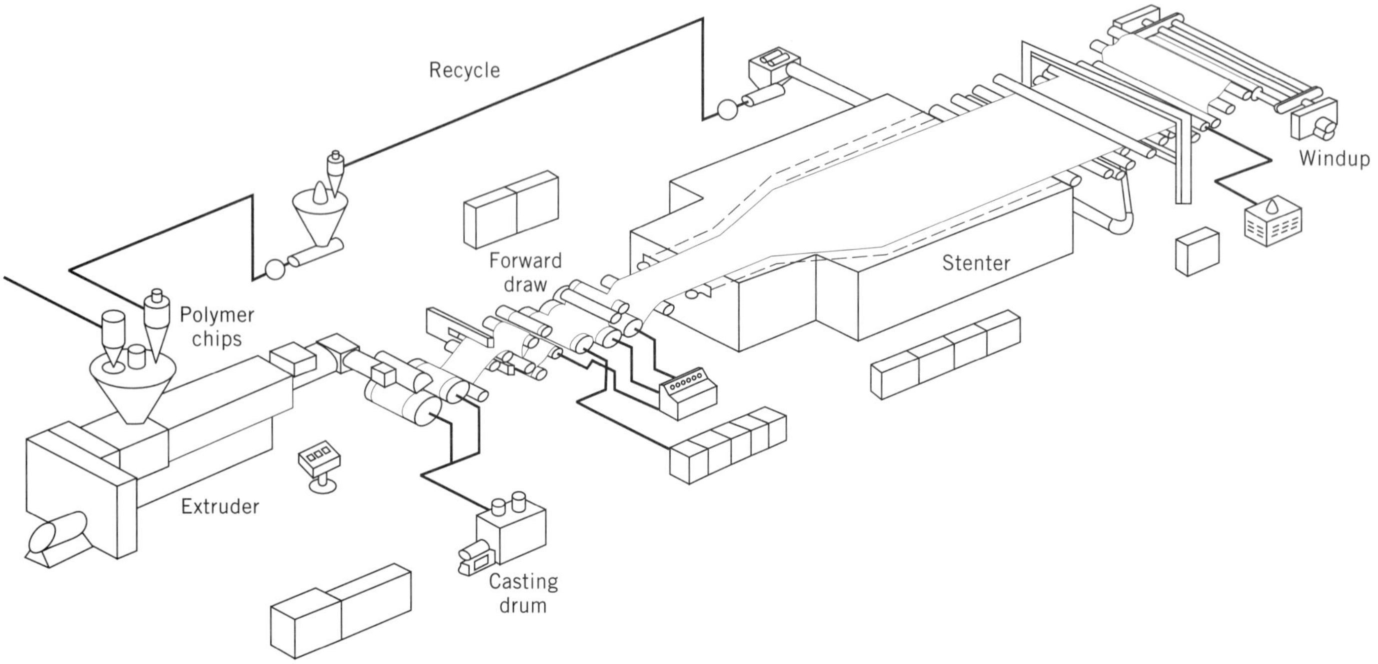

Figure 3. Film processing.

Metallizing. Metallized films provide value to food packagers concerned with package integrity and shelf life. Metallized materials are decorative, and also enhance properties such as moisture, light and gas barrier, flavor protection, protection against static electricity, and excellent machineability. Films other than polyester are now being metallized and used in packaging. However, polyester is still chosen when superior thermal stability and moisture and oxygen barrier are desired. Typical barriers for metallized polyester are 0.03–0.1 for water-vapor transmission and 0.02–0.1 for oxygen. Increasing the metal thickness further improves these barriers (see also Metallizing, vacuum).

PVC. PET can be coated off-line (after the manufacturing process) with polyvinylidene chloride. PVDC-coated PET is used in transparent packages to provide good barrier properties without metallizing. Typical coating thickness are 0.0001 in. (approximately 2 lb per ream). A major application for PVC-coated PET is in packaging processed meats such as hot dogs and luncheon meats (see also Vinyledene chloride polymers).

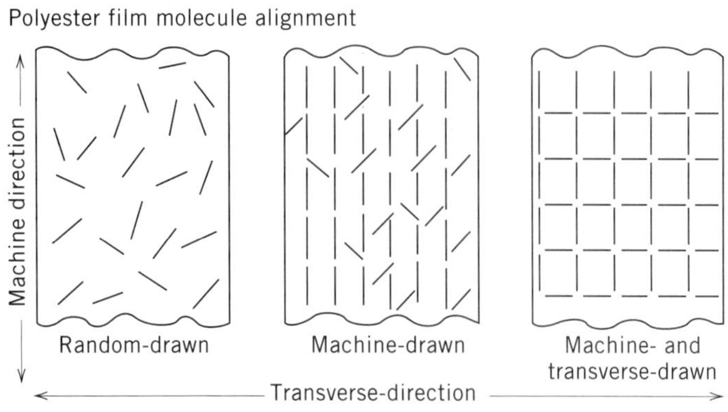

Figure 4. Polyester orientation.

Flavor scalping. PET film is an excellent flavor and odor barrier for packaging products that contain strong flavors, smells, and migratory agents. The sealant that is used in a package is important if flavor scalping (the loss of flavor from a product to the laminate or environment) is an issue. A coextruded, heat-sealable PET used as the sealant layer scalps significantly less flavor from products than a PE sealant or an ionomeric (surlyn) sealant layer does (see Fig. 9).

The sealant (in conjunction with the laminate chosen) can either adsorb the flavor or allow permeation through the sealant and perhaps the laminate. PET film as a barrier web and coextruded, heat-sealable PET as a sealant layer each provide superior flavor barriers in comparison with other packaging films and sealants.

Surface modifications of PET films

Corona treating. Corona treatments use an electrical charge to raise the surface tension of film and enhance its wetout properties. Corona treatment more closely matches the energy of the coating surface to the energy of the substrate surface. This promotes the polyester film's adhesion to various materials used in different phases of conversion. Corona treating has been an accepted method for over 30 years. Its use is limited, however, because the treatment level dissipates with time.

Pretreatments. Polyester can be pretreated in-line for a variety of surface properties, making the coating an integral part of the film. With pretreatment, the coating is so thin that it is virtually invisible. The specific coating to be used depends on customer needs and processes, including the inks or solvents to be used, the processing equipment, and the demand of the end use. Pretreatments can improve handling by

	Property		Gauge	Typical values	Units	Test methods
Physical	Nominal yield		48	42,200	in.^{2}/lb	
	Tensile strength	MD	48	32,000	psi	ASTM DB82A
		TD		39,000		
	Elongation	MD	48	110	%	ASTM D882A
	at break	TD		70		
	Coefficient of friction	Dynamic	48	0.4		ASTM D1894
	(in/out)	Static		0.5		
	Density		48	1.40	g/cc	
Optical	Total luminous transmission		48	88.5	%	ASTM D1003
	Gardner haze		48	3.6	%	ASTM D1003
Thermal	Typical shrinkage	MD	48	3.6	%	Unrestrained at
		TD		4.0		374°F for 5 min
Barrier Unmetallized	MVTR (24 h at 100°F and 90% rh)		48	2.8	g/100 in.^{2}	ASTM F372
	Gas permeability (24 h at 77°F and 75% rh at 1 atm)		48	O_2 N_2 CO_2 6.0 1.6 31.0	cc/100 in.^{2}	ASTM D1434
Metallized	MVTR (24 h at 100°F and 90%rh		48	0.05	g/100 in.^{2}	ASTM F372

Figure 5. Typical properties of polyester film.

affecting slip characteristics on one or both sides. Some pretreats create a film surface that is receptive to a broad range of coatings and inks. Antistat pretreatments reduce static, which can cause handling problems and compromise worker safety during processing. Other pretreats produce superior metal adhesion for certain end-use applications.

Technology	Film properties affected
Manufacturing process	Tensile strength Flatness Shrinkage Abrasion resistance Roll formation
Polmer modification (eg, additives or alloying)	Gloss Clarity Surface roughness Slip Abrasion resistance Flexcrack Adhesion Heat seal Barrier
Surface modifications (eg, Treating or coating)	Adhesion Slip Barrier Heat seal
Coextrusion	Adhesion Slip Heat seal Barrier Surface roughness Gloss

Figure 6. Technologies available that affect film properties.

Coextrusion. Coextrusion is another method for changing film properties. This process involves casting together two or more layers of polyester film to produce desired surface properties (see Fig. 10). For instance, one layer of clear film can be coextruded with a layer of matte film, or a film that handles well can be coextruded with a layer of heat-sealing polyester, or resins of different colors can be coextruded. The first commercially available coextruded PET film imparted gloss and resisted glass fiber bloom when used as a construction material for pool enclosures and glazing. Today, one of the major applications for coextruded PET is metallized snack-food laminations, where it provides an absence of microcracking, a broad seal range, temperature resistance, excellent barrier properties, and improved handling.

Other surface modifications. Techniques for producing high-barrier transparent coatings have been available for many years. However, they were not economically viable. Sputtering techniques using indium tin oxide and aluminum oxide have produced films with oxygen transmission rates equal to those of metallizing. More recently, plasma treatments using silicon oxide products have also yielded barrier properties equal to metallized polyester (see Figs. 7 and 8). These products still cannot be produced today at the costs associated with metallizing. However, companies are striving to narrow the gap. Techniques for producing better high-barrier transparent polyester structures at economical prices are expected to be available in the not-too-distant future. Methods are also being developed for producing nonmetallic high-barrier substrates (see also Surface treatments).

Types and Applications

Flexible packaging is a large and ever-changing source of demand for oriented PET film. Applications continue to grow as

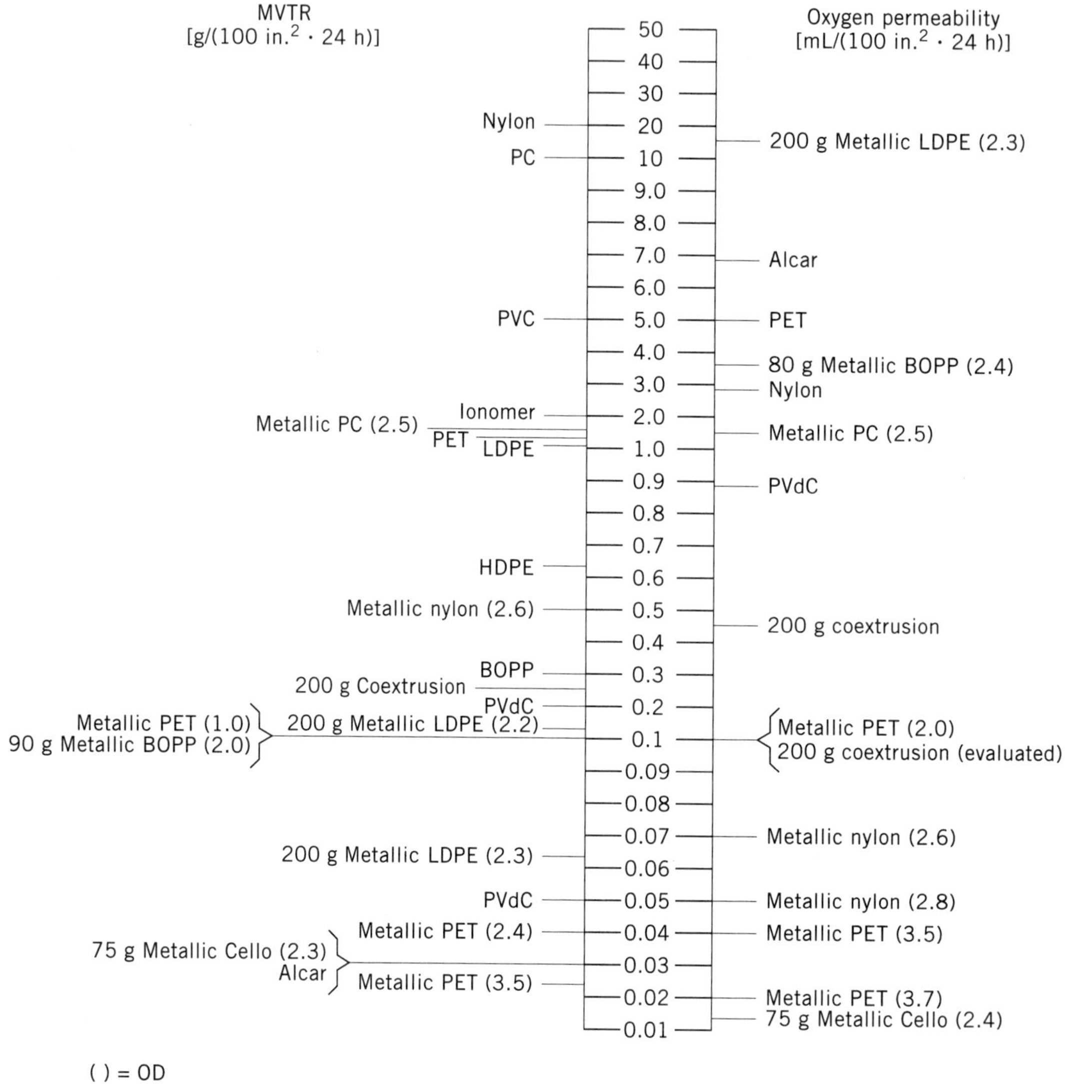

Figure 7. Barrier comparison. (*Base substrates 1 mil/metallic substrates 48 g unless noted otherwise.)

manufacturers change processes and properties to meet customer needs. In the past, polyester was considered the premier packaging material because of its unique characteristics. However, many companies continued to look for more economical ways of packaging. In many cases, they were able to maintain product integrity while using a less expensive package.

In this day of downsizing and consolidation, however, companies are investing in equipment that allows them to run faster and use fewer people, and polyester is regaining its cost-competitiveness. To run equipment at full potential, polyester is preferred by converters because of its excellent ability to maintain registration (temperature stability) and because of its outstanding uniformity across the web and from point to point. The companies that use these converted products are also looking to reduce their costs by running faster. At one time candy bars were wrapped at a rate of 100 per minute; today they exceed 400 per minute. Companies using polyester for packaging can obtain these higher speeds without distortion.

Coffee. Coffee has long been packaged in metallized polyester structures for the fractional pack market. Typically, it has been used for restaurants and offices. More recently, PET in conjunction with aluminum foil and metallizing has taken a substantial portion of the coffee-packaging market previously held by metal cans. The brick-pack concept offers better economics and reduced shelf space.

Boil-in-bag. Boil-in-bag applications typically use a 0.5-mil PET film with 2-mil medium-density PE adhesive lamination. A two-part adhesive system is used. The PET film is necessary for good dimensional stability in fabricating and sealing the pouch. It also provides heat-dimensional stability during the boiling operation.

Bag-in-box. Typical structures are polyethylene/metallized PET/polyethylene. Metallized PET is used because of good oxygen- and moisture-barrier properties and resistance to flavor scalping. Wine was the first application for this type of structure. Today it is also used for bulk packaging of condiments

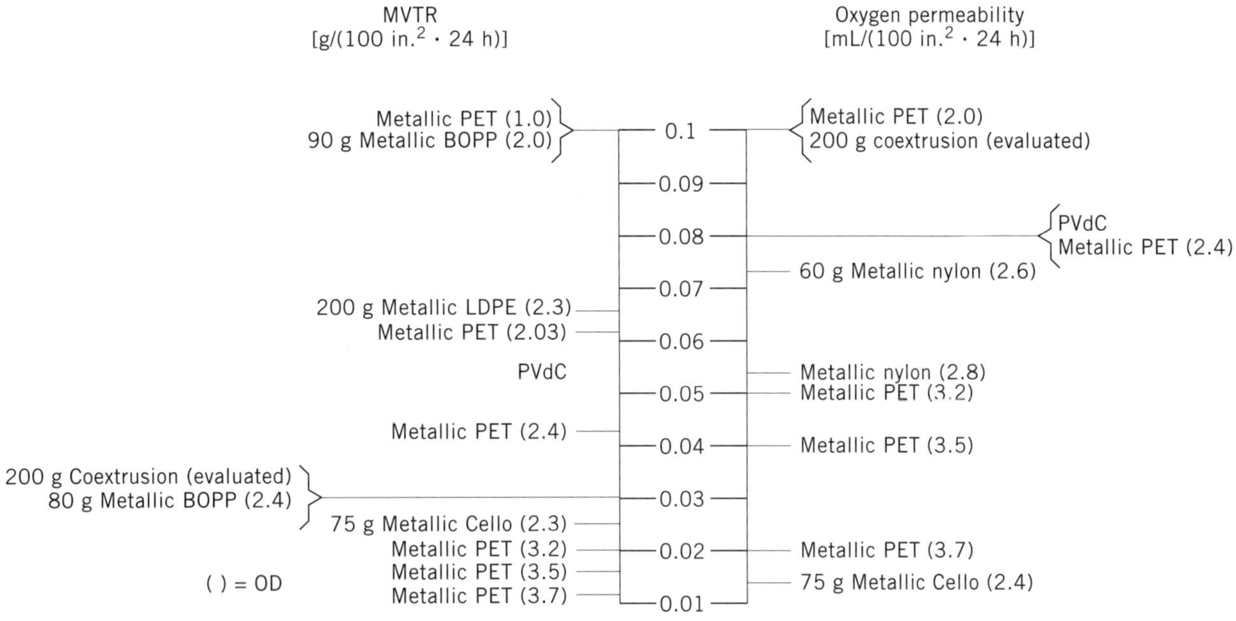

Figure 8. Barrier comparison. (*All substrates 48 g unless noted otherwise.)

such as ketchup and mustard, and for foods such as tomato sauce (see Bag-in-box-liquid product).

Retort. Retort packaging applications are commonly used in Europe, but have not been a large-volume application in the United States. The term *retort* refers to the process of sterilizing (cooking) a product after it is already packaged. A typical structure contains PET/aluminum foil/high-temperature adhesive/cast polypropylene (as a sealant layer). This type of packaged provides flavor that is closer to fresh vegetables in comparison with canned vegetables.

Dual ovenable lidding. PET is typically used as a lid to cover frozen single-serve dinners packaged in CPET trays. PET film can withstand the temperature differential of the freezer-to-oven transition, for both convection and microwave ovens. This market has a unique requirement—the film must seal to the tray, but must be peelable after heating.

Standup pouches. While standup pouches have enjoyed a great deal of success in Europe in the last decade, the conversion from rigid containers to standup pouches in the United States is just beginning. Although the new packaging is touted for source reduction, the conversion will be driven by economics. The cost of producing, transporting, and storing unfilled flat pouches is less than that of rigid containers. Also, the pouches easily accept high-end graphics to create bolder, brighter packages on the shelf. These structures are typically thicker laminates (5–7 mil) to provide the stiffness necessary to stand on the shelf. PET, when reverse-printed on the out-

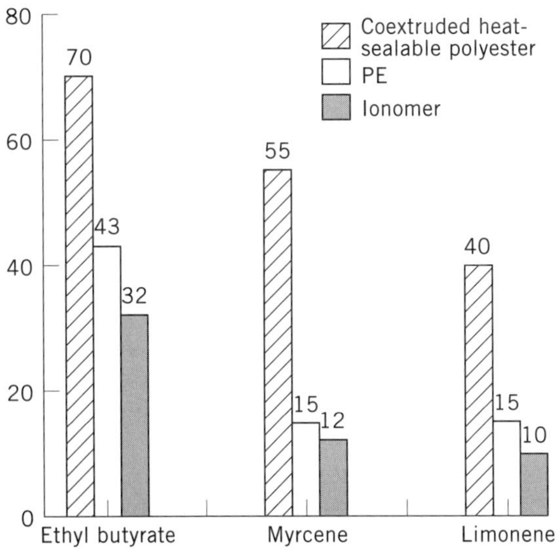

Figure 9. Percent of flavor compound retained. This graph indicates the percentage of flavor compound retained in orange-flavored Kool-Aid powder. Ethyl butyrate, myrcene, and limonene are the three key flavor components in the orange-flavored powder drink mix. The laminate structures were paper/poly/foil/PE, paper/poly/foil/ionomer (surlyn), and paper/poly/metallized heat-sealable PET.

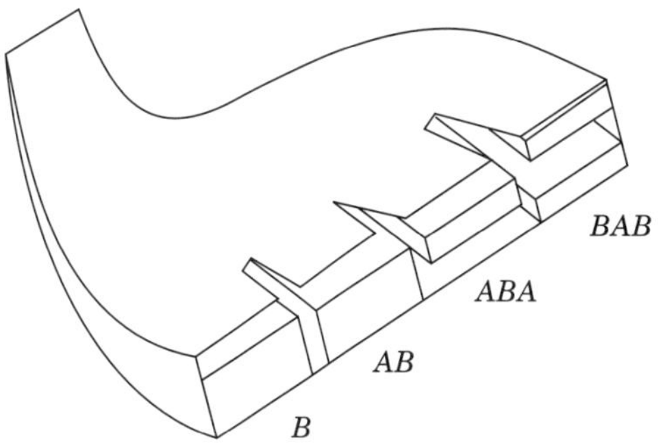

Figure 10. Coextrusion.

side of these packages, provides the thermal stability necessary to heat-seal through the thickness of the laminates (see Standup pouches).

Medical applications. Oriented polyester film is ideal for packaging of medical and surgical devices because of its clarity (visibility) and temperature stability during sterilization. Untreated polyester has no migratory products. The PET film is combined with special paper and spun-bonded polyolefin, which allow gases to permeate both ways and restrict unwanted bacteria. Ethylene oxide is the preferred sterilization method today. Other techniques include steam and gamma radiation. The properties of polyester film are not affected at the highest levels of γ radiation.

Shrinkable PET. Historically, both tamper-evident sleeves and labels have been made of vinyl. Over the last few years, Europe has begun to insist on products other than vinyl because of environmental concerns. Now being offered in Europe is a polyester that provides 50% shrink in the transverse direction and 5% in the machine direction. This trend is expected to spread to other parts of the world in the near future.

Hot-stamping foils. The preferred substrate for hot-stamping foil has, for many years, been polyester film. The typical thickness is 48 gauge. Most of the film used is heat-stabilized to allow for better quality in processing and reproduction. In the normal hot-stamping process, heat and pressure are applied to produce a decorative image. Metallic hot-stamping foils are used to decorate boxes and bottles for the cosmetics and personal-care industries. A pigmented stamping foil is used to code or date products.

Holograms. In the past, holograms were used primarily as security devices on credit cards. Polyester was initially chosen as a substrate for use in this application because of its high-temperature stability and dimensional stability, and its ability to be produced in thin gauges (48 gauge). The ability to use continuous processes with wider film widths has led packagers to begin incorporating holograms into packaging. New polyester films are being developed that allow holograms to be produced at faster speeds without coating. These structures will be available in the near future with a heat-sealable surface. Using this process, the production of packages with holograms should be increasingly feasible (see also Holography).

Labels. The label industry uses many substrates, including paper, PVC, styrene, polypropylene, and polyester. Polyester plays many roles in this industry. It is sometimes used as a base stock that receives printing or bar codes. It can also be used as a protective cover lay or as a release liner. It is available either in transparent or white. The ease with which polyester can be metallized makes it suitable for high-quality decorative applications such as personal-care products, cosmetics, and pharmaceuticals.

Environmental Benefits of PET

PET film has a number of opportunities to stand out against other flexible-packaging films as more and more environmental pressure is placed on the packaging industry. Polyester is a good environmental choice because it is inherently more recyclable than other plastics and because of its capacity for including recycled content.

The biggest impact on the environment comes from reducing the amount of overall material used in packaging. Source reduction is driven by economic forces as well. Rigid containers (glass, metal cans, rigid plastics, etc) are being converted to flexible standup pouches where possible. Flexible standup pouches can reduce the amount of packaging material by 70–80% when compared to a rigid container. Empty pouches occupy 70–90% less landfill volume. Rigid containers (plastic, cardboard, etc) are also being source-reduced to the minimum technical requirements for their particular product.

PET film can also accommodate postconsumer recycled (PCR) feedstock within its manufacturing process. The feedstock can be mechanically recycled, which requires a relatively pure PET feedstock. Mechanically recycled PET soda bottles can create a feedstock pure enough for FDA compliance. The feedstock can also be chemically recycled by either methanolysis or glycolysis process, and remain in FDA compliance. PCR feedstock is more expensive than virgin PET resin because of the additional processing involved (see also Recycling).

Future Trends

A few of the applications listed above, such as standup pouches, holograms for packaging, shrinkable PET, and clear barrier films (as described under "Surface of PET Films" modifications), are in their infancy in the United States. Their potential has yet to be realized. Listed below are several other trends the industry sees for the future.

Metal lamination. A new and exciting application for heat-sealable polyester is its use on the inside and outside of metal-formed cans. Polyester eliminates the need for coatings on the inside of the cans and enhances graphics and eliminates rust on the outside of steel cans. The product (coextruded or amorphous) allows excellent bonding to the steel or aluminum prior to drawing of the two-piece cans. The opportunities in this market throughout the world include

Japan. The target market is beverage cans for noncarbonated beverages such as tea and coffee and for carbonated soft drinks. The 1995 capacity was 10,000 tons of laminate per month.

Europe. The target markets are baby-food cans, pet-food cans, and aerosol-can ends.

United States. The market is under development.

Other rigid containers. There are numerous other rigid containers that will contain polyester in the future. Since polyester does not absorb flavors, it would be an ideal product to use on the inside of containers for various liquid and dry products. If these containers require heat to seal them, polyester would be desirable on the exterior, since it will not deform with heat.

PEN films. Opportunities in the packaging market are being created by the introduction of PEN film and resin for bottles. PEN (polyethylene naphthalate) is a member of the polyester family with a higher glass-transition temperature

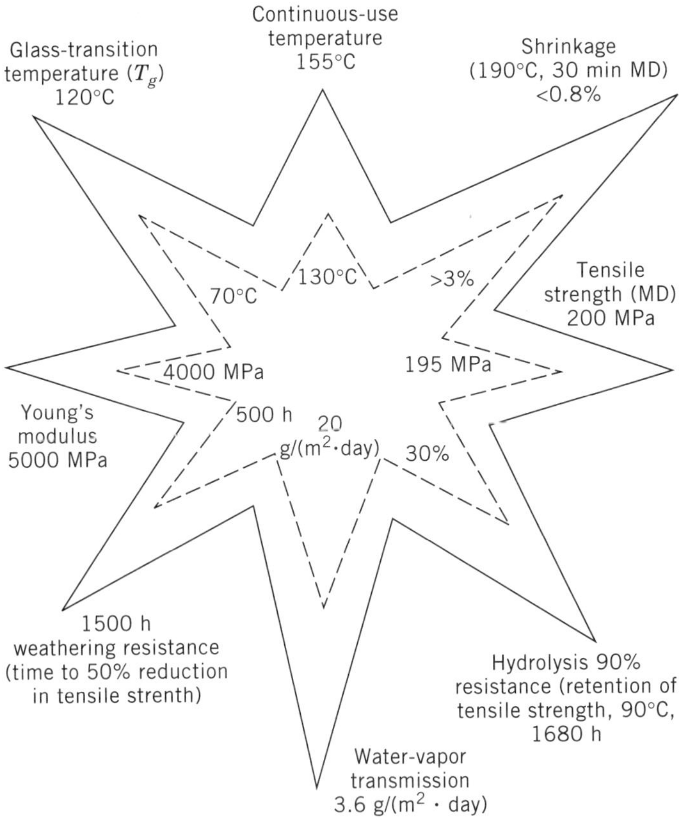

Figure 11. Pen films.

that delivers improved performance characteristics. Among the most important is PEN's improved barrier. Compared to PET, PEN provides approximately five times the barrier for carbon dioxide, oxygen, or water-vapor transmission (see Fig. 11). PEN also provides superior performance at higher temperatures than PET. When used as a resin for rigid-container packaging, PEN allows them to be hot-fillable and both returnable and rewashable. As a film, PEN provides 25% greater stiffness, allowing thinner designs and greater barriers simultaneously.

Historically, the price of PEN film and resin have been 10 times that of PET. However, with the emergence of a commercially available supply of NDC monomer and the launch of major applications that specify PEN, the costs of PEN film and PEN resin are falling to about three to four times that of PET.

John Newton
ICI
Wilmington, Delaware

FILM, ORIENTED POLYPROPYLENE

The first oriented polypropylene (OPP) film was made commercially by Montecatini of Italy in the late 1950s. Shortly after, Imperial Chemical Industries (ICI) of the United Kingdom began OPP film production, Kordite (the predecessor of Mobil's Commercial Films Division), Hercules (now AET), Olin, and DuPont started production in the early 1960s. In 1995, about 4×10^9 lb (1.8×10^6 tons) of effective annual capacity was in place worldwide. OPP film shipments are estimated at 3.8×10^9 lb (1.7×10^6 ton). Consumption of this high-performance thermoplastic film is expected to continue to grow 6–8% per year through 2005. OPP continues to find new markets and applications and is a real workhorse for many products as it displaces PET and various paper grades.

Oriented polypropylene film is made by approximately 100 companies around the world. The two largest producers in North America are

- AET (Applied Extrusion Technologies, Inc.) with plants in the United States, Canada, United Kingdom, and Brazil
- Mobil Chemical Co., Films Division, with plants in the United States, Canada, Belgium, Netherlands, Italy, and a joint venture in Thailand.

Other North American manufacturers include InterPlast-Amtopp (Formosa Plastics), Simpro, Toray America, AEP Industries (Borden), QPF (Quantum Performance Films), BCF (Bemis-Curwood Films), Central Products Co., 3M, Intertape Polymer, General Electric, Steiner Films, Vifan Canada, Bi-Ax International, AltoPro, MasterPak Inc., and Celanese-Hoechst.

The two largest manufacturers in South America are Votorantin in Brazil and Biofilm in Colombia with more than ten other producers.

In Europe, Mobil Plastics Ltd. was the largest producer until Hoechst-Kalle and Courtaulds merged in 1995. While these two companies are a close number 1 or 2 in Europe, Mobil is still the largest producer of OPP worldwide. Other producers in Europe are Moplefan, Vifan, Manuli, Wolf-Walsrode, Polinas, Bimo, Radici, UCB, ICI, Derprosa, Stirosir, and about 12 other small producers spread from Finland to the old Eastern Block countries to Greece to Spain.

The Japanese suppliers include Futamura-Shansho, Honshu Seishi, Toray, Toybo, Tocello, Gunze, Daicel, Tokayama, Toyo Rayon, and Showa Poly.

Several other producers are in Taiwan, South Korea, China, Indonesia, Malaysia, Turkey, India, Egypt, Israel, Australia, and South Africa.

Raw Materials

Four resin categories are used for OPP films: homopolymer, copolymer, terpolymer, and modified resins (see Polypropylene). Homopolymer resins are made by polymerizing propylene monomer with a variety of catalysts. Copolymers incorporate 0.1–20% ethylene as a comonomer with propylene. Copolymers can be either random or block copolymers. The random copolymers at less than 10% ethylene are generally used as heat-sealant skins in a coextrusion. The block copolymers are not heat-sealant resins and are rarely used. How much comonomer is polymerized into the resin depends on the resin's end use. Terpolymer production uses butene along with the ethylene and propylene feeds. Terpolymers are also used in coextrusion as a sealant skin resin. Polypropylene homopolymer resins are also modified with terpenes to give them heat-sealing properties (see Sealing, heat). Modified resins are also used to improve the moisture-barrier properties of OPP.

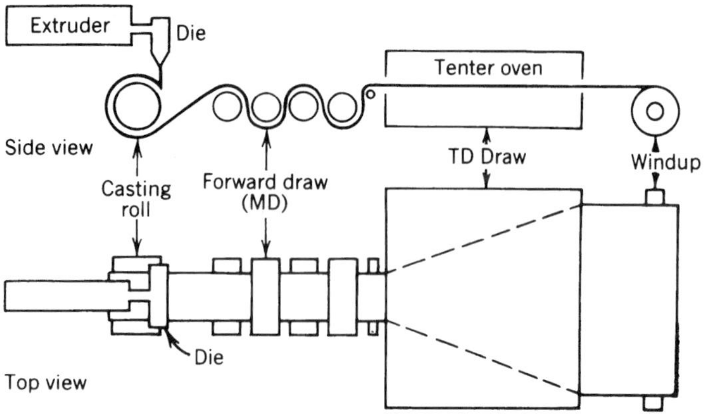

Figure 1. Tenter-frame process.

Manufacturing Process

When a film is biaxially oriented, it is mechanically stretched in perpendicular directions. This aligns the film's molecules in the machine direction (MD) and in the transverse direction (TD). Once biaxial orientation has taken place, the film exhibits a marked improvement in optics, strength, moisture barrier, and low-temperature durability over cast polypropylene film (see Film, nonoriented polypropylene). The increase in strength allows OPP film to be made as thin as 0.45 mil (11 μm) and still function as a laminating substrate for packaging applications (see Laminating; Multilayer flexible packaging). Films as thin as 0.25 mil (6 μm) are routinely used for capacitors and under special conditions films as thin as 0.04 mil (1 μm) can be made. Thicker films [0.60–1.3 mil (15–32 μm) are readily available. Cavitated films, which are opaque, are available up to 3 mils (75 μm) thick. Thicker films up to $\leq$10 mils (250 μm)] can be produced, but they currently do not have any large commerical applications.

Oriented polypropylene film is made by three methods: tenter frame (Fig. 1), blown tubular (Fig. 2), and simultaneous processes. The tenter-frame process is a sequential MD–TD process that uses machine-direction roll-speed differentials to impart the machine-direction orientation and a tenter frame with clips on a chain that diverges to impart the transverse-direction orientation. This MD–TD process generally produces an unbalanced orientation. The blown tubular process casts an annular tube that is reheated and blown in a quasi-simultaneous process that starts TD and finishes MD. Balanced orientation, ie, when the tensile properties are approximately equal to each other (within 3000 psi/20 MPa), is easier to obtain in the blown tubular process. The simultaneous process requires a complex mechanism to impart the simultaneous MD–TD orientation inside an oven. The properties of a true simultaneous orientation are not widely known as this process is expensive to operate and its advantages, if they exist, have not been recognized by the industry. The OPP industry gravitated to the tenter-frame process as the process of choice because this process has better economies of scale and balanced tensile property is not a characteristic that determines fitness for use.

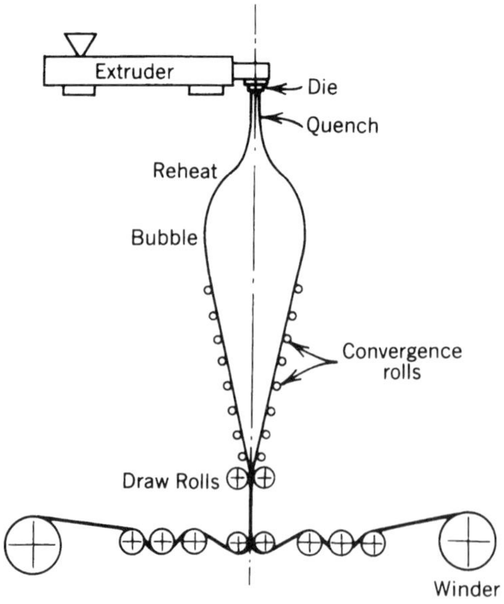

Figure 2. Bubble process.

In the 1970s the blown-film process favored thin film products and the tenter-frame process favored thick-film products. The ranges were 0.5 to <1.0 mil for blown film and 0.65–1.3 mils for tentered film. Today both processes offer packaging films over the range of 0.45–3 mils. Tenter frames also make capacitor films in the <0.3-mil range. The tenter frame has an inherent advantage when producing heavy-gauge films.

The explanation of what happens to polypropylene as it is being oriented requires an understanding of polypropylene's morphology. This morphological understanding allows one to understand what happens to the polymer during the orientation process.

Polypropylene Morphology

In understanding PP morphology, we are concerned with the following areas: chain conformation, helix formation, lamellae, and spherulites.

Chain conformation. During polymerization, PP can take on an isotactic, atactic, or syndiotactic conformation. Tacticity defines the stearic placement of the methyl side groups about the main carbon–carbon polymer bonds. Isotactic chains have identical tetrahedral bond configurations. Atactic chains have a random bond configuration. Syndiotactic chains have an alternating tetrahedral bond configuration. These configurations are shown in Figure 3.

Isotactic PP is the main component (~90–98%) of standard PP resins. The balance is atactic PP. Syndiotactic PP is now commercially available and has not yet found a use in the manufacture of OPP.

Helix formation. Isotactic PP tetrahedral bond configuration results in a stearic hindrance that causes it to form a helix on crystallization. This helix is designated as a 2*3/1 helix. This means that two atoms are along the chain axis for a repeating unit and that it takes three monomer units for a repeating structure. One turn about the helix axis completes the repeating helix structure, as shown in Figure 4. Polyethylene does not have the stearic hindrance and forms a 1*2/1 repeating unit, which is a planar zigzag.

Lamellae. As PP helixes solidify, they crystallize to form lamellae, which are the basic PP crystal structures. PP lamel-

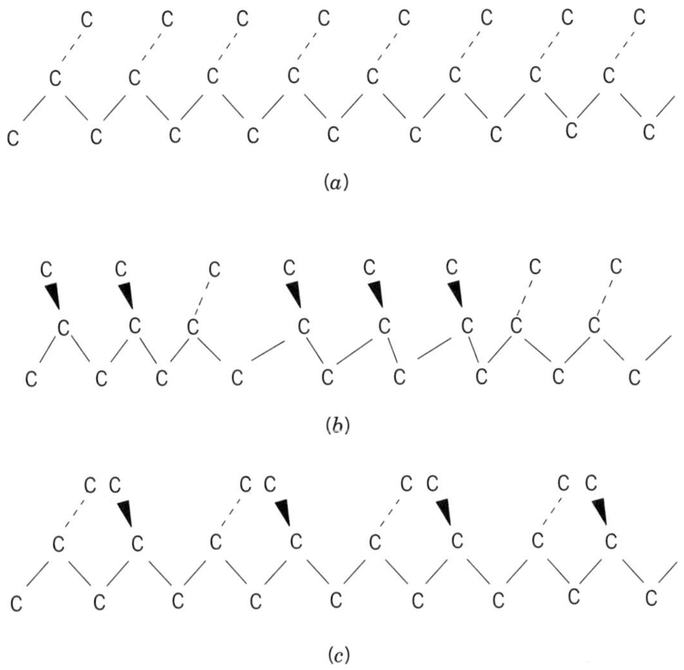

Figure 3. Polypropylene bond configurations: (*a*) isotactic—identical placement of methyl groups; (*b*) atatic—random placement; (*c*) syndiotatic—alternative placement of pendant methyl group. Tacticity designates the bond configurations. Shown are the carbons of the main polymer chain. The hydrogen atoms on the pendant methyl carbon atoms and on the main-chain carbon atoms have been omitted for clarity.

lae come in three forms: monoclinic (form I), trigonal (form II), and triclinic. Only the monoclinic and trigonol form are of interest in biaxially oriented PP production. These lamellae can be thought of as bricks with angles slightly greater or less than 90°. As the helix makes up the basic structure of the lamella, the lamella makes up the basic structure of the spherulite.

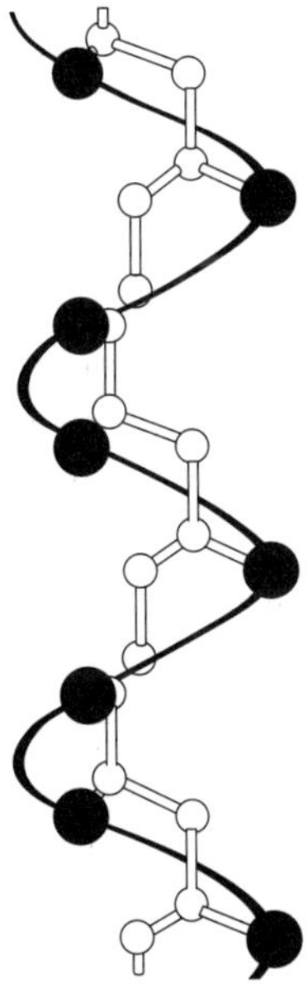

Figure 4. Perspective drawing of isotactic helix of type 2 * 3/1. [Drawn after Natta and Corradini (1960)]

Spherulites. A spherulite consists of folded-chain lamellae that developed radially outward. X-ray studies show that the B-crystalline direction coincides with the radial direction. Spherulite sizes range from <1 μm to more than a hundred micrometers depending on the nucleation density and cooling rate. Under a microscope with crossed poloroids, Maltese crosses are observed. Because spherulites are larger than the wavelength of light, they are responsible for the base-sheet opacity observed. The spherulites are destroyed during the stretching process, leaving the lamellae as the largest crystal structures. Because lamellae sizes are less than the wavelength of light, biaxially oriented PP films are highly transparent.

Orientation Process

Following is a description of the OPP orientation process using the tenter-frame process as the model. The tenter-frame process has the following major sections:

- Extrusion
- Casting
- MD orientation
- TD orientation
- Annealing
- Treatment
- Winding

The extrusion section melts the polymer and provides the pressure to force the molten polymer through a sheet die to the casting section. The casting section solidifies the molten PP sheet and forms its morphology. The MD-orientation section reheats the sheet and stretches it between differential speed rolls. A short annealing section relieves some stresses, and then the MD-oriented sheet passes into the TD-orientation section. Here the sheet is clamped between two moving chains and is passed into a convection oven. The web is reheated to its orientation temperature, then the chains diverge and impart the TD stretch. Annealing and cooling take place before the film leaves the TD-oriented chain. The edges with the unstretched tapes are slit off for recycling, and the web is passed on to be treated (ie, surface-oxidized) if necessary, and then wound as a mill roll.

The speed differential between the casting speed and the MD-orienter speed controls the MD stretch ratio. The ratio of the starting base-sheet width to the film width controls the TD stretch ratio. The combination of stretch ratios and stretching temperatures creates the actual molecular orientation. Higher stretch ratios increase molecular orientation. Higher orientation temperatures reduce the orientation efficiency as a result of higher viscous flow vs molecular orientation.

Extrusion. The extrusion section melts the solid resin pellets and transforms them into a homogeneous stream. As many OPP films are multilayer films, the extrusion section consists of multiple extruders and a coextrusion die to combine the different polymers in the desired layer ratios and configuration.

Casting section. The casting section forms the basic crystal structure, starting with lamellae formation. The lamellae form into a regular structure that defines the spherulites. The crystallization starts at a nucleation point and grows radially outward until it hits another growing region. Inside this spherical structure there exists the lamellae stacked in an expanding radial pattern. The spaces between the lamellae are filled with amorphous material. This amorphous material is the tie chains or glue that holds the individual crystals together.

Also, side-to-side differential base-sheet cooling can be imposed on the base sheet depending on the particular equipment used. The cooling conditions can cause regions of different sized spherulites.

MD-orientation section. In this section the base sheet is preheated over a series of rolls and then stretched between differential speed rolls. This first-direction stretch initiates spherulite breakup by streching the tie molecules in the amorphous regions between the lamellae.

TD-orientation section. In this section the MD-oriented sheet is conditioned and stretched in the second direction. This second stretch completes the spherulite breakup by stretching the previously unstretched TD tie molecules and partially reorienting some MD-oriented chains.

Annealing section. In this section, after the orientation is complete, the film is held for an appropriate time and temperature to relieve residual stresses by allowing viscous flow to occur, thus creating a stabilized, heat-set packaging film. If no annealing is done, then the film is a shrink film.

Proper design and operation of the annealing section provides a heat-set OPP film that can withstand brief contact with sealing mechanisms that are 300°F (150°C) or slightly above, without surface distortion in the seal areas. Sealing mechanisms above this temperature do create progressively greater distortion potential until at 310°F (155°C) the OPP film attempts to return to its unoriented dimensions. Because of this reaction to very high temperatures, OPP film's sealing range has an upper limit of 300–305°F (149–152°C).

Treatment section. In this section the surface of the film is oxidized to change the polypropylene surface from a nonpolar to a polar character. This is required to allow coating, inks, and laminating agents such as adhesives and polyolefin extrudates to adhere to the OPP surface.

Winding section. In this section the film is wound into mill rolls so it can be transferred to a coater where heat sealant or barrier coating can be applied. If the film is ready for shipment to a customer, it is transferred to a slitter where it is slit to customer specifications for width, outside diameter, and core size.

General Comments

Heat-set OPP film 1.00 mil (25.4 μm) thick, when subjected to hot air for 7 min at 250°F (121°C), exhibits an unrestrained shrinkage of 0.5% in the TD and 2.5% in the MD. As a result, heat-set OPP films will show some shrinkage if they are exposed to a very hot environment such as is found in retort packaging [250°F (121°C) at 15 psi (103 kPa)]. However, below 230°F (110°C) OPP film shows very little shrinkage, and a package wrapped in this material maintains its dimensional integrity.

In the early days of OPP film production, blown bubbles were 120 in. (3 m) in circumference and tenter frames were 60 in. (1.5 m) wide. Today, 240-in. (6 m)-circumference bubbles and 360-in. (8 m) wide tenter-frame lines are in operation around the world. At present, 10-m-wide tenter-frame lines are being introduced. This movement to wider lines has greatly increased the OPP film producers' productivity and has also significantly reduced the generation of narrow offcut rolls in slitting operations. Both factors have contributed to OPP film's reputation of being the most economical of all highly engineered, high-performance flexible-packaging materials.

The very low density of polypropylene (0.905 g/cm^3) allows OPP film to have the greatest area-coverage yield per unit of weight of any commercially significant thermoplastic or cellulose-based sheet. This high area yield and ability to be made into very thin films are important contributions to its overall favorable economics.

There are few physical property differences between films made by the two processes (see Table 1). Tensile strength and elongation of blown films are about the same in both directions. For tenter-frame films, TD strength is higher than MD strength and MD elongation is higher than TD elongation. Tenter-frame film has a generally superior heat stability over blown film because of better annealing conditions that can be practiced in the tenter-frame process. Blown-tubular OPP films have better impact strength. Only in special packaging applications would these relatively small differences become significant. Both films possess excellent moisture-barrier properties. This value increases and decreases proportionately with the thickness of the film under consideration.

Product Developments

The first two OPP films produced were homopolymer shrink film, and heat-stabilized film that could not be heat-sealed. The former had limited application because of its need for hot-wire sealing and tremendous shrink energy that originally precluded it from wrapping high-profile products. The introduction of ethylene-propylene copolymers allowed the widespread development of propylene shrink films that today have replaced much of the original PVC shrink films and compete strongly with crosslinked polyethylene shrink films. Additionally, the development of special polymeric coatings and polypropylene copolymers and terpolymers during the 1960s, 1970s, and 1980s, respectively, has resulted in improved sealability in many applications.

The original heat-set OPP film gained wide acceptance as a laminating substrate when combined with heat-sealing polymer-coated cellophane (see Cellophane) and glassine papers. The OPP film contributed strength, moisture barrier, and high surface gloss to the lamination. However, before acceptance was forthcoming from the marketplace, OPP film producers had to improve the film's slip characteristics over the high coefficient of friction (COF) that is natural to the film and improve the wettability. The addition of internal surface-blooming slip agents reduces the film-to-film COF to 0.2–0.4. This COF provides excellent machinability on automatic bag-forming machinery used in the snack-food and candy indus-

Table 1. Tensile Properties of Blown- and Tenter-Frame OPP Films

Typical Film Property	Direction	Blown OPP Film	Tenter-frame OPP Film
Tensile strength, kpsi (MPa)	MD	27–30 (186–207)	18–24 (124–165)
	TD	27–30 (186–207)	30–35 (207–241)
Elongation (%)	MD	70–90	110–150
	TD	70–90	30–60
Modulus 1% secant, kpsi (MPa)	MD	350 (2415)	250–350 (1725–2415)
	TD	350 (2415)	400–600 (2760–4140)

Source: ASTM Test D 882.

tries, while surface oxidation improved wettability. Converters developed the hot-melt adhesive based, heat-sealing thermal stripe that allowed the production of lap back seals on vertical form/fill/seal machines (see Form/fill/seal, vertical). This technology widened the use of OPP film laminations for snack foods, candy, and pasta products. Another application for this non-heat-sealing-film is the inner liner of paper bags for cookies and pet foods.

Since the beginning, the history of further OPP product development has been the creation of improved barrier technologies and the development of surfaces with special attributes to permit the use of the base-sheet barrier properties in packaging and industrial applications. The primary protection that OPP inherently supplies is a moisture barrier. Additional barriers have been added during the development of new products such as oxygen, aroma (chemical) barriers, and light barriers. However, it is the moisture-barrier property, combined with the unusually low polymer density, which results in the maximum product coverage with one of the lowest moisture permeabilities. When these two features were combined with special surface attributes such as sealability, printability, laminatability, coatability, metallizability, and machinability, a high-quality, low-cost packaging film emerged that replaced most of the cellophane then in use in snack-food packaging. Besides cellophane replacement, OPP has made further inroads into glassine paper and some foil markets as well. In order to reach this point several weaknesses of OPP film relative to packaging in general and cellophane in particular, had to be addressed. The inherent limitations to homopolymer OPP film are no sealability, high COF surfaces, low surface wettability, static generation, and a lower use temperature range than cellophane.

To overcome these weaknesses in the product, both film and packaging machine changes were required. The temperature limitations were overcome by improvements in packaging machine design and temperature control of sealing jaws plus the use of special acrylic and PVDC polymer coatings and the use of low-melting copolymers and terpolymers as film-sealing surfaces. Today the machine evolution is complete, and packaging machines are designed with OPP film as a primary wrap. Static generation occurs at all surfaces and, because of the high dielectric strength of OPP, must be eliminated at each point where static interferes with packaging machine performance. Product formulation approaches have been hampered by a lack of FDA-approved additives and the relatively high additive levels that would be required. Surface wettability is overcome by coating with acrylic polymers. For uncoated films, corona or flame treatment surface-oxidizes the film to provide the wettability. This treatment can be applied after the tenter or at subsequent converting steps. The development of multilayer coextrusion systems has also permitted the incorporation of nonpropylene skins to be used, further broadening the range of product developments. The development of low-COF surface modifications has relied almost exclusively on film-formulation developments and has seen many developments over the years. The original surface modification for COF was a formulated acrylic coating that combined excellent printability, low COF, and broad sealability. The second development was the use of fatty-acid amides, which, when added to the polypropylene, would diffuse to the film surface (called "slip bloom") with time and heat treatment after film orientation. A lowered film COF is difficult to control, and the additive often interferes with printing ink adhesion and lamination bonds. Today this slip-blooming technology has been replaced by using mineral and polymeric particles dispersed in the film surface layers and the use of silicone oils for higher temperature applications. This results in a film COF of 0.25–0.5. While these COF values appear high by the original standard, the films all exhibit a low but uniform force to transport the film over the forming collar and through packaging machines. It is this low film-transport force that is the critical property for proper machinability and not the film-to-film COF.

Today OPP film has found wide acceptance and continues to grow as new markets are developed making use of its special product characteristics. Many technologies developed to permit the use of OPP have gone through several developmental generations as new resin and additive technologies were invented to overcome the limitations of the first-generation film modifications. As this article is being written, a new revolution in polymer metallocene catalysis is occurring that is generating new propylene resins that will have a profound impact on future growth in new OPP packaging film modification technologies.

Sealability. The quest for functional heat sealability with conventional heat-sealing mechanisms installed on most packaging machines led OPP film producers down four avenues of development. The first (avenue 1) was the modified polymer route. This film uses a polypropylene modified with terpenes. The film is heat-sealable, although its sealing range is narrow (25°F or ~14°C), and it does not have good hot-tack-seal strengths needed for form/fill/seal packaging. In overwrapping applications where the product to be packaged is contained in a carton or another self-supporting structure, this OPP film has found good acceptance. Because this film is only partially heat-stabilized, it can be snugly tightened around the carton or bundle by using a heat tunnel. One version is used today as a cigarette-packaging material. However, this technology has been supplanted largely by surface sealability modifications.

The second method (avenue 2) of imparting sealability to OPP film is by addition of an off-line heat-sealable coating based on an acrylic polymer (see Acrylics). This coating adds flavor and aroma barriers, but no moisture- or oxygen-barrier properties to the OPP film. However, it does provide a wide sealing range of 80°F or ~45°C, and adds sparkle and good machine slip characteristics. It has a film-to-film COF of 0.2–0.3. It has high-volume use in horizontal form/fill/seal operations (see Form/fill/seal, horizontal) for items such as baked foods (cookies in particular), and candy. As an overwrap film, it is used for pet foods and various tobacco products such as cigarettes, cigars, and pipe-tobacco cartons. Its hot-tack-seal strength characteristics are such that it is used extensively for VFFS (vertical form/fill/seal) packaging of lightweight products, such as snack foods, but is not generally proposed for similar packaging of heavy products. These acrylic coatings are compatible with many inks (see Inks) and adhesives (see Adhesives) used by converters and are also heat-sealable with the PVDC coatings used on cellophane, glassine, and other OPP films (see Vinylidene chloride copolymers).

The third method (avenue 3) to obtain heat sealability is to coat the OPP film off-line with a PVDC coating, which also addresses the gas-barrier deficiency. This is required because oriented polypropylene film in its uncoated form is not considered a gas-barrier film (see Table 3). Its oxygen permeability is high at 73°F (23°C) for 1.00-mil (25.4 μm)-thick film (see Table 3). PVDC-coated films fall into four categories:

1. The first category is readily machinable on heat-sealing, automatic packaging equipment. The sealing range here is wide (60–80°F or 33–45°C), and the oxygen-transmission rate is improved (see Table 3). These films are available with one or two sides coated. The former is a laminating substrate for use with other OPP films and cellulosic materials, to package snack foods and candies. The two-side-coated product is used in unsupported form for bags and carton overwraps. In these applications, the PVDC supplies an aroma barrier, which is generally wanted by the end user.

2. The second category of PVDC-coated OPP film has a superior gas barrier, with an oxygen-transmission rate of 8–12 $cm^3/(m^2 \cdot day)$. However, because its sealing range is no more than 50°F (28°C), this film is not used for its heat sealability. Rather, it is used as a laminating substrate with various other materials for the controlled-atmosphere packaging of coffee, natural cheese, and processed meats (see Controlled-atmosphere packaging).

3. The third category combines an acrylic coating on one side and a PVDC coating on the other side. The advantage cited for this film is that it provides a degree of gas and improved aroma barrier, is lap-sealable, machines well, and costs less than two-sided, heat-sealable, PVDC-coated OPP films.

4. The fourth category of PVDC coated films is the combination of the high-barrier PVDC coating with a broad-seal-range, coextruded sealant skin. This film is used as a laminating film with the high-barrier PVDC buried in the structures and makes use of the excellent seal properties of the coextruded skins (see Table 2).

Table 2. Coextruded OPP Film-Sealing Temperature Ranges

OPP Film	Sealing Temperature Range
Copolymer-group A (0.5–3.0% ethylene)	30–50°F (16–28°C)
Copolymer-group B (≥4% ethylene)	60–70°F (33–39°C)
Terpolymers (ethylene and butene)	85–95°F (47–53°C)
Ionomer	95–105°F (53–58°C)

The fourth avenue of development of heat sealability of OPP films is through coextrusion. Although coextruded, multiply (multiple-ply), heat-sealing OPP films have been available since 1965, this type of film came into large-volume usage during the late 1970s and early 1980s. This was due in large part to the development of ethylene propylene copolymers (EP copolymer) and later by ethylene, butylene propylene terpolymers (EBP terpolymer) resins, and advances in the technology of coextrusion die design. Coextrusions are made using several basic approaches. In one approach, two or more extruders feed their individual molten polymer streams to a single-manifold extrusion die (see Coextrusion, flat: Coextrusion, tubular). In the upstream section of the die is a round or rectangular "feed block" mechanism that brings these polymer flows together into layers before passing them to the single cavity (manifold), where they are transformed into a single wide, thin rectangular sheet as it passes out of the die lips. Another method is termed "tandem" coextrusion. With this procedure the film's core material is extruded from its own extruder and die. This extrudate is then extrusion-coated (see Extrusion coating) from one or more extruders and dies, or it is hot-nip-laminated to other compatible, independently manufactured films. In recent years this "tandem" coextrusion has been largely replaced with multimanifold coextrusion technology. In this OPP technology, generally three separate die manifolds are contained together in the same die body. As the polymer flows toward the die exit, the separate layers are combined in layers and then exit the die lips together. This technology has permitted the combination of resins that cannot be coextruded by the feed-block method or "tandem" methods described above. More recent developments have combined the feed block with the multimanifold dies to produce five and more layers in a single sheet. In all coextrusion cases the multilayered sheets are moved into the orientation units of the line. Coextrusions can be made by both the blown-bubble and tenter-frame processes.

Polypropylene does not anchor well to many other polymers. As a result, great care must be taken in the choice of materials use in OPP film coextrusions or interply bonds may be broken during orientation or later by the stresses of package-forming techniques and distribution. Other demands on outer ply materials include hot-slip characteristics that allow easy film movement over hot-packaging machine parts and surface-slip characteristics for good overall trouble-free automatic high-speed machining. The ply destined to be printed or laminated must also be corona or flame treated.

Today most coextruded OPP films consist of from two to five layers. Most of the film's mass is in the core ply, which is usually made with virgin homopolymer and some regrind. One or both sides of the core may be covered with a polypropylene-rich copolymer or terpolymer, an ionomer (see Ionomers), or a vinyl acetate–modified polyethylene layer. The

Table 3. Properties of Oriented Polypropylene Films, 1 mil (25.4 μm) Thick [a]

Typical Film Properties	ASTM Test	Blown OPP Film	Tentered OPP Film	Slip-modified Film	Acrylic Coated Film	Low Temperature Heat Seal Coated Film	PVDC Coated Film	PVOH Coated Film	Coextruded Film	White Opaque Film	Vacuum Metallized Film
Specific gravity	D1505	0.905	0.905	0.905	0.91	0.91			0.91		0.905
Yield in.^{2}/lb at 1 mil	D2673	30,600	30,600	30,600	30,500	30,500	27,300	29,900	30,500	~1.4 ×	30,600
m^2/kg at 25 μm		43.5	43.5	43.5	43.3	43.3	38.7	42.5	43.3	OPP	43.5
Haze, %	D1003	2.0	2.0	2.0	1.1	1.5	1.5	1.0	2.0		
Light transmission, %										15–60[b]	0.5–2.5[c]
Tear strength, gf/mil (N/mm), propagating	D1922	4–6 (1.5–2.3)	4–6 (1.5–2.3)	4–6 (1.5–2.3)	4–6 (1.5–2.3)						
Heat seal range, °F (°C)					70 (39)	135 (76)	50 (28)		40–110 (22–61)		
Heat-seal temperature, °F (°C), min–max					230–300 (110–150)	165–300 (74–150)	250–300 (121–150)		190–300 (88–150)		
Oxygen permeability (OTR) at 73.4°F (23°C), 0% rh (cm^3 · mil) (100 in.^{2} · day · atm)	D3985	100–150	100–150	100–150	100–150	100–150	0.6–3	0.02	125–175	100–150	0.03–10
(cm^3 · 25 μm)/m^2 · day · atm)		1500–2300	1500–2300	1500–2300	1500–2300	1500–2300	10–45	0.3	2000–2700	1500–2300	0.45–150
Water-vapor transmission rate (WVTR) at 100°F (38°C), 90% rh (g · mil)/(100 in.^{2} · day)	F372	0.25	0.3	0.3	0.3		0.16–0.3	0.3	0.35	0.3–0.6	0.03–0.3
(g · 25 μm)/m^2 · day)		4	5	5	5		3–5	5	5.5	5–9	0.45–5
COF at 73°F (23°C), 50% rh	D1894	0.4	0.4	0.2–0.4	0.25	0.23	0.2–0.4		0.2–0.4		
Face-to-face				0.2–0.4	0.25		0.2–0.4		0.2–0.4		
Back-to-back											

[a] *Note:* For coated films barrier properties are a function of the coating, not the OPP substrate gauge.
[b] Tappi Opacity T 425.
[c] Calculated from Optical Density mesurements [OD = $\log_{10}$(100/% Light Transmission)].

choice of a polymer depends on the purpose for which the finished film will be used. Polypropylene-rich copolymers of propylene and ethylene are used for their good hot-tack seal strengths and excellent finished seal strengths. Terpolymers of propylene, ethylene, and butene are used to improve the film's breadth of sealing range, and today's terpolymers have been developed to give excellent hot-track strength as well. Vinyl acetate–modified polyethylene plies are used for their anchoring potential in off-line coating, laminating, and metallizing operations (see Metallizing). An ionomer resin is used for its wide sealing range and ability to repel grease and oils. Handling it in the orientation process is difficult, but once oriented and heat-stabilized, it provides lower-level seal strengths of about 300 g/in. (116 N/m), which can be used advantageously in packages that demand an easy-open feature. Ionomer skinned films are not generally made by the tenter process because of its aggressive adhesion to hot-metal surfaces used in the process and are only produced on bubble equipment. These films are therefore at risk of disappearing as lower productivity bubble lines are shut down in favor of higher-productivity tenters.

Coextrusion sealing temperature ranges vary greatly depending on the sealing ply materials. General categories are listed in Table 2.

Thin coextruded OPP films, 0.57–0.70 mil (14–18 μm) thick, are often laminated to other substrates such as polymer-coated cellophane, glassine, paper, and other OPP films for snack-food, candy, processed-cheese, and pasta-product packaging. In unsupported form they package lightweight products such as single-service packages of crackers used in restaurants. Medium-weight films [0.80–0.90 mil (20–30 μm) thick] are used for carton and tray overwraps by the bakery and candy industries, as well as for form/fill/seal single wall pouches by pet-food producers. A specially designed medium-weight material is used as a cigarette package overwrap. Heavy, thick coextruded OPP films, over 1.00 mil (25.4 μm) thick, are normally used in unsupported form to package cookies, candies, snack foods in small bags, and other products that need the stiffness, strength, an excellent moisture barrier, and a rich feel offered by these films.

Opaque films. White opaque film is one of the fastest-growing OPP film developments in the United States and western Europe. The most widely accepted film products of this type are made by the tenter-frame process. Homopolymer resin is evenly mixed with a small amount of foreign particulate matter. In one product, when the thick filled sheet is oriented, the polypropylene pulls away from each particle creating an air-filled void (closed cell). After heat stabilization the OPP film is similar to a micropore foamed product. In the second product, the material is produced as a filled film without voids. The opacity is a direct result of the amount of particulate material included in the film.

In the film with air-filled voids, the imparted opacity and whiteness are created to a small degree by the encapsulated particulate matter. However, the primary opacification is caused by light rays bouncing off the polypropylene cell walls and the air within each cell. The refractive index of the air is less than that for polypropylene. This refraction difference results in a TAPPI opacity of about 55%. The light diffusion gives the film the visual effect of pearlescence. Brightness values are calculated at 65–75% by the GE method. The actual gauges of this white opaque film are deceptively thick when their area-coverage yield per unit of weight is considered. A 1.5-mil (38-μm)-thick white OPP film has an area-coverage yield of 30,000 in. ^{2}/lb (427 cm ^{2}/g), which is about the same yield provided by a 1.0-mil (25.4-μm)-thick transparent OPP film.

Coextrusion and out-of-line coating techniques have greatly expanded the market acceptability of white opaque OPP film. By using these approaches to film manufacture, the film can be made one- or two-side heat-sealable. These steps can also increase the film's moisture, oxygen, and aroma barriers. White opaque OPP films are finding growing volumes in snack-food packaging, candy-bar overwraps, beverage-bottle labels, soup wrappers, and other applications that traditionally have used specialty paper-based packaging materials.

Metallizing. The use of OPP film as a metallizing substrate is another area of new market growth (see Metallizing). In this process, aluminum is vaporized onto one side of a film in a high-vacuum environment. The speed of the substrate's passage over the vaporization boats or crucible, the wire feed rate, and the vaporization temperature control how much metal is deposited on the film. If the metallized surface is to be used primarily for decorative purposes, a light coating of approximately 4 Ω per square (1.6 optical density OD) can be applied. This gives OPP film, which was originally transparent, a light-transmission rate of about 2.5%. If the metal coating is used to improve the finished film's barrier properties, a heavier disposition of 2 Ω per square (2.2 OD) is employed. This gives a light transmission of about 0.6%. Moisture and oxygen barriers are improved dramatically (see Table 3). In packaging, metallized OPP films are widely used in snack-food laminations and for confectionery wraps. Today, product design advances are available that lower the moisture-barrier properties of metallized OPP below that of metallized polyester (PET) films while approaching the oxygen-barrier properties of metallized PET. This is accomplished by the incorporation of non-polypropylene layers for the metallized surface. Further developments have produced metallized OPP that have moisture and oxygen barriers significantly better than metallized (PET) (Table 3).

New coating developments. As the product line continues to evolve, the heat-seal range of acrylic-coated films has been widened with the development of a low-temperature heat-seal-coated (LTSC) film. This new product increases the heat-seal range to 145°F (76°C) from 70°F (40°C). Along with the increased heat-seal range, the seal integrity has been improved. As oxygen barrier has become more important, a line of poylvinylalcohol (PVOH)-coated films has been commercialized. This product provides excellent moisture barrier at relative humidities of <60%. At 60–80% rh the oxygen permeability increases to that of PVDC and is generally greater than PVDC above 80% rh (see Table 3 for OTR at 0% rh). As both of these coatings have a specific function, they are only one-side-coated onto OPP. The other side is coated with standard acrylic coating to provide machinability characteristics to the film.

ELDRIDGE M. MOUNT III
JOHN R. WAGNER, JR.
Mobil Chemical Company
Macedon, New York

FILM, PLASTIC

Films are continuous membranes that can separate one area from another. These membranes can vary in thickness, ranging from less than that for rigid containers, to sheeting, to film, and even thin coatings. Usually films are considered self-supporting and less than ~ 10 mils (250 μm) in thickness. No definition is given for a minimum thickness, which can vary depending on the material of construction of the film; however, thickness of < 1 mil (25 μm) are common.

An important feature of most of the films discussed in this article is *heat sealability,* which refers to the thermoplastic property of the film, or coating on the film, which allows it to be fashioned into pouches, bags, and overwrapped packages by virtue of its ability to make a hermetic seal to itself. Heat sealing is accomplished by heating up the film areas, then applying the hot areas to each other under pressure. Sometimes these operations occur simultaneously. During heat sealing the polymer molecules become entangled; the better the intermingling, the stronger the seal. The time allowed for heat sealing in a typical high-speed food-packaging machine is less than one second, during which time heat-seal strengths of very high values can be achieved, ie, several pounds per inch of seal width.

Film Uses

Because of their ability to keep two areas (or volumes) separated from each other, films have come to have a multitude of commercial uses over the years, particularly in the area of food packaging. Films do provide protection to the foodstuffs being displayed, and in doing so can provide a multitude of other functional attributes. They can function as barriers for gases, vaporous flavor components, or moisture from escaping from the food, thus preserving its freshness. They can also protect the foodstuff from attack by undesirable outside agents such as air (oxygen), moisture, or sunlight, thus ensuring that the food does not become rancid or soggy. The film material can serve as a partial gas barrier, thus allowing some gases to escape but not others; thus the choice of packaging film to protect fresh vegetables and fruit can be very selective, especially if the vegetables and fruit are already cut or sliced for customer convenience. For fresh produce the film must allow CO_2 to easily leave the package while allowing only low levels of air (oxygen) to enter the package for longer shelf life.

Films also provide a billboard for information about the contained product, directions on how to open the package, recommended dose levels, warnings about toxic contents, nutritional information, and UPC symbols for pricing and inventory control. This information can be displayed in a multicolor and aesthetically pleasing way on the packaging film so that it "sells" itself and acts as an impulse-buying aid. In addition, the transparency and high clarity of the film can provide a window in the package to allow the consumer to judge desirability of the product before buying or to judge remaining quantity in the package after purchase and as it sits in the consumer's home.

Film applications can be quite large, as in the case for some construction uses. Films or sheeting materials are used to wrap houses and buildings under construction, thus serving as moisture and wind barriers in the finished construction. Film products can be used in highway construction, as leakage barriers in municipal water reservoirs and as protective liners for municipal solid-waste landfill sites. Film applications can be quite small, such as the push-through blister pack for individually packaged medical pills. Packaging film uses can be categorized as follows.

Overwrap. This type of packaging usually involves unwinding a roll of flat film into the packaging machine, where it is folded around the object, the film sealed to itself, and the finished package removed by cutting the film. Typical products packaged by this method include small packs of snack crackers and cookies. The machine is commonly called a "horizontal make-and-fill machine." Products such a potato chips, corn chips, and bags of candy are packaged in vertical form/fill/seal machines in which a roll of flat film feeds into the machine over a forming collar where the film is formed into a tube. This tube is sealed along the back with a vertical heat seal, and then the bottom cross seal is made. The open-top tube is filled with the product and indexed down, and the top cross seal made simultaneously as the bottom seal is formed for the next pouch to be filled and as the filled pouch is removed from the packaging machine. Coated cellophane, coated oriented polypropylene (OPP) film, and laminated films typically are used for this packaging application.

Skin packaging. As the term implies, this type of packaging involves forming a skin of film over the object being packaged and sealing the skin film tightly to a heat-seal-coated display card. This is accomplished by placing the object to be packaged, usually a hardware item, onto the coated display card, then draping a hot sheet of film over the entire assembly, and applying vacuum to the backside of the permeable display card. The vacuum pulls the cooling hot film down over the object and applies pressure to assist in making the seal to the display card. Typically ionomer, LDPE, or PVC resin films are used for this packaging applications, and the board coating is LDPE or EVA. (See also Skin packaging.)

Blister packaging. Preformed blisters are filled with the object, the coated paperboard is placed over the filled blister, and the assembly is heat-sealed. Small-hardware items and small notions are commonly packaged in this manner.

Shrink packaging. This type of packaging refers to the use of a film manufactured in such a way that when it is heated it will contract in both directions, reducing its surface area. When this type of film is wrapped around an object and sealed around its edges, then sent through a shrink tunnel where large volumes of heated air wash over it, the film will react to the heat and contract down to the object, thus making an attractive skintight package. A vent hole must be cut into the film to allow the trapped air to escape from the package as is undergoes film shrinkage in the hot air tunnel. Shrink-film resins are polyethylenes, polypropylenes, PVC, polyesters, and some coextruded structures. Articles packaged in this manner are numerous and can include frozen poultry, pizzas, stationery, toys, cassettes, and compact disks.

Stretch packaging. This packaging technique is useful for bundling objects together for convenience in shipping or handling. Resins used in this application include LLDPE, LDPE,

EVA, PVC, and some copolymers of polypropylene. Typically when stretch wrap is used for pallet bundling, the tacky film is wound around the pallet many times to ensure tightness, the film cut, and the film end stuck down on the wrapped pallet. There is no need to heat-seal or use mechanical fasteners with this packaging technique.

Plastic–Film Resins

It would not be possible to cover all resins that have been fabricated into film in this article. What are presented here are typical resins used over the years as packaging films.

Food packaging has successfully employed films for many years, starting in the 1920s in the United States with the introduction of cellophane film, which was initially used to package loaves of bread that had been previously overwrapped in paper. Because it was transparent, cellophane allowed the bread to be viewed by the customer before purchase. Uncoated cellophane is composed of regenerated cellulose, a softening agent, and water as a plasticizer. While it is not moistureproof, it is impermeable to gases when the film is kept dry. Additional slip, antistatic, and antifog properties can be provided by special formulations applied as coatings to the base sheet. Since cellophane is not a thermoplastic and does not have a melting point, heat sealability is possible only by applying a heat-seal coating such as nitrocellusose (NC) or polyvinylidene chloride (PVDC) to the base sheet.

By the early 1950s low-density polyethylene (LDPE) resin was developed and began serving as the first major human-made (synthetic) thermoplastic packaging resin. Because this resin has a melting point and is thermoplastic, the packaging industry developed thermal extrusion equipment to extrude thin films of this new resin for packaging. Polypropylene (PP), another popular packaging resin, was first produced during the 1950s. However, it did not develop commercially until the 1970s, when new high-efficiency catalysts were discovered for its production. PP competed effectively for many of the cellophane packaging applications. While cellophane had attractive properties of clarity, stiffness, gas barrier, and heat resistance, it also had disadvantages of limited shelf life due to loss of volatile plasticizer. This loss of plasticizer caused the cellophane to embrittle to the point where the packaged product was no longer protected. PP has the inherent attributes of clarity, toughness, thermoplasticity, and heat sealability after being coated by PVDC. PP also has the advantage of having a density lower than that of cellophane; thus more square inches of packaging film are possible from a pound of PP than from a pound of cellophane film—a definite economic advantage. As PP is a thermoplastic resin it did not have the thermal resistance to provide fail-safe high-temperature sealing protection against substrate burn-through on the packaging machine. Additionally, the lower-modulus PP could not be easily fed through the push-feed packaging equipment developed years earlier for the stiffer cellophane. Feeding problems for PP were exacerbated by its greater tendency to generate and hold static charge, which caused clinging and feeding jams in the packaging machine. These serious packaging problems were eventually eliminated by machinery redesign and by improvements in PP base-sheet and coating technology. Plastic film resins can be categorized by the chemical process used to form the molecules in the polymer resin backbone.

Addition polymerization (homopolymers). Polymers in this category include LDPE, HDPE, PP, polybutylene (PB), and polystyrene. During the polymerization or building of the polymer molecules, the individual monomer units (ethylene, propylene, butylene, respectively) are chemically connected together in the pressure reactor in the presence of high temperature and an appropriate catalyst. By control of the residence time in the reactor, the number of these monomer connections can be varied. Long reactor time, longer polymer chain length, and higher molecular weight result. By controlling molecular weight, the manufacturer can control polymer resin properties such as viscosity, which, in turn, can control final film properties such as toughness. These resins, called *aliphatic resins,* are prepared from monomers consisting of only carbon and hydrogen atoms. The characteristic chemical structure in the polymer backbone is a carbon to carbon bond ($-CH_2-CH_2-$) linking the individual monomer units together. Aliphatic ethylene and propylene homopolymer resins are nonpolar, have good clarity, relatively low melting points ranging from $\sim$ 105°C for LDPE to 125°C for HDPE, and 165°C for PP. Polystyrene does not have a melting point but does exhibit a Vicat softening point at $T \leq$ 106°C. Because their monomers are relatively easily derived from petroleum feedstocks and the corresponding resins are used in large-volume packaging applications, resin cost/lb. is not high, particularly for LDPE, HDPE, and PP. LDPE resins can be found, for example, as premade flexible bags and pouches and other large-volume packaging applications. HDPE, which is, stiffer and tougher, is used as grocery bags and multiwall bags and bag liners. PP, as described above, has replaced cellophane film in many food-packaging applications. Some of these applications include overwrap for cigarette packs, and snack-food packs, and pouches and bags for potato and corn chips. PB has found some applications as a meat-packaging film and as an additive for hot-melt adhesives.

Addition polymerization (copolymers). Polymers prepared from combinations of two or more monomers can give rise to plastic resins having a wide range of properties not possible from homopolymers. Resins of E/P (ethylene/propylene), E/B (ethylene/butylene), E/H (ethylene/hexene), and E/O (ethylene/octene) are possible combinations. The last three mentioned resins are gaining popularity as new aliphatic polymers having a much broader range of thermal and mechanical properties than previously possible. These resins, designated metallocene polyethylenes (mPEs) are prepared using the relatively new constrained geometry catalysts that allow the tacticity and branching of the polymer molecule to be better controlled, yielding polymers with narrower molecular distributions and lower densities. mPE resins are aliphatic copolymers somewhat endowed with the properties of the higher-polarity copolymers described below.

Addition copolymers can also be prepared with ethylene and a polar second monomer. These polar monomers can include acrylic acid (AA), methacrylic acid (MAA), ethyl acrylate (EA), and vinyl acetate (VA). Because these monomers contain oxygen atoms in addition to carbon and hydrogen and because oxygen is a heavier atom and rich in electrons, polymerization of these polar monomers results in a copolymer resin having higher polarity than carbon or hydrogen containing aliphatic resins. Resins of E/AA, E/MAA, E/EA, and

E/VA are characterized not by an unusual chemical linkage between the repeated monomer units, but rather by the polar nature of the copolymer in the polymer chain with the ethylene monomer. The E/AA and E/MAA resins, also called *acid copolymer resins,* have good clarity, low haze, and lower melt and sealing temperatures, and are able to adhere strongly to polar substrates such as paper, foil, and some highly polar film resins. Thus they have found use more in specialty packaging applications and in association with multilayer structures for meat, cheese, snack foods, and medical items. E/VA resins typically contain between 5–40% VA comonomer, with VA contents above about 25% used largely in hot-melt adhesive applications. Packaging-film applications employ resins having lower VA contents ranging from 5 to about 18%. These compositions are not too sticky, and films of these compositions can be handled on typical packaging machines; however often slip and antiblock additives must be incorporated into the resin to ensure the proper slip level. Because of the low crystallinity, toughness properties at low temperatures, and low melting points, E/VA resins are used as a poultry and meat wrap, bag-in-box for liquid packaging, stretch film, and as ice bags.

Because E/AA and E/MAA resins contain free-acid groups, they can be partially neutralized to form a class of resins called *ionomers.* The ionomers serve some of the same market applications and have excellent hot-tack sealability, the ability to seal through contamination, and grease/oil resistance. These polar copolymer resins find use in packaging meat, cheese, breakfast cereals, and a number of other wet and dry products. They also are used as skin-packaging films for hardware products.

Condensation polymerization. Some chemically reactive polar monomers can react with each other or with a second monomer to form a polymer chain and a small volatile second molecule. Many times that second molecule is water, thus water seems to "condense" from the reaction. Removal of this second molecule leaves the relatively clean polymer behind. Resins made by this condensation polymerization route include polyamides or nylon, and polyester. Nylon resins are made from diamines and dibasic acids and are characterized by the amide group (—CONH—) in the polymer backbone. Nylon resins are identified by the number of carbon atoms in the monomers. Thus homopolymer nylon 6 can be prepared from a monomer having six carbons in a straight line with reactive end groups on each end. Also nylon 6/6, 6/10, and 6/12 can be prepared from two monomers with the differing number of carbon atoms. Polyamides are regarded as semicrystalline resins with high melting points (175–275°C). As packaging films, they offer toughness, chemical resistance, resistance to oils and greases, and moderate gas barrier. Food products packaged in polyamide films include meats and cheeses.

Polyesters similarly are formed by reaction between an aromatic diacid such as terephthalic acid or the dimethyl ester of terephthalic acid and a polyalcohol such as ethylene glycol. These resins are characterized by the ester group (—CO—O—) linking the monomer units in the polymer backbone. Thus these resins contain a stiff aromatic molecule within the polymer backbone as well as a flexible aliphatic portion. This stiff resin can be extruded into oriented high-tensile-strength film, metallized, and heat-seal-coated to yield a packaging film in very thin gauge. In combination with other sealant coatings or laminations, polyester films are used as boil-in-bag pouches and for processed-meat packaging.

Chlorinated vinyl addition polymerization. Polyvinyl chloride resin for film applications is prepared by suspension polymerization and employs the vinyl chloride (CH_2=CHCl) monomer. A homopolymer resin is polymerized from this monomer by a free-radical process. After purifying and removal of excess water, this rigid resin must be blended with softeners or plasticizers in order to reduce its glass-transition temperature and render it a usable resin. For a more detailed discussion on how to modify melt temperature and glass-transition temperature in polymers, refer to Nielsen (1). Other additives such as lubricants and thermal stabilizers are blended into the resin as well. Packaging applications for PVC film include fresh meat, shrink film, and blister packs. For more details on how packaging film choices are made for different foodstuffs, refer to Jenkins (2).

Multilayer Films

Laminated film structures refer to multilayered structures composed of a number of monofilm layers. By effecting adhesion between the layers, much more sophisticated film properties can be obtained in a single film. Heat-sealable films can provide the outside layers and can cover interior layers of nonthermoplastic materials such as foil, metallized film, or paper. Previously made films containing light blocking levels of TiO_2 can be printed before being laminated with a clear glossy, heat-sealable film, thus yielding an attractive multilayered film capable of being heat-sealed, providing protection of the product from UV degradation, and exhibiting striking printed panels. Since each monofilm of the laminate must have been previously prepared individually as a self-supporting film, the final multilayered laminates can end up quite thick and costly.

Coating of monofilms also provides a higher level of property sophistication to monofilms. Frequently solvent coatings are applied to films to provide heat sealability or enhanced gas and moisture barrier to the film. Vacuum metallization can be considered a form of coating. This technique enhances light opacity and gas- and moisture barrier-properties of the film. Individual coating thicknesses are usually much less than 1 mil (25 μm); thus coating a monofilm does not substantially add to its thickness.

Coextrusion describes a process in which a multilayered film is prepared by adhering several individual film layers within the body of the extrusion die. This process can prepare as many as nine separate layers of film within the same final film. Each resin may have its own extruder feeding a melt stream into the complex feed block attached to the die. Sometimes one extruder will feed melt into two or more separate flow streams by the use of a stream divider within the feed block. This practice is fairly common for the extruder supplying the tie layer resin, the resin used to adhere two dissimilar resin types in the final multilayered film. Many factors must be considered when choosing the tie layer resin for a coextruded structure (3). Obviously, the initial cost for coextrusion equipment is very high. However, this technique is quite popular and is able to provide outstanding film property combinations at a reasonable cost.

Basic Film-Forming Processes

Thermoplastic resins such as ethylene and propylene homopolymers and copolymers, and the condensation polymers described above, by definition have a melting temperature above which they are not rigid solids but rather viscoelastic materials. As such, they are soft plastic melts capable of being formed into shapes convenient for packaging and other commercial uses. There are several typical thermal processes for effecting this conversion from the small pellet supplied in the resin bag, box, hopper truck, or hopper railroad car to the final film structure. The following descriptions describe the preparation of monofilms; however, keep in mind that the coextrusion process is also possible for all these extrusion processes. Refer to Figure 4, which relates typical EVA resin melt flow requirements to film manufacturing process type.

Extrusion–cast film. In this process the plastic resin pellets are introduced into the feed hopper of the extruder, where they are funneled down into the extruder barrel and onto the rotating screw (see Fig. 1). As the screw rotates, it drives the pellets deeper along the hot barrel, compressing and heating them and applying force or shear so that the air spaces between the pellets are driven out and the pellets are heated to the melt temperature or softening temperature of the resin. The screw, which is driven by a powerful electric motor, is typically composed of three separate regions: (1) the *compression zone,* which is closest to the feed hopper; (2) the *transition zone;* and (3) the last region, the *metering* or *blending zone.* Each zone occupies about one-third of the screw length. The compression zone heats up the pellets driving out the air, the transition zone describes the length of the screw where the transformation from solid pellet to melted resin takes place, and the metering zone allows the melt pool to become homogenized so that a uniformly mixed melt stream is presented to the heated die. The die serves the purpose of fixing the dimensions of the final shape of the polymer. Thus most dies have a circular cross section to the melt inlet, which then gradually transforms to the thin and wide cross section of the film. As the flat, thin sheet of polymer melt exits the die, it is quenched to a lower temperature below its melting point, thus solidifying it into the final and desired shape. Often the quenching is accomplished with water-cooled metal drums or rolls having a high-gloss chrome-plated and smooth finish so that the quenched surface on the film is very flat and blemish-free. By controlling the takeoff speed of the quench roll at a faster rate than what the melt exits the die, film thickness can be reduced and controlled. This differential in speed also impacts some final film properties such as toughness, stiffness, and film clarity or haze. Basically, this stretching of the film melt as it is cooling causes the polymer molecules to become better aligned or oriented in the long direction of the film, machine direction (MD). This molecular alignment is responsible for enhance stiffness and toughness properties. The film is then accumulated as rollstock on the windup turret of the cast-film line. Film widths of ≤120 in. are possible by this process. The length of the film is defined by how much is wound on the roll and usually as one film roll is being completed the windup turret can be indexed around so that a new roll can be started without interrupting the continuous resin flow through the extruder and die. The rolls are then removed from the windup turret and slit to the width desired by the customer. Property enhancing lubricants, stabilizers, or colorant additives can be added to the resins pellets at the feed hopper. They will become mixed with the melt stream by the extruder screw. These additives can be added as powders to the feed hopper or more likely will be added as a preblended concentrate in pellet form.

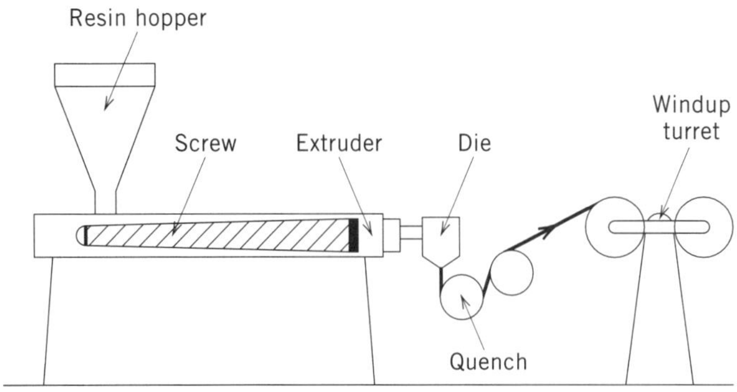

Figure 1. Cast-film line.

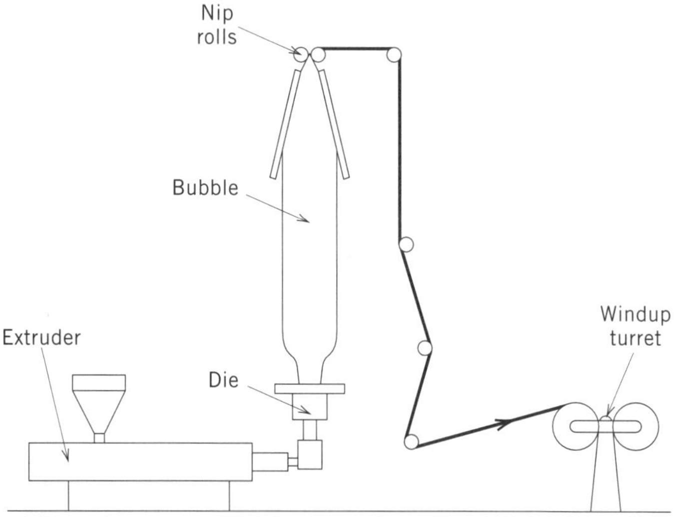

Figure 2. Blown-film line.

Extrusion–blown film. The blown-film process (Fig. 2) is somewhat similar to the cast-film process in that a plasticating screw extruder is employed to convert the solid resin pellets into a uniform plastic melt. The blown-film process employs a circular or annular die to form the film, which is then stretched over and around a captured bubble of air. The film is usually air-quenched, collapsed as a flat tube, and wound up on the windup stand. Sometimes the bubble is slit in the vicinity of the nip rolls so that two sheets of film may be fed down to the windup stand. This bubble process can be oriented so that the film extrudes upward, downward, or in the horizontal direction. The bubble may be very long but typically is 20–30 ft long. Bubble diameters may be 6 ft or more. The resin used in a blown film process must have a high melt strength in order to sustain the relatively high hoop stress and MD tension. This process requires that the melt emerge from the die, then quickly flare out or inflate around the captured air bubble while it is being cooled. Thus the polymer

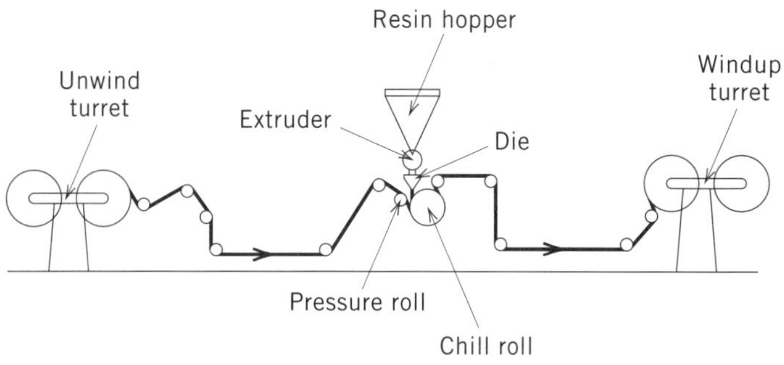

Figure 3. Extrusion coating line.

melt is being cooled and biaxially oriented at the same time. The ratio between the annular die diameter and the final bubble diameter defines the "blowup ratio" (BUR) for the film from this process. The die gap opening and BUR define the final film thickness. Because the blown film process requires such high polymer melt strengths, higher-molecular-weight polymers must be used and are reflected in the typical melt index values of <1–5 dg/min for polyolefin resins (see Fig. 4). Coextrusion blown film processes are useful for preparing multilayer films. Care must be taken to ensure that the tacky heat-seal layer ends up as the outside of the bubble and not the bubble inside, where it might stick to itself as the warm bubble is collapsed at the nip. One advantage of the blown-film process is that it yields a film tube and not a flat film with thick edge beads. Thus blown-film products can be more efficiently used since the thicker-edge beads are not present and do not need to be removed or recycled.

Extrusion coating. As this name implies, the extrusion coating process is employed to apply a thin layer of polymer film onto an existing film structure (see Fig. 3). Thus it is not usually used to make structural mono or coextruded films. It is often used to apply a thin sealant layer to paper, foil, or plastic film. As Figure 3 shows, this process resembles a cast-film process. Extrusion coating requires that the exiting melt curtain quickly contact and adhere to the substrate film. Thus the proper compatibility and adhesion between the substrate and the extruded resin is important as is the need for the resin to have a relatively high melt flow value (see Fig. 4).

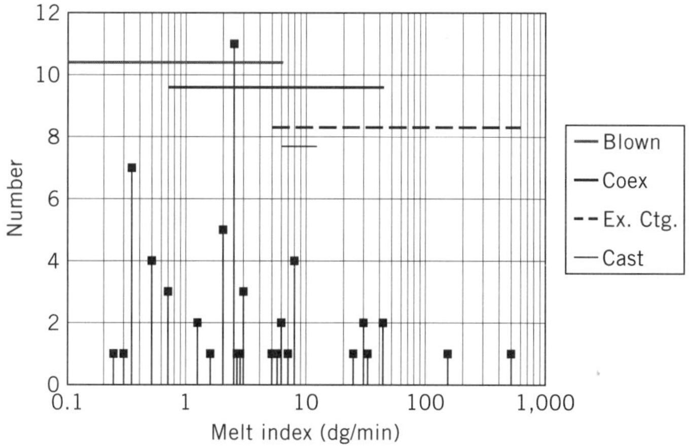

Figure 4. Resin meet flow vs extrusion process.

Typical extrusion coating resin melt flow values can range from about 8 dg/min up to values of several hundred decigrams per minute. The adhesion between the melt curtain and the substrate is aided by the pressure on the nip roll. Since the high-melt-flow polymer is hot and quickly contacts both the coated stock and the quench roll, the quench-roll surface is usually machined or treated to reduce adhesion. Chill-roll release additives can also be incorporated into the resin to prevent sticking (see also Extrusion coating).

BIBLIOGRAPHY

1. L. E. Nielsen, *Mechanical Properties of Polymers,* Reinhold, New York, pp. 27, 31.
2. W. A. Jenkins, *Packaging Foods with Plastics,* Technomic Publishing, Lancaster, PA, 1991.
3. B. A. Morris, *TAPPI J.* **75**(8), 107–110. (Aug. 1992).

General References

J. H. Briston, *Plastic Films,* 2nd ed., Godwin, London, 1983.

F. W. Billmeyer, Jr., *Textbook of Polymer Science,* Wiley, New York, 1965.

Allen G. Kirk
E. I. Du Pont de Nemours, Inc.
Wilmington, Delaware

FILM, RIGID PVC

Poly(vinyl chloride) (PVC) is extraordinarily adaptable to custom compounding for desired performance. This diversity of physical properties at relatively low cost has been the driving force behind its popularity in the packaging industry.

Homopolymer PVC Resin

Commercial PVC resin is a dry free-flowing powder produced by the polymerization of vinyl chloride monomer (see Poly(vinyl chloride).

Molecular weight and viscosity. The fundamental property of PVC is its molecular weight: a measure of its means polymer chain length and a parameter proportional to the resin's viscosity. In general, higher-molecular-weight (higher-viscosity) resins require higher processing temperatures and yield film or sheet with higher heat-distortion temperatures, impact resistance, and stiffness than do resins of lower average molecular weight. The processing equipment for generating film and sheet for the packaging industry demands low- to medium-viscosity PVC resins with relative viscosities of 1.75–2.10, in contrast to extrusion pipe and construction markets that typically use medium- to high-molecular-weight resins.

Heat stability. All PVC resins are subject to thermal degradation during processing and must be compounded with appropriate heat stabilizers to minimize discoloration. These stabilizers serve to scavenge free radicals that perpetuate degradation, as well as hydrogen chloride, the principal degradation product. The highly complex thermal degradation mechanism proceeds by an "unzippering" process whereby liable allylic chlorides act as reaction sites for the liberation of hydrogen chloride and the formation of conjugate-bonding systems. When the conjugation exceeds six bonds in length, color development begins, and if unchecked by ample and appropriate use of heat stabilizers, will progress from a very subtle yellow tint to amber, and finally to black. Commerical processing methods occasionally generate "burned" material that has reached these initial stages of degradation. It is important to recognize that such thermal degradation is both time- and temperature-dependent and that although stabilizers retard the rate of degradation during processing, they do not prevent it. The stabilizers also help protect the film or sheet during subsequent processing (eg, thermoforming), and during the lifetime of the package itself.

Compounding for Properties

Among all polymers used in the packaging industry, PVC is widely regarded as the most versatile and suitable for custom compounding to deliver special properties (see Additives, plastics). It may be compounded for high clarity and sparkle or for maximum opacity and it accepts a full range of custom colorants (see Colorants). Properly compounded, PVC film and sheet are approved for food and drug contact and are available with residual vinyl chloride monomer (VCM) levels below 10 ppb. Examples of what custom compounding can produce are shown in Table 1.

After identification of such desired properties for the specific packaging application, the compounder selects a suitable resin viscosity. If the film is to be approved for food or drug contact, the resin's VCM level must be low enough before processing to ensure that the resulting film will meet all customer requirements on residual VCM. A heat stabilizer must then be selected. Tin mercaptides are frequently chosen because of their high efficiency, excellent early color, good light stability, and excellent crystal clarity in the product (1). Some of these stabilizers (octyl tins) are cleared for food and drug contact. Uncleared options include lead stabilizers, which are limited to opaque systems, and combinations of barium, cadmium, and zinc. Although there are a few calcium/zinc systems in limited use in food packaging, the tin stabilizers dominate the packaging field. Octyl, tins are the principal systems used in food and pharmaceutical packaging (2). Stabilizers are also available that impart improved uv resistance. All of the stabilizers mentioned above are used in rigid PVC at only very small loadings.

Table 1. Examples of Custom Compounding

Higher heat-distortion temperatures for hot-fill packaging
Ethylene oxide (ETO) sterilizable film without water-blush for medical devices
Improved low-temperature impact resistance for drop tests of shipping cartons
Improved uv resistance
Improved outdoor weatherability
Improved sealability (impulse heat. RF. ultrasound)
Denesting formulations for machine-fed blisters
Static-resistant formulations
Optimum performance in laminating to other materials (PVDC. PE, etc)
Formulations for vacuum metallizing
Improved printability
Absence of "white break" or crease whitening

In contrast, impact modifiers may be present up to 15% of the product's weight. As a result, the proper selection and loading of impact modifier is an important compounding decision. Clear packaging films typically contain MBS impact modifiers because of their superior clarity, heat stability, and room-temperature efficiency. ABS modifiers are good for opaque products, and chlorinated polyethylene (CPE) and acrylics (see Acrylics) are often selected for outdoor applications and/or low-temperature environments in opaque systems. Pigments may then be added to provide custom color, and titanium dioxide is generally used at levels as high as 15% to provide the desired level of opacity. Fillers may be used for cost reduction in opaque systems, and in many cases, to improve such physical properties as impact strength, stiffness, and heat-distortion temperature. Present in very low levels are a variety of proprietary lubricants and processing aids that are necessary to facilitate processing and to provide desirable properties such as slip, denesting, and improved thermoformability for the film processor/packager. Flame retardants, antioxidants, coupling agents, antistatic agents, phosphite stabilizers, and a host of additional additives may be included if necessary. Because of this tremendous facility for custom compounding, and the variety of products that result from it, Table 2 must be considered only as a general guide to typical rigid PVC properties.

Film and Sheet Production Methods

Extrusion (see Extrusion) and calendering are the principal methods of producing rigid PVC for the packaging industry. Extrusion is used to produce very thin blown films (see Film, flexible PVC) as well as heavy-gauge sheeting nearly 1 in. (2.54 cm) thick produced by sheet-die methods. Calendering requires a much greater capital investment, but it offers much greater production rates, superior gauge control (cross direction and machine direction) (±5%), superior cosmetic quality including clarity, and much wider versatility in accomodating gauge and width changes. Calendered film and sheet generally has better dimensional stability, which provides thermoforming consistency throughout a given lot. Rigid calendered PVC is available in thickness of 2–45 mil (51–1143 μm) with gloss, matte, or embossed surfaces, either in rolls or in sheets up to about 60 in. (1.5 m) wide. Calendering is the principal means of processing rigid PVC film for packaging.

Calendering. In a modern rigid PVC calendering operation, compounding is done by computer-controlled electronic scales that supply precise amounts of each ingredient to a high-intensity mixer designed to incorporate all liquids into the resin particles and to secure uniform distribution of all powdered ingredients. Blending is generally done for a specific time period and to a specific temperature. The still-dry, free-flowing blend is then charged to a feed hopper where it is screw-fed into a continuous mixer such as an extruder or kneader. Under the action of this mixer's reciprocating screw in the confined volume of the mixing chamber, the blend begins to flux or masticate into the plastic state. It is then forced out of the barrel of the mixing chamber. The continuous strand may be chopped into small fist-sized buns of hot material or simply exit as a continuous rope. This material may then be directly conveyed to the calender, or it may first pass through a two-

Table 2. Typical Physical Properties of Rigid PVC (Clear)

Property	Test method[a]	Units	Values
Specific gravity[b]	D1505		1.30–1.36
Yield (1.30 sp gr)	D1505	in.^{2} lb (cm^2/g)	
7.5 mil (0.19 mm)			2850 (40.5)
10.0 mil (0.25 mm)			2130 (30.3)
12.0 mil (0.30 mm)			1780 (25.3)
15.0 mil (0.38 mm)			1420 (20.2)
20.0 mil (0.51 mm)			1070 (15.2)
Tensile strength (yield)	D882	psi (MPa)	6500–7800 (44.8–53.8)
Tensile modulus	D882	psi (MPa)	2.5–4.0 × 10^5 (1723–2757)
Elongation (break)	D882	%	180–220
Izod impact (1/4 in. or 6.4 mm)	D256	ft · lbf in (J/m)	0.5–20.0 (26.7–1068)
Gloss, 20°	D247		120–160
Heat-distortion temperature (264 psi or 1.82 MPa)	D648	°F (°C)	158–169 (70–76)
Cold-break temperature	D1790	°F (°C)	14 to −40 (−10 to −40)
WVTR (38°C, 90% rh)	DIN53122	g (100 in.^{2} · 24 h) [g (m^2 · day)]	
7.5 mil (0.19 mm)			0.30 [4.7]
10.0 mil (0.25 mm)			0.20 [3.1]
Surface resistance	DIN53482	Ω	10^9–10^{13}
Specific resistance	DIN40634	Ω · cm	10^{13}–10^{15}
Dielectric strength	DIN40634	kV/mm	60–70
Specific heat (20°C)		kJ/(kg · K)	0.8
Thermal conductivity		W/(m · K)	0.16
Linear thermal expansion		K^{-1}	7.0–8.0 × 10^{-5}
Infrared absorption[c] (3–18 μm)			Various intensities

[a] Ds are ASTM test methods, and DINs are German (Deutsche) Industrial Norm test methods.
[b] Indirectly related to amount of impact modifier. Increased opacity may raise to 1.40.
[c] 20 mil (508 μm) unmodified.

roll mill. The calender is a large unit, typically consisting of four or five heated rolls designed to process masticated PVC buns into a continuous web of designated width and thickness (see Fig. 1).

Figure 2 illustrates the typical "L" and "inverted L" configurations generally used for rigid and flexible production, respectively (3). The calender rolls have separate temperature and speed controls as well as roll bending and crossing capabilities to control profile across the web. Proper use of these controls, as well as speed and stretch in the takeoff train, allow the production of an extremely flat sheet with a profile tolerance of less than ±5% across and down the web. Such control is maintained by continuous beta scanning equipment that traverses the web constantly and calls for adjustments in nip openings, skew, and/or roll bending. Such constant in-process monitoring and continuous profile adjustments is a significant advantage of calendering over other processing methods. By using special grit-blasting techniques, the third and fourth calender rolls may be custom-surfaced to generate a uniform two-sided matte product. Alternatively, one or more downstream embossing stations may be utilized to produce a custom surface on one or both sides of the film. Antistatic and or denesting slip agents may be applied to the surface(s) of the web after separating from the last calender roll. Finally, after the cooling section, the web is cut in-line into finished sheets or wound about a core into a master roll for subsequent custom slitting. Typical slit widths are made to the nearest ⅓₂ in. (0.8 mm) on 3- or 6-in. (7.6- or 15.2-cm) cores with roll diameters of 14–40 in. (36–102 cm).

Figure 1. PVC calender.

Package Production by Thermoforming

Most commercial PVC packages are the result of thermoforming rollstock into custom blisters. In those cases where further enhancement of PVC's own oxygen and/or moisture barrier properties are required, barrier materials (see Barrier polymers) such as PE, PVDC, or fluoropolymer film (see Film, fluoropolymer) may first be laminated to the PVC web prior to thermoforming. Thermoforming processing conditions are generally dictated by the PVC material itself regardless of

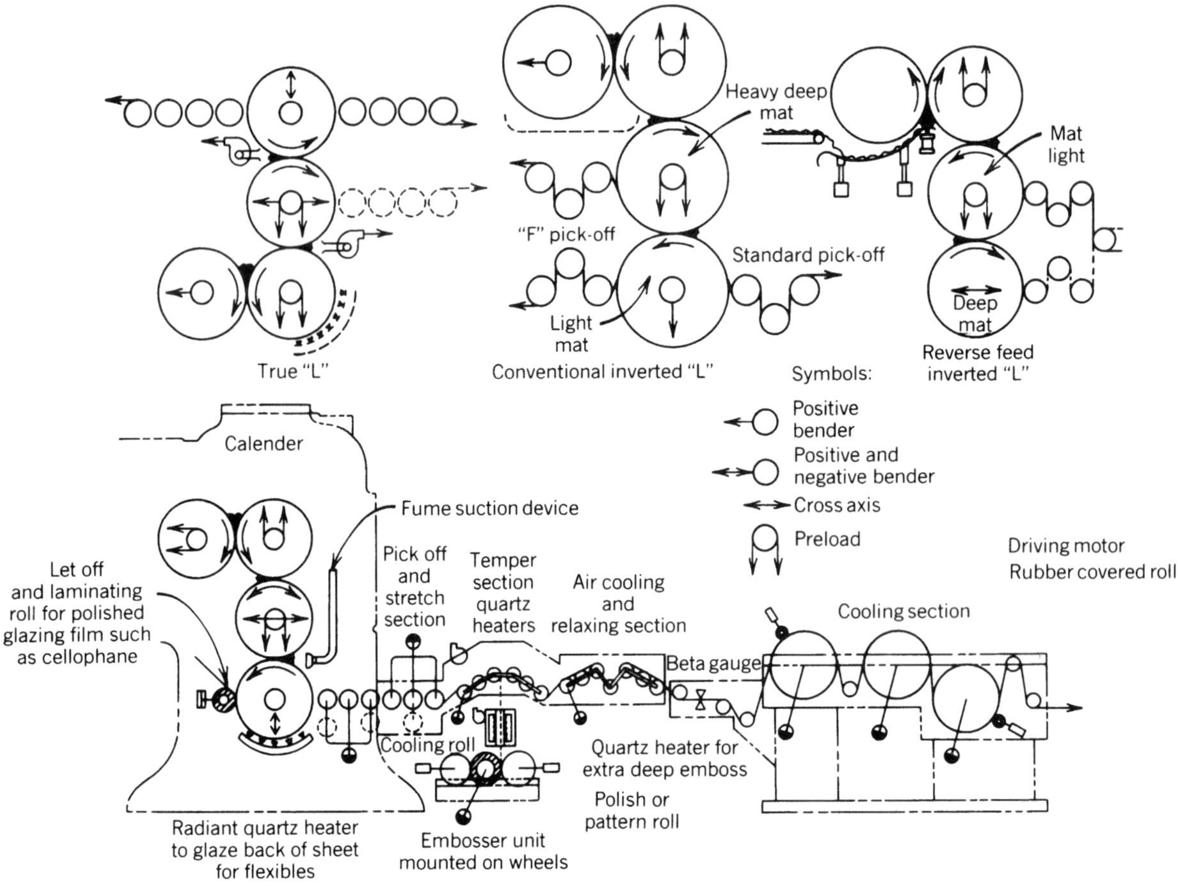

Figure 2. PVC calender operation.

lamination. Since PVC is an amorphous material (see Polymer properties), it softens over a large temperature range and has no sharp melting point. There are two temperatures ranges in which rigid PVC can most readily be formed (see Fig. 3). It must be emphasized that these temperatures are actual film temperatures that must be measured with thermocouples located directly on the surface of the film (heating-element temperatures are very much hotter).

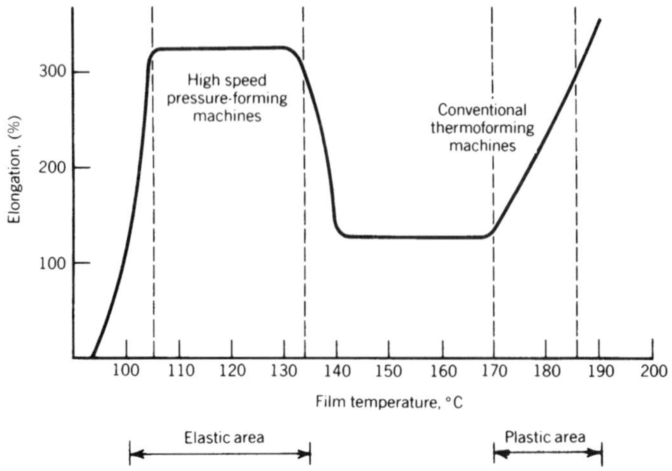

Figure 3. Temperature ranges for PVC thermoforming.

The first plateau at 221–275°F (105–135°C) is the elastic area best for most high-speed form/fill/seal pressure-forming machines (see Thermoform/fill/seal). In this area, the film has sufficient hot strength to elongate or stretch to the conformation of the mold. Between 275 and 338°F (135 and 170°C) is a region of inadequate elongation for proper forming, and attempts to process in this area may result in blowing holes, tearoffs, and poor definition. Failure to respect this "no man's land" is among the most common reasons for thermoforming problems seen in the field by PVC technical service representatives. The elongation quickly increases in the film at 338°F (170°C), and excellent forming is possible up to about 365°F (185°C). This is the range for optimum performance for conventional commercial thermoforming machines (see Thermoforming). Beyond 365°F (185°C), the material will sag excessively, resulting in webs, wrinkles, holes, or thin areas. In general, rigid PVC film thermoforms best when the film temperature is slightly above 338°F (170°C), with relatively cool molds (as low as 50–60°F or 10–16°C), and when plug assist temperatures are about 194°C (90°C). Mold design should avoid exceeding a 1:1 draw ratio and few extra degress of draft and extra radius on corners will help avoid problems.

Film shrinkage for optimum forming. Since distinct types of equipment are used in each of the two forming regions, it is particularly important that the PVC-film supplier have a

Table 3. U.S. Markets for Calendered Rigid PVC Film and Sheet[a]

	Market 10^6 lb (10^3 tons)
Packaging (FDA-grade)	
Food	25 (11.4)
Pharmaceuticals	30 (13.6)
Medical devices and supplies	10 (4.5)
Total packaging (FDA grade)	65 (29.5)
Packaging (general-purpose)	
Thermoforming—blisters, clam shells, etc	225 (102.3)
Vinyl boxes and lids	35 (15.9)
Static control thermoforms	5 (2.3)
Total general-purpose and specialty packaging	265 (120.5)
Specialty and industrial applications	
Printing and stationery	90 (40.9)
Cooling tower and wastewater fill media	40 (18.2)
Artificial Christmas trees	30 (13.6)
Furniture lamination	10 (4.5)
Construction and flooring	10 (4.5)
Floppy disks	8 (3.6)
Litho mask	5 (2.3)
Total specialty and industrial applications	193 (87.6)
Total packaging (61.5%)	330 (150.0)
Total nonpackaging (36.5%)	193 (87.7)
Total rigid PVC film and sheet	523 (237.7)

[a] *Source:* Klöckner-Pentaplast of America, Inc.

thorough understanding of the appropriate shrinkages to be put into the feedstock for each type of machine. PVC destined for pressure-forming over the lower temperature region should have its shrinkage controlled by the producer at 284°F (140°C, a temperature which will release all plastic memory relevant to the processing range of 221–275°F (105–135°C). Typically best results are achieved with PVC film having 284°F (140°C) shrinkages of 2–5% in the machine direction and 0–1% in the transverse direction. Such very slight growth in the transverse direction is desirable to compensate for the repeating necking in that may otherwise occur across each line of blisters such as those used for press-through packaging of ethical drugs.

In the higher temperature range for conventional thermoforming, film shrinkages should be controlled by the film supplier at 350°F (177°C), a temperature that will release all plastic memory relevant to this higher processing range. Best results are typically achieved with PVC film having 350°F (177°C) shrinkages of 4–8% in the machine direction and −1 to +1% in the transverse direction. Excessive shrinkages may cause the film to pull out of the chains, clips, or frame. Insufficent machine direction shrinkage and/or excessive transverse growth may led to webbing. Thermoforming machine operators often respond to such webbing problems by reducing operating temperatures to prevent excessive sag, but this step can result in blowing holes and poor wall distribution because it takes the film out of the optimum thermoforming temperature range and elongation falls off rapidly. Careful shrinkage control is critical to successful thermoforming of rigid PVC film with all types of forming equipment. Proprietary process controls are available in modern PVC calendering technology to custom produce film with the proper shrinkages for optimum thermoformability within each temperature region and on each design of forming equipment.

Packaging Market

U.S. consumption of calendered rigid PVC film and sheet in 1994 is estimated at 523×10^6 lb (237,000 tons), not including an additional 30% for extruded sheet. About 63% of this calendered production was used in packaging (see Table 3). Total 1994 consumption of calendered rigid PVC film and sheet for packaging was 330×10^6 lb (150,000 tons).

BIBLIOGRAPHY

1. A. A. Schoengood, ed., *Plast. Eng.* **32**(3), 25 (March 1976).
2. M. McMurrer, ed., "Update: PVC Heat Stabilizers," in *Plastic Compounding,* Resin Publications, Cleveland, OH, July/Aug. 1980, pp. 83–90.
3. L. R. Samuelson, *Plast. Des. Process.* **21**(8), 14 (1981).

General References

P. Bredereck, *J. of Vinyl Technology* **1**(4), 218–220 (Dec. 1979). *Guide to Plastics,* McGraw-Hill, New York, 1979, p. 27.

M. McMurrer, ed., *Plastic Compounding 1984/85 Redbook,* Vol. 7, No. 6, Resin Publication, Cleveland, OH. 1985.

M. McMurrer, ed., "Update—U.V. Stabilizers," *Plastic Compounding,* Resin Publications, Cleveland, OH, Jan.Feb. 1985, pp. 40–57.

Thermal Stabilization of Vinyl Chloride Polymers, Technical Report 3250, Rev. 7/68, Rohm & Haas, Philadelphia, 1968.

L. R. Samuelson, *Plastics Design and Processing* **21**(8), 13–15 (1981).

V. Struber, *Theory and Practice of Vinyl Compounding,* Argus Chemical Corp., New York, 1968.

H. J. G. Van Beek
R. G. Ryder
Klöckner-Pentaplast of America, Inc.,
Gordonsville, Virginia

FILM, SHRINK

Shrink film is a name given to a unique family of films that are distinguishable by their attributes, production processes, and end uses. They are composed of two basic categories: polyvinyl chlorides (PVCs) and polyolefins. Although most plastics exhibit some amount of free shrink and shrink force at elevated temperatures, true shrink films (sometimes known as *heat-shrinkable films*) provide a high degree of free shrink with a controlled level of shrink force over a broad temperature range.

Shrink force must be controlled to prevent crushing or deforming the product being packaged. Providing the proper level is important to the marketing of toys; games; cards, calendars, and other paper products; hardware; food; and a variety of merchandise where a tight, glossy package appearance is essential.

Shrink films are produced by uniaxially or biaxially orienting a sheet or tube of film by imposing a draw force at a temperature where the film is softened but kept below its melting point, then quickly cooled to retain the physical properties generated during orientation. It is important to note that the

Table 1. Glass-Transition Temperatures and Vicat Softening Points of Common Polymers

Polymer	T_g (°F)	Vicat (°F)
Polystyrene	181–201	208
LLDPE	−188 to −5	177–220
Polypropylene	7–41	307
PVC	158–176	183

orientation temperature occurs between the vicat softening point (ASTM D1525) and the melting point, but is not directly related to the glass-transition temperature (T_g) as some literature would indicate. This can be seen in Table 1, which compares glass-transition temperatures to vicat softening points for several polymers.

Prior to orientation, the molecules of the film are randomly intertwined, exhibiting no particular alignment. However, when a draw force is imposed, the amorphous regions are straightened and oriented to the direction of force. By applying proper cooling, the molecules will be frozen in this state until sufficient heat energy is applied to allow the chains to shrink back. One can visualize this phenomenon by stretching a rubber band and dipping it into liquid nitrogen. It will remain stretched as long as it is kept at sufficiently cold temperatures. However, when enough heat energy is applied, the rubber band will shrink back to its original relaxed state.

Orientation on a commercial scale can be achieved using either of two methods: a tenter-frame or a bubble process. Tenter-frame technology produces a variety of "heat-set" products, of which biaxially oriented polypropylene (BOPP) is the most common. Heat setting is a process whereby a film is reheated in a constrained state such that the shrink properties are destroyed. Other important characteristics derived from orientation (optics, tensile strength, and modulus) remain intact, however. Current tenter-frame technology does not allow the production of materials with a high degree of free shrink and shrink force due to the mechanics and thermodynamics of the process.

In the tenter-frame process, a flat sheet is produced and cooled on a chill roll, which is usually immersed in a waterbath. The sheet then proceeds through a machine-direction orientation (MDO) unit, where it is heated and stretched to the desired ratio. On exiting the MDO unit, the sheet enters the transverse-direction orientation (TDO) unit, where it is reheated and stretched. BOPP is commonly stretched 700–800% in both the machine and transverse directions. After exiting the TDO, the material is wound into large mill rolls for aging (aging allows secondary polymer crystallization, ensuring film flatness and roll uniformity) and converting. Some tenter-frame systems have additional downstream equipment for retensilization and further heat setting or annealing. The primary suppliers of tenterframe systems are Brückner (Germany), Mitsubishi (Japan), and Marshall & Williams (United States).

The second commercial method is the bubble process, sometimes referred to as a *tubular process.* A primary tube is produced by blowing the film onto an external mandrel or casting it onto an internal mandrel. Water is used to help cool the tube at this point.

After it has been cooled, the tube is reheated and air is used to inflate it into a bubble. On inflation, it is oriented in both directions simultaneously, typically 700–800% for BOPP. Other films are oriented 200–1000% in either direction. After orientation, the bubble is cooled using an air ring. The tube is slit and separated, then each half is wound onto mill rolls for aging and converting. Additional in-line processing, such as corona treatment for materials that will be printed at a later time, is often used with this technology. Commercial tubular orientation equipment suppliers include Prandi (Italy) and Gloenco (England). ICI, the developer of much of the early technology, exited this market in the mid 1980s.

Shrink films were limited to monolayer constructions until recently. Technological advances over the past decade have led to the development of multilayer coextrusions. These process improvements have enabled films to be designed with greater control over desired characteristics to meet a much broader range of packaging requirements. Table 2 lists the major shrink films offered in the U.S. market.

The key attributes that are important to shrink films include shrink, sealability, optics, toughness, and slip. Each of these attributes is composed of several facets. For shrink properties, they are onset temperature, free shrink, shrink-force, shrink temperature range, memory, and overall package appearance. For sealing properties, one must also consider ease of trim sealing, trim seal strength, trim seal

Table 2. U.S. Shrink-Film Offerings

Polyolefins	
Polyethylene monolayer	DuPont Clysar HP, LLP Cryovac D601
Polyethylene multilayer	Cryovac D955, D940, RD106, D959
Polypropylene/monolayer ethylene–propylene copolymer	DuPont Clysar EHC, CHS, RSW Okura Vanguard 100 Gunze PSS
PP/EPC multilayer	Cryovac MPD2055, MPD2100, J960, J961 Okura Vanguard 501 Intertape Exlfilm IP-33 DuPont Clysar EZ
Polyvinyl Chloride (PVC)	
Biaxially oriented	Reynolds Reynolon 1044, 2044, 3044, 4044, 5044 Allied Krystaltite T111, T122, T133, T144, T15, R11, R22, R44 Gunze Fancywrap
Preferentially oriented	Reynolds Reynolon 3023, 4061, 5032, 7052 Allied Krystaltite PT152
Specialty Films	
Aroma and oxygen barrier	Cryovac BDF2001, BDF2050
Moisture barrier	Cryovac BDF1000
Moisture, gas, and aroma barrier	Cryovac BDF3000

Table 3. Typical Shrink-Film Properties

Film Type	Unrestrained Shrink (% at 260 °F)	Clarity (%)	Ball Burst (cm/kg)	Coefficient of Friction
Polyethylene monolayer	70–80	40–65	16–24	0.11–0.14
Polyethylene multilayer	70–80	80–90	21–28	0.22–0.32
Polypropylene or EPC monolayer	25–45	75–82	4–10	0.26–0.38
PP/EPC multilayer	50–60	75–85	10–20	0.30
PVC, biaxial	30–55	75–80	6–10	0.20—blocked

appearance, static lap sealability, and thermal lap sealability. For optics, it is important to consider clarity, gloss, and haze.

There are three aspects to toughness: impact strength, slow puncture resistance, and tear resistance. Impact strength measures how well a material resists a sudden force, such as when a box is dropped from a certain height. Slow puncture resistance measures how well a material resists a gradual increase in tension, such as when someone tries to poke a finger through a piece of film. Tear resistance is the measure of how well a film resists tearing once it has been nicked or abraded.

Finally, for slip, it is important to consider both hot slip (such as a warm package being placed in a carton during packoff) and cold slip (important in removing a package from a carton). Many other attributes must be considered for any given application, depending on the nature of the product being packaged. Film properties along with potential advantages and disadvantages are shown in Tables 3 and 4.

Packaging equipment for shrink films are available in a variety of models and price ranges depending on the features desired (see Wrapping machinery, shrink film). An important attribute for equipment is the sealing method. Until recently, most sealing systems for typical shrink-film applications used a hot wire in order to seal and cut the film at the same time. More recently, hot-knife systems have gained popularity on many models because of their increased durability.

The wire or knife seals the film against a pad that is covered with Teflon tape. Impulse seals are also employed for a number of applications. Time, temperature, and pressure are the three variables that must be balanced to optimize seal quality. If the time that the sealing head is held on the film is not long enough, the temperature is too low, or the pressure is insufficient or uneven, the seal quality will be poor. Conversely, if the sealing temperature or the pressure is too high, the result may be a weak spot just behind the seal area, causing failure during distribution.

To prevent seal failure problems, there must be an appropriate amount of time for cooling of the seal prior to any force being applied to the film. This cooling time allows the molten seal to solidify and prevents it from wrinkling or bunching.

Table 4. Advantages and Disadvantages of Various Film Types

Film Type	Advantages	Disadvantages
Polyethylene monolayer	Low cost High slip Strong trim seals More dimensionally stable than PVC Printable	Low modulus Narrow shrink temperature range
Polyethylene multilayer	Excellent optics Low cost Strong trim seals Broad shrink temperature range High abuse resistance Printable	Low modulus
PP or EPC monolayer	High modulus High gloss Low cost High shrink force	Low abuse resistance Easy tear High shrink temperature required Low seal strength
PP/EPC multilayer	High gloss Low cost Strong trim seals Broad shrink temperature range High abuse resistance High modulus Printable	Low abuse resistance Easy tear High shrink temperature required Low seal strength
PVC, biaxial	Low shrink temperature High modulus Low shrink force	Weak seals Noxious fumes created during sealing Corrosion problems Poor machinability Poor slip

PVC shrink films can exhibit another sealing problem. They form a carbon buildup on the seal head, which must be cleaned regularly or it will create a charred seal on the package. Some of the more common equipment manufacturers include Doboy, Great Lakes, Hanagata, Ilapack, Omori, Shanklin, and Weldotron (see also Sealing, heat).

One other piece of equipment needed to provide a sharp, tightly finished package is a shrink tunnel. In order for adequate shrinkage to occur, the package must be exposed to the correct temperature and airflow for the proper amount of time. The ultimate goal is to have a package that is tightly shrunk with no excess material or "ears" on the corners of the finished package. PVC films do not require the same level of temperature control as do polyolefin films, since they shrink more easily when exposed to heat. Shrink tunnels are generally purchased in conjunction with packaging equipment to ensure correct sizing for speed and product application.

BIBLIOGRAPHY

General References

J. L. Throne, *Plastics Process Engineering,* Marcel Dekker, New York, 1979.

R. J. Young and P. A. Lovell, *Introduction to Polymers,* 2nd ed., Chapman and Hall, London, 1991.

E. C. Bernhardt, *Processing of Thermoplastic Materials,* Krieger, Malabar, FL, 1974.

CHARLES R. JOLLEY
GEORGE D. WOFFORD
Cryovac Division, W. R. Grace & Co.-Conn.
Duncan, South Carolina

FILM, STRETCH

Unitization with elastic plastic films (stretch film) is a U.S. innovation that began in the early 1970s and has spread throughout the world to become the most widespread means of containing unit loads. This article focuses on the selection, use, and properties available for stretch packaging unit loads and individual products. Little reference is made to mechanical equipment systems available to accomplish stretch-film unitization.

Saving with Stretch Film

Stretch film is a product thousands of companies throughout the United States and abroad used every day in tremendous quantities. And, although acceptance and audience of stretch film continue to grow, the majority of stretch-film users know little of its true potential or how it has developed over the years.

Introduction

Stretch film is such an effective product that it is often easy to take all that it does for granted. This section is intended to serve as a "reminder" of the benefits of unitizing, and the additional benefits that you reap by unitizing with stretch film. As part of the discussion on the benefits of stretch film, the testing of packages and unit loads are addressed. Finally, in order to give you background knowledge on stretch film, this section also reviews the development and refinement of stretch film over the years. All these issues are covered under the following sections:

- Why unitize?
- Why unitize with stretch film?
- Testing packages and unit loads
- The evolution of stretch film

Why unitize. *Unitization* is defined as a process in which smaller individual items are gathered and made into larger unit loads.

The benefits of unitizing products are many. Although unitization can be achieved with other packaging materials, none does it as effectively and cost-efficiently as stretch film. But before discussing the merits of stretch film, it is important to first understand why its necessary to unitize. Here are the reasons.

Lower handling costs. In general, whenever a load of packaged products increases in quantity, size, and weight, the cost per unit handled becomes lower. Unitizing products, therefore, significantly reduces handling costs.

Labor savings. Without exception, the handling of loose boxes is more time-intensive than handling a unitized load. Thus, another benefit of unitization is the amount of labor hours saved in the movement and shipment of packaged goods. The end result is a quicker flow of goods throughout factories, warehouses, and cross-dock operations while utilizing fewer hands.

Transportation savings. Since unitizing accelerates the speed at which goods are moved, carrier vehicles spend less time at unloading/loading docks. This reduction in unloading/loading time provides valuable cost savings to the company that unitizes and the customers receiving its prodcuts.

Protection. The protection that unit loads afford an operator are threefold:

1. Unit loads reduce incidents of pilferage and theft because packages from an enclosed load are difficult to remove. Moreover, unit loads are much easier to track than individually packaged products, making the "disappearance" of goods more detectable.

2. Unit loads, which are placed on pallets and moved by lift truck, are less likely to suffer damage in transit than loose boxes carried by hand, because lift trucks keep pallets closer to the ground. For instance, whereas small boxes ($\leq$20 lb) may often receive impacts from 30 in. or higher when accidentally dropped, a unit load on a pallet will seldom receive 12 in. in drop.

3. It is less likely in LTL (lift-truck load) shipments that dense, heavy items will be parked on top of palletized unit loads. Hence, your goods stand a greater chance of not being damaged. This benefit, however, does not hold true for cartons shipped individually.

Inventory control. Unitizing products facilitates improvements in control of inventory since large unitized loads (as opposed to parcels of smaller ones) can be identified, counted, and managed more easily.

Customer service. Customers appreciate the effects of unitizing because it allows them to unload their trailers and move goods through their warehousing systems more efficiently.

Why unitize with stretch film? From the advantages listed, the importance of unitizing is clearly evident. What is equally important to realize is that stretch film accomplishes each of these unitizing functions better than any other packaging material. Also, stretch film performs additional functions other unitizing materials simply cannot:

- *Low Supply Cost.* Compared to other packaging and bundling materials, stretch film is the lowest in cost to use because it offers the most yield for the dollar.
- *Protection from Moisture, Dirt, and Abrasion.* Stretch film protects against these elements, safeguarding the goods it wraps.
- *Reliable Performance.* With stretch film, you can quickly see the holding performance and protection it offers. The holding power, however, of other packaging materials such as adhesives is not always immediately noticeable since they can unravel, loosen, or come undone without being noticeable a few minutes or hours after observation. Stretch film, manufactured by a company committed to quality, will produce consistent results every time it is used.
- *Automation.* Stretch-film machines are either partially or fully automated, which improves the productivity of the packaging process, reduces labor, and alleviates back-breaking and time-consuming work.
- *Scan-through Optics.* The clarity of the stretch film enables UPC codes to be quickly read and scanned while protecting labels and lot codes printed underneath the film.
- *Ease of Removal.* Stretch film only clings to itself and not surfaces, making its removal fast, clean, and easy.
- *Elimination or Reduction of Other Packaging Materials.* The superior load-holding capability of stretch film either eliminates the need for or helps scale down the bulk of other packaging materials (eg, in some cases, stretch film may enable thinner cardboard cartons to used for packaging goods).
- *Recycling.* Stretch film is recyclable. In fact, several stretch-film manufacturers will pick up and recycle your stretch film if you qualify for their recycling programs.

Testing packages and unit loads. Stretch film saves you money. By offering superior unitizing and protective capabilities, it drastically reduces product damage, which contributes to your bottom line. But you do not have to take anyone's word for it. There are two in-plant tests you can perform to get an initial feel for the durability of the unit loads you wrap in stretch film before sending them off to your customers.

The first preshipment test is "test course handling." In this test a lift truck equipped with forks or slip sheet attachment approaches the unitized load, picks it up, accelerates, corners, decelerates, and sets the load down. The lift truck then repeats this same test once again, except this time moves in reverse. A typical test sequence includes two to seven repetitions, with an average of four repetitions for the assurance required by most users.

The second preshipment test is the "free-fall drop." This test does exactly what its name implies. In this test, one end of a unit load is raised to a specified height of 3–12 in. and released to fall flat on the floor. The drop is then repeated on the opposite end of the unit load. Drop height will depend on the weight of the unit load and the assurance level that is desired.

Further shake-table or vibration-table test procedures will provide you with stretch-film load-force data for your products. It is recommended that you work closely with your local stretch-film packaging sales professional when packaging new products or considering stretch-film changes.

The evolution of stretch film. Although many people enjoy the benefits of stretch film today, stretch film was not developed overnight. The creation and development of stretch film has been a gradual process, one that continues to undergo refinements to this day. For a better idea of where stretch film has been and where it's headed, let's take a look at this timeline:

Year	Machine Evolution	Film Development	Volume and Cost Per Average Unit Load Wrapped
Early 1960s		Shrink-wrap use begins to grow in Europe	
1965		Shrink wrap adopted in United States	30 oz/$1.50 –2.00
1973	Lantech develops first U.S. stretch-film wrapping equipment	Mobil develops the first stretch film	
		Introduction of Mobilrap "C" (LDPE) 30–50% stretch	18 oz/$1.10
1974		PVC stretch films introduced	
1976		EVA stretch films introduced	
1978		First LLPDE film introduced with much higher stretch levels	10–12 oz/50–80¢
1980	Prestretch introduced	Hand-wrap market begins to grow	

Year	Machine Evolution	Film Development	Volume and Cost Per Average Unit Load Wrapped
1981	Powered prestretch equipment introduced 150–250%	Stretch film cling at high stretch a challenge Convenience hand wraps introduced	8 oz/50¢
1983	Machinery and film advances allow prestretch up to 300%	Cling consistency improves with better control by film manufacturing Nonmigratory cling introduced	
1987		Higher-load-force cast films introduced Thinner gauge (50, 60) become available	4–6 oz/25–40¢
1991	Bi-stretch (trademark) introduced	Mobile introduces Stretch-film recycling Program	4 oz/25¢
1992		Postconsumer recycled content (PCM) hand wrap introduced	

New Packaging Solutions

When you mention stretch film, most people conjure up the image of a load of boxes or goods wrapped on a pallet. For good reason. This application is the most popular one for stretch film. But an "on pallet" application is just one of hundreds that stretch film is capable of performing. All it takes is a little imagination and ingenuity to customize stretch film for your own uses. An increasing number of companies are looking to stretch film to bundle and package goods once wrapped in other materials such as tape, twine, strapping, and corrugated boxes. In this section, some of these innovative applications and the benefits derived from them are shared with you under the following headings:

- Stretch-film advantages
  - Versatility
  - Protection
  - Productivity
- Cost benefits

Stretch-film advantages. Think about all the areas in your operation where you currently use tape, twine, strapping, etc, and chances are stretch film can do a better job. Stretch film surpasses other bundling materials for the following reasons:

Versatility

Stretch film is able to fit the contours of any product. Door and window manufacturers have switched over in increasing numbers to stretch film. Stretch film, because of its stretchability, has largely replaced the custom-size boxes that were once needed to package the different-size doors and windows being produced.

Stretch-film clarity makes products easier to identify. Labels and lot and UPC codes can be read as well as scanned because of the transparency of stretch film. Food wholesalers and distributors have taken advantage of this feature in implementing and expediting their cross-docking practices.

Protection

Stretch film provides better holding force than do other bundling and packaging materials. Because stretch film spreads across the entire surface area of a load, its holding force is dispersed, not concentrated at only a few points like strapping. This attribute enables stretch film to maintain exceptional load integrity. The furniture industry has come to recognize the ability of stretch film to hold even the most unstable of loads together, which is why an increasing number of component manufacturers have turned to stretch film to bundle loose furniture parts, such as table and chair legs and stair banisters.

Because stretch film clings only to itself, it does not leave any sticky or messy residue on product surfaces. This feature has made stretch film a favorite of the furniture industry. With stretch film, there are no glues or adhesives that can damage fragile finished wood. As a result, furniture manufacturers use stretch film for a wide variety of tasks, some of which are to keep furniture drawers shut so that they do not open in transit, attaching accessories to furniture pieces and to wrapping-desk and table tops.

Unlike other bundling and packaging materials, stretch film protects the products it wraps from dirt, moisture, and abrasion. Carpet and textile manufacturers depend on the protective attributes of stretch film to safeguard their rolls of carpet and fabric.

Productivity

Stretch film increases productivity. Thanks to automated stretch-film machines, newspaper companies have been able to bind stacks of newspapers together more time-efficiently and cost-effectively by reducing much of the manual handling involved.

Cost benefits. Stretch film costs less than most conventional bundling and packaging materials. But the savings you earn with stretch film goes way beyond just a lower unit cost. Stretch film also saves you money in the following ways.

Stretch film offers more yield than do other bundling materials. Because stretch film has the ability to stretch, you get more product per square inch than with any other bundling material.

Stretch film can eliminate or reduce the need for secondary packaging materials. One example of a stretch packaging cost savings is that of a national T-shirt manufacturer who used to package its T-shirts for shipment in expensive corrugated boxes. However, once it was seen that stretch film could do the same job at a fraction of the cost, the company switched to stretch film. Today, the company is saving $5000 a day in packaging and shipping costs.

Stretch film eliminates or reduces the need for tertiary packaging. Because of superior load-holding force of stretch film, money once spent on tertiary materials such as strapping, twine, and kraft paper can now be saved and better spent.

Stretch film saves on shipping costs. Since products bundled in stretch film are usually more compact—this reduction in size is achieved by eliminating the need for extrabulky packaging materials like boxes—they take up less space in freight trucks. More space in the back of a truck means additional space for more of your products—and that saves you money!

Selecting the Right Film for the Right Job

There is a wide array of stretch films available on the market, each designed to meet different needs. Choosing the right stretch film(s) for your operation can be a bit confusing. In this section, the variables to choosing the proper stretch film for your particular needs are explained. Finally, this section discusses the characteristics of the different types of stretch film and explain how they are made. All these issues are covered under the following headings:

- Proper film selection
- How stretch film is made
  - Extrusion
  - Coextrusion
  - Cast fabrication process
  - Blown fabrication process
- Process influence on stretch-film properties
- Opaque and color tint films

Proper film selection. Whenever selecting film, the following variables should be taken into account.

Type of product. Stretch-film type is dependent on what you are wrapping. Lighter, more fragile goods require a lower gauge film to avoid being crushed than do heavier, more solid goods, which can withstand the pressure of higher gauges.

Type of load. There are two factors that determine load type: weight of the load and uniformity of its outside surface. In stretch film, load types are distinguished by the first three letters of the alphabet. A loads have no irregularities, B loads have protrusions of ≤ 3 in., and C loads have protrusions of >3 in.

Type of equipment. Machine stretch levels will help determine what type of film is necessary to wrap a load. Whereas higher gauges were once required for machines with high prestretch levels, today, Mobil is manufacturing thinner, tougher films that are able to endure greater levels of stretch while maintaining their strength.

Film thickness. Commonly available gauges for stretch film range from 0.5 to 3.0 mil. Load type, weight, and height are important determinants in selecting film thickness. As prestretch increases, it is extremely important that the stretch film not lose its strength.

Light gauges are excellent dust covers for already secure and strapped loads. Heavy gauges or new-technology metallocene-based films are excellent for loads that require a high degree of containment.

Film width. Most stretch films range from 2 to 70 in., but can be made up to 100 in. by request. Rotary spiral equipment usually uses 20-in.-width film, but can also use 25- and 30-in. film. Full-web equipment uses a wide variety of widths depending on load height.

To determine which film widths are best for your individual applications, contact a packaging sales professional for assistance.

How stretch film is made. Understanding how stretch film is made, and the resulting properties that each type of film possesses, can help you when it comes time to purchase the proper film for your operation. The following industry terms, accompanied by diagrams, describe the various manufacturing processes of stretch film and the types of film that are produced as a result:

Extrusion. This is the first step in the production of stretch film in which plastic (polyethylene) resin pellets are converted into melted plastic. For this melting to occur, the resin pellets are loaded into the hopper funnel of the extruder and dropped by gravity into the barrel where the mixture is heated to its melting point. This liquefied plastic is then pushed forward by a continuously turning screw. In the final phase of extrusion, the molten material (or extrudate) is pushed through an opening called a *die,* which shapes the plastic for film formation. (See Fig. 1.)

Coextrusion. Coextrusion is a form of extrusion using multiple extrudates and dies to create stretch film with multiple layers. Coextrusion involves the simultaneous extrusion of two or more layers of plastic that are brought together while still melted and then cooled to form a multilayer film. The layers of this film may or may not be different materials and may or may not exhibit different properties. (See Fig. 2.)

Cast-film fabrication. After the extrusion process melts the plastic, this plastic is pushed through a slit-shaped die. The plastic is then formed into a sheet as it falls into a roller and is pulled vertically downward, orienting the polymer in that direction. This sheet is cooled by passing it through chill rolls, which are kept cool by circulating water inside them. Finally, the sheet is cut into the proper combination of widths and wound into finished stretch film rolls. (See Fig. 3.)

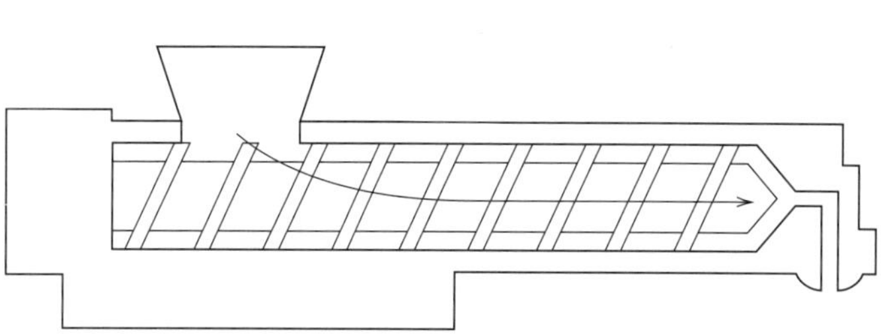

Figure 1. Extruder.

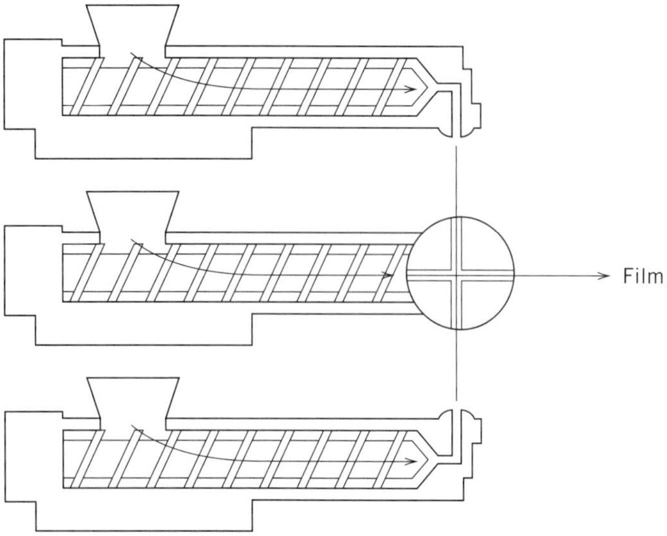

Figure 2. Coextrusion.

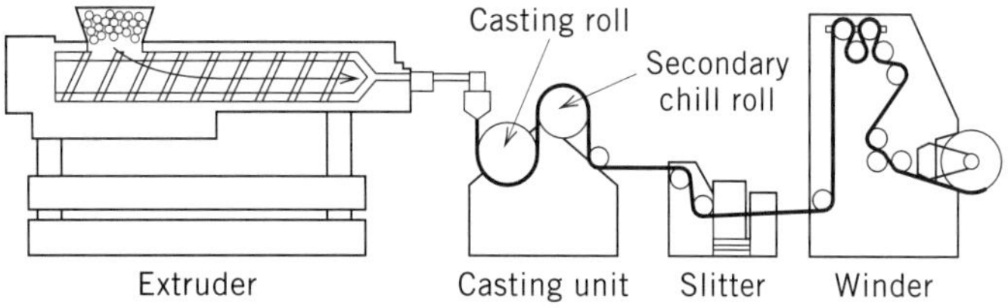

Figure 3. Cast-film fabrication.

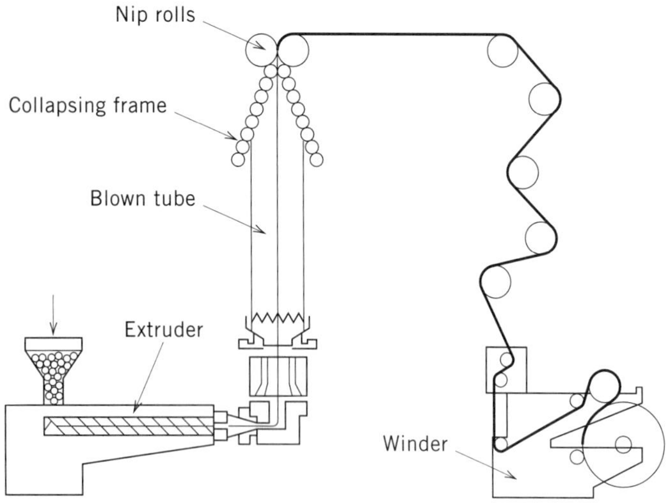

Figure 4. Film-blowing equipment.

Blown fabrication process. In the blown fabrication process, the extrudate is pushed through a round-shaped die. This action forms a continuous hollow tube from the extrudate, which is blown vertically upward. During this phase, the film is pulled or oriented in both the pulling (machine) and sideways (transverse) direction. High above the die, the tube is cooled and squeezed shut by a collapsing framework and a set of nip rolls. The flattened tube is then slit and wound into rolls. (See Fig. 4.)

Process influence on stretch-film properties. Cast films are typically different from blown film because of the difference in the way the plastic is oriented and cooled.

In the cast process, film is cooled more rapidly than in the blown process. This faster cooling, which is induced by the chill rollers, causes single-direction orientation in the film. In other words, the polyethylene molecules are cooled so quickly that the molecular structure is formed longitudinally in long, bidirectional branches.

By contrast, blown film is cooled slowly by air while the tube of plastic is drawn upward. This slower cooling allows for the polyethylene molecules to become oriented in all directions, producing a film with tougher puncture resistance.

The differences in these two manufacturing processes produce the following differences in cast and blown stretch films:

Cast Film	Blown Film
Clear	Hazy
High gloss	Dull
Lower modulus—stretches easily	Tough
Quiet unwind	Noisy unwind
	Stiffer film—high load containment
Tear resistance (cross direction)	
Extensibility	
Nonmigratory cling	Migratory cling
Good-to-excellent gauge control	Poorer gauge control
Good puncture resistance	Excellent puncture resistance
Gauge band potential	Gauge spread across roll (rotating die)

Although these are textbook examples of the differences between cast and blown films, it is important to note that through use of various polymers, additives, or process alterations, these differences can be reduced and even reversed. This ability to manipulate the manufacturing process has enabled the stretch-film manufacturer to produce films that combine the best properties of blown and cast films into one film.

Opaque and color-tint films. Opaque and color-tint films are available from film suppliers as special production runs. Because these requests are customized, opaque, and color-tint films have longer lead times and higher costs than do standard stretch films and must meet sizable minimum-order requirements (see Colorants).

Measuring Stretch-Film Performance

To most people, if a unitized load maintains its integrity as it moves from point A to point B, it has performed well. However, numerous tests can be completed to thoroughly analyze and measure film performance. Many of these tests can be conducted and evaluated only in a laboratory, but there are a few that any operator can complete with some very simple tools.

This section answers the following stretch-film performance questions:

- What properties are important?
- How can the properties be measured?
  - In the factory or warehouse
  - In the laboratory
- What effect does my machine have on film performance?

What properties are important? Because there are hundreds of different stretch films now available, knowing how to evaluate film is an important key to success. Equally important is how to use this information to determine load-wrapping cost.

The following is a list of the top six attributes most commonly evaluated:

1. *Tensile Property.* How much even force can be used before the film breaks?
2. *Puncturing.* How much pointed force can the film take before puncturing?
3. *Tearing.* How much force, after a cut has been made, can the film take before tearing?
4. *Optical Scanning.* How much can be seen and read through the film?
5. *Cling on Slip.* How much does the film stick to itself or other surfaces?
6. *Load Force.* How much actual force does the film place on the load?

It is important to note that all of these film attributes are related. Depending on the specific film application, the importance of each attribute will vary. Also, be aware that each film has a balance of attributes and that producing a film that meets maximum performance levels with each attribute would yield an overall weaker film because you cannot affect just one attribute without affecting all of them.

How can the properties be measured?

In the factory or warehouse. A knowledgeable stretch-film sales professional will employ effective methods to measure film performance on-site. One such method tests the actual stretch level of film after being placed around a load. Using a stretch wheel and a tape measure, the percentage of stretch your machine delivers is calculated in order to ensure that you are not overstretching or underutilizing your stretch film.

Force-to-load is another test that should be performed. The point of this exercise is to measure the pressure the stretch film places against a load. By applying a force-to-load gauge to the stretch film, the apparatus reveals whether the film is administering the appropriate amount of force to contain a load of goods. Snapback is yet another demonstration test used to measure film performance. In the snapback test, you ascertain film recovery; ie, how much has the stretch film returned to its original shape after its been stretched? Film with poor recovery has been stretched beyond its limits and will cause the load to loose integrity over time.

Finally, the "cut & weigh" is a test used to demonstrate the actual cost savings of a particular film in comparison to other brands. In the cut & weigh, a load wrapped in stretch film is cut, removed entirely from the load, and measured on a scale. The same load is then wrapped in another manufacturer's brand of stretch film and is also cut, removed, and weighed. The two weights are then compared.

To see these film performance measurements done firsthand, contact your local packaging sales professional.

In the laboratory. The following properties are routinely tested in the lab:

Tensile properties are determined by measuring film performance as it is elongated at a constant rate. By measuring the force required to elongate the film, the following properties can be determined: *yield strength*—the point where a film permanently deforms as a result of force applied during elongation; *ultimate tensile strength*—the maximum force a film can withstand in a tension test measured in psi (pounds per square inch); and *elongation*—the percent increase in film length at its break point.

Puncture and tear properties are easy to visually identify while the film is stretched around a pallet load. A film that has poor tear and puncture characteristics will usually fail, forcing an operator to compensate by increasing the number of pallet wraps or loosen the tension. This is an expensive way to compensate for having the wrong film or a film with poor quality. In the lab, puncture resistance is measured by the force needed to puncture the film. This is accomplished by pushing a pear-shaped probe through a tightly secured piece of film. It is measured in inch-pounds or in pounds per mil if film-gauge correction is required. Tear tests or an "Elmendorf test" is simply the force required to propagate a precut slit in a film sample. Another common name for film tearing is "zippering." The Elmendorf test is measured in grams or grams per mil if gauge correction is required. Tears can occur in both vertical (machine) and horizontal (transverse) directions. See Figure 5 for clarification.

Optical properties describe the film's ability to transmit light (haze) or reflect light (gloss). Both of these properties are significant in determining film clarity.

Haze is a quantitative measure of film transparency, which is very important for scanning labels and identifying pallet-load contents.

Gloss is a measure of surface finish and the films ability to reflect light shine. The higher the percent gloss, the better the shine. Cast-manufactured films tend to exhibit better clarity than do blown films and have higher gloss.

Load-force properties describe the amount of force a film can provide after it is stretched wrapped to a pallet of goods. As mentioned earlier, this test is done in the field as well as in the lab. The load-force property is very important because

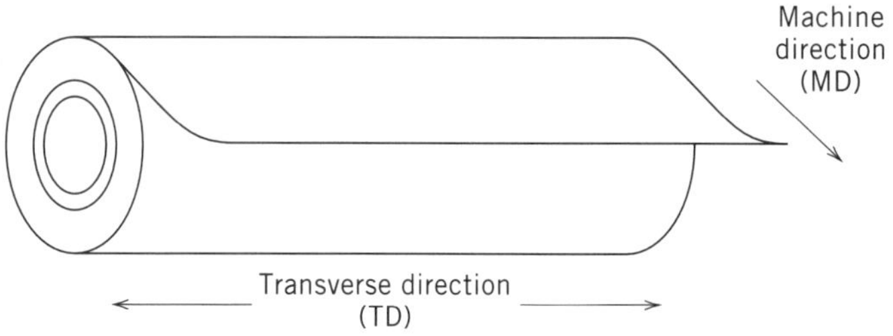

Figure 5. Full-dimensional orientation.

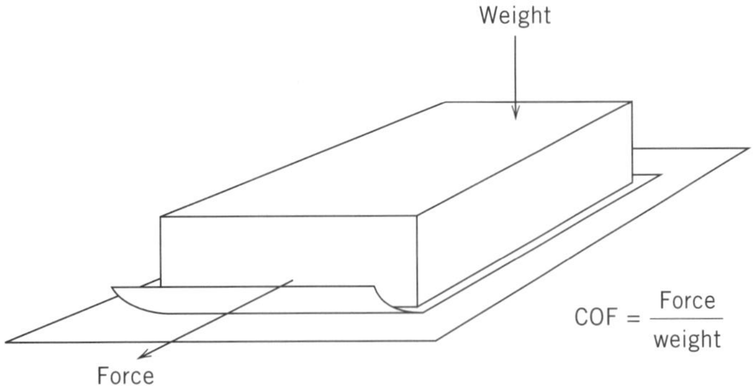

Figure 6. Measurement of coefficient of motion.

it is a key to keeping pallets loads tightly unitized. The best way to measure load force is to use a force-to-load gauge. This gauge will show the actual force (pressure) being placed on the load by the film.

Cling and slip properties are measured with an Instron apparatus. Both cling and slip is measured in grams. Cling is simply the ability of a film to stick to itself or another smooth surface. Cling is affected by the amount of tackifier additives, such as polyisobutylene (PIB) in blown films, which are added or by the smooth outer surface of cast films. To determine slip, a film is measured for its *coefficient of friction* (COF), which is a measure of the amount of force required to move a film sample with no cling out from underneath a standard unit of weight. (See Fig. 6.)

What effect does my machine have on film performance? Film performance is critical when determining which film can offer the best load protection at the lowest cost. Often, the equipment used to apply film has a dramatic impact on the film. A film that performs well on one machine may not offer the same results on another system; for instance, stretch levels may differ by machine. Even equipment operators can affect film performance by mishandling the product, not keeping equipment clean and running efficiently, or not stretching the film to its maximum potential.

The challenge is look beyond a "no complaints" or "it's not broken, don't fix it" mentality to evaluate true film perfomance. In other words, how can a user methodically evaluate film performance levels to discover the best film for the operation under consideration? By employing the film performance measurements outlined in this section, you can make sure that your stretch film is performing up to its full potential.

Inspecting and Handling Stretch Film

Because stretch film exhibits toughness when wrapped, most people would think that stretch film rolls are not fragile. But they are. Caution must be exercised when handling stretch-film rolls and loading them onto the wrapping machine to avoid edge damage, which will cause tearing during prestretch. This section will teach you proper roll-handling techniques and enable you to spot roll defects by covering the following topics:

- Stretch film handling and storage
- Manufacturing defects

Stretch-film handling and storage. Stretch-film rolls, especially their edges, are sensitive to damage. Bumps, nicks, and cuts on the outer edges will easily render the roll unusable. As a result, the rolls must be handled with care. The following are some recommended tips on handling stretch film rolls:

- *Step 1—Minimize Roll Handling.* The packaging that holds a stretch film pallet together has been designed to be opened as close as possible to the wrapping machine. For exactly that reason, it is recommended that pallets of stretch film be stored as close as possible to the wrapping machine to minimize roll handling and prevent damage that could be caused by dropping the rolls.
- *Step 2—Inspect Rolls.* After opening your pallet of stretch film, the rolls should be inspected for shipping damage that can occur if they have shifted in transit. This damage can take the form of dents, product rubs, and gouges, nicks, and tears. Should these exist, try to remove them by unraveling the first few layers of the roll before loading it onto the machine.
- *Step 3—Keep Film Away from Surfaces.* Stretch-film rolls are designed with a $\frac{1}{4}$-in. core protrusion on each end. When either placing stretch film on the floor for storage purposes or on a machine roll carriage, always make sure that the core, and not the film, is touching the ground or machine surface. The less the film comes in direct contact with any surface, the less likely that your roll edges will become damaged.
- *Step 4—Keep Film Away from Excessive Heat.* To prevent film damage, keep rolls away from heating vents or machines that give off an inordinate amount of heat.

Manufacturing defects. Several defects can occur in the manufacture of stretch film. Users of stretch film should be aware of these defects (highlighted in italics in legends to Figs. 7–13) which may adversely affect film performance.

The best way to avoid manufacturer defects is to pick a manufacturer who places extremely high emphasis on quality, consistency, and performance.

Maintaining consistent stretch levels. To measure the degree of stretch your film achieves, you need only three tools: a

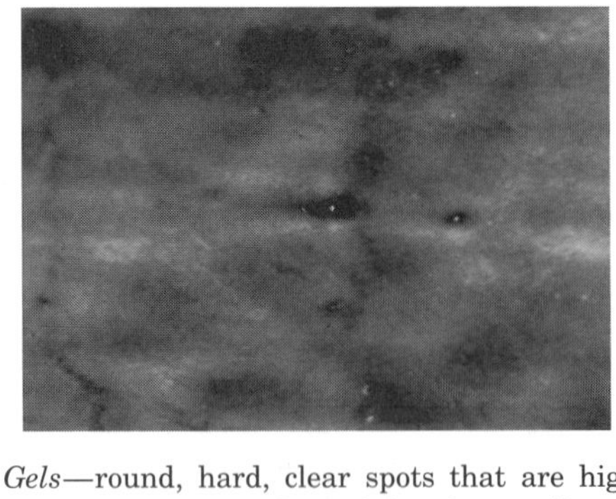

Figure 7. *Gels*—round, hard, clear spots that are high molecular weight or unmelted polymer. Caused by abnormalities in resin or poor mixing. Large gels can cause holes to occur during the wrapping process which may result in film breaks.

Figure 8. *Black specks*—degraded polymer or other film imperfections caused by a foreign material. Not a common problem. Gives appearance of poor film quality.

Figure 9. *Roll telescoping*—the sliding of inner layers of the film in a roll or the core out of a roll, increasing roll width and making the roll unusable. This is typically caused by tackifier levels and winding conditions.

Figure 10. *Feathered edges*—extension of a layer or layers of film past the end of a film roll. Caused if the film wanders back and forth during winding. If these loose edges fold over they may cause spiral tears during film unwinding.

Figure 11. *Scalloped edges*—sawtooth-appearing film edges caused by poor slitting. Can cause web failure from edges.

Figure 12. *Gauge bands*—hard or soft rings on a roll of film, caused by gauge (thickness) variations in the film. Normally, gauge bands do not result in any wrapper failure. If a failure does occur, it will come from a wrinkle in the band area. It will appear as a slit starting between the prestretch rolls.

Figure 13. *Tails–overwind*—material that is disrupted by the automatic index or transfer of the winder. During the transfer from a full roll to any empty core, the web tends to draw in and wrinkle. This allows additional air to be trapped and results in a disrupted film appearance.

stretch wheel, a tape measure, and knowledge of a simple mathematical formula.

A stretch wheel is 10 in. in circumference, rotates on an axis, and is held in place by a handle. Jutting out from this wheel is an ink mark that makes a black mark every 10 in. on a roll of unstretched stretch film when the wheel is allowed to rotate freely against the film.

To measure exactly how much a sheet of film has stretched, roll your wheel horizontally against the outer layer of film while it is on the spindle of the machine. Make sure that at least four black marks have been made. Initiate the wrap cycle so that the film is stretched. Measure the distance between the marks. (*Note:* Make sure that you measure the marks that do not cross corners, but are on the flat side of the pallet.)

Now, using your mathematical formula, subtract 10 in. from your measured distance and multiply that difference by 10. The answer is the percentage of stretch that film has endured.

Example. After a sheet of film has been stretched, the distance between the two black wheel marks reads 27 in. Subtract 10 from this number ($27 - 10 = 17$), and then multiply 17 by 10 ($17 \times 10 = 170$). (*Answer:* The film was stretched 170%.)

Film Problems

There are some problems that can occur when using stretch film. As seen in the preceding section ("Inspecting and Handling Stretch Film") some of these problems are manufacturer-related and some are shipping related. Stretch-film problems can also stem from equipment conditions, wrapping speeds, and operator ignorance. However, as long as you know how to identify these problems, most have correctives.

This section focuses on how to spot common stretch-film problems and what you can do to rectify them when they arise. Stretch film failures and solutions will be covered under

- Film breaks
- Film "tails"
- Tackifier buildup

Film breaks. There are three common film breaks that can be identified by the way the stretch film tears. These breaks and the solutions required to repair them are described in the following paragraphs.

Edge. An edge tear starts at the top or bottom of a roll and tears in a diagonal direction. It is usually caused by a nick in the film edge. To fix the problem, simply unwind the film until the imperfection is gone. (See Fig. 14.)

Ultimate break. An ultimate break is a straight-line break with pointed edges. The break occurs when the film has been prestretched beyond its limits. To avoid this type of break, reduce the percentage of prestretch on your machine. (See Fig. 15.)

Gel. A gel is a V-shaped break in the film caused by resin imperfection or some other manufacturing problem. Before replacing this roll of film, first try to correct the problem by lowering the F2 and/or prestretch percentage on your machine. (See Fig. 16.)

Film "tails." A stretch-film "tail" is a loose end piece of stretch film that hangs off a wrapped pallet.

Tails loosen or fall down on wrapped pallets if cling is too low or there is too much dust on the film. But tails occur primarily because film is cut off under tension. To avoid creating tails, the final wrap should be made with a reduced stretch percentage and lowered F2 so that the film elasticity does not pull the tail free of the load.

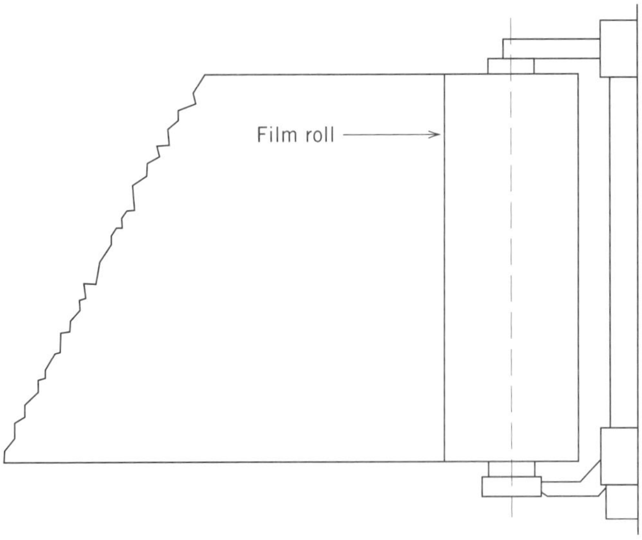

Figure 14. Film break at edge.

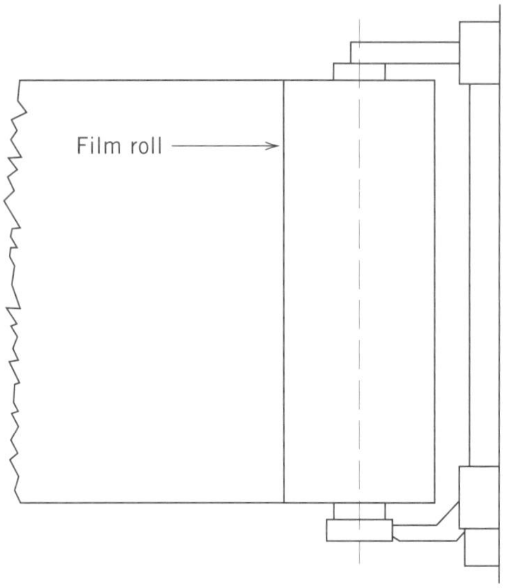

Figure 15. Ultimate break.

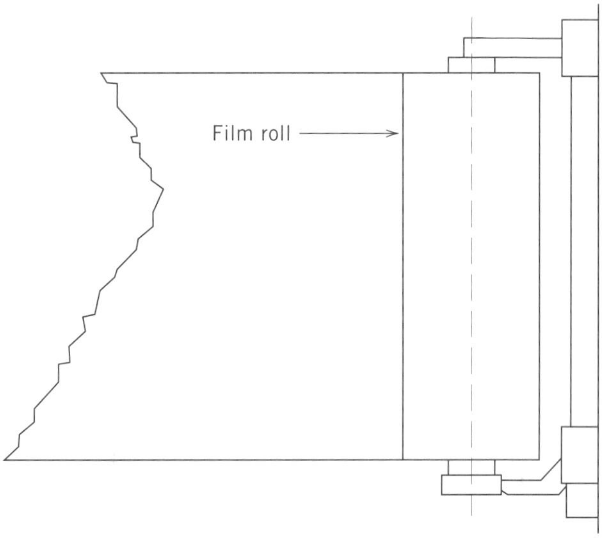

Figure 16. Gel or V-shaped break.

Tackifier buildup. Tackifier, the most common of which is PIB, is an additive responsible for giving blown stretch film its cling. Over time, tackifying agents will build up on machine rollers. If allowed to accumulate, tackifier residue can cause stretch film to stick to the rollers and to itself.

To avoid this scenario, machine rollers should be cleaned on a regular basis. It is advised that a mild detergent (isopropyl alcohol makes a good cleaner), warm water, and a nonabrasive cloth be used for this purpose because of the sensitivity of the machine rollers.

Environmental Effects on Film

The strength and protection stretch film affords the packages it wraps has made it the unitizing material of choice among thousands of businesses worldwide. But this is not to say that stretch film is without its limitations. There are certain environmental and physical conditions that can impair regular stretch film's performance. This section reveals what these conditions are so that you can avoid them and get the most out of your stretch film. The topics discussed are

- Outside storage
- Static discharge
- The effect of temperature on film
- The effect of dust and grit on film
- Stacking pallets

Outside storage. Regular stretch film is not designed for prolonged outside storage. Stretch film is subject to photodegradation from the ultraviolet rays in normal sunlight, which substantially weakens the film.

If wrapped pallets are to be stored outside for more than several weeks, they should be covered. Otherwise, for prolonged outdoor applications, specially formulated ultraviolet-irradiation (UVI)-stabilized films should be used. With UVI film, uv protection is optimized when used in conjunction with opaque tinting.

Static discharge. Stretch films will support a static discharge and, therefore, should not be used in explosion-prone environments. Since stretch film can induce a static charge, electronic devices packaged in stretch film run the risk of suffering damage. Static buildup, however, can be dissipated through use of Christmas garland or static eliminators.

Effect of temperature on film. Generally, higher temperatures make film softer and easier to deform. Lower temperatures make film stiffer and harder to stretch and deform. At higher temperatures, where film is softer, machine-direction tensile strength, tensile yield, puncture force, and transverse-direction tear resistance all decrease. Machine-direction ultimate elongation increases.

At lower temperatures, the opposite is true. Machine-direction tensile strength, tensile yield, puncture force, and transverse direction tear all increase. Machine-direction elongation decreases. If temperature plays a role in your stretch-film requirements, contact your packaging sales professional, who can recommend films that are best suited for your particular needs.

Effect of dust and grit on film. Dust has an adverse effect on film, causing it to lose its cling. In high-dust and high-grit environments, debris can build up on the prestretch and metal rollers and cause failures in film performance. As mentioned earlier in this section, the debris that collects on the rollers is often the result of a buildup of tackifying agents. PIB, the most common cling additive, can accumulate over time, causing the film to wrap and/or sharp objects to be embedded in the rubber rollers. Therefore, it is of utmost importance to keep the rollers clean and free of buildup.

Stacking pallets. Too much pressure may be placed on pallets by stacking them too high. Two is the recommended limit.

Glossary of Terms

Antiblock. Substances added into plastic materials that retard or prevent film layers from sticking together.

Blowup ratio. The ratio of "bubble" or "tube" diameter to extrusion die diameter. A primary consideration in determination of transverse-directional (TD) shrinkage and a partial determinant of machine-directional (MD) shrinkage.

Blown film. Film that is fabricated by continuously pumping the polymer through a circular orifice (die). Once through the die, this polymer is then drawn upward and filled with air, creating a tube. The volume of air contained within the tube stretches the tube out to the desired width and, in conjunction with rate at which it is being pulled away from the die, the desired thickness is created.

Caliper. The thickness of a film or bag, normally expressed in mils or as "gauge" (0.9 mil = 90 gauge = 0.0009 in.).

Cast film. Film that is fabricated by continuously pumping the polymer through a straight slot orifice (die), then chilling this hot sheet of plastic immediately afterward by contact with a colled roll. Film width is determined by the length of the slot. Thickness is determined by how fast the casting roll pulls the plastic away from the slot (or die).

Cling. The characteristic of stretch film that makes it stick to itself or other clean objects.

Coefficient of friction. A dimensionless numerical representation of the ease with which two objects or surfaces will slide against each other. Low numbers slide easily. Total range due to the test definition is 0–1.

Coextrusion. Fabrication of a multilayer film by pumping the various materials through separate extruders and then merging the extrudates into a common die assembly. These die assemblies are constructed to maintain distinct material layers with fusion occurring at the boundaries, due to the pressures and temperature of the extrudates.

Copolymer. A plastic or polymeric whose chemical composition consists of more than one basic hydrocarbon type. The materials are chemically bonded and remain in combination, thus influencing the performance by their own distinct characteristics.

Crosslinked PE. A polyethylene that has been specifically treated by chemical or physical means to cause intrachain bonding.

Crystallization. The formation of distinct, ordered, and repeated molecular groups. Polyethylene forms varying amounts of crystalline structure depending on process conditions and original molecular makeup. The remaining portion is amorphous (which is the antonym of crystalline). Ice is crystalline; water is amorphous.

Dart drop. A method of measuring a film's impact strength or dynamic toughness. A hemispherical shaped, weighted "dart" is dropped onto a film sample. The weight at which 50% of the samples are punctured is considered the WF_{50} reading.

Die. A device used in extrusion processes to shape the extrudate. Circular dies are used for blown products, and slot dies are used for cast products.

Drawdown ratio. Commonly expressed as the ratio of the width of the extrusion die orifice to the thickness of a particular film. Frequently, however, and more correctly, it is a ratio involving die orifice, blowup ratio, and final film thickness as follows: die gap/(film gauge × blowup ratio).

Eldastic recovery. The ability of a material to return to its original shape or size after having been deformed or subjected to strain.

Elmendorf tear. A testing method used to quantify a material's resistance to tearing forces. Generally a sample is slit (initiated tear), then the force required to tear apart the sample is measured and given as the tear value.

Elongation at break. The strain or deformation required to break a sample. Generally expressed as a percentage by dividing the strain (distance) at which a sample breaks by the original sample length.

EVA. Abbreviation (acronym) for the copolymer ethylene vinyl acetate. Small percentages (1–18%) of vinyl acetate monomers are frequently polymerized with ethylene to provide greater extensibility or greater low-temperature strength or to improve sealability.

Extrudate. Molten polymer.

Extrusion. The process of feeding, melting, and pumping a material such that a desired shape or configuration can be created. It is a continuous process and utilizes a device similar to a meat grinder.

Film yield. Generally expressed as the square inches of a film (area) that weighs one pound if the film is 1 mil (0.001 in.) thick. Coverage is on a per pound (lb^{-1}) basis. This should not be confused with yield point, yield strength, or tensile yield, which are strength measurements. Polyethylene yields 30,000 $in.^2/lb$ at 1 mil.

Gauge. Used as a synonym for film thickness or film caliper. Sometimes expressed as 80- or 100-gauge, which equates to 0.8 or 1 mil, respectively or 0.008 or 0.001 in., respectively.

Gauge band. A conformation irregularity found in rolls of material. A thick area in a film will produce a raised or elevated ring in a finished roll of product. Conversely, a thin area will result in a soft ring in a finished roll.

Gloss. The shine or sparkle of a surface. In LDPE film, gloss is described as the amount of light reflected from the surface. Standard technique places a light source and a receiver at 45° angles from the surface. The number value produced is roughly the percentage of light reflected from the source into the receiver.

Haze. The lack of film transparency. It can be induced by process considerations, inherently due to molecular configurations or created by pigmentation. It is measured by determination of the percentages of light not transmitted through a film sample.

Homopolymer. A plastic resin or polymer whose entire chemical molecular structure is of a singe hydrocarbon group. It is made by feeding a single monomer to the reactor.

Impact strength. The ability of an object or material to resist rapidly applied destructive forces. Refer to "Dart drop."

LDPE. Low-density polyethylene—any polyethylene homopolymer whose density is between 0.913 and 0.925 g/cm^3 (g/mL).

Modulus. Short for modulus of elasticity, which is a numerical value reflecting a material's resistance to deformation. A film with a high modulus is hard to stretch or elongate.

Monomer. The incremental or elemental chemical elements before polymerization. Ethylene is a monomer.

Neckdown. The "narrowing" tendency of a film when it is being stretched or pulled. Occurs when film is stretched in the machine direction, resulting in decreased transverse direction width.

Opaque. Film that is impervious to light: 100% haze level.

Polyisobutylene (PIB). Tackifier additive in blown film that gives it cling.

psi. Abbreviation for pounds per square inch ($lb/in.^2$). A unifying statistical measurement by which various thickness of material can be tested, equated, and compared without regard to actual specimen thickness.

Puncture performance. The relative comparison of a material's resistance to failures caused by penetration; eg, how easily your finger pokes through a film sample.

PVC. Abbreviation for poly(vinyl chloride). In film form, it is used as a meat or produce wrap, as stretch film, and as a high-clarity shrink wrap for consumer or retail packaging.

Polyolefin. Polymers of basic unsaturated hydrocarbon chains containing at least one double bond (eg, polyethylene, polypropylene, and polybutene). Not PVC or PVA.

Polymer. Means "many members." A structure generated by the repetitious joining of many of the same elementary units. Natural polymers are cotton, wood, or protein. Synthetic polymers are polyethylene, polystyrene, nylon, etc.

Reactor. The equipment used to transform ethylene gas to polyethylene.

Stress relaxation. Primarily it is the phenomenon of force decay as a function of time. It means that the "rubber band" tying a stretch load together exerts less force as time passes. This rate of decrease varies greatly with different polymers. PVC has an early decay that is much greater and more rapid than that of LDPE.

Stretchability. A combination of factors related to the ability of a material to be stretched or elongated; eg, how easily and to what extent a film stretches and whether it will return to its original length.

Tack. An adjective used to describe a film's resistance to slide against itself or another surface, or its resistance to separation from itself. Similar terms are *surface adhesion* or *cohesion, surface seal, wetting of film layers, blocking,* or *cling.*

Tackifiers. Chemical substances added to increase the tack of the parent or base material.

Tear resistance. The resistance of a film to be torn. This is quantified by Elmendorf tear testing, and is measured as the force required to propagate an initated tear in the MD or TD direction.

Tensile yield. This is a stress level, measured in psi, beyond which permanent deformation occurs. "Tensile" indicates that the specimen is pulled or "tensioned" rather than compressed. Up to this stres or applied force level, a release of the force will result in the specimen returning to its original size. Pulling with a higher force will result in an elongated sample, even after the force is removed.

Tensile ultimate. This is a stress level, measured in psi, beyond which the specimen will break. It is the laboratory-measured, maximum stress (applied force) that the material will withstand. Tensile indicates that test is conducted with a pulling or tensioning type of loading.

Tint. A slight coloring that allows light transmission and relatively good clarity.

Toughness. An overall strength measurement that takes into account both the amount of pull and the amount of elongation a sample can withstand. A tough material will resist breakage by both resisting force and by elongating. A brittle material (eg, glass) will resist force, but will break instead of deforming; hence it has low toughness. LDPE, which is plastic or tough, both resists and elongates; hence is tougher than glass. The rate of force application has significant effects on toughness, so comparisons must include testing data.

Ultimate Strength. Sames as "Tensile ultimate."

Wide web. Wrapping with a roll of film approximately the size of the load.

Wrap. One revolution of a machine turntable.

Yield stretch. Same as "Tensile yield."

Zippering. Lack of resistance to tear propagation of an initiated TD cut or tear. Once the film has been torn or cut, the resulting slit rapidly opens and completes a full-web break.

THOMAS B. BRIGHTON
Tenneco Packaging
Deerfield, Illinois

FILM, TRANSPARENT GLASS ON PLASTIC FOOD-PACKAGING MATERIALS

Introduction

Commonly used polymeric packaging materials such as polyethylene terephthalate (PET), oriented polypropylene (OPP), and biaxially oriented nylon (BON) do not have sufficient oxygen or water-vapor barrier properties for many packaging applications. To resolve this, multilayer material structures incorporating aluminum foil and polymeric films, aluminum metallized films, PVDC-coated films, or coextruded EVOH films are routinely applied. However, environmental considerations and the push toward source reduction have created a market need for a thin-film transparent oxygen or water-vapor barrier coating that is friendly to the environment and convertible into usable packaging materials while at the same time capable of meeting the cost requirements of the food-packaging industry. The coating should also utilize a low temperature process to coat heat sensitive polymeric films such as OPP and BON.

The goal of the scaleup process for the QLF transparent barrier coating was to design and manufacture equipment to provide SiO_x-coated material that is environmentally friendly, recyclable, microwavable, crystal-clear in transparency, excellent in adhesion to the substrate, and able to allow metal detection. In addition, the coating must withstand mechanical stress and the high temperatures generated during extrusion lamination and heat seal. It should be printable, able to run at a high speed on standard converting equipment, and be cost-competitive compared with existing barrier materials. For commercial success, the coating process must be capable of producing high throughputs to meet the low coating cost demands of the packaging industry.

Why Plasma-Enhanced Chemical Vapor Deposition (PECVD)?

Coatings of SiO_x can be produced by a number of methods including sputtering, evaporation and PECVD.

Extensive R&D programs have been completed to investigate each technology to determine the most effective method for creating a SiO_x coating to meet scaleup process goals. At the same time, a prototype machine was built capable of coating rolls 660 mm wide to prove the production and economic viability of whichever process was chosen. This system had the capability to utilize all three technologies. Table 1 shows the results of the study that led to the focus on the PECVD process. Sputtering was disqualified fairly early, due to large power requirements and low deposition rate. The process was not economical for food packaging.

In contrast, the main advantage of evaporation was the *potential* to run at very high line speeds, comparable to those used in vacuum metallizing of aluminum. This potential has not been realized. SiO cannot be evaporated at as high a rate as aluminum. If it is evaporated at a power level to give the same rate as aluminum, molten SiO is spattered onto the film, causing holes in the film. Also, the thickness of the coating needed to gain a suitable barrier is much greater than that for aluminum metallizing (1500–3000 Å vs 200 Å).

On the basis of the coating work carried out, the PECVD process showed much more promise than evaporation in terms of economics and coating performance. The economics for the PECVD process are better than those for evaporation because the vacuum requirements are less stringent, meaning that the time to change rolls should be less; the power requirement per unit area is less; cooling requirements are less; raw material is inexpensive and available; source utilization is higher, mainly because of patented equipment design; and coating thickness is less.

Scaleup of Manufacturing Systems

The SiOx deposition process described is a low-pressure, low-frequency, and low-temperature PECVD process using a mixture of helium, oxygen, and the organosilicon compound hexamethyldisiloxane (HMDSO), as shown in Figure 1. The plasma source uses a multipole magnetic field to confine the plasma in order to obtain enhanced deposition rates at low

Table 1. Comparison of Glass-Coating Technologies

Feature	Sputtering	Evaporation	PECVD
Vacuum (torr)	10^{-3} (proc.)	10^{-6}	10^{-2}
	10^{-6} (base)	10^{-6}	10^{-2}
Power (kw)	Thousands	Hundreds	Tens
Process temperature	High	High	Low
Source	Si + O_2	SiO/SiO_2	Monomer
Source cost	High	High/medium	Low
Source utilization	Roughly 40%	Roughly 25%	>50%
Coating thickness	400–500 Å	1500–3000 Å	150–300 Å
Film color	Yellow	Yellow	Clear
Coating bond	Mechanical	Mechanical	Chemical
Barrier (unlaminated)	Medium	Medium	High

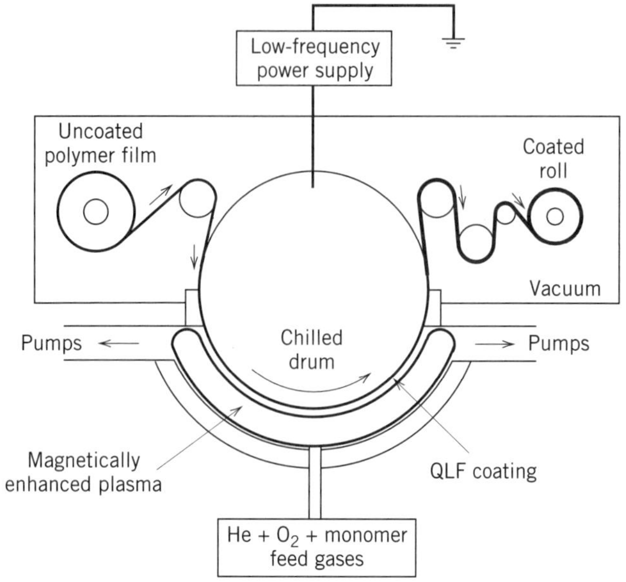

Figure 1. Proprietary plasma-enhanced chemical vapor deposition.

pressures. The resulting silicon oxide is the QLF barrier coating.

The three coaters used in the course of the maufacturing scaleup effort are shown schematically in Figure 2. The QLF coating was initially developed in a small research roll coater. Because of technical risks and the capital cost required to build a 1.5- or 2-m-wide vacuum roll coater, the strategy involved incremental steps in scaling up the coating process from a web width and speed of 0.3 m wide at 0.3 m/min, to 0.66 m wide at 100 m/min, then 1.5 m at >100 m/min and finally to 2 m at >100 m/min. The challenge to incrementally increase the web width and speed was recognized due to the need to develop hardware capable of running under the process conditions required to achieve the target barrier performances and line speeds. BOC Coating Technology has systematically scaled up the process using statistically designed experiments that optimized the parameters through different treatment conditions as well as using previous experience from other vacuum coating machines.

The first scaleup phase was completed in a development coater that can mechanically handle 0.66-m-wide webs at speeds from 0.1 to 150 m/min. The second scaleup phase was conducted in a production-scale coater that can handle 1.5-m-wide webs at speeds from 10 to 300 m/min (Fig. 3). The process scaleup in this coater was completed using HMDSO as the monomer. The third scaleup phase is a production coater that can handle 2-m-wide webs at speeds from 30 to 300 m/min (Fig. 4). This is a modular design production coater with three isolated process zones that provide flexibility for pretreatment and/or posttreatment or for deposition with different chemistries.

Barrier Performance

The *unlaminated* oxygen transmission properties achieved to date for PET and OPA are listed in Figures 5 and 6. A typical oxygen transmission rate (OTR) for uncoated PET is 115 $cm^3/(m^2 \cdot 24\ h)$. With QLF coating applied at a speed of 100m/min the OTR is reduced to 1.1 $cm^3/(m^2 \cdot 24\ h)$. Increasing the line speeds to 200 m/min will result in an OTR reading of 2 $cm^3/(m^2 \cdot 24\ h)$. The opportunity for high barrier performance on other substrates is also possible. Production rolls of OPA with QLF coating run at line speeds of 300 m/min demonstrate an OTR of 7 $cm^3/(m^2 \cdot 24\ h)$. These results represent the flexibility a roll coater can offer. *Whether a customer is targeting a high or medium barrier application a roll coater can satisfy their requirements.*

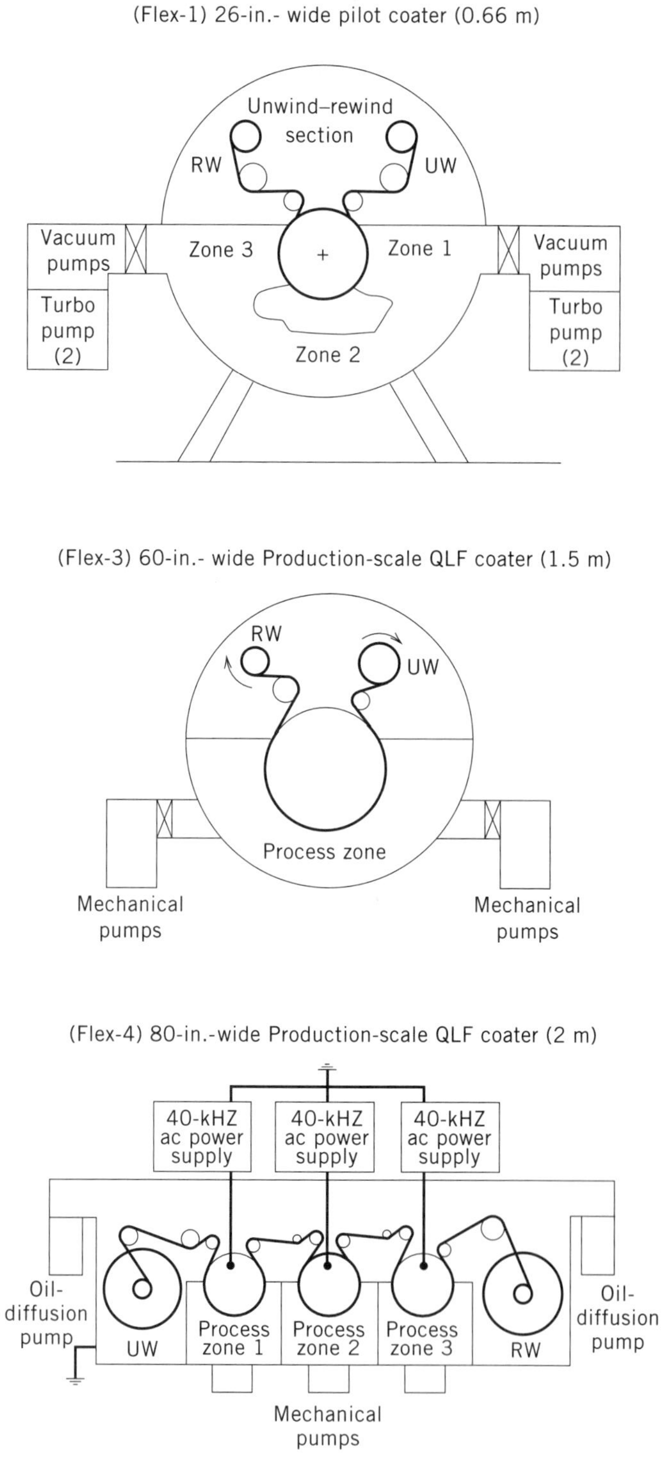

Figure 2. Flex-1, Flex-3, and Flex-4 schematic.

Figure 3. Flex-3: production-scale QLF roll coater.

Initially, the technology was focused on an effort to optimize the oxygen barrier primarily for liquid packaging. Further development was targeted toward upgrading and optimizing the process for water vapor (WVTR) as well. As indicated in Figure 7, the WVTR for PET with QLF coating has been reduced from a level of 5.6 g/($m^2 \cdot 24$ h) in August 1995 to 1.4 g/($m^2 \cdot 24$ h) in December 1995. Development efforts are now focused on increasing line speeds and introducing a range of different substrates to the PECVD process. (See also Barrier Polymers)

Converting

Successful large-scale acceptability of QLF-coated materials is dependent on the ease of convertibility as well as their cost and barrier properties. Extensive research has been dedicated to determining how "robust" QLF coatings are in varying converting processes. Because of the strong chemical bond that exists between QLF coatings and PET, converting results have been encouraging. Several packaging converters have successfully adhesive- and extrusion-laminated QLF-coated PET, maintaining barrier properties and achieving excellent bond-strength results. Good ink adhesion was also achieved directly on the QLF-coating surface. BOCCT can supply a recommended list of inks and adhesives on request.

Gelbo flexing, which is a test commonly used to simulate stresses encountered in finished-packaged shelf life in the marketplace, is also being investigated. Earlier papers (1,2) have already reported that the finished-packaging structures with the QLF coating survived the tests when properly laminated.

Commercialization of QLF Barrier Coating

The transparent oxide barrier market has been plagued by high cost and/or difficulties in converting or printing on the barrier coating surface without losing the barrier properties. One of the most crucial success factors in the transparent high-barrier market is the ability to provide a better, or comparable, material at a lower cost. QLF coatings are not only the most "robust" transparent oxide coatings available in the marketplace—they also provide the most competitive economics.

Today, BOCCT roll coaters are built to offer the converter or film manufacturer the flexibility to target both the medium-barrier and high-barrier markets. Depending on the barrier level one is looking to achieve, QLF coating costs can range from $<$ \$0.03 to $<$ \$0.045/m^2. These costs will continue to lower as further advances in line speeds and barriers are achieved.

The use of commercial size coaters to improve the QLF coating performance is continuing. QLF coatings are FDA approved, provide a better barrier than polyvinylidene chloride (PVDC) coatings, and are insensitive to humidity changes as opposed to ethylene vinyl alcohol (EVOH). Replacement of these coatings on substrates such as PET and BON are the initial targets for large-scale commercialization of QLF coatings.

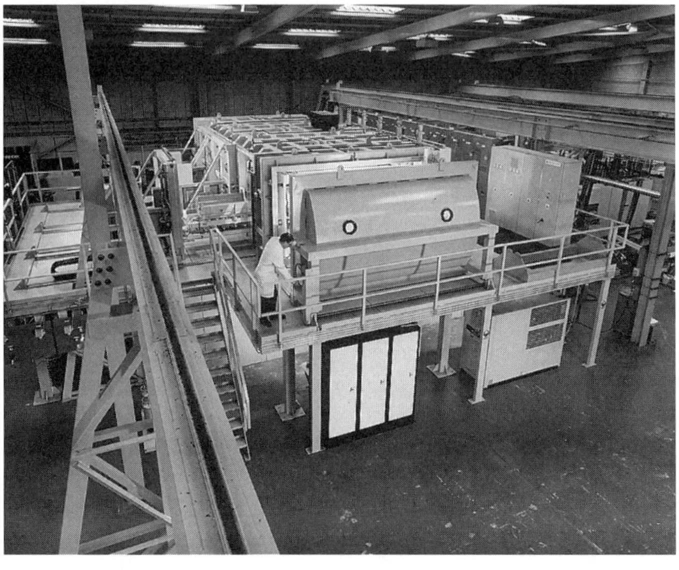

Figure 4. Flex-4: production-scale QLF roll coater.

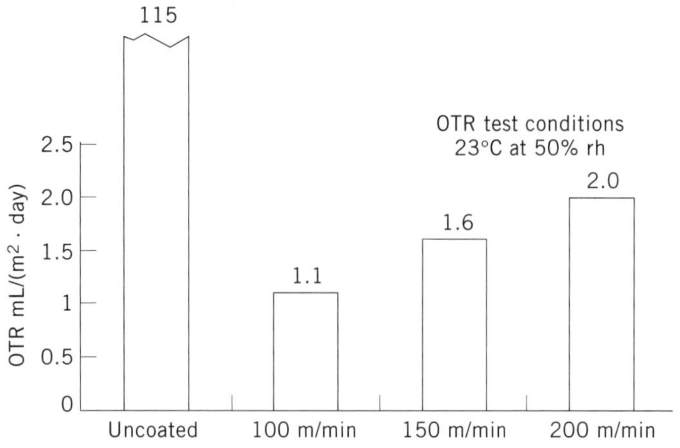

Figure 5. OTR of 12-μm PET with QLF coating.

Conclusion

QLF barrier coatings have been scaled up from a 0.3-m-wide web laboratory coater at a line speed 0.3 m/min to a 2-m-wide web production coater capable of 300 m/min. BOC Coating Technology is committed to continued improvements in both line speeds and barrier performance. Currently, technology-driven converters and film manufacturers are forming partnerships with BOCCT in order to make QLF coatings a commercial reality.

QLF coatings applied to PET substrates can now perform with barriers and line speeds that exceed even the most demanding expectations of the packaging industry. Meeting the demand in the packaging industry for an environmentally friendly, transparent barrier coating is the goal of BOC Coating Technology.

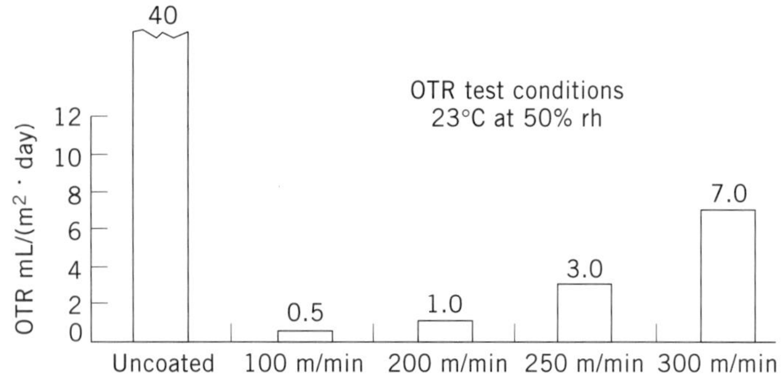

Figure 6. OTR of 15-μm OPA with QLF coating.

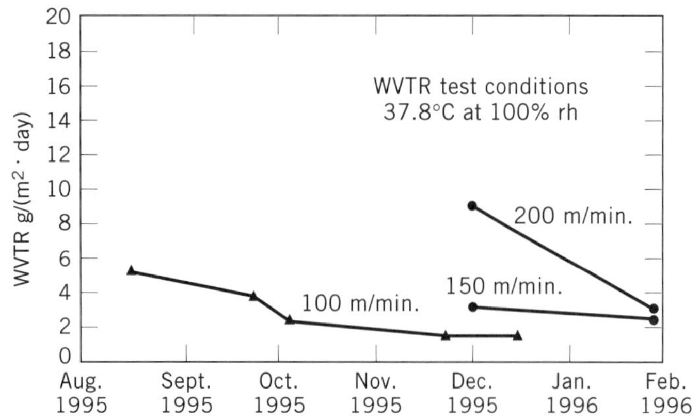

Figure 7. WVTR of 12-μm PET with QLF coating.

BIBLIOGRAPHY

1. E. Finson and J. Felts, "Transparent SiO_2 Barrier Coatings: Conversion and Production Status," *TAPPI J.* **78**(1), (Jan. 1995).
2. E. Finson, and R. J. Hill, "Glass-Coated Packaging Films Ready for Commercialization," *Packag. Technol. Eng.,* 36 (June/July 1995).

RUSSELL J. HILL
BOC Coating Technology
Concord, California

FLEXIBLE INTERMEDIATE BULK CONTAINERS

Flexible intermediate bulk containers are defined as an intermediate bulk container, having a body made of flexible fabric, which (*1*) cannot be handled manually when filled; (*2*) is intended for the shipment of solid material in powder, flake, or granular form; (*3*) does not require further packaging; and (*4*) is designed to be lifted from the top by means of integral, permanently attached devices (lift loops or straps).

Flexible intermediate bulk containers (FIBCs), also known as "big bags," "bulk bags," and "bulk sacks," were first manufactured in the late 1950s or early 1960s. There is some controversy as to where the first FIBCs were made; however, it is known that FIBCs were made in the United States, Europe, and Japan during the time period mentioned above. The first FIBCs were constructed with heavy-duty PVC-coated nylon or polyester where the cut sheets are welded together to form the FIBC. These FIBCs were made with integrated lift slings around the container, or attached to a specially made pallet, or a metal lifting device that the container sat on. These handling devices allowed the container to be filled from the top and discharged from the bottom.

The initial cost of these heavy-duty containers is high; therefore, they are designed to be reused many times in a closed-circuit system, where problems of control logistics, prevention of contamination, cleaning, and liability for loss or damage can be agreed on by the shipper and receiver of the product.

Flexible intermediate bulk containers manufactured with polyolefin fabrics were experimented with in England, Japan, Canada, and the United States all at about the same time in the late 1960s to the early 1970s. It was the development of these high-strength light-weight fabrics (ie, polypropylene) that spurned the growth of the flexible bulk bags that is universally used today.

The rapid growth in Europe in the manufacture of FIBCs occurred in the mid 1970s during the oil crisis. The oil-producing countries building program required large quantities of cement. The demand for cement was shipped in FIBCs at the rate of 30,000–50,000 metric tons per week from Northern Europe, Spain, and Italy to the Middle East.

The demand for bulk bags in the United States grew slower than in Europe until 1984, when the U.S. Department of Transportation (DOT) agreed to grant exemptions for the shipment of hazardous products in FIBCs. Performance standards for FIBCs were established and issued by the Chemical Packaging Committee of the Packaging Institute, USA under T-4102-85. These standards were used to obtain exemptions until DOT included flexible containers with the other types of IBCs in the regulation HM181-E for hazardous products.

The flexible bulk container offers many features that are unique to this package. It can be folded flat and baled for shipment to the user. The weight of a bulk bag to ship one metric ton of product weighs 8–10 lb, offering a low package : product weight ratio. The cost of FIBCs is competitive with other forms of packaging as it is usually utilized with out pallets. They are easy to store and handle in warehouses with standard equipment. When shipping by boat the FIBCs are gang-loaded with up to 14 bulk bags on a spreader bar, and are shipped as break bulk.

The standard diameter of FIBCs is 45–48 in., designed to fit two across in a truck or a shipping container. Special configured containers are made to meet specific requirements of the container user.

FIBCs generally are manufactured to meet specific requirements of the container users. The height of the container, the diameter and length of the spouts, coated or uncoated fabric, and whether a polyethylene liner is necessary will be specified according to the type of product that will be shipped.

When hazardous products are shipped in FIBCs, the UN mark for the product must be printed on the container body. In the United States the manufacturer of the container is self-certifying and marks the container according to the regulations HM 181E. All other countries require the container manufacturer to submit containers to third-party testing for certification.

FIBCs containing nonhazardous or nonregulated product when shipped export from the United States must have performance testing certification if destined for a country that require performance standards for bulk bags.

The following regulation numbers apply for performance standards for containers with regulated products: United States, HM 181E; United Nations, Chapter 16; England, B.S. 6939; Canada, CGSB 43.146.

BIBLIOGRAPHY

General Reference

Flexible Intermediate Bulk Containers / Big Bags, Loadstar Publications, London.

BRUCE CUTHBERTSON
Flexible Intermediate Bulk
Container Association

FOAM, EXTRUDED POLYSTYRENE

Polystyrene foam sheet is a reduced-density sheet made from polystyrene by the extrusion process. Produced in many forms, it is most easily classified by its density and thickness. This article deals with sheet that has densities of 3–12 lb/ft^3 (0.05–0.19 g/cm^3) and thickness 0.012–0.250 in. (0.3–6.35 mm).

The first extruded foam sheet was produced in 1958 by extruding expandable polystyrene beads. In the early 1960s, the direct-gas-injection extrusion technology was developed and is by far the most widely used system today. The growth of the industry in the United States has been extremely fast, growing from about 15 million lb per year in 1966 to over 550 million lb in 1993. In the late 1980s and early 1990s, there was a decrease in the volumes of foam produced due to the environmental issues related to the blowing agents being used and disposal of the foam. The use of new and more ecologically acceptable blowing agents and increased recycling of the products has reduced these concerns, and the foam industry is now growing again. Most of this is used for disposable packaging such as meat and produce trays, egg cartons, containers and trays for carryout meals, and disposable dinnerware. It is also used for drink cups, bottle labels, miscellaneous cushion packaging, and applications where it is laminated to paper or films. A large percentage of all foam sheet produced is thermoformed to manufacture the finished product. When it is laminated with paper, it is usually fabricated by die cutting and scoring to produce folding containers.

There are four direct gas injection processes to produce polystyrene foam sheet. All four have the same objectives: to melt the polymer, uniformly mix in the blowing agent and nucleator, cool the melt, and allow the mix to expand into a biaxially oriented sheet. The four processes employ large-diameter or long *L/D* extruders, twin screw extruders, the Winstead system using a single screw extruder and its cooling system, and the two single-screw extruder tandem system. Of these processes, the tandem system is most successful and widely used (see Fig. 1).

The tandem extruder system utilizes two single-screw extruders: a primary extruder for melting, mixing, and feeding the extrudate to a secondary extruder for cooling and additional mixing of the extrudate prior to exiting an annular die. Most tandem systems use 4.5-in. (11.4-cm) primary extruders and 6-in. (15.2-cm) secondary extruders with output rates of 500–1200 lb/h (227–544 kg/h). As producers strive for higher output rates, 6-in. (15.2-cm) primary and 8-in. (20.3-cm) secondary extruders with output rates of 900–2000 lb/h (408–909 kg/h) are becoming more popular. Tandem foam lines are supplied by several extruder manufacturers (see also Extrusion).

In addition to the rather unique extrusion system, the die–mandrel–takeoff system must be carefully designed. As mentioned above, an annular die is generally used because of the three-dimensional expansion of the foam. The foam is drawn over a cooling sizing mandrel to cool the foam and provide the desired sheet width. Blowup ratios, the die diameter : mandrel diameter ratio must be calculated according to the density and desired orientation of the foam to be produced. Ratios of 3 : 1 to 4.5 : 1 are common. This tubular sheet is slit into the required widths as it is drawn off the mandrel and is then wound into rolls.

The ingredients required to produce foam sheet are resin, nucleator, and blowing agent. The resin is normally a high-heat general-purpose polystyrene such as Dow STYRON* 685D. Nucleators, such as talc or a citric acid–sodium bicarbonate mixture are added to provide foaming sites to obtain the desired cell size and uniformity. The blowing agents used have changed dramatically in the past few years in response to environmental issues. Fully halogenated chlorofluorocar-

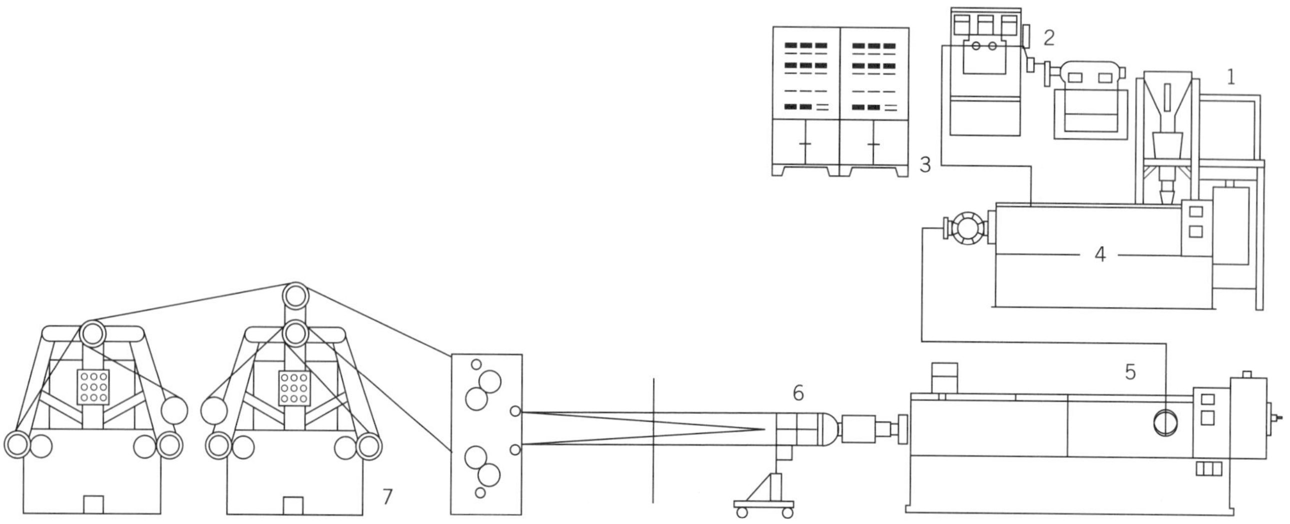

Figure 1. A two-extruder tandem system for the production of foamed sheet: (**1**) continuous feeding and blending system; (**2**) volumetric pump system; (**3**) process control; (**4**) primary extruder; (**5**) secondary extruder; (**6**) annular die and cooling mandrel; (**7**) draw rolls and winders.

bons such as CFC-11, CFC-12, and HCFC-22, which were used previously, have been banned for this application. The blowing agents presently being used are light aliphatic hydrocarbons, HFC-152a, carbon dioxide, and blends of these blowing agents. The blowing agent is injected into the primary extruder approximately two-thirds of the way up the barrel. This is accomplished by the use of a positive-displacement volumetric pump. The amount and type of blowing agent control the density of the foam produced. Figure 2 shows the relationships for three of today's most commonly used blowing agents. The foam can also be colored for aesthetics by the addition of color concentrates (see also Colorants).

A large percentage of all foam sheet is thermoformed to produce the final product. To achieve good postexpansion of the sheet, the sheet should be allowed to age for 3–5 days. This aging allows cell gas pressure to reach equilibrium. Matched metal molds are normally used in the thermoforming process. Oven temperature control is critical for consistent thermoforming. Scrap produced from the extrusion and thermoforming process can be reprocessed by grinding the scrap and densifying it in an extruder.

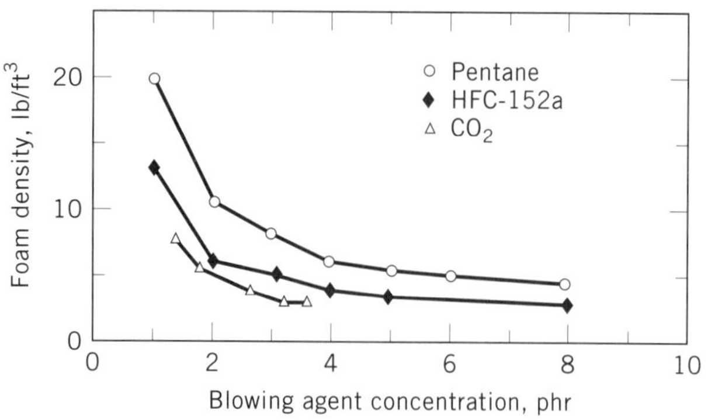

Figure 2. Blowing-agent requirements for polystyrene foam. HFC-152a (Formacel Z-2) data supplied by DuPont. To convert lb/ft^3 to g/cm^3, multiply by 0.0162.

Ecologically, polystyrene foam is very easily recycled, and significant quantities are presently being reused. It also conforms favorably to current forms of disposal, such as landfill and incineration. In landfill foam remains inert, but packs and crushes easily and there is no pollution of underground water streams by decaying material. In incineration the chief products of combustion are water, carbon dioxide, and carbon monoxide, which are typical of organic materials, and the Btu (British thermal unit) value is high.

Various patents have been granted on thin foam sheet and methods for its manufacture, and a freedom to practice the information presented above is not to be implied.

BIBLIOGRAPHY

1. "Controlled Density Polystyrene Foam Extrusion," *SPE J.* (July 1960).
2. "Twin Screw Extrusion of Expanded Polystyrene Sheet," *Modern Plast.* (Sept. 1969).
3. "Inline Process Makes Laminated Paper/PS Foam Board," *Plast. Technol.* (April 1980).
4. "Dual Blowing Agents Lower Cost of Polystyrene Foams," *Plast. Eng.* (June 1986).
5. "Extruding Thermoplastic Foams with Non-CFC Blowing Agent," *Plast. Eng.* (May 1990).
6. *DuPont Alternate Blowing Agents for Thermoplastic Foams,* Bulletin ABA-5.
7. U.S. Pat. 3,479,694 (Nov. 25, 1969), T. W. Winstead.
8. U.S. Pat. 4,532,264 (July 30, 1985), L. C. Rubens.
9. U.S. Pat. 5,250,577 (Oct. 5, 1993), G. C. Welsh.

PHILLIP A. WAGNER
The Dow Chemical Company
Midland, Michigan

FOAM PLASTICS

Foamed plastics, also known as *cellular plastics* or *plastic foams,* have been important since primitive humans began to use wood, a cellular form of the polymer cellulose. Cellulose is the most abundant of all naturally occurring organic compounds, representing approximately one-third of all vegetable matter in the world. The high strength : weight ratio of wood, good insulating properties of cork and balsa, and cushioning properties of cork and straw have contributed to the development of the board range of cellular synthetic polymers in use today.

Cellular polymers have been commercially accepted for a wide variety of applications since the 1940s. The total usage of foamed plastics in the United States rose from about 2.1×10^6 tons in 1993. It is expected to rise to about 2.9×10^6 metric tons in 1998 (1).

Nomenclature

A *cellular plastic* is defined as a plastic whose apparent density is decreased substantially by the presence of numerous cells disposed throughout its mass (2). In this article the terms *cellular polymer, foamed plastic, expanded plastic,* and *plastic foam* are used interchangeably to denote all two-phase gas–solid systems in which the solid is continuous and composed of a synthetic polymer or rubber.

Theory of the Expansion Process

Foamed plastics can be prepared by various methods. The most widely used, called the *dispersion process,* involves dispersion of a gaseous phase throughout a fluid polymer phase and the preservation of the resultant state. Other methods of producing cellular plastics include sintering dispersed small particles and dispersing small cellular particles in the plastic. The latter processes are relatively straightforward techniques of lesser commercial importance.

The expansion process has been the subject of extensive investigation because it is the foundation of foamed plastics (3–6). In general, the expansion process consists of three steps: creation of small discontinuities or cells in a fluid or plastic phase, growth of these cells to a desired volume, and stabilization of the resultant cellular structure by physical or chemical means.

Bubble initiation. The development of bubbles within a liquid or polymer solution is generally called *nucleation,* although the term actually refers only to those bubbles that separate from the supersaturated liquid or polymer solution in the presence of an initiating site such as a surface irregularity. This process has several sources: (1) dissolved gases that are normally present in the liquid or polymer solution and forced into supersaturation by increased temperature; (2) low-boiling liquids that are incorporated into the system as blowing agents and forced into the gas phase by increased temperature and decreased pressure; (3) gases produced as blowing agents, eg, by the water–isocyanate reaction used for CO_2 production in polyurethane foams; and (4) chemical blowing agents that decompose thermally to form a gas.

Bubble nucleation is affected by a number of conditions. Physically, the effects of temperature, pressure, and in some cases humidity are fairly obvious. Other important parameters are surface smoothness of the substrate, surface characteristics of filler particles, presence and concentration of certain surfactants or nucleators, size and amount of second-phase liquid bubbles, and the rate of gas generation.

In many cases bubbles of gas and other contaminants are already present in the liquid or polymer solution, and these serve as sites into which the gas may diffuse. The number and size of these gas bubbles may be another important factor in bubble development.

Bubble growth. The initial bubble is ideally a sphere that grows as a result of the interaction of the differential pressure (ΔP) between the inside and outside of the cell and the interfacial surface tension (γ). The radius (r) of the bubble at equilibrium is related to these factors as shown in Eq. (1):

$$\Delta P = \frac{2\gamma}{r} \quad (1)$$

The differential pressure is larger for a small bubble at a fixed surface tension. Accordingly, smaller bubbles tend to equalize these pressures by growing faster, breaking the wall separating the cells, or by diffusion of the blowing agent from the small to the large cells as indicated by Eq. (2):

$$\Delta P_{1,2} = 2\gamma \left(\frac{1}{r_1} - \frac{1}{r_2}\right) \quad (2)$$

where $\Delta P_{1,2}$ is the difference in pressure between cells of radius r_1 and r_2. Therefore, the rate of growth of the cell depends on the viscoelastic nature of the polymer phase, the blowing-agent pressure, the external pressure on the foam, the cell size, and the permeation rate of blowing agent through the polymer phase.

Bubble stabilization. Bubble stability during growth is a function of the type and concentration of surfactant, rate of bubble growth, viscosity of the fluid medium, differential pressure variations, and presence of cell-disrupting agents such as solid particles, liquids, or gases. As the cell walls are squeezed into polyhedra, a wall-thinning effect takes place, and liquid is drained from cell-wall faces into the lines of cell intersections to form ribs or struts, which are typically triangular in cross-section. This cell-wall membrane thinning can continue to the point where the cell walls collapse and the cells open. This becomes a very important characteristic of most plastic foams and affects properties such as thermal conductivity, moisture absorption, breathability, and load bearing.

Ultimate stabilization occurs as a result of either chemical reaction continuing to the point of complete gelation or the physical effect of cooling below the second-order transition point to prevent polymer flow. As the final solidification is approached, the previously formed bubbles may be distorted by the system flow or gravity, thereby producing anisotropy in the cellular structure. This effect must be considered when evaluating physical properties of plastic foams by obtaining samples oriented in specific directions to the process flow.

Properties

The data in Table 1 show the broad ranges of properties of rigid foamed plastics (7); the manufacturer should be con-

Table 1. Physical Properties of Commercial Rigid Foamed Plastics

Property	ASTM Test	Cellulose Acetate	Epoxy	Phenolic	Polystyrene: Extruded Plank		Polystyrene: Expanded Plank			Polystyrene: Extruded Sheet		PVC		Polyurethane: Polyether		Polyurethane: Isocyanurate Bun	Polyurethane: Isocyanurate Laminate	Urea–formal-dehyde
Density, kg/m³[a]		96–128	32–48	32–64	35	53	16	32	80	96	160	32	64	32–48	64–128	32	32	13–19
Mechanical properties, compressive strength, kPa[b] at 10%	D1621	862	138–172	138–620	310	862	90–124	207–276	586–896	290	469	345	1,035	138–344	482–1896	210	117–206	34
Tensile strength, kPa[b]	D1623	1,172		138–379	517		145–193	310–379	1020–1186	2070–3450	4,137 6,900	551	1,207	138–482	620–2000	250	248–290	
Flexural strength, kPa[b]	D790	1,104		172–448	1138		193–241	379–517				586	1,620	413–689	1380–2400			
Shear strength, kPa[b]	C273	965		103–207	241			241				241	793	138–207	413–896	180	117	
Compression modulus, MPa[b]	D1621	38–90	3.9		10.3			3.4–14				13.1	35	2.0–4.1	10.3–31			
Flexural modulus, Mpa[b]	D790	38			41			9.0–26				10.3	36	5.5–6.2	5.5–10.3			
Shear modulus, MPa[b]	C273			2.8–4.8	10.3			7.6–11.0				6.2	21	1.2–1.4	3.4–10.3		1.7	
Thermal properties, thermal conductivity W/(m · K)	C177	0.045–0.046	0.016 0.022	0.029 0.032	0.030		0.037	0.035	0.035	0.035	0.035	0.023		0.016–0.025	0.022–0.030	0.054	0.019	0.026 0.030
Coefficient of linear expansion, 10^{-5}/°C	D696		1.5	0.9	6.3	6.3	5.4–7.2	5.4–7.2						5.4–7.2	7.2	7.2		
Maximum service temperature, °C		177	205–260	132	74		74–80	74–80	74–80	77–80	80			93–121	121–149	149	149	
Specific heat, kJ/(kg · K)[c]	C351				1.1									~0.9	~0.9	~0.9		
Electrical properties					1.19–1.20	<1.05	<1.05	1.02	1.02	1.02	1.27	1.28			1.05	1.1	1.4	
Dielectric constant	D1673	1.12			0.028–0.031	<0.0004	<0.0004	0.0007	0.0007	0.0007	0.00011	0.00014			13	18		
Dissipation factor		20																
Moisture-resistance water absorption, vol%	C272	4.5		13–51	0.02	0.05	1–4	1–4	1–4									
Moisture-vapor transmission, g/(m · s · GPa)[b]	E96		58		35		<120	35–120	23–35	86	56	15		35–230	50–120		230	1,610–2,000

[a] To convert kg/m³ to lb/ft³, multiply by 0.0624.
[b] To convert kPa to psi, multiply by 0.145.
[c] To convert kJ/(kg · K) to Btu/(lb · °F), divide by 4.184.
Source: Suh and Webb (7).

sulted for the properties of a particular product. The properties of some commercial flexible foamed plastics are given in Table 2 (7). These values depend on several structural variables and should be used only as general guidelines (see also Polymer properties).

The properties of a foamed plastic depend on composition and geometry, often referred to as *structural variables*. Furthermore, they are influenced by the foam structure and the properties of the parent polymer. The polymer phase description must include the additives present.

Cells. A complete knowledge of the cell structure of a particular polymer would require the size, shape, and location of each cell. Because this is impractical, approximations are employed. Cell size has been characterized by measurements of the cell diameter and as a measurement of average cell volume. Mechanical, optical, and thermal foam properties depend on cell size.

Geometry. Cell shape is governed predominantly by final foam density and the external forces exerted on the cellular structure before its stabilization in the expanded state. In the presence of external forces, the cells may be elongated or flattened. Cell orientation can influence many properties.

Fraction open cells. An important characteristic of the cell structure is the extent of communication with other cells. This is expressed as *fraction open cells*.

Gas composition. In closed-celled foams, the gas phase in the cells can contain blowing agent (so-called captive blowing agent), air, or other gases generated during foaming. Thermal and electrical conductivity can be profoundly influenced by the cell-gas composition.

Mechanical properties. In mechanical properties, rigid foams differ from flexible foams. The tests used to characterize them are therefore different, as are their application properties. In the last two decades, a separate class of high-density, rigid cellular polymers called *structural foams* (density >0.3 g/cm^3) have become commercially significant.

Structural variables that affect the compressive strength and modulus of a rigid plastic foam are, in order of decreasing importance, plastic-phase composition, density, cell structure, and plastic state. The effect of gas composition is minor with a slight effect of gas pressure in some cases.

Structural foams or foams with integral skins are usually produced as fabricated articles in injection-molding or extrusion processes. They have relatively high densities, and cell structures are composed primarily of holes in contrast to a pentagonal dodecahedron structure in low-density plastic foams. Because structural foams are generally not uniform in cell structure, they exhibit considerable variation in properties with article geometry.

In flexible foams, which are used in comfort cushioning, packaging, and wearing apparel, different mechanical properties are emphasized than for rigid foams. The compressive nature of flexible foams, both static and dynamic, is their most significant mechanical property. Other important properties are tensile strength and elongation, tear strength, and compression set.

Thermal properties. The thermal conductivity of cellular polymers has been thoroughly studied in heterogeneous materials and plastic foams (8). Heat transfer can be separated into its component parts as follows:

$$k = k_s + k_g + k_r + k_c \tag{3}$$

where k = total thermal conductivity and k_s, k_g, k_r, and k_c represent solid conduction, gases conduction, radiation, and convection, respectively.

As a first approximation, the heat conduction of low-density foams through the solid and gas phases can be expressed as the product of the thermal conductivity of each phase and its volume fraction. In most cellular polymers the conduction through the solid phase is determined primarily by the density and the polymer-phase composition. Although conductivity through gases is much lower than through solids, the amount of heat transferred through the gas phase in a foam is usually the largest component of a total heat transfer because of the large gas-phase volume. Ordinarily, convection cannot be measured in cells of diameter less than ~4 mm (8). Since most cellular polymers have cell diameters of <4 mm, convection can be ignored. Radiant heat transfer through cellular polymers has also been studied (8). The increase in k at low densities is due to increased radiant heat transfer; at high densities, to an increasing contribution of k_s.

The thermal conductivity of most materials decreases with temperature. When foam structure and gas composition are not influenced by temperature, the k of the cellular material falls with decreasing temperature. The thermal conductivity of a cellular polymer can change on aging under ambient conditions if the gas composition is influenced by aging. This is the case when oxygen or nitrogen diffuses into cellular foams that initially have an insulating fluorocarbon blowing agent in the cells (8). The thermal conductivity of foamed plastics varies with thickness (9). This has been attributed to the boundary effects of the radiant contribution to heat transfer.

The specific heat of a cellular polymer is the sum of the specific heats of each of its components. The contribution of the gas is small and can often be neglected.

The coefficients of linear thermal expansion of polymers are higher than those of most rigid materials at ambient temperatures because of the supercooled liquid nature of the polymeric state. This large coefficient is carried over directly to the cellular state. A variation of this property with density and temperature has been reported for foams in general (10).

Moisture resistance. Plastic foams perform better than do other thermal insulation when exposed to moisture, particularly when subjected to a combination of thermal and moisture gradients.

Moisture absorption and freeze–thaw resistance of various insulations and the effect of moisture on thermal performance show that in protected-membrane roofing applications the order of resisting moisture pickup is extruded polystyrene ≫ polyurethane > molded polystyrene (11). Water absorption values for insulation in use for 5 years were 0.2 vol% for extruded polystyrene, 5 vol% for polyurethane without skins, and 8–30 vol% for molded polystyrene. These values correspond to increases in k of 5–265%. Far below-grade applications, extruded polystyrene was better than molded polystyrene or polyurethane without skins in terms of moisture and thermal resistance.

Table 2. Physical Properties of Commercial Flexible Foamed Plastics

Property	ASTM Test	Expanded Natural Rubber		Expanded SBR	Latex Foam Rubber		Polyethylene			Polypropylene		Polyurethane		Poly(vinyl Chloride)		Silicone	
							Extruded Plank	Extruded Sheet	Crosslinked Sheet	Modified	Sheet	Standard Cushioning	High-Resilience Type			Liquid	Sheet
Density, kg/m³[a]		56	320	72	80	130	35	43	26–28	64–96	10	24	40	56	112	272	160
Cell structure		Closed	Closed	Closed	Open	Open	Closed	Closed	Closed	Closed		Open	Open	Closed	Closed	Open	Open
Compression strength 25%, deflection, kPa[b]	D33574 D3575						48			206	4.8	5.7	4.6				
Tensile strength, kPa[b]	D33574		206	551	103		138	41		344		118	103	10.3	24	36 at 20%	
Tensile elongation, %	D33574				310		60	276	276–480	1380	138–275	205	160			227	310
Rebound resilience, %	D3574				73			50		75		40	62				
Tear strength, (N/m)[c] × 10²	D3574						10.5	26				4.4	2.4				
Maximum service temperature, °C		70	70	70			82	82	79–93	135	121					350	260
Thermal conductivity, W/(m · K)	C177	0.036	0.043	0.030		0.050	0.053	0.040–0.049	0.036–0.040	0.039	0.039			0.03 5	0.040	0.078	0.086

[a] To convert kg/m³ to lb/ft³, multiply by 0.0624.
[b] To convert kPa to psi, multiply by 0.145.
[c] To convert kJ/(kg · K) to Btu/(lb · °F), divide by 4.184.
Source: Suh and Webb (7).

Electrical properties. Electrical insulation takes advantage of the toughness and moisture resistance of polymers along with the decreased dielectric constant and dissipation factor of the foams. In addition, because of the low dissipation factor and rigidity, plastic foams are used in the construction of radar domes.

Environmental aging. The response of cellular materials to light and oxygen is governed almost entirely by the composition and state of the polymer phase. Expansion into the cellular state increases the surface area; reactions of the foam with vapors and liquids are correspondingly faster than those of solid polymer. All cellular polymers deteriorate under the combined effect of light or heat and oxygen; this may be alleviated by additives.

Other properties. Cellular polymers by themselves are poor materials for reducing sound transmission. They are, however, effective in absorbing sound waves of certain frequencies. Materials with open cells on the surface are particularly effective in this respect. The presence of open cells in a foam allow gases and vapors to permeate the cell structure by diffusion and convection, thus yielding high permeation rates. In closed-cell foams, the permeation of gases or vapors is governed by the composition of the polymer phase, gas composition, density, and cellular structure. The penetration of visible light through foamed polystyrene has been shown to follow approximately the Beer–Lambert law of light absorption (10).

The resistance to rot, mildew, and fungus is related to moisture absorption. Therefore, open-cell foams support such growth better than do closed-cell foams.

Manufacturing Processes

Cellular plastics and polymers have been prepared by processes involving many methods of cell initiation, growth, and stabilization. The most convenient method of classifying these methods appears to be based on cell growth and stabilization. According to Eq. 1, the growth of the cell depends on the pressure difference between the inside of the cell and the surrounding medium. Such pressure differences may be generated by lowering the external pressure, ie, decompression; or by increasing the internal pressure in the cells, ie, pressure generation. Other methods of generating the cellular structure are by dispersing gas or solid in the fluid state and stabilizing this cellular state, or by sintering polymer particles in a structure that contains a gas phase (see Table 3).

Foamable compositions in which the pressure within the cells is increased relative to that of the surroundings are called *expandable formulations.* Both chemical and physical processes are used to stabilize plastic foams for expandable formulations. There is no single name for the group of cellular plastics produced by the decompression processes. The operations used are extrusion, injection molding, and compression molding. Physical or chemical methods may be used to stabilize the products.

Expandable formulations

Physical stabilization. Cellular polystyrene, poly(vinyl chloride), polyethylene, and copolymers of styrene and acrylonitrile can be manufactured by this process.

Chemical stabilization. This method has been used successfully for more materials than the physical stabilization process. It is generally more suitable for condensation polymers than the vinyl polymers because of the fast yet controllable curing reactions. Chemical stabilization can be used for foam formation from polyurethane, polyisocyanurate, phenolic, epoxy, and silicone resins.

Decompression expansion processes

Physical stabilization. Cellular polystyrene, cellulose acetate, polyolefins, and poly(vinyl chloride) can be manufactured by this process.

Chemical stabilization. Cellular rubber and ebonite are produced by chemical stabilization processes.

Dispersion. In several techniques for producing cellular polymers, the gas cells are produced by dispersion of a gas or solid in the polymer phase followed, when necessary, by stabilization and treatment of the dispersion. In frothing, a quantity of gas is mechanically dispersed in the fluid polymer phase and stabilized. Latex foam rubber was the first cellular polymer produced by frothing. In another method, solid particles are dispersed in a fluid polymer phase, the dispersion stabilized, and the solid phase dissolved or leached, thus leaving the cellular polymer. Cellular polymers called *syntactic foams* are made by dispersing an already cellular solid phase in a fluid polymer and stabilizing the dispersion.

Uses

Concern over energy conservation and safety has stimulated growth in applications for insulation and cushioning in transport. A healthy economy is also expected to increase the demand for cushioning in furniture, bedding, and flooring as well as for packaging. Structural foams are widely used as substitutes for wood, metal, or unfoamed plastics. Table 4 shows the demand for plastic foams in the United States.

Cushioning. The properties of significance in the cushioning applications of cellular polymers are compression–deflection behavior, resilience, compression set, tensile strength and elongation, and mechanical and environmental aging. The broad range of compressive behavior of flexible foam is one of the advantages of cellular polymers because the needs of almost any cushioning application can be met by changing either the chemical nature or the physical structure of the foam. Flexible urethanes, vinyls, latex foam rubber, and olefins are used to make foamed plastic cushioning for automobile padding, seats, furniture, flooring, mattresses, and pillows.

Thermal insulation. Thermal insulation is the largest application for the rigid materials because of their thermal conductivity, ease of application, cost, moisture absorption, and transmission permeance. Plastic foams containing a captive blowing agent have much lower thermal conductivities than do other insulating materials. The low-thermal-conductivity polyurethane foams are used in refrigerators and freezers. Polystyrene foam is popular where cost and moisture resistance are important and polyurethane foams are used in spray applications.

Extruded polystyrene foam is found in residential con-

Table 3. Production of Cellular Polymers

Polymer	Extrusion	Expandable Formulation	Spray	Forth Foam	Compression Mold	Injection Mold	Sintering	Leaching
Cellulose acetate	X							X
Epoxy resin		X	X	X				
Phenolic resin		X						
Polyethylene	X	X			X	X	X	X
Polystyrene	X	X				X	X	X
Silicones		X						
Urea–formaldehyde resin				X				
Urethane polymers		X	X	X		X		
Latex foam rubber				X				
Natural rubber	X	X			X			
Synthetic elastomers	X	X			X			
Poly(vinyl chloride)	X	X		X	X	X		X
Ebonite							X	
Polytetrafluoroethylene							X	

Table 4. U.S. Cellular Polymer Market (million lb)

Foam	1990	1992	1993	1998[a]	Annual Growth, % 1993–1998
Polyurethane	2681	2752	2848	3509	4.3
Polystyrene	1417	1366	1416	1624	2.8
PVC	320	340	350	395	2.5
Polyolefins	100	105	110	135	4.2
Other	62	67	70	80	2.7
Total	4580	4630	4794	5743	3.7

[a] Projected.
Source: Courtesy of Business Communications Co., Inc.

struction as sheathing, perimeter, and foundation insulation. Both polystyrene and polyurethane foams are highly desirable roof insulators.

Packaging. The entire range of cellular polymers from rigid to flexible is used for packaging because of their low cost, ease of application or fabrication, moisture susceptibility, thermal conductivity, consumer appeal, and mechanical properties, especially compressive properties.

Extruded polystyrene foam sheet. Polystyrene foam sheet is made from a combination of polystyrene resin, blowing agents, nucleating agents, and pigments in an extrusion process (12,13). Polystyrene foam sheet is used in many shapes and can usually be characterized by its stiffness and low density, generally 2–12 lb/ft^3 (32–190 kg/m^3). Sheet products are typically 0.015–0.150 in. thick.

The major uses for foam sheet are for disposable applications such as carryout cups and clam shells, plates, bowls, egg cartons, and meat and poultry trays. New market growth has come in the applications of polystyrene foam sheet for use in art boards, insulated boxes, and bottle wraps. For applications requiring extra stiffness or modulus, films or paper may be laminated to the foam. Foam sheet can be fabricated by die cutting and scoring, in addition to thermoforming to shape and bend the foam.

Typically, the foam-sheet process uses an annular die in the production of thin sheets. Single-screw, twin-screw, long-barrel single-screw, and tandem-extruder systems can be used with the latter being most common.

Expanded polystyrene beadboard. Expanded polystyrene (EPS) beadboard is produced with expandable polystyrene beads. These beads are impregnated with 5–8% pentane and sometimes with flame retardants during suspension polymerization. The beads are preexpanded by fabricators with steam or vacuum and aged. Then they are fed to steam-heated block molds where expansion and fusion of beads continue. The molded blocks are sliced after curing. Block densities range from 13 to 48 kg/m^3, with 24 kg/m^3 most common for cushion packaging and 16 kg/m^3 for insulation.

Expandable polystyrene bead-molding products account for the main portion of the drink-cup market and are used in packaging materials, insulation board, and ice chests.

Extruded polystyrene loose-fill packaging. Loose-fill packaging particles are used in void-filling packaging applications. The foamed particles provide cushioning for lightweight, often fragile products. The foamed particles usually resilient and are very low in density [bulk density properties under <0.30 lb/ft^3 (4.8 kg/m^3)] and can be reused many times. Polystyrene loose-fill particles can be recycled. Growth of this market has been affected by the development of starch-based extrusion products.

Expandable polystyrene beads for loose-fill packaging are produced by the extrusion process. Virgin and recycled polystyrene with nucleating agents and additives are fed into the extruder, and blowing agents are usually fed downstream into the extrusion process. The blowing agents used have progressed from CFCs to HCFCs to hydrocarbons or blends of hydrocarbons with carbon dioxide. Some extrusion processes can use expandable beads (pentane/isopentane blowing agent) for feedstock.

Loose-fill particles are produced by a foam-at-the-die process, or by an expandable-bead process. In the foam-at-the-die process, the molten gel is allowed to foam to a density of approximately 1–2 lb/ft^3. The foam is then cut at the die face, or downstream after some cooling has occurred. Subsequent density reduction is accomplished by steam treatment, followed by air aging. The foam may be steamed and aged 2–3 times to lower the density down to below 0.25 lb/ft.[3]

The expandable-bead process produces an unfoamed particle that has the blowing agent incorporated into the bead. The expandable beads are produced at a central extrusion location and then shipped to local expanders. Local expansion facilities are needed because of the lightweight nature of the final product. The expandable beads are exposed to steam as done with the foam-at-the-die process, but will require an extra expansion step.

Polyolefin foam sheet and plank. Polyolefin foams are closed-cell foams that are tough, flexible, and resilient and are used in applications such as cushioning, flotation, furniture overwrap, water sports, surface protection, and many other uses. The energy-absorbing characteristics of polyolefin foams make them excellent foams for use in cushioning applications. Polyethylene and polypropylene foams are low-density products, generally less than 10 lb/ft^3 (160 kg/m^3), that are made by the extrusion process. These foams are made in many forms—plank, sheet, rounds, multiple strands—and can be made with crosslinked or noncrosslinked formulation.

Polyethylene and polypropylene foams are usually produced from a combination of virgin and recycled resins, blowing agents, nucleating agents, and other additives such as colorants, antistatic additives, and flame retardants. Both polyethylene and polypropylene can be made into sheet-type products, typically 0.06–0.625 in. thick (2–15 mm) by up to 72 in. (1800 mm) wide. Because of the low melt strengths typical of polypropylene, only polyethylene is used in the manufacture of large cross sections of foam of 2–4-in. (50–100-mm) thickness and 24–48-in. (600–1200-mm) width. Planks of foam are often fabricated into specific shapes to meet the packaging requirements of electronic components such as computers and their monitors, the handling requirements for water sports uses such as boogie boards, or the part separation requirements of multiple-use industrial applications.

Noncrosslinked polyolefin sheet foams can be made on single-, dual-, and twin-screw extrusion lines that are designed to mix in large quantities of blowing agents and then extruding the supersaturated gel through an annular die at the proper temperature, usually very near the crystalline melting point. Figure 1 shows the theoretical relationship between the blowing-agent content and the expected foam density. To produce a foam with a density of ~2 lb/ft^3 (32 kg/m^3), the extrusion system must be able to mix in 6–16 lb of blowing agent per 100 lb of resin, depending on the blowing agent selected. Gel pressure at the die is also very important. If the pressure is too low, the blowing agent will begin to come out of solution and will allow the molten gel to begin foaming prior to exiting the die lips. If this occurs, severe disruption of the skin surface will occur and a poor-quality foam will result. Blowing agents are injected into the molten resin gel downstream from the hopper. The exact location will depend on the type of equipment being used. To extrude sheet products that are

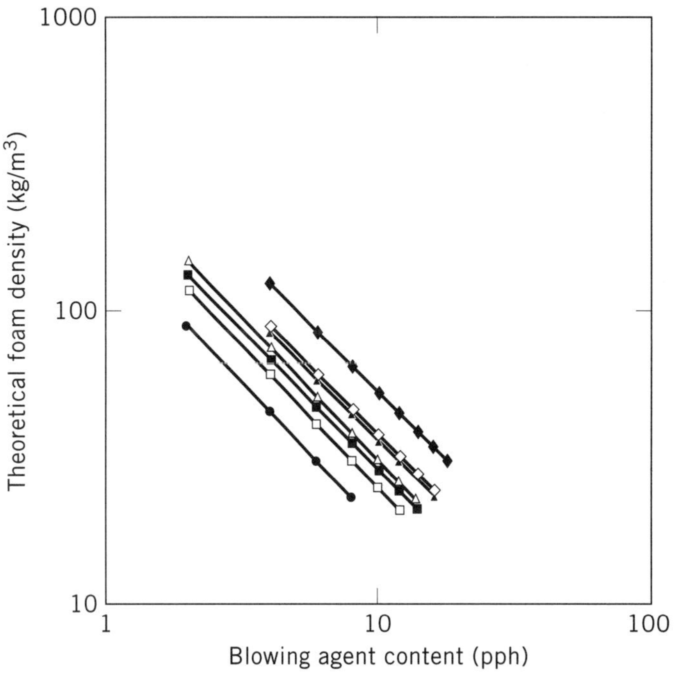

Figure 1. Effect of blowing-agent concentration of theoretical foam density: ◆ CFC-12, ◇ HCFC-22, ▲ HFC-134a, △ pentane, ■ HFC-152a, □ butane, ● CO_2

very wide and very thin, an annular die and forming mandrel is used. The proper-size forming mandrel allows the extruded tube to be stretched both radially and downstream at the same time, and the mandrel provides a means to cool the inner surface of the foam tube. The cooled tube is then slit at one or two locations and subsequently wound into rolls. Plank extrusion is usually through a more rectangular shaped die onto a conveyor belt or some other means to move the foam away from the die. Plank extrusion is usually through a more rectangle-shaped die onto a conveyor belt or some other means to move the foam away from the die. Plank extrusion usually requires higher rates than sheet because of the larger die opening and the need for maintaining sufficient pressure to prevent prefoaming.

Polyolefin foam manufacture generally uses physical agents, materials that are a gas at the foaming temperature (usually at the resin melting point). Chemical blowing agents are rarely used in noncrosslinked applications. Blowing agents for polyolefin foam extrusion are now typically hydrocarbons (14) or a blend of hydrocarbon and carbon dioxide. Prior to the Montreal Protocol agreements, CFC-114 was the main foaming agent used in polyolefin foams. CFC-114 was an excellent foaming agent because of its solubility and permeability characteristics. Another very strong attribute was its nonflammable characteristic. With the Montreal Agreement, use of CFCs was prohibited after 1991 in noninsulating applications. The CFC-114 replacement list for foaming agents was not very long. Hydrocarbons (propane, butane, and isobutane), HCFCs (22,142b), HFCs (152a,134a), and carbon dioxide were considered. Most of the replacement options included an increase in flammability and an increase in cost to the manufacturing operations. The use of carbon dioxide is attractive in terms of cost, flammability, and environmental, but the processing aspect makes total substitution very difficult. Because of the high vapor pressure of this physical gas, maintaining the proper resin-blowing-agent solution characteristics at the die is very difficult. The Clean Air Act of 1990 (CAA) has since further reduced the blowing-agent alternatives. The CAA has prohibited the use of HCFCs, leaving only hydrocarbon and their blends, HFCs (15), and carbon dioxide, or blends thereof.

Crosslinked foams offer higher temperature stability, more flexibility, finer cell size, and better thermoforming properties than do noncrosslinked foams. Crosslinking of the olefin resin is accomplished by chemical crosslinking (such as dicumyl peroxide, or silanes) or radiation crosslinking via X-ray or electron-beam exposure (7). Typically, the polyethylene resin, additives, crosslinking agents, and chemical blowing agents (such as azodicarbonamide) are mixed together at temperatures below the activation temperature of the blowing agents, and then extruded into a flat sheet that can be rolled up, or into some other profile, prior to expansion into a foam product.

Crosslinking occurs before the foam-expansion step. Expansion is done by exposure of the crosslinked sheet to hot (~200°C) air. Generally, chemical crosslinking is used in the production of thick products while radiation crosslinking is used for thinner foams.

Health and Safety Factors

Flammability. Plastic foams are organic and therefore combustible. All plastic foams should be handled, transported, and used according to manufacturers' recommendations and local and national regulations.

Virtually all plastic foams are blown with inert gases, chemical blowing agents that release inert gases, hydrocarbons containing three to five carbon atoms, chlorinated hydrocarbons, and hydrochlorofluorocarbons such as HCFC-22, HCFC-141b, and HCFC-142b.

Atmospheric emissions. Certain organic compounds generate smog photochemically. Interaction with the total environment must be considered in developing environmentally acceptable blowing agents. The products of combustion of plastic foams are usually carbon monoxide and carbon dioxide with smaller amounts of many other substances.

The presence of additives or unreacted monomer in certain plastic foams can limit their use where food or human contact is anticipated. The manufacturers' recommendations and existing regulations should be followed.

BIBLIOGRAPHY

1. H. Kibbel, P-120R, *Polymeric Foams,* A. Huge Sub-Industry, Business Communications Company, Inc., Norwalk, CT, 1994.
2. ASTM D883-80C, *Definitions of Terms Relating to Plastics,* American Society for Testing and Materials, Philadelphia, 1982.
3. K. C. Frisch and J. H. Saunders, *Plastic Foams,* Vol. 1, Parts 1 and 2, Marcel Dekker, New York, 1972 and 1973.
4. C. J. Benning, *Plastic Foams,* Vol. 1 and 2, Wiley-Interscience, New York, 1969.
5. N. C. Hilyard and co-worker, *Mechanics of Cellular Plastics,* Macmillan, New York, 1982.
6. D. Klempner and K. C. Frisch, *Handbook of Polymeric Foams and Foam Technology,* Hanser Publishers, New York, 1991.
7. K. W. Suh and D. D. Webb, *Encyclopedia of Polymer Science and Engineering,* Vol. 3, 2nd ed., Wiley, New York, 1985, pp. 1–59.
8. R. E. Skochdopole, *Chem. Eng. Prog.* **57**(10), 55 (1961).
9. B. Y. Lao and R. E. Skochdopole, paper presented at 4th SPI International Cellular Plastics Conference, Montreal, Canada, SPI, New York, Nov., 1976.
10. J. D. Griffin and R. E. Skochdopole, "Plastic Foams," in E. Baer, ed., *Engineering Design for Plastics,* Reinhold, New York, 1964.
11. F. J. Dechow and K. A. Epstein, ASTM STP 660, *Thermal Transmission Measurements of Insulation,* ASTM, Philadelphia, 1978, p. 234.
12. U.S. Pat. 5,250,577 (Oct. 5, 1993), G. C. Welsh (to The Dow Chemical Company).
13. U.S. Pat. 5,364,696 (Nov. 15, 1994), P. A. Wagner (to The Dow Chemical Company).
14. U.S. Pat. 4,640,933 (Feb. 3, 1987), C. P. Park (to The Dow Chemical Company).
15. U.S. Pat. 5,411,684 (May 2, 1995), M. H. Tusim and C. P. Park (to The Dow Chemical Company).

K. W. SUH
M. H. TUSIM
The Dow Chemical Company
Midland, Michigan

FOIL, ALUMINUM

The Material

Aluminum foil is a thin-rolled sheet of pure or alloyed aluminum, varying in thickness from about 0.00017 (4.3 μm) in. to a maximum of 0.0059 (150 μ) in. (1). By industry definition, rolled aluminum becomes foil when it reaches a thickness less than 0.006 in. (152.4 μm) (see Table 1).

Aluminum, from which the foil is made, is a bluish silver-white trivalent metallic element that is very malleable and ductile. Noted for its light weight, good electrical and thermal conductivity, high reflectivity, and resistance to oxidation, aluminum is the third most abundant element in the earth's crust (1).

Aluminum always occurs in combination with other elements in mineral forms such as bauxite, cryolite, corundum, alunite, diaspore, turquoise, spinel, kaolin, feldspar, and mica. Of these, bauxite is the most economical mineral for production of aluminum. It can contain up to 60% alumina, which is hydrated aluminum oxide. It takes about 4 kg of bauxite to produce 1 kg of aluminum (2).

Alumina is converted into aluminum at a reduction plant or smelter. In the Hall–Héroult process, the alumina is dissolved in a molten salt called *cryolite.* The action takes place in steel boxes lined with carbon called *pots.* A carbon electrode or anode is lowered into the solution, and electric current of 50,000–150,000 A flows from the anode through the mixture to the carbon-cathode lining of the steel pot. The electric current reduces, or separates, the alumina molecules into aluminum and oxygen. The oxygen combines with the anode's carbon to form carbon dioxide. The aluminum, heavier than cryolite, settles to the bottom of the pot from which it is si-

Table 1. Physical Properties of Aluminum Foil

Property	Value
Density	0.0976 lb/in.^{3} (2.70 g/cm^3)
Specific gravity	2.7 (approx.)
Melting range	1190–1215°F (643–657°C)
Electrical conductivity	59° IACS. vol., 200% IACS (approx.), weight
Thermal conductivity	53 W/(m · K) at 25°C
Thermal coefficient of linear expansion	13.1 × 10^{-6} per °F, 68–212°F (23.6 × 10^{-6} per °C, 29–100°C)
Reflectivity for white light, tungsten filament lamp	85–88%
Reflectivity for radiant heat, from source at 100°F (37.8°C)	95% (approx.)
Emissivity, at 100°F (37.8°C)	5% (approx.)
Atomic number	13
Atomic weight	26.98
Valence	4
Specific heat at 20°C	0.21–0.23
Boiling point	3200°F (1760°C)
Temperature coefficient of resistance (representative values per °C)	
at 20°C	0.0040–0.0036
at 100°C	0.0031–0.0028
Low-temperature properties—aluminum increases in strength and ductility as temperature is lowered, even down to −320°F (−195.6°C)	

phoned into crucibles. The molten aluminum is eventually processed into products.

Foil

One of aluminum's most common uses is as foil. Aluminum foil is generally produced by passing heated aluminum-sheet ingot between rolls in a mill under pressure. Ingot is flattened to reroll sheet gauges on sheet and plate mills and finally to foil gauges in specialized foil-rolling mills.

A second method for producing aluminum foil, rapidly gaining popularity, involves continuous casting and cold rolling. This method can eliminate the conventional energy-intensive and costly steps of casting ingot, cooling, transporting to rolling plants, and then reheating and hot rolling to various gauges.

First produced commercially in the United States in 1913, aluminum foil became a highly marketable commodity because of its protective qualities, economic production capability, and attractive appearance. The first aluminum foil laminated on paperboard for folding cartons was produced in 1921. Household foil was marketed in the late 1920s, and the first heat-sealable foil was developed in 1938.

World War II established aluminum as a major packaging material. During the war, aluminum foil was used to protect products against moisture, vermin, and heat damage. It was also used in electrical capacitors, for insulation, and as a radar shield.

After the war, large quantities of aluminum foil became available for commercial use. Its applications boomed with the postwar economy. The first formed or semirigid containers appeared on the market in 1948. Large-scale promotion and distribution of food service foil in 1949 quickly expanded the market (1).

Table 2. Principal Aluminum-Foil Alloys (Non-heat-treatable)

Alloy and Temper (Aluminum Association Number)	Aluminum, %	Principal Other Elements,[a] %
1100–H19	99.00	0.12 Cu
1145–H19	99.45	
1235–H19	99.35	
1350–H19	99.50	
3003–H19	97.00	0.12 Cu, 1.2 Mn
5052–H19	96.00	2.5 Mg, 0.25 CR
5056–H19	93.6	0.12 Mn,
5056–H39		5.0 Mg, 0.12 Cr
	Heat-treatable	
2024–T4	91.8	4.4 Cu, 0.6 Mn, 1.5 Mg

[a] Nominal compositions.

Aluminum foil's compatibility with foods and health products contributes greatly to its utility as a packaging material. A concise guide to the behavior of aluminum with a wide variety of foods and chemicals is a reference entitled *Guidelines for the Use of Aluminum with Food and Chemicals,* published by the Aluminum Association (3).

Standard aluminum foil alloying elements are silicon, iron, copper, manganese, magnesium, chromium, nickel, zinc, and titanium (see Table 2). These elements constitute only a small percent (in most cases, no more than 4%) of aluminum-foil alloy composition.

Properties

Chemical resistance. Resistance of aluminum foil to chemical attack depends on the specific compound or agent. However, with most compounds, foil has excellent to good compatibility.

Aluminum has high resistance to most fats, petroleum greases, and organic solvents. Intermittent contact with water generally has no visible effect on aluminum otherwise exposed to clean air. Standing water in the presence of certain salts and caustics can be corrosive (3).

Aluminum resists mildly acidic products better than it does mildly alkaline compounds, such as soaps and detergents. Use with stronger concentrations of mineral acids is not recommended without proper protection because of possible severe corrosion. Weak organic acids, such as those found in foods, generally have little or no effect on aluminum. A clear vinyl coating, however, is recommended for use with tomato sauce and other acetic foods.

Temperature resistance. Since aluminum foil is unaffected by heat and moisture, it is easily sterilizable and is actually sterile when heat-treated in production. Unlike many packaging materials, aluminum foil increases in strength and ductility at lower temperatures. Its opacity protects products that would otherwise deteriorate from exposure to light (see Table 3.

Table 3. Functional Properties of Aluminum Foil

Form	Continuous rolls and sheets	Hygienic	Sterile when heat-treated in production; smooth metallic surface sheds most contaminants and moisture of sterilization
Thickness	0.00017–0.0059 in. (4.3–150 μm)		
Maximum width	68 in. (1.7 m) for pack-rolled lighter gauges; 72 in. (1.8 m) for gauges 0.001 in. (25.4 μm) and heavier, single-web rolled	Sterilizable	Metal unaffected by heat and moisture of sterilization (except for staining in some cases)
Impermeability	(WVTR)[a] 0.001 in. (25.4 μm) and thicker is impermeable; 0.00035 in. (8.9 μm) has a WVTR of ≤0.02 g/100 in.^{2} (0.065 m^2); 24 h at 100°F (37.8°C)/100th—WVTR drops to practically zero when 0.00035-in. (8.9-μm) foil is laminated to appropriate film	Nontoxic	Inert to or forms no harmful compounds with most food, drug, cosmetic, chemical, or other industrial products
		Tasteless, odorless	Imparts no detectable taste or odor to products
Corrosion resistance	Aluminum's natural oxide shielding, which is maintained in the presence of air, renders it substantially corrosion resistant	Opacity permanence	Solid metal, transmits no light Highly corrosion-resistant in most environments
Compatibility with food, drugs, and cosmetics	Nontoxic; corrosion-resistant to many compounds in solution	Sealability	Excellent dead fold and adhesion to a wide variety of compounds
Formability	Dead fold	Insignificantly magnetic	Provides excellent electrical, nonmagnetic shielding
Nonabsorptivity	Proof against water and wide variety of liquids	Nonsparking	The leading metallic material for applications with volatile, flammable compounds
Greaseproof	Nonabsorbent		

[a]WVTR = water-vapor transmission rate.

Mechanical properties. The addition of certain alloying elements strengthens aluminum. The alloys produced from these compositions can be further strengthened by mechanical and thermal treatments of varying degree and combinations. For this reason, the mechanical properties of aluminum foil are significant to an understanding of its versatility.

The lowest, or basic, strength of aluminum and each of its alloys is determined when the metal is in the annealed, or soft, condition. Annealing consists of heating the metal and slowly cooling it for a predetermined period of time. Rerollstock from which foil gauges are produced is annealed (a process of heating and cooling) prior to the foil-rolling operations to make it softer and less brittle.

All alloys are strain-hardened and strengthened when cold-worked, as in foil rolling. When the product is wanted in the soft condition, it is given a final anneal (1).

Converting

Aluminum foil is converted into a multitude of shapes and products (see Table 4). Processes involved may include converting mill rolls of plain foil by rewinding into short rolls, cutting into sheets, forming, laminating, coloring, printing, coating, and the like (4) (see Slitting and rewinding; Laminating; Printing; Coating equipment).

Packaging products account for about 75% of the market for aluminum foil (5). Aluminum foil is also used in households and institutions as a protective wrap, and for decorative and construction purposes.

Among packaging end uses, aluminum foil is formed into semirigid containers produced from unlaminated metal for frozen and nonfrozen foods, as well as caps, cap liners, and closures for beverages, milk, and other liquid foods (see Closure liners; Closures). It is also formed into composite containers with films and plastics to package powdered drinks, citrus and other juices, and motor oil and other auto supplies (see Cans, composite).

Aluminum foil combined with other materials such as paper or plastic film can be used to package a host of basic nonfood products, including tobacco, soaps and detergents, photographic films, drugs, and cosmetics.

When water-vapor or gas-barrier qualities are critical to the success of a packaging material, aluminum foil is usually considered. Aluminum-foil containers are odorless, moistureproof, and stable in hot and cold temperatures. They were originally developed to supply bakers with cost-cutting disposable pie plates and bake pans.

A major reason for the wide use of aluminum foil in packaging is its versatility. It is adaptable to practically all converting processes and can be used plain or in combination with other materials. Aluminum foil can be laminated to papers, paperboards, and plastic films. It can be cut by any method and can be wrapped and die-formed into virtually any shape. Foil can be printed, embossed, etched, or anodized.

Table 4. Classifications for Converted and Nonconverted Aluminum-Foil End Uses[a]

Packaging End Uses

- Semirigid foil containers (including formed foil lids) produced from unlaminated metal for
  - Bakery goods
    - Frozen
    - Nonfrozen
  - Frozen foods, other than bakery
- Caps, cap-liners, and packaging closures for
  - Beverages and milk
  - Foods
- Composite cans and canisters (including labels and liners for composite cans) for powdered drinks; auto supplies; food snacks; citrus and other juices; refrigerated dough; other refrigerated and frozen products
- Flexible-packaging end uses (including labels, cartons, overwraps, wrappers, capsules, bags, pouches, seal hoods and overlays for Semirigid foil containers)
  - Food products
    - Dairy products—cheese, butter, milk, milk powder, ice cream dried and dehydrated food products—fruit, vegetables, potato products, soup mixes, yeast
    - Baked goods—bread, cookies, crackers
    - Cereals and baking mixes—cereals, rice, cake mixes, frosting mixes, macaroni products
    - Powdered goods—coffee, tea, gelatins, dessert mixes, drink powders, cocoa, dry concentrates, sugar, salt
    - Meat, poultry, and seafoods (fresh, frozen, irradiated, dried, retorted)
    - Frozen prepared foods
    - Confections—candy, mints, chewing gum, chocolate bars (converted foil only)
    - Dry snack foods—potato chips, popcorn, including coated popcorn
    - Beverages—soft drinks, beer, distilled liquors, wines
    - Food products, NEC[b] (including pet foods)
- Nonfood products
  - Tobacco—cigars, cigarettes
  - Soaps and detergents
  - Photographic film and supplies
  - Drugs, pharmaceuticals, cosmetics, toiletries, kindred products
  - Nonfood products, NEC[b]
  - Military specification packaging[c]

Selected Nonconverted Foil Products[d]

- Packaging end uses (unmounted, unconverted foil stock sold to end users for candy and gum wraps)

[a] U.S. Department of Commerce.
[b] Not elsewhere classified.
[c] On direct government orders only.
[d] Reported by foil producers (rollers) only.

Foil lamination. In many laminations, light-gauge foil is the primary barrier against water-vapor transfer. While creasing can create pinholes or breaks in this barrier, problems can be minimized by proper lamination (see Laminating machinery).

Laminations of foil and waxed paper have been popular as overwraps and liners for cereal packages for more than 40 years. Snack foods that once presented a problem because of their high oil content are now wrapped by specially formulated foil-paper and foil-film laminates.

Aluminum foil laminates have been designed to meet exacting requirements of drugs used in transdermal medication systems that deliver medication through the skin at a constant rate over a specific period of time.

Peelable foil-laminated pouches protect transdermal drugs. Space-shuttle astronauts wore a U.S.-quarter-sized transdermal patch behind the ear to help prevent motion sickness.

Printing. Either side of foil may be printed directly, or the foil may be laminated to a reverse-printed clear film to provide attractive designs with accents (6). Use of transparent inks through which the foil can be seen produces pleasing, metallic colors with no loss in brightness or sparkle.

The same presses and the same types of processes used for printing paper and plastic films are used for printing on foil and foil-paper laminations. Processes include rotogravure, flexography, letterpress, lithography, and silk screen (1) (see Printing; Decorating). Inks for all of these processes are readily available in formulations expressly made for printing on aluminum (see Inks).

To provide anchorage for the inks, prime or wash coatings are nearly always used on foil to be printed. The coatings also provide a barrier that prevents offsetting of undesirable materials from the paper to the foil surface.

Embossing. In-line or out-of-line printing and embossing units allow designs of limitless combinations of color and form. Any embossing pattern in foil produces two basic visual effects, namely, three-dimensional patterns or illustrations and continual reflective contrasts.

Foil embossing often is performed in continuous roll form by passing the web or sheet through a roll stand equipped with one engraved steel roll and a soft matrix roll of paper. The pressure for embossing may be obtained by maintaining the paper roll with the axis in a fixed position and using only the weight of the steel roll to depress the negative pattern into the paper roll (see Fig. 1).

Flexible Foil Packages

Flexible foil-containing packages are extremely popular for food packaging because they provide superior flavor retention and longer shelf life than packages formed from other flexible materials. Flexible foil packaging is impervious to light, air, water, and most other gases and liquids. The packages protect contents from harmful oxygen, sunlight, and bacteria (see Fig. 2) (see Multilayer flexible packaging).

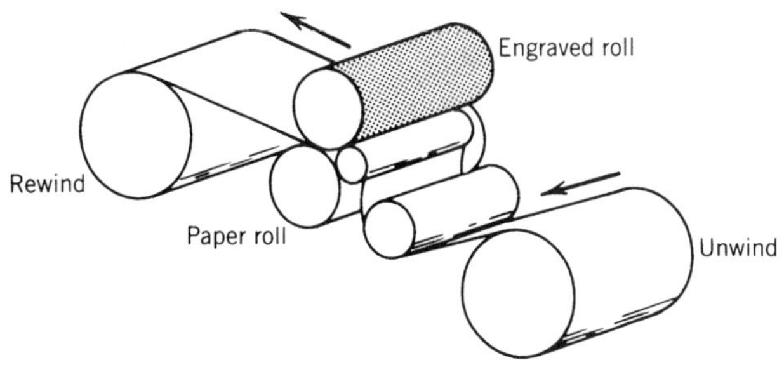

Figure 1. Foil embossing unit. An engraved steel roll, usually the top roll, carries the design, and a matching paper roll becomes the matrix. This matrix roll is constructed of layers of paper (wool-rag) rings compressed solidly into one continuous mass and mounted on an appropriate core.

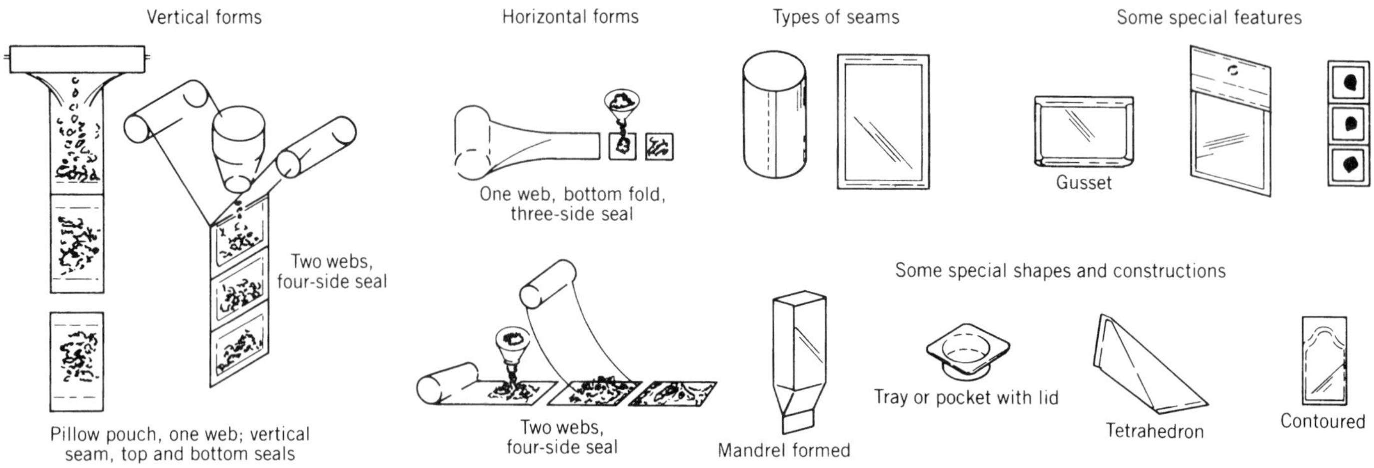

Figure 2. Features and constructions of pouches.

One of the most important flexible-packaging applications of aluminum foil is the form/fill/seal pouch (1). Pouches are formed continously from roll-fed laminated material and are filled and sealed immediately as formed (see Form/fill/seal, horizontal; Form/fill/seal, vertical; Pouches). The pouch is one of the oldest automated-package forms. The retort pouch, developed in the early 1950s, represented a significant advance in food packaging. It is a flexible package made from a laminate of three materials: an outer layer of polyester for strength; a middle layer of aluminum foil as a moisture, light, and gas barrier; and an inner layer of polypropylene as the heat seal and food-contact material (7) (see Retortable flexible and semirigid packages).

The Q-Pouch is a new foil-lined paperboard pouch that can be turned into a drinking cup. A user of the package simply rips the top edge along perforations and squeezes the sides of the packet. Scored paperboard opens into a hexagon-shaped cup (8).

Aluminum foil is also used to some extent for bag-in-box liquid packaging (9) (see Bag-in-box, liquid).

Foil Lidding

Another popular packaging application is the flexible closure or flexible lid. The flexible closure got its start with single-service dairy creamers, yogurt, and cheese dips. It is also widely used in the healthcare industry, especially in hospitals, where the trend is toward disposable packaging (see Healthcare packaging). Liquid medications are also packaged in single-dosage containers, which are convenient and provide easy evidence of tampering (10) (see Pharmaceutical packaging; Tamper-evident packaging).

Flexible lids were introduced in 1966 when a U.S. Health Service regulation required that the pouring lip of dairy containers be covered during shipment and storage. Besides covering the pouring lip of the container, a heat-sealed foil lid offered a more reliable seal, longer shelf life, and greater protection than other lidding materials.

The use of inductive heat-sealing equipment instead of conductive heat sealing is broadening the applications for foil lidding. Inductive sealing does not heat the lid by direct contact. Heat is brought into the lidding stock by a magnetic field through which the container and lid pass. The magnetic field heats the foil in the lid and an effective, hermetic seal forms between the lid and the container. Flexible foil lidding can be used on glass, plastic, or composite cans. This key development enables the aluminum foil industry to provide tamper-evident packaging (see Tamper-evident packages).

Regulated Packages

Aluminum foil is an integral part of many tamper-evident packages. Although the term came into prominence only in the 1980s, this kind of packaging has been on the market before then. As now defined by the *Proprietary Association,* a tamper-evident package is "one which, if breached or missing, can reasonably be expected to provide visible evidence to consumers that the package has been tampered with or opened" (11).

Of 11 methods listed by the Proprietary Association that conformed to the regulation when it was first pronounced, six featured aluminum foil as part of the tamper-evident system.

For a system to be tamper-evident, its package cannot be removed and replaced without leaving evidence. It is extremely difficult to mend or repair a foil-membrane seal after it is broken or to return an aluminum rollon cap to its original state once it has been removed.

One of the first major laws that required protective packaging was the Poison Prevention Packaging Act of 1970, which essentially said that drugs and dangerous substances must be packed in child-resistant packages. This type of package includes blister packs of plastic-foil-paper combinations; tape seals of paper or foil which adhere to the cap or shoulder of a bottle; and pouches which can be tighly sealed.

Aseptic Packaging

Most aseptic packages are laminations of paperboard, plastic, and aluminum foil. In aseptic packaging, the product is sterilized with high heat and then cooled. The package is sterilized separately, and then the sterile product and packaging are combined in a sterile chamber (12) (see Aseptic packaging).

Foil provides superior adhesion for the containers which, in the main, hold highly acidic products. In almost all methods of aseptic container manufacturing, aluminum foil is used as a barrier to light and oxygen (12).

Foil and Microwave Ovens

In 1984, Underwriters Laboratories undertook a study on the use of aluminum foil trays to reheat food in microwave ovens. Using various models of microwave ovens and standard types of aluminum trays containing frozen foods, the study concluded that (13):

The power density of microwave radiation emissions did not exceed maximum allowable limits.

There was no significant change in heat going into various liquids and frozen foods in aluminum containers as compared to other containers.

The temperatures in the foods and liquids being tested were generally comparable in the various containers.

The foil trays containing different amounts of water and empty trays produced no fire emissions of flaming or molten metal.

The test trays used for heating frozen foods in accordance with specific instructions did not increase the risk of radiation, fire, or shock hazard.

Semirigid Packaging

Other classifications of aluminum packaging include semirigid and rigid containers. Semirigid containers include those that are die-formed, folding cartons, and collapsible tubes. Die-formed containers are one-way, disposable devices such as pie plates, loaf pans, and dinner trays. Folding cartons come in many sizes and shapes and hold such products as dry cereals, eggs, and milk or other liquids (see Cartons, folding; Cartons, gable top). Collapsible tubes were first made of lead and used by artists more than 100 years ago. In modern times, a proprietary aluminum foil/film version of the collapsible tube holds toothpaste and products of similar consistency, including hair coloring and depilatories (1) (see Tubes, collapsible).

Aluminum foil provides the barrier to permeation of the oils used in most products and compounds that go into the tubes.

Aluminum tubes have the advantage of providing light weight, high strength, flexibility, and good corrosion resistance. Aluminum tubes also have low permeability and offer quality appearance. Traditionally, aluminum collapsible tubes were impact-extruded, but a new process forms the laminated rollstock into continuous tubing by use of heat and pressure. The tubing is cut into individual sleeves and is automatically headed by injection molding.

As opposed to flexible containers that conform to the shape of the product, semirigid containers have a shape of their own. They can be deformed from their original shape either while they are emptied (as in the collapsible toothpaste tube), or before they are filled (as with the folding carton).

Die-formed aluminum containers are among the most versatile of all packages. They easily withstand all normal extremes of handling and temperature variation. A product in an aluminum container can be frozen, distributed, stocked, purchased, prepared, and served without soiling a single dish.

Bare foil is used for most formed aluminum-foil products but protective coatings are used on the containers for some foods and other products. Although the frozen-food tray is the most common, the aluminum-formed container is available in scores of shapes and sizes from the half-ounce portion cup to full-size steam-table containers for institutional feeding. Closures for these containers vary from laminated hooding to the hermetically sealed closure according to the amount of protection needed for the product.

Folding cartons offer protective and display characteristics unique in packaging and have some of the advantages of both the flexible and the rigid container. Before use, when it is folded flat, the folding carton offers the storage economy of the flexible bag. When it is filled, it offers much of the protection of the setup box.

Rigid Containers

Composite cans and drums feature aluminum foil combined with fiber. They are widely used for refrigerated dough products, snack foods, pet foods, and powdered drink mixes.

In some processes, a composite can is made up of paper-polyethylene-aluminum foil-polyethylene laminated stock with a foil-membrane closure (14). Cans are produced from preprinted gravure rolls of bottom aluminum stock (see Cans, composite).

BIBLIOGRAPHY

"Foil, Aluminum" in *The Wiley Encyclopedia of Packaging Technology,* 1st ed., by Foil Division of the Aluminum Association, Inc., pp. 346–351.

1. *Aluminum Foil,* the Aluminum Association, Washington, DC, 1981.
2. *The Story and Uses of Aluminum,* the Aluminum Association, Washington, DC, 1984.
3. *Guidelines for the Use of Aluminum with Food and Chemicals,* the Aluminum Association, Washington, DC, 1984.
4. *Aluminum Foil Converted,* U.S. Department of Commerce, Washington, DC, 1981.
5. *Aluminum Statistical Review for 1982,* the Aluminum Association, Washington, DC, 1983.
6. W. C. Simms, ed., *1984 Packaging Encyclopedia,* Cahners Publishing, Boston.
7. *Food Technology,* Institute of Food Technologists, June 1978.
8. *Food Eng.* **55,** 60 (Feb. 1983).
9. *Packaging Newsletter,* the Aluminum Association, Spring 1981.
10. *Packaging Newsletter,* the Aluminum Association, Fall 1983.
11. News release, the Proprietary Association, Washington, DC, Oct. 14, 1982.
12. *Food Eng. Internatl.* (Jan./Feb. 1984).
13. Underwriters Laboratories, letter of July 23, 1984.
14. *Packaging Newsletter,* the Aluminum Association, Fall 1979.

FOOD PACKAGING. See Packaging of Food.

FORENSIC PACKAGING

Forensic packaging is a legal term describing the science of determining the cause of package function as intended. Most people think of forensics as being associated with criminal investigation (especially homicide), when in reality there are many areas of forensic specialization. Forensic medicine involves the cause of disease, forensic pathology is related to the cause of death, forensic psychiatry seeks the cause of mental disorders, forensic engineering is used to determine the cause of accidents such as train or automobile, forensic accounting identifies procedures used to hide illegal accounting such as embezzlement, and forensics in law enforcement refers to scientific physical evidence such as latent prints, ballistics, handwriting, fiber analysis, and similar sciences.

Walter Stern, a packaging consultant in Wilmette, Illinois, was the first person known to use the term *forensics* in defining the science of determining the cause of a package to fail to function that resulted in injury.

Determining the cause of package failure may be as simple as discovering that liners are missing from a closure or as complicated as duplicating actual environmental factors such as temperature and humidity present during the time the package was in transport and storage and measuring the effect the conditions have on the subject package. Determining the cause of package failure may fall on the shoulders of any person involved in the sale, manufacture, or use of the package. If the salesperson is unable to readily identify the cause, production may be asked to determine the cause. If production determines the package was produced in spec (specification), the packaging engineer may analyze whether the spec was sufficient to meet the objectives of the package under the changing conditions. If the engineer is unable to determine the cause, a specialist in forensic packaging may be needed to identify all factors related to the package and conduct extensive tests to determine the cause and recommend solutions to the problem.

The concepts of forensic packaging are most useful in identifying weaknesses in the package design, so that a better package or component can be made and the possibility of package failure reduced. Improvements resulting from proper application of forensic packaging concepts are not limited to materials or designs but may also involve manufacturing processes.

The science of forensic packaging requires an objective analytical process, where the results can be replicated by others using the same procedure. Many of the procedures used are found in ASTM 15.06, such as vibration, torque retention, and drop tests. Others may have been developed by various manufacturers, packaging schools, or independent laboratories. The source of the procedure is not as important as its relationship to the problem, reliability of its results, and ability to be replicated by others.

Many improvements in packaging have been the result of identifying the cause of a package failure such as new media for corrugated cartons to increase stacking strength, easier-to-open child-resistant closures, improved lettering on labels, and better barrier properties in films. Identifying and correcting any weakness in a package before the package fails results in savings in tooling, fewer claims of defective products, higher customer satisfaction, and improved profits.

In today's litigious atmosphere, forensic packaging becomes most useful in legal actions. The science of forensic packaging, when properly applied and explained, can be used to objectively explain the process used in developing a package so that it will function as intended, that any weaknesses known would not affect package performance, or that the cause of package failure was not foreseeable and reasonable for the package to encounter. It is reasonable to expect that the package would survive transport and storage in the high temperatures of the summer, but is it reasonable to expect it to withstand abuse by the consumer? The level of abuse the package would withstand should be determined by testing, including normal use and extreme abuse.

In some cases the package was designed with several safeguards, functioned as intended, but the consumer ignored indications of prior opening and died after consuming an adulterated product that resulted in severe injury or toxicity. Should the company have been liable for the failure of the consumer, when the physical evidence of the package remains indicated that it functioned properly?

A qualified forensic packaging consultant or scientist can explain in either technical or lay terms the cause of and solutions to any problem so that the problem can be understood. The results of all tests performed by the consultant or in-house scientist should be verifiable and replicable by others. While the consultant's findings may be based on experience, results based on objective scientific reasoning may carry more weight. The consultant should be capable of developing evidence for use in litigation and testifying in court if necessary.

In forensic engineering it is possible to determine whether a turn signal was on at the time of impact even though the light is broken when examined. In forensic packaging similar results can be obtained by examining the remaining components, duplicating environmental and physical conditions, and conducting tests related to the function of the package. Through the use of objective test-duplicating the package components and conditions, it is possible to determine the probability that a package would or would not leak under similar circumstances. The same analysis combined with variations of the package can identify the cause of the package failure.

What is the value of forensic packaging to management? Proper use of the principles of forensic packaging can prevent the use of manufacture of a package that is likely to fail to function and thereby result in injury to the consumer. It can also assure that the package is the most effective for the specific application, provided accurate information on intended use and environmental conditions are known.

BIBLIOGRAPHY

General References

Books

J. L. Rosette, *Improving Tamper-Evident Packaging,* Technomic Publishing, Lancaster, PA, 1992.

J. L. Rosette, *Product Tampering Detection,* Forensic Packaging Concepts, Inc., Atlanta, 1993.

Publications

American Society for Testing and Materials (ASTM) 15.06

Food & Drug Packaging, published monthly.

Packaging, published monthly.

Packaging Digest, published monthly.

Packaging Technology & Engineering, published eight times per year.

Articles and technical papers

"Package Failure and the Courts," W. Stern presentation to Packaging Institute/International, Lisle, IL, Sept. 1986.

"Tamper-Resistant Packaging," J. Sneden, H. Lockhart, and M. Richmond, Michigan State University, 1983

JACK L. ROSETTE
Forensic Packaging Concepts, Inc.
Fort Mill, South Carolina

FORM/FILL/SEAL, HORIZONTAL

A variety of packages can be made on horizontal form/fill/seal equipment. This article deals primarily with pouch making, but similar concepts are applied to thermoform/fill/seal (see Thermoform/fill/seal) and bag-in-box packages (see Bag-in-box, dry product). Pouch styles include the following: three-sided fin seal; four-sided fin seal; single gusset; double gusset; pillow pouch (lap seal/fin seal); and shaped seal. The package is usually made from rolls of film (see Films, plastic) or other "webs" (see Multilayer flexible packaging). The sides are normally heat-sealed (see Sealing, heat), but other methods such as ultrasonic, laser, or radiofrequency welding can be used to meet specific requirements.

Filling can be done in a number of ways, depending on the characteristics of the product. Fillers include liquid fillers (see Filling machinery, still liquid), paste fillers, augers, pocket fillers, vibratory, orifice-type, and gravimetric units (see Filling machinery, dry product). Accessories can be provided with most form/fill/seal equipment to provide registration (feed-to-the-mark or stretch hardware); web splicing ("flying splice" automatic or manual equipment); in-line printing (see Printing); coding (noncontact, hot-leaf stamping, or printing units) (see Code marking and imprinting); embossing; perforating; notching; vacuum or inert-gas packaging (see Controlled atmosphere packaging); aseptic packaging (see Aseptic packaging); tear string; cartoning or bagging.

Some machines are small and relatively portable; others are massive and fixed. Some versatile machines can be readily changed over within limits. Other equipment is dedicated to a given size, and any change requires extensive and costly modifications. Cost of equipment varies widely depending on the output required and the extent of the system purchased, and on accessories specified and design requirements (eg, sanitary criteria and environmental considerations).

This discussion covers pouch, thermoform, and horizontal bag/box machinery according to the outline shown in Table 1. The equipment is classified by type and functional sequence, and further subdivided by design parameters. These are all horizontal form/fill/seal machines, even though the pouch may be horizontal or vertical (see Form/fill/seal, vertical). It would be impossible to mention all the machinery available. Representative equipment is mentioned for clarity, not as endorsement.

Table 1. Horizontal Form/Fill/Seal Equipment

Pouch form/cut/fill/seal, pouch vertical
- In-line equipment
  - Single-lane, intermittent, and continuous motion
- Rotary equipment
  - Single-lane, intermittent motion

Pouch form/fill/seal/cut, pouch vertical
- Single-lane, continuous motion
- Multilane, continuous motion
- Single-lane, intermittent motion

Pouch form/fill/seal/cut, pouch horizontal
- Single-lane, intermittent, and continuous motion
- Multilane, intermittent, and continuous motion

Thermoform/fill/seal equipment
- Multilane, intermittent, and continuous motion

Horizontal form/fill/seal bag-in-box equipment
- Single-lane, intermittent motion

Pouch Form/Cut/Fill/Seal, Pouch Vertical

In-line, single-lane, intermittent motion. The basic "workhorse" of the pouch machinery group is the single-lane, in-line vertical pouch form/cut/fill/seal intermittent motion machine. It requires some web stiffness to transfer from the cut-off knife to the bag clamps. Size changes, within limits, can be accomplished by adjustments to the machine. Its flexibility includes the ability to handle a wide variety of pouch materials, including most self-supported heat-sealable materials like polyesters, polyethylenes, foils, cellophanes, and paper. Liquids, creams, pastes, granular materials, pills, tablets, and small hard goods, or a combination of these, may be packaged on this type of equipment. Fillers include piston fillers; auger fillers; and vibratory, volumetric, gravimetric, count, and timed-cutoff units. Combination of fillers can be used to permit formulation in the pouch. Gassing and sterilization systems can be provided. The major factor in the selection of this equipment is versatility.

Bartelt Machinery Division of Rexham builds equipment of this type (see Fig. 1). Pouch styles that can be produced on this equipment are three- and four-sided fin seal, bottom gusset pouches for greater volume, wraparound pouches, multiple compartment pouches, and die-cut pouches as well as some combinations of these. Bartelt has been a pioneer in the

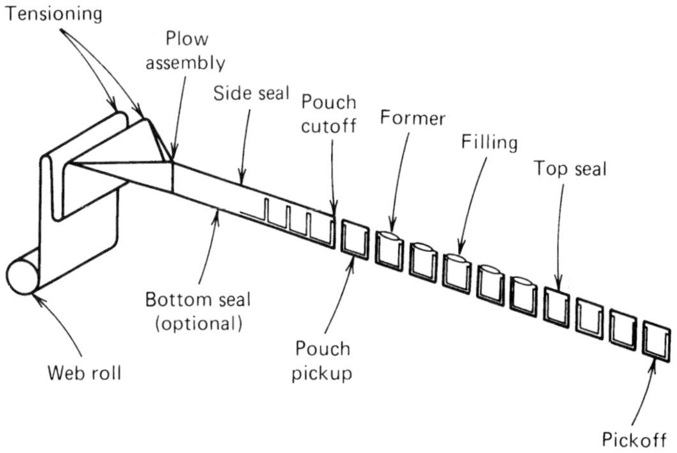

Figure 1. Material-flow diagram of typical Bartelt intermittent bottom horizontal form/fill/seal/machine. Filling of individual pouches allows maximum versatility for multiple component fills and easy pouch-size change. The machine primarily utilizes supported paper, foil, or laminated film structures.

production of retort pouches, and they can supply automated retort systems from rollstock through cartoning.

Costs for equipment of this type vary widely. For instance, N and W Packaging Systems of Kansas City, Missouri, makes a relatively simple unit that produces 55 packages per minute. Bartelt makes complete systems that take packages from several machines with an "on-demand" feature and cartons them on high-speed equipment.

Hassia, a division of IWK, Germany, builds a flat pouch unit that operates at rates of up to 120 packages per minute. Another German firm that produces in-line intermittent-motion pouch equipment is Hamac-Hoeller, a division of the Bosch Packaging Group.

In-line, single-lane, continuous motion. The sequence of form/cut/fill/seal on a continuous-motion machine entails running a cut pouch into clips attached to a continuously moving chain. Filling is usually accomplished by funnels that move with the pouch to allow time for the fill cycle. Sealing is done either with moving heat sealing bars (see Sealing, heat) or band or contact heaters with squeeze rolls at the discharge of the heater elements. Continuous-motion equipment is more complex and expensive than intermittent-motion machines. They are relatively fixed in the pouch size and do not provide the flexibility of intermittent units, but they operate at significantly higher rates.

Delamere and Williams, a subsidiary of Pneumatic Scale Corporation, builds a Packetron Pouch Maker that will run at speeds of ≤500 pouches per minute. This machine can package products like instant cocoa mix, oatmeal, coffee, seasonings, dairy mix, powdered beverages, and most free-flowing food products. Sealing can be accomplished on three or four sides. A gusseted bottom can be supplied to provide an increased pouch volume.

Rotary single-lane, intermittent motion. In this design concept the pouch sides are sealed, then the pouch is cut and transferred to an indexing wheel. Pouch opening, filling, and top sealing take place at various stations around the circumference of the indexing assembly.

Hamac-Hoeller manufactures a unit of this type called the BMR-100. This machine can run up to 120 pouches per minute and produce three- and four-sided fin seals, bottom gusset, standup, and shaped seal pouches. The BMR-200 machine handles two pouches at a time. This unit has a top rate of 200 packages per minute and makes the same style pouches as the BMR-100 plus a twin pouch with a central seal. These machines can be fitted with various type fillers and auxiliaries to meet specific needs.

The Wrap-Ade Machine Company of Clifton, New Jersey, makes a more economical model of a single-lane rotary pouch machine. This unit produces pouches at a rate of 15–60 pouches per minute. It is a 12-station indexing system with seven possible loading stations.

A somewhat related system is available from Jenco (Germany). This machinery produces a standup bag of relatively large volume. The system can be provided with gas flushing, postpasteurization, or sterilization. Jenco offers complete systems from bagmaking to cartoning. System rates are up to 300 bags per minute.

The machines described above produce relatively small pouches. Horizontal form/fill/seal intermittent motion equipment to make and fill large bags is described below.

Pouch Form/Fill/Seal/Cut, Pouch Vertical

This area of pouchmaking has advanced significantly. This concept takes the web continuously from a roll, runs it over a forming plough, makes the side seals on a rotary drum (and bottom seal if required), then forms and fills the pouch while it moves around a rotary filling hopper. The top of the web is stretched, and the top seal is usually accomplished by running the web through contact heaters and squeeze rolls. A rotary knife is used to accomplish the pouch cutoff. This type of machine can produce packages at very high rates, up to 1300 packages per minute on a single-lane machine. It is an inflexible system and is normally dedicated to one size.

Single-lane, continuous motion. R. A. Jones & Co., Inc., Cincinnati, Ohio, makes a machine of this type called a Pouch King (see Fig. 2). A Japanese firm, Showa Boeki Company Ltd., makes a slower unit called the Toyo type R-10.

The Jones unit was originally developed by the Cloud Machinery Company of Chicago for use on sugar pouches with a paper–poly film (see Multilayer flexible packaging). The current capability of the equipment has been greatly expanded. The machine runs a wide range of sizes and flexible heat-sealable structures. These include glassine, cellophane, foil, polypropylene film, and other materials with a variety of sealant materials.

The original machine could fill only relatively small product volumes relative to pouch size, but Jones has developed a system called the "tucked-up-bottom" that holds a significantly larger product volume.

Various types of volumetric fillers that handle free-flowing products can be used on this equipment, including augers, orifice metering, vibratory, and pocket fillers.

Jones now has six models that run a wide range of products including salt, pepper, sugar, beverage mixes, instant coffee, roasted coffee, cocoa mix, and many other free-flowing products. Output is 750–1300 packages per minute.

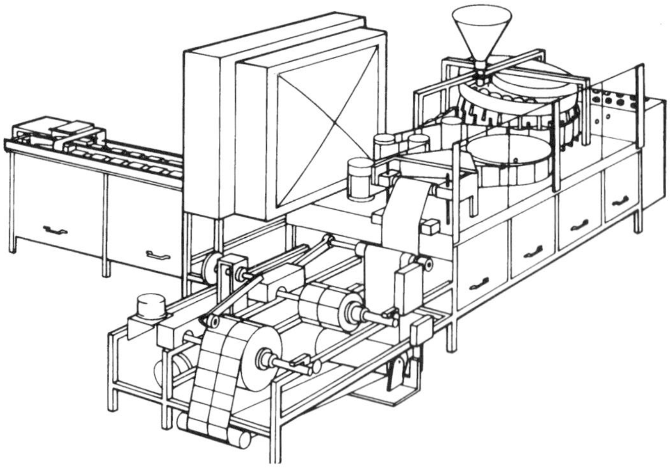

Figure 2. High-speed (700–1200 pouches per minute)—horizontal/fill/seal machine is all-rotary motion to provide long filling and sealing time. Pouches are automatically collated in the count and pattern required. (Courtesy of R. A. Jones & Co., Inc.)

Multilane continuous motion. An example in this category is the Matthews Industries (Decature, Alabama) "Ropak" machine. This equipment was also originally designed to run sugar pouches at rates up to 2000 packages/min. This rate is obtained by running two lanes at 1000 packages/min per lane. Previously, the only structure run on this unit was a paper-poly combination, but a paper–poly–foil–poly web is now being successfully packaged. Machines are currently running sugar, pepper, and instant coffee. The majority of the product is bulk packed in drums or cases for institutional use. The machine is equipped with a predetermined counter-and-swing spout for bulk packing by count.

Single-lane, intermittent motion. Simplicity, economics, versatility, and space saving were the criteria for this approach to form/fill/seal pouches. Semirigid packaging material is drawn from rollstock across a vertically mounted tension control. It then passes over an adjustable guide roller into a horizontal and flat position. At this point, the packaging material is folded in half vertically by passing through two round guide bars. In its horizontal direction of travel, the package material passes the bottom seal station (not required for a three-sided seal pouch). A pouch-opening device separates the folded packaging material for entry of the filling funnels before the side seals are completed. After the product is filled, the packaging material passes the top sealing station, followed by the pouch-cutting device. The final station is the packaging material transport station. The machine is adjustable to a maximum pouch of 5⅛ in. × 5⅛ in. (130 mm × 130 mm) at output speeds of ≤120 pouches/min (240 for duplex operations). The machine is manufactured by Kloeckner Wolkogon, a division of Otto Haensel (Germany).

Pouch Form/Fill/Seal/Cut, Pouch Horizontal

In this wrapper-type equipment the product is fed horizontally into a web that is wrapped around it and sealed. The product can be fed onto the moving web or the web can be formed around the product being carried on an indexing conveyor. The web is longitudinally sealed to form a tube around the product and the ends are sealed and cut off. The machinery is normally single lane but dual lanes can be run on some equipment. Machines are available that utilize automatic product feed, manual product feed, run registered web, and they can be equipped for inert gas packaging. A wide range of films and sizes can be run on this equipment. Products such as cookies, candy bars, cheese, and other rigid-type products can be packaged. The pouch is basically a pillow pouch style with a lap- or fin-longitudinal seal. Gussets can be added to handle increased product volumes. Four-sided fin-seal packages can also be produced on this equipment. Machinery can be supplied that operates on either an intermittent or a continuous basis. Speeds of ≤300 packages per minute are available. The equipment is relatively flexible and changeovers are normally made by adjustments. An interesting innovation offered by Weldotron Corporation on its wrapper is a computer-controlled changeover system. Simply by pressing a button, the machine will adjust itself to any one of six preset sizes. This system was developed by Omori Machine Co. of Tokyo and is distributed in the United States by Weldotron's OMC Packaging Division.

Machinery of this type is supplied by Hayssen Package Machinery, Sig, Bosch, Doboy and Oliver, to name a few. Many of these companies specilize in wrapping a given type of product.

Another category of horizontal form/fill/seal equipment is the pouch strip-packaging unit. This machine is basically used to package low-profile products such as flat candy bars; tablets; hardware; and medical, and novelty items. The machinery is capable of running two different webs and can run multiple rows of the same or different products.

Thermoform/Fill/Seal Equipment

Thermoform equipment takes the horizontal-packaging concept further. In this machine concept a thermoplastic web is heated and formed. The cavity is filled, lidded, and cut from the web. In some machines, a pressure-forming die can be substituted for the heat-form station to form a soft aluminum tray (see Thermoform/fill/seal).

A wide variety of packages can be made on intermittent and continuous motion machines of this type. Liquids, solid foods, pharmaceuticals, medical devices, hardware, and beauty aids can all be packaged on thermoform machines. Packages can range from small blisterpacks to deep drawn cups. The product can be gas, vacuum, or aseptically packaged. Films are available today that provide excellent forming characteristics and barrier protection for sensitive food products.

There is a wide range of thermoforming machines available. The method of transporting the web through the machine depends on the characteristics of the package and product requirements. There are die machines where the film is formed into a die and the die train moves through the machine, supporting the web at every station. By far the most prevalent is the dieless clip machine where the film is carried on its sides by means of clips. Some machines do not use clips but rely on the strength of the web to pull the packages through the machine. The film is heated, prior to forming, by contact or radiant heaters, then formed at the same station or indexed to a separate forming station. Accessories for film forming such as plug or pressure assists can be supplied to obtain deep draws with a uniform wall thickness.

Horizontal Bag-in-Box Form/Fill/Seal Equipment

There are several types of equipment that produce this type of package. One is a system that uses a rotary indexing mandrel to form the inner and outer container. The inner container is usually a flexible film with barrier requirements to meet the needs of the product. The outer packaging material is usually for package appearance or structural requirements and can be either a printed web or a box. Another system forms the box in-line with the rest of the system, then the inner liner is formed and inserted in the box and the container is filled. A variation on this is that the container is filled while the inner liner is being inserted in the box (see Bag-in-box, dry product).

Some of the major manufacturers of this type of equipment are Pneumatic Scale Corp., Hesser, and Sig Industrial Company. Only Sig and Hesser provide a vacuum system. They also provide a package with a valve system that will allow a product such as coffee to outgas without causing the package to rupture or balloon (see Vacuum coffee packaging).

Hesser also makes a system that produces a container similar to a composite can. This machine takes a laminate from rollstock and forms a rectangular body. It then attaches one end, fills the container and seals a lid to it. This equipment has the capability to make an aseptic package.

Horizontal form/fill/seal is an extremely dynamic segment of the packaging industry, for materials and equipment are continually improving, presenting new opportunities. New coextruded structures offer barrier and machining possibilities that are expanding the range of food products that can be packaged in flexible film (see Coextrusions for flexible packaging; Coextrusions for semirigid packaging). Aseptic packaging is another growth area that will add new dimensions to food packaging. In conjunction with the advance in these technologies are equipment developments that will provide a basis for expansion and cost reduction.

BIBLIOGRAPHY

"Form/Fill/Seal, Horizontal" in *The Wiley Encyclopedia of Packaging Technology,* 1st ed., by R. F. Bardsley, Bard Associates, pp. 364–367.

General References

Packaging Encyclopedia and Yearbook 1985, Cahners Publishing.

PMMI Packaging Machinery Directory, Annual, Packaging Machinery Manufacturers Institute, Washington, DC.

J. H. Briston, L. L. Katan, and G. Godwin. *Plastics Films,* Longman, New York, 1983.

F. A. Paine and H. Y. Paine, *A Handbook of Food Packaging,* Blackie and Son Ltd., Glasgow, 1983.

FORM/FILL/SEAL, VERTICAL

The term form/fill/seal means producing a bag or pouch from a flexible packaging material, inserting a measured amount of product, and closing the bag top. Two distinct principles are utilized for form/fill/seal packaging; horizontal (HFFS) (see Form/fill/seal, horizontal) and vertical (VFFS). Generally, the type of product dictates which machine category applies. This article deals specifically with VFFS equipment, which forms and fills vertically. It is used to produce single-service pouches for condiments, sugar, etc, as well as bags for retail sale and institutional use. The range of products and sizes is very large.

Package Styles

VFFS machines can make a number of different bag styles (see Fig. 1):

- A pillow-style bag with conventional seals on the top and bottom, and a long (vertical) seal in the center of the back panel from top to bottom. The long seal can be a fin seal or a lap seal (see Fig. 1**a**,**b**).
- A gusseted bag with tucks on both sides to make more space for more product and maintain the generally rectangular shape of the filled bag (see Fig. 1**c**). This style is used inside folding cartons for cereal and other dry products (see Bag-in-box, dry product).
- A three- or four-sided seal package is similar to those made on HFFS machinery (see Fig. 1**d**).

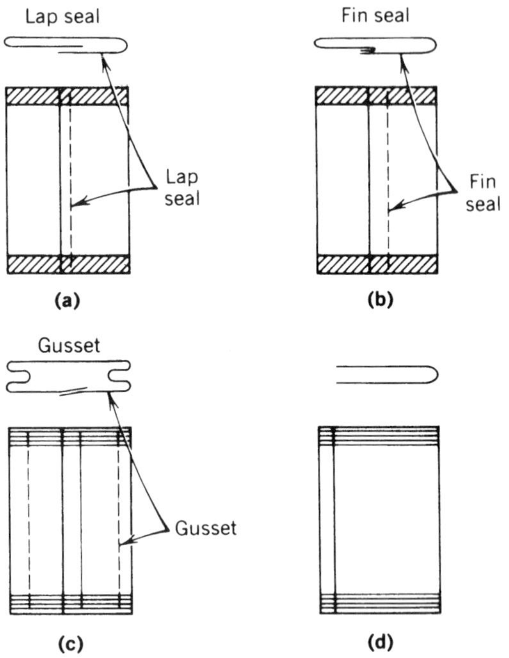

Figure 1. Selected package styles on VFFS machinery: (**a**,**b**) pillow style; (**c**) gusseted style; (**d**) three-sided seal.

- A standup bag (flat-bottom, gabletop) of the type that used to be common for packaging coffee.
- Other special designs such as tetrahedrons, parallelograms, and chubs (see Chub packaging).

A flat-bottom bag needs a relatively stiff material to hold its desired shape, but any type of machinable material can be used to make a pillow-style bag. Various options are available, such as a hole punch for peg-board display, header labels that are an extension of a standard top of a bag, carry handles for large consumer-type packages, and special sealing tools for hermetic seal integrity.

Materials

Two types of packaging materials are suitable for VFFS: thermoplastic and "heat-sealable" materials. Polyethylenes (thermoplastics) require a special bag-sealing technique. Polyethylene films must be melted under controlled conditions until the areas to be attached to each other are fused. The operation is analogous to welding metals. Heat is applied to fuse the materials and then a cooling process allows the seal to set. The sequence for making good seals requires careful control in order to get quality-seal integrity. Impulse sealing is used to seal thermoplastics on VFFS machines. A charge of electricity is put into a Nichrome wire that heats to a preestablished temperature (governed by material thickness) that will melt and fuse the materials. Since thermoplastics become sticky when melted, the Nichrome wire is covered by a Teflon (DuPont Company) sheath. The principle of impulse sealing does not require any specific tooling pressure.

Thermoplastic materials are generally used when a high degree of product protection is not required and low material cost is important. Polyethylene materials have some porosity and are not ideal for applications where hermetic seals are necessary for good shelf life, product freshness, gas flushing, etc. They are used, for example, for frozen foods, chemicals, confectionary items, fertilizers, and peat moss.

The class of "heat-sealable" materials or "resistance seal films" includes paper and cellophane as well as some coextrusions and laminations. Because these materials do not melt at sealing temperatures, or do not melt at all, they require a heat-seal layer that provides a seal with the right combination of time, temperature, and pressure. The sealant layer can be on one or two sides of the web, depending on the desired package configuration (see Multilayer flexible packaging).

A fin seal (see Fig. 1) can be made of materials with sealing properties on one side only, because the "heat sealable" surface seals to itself. This seal is effective for powder products that need the seal to eliminate sifting. It is also a good seal if hermetic-seal integrity is important, as in gas-flush packaging. A lap seal uses slightly less material, but it requires sealing properties on both sides because the lap is made by sealing the inner ply of one edge to the outer ply of the other edge.

Machine Operation

A VFFS machine produces a flexible bag from flat roll-stock. Material from a roll of a given web dimension is fed through a series of rollers to a bag-forming collar/tube, where the finished bag is formed (see Fig. 2). The roller arrangement maintains minimum tension and controls the material as it passes through the machine, preventing overfeed or whipping action. The higher the linear speed of the film, the more critical this handling capability becomes.

The bag-forming collar is a precision-engineered component that receives the film web from the rollers and changes the film travel from a flat plane and shapes it around a bag-forming tube. The design of the bag-forming collar can be engineered to get the optimum efficiency from metallized materials, heavy paper laminates, etc. As the wrapping material moves down around the forming tube, the film is overlapped for either the fin or lap seal. At this point, with the material wrapped around the tube, the actual sealing functions start. The overlapped material moving down (vertically) along the bag-forming tube will be sealed. The packaging material/film advances a predetermined distance that equals the desired bag-length dimension. The bag length is the extent of the material hanging down from the bottom of the tube. The bag

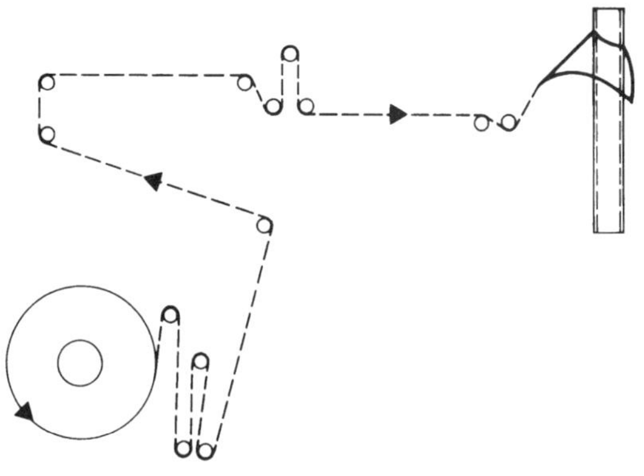

Figure 2. Typical film feed path through a vertical form/fill/seal machine.

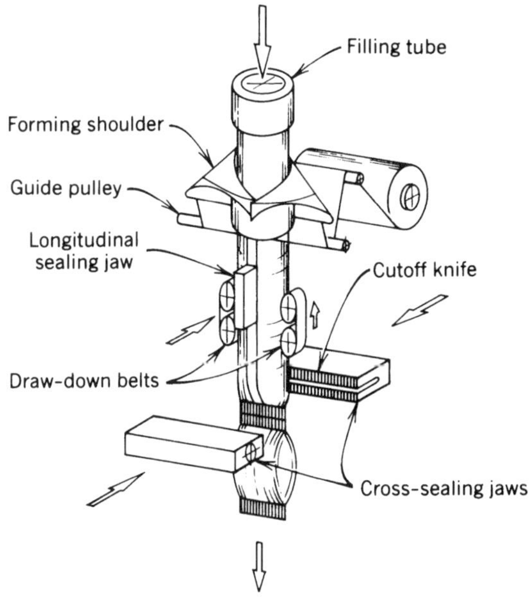

Figure 3. Typical VFFS configuration.

width is equal to ½ of the outside circumference dimension of the tube. After the film advance is completed, the bag-sealing and -filling completes the remainder of one cycle (film advance/fill/seal). There are two sets of tooling on the front of the machine. One of the sealing tools, the vertical (longitudinal or back) seal bar, is mounted adjacent to the face of the forming tube. Its function is to seal the fin- or lap-longitudinal seal which makes the package material into a tube.

The other set of tooling, the cross (end) seal, consists of a front and rear cross-sealing jaw that combines top- and bottom-sealing sections with a bag cutoff device in between. The top-sealing portion seals the bottom of an empty bag suspended down from the tube, and the bottom portion seals the top of a filled bag. The cutoff device, which can be a knife or a hot wire, operates during the jaw closing/sealing operation. This means that when the jaws open, the filled bag is released from the machine. All vertical bag machines utilize this principle to make a bag (see Fig. 3).

Machine Variations

Film transport. Two distinct machine designs are used for transporting the packaging material/film through the machine. The traditional design clamps the material with the cross-seal jaws and advances the material by moving the cross-seal jaws down. This is called a "draw bar" (reciprocating up-down cross-seal jaws). The other is a drive-belt principle for film advance, which leaves the cross-seal bars in a fixed horizontal position with only open-close motion. The belt-drive film-advance principle has been shown to be the most versatile design for high speed packaging and simplicity of operation, and a number of companies have converted to this principle.

Power. There are several approaches to providing power for material/film transport and the filling and sealing operations: all electromechanical; electromechanical/pneumatic; and electromechanical/pneumatic/vacuum.

The electromechanical vertical-bag machine incorporates a cam shaft with a series of cams to operate the various functions. The package material/film drive motion works off a motor/reducer/clutch/brake arrangement. The long-seal and the cross-seal tooling are operated by cams. This allows cycle-to-cycle repeatability of machine settings. This is a very basic principle that has fixed timing on all components and is generally accepted as a heavy-duty, low-maintenance design.

The most common VFFS design incorporates electromechanical power and pneumatics. This combination offers a manufacturing cost advantage in a highly competitive industry; but the tradeoff is the need to control air supplies carefully in order to keep the performance of the machine up to its top efficiency. The use of solenoid valves, pneumatic valves, flow controls, and air lines increases the maintenance requirements somewhat.

The electromechanical/pneumatic/vacuum principle is quite unique. The design is similar to the types utilizing air, with the addition of vacuum material/film transport belts. This principle locks the film to perforated drive belts by means of a vacuum pump. It works quite well, but generally imposes limitations on speed and minimum bag-width dimensions due to the design requirements for utilizing vacuum draw-down belts. Also, the vacuum pump adds another power requirement and noise factor.

Bag-Filling Factors

The product being packaged is generally the limiting factor regarding production rate capabilities on any of the machine designs. Machine operation is affected by product characteristics such as dust, fines, and stickiness, as well as by piece size, piece weight, and product volume. Some products create a piston effect when dropped down inside a bag-forming tube, by pushing air down into the sealed end of the packaging material. This air must escape somewhere. There are various controls such as inner fill tubes and snorkel tubes, to release air pressure before the product drops down the tube.

All of these factors affect the end result. Too often, cycle capabilities of the bag machine and of the product measuring system are calculated independently, without considering what happens when the product moves from the measuring system down through the tube and into the bag. Achievable production rates are based on the compatibility of the three components: bag machine, filler, and product(s). Users of VFFS machinery should supply complete information concerning the products to be packaged to the manufacturers of the equipment so that they can factually evaluate the achievable speed, weight accuracy, and efficiency capabilities.

Product Fillers

There are several different choices of measuring/filling equipment: net weigh scales, auger fillers, volumetric fillers, counters, bucket elevators, and liquid fillers.

Net weight scales provide the most accurate means of measuring products for packaging. The invention and marketing of the multiple-head computer scale system in the past 25 years has literally revolutionized the product-weighing industry. Package weight controls can be held to ± 1 g regardless of the size of the piece or particle being packaged.

Next in line for accuracy is the *auger filler*. This is applicable to products that are powdery in form and can be handled through a screw contained inside a tube. Most chemicals, baking products, and other powdery forms use an auger filler. The accuracy is dependent on bulk density control of the product throughout the augering system as well as the cycle repeatability of the chosen auger filler.

A volumetric cup filler fills by volume and is generally used for inexpensive products where high production rates are desirable and product overweight giveaway is unimportant. Counters (see Filling machinery, by count) apply to applications such as hardware, confectionary items, and other items that must be packaged by count. The counter(s) can be mounted directly over the bag machine, or they can work in conjunction with a bucket elevator.

The *bucket elevator* can be an intermediate between any of the other product fillers, but it should be utilized only where space restraints or other impractical reasons dictate needs. The shortest distance between two points is the rule of thumb on VFFS systems, so mounting the filler directly above the vertical bag machine is the best and should be first choice where applicable.

The major VFFS equipment suppliers to the United States market are General Packaging Equipment Corporation, Hayssen Manufacturing Company, Package Machinery Company, Pneumatic Scale Corporation, Rovema Packaging Machines, Inc., Triangle Package Machinery Company, The Woodman Company, Inc., and Wright Machinery Division.

BIBLIOGRAPHY

"Form/Fill/Seal, Vertical" in *The Wiley Encyclopedia of Packaging,* 1st ed., by George "Rocky" Moyer, Rovema Packaging Machines, pp. 367–369.

General References

E. Dierking, F. Klien Schmidt, and J. Hesse, *Packaging on Form, Fill, and Seal Flexible Bag Machines,* Massuers Wolff Walsorde, Germany (formerly FRG), 1975; discusses the importance of the bag-forming collar with regard to package quality as well as the relationship of material friction on machines utilizing the belt-drive principle.

"Extending the Shelf Life of Foodstuffs by Using the Aroma Perm Protective Gas Flushing System for Flexible Bags," *Verpack. Rundesch.* **4,** 412–414 (1979).

The PMMI Packaging Machinery Directory, The Package Machinery Manufacturers Institute, Washington, DC, annual.

G

GAS PACKAGING.

See CONTROLLED-ATMOSPHERE PACKAGING.

GABLETOP CARTONS.

See CARTONS, GABLETOP.

GLASS CONTAINER DESIGN

Glass has many advantages as a packaging material. It is rigid, transparent, inert, impermeable, and odorless, and a wide variety of shapes, sizes, and colors provide customer appeal. An abundant supply of inexpensive raw materials makes glass a natural packaging material. An effective glass container design considers many factors that relate to shape, manufacturing, marketing, and strength.

Container Shape and Dimensions

The basic container shape is determined primarily by the type and quantity of the product. Each product group, such as beverages, wine, liquor, food, pharmaceuticals, cosmetics, and chemicals, has a few characteristic container shapes. A sampling of these shapes is shown in Figure 1. In general, liquid products have small-diameter finishes for easy pouring. Some food containers need larger finishes for filling and removing the contents. A specific product may be available in a family of similar shapes to provide a range of desired capacities. Certain products have practical size limits; for example, perfume containers are much smaller than wine bottles.

The overall container shape can be further defined by three methods. First, the shape is defined by the Glass Packaging Institute (GPI). If several glass manufacturers supply containers to a few large-volume fillers, certain critical container dimensions are agreed on to ensure that the container will be compatible with labeling, filling, packaging, and shipping operations. Second, the shape can be specified from the manufacturer's drawing of a stock job. Container manufacturers have a catalog of stock jobs for each product type. Third, a distinctive shape can be specially designed for a customer.

Finishes. After the general container shape has been defined, some very specific dimensions are required to ensure finish compatability with the closure (see Fig. 2 for container nomenclature). Finishes can be broadly classified by size, sealing method, and special features. The size, expressed in millimeters, refers to the nominal finish diameter. A closure size of approximately 1.38 in. (35 mm) is the dividing line between narrow-neck (blow-and-blow process) and wide-mouth (press-and-blow process) finishes (see Glass container manufacturing). Common sealing methods are continuous-thread, cork, crown, threaded crown, rollon, and lug (see Closures). Special features include handle, pourout, sprinkler, and snap caps. Selected examples are shown in Figure 3.

Headspace. The product capacity and headspace determine the container's interior volume. Headspace is needed to accommodate any thermal expansion of the product and to assist the filling operation.

Container stability. The center of gravity and bearing surface diameter determine a container's stability. Stability contributes to forming and filling-line efficiency and minimizes customer-handling problems.

Manufacturing Conditions

Dimensions. Forming-machine requirements also effect container shape and dimensions. Bottle-forming machines have inherent height and diameter restrictions. The popular IS machine (see Glass container manufacturing) can make heights of roughly 1–12 in. (25.4–305 mm). IS-machine diameter limits depend on whether 1, 2, 3, or 4 bottles are made

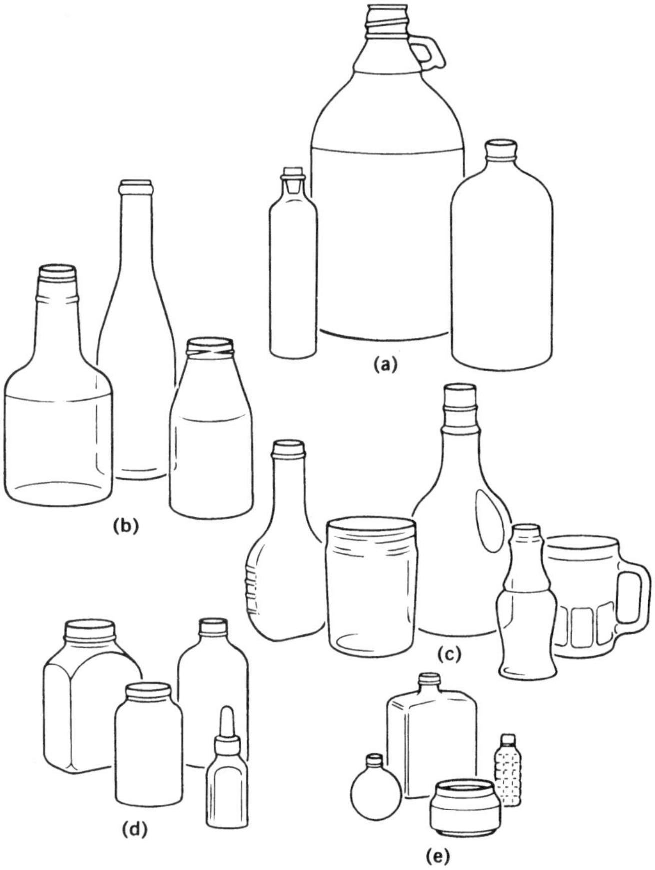

Figure 1. Typical glass container shapes: (**a**) laboratory (left to right)—sample oil bottle, amber gallon jug, amber Boston round; (**b**) beverage (left to right)—liquor round, wine round, single-serving juice bottle; (**c**) food (left to right)—dressing/sauce bottle, peanut butter jar, handled round sauce/oil decanter, spice bottle, specialty mustard mug: (**d**) drug and pharmaceutical (left to right)—amber shelf-pack, amber wide-mouth packer round; amber Boston round, amber iodine dropper bottle; (**e**) toiletries and cosmetics (left to right)—round nail-polish bottle, oblong cologne bottle, opal cream jar, and crystal-cut perfume square. (Courtesy of Continental Glass & Plastic, Inc., Chicago.)

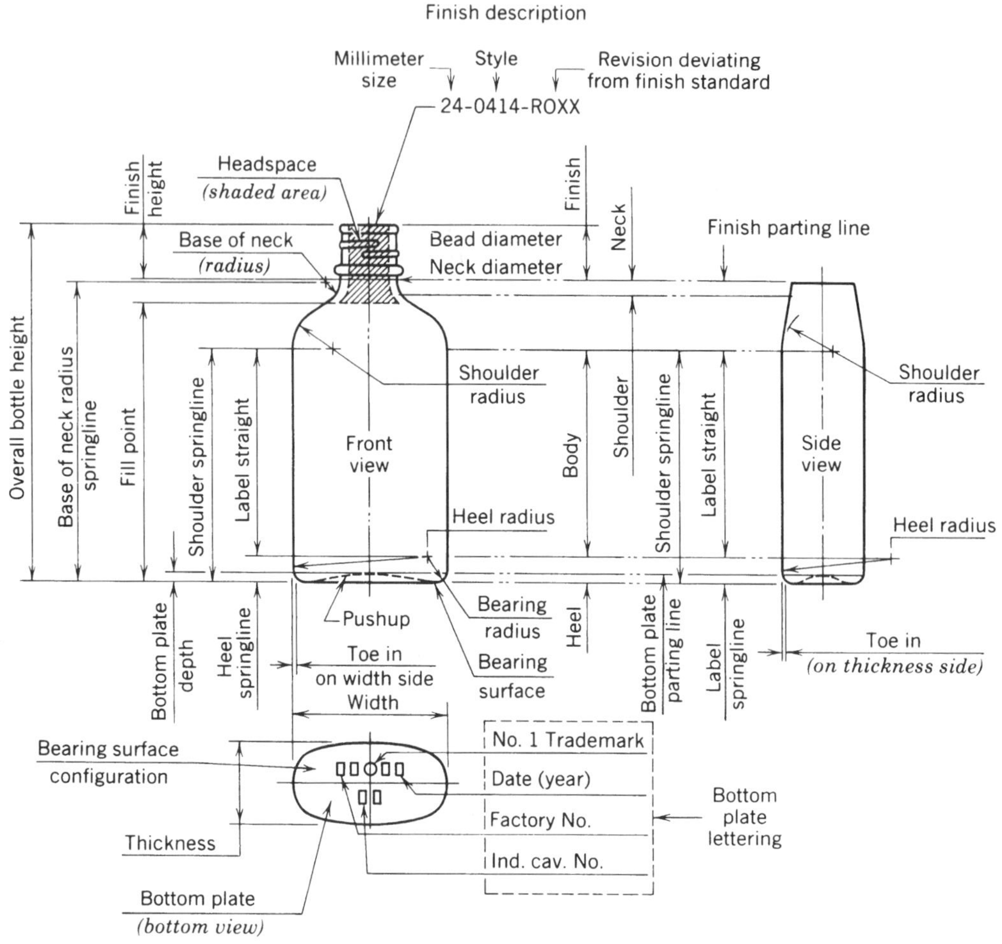

Figure 2. Glass container nomenclature.

on a single section. The widest diameter range is approximately 0.5–6 in. (12.7–152 mm). Other glass-forming machines may have different limitations.

Contours. Hot viscous glass flows more easily into a mold with a smooth, rounded shape. Containers with sharp corners or square heels are more difficult to manufacture than those with rounded profiles.

Tolerances. Acceptable limits or tolerances on container dimensions are necessary because forming-process variables prevent exact container specifications. The Glass Packaging Institute standard tolerances apply to capacity, weight, height, and diameter. The capacity tolerance varies from 15% for very small containers to less than 1% for large bottles. The weight tolerance is approximately 5% of the specified weight. Height limits range from 0.5 to 0.8% of the overall height. Diameter tolerances vary from 3% near the minimum 1-in. (25-mm) diameter limits to 1.5% at the maximum 8-in. (200 mm) diameter size.

Marketing and Sales Factors

Glass containers enhance product customer appeal with distinctive shapes, attractive colors, graphic labels, and special decorations.

Shape. Distinctive container shapes have instant manufacturer or brand recognition. Customers can identify certain brands of beer, wine, or food by the container shape. An attractive shape, especially for cosmetic items, promotes increased sales. Sometimes the container has more sales appeal than the product.

Color. Flint glass (colorless) provides the customer with a clear view of the container's contents. In addition to their aesthetic appeal, amber and green glass protect certain foods and chemicals by limiting the transmission of ultraviolet light (see Colorants). Blue and opal glass can be used for decorative effects. Many shapes of each glass color can be produced to suit the customer's preference.

Labels. Container labels are both attractive and informative (see Labels and labeling). They can be applied by firing an enamel, gluing on paper, or shrinking a preprinted plastic foam. Foam labels also provide cushioning and abrasion protection. Labels and labeling require consideration during container design. Flat areas or areas with only slight curvature must be provided when a label is to be glued or wrapped on the container. Designs often include an indented label panel to protect the label from abrasion.

Decoration. Decorating techniques include cut design, frosting, decal application, and fired enamels. Geometric pat-

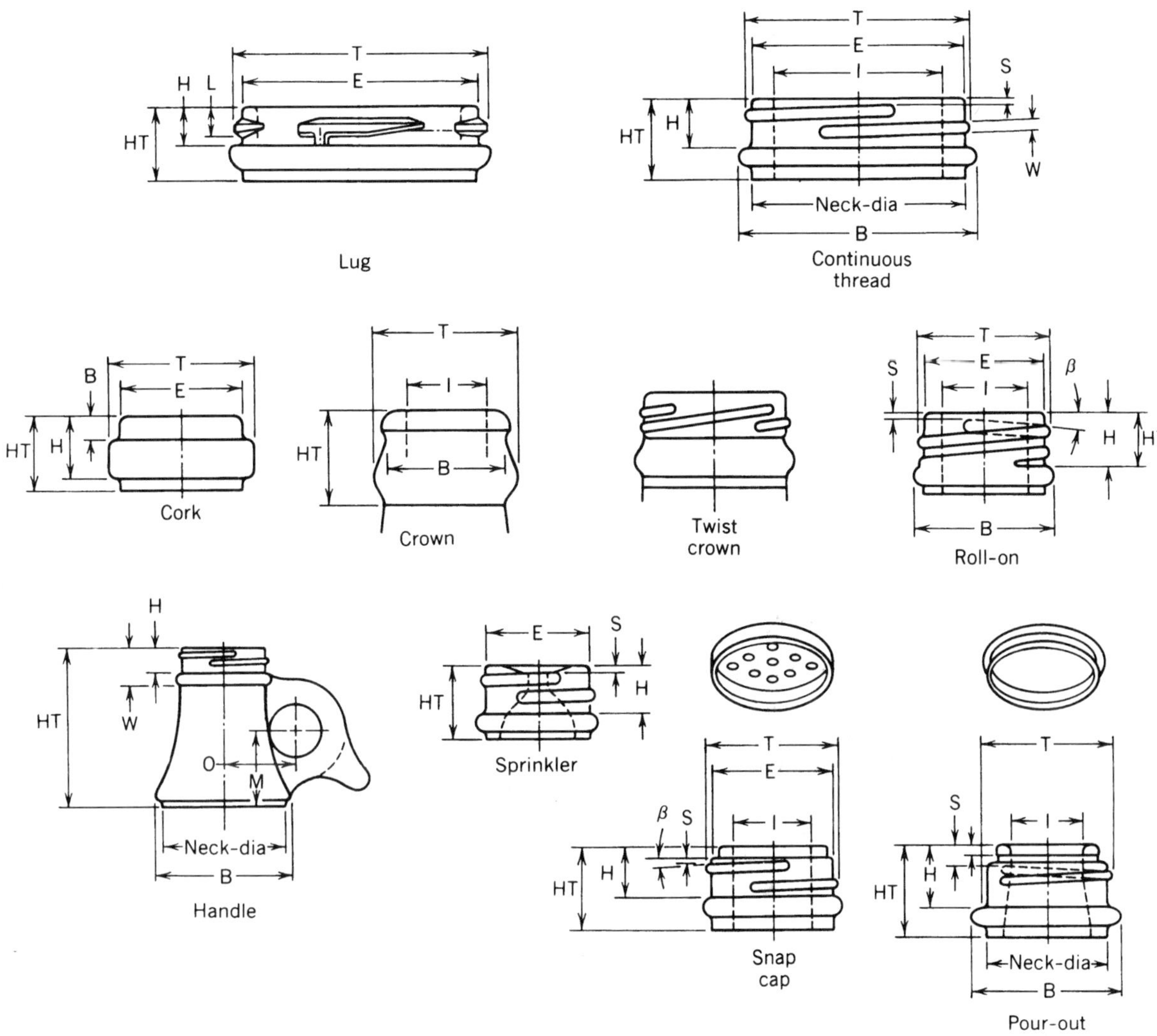

Figure 3. Typical glass container finishes: B = bead diameter; β (beta) = angle of thread helix; E = wall diameter; H = intersection of "T" with bead or shoulder; HT = finish height; I = inside opening through finish; S = start of thread; T = diameter of thread; W = width of thread. (Thread must make a minimum of one complete turn from centerline to centerline of cutter.)

terns machined in blow molds provide a high-quality image at a low cost. A frosted appearance can be achieved by sandblasting or hydrofluoric acid etching the exterior surface. Both organic and inorganic enamels can be applied by spraying or screening, and then they are heated to fuse the coating on the glass.

Container Strength Factors

A glass container's strength is controlled by shape, surface condition, applied stresses, and glass weight. The following general statements on container strength assume that all other strength factors are constant. This is valid for discussion purposes. but the strength of a specific container is governed by a combination of these factors.

Shape. A container shape with a smooth vertical profile is usually stronger than one with sharp transitions. Similarly, a round-cross-section container is usually stronger than a rectangular one. High stress concentrations in sharp transition regions tend to influence container strength. Computer stress analysis of many container shapes shows that a balanced design of the shoulder, heel, and bottom regions can contribute to a container's strength.

Surface condition. A glass surface normally contains small surface imperfections. These imperfections are inherently formed during the manufacturing process and subsequent handling operations as a result of surface contact. The number, magnitude, and effect of these imperfections are controlled by good melting, forming, and handling procedures. Container design, surface coatings, or wraps also reduce the opportunity for formation of imperfections due to surface contact.

Containers can be designed with specific contact areas that concentrate abrasions where they will have minimal effect on glass strength. One example is the use of knurling on the bearing surface. The knurls (small protrusions) are the contact points between the containers and conveyors or other surfaces. Abrasions are concentrated on the tips of the knurls. Since the tips of the knurls are under less stress than the underlying glass, the abrasions have less influence on the container strength than they would otherwise.

Very thin, invisible coatings or surface treatments are often applied to a newly formed container. These treatments lubricate the glass surface and reduce surface abrasions. Bottles with a combination hot- and cold-end coating can be several times stronger than uncoated containers. Hot-end coatings, applied before annealing, include the oxides of tin and titanium. Possible cold-end coatings, applied after annealing, include polyethylene, stearates, oleic acid, silicones, and waxes.

Applied stresses. Container design must consider the type and magnitude of forces applied during its intended use. The most common forces applied to a container are internal pressure, vertical-load, impact, and thermal shock. Tension stress, created by these forces, influences glass strength more than compression stress.

Internal pressure. Internal pressure stresses are developed from a carbonated liquid, or a vacuum-packed food product. The internal pressure of a soft-drink container may reach 50 psi (0.34 MPa). During pasteurization, internal pressure in a beer bottle may be as high as 120 psi (0.83 MPa) (see Carbonated beverage packaging).

Internal pressure in a close container generates predominately circumferential and longitudinal stresses. In the cylindrical part of a typical bottle, the circumferential stress S depends on the bottle diameter d, the glass thickness t, and pressure p as approximated by the following equation:

$$S = \frac{pd}{2t}$$

The longitudinal stress in this part of the bottle is one half of this circumferential value. In the noncylindrical parts of a bottle, this equation does not accurately describe the stress. The rapidly changing curvature and wall thickness in the shoulder, heel, or bottom make it necessary to use much more complicated equations or numerical methods to calculate stress. Finite-element computer programs can be used to calculate the stress in these regions. These programs make it possible to evaluate the stress conditions in alternative designs without actually manufacturing bottles. Figure 4 shows the highly magnified deformation of a container, caused by internal pressure.

Vertical-load stresses. Vertical-load stresses are generated by stacking containers on top of each other or by applying a closure. These compressive forces produce tensile components in the shoulder-and-heel region, which may approach 1000 psi (6.9 MPa). Vertical-load stresses can be lowered by reducing the compressive force or modifying the container design to withstand these forces. Computer stress analyses show that vertical-load stresses can be lowered by decreasing the diameter difference between the neck and body, increasing the shoulder radius, and reducing the diameter difference between the body and bearing surface. The vertical-load stress reduction from the latter case is shown in Figure 5.

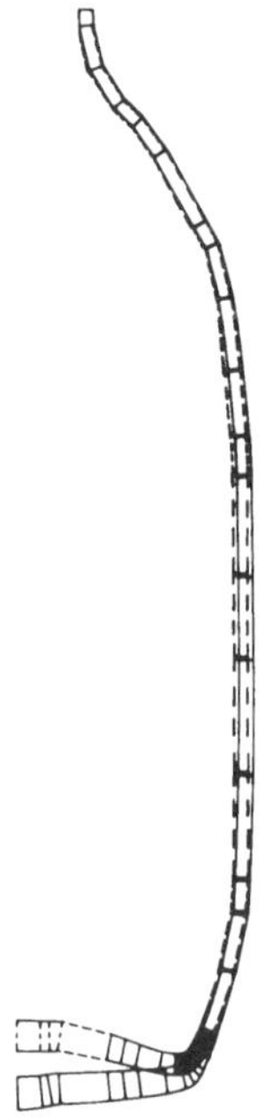

Figure 4. Magnified bottle deformation from internal pressure: ----, displaced profile; ——, original profile.

This graph also shows the associated higher internal pressure stress for this heel-shape change. It also illustrates that a particular shape may be better for one type of loading and less favorable for another type. The designer must balance the design to accommodate all the requirements.

Impact stresses. Impact stresses are produced when a container contacts another object. These contacts can occur during manufacturing, filling, transporting, and consumer handling. Figure 6 shows the three main stresses created during a container impact: contact, flexure, and hinge. The contact and flexure stresses are relatively high compared to the hinge stress, but localized to a small area. The lower hinge

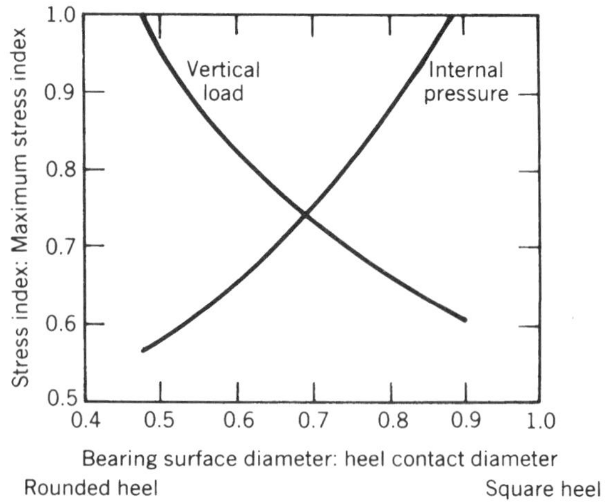

Figure 5. The effect of heel shape on bottle stress for vertical loading and internal pressure.

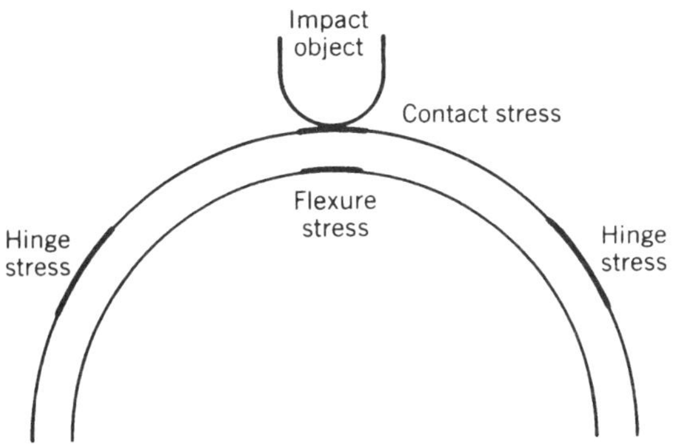

Figure 6. Stresses produced during impact (1).

stress is important because it acts on a large area on the outer surface.

Impacts produce a very complex and dynamic event. The stresses are related to the impact location on the container and the impact material. Also, the stresses vary depending on whether the container is empty or full, stationary or moving, or supported or free-standing. All these factors make a dynamic impact analysis exceedingly difficult, but a steady-state finite-element model of a side load can provide useful design guides. These analyses show that higher hinge stresses are developed by impacts in the midbody than in the shoulder and heel regions.

Thermal shock. Thermal shock stresses are developed from rapid temperature changes during pasteurizing or filling a hot or cold product. The thermal-expansion differential between the hot and cold surfaces generates these thermal stresses. Tension stresses are produced on the cold side, and compression stresses are developed on the hot surface. This stress pattern is complicated by bending stresses generated by expansions and contractions in the bottle. Thermal stresses can be reduced by minimizing the temperature gradient from the hot to cold side, decreasing the glass thickness, and avoiding sharp corners, especially in the heel. A finite-element analysis can be used to evaluate a container design for thermal shock.

Glass weight. The intended use and expected stress conditions are important factors governing the glass weight. For example, a single-trip beverage container is subject to much less abuse than a multiple-trip container, so the single-service bottle need not be as heavy. Conversely, a high-internal-pressure champagne bottle must be heavier than a noncarbonated juice container. Use of surface treatments or coatings on the external surface of glass containers can reduce the amount of glass required to make the containers. Special-shape cosmetic containers are a category where the lightest weight may not be important. In this case, the aesthetic appeal may override glass-weight factors. For a new container design, the glass weight is determined by a combination of computer strength analyses and a comparison to similar jobs that have performed well. For a long-running job, weight reduction may be possible because the design and forming process can be fine-tuned.

Computer Uses in Glass Container Design

State-of-the-art computer-aided design (CAD), computer-aided manufacturing (CAM), and computer-aided engineering (CAE) systems are used for container development. Computers have reduced design cost and improved container performance. Container shape definition can be modified on a computer to achieve the desired image. Tedious volume calculations that previously required hours are now completed in seconds. Clearer and more accurate bottle and mold-part drawings are plotted from computer-defined containers. Exact container geometry is supplied for numerical-control mold-part manufacturing. Computer-forming process simulation programs assist mold designers to achieve better parison shapes. Finite-element analysis of containers provides quantitative stress predictions that were previously unavailable. All these factors have played an important part in lightweighting glass containers.

BIBLIOGRAPHY

"Glass Container Design" in *The Wiley Encyclopedia of Packaging Technology,* 1st ed., by D. L. Hambley, Owens-Illinois, pp. 370–374.

1. R. E. Mould, *J. Am. Ceram. Soc.* **35,** 230 (1952).

General References

B. E. Moody, *Packaging in Glass,* Hutchinson, London, 1963.

G. W. McLellan and E. B. Shand, *Glass Engineering Handbook,* McGraw-Hill, New York, 1984.

J. F. Hanlon, *Handbook of Package Engineering,* McGraw-Hill, New York, 1971; 2nd ed., 1984.

F. A. Paine, *Packaging Materials and Containers,* Blackie & Son, London, 1967.

F. V. Tooley, *Handbood of Glass Manufacture,* Ogden Publishing, New York, 1961.

S. M. Budd and W. F. Cornelius, *Glass Technol.* **17,** 54 (1976).

P. W. L. Graham, "Lightweighting, Strengthening, and Coatings," *Glass Technol.* **25,** 7 (1984).

GLASS CONTAINER MANUFACTURING

Glass containers compete for market share with metal cans and plastic bottles in a packaging segment often called *rigid containers.* For the past 35 years, shipments of all rigid containers have increased at an annual compound rate of close to 3%. In 1960, of the total of 68.4 billion rigid container units shipped, glass had about 33%; metal cans, about 66%; and plastic bottles, less than 1%. The current 1995 share of an estimated total of 200 billion units is about 62% metal, and 19% for both glass and plastic containers. In 1995, the glass container industry in the United States shipped 38.5 billion units, with an aggregate weight of 10.7 million tons of glass, and a sales volume of about five billion dollars. Table 1 shows the glass container shipments in the recent past as compiled by the DOC (U.S. Department of Commerce).

It can be seen that, although the overall growth rate of glass containers has been about 2% annually, there are major fluctuations in the several market segments, where the

Table 1. Millions of Glass Containers

Year	Food	Beer	Soda	Liquor	Wine	Others[a]	Total
1960	9,189	2,377	1,656	1,421	727	6,905	22,274
1965	11,024	5,203	2,921	1,703	784	6,578	28,213
1970	12,080	7,578	10,059	2,015	1,030	5,547	38,309
1975	12,176	11,064	9,276	2,142	1,229	4,293	40,180
1980	12,857	17,666	8,330	2,274	1,265	4,562	46,954
1985	12,054	12,064	8,652	1,719	2,021	2,776	39,286
1990	13,504	12,751	8,928	1,613	2,174	2,068	41,038
1995	14,757	15,969	2,358	1,538	2,010	1,807	38,439

[a] "Others" include medicine and health, household and industrial chemicals, and drugs and cosmetics.

growth rate has been higher or lower, indicating gains or losses to metal cans and plastic bottles.

Glass container manufacturing is an industry with high capital cost, low material cost, and relatively high energy and labor cost. It is an industry that starts with basic raw materials that come from mines or quarries, and smelts or chemically reduces these raw materials to their oxides, at temperatures of ≤2700°F; and, in a continuous operation, forms the molten glass into a container. From an analytical standpoint, it is best viewed as two major manufacturing systems operated in line: the making of the molten glass, often called the *chemical phase;* and the forming of the molten glass into a container, often called the *mechanical phase.*

Chemical Phase

The glass container manufacturer is interested in obtaining a fairly exact metal oxide composition at the lowest material cost; and often uses a wide variety of raw materials to attain this objective. Table 2 shows a typical glass container raw material batch. It can be seen that a one ton batch starts with 2303 lb of raw materials. The silica sand is mined in the form of an oxide, and the feldspar is made up of oxides. The other materials are found as carbonates, and must be reduced to oxides as part of the glassmaking process, releasing 303 lb of carbon dioxide as part of the chemical reaction. This 15% is generally referred to as the *fusion loss.* The formula in Table 2 assumes that the sand has trace amounts of alumina, and maximizes the use of feldspar for its sodium and potassium content (limited by the amount of alumina). It uses a very pure form of calcium carbonate called aragonite, which is found in the Bahamas and available at East Coast and Gulf Coast ports. Certain minor ingredients are used as aids in melting; or to decolorize flint glass, where the sand contains iron oxide; or to add color such as to make amber glass used for beer bottles, or green for lemon flavored soft drinks. In 1995, the raw-material cost to yield a ton of glass varied throughout the United States from about $50.00 to $55.00 delivered to the glass plant.

Recycled glass, and self-generated cullet are added to a raw-material batch in amounts ranging from about 15% to over 50% of its weight, depending on the availability and quality of the recycled glass. Postconsumer recycled glass is currently collected and returned to glass plants for reuse at a national rate of about 35% of the amount entering the consumer marketplace. A much higher percent is collected in states with container deposit laws. The delivered cost of recycled glass in 1995 ranged from $50.00 to $60.00 per ton.

The most significant developments that have taken place in the past 30 years in the chemical phase of glass making has been the improvement in the efficiency of fuel usage, and the extension of furnace life through the use of improved ceramics in furnace construction. The industry average for fuel consumption in the sixties was about 13 decatherms per ton of glass. Today's modern furnaces are able to produce a ton of glass with fuel expenditures as low as four decatherms, because of improved firing systems and improved furnace insulation. Further improvements are to be expected as oxygen replaces air in future glass melting furnaces and furnace rebuilds. Furnace life has been about doubled to 10 years and longer, with increased use of fused-cast ceramics in furnace floors and sidewalls.

Capital costs in the chemical phase. A mimimum cost-effective glass containermaking operation would be scaled at 500 tons of glass containers per 24-h day. Such a plant would normally have a minimum of two furnaces rated at 280–300 tons

Table 2. One-Ton Glass Container Raw Material Charge

		Metallic Oxides Formed				
Material Charged	Pounds	Silicon	Aluminum	Calcium	Sodium	Potassium
Silica sand	1408	1404	4		—	
Feldspar	200	136	36	—	16	10
Aragonite	261	—	—	146	—	—
Soda ash	434	—	—	—	248	—
Total charge	2303					
Total yield	2000	1540	40	146	264	10
Composition, %		77.0	2.0	7.3	13.2	0.5

per day, and would represent a capital investment of about $20 million. If an oxygen fired furnace were installed on a rebuild, the total capital cost would likely be up to 10% less, and the capacity would increase to more like 350–400 tons per day for each furnace.

Energy. Natural gas is the preferred energy source for glassmaking, combined with electric boosting, used to improve productivity from a given furnace size. With the exception of Oklahoma, Texas, and Louisiana, where gas is found in abundant supply, natural gas is often subject to interruption in many areas of the country during severe cold spells. Glassmakers normally have alternate fuels available in the form of oil and propane to use during these emergency periods. The energy required to produce a ton of glass, when the gas is burned with air, is about 4.0–4.5 decatherms. (one thousand cubic feet MCF of natural gas contains about one decatherm of energy)

In recent years, partly spurred on by environmental regulations calling for the reduction of NO*x* emissions from glass container furnaces, various systems have been introduced using natural gas and oxygen as the furnace fuel. Air, of course, is made up of 80% nitrogen, and when subjected to very high temperatures, can form various nitrogen oxides. The use of oxygen and natural gas as the furnace fuel removes the nitrogen and eliminates fuel firing as a source of nitrous oxides. Moreover, when using natural gas and air, 10 volumes of air are required to burn 1 volume of natural gas. When using oxygen, only 2 volumes are needed. This improvement in efficiency reduces the total energy requirement by up to one-third, improving the furnace output by up to one-third, and does away for the need of the relatively inefficient regenerators now widely used, replacing them with recuperators and other more efficient heat-conservation systems. Furnaces must be rebuilt on an 8–10-year schedule. At the cost of oxygen continues to drop, it is believed that future furnace rebuilds will incorporate oxygen-firing systems, at significantly lower capital costs per ton of daily capacity, tending to offset any increased operating cost.

Labor. The chemical phase of glassmaking is fully automated. The direct-labor consists of an average of four workers per shift, including maintenance workers, or a total of 16 workers for the four-shift operation. This is a continuous operation, and the employees normally work up to 2190 h per year.

Supplies and maintenance. An allowance must be made for miscellaneous factory supplies and maintenance of equipment used in the chemical phase. These costs tend to average 5% of the equipment capital costs per year.

Glass pack : pull ratio. Not every gob of glass that is melted ends up as a shippable container. Some are rejected by the automated inspection equipment for a variety of defects, some gobs are lost while molds are changed for maintenance, and so on. All this rejected glass, generally called *cullet,* is automatically fed back into the furnace, where it is remelted. There are three standard pack : pull ratios commonly used in the industry, depending largely on the length of run of the particular mold. They are 95, 90, and 85% pack : pull.

The Mechanical or Forming Phase

Glass forming has been aptly described as a controlled cooling process. Multiple gobs of glass (from one to four) are formed at the furnace feeder and delivered by dropping through chutes to the forming machines below. The forming machine, in turn, normally contains 6–10 sections, or 6–40 molds. There are also several triple-gob 12-section machines, and triple-gob 16-section machines in operation in the United States at this time. When a forming machine has more than 10 sections, two feeders are used to supply glass gobs to the machine. The bulk of the equipment in the field consists of double- and triple-gob 8s, and double and triple gob 10s.

The most significant improvements that have taken place in the last 30 years in the mechanical phase of glassmaking has been the adaptation of electronic controls to improve the timing of the feeders, forming machines, conveyors, etc and the use of electric eyes and other electronic devices to automatically inspect the quality of the containers as they are made. Initially these devices were installed to monitor the performance of various mechanical operations, with adjustments made manually. Now, virtually any mechanical movement than can be electronically monitored is also able to be automatically controlled by using electronic feedback to make the proper adjustments to the feeder and the forming machine.

Capital costs. A minimum-size plant, with a packed capacity of 500 tons/day (tpd), or a melting capacity of about 575 tpd, would have two furnaces of about 280–300 tpd each. There would be six glass-forming lines. Total capital costs for such an installation would be about $28 million.

Energy. The forming machine is a pneumatic machine, operated by compressed air supplied by a bank of air compressors. Modern-day glass lines require about 50 horsepower (hp) per section, equivalent to about 7152 kWh for an 8-section, and 8940 kWh for a 10-section machine.

Annealing energy formerly required about 1.1 decatherms of heat per ton of glass. Modern annealing lehrs are fully insulated, and are, essentially, a closed energy system, utilizing low-output infrared heaters after an initial start-up with portable propane heaters. Daily electric consumption is estimated at 2500 kWh per line.

An additional 100 hp per line is required for conveyors and other support equipment. This equates to about 1800 kWh per line per day.

Labor. Hourly monitor in a glass plant tends to become fixed at the minimum number of people required to operate the line, without reference to the productivity of the line. The present industry average is about 30 people per line, spread over four shifts. The weekday, daytime monitoring, when most of the shipping and receiving takes place, is normally nine per line, with the other three shifts, using seven per line. There are a few plants, using automatic palletizers and other labor-saving devices, where the monitoring has been reduced to about twenty per line. A typical plant would also require a total of 27 salaried forepersons, supervisors, quality-control people, plant security, etc spread over the four shifts, plus an additional five people in plant management.

Other forming costs. Mold costs are, in theory, advanced by the customer, and are rebated on a per gross basis until the purchase costs are recovered. A full set of molds for a triple-gob, 10-section machine can cost up to $150,000. In practice, however, the customer normally advances payment only for the experimental unit molds (about $10,000), and then only if the container is of an unusual or proprietary design. In any event, mold costs are an operating cost for the glassmaker to buy and maintain. In order to cover mold costs and mold maintenance costs, an allocation of about $15.00 per day per mold cavity is made. Other forming machine and support equipment maintenance, including operating supplies and services, is calculated at 5% per year, of the initial machinery capital cost.

Operating mode. Some glass lines operate for extended periods, several months, for example, producing the same or similar size, weight, and shape containers, such as beer or soft-drink bottles. Mold changes are made mainly for routine mold maintenance, and there is essentially no downtime on the machine. These are often referred to as "dedicated" lines. Other short-run lines operate on a single mold for runs only 5–10 days, where there is considerable downtime associated with changing molds, adjusting the forming machine timing sequence, and adjusting the feeder to a different glass weight.

Glassmakers have established operating standards for every container they produce, on each piece of equipment that is likely to be used. They use these standards to measure their performance against their operating plan. These standards normally measure only the variable costs that can be controlled by the plant management. For convenience in analysis, four operating levels for classifying glass forming machine operations have been developed, as follows:

Mode I	Long-run dedicated line	No downtime	95% pack
Mode II	Long-run non-dedicated line	10% No downtime	90% pack
Mode III	Short run non-dedicated line	15% No downtime	85% pack
Mode IV	Seasonally operated line	25% No downtime	85% pack

Carton costs. Some glass is sold and delivered on bulk pallets without cartons. However, most glass containers are priced to include the corrugated reshipper carton. Glass companies contract with corrugated suppliers for the supply of printed cartons on the basis of cost per thousand square feet. Current carton costs are about $46.00 per thousand square feet for a 175 lb test RSC (regular slotted corrugated).

Shipping costs. Glass selling prices normally include a shipping cost based on $0.75 per cwt (hundredweight) for beer and soft drink bottles, and $1.00 per cwt for all other containers.

Administrative and selling costs. There is no universally accepted method that can be used to accurately allocate administrative and selling costs to a specific container. The most widely used method is to allocate these costs as a percentage of dollar sales for each market segment. Sales organizations are often made up of a combination of geographic and market areas, making it possible to segregate sales costs by markets.

Administrative expenses cover such items as insurance, real property taxes, accounting and other overhead costs, office rentals, and telephones. An average percentage of 5% of sales revenue is often used to cover these costs.

The Manufacturing Process

Glassmaking raw materials. These are received at the plant by bulk railcars or trucks and are unloaded under the control of a computerized materials-handling system and directed into large silos adjacent to the batch house, and close to the furnace area. The same computer system automatically weighs and mixes the raw materials in an area of the plant called a "batch house," and delivers the mixed batch to the furnace batch feeder, located adjacent to the furnace. The mixing and feeding system is automatically controlled to maintain a constant level of molten glass in the furnace.

Glass furnaces. These are constructed of refractory bricks, using different types of bricks for different areas of the furnace. There are two distinct areas of a glass furnace. The larger area of the furnace, called the *melting area,* is normally constructed of chemically resistant fused-cast blocks, up to 18 in. thick, that form the sidewalls and furnace floor. The melting area is separated from the smaller refining area by a solid wall rising from the furnace floor to a level above the molten glass, and continuing as an open checkerwork to the furnace roof or crown. The two areas are connected by means of a depressed throat, allowing the molten glass to flow freely from the melting area to the refining area. The refining area is also constructed of a high-temperature fused-cast refractory material. The furnace crown, which is in the shape of a parabolic curve in order to radiate the heat uniformly to the molten glass surface, is made of pure silica brick.

Two methods are used to fire glass furnaces: (*1*) *end-port furnaces,* which are fired from the end of the furnace down one side of the melter and refiner and exhausted back the other side to one of a pair of regenerators located at the rear of the furnace (see Fig. 1); and (*2*) *side-port furnaces* (see Fig. 2), which are fired from a series of ports located on the sides of the furnace across the furnace to one of a pair of regenerators located on the opposite sides of the furnace. In each type of furnace, the firing is done from one direction for a period of 15–20 min, after which time the firing direction is switched, and combustion air is drawn from the regenerators, which have been heated up from the previous firing cycle. The regenerators are constructed of refractory brick in an open-checkerwork pattern (regenerators are often referred to as *checkers*). The upper layers of brick are made of higher-temperature resistant brick, and lower-cost and lower-temperature resistant brick is used in the bottom of the regenerators. The purpose of a regenerator is to extract heat from the spent-fuel gases, before they are exhausted to the atmosphere. When one regenerator has heated up, the direction of firing is reversed and the hot checkers are used to preheat the combustion air before it is mixed with the fuel and ignited. The regenerators are a costly, high-maintenance portion of the furnace. As oxygen-fired furnaces are introduced, the regenerator is expected to be replaced with a lower-cost

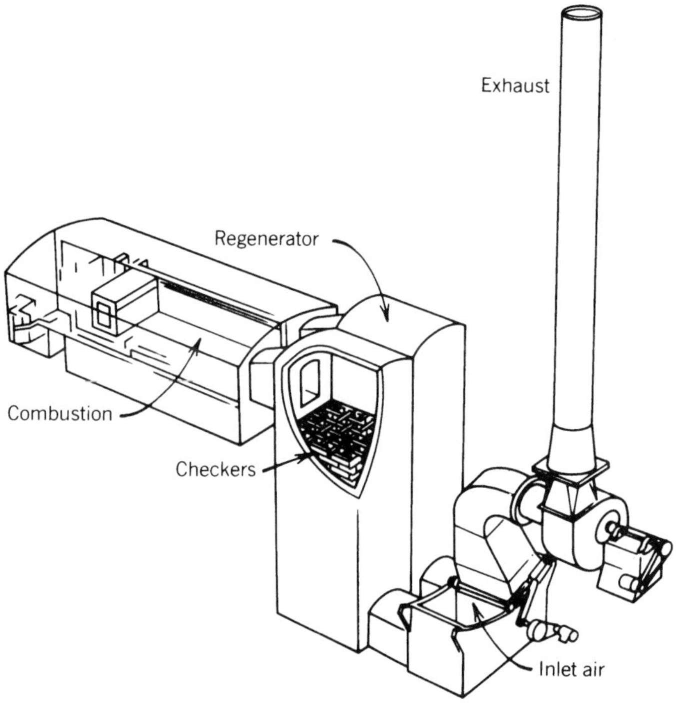

Figure 1. End-port furnace with forced-draft stack.

recuperator system of heat conservation. Glass furnaces are generally well insulated to improve thermal efficiency.

During the time when the glass is in the refiner, it reaches a level of thermal and chemical homogeneity desirable for container glassmaking. The molten glass continues to flow from the refiner, through a shallow alcove into the forehearth. The forehearth is a refractory lined channel up to 3 ft wide, about 8 in. deep, and up to 35 ft long, where the glass is cooled from the furnace temperature to the desired lower forming temperature. The forehearth narrows at the end, where the glass flows into a bowl-shaped device called the feeder (see Fig. 3).

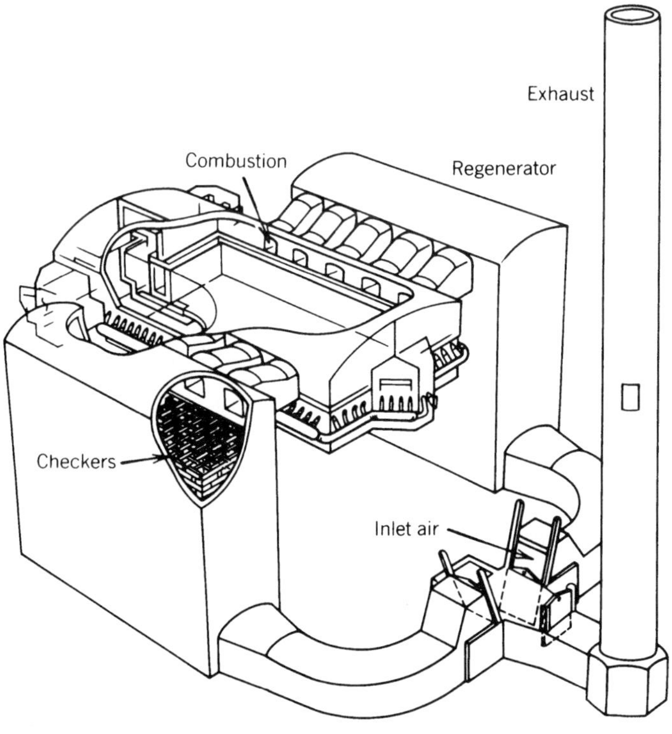

Figure 2. Side-port furnace with natural-draft stack.

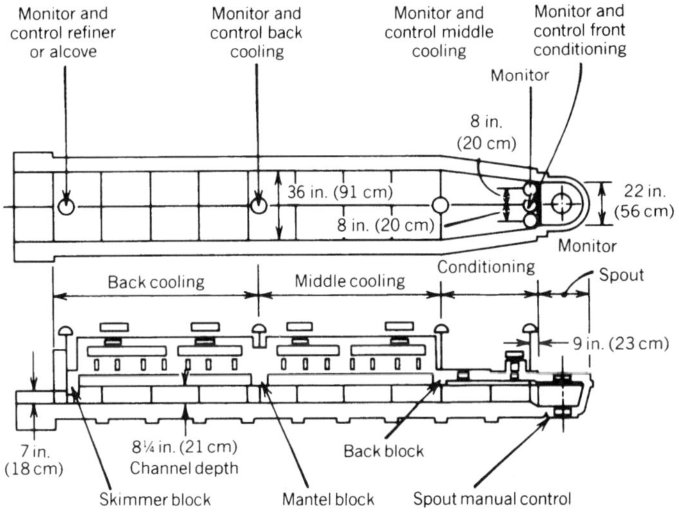

Figure 3. Top and side view of a forehearth.

The feeding–distribution system. This consists of one to four ceramic orifices located at the bottom of the feeder, depending on how many glass molds are in each section of the forming machine. Glass gobs are formed when the vertical motion of the plungers alternately are lifted to permit glass to flow through the orifice, and then lowered to cut off the flow. Discrete gobs are formed when shears cut the glass gobs coincident with the downward motion of the plunger. The gob or gobs fall into a distributor, which directs them through a scoop, trough, and deflector, to the forming machine (see Fig. 4).

The forming machine. This is designed to convert from one to four solid, cylindrical gobs of glass into hollow containers. There are two systems used to accomplish this task in two stages, first shaping a preform or blank mold, also called a *parison,* and then forming the finished container in a second mold. The two systems are called "blow and blow" (B&B) and "press and blow" (P&B). In the B&B process (see Fig. 5), both the preform and the finished container are formed by blowing air into the molds. In P&B, the preform or parison is formed by inserting and pressing a plunger into the molten glass in the preform mold, similar to the injection process used in the plastic industry, to form an intermediate shape, and then form the final shape in the second mold by blowing air (see Fig. 6). Formerly, the P&B process was used only for making wide-mouth containers, but in recent years, a narrow-neck P&B process has proved very successful for narrow-neck bottles for beer and soft drinks, particularly in its ability to produce a lightweight bottle with uniform sidewall distribution.

After the gob has been loaded into the blank mold, the baffle is seated on top of the mold funnel and a settle blow of air

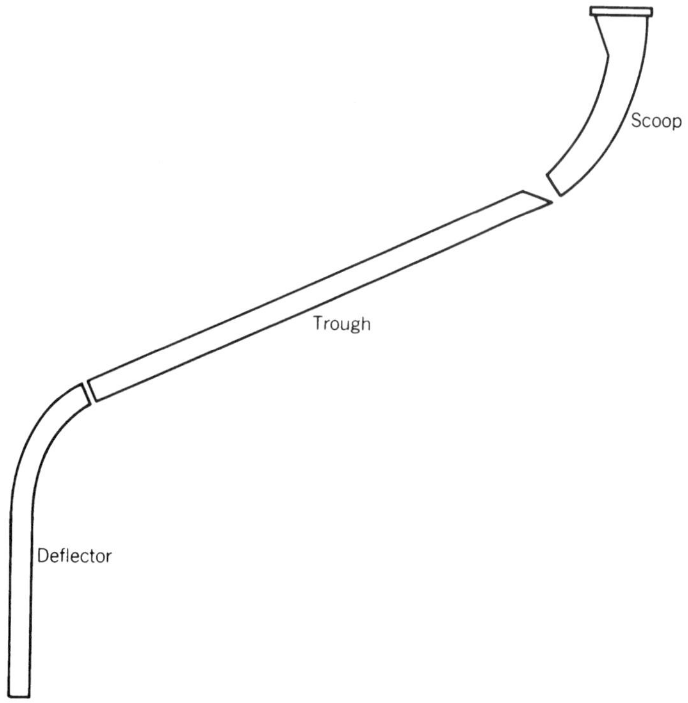

Figure 4. Delivery system for individual-section (IS) machine.

is applied to force the glass down into the finish area of the container (the area from the neck ring to the sealing surface as shown in Fig. 5). Before the application of the settle blow, a small plunger assembly is inserted into the neck ring. This plunger assembly forms the top and inside surface of the finish.

In the *blow-and-blow* (B&B) process, after the settle blow, the plunger is retracted, the funnel is removed, and the baffle is reseated directly on the blank. When the baffle is seated, counterblow air is applied to the gob from the bottom, in the void left by the plunger assembly. The counterblow air enlarges this bubble until the glass is forced out against the sides of the mold and the baffle. The resulting hollow preform is called a parison.

In the *press-and blow* (P&B) process, the baffle is seated directly on the blank after loading the gob. The plunger then moves up and presses the glass against the blank mold walls and the baffle, ensuring that glass is also forced into the finish area of the parison.

From this point on, both systems follow a similar pattern. The blank opens, and the parison is transferred by the invert mechanism of the forming machine, from a neckdown position to an upright position in the final mold. Contact with the metal surface of the blank mold has formed a thin solid skin on the surface of the parison. Before the final blowing operation is started, the parison is held to enable it to reheat, allowing the skin to remelt, and the parison to elongate. This is essential to ensure uniformity of glass distribution throughout the container. Heat is rapidly removed from the surface of the glass by contact with the final mold.

The takeout mechanism. This device now transfers the container from the mold and holds it over the deadplate for a brief time to allow cooling air to flow up through the deadplate and around the container to further the cooling of the container. Finally, the takeout jaws open and deposit the container on the deadplate, from which it is pushed onto the machine conveyor and transported to the annealing lehr.

Glass forming is a controlled cooling process. The total time between release of the gob from the feeder, to the pushout to the machine conveyor, is normally about 12 s for a typical 12-oz beer bottle, weighing 7 oz. However, the operations of shearing, delivery, parison formation, final blow, and takeout overlap each other, so that the cycle time of the machine, or the time between the delivery of a container from each mold, principally limited by the time spent in the final blow mold, is more like 4.5–5 seconds. A good estimate of the cycle time can be obtained using the following formula:

$$(0.20 \times \text{wt in ounces}) + 3.15 = \text{cycle time in seconds}$$

$$\text{using a 7-oz wt } (7 \times 0.20) + 3.15 = 4.55 \text{ s}$$

With a cycle time of 4.55 s, each mold produces one container every 4.55 s, and a triple-gob, 10-section machine, having 30 molds, has $30 \times 60 = 1800$ mold seconds available per minute, and would be expected to make $1800/4.55 = 396.5$, or approximately 400 bottles per minute.

The fact that more than twice the time is spent in the final mold than in the parison mold has led to several attempts to develop forming machines with two final molds for every parison mold. As of this time, none of these machines have received wide commercial acceptance.

IS machines. IS machines are powered by compressed air, with the timing of the machine sequences, formerly controlled by "buttons" on rotating drums actuating various air valves. These "buttons" have been replaced in recent years by electronic timing devices, actuating various solenoids, which are more precise and easier to adjust. Electronic timing, coupled with improved machine mechanical tolerances, has resulted in a smoother overall operation and the reduction of the jerky operation associated with IS machines. The advent of the use of electronics has been extended to include devices that weigh each container periodically, and, by feedback, controls the plunger mechanism in the feeder, to maintain a desired gob weight.

Mold materials. Molds are normally made from gray, or ductile, cast iron. Copper–nickel–aluminum alloys are often used to fabricate neck rings to take advantage of the greater thermal conductivity of such alloys. Baffles and plungers are sometimes made of a nickel–boron alloy to take advantage of its high-temperature hardness and excellent glass-release properties. Blank molds are periodically "swabbed" with graphite-based compounds to improve release properties. Compounds have been developed that can be applied to the blank molds before they are put into service on the forming machines, thus reducing the need for swabbing, and increasing the time that the blanks can remain on a machine before removal for cleaning.

Mold cooling. Molds were cooled by blowing air or "wind" on the outer surface of the molds, through a wind stack, using ambient air. This method proved to be inadequate as machine speeds increased. Now, air is also blown through holes drilled

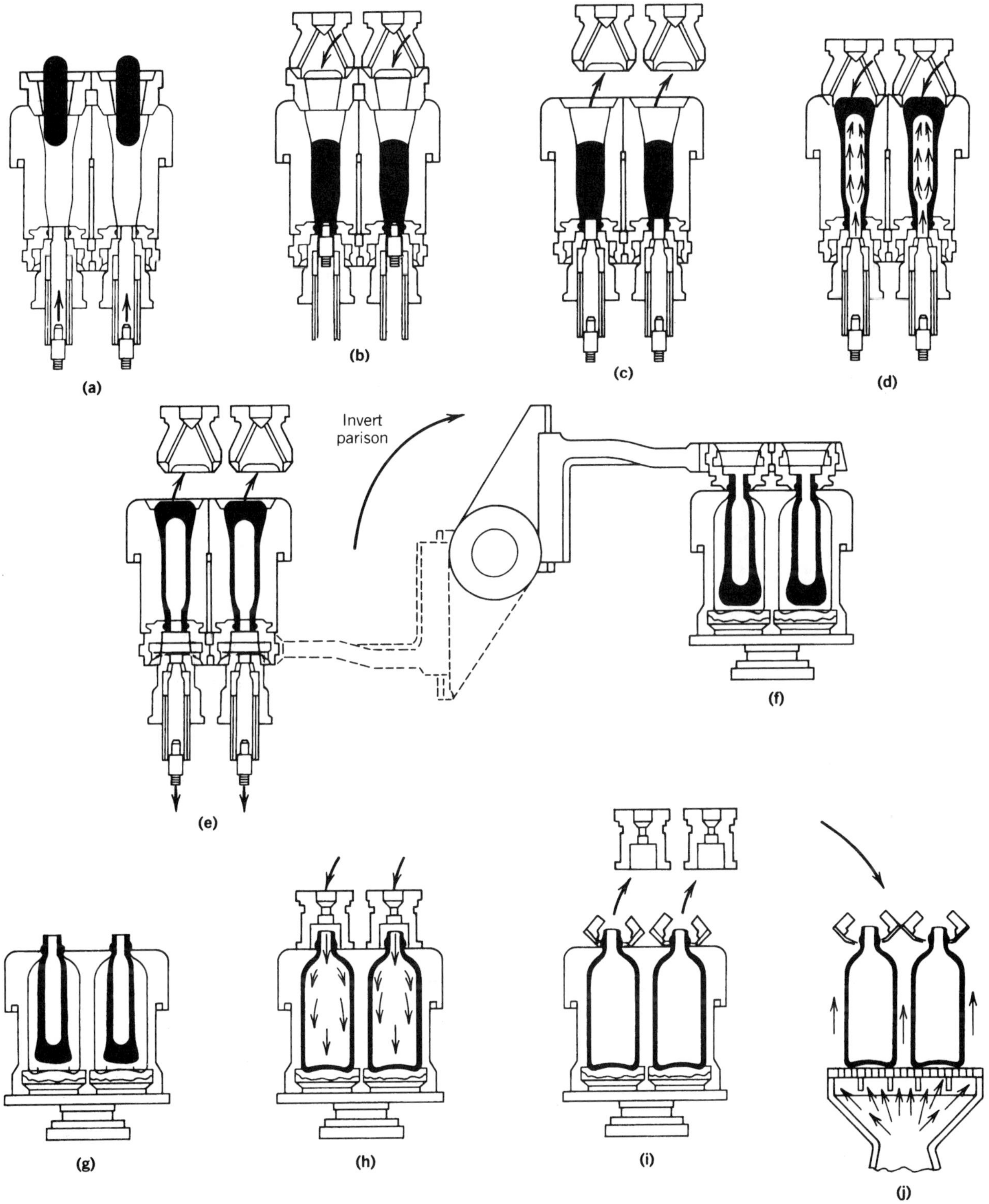

Figure 5. A functional diagram of parison and bottle-forming by the blow-and-blow process: (**a**) gob delivery through funnels into closed blank molds; (**b**) baffle on, settle blow pushes glass downward onto plunger and into finish; (**c**) corkage reheat, plungers down, baffles and funnels off—bubble inside finish reheats; (**d**) baffles down, counterblow air forms parison; (**e**) baffles off, thimbles down, blanks open, parison reheat starts; (**f**) blow molds close and neck rings open, parison released into blow molds; (**g**) reheat continues as parison elongates; (**h**) blowheads down and blow air on—bottles formed to blow mold; (**i**) blow-heads off and blow molds open—bottles taken out with takeouts; (**j**) bottles held in takeout jaws over deadplates and then released and swept onto conveyor.

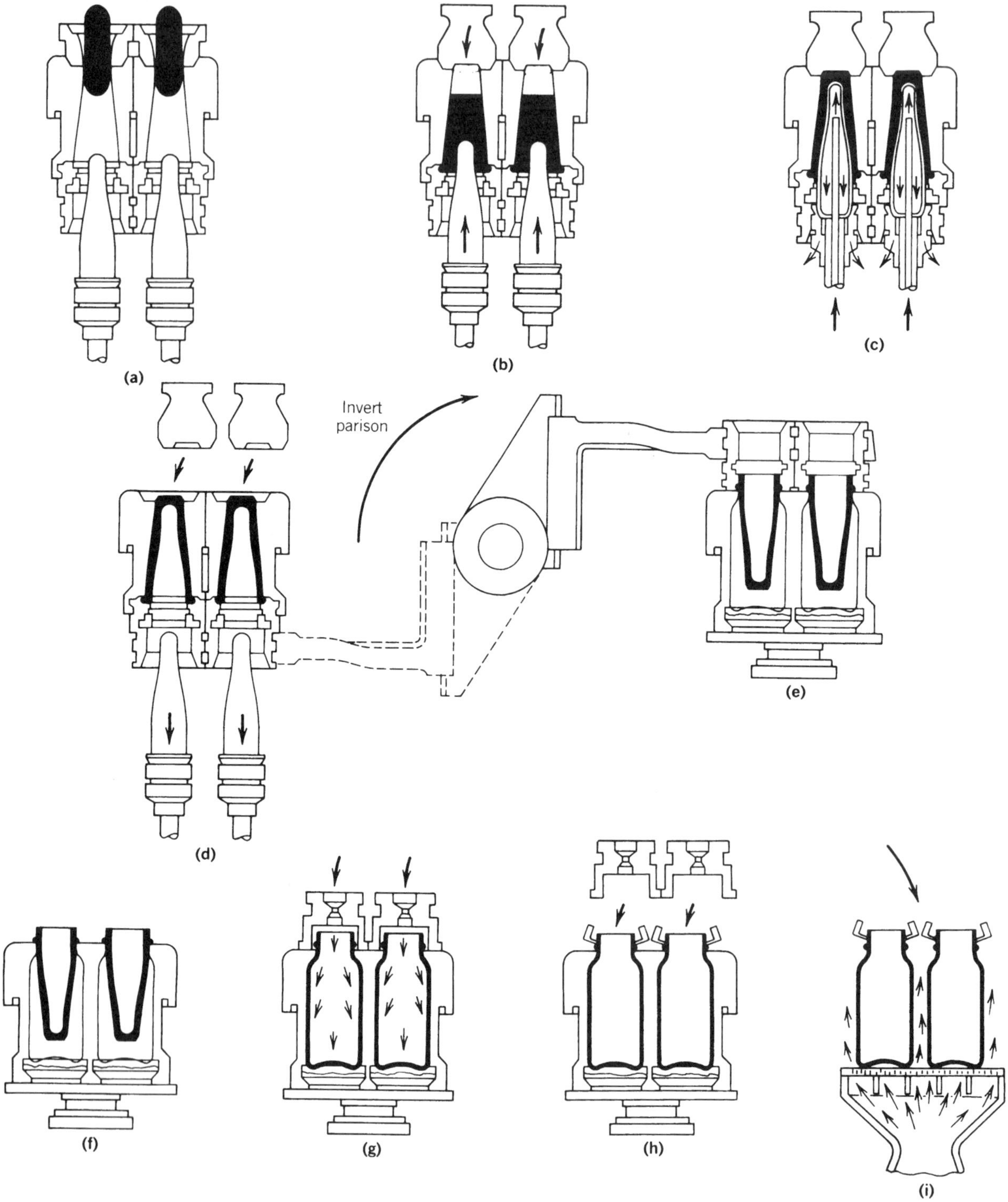

Figure 6. A functional diagram of parison and bottle forming by the press-and-blow process: (**a**) gob delivery through funnels into closed blank molds—plungers in position for loading; (**b**) baffles on and plungers move up starting to press glass; (**c**) pressing of glass into parison shape and into finish is completed; (**d**) plungers down, baffles off, and blanks open—parison reheat—begins; (**e**) blow molds close and neck rings open, parison released into blow molds; (**f**) reheat continues as parison elongates; (**g**) blowheads down and blow air on—bottles (jars) formed to blow mold; (**h**) blowheads off and molds open—Bottles taken out with takeouts; (**i**) bottles held in takeout jaws over deadplates then released and swept onto conveyor.

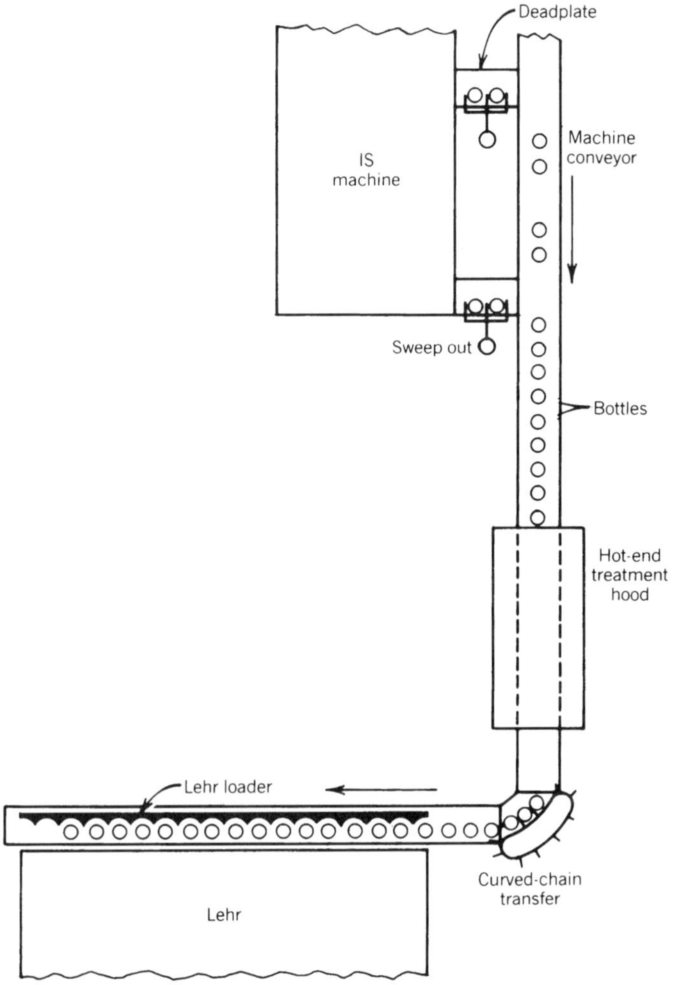

Figure 7. Hot-end handling process.

vertically through the blank and blow molds to improve cooling efficiencies.

Hot-bottle handling. This involves transferring the newly formed container from the deadplate on each machine section, to the machine conveyor for delivery to the annealing lehr. In a typical plant layout, this involves three 90° turns for the container (see Fig. 7). The most critical mechanism is the sweepout, where the containers must be swept off the deadplate in precise timing to match the speed of the machine conveyor. The sweepout mechanism consists of a retractable, rotating arm with fingers designed to make three-sided pockets for the containers. After the containers have been released from the takeout mechanism, the sweepout arm extends, inserting the fingers between each of the containers now resting on the deadplate. The sweepout mechanism then rotates approximately 90° dragging the containers from the deadplate to the moving machine conveyor, sometimes called the "flight" conveyor. Electronic controls are now used to control the timing of this critical interaction.

At the end of the machine conveyor, the containers make a second 90° turn by means of a curved-track ware transfer that moves the containers to the cross-conveyor at the feed end of the annealing lehr. At this point, when the cross-conveyor is filled with containers, a push-bar stacker, or lehr loader moves the line of containers, changing their direction again by 90°, and placing them on the moving-chain belt conveyor, often called the *lehr mat,* of the annealing lehr (see Fig. 7). This sequence of movements takes up to 30 s, during which time the containers have normally cooled to slightly below their annealing temperature.

Annealing. This is required to relieve any residual stresses that may be in the container as a result of the forming and hot-bottle-handling operations. This is done in an annealing lehr, by raising the container temperature to slightly above the annealing point (~1300°F) and holding at that temperature for several minutes. The container is then slowly cooled for several more minutes, to below the annealing temperature, and then rapidly cooled through the remainder of the annealing lehr.

The lehr is normally ≤125 ft long and ≤16 ft wide. The lehr mat is made of a continuous chain-link belt, and moves slowly with its load of containers from the hot end to the cold end, taking 30–45 min.

Surface treatments. These generally consist of two types:

A Hot-End Treatment. This consists of an application of tin or titanium. The application is made by passing the containers through a hood, located on the machine conveyor, and into which an atmosphere of the tin or titanium tetrachloride has been formed. This results in the formation of a unimolecular film of the oxide of the metal on the outer surface of the container. This layer strengthens the surface and provides for improved adhesion of any cold-end treatment.

Cold-End Treatments. This generally consists of a spray application of water emulsions of waxes and or polyethylene to reduce the coefficient of friction (COF) on the outer surface of the container. The application is made in the cooling section of the annealing lehr, at a point where the container temperatures are about 200°F.

Inspection of the containers. This is the next step. Until about 30 years ago, the inspection of glass containers for manufacturing defects was dependent on visual inspection of the containers, generally by women, who sat in front of lightboxes, and watched the containers pass by on a conveyor. They were charged with picking out the containers that were defective, and manually discarding them. A defective container may contain stones, blisters, or checks in the glass; may not meet dimensional specifications or the wall thickness may not be uniform; and a variety of other requirements peculiar to a specific container may be inspected for. As line speeds increased, visual inspection became impossible. Various mechanical and electronic devices were developed that now perform 100% inspection of containers for a variety of possible defects.

Squeeze testers. These are mechanical devices that pass the container between two rollers, causing stress on the container that simulate what may be expected when the container is used. *Plug gauges* mechanically check the height and perpendicularity of the container and the inside and outside dimensions of the finish. *Optical inspection devices* are used to locate stones, blisters, blemishes, and checks by rotating the container in the field of several photocells. Defects will interrupt the circuit and cause the container to be automatically rejected. *Thickness gauges* measure the thickness of the glass by using a capacitance device and comparing the reading to known standards. *Mold-cavity identification devices* are optical devices that read mold numbers and are used to identify and record which molds are giving trouble so that corrective action can be taken.

While each of these inspection machines can be installed separately, equipment manufacturers have developed total inspection machines that combine the inspection functions in a single machine, capable of handling round containers at speeds of 400 containers per minute.

Decorating. Returnable and refillable soft-drink bottles once represented a major market for ceramic decorated glass bottles. Two to four "silk screen" applications of colored glass frit were mechanically applied to glass bottles using a heated fine-wire mesh in place of a silk screen. The finely ground colored frit was suspended in a heavy mineral oil, forming a fluid ink able to be applied by a squeegee, whereupon the ink solidified on contact with the cold bottle surface. After the frit was applied, the bottle proceeded to a decorating lehr, where the oil was burned off, and the frit melted and adhered to the glass surface. Because of the high initial cost of this decorated bottle, this system depended on the bottle making many round trips to be economical. When the nonrefillable metal can and lighter and lower-cost nonrefillable glass and plastic bottles were introduced, their convenience and ultimate economics doomed the decorated refillable soft-drink bottle. Attempts were made to reduce decorating costs by substituting organic inks in place of frit, and using lower baking temperatures, but the market was lost to the nonrefillable container. Both ceramic frit and organic cold color decoration is still used in the cosmetics and other specialty applications (see also the articles on Decorating).

Labeling. Because operating speeds at glass plants are normally slower than at bottle-filling locations, it can be shown that labeling at the glass-forming line is usually less costly overall than at the filling line. At the same time, a full wraparound label, which is normally applied at the glass plant, tends to protect the bottle from glass-to-glass contact and damage from abrasion. This, in turn, makes it possible to further lightweight a labeled container. Also, because the glass-factory-decorated soft-drink bottle once dominated the soft-drink industry, soft-drink bottlers generally never installed labeling equipment on their lines. As a result, most glass bottles for soft drinks, bottled water, iced teas and many fruit drinks are labeled at the glass plant. They are normally wraparound labels applied with roll-fed labelers. The label construction consists of a wide variety of materials such as foamed polystyrene, which is normally shrunk in place as part of the labeling operation; and other nonshrunk compositions such as polypropylene laminates with paper or other plastic substrates. The labels may be printed by flexography, gravure, or silk screening. Most labels are applied using hotmelt or other adhesive systems, and some labels are pressure-sensitive.

Packing. After the containers have been inspected, and after any decoration or labeling operation, they are automatically packed in corrugated containers (called *reshippers*) and placed on pallets, or bulk-palletized for shipment to the customer. The pricing schedule for glass containers normally includes the cost of the reshipper and delivery within a certain shipping radius.

GLOSSARY OF TERMS

Blank mold. The mold used to shape the parison.

Blister. A glass bubble larger than 0.06 in.

Blow mold. The mold used to form the final container shape.

Cavity. The blank-and-blow mold combination used to produce one container.

Check. Small crack in the container.

Cullet. Crushed glass.

Finish. The top portion of the container over which the closure is applied.

Forehearth. The channel-like system used to deliver molten glass.

Frit. A mixture of finely ground glass and inorganic coloring material. Glass from the refiner to the feeder.

Gob. The quantity of molten glass used to make one container.

IS machine. Individual-section glass-forming machine.

Line. The equipment associated with one forming machine.

Parison. The preform or blank from which the container is blown.

Seed. A bubble in glass smaller than 0.06 in. in diameter.

Stone. Particle of unmelted batch or refractory material in a container.

BIBLIOGRAPHY

General Reference

F. V. Tooley, *The Handbook of Glass Manufacture,* Ashlee Publishing, New York, 1974.

Joseph Cavanagh
Joe Cavanagh Associates
Bay Shore, New York

H

HACCP

Origin of HACCP

The "hazard analysis critical control point" (HACCP) system was developed by the Pillsbury company in response to the food safety requirements imposed by NASA for "space foods" produced for manned space flights beginning in 1959. The concept was presented to the public at the 1971 National Conference on Food Protection. While interest was strong in the early 1970s HACCP was not widely incorporated into the food industry. However, in 1985, a Subcommittee of the Food Protection Committee of the National Academy of Sciences (NAS) issued a report on microbiological criteria that included a strong endorsement of HACCP. As a result of this recommendation, the National Advisory Committee on Microbiological Criteria for Foods (NACMCF) was formed in 1988 to advise the Secretaries of Agriculture, Commerce, Defense, and Health and Human Services. Part of the mission of the NACMCF was to encourage adoption of the HACCP approach to food safety, and the NACMCF has been the leader in developing HACCP principles and applying them to food processing operations.

HACCP Concept

HACCP is a preventive system for assuring the safe production of food products. It is based on a common-sense application of technical and scientific principles to the food production process from field, farm, or boat to table. The principles of HACCP are applicable to all phases of food production, including basic agriculture, food preparation and handling, food processing, food packaging, food service, distribution systems, and consumer handling and use.

The most basic concept underlying HACCP is prevention rather than inspection, detection, and correction. A grower, processor, or distributor should have sufficient information concerning that segment of the food system, to be able to locate where and how a food safety problem may occur. If the "where" and "how" are known, prevention is simplified. A HACCP program deals with control of safety factors affecting the ingredients, product processing, and packaging. The objective is to make the product so that it is safe to consume *and* be able to prove that the product presents minimum risk to consumers. The where and how are the *HA* (hazard analysis) part of HACCP. The proof of the control of processes and conditions is the *CCP* (critical control point) part. Flowing from this basic concept, HACCP is simply a methodical and systematic application of the appropriate science and technology to plan, control, and document the safe production of foods. HACCP covers all types of potential food safety hazards that could lead to a health risk—biological, physical, and chemical—whether they are naturally occurring in the food, contributed by the environment, or generated by something in the manufacturing process. Chemical hazards will be of primary interest to packagers, although, of course, chemicals are used at numerous points in the food production chain (see Table 1). While these types of chemicals do not represent health hazards when used properly, some of them are capable of causing illness or even death when used improperly.

Principles of HACCP

HACCP is a systematic approach to food safety consisting of seven principles:

Principle 1: Conduct a hazard analysis. Prepare a list of steps in the process where significant hazards occur and describe the preventive measures. The steps that precede the development of a hazard analysis are 1) assemble the HACCP team, 2) describe the food and its method of distribution, 3) identify the intended use and consumers of the food, 4) develop a flow diagram which describes the process, and 5) verify the flow diagram. The HACCP team then conducts a hazard analysis and identifies the steps in the process where hazards of potential significance can occur. Table 2 enumerates examples of packaging questions that may be considered during the hazard analysis. For inclusion in the hazard analysis list, the hazards must be of such a nature that their prevention, elimination, or reduction to acceptable levels is essential to the production of a safe food. Hazards that are of a low risk and not likely to occur would not require further consideration when developing the HACCP plan. Low-risk hazards, however, should not be dismissed as insignificant and may need to be addressed by other means, for example, by the total quality management (TQM) team. The HACCP team must then consider what preventive measures, if any, exist that

Table 1. Chemicals Used in Food Processing

Point of Use	Types of Chemicals
Growing crops	Pesticides, herbicides, defoliants
Raising livestock	Growth hormones, antibiotics
Production	Food additives, processing aids
Plant maintenance	Lubricants, paints
Plant sanitation	Cleaners, sanitizing agents, pesticides
Packaging	Adhesives resins, surfactants, defoaming agents, slimicides, polymers

Table 2. Examples of Packaging-Related Questions to be Considered in a Hazard Analysis

Does the method of packaging affect the multiplication of microbial pathogens and/or the formation of toxins?
Is the package clearly labeled "Keep refrigerated" if this is needed for safety?
Does the package include instructions for the safe handling and preparation of the food by the end user?
Is the packaging material resistant to damage, thereby preventing the entrance of microbial contamination?
Are tamper-evident packaging features used?
Is each package and case legibly and accurately coded?
Does each package contain the proper label?
Could the packaging material contribute any chemical compound to the product that may make the product unsafe to consume?

can be applied for control of each hazard. Preventive measures are physical, chemical, or other factors that can be used to control an identified health hazard. More than one preventive measure may be required to control a specific hazard. More than one hazard may be controlled by a specified preventive measure. Chemical hazards applicable to packaging that the HACCP team may consider during the analysis include the inks, indirect additives, and prohibited substances in packaged ingredients and packaging materials if the intended use is food packaging. The points of control are prior to receipt, and the control measures may be compliance with specifications, letters of guarantee, vendor certification, and/or approved uses listed in the Code of Federal Regulations. In a food plant producing canned salmon, the hazard is biological in the form of pathogens in the salmon. Retorting (a point of control) using the appropriate scheduled process is a preventive measure that can be used to control the pathogen hazard. In the packaging step (a point of control), proper double seaming will prevent the identified hazard of recontamination and the preventive measure is forming a proper double seam to prevent leakage.

Principle 2: Identify the CCPs in the process. A critical control point is defined as a point, step, or procedure at which control can be applied and a food safety hazard can be prevented, eliminated, or reduced to acceptable levels. An ideal CCP has the following characteristics: (1) critical limits that are supported by research and the technical literature; (2) critical limits that are specific, quantifiable, and provide the basis for a yes/no decision on acceptability of product; (3) technology for controlling the process at a CCP that is readily available and at reasonable cost; (4) adequate monitoring (preferably continuous) and automatic adjustment of the operation to maintain control; (5) historical point of control; and (6) a point at which significant hazards are prevented, eliminated, or reduced to acceptable levels. All significant hazards identified by the HACCP team during the hazard analysis must be addressed.

Examples of CCPs may include cooking, chilling, packaging of processed foods, specific sanitation procedures, product formulation control, prevention of cross-contamination, and certain aspects of employee and environmental hygiene.

Principle 3: Establish critical limits for preventive measures associated with each identified CCP. A *critical limit* is defined as a criterion that must be met for each preventive measure associated with a CCP. Each CCP will have one or more preventive measures that must be properly controlled to assure prevention, elimination, or reduction of hazards to acceptable levels. Each preventive measure has associated critical limits that serve as boundaries of safety for each CCP. Critical limits may be set for control measures such as temperature, time, physical dimensions, sealing conditions, humidity, moisture level, water activity (a_w), pH, titratable acidity, salt concentration, available chlorine, viscosity, presence or concentration of preservatives, and occasionally sensory information. Critical limits may be derived from sources such as regulatory standards and guidelines, literature surveys, experimental studies, and experts. The food industry is responsible for engaging competent authorities to validate that the critical limits will control the identified hazard.

An example of a critical limit for packaging can be seen from canning operations. The Low Acid Canned Food Regulations (Title 21, Code of Federal Regulations Part 113) require that cans should be examined for presence of visual defects, and cans should be periodically removed from the line and torn down to verify that can seam measurements are within specifications. If we accept the fact that proper seams on cans are necessary to protect consumers from harm, then a CCP would be identified at the sealing machine. The monitoring activities are the visual observation and the seam teardown and double-seam measurements. The seam measurement would be compared to the minimum acceptable dimensions (critical limits) specified by the container supplier. A processing example of a critical limit is the time and temperature for the retorting of canned salmon. The process should be designed to eliminate the heat-resistant spore-forming pathogen, *Clostridium botulinum,* which could reasonably be expected to be in the product. Technical development of the appropriate critical limit(s) requires accurate information on the operation of the retort, heating rate of the product, and heat resistance of *Clostridium botulinum.* A partial hazard analysis for canned salmon is shown in Table 3. The diagram illustrates how the first three principles of HACCP may be documented by the team.

Principle 4: Establish CCP monitoring requirements. Establish procedures for using the results of monitoring to adjust the process and maintain control. Monitoring is a planned sequence of observations or measurements to assess whether a CCP is un-

Table 3. Partial Hazard Analysis for Canned Salmon

CCP	Hazard	Prevention	Limits
Empty containers, three-piece only	B[a]	Acceptable seams	Meets specifications
Filled containers	B[a]	Acceptable seams	Meets can specifications
Processing—processes designed to provide commercial sterility	B[a]	Vent time and temperature, or other parameters	Meets vent specifications as per process authority
	B[a]	Process temperature	Specifications by process authority
	B[a]	Process time	Specifications by process authority
	B[a]	Initial temperature	Specifications by process authority
	B[a]	Other critical factors as specified by processing authority	Specifications by process authority
Cooling—applies to water-cooled cans only	B[a]	Chlorination	Measurable residual at discharge

[a] Biological hazard.

Table 4. HACCP Master Sheet

Critical Control Point (CCP)	Hazard	Critical Limits of the Preventive Measures	Monitoring				Corrective Action	Records	Verification
			What	How	Frequency	Who			

Source: Copyright Food Processors Institute.

der control and to produce an accurate record for future use in verification. Monitoring serves three main purposes:

1. Monitoring is essential to food safety management in that it tracks the system's operation. If monitoring indicates that there is a trend toward loss of control, ie, exceeding a target level, then action can be taken to bring the process back into control before a deviation occurs.
2. Monitoring is used to determine when there is loss of control and a deviation occurs at a CCP, ie, exceeding the critical limit. Corrective action then must be taken.
3. Monitoring provides written documentation for use in verification of the HACCP plan.

Examples of measurements for monitoring include

- Visual observations
- Temperature
- Time
- pH
- Moisture level
- Can teardown and seam dimensions

Principle 5: Establish corrective action to be taken when monitoring indicates that there is a deviation from an established critical limit. The HACCP system for food safety management is designed to identify potential health hazards and to establish strategies to prevent their occurrence. However, ideal circumstances do not always prevail, and deviations from established processes may occur. For instance, where there is a deviation from established critical limits, corrective-action plans must be in place to (*1*) determine the disposition of noncompliance product, (*2*) fix or correct the cause of noncompliance to assure that the CCP is brought under control, and (*3*) maintain records of the corrective actions that have been taken. Because of the variations in CCPs for different foods and the diversity of possible deviations, specific corrective-action plans must be developed for each CCP. The actions must demonstrate the CCP has been brought under control. Individuals who have a thorough understanding of the process, product, and HACCP plan are to be assigned responsibility for taking corrective action. Corrective-action procedures must be documented in the HACCP plan.

Principle 6: Establish effective recordkeeping procedures that document the HACCP system. The approved HACCP plan and associated records must be on file at the food establishment. Generally, the records utilized in the total HACCP system will include the following:

1. The HACCP plan
   a. Listing of the HACCP team and assigned responsibilities.
   b. Description of the product and its intended use.
   c. Flow diagram for the manufacturing process indicating CCPs (see example in Table 3).
   d. Hazards associated with each CCP and preventive measures.
   e. Critical limits.
   f. Monitoring system.
   g. Corrective-action plans for deviations from critical limits.
   h. Recordkeeping procedures.
   i. Procedures for verification of the HACCP system.

   In addition to listing the HACCP team, product description and uses, and providing a flow diagram, other information in the HACCP plan can be tabulated as in Table 4.

2. Records obtained during the operation of the plan, especially those records of monitoring and verification activities.

Principle 7: Establish procedures for verification that the HACCP system is working correctly. The National Academy of Sciences (1) pointed out that the major infusion of science in a HACCP system centers on proper identification of the hazards, critical control points, critical limits, and instituting proper verification procedures. These processes should take place during the development of the HACCP plan. There are four processes involved in verification. The first is the scientific or technical process to verify that critical limits at CCPs are satisfactory. This is sometimes referred to as *validation* of the HACCP plan. The second process of verification ensures that the facility's HACCP plan is functioning effectively. The third process consists of documented periodic validations, independent of audits or other verification procedures, that must be performed to ensure the accuracy of the HACCP plan. The fourth process of verification deals with the government's regulatory responsibility and actions to ensure that the establishment's HACCP system is functioning satisfactorily.

Summary

The HACCP concept is intended to provide a systematic, structured approach to assuring the safety of food products. However, there is no universal formula for putting together the specific details of a HACCP plan. The plan must be dynamic, allowing for modifications to production, substitution of new materials and ingredients, and development of new products. The plan is participatory at all levels of management, both in formulating and managing the plan. The strength in a HACCP program is in providing a system which a company can use effectively to organize and manage the safety of the products which are produced.

BIBLIOGRAPHY

1. *An Evaluation of the Role of Microbiological Criteria for Foods and Food Ingredients.* National Academy of Sciences, National Academy Press, Washington, DC, 1985.

General References

DHEW, *Proceedings of the 1971 National Conference on Food Protection,* U.S. Dept. Health, Education, and Welfare, Public Health Service, Washington, DC, 1971

FDA, *Thermally Processed Low-Acid Foods Packaged in Hermetically Sealed Containers,* Title 21, Code of Federal Regulations, U.S. Government Printing Office, Washington, DC, 1992.

NACMCF, *HACCP Principles for Food Production,* USDA, FSIS, Washington, DC, 1989.

NACMCF, "Hazard Analysis and Critical Control Point System," *Internatl. J. Food Microbiol.* **16,** 1 (1992).

Merle D. Pierson and Donald A. Corlett, Jr., eds., *HACCP, Principles and Applications,* Van Nostrand Reinhold, New York, 1992.

Pillsbury Company, *Food Safety through the Hazard Analysis Critical Control Point System,* Contract No. FDA 72-59, Research and Development Department, The Pillsbury Company, Minneapolis, MN, 1973.

Kenneth E. Stevenson and Dane T. Bernard, eds., *HACCP. Establishing Hazard Analysis Critical Control Point Programs, a Workshop Manual,* The Food Processors Institute, Washington, DC, 1995.

BARBARA BLAKISTONE
DANE BERNARD
National Food Processors Association
Washington, D.C.

HEALTHCARE PACKAGING.

See MEDICAL PACKAGING.

HEAT TRANSFER LABELS.

See DECORATING; LABELS AND LABELING MACHINERY.

HEAVY-DUTY PLASTIC BAGS.

See BAGS, HEAVY-DUTY PLASTIC.

HOLOGRAPHIC PACKAGING

Holographic Packaging

Holograms have been commercially available for about 15 years. The unique optical properties of a hologram, parallax (the ability to see "around" or "behind" the image) and depth, were fascinating and attracted much attention. Many of the earliest uses were as novelty stickers where price is not a major consideration. Holograms were also used for promotional purposes such as the 1983, 1984, and 1988 covers of the *National Geographic* magazine. The first significant value-added use was in security and authenticating applications, eg, credit cards (circa 1983). Reasons for this were that holograms were difficult to counterfeit or simulate (could not be replicated by photography, xerography, or printing), and the high cost of production limited their use to a relatively small size and high-value products. Early packaging applications included pressure-sensitive labels and hot-stamped images where a relatively small and expensive hologram could be applied to a much larger package or carton, such as Ralston "Ghost Busters" and Kellogg "Fruit Loops" cereal boxes. The first major all-hologram folding carton was produced in 1987 for Bacardi in which the entire carton was overlaminated with predominantly holographic film, then printed. Use of holography in the packaging industry is still in its infancy, but is beginning to grow rapidly (see Fig. 1). The worldwide holographic market in 1995 was estimated to be between \$250–\$300 million (hologram only, not converted value), of which an estimated \$40–\$50 million is in packaging (1).

Holograms are of two basic structures: embossed and volume. Volume holograms, under proper lighting conditions, are far superior in appearance and efficiency to embossed holograms and are the medium of choice for art holography and scientific applications. The cost to produce them, however, has precluded their use in mass marketing and packaging other than as labels. Volume holograms are created in a manner similar to a photograph in that each hologram is exposed on a photosensitive medium, then developed. The holographic information is recorded within the photosensitive material. By packaging standards, the substrate on which the image is exposed is very thick and expensive, the process is relatively slow, and, at this time, is produced only on rather narrow webs.

Embossed holograms, on the other hand, once mastered, can be replicated through various types of embossing processes. This is because the holographic information is recorded on the *surface* of the photosensitive mastering material. This surface can then be replicated via nickel electroforming. A number of proprietary techniques are employed to actually accomplish the embossing process; however, in the simplest terms it can be viewed as similar to a web printing operation in that a continuous web of film (polypropylene, polyester, PVC, etc) passes between nip rollers, on which the embossing plate is mounted. The holographic information, sub-micron in size, is imparted into the film or coating on the film, with heat, pressure, and/or radiation—either ultraviolet or electron beam. Unlike printing, however, no ink is employed; the colors seen in a hologram are derived from diffraction of the light from the surface. "Diffractive" patterns are similar to holograms in that they diffract light into a rainbow of colors, are embossed, and in some cases, are originated by much the same process as holograms. Diffractive patterns are often used as backgrounds for graphics and are generally considered interchangeable with holograms, at least in packaging applications. They are seldom three-dimensional, nor are they image-specific as are most holograms.

A detailed discussion of the holographic origination process is beyond the scope of this article; however, it is beneficial to understand the different types of holograms and the characteristics of each. All of these variations are "white light"-viewable, meaning that they can be viewed under traditional light sources as opposed to narrowband coherent laser or light and are "rainbow" holograms in that they diffract light into a spectrum of colors. Traditional 3D holography is probably the

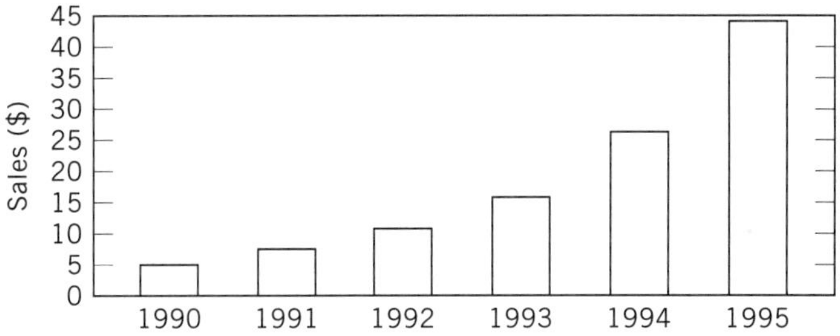

Figure 1. Holographic packaging growth.

one most people think of in connection with holography. This process employs an actual-size model to create the hologram, which exhibits parallax and depth (commonly referred to as 3D). A two-dimensional (2D) hologram can be produced by using a mask instead of a model; the image can be at the "surface," behind it, or in front of it. A 2D multiplane hologram (sometimes called 2D/3D) simply uses two or more 2D images of differing depths to create parallax and the illusion of three dimensions. Stereograms are made from a series of computer generated views or photographs taken from 180 or more closely spaced angles around the subject. Animation can be achieved as well as true color at a specific angle. A relatively new holographic technique known as "dot matrix" employs a computer-generated array of diffractive pixels (picture elements) that combine to form a very bright pattern or image. There are numerous variations in names and techniques for these last two processes. Similar in brightness to dot-matrix holograms are ruled gratings, where a microscopically fine diamond is used to etch a diffractive pattern. Likewise, very bright 2D surface gratings can be achieved by using photomasks. The latter method has an advantage in that it can be done in full color. Deep 3D holograms and stereograms, because of the depth generally employed by them, have the most critical lighting requirements for proper viewing. They require a point light source and would look blurred or even indistinguishable in the uniform fluorescent lighting typical in most supermarkets. Shallow 3D, 2D, and 2D multiplane holograms avoid this problem as long as the image is not too far from the "surface." Dot-matrix and diffractive gratings are viewable and eye-catching under uniform illumination found in most stores and supermarkets.

The hologram industry in general got off to a rocky start in the 1980s because it was a patent minefield. The rights to several very strong patents were licensed by a single company. These patent rights were the subject of legal challenges for many years; however, the net result is that the industry as a whole suffered because most large corporations chose to steer clear of the entire industry for fear of legal entanglement. As a result, the hologram industry was composed of numerous relatively small participants with "niche" markets. Expiration of these powerful patents and settlement of several significant lawsuits have opened the field to more competition, which will benefit the packaging industry as well as others who employ holography.

Several processes can be used to incorporate holograms into packaging: flexible, folding carton, or bottles. Each has advantages and disadvantages that must be considered for the specific application. These methods include pressure-sensitive-, shrink-, or glue-applied labels; hot stamping; web transfer; and lamination. Some methods lend themselves naturally to specific types of packaging; for example, returnable beverage bottles must utilize a label construction that conforms to industry standards for removability.

Labels. Labels, either PSA (pressure-sensitive-adhesive), glue-applied, shrink, or in-mold, have always been the medium of choice for bottles or cans. Holographic labels can be produced on paper or film substrates suitable for pressure-sensitive and glue-application and have been the most widely used method to incorporate holograms in packaging. There have been a large number of successful holographic label applications in both the United States and Europe on beverages, petroleum products, cereals, confectioneries, frozen foods, and toys, to name a few. In the case of bottle labels, no ancillary equipment is needed, and the addition of a hologram has no adverse effect on application cost or throughput. On a folding carton, the addition of a label requires application machinery and can significantly affect productivity.

Hot stamping. Holographic hot-stamp foil has been available since the early 1980s. It is widely used as a security feature on credit cards and documents and has often been used in the publishing industry. Use of holographic hot stamping in packaging is less common, primarily because of application limitations of hot stamping (inability to incorporate the hot-stamping process in tandem with other converting operations), substrate surface requirements (smoothness, adhesion), cost, production rates, and manufacturing logistics. Holographic hot-stamp foils are readily available around the world and, like labels, have been traditionally produced on narrow web embossers. Unit area cost is relatively high (see also Decorating, hot stamping.)

Foil transfer. Foil transfer as a concept has been around for many years but has only recently found commercial success in packaging. It can take several forms. One is similar to a hot-stamp foil, in which a releasable embossable coating is applied to a polyester carrier. The holographic image is embossed into this coating either before or after vacuum metallizing. A transfer adhesive is then applied to the metal surface. In use, the film is hot-nipped to the substrate (board, paper, etc.), then the carrier is peeled away, leaving the holographically embossed coating behind, ready for printing. In the Aluglas process, the hologram is embossed directly into the carrier, usually a low-surface-energy polypropylene. This surface is then vacuum-aluminized and coated with a transfer adhesive. After the transfer adhesive is nipped to the substrate, the embossed carrier is peeled away, leaving the embossed aluminum behind. This exposed surface is then coated with a protective printable top coat. In the Aluglas process, the embossed carrier can be reused a number of times. Both processes have the advantage of easy recyclability, since the finished structure does not contain a plastic film.

Lamination. This is the process that has seen the greatest growth in packaging. The holographic image is embossed into a coated paper or film substrate such as polypropylene or polyester. The hologram side is then vacuum-metallized. In the case of flexible packaging, the holographic film is laminated to a reverse-printed clear film, making the hologram an integral part of the package. In folding cartons, the holographic film is laminated to the board, hologram side down, then surface-printed. If paper is used, it is laminated hologram side up. The key to the growth in laminated holograms is the relatively recent availability of high-quality, wide-web product at prices acceptable to the end user.

Holographic packaging in 1996 appears to be on the verge of explosive growth. A number of limitations in the use of holograms have been overcome, opening the door to much wider use. Some of these limitations have been quality and consistency of the image as well as the substrate material, actual embossing capacity, web width, and cost. Needless to say, some opportunities have arisen in spite of these limitations where either the job was small or cost was not a major consideration. However, to make significant impact in the industry, all of these issues must be reconciled.

In the case of flexible packaging, the package must perform a number of functions. It identifies the product, the manufacturer, ingredients, and nutritional data; provides barrier against light, moisture, oxygen, and microorganisms; and contains the ingredients through the physical properties of the film and sealing layers. When considering a hologram on a flexible package, it becomes apparent that the hologram substrate will itself become part of the package, and consequently should possess properties needed for the particular application. Polypropylene is widely used in flexible packaging and is often the substrate of choice for the hologram. For folding cartons and other board packaging, the hologram and its carrier are usually ancillary to the package itself; hence a wider variety of substrates can be employed such as polyester, paper, PVC, or foil-transfer holograms. The ability to supply large-diameter rolls with good formation is crucial to most packaging converters. This is true in spite of the fact that the hologram material will have undergone three operations (embossing, metallizing, and rewinding) before converting. Sixty thousand foot rolls are not uncommon.

Holographic packaging is still significantly more costly than conventional printing, but is finding increasing application where the customer recognizes the value of on-package advertising, eye appeal, and impulse buying. A 1995 Survey by the Point-of-purchase Advertising Institute now pegs in-store impulse decision as being responsible for 70% of all branded product purchases in supermarkets (2). With the wide variety of similar products displayed on supermarket shelves, the decision on which product to select is often made at the last second. Holography offers a new dimension to package designers in vying for that critical last-second decision. It has also been demonstrated that holographic packaging can be more a cost-effective promotional tool than discounting price. Use of a hologram in flexible packaging or folding-carton packaging, in which the hologram is one of several component layers, can be expected to approximately double the cost of the package. In applications where the hologram is the major component, the cost increase may be significantly higher (see also Point-of-purchase packaging).

In addition to the obvious issues of quality and cost, a more subtle consideration is manufacturing capacity. It is not uncommon for a single flexible-packaging order to be several hundred million square inches (a recent holographic soft-drink promotion was well over 8 *billion* square inches). The ability of the hologram manufacturer to supply large quantities of wide-web embossed film in an acceptable time frame is crucial to winning may orders. Widths of 40–60 in. are common in flexible-packaging and folding-carton industries. Many hologram manufacturers of the late 1980s and early 1990s were restricted to one meter or less. This was mostly due to the difficulty in embossing a quality image across a wide web. Today, there is a growing number of manufacturers with wide-web capability.

Successful use of holograms in packaging requires close teamwork between all participants; holographer, holographic manufacturer, package converter, the customer's art, marketing, purchasing, and production departments and design or advertising agencies. Incorporating holograms or diffractive patterns into a package design can be relatively simple in the case of random backgrounds. The major consideration will be the pattern choice, amount of ink coverage, opacity of inks, and, of course, the physical properties of the hologram substrate as it relates to the overall package. As the holographic pattern becomes more custom, complexity increases. Consideration must be given to image orientation, registration, type of holographic mastering, and lighting conditions under which the product will be viewed, as well as the printed graphics. One of the more alluring aspects of holography is the three-dimensionality, depth, and movement available in the image. These characteristics require a point light source to be most visible. Unfortunately, supermarkets and stores where most of the flexible- and folding-carton packaging is displayed, are fluorescent-lighted, causing these "deep" images to appear blurred and unimpressive. Viewing angle is also important. When holograms are mastered, there is a specific relationship between the angle of the light source, the plane of the hologram, and the viewer. An image that is brilliant at one angle is totally "dead" at most others. The use of dot-matrix mastering permits the generation of images that diffract light at many angles, which in many cases is preferred. The tradeoff, however, is that the image is on one plane and lacks the depth of traditional holograms. New techniques continue to be developed, allowing package designers more options in the use of holography. Because of the higher cost, holography is often used in special promotions. One of the largest holographic packaging promotions to date was undertaken by Pepsi Cola in November 1995, when 10 million, 24-can "cubes" (see Fig. 2) of Pepsi Cola and Mountain Dew were released nationally. This single promotion consumed more than 350,000 lbs of embossed holographic film, and 2200 tons of paperboard. A promotion of this size should not be undertaken without thorough planning. In this case, every step in the process, including embossing, metallizing, lamination, printing, die cutting, gluing, filling, and distribution, was tested with several trial runs before the final order was placed.

Figure 2. Hologram "cube."

Use of holography as primary labeling is still relatively uncommon, because of cost and also because of limited sources. There are, however several examples such as Molson's "Ice Beer," Brock and Brach's "Power Rangers," "Batman," and "Street Sharks" candies, SmithKline Beecham's "Aquafresh Whitening" toothpaste, and Lifesaver's "Icebreaker" gum. Look for more holographic primary labeling in the future.

The Future

Holography, as a packaging element, will find increasing application as cost is reduced and integration with the converting operations becomes more transparent. To this end, holographic manufacturers will form strategic alliances or partnerships with major packaging converters, and more converters will invest in captive hologram manufacturing. Registration of the holographic image to the printing will improve. Were it not for the need to metallize the holographic image before printing or lamination, the embossing and printing operations could be performed in tandem, reducing cost, and lead time, while improving registration and efficiency. Holography, because of—and in spite of—its unique properties, has the potential to become the next printing process. For more information on holography and sources of holographic materials, the following directories may be of help:

Holography Marketplace, 5th ed., Ross Books, Berkeley, CA 97904.

Holo-pack-Holo-print Guidebook, Reconnaissance Holographics Ltd., Egham, Surrey, TW20 9BD, England.

BIBLIOGRAPHY

1. Sixth International Conference on Holo-pack/Holo-print 95, Reconnaissance Holographics, Nov. 6–8, 1995.
2. Point-of-Purchase Advertising Institute, "1995 Consumer Buying Habits Study," conducted by Myers Research Institute Center of New York.

DONALD W. MALLIK
Virtual Image, a Printpack Company
Roswell, Georgia

HOT-FILL TECHNOLOGY

Food processors use a variety of processing, formulation, and storage methods to assure that food products do not spoil, due to microbiological activity, throughout their shelf life. Products such as cereals, breads, and confectionaries are formulated with sugars and/or salts to lower water activity (a_w). Water activity below ~0.90 will not support the growth of most food-spoilage bacteria (1). Chemical preservatives, such as benzoate or sorbate, are added to many products to inhibit microbiological growth. Many dairy products, fruit juices, fruits, vegetables, meat, fish, and prepared meals are distributed under refrigeration or frozen conditions. Shelf-stable, low-acid foods, such as vegetables, meat, and fish, with a pH ≥4.6, rely on a thermal process. Thermal processing of low-acid foods is achieved in a retort at pressures of 10–20 psi (7–14 kg/m^2) and temperatures of 240–260°F (116–127°C) (2,3). Food products with a pH ≥ 4.6 will support bacteria, yeast, and mold growth. Commercial sterility is defined as the elimination of all microorganisms that can grow and metabolize in a product. To achieve commercial sterility in foods with a pH ≥ 4.6, bacteria must be destroyed as well as yeast and mold. The destruction of many bacteria requires temperatures in excess of 212°F (100°C). To achieve temperatures above the boiling point, overpressurization in a retort process is necessary. The high temperature achieved not only destroys microorganisms in the food but also sterilizes the package. Packages historically used for retorted low-acid foods include metal cans, glass jars, and, to a lesser extent, retortable pouches.

High-acid foods with a pH <4.6 support the growth of yeast and mold but, unlike low-acid foods, only a limited number of bacteria. Yeast, mold, and the types of bacteria that can metabolize in high-acid foods are more temperature-sensitive than those that grow in low-acid foods. Temperatures in the range of 170–200°F (77–90°C) are usually adequate to destroy these microorganisms. Commercial sterility of high-acid food products and the sterilization of the containers, in which they are packed, is achieved with temperatures of <212°F (100°C). Therefore, retorting with overpressure is not required. High-acid foods are thermally processed in one of three different ways: hot filling, postfill pasteurization, or pasteurization with aseptic filling. This aritcle focuses on hot filling.

Hot-Fill Processing

Foods such as fruit juices, fruit sauces, jams, jellies, tomato sauces, ketchups, and barbeque sauces are often hot-filled. The hot-fill process consists of heating the food product to an adequate temperature to destroy the yeast, mold, and limited types of bacteria that can grow in the food, and then holding the food at that temperature for an adequate time to destroy these microorganisms. The temperatures required range from 170 to 212°F (77–100°C), and the hold times vary from 30–60 s depending on the food product. The hot product is then filled directly into the container or package. The headspace in the package may then be nitrogen- or steam-flushed to reduce headspace oxygen. The package is then hermetically sealed and inverted to sterilize the lid or closure. The hot product sterilizes the package. To accomplish package sterilization, the product must be at or above the minimum temperatures [170–200°F (77–93°C)] for an adequate time (usually 1–3 min) (4). Thus, the control points for a hot-fill process are the *minimum fill temperature* and the *minimum hold time* (see Fig. 1). Depending on the heat sensitivity of the specific food, the products may or may not be quickly cooled after the minimum hold time is achieved. Cooling is usually accomplished in a water-spray tunnel or immersion bath. Fruit juices and preserves are examples of products requiring cooling directly after hot filling to protect product color, flavor, and nutrient content (vitamin C). Barbecue sauce is an example of a product that is not heat-sensitive and can be allowed to cool under ambient conditions after being packed in corrugated shipping containers.

Historically, high-acid products have been packaged primarily in metal cans and glass jars. However, in the 1980s and 1990s a number of rigid and flexible plastic packages were developed for hot filling.

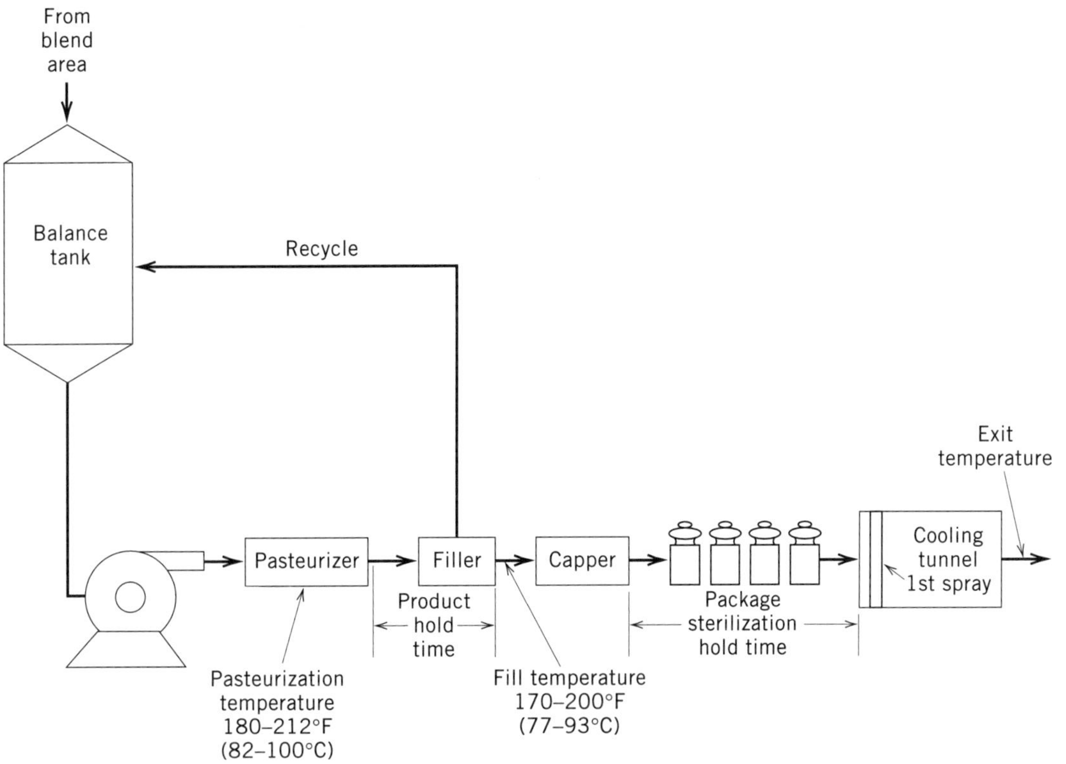

Figure 1. A typical hot-fill process for high-acid food or beverage products.

Tinplate Cans

The traditional metal cans used for hot-filled foods are made of tinplate. Some are enameled inside to protect the food from the migration of iron or tin from the can into the food, which may cause the loss of color or off-taste. Other tinplate cans are not enameled inside; these are referred to as plain tinplate cans. The lack of an inside enamel allows the tin to migrate into the product. The color of some canned products such as white grapefruit juice is preserved by tin salts. Tinplate cans are also enameled to protect the can itself from the corrosive effects of some high-acid products. High-acid products, such as tomatoes and cranberries, can corrode tinplate cans, resulting in perforations and leaking cans. The hot filling of tinplate cans consists of hot-water washing the cans to remove dust or other foreign contamination, hot filling the cans with the high-acid food, flushing the headspace with steam or nitrogen, double-steaming a metal lid onto the can, inverting the can to sterilize the lid, and conveying the hot can for an adequate distance to provide the necessary hold time prior to entering a water-spray cooler. The cans are cooled to approximately 90–110°F (32–43°C). The maximum cooler exit temperature is specified according to the product's heat sensitivity. The minimum cooler exit temperature is established to assure that the cans are not too cool to prevent the evaporation of remaining water beads. If the can is not warm, rust will occur, staining the label and distracting from the general appearance of the can. If the can is too cool, labeling may be difficult as the hot melt used in labeling will set up too quickly, causing misapplied labels. A vacuum is generated in the can as it cools. The vacuum at the exit of the cooler ranges from approximately 16 to 25 in. of mercury (41–64 cm Hg) depending on the headspace, product density, and size of the can. The vacuum is formed by a combination of the condensation of the steam in the headspace and the contraction of the product as it cools. The vacuum is important as it provides integrity to the can by pulling the ends inward and not allowing them to flex during distribution. The concavity of the end caused by the vacuum also provides visual evidence to the consumer that the can is properly sealed.

Lightweight Aluminum Cans

In the 1980s a new hot-filling process was developed that allowed the hot filling of lightweight aluminum cans. This technology is now used extensively for hot-filled fruit juices and juice drinks in $11\frac{1}{2}$-fl-oz cans. Lightweight aluminum cans could not be hot filled with traditional processes because the aluminum cans would panel as a result of the vacuum created when the product cooled. The new hot-fill process includes dosing the filled cans with liquid nitrogen immediately prior to applying and double-seaming the can end. The cans consist of a drawn and ironed aluminum body with an E-Z open aluminum end. Directly following the dosing of the liquid nitrogen and the application of the end, the nitrogen vaporizes and creates internal pressure. In the case of hot fruit juices, this nitrogen dosing creates an initial pressure of approximately 45–50 psi (31.6–35.1 kg/m^2) prior to cooling and 25–35 psi (17.6–24.6 kg/m^2) after cooling. Thus, the nitrogen overpressure more than compensates for the vacuum created by the product as it cools, and the final package is overpressured to compensate for its lightweight construction. The hot-fill nitrogen dosing process for juice and juice drinks consists of cleaning the cans with hot water or air, hot filling with juice, leaving a headspace of approximately 0.3–0.4 in. (7–10 mm), dosing liquid nitrogen, applying the can end, double-seaming the can end to the can body, and finally cooling the product to approximately 90–110°F (32–43°C). This hot-fill process allows the use of lightweight aluminum cans for fruit juice and offers the potential for hot filling a variety of other food products. (See also Cans, aluminum.)

Glass Containers

Hot-filling technology has been used for many years for not only tinplate cans but also for vacuum-packed glass jars and bottles. The hot-fill process used for glass packages is very similar to that used for tinplate cans. One main difference is that glass packages must be preheated prior to hot filling to prevent breakage due to thermal shock. This problem is particularly troublesome in the winter, when glass is either stored in cold warehouses or filled directly off delivery trucks. It is generally believed that thermal-shock breakage is not a problem if the glass temperature after preheating is no more than 75°F (24°C) less than the hot product. Thus, if the product fill temperature is 185°F (85°C), the glass must be preheated to at least 100°F (43°C) before filling. The preheating process is usually accomplished with hot water and also serves the function of cleaning the jars or bottles. Cleaning is particularly important with glass as glass is often received in corrugated shipping and thus may be contaminated with carton dust. After cleaning and preheating, glass containers are filled with hot product, ranging from approximately 170 to 200°F (77–93°C). A small headspace is maintained. The headspace is usually flushed with steam in the capper. Metal twist-on lug-style caps are usually used. These caps contain a gasket compound used to seal the jar or bottle. The caps are preheated in the cap shoot to soften the compound. A small initial vacuum is generated directly following the capper by the condensation of the steam in the headspace. This initial vacuum varies depending on the size of the headspace and the amount of steam in the headspace. An initial vacuum of at least 4–5 in. Hg (10.2–12.7 cm of Hg) is necessary prior to cooling to assure that the seal between the glass jar or bottle and the cap compound is secure prior to generating the larger vacuum created by product contraction as it cools. Conveyors are designed to allow for at least a 60-s hold after capping prior to entering the cooler. This provides time for the hot product to sterilize the jar or bottle. The first zone of the cooler uses a preheated water spray to prevent glass breakage due to thermal shock. The 75°F (24°C) temperature differential, important in the preheating operation, is also critical in cooling. The water-spray temperature should be no more than 75°F (24°C) less than the product temperature. Often the cooler is designed such that cold water enters the discharge end of the cooler and is transported to the infeed end. Thus the water is warmed as it cools the product throughout the cooler, which prevents thermal-shock glass breakage while minimizing energy cost. The product temperature at the end of the cooler is usually 90–110°F (32–43°C). The final vacuum depends on the viscosity characteristics of the product and the size of the headspace, but usually is 16–22 in. Hg (41–56 cm Hg).

As mentioned previously, barbecue sauce is not cooled after filling. Products such as this are packed in corrugated shipping containers, palletized, and cooled over several days in normal warehouse storage. As containers in the center of the pallet are insulated by the product stacked on the outside and top and bottom, they may cool at a much slower rate than product on the outside. To reduce the variation in cooling rates, pallet patterns with ventilation chimneys are used to facilitate heat removal from the center of the pallet.

As hot-filled glass jars and bottles contain a vacuum, they must be handled carefully after cooling. Glass containers, particularly those filled with highly viscose products, break as a result of a phenomenon commonly known as "water hammer" breakage, which occurs when a glass jar or bottle, containing a vacuum, is dropped or otherwise impacted. On impact, a vacuum bubble forms and collapses quickly. This results in the creation of a hydraulic action that concentrates forces on any defect in the container causing breakage (5).

Care must be taken to properly adjust drop case packers to prevent water-hammer breakage. Corrugated shipping containers are designed to cushion the bottom of glass jars and bottles to minimize the impact of dropping. Partitions are used to minimize glass-to-glass impact. See also Glass container design.

Plastic Packages

Historically, the hot-fill process has required packages that can withstand the hot-fill temperatures and are sufficiently rigid to withstand the vacuum developed after cooling without paneling or otherwise distorting. Until recently, these packaging material requirements have limited the use of materials for hot-fill packages to metal and glass. During the 1980s packaging material and hot-fill technology developments occurred that allowed the use of not only lightweight aluminum cans, as already discussed, but also plastics (6).

A number of plastic package technologies were developed in the 1980s that resulted in the commercialization of several hot-fill plastic packages that replaced traditional glass or tinplate packages. Those included plastic squeeze bottles with dispensing closures for ketchup and barbeque sauce, plastic table-ready bowls with peelable membrane seals for fruit sauces, plastic cans with double-seamed metal ends for a variety of products, and clear plastic bottles for fruit juices, juice drinks, and isotonic sport drinks (see also Blow molding).

Poly(ethylene terephthalate). The late 1970s and most of the 1980s was an era of rigid plastic food-packaging development. Nearly all the major U.S. packaging companies were actively developing plastic packaging technologies. Several Japanese and several European packaging companies were also focusing their research and development activities on rigid plastic package developments for food and beverages. In the mid- to late 1970s, polyethylene terephthalate (PET) was commercialized for carbonated soft-drink bottles. The initial application was the 2-L carbonated soft-drink bottle. Carbonated soft drinks contain preservatives and thus are cold-filled. They require a high gas-barrier to retain CO_2 but do not require the high heat resistance or physical strength necessary for hot filling. PET, as used in soft-drink bottles, softens at temperatures too low for hot filling or retorting. However, the success in the market of PET soft-drink bottles proved the consumer acceptance of plastic, and it confirmed the business potential that supported other major plastic food and beverage R&D programs.

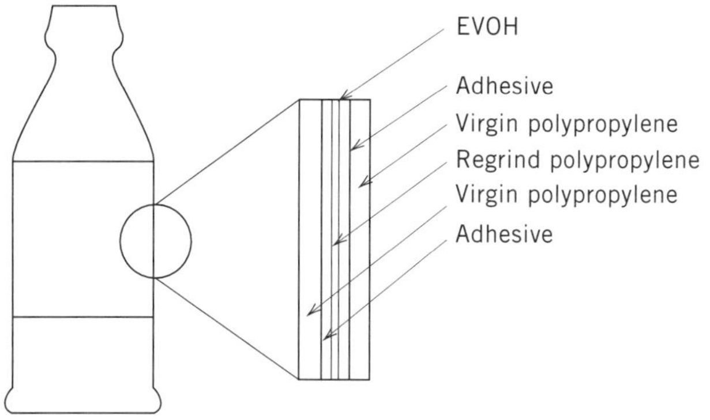

Figure 2. A typical hot-fillable polypropylene/EVOH/polypropylene bottle.

Acrylonitrile. At the same time as PET was being developed for cold fill, another plastic resin—acrylonitrile (AN)—was under development for food and beverage packaging. While AN, like PET, was clear and provided an adequate gas barrier for carbonated soft drinks and oxygen-sensitive foods and beverages, it had a significantly higher melting point and thus could be hot-filled and potentially even retorted. In the 1970s AN was test-marketed for carbonated soft drinks. In addition, several hot-fill containers were developed and test-marketed. A squeeze bottle with a dispensing closure and an induction aluminum-foil seal was test-marketed with barbeque sauce. At least two fruit-juice processors introduced into test markets apple- and cranberry-juice products in $5\frac{1}{2}$-fl-oz. (162.6-mL) AN cans with seamed-on metal ends. A number of other developments were under way. In 1977 the FDA banned the use of nitrile polymers in direct contact with beverages because of the potential migration of AN monomer from AN packages into beverages and the possibility that it could be a carcinogen. AN was reapproved in 1984 with limitations (8).

Polypropylene. Polypropylene containers were also being developed for food and beverage applications in this period. Polypropylene can be thermoformed or extrusion blow-molded and thus made into bowls, tubs, and bottles. It has a high melting point and thus is well suited for hot-fill applications. Polypropylene is not crystal-clear, like PET and AN, but it is semiclear (contact-clear). The oxygen-barrier properties of the polypropylene are relatively poor. This limited polypropylene's application to foods and beverages, until coextrusion technology was developed in the 1980s to allow the addition of small amounts of very high-barrier materials to polypropylene packages. Coextrusion technology resulted in the production of multilayer extrusion blow-molded bottles and thermoformed containers containing five to seven layers. Figure 2 shows a typical six-layer hot-fillable, high-oxygen-barrier polypropylene container. The inner and outer layers consist of virgin polypropylene. Ethylene–vinyl alcohol (EVOH) is used in the middle layer to provide a high-oxygen barrier. Polypropylene regrind is incorporated to use the scrape generated in the extrusion blow-molding process. Two adhesive tie layers are required for proper bonding of the EVOH to the polypropylene. Containers made of this material are hot-fillable, and abuse-resistant and provide a high-oxygen barrier and contact clarity. In the 1980s a number of food processors modified traditional hot-fill processes to allow conversion to polypropylene/EVOH containers from glass and tinplate. Examples include 4-oz apple-sauce cups, 12–16 oz cranberry-sauce containers, ketchup and barbecue sauce bottles, and some fruit-juice containers. Most of these hot-fill packages were sealed with heat-sealed or induction-sealed aluminum-foil laminates.

Heat-set PET. Polypropylene/EVOH co-extrusions provide an excellent hot-fill food and beverage packaging material with the possible shortcoming, for some products, of a lack of glasslike clarity. While PET provides the clarity of glass, it was not hot-fillable until advances in injection blow-molding technology were made in the mid-1980s. Heat-setting technologies were pioneered, primarily by the Japanese, in the early to mid-1980s. Further advances were made shortly thereafter by U.S. PET bottle manufacturers in conjunction with global injection and blow-molding equipment suppliers. The first commercial application of hot-fillable PET bottles, technologically possible because of the heat-setting process, were for fruit juice and juice drinks. In September 1985, (64-fl-oz.) (1.89-L) heat-set PET bottles were commercially introduced with a line of cranberry-juice drink products using the Japanese technology. Shortly thereafter, U.S. bottle manufacturers commercialized alternate thermally stable technologies for PET, and the commercial use spread beyond fruit juices and juice drinks to isotonics. Ketchup also converted from contact-clear polypropylene/EVOH bottles to hot-fillable PET/EVOH.

Heat-set, or otherwise thermally stable, PET provides an adequate oxygen barrier for many products, excellent abuse resistance, good recyclability characteristics, hot fillability up to 180–185°F (82–85°C), and the clarity of glass. By the mid 1990s, hot-fillable PET became a prevalent and popular packaging material for several categories of hot-filled foods and beverages (eg, fruit juices, juice drinks, isotonics, and ketchup) which traditionally were packed in glass.

Summary

In summary, hot-fill processing and packaging is a basic traditional method of preserving high-acid foods and beverages. It has been practiced since the early days of commercial food processing and before that in home canning. It is one of the food-processing technologies that has allowed nutritious foods to be available year-round in all parts of the nation. Historically, its application has been limited to rigid metal and glass containers.

However, technological development of new materials, new package manufacturing processes, and hot-fill methods in the 1980s have resulted in the commercialization of a variety of new hot-fill packages that provide the product protection benefits of metal and glass with additional consumer benefits.

BIBLIOGRAPHY

1. N. N. Potter, *Food Science,* 4th ed., AVI Publishing, Westport, CT, 1986, pp. 297–298.
2. *Ibid.,* pp. 172–189.
3. *Thermal Processes for Low-Acid Foods in Metal Containers,* Bulletin

26-L, 12th ed., National Food Processors Assoc., Washington, DC, 1982, pp. 22–56.
4. Ref. 1, p. 197.
5. J. F. Hanlon, *Handbook of Package Engineering,* 2nd ed., Technomic Publishing, Lancaster, PA, 1992, pp. 9-10, 9-11.
6. R. A. Bourque, "Hot Fill and Aseptic Packaging," *Food PLAS VI-89 Conf. Proc.,* Plastics Institute of America, Stevens Institute of Technology, Hoboken, NJ, 1989.
7. R. Juran, *Modern Plastics Encyclopedia,* Vol. 64, No. 10A, McGraw-Hill, New York, 1988, pp. 34.

General References

M. Bennion, *The Science of Food,* Harper and Row, New York, 1980.

O. R. Fennema, M. Karel, and D. B. Lund, *Principles of Food Science,* Part II, *Physical Principles of Food Preservation,* Marcel Dekker, New York, 1975.

Raymond A. Bourque
Ocean Spray Cranberries, Inc.
Lakeville, Massachusetts

INDIA PACKAGING

Background

The packaging industry in India is a heterogeneous mix of both organized and unorganized sectors, including a number of manufacturers of basic materials, package converters, machinery manufacturers, and ancillary units. While the basic material manufacturers are by and large in the organized sector, others invariably fall into the small-scale sector. Whereas the industry remained in the background for a long time, the recent years have witnessed a revolutionary change.

India is influenced by an agricultural, horticultural economy, and a high percentage of the population is still rural-oriented. The packaged-product market is thus spread in value in the urban sector but in quantity in the rural sector. The wealth of the packaging industry therefore is evenly distributed, and even a small shift in and penetration of value-added packaging among rural consumers could elicit a major—and favorable—change for the packaging industry. Such a shift could also influence industrialization and generate employment.

Industry Setup

The Indian packaging industry is mostly domestic, and comparatively less dependent on imports except for a few specific areas. Whereas the conversion industry progressed satisfactorily, development of the packaging-machinery industry was rather slow. Recent trends, however, are a clear index of the potential growth and prosperity of this industry.

Basic Materials

Most of the basic materials required by the Indian packaging industry are met through indigenous sources. Plastics resins and tinplate, however, are the two primary media imported in appreciable quantities to meet domestic needs and export needs in converted forms. Taking the normal industry growth rate, the requirements assessed would be as illustrated in Table 1.

Package Conversion

The Indian package-conversion industry uses both indigenous and imported machinery and technology. It is estimated that over 15,000 units are engaged in the business of package conversion and almost 95% of them are in the small-scale sector. The total output of converted material during 1992/93 was placed at 5.3 million tons.

The level of conversion technology adopted varies considerably. Often this is considered to be the reason for the indifferent growth and as a constraint for research and development. The other factor cited is the limited infrastructural facilities. The situation is expected to take a turn soon, particularly because of upgrades in indigenous technologies.

Table 1. Packaging Media: Demand/Consumption Pattern

	Demand/consumption Pattern (1000 tons)		
Packaging Medium	1991/92	1995/96[a]	1999/2000
Tinplate (containers)	465	466	637
Glass (Bottles)	900	850	1340
Paper and Boards	1350	1250	2160
Plastics (semirigids/ rigids) and films	465	405	952
Jute	1170	1300	1300
Corrugated board	584	700	1000
Flexible materials	155	200	420

[a] Others for 95/96 = 621.

Ancillary Materials

The packaging industry uses a variety of ancillary media, such as printing inks, adhesives, lacquers, varnishes, caps and closures, wads, cushioning, and reinforcements. The requirements are generally met from domestic sources, except for certain components and ingredients such as adhesives and coating media. Considering the current and anticipated developments, there is considerable scope for new-product introduction, diversification, and growth. The developments are envisaged to include aqueous and alcohol-based inks, hot melts, acrylic and other resistant coatings, solventless lamination media, UV inks, and the like.

Machinery for the Packaging Industry

The design, development, and production of packaging-industry machinery in India have noticeably progressed within the past decade or so. There are about 700 units of packaging machinery in operation, about 40 of which benefit from joint-venture expertise or technical knowhow and belong to the organized sector. Although the engineering industry as such has progressed at a good pace in India, the packaging machinery sector has not. This could be due to the uncertainty of demand levels. The growth of the packaging industry has now augmented interest in the machinery industry and has also attracted the involvement of a large number of overseas leaders.

Packaging Awareness

The implementation of various national programs has resulted in many positive benefits, primarily in the areas of health and hygiene, increased lifespan, higher literacy rates, satisfactory foodstocks, and increasing industrialization and exports. The personal purchasing power is estimated at U.S. $1250. The consumerism has in its wake raised the living standard and way of life. Convenience is a major sought-after factor. The majority of the general population realized the "wealth of health" and raised the demand for products in a packaged form. Awareness of packaging and realization of its importance are growing rapidly, as is reflected by the increasing number of department stores and supermarkets opening throughout the country, even in small towns. The fact that courses on packaging are now included in the curricula of many educational programs is yet another index of the grass-roots approach seen in the country toward creation of packaging awareness.

Concept Change

The growth of the packaging industry in India could be attributed to certain socioeconomic changes in the country. This has also led to the introduction of many newer packaging media and systems. There is a significant trend in the distribution of products from bulk to retail packaging and from loose (nonprepackaged) to prepackaged product sales. Even at retail selling vendor points, the products are dispersed in plastic bags. Consumers today avoid carrying their own bags because most retailers provide bags with purchase. The resultant feature is a high quantity of synthetic bags. Other significant changes noticed are shifts from conventional to newer packaging systems, with the use of more durable packaging materials; streamlining of package production and distribution methods; improving package design to meet the market demand for competitive, attractive, and aesthetically appealing packages; and producing packages modular to the modern concept of ULD and the containerization system of cargo movement. The emphasis has shifted to productivity-oriented systems, with packages that would be amenable to new environmental and ecosystem concerns. Improvements in storage and handling systems are clearly visible with the use of pallets and materials-handling equipment.

Growth Potential

The projected growth rate of consumption and demand for packaging in India is 10%. By the year 2000, the production levels of the following major packaging media are expected to increase: tinplate, by 125%; plastics, 200%; paper and board, 135%; flexible materials, 250%; and glass, 50%. Thus the total demand/consumption pattern projected for 2000 A.D. indicates that the total requirement of all packaging combined would be about 9.3 million tons, representing an increase of 170% from the present level of about 5.5 million tons.

However, current fiscal and industrial policies toward industrialization and globalization, various market shifts and trends, current thrusts in food-processing industries, adoption of new technologies, and exports, as well as consumer awareness and increasing purchasing powers, are expected to change the scenario further in favor of packaging, resulting in a considerably increased demand level. Some factors that could significantly increase the demand for packaging media are briefly

- Packaging of commodities made available to the general population through a public distribution system (PDS). Daily, staples such as wheat, rice, sugar, and edible oils are currently provided in loose (nonprepackaged) form, and this is a potential area for improvement. Considering optimum packaging media, the quantity for PDS needed by 1995 is estimated at 0.64 million tons and by 2000, 1.20 million tons.
- A key area presently addressed is processed food. In India, the current procesing level is less than 1%; even if a minimum of 5% of fruits and vegetables and 10% of cereals and grains are processed, estimated the quantities of packaging required is 3.64 and 7.28 million tons for 1995 and 2000, respectively.
- Adoption of modern technology for food processing and packaging, the expanding Indian real estate market (more modern, spacious houses), and the influx of foreign nationals and influences in India, and unanticipated growth of confectionery, dairy, and other food-industry sectors could increase packaging-media requirements by a further 2 million tons for 1995 and 4 million for 2000.

These new developments would take the total demand of packaging media to 21.86 million tons by 2000, in contrast to 9.3 million tons in the normal growth state.

New Issues

Analysis of the demand and potentials for packaging applications indeed reveals highly encouraging statistics, but at the same time would seem to raise a number of issues besides considerable inputs. The task of implementation is enormous and would have to include

- Construction of infrastructural facilities for package conversion and packaging operation, designed to facilitate all activities associated with the packaging industry and package movement
- Quality-control and quality-assurance programs
- Qualified and skilled workforce at all levels
- Adoption of ULD and containerized distribution systems
- Implementation of statutory and other regulations as well as development of appropriate implementation and monitoring mechanisms
- R&D facilities and programs for constant upgrades in all related sectors
- Reducing the source of solid wastes as well as solid-waste management

The packaging industry in India has a promising future and should be ranked as one of the core industries with a potentially significant contribution to the national economy.

P. V. NARAYANAN
Indian Institute of Packaging
Bombay, India

INDICATING DEVICES

One summer in the 1930s, Eastman Kodak Co. received a telegram from the Kitt Observatory in Arizona stating that a shipment of photographic plates had no light sensitivity. Each of these plates was developed completely and was uniformly black. A new set of plates was carefully packed and shipped to the remote observatory. Again, all of the plates were black when developed. A well-respected Pinkerton guard accompanied the next shipment. He never let the box of plates out of his sight, and the mystery was solved at a small siding in Arizona. The rail clerk received the box clearly marked "FRAGILE GLASS PLATES," and methodically opened the box, took out each plate in the bright afternoon sun, removed the black paper cover, and examined each one. "All in perfect shape," he said to the guard as he rewrapped the plates and returned them to the carton.

A shipper must know what happens to a package after it leaves the shipping room, but it is impossible to send a secu-

Table 1. Biologics, Shipping Temperatures[a]

Product	Maximum Temperature, °F (°C)
Cryoprecipitated antihemophilic factor[b]	<0 (−18)
Measles, mumps, and rubella virus vaccine, live	<50 (10)
Measles and rubella virus vaccine, live	<50 (10)
Measles-smallpox vaccine live	<50 (10)
Measles virus vaccine, live, attenuated	<50 (10)
Mumps virus vaccine, live	<50 (10)
Poliovirus vaccine, live, oral, type 1	<32 (0)
Poliovirus vaccine, live, oral, type 2	<32 (0)
Poliovirus vaccine, live, oral, type 3	<32 (0)
Poliovirus vaccine, live, oral, trivalent	<32 (0)
Red blood cells,[b] frozen	<−85 (−65)
Red Blood cells,[b] liquid	34–50 (1–10)
Rubella and mumps virus, live	<50 (10)
Rubella virus vaccine, live	<50 (10)
Single-donor plasma,[b] frozen	<0 (−18)
Smallpox vaccine, liquid	<32 (0)
Source plasma[b]	23 (−5)
Whole blood[b]	34–50 (1–10)
Yellow fever vaccine	<32 (0)

[a] Ref. 1.
[b] Human.

rity guard with every carton. Instead, a variety of indicating devices have become the security guards for the packages.

Perishable products may speak for themselves by their physical condition or their odor. Damage to other products from inappropriate handling may not be evident until actual use. As an example, whole blood stored below 50°F (10°C) is stable for months. If the temperature rises above that point for 20 min, however, enzymatic changes could occur that would make it life-threatening if transfused. The potency of vaccines is lost by exceeding critical temperatures; an even greater hazard may result by presuming that it is effective. Because microcircuits are sensitive to mechanical shock and electrical potential, individual components and wired circuits must be monitored.

Time and Temperature Indicators

The fully integrating monitoring device can indicate the temperature gradient to which it has been exposed, as well as the amount of time that it has been at that temperature. Many inventors and their companies have focused on developing such devices, which operate on the following principles: physical change, chemical change, electrochemical indication, electromechanical indication, electronic readout, and others. The devices monitor perishable foods such as fish, fruit, and vegetables, as well as pharmaceuticals and vaccines. Table 1 lists the FDA temperature limits for various biological preparations.

Indicators that show time only are also available in many styles. These are used primarily on shipping containers, not on consumer packages. Life-dated products, such as photographic film and some foods, can have greatly extended salable time by refrigerated storage. A time indicator would benefit the manufacturer and retailer as well as the consumer.

The Andover Laboratories manufactures a time–temperature integrator called the Tempchron, formerly known as the Ambitemp (see Fig. 1), which functions with a fluid that has a specific melt temperature for the product to be monitored. This device can be described as an integrator because it provides information on the multiple of the time and temperature, giving a readout in degree minutes that can be interpreted from a chart. A wide range of temperatures can be monitored by selection of the liquid that is to be frozen in the tube.

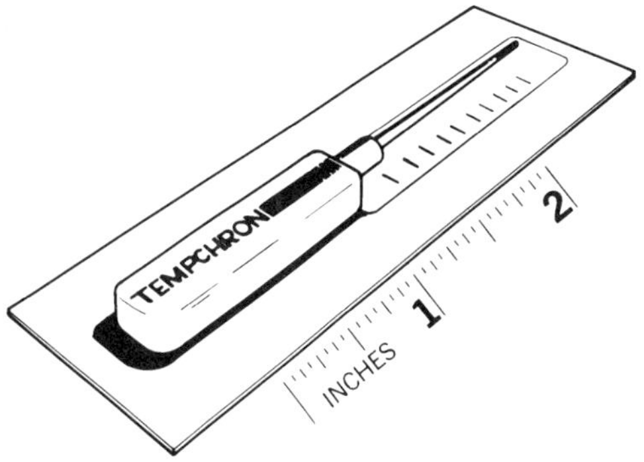

Figure 1. Integrated time–temperature indicator. The Tempchron (Andover Laboratories) utilizes the melt temperature of the product to be measured. To convert in. to cm, multiply by 2.54.

Figure 2 shows an interesting concept of a time–temperature integrator that was at one time manufactured by Honeywell Corp. This device, the TTI, comprised an electrolytic battery that was activated by breaking an ampul of electrolyte between copper and cadmium strips. An electrochemical color reaction that was accelerated at higher temperatures ensued.

Kokum Chemical in Malmo, Sweden, manufacturers a time–temperature monitor that utilizes an enzyme to bring about a color change. This product (see Fig. 3), the I-Point TTM, is intended to match the enzyme reactions of the product it is monitoring. A progression of color changes occurs.

Another time–temperature integrator functions by the use of a selective gas-absorption plastic membrane and leuko (colorless) dyes that are sensitive to oxygen (2). The leuko dye is converted to the colored form inside the pouch in proportion to the rate of oxygen penetration through the membrane and the time. The rate of penetration of the oxygen through the membrane is also related to the temperature. The storage life of foods for emergency survival has been studied as a function of the temperatures at which they were stored. With the accu-

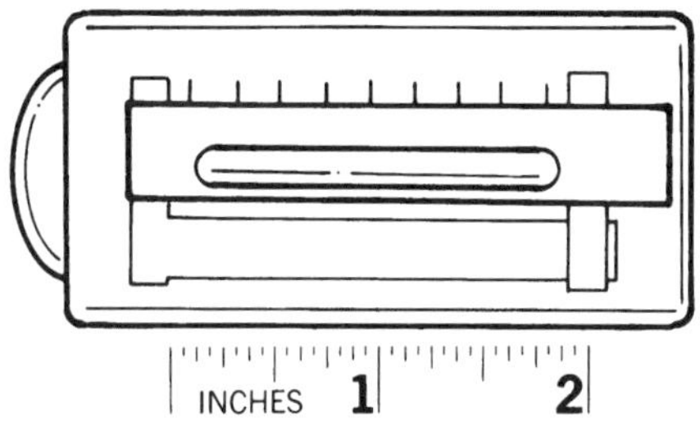

Figure 2. Integrated time–temperature indicator. The TTI (Honeywell Corp.) used an ampul of electrolyte between metal plates. To convert in. to cm, multiply by 2.54.

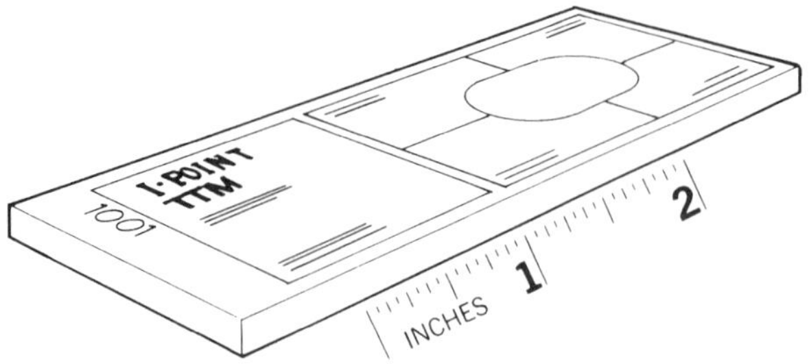

Figure 3. The I-Point TTM (Kokum Chemical) is an integrating time–temperature monitor utilizing enzyme-produced color change. To convert in. to cm, multiply by 2.54.

rately calibrated barrier films (see Barrier polymers) now available, this may prove a valuable area of investigation.

Allied Chemical Co. has a patent (3) on work in which combinations of conjugated acetylene compounds irreversibly change color when heated. After finalizing the above design, the Program for Appropriate Technology in Health (PATH) in Seattle, Wash., has marketed a product called the PATHmarker, which indicates the change in color.

There are two patents (4,5) for time–temperature integrators under the name Monitor Mark (see Fig. 4**a**). In this product, a dyed meltable solid travels down a porous wick when heated above its melting point. The time interval is a function of the travel of the color down the wick. The device is activated by removal of a barrier film that is positioned between the wick and the meltable solid (see Fig. 4**b**). The product is available in several response temperature ranges.

The Therma-Gard recorder (Impact-O-Graph Corp.) monitors temperatures on a cassette and functions for 30 days. The TSI International Corp. has introduced a recorder that operates for up to 90 days in the indicating range of −20 to 100°F (−29 to 30°C).

Two workers (6), following the lead of another (7), have fabricated two styles of electronic recorders with built-in memories (see Fig. 5); however, they are not commercially available. Workers at 3M Co. have invented a Thermal History Indicating Device (THID) (8). Another product that interfaces with a computer for its readout is the Tattletale thermograph (Onset Computer Co.) (see Fig. 6). This thermograph reports temperature transients to a memory and can be programmed to record at 5-s intervals for 5.5–6-h intervals for 32 mo. The functional temperature range is −41 to 185°F (−41 to 85°C).

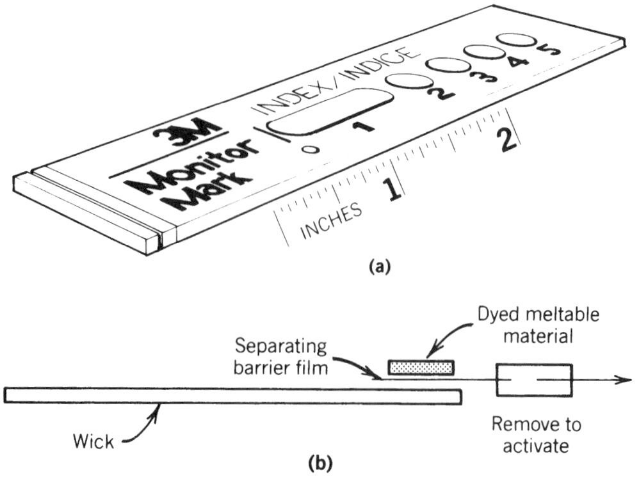

Figure 4. (**a**) The Monitor Mark Time Temperature Tag (3M Co.), an integrator available in many temperature ranges. To convert in. to cm, multiply by 2.54, (**b**) Cross section.

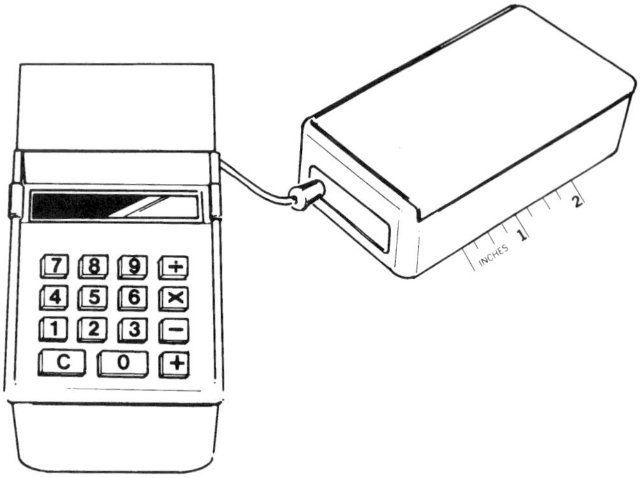

Figure 5. Electronic recorder (6,7). To convert in. to cm, multiply by 2.54.

Temperature Indicators

Temperature-measuring devices are numerous. In the simplest form, a sphere of ice is frozen with one half clear and the other red (see Fig. 7). If the product thaws, the sphere

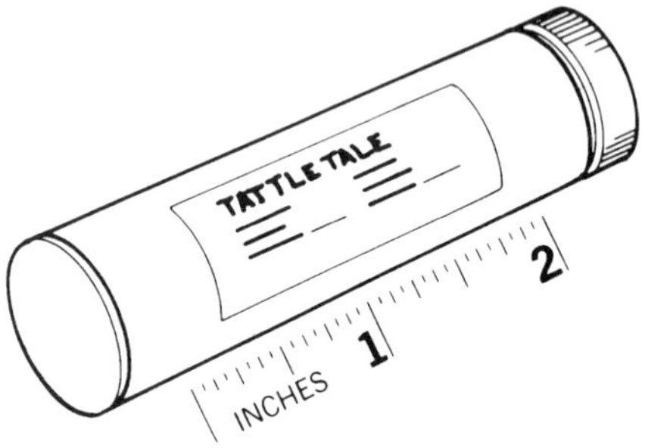

Figure 6. Tattletale thermograph electronic recorder (Onset Computer Corp.). To convert in. to cm, multiply by 2.54.

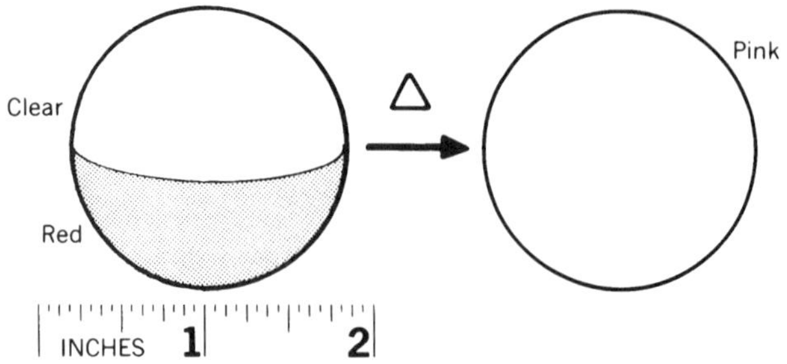

Figure 7. Thaw indicator. Half-colored sphere of ice becomes homogeneous when thawed. To convert in. to cm, multiply by 2.54.

Table 2. Color-Change Temperature Indicators

Product Name	Company and Location	Temperature Range, °F (°C)
Celcistrip	Solder Absorbing Technology (Agawan, MA)	104–465 (40–240) 105–500 (40–260)
Thermomarkers	W. H. Brady Co. (Milwaukee, WI)	100–500 (40–260)
Tesa-Temperatur Indikatoren, irreversible	Biersdorf AG (Hamburg, Germany)	175–360 (80–180)
Thermolabel	Paper Thermometer Co. (Greenfield, NH)	140–180 (60–81)
Tattleherm	Everest Interscience (Tustin, CA)	100–500 (40–260)
Reatec	MRC Corp. (Wayne, PA)	100–420 (40–213)
Template	Wahl Corp. (Los Angeles)	100–1000 (40–590)
Telatemp	Telatemp Corp. (Fullerton, CA)	100–350 (40–155)
Thermo Strip	Archie Soloman and Associates (Roswell, CA)	219–435 (103–220)
Thermindex	Thermindex Chemicals and Coatings, Ltd. (Missisauga, Ontario, Canada)	100–500 (38–260)
OmegaLabels	Omega Engineering (Los Angeles)	100–400 (40–202)
Hermet	Markal Co. (Chicago)	100–500 (40–260)
Temp Tabs	Jardine Engineering (Hong Kong)	100–900 (40–500)
Celsipoint	Signalarm, Inc. (Springfield, MA)	100–500 (40–260)
Templilabel Temp-Alarm Templistik	Big Three Industries, Inc. (South Plainfield, NJ)	125–750 (50–395)
Thermax	Thermagraphics Measurements, Ltd. (Chicago)	150–400 (65–200)
T-Dot	Westemp Instruments Co. (Cardiff, CA)	100–400 (40–200)

becomes pink. High-technology temperature indicators include thermistors, circuits, and liquid-crystal displays (LCD).

Most temperature indicators depend on a chemical color change. These compact units provide direct, reliable information on the temperature to which the product was exposed. They are widely used for materials with critical temperatures such as pharmaceutical and biological preparations (Table 1) and fresh and frozen fish. Table 2 lists some of these indicators; Figure 8 is representative of these devices.

There are two other types of indicators not listed in Table 2: the Criti-Temp (Schobl Enterprise, Inc.), which is a bimetallic spring-loaded monitor, and the Precision Digital Thermometer (TSI International Corp.), which functions from a thermistor to give a temperature display.

Freeze and Thaw Indicators

The main purpose of the indicators listed in Table 2 is to show a response to a rising temperature well above room temperature. Devices to indicate temperatures near the freezing point of water require special design since they must be unbroken before use and function at low temperatures.

Akzo N.V. of the Netherlands has exerted a strong influence in the monitor area through its subsidiary, Organon, heir to the BMS disposable-thermometer technology, and through its Info-Chem Protective Products Division. Info-Chem manufactures the Thaw-Watch and the Freeze-Watch. These products utilize an ampul filled with a colored fluid. Freezing breaks the ampul, which spills the contents onto a paper indicator (see Fig. 9).

Biosynergy, Inc. manufacturers a liquid crystal Hemo-Temp II for use on blood-collection bags. This device, after being frozen, indicates the temperature of the blood (see Fig. 10).

The Check-Spot (Check-Spot, Inc.) (9) functions at 3–32°F (−16 to 0°C) and is based upon a solid emulsion (see Fig. 11).

The Monitor-Mark Button (3M Co.) (see Fig. 12**a**) operates by means of a meltable, dyed compound contained in a porous reservoir. In the inactivated form, a domed indicator paper is separated from the reservoir by a small distance. When the dome is pressed, the two materials come in contact, allowing wicking to occur when the melt temperature is reached (see Fig. 12**b**). This product has many different applications, including irreversible monitoring of bags of whole blood. The specifications for blood banking have been outlined (10).

Figure 8. Temperature indicator representative of devices listed in Table 2. To convert in. to cm, multiply by 2.54.

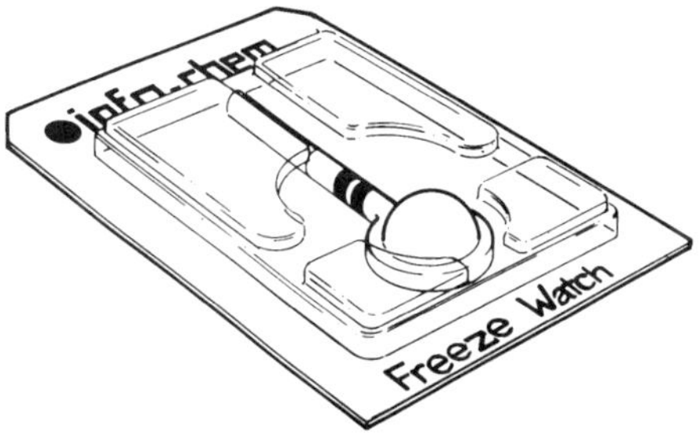

Figure 9. Freeze-Watch (Info-Chem Protective Products).

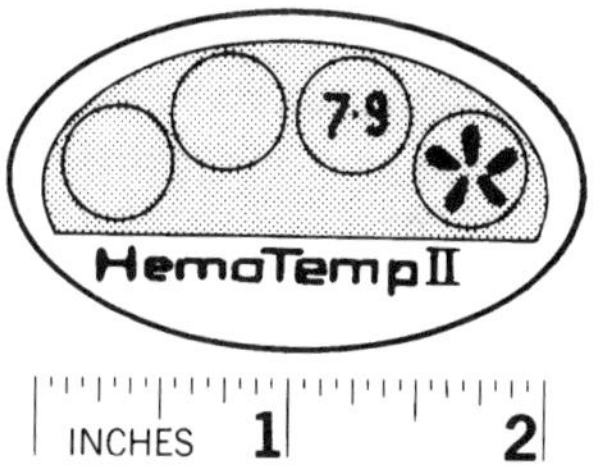

Figure 10. HemoTemp II (Biosynergy, Inc.). To convert in. to cm, multiply by 2.54.

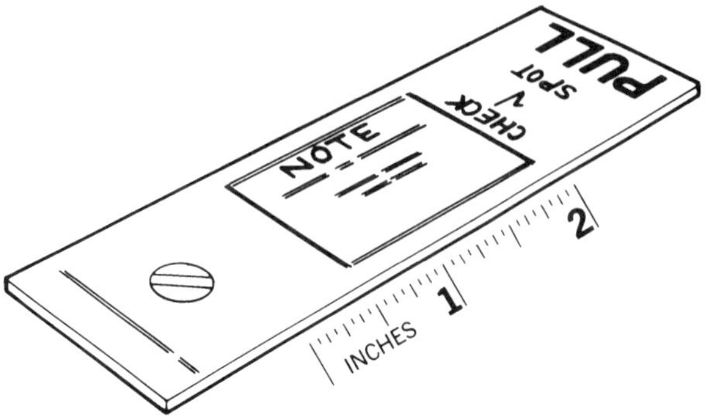

Figure 11. Check-Spot (Check-Spot, Inc.). To convert in. to cm, multiply by 2.54.

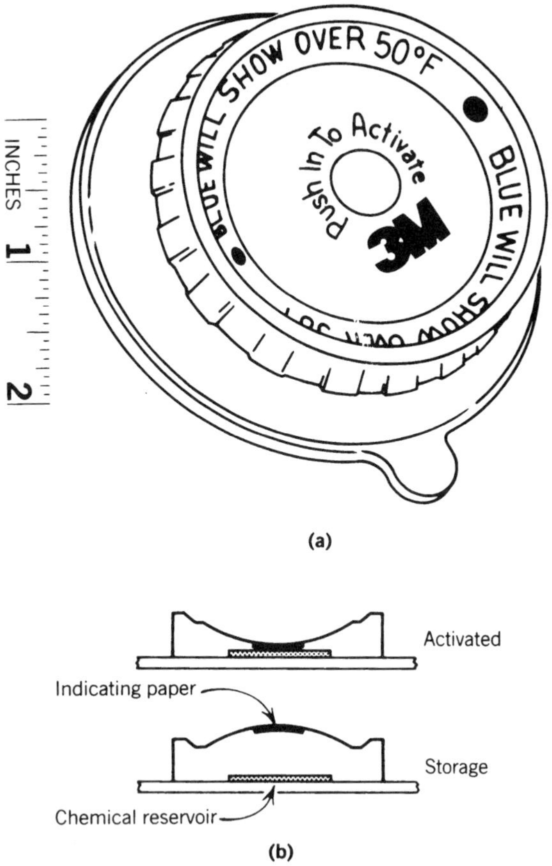

Figure 12. (**a**) Monitor-Mark Button (3M Co.). To convert in. to cm, multiply by 2.54. (**b**) Cross section.

Other styles of freeze indicators are manufactured by 3M Co. (11–13), including the Monitor-Mark Cold Side Indicator, model 32F (see Fig. 13). When temperatures fall below 32°F (0°C), there is an irreversible color change. In the past, the IWI, or irreversible warmup indicator, was manufactured by Artech Corp. (14).

Humidity Indicators

Many items are sensitive to moisture in ways that are irreversible. Simply drying out or humidifying does not restore the original function, and the damage that may have occurred during shipping or storage may not be evident to the recipient. To protect the manufacturer and the customer, humidity changes must be monitored. A humidity monitor based on a color change is available from Herrmann Chemie and Packmittel (see Fig. 14).

A U.S. patent has been granted for a time/history humidity indicator (15). This device comprises a salt that absorbs moisture from the air, a water-soluble dye, and an absorbent material. After activation by removal of a barrir film, the dissolved dye migrates into the absorbent material as the humidity increases. This product resembles the product shown in Figure 4.

Gravitational-Force Indicators

Gravity is a useful force in packaging because it keeps containers in place. Many times the normal force of gravity may

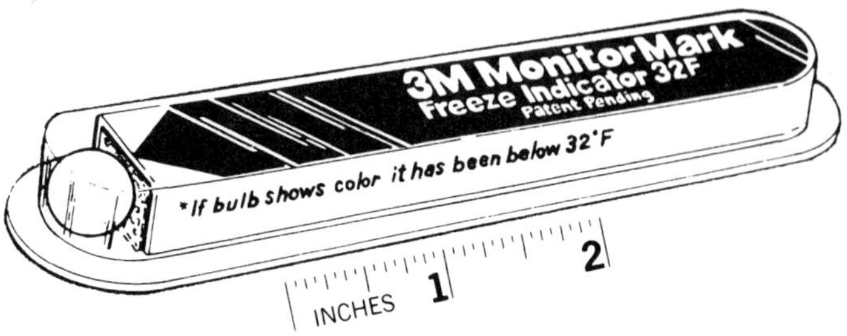

Figure 13. Monitor-Mark Freeze Indicator 32F (3M Co.). To convert in. to cm, multiply by 2.54.

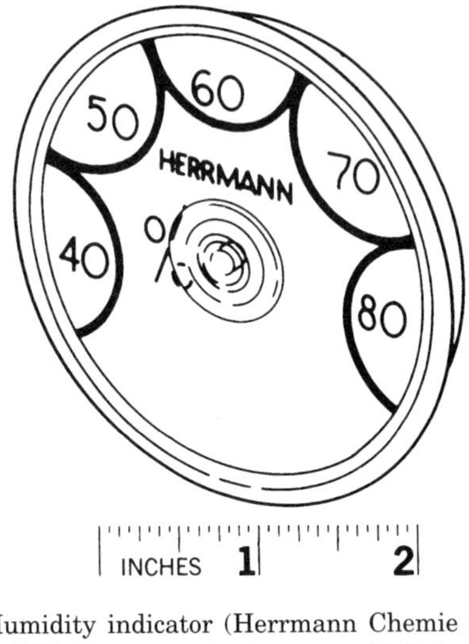

Figure 14. Humidity indicator (Herrmann Chemie and Packmittel of Sod-Chemie AG). To convert in. to cm, multiply by 2.54.

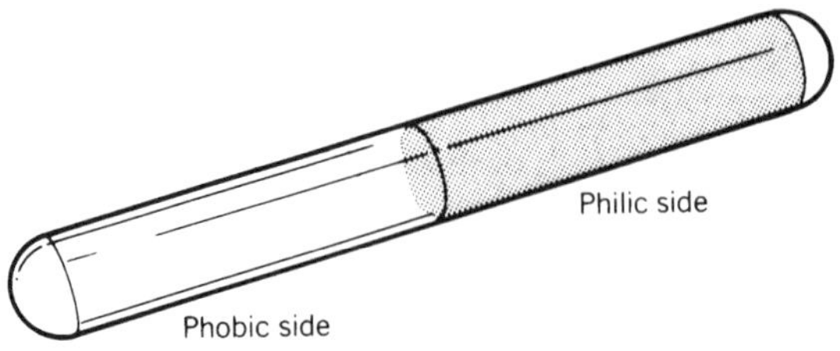

Figure 15. Capillary tube of Shockwatch.

be applied to a product in shipping and handling. This abusive treatment can cause hidden damage. The Hump-Gard (Impacto-Graph Corp.) is a 30-day monitor for the high G forces encountered in railroad shipping.

The Shockwatch is a compact device for measuring high gravitational forces (16). The active element is a capillary tube with surface-energy differences in each end (see Fig. 15). A colored fluid is placed in the -philic end, and when forces exceed a predetermined value, the liquid is forced into the -phobic end where it is visible. Once distributed by 3M Co., it is now available from Media Recovery Co.

This article has discussed only a few of the hundreds of indicating devices availabe (17). A very low cost indicator is still needed, preferably one that could be imprinted as part of the label of consumer products such as frozen foods.

BIBLIOGRAPHY

"Indicating Devices" in *The Wiley Encyclopedia of Packaging Technology,* 1st ed., by Dee Lynn Johnson, 3M Company, pp. 400–406.

1. *Fed. Reg.* **39**(219) (Nov. 12, 1974).
2. K. H. Hu, *Food Technol.* **26,** 56 (Aug. 1972).
3. U.S. Pat. 3,999,946 (Dec. 28, 1976), G. N. Patel (to Allied Chemical Co.).
4. U.S. Pat. 3,954,011 (May 4, 1976), W. Manske (to 3M Co.).
5. U.S. Pat 3,962,920 (June 15, 1976), W. Manske (to 3M Co.).
6. A. J. C. Carselidine and R. R. Weste, *CSIRO Food Res. Q.* **36,** 41 (1973).
7. M. J. M. Van't Root, *Proceedings of the 13th Congress on Refrigeration,* Washington, DC, Vol. 4, pp. 445–451.
8. The author and F. R. Parham, 3M Co., Saint Paul, Minn., have been active in microprocessor monitors.
9. U.S. Pat. 2,971,852 (Feb. 14, 1961), J. Schulein (to Check-Spot, Inc.).
10. B. A. Myhre, *Quality Control in Blood Banking,* Wiley, New York, 1976, pp. 168, 169.
11. U.S. Pat. 4,132,186 (Jan. 7, 1979), W. Manske (to 3M Co.).
12. U.S. Pat. 4,457,252 (July 3, 1984), W. Manske (to 3M Co.).
13. U.S. Pat. 4,457,253 (July 3, 1984), W. Manske (to 3M Co.).
14. Product literature, Artech Corp., Falls Church, VA.
15. U.S. Pat. 4,098,120 (July 4, 1978), W. Mansek (to 3M Co.).
16. U.S. Pat. 4,068,713 (Jan. 17, 1978), U.R. Rubey (to Detectors, Inc.).
17. W. Manske, 3M Co., Saint Paul, Minn., has a working file of over 255 U.S. patents on Indicating Devices.

INJECTION MOLDING

Injection molding, one of the principal methods of forming thermoplastic materials, involves feeding plastic resin to a rotating screw in a heater barrel. There the plastic is melted and mixed. The resulting hot plastic is then injected at high pressure into a closed mold that has one or more cavities in the shape of the desired part. The mold is cooled, and when the plastic solidifies, the mold is opened and the part is ejected. The process, defined as a "cycle," can then repeat itself.

Typical injection-molded products (see Fig. 1) for packaging are polyethylene terephthalate (PET) preforms (see Blow molding), thin-walled yogurt and margarine containers, 5-gal (19-L) pails and other industrial containers (see Pails, plastic), and threaded closures (see Closures, bottle and jar).

Advances in technology continue to impact-injection molding for the packaging industry. Increasing resin costs, ongoing introduction of new resins, and competition from traditional packaging (cardboard, metal, etc), as well as other processes (see Blow molding; Thermoforming), continue to drive technological developments. To effectively compete under these conditions, the cost of injection-molded parts has to be reduced while maintaining adequate part strength and quality. Resin producers have developed high-flow polyethylene (see Polyethylene, low-density; Polyethylene, high-density) and polypropylene (see Polypropylene) resins, in addition to other high-flow resins, to allow the molding of parts with significantly thinner walls and less weight, and equipment suppliers continue to develop high-performance injection-molding machines for the packaging industry.

The Injection Mold

The mold design has a tremendous impact on system productivity and product quality, and therefore on the overall economics of the injection-molding operation. The design of a mold is heavily influenced by characteristics of the part. Most injection-molded parts for packaging have relatively large length-to-thickness ($L:T$) ratios. The length is defined as the maximum flow length in the cavity and is measured from the point where plastic enters the mold to the furthest point it travels. The thickness is the average wall thickness of the part. Early molds had $L:T$ ratios of up to 200 : 1, today they can be as high as 500 : 1. For such thin-wall parts, design emphasis must be placed on the hot-runner system, part ejection, mold cooling, alignment, and mold material selection. A typical mold for thin-walled containers is shown in Figure 2.

Figure 1. A selection of injection molded closures, lids, containers, and bottle preforms for blow-molded polyethylene terephthalate (PET) soft-drink bottles.

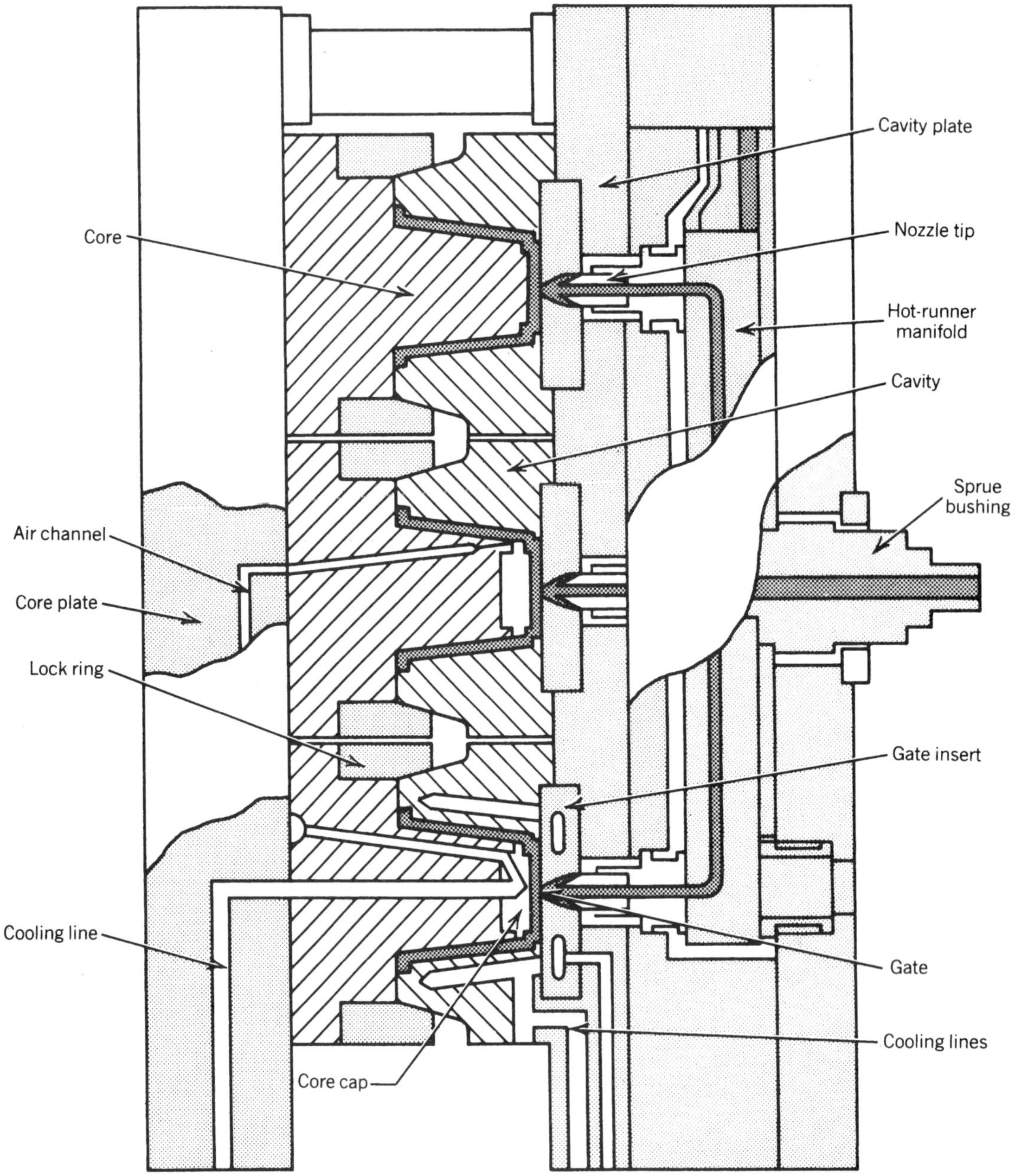

Figure 2. Hot-runner multicavity mold with core lock and air ejection.

Operation. As the mold closes, the cores and cavities are aligned by some form of tapered mating interface. Plastic melt, ie, melted resin, is injected by the injection unit of the molding machine into the sprue bushing and forced through a network of flow channels called the runner system. In Figure 2, the runner system is heated by electric heaters, and hence the term "hot runner." The hot-runner system maintains and controls the temperature of the melt right up to the gates by use of a sprue heater, manifold heaters, and nozzle heaters. Past the gates, the melt flows into the cavities. As the molten plastic solidifies during cooling, the parts typically shrink 1–2% of their diameter causing them to stick to the cores. When the mold opens, parts are ejected from the cores and drop between the mold halves.

Runner system. There are basically two types of runner systems: cold and hot. In cold-runner systems, the melt in the flow channels between the sprue bushing and the gates is allowed to cool and solidify during each cycle and must be ejected along with the parts before the next cycle begins. Cold-runner systems are simpler to design and less expensive than hot-runner systems. A disadvantage is the plastic of the cold runner must be reprocessed or scrapped. The introduction of scrap material to virgin resin can adversely affect process repeatability and part quality. Also, the cold runner can significantly slow the cycle at which the mold operates, since the cold runner must be cooled before ejection. For these reasons, hot-runner systems are extensively used in packaging applications.

To ensure consistent quality, the hot-runner system should be balanced so that it supplies melt to each cavity at the same pressure and temperature. Flow balance is maximized by ensuring that each flow channel from the sprue bushing to the cavity gates has an equal length and an equal number of turns. Temperature balance is achieved through uniform heating and insulation. Hot-runner molds with up to 128 cavities operate reliably with this balanced approach. The hot-runner system should also have channels with smooth corners to reduce friction and pressure drop. There should be no dead spots where plastic can be trapped and degrade.

Cooling. Cooling time is a key variable in the total molding cycle. To decrease cooling time and thus increase productivity, cooling channels should be located near the molding surface. High coolant flow rates are necessary to achieve high heat-transfer rates. In addition, metals with excellent heat-conductivity properties such as beryllium copper can be used for inserts to improve heat transfer where high heat is generated, such as around the gates. Beryllium copper has approximately 10 times the coefficient of thermal conductivity of steel. With a well-designed cooling system, typical thin-walled parts can be cooled rapidly. Total cycle times as short as 2 s are possible. It is also important for the cooling system to maintain the cores and cavities at approximately the same temperature. This prevents excessive wear between interlocking faces due to differences in thermal expansion between the mold halves.

Part ejection. There are two basic methods of part ejection: mechanical and air. Mechanical ejection commonly uses stripper rings or pins surrounding each mold core, or a moving core cap to physically push the parts off the cores. The stripper rings can be activated by hydraulic cylinders attached to the machine platen, air cylinders in the mold, or a mechanical linkage tied to the motion of the machine platen. Air ejection uses blasts of air to loosen and blow the parts off the cores. Air ejection is the more popular method because it involves fewer moving parts and, hence, less maintenance. The mold can also be more compact.

Alignment. The proper alignment of core and cavity is critical in thin-wall molding. Slight misalignment leads to preferential filling of the thicker part of the cavity, which causes a pressure imbalance around the core and cavity, thus further shifting the core. The overall result is uneven wall thickness in the molded part. In extreme cases, it can result in incomplete filling of the cavity, ie, a short shot, and rejected parts. To achieve satisfactory part quality, center-to-center alignment of the core and cavity within ± 0.2 mil (± 0.005 mm) may be necessary.

There are different methods of aligning the cores and cavities of packaging molds. One common method is a stripper ring that uses tapers on the stripper ring to align the core and cavity. This popular method of alignment has disadvantages for thin-wall molding, where alignment forces cause relatively rapid wear of the stripper rings. Since the stripper rings form part of the molding surface, excessive wear in this area reduces part quality. To avoid a drop in part quality, frequent mold maintenance is required. Wear is not critical if the locking rings are not part of the moldign surface. Either air or mechanical ejection can be used with this approach.

The floating-core method is a more recent design that is suitable for some packaging applications. The cores and locking rings are not rigidly fixed to the core plate, but are allowed to float to more easily align themselves with the cavities. Easier alignment results in less wear on the aligning tapers. This method can also be used with either air or mechanical ejection.

Mold materials. High-quality mold materials are of the utmost importance. Most mold components are made of through-hardened high-quality tool steel. Hardness ranges from Rc 49–51 (Rockwell C scale) for the cores and cavities to Rc 30 for the mold plates. Only periodic rebuilding of wear items, such as stripper rings and leaderpin bushings, is required.

Stack molds. A two-level stack mold typically has identical sets of cores and cavities on each mold face, which are then stacked together back-to-back. In the case of a family stack mold, the cores and cavities differ on between the two faces, allowing similar or matched pieces (compact-disk jewel case, floppy-disk shell, etc) to be produced each cycle. Use of a stack mold almost doubles the productivity of a machine. For these reasons, stack molds are increasingly popular. Four-level stack molds have recently been introduced for thin-wall lids and containers to further increase machine productivity.

The Injection-Molding Machine

An injection molding machine consists of three main elements: injection unit, clamp, and controls (see Fig. 3). The injection unit plasticizes (ie, melts) and injects the resin at high pressure into the closed mold. The clamp unit supports the mold halves, closes and clamps them together during injection, and opens them for ejection. The controls consist of electrical, electronic, and hydraulic systems for machine operation. All of these elements are mounted on the machine base, which can be of either a single- or split-base design. Modular machine designs allow a wide variety of combinations of injection units, clamps, and controls to meet the individual requirements of specific applications.

Injection unit. Two types of injection units are used in packaging applications: reciprocating-screw and two-stage. Figure 4 shows a typical design for a reciprocating-screw unit. As the motor rotates the screw, the shearing action on the resin generates most of the heat needed to plasticize the resin. Only about 20% of the heat requirement is from conductive heaters around the barrel. As the screw rotates, it retracts and feeds melt to the front of the screw. When the required volume of melt has accumulated at the front of the screw, the screw stops rotating and moves forward to inject the resin into the mold. Injection takes place at a predetermined speed and pressure for a set period of time to fill the part and then hold plastic under pressure to compensate for shrinkage of the part as it cools in the mold.

A two-stage unit is shown in Figure 5. This type of unit has separate components for plasticizing and injection. The extruder screw feeds the melt at low pressure to a separate shooting pot, from which the melt is injected into the mold by a simple piston.

In thin-wall molding, injection unit design emphasis must

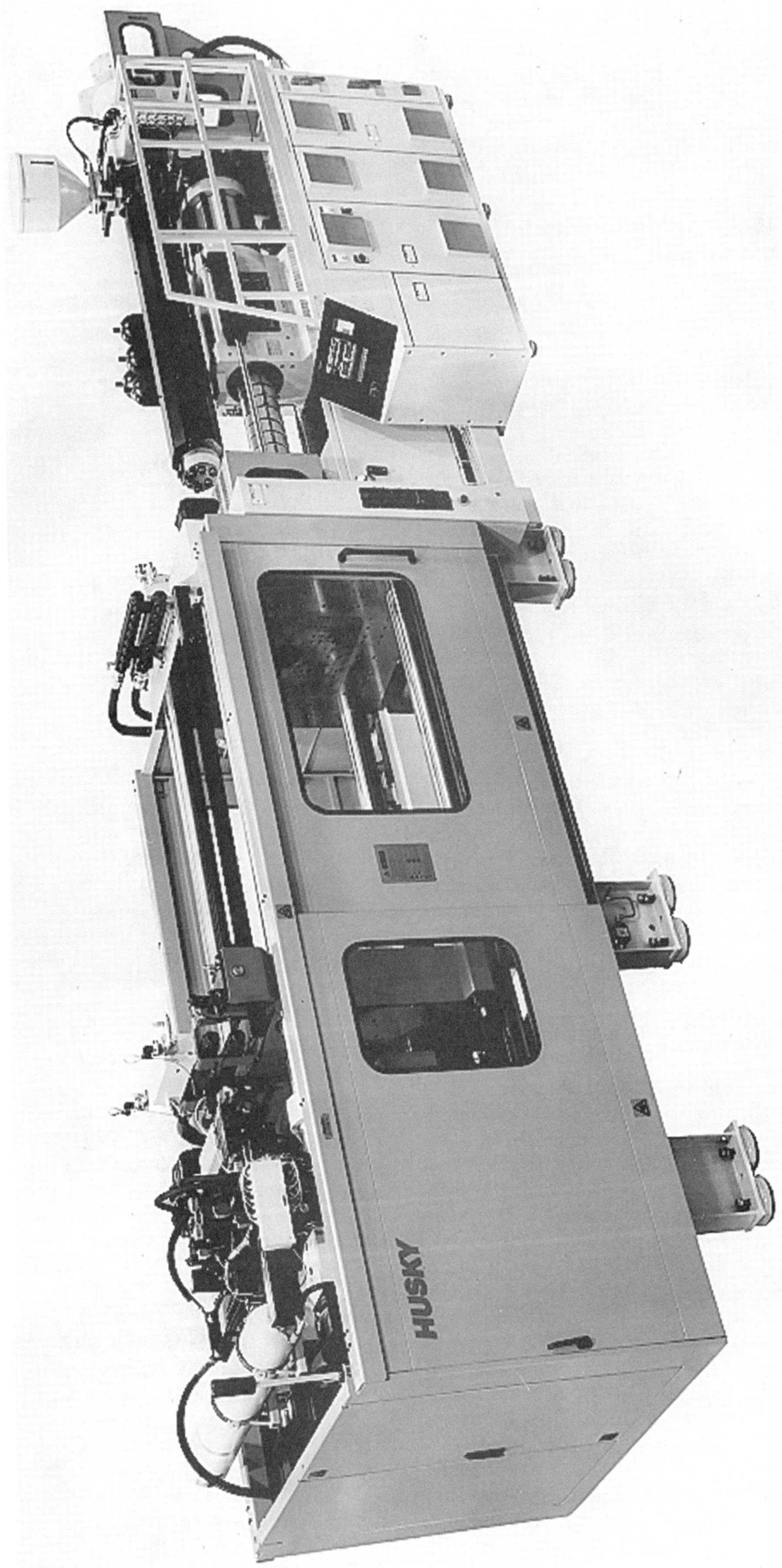

Figure 3. Hydromechanical injection-molding machine with two-stage injection unit.

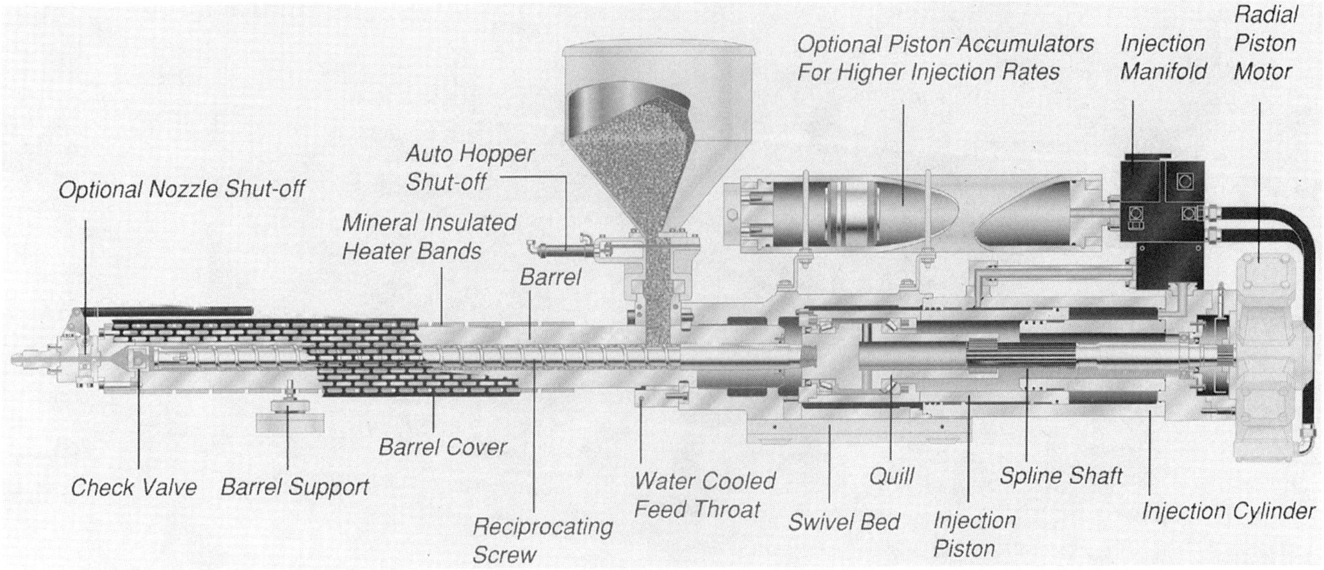

Figure 4. Reciprocating-screw injection unit.

be placed on fast injection speeds, high injection pressures and plasticizing rates, and shot-to-shot repeatability. Injection speed is critical. Molten resin moving through a thin cavity over a long distance (high $L:T$ ratio) tends to cool quickly and freeze off before completely filling the mold. To prevent this, injection speed must be very rapid. With today's multi-cavity molds, this requires high injection rates. Injection rates of 4.4 lb/s (2 kg/s) or more are not uncommon in thin-wall molding.

High injection pressures are mandatory to achieve the high injection rates into thin-walled cavities. Injection pressures of 29,000 psi (200 MPa) are generally used for typical

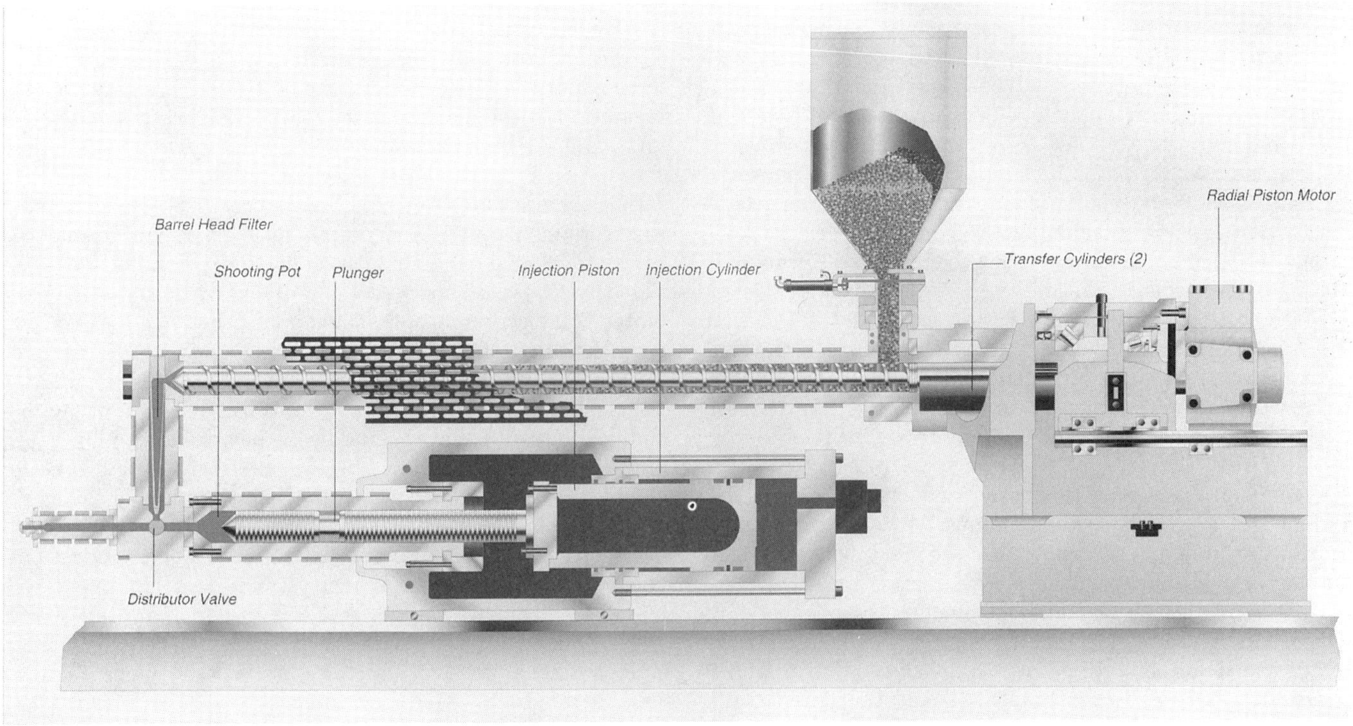

Figure 5. Two-stage injection unit.

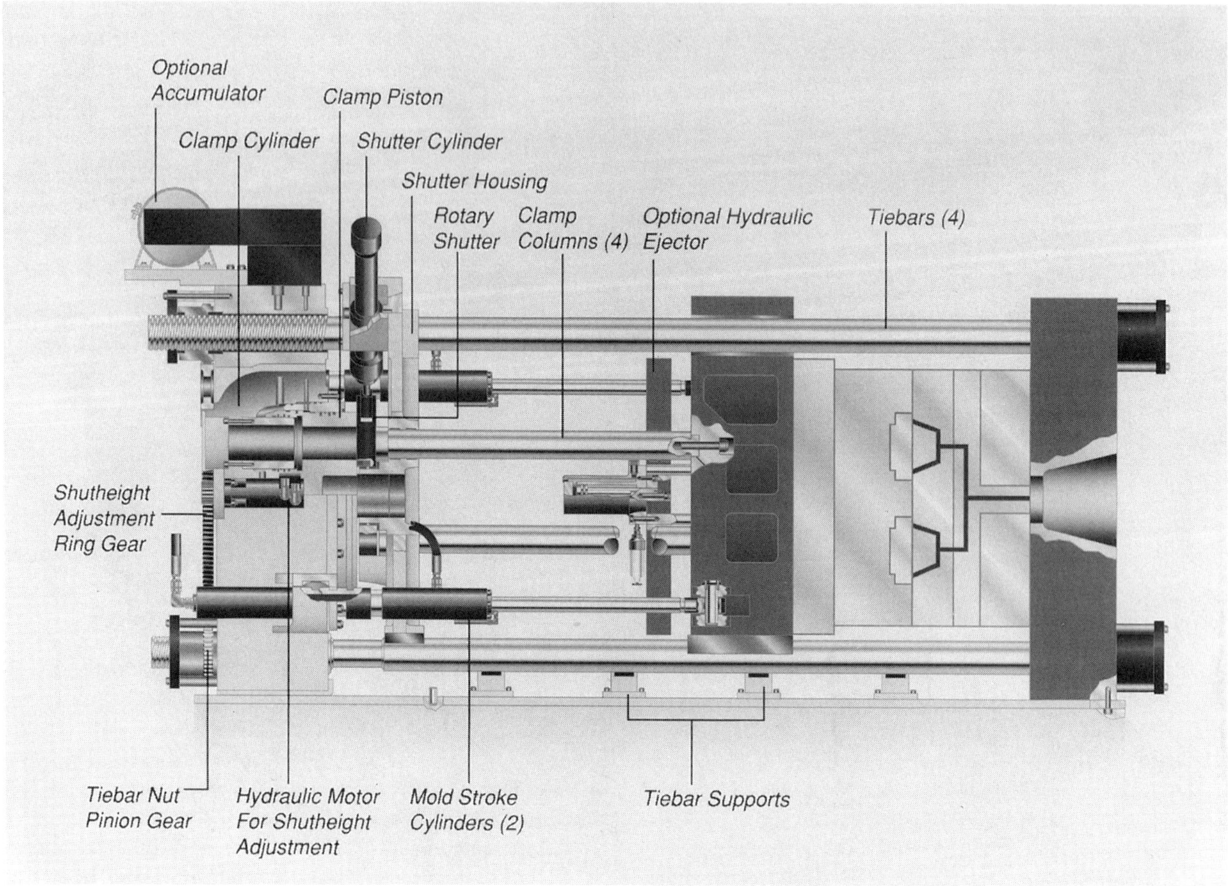

Figure 6. Hydromechanical clamp.

thin-wall applications. Hydraulic accumulators can provide the instantaneous power necessary for injection without the need for oversized hydraulic pumps and motors. High plasticizing rates are also needed with the short cycle times characteristic of packaging applications. An advantage of two-stage injection is that the extruder screw can continue to turn without interruption during both the injection and hold phases.

Shot-to-shot repeatability is essential for consistently high-quality parts and for keeping part weight at the desired minimum. The two-stage design shown in Figure 5 can offer ±0.15% shot-to-shot repeatability in injection volume. Generally, the accuracy of the reciprocating screw unit is less than that of a two-stage unit because of the time delay for the ring-check valve to sell off during injection. Also, accumulators dedicated solely to the injection unit can ensure that the same level of hydraulic energy is stored and ready for every shot.

Clamp unit. There are two broad categories of clamps: hydraulic and toggle. Hydraulic clamps typically use a single hydraulic cylinder to open and close the mold, whereas toggle clamps use mechanical linkages.

One type of hydraulic clamp, commonly described as a hydromechanical clamp, is shown in Figure 6. Two large platens are provided for mounting the mold halves. One platen is firmly attached to the machine base, and the other can be moved to open and close the mold. A small-diameter-clamp stroke cylinder moves the moving platen at high speeds with minimum oil consumption. The machine's electronic control system provides consistent slowdown before the mold faces contact by gradually closing the hydraulic valve, allowing oil to flow to the stroke cylinder. After the mold faces are brought together and the mold is closed, shutters block the clamp column, and the high clamping force is then applied by the large clamping piston. The large clamping piston also provides high initial opening force for mold breakaway. Tiebars, which join the stationary platen and the clamp mechanism, absorb the reaction forces during application of clamp tonnage and guide the moving platen to ensure alignment with the stationary platen. An ejection mechanism supplies the force necessary to strip the parts off the mold cores.

Another type of clamping unit is the toggle clamp (Fig. 7). Toggle linkages inherently provide high speed and low mechanical advantage at the middle of the stroke, and low speed and high mechanical advantage at the end of the stroke for clamp-up. The toggle clamp system can be actuated either hydraulically or electrically.

To design a clamp unit suitable for packaging applications, one must consider several factors. The fast molding cycles possible with thin-walled containers demand high clamp speed. The need for consistent wall thickness demands clamp rigidity and alignment,and the high cost of molds requires reliable mold-protection mechanisms.

Fast speed can be achieved with a hydraulic clamp by using a relatively small clamp stroke cylinder. However, the oil

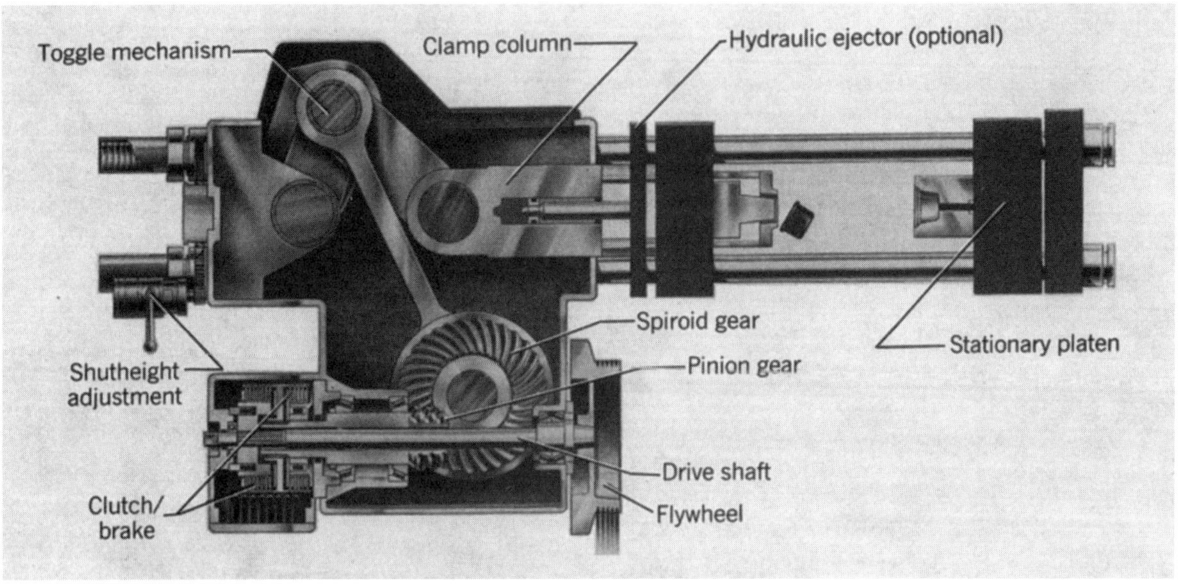

Figure 7. Toggle clamp.

required to rapidly open or close the clamp still places a peak demand on the hydraulic system. To avoid over sizing the hydraulic pumps to meet this demand, hydraulic accumulators can be used. With accumulators, speeds of 5 ft/s (1.5 m/s) are economically feasible. Toggle clamps are generally capable of higher speeds than hydraulic clamps and provide smooth acceleration and deceleration.

To ensure clamp rigidity, massive platens are necessary to limit deflection to under 5 mil (0.13 mm) at full clamping pressures. Also, a large centrally positioned platen backup is important to transmit the clamping force evenly from the column to the moving platen. Excessive platen deflection must be eliminated to avoid flashing (ie, separation of the mold parting line during injection and resultant flow of melt) and uneven wall thickness of the molded part.

Center-to-center alignment and parallelism of the platens is critical. For example, the lower tiebars that guide the moving platen (see Fig. 6) should be supported by the machine base to prevent tiebar droop and resultant misalignment. Also, the clamp unit should be designed and built with close tolerances to ensure parallelism between platens.

Mold protection mechanisms prevent the mold halves from slamming together during clamp closing. They also prevent damage to cores and cavities when parts or other foreign objects become accidentally jammed between the mold faces. Mold protection can be provided by a variety of machine features. In addition to a hydro-mechanical cushion, the pressure in the clamp stroke cylinder can be automatically reduced before closing the mold. The shutter design can also prevent application of full locking force unless the mold is fully closed and the shutters have moved in.

Controls. Injection molding machine controls can be divided into two categories: sequence controls and process controls.

Sequence controls. They provide the logic signals to lead the machine through its normal operating sequence, including steps such as closing the mold, injection, starting of screw rotation, etc. Control design is based on programmable logic, which allows the flexibility of changing the operating sequence and timing to suit different molding jobs and to incorporate downstream parts handling equipment into the machine operating sequence.

Process controls. They are used to control the variables affecting the molding process and therefore the quality of the molded parts. Examples of process control variables are injection speed, injection pressure, screw speed, and resin temperature. Process controls can be divided into two main types: open-loop and closed-loop. In open-loop control, no feedback is available to determine if a process variable has reached the desired setting. It is assumed that it has. In closed-loop control, feedback is provided and automatic corrections are made to ensure that a process variable reaches and remains at the desired setting.

Machine controls must provide the accuracy and repeatability required by the injection and clamp units. Accuracy and repeatability are achieved by a well matched combination of electronics, hydraulics, and transducers. The machine's mi-

Figure 8. Thin-walled container stack mold equipped with servo swing chutes.

Figure 9. Free-standing, positive-takeout side-entry robot: (**a**) the servo-driven, multiposition robot moves the end-of-arm-tooling (EOAT) into the molding area; (**b**) parts are ejected from the mold into a set of water-cooled takeout tubes, where they are held by vacuum for three cycles; (**c**) the robot moves out of the molding area each cycle, allowing the EOAT to rotate and release a set of cooled parts onto a soft conveyor.

croprocessor can store set-up parameters, monitor system performance, and provide management information. The hydraulic package should include accurate and fast response hydraulic valves. Valve leakage, which can destabilize machine functions, must be minimized. Cartridge valves are most suitable for these requirements. The transducer package must include accurate sensors for clamp and screw position. Noncontact digital transducers are available today with an accuracy of ±0.4 mil (±0.01 mm).

Product handling. Manual product handling can account for up to 40% of the total processing cost of the product (excluding resin cost). Adding to this is the problem of work related injuries from repetitive motion and a need to offer challenging jobs for the labor force. As a result, many molders are reducing costs and increasing product quality with automated product handling methods.

Product handling begins with ejection of the parts from the mold. The three basic categories of automated handling are as follows:

1. Free drop and unscrambling: after parts drop from the mold, they are unscrambled and reoriented to meet

down stream operation requirements, such as stacking, assembly, printing, and packaging.

2. Controlled drop: guide pins or guide rails are built into the mold to control part orientation as the parts fall by gravity into chutes located beneath the mold; the orientation of the part established in the mold is never lost. Swing arms can also be used to position parts into chutes.
3. Positive take out robots: parts are mechanically removed from the mold and transferred to product handling equipment located beside the molding machine without losing part orientation.

To further minimize cycle times and provide oriented parts handling for high speed thinwall container applications, innovative product handling systems such as servo swing chutes have been developed. Servo swing chutes consist of arms which swing 90° in front of each core during mold opening to remove the parts. Suction cups on the arms hold the parts which are either physically stripped or air ejected from the cores. The servo drive motors swing the arms out of the molding area before the mold reaches the fully open position allowing mold closing to begin immediately. Once the mold is closed, the vacuum is released allowing the parts to slide down the chutes in an oriented manner. Swing chutes can be used on either single face or stack molds to provide faster cycles and oriented parts handling for a variety of thinwall injection molding applications (Fig. 8).

Robots are highly reliable, can handle deep parts with various shapes, and virtually eliminate parts handling damage. A typical robot application is shown in Figure 9, handling PET bottle preforms. Positive take-out robots are becoming increasing popular for parts with relatively thick walls and longer cycles. Part quality is increased by both improved handling and a more consistent processing cycle.

Changing market demands for product applications combined with new materials will continue to drive innovation in the thin-wall packaging industry. Providing the highest-quality parts at the lowest cost will remain a driving factor in the introduction and acceptance of new technology.

BIBLIOGRAPHY

1. F. Johannaber, *Injection Molding Machines—a User's Guide,* Carl Hanser Verlag, Munich, 1985.
2. H. Rees, *Understanding Injection Molding Technology,* Carl Hanser Verlag, Munich, 1994.

Richard Carter
Husky Injection Molding Systems Ltd.
Ontario, Canada

INK JET PRINTING.

See Code Marking and Imprinting.

INKS

Printing inks are used to decorate the exterior of virtually every package and substrate used in packaging. Since there is a variety of substrates, a number of different printing methods are required to satisfy the needs of this entire market (1). In general, the inks required may be subdivided into two classes: liquid inks and paste inks. The liquid inks include those printed by the flexographic, rotogravure, and screen process methods; the paste inks include those printed by lithographic offset, letterset, and letterpress (see Decorating; Printing).

Table 1. Ink Viscosity and Film Thickness by Process

Process	Viscosity, poise (Pa · s)	Printed film thickness, μm
lithography	50–500 (5–50)	1–2
letterpress	20–200 (2–20)	3–5
flexography	0.1–1 (0.01–0.1)	6–8
gravure	0.1–0.5 (0.01–0.05)	8–12
letterset	30–300 (3–30)	1.5–3
screen	1–100 (0.1–10)	20–100

Basically, all printing inks consist of a colorant (see Colorants), which is usually a pigment but may be a dye, and a vehicle that acts as a binder for the colorant and a film former. Vehicles generally consist of a resin or polymer and a liquid dispersant, which may be a solvent, oil, or monomer. Many other additives are used to provide some specific property or function to the ink or ink film. The ink constituents are discussed in detail in the following sections. Approximate viscosities and film thicknesses used in different printing processes are given in Table 1.

Liquid Inks

Flexographic ink. Flexographic printing utilizes a rubber or plastic printing plate and is essentially a typographic process, printing from a raised area. The inks used are very low in viscosity (see Table 1) and dry rapidly because of their relatively high volatility. Since the printing press used has a very simple ink distribution system, these volatile inks do not cause problems in drying on the inking rollers. The thickness of the ink film is controlled by the depth of engraving used on the anilox inking roller. The commercial speeds used for this type of printing range from 500 to 1000 ft/min (152–305 m/min). Flexographic ink accounts for more than half of the ink utilized in decorative packaging.

Flexographic inks may be subdivided into two general classes: solvent-based and water-based. The volatile solvents selected for the formulation must be chosen with care, since they are in constant contact with the plate elastomer. It is important to screen these ink solvents with the plate and roller elastomers to ensure that no swelling or attack of the surface takes place. In most cases, alcohols are the material of choice for solvent-based flexographic inks, with additions of lower esters and small amounts of hydrocarbons. These are used as needed to achieve solubility of the vehicle resin and proper drying of the ink film at press speed. Solvent blends are required to achieve the best balance of viscosity, volatility, and substrate wetting, and must also take into account EPA (Environmental Protection Agency) requirements.

Water-based inks are increasing in popularity because of air pollution concerns and are being used in increasing amounts for both absorbent and nonabsorbent substrates. The solvent used in these inks is usually not 100% water.

They may contain as much as 20% of an alcohol to increase drying speed, supress foaming, and increase resin compatibility. The wetting of plastic substrates is also greatly aided by the lower surface tension of the alcohol–water mixture.

The resins and polymers used in the flexo-ink vehicles cover a wide range of chemistry and are selected to achieve adhesion to various substrates or to confer resistance properties to the dried ink film. A number of resin classes and the substrates upon which they are generally used are shown in Table 2. Note that the water-based vehicles are listed in a separate portion of the table, since they are generally emulsions or colloidal dispersions rather than true solutions.

A typical formulation for a white solvent-based flexographic ink used in high-speed printing of polyethylene film for bread bags illustrates the types of constituents present:

Titanium dioxide pigment	40%
Alcohol-soluble polyamide resin	20%
Nitrocellulose varnish	5%
Normal propyl acetate	5%
Ethanol	26.7%
Slip additive	0.3%
Wax	1%
Plasticizer	2%

Gravure ink. Gravure printing utilizes an etched or engraved cylinder to transfer ink directly onto the substrate. Because of the mechanics of filling the very small cells, the viscosity of gravure inks must be relatively low (see Table 1). This is also required in order to transfer ink from the engraved cells at high speeds. The rotogravure press uses a simple inking system, where ink is applied directly to the gravure cylinder and excess ink is scraped off with a doctor blade. The ink-film thickness is determined by both the depth of engraving and the area of the cells, as these two factors determine the volume of ink transferred. Commercial speeds vary from 800 to 1500 ft/min (244–457 m/min) or more. Gravure ink is the second-largest category of packaging inks. Together with flexographic ink, these two liquid inks account for more than 80% of all packaging ink.

Rotogravure inks can also be classified as solvent-based or water-based types. The solvent-based inks are not as restricted by plate compatibility problems as are the flexographic inks because the plate is metal and resistant to nearly all solvents. Since a wider variety of solvents can be utilized in manufacturing a rotogravure ink, a much wider range of resin types and resin molecular weights can be utilized in achieving the desired ink properties. This means that gravure inks for packaging can be tailored for good adhesion to the widest selection of substrates. The classes of solvents that may be found in these inks include the following types: aromatic hydrocarbons; aliphatic hydrocarbons; alcohols; esters; ketones; chlorinated solvents; nitroparaffins; and glycol ethers. Even uncommon solvents such as tetrahydrofuran can be utilized in the manufacture of inks for certain speciality applications.

The resins and polymers used in packaging gravure vehicles also cover a wide range of chemistries and are chosen to achieve adhesion to various substrates or to confer specific properties to the finished ink film. These are listed in Table 2.

Water-based gravure inks are used widely for printing paper and paperboard substrates, and their use for nonpaper substrates is growing because of concern about air pollution. The formulation of these inks is very similar to those used in water-based flexography and the polymers are also very similar. These are listed in a separate section of Table 2. A significant problem that can occur with water-based inks in gravure is the drying of the vehicle polymer or resin in the engraved cells. This can occur during shutdowns of the press. Some of these materials are not readily resoluble once they have dried because they are usually emulsions or colloidal dispersions in water. To assist in the drying of water-based inks, the gravure cylinders are usually engraved or etched with shallower cells. A thinner ink film is applied, which contains less water to evaporate. This also means, however, that the amount of pigment in the press-ready ink must be higher in concentration to achieve the same relative printing density as the solvent-based ink. A typical formulation for a water-based packaging ink such as used for printing of paperboard cartons (see Cartons, folding) is given below:

Organic pigment	16.5%
Clay extender (wiping aid)	5%
Acrylic emulsion	40%

Table 2. Liquid Ink Applications for Packaging: Flexographic and Gravure

	Paper/paperboard	Foil	PE	PP	Vinyl	PET	Cellophane
Solvent ink systems							
NC–maleic	X	X	X	X		X	
NC–polyamide	X	X	X	X		X	
NC–acrylic		X		X			X
NC–urethane				X		X	
NC–melamine	X	X					
chlorinated rubber	X	X					
vinyl		X			X		
acrylic	X	X	X		X		X
ASP–acrylic			X	X			
CAP–acrylic			X	X			
styrene	X						
Water ink systems							
acrylic emulsion	X	X	X	X	X	X	X
maleic resin dispersion	X	X					
styrene–maleic anhydride resins	X	X	X	X			

NC = Nitrocellulose
ASP = Alcohol Soluble Propionate
CAP = Cellulose Acetate Propionate

Plasticizer	2%
Isopropyl alcohol	7%
Wax	2.5%
Morpholine (pH controller)	1%
Water	26%

Screen ink. Screen printing, which accounts for a relatively small percentage of ink used in the packaging market, is used on low volume specialty items or where very thick films are desired (see Table 1). Screen ink is included in the liquid-ink section because of its paintlike rheology and chemistry.

Screen inks are applied by squeegeeing the ink through a stencil screen with a rubber blade. The inks must have adequate flow to pass through the screen, but must also have enough body to resist dripping and stringing when the screen is lifted. The two largest classifications of screen inks are solvent types and plastisol types, although water-based, radiation-curing, and two-part catalytic systems are also available. A solvent-based formulation for printing on vinyl is given below:

Organic pigment	10%
Titanium dioxide	20%
Vinyl resin	18%
Acrylic resin	8%
Glycol ether solvent	29%
Ketone solvent	15%

Paste Inks

Offset lithographic ink. Lithographic printing uses a planographic plate in which the ink-receptive image is chemically differentiated from the nonimage area. Since these plates are constantly wetted with a dampening solution, the inks must resist the chemicals contained in these solutions without changing in their printing characteristics. The thickness of the ink film in this process is 1–2 μm which is the thinnest film in any commercial process. Because of this, the colorant concentrations in lithographic inks are generally higher than those found in inks for other processes. Lithographic inks generally have relatively high viscosities due to the ink-distribution systems used on the press equipment for this process. It is also common to find a gelled consistency in the body of these inks because of the need to obtain high printing resolution and faithful reproduction of the plate image. Image quality of half-tone reproduction is extremely high.

Sheetfed offset lithography. This process is widely used for printing on packaging board, paper, metal, and plastic sheets. The use of the offset blanket permits excellent reproduction even on surfaces that are not entirely flat owing to the compressibility of the blanket material. The inks used for most conventional sheetfed printing are dried by an oxidative process and may also be accelerated by the use of infrared radiation. These inks set rapidly in a matter of seconds but are not truly dry for several hours. In the case of metal, the inks are usually reactive only at high temperatures of ≥300°F (≥149°C), which are achieved by passing the metal sheets through a heated oven after removal from the press.

Drying of sheetfed lithographic inks can be done with ultraviolet radiation, which is applied by means of high-powered mercury arc lamps immediately after printing. This process is widely used in the production of packaging that must be die cut and finished in-line, such as cartons for cosmetics and alcoholic beverages. Ultraviolet drying of the ink offers significant energy savings in metal decorating where it replaces long, energy-consuming ovens.

The formula given below is for a typical sheetfed offset ink for paperboard:

Organic pigment	20%
Phenolic–hydrocarbon resin varnish	40%
Drying oil alkyd	10%
Hydrocarbon solvent (500–600°F or 260–316°C)	25.5%
Wax	2.5%
Cobalt drier	1%
Manganese drier	1%

A formula for a simple radiation-curing (see Curing) metal-decorating ink is

Organic pigment	15%
Acrylate oligomer	40%
Acrylate monomer	30%
Photoinitiator and sensitizer	8%
Wax	3%
Tack reducer	4%

Web offset lithography. In web offset, the same lithographic principles that are used in sheetfed lithography are applied to the printing of the substrate in the form of a web. The web of the substrate is fed into the printing press for decoration from a large roll. The printed substrate, upon exiting the press, must be dried immediately so that the finished product can either be rewound, sheeted, or finished in-line. The most common method to dry the printed ink immediately is the utilization of a high-temperature oven that employs recirculated hot air at a temperature of about 250–350°F (121–177°C). For this reason, the handling of plastic substrates is generally not possible because of the distortion of the substrates at these high temperatures. Drying inks with ultraviolet or electron beam radiation allows the printing of temperature-sensitive substrates. Generally, web offset printing in packaging is confined to paper and board printing. However, drying technology utilizing radiation curing promises to offer the packaging market the advantages of high-speed printing, the quality of offset lithography, and the lower cost of offset plates compared to gravure cylinders. Typical speeds for web offset printing are 800–1200 ft/min (244–366 m/min).

A formula for an electron-beam-curable web offset ink is given below:

Organic pigment	16%
Acrylate oligomers	40%
Acrylate monomers	30%
Extender resins	10%
Wax	2.5%
Stabilizer	1.5%

Letterset ink. This printing process, formerly known as dry offset, is a combination of letterpress and offset in that the printing plate uses raised images, but the printing on the substrate is accomplished with a rubber blanket. Therefore, the inks used in letterset generally have the viscosity and body of a letterpress ink. The predominant use of letterset printing in packaging is for the decoration of two-piece metal cans (see Metal cans, fabrication). Plastic preformed tubs and containers are also printed utilizing the letterset process on a mandrel press. Two-piece can printing is accomplished on special presses that produce at the rate of 1200 cans per minute with up to five colors and a clear varnish. The inks for beverage cans are generally dried by heat. Ultraviolet drying is used by several metal-decorating printers, primarily for beer cans. Ultraviolet is widely used for curing plastic containers where thermal sensitivity is a serious problem and heat curing cannot be used. The metal-decorating ovens used for thermal curing have recently gone to short cycles that use temperatures of 600°F (316°C) for only a few seconds to cure the inks. A formula for a typical heat-curing two-piece can ink for short cycle ovens is given below:

Titanium dioxide pigment	55%
Melamine resin varnish	13%
Polyester resin varnish	26%
Acid catalyst	1%
Slip additives	2%
Tack reducer	3%

Letterpress inks. These inks are used primarily for printing corrugated packaging (see Boxes, corrugated), although small amounts are still used for folding-carton and multiwall-bag (see Bags, paper) printing. Since letterpress uses raised images, the inks are fairly heavy in body (Table 1), and the primary mechanism for drying is oxidative or absorptive. The inks are generally formulated in a manner similar to sheetfed offset inks but with a slightly lower viscosity. This printing method has been losing market share to the other printing processes for a number of years because of high preparatory and labor costs. This is particularly true in corrugated printing, which is going primarily to water based flexography.

BIBLIOGRAPHY

R. Leach, ed., *The Printing Ink Manual,* Van Nostrand Reinhold, Berkshire, UK, 1988.

General References

L. Larsen, *Industrial Printing Inks,* Reinhold Publishing Corp., New York, 1962.

H. Wolfe, *Printing and Litho Inks,* 6th ed., MacNair-Dorland Co., New York, 1967.

The Printing Ink Handbook, National Association of Printing Ink Manufacturers, Hasbrouck Heights, NJ, 1988

Ultraviolet / Electron Beam Curing Formulation for Printing Ink, Coating and Paints, 3 Vols., SITA Technology, London, 1991.

R. W. Bassemir
A. J. Bean
Sun Chemical Corporation

IN-MOLD LABELING. See Decorating.

INTEGRATED PACKAGING DESIGN AND DEVELOPMENT

Many factors drive the performance of a company in today's global marketplace. Among the most important of these factors is the manner in which the company manages information, strategy, people, and time. If an organization can efficiently utilize its people and information to design and launch products within a very short period of time, that company will dominate its competitors. It should be recognized that speed alone is not enough. This accelerated development cycle must be supported by a customer-driven strategy that incorporates current and future requirements in terms of quality, costs, lead time, flexibility, variability, and service. "Great!" you might say, "But how is this accomplished?"

Recent management trends have advocated focusing on corporate culture to drive toward better performance. "Empowerment groups," "self-directed work teams," "quality circles," etc. have all focused attention on the need for eradicating barriers to ideas and suggestions for improvements. Some advocate implementation of "total quality management" (TQM) to refocus energies and radically change the culture of the organization. Still newer concepts such as "reengineering the organization" continually point to the need for sweeping change in the corporate culture, and ultimately the design strategy, of U.S. companies.

In addition to efforts directed toward changing corporate culture, some experts have advocated using various "operations management techniques" to focus on needed change. Just-in-time scheduling, MRP II (manufacturing resource planning), KanBan scheduling, etc have focused on the need to control the resources and materials needed for production in order to reduce waste. Plant layout techniques have been designed to minimize materials-handling and transportation costs.

Other experts advocate more technology-oriented approaches. Management information systems (MISs) are recommended to coordinate information flows within a corporation. Expert systems and other artificial intelligence (AI) techniques are advocated to capture or reproduce behavior and capabilities of valuable human workers. CAD/CAM, robotics, flexible manufacturing systems (FMSs), etc are offered as solutions to various production problems. Even the extensive scope of computer-integrated manufacturing (CIM) and the ideal of the computer-integrated enterprise (CIE) often serve to point out just how many areas of importance can be (or have been) neglected in a business enterprise.

These models, tools, and/or techniques can often individually help to accomplish many of the goals stated initially. However, as these methods become more and more powerful, they also become increasingly difficult for most people to understand, let alone use effectively. This confusion has led to efforts to develop various design concepts and methods that are intended to help solve design and manufacturing problems. These solutions often focus only on specific areas and are capable of providing significant assistance only during a specific stage in the life of a product. In order to be truly useful, it is necessary that these methods and concepts be merged into a larger methodology that is structured to ensure that adequate focus is given to each concept at the appropriate time.

For instance, design for assembly (DFA) and design for manufacturability (DFM) are well-known concepts that advocate consideration of manufacturing and assembly requirements during the product design effort. Few will argue with these concepts, and many companies actively pursue this approach to design. However, DFA/DFM is not enough. While concepts such as DFA/DFM and others would include consideration of the packaging of the finished product if they were interpreted in their most comprehensive meanings, this is seldom how they are applied. Very little emphasis has been placed on integrating the product design process with the packaging design process to achieve better end results. As a result of this, many manufacturing and distribution problems can be traced to poor product and packaging interactions that are caused by an inadequate design process.

It has been noted that "quality cannot be built-in unless it is designed in" and "about 40 percent of all quality problems can be traced to poor design" (1). It is also generally recognized that conventional design and manufacturing procedures can often result in long lead times, high costs, and low process efficiencies. Knowing these things, it is easy to see that a better design methodology is needed. Only proper product and packaging design can ensure product quality from the factory to the point of use. What is needed, therefore, is an integrated-design model that can be used by integrated product and packaging design teams as a prescriptive guide to seek a combined solution for both the product and its packaging.

Challenges

The traditional design methods of "craft evolution" and "design by drawing" are no longer adequate in the complex global marketplace that exists today. Even the best of the current design models and methods cannot promise to solve all product or package design problems. Furthermore, it is likely that there will always be opportunities for research in the development of design methodologies (2).

Traditionally, the product, instead of both the product and its packaging, is the major focal point in design and manufacturing. Product design has gradually developed into a process that can involve sophisticated technology and overwhelming complexity. In many companies, design has become too complicated for a single person to complete and control the entire design without assistance.

Similar problems are experienced by engineers all across the typical corporation, in product design, packaging design, and production engineering alike. Usually, most designers lack sufficient information, experience, or knowledge to make design decisions in fields outside their own expertise. Even those engineers familiar with the purpose of the product and the manufacturing systems that will create it cannot be expected to also be informed about scheduling considerations, capacity constraints, environmental hazards, marketing preferences, etc. Even if they are assumed to have access to experts or outside resources, communication problems can still result in errors in interpretation of advice. Most of these limitations could be overcome, given enough time and adequate resources. However, time is a precious commodity and cannot be "wasted" without serious repurcussions. Additionally, the likelihood of imitation by competitors, the possibility of unsuccessful marketing strategies, and the virtual certainty of quickly changing customer preferences all combine to require that products and/or their packages be redesigned or abandoned frequently.

A frustrating element in this problem is that package design is seldom a technology-based part of a product's development plan. Insufficient analytical consideration of the package itself is not an aberration, but rather the norm. Perhaps this is because decisionmakers do not understand the importance of truly integrated design—including packaging. To assist decisionmakers in understanding the place of packaging engineering in modern manufacturing, the following section briefly discusses some of the current technologies, integrated-design models, and management methods that exist. After providing this context, the remaining sections will then present a comprehensive model/methodology that is intended to facilitate truly integrated design.

Integrated-Design Tools, Technologies, and Methodologies

DFA/DFM/DFE. A large number of authors have created various "design for" acronyms. Some of the more popular ones are discussed here.

Design for assembly. Boothroyd and Dewhurst (3,4) are perhaps the best-known advocates of DFA, and have written multiple articles describing their basic approach. In addition to articles, they (and others) have created software packages to assist in the design effort. Most of these packages consist of a structured design review that is intended to guide the user through a "checklist" of common design errors and recommendations. The guidelines provided for DFA offer guidelines for both manual and automatic assembly. Some of the specific recommendations include reduce the use of fasteners, modularize assembly, and design for top–down assembly.

Design for manufacturability. DFM focuses primarily on considering various tradeoffs related to "ease of processing" versus "form, fit, and function." This includes a stage of process identification in which various manufacturing process options are considered and compared. Once the likely manufacturing processes are identified, the next stage consists of modification of the product design to make the parts easier to manufacture. In some cases, several feasible manufacturing processes can be identified, and it is up to the design team to determine the most cost-effective process that provides the best design solution for their specific company.

This design paradigm (DFM) was developed, in part, to eliminate the likelihood of unmanufacturable parts being designed by product designers. DFM efforts are intended to help break down the proverbial "wall" between design engineering and production.

Design for the environment. Recent concerns about the environment have led to efforts to encourage designers to look at the entire life cycle of a product. Issues such as recyclability, ease of disposal, and clean operating characteristics have been focused on as important design criteria. The progress in this area has been slow, and limited mostly to efforts to gain "marketing advantages" by claiming "green" products. However, regulations and litigation are driving most companies to pursue "environmentally friendly" designs. This area obviously impacts the packaging industry, and many of the consumer pressures have been directed toward consumer product packaging.

Design issues in this area encompass more than simply ensuring that recyclable materials are used. Many plastics are recyclable, and industry (eg, the automobile industry) has dramatically increased the amount of plastic used in their products. This has not resulted in significant amounts of "automobile recycling," though. The problem is that, while the individual parts used in the vehicle are often made of recyclable material, the parts are not all made of the same materials.

The difficulty of separating these materials has made them uneconomical to recover.

In addition to the material selection and disposal issues, environmental design demands an awareness of the consequences of specific manufacturing processes. Knowing that a process will provide certain desirable characteristics to a product must be balanced with whatever use of solvents and chemicals might be necessary, as well as with the waste that will be generated by the process. Even something as seemingly "innocent" as specifying something as a "painted" surface, has potential ramifications on the environmental impact of the proposed product.

DFX (design for "fill in the blank"). In addition to the three design issues discussed above, it is possible to find articles on many additional "design for" acronyms. Sometimes it seems that everyone has a "pet emphasis" that the feel should be incorporated into the engineering design curriculum. These methodologies have provided what has become a cottage industry for consultants, focused on promoting various design criteria. Recent years have seen such methodologies as design for maintenance, design for modular replacement, and design for inspection. Despite the importance of many of these concepts, it is not really reasonable to expect designers to know "everything", even if they claim too!

Simultaneous or concurrent engineering If you cannot expect individual designers to know it all, bring together teams of specialists and have them work together. This is the concept behind "simultaneous" or "concurrent" engineering. The advantages of this approach include the increased likelihood that the team will be making good decisions early in the design process. According to Fabrycky (5):

> Fully two thirds of the life-cycle cost of a product or system is committed by the time conceptual and preliminary design is completed. This committment reaches about 80% upon completion of detail design and development.

If production requirements, ease of assembly, serviceability, and likely disposal or reuse have been considered during design, it is relatively easy to believe that the resulting product will be more cost effective. The possible downfalls come in the additional resources required to assemble and support the design team. Results of simultaneous engineering can be dependent on the personalities and strengths of the team members. Simply being together doesn't mean that good designs will result. Care must be taken to ensure that team members are properly trained in interpersonal communications, and that the team "management" is skilled and able to motivate the members. If properly implemented, simultaneous engineering can shorten development cycles and improve manufacturability, but it requires significant support and very dedicated management.

CAD. One of the most widely accepted information automation applications has been the effort to move design off of the drawing board and into the computer via computer-aided design (CAD). "Survey results show that CAD is by far the technology with the greatest degree of use" among industrial automation tools (6). Extensions beyond simple drawing packages have made possible numerous tasks that simply could not be done before, or required tremendous manual calculation efforts. Some examples would include finite-element analysis, simulation of metal flows within casting molds, off-line programming of robots, and many other valuable modeling tasks.

Nonetheless, the lack of universal standards and competition between different vendors and hardware platforms has led to a "Tower of Babel" in actual practice. Most users feel that the benefits of CAD outweigh the problems, but full benefits are not being realized except within vertically integrated proprietary systems.

CIM. Computer-integrated manufacturing (CIM) was one of the most popular buzzwords of the 1980s. No precise definition of CIM was ever accepted, but most authors agreed that the definition incorporated a focus on using computer technology to make information broadly available across all departments of a company. The emphasis on information management, and the inclusion of "nonproduction" departments within the network was one of the hallmarks of early CIM papers. Although no longer viewed as the top priority of most companies, CIM is still considered a worthwhile goal.

The CIM concept of totally open and available information resources is not really feasible with today's commercially available software. Custom software development work or vendor-specific hardware implementations can approach it, but these approaches make the customer company very dependent on a specific hardware platform and/or software approach. So-called open-systems vendors have tried to remove some of the barriers, but have done so primarily in order to combat dominance by large individual corporations rather than any true intention to "open up communications."

CIE. Computer-integrated enterprise (CIE) is the attempt to "out-CIM" CIM. This concept basically reemphasizes the automation of information management in the "nonmanufacturing" or "business" portions of the company. Actually, the "business portions" in most companies were automated fairly early (7). Automation of information management closer to the shop floor is perhaps more difficult, and therefore is taking longer to realize.

The key to understanding CIE is knowing that American industry has only recently rediscovered the importance of outside forces on how a company is run. This recognition of the importance of interacting with customers, suppliers, regulators, and investors has resulted a desire to make electronic links to as many areas as possible, and ensure that information is freely available. The basic argument for the "name change" from CIM came from a assertion that "manufacturing" was only one portion of the company, and the "computer integration" should be more comprehensive. The counterargument says that manufacturing describes the entire scope of industries that this concept affects, and that switching to "enterprise" is only a superficial change.

QFD. By definition, quality function deployment (QFD) is a method or philosophy of

> mapping, the elements, events and activities necessary throughout the development process to achieve customer satisfaction. A techniques-oriented approach using surveys, reviews, analyses, and robust designs all centered on the theme of translating

the . . . Voice of the Customer . . . into items that can be measured, assessed, and improved (8).

The roots of QFD can be traced to efforts by Mitsubishi Heavy Industries, Ltd. at the Kobe Shipyards in Japan in 1972. The Ford Motor Company began using QFD in the early 1980s, and it is used extensively today across a wide range of industries, prodcuts, and services.

Essentially, QFD is an integrated matrix-based approach applied toward product development. At its simplest, QFD seeks to serve as the "voice of the customer" by translating their desires or "customer voice" (also called "wants" and "needs") into specific requirements that are addressed throughout the entire design process. These customer "wants," and the engineering characteristics, or "how's", are often built into what is referred to as a "house of quality" (HOQ). The HOQ utilizes customer perceptions, relative relationships, and relative measures of importance to ultimately generate quantitative measures of customer wants, or design requirements. This method can be repeated to identify other specifications such as part characteristics, manufacturing operations, and production requirements. Conceivably the methodology could be applied with a "cradle to grave" approach for defining customers, thus generating distribution, disposability, or recyclability requirements. Obviously, packaging design could be incorporated in this "expanded view" of QFD.

The QFD framework provides a total systems approach often leading to a reduction of production start-up problems, a reduction in the length of the total design process, and an increase in the use of engineers as system planners rather than simply problem solvers. It also serves as a mechanism for maintaining and translating institutional knowledge to new design teams. Most importantly, however, is the fact that the "voice of the customer" is "listened to" throughout the process. QFD is not the final solution; it also has its limitations. It can be time-consuming, costly, and initially a very complex endeavor. However, the most mundane aspects are readily accomplished through the use of specialized computer software, and the method is fairly widely accepted.

AI methods—neural nets, knowledge-based systems. Among all of these discussions of management approaches and efforts to identify important design considerations, authors will periodically recommend and make use of various advanced computing technologies to implement their particular system. These technologies are useful because, even with a team of engineers simultaneously designing a product or system, some areas of expertise will be missing or under represented.

A possible solution is to gather, structure, and automate information from recognized experts and other sources. Automated "helpers" and "advisors" are becoming more and more prevalent, and currently have the ability to check for basic flaws and make general recommendations. These advisors are still somewhat primitive. Some CAD software will ask if you "realize" that you have extra material that is "unneeded" to meet the defined strength specifications for a mechanical part and offer to take it away. While this is an important feature, and a dramatically improved tool for part designers, the human designer still makes the decisions. Other software tools can automatically route wires on a circuit board or pipes within a building, but still must be "sanity checked" by the human designer.

The current maturity of AI-driven assistants and advisors for designers is low. They are limited to generalizations and special cases, and need to be balanced by the creativity and experience of the designer. Automatic toolpath generation software, for instance, results in feasible numerical control code, but not necessarily optimal toolpaths. Picking the wrong "angle" in a pocketing routine can result in dramatically wasteful toolpaths. Some AI methods are more directly usable than others, and they each have different capabilities and strengths. Most likely AI-driven advisors will need to be used as embedded systems within other user applications (such as CAD systems) rather than as standalone systems.

Packaging-design methodologies, tools, and techniques. Generalized integrated packaging design techniques are much less prevalent in the literature than those typical of manufacturing, product, or process design. This is not meant to imply that individual companies do not practice packaging design in an integrated fashion; rather, that if they do, the methods or techniques are not typically found in the public domain. However, some packaging-design techniques can be found in the literature. For instance, the six-step method for protective-packaging design (9), and the four-step packaging design technique (10) for consumer goods, are well documented in the literature. In theory they also seek to design packaging concurrent with the product design stage. These methods appear to be more suitable for distribution packaging rather than primary and/or secondary packaging.

Packaging tools or software also appear to be less frequent than tools typical for product and manufacturing design. Optimization software packages which seek to optimize cubic space for primary, secondary and tertiary packaging, such as CAPE (11) or TOPS (12), provide tools that could be useful in an integrated-design method. However, they generally are not used in that fashion. Simulation software for modeling manufacturing and packaging lines, such as Pristker Corporation's *Packaging* (13), provides a more specific design technique. However, it is commonly used to review and analyze existing lines rather than as an integrated "concurrent design" technique. It also focuses primarily on the machinery details, an important but restricted viewpoint.

Within the literature, there are also flowcharts and descriptive methods that, while they may qualify as a tool or technique for packaging design and development (14,15), are not typically presented as an integrated-design methodology. Examples of more specific tools that ask general questions such as the Institute of Packaging Professionals' "packaging recycling guidelines," also exist.

Each of the tools, techniques, and methods presented above can be beneficial within the overall design activity of a firm, and can help to achieve some or all of the goals described initially in this paper. However, there appears to be no one tool, technique, or method that incorporates the multitude of seemingly diverse design activities—product design, process design, and packaging design. The following section provides a descriptive model that views each design area as equally important in the process and provides a roadmap for integrated as well as concurrent design.

An Integrated-Design Model

Design models are always rather abstract things. They describe a framework wherein simple ideas and relationships can sometimes seem very complex or confusing. Yet the goal of a good design model is to illustrate the connections between different processes and stages in the design effort and make it easy to track the progress of a given design project. Figure 1 presents an integrated-design model that can be used to guide and enhance the design process. This model emphasizes the need for a comprehensive consideration of both the manufacturing and packaging systems that will be used to produce the product and the product's own unique functional characteristics.

Specific to packaging, the model leads to a rather obvious insight. That insight consists of the following:

> If a product is viewed as consisting not only of the components that will be removed and used by the end user, but also of the delivery system which preserves and presents the components to the end user, then "packaging design" is no longer a separate entity. It is, and should be an integral part of the product design.

Clearly, a properly designed package can enhance a product's attractiveness in order to win the initial sale. Its structural and functional strengths can ensure an undamaged product so that people will buy the product again. It can assist a firm in distributing their products economically and profitably.

The integrated-design model focuses on simplification of the design process via concurrent communication and proper utilization of the various external support systems currently available. Figure 1 depicts a process designed to coordinate the seemingly diverse functions within an organization simultaneously and at an early stage in the product research and development process through the use of systems analysis. Requirements and capabilities are then generated and subsequent preliminary efforts commence. At this point, it is important that each member of the design team have some knowledge of the other team members' areas. Ideally, each team member could be required to have expertise in other disciplines. In reality and practice, that can be very difficult to achieve, considering time constraints and personnel limitations within the firm. In order to mitigate these problems, the integrated-design model employs external supporting systems. Table 1 lists some of the more common external support systems that may be used in the process. External support systems are also used in the next step to reach final criteria and considerations.

Table 1. External Support Systems

DFM/DFA	QFD
ABC costing	CIM/CIE
Computer-aided packaging design	Design optimization
SPC	TQM
CAD/CAM/CAE	Simulation
Group technology	CAPP
Artificial intelligence (expert systems, neural nets)	Logistics
Reliability and maintainability	GDSS
Value engineering and analysis	Systems analysis

This model is clearly not rocket science, but more appropriately common sense. However, common sense and common practice are not always the same thing. Therefore, this method is depicted in this chapter to help meet the needs of industry, and was derived from discussions and surveys conducted within industry (16).

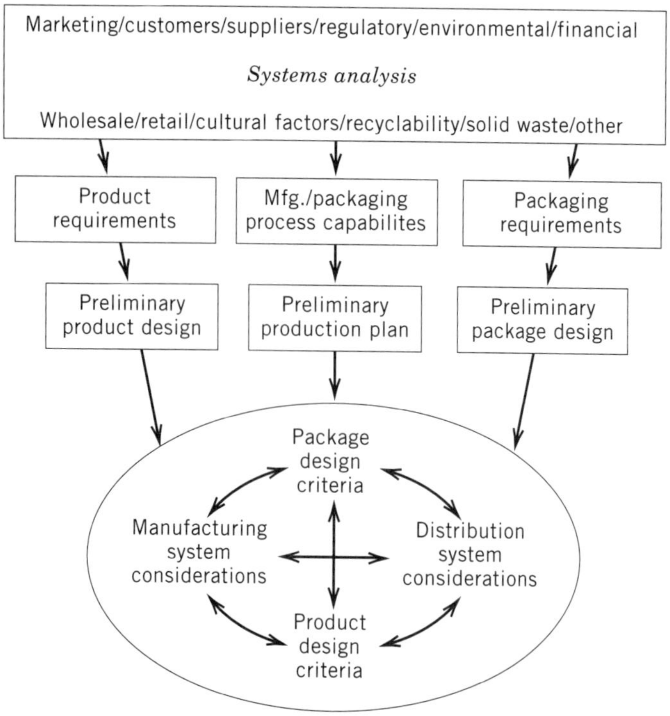

Figure 1. Integrated-design model.

Conclusions

Integrated design is no longer an option. It is a necessity. Meeting constantly expanding customer expectations within ever-shrinking time periods, while still achieving both profit and quality objectives, is a daunting task. But it need not be insurmountable. Old ways of doing things must cease, while new methods must be embraced and implemented. This article has pointed out the potential benefits of integrating all of an organization's design activities in a concurrent or simultaneous fashion while briefly describing some of the most common integrated process and product design methods. It has pointed out several typical packaging design tools, techniques, and methods. Finally, an integrated design methodology has been presented. What remains to be done is for the professional packaging community to recognize their important role in this process, and rise to the challenge of integrated design. This simple method can be used by any organization wishing to integrate its seemingly dissimilar design activities. It places packaging design in its proper peer relationship to product and process design, and provides a framework for incorporating modern packaging science.

BIBLIOGRAPHY

1. R. Suri and M. Shimizu, "Design for Analysis: A New Strategy to Improve the Design Process" in *Research in Engineering Design*, Springer-Verlag, New York, 1989, pp. 105–120.

2. J. C. Jones, *Design Methods: Seeds of Human Futures,* Wiley-Interscience, New York, 1970.
3. G. Boothroyd and P. Dewhurst, "Design for Assembly: Selecting the Right Method," *Machine Design* pp. 94–98 (Nov. 10, 1983).
4. G. Boothroyd and P. Dewhurst, "Product Design for Manufacture and Assembly," *Mfg. Eng.* pp. 42–46 (April 1988).
5. W. J. Fabrycky, "Engineering and System Design: Opportunities for ISE Professionals" in Joe Mize, ed., *1989 IIE Integrated Systems Conference Proceedings, Guide to Systems Integration,* Institute of Industrial Engineers.
6. K. K. Boyer, "Balancing Investments in Advanced Manufacturing Technologies and Infrastructure Improvements," *SME Blue Book Series,* Society of Manufacturing Engineers, May 1995.
7. E. D. Bennett, B. M. Fossums, R. D. Harris, S. A. Reed, and J. Franklin Skipper, "Why Strategy Drives Integration" in Joe Mize, ed., *CIM Review, Guide to Systems Integration,* Institute of Industrial Engineers,
8. *Quality Function Deployment: Translating the Voice of the Customer,* ASQC Dayton Section presentation to Sinclair Community College, Dayton, OH, Oct. 27, 1989.
9. Lansmont Corporation, 6-Step Method For Cushioned Package Development, Lansmont Corporation, Oct. 1988.
10. F. C. Bresk, "Using a Transport Test Lab to Design Intelligent Packaging for Distribution," *Test Eng. Manage.* pp. 10–17 (Oct./Nov. 1992).
11. CAPE Systems, Inc., 2000 N. Central Expressway, Suite 210, Plano, TX 75074.
12. TOPS Engineering Corporation, 1721 West Plano Parkway, Suite 209, Plano, TX 75075.
13. Pritsker Corporation, P.O. Box 2413, West Lafayette, IN 47906.
14. W. Soroka, *Fundamentals of Packaging Technology,* Institute of Packaging Professionals, Herndon, VA, 1995, Chapter 20.
15. S. A. Raper, *Packaging Management Organizations: A Case Study,* master's thesis, The University of Missouri-Rolla, Rolla, MO, 1987.
16. M.-R. Sun, *Integrating Product and Packaging Design for Manufacturing and Distribution: A Survey and Cases,* Ph.D. dissertation, The University of Missouri-Rolla, Rolla, MO, 1991.

Stephen A. Raper
Robert Borchelt
University of Wisconsin
Milwaukee, Wisconsin

INTERMEDIATE BULK CONTAINERS

Intermediate bulk containers (IBCs) are a rapidly growing class of packaging systems designed initially for the transport of hazardous materials (but also currently used with nonhazardous ladings) in quantities intermediate between those packaged in drums and in tank trucks. The growing popularity of IBCs (at the expense of drums) can be attributed to their efficient handling (typically one IBC, handled by forklift truck, carries the amount of product of five or six drums in the space taken by four) and to simplification of disposal problems (only one package, rather than five or six, to be disposed of). In addition, many styles of IBC are designed for reuse. Many shippers use a closed-loop system, wherein IBCs shuttle back and forth, for example, between a chemical manufacturer and its customer, many times before the IBC is worn out.

The packaging of many hazardous materials (or dangerous goods, in international parlance) in IBCs has been sanctioned by both national and international regulatory agencies, such as the U.S. Department of Transportation (DOT) in its "Hazardous Materials Regulations" (HMR) (1), the International Maritime Organization (IMO) in its *International Maritime Dangerous Goods Code* (IMDG) (2), and regulatory agencies of many other nations that have based their national hazardous materials transport regulations on the UN's *Recommendations on the Transport of Dangerous Goods* (3).

While the IBC's forebears were actually developed in the United States (types 52, 53, 56, and 57 metal portable tanks), IBCs have grown to maturity, both commercially and with respect to transport regulations, in Europe. Thus while DOT's IBC regulations (HM-181E) were promulgated in 1993, Chapter 16 of the *UN Recommendations* came out in the mid-1980s, and Chapter 26 of the IMDG went into effect in the late 1980s.

Definition

An IBC, as defined in the recently promulgated DOT regulations, is "a rigid or flexible portable packaging, other than a cylinder or portable tank, which is designed for mechanical handling." The DOT regulations specify that the minimum capacity shall be 450 L (119 gal) and the maximum capacity shall be 3000 L (793 gal). (Under the UN Recommendations there is no minimum size prescribed, but 3000 L is the maximum.) Beyond this, the regulations allow a wide variety of sizes, shapes, and constructions. IBCs are generally mounted on pallets, for ease of handling by forklift truck. If they are intended for the packaging of liquids, they will generally be designed to empty almost completely through a valve at the bottom, and will usually have a 6–8-in.-diameter fill port with a screw cap on the top.

IBCs may be described as metal, rigid plastic, composite (a rigid plastic bottle or flexible plastic bag within a protective framework), fiberboard, wooden, or flexible (with body constructed of film, woven plastic, woven fabric, paper, or a combination thereof). Any of these IBCs may be used for the shipment of hazardous solids, but only the metal, rigid plastic, and composite IBC types may be used to ship hazardous liquids.

Containers fitting the description of any of the IBC types may, of course, also be used for the packaging of nonhazardous products, but the terminology and definitions pertaining to IBCs arose from the regulatory community. IBCs are performance-defined packagings, whereas most other packagings have been developed and defined according to construction standards, and by industry, rather than by government regulators. This perhaps accounts for the great diversity of container styles that legitimately fall under the heading "IBC." The performance tests that IBC's must pass in order to be certified for the carriage of hazardous materials include a drop test; a bottom or top lift test, depending on how the IBC is intended to be handled; a leakproofness test and a hydrostatic test, for IBCs intended for the packaging of liquids; a stacking test for IBCs intended to be stacked; a topple test, a righting test, and a tear test for flexible IBCs; and, only in the United States, a vibration test.

The UN Code

The regulations provide a complex coding scheme (the UN code) to identify the various types of IBC, to indicate the per-

A typical marking for a composite or rigid plastic IBC intended for liquids but not designed for filling or discharge under pressure would be as follows:

(u n) 31HA1/Y/03 88/USA/MFR/7620/2540
1250/235KG/100/09 09/09 90

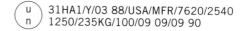

Note: Additional markings would be required for an IBC designed to be filled or discharged under pressure.

The elements in this marking have the following meanings:

(u n) 31HA1/Y/03 88/USA/MFR/7620/2540

- Maximum permissible gross weight (in kg)
- Stacking test load (in kg)[a]
- Name and address or registered symbol of manufacturer or approval agency
- Authorizing State (country)
- Month and year of manufacture
- Packing group to which IBC has been tested
- Code designating the type of IBC[b]
- United Nations packaging symbol

1250L/235KG/100/09 90/09 90

- Month and year of last inspection
- Month and year of last leakproofness test
- Test (gauge) pressure (in kPa)[c]
- Tare mass (in kg)[c]
- Reted capacity in liters[c]

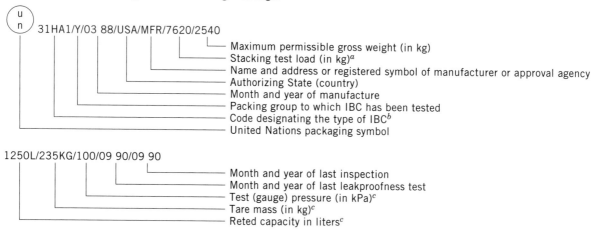

[a] For an IBC not designed for stacking, the figure "O" must be inserted in this position.

[b] The code shown is that designating a composite packaging consisting of a metal (sheet or wire) outer casing and rigid plastic inner receptacle. The corresponding code for a rigid plastic IBC intended for liquids and not designed for filling or discharge under pressure would be "31H2" or, in case of an IBC fitted with structural equipment designed to withstand the whole load when IBCs are stacked, "31H1."

[c] The unit measurement must be indicated in the marking.

Figure 1. UN marking code for IBCs.

formance levels at which design prototypes of the IBCs were tested, and to provide assurance that IBCs so marked meet the performance requirements of DOT and other regulations. A prototypical example of a UN code for a composite IBC with a rigid plastic inner receptacle and steel outer packaging is shown in Figure 1 (4).

Since there are so many different types of IBCs available for the packaging of different kinds of product, the first alphanumeric grouping of the UN coding scheme is used to identify the category of IBC. It consists of a two-digit number, followed by one or two capital letters, followed in some cases by a single numeral. The initial two-digit number provides information on the physical state of the hazardous material that may be packaged in the IBC, on whether the IBC is rigid or flexible, and, for solid ladings, on whether the lading is discharged by gravity or under pressure. Table 1, taken from the *DOT Regulations,* defines this part of the code. Next, for most IBC types, a single capital letter specifies the material of construction of the IBC. For composite IBCs, two capital letters are used: the first capital letter, H, specifies that the material of construction of the inner receptacle is plastic; a second capital letter identifies the material of construction of the outer, structural, packaging. Table 2 lists the identifying letters for the various materials of construction. The final numeral of this code section, if present, provides other information such as (*1*) for rigid plastic IBCs, whether it is intended to be stacked or free-standing; (*2*) for composite IBCs, whether the inner receptacle is rigid or flexible; and (*3*) for flexible IBCs, more specific information about the nature of the material of construction.

As shown in Figure 1, the other sections of the *UN Code,* which are separated by slashes (/), identify such information as the performance standard (X, Y, or Z) to which the IBC has been tested, as well as the date of manufacture, the authorizing state, the identity of the manufacturer, the maximum permissible load, the volumetric capacity, etc. Thus the

Table 1. IBC Code-Number Designations

	For Solids, Discharged		
IBC Type	By Gravity	Under >10-kPa (>1.45 psi) Pressure	For Liquids
Rigid	11	21	31
Flexible	13		

Table 2. IBC Code-Letter Designations

Code Letter	Material of Construction
A	Steel
B	Aluminum
C	Natural wood
D	Plywood
F	Reconstituted wood
G	Fiberboard
H	Plastic
L	Textile
M	Paper, multiwall
N	Metal (other than steel or aluminum)

UN Code for IBCs contains much more information than is found on other types of container intended for the packaging of hazardous materials.

Applications

There are many applications of IBCs that involve their use in the shipment of nonhazardous materials. These may range from the shipment of plastic pellets in "Gaylords" and dry chemicals in flexible IBCs to latex adhesives, pharmaceuticals, and liquid foods, such as vegetable oil, honey and hot sauce, in composite IBCs.

For determining the viability of potential applications for the shipment of hazardous materials in IBCs, the DOT regulations must be taken into account. To determine the appropriate packaging for the shipment for a hazardous material, one must first identify the hazard classification and then look the material up by its "proper shipping name" in 49 CFR 172.101, the DOT "Hazardous Materials" table. Column 8C of the table lists "Packaging Authorizations for Bulk Packaging," which includes authorizations for IBCs. Depending on the physical state of the lading and the severity of the hazard, Column 8C will direct the shipper to the appropriate paragraph in Part 173, Sections 173.240–173.245. If the reference in Column 8C is to Paragraph 240, ie, 173.240 (low-hazard solids), 241 (low-hazard liquids and solids), 242 (medium-hazard liquids and solids, including solids with dual hazards), or 243 (high-hazard liquids and dual-hazard materials that pose a moderate hazard), then an IBC of one kind or another may be used for shipping the product. If the reference is to paragraphs 173.244 or 173.245, then the type of hazard is such that DOT currently considers the use of IBCs to be inappropriate. In addition, Column 7, "Special Provisions," must be consulted. The Special Provision numbers listed in this column direct the shipper to notes that may restrict for a specific packaging type the general authorizations provided in Column 8C.

Some examples of hazardous materials that may be shipped in IBCs of the appropriate type are; lubricating oils, oil-based paints, paper- and water-treatment compounds, firefighting chemicals, agricultural chemicals, and various resins.

BIBLIOGRAPHY

1. "Hazardous Materials Regulations of the U.S. Department of Transportation," *Code of Federal Regulations,* 49 CFR, Parts 171, 172, 173, 178 and 180, as promulgated in Docket HM-181E in the *Federal Register,* 38040–38080 (July 26, 1994).
2. *International Maritime Dangerous Goods Code,* Amendment 26-91, International Maritime Organization, London, 1991, Chapt. 26.
3. *Recommendations on the Transport of Dangerous Goods,* 8th rev. ed., United Nations, New York, 1993, Chapt. 16.
4. G. Rousseau, Executive Director of RIBCA, the Rigid Intermediate Bulk Container Association, 1994, private communication, Figure 1.

GERALD A. GORDON
Industrial Container Division
Sonoco Products Company
Lombard, Illinois

INTERNATIONAL STANDARDS

International standards are developed in a variety of ways. Formal preparation by official international standards organizations is the most obvious route, but regulatory requirements often create standards, directly or indirectly. The purchasing policies of companies, organizations, government agencies, or even international bodies frequently involve compliance with various standards.

What Are Standards?

A standard has been defined as

> A technical specification or other document available to the public, drawn up with the cooperation and consensus or general approval of all interests affected by it, based on the consolidated results of science, technology and experience, aimed at the promotion of optimum community benefits and approved by a body recognized on the national, regional, or international level (1).

A voluntary standard can be incorporated into a contract or regulation and its provisions thus become mandatory.

The aims, or purpose, of standardization have been usefully summarized (2):

1. Provision of means of communication among all interested parties
2. Promotion of economy in human effort, materials, and energy in the production and exchange of goods
3. Protection of consumer interests through adequate and consistent quality of goods and services
4. Promotion of quality of life; safety, health, and the protection of the environment
5. Promotion of trade by removal of barriers caused by differences in national practices

In short, standards are intended to promote communication, efficiency, quality, safety, health, environmental protection, and free trade.

The International Organization for Standardization (ISO)

Established in 1947, the International Organization for Standardization (ISO) is a worldwide federation of national standards bodies (ISO member bodies) formed "to facilitate international coordination and unification of industrial standards" (3). A *member body* is the national body "most representative of standardization in its country," and thus only one such body for each country is accepted for membership in ISO. Member bodies are entitled to participate and exercise full voting rights on any technical committee of ISO. A *correspondent member* is normally an organization in a developing country that does not yet have its own national standards body. Correspondent members do not take an active part in the technical work, but are kept fully informed about the work. Table 1 lists the acronym of each member body and its country.

The work of preparing ISO standards is normally carried out through technical committees (TCs), their subcommittees (SCs), and their working groups (WGs). Each member body interested in a subject for which a technical committee has been established has the right to be represented on that committee. International organizations, governmental and nongovernmental, in liaison with ISO, also take part in the work. Each TC and SC has a secretariat assigned to an ISO member body. Details of each TC, including its secretariat and scope (determined by the ISO Council), are set out in the annual *ISO Memento.* The ISO technical committees directly concerned with packaging and the distribution of goods are shown in Table 2. It is important to note, however, that the work of other TCs often relates to packaging; for example, TC 6 "Paper, board and pulps," and TC 61, "Plastics," for materials standards and associated test methods. The scope of TC 34, "Agricultural Food Products," and TC 126, "Tobacco and Tobacco Products," includes packaging, storage, and transport of such products but is, in practice, limited to general, practical advice.

Because international standardization is a lengthy and expensive process, proposals for new standards must be justified by identifying the need, the aim(s) of the proposed standard, and the interests that may be affected. Once a document is developed, it is listed under the relevant Technical Committee in the annual *ISO Technical Programme* (4) as draft proposals or draft international standards (DISs) in increasing degrees of formality. When a technical committee agrees on a draft standard, it is proposed for approval by all ISO mem-

Table 1. ISO Member Bodies[a]

Country	Member Body
Albania	DSC
Algeria	INAPI
Argentina	IRAM
Australia	SAA
Austria	ON
Bangladesh	BSTI
Belarus	BELST
Belgium	IBN
Brazil	ABNT
Bulgaria	BDS
Canada	SCC
Chile	INN
China	CSBTS
Columbia	ICONTEC
Costa Rica	INTECO
Croatia	DZNM
Cuba	NC
Cyprus	CYS
Czech Republic	COSMT
Denmark	DS
Egypt	EOS
Ethiopia	ESA
Finland	SFS
France	AFNOR
Germany	DIN
Greece	ELOT
Hungary	MSZH
Iceland	STRI
India	BIS
Indonesia	DSN
Iran, Islamic Republic of	ISIRI
Ireland	NSAI
Israel	SII
Italy	UNI
Jamaica	JBS
Japan	JISC
Kazakhstan	KAZMEMST
Kenya	KEBS
Korea, Democratic People's Republic of	CSK
Korea, Republic of	KBS
Libyan Arab Jamahiriya	LNCSM
Malaysia	SIRIM
Mauritius	MSB
Mexico	DGN
Mongolia	MNISM
Morocco	SNIMA
Netherlands	NNI
New Zealand	SNZ
Nigeria	SON
Norway	NSF
Pakistan	PSI
Panama	COPANIT
Philippines	BPS
Poland	PKN
Portugal	IPQ
Romania	IRS
Russian Federation	GOSTR
Saudia Arabia	SASO
Singapore	SISIR
Slovakia	UNMS
Slovenia	SMIS
South Africa	SABS
Spain	AENOR
Sri Lanka	SLSI
Sweden	SIS
Switzerland	SNV
Syrian Arab Republic	SASMO
Tanzania, United Republic of	TBS
Thailand	TISI
Trinidad and Tobago	TTBS
Tunisia	INNORPI
Turkey	TSE
Ukraine	DSTU
United Kingdom	BSI
Uruguay	UNIT
United States of America	ANSI
Uzbekistan	UZGOST
Venezuela	CONVENIN
Vietnam	TCVN
Yugoslavia	SZS
Zimbabwe	SAZ

[a] Full names, postal and telegraphic addresses, and telephone and telex numbers are contained in Refs. 3 and 5.

Table 2. ISO Technical Committees Concerned with Packaging

TC 51: pallets for unit load method of materials handling	
Secretariat:	British Standards Institution (BSI) 389 Chiswick High Road GB-London W4 4AL Tel. +441 81 996 90 00
Scope:	Standardization of pallets in general use in the form of platforms or trays on which goods may be packed to form unit loads for handling by mechanical devices
TC 52: light-gauge metal containers	
Secretariat:	Association française de normalisation (AFNOR) Tour Europe F-92049 Paris La Défense Cedex, France Tel. +33 1 42 91 55 55
Scope:	Standardization in the field of light-gauge metal containers with a nominal material thickness of ≤0.49 mm
TC 63: glass containers (this TC has been dissolved)	
Secretariat:	Slovak Office of Standards, Metrology and Testing (UNMS) Štefanovicova 3 814 39 Bratislava, Slovakia Tel. +42 7 49 10 85
Scope:	Standardization of glass containers used as a means of packaging; excludes containers made from tubular glass
TC 104: freight containers	
Secretariat:	American National Standards Institute (ANSI) 11 West 42nd Street New York, NY 10036 Tel. +1 212 642 49 00
Scope:	Standardization of freight containers, having an external volume of ≥1 m^3 (≥35.3 Ft^3), with regard to terminology, classification, dimensions, specifications, test methods, and marking
TC 122: packaging	
Secretariat:	Türk Standardlari Enstitüsü (TSE) Necatibey Cad. 112 Bakanliklar 06100 Ankara Turkey Tel. +90 312 417 83 30
Scope:	Standardization in the field of packaging with regard to terminology and definitions, packaging dimensions, performance requirements, and tests; excludes matters falling within the scopes of particular committees (eg, TC 6, 52, 104)

bers. If 75% of the votes cast are in favor of the DIS, it is sent for acceptance to the ISO Council. This provides an assurance that no important objections have been overlooked.

There is no requirement for ISO member bodies to publish an ISO standard as a national standard, although it is through such action that ISO standards are put into use throughout the world. On occasion, publication of an ISO standard as a national standard is not considered appropriate.

All ISO standards are listed in the ISO Catalogue (5) and on the *ISO Online* (Internet World Wide Web site at http://www.iso.ch/welcome.html). Alternative methods of access to the large number of ISO standards include handbooks covering specific technical fields, and the ISO KWIC Index of international standards, which is prepared in keyword-in-context format. ISO standards can frequently be purchased from national standards bodies (ISO member bodies) or directly from ISO at:

International Organization for Standardization (ISO)
1 rue de Varembré
Case postale 56
CH-1211 Genève 20, Switzerland
Tel. +41 22 749 01 11; fax +41 22 734 10 79
Internet e-mail: sales@isocs.iso.ch

ISO Packaging Standards

Five technical committees have developed ISO standards directly concerned with packaging.

TC 51: pallets for unit load method of materials handling. Tasked with standardizing pallets for general use in the form of platforms or trays on which goods may be packed to form unit loads, TC 51 has currently developed nine standards listed in Table 3.

TC 52: light-gauge metal containers. Through the work of TC 52, the number of different sizes of open-top, general-purpose food cans has been reduced from over 2000 in 1947, to a total of 35 by the time ISO 3004-1:1986 was issued. This still may be too many, but it represents the best possible compromise at the present time. The range of standards for light-gauge metal containers, following the current program of work, is listed in Table 4.

TC 63: glass containers. Concerned with standardizing glass containers used as a means of packaging (excluding containers made from tubular glass), TC 63 developed 15 standards, listed in Table 5. With no further work item under study or foreseen, TC 63 has since been dissolved.

TC 104: freight containers. The standards relating to freight containers (Table 6) are perhaps the best possible example of international standardization, because without them, the

Table 3. ISO Standards for General-Purpose Pallets (TC 51)

ISO 445	*Pallets for Materials Handling—Vocabulary*
ISO 6780	*General-Purpose Flat Pallets for through Transit of Goods—Principal Dimensions and Tolerances*
ISO 8611	*General-Purpose Flat Pallets for through Transit of Goods—Test Methods*
ISO/TR 10232	*General-Purpose Flat Pallets for through Transit of Goods—Design Rating and Maximum Working Load*
ISO/TR 10233	*General-Purpose Flat Pallets for through Transit of Goods—Performance Requirements*
ISO/TR 10234	*General-Purpose Flat Pallets for through Transit of Goods—Phytosanitary* (plant health) *Requirment for Wooden Pallets*
ISO/TR 11444	*Quality of Sawn Wood Used for the Construction of Pallets*
ISO/TR 12776	*Pallets—Slip Sheets*
ISO 12777-1	*Methods of Test for Pallet Joints—Part 1: Determination of Bending Resistance of Pallet Nails, Other Dowel-Type Fasteners and Staples*

Table 4. ISO Standards for Light-Gauge Metal Containers (TC 52)

ISO 90-1	*Light-Gauge Metal Containers—Definitions and Determination Methods for Dimensions and Capacities—Part 1: Open-Top Cans*
ISO 90-2	*Light-Gauge Metal Containers—Definitions and Determination Methods for Dimensions and Capacities—Part 2: General-Use Containers*
ISO 1361	*Light-Gauge Metal Containers—Open-Top Cans—Round Cans—Internal Diameters*
ISO 3004-1	*Light-Gauge Metal Containers—Capacities and Related Cross Sections—Part 1: Open-Top Cans for General Food*
ISO 3004-2	*Light-Gauge Metal Containers—Capacities and Related Cross Sections—Part 2: Open-Top Cans for Meat and Products Containing Meat for Human Consumption*
ISO 3004-3	*Light-Gauge Metal Containers—Capacities and Related Cross Sections—Part 3: Open-Top Cans for Drinks*
ISO 3004-4	*Light-Gauge Metal Containers—Capacities and Related Cross Sections—Part 4: Open-Top Cans for Edible Oil*
ISO 3004-5	*Light-Gauge Metal Containers—Capacities and Related Cross Sections—Part 5: Open-Top Cans for Fish and Other Fishery Products*
ISO 3004-6	*Light-Gauge Metal Containers—Capacities and Related Cross Sections—Part 6: Open-Top Cans for Milk*
ISO/TR 8610	*Light-Gauge Metal Containers—Round Vent-Hole Cans with Soldered Ends for Milk and Milk Products—Capacities and Related Diameters*
ISO/TR 10193	*Round General-Use Light-Gauge Metal Containers—Nominal Filling Volumes and Nominal Diameters*
ISO/TR 10194	*Nonround General-Use Light-Gauge Metal Containers—Nominal Filling Volumes and Nominal Cross Sections*
ISO 10653	*Light-Gauge Metal Containers—Round Open-Top Cans—Cans Defined by Their Nominal Gross Lidded Capacities*
ISO 10654	*Light-Gauge Metal Containers—Round Open-Top Cans—Cans for Liquid Products with Added Gas, Defined by Their Nominal Filling Volumes*
ISO/TR 11761	*Light-Gauge Metal Containers—Round Open-Top Cans—Classification of Can Sizes by Construction Type*
ISO/TR 11762	*Light-Gauge Metal Containers—Round Open-Top Cans for Liquid Products with Added Gas—Classification of Can Sizes by Construction Type*
ISO/TR 11776	*Light-Gauge Metal Containers—Nonround Open-Top Cans—Cans Defined by Their Nominal Capacities*
ISO 11944	*Round General-Use Light-Gauge Metal Containers—Nominal Diameters for Cylindrical and Tapered Cans up to 10,000-mL Capacity*

"container revolution" could never have taken place. Standardized dimensions, maximum gross masses, and above all, standardized methods of lifting (twist locks and corner fittings) have enabled rapid, safe transfer of the containers from one transport mode to another. These type standards relate not only to box containers but also to tank containers for liquids, gases, and solids.

TC 122: packaging. The committee with the broadest scope, TC 122 works to standardize packaging terminology and definitions, as well as packaging dimensions, performance requirements, and tests; but excludes matters falling within the scopes of other committees (eg, TC 51, TC 52, TC 63, and TC 104). These standards are listed in Table 7.

ISO 9000: Quality Systems Standards

In 1979, the ISO formed Technical Committee ISO/TC 176 on Quality Management and Quality Assurance to develop a single, generic series of quality system standards that could be used for external quality-assurance purposes. In 1987, on the basis of the work of TC 176, the ISO published the ISO 9000 Standard Series on quality management and assurance. No other standards developed by the ISO have had as much impact on a global scale as this series.

The ISO 9000 series was developed by researching a number of national standards from several ISO member countries including Great Britain's BS 5750 and BS 4891, France's AFNOR Z 50-110, Germany's DIN 55-355, the Netherland's NEN 2646, Canada's Z-299, and the United States' ANSI/ASQC Z-1.15, MIL-Q-9858A, ANSI/ASQC C-1, and ANSI/ASME NQA-1.

Each representative member country participating in the ISO has taken the ISO 9000 series back to their respective country and, after translation, issued it a designation unique to that country. In the United States, the ISO 9000 series is published by the American National Standards Institute (ANSI), in conjunction with the American Society for Quality Control (ASQC), and is referenced as the ANSI/ASQC Q9000 series (this designation was changed with the 1994 revision from the ANSI/ASQC Q90 Series); in Great Britain, as BS EN ISO 9000 (previously BS 5750); and in the European Union

Table 5. ISO Standards for Glass Containers (TC 63)

ISO 7348	*Glass Containers—Manufacture—Vocabulary*
ISO 7458	*Glass Containers—Internal Pressure Resistance—Test Methods*
ISO 7459	*Glass Containers—Thermal Shock Resistance and Thermal Shock Endurance—Test Methods*
ISO 8106	*Glass Containers—Determination of Capacity by Gravimetric Methods—Test Method*
ISO 8113	*Glass Containers—Resistance to Vertical Load—Test Method*
ISO 8162	*Glass Containers—Tall Crown Finishes—Dimensions*
ISO 8163	*Glass Containers—Shallow Crown Finishes—Dimensions*
ISO 8164	*Glass Containers—520-mL Euro-form Bottles—Dimensions*
ISO 9008	*Glass Containers—Verticality—Test Method*
ISO 9009	*Glass Containers—Height and Nonparallelism of Finish with Reference to Container Base—Test Methods*
ISO 9056	*Glass Containers—Series of Pilferproof Finish—Dimensions*
ISO 9057	*Glass Containers—28-mm Tamper-Evident Finish for Pressurized Liquids—Dimensions*
ISO 9058	*Glass Containers—Tolerances*
ISO 9100	*Wide-Mouth Glass Containers—Vacuum Lug Finishes—Dimensions*
ISO 9885	*Wide-Mouth Glass Containers—Deviation from Flatness of Top Sealing Surface—Test Methods*

Table 6. ISO Standards for Freight Containers (TC 104)

ISO 668	*Series 1 Freight Containers—Classification, Dimensions, and Ratings*
ISO 830	*Freight Containers—Terminology*
ISO 1161	*Series 1 Freight Containers—Corner Fittings—Specification*
ISO 1496-1	*Series 1 Freight Containers—Specification and Testing—Part 1: General Cargo Containers for General Purposes*
ISO 1496-2	*Series 1 Freight Containers—Specification and Testing—Part 2: Thermal Containers*
ISO 1496-3	*Series 1 Freight Containers—Specification and Testing—Part 3: Tank Containers for Liquids, Gases, and Pressurized Dry Bulk*
ISO 1496-4	*Series 1 Freight Containers—Specification and Testing—Part 4: Nonpressurized Containers for Dry Milk*
ISO 1496-5	*Series 1 Freight Containers—Specification and Testing—Part 5: Platform and Platform-Based Containers*
ISO 3874	*Series 1 Freight Containers—Handling and Securing*
ISO 6346	*Freight Containers—Coding, Identification, and Marking*
ISO 8323	*Freight Containers—Air/Surface* (intermodal) *General-Purpose Containers—Specification and Tests*
ISO 9669	*Series 1 Freight Containers—Interface Connections for Tank Containers*
ISO 9711-1	*Freight Containers—Information Related to Containers on Board Vessels—Part 1: Bay Plan System*
ISO 9711-2	*Freight Containers—Information Related to Containers on Board Vessels—Part 2: Telex Data Transmission*
ISO 9897-1	*Freight Containers—Container Equipment Data Exchange (CEDEX)—Part 1: General Communication Codes*
ISO 9897-3	*Freight Containers—Container Equipment Data Exchange (CEDEX)—Part 3: Message Types for Electronic Data Interchange*
ISO 10368	*Freight-Thermal Containers—Remote Condition Monitoring*
ISO 10374	*Freight Containers—Automatic Identification*

(EEC), as the EN ISO 9000 series (the standard had been published previously by the European Community as the EN 29000 Series).

ISO 9000 is a generic, baseline family of quality standards written with the intent of being broadly applicable to a wide range of varying nonspecific industries and products. With this objective, they establish the basic requirements necessary to document and maintain an effective quality system. These quality system requirements are intended to be complimentary to specific product requirements. This broad objective is ISO 9000's strength. However, the standard is written so generically that users often have difficulty interpreting and adapting the standard to their specific industry application. To meet this need, a series of supporting documents were developed. The ISO 9000 family consists of both "models," which define specific minimum requirements for external suppliers, and "guidelines" for development of internal quality systems. The ISO 9000 series is also supported by several ISO 100000 series standards consisting of both guidelines and industry specific requirements. (See also ISO 9000 opens gateway to new opportunities.)

Guidelines for selection and use. The cornerstone of the ISO 9000 family is *ISO 9000-1: Quality Management and Quality Assurance Standards—Guidelines for Selection and Use.* The purpose of this document is to

- Clarify the principal quality-related concepts and the distinctions and interrelationships among them
- Provide guidance for the selection and use of the ISO 9000 family of International Standards on quality management and quality assurance.

The ISO 9000 standards are intended to be used in four situations:

Table 7. ISO Standards for Packaging in General (TC 122)

ISO 780	*Packaging—Pictorial Marking for Handling of Goods*
ISO 2206	*Packaging—Complete, Filled Transport Packages—Identification of Parts When Testing*
ISO 2233	*Packaging—Complete, Filled Transport Packages—Conditioning for Testing*
ISO 2234	*Packaging—Complete, Filled Transport Packages—Stacking Tests Using Static Load*
ISO 2244	*Packaging—Complete, Filled Transport Packages—Horizontal Impact Tests* (horizontal or inclined plane test; pendulum test)
ISO 2247	*Packaging—Complete, Filled Transport Packages—Vibration Test at Fixed Low Frequency*
ISO 2248	*Packaging—Complete, Filled Transport Packages—Vertical Impact Test by Dropping*
ISO 2873	*Packaging—Complete, Filled Transport Packages—Low-Pressure Test*
ISO 2875	*Packaging—Complete, Filled Transport Packages—Water-Spray Test*
ISO 2876	*Packaging—Complete, Filled Transport Packages—Rolling Test*
ISO 3394	*Dimensions of Rigid Rectangular Packages—Transport Packages*
ISO 3676	*Packaging—Unit-Load Sizes—Dimensions*
ISO 4178	*Complete. Filled Transport Packages—Distribution Trials–Information to Be Recorded*
ISO 4180-1	*Complete. Filled Transport Packages—General Rules for the Compilation of Performance Test Schedules—Part 1: General Principles*
ISO 4180-2	*Complete. Filled Transport Packages—General Rules for the Compilation of Performance Test Schedules—Part 2: Quantitative Data*
ISO/TR 8281-1	*Packaging—Estimating the Filled Volume Using the Flat Dimensions—Part 1: Paper Sacks*
ISO 8317	*Child-Resistant Packaging—Requirements and Testing Procedures for Reclosable Packages*
ISO 8318	*Packaging—Complete, Filled Transport Packages—Vibration Tests Using a Sinusoidal Variable Frequency*
ISO 8474	*Packaging—Complete, Filled Transport Packages—Water Immersion Test*
ISO 8768	*Packaging—Complete, Filled Transport Packages—Toppling Test*
ISO 10531	*Packaging—Complete, Filled Transport Packages—Stability Testing of Unit Loads*
ISO 11683	*Packaging—Tactile Danger Warnings—Requirements*
ISO 12048	*Packaging—Complete, Filled Transport Packages—Compression and Stacking Tests Using a Compression Tester*

Table 8. ISO 9000 Series Quality Systems Models

ISO 9001	*Quality Systems—Model for Quality Assurance in Design/Development, Production, Installation and Servicing*
ISO 9002	*Quality Systems—Model for Quality Assurance in Production, Installation and Servicing*
ISO 9003	*Quality Systems—Model for Quality Assurance in Final Inspection and Test*

1. Guidance for quality management
2. Contractual, between first and second parties
3. Second party approval or registration
4. Third-party certification or registration

ISO 9000-1 provides general guidance addressing each situation.

The ISO 9000 models, guidelines and supporting documents. The ISO 9000 "models" are intended to define specific minimum requirements for external suppliers. When a customer requires a vendor to meet the requirements of an ISO 9000 standard, the customer will reference one of the models listed in Table 8.

ISO 9001 is the most comprehensive model. ISO 9002 is identical to ISO 9001 with the exception of the elimination of the sections addressing design and development. ISO 9003 can be differentiated from ISO 9002 in that the sections addressing production and installation are absent, while other sections are abbreviated. ISO 9003 is the least comprehensive of the series, and is intended primarily for final inspection and test applications. All of the models, ISO 9001, 9002 and 9003 include requirements for: an effective quality system, contract review, and effective document and data control, ensuring that measurements are valid and that measuring and test equipment is calibrated regularly, having an adequate isnpection and testing system as well as a process for dealing with nonconforming product, maintaining adequate record keeping, the use of appropriate statistical techniques and adequate packing, packaging, and marking processes.

ISO 9000 series guidelines are intended to assist the user in expanding the scope of the model selected, and in considering such factors as the market being served, the nature of the product, the production, the processes, and the consumer's needs. These guidelines are listed in Table 9.

Table 9. ISO 9000 Series Guidelines

ISO 9000-1	*Quality Management and Quality Assurance Standards—Part 1: Guidelines for Selection and Use*
ISO 9000-2	*Quality Management and Quality Assurance Standards—Part 2: Generic Guidelines for Application of ISO 9001, ISO 9002, and ISO 9003*
ISO 9000-3	*Quality Management and Quality Assurance Standards—Part 3: Guidelines for the Application of ISO 9001 to the Development, Supply, and Maintenance of Software*
ISO 9000-4	*Quality Management and Quality Assurance Standards—Part 4: Application to Dependability Program Management*
ISO 9004-1	*Quality Management and Quality System Elements—Part 1: Guidelines*
ISO 9004-2	*Quality Management and Quality System Elements—Part 2: Guidelines for Services*
ISO 9004-3	*Quality Management and Quality System Elements—Part 3: Guidelines for Processed Materials*
ISO 9004-4	*Quality Management and Quality System Elements—Part 4: Guidelines for Quality Improvement*

When the ISO 9000 family was still relatively new, TC 176 recognized the need for supporting documents containing either guidelines or additional requirements. Currently, several supporting standards have been published in the ISO 10000 series (as listed in Table 10). Many others are still being developed.

Throughout the ISO 9000 family of standards, emphasis is placed on

- Satisfaction of the customer's need
- The establishment of functional responsibilities
- The importance of assessing (as far as possible) the potential risk and benefits

All of these aspects should be considered in developing and maintaining an effective quality system.

International Codes, Recommendations, Regulations, and Technical Instructions

Several other international organizations have developed codes, recommendations, regulations and technical instructions that should also be mentioned here:

1. *The United Nations Recommendations on the Transport of Dangerous Goods* ("Orange Book") prepared by the United Nations Committee of Experts on the Transport of Dangerous Goods (ST/ECA/43-E/CN.2/170). Reference United Nations Publication Sales Number E.93.VIII.1. For North America, Latin America, Asia and the Pacific this document may be ordered from
United Nations Publications
2 United Nations Plaza
Room DC2-853, Dept. 007C
New York, NY 10017
Tel. (212) 963-8302; (800) 253-9646; Fax (212) 963-3489
Internet e-mail: PUBLICATIONS@UN.ORG

2. *The International Maritime Dangerous Goods Code* published by
International Maritime Organization
4 Albert Embankment
London, England SE1 7SR
Tel. +441 71 735 76 11; Fax +441 71 587 32 10

Table 10. ISO 10000 Series Supporting Documents

ISO 10005	*Quality Management—Guidelines for Quality Plans*
ISO 10007	*Quality Management—Guidelines for Configuration Management*
ISO 10011-1	*Guidelines for Auditing Quality Systems—Part 1: Auditing*
ISO 10011-2	*Guidelines for Auditing Quality Systems—Part 2: Qualification Criteria for Auditors*
ISO 10011-3	*Guidelines for Auditing Quality Systems—Part 3: Managing Audit Programs*
ISO 10012-1	*Quality Assurance Requirements for Measuring Equipment—Part 1: Metrological Confirmation System for Measuring Equipment*
ISO 10013	*Guidelines for Developing Quality Manuals*

3. *The Dangerous Goods Regulations* available from
International Air Transportation Association (IATA)
IATA Building
2000 Peel Street
Montreal, Quebec, Canada H3A 2R4
Tel. (514) 844-6311; Fax (514) 844-9089 (Publications Department)

4. *The Technical Instructions for the Safe Transportation of Dangerous Goods by Air published by*
International Civil Aviation Organization (ICAO)
1000 Sherbrooke Street, West
Montreal, Quebec, Canada H3A 2R2
Tel. (514) 285-8221; Fax (514) 286-6376

BIBLIOGRAPHY

1. *ISO/IEC Guide 2:1991, General Terms and Their Definitions Concerning Standardization and Related Activities,* International Organization for Standardization, Geneva, Switzerland.
2. *A Standard for Standards. BSO: Part 1. General Principles of Standardization,* British Standards Institution, London, UK.
3. *ISO Mememto,* International Organization for Standardization, Geneva, Switzerland, annual publication.
4. *ISO Technical Programme.* International Organization for Standardization, Geneva, Switzerland, annual publication.
5. *ISO Catalogue,* International Organization of Standardization, Geneva, Switzerland, annual publication.
6. *ISO 900-1: 1994, Quality Management and Quality Assurance Standards—Guidelines for Selection and Use,* International Organization for Standardization, Geneva, Switzerland, 1994.
7. M. Breitenberg, *Questions and Answers on Quality, the ISO 9000 Standard Series Quality System Registration, and Related Issues,* U.S. Department of Commerce, National Institute of Standards and Technology. Gaithersburg, MD (Nov. 1991).
8. R. Randall, *Randall's Practical Guide to ISO 9000,* Addison-Wesley, Reading, MA, May 1995.

RICHARD C. RANDALL
President
Randall's Practical Resources
Denver, Colorado

IONOMERS

Ionomers are premier heat seal polymers that have found use in demanding packaging applications requiring excellent seal reliability, formability, and toughness. The term *ionomers* was coined by the DuPont Company in the early 1960s to describe polymers having both ionic and covalent linkages (1,2). In particular, ionomers was used to describe the polymers originally sold under the registered trademark of SURLYN. These polymers are made by high-pressure free-radical catalyzed polymerization of ethylene with an unsaturated organic carboxylic acid and then partially neutralized with a metal ion (3). The acid groups in these polymers are incorporated randomly along the polymer backbone. While DuPont ionomers are made from ethylene–methacrylic acid copolymers, Exxon Chemical Company ionomers are made from ethylene–acrylic acid copolymers. Both companies supply grades neutralized with either zinc or sodium ions. The ionomers above are composed of ethylene–unsaturated carboxylic acid–carboxylate salt. In addition, DuPont offers some grades that contain a termonomer, isobutyl acrylate. Isobutyl acrylate acts to reduce the polymer's modulus and melting point. It also increases toughness and improves adhesion to certain substrates. Mitsui-DuPont in Japan offers ionomers similar to SURLYN under the registered trademark of HI-MILAN.

Recently, Chevron Chemical Company presented information on ionomers they produced by saponification of ethylene–alkyl acrylate copolymers (4), such as ethylene–methyl acrylate. These ionomers differ from those above in that all of the acid groups in the polymer are neutralized with sodium. The product is an ionomer of ethylene–methyl acrylate–sodium acrylate. The authors note similarities to the ionomers made by neutralization, and also differences such as lower tensile strength, stiffness, notched tear, and lower seal initiation and hot tack temperatures.

Other polymeric materials are also included in the classification of ionomers (5,6) but the discussion here will be limited to those based on ethylene copolymers that are used in packaging applications. Structure, physical properties, processing, and selected end uses are covered below.

Ionomer Structure

The polymer that can be considered the "parent" of these ionomers is low-density polyethylene (LDPE). The structure of LDPE consists of crystalline and amorphous phases, but ionomers contain a third ionic phase composed of metal ions and neutralized carboxylate groups. The structural units of this third phase are called *ionic clusters*. They possess an internal order, the exact nature of which depends on the type of ion present. These clusters are responsible for the unique set of properties of ionomers.

For instance, ethylene–unsaturated acid copolymers exhibit improved toughness compared to LDPE because of hydrogen bonding between carboxylic acid groups on different polymer chains. Ionomers made by neutralizing some of the acid groups in such polymers have not only hydrogen-bonding forces holding polymer chains together but also the much stronger ionic forces from clusters that interconnect polymer chains. The attractive forces, called *ionic crosslinks,* weaken with increasing temperature, making the polymer more readily processable than an LDPE polymer having comparable melt flow index.

Although the order within the clusters decreases with increasing temperature, the clusters themselves are not destroyed even at temperatures as high as 572°F (300°C). When heated polymer is allowed to cool, reordering of the structure within the clusters and crystallization of short polyethylene segments causes a substantial increase in the modulus of the polymer. The modulus of the cooled polymer and its rate of attaining ultimate modulus depends on the processing temperature, the storage temperature, and relative humidity.

Physical Properties

The amount of acid incorporated into the backbones of ionomer chains may range from 7 to 30 wt% for commercial products. These acid groups are usually neutralized from about 15 to 80%. Even a relatively low amount of neutralized acid provides significant changes in physical properties compared to LDPE. Increasing acid content and percentage neutralization causes the following properties to increase: tensile strength, modulus, toughness, clarity, abrasion resistance, resistance to both mineral and vegetable oils and fats, moisture absorption, and melt strength. Properties that decrease are melting point, haze, surface tack, and notched tear resistance.

Adhesion to polar materials such as aluminum foil, nylon, and paper in coextrusion or extrusion coating is a direct function of the amount of unneutralized acid in the ionomer. Adhesion to nonpolar polyethylene is decreased by increasing amounts of acid and carboxylate salt. Zinc ionomers tend to have a broader range of adhesion than do sodium ionomers and are less hygroscopic. Sodium ionomers offer superior grease resistance and optical properties.

Processing

Unlike LDPE, ionomers are hygroscopic. Manufacturers supply them in a dry state, and they need to be kept dry for proper processing. Ionomers can be processed in typical extrusion and coextrusion equipment that is suitable for extruding LDPE, with the exception that corrosion-resistant equipment needs to be used. For blown-film extrusion, ionomer grades having melt-flow indexes ranging from fractional to about 5 dg/min are suitable, although grades having melt flows as high as 14 can be extruded because ionomers possess exceptional melt strength. The melt strength of ionomers is sometimes used to help carry other layers in coextrusion. For extrusion coating, grades having melt-flow indexes ranging upward from about 2.5 dg/min are suitable. Because ionomers have affinity for metal surfaces, grades for extrusion coating benefit from the use of matte chill rolls. Use of a chill-roll release additive also can be helpful, especially if a gloss chill roll is required. Cast-film processing is also suitable for ionomers.

Unlike LDPE, the extrusion temperature should not exceed 590°F (310°C). Good practice calls for maintaining a slow flow of ionomer through the equipment during nonproduction periods. Ionomer needs to be purged with a polyolefin resin from the extruder and die prior to extended shutdown.

Applications

Vertical form/fill/seal packaging. Many of the uses of ionomers in packaging depend on the melt-strength, melt-flow, and adhesion characteristics of this polymer family. As a heat-seal layer especially for vertical form/fill/seal (VFFS) packaging, a property that is extremely important is called "hot tack," which is the capability of a newly formed (still molten) heat-seal bond to resist an opening force. In the case of VFFS packaging, the force is exerted on the end seal by product being dumped into the package. If the hot tack of the seal layer is insufficient, the seal may open partially or completely. Ionomers excel in melt strength, and seal widths of packages can be minimized by their use as seal layer (see also Form/fill/seal, vertical.)

A comparison of the hot tack profiles of a low-performance zinc ionomer, an acid copolymer, and LDPE is shown in Figure 1. For a given hot-tack strength, say, 750 g/in., the operating-temperature window of the ionomer is greater than 75°F, that of the acid copolymer is 40°F, and the LDPE polymer has insufficient hot-tack strength to meet the requirement. The width of the hot-tack window for ionomers contributes to seal reliability and reduced dependence on highly accurate temperature control of the sealing operation.

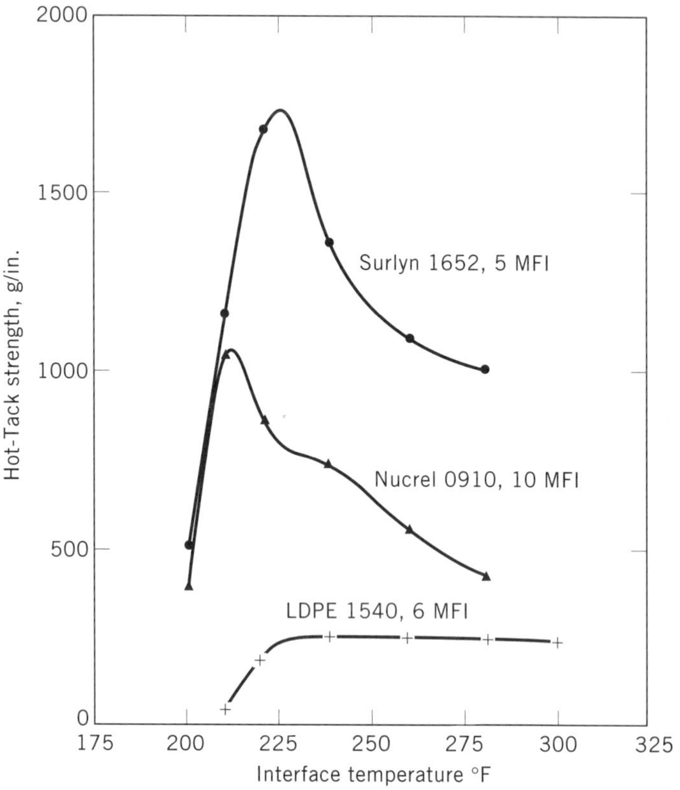

Figure 1. Hot-tack strength comparison, DuPont spring test.

Ionomers also have low seal initiation temperatures because the crystalline melting point of the parent LDPE has been reduced by copolymerization of acid and/or ester. Since their seal strength is maintained at the high-temperature end of the sealing range, this means that the overall heat-seal operating window is widened, and contributes to increased packaging line speeds and seal reliability. The melt-flow characteristics of ionomers also allows them to flow and seal around particulate contaminants that may be in the seal area, also increasing seal reliability. Abrasion resistance of ionomers is also a benefit in packaging food products that tend to be abrasive.

Visual carded display packaging. This type of packaging, also called "skin packaging," is used both for retail hanging card displays and for industrial packaging to hold products to a board for shipping. The toughness, optical clarity, and minimal board-curling properties of ionomers make them particularly suited for this use.

Thermoformed packages. The melt strength of ionomers allows deeper draw depths of packaging webs utilizing an ionomer layer compared to use of other polymers. Coextrusions of ionomers with nylon have been useful in packaging processed meats and cheese. Methacrylic acid-based ionomers also have the property of providing adhesion to the protein of meats, and reduce the amount of purge, or liquid exudation formed during meat processing. Because of their melt-flow properties, ionomers also have superior properties compared to other polymers for sealing through contamination, as from meat juices. This property gives webs having ionomer seal layers a reduced amount of package leakage, reducing waste and cost.

A specialized type of thermoformed packaging is called "Stretch Pak," in which a film attached to a windowed card is thermoformed, and the product is placed in the cavity. The packaged object is held snugly in the custom-shaped cavity when the film is shrunk by passing the package through a heated shrink tunnel.

Aluminum-foil-based packaging. Ionomers containing unneutralized acid groups have found wide utility as the seal layer in extrusion-coated aluminum foil structures. The acid groups provide adhesion to the foil, while the superior sealing and chemical-resistance characteristics of the ionomer permit packaging a wide variety of products.

BIBLIOGRAPHY

1. R. W. Rees, *Modern Plast.* **42,** 209 (Sept. 1964).
2. R. W. Rees and D. J. Vaughan, *Polymer Preprints,* Vol. 6, American Chemical Society Division of Polymer Chemistry, 1965, p. 287.
3. U.S. Pat. 3,264,272, R. W. Rees.
4. G. L. Baker and L. R. Compton, "Introduction of Acrylate Copolymer Based Ionomer Resins for Packaging Applications," *1995 TAPPI Polymers, Laminations & Coatings Conference, Preprints,* p. 625.
5. R. W. Rees, "Ionomers" in J. Kroschwitz, ed., *Kirk-Othmer Encyclopedia of Chemical Technology,* 4th ed., Vol. 14, Wiley, New York, 1995.
6. R. Longworth, "Ion Containing Polymers: Their Structure and Application," *Plast. Rubber,* 75 (Aug. 1978).

GEORGE HOH
DuPont Packaging and Industrial Polymers
Wilmington, Delaware

ISO 9000 OPENS GATEWAY TO NEW OPPORTUNITIES

Customer satisfaction.
Increased productivity.
Reduced waste.

A windfall of business opportunities could be standing just outside your doorway. If you are ready to unlock the gateway to a wealth of possibilities, ISO 9000 may be your key to success.

In the last decade, ISO 9000 has seized hold of the world's attention. Industries around the globe are demanding quality, and the number of businesses seeking ISO 9000 registration is soaring to unprecedented heights.

According to a 1995 survey conducted by Mobil Europe Ltd., an astounding 86 countries have adopted the international standard for quality. As of March 1995, a total of 95,476 companies worldwide have become ISO 9000–registered.

In the European Union (EU; formerly EEC), a single market comprising 15 European countries and more than 370 million people, ISO 9000 plays a critical role. When the EU was established in 1992, internal borders to all 15 member countries were dismantled to form a single market, based on intergovernmental cooperation and economic unity.

As a means of ensuring confidence in the quality of products that are circulated throughout the EU, the ISO 9000 series has been adopted in the EU as part of a conformity assessment plan to establish uniform systems for product certification and quality systems registration.

Today, the EU represents one of the world's largest trading powers. And though ISO 9000 is not an absolute requirement in the EU, more and more U.S. companies are realizing that if they want to gain a competitive foothold in this leading market, then ISO 9000 registration is an important first step.

Clearly, ISO 9000 is revolutionizing business, and as global acceptance of the ISO 9000 standards continues to flourish, industry experts say the move toward registration will skyrocket to new heights in the coming years (see also international standards).

What is ISO 9000?

ISO 9000 is a series of quality assurance and quality management standards that were created by the International Organization of Standardization, based in Geneva, Switzerland. The organization is a consortium of virtually all the world's industrialized nations. As of this publication, there are over 110 members. The United States representative to the worldwide federation is the American National Standards Institute (ANSI), a very familiar name in American industry.

The International Organization of Standardization was established in 1946 with the purpose of developing industrial standards that would facilitate the international exchange of goods and services. That goal has been realized through ISO 9000. The standards, first published in 1987, are generic in nature so that they can be applied to any company, of any size, anywhere in the world.

To achieve this generic state, ISO 9000 refrains, to the greatest extent possible, from mandating specific methods, practices, and techniques. Instead, it emphasizes principles, goals, and objectives.

The series incorporates three contractual quality-system models (ISO 9001, 9002, and 9003) to which organizations become certified, and several guidance documents, providing strategies for implementing ISO 9000 quality systems, developing quality manuals, managing audit programs, and so on.

ISO 9001 is the most comprehensive quality system model, covering design, manufacturing, installation, and servicing systems. It contains 20 sections describing various elements of the quality system. ISO 9002 covers production and installation. It contains all the elements found in ISO 9001, excluding design. ISO 9003 covers only final product inspection and test. It contains 16 of the 20 elements outlined in ISO 9001.

Which model a company chooses to implement depends on its scope of operation. A company that designs its own product, for example, would consider ISO 9001. If, on the other hand, the company only manufactures the product (the design is subcontracted), then ISO 9002 would be the most appropriate model. ISO 9003 would be suitable for a company that neither designs nor manufactures.

There is a total of 137 requirements an organization must satisfy in order to become certified to ISO 9001. Compliance with these requirements ensures the quality of the product or service delivered. Most of these requirements focus on

- Document control
- Quality records
- Process control
- Inventory control
- Shipping and receiving
- Purchasing
- Inspection
- Testing
- Training

ISO 9000 is not a product standard, but a quality system standard. It is not applied to particular products, but to the processes that create them. An ISO 9000 quality system that is well designed, well implemented, and carefully managed will provide confidence that the product or service will meet customer expectations and requirements. It is aimed at providing that confidence to three audiences:

- Customers directly
- Customers indirectly (via third-party assessments and quality system registration)
- Company management and staff

It does so by requiring that every business activity affecting quality be conducted in a three-part never-ending cycle: planning, control, and documentation.

- Activities affecting quality must be *planned* to ensure that goals, authority, and responsibility are defined and understood
- Activities affecting quality must be *controlled* to assure that specified requirements (at all levels) are met, problems are anticipated and averted, and corrective actions are planned and carried out
- Activities affecting quality must be *documented* to ensure understanding of quality objectives and methods, smooth interaction within the organization, feedback for the planning cycle, and objective evidence of quality system performance for those who require it, such as customers or third-party assessors

ISO 9000 is not new or radical. It is good, hard-headed, common business sense in codified, verifiable, and easily adapted form. It has strong commonalities with other quality schemes, such as MIL-Q, Deming's 14 points, total quality management, and the Malcolm Baldrige National Quality Award standards. The main difference is that one can *register* to ISO 9000.

Why Implement an ISO 9000 Quality System?

Invariably, companies that are not familiar with the ISO 9000 standards want to know what's in it for them. Why should they invest the time, money and effort to implement an ISO 9000 quality system?

The answer to that question can be found by talking to people who have actually implemented the system and attained registration status. For many firms, the answer is obvious. They know for a fact that without ISO 9000 they will lose business. Others will tell you ISO 9000 provides strategic and tactical methods for assuring that their products or services meet their customers' quality requirements, while meeting return-on-investment goals. This equates to good, common business sense.

Others still will tell you that ISO 9000 provides access to markets. It enables them to maintain or create customer relationships in situations in which ISO 9000 certification is required.

Everyone will tell you that an effective quality system is the philosophical and procedural "glue" that unites all elements of the facility—including employees, plant, equipment, and procedures—with *suppliers* at the front end and *customers* at the output end.

Facilities that operate quality systems tend to exhibit the following attributes:

- *Philosophy of prevention* rather than detection
- *Continuous review* of critical process points, corrective actions, and outcomes
- *Consistent communication* within the process, and between facility, suppliers, and customers
- *Thorough recordkeeping* and efficient control of critical documents
- *Total quality awareness* by all employees
- *High level of management confidence*

These attributes inevitably lead to the following tangible benefits:

- Informed, competent management decision-making
- Dependable process input (supplier control)
- Control of quality costs
- Increased productivity
- Reduced waste

In summary, a company with a well-designed and well-implemented quality system has a process that tends to be lean, sensitive to customer needs, highly reactive, efficient, and positioned at the leading edge of its marketplace. Quality needs are continuously satisfied, securing greater customer loyalty.

ISO 9000 is an ideal quality system for facilities that are serious about quality. It is an emerging imperative for any facility that has, or expects to have, commerce with EU nations. These, along with the quality system benefits cited above, make achievement of ISO 9000 standards a powerful strategic tool, regardless of whether the facility goes the certification route.

How Does ISO 9000 Work?

Implementing an ISO 9000 quality system can be highly effective because it provides a means for companies to function in a planned and systematic way. The basic elements of ISO 9000 are

1. *Say what you do.* Describe, through written procedures, the practices of your organization. In doing so, ask yourself if the best methods are being used. Are there ways in which work processes could be better carried out? Adopt the best practice.

2. *Do what you say.* Follow these same procedures to make sure that jobs are performed correctly and consistently.
3. *Prove it.* Keep proper records, as evidence, to demonstrate that your organization is doing what it says it is doing.

By adhering to these three basic steps, many common quality problems can be avoided because you're implementing a system that encourages checking and good communications. For example, because work processes must be documented, employees gain a better understanding of their job responsibilities and objectives. Job-related stress levels are reduced because ISO 9000 requires everyone to know what their job is and how to do it. Furthermore, current versions of these documents must be located at all relevant workstations. This ensures that obsolete specifications are not used, which, in turn, prevents faulty products from being produced.

ISO 9000 does not guarantee excellence, but it does provide a means for gaining consistent quality.

As mentioned earlier, the ISO 9001 quality system model consists of 20 basic elements. Let's take a look at the various quality system requirements:

4.1 *Management Responsibility.* The responsibility for quality management belongs to top management. Executive management must define its quality policy, execute it, and participate actively in the quality system by conducting verification and review activities. Additionally, a *management representative* must be appointed to oversee the day-to-day responsibilities of the quality system.

4.2 *Quality System.* A quality system must be established, documented, and maintained to meet specified requirements, policies, and objectives. The quality system is the organizational structure, procedure, processes, and resources needed to implement quality management. The system should satisfy customer needs and place emphasis on problem prevention.

4.3 *Contract Review.* Procedures must be maintained for evaluating customer requirements and comparing them with the facility's capabilities. Before accepting a contract or order: requirements must be adequately defined and documented; requirements must be within the facility's capabilities; and all discrepancies must be resolved.

4.4 *Design Control.* Procedures for controlling the release, change, and use of designs must be established. The design function must be aware of its responsibilities for quality. The chief goal of the design process is to translate customer's requirements into technical specifications for output.

4.5 *Document and Data Control.* Documents relating to the quality system—drawings, specifications, blueprints, procedures, electronic media, records, etc—must be created. All are intended to document the quality level of output, as well as the overall performance of the quality system. Personnel must have access to current editions of documents pertinent to their quality-related functions.

4.6 *Purchasing.* A close working relationship and open channels of communication must be established with each subcontractor. Subcontractors must be chosen on the basis of their ability to meet quality requirements, which should be clearly defined and understood. There must be a mutual agreement on the degree of quality assurance the subcontractor is to provide; methods for verifying conformance must be agreed on; and procedures must be in place for dealing with matters—particularly disputes—affecting quality.

4.7 *Control of Customer-Supplied Product.* Facilities that receive products (or services) from a customer must monitor and secure such at each point during which it is under the facility's control.

4.8 *Product Identification and Traceability.* The facility should operate a documented system for identifying and tracing output, as appropriate, from receipt and throughout all stages of production, delivery, and installation. Input (supplied products or services) should be tracked all the way through the process; specific operations (ie, equipment or personnel) must be tracked to monitor effectiveness; and the ultimate destination of output must be traceable to ensure long-term customer satisfaction.

4.9 *Process Control.* Production should be carried out under planned, controlled, and documented conditions to ensure that quality requirements are met. Process capability studies should be conducted to determine the potential effectiveness of a process. Processes that are important to product quality should be closely monitored to ensure that quality goals are met.

4.10 *Inspection and Testing.* Procedures for inspection and testing activities must be established and maintained to verify that specified requirements for the product or service are met. Incoming product cannot be used until it has been inspected and verified as conforming to specified requirements. In cases where product is released for urgent production prior to verification, it must be positively identified and recorded to permit immediate recall and replacement.

4.11 *Control of Inspection, Measuring, and Test Equipment.* Devices used to assess output conformance to specified requirements should be selected and controlled so as to preserve confidence in its accuracy and precision. Devices must be calibrated to prescribed intervals, and properly handled, preserved, and stored.

4.12 *Inspection and Test Status.* The inspection and test status of product shall be identified by suitable means. Products that have not been verified, or that are found to be in conformance or nonconformance, should be identified via stamps, tags, notations, inspection records, etc. The identification should include the unit responsible for verification.

4.13 *Control of Nonconforming Product.* Procedures should be in place to prevent customers from inadvertently receiving nonconforming product. Nonconforming product must be identified, documented, evaluated, and segregated. Nonconforming product can be reworked, accepted (with or without repair) by concession, regraded, or scrapped.

4.14 *Corrective and Preventive Action.* When a quality-related problem is detected (via audits, management reviews, customer complaints, etc), measures must be taken to eliminate or minimize the recurrence of the problem. The cause of the nonconformity must be investigated, a plan for corrective action must be determined and initiated, and controls must be applied to ensure that the corrective action is effective.

4.15 *Handling, Storage, Packaging, Preservation, and Delivery.* Product quality must be protected and safeguarded throughout all process activities—from conception through final delivery. Product must be handled and stored to prevent damage or deterioration; packing, packaging, and marking processes must be controlled to ensure conformance to specified requirements; product must be preserved while under the supplier's control; and arrangements must be made to protect the quality of product after final inspection and test and through delivery.

4.16 *Control of Quality Records.* Quality records must be established and maintained to demonstrate conformance to specified requirements and the effective operation of the quality system. Records must be legible, protected from damage and deterioration, and stored in such a way that they are readily retrievable.

4.17 *Internal Quality Audits.* Regular internal quality audits must be carried out to determine the effectiveness of the quality system and to verify whether quality activities and related results comply with planned arrangements.

4.18 *Training.* Appropriate training should be provided to all levels of personnel performing activities that affect quality.

4.19 *Servicing.* Procedures must be established and maintained to ensure that contractually-required servicing is performed according to requirements.

4.20 *Statistical Techniques.* The quality system should include procedures for identifying statistical techniques (design of experiments, statistical sampling, quality-control charts, etc) used in assessing process capability and product characteristics.

(*Note:* All procedures must be documented.)

Registering to the Standard

Implementing an ISO 9000 quality system can instill efficiency in a business, and it can secure a high level of quality in the production of goods and services. But implementation alone is not enough to reap all the benefits that are inherent with ISO 9000. You must register your quality system to the standard to experience all of its advantages.

Because ISO 9000 certification carries great weight among industries all over the world, attaining registration status can open your company's doors to a vast, global marketplace. This alone is a powerful reason to pursue registration, but there are many other reasons to consider as well.

ISO 9000 registration provides objective evidence that a company is meeting all of its required quality levels. Proof of certification also provides assurance to the customer that a company's products or services really are of the highest standards to which the manufacturer is claiming. ISO 9000 is customer-driven. When implemented correctly, the elements of the ISO 9000 standard work meticulously together to ensure that the customer's needs are satisfied.

Furthermore, because registration requires an independent party (a registrar) to assess the effectiveness of a company's quality system, customers can rest assured that the company is being monitored regularly and that it is performing at its highest level.

How Does a Company Become ISO 9000–Registered?

Before a company can be considered for registration, several preliminary steps must be taken.

The first is to implement a quality system that incorporates all the technical requirements of the appropriate model: ISO 9001, 9002, or 9003, depending on the scope of your operations.

The quality system must be up and running long enough for employees to become familiar with the system, and to develop a sufficient evidentiary trail of documents that can be audited (3–6 months is the general rule of thumb).

In addition to implementing a quality system, a company must create the appropriate documentation before it can apply for registration status. This should include a *quality policy* that describes the company's overall intent and commitment to quality objectives. A *quality manual* (or equivalent document) that describes what your company does in the form of procedures and work instructions. *Quality plans* that explain how quality will be managed for individual projects, contracts, or products. And *quality records*—things such as design charts, files, inspection and testing records, and any other records—that provide supporting evidence that your quality system is in conformance to specified requirements.

When a company has successfully implemented a quality system and created enough supporting documentation for auditors to measure the effectiveness of the system, a company is ready to file a request for registration.

Filing an Application

When a company is ready to file for registration status, normally the first step is for the company to complete an application or questionnaire describing its scope of operations. (Some registrars charge an associated application fee.) Some typical questions a company can expect to answer are

What is your desired time frame for registration?

Describe your business and list any applicable SIC codes.

What is the size of your company (number of facilities and employees)?

What is the status of your existing quality system and related documentation?

Using this information, the registrar will prepare a price quote and an estimate of time required for completing the registration assessment. When a company has selected the registrar it wants to do business with, a contract will be drafted.

Documentation Review

When the registration audit officially begins, the first thing the registrar will request is a controlled copy of the company's quality manual (or equivalent document). The registrar will review the manual to determine whether it meets all the requirements of the ISO 9000 quality system model (ISO 9001, 9002, or 9003) the company has implemented.

After the quality manual has been reviewed, the registrar will submit a report to the company. If the company's documentation fails to meet all the criteria stipulated in the applicable ISO 9000 standard, the nonconformances will be identified in the report and the company will be required to take corrective action.

If the registrar determines that the company's quality manual is satisfactory, final arrangements will be made to conduct a full, on-site registration assessment of the company's quality system. During this assessment, the registrar's audit team will verify that

1. The company's quality system meets all the requirements of the applicable ISO 9000 standard.
2. The quality system is designed to achieve, and is achieving, continuous improvement.
3. The organization is complying with its own policies and procedures.

The Registration Assessment

During the final registration assessment, the audit team will walk through various areas of the company to observe activities. Auditors may conduct one-on-one interviews with employees, they may inspect documents and records, and they may examine equipment and products. Throughout the audit, they will be seeking objective evidence—statements, documented procedures, written policies, etc—to support their observations.

The auditor(s) will be looking for answers to the following questions:

- Is the documented system consistent with the standard? (Do you describe what you do?)
- Are activities consistent with the documented system? (Do you do what you say you do?)
- Is the quality system effective, and is the company meeting its objectives and goals? (Can you prove that your quality system is functioning effectively?)

If any nonconformances are found, the auditor will record them on a "noncompliance report" and bring them to the attention of the management representative. In the report, the auditor will specifically describe what the nonconformance is and classify it as either major or minor.

A *major* nonconformance is a total lapse of a required element (ie, lack of quality system documentation or a major deficiency in the quality system, product, or service) that can block the possibility of immediate registration. A major nonconformance can also be numerous minor nonconformances, which, when taken together, indicate a total breakdown of a required element.

A *minor* nonconformance is something that does not directly affect the quality of a product or service (ie, instrumentation out of calibration or drawings marked up with unauthorized design changes) and, in most circumstances, can be easily remedied. However, the registrar cannot recommend certification until it can verify that all nonconformances—major or minor—have been corrected.

When all corrective actions have been closed, the company's application, audit results and other documentation will be forwarded to the registrar's executive committee; this is an independent, decision-making body of the registrar. The executive committee will review the information and decide whether to grant registration.

If the committee determines that a company has met all the requirements for ISO 9000 registration, a certificate of registration will be prepared. The certificate will bear the seals of the accreditation agency(s) of the registrar, as well as the registrar's own logo. Companies can display the registration mark in advertising, promotional literature, and stationary to show customers its commitment to quality.

Maintaining Registration

Once a company has attained ISO 9000 registration status, it will be subjected to periodic surveillance by the registrar. The registrar will reinspect the company's quality system to look for objective evidence that it is still in compliance with the applicable standard, and that the company is continually working to improve and maintain the system.

Most registrars will conduct surveillance visits once a year. During these visits, the registrar will send an auditor to the company's facility to carry out a sampling of its quality system, not a full re-audit.

The standard validity of a registration certificate is three years, after which a full re-audit must be performed for renewal. Some registrars offer a "continuous assessment" option whereby a company's certification never expires, as long as certain requirements are met and surveillance visits reveal continuing compliance to the standard.

ALICIA GARRISON
Perry Johnson, Inc.
Southfield, Michigan

ISO 14001 ENVIRONMENTAL MANAGEMENT SYSTEM

ISO and 14000 together represent a new international corporate environmental management standard. It is a set of generic standards that provides business managers a systematic method for controlling and monitoring the complex and diverse environmental activities of a company while lifting another barrier to international trade. The International Organization for Standardization (ISO) has worked with the United Nations and organizations from over 100 countries around the world to create this common uniform international standard for basic environmental management. ISO is not a new law, act, or regulation. It is for voluntary participation as an internal management tool to effectively manage environmental risks and is advisory in nature. Companies around the world are encouraging certification under the ISO standard. Examples of *Fortune* 500 companies that are actively pursuing certification include Anheuser Busch, Union Camp, AT&T, OKI Telecommunications, Chrysler Automotive, and 3M. However, it can apply to manufacturing organizations of any size that are implementing, maintaining, or improving environmental affairs. Many companies have started to either implement the principles of ISO 14000 standards or are evaluating the differences between their existing environmental programs and the ISO standards.

According to a recent survey conducted by Arthur D. Little (1), 61% of respondents indicated that ISO 14000 will have an impact on international trade. After final publication of the first portion of ISO 14000 referenced as ISO 14001 in mid-1996, it will become "the international standard" that companies and their environmental managers uphold. In the ISO 14000 Implementation Survey (2), 33% of respondents identified "customer demand" as the most compelling reason to pursue certification. Companies that receive ISO 14000 certification and that do business internationally will receive some distinct competitive advantages over those that do not participate. They will build customer confidence through evidence that it employs environmentally friendly production, which will enhance product image. They will have access to more export markets around the world as countries with different environmental regulations and redundant levels of enforcement begin accepting the ISO 14000 guidelines. Domestic companies with ISO certification will have more access to global companies who demand ISO certification from their vendors as a condition of doing business. Since current estimates indicate that the environmental regulatory burden on the U.S. gross national product is as great as 8% (3), ISO will also provide improved cost control, more efficient management of environmental resources and reduced environmental liability. It will facilitate financing through reduced environmental liability and demonstration of due diligence. Through identification of lower cleanup liabilities, insurance costs can be reduced. It will improve relationships with government regulators because it can show them that noncompliance incidents may occur less frequently. In addition, many U.S. states (e.g., Colorado, Massachusetts, New York, Washington, and Wisconsin) are considering incorporating ISO 14000 in place of certain rules or guidance documents.

There is a broad range of separate disciplines within ISO 14000, including the basic environmental management system, environmental auditing, environmental performance evaluation, environmental labeling (ie, green labeling), life-cycle assessment, and product standards. There are separate ISO designations assigned to each discipline from 14001 to 14025 that is in various stages of consideration or adoption. The two broad and separate categories for these designations are guidance and specification. All standards are guidance except for ISO 14001. The designation closest to ratification is ISO 14001, the environmental management system portion of ISO 14000. It is a comprehensive environmental management system (EMS) with five guiding principles outlined in the August 1995 Draft International Standards (DIS). The general elements within 14001 are (1) environmental commitment and policy, (2) planning, (3) implementation, (4) measurement and evaluation, and (5) review and improvement. Each element is described in order to better understand the elements within ISO 14001 scheduled for ratification in mid-1996.

Environmental Commitment Policy

Environmental commitment and policy is a commitment from top management that establishes an overall sense of direction and sets the principles of action in a formal written statement. It sets an overachieving goal as to the level of overall environmental responsibility and performance required by the organization against which all subsequent actions are judged. The corporate environmental policy considers the following: the organization's mission, vision, core values, and beliefs; requirements of and communication with interested parties; continual improvement; prevention of pollution; guiding principals; coordination with other policies such as quality and occupational health and safety; specific local and regional conditions; and compliance with relevant environmental regulations and laws. Communication throughout the organization and other interested parties makes for an effective policy.

Planning

Environmental planning is the formal plan to fulfill the environmental policy. The plan is based on knowledge of the environmental aspects and significant environmental impacts associated with the organization as well as the legal (regulatory) implications and internal performance. The aspect is an element of an organization's activity, product, or service that can have a beneficial or adverse impact on the environment. It must be large enough to have meaning and small enough to be understood. For example, it could involve a discharge, an emission, consumption, or reuse of a material. An *impact* refers to the change that takes place in the environment as a result of the aspect. Examples of impacts might include contamination of water or depletion of a natural resource.

Implementation

Effective implementation is developing the capabilities and support mechanisms necessary to achieve its environmental policy, objectives, and targets. This includes resources, training, communication, and recordkeeping.

Resources. The organization needs the people with the knowledge skills, and physical, technological, and financial resources to implement the environmental management system.

Training. Appropriate training is relevant to the achievement of environmental policies, objectives, and targets to personnel within the organization. Training maintains the methods and skills required to perform tasks in an efficient and competent fashion and provides knowledge of the activities that can have an impact on the environment.

Communication. An effective communication process provides for two-way reporting internally and, where desired, externally on the environmental activities of the organization. It transfers management's commitment to the environment and deals with concerns and questions about the environmental issues of the organizations' activities, products, and services. Communication raises awareness of the organization's environmental policies, objectives, targets, and programs.

Recordkeeping. Written environmental operating procedures and protocols are essential to effective functioning of the environmental management system. Written documentation in organized systems demonstrates compliance with requirements of the environmental management system and records the extent that environmental objectives and targets have been met. Recordkeeping provides direction and descriptions of key elements of the environmental management system.

Planning is a demonstration of a commitment to provide adequate employment for people and equip them with the tools such as computers and reference information along with resources for the environmental management system.

Measurement and Evaluation

Measurement and evaluation are the periodic monitoring measuring and evaluation of the key characteristics of the environmental management system operations and activities. This provides for the performance evaluation of compliance with relevant environmental legislation and regulations. It provides an opportunity to periodically analyze success and failures through audits and identify corrective actions.

Review and Improvement

The last element is review and improvement. This is the continual review for suitability and effectiveness of the organizations environmental management systems. It consists of reviewing the environmental dimensions of all activities, products, and services of the organization after considering changes in legislation, products, and advances in science and their impact on financial performance. It considers findings of EMS audits and evaluates the effectiveness of the EMS. The ongoing review provides opportunities for continually improving the EMS and environmental performance.

This provides a framework that identifies the minimum standards required by the ISO 14001 draft international standards for environmental management systems in companies. A company can take advantage of the DIS by taking steps to become certified. The company can incorporate strategic planning that can then estimate implementation schedules, resource needs and budget requirements in order to upgrade current environmental programs and become an ISO 14001–certified company. Companies not desiring to become certified under ISO can use the standard to design or improve management of their environmental program in order to gain economic and business advantages. Proactive environmental management systems make good business sense.

BIBLIOGRAPHY

1. Survey, Arthur D. Little, *Financial Times* (Oct. 13, 1995).
2. Strategies, Inc., M. L., "ISO 14001 Implementation Survey," Feb. 26, 1996.
3. M. Barcaskey, "Get Ready . . . ISO 14000 Is Just Around the Corner," *The Source,* Georgia Pollution Prevention Assistance Division, Summer 1995.

General References

R. T. Brody, "Case Study of an ISO 14001 Readiness Assessment," Georgia Water and Pollution Control, Industrial Pollution Control Conference, 1996.

International Environmental Systems Update, 1995 and 1996; an ISO 14000 Information Service.

ISO/DIS Ballot—ISO 14001 "Environmental Management Systems—Specification with Guidance for Use" and ISO/DIS Ballot—ISO 14004 "Environmental Management Systems—General Guidelines on Principals, Systems and Supporting Techniques," ASQC Committee Correspondence, Sept. 5, 1995.

Kilpatrick and Cody, *Environmental Review,* Winter/Spring 1996.

B. Lawrence, "Environmental Management Systems and ISO 14000," Georgia Water and Pollution Control, Industrial Pollution Control Conference, 1966.

G. K. Nestel, "The Road to ISO 14000," Times Mirror Higher Education Group, Inc., 1996.

ROBYN T. BRODY
Roy F. Weston, Inc.
Norcross, Georgia

L

LABELS AND LABELING MACHINERY

Labels can be affixed to almost anything to indicate its contents, nature, ownership, or destination. Labeling is the art of applying or attaching the label to a particular surface, item, or product (see also Tags).

Originally, labels were used merely to identify a product or to supply information about the properties, nature, or purpose of the items labeled. Today their use is often required by legislation (1), and they are seen as sales and marketing aids in the total design of a package or product. In some cases, the labels have become products in themselves.

Types of Labels and Materials

The range and variety of labels used, and the markets and applications for them, are increasingly diverse. Materials include paperboard, laminates, metallic foils, paper, plastics, fabric, and synthetic substrates. These, in turn, may be adhesive or nonadhesive. Even then, the range of label materials can be further subdivided into a variety of different types: coated or uncoated; pressure-sensitive or heat-sensitive; conventional gummed or particle gummed (2). Figure 1 shows the main types of labels and materials.

Of all the different types of labels, nonadhesive plain paper labels applied by wet gluing tend to dominate the total user market throughout the world. In recent years, however, there has been a marked trend toward self-adhesive pressure-sensitive labels. Newer methods of labeling such as shrink sleeve, in-mold, and heat transfer have also gained market acceptance (see Bands, shrink; Decorating).

Overall, the U.S. pressure-sensitive (self-adhesive) roll label market is growing at about 10–15% per year, with some printers and market sectors showing growth of 20% or more. The two main markets showing above average growth are electronic data processing (EDP) labels and prime labels.

Plain-paper labels. The conventional wet-glue plain-paper label is widely used, particularly for large-volume items such as beer, soft drinks, wines, and canned foods, where high label application speeds, sometimes in excess of 80–100,000 labels per hour, are required.

Most of the papers used in this type of labeling are one-sided coated grades, which make up the familiar can and bottle labels, but a fair volume of uncoated grades is still used. These tend to be limited to special effects, such as colored papers, embossed papers, or antique laids for wine labels, where particular characteristics are required (3).

Gummed-paper labels. Gummed-paper labels have been declining in usage in recent years and are now limited mainly to applications such as address labeling for cartons, point-of-sale displays, and any labeling applications where automation is either difficult or unnecessary. Two main types of gummed labels are in use: conventional gummed and particle gummed (4).

Conventional gummed labels are those that are made up of a base support of paper coated on the reverse side with a film of water-moistenable gum. On particle gummed papers, the adhesive is applied in the form of minute granules. This avoids problems of curling often associated with conventional gummed papers. Most of the newer applications for gummed papers require the properties of high-tack particle gummed papers, which offer many advantages in terms of processing efficiency and security of application.

Self-adhesive labels. Self-adhesive label materials (see Fig. 2) today range from permanent to removable adhesive types, as well as a whole range of special adhesive materials (see Adhesives).

Permanent adhesives are those that are required to stay in position for a long time or for application to surfaces that are round, irregular, or flexible. Because of the strong adhesive "grab," the label normally becomes damaged or defaced if attempts are made to remove it.

Removable adhesives are those that can be removed after a specified time without damage to the surface on which they have been applied. Uses include short-term food-packaging applications, point-of-sale or advertising stickers, china, and

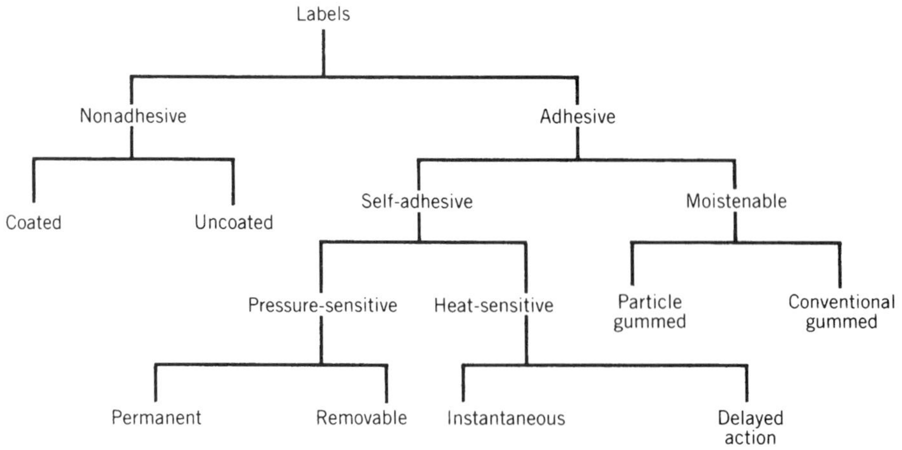

Figure 1. The main types of labels and materials.

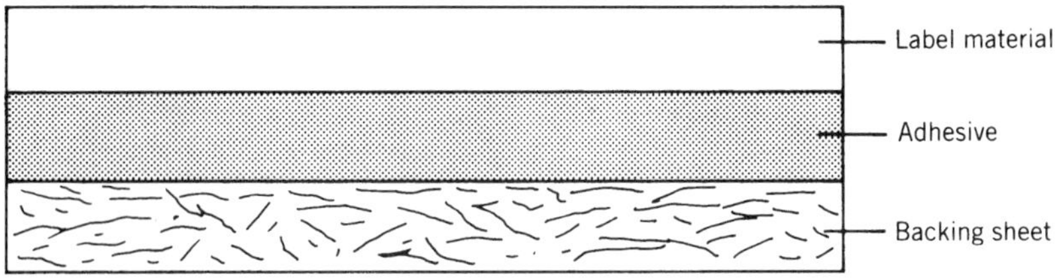

Figure 2. Self-adhesive labels consist of three main layers: the label-face material itself, the pressure-sensitive adhesive coating, and a release coating on a backing material to prevent the adhesive from sticking to the face material in the reel or sheets.

ovenware, where the label should be easy to remove before use.

Within the broad categories of permanent and removable self-adhesive labels there is a wide variety of special adhesive types to cover requirements such as water removability, low- or high-temperature adhesion, ultraviolet-light resistance, and high- and low-tack properties. Adhesives, which at one time were largely solvent-based, have moved rapidly in recent years to hot-melt or water-based acrylic adhesives (5) (see Acrylics).

For all normal self-adhesive labeling requirements (food and supermarket labels, retail/price marking labels, etc) surface papers in a variety of weights and finishes, colors, and radiants are available. For special applications, such as outdoor labels, instruction and nameplates for appliances, underwater labels, ampules and vials, and luxury-labeling of cosmetics and toiletries, there is a wide range of nonpaper materials. These include metallic foils and plastic films.

One of the most recent developments in self-adhesive materials is a range of thermal imaging label facestocks, which have become popular for in-store supermarket labeling of meat and fish and fresh produce (fruit and vegetables). These materials contain a surface coating of chemicals that darken under the action of a heated printhead (see Fig. 3) (6), thereby providing clear price and weight information and accurate scannable bar codes (7) (see Codebar; Code marking and imprinting).

Almost all self-adhesive label materials are available in both sheet form, from paper merchants and suppliers, or, for reel-fed conversion, from manufacturers and specialist suppliers in roll form.

The main markets for self-adhesive labels in the United States are basically segmented into three areas: industrial labels (or primary and secondary labels used in retail, wholesale, and industrial packaged goods), which make up about 80% of the U.S. self-adhesive label market; EDP labels (those used for marking/imprinting variable information); and price-marking and imprinting (which addresses label applications through handheld dispensers and the whole range of imprint technology). Split into some of the main market application areas, this includes promotional labels, price labels, functional labels, office/retail labels, nameplates, stickers, EDP labels, and prime labels.

Heat-sensitive labels. There are two basic types of heat-sensitive or heat-seal labels: instantaneous and delayed-action. In the former, heat and pressure are applied to the label to fix it directly on to the product, whereas with delayed action, applied heat turns the product into a pressure-sensitive item. No direct heat is applied to the goods and this clearly is vital to some products, such as food.

Typical market applications for delayed-action heat-sensitive label papers include pharmaceutical, price/weigh, some glass-bottle, and rigid or semirigid plastic container labeling. Instantaneous materials are used for applications such as end seals on biscuit or toilet tissue packs, labels on "pleat wrapped" articles (eg, disinfectant pads, pies, cakes), and for some banding applications (8).

Label Printing

The main printing processes in use for label production are flexography, letterpress, gravure, lithography, silkscreen, and to a lesser extent, hot-foil stamping. This applies for most types of labels whether printed flat or from a reel (see Decorating; Printing).

Although the principles of each process remain the same for reel- or sheet-fed printing, the selection of the particular process and the cost factors in that decision are different for each particular section of the label industry. For example, printing of sheet labels is done primarily by lithography or gravure; but letterpress and flexography are major processes for the production of reel-fed self-adhesive labels.

The fastest-growing technique of label printing is the nar-

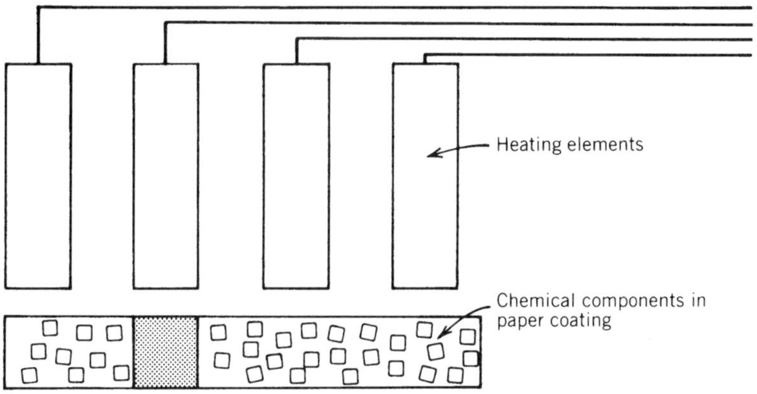

Figure 3. The chemical components in the thermal label paper coating combine and darken when activated by heat.

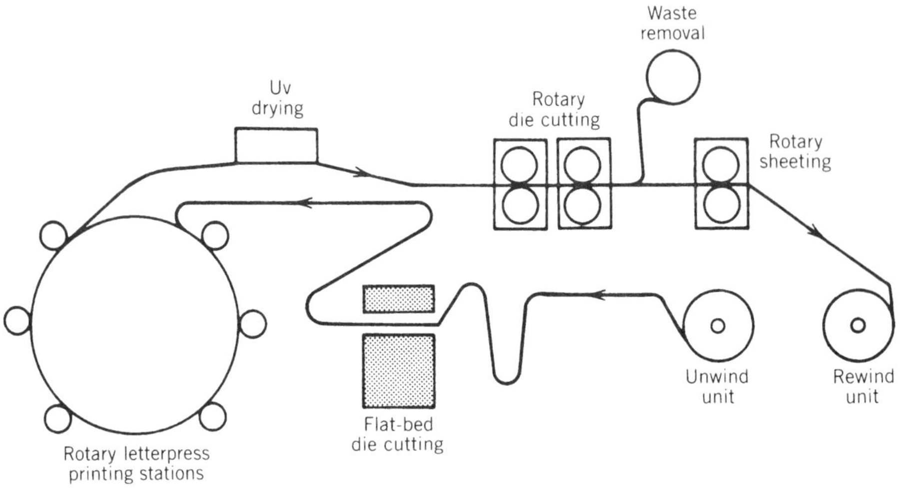

Figure 4. Rotary letterpress roll label press.

row-web roll label conversion of self-adhesive materials using presses that may be printing and converting webs as narrow as 2 in. (50 mm), ranging up to 250, 300, or 400 mm wide (9). These presses not only print substrates up to 6, 8, 10, or more colors in one press pass but also undertake flat-bed or rotary die cutting of the labels, and possibly uv varnishing, over-laminating with clear synthetic films, removing matrix waste, slitting, fan-folding, sprocket-hole punching, etc, and finally, rewinding of the printed reel of labels ready for automatic application (10). A schematic diagram of a typical roll label printing and converting press is shown in Figure 4.

Printing Methods

Flexography. The main advantage of flexography, the most widely used process for reel-fed label printing, is its output speed, which is due to full rotary printing and machine die cutting. In addition, spirit-based inks evaporate quickly, allowing printing on a wide variety of substrates other than paper (see Inks).

Owing to speed of output, flexo labels generally cost less than labels printed by the other major processes. Because presses can be purchased at a relatively low capital cost; many of the smaller label printers use the process. The presses do, however, vary enormously in complexity. Equally important, the quality of the labels produced depends greatly on the skill of the operator.

Letterpress. Although letterpress has declined in usage in the general printing industry, it is a major process in the label industry and gaining in popularity. It is the second most popular process in the roll label industry after flexography, and in some countries is almost the only process used.

The process has adapted well to the needs of the label industry, giving good-quality reproduction and relatively low-cost origination. Presses are available in flat (platen) form, semirotary, or full-rotary letterpress (9).

The main advantage of letterpress is superior quality and ability to hold accurate print definition throughout the run. Solid- and fine-line printing are no problem, as the inks are fully pigmented with excellent lightfastness and color matching. Printing plates can be metal or photopolymer.

The development of uv-cured inks has meant that previously difficult or even impossible to print substances such as vinyls and polyesters can now be "cured" using polymer inks and uv lamps.

Gravure. The gravure process gives excellent quality and print definition. With a very fast throughput it is best suited to long runs where the high cost of originating printing cylinders can be recovered.

Silkscreen. The unique method of applying ink through a screen gives screen process an ink coverage that cannot be completely matched by any other printing method (9). It will also print on virtually any substrate.

The presses are made as complete screen printing units, or a series of screen printing heads can be added to suitable letterpress machines, giving a combination of benefits.

Hot-foil stamping. Hot-foil stamping is generally done in one of two ways: by complete hot-foil presses in their own right, and hot-foil units set in line on to an existing label press. The process gives striking effects when combined with other processes and is very much a designer's medium (9).

Labeling

Once a label has been produced, it must be applied in the correct position to a particular surface or product. The label must be applied securely enough for it to remain fixed in that position throughout the useful life of the product or container. It must stand up to whatever conditions it is likely to be exposed to, and should also maintain a good appearance. In the case of returnable bottles, however, the label must be able to be removed easily during the washing operation.

The labeling operation must also be able to meet production requirements. On packaging lines, the labeling machine must be able to keep pace with the line, often at high speeds, having minimum stoppages and down time (11).

Some labels are applied to products or containers with a simple handheld gun or dispenser; others are applied by a semiautomatic or fully automatic labeling machine suited to the particular type of label: wet-glue, pressure-sensitive, or heat-seal (heat-sensitive).

Wet-glue labeling machinery. Wet-glue labeling is the least expensive labeling system in terms of label costs. There are many machines on the market from the simple semiautomatic to the high-speed advanced models for speeds of up to 600 containers per minute. Straight line or rotary, vacuum transfer of the label or transfer by picker plate, and all-over gumming or strip gumming all have a place and are used in many industries, particularly in food, wine, and spirits plants.

New types of adhesives make glue labeling important for plastic containers as well as glass. Hot-melt adhesive can be used. Large labels are probably cheaper to apply with wet glue than with a heat-seal or pressure-sensitive system, and wet-glue labels are probably superior for applying multiple labels to a range of containers, such as liquor bottles.

Labels may be applied by hand or by semiautomatic or automatic machines (see Fig. 5). There are many different types of wet-glue labeling machines, but they all have to perform the following functions:

1. Feed labels one at a time from a magazine.
2. Coat the labels with adhesive.
3. Feed the glued labels on to the articles to be labeled in the correct position.
4. Ensure that the articles are held in the correct position to be labeled.
5. Apply pressure to smooth the label on to the article and press it into good contact.
6. Remove the article when it has been labeled.

If any of these operations are not performed at the right times, with everything positioned accurately, the results will be unlabeled articles, badly positioned, skewed, or torn labels, and machine stoppages. Machines vary mainly in the way they perform the various functions. For example, some machines may use suction to remove labels from the magazine, and others use the tack of the adhesive itself (11).

The adhesives used for wet labeling fall into five main classes: dextrin-based, casein-based, starch-based, synthetic-resin dispersions, and hot melts.

Apart from the hot melts, all these adhesives are waterborne. The speed at which they set depends, therefore, on the rate at which the water phase can be removed by the absorbency of the label stock. If the water cannot get away, they will not set.

Synthetic resin dispersions, of which the most widely used are those based on poly(vinyl acetate) (PVA), have the major advantage of faster setting, owing to the ability of the polymer particles to draw together to form a continuous film for the loss of much less water than is the case with those adhesives based on natural materials. They are used particularly in the labeling of plastic bottles or coated glass, where normal dextrin, casein, and starch adhesives have difficulty in producing a permanent bond. They are limited to nonreturnable bottles, however, as the dried film has considerable resistance to water and cleaning fluids. One advantage is a low initial tack, which restricts their application to certain types of machines with sufficient brushing-on capacity.

Hot melts are 100% solids, melting when heated and setting almost instantaneously on cooling. They have high initial tack, thus labels adhere to surfaces such as PVC and polyethylene (PE) at high speeds, but are not suitable for wet bottles. To select the correct adhesive for a particular application every factor of the operation must be taken into account: the operating conditions, type, and condition of the articles to be labeled, nature of the label papers, transport and storage conditions, and any particular usage requirements (12).

Pressure-sensitive labeling machines. A pressure-sensitive adhesive is one that remains permanently and aggressively tacky in the dry form and has the ability to bond instantaneously to a wide variety of materials solely by the application of light pressure. No water, solvent, or heat is needed to activate these materials.

Because the adhesive is permanently tacky, it would stick to anything it contacted, so a backing sheet is required to protect the adhesive layer until it contacts the article to be labeled. The backing sheet is coated with a release coating to prevent the label from sticking too firmly to it (13).

There are many different types of applicators, but they all have one thing in common: a means of peeling the labels away from the backing (see Fig. 6). This is usually accomplished by unwinding a reel of die-cut labels, and pulling them under

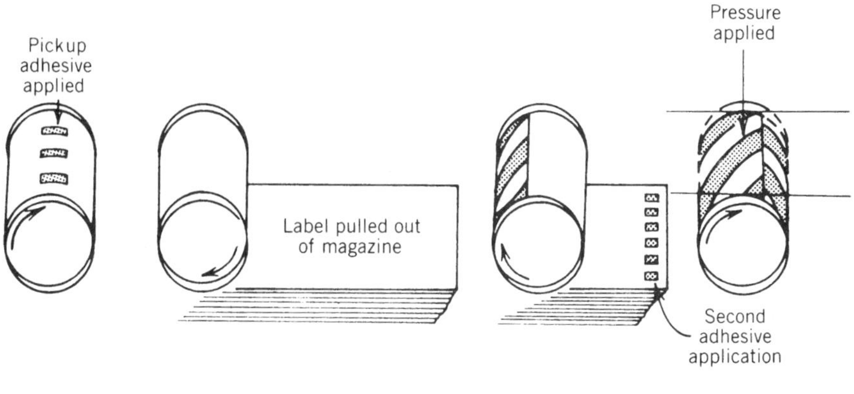

Figure 5. Stages in the labeling of cans. Adhesive is first applied to the container. As the container rotates over the label magazine it picks up its own label, which is then rolled around it. A second adhesive application to the trailing edge of the labels, followed by pressure, completes the labeling operation.

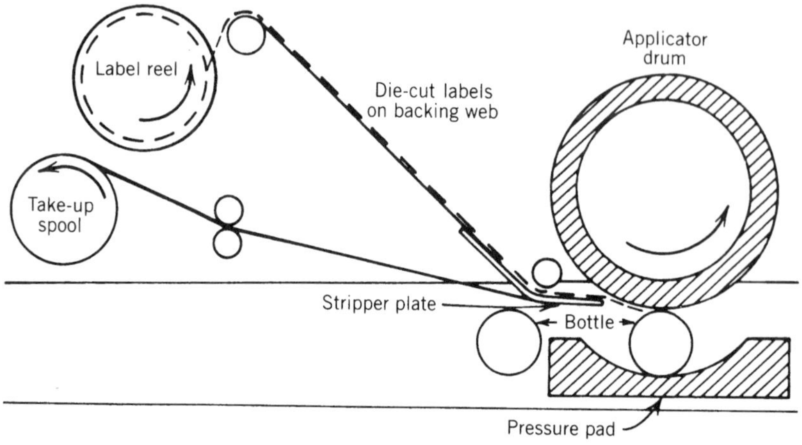

Figure 6. A common type of self-adhesive label applicator.

tension around a stripper plate. As the backing is bent around a sharp angle, the front edge of the label peels away. Once the labels have been detached from the backing, there are various ways of feeding them forward and pressing them onto the containers in the correct position. Containers may be fed forward to an applicator drum where the label is transferred to it under light pressure created by an applicator drum and pressure pad. Alternatively, the label may be held in position on a vacuum box or drum and released onto the article when it is in the correct position, or it may be blown onto the container by releasing the vacuum and applying air pressure. The backing paper is then rewound on a take-up spool (12).

Comparatively simple systems have been devised to apply the label to the product. The choice ranges from a dispenser that releases a label for hand application up to high-speed automatic labelers. Labeling machines for applying pressure-sensitive labels vary in complexity according to the nature of the package to be labeled, the number of labels to be applied at one time, and the speed of application. There are, of course, semiautomatic labelers for single-product medium-volume application. Although pressure-sensitive labelers are available as standard machines, they are usually readily customized to meet special requirements. Generally, they are only about one-third to one-half the cost of most competitive labeling systems.

The advantages of pressure-sensitive labels are said to be numerous. They are cleaner than other methods, less wasteful than wet glue, and more easily controlled because they come in roll form so that labels cannot be mixed up. A wide variety of adhesives for different surfaces is available, and clear acetate, vinyl, or polyester film can be used for "see-through" labels.

Heat-seal labeling machines. Heat-seal labeling has most of the advantages of pressure-sensitive labels but generally offers a lower label cost. Heat-seal describes a plasticized paper which, when heated by the machine, becomes sticky on its underside. Labeling equipment for heat-seal ranges from simple aids to automatic high-speed machines, giving the unique advantage of being able to apply to many varied types of surface labels that can be precision-placed, give all-over adhesion, and be perfectly clean when applied.

The machines free the user from mixing and adding glue, selecting grades, controlling viscosity, and so on. The absence of glue eliminates cleaning the machine, and these savings give substantial increases in productivity and reductions in maintenance expenses. The labels offer a high degree of adhesion security that is useful, for example, for bottles subjected to high humidity, or steam of water saturation (14).

Of the two types of heat-seal paper, the delayed-action type is best suited for machine operation. The ability of the adhesive to remain tacky after removal of the heat source allows the machine designer to separate the heat source from the pressure application of the label. Thus the heating plates can be simple whereas the pressure-applying devices can be of complex shapes, enabling a diverse range of articles to be labeled, varying from flat to round or irregular shapes (10).

Label Overprinting Machinery

For certain label applications there is a requirement to have a plain paper or preprinted label or labels overprinted with price, price/weight, bar code, or other variable information just prior to or at the point of application. A variety of mechanical, electronic, or computer-based overprinting systems are available for such operations. These are now widely used by department stores, the retail and wholesale trade, industry, and even some label producers.

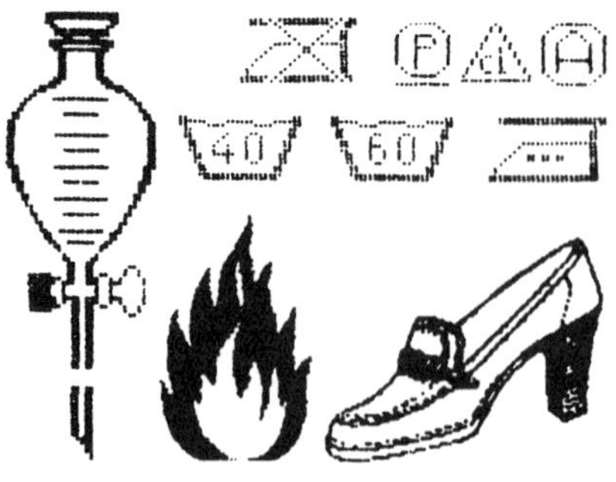

Figure 7. Electronically produced graphics are available on electronic label overprinters.

Mechanical overprinting systems range from small hand-operated machines wtih one or two wheelsets and two type channels, to hand-operated and electronic versions used for price/description labeling with either conventional or EDP-readable fonts (10).

The current trend in overprinting is away from mechanical to totally electronic overprinters, which can now print labels and tags without any form of marginal or sprocket-hole punching. Free formating and graphics, as well as alphanumeric information and bar codes are now possible using such overprinting technology, the graphics being keyboard entered and edited. Graphics suitable for reproduction include product outlines, such as shoes, laboratory equipment, small hardware items, safety hazard designs, and laundry symbols. Figure 7 shows examples of such electronically produced and over-printed graphics.

BIBLIOGRAPHY

"Labels and Labeling Machinery" in *The Wiley Encyclopedia of Packaging Technology*, 1st ed., by M. C. Fairley, Labels and Labelling Data and Consultancy Services Limited, pp. 424–430.

1. *A Guide to Legislation and Standards on Labelling* (UK), Labels & Labeling Data and Consultancy Services Ltd, Potters Bar, Herts, UK, 1984.
2. D. G. N. Alder, "Label Papers and Their Uses," *Labels & Labelling,* **14,** 16 (Jan. 1979).
3. "The Most Important Types of Paper for Beverage Labels," *Krones Manual of Labelling Technology,* Hermann Kronseder Maschinefabrik, Neutraubling, FRG, 1978, pp. 118–121, 1978.
4. A. Hildrup, "Review of the (UK) Gummed Paper Market," *Labels & Labelling Yearbook,* 23 (1981).
5. M. Fairley, "The challenge for water-borne adhesives," *Labels & Labelling*, 29 (Sept. 1984).
6. M. Fairley and R. Brown, *Thermal Labelling,* Labels & Labelling Data and Consultancy Services Ltd., Potters Bar, Herts, UK, 1984.
7. M. Fairley, *Bar Coding,* Labels & Labelling Data and Consultancy Services Ltd., Potters Bar, Herts, UK, 1983.
8. N. Henderson, "Trends and Developments in Heatseal Label Materials," *Labels & Labelling Yearbook*, 31 (1981).
9. M. Fairley, *Label Printing Processes and Techniques,* Labels & Labelling Data and Consultancy Services Ltd., Potters Bar, Herts, UK, 1984.
10. *Labels—A Product Knowledge Book,* National Business Forms Association, Alexandria, VA, 1983.
11. *Labelling Operations*, Pira, The Research Association for the Paper and Board, Printing and Packaging Industries, Leatherhead, Surrey, UK, 1981.
12. E. Pritchard, "Labelling, and Labelling Systems," *Labels & Labelling,* 14, 16 (March 1982).
13. *Labelling—Basic Principles,* Pira, The Research Association for the Paper and Board, Printing and Packaging Industries, Leatherhead, Surrey, UK, 1980.
14. D. Miles, "Pre-adhesed Labelling," *Labels & Labelling,* 20 (Jan. 1981).

LAWS AND REGULATIONS, EUROPE

There is no separate branch of law that may be conveniently classified as packaging law; statutes in all the European countries affect packaging. Legislation on the sale of goods, trade descriptions, transport, weights and measures, food and drugs, food safety, environmental issues, and waste management are all concerned with packaging.

The European Union (EU) has had a major impact on the legislation that is intended to provide a uniform practice across all countries of the Union. However, in practice some countries may deviate from this either because legislation is already in place, on environmental grounds or where exemptions have been granted by the European Commission. For example, Greece is not required to meet the used packaging recovery rates set for the Union because of its geographic spread over 2000 islands. It is therefore important for exporters, particularly those from non-European Union countries, to check both the European Union law and those of the country concerned. Alternatively, consult an expert who has the necessary contacts and experience of the EU countries and is skilled at finding out the precise requirements.

In 1944 the European Union consisted of 12 countries: France, Germany, Netherlands, Belgium, Denmark, Luxembourg, Spain, Portugal, the United Kingdom, Greece, Italy, and Ireland. As from January 1, 1995, Austria, Finland, and Sweden joined the Union. The collpase of communism has opened up new opportunities in the former Eastern Bloc European countries, and although these have had very little packaging legislation in the past, some of them, notably Hungary, Poland, the Czech and Slovak Republics, and Romania, are adopting the European Union (EU) legislation in preparation for membership around the year 2000. In Poland, it should be noted, in addition to following EU legislation there is already a law in place that requires all packaging to be approved by the government authority before it can be placed on the market.

How the European Union Functions

The member states. The European Union is a group of 15 countries bound together by three major international treaties and a number of smaller ones, the most important of which is the Treaty of Rome.

The institutions. There are four main EU institutions: the Commission, the Council, the Parliament, and the Court of Justice.

The Commission. The Commission proposes EU policy and legislation; executes the decisions taken by the Council of Ministers and supervises the day-to-day running of EU policies; acts as the "guardian of the Treaties" and can initiate action against member states who do not comply with EU rules; and has its own powers under the Treaties on some areas, notably competition policy and the control of government subsidies.

There is a commissioner in charge of each area of policy; all commissioners formulate proposals within their area of responsibility aimed at implementing the Treaties.

The Council of Ministers. The Council is the EU's principal decisionmaking body. In most cases it adopts legislation on the basis of proposals from the Commission. Council meetings take place at several levels:

Council or Summits. These take place at least twice a year and are attended by heads of state or government, foreign ministers, and representatives from the Commission who meet to discuss broad areas of policy.

Specialist Councils. These deal with particular areas of policy. They are attended by the relevant ministers from member states and by the Commission. Policy initiatives are discussed and legislation is formally adopted at these meetings.

COREPER I and COREPER II (Committee of Permanent Representatives). These meet weekly and consist of permanent representatives or deputy permanent representatives of each member state. They pave the way for political decisions to be taken by ministers.

Council Working Groups. These are attended by officials from member states and from permanent representatives in Brussels. These groups are convened as necessary and there are usually about 150 working groups in operation during any given presidency. They examine the issues in detail. Council meetings are chaired by the member state holding the presidency of the Union. Each member state holds this office in turn for a 6-month period.

The Parliament. The European Parliament is a directly elected body of 626 members. Its formal opinion is required on most proposals before they can be adopted by the Council. Most of the detailed work in Parliament is done by its specialist Committees, which examine and draw up reports on Commission proposals before they are put to the Parliament as a whole.

The European Court of Justice. The European Court of Justice (ECJ) rules on the interpretation and application of Union (EU) laws. A Court of First Instance has been established to relieve the ECJ of some of its excessive workload.

Other bodies. There are three other main bodies: the Economic and Social Committee—an advisory body representing employers, trade unions, and other interest groups (such as consumers), which has to be formally consulted by the Commission on proposals relating to economic and social matters; the Court of Auditors, which examines and assesses the Union's revenue and expenditure; and the European Investment Bank—the European Union's bank, which lends money to finance capital investment projects.

Types of legislation. Under the Treaties, the Council and the Commission may make Regulations, issue Directives, take Decisions, make Recommendations, or deliver opinions.

Regulations have general application and are directly applicable in all member states. They do not have to be confirmed by national parliaments in order to have binding legal effects. If there is a conflict between a Regulation and existing (or future) national law, the Regulation prevails.

Directives are binding on member states as to the result to be achieved within a stated period, but leave the method of implementation to national governments. In itself, a Directive does not have legal force within member states, but may take direct effect if the Directive is not duly implemented.

Decisions are specific to particular parties and are binding in their entirety on those to whom they are addressed, whether member states, companies, or individuals. Decisions imposing financial obligations are enforceable in national courts.

Recommendations and opinions have no binding force but merely state the view of the institution that issues them.

The legislative process. Union legislation is the result of a complex and often lengthy process of consultation and negotiation. Before legislation is formally proposed, the Commission will often discuss its ideas informally with national experts, professional and business organizations, and interested parties.

The formal process usually starts with proposals made by the Commission. The Council can, however, request the Commission to undertake studies or submit appropriate proposals. The proposals are then submitted to the Council, which must usually consult the Parliament and the Economic and Social Committee.

Most single market proposals are subject to the cooperation procedure under which the Parliament gives two readings, once when the Commission proposal is submitted to the Council and again after the Council has reached an agreement in principle (a common position). The Parliament can propose amendments at both stages.

The Council can formally adopt the Commission proposal as drafted, request the Commission to amend it, amend it itself, reject it or simply take no decision.

There are three methods of decision taking in Council depending on the nature of the proposal and the Treaty article on which it is based: unanimity, simple majority voting (at least seven member states in favor), and qualified majority (weighted voting based on the relative size of the member states by population). Most single market measures are subject to qualified voting.

A cooperation procedure was introduced as part of the Single European Act, allowing the European Parliament to slow down but not to take the final decision on legislation. However, the Maastrict Treaty gave the Parliament the power of veto and set up a "codecision" procedure (see Fig. 1).

When the Commission draws up a proposal, it must be well informed about the subject in question. To this end it consults with national authorities who have formed working parties and subgroups to discuss the subjects being treated. In the food area, the Commission consults with the EU Scientific Committee for Food, and it also consults professional groups such as trade associations and consumer organizations. Because the opinions of these organizations vary, they are represented on an Advisory Committee for Foodstuffs, where they can compare and discuss their different views and approaches.

Once a Directive is adopted, it is published in the *Official Journal* (*OJ*) of the European Union and sent to each member state with a request to amend their laws within the prescribed time.

When violations of any Union law is reported to the Court of Justice, it has a duty to report the case and interpret the Union law. Its decisions are binding on all parties (see Fig. 2).

Food-Package Compatibility

Few materials suitable for food packaging are completely inert to food, and those materials that are inert often have to

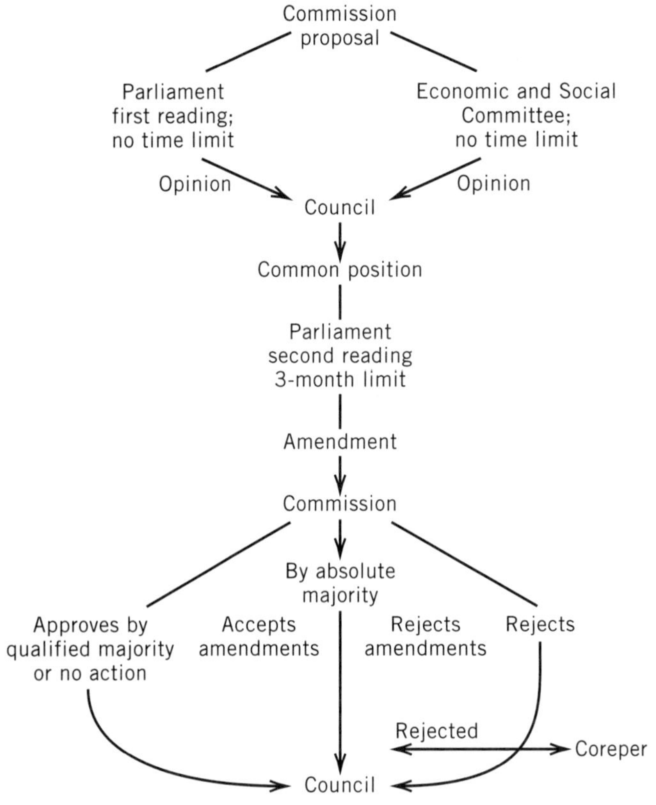

Figure 1. Codecision procedure.

be used in conjunction with others that are not. The care taken to produce wholesome and attractive food must be matched by the care taken in the production of the packaging used to contain and protect them. Recognizing this, responsible suppliers of packaging materials and food manufacturers have for many years worked together for the benefit and protection of the consumer. In UK this has culminated in the preparation of an advisory code of practice for manufacturers of food packaging.

The fitness of packaging materials for this purpose was governed by the requirement that food shall be of the substance and quality demanded and any packaging that did not adversely affect the flavor or olfactory properties of the food

Figure 2. Binding-Decision process.

was allowed. As most food is packaged, it is necessary to know the extent of any interactions between foods and the materials or containers in which they are packaged. Clearly, any interactions must be small. Some small interactions can be detected by the sense of taste or smell, but these are not necessarily definitive, and modern analytical methods are applied to packaging that is to be used for food packaging.

The EU has in recent years introduced extensive legislation in this area. At present it is concerned mostly with plastics in contact with foodstuffs, but it is intended that the principles should be applied to all packaging materials.

The EU has published *Practical Guide NI,* which is a practical guide for users of EU Directives on materials and articles intended to come into contact with foodstuffs. It states that the general requirements for toxicological studies should be aimed at obtaining the maximum amount of information from the minimum number of animals. The essential core set of tests cover a 90-day oral study; three mutagenicity studies; a test for gene mutation in bacteria; a test for chromosomal aberrations in cultured mammalian cells; a test for gene mutations in cultured mammalian cells under special circumstances; studies in absorption, distribution, metabolism, and excretion; data on reproduction; data on teratogenicity; and data on long-term toxicity or carcinogenicity.

These studies should be carried out according to EU Directives 84/449/EEC, April 25, 1984 and Directive 87/302/EEC, December 18, 1987; Directive 87/18/EEC, December 18, 1987; Directive 87/18/EEC, December 18, 1986; Directive 88/320/EEC, June 9, 1988; Council Decision 89/569/EEC, July 28, 1989; and Directive 90/18/EEC, December 18, 1989.

Once a plastic material is approved for food use, it is up to the packaging manufacturer and the user to show due diligence and routinely test the migration levels of these materials.

Food-Contact Legislation

In 1970 a "framework" Directive on materials on articles intended to come into contact with foodstuffs was introduced. The preparation of the specific EU legislation and regulations for the various materials used for food packaging—plastics, paper and board, glass, etc—has up to now been slow, but the first "Plastics Directive" has now been adopted and a target date of end 1995 has been set for the next major Directive on additives used in plastics.

Any substance that migrates from packaging into food is of particular concern if it could either (*1*) be harmful to the consumer of the food or (*2*) have an adverse effect on the organoleptic properties of the food—taste and aroma. Even if the migrating substance neither is harmful nor has adverse organoleptic properties, its presence in the food as an "inert" contaminant is undesirable.

Until the late 1970s food packaging was covered in the UK by the "Food Laws," which make it an offense to sell food unless it is safe, fit to eat, and of the nature and substance demanded by the consumer.

In 1979, Statutory Instrument (SI) No. 1927 (1978) was introduced, specifically aimed at food packaging and other "food contact" materials. This SI-implemented EU Directive 76/893/EEC on the approximation of the laws of the member states relating to materials and articles intended to come into contact with foodstuffs. These laws were, in essence, framework regulations, requiring that materials and articles be manufactured in accordance with good manufacturing practice, so that, under their normal or foreseeable conditions of use, they do not transfer their constituents to foods in quantities that could endanger human health or bring about a deterioration in the organoleptic characteristics of such food or an unacceptable change in its nature, substance, or quality.

To supplement the "framework" regulations, the "food-packaging industry" has relied on codes of practice. An example is that published jointly by the British Plastics Federation and the British Industrial Biological Research Association entitled *Plastics for Food Contact Applications.*

Statutory Instrument 1979 No. 1927 was consolidated with two amendments issued in 1980 and 1982, that dealt with poly (vinyl chloride) (PVC) plastics and vinyl chloride monomer. Both of these amendments had originated from EU Directives. In addition, two EU Directives on regenerated cellulose films (83/229/EEC and 86/388/EEC) were implemented and an EU Directive determining the symbol that may accompany any material or article intended to come into contact with foodstuffs. The regulations also specify requirements for the labelling of materials and articles sold. The new Statutory Instrument—1978 No. 1523—retains the title *Materials and Articles in Contact with Food Regulations.*

In these current regulations only the two plastics mentioned—PVC and regenerated cellulose films—are specifically "controlled." With PVC, control is limited to the vinyl chloride monomer, which should not be present in the plastic at a level exceeding 1 mg/kg (1 ppm), and should not transfer to foods in any quantity greater than 0.01 mg/kg of food (10 ppb). The reason for controlling this particular plastics monomer is because it is a known carcinogen. The controls on the regenerated cellulose films are more comprehensive, with positive lists of substances that may be used in the manufacture, plus a restriction on coated regenerated cellulose film that states that the film should not transfer the softening agents, monoethylene glycol and diethylene glycol, into food in a quantity exceeding 50 mg/kg of food. Further amendments to the Directives on regenerated cellulose films have now been adopted introducing changes considered necessary from recent data.

In 1989 the EU introduced a new "framework" Directive for "food-contact materials and articles" to replace 76/893/EEC. This Directive (89/109/EEC) is essentially the same as 76/893/EEC but makes provision for specific Directives for the various categories of "food contact" materials and their adoption through the Commission's Standing Committee on Foodstuffs, on which all member states are represented: decisions may be taken by a qualified majority. Each member state has an allocated number of votes ranging from 2 to 10, depending on its size. There are a total of 87 votes, 10 of which were UK votes. A qualified majority requires 54 (71%) in favor. Where the Committee cannot reach a decision, the matter is referred to the Council. The aim of this new procedure is to reduce the time to reach decisions.

Article 2 of the Directive restates the fundamental requirement that "materials and articles in their finished state, must not transfer their constituents to foodstuffs in quantities which could endanger human health or bring about an unacceptable change in the composition of the foodstuffs."

The various classes of materials that are listed in the Directive, and which will be eventually covered by specific Commission Directives, are plastics, including varnish and coatings; regenerated cellulose; elastomers and rubber; paper and board; ceramics; glass; metals and alloys; wood, including cork; textile products; paraffin waxes and microcrystalline waxes.

Regenerated cellulose films have already been dealt with in two specific Directives, as mentioned above, with further amendments. The main current effort is being directed toward producing Directives for plastics.

EU Directive 89/109/EEC has not yet been implemented by UK legislation.

The first "specific" Commission Directive relating to plastics materials and articles intended to come into contact with foodstuffs (90/128/EEC), was adopted by the Standing Committee on Foodstuffs in November 1989, and published in the *Official Journal* in February 1990. This Directive contains a positive list of monomers and other starting substances used in the manufacture of plastics, but not the other classes of components used in plastics, such as additives, catalysts, and colorants.

The Directive also includes a limit on overall migration.

An important aspect of these regulations is that they apply to the finished package or packaging material—films, bottles, pots, containers, etc—and responsibilities will extend from the plastics manufacturer to the food retailer.

As there was a large number of mistakes and errors in the Directive, a corrigenda was published in the *Official Journal* in December 1990. A number of changes to this Directive have, however, already been decided and amendment Directives in 1992 and 1993 have been adopted. The changes are mainly to the lists of monomers and regulation restrictions.

Not all the monomers in the positive, or permitted, list of Directive 90/128/EEC were fully toxicologically assessed by the Scientific Committee for Food (SCF)—the expert advisory body to the EU, due to lack of data. They were therefore categorized into two groups, Annex I and Annex II. Annex I contains all those monomers for which a definite decision was possible; and Annex II, those monomers for which a definite decision could not be made because of insufficient toxicological data. In the Directive, the SCF Annex I monomers have been placed in Section A, and the majority are listed without any restrictions, but a number, because of their toxicological properties, have been assigned the following restrictions: a limit of the quantity that is allowed to migrate into foods − a specific migration limit = SML, a limit on the residual quantity in the plastic − a compositional limit = QM.

For some monomers, only one of the restrictions applies, but for other monomers, both restrictions apply.

For those monomers in Section A that have well-established toxicological properties, the limits on the quantities that will be allowed to migrate into foodstuffs will be very low. For example, for the monomer acrylonitrile, used in the manufacture of plastics for ABS margarine tubs, the restriction is that it should not be detectable in foods when analyzed using a method with a detection limit of 0.02 mg/kg (20 ppb).

For other monomers considered to have less hazardous

properties, the limits on the quantities allowed to migrate into foods are higher. A monomer in this category is terephthalic acid—a principal monomer used in the manufacture of PET plastics—with a specific migration limit (SML) of 7.5 mg/kg (7.5 ppm).

Restrictions on the residual quantity of the monomer allowed in the plastics are applied to those monomers that have a tendency to hydrolyze readily. Monomers in this category include the isocyanates used in the manufacture of polyurethane and certain acrylates. The (QM) limits typically range from 1 to 5 mg/kg.

Monomers listed with both types of restrictions include vinylidene chloride. The limits for this monomer are 5 mg/kg (5 ppm) as a residue in the plastic (QM), and 0.05 mg/kg (50 ppb) as a limit on the migration into foods (SML).

The SCF Annex II monomers, which represented about 70% of the total number of monomers in the positive list of Directive 90/128/EEC, were placed in Section B of the Directive. Substances can be transferred from Section B to Section A when satisfactory additional information has been provided.

This additional information ideally should be toxicological, but it has been recognized that in view of the considerable volume of experimental work that would be necessary to provide a basis for the toxicological assessment of each substance, priorities have to be set. It has also been indicated that the criteria for setting these priorities should include data on exposure, such as usage, residual monomer content, and extent of migration.

Of these data, those on migration and residual monomer content in the plastics are of prime importance. If it can be shown that the levels migrating from the plastic into foods or food simulants are very low (eg, in the low-ppb range), then only limited toxicological data are required. For example, where migration of the monomer is <0.05 mg/kg of food (food simulant), only three mutagenicity test results are required.

Where full toxicological data are supplied for the substance, ADI (acceptable daily intake) and TDI (tolerable daily intake) values are calculated. These ADI and TDI values are then used to calculate any specific migration limits (SML). The SML is determined by multiplying the ADI or TDI value by 60. For example, the ADI value for terephthalic acid is 0.125 mg/kg body weight (bw), and multiplying by 60 gives the SML for terephthalic acid of 7.5 mg/kg.

For any substance for which the ADI or TDI value is 1 mg/kg bw, no specific migration limit is assigned. This is because the calculated specific migration limit would be ≥ 60 mg/kg, and Directive 90/128/EEC also contains an overall migration limit of 60 mg/kg. This overall migration limit restricts the total quantity of substances that may migrate from the plastics material into the food.

Amendments to Directive 90/128/EEC contain changes to the lists of monomers. A number of substances have been added to Section A, and others have been transferred from Section B to Section A. The largest change is the deletion of about 200 substances from Section B.

Measurements to determine the migration of monomers into foods can be quite difficult, particularly where the specific migration limit is very low. Tests on plastics food packaging to determine whether any specific migration of a monomer is within the specified limit are often performed with food simulants instead of foods. These food simulants are simple liquids, making the analytical determinations less difficult.

The work on the plastics additives list is well advanced, and its completion was expected toward the end of 1994. The large number of substances for the positive additives list, in excess of 1000, will be dealt with in a similar manner to the monomers, but the Directive is expected only to have an "A" list of fully approved substances. Plastics additives include antioxidants, stabilizers, slip agents, and antistatics.

The latest information on plastics additives is contained in the EU draft Synoptic Document N6. This document also contains monomers and additives for coatings used on plastics substrates and other substrates such as board and metals.

Even if substances that migrate from packaging into the food are neither harmful nor have unacceptable organoleptic properties, their presence in food is undesirable, particularly if there is significant contamination. The overall migration limit in EU Directive 90/128/EEC will prevent food being excessively contaminated.

Overall migration, or global migration, as it is sometimes called, is the total migration of substances from the plastic into a food simulant. No attempt is made to identify the nature of the substances. It is intended as a measure of the inertness of the plastic and a means of preventing unacceptable contamination of the food. The value of the overall migration limit has been strongly disputed by some member states, including the United Kingdom, and also by some trade associations. However, the points put forward in favor of the overall migration limit, in addition to those above, are *(1)* specific migration limits, or other restrictions, are avoided for those substances that at concentrations above the overall migration limit, are regarded as unsafe; and *(2)* the health risks are reduced from migration of those substances whose toxicological effects are not fully known because of incomplete data.

In the Commission Directive 90/128/EEC on the approximation of the laws of the member states relating to plastic materials and articles intended to come into contact with foodstuffs, Article 2 states:

> Plastic materials and articles shall not transfer their constituents to foodstuff in quantities exceeding 10 milligrams of substance(s) per square decimetre of surface area of material or article (overall migration limit). However, this limit is 60 milligrams of substance(s) released per kilogram of foodstuff in the following cases: containers or articles which are comparable to containers, which can be filled with a capacity of not less than 500ml and not more than 10 litres; articles which can be filled and for which it is impracticable to estimate the surface area which is in contact with foodstuff; caps, gaskets, stoppers, or similar devices for sealing.

These are not two different overall migration limit values as they are directly related.

The relationship is a follows. A cube, with each of the six faces 1 dm^2 in area, has an internal volume of 1 L. If each of the six faces releases 10 mg of substance into the internal volume of the cube, then the total quantity of substance is 60 mg. If the internal contents of the cube are assumed to have a specific gravity of 1, then the concentration of substance from the six faces can be expressed as 60 mg/kg.

Overall migration tests are carried out with the food simulants that are laid down in Directive 82/711/EEC: distilled water; 15% v/v ethanol; 3% w/v acetic acid; and olive oil, sunflower oil, or synthetic triglyceride. These food simulants are also used for specific migration tests for monomers, mentioned earlier.

Directive 82/711/EEC also lists the test conditions relative to the intended conditions of use of the plastic, eg, time and temperature. A further Directive (85/572/EEC) relates classes of foods to the food simulants to be used for the migration tests.

An amendment to Directive 82/711/EEC now contains high-temperature test conditions. Directive 82/711/EEC covered conditions of use only up to 121°C.

Standard overall migration test methods and specific migration methods for monomers are in preparation under the European Commission for Standardisation (CEN). The overall migration draft standards have recently been completed and cover methods for all four food simulants with testing by total immersion, single-side contact with cell and pouch, and article filling. They are published as Standard ENV1186.

For the three aqueous food stimulants, the test procedure is relatively simple. At the end of the test period the overall migration is determined gravimetrically, after evaporating the food simulant to dryness. This method obviously cannot be used for olive oil or other oils. With these simulants, the test specimen is weighed before and after contact with the oil, and the overall migration is calculated as a weight loss after allowing for absorbed oil. This is determined separately by gas chromatrography after extraction, hydrolysis to the fatty acids, and conversion of these acids to methyl esters. An involved procedure such as this can easily produce unreliable results, particularly with plastics from which it is difficult to extract all the absorbed olive oil, and plastics that are moisture-sensitive.

The European Commission has not yet started work on a Directive for "food contact" paper and board materials.

In the current UK Materials and Articles in Contact with Food Regulations (SI No. 1523), there is no specific reference to paper and board products. Also, there is no code of practice similar to that for plastics published by BPF/BIBRA.

Paper and board products for food packaging must of course comply with the main basic requirements of UK SI No. 1523 and the EU Directive 89/109/EEC, which has already been stated.

In the absence of specific regulations or code of practice, manufacturers and users of paper and board products for food packaging and other food-contact applications have in the past relied on regulations and codes of practices of other countries, such as the American FDA and Dutch regulations, and the German BGA recommendations.

Recently, the Council of Europe has been preparing a Resolution on "food contact" paper and board materials, with Sweden and the United Kingdom being mainly responsible for the work. It is expected that the resolution will be taken up by the European Commission and progressed for a future Directive. The basis for this resolution has been a document produced by the European Confederation of Pulp, Paper and Board Industries (CEPAC) in collaboration with the International Committee of Paper and Board converters in the Common Market (CITPA). The draft resolution contains a preamble defining the field of application of the measures recommended depending on the nature of the packaging and the food contact; positive lists of substances for the manufacture of the paper and board products, to include uses for baking and hot-water filtration; and extraction and/or purity limits on listed substances, where considered necessary from toxicological data.

There is no provision for an overall migration limit and overall migration tests with the four standard food simulants, as applied to "food contact" plastics. Such a limit is considered inappropriate as paper and board materials are not used in direct contact with liquid foods. Likewise, specific migration tests in most cases are considered inappropriate and, as indicated above, controls on individual substances will be by extraction or purity limits.

A reduced testing schedule will apply to paper and board materials intended for contact with dry foods only.

Substances controlled by extraction and/or purity limits will include both additives and also contaminants, including toxic heavy metals, pentachlorophenol, and PCBs.

It is proposed that paper and board made with recycled fibers should be allowed, provided the recycled fibers originate from specified qualities of reclaimed paper and/or board that has been subjected to appropriate processing and cleaning. The finished materials, of course, must also fully comply with the specifications in the Resolution.

It is also proposed that printed surfaces should not come into contact with foods, but it is stated that inks used should comply with a complementary draft resolution on printing inks.

The possibility of dioxins being present in paper and board materials made from pulp bleached with chlorine is mentioned. Although no limits are proposed for these highly toxic substances, there is a direction in the Resolution that manufacturers of paper and board for food-contact applications should use raw materials produced by processes that reduce the amount of dioxins to the lowest possible levels.

The extensive tests carried out on paper and board food-packaging materials have shown that the levels of dioxins present are only just detectable, even using the most sensitive analytical techniques currently available. The general conclusion is that any dioxins present in paper and board food packaging do not present a health hazard to consumers of the packaged foods.

Although a European Commission Directive on "food contact" paper and board is some years away, work is currently in hand by a Technical Committee of the European Committee for Standardisation (CEN) developing test methods based on the proposed limits in the Council of Europe Resolution.

Metals and glass are on the list of materials for which the European Commission intends to produce specific Directives. However, there are no plans for the immediate future to prepare these Directives.

Many metal cans, either steel or aluminum, often have internal lacquer coatings to act as a barrier between the metal and the foodstuff. These lacquers are polymer-based and sometimes similar in composition to plastics that are covered by the "plastics" Directive 90/128/EEC and UK SI No. 3145. This Directive and the corresponding UK SI, however, apply only to materials and articles made totally of plastics and not to articles where the "plastics material" is in combination with another material.

The Commission's Synoptic 6 document contains the draft lists of monomers and additives for "coatings." Work is progressing well with the examination of these substances, and a Directive on "coatings" is expected to be completed in 1996.

The future migration and other restrictions on monomers and additives used in polymeric coatings are expected to be similar to those types of substances in the "plastics" Directive 90/128/EEC.

A European Committee for Standardisation Technical Committee (TC194/Sub-Committee 1/Working Group 5) has also been set up to establish the necessary migration and other test methods that will apply to lacquer-coated food cans. In many cases the migration tests developed for plastics will be directly applicable to the polymer-coated food cans. With some migration tests problems have already been encountered, and special tests will have to be developed.

In addition to EU rules on migration some countries in Europe have bans on specific plastics. In Sweden there is an agreement between industry and the government that PVC should not be used for packaging of fatty foods. In practice this has become a ban on the use of PVC. One Dutch supermarket also banned the use of PVC but then rescinded the ban after protests. Although PVC is an excellent packaging material, some emotion—rather than fact-based criticism of its use has arisen, and therefore anyone planning to export to the EU should first ensure that the material and its additives they plan to use are acceptable. Some materials, such as PVC bottles in Switzerland or PET for nonreturnable bottles for carbonated drinks in Germany, are restricted on environmental grounds rather than failure to meet toxicity or migration data.

Labeling

The labeling Directive, EU Directive 79/112 Labelling, Presentation, and Advertising of Foodstuffs for Sale to the Ultimate Consumer, was adopted on December 18, 1979. It was then incumbent on member states to permit trade in goods labeled according to the Directive 2 years later (Dec. 18, 1981) and to prohibit goods not labeled according to the Directive 4 years later (Dec. 18, 1983). To accommodate nationally established practices, the final Directive allowed member states to be selective sometimes, and also permitted more detailed national labeling controls to be imposed, provided a notification procedure was followed. There are about 30 options or derogations in the Directive, from which member states can choose. Those chosen differ from one country to another so that the 10 sets of regulations stemming from the same Directive each differ in some way.

Judging by UK experience, member states have faced problems in incorporating the provisions of the labeling Directive into national legislation. The Directive is based on the principle that labeling methods used must not mislead purchasers as to the nature, identity, properties, composition, quantity, durability, origin or provenance, methods of manufacture, or production of a food, by attributing to it properties that it does not possess or by suggesting that it has special characteristics when all similar foodstuffs also have them.

Leaving the exemption and derogations aside, the Directive requires the following particulars to be shown on food labels: the name under which the product is sold; a list of ingredients; the net quantity; the date of minimum durability; any special storage conditions or conditions of use; the name or business name and address of the manufacturer or packager, or of a seller established within the Union; the place of origin, where failure to give such information could mislead the consumer as to where the foodstuff came from; and instructions for use when it would be impossible to make proper use of the foodstuff if these were not supplied.

Member states may retain national requirements for an indication of the factory or packaging center where domestic production is involved. Member states may also apply more extensive provisions regarding weights and measures.

Product naming must follow a set pattern. First, if the name is prescribed by Union or national law or by administrative rules, then that is the name that must be used. If a prescribed name does not exist, a customary name may be used, or the product must be described precisely to inform the purchaser of its true nature and to distinguish it from products with which it could be confused. A customary name is a name that is customary in the member state in which the product is to be sold. Some customary names are used in many countries (eg, spaghetti and frankfurter), but generally speaking, they are more local.

Recent (1993/4) cases in the European Court of Justice prohibit the use of certain names for products that do not come from specific regions; for instance, champagne can be so called only if it comes from the Champagne region of France. If produced anywhere else, it can be described only as Champagne type or sparkling wine.

The Directive prohibits the substitution of a trademark, brand name, or fancy name for the product name. Hence, although brand names such as Pepsi Cola and Coca Cola are internationally recognized, they must be more fully described.

A product name must also include or be accompanied by details as to the physical condition of the foodstuff, or of a treatment it has undergone, if the omission of such information could create confusion for the consumer. If it is not readily apparent that a food is powdered, freeze-dried, concentrated, cured, etc, then this must be made clear in the product name. By this, it is presumed that a product name should reflect, for example, the polyphosphate treatment of chilled or frozen chicken, as well as mentioning whether the product is chilled or frozen.

This is particularly important where food has been irradiated to preserve it. Although the sale of such food is permitted, many large retailers are refusing to stock such products because of the emotional impact radiation has with the public.

"Ingredients" means any substance, including additives, used in the manufacture or preparation of a food and still present in the finished product, even if it has altered in form. Additives must be listed, together with either their chemical name or the EU serial number. However, additives used only as processing aids need not be declared, nor those additives that are there because they were present in one or more of the ingredients, as long as the levels of carryover are insufficient for them to have a technological function in the finished product.

Ingredients must be listed in descending order of weight, determined at the mixing stage. The only exceptions are volatile products and added water, which must be listed according to the amount present in the finished product. This allows for cooking losses. Less than 5% of added water need not be declared.

The names of ingredients must be those used when the ingredients are sold separately. Some can be described generically, such as fish, poultry, meat, and cheese. Whether oils and fats are of animals or vegetable origin must be stated, but there is no provision to permit manufacturers to indicate the type of oil or fat likely to be present, in order to allow more flexibility in the composition of the fat blend used and to accommodate fluctuations in the supply and world market prices of specific oils and fats.

The foregoing is not a comprehensive account of the provisions in the Directive covering ingredient listing, but it does identify the main points. Where the labeling of a food places emphasis on the presence or low content of one or more of its ingredients, or where the description of the food has the same effect, the minimum or maximum percentage, as appropriate, used in its manufacture, must be stated.

When considering the quantity marking of goods, regardless of whether these are foodstuffs, other European Union legislation must be taken into account. For example, there is a Directive concerned with the packing of goods in prescribed quantities (80/232); another Directive specifies the prepacking of solids on a qualified average-quantity basis (76/211); and similar control is imposed on prepackaged liquids (75/106). Other relevant Directives are 86/96 and 87/356.

Foods prepackaged in quantities greater than 5 g or 5 mL must have the quantity marked. There is some freedom for member states to increase the 5-g or 5-mL threshold in exceptional cases and also to make national provisions in specific circumstances, to derogate from the need to quantity mark at all.

Briefly, goods packed according to EU weights and measures legislation must contain, on average, at least the stated quantity. Variation from the average must be within specified tolerances, which are related to the quantity in the pack. Packers must keep adequate, statistically based, quantity-control records, and inspectors must carry out reference tests in a prescribed manner. The symbol "E" alongside a quantity mark on a label means that it conforms to EU quantity-control standards and has been checked by the member state concerned. Control of the volumes of liquids packed in bottles is covered by a Directive (75/107) that defines the accuracy to which the volume of bottles must conform if the bottles are used as measuring containers. The checking of fill levels is the basis for quantity measurement.

In all EU countries except the United Kingdom and Ireland, products are sold in metric units. Legislation is now being introduced in the United Kingdom that will phase out imperial units under the Weights and Measures Act 1994. From October 1, 1995 all goods sold by quantity, including food prepacked in variable weights such as cheese and meat that are not already traded in metric will have to be so traded. The principal exception is in respect of goods sold loose from bulk by imperial units—primarily foods such as meat, poultry, cheese, fish, fresh fruit, and vegetables that will not have to switch to metric until January 1, 2000. Retailers who price food sold loose from bulk during the transitional periods will be required to display a price conversion chart or to dual-price items. The legislation provides that the doorstep pint of milk and the pint of draught beer or cider will be allowed without time limit.

The principle on which EU date marking is based is minimum durability. This is defined as the date until which a foodstuff retains its specific properties when properly stored. The mode of marking is specified in the Directive. In most instances it takes the form of "Best before" followed by the minimum durability date. In the case of foodstuffs that are highly perishable from a microbiological point of view, member states can adopt the words "Use before" or some equivalent phrase, instead of "Best before." The equivalent chosen by the United Kingdom is "Use by (date)."

A number of foodstuffs are not required to have open date marking. Fresh fruits and vegetables, wines, beverages with more than 10% alcohol, vinegar, and cooking salt are examples. Also, member states may exempt long shelf-life foods (those remaining in good condition for longer than 18 months).

To reduce practical problems in applying data marks, the "Best before" statement can be followed either by the actual date or an indication of where the date is placed on the pack, eg, "Best before: (see date on can end)." Remember, however, that if particular storage conditions are needed to keep the product for the specified time, then these particulars must follow the "Best before" indication; for example, "Best before (date): Keep in refrigerator." Canned food products now carry "Best before" dates, and some nonfood products such as perfumed bath salts where the fragrance of the product may be lost are carrying best before dates.

The Directive specifies that instructions as to the method of using a foodstuff shall be provided to ensure that it is used appropriately. UK labeling regulations have always required that if additions to the food are needed, this must be clearly stated. The Directive adopts the same principle so that if, for example, a prepacked cake mix requires the addition of an egg or other ingredient, this is made clear on the label in close proximity to the product name.

Manner of marking. The Directive does not demand an information panel, nor does it specify type size requirements to determine how the required particulars must be marked on the label. It only specifies that they be easily understood, be placed conspicuously, and be easily visible, clearly legible, and indelible. Furthermore, they must not be hidden, obscured, or interrupted by other written or pictorial matter. The product name, quantity, and date must all appear in the same field of vision.

Nutritional labeling is not a mandatory requirement in any EUmember state, although when such claims are made, nearly all countries require them to be capable of substantiation and to be supported by the appropriate details on the label. The nearest approach to nutritional labeling is provided in the Directive on Food Particular Nutritional Uses (77/94). This is concerned specifically with foods that are nutritionally balanced, such as slimming foods and infant and baby foods, and other foods that are designed to meet a specific need, eg, for diabetics. The Directive requires that these foods be suitable for their use, claims substantiated, and certain labeling provisions met. Specifically, the manufacturer may not claim that the food can prevent, treat, or cure human diseases, except in exceptional and clearly defined cases provided for under national legislation. This

Directive is concerned solely with claims related to dietary and dietetic foods, and they are still subject to the wider requirements of the Labeling Directive and any national labeling rules permitted by it.

The EU Labeling Directive provides only a starting point for determining labeling requirements within the countries of the EU. Member states use the permissible exemptions and derogations in the Directive extensively, and this means that national regulations based on the Directive vary in a number of respects. Labeling requirements in any specific country are not confined to those of the Directive. Hence, exporters, particularly those operating from non-EU countries, wishing to label products to be sold in one or more EU countries correctly, are advised to consult a labeling expert in the country of sale. Alternatively, consult an expert who has the necessary contacts in EU countries and is skilled in finding out the precise requirements.

Eco Labeling

The Directive on eco labeling outlines an environmental labeling scheme for the identification of "green" or environmentally acceptable products. It is a European-wide scheme that operates in parallel with national schemes, such as the "Blue Angel" in Germany. It is intended that eventually the EU ecomark would replace the natural ones to give uniformity throughout the Union.

At present some products are excluded from the scheme. These include drugs, food, drink, and packaging, although some of the National Working Groups wish to include the latter. The United Kingdom has been the first member state to process applications and the first products allowed to carry the symbol (a flower made up of the stars of the EU surrounded by the European "E" standard) are dishwashers and washing machines.

Speciality Labels Commission Regulations (EU) 2515/94

This permits the producer of foodstuffs and agricultural products that have a specific character (specific to a region) to label their product with the appropriate Union symbol. A blue circle with the stars of the member states around the circumference surrounded by the legend "Traditional speciality guaranteed."

Trademarks

The UK Trade Mark Act 1994 brought the EU Trade Mark Directive into force, provides for the Union Trade Mark, which will be available from 1996 and enables companies to register one mark for the whole EU. Trademarks may comprise words, personal names, design, letters, numerals, or the shape of goods or their packaging. This reverses previous English case law that held that packaging could not be registered as a trademark.

Child-Resistant Closures/Tactile Warnings Directive 90/35/EEC

Containers containing certain categories of dangerous preparations offered or sold to the general public are required to be fitted with child-resistant closures and/or carry a tactile warning of danger. The products covered by the Directive are listed in Article 21 of Council Directive 67/548/EEC as last amended by Directive 88/490/EEC. Containers labeled as toxic or corrosive must be fitted with child-resistant closures and a tactile warning of danger. Containers labeled as harmful, extremely inflammable, or highly inflammable must carry a tactile warning of danger. The warning must conform to the specifications in Part B of Annex IX to Directive 67/548/EEC ISO 8317.1989 standard lays down the procedure for testing child-resistant closures.

Marking and Labeling Transport Packaging

In order to reduce losses in transit, it is essential that transport packaging be marked in such a way that the requirements are easily understood in every port around the world. The regulations and legislation on transport modes and country vary so much. However, the International Cargo Handling Co-ordination Association, 71 Bondway, London SW8 1SH, UK, publishes a manual on marking and labeling of transport packaging. Although not mandatory, it is an invaluable document for all those involved with the transport of goods.

The Transport and Packaging of Dangerous Goods

In January 1993 the United Nations published recommendations on the packaging and transport of dangerous goods (usually referred to as "hazardous" in the United States). Known as the "Orange Book," it does not have the force of law and its recommendations can only become legally binding if adopted by a national government. Today the Orange Book forms the basis of all the international regulations. There are still modal differences, some of which are significant but far fewer than there used to be, probably a number will remain in place for many years because they reflect the different hazards between transport modes but may well disappear by the year 2000. These regulations apply to the transport of dangerous goods in Europe and now form part of Directive 94/55/EEC, November 21, 1994, which sets out the laws regarding the transport of dangerous goods by road in the Union.

BIBLIOGRAPHY

1. *UN Recommendations on the Transport of Dangerous Goods, published by the United Nations.*
2. *The Transport of Dangerous Goods,* M. Castle, published by Pira International, Leatherhead, UK, ISBN 1 85802 031X

Environmental Directives

There are Directives on several environmental aspects that affect packaging indirectly mostly in manufacturing or waste disposal. Manufacturing directives on air pollution, water quality, toxic waste, and waste disposal are relevant. In addition, the COSH regulations concerning health and safety at work and exposure of personnel to solvents and compounds such as isocyanates are relevant. Urethane-foam materials are also required to have a fire retardant added to them.

After many years of negotiation, the EU published Directive 94/62/EEC on Packaging and Packaging Waste in December 1994. The Directive, which must be implemented by all member states by June 30, 1996, aims to harmonize natural measures concerning the management of packaging and packaging waste to provide a high level of environmental protection and to ensure the functioning of the internal market and to avoid obstacles to trade and distortion and to ensure the functioning of the internal market and to avoid obstacles to trade and distortion and restriction of competition within the Union.

The first priority is to prevent the production of packaging waste and at reusing packaging at recycling and the other forms of recovering packaging waste and reducing the final disposal of such waste. The scope of the Directive covers all packaging placed on the market in the Union and all packaging waste, whether it is used or released at industrial, commercial, office, shop, service, household, or any other level, regardless of the material used.

The Directive gives definitions of packaging, packaging waste, reuse, and other terminology used. It lays down that member states must implement schemes to prevent packaging waste and to encourage the reuse of packaging as well as the collection and recycling of packaging waste. Burning of waste with energy recovery is also allowed, but the Directive will not set out a hierarchy between reusable, recyclable, and recoverable packaging until life-cycle assessments have been completed.

Five years from the date that the Directive must be implemented (June 30, 1996), recovery targets for packaging waste must be implemented: a 50% minimum and 65% maximum by weight within the general target and a 25% minimum and 45% maximum by weight of the totality of packaging materials contained in the packaging waste will be recycled with a minimum of 15% by weight for each packaging material.

Within 10 years of implementation of the Directive, revised levels for recovery and recycling will be set by the European Council. These will be based on the practicality and degree of success of the systems used in the first 5 years. Temporary (5 years) exemptions are granted for Greece, Ireland, and Portugal because of the geographic nature, presence of rural and mountain areas, and the current low level of packaging consumption, but they are required to achieve 25% recovery levels. Member states who have introduced or plan to introduce higher targets in the interest of a high level of environmental protection, provided these measures avoid distortions of the internal market and do not hinder compliance of the Directive by other member states.

Governments are required to set up the appropriate collection schemes and may introduce appropriate economic instruments to promote implementation. In order to identify materials used in the manufacture of packaging, a numbering system and/or abbreviation for the relevant material may be used in the center or below the graphic marking indicating the reusable or recoverable nature of the packaging.

By 1999, packaging may be placed on the market only if it complies with all essential requirements of the Directive. Member states are also required to publish standards and the Commission will promote the preparation of European standards relating to the essential requirements on the composition and the reusable and recoverable, including recyclable nature of packaging set out in Annex II.

Heavy-metal concentration (lead, cadmium, mercury and hexavalent chromium) present in packaging must not exceed

600 ppm by weight by June 30, 1998
250 ppm by weight by June 30, 1999
100 ppm by weight by June 30, 2001

(Packaging made from lead crystal is exempt.)

Governments are also required to set up data collection systems on packaging and packaging waste. Users of packaging are also to be provided with information on the recovery, recycling systems available to them, and their role in contributing to reuse, recycling, etc. Member states are not allowed to ban packaging that meets the conditions of the Directive.

Other Environmental Legislation

Most of the Union member states have introduced some form of collection–recycling scheme for used packaging. Some of this conflicts with the EU Directive. However, it is likely that in the short term some of these measures will continue after the full implementation of the EU Directive.

Austria. Collection scheme in force, Alstaff Recycling Austria (ARA) has set up a nationwide collection scheme. It is funded through licence fees paid by companies placing packaging on the market calculated on a material/weight basis. There are also regulations on collection and reuse of beverage containers.

Belgium. Proposed taxes on beverage containers and other disposable items. A voluntary scheme for collecting used packaging is currently not well supported by industry because manufacturers are not obliged to take responsibility for collection and recovery of their waste packaging until agreement on a common decree has been reached by Belgium's three regions.

Denmark. The 1991 Environmental Protection Act provides a legal framework that obliges manufacturers and importers to ensure that packs have as long a life as possible, can be recycled to the greatest possible extent, and do not have an adverse environmental impact at time of disposal.

Total ban on cans for beverages is in force and refillable bottles are compulsory for beer and carbonated soft drinks.

A reuse target of 50% of packaging waste by 2000 has been set, and industry has agreed to reduce the amount of PVC packaging to 85% of the weight consumed in 1987.

A draft voluntary agreement on the recovery of transport packaging could be made compulsory and extended to cover all industrial companies.

Manufacturers, fillers, and retailers operate a mandatory deposit scheme for liquid containers and local authorities are responsible for collection and recycling of packaging waste and finance recovery through local taxes.

Finland. There is a packaging tax on nonrefillable beverage containers, and the government intends to introduce legislation to encourage recycling and energy recovery rather than using landfill for waste disposal.

France. Industry has a voluntary objective to recovery 75% of packaging waste by 2003 and incineration with energy recovery. Consumer goods manufacturers, importers, and producers are responsible for the recovery of consumer sales packaging.

Germany. Probably the most draconian legislation in Europe. The 1991 Waste Avoidance Ordinance requires companies to take back all sales and transit packaging. The recovery system, Duales System Deutschland (DSD), operates a national scheme for collection and sorting of waste. DSD members are manufacturers and distributors who finance the operation with a charge based on the weight and type of material of packaging placed on the market. Members use a green dot on packs to signify that they are part of the scheme.

Greece. Allowed to delay introduction of the EU Directive but some pilot collection schemes are operating.

Ireland. Allowed to delay introduction of EU Directive but some legislation under discussion.

Italy. Legislation covering liquid packaging and plastic and paper carrier bags requires manufacturers to work with local authorities to operate collection and recycling schemes. Separate consortia have been established to collect and recycle glass, aluminum, and plastics.

Luxembourg. A convention between government and industry designed to promote refillable packs for liquids.

Netherlands. A Packaging Convenant in which industry accepted targets to reduce packaging and set up collection and recovery systems. Some forms of packaging such as aluminum-foil-lined cartons for spirits are banned.

Portugal. Allowed to delay implementation of EU Directives.

Spain. Some regional legislation requiring local councils to collect and requiring local councils to collect and recycle waste.

Sweden. Producers responsible for meeting government recycling targets.

Switzerland. Deposits on beer and soft-drink containers. PVC containers banned.

United Kingdom. Legislation under discussion between government and industry.

The legislation covering collection and recycling is an ever-changing scene, and anyone wishing to put products on a particular market should consult with a local expert beforehand.

List of UK Statutory Instruments, EU Directives, and other EU documents

1. Statutory Instrument No. 1927 (1978), *The Materials and Articles in Contact with Food Regulations.*
2. Statutory Instrument No. 1523 (1987), *The Materials and Articles in Contact with Food Regulations.*
3. Statutory Instrument No. 3145 (1992), *The Plastic Materials and Articles in Contact with Food Regulations.*
4. EU Directive 76/893/EEC (first framework Directive) on the approximation of the laws of the member states relating to materials and articles intended to come into contact with foodstuffs; see *OJ* (*Official Journal*) **L340,** 19–24 (Dec. 9, 1976).
5. EU Directive 89/109/EEC (second framework Directive) on the approximation of the laws of the member states relating to materials and articles intended to come into contact with foodstuffs; (see *OJ* **L40,** 38–44 (Feb. 11, 1989).
6. EU Directive 78/142/EEC on the approximation of the laws of the Member States relating to materials and articles that contain vinyl chloride monomer and are intended to come into contact with foodstuffs; see *OJ* **L44,** 15–17 (Feb. 15, 1978).
7. EU Directive 80/766/EEC on laying down the Union method of analysis for the official control of the vinyl chloride monomer level in materials and articles that are intended to come into contact with foodstuffs; see *OJ* **L213,** 42–46 (Aug. 16, 1980).
8. EU Directive 81/432/EEC on laying down the Union method of analysis for the official control of vinyl chloride released by materials and articles into foodstuffs; see *OJ* **L167,** 6–11 (June 24, 1981).
9. EU Directive 80/590/EEC on determining the symbol that may accompany materials and articles intended to come into contact with foodstuffs; see *OJ* **L151,** 21–22 (June 19, 1980).
10. EU Directive 83/299/EEC on the approximation of the laws of the member states relating to materials and articles made of regenerated cellulose film intended to come into contact with foodstuffs; see *OJ* **L213,** 31–39 (May 11, 1983).
11. EU Directive 86/388/EEC amending Council Directive 83/388/EEC on the approximation of the laws of the member states relating to materials and articles made of regenerated cellulose film intended to come into contact with foodstuffs; see *OJ* **L228,** 32–33 (Aug. 14, 1986).
12. EU Directive 92/15/EEC amending Council Directive 83/388/EEC on the approximation of the laws of the member states relating to materials and articles made of regenerated cellulose film intended to come into contact with foodstuffs; *OJ* **L102,** 44 (April 16, 1992).
13. EU Directive 93/10/EEC relating to materials and articles made of regnerated cellulose film intended to come into contact with foodstuffs; see *OJ* **L93,** 27–36 (April 17, 1993).
14. EU Directive 82/711/EEC laying down the basic rules necessary for testing migration of the constituents of plastic materials and articles intended to come into contact with foodstuffs; see *OJ* **L297,** 26–30 (Oct. 23, 1982).
15. EU Directive 93/8/EEC amending Council Directive 82/11/EEC laying down the basic rules necessary for testing migration of the constituents of plastic materials and articles intended to come into contact with foodstuffs; see *OJ* **L90,** 22–25 (April 14, 1993).

16. EU Directive 85/572/EEC laying down the list of simulants to be used for testing migration of constituents of plastic materials and articles intended to come into contact with foodstuffs; see *OJ* **L372,** 14–21 (Dec. 31, 1985).
17. EU Directive 90/128/EEC relating to plastics materials and articles intended to come into contact with foodstuffs (Corrigenda); see *OJ* **L349,** 26–47 (Dec. 13, 1990).
18. First amendment to EU Directive 90/128/EEC–EU Directive 92/39/EEC; see *OJ* **L168,** 21–29 (June 23, 1992).
19. Second amendment to EU Directive 90/128/EEC–EU Directive 93/9/EEC; see *OJ* **L90,** 26–32 (April 14, 1993).
20. Synoptic Document N6 Draft of provisional list of monomers and additives used in the manufacture of plastics and coatings intended to come into contact with foodstuffs (updated to April 2, 1993): CS/PM/2064.
21. First Supplement to Synoptic 6, Brussels, October 22, 1993.
22. Synoptic Document N7 Draft of provisional list of monomers and additives used in the manufacture of plastics and coatings intended to come into contact with foodstuffs (updated to May 15, 1994); CS/PM/2356.
23. EU Directive 84/500/EEC on the approximation of the laws of the member states relating to ceramic articles intended to come into contact with foodstuffs; see *OJ* **L277,** 12–16 (Oct. 20, 1984).

R. M. WHITE
P. A. TICE
Pira International
Surrey, United Kingdom

LAWS AND REGULATIONS, UNITED STATES

There is no single body of law or regulations pertaining to all aspects of the manufacture, distribution, and use of packaging in the United States. Many different types of federal, state, and even local laws and regulations affect the packaging industry. These laws and regulations can be separated into two broad categories: (1) those that pertain directly to the packaging itself and (2) those that pertain generally to manufacturing and commercial transactions, including packaging among many other products.

This article focuses on the former category of legal requirements pertaining most directly to packaging. These include health and safety regulations of packaging materials, labeling requirements, and environmental laws and regulations relating to the disposal of packaging.

FDA's Regulatory Framework for Packaging

The Food and Drug Administration (FDA) has primary regulatory authority in the United States over the safety of packaging used for food, drugs, cosmetics, and medical devices to the extent the products are in interstate commerce (which they almost always are) (1). The packaging for these products is regulated by FDA so that it does not cause the products themselves to become contaminated or "adulterated" in violation of the Federal Food, Drug and Cosmetic Act.

The most extensive of FDA's packaging regulations are the "food-additive regulations," which provide official permission for the use of substances as food ingredients (direct additives), and as components of food packaging or other food-contact materials (indirect additives). FDA does not have regulations clearing specific materials for use in packaging for drugs, medical devices, or cosmetics. In the case of drugs and medical devices, packaging materials are reviewed and approved by FDA in the context of approving the drug and medical-device products themselves, if such approval is necessary. FDA does have a general requirement for tamper-resistant packaging for over-the-counter (OTC) drugs. Suppliers of packaging materials to drug and medical-device manufacturers typically assist those manufacturers in obtaining FDA approval by providing information on the packaging to FDA through a confidential drug or device master file. Neither cosmetic products nor their packaging are subject to premarket clearance by FDA; it is up to the cosmetics manufacturer and its packaging suppliers to ensure that the products and packaging are safe for their intended use.

Definition of food additive. FDA's regulation of food packaging is based on Section 201(s) of the Federal Food, Drug and Cosmetic Act. Section 201(s) of the Act defines a food additive, in relevant part, as follows:

> [A]ny substance the intended use of which results or may reasonably be expected to result, directly or indirectly, in its becoming a component . . . of any food . . . if such substance is not generally recognized . . . to be safe under the conditions of its intended use; except that such term does not include . . .
>
> (4) any substance used in accordance with a sanction or approval granted prior to the enactment of this paragraph pursuant to this Act (4)

Section 409 of the Act requires that a food additive, as defined above, must be the subject of an applicable food-additive regulation (2). Thus, substances used in food-contact applications that, under their intended conditions of use, may reasonably be expected to become components of food are "food additives" and can be used only in accordance with an applicable food-additive regulation unless they are either *(1)* the subject of a prior sanction or approval granted by FDA or the U.S. Department of Agriculture (USDA) prior to the enactment of the Food Additives Amendment of 1958 or *(2)* generally recognized as safe (GRAS). If the substance is not reasonably expected to become a component of food under the intended conditions of use, however, it is not a food additive within the meaning of Section 201(s) of the Act, and may be used as intended without the need for consultation with or prior action by FDA.

The food-additive regulations. FDA's food-additive regulations, found in Title 21, Parts 170–197 of the *Code of Federal Regulations* (*CFR*), cover both direct and indirect additives. The term "direct food additive" commonly refers to materials directly and intentionally added to foods to perform a functional effect in the food. The term "indirect food additive" generally refers to substances that are not intended to, but nevertheless become, components of food as a result of use in articles that contact food (eg, substances used in packaging materials).

Over the years, indirect food-additive regulations have been promulgated in response to individual food-additive petitions with no overall structural scheme. This has led to the existence of three general types of indirect food-additive regulations: those that clear specific polymers (eg, Section 177.1520, applicable to polyolefins), those that clear substances categorized by function (eg, Section 178.3400, applicable to emulsifiers and surface-active agents), and those that list substances that may be utilized in specific types of packaging (eg, Section 176.170, applicable to components of paper and paperboard). For this reason, a great deal of overlap among different regulations has occurred.

A major limitation on the types of substances that may be listed in the food-additive regulations is the so-called Delaney Clause of the Federal Food, Drug and Cosmetic Act. The Delaney Clause prohibits FDA from issuing regulations permitting the use as food additives of substances that are "found to induce cancer when ingested by man or animals" (3).

FDA has mitigated the impact of the Delaney Clause, however, by interpreting it to apply to the finished food ingredient or food-contact material itself, not to constituents or components of the food additive (eg, the Delaney Clause applies to polymers used to make food packaging, but not to monomers used to make the polymers). This "constituents policy" is not set forth in any law or regulations; but is applied by FDA in evaluating food additive petitions, was described by FDA in an Advance Notice of Proposed Rulemaking, and has been upheld by the courts (4).

Prior-sanction exemption. The Federal Food, Drug and Cosmetic Act specifically exempts from the definition of a food additive any substance that is "prior-sanctioned," meaning any substance permitted for its intended use by an FDA or USDA letter or memorandum written prior to the Food Additives Amendment of 1958. Unfortunately, no authoritative master list exists for the materials covered (5). If a company has any such letters in its files, or knows of any relevant ones elsewhere, they can be relied on, but FDA interprets such letters narrowly.

"Generally recognized as safe" (GRAS) exemption. Another exemption from the food additive definition pertains to GRAS substances. Section 174.5(d)(1) of the *Food Additive Regulations* states that substances generally recognized as safe "among experts qualified by scientific training and experience to evaluate their safety" are permitted to be used as components of articles that contact food. Parts 182, 184, and 186 of the *Food Additive Regulations* provide a listing of some substances that are considered by FDA to be GRAS, but the substances actually listed in the regulations by no means constitute all substances that are GRAS (6). Section 182.1 of the *Food Additive Regulations* makes it clear that the GRAS listings are by way of illustration and do not represent an all-inclusive list. Furthermore, the absence of a substance from the list does not preclude a company from making a self-determination that its products are GRAS, where such a determination is supported by the available information.

FDA has codified requirements for the classification of food additives as GRAS under 21 *CFR* 170.30(a). As stated in that provision, general recognition of safety must be based on (*1*) scientific procedures or (*2*) in the case of a substance used in food prior to January 1, 1958, through experience based on common use in food. General recognition of safety requires a "common knowledge" about the substance throughout the scientific community knowledgeable about the safety of substances directly or indirectly added to food. For substances not widely used in food prior to 1958, general recognition of safety requires the same quantity and quality of scientific evidence as is required to obtain approval of a food additive regulation for the ingredient. Unlike a food-additive petition, however, general recognition of safety is ordinarily based on published studies that may be corroborated by unpublished studies and other data and information (7).

The exclusion of GRAS substances from the definition of the term "food additive" means that such substances do not require premarket clearance by FDA. Any manufacturer who determines that a particular substance is GRAS is free to market the substance without notification to, or approval by, FDA. Obviously, if FDA should consider the manufacturer's determination of GRAS status to be erroneous, the Agency can take appropriate regulatory action; in such a case the burden of proof would fall on FDA to demonstrate that the substance is not GRAS. This is an unlikely consequence unless a real public-health problem is presented.

A manufacturer who pursues a self-determined GRAS position for a substance can also file a GRAS Affirmation Petition (GRASP) with FDA seeking the Agency's concurrence. In contrast to a Food Additive Petition, if a GRASP is accepted by FDA for filing, the petitioner may market the product while the petition undergoes technical review since FDA has been informed of the manufacturer's determination that the product is GRAS and has raised no objection. However, unlike a Food Additive Petition, all materials submitted in support of the GRASP become immediately available in full to the public, and even manufacturing details cannot be kept confidential. The requirements and type of information a petitioner must supply for a GRASP are spelled out in 21 *CFR* § 170.35.

The "no-migration" exemption. One of the most important means of establishing satisfactory FDA status for an indirect additive is to establish a rational basis on which to conclude that there is no reasonable expectation of the substance becoming a component of food. If a substance is not reasonably expected to become a component of food under its intended conditions of use, it is not a food additive by definition and, therefore, may be so used without obtaining any FDA "approval." Unfortunately, FDA has not provided definitive, objective criteria for determining when a substance in a food-packaging material may reasonably be expected to become a component of food. Nevertheless, reliable guidance is available from at least two different sources:

1. *Ramsey Proposal.* The first source of guidance is a proposal circulated by FDA as its response to widespread criticisms by the food-packaging industry at the National Conference for Indirect Additives held in Washington, DC in February 1968. This so-called Ramsey proposal would have acknowledged in a regulation the propriety of the use, without the prior promulgation of a food-additive regulation, of substances that will transfer from packaging or other food-

contact articles to food at levels no higher than 50 parts per billion (ppb); components of articles used in contact with dry, nonfatty foods; and components of articles intended for repeated use in contact with bulk quantities of food. This would have applied to all substances except those known to pose some special toxicological concern, such as a heavy metal, a known carcinogen, or something that produces toxic reactions at levels of 40 parts per million (ppm) or less in the diet of humans or animals. Although never formally adopted, the standards in the Ramsey proposal were deemed scientifically acceptable and have been generally honored by FDA either explicitly or by its having taken no enforcement action in cases where the criteria are met.

2. *Monsanto v Kennedy.* The second source of clarification of the "food additive" statutory definition is provided by a U.S. Court of Appeals decision. In *Monsanto v Kennedy,* 613 F.2d 947 (D.C. Cir. 1979), FDA argued that any contact of a packaging material with food must result in some transfer of the packaging constituent to the contained food. The Court rejected this argument and stated

> Congress did not intend that the component requirement of "food additive" would be satisfied by . . . a mere finding of any contact whatever with food . . . . For the component element of the definition to be satisfied, Congress must have intended the Commissioner to determine with a fair degree of confidence that a substance migrates to food in more than insignificant amounts (8).

Role of extraction studies. On the basis of the Ramsey proposal and the *Monsanto* precedent, it is widely accepted that, if extraction testing properly simulating the intended conditions of use for a potential additive does not yield detectable migration of the additive or its components at an appropriate analytical sensitivity, the substance is not reasonably expected to become a component of food under these conditions and, thus, is not a food additive within the meaning of Section 201(s) of the Act. Industry generally considers a finding of "nondetected" in a properly conducted migration study utilizing methods sensitive to the equivalent of 50 ppb of the substance in contacted food to be a sound basis for concluding that a substance is not a food additive. However, in some cases, it is necessary to make use of analytical methodology that can detect as little as 10 ppb or even 1 ppb to support a "no-migration" determination because of the sensitive nature of the material being used, such as a heavy metal or a substance for which there will be a high level of use, such as milk or soda containers.

General adulteration provision. If the packaging material is not reasonably expected to become a component of food and is, therefore, not a food additive, it is still subject to the nonadulteration provision of the Act. FDA sometimes refers to this as the general safety clause, contained in Section 402 of the Act. This general prohibition against adulteration is the basis for FDA's "good manufacturing practice" regulation, requiring that "[a]ny substance used as a component of articles that contact food shall be of a purity suitable for its intended use" (9). Fortunately, the same analysis that demonstrates that a material will not become a component of food will also satisfy the Act's general safety requirements for food-packaging material, specifically, that the packaging will not adulterate the food by rendering it injurious to health.

The "basic resin doctrine" exemption. A significant characteristic of the *Food Additive Regulations* is that FDA clears substances on a generic rather than a proprietary basis. In the case of resins and polymers, as long as the basic resin is listed in a regulation, is manufactured in accordance with good manufacturing practices, and complies with applicable limitations such as stated extraction requirements, it is covered by that regulation even though different manufacturers may make the resin by different processes and employ different catalysts, reaction control agents, and the like.

A "basic polymer" is the material that comes out of the polymerization "kettle" or reactor, that is, the product that results when the polymerization process has been carried to commerical completion. Substances such as catalyst residues, chain-transfer agents, very minor comonomers, and other materials required to produce the basic resin are considered a part of it and are *not* subject to independent regulatory consideration. FDA's basis for this principle is that, where a substance is used only in a small quantity and either becomes part of the resin during polymerization, or is washed from the resin at the conclusion of polymerization, its potential for significant migration is minimal. In other words, there is no reasonable expectation of migration, and therefore the substance is not considered an independent food additive.

The basic polymer doctrine also reflects the practical reality that FDA cannot write generic regulations for food-packaging materials that describe and specifically clear every substance that might properly be a trace component or contaminant of the packaging material as a result of every manufacturing process that yields a suitable resin. Since trace quantities of these "unregulated substances" are not perceived to present a public-health hazard, FDA has wisely chosen not to subject these substances to the burdensome preclearance provisions of Section 409 of the Act that apply to food additives.

Generally speaking, it is reasonable to consider catalysts and minor comonomers to be covered by the basic polymer doctrine when used at levels of up to approximately 0.5%; between this level and 1%, the substance and use in question must be evaluated on a case-by-case basis. At levels above 1%, the basic polymer doctrine is not considered applicable.

Even if a component of a food-contact material falls within the basic-polymer doctrine, the finished food-contact material must still satisfy the necessary purity requirements of the *Food Additive Regulations;* that is, it must be of a purity suitable for its intended use, as required under 21 *CFR* § 174.5(a)(2). Obviously, if a particular catalyst being used were to render a finished food-contact material unsafe or unfit for its use, that material would not satisfy the purity requirements.

The housewares exemption. The components of eating utensils, receptacles, paper towels, or other kitchenware, sold as such, are outside the scope of coverage of the Food Additive Amendments of 1958, and therefore do not require premarket clearance by FDA. This has come to be called the "housewares" exemption. This exemption is grounded on a declaration of intent in the legislative history of the 1958

Food Additives Amendment. The following statement appears in the official record of the debate on the Amendment and was made by the floor manager of the bill, The Honorable John Bell Williams, Chairman of the House Subcommittee on Health and Science:

> I have been asked since the Committee report of the bill what is meant by a substance "holding" food, as mentioned in the bill. An example of what is meant by this would be a plastic film or paper wrapper which surrounds the package of food. This bill is not intended, for example, to give the Food and Drug Administration the authority to regulate the use of components of dinnerware or ordinary eating utensils (10).

FDA has written many letters affirming this exclusion. The exclusion of housewares from the need for filing food-additive petitions is a recognition that such products generally do not give rise to any public-health concern. However, FDA does take action against products falling within the housewares exemption in instances involving migration of a substance, such as lead, which has been proven to pose a true public health concern.

G. Threshold of Regulation. The exemptions from FDA's premarket clearance authority discussed so far all have one important characteristic in common—a company is entitled to determine for itself whether the exemptions apply, and whether no consultation with or concurrence by FDA is required. FDA recently has adopted an additional exemption that only the Agency itself may apply, the Threshold of Regulation.

In 1995, FDA published the final rule establishing the Threshold of Regulation, which is FDA's "process for determining when the likelihood or extent of migration to food of a substance used in a food-contact article is so trivial as not to require regulation of the substance as a food additive" (11). The "threshold" level is defined as a dietary concentration of 0.5 part per billion (ppb) or less. The threshold for substances that have been cleared by FDA as direct additives is that the indirect or packaging use must be limited to 1% or less of the acceptable daily intake established by the Agency for the direct use of the substance. To be eligible for the exemption, the food-contact material must not be a carcinogen and must not contain carcinogenic impurities that are more potent than a certain level defined in the regulation (the impurity cannot have a TD_{50} value of <6.25 mg/kg bw (body weight) per day, with TD_{50} defined as the feeding dose that causes cancer in 50 percent of the test animals.

FDA has reserved to itself the authority to grant official exemptions from regulation under the Threshold of Regulation Rule. However, in proposing the Rule, FDA explicitly recognized the continuing right of a company to determine on its own "that a particular use of a substance does not meet the definition of a food additive" (12). Therefore, companies continue to be entitled to determine for themselves whether the "no migration" exemption or any of the other exemptions from the definition of "food additive" apply to a particular intended use of a food-packaging material.

The food-additive petition process. If none of the exemptions from FDA's premarket food-additive regulatory authority is available for a packaging material, regulatory clearance must be obtained from FDA through a food-additive petition. Section 409(b) of the Federal Food, Drug and Cosmetic Act permits any person to petition FDA for a regulation prescribing the conditions under which an additive may be safely used. The Act delineates generally the type of information to be submitted in support of a petition and the procedures and substantive parameters to be followed by FDA in considering a petition. Additional detail on these information requirements and procedures is provided in Part 171 of the *Food Additive Regulations.*

Reduced to its fundamentals, a food-additive petition must describe the substance to be cleared, provide an estimate of the quantity that will enter the diet, and demonstrate that this quantity will be safe. The Act requires FDA to act on a petition within ninety days, with one extension for an additional 90 days available to the Agency. In practice, very few petitions are acted on in 18 months or less, and most are under review for at least 2–3 years.

In 1995, several industry groups representing food manufacturers and their packaging suppliers revived their efforts to reform the legal system for food-additive regulation by FDA. The basic goals of this effort are to replace the Delaney Clause with a negligible risk standard based on sound science and to require FDA to accelerate the process of reviewing food-additive petitions. With respect to packaging materials, the proposed reforms would replace the petition process with a premarket notification system. FDA would have 120 days to review the data on a new packaging material. The material would be permitted for use at the end of the 120-day period unless FDA objected on the basis of real safety concerns.

FDA regulation of recycled materials for food packaging. A very specialized area commanding much attention in the early 1990s has related to the recycling of postconsumer waste materials, including their use in some cases to make food packaging. There is no legal requirement to obtain any special or new FDA "approval" for the use of recycled material in food packaging. FDA has no special regulations governing recycled materials, with the exception of a regulation on recycled paper that merely restates the universal FDA requirement that all food-contact materials must be suitably pure for their intended use (13). The Agency generally regulates food-contact materials on the basis of their composition, not on the specific process by which they are manufactured or the source of the raw materials. Any food-contact material (whether plastic, glass, metal, or paper and whether virgin or recycled) must meet the safety requirements of the Act and the specifications and limitations of any food-additive regulations applicable to the generic type of material (14).

Although no FDA approval is required for recycled content in food packaging, the requirements of food companies for assurance that recycled materials are safe has resulted in a common industry practice of seeking an official blessing from FDA, particularly with respect to recycled plastics. These blessings take the form of so-called no-objection letters from FDA. The Agency generally applies the "Threshold of Regulation" analysis to requests for these letters, requiring a showing that any potential contaminants in the recycled material have been reduced to the "threshold" dietary concentration of 0.5 ppb or below.

Environmental Laws and Regulations

The federal role: Environmental Protection Agency. The Environmental Protection Agency (EPA), the primary federal agency responsible for regulating waste disposal, has broad authority over the management of solid waste under the Resource Conservation and Recovery Act (RCRA) (15), as well as the Clean Air Act (16), the Clean Water Act (17), and the Comprehensive Environmental Response, Compensation and Liability Act (18) (commonly referred to as "Superfund"). This authority includes power to set standards for the design and operation of landfills and incinerators and other modes of waste disposal.

In general, EPA's modus operandi for dealing with solid-waste management issues, particularly issues concerning packaging, has been to make recommendations, such as on preferred options for waste disposal, but to leave much of the actual regulation of waste to state and local officials (19). At this time, EPA has no specific rules governing the disposal of postconsumer packaging waste.

State regulation. A few states have adopted bans or other limitations on packaging that does not meet certain standards of "environmental acceptability." The criteria of environmental acceptability in these proposals are typically that the packaging be recycled at a specified rate, be reusable, or be made from recycled materials.

California enacted a law in October 1991 that purported to ban, as of January 1, 1995, rigid plastic containers that (1) are not composed of 25% postconsumer material; (2) do not have a recycling rate of 25% (55% in the case of PET containers, if measured separately); (3) are not source-reduced by 10%; or (4) are not reusable or refillable five times (20).

The statute contains significant waiver and exemption provisions. The state will waive the postconsumer material content requirement if it finds that (1) the containers cannot meet the content requirement and remain in compliance with FDA regulations or other state or federal laws or regulations or (2) the use of containers meeting the content requirement is "technologically infeasible." The statute also exempts containers for food, cosmetics, drugs, medical devices, medical food, or infant formula.

Oregon has a statute nearly identical to California's (21). The Oregon law has a permanent exemption for food packaging, except for beverage containers.

Wisconsin has a law purportedly requiring 10% recycled content in plastic containers (22). However, this law will go into effect for FDA-regulated products only if FDA has "approved" the use of recycled content through a "formal" procedure. So far, FDA's "no-objection letter" approach to permitting recycled materials has not been recognized by Wisconsin as a "formal" procedure. Wisconsin also has banned recyclable packaging from landfills and incinerators (23), but has been granting a series of 1-year waivers for certain plastic packaging that is not widely collected for recycling.

Nine states have deposit laws that place a surcharge on certain beverage containers, and provide a refund of the surcharge to consumers who return the containers for recycling (24). California has its own modified form of deposit law, known as AB (Assembly Bill) 2020, under which, in addition to collection and refund of deposits, industry may be assessed fees on the basis of costs of recycling.

Control of alleged toxic substances in packaging

State restrictions on heavy-metal content of packaging. Eighteen states (25) have enacted statutes restricting lead, cadmium, mercury, and hexavalent chromium in inks, dyes, pigments, adhesives, stabilizers, and other components of packaging. These laws are based on model legislation developed by the Coalition of Northeastern Governors (CONEG).

California's Proposition 65. Perhaps the most far-reaching state environmental law directly affecting packaging and other consumer products is California's Safe Drinking Water and Toxic Enforcement Act (26), popularly known as "Proposition 65." Proposition 65, approved by California voters as an initiative in 1986, is a right-to-know law that requires companies to either establish that their products are not likely to expose any individual to a "significant" amount of any of over 400 chemicals, or to provide a "clear and reasonable" warning that the product contains a known carcinogen or reproductive toxin. To avoid the warning requirement, a manufacturer must establish that exposure to a chemical listed under Proposition 65 from a particular package or other product presents "no significant risk." The state has established "no significant risk levels" for some, but by no means all, of the substances listed under Proposition 65.

Labeling

For most packaging, the requirements pertaining to labeling will depend on the nature of the product contained within the package. Many products require only that the labeling not be false or deceptive. This standard is enforced on a nationwide basis by the Federal Trade Commission (FTC) (27) and by appropriate authorities in each state, enforcing the so-called "Little FTC Acts."

Certain products, however, are subject to special labeling laws and regulations, frequently for health or safety reasons.

FDA regulation. FDA regulates the labeling on packaging for foods, drugs, cosmetics, and medical devices. The Federal Food, Drug, and Cosmetic Act prohibits the "misbranding" of FDA-regulated products. A product is "misbranded" if its labeling is "false or misleading," or if it violates other specific statutory or regulatory requirements.

FDA has issued separate regulations pertaining to the labeling of foods, drugs, cosmetics, and medical devices. The food labeling regulations are the most extensive of the group, and now include a requirement for nutrition labeling on most food packages (28).

Packaging and labeling of hazardous materials. The U.S. Department of Transportation (DOT) regulates the packaging and labeling of hazardous materials for interstate shipment (and has proposed extending its regulations to intrastate shipments as well). The authorized types of packaging and the required labeling vary according to the degree of hazard presented by a particular material (29).

Environmental advertising and labeling

Federal and state guidelines. Labeling and advertising claims about the environmental attributes of packaging or products are not subject to special laws or regulations at the federal level or in most states. The permissibility of particular claims generally is tested according to the "false or deceptive" standard of the Federal Trade Commission Act and similar state laws.

Both federal and state authorities have issued guidelines for applying the "false or deceptive" standard to environmental labeling and advertising. In July 1992, the FTC released its *Guidelines for Environmental Marketing Claims.* The *Guidelines* address eight categories of claims: general claims such as "environmentally friendly," degradable claims, compostable claims, recyclable claims, recycled content claims, source-reduction claims, refillable claims, and ozone-safe/ozone-friendly claims. The FTC is currently reviewing its *Guidelines* to determine whether they should be modified to be more effective. The *Guidelines* do not have the force of law, but are considered authoritative statements of the way in which the FTC will exercise its enforcement authority. A task force of state attorneys general also issued guidelines in 1991, known as the *Green Report II,* addressing the same types of claims covered by the FTC *Guidelines,* but taking a more restrictive approach. In the absence of formal regulations issued by the FTC, states can enforce stricter limitations on environmental claims than would the FTC.

State regulation. A few states have actually adopted laws or regulations limiting environmental claims for products and packaging. Two states, California and Indiana, enacted laws governing environmental advertising, including use of such terms as "recycled" and "recyclable" (30). California's law was challenged in court as an infringement on commerical free speech (31). Although most of the law was left intact when the United States Supreme Court declined to hear the case, California subsequently repealed the essential provisions of the law. Indiana's law as enacted was similar to California's, but the state now defers to the FTC's *Guidelines.*

Some states, such as New York and New Hampshire, specifically restrict use of the terms "recycled" and "recyclable" (32). Both New York and New Hampshire restrict the use of the term "recycled" to products containing a certain minimum percentage of postconsumer material. They also require significant levels of recycling for a package or product to be labeled "recyclable."

BIBLIOGRAPHY AND NOTES

1. The U.S. Department of Agriculture (USDA) technically has authority over packaging used in federally inspected meat and poultry processing plants. However, in practice, USDA does not independently evaluate the safety of packaging materials, deferring to FDA on this issue.
2. Food that contains an uncleared food additive is considered "adulterated" and would be subject to adverse FDA regulatory action.
3. 21 *United States Code (USC)* § 248 (c) (3) (A).
4. 47 *Federal Register* 14464 (April 2, 1982); *Scott v Food and Drug Administration,* 728 F.2d 322 (6th Cir. 1984).
5. See *Food Chemical News Guide,* published by Food Chemical News, Inc. The *Food Chemical News Guide,* commonly known in industry as the "Knife and Fork" book, contains listings of both direct and indirect food additives and their status under the Food and Drug Regulations; the "Knife and Fork" book also documents public "prior sanctioned" status of substances. The *Guide* is regularly updated.
6. Idem. The "Knife and Fork" book also documents filings and status of petitions to FDA for listing of substances as GRAS in the regulations.
7. See 21 *Code of Federal Regulations (CFR)* § 170.30 (b).
8. *Monsanto v Kennedy,* at 948; see also, *C & K Manufacturing & Sales Co. v Clayton Yeutter,* 749 F. Supp. 8 (D.D.C. 1990).
9. 21 *CFR* § 174.5.
10. 104 *Congressional Record* 17,418 (1958).
11. 60 *Federal Register* 36582 (July 17, 1995).
12. 50 *Federal Register* 52720 (1993).
13. 21 *CFR* § 176.260.
14. See "FDA's Current Thinking on Recycled Polymers for Food-Contact Use," by Dr. Alan Rulis, then Director of FDA's Division of Food and Color Additives, presented at the GMA Environmental Issues Conference on Solid Waste, Washington, DC May 1, 1991.
15. 42 *USC* § § 6901 et seq. Under the complex RCRA statute, Congress has divided waste management into two discrete universes: hazardous waste, which is regulated under subtitle C of the statute, and all other waste, which is regulated under subtitle D. Most packaging waste is subject to subtitle D, which governs MSW. During the past decade, most of EPA's resources have been directed toward toughening standards for land disposal of subtitle C hazardous waste to make them commensurate with pollution controls already in place for air emissions and water discharges under the Clean Air Act and Clean Water Act.
16. 42 *USC* § § 7401 et seq.
17. 33 *USC* § § 251 et seq.
18. 42 *USC* § § 9601 et seq.
19. EPA advocates, but does not mandate, an integrated waste management approach based, in order of preference, on (*1*) source reduction, (*2*) recycling and composting, (*3*) incineration, and (*4*) land disposal (EPA, 1989).
20. California Public Resources Code § § 42300-42340.
21. Oregon Revised Statues § § 459A.650-459A.665.
22. Wisconsin Statutes Annotated § 100.297.
23. Wisconsin Act 335 (1989).
24. The nine states with deposite laws are Connecticut, Delaware, Iowa, Maine, Massachusetts, Michigan, New York, Oregon, and Vermont.
25. The 18 states that have adopted versions of the CONEG heavy-metals restrictions are Connecticut, Florida, Georgia, Illinois, Iowa, Maine, Maryland, Minnesota, Missouri, New Hampshire, New Jersey, New York, Pennsylvania, Rhode Island, Vermont, Virginia, Washington, and Wisconsin.
26. California Health & Safety Code § 25249.5-.13.
27. The FTC regulates false, deceptive, or misleading advertising pursuant to authority granted the Agency under section 5 of the Federal Trade Commission Act [15 *USC* § 45(a) (1)], which prohibits unfair or deceptive acts and practices affecting interstate commerce. The Commission has interpreted the act as essentially requiring companies to be able to substantiate the truthfulness of both express claims and any inferences a reasonable consumer is likely to draw from the express claim.
28. The nutrition lableing requirement was added to the Federal Food, Drug and Cosmetic Act by the Nutrition Labeling and Education Act of 1990. (Public Law Number 101-535, now found in 21 *USC* § 343).

29. See 49 *CFR* § § 172.400-407 for the labeling requirements and § 173.24 for the packaging requirements.
30. *California Business and Professions Code* § 17508.5; Burns Indiana Code Annotated § 24-5-17-1.
31. *The Association of National Advertisers, Inc., et al. v Lungren,* Docket No. C-92-0060, United States District Court for the Northern District of California.
32. New York Environmental Conservation Law § 27-0717; New Hampshire Revised Statutes Annotated § 149-N : 1.
33. A. Schmidt, "Ideas and Reality in Actual Regulatory Experience," Regul. Toxicol. Pharmacol. (Sept. 1988).
34. J. Heckman, "Fathoming Food Packaging Regulations, " 42 Food Drug Cosmetic Law J. 38 (Jan. 1987).
35. Food and Drug Administration, *A Food Labeling Guide,* Sept. 1994.

RALPH A. SIMMONS
Keller and Heckman
Washington, D.C.

LEAK TESTING

Introduction

The function of a package is to protect products from changes brought about by the outside environment. Prior to the development of packaging, food born diseases caused by pathogenic microorganisms were one of the most serious problems faced by humans. Microorganisms excrete enzymes that break down materials releasing nutrients that they can absorb. Packaging made from synthetic materials are generally resistant to microbial enzymes and form a barrier preventing reinfection. During prolonged dry periods many microorganisms survive inactively as spores. Bacteria, yeast, and mold require water to feed grow and reproduce, so packages need to be maintained in a clean and dry condition. Packages are inspected before they are filled for evidence of water stains and contamination. Then packages may be washed or blown with compressed air to remove dust-containing spores. Thermal processing and chemical preservatives kill vegetative microorganisms and many, but not all, spores. During filling and closure care must be exercised to assure minimal contamination of package seals that would create defects. Package defects form a pathway for reinfection by microorganisms. The quality control aspects of packaging by processors focus upon sanitation and closure integrity (see also Testing, permeation and leakage).

Available Methods

There are many ways to measure the hermetic integrity of packages. Test methods recommended by the National Food Processors Association are shown in Table 1.

Descriptions of Test Methods for Package Integrity

For detailed procedures to conduct these tests, see *Bacteriological Analytical Manual,* 7th ed.

1. Air-leak testing involves pressure decay or measurement of airflow rates.

2. Biotesting uses solutions with concentrations greater than one million nonpathogenic microorganisms per milliliter. Individual packages are immersed and pressure, vacuum, or mechanical flexing is applied to distort the package. Alternatively, packages contained within a chamber may be sprayed with an aerosol mist containing microorganisms.

3. Burst testing may be used as a dynamic test where the maximum force required to explode a package is used as an index of seal strength, or as a static test where packages are held at a fixed pressure to determine if the package will maintain pressure for a specific period of time.

4. Chemical etching separates or dissolves overlaying materials to expose the sealant layer of packages. This is useful for visual examination of contaminated seals of plastic and paperboard packages.

5. Compression testing is a burst test using pressure applied to the outside of a package. Weak packages will crush or burst, and stronger packages should not.

6. Distribution testing is a simulation of forces that are likely to be incurred by packages during storage distribution. Tests include vibration, drop testing, and compression in a laboratory having controlled temperature and humidity. Simulated abuse testing is useful for side-by-side comparison of two or more package designs in order to determine if one package design is more fragile than another. Simulation testing is preferred over shipping tests because there are fewer variables and results from simulated distribution testing are generally reproducible.

7. Dye testing is used to determine if holes in packages are present or not present. Water soluble dyes should be used with plastics or paperboard packages and solvent-based penetrating dyes should be used with glass or metal packages.

8. Electesting is a nondestructive method for liquid foods which change viscosity as they spoil. A shock wave is created by shaking the package. The amplitude and wavelength of the shock wave is a characteristic of the viscosity of liquids. When the device detects a signal response outside the normal range, the package is rejected. Other tests may be used to examine rejected packages for leaks.

9. The electrolytic test uses electricity to determine whether holes are present in plastic films, coatings, and seals. The detector is an open circuit that uses electrolyte or metal components of the package to conduct electricity. Holes are indicated when the circuit closes activating alarms. Liquid products containing at least one percent salt, or a solution containing 1% table salt (NaCl) in tap water may be used to conduct electricity through very small holes in packages having plastic layers. Breaks in the coatings of metal cans may be located using this method. Porous coatings and breaks in coatings expose metal to the product and these are the locations where corrosion can occur.

10. Gas leaks may be detected by mass spectrophotometer or a gas chromatograph. Helium is a useful tracer gas, because it moves readily through small holes in packages. Sulfur hexafloride is inert like helium and may be used to detect, locate, and measure small holes in packages in a nondestructive manner.

11. Incubation is a test that encourages the growth of microorganisms that are present within packages. Packages that swell as a result of microbial fermentation of the product may be identified by visual inspection. The test can be per-

Table 1. Recommended Test Methods for Food Packages[a]

Package descriptions
Paperboard packages
Flexible pouches
Plastic packages with heat-sealed lid
Plastic cans with double-seamed metal end
Metal cans
Glass bottles and jars

Test Method	1	2	3	4	5	6
1. Air-leak testing	O	O	O	O	O	O
2. Biotesting	O	O	O	O	O	O
3. Burst testing	O	X	X	O	NA	NA
4. Chemical etching	O	O	O	NA	NA	NA
5. Compression, squeeze testing	X	X	O	O	NA	NA
6. Distribution (abuse) testing	O	O	O	O	O	O
7. Dye penetration	X	O	X	O	X	X
8. Electroconductivity	O	NA	NA	O	O	O
9. Electrolytic	X	O	O	NA	NA	NA
10. Gas-leak detection	O	O	O	O	O	O
11. Incubation	X	X	X	X	X	X
12. Light	NA	O	O	O	O	O
13. Maching vision	O	O	O	O	O	O
14. Proximity tester	O	O	O	X	X	X
15. Seam scope projection	NA	NA	NA	X	X	X
16. Sound	X	NA	X	X	X	X
17. Tensile (peel) test	NA	X	X	O	O	O
18. Vacuum testing	NA	O	X	O	X	X
19. Visual inspection	X	X	X	X	X	X

[a] X = recommended method; O = optional method; NA = not appropriate method.

formed on large numbers of packages while they are held in a warm warehouse.

12. Light may be used to spot holes in packages. Visible light is used to backlight packages and reveal small holes. Polarized light can reveal areas within transparent plastic heat seals that are not fused or have different amorphous or crystaline compositions. X-rays are used to detect foreign materials such as stones and glass shards within sealed packages. Lasers are used to measure very small differences in packages dimensions that indicate that a leak has allowed packages to change shape. Laser shearography may be used to measure the effect of stress on packaging materials.

13. Machine vision may be used to measure critical dimensions of packages traveling rapidly along conveyors. Strobe lights stop the action and the video image is digitized. The picture is compared to images stored in computer memory. Parallel networks enable rapid comparison of images to standards. Nonconforming packages are side-lined for inspection by technicians using other tests for integrity.

14. Eddy-current meters sense the position of metal in packages. Concave ends indicate retained vacuum. Flat ends indicate packages have lost vacuum. The profile of container ends is consistent when the package, product, and filling conditions are rigorously controlled. Proximity sensors may also be used to measure the position of metal components of packages when a force is applied. Those packages that respond differently under stress may be identified by this method.

15. Seam scopes are large-format low-power microscopes. Mechanical or electronic measuring devices may be used to measure critical dimensions. Specialized devices are designed for metal-can double seams. The dimensions of very small holes may be measured using a seam scope as a microscope.

16. Sound is a novel nondestructive testing method. Rigid packages with internal vacuum may have ends tight, like a drum. Tapping on a can end reveals a difference in sound between a normal can and one that has lost vacuum. Electromagnets may be used to pull up can ends, which then snap back. A transducer can register the sound for analysis. Ultrasonics and laser acoustics may be used to measure pressure within packages without contact.

17. *Tensile testing* is a term applied to laboratory methods that stress packages or materials by pulling to create tension. Plastic ends and metal pull tabs on packages may be peel-tested. Composite materials may be delaminated and the strength of adhesion and cohesion measured. Glue bonds may be evaluated using tensile tests. Pieces of metal or film may be pull tested to determine elasticity, stretching, and break strength.

18. Vacuum may be used to test packages in many ways. Packages dipped in soapy water to coat the surface before being placed within a vacuum jar will exhibit bubbles revealing the location and approximate size of a leak. If the vacuum chamber is filled with water and the package held below the surface a steady stream of bubbles in the water may reveal the location of the leak on the package. Vacuum decay within a closed vessel is a useful method for measuring small leaks over a period of time.

19. Visual inspection is the most common method of pack-

Table 2. Nondestructive On-line Testing Methods for Packages

Method	Nondestructive	Application
1. Air-leak testing	Yes	Empty rigid and flexible containers and sealed packages
2. Biotesting	Yes	Recommended only for testing new package designs
3. Burst testing	Yes	Heat seals of flexible and semirigid packages
4. Chemical etching	No	Only a laboratory method for samples
5. Compression testing	Yes	Sealed flexible packages
6. Distribution simulation	Yes	Occurs daily on trucks laboratory simulation is semiautomated
7. Dye testing	No	A laboratory test
8. Electester	Yes	Possible with a pick-in-place (robotic) device to load testing machine
9. Electrolytic	Yes	Possible, not yet commercial
10. Gas-leak detection	Yes	Many gasses and test devices
11. Incubation	Yes	Requires weeks of storage
12. Light	Yes	Empty formed packages
13. Machine vision	Yes	Many commercial devices
14. Proximity tester	Yes	Metal cans, metal jar lids, and flexible films containing metal foils
15. Seam scope	No	A laboratory method
16. Sound	Yes	Many commercial devices
17. Tensile testing	No	A laboratory method
18. Vacuum testing	Yes	Both empty and sealed packages
19. Visual inspection	Yes	Most common method

age inspection. Holes as small as 100 μm may be seen with unaided normal vision under ideal lighting conditions. Unfortunately, inspectors quickly fatigue and the best visual inspection is only 85% effective.

Selecting the Right Test Method

Some test methods are specific to materials and others are specific to the design of the package closure. No single test method will work for all applications.

First, begin by collecting as many different package defects as possible and group them according to their public-health significance. A three-class system for package defects is used in the United States:

1. Critical defect
   a. Loss of the microbial barrier
   b. Evidence of microbial growth within package holes, fractures, punctures, swelling, leakage
2. Major defect
   a. Might leak, but has not yet leaked
   b. Testing is needed to determine whether packages will leak
   c. Weak seals, deformed double seams, deep scratches, cracks, creases
3. Minor defect
   a. Do not leak
   b. Are of economic, not public-health, concern
   c. Flexcracks, dents, abrasions, convolutions

Classification of defects has been completed for packages which contain food products that may support the growth of *Clostridium botulism,* the bacteria causing botulism. Sources of information are listed in the Bibliography at the end of this article.

Second, make a drawing of the package and copy it so you have one drawing for each type of defect. Mark the drawings to show the location of the similar defects using one mark for each defective package having that defect. This graphic summary will reveal where defects occur most frequently on the package.

Third, select the test method most likely to reveal the defect at the point on the package where it occurs. Methods that test the entire package are preferred to methods that test only a portion of the package. With most test methods for package integrity the design of fixtures to hold packages during testing is important. Dynamic test methods apply stress to packages and changes may be measured over a range. Static tests apply pressure at a fixed level and measurement is made after the package has been under this condition for a specified period of time. Evaluation of packages is needed to determine when a static or a dynamic method provides better results.

Fourth, follow the path taken by packages starting at the point of manufacture, through filling, processing, warehouse and distribution, all the way to the consumer. Draw this pathway as a flow chart and identify where defects are first detected. The cause is often just upstream from this point. Go to that location and observe for the occurrence of defects on the packages. Proceed upstream until defects are no longer detected. Search for actions which contact the package at the point where defects commonly appear.

Fifth, determine what physical action might form each defect. This is easier to visualize if you are able to locate the source of the defects and observe their formation during production or handling. There may be more than one cause for the same defect and some defects may be nothing more than different stages of the same defect. Use normal packages to try to reproduce the defects to verify the possible causes of damage. Often testing in a packaging laboratory can help to verify damage mechanisms.

Sixth, conduct validation testing of the method to determine the threshold sensitivity for small holes. Some test methods may be used on-line in a nondestructive manner.

Nondestructive Testing Methods

Most package testing methods may be automated for laboratory testing or for use in manufacturing. The ideal testing method will be nondestructive and fast enough to be performed on the production line. On-line nondestructive testing provides the opportunity to individually test packages for hermetic integrity. Table 2 lists some current applications and limitations of on-line nondestructive test methods for packages.

Many of the on-line nondestructive testing methods available today use sensors that do not contact the package. Examples include light leak detectors, pressure-differential sensors, proximity detectors, and sound and machine vision systems. The ability to perform nondestructive testing to packages using other methods that contact packages is limited by the design of fixturing capable of coupling the package to the sensor at production line speeds.

BIBLIOGRAPHY

General References

AOAC/FDA, "Classification of Visible Can Defects (Exterior)" in *Bacteriological Analytical Manual,* 6th ed., Food and Drug Administration/Association of Official Analytical Chemists, Arlington, VA, 1984.

AOAC/FDA, "Examination of Flexible and Semirigid Packages for Integrity" in *Bacteriological Analytical Manual,* 7th ed., Food and Drug Administration/Association of Official Analytical Chemists, Arlington, VA, 1992.

FPI, *Canned Foods: Principles of Thermal Process Control, Acidification, and Container Closure Evaluation,* Food Processors Institute, Washington, DC, 1988.

NFPA, *Flexible Package Integrity Bulletin,* Bulletin 41-L, National Food Processors Association, Washington, DC, 1989.

J. A. G. Rees and J. Bettison, *Processing and Packaging of Heat Preserved Foods,* Van Nostrand Reinhold, New York, 1990.

GEORGE W. ARNDT, JR.
Pet Inc.
St. Louis, Missouri

LIDDING

Lidding is a very specialized aspect of flexible-packaging technology. The advent of portion packaging and dispensing packages created a need for flexible-packaging lidding materials, and liddings are frequently used to seal other types of packages, including semirigid containers. Lidding materials are rarely composed of just one layer. One or more layers provide physical properties, and other layers provide sealability. Generally, the ideal lid is one that is easily peelable, leaves no traces of sealant residue, and is tamper-evident (see Tamper-evident packaging). The sealant should melt at a relatively low temperature unless heat resistance is necessary for sterilization of the contents or reheating of a food product. Typical examples of lidding applications are shown in Figure 1.

The first considerations in the choice of a lidding must be the intended use. Some of the questions that should be asked follow:

Must the lid prevent contamination or aid in dispensing the product by being peelable or having push-through properties?

What are the requirements for gas, moisture-barrier and light protection?

Should the lid be fusion-sealed or peelable?

What temperature resistance is required of the lidding? Is the product to be packaged hot? Will retained heat be a problem?

Should the lidding be heat-sealable or pressure-sensitive?

Will the use of cold-seal adhesives be advantageous?

Should the lidding be tamper-resistant?

What type of container will the lidding cover, and how will it be sealed to the container?

Will the lidding be left on during a temperature cycle, (eg, sterilization)?

Must it also be resistant to electron-beam or nuclear sterilization? If a lidding is used on a "cook-in-tray," will it

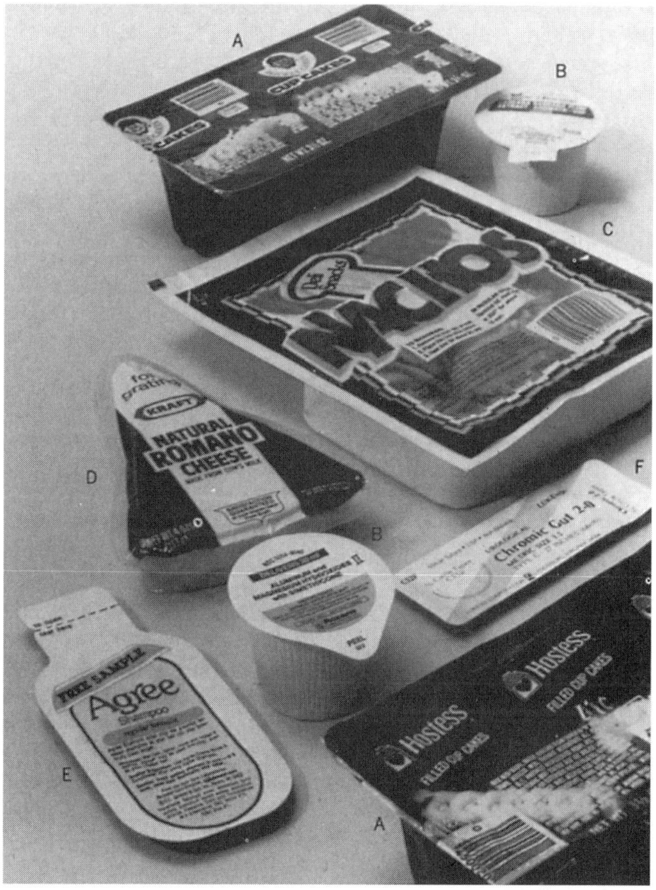

Figure 1. Typical examples of liddings used in packaging applications: A, transparent lidding sealed to opaque preformed cupcake tray; because of rapid turnover, this type of product requires only the minimal barrier properties of a peelable lidding; B, peelable lidding sealed to a vinyl-coated aluminum cup for liquid unit-dose drug applications; this type of lidding should be easily peelable, but also requires tamper evidence; C, snack package utilizing a flexible, peelable lidding with good gas- and moisture-barrier properties; D, peelable cheese lidding with good gas- and moisture-barrier properties; E, sample package of shampoo utilizing a flexible lidding sealed to a tray formed from polyethylene-coated PVC sheet—this is an example of a fusion seal; F, lidding for a PVC preformed tray containing sutures packed in alcohol.

be left on during heating in a microwave or conventional oven? Liddings for cook-in-trays must also have good adhesion to the tray while the product is in a refrigerated or frozen state.

If the lidding is for an industrial application, it might be very easy to find an appropriate material. A typical example would be a polyethylene-lined cannister filled with lubricating oil. An excellent lidding would be nylon/LLDPE (see Polyethylene, low-density). However, if the application is for a food product, the parameters of the barrier properties, compatibility of food and container, storage conditions, and when and how the lid is to be removed must be evaluated. Meeting applicable FDA regulations is another important consideration. With medical products, the problems of contamination of the packaged item also have to be very carefully considered.

How a lid is removed is of prime importance on many packages. A tab, or something to hold onto, is highly desirable. Ideally, the lid should peel off in one piece. A residue of sealant should be avoided if possible, but sometimes the peelability of the lid is designed to come about by separating the coating from the base stock. Examples of good bases for lidding stocks are paper/polymer/foil (see Multilayer flexible packaging) or oriented polyester film (OPET), which also might have a barrier coating for improved protection.

In a paper/polymer/foil base stock, the polymer is generally polyethylene, but it can also be an ionomer (see Ionomers) or ethylene–acrylic acid (EAA) to meet specific requirements. An ionomer might be used to improve toughness or tear resistance. The next consideration is the protection of that side of the foil that will face the container from the product being packaged. If the product is inert to foil, protection is not necessary, but because few products are chemically neutral, a barrier layer next to the foil is generally essential. This can be as simple as a vinyl coating, a vinyl film, a polyolefin film, or a polyester film, depending on the product being packaged and the type of protection needed. The heat-seal material is then applied over the protective layer (see Sealing, heat).

One of the simplest overwraps is nothing more than a corona-treated (see Surface modification) OPET film (see Film, oriented polyester). Polyester film is normally not heat-sealable, but corona treatment changes the surface so that it can be heat-sealed to itself. The seal must occur at a temperature near the polymer's melting point with high seal pressures. The seal is a fusion-type seal, but it is brittle and tears open easily. The chief use of this type of material is to overwrap school lunch trays or sandwiches. The physical protection is minimal, but it does act as a cover to prevent direct contamination. Another use is as a wrapper around frozen pizza. It is one of the most inexpensive liddings that can be sealed to a polyester-coated tray, but it is normally not used in "ovenable" applications.

Another important group of materials for tray liddings is OPET film coated with ethylene–vinyl acetate (EVA) applied from a solvent system or as an extrusion coating (see Extrusion coating). Normally, lower coating weights can be applied from a solvent system. The coating makes the material not just sealable to itself, but to a variety of other materials. It also seals at relatively low temperatures. EVA coatings do not form fusion seals, but the seals are peelable and usually removable in one piece. This type of lidding is very popular for polystyrene sandwich trays or other food trays. Tray packs of cheese and luncheon meat often use a PVDC-coated OPET film with an EVA coating.

Ovenable liddings are usually solvent-based polyester coatings applied to a polyester base film. The coating is used to provide heat sealability, and by proper selection of polyester resins used in the coating formulation, different seal ranges can be obtained and the degree of peelability regulated. Another route to obtaining peelability is to incorporate inert fillers into a coating which normally makes fusion seals. Polyester coatings, in addition to sealing to polyester materials, usually also seal to vinyl materials. For example, a polyester-coated OPET lid seals very well to a semirigid PVC blister.

PVC (see Poly(vinyl chloride)) films and solvent coatings are used as sealants on liddings where fusion seals to PVC semi-rigid stocks are required. The coating can be modified to provide peelability. Liddings for orange-juice portion packs have traditionally been aluminum foil with vinyl-type coatings which seal to a vinyl cup. The vinyl coating is inert to the acidic juice and is also good film-former to protect the foil from corrosion. Other liddings are foil/film laminations (see Laminating machinery) coated with a peelable heat-sealed coating. Similar liddings are used for yogurt, but they also require good barrier properties to extend the shelf life of the product.

Medicinal products in pill form are sometimes packed in PVC trays with a push-through-type lidding for ease in dispensing (see Pharmaceutical packaging). The tray is designed with wide flanges around each pill so that every pill is fusion-sealed in its own compartment and kept free of contamination. The lidding is usually a vinyl-coated aluminum foil, at least 0.001 in. (25.4 μm) thick for good barrier properties, which is sealed to a tray formed from semirigid PVC. The other popular pill package is the strip package. The strip package generally incorporates a peelable lidding having several plies. The outer layer is usually paper (to provide a good printing surface) which is then mounted to foil either by extrusion coating or adhesive lamination. The sealant side of the lidding can be a film or coating or a combination. The actual construction depends on the barrier requirements. If an extremely good barrier is necessary for very long shelf life, the structure can contain Aclar (Allied Corp.), which has exceptional barrier properties (see Film, fluoropolymer).

Other medical uses for liddings are safety seals on bottles to prevent tampering. These are combinations of materials that form fusion seals and are destroyed, ie, delaminated, when opened so that resealing is difficult. A safety seal is often fabricated with aluminum foil and mounted to a bottle-cap liner stock (see Closure liners) with wax. The combination of safety seal and cap liner is die-cut and placed in the bottle cap. The filled and capped bottle is passed through an induction sealer that fuses the safety seal to the bottle and melts the wax adhesive layer so the two parts separate when the bottle is opened.

For medical devices (see Medical packaging) that are to be ethylene oxide (ETO) sterilized, a popular packaging technique is to use a PVC tray with a Tyvek-coated lid. Tyvek (DuPont) medical-grade materials are porous to ETO gas, but not to bacteria. The sealant requires only limited heat resistance, but the web must be porous to allow the ETO to penetrate the Tyvek. Tyvek can be coated with a sealant in a pattern that does not change the porosity of the lidding. Another variation is to use a coating that is heat-sealable but not fused in drying so that it does not form a continuous film and therefore maintains porosity. Yet another option is to put the sealant material on the forming web so the Tyvek does not have to be coated.

Aluminum foil is normally used in a lidding if excellent barrier properties and protection from light are required. If the package is to be microwave-heated or requires transparency, foil is normally replaced with a PVDC-coated film which also gives very good barrier properties. Sealing methods are usually conductive-type heat seals. Important exceptions to this are safety seals in conjunction with a bottle-cap liner which use induction-type sealing equipment.

Further refinements of lidding technology can be expected as part of the current focus on semirigid replacements for metal cans (see Retortable flexible and semirigid packages).

BIBLIOGRAPHY

"Lidding" in *The Wiley Encyclopedia of Packaging Technology*, 1st ed., by W. R. Sibbach, Jefferson Smurfit Corporation, pp. 440–442.

General References

C. J. Benning, *Plastic Films for Packaging,* Technomic Publishing, Lancaster, PA, 1983.

R. C. Griffin, Jr., S. Sacharow, and A. L. Brody, *Principles of Package Development,* 2nd ed., AVI Publishing, Westport, CT, 1985.

R. C. Griffin, Jr., and S. Sacharow, *Food Packaging,* AVI Publishing, Westport, CT, 1981.

A. R. Endress, "Heat-Seal Coatings—Water and Solvent Based For Paper, Foil and Film," *Paper Synthetics Conference, Atlanta, GA, Sept. 27–29, 1976,* TAPPI, Atlanta, 1976, pp. 55–57.

A. R. Endress, "Water Base and High Solids Coatings For Complying with E.P.A. Regulations," *Paper Synthetics Conference, Cincinnati, Ohio, Sept. 15–17, 1980,* TAPPI, Atlanta, 1980, pp. 205–207.

R. B. Schultz, "Heat Seal Coatings For Disposable Medical Device Packaging," *Paper Synthetics Conference, Atlanta, Sept. 13–15, 1982,* TAPPI, Atlanta, 1982, pp. 13–15.

F. R. Solenberger, "Health Care Packaging With Ionomers and EVA," *Paper Synthetics Conference, Atlanta, Sept. 13–15, 1982,* TAPPI, Atlanta, 1982, pp. 65–69.

R. Pilchik, "Lidding Materials for Formed-Filled-Sealed Packages of Medical Devices," *Paper Synthetics Conference, Lake Buena Vista, FL, Sept. 26–28, 1983,* TAPPI, Atlanta, 1983, (1), pp. 221–224.

LIFE-CYCLE ASSESSMENT: AN ENVIRONMENTAL MANAGEMENT TOOL FOR PACKAGING

Consumer product companies are facing new challenges in the packaging arena today. The packaging they use must be attractive, keep the product fresh and usable, and provide for the safety of users by being tamper-resistant, childproof, or unbreakable. Now packaging must also be environmentally friendly. Although it is more difficult to claim environmental friendliness than it used to be, the environmental effect of packaging materials is a primary area of concern for many companies. *Life-cycle assessment* (LCA) is a tool that can be used to address this concern by evaluating the potential environmental effects of packaging options. LCA can show companies whether a packaging change is likely to cause more or less environmental burdens than what is in current use.

LCA Defined

Life-cycle assessment is the quantitative determination of the resource and energy use, and environmental burdens of a given product or process over its entire life cycle. It is often called a "cradle to grave" analysis, because it takes into account the resource and energy use, and environmental burdens of a given product or package system from cradle (raw-material acquisition), to grave (final disposal by the consumer). Life-cycle assessments are generally comparative in nature. Typically there is a comparison of one or more products or packages that provide the same functional use. The benefit that LCAs provide is that they look beyond the resource and environmental burdens of the immediate production and disposal of the product or package being studied. This is done by including the indirect effects caused by the production processes leading up to the final product or packages. Some of these indirect effects are the mining of raw materials, the energy and emissions resulting from transportation, and the energy and emissions associated with the production of electricity. These added "indirect" effects often result in substantial added environmental burdens of a product.

The first life-cycle study was done on packaging (in 1970), and packaging has been the focus of a high percentage of the LCAs that have been conducted since then. The use of LCAs in this industry has continued to increase over the past 5 years, even as broader uses of LCA are growing in importance. Typically, LCA is viewed as an environmental management tool. More specifically, it has been used as a green design tool, especially in the packaging industry. "Green design" is defined as a product or package design that is compatible with environmental goals.

Packaging Choices: Choosing Optimal Size Based on Environmental Evaluation

For example, a beverage bottler may want to determine what bottle size produces the least environmental burdens for delivery of an equivalent amount of beverage to their customer. The choices in this example are 1-, 2-, or 3-L PET bottles. Each bottle requires the manufacture of a cap, a thin PET bottle wall and bottom, and a thicker bottle neck. The manufacture of the bottle neck and cap requires substantially more resources than the manufacture of the thin PET bottle wall and bottom. All three bottles use the same size cap. A logical review of these choices may lead one to believe that the 3-L bottle would be the optimum choice. This would seem to be an obvious conclusion because the delivery of 6L of beverage requires six 1-L bottles, three 2-L bottles, or two 3-L bottles and the manufacture of their associated caps and necks (putting them on an equivalent basis). Thus, one would assume that the 3-L bottle would deliver the desired quantity of the beverage with relatively smaller environmental burdens. As expected, the 2-L bottle uses less energy and produces less emissions to the air, water, and land than does the 1-L because of the reduced number of bottle tops manufactured.

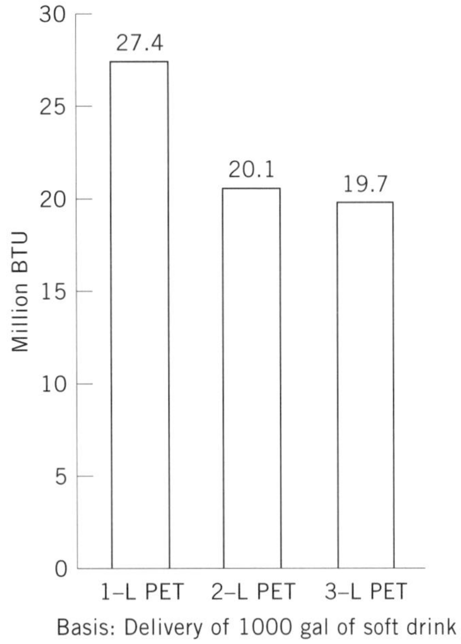

Figure 1. Energy requirement for PET soft-drink containers.

However, this same logic does not hold for the 3-L bottle. As an LCA would indicate, the 3-L bottle, because it holds larger quantities of liquid, requires a greater rigidity and wall thickness than do the 1- and 2-L sizes. This extra requirement is enough to make the environmental burdens of the 3-L bottle equivalent to that of the 2-L bottle (1). This is illustrated in Figures 1 and 2, which show the energy and solid-waste totals generated by a computer model for the delivery of an equivalent quantity of soft drink. Figure 1 shows the total energy usage for the three soft-drink container sizes. The graph illustrates the extra energy necessary for the 1-L bottle, and shows that this same advantage between the 2- and 3-L bottles does not exist. This same relationship holds for solid-waste totals, illustrated in Figure 2.

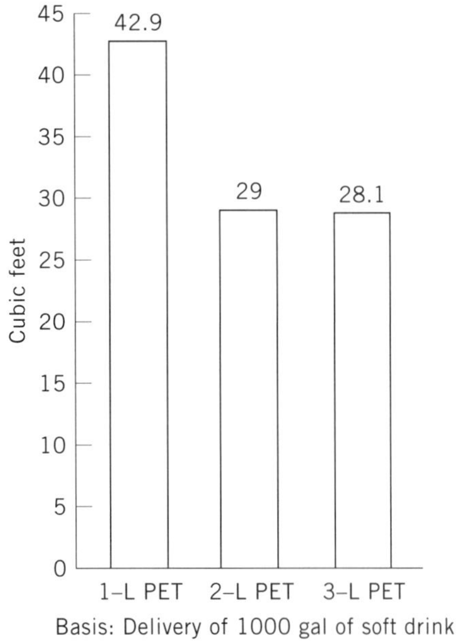

Figure 2. Solid waste for PET soft-drink containers.

So, in actuality, the optimal beverage delivery size is either the 2- or 3-L sizes, and not the 3-L, as logic would indicate. This is not to say that the 1-L bottle is without merit; in fact, all three sizes are strong in the marketplace. For example, the 1-L (or smaller) container is an individual serving size. As this example shows, the value of an LCA is that it takes into account the products entire life cycle, and provides a comparison to the other packaging options. Often this cradle-to-grave comparison provides results that are contrary to common opinion.

Uses of LCA

Although LCA is now a tool for green design, its uses have continued to extend into other applications. The areas of LCA usage can be broken down into four basic categories: (*1*) environmental management, (*2*) ecolabeling, (*3*) regulatory purposes, and (*4*) information and education (2–4). The first category is using LCAs as an environmental management tool for internal use within an industry and/or company. Some of these uses include product design (more specifically green design), assistance in meeting ISO 14000 guidelines, and cost-effectiveness. The second category of LCA uses also has economic implications, and that is for ecolabeling. Typically ecolabeling may be used to sway public opinion regarding a product or give environment sanction to a product based on some stated criteria. The label can sway public opinion to increase market share. Ecolabeling based on LCA principles is a very active concept in Europe. However, the ecolabeling approach is fraught with complexities that are not resolved. In the United States, LCA is not viewed widely as a public policy tool for labeling or for environmentally preferable products or packages. Nonetheless, this type of initiative is a proposal under public review.

The third category of LCA uses is providing information for regulatory purposes. Both the FDA and EPA are requiring more environmental information when applying for product approval. For example, applications to the FDA for approval of new packaging that will be in contact with food now require information on the environmental consequences of the new packaging. This information must include a comparison of the new packaging to the packaging it will replace or compete with. This additional information is typically outlined well by an LCA.

The final category of LCA uses is also for information and education. This includes providing information that will be used internally to educate the members of an industry of the extent and complexity of a products production. This same information can be used externally to educate the public or regulatory agencies about an identified product. All of these uses show that LCA utility extends beyond the world of environmental applications. As its utility continues to expand, knowledge of LCA becomes more and more important for those involved in the packaging industry.

LCA Components

The LCA can be broken down into four major components: goal definition and scoping, life-cycle inventory (LCI), impact assessment (now defined as inventory interpretation), and improvement assessment (2,3). To this date most life-cycle

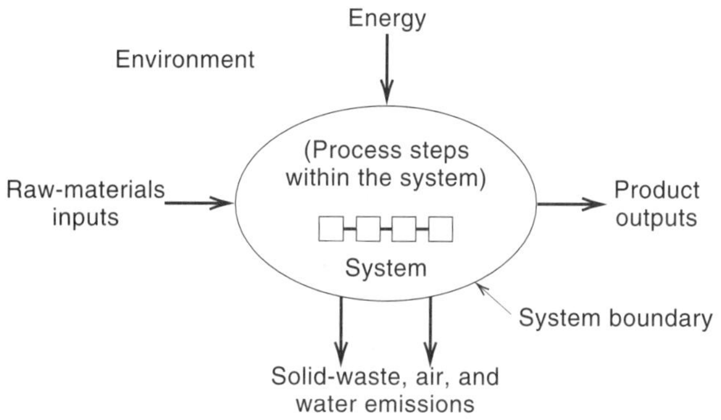

Figure 3. Definition of system in LCA.

studies have included life-cycle inventories and some improvement assessment. There has been limited success in completing impact assessments because of the lack of quality data and a consistent, accepted methodology.

Goal definition and scoping. "Goal definition and scoping" is the task of identifying the purpose and basis of the study. The purpose of the study should be a clear and unambiguous statement of why the study is being pursued. This statement helps to define the system and the processes within the system that need to be analyzed. The "system" under study is defined in the same manner as it is defined in classic thermodynamics.

Thermodynamics breaks a process down into two components: the system and the surrounding environment. The surrounding environment acts as a source of all inputs and a "sink" for all outputs. The description of the system is a description of all flows that go across the system's boundaries. For example, the surroundings provides the raw materials for the system that enter across the boundaries. Conversely, all air, water and solid-waste emissions along with the desired product cross the boundaries into the surroundings. This is illustrated in Figure 3. Thus an LCA takes into account all materials and energy that cross or are held within the system.

The boundaries for an LCA include the acquisition of raw materials, manufacture of intermediate materials and final product, use of product, and final disposition of the product. Also included in the boundaries are any use, reuse, or recycling that occurs; the energy and emissions for each processing step; the energy and emissions to transport from one step to the next; and any energy and emissions necessary to process energy sources into usable fuels. The energy and emissions from extracting the fuels that are consumed often contribute significantly to the LCA results. Therefore, the fuel extraction steps, including spills and losses, should be included with the data for fuel combustion. In fact, relatively simple flow diagrams such as Figure 3 turn into a complex network of processes when an actual product or package is evaluated. Also, whether the product or package is reusable can be a very important consideration. A good example is a reusable glass or ceramic cup. Each reuse requires the washing of the cup with hot water and detergent. Over the lifetime of a cup, it may be washed hundreds of times, or it may break and thus have a short life. The burdens from the generation of the hot water and detergent for the reuse of the cup could outweigh the burdens from the generation of the cup from raw materials. In other words, the reuse cycle may create unanticipated burdens compared to a single-use item such as a paper or plastic cup.

One of the most important parts of defining the scope and boundaries of a study is the definition of the functional unit. This identifies the unit of product that is going to be studied and compared. Some examples are the unit surface area covered by paint, the packaging used to deliver a given volume of beverage, or the amount of detergent used for a standard household wash (2). So with these boundary definitions the product under study is examined from cradle (raw-material acquisition) to grave (final disposition of the product).

Life-cycle inventory (LCI). The boundaries and scope provide direction into completion of LCA. The LCI is the next stage of the LCA and is the most frequently used and best developed stage. The LCI is often used separately from the other components of the LCA, because the inventory portion of the LCA is more clearly defined. This clearer definition is due to the fact that the methodology of LCI is based on scientific principles. LCI is essentially based on the law of conservation of mass, the conservation of energy, and thermodynamic definitions of systems. Using these scientific principles, the basis of LCIs can be developed and applied consistently with each LCI regardless of the product analyzed. The LCI consists of seven unique steps: (*1*) definition of scope and boundaries, (*2*) identification of unit operations, (*3*) development of systems and network diagrams, (*4*) data gathering, (*5*) creation of a computer model, (*6*) analysis and reporting of study results, and (*7*) the interpretation of results leading to specific conclusions.

The definition of scope and boundaries is directed by the scope and boundaries identified by the full LCA. If the LCI is used separately from an LCA, the scope and boundaries for the LCI need to be defined in a similar manner as was described for the full LCA.

Once the functional unit and the system to be studied are identified, the system must be broken down into individual unit operations from the beginning to the end of the life cycle. This process serves as a blueprint for the process of gathering data (5). For example, in order to make ethylene from petroleum, each step in the life cycle, from the drilling of crude oil and natural gas, to producing ethylene, needs to be broken down into discrete steps. For each of these discrete steps a

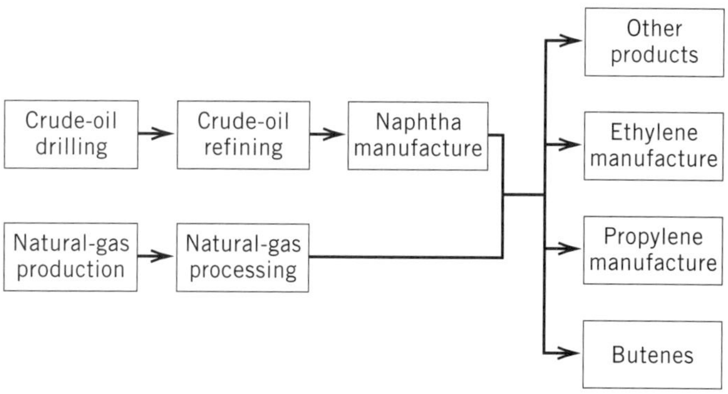

Figure 4. Steps in the production of ethylene.

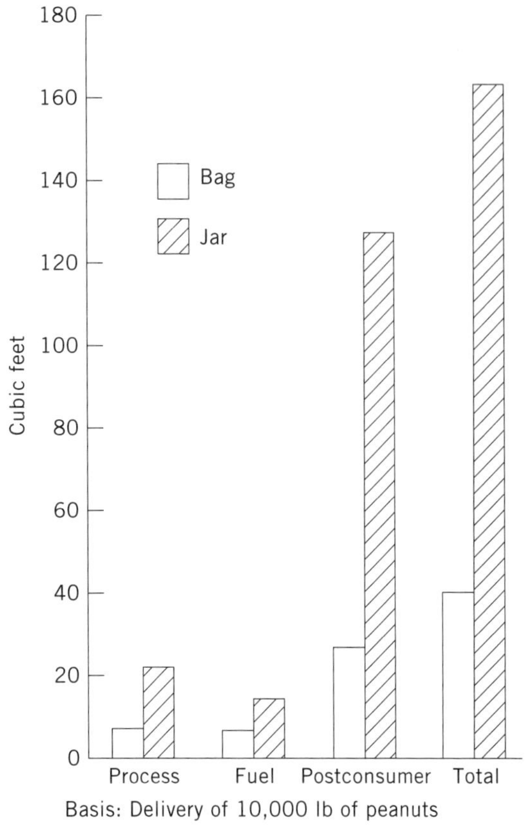

Figure 5. REPAQ comparison of solid waste.

system boundary must be identified, with all the inputs and outputs that cross the boundaries identified as well. For the preceding example, the process of making ethylene would include the inputs of natural gas and/or naphtha as raw materials and energy resources (6). Outputs would include ethylene, propylene, and butenes (coproducts), air, water, and solid-waste emissions to the environment. This process as part of the whole life cycle is shown in Figure 4 (see page 565). Once the inputs and outputs for each individual process are identified in this manner, the entire process is linked according to the functional unit defined in the scope of the study. The difficulty lies in obtaining the data for each individual step within the life cycle.

Data gathering is a resource-consuming process, but its importance cannot be overemphasized. The entire LCI/LCA is dependent on the quality of the data gathered. Each unit operation should have a complete breakdown of all inputs and outputs, both material and energy. With the advent of computer technology, the quantity of data used and the accuracy and complexity of LCIs that can be undertaken have increased tremendously. With the use of spreadsheets or computer programs, the air, water, and solid-waste emissions, as well as the energy usage for each individual unit operation, can be totaled with respect to the functional unit selected in the scoping of the project. The computer model can then be modified to determine how changes in the functional unit, or the process conditions, will affect the overall results. For example, the delivery of peanuts could have two packaging alternatives: a glass jar or a composite aluminum and plastic bag. Once the two packaging alternatives are put on an equivalent basis, an LCA can identify the environmental burdens and resource consumption for each packaging alternative. From the LCA comparisons of the two packages, a choice can be made as to which of the two alternatives produces the least potential environmental burden.

Various LCI software packages are available to companies that do not wish to collect data or develop their own computer models. One such software package is REPAQ (trademark), developed specifically for screening packaging options. With the availability of LCI software such as REPAQ, the packaging engineer can now see how modifications in packaging systems (shape, size, weight, and composition) can change the overall outcome of the LCI results (7). Figures 5–7 illustrate a partial output from REPAQ. The figures illustrate the comparative solid waste, atmospheric emissions, and energy usage of the jar and the bag from the example above. A typical computer output will consist of a comparative listing of 20–30 atmospheric emissions, 20–30 waterborne emissions, and a breakdown of solid waste and energy. The advantage of computer modeling that is the size, weight, and compositions of the two options can be modified, or secondary packaging can be added or removed, and the graphic summary will give immediate feedback on how the LCI results would change. With this type of information, a packaging system that helps meet environmental goals can be quickly designed.

Impact Assessment (Inventory Assessment)

The life cycle inventory provides a great deal of information about the energy use and emissions associated with the life

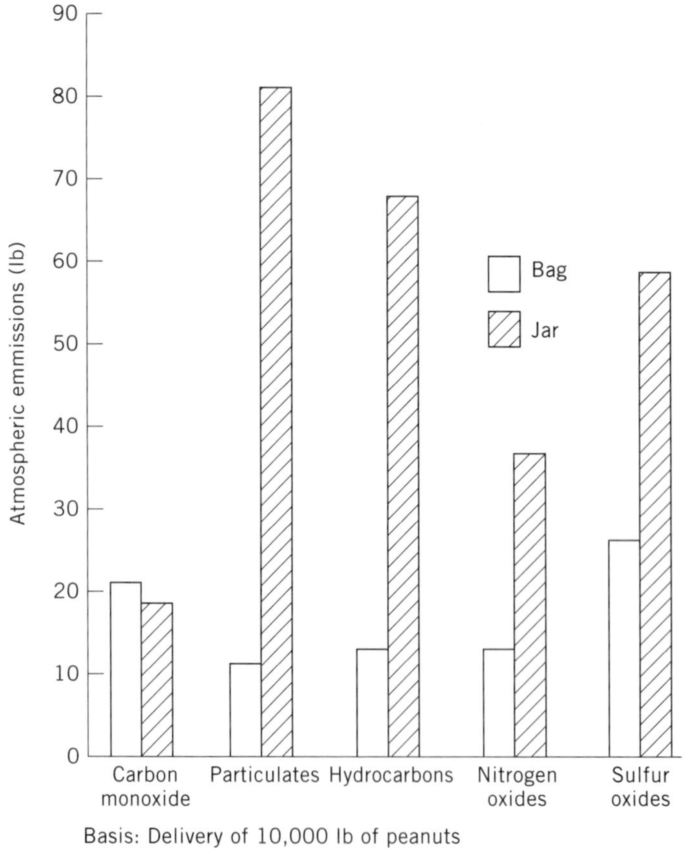

Figure 6. REPAQ comparison of atmospheric emissions.

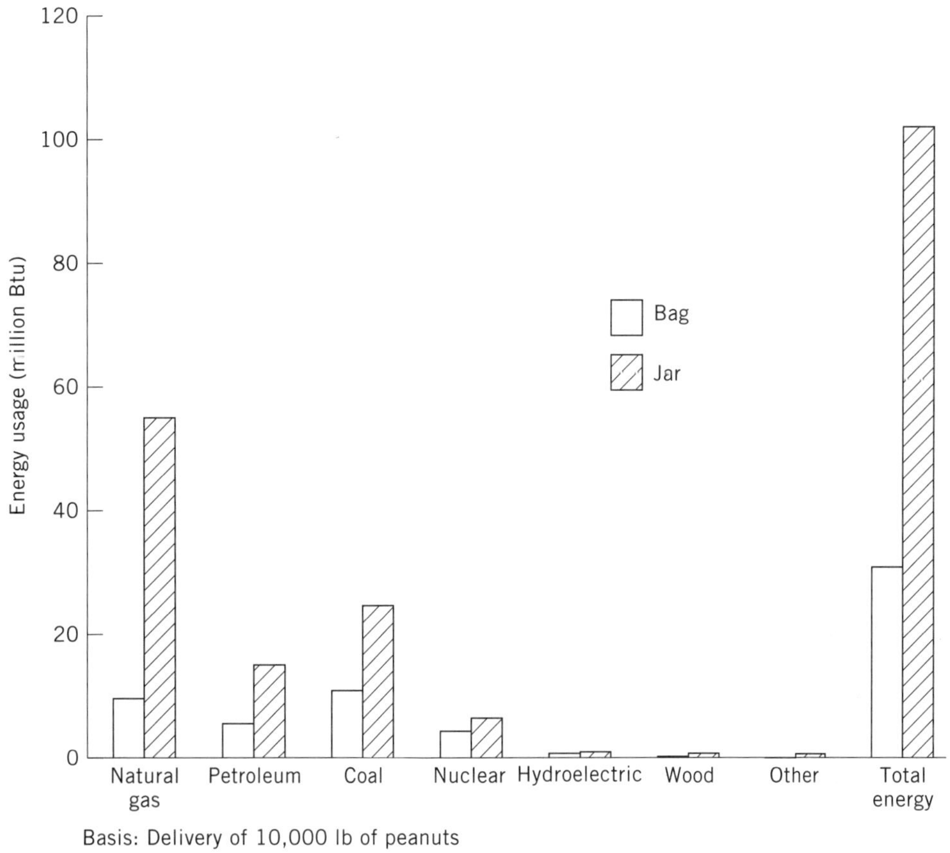

Figure 7. REPAQ comparison of energy usage.

cycle of a given product. This information, however, cannot be generally applied to the product's effect on human health, ecological quality, and natural resource depletion. This is simply because one pound of a given emission may provide substantially different effects on human health or the environment than a pound of a different emission. Impact-assessment methodology helps to categorize the LCI information into sets of common impact measures such as increased mortality, habitat destruction, or global warming that allows interpretation of the total environmental effects of the system being evaluated (8).

According to the definition of impact assessment put forth by SETAC and the U.S. EPA, the impact assessment can be broken down into three steps: classification, characterization, and valuation (2,3,9). *Classification* is the process of assigning the initial aggregation of LCI data into relatively homogenous impact groups. There are four general impact categories: environmental or ecosystem quality, quality of human life (including health), natural resource utilization, and social welfare. These categories are then broken down into even more specific impact groups. Two examples of these groups are global warming and ozone depletion. Each emission outlined in the inventory is placed into one or more of these impact groups. The selection of impact groups is somewhat related to what emissions are being released by the systems under consideration. The selection process is twofold, with the analyst first asking: "What environmental problems could these emissions potentially cause?" The second question the analyst asks is:

	Eliminate packaging	Recycled content	Lightweight material	Material substitution	Redesign packaging	Redesign product
Paperboard box						
HDPE bottle						
Polypropylene cap						
Corrugated container						

Figure 8. Improvement options for packaging a consumer product.

"What environmental impacts are important or of interest to the audience of the study?" Answering both questions results in a comprehensive list of impact groups to evaluate.

Once the inventory emissions are classified into impact groups, the process of characterization begins. *Characterization* is the selection of the actual or surrogate characteristics to describe impacts. For example, acid equivalents could be used to characterize emissions classified as having acid-rain potential. This provides an assessment of the relative magnitude of the given impact on the more general categories such as human health. The difficulty in the characterization process comes with the fact that scientific factors have not been developed for all the impact categories, and thus coming up with a common descriptor for each group is subjective at best.

The third and final step of impact assessment is *valuation,* the assignment of relative values or weights to different impacts. This step essentially allows for the evaluation of the different emissions across all impact groups. With this composite information, decisionmakers can then decide the total impacts of a product system. Valuation is extremely subjective and is dependent on the value system of those assigning relative weights to the impact groups.

Beyond the classification step, impact assessment quickly becomes subjective and value-laden. The lack of a well-accepted and nonsubjective model for impact assessment limits its applicability to industry. However, development of nonsubjective models, or models that limit subjectivity, is ongoing.

Improvement Assessment

Similar to impact assessment, improvement assessment has not undergone the consensus examination of the methodology that life-cycle inventory has undergone. However, improvement assessment based on life-cycle inventories is common practice. The purpose of the improvement assessment is threefold: (*1*) it evaluates LCA results for their relevance to the predetermined LCA goals and objectives, and identifies opportunities for reducing the environmental impacts of a product system; (*2*) it translates all results and their determined importance to the LCA audience in a clear and concise manner; and (*3*) it uses the impact assessment or the LCI as a baseline to determine the effect, if any, of the instituted improvements.

Improvement assessment identifies and evaluates any opportunities for reducing environmental burdens. These opportunities may be uncovered by the inventory portion of the LCA by identifying areas of large energy usage or environmental emissions. The improvement assessment can then identify those areas where increased efficiency or modifications would be appropriate. Figure 8 (see page 567) is a table of some areas for improved efficiency that could be identified in an improvement assessment. The packaging of over-the-counter (OTC) medication is used for this example. The package components are listed in the first column, and possible modifications to the package system are listed in the top row. With this sort of grid, possible improvements can be assessed. For example, could some of the packaging, such as the paperboard box, be eliminated? Instead of packaging the bottle in a box, the bottle could be placed directly on the shelf. Could the packaging makeup be changed by using recycled content, making it lighter, or making it from a different resin? Could the package or the product be redesigned? For example, could the bottle be filled more, or could some filler be removed from the product? Each of these options can be reviewed to see how modification would affect the environmental burdens of the packaging as a whole.

Similarly, the improvement assessment can use information provided by the impact assessment. This information takes the inventory results one step further by identifying not only the quantities of energy used and environmental emissions, but also their potential impacts. The improvement assessment takes the inventory and impact information and condenses it to a few clear and concise conclusions. These conclusions drive the process or product change that will result in reducing the environmental impact of the system being studied. Once these changes are made, the impact and inventory portions of the LCA can be used as a baseline to determine the effects of any improvements made to the system.

Limitations of LCA

Beyond the obvious limitations of the lack of consensus for impact and improvement assessments methodologies, LCAs do have limitations. Often LCA results—or, more specifically, LCI results—are extrapolated beyond what is supported by the study. Most studies have very specific parameters that describe a given comparison. These parameters can include weight of the product, delivery of a quantity of useful product, time period, and specific location of the comparison. If any of these parameters are changed, the relative comparison may or may not be accurate. Thus the study results are limited to the specific study conditions. Second, there has been a desire to use LCAs as public policy instruments to determine whether a specific material or product is "good" or "bad" for the environment. Although LCAs and LCIs provide a foundation of useful information on resource usage, environmental emissions, and potential impacts, they do not quantify all environmental consequences of a product or process. Thus, LCA is an effective tool for evaluating environmental tradeoffs and consequences of specific actions. It is not a tool for measuring the magnitude of environmental "goodness" for specific materials or products. As processes and products change, their LCA profile also changes relative to other alternative processes and products.

Life-cycle assessments have been used by the packaging industry for 25 years to help meet both financial and environmental goals. They provide a strong basis for making design decisions as well as environmental planning. As the applications and knowledge of LCAs continue to grow, the value to the packaging industry will continue to grow as well.

BIBLIOGRAPHY

1. Franklin Associates, Ltd., *Comparative Energy and Environmental Impacts for Soft Drink Delivery Systems,* March 1989.
2. Society of Environmental Toxicology and Chemistry, *A Technical Framework for Life-Cycle Assessments,* workshop report for Society of Environmental Toxicology and Chemistry, August 1990; Society of Environmental Toxicology and Chemistry and SETAC Foundation for Environmental Education, Inc., Washington, DC, 1991.
3. Society of Environmental Toxicology and Chemistry, "Guidelines for Life-Cycle Assessment: A 'Code of Practice'," from the SETAC

Workshop held at Sesimbra, Portugal, March 31–April 3, 1993, Society of Environmental Toxicology and Chemistry (SETAC), 1993.

4. The Nordic Council, *Product Life Cycle Assessment, Principles and Methodology,* Nordic Council of Ministers, Copenhagen, 1992.
5. T. K. Boguski, R. G. Hunt, J. M. Cholakis, and W. E. Franklin, "Environmental Life Cycle Assessment," in *LCA Methodology,* M. A. Curran, ed., McGraw-Hill, in press.
6. I. Bousted, *Eco-balance, Methodology for Commodity Thermoplastics. A Report for The European Center for Plastics in the Environment,* Brussels, Dec. 1992.
7. REPAQ (trademark): a software package designed to evaluate and design optimum packaging alternatives; Franklin Associates, Ltd. Prairie Village, KS.
8. Franklin Associates, Ltd., *Life Cycle Assessment of Ethylene Glycol and Propylene Glycol Based Antifreeze,* Aug. 1994.
9. U.S. Environmental Protection Agency, *Life-Cycle Impact Assessment: A Conceptual Framework, Key Issues, and Summary of Existing Methods,* EPA-452/R-95-002, July 1995.

General References

T. K. Boguski, R. G. Hunt, and W. E. Franklin, "General Mathematical Models for LCI Recycling," *Resource, Conserv. Recycl.* **12:** 147–163 (1994).

T. Boguski, J. Wood, R. Hunt, and W. Franklin, *Life Cycle Analysis and Green Design,* Franklin Associates, Ltd., Prairie Village, KS.

R. Hunt, J. Sellers, and W. Franklin, "Resource and Environmental Profile Analysis: A Life Cycle Environmental Assessment for Products and Procedures," *Environ. Impact Assess. Rev.* (Spring 1992).

William E. Franklin
Terrie K. Boguski
Paul Fry
Franklin Associates, Ltd.
Prairie Village, Kansas

LIQUID-CRYSTALLINE POLYMERS, THERMOTROPIC

Main-chain thermotropic or melt-processible liquid-crystalline polymers (TLCPs) are a class of high-performance condensation polymers represented mainly by *para*-extended aromatic polyesters (1). Although these polymers were invented in U.S. industrial research laboratories in the 1970s, development was inhibited until the 1980s by high monomer costs and the lack of attractive markets. With the emergence of a market niche in molded electronics packaging and enhanced prospects for lower monomer costs, rapid 20–25% annual growth of the TLCP market presently estimated at 10–15 million lb is projected for the remaining 1990s (2).

TLCP property advantages include high thermal, electrical, and solvent resistance combined with exceptional flow (low-melt viscosity) in the 300°C (572°F) range, precision moldability, and dimensional stability; these features are ideal for injection-molded, close-tolerance applications such as thin-walled electrical connectors that constitute about 90% of the present market. In film form, the dense uniaxial LC morphology affords excellent barrier properties, but development in packaging applications has been inhibited by high resin costs and the need for special technology to prepare balanced film. With lower resin costs and recent progress in balanced and coextruded film technology most notably using counter-rotating dies (3), medical and food-packaging applications are expected to emerge rapidly. In addition, the unique post-heat-strengthening capability of these materials by solid-phase polymerization of the final part results, for example, in high-strength fibers with the mechanical properties of Kevlar aramid fiber (4).

Fundamentals

TLCP chemistry is relatively simple and involves reaction of an acetylated aromatic diol with an aromatic dicarboxylic acid, as illustrated in Figure 1. The diol is usually acetylated *in situ* with acetic anhydride at 150–200°C (300–392°F) under nitrogen in the presence of the diacid, and the melt polymerization finished under vacuum at higher temperatures (300°C or 572°F range) over a period of several hours. Acetic acid is the principal reaction byproduct. Principal monomers used in commerical materials shown in Figure 2 are *para*-oriented, phenyl or naphthyl rings preferred, but a certain amount ($\leq$40 mol%) of non-*para* or "soft" comonomers (eg, isophthalic acid, ethylene glycol) can be tolerated without destroying liquid crystallinity. Other linkages or colinkages are also feasible, eg, ester–amide, ester–imide, carbonate, and azomethine, but the predominant linkage is ester.

The rodlike character of the extended chain molecules results in a preference for parallel packing in the melt, driven by entropy, much like "logs in a stream." Packing occurs in the form of optical-size birefringent domains that contain uniaxially oriented polymer chains, but which are randomly oriented with respect to one another. At flow conditions, the domains easily orient with respect to each other, which frequently results in anisotropic properties in the final part. This can be an advantage in some structural materials (eg, fibers), but a disadvantage in others requiring balanced properties (eg, films). Differences in flow behavior of liquid crystalline and conventional or flexible chain polymers are illustrated in Figure 3. This type of simple uniaxial ordering is referred to as a "nematic" *meso*phase in LC terminology, and is predominant for TLCPs. The use of optically active comonomers can also lead to chiral stacking of domains and the "cholesteric" mesophase in the melt (5).

Commercial Sources

Major commercial TLCPs include Vectra (Hoechst-Celanese), Xydar (Amoco) and Zenite (DuPont). All are high-temperature aromatic polyesters with heat-distortion temperatures in the 240–270°C (464–518°F) range and are glass-filled, geared mainly to injection-molded applications in electronics. Vectra is a 73/27 copolymer of *p*-hydroxybenzoic and hydroxynaphthoic acids, Xydar a 50/25/25 copolymer of *p*-hydroxybenzoic acid, 4,4′-biphenol and terephthalic acid (and optionally a small amount of isophthalic acid) developed initially by Carborundum in the 1970s and Zenite, an all-aromatic polyester. Vectran is a high-performance TLCP fiber based on the Vectra composition manufactured by Kuraray in Japan under license from Hoechst-Celanese. Worldwide capacity of these materials is estimated at 15–20 million lb; Hoechst-Celanese/Polyplastics and DuPont have recently announced new plants in Japan and Tennessee, respectively, onstream in 1995. Ro-

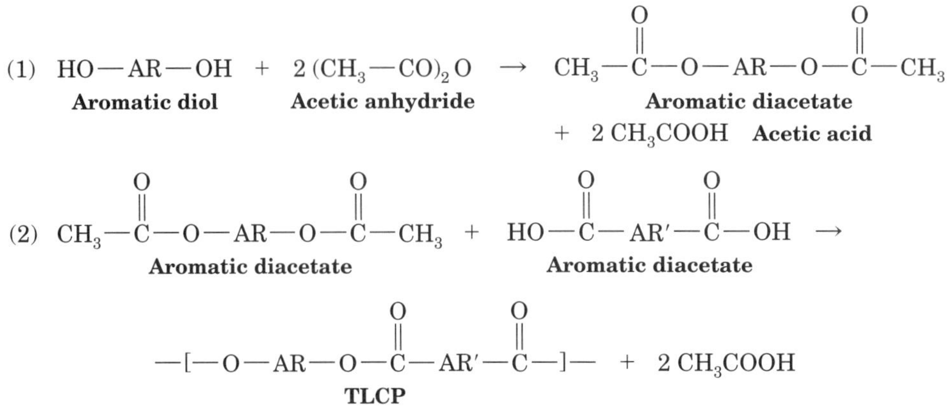

Figure 1. TLCP Melt Polymerization.

drun, a lower-temperature TLCP based on PET (20–40%) and *p*-hydroxybenzoic acid developed initially by Eastman in the 1970s, is also available in limited quantities from Unitika–Japan.

TLCP cost is presently high: \$5–\$11/lb for glass- or mineral-filled molded grades and \$12–\$22/lb for extrusion grades; resin cost, however, is expected to approach the \$3–\$5/lb range over the next several years (2) owing to lower prices projected for key monomers such as 2,6-naphthoic acid, also a monomer for PEN and now available from Amoco at \$1.50/lb, and economies-to-scale price reductions expected for *p*-hydroxybenzoic acid (to \$1.00/lb range from \$2.50/lb present) and 4,4′-biphenol (to \$2.00/lb from \$5–\$7/lb).

Films

Processing. TLCP ease-of-flow orientation results in unbalanced films with poor lateral properties by conventional film extrusion that are generally not poststretchable. Kuraray has reported reasonably balanced Vectra-type films by conventional film blowing at high shear rates (6), but the best results have been obtained by Superex using a combination of film-blowing and counter-rotating extrusion dies (3). In the latter case the annular die consists of counter-rotating inner and outer mandrels that impart an opposing biaxial or "crisscross" orientation to the respective inner and outer film surfaces. Stretch is controlled by gas blown through the center and attenuation/takeup speed. The latter technology, also applicable to tubing, is available from Superex by special licensing. Another approach reported in patents by Nippon Oil involves conventional extrusion of cholesteric TLCP film with improved property balance through the use of chiral (optically active) comonomers that results in a biaxial twist or chirality to the morphology in the film plane (7).

Properties. TLCP films generally possess exellent permeability resistance to gases (eg, oxygen, water vapor, and carbon dioxide), which is attributed to the highly compact uniaxial morphology. A comparison of permeability of various polymer films in Figure 4 illustrates the unique barrier resistance of TLCPs to both oxygen and water vapor, far exceeding other packaging resins. Film studies attribute low permeability to low solubility indicative of a highly ordered substrate (8); moderate reductions in oxygen permeability were observed (9) with orientation (45%) and solid-phase polymerization (10–20%). Larger reductions in oxygen permeability (200–300%) have been observed by Kuraray in

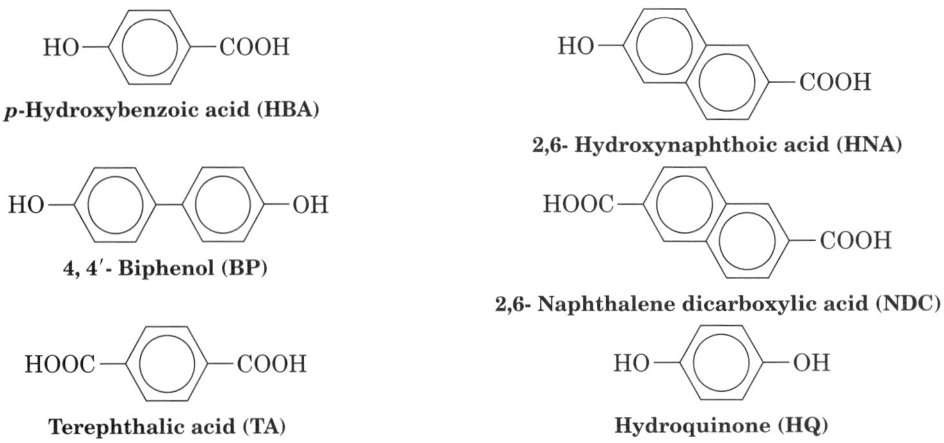

Figure 2. TLCP Monomers.

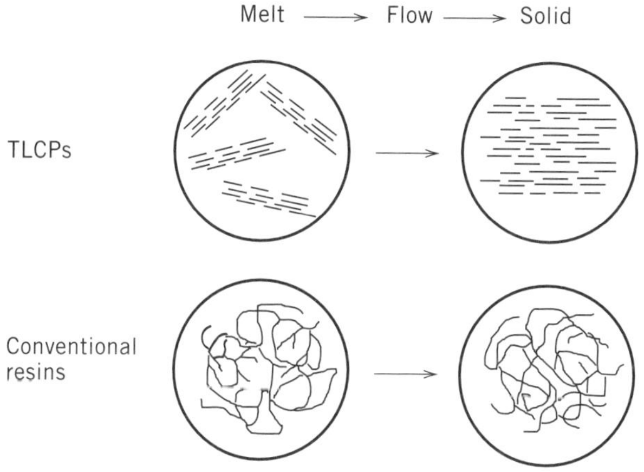

Figure 3. Flow behavior of TLCPs vs conventional resins.

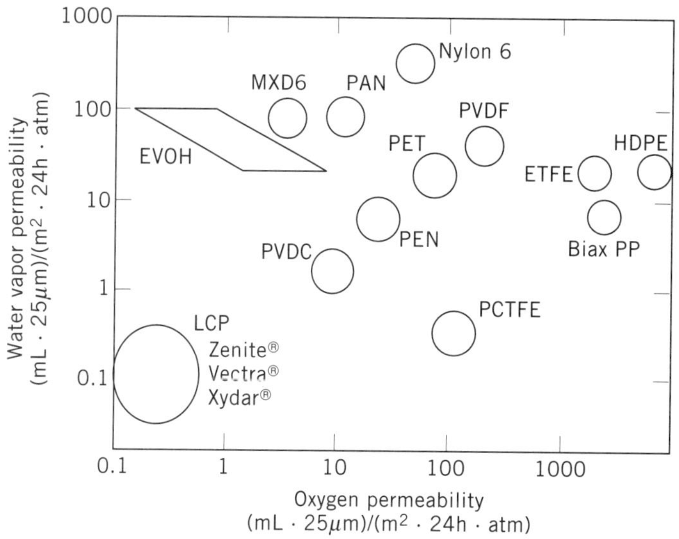

Figure 4. Permeability of selected polymer films. (Courtesy of Superex Polymer, Inc.)

studies of aliphatic–aromatic TLCPs based on PET and PEN copolymers with HBA and HNA (10), whereby permeability is found to decrease with increasing naphthyl (vs phenyl) content, probably due to decreased solubility. These partially aliphatic materials were also found to be poststretchable biaxially and, in some cases, transparent. Wholly aromatic TLCP films (eg, Vectra), in contrast, are not poststretchable and are opaque, except in the case of ultra-high-oriented ("monodomain") film prepared from uniaxial layups of TLCP fiber (11).

Applications. A comparison of the general properties of balanced TLCP Vectra film with other high-performance polyester (PET, PEN) and polyimide (Kapton, Upilex) films is shown in Table 1. In contrast to polyimide, TLCP appears to offer similar high-temperature and electrical properties with lower moisture absorption and lower cost, for electrical applications in flexible-film substrates. Compared to polyester, TLCP offers higher thermal, mechanical, and barrier properties for medical and food-packaging applications requiring sterilizing, retorting and long shelf life. In the latter case, even at a resin cost of \$20/lb, TLCP offers about five times better cost-barrier performance (oxygen, water vapor) vs PET and PEN, assuming resin costs of \$1 and \$4/lb for the latter polymers (12). Also noted is the high-use temperature capability of TLCP film [~200°C (~392°F)] despite relatively low glass temperature [110–125°C (230–255°F)], attributed to the crystalline uniaxial morphology. TLCPs with glass temperatures in the 200°C (392°F) range that may have future application in packaging have recently been disclosed (13).

A TLCP drawback has been the lack of availability of balanced film in the marketplace, but this may change very soon. Superex has recently announced a number of consortium activities targeted for the marketplace, including development of a copper-clad balanced film for electronics multilayer circuit boards and multichip modules (14), and coextruded TLCP and PET film and tubing for electronics, medical, and food applications requiring high temperature [205°C (400°F)] resistance (15). The latter features a tie layer probably of a PET/TLCP blend to improve heat-sealing and formability. A 1-million-pound production level is expected by 1997, and cost will be competitive with EVOH and PVDC. A possible application is a shelf-stable, retortable food tray. Superex has also

Table 1. Properties of Balanced Films

	Vectra TLCP	PET(ICI) MD/TD	Kaladex PEN	Kapton PI	Upilex S PI
Tensile strength, MPa	345	165/228	220	90	413
Tensile modulus, GPa	8.3	4.1/4.8	6.0	2.6	8.9
T_g/T_m,°C	110–125/280	72/269	120/262	385/none	>500/none
Upper use temperature, °C	180–200	130	150	200	260
CTE, ppm/C	−5 to +9	23/18	20	22	2.6
Moisture absorption, %	0.02	<0.1		2.1	1.0
Solvent resistance	Excellent	Good	Good	Good	Excellent
Relative permeation, O_2/H_2O	1/1	100/100	25/30		
Dielectric constant at 1 kHz	3.3	3.2		3.4	3.5
Dissipation factor at 1 kHz	0.003	0.005		.002	0.0013
Resin/film price, \$/lb	12–22/44	1/2–6	4/10–19	–/55	–/60

Source: Courtesy of The PST Group.

announced consortium development of a PET/TLCP alloy with twice the oxygen- and moisture-barrier resistance of PET at 10% TLCP loading (16). Flexible-to-rigid monolayer or coextruded packaging that is both recyclable and reusable should be feasible. Projected price is $3/lb with up to 50% package downgauging possible. A key to the PET alloy technology is the preservation of the TLCP barrier phase using special compatibilizers to control ester interchange. In addition to film packaging, Superex TLCP biaxial tubing is now being manufactured under license by ACT Medical for endoscopic surgical instruments (17). Advantages include five times higher stiffness vs conventional tubes and 50% cost savings over fiber-reinforced composite tubes. Superior electrical properties and broad sterilization capabilities are also exhibited. Superex' biaxial TLCP tubing was a 1995 Award Winner selected by *R&D Magazine* as one of the top 100 technologically significant products of the year.

Neste Oy (18) has also announced development of coextruded LDPE/TLCP barrier film based on a new low-melting TLCP [210–260°C (410–500°F)] with heat-distortion temperature of 148°C (300°F). The blend includes a compatibilizer to improve heat sealing and processibility. Multilayer coextruded and paper-coated constructions with LDPE have been prepared by conventional processing. Special biaxial processing is not required, apparently because films are less anisotropic because of the higher flexibility of the TLCP (perhaps due to aliphatic linkages). Three- and five-layer constructions with LDPE far exceed oxygen-barrier cost performance of similar PET constructions and are competitive with those of EVOH, with the added advantage of recyclability. The LDPE/TLCP materials are intended as EVOH and aluminum-foil replacement, particularly in aseptic packaging.

SUMMARY

TLCPs are a class of high-performance engineering polymers represented mainly by aromatic copolyesters with exceptional processibility and high thermal, electrical, and barrier properties. About 90% of the present TLCP market of 10–15 million lb consists of injection-molded electronics connectors, but rapid development of biaxial film and tubing technology (Superex) and coextruded films and blends with conventional packaging resins (eg, PET, PEN and LDPE) should expand opportunities near-term to applications in electronics, medical, and food packaging. With resin costs expected to drop to the $3–$5/lb range over the next several years because of reduced monomer costs, rapid 20–25% annual growth is forecast for the remaining 1990s to a 40–50-million-lb market by the year 2000. It is expected that food and medical packaging will constitute a significant share of this market driven by improved thermal stability, longer shelf life, and ease of recycling afforded by TLCPs.

BIBLIOGRAPHY

1. W. J. Jackson, Jr., *Br. Polym. J.* 154–162 (Dec. 1980).
2. D. J. Williams, "Applications and Markets for Thermotropic Liquid Crystalline Polyesters (TLCPs)," paper presented at Conference on High Temperature Polymers, Executive Conference Management, Clearwater Beach, FL, Feb. 20–21, 1995; in *Applications of High Temperature Polymers,* R. R. Luise, ed., CRC Press, Boca Raton, FL, 1996.
3. *Plast. Technol.,* 35 (Feb. 1991).
4. R. R. Luise, "Advances in Heat-Strengthened Thermotropic Liquid Crystalline Polymer Materials" in *Polymeric Materials Encyclopedia,* J. C. Salamone, ed., CRC Press, Boca Raton, FL, 1996.
5. E. Chiellini and G. Gallo, "Chiral Thermotropic Liquid Crystal Polymers" in *Recent Advances in Liquid Crystalline Polymers,* L. Chapoy, ed., Elsevier, London, 1985.
6. EP Pat. Appl. 346 926 (Dec. 20, 1989), T. Ishii and M. Sato (to Kuraray).
7. U.S. Pat. 4,698,397 (Oct. 6, 1987) and U.S. Pat. 4,891,418 (Jan. 2, 1990) T. Toya et al. (to Nippon Oil).
8. J. S. Chiou and D. R. Paul, *J. Polym. Sci., B, Polym. Phys.* **25,** 1699 (1987).
9. D. H. Weinkauf and D. R. Paul, *J. Polym. Sci., B, Polym. Phys.* **30,** 817 (1992).
10. EP Pat. Appl. 466 085 (Jan. 15, 1992), T. Kashimura and M. Matsumoto (to Kuraray).
11. R. R. Luise, "Liquid Crystalline Condensation Polymers" in *Integration of Fundamental Polymer Science and Technology,* Vol. 5, P. J. Lemstra and L. A. Kleintjens, eds., Elsevier, London, 1990.
12. M. A. Kirsch and D. J. Williams, "Understanding the Thermoplastic Polyester Business," *Chemtech,* 40 (April, 1994).
13. WO Pat. Appl. 94/28069 (Dec. 8, 1994), R. R. Luise (to DuPont).
14. T. Knoll, "Controlled Thermal Expansion Printed Wiring Boards on TLCP Dielectrics," *Technical Conference Proceedings,* IPC, Boston, April 1994.
15. *Plast. Technol.,* 104 (July 1994).
16. *Modern Plast.,* 25 (July 1995).
17. R. W. Lusignea, "Fabrication of LCP Tubing for Medical Applications," paper presented at Specialty Polyesters '95, Schotland Business Research, Brussels, Belgium, June 27–28, 1995.
18. T. Heiskanen and E. Suokas, "A New Low Processing Temperature LCP with High Mechanical Properties and Suitability for Blending," paper presented at Specialty Polyesters '95, Schotland Business Research, Brussels, Belgium, June 27–28, 1995.

ROBERT R. LUISE
Hytem Consultants, Inc.
Boothwyn, Pennsylvania

LITHOGRAPHY. See Printing

LOGISTICAL/DISTRIBUTION PACKAGING

Logistical packaging affects the cost of every logistical activity, and has a significant impact on the productivity of logistical systems. The objective of logistics is to deliver finished inventory, work-in-process (WIP) inventory, and raw materials to the right place, at the right time, and at the lowest total cost.

> It is through the logistical process that materials flow into the vast manufacturing capacity of an industrial nation and products are distributed through marketing channels for consumption. The complexity of logistical systems is awesome. In the United States alone, channels of marketing contain over 2.2 million retailers and more than 350 thousand wholesalers (1).

Transport and storage costs are directly related to the size and density of packages. Handling cost depends on unit-loading techniques. Inventory control depends on the accuracy of manual or automatic identification systems. Customer service depends on the protection afforded to products as well as the cost to unpack and discard packing materials. And the packaging postponement/speculation decision affects the cost of the entire logistical system. An integrated logistics approach to packaging can yield dramatic savings.

Logistical packaging facilitates product flow during manufacturing and distribution. Logistical packaging includes shipping containers for consumer goods (which are almost always also in consumer packages), industrial packaging for factors of production ranging from automobile parts to food ingredients, and institutional packages. There is also a packaging aspect to vehicle loading and unloading, as well as to intermodal containerization. Every factory and/or logistical operation receives and ships logistical packaging; most operations unpack, reconfigure, and repack products, as well as purchasing and disposing of packaging materials.

This article deals with the role of packaging in logistics systems. Readers who seek more general information about logistics are referred to all references on business logistics listed at the end of the article.

Logistical Packaging Forms

Since the early 1900s, common carriers in the United States have tried to "regulate" the packages they transport. The American Association of Railroads and the American Trucking Association publish packaging material "requirements" in their freight classification books (2,3). Most of the requirements were developed with the help of the Fiber Box Association (or its predecessors) and use more corrugated fiberboard material than may be required for adequate performance.

The justification for these requirements has been that carriers are responsible for paying in-transit damage claims. One of the carriers' common law liability defenses is "an act of the shipper," including insufficient packaging, and carriers have tried to maintain that only the packages in the "Classification" are "sufficient."

These requirements have traditionally been a barrier to packaging innovation. The existence of the "Cardboard Rules" has caused many firms to make logistical packaging simply a purchasing responsibility, rather than an area for proactive management (4).

The barrier has fallen, however; since transportation was deregulated in 1980, carriers have been able to exercise much less control over logistical packaging. After all, carriers are responsible only for in-transit damage, and there are nonapproved packages that are at least as protective as those in the Classification.

The logistical channel members who buy, own, and sell products have a much bigger stake in preventing damage and controlling other packaging-related costs. The underlying logic of integrated logistics management is the ability to control such systemwide costs.

Furthermore, transportation deregulation has caused less freight to be subject to common carriers' packaging rules since much less freight pricing is governed by the Classification and more contracts are negotiated. More freight today is shipped by full truckload, never handled by the carriers' workers, and is therefore less subject to abuse in transit. Today, the Classification "Cardboard Rules" apply only to less-than-truckload (LTL) common-carrier freight. And even LTL carriers will accept a packaging exception to the rules if it is certified as passing a series of performance tests (5).

This loosening of "rules" has triggered a renaissance in logistical packaging. Shippers are questioning traditional packaging materials and forms and are experimenting with new, less costly, packaging systems. For example, some appliance shippers have reduced the cost of their packaging by 50%, by replacing corrugated fiberboard with film-based (shrink- or stretch-wrap) packaging systems. Some furniture manufacturers have reduced their package purchase and disposal costs to almost nothing by utilizing "blanket wrap" (moving van) trucking service. These innovations have also been shown to dramatically reduce damage (6).

Most of the new packaging systems are simply variations of traditional packaging forms, using traditional materials in new and innovative ways. They all have two things in common: they are tailored to perform for specific products and logistical systems; and they minimize packaging purchase and solid-waste disposal costs.

The traditional logistical packaging materials and forms are

- Wood pallets, crates, blocking, and bracing
- Corrugated fiberboard boxes, dividers, inserts, and dunnage
- Solid fiberboard slipsheets and boxes
- Plastic and multiwall paper bags and drums
- Intermediate bulk containers, bulk bags
- Steel cans, pails, drums, and straps
- Steel racks and cages
- Fabric (burlap and woven plastic) bags and blankets
- Plastic film shrink wrap and stretch wrap
- High-density plastic boxes, slipsheets, and pallets
- Blanket-wrapping, decks, and restraints
- Plastic strapping
- Plastic foam cushioning and dunnage for fragile or irregular shapes

Descriptions, properties, and specifications of most of these materials can be found elsewhere in this *Encyclopedia*. For more information regarding the traditional packaging forms, the reader is referred to the publications listed in General References list at the end of this article.

Logistical Packaging Performance

Logistical packaging only adds value as it functions in a logistical system to provide protection, utility, and communication. Performance specification and value analysis, which can be used to guide packaging innovation, employ a functional approach.

The amount of protection that a package must provide depends on the characteristics of the product and conditions in the logistical system. The relationship can be conceptualized as

Product characteristics + logistical hazards = package protection

Damage is a symptom of a system out of adjustment, resulting from product problems, handling problems, or packaging problems. Packaging is generally responsible for the product quality that is maintained after the name goes on.

The relevant *product characteristics* are those that can be damaged during distribution. Examples include the propensity of food to deteriorate as a result of temperature, oxygen, or moisture; the tendency for furniture to rub during transit vehicle vibration; and the fragility of products and packages that can break when dropped during material handling operations.

The *hazards* of a logistical system depend on the types of transportation, storage, and handling used. For example, full truckload (TL) transportation generally causes less damage than less-than-truckload (LTL) transportation where packages are handled repeatedly during transloading operations and have many more chances to be dropped or loaded beneath or beside damaging cargo. Railroad shipment often incurs damage as a result of railroad switching and coupling operations.

The hazards of international transportation range from flatbeds to bikecarts. Intermodal containerization for international shipment has reduced impact and moisture damage, compared to "breakbulk" shipping. But even when products are shipped containerized, they will probably be handled in a breakbulk operation inland overseas and require adequate packaging performance.

In general, moisture protection and protection from oxidation are obtained by using barrier materials—which keep water and water vapor, oxygen and other destructive contaminants away from the product, and conserve flavor components in food—or by enclosing desiccant or oxygen-absorbing packets. Shelf life can be lengthened by the use of barrier materials and/or controlling the atmosphere inside food packages. Temperature protection requires insulated containers.

Protection against impacts can be obtained through the use of cushions for shock-sensitive products. Protection to prevent impacts from bursting a package wall requires improving the walls' tensile strength and puncture resistance. To prevent products from shifting during railroad switching and coupling impacts or as an intermodal container is placed onto or removed from a flatbed railcar, trailer chassis, or ship, loads are blocked and braced to the boxcar or container floor (7,8).

A basic rule for improving protection against impacts, vibration, and compression is to always pack products tightly so there is no headspace, no room to rattle and less reliance on box walls for stacking strength. Vibration or abrasion protection may also require changing package surface or spring-mass characteristics. To improve stacking, compression strength may need to be added to the package walls.

The relative protectiveness of alternative packaging systems can be evaluated and compared in laboratory and field tests. Standardized tests are available (American Society for Testing and Materials and International Safe Transit Association), and they can be used to develop more specialized tests to evaluate specific properties. The best tests target the package's specific damage characteristics (eg, bags' ends burst in side drops) and provide a measure of the difference between alternative packages' performance (how much energy is required to burst), rather than simple pass/fail tests.

The more liable the product is to damage and the more hazardous the logistical system's operations, or the higher the cost of damage, the more packaging protection is required. It is important to note, however, that the amount of protection is not directly related to the cost of packaging. In most cases it is possible to improve protection and reduce packaging cost at the same time by simply choosing more appropriate materials and methods. For example, the U.S. Food for Peace program has been able to reduce the number of paper piles in multiwall bags and improve protection by simply changing the bag sealing method (9).

Accurate *measurement of damage* can provide the most valuable information for directing attention to problems. But packaging decisions regarding damage prevention are often made without sufficient information. Typical reports of damage problems are unspecific as to the amount, types, and causes. It is difficult to quantify costs for damage, because they are scattered among firms in a logistical system (10). Since carriers are responsible for the damage they cause, documented in-transit damage claims are the first place to look for damage trends (11,12). Damage that is caused by wholesalers and retailers is more elusive and their cooperation is required to report and evaluate losses.

In order to reduce waste, improve customer service and control packaging costs, some firms have developed systems to track and manage product quality during distribution (13). Such logistics quality-control systems resemble systems for managing product quality during manufacturing and include the following steps (14):

- Setting standards for critical defects
- Appraising conformance to standards, including
- Monitoring and collecting data (eg, electronic data interchange)
- Reporting and evaluating data (Pareto analysis)
- Corrective action (modify the package, product, or handling

Once a firm knows about its logistical damage, it can target initiatives where they will do the most good. For example, some furniture companies target specific packages and products for improvement when damage quantity and/or costs are higher than normal (15). Initiatives may include changing the product, changing the logistical hazards, or changing the package. Specific information about damage is particularly useful when developing package tests and performance levels for alternative packages. For example, some food manufacturers have learned that a most common source of damage is box-cutter damage. In response, they have developed easy-open display-ready packages (16).

In many cases, it costs much less to reduce the hazards than to "improve" the packaging. Examples include

- The use of refrigerated transportation and storage to protect fresh produce
- The use of storage racks in warehouses to prevent compression damage

- Good sanitation practices during distribution to reduce the need for packaging to prevent vermin infestation

In other cases, the lowest-cost solution is to design products to better survive shipment hazards (17). Product design changes range from reducing fruit bruising through genetic modification, to improving the impact resistance of an electronic product's circuit-board fasteners. An impact-resistant product requires little cushioning, which minimizes cube and trash, and is more reliable in use than a fragile product.

Packaging must also protect people in the logistical system from injury and accidents. Workers' ergonomics and safety problems can be solved through better packaging, especially since there are so many possibilities for injury while working with packages. Routine package handling can be linked to chronic stress injuries, and material-handling accidents can kill people.

Routine manual handling of packages has always been taken for granted, but it is traditionally the source of many back injuries. OSHA guidelines answer the question "How heavy should manually handled packages be?" with an answer that depends on how far and how often the lifting is done. For example, for "continuous high-frequency lifting, variable tasks," such as manual receiving operations, weights below 17 lb are acceptable and represent nominal risk, weights between 17 and 50 lb are acceptable only with administrative or engineering controls, and packages weighing over 50 lb are unacceptable (18).

Accidents happen. Personal-injury lawsuits involving packaging materials and methods are increasing. The liability judgment generally depends on whether the accident could have been foreseen and whether adequate packaging and training procedures were followed. For example, when a longshoreperson is injured by a bag that slips from a stack, the fabric is accused of not complying to standards, the unit-loading technique is examined (maybe it should have been stretch-wrapped), and the port practices are questioned. It is especially important for firms to evaluate packages that are involved in previous accidents, in order to prevent future occurrences.

The packaging function of utility relates to how packaging affects the productivity and efficiency of logistical operations. All logistical operations are affected by packaging utility—from truckloading and warehouse-picking productivity, to transportation and storage cube utilization, to customer productivity and packaging waste reduction.

Productivity is the ratio of real output to real input:

$$\text{Productivity} = \frac{\text{number of packages output}}{\text{logistics input}}$$

Logistical productivity is the ratio of the output of a logistical activity (like loading a truck) to the input (like labor and forklift time required). Most logistical productivity studies center around making the input, particularly labor, work harder.

But packaging unitization and size reduction initiatives easily increase the output of logistical activities. Almost all logistical activity outputs are described in terms of number of packages. Some examples include number of cartons loaded per hour into a trailer, number of packages picked per hour at a distribution center, and number of packages that fit into a cubic foot (ft^3; "cube utilization") of vehicle or warehouse space (19). Testing alternative packages for their logistical efficiency is a simple matter of productivity measurement.

For example, palletization and/or unitization improves the productivity of most handling operations. Warehouse-order-picking productivity can be improved by making packages easy to find and recognize, by packing items in order picking quantities (eg, dozens), and by making it easy to repack the heterogeneous order.

Cube utilization can be improved by reducing package size. Package size can be reduced by concentrating products (eg, orange juice and fabric softener), or by eliminating space inside packages by shipping items unassembled, nested, and with minimal dunnage. In most cases, the amount of dunnage materials (eg, expanded polystyrene loose-fill) can be minimized simply by reducing box size. Experts believe that improving cube utilization is packaging's greatest opportunity, and predict that, in general, packaging cube can be reduced by 50%, doubling transportation efficiency (20).

Cube minimization is most important for lightweight products (like assembled furniture) that "cube out" a transport vehicle far below its weight limit. Trailer sizes vary; most in the United States are 48 ft long × 102 in × 102 in.; and cargo weight limits are generally about 40,000 lb. On the other hand, heavy products (eg, liquid in glass bottles) "weigh out" a transport vehicle before it is filled. Weight can be reduced by changing the product or the package. For example, substituting plastic bottles for glass significantly increases the number of bottles transported in a trailer.

Packaging utility also depends on the compatibility of packages with pallets and other material-handling equipment used in a logistical system. Pallet-load efficiency can be improved by designing shipping containers to better fit the footprint: most grocery pallets in the United States are 40 ft × 48 ft (21); most in Europe are 1200 mm × 800 mm. In some logistical systems, especially internationally, pallets are not used, and packages may need the ability to be efficiently manually handled (or by the longshoreperson's hook) repeatedly.

Packaging can also add utility for logistical customers. The cost of unpacking is seldom considered by the firm that packs products. But easy-opening features improve retailer's direct product profitability by reducing customers' costs of opening and displaying or installing products.

Packaging disposal is also borne by logistical customers. Besides the environmental impact, logistical packaging trash disposal costs money and can severely reduce a customer's productivity, particularly in nations such as Canada and Germany, which have enacted legislation to reduce packaging waste.

Packaging solutions to disposal problems include reducing, reusing, and recycling. Reduction of packaging also saves on packaging purchase costs. Packaging reuse generally adds some costs for sorting and return transportation. Recycling is growing in more widespread use and gaining popularity, thus reducing costs and increasing markets.

Recycling is a good disposal method for most logistical packaging waste, since logistical packaging waste naturally collects in large heterogeneous piles. Manufacturers, warehouses, and retailers discard large amounts of pallets, corrugated fiberboard, polyethylene film, plastic foam, and strapping. Recyclers appreciate such concentrated and relatively clean sources (compared to sorting and cleaning curbside and food-service wastes). Likewise, purchasing packages made from recycled material encourages the growth of a recycled products market and infrastructure (22). Recycling is sometimes called "reverse logistics" (23).

One of the most popular trends is the use of more *returnable packaging* for products such as assembly parts, food ingredients, and snack chips, from retail warehouse to store totes and intraplant shipments. Most reusable packages are steel or plastic; some firms reuse corrugated fiberboard boxes and pallet boxes. There are also flexible reusable systems: blanket-wrapping and intermediate bulk containers such as bulk bags.

The growing returnable-packaging applications all have one thing in common: a short vertical marketing system. A vertical marketing system (as contrasted to a "free-flow" marketing system) is one in which the primary participants are either corporate ownership, contracts, or administration under the control of one firm (24).

A vertical marketing system is important because of the need to control the movement of returnable containers and the need to share the investment benefits. All partners in a returnable system must cooperate to maximize container use, and an explicit relationship is required for coordination and control. Otherwise, containers are easily lost or misplaced. Alternatively, deposit sytems may be necessary in free-flow marketing systems (like groceries), where the channel members are linked only by transactions. Deposit systems have been used for beverage bottles, pallets, and steel drums.

The shipment cycle should be short, in terms of time and space, in order to minimize the investment in the container "fleet" and to minimize return transportation costs. The size of the investment depends on the number of days in the cycle.

Deciding to invest in a returnable-packaging system is a very different task from purchasing "expendable" containers. The decision involves considering explicit relevant costs—purchase cost, including number of packages in the cycle and transport cost—versus the purchase and disposal cost of expendables. Intangible benefits such as improved factory housekeeping, improved ergonomics, and decreased damage are often considered. Generally, there are a number of unexpected costs, including sorting, tracking, and cleaning. Most financial analysis of returnable systems has been limited to "payback period" justification rather than net present value calculations, which would demonstrate the strategic and profit potential of a returnable-packaging system investment (25).

The function of communication is becoming more important for logistical packaging, as logistical management information systems become more comprehensive. For all practical purposes, the package symbolizes the product throughout logistical channels. Correct identification of stockkeeping units (including name, brand, size, color), counting, special shipping instructions (eg, "Hazardous") and address are critical to quality logistical information management. International shipments require the language of origin, destination, and intermediate stops, as well as international markings for handling instructions.

For example, automatic identification (eg, bar coding) highlights the communication function, and can interface with information systems throughout a logistical system if symbology is standardized. But even manually readable packaging must be clearly legible to interface with logistical management information systems; workers must be able to quickly recognize a package from its label. Inventory control, shipping and receiving, order picking, sorting, tracking—almost every logistical activity entails reading the package and recording or changing its status in an information system. Automatic identification technologies such as bar coding and radiofrequency identification enable a systems approach to managing logistical information where every input is standardized and errors are reduced.

The functions of packaging—protection, utility, and communication—are the basis for packaging performance specifications. Performance specifications outline what the package must do (eg, be able to survive an impact). The specification of performance guides packaging changes that add value and reduce costs.

Many firms and government agencies have adopted packaging performance specifications for logistical packaging. For example, the U.S. Department of Transportation (DOT) has replaced hazardous-material packaging specifications with "Performance Standards" that do not specify material, but do specify tests (eg, impact, permeability, or compatibility tests) for specific product/package/hazard conditions (26).

Performance specifications encourage innovation, and innovation results in lower packaging costs. Material specifications are necessary, of course, for routine packaging purchases, but adding a performance specification offers two benefits: (*1*) a firm that examines its current and expected levels of packaging performance will always uncover opportunities to cut packaging-related costs; and (*2*) suppliers who are invited to compete on a performance basis will always suggest lower-cost packaging solutions.

Typical Logistical Packaging Problems

There are three types of logistical packaging problems: engineering problems, cost problems, and logistical problems with packaging solutions. Generally, logistical packaging problems require on-site investigations—the best way to find out more is to go and look.

Packaging professionals use *engineering* principles of material science to develop new packages and solve protection and productivity problems. The most common material properties that can be scientifically determined are cushioning (for fragile products), impact and vibration resistance of filled packages (27), shelf life and permeability for oxygen- or moisture-sensitive products (28), and compression strength for hard-to-stack products (29). Packaging professionals also use some industrial engineering principles to configure packing operations and choose packaging machinery.

Management of logistical packaging, however, involves more than "package engineering." Most packaging problem solving is not associated with developing new packages for new products. In fact, most "new" logistical packages are simply variations of packages used for similar products and require very little "engineering."

The second principal activity of a packaging professional is to *reduce the cost* of purchasing packaging materials and the cost of packing operations. The most common areas for cost reduction are reducing package weight or size, material substitution, size reduction, automation of manual tasks, and standardization. Innovative incremental packaging solutions, that reduce cost and improve performance, abound in the packaging supply industry. Manufacturers of packaging materials and equipment are sources of a great amount of free, if potentially biased, consulting advice, so it is wise to consult more than one. Other good references include trade journals such as *Packaging, Materials Handling Engineering, Modern Materials Handling,* and *Transportation and Distribution.*

The third type of packaging problems are *logistical problems* and *opportunities with packaging solutions.* Packaging can reduce the cost of every logistical activity: transport, storage, handling, inventory control, and customer service. It can reduce the cost of damage, safety, and packaging disposal. Integrated management of packaging and logistics is required, if a firm is to realize such opportunities (30–32).

When packaging and logistics are integrated, it is reasonable to consider where and when packaging operations fit into a logistical channel. Packaging postponement can dramatically reduce the cost of an entire logistical system.

Postponement and *speculation* are logistical concepts that question where and when to add value (time, place, and form utility) in distribution channels in order to reduce costs. The principle of speculation says that manufacturing and shipping should occur at the earliest possible time in the marketing flow in order to achieve economies of scale (33). The opposite principle of time and form postponement reduces risk by requiring that changes in form and identity and inventory location occur at the latest possible point in the marketing flow, since every differentiation that makes a product more suitable for a specified segment of the market makes it less suitable for other segments (34). The traditional inventory tradeoff is to reduce risk by postponement (just-in-time logistics) or to reduce cost with economies of scale (forward buying).

Traditionally, also, the packaging point for bulk products has been postponed or speculated, on the basis of product and market conditions (35). For example, produce is canned in the fall and stored in "bright" cans; labeling is postponed until food-marketing companies buy and brand it throughout the year. Labeling postponement minimizes the risk of inventory misforecasts and economizes on canning production during a busy season. On the other hand, processing and packaging of frozen fruit concentrates, such as orange or grapefruit juice, is an example of speculation, resulting in low costs because weight is eliminated before shipping and homogenous waste at the orchard is processed for animal food.

Postponement and speculation also apply to assembly and packaging of more sophisticated products in a global market with global sources. When products are similar, but differentiated for local markets, assembly and packaging postponement offers opportunities for customizing packaging for the local markets while minimizing long-distance transport costs and the amount of differentiated inventory in the pipeline.

For example, one of the largest cost-savings initiatives in packaging history (tens of millions of dollars per year) is a case of international packaging postponement. To reduce high transport and inventory management costs, a major electronics manufacturer has postponed the packaging for products shipped overseas. The products are shipped in "bulk packs," palletized and stretch-wrapped, with minimal interleaving between layers. Once there, products are packaged to order. Transport costs have been cut in half, since differentiation and cushioning are postponed until they are required (36).

Innovative Logistical Packaging

This article has approached the field of logistical packaging from a nontraditional perspective. The reason for this approach is to encourage innovative packaging solutions to logistical problems. Most traditional packaging forms date from the early 1900s and are the most costly packaging systems available. Innovation represents a dramatic potential for savings.

Packaging innovation is a process of problem identification, finding and testing potential solutions, deciding which action to take, and following through to implementation. It requires proactive management by a project champion—usually a packaging professional. It also requires a systems and team approach, since packaging affects so many other functional areas such as marketing, operations, plant and product engineering, logistics, accounting, and finance.

Identification of packaging-related problems is the first step. Problems of cost, protection, utility, and communication may be identified by customers, salespeople, logistical managers, or packaging professionals. The search for packaging solutions may be formal or informal, but the innovation process requires a potential solution before it can proceed. Sometimes problems languish for years before someone thinks of a good solution. Ideas for potential solutions can come from suppliers or consultants, or from the advice of colleagues. Alternative packages are judged, subjectively or in lab tests, for their protection, utility, and communication performance, as well as systemwide cost implications. Next, the "best" solution packages are tested in the field (test shipments) and for market acceptance. The early rollout is a period of fine-tuning. Problems with the new package are discovered and corrected. Sustained implementation returns to a new level of business as usual—generally with substantial cost savings.

Conclusion

This chapter has shown that, beyond its purchase cost, packaging affects the cost of every logistical activity. Package size affects transportation cost; unitization methods affect order picking and handling costs; identification techniques affect inventory control; and customer service depends on damage control and the cost to unpack and discard packaging. Firms that manage packaging from an integrated logistical perspective, rather than from a traditional purchasing perspective, find many opportunities for improving their profitability.

BIBLIOGRAPHY

1. D. J. Bowersox, D. J. Closs, and O. K. Helferich, *Logistical Management,* Macmillan, New York, 1986.

2. National Classification Board of the American Trucking Association, *National Motor Freight Classification,* National Classification Board, Washington, DC, revised annually.
3. Uniform Classification Committee of the Western Railroad Association, *Uniform Freight Classification,* Uniform Classification Committee, Chicago, revised annually.
4. M. A. McGinnis and C. J. Hollon, Packaging Organization, "Objectives, and Interactions," *J. Business Logistics* **1,** 45–62 (1978).
5. International Safe Transit Association, *Pre-Shipment Test Methods,* ISTA, Chicago, revised periodically.
6. D. Twede, "The Process of Logistical Packaging Innovation," *J. Business Logistics* **13**(1), 69–94 (1992).
7. Association of American Railroads, *Loading Methods for Closed Cars,* AAR, Chicago, various years.
8. J. Agnew and J. Huntley, *Container Stowage: A Practical Approach,* Container Publications, Dover, England, 1972.
9. D. Twede, B. Harte, J. W. Goff, and S. P. Miteff, "Breaking Bags: A Performance Specification Philosophy Based on Damage," *J. Packag. Technol.* **4,** 17–21 (Nov./Dec. 1990).
10. J. L. Cavinato, *Analysis of Loss and Damage in a Procurement Distribution System Using a Shrinkage Approach,* Ph.D. dissertation, Pennsylvania State University, 1975.
11. W. Augello, *Freight Claims in Plain English* in Transportation Claims and Prevention Council, Huntingon, NY, 1982.
12. S. Smith, A. H. McKinlay, and W. J. Augello, *Freight Claim Prevention in Plain English"* in Transportation Claims and Prevention Council, Huntington, NY, 1989.
13. D. Twede, "Distribution Damage Measurement, Analysis and Correction," *Packag. Technol. Sci.* **4,** 305–310 (1991).
14. R. A. Novak, "Quality and Control in Logistics: A Process Model," *Internatl. J. Phys. Distribution Mater. Manage.,* 2–44 (1989).
15. T. J. Thomas, "Quality Control of Products During Distribution," in *Logistical Packaging Innovation Proceedings,* Council of Logistics Management, Oakbrook, IL, 1991.
16. W. Thompson, "Improving Customer Service "Packages: Direct Product Profitability" in *Logistical Packaging Innovation Proceedings,* Council of Logistics Management, Oakbrook, IL, 1991.
17. E. Maezawa, "Product Modification to Reduce Distribution Costs," *SPHE J.,* 21–22 (1987).
18. American Industrial Hygiene Association, *Work Practices Guide for Manual Lifting,* American Industrial Hygiene Association, Akron, OH, 1987.
19. A. T. Kearney, Inc., *Measuring and Improving Productivity in Physical Distribution,* Council of Logistics Management, Oakbrook, IL, 1984.
20. J. Goff, "Packaging-Distribution Relationships: A Look to the Future," *Logistical Packaging Innovation Proceedings,* Council of Logistics Management, Oakbrook, IL, 1991.
21. Food Marketing Institute, *Pallet Patterns and Case Sizes for Dry Grocery Shipping Containers and Voluntary Industry Guidelines for Dry Grocery Shipping Containers,* FMI, Washington, DC, 1988.
22. S. M. Selke, *Packaging and the Environment: Alternatives, Trends and Solutions,* Technomic, Lancaster, PA, 1990.
23. R. Kopicki, M. J. Berg, L. Legg, V. Da Sappa, and C. Maggioni, *Reuse and Recycling—Reverse Logistics Opportunities,* Council of Logistics Management, Oak Brook, IL 1993.
24. L. P. Bucklin, "The Classification of Channel Structures," in *Vertical Marketing Systems,* Louis P. Bucklin, ed., Scott, Foresman, Glenview, IL, 1970, pp. 16–31.
25. W. V. Uxa, *Returnable/Reusable Containers in the Automotive Industry and the Related Capital Budgeting Investment Decision,* M.S. thesis, Michigan State University, 1994.
26. U.S. Government Printing Office, *Code of Federal Regulations, Title 49,* Parts 171–179, updated annually.
27. R. K. Brandenburg, and J. J. Lee, *Fundamentals of Packaging Dynamics,* MTS Systems, Minneapolis, MN, 1985.
28. W. A. Jenkins and J. P. Harrington, *Packaging Foods with Plastics,* Technomic, Lancaster, PA, 1991.
29. G. G. Maltenfort, *Performance and Evaluation of Shipping Containers,* Jelmar, Plainview, NY, 1989.
30. D. B. Carmody, *Packaging Role in Physical Damage,* American Management Association, New York, 1966.
31. C. W. Ebeling, *Integrated Packaging Systems for Transportation and Distribution,* Marcel Dekker, New York, 1990.
32. J. L. Heskett, N. A. Glaskowsky, Jr., and R. M. Ivie, "Packaging," in *Business Logistics,* Ronald Press, New York, 1973, pp. 472–602.
33. L. P. Bucklin, "Postponement, Speculation, and the Structure of Distribution Channels," *J. Market. Res.* (Feb., V.2 pp. 26–32, 1965).
34. W. Alderson, *Marketing Behavior and Executive Action,* Richard D. Irwin, Homewood, IL, 1950.
35. W. Zinn, "Planning Physical Distribution with the Principle of Postponement," *J. Business Logistics* **9**(2), 117–135 (1988).
36. K. A. Howard, "Packaging Postponement Lowers Logistical Costs" in *Logistical Packaging Innovation Proceedings,* Council of Logistics Management, Oakbrook, IL, 1991.

General References

American Society for Testing and Materials, *Annual Books of Standards,* ASTM, Philadelphia, revised annually.

R. H. Ballou, *Business Logistics Management,* Prentice-Hall, Englewood Cliffs, NJ, 1992.

D. J. Bowersox, D. J. Closs, and O. K. Helferich, *Logistical Management,* Collier Macmillan, New York, 1986.

D. J. Bowersox, P. Daugherty, C. Droge, R. Germain, and D. Rogers, *Logistical Excellence: It's Not Business as Usual,* Digital Press, Burlington, MA, 1992.

M. Christopher, *Logistics and Supply Chain Management,* Irwin, Burr Ridge, IL, 1994.

W. F. Friedman and J. J. Kipnees, *Distribution Packaging,* Robert E. Krieger, Huntington, NY, 1977.

N. A. Glaskowsky, D. R. Hudson, and R. M. Ivie, *Business Logistics,* Harcourt, Brace Jovanovich, Fort Worth, TX, 1992.

J. F. Hanlon, *Handbook of Package Engineering,* McGraw-Hill, New York, 1971.

International Journal of Physical Distribution and Management, MCB University Press, Bradford, West Yorkshire, UK, published monthly.

Journal of Business Logistics, Council of Logistics Management, Oak Brook, IL, published biannually.

B. LaLonde, ed., *Bibliography on Logistics and Physical Distribution Management,* Council of Logistics Management, Oak Brook, IL, updated annually.

D. Lambert and J. R. Stock, *Strategic Logistics Management,* Irwin, Homewood, IL, 1993.

E. A. Leonard, *Packaging Economics,* Books for Industry, New York, 1980.

E. A. Leonard, *Packaging Specifications, Purchasing and Quality Control,* Marcel Dekker, New York, 1987.

G. G. Maltenfort, ed., *Corrugated Shipping Containers: An Engineering Approach,* Jelmar, Plainview, NY, 1990.

Material Handling Engineering, Penton, Cleveland, OH, *Material Handling Engineering Reference Guide,* Penton, Cleveland, OH, revised annually.

M. A. McGinnis and C. J. Hollon, "Packaging: Organization, Objectives, and Interactions," *J. Business Logistics* **1,** 45–62 (1978).

Modern Materials Handling, published monthly. Modern Materials Handling Casebook Reference Issue, Cahners, Denver, revised annually.

National Safe Transit Association, *Procedures for Pre-Shipment Testing,* NSTA, Chicago, 1990.

Packaging Digest magazine, Cahners, Chicago, published monthly.

F. A. Paine, ed., *The Packaging User's Handbook,* Blackie, Glasgow and in AVI imprint of Van Nostrand Reinhold, New York, 1991.

C. A. Plaskett, "Principles of Box and Crate Construction" in *Technical Bulletin No. 171,* U.S. Department of Agriculture, Washington, DC, 1930.

J. F. Robeson and W. C. Copacino, eds. *The Logistics Handbook,* The Free Press, New York, 1994.

W. Stern, "Safer Packaging through Structural Improvements," *IoPP Tech. J.* 19–24 (Winter 1989).

Transportation and Distribution, Penton, Cleveland, OH, published monthly.

DIANA TWEDE
Michigan State University
School of Packaging
East Lansing, Michigan

M

MACHINERY, CARTONING, END-LOAD

The cartoning machine, located near the end of the packaging line, produces a package that in most cases brings the product to the consumer. For this reason the package must be attractive and adequately protect the product. Cartoning machines handle an almost infinite variety of products; solid objects like bottles, jars, and tubes, and also breakfast cereals, rice, pasta products, and similar free-flowing items which may be filled into the carton by scales or volumetric means. This article is devoted primarily to horizontal and vertical cartoners handling solid products.

There are two basic types of cartoning machines: semiautomatic and fully automatic. By definition, a semiautomatic cartoner is one with which the operator manually places the product into the carton; a fully automatic cartoner is a machine that automatically loads the product into the carton even though an operator may place the product in a bucket or flight of the intake conveyor (1). Both types are available in horizontal or vertical modes, that is, the carton is carried through the machine lying horizontally or standing upright in the conveyor carrying the carton.

Semiautomatic Vertical Cartoners

In most cases the semiautomatic verticle cartoner is arranged as shown in Figure 1. The tubular carton is fed from a horizontal magazine, expanded and transferred into the carton conveyor. The bottom of the carton is closed by tucking or gluing and then conveyed past one or more operators who manually place the product in the carton. The machine then closes the top of the carton by tucking or gluing.

The semiautomatic vertical cartoner is usually used for products with low production volume where changeover to another size is frequent. It is also well suited for packages containing a variety of different items. A semiautomatic vertical cartoner usually has a wider size range than a fully automatic machine. Common operating speed is up to 120 packages per minute, although many machines of this type operate at much slower rates because of the need to place the product manually into the moving carton.

The semiautomatic cartoner can be equipped with a number of attachments such as a leaflet feed mechanism, code impressor or printing mechanism for lot numbers, expiration dates, or prices. On-machine printing is usually located on the end panel or tuck flap of the carton rather than on any of the body panels because adequate backup pressure is required in order to obtain a good, legible impression (see Printing).

Many packages require a leaflet or coupon to be placed inside the carton with the product. This may be accomplished by an operator stationed along the carton conveyor who can place a prefolded leaflet into the carton next to the product. For marketing reasons this is not always acceptable because the product can be removed by the consumer without removing the leaflet. To overcome this, the prefolded leaflet is often folded around the product in a U-shape. If the leaflet is to be readily visible when the carton is opened, the ends of the leaflet must be inserted first. It is much easier to wrap the leaflet around the product and insert the assembly into the carton with the ends of the leaflet trailing; consequently, many cartons are run through semiautomatic vertical cartoners upside down so the cap of the bottle (or top of the bottle) with the leaflet is at the bottom of the carton as it runs through the machine.

An automatic mechanism can be installed to feed a prefolded leaflet from a magazine and partially insert it into the carton. When the product is loaded by the operator, the leaflet is pushed down to the bottom of the carton. A pick-and-place unit may be used to transfer a prefolded leaflet or coupon from a magazine into the carton, but generally such a unit does not orient the coupon in the carton; it merely drops it into the moving carton, usually before the product is inserted.

Semiautomatic Horizontal Cartoners

The semiautomatic horizontal cartoner is similar to the vertical cartoner except that the carton is carried through the machine lying on its back panel. Because the product is inserted into the carton manually, it is classified as semiautomatic. Some models have the carton conveyor arranged in such a way that the carton is inclined downward away from the operator, enabling the product to be loaded without lifting, that is, just slid into the carton by gravity. Some of the slant-type horizontals can be adjusted by a crank so the carton is carried anywhere from the horizontal to 20° to suit the type of product being handled. The semiautomatic horizontal machine, like the vertical, can be equipped with many different attachments, code impressors, printers, and hot melt adhesive systems.

Fully Automatic Horizontal Cartoners

In general, this class of packaging machine is similar to Figure 2. It consists of a product infeed conveyor, carton feed, carton conveyor, loading mechanism, and closing system. Fully automatic machines can operate at speeds from 50 packages per minute to well over 600 packages per minute for certain items, although most are designed for operation in the range of 150–300. This type of cartoner usually has less flexibility in size range than the semiautomatic vertical machines but has the advantage that all operations are automatic; that is, no operators are required to place the product into the carton. Each mechanism is critical to the efficient operation of the machine, but probably the most important is the carton feed.

Carton feeds. There are many types of carton feeds used on cartoning machines, the selection dependent upon the speed of operation required, the size of the carton to be handled, and the style of carton (ie, preglued tubular style or flat blanks). The tubular carton, preglued by the carton manufacturer, is the most common although flat blanks also have advantages.

The basic carton feed for tubular cartons consists of a magazine to hold the supply of unexpanded cartons, a vacuum head, and a transfer system to place the open carton into a

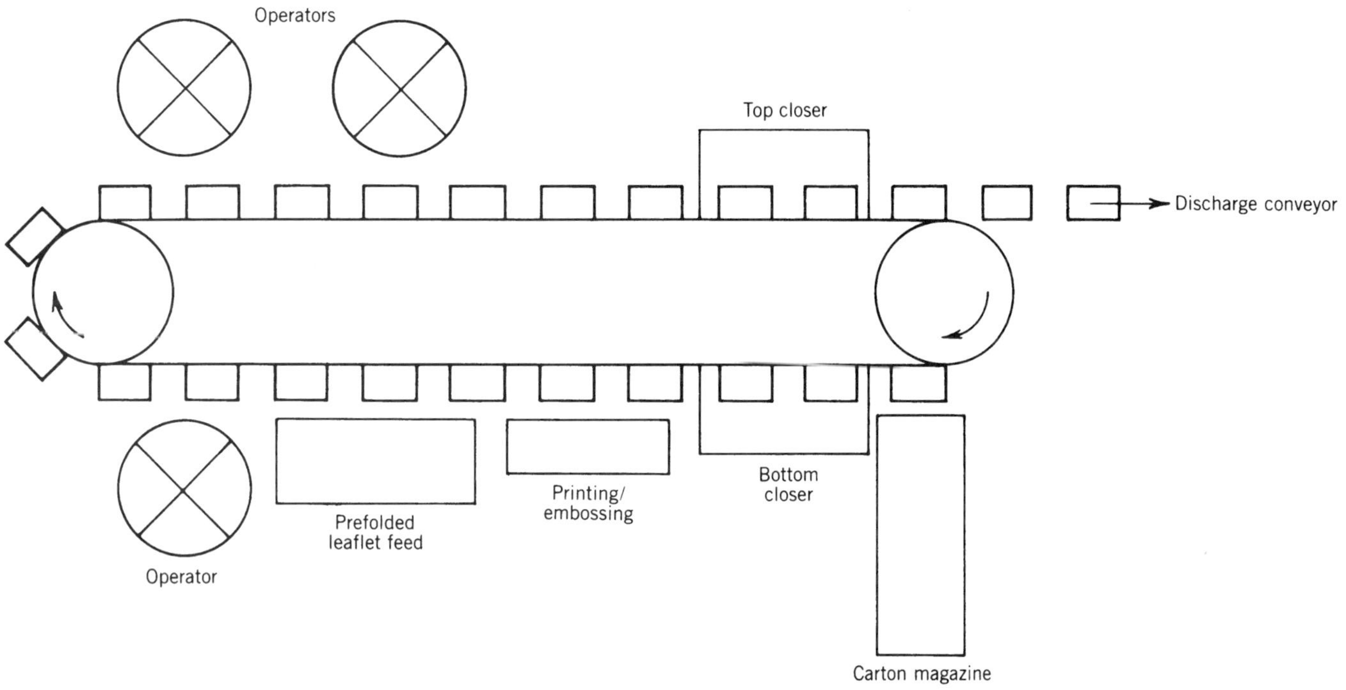

Figure 1. Semiautomatic vertical cartoner.

conveyor. The magazine can be vertical, horizontal, or inclined. In most cases the cartons are retained in the magazine by ledges, which are small projections from the side of the magazine, usually located in the cut-out areas of the carton flaps so the movement of the carton by the vacuum head releases the carton from the magazine. The carton must be pulled free of the stack before expansion can start (2). A simple carton feed is illustrated in Figure 3. The suction head, moving downward, pulls the carton from the ledges, and expansion begins when the carton contacts the beveled expander block. As the vacuum head reaches the bottom of its stroke, it straddles or goes between the carton conveyor chains, and as the vacuum is released, the carton is transferred into the conveyor lugs which are moving around the chain sprocket. Expansion is completed as the lugs level out. The flaps on the loading side of the carton are usually guided outward so they mesh with the mouth of the bucket carrying the product to create a funnel for loading the carton.

Another type of carton feed bows the bottom carton from under the stack in the magazine to provide space for a blade to enter between the bottom carton and the one above. This blade supports the weight of the stack to enable the bottom carton to be pushed by reciprocating fingers into the expansion position. Expander blades are inserted inside the carton, frequently from both ends of the carton, to start the expansion. An operated finger partially expands the carton by pressing down against the side panel while the carton is still held by the blades. Forming of the cartons is completed by the lugs of the carton conveyor as they come around the sprocket.

When a machine is to handle large, flat cartons such as those required for pizza pies, the carton may be transferred from the magazine in a manner similar to that described above, but expansion is performed by top and bottom vacuum heads. The bottom panel may be held in position by short-stroke vacuum heads moving up and down. When firmly held by vacuum, a second pivoted vacuum head may contact the top panel and expand the carton by moving in an arc corresponding to the height of the carton as shown in Figure 4. In some cases the top vacuum head may move almost 180° in an arc to completely overexpand the carton, which is advantageous when handling wide cartons because it breaks the score lines and produces a neat, square carton. This type of feed requires more time to expand the carton so the operating speed is somewhat lower than that for a smaller size.

Rotary motion carton feeds are used extensively for high speed operation. A typical feed is illustrated in Figure 5. The only reciprocating motions are a short-stroke vacuum head and stripper fingers. The carton is bowed by a vacuum cup and pushed downward a very short distance so the flat carton can be nipped by a set of feed rolls which drive the carton down in time with a lugged pair of wheels. These wheels mesh the carton with a second wheel carrying a number of vacuum heads, one for each pitch of the wheel. While the carton is firmly gripped by the vacuum cups an expander plate, controlled by a fixed cam, gradually expands the carton over several positions. This system allows the carton to be overexpanded to take the "fight" out of the score lines. While still on the vacuum wheel, the leading side flap is operated forward, the trailing side flap plowed backward, and the end panels guided open; as the carton meshes with the carton conveyor the flaps spread around the end of the product bucket, creating a natural funnel for entry of the product into the carton.

Air expansion of cartons is an excellent system, particularly for large cartons, although it also has been used successfully on very small cartons. The unexpanded tubular cartons are usually placed in a horizontal or slightly inclined magazine. In some feeds a metering wheel assisted by air flow from the top brings the flat carton to the horizontal position, where

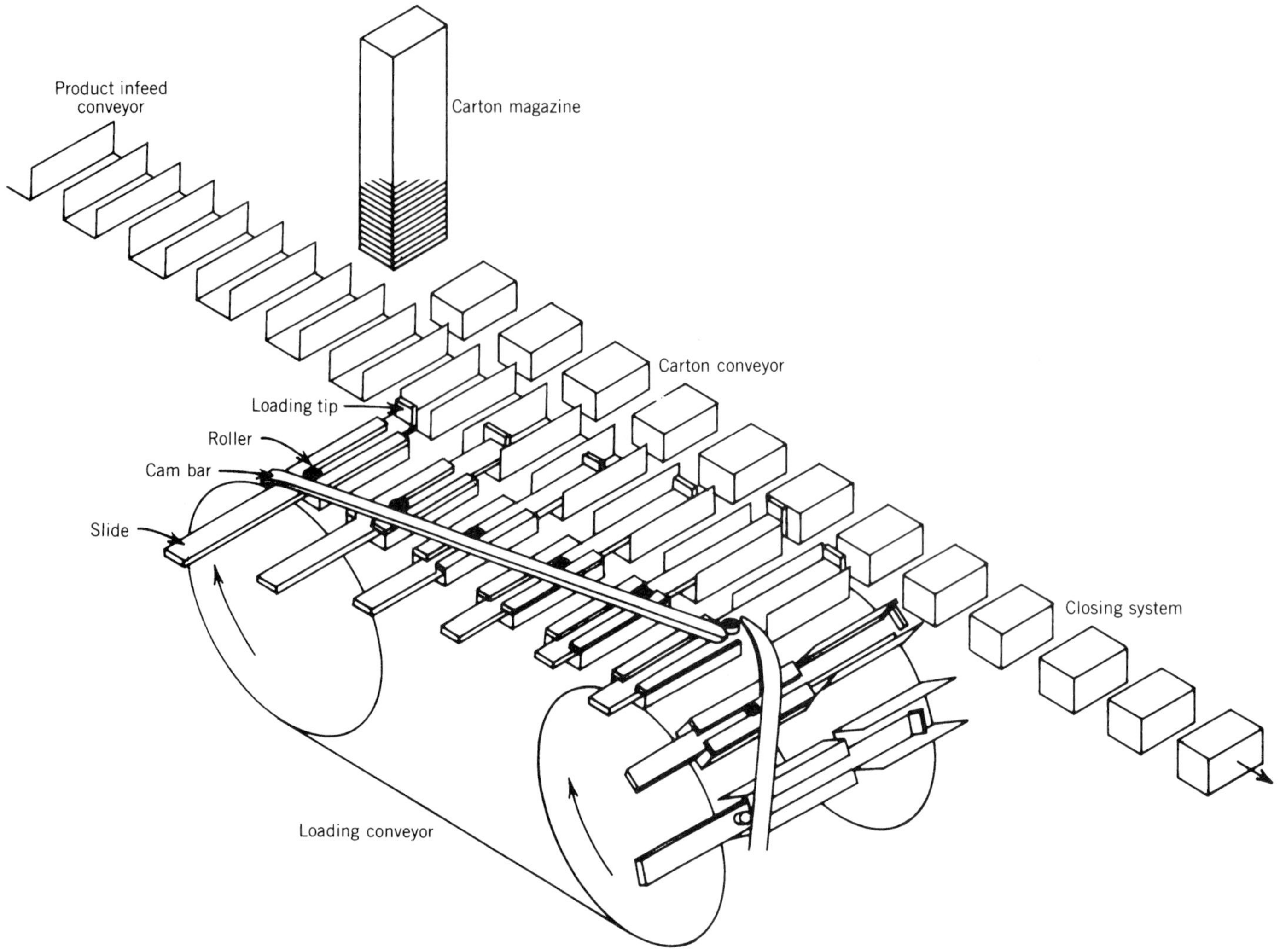

Figure 2. Fully automatic horizontal cartoner.

it is picked up by feed chain lugs which transport it to the opening station. The end flaps are guided slightly apart, and the carton is carried past air manifolds which pass a high volume of air at low pressure into the carton. The volume of air expands the carton completely while still in the transport chain conveyor. This style of carton feed has no reciprocating motions, vacuum pump, vacuum lines, or cups to wear or clog.

Side Seam Gluing

Most cartoning machines in operation handle tubular cartons with the side seam or manufacturer's joint preglued on high speed in-line gluers at the carton manufacturer's plant. The carton must be carefully packed on edge in the corrugated shipper by the manufacturer with each tier in the shipper separated by a piece of carton board. If cartons are packed too tightly, they tend to lose their prebreak or become warped. Preglued cartons tend to lose their prebreak in direct proportion to the time they are stored in the shipping container and develop a "set", making the carton more difficult to open on the cartoning machine.

Many cartoners can be equipped to handle flat, unglued carton blanks with the cartons shipped in stacks on a pallet. The elimination of pregluing by the carton manufacturer and the cost of the corrugated shipper results in direct savings. Indirect savings to the packager result from savings in storage space, usually no additional labor required on the cartoner, and improved cartoner efficiency due to better score breaking (3).

There are two basic systems for on-machine side seam gluing. A separate gluing machine located at the cartoner can feed flat blanks from a magazine, break the scores, apply hot melt adhesive to the side seam, fold and compress the carton joint, and discharge the carton into the magazine of a conventional cartoning machine. This system can handle a substantial size range and has the advantage that it can be added to a cartoning machine at a later date.

The second major system is the wrap around style which feeds a flat carton blank from a magazine, forms it, glues the seam around a three-sided mandrel carrying the product, and transfers the filled and formed carton from the mandrel into a second conveyor where the ends of the carton are closed. Because the carton is formed and glued around the mandrel, the package is extremely square. The product is already inside the carton when the carton is formed, so there is no

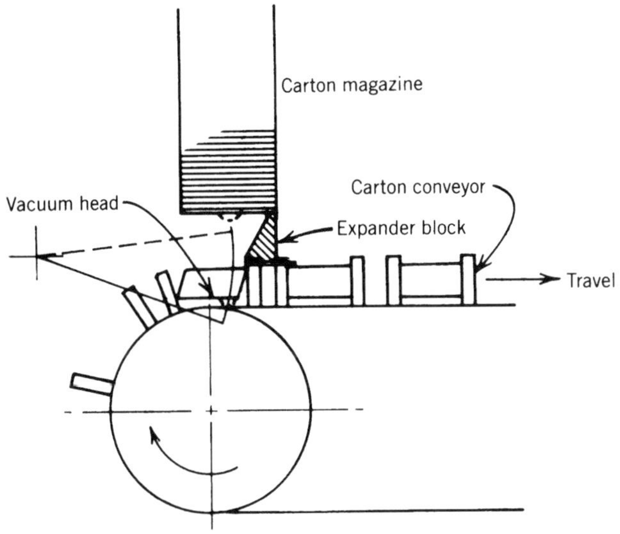

Figure 3. Basic carton feed for tubular cartons.

transfer of the product into the carton as there is on conventional cartoners. This system requires less clearance between the product and the carton, resulting in the greatest box capacity utilization of any cartoning system. The carton blanks require fewer carton converter operations, larger quantities can be shipped on a pallet requiring less storage space in the packager's plant, a slightly smaller carton can be utilized, and quite often a lighter caliper of board can be handled than on conventional cartoners running tubular cartons. Savings of as much as 25% can be obtained in some cases.

Carton Loading

The conventional method of loading a carton on a fully automatic horizontal cartoner is illustrated in Figure 3. The loader consists of a pair of parallel chains running at the same speed as the carton conveyor. A series of slide bearings is fastened to the chains on the same pitch as the carton. On each bearing is a slide with a suitable loading head to gradually insert the product into the carton as a roller on the slide rides against a fixed cam bar. The angle of the cam bar usually does not exceed 38–40°, which provides a smooth transfer of the product from the product bucket into the carton. Loading is accomplished over several positions and therefore at a fraction of the machine operating speed.

Product Infeeds

There are a host of different automatic infeeds available, all designed for a specific type of product, such as bottles, jars, tubes, pouches, stacks of tissues, spaghetti products, pencils, and shotgun shells. Because of space limitations only a few of the most common will be described.

Bottle infeeds. Bottles are usually brought into the cartoner on a platform chain, timed by a star wheel or screw, and transferred into a pocketed wheel and then into the product

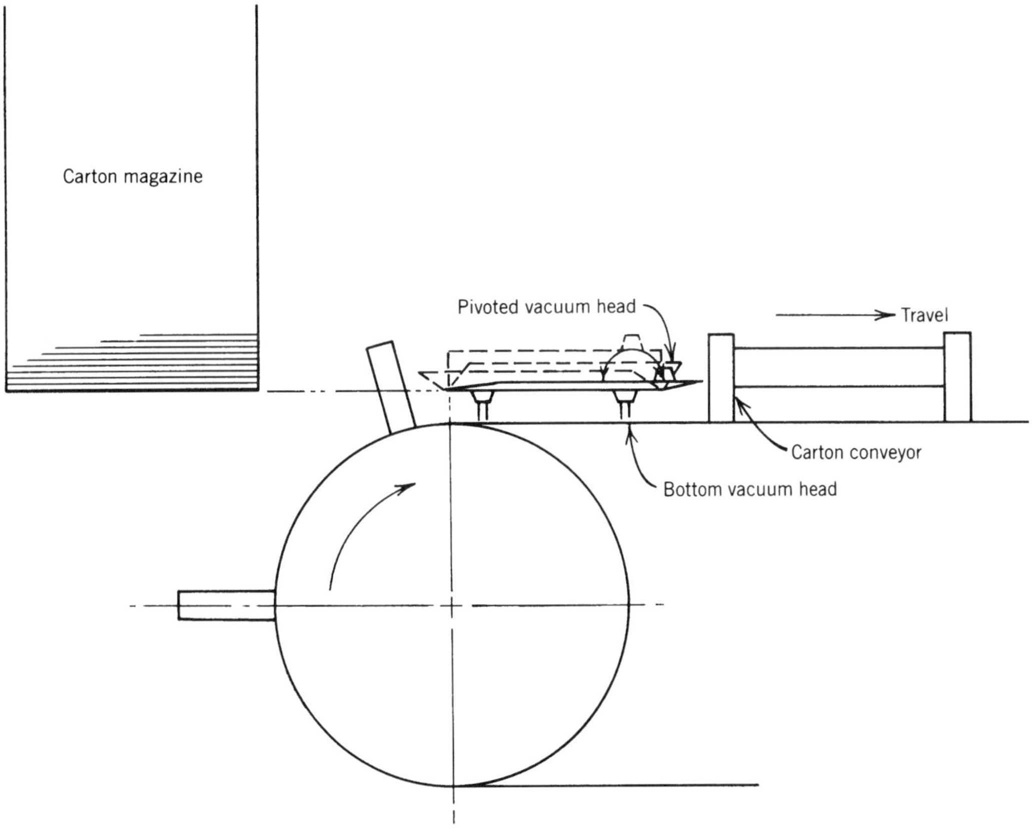

Figure 4. Carton feed for large flat cartons.

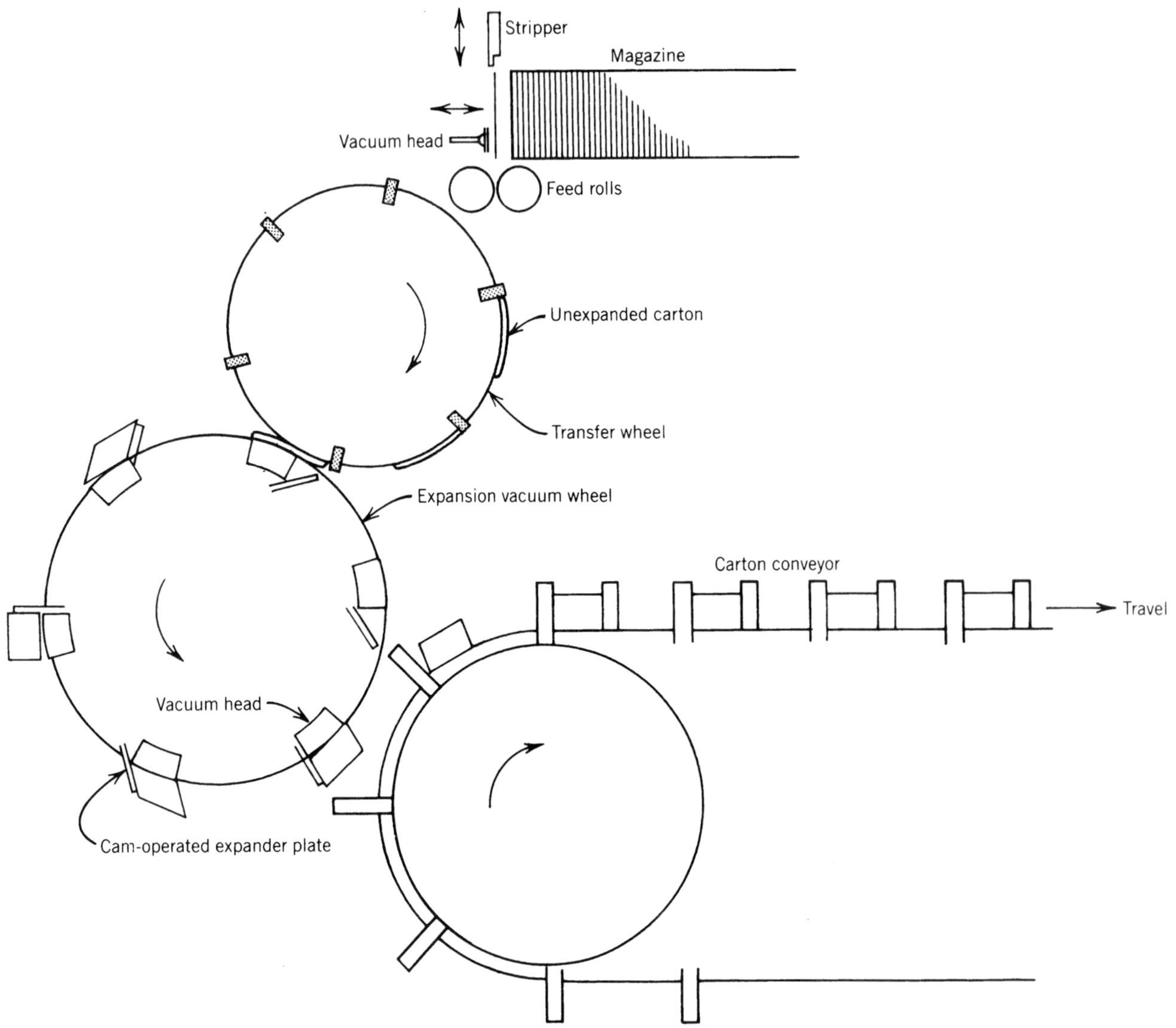

Figure 5. Rotary motion carton feed for high speed operation.

bucket with the bottle still standing on end. The trailing wall of the product bucket pushes the bottle which usually rides up on a wedge-shaped guide tipping the bottle against guides which gently lay it horizontally in the bucket. Bottle transfers are usually completely adjustable for the entire range of bottles to be run, although it may be necessary to use a different timing wheel or screw for some sizes.

Tube transfers. A tube filling machine may be driven by the cartoner to obtain the necessary timing between the two machines, although it is now common to drive the machines separately and time the tubes on infeed belts by means of clutches driving the belts. Tubes are frequently filled in tandem so the cartoner infeed is also in tandem, accurately delivering each tube into its product bucket despite the small difference in timing between the two tubes. Usually the transfer into the bucket is by means of a shuttle sweeping the tubes off a dead plate.

Bags-in-box. Bags are usually made and filled on a vertical form/fill/seal machine, frequently at the rate of two bags per cycle (see Form/fill/seal, horizontal and vertical). As with tube fillers, the form/fill/seal machines may be driven by the cartoner but normally are synchronized electrically with the cartoner. The bags drop onto an inclined conveyor and may be conditioned while on the conveyor to distribute the product within the bag. The conveyor elevates them to the level of the carton conveyor buckets where they will rest on a dead plate. An overhead conveyor, driven in time with the buckets, sweeps each bag into its bucket. At the loading position of the cartoner the bags may be confined within the carton dimensions on the leading side and the top by an overhead conveyor. To obtain high speed it is not uncommon to have two or even more form/fill/seal machines feeding one cartoner.

Multipackers

Another form of cartoner is the multipacker, a machine which accumulates a number of identical products and loads them into a carton which frequently is used as a display as illustrated in Figure 6. Cartons used for multipacks can be set-up trays, trunk-style top-loaded cartons, or end-loaded tubular cartons as shown in Figure 6. The latter has the advantage of excellent control of the assembled unit packages during loading, it results in a strong package to protect the contents, and by folding the end flaps with a special folding sequence, it results in a trunk-style carton for display purposes. Figure 6 shows the sequence of folding and the application of glue. The inner end panel is folded first, then the side flaps, and finally glue is applied to the outer end panel so the panel adheres to the side flaps. For display purposes the "ears" can be removed by tearing along the perforated score line or can remain to provide a wider display flap. The side seam (manufacturer's joint) of the multipack carton is spot glued by the carton supplier, providing enough strength to hold the carton together during expansion. The joint can be broken on the multipack machine by a slitting mechanism or, if only a few spots of adhesive are required to provide the necessary strength, by the machine which exerts pressure along the top panel. The length of the glue flap is less than the carton depth which provides room for the flap to be depressed enough to break the glue spots.

The multipacker operates at relatively low speeds except for the accumulator which may take the output of more than one machine. The motion of the cartoner may be intermittent or even cyclic, operating only when the assembly of packages has been collated. Because the multipacker takes the output of one or more machines it must be very efficient; when the multipacker is down the entire line must stop. If the multipacker is designed to take the output of more than one machine, it must be designed to operate at reduced speed if one of the machines feeding it goes down. A separate collating mechanism may be built to take the discharge of each machine feeding the multipacker with each collator feeding the complete assembly into a sequenced pocket of the cartoner infeed. If one of the collators goes down, the pocket for the inoperative collator will be empty but the cartoner will merely not feed or skip a carton for that bucket and will continue to run. On the other hand, one collator may take the output of two or more machines. As an example, let us assume four machines are feeding an accumulator and four products are required to be placed in the multipack carton, one from each machine. If one of the machines goes down, only three products would be available for the multipacker. To overcome this it is necessary to combine the total output into a single line, but this could result in very high line speed on the incoming line to the collator. For this reason a multipacker may be equipped with more than one accumulator.

Leaflet Feeds

One of the most important attachments on a cartoning machine is a mechanism to place a leaflet or coupon into the carton with the product. The leaflets may be for advertising

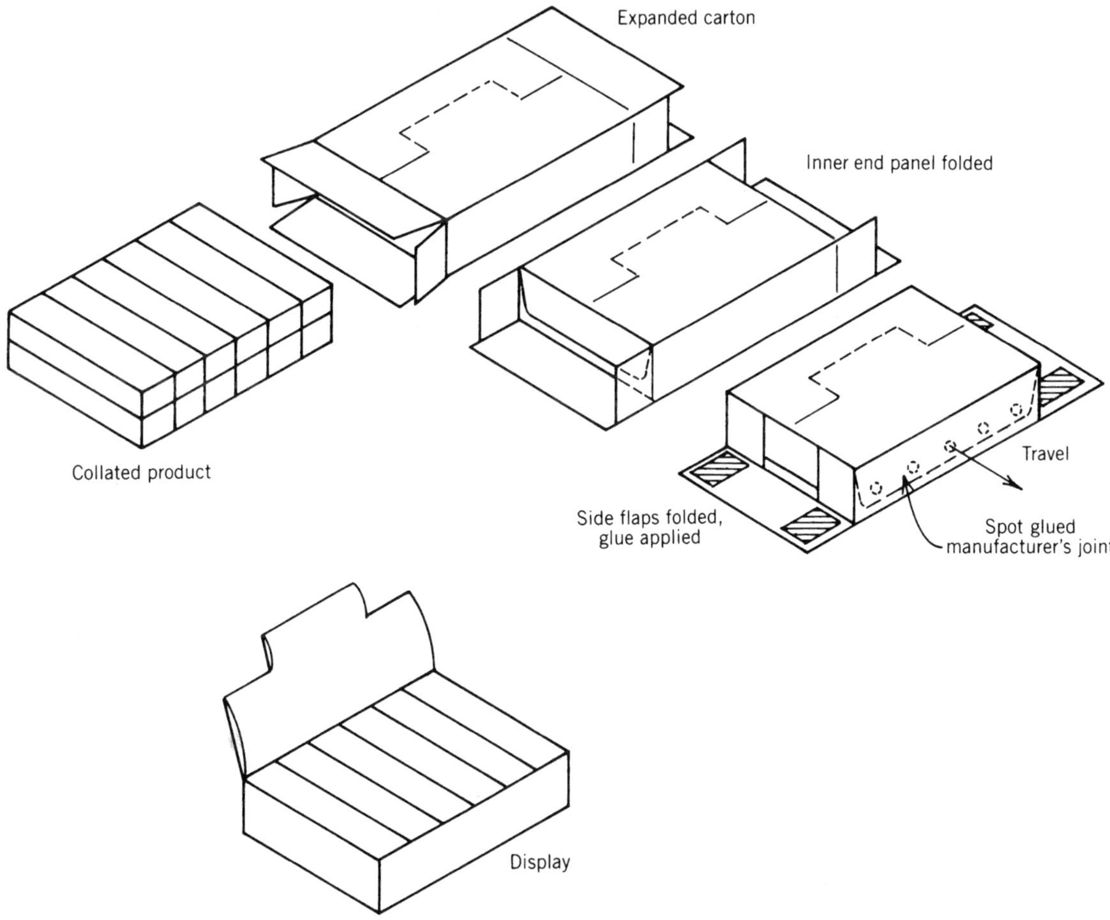

Figure 6. Multipacker.

purposes or may be essential to provide the technical information regarding the product (eg, pharmaceuticals). Feeds are available to handle prefolded leaflets which are frequently used with semiautomatic vertical cartoners (see Fig. 2), die cut sheets which the mechanism folds prior to insertion, or leaflets from roll stock, which are cut to length, folded, and then inserted into the carton with the product. Leaflets from roll stock have some decided advantages over die cut sheets: higher speeds can be obtained since the web is in continuous motion, and since the leaflets are on rolls, the chance of an incorrect leaflet on a given run is eliminated.

A typical mechanism to feed and fold die cut sheets is illustrated in Figure 7. The sheets are stacked in a magazine with the weight of the stack usually supported by freely turning wheels and the rear of the stack resting on a pair of very sharp needles which penetrate several leaflets. A set of start rolls with rubber segments pulls the bottom leaflet from under the stack, the sheet above being held back by the needles. The grain of the paper is parallel to the direction of feed so the tears from the needles are very small. Generally the start rolls are driven with a variable driving mechanism which enables the leaflet to be picked up at a slower speed. As the leading edge of the leaflet reaches the outside of the magazine, it is usually nipped by a set of feed rolls which are running faster than the start rolls, pulling it ahead. The folding of the leaflet is accomplished by operated blades which usually move just enough to drive the folded leaflet into another set of feed rolls.

The folding of the leaflet may be done by buckle folding, eliminating the need for folder blades. The leaflet enters a slot until it contacts an adjustable stop. As the feed rolls are still driving, the leaflet is forced ahead, which causes the paper to buckle and enter another slot leading to a set of feed rolls at the desired location. Multiple folds can be made easily with this type of unit. If a certain fold is not required, the stop in that slot is located forward to block off the slot to prevent entry of the leaflet.

Roll-fed leaflet mechanisms are equipped with either a blade folding unit or a buckle folder. Cross folds can be made if required, usually by folding the web parallel to the direction of flow by passing it over a plow before cutting the sheet to length. Cross folds can also be made on a blade folder mechanism.

After the leaflet is folded it must be transferred into the carton with the product. The most common system used with the horizontal cartoner is to insert the folded leaflet into the bucket, carrying the product either in a slot underneath the bucket, clamped to the bottom of the bucket with a chain synchronized with the bucket, or in some cases, held in position by a clamp which is an integral part of the bucket. While held under the bucket the projecting end of the leaflet is folded up and back over the product and held in this position by guides until the loading conveyor transfers the assembly of product and leaflet into the carton. In most cases the cap of the bottle leads as it enters so the folded leaflet is readily visible when the package is opened.

When handling leaflets, particularly with pharmaceutical products, it is essential that the leaflet is carried into the carton with the product. As the loading tip contacts the base of the product, the clamp or chain holding the leaflet to the product bucket releases and the product pushes the leaflet into the carton. However, if the sides of the leaflet are of unequal length the leaflet could slip alongside the product and not enter the carton properly. To overcome this, some horizontal cartoners can be equipped with a clamp conveyor which is similar to the loading conveyor described earlier, but mounted on the rear of the machine. The slide enters the open carton from the rear and passes completely through the carton until the spring-loaded tip contacts the leaflet and clamps it against the leading surface of the product. The slide is withdrawn at the same rate the loading slide pushes the product into the carton, keeping the leaflet securely clamped to the product. Because the clamp conveyor slide must pass through the carton, it enters the carton several positions ahead of the loading position. This type of mechanism is usually designed for the full range of carton length, providing the same length of stroke whether the carton being handled is short or long. A similar mechanism is sometimes used for certain products that may tend to topple when loaded, such as a group of wrapped chocolate bars stacked on edge.

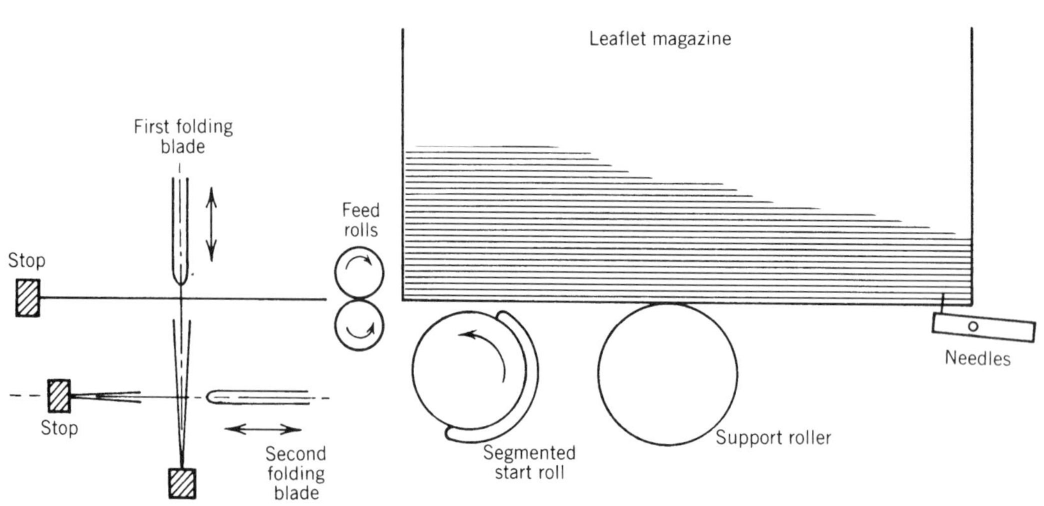

Figure 7. A typical mechanism to feed and fold die cut sheets.

Carton Closing

Tucking. The tucking operation is accomplished over a number of positions on the machine: side (dust flap) folding, prebreak of the tuck flap score line, possible slitting of the dust flaps to assure clearance if lock slits are used, alignment with the carton, first and usually second tucking, followed by a final contact to seat the lock slits. Usually these operations are done with the tucking parts on a bar carried by parallel cranks with each tucking position on a separate pitch of the machine. Tucking by means of belts is also done successfully, eliminating the need for cranks, although the trailing side flaps must still be folded over by a separate mechanism prior to tucking. The final tuck for locking the lock slit is accomplished by a separate rotary mechanism.

Gluing. A glue end carton may be closed by single or double gluing. When single gluing, the two outer end flaps are glued together; when double gluing, the inner end flap is glued to the side flaps and then the outer end flap is glued to the inner. Single gluing results in a slight crack between the glued end flaps and the folded side flap which may be acceptable for a bottle or other solid object. Cartons for food products or facial tissue are usually double glued to protect the contents.

There are two general types of adhesive used, cold glue or hot melt. Cold glue may be either dextrin or resin base, both obtaining their set by evaporation of the water in the adhesive. In general, dextrin glues require approximately 30 seconds to set and resin 20 seconds, but in either case a compression conveyor is required to keep the carton under pressure until the glue has set. High packaging speeds result in long compression conveyors, particularly when running thick cartons such as facial tissue boxes.

The use of hot melt adhesives has greatly reduced floorspace requirements because the glue sets by temperature reduction instead of water evaporation. A compression system is still required, but usually the carton can remain in the carton conveyor since only a very few seconds are required for the glue to set, just enough time to square up the flaps.

Hot melt adhesive is applied by guns or intaglio rollers (see Adhesive applicators). When using guns there are three definite heated components, that is, the glue reservoir, the hose, and the gun itself, each controlled by individual thermostats to maintain precise temperatures. The glue gun can apply a strip of glue, an interrupted strip, or dots. For single gluing (ie, gluing the outer end flaps together), a single strip of glue is usually applied, but if a thick carton is being run a multiple orifice nozzle may be used to apply two or even more lines simultaneously. Obviously a separate gun is required for each end of the carton although only one reservoir is needed. When double gluing a carton, two guns are required at each end of the carton, one to fire an interrupted line to apply glue to the inside of the inner end panel to glue to the side flaps and the other to apply a continuous line to glue the outer end panels to the inner. Intaglio rollers can apply the dots of adhesive in the required pattern for double gluing. In this case the glue tank and the intaglio roller are heated.

Detectors

Cartoning machines, particularly the fully automatic horizontal cartoner, are usually equipped with a variety of detection systems. Some of the many detectors available are listed below:

Skip carton. The carton feed does not feed the carton for an empty product bucket; instead, it "skips" a carton.

No carton. If a carton fails to feed or expand for a properly filled article bucket the machine will stop. In some cases, usually when the carton conveyor consists of buckets instead of lugs, the product can be transferred into the carton bucket and discharged at the end of the carton conveyor, which enables the cartoner to continue to run.

Skip load. If a carton is not present in the carton conveyor the loading slide is not cammed forward and the contents of the product bucket remain in the bucket and are usually discharged from the bucket as it goes around the loader sprocket. The machine can continue to run.

Missing leaflet. If a leaflet is not placed properly in the product bucket the machine will stop. In some cases the presence of the leaflet can be detected in the carton. This is a more advantageous system because a leaflet may be lost during the transfer into the carton even though it was in the product bucket. However, it is much more difficult to sense the leaflet after it has been placed in the carton.

Low product level. If the incoming products reach a low point the cartoner can be arranged to stop. When handling glue end cartons, particularly with hot melt adhesives, the machine will come to a timed stop, a position where the carton just glued is held under compression.

Overloads

To protect the machine most cartoners are equipped with several overload devices, frequently merely a detent that contacts a microswitch if an overload is sensed. Most bottle feeds have an overload device on the timing wheel and sometimes also on the transfer wheel. Another common overload is a detent in the side-flap-folding drive mechanism. The operated folder or star wheel may contact the product if it did not enter the carton completely or, as sometimes occurs when handling a leaflet with many folds, is pushed out of the carton by the springiness of the leaflet. The overload will stop the machine before the product or machine parts are damaged. Some cartoners have an overload mechanism on the main drive to protect the machine in the event of an overload in any portion of the machine not already protected. This type of safety is an excellent feature although it must be set in such a way that it does not throw out because of the starting torque of the machine yet will sense an overload during normal running.

Many cartoners equipped with detectors and overloads have indicator lights on the control panel to show the operator which detector stopped the machine. Usually the light operates through a holding relay to keep it on as the machine may coast past the microswitch as it comes to a stop.

Microprocessors

Centralized microprocessors are frequently used to control and monitor the operation of cartoning machines. Encoders, working with programmable limit switches, have greatly simplified operating, set-up, and maintenance procedures. They have replaced electromechanical sequencers and shaft driven cams which have been common on cartoners. Programmable devices can be adjusted remotely, can be set up and changed very quickly, provide beter resolution, and better repeat accuracy, and generally can operate at higher speeds.

BIBLIOGRAPHY

"Cartoning, Machinery, End-Load, from the *Wiley Encyclopedia of Packaging Technology*, 1st ed., by W. F. Dent Ansell, Incorporated, pp. 132–142.

1. S. Knapp, *The Packaging Encyclopedia,* Cahners Publishing Co., Boston, Mass., 1984, pp. 218–222.
2. C. Glenn Davis, *Packaging Machinery Operations—Cartoning,* Packaging Machinery Manufacturers Institute, Washington, D.C., 1980, pp. 2/2–2/7.
3. P. A. Toensmeier, *Impact and Future: The Packager Side Seam Gluing of Folding Paper Cartons,* A Packaging Technology and Marketing Study/Report, Patrick A. Toensmeier, Hamden, Conn., 1979, pp. 11–15.

MANAGING THE PACKAGING FUNCTION

There are ongoing disagreements and/or confusion as to where in the organization packaging best "fits" and, even worse, a general lack of understanding on the part of senior management, and other corporate groups, as to what packaging can or should be doing to best serve that business.

Location in the Organization

Where is the packaging function best positioned to serve the company? Granted, there is no easy answer to this question. The close relationship with suppliers that are typically desired or required directs one to elect location within the purchasing department as optimum. Especially so when you consider the increasing importance of cost control, cost reductions, specification negotiation, and supplier problem resolution—especially quality problems.

There are cases in which perhaps the manufacturing function has a dominant need; and in all cases, packaging must assure good working relationships with production management *and* line personnel.

In essence, there is no one best answer for all, *but* the wrong or inappropriate corporate location of the packaging function assures hampered performance and limited professional growth or recognition of value.

Packaging Responsibilities: Known and Understood

Perhaps the prime responsibility of a packaging manager is to clearly understand all that is *currently* and appropriately expected of the function; and likewise, to continually educate upper management, as well as their peers or "customers" in other company functions, as to what the department's capabilities and expected productivity are. The latter is basis staffing, training thereof, budget, and assigned priorities.

The emphasis on *current* expectations is to asure that the department manager keeps aware of ongoing changes in the company and/or industry that continually impacts respective responsibilities and priorities. Those changes must also be quickly communicated to the packaging staff.

Staffing

One must expect the typical and likely ongoing understaffing that exists today, has existed for years, and will likely continue in the future. This is due partially to failure to communicate to senior management the needs, benefits, and cost-effectiveness possible within the packaging function. "Cost-effectiveness" is a combination of cost reductions and avoiding major costly claims or mistakes that impact other corporate functions. This is covered in a later paragraph.

Appropriate staffing is usually a matter of budget restraints and the collective expectations of the packaging department. Clearly delineated project assignments and regular progress reports help emphasize possible understaffing. If certain engineers have more work assigned than they can reasonably handle, the resultant tardy results and stress will quickly magnify the need for added staffing. One cannot emphasize enough the need for establishing a project priority system as the best measure of "appropriate" staffing. In essence, continuing failure to meet objectives and particularly timelines (schedules) will be a failure of *management.* That could include department managers and their superiors! If added staff is not made available, then a reduction in priority assignments is in order if timelines are to be met.

Training

Because packaging as a whole is both complex and ever changing, ongoing training programs are of paramount importance, including both technical and nontechnical. Good examples of the latter include total quality, teamwork, and the development of leadership thereof, and interpersonal relations and communications.

There is always a need for prompt and optimum decisionmaking. Therefore, *everyone* in the department should attend a course in Problem Analysis and Decisionmaking. They last throughout your career!

Finally, department managers are responsible for training their successors. The earlier, the better.

Budget

The woes of an inappropriate budget are a great teacher for future budgeting. Properly budgeted departmental costs, as well as cost reductions, are a key to avoiding monthly or quarterly stress through the year. Department managers must take the time to fully understand what is expected of them in the year ahead, both of themselves and all others. This helps establish a budget which at least *allows* for a good to excellent performance. A constant review is essential through the year as "surprises" come or situations change. The resultant changes in priorities or workload can be deafening (which refers to the screams from those who expect a certain level of performance from the packaging functions).

Educating Associates and Senior Management

It can be enlightening to find out how little your peers and/or senior management understand about packaging! The better packaging is understood, the more likely you can expect proper recognition and understanding of *your* problems in day-to-day affairs.

It is clearly worthwhile to prepare a rather brief, but thorough, description of what you do and for whom. Also, consider the benefits of a meeting where you can explain in some detail, and others can ask questions to learn, the complexities and potential of the packaging functions *in your company.*

Cost Reductions and Crisis Avoidance ($ and $)

These are two critical aspects of a packaging department which impact staffing, growth and respect. Every packaging person in the department should have some cost reduction budget dollars assigned annually, with a total cost reduction budget of 1–3% annually on the basis of the total package material expenditures for that year. Cost reductions affect the corporate "bottom line" and should get attention at the highest levels, especially if there is a year-end summary with wide circulation!

Equally important is the avoidance of major packaging crises, especially claims due to unsalable merchandise. These can and have mushroomed into millions of dollars. Then packaging is in the spotlight negatively! Good packaging specifications and the enforcement thereof, along with participation in vendor screening or selection, are the responsibility of packaging. There are many horror stories about packaging and product claims that serve to remind us how important the proper package and package quality are to the marketplace.

Professional Growth and Associations

Packaging managers should assume responsibility for their own and the department's growth professionally speaking. In addition to training programs previously mentioned, there are opportunities to participate in the local chapter of the Institute of Packaging Professionals (IoPP). Further, it allows you to become a Certified Packaging Professional (CPP) through the IoPP exam. The CPP program has been modified to allow more alternatives to qualify for the CPP certification. This is truly a matter of personal pride and industry recognition of your interest in being a professional member.

Project Control

The packaging manager should assume full responsibility for projects assigned to the department and their successful completion on a timely basis. The number of projects accepted, and their comparative or assigned priority, should be a direct responsibility of the department manager, and clearly understood in total by the manager and the assigned engineer.

This is a critical basis for good interrelationships between packaging and other departments, especially new-product development, marketing, and manufacturing. Such activity often calls on the manager's diplomatic skills in explaining current workloads and any new assignments' comparative priority.

Self-Directed Projects

One major means of helping the staff engineer gain and retain pride in their work is to encourage their self-initiated projects. This could, if not should, involve up to 10% of their time. It certainly develops a sense of personal pride and professional growth.

This might be a good time to emphasize the benefit of motivation in the individual engineer. A lack of motivation to excel and/or improve is a sure sign that something is awry and needs attention from the department manager. Such lack of motivation can and usually does result in less than quality efforts, all the way to sloppy or careless work. The longer it continues unabated, the worse the results become.

Global Outlook

Today's packaging manager must have a global interest and awareness. Matters such as new products, materials or equipment and cost impact due to impending worldwide shortages or market glut allow for constructive planning and activity. Whether this knowledge is primarily from trade magazines, meetings or seminars, supplier presentation, or whatever, the point is that there must be sources of global packaging knowledge for packaging department managers. They, in turn, can sort out the likely impact on their company products, or industry, and keep the management appropriately informed.

Packaging Specifications

A first step in developing a worthwhile specification program might be "descriptive" specifications. By definition, these describe the package structure and size. They lack any assurance of quality, or better-controlled quality, from the vendor. This benefit comes only from "performance" specifications that are complete and negotiated with the vendor to assure that they are realistic. They should be signed off by the supplier and your purchasing department as recognized operating documents. There are many historical examples, in all industries, of how performance packaging specifications prevented claims and/or market disasters. Today, with the constant emphasis on quality, there is no better way to assure that the package is what was planned, tested, and expected from the package development effort.

Conclusion

In summary, the packaging manager has the opportunity to hire motivated engineers and technicians, implement an ongoing training program for their growth on the job, and help assure optimum department location in the corporate structure that improves effectiveness, and improves the company profits. The latter is the result of well-executed cost reductions and claim avoidance, along with carefully prepared performance packaging specifications. Truly, a professional effort, conducted by a Certified Packaging Professional (CPP).

AL CORNING
A Corning Consulting, Inc.
Lombard, Illinois

MARINE ENVIRONMENT AND EXPORT PACKING

Purpose and Scope

Before there can be any meaningful discussion about the preparation of goods for export shipment, there must first be an understanding of the terms preservation, packaging, and packing. These three terms constitute a total system that when properly implemented serves

- To protect cargo from the elements and the normal rigors encountered in export transportation
- To facilitate handling
- To provide a degree of rigidity to the cargo
- To provide continued protection to the cargo in short- and long-term storage

This article is intended to offer the shipper a ready reference on how to prepare a product for export shipment. The fundamentals of preservation, packaging, and packing, including the selection of materials, are discussed. Proven methods of construction of skids, crates, and boxes are presented. The problems associated with marine transportation and with the use of sea containers are given special attention.

Definitions

Preservation is the operation by which goods are protected from the elements, ie, fresh- and saltwater, heat, cold, corrosive atmospheres, and humidity. Preservation usually consists of the direct application of a coating specifically designed to protect against a known hazard such as rust. Other types of preservatives are vapor-corrosion and vapor-phase inhibitors (VCI and VPI), which are additives in the form of powder or chemically treated paper that serve to inhibit the formation of water vapor and the rust resulting therefrom. Desiccants are also used as preservatives because they absorb moisture.

Packaging is an operation consisting of wrapping or boxing an item as a preparatory step prior to placing the item into the final shipping container. The nature of an export shipment, ie, by marine transportation, is such that the cargo is handled many more times than a domestic shipment, and for that reason, greater consideration must be given to known or assumed levels of shock and vibration. The type and extent of packaging depends on these considerations as well as on marketing requirements and segregation requirements. It is usually performed as an intermediate step in the protection of the item from the rigors of transportation.

Packing is the final step in the preparation of goods for shipment. Packing consists of the outer or external shipping container, whether a box, crate, or sea container. Cushioning, blocking, bracing, and anchoring are additional protective measures taken within this outer shipping container.

If shippers would view the preparation of their goods for shipment as a system consisting of preservation, packaging, and packing and would conscientiously evaluate their goods for susceptibility to the rigors encountered in marine transportation, their goods would stand a substantially better chance of arriving at the ultimate destination in sound condition.

VCIs and VPIs, when incorporated in waterproof packaging, can add a substantial measure of protection against moisture.

Desiccants are usually moisture-absorbing powders contained in packages such as boxes or cloth bags. When placed in relatively watertight or waterproof voids, these desiccants will absorb excess moisture. They cannot, however, be expected to continue to absorb water or moisture beyond their absorption capability; therefore, the amount of desiccant used must be carefully determined considering the expected temperature and humidity levels and the volume of the spaces that they are to protect.

Marks and numbers are extremely important. The outer or external shipping container must clearly show certain critical information essential to the safe handling and expeditious delivery of cargoes to their final destination: The name and full address of the consignee; an identifying number or code relating to the particular item being shipped; full dimensions and gross and net weight in both metric and British systems; and the appropriate international shipping symbols such as "Keep dry," "Fragile," "This side up," and "Lift here" or "Sling here," are all part of the labeling requirements (see also Fig. 1).

Unitization is the assembly of cargoes of similar size and configuration into a single bundle, group, platform, or pallet. The purpose of unitization is to reduce cubage, to facilitate handling, and usually to provide a greater degree of rigidity

Figure 1. International shipping symbols.

to the unitized cargoes. The bundling of pipe, coiled wire, lumber, and structural steel members are examples of unitization.

Palletization is a special case of unitization and consists of placing one or more similar units of cargo on wooden pallets and securing the cargoes to the pallets usually by means of steel or synthetic fabric tension bands. If required, overwraps may be applied to the palletized units to provide either watershed or waterproof protection. Other common methods of restraining and securing palletized cargoes are shrink wrap and tension wrap (polyethylene or polypropylene films).

The term *protective materials* generally refers to contact preservatives such as paint, grease, and oil or to the watershed or waterproof liners and wraps applied to protect the cargo from the elements. Protective materials can also include cushioning materials and blocking and bracing.

Skids are wooden timbers or alternative materials used either as a base to facilitate handling for heavy cargoes where cargoes possess inherent rigidity; or serve alone as a base for cargoes lacking inherent rigidity, protecting them and facilitating handling. Skids are also major structural members in crate construction. It is important to note that if skids are used alone as a base to carry a cargo with or without inherent rigidity, attention must be given to the type of load (ie, uniform or concentrated), and the skid must be designed accordingly. In addition, for economic reasons, skids used as structural members in true crates can be significantly smaller in cross section than the skids used as support bases because crate side and top and end panels contribute significantly to the overall strength of the crate.

A *wood crate* (see Figs. 2–7 for examples of open-crate design) is a structural framework of members fastened together to form a rigid enclosure that will protect the contents during shipping and storage. This enclosure is usually of rectangular outline and may or may not be sheathed. A crate differs from a nailed wooden box in that the framework of members in the sides and ends provides the basic strength, whereas a box must rely for its strength solely on the boards of the sides, ends, top, and bottom. Crate framework can be considered to be similar to a type of truss used in bridge construction. It is designed to absorb most of the stresses imposed by handling and stacking.

A *wooden box* is simply a rectangular enclosure consisting of wooden boards or planks nailed along their ends or sides to adjacent boards or planks. Boxes generally do not employ framework such as that used in crates. However, boxes can and should employ *cleats* and *battens,* wooden members that act as nailers and stiffeners.

The *wood species* most commonly used in crate construction are divided into four groups, largely on the basis of density. In general, it is good practice to use species in the same group for similar parts. Group I consists of the softer woods such as pine, spruce, and chestnut. Group II consists of heavier woods such as Douglas fir and southern yellow pine. Group III woods are generally hard woods of medium density and possess relatively good nail-holding capacity. Examples of these woods are ash, soft elm, and soft maple. Group IV woods are heavy hardwood species, the heaviest and hardest domestic woods. They have the greatest capacity to both resist shock and hold nails. They are difficult to nail and tend to split but are especially useful where high nail-holding capacity is required. Oaks, hickory, birch, and white ash are examples of Group IV wood.

Wood defects should be carefully considered when selecting wood for skid, crate, and box construction. Seasoned wood is preferable to green lumber. Knots and other various defects such as slope of grain, decay, wane, shakes, checks, splits, and warping are defects that affect the strength and long-term serviceability of wood. Guidelines for selecting wood with defects are contained in great detail in the USDA *Wood Handbook* (1) and *Wood Crate Design Manual,* Agricultural Handbook No. 252 (2).

Rub strips or runners are lumber strips secured to the underside of skids and skid-based crates to facilitate handling.

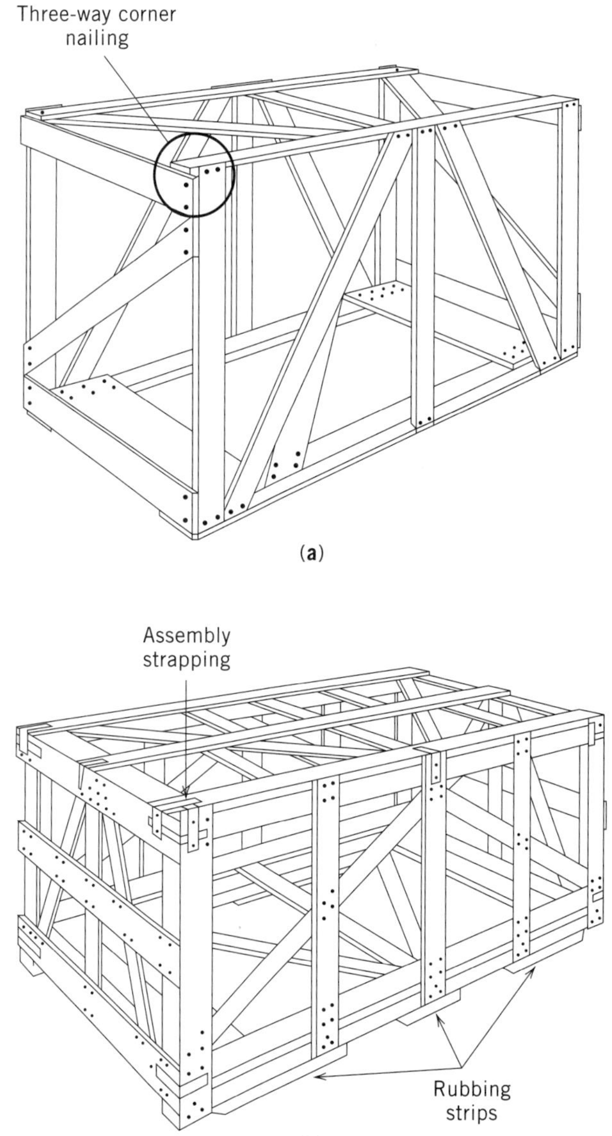

Figure 2. Open-crate designs: (**a**) light-duty open crate (for loads to ≤200 lb) with three-way corner nailing, the key to proper assembly nailing of crates of this type; (**b**) limited military-type open crate (for loads ≤2500 lb) designed for the shipment of light, bulky items that do not require a waterproof container (2).

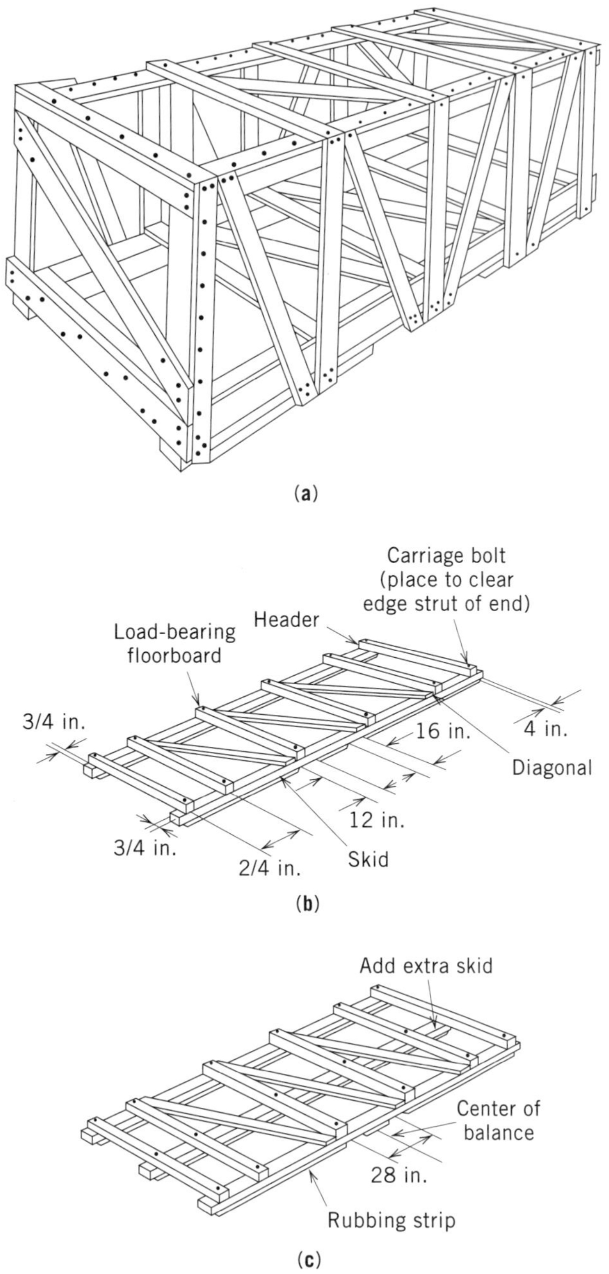

Figure 3. (**a**) Typical open crate; (**b,c**) bases for light-duty open crates [(**b**) ≤42 in. wide; (**c**) >42 in. wide] (2).

Rub strips are an essential part of skid assemblies and crates and in box construction for loads in excess of 200 lb.

Tension bands are metal or fabric bands or straps used to secure or bind cargoes to pallets or skid bases or as an added measure of binding the external members of crates and boxes. Tension bands are manufactured in a variety of dimensions and are applied by means of mechanical or hydraulic tensioning equipment. Tension bands are also an essential part of good packing.

Reinforcing straps (see Fig. 8) are mild steel metal bands fitted with nail holes and can be affixed at the corners of crates and at the junction of side-panel vertical members and top-panel horizontal members. Reinforcing straps should be applied in two directions, where possible. These serve, as an added measure, to prevent the separation of joints.

Headers are timbers or alternative materials which are bolted across the end of skid assemblies. Generally, they are to be of at least the same cross section as the skid members they join. Headers are an essential part of skid assemblies in skids acting alone as well as skids serving as part of a crate.

Load-bearing floorboards are wooden members or alternative materials that are fastened transversely across the tops of skid assemblies and carry the load of the item for which the skid assembly is intended.

Splicing refers to the joining, end to end, of two (2) lengths of timbers of similar cross section in order to make one longer section with approximately the same longitudinal strength as a single continuous piece. The ends to be spliced are to be notched to half the timber thickness, and through-bolts are to be placed vertically through the joining notched sections. Lapped along each side of the splice is to be a $\frac{1}{4}$-in. steel plate of length no less than three (3) times the thickness of the members joined and through-bolted, side to side, with at least two (2) bolts in each half of the splice.

A *uniform load* refers to a load that is distributed more or less evenly over the entire unsupported length of the member on which it rests. As was pointed out above, consideration must be given to the type of load, ie, uniform or concentrated, when determining the size of members to be used as skids and load-bearing floorboards.

A *concentrated load* is a load that is not distributed evenly over the unsupported length of the member on which it rests. Rather, it is centered over a shorter segment of the member on which its rests, resulting in a bending moment higher than that produced by a uniform load of similar magnitude.

Rigidity refers to a strength property of materials that allows them to resist deformation such as longitudinal bending or racking. For example, a 20-in.-diameter pipe with $\frac{1}{2}$-in. wall thickness and length of 20 ft may be able to be picked up by a sling at its midlength without any noticeable deformation, while a 2-in.-diameter pipe with wall thickness of $\frac{3}{16}$ in. and length of 20 ft would droop noticeably if lifted with a sling at its midlength. Items with no inherent rigidity, which are intended for export shipment, should be skidded or crated in order to facilitate handling and prevent damage.

Fragility is a term used to describe the degree of susceptibility to damage from the normal rigors of transportation, ie, shock and vibration.

Bending moment, as used within the scope of this article, refers to the product of a load acting over a distance to produce deformation of a structural member that is subjected to the load at some point along its length. The load may be uniform or concentrated as described above. As an example, consider a beam supported only at its extreme ends and subjected to a load at its midlength such that the resultant of the load acting downward and the reaction forces at the extreme ends acting upward is a sagging of the beam. What actually happens here is that the fibers of the beam in contact with the load are in compression while the fibers on the underside of the beam opposite the load are in tension. The bending moments described above can be written in the form of the mathematical expression

$$\text{BM} = \frac{WL}{4} \quad (\text{ft·lb})$$

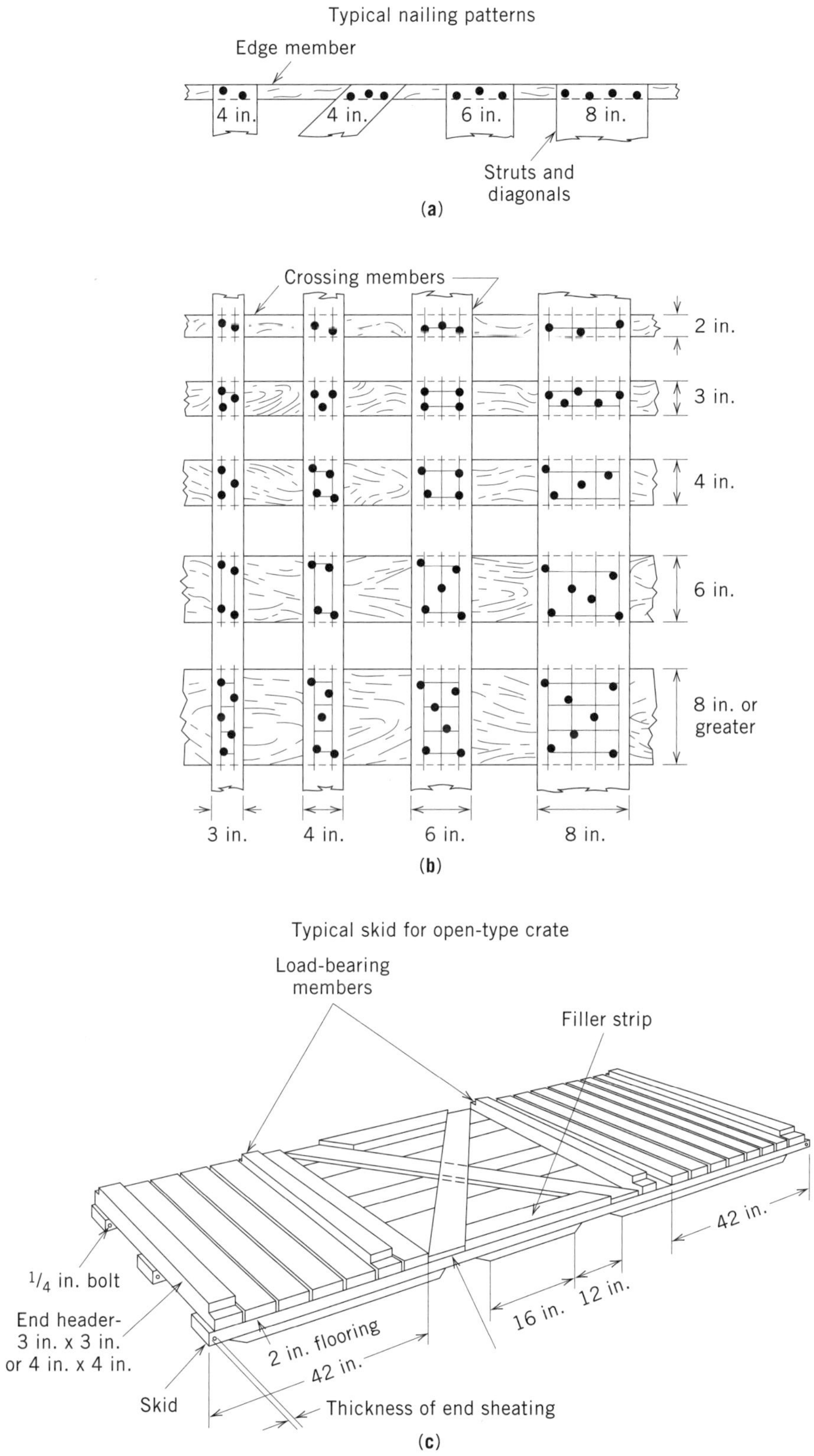

Figure 4. (**a,b**) typical nailing patterns; (**c**) typical skid for open-type crate (2).

where BM is the bending moment, W is the applied load, L is the unsupported length of the beam, and the denominator 4 takes into account that W is a concentrated load. Similarly, the mathematical expression for the bending moment resulting from the application of a uniformly distributed load on the same beam

$$\mathrm{BM} = \frac{WL}{8} \quad (\mathrm{ft{\cdot}lb})$$

where the denominator 8 accounts for the fact that W is a uniform load. In some calculations for determining bending moments on beams where the applied loads are neither

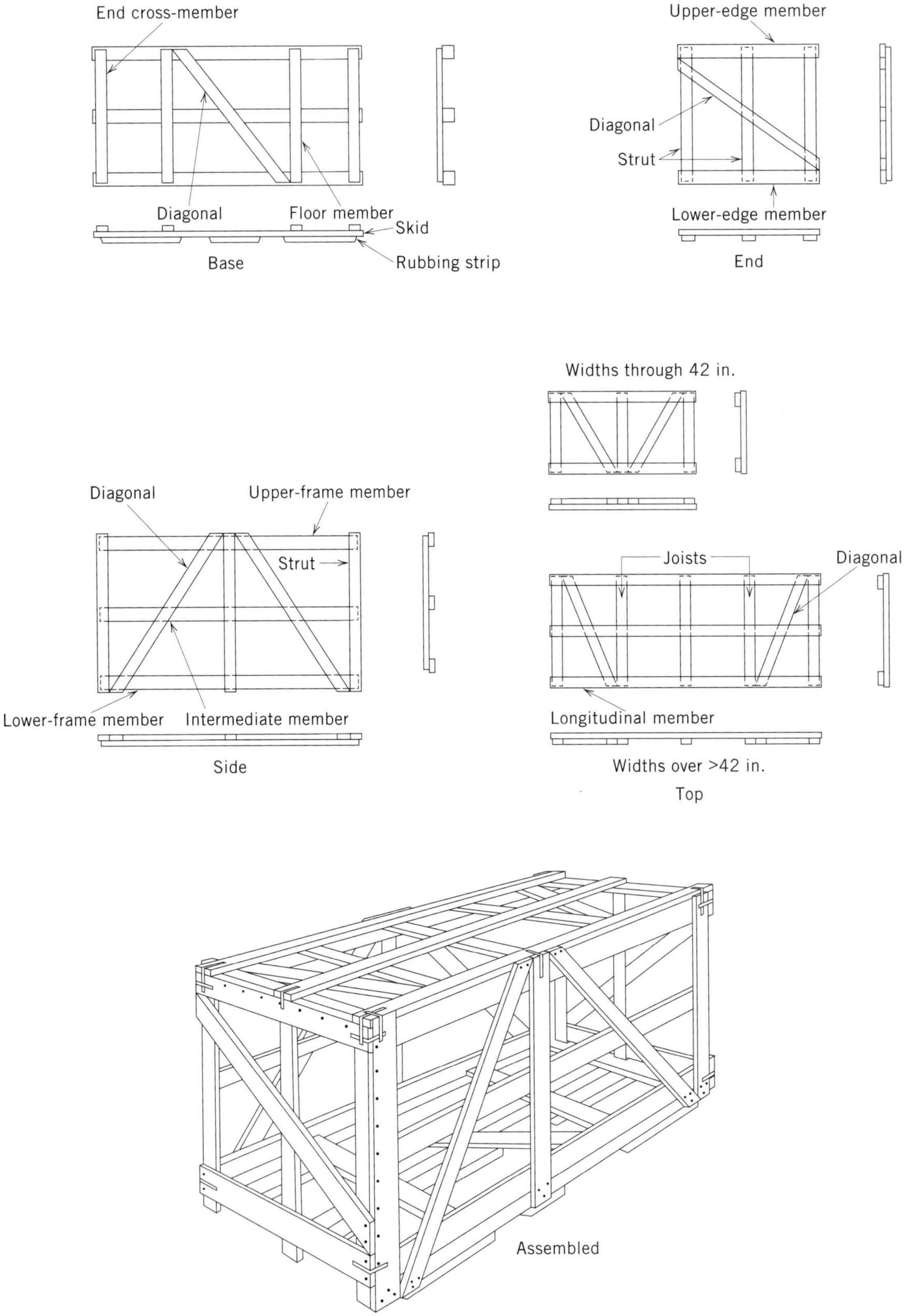

Figure 5. Crate components and final assembly (2).

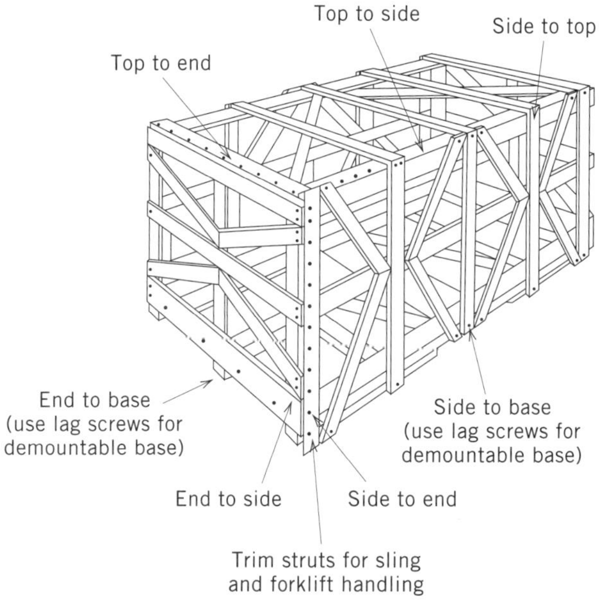

(a)

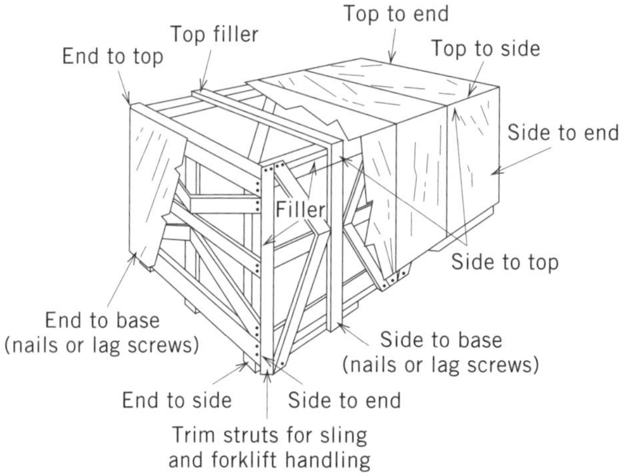

(b)

Figure 6. Assembly of (**a**) light-duty open crate; (**b**) covered-style light-duty open crate (2).

strictly uniform nor strictly concentrated, it is convenient to use the denominator 6.

Stress is directly related to the bending moment. Mathematically

$$\text{Stress} = \frac{\text{BM}}{\text{SM}} \quad [(\text{in.}\cdot\text{lb})/\text{in.}^3] = \text{lb/in.}^2$$

where SM (section modulus) denotes a property of a particular beam cross section. It is defined as the moment of inertia I (in.^{4}) divided by the distance of the most remote fiber from the neutral axis, and is denoted by the symbol Z or SM. A simple mathematical expression for section modulus is $(bd^2)/6$, where b is the width (inches) and d is the depth (inches) of the cross section. It is clearly seen that section modulus is extremely helpful in determining the actual stress to which a beam is subjected under a known bending moment.

In this section, the term "engineered" means that an analysis must be made of the types of loads and their magnitudes on the various structural members of crates so that their required dimensions may be correctly determined by calculation.

Marks and Numbers

The purpose of marks and numbers is to identify the shipper, consignee, and purchase order; aid in safe handling; provide general information; and identify hazardous materials. Never identify the contents unless it happens to be hazardous materials.

Marks and numbers should be made in large (2-in. minimum height) block letters and numbers. All numbers such as gross and net weight and dimensions should be in both British and metric units. Marks and numbers should be placed on three (3) sides of boxes and crates to ensure that they can be seen, and these should be accompanied by the international symbols such as the umbrella denoting "Keep dry," the wind glass denoting "Fragile," and the link chain denoting "Sling here." Marks and numbers should be in permanent ink or paint that contrasts with the background.

Preservation

Some items, by their method of manufacture, are already preserved. Others require protective coatings such as oils, grease, paint, or some other contact preservative that may be specified by manufacturing, engineering, quality control, or customer requirements. Some items, in addition to contact preservatives, require a wax-paper wrapping or some other protective film around the item.

Certain items may require, besides contact preservatives, additional protection from the elements, especially water, such as shrouds, liners and linings, wraps, VPI or VCI (moisture inhibitors), or desiccant materials. Wraps applied over a contact preservative must be greaseproof and require a means of assuring that they remain in place during handling and transportation. Liners and linings are waterproof barrier materials applied between framing and sheathing to provide watershed and diversion protection for the items crated. When shrouds are applied, all projections in sharp corners must be cushioned to prevent penetration or abrasion of the shroud from either internal or external sources. The bottom edges of all shrouds should be fastened down with batten boards and additional restraints provided to prevent ballooning or flapping of the shrouds in transit. Adequate bottom ventilation must be provided either by spacing floorboards ($\frac{3}{8}$-in. minimum) or by drilling one 1-in.-diameter hole for each 5 ft^3 of shrouded volume. When VPI/VCI materials are used, a sufficient quantity of treated paper must be used to completely enclose the item.

For general watershed and diversion protection, polyethylene films (4 mils minimum thickness), asphalt-laminated kraft paper, or filament-reinforced polyethylene- or polyethylene-coated kraft paper may be used. When applying water-vapor-resistant liners or lining, such as asphalt-laminated kraft paper or polyethylene films (6 mils minimum thickness), it is essential that they be sealed tight. Whenever VPI/VCI-treated papers are used, they must be inserted within a tight

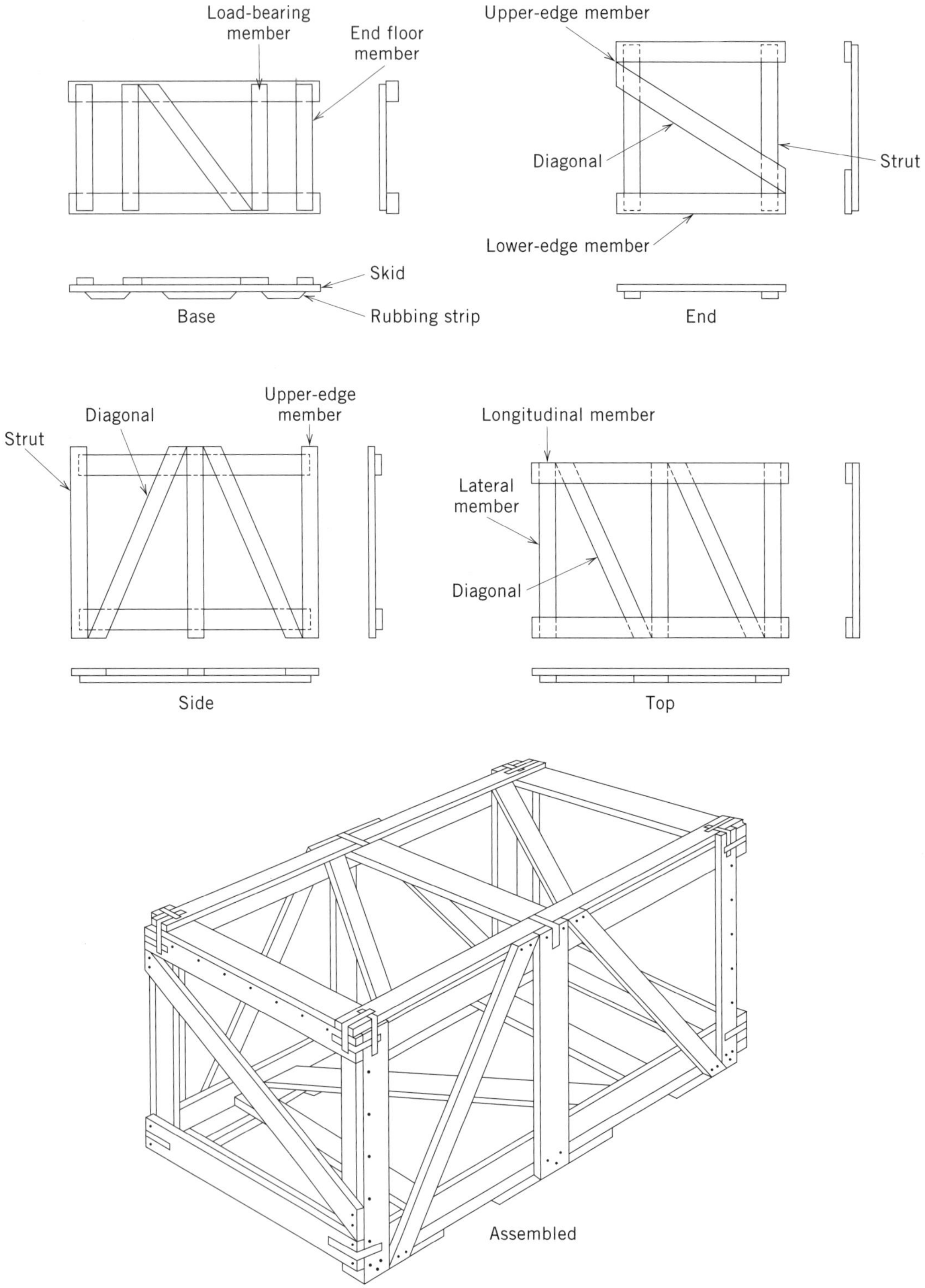

Figure 7. Style A, limited military-type open crate (2).

enclosure, and the linear or lining must be sealed with waterproof tape. This is also true of desiccants.

Skids, crates, and boxes. The materials used for skids must be Group IV (hard) woods such as beech, birch, hard maple, hickory, oak, rock elm, and white ash. These woods have the greatest nail-holding power and the highest overall strength and resistance to shock. However, they are very susceptible to splitting. These woods should be used for skids, headers, load-bearing floorboards, and critical joists.

Group II and III are generally acceptable for exterior box and crate construction. Group II woods include Doublas fir,

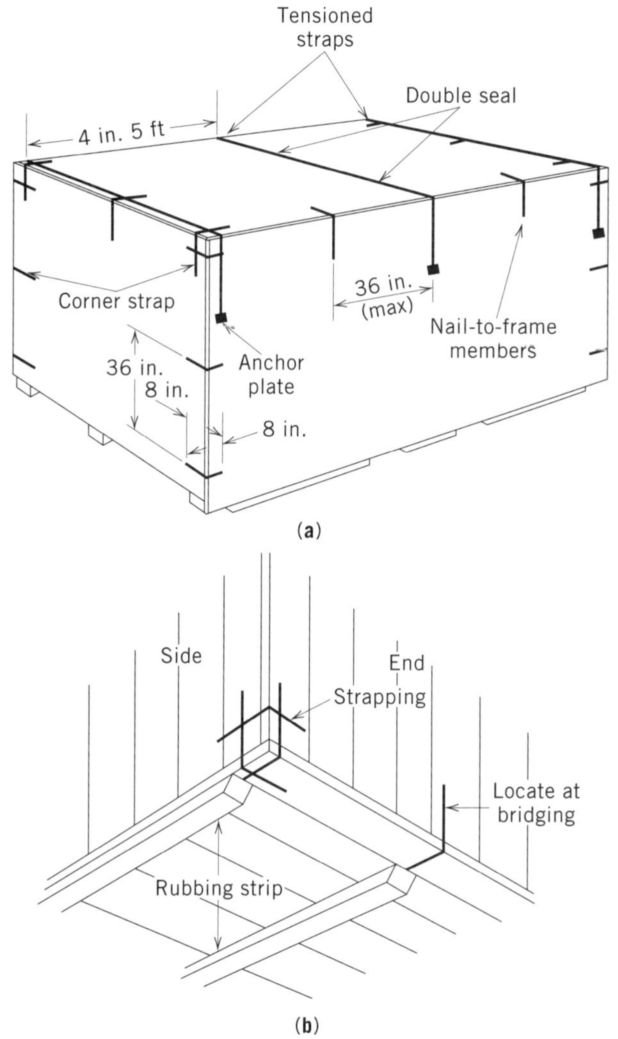

Figure 8. Strapping for crates: (**a**) sheathed crate with skid base; (**b**) additional strap for sill-type base (2).

hemlock, Southern yellow pine, Tamarack, and Western larch. These woods are harder soft woods. They possess greater nail-holding power, greater strength, and greater shock-resisting capacity than do Group I woods. They are more inclined to split, and their grains often deflect nails. Examples of Group III woods are ash (except white), cherry, soft elm, soft maple, sweet gum, sycamore, and tupello. These woods are similar to Group II in nail-holding power and beam strength but have less tendency to split and shatter under impact.

When using plywood as sheathing or floorboard material, type C–D interior plywood with exterior glue is the best economical selection.

Lumber selection and defects. A great deal of consideration must be given to a standard of quality of the wood selected for skids, crates, and boxes. Often the cheapest grade of lumber will contain a large number of defects that are undesirable. Therefore, it may be necessary to select a higher grade of lumber to eliminate the majority of disqualifying defects. Generally, knots are not to exceed one-fourth of the width of the member. No knot, regardless of size, will be permitted on the edge of any member. Obviously, this is because the member may be subjected to bending and a knot may fall in an area subjected to compression or tension. Cross-grain or slope of grain, disregarding slight local deviations, along the general direction of the grain as related to the longitudinal axis of the wood member, should not be steeper than 1 in. in 15 in. of length. Wane is either bark or lack of wood on the edge or corner of a piece of lumber. Generally, wane is permitted on one edge only, but should not exceed one-sixth of the width, thickness, or one-third of the length of the member. Checks and splits, lengthwise openings from separations during seasoning, may reduce wood's resistance to shear. Checks and splits that extend through the entire thickness of the piece are not permitted. Shake is a separation along the grain, largely between the growth rings, which occur while the wood is seasoning. Shakes in members subjected to bending reduce the resistance to shear and, therefore, should be closely limited in structural members. Decay, a disintegration of wood, results from the action of wood-destroying fungi. It seriously affects the strength properties of wood and its resistance to nail withdrawal. However, if it is determined that the total amount of decay beneath the surface does not extend beyond the surface outline, the dimensional limitations for knots apply. Warp along the longitudinal axis of the lumber should not be more than 1 in. in 8 in. of length. Lengthwise, warp should not exceed $\frac{1}{8}$ in. in 4 in. of width. The acceptable range of moisture content in lumber used for skids, crates, and boxes should lie between 12% and 19%.

No opportunity should be overlooked to utilize materials besides wood or plywood for crate, box, or skid construction where practically and economically feasible. Steel alternatives for wood skid members may be the simplest substitution. For example, a 4-in. × 4-in timber may be replaced by a 3-in. × 4.1-lb steel channel iron or by a 3 in × 5.7-lb I-beam or by a 3-in × 3-in. × $\frac{1}{2}$-in.-angle iron. Wherever material substitutions are contemplated for heavy loads, they should be engineered for the specific purpose intended.

Design

A proper analysis of forces to which skids, crates, and boxes may be subjected must include consideration of compression, lateral thrust, impacts, repeated handlings, abuse, tension, and inertial forces on the structure and the cargo. In this regard, it must be remembered that inertial loadings encountered aboard ship often exceed those encountered in the other modes of transportation.

Basic design criteria. Skids must be of Group IV woods and engineered for uniform or concentrated loads. Headers are to be bolted using washers and double nuts or upset threads. Skids are to be spaced not more than 48 in. apart and are to be of single-piece construction or, if over 12 ft long, spliced according to an approved method. The ends of skids are to be chamfered and are to rest on chamfered rub strips. Cargo is to be bolted or tension banded to skids using engineering design methods for determining the sizes of bolts and tension bands. For very heavy loads, consideration must be given to the crushing strength of the wood or alternative material. Load-bearing floorboards can be analyzed as beams using the formulas for bending moment to determine the required section modulus. The allowable stress used in the formula for section modulus can be taken as 1000 psi. For hard woods, this allowable stress provides an adequate factor of safety. Where the distance between skids might exceed 48 in., it will be necessary to add an additional skid. The bending moment on the load-bearing floorboards, then, should still be calculated on the basis of the length between outside skids.

For example, suppose that a 15,000-lb load is to be supported on skids whose length has been determined to be 15 ft. Calculate the required timber size. Using the formula BM = (WL/6, the bending moment will be 37,500 ft · lb. Then calculate the required section modulus = BM allowable stress = 450 in.^{3}. Remember to multiply the bending moment by 12 in order to convert foot-pounds to inch-pounds. Since section modulus = (*bd*)/6, you can solve by trial and error by substituting cross-sectional dimensions as follows. Try a 10-in × 10-in. timber. This yields a section modulus of 166.7 in.^{3}. Two (2) skids of these cross-sectional dimensions yield a total of 333.4 in.^{3}. It is obvious, then, that two (2) 10-in. × 10-in. timbers do *not* provide the required section modulus (450 in.^{3}) for a quasi-uniform load. The next logical choice would be two (2) 12-in. × 12-in. timbers. Note, however, that if the load is truly uniform (BM = *WL*/8), the required section modulus is only 337.5 in.^{3}. The two (2) 10-in. × 10-in. timbers then come very close to making the required section modulus, and with the factor of safety provided by using the allowable stress of 1000 psi, the two (2) 10-in. × 10-in. timbers would serve satisfactorily for a strictly uniform load. For most hard woods, the allowable stress of 1000 psi provides a comfortable factor of safety of approximately 1.5. In calculating the basic stresses for structural lumber, impact loading is generally ignored, but long-time loading and a safety factor are considered. A piece of wood will carry less load for a long time than it will for a short time. Consideration should be given to the expected interval of storage, ie, short time versus long time. In addition, if research or historical data indicate that a particular port of loading and/or a particular port of discharge have a higher frequency of cargo claims due to rough handling, timber sizes may be increased by applying an appropriate factor of safety or by requiring that the actual timber size be one full unit larger than derived by calculation. Remember that rub strips must be applied to the skids. An added precaution against the accidental loss of a rub strip is the use of reinforcing straps from the rub strip to the underside of the skid members.

Crates. In general, Groups II and III woods are to be used in crate construction except for skids, headers, and occasionally load-bearing floorboards that should be Group IV woods. All nails used are to be cement-coated. Barbed, screw, or serrated nails are occasionally used, but for practicality and economy, cement-coated nails make the best selection.

All skid members in crate construction generally follow the same construction requirements as skid members used as skids alone. However, skid members sizes in crates can be substantially smaller in cross-sectional area because, in crate design, it is assumed that a large part of the load imposed by the contents is carried by the side panels acting as trusses. Therefore, large skids are not necessary as load-carrying members because the sides act integrally with the skids in this function. While this assumption results in smaller size skids, it does not permit handling and moving a loaded base alone without the sides and ends fastened in place.

Crates may be either open or closed (sheathed). It is not always necessary to sheathe a crate if the contents do not require a degree of protection from the elements. Opened crates require less lumber and, therefore, if properly designed, can be less expensive to fabricate. This may not always be true, and, to achieve greater economy, it is necessary to make an engineering economy analysis to determine whether an open crate or a sheathed crate is more economical in terms of labor and cost of materials. This article discusses the design and construction of open crates since the contents impose loadings on structural members that must be determined by either graphic analysis or engineering methods of resolving forces. The sides and ends, as well as some members of the base, are considered as part of a bridge truss. The Howe truss, with its parallel upper and lower chords and its vertical and diagonal members, has the same general pattern as the frame members of the side of a wooden crate. Since the truss is a framed structure composed of straight members, the stresses in the members due to loads must be either compression or tension. The magnitude and character of the stress in each member can be determined. The required size of all tension members in the truss is determined by the formula $A = P/f$, where A is the required cross-sectional area of the member in square inches, P is the total load in pounds (as determined from the resolution of forces or from graphic analysis), and f is the working stress in pounds per square inch (in tension parallel to the grain). With support at each end, the major stresses in strut and lower-frame members are tension stresses and those in the diagonals and upper-frame members are compression stresses.

Compression members must be designed as columns, and column formulas must be used to determine their sizes. In crate design, there are essentially three column formulas. The selection of the proper formula depends on the unsupported length of the column and, columns are therefore classified as short, intermediate, and long. It is beyond the scope of this article to discuss the various factors and the application of these in the appropriate column formulas. The *Wood Crate Design Manual* is an excellent reference text for the more determined crate builder. It must be remembered that it is most probable that crates of similar dimensions will be stacked vertically (one on top of the other) aboard ship or in storage. If crates are of exactly the same overall dimensions, the corner posts will carry the superimposed loads as columns. These anticipated superimposed loads must be included in the strength analyses. If superimposed loads consists of smaller crates or other types of loads that do not rest on the corner posts, the structural members of the top panels will be subjected to bending as well as to compressive loadings resulting from handling, especially with slings or chains. Therefore, these structural members must be analyzed as beams subjected to bending and axial compression. If highly concentrated loads (in the center third of the crate) are to be crated, the skid depth must be increased. This, however, will be apparent if the side panels are properly analyzed.

Crates should not be higher than they are wide (even well-constructed, durable shallow crates are acceptable; see examples in Fig. 9). Side panels are to be divided by a suitable number of vertical members so that diagonals are as close to a 45° angle as possible. Top and bottom horizontal and vertical

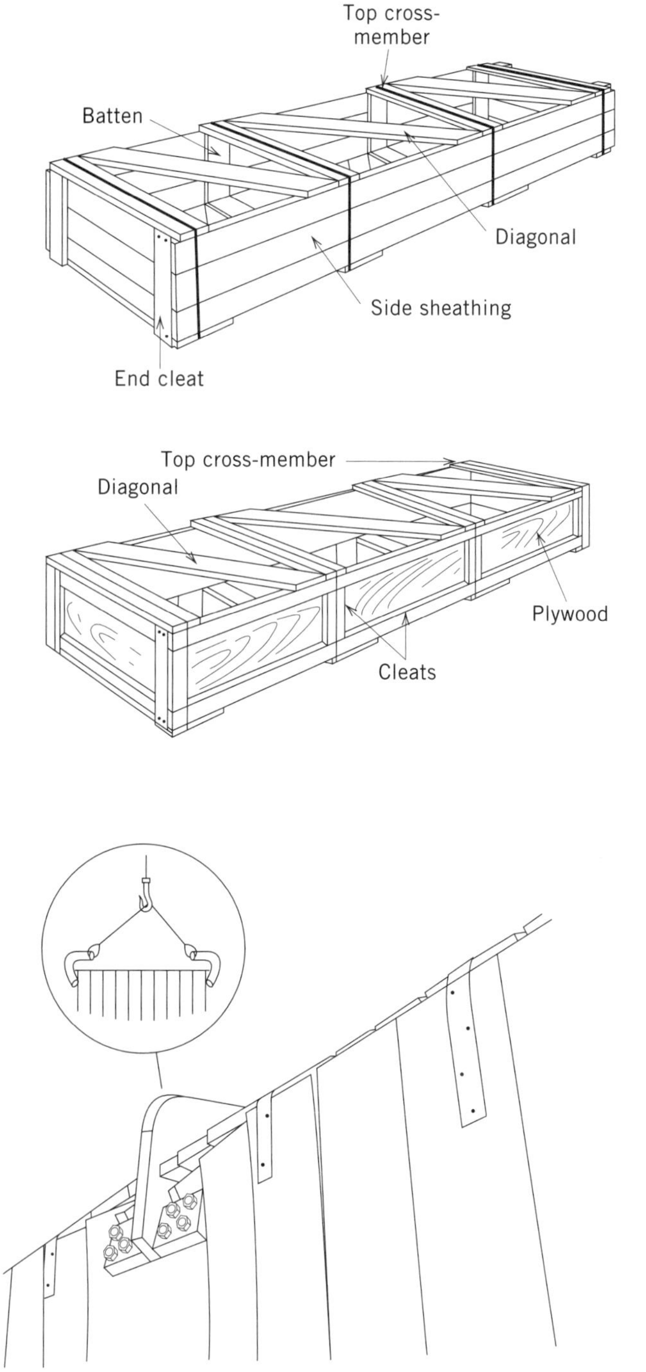

Figure 9. (**a,b**) Typical shallow crates [(**a**) lumber; (**b**) plywood]; (**c**) crate damage caused by a grabhook when there was insufficient joint support in the top of the crate (2).

members are to be through-members (extend to full extent of length and height dimensions). All end panels are to have one or more diagonals. Avoid splicing members less than 16 ft long. Splicing, where required, must be accomplished using an approved method. Intersections of diagonals are to be reinforced using plywood gussets or lumber bridge.

If the crate height exceeds 72 in., a through intermediate horizontal member is required at midheight of the crate. Side-panel width (ie, between vertical members) must not exceed 42 in. The geometry of the panels must be altered to add an additional panel so that panel widths do not exceed 42 in. Lower horizontal members are to rest on the ends of the floorboards with notches cut in headers and load-bearing floorboards to accept these lower horizontal members. Vertical and diagonal members must be in the same plane and to the outside of the horizontal members. Side-panel diagonals should slope from the top of a vertical member down to the skid base.

End-panel design should not exceed 42 in. in width. End panels for crates having a skid width in excess of 42 in. will require intermediate vertical members. End panels for crates having more than two (2) skids will require a vertical member above each skid. Filler strips will be required to allow end panel vertical members to rest across the ends of skids. The filler strips will fill the void between the headers and the lower horizontal members.

Top panels are critical since most members will be in compression under the influence of various types of materials-handling equipment and superimposed loadings. Top-panel lateral and diagonal members will be in the same plane and rest on the longitudinals and extend to their edges.

Where protection from the elements is essential or to meet customer requirements, a sheathed-crate, either $\frac{3}{4}$-in. lumber or $\frac{1}{2}$-in. plywood (C–D, exterior glue) sheathing can be used. Reinforcement strapping is required on all corners and at all intermediate, horizontal, and vertical member junctures after the crate is assembled.

Where crate items require additional protection from the elements, as discussed earlier in this article, appropriate steps must be taken in the early stages of crate construction to apply the required crate liner between internal framing and external sheathing. If the crate is not to be watertight, sufficient ventilation must be provided. If the contents are to be provided with waterproof protection, the item must be properly and totally wrapped with either asphalt-laminated kraft paper or polyethylene film and sealed with waterproof tape in addition to the crate liner material.

The strength and rigidity of crates are highly dependent on the fastenings. Nails, lag screws, bolts, screws, and metal connectors are the most important fastenings in crate construction. This article briefly covers nails and nailing rules. Refer to Anderson's treatise on nailing better wood boxes and crates (3) for a more complete discussion of nails and nailing rules. Cement-coated nails are being more widely used today than ever before, primarily because many crate builders use automatic nailing equipment more readily amenable to the use of belted cement-coated nails. Nails should be driven through the thinner member into the thicker member where possible and should penetrate both members, leaving a minimum of $\frac{1}{4}$ in. for clinching. Clinching is one of the best methods of increasing effectiveness of nails. It is used almost entirely in the fabrication of crate panels, except when frame members or other crate parts are >2-in. thick. Clinched nails have 50–150% greater withdrawal resistance than unclinched nails when driven into drywood. Predrilling the wood before the nails are driven may be necessary to prevent splitting in very dense woods or with nails of large diameter. Nails should not be overdriven. They should be positioned no less than the thickness of the piece from the end or half the thickness of the piece from the side edge. When members (both of which have ≥3-in. thickness) are joined together, bolts are to be used. Nails should be staggered to prevent any two nails from entering the same grain line of any board. Follow approved nailing techniques as outlined in the handbook described above.

Where practical, the use of tension bands will provide an added degree of a reinforcement to the crate assembly. Tension bands should be tight enough to cause a light crushing of the ends of the main structure over which they pass. Tension bands should be stapled to prevent their movement during transportation and handling.

A crate can be assembled before or after loading the item on the skid base, whichever is appropriate, but internal blocking and bracing of the item in the crate must be done at the time the item is set on the skid base. The blocking and bracing must be of substantial design and construction to prevent the free movement of the item within the crate caused by the forces encountered in transportation and handling. Blocking and bracing are not to be taken lightly. The inertial loading on the crate and its contents aboard ship often exceeds twice that encountered in the other modes of transportation.

Boxes. Boxes are not to be used for loads in excess of 5000 lb or lengths in excess of 16 ft. Groups II and III woods are to be used. All nails are to be cement-coated. Boxes that exceed 200 lb gross weight will require 2-in × 4-in. rub strips. When any unsupported span of top, end, or side lumber exceeds 36 in., additional interior battens or cleats will be required. Cleated plywood boxes for loads over 1000 lb must have load-bearing bases with properly sized skids, headers, and rub strips and are to be nailed in accordance with the preceding nailing rules. All boxes require tension bands sized according to the load and arranged in two directions passing over cleats or battens.

Bands should be stapled to prevent movement and should be tight enough to cause light crushing of the corners of the members over which they pass. If a boxed item is of large mass (weight), appropriate filler or cushioning materials filling the voids around the item may not be effective. Therefore, dunnage or blocking and bracing, where practical, should be used. Where the boxed item requires protection from the elements, the internal surfaces of the box may be lined as described under crates and the item itself should be properly preserved.

Unitization and Palletization

Unitization provides a more economic package than individually shipped pieces. It also provides ease of handling and storage and provides a greater degree of protection than individual units. Pipe and conduit are easily bundled. They may require special plastic end protectors to protect internal or external threads. The ends of bundled pipe can be boxed or completely covered with double- or triple-wall fiberboard secured with tension bands. In addition, long bundles may be bucked. *Bucks* are wooden frames that surround the girth of the bundle. They are to be nailed at the right angle junctures and tension-banded. Tension bands are to be stapled to prevent accidental loss of the bands. Bucks should be located at 5-ft maximum intervals.

Lumber and paneling are usually bundled. Care must be taken to avoid overtensioning bands that will mar their edges. Paneling should be covered with water-resistant paper or polyethylene film.

Coiled wire can be bundled. Because of the historically poor bundling methods used for wire, substantial losses have occurred. Consignees frequently use coiled wire to make nails, screws, and bolts. Kinked or crooked wire will interfere with the extrusion process and substantial claims may arise.

Lumber and paneling can also be bundled. Care must be taken to ensure that tension bands do not damage the items bundled. Coiled wire can be bundled. The usual problem encountered here is when tension bands or wrapping wires comes off, allowing the coils to become disarranged or crushed. The consignee usually has a specific machine set up for manufacturing items such as bolts, screws, and rods. Hence, damage to the coils can result in a claim.

Palletization is best for similar-size fiberboard cartons, drums, and coiled sheet steel. Do not overload or stack the items too high. Overloaded pallets break. An excessively high stow can result in crushing of the cartons in the lower levels. Then further damage occurs when the otherwise neat stacking falls apart. In-transit handling of broken-down pallets usually results in further abuse. Fabric or steel banding is usually one of the best methods of securing an entire unit load. It may be advisable to employ vertical and horizontal corner protectors to eliminate banding damage. Shrink wrapping provides a high-strength securement to a palletized load. It can also be applied to protect against contact with water.

Palletized drums must be fitted with partially framed tops to allow for safe utilization of tension bands. Tensioning of the bands must not cause damage to the pallet load-bearing boards or to the cargo. Fiberboard sheets across the top and partially down the sides of drums will serve to keep the tension bands positioned and provide a level surface for stacking other pallets.

In general, old or badly worn pallets should not be used. Pallets can be repaired. Some companies use secondary reusable pallets with good success. Still, old wood and loose nails can result in broken pallets that invite further abuse in handling.

Container Problems

Containers are subject to the same shock and vibration forces as are over-the-road trailers and railroad cars. In addition, very high inertial loadings are experienced by the contents as a result of swaying, pitching, rolling, heaving, yawing, and surging of a ship at sea.

The major causes of damage to containerized cargoes are improper stowage, inadequate dunnage, lack of security, overloading, and poor weight distribution. Any assumption that the container is a substitute for safe stowage and handling is an invitation to disaster. A good rule to adopt is to, stow and secure for the worst conditions.

Containers may travel to seaport by railroad. Constant vibration and occasional sharp humping forces must be taken into consideration. When loading a container for sea transport, always consider the six basic ship motions; yaw, heave, sway, pitch, roll, and surge. Occasionally heavy seas will have a pronounced effect on cargoes in containers.

Blocking and bracing is essential to the safety of the cargo.

Materials commonly used for securement are lumber, plywood, strapping, fiberboard, and inflatables. Lumber can be used as a filler, for decking, blocking and bracing, and constructing partitions. Plywood can be used as a partition, divider, and auxiliary decking. Heavy-duty strapping can be used to separate cargo and to tie down heavy and awkward items. The strapping must be firmly anchored and properly tensioned for the greatest effectiveness. Fiberboard is available in sheets, rolls, and structural shapes and can be used for light-duty bracing, as dividers, decks, and partitions. Inflatables are available in paper or rubber and may be reusable or disposable. They are expensive and not recommended for voids in excess of 18 in. Inflatables are used for light- and medium-duty bracing.

Condensation is another major cause of damage to cargo. It is not generally known that fiberboard boxes and crate or dunnage lumber contribute to the generation of water vapor in sea containers. Furthermore, in changing temperature and humidity conditions, condensation can form within containers and drip down from the overhead onto moisture-sensitive packaging and cargoes. These conditions can be remedied, in part, by using seasoned lumber for dunnage and crate construction. Adequate (prepackaging) preservation of moisture-sensitive items can be the best method of protecting against water damage. Overwraps such as canvas or polyethylene sheeting will serve to divert condensation that drips from the overhead. With some types of cargoes, it may be necessary to provide completely waterproof liners or wraps constructed VCI/VPI or desiccant materials. Cargoes are more likely to be subjected to condensation and water damage over a long time interval such as long transportation distances or long-duration storage. Therefore, it is incumbent on the shipper to consider all these factors when preparing cargo for export.

The annual report of the Cargo Loss Prevention Committee of American Institute of Marine Underwriters contained valuable advice on the use of containers in shipping. Extracts from the report follow:

> Normal reaction is for a shipper to use domestic packaging and consider the container as an additional protection. It is our experience that containers may leak and the cargo must be protected against moisture and water damage. Additional emphasis is needed on the use of dunnage and tight packing in containers.
>
> Several offices report that losses have been experienced due to defective containers. Many shippers do not employ adequate container inspection procedures. It appears that some insureds think that because their goods are containerized the packing can be of a minimal nature. Our experience continues to show that such minimally packed goods are likely to be damaged by rain and/or seawater when shipped in poorly maintained containers.
>
> Many shippers continue to use containers as a means of packaging and claims due to crushing, water damage and improper or inadequate stowage, especially of heavy machinery, continue to be a problem.
>
> Losses have also been experienced where shipments of high value electronic units have been made in containers. The units are secured on lumber bases which in turn are then secured to pallets with the units being entirely shrouded in polyethylene sheeting which also encompasses the wooden base to which the units are attached. It was found that these wooden bases were constructed of green or improperly dried lumber, which gave off moisture, which became trapped in the polyethylene shrouding. During the course of transportation, fluctuations in temperature caused this moisture to condense on the exposed metal areas of the electronic unit, causing serious damage.
>
> We have also observed that machinery which is top heavy has on occasion been stowed in containers without adequate bracing and precautions to prevent the machine from tipping. On one occasion, a container loaded with tractor engines arrived at destination with every engine damaged, as a result of their center of gravity being high. Although the engines were braced along the floor of the container, and would prevent lateral or fore and aft movement, there was no bracing up higher and with their relatively high center of gravity they tipped, coming in contact with one another, causing damage. Shippers should be cautioned to be aware of the center of gravity of machinery and other heavy or odd sized pieces being shipped in containers and adequate bracing to prevent tipping should be done.
>
> Although containerization of cargo continues to grow worldwide, the packing of goods into house to house containers by shippers, as well as securing of the goods in the containers, continues to be a cause for concern. We find in many instances cargo to be damaged by condensation within the container and/or by extreme heating or freezing. Once a container is loaded and the doors securely shut, the only way in which the outside weather can affect the cargo is through changes in the temperature.
>
> Unless the internal temperature of the container is controlled by mechanical or other means, the temperature of the air inside the container will follow the temperature of the air outside. Extreme fluctuations in the air temperature can cause condensation to settle on the cargo; if the container is exposed to very high temperatures, the cargo within the container may sustain damage or, if the container is exposed to extremely low temperatures, the cargo can be damaged, such as change in the chemical state of some goods (eg, drugs) making them useless and sometimes dangerous, or freezing of bottled liquids with subsequent bursting of bottles.
>
> Taking into account the anticipated voyage, its length, and temperatures normally expected to be encountered, consideration should be given to goods which are susceptible to damage by extreme heat or cold by shipping them in insulated or refrigerated containers where the temperature inside the container can be controlled to some degree.
>
> In other cases, certain precision machinery or electronic equipment may require the use of VCI, VPI or powder desiccants. It must be pointed out, however, that in order for these moisture control measures to take full effect, the packaging and packing must be *totally* sealed.

Shipping Losses and Insurance

Under normal conditions of domestic rail shipments, material loaded in open cars is inspected by railroad representatives before the car is moved, and the railroad assumes responsibility. In closed cars, however, the Association of American Railroads rules are not mandatory and are used only for guidance. Closed cars are not ordinarily inspected, but if damage occurs to the contents, they are inspected at the destination to determine the cause before a settlement is made.

Export shipping companies are specifically exempt from most forms of liabilities under the laws of many countries. The exceptions to this usually include loss or damage due to negligence in proper loading, custody, or delivery of the goods.

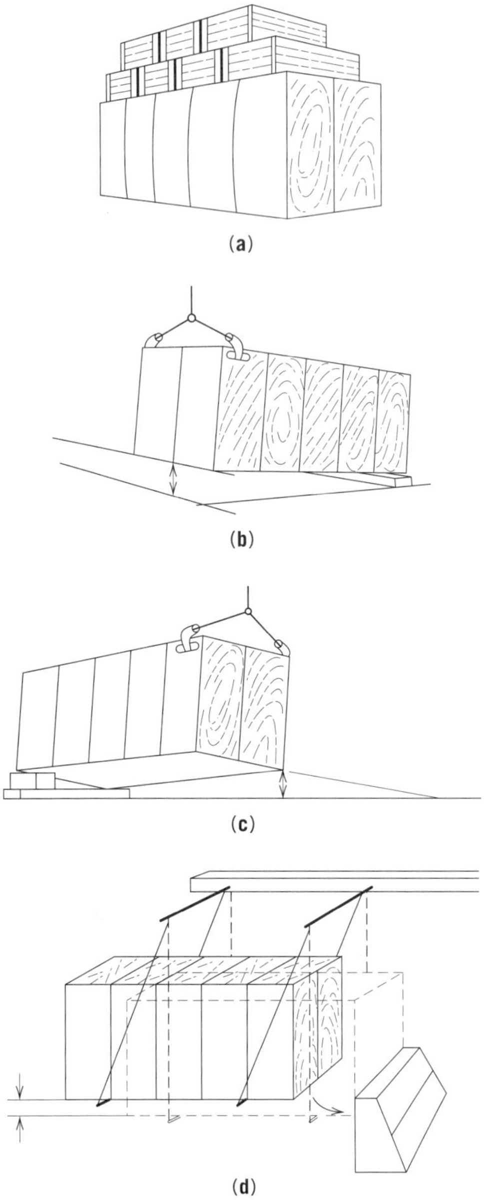

Figure 10. Rough-handling tests for crates; (**a**) superimposed load; (**b**) edgewise drop; (**c**) cornerwise drop; (**d**) pendulum impact (2).

The shipper or consignee must assume responsibility for all remaining risks during the shipment.

To prevent loss to the shipper, a form of marine insurance covers these losses. Marine insurance may be obtained to cover such perils as pilferage, theft, and leakage, as well as loss or damage if the ship should sink, burn, or be involved in a collision. However, the more hazards covered by the policy, the higher the rate, so it is not economical to pay for broader protection than is actually required.

Rates in marine insurance are rather complex and are not fixed. They depend on, among other things, the type of vessel, the route, the perils insured against, the type of packing used, and the loss record of the shipper. This latter factor reflects the type of container and the method of cushioning, blocking, and bracing used by the shipper, because well-constructed and well-packed crates will normally receive little damage during the voyage. A shipper who uses adequate containers pays lower rates. Underwriters keep statistical records of shippers with whom they deal and allow lower rates for those with good records.

Damage and Claims

Too frequently one or more of the three steps in preparation for shipment are less than adequate, and this will inevitably result in damage of one type or another to the cargo. Precision machinery shipped long distances or subjected to long-term storage without adequate preservation will rust. Electronic equipment shipped without serious consideration being given to adequate protection from vibration and shock will sustain physical damage. Items requiring a degree of dimensional stability (rigidity) will be damaged if the packaging and packing is not properly engineered and constructed. One way to ensure greater construction durability is by preshipment damage testing of crates or other containers (see Fig. 10).

With regard to containerization (sea containers), a tremendous number of cargo losses occur each year because of ignorance of the proper use of these containers, specifically the requirement for adequate blocking and bracing and protection of the cargo against condensation. It is imperative that the shipper understand that the use of sea containers does not guarantee that the cargo will be received by the consignee in sound condition.

A major marine insurance company has made an analysis of principal causes of losses and found that, in the "handling and stowage" category, 39% of all losses are due to container damage including breakage, leakage, and crushing. In the "water damage" category, 10% of all losses are due to freshwater, sweat, and saltwater damage. Nevertheless, it concludes, 70% of all cargo losses are preventable.

Conclusion

In summary, shippers would do well to have, as their first concern, getting the goods to their customers in sound condition, and to that end, the author hopes that this article is a valuable aid.

BIBLIOGRAPHY

1. U.S. Department of Agriculture, *Wood Handbook,* Agricultural Handbook No. 72, Forest Products Laboratory, Madison, WI, 1974.
2. U.S. Department of Agriculture, *Wood Crate Design Manual.* Agricultural Handbook No. 252, Forest Products Laboratory, Madison, WI, 1964.
3. L. O. Anderson, *Nailing Better Wood Boxes and Crates,* Agricultural Handbook No. 160, USDA, 1959.

General References

Code of Federal Regulations, 49 *CFR* Parts 100–199, Subparts D and E, Office of the Federal Register, National Archives and Record Service, Oct. 1978.

J. P. Colletti, *Export Packing in the Marine Environment,* U.S. Department of Commerce, *Damage Prevention in the Transportation Envi-*

ronment, Mechanical Failures Prevention Group, National Bureau of Standards, Oct. 1981, Gaithersburg, MD, pp. 8–37.

Departments of the Army, Navy, and Air Force, *Packaging of Materiel: Preservation,* Vol. 1, Manual DSAM 4145. 2, 1976.

W. Friedman, and J. Kipnees, *Distribution Packaging.* Krieger, New York, 1977.

J. F. Hanlon, *Handbook of Package Engineering,* McGraw-Hill, New York, 1971.

U.S. Department of Transportation, *A Shipper's Guide to Stowage of Cargo in Marine Containers,* Maritime Administration, Washington, DC, 1982.

JOHN P. COLLETTI
John P. Colletti and Associates
Pittsburgh, Pennsylvania

MATERIALS HANDLING

Introduction

The Industrial Revolution of the late 18th–early 19th century saw the introduction of powered machines to replace manual labor in accomplishing the same or similar tasks. Consequently gains in labor and productivity per worker-hour of individual workers were accomplished. Moreover, this time period saw the transition from a "domestic system" to a "factory system." Concurrently the need for materials-handling systems increased to meet the needs of the more "productive" factory system. Initially this need was for labor-saving devices to handle material in manufacturing and distribution (1,2). However in today's competitive and global marketplace the rationale for materials-handling systems is multifaceted.

The "cost" estimates for materials handling vary in nature. Generally materials handling does not add value to the product, but it can and does add to the final product costs. Estimates of this cost range from as little as 20% up to 90% in some cases. Conversely, however, efficiently designed materials-handling systems have been known to reduce a plant's operating cost by 15–30% (3). This figure could vary depending on the industry and product categories. For example, the food-processing industry would have a fairly high per unit cost whereas an industry utilizing tanks and pipes would have a significantly lower per unit cost.

The potential costs and benefits associated with materials-handling systems are significant and thus impact our domestic and global competitiveness. This article provides an overview of materials-handling principles and guidelines, and relates these to the field of packaging. Generalized definitions, objectives, principles, and examples of materials-handling checklists are included. Brief discussions on equipment are also presented, as well as approaches toward design of materials-handling systems. Because of the strong relationship between facilities layout and materials handling, a brief section on techniques for facilities layout is also included. A final section showing the relationship between packaging and materials handling concludes the article.

Definitions

Simplistically and intuitively, *materials handling* is defined as handling material. However, various authors define this term in increasingly comprehensive manners. Meyers (4) indicates that materials handling is "the function of moving the right material to the right place, at the right time, in the right amount, in sequence, and in the right position or condition to minimize production costs." He further states that materials handling involves the handling equipment, the storage facilities, and the control apparatus. Tompkins and White (5) similarly state that materials handling uses the right method to provide the right amount of the right material at the right place, at the right time, in the right sequence, in the right position, in the right condition, and at the right cost. Again, the emphasis is more than just handling materials. Storage and control are also important. Time utility—the right material at the right time—and place utility—the right material at the proper place—are also emphasized. They also point out that materials handling can be described in terms of quantity, position or orientation, condition, space, profit, quality, safety, and productivity, which are also significant factors relative to packaging.

Gelders and Pintelon (2) offer the following definition from an engineering point of view: "the art and science involved in picking the right system, composed of a series of related equipment elements or devices designed to work in concert or sequence in the movement, packaging, storage and control of material in a process or logistics activity." Sims (6) states that materials handling includes the movement of liquids, bulk solids, pieces, packages, unit loads, bulk containers, vehicles, and vessels. He further defines a materials-handling system as a series of related equipment elements or devices designed to work in concert or sequence in the movement, storage, and control of material in a process or logistics activity, where each system must be custom designed for each unique application. Finally, Sims (6) states that materials handling is the portion of the business and economic system that affects the physical relationship of material, products, and packaging to the product, process, facility, geography, or customer without adding usable worth or changing the nature of the products.

Clearly, materials handling is more than "handling materials." It is a complex, multifaceted activity that extends beyond the physical confines of a manufacturing facility. Properly designed materials-handling systems, while not adding value to a product have the potential to minimize overall costs, and even in the case of product/package-delivery systems, impact the end-use customer or consumer.

Objectives of Materials Handling

It is generally accepted that the primary goal of material handling is to have the least possible handling, ie, less handling at less cost leading to reduced costs of production. However, a more specific and comprehensive set of objectives that more accurately reflects the diverse nature of materials handling, as stated by Gelders and Pintelon (2), are as follows:

- To increase the efficiency and effectiveness of materials flow
- To increase productivity in manufacturing (plant) or in distribution (warehouse)
- To increase space and equipment use
- To improve safety and working conditions
- To reduce materials-handling costs

- To avoid high capital requirements
- To ensure a high level of systems flexibility, reliability, availability, and maintainability
- To improve integration between materials and information flow
- To smooth the flow of materials through the logistics pipeline (from supplier to final customer)

How can these objectives be achieved via appropriate materials-handling system design? The following section provides guidance on basic principles, checklists, and critical questions that may be used to meet one or more of the above objectives.

Basic Principles of Materials Handling

In order to meet the primary goal of materials handling, as well as the expanded objectives listed above, a list of principles based on accumulated knowledge and practice of experts in the field has been developed. The list of 20 principles adopted by the Materials Handling Institute Inc., provides rule-of-thumb advice that facilitates optimized materials-handling system design, and also serves to help develop checklists relevant to the system under focus. A slightly modified version (2) is shown in Table 1. These 20 principles provide an initial frame of reference to begin design of a new materials-handling system, make modifications to an existing system, or analyze a system to see if it is meeting predetermined objectives.

Table 1. Basic Principles of Materials Handling

1. *Planning:* Study the problem thoroughly to identify potential solutions and constraints and to establish clear objectives.
2. *Flow:* Integrate data flow with physical materials flow in handling and storage.
3. *Simplification:* Try to simplify materials handling by eliminating, reducing, or combining unnecessary movements and equipment.
4. *Gravity:* Use gravity to move materials wherever possible, while respecting limitations concerning safety and damage.
5. *Standardization:* Standardize handling methods and equipment wherever possible.
6. *Flexibility:* Use methods and equipment that can perform a variety of tasks.
7. *Unit load:* Handle products in as large a unit load as possible.
8. *Maintenance:* Plan maintenance carefully to ensure high system reliability and availability.
9. *Obsolescence:* Make a long-range plan, taking into account equipment life-cycle costs and equipment replacement.
10. *Performance:* Determine the efficiency, effectiveness, and cost of the materials-handling alternatives.
11. *Safety:* Provide safe materials-handling equipment and methods.
12. *Ecology:* Use equipment and procedures that have no negative impact on the environment.
13. *Ergonomics:* Take human capabilities and limitations into account while designing a materials-handling system.
14. *Computerization:* Consider computerization wherever viable for improved materials and information control.
15. *Utilization:* Try to obtain a good use of the installed capacity.
16. *Automation:* Consider automation of the handling process to increase efficiency and economy.
17. *Operation:* Include operating costs (energy) in the comparison of materials-handling alternatives.
18. *Integration:* Integrate as much as handling and storage activities into one coordinated system, covering receiving, inspection, storage, transportation, production, packaging, warehousing, and shipping.
19. *Layout:* Keep in mind that layout and materials handling are closely linked and that an interactive procedure is often needed to obtain their best coordination.
20. *Space use:* Choose the material handling equipment so that effective use is made of all (cubic) space.

Source: Reprinted with permission from Materials Handling Institute, Inc.

Methods of Analysis

Analysis of existing or proposed materials-handling systems can be approached in either a qualitative or quantitative manner. In order to facilitate the analysis of a materials-handling system in general, Tompkins and White (5) propose modifying the well-known engineering design process as follows:

1. Define the objectives and scope for the materials-handling system.
2. Analyze the requirements for handling, storing, and controlling material.
3. Generate alternative designs for meeting materials-handling system requirements.
4. Evaluate alternative materials-handling system designs.
5. Select the preferred design for handling, storing, and controlling material.
6. Implement the preferred design, including the selection of supplies, training of personnel, installation, debug and start-up of equipment, and periodic audits of system performance.

Several other authorities in the field also suggest using this approach. Utilizing the expanded list of objectives and the principles of materials handling serves to address step 1 of this process. The next four steps can be approached qualitatively, or quantitatively. A qualitative analysis may take a questioning approach with yes/no or scaled/weighted answers. Basic questions to consider asking when addressing the system under study include "why," "what," "where," "when," "how," and "who," and "which." More specifically, and at a *minimum,* Tompkins and White suggest asking and answering a set of questions as shown in Table 2 (see also Fig. 1).

This "questioning attitude" leads to the basic equation of developing the best materials, moves, and methods in order to arrive at the preferred system. Some issues to consider in this equation include the type of materials and their physical characteristics, the quantities of the materials to be moved, frequency of moves, sources of moves, and methods by which this may occur.

Numerous checklists are available in the literature that aid in qualitative analysis and help to answer the basic questions listed above (2,4–6). It should be noted that any checklist may have to undergo a certain amount of modification to fit the system at hand. Certain aspects of the checklist may

Table 2. Basic Materials-Handling System Design Questions

Why
Is handling required?
Are the operations to be performed as they are?
Are the operations to be performed in the given sequence?
Is material received as it is?
What
Is to be moved?
Data are available and required?
Alternatives are available?
Are the benefits and disbenefits (costs) for each alternative?
Is the planning horizon for the system?
Should be mechanized or automated?
Should be done manually?
Shouldn't be done at all?
Other firms have related problems?
Criteria will be used to evaluate alternative designs?
Exceptions can be anticipated?
Where
Is materials handling required?
Do materials-handling problems exist?
Should materials-handling equipment be used?
Should materials-handling responsibility exist in the organization?
Will future changes occur?
Can operations be eliminated, combined, simplified?
Can assistance be obtained?
Should material be stored?
When
Should material be moved?
Should I automate?
Should I consolidate?
Should I eliminate?
Should I expand?
Should I consult vendors?
Should a postaudit of the system be performed?
How
Should materials be moved?
Do I analyze the materials-handling problem?
Do I sell everyone involved?
Do I learn more about materials handling?
Do I choose from among the alternatives available?
Do I measure materials-handling performance?
Should exceptions be accommodated?
Who
Should be handling materials?
Should be involved in designing the system?
Should be involved in evaluating the system?
Should be involved in installing the system?
Should be involved in auditing the system?
Has faced a similar problem in the past?
Which
Operations are necessary?
Problems should be studied first?
Type equipment (if any) should be considered?
Materials should have real-time control?
Alternative is preferred?

Source: Reprinted with permission from Tompkins and White (5).

not be appropriate, or additional aspects may be required. However, the previous stated objectives as well as the principles of materials handling can facilitate selection or development of checklists for use as a qualitative analysis tool. One such example of a recent checklist as compiled by Gelders and Pintelon (2) is shown in Table 3.

Quantitative analysis can provide an objective basis to evaluate MHSs as compared to the more subjective qualitative techniques. This may include computing measures such as *efficiency,* which can be defined as the theoretical computed input divided by the actual consumed output; *effectiveness,* which is the actual achieved output divided by the theoretical expected output; and *productivity,* which is defined as output over inputs. Gelders and Pentelon (2), among others, refer to

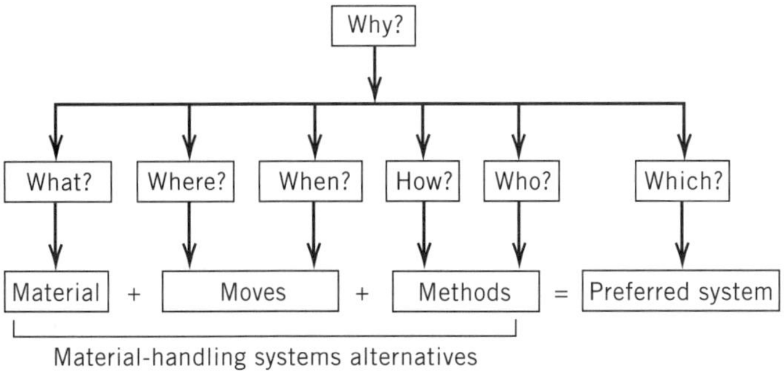

Figure 1. Basic equation for materials-handling system design. Reprinted with permission from Tompkins and White (5).

these as performance indicators (a related discussion of these issues can be found in the article by Zepf in this Encyclopedia). For instance, throughput time (cycle time) is one measure of effectiveness, utilization of warehouse space, or pallet space (unit load) is a measure of efficiency, and productivity may be measured in terms of labor components relative to the materials-handling function. Comparing these or other measures against predefined targets may help to direct modifications to the entire MHS, or only portions of the same.

Overall MHS analysis may be accomplished through the use of simulation software. The software can be used to compare various alternatives against specified criteria, and perform what-if analysis (7–9). The majority of current simulation software packages offer animation capabilities that allow the designer to see dynamic representations of the system under study. Specialized software such as CAPE (10) or TOPS (11) may be used to focus on palletization and the unit-load concept as well as optimized package design. Simple spreadsheet analysis utilizing macro programming may also be used for this purpose. Beyond ones own operations, benchmarking, that is, comparing one's own strengths and weaknesses to a competitor, may also be applied relative to materials-handling analysis.

One of the earlier structured approaches to MHS analysis was developed by Muther and Haganas (12). Their method, *systematic handling analysis* (SHA), was developed from the premise that any analysis must depend on material, moves, and methods. They describe the approach as "an organized, universally applicable approach to any materials-handling project," consisting of a "framework of phases," "pattern of

Table 3. Checklist for Materials-Handling System (MHS)

General questions
Topic: Global aspects of the MHS
Examples: Is the MHS flexible enough to cope with changing product volumes and mixes, throughput and service needs, technology options, customer requirements, etc?
Is the MHS integrated with the production system? Are interfaces with other business functions working okay?
Is standardization a main concern while investing in new equipment?
Has the company clear objectives concerning the degree of automation to apply to the MHS?
Specific questions
Topic: Different areas of (and interfaces with) the MHS
Examples:
Receiving, including dock operations and inspection:
Does the dock equipment match the types of warehouse help to reduce energy costs?
Could doors between dock and warehouse help to reduce energy costs?
Do quality control and inspection reports satisfy the needs of the customer, ie, the manufacturing process?
Storage
Are there any unnecessary packing-unpacking operations needed due to lack of standardization or choice of the wrong unit load?
Is the storage equipment sized correctly for the material stored?
Is the tradeoff between storage density and selectivity optimized?
Handling operations in the manufacturing process
Is there any backtracking in the flow path?
Are there any areas with traffic congestion?
Is production work delayed owing to poorly scheduled delivery and removal of material?
Other picking
Are travel distances between picks minimized where possible?
Would a part-to-picker instead of a picker to part system improve the picking operations?
Are low-activity items out of the way but accessible to pickers when necessary?
Packaging
Are mispicked orders corrected in a timely manner?
Is volume sufficiently high to justify addition of automatic packing equipment to replace manual operations?
Did you make the right choice between strapping, wrapping, etc?
Shipping
Are receiving and shipping operations separated so that they do not interfere?
Are the shipping operations well planned?
Would additional equipment speed up the loading operations?
Information flow and data collection
Topic: The amount and type of data collected, the data processing, and information flow accompanying the material flow
Examples:
Does a computerized database of inventory information exist?
Does each stockkeeping unit have a unique identification number?
Is there a satisfactory performance reporting?

Source: Reprinted with permission from Gelders and Pintelton (2).

procedures," and "set of conventions." They further identify four phases as "external integration," "overall handling plan," "detailed handling plan," and "installation." The total methodology makes extensive use of charts, checksheets, flowcharts, and detailed procedures. In short, the method provides a systematic approach to determine solutions to materials-handling problems. When used in conjunction with the basic principles of material handling, it can provide a powerful framework for identifying good solutions to material handling problems.

Plant Layout and Facilities Layout

Materials handling is an integral part of plant layout. Conversely stated, plant layout is an integral part of materials handling (the proverbial chicken and egg). In short, both MHS design and plant layout design must go hand-in-hand. A cursory review of the stated objectives of materials handling and the MHS principles reveals commonalties, for instance, increase space and equipment use and improve safety and working conditions (objectives), flow, utilization, and space use (principles). These objectives and principles, to name just a few, are also pertinent to effective facilities and plant layout.

Facilities layout, much like material handling is a voluminous subject in and of itself. However, for the purposes of this chapter, some methods of analysis are briefly discussed. The techniques used may be simple or complex, and either quantitative or nonquantitative. Tompkins and White (5) provide a comprehensive discussion of the subject matter. The most basic method to determine layout design is by way of scaled templates of the facility, machinery, material, and people. The templates can be arranged using heuristics to determine the most appropriate layout. This can be accomplished with the aid of computer-aided-design (CAD) packages. When areas of a facility can be departmentalized, matrix techniques such as from–to charts, or "closeness" charts may be used. If some objective such as minimizing materials-handling costs, or minimizing distances traveled can be identified, several optimization models may be used (13).

A complete methodology, *systematic layout planning* (SLP), developed by Richard Muther, and similar in nature to systematic handling analysis (SHA) also is effective for use in existing or new facilities. Similar to SHA, it consists of a framework of phases, a pattern of procedures, and a set of conventions. In general the process includes four overall phases of establishing a location, planning the general overall layout, preparing detailed layout plans, and finally installation at the facility. Also like SHA, the method is logical in nature, includes detailed graphic procedures, and leads to good solutions, based on the initial inputs.

Numerous computer software programs have been developed which address facilities layout. Three such programs are ALDEP (automated layout design program), CORELAP (computerized relationship layout planning), and CRAFT (computerized relative allocation of facilities technique). (13) More recently, *factory CAD, factory plan and facility plan flow* suite of facilities-planning software (14) has been introduced based largely on the procedures pioneered by Muther. Programs such as these are useful when problems are large and complex. However, they are not guaranteed to provide optimum solutions. Experience, judgment, and intuition are still necessary when utilizing this or other kinds of software.

Materials-Handling Equipment, Material, and Methods

There is an abundance of materials-handling equipment that exists for various applications. Usually there are multiple choices for each application. The literature for materials handling contains guidelines, tables, and classification schemes to help in selecting the right equipment for the right purpose (1,3). Furthermore proper use of the principles of materials handling and appropriate checklists, in line with stated objectives, can also serve to identify the right equipment for the given application. In general, the three broad categories of materials-handling equipment are trucks, conveyors, and cranes and hoists. In the total systems concept, storage devices such as pallet racks and bins, automated storage and retrieval systems, are also included, and pallets, slipsheets, and returnable containers are examples of material and methods.

Materials Handling and Packaging

The previous sections of this article discussed general principles, objectives, and techniques of materials handling in a broad sense without reference to a particular industry or manufacturing application. However, as the most comprehensive definition (6) stated earlier implies, the issue of material handling is usually not singular in nature but is indicative of a more complex and comprehensive system. Consequently, materials handling can be approached from a systems perspective. Figure 2 illustrates the extent to which a systems view can be taken. This view shows an inclusive life-cycle approach that includes the supplier of the basic raw material and extends to disposal of the product and package. Decisionmaking criteria related to material handling, therefore should extend beyond the confines of the "manufacturing" facility, or process, to include issues related to packaging, distribution and warehousing, wholesale or retail, the consumer, and solid waste.

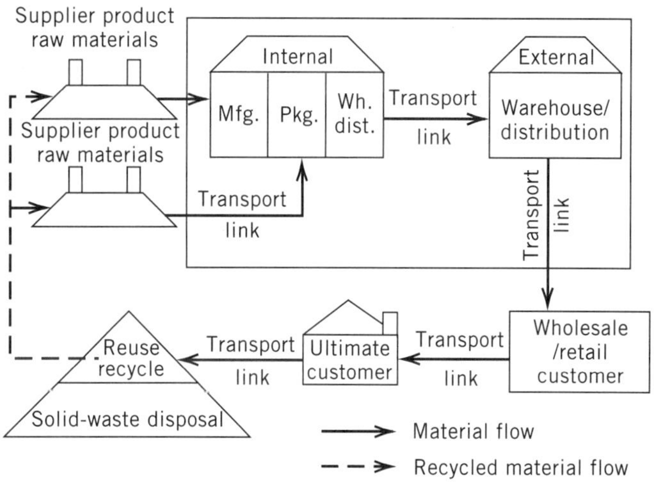

Figure 2. Materials handling in a system context.

Packaging, like materials handling, can also be viewed from a systems approach as shown in Figures 3 and 4. Figure 3 shows the broad, interdisciplinary nature of packaging, and identifies the functions it must address, ie, containment, protection, communication, and performance. It also illustrates that a package can be primary, secondary, tertiary, or quaternary in form and function. This view also indicates packaging extends beyond the confines of the "manufacturing" facility, and that a life-cycle approach toward the development of packaging is appropriate. The complexity of the systems view is influenced by the product itself and the particular industry classification. For instance, a consumer foods or pharmaceuticals company would have a very complex system that may need to take into account product requirements, manufacturing requirements, distribution and warehousing requirements, marketing and consumer requirements, and solid-waste requirements relative to packaging development. The range of packaging material, processes, and equipment in this instance could be quite wide. However, the packaging for the manufacturer of a small electric motor for use in another product may be limited to corrugated boxes that are hand-packed and hand-palletized.

Figure 4 focuses more specifically on the point where the product, the package, and the package machinery meet and are united, influenced by human factors (labor, management policies). This view is relevant at the initial "marriage" between the product and the primary package as well as subsequent "marriages" when the primary filled package (becomes product) is united to a secondary package, and so on. This view is typically confined to the "manufacturing packaging" facility, and does not usually extend beyond those boundaries.

Although it is beyond the scope of this article to identify every packaging issue that influences materials-handling design, or vice versa, several examples may show that synergism does exist between material handling and packaging. In the systems view of both materials handling and packaging, the concept of the unit load—handle products in as large a unit load as possible (principle number 7)—is relevant to each. As previously mentioned, specialized software (10,11) exists that can be used to optimize unit loading on a pallet, for single and mixed loads, or for optimized loading in the

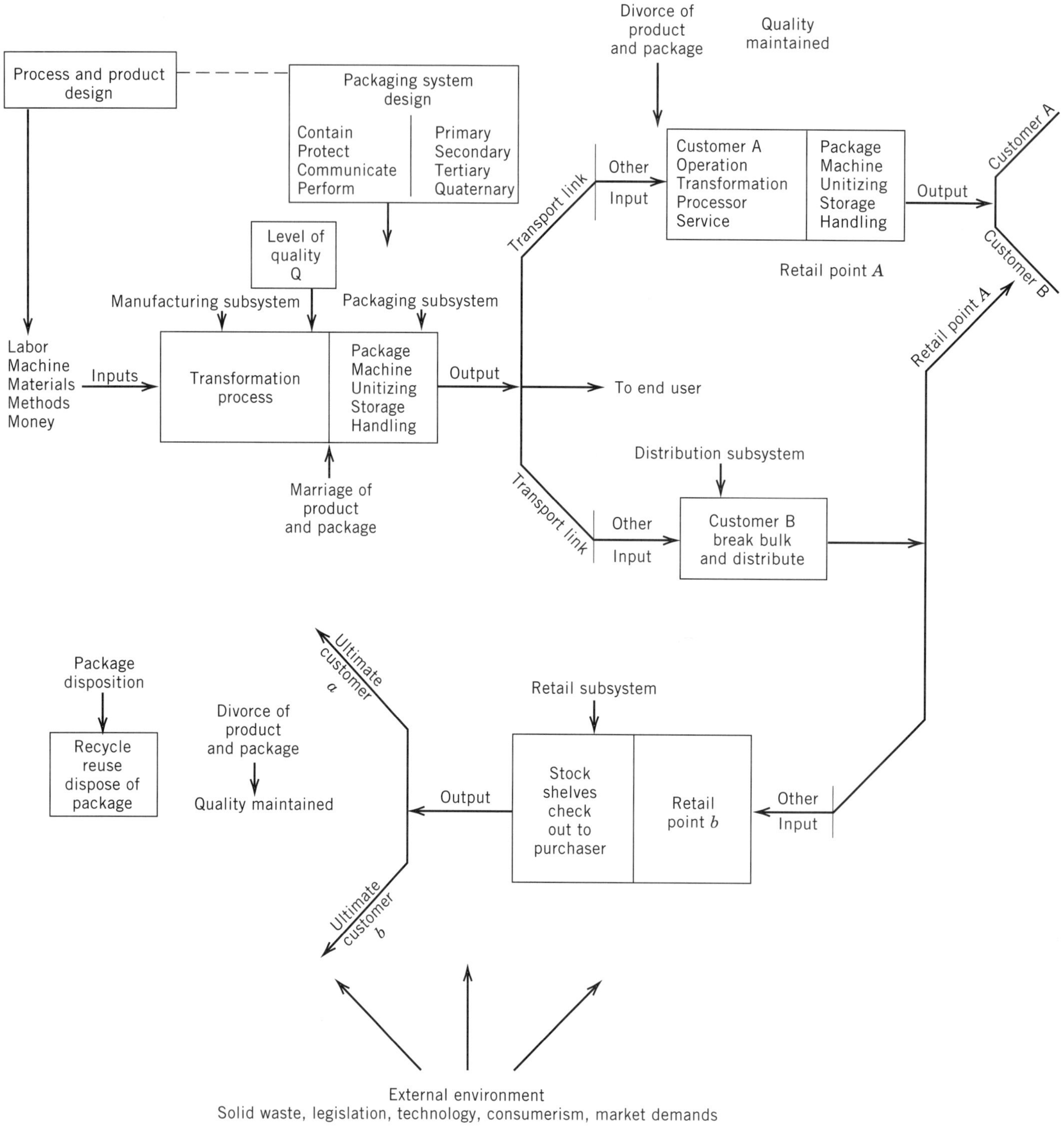

Figure 3. A systems representation of packaging.

transport vehicle. Addressing this concept can help to meet objectives such as reducing materials-handling costs and packaging material costs, and also to increase productivity in manufacturing and distribution. In addressing the pallet, choices exist depending on a number of factors. Choices can be made between one-way pallets, or closed-loop pallets made from wood or plastic, one-way or closed-loop pallets made of recycled material, slipsheets, or no pallet at all. Those choices, in turn, impact the handling equipment such as fork-lift trucks, automated guided vehicles, etc. Warehouse storage options such as rack storage or open-bay storage may also be of concern. Choices on how to maintain the optimized unit load may also exist. Current choices include stretch or shrink film, or spot gluing (15), strapping, etc, which, in turn, influences equipment options. Using closed-loop pallet systems, pallets made from recycled material, or no pallet at all help

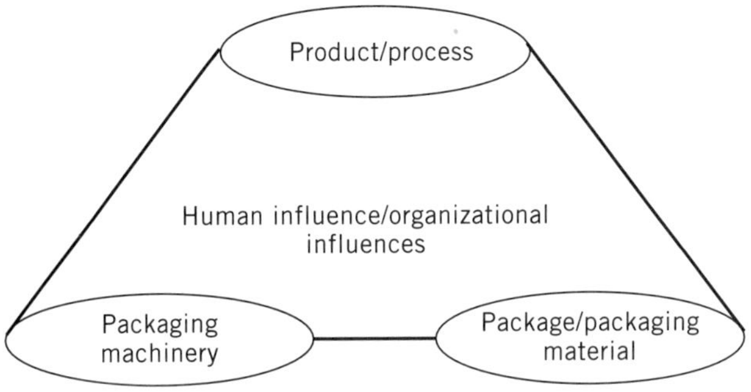

Figure 4. Marriage of product and package.

to address material handling principle 12—ecology—which advocates using equipment and procedures that have no impact on the environment.

In the more restricted view of packaging shown in Figure 4, the packaging line is the primary issue of focus. Typically the packaging line will consist of specific packaging machinery (filling, sealing, labeling), handling equipment (conveying, accumulation, etc), inspecting equipment, coding equipment, ect, usually beginning at the depalletizer for empty containers and ending at the palletizer for filled and "packaged" products. The level of automation of the packaging line may be manual, semiautomatic, fully automatic, or a fully computer integrated operation.

The integration of the packaging line from a "macro" level through the proper use and selection of various materials-handling equipment can help to address objectives such as increasing productivity, increasing space and equipment use, and ensuring a high level of systems flexibility, reliability, availability, and maintainability, among others. For example, the choice of the appropriate conveying and accumulation equipment becomes extremely important depending on a range of factors specific to the package or packaging material, the speed of the packaging lines, how flexible the lines must be, etc. For instance, handling requirements for empty lightweight plastic bottles are very different from those for comparable heavier glass containers (16), thus potentially influencing productivity and other objectives (see Conveyors, accumulation, palletizing, depalletizing). Other examples may include the use of robotic palletizers (see Palletizers) in order to increase the flexibility of the packaging line, or the use of vertical versus horizontal accumulation to optimize space and overall facilities layout. Objectives can also be influenced at a "micro" level by proper design and selection of in-feed mechanisms (star wheels, worm screw), transfer points (dead plate, power roll), right angle turns, etc.

In short, there is much similarity between a systems view of materials handling and a systems view of packaging. In designing or evaluating packaging systems, issues relative to materials handling should be addressed. Correspondingly, when designing materials-handling systems, the requirements of the packaging system must be addressed as well. This is particularly true for industries such as consumer foods and products, pharmaceuticals, and beer and beverages. This is also true for the suppliers of packaging materials and equipment.

Conclusions

Materials handling is more than handling material. It is, in fact, a complex, multifaceted activity that extends beyond the boundaries of the "manufacturing" facility. This article has presented definitions, objectives, checklists, questions, and techniques that help to better understand the complexities surrounding materials handling. Similarly, packaging is "more than just a box." This article briefly shows that packaging can also be viewed from a systems perspective, and that synergies exist between material handling and packaging. This article mentioned only a few examples of the commonalties that exist between materials handling and packaging, but an entire article could be devoted to the topic. In conclusion, it is apparent that optimum design for both can depend in large part on understanding and analyzing each from a systems perspective.

BIBLIOGRAPHY

1. T. H. Allegri, Sr., *Materials Handling: Principles and Practice,* Van Nostrand Reinhold, New York, 1984.
2. L. F. Gelders and L. M. A. Pintelon, "Material Handling" in *Handbook of Design, Manufacturing and Automation,* Richard C. Dorf and A. Kusiak, eds., Wiley, New York, 1994, Chapter 17.
3. D. R. Sule, *Manufacturing Facilities: Location, Planning and Design,* 2nd ed., PWS-Kent Publishing, Boston, 1994.
4. F. E. Meyers, *Plant Layout and Materials Handling,* Regents/ Prentice-Hall, Englewood Cliffs, NJ, 1993.
5. J. A. Tompkins and J. A. White, *Facilities Planning,* Wiley, New York, 1984.
6. R. E. Sims, Jr., "Materials Handling Systems," in *Handbook of Industrial Engineering,* Gavriel Salvendy, ed., Wiley, New York, 1982, Chapter 10.3.
7. K. J. Musselman, "Simulations Spectrum of Power In Manufacturing" in *Proceedings, 1990 IIE Integrated Systems Conference & Society for Integrated Manufacturing Conference,* San Antonio, TX, Oct. 28–31, 1990, pp. 65–70.
8. W. S. Haider and S. Rajan, "Effective Analysis of Manufacturing Systems Using Appropriate Modeling and Simulation Tools" in *Proceedings, 1990 IIE Integrated Systems Conference & Society for Integrated Manufacturing Conference,* San Antonio, TX, Oct. 28–31, 1990, pp. 94–99.
9. A. M. Law and D. W. Kelton, *Simulation Modeling & Analysis,* 2nd ed., McGraw-Hill, New York, 1991.
10. CAPE—Computer-Assisted Packaging Evaluation, CAPE Systems Inc., Plano, TX 75074.
11. TOPS—Total Optimization Packaging Systems, TOPS Engineering Corporation, Plano, TX 75075.
12. R. Muther and K. Haganas, *Systematic Handling Analysis,* Management and Industrial Research Publications, Kansas City, MO, 1975.
13. W. J. Stevenson, *Production / Operations Management,* 4th ed., Irwin, Boston, 1993.
14. Cimtechnologies Corporation, *Factory CAD, Factory Plan & Factory Flow,* Release 3.0 Technical and Reference Manuals, Ames, IA, 1992.
15. "Palletizing System Brews up Savings for Miller," *Packag. World,* 36 (Dec. 1994).

16. P. Reynolds, "Empty-Bottle Handling Is No Picnic," *Packag. World* 42, 46 (Jan. 1995).

STEPHEN A. RAPER
The University of Missouri-Rolla
Rolla, Missouri

COLIN BENJAMIN
Florida A&M University

MEDICAL PACKAGING

Introduction

This topic primarily addresses packaging of sterile disposable medical devices. Medical devices have a number of criteria that differentiate their packaging needs from those typically identified with pharmaceuticals packaging. However, growth of single-dose pharmaceuticals packaging has led to some parallels with medical devices in materials and process developments.

Rapid expansion of the disposable medical-device industry has created a specialty market for high-performance, high-integrity single-use packaging. The global market for primary disposable-device packaging is estimated to be in excess of $1,000 MM per year, exclusive of secondary packaging such as cartons, cases, etc. This is small by comparison to industrial and consumer packaging. However, it is a market that has carved out its own niche of packaging criteria for materials, processes, and performance. Growth of this market is expected to be in the 5–10% annual rate throughout this decade.

Significant driving forces affecting the rapid growth and diversity of medical packaging include the following:

1. Almost universal acceptance of single use or disposable devices in hospitals
2. High growth rate of implantable devices
3. Expansion of healthcare services, to include home-care applications
4. Major shifts in methods of sterilization, creating opportunities for new packaging materials
5. Safety and conservation concerns, requiring more environmentally friendly materials and designs
6. Cost control and continuous improvement programs imposed by increasing competitiveness of single-use products.
7. Growth of hybrid drug/device products, such as transdermals, solutions, and prefilled syringes.

Defining the Package Function

Every sterilized medical package is expected to provide a common set of performance functions. These include

1. Product protection throughout the sterilization process, shipping, storage and handling of the product see also Package Sterilization Issues for Healthcare Packaging. Protection takes many forms, and may be tailored to the particular characteristics of the device.

a. Sterilization. Once sterilized, the function of sterility maintenance becomes the most important role of the package. Microbial barrier is an inherent material property required of all primary medical packaging materials. This means pinhole freeness of barrier materials, and controlled pore characteristics of breathable materials used in ETO, steam, or chemical sterilization processes.

b. Seal Integrity. This is the second most important characteristic in product protection. Seals not only need to be completely free of voids or channels, but must withstand the rigors of sterilization, and impact of vibration factors in transit.

c. Puncture and Abrasion Resistance. These have particular significance, depending on the package style and physical shape, or surface characteristics of the device. In addition to the basic protective criteria, special properties may need to be present, such as moisture, oxygen, light, or chemical barrier. Other protective properties can be designed into the package, particularly if it is a thermoformed, by nesting, to keep product from shifting, or kit components separated. Even flexible packages may be compartmented to separate components until time of use.

d. Product Identification. Many medical packages today are preprinted, and provide important information about the product. In addition to product name and supplier, information may include size, product codes, instructions for use, date codes, precautions, and barcoded information. It is becoming more common to apply multilingual copy, and in cases of limited space, printing may also appear on the inside of the package. Because the printing is usually an integral part of the packaging material, it requires low-toxicity inks. Stringent print quality standards must be met, to ensure accurate copy, clear legibility, color control, and print registration.

e. Processability. Packaging materials must be versatile enough to run on a wide range of equipment, both in converting operations (coating, slitting, sheeting, die cutting, printing, or thermoforming) and in packaging operations. Materials have to be designed to generate minimal amounts of particulate matter when put through various mechanical operations. Adhesive particles, loose fibers, slivers, or ink pickoff are all objectionable defects in materials converting. The most important function in the packaging operation is sealing or closure of the package. In the case of heat sealable packages, the primary variables influencing seal quality are temperature, pressure, and dwell time. Consistency of materials, particularly the heat-seal coatings, plays a key role in validation of packaging processes. Adhesives with wide sealing parameters are more adaptable to different equipment, and can sometimes compensate for variation in the sealing process.

f. Ease of Use. In the case of sterile medical devices, it is important not only for the package to be opened easily but to provide sterile delivery of the product. Package design plays a key role in the opening function. It is important for the seal area to be designed to permit the peeling force to be transmitted smoothly around the package. Shape of the seal, and design and location of the peel tabs can influence the relative ease of opening. For example, it is highly undesirable to have a narrow peel tab midway in the side of a tray seal. Ideally, peeling is done from the corners of rectangular tray packs. Lids should also extend slightly over the seal areas, to prevent edge tear when peeled. Round package lids are often tear drop shaped for optimum distribution of peel force. Several pouch seal designs may be used to allow for easy opening, but most are based on the classic "Chevron seal" or "corner peel" designs. Functionality of a specific package is usually determined by various characteristics of the device being packaged. Therefore, in the early stages of package selection, it is important to define, as completely as possible, the following:

(*1*) What is the product; what are its critical characteristics?
(*2*) What type of protection is required of the package design and materials?
(*3*) How will it be sterilized?
(*4*) Where and how will the product be dispensed?

2. Generally, medical devices may be categorized as follows:
a. Solid, dry, require no special environmental protection, soft, low profile, no sharp edges or protrusions.
b. Same as above, but may have protruding parts, high profile, or heavy weight.
c. Dry solid, but require environmental protection, either from light, oxygen, moisture, temperature or chemical gases.
d. Wet solid or liquid, migratory, needs high-barrier package for containment.

3. After the basic characteristics are known, and the type of protection required is understood, other functional requirements of the package can be taken into consideration. The style of package, whether flexible (pouch) or rigid (tray lid), will play a big part in the integrity of the package. For example, a high-profile, irregularly shaped device would be more securely packaged in a preformed plastic tray than in a flexible package. When the product needs and package style issues have been resolved, the materials selection process can generally proceed more logically.

Materials Selection

Materials used in sterile packaging fall into three general categories: flexible, semirigid, and sealants.

Flexible materials. Used in pouch and bag construction, or as lidding materials, either in die-cut form or as rolls. Form/fill/seal packaging, one of the major growth areas of medical packaging, typically uses a top web and a formable bottom web. Bottom web materials may be either semirigid or flexible. Most widely used flexible materials are paper; Tyvek: various films; and laminations of aluminum foil with films, papers, or extrusion coatings.

Papers. For a long time the traditional single-use packaging medium, papers continue to play a key role in device packaging, particularly for high-volume, cost-sensitive, relatively easy-to-package devices. Examples are dressings and bandages, surgical gloves, syringes and needles, sutures, and in-hospital packaging of reusable devices. Papers are supplied in several basis weights, ranging from 25 lb per ream (3000 ft^3) up to board stock. They are used either in uncoated form, sealed to polyethylene surfaces, or coated with heal seal, cold seal, or polyethylene. Typical grades used in pouches and lidding fall into three categories:

Bleached Surgical Kraft. High-strength virgin pulp, with minimal ingredients that may serve as nutrients to microorganisms. Paper formation is closely controlled to provide optimum porosity and effective bacterial barrier.

Latex-Reinforced Surgical Kraft. Higher strength for improved tear and puncture resistance. Latex binders also improve wet strength, peelability, and dimensional stability.

Other Papers. These are used as the printable outer layer of several laminations, where the strength and bacterial properties are provided by non-permeable layers of plastics or foil.

Several factors continue to contribute to paper role in medical packaging, including cost, disposability, steam resistance, printing and laminating versatility, and wide range of product grades available.

Tyvek. The premium packaging material for high performance medical packaging, Tyvek has experienced continuous growth in its approximate 25 years in this market. Tyvek is a fibrous web material composed entirely of extremely fine, continuous strands of high-density polyethylene. Since the material contains no separate binders, fillers, or sizing agents, it relies on its own thermoplasticity to hold the fibers together. This is achieved by compressing the "spun" web—so called because it is extruded from spinnerets under heat and pressure until the individual fibers begin to adhere to each other, hence the term "spun-bonded." Tyvek is made in a number of physical styles, ranging from relatively stiff, paperlike grades, to very soft and comfortable grades, which may also be embossed, needled, or perforated. The type used in packaging (type 10) are the paperlike grades and it is to these that we refer.

The composition and structure of Tyvek provide its outstanding characteristics as a medical packaging material. High strength is the first characteristic that distinguishes it from other fibrous packaging materials. Tyvek's strength, combined with its high porosity and effective bacterial barrier, gave the medical device industry its first truly nonparticulating, convenience packaging material suitable for ethylene oxide sterilization processes. In addition to these properties, Tyvek provides several other attributes useful to package integrity and aesthetics:

Water Repellancy. The grade of Tyvek used in medical packaging shows a strong repellancy to liquid water. Since the surface is untreated, the inherent property of the polyolefin is to shed water. This has benefit in that Tyvek is unaffected by limited spillage or by condensate during sterilization. It is also insensitive to changes in relative humidity. However, Tyvek is porous to moisture vapor, just as it is to air or sterilizing gases.

Chemical Resistance. Tyvek resists the usual agents of age degradation (eg, moisture, oxidation, rot, mildew, and many inorganic chemicals).

Radiation Stability. When subjected to radiation levels commonly used for sterilization (2.5–5.0 Mrad), Tyvek is relatively unaffected.

Low-Temperature Stability. While many plastics embrittle, Tyvek retains strength and flexibility even at subzero temperatures.

Aesthetic Qualities. The bright, white, clean appearance of Tyvek makes it a strong candidate for packaging that has to convey a "cleanroom" image.

Although Tyvek has gained almost universal acceptance in sterile device packaging, some of its properties may limit its use in specific applications. These include static properties, swelling by certain oily substances or solvents, ink adhesion limited on untreated surface, and limited heat resistance, although Tyvek is enjoying new growth in controlled steam sterilization processes where temperatures are maintained between 250 and 260°F (121 and 127°C).

Films, laminations, and coextrusions. The combination of structures possible with flexible plastics is almost limitless. The most common structures used in pouch packaging for several years have included:

Polyester/Pe—strong, heat-sealable, pinhole-free, clear, suitable for ETO and radiation sterilization. Usually the Pe is EVA-modified.

Polyester/PP—similar to above, but used with paper for steam sterilizable pouches.

Nylon/Pe—puncture- and abrasion-resistant, also used as forming web in form/fill/seal.

In recent years lamination and coextrusion technology has kept pace with plastics developments, so that now custom structures may be designed to satisfy specific packaging criteria. Furthermore, films can be either adhesive laminated or extrusion-laminated with foil or paper to form high-barrier structures. All plastic barrier films continue to improve, as glass (SiOx)-coated polyesters and fluorocarbon (Aclar) films are finding uses in see-through medical packaging. Special film blends are now available that form peelable bonds when sealed face to face. This feature opens the door for economical all plastic packaging for radiation-sterilized packaging. These films may also find applications in liquid tight peelable packaging for difficult materials such as saline-, iodine-, or petrolatum-based compounds. Heat-seal-coated films are also available for blister lidding and pouching with similar plastics. Heat-seal coatings provide adhesive versatility, wide seal range, and visible seal quality features. Multilayer coextruded films are widely used as flexible form/fill/seal (FFS) bottom webs for products not requiring package rigidity. Flexible webs offer economy over rigid materials, and when combined with a coex top web or Tyvek, they can provide a completely recyclable package without separation of components.

Semirigid materials. Several thermoformable plastics are used for preformed trays and FFS bottom webs. The versatility of thermoplastics makes possible an almost unlimited variety of three-dimensional packages for medical devices. It has become possible to tailor make packages to fit the contours of a product or to hold several components in one unit. Features may also be designed right into the thermoformed container to perform such functions as denesting, stocking, compartmenting, hinges, gripping, and product locating. Other built-in features hold products in place in the container, reinforce the structure, or provide code or manufacturer's information. The selection of a specific material may be dependent on several factors.

Once it has been decided that a particular device or line of products is to be packaged in a thermoformed container, much attention should be given to finding the plastic material that will best complement the design and functionality of the product. In screening materials, the first step is to identify the package requirements. As many package criteria as possible should be listed and then evaluated against the various characteristics of different materials. The material of choice will be the one that most closely fits the package criteria.

Several plastics are used for trays, depending on specific functional or cost requirements. Properties that enter into the plastic selection process include clarity, ease of thermoforming, temperature resistance, compatibility with sterilization method, impact resistance, cleanliness in die cutting, and ease of sealing.

Polyester (PETG) has shown most growth in recent years, while other plastics such as PVC and XT polymer have declined in their share of the packaging market. Yield (density) of tray plastics may affect tray costs, since lower-density plastics provide more trays per pound than do higher-density materials (see Table 1). A small, but growing market exists for high-temperature thermoplastics used in steam-sterilizable tray packaging, with Tyvek lids. Polycarbonate (see barrier properties in Table 2) has been the most widely used plastic for this application, but interest in polypropylene is increasing, for obvious cost reasons.

Table 1. Typical Density and Yield Factors of Various Thermoplastics[a]

Plastic	Typical density (g/cm^3)	Typical yield [in. ^{2}/(lb · mil)]
ABS	1.03	26,885
Acrylic multipolymer (XT)	1.10	25,175
Acrylonitrile (Barex)	1.10	25,175
Butadiene–styrene (K)	1.03	26,885
Polycarbonate (Lexan)	1.20	23,075
Polyester (Kodar)	1.20	23,075
Polyethylene		
High-density	0.95	29,145
Low-density	0.92	30,098
Polypropylene	0.90	30,765
Polystyrene		
Crystal	1.05	26,370
High-impact	1.03	26,885
Polyvinyl chloride	1.30	26,885
Propionate (CAP)	1.20	23,075

[a] *Modern Plastics Encyclopedia,* McGraw-Hill, New York.

Sealants (Adhesives). Sealants are the "glue" that holds packages together, particularly those made from dissimilar materials that will not seal by themselves. More importantly, with peelable packaging, sealants (adhesives) provide con-

Table 2. Barrier Properties of Plastic

Plastic	Permeability	
	Oxygen (O_2)	Water (H_2O)
Low-density polyethylene (LDPe)	300–400	1.0–1.5
High-density polyethylene (HDPe)	100–200	0.3–0.5
Polypropylene (PP)	150–200	0.2–0.5
Polystyrene	300–400	5–10
Polycarbonate	200–300	3–8
Nitrile	0.8	3–5
PVC	5–10	0.9–2
Polyester (PET)	10	0.9
Copolyester (PETG)	25	1.2
Nylon	1–3	6–22
PVDC	0.1	0.01
EVOH (dry)	0.01	6

trolled bond strengths, making it possible to seal various tray or pouch materials and get consistent peel values without tearing or rupturing the package (see also Adhesives).

Most peelable medical packaging is achieved by coating materials such as Tyvek, papers, films, and foil laminations with specially formulated adhesives. Sealing of most device packaging is done by a combination of heat and pressure. Cold seals require pressure only to effect a seal, but require adhesive on both surfaces being joined. Since heat sealing is the most widely used method of medical package closure, we will deal with various factors affecting heat-seal/peel package performance. Heat seals can be made by various methods such as platen or bar, roller, impulse, hot-wire, induction, and ultrasonic. The two methods most commonly used for peelable sterile packaging are platen and roller (rotary). Optimum seal conditions are determined by finding the best combination of pressure, temperature, and dwell to give desired seal characteristics. Since most heat-seal coatings are applied uniformly over one surface, they become an internal part of the package. Therefore, they have to be able to satisfy a variety of criteria in addition to holding the package together. The following characteristics are essential for a seal/peel adhesive coating on lidding or pouch materials.

Sealability. Lidding materials, particularly, might be required to seal to a variety of tray plastics [eg, PVC, polystyrene, acrylics (XT), polyester, (PETG), acrylonitriles (Barex), cellulosics (CAP, CA, CAB), polypropylenes, polycarbonates, and various polyethylenes].

Visible Seal Indication. Seals should be clearly distinguishable through transparent packaging materials. Furthermore, when peeled, it is desirable for the seal to split internally, leaving surfaces with a high degree of light diffraction in the previously sealed area. This gives a visual effect of a highly contrasting seal image. By evaluating the quality and continuity of this seal image, a trained inspector can determine the relative quality of seals.

Seal Range. An ideal adhesive will effectively seal to this variety of plastics in a broad temperature range that will not distort the plastics.

Uniform Peelability. Peel strength should fall in a range that allows convenient opening, without compromising package integrity. Depending on substrate, size of package, and seal area, peelable seal strengths may be achieved in the following ranges:

Paper:	0.75–2.0 lb/in. (340– 900 g/in.)
Tyvek:	1.0–3.0 lb/in. (454–1360 g/in.)
Films:	1.0–3.0 lb/in. (454–1360 g/in.)
Foils:	1.0–5.0 lb/in. (454–2270 g/in.)

An ideal adhesive is the one that will peel in approximatcly the same range regardless of the plastic to which it is sealed. This can be achieved if the adhesion to the plastic exceeds the internal strength of the adhesive (cohesion). This can enable a packager to interchange plastics with very little difference in adhesive performance.

Porosity. Important for ETO and steam sterilization, this characteristic is often a combination of coating and substrate characteristic. In cases where the coating is inherently nonporous, patterned grid, dots, or perimeter (zone) coating may be employed.

Creep Resistance. This phenomenon is experienced in some ETO and steam sterilization cycles. Seals should be able to withstand the effects of temperature, moisture, pressure changes, and any solvating effects of sterilant gases.

Radiation Resistance. The adhesive must be resistant to radiation within commerical dose levels of sterilization.

Hot Tack. This is defined as the adhesive strength of the seal immediately after sealing. Stresses in packages or movement in the packaging line can cause seals to fail if they are not sufficiently strong in the hot state.

Cling Resistance. Coatings in contact with product should not activate and adhere to the product during or after sterilization.

Compatibility. Adhesives should not react with device materials, such as rubber compounds or plasticized plastics.

Mechanical Processability. Coatings should be smooth, be well anchored to the substrate, be scuff- and scratch-resistant, printable, resist blocking, resist ink off-setting, possess good aging qualities, and resist discoloration and odor formation.

Low-Toxicity Materials Composition. Materials must be certified to be acceptable for direct food contact under FDA regulations.

Adhesives can also be applied in zoned patterns, stripes, dots,or grids to meet certain criteria not achievable by overall coating. Attributes of discontinuous coatings might be

- Improved porosity
- No adhesive contact with device
- More controllable peel on certain substrates

Perimeter zone coatings, used for several years in glove packaging, are finding increasing use in sensitive applications such as implant packaging and steam-sterilizable lidding.

Table 3. Packaging Materials Problems

Defects	Causes	Typical Effects	Probable Solutions
Faulty materials	Nonuniform substrates (eg, papers and films)	Package punctures, tears leaks	Stronger, more tear-resistant materials
	Pinholing	Loss of sterility barrier	Heavier material or better-quality source
	Loose matter in packaging	Particulate contamination	Stronger substrates; peelable adhesives
	Inadequate adhesive strength or coverage	Seal voids, open packages	Heavier coating, or modify seal conditions
	Materials incompatibility	Blocking, embrittlement, discoloration, haze, migration, transfer of printing	Screen product/material variables; alternative materials
	Unstable ink system	Fading, discoloration, offsetting of print	Investigate pigments; improve ink adhesion
Faulty package design	Size of package	Package stresses, separation vibration damage	Stronger materials, double-pack, cushion
	Weight : strength ratio	Impact failures, package distortion	Double-pack, thicker, better-impact materials
	Use and storage testing	Failure at low or high temperatures	Improve flexibility and heat resistance; alternative materials
	"Economy" materials	Production setbacks, waste, reruns	Upgrade materials
	Graphic design, copy layouts	Legibility, inaccessibility of information	Change print method, plate materials; reevaluate copy
	Location and type of opening texture	Package tearing, product damage or loss	Package redesign, stronger materials
Faulty processes	Control of sealing conditions	Weak packages, sterility failures	Seal study; may need stronger adhesive
	Sterilizer conditions	Deteriorating seals, product reactions with package, package distortion	Compatibility studies; need more heat-resistant materials
	Uncontrolled storage	Temperature, moisture, chemical change	Change storage conditions; more durable package materials
	Equipment maintenance	Wrinkling, scuffing, tearing, contamination of package	Smooth out operation, tougher material

Trends in the Medical Industry that Affect the Selection and Use of Packaging Materials

Sterilization methods. ETO, once the dominant sterilization method, has been steadily losing market share. However, about 50% of all devices are still sterilized by ETO, making it still an important process. Faster and higher-temperature ETO cycles tend to put more stress on package seals. Lidding adhesives have been improved to provide better porosity and temperature resistant to withstand higher sterilizer stress.

Radiation (gamma and electron beams) is currently the fastest-growing process approaching the volume of ETO. Since radiation processes do not require gas porous packaging, they open up possibilities for all-plastic film packaging. Nevertheless, Tyvek is still a key material in radiation packaging, because of its other high-performance characteristics and radiation resistance.

Alternate methods of sterilization, such as steam, dry-heat, and various chemical and plasma methods, still represent a very small share of the EOM market (probably less than 5%). However some of these methods offer efficiency, safety, and cost benefits, which may make them attractive for certain segments of device sterilization. This is an area where packaging development can help to open up new opportunities for alternate sterilization methods.

Environmental responsibility. This has become a worldwide issue that affects the medical industry's treatment of disposable products and packaging. From a packaging perspective, the focus will be on *materials, process,* and *design,* to reduce the mass of materials consumed, make them more easily recycled or otherwise disposed of.

Materials. This involves the practice of source reduction. By reducing the variety of materials used, recycling would be made easier. Use of stronger, lighter materials would enable downsizing of packages, reducing total usage. Materials that can be easily recycled, can be incinerated cleanly for energy recovery, or produce nontoxic landfill will be more in demand for future packaging. (See also Table 3).

Process. Waste can be minimized through process improvements and higher-performance packaging materials, resulting in fewer rejects, damage of product or loss of sterility. Cleaner sterilization processes, such as steam or vapor-phase plasmas, can reduce process and validation costs, and use of expensive, toxic byproducts.

Design. Downsizing and downgauging of materials can have a considerable effect on the volume of materials used. However, changes of this nature need to be evaluated carefully through all phases of the product life cycle, to ensure safe and sterile delivery of the product at its end use.

Summary

While the basic forms of medical packaging are expected to undergo relatively little change, processes and materials will continue to go through steady evolution, driven primarily by the need to supply defect-free sterile products at more cost-effective levels.

BIBLIOGRAPHY

General References

K. Beagley, "Hospital Packaging Trends," *Pharmaceut. Med. Packag. News* (May 1994).

K. Beagley, "Packaging for the Operating Room," *Pharmaceut. Med. Packag. News* (Feb. 1995).

D. Dyke, "How Design Control Affects The Packaging Development Process," *Med. Device Diagn. Mag.* (Oct. 1994).

R. Hall, "Developing a Medical Device Package," *Med. Device Technol. Mag.* (Jan./Feb. 1993).

R. Hall, "A Study of Medical Device Packaging," *Med. Device Technol. Mag.* (March 1993).

C. Marotta, "Medical Packaging Materials" in *Encyclopedia of Polymer Science and Engineering,* Vol. 10, 2nd ed., Wiley, New York, 1987.

C. Marotta, "Medical Device Packaging Handbook" Chapter 3—"Packaging Materials" Copyright 1990 Marcel Dekker, Inc.

R. Pilchik, "Lidding Materials for Form–Fill–Seal Packaging of Medical Devices," *J. Packag. Technol.* (Feb. 1988).

P. Schneider, "Low Temperature Sterilization Alternatives in the 1990's," *TAPPI J.* **77**(1), (March 1994).

M. Spaulding, "Medical Device Paths Change with the Market" *Packag. Mag.* (Dec. 1992).

J. Spitzley, "Implementing Packaging Changes to Meet Ecological and Consumer Concerns" *Med. Device Diagn. Mag.* (Feb. 1992).

J. Spitzley et al. "Toward a New Consensus on Sterile Device Packaging," *Med. Device Diagn. Mag.* (Jan. 1993).

J. Wagner, "Surviving Sterilization: A Packaging Guide," *Pharmaceut. Med. Packag. News* (June 1994).

J. Wagner, "Lidding and Lidstock Trends," *Pharmaceut. Med. Packag. News* (March 1995).

CARL MAROTTA
Tolas Health Care Packaging
Feasterville, Pennsylvania

METAL CANS, FABRICATION

The metal can is one of the oldest forms of packaging preserved food for long periods. In the same way it has proven to be an adequate container for food stuffs, beverages, and industrial products. Thousands of different products of all kinds have been packed in metal cans. The traditional method of manufacture is to start with a rectangular sheet of tinplate or canstock that have special surface treatments. Blanks, or bodies, are cut from the sheets, flexed and rolled into a cylinder, notched and hooked so as to form a locked side seam which results in a longitudinal joint line bonding both the lateral cut edges of the blank.

Bodies are beaded for increased resistance against implosion of the can. More and more can bodies are necked in order to use ends of reduced diameters or/and thinner and more economical end stock and also to make cans stackable, etc.

Ends, ie, the closures at the bottom and the top of the cylindrical or any other geometrical section of the can body, are tightly secured to the body by a double-seam, (called so because they are made in two operations) the first and the second ones. Both extremities of the can bodies are flanged so as to create the body hook which engages with the end hook so as to form a tight, compact, interlocked closure. One end, called the maker's end, is fitted by the can manufacturer. The other, fitted by the filler or the packer, upon filling the "open top" can, is known as the packer's end.

Since the container is made from three separate elements, it is known as a three-piece can. Its construction had remained basically unchanged for over 150 years. Advances have been made in engineering, automation and speeding up of the original manual canmaking processes. The metal input has been gradually and constantly reduced through more sophisticated design geometry, and more recently, by changing the methods of making the side seam from tin/lead alloy soldering to welding (see Fig. 1).

Since the early 1970s, a different concept of canmaking has gained acceptance in commercial production. In this, the body and one of the ends, are formed in one entity from a flat circular blank by press forming technique (1). The open top end is sealed with the usual packer's end. It is known as two-piece can. The methods of forming are identified: drawing and ironing (D&I in the United States, DWI in Europe) and draw and redraw (DRD) (2). D&I, for instance, was used in World War I for making shell cases. What distinguishes them in canmaking is the use of ultrathin metal in very high speed production to yield outputs counted in billions of cans per year.

Can Types

All processes convert flat sheet material into finished cans, supplied with a loose end for the packer or filler, according to this basic scheme: prepare plate matching the products to be packed, their filling and processing conditions as well as market conditions (climates, shelf lifes, presentation of cans, sanitary regulations, etc); make bodies and ends; apply finishes (decorative or/and protective barriers). The order may vary, depending on the process used. The manufacture of three-piece and two-piece DRD (Draw and Re-

Figure 1. Three-piece can side seams (**a**) Soldered; (**b**) cemented; (**c**) welded. Courtesy of *Proceedings, 3rd International Tinplate Conference,* 1984.

draw) cans starts with the finishing step. Some cans are resprayed a top-coat for increased corrosion resistance in a final stage of fabrication.

Coils are usually cut into sheets, or scrolled strips (technique Littell) if coil coated stock is used. Sheets are coated on one or both sides and decorated if appropriate. The coatings are called enamels in the United States and lacquers in the United Kingdom. Decorations are always protected by over- or finishing varnishes to make them scratch proof and to add gloss. In case of processed cans, decorative or barrier coats have to resist the applied heat treatments in autoclaves.

If the starting point is a circular blank, as for DRD cans or ends, the cut edges of the sheets are scrolled for economy of metal usage (Fig. 2). Alternatively, precoated coil stock may be fed directly into the cupping press for blanking and drawing or into the multi-die end stamping press. In the manufacture of two-piece D&I cans, plain coil, as supplied from the mill, is the starting point.

Three-piece can manufacture is readily adaptable to making cans of any diameter and height. The production equipment is amendable to changes in size and is capable of production speeds of up 700+ cans per minutes when cans are made as singles. In case of multi-high canmaking, bodies are scored to final individual height prior to rolling and welding and subsequently separated outputs of 1200 to 1500 cans per minute are achieved on appropriately designed lines. Where the use of lead/tin solder is no longer acceptable, the change to welding can be made at minimal cost on existing lines, since only the equipment which makes the cylinder (the bodymaker) needs to be changed.

Three-piece manufacture is the choice of the small to medium sized operation requiring flexibility, for producing can sizes required in relatively modest quantities or to suit a variety of fill products requiring changes in coating specifications. This is the salient asset of three-piece canmaking facilities. It allows further the manufacture of various can types, ie, food,

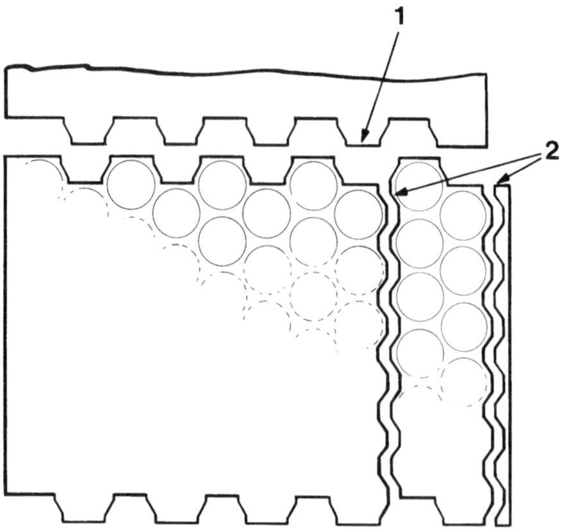

Figure 2. Scrolled sheet showing layout of blanks and scrolled edges for material utilization. 1, Primary scrolling used when cutting coil into sheets for coating. 2, Secondary scrolling used for cutting sheets into strips for feeding into the shell press.

flat and/or carbonated beverages, aerosols, general line cans (industrial and dry products) by adding the adequate complementing equipment at limited cost, to the existing line as long as the bodymaker can cope with the range of cans to be made.

Two-piece can manufacture is basically suited to a single can size, requiring outputs of at least 150 to 300 million cans per year. Yearly productions of modern lines are usually much higher, ie, in the range of 400 to 500 million per annum. DRD, using precoated sheet or coil material, is used for food cans and predominently in the shallower sizes ($h/d < 1$). The process is being used, however, for food cans with an h/d ratio of 1.5 in the popular 3 inch (7.6 cm) diameter, eg, 300 × 406 in the United States (for explanation of can dimensions: the first digit stands for inches and the two last ones, for ¹⁄₁₆ths of inches, the measures being taken on the finished can, overall in length and over the double seems in diameter).

The application of enamel to both sides permits the use of ECCS (electrolytically chromium coated steel) which has a surface that is too abrasive to be used uncoated (plain). It does also not offer any corrosion resistance. Because economics demand the use of the thinnest possible can stock, the high strength needed dictates the use of double-reduced (DR) grades. Further additional strength is provided by beading, or reforming the can bodies, and adapting the profiles of the ends for increased flexibility in the autoclave, ie, providing a maximum of expansion volume.

Generally utilizing two presses with a minimum of peripheral equipment, DRD provides a compact installation of relatively low capital cost for the achievable outputs. For that reason, it appears to be a preferred method for self-manufacture (captive canmaking) of relatively shallow containers by packers (eg, tuna fish cans). However the material property requirements, its constance of drawability, are high and the standard of enameling is critical to maintain integrity as the metal is formed.

D&I converts plain coil into a fully finished can in a totally integrated, fully automatic process. The high capital cost of the full range of equipment needed implies a high volume, nonfluctuating and constant ensured market. This together with the desirability of keeping the line running once harmonious operation has been achieved, results in a 24-hour per day, 7-day per week operation with annual production up to planned business feasibility studies. Basically a single-can-size process, it is best suited for the production of beverage cans, which are made worldwide in a few standard sizes for use of high speed filling-closing lines. D&I thin walled cans are pressure packs (4 to 6 kg/cm^2 constant pressure in carbonated drinks cans and much higher in D&I aerosol cans). Their rigidity derives from the high inside pressure. Such pressures are currently generated in noncarbonated fills by inoculating a specific quantity of liquid nitrogen into the headspace of the filled cans just prior to closing the cans. This offers more and vast opportunities for lightweight two-piece D&I cans.

Either tinplate or aluminium is used, usually on a dedicated line. In recent years, due to the uncertain economics relating to metal costs, new lines are built as "swing lines", basically capable of handing either metal. It does nevertheless require some downtime to make the change-over and lines normally operate continuously on the one metal.

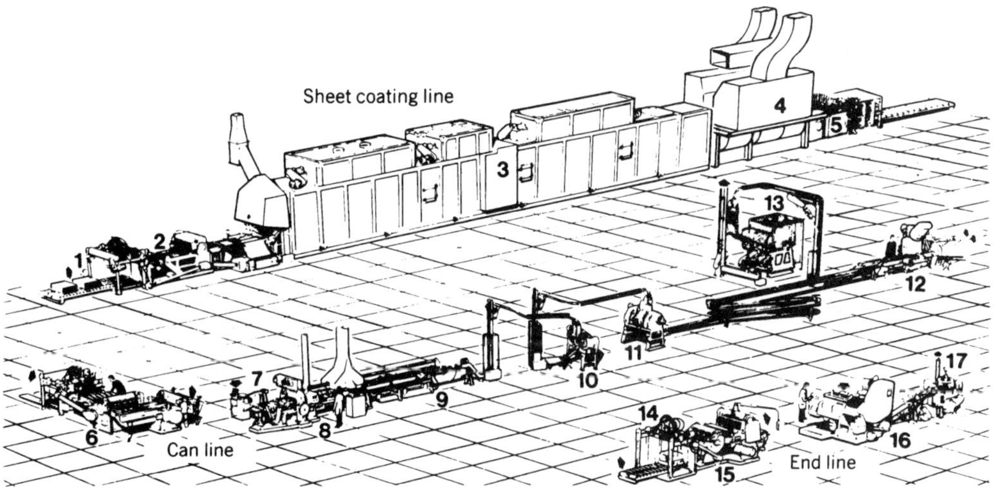

Figure 3. Three-piece can manufacture. Sheet-coating line: 1, sheet feeder; 2, roll coater; 3, oven; 4, cooler; and 5, unloader and sheet stacker. Can line: 6, slitter; 7, blank feed; 8, bodymaker; 9, side seamer; 10, flanger; 11 beader; 12, end seamer; and 13, tester. End line: 14, sheet feeder; 15, secondary scroll shears; 16, shell press; and 17, compound liner. Courtesy of Metal Box Ltd.

Three-Piece Can Manufacture

The manufacture of three-piece cans is as follows: cut up coil stock into rectangular sheets; coat and/or decorate sheets; slit into rectangular blanks; form cylinders and side seam (soldered or nonsoldered, or welded side seams); inside and/or outside side seam striping (lacquer or powder application and curing); separate cylinders in case of multi-high cylinders; neck and or bead if appropriate; expand or reshape or reform cylinders if appropriate; form flanges at both extremities; fit maker's end; organized visual inspection or/and automatic leak testing; casing or palletizing (see Fig. 3).

Cans for very corrosive or delicate products (carbonated beverages or red fruit, etc) are internally spray-coated and cured prior to applying the maker's end. This coat is called "top coat" whereas the "bottom coat" is applied on the flat sheet in roller coaters and cured in wicket ovens. Both bottom and top coats can be spray coats with or without intermediate curing (spray–spray–cure or spray–cure–spray–cure).

Slitting. Large rectangular sheets of can body stock, precoated if required on one or both sides by roller coating techniques, are slit by pairs of circular knives (3) on tandem slitters, first logitudinally into strips which width corresponds with the development of the can bodies, and then transversely into rectangular blanks of the appropriate size corresponding with the can height, respectively with the allowance to form the welding overlap (or the lock side seam in case lap-locked side seamed cans are made, soldered, or nonsoldered or cemented).

Because welded cans have considerably less overlap (0.4 to 0.8 mm) at the side seam than soldered cans (side seam allowance), slitting demands more rigorous precision (dimensions and right angles) than is needed for soldered can blanks. Circular slitting cutters can be positioned very accurately by applying internal pressure at the bore of the cutter hubs to expand them so as to move freely on the slitter shaft (4). The said hubs can also be blocked on the shafts by screw jammed wedges or by application of hydraulic pressure (compressed grease) inside of the hub. Any of these proven methods must apply the needed jamming pressure perfectly symetrically in order not to distort the setting (very narrow slitting gap, usually 5% of the body stock thickness) or/and cause wabbling of the slitter knives' cutting edges.

Bodymaking, soldered cans. The solder alloys commonly used consist of 98% lead and 2% tin. As more and more countries impose lower limits of lead content in food, the soldered can is being phased out in Western countries, if not being banned altogether as in the United States. Nevertheless, many soldering lines around the world will continue in operation for some time. The machine consists of a "bodymaker" which feeds single blanks from the feed hopper, scores (in case of multi-high can bodies), flexes and forms the blanks into cylinders upon slit and clip notching and hooking. The bodymaker is coupled to the side seamer ie, the soldering attachment where the joint is soldered and wiped off the excess of solder accumulated at the outside of the side seam.

Blanks are still commonly transferred manually from the slitter blank collecter although equipment exists for automatic transfer. In the first case, a visual check for quality and eventual rejects is possible whereas sophisticated devices check automatically any miscut blanks and trigger the programmed intervention. The prepared hooks, or edges, as prepared and formed in the hooking station(s) of the bodymaker are interlocked on the forming horn and compressed in the horn spline groove by the bumping steel. For high speed operation, the body cylinder is usually formed by a pair of wings. (The term "wing form bodymaker" contrasting with "roll form bodymaker" which is a former American Can Co, technique.) In this technique bodies are rolled and consecutively notched and hooked in the rolled shape.) Liquid flux is applied on one or both hooks prior to engage and bump the side-seam on the forming horn.

The lap-lock assembled bodies are fed out of the bodymaker and picked up synchroneously and individually by the clamping pawls of the soldering chain. The side seam area is preheated by ribbon gas burners before passing over a longitudinal solder roll where solder is applied. The presolder ribbon burners preheat the side seam and by appropriate setting of the "bow and counter bow" burners the side seam is kept as straight as possible. Upon applying solder, post-solder ribbon burners ensure that the solder is "sweat" into the capillary gaps of the side seam after which a rotating wiper mop removes excess solder, mainly in the form of blobs. Any solder splashes towards the inside of the can bodies have to be avoided. Rotary solder splash shields provide for this.

The can is then cooled by air jets before conveying to the downstream finishing operations. As lead contamination has become an increasing concern, the wiper mop unit has been the subject of considerable improvement to contain the lead dust generated by its operation, particularly at high speed canmaking operation. By its very nature, the process requires the use of tinplate, although blackplate had been soldered during World War II in the United States. In order to ensure sound soldering, a minimum "solderability" has to be ensured: rugosity of the steel base, a minimum of alloyed and free tin, oxidation, etc. The line operator's skill and experience will overcome most of the material inherent drawbacks, the major one of which is the fluctuation of stiffness of the steel base of the blanks (open or thick laps).

Bodymaking, welded cans. The side seam is made by a resistance-welding process using the "lost-wire-electrode" principle (5) as well as *The Canmaker* June 1996 issue Evolution of a bodymaker by Sigfried Frei, Frei AG Switzerland. Body blanks are fed from the bottom of the feed hopper, transferred through double blank detectors, scoring for eventual multi-high can bodies, flexing, into the forming rolls (See Fig. 4).

The two laps of the rounded cylinder are butted in the grooves of the Z-bar, and the cylinders pushed along the Z-bar by one or more driven chains provided with pawls. Upon perfect radial centering in the Z-bar, the cylinders are introduced into the welding rolls at the same speed as the said rolls are driven at. The actuating mechanism of the reciprocating introduction pusher pawls are designed so as to guarantee that pushing speeds are accurately matching the peripheral roller electrodes speed. Only the outer electrode is driven.

Overlap accuracy over the full length of the side seam, which is also dependent on gauge and temper of the plate, has to be controlled as to avoid bodymaker jams and wrecks or mainly irregularly welded side seams. The two overlapping edges of the cylinders are bonded by a-c resistance continuous nugget welding using approximately 4000 to 7000 A at 8 to 5 V. Both overlapping edges must be free of contamination, each one on both sides (lacquer splashes or traces) to eliminate variations of resistance which would lead to welding faults as well as eventual copper wire ruptures.

A significant amount of energy is lost in heating other parts of the welder such as welding arms and electrodes, which need water cooling. In high humidity, this can lead to problems with condensation. Thermostatically controlled cooling media, even on upper acceptable limits of operation temperature, should avoid reaching dew-points.

Each resistance welding spot, called a nugget, is achieved by one half of the a-c wave cycle. Welding current supply to the electrodes and welding speed are limited because the nuggets should overlap longitudinally to ensure a homogenous side seam over the full length of the side seam. To achieve higher welding speeds of up to 120 + meter per minute, higher welding current frequencies have to be generated via an alternator or static transformers. Other than sinuosoidal waveforms are applied and contribute to reaching high welding speeds coping with different canstock surface conditions.

Earlier systems used a large overlap (2- to 3- mm) and raised the steel temperature to the melting point by applying welding roller pressure to forge-weld the metal. The latest welders use a small overlap as mentioned (0.4 to 0.8 mm) with metal temperatures just below the melting point and increased roll pressure to forge the two laps together. To ensure reproducible welding conditions over the full length of all double seams produced, the electrode contact is made by endless copper wire wrapping around both welding rolls and moving the cylinders at the preset welding speed. Any contamination of the welding electrodes by tin pick-up is thus continuously removed from the contact area. After use on both sides of the profiled wire, it is either chopped or rewound for recycling.

Having dealt effectively with the problem of tin contamination of the copper wire electrodes, the system paradoxically requires a minimum of tin coating on the can stock, around 0.09 lb/bb on both sides (1.2 g/m^2 on one side.) Table 1 shows the comparison of properties of various materials wherein nr 25 and nr 10 stocks have respectively tin coatings of 2.8 and 1 g/m^2 on one side. TFS or ECCS materials, as well as blackplates, are poorly weldable, if at all under acceptable production conditions. They have to be "edge cleaned", ie, the oxide films have to be abrased from the four sides of the overlapping edges of the side seam. Edge cleaning has however never found reliable solutions apart from edge cleaning by "edge milling" as practised by the Continental Can Company in their Conoweld Technique. This system used welding rolls without an intermediate copper wire, but it is now of less importance in high speed canmaking, mainly because of the frequent need for changing the electrode rolls.

The integrity and quality of the seam weld is usually tested by visual and mechanical means (eg, Ball test). For a more detailed examination, weld cross and longitudinal metallographic inspection will reveal any sign of separation between laps, cavities, etc. Radiographical examinations are also used for quality inspections. Welders have been fitted with "weld monitors" to continually monitor welded seam quality. Usually these rely on measurements of voltage or current fluctuations between the welding electrodes. Welded seams as well as single nuggets, made outside preset limits are detected and the faulty cans ejected. Other monitoring systems have been explored for enhanced performance and were based on weld temperature or on the final thickness of the forge welded overlapped side seam.

Renewed effort went into welding the side seam by means of a laser beam (6). The principle of can welding by laser was demonstrated in the late 1970s, but welding speeds appeared to be too low to justify commercial exploitation. The technique was then discarded. Positive results of other ongoing R&D work on the subject in the U.S., Europe, and Japan, are not known. Apart from the elimination of costly copper wire, the

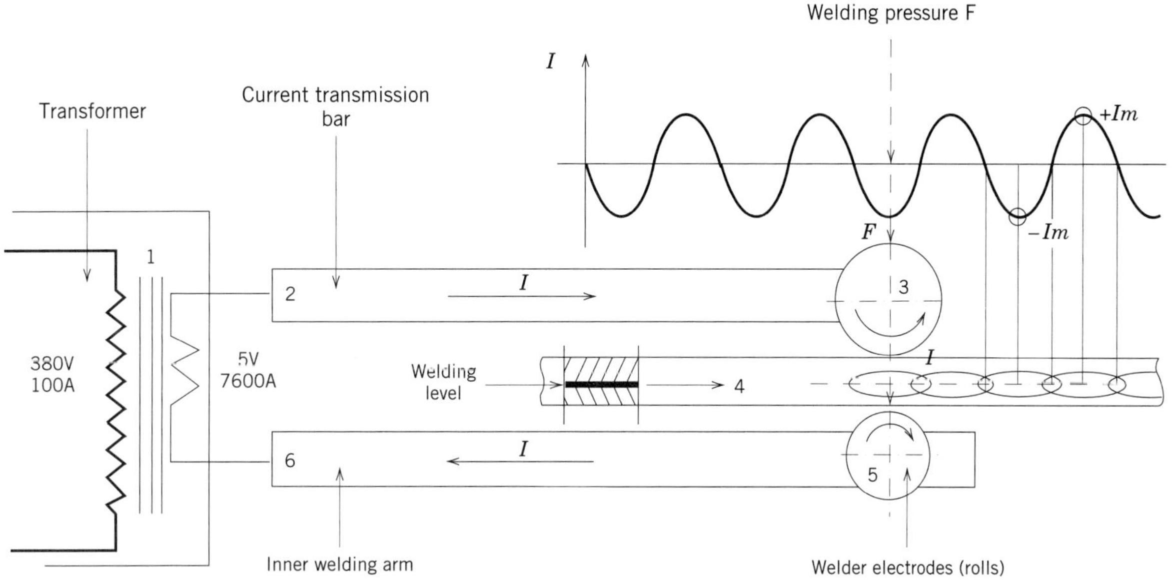

(a)

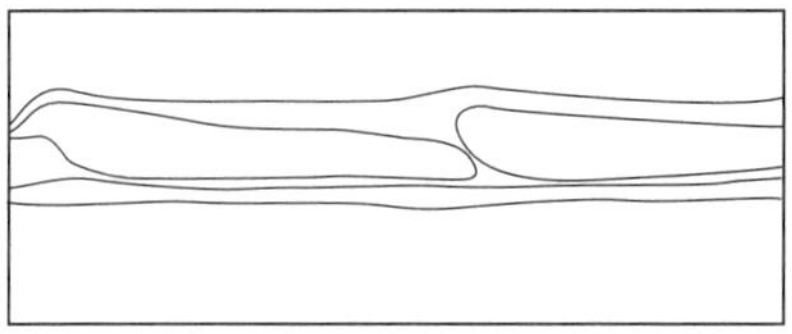

(b)

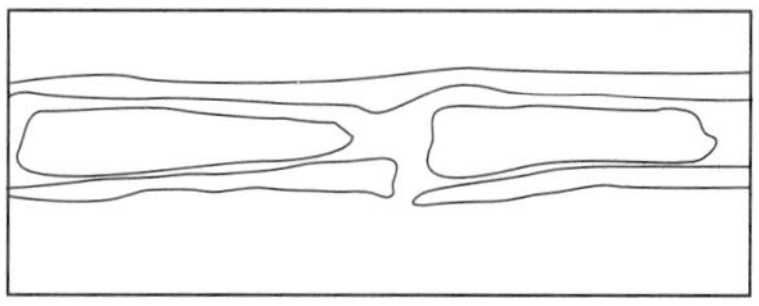

(c)

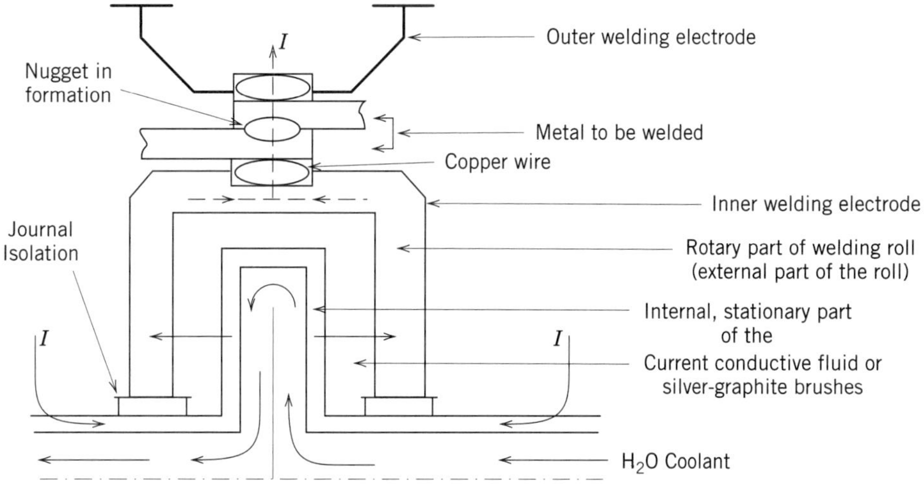

(d)

Figure 4. (**a**) Fundamentals of electric canbody welding. (**b**) Longitudinal section of a welded seam showing overlapping weld nuggets (40 ×). (**c**) Longitudinal section of a welded seam showing separated weld nuggets (40 ×). (**d**) Section through elements involved directly at the welding station.

Table 1. Comparison of Properties of Various Materials[a]

Material	Section structure	Properties[a]			
		Weldability	Lacquer adhesion	Iron Pickup value(IPV)	Filiform corrosion (FFC)
LITEWEL-N	Cr^{ox} Cr^{o} Sn Ni-Sn-Fe Base steel	○	○	●	●
#25 Tin plate	Cr^{ox} Sn $FeSn_2$ Base steel	●	△	●	●
#10 Tin plate	Cr^{ox} Sn $FeSn_2$ Base steel	○	△	△	○
TFS	Cr^{ox} Cr^{o} Base steel	x	●	○	●
Low nickel plated steel	Cr^{ox} Cr^{o} Ni_2 Base steel	△	○	○	○

[b] ● Excellent ○ Good △ Normal x Poor Cr^{ox}: Chromium oxide Cr^{o}: Chromium metal

[a] *Source:* NKK Technical Bulletin- The Steel Canstock Guide by Elizabeth Parr, Sheffield, UK.

method offers pure butt welding with advantages to double seaming of the ends, necking as well as versatile decoration such as wrap-around printing. The copper wire welded seam and the plain welding margin in case of lacquered insides of the cylinder, interrupt an otherwise smooth inside enameled (or plain) surface. The welded seam contains exposed iron and iron oxide as well as eventual traces of copper besides tin at either side of the weld. To protect the product from contamination and/or the weld from attack by the product which still results in contamination of the product, the side seam needs to be coated in most cases.

Formation of iron oxide can be avoided by neutral gas welding in which case the weld area is isolated from the atmosphere by neutral gas jets such as nitrogen and a small percentage of hydrogen. This can be done at the inside and/or at the outside of the side seam.

Bodymaking, cemented cans. Cans used only for dry or otherwise neutral products such as powders of specific natures, mineral or/and vegetable oils as well as many industrial products. Melted organic cements are injected into the inner hook of lap-locked side seams and, upon engaging and flattening the side seam, they act as sealents. Before the almost universal use of two-piece cans for high output beverage can production, one method for making them was to cover the longitudinal edges of the blanks with a nylon strip which was fused after forming the cylinder. The process has a number of proprietary names (eg, Miraseam, A-seam). An advantage was the complete protection of the raw cut edges of the blank. It could only be used with precoated TFS since the melting point of tin is close to the fusion temperature of the plastic. Even tapered corned beef cans were made in this way. When tin prices were a multiple of the present ones, using TFS was most cost attractive a solution. Cost cutting in metal packagings is therefore not an exclusive present day problem. The bodymaker used to make the nylon-bonded cans was an adaptation of the soldering configuration.

Completing the body. The plain cylinder must be furnished with a flange at each extremity of the cylinder or otherwise odd shaped body for attachment of the closure. For food processed in and with the can, where the can may be subjected to external pressure especially at the cooling phase of the autoclaving cycle, or remain under internal vacuum during storage, the cylinder wall may be ribbed or beaded for radial strength. Beading profiles and distribution along the can body are to be designed so as to reach the desired proportion of radial against axial resistance of the cans.

Cylinders for shallow containers, especially those which can not be made on bodymakers because of limited minimum cylinder height, may be made "multi high", ie, in a length suitable for two or three cans to obtain maximum efficiency from the forming machine. The welding speed remains unchanged thereby whereas the mechanical motions are divided by 2 or 3. Two techniques are used for parting the multi-high bodies into their double or triple components by breaking the score line or by cutting peripherally the unscored bodies.

Two-Piece Can Manufacture

Metal forming methods. Both methods of making two-piece cans use metal forming principles that depend on the properties of metal to "flow" by rearrangement of the crystal structure under the influence of compound stresses without rupturing the metal however.

Drawing. In drawing, as applied to can manufacture, a flat sheet is formed into a cylinder, or other section such as oblong, eg, fish cans, by the action of a punch drawing it through a circular, or different shaped, die. See Fig. 5 showing the basic tool parts involved and the stresses and strains occuring in the material. Some thickening towards the upper part of the drawn element is inevitable, but the process is essentially one of diameter reduction at constant metal thickness and volume so that the surface area of the drawn part is practically equal to the surface area of the blank from which is has been formed. This factor forms the basis of design, in particular the calculation of metal utilization. The amount of diameter reduction achievable, ie, from blank to cup diameter, is governed by the properties of the material, the surface friction interactions between the tooling and the material influenced by tool and material surface conditions and lubrication. The blankholder force is an additional factor which adds to the drawing force which the material has to stand between punch and draw radii. Drawing has been the subject of much research. One of the classical relationships is shown in Figure 6.

The diameter of the cup produced in the initial draw may be further reduced by a similar redraw operation with a draw sleeve fitting between punch and the inside diameter of the cup. The said draw sleeve is then acting as a blankholder. The rule of constant area and volume determines that the reduction in diameter is accompanied by a corresponding increase in height (see Fig. 7). The redraw operation can be repeated in several drawing stages, provided the progressive reductions fall between definite limits so as to avoid metal failure. Draws can be made "through the die" and the drawn part be stripped from the punch at the end of the press stroke below the die or the press stroke can be limited so as to leave the drawn part with a flange, in which case the drawn part is extracted from the die, being retained by the punch (friction and vacuum), and ultimately stripped from the punch for transfer or evacuating from the press. The stroke of the press has to be adapted accordingly. In the latter case stated it is usually equal to 2.5 times the cylinder height as drawn. Fig 8 shows (**c**) typical drawn can provided with a stacking or/and label retaining bead if canstock is not decorated, (**b**) the sequence of operations from the blank to the finished can, (**a**) a typical schematic lay-out of a DRD cans production line of medium production capacity.

Wall ironing. In pure ironing, as used in the D&I process, a cylindrically drawn cup is redrawn by the ironing punch and does therefore precisely fit the said punch diameter. The redrawn cup is then forced axially through a set of ironing die rings, whose diameters, which are progressively decreasing, create a gap with the punch, which is smaller than the wall thicknesses of the redrawn cup and the consecutive wall thicknesses from t1 to t4, see Figure 9(**a**) whereon the stated wall thicknesses are related with the punch and rings radii differences.

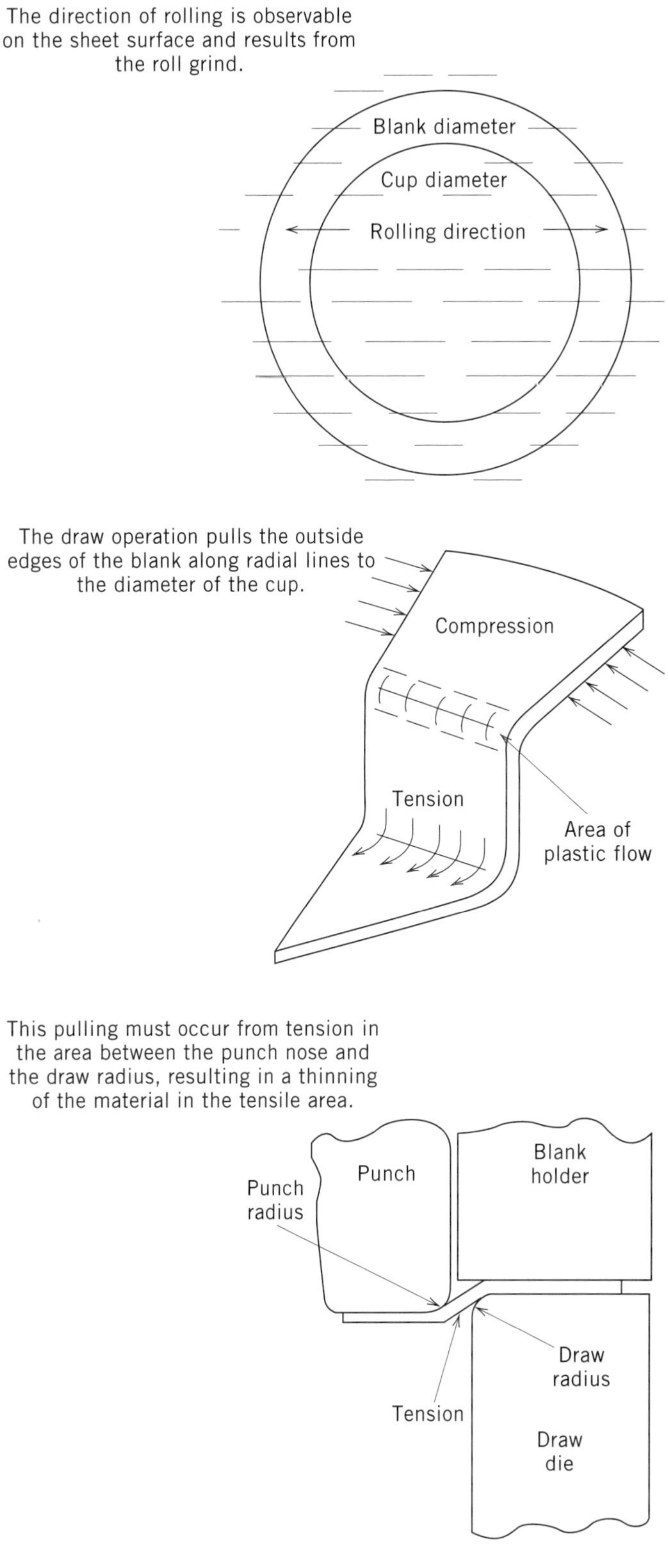

Figure 5. Basic stresses and strains. Courtesy of ALCOA Forming Aluminum.

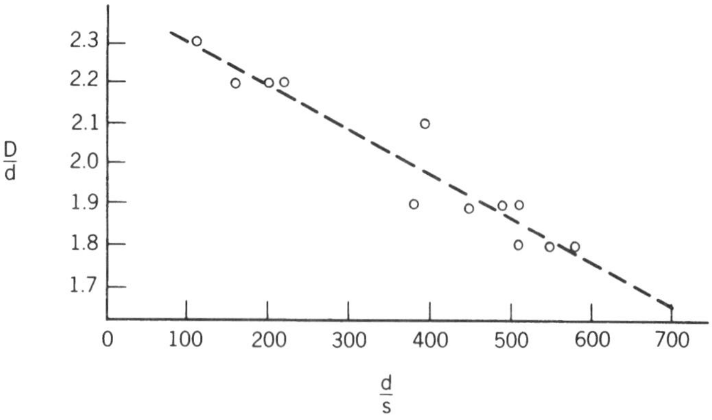

Figure 6. Limiting draw ratio. The maximum blank diameter which can be drawn into a cup without metal failure. The relationship illustrate the significance of sheet thickness (gauge). D = blank diameter; d = punch diameter; s = sheet gauge. Courtesy of International Tinplate Conference, 1976.

The process, similar to impact extrusion, thus results in a reduction in wall thicknesses at a constant diameter excepted for t3 where the punch diameter is recessed so as to form the "thick wall" whereas t4 is the "thin wall" which is reduced from 0.22 mm for t0 down to some 0.07 mm. Here too, the governing principle is the constant metal volume. In other words, the volume of the metal in the ironed can body (prior to trimming) is equal to the ingoing cup and consequently to the original blanked disc. Note the sequence of operations in Figure 10. The amount of reduction at each stage is determined by the material properties, governed by the needs to avoid metal failure. The very high friction, under extreme surface pressure, mainly at the outer surface of the ironed can body, makes special demands upon lubrication, which is combined with copious flood cooling to maintain the critical punch-to-die gap.

A similar effect can be obtained in drawing if the gap between the draw die and the punch is less than that of the metal being drawn. It is common practice to keep this gap equal to the nominal thickness of the canstock to control thickening due to diameter reduction. This process is called sizing. The drawing gap can be further reduced to produce a definite thinning of the wall, relative to the base material as a combination of drawing and ironing, or thinning. Special lacquering is primordial in this case.

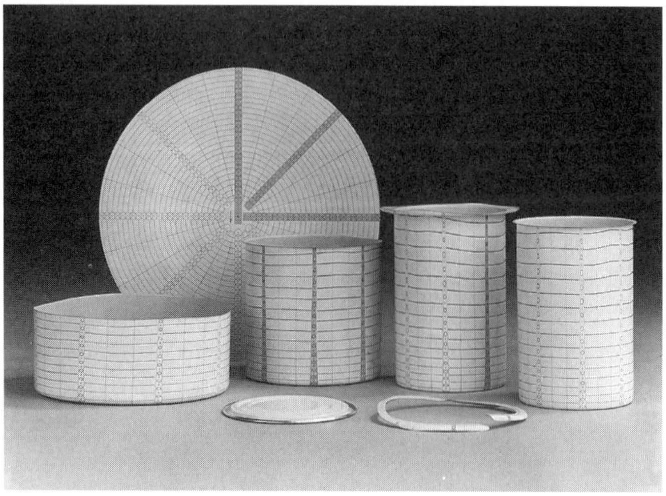

Figure 7. Reduction in diameter is accompanied by a corresponding increase in height. Courtesy of Styner O Bienz, Nieder Wangen/Bern Switzerland.

D&I can manufacturing. The procedure of producing D&I cans is schematically as follows: unwind plain coil; lubricate; blanking and cup drawing; redraw; wall ironing; dome forming; trim body to correct height; wash and surface treat (if appropriate as for aluminum); dry or wash coat and cure. Then for *beverage cans:* base coat outside and cure (if appropriate; decorate and cure; coat inside and cure (eventually a double coat); neck and flange open end (in case of tinplate, the order of the last two operations may be reversed). For *food cans:* wash coat outside; neck and/or flange as appropriate and if body is not beaded; body beading; coat inside (in some lines the order of the two last operations is reversed). The *closures,* or *ends,* are made from precoated sheets or coil on multi-die presses, profiled in such a way that internal pressure will not cause any permanent deformation (peaking) in the chuck wall radius area. Their turned down edge of the flange (crown shaped or plain) is curled so as to allow stacking and automatic handling and assist the formation of the double-seam. A sealing compound is nozzle injected onto the spinning ends, placed and centrifuged so as to form a gasket in the finished double-seam (see Fig. 11).

Compounds are either solvent-based (hexane) or water-based, which are respectively, drying by evaporation of the solvent or by heat drying (hot air or various radiations).

The components of a D&I line are shown in Figure 12. Coils are shipped with the axis vertical, for safety and to avoid damage to the laps. A down-ender is used to bring the axis horizontal and transfer the coil to a coil car for distribution to one of a number of dereelers (unwind stands), each feeding one cupping press, or cupper. Dereelers may be dual and engineered in such a way that coil changes are rapid either when coils are finished or if one has to be removed for being found defective. The coil is passed through a lubricator where lubricant is applied by dipping in a tank, the excess being removed in a couple of rubber covered rolls. The lubricant is constantly recirculated for filtering, temperature control, bacteriological control. It is essential that lubricants, coolant-lubricants and washing chemicals be compatible and form a system possibly supplied by the same manufacturer. Steel coils are additionally and continuously inspected for detection of pinholes, surface defects (visual inspection) and monitoring of gauge thickness, interlocked to stop the cupper when out of specification material is detected. The cupper (7) is a double action press so as to blank and draw the cups under a perfectly controlled blankholder force. Blanking the disc and drawing the cup are two consecutive phases in one stamping cycle or press stroke.

Depending on coil width, as many as 12 cups may be produced in one working stroke. Cuppers can be designed with top drives (Minster) or underdrive (Standun). Cups are produced at speeds of up to 200 strokes per minute.

The cupping process demands great precision due to (*1*) extremely narrow drawing gaps of 10% in excess of the

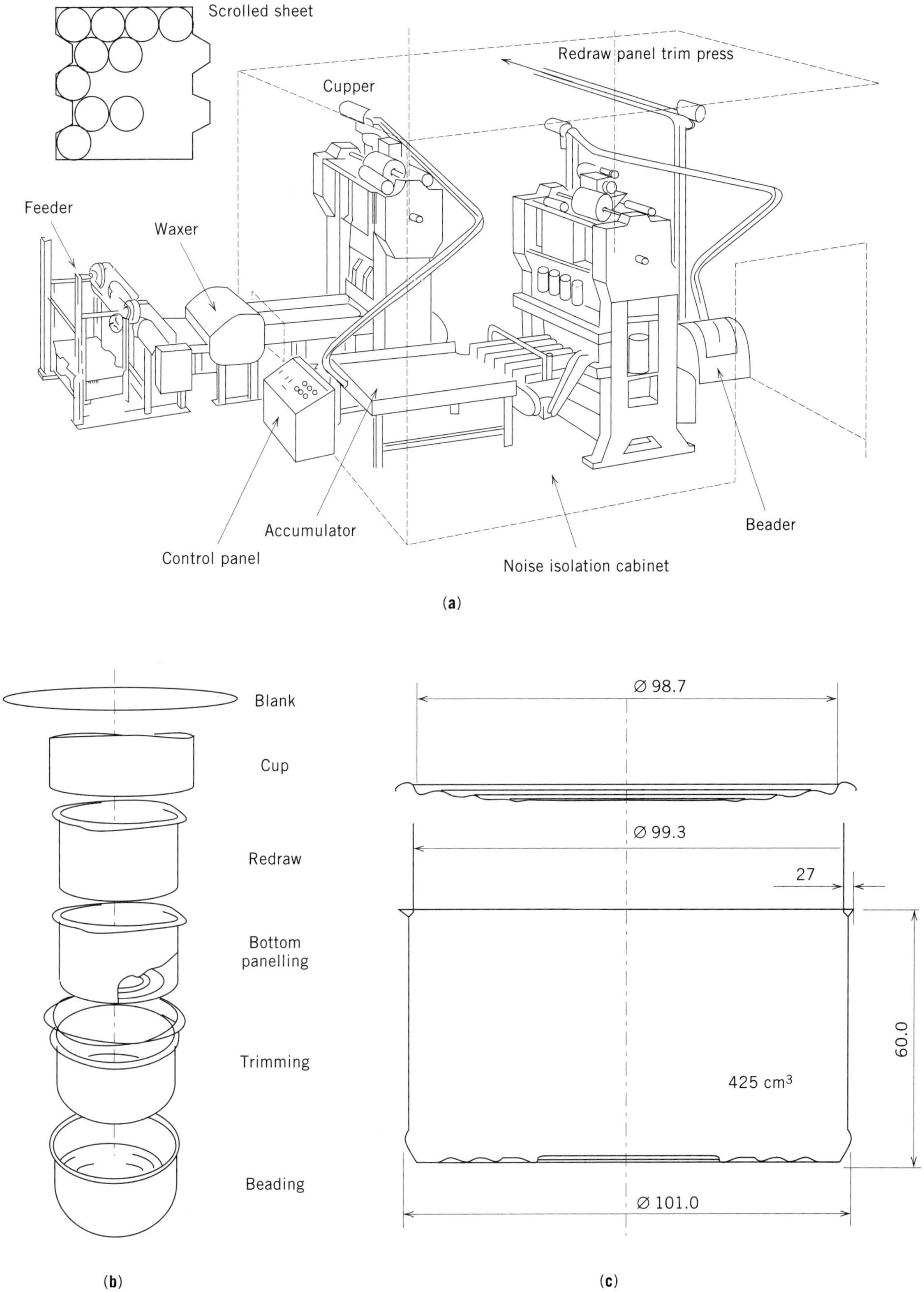

Figure 8. (**a**) Typical lay out of medium production of DRD cans. Courtesy of Ferembal Nancy/ France. (**b**) Sequence of DRD fabrication steps (**c**) Typical 401 × 206 DRD can with end (for ref. only).

material thickness, between punch and die. Materials currently used have a thickness of 0.009" (0.24 mm) and canstock will undoubtedly become thinner still in the future. (*2*) Shreds (intervals) between two adjacent cuts of blanks is about 0.04" (1 mm). (*3*) All press and tool parts must be perfectly aligned, flat, parallel and/or concentric. Cups should show no wrinkling or measurable or visual differences all around the walls. This is fundamental for successful ironing. The wall ironer (bodymaker in the United States) (8, 9), converts the cups into a cylinder with the correct preestablished and correct thickness distribution along the wall and dome shapes

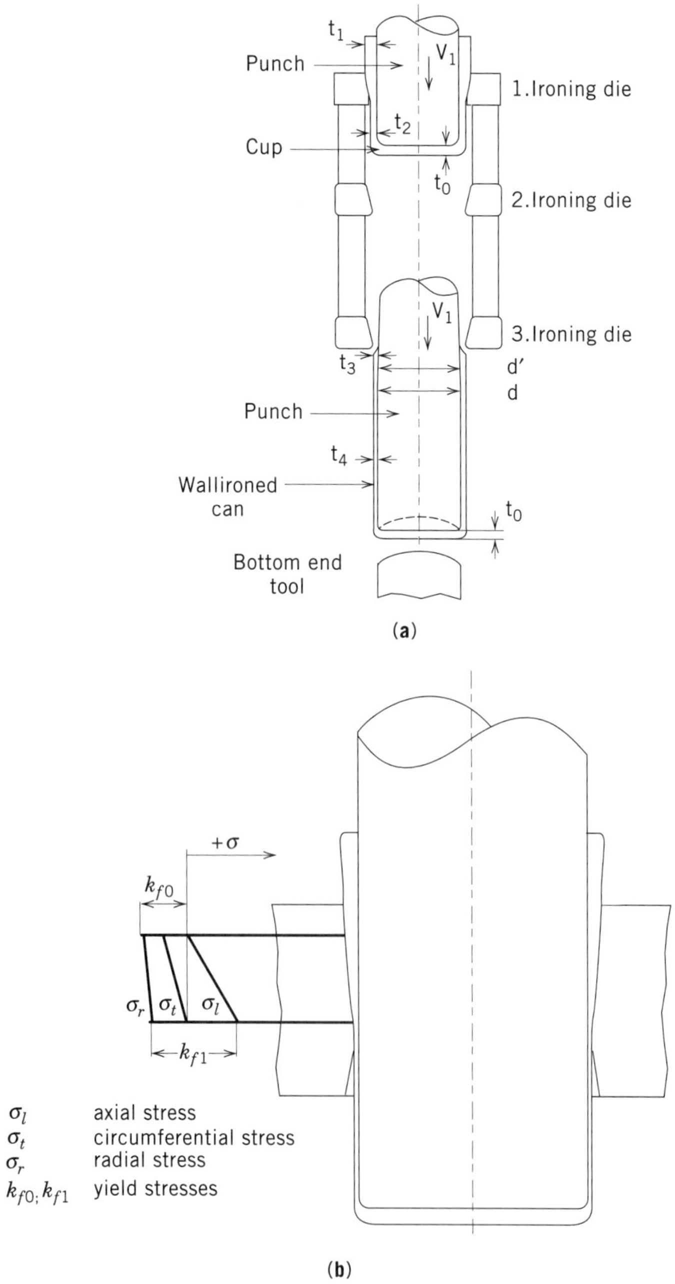

Figure 9. (**a**) Schematic of wall ironing principal Courtesy of: Tinplate and modern canmaking technology by E. Morgan (**b**) Ironing DWI can body and stresses on the material. Courtesy of Rasselstein AG Germany.

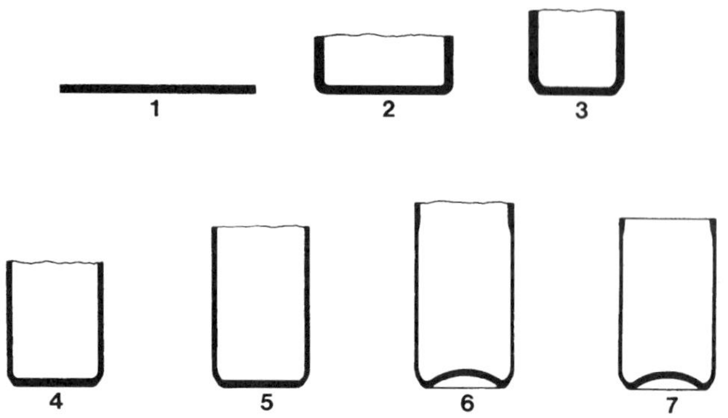

Figure 10. Stages in forming a drawn and ironed (D&I) body. 1, Circular blank; 2, cup; 3, redraw; 4, first ironing stage; 5, second ironing stage; 6, third ironing and dome forming (note thickening at top where flange is to be formed); and 7, trimmed body.

(the base) so as to resist elevated internal pressure. Figure 9(**a**) shows the arrangement of the ironing tool set as it is basically used in all ironers. At the end of the stroke, at the moment the punch reciprocates after the front dead point of the ram, spring loaded stripper fingers, assisted by compressed air fed through the ram to the punch nose, strip the can body from the punch into a discharge conveyor which transfers the cans sideways to the trimmer. For the alledged base of constance of metal volume the length of the cans will vary with the tolerances of the material and toolings. The top end of the cylinder will be more or less wavy, earing, which is due to specific characteristics of the canstock as long as toolings and alignments, as well as ironer stability, cannot be incriminated.

Ironers run at about the same speed as the cuppers, so that one wall ironer is needed for every tool, or "out" in the cupper. In practice, one ironer was added as stand-by to bridge short interventions on the machine or/and clear jams, etc. Newer generations of ironers run faster than the cupper and twin-ram machines allow to half the number of ironers in the canmaking line. Stand-by must still be available as the loss of output is doubled during down-time in case of twin-

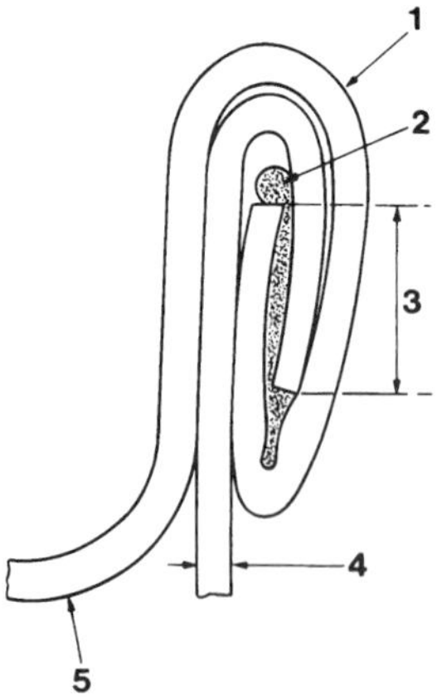

Figure 11. Double-seam closure. Showing fixing of end to cylinder, and function of lining compound: 1, end curl, folding round flange on cylinder, 2, lining compound; 3, seam overlap; 4, cylinder; and 5, end.

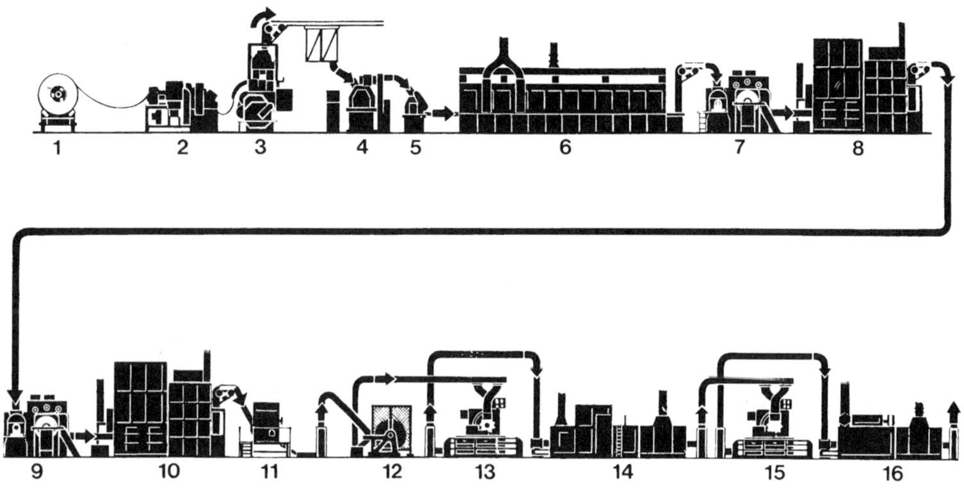

Figure 12. D&I can manufacture. Diagram showing all the equipment that may be required. 1, Uncoiler (dereeler); 2, lubricator, and optional coil inspection; 3, cupping press; 4 wall ironer (bodymaker); 5, trimmer, 6, can washer; 7, external coater; 8, pin oven, 9, decorator; 10, pin oven, 11, necker–flanger; 12, tester; 13, internal spray machines; 14, curing oven (IBO); 15, optional second spray; and 16, IBO. Courtesy of Metal Box Ltd.

ram machines. The tooling uses inserts of carbides or/and ceramics in the ironing dies together with carbide or ceramic punches when steel cans are fabricated. Good quality tool steels are adequate for aluminum cans. Associated with every ironer is a trimmer to which the eared bodies are transferred positively held by the base either magnetically or by vacuum. Rail (sickel knife) or roller trimmers are used.

In rail trimmers, the end of the canbodies is trimmed between a mandrel mounted on a rotating turret and a stationary rail. A certain number of mandrels are fitted and the turret is in continuous motion.

In the roll trimmer, cans are indexed to a position where the end is inserted between two rolls and rotated about their axis while the rolls are closed, one being pivotly mounted, to perform a peripheral trimming, or cutting, action. In one widely used machine, the cutting rolls are mounted in an easily removable cartridge for refurbishing and precision-setting in a toolroom environment.

Upon trimming the canbodies to their specified height they are mass conveyed in bulk, in upright position, to a washer. Cans are conveyed on a flatbed open mesh belt, open end down, through a series of compartments fitted with spray nozzles above and below. Top conveyors are used in each spray compartment to restrain the cans from falling over. After prewashing with detergents for removing drawing and ironing lubricants, cans are washed, prerinsed and rinsed with demineralized/deionized water. Aluminum cans commonly receive an etching treatment to make the surface receptive to organic coatings applied in the finishing operations. Deionized water rinse ensures stain free drying.

Alternatively, the final rinse may contain an organic coating to protect the outside surface of tinplate food cans underneath the paper label applied by the filler. The washed and dried cans are now ready for receiving internal and external finishes, both structural (necking/flanging, eventual reforming) and protective (coatings and decoration). Food cans are usually beaded for the thin wall to stand buckling and/or implosion in the same way as with three-piece cans. The machinery however is differing in detail due to the presence of the integral bottom and the absence of side seam extra thickness as on three-piece cans. Spin flanging is invariably used to avoid split flanges in the axial metal grain structure caused by ironing although excessive work-hardening does not occur. The top of the beverage cans is necked-in to yield metal cost savings through the use of smaller diameter, and therefore thinner ends material. The first necking steps may be die necking operations followed by spin necking/flanging. This method is finding increasing application (10). When first cans were die-necked and spin-flanged from 211 to 209 diameter, they are now necked down to 202. It creates an aesthetically pleasing neck profile and contour when cans are spin necked and this with minimal strain on the split-prone flange and very low axial force on the can. For the smaller necks, such as 211/202, several die prenecks are necessary when die necking is applied (11). Single stage prenecks are currently practised by spinning. The final spin-neck/flanging operation is done either before or after internal protection depending on local circumstances. In general, aluminum requires an enameled surface to prevent pick-up (galling) on the tooling. In some way, the enamel acts as a lubricant.

DRD can manufacturing. The manufacturing steps are as follows: cut up coil into scrolled sheets (eventually coil coated stock); coat and/or decorate sheets; blank discs and form cups; redraw once or twice depending on the h/d ratio of the can; form base (or panel can bottom); trim flange to applying width; bead (if appropriate); spray coat (either single or top coat on roller coated material according to pack) and cure; test and palletize. See typical production set-up in Figure 8. As in D&I, a multi-tool cupper (7) (Fig. 8) is used to cut blanks from wide coils or scrolled sheets and form them into shallow cups. Their D/d ratio (blank to cup diameter ratio) as chosen will influence the performance of the subsequent draws if cups are reformed by one or two further draws which

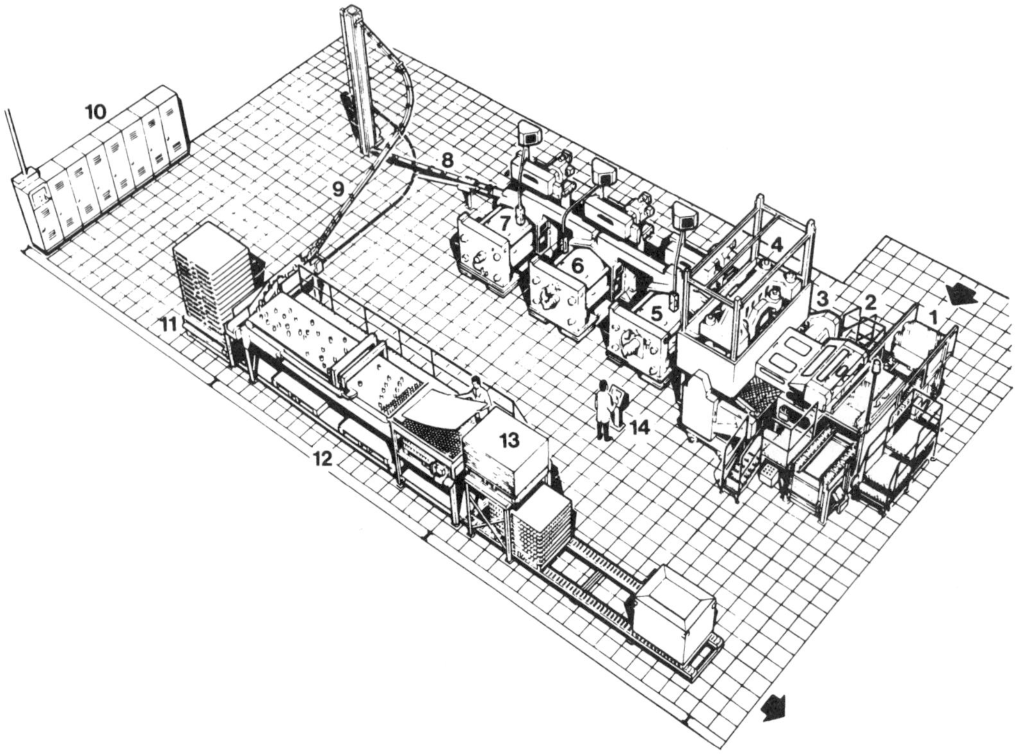

Figure 13. Rotary Press DRD line. 1, Infeed conveyor (for stack of coated sheets); 2, lubricator; 3, sheet feeder, 4, cupping press,; 5 first redraw press; 6, second redraw press; 7, trimming and base forming press; 8, can sampling chute; 9, can conveyor (a tester is usually inserted here); 10, electrical controls, includes press synchronization; 11, pallet feed; 12, palletizer; 13, layer pads; and 14, master control station. Courtesy of Metal Box Ltd.

progressively reduce the diameter and increse the height. By stopping the final draw at an appropriate point, a flange is left on the can, but as with all metal-forming operations, the eared and irregular collar has to be subsequently trimmed. Prior to this, the can bottom, or base, has to be profiled, or paneled, see Figure 8, so as to withstand processing requirements. All these operations are usually carried out on multilane, multistage transfer press. Set-ups have to be coherent with regards to their production capacity. Each follow-up machine should have a capacity of 0.5 to 1% in excess of the foregoing one and controlled infeed should ensure that no jams occur at stops and starts of the infeed whereas the machines should not be stopped. However, due to the critical nature of the die-trimming operation, separate machines employing different principles can be used, such as rotary trimmers. When simultaneously drawing and ironing, it is convenient to draw the body right through the die in order to achieve thickness control in the flange area. In this case, the body is trimmed and flanged in conventional D&I equipment. For taller cans, the walls may then be beaded in the usual machine.

Variants in metal forming equipment are the Metal Box rotary press line (12) (Fig. 13) and the Standun opposed-action press. The latter features a split blankholder with separate pressure controls, which is claimed to permit greater draw reductions and provide control of wall thickness through stretching rather than ironing, which is said to be less severe on the coating. Equipment of various origins are on the market for small to high production. More intelligently designed lines are engineered together by canmaking experts on proven presses, machines and reliable toolings. Figure 8 represents such a set-up.

Coating

With the exception of three-piece cans for certain products, organic protective coatings are applied to both in-and/or outside surfaces of the cans. Internally they provide a protective barrier between the product and the metal whereas externally they may protect from atmospheric corrosion. Outside decoration provides product identity as well as protection especially if inks are protected by an over-varnish. Coatings are applied "wet", ie, the resins are suspended in a carrier for ease of application and allowed to coalesce, flow out or extend for proper film forming. The coating is then baked, or stoved, or cured, according to countries, first to remove the carrier, which may be an organic solvent or a mixture, predominantly water (waterbased lacquers) and then to polymerize (cure) the resins. The methods of application are in general commercial use: roller coating in case of sheet stock or on the outside of two-piece cans, and spraying on the inside of formed bodies. Spraying is predominantly used for the inside of two-piece D&I cans, but DRD cans are sometimes sprayed as well. Certain categories of three-piece cans are sprayed for carbonated beverages or red fruit cans, the spray being a top coat applied upon bodymaking and side seam striping, the material having been roller coated. Alternative methods of application and curing are available such as electrophoretic deposition. This method is also used widely for repair-coating Easy-Opening-Ends upon conversion, ie, scoring, riveting, and securing the pull tab. Drying and/or curing (polymerizing) are normally carried out by forced convection using hot air. To a lesser extent inks and some lacquers are cured by UV-radiation, electron-beam curing, infrared, or high frequency energy. The most diffused nonconventional inks and lacquers are ultraviolet curing products and they permit applications of virtually solvent-free "wet" coatings by conventional means. It requires the use of specially selected materials amenable to this form of polymerization which demand care in storage and handling and complete polymerization in the interest of health and safety. Its attraction lies in the absence of solvents in difficult environmental situations, and extremely short curing times coupled with relatively low energy demand. It is thus used for very high speed operations or where space is limited as for instance between two or more print color applications ("Interdeck-Drying").

Powder coating. When resins are applied "dry" in the form of fine powder, powder coating is directed in many cases by creating an electrostatic field, charging the powder particles on the negative pole for being deposited by the positive pole, the latter being the element of the can, or the entire canbody, which has to be powder coated. In three-piece canmaking, it is specifically the weld and the weld margin (lacquer free) area which are coated. The absence of solvents avoids the otherwise excessive cost of drying and eliminates eventual blistering caused by trapped solvents, which cause porosity, or boiling off because of solvent based lacquers being applied on too hot welds. Curing, which is a simple fusion of the applied powder layer, is usually done by infrared radiation or high frequency induction heating, as hot convected air could disturb the uncured coating. In a great many cases, the same curing ovens are used however for powders as well as for lacquers. Only their length is different as curing of the different products used for side seam striping need more or less curing time: polyester powder, 8 s; epoxy powders, 10 s; airspray lacquers, 12 s; airless sprayed lacquers, 14 s; and roller coated lacquers, 17 s. In case both powder or lacquer have to be applied, the oven length should fit the longer curing time. *Electrophoretic deposition* provides a means for electrically depositing a resin film on a metallic substrate from an acqueous suspension (13). Originally used at low speed for protecting automobile bodies, it has now been developed to be applied at can production speeds with practicable voltage and current demands. Two-piece and three-piece cans are currently coated on high production lines on both in- and/or outside of the bodies. Decorated cans of either type can be washed, without affecting the prints, and inside coated in the most positive way. A system of process cells is described in refs. 14 and 15.

It must be stressed that for food and drinks cans, only the anodic system is applicable. Compared to spray application it gives far more even distribution of coating over the full can wall and base, providing a saving of the quantity used without sacrificing the required minimum coating thickness. Its throwing power enables it to coat regions inaccessible to spray, useful for the severe profiles which have to be used to obtain adequate container strength with thin plate. The system provides practically porefree films and practically near to zero mA curent permeability tests are obtained, even when particularly severe tests are applied. Electrodeposition was initially used (American Can Co.) to repair-coat converted Easy Open ends, thus protecting scores, rivets and cut edges of the tabs, as well as other possible damages to the roller-coated lacquer film applied prior to stamping shells and converting. This equipment is available on the market as an improved version, offered by CORIMA in Tresigallo/Italy. *Thermal transfer* of the complete design from printed paper to a plain coating on the can provides a method of obtaining very high quality decoration in two-piece cans if the extra cost is justified (16). It is used in the UK under the trade name Reprotherm (Metal Box Ltd) for promotional designs or single-service beer cans.

Coating Equipment

Offset coating. Offset coating is based on the offset printing principle, whereby a metered quantity of coating is applied to a rubber or polyurethane covered roller, to be transferred therefrom to the metal substrate (see Fig. 14). In roll coating, metering is accomplished by a series of steel or rubber rollers which pick-up the coating from a trough and ensure even distribution and weight control through a combination of relative surface speed and the pressure or gaps between them. Another method, gravure coating, employs a pattern of cavities etched into a steel roll, which are filled as the roll dips into a trough. The excess is removed, and the precise amount filling the cavities is then given up to the transfer roll for offset application onto the metal substrate.

Rotary screen lacquering. A third method is a development which uses the old established principles of silk-screen printing. The screen is in the form of a cylindrical thin, narrow meshed sieve like metallic screen, fed internally with the coating material, which is forced through the screen by an internal doctor blade onto the transfer roll. Patterns governing shape (cutting diagrams for blanks and/or ends) and weight of the applied coating by photomechanic methods.

Roll coating. This is the most common method for coating in the flat, used for both coil and sheets. In sheet coating it is possible to cut a pattern (stencil) into the transfer roll (which would be the blanket cylinder on an offset press) and coat only the areas to be used for blanks. In a large scale operation this could yield significant savings in material, but requires the sheet to be accurately registered with the transfer roll. Most of the present day coaters are provided with front lays which ensure precise registering and magnetic or vacuum cylinders avoid irregular travel of the sheets while being transferred by the pressure contact between the transfer roll and the compression cylinder.

Rotary-screen coating permits a better definition of the pattern in the applicator itself without cutting (or stenciling) the transfer roll, thus leaving a sharper, cleaner edge between coated and uncoated areas. Other advantages claimed are the ability to apply double the coating, possibly saving one pass, ie, the coating operation and curing, as well as applying differential coating weights on adjacent areas of the sheet or on individually coated spots. Thus the coating pattern can be identical to an end-cutting diagram whereon all geometrical

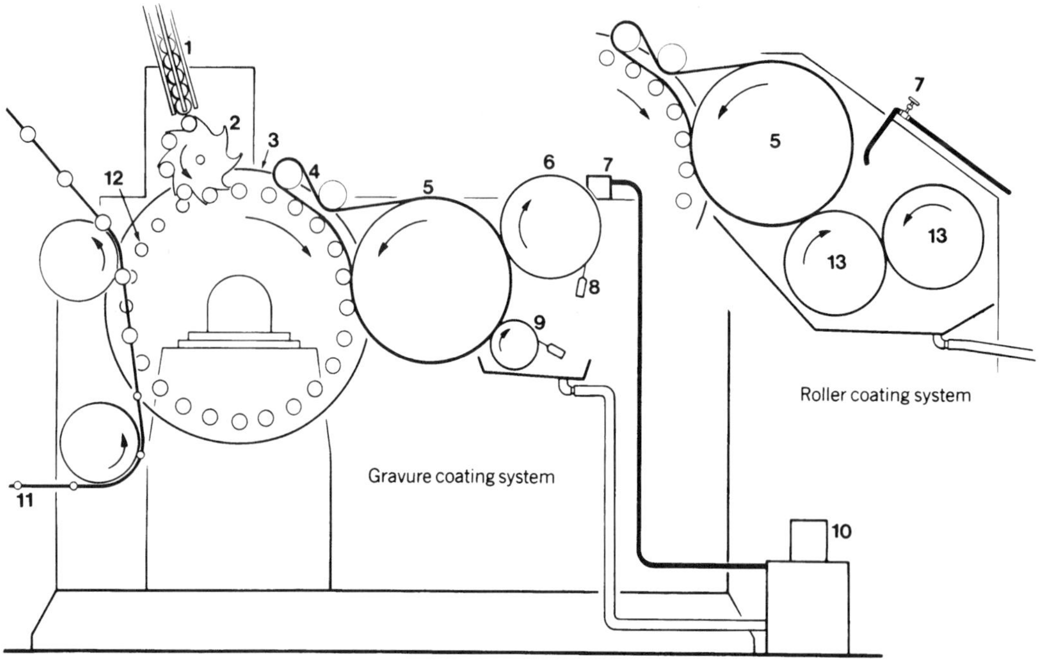

Figure 14. Offset coating, showing application to two-piece can bodies. 1, Body infeed; 2, star-wheel; 3, mandrel wheel; 4, pre-spin; 5, offset cylinder; 6, gravure roll; 7, coating material feed; 8, doctor blade; 9, stabilizer roll; 10, coating material tank, with pump; 11, oven pin chain; 12, mandrel; and 13, metering rolls.

losses (shreds, skeleton) remain uncoated. In this way the flange area of any end, receiving the sealing compound can receive a low weight lacquer film whereas the curl portion can remain totally unlacquered. There is a twofold advantage: water based compound adheres better on uncoated endstock and the blank cut in the end stamping press is made on a lacquer free area. Cured resins, especially when they "feather" act as emery in end tools, adhere to the tool parts and scratch other ends causing downtime for cleaning. The uncoated curl area is covered by compound, at least to 1/64th" from the cut edge of the end, and does therefore disappear in the double seam end hook.

Roller coating is similarly used, although in reduced dimensions as compared with sheet coaters, for the outside of two-piece cans. For this application however, the more precise method of gravure coating is gaining favor. Decorators operate in a similar fashion (17) but are considerably more complicated, since they apply four, and sometimes up to six colors with extreme accuracy of registration to build up the total picture. Thus a can decorator uses the same can-handling system as the coater, replacing the coating head by a reduced size offset press printing equipment for each colour, all of them four to six, transferred onto one and the same blanket cylinder which transfers the total label onto the can envelope in one rotation of the can. Cans (in the same way as collapsible tubes) are stripped from the printing mandrels and transferred onto the pins of the pin oven chain for curing.

Spraying. The method normally employed is the airless spray system, where atomization of the enamel into fine droplets of appropriate size is achieved by the use of high pressure on the enamel, about 650–850 psi (45–60 kg/cm^2). A spray gun consists of a spray nozzle designed to give the desired spray pattern, a needle valve to cut off the flow without dripping, and the means to activate the valve at high speeds, usually a solenoid.

Since viscosity of the enamel has a decisive influence on the spray pattern, discharged quantity and dripping, it is heated and continuously circulated through the gun. The quantity of lacquer deposited in the can is determined by precise timing of the valve. A typical spray time is approximately 100 milliseconds. In modern practice, the gun is at a fixed position at the mouth of the can although machinery exists for moving it axially inside the can during spraying. This is called a lancing gun. The can is indexed to a position in front of the gun and rotated at about 2000 rpm, to ensure an adequate number of rotations while spraying is in progress. It is practically compulsory to keep rotating the can on its way from the spray coater to the oven to assist flow-out (extension and full wetting for poreless film formation and avoid rundown) and even distribution. The pattern of coverage and number of coats depend on the type of product to be packed (beer or carbonated soft drinks), the can material (steel or aluminum) and the can size (h/d ratio). Coverage can be total, where one spray covers the total area, or zonal, where one spray is directed towards the base and lower side wall of the can, and a second one covering the upper side wall of the can up to its open end. Two coats may be applied in consecutive indexing stations on the same machine (wet on wet) or in separate machines with intermediate drying (wet on dry).

Internal side-seam protection for welded cans. The bare metal that exists in the weld area must be protected in many cases. Roller coating, spray, electrostatic powder coating, are used.

The applicator is mounted on an extension of the welding arm through which the coating material is supplied from a connexion fitted before the can cylinder is formed around it. Roller coating permits a low pressure fluid supply, but lack of space makes the applicator components extremely small as they must be contained inside the sometimes quite small internal can diameter, ie, 202 cans. Spray application requires a high pressure fluid supply. Although good initial coverage can be achieved, liquid enamel tends to retract from sharp edges and eventual "splashes", so that high application weights are needed for adequate coverage of the cut edge of the blank. The overspray, which escapes from the gaps between the cans, poses exhaust problems, especially in avoiding external contamination through drips. Powder is fluidized with air for conveyance through the welding arm and electrostatically charged to achieve deposition on the welded area. This method undoubtedly provides the best protection, but is expensive in material and must be run carefully in production. Wherever epoxy powders can be applied instead of polyester ones, the material cost is substantially reduced.

BIBLIOGRAPHY

1. Brit. Pat. 621,629 (June 16, 1949), J. Keller.
2. U.S. Pat. 760,921 (May 24, 1904), J. J. Rigby (to E. W. Bliss Company).
3. U.S. Pat. 2,355,079 (Aug. 8, 1944), L. L. Jones (to American Can Company).
4. Brit. Pat. 1,574,421 (Sept. 10, 1980), J. T. Franek and E. W. Morgan (to Metal Box Ltd).
5. Brit. Pat. 910,206 (Nov. 14, 1962), Soudronic AG.
6. Fr. Pat. 2,338,766 (Aug. 19,1977), E. E. V. V. Saurin and E. V. Gariglio.
7. Brit. Pat. 1,256,044 (Dec. 12, 1971), E. Paramonoff and H. Dunkin (to Standun Inc.).
8. U.S. Pat. 3,270,544 (Sept. 6, 1966), E. G. Maeder and G. Kraus (to Reynolds Metals Company).
9. U.S. Pat. 3,704,619 (Dec. 5, 1972), E. Paramonoff (to Standun Inc.).
10. Brit. Pat. 1,534,716 (Dec. 6, 1978), J. T. Franek and P. H. Doncaster (to Metal Box Ltd).
11. Brit. Pat. 2,083,382B (Mar. 24, 1982), J. B. Abbott and E. O. Kohn (to Metal Box Ltd.).
12. Brit. Pat. 1,509,905 (May 4, 1978), J. T. Franek and P. Porucznik (to Metal Box Ltd.).
13. Brit. Pat. 455, 810 (Oct. 28, 1936), C. G. Sumner, W. Clayton, G. F. Morse, and R. I. Johnson (to Crosse & Blackwell Ltd.).
14. U.S. Pat. 3,922,213 (Nov. 25, 1975), D. A. Smith, S. C. Smith, and J. J. Davidson (to Aluminum Company of America).
15. Brit. Pat. 1,604,035 (Dec. 2, 1981), T. P. Murphy, G. Bell, and F. Fidler (to Metal Box Ltd.).
16. Brit. Pat. 2,101,530A (Jan. 19, 1983), L. A. Jenkins and T. A. Turner (to Metal Box Ltd.).
17. Brit. Pat. 1,468,904 (Mar. 30, 1977), (to Van Vlaanderen Container Machinery Inc.).

General References

A. L. Stuchbery, "Engineering and Canmaking," *Proceedings of the Institution of Mechanical Engineers,* **180,** 1, 167–1193 (1965–1966).

J. T. Winship, *Am. Machinist,* Special Rep. No. 721, 155 (Apr. 1980). Explains how metal containers are made.

The Metal Can, Open University, Milton Keynes, UK 1979, 52 pp.

C. Langewis, Technical Paper MF80–908, Society of Manufacturing Engineers, Detroit, Mich., 1980.

E. Morgan, *Tinplate and Modern Canmaking Technology,* Pergamon Press, 1985. Contains a broad summary of printing processes.

W. A. H. Collier in *Proceedings of Cold Processing of Steel,* The Iron and Steel Institute and the Staffordshire Iron and Steel Institute, Bilston, U.K., October 1971 and Mar. 1972, pp. 70–77. Describes drawing, forming, and joining of steel containers.

J. D. Mastrovich, *Lubrication* **61,** 17 (Apr./June 1975). Describes aluminum can manufacture.

Proceedings of 1st International Tinplate Conference, International Tin Research Council, London, Oct. 1976. Full set of papers and discussions issued in book form which provide the best source of reference for can-making technology. Papers include G. F. Norman, "Welding of Tinplate Containers—An Alternative to Soldering," Paper 20, pp. 239–248; J. Siewert and M. Sodeik, "Seamless Food Cans Made of Tinplate," Paper 13, pp. 154–164; and W. Panknin, "Principles of Drawing and Wall Ironing for the Manufacture of Two-Piece Tinplate Cans," Paper 17, pp. 200–214.

Proceedings of 2nd International Tinplate Conference, International Tin Research Council, London, Oct. 1980. Papers include G. Schaerer, "Food and Beverage Can Manufacture," Paper 17, pp. 176–186.

Proceedings of 3rd International Tinplate Conference, International Tin Research Council, London, Oct, 1984. Papers include W. Panknin, "New Developments in Welding Can Bodies;" G. Schaerer, "Soudronic Welding Techniques—A Promoter for the Tinplate Container;" and R. Pearson, "Side Seam Protection of Welded Cans."

Developments in the Drawing of Metals, Conference organized by The Metals Society, London, May 1983. Section on deep drawing and stretch forming, pp. 76–125. Papers include P. D. C. Roges and G. Rothwell, "DWI Canmaking: The Effect of Tinplate and Aluminum Properties," E. O. Kohn, "The Use of Spinning for Re-forming Ultra-thin Walled Tubular Containers;" and three papers on aspects of deep drawing.

A. M. Coles and C. J. Evans, *Tin and Its Uses,* International Tin Research Institute, No. 139, 1984, pp. 1–5.

F. L. Church, *Mod. Metals* **37,** 18 (July 1981).

N. T. Williams, D. E. Thomas, and K. Wood, *Metal Constr.* **9,** 157 (Apr. 1977); **9,** 202 (May 1977).

A. G. Maeder, *Aerosol Age* **18,** 12 (Nov. 1973). Still the best first introduction to the subject.

D. Campion, *Sheet Metal Ind.* **57,** 111 (Feb. 1980); **57,** 330 (Apr. 1980); **57,** 563 (June 1980); **57,** 830 (Sept. 1980).

S. Karpel, *Tin Int.* **54,** 208 (June 1981).

Metal Decorating and Coating, Annual Convention of National Metal Decorators Association (NMDA), October, annual. Reviewed in *Modern Metals.*

E. A. Gamble, *J. Oil Colour Chem.* **10,** 283 (1983).

JEAN SILBEREIS
SLCC Can Consultants
La Tour de Peilz, Switzerland

METALLIZING, VACUUM

Introduction

Vacuum metallizing, which is a physical vapor-deposition process, produces coatings by thermal evaporation of metals un-

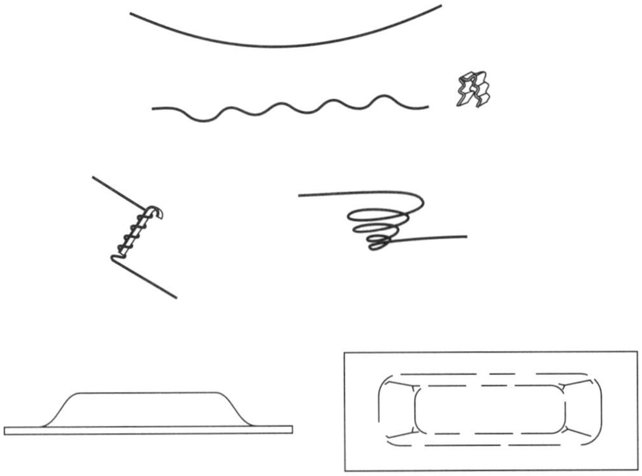

Figure 1. Resistance-heated wire sources (*top*); refractory metal boat source (*bottom*).

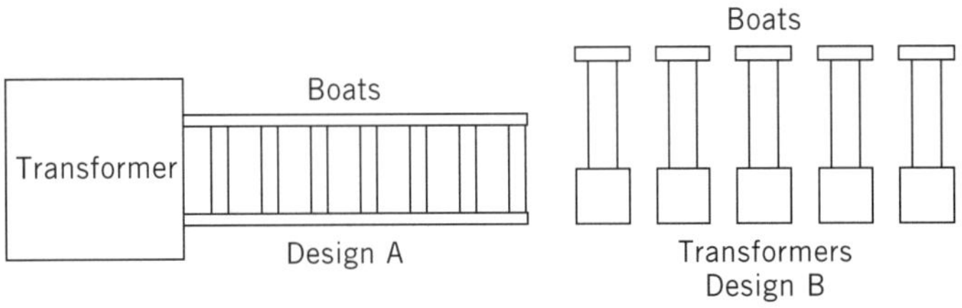

Figure 2. Two modes of powering resistance boats.

Table 1. Some Factors Influencing Boat Life

Electrical contact[a]
Initial heating[a]
Initial wetting of the cavity[a]
Operating temperture[a]
Utilization of cavity area[a]
Vacuum level
Number of cycles (length of rolls)
Purity of aluminum wire
Correct resistivity
Age of unused boats
Cooling of boats[a]
Boat defects

[a] Operator-controllable.

Table 2. Advantages of Induction Heating

Constant evaporation rate
Stable control on thickness uniformity using multiple evaporation source
No splaches; pinhole-free film can be coated
Automatic control of evaporation rate is easy and simple to operate
Aluminum purity can be 3N, instead of 4N or 5N
Aluminum blocks used can be easily available (10–300 g/piece)
Cheap running cost of evaporation source
Crucible life is long, and can be used for more than 40 batches
Simple evaporation source mechanism
Dense evaporated film structure
Higher yield of evaporation source: 95–98% for crucible and 90–93% for boat

der conditions that permit and enhance this deposition on a wide range of substrates. It is a process for the manufacture of a great variety of products, including a diversity of polymeric webs of interest to the packaging industry.

The key here is the thermal evaporation of metals. Changes in needs and demand for new products have also led to the application of this process for coating of these webs with a variety of ceramics.

The control of the evaporation process and the evaporant depends on the thermal source, the properties of the evaporant, the vessel from which one evaporates, the manner of holding and supplying of the evaporant, and, last but not least, the prevailing process pressure and/or environment.

	Dimension	EH 60	EH 100
Beam Power	kW	60	100
Accelerating Voltage	kV	30	33
Beam Current	A	2	3
Life Time of Cathode	h	≥120	≥100
Magnetic Lenses	pcs	2(1)[1)]	2
Maximum Angle of Deflection	deg	±40	±30
Limiting Frequency (Standard)	Hz	1000	1000[3),4)]
Min. Spot Diameter[5)]	mm	about 10	about 10
No. of Chambers	pcs	2(1)[1)]	2
Plate Valve		yes (no)[1)]	yes
Max. Pressure in Process Chamber	mbar	$5 \times 10^{-2} (3 \times 10^{-4})$[1)]	5×10^{-2}
Cooling Water Consumption	m^3/h	0.5	0.5
T-Meter for Beam Control		yes	yes
Height	mm	820 (610)[1)]	900
Weight	kg	120 (95)[1)]	200

1) Variant of EH 60 in Parenthesis
2) higher frequency on request
3) Optional with a Super Deflection Systems (SDS-System) with limiting frequency of >10,000Hz
4) Deflection system with lower frequency on request

Figure 3. A 100-kW von Ardenne Anlagentechnik gun and parameters.

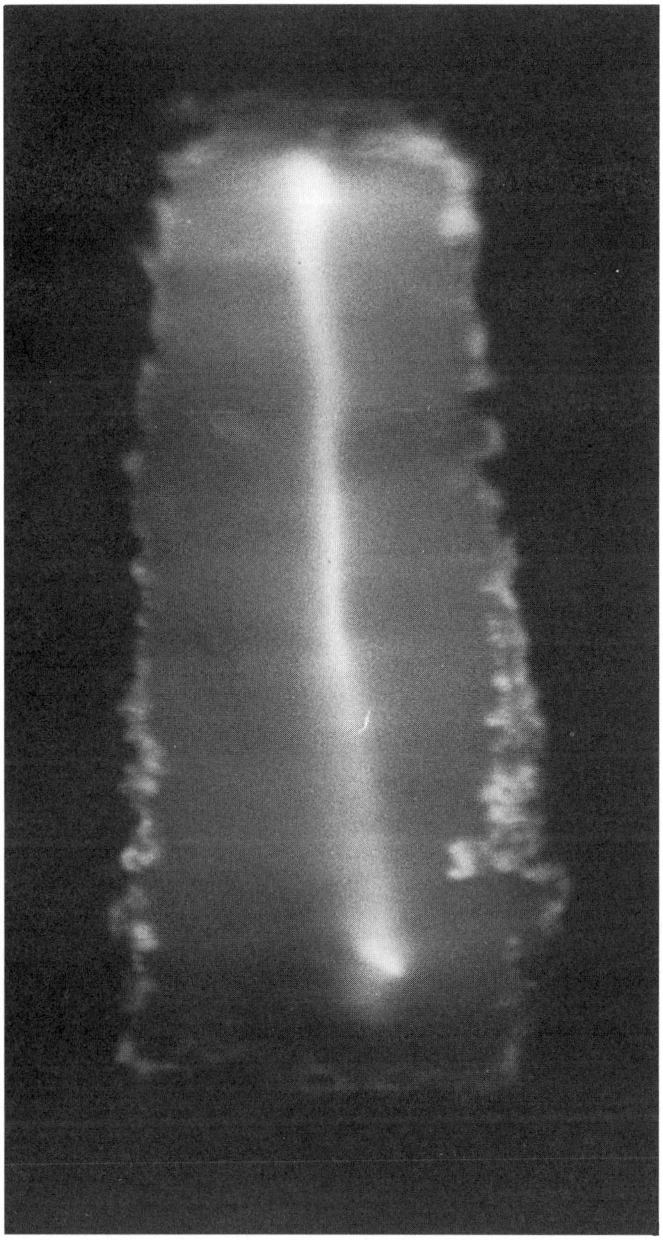

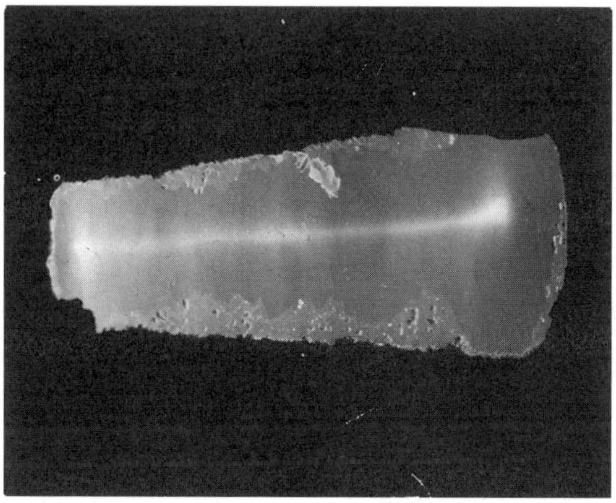

Figure 4. Evaporation line of an electron-beam line evaporator for 1000-mm web width.

In the equipment where this coating is performed, ie, the vacuum metallizers, variables such as geometry, distance between the evaporant source and web, and presence or absence of chill rolls, a system component that determines the substrate temperature and coating velocity, all affect the properties of the coating.

While sputter technology, another PVD process, could also be considered and in fact is a vacuum metallizing process, it is not referred to here. The process is considerably slower, and its high costs truly eliminate it from consideration in production of coatings for the packaging industry that this encyclopedia serves.

Thermal Sources

Today three basic heat sources are used in vacuum metallizing: resistance heating, induction heating, and electron-beam heating.

Resistance heating. Source development began with resistance heating using refractory metals. The successful resistance source is an electrical conductor with resistivity that allows it to reach the evaporation temperature with reasonable current flow. This material must have a higher melting point and lower vapor pressure than the evaporant. It also must be chemically inert to the evaporant, and it certainly should not form byproducts that could degrade the properties of the films produced. The source should be so designed to be able to contain the evaporant. Ease of handling and advantageous cost should also be integral characteristics of this source. The refractory metals—tungsten, molybdenum, and tantalum—have these characteristics, although tungsten is by far the preferred source. A range of configurations for braided wires to refractory metal boat sources are to be found in service. Usually when complexly shaped containers are needed, one will use tantalum. Resistance-heated wire

Figure 5. 1000-mm crucible and magnetic traps in a roll-coating plant.

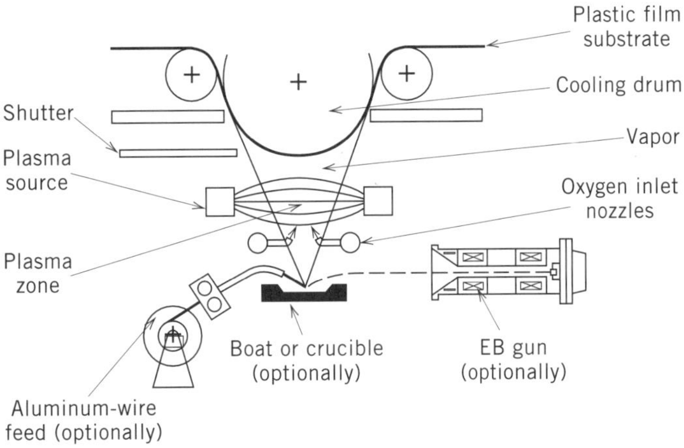

Figure 6. Schematic of the aluminum oxide deposition process by plasma-activated reactive evaporation of aluminum.

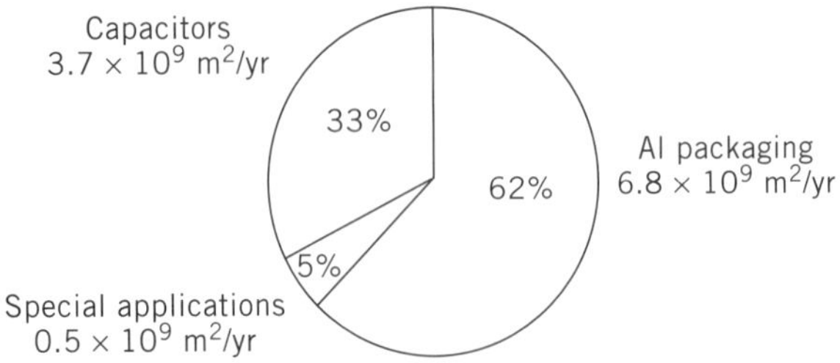

Figure 7. Estimates of worldwide 1993 metallizing production capacity.

sources and refractory metal boats sources are shown in Figure 1 (1). The refractory metal resistance source remain in use in small installations, but when we speak of industrial applications, the intermetallic boat is now the mainstay. Initial success with resistance-heated ceramic boats were recorded in the 1960s (2).

Let us now look at these. Boron nitride (3) with a filler of aluminum nitride, which permits the materials to be shaped in virtually any desired shape, appears to have wide acceptance for boats. This material has zero resistance when cold and an increasing resistance as the temperature increases.

Table 3. Comparative Theoretical Evaluations between EB and Resistance Heating

General assumptions
- 1000-mm coating width
- 400-A Al layer
- 10-ms^{-1} coating speed
- 12-μm PET
- 645/1000-mm roll diameter
- 3-shift operation
- 85% machine availability

	Resistance-heated boats	EB
Cost calculation basis		
Number of boats/crucible	11	1
Price per boat ($)	35	400
Boat/crucible lifetime (h)	17	100
Pumpdown time/conditioning (min)	5	20 (15)[a]
Vent time (min)	5	20 (5)[a]
Investment costs (percent)	100	112 (120)[a]

	Resistance-heated boats	EB	EB with crucible lock chamber
Coating profitability		645-mm roll diameter	
Annual productivity (tons/yr)	1502	1141	1402
Coating costs ($/kg)	0.47	0.62[2]	0.53[b]
		0.57[3]	0.50[c]
		1000-mm roll diameter	
Annual productivity (tons/yr)	2093	1799	2036
Coating costs ($/kg)	0.37	0.45[b]	0.43[b]
		0.40[c]	0.38[c]

[a] With crucible lock chamber.
[b] EB evaporator with 50% higher Al consumption.
[c] EB evaporator with same Al efficiency as intermetallic boats. The calculation shows that in case of aluminum coating of PET, the EB-coating costs are definitely higher mainly because the inherently higher possible evaporation speed of the EB evaporator cannot be utilized.

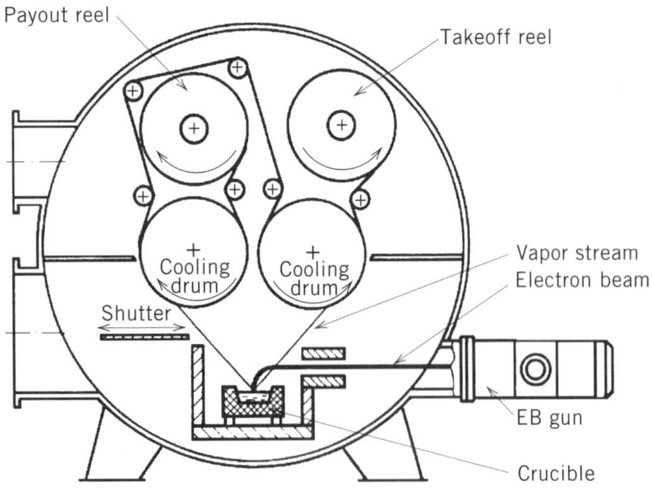

Figure 8. Outline of a vacuum roll coater typical of smaller machines.

Table 4. Maximum Evaporation Rates and Theoretically Derived Deposition Rates for Reactive Aluminum Evaporation[a]

		Boat Evaporator		EB Line Evaporator per 1000 mm
		per 100 mm (1 boat)	per 1000 mm (10 boats)	
Maximum evaporation rate (Al)	$\frac{g}{min}$	12 (cavity 30 cm^2)	120	≈200
Derived deposition rate (Al)	$\frac{nm}{s}$	2400 (200 nm/s per 1 g/min)	2400	≈4000
Derived deposition rate (Al_2O_3)	$\frac{nm}{s}$	75%–150% of Al deposition rate depending on the location of O_2 inlet		
Limitation for Al evaporation rate		Size of boats (max. 0.4 g/min per cm^2 cavity)	Number of boats per 1000 mm (10 boats per row)	Splashing (max. 2 g/min per cm^2)

[a]*Precondition:* No other limitation for deposition rate.

Table 5. Advantages and Limitations of EB Evaporation

Advantages	Limitations
High deposition rates for metals and dielectrics	Low source efficiency
Direct deposition of oxides, fluorides, sulfides, and magnetic materials	Multisource control
Coating material change over flexibility	Source inventory
	Thermal load to substrate
	Source orientation

Because of this, sophisticated controls of the heat input are needed for these boats. Three-phase boats consisting of boron nitride, aluminum nitride, and titanium diborate are also used. Here the A1N is added for its thermal conductivity and electrical resistivity properties. These boats operate at voltages of the order of 20 V with programmable current input that limits the power to the boat when it is cold in order to avoid cracking. There is a variety of ideas for the most efficient way to power the boats, two of these are shown in Figure 2 (4). A, which shows a system where a single transformer is used to power all boats, has two main disadvantages. The first is that there is a voltage drop from the first to the last boat, making it virtually impossible to control the individual boats. The alternative approach, shown in part **b,** is the more common and preferred method used. It permits individual boat control, which makes possible setting of uniform heating despite resistivity changes due to aging of the boats.

In addition to this, the design of boats clamping must be properly engineered to permit good electrical contact and avoid mechanical stresses due to thermal expansion. Various designs are possible. The size of the evaporation boats is dependent on the individual machine, and there is a wide range of opinions related to boat size. Boats are habitually wire-fed with a wide variety of feed drives to be found in the field. Overall resistance heating is rapid, and provides efficient deposition rates. This source also has lower outgassing rate. The choice of boat material is determined for each installations based on machine parameters, operation experiences, and end-product requirements. Boat life varies extensively. Table 1 (5) shows some of the factors that influence it.

Induction heating. Induction heating is the second heat source used here. While in the past it used to be a very popular source, it appears that U.S. and European producers of metallizers have abandoned it in favor of resistance heating. Their reason for the move in this direction was their feeling, based on experience, that this source was difficult to operate, that it could not be feed continuously, and that it had unfavorable economics.

By contrast, it appears that ULVAC, the premiere Japa-

Table 6. Present and Future Applications of Roll Coaters

Application	Products
Gold and silver yarns	Obi(sash), sash clip, bag, tie, sandals, curtain, towel, shawl, socks, slippers, embroidery, table cloth, lace, stage costome, tatami brim, etc
Ornaments	Interior decorations, Christmas-tree decorations, nameplate, case, ballom, cosmetic package, etc
Stamp foils	Pencil, book cover, postcard, label, Christmas card, poster, cigarette package, picture frame, lighter, cassete case, kimono, kimono sash, etc
Packages	Medicines, confectionary, rice, cigarette, food, etc
Building materials	Parasol, sunshade, greenhouse shade, interior wall, heat-insulation wall, solar film, firefighter's suite, curtain, space suit, etc
Farming materials	Reflective sheet, beach mat, cooler bag, scarecrow, etc
Reflection sheet	Traffic sign, number plate, mirror, guidance sign (airplane), etc
Food package	Lunchbox, tableware, etc
High added value products	Magnetic tape, copying film (substitute of copy drum), aerial photograph film, optical filter, multilayer flexible circuit board with polyimide, etc
Precision capacitor	Logic circuit (communication device, personal computers, etc); bypass capacitor (TV, radio, VCR, etc)
Phase advance capacitors	High industrial power unit, etc; low power—washing machine, motor, Pachinko machine, home bakery, fluorescent lamp

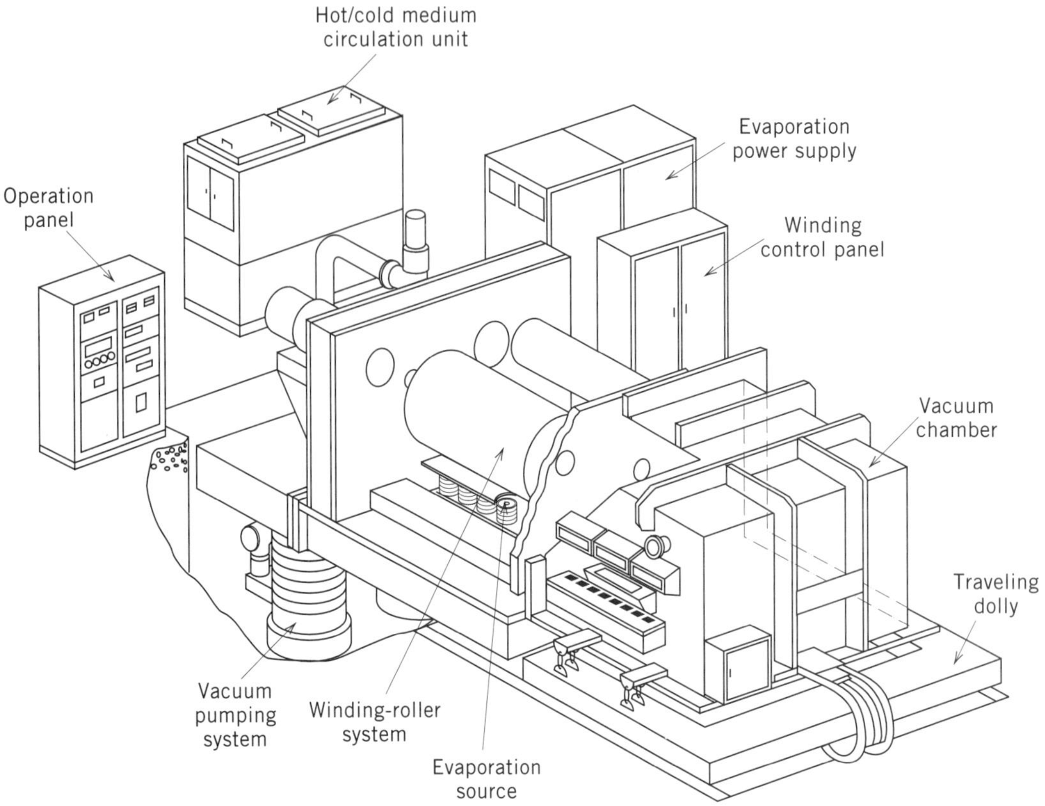

Figure 9. Schematic of a coater.

nese metallizing equipment producer, has remained loyal to induction heating (6) for its metallizers. ULVAC feels quite strongly about it, listing the advantages in Table 2 (7).

Electron-beam heating. In electron-beam heating the heat generated is a result of an energy conversion process. Here the kinetic energy of the highly accelerated electrons is converted to heat on impact with the workplace. The *electron gun* is a device that generates electrons and, after columniating them, directs them to the workplace. Electron-beam heating has no upper limit, making it possible to evaporate virtually any material. Today transverse and axial electron guns are to be found in metallizing systems. The transverse guns generate the electron beam, and after a 270° bending direct it to the workpiece. The reason for this path is mandated by the need to maintain the area where the electrons are generated free of the high-density material vapors that habitually are found on top of the evaporant and which can and will defocus the beam reducing its efficiency. The axial guns generate the beam and axially directs them to the workpiece.

With these guns the beam is magnetically deflected to effect desired coverage of the evaporant filled crucible. A typical axial gun is given in Figure 3. It can produce a beam that, by suitable deflection, can cover the crucible containing the evaporant with virtually any desired configuration. This includes the beam in a line configuration that was first introduced in 1984 (8), making possible the generation of a uniform vapor curtain (see Fig. 4).

When using electron-beam heating, one must assure that virtually no secondary electrons impinge on the polymeric web, Unless this is accomplished, their thermal load will lead to severe damage and even complete destruction of the web. Figure 5 shows 1000-mm crucible and magnetic trap for an electron-beam line evaporator in a roll-coating plant. Note, in electron-beam heating, that the evaporant, depending on its nature, is evaporated either from a cold crucible (ie, water-cooled copper) or from a hot crucible vessel that can be made of a variety of ceramic materials. The specific selection is based on compatibility with evaporant. Table 3 shows a theoretical comparison between resistance and electron-beam heating for aluminum coating (9). As you can see from this table, the electron-beam heating, although more expensive, is notably simpler.

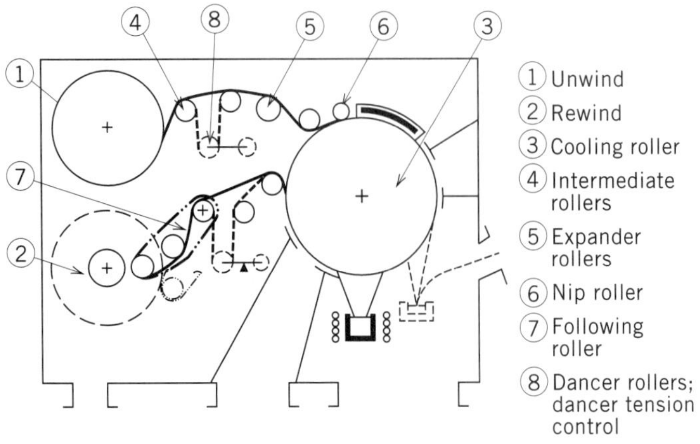

Figure 10. Internal roll assembly of water shown in Fig. 9.

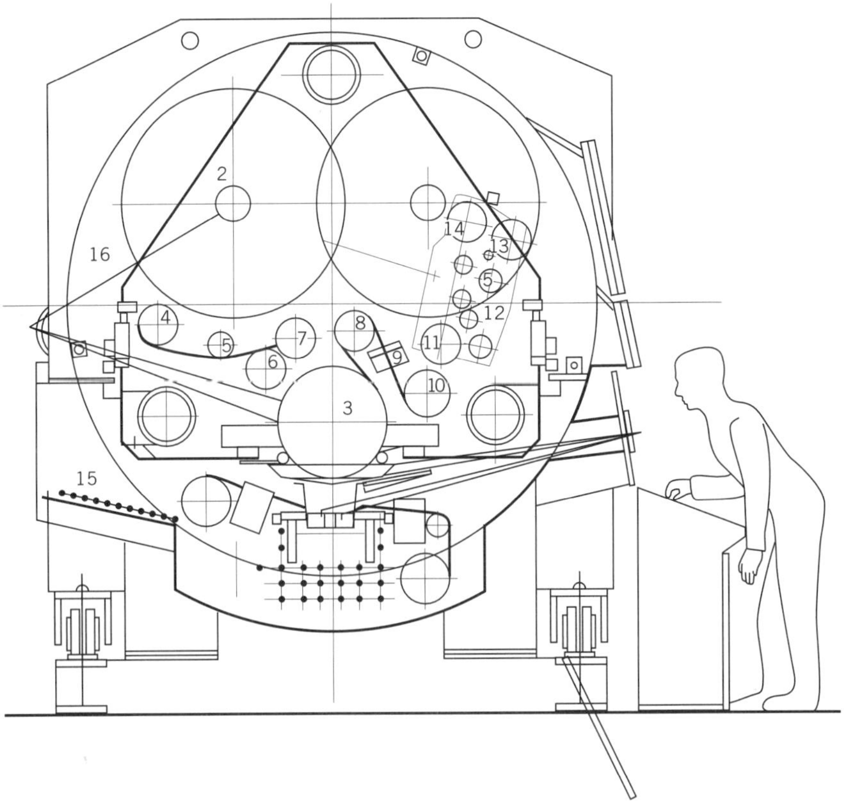

Figure 11. Cross section of typical high-speed large machine.

From the heat sources only electron-beam heating can be used for direct evaporation of ceramics. The electron beam can also be used for a deposition process referred to as plasma activated reactive evaporation of aluminum (10). The geometry of the process is shown in Figure 6 (10). This process can be carried out equally well with a resistance-heated boat evaporator. Table 4 (10) shows maximum rates obtained by reactive evaporation. We are likely to see considerably more reactive evaporation applications in the future. Reference should also be made to Table 5 for the advantages and limitations of the electron-beam evaporation (11).

Metallizing Equipment

As with equipment to carry out any other process, so with metallizing equipment the details of the system are governed by the process to be used, and the product that it must produce. Once details are fixed, the ability to carry out the process in the system depends on the competence and the skill of the equipment producer. By saying this, one could say that the estimated 500 metallizers operating in the world today are, in fact, 500 different pieces of equipment. The estimate of their production is given in Figure 7 (12). While in the true

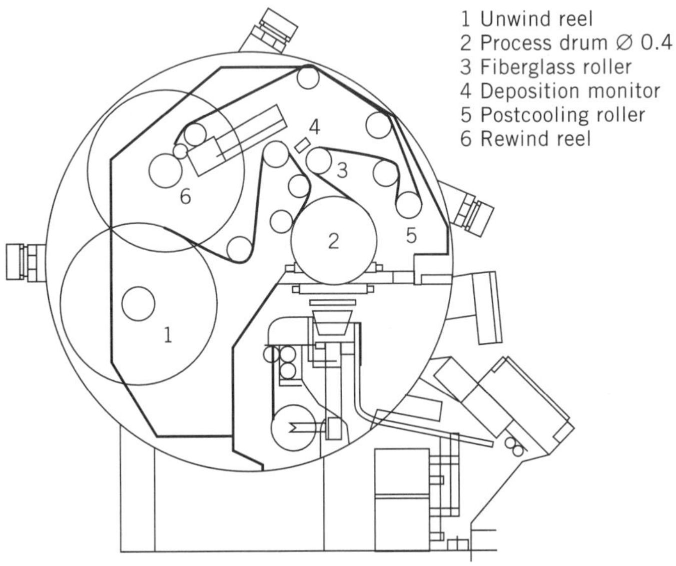

Figure 12. Cross section of typical high-speed large machine.

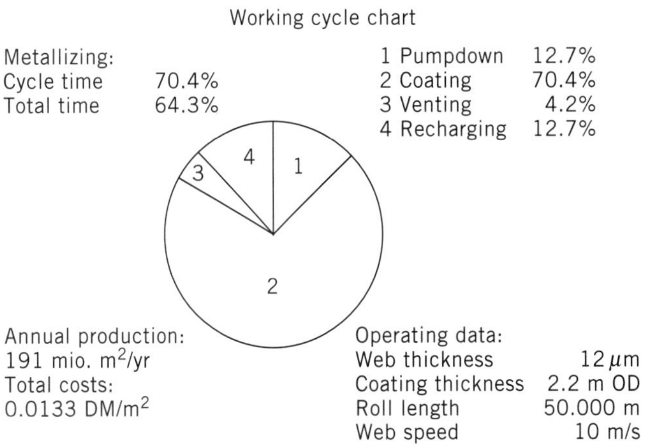

Figure 13. Representative working cycle.

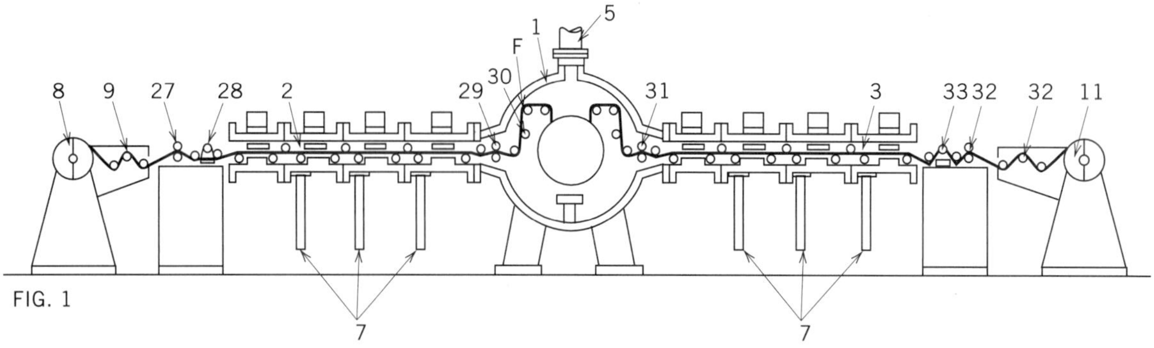

Figure 14. Schematic of a Hitachi air-to-air metallizer.

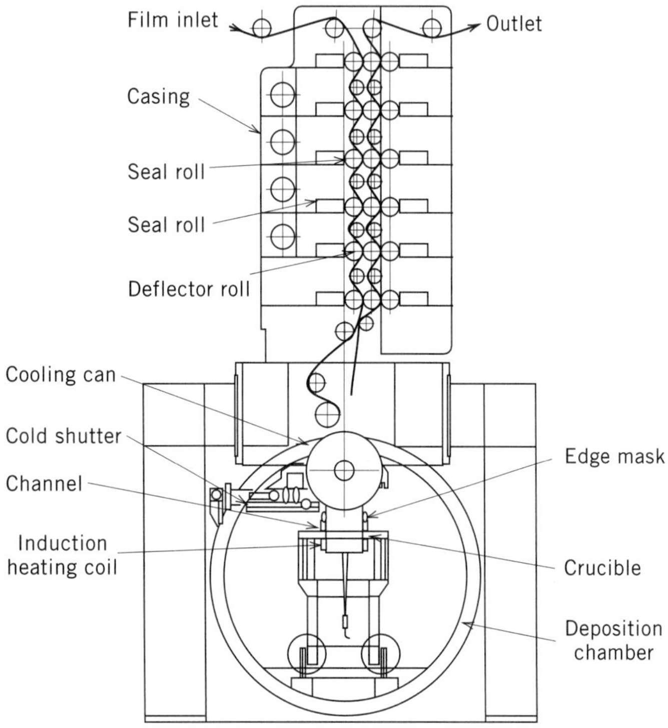

Figure 15. Schematic of an Mitsubishi air-to-air machine.

sense of the word this might be correct, I will only refer to a few typical systems. The metallizers, or roll coaters, if you will, accomplish a wide range of tasks far and beyond the needs of the packaging industry, which is the largest consumer of polymer films coated with aluminum. It is thus appropriate to refer to Table 6 (13), which refers to most, if not all, products that can be made in metallizers. However, many of these are actually products of sputter technology.

Increased productivity and lower costs have been the engine driving this technology since its introduction. They have been responsible for (1) the widening of the metallizers, with ability to coat today's widths, exceeding 125 in.; (2) the increases in coating speed, today getting ready to exceed 3000 ft/min; and (3) numerous quantum leaps from the manual controls of yesterday, to fully computer-controlled systems of today (14), including the use of artificial intelligence (AI) techniques to provide real-time process control and troubleshooting advice (15).

Let us now look at schematics of few of the large variety of web coaters built over the years for the purpose of illustrating the equipment that operates in industry. Figure 8 (16) shows cross section of a coater representing smaller-type machines. In this schematic electron-beam heating is indicated, although resistance heating for such systems is much more common. Figure 9 (17) shows a schematic of a coater, and Figure 10 (17) shows its internal rolls assembly; note both electron-beam- and induction-heated sources. Figures 11 (18) and 12 (19) shows cross sections typical of a high-speed, large machine. All these are referred to as *batch* or *semicontinuous coaters.* The production cycle of these machines habitually consists of four steps as follows: (*1*) precoating, (pumpdown), (*2*) coating, (*3*) venting, and (*4*) recharging. These vary from machine to machine. Figure 13 (20) shows an example of such a cycle. One could look at it as somewhat representative of batch systems. However, one should note that there is considerable variation in the production cycle from machine to machine.

To further increase productivity, air-to-air machines have been built. The producer of one of these (21) claims that output here can be increased up to 50% in relation to a batch-type machine. Figures 14 (20) and 15 (21) show schematics of air-to-air metallizers. Cost of the equipment, however, are notably increased. While there are a limited number of air-to-air machines serving the packaging field, this type of installation remains primarily the province of metal strip coaters. To round up the discussion on metallizers one should look at a couple of 1995 vintage systems serving the packaging and related industries. Figure 16 (*left*—machine; *right*—schematic) shows the Leybold electron-beam-powered "top beam" 2-m-wide coater for the packaging industry, while Figure 17 shows the resistance-powered "mega" 3-m-wide Galileo metallizer.

One can look at the equipment discussed above as the classic metallizing equipment. Efforts to develop competitively priced transparent barrier coatings, in progress for some time, are continuing. With minor exceptions, products developed from these activities to date have yet to make a major impact on the coated products for the packaging market. Numerous reports on this research and development and on specific systems for production of such coatings can be located in the proceedings of the International Vacuum Web Coating Conference beginning in 1987 and those of the Society of Vacuum Coaters beginning in 1986. A number of technologies for applying transparent barrier coatings are referred to in these publications. Specifically for coatings of SiOx, one can elec-

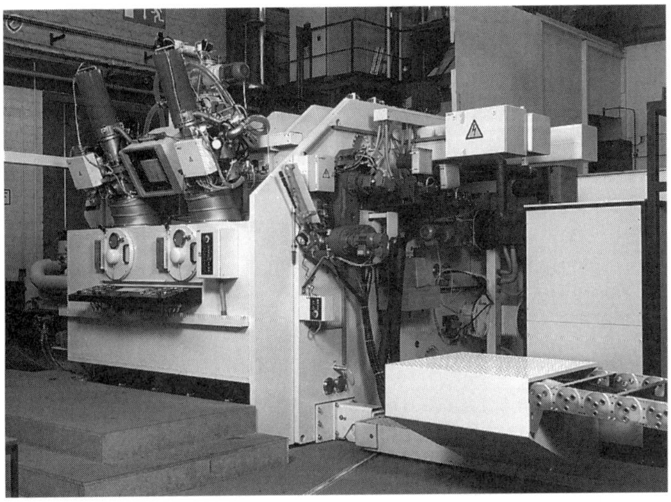

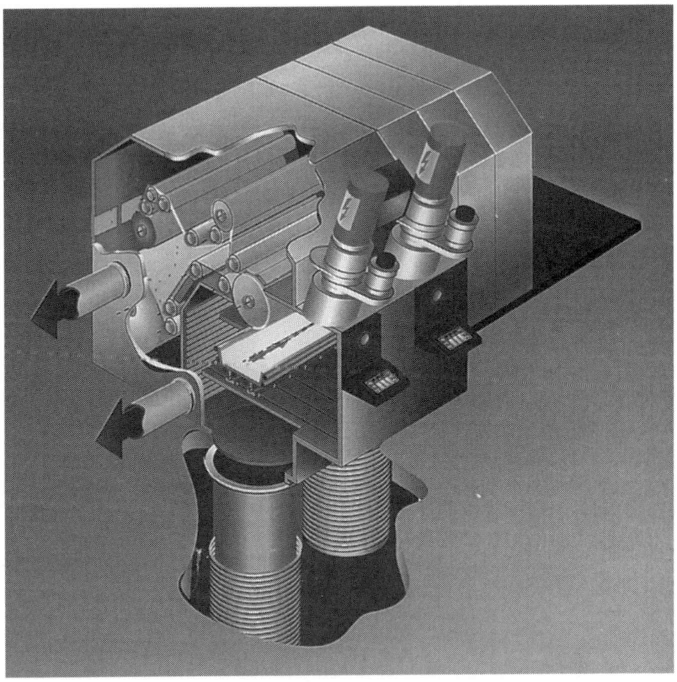

Figure 16. "Top-beam" coater.

tron-beam-evaporate silica. An alternate way to produce coatings of the same compound is through the reactive evaporation of silicon monoxide. SiO_2 coatings can be produced by PECVD of organosilenes. Al_2O_3 coatings can be produced by anodization of aluminum coatings, which is entirely impractical, and by electron-beam evaporation of Al_2O_3. AlOx coatings can be produced by reactive evaporation of aluminum.

Conclusion and a Look at the Future

The field of vacuum metallizing serving the packaging industry has been one of steady-state growth. Advances in equipment have lead to price reductions of vacuum-metallized products and an ever-increasing market for these. The last decade has seen tremendous interest and activities in development of transparent barrier coatings. The technology to bring prices of these coated materials to levels acceptable to the packaging industry has prevented real acceptance of these products except in special situations. This effort undoubtly will lead to a breakthrough in the technology that will enable the suppliers to meet the pricing targets of this industry. Only then would transparent barrier-coated materials begin a real penetration in the market dominated by vacuum-metallized aluminum in the packaging industry.

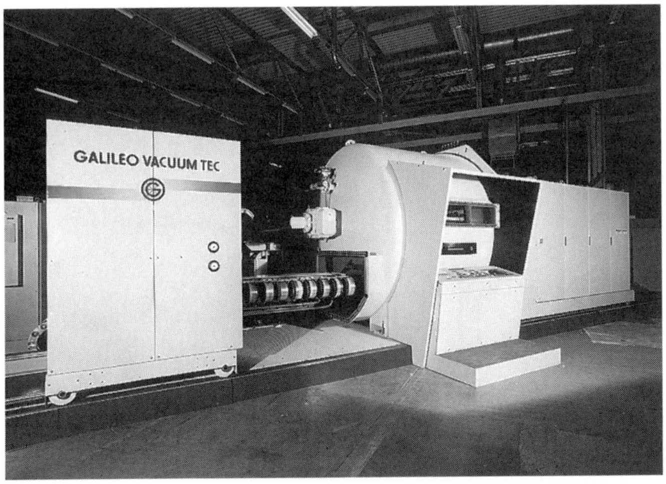

Figure 17. "Mega" line.

BIBLIOGRAPHY

1. K. M. Anetsberger, "Metallization of Plastics with Resistance Heated Sources" in *Metallized Plastics,* K. L. Metal and J. R. Susko, eds., Plenum Press, 1989, pp. 29–44.
2. H. Zoelner, "Resistance Heated Ceramic Evaporation Boats" in *Proceedings of the 27th Annual SVC Conference,* 1984, pp. 1–4.
3. I. Watts, "20 Years Resistance Source Development" in *Proceedings of the 34th Annual SVC Conference,* 1991, pp. 118–123.
4. F. Casey and A. Bloomfield, "Recent Advances in Source Design in Resistive Evaporation Web Coaters" in *Proceedings of the 34th Annual SVC Conference,* 1991, pp. 120–124.
5. C. Gibson and K. Kohnken, "Intermetallic Evaporation Boats" in *Proceedings of the 3rd International Vacuum Web Coating Conference,* R. Bakish ed., BMC, 1989, pp. 121–130.
6. I. Tada, A. T. Yamamori, K. Mutsumory, and Y. Yoneda, "Comparison of Evaporation Sources for Web Coaters" in *Proceedings of the 32nd SVC Conference,* 1989, pp. 131–149.
7. Y. Toneda, T. Chiba, and O. Ohkubo, "Recent Advances in Vacuum Roll Coating" in *Proceedings of the 30th Annual SVC Conference,* 1987, pp. 105–110.
8. S. Schiller, M. Neumann, and R. Bakish, "Electron Beam Line Evaporators for Roll Coating" in *Proceedings of the 27th Annual SVC Conference,* 1984, pp. 192–210.

9. A. Feuerstein, "Latest Technology High Performance Electron Beams" in *Proceedings of the 29th Annual SVC Conference,* 1986, pp. 118–135.
10. S. Schiller, M. Neumann, H. Morgner, and N. Schiller, "Reactive Aluminum Oxide Coatings of Plastic Films" in *Proceedings of the 38th Annual SVC Conference,* 1995, pp. 18–27.
11. A. S. Matteucci, "Electron Beam Evaporation and Magnetic Sputtering in Roll Coating" in *Proceedings of the 30th Annual SVC Conference,* 1987, pp. 91–104.
12. E. K. Hartwig, "Future of Al Vac Web Coating Considering Environmental Legislation and Non Metallic Vacuum Coating" in *Proceedings of the 7th International Vacuum Web Coating Conference,* R. Bakish, ed., BMC, 1993, pp. 2–10.
13. C. Hayashi, "Advances in Vacuum Roll Coating" in *Proceedings of the 1st International Vacuum Web Coating Conference,* R. Bakish, ed., BMC, 1987, pp. 2–10.
14. C. Beccaria, A. Pasqui, and F. Remedotti, "Integral Computer Control of Roll to Roll Coater" in *Proceedings of the 35th Annual SVC Conference,* 1992, pp 100–105.
15. I. M. Boswarva, V. Lee, and G. Duchame, "The Use of Artificial Intelligence Techniques to Provide Real Time Process Control and Trouble Shooting Advise in Vacuum Roll Coaster Operations" in *Proceedings of the 7th International Vacuum Web Coating Conference,* R. Bakish, ed., BMC, 1993, pp. 80–84.
16. S. Schiller and M. Neumann, "Application of EB Line Evaporator for Web Coating" in *Proceedings of the 1st International Vacuum Web Coating Conference,* R. Bakish, ed., BMC, 1987, pp. 113–128.
17. T. Yamamori, "Vacuum Roll Coater for the Production of Decorative and Magnetic Tapes" in *Proceedings of the 28th Annual SVC Conference,* 1985, pp. 90–100.
18. A. Paqui, F. Grazzini, and A. Carleti, "High Rate Deposition on High Speed Aluminum Roll Coaters" in *Proceedings of the 36th Annual SVC Conference,* 1993, pp. 185–190.
19. I. K. Baxter, "Effective Film Temperature Control for Vacuum Web Coater" in *Proceedings of the 35th Annual SVC Conference,* 1992, pp. 106–121.
20. E. Hartwig, "Air to Air Vacuum Web Coating" in *Proceedings of the 34th Annual SVC Conference,* 1991, pp. 152–160.
21. T. Taguchi, S. Kamikawa, Y. Ito, M. Mitarai, and K. Matsuda, "Air to Air Metallizer Design and Operational Data" in *Proceedings of the 36th Annual SVC Conference,* 1992, pp. 135–140.

Robert Bakish
Bakish Materials Corporation
Englewood, New Jersey

METRICATION IN PACKAGING

The complete expression of any quantity requires both a *number* and a *unit.* While the unit tells us what the quantity is about (and the standard by which is measured), the number indicates how many of those units are considered. Units are part of the daily framework of reference used by a society; when quantities are expressed in familiar units, they are easily understood. In the worlds of commerce, science, industry, and everyday life, we need to express quantities within one group of widely recognized coherent units, called a *system* of units. Nevertheless, because of the relative predominance of customary units in the United States, many people still feel uncomfortable when operating with quantities expressed in metric units. In an effort to glean the benefits of using the metric system, the United States has been making slow but steady progress in replacing the customary system of units with the metric one.

Table 1. SI Dimensions, Units, and Symbols

Dimension	Unit	Symbol
Basic units		
Length	meter	m
Mass	kilogram	kg
Time	second	s
Temperature	kelvin	K
Amount of substance	mole	mol
Luminous intensity	candela	cd
Electric current	ampere	A
Supplementary units		
Plane angle	radian	rad
Solid angle	steradian	sr

Because the metric system is simpler and easier to use than the customary system, it has become the internationally accepted system of units. The international metric system is a rational and coherent group of measurement units based on a decimal scale. Like the decimal money system (cent, dime, dollar, etc), the metric system is organized in factors of tens: 1, 10, 100, etc. The metric system is based on seven basic units: kilogram (mass), meter (length), degree Celsius (temperature), liter (volume), second (time), candela (luminous intensity), and ampere (electric current). In this article, a brief historical note of the metric system, its characteristics, and common conversion units used in packaging are presented.

Historical Background

While all industrialized countries have only one system of units, in the United States there are two major systems of units in use: the U.S. Customary System (USCS) and the International System, also denoted as the *metric system.* The USCS is derived from the British Imperial System and has been customarily used since colonial time, although the current USCS is now somewhat different from the old British Imperial System. By action of state and federal legislation, the USCS has become the fundamental system of units in the United States (1). The metric system, on the other hand, is

Table 2. Prefixes

Multiples	Prefix	Symbol
10^{18}	exa	E
10^{15}	peta	P
10^{12}	tera	T
10^{9}	giga	G
10^{6}	mega	M
10^{3}	kilo	k
10^{2}	hecto	h
10^{1}	deka	da
10^{-1}	deci	d
10^{-2}	centi	c
10^{-3}	mili	m
10^{-6}	micro	μ
10^{-9}	nano	n
10^{-12}	pico	p
10^{-15}	femto	f
10^{-18}	atto	a

Table 3. Selected Units with Special Names in SI

Quantity	Unit Name	Dimensions	Symbol
Absorbed dose	gray	Gy	J/kg
Activity (radioactivity)	becquerel	Bq	s^{-1}
Capacitance	farad	C/V	F
Electricity (amount of)	coulomb	A · s	C
Electrical resistance	ohm	V/A	Ω
Energy	joule	N · m	J
Enthalpy	joule		J
Force	newton	kg · m/s^2	N
Frequency	hertz	s^{-1}	Hz
Heat (amount of)	joule		J
Illuminance	lux	lm/m^2	lx
Luminous flux	lumen	cd/sr	lm
Mass	metric ton	1000 kg	t
Potential difference	volt	W/A	V
Power	watt	J/s	W
Pressure	pascal	N/m^2	Pa
Stress	pascal		
Volume	liter	dm^3	L
Work	joule		

the only system of units that has ever been explicitly approved by the U.S. Congress. In 1866 Congress voted a law making it "lawful through the United States of America to employ the weights and measures of the metric system." In 1875 the United States was an original signatory part to the Treaty of the Meter, which established the General Conference of Weights and Measures, the International Committee of Weights and Measurements, and the International Bureau of Weights and Measures.

In 1964 the National Bureau of Standard made its policy to use the International System, "except when the use of these units would obviously impair communication or reduce the usefulness of a report." In 1975 Congress approved the Metric Conversion Act, amending it in 1988, and declaring the metric system to be the preferred system of measurement for U.S. trade and commerce. Furthermore, the Executive Order 12770, "Metric Usage in Federal Government Programs," was signed in 1991. In spite of all the legislation, the United States has not yet embraced use the metric system for the bulk of its commercial and standard activities. Although the use of the metric system is mandatory in federal agencies, its use by the industry is voluntary. In an effort to promote metrication (the transitional process for converting to the use of the metric system), the Metric Program at the National Institute of Standard and Technology (NIST) has been established (2).

Benefits of the Metric System

There are many commercial, educational and technical benefits in using the metric system. The metric system is the measurement system for products, processes, and information in international commerce (3). The European Union regulations will require by the year 2000 that all products sold in their boundaries be labeled in metric units and all the corresponding documentation be expressed in metric system. Japan opposes the import of products with nonmetric units. Other countries and regions, like England, Canada, Australia, China, Mexico, and South and Central America, have already converted their system of measurements to metric units.

Table 4. Selected Derived Units in Terms of SI Units

Quantity	Name	SI units
Acceleration	meter per square second	m/s^2
Area	square meter	m^2
Concentration	mole per cubic meter	mol/m^3
Density	kilogram per cubic meter	kg/m^3
Luminance	candela per square meter	cd/m^2
Specific volume	cubic meter per kilogram	m^3/kg
Velocity	meter per second	m/s
Volume	cubic meter	m^3
Wave number	reciprocal meter	m^{-1}

Table 5. Units Accepted in the Metric System

Names	Symbol	Value in SI
liter	l or L	$1\ L = 10^{-3}\ m^3$
day	d	1 day = 86,400 s
hour	h	1 h = 3600 s
minute (time)	min	1 min = 60 s
degree (angle)	°	$1° = (\pi/180)$ rad
minute (angle)	′	1′ = 1/60°
second (angle)	″	1″ = 1/60′
electronvolt	eV	$1\ eV \approx 1.602\ 18 \times 10^{-19}\ J$
hectare	ha	$1\ ha = 1\ hm^2 = 10{,}000\ m^2$
metric ton	t	1 t = 1000 kg
unified atomic mass	u	$1\ u \approx 1.602\ 18 \times 10^{-27}\ kg$

With the steady increase in world trade and international interdependence, it is expected that the use of the metric system in the United States will necessarily increase. The use of the metric system by industry is also expected to benefit the consumer because it will promote standardized and simplified product packaging. This will reduce the vast amount of packaging, facilitate price–quantity of product comparison, and standardize shipping and distribution conditions, including items such as boxes, containers, and pallets. Parallel to this, students learning mathematics will benefit since less time will be spent on cumbersome and distractive conversions.

International System (SI) and the Metric System

In 1960, the General Conference of Weights and Measures, made up of the signatory nations to the Treaty of the Meter, approved the modern version of the metric system, the International System of Units. The International Systems is officially known by the French title *Systeme International d'Unites* (SI), and is referred as SI. In this respect the SI and the metric system are equivalent. Nevertheless, it is accepted that products standards and preferred sizes (like liter) used by industries and governments through the world are considered part of the metric system. For instance, the Metric Conversion Act of 1975 indicates that "the system of measurement means the International System of Units . . . and as interpreted or modified for the United States by the Secretary of Commerce." Thus, although the metric system and SI are equivalent, the metric system, also refers to other units related to SI.

The basic units of the SI are the seven units listed in Table 1 (4). Since the numbers in the quantities expressed in these units can be very large or very small, prefixes have been in-

Table 6. Customary and Selected Commonly Used Units into the Metric Units

Quantity	To Convert Customary or Commonly Used Unit	Multiply by	To Obtain Metric Unit
Acceleration	ft/s^2	3.048 E − 01[a]	m/s^2
		3.048 E + 01[a]	cm/s^2
	in./s^2	2.54 E − 02[a]	m/s^2
Amount of substance	lb · mol	4.535 E + 01	kmol
	STP m^3 (0°C, 1 atm)	4.461 E − 02	kmol
	22,414 STP mL (0°C, 1 atm)	1.000 E + 00	mol
Area	yd^2	8.3613 E − 01	m^2
	ft^2	9.290304 E − 02[a]	m^2
	in.^{2}	6.4516 E + 00[a]	cm^2
	100 in.^{2}	6.4516 E − 02[a]	m^2
Area factor, yield	ft^2/lb	2.0485 E − 04	m^2/g
		2.0485 E − 01	m^2/kg
	yd^2/lb	1.8437 E − 03	m^2/g
Basis weight, grammage, substance	lb/1000 ft^2	4.882 E + 00	g/m^2
	lb/3000 ft^2	1.627 E + 00	g/m^2
Bending moment/length	(lb$_f$ · ft)/in.	5.338 E + 01	(N · m)/m
	(lb$_f$ · in.)/in.	4.448 E + 00	(N · m)/m
Caloric value, enthalpy per mass unit	Btu/lb	2.326 E + 00	kJ/kg
	cal/g	4.184 E + 00	kJ/kg
	cal/lb	9.224 E + 00	J/kg
Caloric value, enthalpy per mole unit	kcal/g · mol)	4.184 E + 03[a]	kJ/kmol
	Btu/(lb · mol)	2.326 E + 00	kJ/kmol
Concentration	wt%	1.0 E − 02[a]	kg/kg
		1.0 E + 01[a]	g/kg
	ppm (wt/wt)	1.0 E + 00[a]	mg/kg
	ppm (vol/vol)	1.0 E + 00[a]	cm^3/m^3
	ppm (wt/vol)	1.0 E + 00[a]	g/m^3 = mg/L
Corrosion rate	in.^{3}/yr	1.6387 E + 01	cm^3/year
	in./yr	2.54 E + 01[a]	mm/year
	mil/yr	2.54 E − 02[a]	mm/year
Density	lb/ft^3	1.601 E + 01	kg/m^3
		1.601 E + 04	g/m^3
	g/cm^3	1.0 E + 03[a]	kg/m^3 = g/L
Diffusion coefficient	ft^2/s	9.290 E − 02	m^2/s
	cm^2/s	1.000 E − 04[a]	m^2/s
Edge crush	lb$_f$/in.	1.7513 E − 01	kN/m
Energy, work	Btu	1.055 E + 00[a]	kJ
	kcal	4.184 E + 00[a]	kJ
	cal	4.184 E − 03[a]	kJ
	J	1.000 E − 03[a]	kJ
	cal	4.184 E + 00[a]	J
	erg = dyn · cm	1.000 E − 07[a]	J
	ft · lb	1.3557 E + 00	J
Energy density	in. · lb$_f$/in.^{3} = psi	6.8948 E + 00	kPa
		6.8948 E + 03	J/m^3
Flow rate (mole basis)	(lb · mol)/s	4.535 E − 01	kmol/s
	(lb · mol)/h	1.259 E − 04	kmol/s
Flow rate (volume basis)	ft^3/min	4.719 E − 01	dm^3/s = L/s
	ft^3/s	2.831 E + 01	dm^3/s = L/s
	cc/min	1.667 E − 05	dm^3/s = L/s
Force	U.K. ton$_f$	9.964 E + 00	kN
	U.S. ton$_f$	8.896 E + 00	kN
	kg$_f$ (kp)	9.80665 E + 00[a]	N
	lb$_f$	4.448 E + 00	N
	dyn	1.0 E − 05[a]	N
Length	mi	1.609344 E + 00[a]	km
	yd	9.144 E − 01[a]	m
	ft	3.048 E − 01[a]	m
		3.048 E + 01[a]	cm
	in.	2.54 E + 01[a]	mm
		2.54 E + 00[a]	cm
	mil = point	2.54 E + 01[a]	μm
	gauge = 0.01 mil	2.54 E − 01[a]	μm
	caliper = 10 mil	2.54 E + 02[a]	μm
Mass	short ton	9.07185 E − 01	Mg (metric ton)
	lb	4.535 E − 01	kg
	oz (troy)	3.011 E + 01	g
	oz (av)	2.835 E + 01	g

Table 6. (*Continued*)

Quantity	To Convert Customary or Commonly Used Unit	Multiply by	To Obtain Metric Unit
Power	hp (electric)	7.46 E − 01[a]	kW
	cal/h	1.1622 E + 00	W
	Btu/h	2.9307 E − 01	W
	ft · lb/min	2.2597 E − 02	W
Pressure, vacuum, bursting strength	atm = 14,696 psi	1.01325 E − 01[a]	MPa
		1.01325 E + 02[a]	kPa
		1.01325 E + 0[a]	bar
	bar	1.0 E + 02[a]	kPa
		1.0 E + 05[a]	Pa
	psi	6.8948 E + 03	N/m^2
		6.8948 E + 00	kPa
	mmHg(0°C) = torr	1.33324 E − 01	kPa
		1.33324 E − 03	bar
	in. Hg (60°F)	3.3769 E + 00	kPa
	in. H_2O (4.0°C)	2.4908 E − 01	kPa
Puncture resistance	in. · oz/in.	3.0E − 02	J
Rotational frequency	r/min (rpm)	1.666 E − 02	r/s
		1.047 E − 01	rad/s
Specific-heat capacity (mass basis)	Btu/(lb · °F)	4.1868 E + 00[a]	kJ/(kg · K)
	kcal/(kg · °/C)	4.184 E + 00[a]	kJ/(kg · K)
Specific-heat capacity (mole basis)	cal/(g · mol · °C)	4.184 E + 00[a]	kJ/(kmol · K)
	Btu/(lb · °F)	4.1868 E + 00[a]	kJ/(kmol · K)
Stiffness	g_f · cm (Taber unit)	9.807 E − 05	N · m
Stress	$lb_f/in.^2$ (psi)	6.895 E − 03	Mpa
	lb_f/ft^2 (psf)	4.788 E − 02	kPa
	dyn/cm^2	1.000 E − 01[a]	Pa
Surface tension	dyn/cm	1.000 E + 01[a]	mN/m
Surface energy, tensile energy absorption (TEA)	erg/cm^2 = dyn/cm	1.000 E + 00	mJ/m^2
		1.000 E + 03	J/m^2
	psi · in.	1.7513 E + 02	J/m^2
Tear resistance	Elmendorf unit/ply	1.5696 E + 02	mN
Temperature	°R	5/9	K
	°F	5/9 (°F − 32)	°C
	°F	5/9 (°F − 32) + 273.18	K
Thermal conductivity	(cal · cm)/(s · cm^2 · °C)	4.184 E + 02[a]	W/(m · K)
	(Btu · ft)/(h · ft^2 · °F)	1.731 E + 00	W/(m · K)
		6.231 E + 00	(kJ · m)/(h · m^2 · K)
	(kcal · m)/(h · m^2 · °C)	1.162 E + 00	W/(m · K)
	(cal · cm)/(h · cm^2 · °C)	1.162 E − 01	W/(m · K)
Thermal resistance	(°C · m^2 · h)/kcal	8.604 E + 02	(K · m^2)/kW
	(°F · ft^2 · h)/Btu	1.761 E + 02	(K · m^2)/kW
Torque, moment, TIP	kg_f · m	9.80665 E + 00[a]	N · m
	lb_f · ft	1.356 E + 00	N · m
	lb_f · in.	1.130 E − 01	N · m
Velocity (linear), speed	mi/h (mph)	1.609344 E + 00[a]	km/h
	ft/s	3.048 E − 01[a]	m/s
		3.048 E + 01[a]	cm/s
	ft/min	5.08 E − 03[a]	m/s
	ft/h	8.4667 E − 02	mm/s
	in./s	2.54 E + 01[a]	mm/s
	in./min	4.233 E − 01	mm/s
Viscosity	(kg_f · s)/m^2	9.806 E + 00	Pa · s
	(dyn · s)/cm^2	1.0 E − 01	Pa · s
	cP (centipoise)	1.0 E − 03	Pa · s
Viscosity (kinematic)	$in.^2/s$	6.451 E + 02	mm^2/s
	ft^2/h	2.580 E − 05	m^2/s
	cSt (centistoke)	1	mm^2/s
Volume	ft^3	2.831 E − 02	m^3
	U.K. gal	4.546 E − 03	m^3
		4.546 E + 00	L = dm^3
	U.S. gal	3.785 E − 03	m^3
		3.785 E + 00	L = dm^3
	qt (quarts)	9.464 E − 01	L
	pt (pint)	4.732 E − 01	L
	c (cups)	2.4 E − 01	L
	fl oz	2.957 E + 01	mL
	$in.^3$	1.638 E + 01	mL
	Tbsp (tablespoon)	1.5 E + 01	mL
	tsp (teaspoon)	5.0 E + 00	mL

[a] Indicates exact value

Table 7. Permeability Units Conversion Table

To Convert From	To	Multiply by
$\frac{kg \cdot m}{m^2 \cdot s \cdot Pa}$	as[a]	1
$\frac{kg \cdot m}{m^2 \cdot s \cdot Pa}$	$\frac{kg \cdot \mu m}{m^2 \cdot day \cdot kPa}$	8.64×10^{13}
$\frac{g \cdot mil}{m^2 \cdot day}$ at 100°F, 90% rh	$\frac{g \cdot \mu m}{m^2 \cdot day \cdot kPa}$	4.264
$\frac{kg \cdot \mu m}{m^2 \cdot day \cdot kPa}$	$\frac{g \cdot mil}{m^2 \cdot day\ atm}$	3.989×10^3
$\frac{kg \cdot m}{m^2 \cdot s \cdot Pa}$	$\frac{g \cdot mil}{100\ in.^2 \cdot day \cdot atm}$	2.224×10^{16}
$\frac{g \cdot mil}{100\ in.^2 \cdot day}$ at 100°F, 90% rh	$\frac{g \cdot \mu m}{m^2 \cdot day \cdot kPa}$	66.09
$\frac{kg \cdot m}{m^2 \cdot s \cdot Pa}$	$\frac{m^3 (STP) \cdot m}{m^2 \cdot s \cdot Pa}$	$\frac{22.414}{MW^b}$
$\frac{m^3 (STP) \cdot m}{m^2 \cdot s \cdot Pa}$	$\frac{mL (STP) \cdot mil}{100\ in.^2 \cdot day \cdot atm}$	2.224×10^{19}
$\frac{mL (STP) \cdot mil}{100\ in.^2 \cdot day \cdot atm}$	$\frac{mL (STP) \cdot \mu m}{m^2 \cdot day \cdot kPa}$	3.883
$\frac{mL (STP) \cdot mil}{100\ in.^2 \cdot day \cdot atm}$	$\frac{mL (STP) \cdot mil}{m^2 \cdot day \cdot atm}$	15.5
as	$\frac{mL (STP) \cdot mil}{m^2 \cdot day \cdot atm}$	$\frac{1936.57}{MW}$
as	$\frac{mL (STP) \cdot mil}{100\ in.^2 \cdot day \cdot atm}$	$\frac{498.47}{MW}$

[a] as = 1 atto second = 1×10^{-18} s.
[b] MW = molecular weight.

cluded as part of SI. A list of common prefixes is presented in Table 2. Additionally, derived units with special names are officially recognized in the SI and are presented in Table 3. Selected derived units in terms of SI units are shown in Table 4. Table 5 lists selected units accepted in the metric system. A list of conversion factors to convert units of the USCS into the SI, and often used in packaging, was compiled in Table 6. Since the permeability units are more complex, they are presented separately in Table 7.

Converting and Rounding

Conversions should follow the simple rule of error propagation for multiplication (or division) of two numbers; figures that imply more accuracy than justified by the original data should not be included. For instance, 24 in. converted to cm (24 × 2.54) should be written as 61 cm, not 60.91 cm. Only two significative figures are given in 24 in. Similarly, 2.7 lb/ft^3 converts to 43 kg/m^3, and from Table 7, 0.3 mL (STP)·mil/(100 in.^{2} · day · atm) converts to 1×10^{-20} m^3(STP) · m/(m^2 · s · Pa).

Naming the Metric System

All the metric system units have a name and a symbol; for instance, the symbol for the unit named liter is L. Important rules referring to the use of name, symbols, and prefixes in the metric system are as follows (1) and (5):

1. The names of all units begin with a lowercase letter (unless at the beginning of the sentence). There is one exception to this rule: degrees Celsius, °C. The unit "degrees" in lower case but the modifier "Celsius" is capitalized. For instance, the glass-transition temperature of PET is 78°C. For other units, see Tables 1, 3, and 5.
2. Unit symbols are written in lower case (eg, m, kg). However, the following exceptions apply: liter is written in capital L (to avoid confusion between "l" with the numeral "1"), as well as units honoring a person, like W for watt, K for kelvin, Pa for pascal, etc.
3. Symbols of prefixes indicating a million or more are capitalized (see Table 2); for example, M for mega (10^6). Symbols indicating less than a million are in lowercase: for kilo (10^3) is k, and milli (10^{-3}) is m. This may create some confusion, and attention should be given to distinguish for example, mN (millinewton) from N·m (Newton-meter)
4. When the preceding numerical value is more than 1, the name of the unit is made plural, and singular is used in the case of 1 or less (eg, 50 μm, 105 kP, 0.25 g, 0.5 L). Still, "degrees" is always plural when not referring to one; we say 1 degree Celsius, but 0.5 degrees Celsius. The symbols are never written in plural: 13.5 liters is 13.5 L, 67.3 megapascals is 67.3 MPa.
5. A prefix, however common, must not be used to replace a unit. For example, "kilo" (which is a prefix for 1000) should not be used for "kilogram." "Micron" is no longer accepted to indicate 1×10^{-6}m; the correct term is micrometer (μm). The unit of temperature is degree Celsius, not degree centigrade (the latter term is no longer accepted).
6. Numeric values with five or more digits are sometimes separated by spaces, not commas (eg, 7 689 734 Pa or 0.047 65 cm). For values of up to four digits, spacing is optional.

BIBLIOGRAPHY

1. Anonymous, National Institute of Standards and Technology, LC 1136, May 1992.
2. *Metric Program,* National Institute of Standards and Technology, U.S. Department of Commerce, Gaithersburg, MD.
3. G. P. Carver, *The Global Path to Global Markets and New Jobs: A Question-and-Answer and Thematic Discussion,* NISTIR 5663, National Institute of Standards and Technology, June 1990.
4. B. N. Taylor, *Interpretation of the SI for the US and Metric Conversion Policy for Federal Agencies,* NIST Special Publication 814, October 1991.
5. *Federal Register,* 55 *CFR* 52242-522450, Dec. 20, 1990.

RUBEN J. HERNANDEZ
School of Packaging, Michigan State University
East Lansing, Michigan

MICROWAVABLE PACKAGING AND DUAL-OVENABLE MATERIALS

Introduction

The 1980s saw the prolific growth of the consumer microwave oven resulting in a new category of food products. Development of the new products spawned a new market for packaging materials, some of which were capable of being rethermalized in a conventional oven as well as the microwave oven (known as *dual-ovenable* products and materials), while still other new products and their packages evolved that were developed strictly for heating in the microwave oven. The penetration of microwave ovens into U.S. homes and lifestyles grew rapidly from a mere 15% of households owning a microwave oven in 1980 to an impressive 78% by 1989 (1). Consumers were hungry for food products designed for cooking and rethermalization in the microwave oven and a new category known as "microwave foods" developed seemingly overnight.

The new category was bolstered by microwave popcorn, a product that was developed specifically for preparation in the microwave oven. The standard was now set for other new food products that attempted to cater to consumers' desires for foods offering quickness and convenience in their preparation. In the mid-1980s we saw a wave of premium priced frozen dinners, positioned as a higher-quality "TV dinner," that were now available on thermoset polyester plates, capable of being heated in the conventional-oven environment (up to 400°F) or the microwave. The new category of frozen dinners saw active growth and entry by a number of companies who were keen on capitalizing on the consumers' desire for foods that could be conveniently heated in the microwave oven. In many cases, the feature of a product that could be prepared as packaged in either a conventional oven or a microwave oven was important to food marketers, even when their research showed that 85% of their consumers were using the microwave oven.

The new microwave-foods category was growing at such a rate that was difficult to ignore. New departments and teams were assembled in food companies to develop new products for the burgeoning new category. *Gorman's New Product News* tracked the number of product introductions, which increased from 278 in 1986 to nearly 1000 in 1988. Often, these new foods were merely traditional food products that were packaged for microwave-oven preparation or traditional foods that now included preparation directions for the microwave oven. As a greater understanding of the effect of microwave energy on different foods and packaging materials grew, so, too, did a desire to improve the quality of the food as eaten to the same level seen with conventional-oven preparation. Greater focus on foods engineered specifically for the microwave oven led to packaging materials that performed strictly in the microwave oven. The following is a list of materials that are currently available, or were used at some time for use as packaging components for foods to be heated in the conventional oven, the microwave oven, or, in some cases, both oven environments.

Dual-Ovenable Materials

Coated paperboard. Paperboard is available as a dual-ovenable material in the form of trays, plates, and cartons. Generally, the paperboard used is solid bleached sulfate (SBS) but solid unbleached sulfate (SUS), also referred to as *natural* or *kraft paperboard,* can be used; however, market applications are strongly skewed to SBS use. To make an ovenable material, the paperboard is coated, generally with a fine clay on one side to impart higher surface gloss and a proper surface for printing, along with an extrusion coating of a film made of polyester (PET), or in some cases 4-methylpentane-1 copolymer (TPX). The thin layer of a high-temperature plastic provides a relatively inexpensive method to produce containers that capitalize on the structural strength and economics of the paperboard component while adding the barrier required to keep fats and moisture in the food from entering the paperboard. Other plastics are also extrusion-coated onto paperboard, notably low-density polyethylene (LDPE) and polypropylene (PP); however, the maximum temperature resistance of LDPE is 215°F and 260°F for PP. The upper temperature limit for LDPE- and PP-coated paperboard will dictate the packaging application. With a maximum temperature limit of 215°F, LDPE-coated paperboard provides excellent properties for milk cartons and cold and hot-drink cups; however, its use in the microwave oven will be limited to products that will not get hot, such as a frozen dessert that may be microwaved to only soften the contents. Similarly, PP-coated paperboard is available for microwave only packages in which there is no risk that the hot spots within a package can reach a temperature where the PP softens, causing structural changes in the package or causes the PP to break down, allowing some of its constituents to enter the food.

The dominant ovenable paperboard is PET-extrusion-coated. With a maximum temperature use of 400°F, PET-coated paperboard is well suited for forming containers to be used in dual-ovenable containers for a variety of food products. TPX-coated paperboard is preferred in baking applications because of its higher temperature resistance and also its release characteristics for sugars that may become caramelized during the cooking cycle. TPX-coated paperboard is more expensive than PET-coated paperboard.

The clay coating is put on the paperboard either in-line on the paperboard-making machine or at a remote station. The clay coating in both cases is done on mill-size rolls of paperboard that are about 200 in. wide. The coated mill roll can then be slit to widths more appropriate for laminating equipment and forming and carton-making machines.

To form a tray from a roll of PET coated paperboard, the paperboard is first moistened with water to a level of 8–11%. This softens the paperboard to allow pressure forming. The roll of paperboard is first cut into blanks that have the flange dimension of the finished container. The blank is then indexed into a heated matched metal mold that, when closed, forms the blank into the mold shape, ie, the shape of the container. The heat in the mold dries the paperboard. Since the paperboard does not stretch, this type of forming will produce a square container with small creases in the corners or creases around the base of a round or oval container.

Molded pulp, PET-film-laminated. PET-film-laminated molded pulp trays are cellulose-based containers most commonly seen as plates and trays. The PET film is required to provide the barrier resistance to fats and oils to go along with the structural characteristics of the lower-cost pulp. The cellulose fibers are suspended in a slurry (which may contain other components such as sizing and treatments to provide better barrier properties) and pumped to the compression molds, which form the pulp into the desired container shape. The slurry is held into the mold by vacuum, which also draws most of the water out through a screen. Pressure and heat are applied as the mold closes to form the pulp into the container shape and draw out the remaining water. An advantage to this type of molding is the flexibility in the shape of the finished container, which may include divided compartments and areas with varying dimensional thickness for strength. Following the molding cycle, the containers are laminated with a cast PET film, which is then trimmed of any excess and packaged.

PET-laminated molded pulp trays will withstand oven temperatures of 400°F with good stiffness and structural characteristics.

CPET. Crystallized polyester (polyethylene terephthalate) is a rigid plastic material that is able to be thermoformed into containers, generally shallow plates and trays. To become dual-ovenable, the PET must be crystallized during the thermoforming process. The PET contains nucleating agents that assist in the molecular crystallization. A key factor to consider when thermoforming CPET is the intrinsic viscosity (IV) of the material. The amount of crystallization and the IV will determine the balance between the container's stiffness at low (−40°F) and high (400°F) temperatures. Generally the crystallinity of the finished container will be 28–32% and the IV will range from 0.85 to 0.95.

Prior to extruding, the PET must be thoroughly dried to a level of 0.003% to remove inherent water. For thermoforming, great care must be given to temperature control to ensure consistency. The ovens used to heat the sheet on the thermoformer prior to forming must heat the sheet evenly across its dimensions. CPET is considered to be a difficult material to work with because of its toughness and narrow window of operating temperatures, so proper mold design is a consideration. Aluminum molds are used to promote even thermal conductivity during forming. Female molds are used and the design should allow for generous radii and minimize undercuts. Often a second stage used in the molding process is a cooling mold that assists in shortening the cycle time, helps to stabilize the material after it is formed in the heated mold, and makes trimming easier. Because of its toughness, CPET is difficult to trim. Matched metal dies are used and should be sharpened periodically. Additionally, heavy-duty trim presses with quick cycle times should be used.

CPET has a temperature resistance of 400°F, has a high gloss, has a hard surface, and can be colored with pigment effectively, although the preferred colors in the market are black, white, and ivory.

PCTA. Another material in the polyester family that has higher temperature-resistance properties than CPET is a copolyester resin composed of a polymer of cyclohexanedimethanol and terephthalic acid (PCTA), often referred to by Eastman Chemical Company's trade name, Thermx. PCTA is a thermoformable material capable of withstanding temperatures in the range of 425–450°F. Processing is generally considered to be more difficult than CPET because of the higher temperatures required for extrusion and thermoforming and greater cooling requirements. A special nucleating agent is required; however, equipment specified for running CPET will generally be able to run PCTA with the proper adjustments. PCTA, like CPET, is able to be marked with the Society of the Plastics Industry (SPI) code as number 1—PETE for recycling purposes.

Foamed CPET. Shell Chemical Company has developed a method of making foamed CPET that they market under the trade name PETLITE. The objective of this material is to produce containers with 35–40% less material than conventional CPET. Extrusion equipment used for CPET must be modified for running foamed CPET; however, a single-screw extruder can be used. The blowing agent used for the expansion is an inert gas. Generally, processing temperature for extrusion and thermoforming are comparable to CPET as are pigmenting and trimming requirements.

PETLITE containers have a temperature resistance of 400°F. Currently, commercial applications include containers for baked goods such as muffins and cakes.

Borden Global Packaging, UK has recently announced the availability of their version of foamed CPET under the trade name Ovenex Lite. Trays made of Ovenex Lite contain 25–33% less material than their solid counterparts. Commercial availability in stock sizes are available from Borden in mid-1996. Borden uses their own process for foaming CPET for which a patent is pending.

Thermoset polyester. Thermoset polyester plates were the first commercial application of dual-ovenable materials when used for frozen meals in the mid-1980s. The compound used is an unsaturated polyester that is highly filled with minerals such as talc and calcium carbonate, along with glass fibers and catalyst materials to produce the chemical reaction to convert the compound into material that is irreversibly set. The polyester compound is mixed and extruded to form logs that are cut to the proper size and weight for the finished container. The material is placed into a heated mold in a hydraulic press that closes the mold. The pressure causes the material to flow into the shape of the mold, and the heat cures the compound while under pressure into the finished and irreversible material. Typically, the container must be sanded to remove any flashing around the edges and often is run through a conveyor oven for a postbake cycle to drive off any residual uncured compound, and washed to remove any dust from sanding. This process produces containers that are very strong and stiff, even at high temperatures (425°F), and are heavy with a chinalike feel and appearance. Unlike thermoforming from a sheet of material with consistent thickness, thus producing containers having essentially the same material thickness throughout, compression molding permits the finished containers to have varying degrees of thickness throughout the dimension to add strength or design features where desired. Because of the amount of material used, the multiple steps in the manufacturing process, and the relatively low output per machine cycle for compression molding presses, the price for a thermoset polyester plate is proportionally higher than competitive thermoformed plastic materials that can be produced at much higher rates with less material and cellulose-based containers that have a lower material cost.

Nylon 6/6. Mineral-filled nylon (or polyamides) is a dual-ovenable material that today is no longer commercially available. Developed by DuPont Canada during the mid-1980s, mineral filled nylon was also converted by DuPont Canada by injection molding into containers. Mineral filled nylon plates have a higher temperature resistance and stiffness than (500°F) does CPET and were priced between CPET and thermoset polyester containers. Although the material was successfully introduced commercially, the use was limited because of the higher cost than CPET; also, the hydroscopic nature of filled nylon plates sometimes caused performance problems in the microwave oven. Nylon has good resistance to oils and fats, however, within the moist operating environment of the microwave oven, some of the water present in the food would be absorbed into the nylon plate, resulting in a loss of dimensional stability. Nylon plates' strength was in the conventional oven, where, unlike some competitive thermoplastic materials, it retained its rigidity.

Nylon plates were injection-molded, a process that generally has higher tooling costs and higher operating costs when compared with thermoforming comparable unit volumes.

Polyetherimide. Polyetherimide (PEI) is a high-temperature thermoplastic resin available from General Electric Plastics under their trade name Ultem.

PEI provides a dual ovenable material for applications of about 350°F. PEI is available only in injection-molding grades, and with its high price per pound, is generally better suited for multiple-use versus single-use applications. PEI also has good chemical and stain resistance, allowing for it to effectively be cycled through commercial dishwashing systems.

PEI has been used for plates and containers in institutional feeding programs and airline meals.

Polysulfone. Polysulfone (PSO) is an amorphous thermoplastic with good rigidity and toughness. PSO is capable of withstanding oven temperatures below 325°F, making it well suited to low-temperature and microwave applications. PSO is available in transparent form, lending its use in appliances as a replacement for glass and other multiple-use applications.

Because of its high price, PSO has not been used in single-use applications.

Liquid-crystal polymer. Liquid-crystal polymers (LCP) offers very good high-temperature resistance up to about 500°F in some grades. LCP can be injection-molded and is generally pigmented from its natural beige color (see also Liquid crystalline polymers).

LCP is transparent to microwave energy; however, the high price of this resin limits its use to specialized applications. At one time, LCP was in commercial use as dual-ovenable cookware marketed by Tupperware.

Aluminum. Prior to the explosion of microwavable foods during the mid-1980s, aluminum trays dominated the prepared- and frozen-meals market as well as food-service applications such as school-lunch programs. As a package material for use only in conventional ovens, aluminum was ideal. It was capable of withstanding very high temperatures for long times—certainly exceeding the temperature limitations of the food, was usually available at attractive prices, and could be run at high speed on packaging equipment.

During the early stages of the microwave oven, arcing occurred when metal objects were placed in the oven cavity during operation, sometimes disabling the unit. This was largely corrected when the electronics were improved so that energy could not be reflected back into the magnetron. Even though it was safe for metal objects to be used in the microwave oven, consumers did not want to take the risk.

In the mid-1980s, Alcoa developed a plastic-coated aluminum tray that was formed without the typical wrinkled corners. The vinyl/epoxy coating was often pigmented to appear more like plastic and allowed the tray to work in both conventional and microwave ovens. Performance in the microwave oven is different from that of plastic- or cellulose-based materials since the aluminum shields microwave energy, often leading to longer heating times than for similar trays of competitive materials. This lends the design of the tray shape to be shallow to lessen the amount of microwave energy shielded. Coated aluminum foil trays were used commercially for pot pies where they were able to maintain high filling line speeds.

Polycarbonate. Polycarbonate (PC) is an amorphous thermoplastic resin that is capable of withstanding temperatures above 400°F. PC can be injection-molded, blow-molded, and thermoformed.

PC has been used in applications for multiple-use products such as microwavable cookware. PC is virtually unbreakable, making it a good replacement for glass. PC was used in a commercial package during the late 1980s for Stouffer's dual-ovenable meals. The structure used was thermoformed from a three-layer coextruded sheet; however, because of its high price per pound, it was replaced by competitive materials that offered acceptable performance for a much lower cost.

Microwave-Only Materials

Polypropylene. During the late 1980s, polypropylene grew as a microwave material because of its flexibility and relatively low cost. Polypropylene (PP) use in packaging was bolstered by the rapid growth of shelf-stable meals such as Lunch Buckets (registered trademark) and the like, in which ethyl vinyl alcohol (EVOH) is used to improve the oxygen-barrier properties required to safely preserve the cooked food at ambient temperatures.

Microwavable containers of PP can be thermoformed or injection-molded, generally in two different types: homopolymer PP and random copolymer PP. Homopolymer PP is produced using propylene monomer without the addition of other monomers. Random copolymer PP is similar in polymeric structure to homopolymer PP but also includes the random addition of ethylene to a polypropylene chain as it grows. Random copolymer PP gains some molecular orientation, providing certain advantages over homopolymer PP such as improved impact strength and much better clarity.

Homopolymer PP can also be filled with minerals such as talc or calcium carbonate at levels of 20–40%. Filled homopolymer PP will have a slightly higher end-use temperature and greater stiffness than its unfilled form.

Polyphenylene oxide/polystyrene. Polystyrene (PS) by itself does not have a sufficiently high temperature resistance (about 180°F), but when blended with polyphenylene oxide (PPO), the temperature-resistance properties are increased depending on the ratio of PS to PPO. For temperature resistance in the range of 212–230°F, a blend of 25% PPO and 75% PS is recommended.

PPO has a low resin flow and is therefore difficult to form;

however, when it is blended with PS, the flow characteristics and processing requirements are improved. PPO/PS can be thermoformed on equipment used for PS forming with only minor modifications. It is important to have accurate blending during the extrusion process, therefore, a high-intensity mixing screw is required. Additionally, PPO/PS is able to be foamed (much like expanded polystyrene) by extruding with tandem extruder systems and blowing agents used for polystyrene such as pentane.

PPO is available from General Electric Plastics under their trade name Noryl.

Polyethylene. High-density polyethylene (HDPE) is an acceptable thermoplastic resin for some microwave applications. HDPE starts to lose its rigidity at temperatures above 200°F, resulting in distortion of the tray or container, so care must be given in selecting this material for food products that will not exceed this temperature. This means that foods that have a high fat or oil content or those that generate steam will not be good candidates for HDPE. Generally, applications for HDPE are for foods that do not have a long heating cycle and have a homogenous texture to balance heating throughout the food, thereby eliminating hot spots.

Advantages of HDPE are its relatively low cost in comparison to other resins, its processing ease, and its good impact properties at frozen temperatures. HDPE for food trays is most commonly thermoformed but also can be injection-molded (see also Polyethylene, high density).

Glass. Although glass usage as a packaging material has declined steadily, it is a material that is able to withstand the rigors of microwave heating. Because of the advantages plastic has over glass in consumer safety, transportation costs and design flexibility, there are very few applications where glass is selected as a packaging material because of its ability to be used in the microwave.

BIBLIOGRAPHY

H. A. Rubbright and N. O. Davis, The *Microwave Decade,* Packaging Strategies, West Chester, PA, 1989, p. 6.

George J. Huss
Rubbright·Brody, Inc.
Eagen, Minnesota

MICROWAVE PASTEURIZATION AND STERILIZATION OF FOODS

In the modern food-processing industry, a variety of preservation technologies are used to extend the shelf life of foodstuffs and packaged retail food products. This extension of shelf life is an important factor in the delivery of safe, wholesome, and high-quality food products to the consumer. Typical preservation techniques include salting, smoking, canning, freezing, drying, and the use of chemical additives. These methods, while traditional and highly accepted by consumers, have the drawback of delivering food products with decreased perception of quality and freshness.

There is an increasing desire among food processors to deliver fresh, wholesome, preservative-free foods to consumers in a convenient form. This has been driven by observed changes in lifestyle and income patterns, which put a premium on time availability. These market dynamics have led to developments in the areas of aseptic processing and packaging, irradiation technology, modified atmosphere packaging, and microwave technology. The consumer in the United States and western Europe has become familiar with microwave technology through the development and widespread use of the microwave oven as an appliance.

Unit operation	Objective
Blanching	Inactive spoilage enzymes
Cooking	Modify flavor and texture
Drying	Reduce moisture
Pasteurization	Destroy vegetative microbes
Sterilization	Inactivate spores
Tempering	Raise temperature below freezing

Figure 1. Major new operations in microwave food processing.

Microwave technology has been employed in the food industry over the past 20 years to perform a variety of unit operations (Fig. 1). Using microwaves to pasteurize or to sterilize prepared food products has given food processors an opportunity to deliver against increasingly challenging consumer demands for food products of high quality that are easy to prepare. Microwaves offer the food processor many advantages over other thermostabilization technologies in that inactivation of enzymes and undesirable microorganisms can be achieved with minimal deterioration of organoleptic qualities.

We must be careful to define pasteurization and sterilization in the context of food processing. These definitions are accepted by the regulatory bodies overseeing the food industry in the United States. When we talk of *pasteurizing* a food, this refers to *the application of heat to a food product sufficient to destroy all vegetative bacteria, yeast, and mold.* This product must be stored and distributed at refrigerated temperature conditions of <40°F (4.5°C), until consumption. In the context of this subject, *pasteurization* refers to this operation being performed to a food product *after* it has been placed in the package. *Sterilization* refers to *the application of heat to a food product sufficient to destroy all spore-forming bacteria and render the product commercially sterile.* This product can be stored and distributed at ambient temperatures, and will remain safe as long as the package and seal are not compromised. This operation will be performed on a food product *after* it has been placed in the package. It is important to understand that in order to perform either pasteurization or sterilization on a food product, three distinct operating steps are necessary:

1. The food product must be heated to the sterilization or pasteurization temperature. In the case of pasteurization, this is normally in the range of 160–195°F. In the case of sterilization, temperatures are normally in the range of 250–290°F.

2. The product must be held at the appropriate temperature for a specified time to allow for lethality of the targeted organisms.
3. The product must be cooled to temperatures appropriate for storage and distribution.

Why Use Microwaves

Conventional means of heating food products using heat exchangers work well on items that are easily pumpable. This implies that the food is either in liquid form (milk, juice, beverages) or has a liquid such as a sauce as a major ingredient. Additionally, in the case of pumpable food items, particulate integrity is not of significant importance. When it comes to processing food items with large particulates such as slices of meat, whole loaves of bread, filled pasta items, or prepared meals with separate components, conventional means of imparting sufficient heat to these food products are no longer appropriate. Microwaves offer the food processor a means to *rapidly* heat a large particulate food item or prepared multicomponent meals to process temperatures without the need for liquid as a heat-transfer medium. This rapid rise to process temperature offers the possibility of ensuring higher-quality finished products, due to a reduction in product degradation from long exposures to high temperatures. Additionally, with recent advances in susceptor technology, it may be possible to "focus" the microwave energy to specific areas of a meal or food product by using carefully designed microwave susceptors in the packaging materials. This could achieve a desired effect in one portion of a meal without excessive thermal abuse to other portions.

The use of microwave pasteurizing systems for extending the shelf life of bread has allowed bakers to eliminate preservatives from their recipes. Microwave pasteurization of ready-to-eat meals has allowed food processors to deliver fresh-tasting, refrigerated items to consumers in convenient packages. These meals are easy to heat in the home (or workplace) microwave oven.

Current Uses

While the possibility of sterilizing food products using microwave has been demonstrated many times, this use of microwaves is still in the development phase. Attempts to commercialize in both Europe and in the United States have been unsuccessful. The reasons for this relate to regulatory issues in the United States and to economic issues in Europe. Key challenges to overcome in the development of commercial microwave processes include the development of continuous microwave ovens capable of overpressure sufficient to allow achievement of elevated temperatures, the "taming" of microwave energy to ensure uniform heating, and the development of appropriate sensor technology to allow accurate measurement and characterization of microwave sterilization processes. Much work has been done to overcome these challenges and significant progress has been made in preparation for commercial application when the regulatory requirements are met and appropriate market economics are achieved.

In contrast, over 100 commercially successful microwave pasteurization installations exist in both Europe and the United States. Systems manufactured by Omac of Italy, and Berstorff of Germany (Fig. 2) are in use in Europe to pasteurize bread loaves and a variety of filled and unfilled pasta products to extend their shelf life. An article published in *Packaging Magazine* (1) describes a commercial production line manufacturing high-quality prepared meals for refrigerated distribution. This facility is located in Amsterdam and, because of its commercial success, a sister facility has been constructed in Canada. Similar installations exist in Belgium, the United Kingdom, and France.

Vendor	Country	MHz
Magnetonics (APV)	UK	915
Berstorff	Germany	2450
Omac	Italy	2450
Alfastar (Tetra Laval)	Sweden	2450
Calorex	Sweden	2450
Schrade	Germany	2450
SFAMO	France	2450
Microdry	USA	2450
Cober	USA	2450
Raytheon	USA	915
Toppan	Japan	2450
Oshikiri	Japan	2450

Figure 2. Worldwide microwave equipment vendors.

System Characteristics

Industrial microwave systems are available from manufacturers in both batch and continuous design configurations, and use magnetrons that develop either 915 or 2450 MHz. Frequency allocations for industrial, scientific, and medical use have been specified by the World Administrative Radio Conference (WARC). These allocations represent unused frequency ranges in the telecommunications industry. Comparisons of the two frequencies reveal some significant differences in overall characteristics of microwave systems. Microwaves of 915 MHz will penetrate up to three times the depth of 2450 MHz. Also, studies have concluded that more uniform temperature distributions are achieved using 915 MHz. The conversion of electrical power into product heating is in the 80–90% range with 915 MHz, which is double the 40–48% conversion observed with 2450 MHz. This makes the 915-MHz power sources cheaper to operate. Finally, 915 MHz can be delivered by magnetrons up to 60 kW in size, while the largest (2450-MHz) magnetron is only 10 kW. These differences can and should be used to advantage by the engineers designing systems for use on specific food types. Large doses of high-powered energy should be delivered by 915-MHz units where food thickness is a factor. Small and frequent doses of low-powered energy to achieve more "focused" energy can be delivered by 2450-MHz power sources. Manufacturers have made advances in the use of waveguides, antennas, and other means to help distribute,

	Continental Europe	USA	U K
915 MHz	X	√	√
2450MHz	√	√	√

Figure 3. Operating frequencies.

control, and direct microwave energy to specific target areas within food products. It is interesting to note (Fig. 3) that the 915-MHz frequency is not permitted in continental Europe.

Packaging

As discussed, the process of sterilization or pasteurization takes place with the product in the package. The package is therefore treated to the same conditions as the product and is effectively pasteurized or sterilized along with the product. It is important that the package is able to maintain the microenvironment within until the product is prepared and consumed. The package and seal must remain intact during the stresses incurred in heating and cooling. Materials must be chosen that are compatible with microwaves and are able to protect the product from moisture and oxygen transfer. These characteristics are essential for extended shelf life. A variety of containers and lid stock materials have been developed specifically for use with microwave systems that offer dual ovenability (microwave and convection) and easy-to-remove, peelable seals. There has also been development in the area of susceptors that enable multicomponent meal packages to direct heating to specified portions of the meal.

Microwave technology has been successfully incorporated into a food-processing system to derive many unique and beneficial results. When combined with advanced package designs, this technology has provided opportunities for high-quality, safe, and convenient food products. The full potential of this technology has not yet been realized.

BIBLIOGRAPHY

1. Bruce Holmgrin, "High Volume Line Turns out Meals with Fresh Look," *Packag. Mag.* (Aug. 1988).

General References

A. Brody, "Microwave Processing for Micro-ready Foods: Packaging Requirements" in *Micro '89 Conference,* Schotland Business Research Inc., 1989.

R. V. Decerau, *Microwaves in the Food Processing Industry,* Academic Press, Orlando, FL, 1985.

R. V. Decereau, "Microwave Uses in Food Processing" in *Symposium Proceedings: The Application of Microwaves to Food Processing in Australia,* Sydney, Aug. 1993.

J. Giacone, and S. Sommerfeld, "The Food Industry in the Year 2000," *Chem. Eng. Prog.* (May 1988).

Joseph Giacone
Kraft Foods
Tarrytown, New York

MILITARY PACKAGING

NEW WAY OF DOING BUSINESS

On June 29, 1994, Secretary of Defense, William Perry directed the implementation of a memorandum titled, "Specifications and Standards—A New Way of Doing Business," which directed the implementation of initiatives related to military specifications and standards in the Department of Defense. The memo stated, in part, that "moving to greater use of performance and commercial specifications and standards in one of the most important actions the DoD must take to ensure we are able to meet our military, economic and policy objectives in the future." A complete copy of the memorandum appears at the end of this paragraph. This policy has had a significant effect on all military requirements, including packaging. Military packaging requirements are being converted from detailed to performance oriented requirements and, for acquisition purposes, will appear in the government contract or order in lieu of the item specification. Two new or revised documents (1,2) are available that provide more detail on content and format of performance specifications.

MEMORANDUM: SPECIFICATIONS AND STANDARDS—A NEW WAY OF DOING BUSINESS, FROM THE SECRETARY OF DEFENSE, WILLIAM PERRY

To meet future needs, the Department of Defense must increase access to commercial state-of-the-art technology and must facilitate the adoption by its suppliers of business processes characteristic of world class suppliers. In addition, integration of commercial and military development and manufacturing facilitates the development of dual-use processes and products and contributes to an expanded industrial base that is capable of meeting defense needs at lower costs.

I have repeatedly stated that moving to greater use of performance and commercial specifications and standards is one of the most important actions that DoD must take to ensure we are able to meet our military, economic, and policy objectives in the future. Moreover, the Vice President's National Performance Review recommends that agencies avoid government-unique requirements and rely more on the commercial marketplace.

To accomplish this objective, the Deputy Under Secretary of Defense (Acquisition Reform) chartered a Process Action Team to develop a strategy and a specific plan of action to decrease reliance, to the maximum extent practicable, on military specifications and standards. The Process Action Team report, "Blueprint for Change," identifies the tasks necessary to achieve this objective. I wholeheartedly accept the Team's report and approve the report's primary recommendation to use performance and commercial specifications and standards in lieu of military specifications and standards, unless no practical alternative exists to meet the user's needs. I also accept the report of the Industry Review Panel on Specifications and Standards and direct the Under Secretary of Defense (Acquisition and Technology) to appropriately implement the Panel's recommendations.

I direct the addressees to take immediate action to implement the Team's recommendations and assign the Under Secretary of Defense (Acquisition and Technology) overall implementation responsibility. *I direct the Under Secretary of Defense (Acquisition and Technology) to immediately arrange for reprogramming the funds needed in FY94 and FY95 to efficiently implement the recommendations.* I direct the Secretaries of the Military Departments and the Directors of the Defense Agencies to program funding for FY96 and beyond in accordance with the Defense Planning Guidance.

Policy Changes

Listed below are a number of the most critical changes to current policy that are needed to implement the Process Action Team's recommendations. These changes are effective immediately. However, it is not my intent to disrupt on-going solicitations or contract negotiations. Therefore, the Component Acquisition Executive (as defined in Part 15 of DoD instruction 5000.2), or a designee, may waive the implementation of these changes for ongoing solicitations or contracts during the next 180 days following the date of this memorandum. The Under Secretary of Defense (Acquisition and Technology) shall implement these policy changes in DoD Instruction 5000.2, the Defense Federal Acquisition Regulation Supplement (DFARS), and any other instructions, manuals, regulations, or policy documents, as appropriate.

Military specifications and standards. Performance specifications shall be used when purchasing new systems, major modifications, upgrades to current systems, and nondevelopmental and commercial items, for programs in any acquisition category. If it is not practicable to use a performance specification, a non-government standard shall be used. Since there will be cases when military specifications are needed to define an exact design solution because there is no acceptable non-governmental standard or because the use of a performance specification or non-government standard is not cost effective, the use of military specifications and standards is authorized as a last resort, with an appropriate waiver.

Waivers for the use of military specifications and standards must be approved by the Milestone Decision Authority (as defined in Part 2 of DoD Instruction 5000.2). In the case of acquisition category ID programs, waivers may be granted by the Component Acquisition Executive, or a designee. The Director, Naval Nuclear Propulsion shall determine the specifications and standards to be used for naval nuclear propulsion plants in accordance with Pub. L. 98-525 (42 U.S.C. §7158 note). Waivers for reprocurement of items already in the inventory are not required. Waivers may be made on a "class" or item basis for a period of time not to exceed two years.

Innovative contract management. The Under Secretary of Defense (Acquisition and Technology) shall develop, within 60 days of the date of this memorandum, Defense Federal Acquisition Regulation Supplement (DFARS) language to encourage contractors to propose non-government standards and industry-wide practices that meet the intent of the military specifications and standards. The Under Secretary will make this language effective 180 days after the date of this memorandum. This language will be developed for inclusion in both requests for proposal and in on-going contracts. These standards and practices shall be considered as alternatives to those military specifications and standards cited in all new contracts expected to have a value of $100,000 or more, and in existing contracts of $500,000 or more having a substantial contract effort remaining to be performed.

Pending completion of the language, I encourage the Secretaries of the Military Departments and the Directors of the Defense Agencies to exercise their existing authority to use solicitation and contract clause language such as the language proposed in the Process Action Team's report. Government contracting officers shall expedite the processing of proposed alternatives to military specifications and standards and are encouraged to use the Value Engineering no-cost settlement method (permitted by FAR 48.104-3) in existing contracts.

Program use of specifications and standards. Use of specifications and standards listed in DoD Instruction 5000.2 is not mandatory for Program Managers. These specifications and standards are tools available to the Program Manager, who shall view them as guidance, as stated in Section 6-Q of DoD Instruction 5000.2.

Tiering of specifications and standards. During production, those system specifications, subsystem specifications and equipment/product specifications (through and including the first-tier references in the equipment/product specifications) cited in the contract shall be mandatory for use. Lower tier references will be for guidance only, and will not be contractually binding unless they are directly cited in the contract. Specifications and standards listed on engineering drawings are to be considered as first-tier references. Approval of exceptions to this policy may only be made by the Head of the Departmental or Agency Standards Improvement Office and the Director, Naval Nuclear Propulsion for specifications and drawings used in nuclear propulsion plants in accordance with Pub. L. 98-525 (42 U.S.C. §7158 Note).

New Directions

Management and manufacturing specifications and standards. Program Managers shall use management and manufacturing specifications and standards for guidance only. The Under Secretary of Defense (Acquisition and Technology) shall develop a plan for canceling these specifications and standards, inactivating them for new designs, transferring the specifications and standards to non-government standards, converting them to performance-based specifications, or justifying their retention as military specifications and standards. The plan shall begin with the ten management and manufacturing standards identified in the Report of the Industry Review Panel on Specifications and Standards and shall require completion of the appropriate action, to the maximum extent practicable, within two years.

Configuration control. To the extent practicable, the Government should maintain configuration control of the functional and performance requirements only, giving contractors responsibility for the detailed design.

Obsolete specifications. The "Department of Defense Index of Specifications and Standards" and the "Acquisition Management System and Data Requirements Control List" contain outdated military specifications and standards and data requirements that should not be used for new development efforts. The Under Secretary of Defense (Acquisition and Technology) shall develop a procedure for identifying and removing these obsolete requirements.

Use of non-government standards. I encourage the Under Secretary of Defense (Acquisition and Technology) to form partnerships with industry associations to develop non-government standards for replacement of military standards where practicable. The Under Secretary shall adopt and list in the "Department of Defense Index of Specifications and Standards" (DODISS) non-government standards currently being used by DoD. The Under Secretary shall also establish teams to review the federal supply classes and standardization areas to identify candidates for conversion or replacement.

Reducing oversight. I direct the Secretaries of the Military Departments and the Directors of the Defense Agencies to reduce direct Government oversight by substituting process controls and non-government standards in place of development and/or production testing and inspection and military-unique quality assurance systems.

Cultural Changes

Challenge acquisition requirements. Program Managers and acquisition decisionmakers at all levels shall challenge requirements because the problem of unique military systems does not begin with the standards. The problem is rooted in the requirements determination phase of the acquisition cycle.

Enhance pollution controls. The Secretaries of the Military Departments and the Directors of the Defense Agencies shall establish and execute an aggressive program to identify and reduce or eliminate toxic pollutants procured or generated through the use of specifications and standards.

Education and training. The Under Secretary of Defense (Acquisition and Technology) shall ensure that training and education programs throughout the Department are revised to incorporate specifications and standards reform.

Program reviews. Milestone Decision Authority (MDA) review of programs at all levels shall include consideration of the extent streamlining, both in the contract and in the oversight process, is being pursued. The MDA (ie, the Component Acquisition Executive or his/her designee, for all but ACAT 1D programs) will be responsible for ensuring that progress is being made with respect to programs under his/her cognizance.

Standards improvement executives. The Under Secretary, the Secretaries of the Military Departments, and the Director of the Defense Logistics Agency shall appoint Standards Improvement Executives within 30 days. The Standards Improvement Executives shall assume the responsibilities of the current Standardization Executives, support those carrying out acquisition reform, direct implementation of the military specifications and standards reform program, and participate on the Defense Standards Improvement Council. The Defense Standards Improvement Council shall be the primary coordinating body for the specification and standards program within the Department of Defense and shall report directly to the Assistant Secretary of Defense (Economic Security). The Council shall coordinate with the Deputy Under Secretary of Defense (Acquisition Reform) regarding specification and standards reform matters, and shall provide periodic progress reports to the Acquisition Reform Senior Steering Group, who will monitor overall implementation progress.

Management Commitment

This Process Action Team tackled one of the most difficult issues we will face in reforming the acquisition process. I would like to commend the team, composed of representatives from all of the Military Departments and appropriate Defense Agencies, and its leader, Mr. Darold Griffin, for a job well done. In addition, I would like to thank the Army, and in particular, Army Materiel Command, for its administrative support of the team.

The Process Action Team's report and the policies contained in this memorandum are not a total solution to the problems inherent in the use of military specifications and standards; however, they are a solid beginning that will increase the use of performance and commercial specifications and standards. Your leadership and good judgment will be critical to successful implementation of this reform. I encourage you and your leadership teams to be active participants in establishing the environment essential for implementing this cultural change.

This memorandum is intended only to improve the internal management of the Department of Defense and does not create any right or benefit, substantive or procedural, enforceable at law or equity by a party against the Department of Defense or its officers and employees.

Signed by William J. Perry

BIBLIOGRAPHY

The government documents listed below can be obtained from the Standardization Document Order Desk, Building 4D, 700 Robbins Avenue, Philadelphia, PA 19111-5094, or by calling (215) 697-2667/2179.

1. *MIL-STD-961D, Department of Defense Standard Practice for Defense Specifications,* March 22, 1995.
2. AMC-P-715-17, *Guide for Preparation and Use of Performance Specifications,* March 15, 1994.

RAYMOND T. MANSUR
SSCOM (NRDEC)
Natick, Massachusetts

MODIFIED ATMOSPHERE PACKAGING

Modified atmosphere packaging (MAP) is a term applied to a range of food packaging technologies that rely on mixtures of the atmospheric gases oxygen (O_2), carbon dioxide (CO_2), and nitrogen (N_2), in concentrations different than those in air, to retard deterioration processes in foods. Such atmospheres, sometimes with the addition of small amounts of other gases such as carbon monoxide (CO), ethanol (EtOH), sulfur dioxide (SO_2), or argon (Ar), maintain foods in a "fresh" state for periods of time necessary to move them through extended distribution and marketing chains.

The majority of these technologies rely on a combination of MAP and rigorous refrigeration to forestall microbial and chemical deterioration. The key to the technologies lies in the different concentrations of the common atmospheric gases and the beneficial effects conferred by specific concentrations of various gases. The technologies, thus, rely on gases that are safe, common, cheap, readily available, and usually not considered chemical additives. Different combinations of these gases are appropriate for different foods, package types and situations. The proper marriage of food, gas mixture and package type has been the subject of most of the developmental efforts in the area.

Vacuum packaging (VP) is a form of MAP that establishes an atmosphere by drawing a partial vacuum to remove the ambient gases inside a gas barrier package and then sealing the package with the purpose of excluding atmospheric gases, principally O_2. A complete vacuum is never established under practical conditions but the exclusion of gases is the aim.

MAP/VP are employed to delay deterioration of foods that are not sterile, and whose enzymatic systems may still be operative. Thus, these techniques differ from traditional methods of food preservation, such as cooking, canning, drying or freezing, but rather are methods to maintain foods in a "natural" condition while slowing specific deterioration processes. Rigorous temperature control is often necessary for the packages to work properly. For this reason, the capabilities of the distribution system define the degree to which MAP can be used. These packaging technologies have fit very well into the vertically integrated, geographically restricted distribution systems in Western Europe. They have been more difficult to implement in the more fragmented and longer distance distribution systems of North America.

Gases Used In MAP

Oxygen (O_2). Most of the reactions with food constituents involving oxygen are degradation reactions resulting in the oxidative breakdown of foods into their constitutive parts. Oxygen combines readily with fats and oils and causes rancidity. In addition, most spoilage microorganisms require O_2 to grow and will cause off odors in the presence of sufficient oxygen. Oxygen is necessary to the normal respiratory metabolism of fresh fruits and vegetables and normal atmospheric concentrations of O_2 contribute to senescence and degradation of produce quality. Oxygen is implicated in staling of bakery goods and pasta. It oxidizes the pigments of red meats to an undesirable brown color. In the absence of O_2 meats will take on a purplish color that some consumers find objectionable.

Oxygen permeates through plastic polymers at rates depending on the polymer, but it generally permeates more slowly than carbon dioxide. The permeability rate of oxygen (and all gases) in plastics increases as temperature increases. Similarly, the chemical reactivity of oxygen with food constituents increases as temperature increases.

Carbon dioxide (CO_2). Carbon dioxide is present in the atmosphere at low levels, (~0.03%) but is a product of combustion and so is easily produced. It is very soluble in water, especially in cold water, (179.7 cm^3/100mL at 0°C), and will thus be readily absorbed by high moisture, refrigerated foods. Carbon dioxide is soluble in fats and oils, as well as in water, and this solubility can lead to package collapse, off flavors, excess purge by muscle foods, and discoloration of fresh produce. Carbon dioxide permeates most packaging materials more rapidly than other atmospheric gases.

When CO_2 dissolves in water it has an acidifying effect. This acidification, as well as direct antimicrobial effects of $CO_2 > 10–15\%$, can suppress the growth of many spoilage microorganisms and for this reason it is an important component of MAP.

Carbon dioxide also suppresses the respiration of some fresh fruits and vegetables and thus can help extend their shelf lives. In addition, CO_2 concentrations above about 1% can render many plant tissues insensitive to the ripening hormone ethylene and thus slow their senescence and deterioration. However, too much CO_2 can be damaging to plant tissues and individual fruits and vegetables differ in their tolerance of CO_2.

Nitrogen (N_2). Nitrogen is the most abundant component in air (~78%) and can be used in either gaseous or liquid form. It is physiologically inert in its gaseous and liquid forms and is used in packaging primarily as a filler and to exclude other more reactive gases. It is sparingly soluble in water (2.33 cm^3/100mL at 0°C).

Carbon monoxide (CO). Carbon monoxide is a colorless, odorless, tasteless, very toxic gas which is effective as a browning and microbial inhibitor. In concentrations as low as 1%, CO will inhibit the growth of many bacteria, yeasts and molds. It can also delay oxidative browning of fruits and vegetables when combined with low O_2 concentrations (2–5%) and has found limited use commercially for this purpose. However, due to the toxicity of the gas, and its explosive nature at 12.5–74.2% in air, CO must be handled using special precautions.

Sulfur dioxide (SO_2). Sulfur dioxide has been used to control growth of mold and bacteria on a number of soft fruits, particularly grapes and dried fruits. It has also found use in the control of microbial growth in fruit juices, wines, shrimp, pickles, and some sausages. Sulfur dioxide is very chemically reactive in aqueous solution and forms sulfite compounds, which are inhibitory to bacteria in acid conditions (pH <4). However, a significant minority of people display hypersensitivity to sulfite compounds in foods and the use of sulfites has come under public and regulatory scrutiny in recent years.

Ethanol (EtOH). Ethanol has antifungal activity and has been used with some baked goods in Japan to reduce microbial spoilage. In addition, research results have shown ethanol to enhance firmness of tomatoes and may act as a flavor enhancer for other fruits.

Argon (Ar). Argon, a noble gas, is not known to have any chemical or biological activity. Nevertheless, reports from one company suggest that it may have antimicrobial effects. Argon comprises 0.9% of the atmosphere and so is relatively abundant.

Packaging Materials

Most MAPs rely on flexible monolayer or multilayer plastic films and/or plastic and composite trays to maintain product and atmosphere integrity. Plastic-paper and plastic-foil films, as well as metal cans and glass jars are also used in such packaging. Many kinds of plastic and other materials are available for these uses and more are appearing every day. Plastics probably comprise upwards of 90% of the materials used in MAP with paper, paperboard, aluminum foil, metal and glass accounting for the remainder. However, for each product category, relatively few kinds of materials are used. The selection of materials depends on technical considerations of performance, strength, sealability, printability, and recyclability as well as on safety and cost considerations.

The most commonly used plastics are polyethylene (PE), polypropylene (PP), polyester, polycarbonate, polystyrene (PS), nylon, ethylene–vinyl alcohol (EVOH), polyvinylidene dichloride (PVDC), cellophane, rubber and butadiene. In addition, sealant layers of ionomer or other plastic materials are sometimes used.

Barrier properties. The function of many MAPs is to exclude O_2 and moisture from the packaged food and thereby slow oxidative rancidity (baked goods, *Sous Vide,* and pasta), retard growth of spoilage microorganisms (meats, pasta, baked goods), maintain crispness (baked goods and snack foods), and maintain proper color (red meat). In other packages, the aim is to prevent egress of CO_2 to retard growth of microorganisms (meat, baked goods and pasta). These packages require the use of gas barrier materials which allow very litle permeation of gases.

MAPs for fresh produce, on the other hand, must allow entry of some atmospheric O_2 to maintain the aerobic metabolism of the product. In addition, some CO_2 must exit the package to avoid buildup of injurious levels of the gas. These packages rely on the use of plastic films with relatively high gas permeability characteristics.

The barrier properties of plastic films depend primarily upon the rates of diffusion processes. In some cases, where small holes or pores are incorporated in the film structure, mass flow of gases may also play a role. The movement of gases across films depends on several physical factors that are related through Fick's Law as follows:

$$J_{\text{gas}} = \frac{A \times \Delta C_{\text{gas}}}{R}$$

Where:

J_{gas} = total flux of gas (cm^3/s)
A = surface area of the film (cm^2)
ΔC_{gas} = concentration gradient across the film
R = resistance of the film to gas diffusion (s/cm)

The gas flow across a film increases with increasing surface area and with increasing concentration gradient across the film. The gas flow across the film decreases with increasing film resistance to gas diffusion. A barrier film has high resistance to gas diffusion.

Gas barrier properties of plastic films are sensitive to temperature and, in a few cases such as for EVOH or Nylon, relative humidity. As most film permeability testing occurs at room temperature and low relative humidity, it is often difficult to predict barrier performance under the high humidity refrigerated conditions typical of extended shelf life fresh foods. As a rule, both WVTR and gas transmission rates increase with increasing temperature (see also Barrier Polymers).

Active Packaging

Besides gas flushing, gas injection and wrapping in packages that mediate or obstruct gas movement, there are a growing number of supplementary materials that can be added to MAPs or used with those packages to further alter and control the package environment. These include absorbers, emitters, scavengers, scrubbers, and desiccants that, when added to a package, alter the package structure, function, or properties. Their purpose is to extend product shelf life, usually through the control of atmospheric gases and moisture, without the use of "preservatives." These materials are diverse and their claims in the marketplace may sometimes outpace their capabilities. But there is no question that this is a rapidly growing area of packaging that is likely to move into the mainstream of food packaging in the near future (1).

Oxygen absorbers. For packaging applications that involve oxygen sensitive products, gas flushing with nitrogen or carbon dioxide often is insufficient to remove residual oxygen. Because low concentrations of oxygen can participate in the degradation of fatty foods, accelerate staling of baked goods, and facilitate other forms of deterioration, the use of oxygen absorbers as an adjunct to gas flushing has become an important methodology. Most commercially available O_2 absorbers use metallic reducing agents such as powdered iron oxide, ferrous carbonate or metallic platinum. Iron powder is the most common active ingredient. These products often utilize powdered FeO which becomes Fe_2O_3 and Fe_3O_4 and their hydroxide forms after absorption of O_2. The lower the temperature, the lower the O_2 absorption speed (2,3).

Reduced iron is listed as GRAS (generally regarded as safe by the FDA) with no limitations on its use other than good manufacturing practices. It should be pointed out that the use of O_2 scavengers reduces O_2 to levels that may be conducive to the growth and toxigenesis of certain anaerobic pathogens, notably *Clostridium botulinum,* and thus should be used with great care with nonsterile fresh foods.

Carbon dioxide absorbers. Some of the absorbents that are currently being used to remove excess CO_2 from CA storage rooms could be adapted for their utilization in MAP. They include lime (calcium hydroxide [$Ca(OH)_2$] + sodium hydroxide or potassium hydroxide), activated charcoal and magnesium oxide (3).

Moisture regulators. The stability and shelf life of many food products depends, in part, on the maintenance of an appropriate water activity inside the package. Inappropriate moisture levels can lead to shortened shelf life, reduced quality and, in the worst case, unwanted microbiological activity. While many films are available that will provide an excellent moisture barrier to prevent the ingress or egress of unwanted moisture, often the food product itself can contribute unwanted humidity within a package. Several products are available that will aid in the regulation of proper moisture within the package. Most rely on physical absorption of moisture, though a few rely on actual chemical bonding to the water molecules. The most common desiccants currently used are based on silica gel, montmorillonite clay, calcium oxide, activated carbon or glycerol. Of these, silica gel is the most common.

Silica gel is a partially dehydrated colloid of silicic acid. Its structure is such that it forms interconnected micropores that form a vast surface area that can attract and hold water through adsorption and capillary condensation. Silica gel is most efficient at temperatures below 25°C where it can adsorb up to 40% of its weight in water. Its efficiency falls off at higher temperatures. Silica gel is nontoxic and noncorrosive. Montmorillonite clay is less efficient at moisture adsorption than silica gel and also slower.

Alternatively, saturated salt solutions can be employed to maintain a desired equilibrium relative humidity within the package. Individual salts, if present in sufficient quantities, will maintain a characteristic relative humidity (RH) at a given temperature. For example, NaCl, at 5°C, will maintain a RH of 75%. For a higher equilibrium RH, K_2SO_4, at 5°C, will maintain an RH of 98.5% (4,5). Selection of the proper salt can result in almost any desired RH.

Absorbers/emitters of ethylene and other volatiles. Because ethylene gas is a plant hormone with powerful physiological effects at very low concentrations it is sometimes removed from the environment of sensitive commodities. Ethylene gas (C_2H_4), in part per million concentrations, can induce rapid ripening and senescence in many fruits, yellowing of leafy vegetables, and physiological injury of selected produce items. Potassium permanganate has long been used to remove ethylene from controlled atmosphere apple storage rooms. The same compound, adsorbed on various inert substrates, is available commercially in sachets, on pads, or in granular form. Many other compounds have ethylene adsorbing capacity, including various hydrocarbons, silicones, glycols, and clay materials (6).

In addition, many other volatile compounds can be responsible for off odors in foods but can be reduced through the use of absorbers. Potassium permanganate and activated charcoal are the most important and can absorb many volatile gases other than ethylene.

Active films. While the above methods of active package modification have their virtues, they all suffer because they have to be utilized in addition to the packaging itself. As such, they represent an added expense, and, in some cases, the addition of a potentially hazardous material. A number of innovative approaches to avoiding these problems have emerged in the past few years. There are now several plastic films that incorporate some of the functions described above into the film itself. Thus, the package itself may absorb O_2, ethylene or CO_2, or control decay (1).

There are several new films, mostly from Japan, that are reputed to have ethylene adsorbing capabilities to improve the shelf life and freshness of fruits and vegetables. These films are typically made of common plastic polymers (usually low density polyethylene) with one or several clay materials (zeolites, cristobilites, etc) embedded in the film matrix. These ceramic materials have putative ethylene-adsorbing capacity. However, investigators have found little ethylene adsorbing capacity by porous ceramic materials (6,7). Urushizaki (8) tested one such ceramic material and detected adsorption of less than 1μl/gm of ceramic material after exposure to 500 ppm of ethylene. Such adsorptive capacity is insufficient to be of benefit in a modified atmosphere package of fresh produce.

Some films containing ceramic materials are claimed to aid in the preservation of produce through the emission of far-infrared radiation. Such radiation normally is effective mainly in heating an absorbing body, in this case the fresh produce. This is antithetical to the usual aim in preserving produce. The reputed beneficial effect is that the infrared radiation excites the plant cell membranes thereby rendering the cells more resistant to microbial invasion. Such claims are as yet unsubstantiated in the scientific literature.

Films that will pass ultraviolet radiation have been suggested for reduction of ethylene and for control of microorganisms. Ultraviolet radiation reacts with oxygen to form ozone which scavenges ethylene and also kills microorganisms. Unfortunately, ozone is also very damaging to most plant tissues. In addition, the reaction may not be very efficient in low oxygen environments such as those found inside MAP.

A promising approach to reduction of oxygen content inside a package is based on imbedding in the plastic film a photosensitizing dye and an electron-rich oxidizable compound termed a singlet oxygen acceptor. When irradiated with ultraviolet, visible or near infrared radiation of appropriate wavelengths the excited dye molecules sensitize oxygen molecules by converting them to the singlet state. Such oxygen molecules then more readily react with oxygen acceptor molecules that are also embedded in the film (9).

Other workers are making progress in incorporating antimicrobial agents into plastic films so that the package itself controls bacteria and fungi. However, the incorporation of pesticides in films may meet resistance in the current regulatory climate.

MAP of Red Meat

Deterioration of fresh meat. Fresh meat undergoes both oxidative color deterioration and microbiological deterioration. The color of red meat is determined by the oxidation state of the pigment myoglobin. The reduced form, deoxymyoglobin, is a purple color. Oxygenated oxymyoglobin is a bright red color associated with freshness in the minds of many consumers. Atmospheric O_2 slowly converts myoglobin to the dull brown metmyoglobin which many consumers associate with poor quality. Deoxymyoglobin is more susceptible to oxidation than oxymyoglobin, so metmyoglobin forms more readily at low O_2 concentrations than at higher concentrations (10).

Fresh meat is contaminated on the surface with spoilage bacteria from the slaughtering process. The presence of O_2 encourages the growth of aerobic spoilage bacteria, particularly species of *Pseudomonas,* which will produce putrid, malodorous compounds.

Vacuum packaging. The exclusion of atmospheric O_2 prevents the oxidation of myoglobin to metmyoglobin as well as suppressing the growth of malodorous *Pseudomonas* bacteria.

The delay of these two primary forms of deterioration through packaging can significantly extend shelf life of fresh red meat.

Vacuum packaging has been applied to 2–9 kg primal and subprimal cuts of meat for distribution to stores. Appropriate films for such packaging must have good gas barrier properties, generally admitting <50 cc $O_2/m^2/24h/atm$ (10). In addition they must resist puncturing by protruding bone and must be flexible enough to form a tight "skin" over the meat and prevent pockets of air from forming where aerobic deterioration could rapidly occur. Films made of polyamide–polyethylene laminates and EVA copolymer–PVC/PVDC copolymer laminates, and ionomer–polyamide–EVA copolymer, among others, have been used for this purpose (13).

When vacuum packaged meats reach the store they are unpacked, cut into appropriate consumer units and placed in polystyrene foam trays or PVC trays and overwrapped with O_2 permeable films. The ingress of O_2 causes the deoxymyoglobin to "bloom" into red oxymyoglobin.

Vacuum packages for meat are formed in four basic ways: *1.* Through heat-shrinking a flexible bag around the primal cuts, *2.* Using a preformed plastic bag in an evacuation chamber, *3.* Thermoforming trays in-line from a base web, or *4.* Vacuum skin packaging in which the product acts as the forming mold. Vacuum packaged beef of normal pH and kept at chill temperature can be stored for 10–12 weeks. Lamb, because it tends to have more neutral pH exterior adipose tissue than beef, will keep 6–8 weeks in vacuum packages (11).

High O_2 MAP. High O_2 MAP employs 40–80% O_2 to extend color stability as well as 20–30% CO_2 to delay microbial spoilage. However, the presence of O_2 can lead to off odors and rancidity and so high O_2 MAP has not been suitable for prolonged storage of red meat.

Low O_2 MAP. Combinations of low O_2 and high CO_2 have been achieved through gas-flushing packages with N_2 and CO_2. The absence of O_2 retards formation of brown metmyoglobin and the CO_2 suppresses growth of spoilage bacteria. However, because of CO_2 is very soluble in both water and fat, excess CO_2 must be added to the package to allow for the solubility. Package collapse can result unless N_2 in the headspace of the package significantly exceeds the amount of CO_2 that solubilizes. Such packages have resulted in shelf lives of 3–4 months for pork and lamb.

MAP of Poultry

Poultry is usually contaminated with a large bacterial population that consists primarily of spoilage bacteria. Poultry is high pH compared to red meat and so provides a good environment for growth of these bacteria. Vacuum packages are difficult to form around poultry because of the irregular shape and sharp edges commonly encountered. Consequently, shelf life in vacuum packages is usually short, limited to about 2 weeks before putrid odors develop.

Low O_2 combined with high CO_2 has been extensively used with bulk packaged poultry. Oxygen is removed by drawing a vacuum and CO_2 and N_2 are introduced. Despite the presence of some residual O_2, shelf lives of 2–3 weeks can be achieved (10).

MAP of Fresh Fish

Fish encompasses a great diversity of species and habitats. In general, fish differ from terrestrial foods in that the interior of their muscles are not sterile, they undergo rapid enzymatic breakdown of their proteins even in low O_2 environments, and they are often not raised in controlled environments. Fish are only as clean as the water they were taken from. Thus, fish can be carriers or several pathogens such as *Clostridium botulinum* nonproteolytic types E and B, *Vibrio parahaemolyticus, Listeria monocytogenes,* and others (12).

Fish rapidly spoil due to the activity of gram-negative spoilage bacteria and they undergo enzymatic breakdown. Vacuum and MAP employing high CO_2 have been used to extend the normally limited shelf lives of various kinds of fish. Gas mixtures of 30% O_2, 40% CO_2 and 30% N_2 have been used for nonfatty fish while, for smoked and fatty fish, 40% CO_2 and 60% N_2 have been used (13).

The use of MAP and vacuum packaging are not capable of preventing growth and toxin production by *C. botulinum* (14). Maintaining the temperature below 3°C throughout distribution is the only barrier to the growth of this pathogen. While commercially packaged raw fish have not been implicated in food poisoning incidents, MAP of fresh fish should only be undertaken with extreme caution.

MAP of Bakery and Pasta Products

Pasta and baked goods are subject to moisture loss, oxidative rancidity and microbiological breakdown, primarily due to growth of molds. MAP has been used extensively in Europe and, less so, in the United States to prevent deterioration and extend shelf life. Vacuum packaging has been used for some baked goods to remove headspace O_2 and inhibit oxidative rancidity and growth of aerobic spoilage microbes. Packaging in a combination of N_2 and CO_2 has been used more commonly to replace O_2 and suppress growth of bacteria and molds due to the presence of the CO_2. Oxygen absorbers, CO_2 generators and ethanol emitters have been used to extend the shelf life of baked goods in Japan and other parts of Asia (15).

Because the exclusion of O_2 is so crucial to MAP of pasta and baked goods, packaging materials must provide a good barrier to O_2. Rigid trays with nylon/LDPE or PVC/PVDC lidstocks have been used. Alternatively, laminated films made from nylon/polyethylene, nylon/PVDC/PE, or nylon/EVOH/PE have also been used. All maintain a reasonable moisture vapor as well as an O_2 barrier.

Baked goods have been packaged in atmospheres of 50–100% CO_2 plus 0–50% N_2 which can result in shelf lives up to several months for some products. Fresh pasta will stay fresh in MAP up to 2 weeks if not pasteurized or, in some cases, 3 months if pasteurized (16).

MAP Of Prepared Foods

Chilled, prepared foods are perhaps the fastest growing area of all food categories, and many of these foods are being shipped and marketed in MAP. As with all MAP applications, use of top quality ingredients, strict adherence to temperature control and careful sanitation are prerequisites to entering this market. The most rapidly growing areas of this sector include such prepared foods as pasta, pizza, precooked meats, sandwiches, precooked French fries, and complete prepared dishes such as "*Sous Vide*".

Most chilled, prepared foods are packaged in combinations of N_2 and CO_2 with little or no O_2 in the package. Such atmospheres confer several benefits including reductions of oxidative rancidity, lack of growth of aerobic spoilage microorganisms, suppression of growth of molds by CO_2, little moisture loss through the film package, and reduced oxidative breakdown of flavor and aroma volatiles.

The packages for these products are generally high-barrier plastic laminations, either bags or lidding material sealed onto rigid trays. Most packaging is performed on thermoform/fill/seal machines using PVC/PVDC/LDPE or PET/PVDC/LDPE for the lidding material.

Sous Vide involves vacuum packaging of foods, usually in multilayer plastic pouches, cooking the vacuum packaged product in a water bath, moist steam or pressure cooker, cooling rapidly in cold water and then storing under refrigeration. These products will have a shelf life of 2–3 weeks under refrigeration. Cooking under vacuum protects the flavors from oxidative breakdown as well as preventing the growth of aerobic spoilage microorganisms (17).

While the food is not sterile, it will have low microbe counts and refrigeration should prevent growth of those few microbes that survived the cooking process. Because the cooking process will not kill spore-forming bacterial pathogens, the inclusion of barriers to pathogen growth, such as low pH, reduced water activity, or the introduction of lactic acid bacteria, in addition to refrigeration, has been suggested. *Sous Vide* was developed in France and has found widespread acceptance there. Acceptance has been slower in the United States but continues to grow, particularly in the food service sector.

MAP of Fresh Fruits and Vegetables

The primary effects of MAP of fruits and vegetables are based on the often observed slowing of plant respiration in low O_2 environments. Reduced respiration leads to reduced depletion of carbohydrate reserves, slower ripening of fruits, and longer shelf life. This suppression of respiration continues until O_2 reaches about 2–4% for most produce. If O_2 gets lower than 2–4% (depending on product and temperature), fermentative metabolism replaces normal aerobic metabolism and off flavors, off odors and undesirable volatiles are produced. Low O_2 can also reduce enzymatic browning of injured plant tissues. As CO_2 increases above the 0.03% found in air, a suppression of respiration results for some commodities. Reduced O_2 and elevated CO_2 together can reduce respiration more than either alone. In addition, elevated CO_2 suppresses plant tissue sensitivity to the effects of the ripening hormone ethylene. For those products that tolerate high concentrations of CO_2, suppression of the growth of many bacteria and fungi results at >10% CO_2 (18).

Although package O_2 permeability increases somewhat as temperature increases, product respiration (demand for O_2) increases much faster. Thus, O_2 will rapidly be depleted if package temperature increases. Because package performance can only be specified within a particular temperature range, it is important that packages be used within that range. If the temperature gets outside of that range, product quality will suffer (5).

The rapid growth of the fresh-cut produce industry, including retail salads, baby peeled carrots and many fruits and vegetables produced for food service, has been possible largely due to the improved quality and shelf life of cut produce in MAP. Most products are packaged in laminated or coextruded copolymers of LDPE, PS, and PP. Other additives and comonomers are added to facilitate sealing, printing and to add antifog properties (5).

Package atmospheres can be created passively by sealing the bag and allowing the product respiration to decrease the O_2 concentration and increase the CO_2 concentration until equilibrium is reached. Alternatively, the atmosphere can be rapidly modified by reducing the package headspace by drawing a vacuum and, sometimes, by injecting a desirable gas mixture into the bag. In any case, the equilibrium package atmosphere is determined by the respiration rate of the product and the gas permeability properties of the package, not by the initial atmosphere in the bag.

BIBLIOGRAPHY

1. M. L. Rooney, "Active Packaging in Polymer Films" in M. L. Rooney, ed., *Active Food Packaging,* Blackie Academic & Professional, 1995 pp. 74–110.
2. J. P. Smith, J. Hosohino, and Y. Abe. "Interactive packaging involving sachet technology" in Ref. 1, pp. 143–173.
3. J. P. Smith, Y. Abe, and J. Hoshino, "Modified Atmosphere Packaging—Present and Future Uses of Gas Absorbents and Generators," in J. M. Farber and K. L. Dodds, eds., *Principles of Modified-Atmosphere and Sous Vide Product Packaging.* Technomic Publishing Co., Inc. Lancaster, 1995, pp. 287–323.
4. P. W. Winston and D. H. Bates. "Saturated Solutions for the Control of Humidity in Biological Research, *Ecology* **41**(1), 232–237 (1960).
5. D. Zagory, "Principles and Practice of Modified and Atmosphere Packaging of Horticultural Commodities," in Ref. 3, pp. 175–206.
6. D. Zagory, "Ethylene-Removing Packaging" in Ref. 1, pp. 38–54.
7. D. Faubion and A. A. Kader, Evaluation of Two New Products with Claimed Ethylene Removal Capacity," *Perishables Handling,* **86,** 27–28 (1996).
8. S. Urushizaki, "Development of Ethylene Absorbable Film and its Application to Vegetable and Fruit Packaging, *Autumn Meeting of Japan Soc. Hort. Sci. Symposium: Postharvest Ethylene and Quality of Horticultural Crops,* University of Kyushu, Oct. 8, 1987.
9. M. L. Rooney, R. V. Holland, and A. J. Shorter, "Removal of headspace oxygen by a singlet oxygen reaction in a polymer film," *J. Sci. Food Agric.,* **32,** 265–272 (1981).
10. C. O. Gill, "MAP and CAP of fresh, red meats, poultry and offals," in Ref. 3, pp. 105–136.
11. A. L. Brody, "Modified atmosphere/vacuum packaging of meat" in A. L. Brody, ed., *Controlled / Modified Atmosphere / Vacuum Packaging of Foods,* Food & Nutrition Press, 1989, pp. 17–37.
12. D. M. Gibson and H. K. Davis, "Fish and shellfish products in Sous Vide and modified atmosphere packs" in Ref. 3, pp. 153–174.
13. G. Robertson, *Food Packaging, Principles and Practice.* Marcel Dekker, Inc., New York, 1993, 676 pp.
14. L. S. Post, D. A. Lee, M. Solberg, D. Furgang, J. Specchio, and C. Graham, "Development of botulinum toxin and sensory deterioration during storage of vacuum and modified atmosphere packaged fish fillets," *J. Food Sci.,* **50,** 990–996 (1985).
15. J. P. Smith and B. K. Simpson, 1995. "Modified atmosphere packaging of bakery and pasta products" in Ref. 3, pp. 207–242.

16. F. Castelvetri, *Proceedings of the Fourth International Conference on Controlled/Modified/Vacuum Packaging,* December 2–6, 1988. Glen Cove, New York, 1989 pp. 43–62.
17. T. Martens, "Current status of *Sous Vide* in Europe," in Ref. 3 pp. 37–68.
18. A. A. Kader, D. Zagory, and E. L. Kerbel, "Modified Atmosphere Packaging of Fruits and Vegetables." *CRC Critical Rev. Food Sci. Nut.,* **28**(1), 1–30 (1988).

DEVON ZAGORY
Devon Zagory and Associates
Davis, California

MODIFIED-ATMOSPHERE PACKAGING MARKETS, EUROPE

Information emanating from this Research Association is given after the exercise of all reasonable care and skill in its compilation, preparation, and issue, but is provided without liability in its application and use.

Introduction and MAP Market Information

In recent years there has been a dramatic growth in the chilled foods market in Europe. This growth has been stimulated by an increasing consumer demand for fresh convenient and additive-free foods. This consumer demand has led to the growth of modified-atmosphere packaging (MAP) as a technique to improve product image, reduce wastage, and extend the shelf-life quality of a wide range of foods. (See also Modified atmosphere packaging).

The United Kingdom leads the world in the expanding chilled-food market, with new products being launched at an increasing rate. Proportionally more retail area is being devoted to chilled foods at the expense of frozen and ambient-stable foods. The main reason for this development is that the concentration of retail power in the United Kingdom is the highest in the world, with Sainsbury, Tesco, Safeway, Asda, Somerfield, Waitrose, and Marks & Spencer accounting for well over 50% of all foods sales and over 80% of chilled-food sales (1). Their operations are increasingly being focused on centralized distribution and a quality image that they feel is enhanced by an expanding chilled-foods sector (2).

Over the past few years, the development and commercialization of MAP for fresh chilled food products has been most rapid in the United Kingdom and France, which hold about 40% and 25% of the European market, respectively. Although reliable market figures for the total European MAP market are notoriously difficult to compile, Marketing Strategies for Industry (MSI) have recently estimated that in the United Kingdom, 2.191 billion MA packs were sold in 1994, with growth predicted to reach 3.249 billion MA packs in 1999 (3). Figure 1 illustrates the food product segmentation for the UK MAP retail market.

MSI have also estimated that in France, 1.39 billion MA packs were sold in 1994, with growth predicted to reach 2.428 billion MA packs in 1999 (4,5). Figure 2 illustrates the food product segmentation for the French MAP retail market.

This article is an updated version of a recent article (6). It includes background information on MAP, the market for MA packed foods, and selected recent developments in MAP along with brief comments on new-product applications, food legislation, food-safety issues, and guidelines for the food manufacture and handling of MA packed foods. In addition, the implications of a recently published UK government report on vacuum packaging and associated processes are discussed.

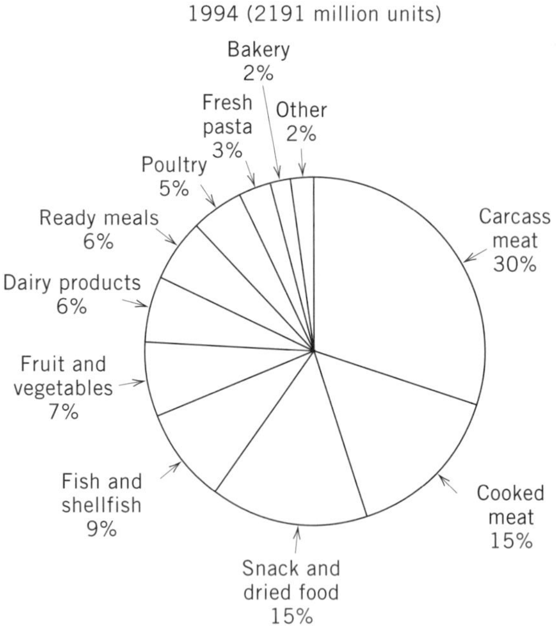

Figure 1. Segmentation of the UK MAP retail market, by food-product sector, 1994.

Background Information

Modified-atmosphere packaging (MAP) is an increasingly popular food-preservation technique whereby the composition

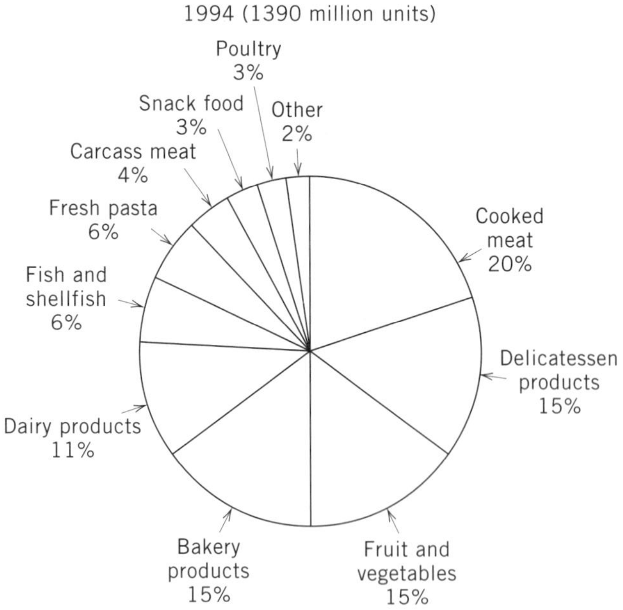

Figure 2. Segmentation of the French MAP retail market, by food-product sector, 1994.

of the atmosphere surrounding the food is different from the normal composition of air. In many respects, vacuum packing (VP) is similar to MAP, but in VP there is the removal of most of the air within a package without deliberate replacement with another gas mixture. Active packaging is another technique similar to MAP since certain additives incorporated into packaging film or within packaging containers are capable of altering the atmospheric composition surrounding the food. Controlled-atmosphere storage (CAS) is yet another food-preservation technique similar to MAP, but unlike MAP, where there is no way of controlling atmospheric constituents at specific concentrations once a package has been hermetically sealed, in CAS the atmospheric constituents are precisely adjusted to specific concentrations throughout the storage and/or distribution of perishable foods. CAS is used primarily for the warehouse storage of whole fruit and vegetables and the road or sea-freight container transport of perishable foods (7).

Gases used for MAP. The gas mixture used in MAP must be chosen to meet the needs of the specific food product (7), but for nearly all food applications this will be some combination of carbon dioxide (CO_2), oxygen (O_2), and nitrogen (N_2). Other gases such as carbon monoxide, ozone, ethylene oxide, nitrous oxide, helium, neon, argon, propylene oxide, ethanol vapor, hydrogen, sulfur dioxide, and chlorine have been used experimentally or on a restricted commercial basis to extend the shelf life of a number of food products. For example, carbon monoxide has been shown to be very effective at maintaining the color of red meats, maintaining the red stripe of salmon and inhibiting plant tissue decay.

Also, argon has been shown to inhibit the tissue fermentation of sliced tomatoes, inhibit the protein breakdown of seafood, and extend the shelf life of prepared fruit (8). The commercial use of most of these gases is severely limited in the United Kingdom because of regulatory constraints, safety concerns, negative effects on sensory quality, or economic factors (7). However, argon (E938), helium (E939), nitrous oxide (E942), and sulfur dioxide (E220) are now permitted for food use in the EU (9).

Products packed under MA. The United Kingdom is undoubtedly leading the world in terms of the range of food products packed under MA (6). These product developments have been prompted by the giant retail chains, which wield considerable power in the British food industry (5). Established products packed under MA include red meats, seafood, pasta, offal, cheese, bakery goods, poultry, cooked and cured meats, pizza, quiche, ready meals, dried foods, herbs, fruit, and vegetables. Many new value-added products packed under MA have recently appeared in UK retail chill cabinets. These include meat and gravy, steak and flavored butterpats, pasta and meat sauce, ready-to-cook recipe kits, prepared vegetables and dressing, medallions of salmon, tropical fruit salads, and fresh herbs. In addition, predicted future commercial MAP applications include casseroles, chilli, sandwiches with meat fillings, meat-based salads, and a variety of cook–chill products (10).

The use of N_2 and/or CO_2 is also widely used for many beverages such as beer, lager, fruit juices, and carbonated soft drinks. N_2 and/or CO_2/N_2 sparging has been promoted for improving the quality of beer, lager, wine, and liquid dairy products (11–13).

Although retail products have tended to attract most of the publicity in relation to MAP developments, bulk MA technology is currently being used for products such as pork primals, poultry, and bacon (6). In addition, this technology is presently being used to gas-flush conventionally packaged retail products, such as overwrapped trays of red meats, into large bag-in-box master packs. Excellent opportunities also exist for using MAP technology for supplying the catering, food-processing, and wholesale industries with bulk packs of many dried commodities, such as tea, coffee, cocoa, herbs, spices and nuts, as well as chilled perishable foods (6).

MAP Developments

Vacuum-skin packaging (VSP)/MAP three-web packaging. A new development that has been introduced to the chilled-food market combines the advantages of both VSP and MAP for fresh-red-meat packaging and other food applications. This is done by producing a standard VSP pack for fresh meat but with a permeable-top web skin. After this process, the product enters a second sealing station where a third web is used to place a lid on the pack. The space between the top web skin film and the lid is gas-flushed with an O_2 rich mixture and, since the top web skin is oxygen permeable, the meat remains bright red in color (14). Another application for VSP/MAP three-web packaging is seafood (14).

High-permeability films for produce. Unlike other foods that are MA packed, fresh fruit and vegetables continue to respire after harvesting and consequently any subsequent packaging must take into account this respiratory activity. The depletion of O_2 and the accumulation of CO_2 are natural consequences of the progress of respiration when fresh fruit or vegetables are stored in a hermetically sealed package. Such modification of the atmospheric composition results in a decrease in the respiration rate, with a consequent extension in the shelf life of fresh product. However, packaging film of the correct permeability must be chosen to realize the full benefits of MAP for fresh produce.

Typically, the key to successful MAP of fresh produce is to maintain an equilibrium MA (EMA) containing 2–10% O_2/CO_2 within the package. For highly respiring produce such as mushrooms, beansprouts, leeks, herbs, peas and broccoli, traditional films such as LDPE, PVC, EVA, and OPP are not sufficiently permeable. Newly developed highly permeable microperforated films for highly respiring produce are most suitable at the present time, but further new developments may change this picture soon (9). Two novel temperature compensating films for produce have been reported recently (15,16), but the likely high cost of these novel films is liable to hamper future commercialization (17).

Novel MAP for fresh-prepared produce. Information gathered by the author via verbal communication during the last 4 years has confirmed that a handful of UK packers of fresh-prepared green vegetables have been experimenting with high O_2 mixtures (eg, 70–100% O_2) with surprisingly beneficial results. To test this radically new approach to MAP of fresh-prepared produce, limited experimental trials were undertaken at CCFRA on high-O_2 MAP of prepared iceberg lettuce and tropical fruits. These experiments confirmed that such an approach could overcome the many disadvantages of current low-O_2 MAP of fresh produce. This so-called oxygen-shock or gas-shock treatment, which can be considered somewhat analogous to hyperbaric controlled-atmosphere storage, has been found to be particularly effective in inhibiting enzymic discoloration, preventing anaerobic fermentation reactions, and inhibiting aerobic and anaerobic microbial growth (9).

An industrially funded research club has recently been set up at CCFRA to study the interesting effects of high-O_2 MAP in more detail. The key objectives are to establish safe commercial applications for high-O_2 MAP and to investigate the underlying microbial and biochemical spoilage mechanisms that are inhibited. Matching funds from the European Union FAIR programme has also been secured for more extensive underpinning research on high-O_2 MAP. In addition, argon MAP and, to a minor extent, nitrous oxide MAP will be investigated along with functional alternatives to sulfite dipping (9).

Valle Spluga two-phase MAP. Pack collapse is a common problem encountered when high-moisture, high-fat foods such as meats, poultry, and seafood are packed under MAs containing a high proportion of CO_2, because of the solubility of CO_2 into such foods, especially at chilled temperatures. In conventional retail MA packs, pack collapse is minimized by limiting the proportion of CO_2 to less than 40% and having a large headspace volume above the food. However, these procedures limit the shelf-life extension potential since lower levels of bacteriostatic CO_2 are maintained within MA packs. Additionally, the larger headspace volume required results in a decreased packing density, with associated higher costs of production, storage, and distribution (6).

An innovative process for solving the problem of pack collapse has been patented by Valle Spluga (Gordona, Italy). This two-phase process is based on the use of solid and gaseous CO_2 in MA packs (6). A weighed tablet of solid CO_2 is dispensed into a thermoformed tray containing the food product, just before gas flushing and sealing. The solid CO_2 sublimes to gaseous CO_2 after sealing and causes the semirigid package to swell. However, after a few hours, an equilibrium is established between the gaseous CO_2 in the headspace and the CO_2 absorbed by the food, and hence the package reverts back to its original shape. Valle Spluga has successfully used this process to market spring chickens in MA packs that maintain high levels of CO_2 with a minimum headspace (6).

Thermoformed pillow packs. Pillow packs are normally produced on horizontal or vertical form/fill/seal machines. Multivac (Swindon, UK), however, have recently developed a fully automated MAP operation that partially thermoforms both top and bottom films to produce a balanced pillow pack. Beni Foods (Milton Keynes, UK) use this type of pack for their wafer-thin sliced-meat products Easy-tear slit openings ensure that the pack is consumer-friendly, and each pack has a hanging feature that allows retail display on either single or double-peg systems (6).

Easy-opening/reclosing systems. Several developments in these types of systems, which prevent food products from drying out after opening, are finding applications in the European MAP market. For example, ESS Food Danepak have used an easy-peel reclosable MA pack developed by LM Smith Brothers (Whitehaven, Cumbria, UK) to market rindless bacon (6).

A French company, Sodebo (St. Georges due Montaigo), is using a Zipseal on its thermoformed MA packs for its "Le Fumay" smoked ham. The pack developed by Multivac is claimed to be the first time that Zipseal grips have been used on a thermoformed semirigid MA pack (6).

Bulk MAP systems. Apart from the bag-in-box applications mentioned previously, other bulk MAP systems are presently being utilized in the European chilled-food market. For example, a French development that was introduced in the late 1980s by Socar (St. Mandé, France) consists of a large thermoformed plastic tray inside a corrugated cardboard box. It has been used primarily for the MAP of catering portions of meat, but by using an appropriate permeable lidding film, it could easily be adapted for fresh-prepared produce (6).

Yet another interesting development used by Sodebo is the "Fresh Container," which is a stackable system that can be used for a wide range of products transported in bulk. Sodebo utilizes a Multivac R7000 to create this new-style pack, which contains six pizzas on the base web with a further six pizzas on the top, separated by a film interleave. The "Fresh Container" features special "pillar" supports designed into the formings that are capable of withstanding weights of up to 150 lb when stacked on pallets (6).

BDF/Delta P system. The first commercially available barrier shrink film for MAP of poultry and meat products has been introduced by the Cryovac Division of W. R. Grace Ltd (Duncan, SC, USA), and is now being successfully used in Europe. The Cryovac BDF 250 film is a 25-μm crosslinked, multilayered, coextruded polyolefin with antifog formulation and low-oxygen and aroma-barrier properties. The film has been developed to be used in conjunction with an Ilapak Delta-P HFFS machine, which is equipped with a gas-flush system and a presser pad for headspace reduction. The food product is hermetically sealed on a suitable tray (foamed EPS, injection-molded polystyrene type) that has sufficient stiffness to avoid deformation after shrinking in a hot-air tunnel. As the packages are shrunk, all seals are drawn down to the underside of the package, and hence the appearance is very similar to a traditional stretch overwrapped PVC package.

The BDF-250/Delta-P system is an alternative to the bulk masterpack system. The disadvantages of the masterpack system, such as the dilution effect on nonflushed trays on the equilibrated headspace in the bag and the loss of MAP once the bag is opened, are circumvented by the BDF-250/Delta-P system. To date, development work has explored the options of using the system for fresh poultry items, red meats, kebabs, pizzas, cheeses, and prepared meals (6).

Barrier EPS trays. Multivac have introduced an in-line thermoformed high-barrier version of the familiar EPS foam tray, which can then be lidded with a conventional-barrier top web film for MAP applications. The gas-barrier qualities of this combination provide a much longer shelf life than do the familiar EPS tray and overwrap currently used widely by instore butchery departments of supermarkets (6).

Guidelines for the Manufacture and Handling of MA Packed Foods

MA-packed chilled foods have been marketed in Europe for many years, and during this period they have maintained an excellent safety record, providing the consumer with high-quality, safe, and fresh products. It is important, however, for all personnel involved in the manufacture and handling of MA-packed chilled foods to be vigilant about possible food-safety hazards, especially in the light of increased regulatory and consumer concern about the perceived rise in food-poisoning incidents. It is imperative that the food safety of such products is not compromised by complacency and poor manufacturing or handling practices. Such a situation would seriously compromise the safety of MAP technology and hence its further application as a preservation technique for chilled foods (7).

With these issues in mind, 6 years ago the Campden & Chorleywood Food Research Association set up a MAP Club with representatives from the retail, food-manufacturing, packaging-film/machinery, and gas-supply industries. The ultimate objective of the MAP Club was the provision of a guidelines document that provides the food and related industries with advice and recommendations to ensure the safety and extended quality shelf life of all MA-packed foods. This guidelines document has now been published and is available from CCFRA (7).

Report on Vacuum Packaging and Associated Processes

The UK Advisory Committee on the Microbiological Safety of Food (ACMSF) have published a report on the potential hazards of vacuum packaging and associated processes such as "sous vide" and MAP (18). Emphasis is placed on the safety aspects of chilled foods, with particular reference to the risks of botulism. Preventative measures have been identified, and possible mechanisms for control have been detailed. Among the many recommendations is one that states that chilled prepared food packed under reduced O_2 levels with an assigned shelf life of more than 10 days should contain one or more controlling factors in addition to chill temperatures to prevent growth and toxin production by *Clostridium botulinum*. Controlling factors include heat treatment, acidity and salt levels, and water activity. The UK government endorses this recommendation and has drawn it to the attention of appropriate trade and professional bodies (18).

Conclusions

MAP is one of the most exciting and innovative areas of the packaging industry. New developments in both packaging-materials/machinery and food-product applications seem to be opening up at an increasing rate. The MAP chilled-food market in Europe is substantial and has enjoyed considerable growth in recent years because of the important benefits it provides to food manufacturers, retailers, and consumers alike. Although MAP has been used primarily for red meats, tremendous opportunities exist for the MAP of other foods. The success of MA-packed products in the UK and France is expected to stimulate future growth in other European nations, especially in the underdeveloped markets of Germany, Italy, and Spain. The establishment of an integrated, trade barrier–free European Union can only help such market growth.

BIBLIOGRAPHY

1. B. P. F. Day, "Extension of Shelf-Life of Chilled Foods," *Eur. Food Drink Rev.* **4,** 47 (1989).
2. R. Pearson, "The Quiet Revolution. Imperial Chemical Company," *Plast. Today* No. 29 (1988).
3. MSI, *Data Report: Modified Atmosphere Packaging: UK,* Marketing Strategies for Industry (UK) Ltd., Chester, UK, 1994.
4. MSI, *Data Report: Modified Atmosphere Packaging: France,* Marketing Strategies for Industry (UK) Ltd., Chester, UK, 1994.
5. B. P. F. Day, and L. G. M. Gorris, "Modified Atmosphere Packaging of Fresh Produce on the Western European Market," *ZFL* **44(½),** 32. (1993).
6. B. P. F. Day, "Recent Developments in MAP," *Eur. Food Drink Rev.* **2,** 87. (1993).
7. B. P. F. Day, *Guidelines for the Manufacture and Handling of Modified Atmosphere Packed Food Products,* CCFRA Technical Manual 34, Chipping Campden, Glos., UK, 1992.
8. K. C. Spencer, "The Use of Argon and Other Noble Gases for the MAP of Foods," paper presented at International Conference on MAP and Related Technologies, CCFRA, Chipping Campden, Glos., UK, Sept. 6–7, 1995.
9. B. P. F. Day "Novel MAP for Fresh Prepared Produce," *Eur. Food Drink Rev.* **1,** 73 (1996).
10. A. L. Brody, "MAP—a Vision of the Future" in *Proceedings of the International Conference on MAP,* CCFRA, Chipping Campden, Glos., UK, 1990.
11. M. J. Butterworth, "Uses for Nitrogen," *Brewer* **69**(828), 407 (1983).
12. N. Falkenberg, "Practical Application of Mixed Gases in Wineries," *Austral. Grapegrower Winemaker* No. 268, 55 (1986).
13. U.S. Pat. 4,935,255 (1990) D. L. Anderson, D. J. Keller, and P. J. Streiff.
14. Anonymous, "New Three Web Pack Innovation," *Multivac Bull.* **1,** 3 (1990).
15. Anonymous, "Temperature Compensating Films for Produce," *Prepared Foods* (Sept. 1992).
16. Internatl. Pat. Appl. PCT/GB92/00537 (1992) A. A. L. Challis, and M. J. Bevis.
17. B. P. F. Day, "Commercial and High Oxygen Modified Atmosphere Packaging (MAP) of Fresh Prepared Produce," paper presented at International Conference on MAP and Related Technologies, CCFRA, Chipping Campden, Glos., UK, Sept. 6–7, 1995.
18. ACMSF, *Report on Vacuum Packaging and Associated Processes,* Advisory Committee on the Microbiological Safety of Food, HMSO, London, 1992.

BRIAN P. F. DAY
Campden & Chorleywood Food Research Association
Chipping Campden, Gloucestershire, UK

MULTILAYER FLEXIBLE PACKAGING

Multilayer flexible packaging offers packaged-goods manufacturers the opportunity to match the functional needs of their products precisely to the materials used to package them. These functions, broadly defined, are (1) package appearance, (2) package barrier, and (3) product containment.

Multilayer flexible packaging is any combination of metal, plastic, or cellulosic substrates (foils, films, and paper), individually ranging in thickness from 0.0001 in. (2.5 μm) to about 0.005 in. (125 μm). The aggregate thickness of the combination is defined by convention to be less than 0.01 in. (250 μm). (Thicker composites are typically considered "sheet" materials.) This material is fashioned into a container (pouch, bag, wrapper, etc) that holds a product.

The manufacture of multilayer flexible packaging is known as *converting*. In producing material for a particular application, specific text and related graphic information (eg, pictorials and use suggestions or bar codes) are typically printed onto one of the substrates. Several processes—*coating, laminating,* and *coextrusion*—then combine this printed layer with others to provide the overall material functionality.

Packaged-goods manufacturers usually buy the composite directly in roll form. They use various types of integrated packaging machinery to form the material into a pouch or bag-like container, fill it with their product, and finally seal it. Alternately, the supplier will fabricate bags or pouches from the composite material. The packaged-goods manufacturer then fills and seals these containers using a variety of manual or automated means.

Package Appearance

Consumers expect various visual cues and information on packages of the products they buy. Multilayer flexible packages must communicate to consumers specific information about the product within: product identification (including ingredients and nutritional data in the case of retail food products), product quantification (eg, count, volume, or net weight), manufacturer's identity and address, and warnings about any hazards presented by the product. In addition, manufacturers will usually provide product-use information (with text and/or pictorials) on the package. If the product is branded, the package must effectively communicate trademarks, logos, brand colors, and so forth. The appearance of a multilayer flexible package may also be used to communicate a variety of marketing claims for the product (eg, new formulations, nutritional positioning, compliance with certain religious requirements), promotional efforts of the manufacturer intended to stimulate sales, or linkages to other products. Multilayer flexible packaging provides abundant options to fill these needs.

Materials options. Multilayer flexible packaging ranges in opacity from totally opaque to brilliantly transparent. In the opaque form, it serves to deliver light-sensitive photographic and X-ray films to users. When transparent, it can merchandise a variety of food products, allowing the consumer to directly view them before purchase. Aluminum foil historically served as the most effective material for blocking light. With mirrorlike spectral reflecting properties, composite materials with foil have come to connote product quality and freshness (1). Metallized plastic films with vacuum-deposited aluminum coatings have been effective in imitating this shelf appearance, but they typically do not provide the light barrier or other keeping qualities of foil.

A variety of plastic films combine to make transparent multilayer flexible materials. Cellophane, the earliest transparent film, was a cellulosic product, however (2). This material was first used as an unsupported (single-layer) packaging material. In fact, much of the product–process development in multilayer flexible packaging over the past 50 years derives from the need to overcome various use limitations of cellophane (eg, brittleness from cold-temperature or low-humidity storage). These same transparent plastic films are often pigmented with white, black, or various hues to complement colors printed on the packaging.

Printing options. Multilayer flexible-packaging materials are printed using primarily rotogravure and flexographic processes. These high-quality, high-speed methods provide a full range of text, machine-readable, and pictorial information. Some materials are printed on their outside surface. In such cases, a protective coating is often applied over the ink to protect it from scuffing, chemical destruction, or other abuse. In contrast, the most common practice for multilayer flexible materials is to print on the inner surface of an outer, transparent, layer of the multilayer flexible material. The surface is then laminated to the other layers. The outside layer itself then serves to protect the ink from abuse. It also usually presents a very glossy appearance to the package as it is displayed (see also Printing).

Package Barrier

Multilayer flexible-packaging materials provide documented levels of barrier to environmental factors: light; moisture vapor; and oxygen (Table 1). These factors interact with many of the food products for which multilayer flexible films serve as protective-packaging materials. The materials are generally considered impervious to microorganisms, although packaging shelf-stable processed foods in them requires confirmation of this barrier function down to the level of a submicrometer void (3).

Oxidation of oils and fats in foods can be retarded by barrier multilayer flexible materials if the oxygen-rich atmosphere within the package is replaced with an inert gas such as nitrogen or evacuated, packaging and product under vacuum. Interaction with water vapor can be more dynamic. Dry-food (moisture content <3%) and hygroscopic granular products need protection from moisture migrating from a humid exterior into the package. Moist products, including aqueous solutions, need moisture-barrier materials that will prevent the loss of product moisture into a desiccating exterior. Recent advances in controlled and modified atmosphere packaging systems utilize the barrier of flexible multilayer packaging materials to nitrogen and carbon dioxide as well. These properties and standard methods for quantifying them are not generally established in the industry.

Extended shelf life. Flexible multilayer packaging is successfully used for retort processing of various food formulations. These "retort pouches" are capable of preserving their contents for 24 months. The primary barrier properties are provided by a layer of aluminum foil or, in a few cases, by a transparent layer of saran (polyvinylidene chloride, "PVDC") film. Industry-standard practices are established for the safe and reliable production of these packaged products (5).

Table 1. Typical Transmission Rates of Selected Flexible Materials[a]

Gauge (mil)	Material	Optics: haze or optical density	Transmission rate	
			Moisture vapor	Oxygen
1	LDPE	35% (haze)	0.8	>100
1	HDPE	6% (haze)	0.35	>100
0.7	Oriented polypropylene	1.6% (haze)	0.49	90
0.5	Oriented polyester	2.5% (haze)	2.8	6
0.5	Oriented nylon	3.0% (haze)	24	4
0.7	Metallized oriented polypropylene	2.3 (OD)	0.02	8
0.5	Metallized oriented polyester	3.0 (OD)	0.05	0.08

[a] Values as supplied by various commercial film suppliers. Moisture-vapor values g/(100 in.^{2} · 24 h) at 100°F and 90% rh; oxygen values mL/(100 in.^{2} · 24 h) at 100°F and 90% rh.

Shelf-life extension is also achieved in practice by aseptic packaging of sterile products. In this process, a sterilizing technique, such as a steam of hydrogen peroxide bath, or exposure to an electron beam, is used to clean the food-contact surfaces of the flexible multilayer material in a sterile environment. Then, while still in a sterile environment, a commercially sterile product is packaged in this material and hermetically sealed.

The barrier properties of flexible multilayer packaging materials are typically not as dependable as the rigid packages (eg, metal cans and glass bottles) traditionally used for extended shelf-life requirements. This is the result of (*1*) imperfections in the thin layer of materials involved, (*2*) measurable transmission of oxygen through fabricated multilayer flexible containers, and (*3*) voids in heat-sealed (ie, welded) areas of the fabricated containers. Ongoing research is directed toward developing reliable, nondestructive testing of 100% of flexible pouches containing long-shelf-life processed foods (5) (see also Shelf life).

Material barrier properties. The barrier presented by a plastic film to migrating gaseous molecules is a complex property of that film (see also Barrier Polymers). It is considered to be a mas-transport phenomenon in which the gas must first dissolve into the polymer on the side of higher gas concentration; next, diffuse through the polymer in response to the concentration gradient; and finally, desorb and evaporate from the polymer on the side of lower gas concentration (6). In practice, the industry uses an equilibrium (isostatic) technique in which a constant concentration gradient for the permeating gas is maintained across the film. In this steady-state condition, transmission of the permeant is expressed as a mass (or standard volume) of the permeant per unit area in a 24-h period. Ambient temperature and film thickness are also reported (7). Table 1 provides a summary of barrier performance for commonly used films.

The barrier to migrating gaseous molecules presented by aluminum foil is not characterized by this plastic film model. Foil, free from defects, is a perfect barrier to migrating molecules. With the inevitable pinholes in foils (representing <0.001% of the surface area; however, the transport mechanism is one of flow through orifices (8). At this point, the polymeric model applies, assuming that the foil is, in fact, laminated to a plastic film.

Barriers to product volatiles. Food and nonfood products often contain volatile components critical to the utility and purpose of that product. Inside a sealed flexible multilayer package, these components can volatize into the headspace of the container. In this mobile form, they can dissolve and diffuse through a polymer film, much like environmental gases. If the solubility and transmission rates are of sufficient magnitude, product shelf life can be significantly decreased. Similarly, volatiles from the packaging materials themselves can desorb into package headspace and contaminate the products inside. While a variety of proprietary studies have been reported, attempts to develop standardized methods and measures for typical volatiles have gained the support of industry participants only very recently.

Product Containment

This function of multilayer flexible packaging is the most complex, and perhaps the most critical. If the material fails to contain its product properly, its intended appearance and barrier performance are quickly compromised. Containment relies on the physical strength of the composite material to deliver a product securely to its intended market. It also depends on the ability to "seal" packages by using heat and/or pressure to weld opposing surfaces of a two-dimensional leaf of material into a three-dimensional container (eg, pouch, bag).

Physical strength. Standard physical measures of material strength are used to characterize films, foils, and papers. Table 2 summarizes typical properties of interest. The specific product to be packaged dictates the level of strength, as measured by any particular physical property, that a composite flexible material must have. For example, dried soup mixes and ground coffee contain small, sharp, granular particles that can readily puncture or rub through a material unless "puncture resistance" is adequate. The actual environmental contexts within which punctures take place are complex and dynamic—so much so that traditional strength of materials

Table 2. Materials Properties of Interest

Property	Reference
Ultimate tensile	ASTM D882
Tensile at elongation	ASTM D882
Elongation at break	ASTM D882
Secant modulus	ASTM D882
Density	ASTM D1505

Table 3. Specialized Materials Properties

Property	Reference
Tear strength	TAPPI T414
Dart drop impact	ASTM D4272
Puncture	ASTM F1306
Burst	TAPPI T403
Flex durability	ASTM F392
Bond strength	ASTM F904

measurements taken alone are not sufficient predictors of success in a given application. The pouch of dried soup mix, for example, is typically sold as a multipack inside a paperboard carton. Through distribution and storage, any pouch face in direct contact with the paperboard will experience abrasive forces greater than ones rubbing against other pouches. To satisfy the need to specify materials without the time and expense of dynamic simulation or actual field testing, a variety of specialized tests with associated testing apparatus have been developed to discriminate among the relevant performance levels of various materials (Table 3). Many of these have been adapted from the paper industry where analogous needs for distinguishing among materials in nonpackaging uses has long existed.

The growing interest in packaging relatively large volumes of liquids (ie, >200 ml) demands a new level of strength in multilayer flexible-packaging materials. Linear low-density polyethylene (a copolymer of ethylene and longer-chain α olefins, typically octene) is the material of choice for such applications. Its tensile strength and elasticity can withstand the surges of hydraulic pressure as liquids in pouches are compressed during shipping and handling. Laminating this film to oriented-polyester film with its high tensile strength provides a reinforced composite material with the ability to contain several liters of liquid.

Heat-seal strength. Obviously, the container made of a multilayer flexible material can be no stronger than the seals that hold it together. Seals are typically made by applying pressure with heated surfaces to opposing faces of the composite materials, melting the thermoplastic material of the contacting surfaces. The two surfaces are then welded together as the thermoplastics cool.

Cohesive coatings are the major exception to heat sealing. Such coatings are similar to pressure-sensitive ones in that they will adhere when they are applied with pressure to another surface. Unlike pressure sensitives, cohesives adhere only to surfaces coated with similar coatings. In practice, cohesive coatings are applied as a perimeter pattern on the inside surface of a packaging film. The cohesive property allows this pattern-coated film to be unwound from roll form on a packaging machine. The package is formed around a product using only pressure to adhere the opposing cohesive surfaces. Heat-sensitive products, such as chocolate bars, can be wrapped in this manner without melting. Packaging line speeds can also be increased as the sealing pressure can be transmitted to the interface essentially instantly (9). Heat sealing, in contrast, requires time to raise the interface temperature to the melting point because the heated surfaces can contact only the opposite side of the surface to be sealed.

The process of making heat seals relies on the basic thermoplastic properties of the polymers on the inside of a multilayer flexible package (the "seal" layer) (10). In theory, this welding process involves a dynamic fluid material moving under pressure. However, heat-sealing practice for the most part ignores the rheological properties of the molten polymer and uses a static set of temperature, pressure, and time variables to control the heat-sealing process on packaging machinery. A matched pair of sealing "jaws" is maintained at constant temperature and brought together with the packaging material between them. The thickness of the material and the minimum distance between the jaws dictate the pressure at the sealing interface. The "dwell" time during which maximum pressure is maintained is dictated by the line speed of the machinery. The process assumes that the outer layers of the multilayer flexible-packaging material are sufficiently heat-resistant themselves to avoid melting. Paper and cellophane can easily withstand sealing temperatures, typically 250–350°F. Heat-set oriented-plastic films of nylon, polyester, or polypropylene are the usual thermoplastic options. Pressure effects at the interface of seal layers can be enhanced by using serrated or ridge surfaces on the seal jaws that mate with complementary patterns in the closed position. (In theory, such shapes can be used to impart shear forces to the molten polymers. This would effect stronger welds more quickly.) Various preheating techniques (eg, infrared lamps or heated guide surfaces) have been developed to speed up the process of melting the interface layers.

While the integrity of the package itself depends on reliable seals, consumer access to the packaged products can require that the seals be broached with relatively little effort. Such "easy-open" seals have been accomplished by a variety of means (Fig. 1). One major technique ("delamination seals") involves using a thin layer of low melt point polymer as the seal layer. This is welded to itself in the sealing process. As the consumer pulls on the seal to open the pouch, the thin layer breaks away from the other layer(s) of the multilayer material. This allows access to the interior of the package past a layer that was previously buried within the composite material. Alternatively, the seal layer can be contaminated with an immiscible material that is blended into the polymer that will actually melt to form the weld. This can be either an inorganic material or a compatible polymer with a higher melt point. When the seal is made, the welded area is decreased by the amount of interface area composed of the contaminate. This is designed to be enough area to allow access

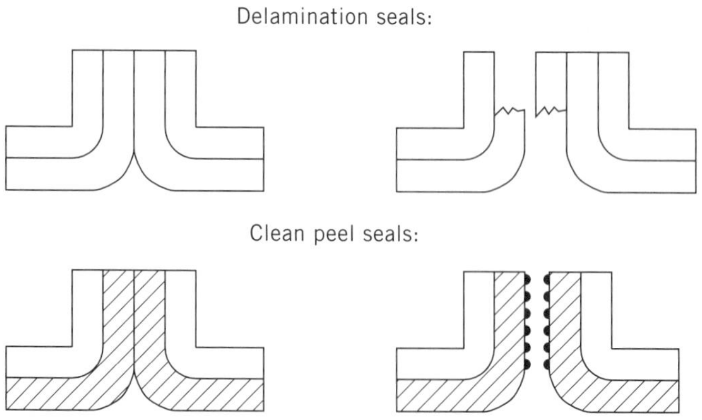

Figure 1. Easy-open seal types.

to the package interior by using a specified force to separate the original seal interface ("clean-peel seals").

Reclosing flexible packages remains an industry challenge. Dry foods (eg, cereals, snacks, cookies, and crackers) can quickly lose their crispness in an opened package. And moist foods, such as shredded cheese, can dry out, particularly in refrigerated air. Pressure-sensitive tapes and interlocking plastic ridges (plastic "zippers") are used in several applications at present. Widespread adoption of recloseable features for flexible packages awaits a reliable, low-cost technique.

Manufacturing Multilayer Flexible Packaging

The converting operations that manufacture multilayer flexible packaging are printing, laminating, and finishing processes. Such operations are designed and equipped to convert a variety of raw materials into many kinds of customized products. Even within a specific category of packaged product, the actual multilayer flexible packaging specified by one customer may differ dramatically from that purchased by another. Differences in product formulations, packaging machinery, and distribution systems combine to require different levels of appearance, barrier, and containment. These functional differences, in turn, dictate differences in specified raw materials.

Printing. Flexible-packing printing involves both flexographic (images on rubber plates that stamp ink onto a substrate) and rotogravure (images etched into metal cylinders that release ink to a substrate from small engraved cells) printing processes. The choice between processes is now primarily one of customer preference. Historically, converters specializing in foil and opaque packaging were rotogravure printers. Until the 1980s, this process was the only one capable of reproducing quality continuous tone (pictorial) graphics. Flexographic printers were more common among cellophane and plastic-film converters, where transparent packaging required only basic text printing. Flexographic continuous tone printing has become much more acceptable for high-quality graphics over the past 10 years. With consolidation of converters within the industry and internal diversification by the remaining suppliers, this distinction of capabilities along opaque and transparent substrate lines has virtually disappeared.

Both printing processes use liquid inks comprised of colored pigments, a polymeric "vehicle" (binder), and a volatile solvent. The solvent (an organic liquid or water) is evaporated from the printed substrates as the vehicle solidifies and adheres to the printed substrate. Each of these three ink components must be selected to meet various functional and environmental requirements (11).

When the printed surface is to be the outside of the outer layer of the multilayer flexible package, a protective layer, or overprint varnish, is often applied over the ink to prevent scuffing and to provide a high-gloss effect. This coating must also be resistant to the hot-seal jaws that will contact it as heat seals are made. In contrast, when the printed surface is on the inside of the outer layer, the ink must be formulated to adhere to the material that will be used to bond it to the next layer. It must also have an internal strength sufficient to support the overall strength requirements of the entire multilayer material. These properties must be maintained at the temperatures that the ink will experience during heat sealing.

Adhesive laminating. The majority of adhesive laminations in multilayer flexible packaging are manufactured using the dry-bond process. In this technique, a liquid adhesive is applied to one substrate. Any one of a number of web-coating methods (eg, direct, offset, or reverse gravure, wirewound rods, air knife) is used to meter the adhesive onto the substrate. The adhesive coating is then dried with hot air. This dried surface can be adhered to a second substrate using heat and pressure at a nip point.

The adhesive formulations themselves represent a reactive chemistry (typically urethanes or acrylics) that is chosen to withstand the processing and distribution environment of the filled product. This can be high temperatures of retort processing (280°F), migrating volatile organics that can redissolve adhesive solids, or other packaging components that can interact to discolor components of the lamination. Because the adhesives are reactive chemicals that are expected to polymerize and/or crosslink when coated, government food-additive regulations control the presence of any unreacted residuals in the composite materials (12). (See also Adhesives).

Extrusion lamination. Extrusion lamination involves using a thin (as low as 12-μm) layer of plastic (typically low-density polyethylene) to bond together two layers of film, paper, and/or foil. The method has the advantage over adhesive laminations of adding substantial thickness to a multilayer lamination as well as contributing to the overall strength of the material. It grew out of the business of coating paperboard with low-density polyethylene (an alternative to wax coatings for such packages as milk cartons) (13). At present, the process accounts for the majority of printed multilayer flexible packaging.

The process enjoys significant cost advantages over the alternative of producing separate thin layers of polyethylene and using adhesives to bond the separate layers together. It has no environmental emissions and permits the processing of thinner layers of plastic than can be handled otherwise. The intrinsic adhesive affinity of low-density polyethylene (LDPE) toward other substrates is limited. However, a variety of LDPE copolymers, adhesion primers, and processing aids have been developed to broaden the applicability of the method (14).

Coextrusion. The cost advantages of extrusion laminations are exploited in the extreme in the coextrusion process. This method entirely eliminates the use of any separately manufactured substrates. It simultaneously extrudes several layers of molten plastic into a single multilayer material. Each plastic maintains its identity as a separate layer in the film and can contribute various functions accordingly. For example, a coextruded film of nylon, ethylene vinyl alcohol, and ionomer provides, respectively, heat resistance, oxygen barrier, and low-temperature sealing for the material used to vacuum-package-processed meats. Of course, a pure coextruded film precludes the use of reverse-printed substrates, and paper and foil entirely. However, hybrid processes in which a multilayer extruded curtain is used to extrusion-laminate traditional substrates have been developed.

Table 4. Flexible-Packaging Source Reduction

Packaging system		Reduction (%)	
Baseline	Reduced	Weight	Volume
Coffee can	Brick pack	70	55
Detergent bottle	Concentrate pouch	85	84
Diaper carton	Film bag	85	86
Soup can	Soup pouch	93	97
Glass bottle	Flexible pouch	96	82
Glass bottle	Drink box	90	70

Source: Flexible Packaging Association.

Coextrusion has proved to be a powerful option for achieving a mix of functional features in ultra-thin-base films. In effect, base-film suppliers can mix and match materials in a manner analogous to a converter's lamination processes. For example, the current standard seal layer for snack food packaging is a coextruded oriented-polypropylene film comprised of a ~1-μm bonding layer for metal adhesion, a core layer of ~15-μm of homopolymer polypropylene, and a seal layer of ~2 μm of propylene–ethylene copolymer. This film is vacuum-metallized and extrusion-laminated to a reverse-printed oriented-polypropylene film to provide a strong, easy-open package with excellent oxygen, light, and moisture barrier.

Finishing. The converter will print and laminate multilayer flexible-packaging materials in widths of 40–60 in. (1–1.7 m). These are usually wider than the rolls that a packaging machine can form, fill, and seal. The converter then slits the wide rolls into several narrower lanes suitable for the customers' machines. This product is then sold in roll form to the customer in very compact form. A typical 0.003-in. (75-μm)-thick multilayer material wound on a core 3 in. (75 mm) in diameter can be delivered in 18-in. (450-mm)-diameter rolls containing about 7000 packages 12 in. (300 mm) long.

The alternative to multilayer flexible packaging in rolls is the premade pouch form. For this product, the front and back of the pouch are printed side-by-side. This single package is folded longitudinally and heat-sealed inside-to-inside along the fold and two of the other three sides. These flat pouches are sent to the customer, who fills them with product and seals shut the fourth side.

Current Trends in Multilayer Flexible Packaging

Significant development resources are being devoted at present to commercializing flexible films and composites that improve on the functionality in essentially all of the areas discussed here (15). In addition, two recent issues, source reduction and standup flexible packages, have emerged to further drive the continued evolution of flexible packaging.

Source reduction. By nature, multilayer flexible packaging contains many materials intrinsically bonded together. This fact mitigates against recycling the component materials, as has been possible for many monomaterial packaging systems (eg, HDPE merchandise bags and milk bottles, polyester beverage containers, aluminum cans, and glass bottles; note that some progress has been made in separating at least some components of multilayer materials for economic recovery) (16). Rather than minimizing packaging residuals after use by recycling, the flexible-packaging industry has been urging consumer and regulatory consideration of the advantages of using less packaging material for a product as it is put into commerce (17). Flexible-material packaging systems offer significant reduction of both waste volumes and weights when compared to traditional systems. A few of these are summarized in Table 4.

Standup flexible pouches. The improved materials and composite functionality of multilayer flexible packaging has expanded the universe of possible applications for these systems to include products traditionally packaged in rigid packages such as metal cans and glass or plastic bottles. The standup pouch concept has been used for hot-fill juice containers for over two decades (18). European and Japanese consumers are exposed to many other commercial examples at present. Only a few domestic market trials have been conducted, and these have been largely unsuccessful for a variety of reasons. In spite of these experiences, enthusiasm for novel stand up flexible pouches remains high (19). Hopes for a major new market for multilayer flexible packaging depend on the concurrent development of both materials and machinery.

BIBLIOGRAPHY

1. Aluminum Association, *Aluminum Foil,* 2nd ed., Washington, DC, 1981.
2. W. A. Jenkins and J. P. Harrington, *Packaging Foods with Plastics,* Technomic Publishing, Lancaster, PA, 1991, p. 34.
3. B. Blakistone, "New Developments in Plastic Packaging Seal Integrity Testing: One Key to the Future of High Speed Plastic Packaging" in *Proceedings of the 1994 IOPP Packaging Technology Conference,* Institute of Packaging Professionals, Herndon, VA, 1994.
4. ASTM, F1168, *Standard Guide for Use in the Establishment of Thermal Processes for Foods Packaged in Flexible Containers,* American Society for Testing and Materials, Philadelphia, 1988.
5. Blakistone (Ref. 3).
6. Michigan State University, School of Packaging, "Product Storage Stability Based on the Permeability of the Package Storage System," Course Notes, July 10–13, 1984, East Lansing, MI, 1984.
7. ASTM, D3985, *Standard Test Method for Oxygen Gas Transmission Rate through Plastic Film and Sheeting Using a Coulometric Sensor* and F1249, *Standard Test Method for Water Vapor Transmission Rate through Plastic Film and Using a Modulated Infrared Sensor,* American Society for Testing and Materials, Philadelphia, 1988.
8. British Aluminium Foil Rollers Association, *Barrier Properties of Aluminum Foil: A Handbook,* London, 1973.
9. R. J. Ginsberg, "Cohesive Coatings Offer Food Field Cost Advantages, Higher Speeds," *Packag. Devel. Syst.* 23–35 (July/Aug. 1979).
10. T. J. Dunn, "Use of Differential Scanning Calorimetry in Developing and Apply Films for Flexible Packaging" in M. L. Troedel, ed., *Current Technologies in Flexible Packaging,* ASTM STP912, American Society for Testing and Materials, Philadelphia, 1984.
11. T. J. Dunn, "Flexible Packaging and Environmental Control in the 80's and 90's in *Proceedings of Envirocon '89,* The Packaging Group Inc., Princeton, NJ, 1989.

12. U.S. Food and Drug Administration, "Indirect Food Additives: Adhesive Coatings and Components," *Code of Federal Regulations,* Title 21, Part 175.
13. T. Bezigian, 1992, "Extrusion Coating and Laminating—the Growth of an Industry," *Converting Mag.* (Part I, p. 48ff., Jan. 1992; Part II, p. 30ff., Feb. 1992).
14. R. Isbister, "Chemical Priming for Extrusion Coating," *TAPPI J.* **71** (5) 101–104 (1988).
15. T. J. Dunn, "Flexible Packaging 1994: Issues, Trends, and Life Cycle Analysis" in *Proceedings of the TAPPI Internatl. Packaging Symposium,* Chicago, July 1994.
16. E. Klein, "The Greening of Coated Paperboard," *Proceedings of the TAPPI Polymers, Laminations, & Coatings Conference,* Nashville, TN, Aug. 1994.
17. T. J. Dunn, "Source Reduction Needs Defining to Have Maximum Impact," *Packag. Technol. Eng.* **3**(4), 44–47 (1994).
18. A. Brodie, "The Source Reduction Pros & Cons of Liquid Stand-up Pouches" in *Proceedings Green Packaging '94, Conference,* Packaging Strategies, Washington, DC, June 1994.
19. M. Kapel, 1994, "Flexible Stand-up Pouch Technology for Liquids and Solids: 'Are we on the verge of marketing and technical breakthroughs?'" in *Proceedings of the 1994 IOPP Packaging Technology Conference,* Institute of Packaging Professionals, Herndon, VA, 1994.

THOMAS J. DUNN
Printpack Inc.
Atlanta, Georgia

MULTILAYER PLASTIC BOTTLES. See BLOW MOLDING; BARRIER POLYMERS; COEXTRUSION.

N

NETTING, PLASTIC

Today there are two popular manufacturing methods to produce plastic netting: knitting and extrusion. Both methods produce a tubular or sheeted netting product in a wide variety of mesh sizes, diameters, widths, and colors. Plastic netting is purchased as individual pieces, in continuous rope form, or sheared onto cardboard mandrels.

The primary materials used to produce plastic netting are high-density polyethylene (HDPE) and polypropylene. Ultraviolet inhibitors can be added to the resin to give the netting extra durability in direct sunlight applications. HDPE has a high elongation value and must undergo the process of orientation to achieve its maximum tensile strength. The process of orientation is when the material is heated and then stretched to the point just before breakage. This process also keeps the netting from drooping when exposed to heat applications. The melt index of HDPE is 208°F. Polypropylene has a high melt index (325°F), and is used when temperature requirements exceed 208°F. Polypropylene tensile strength is almost twice that of HDPE with the equivalent weight ratio. With the heat and strength attributes, polypropylene netting is widely used in the meat–poultry industry.

The process of extruding plastic netting was invented in the mid-1950s. Extruded netting is produced through counterrotating dies. As the inner and outer die rotate, small strands of molten plastic overlap each other, bonding themselves together where they overlap. The extruded plastic is then cooled by water; this material is called a *cast*. The next process is orientation. The cast is heated and stretched to the point just before breakage. After the cast is oriented and has become the final product, the net is transverse-wound onto large spools (1).

The manufacturing process of knitted plastic netting offers the most versatile and variety of netting products. Knitting machines offer simple and sophisticated stitch patterns. The first knitting machine was invented by Reverend William Lee, an Englishman, in 1589. The first power knitting machine was introduced in 1832, in Cohoe, New York by Egberts and American (2). The knitting process is complex and varies by different types of machinery. To manufacture plastic netting via knitting, the plastic resin must be manufactured into yarn or tape. The yarn is then put onto a creel, with several individual packages consisting of a single thread, or beam, and several threads wound onto a large spool. The basic concept of knitting is to take the yarn or plastic tape from the creel or beam, and thread the set of needle guides in the knitter. A bed of latch hooks move up and down (vertically) as the needle guides move right and left (horizontally). The needle guides are directed by a pattern chain that controls the different mesh patterns; hexagon, diamond, or countless other configurations. Knitting allows the manufacture to choose the style and the material to be used, allowing a multitude of products.

Plastic netting is used in an array of applications. As a flexible material it conforms to irregular products (toys, houseware products, etc). Netting provides an excellent packaging forum as a decorative and protective overwrap. It provides the necessary air circulation for products such as plants (tropical and flowering) and various types of fruits and vegetables. Because of plastic netting's flexibility it can be looped, making a convenient carrying handle for the consumer (fresh or frozen whole turkey and ham products). In the meat–poultry industry netting is used as a transportation vehicle. The meat and poultry items are encased in the netting in the production area and then hung on smoke racks ready to go into the smoke house. The netting allows the meat and poultry products to receive smoke for flavoring and coloring during the cooking process. The netting leaves an attractive pattern on the products, giving them the "old world" look. The netting can be left on the product as a marketing tool for product identity.

Netting has a wide variety of uses in sheet form. It can be used to wrap pallets as an alternative to stretch wrap. The netting can be put up the same way as stretch wrap, to utilize modern equipment and hand applicators that are available. This is an important tool where ventilation or heat dissipation is needed (produce, bagged mulch, flowers, flour etc). Because of its exceptional strength, netting can be used to wrap heavy loads such as bricks. In sheet form, netting has an abundance of uses. It makes a great fence for construction or snow applications. It can be used as a safety net in construction to catch debris and increase the safety of workers. It is widely used in the horticulture industry as shade cloth protecting plants from the harsh sun. It is also used as bird net, protecting fruit trees, berry bushes, and fish ponds from predator birds.

By using the knitted process to manufacture plastic netting, different-color strings or threads can be combined to form a multicolored netted product. An example is Candy-Cane (trademark) red and white netting used to package Christmas trees, (3). This netted package allows the tree to be easily transported from the retail lot to the consumer's tree stand at home. Netting is also used by Christmas-tree growers to compress the tree. This gives the grower the ability to ship more trees, saving in transportation cost. The netting also protects the tree from damage, and makes the tree much easier to handle. This packaging process also applies to other types of trees (white oak, maple, bamboo, etc).

Knitted netting is soft, durable, and strong; making it an ideal packaging medium. The produce industry, which uses extruded netting primarily for consumer-sized packages, is seeing a trend moving toward knitted netting. Both styles of netting offer adequate ventilation needed for fruits and vegetables to reduce spoilage. The soft texture of knitted netting does not damage the delicate produce skin and makes an eye-pleasing package. A draw string can be automatically inserted during the knitting process making a convenient closing mechanism as well as a carrying handle. Potato sacks are a good example of this process.

Plastic netting is a valuable packaging tool. It is resilient, strong, and flexible. Netting can be frozen then heated, or vice versa, keeping its strength and flexibility. Plastic netting is very cost-effective compared to other forms of packaging and is recyclable. The applications and areas of usage keep growing as creative minds keep developing new and innovative ways to use plastic netting.

Some manufactures are

Tipper Tie-Net, 390 Wegner Drive, West Chicago, IL 60185

Polly Net Inc., P.O. Box 27, Three Rivers, MA 01080

C&K Manufacturing, 28025 Ranney Parkway, Westlake, OH 44140

BIBLIOGRAPHY

1. J. Scoba, Polly Net Inc., P.O. Box 27, Three Rivers, MA 01080.
2. *The World Book Encyclopedia,* Field Enterprises Educational Corporation, 1970, pp. 279–280.
3. "Candy Cane," manufactured by Tipper Tie-Net, 390 Wegner Drive, West Chicago, IL 60185.

JIM OHLINGER
Tipper Tie-Net
West Chicago, Illinois

NIPPON'S PACKAGING (JAPAN'S PACKAGING)

Packaging Culture and Commerce

The Japanese have an unusually close relationship to packaging which represents an historic part of their culture, as well as being an important component in their economic success.

Japan's packaging manufacturing industry is relatively fragmented, with about 100 companies accounting for 85% of the market. Of these Toyo Seikan has a 10.7% share of the total market, with 5 others accounting for a further 40%. Unusually, there is a large degree of vertical integration. Ajinomoto (the name means monosodium glutamate) is a substantial food producer yet is also a leading edge packaging company. Although a dynamic economic force, Japan is still essentially a traditional country where close family values, individual status and personal dedication are paramount. These cultural aspects are reflected in their packaging both as the consumer sees it and the manufacturing and technology drives it.

Product and packaging manufacturing companies expect and receive commitment and respect from all their employees; in practical packaging development terms this means that 20 assorted managers representing a complete spectrum of different disciplines can adapt their skills and resources to a single goal in complete synergy. In westernized societies it is often difficult to get commonality of goal let alone implement it. This is but one example that differentiates Japan as a powerful force in packaging technology and its market exploitation. Additionally, for most of the last 50 years Japan has typically invested two to three times that of European companies in new product and packaging development in its cultural, as well as commercial search for packaging innovation and excellence.

However, it is the gift concept that is most striking. Nearly all packaging is regarded by the consumer as a gift, where there is subliminal pleasure in recognizing, exploring, opening, etc, with the pack. This can be more important than the product itself. A trusted proverb "To travel (the unwrapping) is as important as to arrive (the product)." This is as true for the most elaborate gift wrapped box right down to a humble can of beans or roll of sweets. It is the demonstrative statement of care that is critical to successful packaging and is also so recognizable in the Japanese products that we use.

This special relationship may account for the US$326 per capita spend on packaging. A total of US$40.3 billion (1992). Despite the quantity and investment, exports are not high (see Figure 1).

Key Trends and Technologies

Japan is a world trend-setter and the source of many successful everyday packaging technologies: shrinksleeving, microwaveable food packaging, PET bottles, high definition printing processes, high barrier flexibles and closures in pouches are examples.

The environmental agenda arrived late in Japan (early to mid-1990s in practice) but response has been swift, with initiatives following the "Reduce, Recycle, Reuse" principles. This has caused a renaissance in cartonboard with one major manufacturer, Toppan Printing, publicly committing to board as a significant part of its future. Board technologies judged to be important include the development of the cartonboard can, an aseptic fill system in use for fruit juices, wine and sake. This pack is designed with a lift-up aperture for a one-shot drink. Another approach is a clever rip-pull to a card drinking cup with a sealed lid. This system is for short shelf life in the chiller and is the intellectual property of food giant Meiji (see Figure 2).

Flexible packaging has had a larger and more widely applied use in Japan than anywhere else in the world. Environmental pressures have accelerated the diversity of application. Formats such as steel cans have lost out to pouches in chopped fruit, fish and meats.

There are several critical packaging technologies evolving in flexibles. The addition of injection molded closures has moved the pouch from a refill to a primary pack. Of special significance is the ability of equipment to fix the closure in various locations on a pack, as demonstrated in Figure 3. The majority of pouches are premade, but reel fed form–fill–seal is penetrating the market. Part of this achievement is the ability to achieve heat seals that can resist the stolic pressure of liquid products and the increased pressures in post filling sterilization processes, like retorting. A leading exponent is Komatsu Manufacturing. Materials in the pouch market have conventionally involved multilayers including PET, which provides stiffness and gloss as well as barrier but there is a trend to single substrate or simple multi-layers without PET. Figure 3 includes form–fill–seal with a closure on the side of an orientated polypropylene pouch, plus other more conventional approaches.

Pouch making machines are now moving into shaping of a pouch to create marketplace and branding differentiations. There are two distinctively different technical approaches. First, the 2D shaped pouch where a perimeter heatseal is created that has an outline of the desired shape (see Figure 4). Heatseal width is often greater than conventional and it is necessary to use a stiff flexible substrate, often PET. The technology lends itself to providing important easy open devices. Figure 5 shows examples now in the marketplace.

The alternative is the emergence of 3D shaped pouches.

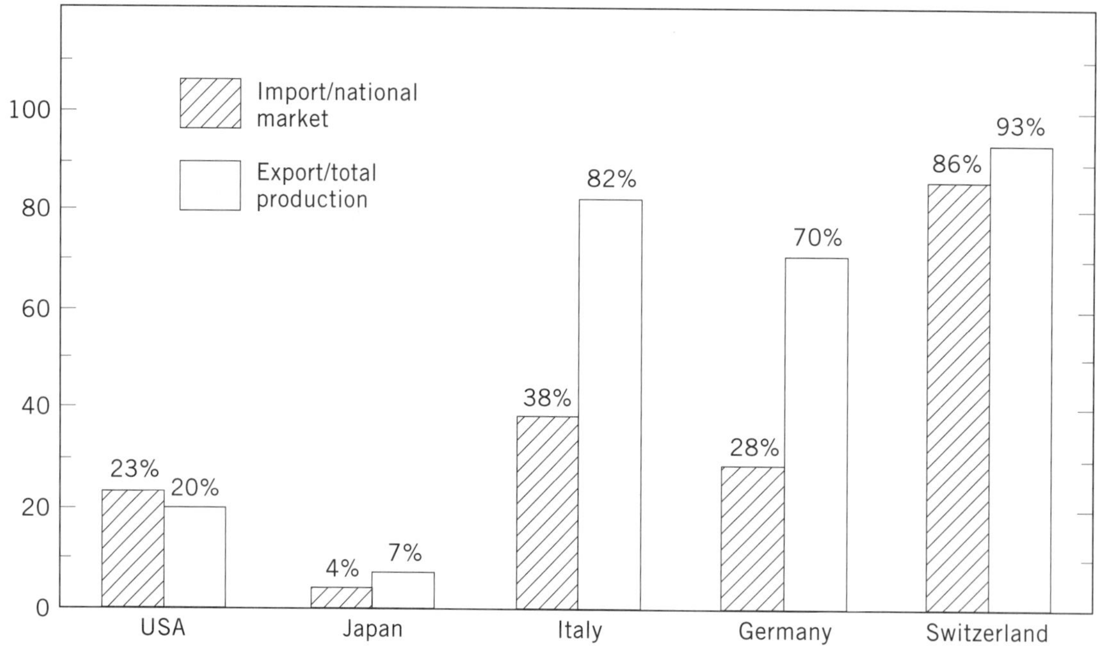

Figure 1. Export trade in selected countries of packaging machinery, 1993.

These developments rest firmly with the pouch machine manufacturers and challenge head on both glass and blow molded plastics bottles. Providing shape involves two factors. The original cut perimeter or joining adjacent film faces with a shaped heat seal form; this is coupled with multilayer films that have strategically placed reinforcing beads running through them to provide stiffness to front and back panels, to sustain shape especially in the vertical plane. Sumitomo Bakelite have been championing this technology under the brand name Poucher and have household and industrial cleaners in this format as well as mineral water. Pouch sizes range from 250 ml up to 3 liters. Figure 6 demonstrates Poucher examples and other approaches.

A third important force in flexibles is the application of silicon dioxide (in simple terms glass) as a barrier coating to flexible packaging. It is applied to the film web by vaporizing the silicon dioxide through extreme heat and under vacuum, allowing the vapor to deposit on the surface of the film. The coating is very thin, at a few microns. The technology has been both difficult to achieve and make at a commercially attractive price. By 1996 about 60% of Toppan Printing's vacuum depositor capacity is silica (the other 40% being the traditional aluminum depositing). The benefits of silicon dioxide coating is a very high barrier almost equivalent to aluminum foil (in some instances better); it is less harmful on disposal, especially where energy recovery through incineration is undertaken as it is Japan; and creates the marketplace opportunity to see the product inside the pack where, in the past this was not possible with foil barrier or maybe even before that, a steel or aluminum can.

Surface decoration technologies are a Japanese strength. Communication enhancement is possible by fibers embedded into the surface of films (polypropylene). Other textural effects can be achieved with embossing and inks containing ground leather to give a matt cum sensual finish. Temperature sensitive inks are well established providing fun, but

Figure 2. A range of board cups for one shot drinks.

Figure 3. The ability to apply closures in a variety of positions.

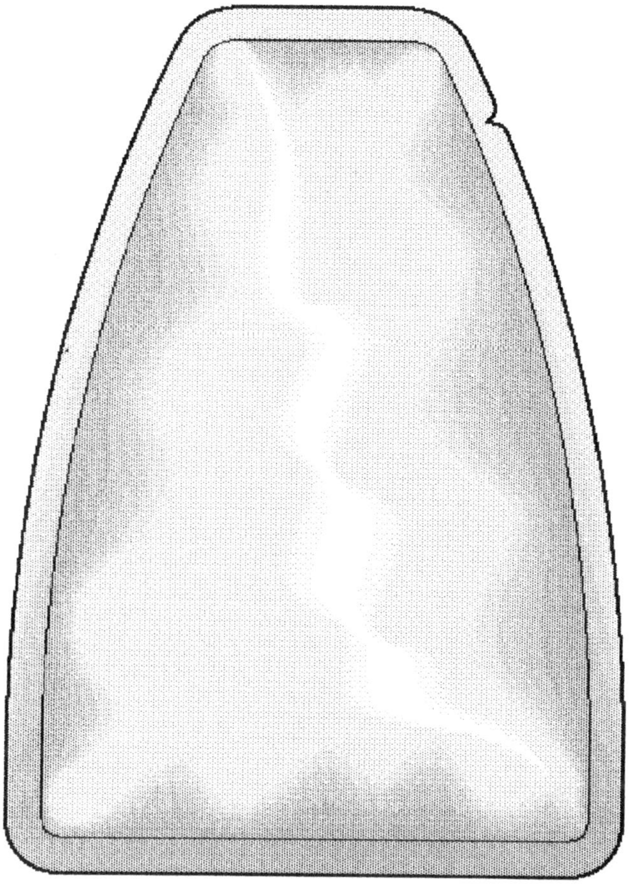

Figure 4. The ability to apply perimeter shaping.

also the important message of being at the right temperature to consume the product. To come will be light emitting surfaces and more advanced 3D holographic effects.

Glass is enjoying revival, with growth in surface treatments especially cold end surface sprays and post curing to create an acid etched effect. Steel cans have received a boost from Toyo Seikan with the development of the TULC can, a lightweight, two piece deep drawn can with a protective coat of PET rather than lacquer.

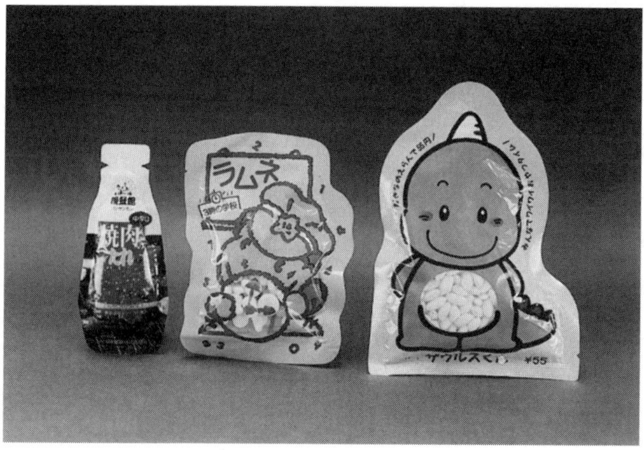

Figure 5. The incorporation of shape and easy open in shaped pouches.

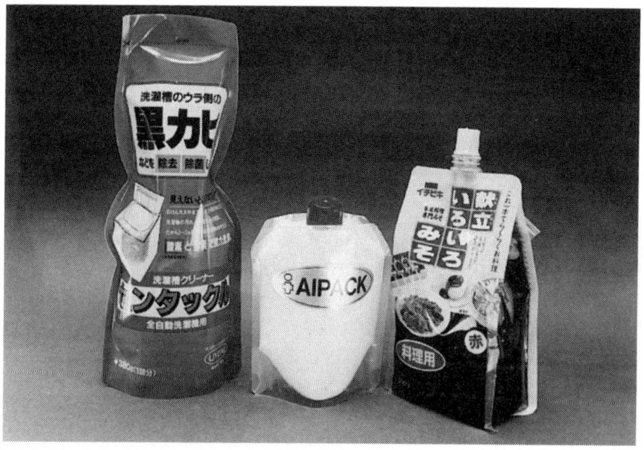

Figure 6. Shaped pouches in a variety of product sectors.

Technologies are transferred from format to format more freely in Japan than elsewhere. For example, matt and textured inks are found on shrinksleeves, cans, can labels, films, plastics containers and cartons.

Improving food shelf life and real freshness is being achieved through oxygen absorbers, also known as scrubbers. Traditional materials were ferrous based and came in small, select forms, boldly marked "Do Not Eat." They remain in wide use and are very effective. There is concern as to the application of metal with food and an important emerging technology is plastics based oxygen absorbers which have the benefit of being an integral part of the packaging structure. Mitsubishi and Idemitsu have both produced plastic food trays and Mitsubishi have experimental oxygen absorbing films.

The next strategic development awaited is a PET or PEN bottle with an integral handle. If it is a possibility, Japan will be at the forefront. This achievement will have a dramatic effect on consumer care and open up new design and shaping opportunities. It will also present a powerful threat to glass bottles and accelerate a switch in usage.

Japan's packaging influences North America and Europe, as well as its neighbors of the Pacific Rim. Its continuing success though is in placing the consumer first and applying single-mindedly innovative packaging technology vertically through industry, to enhance consumer care in all respects in a demonstrable way to the consumer. Packaging in Japan offers both technology and cultural benchmarks for the world market.

Material drawn from Pack-Track-Japan research programme (1988–1996) operated by CPS International.

Andrew Streeter
CPS International
Essex, United Kingdom

NITRILE POLYMERS

Nitrile polymers are generally those which contain the cyano ($C{\equiv}N$) functional group, also called the nitrile group. The commercial development of these materials was due in large part to the 1957 discovery by BP Chemicals of a low-cost one-step process for acrylonitrile (AN) production (1). The pure nitrile polymer, polyacrylonitrile (PAN), is 49% nitrile. It is an amorphous, transparent polymer with a relatively low glass-transition temperature (T_g = 87°C) (see Polymer properties) that provides an outstanding barrier to gas permeation and exceptional resistance to a wide range of chemical reagents. Unfortunately, its combination of properties is not of commercial value in the packaging industry. Its primary deficiency is that it is not melt processable. It degrades at 428°F (220°C), which is below that required for melt processing. To overcome this handicap, nitrile polymers are produced using acrylonitrile ($CH_2{=}CHCN$) as the monomer with other comonomers that impart melt processability. Through this copolymerization process, the desirable properties can be retained and the undesirable properties can be suppressed.

Copolymers

Styrene–acrylonitrile (SAN). Typical SAN polymers are made using a 3:1 ratio, by weight, of styrene to acrylonitrile. The copolymer has a combination of properties which reflect the processability of the styrene component and the chemical resistance of the AN. Gas-barrier properties are low because of the relatively low AN concentration. SAN is of relatively minor significance in packaging, used in applications where PS would suffice with an added measure of chemical resistance.

Acrylonitrile–butadiene–styrene (ABS). ABS is a graft copolymer of SAN onto a polybutadiene backbone. The SAN forms a matrix phase; the polybutadiene a discrete (dispersed) phase. A tough impact-resistance thermoplastic is produced by using the grafting mechanism to compatibilize the two phases. Although the polybutadiene is an excellent impact modifier with low T_g (≅ 85°C), its refractive index is different from the SAN matrix polymer. Therefore, in contrast to PS (see Polystyrene), PAN, and SAN are not transparent. ABS is a major commercial thermoplastic, but it is rarely used in packaging because of its opacity, lack of gas-barrier properties, and economics relative to other commodity resins. However, thermoformed ABS (see Thermoforming) has been used, eg, to produce margarine tubs. This is an application in which PS would suffice if not for its limited resistance to stress cracking. A wide range of SAN ratios is used to achieve properties of value in nonpackaging applications.

Acrylic multipolymers. Some acrylic multipolymers are produced using AN as a comonomer with methyl methacrylate (MMA); for example, XT Polymer (Cyro Industries). Refractive index matched rubber modifiers are incorporated to combine toughness with transparency. These materials are used in health-care packaging (2) (see Medical packaging) and in some food packaging applications. Because the AN concentration is low, these materials do not have exceptional gas-barrier properties [18.7 cm^3 mil/100 in.^{2} d atm at 73°F (23°C)].

High-nitrile resins (HNR). As noted above, SAN is a styrene–acrylonitrile copolymer. Its gas-barrier properties are limited, however, by the low nitrile content. High barrier melt processable copolymers can be produced by raising the nitrile content above 25%. Over this threshold, copolymer properties begin to resemble those of PAN, particularly with respect to gas barrier and chemical resistance.

Acrylonitrile–styrene copolymers (ANS). High-nitrile copolymers can be produced by combining acrylonitrile and styrene in a 70:30 ratio. This was the approach taken by Monsanto Company (Lopac) and Borg-Warner (Cycopac) during the 1960s and early 1970s in their development of resins for carbonated-beverage packaging. As serious commercial development began, toxicological problems with AN surfaced (3). The FDA banned the use of HNR in beverage packaging because of concern for potential AN-extraction from the bottle into the beverage. Since beverages are a major component of the diet, their treatment by FDA was most severe. The FDA continued to permit the use of HNR for direct and continuous nonbeverage-food-contact applications (4) with filling and storage temperatures less than 150°F (65.6°C); but because the principal commercial significance of ANS polymers was in carbonated beverage packaging, all commercial production of ANS polymers was eventually discontinued.

In 1984, the FDA amended its position on HNR-beverage applications (5). Monsanto Company had petitioned on behalf of their ANS high nitrile resin for approval. As described in their process patent (6), the bottle preform is irradiated with an electron beam prior to blowing of the bottle. Monsanto claimed that this process resulted in a bottle that, because of the thermodynamics of the extraction process, would have essentially no extraction of AN by the contained beverage. The FDA did not accept this claim, but it did decide that the extraction would be below the detection limit of 0.16 ppb. The agency ruled that AN concentrations at or below this detection limit would be considered acceptable, and limited the residual AN content of the finished container of 0.1 ppm.

Rubber-modified, acrylonitrile–methacrylate copolymers (AN/MA). High barrier properties can also be achieved by copolymerizing acrylonitrile and methacrylate in a 75:25 ratio onto a nitrile rubber backbone. This is the approach taken by BP Chemicals (formerly Sohio Chemical Company) in the production of Barex (registered trademark) resins. This family of resins can be processed on almost all conventional plastic processing equipment including, but not limited to, injection, injection blow molding, injection stretch blow molding, extrusion, extrusion blow molding, extrusion stretch blow molding, profile extrusion, pipe extrusion, tubing extrusion and thermoforming (see individual processing sections). Transparency is retained through the use of a refractive-index-matched rubber modifier.

The Barex family of resins are the only high nitrile resins in commercial production today and are in full compliance with all applicable FDA regulations for direct food contact and meets all the known requirements of a USP Class VI plastic.

Properties

Among the commercial packaging polymers that have the physical properties required for monolithic structures, HNR offer the greatest gas barrier [0.8 cm^3 mil/100 in.^{2} d atm or 3.0 cm^2 μm/m^2 d kPa] and chemical resistance. Their gas-barrier properties (see Barrier polymers) are surpassed only by EVOH (see Ethylene–vinyl alcohol) and PVdC (see Vinylidene chloride copolymers), which are used as components of

Table 1. Properties of High Nitrile Resins

Property	ASTM Test	Units	Barex 210 Resin	Barex 218 Resin
Specific gravity	D1505		1.15	1.11
Yield		$in.^2$/lb at 1 mil	24,080	24,950
Brabender torque		meter-grams	950	1050
Notched Izod	D256	ft-lb/in.	5.0	9.0
Flexural modulus	D790	psi	490,000	400,000
Oxygen T. R. (day)	D3985	cc-mil/$100in.^2$	0.8	1.6
CO_2 T. R. (day)	D3985	cc-mil/$100in.^2$	1.2	1.6
Nitrogen T. R. (day)	D3985	cc-mil/$100in.^2$	0.2	0.4
WVTR (day)	F1249	g-mil/$100in.^2$	5.0	7.5
Heat deflection temperature	D648	°F	160	170
Heat seal temperature		°F	250–375	275–350

multilayer structures (see Coextrusions for flexible packaging; Coextrusions for semirigid packaging. Because of the polarity that the nitrile group imparts to the molecule, HNRs show an affinity for water. Water-vapor barrier is lower than that of the nonpolar polyolefins (e.g., polyethylene, polypropylene); but that same polarity imparts resistance to nonpolar solvents. The relatively high flexural modulus (combined with the lower specific gravity) means that for structures with identical geometry, the HNR parts can be source reduced (downgauged) and therefore designed with less material for equivalent stiffness compared to the polyolefins, PVC, PET, PETG and many other plastic polymers. A summary of the properties of high nitrile resins is shown in Table 1.

Applications. HNR are used in packaging in a variety of physical forms. These include: film, semirigid sheet, and injection molded (see Injection molding) and blowmolded containers. Blown film (see Extrusion) is used in polyolefin-container structures to provide formability, chemical resistance, and gas barrier. Spices, medical devices, and household chemical products are example of such applications. Laminations with polyolefins and aluminum foil are used in applications ranging from food packaging to oil-drilling core wraps (7). These structures have exceptional barrier properties, as well as sealability and chemical resistance. Indeed their use in sachet packaging is growing rapidly as individual dose and unit packaging become more popular.

The semirigid sheet market for HNR is primarily meat and cheese packaging in thermoformed (see Thermoforming) blister packages. With its excellent gas barrier, clarity, and rigidity, the HNR are the premium packaging material. Of increasing importance in semirigid applications, however, is disposable medical device packaging (see Medical packaging). Here HNR can be sterilized by either ethylene oxide (ETO) or gamma radiation and is unaffected by plasticizers present in many devices, making it ideal for many medical packaging applications. As a result of its unique combination of properties, including source reduction and superior thermoforming, usage has been steadily increasing in this market area.

Blow molding applications for HNR are dominated by chemical resistance requirements. Injection blow molding is the most widely used method for manufacturing small containers; for example, bottles for correction fluid, nail enamel, and other cosmetics. Larger bottles are generally extrusion blow-molded. Some important applications include: pesticides, herbicides and other agricultural chemicals, fuel additives, and hard to hold household chemicals. Extrusion stretch blow molding is gaining increasing acceptance for bottles 16 ounces (473 mL) or larger. The orientation achieved during stretching greatly increases the drop-impact performance, and the walls can be relatively thin.

HNR has been coextruded with many different polymers, but the polyolefins have been of greatest commercial significance. Coextrusions are available in sheet, film, and bottle form. They typically gain gas barrier or chemical resistance from the HNR and water vapor barrier and economics from the polyolefin. In structures with polypropylene, the heat-deflection temperature (HDT) of the structure is increased by the higher HDT of the polypropylene. This permits the use of HNR in high temperature environments such as microwave ovens.

The adhesive used to combine the layers in an HNR coextrusion are typically styrene–isoprene or styrene–butadiene block copolymers (see Multilayer flexible packaging). Scrap is reusable in the polyolefin layer if the nonolefin percentage in that layer is well dispersed and of lower concentration than about 15%.

Chemical-resistant coextruded bottles containing HNR are now being commercialized. HNR is the inner contact layer enclosed by adhesive and polyolefin, typically HDPE. Use of a three-layer structure limits the cost and complexity of the machinery. It also places the solvent-resistant polymer in contact with the chemicals. Five layer or laminar structures (see Surface modification; Nylon) using other barrier resins place the polyolefin in direct contact with the aggressive contents of the container. The three layer structure also allows visual inspection of the barrier layer and maximizes the sealing area of the barrier layer at the pinchoff of the bottle. HNR coextruded containers offer high performance with economics superior to the other packaging alternatives (8).

Alloying and blending also offer other unique properties that can be imparted to the base resins, typically polyolefins. Here they can be used to reduce ESCR as well as provide increased flexural modulus (stiffness) to polyolefinic resins (9).

The Barex family of resins offers a unique combination of properties to the packaging industry. Developing layer-combining technologies, increasing consumer acceptance of plastic packaging, and the changing FDA status bode well for HNR. Displacement of metal and glass, as well as other plastics, in packaging applications by HNR should be accelerated by the new processing technologies. As an example, sheet and film applications are expected to show increased growth in both medical disposable and food market areas because of economic advantages from source reduction and processing via calendering.

BIBLIOGRAPHY

1. U.S. Pat. 2,904,580 (Sept. 15, 1959), J. D. Idol (to BP Chemical, OH).
2. J. M. Lasito, "Acrylic Multipolymers in Medical Packaging," *Proceedings of the TAPPI* **71,** 74 (Sept. 15, 1982).
3. Report to the FDA by the Manufacturing Chemists Association, Jan. 14, 1977.
4. *Fed. Regist.* 41 23,940. Title 21, Part 177.1480 (June 14, 1976).
5. *Fed. Regist.* 49 36,637. Title 21, Part 177 (Sept. 19, 1984).
6. U.S. Pat. 4,174,043 (Nov. 13, 1979), M. Salome and S. Steingher (to Monsanto).
7. U.S. Pat. 4,505,161 (Mar. 19, 1985), P. K. Hunt and S. J. Waisala (to BP Chemical, OH).
8. J. P. McCaul, "The Economics of Coextrusion," *Proceedings of the High Technology Plastic Container Conference,* SPE, Nov. 11, 1985.
9. P. R. Lund and co-workers "High Nitrile Polymer/Polyolefin Blends: A Low Cost, High Performance Alternative," *Proceedings of the 1995 Annual Technical Conference,* SPE, May 7–11, 1995.

Paul R. Lund
Joseph P. McCaul
The BP Chemical Company
Cleveland, Ohio

NONWOVENS

Definition

"Nonwovens are fabric-like materials consisting of a conglomeration of fibers that are bonded in some way or other" (1). The total world wide sales of nonwovens amounts to 7.1 billion dollars and the ten top producers in order or predominance are listed in Table 1. Nonwovens may be sold by a yield or area basis, weight basis or product unit basis. Because of their unique properties and high production rates, nonwovens offer a substantially high performance/price ratio." However, in order to appreciate the full range of products, properties and end uses of nonwovens, it is necessary to consider the fibers, the web-formation method, and the bonding methods used to make them.

Table 1. Top 10 International Nonwoven Roll Goods Companies[a]

1995 Ranking	Company	Worldwide Sales (Million $)
1	Freudenberg	1117
2	DuPont	800
3	PGI	600
4	BBA Group	500
5	Kimberly-Clark	482
6	Veratec	300
7	Japan Vilene	295
8	Dexter	284
9	Hoechst	189
10	Asahi	172
Total Other Company Sales		2395
Total Combined Sales		7134

[a] From Ref. 2.

Fibers

The primary ingredient of nonwovens are the fibers used to produce them. A fiber is substantially longer in length than diameter. In nonwovens, fibers may vary in length from 1 mm to a continuous length. Short (1-mm) fibers are used in air-laid or wet-laid nonwovens. Whereas spunbonded nonwovens use continuous-length fibers in their fabrication. The length of the fibers will affect both the uniformity of fiber distribution and the strength of the nonwoven. Short fibers tend to give good fiber distribution, but longer fibers tend to produce greater strength. However, the fiber length actually utilized is dependent on the nonwoven-web-formation process and the type of product being manufactured.

The diameter of the fiber affects the properties of the nonwoven considerably. If it was desired to produce a soft nonwoven facial wipe, baby wipe, or polishing cloth, a fine fiber should be used. If, on the other hand, it was desired to produce a scrubbing pad of the required stiffness for a frying pan, large-diameter fibers should be used. The latter would also give high incompressibility or resilience.

While many fibers are perfectly round, some are triangular, square, hollow, and of other shapes. In addition, they may consist of one or more generic type of materials in their makeup.

One major variable in nonwovens is the type of fiber used in its construction. Virtually any fiber can be utilized to make nonwovens. The principal natural fibers found in nonwovens are cotton, jute, and especially wood pulp. The major man-made fibers used are polypropylene, polyester, rayon, and glass. Because of their far-reaching properties, the range of fiber types available offer great latitude in the performance characteristics of the final nonwoven. For example, since moisture absorbency is desired in diapers, incontinence pads, and wipes, rayon, cotton, and wood pulp are prime contenders. For oil absorbency in shop wipes and oil scavengers, polypropylene, polyethylene, or polyester can be used. Because of their high hydrophobic properties, moisture-barrier properties are provided by polypropylene, polyethylene, polyester, and PTFE. High tensile strength may be obtained using Kevlar aramid, Spectra olefin, nylon, polyester, silk, and glass. Low elongating properties for gaskets can be obtained with Kevlar aramid, Spectra olefin, and glass. High-temperature resistance or low flammability can be obtained with carbon, glass, Kynol novoloid, Kevlar and Nomex aramid, modacrylic, PBI, PTFE, ceramic, and metallic fibers. For composites, glass, carbon, Kevlar, and Nomex aramid may be used. Chemical resistance may be obtained with PTFE, glass, carbon, acrylic, olefin, and some metallic fibers. It should be apparent, therefore, that proper fiber selection is essential in order to assure desirable end use performance characteristics of the nonwoven.

Nonwoven Web-Manufacturing Methods

Nonwoven webs can be produced in four primary ways: carding, air laying, wet laying, or spunbonding (3). Within each one of these, the process and resulting product will often vary considerably.

A carding machine feeds a mass of crimped fibers, opens, disentangles, and drafts the fibers and then delivers a thin wide web of uniformly distributed fibers to some form of delivery apron. The fibers used to produce carded nonwovens must have crimp and will usually have a fiber length ranging between 34 and 152 mm. The fibers exiting the card tend to be aligned in the machine direction, and, if used in this manner, the product is referred to as a *parallel-laid nonwoven.* Often it is necessary to combine the web from several carding machines to achieve the desired nonwoven weight. A parallel-laid nonwoven may be up to 11 times stronger in the machine direction or fiber-oriented direction than it is in the cross-machine direction. Also, the elongation in the machine direction is substantially less than that found in the cross-machine direction.

The web exiting the card may be laid back and forth across a moving apron to produce what is known as a *cross-laid nonwoven.* Because systems of fibers cross each other, the resulting nonwoven has more uniform strength and elongation properties than found in parallel-laid nonwovens.

Sometimes a parallel-laid web is combined with a cross-laid web to produce what is termed a *composite carded nonwoven.* The latter has the greatest uniformity in strength and elongation properties of carded nonwovens.

A recent innovation is *randomized carded nonwovens.* The card is similar to that used for parallel laying, but the doffer roll is run in a direction counter to that of the main cylinder at the point of near-wire contact. Such a configuration deparallelizes the web, permitting more loft and greater uniformity of strength and elongation in the machine direction and cross direction.

Because of the great latitude in weight, thickness, fiber orientation, strength, elongation, and fibers utilized, carded nonwovens have a vast number of product end use applications. This might include apparel interfacing, shoe components, thermal insulation, blankets, computer disk liners, wipes, and papermaker felts.

Air-laid nonwovens utilize fibers having a length ranging between 1 and 76 mm. The fibers are first opened, then conveyed with an air stream against either a condensing cylinder or a screen to form the web. The product produced tends to have an *x, y, z*-fiber orientation, and as a result the webs are soft and lofty. Such a web is especially useful in padding, insulation, filtration, and wipes.

Wet-laid nonwovens utilize papermaking equipment and technology. The fiber length ranges between 1 and 38 mm. The longer fibers are either glass for roofing shingles and felt or carbon for composites. Shorter natural and man-made fibers are used in tea bags, automotive air and oil filters, coffee filters, bunting, hang tags, and composites. While flat tabled fourdrinier machines have been used to make some nonwovens, the best machine for producing wet-laid nonwovens is the inclined-wire machine since long fibers are held in suspension in water at the time of formation on the moving wire. Often multiple headboxes are used to produce layered or stratified high-efficiency air filters and other products.

Spunbonded and melt-blown nonwovens (see Figure 1) are produced at the time the fiber is extruded from the spinnerets. Spunbonded nonwovens are produced with continuous filament fibers that are drafted with either multiple drafting rollers or high-velocity air used in combination with a venture to achieve molecular fiber orientation. Most spunbonded nonwovens are produced with polypropylene. However, polyethylene, polyester, nylon, acrylic, and rayon are also used. End uses include wallpaper, bailing, hang tags, bagging, filters, geotextiles, mattress-spring assemblies and covers, furniture construction, moisture-barrier fabrics, and medical products.

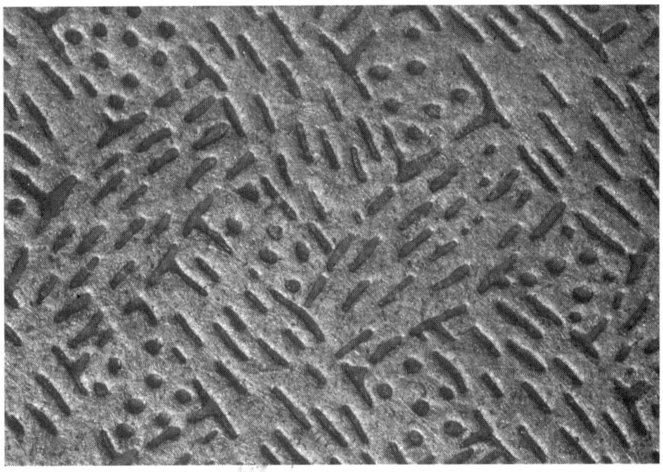

Figure 1. Illustration of a polypropylene melt blown nonwoven that is bonded using a heated pattern engraved calender roll. Such a product is manufactured using a fine oil loving fiber and is excellent for use as a shop towel or wipe.

Melt-blown nonwovens are similar to spunbonded nonwovens but instead use a spinnerette in which both the polymer and hot air exit the spinnerette from different orifices simultaneously. The velocity of the air is sufficient to draft the polymer into very fine discontinuous fibers. Melt-blown nonwovens are used for thermal insulation, filters, padding, personal products, synthetic leather, and other products.

Bonding Methods

There are four principal methods of bonding nonwovens: 51% chemical, 43% mechanical, 5% thermal, and 1% inherent or self-bonding. Chemical bonding may rely on binders, solvents, or hydrophilic fibers. Of these, binders are by far the most important. Binders act like adhesive and include acrylic, phenol, urea and melamine formaldehyde, polyvinyl acetate, styrene–butadiene rubber, vinyl acetate–ethylene, polyvinyl chloride, nitrile rubber, polyvinyl alcohol, and other lesser binders (4). Such binders vary considerably in cost, properties, and performance characteristics.

Mechanical bonding includes needle punching, stitch-through, and hydroentangling. Needle punching may use either barbed or forked needles. Barbed needles interlock the fibers to provide mechanical strength to produce such products as papermaker's felts, blankets, geotextiles, synthetic shoe leather, thermal lining, and automotive trunk liners. The forked needles can produce velour and patterned products for use as institutional and automotive carpeting, wallscaping, and throwrugs.

Stitch through technology is a knitting-through process that may be accomplished either with or without yarns to stitch the nonwoven web together. The Mali and Arachne machines are most common and produce products such as carpets, upholstery, drapery, apparel, blankets, interlining, wallscaping, geotextiles, and automotive products.

The hydrogenating process was invented by E. I. DuPont and was first referred to as the *spunlace process.* It utilizes very fine water jets to mechanically entangle fibrous webs. The materials produced have sufficient integrity to be used as polishing cloths, wipes, headrests, curtains, and even in the manufacture of composites.

Thermal bonding may include the use of thermoplastic fibers, yarns, powders, or films. Often bicomponent fibers having differential melting temperatures in either a core-and-sheath or a side-by-side configuration, may be utilized. Bonding is facilitated by through-air heaters, and smooth or patterned calender rolls.

Inherent or self-bonded nonwovens are bonded at the time they are produced. This might include film splitting, blow extrusion, or static-charge extrusion. However, because of their limited properties and aesthetics, nonwovens produced in this manner have limited uses.

Because of the fiber types available, the differences in web-manufacturing processes possible and the various bonding methods, a cost-effective nonwoven product can be produced with properties that will perform in almost any application. However, adequate performance characteristics are assured only if a nonwoven incorporates the proper fibers, web manufacturing process, bonding method, and proper finishing and fabrication to produce the final product for the end use application desired.

BIBLIOGRAPHY

1. J. R. Wagner, *Nonwoven Fabrics,* Norristown, PA, 1982.
2. *International Top 40 Roll Goods Companies, Nonwovens Industry,* Rodman Publications, Inc., September 1995, p 39.
3. J. R. Wagner, *Introduction to Nonwovens,* Home Study Library, Vol. 9, TAPPI Press, Atlanta, 1985.
4. J. R. Wagner, *Nonwoven Notes,* Norristown, PA, 1989.

J. Robert Wagner
Philadelphia College of Textiles and Science
Philadelphia, Pennsylvania

NUTRITION LABELING

The regulatory issues associated with nutrition labeling of food products are complex and extensive. Specific nutrition information must be included on most food products following particular design requirements. The regulations also delineate what nonmandatory nutrition information may be included and how it can be presented. Professionals involved in food processing, manufacturing, marketing, and packaging need to be aware of nutrition labeling regulations and understand how they affect food packaging and marketing.

The purpose of this article is to provide a broad understanding of the nutrition-labeling regulations, explain why they were developed, and describe some of the approaches that may be used for developing nutrition labels. A heavier emphasis will be placed on the issues related to design and format, rather than those pertaining to the content of the nutrition label. Because the regulations are extensive and detailed, it is not possible in the scope of this article to cover all the issues necessary to make this a "how to" reference. In order to apply the regulatory requirements to specific packaging issues, it is necessary to consult with regulatory source documents, regulatory guidebooks, and/or legal or regulatory advisors.

History of Nutrition Labeling

Although the history of food labeling laws reach back as far as 1906 with the Pure Food and Drug Act, it was not until after World War II that consumer interest in and need for nutrition information became evident (1). During the war, food preparation shifted from the household to large-scale food processing. The extended shipping times for delivery of goods needed in war encouraged the development of additives to extend shelf life. Wartime health problems led to the recognition of the relationship between nutrition and diseases, spawning interest in nutrition research and food labeling. Scientists recognized the need for vitamins and minerals to prevent certain diseases such as beriberi and pellagra.

By 1969, there was significance public interest in the relationship between diet and health. As a response to this interest, President Nixon convened the White House Conference on Food, Nutrition, and Health, addressing malnutrition in America (2). He told Congress of the need to ensure that the private food industry serves all Americans well and that people are educated in choosing proper foods. As an outcome of this conference, and Congress' effort to mandate nutrition labeling, the Food and Drug Administration adopted a voluntary nutrition labeling program in 1973 that required the labeling of nutrition information whenever a nutrient claim was made or a product was fortified (3). At this time, the voluntary nutrition label focused on key vitamins or minerals known to cause deficiency disease.

Proponents of mandatory nutrition labeling were not satisfied with the limitations of the voluntary program—typically, only products with a good nutrition story included nutrition labeling. Throughout the 1970s, mandatory nutrition labeling continued to be addressed by legislators, in particular, Senator George McGovern, who chaired the Senate Select Committee on Nutrition and Human Needs. This Committee investigated a number of nutrition issues affecting Americans and was responsible for issuing the U.S. *Dietary Goals,* the first government report setting prudent dietary guidelines for Americans (4). In 1978, McGovern initiated a series of hearings to explore nutrition labeling and information, where he is quoted: "It appears from all I have heard and read to date that the present labeling system is not useful or appropriate. Therefore we must determine what nutrition information the public wants and needs, and how best to convey that information" (5). Although the hearing displayed the bipartisan nature of the issue, it took Congress another 13 years before the Nutrition Labeling and Education Act of 1990 (NLEA) was passed.

With the election of President Reagan in 1980, the thrust to adopt mandatory nutrition-labeling legislation came to an abrupt halt (1). Deregulation limited available resources, causing FDA to focus on food-safety issues and to deemphasize economic issues such as food labeling. Despite this shift in legislative priorities, there were major advances in science that documented the link between nutrition and health. Several reports issued by the federal government (6–8), in addi-

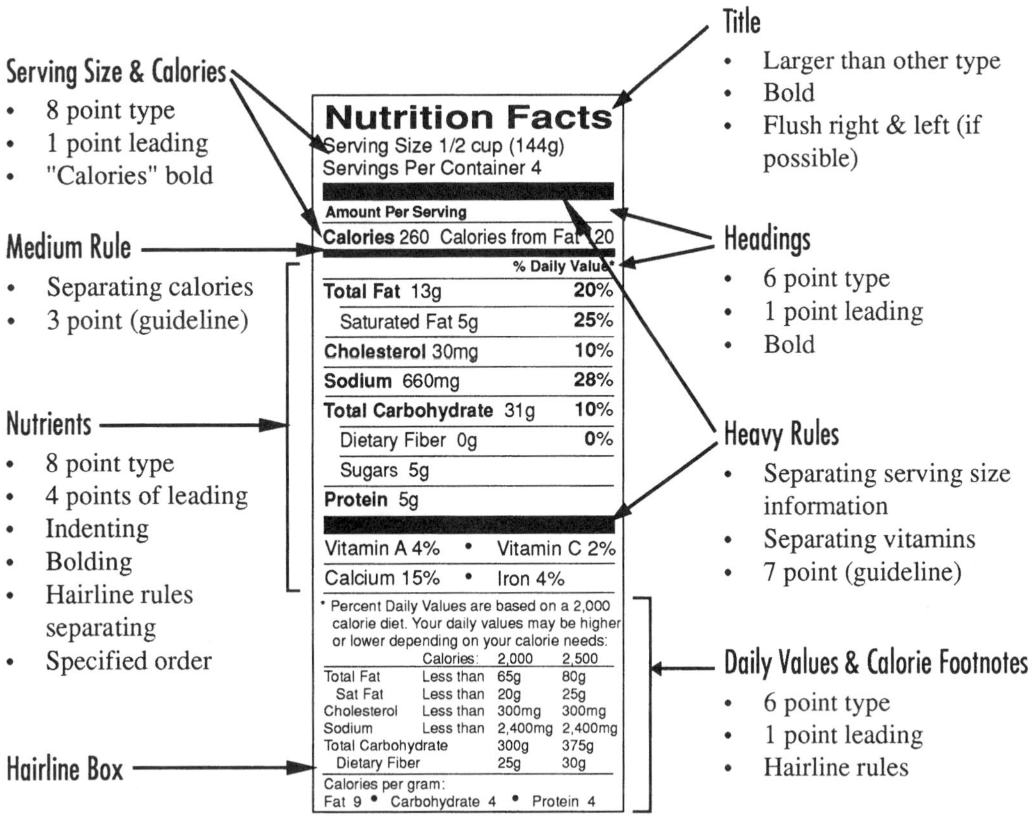

Figure 1. Design features of the Nutrition Facts panel.

tion to reports released by nonprofit agencies, such as the American Cancer Society (9) and the American Heart Association (10), clearly showed the growing consensus about the relationship between dietary imbalances and chronic diseases, specifically the overconsumption of calories, fat, saturated fat, cholesterol, and sodium and the underconsumption of dietary fiber. As a result of extensive media coverage of these findings, consumers began to demand healthier food products and more information about the fat, cholesterol, and fiber content of foods. Food manufacturers responded by introducing nutrient-focused products—products lower in fat, cholesterol, and sodium and high in fiber proliferated. Unfortunately, there were no guidelines for defining "low" or "high" or for labeling nutrients such as dietary fiber, saturated fat, or cholesterol. As a result, manufacturers' efforts to produce and market such products without consistent rules led to incidence of abuse, and consumer groups complained of food-labeling deceptions. By the end of the 1980s, the food industry, as well as regulators and consumers, were frustrated by the contradictions that prevailed on grocery-store shelves.

It was not until President Bush took office in the late 1980s that legislative and regulatory efforts again focused on nutrition labeling. Senators Metzenbaum, Kennedy, and Hatch and Representatives Waxman and Madigan renewed the spirit of the 1970s with nutrition labeling regulations that ultimately resulted in mandatory labeling, nutrient content and health claim definitions, and federal preemption, now known as NLEA. While these legislators were lobbying for mandatory nutrition labeling laws, the National Academy of Sciences' Institute of Medicine (IOM) issued a report entitled *Nutrition Labeling: Issues and Directions for the 1990s* (11). The IOM report recommended changes in food labeling to assist consumers in implementing the recommendations of the Surgeon General (7) and the National Research Council (8). FDA responded by working with all concerned parties, including industry, consumer groups, and the states, to define regulations that would meet the goals set out by the IOM report.

On November 8, 1990, President George Bush signed into law the Nutrition Labeling and Education Act (NLEA) of 1990 (12), which has dramatically changed the way food products are labeled in the United States. The NLEA represents a comprehensive mandatory nutrition-labeling system, designed to help consumers meet the U.S. *Dietary Guidelines* (6) and reduce their risk of chronic diseases. The regulations, although not perfect, provide the guidance to food manufacturers and package designers needed to ensure the consistent presentation of nutrition information and to prevent erroneous nutrient content and health claims.

Regulatory Agencies

Regulation of food labeling falls primarily under the jurisdiction of two federal agencies: the Food and Drug Administration (FDA) and the United States Department of Agriculture (USDA) (13). USDA's Food Safety and Inspection Service (FSIS) oversees food labeling of products containing meat or poultry (≥2% or more cooked, ≥3% raw). All other food prod-

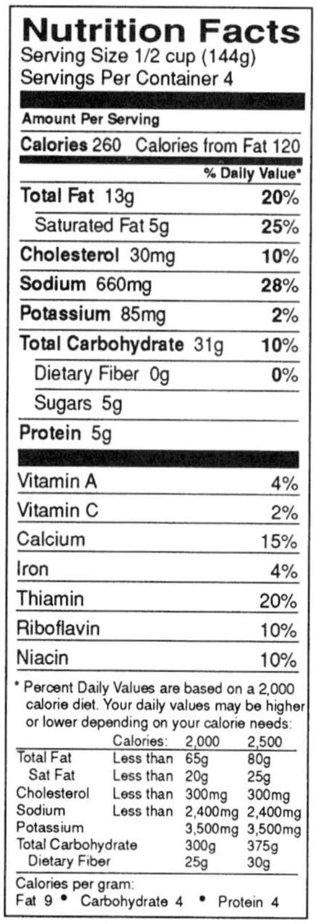

Figure 2. Extended Nutrition Facts format.

ucts fall under the jurisdiction of the Center for Food Safety and Applied Nutrition (CFSAN) of FDA. FDA regulations are governed by the Federal Food, Drug, and Cosmetic Act (FDCA) and the Fair Packaging and Labeling Act (FPLA). [The Nutrition Labeling and Education Act of 1990 (NLEA) is an amendment to the FDCA]. USDA's regulatory role and responsibilities are defined by the Federal Meat Inspection and Poultry Products Inspection Acts. Because the two agencies are governed by separate laws, they have different missions, philosophies, and approaches to food labeling. This results in subtle differences throughout the regulations.

USDA's primary regulatory role has been to prevent public health hazards resulting from improper handling of meat and poultry during production and packaging. USDA's network of field offices are responsible for conducting frequent plant inspections, and the national office administers a label preapproval program. Through these mechanisms, USDA tightly controls food labeling during the production process. Unlike USDA, FDA has not been funded to function as an inspection service; therefore, the agency does not have the staff resources to play a hands-on role in label development and approval. FDA relies on postmarket surveys to enforce the regulations.

As discussed above, Congress passed the NLEA in 1990, mandating FDA to initiate extensive changes in the content and format of the nutrition label. FDA responded by issuing final rules in January 1993 (14). Nutrition labeling of USDA products was not included in the Congressional mandate; however, USDA decided to follow suit in order to prevent consumer confusion arising from two different nutrition labels (15). In most of the significant aspects of nutrition labeling, the USDA regulations mimic the FDA's; however, the regula-

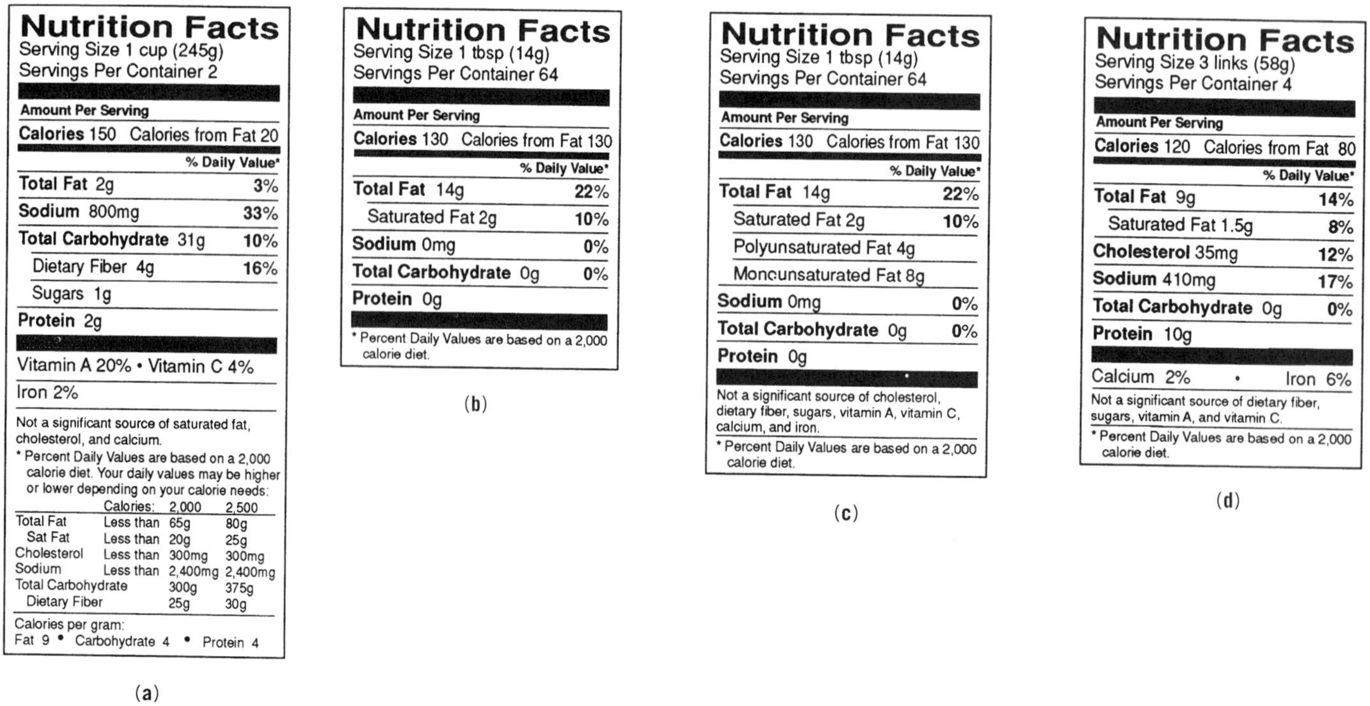

Figure 3. Abbreviated Nutrition Facts format: (**a**) Shortened format; (**b**) FDA Simplified format; (**c**) FDA Simplified Extended format; (**d**) USDA Simplified format.

(a)

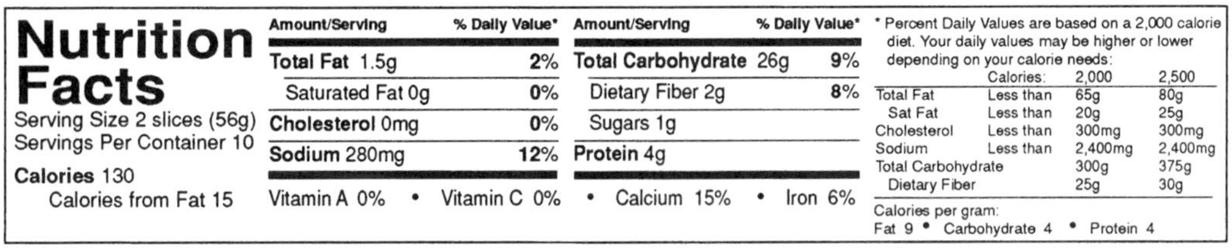

(b)

Figure 4. Modified (**a**) and Tabular (**b**) displays.

tions diverge in a few areas. For example, USDA has different exemption criteria, more relaxed rules for use of the "Simplified" format, and less extensive requirements for statements that accompany nutrient content claims.

Nutrition Facts Panel

The Nutrition Facts panel presents the nutrient profile of a food product on a per serving basis. Since a goal of NLEA was to make nutrition information easy for consumers to locate, read, and understand, both the content and layout of the Nutrition Facts panel were studied extensively before the regulations were finalized. The Nutrition Facts panel focuses on the nutrients important to the health of Americans (eg, calories, fat, saturated fat, cholesterol, sodium, fiber). The nutrient content of the food is presented in absolute terms (eg, grams, milligrams), and it is also presented as a percentage of the Daily Value (% Daily Value), which provides a method for comparing the nutrient profile of the food to recommended nutrient intakes. Larger food packages are also required to include a list of the Daily Values at two calorie levels (ie, 2000 and 2500) in a footnote at the bottom of the Nutrition Facts Panel. The serving size information has been standardized to make it easier to compare similar products. Serving sizes are presented in common household measures (eg, cups, tablespoons) to help consumers visualize portion sizes, as well as metric units (eg, grams), to encourage familiarity with the International System of Measures (13).

To maintain a consistent look from package to package, both FDA and USDA regulations explicitly define the layout of the Nutrition Facts panel (Fig. 1). A hairline box surrounds the nutrition information, which is presented in a required order. The type size and leading (spacing between lines of type) of each line is specified. Boldface type is used to enhance certain nutrients, and other nutrients are indented. Heavy-, medium-, and lightweight rules are incorporated in specified places.

To provide some flexibility in tailoring the Nutrition Facts panel to various food packages, the regulations allow for several variations of the Nutrition Facts layout pictured in Figure 1. These variations are not a random choice; they each have explicit usage criteria (13). Format options are based on the nutrient profile of the food. Display options are available to accommodate different package shapes and sizes, as well as special situations (eg, variety packages, foods for children, bilingual labels). In addition, the regulations allow for modification of the Nutrition Facts on packages which have limited space for labeling.

Format options. The format of the Nutrition Facts panel refers to the nutrients listed on the panel (13). The regulations define 14 mandatory nutrients for inclusion in the Nutrition Facts panel; the "Full" format includes these 14 nutrients (Fig. 1). In addition, the regulations allow for inclusion of certain nonmandatory nutrients, on either a required or voluntary basis. The Extended format accommodates the inclusion of non-mandatory nutrients in a specified order. An example of the Extended format is presented in Figure 2.

When a food product has one or more mandatory nutrients at an insignificant level (eg, nutrient declaration of 0), an abbreviated format may be used. The "Shortened" format is

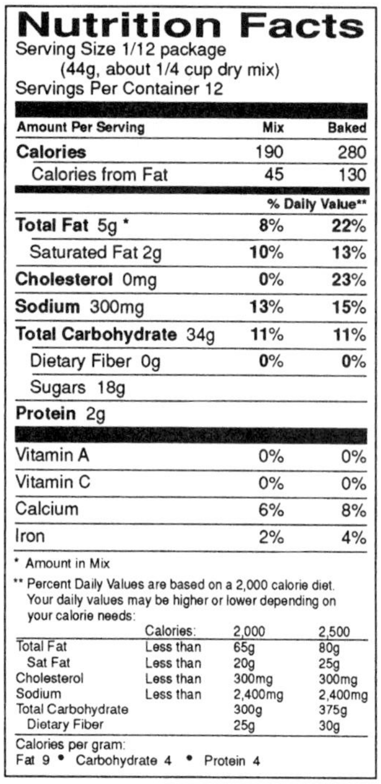

(a)

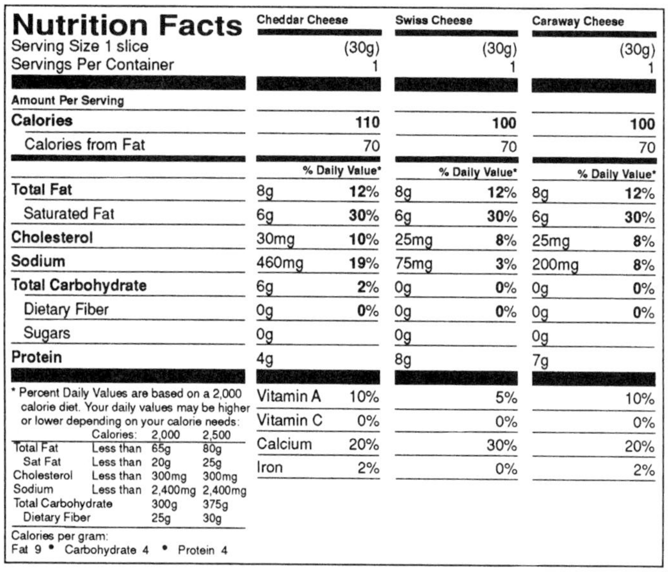

(b)

Figure 5. Dual Column (**a**) and Aggregate (**b**) displays.

available for FDA products that have between one and six insignificant nutrients. With this format, insignificant nutrients may be listed in a footnote stating, "Not a significant source of [list of insignificant nutrients]." FDA products with seven or more insignificant nutrients may bear the Simplified format that allows insignificant nutrients and the Daily Values table in the footnote to be omitted. However, if nonmandatory nutrients are declared, then a variation, called the "Simplified Extended" format, is used. USDA employs a different approach to the Simplified format. The Simplified format is available for products that have one or more noncore, mandatory nutrients at insignificant levels. With all these abbreviated formats, five core nutrients (viz, calories, total fat, sodium, total carbohydrate, protein) must be listed, even if they are insignificant. Figure 3 presents examples of the abbreviated formats.

Table 1. Display Options Based on Package Size

Package size	Display options	Modifications
>40 in.^{2}	Standard Modified (if limited vertical space) Tabular (if limited vertical space) Dual Column Aggregate Bilingual Child	None available
12–40 in.^{2}	Standard Tabular Dual Column Aggregate Bilingual Child Linear (if no other displays will fit)	Daily Value table omitted from footnote Abbreviations permitted
<12 in.^{2}	Linear	Address/phone number (if no claims or nutrition information) Daily Value table omitted from footnote Abbreviations permitted

Table 2. Types of Nutrient-Content Claims

Type of claim	Definition	Examples
Absolute	Statement about the nutrient level in food without referencing or comparing to another product	Free Low Very low High Source of Healthy Lean Extralean
Comparative	A claim comparing the level of a nutrient in one product to the level of that nutrient in another product or class of foods; also called a "relative claim"	Light or "lite" Reduced Less More
Implied	Statement that leads a consumer to assume that a nutrient is absent or present in a certain amount or that the food may be useful in achieving dietary recommendations	"High in oat bran" implies "high in dietary fiber"
Expressed	Any direct statement about the level or range of a nutrient in a food	Low fat Low cholesterol Low sodium High in dietary fiber

Nutrition Facts	
Serv. Size 1 slice (28g)	
Servings 12	
Amount Per Serving	
Calories 110 Fat Cal. 70	
	% DV*
Total Fat 8g	**12%**
Sat. Fat 6g	**30%**
Cholest. 30mg	**10%**
Sodium 460mg	**19%**
Total Carb. 6g	**2%**
Fiber 0g	**0%**
Sugars 6g	
Protein 4g	
Vitamin A 10% • Vitamin C 0%	
Calcium 20% • Iron 0%	
* Percent Daily Values (DV) are based on a 2,000 calorie diet.	

Nutrition Facts	Amount/Serving	% DV*	Amount/Serving	% DV*
Serv. Size 1/3 cup (56g)	**Total Fat** 1g	**2%**	**Total Carb.** 0g	**0%**
Servings 3	Sat. Fat 0g	**0%**	Fiber 0g	**0%**
Calories 80	**Cholest.** 10mg	**3%**	Sugars 0g	
Fat Cal. 10	**Sodium** 200mg	**8%**	**Protein** 17g	
* Percent Daily Values (DV) are based on a 2,000 calorie diet.	Vitamin A 0% • Vitamin C 0% • Calcium 15% • Iron 6%			

Figure 6. Modification of Standard and Tabular displays for packages with <40 in.^{2} "available space to bear labeling."

Display options. The "Standard" display (vertical column) pictured in Figure 1 is the "model" Nutrition Facts layout from which all others are derived (13). On packages with limited vertical space, the "Modified" (footnote to the side) display may be used. If neither the Standard or Modified displays will fit, then the "Tabular" (horizontal) display may be used. Figure 4 presents the Modified and Tabular displays.

To accommodate special packaging and labeling situations, other displays are allowed. The "Dual Column" display provides for listing nutrients on an "as prepared" as well as an "as packaged" basis. The "Aggregate" display allows for presentation of nutrition information for a variety of products contained within one package (eg, variety package of cereal). Figure 5 pictures the Dual Column and Aggregate displays.

A "Bilingual" display accommodates inclusion of a second language in the Nutrition Facts panel. Special labeling requirements apply to foods intended for young children, so the Nutrition Facts panel is modified for these products.

Package-size considerations. For packages with <40 in.^{2} of "available space to bear labeling," any of the displays can be modified to create a smaller Nutrition Facts panel. Specifically, the Daily Values table may be omitted from the footnote and abbreviations may be used. Figure 6 illustrates these modifications of the Standard and Tabular displays.

On packages with <12 in.^{2} of available space that contain no other nutrition information and make no claims, the Nutrition Facts panel may be replaced with an address or phone number where nutrition information can be obtained. The "Linear" (paragraph) display is provided to accommodate nutrition labeling of these small packages when claims or nutrition information is present (Fig. 7). (A larger display may be employed at the manufacturer's discretion.) If no other display will fit, then the Linear (paragraph) display also may be used on packages with 12–40 in.^{2} of available space.

Nutrition Facts Serv. size: 1 package (30g), Servings: 1, Amount Per Serving: **Calories** 45, Fat Cal. 10, **Total Fat** 1g (2% DV), Sat. Fat 1g (5% DV), **Cholest.** 0mg (0% DV), **Sodium** 50mg (2% DV), **Total Carb.** 8g (3% DV), Fiber 1g (4% DV), Sugars 4g, **Protein** 1g, Vitamin A (8% DV), Vitamin C (0% DV), Calcium (4% DV), Iron (2% DV). Percent Daily Values (DV) are based on a 2,000 calorie diet.

Figure 7. Linear display.

Table 1 summarizes what display options are available to various package sizes and the modifications that can be made.

Claims

The regulations define two categories of claims: nutrient content (or descriptors) and health claims (13). Nutrient-content claims are statements about the level of a nutrient in a food. Health claims, on the other hand, link the nutrient profile of a food to a health or disease condition. Nutrient content claims are used widely by the food industry. Since the regulations governing health claims are more complicated and restrictive, health claims are not as common.

Nutrient-content claims. Nutrient-content claims (sometimes referred to as "descriptors") characterize the level of a nutrient in a food. Only defined terms can be used on the label to describe a food's nutrient content. When these terms are used, a product must meet specific criteria. Nutrient-content claims may be implied or expressed, comparative or absolute (Table 2).

When a product bears a nutrient content claim (whether it is implied or expressed), certain information must be incorporated into the design of the package. Depending on the type of claim and whether the product is governed by FDA or USDA, the requirements for claims-related information will vary. FDA mandates more extensive information than USDA (ie, the inclusion of either a Referral or Disclosure Statement), and both agencies require additional information on products bearing comparative claims. All required statements must adhere to placement and typesetting requirements that are defined by the regulations. Table 3 summarizes the required statements that must accompany nutrient-content claims.

All FDA products bearing nutrient-content claims require either a Referral or Disclosure Statement to be included on each panel where a claim appears, except the Nutrition Facts panel. A Referral Statement, appearing next to the largest claim on each panel, directs the consumer to the Nutrition Facts statement. When a product contains excessive levels of key nutrients that are associated with health risks, a Disclosure Statement that flags the nutrient(s) of concern and directs the consumer of the Nutrition Facts panel is used instead of the Referral Statement.

Table 3. Required Statements for Nutrient Content Claims

	When to include		
	Absolute claims	Comparative claims	What to state
Referral Statement or Disclosure Statement	FDA products	FDA products	"See ——— panel for nutrition information." "See ——— panel for information on [*nutrient(s)*] and other nutrients."
Nutrient Claim Clarification Statement		FDA & USDA products	Identity of comparison food and percent (or fraction) by which the nutrient differs Example: "$\frac{1}{3}$ fewer calories than comparison food]."
Quantitative Information		FDA & USDA products	Comparison of claims-related nutrient(s) in product and comparison food on per serving basis. Example:

Product	Fat[a]	Calories[a]
Light	4g	200
Regular	8g	300

[a] Per labeled serving.

Source: J. Storlie, *Food Label Design: A Regulatory Resource Kit,* Institute of Packaging Professionals, Herndon, VA, 1996. Used with permission.

Both FDA and USDA product labels that bear a comparative claim must include a Nutrient Claim Clarification statement and Quantitative Information. On FDA products, this information is required in addition to a Referral or Disclosure Statement. The Nutrient Claim Clarification statement identifies the comparison food and states the percentage (or fractional) difference in the subject nutrient(s) between the product and its comparison food [eg, 50% less fat than (comparison food), ~33% fewer calories than (comparison food)]. The Quantitative Information provides the absolute amounts of the subject nutrient(s) in the product and the comparison food. This required information can be presented in a chart or narrative format.

Health claims. FDA has defined nine health claims and created very strict criteria for use of these claims. No other nutrient-disease associations can be made in food labeling. Written statements, third-party references (eg, "American Heart Association"), use of certain terminology in a brand name (eg, "heart"), symbols, and vignettes may be considered a health claim if the context in which they are presented either suggests or states a relationship between a nutrient and a disease. When a statement, symbol, vignette, or other form of communication suggests a link between a nutrient and a disease, it is considered an implied health claim, and it is subject to all the requirements for health claims. USDA has proposed regulations that parallel FDA's.

The regulations governing health claims are very complex. When a product bears a health claim, very specific language must be used on the label. The regulations pertaining to each health-claim outline the assertions that can be made and any additional required statements. When a claim is implied through graphic representations, a complete claim statement must be included on the label.

Since the required statements can be quite lengthy, the complete claim may appear on a back or side panel. When this approach is taken, a reference statement may be placed on the front panel, flagging the claim and directing the consumer to the location of the claim (eg, "See ——— for information about the relationship between ——— and ———." The first blank contains the location of the health claim (eg, back panel, attached pamphlet); the second blank should state the nutrient; and the third blank, the disease or health-related condition. Products bearing health claims must undergo careful review by legal and/or regulatory experts to make certain that the language is accurate and any graphics are acceptable.

Nutrition-Labeling Resources

Although this article summarizes the evolution of nutrition labeling and provides an overview of the nutrition-labeling regulations, it is not a sufficient resource for implementing the regulations. Understanding and complying with nutrition-labeling regulations requires a detailed working knowledge of the regulations and continuous monitoring of regulatory changes. Regulatory guidebooks, such as those listed in the Bibliography at the end of this article, can be expedient; however, consulting the regulatory source documents is the most accurate approach.

All federal regulations are compiled and updated annually in the *Code of Federal Regulations* (*CFR*). Each federal agency is assigned a numerical title of the *CFR* in which the agency's regulations are published. The FDA regulations are published in Title 21, which includes eight volumes of regulations, pertaining to foods, drugs, and other products administered by the FDA. The first three volumes contain the regulations related to foods (Part 1 through 99, Parts 100 through 169, and Parts 170 through 199). USDA's food labeling regulations can be found in the second volume (Parts 200 to end) of Title 9 of the *CFR*.

Since the *CFR* is updated only on a annual basis, other sources must be consulted to identify recent regulatory changes. The *Federal Register* releases proposed and final regulations for all federal regulations on a daily basis. Regular monitoring of the *Federal Register* is the most accurate method of staying current. Many trade publications include columns that summarize regulatory updates of importance to their readers. *Food Labeling News,* a weekly newsletter, is another source for obtaining current information on regulatory aspects of food labeling, advertising, and packaging.

BIBLIOGRAPHY

1. Frank Olsson, and P. C. Weeda, *Complying with the Nutrition Labeling and Education Act,* The Food Institute, Fairlawn, NJ, 1993.
2. *White House Conference on Food, Nutrition, and Health. Final Report,* U.S. Government Printing Office, Washington, DC, 1970.
3. *Fed. Reg.* **38**, 2125–2132 (Jan. 19, 1973).
4. *Dietary Goals for the United States,* U.S. Government Printing Office, Washington, DC, Stock No. 052-070-04376-8, 1977.
5. Nutrition Labeling and Information: Hearings Before the Subcommittee on Nutrition of the Senate Committee on Agriculture, Nutrition, and Forestry, 95th Congress, 2nd Session, 38, 1978.
6. U.S. Department of Agriculture/Department of Health and Human Services, *Nutrition and Your Health: Dietary Guidelines for Americans,* Consumer Information Center, Department 622N, Pueblo, CO, 1985.
7. *The Surgeon General's Report on Nutrition and Health,* USDHHS (PHS) Publication No. 88-50211, 1988.
8. National Research Council, *Diet and Health: Implications for Reducing Chronic Disease Risk,* National Academy Press, Washington, DC, 1989.
9. American Cancer Society Special Report, *Nutrition and Cancer: Cause and Prevention,* American Cancer Society, 1984.
10. American Heart Association, "Dietary Guidelines for Healthy American Adults: A Statement for Physicians and Health Professionals by the Nutrition Committee, American Heart Association," *Circulation* **77**, 721A–724A (1988).
11. Institute of Medicine, *Nutrition Labeling: Issues and Directions for the 1990s,* National Academy Press, Washington, DC, 1990.
12. Public Law 1001-535.
13. J. Storlie, *Food Label Design: A Regulatory Resource Kit,* Institute of Packaging Professionals, Herndon, VA, 1996.
14. *Fed. Reg.* **58**, 2066–2956 (Jan. 6, 1993).
15. *Fed. Reg.* **58**, 632–691 (Jan. 6, 1993).

General References

9 *Code of Federal Regulations,* Parts 200 to end, Superintendent of Documents, U.S. Government Printing Office, Washington, DC (for USDA food labeling regulations Title 9).

21 *Code of Federal Regulations,* Parts 1–99, 100–169, Superintendent of Documents, U.S. Government Printing Office, Washington, DC (for FDA food labeling regulations).

Food Labeling News, CRC Press, Washington, DC.

J. L. Vetter, *Food Labeling: Requirements for FDA Regulated Products,* American Institute of Banking, Manhattan, KS, 1993.

JEAN STORLIE
Nutrition Labeling Solutions
Ithaca, New York

KAREN HARE
Nutrition Services, Inc.
Crystal Lake, IL

NYLON

Nylons are selected for applications in the packaging industry mainly for their functional contributions. In general, they offer clarity, thermoformability, high strength and toughness over a broad temperature range, chemical resistance, and barrier to gases, oils, fats, and aromas. For most packaging applications, nylons are used in film form, as single components in multilayer structures (see Films, plastic; Multilayer flexible packaging).

Nylons, or polyamides, are thermoplastics characterized by repeating amide groups ($-CONH-$) in the main polymer chain. The various types of nylon differ structurally by the chain length of the aliphatic segments separating adjacent amide groups. The combination of hydrogen bonding of amide groups and crystallinity yields tough, high-melting thermoplastic materials. The flexibility of aliphatic chains permits film orientation to further enhance strength (1). As the length of the aliphatic segment increases, there is a reduction in melting point, tensile strength, water absorption, and an increase in elongation and impact strength. Copolymerization also tends to inhibit crystallization by breaking up the regular polymer chain structure and likewise yields lower melting points than the corresponding homopolymers (2). The selection of a particular nylon for an application involves consideration of specific physical requirements (mechanical properties, barrier properties, dimensional stability), processability (melting point), cost, etc. Table 1 lists comparative properties of commercial nylon films.

As noted above, polyamides can be described as long-chain molecules with amide functionalities ($-CONH-$) as an integral part of the repeat unit. Film-forming nylons are usually linear and conform to either of two general structures.

$$NH_2-R-\overset{O}{\overset{\|}{C}}\left[-N-R-\overset{O}{\overset{\|}{C}}-\right]_n N-R-\overset{O}{\overset{\|}{C}}OH$$
$$\underset{H}{} \qquad \underset{H}{}$$

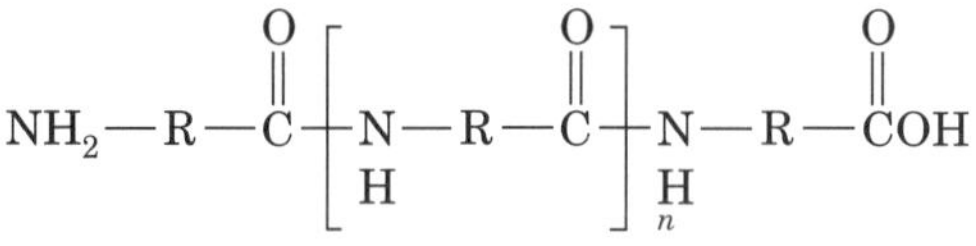

(1)

$$NH_2-R'-\underset{H}{N}\left[-\overset{O}{\overset{\|}{C}}-R''-\overset{O}{\overset{\|}{C}}N-R'-\underset{H}{N}-\right]_n \overset{O}{\overset{\|}{C}}-R''-\overset{O}{\overset{\|}{C}}OH$$

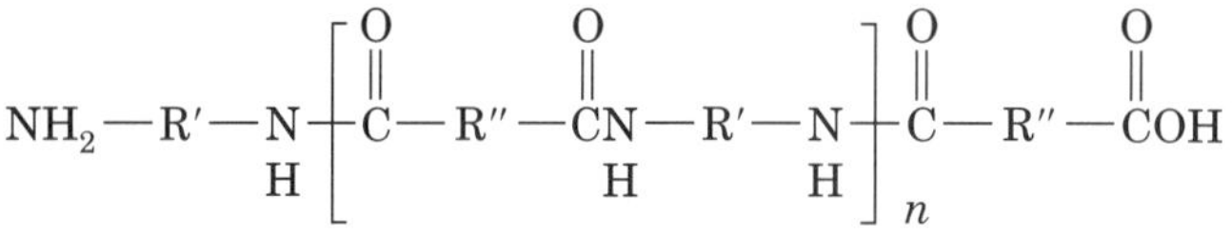

(2)

Examples of the first polymers are nylon 6 [$R = (CH_2)_5$, nylon 11 [$R = (CH_2)_{10}$] and nylon 12 [$R = (CH_2)_{11}$]. Here, the nylon type corresponds to the total number of carbon atoms in the repeat unit. Examples of the second-type nylons are nylon 6,6 [$R' = -(CH_2)_6-$, $R'' = -(CH_2)_4-$], and nylon 6, 10 [$R' = -(CH_2)_6-$, $R'' = -(CH_2)_8-$]. In the case of the second-type polymers, R' refers to the number of methylene (CH_2) groups or carbon atoms between the nitrogen atoms and the R'' refers to the number of methylene groups or carbon atoms between the CO groups. The n in the formula is the degree of polymerization and its value determines the molecular weight of the polymer.

Nylons can also be prepared as copolymers. For example, a nylon 6/6, 6-copolymer might have the formula

$$NH_2-R-\overset{O}{\overset{\|}{C}}\left[-\underset{H}{N}-R'-\underset{H}{N}\overset{O}{\overset{\|}{C}}-R-\overset{O}{\overset{\|}{N}C}-R''-\overset{O}{\overset{\|}{C}}-\right]_n \underset{H}{N}-R'-NH_2$$

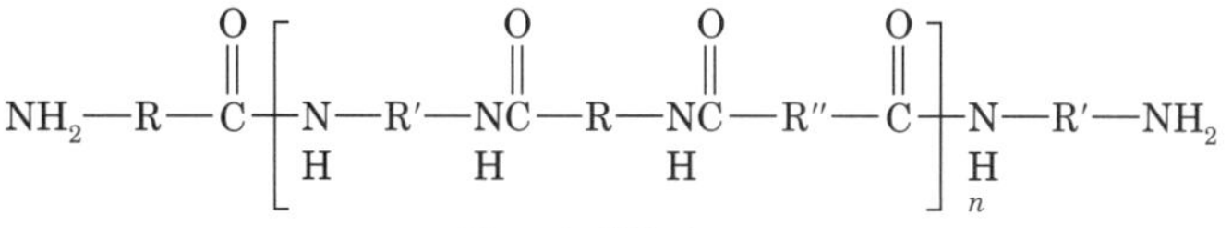

$$R = -(CH_2)_5-$$
$$R' = -(CH_2)_6-, \qquad R'' = -(CH_2)_4-$$

If the amounts of the nylon 6 and nylon 6,6 monomers are varied, many combinations of the comonomers are possible. This would give rise to a series of nylon 6/6,6 copolymers with different properties (3).

Table 1. Properties of 1-mil (254.4-μm) Nylon Films

		Value				
Property	ASTM test	Nylon 6	Nylon 6,6	Nylon 6/6,6	Nylon 6/12	Nylon 6, biaxially oriented
Melting point, °F (°C)		424–428 (218–220)	510 (266)	388–395 (198–202)	386–392 (197–200)	424–428 (218–220)
Specific gravity	D1505	1.13	1.14	1.11	1.10	1.15
Yield, in.^{2}/(lb·mil) [m^2/(kg·mm)]		24,500 [1,372]	24,300 [1,361]	25,000 [1,400]	25,200 [1,411]	24,000 [1,344]
Haze, %	D1003	1.5–4.5	1.5	2.0		1.3
Light transmission, %						
Tensile strength, psi (MPa) MD	D882 12,000 (82.8)	9,000 (62.1)	16,000 (110.3)	4,640 (32.0)	32,000 (220.7)	
XD		10,000 (69.0)	9,000 (62.1)	16,000 (110.3)	4,500 (31.0)	32,000 (220.7)
Elongation, %	D882					
MD		400	300	400	400	90
XD		500	300	400	500	90
Tensile modulus, 1% secant, psi (MPa) MD		100,000 (689.7)	100,000 (689.7)	100,000 (689.7)		250,000 (1724.1)
XD		115,000 (793.1)	100,000 (689.7)	100,000 (689.7)		250,000 (1724.1)
Tear strength, gf/mil (N/mm)						
initial	D1004	500 (193)	600 (232)	500 (193)		200 (77.2)
propagating	D1922	35 (13.5)	35 (13.5)	70 (27)		10 (3.9)
Bursting strength, 1 mil (25.4 μm), psi (kPa)	D774	Does not burst, 10–18 (69–124)	18 (124)			
Water absorption, 24 h, %	D570	9	8	9	3	7–9
Folding endurance cycles	D2176	>250,000				
Change in linear dimensions at 212°F (100°C) for 30 min, %	D1204	<2		1		<2.5
Service temperature, °F (°C) range		−50–250 (−46–121)	−100–450 (−73–232)	−50–240 (−46–116)		−76–266 (−60–130)
Heat-seal temperature, °F (°C) range		410–420 (210–216)	490–500 (254–260)	375–385 (191–196)	360–375 (182–191)	410–420 (210–216)
Oxygen permeability cm^3·mil/(100 in.^{2}·day·atm) [cm^3·μm/(m^2·day·kPa)]	D1434					
23°C, 0% rh		2.6 [10.1]	3.5 [13.6]		2.0 [7.8]	1.2 [4.7]
23°C, 50% rh				5.0 [19.4]		
23°C, 95% rh		3.5 [13.6]	16.0 [62.2]		2.2 [8.5]	3.5 [13.6]
Water-vapor transmission rate, g·mil/100 in^2·day) [g·mm/(m^2·day)], 100°F (38°C), 90 % rh		18 [7.1]	19 [7.5]	31 [12.2]	7 [2.8]	17 [6.7]
COF, face-to-face back-to-back	D1894	0.25–0.65	0.45			

Processing Methods

Extrusion. Nylons are melt-processable via conventional extrusion, but some parameters differ from those used in extruding other resins (see Extrusion). Since the quality of extruded film is sensitive to raw-material defects, nylon resins should be clean and free of gel particles. Selection of resin viscosity depends on use. Low-viscosity resins are used in extrusion coating to allow rapid drawdown rates, whereas medium- to high-viscosity resins are preferred for film production. Because nylons are hygroscopic, special care is required to assure sufficiently low water content (<0.10 wt%) for the extrusion process. Unless properly packaged and stored, nylon resins absorb moisture from the atmosphere and inevitably pose processing difficulties (4). In any extrusion operation, continuous production of uniform-quality film as well as

Table 2. Typical Temperature Profiles for Nylon Extrusion, °F(°C)

	Nylon 6	Nylon 6,6	Nylon 6/6,6[a]
Feed zone	446–482 (230–250)	500–554 (260–290)	401–464 (205–240)
Transition zone	437–500 (225–260)	500–545 (260–285)	401–482 (205–250)
Metering zone	396–527 (220–275)	500–545 (260–285)	392–491 (200–255)
Head	437–518 (225–270)	500–545 (260–285)	401–482 (205–250)
Die	419–518 (215–270)	494–563 (255–295)	392–482 (200–250)
Melt temperature	437–518 (225–270)	500–512 (260–300)	401–482 (205–250)

[a] Nylon 6/6,6 temperature profile reported by Allied Corporation for XTRAFORM resin.

maintenance of a safe work environment is best achieved by monitoring and tightly controlling all machine variables (temperatures, head pressure, extruder drive load, screw rpm, etc) (5,6).

Many film converters have been successful in producing nylon film on conventional polyethylene extruders, making only small modifications. The most important factors to consider are temperature control and screw design. The extruder must be equipped with adequate heaters for the required processing temperatures (5,7). In addition to heating capability, temperature control should permit little fluctuation (±3°C) to assure delivery of a proper and consistent melt. Typical temperature profilies are shown in Table 2.

Although a variety of screws have been used successfully for processing nylon resins, not all screws are optimum for the nylon being processed. In general, a nylon screw is thought of as a rapid-transition metering-type screw. A compression ratio (volume of a feed flight relative to the volume of a metering flight) of 3 : 1 to 4 : 1 is acceptable for most nylons. Most nylon screws are 40% metering, 3–4 turns transition, and the remainder feed zone. Length to diameter ($L:D$) ratios of 20 : 1 and 24 : 1 are common and acceptable. In designing a screw for a specific nylon, factors that must be considered include melting point of the resin, melt viscosity characteristics, machine type, extrusion rate, etc (4).

Film manufacture. Nylon film can be produced by either the cast-film process or the blown-film process. Semicrystalline polymers like the nylons can be made "amorphous" by rapidly quenching the melt via the cast-film process. That is, the polymer chains are prevented from aligning and organizing into regular, three-dimensional "crystalline" structures. Thus, by varying quenching rates, nylons are capable of existing in a low-order or amorphous state and in a highly crystalline state. The properties of the final film are highly dependent on the crystalline state of the polymer. By a rapidly quenching the melt, favorable properties of transparency and thermoformability are induced.

In producing nylon film by the blown-film process, the cooling rate is much slower than for the cast-film process: the film is allowed to crystallize and film clarity is generally sacrificed. Blown films are used in applications requiring tubular film, or films of higher strength or better gas-barrier properties than yielded by the cast process. A chrome-plated bottom-fed die with a spiral mandrel is generally recommended for blown-film processing to minimize weld lines and polymer degradation due to stagnant hold-up of melt in the die. Because of their stiffness, nylons pose wrinkling problems in the bubble-collapsing phase of production. Special care is required to properly align the collapsing frame and nip rolls, minimize gauge variations, and limit drag force in the collapsing assembly. It has been suggested that the lower-density nylons show less-severe wrinkling problems (4).

Oriented nylon films. The high strength and toughness properties of nylon films can be further enhanced by orientation. The increased alignment and tighter packing of polymer chains resulting from the process also yields improved barrier properties. Table 3 lists property comparisons for unoriented vs biaxially oriented films. Preferential orientation, typically machine-direction, improves strength and toughness in the direction of orientation. Biaxial orientation yields films with balanced properties. Market development efforts in this area are concentrated on critical packaging applications requiring soft films that permit tight package conformation and offer improved impact strength and reduced pinholing, superior burst strength, and flexcrack resistance. As with most other flexible substrates, oriented films permit further conversion, (eg printing, lamination, metallization, etc).

Three processes are used to manufacture biaxially oriented nylon film.

One-step tenter frame. This process simultaneously draws cast nylon film in both the machine and transverse direction (8).

Two-step tenter frame. This is a two-step orientation process in which a nylon film that has been modified with a plasticizer is first drawn in the machine direction and then drawn in the transverse direction (9).

Blown bubble. Nylon film extruded from a circular die is oriented in the transverse direction by controlled internal air pressure, and oriented in the machine direction by regulating the bubble takeoff speed (10).

Coextrusion. Nylons used in film extrusion are often combined with other plastic materials via coextrusion. In most cases, polyolefins are used as coextrusion partners for nylon to provide heat sealability, moisture barrier, and good economics. As in single-layer-film extrusion, nylons can be processed by cast-film coextrusion and by blown-film coextrusion (see Coextrusion, tubular). The combining-adapter technology is generally the preferred method for joining layers in cast film coextrusion of nylon, although multimanifold-die systems are also used. Special care in matching resin viscosities is required when using the combining-adapter system in order to produce films of uniform layer profile. For both systems, nylon 6 is the most common and preferred polyamide used in cast film coextrusion.

Because the blown-film coextrusion process employs air as

the cooling medium, the melt is cooled slowly, permitting spherulite formation in semicrystalline nylon homopolymers (ie, nylon 6, nylon 6,6) and film transparency is sacrificed. For this reason, less-crystalline nylon copolymers such as nylon 6/12 and nylon-6/6,6 are gaining acceptance in blown-film coextrusion.

Applications for nylon/polyolefin coextruded films include vacuum packaging of meat products, cheese-ripening pouches, consumer packaging of cheese and fish products, and several nonfood packaging applications including containers for chemicals, fertilizers, and animal foods.

Extrusion coating. Nylons with lower viscosity permit rapid drawdown rates for extrusion coating (see Extrusion coating). Typical substrates range from heavy-duty paperboard to intermediate aluminum foils and papers, to thin polyethylene films. Published literature describes the nylon extrusion coating process in detail (4,11). As in cast-film production, annealing the nylon coating is necessary to impart dimensional stability when a flexible substrate is used.

Blow molding. Nylons of ultrahigh viscosity are commonly used in blow molding (see Blow molding). Because of their excellent impact strength, chemical resistance, toughness, and wide temperature use properties, nylons are ideally suited for several blow-molded applications, including industrial containers, moped fuel tanks, and automotive oil reservoirs. A recent development for nylon in blow molding has been in the area of blow-molded hydrocarbon-barrier containers. The process employs a blend of a modified nylon barrier resin and a polyolefin, extrusion blow molded under controlled mixing conditions to produce a barrier container competitive with surface-treated HDPE bottles (1,2,11). Packaging applications include containers for charcoal lighter fluids, general-purpose cleaners, waxes, polishes (see Surface modification) and those described in several patents (12).

Secondary Conversion

Thermoforming. Nylon films are readily thermoformed by conventional methods (see Thermoforming.) Ease of formability is affected by nylon type (melting point), molecular weight, degree of crystallinity, and machine variables. In general, nylon films offer excellent thermoformability due to their high elongation. Further, the high elongation facilitates deep draw, and flex- and stress-crack resistance of nylons minimize film breaks during and after forming. Current applications include vacuum and gas packaging of meats and cheeses, and thermoform/fill/seal packaging of disposable medical devices (see Controlled Atmosphere Packaging; Healthcare packaging; Thermoform/fill/seal.)

Heat sealing. Because of their high melting points, nylons are typically coextruded with, laminated to, or coated with a polyolefin heat seal layer (PE, EVA, EAA, ionomer, etc). However, unsupported nylon films are heat sealed for applications that require heat-seal integrity under high-temperature exposure (eg, oven "cook-in" bags). By properly balancing the variables of time, temperature, and pressure, unsupported nylons can be heat-sealed at relatively low temperatures. For most commercial applications, however, it is necessary to heat the films to temperatures that closely approach their melting points. This factor, compounded by the relatively narrow melting range of nylons, necessitates precise temperature (±3°C) and pressure control. By making necessary modifications to conventional machinery, nylons are successfully heat-sealed by thermal impulse and constant-heat techniques.

Adhesive lamination. Nylon films are combined with other flexible materials via adhesive lamination to produce multiple structures, each ply contributing to the requirements of the end product (see Laminating; Multilayer flexible packaging). Typical substrates include sealant webs (PE, EVA, ionomer,

Table 3. Comparative Properties of Unoriented, Uniaxially Oriented (MD), and Biaxially Oriented Nylon-6 Films (1-mil or 25.4-μm Filims)

Property		Values		
		Nylon 6 film, unoriented	Nylon 6 film, MD-oriented	Nylon 6 film, biaxially oriented
Specific gravity		1.13	1.14	1.15
Tensile strength, psi	MD	12,000 (82.8)	50,000 (344.8)	32,000 (220.7)
(MPa)	XD	10,000 (69.0)	10,000 (69.0)	32,000 (220.7)
Elongation, %	MD	400	60	90
	XD	500	450	90
Tensile modulus psi (MPa)	MD	100,000 (689.7)	300,000 (2069)	250,000 (1724.1)
	XD	115,000 (793.1)	100,000 (689.7)	250,000 (1724.1)
Initial tear strength, gf/mil (N/mm)	MD	500 (193)	650 (251)	200 (77.2)
	XD	500 (193)	1,300 (502)	200 (77.2)
Propagating tear strength, gf/mil (N/mm)	MD	35 (13.5)	40 (15.4)	10 (3.9)
	XD	35 (13.5)	100 (38.6)	10 (3.9)
Oxygen permeability, [$cm^3 \cdot \mu m/(m^2 \cdot day \cdot kPa)$] $cm^3 \cdot \mu m/(100\ in.^2 \cdot d \cdot atm)$		2.6 [10.1]	2.4 [9.3]	1.2 [4.7]
WVTR, $g \cdot mil/(100\ in.^2 \cdot day)$ [$g \cdot mm/(m^2 \cdot day)$]		18 [7.1]	18 [7.1]	17 [6.7]

etc), and aluminum foils. Converters may choose to laminate rather than coextrude nylon to minimize scrap losses for short runs. Lamination is also necessary when combining non-coextrudable or incompatible plastics. The lamination process is described in detail in published literature (13). The adhesive is generally applied to the nylon web which has been corona treated (see Surface modification) to assist wettability, as the high melting point of nylon provides suitable stability in solvent-drying ovens. Adhesive systems are typically two-component types that vary for nylon types and substrates. Nylon film suppliers recommend adhesives for specific substrates (see Adhesives).

Vacuum metallizing. Applications for metallized oriented nylon films are expanding in the packaging industry. Metallization offers functional contributions of improved moisture, oxygen, and light barriers and unique aesthetic appeal at the consumer level (14). The resultant film offers excellent flexibility, oxygen barrier, flex-crack resistance, antistatic properties, and printability. These properties meet the necessary requirements for use in such applications as institutional coffee pouches, metallized balloons, and liquid-box containers (see Bag-in-box, liquid product; Metallizing).

Packaging Applications

Although nylons are not generally considered commodity-packaging resins, the added material cost is easily justified in specific demanding applications where the nylons' physical properties provide added protection, extended shelf life, or reduced losses of expensive contents. The combination of excellent thermoformability, flexcrack resistance, abrasion resistance, gas-, grease-, and odor-barrier, and tensile-, burst- and impact-strength over a broad temperature range make nylons well suited for many packaging applications. For most applications, nylons are combined with other materials that add moisture barrier and heat sealability, such as low-density polyethylene, ionomer, EVA, and EAA.

Most nylon-containing packaging films are used in food packaging, principally in vacuum-packing bacon, cheese, bologna, hot dogs and other processed meats (15). A variation of this package includes a carbon dioxide flush prior to heat sealing to remove traces of oxygen, thus prolonging shelf life for foods such as poultry, fish, and fresh meat (16). Two recent developments are nylon composites used in vacuum-packing cooked whole lobster, boasting a 2-year shelf life and canless "canned" ham, where the product is vacuum-packed and cooked in its package (17). Poly(vinylidene chloride) (PVDC) coatings are offered for improved oxygen-, moisture vapor-, or uv light-barrier properties (see Vinylidene chloride copolymers).

Nylon 6 is the nylon resin used most frequently for packaging applications because of the balance of cost, physical properties, and process adaptability. For blown or cast extrusion as well as cast coextrusion, nylon 6 resins are favored by most converters, while lower-melting nylon copolymers (nylon 6/6,6 or nylon 6/12) have been developed primarily to aid blown-film coextruders (lower melting points permit lower process temperatures for faster melt quenching). In addition to lower melting points, the nylon copolymers are less crystalline than their corresponding homopolymers and provide better clarity and thermoformability. On the high end of the melting-point scale, nylon 6 and nylon 6,6 resins are appropriate for use in oven-cooking bags, where high-temperature tolerance is a key requirement.

Medical-packaging applications, such as packaging of hypodermics and other medical devices, are a relatively new and expanding area for the nylons. The combination of toughness, puncture resistance, impact strength, abrasion resistance, and temperature stability make nylons appropriate for protecting sterile devices during shipping and storage. Although ethylene oxide and steam have always been appropriate means of sterilization for nylons, modified-nylon resins have recently been introduced that permit radiation sterilization as well (18).

Biaxial orientation of nylon films provides improved flex-crack resistance, mechanical properties and barrier properties. These films have new applications in packaging foods such as processed and natural cheese, fresh and processed meats, condiments, and frozen foods. They are used in pouches and in bag-in-box structures (see Bag-in-box, dry product; Bag-in-box, liquid product) (19). In other areas (eg, cooked meats, roasted peanuts, smoked fish) the nylons compete with biaxially oriented polyester (see Film, oriented polyester). Although oriented nylons offer better gas barrier, softness, and puncture resistance, oriented polyester offers better rigidity and moisture barrier.

Other uses for nylon film include a composite pouch for wine (20) a nylon 6 shrink film (see Films, shrink) for meat and fresh-vegetable packaging (21), a nylon composite film used in a system to produce greaseless fried chicken (27), a uniaxially oriented nylon-6 film for food packaging (23), and a nylon 6 film with improved thermoformability (24).

BIBLIOGRAPHY

"Nylon" in *The Wiley Encyclopedia of Packaging,* 1st ed., by M. F. Tubridy and J. P. Sibilia, Allied Corporation, pp. 477–482.

1. R. D. Deanin, *Polymer Structure, Properties and Applications,* Cahners Books, Division of Cahners Publishing Company, Boston, 1972, pp. 455–456.
2. K. J. Saunders, *Organic Polymer Chemistry,* Chapman and Hall London, 1976, pp. 175–202.
3. U.S. Pat. 4,417,032 (Nov. 22, 1983), Y. P. Khanna, E. A. Turi, S. M. Aharoni, and T. Largman (to Allied Corporation).
4. R. M. Bonner, "Extrusion of Nylons", M. I. Kohan, ed., *Nylon Plastics,* Wiley New York, 1973.
5. E. C. Bernhardt, *Processing of Thermoplastic Materials,* Reinhold, New York, 1959.
6. J. M. McKelvey, *Polymer Processing,* Wiley, New York, 1962.
7. R. J. Palmer, "Polyamide Plastics" in J. Kroschwitz, ed., *Kirk-Othmer Encyclopedia of Chemical Technology,* Vol. 17, 4th ed., Wiley Interscience, New York, 1996, pp. 559–584.
8. U.S. Pat 3,794,547 (Feb. 26, 1974), M. Kuga and co-workers (to Unitika, Ltd.)
9. U.S. Pat. 26,340 (reissued Aug. 2, 1977), K. Khisha (Orig); U.S. Pat. 3,843,479 (Oct. 24, 1974), I. Hayashi and K. Matsunumi (to Toyo Boseki, Ltd.)
10. U.S. Pat. 3,499,064 (March 3, 1970) K. Tsuboshima and co-workers (to Kohjin Co., Ltd.)
11. S. M. Weiss, in J. Abranoff, ed., *Modern Plastics Encyclopedia,* Vol. 16, No. 10, McGraw-Hill Publications, New York, 1984, pp. 199–202.

12. U.S. Pat. 4,410,482 (Oct. 18, 1983) P. M. Subramanian; U.S.Pat. 4,416,942 (Nov. 22, 1983) R. C. Di Luccio; U.S. Pat. 4,444,817 (Apr. 24, 1984) P. M. Subramanian,
13. J. A. Pasquale, in J. Agranoff, ed., *Modern Plastics Encyclopedia,* Vol. 61, No. 10, McGraw-Hill Publications, New York, 1984. pp. 284–286.
14. W. Goldie, *Metallic Coating of Plastics,* Electrochemical Publications Ltd., Hatch End, Middlesex, 1969.
15. E. C. Lupton, *Modern Packag.* **52**(12), 26 (Dec. 1979).
16. M. Gilbert, *Plast. Eng.* **34**(10), 41 (Oct. 1978).
17. *Mod. Plast. International* **12**(7), 21 (July 1982).
18. Allied Corporation, *Modern Plast.* **60**(9), 114, 119 (1983).
19. D. May, *Pap. Film Foil Converter* **56**(1), 46 (Jan. 1982).
20. *Mod. Plast.* **59**(12), 28 (1982).
21. *Plast. Ind. News* **29**(10), 148 (1983).
22. *Food & Drug Packag.* **46**(12), 30 (1982).
23. *J. Commerce* **353**(25,263), 22B (July 20, 1982); *Plastics World* **40**(8), 76 (1982).
24. *J. Commerce* **345**(24,747), 5 (July 2, 1980).

General Reference

M. I. Kohan, *Nylon Plastics Handbook,* Carl Hanser Verlag, Munich, Germany 1995. An excellent reference book covering all aspects of nylon technology.

O

OFFSET PRINTING. See PRINTING; DECORATING

OXYGEN SCAVENGERS

Oxygen scavengers are the most effective commercially available means of reducing or essentially eliminating the oxygen in a package's headspace. Oxygen is a well-known causative agent in many types of food spoilage, and its reduction in the package can lead to control of aerobic microorganisms such as mold, and control over detrimental color and flavor changes (1–3). Although food preservation is the predominant commercial use, there are pharmaceutical and industrial uses as well. Oxygen scavengers are most commonly a packet or sachet containing a food-safe sacrificial compound that is formulated to rapidly corrode or otherwise oxidize and thereby chemically eliminate free oxygen in a hermetically sealed package. While the sachet is the most common format, there are also oxygen-scavenging pressure-sensitive labels, and oxygen-scavenging cap liners, and recent patent activity suggests that there should also be oxygen-scavenging films or sheets in the future. The great majority of all oxygen scavengers in commercial use are based on the oxidation of iron powders, although other oxidizeable media are available. The technology has been extensively patented, especially in Japan.

Numerous chemistries and sizes of oxygen scavengers have been developed for specific applications. Misapplications involving the wrong chemistry or too small a capacity may result in failure to adequately control the oxygen in the headspace with resultant food spoilage, so a great deal of care is necessary in system design. In well-designed oxygen-scavenger/package systems, the attainment of 100 ppm or less of residual headspace oxygen should be the norm. At 100 ppm, effective control of the detrimental effects of oxygen has been demonstrated (4). Oxygen scavengers are generally used in conjunction with modified atmosphere as a further enhancement to a gas flush. The gas flush is a more economic means of reducing the oxygen down to as little as 0.5%, but flushing alone is not capable of reaching the ppm levels of the scavenger. Oxygen scavengers have been used commercially in Japan since 1977 and in the United States since 1984. In 1989 usage in Japan alone was reported at 6.7 billion sachets (5) and by 1993 was estimated at 10 billion equivalent sachets (6). For quantitive tracking, this was based on conversion of all sizes into 100-mL-capacity equivalents.

History

The earliest mention of oxygen scavengers in food packaging is generally attributed to a Dr. Tallgren in Finland in 1938. In that year he was granted patent BP 496,935 for "keeping food in closed containers with water carrier and oxidizeable agents such as zinc dust, iron powder, and manganese dust, etc." (5). Dr. Tallgren was followed by a Mr. Isherwood of the Low Temperature Research Station in Cambridge in 1943 with a patent for "removing oxygen from a container containing vacuum or gas packed food in which a metal absorbs oxygen to form an oxide" (5). This earliest activity was followed by efforts to use powdered metals to purify industrial gasses and the development of a system by American National Can in the 1960s to scavenge oxygen by use of palladium catalysts in a hydrogen-flushed package. In 1969, a Mr. Fujishima attempted to market a hydrosulfite-based oxygen scavenger that reacted very quickly, causing problems with lost capacity, and had a propensity to also react with food volatiles to form noxious sulfur-based gasses. However, none of these early developments gave rise to any significant commercial activity. That distinction was Japanese and belongs to the Mitsubishi Gas Chemical Co. for marketing of their very successful "Ageless"-brand iron-based sachets that began in 1977.

The early success in Japan seems to be closely tied to the tradition of giftgiving and the importance of flawless presentation and quality (5). When the gift was a food item, the quality enhancement and assurance afforded by an oxygen scavenger was a reasonable precaution. In addition, there were several confectionery items given as traditional gifts that suffered from short shelf life due to mold. For items such as "rice cake," the shelf life and quality improvements afforded by a barrier package and an oxygen scavenger changed distribution and made "Ageless" a successful product. Competitors did follow in Japan, and today there are approximately one dozen, although Mitsubishi holds a commanding industry lead. Iron-based oxygen scavengers are not the only choice, however, as Toppan Printing Co. has marketed an ascorbic acid–based absorber since 1979. A catechol-based scavenger has also been marketed in Japan.

Oxygen scavengers were first used in the United States in 1984 when General Foods began to ship ground roast coffee with their "Fresh Lock," a Mitsubishi-supplied oxygen-scavenger formulation. Multiform Desiccants Inc. entered the U.S. market with their "Freshpax" oxygen scavengers in 1988 and an initial application for Meal, Ready-to-Eat as developed by the U.S. Army Natick Research, Development and Engineering Command. The Aquanautics Corp. worked on oxygen-scavenger technology for many years and introduced "SmartMix and SmartCup" technology for compounding into packaging such as bottle-cap liners in the late 1980s (7). In addition, CMB Technology attempted to introduce their "Oxbar" total oxygen-barrier system, which used a cobalt catalyst to scavenge oxygen and failed to win Regulatory approval (8).

No other market has yet experienced the explosive acceptance that has characterized the Japanese industry. Food is less expensive in the United States than in Japan, thus the cost of an oxygen scavenger versus the value of preservation afforded a greater incremental packaging expense. The U.S. food industry has had other reservations, especially regarding the liability of loose sachets in food packages. However, there is growing acceptance in the United States and Europe as the technology is adapted to specific applications that have economic significance in our markets (9).

Addressing Spoilage with Oxygen Control

Use of an oxygen scavenger will obviously assist in food preservation only when an oxygen-related food-spoilage phenomenon is the critical factor in deterioration. The economic significance of the various types of spoilage attributable to oxygen also vary and should be factored into any decision to employ a scavenger. In situations where control of oxygen is warranted, the advantages have included greatly reduced spoilage losses, improvement in consumer confidence and relations, increased distribution radius due to longer shelf life, increased sales (10), and improved plant scheduling due to longer product runs taking advantage of the longer shelf life.

Control of aerobic microorganisms is a primary use of an oxygen scavenger. While both vacuum packaging and modified-atmosphere packaging (MAP) have the capability to aid in control of aerobes, Hoshino and Nakamura (4) have reported that mold colonies may grow in even 0.4% residual oxygen in as little as 10 days. Both vacuum and MAP techniques may not be able to effectively deliver or maintain such low initial residual oxygen concentrations, although an appropriate concentration of carbon dioxide may also effectively suppress mold growth. However, Powers and Berkowitz (11) reported that an oxygen absorber prevented mold growth in specially formulated and packaged bread for up to 3 years, and oxygen scavengers are also reported to control mold growth on crusty rolls for at least 60 days (12). In commercial practice, use of an oxygen scavenger to control mold is a major use in Japan but has been of minor importance elsewhere in the world. Use of barrier packaging for baked goods appears to be more prevalent in Japan than in the United States, and oxygen scavenger use makes sense only where a suitable barrier is employed. However, the MRE, beef-jerky, and fresh pasta industries in the United States have adopted oxygen scavengers to control mold growth on their products during distribution. Permeation of oxygen into the package over time which leads to mold growth is the major concern.

Another oxygen related mode of food spoilage is through oxidation of various pigments in the food itself (2,13). This generally takes the form of a darkening of color, and while it does not in and of itself make the food product inedible, consumers appear to perceive that such darkening is an effective indicator of other spoilage problems that may have occurred simultaneously. Therefore, such darkening of pigments may effectively end the salable shelf life of the packaged product. The smoked/processed-meat industry is especially aware of the problems of discoloration and consumer reluctance to purchase packages exhibiting a color darker than other packages in the same retail case. Oxygen scavengers, more than any other packaging method, excel at reducing headspace oxygen to such a low level that pigment oxidation does not occur for extended periods of time (13,14). Use of oxygen scavengers in the United States and Europe is very often for reasons of color preservation.

Another mode of oxygen related deterioration of food is through oxidation of fats and oils in the food product (4,13,15). This process is commonly referred to as "rancidity" and yields organoleptic changes that are usually considered undesirable even though rancidity is commonly accepted in some food products such as fried fish and nutmeats. The byproducts of this type of oxidation may include peroxides, aldehydes, carbonyls, and carboxylic acids. Use of an oxygen scavenger has been shown to positively control this type of oxidation (16) since peroxide values have remained essentially unchanged over time in packages protected by the scavenger. Conversely, reduction of headspace oxygen to the low-ppm range tends to preserve vitamin and nutrient content of foods. Examples in which flavor change was so economically significant to warrant use of a scavenger are thus far limited to processed nuts and dehydrated foods.

An often overlooked example of a type of food spoilage that is related to oxygen in the package headspace is insect infestation. Many grains and flour products suffer losses due to weevils and the like, and it has been shown that these can be controlled by use of an oxygen scavenger (4). While an infestation may have been suppressed during the storage of the bulk grain, either a later hatch or a reinfestation may occur during packaging or distribution. Insect eggs, larvae, or pupae may be present, but all need some amount of oxygen to survive. Elimination of the headspace oxygen will eventually kill all forms of the infestation and prevent further damage, although it may take as long as 12 days to kill the eggs of some of the more resilient insects. There have been limited uses of this concept in the United States for bird seed and preservation of museum artifacts, but it is accepted practice in Japan.

When the product to be protected is not a food product, scavengers can also provide significant benefits. Some pharmaceuticals are oxygen-sensitive, particularly among the vitamins, and β-carotene is commonly packaged with an oxygen scavenger. Although closely related to the food industry, it should be remembered that pet food and pet snacks are also protected with an oxygen scavenger as rancidity and mold appear to be at least as great a problem for a dog's palette as it is for us. There are also incidences where fine metal parts are protected from corrosion by packaging in a barrier package with an oxygen scavenger to sacrificially corrode. However, the food industry is the dominant worldwide consumer of oxygen scavengers.

Chemistry

Applications of this technology are complex primarily because of the chemistry involved and the interactions between the scavenger and the food itself. Therefore, it aids in understanding of the technology when there is a grasp of the chemistry. Since the iron-powder-based oxygen scavengers are the most prevalent, we will examine those first. As we indicated previously, the technology simply relies on a very rapid oxidation of an amount of sacrificial iron powder to prevent headspace oxygen from reacting with foodstuff components. For oxidation of iron to occur, there must be an electrolyte present and an aqueous interface to facilitate the oxidation. This is exemplified by the variance between rusting of metal at the seashore versus in the desert. The chemistry is significantly complicated by the need to control reaction speed, achieve maximum efficiency in achieving the full oxidative potential of the iron powder, maintain compatibility with various foodstuffs, and ensure food safety. Simply stated, the reaction formula may be expressed as follows:

$$4Fe + 3O_2 + 6H_2O \rightarrow 4Fe(OH)_3 \rightarrow 2(Fe_2O_3\ 3H_2O)$$

Many metals are recognized as being sufficiently oxidative to become candidates for an oxygen-scavenger formulation; however, iron does offer unique advantages that have driven the industry to general use of this medium. Iron has a relatively high affinity to combine with oxygen on a per unit weight basis compared to most choices. Elemental candidates that exceed it have drawbacks such as the odor problems with sulfur and the propensity of aluminum to form an oxidized skin layer that limits further oxidation. Of great importance is the food safety of iron powders, which are commonly used as food enrichments for such products as breakfast cereals and baby foods. Also of great importance is the relatively low cost of iron, especially in comparison to some choices such as palladium. Then there is the ready ability to manipulate iron's reactive capability to adapt to a wide variety of applications for both speed and package headspace moisture. One of the few drawbacks is that the oxidation of iron is a temperature-dependent reaction and normally slows dramatically as the temperature approaches freezing.

Knowing that iron is the main reactive vehicle in the most common formulas, what else is in those chemical formulations? We have already mentioned the need for an electrolyte source and this is commonly supplied by a metallic salt (17). While there are many feasible choices, the practical considerations of food safety have dictated that common table salt is frequently the best choice. Another extremely important factor in oxygen-scavenger chemistry is moisture control in the reaction as there may be more variances in formula components used to supply or control moisture than any other aspect. Numerous silicas, zeolites, and activated carbons are used in various proprietary formulations.

The real divisions in the various oxygen scavenger formulations come about in the various ways that the moisture is pulled into the sachet or is carried as an integral part of the formulation. There are two broad categories, which are commonly called the "moisture-dependent type" and the "self-reaction type." For the moisture-dependent chemistries, either no moisture or insufficient moisture is present in the formulation in the sachet and the chemistry must have the capability to pull moisture from the headspace of the package through a desiccant-type action. A great advantage of these formulations is that they are very stable in normal atmospheres as they are inactive until sufficient moisture has been absorbed to allow corrosion to occur. Of course, the package into which they are inserted must have sufficient moisture to allow this to occur, which limits the uses of these formulations to relatively moist food products. For the self-reaction type, the necessary moisture is compounded right into the sachet on a suitable carrier. These carriers must carry a high percentage by weight of moisture, must bind the moisture sufficiently strongly that the formulation remains free-flowing, yet must release the moisture as needed for the reaction. Unfortunately, this type of formulation will scavenge oxygen as soon as it is exposed to the atmosphere and can exhaust its capacity if exposed for extended periods. A further disadvantage is that the moisture is subject to equilibrate in the package over time, which means that with extremely dry products, the moisture will leave the sachet. Even though there will in all probability be too little moisture to damage the product, the exit of the moisture from the sachet does mean that at some future date the oxygen scavenger will stop working even though there may be unoxidized iron powder.

It should be noted that since oxygen scavengers are routinely used with MAP, there will be situations in which the application dictates use with gas flush containing carbon dioxide. This is an area for caution since many formulations are unsuitable for use with carbon dioxide because they will absorb the CO_2 preferentially over oxygen. Most of these formulations have a capacity for about five times as much CO_2 as O_2 and will thus fail to remove sufficient oxygen to prevent spoilage in such applications. Furthermore, these formulations can also reduce the beneficial effects of carbon dioxide by reducing its intended concentration in the headspace. However, there are specific formulations available from various suppliers that both scavenge oxygen and release carbon dioxide, and one of these should be used when there is CO_2 in the flush.

There are also some oxygen scavengers available that are based not on iron corrosion but on the oxidation of ascorbic acid. This technology has only achieved minor acceptance in the United States but it does have unique advantages and disadvantages. Use of an iron-based oxygen scavenger can create problems with metal detection since there is enough iron in most sachets to set off a metal detector. This is generally overcome by using metal detectors just before the insertion of the scavenger to assert that no extraneous metal is present and then another just after to assert that the scavenger has been added. In situations where this has not been possible, the ascorbic acid scavengers have been used with success. However, this organic technology generally requires a larger sachet and weight of reactant to accomplish the same oxygen removal as for the ferrous technology. This technology also releases carbon dioxide as a byproduct of its oxidation.

Common Applications

Oxygen scavengers are now found in several applications in many U.S. supermarkets. Fresh refrigerated pasta is a major user of oxygen scavengers to maintain quality during its distribution life, and at least one supplier has taken the expedient of using an adhesive to affix the sachet in the bottom of the package where it is relatively unavailable for accidental ingestion. The smoked/processed-meat industry has several segments that rely on oxygen scavengers to maintain color or prevent mold during distribution. Both presliced pepperoni in thermoformed packages and bacon bits in glass jars use an oxygen scavenger for color retention. In Great Britain, sliced deli (delicatessen) meats are often produced with very low levels of preservatives and are thus susceptible to rapid discoloration under retail lighting. These are now increasingly packaged with oxygen scavengers to prevent discoloration for up to one month. Beef jerky is commonly packaged with an oxygen scavenger for reasons of both color and mold depending on the package style. Roasted nuts are often protected from rancidity by scavengers, especially where the product will be used for export or for upscale packages. The previously mentioned Meals, Ready-to-Eat remain a major U.S. application that is being adopted by other military organizations around the world. Freeze-dried products often have an oxygen sensitivity, and items such as backpackers' trail foods are now packaged with scavengers. Pizza crusts are also being packaged for some markets with a scavenger to prevent mold during fresh distribution. The one major Japanese use that has not yet been adopted domestically seems to be for baked goods and confections, although this has become an area of study and interest.

Sizing and Chemistry Selection

Because there are many variables to consider when selecting an oxygen scavenger, some guidelines are in order. The major factors to consider are the water activity (A_w) of the food product or the relative humidity in the package headspace, the total volume of the package, the total weight of the food or other product in the package, the temperatures that the package will experience, whether carbon dioxide will be present in the headspace, the percentage of oxygen initially present in the package, the permeation rate of oxygen into the package, and the total shelf life desired. Water activity, temperature, and presence of carbon dioxide will determine the proper chemistry while the rest determine the proper sizing.

Water activity can be expressed as the "ratio of the vapor pressure of the water in a food to isothermal vapor pressure of pure water" (18,19). This concept is far more important than the total moisture content of the food because water activity is a better indicator of the actual moisture available to be absorbed by the scavenger and thus the ability of the food in a package to activate a moisture-dependent type of oxygen scavenger. The exact limits of a specific formulation's lower tolerance to water activity are published by the manufacturer but should always be verified by product tests. Moisture dependent types can be found with water activity thresholds from 0.65 up to 1.0 A_w. Self-reacting types have been used with products such as freeze dried foods having extremely low A_w. When moisture-activated types are used at their lower A_w limits, the speed of activation may be delayed and the actual reaction rate may be greatly slowed. Once the water activity, the temperatures to which the package will be exposed, and whether any carbon dioxide will be present are known, then the proper chemistry may be selected for a specific application.

The objective of the sizing calculations is to determine actual volume of the oxygen that must be scavenged over time. The volume of the package may either be calculated or determined by displacement. These calculations are best done in metric to take advantage of the expedient of one cubic centimeter (ie, 1 mL) of water weighing one gram, which greatly eases the conversions. Since the density of most food products approximates that of water, the product weight in grams can be subtracted from the package volume in cubic centimeters (milliliters) to give an acceptable headspace volume. This is an especially valuable tool because many food products contain entrained air, which will gradually permeate into the headspace and must be dealt with by the scavenger. Obviously, when the product density does not approximate that of water, then the product volume must be determined. Once the headspace volume is known, then the percentage of oxygen in the initial headspace gas is used to calculate the actual volume of oxygen that must be scavenged. If the product has a relatively short shelf life, removing this initial oxygen may be all that is required. However, for many packages, the oxygen permeating the package over its distribution life is of greater volume than the oxygen initially in the headspace. If this is suspected, then the surface area of the package should be multiplied by the permeation rate of the barrier in use to determine the amount of oxygen ingressing over the total days of shelf life. The initial oxygen and the oxygen permeating over the shelf life would then be added to determine the total capacity necessary for the proper oxygen scavenger for that application. Table 1 lists the chemistry designations, the effective water activities, and notes on best uses for several types of iron-based oxygen scavengers readily available in the United States. It should also be noted that sizing has standardized such that most of the chemistry designations are available in sachets of 20, 30, 50, 100, 200, 300, 500, 1000, and 2000 mL of oxygen capacity. Since oxygen is approximately one-fifth of the atmosphere, these oxygen capacities can be used to deoxygenate five times the stated capacity of air.

It should also be obvious that as better barriers are used, the permeation of oxygen decreases and smaller scavengers are needed to offset the ingress of oxygen. However, there is a tradeoff between the cost and complexity of the barrier and the cost and simplicity of the scavenger. This relationship appears to increasingly favor the scavenger over greater barrier. Some manufacturers have favored a barrier transmitting ≤20 mL of oxygen per square meter per day as entirely suitable for use with a scavenger. Others have favored barriers of ≤15 mL. Numerous packaging materials can provide such barriers as long as they are well and truly hermetically sealed.

Testing

One very common difficulty experienced by many first time users of oxygen scavengers is the difficulty of determining quantitatively just how well the scavenger is working. The ability to accurately measure 1% residual oxygen in a package headspace is quite common. The difficulty arises when there is only 0.01% because the partial pressure of oxygen is far less inside the package than outside, causing atmospheric oxygen to rapidly equilibrate through any tiny aperture and contaminate many samples. Septa and sampling syringes that are entirely adequate at 1% will prove inadequate at 0.01%, as many labs have learned. It is now common to use a gated or valved syringe with tight-fitting gaskets and natural-rubber septa to overcome the propensity for leakage. One manufacturer recommends that the sample be taken, the valve closed while the syringe is still in the package, and thumb pressure be maintained on the syringe plunger as it is withdrawn and until the syringe needle is actually inserted into the oxygen meter, and only then is the valve actually opened. By so doing, the consistency of their samples has improved (20). It is also difficult to obtain accurate repeat samples from the same package unless great care is exercised with the septa. It should also be noted that while there are now several makes of oxygen meters that are capable of the accuracy required for analyzing ppm levels of oxygen, there are also numerous instruments in use that do not have this precision. These are typically found in food-plant quality-control work where a gas flush is used.

Food-Safety and Regulatory Concerns

Oxygen scavengers have been used commercially since 1977 and in great quantities, as has been previously reported, and in all this history the author has not learned of a single report of any physical harm coming to anyone ingesting a sachet. The Mitsubishi Gas Chemical Co. reports in their published literature that they have tested the acute oral toxicity of their formulations in rats and find it (LD_{50}) to exceed 16 g per kilogram of body weight (21). Since the major reactive component is iron powder, a check of the LD_{50} of various food-grade iron powders shows these to also be in the range of values exceeding 5–15 g/kg of body weight. Multiform Desiccants, Inc. reports that their oxygen scavenger components are either

Table 1. Classes of Iron-Based Oxygen Scavengers Used in the United States

Water Activity	Product/Chemistry	Deoxygenation Time	Notes
All	Freshpax[a] "R"	4 h–1 day	For refrigerated or frozen foods
>0.85	Ageless[b] "FX"	0.5–1 day	Moist foods, stable in air
<0.85	Ageless "Z"	1–4 days	For dry foods
<0.7	Freshpax "D"	0.5–4 days	For dry foods
>0.65	Ageless "S"	0.5–2 days	Refrigerated, fast-working
>0.65	Freshpax "B"	0.5–2 days	For moist food, stable in air
>0.65	Keplon[c] "TY"	0.5–1 day	Oil and fat contact
>0.65	Freshpax "M"	0.5–2 days	Use with CO_2 flush
>0.65	Freshmax "M" label	0.5–2 days	Use with CO_2 flush
<0.65	Keplon "TS"	0.5–1 day	Deoxidizes in low humidity

[a] Freshpax/Freshmax = Multiform Desiccants, Inc., Buffalo, NY
[b] Ageless = Mitsubishi Gas Chemical Co. Ltd., Tokyo, Japan
[c] Keplon = Keplon Co., Ltd., Sagamihara-City, Japan

food- or pharmaceutical-grade materials. Therefore iron based oxygen scavengers can be used with a great deal of confidence.

There are some emerging issues of which end users should be aware, however. Iron toxicity has been recognized as a problem among infants in the United States as relates to accidental ingestion of dietary iron-supplement products (22), and poison-control centers can be expected to react quickly and with great caution on reports of iron ingestion by persons of low body weight (23). Because of body weight, it is extremely unlikely that an adult could ingest enough iron powder from commercially available sachets to cause harm. There is also a safety factor in that by the time a food package would reach a consumer, the iron would no longer be in a very reactive elemental form but oxidized. There are some concerns where retail packages having large oxygen scavengers may be available to infants, especially toddlers who are teething. Several strategies are being employed to effectively eliminate this risk. One of these is to always employ an oxygen scavenger having such a low iron content that no harm could come from ingestion by an infant. This is commonly accomplished by using an effective gas flush and the resultant small scavenger. Another strategy is to affix the scavenger onto the package and thereby reduce or eliminate the chances of accidental ingestion (10,15). The use of an adhesive to accomplish this with the fresh pasta and a sachet was mentioned previously. Use of the pressure-sensitive oxygen-scavenging label is another, and the push to develop oxygen-scavenging films is the ultimate answer thus far envisioned.

Of equally great concern is the necessity to prevent an oxygen scavenger being used in an application where anaerobic bacteria may grow. There are a number of precautions that the industry commonly employs. Among these is to only use oxygen scavengers when the water activity of the food product is below the point at which *Clostridium bottulinum* will grow (19,24), or to use them only when the food has sufficient acidity to prevent bottulism, or when appropriate curing salts have been used to prevent bottulism. Temperature control is generally discounted as a bottulism-preventive agent where retail packages are concerned, but HAACP systems have been employed for some products. Specific limits have not been given as they will vary from product to product. There is some evidence that the concern over oxygen scavengers and botulism may be overstated. An examination of the oxidative reductive potential in packages with and without oxygen scavengers has been reported to show that the scavenger does not increase the reductive potential (5), and thus does not enhance the risk of outgrowth.

Troubleshooting

Because this technology is not as prevalent as MAP and is somewhat complex, many users have need of a diagnostic tool to help evaluate the performance of oxygen scavengers. The following points can be helpful for iron-based scavengers:

1. Is the package hermetically sealed? If not, a hermetically sealed barrier package must be used so as to prevent gross leakage of oxygen into the package.
2. Are the contents of the sachet or label rusty, or do they show some agglomeration? These are common visual indicators that the components have at least begun to oxidize or scavenge oxygen. If there is evidence of some activity present, then examine the following factors.
   a. Is the test method for oxygen capable of measuring ppm levels? An incapable method may indicate values of several percent oxygen when the actual values may be significantly lower.
   b. Is there carbon dioxide present in the headspace? If so, use a chemistry compatible with carbon dioxide that will adsorb oxygen instead.
   c. Does the scavenger have enough capacity for the package? Recheck the sizing calculations and use a larger scavenger as indicated.
   d. If the scavenger is a self-acting type, was it left in the atmosphere for too long before being placed in the package? If so, it may have exhausted part or all of its capacity. Limit such exposure if this is the case.
3. If the contents of the sachet or label are not rusty or agglomerated:
   a. Verify that the water activity range of the scavenger chosen matches the water-activity range of the food or other product being packaged. If not, change scavenger specification.
   b. Check to make sure that the scavenger was not blocked off from the headspace gas by being placed or having fell under the food, under a tray, or trapped tightly between the product and the barrier film. If so, alter the scavenger placement.

BIBLIOGRAPHY

1. J. D. Flores, "Controlled and Modified Atmospheres in Food Packaging and Storage," *Chem. Eng. Prog.* 25–32 (June 1990).
2. "Shelf Life of Food," *Food Technol. NZ,* 5–11 (Feb. 1981).
3. Aaron Brody, "Active Packaging 2001," *Meat Internatl.* **2**(9/10), 42–45.
4. J. Hoshino and H. Nakamura, "Techniques for the Preservation of Food by Employment of an Oxygen Absorber" in *Sanitation Control for Food Sterilization Techniques,* Sanyu Publishing, Tokyo, 1983.
5. Y. Abe, *Active Packaging—a Japanese Perspective,* Mitsubishi Gas Chemical Co., Ltd., Tokyo, Japan.
6. Y. Kondoh, "Oxygen Absorbers—Review of Current Worldwide Market Situation-and Emerging New Technologies," paper presented at Future Pak '93, Atlanta.
7. B. Zenner, "SmartCap: Controlling Oxygen Degradation in Food and Beverage Packaging," *J. Packag. Technol.,* 37–38 (Jan./Feb. 1991).
8. R. Felland, "The Oxbar Super-Barrier System: A Total Oxygen-Barrier System for PET Packaging," paper presented at Europak '89.
9. R. C. Idol, "New Methods of Incorporating Absorbers in Packaging," paper presented at the Institute of Food Technologists' Food Packaging Update Short Course, Chicago, July 1993.
10. R. C. Idol, "A Retail Application for Oxygen Absorbers in Europe" in *Proceedings of Pack Alimentaire '93.* June 1993.
11. E. Powers and D. Berkowitz, "Efficiency of an Oxygen Scavenger to Modify the Atmosphere and Prevent Mold Growth on Meal, Ready-to-Eat Pouched Bread," *J. Food Protect.* **53**(9), 767–771 (Sept. 1990).
12. J. P. Smith, B. Ooraikul, W. J. Koersen, E. D. Jackson, and R. A. Lawrence, "Novel Approach to Oxygen Control in Modified Atmosphere Packaging of Bakery Products," *Food Microbiol.* No. 3, 315–320 (1986).
13. Y. Harima, "Free Oxygen Scavenging Packaging," in Takashi Kadoya, ed., *Food Packaging,* Academic Press, San Diego, 1990.
14. H. J. Andersen and M. A. Rasmussen, "Interactive Packaging as a Protection Against Photodegredation of the Color of Pasteurized, Sliced Ham," *Internatl. J. Food Sci. Technol.* No. 27, 1–8 (1992).
15. J. Belcher, "Optimizing the Potential of Oxygen Absorber Technology" in *Proceedings of Pack Alimentaire '93,* June 1993.
16. Y. Abe and Y. Kondoh, "Oxygen Absorbers" in A. L. Brody, ed., *Controlled / Modified Atmosphere / Vacuum Packaging of Foods,* Food & Nutrition Press, Inc., Trumbull, CT, 1991.
17. T. Klein and D. Knorr, "Oxygen Adsorption Properties of Powdered Iron," *J. Food Sci.* **55**(3), 869–870 (1990).
18. R. S. Flowers and D. A. Gabis, "Water Activity," *Scope* (a technical bulletin of Silliker Laboratories) (Fall 1984).
19. J. A. Troller and J. H. B. Christian, *Water Activity and Food,* Academic Press, New York, 1978.
20. *Laboratory Procedure for Oxygen Aanalysis,* Multiform Desiccants, Inc., Buffalo, NY.
21. *Ageless—a New Age in Food Preservation,* Mitsubishi Gas Chemical Co., Ltd., Tokyo, Japan.
22. "Pediatric Fatalities from Iron Poisoning Doubled, FDA Told," *Food Chem. News,* 22 (April 6, 1992).
23. W. Klein-Schwartz, G. M. Oderda, R. L. Gorman, F. Favin, and S. R. Rose, "Assessment of Management Guidelines. Acute Iron Ingestion," *J. Clin. Pediatr.* 316–321 (June 1990).
24. K. Glass and M. Doyle, "Relationship between Water Activity of Fresh Pasta and Toxin Production by Proteolytic Clostridium botulinum," *J. Food Protect.* **54**(3), 162–165 (March 1991).

RONALD C. IDOL
Multiform Desiccants, Inc.
Buffalo, New York

P

PACKAGE HANDLING SYSTEMS. See CONVEYING.

PACKAGE-INTEGRITY ISSUES FOR STERILE DISPOSABLE HEALTHCARE PRODUCTS

Sterile disposable medical products are a relatively new product group, in existence only since the 1960s. The use of this new product group created an increase in healthcare quality by utilizing single use sterile medical products. The development of these one-way products has reduced the risk of contamination from previously used products that were sterilized after each use in the hospital. The advent of this product group, based on the "Kleenex" principle, eliminated the cost and risk of in-hospital sterilization. This product group has been growing, with more products being introduced in the healthcare industry each year. These products have a new set of protective requirements that their packages have had to meet. Because the products are disposable after one use, the packaging cost must not significantly increase the cost of the product, so along with functional protective requirements the packaging is frequently restricted by cost constraints.

Package Requirements in Healthcare Packages

The basic requirements of a package are to protect the product from the producer to the consumer. The definition of "protection" is to prevent any activity from taking place during the distribution channel that could render the product useless or significantly reduce its usefulness. It is beneficial to understand the basic requirements and applications of a package prior to a more thorough discussion of specific functional requirements. The main requirements of a disposable device package include the following:

1. *Maintain Sterile Barrier around the Product.* The first and most important functional demand of a device package is for the package to maintain a sterile barrier around the product. This sterile barrier is accomplished mostly by flexible packages or blister packages. The simple, low-cost devices that require little physical protection are usually packaged in flexible bags or pouches. These simple packages provide excellent barriers to microorganisms when the correct materials are chosen, and they are very material efficient. Complex devices such as implants or multiple channeled IV (intravenous) sets require more protection and often additional product-use functions. These products are often packaged in blister trays that offer more physical product protection and other features for product use, such as separate areas in the package for holding items before or after their specific use (see also Skin packaging). If the sterile integrity is compromised by a hole in the package or a ruptured seal, the product is regarded as contaminated and determined unusable. The potential contamination of a patient from a nonsterile product due to package failure is a very real concern with serious outcomes. The maintenance of a sterile barrier is a function of keeping anything in the external environment from breaching the package and keeping the product from breaching the package integrity. The breach does not have to be more than a few micrometers in size for a microorganism to enter the package. This barrier function relies on material and package seal integrity.

2. *Allow for Sterilization.* The second major functional demand that a package may have to meet is the ability to allow for the product to be sterilized. This demand translates into the package having to survive the stress of a sterilization process and allow for complete sterilization of the product within that package. This is a necessary demand on the package system in most products because the products are sterilized after they have been packaged. Sterilization occurs after the product has been sealed in the primary package and the primary package has been packaged into the corrugated shipper. Four main sterilization processes are used to sterilize medical devices, gamma radiation, ethylene oxide gas, electron-beam radiation, and steam. Gamma radiation and ethylene oxide are responsible for about 85% of the disposable market, with each currently having approximately one half of that group. The electron-beam system is gaining market share because of several new developments in that process that make this sterilization technology particularly beneficial to the medical-device industry. Other sterilization systems are used to sterilize devices, including gas plasma, vapor-phase hydrogen peroxide, peracetic acid, and X-ray, but each has limited applications due to certain factors and limitations in each system (1). Certain products have unique needs that are satisfied by these other sterilization systems, but presently they represent minor applications for device sterilization.

3. *Aid in Aseptic Removal of Product.* Another requirement of these packages is that they must allow for an easy opening of the package by the medical personnel. The opening of the package must not compromise the sterility of the product. The package must facilitate sterile, struggle-free opening without the use of nonsterile opening devices. Low-aggression opening of the package allows for easy removal of the sterile product and reduces the risk of contaminating the product by struggling with the package and touching sterile areas on the product or dropping the product on a nonsterile area. This requirement is particularly problematic in that the package, to accomplish an easy opening, must have a low resistance to force applied to the seal areas while remaining intact. This low resistance to physical stress makes it difficult for the package to remain intact through the physical stresses occurring during sterilization and distribution cycles. The easy-opening feature is generally accomplished through the use of a heat-seal material that has low cohesive strength. Ethylene–vinyl acetate (EVA), a copolymer of polyethylene and vinyl acetate, has a low cohesive strength and allows for a thermoplastic heat seal that gives the required microbial barrier (see Aseptic packaging).

4. *The Package Should Not Add to Airborne Contamination.* The package material should remain intact and not tear and create airborne fiber that can settle down and recontami-

nate the product during opening. This translates to having packaging materials that are higher in cohesive integrity than the heat-seal materials used. When a package is opened, the package material will not tear but the adhesive will come apart. Additionally, heat-seal adhesives should not be prone to particulate generation. The adhesive should separate and remain on the package materials. This particulate concern also includes dust that has settled on the exterior of the package becoming airborne through a vigorous opening action. The dust problem can be solved by low-cohesive-strength heat-seal materials, designing a flat package with few folds to allow for dust collection and the use of smooth materials.

5. *Aid in Product Identification.* The product should be easily identifiable while still in the unopened package. This eliminates the opening of a package, with the consumer erroneously thinking that another product or variation of the product was inside. Ease in product identification is a particularly important feature where numerous products or product variations need to be distinguished. In the case of erroneously opening a package, the entire product would be thrown out. Transferral of product-specific information can occur in many ways in the unopened package. Use of clear packaging materials can allow for product inspection through the unopened package. The package can display an illustration of the product with specific items highlighted on the exterior of the package. The package can be color-coded so that a line of similar products with different specific features could be graphically identified by the colors used for copy.

6. *Evident Opening Features.* The opening feature of the package should be designed to clearly indicate that the package has been opened. There should be no reclosable feature on the package that would allow for the package to be opened, contaminated, and then resealed. The package seals should, when opened, visually alert hospital personnel that the package had been opened and sterility had been compromised.

Package and Material Requirements for Sterilization Procedures

Generally the maximum physical stress to which a package will be exposed is during transportation and distribution, including truck shock and vibration, the drops in handling, and finally the twist and compression of handling the packages by hospital personnel. A medical package will face all these stresses, but the most demanding physical stress to which the package will be exposed occurs during product sterilization. The particular sterilization process used exerts a unique set of stresses on the product and also on the package. Ethylene oxide sterilization, for example, will stress the package integrity more than it stresses the materials. Both radiation and sterilization seem to have the most stressful effect on the material properties. The package and materials must withstand these demands and after these processes remain functionally intact throughout distribution and storage.

Since sterilization occurs after the product has been packaged, the package sees the same contact with radiation, chemicals, moisture, and heat that the product is exposed to during sterilization. The material properties may change as a result of the sterilization exposure, and some low level of change is acceptable. The package properties may change as a result of the material changes, but as a whole the package must maintain a level of microbial barrier performance. What is not allowable is *any* loss of sterility. If the sterile integrity of the package is breached to any degree by seal rupture or stress cracks in the material, the product is considered contaminated and unacceptable.

Each sterilization system used requires that the package perform in some way in order to facilitate sterilization. Autoclave sterilization requires that hot steam be allowed to pass through the package and contact the product. The package must not act as an insulator or be a barrier to the steam. An additional requirement for steam sterilization is for the package to tolerate high temperatures without loosing intended properties. Ethylene oxide (EtO) sterilization requires a certain level of porosity in the package materials for the EtO gas to contact and sterilize the product. Porosity is also necessary for removal of the EtO gas from the package at the end of the sterilization cycle. Gamma and electron-beam sterilization both require a tolerance for low-level energy without loss of properties. These packaging materials must also not shelter the product from energies required for an effective kill.

The stress a package must withstand in this industry may be compounded in two ways:

1. It is possible for a package to see multiple sterilization cycles. This occurs when a batch of product is sterilized and the batch fails to meet sterility requirements. If this occurs, the batch is allowed to be resterilized and retested. The sterilization effects on a package are cumulative, and package design must consider the worst-case challenge.

2. A single package design may be required to satisfy the requirements of two different sterilization systems. This occurs when a company has multiple manufacturing locations and different sterilization facilities are used at each location. The company does not want the product to be on the market in two different packages, so one package must meet the functional needs of two different sterilization procedures.

Steam and (EtO) Sterilization. Steam and EtO sterilization systems are similar in that they both use mechanical systems to deliver a sterilant to the product. In the case of steam, the sterilant is hot moist air; in the case of EtO, it is a warm chemical gas. The mechanical systems designed to deliver the sterilants rely on the differences in pressure created by a vacuum vessel to move the sterilants through the package walls and then contact the product. The packages are placed in a sealed chamber, and then a vacuum is pulled on the entire chamber. The sterilant is then released into the chamber, and because of the pressure difference, the incoming gas with the sterilant replaces the low-pressure areas, moving through the package and contacting the product. This is held for a period of time, and then another vacuum is pulled on the chamber to remove the sterilant. Figure 1 shows the pressure changes in a typical steam-sterilization cycle.

The back-and-forth changes in pressure contribute to package-seal rupture in these sterilization systems. This seal stress is increased by the warm to hot temperatures used during the cycle. The EVA heat seal is a low-melt-temperature thermoplastic, and elevating the temperature softens the material and weakens the seal strength. The amount of vacuum drawn, quickness of the evacuation and repressurization, package porosity, and temperature are the key variables in maintaining package integrity. It would be possible for the

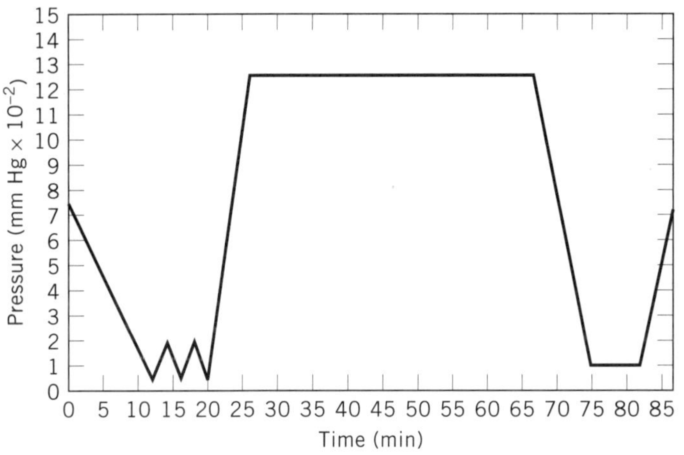

Figure 1. Changes in pressure in a typical steam sterilization cycle.

pressure cycles to be very slow and gradual, but that would increase the processing times and add to the cost. In an EtO cycle, the temperatures may be lowered, but to compensate a longer exposure time may be required. Figure 2 shows the pressure changes that occur during a typical EtO sterilization cycle. Steam sterilization requires a minimum temperature that cannot be reduced. These variables are balanced with the package's ability to survive the sterilization system. If the need for improved package integrity outweighs the cost of a longer cycle, then the cycle may be altered.

The mechanical similarities between the steam and EtO processes allows for some common solutions to the problems caused by the systems. The most serious common problem is the stressing and rupturing of the heat seals during the pressure changes in the sterilization cycles. There are package and package material variables that can be manipulated to alleviate the problem of package seal stress during sterilization:

1. *Porosity Rate of the Package.* The porosity rate of the package impacts on the stresses the seal face because, if the porosity is high, the change in relative pressures occur quickly and the package does not billow out and stress the seals. If the package has a low porosity rate, the package tends to billow out and stress the seal by the pressure exerted on the seal area. This property of the material can be measured by a porosity tester. It is possible to increase the surface area of porous material and by doing so, increase the total package porosity rate.

2. *Seal Aggression.* Another way to improve the seal's ability to survive stress is to increase the cohesive strength of the seal material. The aggressiveness of EVA seal material is designed so that the opening force of the package is low and the hospital worker does not have to struggle to open the package. The drawback of an easy-opening package is that the lower-aggression seals have a hard time surviving sterilizer stresses. It is possible to vary the vinyl acetate content of the copolymer from a 14% to 40%. The higher the vinyl acetate content, the weaker the material's cohesive strength and the lower the seal strength. Another aspect of this seal-strength issue is that if the seal is too aggressive, it may cause the lidstock of the package to tear when opening, increasing the amount of airborne particle contaminants.

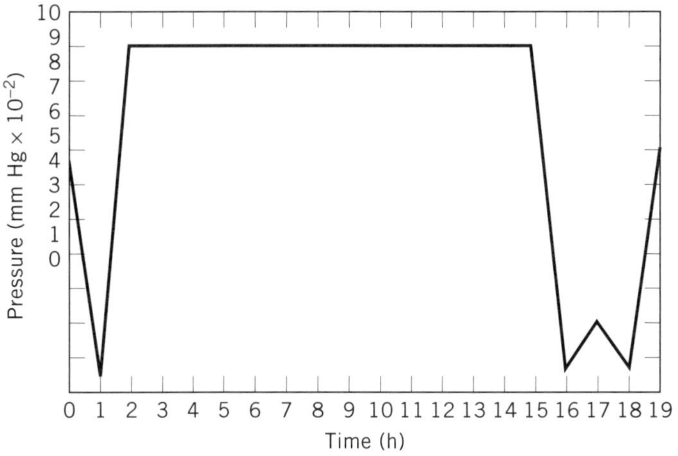

Figure 2. Changes in pressure in a typical EtO sterilization cycle.

Table 1. Steam Sterilization

Advantages	Disadvantages
Highly effective	High heat exposure
Short product release time	Many variables
In-house process	High moisture
Low equipment cost	Package seal stress
	Package porosity requirements
	Labor cost to process

Steam sterilization. The use of steam to sterilize devices is the oldest sterilization method and dates back to the time when devices were reusable and were sterilized in the hospital after each use. The process of using moist heat to kill microorganisms is a very effective system with much data available demonstrating its effectiveness. This process is very difficult to monitor because there are many variables to control, including temperature, moisture content, pressure, and time of exposure. This process is very similar to EtO in that the vacuum vessel is used and the differences in pressure are created to move the sterilizing agent into the packages. In this process the sterilizing agent is very hot steam, with temperatures in excess of 121°C. This process is common in most hospitals, and physicians' and dentists' offices. It is not used for many disposable devices because the temperatures required will soften and weaken most thermoplastic devices. Steam is most useful with glass and metal devices that can tolerate the extreme moist heat. Steam is not a major sterilization process in the disposable-device industry but is more an in-house system used by hospitals and medical offices. The use of steam sterilization as an in-house system provides us with an example of packaging medical devices on a relatively small scale. This is accomplished by the hospitals purchasing premanufactured packages and loading the product into the three-sided pouch and sealing the pouch. This package must then withstand the rigors of a steam cycle and remain intact and provide a sterile barrier until the product is to be used. The advantages and disadvantages of steam sterilization are summarized in Table 1.

Effects on materials. The materials used in packaging

steam-sterilized items must be porous enough to allow for sterilization and be able to withstand high-moist heat. Steam has the ability to weaken papers because of the combination of heat and moisture, so paper packaging must be evaluated to determine the actual loss of material integrity. Steam does not have the problem of toxic residues found in packaging materials after sterilization because the mechanism of lethality to microorganisms is the high temperature itself. The greatest concern with thermoplastic packaging materials is their loss of physical properties on exposure to high temperatures during the cycle. Polyethylene and polystyrene have relatively low melt temperatures, so they soften at too low a temperature for usefulness. Polypropylene and polyesters have higher melt temperatures and may be more suitable for steam cycles.

Effects of package. The package stress stems from exposure to pressure changes during the sterilization cycle. These stresses manifest themselves in the same package-seal problems occurring in EtO sterilization cycles. The difference in temperatures used between steam and EtO accounts for the greater severity of package stress with steam. Steam sterilization requires that the product reach a temperature of 121°C to be considered sterile. This weakens thermoplastic heat seals and packaging films. The required porosity eliminates most plastic films and makes paper a good choice. The one drawback with paper is that it will weaken when saturated with water and must be dried after sterilization.

Ethylene oxide sterilization. Ethylene oxide (EtO) is a chemical that, when in contact with biological systems, disrupts certain cellular activities interfering with the cell function, and in the case of small organisms, the result is death. The use of EtO in sterilizing medical devices is widespread, with approximately 45% of single-use devices using this method (2). This sterilization system is favored over steam for disposables because it employs a warm rather than hot environment and is not as temperature-aggressive to thermoplastics. The process involves, first, the prehumidification of the devices and packages prior to the sterilization cycle. This is done to activate any desiccated microorganisms because the chemical has a more effective kill rate on active microorganisms that are carrying out normal biological activities. This pre-humidification process takes at least 9 h at 40°C and 65–80% relative humidity. After prehumidification, the devices are moved into a vacuum chamber. The products are still in their primary packages and in most high-volume applications, the cases of product are moved into the vacuum chamber still on pallets. In the vacuum chamber a vacuum is drawn and then held until the EtO gas is released into the evacuated chamber. The difference in pressure is the mechanism that draws the gas into the package and into the product where all surfaces that come in contact with the gas are sterilized. After a period of time, depending on the level of bioburden and complexity of the device design, another vacuum is drawn and the EtO gas is drawn out of the chamber. A series of vacuums and air flushes are performed to eliminate the EtO gas residuals from the product and the package. These pallets of products are then removed from the vacuum chamber and allowed to vent while samples are evaluated for sterility before the batch of product is released.

The effectiveness of this sterilization procedure relies on the package having the properties of porosity to EtO gas, heat and moisture resistance, and the ability to maintain a sterile barrier. The maintenance of a sterile barrier depends on the package material and heat seals remaining intact without breaches throughout the repeated vacuum and pressurization cycles.

Table 2. Ethylene Oxide Sterilization

Advantages	Disadvantages
Relatively effective	Toxic residuals
Relatively low heat (compared to steam)	Package porosity requirements
	Many variables to monitor
In-house process	Moderate heat
	Regulatory issues
	Moist process
	Long process time
	Long product release time
	Package seal stress
	Labor cost to process

Effects on materials. The effect of EtO gas on materials is very well documented as this process has been widely used for years. The main concern has been the absorption of EtO and byproducts ethylene glycol and ethylene chlorohydrin by packaging materials (3). Medical papers and nonwoven materials such as Tyvek (manufactured by DuPont) are porous and do not retain EtO because they have ample surface area for the residuals to evaporate and elute. EtO residuals can be absorbed into plastics and exposure of the device manufacturer and healthcare workers to residual amounts of the chemical are a concern. There are differences between materials, but a basic rule is that thicker materials will tend to absorb more EtO than a thinner material and it is tougher for a thicker piece of material to elute EtO residuals.

Polyvinyl chloride will absorb twice the amount of EtO as polyethylene, polypropylene, and polystyrene (3). The longer the sterilization cycle (the longer a material is exposed to EtO), the greater the absorbency. The manufacturers who use EtO have designed aeration cycles during and poststerilization that reduce the residual problem. The residual issue continues to be monitored by the U.S. government regulatory agencies, and this issue has been instrumental in the increasing use of radiation as a sterilization process (2). The advantages and disadvantages of EtO sterilization are summarized in Table 2.

Effects of package. The effects of an EtO sterilization cycle is more apparent on the package than the materials used in that package. The package effects center on the ability of the package, as a combination of materials assembled together, to maintain a sterile barrier:

1. *Package Stress.* The package stress comes from the evacuation cycles during sterilization. The quick changes in pressure have a bellows effect on the heat seals. This effect is to blow the package up and stress the package seal. This stress is compounded by the fact that the plastic used for a heat-seal material is chosen for its low melt temperature and low cohesive force. The package must survive stress on the seals at temperature conditions that have already weakened the heat-seal material. The most challenging part of packaging for EtO-sterilized packages is that the package has to encounter stress when the heat seals are in a preweakened

warm condition. The packaging engineer is caught between creating a strong, aggressive seal to withstand sterilization and creating a weak, easy-open seal to facilitate a sterile removal of the product from the package. The seal weakening from evacuation and pressure cycles is referred to as "CREEP." This is recognized by a scalloped area along the seal that reduces the seal width and also the seal strength at that point. The problem with a seal weakened during sterilization is that during normal distribution the seal may see additional stress that may exceed the total seal strength and cause it to open, compromising the sterile integrity of the package. The scalloped void in the seal can also actually be as wide as the seal, breaching integrity and creating an opening for microorganisms to enter.

2. *Packaging Design to Aid in Penetration.* The effectiveness of EtO sterilization lies in the ability of the cycle to drive the gas into the package and contact the product. It is difficult for the gas to reach sealed chambers in the product and other areas that do not allow for airflow. Also, packages utilizing foam inserts are particularly difficult to sterilize because the volume of air in the foam materials is difficult to remove and impacts on sterilizer effectiveness (4). The package can be designed with large areas of porous material to improve the airflow characteristics of the package. This is necessary with blisters that have only one side that is porous, and "Schuster" or "header" pouches that reduce the surface area of porous materials used in the package (5).

Radiation sterilization. Currently there are two popular methods of generating radiation energy used to sterilize medical devices. Each delivers the same effect, but they differ in mechanism of delivering the energy. It is these differences that drive the specific applications of each system now and in the future.

Gamma sterilization. With gamma sterilization the products are sterilized by action of exposing the packaged product to a decaying isotope, usually cobalt-60 or cesium-137 (^{60}Co or ^{137}Cs). The γ-photon particles bombard the item and excite electrons, causing them to break apart from their existing positions. These ionized particles break other chemical bonds, causing material and biological damage. This damage is what kills microorganisms by disrupting cellular systems. Gamma particles are large, and because of their size and mass, can travel relatively far distances. This means that the source can be located yards away from the product and still receive an effective dose. Gamma sterilization requires a large technical facility requiring a large capital investment. These facilities are generally owned by companies whose business it is to expose items to radiation through this process. The process can effect lethal doses as in the case of medical devices or sublethal in the case of food products (6). The sublethal process is exposure to much lower doses. The number of variables in this process are few and microbial kill effectiveness is high, accounting for high reliability in the process. The sterilization takes place at ambient conditions, so the effects of high temperature and humidity in steam and EtO sterilization cycles are not factors for package integrity. Gamma is characterized by very high penetration, which accounts for its high reliability of kill rate. The time for a sterilization cycle is about 6–9 h at a normal dose rate of 4 kGy/h. The recognized average dose for sterilization is 2.5 Mrad or 25 kGy. A Gamma-cycle will guarantee exposure to at least 2.5 Mrad, but in order to do so, actual doses will exceed that minimum. The advantages and disadvantages of gamma sterilization are summarized in Table 3.

Table 3. Gamma Sterilization

Advantages	Disadvantages
Effective kill	Costly facility
Few variables	Degrades materials
High-volume cost-efficient	Facility location not always convenient
No residuals	Effects are cumulative
No heat	Variability in dose
Short product release time	Long process time

Electron-beam (E-beam) sterilization. E-beam sterilization is a lower-energy system than gamma and has particular advantages in particular applications. This system excites electrons in a electron accelerator and then bombards the package and product with these excited electrons. The free electrons then break other chemical bonds and cause biological damage. The similarity between gamma and E-beam is the use of electrons to effect sterilization, the difference is how the electrons are excited. A gamma particle is much heavier than an electron, so the gamma particle can travel farther and still have the energy left to excite and ionize compounds. Electrons, on the other hand, have very small masses and loose energy quickly, so effective use of excited electrons occurs relatively close to the source. This distance is less than 12 in. for high-energy accelerators (7). The advantage of E-beam sterilization is that a large facility is not needed and consequently does not require as large a capital expense as gamma. The electron accelerator can be placed in a production line with appropriate shields employed.

The use of electron accelerators is not new in the device industry. Johnson and Johnson was one of the first to use this technique as far back as the 1960s (2). Currently the technology is used for such varied applications as dosing polyester tennis strings to improve the tensile strength and curing inks and varnishes. The latter is currently more common in Europe. There are currently electron accelerators capable of producing 2–12 MeV of energy. The lower energy is used mostly to cure inks and varnishes, but this energy level is capable of producing a sterile field within 1 in. of the source. This system has potential for sterilizing very low-profile medical devices (8). At these low energy levels little shielding is required because the energy is lost over small distances. This fact, along with the short time required to dose the product, would make this process a possible in-line operation. The 10–12-MeV accelerators are being used currently in the industry to sterilize cases of products and have a penetration of roughly 1 MeV to 0.8 cm. These higher energy systems require shielding, but this is being accomplished by subtle design changes to the production line so the product can be sterilized in-line. Figure 3 shows an E-beam accelerator being used in-line to sterilize medical devices. The advantages and disadvantages of E-beam sterilization are summarized in Table 4.

Effects on materials. The material changes in packages exposed to radiation vary quite a bit. Some materials have little change even when exposed to large doses of energy, while other materials show profound effect with small energy levels.

Figure 3. High-energy electron accelerator sterilizing cases of medical devices. (Photo Provided by AECL Accelerators, Ontario, Canada.)

The effects of radiation on materials has been fairly well documented by research done by the U.S. government and different private device companies (9). The materials that are used in the majority of device packages are generally low-cost, common plastics and papers, and consequently any method of limiting these effects must be cost-effective.

Plastics are affected several ways when exposed to doses of radiation. The polymer chains exhibit scission and crosslinking, gas elution, and discoloration. The dominant effect depends on the particular material. The effect on materials has been shown to be dose-dependent and cumulative. The effects are increased by repeated exposure from multiple sterilization cycles. Most packaging plastics are very resistant to radiation energies at the levels used for sterilization, but two common materials exhibit significant property changes. It is interesting that two different materials show profound changes at low level energy exposure. Figure 4 shows the effects of different doses of energy on the tensile strength of plastic films commonly used as packaging materials. Polypropylene (PP) and polyvinyl chloride (PVC) are two common materials used to make devices and packages that exhibit acute reactions to low levels of energy. At very low doses polypropylene will turn brittle, loose flexibility, and discolor, turning yellow. Polypropylene shows significant loss in mechanical strength at energy levels used in radiation sterilization. Polyvinyl chloride shows a reaction to radiation at low levels by turning amber and eluting gas. The significant change in mechanical properties in PVC occurs at higher energy doses, around 100 Mrad (10).

In contrast, polyester films show change due to exposure by increasing tensile strength (11). The tensile strength increase is dose-dependent over 0–5 Mrad and then begins to reduce with increased exposure. Even at 12.5 Mrad the tensile strength of the sample is greater than the original, untreated sample as shown in Figure 4. This is a desirable property change in applications where tensile strength is desirable. Polyester tennis strings produced by a company named Gamma Gut are dosed by gamma radiation to increase their tensile strength, which is a quality necessary for tennis strings. The company even identifies the process in the corporate name.

In addition to the mechanical properties changing, the barrier properties of a material are also at risk. PETG, PP, and

Table 4. Electron-Beam Sterilization

Advantages	Disadvantages
Low heat	Degrades some materials
In-line process	Low penetration
Highly effective	High-energy systems require shielding
Few variables	
Quick process time	
Short release time	
Low equipment cost	
Controlled dose	

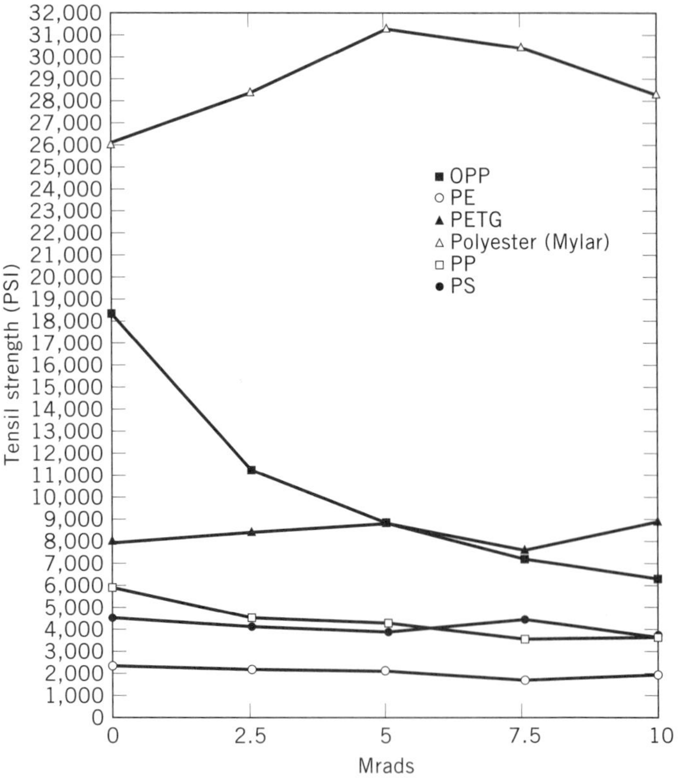

Figure 4. Tensile strengths of packaging films after exposure to low-level radiation.

Table 5. Effects of Radiation on the WVTR of Packaging Materials Tested [g/(100 in.^{2} · 24 h) at 100°F, 90% rh]

Materials (Mrad)	0	2.5	5.0	7.5	10.0
PETG	1.42	1.52	1.57	1.76	1.73
PET/PE	0.46	0.48	0.50	0.55	0.57
PP	0.19	0.21	0.21	0.23	0.24
OPP	0.30	0.33	0.33	0.35	0.37
PE	0.12	0.13	0.15	0.15	0.13
PET	0.25	0.28	0.28	0.29	0.28
Nylon/PE	0.26	0.29	0.34	0.37	0.37

nylon show significant loss of water-vapor-barrier qualities when exposed to sterilizing doses of radiation. These effects have been demonstrated to be dose-dependent, so the greater the dose, the more pronounced the effect (12). It is speculated that the effect comes from the restructuring of the polymer chains, allowing more area for water molecule transport. Table 5 shows the WVTR values of various packaging films at different energy doses.

The stabilization of these polymers to the radiation levels seen in sterilization cycles is being accomplished by the use of various additives. The additives are designed to either react with the energy before it damages the material or absorb the energy before it energizes the material. These additives are specially formulated for the particular application because their compatibility with a product is a key consideration. PVC is modified with lead carbonate or organic tin compounds, and this change reduces the discoloration in the material (13).

Effects of package. The effects on package integrity are secondary to the effects of material integrity. There is no elevated temperature to soften plastics, and there is no evacuation cycle to stress the package seals. The embrittlement of a material is the key issue because if a package material is flexed and cracks, the sterile integrity of the package is compromised and the product is viewed as contaminated and unusable. Stresses encountered during distribution take on greater importance when materials have become brittle, so effective distribution testing is prudent.

Shelf life of device packages. The shelf life of a device package is expressed as the length of time the package will provide a sterile barrier. The shelf life has traditionally been defined as only an event related process where events such as rough handling contribute to premature aging of the package. Current research has indicated that time is also a factor in device shelf life. There has been evidence to suggest that material changes may occur during poststerilization storage. These changes are the same as those that occur immediately after sterilization but seemingly increase over time. One study has shown that aged EVA heatseals on porous packages sterilized by EtO have significantly lower seal strengths than the same packages immediately after sterilization (14). Other observed effects are discoloration and increasing brittleness of materials sterilized by radiation. These few observations make a case for increasing attention on poststorage testing of the packages so that more can be learned about the impact of sterilization method on shelf life (see also Shelf life).

BIBLIOGRAPHY

1. C. Henke, *Med. Device Diagn. Ind.* **14**(12), 41 (Dec. 1992).
2. A. F. Booth, *Med. Device Diagn. Ind.* **17**(2), 64 (Feb. 1995).
3. *Industrial Sterilization,* Duke University Press, Durham, NC, 1973; B-D International Industrial Sterilization Symposium, 1972 Amsterdam, Netherlands.
4. R. R. Reicht and D. J. Burgess, *Med. Device Diagn. Ind.* **13**(1), 124 (Jan. 1991).
5. C. Marotta, *Med. Device Diagn. Ind.* **6**(2), 27 (Feb. 1984).
6. G. G. Giddings, "Irradiation of Packaging Materials and Prepackaged Foods" in *Proceedings from the 4th International Conference on Packaging,* Sept. 1985.
7. J. W. Barnard, *Med. Device Technol.* (June 1991).
8. J. N. Aaronson and S. V. Nablo, "The Application of Electron Sterilization for Devices and Containers," paper presented at TAPPI Polymers, Laminations and Coatings Conference, 1986.
9. J. J. Killoran, *Radiat. Res. Rev.* **3,** 369–388 (1972).
10. W. E. Skeins, *Radiat. Phys. Chem.* **15,** 47–57 (1980).
11. V. Pungthong, M.S. thesis, *Mechanical Properties of Polymeric Packaging Films After Radiation Sterilization,* Rochester Institute of Technology, 1990.
12. R. L. A. Vas, M.S. thesis, *Effects of Low Level Radiation on WVTR of Plastic Packaging Films,* Rochester Institute of Technology, 1989.
13. M. Foure and P. Rakita, *Med. Device Diagn. Ind.* **5**(12) 33, (1983).
14. A. T. Cook, M.S. thesis, *The Effect of Accelerated Aging on Peelable Medical Product Heatseals,* Rochester Institute of Technology, 1994.

Fritz Yambrach
Rochester Institute of Technology
Rochester, New York

PACKAGING OF FOOD

Packaging protects food against a hostile environment. Being biological, food can deteriorate to lose nutritive value; change color, flavor, and masticatory properties; and in some instances can become a toxicological hazard.

Food deteriorates by four vectors: biochemical, enzymatic, microbiological, and physical. The biochemical vector is the result of interaction of food chemicals because of proximity to each other. Enzymatic deterioration is biochemical deterioration catalyzed by enzymes naturally present in food. Microbiological deterioration is the most common food spoilage vector. Microorganisms ubiquitous in food products include yeasts, molds, and bacteria.

Yeasts and molds are generally found in association with high-acid sugary products while bacteria usually are found throughout all foods. Yeast and molds grow best on the surfaces of high-acid, sugary food products but will grow on virtually any food surface exposed to air. Bacteria will grow almost anywhere both within and on the surfaces of foods. Food sterilization is designed to destroy all microorganisms that could grow in or on food products. Water at almost any temperature from 32°F (0°C) to 140°F (60°C) permits the growth of microorganisms. Microbial growth is slowed by reduction in temperature with the rate of reaction changing by a factor of between two and four for every 18°F (10°C) change in temperature.

Bacteria include the largest groups of organisms capable of causing infections and intoxications. Further, low-acid food products with a pH above 4.5, in the absence of oxygen can support the growth of the *Clostridia* organisms capable of producing botulism toxins under anaerobic conditions. Most bacteria, however, do not produce toxic products but rather spoil the food product, reduce its nutritional value, and adversely alter its appearance and flavor.

Damage to food products not associated with biochemical, enzymatic, or microbial spoilage is usually physical, such as gain or loss of water or color. Elevated water activity by dry-food products increases the rate of biochemical reactions. In food products with high water contents such as fresh produce or meat, water loss alters physical characteristics and can lead to conditions for microbiological growth. From an economic standpoint, the water content must be maintained at its original level. Much of the function of food packaging is to ensure against gain or loss of water from the product. Almost all adverse reactions are accelerated by increasing temperatures.

Packaged Food Classification

Approximately half of all packaged products in the United States is food. Few foods are not packaged in some manner, apart from fresh foods such as produce sold at roadside stands.

Fresh-food products include all fresh meats, vegetables, and fruits that are unprocessed except for removal from the original environment and limited trimming and cleaning. Because of spoilage vectors, fresh foods should be consumed as soon as possible and handled in a manner that retards their deterioration, which is relatively rapid at ambient temperature or above. Thus, meats from freshly killed animals are chilled rapidly to below 50°F (10°C). Most vegetables and fruits are generally reduced to below 40°F (4.4°C) by low temperature air, water, or ice.

Partially or minimally processed foods include those which have been altered to help retard deteriorative processes. These include many dairy products which must be refrigerated after pasteurization and cured meats which also must be kept refrigerated to ensure against microbial growth, etc.

Fully processed foods are those intended for long-term shelf life at room temperature, and include almost all heat processed, dried, etc, foods that have traveled through a food-processing plant before packaging.

Fresh Foods

About a quarter of the value of food products in the United States consists of animal meats, includng beef, poultry, fish, mutton, veal, and pork, all of which are susceptible to microbiological, enzymatic, and physical changes.

Meat. The color of red meat depends upon the presence of oxygen. The natural color of myoglobin meat pigment is purplish. The basic color of the red meat is oxymyoglobin or oxygenated pigment. Oxidized myoglobin is the brown color seen when meat is exposed to the air for extended time periods.

To preserve red meat, the objective is to retard spoilage, to permit some enzymatic activity to improve tenderness, to retard weight loss, and to ensure an oxymyoglobin or cherry-red color at consumer level.

In distribution, most red meat is packaged under vacuum in high oxygen–water vapor-barrier flexible packaging materials to retard deterioration. Very little of red meat in the United States is distributed from slaughter to retail use without packaging or with minimum packaging. At the retail level, packaging permits restoration of the bright cherry-red oxymyoglobin color. Oxygen-permeable flexible packaging such as poly(vinyl chloride) (PVC) film permits oxygen into the package while retarding the passage of water vapor. All fresh meats are moved from the slaughter to retail at temperatures below 50°F (10°C) in order to retard deteriorative processes.

Poultry. Poultry is extremely susceptible to microbiological deterioration as it is an excellent substrate for the growth of *Salmonella* microorganisms. Thus, it is vital that the temperature be reduced as rapidly as possible. Many times, poultry is immersed in cold water or ice to reduce the temperature to below the optimum for microbiological propagation. Poultry is shipped in wet or dry ice to retail level. Packaging no longer now occurs at factory level in soft film such as PVC which retards water vapor loss, i.e., poultry is largely centrally packaged and maintained under carefully controlled temperature conditions. Turkey and other poultry consumed seasonally are preserved by freezing after packaging in low oxygen–water vapor-permeable heat-shrinkable films similar to those used for refrigerated distribution packaging of primal cuts of fresh beef.

Fish. Fish are generally taken from cold waters and temperature must be reduced immediately. Fish must be kept so that there will be little or no weight loss. Much fish is marketed as fresh from bulk ice. Significant quantities, however, are frozen and packaged (or packaged and frozen). Packaging generally has low water vapor permeability to permit long-term frozen distribution without freezer burn or surface desiccation. Wrapped or coated paperboard cartons and flexible films are employed to package frozen fish.

Produce. Subtropical and tropical fruits and vegetables such as pineapple, banana, and tomato are susceptible to chill damage at temperatures below 50°F (10°C). At temperatures above 50°F (10°C), normal enzymatic and microbiological deteriorations occur. Products susceptible to chill damage must be reduced in temperature to the lowest level at which chill damage does not occur. The main deterioration vectors are enzymatic, retarded by temperature reduction through immersion in chilled water, air blast, or ice.

Fruits and vegetables should be handled relatively gently because of the ubiquitous presence of microorganisms on the surface. Damage to the product surface provides channels through which the microorganisms can enter to initiate spoilage.

Fruit and vegetable packaging is often in bulk in a variety of almost traditional wooden boxes and crates and corrugated fiberboard cases. In Western Europe, expanded polystyrene foam trays are often used. At or near retail level, bulk produce may be repackaged in oxygen-permeable flexible materials, with or without a tray. Access of oxygen, which may be achieved by punching holes in the package, eliminates the possibility of anaerobic respiration which is another spoilage vector. Significant quantities of pre-cut fruit and vegetables are packaged under modified atmospheres (see Controlled atmosphere packaging).

Partially Processed Food Products

Partially processed food products have received more than minimal processing but still require refrigeration.

Processed meats. Processed meats such as ham, bacon, sausage, bologna, etc, are salted to reduce the water activity, are spiced for flavor, and may also have ingredients to maintain the desired red color. Curing agents include sodium nitrite and sodium nitrate. Because the color is fixed, oxygen is not a vector of color loss for extended time under refrigeration. Thus, cured meats maintained with an absence of oxygen will have extended shelf life (see Shelf life). Most processed meats are packaged under vacuum in thermoform/vacuum seal systems and distributed under refrigeration. Small quantities are packaged in pouches under inert atmosphere such as nitrogen.

Dairy products. Dairy products are derived from milk, which, despite best efforts, usually is not sterile. Pasteurization is a low-heat process to destroy disease microorganisms but not to destroy all microorganisms that could cause spoilage. Pasteurized dairy products must be maintained under refrigeration. The packaging of pasteurized milk must be in clean containers. Nonreturnable packages such as blow-molded high density polyethylene bottles or polyethylene-coated paperboard gabletop packages are most often used for packaging and distributing milk under refrigeration in the United States. Returnable glass bottles are still employed to a minor degree in the United States and to an extent in the United Kingdom. Returnable milk bottles are carefully cleaned on each cycle to remove debris and then sterilized to destroy microorganisms.

Aseptic packaging (see Aseptic packaging) for milk has been used in countries around the world. In aseptic packaging, the milk is sterilized, that is, rendered free of microorganisms. Simultaneously, the high barrier paperboard/foil/plastic-lamination packaging material is sterilized. The two are brought together in a sterile environment and the package is sealed to produce sterile milk in a sterile package. The increased heat required for sterilization of the milk can lead to flavors different from pasteurized refrigerated milk. Aseptically packaged milk may be distributed at ambient temperature.

Both fresh and cured cheeses in the United States employ pasteurized milk into which the enzyme rennin plus pure culture microorganisms are introduced to cause the precipitation of the proteins. With fresh cheese, the protein precipitate is cut and salted slightly for flavor and then packaged in closed containers such as waxed paperboard, thermoformed plastic tubs, foil laminations, or plastic wraps to restrict water loss and retard airborne organisms. For practical purposes, such products are solid versions of milk subject to the same microbiological deterioration. Cheese is protected against water loss to ensure retention of physical and flavor integrity.

Hard cheeses, on the other hand, continue to undergo processing to become less susceptible to microbial deterioration. In a few cases, hard cheeses can be distributed without refrigeration. Hard cheeses include parmesan, mozzarella, and Swiss. Mold cheeses include Roquefort, Stilton, and bleu.

In the past, surface-applied wax retarded moisture loss; today, high oxygen-barrier flexible film coupled with vacuum or inert gas packaging greatly retards water loss and helps suppress microbial activity by retaining an oxygen-free interior.

Yogurt is a partially fermented milk product subject to microbial deterioration. Many yogurts also contain jams, jellies, etc, which may contain spoilage microorganisms. Yogurt products must be maintained under refrigeration in closed waxed paperboard or thermoformed or injection-molded plastic cups or tubs used for packaging.

Ice cream is a frozen aerated emulsion of milk fat with sugar. Ice cream is not subject to microbiological deterioration because the distribution and use temperatures are too low for microbial propagation. Packaging is generally minimal: lacquered or polyethylene extrusion-coated paperboard cartons and molded plastic tubs.

Other products. Fruits and vegetables subjected to natural and/or artificial drying include apples, prunes, raisins, grapes, and apricots. A preservative such as sulfur dioxide retards microorganisms and drying reduces the water content. Dry fruits and vegetables are susceptible to changes in texture from gain or loss of moisture and can be subject to microbial spoilage if the water content increases sufficiently to permit the growth of microorganisms. Although often distributed with minimum packaging, dry fruit and vegetables may be sealed in low water vapor-permeability packaging to maintain the water content.

Fully Processed Foods

Fully processed foods are processed and packaged so that their ambient temperature shelf life can extend three to six months.

Canned foods. Canning is the process of extending shelf life by thermally destroying all microorganisms and enzymes present and maintaining that sterility by hermetic sealing in oxygen- and water vapor-impermeable packaging that excludes microorganisms. The resulting food product has been fully cooked as a consequence of the heat process (see Canning, food).

Whether we use a metal can or a glass jar, the process begins with treating the food product prior to filling. Initial operations inactivate enzymes so that the enzymes will not degrade the product during processing.

The package is cleaned. The product is introduced into the package, usually hot. In general, air that can cause oxidative damage is removed from the interior. Air removal leads to an anaerobic condition which can foster the growth of *Clostridia* organisms. The package is hermetically sealed (see Can closing) and then subjected to heating. The package must be able to withstand heat up to about 212°F (100°C) for high-acid products and up to 260°F (127°C) for low-acid products which must receive added heat to destroy heat-resistant microbial spores. Packages containing low-acid (above pH 4.5) foods must withstand pressure. In glass-jarred products, there must be external overpressure to ensure that the closure stays on the package.

The thermal process is calculated on the basis of the time required for the most remote portion of the food within the package to achieve a temperature that will destroy *Clostridia* microorganisms. After reaching that temperature, the package must be cooled rapidly to retard further cooking.

Since the products contained are biochemically and chemically aggressive, significant interaction between the food product and the internal can coating can occur. Thus, high temperature-resistant inert organic coatings are employed inside the can.

The packaging must contain the product, exclude air, withstand heat conditions, and also maintain hermetic seal throughout its distribution life and ensure that no microorganisms can reenter the package. It must be easy to open and dispense, inert to contents and the environment, shock and vibration resistant, and cost effective.

Attempts have been made to perform the same thermal retorting in a flexible pouch or tray. The retort pouch (see Retortable flexible and semi-rigid packages), has a much higher surface-to-volume ratio and employs a heat seal (see Sealing, heat) rather than a mechanical closure.

The original intent of the retort pouch was to provide a convenient package for military ground forces. This objective was extended to attempting to introduce better quality food by virtue of less heat to effect the thermal sterilization.

Aseptic processing. Aseptic packaging is the independent sterilization of product and package, bringing them together in a sterile environment and sealing so that the contained food may be stored for prolonged periods at ambient temperatures. An important objective is the ability to sterilize the product outside the package, use thin-film heat exchangers, and thus impart less thermal damage to the products. This limited the application of aseptic packaging to liquids such as milk, juices, juice drinks, teas, puddings, etc. Most aseptic packaging is done with laminations of paperboard, aluminum foil, and polyethylene into block shapes; some is being performed in thermoformed cups on thermoform/fill/seal or preformed cup deposit/fill/seal systems.

Frozen foods. Freezing requires reducing the temperature below the freezing point of the water so that microbiological, enzymatic, and biochemical activities are significantly retarded. In freezing, the product temperature passes through the transition from liquid water to ice rapidly so that ice crystals are relatively small and do not physically disrupt food cells.

The product may be frozen inside or outside of the package. In the early days of frozen foods, most freezing was performed with the product inside a waxed paper-wrapped paperboard carton; plate freezers, with direct contact of the cold surface to the package, were employed widely. More recently, however, freezing processes use high velocity cold air or liquid nitrogen to remove the heat from the bulk unpackaged or individually quick frozen (IQF) product. Large quantities of cooking vegetables are frozen usually in IQF form.

The frozen product is then packaged in paperboard cartons or polyethylene pouches. Frozen foods are susceptible to sublimation of ice which can lead to freezer burn.

Probably the most important products frozen today in the United States are precooked, processed entrees in meal-size portions, often in aluminum-foil or paperboard trays with flexible closures, overpackaged in printed-paperboard cartons. Soft baked goods such as cakes, pastries, pies, and breads are frozen after packaging in coated paperboard cartons which are grease- and water vapor-resistant.

Dry foods. By removing all water from food, biochemical activity ceases. Biological activities in dry-food products are related to water activity A_w. Water activity is the ratio of water vapor pressure of the food product to the water vapor pressure of pure water under the same conditions. Thus, a food product with an A_w of 0.6 would have an equilibrium relative humidity of 60%. If the relative humidity outside of the product were to exceed 60%, water would enter the product; if the relative humidity outside of the product is below 60%, water leaves the product. Water activity depends not just on the free water but rather on the active water present and capable of being evaporated.

By water activities, food products may be classified into very dry, with a water content of less than 10%; intermediate water activity or a water content between 10–85%; and completely hydrated with water activity of 0.85 or above. Products with water activities of 0.85 and above are essentially fresh or processed products susceptible to microbiological and rapid enzymatic deterioration and so must be processed for extended shelf life.

Dry products include those dried from liquid or engineered mixes of dried components that become dry products. In the first category are instant coffee, tea, and milk. The liquid is spray, drum, or air-dried because water content above 1% can lead to browning or significant product deterioration as well as particle agglomeration that restricts product rehydration. Engineered mixes include beverage mixes such as sugar, citric acid, color, flavor, etc, and soup mixes which include dehydrated soup stock plus noodles and some fat products, all of which become a liquid on rehydration.

Very dry products are often specially dried with porous surfaces to facilitate rehydration and so are very susceptible to moisture in the air. Oils within the product flavor are susceptible to oxidation, and so many such products are sealed under inert atmospheres to ensure against oxidation. Products with relatively high fat such as bakery mixes or soup mixes must be packaged so that the fat does not interact with the packaging materials. For example, low density polyethylene on the interior of the package can adversely interact with fat products. Further, seasoning mixes that contain herbs and volatile flavoring components can interact with plastic packaging materials. Interaction may be scalping or removal of flavor from the product or the product removing components from the packaging material. The package must be hermetically sealed; that is it must provide a total barrier against access by water vapor, and for products susceptible to oxidation, also exclude oxygen.

Only very dry products with A_w below 10% require the very high barriers. Other dry products are more susceptible to the interaction of the fatty and flavor components of the product with packaging materials and less with the hygroscopicity of the product.

Fats and oils. In general, fats and oils may be classified into those with and without water. Cooking oils and hydrogenated vegetable shortenings contain no water and are extremely stable. Hydrogenated fats and oils are subject to hydrolytic rancidity through reaction at high temperatures or long-term reactions at low temperatures with small quantities of water.

Unsaturated fats and oils are subject to oxidative rancidity; that is, breaking of the fatty acid chain at the double-bond sites to form peroxides and ultimately low molecular weight odorous aldehydes and ketones. Oils are more subject to oxidative rancidity than fats and both are usually packaged under inert atmosphere such as nitrogen. Hydrogenated vegetable shortenings generally are packaged in metal, composite paperboard, or plastic cans with nitrogen to ensure against oxidative rancidity. Oils are packaged in glass and blow-molded plastic such as high density polyethylene or PVC.

Fats and oils containing water include margarine and butter. Such products contain water-soluble ingredients, eg, salt, as well as milk solids which impart flavor and color to the product. Generally, these products are distributed at refrigerated temperatures to retain their quality. Greaseproof packaging such as coated paperboard, aluminum foil/paper and parchment wraps, and plastic tubs are used to contain butter and margarine.

Grain products. The largest volume of food product in the world is in the form of grain, ie, rice, wheat, corn, oats, etc. Water activities are reduced so that the starch and protein contents are stable. Most grain products contain oil subject to rancidity. The presence of water and air renders grains susceptible to biochemical deterioration. Bulk grains also are subject to infestation from insects, rodents, etc. Refined-grain products such as flour have relatively high water activity and so require no special packaging. On the other hand, grain products may contain insect eggs and can be the target of insects and rodents.

Grain products are converted into more edible products. Breakfast cereals contain most of the whole grain in cooked or toasted form plus flavorings, vitamins, and minerals. Cereals have a sufficiently low water content that they are susceptible to moisture absorption. At elevated temperatures, the fat can separate and release liquid fat. Breakfast cereals, therefore, in general require good water vapor- and grease-barrier packaging. Packaging generally should retain the delicate flavors. Breakfast cereals are packaged in glassine, waxed glassine, or, mostly, polyolefin coextrusions in the form of pouches or bags within paperboard carton outer shells (See Bag-in-box, dry product). Sugared cereals are often packaged in aluminum-foil laminations to retard water vapor transmission (see Multilayer flexible packaging).

Soft baked goods including breads, cakes, and pastries are highly aerated structures containing about 45% water content. Soft baked goods are subject to dehydration and staling. Baked goods are also subject to microbiological deterioration as a result of growth of mold and other microorganisms. Equilibrium relative humidity must be maintained for product distributed at ambient temperatures. Because of the short shelf life, fairly good water vapor barriers such as polyethylene-film bags or polyethylene-coated paperboard are used for packaging.

Hard baked goods such as cookies and crackers generally have a relatively low water content and a high fat content. Water can be absorbed, and the product can lose its desirable texture and be subject to rancidity. At elevated temperatures, the contained fat can be exuded and so dry bakery products should be packaged to exclude water vapor and also to eliminate the possibility of fat from the product leaking into the packaging material. Packaging for cookies and crackers includes waxed glassine or polyolefin-coextrusion pouches within paperboard shells, polystyrene trays overwrapped with polyethylene or oriented polypropylene film, etc. Soft cookies are packaged in high water-vapor-barrier laminations containing aluminum foil or metallized plastic film.

Snacks. Snacks include dry grain or potato products such as potato and corn chips, and roasted nuts. Snacks usually have low water content and relatively high fat content. Snack packaging problems are compounded by salt, a catalyst for oxidative rancidity and an ingredient which can abrade the packaging material surface. Snacks are most often packaged in pouches that have low water vapor transmission, relying on rapid distribution to obviate any fat oxidation problems. Some snacks are packaged under inert atmospheres in sealed rigid containers such as pouches or composite cans to permit long-term distribution. Generally, light harms such products, so opaque packaging is usually employed.

Candy. The three basic types of candies of concern are chocolate and hard and soft sugar. Chocolates may be either solid, solid with inclusions such as nuts, or molded with centers. Chocolate is a mixture of fat and nonfat components, infrequently subject to flavor or microbiological change. Inclusions and fillings are susceptible to water gain or loss. The chocolate acts as a water vapor barrier that helps protect the inclusions such as nuts, crisped rice, liquid cherry. Chocolates are packaged in greaseproof materials such as glassine or polypropylene plastic.

Hard sugar candies are amorphous sugar with added flavors and are extremely hygroscopic because of their very low moisture content. Further, sugar candies are hard and fragile and should be protected against physical damage.

Soft sugar products include jellies, marshmallows, etc. These contain a matrix of water which can be lost as a consequence of evaporation, and so these products must be protected against water loss.

Sugar candies are sealed in low-water vapor-transmission packaging.

Spreads. Bread and cracker spreads include jams, jellies, preserves, and peanut butter. A serious defect of jams, jellies, and preserves is browning that takes place as a consequence of interaction of oxygen at the product surface which leads to darkening of the color.

Peanut butter is a fatty product containing ground peanuts and is susceptible to hydrolytic rancidity. The package, of course, must be greaseproof.

Most spreads are packaged in glass with reclosable lids.

Beverages. Beverages may be still or carbonated, alcoholic or nonalcoholic. The largest quantity of packaging in the United States is for two carbonated beverages: beer and soft drinks. (see Carbonated beverage packaging).

Beer and other carbonated beverages contain dissolved carbon dioxide which creates internal pressure within the package. Thus, the package must be capable of withstanding the internal pressure of carbon dioxide up to 90 psi (620 kPa) in some instances. Carbonated beverages are subject to oxidative changes in flavor and so should be protected against oxygen. Further, the packaging material should not contain components that can be scalped or removed by the product. Organically lined metal cans and glass bottles are used for packaging carbonated beverages. Since the 1970s, polyester plastic bottles have been used for soft drinks.

Beer is more sensitive than other carbonated beverages to oxygen, to loss of carbon dioxide, to off flavor, and to light. Further, most U.S. beer undergoes thermal pasteurization.

Most commercial pasteurization is done after sealing in the package, and the internal pressure within the package can build up to well over 100 psi (689 kPa) at 145°F (63°C), the usual pasteurization temperature. Relatively few beers in the United States, except for draft keg beers, are not exposed to heat for stabilization. Beer is sensitive to flavors from packaging materials in which it is stored. Beer and other carbonated beverages are generally packaged at relatively high speeds, which means that the packaging materials must be extremely uniform, free of defects, and dimensionally stable.

About 10% of beer in the United States is sold in returnable glass bottles. Reusable bottles are cleaned and inspected before each use to ensure against chips, cracks, stones, and other foreign objects.

Other liquid beverages packaged in the United States include wine, most of it in glass bottles. Very small quantities of wine are packed in metal cans, and very slight quantities are packed in paperboard–foil composites such as currently used for aseptic packaging of juices.

Liquor packaging requirements are subject to regulatory restrictions. The packages must be resistant to both water and alcohol and must not alter the product flavor or the proof or alcohol content. Alcohol extraction of components from coatings is not unknown. Distilled alcoholic spirits have traditionally been packaged in glass although now liquor is being packaged in lightweight nonbreakable plastic packages.

BIBLIOGRAPHY

General References

Packaging of Horticultural Produce, symposium proceedings, Sprenger Institute, Wageningen, the Netherlands, 1981.

A Processors Guide to Establishment, Registration and Process Filing for Acidified and Low Acid Canned Foods, FDA, HHS Publication 80-2126, U.S. Department of Health & Human Services, 1980.

Retort Pouch Technology Seminar II—Proceedings, Pouch Technology, Inc. Oak Brook, Ill., 1982.

Proceedings of Conference *Aseptic Processing and the Bulk Storage and Distribution of Food,* Food Science Institute, Purdue University, Lafayette, Ind., 1978.

Proceedings *International Conference on UHT Processing and Aseptic Packaging of Milk and Milk Products,* Dept. of Food Science, North Carolina State University, Raleigh, N.C., 1979.

Proceedings of First International Conference on Aseptic Packaging, Schotland Business Research, Princeton, N.J., 1983; Second Conference, 1984.

The Science of Meat and Meat Products, W. H. Freeman & Co., San Francisco, Calif., 1960.

H. M. Broderick, *Beer Packaging,* Master Brewers Association of America, Madison, Wisc., 1982.

A. L. Brody, *Controlled / Modified Atmosphere / Vacuum Packaging of Foods,* Food and Nutrition Press, Trumbull, Connecticut, 1989.

A. L. Brody, *Flexible Packaging of Foods,* CRC, Cleveland, Ohio, 1970.

A. L. Brody, "Food Canning in Rigid and Flexible Packages," *Crit. Rev. in Food Tech.* **2**(2), July 1971).

A. L. Brody, and E. P. Schertz, *New Developments in Meat and Meat Packaging Technology,* Iowa Development Commission, Des Moines, Iowa, 1968.

A. L. Brody and E. P. Schertz, *Convenience Foods: Products, Packaging, Markets,* Iowa Development Commission, Des Moines, Iowa, 1970.

D. Carracher, *Seminar Proceedings Aseptic Packaging,* Campden Food Preservation Research Assoc., Chipping Campden, Glos., UK, April 1983.

C. M. Christensen, *Storage of Cereal Grains and Their Products,* American Association of Cereal Chemists, St. Paul, Minn., 1982.

R. B. Duckworth, *Fruit and Vegetables,* Pergamon Press, Oxford, UK, 1966.

R. Griffin and S. Sacharow, *Food Packaging,* 2nd ed., AVI, Westport, Conn., 1982.

R. S. Harris and E. Karmas, *Nutritional Evaluation of Food Processing,* AVI, Westport, Conn., 1975.

A. C. Hersom, and E. D. Hulland, *Canned Foods. An Introduction to Their Microbiology,* Chemical Publishing Co., New York, 1964.

D. S. Hsu, *Ultra High Temperature Processing and Aseptic Packaging of Dairy Products,* Damana Tech, New York, 1979.

T. P. LaBuza, "Moisture Gain and Loss in Packaged Foods," *Food Tech.* **36,** (April 1982).

R. Lampi, "Retort Pouch. The Development of a Basic Packaging Concept in Today's High Technology Era," *J. Food Proc. Eng.* **4**(1), (1981).

R. A. Lawrie, *Meat Science,* Pergamon Press, Oxford, UK, 1966.

A. Lopez, *A Complete Course in Canning,* The Canning Trade, Baltimore, Md., 1981.

B. J. McKernan, "Developments in Rigid Metal Containers for Food," *Food Tech.* **37,** (April 1983).

C. R. Oswin, "Isotherms and Package Life: Some Advances in Emballistics," *Food Chem.* **12,** 3 (1983).

C. R. Oswin, "The Selection of Plastic Films for Food Packaging," *Food Chem.,* **8**(2), 121–127 (1982).

F. A. Paine and H. Y. Paine, *A Handbook of Food Packaging,* Leonard Hill, London, 1983.

N. Potter, *Food Science,* AVI, Westport, Conn., 1978.

G. Robertson, Food Packaging, N.Y.: Marcel Dekker, 1993.

S. Sacharow and A. L. Brody, *Packaging: An Introduction,* Duluth, MN: 1986.

Aaron L. Brody
Rubbright, Brody, Inc.
Duluth, Georgia

PAD PRINTING. See Decorating.

PAILS, PLASTIC

Plastic pails can be either open-head containers with removable lids or containers produced as a single unit called tight-head containers. The container top or lid can be manufactured with one or more openings for filling and dispensing. The openings, in turn, are designed to be used with a variety of closures. In this article, only the range of container sizes and designs that are normally considered to be shipping containers are discussed in detail. These can be defined as single-wall, heavy-duty containers with a handle. These containers range in capacity from 1 gal (3.785 L) to 6 gal (22.7 L) (see also Drums, plastic).

In the open-head configuration, the containers are usually supplied with a removable lid designed to be used with either liquids or solids. The lids are available with pour fittings of various types. The most commonly used pour fitting for liquids is the flexible polyethylene spout, which either snaps into position in a preformed hole or is a spout that incorporates a metal collar that is crimped onto a formed, ridged opening in the lid. These fitting designs feature the ability to be recessed into themselves and not protrude above the lid surface when in a nonuse position. When positioned for use, however, they extend up and out to form a convenient pour spout. Open-head containers are, with few exceptions, of ta-

Table 1. Industrial Uses of Plastic Pails[a]

Type	Quantity, $\times 10^6$
Printing inks	5
Janitorial products	9
Chemicals	22
Petroleum products	11
Paints	27
Building products (adhesives, cements, joint compounds, etc)	52
Foods	30
Miscellaneous	4
Total	160

[a] Statistics are provided for comparative purposes only.

pered design, which allows the individual pails to nest into each other for efficient storage and transportation, prior to being filled and sealed. These pails are universally injection-molded, as are the matching lids (see Injection molding).

In the closed-head configuration, no nesting advantage exists, but this product has certain advantages in the area of structural integrity arising from the one-piece construction. This construction meets particular government specifications for packaging of hazardous products (dangerous goods). Most closed-head plastic containers manufactured and sold in the North American market are integral one-piece units produced by the blow-molding process (see Blow molding).

The use of plastic pails for the shipment of industrial products in the United States and Canada was estimated to have reached a level of 160 million (10^6) units in 1984 (1). Of the total, less than 25% are closed-head (also called tight-head containers). The uses of these pails and the approximate breakdown of the principal categories of use are shown in Table 1.

Within the subject capacity range, the 5-gal (18.9 L) open-head pail is the predominant package, constituting 60–70% of the total number of pails consumed (2). Other popular packages in the United States include a 4-gal (15.1 L) "food package" and the $3\frac{1}{2}$-gal (13.25 L) open-head pail used in the chemicals industry.

In Canada, since the government-mandated conversion from British Imperial units to metric standards in the early 1980s, the 20-L container size has become the predominant pail size (2). The relative closeness in capacity between the 5-gal (18.9 L) and the 20-L size would seem to indicate that a future U.S. conversion to metric would favor the adoption of the 20-L size in the United States as well.

Raw-Material Considerations

Open-head and closed-head pails are made of high-density polyethylene. The following discussion pertains to the predominant open-head injection-molded pails only (3).

Four basic parameters must be considered in the design of a polyethylene resin for injection molding industrial containers. These are melt index, density, molecular-weight distribution, and type of comonomer incorporated (see Polyethylene, high-density). Adjusting each of these parameters results in a trade-off between the various designed properties. Although such tradeoffs make resin design difficult, there seems to be a fairly small range in each of the above four parameters that gives optimum properties. The injection-molding pail resins from all resin suppliers appear to fit into the ranges listed below.

Melt index is one of the two most important factors. The higher the melt index, the easier the processability of the resin, allowing faster molding cycle times and lower production costs. On the other hand, increasing the melt index reduces all the physical properties, including drop impact strength, top-load capacity, and environmental-stress crack resistance (ESCR). In general, a melt index of 4–8 g/10 min gives an optimum balance of these properties. The smaller the volume of the pail and the thinner the wall, the higher the optimum melt index of the resin should be.

Density is the other most important factor; it is controlled by the amount of comonomer that is polymerized with the ethylene. Lowering the density increases the drop impact strength and the ESCR, but decreases the top-load capacity of the container. Processability is virtually unaffected by density. Most injection-molding pail resins have densities of 0.950–0.955 g/cm^3.

Another factor is the molecular-weight distribution (MWD), which is determined by the type of polymerization process, as well as the type of catalyst used. A narrow MWD means that most of the polymer chains are of similar length, whereas a broad MWD implies that the polymer chains vary in length. A broader MWD aids in processability and generally increases ESCR, but at the expense of some drop-impact strength and top-load capacity. An intermediate MWD seems to be optimum for pail applications.

The type of comonomer also affects the properties of the injection-molded part. A comonomer such as butene or octene is used in the polymerization process to add branches to the polymer chain and thus lower the density. The shortest-chain comonomer available is propylene (C_3), whereas the longest comonomer generally used for polyethylene is octene (C_8). Propylene copolymers possess very poor ESCR. A comonomer with a long-chain, such as octene, results in an increase in drop-impact strength and a large increase in ESCR. However, in the melt index range under discussion, the superior properties of octene can only be realized if the density of the copolymer is below 0.950 g/cm^3. Above this density, so little comonomer is incorporated into the polymer that the octene acts solely as a density modifier and not as a physical property enhancer. As a result, injection-molding pail resins are generally ethylene–butene (C_4) copolymers.

Performance Requirements

Plastic pails are used to package many different products, and many of these products are likely to be shipped across state or national boundaries. As a result of these shipments, the containers themselves are subject to government regulations and specifications. The specifications and regulations that apply to the pails are determined by the nature of the product being transported. Products are categorized as either "Hazardous" or "Nonhazardous." In terms of packaging performance requirements, hazardous products are further split into liquids and solids. All hazardous products shipped in interstate commerce in the United States are regulated by the Department of Transportation (DOT), which defines hazardous materials as "a substance or material in quantity and form which may pose an unreasonable risk to health and safety or property when transported in commerce" (4).

Pails used for the packaging of nonhazardous products are required to meet specifications that are established by the Uniform Classification Committee (UCC) of the Association of American Railroads (for rail transport) and the National Motor Freight Classification Committee (NMFC) (for highway transport).

There are additional regulations for food products. In the United States they are as follows:

1. All containers used for food products must meet the requirements of the Food Additives Law.
2. All containers used for meats and poultry must be approved by the United States Department of Agriculture (USDA).

As a general rule, all ingredients used in the manufacture of pails destined for food packaging are expected to comply with FDA regulations. Users of pails for food-packaging applications typically request certification from their suppliers that this is the case (see Food, drug, and cosmetic regulations).

Specifications

Neither the preceding review of applicable regulations nor the specifications that follow are intended to be a comprehensive and up-to-date list of all specifications and regulations. A number of these regulations and specifications are currently under review. The interested reader is urged to obtain up-to-date information from the appropriate authority. Sources of information can be found in Ref. 5. The primary purpose of this discussion of specifications is to examine the basic parameters for the design of plastic pails that are on the market and in use at this time in North America.

DOT Specifications for hazardous commodities. *Specification 34-Liquids* (6) covers reuseable polyethylene containers for use without overpack, which range in size up to 30 gal (114 L) and have no removable heads. Until recently, in addition to limiting the container size, this specification listed in detail allowable limits of melt index, density, tensile strength, and elongation. All of the above are for polyethylene packagings only. The most recent direction is to eliminate specific raw-material criteria, extend the range of packaging sizes, and in general to increase the emphasis on performance requirements. The test procedures described below have provided the basis for design of current products.

Tests. Space limitations do not allow for a detailed listing of the test procedures; however, they are summarized below.

1. A container filled with liquid conditioned to 0°F (−18°C) is required to withstand a 4-ft (1.22 m) drop on the top diagonal edge, or whichever part is weakest. The container must not leak after the test.

2. A container must withstand a hydrostatic pressure of 15 psi (103.4 kPa) for 5 min without pressure drop.

3. A container must not deflect in excess of 1 in. when filled with water and subjected to a compression load as per Table 2. The load is held constant for 48 h.

Table 2. Compression Load Requirements

Marked (rated) capacity, gal (L)	Compression test, lb_f (N)
2.5–6.5 (9.5–25)	600 (2669)
15 (56.8)	1200 (5338)
30 (113.6)	1800 (8007)

Records of test results must be maintained in current status and retained by each manufacturer at each producing plant. The preceding performance requirements of *Specification 34* are likely to be changed to add a requirement to establish permissible rates of permeation for various hazardous commodities. Also to be included is a test procedure involving time, temperature combinations, and both product weight loss and material testing.

Specification 35-Solids (7) is for nonreusable molded polyethylene drums for use without overpack with removable head required. As in Specification 34, specific parameters of polyethylene are included, together with construction and capacity details. The specifications are likely to change in the future in line with the trends described above, but the existing basic tests are the design parameters used for current DOT specification containers.

Performance requirements are summarized below:

1. In a drop test a container is filled with dry powder, topped with sodium bicarbonate, and closed with lid. Container and product conditioned to 0°F (−18°C) must survive without spillage each of the following drops, with no container being required to submit to more than one test: 4-ft (122-m) drop onto concrete flat on bottom, 4-ft (1.22-m) drop onto concrete on top edge, and 4-ft (1.22-m) drop onto concrete on bail ear (side drop).
2. A vibration test is required, although vibration is not normally a cause of failure.
3. A compression test is required, where container and dry product are conditioned to 130°F (54°C) and are required to withstand without deflection in excess of 1 in. (2.54 cm) a 600-lb (272-kg) top load.

Transport specifications for products classified by DOT as nonhazardous commodities. Rail shipments of products packaged in plastic pails are regulated by the UCC of American Railroads. The most-current applicable reglation is covered under *UFC 6000C, Rule 40, Sect. 7 ¼.* This specification covers open-head pails specifically and in addition to material requirements and construction specifications, there is a brief listing of performance requirements.

Performance requirements. Pails, filled with commodity to marked capacity, must meet the following performance standards without failure. Failure is defined as leakage or spillage of contents. Each test must be performed on a minimum of three sample pails, but no single pail will be required to withstand more than one test.

1. Pails must be conditioned to 0°F (−18°C) for a minimum of 4 h. Drop tests must be performed with the pail flat on its side and also at a 45° angle on the bottom chime onto solid concrete from a height of 48 in. (1.22 m).

2. Pails must be conditioned to 130°F (54°C) for 4 h, stacked three high, and vibrated for 1 h at 1 *g* to a vertical linear motion.

3. Pails filled with commodity to marked capacity must withstand a static load to 600 lb (272 kg) for a period of 48 h without defect or damage.

Tests may be performed using water when the viscosity of the commodity does not exceed 5000 cP (5 Pa · s), or sand when the viscosity of the commodity exceeds 5000 cP (5 Pa · s).

Highway shipments of nonhazardous commodities are regulated by the NMFC. The most current ruling affecting open-head plastic pails is *Classification 100J.* These regulations are similar to those of the railroad regulatory group.

A review of the performance standards listed shows that packaging of nonhazardous commodities is required to undergo much less rigorous testing than that required of pails to be used for hazardous commodities, particularly for liquids. Because the number of open-head pails used for hazardous commodities is relatively small and the emphasis on higher performance requirements for these pails is increasing, some manufacturers of open-head pails do not offer their products for these applications. The market is served by manufacturers of closed-head plastic pails and those with a specialized heavy-duty-design open-head pail package. Recent emphasis by DOT on compatibility testing is a further barrier to be overcome by the HDPE packages that are aimed at the hazardous-commodities packaging markets. Although a number of plastic pail designs currently on the market meet all of the requirements established by the DOT, the majority of product being sold is aimed toward meeting the less stringent nonhazardous-packaging regulations.

Design

Most plastic pails were manufactured and sold without benefit or concern for patent protection (8). In order to meet previously described performance requirements, there are a number of design features that tend to be common to all plastic pails.

These features are related to the ability to withstand impact, without leakage; the ability to withstand compressive loads at elevated temperatures; the maximization of the raw-material component of the package and resin processability; and convenience and aesthetic features.

Design features review. The features listed in the previous section are examined with reference to the designs illustrated in Figure 1.

Impact strength. The fit of lid to pail body and the ability of the combination to withstand deformation is key to maintaining package integrity. Where severe impact inevitably results in deformation, that deformation must occur in such a controlled fashion as to maintain a water-tight seal. As illustrated in Figure 1, pail bead size and design (A), corresponding cover fit (B), together with an adequate reinforcing structure of either radial ribs (C), or a bumper configuration (D) are all contributing factors to the ultimate success or failure of the design to withstand impact.

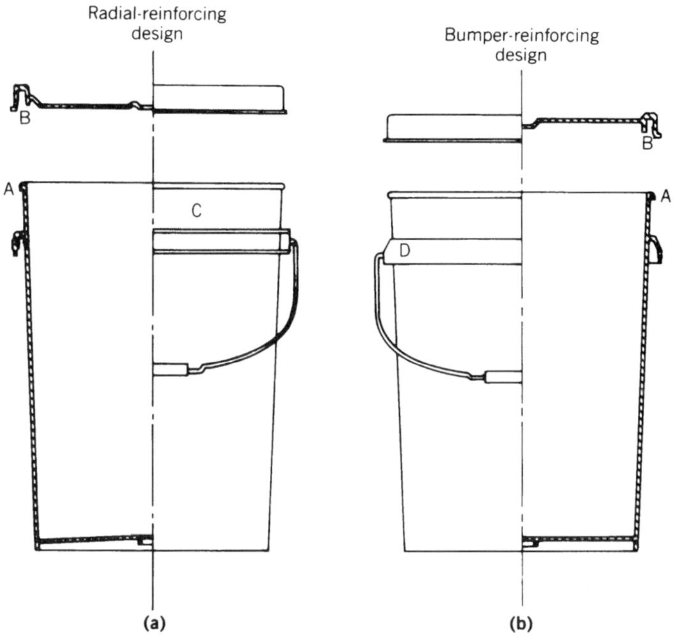

Figure 1. Plastic pail designs: (**a**) radial-reinforcing design; (**b**) bumper-reinforcing design. Parts A–D are described in text.

Compression capability. The following are generally considered important factors in maximizing top-load strength (9): wall thickness and uniformity—no other factor has a greater effect given specific material characteristics of stiffness; number and location of either radial or vertical stiffening structures; lid design; and angle of taper and ratio of container diameter to height.

Processing factors. Design features that relate to processability and effective use of the polymer. These features may include variable wall cross sections of pail walls and lids, and variations in gate sizes and designs.

Convenience and aesthetics. Other areas of design interest are taper or nesting capability; size and texture of the printing-surface area of the container; chime, base ribbing configuration and ear design for bail handle; design of cover for removal and installation ease, pouring, or ability to avoid surface-water pooling; and general ease of cleaning after filling. Ability to be handled with automatic stacking and palletizing equipment.

Mold design is often tailored to product design and *vice versa.* A successful marriage of the two is essential for a satisfactory outcome.

Processing Equipment

The injection-molding process is generally well known, and information regarding the process is readily available from a variety of sources, such as the Society of Plastics Industry (SPI), New York, NY, and the Plastics Shipping Container Institution (PSCI), Chicago, Illinois (see Injection molding). The predominant type of machinery used for pails is the all-hydraulic, horizontal injection-molding press, with 500–700 tons (4.45–6.23 MN) of clamp capability. Development of processing machinery and expansion of the plastic-pail market have resulted in equipment built and targeted for specific uses. These tailored specifications have contributed to substantial improvements in both product quality and output.

The handling, processing, and use of polyethylene is generally considered to be a relatively safe activity. Polyethylene is used widely as a raw material of choice for industrial packaging, and as evidenced by the widespread acceptance of the product in the food industry, can be considered as a safe product in its finished form. Much has been discussed and written with regard to the toxicity of the products of combustion of plastics. Available research (10) indicates that the products of combustion of a typical plastic pail are not greatly different from those of a pine log.

BIBLIOGRAPHY

"Pails, Plastic" in *The Wiley Encyclopedia of Packaging Technology,* 1st ed., by Peter Kirkis, Vulcan Industrial Packaging, Ltd., pp. 483–486.

1. *Sales Statistics,* Vulcan Industrial Packaging, Ltd., Toronto, Canada, 1982.
2. *Vulcan Industrial Packaging Statistics 1983 / 1984,* Vulcan Industrial Packaging, Ltd., Toronto, Canada, 1984.
3. R. Scott, research data, Dow Chemical Canada, Ltd., Canada, 1984.
4. *Title 49, Hazardous Materials Transportation Act,* U.S. Department of Transportation, Washington, DC.
5. DOT, Materials Transportation Bureau, *Hazardous Materials Regulations,* Research and Special Programs Administration-Standards for Polyethylene Packaging, U.S. Department of Transportation, U.S. Government Printing Offices, Washington, D.C.; NMFC-100, Classes & Rules, National Motor Freight Traffic Association Inc., Alexandria, VA; UFC-6000, Ratings, Rules, and Regulations, Uniform Classification Committee, Association of American Railroads, Chicago.
6. *Hazardous Materials Regulations, Specification 34, DOT Requirements, Title 49, Code of Federal Regulations* (49 *CFR*), Sect. 178.19, U.S. Department of Transporation, Washington, DC, 1982.
7. *Hazardous Materials Regulations,* Specification 35, *DOT Requirements, Title 49, Code of Federal Regulations* (49 *CFR*), Sect. 178.16, U.S. Department of Transportation, Washington, DC, 1982.
8. U.S. Pat. 3,510,023 (May 5, 1970), F. E. Ullman, W. Klygis, and M. J. Klygis (to Inland Steel); U.S. Pat 3,516,571 (June 23, 1970), W. Roper, R. E. Roper, R. Roper, C. R. Roper, F. Roper, and R. A. Miller; U.S. Pat. 3,804,289 (Apr. 16, 1974), R. G. Churan (to Vulcan Plastics Inc.).
9. J. E. Boyd and S. B. Falk, *Strength of Materials,* McGraw-Hill, New York, 1950.
10. W. J. Potts, T. S. Lederer, and J. F. Quast, *Combust. Toxicol.* **5,** 412–433 (Nov. 1978).

PALLETIZING

A pallet is a low platform used to stack or accumulate a number of smaller units of product so that they may be conveyed by mechanical means. Pallets may be made of wood (see Pallets, wood), plastic (see Pallets, plastic), or corrugated kraft board (see Pallets, expendable corrugated), and are usually rectangular rather than square. The arrangement of product on the pallet (pallet pattern) is critical to an orderly, compact loading, and distribution system. Patterns correspond to the pallet shape and are designed to afford the maximum load in the minimum space without forfeiting structural strength. Pattern design may be restricted by the primary and secondary product packages, the shipping method and size of conveyance, and warehouse space.

The quantities of pallets and tiers are also determined by the shipping container, shipping method, and distributor space allotment. The second layer, or tier, may be exactly like the first, or rotated 180°, or it may be an entirely different pattern. Unless the cases are heavy and/or large, the pattern of the second layer is changed to form a load which will not break apart. Rotation by 180° is a common solution. The stacking strength of the load must be judged against the space limitations of the pattern. A column stack (each layer identical) with corners one above the other is strongest, but it is most likely to topple since the tiers are not interlocked. Many different pallet patterns are shown in Figure 1.

Palletizing

Once the pattern and number of tiers have been designed, the method of product-to-pallet transfer is determined. Hand palletizing is the most versatile method and is effective when loading is slow. Single-product, single-pattern high level palletization developed from hand palletization. Recent technological advances permit automated palletization of almost every product, but this may not always be economical. The selection of a palletizer demands consideration of performance requirements and space limitations. The palletizer is an element in the total packaging/distribution system, and it should never be a limiting factor. The palletizer should be capable of speeds exceeding normal line speed by 5–10% and change over time from product to product should be minimal. A fully automatic palletizer does not require an operator in attendance full time.

High-level palletizers. High level palletizers (also called moving-pallet palletizers) pick up the product at a level above the height of a full pallet and form the tier pattern on a bed. The pallet is raised to the bed level, and the product is transferred (by a sweep or rake off system) to the pallet. Then, the pallet is lowered one layer, and the next tier is formed in the same manner and placed atop the first. This process continues until the pallet reaches the prescribed height. It is then transferred out to be replaced by an empty pallet that is raised to the bed for a repetition of the process. Numerically controlled tape programs provide simple multipattern capability and are adaptable for a variety of users.

Low-level palletizers. Low-level palletizers (also called "fixed-pallet" palletizers) were developed more slowly than high level palletizers, but with the introduction of programmable logic controllers, they have become highly competitive with high-level units in cost and speed. The low level palletizer operates at floor level (see Fig. 2). The pallet is neither raised nor lowered, and the tier is formed at low level. The transfer bed or apron is raised or lowered to the appropriate level and the tier is transferred to the pallet. The bed returns to the tier-forming position, and the next layer is formed. This

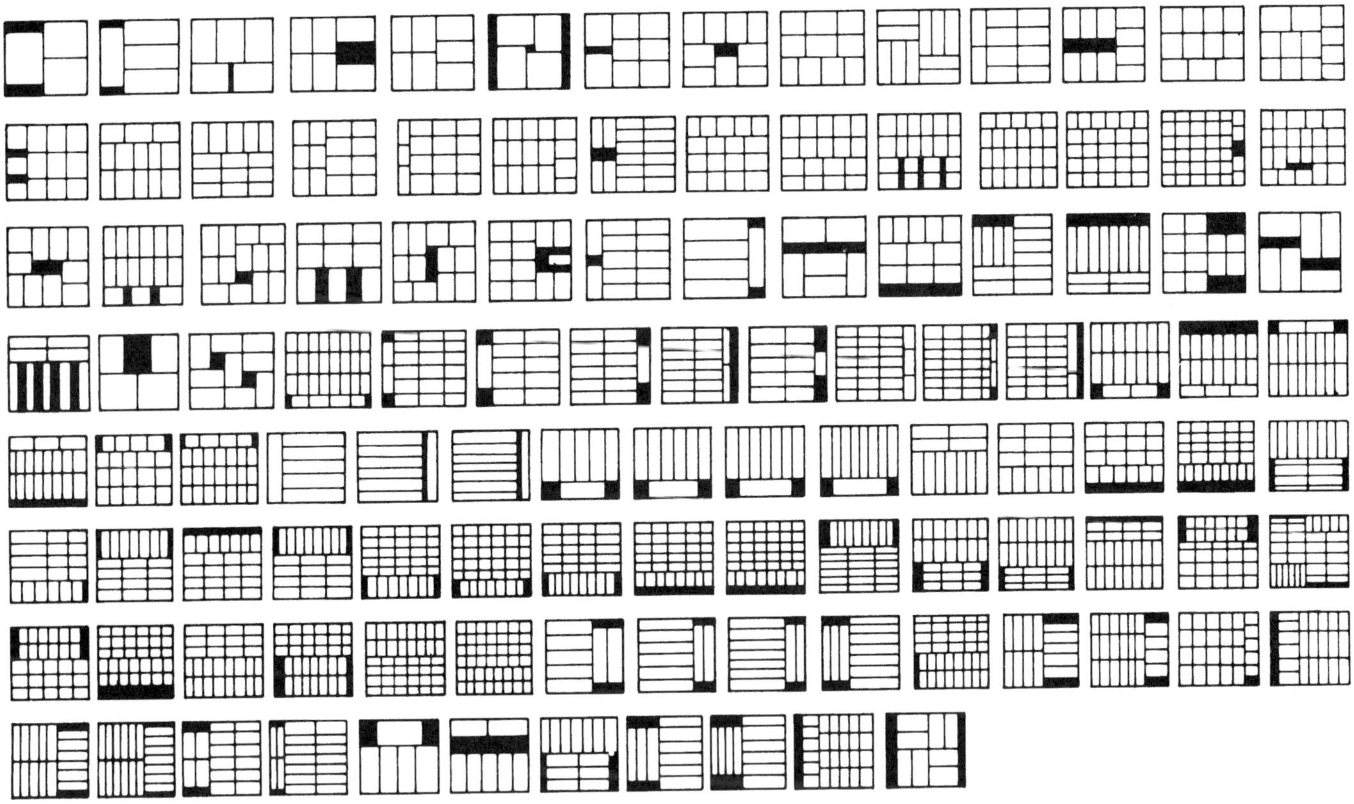

Figure 1. Pallet patterns.

is repeated until the pallet is full. After the full pallet is discharged, an empty one is transferred into the load station, and the process is repeated.

Robotic palletizers. In the past, "pick-and place" systems picked up a product at one point and placed it at a second point repeatedly. At present, robotic palletizers may be thought of as intelligent, discriminating systems capable of picking up several different products and placing them at several different points. These units are usually capable of movement in two or three planes, often on rotational coordinates operating from a fixed point at the center of a circle. They are capable of picking up a variety of products from one point on the circumference of the circle and placing it on one of several pallets located at other points around the circumference. Robotic palletizers are most applicable when a variety of products require different pallet patterns and relatively slow speeds are acceptable.

Bulk palletizers. The shift to bulk handling has reduced the demand for case depalletization of empty containers. There are numerous applications for palletizing without benefit of a case; for example, the transfer of empty cans from the can manufacturer to the filler. Cans are bulk palletized successfully because they are strong enough to withstand relatively rough treatment. Improved handling methods permit bulk palletization of glass containers, and many plastic containers now are being bulk palletized as they gain a greater place in the market.

Nested container-to-container bulk pallets can save up to 15% space for the same number of containers compared to column-stacked cases. That means 15% more containers can be shipped in the same truckload to the user, cutting time and expense. The compact loads also reduce warehouse requirements. In addition, bulk palletizing eliminates the labor and expense associated with uncasing equipment and additional conveyors for empty cases.

Bulk palletization is essentially an adaptation of high- or low-level palletization. The product is accumulated on a table or bed, then transferred by lifting a tier or clamping the perimeter and sweeping the tier onto a pallet. A corrugated sheet the size of the pallet, called a tier sheet, is generally used between layers to provide stability to the load. Further stability can be gained by use of an inverted tray instead of a tier sheet.

Miscellaneous palletizers. Other types of palletizers include drum palletizers, keg palletizers, and bag palletizers. In most cases, these machines address a particular end-use and are concerned only with a narrow portion of the market place.

Depalletizing

The removal of product from pallet depends on the conformation of the product. Bulk depalletizers remove tiers of product from the pallet in much the same manner as bulk palletizers in reverse. In one approach, the tops of the containers are gripped mechanically, pneumatically, or with a vacuum, and the tier is lifted onto a discharge table. In another, the tier is swept onto the discharge table. Removal of the tier sheet or

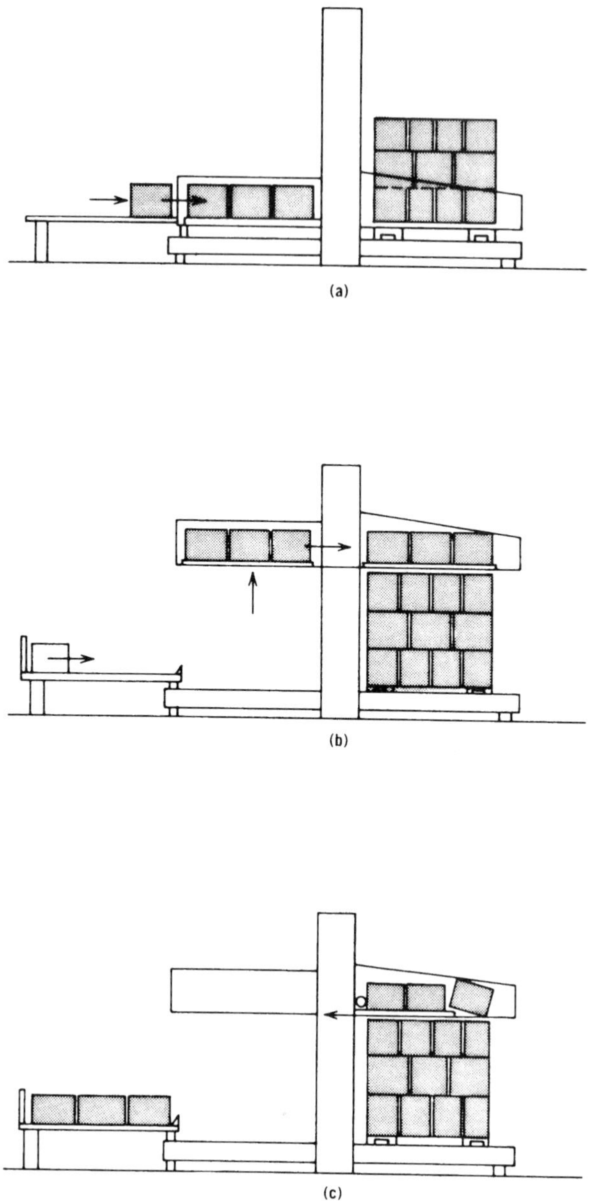

Figure 2. Operation of a low-level palletizer. (**a**) Operation 1. Sealed shipping cases feed in and are oriented to the preprogrammed pattern. When one row is formed, the cases move forward. The next row forms and moves forward, continuing until the layer is complete and the loading plate is filled. (**b**) Operation 2. The layer is lifted to the height of the existing pallet stack. The filled loading plate moves into position just above the pallet stack. (**c**) Operation 3. The loading plate retracts allowing the cases to settle, row by row onto the top of the pallet stack. The pallet is squared by a squaring bar which also assures complete unloading. The loading plate returns to starting position where another accumulated load is ready.

inverted tray is as critical here as in bulk palletization. Product stability is a key factor in all bulk handling operations and the primary determinant of method.

Depalletization of plastic cases or crates may require modified bulk depalletizers or specialized robotic depalletizers. Plastic crates usually have an interlocking feature which requires a tier to be lifted clear of the one below before transfer to the discharge table, thus precluding a sweep system. Most pail depalletizers must handle the products individually in addition to lifting clear of the pail below.

Depalletization of corrugated cases is more difficult than palletization. The flaps on the cases get caught on one another, preventing consistent sweepoff. Corrugated cases do not interlock, so clamping the perimeter causes the center cases in the tier to slip down. The most reliable way to remove a whole tier of corrugated cases is to use tier sheets, but the additional cost discourages wide acceptance. The next-best method is a combination clamping and vacuum system. Some automated warehousing systems remove cases from pallets one at a time in a type of "order-picking" operation. Little effort is being expended today on finding better case-depalletization methods because the shift to bulk handling has shifted research and development work in that direction as well.

BIBLIOGRAPHY

"Palletizing" in *The Wiley Encyclopedia of Packaging,* 1st ed., by S. D. Alley, ABC-KCM Technical Industries, Inc. pp. 486–488.

PALLETS, EXPENDABLE CORRUGATED

Corrugated board can be combined with other materials to produce pallets for lightweight loads. Several manufacturers provide expendable pallets that use corrugated materials for the deck and various other products for the support structure. The deck can be supported by plastic legs, cutdown paper cores, or corrugated buildup material. All provide the features required by materials handling systems. Most expendable corrugated pallets have a load limitation of 1500 lb (680 kg). They can be custom manufactured to fit the exact dimensions of the load placed on them. Typical users include manufacturers of foam products, electrical components, and plastics.

PALLETS, PLASTIC

Plastic pallet construction began during the late 1960s when the low cost of commodity resins such as polystyrene and polyethylene encouraged scores of molders to enter this promising market. In the early 1970s, HDPE was priced at about \$0.16/lb (\$0.35/kg). Still in its infancy, the plastic pallet market was severely curtailed when prices of commodity resins more than doubled at the time of the 1973 oil embargo. A 2 : 1 price differential between wood and plastic pallets quickly became a 4 : 1 price disadvantage.

A number of changes that took place in the 1980s has encouraged and use of plastic pallets:

Packaging is now often considered part of direct production cost instead of fixed overhead. This highlights the savings generated by reusable pallets.

Adoption of the Just-In-Time inventory concept (2) including reusable packaging, inventory reduction, and higher quality.

Figure 1. Typical single-faced plastic pallet.

- Greater use of robots and automated palletizers which require uniform size/weight pallets.
- Increased awareness and regulation of plant sanitation.

Typical single- and double-faced pallets are shown in Figures 1 and 2.

Materials. Most plastic pallets are manufactured from HDPE (see Polyethylene, high-density). Materials such as polystyrene, fiberglass-reinforced plastics (FRP), and polypropylene are used occasionally. Heavy pallet loads and unsupported pallet racking may dictate the use of stiffer polystyrene (see Polystyrene).

FRP (see Thermosets) are used for low-volume custom pallet requirements or prototype pallets. In this situation, low-cost wooden tooling is used with the hand-layup fiberglass technique. Polypropylene (see Polypropylene) has been used to construct structural foam plastic pallets by companies with excess virgin or regrind polypropylene. Polypropylene is not normally used in pallet construction because it requires relatively expensive impact modifiers for cold-weather performance.

Polyethylene is favored for a number of reasons; commodity status (ie, low cost, uniform performance, readily available, wide acceptance); excellent resistance to impact; good performance under a wide range of operating conditions (ie, temperatures of −30 to 150 °F (−34 to 66°C), indoor or outdoor applications, light- to heavy-weight loading); outstanding chemical resistance to most acids and bases; USDA and FDA clearance for use in food and pharmaceutical plants; easy cleaning; and outstanding molding and design flexibility.

Polyethylene's one glaring weakness is its inability to resist deflection (bending) under load. This deflection problem is especially serious in pallet-racking applications. Unsupported racks do not have center supports or decking. In these racks, the pallet must span an open space while maintaining the load. With loads of over 2000 lb (907 kg), plastic pallets are prone to bending (deflection). In addition to the initial deflection, the plastic pallet will continue to bend or creep for up to 2 weeks. Over time, it may become difficult to reenter the pallet with the forks of a lift. Most standard pallet rack, drive-through-rack, and gravity-flow rack is "unsupported."

Figure 3. Two-way entry racking pallet.

In situations where heavyweight racking is a must, steel-reinforced plastic or stiffer polystyrene (PS) are frequently used. Steel reinforcements add expense, and compared to HDPE, PS costs more and offers less chemical and impact resistance. One solution to the racking problem is in the design of rackable pallets. Two-way-entry pallets can rack over 3000 lb (1360 kg) in an unsupported rack (see Fig. 3). Experimental plastic resins are also being tried in an attempt to solve the racking dilemma.

Design and construction. Pallet design and the method of construction greatly influence pallet performance, price, and acceptance. Today plastic pallets are designed and built using several different processing techniques (see Table 1).

Structural foam molding. Most plastic pallets made today are made by structural foam molding. (3,4) This low pressure injection molding process produces parts with a solid skin surrounding a foamed core. Compared to high-pressure injection molding, structural foam molding allows the economic production of heavy wall sections and helps reduce stress points throughout the pallet. The structural foam process provides outstanding design flexibility. Wall thickness of $\frac{3}{16}$–$\frac{3}{4}$ in. can be molded to produce pallets ranging from lightweight single-faced units to super-duty racking pallets. Another benefit is high-speed production, with cycle times as low as 2–3

Figure 2. Typical double-faced plastic pallet.

Table 1. Plastic Pallets, Production Methods

Molding Process	Plastic Pellets Advantages	Plastic Pallets Disadvantages	Ideal Pallet Application	Secondary Applications	Tooling Options	Average Cycle	Wall Thickness
Structural-foam molding	Economic production of heavy-wall sections Short cycle times Good impact resistance Good deflection strength High strength per pound (kilogram) Good weight and dimensional tolerance allows complex shapes	High-cost tooling and processing equipment	Large-volume custom or proprietary pallets with runs of 1000 pallets or more Minimum custom order quantity 3000 units	Manufacture of heavy-duty racking pallets is possible by using filled polyethylene pallets or polystyrene Wall thicknesses of up to 1 in. (2.54 cm) can be used when necessary	Kirksite aluminum steel	2–4 min	$\frac{3}{16}$–1.0 in (4.8–25.4 mm)
Injection molding	Flexible process allows production of lightweight disposable as well as heavy-duty returnable pallets; allows complex geometry	Highest tooling cost Highest equipment costs High energy costs	Largest-volume custom and proprietary pallets	Lightweight disposable pallets can be inexpensively produced by keeping wall sections narrow and cycle times short; heavy-duty racking pallets can be manufactured by using heavier wall sections and a well integrated rib design	Hardened steel	30s–3 min	$\frac{1}{32}$–$\frac{3}{8}$ in. (0.8–9.5 mm)
Rotational molding	Low equipment cost Low tooling cost production of double walled parts	Relatively long cycle times Limited weight and dimensional stability Limited to simpler design (geometry)	Low-volume production of large pallets Custom pallet projects of 1000 units or more are feasible	Manufacture of heavy–duty racking pallet is possible by ecapsulating steel reinforcements into the pallets	Cast-aluminum fabricated metal-plated nickel	3–6 min	$\frac{1}{8}$–$\frac{1}{4}$ in. (3.2–6.4 mm)
Vacuum forming	Low-cost equipment Low-cost tooling	Relatively long cycle times Limited wall thickness Limited depth of draw Limited design complexity	Lower-volume, low-cost, lightweight pallets Pallets projects of 500 units and above are feasible	Heavier loads up to 3000 lb (1361 kg) can be accomodated by using twin-sheet vacuum forming Vacuum-formed pallets are not generally used for heavy-duty racking applications	Metal Plaster Epoxy Wood	3–6 min	$\frac{1}{8}$–$\frac{1}{4}$ in. (3.2–6.4 mm)
Reaction-injection molding	Lighterweight tooling and equipment costs less than injection molding processes Allows complex designs Lower pressures and temperatures afford significant savings (70%) over injection-molding processes	Limited deflection strength Slightly longer cycle times than injection-molding processes Limited dimensional stability	Lighter-duty custom and proprietary pallets Pallet projects of 1000 units and above should be justifiable	Fiberglass-reinforced reaction injection molding is used to increase deflection strength for heavier applications Steel reinforcements can be encapsulated for additional strength	Lightweight steel Aluminum Kirksite Sprayed metal	2–4 min	$\frac{1}{8}$–2.0 in. (3.2–51 mm)

min. Good resistance to impact, high strength per pound (kilogram), and good deflection strength are all positive characteristics that make structural foam a good choice for large-scale production of both custom and proprietary pallets. The chief limitation of structural foam is that relatively high volume (3000 total units minimum) is required to amortize the relatively high tooling cost. When compared to high-pressure injection molding, the tooling for structural foam may be less costly. Most low-pressure foam tools may be built from machined aluminum or Kirksite, which reduces tooling costs by up to 50%. When structural foam tools are built from steel, the cost savings are negligible.

Injection Molding. High-pressure injection molding (see Injection molding), also offers design flexibility. It is used for the production of pallets that range from very lightweight disposables to heavy-duty 60-lb (27-kg) reusables. Injection molded parts generally have narrower wall sections than structural foam, ≤ 0.300 in. (≤ 7.62 mm), and rely on their rib design for structural integrity (5). Injection molding excels in light-weight large-volume production. Cycle times for $\frac{1}{8}$ in. (3.2 mm) injection-molded pallets can be under one minute. Heavy-duty parts with wall section $\frac{1}{8}$ in (3.2 mm) offer high strength and excellent durability. Because high molding pressures require expensive equipment and hardened-steel tooling, high-volume production runs ≥10,000) are generally required to amortize tooling and press costs.

Rotational molding. Rotational molding (3) (see Rotational molding) uses a heated tool into which solid or liquid polymer is placed. This process offers the most economical tooling costs. Myriad sizes and designs of relatively low-quantity (1000–2000 units) can be economically justified. Design innovation, including the molding of steel-encapsulated, smooth-skinned pallets is a feature of rotational molding. Rotationally molded parts offer good resistance to blunt impact and the repair of small puncture damage is possible. Its drawbacks include relatively long cycle times (as high as five min) and relatively narrow $\frac{3}{16}$-in. (4.8-mm) wall thickness (6), limiting rotationally molded pallets to medium-duty applications. Some rotationally molded designs can accommodate the addition of steel reinforcement for heavier loads and pallet racking operations.

Thermoforming. Thermoformed plastic pallets (see Thermoforming) are offered in dozens of low-cost lightweight designs. Inexpensive tooling allows faster amortization of low-volume custom pallets. For example, custom reusable dunnage trays are often thermoformed. Thermoforming, too, has its disadvantages. With cycle times averaging 5 min (5), high-volume projects are sometimes impractical. In addition, relatively narrow wall thicknesses limit these pallets to lighter loads, usually under 3000 lb (1360 kg). They are not often found in heavy-duty racking applications. Twinsheet vacuum forming allows heavier loads with reduced deflection, but it lengthens cycle times and adds cost.

***Reaction injection molding* (RIM).** RIM polyurethane pallets have entered the market. RIM utilizes two or more liquid components (polyol and isocyanate) that are mixed, then injected into a closed mold. These two components react to form a finished polymer, taking on the shape of the tool. The chief advantages of RIM are lower-cost equipment and tooling, especially in building large parts such as pallets. The chief disadvantage of RIM pallets is the lower resistance to deflection. For this reason, many large RIM parts are steel reinforced or manufactured with fiberglass or mineral fillers. These stiffening techniques add cost.

Advantages. Plastic pallets are used primarily in the food, pharmaceutical, textile, high-technology, and automotive industries. Due to the higher cost of plastic pallets, most purchasers use their pallets in-plant, or in closed-loop shipping system. Plastic pallets are almost always found in applications where the user can retrieve most of the pallets after each trip.

Plastic pallets of all types offer certain generic advantages which make them attractive alternatives to other pallet materials. Listed below are some of the plastic pallet's chief benefits.

Long Pallet Life. The relatively expensive plastic pallet must offer a long service life. Many customers experience plastic pallet life of 5–9 years and more (7).

Reduced Load Damage. Smooth molded plastic helps eliminate product damage. There are no broken boards or protruding nails to damage-sensitive loads (7).

Easy Cleanup. Plastic pallets are easy to clean and keep clean (8).

USDA and FDA Clearance. Both polyethylene and polystyrene are acceptable in food and pharmaceutical plants. Pallets made from these materials can be approved on a case by case basis by the on-site inspectors.

Reduced Worker Injury. Smooth construction and consistent weights help to eliminate minor cuts and back strain (7).

Chemically Inert. Polyethylene plastic pallets are highly resistant to acids and bases, and at ambient temperatures, hydrocarbon solvents.

Moistureproof. Plastic pallets will not absorb moisture and soak loads. Plastic pallets will not rust or break down in wet conditions (8).

No Harbor for Pests. Plastic pallets will not harbor or support the growth of worms, eggs, molds, or mildew.

Design advantages of plastic pallets can include

Nestability (single-faced pallets can nest with each other when unloaded). This feature can save over 50% of valuable truck or dock space.

Interstacking (the ability to positively locate one loaded pallet on top of another loaded pallet).

BIBLIOGRAPHY

"Pallets, Plastic" in *The Wiley Encyclopedia of Packaging Technology,* 1st ed., by L. T. Luft, Menasha Corporation, pp. 488–492.

1. *U.S. Industrial Outlook,* United States Department of Commerce, Jan. 1984.
2. J. M. Callahan, "Just-in-Time a Winner," *Automotive Ind. Mag.* **65**(3), 78 (March 1985).
3. *Modern Plastics Encyclopedia,* 1983–1984 ed., McGraw-Hill Publications, New York.

4. "What's Available in Plastics?", *Warehouse Supervisor's Bulletin,* National Foreman's Institute, Waterford, CT, June 25, 1984.
5. "Pallets Take off in All Directions," *Modern Plast. Mag.,* 64–66 (March 1971).
6. "Fitting Plastic Pallets to the Job," *Plast. Design Forum* **9**(3), 57 (May/June 1984).
7. R. F. Ellis, "Plastic Pallets Eliminate Product Damage in Storage," *Modern Mater. Handling Mag.* **39**(8), 87 (June 8, 1984).
8. "Molded Plastic Pallets Solve Odor Transfer Problem," *Food Processing Magazine,* **46**(3), 108 (March 1985).

PAPER

There is no strict distinction between paper and paperboard, particularly in view of the tremendous variation in density possible with current technology. Generally, structures less than 0.012-in. thick (12 "points" or 305-μm) are considered paper regardless of weight per unit area. Except for the use of paper as an overwrap for folding cartons, most packaging papers are used in flexible applications.

The primary intermediate product used to make paper is wood pulp. The properties of an individual paper or paperboard are extremely dependent on the properties of the pulps used. Pulp preparation from deciduous (hardwood) or conifer (softwood) species may be done by mechanical, chemical, or hybrid processes (1–3). These hardwood or softwood pulps may be used unbleached, or they can be bleached to varying degrees by a diversity of techniques.

Mechanical pulps produce papers that are characterized by relatively high bulk, low strength, and moderate to low cost (4). Their use in packaging is very limited.

The kraft (sulfate) pulping process, introduced about 100 years ago, dominates the chemical pulping industry: yields are higher, pulps are stronger, and process chemicals more completely and economically recovered than with any other process. Although the sulfite processes were extremely prominent 75–125 years ago, their chemicals are difficult to recover, the resultant pulps are significantly weaker than those produced by the kraft process, and they produce no unique paper properties. Unbleached pulps are generally stronger, stiffer, and more coarse than their bleached counterparts, but papers made of white, conformable fibers are used in many more applications. General treatments of pulping and bleaching are provided in Refs. 5–8.

The standard ream basis for packaging papers in the United States is 500 sheets cut to a size of 24 × 36 in. (61 × 91.5 cm) (3000 ft^2 or 279 m^2). On this basis, packaging papers normally weigh 18–90 lb/ream (8.2–40.8 kg/ream), but some specialty applications require weights as low as 10 lb (4.5 kg) or as high as 200 lb (90.7 kg)/ream. At any given basis weight, density may typically vary from 0.08–0.16 lb/in.^{3} (2.2–4.4 g/cm^3), providing a very wide range of thickness and strength properties.

The two most general classifications of packaging papers are coarse and fine. Coarse (kraft) packaging papers are almost always made of unbleached kraft softwood pulps. Fine papers, generally made of bleached pulp, are typically used in applications demanding printing, writing, and special functional properties such as barriers to liquid and/or gaseous penetrates.

Kraft Papers

Kraft papers, produced by the kraft process, derive their name from the German word for "strong." Kraft paper is made from at least 80% sulfate wood pulp. It is typically a coarse paper with exceptional strength. Sometimes made with a rough finish to keep bags from sliding off piles, these papers are often made on a fourdrinier machine, and then either machine-finished with a calendar stack or machine glazed by using a Yankee dryer (5,6). The surface of these papers is acceptable for printing by letterpress, flexography, and offset processes (see Printing). In addition to wrapping applications, kraft papers are used for multiwall bags and shipping sacks (see Bags, Multiwall), grocers' sacks, envelopes, gummed sealing tape (see Tape, gummed), butcher wraps, freezer wraps, tire wraps, and specialty bags and wrappings that require economy and strength. Many papers formerly manufactured from sulfite pulps, especially those of tissue weight, are now manufactured with kraft pulps. Unbleached "sulfite" papers are used for products such as oil cans (intermediate liner) and single-service food packages.

Extensible kraft papers have satisfied a special niche in the packaging industry. Although creped papers have long served both decorative and functional roles, other papers capable of absorbing energy at sudden rates of strain have become increasingly important in uses such as shipping sacks. Conventional creping is performed either at the wet press section of a paper machine (wet crepe) or on a Yankee dryer (dry crepe). Dry creping is most commonly used to generate qualities such as softness and absorbency; wet creping is a technique for making tough, flexible papers capable of absorbing tensile energy. A secondary creping operation rewets a dry sheet, and may be done in-line on the paper machine or as an independent manufacturing process. Whereas a standard kraft paper might have a stretch (before breaking) of 3–6%, creped papers generally may be stretched 35–200% of their original length before breaking. Manufacturers of creped, extensible, and other coarse (kraft) papers may be found in Refs. 9 and 10.

Bleached Paper

Packaging applications that place a higher priority on printing, writing, and special functional properties than on economy and strength generally utilize bleached papers. The pulps used to manufacture bleached papers are relatively white, bright, and soft, and they are also receptive to special chemicals necessary to develop many functional properties. Although generically not as strong as unbleached kraft papers, bleached papers can be manufactured to meet simultaneous requirements of both strength and printability. Their whiteness enhances print quality and generates a perception of cleanliness and quality. The aesthetic appeal of bleached packaging papers may be augmented by clay coating one side (C1S) or both sides (C2S). The increasing demand for a combination of functional performance and top quality graphics favors the C1S manufacturer who can satisfy the variety of challenges of this market.

Vegetable Parchment

The process for producing parchment paper was developed in the 1850s, making it one of the grandfathers of special packaging papers. By soaking an absorbent paper in concentrated sulfuric acid, the cellulosic fibers are swollen tremendously and partially dissolved. In this state the plasticized fibers close their pores, fill in voids in the fiber network, and thus produce intimate contact for extensive hydrogen bonding. Rinsing with water causes reprecipitation and network consolidation, resulting in a paper that is stronger wet than dry, lint free, odor free, taste free, and resistant to grease and oils. By combining parchment's natural tensile toughness with extensibility imparted by wet creping, paper with great shock-absorbing capability can be produced. Special finishing processes provide qualities ranging from rough to smooth, brittle to soft, sticky to releasable. Parchment was first used for wrapping fatty substances like butter, but this versatile paper is now also used whenever food is prepared, frozen, packaged, or displayed, and when tough, lint-free, chemically pure surfaces are needed for special packages.

Greaseproof and Glassine

Because cellulose is hydrophilic, it is a good substrate to use for resisting penetration of hydrophobic liquids. As noted above, vegetable parchment performs well as a greaseproof paper because it is essentially pore-free and composed of a hydrophilic material. "Greaseproof" paper is a substrate manufactured to also have an essentially pore-free consolidation; but mechanical refining ("buffing" or cutting) is used in its production instead of swelling with concentrated sulfuric acid. Refining fibrillates, breaks, and swells the cellulose fibers to permit consolidation of a web with many interstitial spaces filled in. Glassine paper is produced by further treating "greaseproof" paper with a supercalendar operation. The supercalendar step involves moist high temperatures (steam), pressure (several hundred pounds per lineal inch or ca 100 kg/cm), and differential hardness (one roll typically cotton or soft rubber, the other roll hard rubber or metal) to polish the surface. Supercalendering a greaseproof paper generates such intimate interfiber hydrogen bonding that the refractive index of the glassine paper approaches the 1.02 value of amorphous cellulose. This indicates that very few pores or other fiber/air interfaces exist for scattering light or allowing liquid penetration.

Greaseproof and glassine papers are frequently plasticized to further increase their toughness. They have a reputation of running well on high speed packaging lines, and have served well for odor and aroma barriers. Like other flexible-packaging papers, they can be chemically modified to enhance functional properties (eg, wet strength, adhesion, release). When waxed, they are standard materials for primary food pouches used to package dry cereals, potato chips, dehydrated soups, cake and frosting mixes, bakery goods, candy and ice cream confections, coffee, sugar, pet food, etc. In addition to their protective functions, these papers heat-seal easily when waxed and reclose well.

Water-, Grease-, and Oil-Resistant Papers

The distinction between greaseproof papers and grease-resistant papers is a fairly subtle one, involving an understanding of the methods of penetration of liquids into surfaces. Because the primary mechanisms involve capillary penetration and/or wetting, it is appropriate to consider the severity of the packaging requirement. The requirement may be for minimal staining by grease, oil, or water under negligible pressure; or it may be for absolute resistance to any penetration of the liquid over long periods of time and/or under substantial pressure; or it may lie between these extremes. Parchment, glassine, and greaseproof papers offer decreasing protection from grease and oil at the more restrictive end of the spectrum. As noted above, their resistance is generated by the lack of capillaries and the oleophobic nature of cellulose.

A consideration in designing primary packages is the economics of using the various barriers available for the job. A bag for a single-service consumable (eg, french-fried potatoes) may require resistance to staining for only several minutes. A lubricating oil package, on the other hand, may require a stain-free barrier for several months. Imparting sufficient resistance to liquid penetration to meet the requirements of the less demanding applications can be done very economically with chemical treatments. If grease and/or oil penetration is the only concern, moderate resistance (package life of minutes to days) can be developed using waxes (see Waxes) and other low surface-energy materials such as fluorocarbons. Fluorocarbon technology in papermaking has expanded from multiwall and consumer bags to labels, coupons, and carry-out food packaging, perishable bakery goods packaging, candy and confection packages, and form-and-fill packages where edge wicking may be an important consideration.

In many applications (eg, carryout food packaging) resistance to both water and oils must be developed for adequate performance. The use of rosin-based chemicals for developing water repellency in paper requires the use of alum ($Al_2(SO_4)_3$) or other multivalent cations that destroy the grease-resisting properties of fluorocarbons. The simultaneous development of both water and oil resistance requires the use of oleophobic chemicals which react with hydroxyl and carboxyl groups on the cellulose fibers, and the use of hydrophobic chemicals such as fluorocarbons, which have also been modified to react with the same cellulose functional groups. Typical oleopholic chemicals currently in widespread use are alkylsuccinic anhydrides and alkylketene dimers. The most successful economic choices are made through close consultation between the paper manufacturer and the user, which allows the careful selection of designs for meeting specific performance requirements. Grease and oil-resistant papers can be made to run well in most any converting and printing processes.

Waxed Papers

Waxed papers are age-old papers that have served the packaging industry well in applications requiring direct contact with food for barrier against penetration of liquids and vapors, as well as heat sealability, lamination, and even printing. Waxing can be performed in-line with the paper-manufacturing process, in-line with printing, converting or lamination processes, or as a discrete process. A great many base papers are suitable for waxing processes, including greaseproof and glassine papers, and water-resistant papers. There are two fundamentally different waxing processes, generating different characteristics for the finished sheet (see Waxes). Wet waxing is an operation in which the wax coating is applied to the surface of the sheet. Surface wax is desirable for heat sealing and lamination, and essential for vapor-barrier development. Dry waxing is performed to absorb wax into the sheet, leaving a surface that often does not look or feel waxy. Penetration of wax allows additional surface treatments for special release applications, or for further lamination. Absence of the continuous film of wax on the surface characteristic of wet waxed paper also allows the dry waxed paper to "breathe" moisture, carbon dioxide, and oxygen.

Waxed papers provide an economical choice for primary food packaging not only because of their versatility, but also because of their safety as tasteless, odorless, nontoxic, and relatively inert materials. Their widespread use in conventional packaging applications includes delicatessen pickup sheets, box liners, cover, scale, and utility sheets, patty papers, sandwich wraps and bags, laminations to other papers and paperboard for food trays, locker papers, carryout cartons, food pails, baking cups, folding cartons, cereal liners, and folding carton overwrap.

Specialty-Treated Papers

Many packaging applications require barrier to substances other than water, grease or oil. Most food-packaging applications are well-served with water- and/or oil-penetration resistance, but a number of products require more elaborate barriers. Meat-wrapping paper demands exceptional strength; resistance to grease, moisture, and blood; easy release with no residual taste or odor; plus "bloom" retention. Freezer paper must remain pliable at low temperatures, and offer puncture resistance, moisture-vapor barrier, exceptional seal integrity, and easy release from frozen or thawed meats. Other products such as chemicals, drugs, cosmetics, personal-care items, and industrial products require package functions such as acid resistance, alkali resistance, alkali solubility, mold resistance, flame retardation, solvent resistance, organic polymer adsorption/absorption resistance or affinity, adhesion, release, tarnish or rust inhibition, heat stability, sterilizability, specific-ion adsorption, conductivity, resistivity, stiffness, or flexibility. To address the general manufacturing techniques or even the functional property classification of such a variety is beyond the scope of this publication. Reference 11 provides a general treatment of some of these products which have transcended proprietary technology. Because many of the manufacturing techniques employed to generate these functional properties are considered trade secrets, the reader is referred to directories of the specialty-packaging paper manufactures (9,10).

Wet-Strength Papers

Conventional papers are not strong when wet. The most predominant fraction of paper's strength is the result of hydrogen bonding between hydroxyl and carboxyl groups of adjacent fibers. The removal of water during the papermaking process generates these bonds, and the process is reversible. The two strategies for manufacturing strong-when-wet papers are (*1*) keep the water out of the paper and (*2*) introduce additional chemical bonds between fibers which are not influenced by the introduction of water.

As mentioned above, paper that has been parchmentized is actually stronger wet than dry, principally because of the loss of individual fiber identity during the gelatinized stage of the process. Through advances in chemical technology, several more economical alternatives exist for generating wet-strength papers. In general, the chemicals used to augment the natural hydrogen bonding are crosslinked during the manufacturing process. Common chemicals for producing wet-strength papers include protein, urea, melamine, resorcinol, and other phenolic or amino resins crosslinked with formaldehyde, and condensation products of polyalkylene polyamines with dicarboxylic acids crosslinked with epichlorohydrin (12).

The conventional tests of wet-strength papers are for tensile, tear, and burst strength, as these are the most useful indicators of use requirements. The degree of wet strength is expressed as a percentage of original dry strength, and is referred to as the percent of strength retention. The typical range for sack or pouch papers is 15–30%. Wet strength may also be generated in a variety of permanence levels, so that the product will either remain tough-when-wet, or eventually disintegrate with soaking time or application of force.

Absorbent Papers

Papers designed for absorption of specific fluids are an important part of the packaging industry. Although they are a member of the class of "Specialty-treated papers," they are distinctive enough to warrant separate discussion. Typically at the low end of the strength spectrum, absorbent papers must not only be exceptionally porous, they must have surface modifications to render affinity to the target liquid. When that liquid is aqueous they must generally also have a definitive level of wet strength. Providing the special affinity to a given liquid is often a proprietary technology involving chemicals that are substantive to cellulose and to the target liquid. This special class of papers currently services industries ranging from fresh-food packaging to industrial-products packaging.

Tissue Papers

Tissues form a special group of fine packaging papers because of their versatile performance. Always fairly thin, tissues range from semitransparent to totally opaque. They can be waxed or treated with any of the specialty treatments (eg, edible oils for fruit wrapping, antitarnish metal protection), or they can be used "as is" for applications ranging from intermediate lamination steps in composite-container construction (see Cans, composite) to gift wrapping. They may be made exceptionally weak for softness, or surprisingly tough in all directions. Tissue papers are generally either machine-finished (MF) or machine-glazed (MG). Machine finishing involves calendering between rolls that are usually constructed of highly polished steel that are hydraulically loaded to squeeze and polish both sides of the paper to similar levels of smoothness. Machine glazing of papers produces a smooth glazed side and a rough back side. The special finish is the result of drying the paper with 60–70% moisture on a Yankee dryer, 6–18-ft (1.8–5.5-m) dia with a mirrorlike surface. This glazing process produces a very smooth surface for printing, adhesion, release, or wet waxing. A physical fusion takes place with the surface fibers, producing a physical barrier similar to a cast film. MG papers may also be machine finished to improve the smoothness on both sides or to produce intermediate rolls which process better through subsequent converting steps. A great many applications for tissue paper today utilize special treatments for adhesion or release in an intermediate package-manufacturing process where the paper's light weight and thinness make it an economical carrier for more costly substances (see Multilayer-flexible packaging).

Coated Papers

Coated papers is a term generally reserved for papers acting as a base for aqueous mixtures of clay and/or other mineral pigments with natural and/or synthetic polymers as pigment binders. The term does not typically refer to papers that have been extrusion- or solvent-coated with organics or plastics, or surface-treated with specialty treatments listed above for functional improvements. Aqueous coatings of paper are performed primarily for market appeal where graphics are important. Several hundred thousand metric tons of coated papers are consumed annually for packaging purposes, not only for labels and multiwall bags but also for lamination or combination with other functional materials in composite structures. Coated papers can be designed for printing by any process from letterpress to ink jet (see Code marking and imprinting). The most valued grades are typically produced in discrete manufacturing processes which vary widely (5,6,8). Detail of coating technology state-of-the-art are contained in annual *TAPPI Coating Conference Proceedings* (13).

Nonwovens

Nonwovens are materials used as cloth substitutes, made entirely or partially from cellulosic fibers. As an industry, nonwovens is dynamic and growing tremendously in medical, healthcare, industrial, food-processing, and consumer- and household-products areas. Differentiated from classical paper, which is formed in water and consolidated with interfiber hydrogen bonds, nonwoven-manufacturing technologies include resin and thermally-bonded carded web process, melt-blown process, and an air-laid process. Because the nonwovens industry is less than 4 decades old, much of the technology is proprietary (see Nonwovens).

BIBLIOGRAPHY

"Paper" in *The Wiley Encyclopedia of Packaging Technology,* 1st ed., by M. Sikora, James River Corporation, Speciality Packaging Papers Group, pp. 497–500.

1. S. A. Rydholm, *Pulping Processes,* Wiley-Interscience, New York, 1965.
2. M. G. Halpern, ed., *Pulp Mill Processes, Developments Since 1977,* Noyes Data Corp., Park Ridge, NJ, 1981.
3. *Proceedings of the Alkaline Pulping Conference, 1981 and 1985,* TAPPI Press, Atlanta.
4. D. R. Allen, ed., *Uncoated Groundwood Papers,* Miller Freeman Publications, San Francisco, 1984.
5. J. P. Casey, ed., *Pulp and Paper Chemistry and Technology,* 3rd ed., Vols. I–IV, Wiley-Interscience, New York, 1980.
6. R. G. MacDonald, ed., *Pulp and Paper Manufacture,* 2nd ed., Vols. I–III, ed., McGraw-Hill, New York, 1969.
7. *Pulp Technology and Treatment for Paper,* Miller Freeman Publications, Inc, San Francisco, Calif., 1979.
8. M. J. Kocurek and C. F. B. Stevens, eds., *Pulp and Paper Manufacture,* 3rd ed., Joint Textbook Committee of The Paper Industry, Atlanta, Ga. 1984.
9. H. Dyer, ed., *Lockwood's Directory of the Paper and Allied Trades,* 108th ed., Vance Publishing, New York, 1985.
10. *Post's Pulp and Paper Directory,* Miller Freeman Publications, San Francisco, 1985.
11. R. H. Mosher and D. S. Davis, *Industrial & Specialty Papers,* Vols. 1–4, Chemical Publishing Co., 1970–1974, 1969–1973, Vol. 1 (*Technology*) 1968, Vol. 2 (*Manufacture*) 1968, Vol. 3 (*Applications*) 1969, Vol. 4 (*Product Development*) 1970.
12. U.S. Pat. 2,926,154 (Feb. 23, 1960), G. A. Keim, (to Hercules Powder Company).
13. *Proceedings of the Polymers, Laminations, and Coatings Conference,* annual publication of TAPPI Press, Atlanta.

General References

B. Wirtzfeld, ed., *The Paper Yearbook,* Harcourt Brace Jovanovich, New York, comprehensive listing of typical uses.

The Competitive Grade Finder, 18th ed., Grade Finders Inc., Pub. Bala-Cynwyd, PA, 1984.

J. Hube, ed. *Kline Guide to the Paper Industry,* 4th ed., C. H. Kline Pub., Fairfield, NJ, 1980.

API annual *Statistics of Paper and Paperboard* for relative shipment and sales volume.

Pulp and Paper North American Industry Fact Book, Miller Freeman Publications, San Francisco, 1982.

The Dictionary of Paper, 4th ed., American Paper Institute, Inc., Pub. New York.

B. Toale, *The Art of Papermaking,* Davis Publications, Worcester, MA, 1986; concise but fairly comprehensive treatment of historical techniques for making contemporary decorative papers.

I. P. Leif, *An International Sourcebook of Paper History,* Archer Dawson, Hamden, CT, 1978.

H. F. Rance, ed., *Handbook of Paper Science,* Elsevier Scientific Publishing, New York, 1982, Vol. 1, *Raw Materials and Processing,* Vol. 2, *Structure and Properties.*

R. P. Singh, ed., *The Bleaching of Pulp,* 3rd ed., TAPPI Press, Atlanta, 1979.

G. R. Hutton, ed., *Phillips Paper Trade Directory-Mills of the World,* Derek G. Muggleton Publishers, Kent, UK, 1985.

M. Bruce Lyne, "Paper," in J. I. Kroschwitz, ed., *Kirk-Othmer Encyclopedia of Chemical Technology,* Vol. 18, 4th ed., Wiley, New York, 1996, pp. 1–34.

1994 Statistics on Paper, Paperboard & Woodpulp, American Forest & Paper Association, Washington, 1994.

PAPERBOARD

Paperboard, often called simply "board," is one of the major raw materials used in packaging. The term includes boxboard, chipboard, containerboard, and solid fiber. Its application covers a wide range of uses from simple cartons (see Cartons, folding) to complex containers used for liquids. It can also be converted into drums for bulk packaging of chemicals or combined with other materials to produce containers large and strong enough for the protection of large, heavy, and often fragile items during transport. In addition to product-protection requirements, paperboard must often have at least

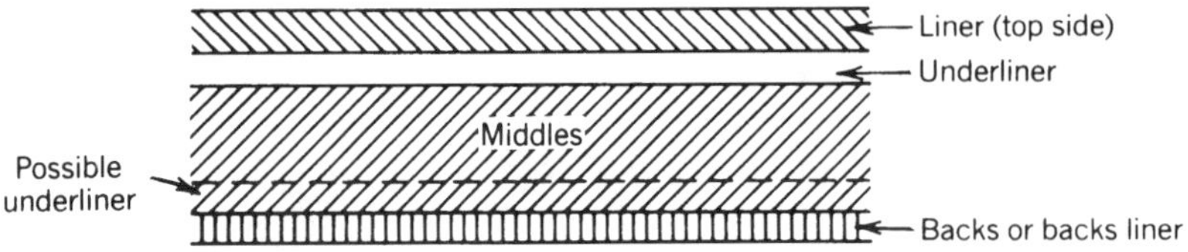

Figure 1. Structure of a typical multi-ply paperboard.

one smooth surface capable of accepting high-quality print (see Printing).

Terminology

"Paper" is the general term for a wide range of matted or felted webs of vegetable fiber (mostly wood) that have been formed on a screen from a water suspension. The general term can be subdivided into paper (see Paper) and paperboard. There is no rigid line of demarcation between paper and board, and board is often defined as a stiff and thick paper. ISO standards state that paper with a basis weight (grammage) generally above 250 g/m^2 (~51 lb/1000 ft^2) shall be known as paperboard, or "board." The definition becomes less clear because in some parts of the world board is classed as such when its caliper (thickness) exceeds 300 μm (~12 mil) [in the United Kingdom, 250 μm (~ 10 mil)]. There are exceptions to the above: blotting papers and drawing papers thicker than 300 μm are classified as paper, and corrugating medium, linerboard, and chipboard less than 300 μm are classified as paperboard.

It is also not possible to strictly define paperboard by its structure or by the type of machine used to produce it. For example, paperboard can have either a single- or multiply structure and can be formed on a fourdrinier-wire part, a single or a series of cylinder molds, or a series of modern formers, or sometimes by means of a combination of one or more of the above (see diagrams and related text). For a small part of the market, paperboard is produced by laminating sheets of paper together. In that case, the product is solid fiber (see Boxes, solid fiber).

The following terms are in general use in paperboard manufacturing and associated converting industries:

Basis weight. This is the weight of a known area; for example, g/m^2 (grammage) or (lb/1000 ft^2), the weight in pounds of a ream (usually 500 sheets) of paperboard cut to its "basic size."

Caliper (thickness). The thickness of the sheet expressed in μm or thousandths of an inch (mil or points) (mil = 25.4 μm).

Size of a sheet. The width and length of a sheet of paperboard. The width, always expressed first, is the dimension cut at right angles to the direction of the sheet. Length is the dimension cut in the machine direction. For example, 20 × 30 in. means 20 in. (50.8 cm) cut across the machine by 30 in. (76.2 cm) cut in the direction of the machine. The first dimension is often termed the cross-direction and the second dimension the machine direction.

Structure. The composition of the web (see Fig. 1).

Ply. A fibrous layer of homogeneous composition.

Topside. The side of the web opposite to the wire side is the normal paper definition for this term. In the case of paperboard, topside can also mean liner side, generally the better-quality face of the web. Some grades of paperboard are known as double-lined, in which case the two faces of the web are both of high quality.

Liner. A ply of good-quality fiber (usually white), on the topside.

Underliner. The layer (ply) of fiber between the external layer (topside/liner) and the middle.

Middle. The layer (ply) of fiber between the two external layers or between the underliner and an external layer.

Backs. The outside layer (ply) of fiber directly opposite to the liner layer.

Duplex (or biplex). This is a board consisting essentially of two layers of different furnish.

Triplex. This is a board consisting essentially of three different furnish layers (external furnish layers may have the same composition).

Multiplex. This is a board with more than three furnish layers. Two or more of the layers can have the same composition. Also known as a multilayer board.

Furnish. The constitution of the various materials that are blended in the stock suspension from which the paperboard plies are made. The main constituents are the fibrous material (pulp or secondary fiber), sizing agent, fillers, and dyes.

Structure and Properties

The structure of a typical multiply paperboard (see Fig. 1) consists mainly of cellulose fibers. The most common source is mechanical and chemical pulps dervied from wood. Secondary fibers are used widely in the cheaper grades. Other sources of cellulose fiber are occasionally used such as straw and esparto.

Examination of the structure shows that it consists of a compact network of the fibers bonded together by mechanical entanglement and chemical links. This structure can be a thick single homogeneous ply or, as is more often the case, of two to eight thinner plies. The multiply construction allows different types of fibers to be used for the different plies. Improved fiber economy is achieved by selected distribution of the fiber types through the web (eg, by using cheaper fibers for the inner layers). Improved characteristics are obtained by using the correct selection of fiber for the individual layers (eg, stronger fibers for the outer plies). A typical example of a multiply construction is a type of paperboard known as whitelined chip. The top ply (liner) is made of bleached chemical wood pulp, which provides surface strength, good appearance, and printing properties. For the other plies, secondary fiber is used.

It is of extreme importance that the individual plies of multilayer paperboard are bonded together. If the bonding is poor, the structure can break up, with resultant deterioration in strength during subsequent processing and use. The level of bonding achieved, which must meet certain requirements, is dependent on having the right balance of mechanical entanglement and chemical bonding. If, for example, a paperboard sheet is subjected to continuous folding (eg, the hinged lid of a cigarette carton) and the bonding level is too high, cracking at the hinge will occur.

Physical Characteristics

Mechanical properties. Because paperboard has the same type of fibrous structure as paper, the strength-to-weight relationship is of the same order as paper. Because of its extra thickness and the bonding between the plies it normally has considerably greater flexural rigidity (stiffness). The choice of furnish for the individual plies influences the stiffness characteristics: increased stiffness is obtained by increasing the strength of the outer plies by using a stronger pulp. During the converting of paperboard, it is often necessary to fold the sheet, and because of the thickness of the paperboard, large internal forces are generated. These can cause structural damage to the outer layers (cracking). In order to carry out the folding operation with minimum damage to the outer layers, the ply bonding is broken down locally by a creasing operation before the paperboard is folded. A paperboard sheet has grain characteristics similar to those of paper, with the sheet being strongest in the machine direction.

Optical properties. Paperboard is normally opaque by virtue of its thickness and only the color (whiteness, etc) and occasionally the gloss of its outside plies are important. For a multiply paperboard with a white liner on darker under plies (eg, when wastepaper is used) liner-ply opacity is important in order to maintain the white appearance of the liner surface. If a dark waste paper is used, an underply (underliner), consisting either of a lighter colored waste paper or a mechanical pulp furnish, is applied. This reduces the showthrough of the dark waste layer and improves the whiteness of the liner surface.

Absorptive properties. The surface of paperboard is often required to have characteristics suitable for printing. To obtain these characteristics the surface layers are sized in ways similar to those used for paper. Paperboard, however, must often be glued during the manufacture of cartons, and for this the surfaces must be absorbent. In a typical case, the back surface has higher absorption characteristics than the liner ply, thus allowing the back ply to absorb some moisture during the gluing operation. Like paper, paperboard changes dimensions if it absorbs moisture. The degree of change depends on many factors such as type of fiber, condition of fiber, and structure of the fibrous network. This effect of dimensional change with change of moisture content presents many problems during converting operations such as printing. For example, in lithography and laminating (see Laminating) serious dimensional change can cause print misregister or curling of the paperboard web.

Paperboard Manufacture

The methods of fiber treatment (beating, refining, cleaning, etc) are essentially the same as those used in manufacturing paper grades. After the sheet has been formed, the methods used for removing excess water and finishing the web (pressing, drying, calendering, etc) are also essentially the same as those in the manufacture of paper. The main difference is found at the forming section of the machine, where the web is formed.

The following are the main forming methods used in the manufacture of paperboard:

Single-ply paperboard: mainly fourdrinier.

Multiply paperboard:

1. Fourdrinier with secondary head boxes.
2. Rotary formers (multiples and using multilayer formers).
3. Twin wire formers (multiples and using multilayer flow boxes).
4. Combination of the above.

Fourdrinier machine. The forming section of a fourdrinier machine is made up of two essential parts: the flow box and the drainage table. The operation of both parts can influence the structure of the resulting paperboard web. It is normally the aim to deliver the fiber suspension, well dispersed, to the moving screen at approximately the same velocity as the screen. The concentration going to the screen for paperboard generally ranges within 0.4–1.2% depending on the furnish and product requirements (Fig. 2).

On early machines and some still used today, preliminary dewatering takes place at the table rolls. The action of each table roll can be considered as a pump, drainage being induced by the suction on the downstream side of each table roll, so that the drainage flow is intermittent. A positive-pressure pulse exists on the upstream side of each table roll, so water is forced up through the screen and the deposited mat. This mechanism has considerable influence on web structure. The magnitude of the positive and negative pulses depends on table roll diameter and screen speed.

Modern practice is to replace the table rolls by stationary foils. These units generate less intense suction pulses than table rolls, and the length of the drainage zone can be considerably extended. These foils are often closely spaced along the Fourdrinier forming table so that there is a gain in available drainage. Further down the fourdrinier table, when concentration is 3–4%, suction flat boxes are used to continue the drainage at a rate controlled by the level of vacuum in the box. It should be noted that flat vacuum boxes can be used at the initial drainage zone, and today a unit is often used that in effect is a combination of foil sections in a vacuum enclosure (wet-suction box). When producing heavy-weight papers and boards, because drainage is more difficult, it is usual to use additional vacuum boxes. This can lead to excessive drag on the screen, resulting in higher power requirements, excessive wear of suction box tops and screen, and may even lead to screen stalling. For the production of multi-ply paperboard, such as linerboard, a secondary flow box is often used. A base ply is formed first and after sufficient dewatering has taken place a second ply is applied by means of the secondary flow box. The associated fibrous suspension is then drained through the base ply and dewatering is carried out essentially by flat vacuum boxes with thickening as the main mechanism.

Rotary forming devices. In these units the forming screen is in the form of a drum and the fibrous suspension is fed to the

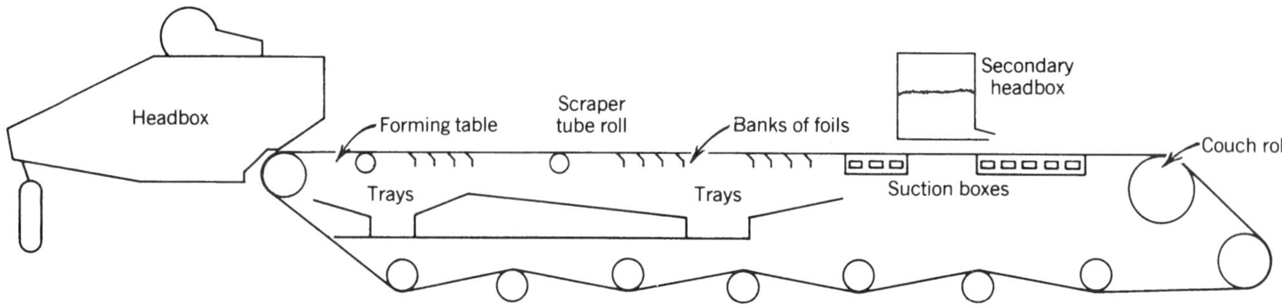

Figure 2. Fourdrinier forming.

screen by various methods ranging from simply immersing the rotating screen in a chamber (vat) containing the fibrous suspension, to the use of a type of flow box.

Cylinder mold machines. There are two basic types: Uniflow and Contra Flow. In the Uniflow machine, the fibrous suspension is fed into the vat at the ingoing face of the cylinder mold (forming screen); in the Contra Flow, the suspension enters at the emerging face (see Fig. 3).

The draining forces are low, typically 1–5 in. of water static head difference. The mechanism of forming is complex, and because of the low-drainage forces, considerable washoff of the fragile newly formed web followed by redeposition takes place in the forming area. The forming zone is obviously too long with continual washing off and redepositioning taking place so that the overall mat deposition is inefficient. Various modifications have taken place in attempts to overcome this problem. Further development has led to a variation known as the rotary former.

Roll formers. Figure 4 shows a typical rotary former. It is easy to see that it consists essentially of a cylinder-forming screen with an associated type of flox box. The forming length has been considerably decreased, and the drainage forces can be far higher than is the case with the previously described cylinder mold units (Fig. 4).

The cylindrical screen can be of relatively simple construction relying on a pressure force in the forming zone to assist drainage or it can be in the form of a suction roll using a series of vacuum boxes to further assist dewatering. Compared to cylinder molds, roll formers have several advantages:

1. They can develop and tolerate higher levels of turbulence in the initial forming zone because the drainage zone is enclosed, whereas the initial forming zone in a cylinder-mold machine starts at a free surface.
2. High pressure in the free suspension and the possibility of using suction on the underside of the forming screen permit a much higher rate of drainage.
3. The rotary former has a more uniform metering of the fibrous suspension onto the forming screen.

Although both types of rotary forming devices can be used to produce paper, the greatest application by far is the manufacture of multiply board. A number of the units are operated in series, progressively building up a multiply web.

Twin-wire formers. The third basic method used for paper and paperboard forming is a relative newcomer, invented in the 1950s and called the twin-wire method In this technique the paper web is formed between two forming screens. The idea dates back to the nineteenth century, but it was only in

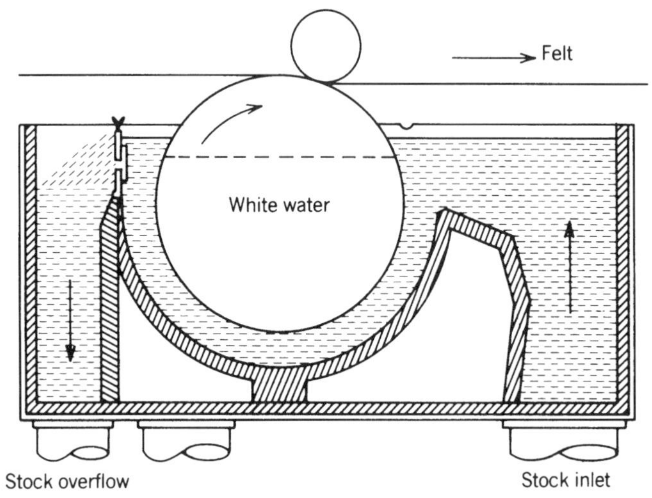

Figure 3. Cylinder-mold forming.

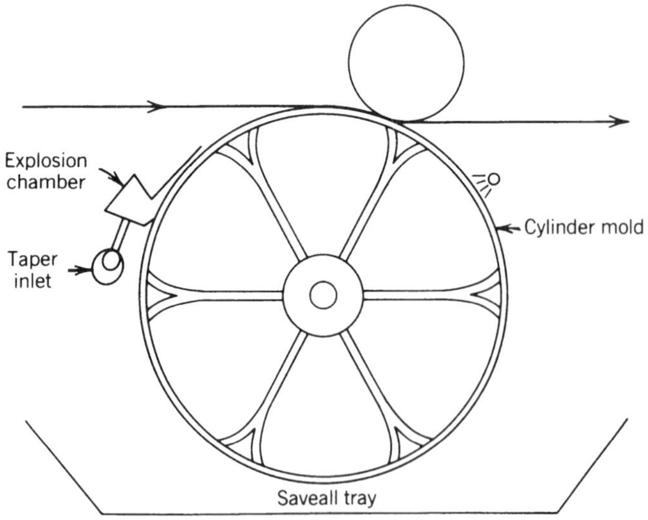

Figure 4. A cylinder former.

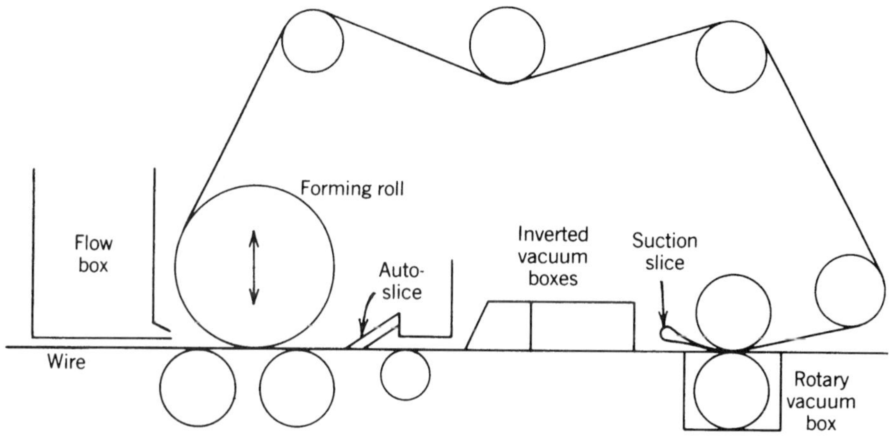

Figure 5. An Inverform (Beloit) twin-wire forming section.

the 1950s that serious development took place, taking advantage of improved ancillary equipment. The process became a commercial reality and a viable contender in some applications. Figure 5 shows a typical twin-wire forming unit, in this case the basic Inverform concept, a UK invention and the first commercial twin-wire system.

In all twin-wire formers, the fibrous suspension is fed into the gap between two converging forming screens by means of a flow box where dewatering and associated web forming takes place. The action related to web structuring in the forming zone are still not fully understood but with regard to dewatering, the twin-wire concept offers the opportunity to carry out symmetrical drainage of the fibrous suspension from both sides. This allows a symmetrically structured web to be formed and provides the opportunity to greatly increase the dewatering potential. Dewatering is assisted by the use of deflectors, which press into the forming screens, and/or vacuum boxes operating on one or both forming screens. A further benefit of the twin wire concept is the absence of a free surface in the forming zone.

In general terms, multiply webs are produced by (*1*) separately forming the individual plies and then combining them together; (*2*) forming onto an existing ply or plies, to form more plies; (*3*) using multilayer (stratified) flow boxes; or (*4*) a combination of *1–3*.

Examples of multiply arrangements in use today are shown in Figures 6 and 7.

Machine finishing. The surface of the paperboard web can be treated during the manufacturing process by various means according to the characteristics required of the finished product. It is not unusual to use as part of the drying process an M.G. (Yankee) cylinder, which imparts a smooth surface to one side of the web without too much densification of the web taking place. In the drier section, there is often a size press where chemicals can be added to the surfaces of the web in order to impart certain characteristics (eg, hard sizing or barrier properties). Paperboard machines often have coaters "in line" with the operation at which one or more layers of coating medium can be applied. At the end of the manufacturing process, right before the windup, one or more stacks of calenders are installed. A stack of calenders consists of a number of horizontal cast iron rolls set one above each other. As the web of paperboard passes through the nips, the calenders increase the smoothness and gloss of the surface of the web.

Types of Paperboard

The simplest types of paperboard are single-ply thick papers used for many nonpackaging purposes (eg, index board for

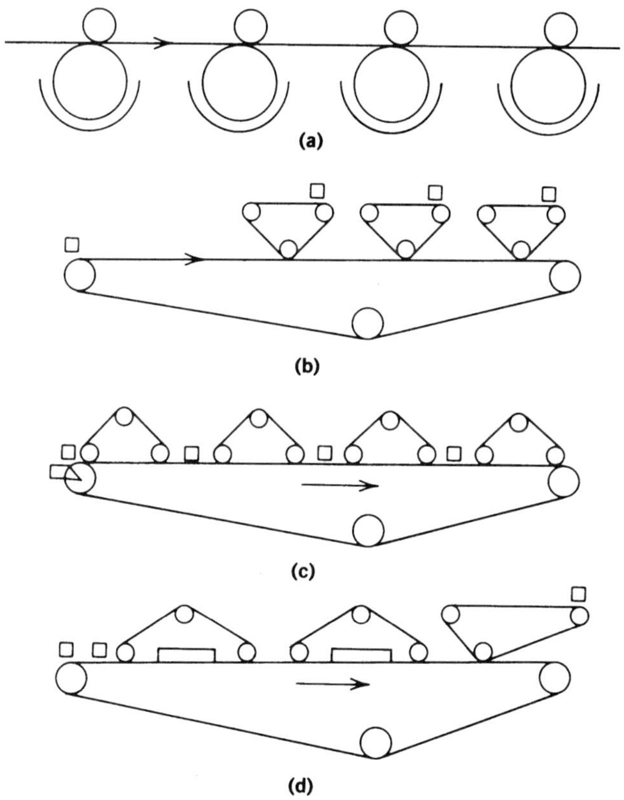

Figure 6. Typical multiply forming arrangements. (**a**) Typical cylinder M/C. Common throughout the world. Can be Uniflow, Contra Flow etc, including mixtures. (**b**) Fourdrinier with "on top" mini-fourdriniers. Examples in Europe and the United States. (**c**) Suction breast roll with Inverform units. Example in the United States. (**d**) Fourdrinier with bel bond units and mini-Fourdrinier. Examples in United Kingdom and Europe.

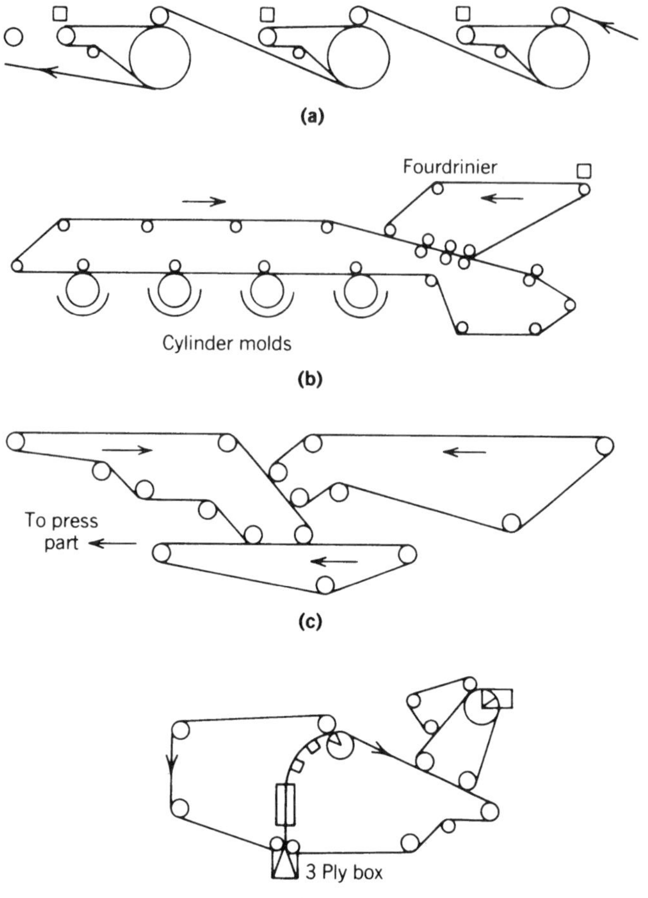

Figure 7. Examples of multiply formers. (**a**) Super ultra former. Examples Japan, the United States, Canada, etc. (**b**) Typical combination M/C. Examples mainly in Europe. (**c**) Typical multiwire M/C. Example Finland. (**d**) Commercial application of contro flow former. Example, Finland. Contro flow former combined with an arcu (rotary) former.

card-filing systems and display mounting). They have a degree of stiffness, an acceptable appearance (eg, a uniform surface). This type of paperboard can also be made from waste paper. The product will have good stiffness characteristics, but not necessarily good appearance. Its main use is as package inserts and envelopes. Paperboard with similar structure, but with improved printing surfaces, has a wide range of uses (eg, high-quality display). The greatest volume of paperboard is used, however, in packaging.

Single-ply paperboard made from 100% bleached-chemical wood pulp is used for food packaging where purity and clean appearance is required together with a degree of strength and a surface of sufficient quality to accept good-quality print. Materials of this type are often coated or laminated with a plastic film to improve barrier properties. Compared to single-ply structures, multiply paperboard can be used over a much wider range of applications because virgin-pulp outer layers with good appearance, strength, and printing properties can be combined with lower-grade middle plies. These paperboards are often combined with other materials such as plastic film or high-quality paper in order to extend the range of application. The general term for this range of paperboard is folding box-board.

Another type of paperboard, known as fiberboard, is used to produce large and strong cases. The materials used for the construction of these containers are made from several layers of paperboard. There are two main types: solid board (two or more boards are laminated together) and corrugated board. In the production of corrugated board, two facings are glued to both sides of the corrugating medium to produce a single-wall corrugated board. For stronger and larger boxes three facings and two media are used (double-wall corrugated). Triple-wall corrugated uses four facings and three media (see Boxes, corrugated).

The types of paperboard mentioned have, in general, a basis weight range from 200–600 g/m^2 (~41–123 lb/1000 ft^2). Thicker and heavier paperboards are used in many applications ranging from building materials to suitcases. They are often produced by laminating many layers of thinner paperboards.

Table 1 shows a typical range of paperboard grades with comments on their furnish, requirements and end usage. The following definitions describe the most common types of paperboard used in packaging.

Body or baseboard. This is a board that is ultimately treated by, for example, a coating or a surface application.

Lined board. A multiply board with a liner ply, usually of high-grade material (eg, white-lined chipboard).

White-lined board (duplex). A board with a bleached pulp liner and the remainder of the board made up of, for example, a mixture of chemical pulp and mechanical pulp. Often used for food packaging.

Kraft-lined chipboard. An unbleached kraft liner on a wastepaper base used for packaging products such as electrical and mechanical components.

Test jute liner. This is a type of kraft-lined chip board. Sometimes the liner is made from a strong kraft waste furnish. This material is combined wtih fluting medium to produce corrugated board.

Double-lined board. A board lined on both sides. For example, the outer surfaces (liner) can consist of bleached pulp and the middle mechanical pulp. It is used for high-quality packaging of foods and cosmetics.

Carton board (folding-box board). Paperboard of various compositions used for the manufacture of folding cartons and set up boxes.

Foodboard. Single- or multiply paperboard used for food packaging. It is hard-sized for water resistance.

Liquid packaging board. Also called special-food board, or milk-carton board, this strong board is usually 100% chemical pulp, often plastic-coated. It is formed into containers for a wide range of liquids (eg, milk, other beverages).

Frozen-food board. Single- or multiply paperboard with high moisture- and water-vapor resistance. It is often single-ply, made from bleached wood pulp with a surface coating for high-quality print.

Kraft linerboard. A strong packaging paperboard with a two-ply construction made essentially from virgin kraft pulp and produced on a Fourdrinier machine. The top ply is added by means of a secondary flow box. It is used in the manufacture of large containers, combined with fluting medium to form corrugated board.

Fluting medium (corrugating medium). A board usually with

Table 1. Examples of Typical Paperboard Grades

Board Grade	Grammage Range g/m² (lb/1000 ft²)[a]	Typical Furnish	Special Requirements (Physical)	Typical Usage
White-lined folding-box board	200–800 (40–160)	Liner, virgin pulp Underliner, mechanical pulp Middles, waste Backs, mixture (mechanical and chemical pulp)	Bending Printing Plybond Stiffness	General packaging cartons
Chipboard	200–800 (40–160)	100% waste	Bending Plybond	Packaging cartons, tubes, and stiffeners
Gypsum board	300–800 (60–160)	100% waste	Plybond	Outer component of plaster-board
Test liner	150–300 (30–60)	Liner, virgin pulp (kraft): rest, waste	Bending Bursting strength Plybond Crush resistance	Outer components of corrugated containerboard
Linerboard	150–300 (30–60)	100% virgin pulp	Bursting strength	Outer components of corrugated board
Foodboard	200–600 (40–120)	100% virgin pulp (single- or multiply)	Bending Printing Plybond Stiffness	All foods, especially frozen foods (high-quality containers)
Liquid packaging	200–400 (40–80)	100% virgin pulp (single- or multiply) with barrier (coating and/or laminate)	Bending Printing Plybond Stiffness	Containers for wide range of liquids including milk
Fluting medium	90–200 (18–40)	Typical—100% waste or semichemical hardwood pulp	Crush resistance	Inner components of corrugated containerboard

[a] g/m² = 0.2 lb/1000 ft².

a basis weight of 100–125 g/m² (20–25 lb/1000 ft²) and a caliper of 225 μm (9 mil) made from semichemical hardwood pulp or waste paper. The material is fluted in a corrugating machine and combined with linerboard to produce corrugated board.

Chipboard. A paperboard made from waste paper, used in low-grade packaging, solid fiber, and bookboard.

Machine-glazed board. Paperboard that in its manufacture has had one face made smooth and glossy by drying on a large polished steam-heated drying cylinder (Yankee cylinder).

Tube board. A paperboard generally unsized and smooth finished. It is slit into narrow widths for winding and pasting into spiral or convoluted mailing tubes, cores, etc.

Can board. Paperboard used for the manufacture of composite cans and fiber drums. The cans are used for packaging a wide range of materials including liquids and powders.

Coated boards. Paperboards of various grades that have been coated on one or both faces to make the surfaces suitable for high-quality printing.

BIBLIOGRAPHY

"Paperboard" in *The Wiley Encyclopedia of Packaging Technology,* 1st ed., by B. W. Attwood, St. Annes Paper and Paperboard Developments Ltd., pp. 500–506.

General References

American Society for Testing and Materials, Philadelphia (ASTM Special Publication No. 60 B), 1963.

British Standard Institution, *Glossary of Paper Terms* (B.S. 3203), London, 1964.

R. Higham, *A Handbook of Paper and Board Manufacture,* Vol. 1, Business Books, Ltd., London, 1968.

C. Klass, *Cylinder Board Manufacture,* Lockwood Trade Journal Co., Ltd., London, 1968.

Dictionary of Paper, American Paper and Pulp Association, New York, 1965.

E. Labarre, *Dictionary and Encyclopedia of Paper and Papermaking,* Oxford, 1952 and 1967 supplements.

Pulp and Paper Manufacture, Vol. III, 2nd ed., McGraw-Hill, New York, 1970.

Paper and Board Manufacture—Technical Division, British Paper and Board Industry Federation, London, 1978.

J. Grant, *A Laboratory Handbook of Pulp and Paper Manufacture,* Arnold, London, 1961.

1994 Statistics on Paper, Paperboard & Woodpulp, American Forest & Paper Association, Washington, 1994.

J. P. Casey, *Pulp and Paper,* 3rd ed., Vols. I–IV, Wiley, New York, 1980–1983.

B. Attwood, "Inverform Developments," *Paper Technol.* **13**(4), 253–258 (Aug. 1972).

B. Attwood, "Multi-Ply Web Forming—Past, Present, and Future" in *Proceedings of the TAPPI Annual Meeting,* Atlanta, Feb. 1980, pp. 229–241.

PAPER, SYNTHETIC

Synthetic plastic paper is a calendered plastic sheet that is a unique mixture of clay (calcium carbonate) and polypropylene resin. This new formulation makes it feel, look, print, and fabricate like paper, but with the durability and tear resistance of plastic. White opaque with a fine matte finish, it is a single layered substrate that offers superior ink adhesion and has excellent bonding characteristics. It is easily printable (no pre-treatment) via: offset, lithography, gravure, letterpress, and screen processing.

Properties

Temperature, abrasion, moisture, and tear resistance. Synthetic paper is dimensionally stable and can withstand temperatures ranging from −60 degrees to +200 degrees fahrenheit. Under any of the above conditions it is resistant to cracking, abrasion, shrinking or any distortion in color or sheet size.

It is 100% waterproof and will not be affected by moisture or humidity and offers superior tear resistance over paper.

Printability. HOP-SYN brand synthetic paper (trademark of Tamerica Products) is formulated with antistatic agents and offers a porous printing surface good layflat and a very close gauge tolerance suitable for high speed printing. Its antistatic surface eliminates double fed sheets when printing, folding, and stacking.

Its porous surface allows ink to be absorbed underneath its surface preventing the printed image from being damaged by scuffing or scratching which makes it easily washable with a damp cloth when removing stains such as grease, oil, foods, soft drinks, soaps, and/or detergents.

Uses

Since synthetic paper is manufactured in a variety of gauge thicknesses from .0033 to .020 in both sheet and roll form it has many new uses. Popular applications include tags, labels, maps, charts, licenses, credit cards, menus, posters, signs, banners, calenders, greeting cards and shopping bags. However, this is only a small sample of applications.

Environmental Considerations

Synthetic paper (in this case HOP-SYN) is made from polypropylene resin and clay (calcium carbonate). These are nontoxic materials that contain no chlorine or other halogens, producing neither dioxins, hydrochloric acids, or other highly toxic materials. It recycles as a plastic, does not drip when burned, emits no toxic fumes and leaves a clean ash when incinerated and can be easily recycled in the manufacturing process.

TAMERICA PRODUCTS
Chino, California

PERMEABILITY. See BARRIER POLYMERS; TESTING, PERMEABILITY AND LEAKAGE.

PERMEABILITY OF AROMAS AND SOLVENTS IN POLYMERIC PACKAGING MATERIALS

In addition to the advantages that polymeric materials provide in food and pharmaceutical packaging, there are corresponding concerns related to product/package interactions (1) (see Pharmaceutical packaging; Packaging, food). This is a broad-base topic associated with the mass transport of gases, water vapor, and low-molecular-mass organic compounds between product, packaging material and storage environment. Mass transport in package systems encompasses a number of phenomena referred to as either *permeability, sorption,* or *migration.* Permeability includes the transfer of molcecules from the product to the external environment through the package, or from the storage environment through the package to the product. Sorption involves the takeup of molecules contained by the product into but not through, the package, while migration is the passage of molecules originally contained by the package itself into the product. The mass transfer process also provides the basis of further physiochemical activities within the package system. Such activities may induce physiochemical changes in the product, as well as physical damage of the package, or both (2). In a package/product system, the mass-transfer processes and the physiochemical activities associated with them are referred to as *product/package interactions.*

There are various terms found in the literature to describe the steady-state permeation of molecules through polymer films of area A and thickness l (see Fig. 1). The following terms are defined to characterize the permeation process:

- Permeant transmission rate:

$$F = \frac{q}{At}$$

- Permeance:

$$R = \frac{q}{At\,\Delta p}$$

- Thickness normalized flow:

$$N = \frac{ql}{At}$$

- Permeability constant:

$$P = \frac{ql}{At\,\Delta p}$$

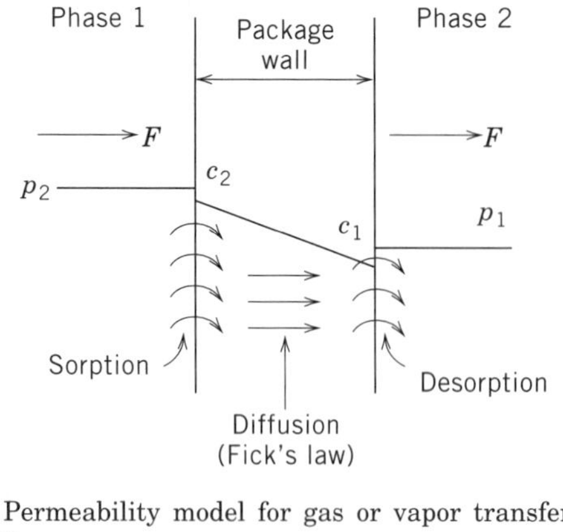

Figure 1. Permeability model for gas or vapor transfer through a package wall.

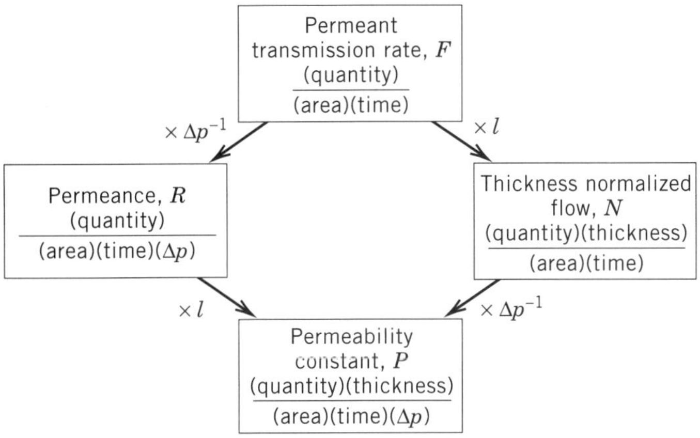

Figure 2. Relationship between permeant transmission rate, permeance, thickness-normalized flow, and permeability constant.

where q is the quantity permeated during time t and $\Delta p = p_2 - p_1$, pressure drop across the film.

The interrelationship between the respective terms is shown graphically in Figure 2.

The permeant transmission rate F is often called the *gas transmission rate* or the *water-vapor transmission rate* (WVTR), to refer to a gas or water vapor, respectively.

Permeation through a polymer film or sheet is a measure of the steady-state transfer rate of the permeant, which is normally expressed as the permeability constant P. The permeability constant is based on two fundamental mass-transfer parameters: the diffusion and solubility coefficients. The diffusion coefficient D is a measure of how rapidly penetrant molecules are advancing through the barrier, in the direction of lower concentration or partial pressure. The solubility coefficient S describes the amount of the transferring molecules retained or dissolved in the film or slab at equilibrium conditions.

The simplest and most common relationship relating P, D, and S is

$$P = D\,S \tag{1}$$

This equation is applicable only for situations where D is independent of permeant concentration and S follows Henry's law of solubility. However, when the permeation process involves highly interactive organic penetrants such as aroma, flavor, or solvent molecules, the diffusion process is more complex than the diffusion of simple gases, and the diffusion coefficient may vary as a function of penetrant concentration and time (3–6). When Fick's second law takes into account the concentration dependency of D, it is written

$$\frac{dc}{dt} - \frac{d}{dc}\left[\mathrm{D(c)}\frac{dc}{dx}\right] \tag{2}$$

for diffusion in a single direction x. Here, c is the concentration of the diffusate in the polymer (7). Where D varies with t, the diffusion is often called *non-Fickian*.

Mears (7) proposed the following expressions for cases where the diffusion coefficient is concentration-dependent:

$$D = D_0(1 + ac) \tag{3}$$

or

$$D = D_0 \exp(bc) \tag{4}$$

Equation 4 is more suitable in cases where the diffusion coefficient is strongly concentration-dependent. Here D is the differential diffusion coefficient, D_0 is the limiting diffusion coefficient, and a and b are constants.

The variables affecting permeation and diffusion processes in a package can be grouped as follows:

1. Compositional variables:
   a. Chemical composition of the polymer and penetrant
   b. Morphology of the polymer
   c. Concentration of the penetrant
   d. Presence of copermeant
2. Environmental and geometric factors
   a. Temperature
   b. Relative humidity
   c. Packaging geometry

While an in-depth treatment of each of the above factors is beyond the scope of this article, selected examples are discussed to illustrate their role in the transport of organic penetrants in barrier polymers. For a more in-depth treatment, the reader is referred to the references listed in the Bibliography and references cited therein.

Chemical Composition of the Packaging Material

The relationship between penetrant transfer characteristics and the basic molecular structure and chemical composition of a polymer is rather complex, and a number of factors contribute to the permeability and diffusion processes; the most important are

- Cohesive-energy density, which produces strong intermolecular bonds, van der Waals or hydrogen bonds, and regular, periodic arrangement of such bonding groups
- The glass-transition temperature (T_g) of the polymer, above which free vibrational motion and rotational motion of polymer chains occur, so that different chain conformations can be assumed.

With respect to the glass-transition temperature of barrier polymer structures, DeLassus (8) reported that glassy polymers have very low diffusion coefficients for flavor, aroma, and solvent molecules at low concentrations. Typically, these values are too low to measure by standard analytical procedures. The diffusion coefficient determines the dynamics of the permeation process and thus the time to reach steady state, which accounts for glassy polymers exhibiting high-barrier characteristics to organic permeants. Polyolefins, being well above their glass-transition temperature, are nonglassy polymers and have high diffusion coefficients for organic permeants, and steady-state permeation is established quickly in such structures.

Polymer-free volume is also a function of structural regularity, orientation, and cohesive-energy density. The aforementioned structure–property relationships all contribute to a decrease in solubility and diffusivity, and thus permeability.

Salame (1) has proposed a relationship between polymer molecular structure and permeability based on an empirical constant (π), or "Permachor" constant, which, when substituted into the Permachor equation, predicts the gas permeability of polymer structures. The correlation parameter or the "Permachor" constant is based on the cohesive-energy density and free volume of the polymer, two major properties of the polymer. Agreement between the Permachor constant and film permeability has been shown to be quite good for oxygen, CO_2 and nitrogen, but not for water vapor and organic vapors.

The equation for relating gas permeability to the Permachor constant is as follows:

$$P = A \exp(-s\pi) \quad (5)$$

where A and s are constants for any given gas at temperature T and π is the Permachor constant of the polymer.

Polymer Morphology

Solid-state polymer chains can be found in a random arrangement to yield an amorphous structure or a highly ordered crystalline phase. Most polymers used in packaging are semicrystalline or amorphous materials (see Polymer properties).

Morphology refers to the physical state by which amorphous and semicrystalline regions coexist and relate to each other in a polymer, and depends not only on its stereochemistry but also on whether the polymer has been oriented, and at which conditions of temperature, strain rate, and cooling temperature, as well as the melt cooling rate.

Fundamental properties that are associated with polymer morphology and will therefore influence the permeability and diffusivity characteristics of the polymer include

- Structural regularity or chain symmetry, which can readily lead to a three-dimensional order or crystallinity. This is determined by the type of monomer(s) and the conditions of the polymerization reaction.
- Chain alignment or orientation, which allows laterally bonding groups to approach each other to the distance of the best interaction, enhancing the tendency to form crystalline materials.

Morphology is thus important in determining the barrier properties of semicrystalline polymers. This is illustrated by the results of permeation studies carried out on biaxially oriented polyethylene terephthalate (PET) films of varying thermomechanical history (9).

Film samples were biaxially stretched at a strain of 350%/s, based on the initial dimensions of 4 in. × 4 in., which corresponded to an orientation rate of 14 in./s biaxially.

The degree of orientation was 400% based on the initial dimensions. The orientation temperature was 90, 100, and 115°C, respectively.

Table 1 summarizes the results of permeability studies carried out with ethyl acetate in PET film biaxially oriented at 90 and 115°C, respectively, and serves to illustrate the effect of thermomechanical history (stretching-temperature values) on the relative barrier properties of PET, for the permeation of ethyl acetate. The percent crystallinity of PET film oriented at 90°C was 22%, while the percent crystallinity of the film sample oriented at 115°C was 31%. As shown, ethyl acetate permeability values decreased by approximately four times by increasing the film orientation temperature from 90 to 115°C.

Table 1. Permeability of Ethyl Acetate through PET Film Biaxially Oriented at 90° and 115°C

Orientation Temperature (°C)	Vapor Activity (a_v)	Run Temperature (°C)	Permeability Constant[a] $P \times 10^{20}$	Lag Diffusion Coefficient[b] $D \times 10^{12}$
90	0.59	30	2.6	1.8
	0.43	37	4.8	2.9
	0.21	54	15.4	11.0
115	0.59	30	0.014[c]	–
	0.21	54	3.6	5.3

[a] Permeability constant units are (kg · m)/(m² · s · Pa).
[b] Diffusion coefficient units are cm²/s.
[c] No permeation after 550 h. Value of P reported represents an upper bound.

Concentration Dependence of the Transport Process

Permeance of limonene vapor through (*1*) oriented polypropylene, (*2*) saran-coated oriented polypropylene, (*3*) two-sided acrylic (heat-seal)-coated biaxially oriented polypropylene, and (*4*) one-side saran-coated, one-side acrylic-coated polypropylene film samples, as a function of penetrant concentration, is presented graphically in Figure 3, where permeance (R) is plotted as a function of penetrant concentration (10). The observed concentration dependency of the permeance values may be attributed to the penetrant/polymer interaction, resulting in configurational changes and alteration of polymer-chain conformational mobility. Zobel (11) reported similar findings for the transport of the penetrant benzyl acetate through coextruded oriented polypropylene and saran-coated oriented polypropylene, at various penetrant concentrations.

Presence of Copermeant

As shown above, organic vapors are capable of exhibiting concentration dependent mass transport processes. Therefore, the type and/or mixture of organic vapors permeating will determine the magnitude of sorption and permeation, as well as the effect of a copermeant on penetrant permeability. The synergistic effect of a copermeant is illustrated by the results of permeability studies carried out on a biaxially oriented polypropylene film. The degree of film orientation was 430% (machine direction) and 800% (cross-machine direction), based on the initial dimensions. Binary mixtures of ethyl acetate and limonene of varying concentration were evaluated as the organic penetrants (12).

Results of permeability studies for selected ethyl acetate/limonene binary vapor mixtures are presented in Figures 4 and 5, respectively. As shown in Figure 4 (ethyl acetate a_v = 0.10 and limonene a_v = 0.18), limonene vapor had a significant effect on the transport properties of the copermeant. A 500% increase in the permeability constant of ethyl acetate was obtained, when compared to ethyl acetate vapor permeability alone, at similar test conditions. However, at this con-

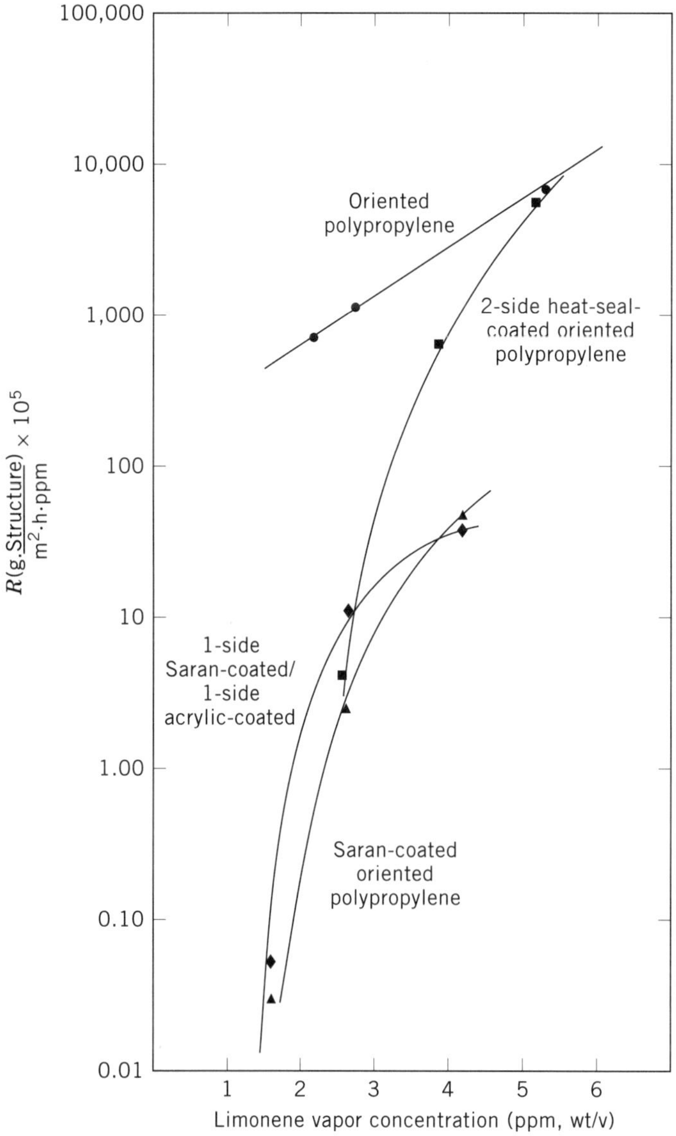

Figure 3. The effect of limonene-vapor concentration on log P for oriented polypropylene and coated oriented polypropylene structures (22° ± 1°C).

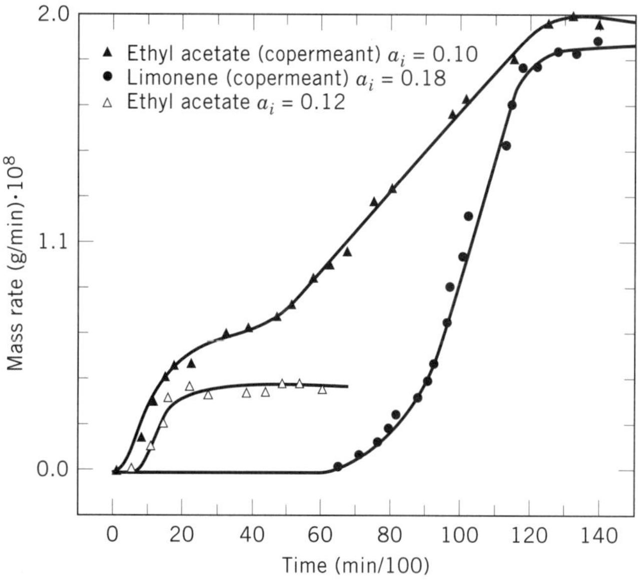

Figure 4. Comparison of the transmission profile of the binary mixture, ethyl acetate a_v = 0.1/limonene a_v = 0.18, with the transmission profile of ethyl acetate (a_v = 0.12) through oriented polypropylene.

centration level, ethyl acetate did not appear to influence the permeation of the limonene vapor. The transmission-rate profile curve for limonene vapor in the binary mixture is superimposed in Figure 4, to provide a complete description of the transmission characteristics of the mixed-vapor system.

For the ethyl acetate a_v = 0.1/limonene a_v = 0.29 binary mixture, a permeation rate 40 times greater than the transmission rate of pure ethyl acetate vapor, of an equivalent concentration, was obtained. This is illustrated in Figure 5, where the transmission-profile plot of the binary mixture is presented, and compared to the transmission-rate profile curve for ethyl acetate vapor alone. Again, at this concentration level, ethyl acetate did not appear to affect the permeability characteristics of limonene vapor.

For studies carried out with the binary mixture of ethyl acetate a_v = 0.48 and limonene a_v = 0.18, the individual components of the mixture were found to have a significant effect on the permeation rates of the copenetrant.

Since the permeability is drastically affected in the sorbate/polymer systems studied, the diffusion coefficient and/or the solubility coefficient must also vary in the presence of a copenetrant to account for the observed increases in the transmission rates for the components of ethyl acetate/limonene binary mixtures. By direct measurement of the equilibrium solubility, Nielsen and Giacin (13) found that the solubility coefficient values were independent of sorbate vapor activity, over the range of activity levels studied, and were not affected by the presence of a copenetrant.

For the ethyl acetate (a_v = 0.10)/limonene (a_v = 0.21) binary mixture, Hensley (12) reported a permeation rate of ethyl acetate 40 times greater than the transmission rate for

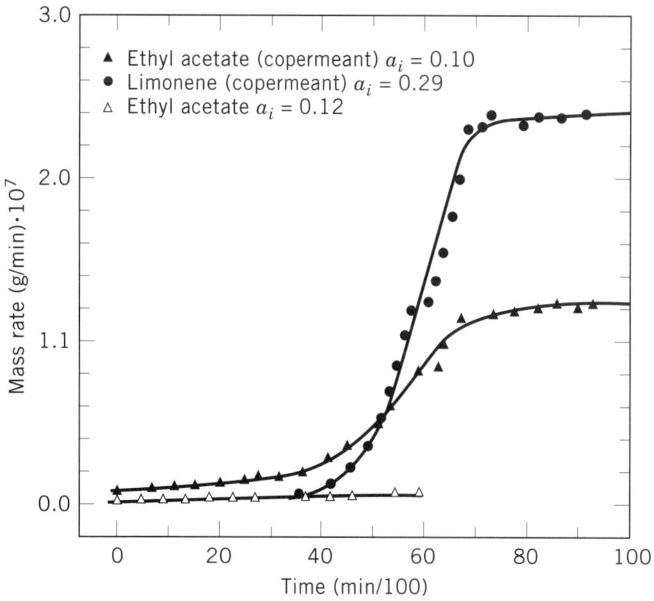

Figure 5. Comparison of the transmission profile of the binary mixture, ethyl acetate a_v = 0.1/limonene a_v = 0.29, with the transmission profile of ethyl acetate (a_v = 0.12).

pure ethyl acetate vapor of an equivalent concentration, while sorption studies showed the solubility coefficient of ethyl acetate to be constant and did not deviate from Henry's law in the presence of a copenetrant. A possible explanation for the dramatic increase in the permeability coefficients for ethyl acetate in the presence of limonene as a copermeant lies with the high copermeant dependence of the diffusion coefficient.

For ethyl acetate/limonene binary mixtures, limonene as a copenetrant appears to have little or no effect on the solubility of ethyl acetate in oriented polypropylene film, while significantly changing the inherent mobility of ethyl acetate within the polymer bulk phase. This accounts for the observed increase in the transmission rates for ethyl acetate through the OPP (oriented polypropylene) film in the presence of limonene.

While not fully understood, the proposed copenetrant dependence of the diffusion coefficient may be due in part to copenetrant-induced relaxation effects occurring within the polymer matrix. The absorption of organic vapors can result in polymer swelling and thus change the conformation of the polymer chains. These conformational changes are controlled by the retardation times of polymer chains. If these times are long, stresses may be set up that relax slowly. Thus, the absorption and diffusion of organic vapors can be accompanied by concentration as well as time-dependent processes within the polymer bulk phase, which are slower than the micro-Brownian motion of polymer chain segments that promote diffusion (7).

There is precedence in the literature in support of such long-time-period relaxation effects occurring in polymer films above their glass-transition temperature (14,15). Thus, there may be copenetrant-induced relaxation effects occurring during the diffusion of ethyl acetate/limonene binary mixtures in the oriented polypropylene film investigated. Such relaxation processes, which occur over a longer time scale than diffusion, may be related to a structural reordering of the free-volume elements in the polymer. Thus, providing additional sites of appropriate size and frequency of formation, which promote diffusion and account for the observed increase in the permeation rate of ethyl acetate in the presence of limonene as a copermeant.

Effect of Relative Humidity

The permeability of acetone vapor through amorphous polyamide (nylon 6I/6T) under dry conditions and in the presence of a humid environment serves to illustrate the effect of water activity or moisture content on the barrier properties of hydrophilic polymer films. Studies were conducted at 60, 75, 85,

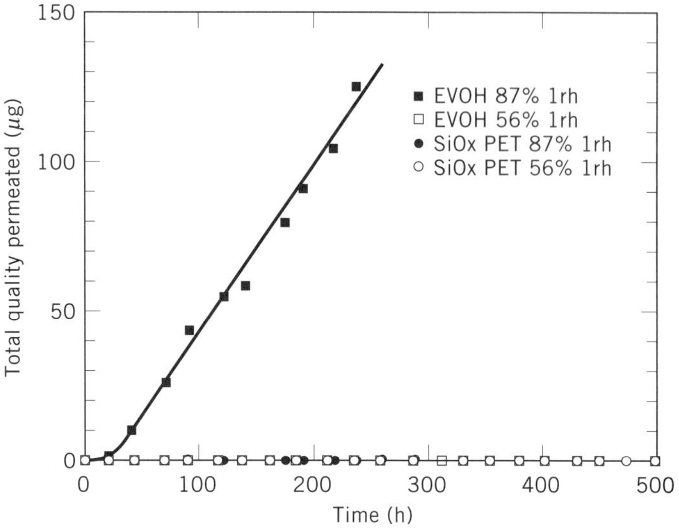

Figure 6. Transmission-rate profile curves of ethyl acetate vapor through SiOx PET and EVAL-F films at 22°C, 190 ppm (wt/v), and 56% and 87%. rh.

and 95°C, at a constant penetrant partial pressure value of 92 mm Hg (0.29 g/L). Water activity (a_w) of the penetrant steam was maintained at 0.7 (70% rh), when measured at 23°C. Different film samples were used for dry- and humid-condition experiments. For an experiment, each run was maintained for a period of 8–14 days after attaining steady state, to ensure that the system was at equilibrium (16).

The respective permeability constant values are summarized in Table 2. As shown, sorption of water vapor resulted in an increase in permeability, as compared to dry conditions, with an increase of approximately 1.5 times being observed.

A further illustration of the effect of water activity on the barrier properties of polymer films is presented in Figure 6, where the total quantity of ethyl acetate permeated is plotted as a function of time, for the permeability of ethyl acetate through SiOx PET and EVAL-F films (17). The test conditions were as follows: temperature 22°C, 190 ppm (wt/v) concentration of ethyl acetate vapor, at 56% and 87% rh, respectively. Fluctuation of relative humidity was ±2%, and fluctuation of organic vapor concentration was ±5%.

For time, >500 h of continuous testing, there was no measurable permeation at 56% rh. The results indicate that both test films were excellent ethyl acetate vapor barriers at 56% rh, and ambient temperature. However, as shown, at 87% rh the EVAL-F film had a significant increase in the permeation rate of ethyl acetate vapor, while the SiOx PET film still showed no measurable increase in the rate of permeation. Ta-

Table 2. Summary of Permeability Constant Values of Acetone through Nylon 6I/6T[a]

Temperature (°C)	$P \times 10^{19a}$ (Dry Condition)	$P \times 10^{19a}$ (Humidified Conditions)
60	3.7	4.9
75	6.5	11.2
85	9.8	17.6
95	11.8	–

[a] Permeability constant units are expressed in (kg·m)/(m²·s·Pa).

Table 3. The Effect of Relative Humidity on the Permeation of Ethyl Acetate Vapor through SiOx PET and EVAL-F Films

Relative Humidity (%)	Sample	Permeance[a] $R \times 10^{17}$
56	SiOx PET	<1.1[b]
	EVAL-F	<2.2[b]
87	SiOx PET	<2.2[b]
	EVAL-F	840 ± 40

[a] Permeance units are expressed in kg/(m²·s·Pa).

Table 4. The Effect of Water Activity on the Permeability of Toluene Vapor through a Two-Side-PVDC Coated OPP Film

Toluene-Vapor Concentration (ppm) (mass/volume)	Toluene-Vapor Pressure (Pa)	Water Activity (a_w)	Transmission Rate[a, b] (F)	Permeance[a, c] (*R*)	Lag Time (h)
40	1203.8	0.50	0.73×10^{-13}	6.0×10^{-17}	94.0
		0.86	0.72×10^{-10}	6.0×10^{-14}	55.0
60	1778.2	0	0.4×10^{-11}	2.2×10^{-15}	82.0
		0.50	0.32×10^{-10}	1.8×10^{-14}	38.0
		0.86	1.6×10^{-10}	9×10^{-14}	29.0
81	2407.6	0	1.1×10^{-7}	1.6×10^{-11}	2.1
		0.50	1.1×10^{-9}	4.6×10^{-11}	2.2
		0.86	1.2×10^{-7}	5.0×10^{-11}	1.9

[a] Average of replicate runs, with a 2σ confidence limit within ± 10%.
[b] Transmission rate units are expressed in kg/(m^2·s).
[c] Permeance units are expressed in kg/(m^2·s·Pa).

ble 3 summarizes the permeance values. Also presented in Table 3 are upper-limit-value estimations for film permeance.

Liu et al. (18) studied the effect of both water activity and permeant vapor concentration on the permeability of toluene vapor through a saran (PVDC)-coated OPP film. Studies were conducted at 22 ± 1°C. Since the film investigated in the present study is a multilayer structure, permeance values are presented to describe the barrier properties of the total structure.

The permeability parameters determined are summarized in Table. 4.

A comparison of the permeance values for toluene, determined as a function of toluene vapor concentration and water activity, showed that the permeance values are highly dependent on both water-activity levels and toluene-vapor concentration. The findings suggested that the sorption of water vapor and/or toluene vapor leads to relaxation in the molecular structure of the saran barrier coating, resulting in a high concentration dependency of the permeance values. Toluene appears to be much more effective than water vapor in promoting relaxation of the saran coatings, since at the highest toluene-vapor concentration level studied (81 ppm), the permeance rate is independent of water-activity or sorbed-water levels. However, at the lower penetrant concentration levels studied, the effect of sorbed water is quite significant, making this polymer structure very vulnerable to the permeation of low concentrations of toluene in the presence of sorbed water.

The results of this study also serve to illustrate the complex interrelationship of the barrier characteristics of the polymer film, penetrant vapor activity, and sorbed-water concentration.

Effect of Temperature on Mass-Transfer Parameters

Permeability, diffusion, and solubility coefficients follow a van't Hoff–Arrhenius relationship as given in Equations 6–8:

$$P = P_0 \exp\left(\frac{-E_p}{RT}\right) \tag{6}$$

$$D = D_0 \exp\left(\frac{-E_D}{RT}\right) \tag{7}$$

$$S = S_0 \exp\left(\frac{\Delta H_s}{RT}\right) \tag{8}$$

where E_p = activation energy for permeation
E_D = activation energy for diffusion
ΔH_s = molar enthalpy of sorption
P_0, D_0, S_0 = constants
R = gas constant
T = absolute temperature

In a recent study Lin (19) determined the permeability of ethyl acetate and toluene for the following commodity films: (*1*) oriented polypropylene, (*2*) high-density polyethylene, (*3*) glassine, (*4*) saran-coated oriented polypropylene, (*5*) acrylic-coated oriented polypropylene, and (*6*) metallized polyethylene terephthalate/OPP laminate.

Permeability studies were carried out at three temperatures to allow evaluation of the activation energies. For each temperature, three vapor activity levels were evaluated. Since several films investigated are multilayer or barrier-coated structures, permeance values are presented to describe the barrier properties of the total structure (see also Barrier polymers).

Permeance values determined at ethyl acetate vapor activity levels of a_v = 0.095, 0.21, and 0.41 for the respective barrier structures are summarized in Tables 5–9. For the metallized polyethylene terephthalate/OPP laminate, no measurable rate of diffusion was detected, following continuous testing for 44 h at 70°C and a vapor activity of a = 0.41.

Table 5. Permeance of Ethyl Acetate through Oriented Polypropylene (OPP) as a Function of Vapor Activity and Temperature[a]

Vapor Activity[b]	Temperature (°C)	Permeance[c] $R \times 10^{12}$
0.095	30	0.6
	40	1.7
	50	4.4
0.21	30	0.7
	40	1.9
	50	4.4
0.41	30	0.9
	40	2.1
	50	4.8

[a] The results reported are the average of duplicate analyses.
[b] Vapor-activity values were determined at ambient temperature (24°C).
[c] Permeance units are expressed in kg/(m^2 · s · Pa).

Table 6. Permeance of Ethyl Acetate through High-Density Polyethylene (HDPE) as a Function of Vapor Activity and Temperature[a]

Vapor Activity[b]	Temperature (°C)	Permeance[c] $R \times 10^{11}$
0.095	30	1.0
	40	1.7
	50	2.7
0.21	30	2.4
	40	2.8
	50	3.4
0.41	30	2.9
	40	3.8
	50	4.8

[a] The results reported are the average of duplicate analyses.
[b] Vapor-activity values were determined at ambient temperature (24°C).
[c] Permeance units are expressed in kg/(m²·s·Pa).

Table 7. Permeance of Ethyl Acetate through Glassine as a Function of Vapor Activity and Temperature[a]

Vapor Activity[b]	Temperature (°C)	Permeance[c] $R \times 10^{12}$
0.095	23	1.1
	30	2.0
	40	4.2
0.21	23	2.3
	30	2.9
	40	5.2

[a] The results reported are the average of duplicate analyses.
[b] Vapor-activity values were determined at ambient temperature (24°C).
[c] Permeance units are expressed in kg/(m²·s·Pa).

Table 8. Permeance of Ethyl Acetate through Saran-Coated OPP as a Function of Vapor Activity and Temperature[a]

Vapor Activity[b]	Temperature (°C)	Permeance[c] $R \times 10^{13}$
0.095	40	1.0
	50	5.0
	60	15.1
0.21	40	1.6
	50	5.2
	60	14.9
0.41	40	2.8
	50	7.7
	60	26.5

[a] The results reported are the average of duplicate analyses.
[b] Vapor-activity values were determined at ambient temperature (24°C).
[c] Permeance units are expressed in kg/(m²·s·Pa).

Determined permeance values for toluene activity levels of a_v = 0.067, 0.22, and 0.44 for the respective barrier structures are summarized in Tables 10–14. There was no measurable rate of diffusion for the polyethylene terephthalate/OPP laminate structure following continuous testing for 44 hours at 70°C and a vapor activity of 0.44.

From equation (6) the temperature dependency of the transport process associated with the respective barrier membranes, over the temperature range studied, was found to follow the Arrhenius relationship. From the slopes of the Arrhenius plots, the activation energy for the permeation process (E_p) was determined for the respective film samples, as a function of vapor activity. The determined activation energy values for both ethyl acetate and toluene are summarized in Tables 15 and 16, respectively.

Numerical Consistency of Permeability Data

In addition to determining the permeability constants of organic penetrants, it is also important to determine the diffusion coefficients and to evaluate the consistency of the experimental data obtained. The numerical consistency of the

Table 9. Permeance of Ethyl Acetate through Acrylic-Coated OPP as a Function of Vapor Activity and Temperature[a]

Vapor Activity[b]	Temperature (°C)	Permeance[c] $R \times 10^{12}$
0.095	50	0.25
	60	1.0
	70	2.6
0.21	50	0.3
	60	1.2
	70	3.6
0.41	50	0.3
	60	1.2
	70	3.3

[a] The results reported are the average of duplicate analyses.
[b] Vapor-activity values were determined at ambient temperature (24°C).
[c] Permeance units are expressed in kg/(m²·s·Pa).

Table 10. Permeance of Toluene through OPP as a Function of Vapor Activity and Temperature[a]

Vapor Activity[b]	Temperature (°C)	Permeance[c] $R \times 10^{11}$
0.067	30	0.5
	40	1.15
	50	1.8
0.22	30	0.75
	40	1.4
	50	2.7
0.44	30	1.4
	40	2.0
	50	3.1

[a] The results reported are the average of duplicate analyses.
[b] Vapor-activity values were determined at ambient temperature (24°C).
[c] Permeance units are expressed in kg/(m²·s·Pa).

Table 11. Permeance of Toluene through HDPE as a Function of Vapor Activity and Temperature[a]

Vapor Activity[b]	Temperature (°C)	Permeance[c] $R \times 10^{11}$
0.067	30	2.1
	40	3.4
	50	5.6
0.22	30	4.0
	40	5.5
	50	8.0
0.44	30	7.7
	40	8.6
	50	9.2

[a] The results reported are the average of duplicate analyses.
[b] Vapor-activity values were determined at ambient temperature (24°C).
[c] Permeance units are expressed in kg/(m²·s·Pa).

permeability data will affect the values of both the diffusion coefficient and the permeability constant and would indicate any variations of the system parameters, such as temperature, or permeant concentration changes during the course of the permeability experiment. Gavara and Hernandez (20) have described a simple procedure for determining the diffusion coefficient and for performing a consistency analysis on a set of experimental permeability data from a continuous-flow permeation study. This procedure was applied to the continuous-flow permeation data obtained, to provide a better understanding of the mechanism of the diffusion and sorption processes associated with the permeation process. The consistency test for continuous-flow permeability experimental data has been described in detail by Gavara and Hernandez (20) and is summarized briefly below.

Table 12. Permeance of Toluene through Glassine as a Function of Vapor Activity and Temperature[a]

Vapor Activity[b]	Temperature (°C)	Permeance[c] $R \times 10^{12}$
0.067	30	2.8
	40	8.9
	50	5.1
0.22	30	3.5
	40	5.3
	50	6.6
0.44	30	4.1
	40	5.5
	50	6.8

[a] The results reported are the average of duplicate analyses.
[b] Vapor-activity values were determined at ambient temperature (24°C).
[c] Permeance units are expressed in kg/(m^2·s·Pa).

Table 13. Permeance of Toluene through Saran-Coated Oriented Polyethylene (Saran OPP) as a Function of Vapor Activity and Temperature[a]

Vapor Activity[b]	Temperature (°C)	Permeance[c] $R \times 10^{13}$
0.067	40	1.0
	50	5.8
	60	21.9
0.22	40	2.5
	50	9.7
	60	25.0
0.44	40	3.1
	50	12.8
	60	28.9

[a] The results reported are the average of duplicate analyses.
[b] Vapor-activity values were determined at ambient temperature (24°C).
[c] Permeance units are expressed in kg/(m^2·s·Pa).

Table 14. Permeance of Toluene through Acrylic-Coated Polypropylene as a Function of Vapor Activity and Temperature[a]

Vapor Activity[b]	Temperature (°C)	Permeance[c] $R \times 10^{13}$
0.067	50	ND[d]
	60	ND[d]
	70	ND[d]
0.22	50	ND[d]
	60	ND[d]
	70	ND[d]
0.44	50	1.4
	60	2.0
	70	3.1

[a] The results reported are the average of duplicate analyses.
[b] Vapor-activity values were determined at ambient temperature (24°C).
[c] Permeance units are expressed in kg/(m^2·s·Pa).
[d] Without detectable response after 44-h test.

Table 15. Activation-Energy Values for the Permeation of Ethyl Acetate through Polymer Membranes

	E_p (kcal/mol)		
Polymer Membranes	$a_v = 0.095$	$a_v = 0.21$	$a_v = 0.41$
OPP	17.7	14.6	14.5
HDPE	9.3	4.4	4.9
Glassine	14.6	9.6	N/A[(a)]
Saran OPP	24.9	23.0	24.5
Acrylic OPP	25.9	27.9	25.0

[a] N/A—data not available.

Table 16. Activation Energy Values for the Permeation of Toluene through Polymer Membranes

	E_p (kcal/mol)		
Polymer Membranes	$a_v = 0.067$	$a_v = 0.22$	$a_v = 0.44$
OPP	12.63	12.53	7.79
HDPE	9.45	6.90	1.90
Glassine	5.78	6.21	5.02
Saran OPP	34.00	34.10	35.41
Acrylic OPP	N/A[a]	N/A[a]	30.25

[a] N/A = data not available.

The value of the permeation rate at any time F_t, during the unsteady-state portion of the permeability experiment varies from zero, at time equal to zero, up to the transmission rate value (F_∞) reached at the steady state. This is described by the following expression (21):

$$\frac{F_t}{F_\infty} = \frac{4}{\pi^{1/2}}\left(\frac{\ell^2}{4Dt}\right)^{1/2} \sum_{n=1,3,5}^{\infty} \exp\left(\frac{-n^2\,\ell^2}{4Dt}\right) \tag{9}$$

Equation 9 is simplified to the following form:

$$\gamma = \left(\frac{4}{\pi^{1/2}}\right)(X)^{1/2}\exp(-X) \tag{10}$$

where γ is equal to the transmission-rate ratio F_t/F_∞ and $X = \ell^2/4Dt$. In Equation 9, D is assumed to be independent of permeant concentration and time.

For each value of F_t/F_∞ (γ), a value of X can be calculated, and from a plot of $1/X$ versus t, the diffusion coefficient (D) can be determined. The authors further described two dimensionless constants, k_1 and k_2;

$$k_1 = \frac{t_{1/4}}{t_{3/4}} = \frac{X_{1/4}}{X_{3/4}} = 0.4405 \tag{11}$$

$$k_2 = \frac{t_{1/4}}{t_{1/2}} = \frac{X_{1/4}}{X_{1/2}} = 0.6681 \tag{12}$$

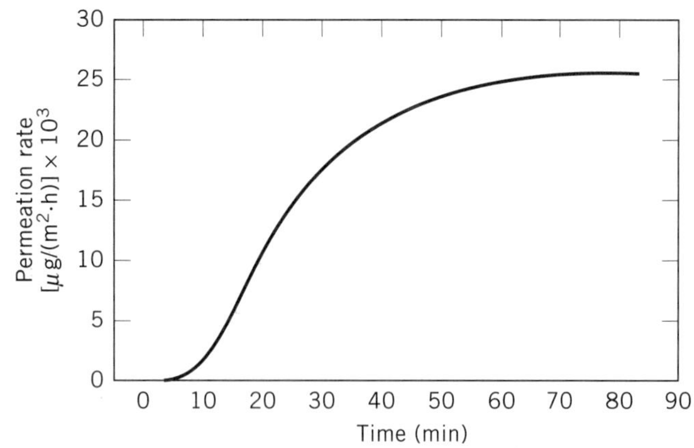

$t_{1/4}$ = 839 (s)
$t_{1/2}$ = 1260 (s)
$t_{3/4}$ = 1936 (s)
$k_1 = t_{1/4}/t_{3/4} = 0.4333$
$k_2 = t_{1/4}/t_{1/2} = 0.6659$
$D_{1/4} = \ell^2/(4\,X_{1/4}t_{1/4}) = 1.62 \times 10^{-13}$ (m²/s)
$D_{1/2} = \ell^2/(4\,X_{1/4}t_{1/4}) = 1.62 \times 10^{-13}$ (m²/s)
$D_{3/4} = \ell^2/(4\,X_{1/4}t_{1/4}) = 1.58 \times 10^{-13}$ (m²/s)
average $D = 1.61 \times 10^{-13}$ (m²/s)

Figure 7. Transmission-rate profile and consistency test for the permeability of limonene ($a_v = 0.4$) through high-density polyethylene at 50°C.

where $X_{1/4}$, $X_{1/2}$, and $X_{3/4}$ denote the numerical values of X when the permeability experiment has reached values of 0.25, 0.5 and 0.75, respectively, for F_t/F_∞, the transmission rate ratio.

The numerical values of the constants k_1 and k_2 as given in Equations 11 and 12, together with the linear relationship of $1/X$ versus t, will provide values of the diffusion coefficient and a criteria to evaluate the consistency of the experimental data.

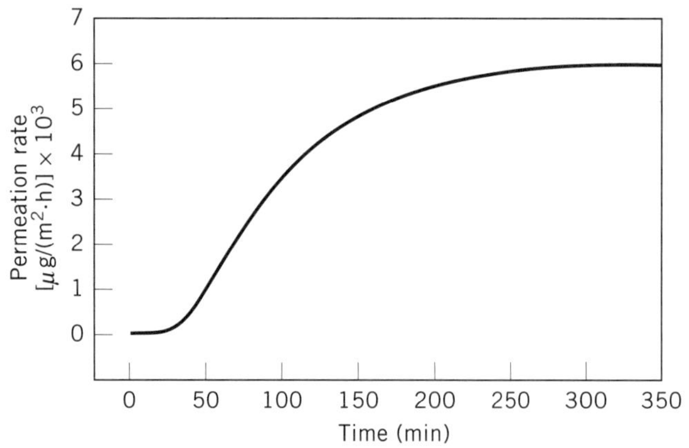

$t_{1/4}$ = 2936.5 (s)
$t_{1/2}$ = 4705.1 (s)
$t_{3/4}$ = 7272.1 (s)
$k_1 = t_{1/4}/t_{3/4} = 0.4037$
$k_2 = t_{1/4}/t_{1/2} = 0.6240$
$D_{1/4} = l^2/(4\,X_{1/4}t_{1/4}) = 0.34 \times 10^{-13}$ (m²/s)
$D_{1/2} = l^2/(4\,X_{1/4}t_{1/4}) = 0.34 \times 10^{-13}$ (m²/s)
$D_{3/4} = l^2/(4\,X_{1/4}t_{1/4}) = 0.41 \times 10^{-13}$ (m²/s)
average $D = 1.61 \times 10^{-13}$ (m²/s)

Figure 8. Transmission-rate profile and consistency test for the permeability of limonene ($a_v = 0.2$) through oriented polypropylene.

The results of a series of permeability studies carried out with limonene vapor are presented graphically in Figures 7 and 8, where the transmission rate is plotted as a function of time, and serves to illustrate the applicability of the consistency test to permeability data obtained for polyethylene and polypropylene (22). The values of k_1 and k_2 calculated from the experimental data and the associated diffusion coefficients, $D_{1/4}$, $D_{1/2}$, and $D_{3/4}$, obtained by substitution into the expressions

$$D_{1/4} = \frac{\ell^2}{4\,X_{1/4}t_{1/4}} \tag{13}$$

$$D_{1/2} = \frac{\ell^2}{4\,X_{1/2}t_{1/2}} \tag{14}$$

$$D_{3/4} = \frac{\ell^2}{4\,X_{3/4}t_{3/4}} \tag{15}$$

are also summarized in the respective transmission-rate profile plots. The values k_1 and k_2 calculated from the experimental data ranged within 4–14% of the theoretical values given by Gavara and Hernandez (20).

BIBLIOGRAPHY

1. M. Salame, J. *Plast. Film Sheet.* **2,** 321 (1986).
2. B. R. Harte and J. R. Gray, *Food Product-Package Compatibility Proceedings,* Technomic Publishing, Lancaster, PA, 1983.
3. E. Bagley and F. A. Long, *J. Am. Chem. Soc.* **77,** 2172 (1958).
4. J. Fujita, *Fortsch-Hochpolym-Forsch.* **3,** 1 (1961).
5. J. Crank, *The Mathematics of Diffusion,* 2nd ed., Clarendon Press, Oxford, UK, 1975.
6. A. R. Berens, *Polymer,* **18,** 697 (1977).
7. P. Mears, *J. Appl. Polymer. Sci.* **9,** 917 (1965).
8. P. T. DeLassus, "Permeation of Flavors and Aromas through Glassy Polymers" in *Proceedings, TAPPI, 1993 Polymers, Laminations and Coatings Conference,* Chicago, 1993, p. 263.
9. R. J. Hernandez, J. R. Giacin, K., Jayaraman, and A. Shirakura, *Proceedings ANTEC '90,* Dallas, 1990.
10. J. R. Giacin and R. J. Hernandez, "Activities Report of the R & D Assoc." in *Proceedings of the Fall 1986 Meeting of Research and Development Associates for Military Food Packaging Systems, Inc.,* 1987, p. 79.
11. M. G. R. Zobel, *Polymer Test.* **3,** 133 (1982).
12. T. M. Hensley, J. R. Giacin, and R. J. Hernandez, *Proceedings of IAPRI 7th World Conference on Packaging,* Utrecht, The Netherlands, 1991.
13. T. J. Nielsen and J. R. Giacin, *Packag. Technol. Sci.* **7,** 247 (1994).
14. A. R. Berens, *Polymer,* **18,** 697 (1977).
15. D. A. Blackadder and J. S. Keniry, *J. Appl. Polym. Sci.* **17,** 351 (1973).
16. S. Nagaraj, *The Effect of Temperature and Sorbed Water on the Permeation of Acetone Vapor through Amorphous Polyamide Film,* M. S. thesis, Michigan State University, East Lansing, MI, 1991.
17. T. Sajaki and J. R. Giacin, *J. Plast. Film Sheet.* **9**(2), 97 (1993).
18. J. J. Liu, R. J. Hernandez, and J. R. Giacin, *J. Plast. Film Sheet.* **7**(1), 56 (1991).

19. C. H. Lin, *Permeability of Organic Vapors through a Packaged Confectionery Product with a Cold Seal Closure: Theoretical and Practical Considerations,* M. S. thesis, Michigan State University, East Lansing, MI, 1996.
20. R. Gavara and R. J. Hernandez, *J. Plast. Film Sheet.* **9,** 126 (1993).
21. R. S. Pasternak, J. F. Schimscheimer, and J. Heller, *J. Polym. Sci.* (Part A-2) **8,** 467 (1970).
22. S. J. Huang, *A Comparison of the Isostatic and Quasi-Isostatic Procedures for Evaluating the Organic Vapor Barrier Properties and Polymer Membranes,* M. S. thesis, Michigan State University, East Lansing, MI, 1996.

J. R. GIACIN
R. J. HERNANDEZ
School of Packaging,
Michigan State University,
East Lansing, Michigan

PHARMACEUTICAL PACKAGING

There are two major categories of pharmaceutical products: ethical products and OTC (over-the-counter) consumer products. Because pharmaceutical companies usually manufacture and package in the same facility, packaging of ethical and OTC drugs is consistent with recognized standards of the pharmaceutical industry.

Ethical Pharmaceutical Packages

The word *ethical* is used to define a product that is sold only on prescription. Federal law prohibits the sale of this drug class in any other way in the United States. (In certain other countries, the ethical or "legend" drug can sometimes be legally obtained without a prescription.) The principles of package design are, therefore, somewhat different from those used in the design of consumer packages. The physician, dentist, nurse, pharmacist, and medical technician are the users of prescription packages.

With the exception of unit-of-use packages and selected injectable products, prescription tablets, capsules, oral liquids, some ointments and creams, and suppositories, are generally purchased by the pharmacist in bulk packages and dispensed according to the physician's directions. The typical prescription is, therefore, not usually sold in the original manufacturer's package, but is transferred to a container supplied and labeled by the pharmacist.

The ethical-prescription package supplied by the manufacturer requires special attention because of the role it plays in the normal distribution process. The primary container and closure (the items in direct contact with the drug) must protect the product both chemically and physically. The selection of primary package material and size is determined by exhaustive scientific study to confirm the product–package compatibility. Testing, as it relates to light protection, water-vapor transmission, gas permeation, potency stability, and product durability under simulated and actual shipping conditions are all important evaluations before the necessary approvals can be requested (see Testing, consumer packages; Testing, packaging materials). The results of development testing also determine the required expiration date of the product. They may rule out certain desirable package features; for example, clarity, shatter resistance, snap cap, plastic syringe, large container, light weight, and low cost.

Product–package testing requires scientific data to substantiate the package choice, and to confirm the requested expiration dating. United States regulations require that these data be submitted to the FDA as part of the New Drug Application (NDA) or an NDA supplement. Written FDA approval of the NDA is then required prior to interstate distribution (see Food, drug, and cosmetic regulations). NDA documentation contains a complete description of the primary package, including desired package sizes. The description of the package material may include manufacture of the material composition (if a compound), the Drug Master File (DMF) number, as well as the type of closure (see Closures), and specifications of cap liner (see Closure liners). After the original NDA has been submitted, one may expect a request for more detail from the FDA examiner. Such demands usually involve primary package data.

All package copy and labeling is also submitted as part of the NDA. This includes the physician's insert or leaflet (see Inserts and outserts). This information folder must be included with every prescription package. Where and how to include the insert with each package is often a challenge. The usual methods include placing the insert and primary container in a folding carton (see Cartoning machinery); automatically locating the insert under the label of the bottle in a way that allows its removal without destruction of the label; placing it under the closure; and taping or banding it to the container.

The Poison Prevention Packaging Act of 1970 (1) and subsequent supplements created new and significant packaging changes for pharmaceuticals. This law meant that all oral prescription products sold in containers designed and intended to be used by the patient or ultimate consumer must have "special packaging," the term in the law to mean child-resistant packages, more commonly referred to as child-resistant safety closures (see Child-resistant packaging; Closures). For those prescription products packaged in containers for bulk dispensing by the pharmacist, the law holds the pharmacist responsible to provide the child-resistant package at the time of the sale to the customer. The law provides that the prescription customer may request a conventional closure, thereby assuming the safety-measurement risk, and the pharmacist is then allowed to sell the product in a package that does not have a safety cap or child-resistant packaging (2).

There are three other types of special-purpose packaging for ethical pharmaceuticals: unit dose; unit-of-use; and parenteral packaging.

Parenterals. *Parenteral* is a medical term for "outside of the intestine." The word is commonly used to mean sterile products injected into the body, either into the muscle (IM) or into the vein (IV). In some cases, the physician can decide whether the product will be given IM or IV. The package design accommodates this choice in many cases and contributes to more convenient and accurate administration. Parenteral or sterile pharmaceutical products are further subdivided into small-volume parenterals (SVPs) and large-volume parenterals (LVPs). The SVPs contain less than 100 mL in a single package. The LVPs contain 100–1000 mL.

Until the late 1970s, parenteral products were always packaged in glass. Such packages took the form of glass ampuls, rubber-stoppered or sealed vials, and rubber-stoppered bottles. The ampul is glass tubing that is product-filled and heat-sealed (see Ampuls and vials). The constricted neck of the container is snapped off to enable withdrawal of the sterile product with a syringe. Glass vials, usually SVPs, permit multidose packaging by allowing a number of syringe withdrawals from the same package. The rubber compounds used to seal and stopper the vials are chosen for their resealing ability, in addition to other physical and chemical demands.

The LVP glass bottle normally contains intravenous (IV) medication. Popular sizes include 250-mL, 500-mL, and the most common 1000-mL bottle. The medication is administered to the patient through an IV needle attached to flexible plastic tubing. The other end of the tubing is connected to a large metal or plastic needle that pierces the rubber-stoppered bottle. The flow rate to the patient is controlled by a pinch clamp on the flexible tubing.

Advancement in plastic packaging is now allowing a limited number of sterile products to be approved in plastic containers. The LVP products are leaders in this early trend. Both rigid and flexible containers store the widely used electrolyte and sugar supplements, and a few pure intravenous drug compounds are now available in "ready-to-use" bottles or bags. Such packaging eliminates the need to add the drug product to the larger infusion (eg, 5% dextrose) container. With a "piggy-back" tubing attachment, the drug product is administered through the same tube that carries the infusion solution. The "ready-to-use" IV packaging system is expected to be widely accepted when FDA clearance is in place.

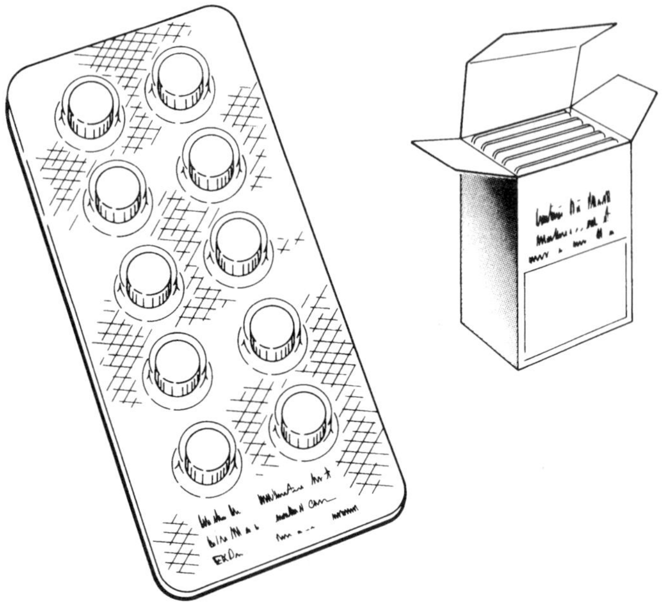

Figure 1. Blister packaging. A push-through vinyl blister with foil back. Either vinyl or foil can be printed.

Unit dose. Another package innovation evolved from hospital use of pharmaceuticals: hospital unit-dose packaging. The product is contained in a package that allows and controls the dispensing and the administration of a prescribed single dosage at the right time with the right product. Hospital unit-dose packaging has significantly reduced hospital medication errors since their introduction, in the form of prefilled disposable syringes, over 30 years ago. In addition to single-dose prefilled syringes, many parenteral drug products became available in single-dose vials.

The most significant advancement for hospital packaging was the introduction of unit-dose oral products. During the early 1950s some tablets and capsules were available in foil-strip packages. Tablets or capsules were packaged in individual pouches that were attached to each other and separable by means of a perforation in the foil strip of 100 tablets or capsules. Each small packet or pouch ws labeled to include the product trade name, the generic chemical name, expiration date, lot number, and name of the manufacturer. From this early concept, the present hospital unit-dose blister package evolved.

Blister packages (see Fig. 1) are recognized as an improvement over the early foil-strip package. The tablet or capsule is visible through the blister side of the unit-dose package. This provides one more safety measure in the hospital dispensing process. Recognition of the product prior to opening the package reduces medication errors, as well as waste of products opened by mistake that must be destroyed. The transparent blister can be made of one of several thermoformable (see Thermoforming) polymers or combinations of polymers that provide improved barrier properties, or the heat-seal capabilities needed to seal the blister side to the lidding stock (see Lidding; Sealing, heat). The selection depends on the chemical and physical barrier demands of the pharmaceutical product. The more moisture-sensitive the product, the better the moisture-vapor transmission barrier properties must be. Available materials range from relatively inexpensive PVC to the expensive PVC/chlorotrifluoroethylene (see Films, fluoropolymer). The choice of film thickness affects both material costs and barrier properties. Other considerations are machinability, production rates, depth of the blister, wall thickness and uniformity of the blister, and sealing properties to the lidding stock. It is obvious, therefore, that expensive and time-consuming testing must take place to verify the package–product acceptance demands.

The reverse side of the hospital unit-dose package is the lidding stock. This usually takes one of two forms: a lamination of aluminum foil/paper (see Multilayer flexible packaging) or preprinted aluminum. The more popular type is the paper/foil combination. The foil side is sealed to the blister containing the product with a heat-seal coating material. The paper side is printed on line with the required label copy. The foil component of unit dose must perform the same protection and sealing functions as the blister. For this reason, a specification of 0.001 in. (25 μm) is the usual standard (see Foil, aluminum). This thickness is considered to be pinhole-free and provides the optimum ratio of cost : product protection.

As with all prescription products submitted to the FDA for approval, complete identification and specifications for hospital unit-dose packaging are included. To assist inventory requirements, a group of products, known as "controlled substances," are sometimes numbered sequentially on each blister. These products include the narcotic compounds, the barbiturates, and other habit-forming substances specified by the FDA (3).

Many hospitals have converted all, or almost all, of their dispensed prescriptions to the unit-dose system for inpatient carer. Because unit-dose packages are not commercially available for all prescription pharmaceuticals, many hospitals have invested in automatic or semiautomatic blister packaging machines. The products are purchased in bulk containers from the pharmaceutical manufacturer and then repackaged into the blister form.

Unit-of-use. Unit-of-use packaging is not the same as hospital unit-dose packaging: unit dose is prescribed for hospital inpatients; unit-of-use is intended for the ambulatory or prescription customer. This package is designed to contain that amount of drug to satisfy the patient's therapeutic requirements for a period of not more than 30 days. The product is intended to be dispensed by the pharmacist in the original package, as supplied by the pharmaceutical manufacturer. The typical pharmacist's label, indicating the dosage prescribed by the doctor, is also attached to the original package. The early unit-of-use packages were probably ointment tubes and oral contraceptive products. Oral pediatric liquid antibiotics and eyedrop preparations are also provided in unit-of-use packaging. It is difficult to package all ethical drugs in this form because of the wide range of dosage requirements due to patient needs, physician judgment, and economics. But the popularity of unit-of-use is steadily increasing in the United States and is now the preferred method in many European areas.

Unit-of-use assures that the product is packaged in an approved container that is labeled by the manufacturer and carries the approved expiration date, as well as the manufacturer's lot or batch number (see Code marking and imprinting). Such measures assure the quality and uniformity of the product as backed by the manufacturer. Like hospital unit-dose packaging, unit-of-use packaging is popular for the hospital outpatient. The outpatient is one who is being discharged to go home or visits the hospital for care and medication without being admitted to the hospital as an inpatient. The convenience of unit-of-use packaging is obvious from the pharmacist's viewpoint, and it gives the physician an easier way to write complicated prescription directions. It also places responsiility for regulatory compliance in the hands of the manufacturer. Proof of child-resistant closure effectiveness is also the responsibility of the drug packager. This is done according to a federal protocol testing procedure (4,5).

The term *secondary packaging* describes the remaining packaging components, which are not in direct contact with the drug product. Folding cartons, paperboard sleeves, corrugated boxes, corrugated dividers, thermoformed trays, and plastic foams are examples of secondary package components used to provide shipping protection. Laboratory testing and actual shipping trials determine the optimum cost-performance relationship between materials-design and product damage.

OTC Packages

Unlike most ethical product packages, OTC packages are researched, designed, and produced for the consumer. They are sold not only in prescription outlets, but also in supermarkets, department stores, small neighborhood convenience stores, and vending machines. The structural design of the package, and its graphics, must appeal to the consumer, since the package frequently sells the product. Focus interviews and other market-research techniques are used to measure this appeal.

Safety-closure reglations apply to selected OTC pharmaceuticals (5). The law allows the buyer to choose one package size without the child-resistant feature. If only one size is available, the safety cap must be part of the package supplied by the drug manufacturer. The pharmacist supplies a conventional cap or package only at the request of the customer. Manufacturers must also comply with the regulation promulgated in 1982 that called for tamper-evident packaging (see Tamper-evident packaging).

Secondary packaging for OTC pharmaceuticals includes display packers that hold 12–36 individual packages, corrugated stands that contain several dozen packages of different sizes, and packages that are designed to be displayed on a wire stand. Trial size packages are recent additions to OTC packaging. The amount of product is intended to acquaint the prospective customer with a sample quantity for taste, size, short-term effect, and price. Trial-size packages are not routine production packages either in terms of volume or machine demand, and on a unit-cost basis, they are relatively expensive.

The FDA governs the use and acceptance of consumer-drug products. A system called the drug monograph regulates those products that are safe and effective and available without prescription. The drug monograph is to OTC products what the NDA is to the ethical products. The FDA is continuing research to determine which prescription products can be authorized for consumer-product status. A number of ethicals have already been changed to OTCs, based on long-term prescription usage and safety. This trend is important in the packaging of OTC products. Because the requirements for ethical and OTC packages are so different, a change in the status of a product presents interesting new package-design challenges.

Production Activities

Pharmaceutical manufacturers use a relatively high number of dissimilar materials and packages, and packaging operations form an intrinsic part of pharmaceutical manufacturing. The federal government controls manufacturing activities in United States plants, and its guidelines are often used in other countries. The regulations are known as CGMPs or Current Good Manufacturing Practices (6). The purpose of the regulations is to ensure that all pharmaceutical products are produced and packaged under specified conditions to assure safety and effectiveness. These rules include periodic inspection of the pharmaceutical-manufacturing facility.

Due to the sometimes large number of package sizes of the same product, and the varied package designs of different products, packaging operations are generally not standardized in large and medium-sized pharmaceutical companies. Long production packaging runs are not routine and frequent machine changeovers are common. This makes it especially important that packaging materials comply with written and issued specifications (see Specifications and Quality Assurance). Successful suppliers of packaging materials and components recognize the importance of consistently providing the pharmaceutical packager with products of the highest quality. Careful incoming inspections are made to ensure that the parts meet the detailed specifications. In most cases, the primary packaging components are examined analytically before production approval. The health industry and medical profession recognizes the importance of pharmaceutical packaging. The quality of the product, the protection and potency of the medication and the accurate means of identification all contribute to modern medical-pharmaceutical success.

BIBLIOGRAPHY

"Pharmaceutical Packaging" in *The Wiley Encyclopedia of Packaging Technology* 1st ed., by H. C. Welch, PharmPack Company, pp. 506–509.

1. *Public Law 91–60 / 5.2162,* Dec. 30, 1970.
2. *Fed. Reg.* **37**(32) (Feb. 1972).
3. 21 USC 801, Comprehensive Drug Abuse Prevention and Control Act, 1970.
4. *Fed. Reg.* **36**(225) ¶295.10 (Nov. 20, 1971).
5. *Fed. Reg.* **38**(151) (Aug. 7, 1973).
6. *CFR,* part 211.

POINT-OF-PURCHASE PACKAGING

Point-of-purchase packaging materials play an important role in marketing efforts of consumer products companies. Displays, fixtures, signs, and related materials at the retail store represent point-of-purchase, or (POP), packaging. Food, drug, specialty, and convenience stores, as well as mass merchandisers and warehouse clubs all make use of POP packaging. The medium is used in almost every category of product from health and beauty aids to household appliances. POP materials are used to increase sales, obtain positioning on the retail floor, and carry the brand name and strategy along with related advertising efforts.

High Visibility versus POP Packaging

A distinction can be made between the primary package that is considered highly visible and packaging that is considered point-of-purchase. Both forms of packaging are an important promotional medium, especially in self-service stores. However, POP packaging is different from high-visibility packaging because in most cases it consists of the promotional materials that surround the primary package. For most consumers, POP packaging implies that the product is new to the market, is offered at a sale price, or comes with another promotional tie-in (1). High-visibility packaging includes blister, clamshell, and skin packaging, as well as packages with windows (see Blister packaging; Carded packaging). These forms of primary packaging display the product's benefits, and they promote impulse purchases. Often, retailers request product manufacturers to replace the standard primary package with a high-visibility package for two reasons: (*1*) sales assistance is not available to help with customer purchase decisions so the package needs to sell itself and (*2*) they request the high-visibility package to have better pilfer resistance than the standard package. Despite the distinction between the two types of packaging, the lines between primary packaging and POP packaging have blurred, and the diversity of POP allows for exceptions. Primary packaging can also function as a point-of-purchase device as seen in Figure 1. These two camera boxes function as the primary package and also as an informational sales tool. Retailers leave these boxes in the open position on a countertop for the customer to read about product features of the professional cameras. The bleached-white laminated E-flute die-cut corrugated box is designed in the style of a counter display (see Board, corrugated; Diecutting). Extra lithographic printing surfaces are available for the selling message as a result of the double wall hinge cover, double sidewalls, double front wall, and the interior shelf-style panel for a clean, finished appearance (see Lithography; Offset printing).

Figure 1. A primary package functions as a point-of-purchase device.

Materials

Because POP packaging is so varied, material selection is based on the intended duration of the promotion. POP is commonly defined as either temporary or permanent. In general, temporary displays are used until the product is depleted or for less than 6 months. Permanent displays have an intended length of use for more than 6 months. A third category called *semipermanent* often marries packaging materials used in both of the categories to give the display an extended shelf life over temporary displays. The most widely used material in temporary exhibits is corrugated board. Paperboard is also popular (see Paperboard). Permanent display construction makes use of extruded plastic sheet fabrications, plastic profile and tubing, plastic molding, wire, metal extrusions, foamboard, or wood. To create semipermanent displays, corrugated board can be used with more durable materials such as wood, plastic, or metal. Solid fibreboard (see Boxes, solid-fiber) or corrugated board of heavy construction could also be used alone. Figure 2 shows a temporary counter display made of bleached-white corrugated board on the left. The permanent counter display on the right is an extruded plastic sheet fabrication.

Temporary POP Categories

Common elements of counter and floor stands. The most widely used display styles for temporary promotions are

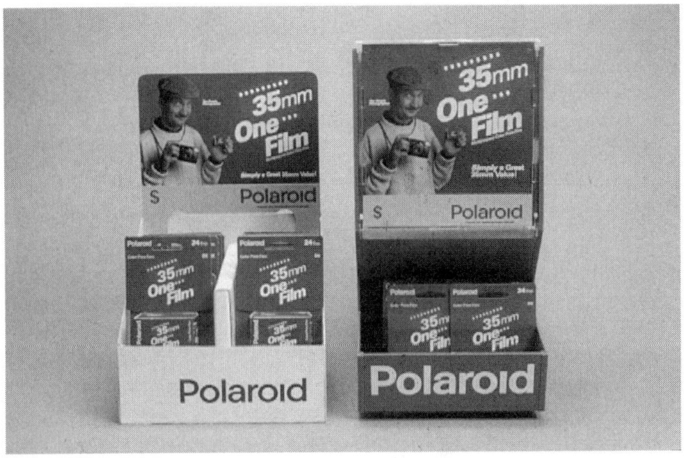

Figure 2. A temporary corrugated-board counter display (*left*) and a permanent counter display made of extruded plastic sheet (*right*).

counter and floor-stand displays. Both styles share commonly used structural design elements. The most visual element that these two styles share is the header portion (also called riser panel) of the display. The header is a display panel that protrudes up from the open side of the display box to advertise the contained product. The header often comes attached to the body of the display as a hinged-cover panel with a front tuck flap. The header folds back for setup. It may also arrive as a separate piece that slides into a slot on the body. The most popular manufacturing technique for header cards is to preprint a lithographic sheet and glue it to a sheet of paperboard or corrugated board. This technique allows for high-quality printing for the sales message while the other parts of the display can be printed at lower cost using flexography or can be left unprinted (see Flexography). The temporary display in Figure 2 has a flexo printed body with a litho sheet header card that is laminated to corrugated board.

The body of counter and floor-stand displays is designed in either an easel style or a tray style. Easels often have cells to partition the product. Figure 3 shows a corrugated-board floor-stand display that uses an easel-style body with cells. This display ships as a prepackaged multipack so that loading of the product is not required by the retailer. The tray style often uses a tiered tray as shown in Figure 4. The tiered-tray floor stand ships on a custom-sized wood skid. This display is intended as an end cap that will abut the end of an aisle. Some tiered trays are offset so that all of the product is displayed at once. The pallet-unit style shown in Figure 4 is commonly referred to as a "stacker." This display is intended for a wholesale or warehouse club store. The large amount of product contained in the display would not be suitable for traditional retailers.

Floor stand displays sometimes require a base to support the display. Figure 3 shows a typical base design. A popular industry practice is to use modular stock display designs from POP manufacturers. The display is then customized with special structural features on the header and the client's graphics. The client incurs a lower cost because no tooling charge for die-cutting exists for the stock elements.

Other counter and floor-stand displays. Counter displays have several unique categories including contest boxes, counter cards, and literature holders. Figure 5 shows a

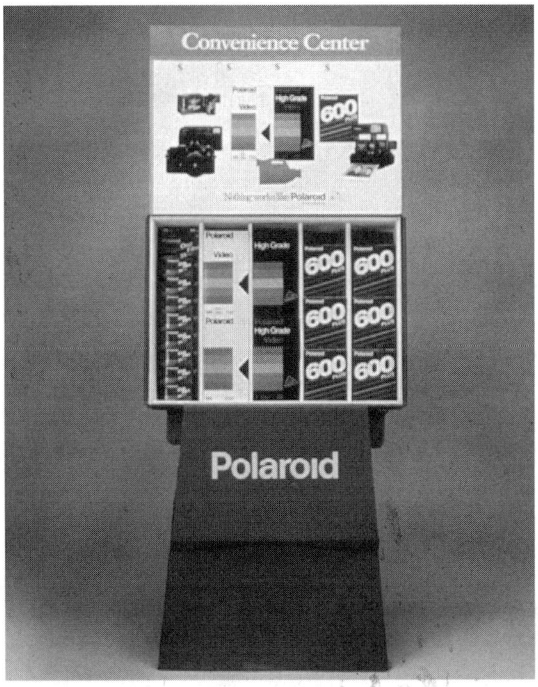

Figure 3. A corrugated-board floor-stand display that uses an easel-style body with cells. The header and base are stock display designs.

counter card with an interactive feature. The camera's audio unit is a talking voice chip with three prerecorded messages. Floor stands include several unique categories, as well. The standee is a standup person or object that is usually two-dimensional with an easel backing for support. Standees are often used when a public figure joins a company's advertising campaign. This visual element triggers recognition and reinforces a positive impression, which leads to consumer loyalty.

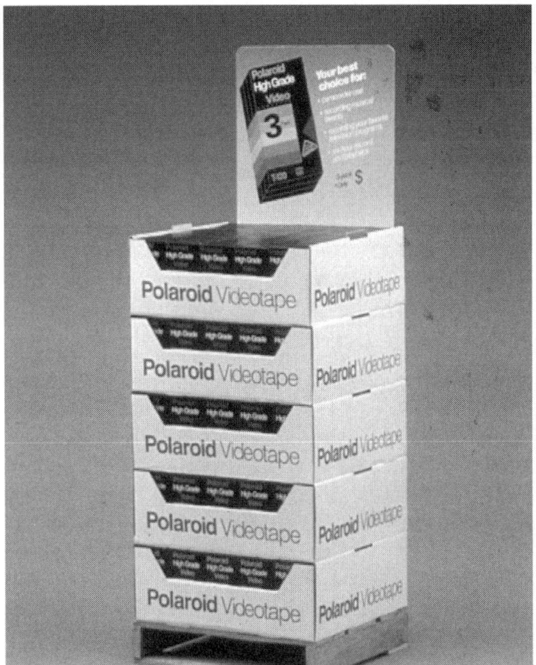

Figure 4. A palletized-unit display called a stacker uses a tiered tray design. A removable header tops the display.

Figure 5. A counter card displays the product by mounting it in the card's panel. Customers can hear three prerecorded messages by pushing the RELEASE button.

Figure 7. Two merchandising strips hold the instant film packs in a vertical display.

Lastly, the dump bin style is also a category used in floor stands.

Miscellaneous temporary POP. Several miscellaneous categories fit under the temporary POP umbrella. Pole displays, overheads, banners, flags, posters, and mobiles are common. Figure 6 includes several examples of POP promotions that are used to complement a display system. The small flags in the photograph are used as "shelf talkers" with plastic clips to secure to the shelf edge. Along with hooks and clips, merchandising strips are popular, especially for use on powerwing or end-cap metal racks. A power wing is the side of an end cap where impulse merchandise is hung, often on hooks. Figure 7 shows two merchandising strips loaded with instant film packs.

The last area that falls under temporary POP is incentive items such as game cards, sweepstakes cards, rebate offers, on-pack coupons, and bottle neck toppers. The $5 mail-in rebate offer shown in Figure 8 has a counter card with a pad of rebate cards that explains the rules of the promotion. Of special note to graphic artists is the packaging graphics redesign of the whole line of instant film shown on the center card. Earlier versions of the film box did not have a unified look for the product line's corporate and brand identity.

Permanent POP Categories

Permanent POP includes many of the categories that have been detailed for temporary displays such as any of the counter and floor stands. Much of permanent POP includes fixtures which are not actually "packaging." However, it is the packaging engineer's job to specify the fixture and design

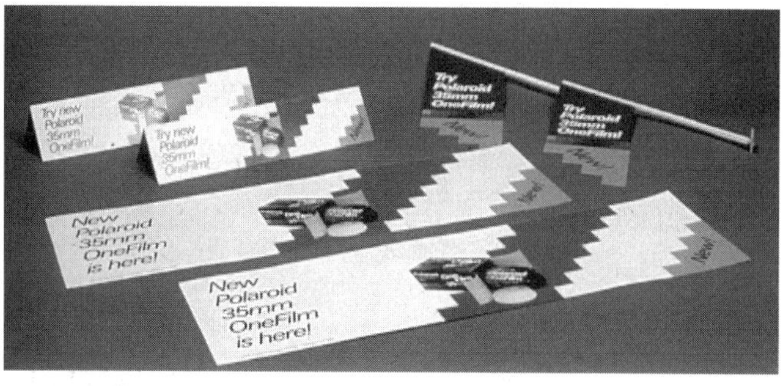

Figure 6. The photo includes banners, signs, and "shelf talker" flags on standout poles that clip onto a shelf edge.

Figure 8. A pad of rebate offers on a counter card. Note the packaging graphics redesign to unify the look of the instant film product line.

Figure 9. A permanent floor-stand spinner rack displays personalized greeting cards. The customer inserts the instant photo she has just taken into the card.

shipping packaging to transport the promotion's fixture and the enclosed product to the retail store. A category called the spinner or spinner rack is most often intended for permanent display. It comes in counter or floor-stand designs. Figure 9 shows a floor-stand spinner rack with wire display pockets. Spinner racks that have hooks or pegs are also popular, and they are for products with hang tabs. Another form of permanent POP is product sampler or tester unit displays for industries such as cosmetics. Other categories usually associated with permanent POP are items such as dangling inflatables and motion displays that are heavily used in the competitive field of beverage and liquor packaging.

The next category of permanent POP is full-line merchandisers, which are category or shelf management systems. They are used to organize a company's offerings for a product line in any department of a retail store. It ensures facings are easily identifiable. Examples are seen in panty hose, wrapping paper and bows, and other product lines with many choices. For a product manufacturer, a full-line merchandiser ensures that the product will not be subject to the whim of a retailer's planogram. Retail stores use planograms—written diagrams with instructions—to lay out departments. The retail store finds planograms useful for cutting in new products, positioning higher-margin products in more visible locations, and periodic cleaning of fixtures. The last permanent category worth mentioning is computerized kiosks with interactive touchpad displays and printers. The kiosks are used in various applications such as multimedia questionnaires or self-service tools in which customers use the kiosk to help identify correct replacement parts for a product.

Trends

Trends in the industry include cross-merchandising where two marketers join together in a promotion program. It is also called cobranding. For instance, a soft-drink manufacturer teams up with a candy company to do a joint temporary display. With changing channels of distribution due to increased competition from wholesale clubs and superstores, manufacturer's are feeling a greater retailer influence. Promotions have become more account-specific because retailers are demanding customizing of displays for their use only. Another trend is comarketing. Comarketing teams up a retailer and a marketer who have developed a promotion together to leverage store and brand-name equity in order to build sales for both. A distinction is made between comarketing and "prepackaged" promotions that a manufacturer has thoroughly thought out and then tailored to a specific retail account (2). For more specific data on trends in POP, *P/O/P Times* magazine published a second annual report on trends and growth patterns in its November/December 1994 issue.

For Further Information

The Point-of-Purchase Advertising Institute offers many reference materials for sale through its information center in Washington, D.C. [phone (202) 530-3000]. These materials include written surveys, brochures, books, monographs, yearbooks, and audio cassettes. A helpful handbook is the *POPAI P-O-P Desktop Reference* which provides 1,200 POP industry terms. The General References list in the Bibliography (below) includes the only two books in print on the subject. Periodicals are the industry's major communication tool. Trade magazines that cater to the industry include *P/O/P Times,* a publication of point-of-purchase advertising and display; *P.O.P. & Sign Design,* for high-volume producers of displays, sign, and fixtures; *Creative,* which is published for sales promotion and marketing professionals who manage point-of-purchase display, trade-show-exhibit, and sales-promotion programs; and *Promo,* a magazine about promotional marketing. *Creative* publishes a yearly illustrated supplier directory arranged by category that includes company listings. In addition, several marketing management and packaging periodicals occasionally carry related journal articles. A *Marketing Science* article details a mathematical model for managing shelf space at retail (3). Research published by Percy and Rossiter shows how to apply successful message construction in POP to encourage a consumer purchase decision on the basis of an interplay of customer involvement and motivation (4).

Several trade shows catering to the POP market are held each year. Merchandising and design innovations in POP packaging are promoted yearly in industry achievement award contests. The Point-of-Purchase Advertising Institute, POPAI, has sponsored a competition each year since 1960. Another contest is held at the annual Display, Sign and Fixture Design show.

BIBLIOGRAPHY

1. Editor, "How Shoppers Feel about Displays," *P/O/P Times,* 58 (Feb. 1995).
2. C. Hoyt, "Co-marketing: What It Is And Is Not," *Promo,* 32 (March 1995).
3. A. Bultez and P. Naert, "S.H.A.R.P.: Shelf Allocation for Retailers' Profit," *Market. Sci.,* 211–231 (Summer 1988).
4. L. Percy and J. Rossiter, "A Model of Brand Awareness and Brand Attitude Advertising Strategies" in *Psychology & Market,* Wiley, July/Aug. 1992, John Wiley & Sons, Inc., New York, 1992.

General References

Point of Purchase Design Annual, Rockport Publications, 1994.

Point of Purchase Design Annual: POPAI's 36th Merchandising Awards, Retail Report, POP Advertising Institute Staff, 1994.

Jennifer Griffin
Polaroid Corp.
Cambridge, Massachusetts

POLYCARBONATE

Polycarbonate (PC) is an amorphous resin that does not require orientation to achieve its full property profile. The molten resin can be extruded by the blown or cast processes (see Extrusion), injection-molded (see Injection molding), or blow-molded by the extrusion-blow or injection-blow techniques (see Blow molding). As the polycarbonate cools into film, sheet, or containers, it exhibits excellent dimensional stability, rigidity, impact resistance, and transparency over a wide range of temperatures and loading rates (see Table 1). Because PC is amorphous, its wide softening range and added strength in thermoforming (see Thermoforming) operations provides deep-draw capabilities.

Polycarbonate provides an excellent combination of tensile strength and flexural modulus at high temperatures (see Table 2). Its heat resistance, combined with superior impact resistance at both high and low temperatures, makes it an excellent structural layer in coextruded or laminated packag-

Table 1. Typical Property Values for Polycarbonate[a]

Property	ASTM Test Method	Melt Flow Indexes 22	16	10	6	PC Copolymer
Specific gravity	D792	1.20	1.20	1.20	1.20	1.20
Light transmittance, 0.125 in. (3.2 mm), %	D1003	89	89	89	89	85
Haze, 0.125 in. (3.2 mm), %	D1003	1	1	1	1	1–2
Deflection temperature at 264 psi (1.8 MPa) °F (°C)	D648	260 (127)	265 (129)	270 (132)	270 (132)	325 (163)
Flammability rating,[b] UL 94, at 0.060 in. (1.5 mm)		V-2	V-2	V-2	V-2	HB
Tensile strength, yield, psi (MPa)	D638	9,000 (62)	9,000 (62)	9,000 (62)	9,000 (62)	9,500 (65.5)
Tensile strength, ultimate, psi (MPa)	D638	9,500 (65.5)	10,000 (68.9)	10,000 (68.9)	10,500 (72.4)	11,300 (77.9)
Elongation, rupture, %	D638	120	125	130	135	78
Flexural strength, psi (MPa)	D790	13,500 (93.1)	14,000 (96.5)	14,000 (96.5)	14,200 (97.9)	14,100 (97.2)
Flexural modulus, psi (MPa)	D790	335,000 (2,310)	340,000 (2,340)	340,000 (2,340)	340,000 (2,340)	338,000 (2,330)
Izod impact strength, notched, $\frac{1}{8}$-in. (3.2-mm) thick, ft·lbf/in. (kJ/m)	D256	12 (0.64)	13 (0.69)	15 (0.80)	17 (0.94)	10 (0.53)
Tensile impact, ft·lbf/in.^{2} (J/cm^2)	D1822	180 (37.8)	225 (47.3)	275 (57.8)	300 (63.0)	275 (57.8)

[a] Properties shown are average values that can be expected from typical manufacturing lots and are not intended for specification purposes. These values are for natural color only. Addition of pigments and other additives may alter some of the properties.
[b] This rating is not intended to reflect hazards of this or any other material under actual fire conditions.

ing for hot fill at 180–210°F (82–99°C), retorting at 250°F (121°C), autoclaving at 270–280°F (132–138°C), and frozen-food packaging. In addition, PC can be sterilized with both gamma and electron-beam irradiation with good stability.

Polycarbonate has light-transmittance values of 88–91% as compared with 92% for clear plate glass. It has a haze factor of less than 1%, and maintains these values throughout the temperature scale. Its high gloss and easy colorability and printing contribute to distinctive package design. Polycarbonate has high resistance to staining by tea, coffee, fruit juices, and tomato sauces, as well as lipstick, ink, soaps, detergents, and many other common household materials. Its relatively dense composition makes its immune to odors, and its hard, smooth surface facilitates easy removal of foodstuffs.

Some of the most significant attributes of polycarbonate stem from its very low water absorption. Added weight increase after 24-h immersion at room temperature is only 0.15%. This low absorption level helps account for the resin's excellent dimensional stability and stain resistance. It also indicates that the resin, itself tasteless and odorless, is unlikely to pick up food odors. PC is available in grades that meet FDA and USDA regulations and is recognized as safe for food-contact applications.

Packaging Applications

Refillable bottles. Polycarbonate is the material of choice for use in reusable bottles, particularly 5-gal (19-L) water bottles, which represent the resin's chief packaging application. These bottles take advantage of polycarbonate's toughness (to resist breakage) and clarity (to see the contents). The fact that PC is much lighter than glass provides fuel savings as well as productivity improvements, since several bottles can be carried at once. Systems have been developed to wash polycarbonate and provide clean bottles for reuse with minimum impact on trippage.

Table 2. Tensile Strength and Modulus of Polycarbonate Over a Temperature Range

Temperature, °F(°C)	Tensile Strength, psi (MPa)	Flexural Modulus, psi (MPa)
73 (23)	10,000 (68.9)	320,000 (2,206)
212 (100)	5,800 (40)	233,600 (1,610)
270 (132)	4,000 (27.6)	211,200 (1,456)

Medical-device packaging. Polycarbonate meets many requirements of medical-device packaging (see Healthcare packaging). It is clear, tough, and can be sterilized by commercial sterilization techniques: ethylene oxide (ETO), radiation, and autoclave sterilization (see Radiation, effects of). The development of coextrusion technology has afforded opportunities in all sterilization systems. In thin films, PC can be coextruded with polyolefin heat-seal layers (see Coextrusions for flexible packaging; Coextrusions for semirigid packaging) to produce a cost-effective alternative to laminations based on oriented films (see Films, plastic). Because it is amorphous, heat sealing does not shrink or embrittle the film (see Sealing, heat). This virtually eliminates puckering, which can lead to hairline cracks and shattering upon opening. A soft blister package with good puncture resistance can be produced by thermoforming heavier-gauge film. For increased stability in the autoclave, a polycarbonate copolymer, poly(phthalate carbonate), can be incorporated into the structure. Properties of monolayer PC films are listed in Table 3.

Food packaging. Coextrusions of polycarbonate are being evaluated for use in several segments of the food-packaging market. The snack-food industry, and other users of flexible-packaging materials, are discovering ways to use PC in coextrusions to replace laminated films in a portion of a structure or to replace the entire structure (see Multilayer flexible

Table 3. Polycarbonate Film Properties, 1 mil (25.4 μm)

Property	ASTM Test		Value
Specific gravity	D792		1.20
Yield, in.^{2}/(lb·mil) (m^2/(kg·mm)			23,100 (1,294)
Haze, %	D1003		0.5
Optical clarity	D1746		86–88
Tensile strength, psi (MPa)		MD	10,735 (74)
		XD	10,009 (69)
Elongation, %	D882	MD	91
		XD	92
Secant modulus, psi (MPa)		MD	185,000 (1,275)
		XD	196,000 (1,351)
Tear strength, gf/mil (N/mm)			
Initial[a]	D1004		454 (175)
Propagating[b]	D1922		16 (6.32)
Tensile impact, S type, ft·lbf/in.^{2} (J/cm^2)	D1822		225–300 (47.3–63.0)
Bursting strength, 1 mil, psi (kPa)	D774		27.4 (189)
Water absorption at 24 h and 73°F (23°C), %	D570		0.15
Folding endurance[c], cycles	D2176		11,000
Heat-seal temperature at 40 psi (276 kPa) for 3 s, °F (°C)			400–420 (204–216)
Oxygen permeability at 77°F (23°C) and 0% rh, cm^3·mil/(100 in.^{2}·day·atm) [cm^3·μm/(m^2·day·kPa)]	D1434		240 (933)
Water-vapor transmission rate at 100°F (38°C) and 90% rh, g·mil/(100 in.^{2} day) [g·mm/m^2·day)]	E96-66		6.5 (2.6)
Coefficient of friction			
Static	D1894		0.570
Kinetic			0.542

[a] Graves.
[b] Elmendorf.
[c] Double folds.

packaging). It is polycarbonate's toughness without orientation that makes it a good candidate for coextrusion.

For frozen foods, the resin's low-temperature impact strength adds durability to dual- "ovenable" trays. In coextrusions, it can add toughness to crystallized polyester or polyetherimide trays. In cases where the wall thickness is determined by impact-strength requirements, gauge reductions up to 33% are possible.

In high barrier multilayer containers, polycarbonate offers the rare combination of dimensional stability at retort and hot-fill temperatures, along with crystal clarity (see Retortable flexible and semirigid packages; Multilayer plastic bottles). To produce a 1% distortion in PC requires 3000 psi (20.7 MPa) stress. A similar distortion in polypropylene occurs at less than 500 psi (3.4 MPa) stress. Polycarbonate is about three times as expensive, but dimensional stability can be translated into value by making lighter containers with thinner walls by faster and more reliable closing a hot-fill temperatures, higher retort temperatures with less critical overpressure control, and enhanced container-design flexibility.

An unexpected benefit from the use of PC in retort applications is an increase in the effectiveness of the barrier material EVOH (see Ethylene–vinyl alcohol). In a retort, moisture is driven into all layers of a plastic container. When the outside skin layer is polypropylene, the polypropylene prevents moisture from entering the structure, but is also prevents moisture trapped in the EVOH layer from escaping. Because polycarbonate has poorer moisture-barrier properties than polypropylene, it allows much more rapid drying of the EVOH layer and longer shelf life. As coextruded plastic containers become more readily available and begin commerical penetration, polycarbonate will become a more important material in food packaging.

BIBLIOGRAPHY

"Polycarbonate" in *The Wiley Encyclopedia of Packaging Technology*, 1st ed., by J. M. Mihalich and L. E. Baccaro, General Electric Plastics, pp. 510–511.

POLYESTER FILM. See FILM, ORIENTED POLYESTER.

POLYESTERS, THERMOPLASTIC

Poly(ethylene terephthalate) (PET) was first developed by a British company, Calico Printers, in 1941 for use in synthetic fibers. The patent rights were then acquired by DuPont and Imperial Chemical Industries (ICI), which in turn sold regional rights to many other companies. Polyester fibers have since made a considerable impact on the textile industry. The second principal application of PET was film. In 1966, PET became available for the manufacture of injection-molded and extruded parts.

The amazing growth of PET in beverage packaging began in the early 1970s with the technical development of biaxially oriented PET bottles (see Blow molding) and with the introduction of the first 2-L PET beverage bottle in 1976. Since then, the U.S. PET beverage-container market has grown from 200 million (10^6) units in 1977 to 5 billion (10^9) units in 1985, and continues to mount into higher billions in the 1990s. Table 1 shows world PET consumption by end use in metric tons (1).

PET has the following formula:

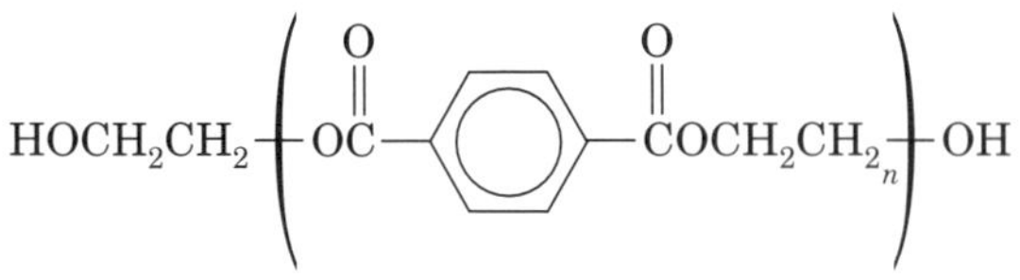

where n = 100–200.

Manufacture of PET

Raw materials. The raw materials for PET (see Fig. 1) are derived from crude oil. *Para*-xylene, one of the two starting materials, is part of the naphtha feedstock which used to be

Table 1. World PET Consumption by End Use[a] 1992, 10^3 t

Country	Carbonated Soft Drink Containers	Hot-Fill Containers	Returnable Containers	Mineral Water Containers	Edible-Oil Containers	Other[b]
Europe	244		21	111	22	92
United States	413	45		13	18	213
Canada	25			4		3
South America	14	1	17	4	5	3
Japan	35	61		4	19	29
Korea	28	10		2	5	11
Taiwan	9	1		2	3	2
China[c]	40			13	2	1
Middle East	20			10	3	9
Africa	10		2	1	1	2
rest of world	42	1	10	19	21	12
Total	*892*	*119*	*50*	*183*	*99*	*377*

[a] Ref. 2.
[b] Includes other drink containers, nonfood packaging, and all other uses.
[c] Includes Hong Kong.

fully available for chemicals because it was a byproduct of limited value to the oil refiner. Now it has become the source of additives that replace lead in unleaded gasoline in many countries. Of the mixed xylenes, as they come from the reformer, only *p*-xylene is suitable for building straight polymer chains, and the straight configuration is necessary to give the polymer its fiber- and film- (or bottle-wall-) forming characteristics and ultimately its high tensile strength. The other raw material for polyester is ethylene, which is contained in the crude oil's gas fraction. It is converted to ethylene glycol by oxidation and hydrolysis. All of these raw materials are now derived from oil, but it is technically feasible to produce them from coal-tar distillation.

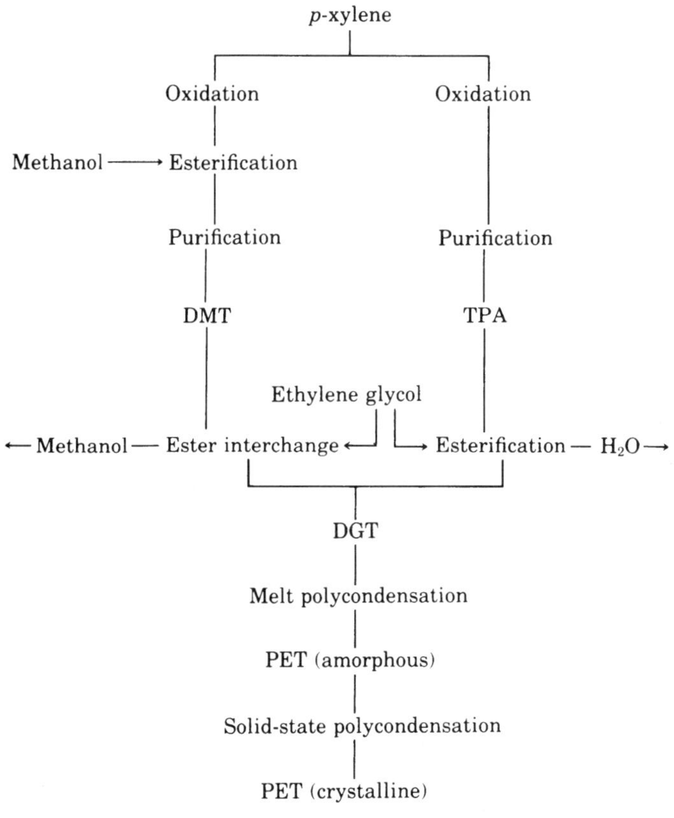

Figure 1. .Manufacture of PET.

Intermediate products. Two different routes are used to manufacture PET: one by way of dimethyl terephthalate (DMT), and the other by way of terephthalic acid (TPA). Both are dibasic acids. Figure 1 shows both routes. The plants for intermediate and end products are shown together, but they are normally physically separated. The process of making DMT is relatively simple: one end of *p*-xylene is first oxidized with air and esterified with methanol to yield a half-ester that is subsequently converted at the other end. The resulting DMT is purified by distillation and repeated crystallization to remove isomers and other impurities. The TPA route is similar. TPA can be produced by oxidation of *p*-xylene in solution and purified by solvent extraction.

Melt polycondensation. Both batch- and continuous-polycondensation processes are used. The continuous process inherently allows better product uniformity; the batch process is preferred for small quantities of specialty resins. On the DMT route, DMT and ethylene glycol (EG) are continuously metered into the ester interchanger, where the methyl end groups of the DMT are replaced by ethyl end groups to form diethylene glycol terephthalate (DGT), the monomer of PET. In this step, EG is consumed while methanol is evaporated and collected to be returned to the DMT plant. On the other route, TPA is esterified with EG to DGT, and water is removed as a by-product. Subsequent polycondensation is the same for both routes. In EG takeoff, excess EG is removed and sent to the distillation plant for recovery. Polycondensation takes place in vacuum reactors designed to evaporate EG, the condensation by-product, thereby shifting the equilibrium toward long polymer chains. The final step includes extrusion of the melt as strands or ribbon, quenching in water, and cutting to the desired chip size.

Solid-state polycondensation. Melt polycondensation produces amorphous PET as used in most fiber and film applications. Unfortunately, this product is not suitable for the injection molding of food containers because the inherently high acetaldehyde level would affect the taste of some foods;

Table 2. PET Polymer Resin Options for Different Manufacturing Processes and Applications

Manufacturing Process or Application (PET Resin Type)	Standard Homopolymer	Standard Bottle Resin	Slow-Crystallizing Copolymer	Fast-Crystallizing Homopolymer
Injection blow-molded bottles (nonoriented)		X		
Injection-stretch blow-molded bottles (biaxially oriented)		X		
Extrusion blow-molded bottles			X	
Sheet extrusion[a]	X		X	X
Film casting[a] (nonoriented)	X		X	X
Biaxially oriented film	X			
Heat-set film				X
Crystallizable PET trays				X
PET coating for paperboard	X			

[a] Resin option depends in desired application.

there are other impurities that might promote degradation during the injection-molding process; and amorphous resin tends to fuse and form lumps in the drying hopper. The polymer must be upgraded by the solid-state polycondensation process. The chips are crystallized to avoid later sticking and then dried to reduce hydrolysis at high processing temperatures. Solid-state polycondensation takes place in the reactor, where the chips are subjected to high temperatures under vacuum (batch process) or in a nitrogen or dry-air stream (continuous process). The product's intrinsic viscosity (IV) is normally between 0.70 and 1.0. (Intrinsic viscosity is a method for the characterization of the average length of the molecule chains in PET.) High-viscosity resin is relatively expensive because its production is lengthy. High-viscosity PET, ie, having longer molecule chains, offers better mechanical properties than the average-viscosity resin. These properties compensate for certain deficiencies of molded articles, eg, excessive volume expansion of beverage bottles can be limited by higher-viscosity resins that creep less under load. The final step is cooling, since the resin should not be exposed to moist air while it is hot.

Homopolymers and Copolymers

PET is a homopolymer made from one part dibasic acid, ie, TPA or DMT, and one part EG. A copolyester (copolymer) is made from more than one dibasic acid and/or glycol. Copolymers remove processing limitations and provide increased physical properties at elevated temperatures. In addition to DMT or TPA, isophthalic acid (IPA) can be used as a comonomer to reduce the rate and degree of crystallization to an extent that depends on its dosage. This broadens the processing parameters of food-container manufacturing machines.

Glycols offer several opportunities for modification. During polycondensation, EG reacts with itself to some extent to form diethylene glycol (DEG). Higher amounts of DEG affect many polymer properties. There are other glycols available as partial substitutes for EG, eg, neopentyl glycol, cyclohexane dimethanol. All these modifications lead to desired polymer property changes, ie, reduction of the crystallization rate, melting point, etc. Cyclohexane dimethanol can react with a mixture of terephthalic and isophthalic acids in order to increase the melt strength of the polymer for extrusion processes (3).

On the other hand, some injection-molding (see Injection molding) and thermoforming (see Thermoforming) applications require accelerated crystallization rates in order to set up crystallization in the article, which prevents physical deformation at elevated temperatures. This objective can be achieved by nucleation, which involves the addition of other ingredients to the polymer. Inert, nonsoluble substances (eg, mica, talc), organic substances (eg, aromatic alcohols), and certain polymers (eg, PP, PE) can be used as nucleation ingredients to increase crystallization rates.

The use of PET and optional added substances for food-packaging applications is governed by FDA Regulation No. 177-1630 of March 16, 1977. Homopolymers and copolymers and additives must conform to this regulation.

Packaging Applications

Homopolymers. By strict definition, most PET resins are modified homopolymers. These homopolymers are used to manufacture containers, ie, bottles, by injection blow molding or injection-stretch blow molding (see Blow molding) (4). About 70–80% of the resin consumption is used for soft-drink bottles (see Carbonated-beverage packaging). PET bottles are also used for liquor, wine, food, toiletries, and pharmaceuticals, and for beer in some countries. Homopolymers cannot be processed by extrusion blow molding because of insufficient melt strength.

Biaxially oriented PET film (see Film, oriented polyester) is usually manufactured by polycondensation and subsequent continuous casting of the film plus direct biaxial orientation, ie, molecule chains of the resin become biaxially oriented. Nonoriented PET film and sheet are manufactured by melting PET resin in an extruder and casting the melt through a flat die with subsequent calendering.

A fast-growing application for PET is "ovenable" trays for frozen food and prepared meals. These trays are thermoformed from cast PET film and crystallized. Crystallization heat-sets the article to prevent deformation during cooking and serving. The main advantages of PET for this application include suitability for both conventional and microwave ovens, light weight, and superior aesthetics (as compared with foil trays).

Copolymers. Commercially available copolymers offer improved melt strength for the vertically positioned extrusion blow-molding process. They are used to some extent for bottles, and to a greater extent as extruded sheet for blister-pack applications. Since the FDA regulation permits certain copolymers for food-contact use, future applications may include packages for noncarbonated drinks, cooking oil, and vitamin preparations.

Polymer options, manufacturing processes, and applications are summarized in Table 2.

The manufacturing process of PBT is very similar to that of PET. Instead of EG, 1,4-butandiol [HO $\leftarrow CH_2 \rightarrow_4$ OH] is used to react with either TPA or DMT. PBT is rarely used in packaging applications.

BIBLIOGRAPHY

1. A. J. East, M. Golden, and S. Makhija, "Polyesters, Thermoplastic, in J. I. Kroschwitz, ed., *Kirk-Othmer Encyclopedia of Chemical Technology,* 4th ed., Vol. 19, Wiley, New York, 1996, pp. 609–653.
2. A. K. Mitchell, *Mod. Plast.* **71**(12), B48–B50 (1994).
3. P. Aspy and co-workers, "Controlled crystallization—PET Copolyester," Paper presented at the Seventh Annual International Conference on Oriented Plastic Containers, Atlanta, March 1983, Ryder Associates, Whippany, NJ, 1983.
4. U.S. Pat. 3,733,309 (May 15, 1973), N. Wyeth (to E. I. du Pont de Nemours & Co., Inc.).

General References

"Polyesters, Thermoplastic" in *The Wiley Encyclopedia of Packaging Technology,* 1st ed., by E. H. Neumann, Hoechst, AG and Ed Sisson, Shell Chemical Company, pp. 512–514.

General Information about Hoechst Thermoplastic Polyester Resin, Technical Bulletin 2, American Hoechst Corp., Somerville, NJ, 1981.

K. D. Asmus, "Polyalkylenterephthalate," *Kunststoffe Handbuch,* Carl Hauser, Verlag, Germany (FRG), 1972.

Chemie-Kompendium für das Selbststudium, Hoechst AG, Frankfurt, Germany (FRG), 1972.

POLYETHYLENE, HIGH-DENSITY

High-density polyethylene is a semicrystalline polymer, conveniently defined by ASTM as having a density of 0.941 or greater, with a typical upper limit of 0.965. For commercial purposes, HDPE is described as a homopolymer when the density is ≥0.960, and a copolymer when the density is below this figure. The polymer is available in a wide range of molecular weights as determined by either MI or HLMI (melt index or high-load melt index).

HDPE is one of the largest-volume plastics used in packaging for the simple reason that it can be successfully employed in numerous transformation processes. Table 1 lists some of the major processes used and the products that can be obtained from them.

Table 1. Examples of Packaging Applications

Conversion Process	Examples of Packages
Blow molding	0.1–5.0-gal juice, water, or milk bottles
	Bottles containing pet food to bleach industrial 55-gal drums
	Bottles for cosmetics and pharmaceuticals
Injection blow molding	Bottles for cosmetics, personal care, pharmaceuticals
Injection molding	Caps and closures
	Pails and buckets
	Thin-walled dairy cups/tubs
	Crates
	Foamed packaging
Blown or cast film	Multiwall bag liners
	Grocery, merchandise or produce bags
	Breakfast-cereal wrappers
	Snack-food wrappers
Thermoforming	Prepackaged-food containers
	Dairy cups and containers
	Tubs for personal care, hygiene items
	Pallets
Sheet extrusion	Foamed sheet for packaging
Rotational molding	Water and chemical tanks and shipping containers

Manufacture

HDPE is commonly sold in pellet form and is manufactured in a two-stage process. Polymerization takes place in a reactor on a continuous basis using either of two types of catalyst: Ziegler or Phillips. The choice of hardware today is limited to the slurry and gas-phase processes, the solution process having largely been bypassed on both cost and environmental counts, except for specialty grades. The slurry (Phillips) process was developed by the Phillips Petroleum Company and was licensed worldwide in the 1960s (1,2). Figure 1 shows a simplified view of the process. Purified ethylene, isobutane (the slurry carrier), activated catalyst, and any comonomer are circulated within loop-style reactors under relatively narrow ranges of pressure and temperature. At periodic intervals, portions of slurry are withdrawn and separated into unreacted materials, carrier, and HDPE. The raw polymer powder is dried and then stored in tanks until compounded in an extruder, together with the requisite additives (stage 2 of the process).

The newer gas-phase process was originally developed by Union Carbide in the 1970s and was named Unipol; BP Chemicals licensed a competing process about 10 years later. The gas-phase hardware dispenses with liquid carrier, and as the name suggests, permits direct polymerization of ethylene using living polymer as a fluid-bed medium, and ethylene as the fluidizing gas. From a distance, a GP reactor looks very much like a water tower because of the elongated bulb sitting on top of the cylindrical tower housing the fluid bed. The bulb is the portion of the reactor where fines deentrainment takes place. New world-scale HDPE plants will be typically rated at 400–500 MM lb/yr and be composed of one or more gas-phase units closed-coupled to twin-screw compounding trains.

Whatever the hardware, the core of the polyolefin manufacturing operation is the catalyst. Ziegler formulations are traditionally based on combinations of aluminum alkyls with titanium compounds, often with proprietary modifiers to improve productivity, comonomer incorporation, or other parameters. The catalyst may be chemically or physically attached to a variety of bases (eg, silica or magnesium chloride) to improve productivity and flowability through controlled particle-size distribution (many catalysts are self-supporting). Ziegler catalysts tend to be unrestricted in molecular-weight (MW) terms, although they produce relatively narrow molecular-weight distributions (MWD). Phillips catalysts, on the other hand, tend to be restricted to fractional melt-index (MI)

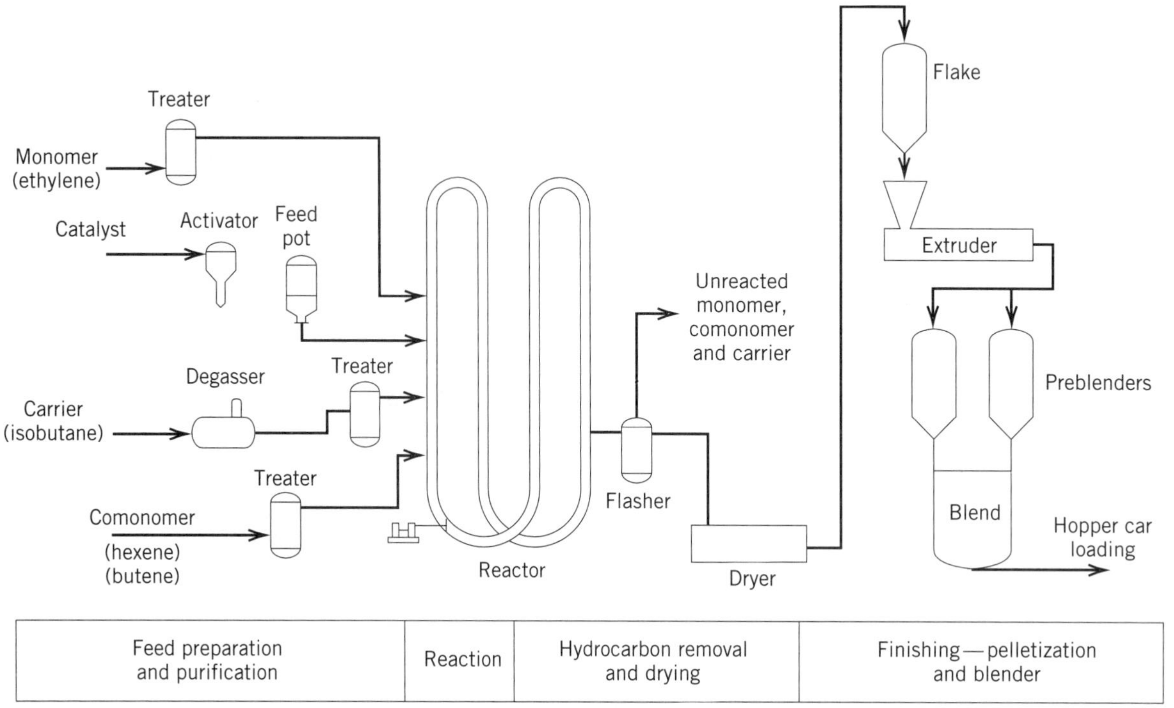

Figure 1. Simplified flow diagram for the Phillips (slurry) manufacturing process.

grades, producing broader MWD HDPE with better environmental stress-crack resistance (ESCR) properties (3,4).

In the realm of catalysts, the most recent development has been the commercialization of the metallocene type. Although currently expensive to make and barely out of the laboratory, many people think that metallocene grades will have a lot to offer in packaging because of the improvements obtainable from single-site catalysis features (5,6).

Molecular Structure of HDPE

There are four principal features of HDPE that affect its packaging and processing properties: density, degree and type of branching, molecular weight, and molecular-weight distribution. As comonomer is introduced into the reactor to control density (typically 1-butene or 1-hexene), short-chain branching (SCB) is incorporated into the backbone of the polymer (catalysts can also create some SCB in the absence of comonomer). The nature and frequency of the SCB induce major disruptions in the HDPE crystallites, with the result that the density and degree of crystallinity are decreased. As little as one ethyl branch per 1000 carbon atoms can produce a density change of 0.01. As implied, the density of HDPE enjoys a rough linear relationship with regard to crystallinity. Commercial polymers exhibit degrees of crystallinity of 70–80% at 0.960, depending on the MW (MI) and type of SCB. A drop of 10% in crystallinity as one progresses from 0.960 to 0.940 is typical. There is increasing evidence that the "blockiness" of the branches is as important as the overall frequency (7). New information has been obtained using *temperature-rising-elution fractionation* (TREF), a relatively new technique for studying SCB that is independent of molecular weight and that fractionates polymers according to crystallinity, composition, and tacticity differences.

As a consequence of SCB, the morphology of HDPE in the solid state consists of three very different regions. The amorphous region contains polyethylene chains that are randomly oriented and acrystalline in nature, while the crystalline regions contain HDPE chains that are tightly packed in folds to yield differing types of geometries. Joining the two regions together is the interface. The crystalline regions imbue PE with its rigidity, tensile strength, and lack of permeability toward small molecules, while the extent of the amorphous areas affect impact resistance and ESCR.

Table 2 lists some of the major properties associated with changes in density, MW, and MWD. An increase in density causes a rise in tensile strength at yield, stiffness, and chemical resistance, while a decrease improves both ESCR and impact resistance. This results in compromises for many applications. For example, a 1-gal milk jug is manufactured from a homopolymer to extract the highest levels of top-load and lightweighting capability, while a bottle containing industrial chemicals (bleach, detergent, etc) is blown from a lower-density resin to take advantage of the higher ESCR. Density is commonly measured by the gradient column method or, more recently, by an ASTM-approved acoustics technique.

The term *molecular weight* colloquially refers to the length of a polyethylene chain. Since all HDPE polymers possess a variety of chain lengths, it becomes necessary to mathematically define some MW terms in order to obtain meaningful information. Thus M_n, the number-average molecular weight, is $\Sigma N_i M_i / \Sigma N_i$, where N = number of polyethylene molecules and M = combined mass of the polyethylene molecules (eg, 100 molecules at 10,000 + 100 molecules at 20,000 = 15,000). The weight-average molecular weight M_w, is $\Sigma N_i M_i^2 / \Sigma N_i M_i$ (eg, 100 molecules at 10,000 + 100 molecules at 20,000 = 16,667). For packaging applications, M_n ranges from about 7000 to 19,000 and M_w between 80,000 and 250,000. When

Table 2. Influence of Density, Molecular Weight, and Molecular-Weight Distribution on Selected HDPE Properties

Property	Density ↑	MW ↑	MWD Broadens
Tensile strength at yield	↑	↑	↑ (sl)
Tensile elongation at break	↓ (sl)[a]	↑	
Flexural modulus	↑	↓ sl	
Stiffness	↑	↑ (sl)	↓ (sl)
Impact strength	↓	↑	↓
Melt strength		↑	↑
Melt viscosity		↑	↓
Die-swell ratio (DSR)		↑	↑
Heat-deflection temperature	↑	↑	↓ (sl)
Permeability	↓	↓	
ESCR	↓	↑	↑
Chemical resistance	↑[b]	↑	
Gloss	↑[c]	↓	↓
Haze	↑	↑	
Shrinkage from melt[d]	↑ (sl)	↑ (sl)	↓ (sl)
Flowability and processability		↓	↑

[a] sl = slightly.
[b] Attack by strong oxidizing agents excepted (eg, hydrogen peroxide, concentrated nitric acid, concentrated bleach).
[c] Highly dependent on polymer type.
[d] Subject to cooling rate, part thickness, etc.

plastics people talk about MW, in reality they are referring to M_w.

Molecular weights are commonly estimated by rheological techniques (capillary/dynamic oscillatory rheometry), solution (intrinsic) viscosity, or the popular size-exclusion method, gel-permeation chromatography (GPC). However, each of these methods does give rise to considerable variation, especially for high-molecular-weight polymers, and any numbers quoted should be used only as a guide.

The most frequently used way of determining a measure of MW is the melt index (MI). For HDPE, this is condition E of ASTM D1238, which uses a well-delineated capillary rheometer, a temperature of 190°C, and a 2.16-kg load to push the extrudate out of the barrel.

The MI is simply the weight of polymer collected during a 10-min period under these conditions. For higher-MW polymers, the load is changed to 21.6 kg and the result referred to as *high-load melt index* (HLMI). The ratio of HLMI/MI represents a crude measure of the molecular-weight distribution and is often the simplest technique for quality-control purposes. Polydispersity, formally M_w/M_n, is a more rigorous gauge of MWD and is usually obtained by GPC. Because this method gives a direct visual representation of the MWD, it can also supply other information that is absent from other techniques. Figure 2 shows typical MWD curves for a variety of polymer types. Note that such features as modality, symmetry, and MWD breadth are easily determined.

From Table 2 it can be seen that both MW and MWD affect virtually all common properties, especially melt parameters, processing, and impact strength.

One other type of branching that we have not yet discussed is long-chain (LCB). These longer side chains to the PE backbone are characteristically 6–12 carbons long. They impact both melt elasticity (important for foamability, large parison stability, etc) and processability (how easily a material can be transformed). In other end uses, LCB sites can be used to initiate crosslinking with a peroxide or other suitable radical generator. LCB can be imaged through elongational or short-die, capillary-die rheometry techniques.

Chemical Properties

HDPE resists most chemicals, including dilute acids and bases, oxidizing agents, hydrocarbons, and some aldehydes/ketones. The major exceptions are chlorohydrocarbons and aromatics, which tend to swell and soften the polymer. Even some of these chemicals can be packaged in HDPE if the interior surface is fluorinated or sulfonated, or if a coextrusion with a barrier layer is incorporated (eg, nylon).

The ability of an agent to degrade HDPE over a period of time is dependent on the ESCR of the polymer. The failure mechanism is the extension of a crack at a focal point, or local area of concentrated stress. On a microscopic level, one might consider this to be a plasticization of the strands of PE that link the crystals of HDPE together in the interfacial zone—in other words, a dissolution of the "glue" that binds the crystalline regions. Both density and MW have a profound effect on this parameter, so the trick is to make a grade of HDPE that has the highest density and lowest melt viscosity compatible with the application.

ESCR is usually measured by the *bent-strip test* (ASTM D1693), which involves taking a strip of HDPE, making a notch in it, and bending it in a jig, so that the notch provides a narrow area of constant strain. The strip is immersed in either 100% detergent (condition A), or 10% solution in water (condition B). The latter is considered a much more aggressive condition. For bottles, it is also possible to employ a more sophisticated, real-world test by applying a low constant pressure to a bottle that contains a small amount of water or other liquid [constant-pressure bottle test (CPBT)]. Failure occurs when either the bent strip snaps or, in the case of the CPBT test, the bottle bursts and leaks liquid. In either case, ESCR is reported in average number of hours to failure based on a statistical approach, or first to fail.

The permeability of HDPE toward small molecules is most dependent on the density, for it is the crystalline regions that present the highest impedance. Longer, more entangled,

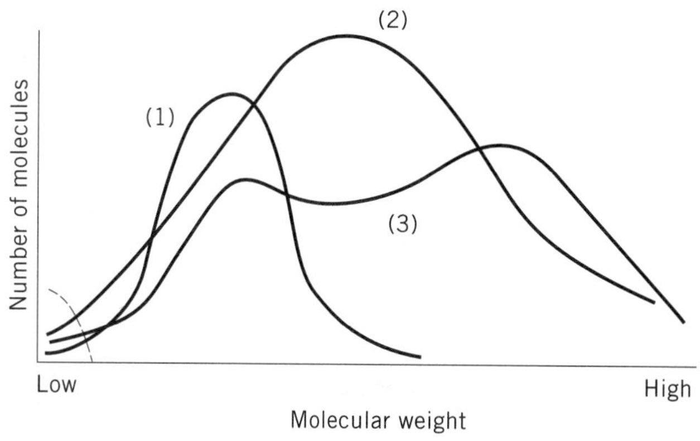

Figure 2. GPC curves for various types of HDPE polymers: curve 1—low-MW, symmetrical, narrow-MWD injection molding resin; curve 2—medium-MW unsymmetrical monomodal blow-molding resin; curve 3—high-MW bimodal film resin. The dashed line in the lower-left corner represents the oligomer region.

higher-MW chains also cause small molecules such as oxygen or water to take a more tortuous path. The transmission rates of gases such as moisture are dependent on the thickness of PE that they have to traverse, as well as the temperature and concentration of the gas. As the thickness of the HDPE is increased, the ability of the layer to prevent small molecules from passing through it decreases exponentially. As barrier polymers go, HDPE is only a fair performer. To get the smallest transmission rates through a film or bottle wall, it may be necessary to make a multilayer construct using coextrusion.

Preparation of the surface of HDPE for decoration is most commonly achieved by treating it with either a gas flame or corona discharge. In the case of either method, time should be taken to optimize the process. The polyethylene on the surface should become sufficiently oxidized so that the surface tension rises from 29 to at least 36 dyn/cm. Simple tests based on water dissipation or dye retention can often be used as a quick check. One should be aware that many additives such as antistats, lubricants, and slip agents can migrate quickly through the plastic and cause print delamination over a period of days. The same result is seen when the oligomer content is high.

Physical Properties

Impact resistance of a bottle or container is one of the most important parameters in packaging. This property is improved by decreases in density and MWD, and an increase in MW. Often the needs for impact resistance are in direct conflict with load-deformation requirements. The degree of impact resistance can be assessed by a variety of tests performed on compression-molded plaques, such as Izod or Charpy, in both notched and unnotched conditions. Dropping heavy weights on plaques in ambient or cold temperatures using the Gardner apparatus is also practised widely (dart impact for films). Many manufacturing companies also drop loaded containers in various attitudes to verify performance and ensure that the design or molding process itself has not introduced a weakness.

Load deformation comes in two flavors: (*1*) instant, or top load, in which the stress is exerted on a bottle or other container during a momentary situation (eg, during filling or capping); and (*2*) constant loading, in which bottles are stacked on top of one another during a long storage period. If the material being contained is nonchemically aggressive, a homopolymer of medium MW should prove sufficient for the application. A more demanding ESCR service will require an HDPE of higher MW, or lower MI to compensate for the reduced stiffness of the lower-density material. The conventional gauge of stiffness is flexural modulus and typical numbers for HDPE range from 250,000 for a homopolymer to half that value for a 0.940-density material.

Recent Issues

Regulatory demands continue to increase. HDPE easily complies with the FDA rules for food-contact use, although many additives compounded into PE do not, and as a result, end users should contact the manufacturer to determine limitations. Many resins are also listed in the Drug Master File, a prerequisite for packaging drugs.

Although many attempts have been made to proscribe heavy metals in HDPE, at present there is no legislation in force limiting the presence of these materials. Nevertheless, efforts continue on a voluntary basis to replace such metals as cadmium, often found in pigments, with organic derivatives. Extremely low levels (eg, hexavalent chromium at or below 1 ppm) will most likely be dealt with by repeal of the Delaney Clause.

Most states have mandatory minimum recycled content programs in place for packaging and HDPE in particular is well suited to fulfulling this need, provided certain practices are followed. A survey in 1995 revealed that over a third of all milk, juice, and water bottles are currently being recycled (8). These homopolymers are typically incorporated into the middle layer of coextrusion applications or are mixed and converted with copolymers into municipal, recreational, or construction applications.

BIBLIOGRAPHY

1. Eur. Pat. 250,169 (Dec. 23, 1987), C. Raufast (to BP Chemicals Ltd.).
2. U.S. Pat. 2,825,721 (March 4, 1958), J. P. Hogan and R. L. Banks (to Phillips Petroleum Co.).
3. V. Chandrasekhar, P. R. Srinivasan, and S. Sivaram, "Recent Developments in Ziegler–Natta Catalysts for Olefin Polymerization and Their Processes," *Ind. J. Technol.* **26,** 53–82 (1988).
4. M. B. Welch and H. L. Hsieh, "Olefin Polymerization Catalyst Technology" in C. Vasile and R. B. Seymour, eds., *Handbook of Polyolefins,* Marcel Dekker, New York, 1993, pp. 21–38.
5. *Proceedings of MetCon '94,* May 25–27, 1994, Houston, TX, Catalyst Consultants Inc., Spring House, PA.
6. *Modern Plast.* **72,** 20–21 (Aug. 1995).
7. L. T. Wardhaugh and M. C. Williams, "Blockiness of Olefin Copolymers and Possible Microphase Separation in the Melt," *Polym. Eng. Sci.* **35,** 18–27 (1995).
8. *Progress Report,* American Plastics Council/Society of the Plastics Industry Inc., Sept. 1995, p. 1.

STEPHEN J. CARTER
Solvay Polymers, Inc.
Deer Park, Texas

POLYETHYLENE, LINEAR AND VERY LOW-DENSITY (LLDPE AND VLDPE)

Polyethylene is a thermoplastic material well suited for packaging applications. It is a large-volume commodity resin. U.S. polyethylene sales for 1994 were estimated to be in excess of 25 billion lb (1). Polyethylene resins are available in a variety of molecular weights and densities tailored to specific end-use markets. Polyethylene can be generally classified into the product types listed in Table 1 on the basis of its density.

History

In the early 1950s, 20 years after ICI pioneered LDPE in the United Kingdom using a high-pressure process, Phillips Petroleum Company commercialized catalysts containing chromium oxide supported on silica. These catalysts were used to produce HDPE, which became the first commercial products of catalytic ethylene polymerization. Almost simultaneously,

Table 1. Commercial Classification of Polyethylene Resins

Product	Density, g/mL
High-density polyethylene (HDPE)	0.94–0.97
Medium-density polyethylene (MDPE)	0.926–0.939
Linear low-density polyethylene (LLDPE)	0.915–0.926
Very low-density polyethylene (VLDPE)	0.89–0.915
Low-density polyethylene (LDPE, produced via free-radical polymerization)	0.915–0.940

K. Ziegler and co-workers (2) discovered a new group of transition-metal catalysts that polymerize ethylene and other α-olefins under mild conditions. The first commercial introduction of LLDPE was made in a solution process by DuPont Canada in 1960 (3) using these catalysts. In 1976, W. Kaminsky and H. Sinn discovered a new family of catalysts for ethylene polymerization using metallocene complexes (4). These catalysts afford the synthesis of ethylene copolymers with a high degree of compositional uniformity. Currently, there are several commercial processes that produce LLDPE and VLDPE as described in the next section (5).

Process

LLDPE and VLDPE resins are commercially produced by several processes, including gas-phase, solution, and slurry polymerization. The first gas-phase process using the fluid-bed process was developed by Union Carbide in 1979 and has been licensed worldwide. Another gas-phase process used globally in one developed by British Petroleum Company. Ethylene copolymers with butene, hexene, and 4-methyl-1-pentene are produced in the fluidized-bed reactors, a highly versatile, economical process that can accommodate various types of catalysts. Although Ziegler–Natta catalysts are most widely used, metallocene-based catalysts can also be used in this gas-phase process (6).

The first solution processes were introduced in the late 1950s. Currently, there are two types of solution processes commonly used for the production of ethylene copolymers. The first process used heavy solvents (C_6–C_{10} hydrocarbons) to solubilize the ethylene and polyethylene at high temperatures and pressures. The process utilizes Ziegler–Natta or metallocene catalysts. The second type of solution polymerization process uses mixtures of supercritical ethylene and molten polyethylene as the medium for ethylene copolymerization reactions. In this case, retrofitted high-pressure reactors that were used previously for producing LDPE are currently being used to produce LLDPE and VLDPE.

The slurry polymerization process is the oldest catalytic polyethylene production technology. The main disadvantage of this technology, however, lies in the fact that the LLDPE resins produced by this process exhibit high swelling even when light solvents are used. The resultant stickiness in the polymer particles significantly limits the polymer density range and production rates that can be achieved via this process. Recent developments in this process have allowed the production of broad-MWD LLDPE using light solvents such as isobutane and isopentane.

Polymer Structure and Properties

Properties of LLDPE and VLDPE are usually specified in terms of density (ASTM method D792) and melt index (ASTM method D1238). Density is a measure of crystallinity, and melt index is related to the polymer molecular weight (see also Polymer properties.).

Density is the most important parameter governing resin properties. Polyethylene is essentially a composite material consisting of a rigid crystalline phase and an elastic amorphous phase. As crystallinity decreases with decreasing density, the product becomes softer and more pliable, clarity increases, and toughness also increases. Density is controlled by the concentration of short-chain branching (SCB), which is introduced in polyethylene resins via copolymerization with α-olefins. Comonomers widely used are 1-butene, 1-hexene, 4-methyl 1-pentene, and 1-octene. The side branches serve to disrupt the polyethylene crystals, channel polymer chains into the amorphous phase, and thus reduce overall density. The comonomer type determines the length of the side branch. The short side branches of a butene copolymer can be partially incorporated into the polymer crystal, whereas those from the higher α-olefins (HAOs) cannot. As a result, the HAO copolymer chain is likely to leave a given crystal and enter another one to form a "tie chain" that helps bind the crystals together. For this reason, HAO copolymers are generally stronger and tougher than butene copolymers.

Another important structural characteristic is the polymer chain length or molecular weight (MW) and molecular-weight distribution (MWD). It is a common practice to characterize MW by means of the melt index (MI), which bears an inverse relation with MW; thus, lower MI values correspond to higher MW. The melt-flow ratio (MFR) is a ratio of the polyethylene viscosity at two different shear rates, and is an indicator of the MWD. In general, long chains are entangled with one another and entanglements contribute to the resin's strength and toughness. However, the desire for higher strength needs to be balanced against processabilty, since MW and MWD have a profound influence on the flow behavior or rheological characteristics of the molten polymer. MW controls the overall viscosity level and MWD controls how viscosity depends on shear rate, ie, its shear-thinning response. Processability, related to flow at high shear rates, is therefore a function of both MW and MWD. Most LLDPE and VLDPEs have relatively narrow MWD and show low degree of shear-thinning (see Fig. 1).

When compared to LDPE (whose shear-thinning is further enhanced by the presence of long-chain branching) of a simi-

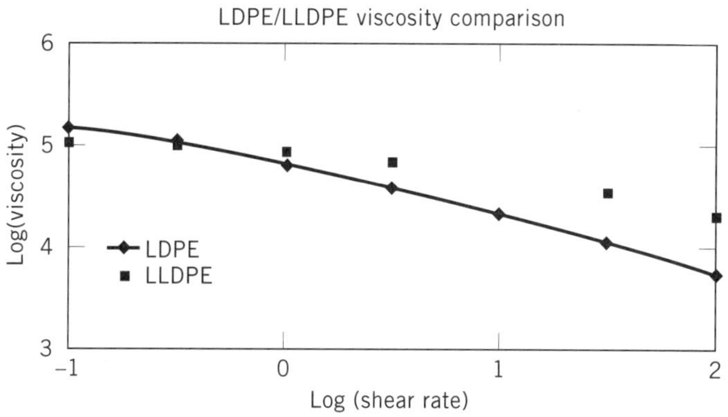

Figure 1. LDPE/LLDPE viscosity comparison.

lar melt index, LLDPE and VLDPE resins tend to build up high stresses during processing at relatively low shear rates. Other consequences of their narrow MWD are low-melt elasticity and "soft" stretching flow behavior. Low-melt elasticity implies rapid relaxation of polymer chains in the melt when subject to stress. For this reason, the tendency to induce molecular orientation during processing is low, and the product physical properties are not as sensitive to processing conditions. The "soft" stretching behavior implies a low extensional viscosity that enables easier drawdown of the molten resin bubble to thin gauges. By the same token, LLDPE and VLDPE resins have low melt strength, and can be sensitive to bubble instability during blown-film extrusion.

A third influential factor in the LLDPE/VLDPE properties and processing is the short-chain branching distribution (SCBD) of the polymers. In general, LLDPE resins are structurally heterogeneous. Not only is there a distribution of MW, but there is also a distribution of SCB. The SCB is not uniform across all polymer chains; some chains have many branches, while others may be essentially linear. This characteristic of heterogeneous SCBD generally contributes to thicker crystallites, higher melting point, higher stiffness, and lower optical clarity in LLDPE when compared to LDPE at the same density. The SCBD also affects the fraction of the resin that can be extracted by solvents such as hexane. This fraction, commonly referred to as the "FDA extractables," is highly branched and low in MW. Both SCBD and MWD are determined by the catalyst system, and to a lesser extent the reactor conditions. Polyethylene produced using metallocene catalysts have a very narrow MWD as well as a much more homogenous SCBD. This leads to very low extractable levels in the final product.

Since VLDPE resins have lower density compared to LLDPE resins, the short-chain branching levels are higher. Conventional VLDPE resins (catalyzed by Ziegler–Natta catalysts), have high extractables, high strength, and clarity. Metallocene-catalyzed VLDPE resins, in spite of the low density, still offer the additional advantage of having very low extractable levels. This is a major advantage in medical applications where high purity is essential. Also due to the homogenous SCBD, the melting point of metallocene-catalyzed resins is significantly lower compared to conventional LLDPE and VLDPE. This leads to improved sealability in the final product.

Applications

LLDPE competes with LDPE for many applications, and has made significant inroads into traditional LDPE markets. This is mainly due to the significantly better downgauging potential, which allows thinner films with enhanced physical properties. On the basis of 1994 usage (1), film applications consume about 3.7 billion lb of LLDPE accounting for about 60% of the total LLDPE consumed annually in the United States. Of the film produced, there is an approximately even split between packaging and nonpackaging applications. Other uses for LLDPE resins include injection and rotomolding, wire and cable, and sheet extrusion. Next to film, injection-molded products account for the largest usage of LLDPE. The 1994 LLDPE sales into this market was about 569 million lb.

Table 2. Typical LLDPE/VLDPE Film Applications

Agricultural rollstock	Ice bags
Bakery bags	Industrial liners and sheet
Diaper backsheet	Medical packaging
Garbage bags, consumer	Newspaper bags
Garbage bags, institutional	Produce bags-on-roll
Grocery sacks	Shopping (retail) bags
Household food bags	Shrink wrap
Heavy duty sacks	Stretch wrap

The improved strength and tensile properties of LLDPE are desirable for several film applications such as those shown in Table 2.

Typically butene LLDPE is used in low-end applications such as generic packaging or liners where premium strength is not required. Other applications such as heavy-duty sacks and consumer garbage bags require the use of HAO-LLDPE. Premium LLDPE, such as Mobil's "superstrength" hexene copolymers or Dow's octene copolymers, are used for applications requiring maximum toughness and tear strength. Metallocene-based LLDPE resins offer excellent clarity and sealability, which makes them especially desirable for applications such as poultry and frozen-food packaging.

VLDPE resins offer an interesting niche market because of their outstanding puncture resistance, low-temperature stability, flexibility, and good sealability. The sealing and flexibility characteristics are similar to ethylene–vinyl acetate (EVA) copolymers, while the physical properties are significantly superior. VLDPE resins are primarily used in coextrusion or in blends with LDPE, HDPE, or LLDPE to take advantage of their properties. Some individual applications include poultry and meat packaging and shrink-film and frozen food packaging. Metallocene-based VLDPE resins offer the additional advantage of having very high clarity, and a very low level of extractable material, which makes them desirable for medical markets. Medical packaging and biohazard bags are currently a small market segment, but are expected to grow rapidly (8).

Over half of the LLDPE used in injection-molding applications is used for housewares (9). LLDPE houseware products are stiffer and more resistant to impact and distortion at elevated temperatures compared to LDPE. The othe large market for LLDPE is in lids for glass and HDPE jars.

LLDPE resins are used in rotomolding applications to produce soft toys and sporting goods. Also, HAO-LLDPE is used to produce tanks and refuse containers. In these applications, the higher impact strength and increased stress-cracking resistance of LLDPE compared to LDPE and HDPE resins is considered desirable.

In sheet and profile extrusion, geomembrane liners are a rapidly growing market for both HAO-LLDPE and VLDPE. Drip/trickle irrigation pipes uses mainly butene LLDPE.

The wire–cable industry used butene and HAO-LLDPE for low- and medium-voltage cables, telephone exchange cables, etc. The use of HAO-LLDPE for fiber-optics is a growing market as new electronic equipment permits more cost-effective use in telephone equipment. Also, metallocene-catalyzed VLDPE resins are especially suited for this application because of their high flexibility, transparency, and strength.

Fabrication processes. As indicated in the previous section, LLDPE and VLDPE are used in a large variety of product types. Since film is the largest market segment, film extrusion is briefly summarized in the following paragraphs (10).

Film extrusion. Polyethylene is converted from pellets into film using one of two techniques: blown- or cast-film extrusion. Depending on the end-use property requirements, the resin is either blown into a tubular film or cast into a thin flat sheet. Compared to cast films, blown films typically have higher haze, more balanced orientation, and good impact strength, and can generally achieve gauge uniformity within 10%. On the other hand, cast film tends to have better clarity and is highly oriented in the machine direction. Excellent gauge uniformity (within 3%) can be achieved by casting the film, which allows film production at very thin gauges (sometimes less than 0.5 mil). The film is quenched much sooner on cast film lines, and significantly higher extrusion rates are achieved. Some basics of film extrusion are explained below.

Blown film. The process of converting pellets into blown tubular film consists of four basic steps: feed/recycle, extrusion, film blowing, and winding.

1. The feed section consists of a hopper in which virgin polymer is loaded. Usually, there is some mechanical means of blending in some recycled material. This could be either postconsumer recycle (PCR), which is mandated in some states and in some applications, or merely scrap film and edge trim that is fed through a grinder and combined into the feed stream with the virgin resin. The feed stream also may contain additives such as pigments and processing aids.

2. The function of the extruder is mainly to compress, melt, and convey the polymer to the discharge end. There are several different types of extruder screws. The original film equipment consisted of screws that were optimized for LDPE resins. When LLDPE was first introduced, it became evident that the LDPE screws would require some modification to process LLDPE, since the power efficiency was very low because of the rheological behavior of these mostly linear resins. The processing requirements of LLDPE were met by designing screws such as the barrier screw or decreasing-pitch screw that could handle a more viscous melt. By far the most commonly used screws for extruding LLDPE or LLDPE/LDPE blends are the barrier screws.

3. After the extruder, the polymer melt is forced into an annular blown-film die for extrusion into a tube. The thickness of the annular opening, the die gap, is generally between 25 and 100 mils. When processed through a narrow die gap, LLDPE resins generally require a processing aid to avoid melt fracture, a film deformity that occurs when a certain critical shear stress is exceeded. Although melt fracture does not cause film property loss (11) except at severe levels, it poses an appearance problem that can be a critical factor in several applications. After the polymer exits the die, it is drawn down in the machine and transverse directions (MD and TD) to achieve the desired film thickness. Film cooling is also an important part of the process. The frost-line height (FLH), where the polymer melt crystallizes, plays an important role in determining final film properties (12).

4. The final step consists of the buddle being flattened by collapsing frames, drawn through nip rolls and over idler rolls to a winder to produce rolls of blown film.

Cast film. The cast-film process is similar in most aspects to the blown-film process outlined above, except in the film formation step. Here, the polymer melt exits the extruder and enters a "T" or coat-hanger die. The orientation of the die can vary depending on processing conditions. Typically, a vertically downward die is preferred. The molten polymer web is cast onto a polished roll that has an internal cooling mechanism to maintain a constant temperature. An air knife is generally used to press the molten polymer against the roll. Usually, two chill rolls are used to provide appropriate tension to the film. After this point, the film goes though nip rolls and into the winder to produce rolls of cast film.

Coextrusion. The reasons for using coextrusion are usually to achieve specific physical properties such as sealability, barrier, and strength. Also, there are economic benefits to using coextrusion including reduced costs (by using an inexpensive polymer in the noncritical layer) and reduced waste (by introducing a recycle layer). Coextruded film can be produced on both blown- and cast-film equipment, and involves multiple extruders feeding a single die. As many as nine-layer lines are now being used commercially, although two and three-layer lines remain the most common.

Physical Properties

Although LLDPE resins are generally considered as having superior physical properties compared to LDPE, the actual impact or tear strength is strongly dependent on the comonomer used, as well as on the melt index and density (13). Some key film properties that are critical for several film applications are explained below:

Dart Impact (ASTM D1709). This test measures the ability of a film to withstand the force of a falling dart, and is an indicator of film toughness. Dart impact is strongly affected by the molecular weight, density, and comonomer type.

Elmendorf Tear (ASTM D1922). The resistance to tear in either the machine direction (MD) or transverse direction (TD) is measured by this test. Typically, LLDPE films tend to have lower MD tear strength compared to TD. Since that becomes the limiting direction, only the MD tear values are often measured and reported, whereas the TD tear is not considered critical in most LLDPE film applications. Ideally, a balanced orientation, or MDT : TDT ratios close to 1.0 are desirable to maximize film performance.

Other Properties. Tensile and yield strength (ASTM D638) of the LLDPE resins are also considered critical end-use properties. Typically, the tensile strength of butene LLDPE is somewhat lower than the HAO-LLDPEs. Yield strength is primarily a function of resin-base density. Hexane and xylene extractables are also measured to ensure that the film meets FDA limits for food-packaging applications. Film blocking, the tendency of a bag to resist opening, is a function of resin-base density and MWD. Other properties such as puncture resistance, puncture propagation tear (PPT), sealability, haze, and

Table 3. Typical Properties of Commercially Available LLDPE Resins[a]

Comonomer	Butene	Standard Hexene	"Super" Hexene	Octane
Melt index, dg/min	1.2	0.9	0.9	1.0
Density, g/mL	0.918	0.917	0.917	0.920
MFR	24	28	25	30
Dart impact, g	85	180	480	250
Elmendorf tear MD, g	80	300	450	350

[a] 3.5 in. Gloucester, 2:1 BUR, 430°F, 100-mil die gap, 250 lb/h, 1.0-mil film.

gloss, are also considered important depending on the application.

Comonomer type plays a strong role in influencing the final film properties. In general, butene copolymers tend to have lower film strength, and are used in low-end applications. Standard hexene copolymers show significantly improved strength properties over butene, but are not considered as strong as octene copolymers. However, a new class of "superstrength" hexene copolymers actually exceed the strength properties of standard octene copolymers. Film strength properties of some commercially available resins are shown in Table 3.

VLDPE resins generally offer much better optical properties and impact strength, due to the increased level of comonomer, or low base density. As in the case of LLDPE, film properties vary with respect to melt index, density, and comonomer type. In general, VLDPE resins exhibit outstanding puncture resistance, flexibility, low-temperature stability, and good sealability.

The use of LLDPE in injection molding offers such advantages as higher stiffness, improved ESCR, and heat-distortion resistance, which can be coupled with higher melt indexes to provide faster cycle times and downgauged articles.

Second-Generation LLDPE

Several companies are accelerating their product development efforts to produce resins that can outperform currently available LLDPE resins. Most notably, efforts are being geared toward using single-site metallocene-based catalysts that provide resins with a very narrow molecular-weight distribution (MWD) and a homogenous short-chain branching distribution (SCBD). LLDPE and VLDPE resins produced using metallocene catalyst offer outstanding toughness and optical properties.

Another class of new-generation LLDPE consists of the broad/bimodal MWD LLDPE. These resins offer excellent processability, and can be extruded on existing HP-LDPE equipment without any modifications.

Safety and Health

Polyethylene is one of the most inert polymers and constitutes no hazard in normal handling (resin suppliers will provide Material Safety Data Sheets on request). It is generally recognized as a safe packaging material by the FDA. Resin suppliers will state which of their products comply with regulations governing polyethylenes used in food-contact applications.

BIBLIOGRAPHY

1. *Modern Plast.* (Jan. 1995).
2. German Pat. 973,626 (1960). K. Ziegler et al.
3. U.S. Pat. 4,076,698 (1978) A. W. Anderson.
4. U.S. Pat. 4,545,762 (1985). W. Kaminsky and H. Hahnsen.
5. B. A. Krensel, Y. V. Kissin, V. I. Kleiner, L. L. Stotskaya, *Polymers and Copolymers of Higher Alpha-Olefins,* Hanser Publishers, in press.
6. A. B. Furtek, *MetCon '93, Proceedings,* p. 125.
7. Peter Sherwood Associates, *LLDPE/LDPE/EVA Film Markets, 1992–1998,* June 1993.
8. Peter Sherwood Associates, *LLDPE/LDPE/EVA Non-Film Markets, 1993–1999,* April 1994.
10. *Film Extrusion Manual,* TAPPI Press, 1992.
11. V. Firdaus and P. P. Tong, *SPE 1993 ANTEC Proceedings,* p. 2550.
12. P. P. Shirodkar, V. Firdaus, and H. Fruitwala, *SPE 1994 ANTEC Proceedings,* p. 211.
13. K. G. Schurzky, *TAPPI 1984 PLC Proceedings,* p. 7.

V. Firdaus
P. P. Tong
Mobil Chemical Company
Edison, New Jersey

POLYETHYLENE, LOW-DENSITY

Polyethylene is a thermoplastic polymer formed from the polymerization of ethylene. It is available in a variety of molecular weights and densities, which have been tailored to specific end-use markets. ASTM has divided polyethylene into four general categories according to density:

Type	Nominal Density, g/mL
I	0.910–0.925
II	0.926–0.940
III	0.941–0.959
IV	≥0.960

In general, the polyethylene industry does not always follow these designations but has broken polyethylene into two broader categories: high-density polyethylene (HDPE), density ≥0.940; and low-density polyethylene (LDPE), density 0.915–0.939. There is also now a type of polyethylene, very low-density polyethylene (VLDPE), with densities of <0.915. Dow Chemical offers VLDPE's under the trade name ATTANE (trademark) (1). Union Carbide sells polyethylene with densities of 0.895–0.913 under the trade name FLEXOMER (registered trademark) (2).

Low density polyethylene was first produced in England in 1933 by Imperial Chemical Industries laboratories, when ethylene gas was compressed to high pressures and heated to high temperatures (3). ICI's development of a commercial process for the manufacture of LDPE was closely followed by the wartime use of the product in critical areas such as high-frequency cables for ground- and airborne radar equipment.

In the early 1950s, another type of polyethylene, HDPE, was commercially introduced by several companies who had developed new low pressure processes for its production (4).

HDPE is a linear polymer, without any of the long-chain branching characteristic of LDPE. Because of its different structure, HDPE possessed properties different from and complementary to those of LDPE and was quickly utilized in many new packaging applications. These two polyethylenes—differentiated with respect to density, properties, and manufacturing processes—coexisted and grew until the early 1960s, when a third type of polyethylene, LLDPE, was introduced. Because the molecular structure of this new type of LDPE was more similar to that of HDPE, and term *l*uinear *l*ow-*d*ensity *p*olyethylene (LLDPE) was coined. Therefore, the polyethylene industry today is composed of three types of polyethylene: HDPE, LDPE (sometimes also referred to as high-pressure LDPE or HP-LDPE), and LLDPE. There is some confusion in the terminology used for LDPE and LLDPE. In some articles and publications, LDPE is used as a generic term for polyethylene below 0.935, thus covering both HP-LDPE and LLDPE. In other cases, LDPE is used to cover only HP-LDPE and LLDPE is a separate category of polymer. In this article, the second type of categorization is used. LLDPE is treated as a separate type of polyethylene and referred to only as LLDPE, while LDPE covers only low-density polyethylene produced by the high-pressure process. When written out in words (eg, low-density polyethylene), both LDPE and HP-LDPE are included. LLDPE has made inroads in many of the markets currently served by LDPE and, in addition, is competing with HDPE in some new applications.

However, by the late 1980s, the penetration of LDPE markets by LLDPE appeared to be leveling off and LLDPE producers looked for new technology to regain momentum. The early 1990s saw an exciting explosion of new low-pressure, low-density PE products and technology designed to address some of the deficiencies of LLDPE compared to LDPE. For example, Phillips Petroleum announced LDLPE (*l*ow-*d*ensity *l*inear *PE*), and claimed that it was easier to process than conventional LLDPE (5). Union Carbide Corp. announced Unipol (registered trademark) II, a technology that would be capable of making low-pressure products which were true "drop-ins" for conventional, commodity LDPE products (6). Their new Unipol II facility will be completed in mid-1995. Himont built a new U.S. Gulf Coast facility to utilize its Spherilene (registered trademark) process, which is claimed to produce polyethylene resins in spherical form directly from the reactor (7). Dow and Exxon both announced the development of single-site catalyst technology that can make polyethylene homo-, co-, or terpolymers from 0.865 to 0.96 density (8,9). By early 1995, most major polyethylene producers had announced their own programs in metallocene or single-site programs and many had formed joint ventures to commercialize the products (10). All this innovative activity certainly challenges the notion and polyethylene is a "mature" technology.

Characterization of Polyethylene

The properties of products made from polyethylene are dependent on some basic characteristics of polyethylene itself. Several of the terms commonly used to describe polyethylene are discussed in the following paragraphs.

Melt index. The melt index (MI) of a polymer is used as an empirical measure of its molecular weight. To measure melt index according to ASTM D1238, a polymer sample is melted and forced through a small orifice of fixed size under a fixed pressure. The weight of polymer that is extruded in 10 min under 44 psi (303 kPa) of pressure is called the *melt index.* When the pressure is increased to 440 psi (3030 kPa), the weight of polymer extruded in 10 min is the *flow index* (FI). Since a polymer with very high molecular weight will be very viscous and resistant to flow, it will not pass through the small orifice quickly and the weight obtained in 10 min (or melt index) will be low. Melt index is therefore inversely proportional to molecular weight. Typical low-density polyethylene melt indices range from 0.2 to over 150 dg/min. In general, products in the lower-melt-index range are used for film extrusion and the higher-melt-index products for molding and extrusion coating. Within any fabrication process, the use of a lower-melt-index resin will result in a stronger product, although usually with some sacrifice in extrudability.

Melt-flow ratio. The melt-flow ratio (MFR) is a rough estimate of the molecular-weight distribution (MWD) of a resin. Since all polymer chains in a given resin are not exactly the same length, a MWD measurement will describe how dissimilar the chains are from each other. Melt-flow ratio is the ratio of the flow index to the melt index. The higher the ratio, the broader the molecular-weight distribution and the more dissimilar the chains are from each other. A polymer with every chain exactly the same length would have a very narrow molecular weight distribution and a very low melt-flow ratio. Melt-flow ratios of commercial low-density polyethylenes vary from about 20 (very narrow) to about 100 (very broad). Polymers with narrow molecular-weight distributions give stronger products but are more difficult to extrude than those with broad molecular-weight distributions.

Density. The density of a polymer is a measure of its crystallinity. Density measured according to ASTM D1505 consists of taking a small sample of polymer that has been molded in a carefully prescribed manner and dropping it into columns with solutions of different viscosities. The position of the unknown polymer is the column is then compared to standard samples of known density. The density of a film or molded article is only partly controlled by the density of the resin used to make the product. The rate at which the product is cooled also plays an important role. The faster a film or molded article is cooled, the less time there is for the polymer chains to crystallize and the lower the density of the final product. For example, the density of a sample cut from a blow-molded polyethylene bottle ws 0.945 g/mL. The same polyethylene resin, when compression-molded into a plaque and cooled according to ASTM D1505, measured 0.954 g/mL. Product properties such as stiffness, rigidity, environmental stress-crack resistance (ESCR), and moisture vapor transmission rate (MVTR) are affected by density. The lower the density of a product, the more limp and flexible it is. A 0.918-g/mL polyethylene product will have better ESCR and higher MVTR than a product with a density of 0.930 g/mL.

LDPE Process

LDPE is made by the high-pressure polymerization of ethylene (4,11). In either a tubular or autoclave reactor ethylene is pressurized to more than 20,000 psi (138 MPa) and heated to >150°C. Small amounts of an initiator, typically oxygen or

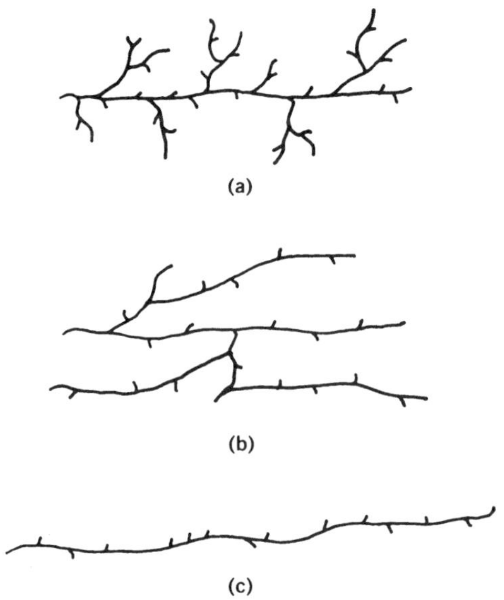

(a) (b) (c)

Figure 1. Molecular structures for LLDPE and HP-LDPE: (**a**) HP-tubular process; (**b**) HP-autoclave processes; (**c**) LLDPE.

peroxide, are added to start the polymerization process. Comonomers such as vinyl acetate or ethyl acrylate can be added to make EVA and EEA copolymers, respectively. Critical molecular characteristics such as molecular weight, molecular-weight distribution, and density are controlled by reaction temperature, ethylene pressure, and the concentration of chan-transfer agents. Constraints on the viscosity of the polymer solution in the reactor limit the rate at which products with high molecular weight and/or high density can be made in the high-pressure process. It has been found that the tubular high-pressure process gives products which differ subtly in the type and degree of branching from products made in the autoclave process. Density depression occurs in all types of polyethylene because of the presence of short branches along the backbone of the polyethylene chain. These short branches, typically one to five carbon atoms long, prevent the long polyethylene chains from folding together and forming crystals. However, early in the development of LDPE, it was found that the presence of these short branches did not totally explain some of the rheological properties of LDPE. It was then hypothesized and proved that LDPE also contains low levels of long-chain branching (12). These long chains, which can be over 1000 carbon atoms in length, have a very small effect on the density of LDPE but have major impact on the processing and properties of LDPE (Fig. 1). The tubular reactor makes LDPE with a large number of long-chain branches. The branches are of relatively short length, however. The autoclave reactor, on the other hand, gives products with low levels of long-chain branches of extremely long length. Because of the difference in long-chain branch type and frequency, some specialization in product applications has occurred with these products. The autoclave reactors produces products that are especially useful in high-speed extrusion coating and in film applications requiring toughness. The tubular reactor gives products with optimal clarity and processing characteristics.

LLDPE Process

Low-pressure processes for the polymerization of ethylene were developed in the 1950s (4,11). These processes used organometallic catalysts to polymerize ethylene to form HDPE at moderate pressures (approximately 300 psi) and temperatures (approximately 100–200°C). Three basic types of low-pressure systems developed: *solution,* where the polymer was completely dissolved in a solvent at high temperatures; *slurry,* where the solid polymer particles were physically suspended in a solvent at lower temperatures; and *gas phase,* where the solid polymer was in contact with only the polymerization gases. Again, because of process constraints in the reactor, molecular weights and densities were thought to be limited, especially in the solution and slurry systems. The density limitations of the low-pressure and high-pressure processes were thought to form incompatible regions, with the low-pressure process making polyethylenes above 0.935 density and the high-pressure process making polyethylene below 0.935 density.

The first commercial production of LLDPE was made in a solution process by DuPont Canada in 1960 (13). A large market for this new polymer did not develop until 1977, when Union Carbide started licensing their gas-phase process for the manufacture of linear low-density polyethylene. Since that time, several other polyethylene manufacturers have announced the conversion of low-pressure HDPE processes or even high-pressure LDPE processes to make LLDPE and some of these processes are available for licensing.

The structure of these new linear low-density polymers is very different from the LDPE made from the high-pressure process. In LLDPE, there is no long-chain branching. Density is controlled by the addition of comonomers such as butene, hexene, or octene to the ethylene. These comonomers give rise to short-chain branches of different lengths, two carbon atoms for butene, four for hexene, and six for octene. The length of the short-chain branches determines some of the strength characteristics of LLDPE. The absence of long-chain branches in LLDPE plays a significant role in the difference in extrusion characteristics between LLDPE and LDPE, as discussed in the sections on specific application areas.

Low-Density Polyethylene Properties

Low-density polyethylene is one of the most widely used packaging materials in the market. Its utility in a variety of different applications is due not to some single outstanding property or characteristic, but usually to a combination of properties. The low price of polyethylene compared to wood, metal, and other polymers has accelerated its penetration into many applications. In addition to the low cost of polyethylene, the excellent toughness, flexibility, moisture barrier, chemical resistance, electrical insulation, and light weight of polyethylene films, bottles, pipes, cables, and other articles make them superior to articles made from conventional material sof construction. Since the introduction of LLDPE into the market, the superior properties of LLDPE have led to its use in new applications for polyethylene as well as the replacement of LDPE and/or HDPE in some areas. Compared to LDPE, LLDPE at the same melt index and density offers better toughness, rigidity, stress-crack resistance, elongation, melting point, and moisture barriers. (See article on polymer properties for more details.) In the 1980s, public concern on the solid-waste issue became a significant concern to the plastics industry. While plastics often offer a low cost, less energy-intensive packaging alternative to paper, glass, and wood, the public was concerned over the lack of plastic recycling and the longevity of plastic in the environment. The plastics industry responded by establishing the American Plastics Council, which sponsors programs in plastics recycling and represents the plastics industry in legal challenges to plastics packaging.

Low-Density Polyethylene Markets

United States. Low-density polyethylene has found utility in a wide variety of applications, ranging from thin-gauge garment film to 500-gal (1892-L) water-storage tanks. The total 1993 production of low-density polyethylene (defined as <0.940 density) in the United States was estimated to be about 12.1 billion lb (5.5 million metric tons) (14). Of this, about 4.8 billion lb (2.2 million metric tons) was LLDPE. This production capacity estimate is becoming increasingly difficult in measure because of the versatility of the various low-pressure processes. It is now relatively easy for a polyethylene producer, who has appropriate manufacturing technology, to switch production from HDPE to LLDPE depending on market situations. As more and more producers convert existing high- or low-density plants or build new plants capable of making both HDPE and LLDPE, the distinction between production capacities of the two polymers will blur and there will be a unified polyethylene market.

In the United States, of the 12.1 billion lb (5.5. million metric tons) of LDPE/LLDPE produced in 1993, approximately 1.7 billion lb (770,000 metric tons) were exported (14). A large quantity of low-density polyethylene, over 1 billion lb (450,700 metric tons), was imported in 1993. Canada represents the country exporting the most polyethylene to the United States by a large margin. Pricing of LDPE/LLDPE is quickly affected by changes in supply and demand and discounting off list price is common. In 1993, the list price of LDPE/LLDPE in the United States varied from 30 to 52 cents per pound, delivered.

In the next few years, the growth rate for LDPE has been predicted to remain low as LLDPE penetrates the traditional markets for LDPE (Fig. 2) and no new investments are made in LDPE facilities. It is expected that by 1995, essentially all LDPE markets will be penetrated to some extent by LLDPE. Growth of the total polyethylene market is expected to stabilize at about 4% per year through 1997 (14). This low growth is reflecting not only the mature state of the market but also the practice of "downgauging" or using less polyethylene in the final product. Downgauging has become very widespread because of the superior properties of LLDPE in film and molding products. It is also being mandated by some state and local legislation aimed at reducing the amount of packaging material in landfill. This overall low growth rate is made up of an almost constant market for LDPE compared to a healthy growth rate of 7% per year for LLDPE (14). This rapid rate of growth by LLDPE is due to its outstanding combination of properties, as discussed earlier.

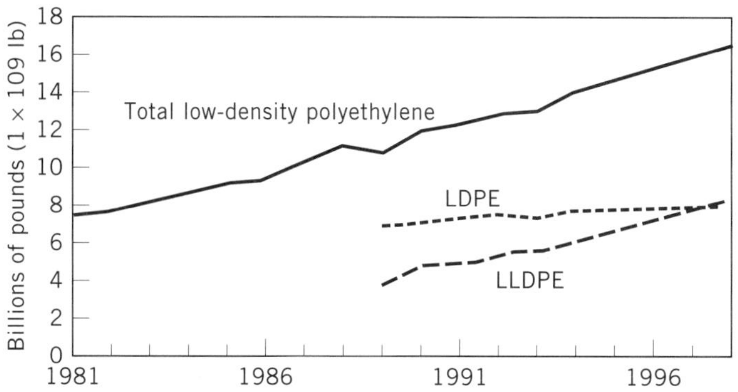

Figure 2. Low-density polyethylene sales and use in the United States (14). To convert 10^9 lb to 10^6 tons, multiply by 0.4536.

By far the largest domestic application for LDPE/LLDPE is in film and sheet. In 1993, over 50% of all low-density polyethylene was used in these applications (14). Other markets, listed in order of size, are:

- Injection molding
- Extrusion coating
- Wire and cable
- Rotomolding

World. The total world demand for LDPE/LLDPE is predicted to be 57.0 billion lb (25.9 million metric tons) in 1997 (15). On the basis of nameplate capacity and announced expansions, the supply of LDPE/LLDPE in 1997 is projected at 83 billion lb (37.8 million metric tons), although a high percentage of that capacity can produce HDPE as well. The rest of the world will lag the United States in incorporating LLDPE into the market. For example, the pentration of LLDPE into all LDPE applications in the United States is estimated to be about 52% and 1997 (14). In contrast, the penetration in western Europe will be only about 30%; and in Japan, about 35%. (Ref. 15, p. III.17). The more rapid penetration of LLDPE in the United States is postulated to be due to the large domestic and institutional markets in trash and garbage bags, where the property benefits and low cost of liner-grade LLDPE provide an economic incentive to the film extruder. The United States is also a more homogeneous market for film and molding articles. There are large plastics fabricators who sell products into all geographic areas and therefore maintain a universal product specification. The success of LLDPE in the United States can also be attributed to the determined efforts of the pioneer LLDPE producers to assist film fabricators in converting their equipment to run LLDPE economically. In Europe, on the other hand, the market is composed of smaller, local companies, who make products to regional specifications that may not be acceptable in another location. It is harder, therefore, for a new material to gain rapid acceptance because of the multitude of different specifications in the various countries. This may change in light of large European polyethylene companies formed by joint ventures, like Borealis (Neste and Statoil) and Polimeri Europa (Union Carbide and Enichem). In addition, the LDPE market in Europe has been characterized by overcapacity and low prices in recent years. There has been very little incentive for a film or bottle producer to look at a new material, such as LLDPE, especially if it requires any equipment modifications at all. An additional barrier to the acceptance of LLDPE was its premium price. In the 1980s, LLDPE was sold in western Europe at a ≤15–20% premium over LDPE (16,17).

Table 1. HP-LDPE Homopolymer Film Resins

Type	HP-LDPE	HP-LDPE	HP-LDPE	HP-LDPE	HP-LDPE	HP-LDPE
Comonomer	None	None	None	None	None	None
Melt index	0.2–0.8	1.5–2.0	1.5–2.0	1.2–2.0	1.0–2.0	5.0–10.0
Density, g/cm^3	0.919–0.923	0.922–0.925	0.930–0.935	0.918–0.924	0.923–0.927	0.917–0.930
Molecular-weight distribution	Broad	Broad	Broad	Broad	Broad	Broad
Process	Blown	Blown	Blown and/or cast	Blown	Cast	Cast
Applications	Shipping sacks; heavy-duty applications	Bread and bakery; general-purpose packaging	Overwrap	General-purpose packaging	Bread and bakery; general-purpose packaging	Extrusion coating
Critical properties	Toughness	Clarity	Clarity; stiffness	Extrudability; toughness	Extrudability; good tear	Drawdown adhesion; pinhole resistance

Applications of LDPE/LLDPE

Film and sheet. About 65% of all the low-density polyethylene used in the world today goes into the film and sheet area. These applications include garbage bags, grocery sacks, garment bags, heavy-duty sacks, shrink film, stretch film, carrier film, pond liners, construction and agricultural film, and food packaging. Polyethylene used in the film area range in melt index from 0.2 to 6 and in density from 0.915 to 0.935. The advantages of LDPE/LLDPE include low cost, flexibility, toughness, and chemical- and moisture-barrier properties (Tables 1–3). This has been the market most rapidly penetrated by LLDPE in the United States. This rapid penetration is due, in part, to the outstanding physical properties of LLDPE film (Table 4). Dart drop and Elmendorf tear measure the toughness of film in a high-speed test and give an indication of the failure behavior of film under catastrophic conditions. Puncture and tensile tests are done at relatively slow speeds and give an indication of intrinsic properties and failure modes that can occur with long-term use.

The differences in molecular structure between LLDPE and LDPE (linear vs branched molecules and molecular-weight distributions) affect the rheology of the two materials. LLDPE is more viscous at extrusion shear rates and requires

Table 2. HP-LDPE Copolymer Film Resins

Type	HP-LDPE	HP-LDPE	HP-LDPE	HP-LDPE
Comonomer	2–5% VA[a]	3–5% VA[a]	7% VA[a]	15–18% EA[b]
Melt index	1.5–2.0	0.2–0.4	0.2–4.0	2.0–6.0
Density, g/cm^3	0.925–0.930	0.923–0.927	0.927–0.945	0.927–0.940
Molecular-weight distribution	Broad	Broad	Broad	Broad
Process	Blown	Blown	Blown; extrusion coating	Blown; extrusion coating
Applications	Frozen food	Ice bags	Sealing layer; liquid packaging	Disposable gloves; ID cards
Critical properties	Clarity; low-temperature properties	Low-temperature properties	Adhesion; low-temperature properties	Low stiffness; adhesion to polar substrates

[a] Vinyl acetate.
[a] Ethyl acrylate.

Table 3. LLDPE Film Resins

Type	LLDPE	LLDPE	LLDPE	LLDPE	LLDPE
Comonomer	Butene	Hexene; octene	Hexene; octene	Hexene; octene	Hexene; octene
Melt index	0.8–2.5	2.0–4.0	2.0–5.0	0.7–1.5	0.8–1.5
Density, g/cm^3	0.917–0.922	0.912–0.919	0.928–0.935	0.924–0.928	0.917–0.923
Molecular-weight distribution	Narrow	Narrow	Narrow	Narrow	Narrow
Process	Blown	Cast	Cast	Blown	Blown
Applications	General-purpose packaging	Stretch wrap	Bread and bakery; overwrap	Grocery sack	Blending; ice bags
Critical properties	Extrudability; toughness	Puncture and tear resistance	Stiffness; moisture barrier	Stiffness; tear resistance	Excellent toughness; Low-temperature properties

Table 4. Blown-Film Properties[a] of LLDPE and HP-LDPE

Property	ASTM test		HP-LDPE	HP-LDPE	LLDPE	LLDPE	LLDPE
Melt index			2.5	0.2	1.0	1.0	1.0
Density, g/cm^3			0.921	0.923	0.918	0.918	0.920
Comonomer			None	None	Butene	Hexene	Octene
Dart drop, gf/mil (N/mm)	D1709		0.75 (29)	185 (71.4)	100 (38.6)	200 (77.2)	250 (96.5)
Puncture energy, in. · lbf/mil (kJ/m)			6 (26.7)	5 (22.2)	16 (71.2)	17 75.6)	
Elmendorf tear, gf/mil (N/mm)	D1922	MD	160 (61.8)	90 (34.7)	140 (54)	340 (131.2)	370 (142.8)
		XD	110 (42.5)	100 (38.6)	340 (131.2)	585 (225.8)	800 (308.8)
Tensile strength, psi (MPa)	D882	MD	2900 (20)	2800 (19.3)	5000 (34.5)	5200 (35.9)	6500 (44.8)
		XD	2700 (18.6)	3000 (20.7)	3800 (26.2)	4700 (32.4)	5100 (35.2)
Tensile impact strength, ft · lbf/in. (MPa)		MD	440 (36.4)	500 (41.4)	1200 (99.3)	1930 (159.7)	
		XD	650 (53.8)	1050 (86.9)	900 (74.5)	1760 (145.6)	

[a] All properties measured on 1.5-mil (38-μm) film produced at 2:1 blowup ratio.

more power to extrude. In addition, it is necessary to use a wide dip gap to avoid melt fracture when extruding 100% LLDPE. Therefore, in the early days of LLDPE introduction, minor modifications to the screw and die gap had to be made in order to extrude LLDPE on extruders designed for HP-LDPE and obtain optimum film properties (18). Most extruders today have been designed to handle the different rheology of LLDPE, and such modifications are not necessary. In addition, there are additives now available that allow the extrusion of conventional LLDPE through narrow die gaps without melt fracture (19). The new generation of LLDPE film products, designed to be "drop-in" for LDPE products, will not require any modifications to extrusion equipment.

Injection molding. This market includes lids, buckets, wash basins, housewares, toys, freezer containers, and general housewares. The advantages of low-density polyethylene include good low-temperature properties, low cost, light weight, and flexibility (Table 5). Low-density polyethylene in the melt-index range of 2–150 and density range of 0.920–0.930 is commonly used in injection-molding applications. LLDPE has made significant penetration into certain markets such as lids and housewares, where property advantages, such as higher stiffness and improved environmental stress-crack resistance and heat-distortion resistance, can be coupled with higher melt indices to result in faster cycle times and down-gauged articles (Table 6) (20). The 1993 consumption of LLDPE for injection molding is estimated at 546 million lb (245,000 metric tons) (14).

Extrusion coating. In this application, polyethylene is used as a coating on another material such as paper, aluminum foil, and cardboard. These coated products are used for liquid packaging, such as milk and juice, and for the new aseptic packages of nonrefrigerated juices. The polyethylene serves as an adhesive, moisture barrier, seal layer, printable surface, and/or barrier to tear. Because of its molecular structure, that is, the type of long chain branching, polyethylene made in tbe high-pressure autoclave reactor is the most successful in this market. Resins with melt indices of 4–10 and densities of 0.920–0.930 are commonly used. Because of its linear structure, LLDPE has not penetrated the extrusion coating market, except as blends with LDPE.

Blow molding. High-density polyethylene is the preferred material for this market because of its rigidity and barrier properties. Low-density polyethylene may be used in those segments where flexibility and excellent stress-crack resistance are required, such as squeeze bottles, toys, and drum liners. Fractional-melt-index polyethylene (density 0.920–0.935) is typically used in this market. LLDPE offers better

Table 5. HP-LDPE and LLDPE Molding Resins

Type	HP-LDPE	HP-LDPE	HP-LDPE	LLDPE	LLDPE	LLDPE
Melt index	6–10	15–25	35–50	0.8–1.2	12–30	50–150
Density, g/cm^3	0.924–0.926	0.914–0.918	0.923–0.925	0.918–0.922	0.920–0.926	0.926–0.935
Molecular-weight distribution	Broad	Narrow	Narrow	Broad	Narrow	Narrow
Process[a]	IM	IM, BM, PE	IM	BM, IM, PE	IM	IM
Applications	Bottle closures	Drug bottles; aseptic packaging	Lids	Drum liners; irrigation tubing; spouts	Industrial containers; Specialty housewares	Lids; housewares
Critical properties	ESCR;[b] stiffness	ESCR;[b] low modulus	Cycle time	ESCR;[b] process-ability	ESCR;[b] low-temperature impact	Cycle time

[a] Key; IM = injection molding; BM = blow molding; PE = profile extrusion.
[b] ESCR = environmental stress-crack resistance.

Table 6. Comparison of LLDPE and HP-LDPE Molded Properties

Type	LLDPE	LLDPE	LLDPE	HP-LDPE	HP-LDPE
Melt index	20	50	100	23	49
Density, g/cm^3	0.924	0.926	0.931	0.924	0.924
Dishpan impact, ft · lbf (J) at −20°C	30 (40.7)	24 (32.5)	3 (4.1)	9 (12.2)	3 (4.1)
Failure mode	Ductile	Ductile	Ductile	Shatter	Shatter
ESCR[a] ($F_{50}h$)[b]	150	3	<2	<1[c]	

[a] Environmental stress-crack resistance.
[b] At 50°C, 100% Igepal, no slit, ASTM D1693.
[c] $F_{100}h$.

stiffness and improved stress-crack resistance compared to LDPE.

Safety and Health

Polyethylene is generally recognized as a safe packaging material by the Food and Drug Administration. Resin suppliers will state which of their resins comply with regulations governing polyethylenes used in food-contact applications. These regulations are covered in the U.S. Food, Drug, and Cosmetic Act as amended under Food Additive Regulation 21 *CFR* 177.1520. Polyethylene is a very stable polymer. However, proper material handling procedures are required to control dust and smoke and to provide adequate ventilation during extrusion. Resin suppliers will provide Material Safety Data Sheets and Materials Handling Guides for polyethylene resins on request.

BIBLIOGRAPHY

1. *Plastics World,* **43,** 92 (Jan. 1985).
2. *Chemical Week,* **135,** 66 (Sept. 19, 1984); *Plastics World,* **42,** 8 (Oct. 1984)
3. Br. Pat. 471,590 (Sept. 6, 1937) and U.S. Pat. 2,153,553 (Apr. 11, 1933), E. W. Fawcett and co-workers (to ICI, Ltd.).
4. A. Renfrew and P. Morgan, eds., *Polyethylene,* Interscience Publishers, New York, 1960, Chapter 2.
5. *Chem. Week,* 12 (May 5, 1993).
6. *Plast. News,* 22 (March 8, 1993).
7. *Chem. Week,* 8 (June 12, 1991).
8. *Chem. Market. Reporter,* 5 (Dec. 23, 1991).
9. *Chem. Eng. News,* 20 (Sept. 18, 1989).
10. *Plast. World,* 33–36 (Jan. 1995).
11. R. A. V. Raff and J. B. Allison, *Polyethylene,* Interscience Publishers, New York, 1956, Chapter 3.
12. J. J. Fox and A. E. Martin, *Proc. Roy: Soc. London Ser.* A **175,** 211, 216 (1940).
13. U.S. Pat. 4,076,698 (Feb. 28, 1978), A. W. Anderson (to E. I. DuPont de Nemours & Co., Inc.).
14. *Monthly Petrochemical & Plastics Analysis,* Chemical Data Systems, Inc., Oxford, PA, July 1994, p. 53.
15. *World Plastics, 1993,* Tecnon, Via Egadi, Milan, Italy, p. 11.9.
16. *Eur. Plast. News* **9,** 13 (Dec. 1982).
17. *Modern Plast. Internatl.* **10,** 6, 8, (Oct.1982).
18. S. J. Kurtz and L. S. Scarola, *Plast. Eng.* **36** (6), 45 (June 1982).
19. *Plast. World* **41,** 52 (Dec. 1983).
20. *Plast. World* **42,** 42 (March 1984); **41,** 37 (March 1983).

General References

J. H. DuBois and F. W. John, *Plastics,* Van Nostrand Reinhold, New York, 1981 (contains comparisons of properties of thermoplastics and thermosets).

T. O. J. Kressler, *Polyethylene,* Reinhold Publishing, New York, 1961 (contains general information on polyethylene applications; has a table containing comparative properties of transparent packaging films of different plastics).

R. A. V. Raff and J. B. Allison, *Polyethylene* (1956) and A. Renfrew and P. Morgan, eds., *Polyethylene* (1960), both by Interscience Publishers, New York (both books contain background information on polyethylene manufacturing and characterization; good source for basic molecular descriptions of polyethylene).

World Polyolefin Industry, 1982–83, Chem Systems, Inc., Tarrytown, NY, June 1983 (good source for economic description of polyethylene industry; also marketing forecasts of supply and demand).

Polyolefins through the 80's—a Time of Change, SRI International, Menlo Park, CA, Aug. 1983 (a four-volume study covering the marketing and production aspects of the polyethylene industry).

NORMA J. MARASCHIN
Union Carbide Corporation
Bound Brook, New Jersey

POLYMER PROPERTIES

The properties of a polymer result from its chemical nature, morphology, formulation, processing, and even use conditions. Intrinsic polymer properties depend primarily on the chemical nature of the polymer, but since most polymers are polymorphic materials, their intrinsic properties may also depend on the polymer's morphology. Morphological changes at room temperature, however, are very slow and highly time-dependent. Formulation (compounding with additives) and processing have a direct impact in the final properties of a polymeric material. Furthermore, storage and use conditions (eg, humidity and aging) may affect a plastic's performance.

Intrinsic properties of a polymer can be classified as either molecular or bulk. Intrinsic molecular properties depend mainly on the chemical structure of the polymer's constitutional units, and to a lesser degree on its macromolecular character. The chemical nature of a constitutional unit, which results from the type and number of atoms, existing side groups, charge distribution, and type of secondary molecular forces, controls important properties. These properties are cohesive energy, molecular packing density, molecular relaxation (including glass-transition temperature), barrier, mechanical strength, frictional forces, surface tension, and adhesion characteristics. In PE, for example, the $-CH_2-$

group yields a set of properties values quite different from the constitutional unit $-OC(CH_2)_5NH-$ of nylon 6. The latter unit is larger and more polar than the first, and it tends to develop strong intermolecular hydrogen bonding.

Intrinsic bulk properties such as stiffness, melting temperature, heat-sealing temperature range, melt-flow index, and viscosity, are largely influenced by the molecular mass and distribution, as well as the architecture of the polymer chain. Molecular mass and molecular distribution include the average molecular mass, dispersion index, and single- or multimodal distribution. The "architecture" of a polymer refers to the monomers layout in the polymer chain. In this respect polymers can be linear, branched, or crosslinked; have different tacticity (eg, atactic, isotactic, and syndiotactic polypropylene); and show various copolymer arrangements such as random, block, alternating, or graft.

Phase morphology affects, in varying degrees, the physical, mechanical, and optical properties of a polymer. Polymers can be isotropic (amorphous); that is, materials that are glassy (hard and brittle) below the glass-transition temperature, and rubberlike (soft and elastic) above it. As temperature increases, amorphous polymers become true liquids without any thermodynamic discontinuity. Polymers that are nonisotropic can crystallize in several arrangements; even two different crystal structures can coexist, depending on the values of temperature and pressure. Sometimes a crystalline polymer remains a liquid below the melting temperature, thus producing a supercooled material. In semicrystalline polymers, crystalline and noncrystalline regions may coexist, forming microcrystalline or paracrystalline regions. Polymers may also have a one or a two-dimensional molecular order in the liquid state that forms thermotropic or lyotropic liquid crystalline mesophases. A simple example of how the morphology affects the properties is illustrated by the density of PE; the density of PE may range from 0.90 to 0.97 g/mL as the percent of crystallinity increases. Besides density, there are other properties proportional to the degree of crystallinity: rigidity, heat resistance, barrier, abrasion resistance, gloss, shrinkage, and parting-line difference. However, stress-crack resistance, clarity, folding endurance, impact strength, and parison sag decrease as crystallinity increases.

Process operation, on the other hand, can alter the morphology of a material. Film orientation, for instance, directly affects the polymer morphology by producing a slightly more compact molecular packing and a more transparent film. Similarly, a rapid cooling process will increase the amorphousness of a supercooled phase, while slow cooling will increase crystallinity. Thus process conditions may affect properties such as the heat of fusion of a material, and may also affect operations such as thermoforming operation and heat sealing. Since transparency (or opacity) is directly controlled by the crystallinity, morphological changes can influence the optical properties of a polymer. When additives are incorporated into a resin, however, they can substantially alter the resin's original properties.

Determination of Polymer Properties

Polymer properties can be measured experimentally (usually according to standardized methods) or be estimated from semiempirical correlations. Many intrinsic properties related to the molar constitutional unit can be estimated from group contributions or increment methods. Some polymer properties that can be estimated by group contribution include density, thermal expansion coefficient, thermal conductivity, specific heat, specific entropy of fusion, melting temperature, glass-transition temperature, cohesive-energy density, solubility parameter, surface tension, viscosity coefficient, dielectric constant, magnetic susceptibility, specific shear modulus, specific bulk modulus, and sound velocity (1).

The most common properties of polymers related to packaging applications are described next. For easy reference they are grouped under these headings: (*1*) density and thermophysical properties, (*2*) mechanical properties, (*3*) solubility and degradation properties, (*4*) barrier properties, (*5*) surface and adhesion properties, (*8*) electrical properties, and (*7*) optical-appearance properties.

Density and thermophysical properties

Density. The density of a plastic is directly proportional to its crystallinity: $d = d_a + C(d_c - d_a)/100$, where d is the density, C is percent volumetric crystallinity, and d_a and d_c are the amorphous and crystalline density, respectively. For example, $d_a = 0.855$ g/mL while $d_c = 1.0$ g/mL for PE (2). Polymer density values vary from 0.87 g/mL for polypropylene up to 1.86 g/mL for PVDC (3). The standard ASTM D1505 (4) describes the density gradient method to evaluate the density of films and resins.

Glass–transition temperature. The glass-transition temperature T_g, which is actually a relaxation temperature, is an important property in amorphous polymers. As a relaxation temperature, T_g is associated with the onset of the rotation and mobility of chain segments involving several monomers. Rather than a single-point temperature, T_g in semicrystalline polymers is a range of temperature values. The broadness of this range depends on the sample morphology. When the polymer is at a temperature below T_g, it becomes glassy and stiff, while above T_g it shows a plastic or rubbery behavior (5). The T_g values for polymers range from about −25°C (−125°C is also commonly reported) in polyethylene to 365°C for thermoplastic polyimide (1,6). Both the T_g and the melting temperature set the application temperature range of a plastic container. For instance, a polypropylene container may become brittle at freezing temperature if its T_g is near 0°C; "crystal" polystyrene is brittle at room temperature because it has a T_g of ~80°C and does not contain plasticizer. On the other hand, a container made of a low-melting-temperature (T_m) plastic, will soften if heat-sterilized. Both differential thermal analysis (DTA) and differential scanning calorimetry (DSC) are used to determine T_m and T_g (these methods are described in ASTM D3418). Van Krevelen (1) showed that polymers have T_g/T_m ratios ranging from 0.5 to 0.76. Most polymers, however, have the T_g/T_m ratio centered at 0.67.

Heat capacity. Heat capacity (or specific heat), c, is the amount of energy needed to change a unit of mass of a material one degree of temperature. The heat capacity of plastics, which are obtained at constant pressure, are temperature-dependent, especially near the glass-transition temperature. In a semicrystalline polymer, the heat capacity of the amorphous phase is greater than the heat capacity of the crystalline phase implying that the c values depend on the percent of polymer's crystallinity. The heat capacity values of polymers at 25°C vary from 0.9 to 1.6 J/(g · K) for amorphous polymers, and from 0.96 to 2.3 J/(g · K) for crystalline polymers. Reliable data of the heat capacity of amorphous and crystalline phases are available for only a limited number of polymers (1,2). Values of c for polymers can be found in the review by Wunderlich (7). The usual techniques to measure c are differential thermal analysis (DTA) and differential scanning calorimetry (DSC).

Heat of fusion. The heat of fusion, ΔHm, is the energy involved during the formation and melting of crystalline regions. For semicrystalline polymers, the heat of fusion is proportional to the percent of crystallinity. Amorphous polymers or amorphous polymer regions do not have heat of crystallization, since amorphous structures have a smooth transition from the liquid amorphous state to the liquid state. Experimental values of crystalline heat of fusion for common packaging plastics vary from 8.2 kJ/mol for polyethylene to 43 kJ/mol for nylon 6,6 (1). ASTM D3417 describes a method for measuring the heat of fusion and crystallization of a polymer by differential scanning calorimetry (DSC).

Melting temperature. The melting temperature T_m is a true transition temperature. This means that at T_m both the liquid and solid phases have the same free-energy value. Most semicrystalline polymers, instead of having a sharp melting point, show a melting temperature range, and for amorphous polymers there is no T_m. Similar to T_g, T_m can be estimated from contribution groups and empirical relationships between T_g and T_m (as indicated above). Values of T_m range from as low as 275 K for polyisobutylene, up to 728 K in polyethylene terephthalamide (1,6). ASTM D2117 and D3418 describe methods for measuring T_m.

Thermal conductivity. Thermal conductivity k, is the parameter in Fourier's law relating the flow of heat to the temperature gradient. In practical terms, k is a measure of the ability of a material to conduct heat. The thermal conductivity of a polymer is the amount of heat conducted through a unit of thickness per unit area, unit time, and degree of temperature. Thermal conductivity values control the heat-transfer process in applications such as plastic processing, heat sealing, cooling and heating a package, and sterilization processes. Plastics have values of k much lower than metals do; for instance, thermal conductivity for plastics ranges from 3×10^{-4} for PP to 12×10^{-4} cal/(s · cm · °C) for HDPE, while for aluminum $k = 0.3$ cal/(s · cm · °C), and for steel $k = 0.08$ cal/(s · cm · °C) (8). Plastic foams show k values much lower than those for the nonfoamed plastics because of the presence of air trapped within the cellular structure. Enhanced by low thermal conductivity of air, foams make excellent insulating materials and good cushioning packaging materials. Plastic fillers may increase the thermal conductivity of plastics. Methods for measuring k are given in ASTM D4351, C518, and C177.

Thermal expansion. The coefficient of linear (or volume) thermal expansion is the change of length (volume) per unit of length (volume) per degree of temperature change at constant pressure: $\beta = (1/L)(dL/dT)$ and $\alpha = (1/V)(dV/dT)$. Units of α and β are K^{-1} or $°F^{-1}$ (reciprocal degrees Kelvin and Fahrenheit). Compared to other materials, polymers have high values of thermal expansion coefficients. While metals and glass have values in the range 0.9–2.2 K^{-1}; polymers range from 5.0 to 12.4 K^{-1} (8). Thermal expansion coefficients can be measured by thermomechanical analysis (TMA). ASTM D696 gives a method using a quartz dilatometer, and ASTM E831 describes the determination of the linear thermal expansion of solid materials. Volume contraction of a container from the molding operation temperature down to room temperature is called *shrinkage,* and its measurement is described in ASTM D955, D702, and D1299-55.

Mechanical properties. The response of polymers subjected to mechanical forces is determined by the polymer's isoentropic elastic and viscoelastic behavior. Ideally elastic behavior should be reversible for small deformations. Two types of small deformation are possible: tensile and shear. Hooke's law describes the response of plastics to tensile forces by relating tensile stress to elongation, while the law of shear deformation relates shear stress to both shear modulus and angle of shear. Viscoelastic behavior is observed in large deformations of solid polymers and polymer melts (9). In a creep experiment, the elongation of solid sample increases slowly as the weight hangs from the sample. When the force is released, the sample partially recovers its shape, decreasing the elongation. This behavior is due to the viscoelasticity behavior of the polymers: it partially recovers because it is elastic, and it creeps because it is viscous. This behavior is due to the continuous-chain molecular rearrangement that takes place at all temperatures. The time lapses associated with molecular rearrangement are much larger below than above the glass-transition temperature. The viscoelastic behavior affects almost all mechanical properties of a polymer and should be carefully considered in packaging design. Polymer strength increases with increasing molecular mass and with increasing intermolecular forces, but it decreases with the presence of plasticizers. For example, at the same molecular mass, polyamides, polyesters, or polyacrylonitriles are stronger than polyolefins, and plasticized PVC is weaker than rigid PVC.

Bursting strength. This is the hydrostatic pressure, usually in pascal (or psi), required to produce rupture of the material when the pressure is applied at a controlled increased rate through a circular rubber diaphragm of 30.48 mm (1.2 in.) in diameter. "Points bursting strength" is the pressure expressed in psi. The measurement of the bursting strength of plastic films is described in ASTM method D774.

Dimensional stability. *Dimensional stability* refers to the capability of a structure to maintain its dimensions under the changing conditions of temperature and humidity. Machine and transverse directions may produce different changes in dimensional stability, which is important in any flexible-material converting process. During printing, for example, even small changes in dimensions may lead to serious problems in holding a print pattern. ASTM D1204 describes a standard method for linear dimensional changes of flexible thermoplastic films and sheets at elevated temperature.

Folding endurance. This is a measure of the resistance of the material to flexure or creasing. Folding endurance is greatly influenced by the polymer's glass-transition temperature and the presence of plasticizers. ASTM D2176 describes the procedure to determine the number of folds necessary to break the sample film.

Impact strength. Impact strength is the material's resistance to breakage under a high-velocity impact. Widely used impact tests are Izod (ASTM D256A) and Charpy (ASTM D256) for rigid materials, and for dart-drop impact (ASTM D4272) and pendulum-impact resistance (ASTM D3420) for flexible structures. A free-falling dart method for polyethylene films is described in ASTM D1709. Unlike low-speed uniaxial tensile tests, the pendulum-impact test measures the resistance of film to impact puncture simulating high-speed end-use applications. Dart drop measures the energy lost by a moderate-velocity blunt impact passing through the film. Both pendulum and impact tests measure the toughness of a flexible structure.

Melt-flow index. Most polymer melts follow a pseudoplastic behavior, which implies that the viscosity decreases with shear rate. In common extrusion processes, the shear-rate value varies from 100 to 100,000 s^{-1} (per second). It follows that a complete rheological description of an extrusion process should include a wide range of shear-rate values at suitable temperature range. The melt-flow index (MFI) is a widely used test to measure polymer flow properties, but it provides flow values at only one shear-rate value. MFI, also called *extrusion plastometer test,* is described in ASTM D1238. The melt-flow index measures the mass of polymer extruded during 10 min. Values of MFI vary between 0.3 and 20 g/10 min, corresponding to shear-rate values ranging from 1 to 50 s^{-1}. Variables that affect the MFI of a resin include average molecular mass, distribution of molecular mass, branching, and temperature (10).

Pinhole flex test. Pinhole flex resistance is the property of a film to resist the formation of pinholes during repeated folding. A related test is the folding endurance. Film having a low value of pinhole flex resistance will easily generate pinholes at the folding line, under repeated flexing. The test is described by the standard ASTM F456.

Poisson's ratio. Poisson's ratio is the ratio of lateral strain to longitudinal strain measured simultaneously in a creep experiment. Its value varies from 0.5 for totally deformed but nonelastic liquids, to zero for pure elastic incompressible solid materials. Typical values for rubber materials range from 0.49 to 0.499; and for plastic, from 0.20 to 0.40. This shows that when a polymer sample is elongated, its volume will increase. Because polymers are viscoelastic materials, the Poisson-ratio value is morphology- and time-dependent (1).

Tensile properties. The mechanical behavior of a polymer can be evaluated by its stress–strain tensile characteristics. The stress is measured in force/area which can be given in pascals or psi, and the strain is the dimensionless fractional length increase. *Modulus of elasticity, E,* is the elastic ratio between the stress applied and the strain produced, giving the material's resistance to elastic deformation. The tensile modulus also gives a measure of the material stiffness; the larger the modulus, the more *brittle* the material. For example, E of LDPE is 250 mPa, while for "crystal" PS, it is 2500 mPa. Comparatively, values of tensile modulus in polymers (1.9×10^3 mPa for nylon), are much lower than for glass (60×10^3 mPa), or mild steel (220×10^3 mPa) (6). *Elastic elongation* is the maximum strain under elastic behavior. *Ultimate strength* or *tensile strength* is the maximum tensile stress the material can sustain. Ultimate elongation is the strain at which the sample ruptures. *Toughness* is how much energy a film can absorb before rupturing, and it is measured by the area under the stress–strain curve. *Brittleness* is the lack of toughness (11). Amorphous and semicrystalline polymers become brittle when cooled below their glass-transition temperature. Tests for tensile properties are described in ASTM D882, for flexural strength in ASTM D790, and for flexural modulus in ASTM D790M.

Tear strength. Tear-strength measurement considers the energy absorbed by the film sample in propagating a tear. Standard methods available are: ASTM standard D1004, which describes the measurement for initial tear resistance; ASTM D1922, and ASTM D1938 which refer to the energy absorbed by a test specimen in propagating the tear that has already been initiated. The value of tear strength in one film depends on the orientation stretching ratio, whether the measurement is performed along the cross-direction or the machine direction (CD or MD).

Viscosity. Viscosity, a fundamental rheological property of fluids, is a measure of the molecular resistance to shear fluid deformation, generated by the action of external forces. When the external action is given by the shear stress and the shear flow deformation given by the rate of strain or shear rate, the *absolute viscosity* is equal to the ratio of shear stress/shear rate. The viscosity of a polymer will increase with the length of the polymer chain, with smaller dispersion index (narrow distribution), and with increasing intermolecular forces. An increase of temperature, however, will decrease the viscosity. The rheological behavior of thermoplastics is, in most cases, Newtonian or pseudoplastic (10). The viscosity of Newtonian fluids is constant at any shear rate value. This behavior is characteristic of fluids with low molecular mass. In pseudoplastic materials, the viscosity decreases with shear rate; that is, the viscosity depends on the value of shear rate. Pseudoplastic behavior is characteristic of polymer melts. The viscosity of polymer melts ranges between 100 and 10^7 N/(s · m^{-2}). Viscosity is also known as either kinematic viscosity or intrinsic viscosity. *Kinematic viscosity* is the absolute viscosity divided by the density. *Intrinsic viscosity* (IV) can be obtained as follows: (*1*) solutions of different concentrations c of the polymer in a solvent (having viscosity η_0) are prepared; (*2*) the viscosity η of the solutions are measured; (*3*) from the plot of viscosity number, $[(\eta - \eta_0)/\eta_0 c]$, versus concentration, the IV is obtained as the value of the viscosity at concentration zero (11). Intrinsic viscosity, therefore, is a measure of the polymer's capacity to enhance the viscosity of the solvent. The units of IV are mL/g. For PE, the relation between molecular mass MW and IV is accepted to be as follows: $MW = 5.3 \times 10^4\ (IV)^{1.37}$ (8). An IV of 5, for example, corresponds to a MW of 500,000. Intrinsic viscosity values of polymers can be estimated by group contribution in terms of molar intrinsic viscosity (1). ASTM has published several methods for measuring viscosity.

Solubility and chemical degradation. Plastics are chemically resistant to many gases, liquids, and solid products within a wide range of pH. Nevertheless, polymers are not inert materials. Given the right conditions (temperature, time, and concentration), polymers can be depolymerized, transformed by chemical reactions, and penetrated by solvents and vapors. They can be affected by environmental agents such as visible and UV radiation, oxygen, microbes, solvents, and organic compounds (12). The extent of the transformation depends on the thermodynamic and kinetics parameters, which, in turn, depend heavily on temperature. In hydrophilic polymers (eg, nylons, EVOH, and PET) water acts as a plasticizer, consequently lowering the glass-transition temperature and changing oxygen permeability and other mechanical properties (13).

Water can dissolve certain polymers such as PVOH. Hydrophobic polymers such as polyolefins, on the other hand, are not affected by water.

Chemical degradation. Chemical degradation of polymers results in the fragmentation of large chains into smaller ones and, eventually, into atomic elements. One important type of polymer chemical degradation is oxidation (12). Often oxidation is promoted or induced by electromagnetic radiation (photodegradation) and by thermal energy at high temperature (thermal oxidation). Polymers such as PE, PP, and PVC are particularly prone to oxidation during melt processing. Tensile strength, elasticity, and impact strength are drastically affected by thermodegradation processes. Polymers may also be degraded by acids and alkalis. Test for resistance to acids, bases, and solvents is described in ASTM D543. Plastic resistance to grease and oil is covered in ASTM D722. The ASTM D756 test covers the effect of atmospheric humidity and temperature on plastics, and ASTM D570 covers the sorption of water of immersed plastics.

Environmental stress cracking. A plastic material may be resistant to a chemical compound in no-stress conditions. Nevertheless, the plastic may crack when subjected to a mechanical stress during exposure to that compound. Almost all plastics can show environmental stress cracking (ESC), when exposed to gases, liquids, or solids while under stress. The mechanism by which a plastic shows ESC is often complex. A practical evaluation of a plastic's resistance to liquids can be done by immersing a polymer sample in the liquid and then checking for the appearance of crazing, cracks, or even total failure of the sample. Also, a plastic container can be tested by filling it with the liquid, causing ESC. Stress conditions applied during testing may be of two types: constant stress or constant strain. The selection of the test method should correspond with the intended use of the container. Common ESC tests for PE bottles include bottle stress crack (ASTM D2561), top-load stress crack, and internal pressure (ASTM D2561). For some applications, the temperatures recommended by ASTM for these tests may be too mild (14).

Flammability. The *flammability* behavior of a material is a broad term related to the easiness to ignite, burn, produce smoke, and endure burning. A flammable material ignites easily and has a rapid flaming combustion process. The initial step is decomposition (or pyrolysis), which is an endothermic process (the material absorbs heat). After ignition, products from pyrolysis are combusted and consequently, heat Q is generated. Some flame-retardant agents tend to minimize Q. The flammability characteristics of a polymer may be predicted from its chemical structure (1), and a method to measure flame resistance is the *oxygen index* (15). ASTM test D3713 describes the response of solid plastics to ignition by a small flame, and the ASTM D4100 method covers the gravimetric determination of smoke-particulate matter produced from combustion or pyrolysis of plastic materials. Testing plastics for practical fire situations is very complex and depends on many variables other than just chemical structure (16,17).

Photodegradation. Sunlight is the source of energy for polymer photochemical degradation. Near-ultraviolet and visible radiation (290–710 nm) carry enough energy to break single covalent bonds such as C—C, C—O, or C—N. The absorption of these radiations is sometimes attributed to impurities in the polymer. Photodegradation processes generate chain ruptures and free-radical formation, resulting in color changes and increasing fragility (because of chain crosslinking and chain scissions). UV light stabilization of PP and other polyolefins can be achieved by the addition of hindered amine light stabilizers (HALSs) (18). These additives reduce the rate of photoinitiation and chain-reaction propagation by their ability to trap free radicals. ASTM D4674 describes the accelerated testing for color stability of plastics exposed to indoor fluorescent lighting and window-filtered daylight.

Solubility parameter. Polymer solubility properties in organic liquids are controlled by the polymer's cohesive energy. The cohesive energy of a solid (or liquid) is the total energy necessary to remove one molecule from the bulk of the solid (or liquid). It is also an indication of the molecular internal pressure. The solubility of a polymer is given by its solubility parameter δ, which is the square root of the cohesive-energy density (CED), $\delta = (\mathrm{CED})^{1/2}$. In addition to cohesive energy density, polymer morphology and molecular mass also strongly affect the tendency of a polymer to dissolve. Crystalline structures, on the other hand, are insoluble below their melting temperature. Polymers have solubility parameter's ranging from 12 to 29 $(\mathrm{J/cm^3})^{1/2}$. Substances that have similar delta values [$< 0.5\ (\mathrm{J/cm^3})^{1/2}$ apart] will tend to mutually dissolve. This substantiates the rule of thumb that "like dissolves like." Solubility parameters can be determined by solvent swelling (19) or intrinsic viscosity methods (20). Values of CED and solubility parameter of a polymer can be estimated by group contribution methods, based on the contribution of intermolecular forces: polar forces, dispersion forces, and hydrogen bonding (1). Extensive tables of δ are presented by Barton (21).

Solubility in organic solvents. Many organic compounds interact with polymers to produce slight swelling or staining, or total dissolution, depending on the polymer molecular mass, crystallinity, and solubility parameter. ASTM D2299 gives a method to assess the susceptibility of plastic staining by incidental contact with organic products. Common solvents for plastics are boiling xylene and trichlorobenzene, for polyolefins, styrene polymers, vinylchloride polymers, and polyacrylates; formic acid for polyamides and polyvinyl alcohol derivatives; tetrahydrofuran for all uncrosslinked polymers; nitrobenzene and *m*-cresol for PET; water for polyacrylamide and polyvinyl alcohol; and dimethylformamide for polyacrylonitrile. In addition, polyvinyl acetate is dissolved by benzene, chloroform, methanol, acetone, and butyl acetate (22). ASTM D543 covers the testing of all plastic materials for resistance to chemical reagents.

Thermodegradation. Thermal degradation of polymers is the breaking of chain bonds by heat, and in absence of oxygen. Heat-stabilizing additives help prevent degradation, but impurities may have the contrary effect. Thermally stable polymers resist thermal degradation and will have high values of bond dissociation energy (1). As the temperature increases, chemical bonds with the lowest values of dissociation energy will be the first to break. Thermodegradation affects PE, PP, and PVC (18). The poor thermal stability of the PVC resin is caused mainly by dehydrochlorination. For each HCl eliminated from the polymer chain, a double bond is created, which, in turn, promotes oxygen chemical degradation. Heat stabilizers include inorganic and organic compounds. In the case of flexible PVC films, barium–zinc and calcium–zinc stabilizers are used, while organotin compounds are commonly used in PVC bottle containers.

Barrier properties. The *barrier property* of a material indicates its resistance to diffusion and sorption of foreign molecules. A high-barrier polymer has low values of both diffusion (D) and solubility (S) coefficients. Since the permeability coefficient P is a derived function of D and S, a high-barrier polymer has low permeability. The diffusion coefficient indicates how fast a penetrant will move within the polymer, while the solubility coefficient gives the amount of the penetrant taken (or sorbed) by the polymer from a contacting phase. Both diffusion and solubility can be applied to the reverse process of sorption, that is, the migration of compounds from the polymer to a surrounding medium. Several factors influence the effective value of the diffusion and solubility coefficients in polymers: (*1*) chemical compositions of the polymer and permeant, (*2*) polymer morphology (since diffusion and sorption occur mainly through the amorphous phase and not through crystals), (*3*) temperature (as temperature increases, diffusion increases while solubility decreases), (*4*) glass-transition temperature, and (*5*) the presence of plasticizers and fillers.

Diffusion coefficient. In Fick's law, the diffusion coefficient D is a parameter that relates the flux of a penetrant in a medium to its concentration gradient. A diffusion-coefficient value is always given for a particular molecule–polymer pair. For solid polymers, the diffusion coefficient values may range from 1×10^{-8} to 1×10^{-13} cm^2/s. The diffusion theory shows that diffusion is an activated phenomenon that follows Arrhenius' law. In addition to temperature, penetrant concentration and plasticizers also affect the value of the diffusion coefficient. Excellent references covering the diffusion process are the books by Crank (23), and Crank and Park (24).

Permeability coefficient. The permeability coefficient (P) combines the effect of the diffusion and solubility coefficients. The barrier characteristic of a polymer is commonly associated with its permeability coefficient value. The well-known relationship $P = DS$ holds when D is concentration independent and S follows Henry's law. Methods for measuring the permeability to organic compounds are described by Hernandez et al. (25). ASTM E96 describes a method for measuring the water-vapor transmission rate (WVTR). ASTM D1434 covers a method for the determination of oxygen permeability.

Solubility coefficient. The solubility coefficient indicates the sorption capacity of a polymer with respect to a particular sorbate. The simplest solubility coefficient is defined by Henry's law of solubility, which is valid at low concentration values. The solubility of CO_2 in PET at high pressure is described by combining Henry's and Langmuir's laws (26). In the case of organic compounds, the Flory–Huggins equation is applicable (24), and for sorption of water in nylons, a special model has been proposed (27).

Surface and adhesion

Adhesive bond strength. Adhesive bond strength between an adhesive and a solid substrate is a complex phenomenon, and is controlled (at least in part) by the values of surface tension, solubility parameters, and adhesive viscosity. To obtain wettability and adhesion between a polymeric substrate and an adhesive, the surface tension of the adhesive must be lower than that of the substrate. Usually, the difference between the two values must be at least 10 dyn/cm. The similarity in solubility parameters between the two phases indicates the similarity of the intermolecular forces between the two phases. For good compatibility, the values of the solubility parameters must be very close. Low viscosity in the adhesive is necessary for good spreadability and wettability of the substrate (28).

Blocking. Blocking is the tendency of a polymer film to stick to itself simply by physical contact. This effect is controlled by the adhesion characteristic of the polymer. Blocking is enhanced by surface smoothness and by pressure on the films present in stacked sheets or compacted rolls. Blocking can be measured by the perpendicular force needed to separate two sheets, and it can be minimized by incorporating additives such as talc in the polymer film. ASTM D1893, ASTM D3354, and Packaging Institute procedure T3629 present methods to evaluate blocking.

Cohesive bond strength. Cohesive bond strength is the force within the adhesive itself when bonding two substrates. The cohesive bond strength depends mainly on the intermolecular forces of the adhesive, molecular mass, and temperature.

Friction. The coefficient of friction (COF) is a measure of the friction forces between two surfaces, and it characterizes a film's frictional behavior. The COF of a surface is determined by the surface adhesivity (surface tension and crystallinity), additives (slip, pigment, and antiblock agents), and surface finish. Cases in which the material's COF values require careful consideration include film passing over free-running rolls, bag forming, the wrapping of film around a product, and the stacking of bags and other containers. Besides the intrinsic variables affecting a material's COF, environmental factors such as machine speed, temperature, electrostatic buildup, and humidity also have considerable influence in its final value. The *static COF* is associated with the force needed to start moving an object. It is usually higher than the *kinetic coefficient,* which is the force needed to sustain movement. Determination of static COF is described in TAPPI standard T503 and ASTM D1894. Thompson (29) has also analyzed the action of additives on the value of COF of polypropylene.

Heat sealing. An important property for wrapping, bagmaking, or sealing a flexible structure is the heat sealability of the material. At a given thickness, heat-sealing characteristics of flexible web material are determined by the material's composition (which controls strength), average molecular mass (controlling temperature and strength), molecular-mass distribution (setting temperature range and molecular entanglement), and the thermal conductivity (controlling dwell time (30). Tests normally conducted to evaluate the heat sealability of a polymeric material are the *cold-peel strength* (ASTM F88) and the *hot-tack strength* (31). Hot tack is the melt strength of a heat seal when the seal interface is still liquid and without mechanical support. The hot adhesivity is associated with the molecular entanglement of the polymer chains, viscosity, and intermolecular forces of the material.

Surface tension. In solids and liquids, the forces associated with inside molecules are balanced because each molecule is surrounded by like molecules. On the other hand, molecules at the surface are not completely surrounded by the same type of molecules generating therefore, unbalanced forces. At the surface, therefore, these molecules show additional free surface energy. The intensity of the free energy is proportional to the intermolecular forces of the material. The free surface energy of liquids and solids, called *surface tension,* can be expressed in mJ/m^2 (millijoules per square meter) or dyn/cm (dynes per centimeter). Values of surface tension in polymers range from 20 dyn/cm for Teflon to 46 dyn/cm for nylon 6,6. The measurement of surface tension by contact-angle measurement is covered by ASTM D2578. Several independent methods are used to estimate surface tension of liquids and solid polymers, including the molar parachor (1).

When two condensed phases are in close contact, the free energy at the interface is called *interfacial energy.* Interfacial energy and surface energy in polymeric materials control adhesion, wetting, printing, surface treatment, and fogging.

Wettability. Adhesion and printing operations on a plastic surface depend on the value of the substrate wettability or surface tension. A measure of the wettability of a surface is given by a material's surface tension as described in ASTM D2578.

Electrical properties. Electrical conductivity, dielectric constants, dissipation factors, and triboelectric behavior are electrical properties of polymers subject to low electric-field strength. Materials can be classified as a function of their conductivity (κ) in $(\Omega/cm)^{-1}$ (reciprocal ohms per centimeter) as follows: conductors, $0–10^{-5}$; dissipatives, $10^{-5}–10^{-12}$; and insulators, $\geq 10^{-12}$. Plastics are considered nonconductive materials (ie, not counting the newly developed conducting materials).

Dielectric constant. The relative dielectric constant of an insulating material (ε) is the ratio of the electric capacities of parallel-plate condenser with and without the material between the plates. A correlation between the dielectric constant and the solubility parameter (δ) is given by the relationship of $\delta \approx 7.0\varepsilon$ (1). There is also a relationship between *resistivity* R (inverse of conductivity) and the dielectric constant at 298 K: $\log R = 23\text{-}2\varepsilon$ (1). Values of ε for polymers are presented in Refs. 1 and 2.

Triboelectric behavior. When two polymers are rubbed against each other, one becomes positively charged and the other negatively charged, depending on the relative electron donor–acceptor characteristics of the polymers. A *triboelectric* series is the listing of polymers according to their charge intensity, which goes from the most negative charge (electron acceptor) to the most positive (electron donor). The charge of polymer films also takes place by friction during industrial operations such as form/fill/seal. The most negative polymers are (value in parentheses is the dielectric constant): PP (ε = 2.2), PE (ε = 2.3), and PS (ε = 2.6); neutral charged polymers include PVC (ε = 2.8), PVDC (ε = 2.9), and PAN (ε = 3.1); and positively charged polymers include cellulose (ε = 3.7), and nylon 6,6 (ε = 4.0). As these values suggest, the tribolectric series of dry polymers can be correlated with the polymer dielectric constant. Since hydrophilic polymers absorb water, they become more conductive, and their dielectric constants increase. Standard methods for measuring the triboelectric charge of films and foams are, at the time of this article, still under consideration by ASTM Committee D10.13 (32).

Optical appearance. Among the most important optical properties of polymers are absorption, reflection, scattering, and refraction. Absorption of light takes place at the molecular level, when the electromagnetic energy is absorbed by atoms or group of atoms. If visible light is absorbed, a color will appear; however, most polymers show no specific absorption with visible light and, therefore, are colorless. Reflection is the light that is remitted on the surface and depends on the refractive indices of air and polymer. Scattering of light is caused by optical inhomogeneities reflecting the light in all directions. Refraction is the change in direction of light due to the difference between the polymer and air refraction indices. Transparency, opacity, and gloss of a polymer are not directly related to the chemical structure or molecular mass; they are determined mainly by the polymer morphology. Optical appearance properties are of two types: optical morphological properties, which correlate with transparency and opacity; and optical surface properties, which produce the specular reflectance and attenuated reflectance (1,33).

Gloss. Gloss is the percentage of incident light that is reflected at an angle equal to the angle of the incident rays (normally 45°). It is a measure of the ability of a surface to reflect the incident light. If the specular reflectance is near zero, the surface is said to be "mat." A surface with high reflectance has a high gloss, which produces a sharp image of any light source and gives a pleasing sparkle. Surface roughness, irregularities, and scratches all decrease gloss. Test method is covered by ASTM 2457.

Haze. Haze is the percentage of transmitted light that, in passing through the sample, deviates by more than 2.5° from an incident parallel beam. The appearance of haze is caused by light being scattered by surface imperfections and nonhomogeneity. The measurement of haze is described in ASTM D1003.

Transparency and opacity. Transmittance is the percent of incident light that passes through the sample and is determined by the intensity of the absorption and scattering effects. The absorption in polymers is insignificant, so, if the scattering is zero, and the sample will be transparent. An opaque material has low transmittance and, therefore, large scattering power. The scattering power of a polymer results from morphological inhomogeneities and/or the presence of crystals. Then, an amorphous homogeneous polymer such as "crystal" polystyrene will have little or no scattering power and therefore will be transparent, while a highly crystalline polymer such as HDPE will be mostly opaque. Transmittance can be determined according to standard ASTM D1003. A transparent material has a transmittance value above 90%.

BIBLIOGRAPHY

1. D. W. Van Krevelen, *Properties of Polymers,* 3rd ed., Elsevier, New York, 1990.
2. J. Brandup, and E. Immergut, eds., *Polymer Handbook,* 3rd ed., Wiley-Interscience, New York, 1989.
3. D. Braun, *Identification of Plastics,* 2nd ed., Hanser Publishers, 1986.
4. American Society for Testing and Materials, Vols. 08.01–08.03, 1988.
5. R. C. Progelhof and J. L. Thorne, *Polymer Engineering Principles,* Hanser Publishers, New York, 1993.
6. *Guide to Plastics Property and Specification Charts, Modern Plastics,* McGraw-Hill, New York, 1987, pp. 46–48.
7. B. Wunderlich, S. Cheng, and K. Loufakis, in J. I. Kroschwitz, ed., *Encyclopedia of Polymer Science and Engineering,* Vol. 16, Wiley, New York, 1989, pp. 767–807.

8. D. Rosato, *Rosato's Plastic Encyclopedia and Dictionary,* Hanser Publishers, New York, 1993.
9. G. Kampf, *Characterization of Plastic by Physical Methods,* Hansen Publishers, New York, 1986.
10. A. Birley, B. Harworth, and J. Batchelor, *Physics of Plastics,* Hansen Publishers, New York, 1992.
11. R. P. Brown, ed., *Handbook of Plastic Testing,* 2nd ed., Pitman Press, Bath, UK, 1981.
12. S. H. Hamid, M. B. Amin, and A. G. Maadhah, eds., *Handbook of Polymer Degradation,* Marcel Dekker, New York, 1992.
13. R. J. Hernandez, *J. Food Eng.* **22,** 495–505 (1991).
14. J. J. Strebel, *Polym. Test.* **14,** 189–202 (1995).
15. C. P. Fenimore, and F. J. Martin, *Modern Plast.* **44,** 141 (1966).
16. F. L. Fire, *Combustion of Plastics,* Van Nostrand Reinhold, New York, 1991.
17. C. J. Hilado, *Flammability Handbook for Plastics,* 4th ed., Technomic, Lancaster, PA, 1990.
18. P. N. Son, "UV Stabilizers," *Modern Plastics Encyclopedia Handbook,* McGraw-Hill, 1994, p. 119.
19. G. M. Bristow and W. F. Watson, *Trans. Fradady, Soc.* **54,** 1731,1742 (1958).
20. H. Mark and A. V. Tobolsky, *Physical Chemistry of High Polymers,* 2nd ed., Interscience, New York, 1950.
21. A. F. M. Barton, *Handbook of Polymer-Liquid Interaction Parameters and Solubility Parameters,* CRC Press, Boca Raton, FL, 1990.
22. D. Braun, *Simple Methods for Identification of Plastics,* 2nd ed., Hanser Publishers, New York, 1986.
23. J. Crank, *The Mathematics of Diffusion,* 2nd ed., Clarendon Press, London, 1975.
24. J. Crank and G. S. Park, *Diffusion in Polymers,* Academic Press, New York, 1968.
25. R. J. Hernandez, J. R. Giacin, and A. L. Baner, *J. Plast. Film Sheet.* **2,** 187–211 (1986).
26. W. R. Vieth, J. M. Howell, and J. Hsieh, *J. Membrane Sci.* **1,** 177 (1976).
27. R. J. Hernandez and R. Gavara, *J. Polym. Sci., Part B* **32,** 2367–2374 (1994).
28. D. E. Packham, *Handbook of Adhesives,* Wiley, New York, 1993.
29. K. I. Thompson, *TAPPI J.,* 157 (Sept. 1988).
30. J. W. Spink, R. J. Hernandez, and J. R. Giacin, *TAPPI Proceedings Polymer, Laminations and Coating Conference,* 1991, pp. 579–587.
31. H. Theller, *J. Plast. Film Sheet.* **5**(1), 66 (1989).
32. C. W. Hall, S. Singh, and G. Burgess, *J. Test. Evaluation,* **21,** 57–61 (1993).
33. M. Born and E. Wolf, *Principle of Optics,* 6th ed., Pergamon Press, Oxford, 1980.

RUBEN J. HERNANDEZ
Michigan State University,
East Lansing, Michigan

POLYPROPYLENE

Polypropylene is an extremely versatile material in the packaging industry. The reason for its adaptability is the ease with which its polymer structure and additive packages can be tailored to meet diverse requirements. Many useful properties are inherent in polypropylene. It has low density (high yield), excellent chemical resistance, a relatively high melting point, and good strength, at modest cost.

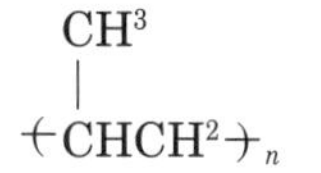

Figure 1. Polypropylene.

General Categories and Definitions

Polypropylene is the result of linking a large number (typically 1000 to over 30,000) of propylene molecules to build long polymer chains (see Fig. 1). Polymers made up only of propylene are called homopolymers. If another monomer is added (typically ethylene), the polymer is called a copolymer. The order and regularity of the monomer units in the polymer control the properties of the product.

One end of a propylene molecule is different from the other. The end with three hydrogens is called the head; a head-to-tail linkage is called stereoregular. If the process links the monomers head-to-head about as often as they are linked head-to-tail, the resulting polymer has little order and does not crystallize (1). At room temperature, this material has a density of about 0.850 g/cm^3. It is called atactic or amorphous polypropylene and is soft, tacky, and soluble in many solvents. Such polymers are useful in hot-melt adhesives (see Adhesives) and several other applications.

If monomers are connected head-to-tail almost every time, the polymer is said to be isotactic and crystallizes (see Fig. 2). Crystallinity is the reason for the solvent resistance, stiffness, and heat resistance of the commercial plastic material. At normal conditions, isotactic polypropylene is usually about half crystalline and has a density of 0.902 ± 0.005 g/cm^3. Normal polypropylene melts at about 329°F (165°C) when heated slowly and contains roughly 5% of atactic material as well as intermediate structures. In this article, the term *polypropylene* (PP) implies the plastic of commerce.

When discussing copolymers and alloys, an additional level of complexity is added. Alloys, also called blends, are mixtures of polymers. Copolymers are the result of polymerizing two or more monomers together. There are many possible types (structures) of copolymers (2), not all of which are useful.

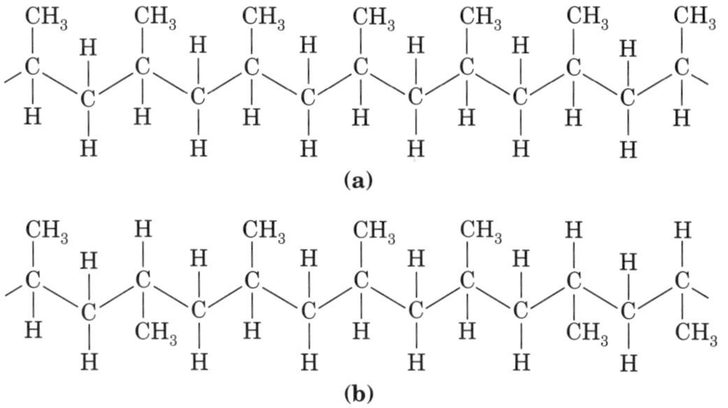

Figure 2. Comparison of the structures of isotactic and atactic polypropylenes: (**a**) isotactic PP–methyl groups iin orderly alignment; (**b**) atactic PP–methyl groups in random alignment.

Table 1. Typical Polypropylene Properties

	Values			
Property	Homopolymer	Clarified Random Copolymer	Impact Copolymer	High Impact Copolymer
Melt flow condition *L*, g/10 min	4	10	4	4
Tensile strength (yield), psi (MPa)	4900 (33.8)	4300 (29.6)	3900 (26.9)	3000 (20.7)
Yield elongation, %	11	9	9	8
Flexural modulus, (1% secant), psi (MPa)	200,000 (1,380)	150,000 (1,030)	175,000 (1210)	130,000 (900)
Hardness Rockwell R (HRC)	86	77	73	70
Heat-deflection temperature (66 psi or 455 kPa), °F (°C)	199 (93)	189 (87)	189 (87)	176 (80)
Notched Izod, 23°C, ft · lbf/in. (J/m)	0.7 (37)	1.4 (75)	1.8 (96)	>10 >500
Application	Injection molding	Injection molding	Injection molding	Injection molding

Random copolymers result when a small amount of comonomer, usually ethylene, is polymerized at random intervals along the PP chain. Typical ethylene levels are from 1 to 5 wt %, and the product has only one phase. These copolymers are relatively clear and have lower and broader melting points than PP homopolymers. The lowering of the melting point is proportional to the randomness of incorporation and the amount of comonomer incorporated. At ethylene levels well above 10%, the product becomes less crystalline and is called EPR, or ethylene-propylene rubber. One cannot make an alloy (polymer blend) that resembles a random copolymer.

Impact copolymers generally contain larger amounts of ethylene monomer, typically 4 to >25%, and are heterophasic. The ethylene may be polymerized in the form of polyethylene and/or EPR. Usually, such products are made by polymerizing a homopolymer and changing the conditions to add ethylene to the polymer chain. Reasonably comparable materials can be made by mechanically blending polypropylene with EPR and/or polyethylene. If crystalline EPR or polyethylene is present, it can be detected by a second melting point at 224–271°F (118–133°C). These products are characterized by lower stiffness, much enhanced toughness at low temperatures, and a relatively opaque appearance (see Table 1).

The molecular weight (related to the average number of monomer units in a chain) and the molecular-weight distribution (MWD) are significant polymer characteristics. High-molecular-weight (HMW) polymers are highly viscous when melted and are difficult to injection mold or push through restrictive dies; but they have high toughness and good "melt strength" (melt elasticity). Melts of low-molecular-weight (LMW) polymers are more fluid. They have less toughness and lower melt strength. In a narrow-MWD polymer, there is less variation in the length of the chains than in a broad-MWD polymer. Narrow MWD allows retention of the toughness of HMW grades (3) in a more-easily processed material. Narrow MWD materials do not generally have high melt strength. They offer advantages in fiber spinning and injection molding.

Substantial modification in properties can be achieved by the use of additives (see Additives, plastics) and fillers. Virtually all commercial grades are stabilized to increase resistance to oxidation on aging or at elevated temperatures. Additives can confer resistance to sunlight, reduce the tendency to retain static electric charges, modify the coefficient of friction, and prevent surface tackiness.

The additives should be selected for the application. In food packaging, the stabilization should be chosen to avoid transfer of taste and odor to the package contents. Antistats reduce static electricity, which attracts dust and makes packages appear dirty. In sunlight, the damaging wavelengths are in the uv range. Uv resistance is important for items that may be stored outdoors. Coefficient of friction and antiblocking properties are important in the winding and handling of film (see Slitting/rewinding). Frictional characteristics are also important in threaded closures (see Closures).

Fillers can greatly increase stiffness, improve processing behavior, confer conductivity, and change the appearance of PP. Commonly used fillers are talc, calcium carbonate (usually powdered marble), glass fiber, mica, carbon black, clays, cellulose fibers, lubricants, and pigments. Many exotic fillers have been used for special purposes. The use of fillers in PP usually reduces toughness and raises density and cost per volume.

Processing

Melt-flow rate (MFR) is one of the key variables in the processing of PP. The American Society for Testing and Materials (ASTM) specifies (4) that PP is to be tested at 446°F (230°C) under a pressure exerted by a nominal 4.4-lb (2-kg) mass in the apparatus specified under ASTM Standard D-238. The result is the weight of material extruded through the standard orifice in 10 min. This test is a crude measure of the melt viscosity of the plastic under low shear rate. Melt viscosity correlates to the weight-average molecular weight of the polymer. Commercial PP grades of interest in packaging span a MFR range of 0.3–40 g/10 min. The low flow grades (up to about 2.5 g/10 min) are used in sheet extrusion (see Extrusion) and blow molding (see Blow molding). Film is manufactured from intermediate-flow grades (MFR 2–15 g/10 min). Injection-molding processes (see Injection molding) ordinarily use PP with MFR of 3–40 g/10 min (see Table 2).

Applications and Markets

In 1984, consumption of polypropylene in packaging applications in the United States was roughly as follows (5):

Applications	1 × 10⁶ lb (1 × 10³ tons)
Containers and lids	
Blow-molded	154 (70.0)
Injection-molded	254 (115)
Thermoformed	43.5 (19.8)
Film	
Oriented	711 (323)
Nonoriented	172 (78.2)
Closures	
Injection molded	583 (265)

Thermoforming. Sheet extrusion for thermoforming (see Thermoforming) is a small, rapidly growing segment of the PP market. Until the late 1970s, few converters were willing to attempt thermoforming of PP. Since that time, thermoforming techniques and grades have been improved. Random copolymers and impact copolymers have a broader temperature "window" and are more easily formed than homopolymers. Several grades now in development promise excellent forming characteristics, even in homopolymers. The several available high-melt strength PP grades ease manufacture via improved material distribution and higher forming rates. The factors that promote the use of PP in thermoforming are low odor and taste transfer, good moisture barrier, chemical resistance, and adequate clarity in thin sections.

Blow molding. In blow molding, most of the resin consumed is for consumer products. High melt strength is required for extrusion blow molding, which is used to produce relatively large containers up to 5.3 gal (20 L) in size. Injection blow molding usually requires MFR of 1–2.5 g/10 min. This process is particularly useful for relatively small containers. The ability to withstand temperatures or aggressive chemicals that would stress-crack or attack other materials is usually the reason to blow-mold PP today. In detergent exposure tests that crack HDPE in a day or two, PP does not fail even after many weeks. Random and impact copolymers are most often used. The development of high-melt-strength resins will assist the growth of PP blow molding. Today, injection stretch blow-molded PP bottles are biaxially oriented, and rival the clarity of PET. If the chains of PP are stretched, they line up to give remarkable strength, clarity, and toughness (even at low temperature). This process uses a preform that is stretched and blown while warm. Random copolymers are used since low temperature toughness is provided by biaxial orientation. Typical resin MFR is 1–3 g/10 min for these applications.

Film. The use of PP in film is very large. Oriented films typically have high toughness and excellent clarity (see Film, oriented polypropylene). They can be produced by a high-expansion bubble process or a tenter process. Product variations are possible based on the amount of transverse- and machine-direction orientation. Nonoriented cast films are usually made by a chill-roll process but there are also water-quench and water-quenched bubble processes in use (see Extrusion; Film, nonoriented polypropylene). Oriented films can have a stiff feel or "hand." They sparkle and tend to "crinkle" audibly. These films are employed as cigarette wrap, candy wrap, and snack-food pouches, often in replacement of cellophane (see Cellophane). Oriented films have been tailored for superior barrier properties (by coating), heat-seal strength (by resin modification or coating), heat-shrink properties (see Films, shrink), printability, and electrical properties. A relatively new addition to the family of oriented films is opaque film. The opacity is produced by a filler and by controlled voiding. A type of decorative ribbon is foamed uniaxially oriented PP with a colorant (see Colorants).

The use of woven slit tape or slit film for heavy-duty agricultural bags is not a major use in the United States, but these bags compete with jute and other natural fiber in carrying much of the world's grain. A related product is strapping, which is made by extruding either sheet or tapes. If sheet is extruded, it is subsequently slit (see Slitting/rewinding). The filaments are then stretched while warm to give tensile strength values of ≥50,000 psi (>345 MPa), ten times that of unoriented PP.

Nonoriented film has a number of growing markets. Compared to oriented film, it is available in thicker gauges and has a softer "hand" at the same thickness. Some applications are release sheets, sanitary products, disposable-diaper layers, bandages, and apparel packaging. The use of PP in composite film and sheet materials is small, but growing rapidly. PP is used in combination with other PP structures, paper, metal foils, fabric (woven and nonwoven), and other plastics. Such composites can be made by coextrusion (see Coextrusion), lamination (see Laminating), or extrusion coating (see Extrusion coating). The motivation for making such structures is often related to barrier properties, temperature resis-

Table 2. Effect of Molecular Weight (Melt Flow) on Homopolymer Properties

Property	Values			
Melt-flow condition *L*, g/10 min	0.4	2	12	35
Tensile strength (yield), psi (MPa)	4800 (33.1)	5000 (34.5)	5100 (35.2)	4600 (31.7)
Yield elongation, %	13	11	10	12
Flexural modulus (1% secant), psi (MPa)	190,000 (1310)	220,000 (1520)	220,000 (1520)	170,000 (1170)
Hardness Rockwell R (HRR)	92	91	90	88
Heat-deflection temperature (66 psi or 455 kPa), °F (°C)	207 (97)	203 (95)	207 (97)	201 (94)
Notched Izod, 23°C, ft · lbf/in. (J/m)	1 (53)	0.8 (43)	0.7 (37)	0.6 (32)
Application	Sheet extrusion Thermoforming Blow molding	Sheet extrusion Thermoforming Blow molding	Injection molding Film	Injection molding

tance, chemical resistance, and cost. The principal market for these products is in food packaging.

Most film processes use PP grades with a MFR in the range of 2–10 g/10 min. Extrusion coating uses materials with flow ranging from 10 to over 60 g/10 min. Selection of the additive package is an important consideration. Printing, winding, static-charge buildup, blocking (sticking together), odor, color, heat-seal strength, and other properties are influenced by the additives.

Injection molding. Injection molding produces many familiar packaging items, including threaded, dispensing, and pump closures (see Closures); aerosol valves and overcaps (see pressurized packaging); wide-mouthed jars, totes, crates (see Crates, plastic), yogurt cups, snuff boxes, cosmetic containers, drug syringes, barrel bungs, dairy tubs, and many others. Most general-purpose PP molding grades are well suited for packaging. In food and drug packaging, one must ensure that the particular grade is suitable for the product to be packaged. Conventional injection-molding grades span a MFR range of 2–>70 g/10 min. Generally, the lower-flow materials are tougher and process less rapidly. New polymer production processes can produce very high-flow PP grades without narrowing the MWD. Although less tough than narrow-MWD grades of similar MFR, these grades are more easily molded. Copolymers with MFR up to 70 g/10 min are being used in thin-walled injection-molded (TWIM) containers, notable for quality appearance and ease of printing or decorating. Thin walls allow high productivity, low part weight, and low container cost. By using narrow-MWD or "controlled-rheology" resins, high MFR grades with good toughness can be made. These grades are appropriate choices for thin-walled moldings and for large multiple-cavity molds. They offer better dimensional control, but are not as stiff as broader-MWD material.

Normal homopolymers are brittle at refrigerator temperatures. The use of copolymers is recommended when shipping or use expose the part to low temperature impact.

Health and Safety Issues

Except for fire-retardant grades containing antimony, PP is generally a nontoxic material. Many grades are available that comply with FDA requirements for food packaging. PP is used in drug packaging and medical devices. It is fiber-spun for use in undergarments, upholstery, sanitary products, and bandages. It is not soluble at normal temperatures in food and beverages. PP is combustible and burns completely when adequate air is available to the flame. As sold by resin manufacturers, PP is usually in a coarse granular form and presents no unusual fire hazard. If it is finely divided (finer than 200 mesh or 74 μm), however, polypropylene can present a combustible dust hazard as do most organic materials. Like other organic materials (eg, wood, wool, flour), the products of incomplete combustion include carbon monoxide and can include a number of unpleasant, partially oxidized pyrolysis products (aldehydes, ketones, etc).

Under most processing conditions, little hazard exists with the use of PP. One should avoid contact with the molten polymer. If the plastic is exposed to air at temperatures above 500°F (260°C), proper ventilation should be used. The autoignition temperature is 675–700°F (357–371°C) for most grades. In summary, there are no unusual risks associated with the use of PP.

BIBLIOGRAPHY

1. M. R. Schoenberg, J. W. Blieszner, and C. G. Papadopoulos, "Propylene," in M. Grayson and D. Eckroth, eds., *Encyclopedia of Chemical Technology,* 3rd ed., Vol. 19, Wiley, New York, 1982, p. 228.
2. J. R. Fried, "Polymer Technology—Part 1: The Polymers of Commercial Plastics," *Plast. Eng.* **38,** 49–55 (June 1982).
3. J. R. Fried, "Polymer Technology—Part 3: Molecular Weight and Its Relation to Properties," *Plast. Eng.* **38,** 27–33 (Aug. 1982) (Parts 4, 5, and 6 followed).
4. *1991 Annual Book of ASTM Standards,* Vol. 8.01, American Society for Testing and Materials, Philadelphia, Pa., pp. 394–402.
5. *Mod. Plast.* **62**(1), 69 (Jan. 1995).

General References

J. R. Fried, "Polymer Technology—Part 2: Polymer Properties in the Solid State," *Plast. Eng.* **38,** 27–37 (July 1982).

J. L. Szajna, "A Supplement to 1980 ANTEC Paper, Functions vs. Economics vs. Aesthetics," paper presented at the 40th Annual Technical Conference and Exhibition of the Society of Plastics Engineers, May 10–13, 1982.

H. Gross and G. Menges, "Influence of Thermoforming Parameters on the Properties of Thermoformed PP," paper presented at the 40th Annual Technical Conference and Exhibition of the Society of Plastics Engineers, May 10–13, 1982.

R. C. Miller
Montell USA Inc.
Wilmington, Delaware

POLYPROPYLENE FILM.

See Film, oriented polypropylene; Film, nonoriented polypropylene.

POLYSTYRENE

Polystyrene is the parent of the family of styrene-based plastics. By copolymerization of styrene with other monomers, a wide range of properties is obtainable. Polystyrene resin is one of the most versatile, easily fabricated, and cost-effective plastics.

General-purpose polystyrene (Table 1) is often called *crystal polystyrene,* which refers to the clarity of the resin. The commercial success of this resin is due to its transparency, ease of fabrication, thermal stability, low specific gravity, high modulus, excellent electrical properties, and low cost. It is commercially produced by two processes. Suspension polymerization is used primarily to produce foam-in-place beads and ion-exchange resin. The bulk of the resin is produced by solution polymerization in a continuous process consisting of one or more vessels followed by volatile removal at high temperatures and high vacuum. There are three common commercial grades of general-purpose polystyrene: easy flow, medium flow, and high heat. The choice of resin depends mostly on the fabrication method. Easy flow and medium flow are used primarily for injection molding and the high-heat extrusion applications.

High-heat polystyrene, such as Styron 685D (trademark of

Table 1. Typical Properties for General-Purpose Polystyrene

Property	ASTM Test Method	Units	High Heat	Medium Flow	Easy Flow
Melt-flow rate (condition G; 200°C/5 kg)	D1238	g/10 min	1.6	7.5	16.0
Vicat softening temperature	D1525	°C	108	102	88
Deflection temperature under load annealed 264 psi	D648	°C	103	84	77
Tensile strength at break	D638	psi	8200	6500	5200
		(MPa)	(56.6)	(44.8)	(35.9)
Tensile elongation at break	D638	%	2.4	2.0	1.6
Tensile modulus	D638	psi	485,000	460,000	450,000
		(MPa)	(3340)	(3175)	(3100)
Izod impact	D256	ft · lb/in.	0.5	0.4	0.3
		(J/m)	(24)	(21)	(16)
Rockwell hardness	D785	M scale	76	75	72
Molecular weight	SEC	Daltons	300,000	225,000	210,000
Molecular number	SEC	Daltons	130,000	92,000	75,000

The Dow Chemical Company), have the highest molecular weight and contain the fewest additives. High-heat resins normally do not contain mineral oil or other flow additives and generally contain a mold release additive or extrusion aid. The higher heat distortion and toughness are offset by low melt-flow rates. The high molecular weight improves the mechanical properties of the resin and also the ability to induce orientation during fabrication. Orientation of the polymeric chains tremendously increases the toughness of resin. For food-packaging applications, low residual levels are desirable to minimize taste or odor transmission to the food.

High-impact polystyrene (Table 2) sometimes called rubber-modified polystyrene, is normally produced by copolymerization of styrene and a synthetic rubber. This produces a two-phase system consisting of a dispersed rubber phase and a continuos polystyrene phase. This system uses a unique feature of polystyrene to allow elongation by the formation of energy-absorbing crazes. The dispersed rubber particles initiate astronomical numbers of crazes without crack formation, thus contributing to the development of very tough products. By varying the amount of rubber, normally 2–15 wt %, and morphology of the rubber-phase physical properties of the resin can be varied considerably. High-impact polystyrene has lower tensile strength and a tremendous increase in elongation and impact strength. The opacity of high-impact polystyrene precludes its use in applications requiring transparency.

Applications

Typical applications for general-purpose polystyrene include packaging products, disposable medical devices, toys, tumblers, cutlery, tape reels, storm windows, light diffusers, appliance parts, and electronic components. It is used in the production of foam products such as egg cartons, meat trays, fast-food containers, building insulation, and cushioning material for packaging. When orientation devices are used in its fabrication, general-purpose polystyrene can be fabricated into clear, tough articles such as lids, salad trays, cookie trays, blister packs, windows for envelopes, sandwich containers, tumblers, bottles, and decorative and lamination films.

Applications for high-impact polystyrene include packaging products, toys, tumblers, housewares, building and construction, and appliances. The superior elongation and impact properties of this resin do not require orientation to allow its fabrication by thermoforming. Therefore the cost of equip-

Table 2. Typical Properties of High-Impact Polystyrene

Properties	ASTM Test Method	Units	Extrusion Resin	Injection-Molding Resin
Melt-flow rate (condition G, 200°C/5 kg)	D1238	g/10 min	3.0	7.5
Vicat softening point	D1525	°C	100	94
Deflection temperature under load annealed 264 psi	D648	°C	92	82
Tensile rupture	D638	psi	2400	2300
		(MPa)	(16.6)	(15.9)
Tensile yield	D638	psi	2600	2700
		(MPa)	(17.9)	(18.6)
Tensile modulus	D638	psi	240,000	260,000
		(MPa)	(1655)	(1793)
Tensile elongation	D638	%	40	40
Flexural modulus	D790	psi	280,000	260,000
		(MPa)	(1931)	(1793)
Flexural strength	D790	psi	5600	4700
		(MPa)	(38.6)	(32.4)
Izod impact notched 73°F	D256	psi	1.5	1.6
		(J/m)	(80.1)	(85.5)
Rockwell hardness	D785	M scale	29	25

Table 3. Polystyrene (PS) Major Markets

Market	Million lb, 1994
Molding (Solid PS Only)	
Appliances and consumer electronics	
Air conditioners	33
Refrigerators and freezers	84
Small appliances	45
Cassettes, reels, etc	310
Radio/TV/Stereo cabinets	203
Other	14
Furniture and furnishings	
Furniture	37
Toilet seats	9
Other	13
Toys and recreational	
Toys	141
Novelties	55
Photographic	67
Other	10
Housewares	
Personal care	87
Other	102
Building and construction	62
Miscellaneous consumer and industrial	
Footwear (heels)	7
Medical	101
Other	22
Packaging and disposables	
Closures	117
Rigid packaging	108
Produce baskets	31
Tumblers and glasses	102
Flatware, cutlery	97
Dishes, cups, bowls	68
Blow-molded items	10
Other injection	119
Total molding	2054
Extrusion (Solid PS Only)	
Appliances and consumer electronics	
Refrigerators and freezers	116
Other	45
Furniture and furnishings	30
Toys and recreational	42
Housewares	67
Building and construction	77
Miscellaneous consumer and industrial	62
Packaging and disposables	
Oriented film and sheet	305
Dairy containers	171
Vending and portion cups	305
Lids	145
Plates and bowls	57
Other extrusion, solid PS	249
Extrusion (foam PS)	
Board	181
Sheet	
Foodstock trays	214
Egg cartons	55
Single-service	
Plates	174
Hinged containers	112
Cups (nonthermoformed)	55
Other foam sheet	40
Total extrusion	2502
Expandable-bead Polystyrene (EPS)	
Billets	
Building and construction	245
Other	45
Shapes	
Packaging	125
Other	59
Cups and containers	185
Loose-fill	90
Total expandable bead	749
Export	350
Other	222
Grand total	5877

Source: Information from *Modern Plastics* (Jan. 1995).

ment for the production of many packaging applications is considerably lower than that required for general-purpose polystyrene. This makes high-impact polystyrene the resin of choice for applications where clarity is not a requirement. Ignition-resistant additives are compounded into impact polystyrene for use in television, radio, and electronic cabinets.

Table 3 contains the markets and approximate pounds consumed by them for 1994.

Fabrication

Processing of polystyrene can be accomplished by most fabricating techniques. Polystyrene resins are among the most widely used thermoplastics for both extrusion and injection-molding applications. Various formulations offer different properties specially suited for specific applications. This also results in slightly different processing conditions. Polystyrene resins are usually processed at melt temperatures of 360–500°F. In extrusion applications, extruders vary significantly in size. Extruder screws can range from small ($\frac{3}{4}$-in.-dia) to large (12-in.-dia) with varying *L/D* ratios from 20 : 1 to 42 : 1. The current trend in extruder design is to 30 : 1 and 36 : 1 ratios using two-stage single screws and vented barrels. Both single- and two-stage screws are used and because of polystyrene's excellent stability compression ratios, between 3 : 1 and 5 : 1, which are used to achieve good mixing and uniform delivery. Screw cooling of the feed section of the screw is recommended on large extruders when operating at high output rates or when high levels of regrind are being used. Because polystyrene is nonhygroscopic, drying of the granules prior to extrusion is usually unnecessary.

An optimum set of operating conditions can be determined only through experience with a particular extruder. Optimum running conditions require the proper balancing of many variables such as screw design, extruder size, extrusion rate, extruder conditions, polymer used, and desired end product. Suggested starting conditions using a typical polystyrene screw are as follows:

Hopper	Adequate Flow of Cooling Water
Feed section	350°F (176°C)
Transition section	400°F (204°C)
Metering section	440°F (226°C)

Table 4. Typical Sizes of Injection-Molding Machines

Nominal Rating of Clamping Capacity, kN[a]	Shot Size, g PS[b]	Rate of Fill, mL/min[c]
1335	170–340	131–262
4450	1360–2155	410–738
8900	4536–5100	1148–1476

[a] To convert kN to tons, divide by 8.9.
[b] To convert g to oz, divide by 28.35.
[c] To convert cm3/min to in3/min, divide by 16.4.

Extruder head pressures range from 1500 to 4000 psi. The die gap is normally set at 10–20% greater than the desired sheet thickness. Polystyrene resins are notably stable materials. Evidence of this is seen in the ability to use regrind and to recycle these resins. Many new packaging applications are being developed through the use of coextrusion. By incorporating layers of other resins such as EVOH and PVDC barrier properties can be achieved, and the ease of polystyrene fabrication and effective cost and performance of polystyrene can be maintained.

In injection-molding applications, machines vary in size and capacity to permit molding very small to very large parts. Machine cost is primarily a function of clamping capacity, which is the force necessary to hold the mold halves together during high-pressure injection. This is directly related to the pressure needed to fill the mold cavity. Typical sizes are listed in Table 4. For best operating conditions, the preferred shot size is usually about 50–75% of the maximum shot size of the machine.

Cooling time is the major portion of a injection molding cycle. The cooling times are influenced mainly by part wall thickness and the heat capacity of the polymer. In the case of crystalline polymers (eg, polypropylene), the heat of crystallization is a large part of the heat that must be removed. The shorter cooling times for amorphous polymers, such as polystyrene resins, allow for higher production rates than are possible with many crystalline polymers. Typical injection-molding conditions for polystyrene are listed in Table 5.

Table 5. Typical Injection-Molding Conditions

Controls	English Units	SI Units
Barrel temperatures		
Zone 1, feed	375–420°F	190–215°C
Zone 2	400–450°F	204–232°C
Zone 3	430–475°F	221–246°C
Zone 4	430–475°F	221–246°C
Zone 5, adapter	430–475°F	221–246°C
Nozzle	425–460°F	218–238°C
Melt temperature	430–475°F	221–246°C
Mold temperature	70–120°F	20–50°C
Injection pressure	1100–2300 psi	8–16 MPa
Injection pressure	Adjust to control part weight and dimensions	
Injection speed	Variable control, slow–fast–slow, adjust to control appearance	
Backpressure	0–150 psi, higher with concentrates	

BIBLIOGRAPHY

1. A. J. Warner, in R. H. Boundy and R. F. Boyer, eds., *Styrene, Its Polymers, Copolymers, and Derivatives,* Reinhold Publishing, New York, 1952.
2. R. F. Boyer, in J. Kroschwitz, ed., *Encyclopedia of Polymer Science and Technology,* Vol. 13, Wiley, New York, 1970.
3. J. Frados, *Plastics Engineering Handbook,* Van Nostrand Reinhold, New York, 1976.
4. Z. Tadmor and I. Klein, *Engineering Principles of Plasticating Extrusion,* Van Nostrand Reinhold, New York, 1970.
5. V. Shah, *Handbook of Plastics Testing Technology,* Wiley, New York, 1984.
6. *Annual Book of ASTM Standards,* American Society of Testing and Materials, Philadelphia.
7. *Thermoforming Process Guide,* Form No. 305-01833-1092SMG, The Dow Chemical Company, Midland, MI.
8. J. D. Griffin and J. Y. Glass, "Polystyrenes" in D. M. Considine, ed., *Chemical and Process Technology Encyclopedia,* McGraw-Hill, New York, 1974, pp. 914–917.
9. C. Harper, *Handbook of Plastics and Elastomers,* McGraw-Hill, New York, 1975.
10. H. S. Gilmore and A. R. Hoge, "Polystyrene," in *Modern Plastics Encyclopedia 1984–85,* McGraw-Hill, New York, 1984.
11. F. L. Burkett, *"Styrene Polymers"* in J. Kroschwitz, ed., *Encyclopedia of Polymer Science and Engineering,* Vol. 16, Wiley, 1989.

PHILLIP A. WAGNER
JOHN SUGDEN
The Dow Chemical Company
Midland, Michigan

POLY(VINYL CHLORIDE)

Poly(vinyl chloride) (PVC) is, in terms of sales volume, the largest member of a group of polymers commonly referred to as "vinyls." These polymers are all based on either the vinyl radical ($CH_2{=}CH{-}$) or the vinylidene radical ($CH_2{=}CR{-}$). Included in this unique and versatile group of polymers are poly(vinyl acetate), poly(vinylidene chloride) (PVDC), poly(vinyl alcohol), poly(vinyl fluoride), poly(vinylidene difluoride) (PVDF), and poly(vinyl butyral). However, homopolymer PVC has the greatest applicability to packaging applications. PVC packaging represented about 7% of total annual North American sales of PVC of 11.0 billion lb in 1994 (see Fig. 1).

When first developed in the 1930s, PVC found little applicability or marketplace acceptance because of its tendency to thermally degrade or dehydrochlorinate (1) when heated. However, scientists soon discovered that additives such a stabilizers and plasticizers could easily be compounded into PVC to make it processable without thermal degradation. More importantly, they found they could also modify PVC's physical properties across a broad spectrum.

As a result, PVC has evolved into one of the world's most versatile polymers and the second-largest volume-produced plastics because of its toughness, relatively low cost, and the ability to modify its physical properties. A wide range of applications today includes pipe and fittings, house siding, windows, electrical wire coatings, credit cards, medical intravenous bags and tubing, as well as packaging applications such as bottles and blister packs.

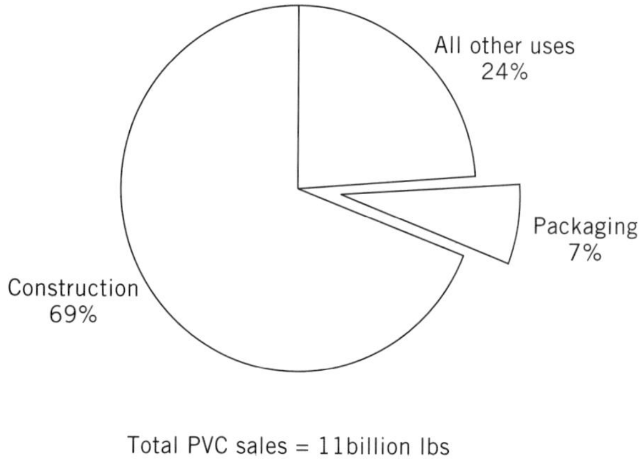

Figure 1. PVC end markets in North America in 1994 (domestic sales in resin equivalents for the United States plus Canada). (*Source:* SPI and The Geon Company.)

How Vinyl is Produced

Poly(vinyl chloride) (PVC), more commonly referred to as *vinyl,* consists of three primary elements: chlorine, carbon, and hydrogen. Chlorine is derived from a chloralkali production process; carbon and hydrogen are derived from ethylene through either a petroleum or natural-gas cracking process (see Fig. 2).

In the chloralkali process, salt is combined with water to form a brine solution. This saltwater ($NaCl + H_2O$) solution is passed through an electric current in a process where chlorine atoms are attracted to the anode and where the sodium ions in the solution are attracted to the cathode electrode. This electrolysis process produces an electricochemical unit (ECU). The ECU is made up of 1.0 parts of chlorine and 1.1 parts of caustic soda (sodium hydroxide).

Ethylene is derived from the cracking of either petroleum or natural gas. In this process the feedstock is put through a catalyst bed at high temperature and pressure to produce ethylene and a number of other coproducts such as propylene and butadiene. The ethylene is further processed to separate it from the coproducts.

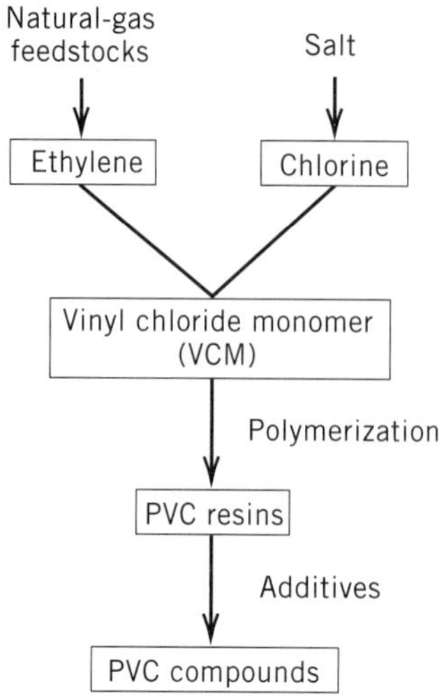

Figure 2. PVC production process.

Ethylene and chlorine are combined to first make 1,2-dichloroethane, ethylene dichloride (EDC), and finally to make vinyl chloride monomer (VCM or $CH_2{=}CHCl$).

PVC is normally polymerized from VCM by one of four processes (suspension, mass, emulsion, and solution) into the following polymer structure:

$$\text{-}\!\!(CH_2CHCl)\!\!\text{-}_n$$

Each process uses peroxide-type initiations to produce free radicals, and the exothermic reaction is normally carried out at 95–167°F (35–75°C) (2). Under different reactor configurations, agitation, and reaction media, these four processes produce PVC resins with uniquely different physical characteristics. These characteristics play important roles in subsequent application processes such as extrusion and calendering.

The suspension and mass polymerization processes account for more than 80% of the PVC produced in North America. The emulsion and solution process produce a much finer particle size for use in dispersion of plastisol applications. Specific applications include coatings, toys, and flooring.

There are presently eight major manufacturers of PVC resins in the United States.

Vinyl Industry Channels

About 75% of all North American PVC resin is sold to captive compounders, which convert the resin to compounds and then fabricate products from these compounds themselves. Custom compounders that make compounds for sale to PVC product fabricators purchase approximately 10% of the vinyl resin production. The remaining 15% is used by PVC resin producers that have their own integrated compounding facilities. These custom compounds are then sold to PVC fabricators in North America and throughout the world.

Structure and Properties

As a basis for discussing PVC's structure or morphology (3–5), some typical suspension-resin characteristics are shown in Table 1.

Molecular weight. The molecular weights of PVC resins produced in the United States are typically 0.55–1.50 inherent viscosity. As molecular weight increases, physical properties such as tensile strength and tear strength increase proportionately (6,7), as does melt viscosity, which affects product processing. The tradeoff in selecting a PVC resin is to choose the minimum molecular weight to meet the end-product physical requirements while minimizing melt viscosity.

Table 1. Typical Characteristics of a PVC Suspension Resin

Molecular weight	
Inherent viscosity, ASTM D1243	0.88–0.98
Weight-average molecular weight	142,000–185,000
Number-average molecular weight	55,000–62,000
Mean particle size	130–165 μm
Apparent bulk density, ASTM D1895	0.450–0.550 g/mL
Porosity, ASTM D2873	0.23–0.35 mL/g
Residual monomer, ASTM D3749 or D3680	<1.0 ppm

Particle size. PVC particles are somewhat spherical as a result of the suspension polymerization process (agitation, suspending agents, etc). The resin particles generally have a size distribution of 70–250 μm in diameter, which results in a mean size of 130–165 μm.

Bulk density. PVC has a specific gravity of approximately 1.40, but the resin's bulk density (8) is significantly less: 0.450–0.550 g/mL. Bulk density is directly related to the particle's morphology and specifically the resin's porosity, particle size and distribution, and particle surface characteristics.

Porosity. A single PVC resin particle from the reactor has a structure that contains many openings in its surface plus a measurable and accessible void within the particle. The amount of free volume within a resin particle is referred to as its *porosity.* Techniques such as mercury intrusion (ASTM D2873) are commonly used to measure a resin's porosity. This unique characteristic allows a PVC resin to absorb liquids such as plasticizer during the compounding operation. The amount of porosity and its accessibility play an important role when considering the amount and viscosity of the liquids added during compounding.

Residual vinyl chloride monomer. During the polymerization of PVC, not all of the VCM is converted to polymer. Today the amount of residual monomer remaining in the dried PVC resin is typically less than 1 ppm. This level establishes safe-worker levels and provides acceptable levels in the product.

Compounding PVC

As discussed earlier, the addition of compounding additives enables PVC to be modified into many useful products. For example, the addition of a liquid plasticizer such as di(2-ethylhexyl) adipate (DEA) permits the production of a flexible film with the oxygen-transmission properties required for meat packaging. The addition of a rubbery polymer such as methacrylate butadiene styrene (MBS) significantly affects the toughness or crack propagation characteristics as measured by impact tests (eg, Gardner or Izod). PVC's tremendous versatility results from the ability to tailor those properties to the requirements of the application. With the proper additives, a rigid PVC bottle can be blow-molded for edible-oil packaging. Other ingredients allow the extrusion of a blown flexible film for produce wrapping. These formulations are generally highly proprietary and very specific to use applications and properties.

Stabilizers. Stabilizers give PVC the ability to withstand the thermal and shear conditions of processing without polymer degradation (9). Stabilizers used in food-contact applications, which must have FDA clearance, include Ca/Zn salts, epoxidized soybean oil, and octyl–tin mercaptides. In other applications, stabilizers such as butyl or methyl tin mercaptides are used.

Plasticizers. There are numerous types of plasticizers, and each imparts a specific set of properties to the final product (10). These liquid or polymeric additives generally reduce the T_g of PVC. At the same time, they reduce tensile strength and increase elongation. Certain plasticizers, such as di(2-ethylhexyl) phthalate (DEP), improve PVC's water-vapor barrier properties; DOZ (dioctyl azelate) significantly improves its low-temperature impact strength.

Lubricants. Generally, lubricants are added to PVC compounds to reduce the frictional properties between the compound and the metal surface of the processing equipment (11). They are also used to reduce the surface friction of the final product or to reduce the product's surface static properties. Lubricants include such families as paraffinic waxes and metal stearates.

Impact modifiers. Many types of impact modifiers have been developed for PVC packaging applications to improve PVC's toughness for transportation and handling. Examples include methacrylate butadiene styrene (MBS), chlorinated polyethylene, and acrylic polymers. The type used depends on application requirements such as clarity, cost efficiency, low-temperature impact, and metal-release properties.

Fillers and pigments. Fillers such as $CaCO_3$ can reduce the raw-material costs of a PVC compound. They have little effect on physical properties at low levels. Some fillers can improve properties such as stiffness or abrasion resistance.

PVC in Packaging

PVC packaging applications fall into three general categories: film and sheet, bottles, and others (including coatings, and cap liners). About 28% of PVC in packaging involves food applications, 31% medical uses, and 41% nonfood end uses.

Film and sheet are the largest applications for PVC in packaging. PVC film generally is flexible and, as shown in Figure 3, is used in food film for packaging meat, cheese, and produce. Among the advantages of PVC for these uses are its clarity, barrier properties, puncture resistance, and cling for good sealability (12).

In medical uses such as blood and intravenous-solution bags, PVC provides excellent characteristics such as heat sealing and ease of sterilization (13). Since replacing glass for critical applications more than 25 years ago, PVC has proved to be a safe and reliable polymer. PVC is also used in tubing for connecting various devices to patients.

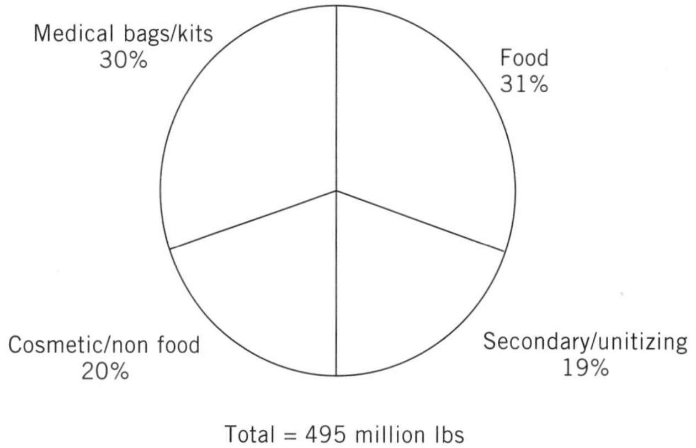

Figure 3. PVC film applications [total–final weight (PVC plus additives)]. (*Source:* The Geon Company.)

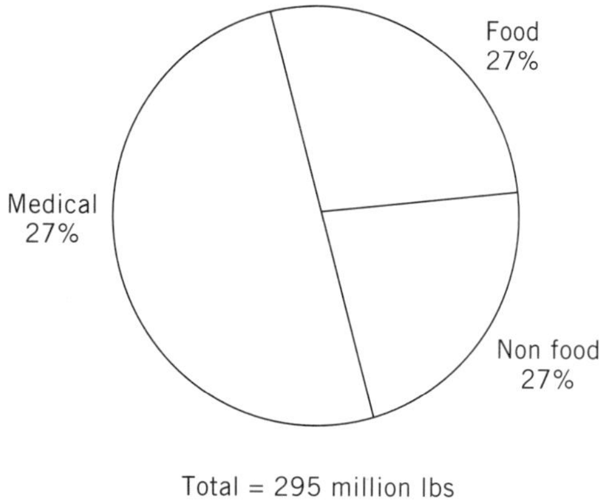

Figure 4. PVC sheet applications [total—final weight (PVC plus additives)]. (*Source:* The Geon Company.)

Table 2. PVC Packaging Market Size[a] (million pounds)

	1991	1992	1993	1994
Film and sheet[b]	407	450	480	547
Bottles	203	191	177	188
Other packaging	75	83	60	80
Subtotal	685	724	717	815
Postrecycled packaging	2	6	6	10
Total	687	730	723	825

[a] All figures in resin equivalents.
[b] Includes both flexible and rigid components.

As Figure 4 depicts, nearly 300 million lb of PVC are used annually to make rigid sheet in thicknesses ranging from 0.010 to 0.150 in. PVC also has proved to be an excellent thermoforming plastic. Because of its ability to hold a form during the thermoforming process, its high impact resistance (even at low temperatures), and excellent clarity, it has become widely used in food, medical, and nonfood applications. One of the single most important growth applications has been in drug blister packaging, including unit-dose packets. PVC's performance and cost-effectiveness have led to its dominance of more than 95% of this market (14).

PVC is used in bottles (see Fig. 5) because of its clarity, impact resistance, ease to formability, chemical resistance, and cost-effectiveness. As a result, the PVC bottle market is primarily for cleaners, chemicals, toiletries, and cosmetics. While the overall use in food packaging has remained flat because of competition from other materials such as PET, PVC is seeing rapid growth worldwide for water bottles. With the limited availability of potable water in developing countries, PVC bottles provide a safe, cost-effective way to transport drinking water.

FDA Status

PVC is used extensively in food-contact applications such as meat wrap. It is prior-sanctioned for use in general food-contact applications by virtue of an article published in July 1951 in the *Journal of the Association of Food and Drug Officials of the United States* by A. L. Lehman of the FDA. In addition, PVC resins are listed in a number of specific FDA regulations relating to food-contact substances.

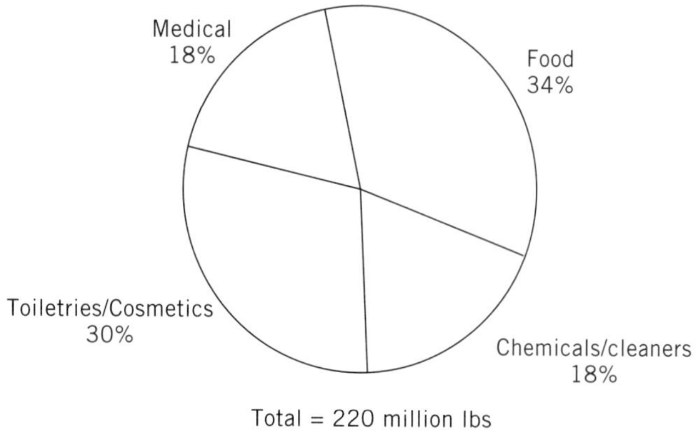

Figure 5. PVC bottle applications [total—final weight (PVC plus additives); includes bottles, jars, other containers]. (*Source:* The Geon Company.)

PVC and Recycling

While PVC is the world's second-most widely used plastic material, it represents only a small percentage (<0.5%) of all solid waste. However, as of 1991, nearly 1200 communities in the United States had access to vinyl recycling through curbside or dropoff center collection programs for mixed plastics (15). In 1991 there were nearly 60 recyclers nationwide processing postindustrial and postconsumer PVC; as economics naturally improve, these figures are expected to grow (see Recycling).

Once collected, vinyl can be processed with other plastic to create "commingled" products such as plastic lumber (16). Or, it can be separated for use in products ranging from new packaging to pipe. Because PVC can easily be reprocessed, it can be used to make new PVC bottles, rigid sheet, or film. Uses for recycled vinyl already exist and, in fact, demand for recycled vinyl far outstrips current supply. As Table 2 depicts, postconsumer vinyl packaging is being recycled at nearly 10 million lb annually.

BIBLIOGRAPHY

1. A. Guyot, M. Bert, P. Burille, and co-workers, "Trial for Correlation between Structural Defects and Thermal Stability in PVC," paper presented at the Third International Symposium on Polyvinylchloride, Case Western Reserve University, Cleveland, OH, Aug. 10–15, 1980.
2. L. F. Albright, *Chem. Eng.,* 145–152 (June 5, 1987).
3. P. R Schweagerle, *Plast. Eng.,* 42–45 (Jan. 1981).
4. N. Berndstein and G. Manges, *J Pure Appl. Chem.* **49,** 597–613 (1977).
5. H. Behrens, *Plaste and Kautschuk* **20**(1), 2–6 (Jan., 1973).
6. J. R. Fried, *Plast. Eng.,* 27–33 (Aug. 1982).
7. S. Kaufman and M. M. Yocum, *Plastic Compound,* 44–46 (Nov./Dec. 1978).
8. M. A. Kauffman and R. S. Guise, *J. Vinyl Technol.* **16,** 39–45 (1994).
9. R. D. Dworkin, *J. Vinyl Technol.* **11,** 15–22 (1989).
10. F. Tomasellik, V. P. Gupta, H. S. Caldiron, and G. R. Brown, *J. Vinyl Technol.* **10,** 72–76 (1988).
11. J. A. Falter and K. S. Geick, *J. Vinyl Technol.* **16,** 112–115 (1994).

12. J. Southus, "PVC Film & Sheet in Packaging," paper presented at the Regional Technical Conference, SPE, Mississauga, Ontario, Canada, Sept. 13–14, 1983.
13. N. Perry, "Flexible PVC Medical Packaging Applications," paper presented at the Regional Technical Conference, SPE, Mississauga, Ontario, Canada, Sept. 13–14, 1983.
14. M. Larson, *Pharmaceut. Med. Packag. News,* 23–27 (Nov./Dec., 1993).
15. W. F. Carroll, Jr., *J. Vinyl Technol.* **16,** 169–176 (1994).
16. F. E. Krause, *J. Vinyl Technol.* **16,** 177–180 (1994).

General References

L. I. Nass, ed., *Encyclopedia of PVC,* Vol. 1 (1976), Vol. 2 (1977), Vol. 3 (1977), Marcel Dekker, New York.

J. K. Sears and J. R. Darley, *The Technology of Plasticizers,* Wiley, New York, 1982.

W. S. Penn, *PVC Technology,* MacLaren, London, 1966.

J. H. Briston and L. L. Katan, *Plastic Films,* Longman, New York, 1983.

D. A. Cocco
The Geon Company
Cleveland, Ohio

PRESSURE CONTAINERS

Several distinct classes of pressure containers exist. All are used to dispense the contents into either open or closed environments. Among the largest are Hortonspheres, with capacities exceeding 1 million gal (3,850,000 L). For transport purposes, the largest pressure containers are "jumbo" tankcars, with capacities of about 31,000 gal (117,000 L). Cylinders down to about 500 gal (1900 L) are designated as bulk tanks. Smaller ones are variously known as "pig tanks," "ton cylinders," or simply as cylinders. Sometimes the intended use serves to identify the general size and shape as in the case of "propane cylinder" or "lecture bottle."

Pressure containers in sizes ranging from 3 to 1000 mL are used to dispense consumer specialties. In two of these, the primary container is pressurized. The most popular is the aerosol, where pressure is generated by a propellant gas. The other uses mechanical pressure from springs or elastomers under tension to propel products from a flexible inner container. A third form is the mechanical pump, such as the finger-pump or trigger-spray dispenser, in which pressure is generated by the user, but only within the valve body. In 1994 an estimated 3.35 billion aerosols were produced in the United States, or about 14 per capita. The worldwide consumption has been estimated at about 9.4 billion for 1994. Based on estimates for the first 6 months, U.S. aerosol production for 1995 should be about 3.4 billion units, equivalent to retail sales of about $6.5 billion. The remainder of this article is restricted to aerosols (see also Aerosol containers).

An aerosol composition normally consists of a combination of concentrate and propellant. The propellant portion can range from about 0.5 to 100%, and at the 100% end the propellant functions as the product itself. The regulatory definition of an aerosol [by the U.S. Department of Transportation (DOT)] is a product that consists of "chemical ingredients pressurized by compressed gas." In a 1991 clarification the DOT further required that the product name reflect the intended use of the concentrate ingredients. This was designed to halt interstate shipments of commodities, under aerosol exemptions, which contained flammable propellants—to which trace amounts of two or more incidential chemicals had been added, with the sole purpose of satisfying the aerosol description in the tariff.

Virtually all aerosols are fitted with valves, for the purpose of dispensing the contents—most often in spray forms, but also as foams, streams, gels, gases, lotions, pastes, ointments, and so forth. Limited numbers of aerosols are packaged with screw-threaded closures, designed to be connected to valve-and-hose assemblies for such purposes as air-conditioner refills. About a dozen firms produce aerosol valves, and they offer a bewildering array of diverse styles, orifices, materials, and associated fitments. Approximately 85% of the U.S. market is currently held by three valvemakers: Precision Valve Corporation, SeaquistPerfect (of Aptar), and Summit Packaging Systems, Inc., and each of these firms is international in scope.

The scope of aerosol products is surprisingly large. Every consumer and institutional and/or industrial product category is included, such as pesticides, household products, pharmaceuticals, drugs, foods, cosmetics, and medical devices. Some liquid streams can be propelled 25 ft (8 m) to reach wasp nests and kill the insects and their eggs. Compressed-gas jets may be used to dedust the surfaces of electronic circuit boards, tapes, and CDs (compact disks). Sterile saline solutions are available for flushing cleaning solution residues from contact eyewear. Cheese and jelly products can be cleanly extruded from special valves onto other food items. Gels can be produced that magically spring into foams when stroked with the fingertip. Unlike pump sprayers, aerosols can produce sprays with an optimum particle size distribution for maximizing their knockdown and cidality to flying insects. Particles are produced that are small enough to remain airborne for long periods, yet not so small that they airflow around the bodies of flying pests. As a result, a maximum number of insects are impacted by lethal doses.

The scope of aerosols has been enhanced by the steadily growing numbers of bag-in-can and piston can products now on the market. The product is partitioned from the propellant by placing it inside a collapsable plastic, metal, or composite laminated bag, or above a piston. The propellant then simply acts to pressurize the system. Nonaerated products, such as lithium grease, toothpaste, gels, artist's pigment dispersions, and standard chemical solutions, can then be dispensed. Product viscosities up to about 2,000,000 cP can be easily handled. Originally restricted to the 2.125-in. (52-mm)-diameter aerosol can, these compartmentalized aerosols are now available in diameters ranging from about 1.378-in. (35-mm) to 2.588-in. (65.7-mm) and in several heights.

Among the myriad aerosol valve designs is the meter-spray type. Originally designed to dispense costly perfumes in dosages of about 0.02–0.05 mL, the primary use has now shifted to the metered-dose inhalents (MDIs). An estimated 150,000,000 of these small packages are sold per year in the United States alone, and the number is steadily increasing as asthmatic problems escalate yearly.

Historical Development

Discounting various crudely designed pressure packs that can be dated back as far as about 1851, the true aerosol concept originated in 1923 when Eric Rotheim of Oslo, Norway developed heavy brass containers fitted with needle valves and used them to spray such products as ski wax and air fresheners. He used butane, dimethyl ether, vinyl chloride, and carbon dioxide as propellants. Because of the cost, nothing was done with the patented Rotheim development until 1943. Lyle Goodhue and William Sullivan, entomologists working for the USDA, found that aerosol sprays in the aerodynamic particle size range of 10–30 μm remained suspended in tranquil air for about 10 to 1 min., respectively. If the spray contained insecticidal ingredients, it became surprisingly effective against flying insects. They patented the discovery, noting that about 85–90% of a recently commercialized propellant (dichlorodifluoromethane, or CFC-12) was necessary to achieve this range of particle sizes. The term *aerosol,* now used to generically describe the packaging form, originally came from the scientific word for a colloidal mixture of liquid droplets in gas, such as a mist or fog.

The Goodhue–Sullivan development was commercialized by the government and used to help members of the armed services combat mosquitoes in the South Pacific theater during World War II. After the war attempts were made to market the product to the civilian population, but the high initial cost, coupled with a need to return the empty can for credit, strongly inhibited sales. The fledgling industry might have foundered, had not a young chemist, Harry E. Peterson, been able to adapt a high-strength beer and beverage can into a low-cost, disposable aerosol container. He also developed an inexpensive valve that was simple to operate. Finally, he convinced several businessmen in Danville, Illinois to have a machine shop build the first aerosol production line and sell it to the newly formed Continental Filling Corporation—the first contract aerosol filler, which went into production in 1947. From this modest beginning the aerosol industry now reaches into all parts of the world and enhances the lives of billions of people.

Aerosol Containers

About 85% of all aerosols produced in the United States during 1994 were steel [actually about 80% electrotinplate (ETP) and 5% electrochrome-coated steel (ECCS)]. Aluminum accounted for about 14%, while glass, plastic-coated glass, stainless steel, and plastic totaled about 1%. Trends now developing in the industry suggest the demise of ECCS cans during the late 1990s. The percentage of aluminum cans should continue to grow very slightly, and plastic containers made principally of biaxially oriented polyethylene terephthalate (B-OPET) should decrease because of reinterpretations of certain DOT shipping regulations.

The percentage of aluminum cans in U.S. aerosol units is perhaps the lowest in the world. Most countries use about 60% steel, 40% aluminum, and virtually no other container materials. The low percentage of aluminum in the United States is considered to be due to three factors: the consumer culture demands less packaging elegance, more large-size utility products are sold, and the prices of aluminum cans are significantly higher. Comparing the net costs to a major marketer or filler, in the case of 52-mm-diameter tinplate and aluminum cans having a 325-mL capacity, the aluminum can will be about 60% higher. The cost increase is largely tied to the much higher price of aluminum, even though the finished can is lighter in weight than tinplate units of the same capacity.

Steel

The usual steel aerosol can is produced from electrolytically tinplated sheet, and consists of three segments: top (dome), body, and bottom (base). In the United Kingdom the top is called the *cone* and the base is known as the *dome,* which has led to some confusion in nomenclature from time to time. The three domestic suppliers are the Crown Cork & Seal Company, the United States Can Company, and Ball Corporation—which purchased the Heekin Can Corporation in 1993.

Perhaps 2% of the large steel aerosol-can market is held by the Sexton Can Company, which produces both one-piece (35- and 38-mm) drawn cans and a line of two-piece cans, where a top shell is sealed to the base by double-seaming. A major portion of these last cans is used for pure HCFC-22 and HFC-134a air-conditioning and refrigerant gases, under two special DOT exemptions, since their pressures exceed the standard aerosol limits.

The final steel aerosol canmaker is Dispensing Containers Corporation, a new firm whose production capabilities are still in the pilot-line stages, although a 240 can/min line is being assembled for 1996 productions. They should produce about 8,000,000 two-piece and possibly 200,000 one-piece steel cans in 1997 and claim to be able to produce the same sizes in aluminum as well. Their containers are unique in that about 92% of the sidewall is drawn and ironed to a thickness of only 0.0038 in. (0.096 mm). This operation acts to elongate the base–body "cup," formed by ordinary extrusion techniques. The cup is then trimmed to height and eventually double-seamed to an aerosol dome. For the small (35- and 38-mm dia) cans, the drawn and ironed bottom shells are trimmed and then formed into a rounded dome and curl by a multistage ("female" only) die-shaping operation.

It is not unusual for the can body of a three-piece steel aerosol can to represent about 65–75% of the total weight, depending on the can height. If the draw–iron process effectively thins some 92% of the body metal to half the normal thickness, the weight of the body will be reduced by some 46% and the weight of the entire can will diminish by about 32%. As a rule, about half the final net cost of the aerosol can is due to the cost of tinplate, so it follows that a 32% reduction in metal will equate to about a 16% reduction in can cost. This economic fact has many marketers interested in testing these new cans. Claims of environmental source reduction, lack of the slightly unsightly side seam and bottom double seam, plus some other stated advantages, serve to enhance interest.

The thickness of tinplated steel plate is specified in terms of the weight per "basis box" (see Cans, steel; Tin-mill products). The *basis box* was an accounting term developed about 1730 in Pontypool, England and was used to note the weight of 112 tinplated steel sheets, 14 in. wide by 20 in. long (35.6 mm × 50.80 mm)—thus having an area of 31,260 in.^{2} (20.2325 m^2) on each side. The Pontypool steel plate averaged 100 lb (45.45 kg) per basis box. After tinplating it weighed about 101–102 lb per basis box, and when crated it constituted a load that was not overly heavy for Welsh longshoremen employed in lading ships bound for canmaking shops in London and continental Europe.

Untinned steel plate weighing 100 lb (45.45 kb) per basis box has an average thickness of 0.0110 in. (0.280 mm). The plate is simply called "100-lb stock." For aerosol cans the minimum plate thickness is often important in meeting the DOT requirements. The steel-mill tolerance is usually ±10% on thin, canmaking plate, but the can companies have been successful in insisting that the 3σ (three-sigma) tolerance be −6% to 10%. For example, the very strong double-reduced, temper 8 (DR-8) 75-lb stock widely used for aerosol can bodies will have a steel thickness range of 0.00776–0.00908 in. (averaging 0.00825 in. or 0.210 mm). The cause of this apparently large variability comes from the upward distortion of the heavy-duty rollers, used to make the plate, causing the center portion of the typical 54-in. (1.37-m)-wide strip to be thicker than at the edges.

The old English basis-box system was introduced to the United States in 1829, when the first tinplate was imported from Wales. It continued to be used in 1858 when the first tinplate was manufactured here. Although the English have (officially) converted from basis box to metric measurements, the United States has yet to do so. However, international pressure is being brought to bear, and it is likely that the transition will begin before the year 2000. For example, 100-lb steel plate would then be called 0.28-mm plate.

Prior to 1937 all tinplate was produced by the hot-dip method, with both sides being coated equally. If 1.00 lb of tin was deposited per basis box, this meant that 0.50 lb was laid down on each side. Technically, such plate could be designated as 0.50/0.50-lb (227/227-g) ETP. When electrodeposition of tin became practical on a large scale, it also became possible to differentially coat the two sides of a given steel sheet. This advance led to some changes in nomenclature. The rule is to designate the tin weight on each side as if the plate were nondifferentially coated. For example, if 1.00 lb of tin were deposited per basis box, but with 0.75 lb on one side and 0.25 lb on the other, the designation would be D-150/50, meaning that the more heavily tinplated side would have the same weight of tin as nondifferential 1.50-lb plate and the light side would have the equivalent of nondifferential 0.50-lb plate. The nominal tin-coating weight on each side would be 0.375/0.125 lb (170/42.5 g). In the metric system, this plate would be designated as having an average of 16.8 g/m^2 of tin on the more heavily plated side and an average of average of 4.2 g/m^2 on the lighter side.

Differential plate is normally marked on the more heavily plated side, sometimes with a "D," but also with various lines, hexagons, triangles, or other figures, imprinted by roll printing a dilute solution of sodium carbonate onto the tin coating, prior to flow melting. This prevents plate reversal problems during canmaking. As a rule, the more heavily tinplated surface is oriented toward the product, to reduce possible corrosion. In a few instances very lightly tinplated plate, such as D-05/15, is placed with the heavily tinned surface facing the atmosphere. An example, is where the product is quite alkaline, such as some oven cleaners or other hard-surface cleaners, which quickly dissolve the amphoteric tin metal. In fact, the 0.05-lb tin coating is so thin and incomplete that the dark-gray $SnFe_2$ alloy layer and gray steel itself can be seen through it. In this case the free tin is typically 1.36 μin. thick and the underlying $FeSn_2$ layer is about 0.19 μin. (0.08 g/m^2). By using the heavier 0.15-lb tin coating on the outside of the can, the white, shiny surface of the tin improves appearance. It even brightens the lithography in some instances.

Table 1. Chemistry of Steel Canmaking Plate

Constituents	Compositions during 1960; Mild Steel, Various Tempers	Compositions during 1995		
		Type MR T-1 to T-4	Type MC T-5	Type D[a] T-8
Carbon	0.04–0.15	0.13 max.	0.13 max.	0.12 max.
Silicon	0.08 max.	0.02 max.	0.01 max.	0.02 max.
Nitrogen	0.001–0.025	0.007–0.012	—	—
Phosphorus	0.01–0.14	0.02 max.	0.06–0.10	0.02 max.
Sulfur	0.015–0.050	0.05 max.	0.05 max.	0.05 max.
Manganese	0.20–0.70	0.60 max.	0.60 max.	0.60 max.
Copper	0.02–0.20	0.20 max.	0.20 max.	0.20 max.
Chromium	—	0.10 max.	—	—
Nickel	—	0.15 max.	—	—

[a] Double-reduced plates.

Different thicknesses of tinplated steel plate are used in the preparation of three-piece aerosol cans. As can size increases, heavier plate must be used to obtain the same pressure resistance. The end sections (dome and base) are typically 60–80% thicker than the can body. They must resist eversion and possible separation, under high-pressure conditions. Even with can bodies as thin as about 0.004 in. (0.10 mm), the so-called "hoop effect" prevents any swelling or bursting, unless the body metal is softened by strong heating—as in a fire.

Plates used for three-piece cans extend from about 60 to 135 lb per basis box. Bodies range from 60 to 90 lb (0.0066–0.0099-in., or 0.168–0.251-mm average thickness), while end sections extend from 90 lb on small 112-diameter (45-mm) cans to 135 lb on the large 300-diameter (76-mm) containers. The use of these plate thicknesses, along with the basic pressure-resistant architecture of aerosol containers, allows even the lowest strength cans to resist pressures to at least 160 psig (lb/in.^{2} gauge) (1.2 MPa) without deformation, and up to at least 226 psig (1.7 MPa) without bursting.

The chemistry of canmaking plate is illustrated in Table 1. Elements such as sulfur and copper are minimized, since they can often sponsor can corrosion when in contact with aggressive formulations. Other elements are sometimes added, such as manganese, to develop certain working properties.

The canmaking plate is also continuously annealed to provide various tempers, or stiffness levels. Annealing also serves to remove certain work-hardening effects, obtain the desired mechanical properties, and achieve an optimum grain structure. The annealing process as fairly rapid. Plate is unreeled into an inert atmosphere oven and heated to about 1250°F (677°C) for about 20 s, then allowed to cool to 900°F (482°C) across the following 25–30 s. The rate of this initial cooling determines temper. The highest tempers are achieved by using the slowest rates of cooling—to allow the formation of a larger number of cementite (Fe_3C) grains per unit volume of the steel. For example, the very stiff temper No. 5 (or T-5) plate contains about 15,000 grains of cementite per mm^3. The various tempered steels and their aerosol canmaking used are illustrated in Table 2.

The thickness and temper of aerosol canmaking plate must meet certain general guidelines. However, there are consider-

Table 2. Tempers of Electrotinplated Steel (USA)

Temper Number	Hardness Range[b]	Aerosol Canmaking Applications
1	46–53	No longer used
2	50–56	Not used for aerosol cans
3	53–60	Sometimes used for can domes
4	58–64	Used for can bodies and commonly used for can domes
4-CA	54–68	No longer used for aerosols
5	62–68	Some slight use for can bodies
5-CA[a]	62–68	Used for can bases; some bodies
6	67–73	Now obsolete for aerosols
DR-7	72–76	Minor use for can bodies
DR-8	N/A	Used for most can bodies
DR-9	N/A	Not used for aerosols as yet
DR-10	N/A	Not used

[a] 5-CA is the successor to the "TU" plate of 1950–1985.
[b] Rockwell tester; R30T range. Not applicable to DR-8, DR-9, and DR-10 plates, since the hardened ball pierces the tough, stiff plate before forming a depression. The preferred method for testing DR plate is by tensile strength. DR-9 and DR-10 are used for three-piece steel beer and beverage cans. Some aerosol testing is in progress. Tempers are inapplicable to impact-extruded and impact-extruded drawn-and-ironed shells.

able variations between canmakers, regarding selection. The data presented in Table 3 can be considered as indicative only.

The plate for aerosol can bodies begins with multistage hot rolling, down to a thickness of about 0.10–0.06-in. (2.5–1.5 mm). The plate is then cold-rolled (temperature below ~500°C) into a thickness range of ~0.0014–0.0065 in. (0.036–0.017 mm). It is then annealed. This produces "CR," or cold-reduced plate of the desired temper. CR plate is still widely used in can bodies. To make "DR," or double (cold)-reduced plate, one starts with CR plate that is ~0.018–0.013-in. thick. Having annealed it, the stock is again cold-reduced, up to 50% in one pass, to produce tinplate with a thickness range of ~0.0094–0.0065-in. This corresponds to 85–60-lb DR tinplate—the gauges normally used in can-body constructions. The advantages of DR plate are that ~10–12% less weight can be used, compared to CR plates, while retaining the same strength. The lower apparent cost is offset by the higher manufacturing cost of the plate—since another operation is involved—a very slightly higher scrap rate, and sometimes a bit more variability in the double seams.

Dimensions

The size of tinplate aerosol cans is still described as the diameter across the double seams, multiplied by the height from the base to the top of the top double seam. These designations nearly always have three digits, unless the can is extremely tall. The first is the inch value, and the next two are the number of $\frac{1}{16}$-in. increments added. Thus the over-the-seams diameter of a standard (non-necked-in) 202-diameter three-piece can would be 2.125-in. This corresponds to 53.99 mm, but this can be misleading, since the International Organization for Standardization (ISO) utilizes the inside diameter of the can for "can diameter," and in this case their figure (expressed to the nearest millimeter) is 52 mm.

Table 3. Thickness and Temper for Tinplate Used for Aerosol Cans

Weight per Base Box, lb (kg)	Average Thickness, in. (mm)	Typical Temper Designation	Typical Application: Component	Typical Application: Can Dia (mm)[a]
55 (25)	0.0061 (0.155)	DR-8	Body	202 (54)
65 (29)	0.0072 (0.183)	DR-8	Body	207.5 (62.7)
70 (32)	0.0077 (0.196)	DR-8	Body	211 (68.3)
70 (32)	0.0077 (0.196)	T-5	Body	202 (54)
75 (34)	0.0083 (0.211)	DR-8	Body	300 (76.2)
75 (34)	0.0083 (0.211)	DR-8 or T-5	Body, DOT Specification 2P	202, 207.5, 211 (54, 62.7, 68.3)
80 (36) 0.0088	(0.224)	DR-8	Body	300 (76.2)
85 (39) 0.0094	(0.239)	DR-8	Body, DOT Specification 2Q	202, 207.5, 211 (54, 62.7, 68.3)
			Body, DOT Specification 2P	300 (76.2)
90 (41)	0.0099 (0.251)	T-5	Body, including DOT Specification 2P	300 (76.2)
107 (49)	0.0118 (0.300)	T-5	Bottoms	113, 202 (30, 54)
112 (51)	0.0123 (0.312)	T-5	Bottoms	207.5 (62.7)
112 (51)	0.0123 (0.312)	T-3	Domes	202 (54)
118 (54)	0.0130 (0.330)	DT-2	Domes	207.5 (62.7)
123 (56)	0.0135 (0.343)	T-5	Bottoms	211 (68.3)
128 (58)	0.0141 (0.358)	T-3	Domes	211 (68.3)
128 (58)	0.0141 (0.358)	T-5	Bottoms, including DOT Specification 2P	211 (68.3)
			Bottoms	300 (76.2)
135 (61)	0.0149 (0.378)	DT-2	Domes, including DOT Specification 2P	300 (76.2)
			Domes, including DOT Specifications 2P and 2Q	211 (68.3)
135 (61)	0.0149 (0.378)	T-5	Bottoms, including DOT Specification 2P	300 (76.2)

[a] 202 = 2 $\frac{2}{16}$-in. (54-mm) dia, etc.

Table 4. Dimensions and Capacities of Standard Three-Piece Cans[a]

Nominal Size (English)	Nominal Size (Metric)	Capacity[b] (mL)
202 × 214	52 × 73	147
202 × 314	52 × 98	192
202 × 406	52 × 111	226
202 × 505	52 × 135	270
202 × 509	52 × 141	290
202 × 700	52 × 178	367
202 × 708	52 × 191	391
207.5 × 413	57 × 122	332
207.5 × 509	57 × 141	389
207.5 × 605	57 × 160	452
207.5 × 701	57 × 179	498
207.5 × 703	57 × 183	508
207.5 × 708	57 × 191	527
207.5 × 713	57 × 198	541
211 × 410	65 × 117	396
211 × 503	65 × 132	438
211 × 600	65 × 152	499
211 × 604	65 × 159	522
211 × 607	65 × 164	539
211 × 612	65 × 171	567
211 × 710	65 × 194	648
211 × 713	65 × 198	657
211 × 908	65 × 241	796
211 × 1008	65 × 266	872
300 × 709	73 × 192	796

[a] Metric measurements are by ISO convention.
[b] Capacities are on a brimfull basis. Subtract about 3 mL for crimped cup volume.

A selection of commercially available, standard (straight-wall) three-piece cans is shown in Table 4.

During the past dozen years or so the "necked-in" three-piece can profile has finally become popular in the United States—some 20 years after it was first introduced by (then) Metal Box, Ltd. in Europe. (*Note:* Metal Box was purchased by Crown Cork & Seal Company in late 1995.)

Several can diameters are now available in the necked-in style that are not available in the older, straight-wall can (eg, 113, 205, and 214). The diameter over the top double seam is less than the diameter of the can body, and this has led to a curious nomenclature, where the top-seam diameter is described, followed by that of the can body, and then the height over the double seams. For example, the smallest three-piece tinplate can is described as a 111/113 × 214 container, and the largest is a 211/214 × 1006 can. The necked-in cans are more aesthetically pleasing than the standard or straight-wall types. They are also lighter. Because there are no protruding double seams, they fit into slightly smaller corrugate or tray/shrink-wrap shippers. An idea of the diverse necked-in can sizes can be seen from Table 5.

Construction

Very briefly, a coil of tinplate, weighing 20,000–25,000 lb (9090–11,360 kg) is unrolled and cut into sheets approximating 24 × 30 in. (610 × 760 mm)—or about enough to produce about 10–15 large-size can bodies. The sheets are lithographed on one side and often lined on the other. Large ovens are used to bake these coatings onto the tinplate. The sheets are then slit into body blanks. The blanks are formed into cylinders, tack-welded, and then side-seam-welded. Inside and outside side-seam stripes are generally applied to protect the raw steel edges. If necked-in cans are desired, the cylinders must be end-formed (necked) at this point, followed by flanging, which is needed to attached the end sections by a double-seaming operation.

The end sections are manufactured from sheets of plain, enameled, and/or lithographed tinplate, which are fed through a scroll shear to produce multiple end-cut strips. The strips (each sufficient to produce about 12–24 ends) are fed to a can-end-forming press to cut out disks and form them to the desired contour. A can-end conveyor curler then bends the outer edge upward, so that the can-end compound-applying machine can pour in an elastomer-based material. This is cured, and the completed can ends move, either directly to the double seamer, or to a can-end-wrapping machine where they are nested into water-resistant paper sleeves for later fabrication onto can bodies.

The double seamers consecutively join the end sections to the flanged can body, using rollers that force the metal into

Table 5. Dimensions and Capacities of Necked-in Three-Piece Cans

Nominal Size (English)	Nominal Size (Metric)	Capacity (mL)
111/113×	43/45×	
214	73	106
312	95	140
410	117	175
508	140	209
608	165	247
200/202×	50/52×	
214	73	145
314	98	190
406	111	224
509	141	288
514	149	303
700	178	365
708	191	389
711	195	398
202/205×	52/57×	
410	117	272
604	159	388
607	164	401
701	179	428
704	184	445
710	194	486
802	206	522
805	211	536
207.5/211×	60/65×	
410	117	395
413	122	419
604	159	518
612	171	563
710	194	647
713	198	671
211/214×	65/67×	
315	100	346
413	122	480
714	200	766
804	210	802
1006	263	999

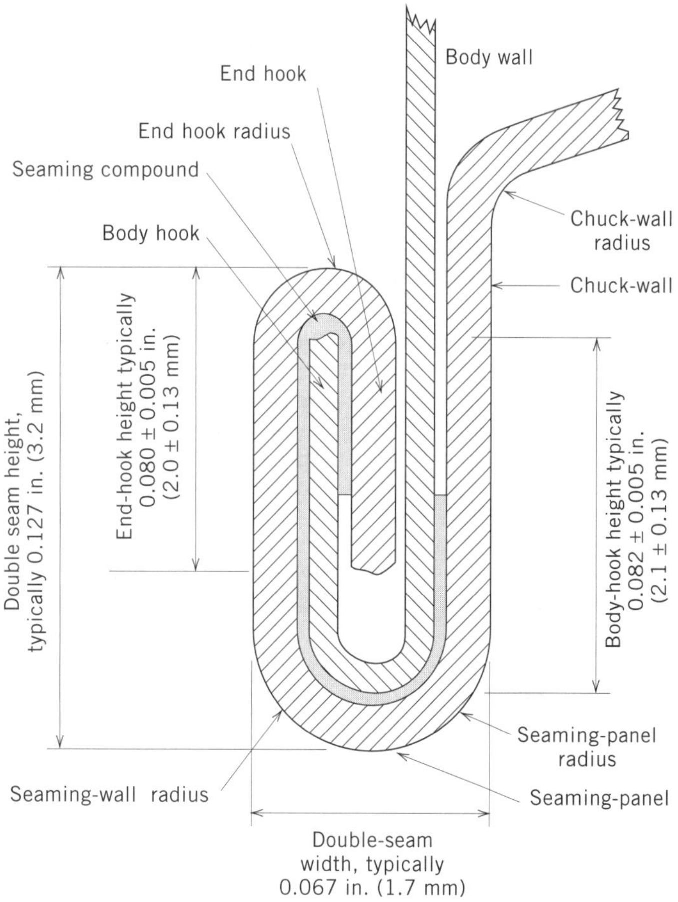

Figure 1. Aerosol tinplate can bottom double seam: terminology and dimensions.

specially shaped dies. The first-operation roll begins the wrapping process, and the second-operation roll sets most of the final dimensions and tightness factors, so important in making the double seam function as an hermetic seal that is essentially impervious to leaching or penetration by any of the vast array of solvents that are used in aerosol formula. A profile of a typical double seam is shown in Figure 1.

Tinplate Can Linings

In the United States, tinplate cans are available plain, or with single-coat, double-coat, or even triple-coat linings. The major type is the epon–phenolic, which combines the excellent product resistance of the phenolic portion with the good fabricability of the epoxy. (Otherwise, the phenolic would craze or crack when the can is bent during forming). Vinyl resins are nearly always used as second coats. They form an excellent barrier to water and alcohols, but are dissolved by acetone, MIBK, methylene chloride, and dimethyl ether. Newer linings include the special organosols, the polyamideimides (PAMs), and a few others. If linings fail to resist aggressive products, the can company should be contacted for advice.

Can corrosion most often occurs at the bottommost 20–40 mm of the side seam, and this reflects a lack of protective tinplate, exposure of the raw steel cut edge, crevices under that edge in some cases, and incomplete thermal curing of the side-seam stripe. Quite often enamel blisters are seen, enclosing "concentration cells" that may have much higher levels of acidity and/or chloride ion than the general solution. These corrosion promoters often lead to severe pitting and eventual perforation. Corrosion inhibitors should be considered whenever an aqueous solution or dispersion is packed in tinplate. Typical ones include sodium nitrite, sodium benzoate, and various amines.

Pressure Resistance

The 70°F pressure range of aerosols is generally about 20–142 psig (320–1100 kPa), although most fall in the 30–100-psig (360–791-kPa) range. However, all aerosols having a capacity of more than 4.00 fl oz (118.2 mL) are regulated by the DOT Dangerous Goods Transport Tariff. Cans are divided into four groupings. The "nonspecification" can may be used for products having pressures of ≤140 psig (1.07 MPa) at 130°F (54.4°C). They must hold the product, when heated to 130°F, without leakage or permanent distortion, and must also tolerate a pressure one and one-half times that high without bursting. This can probably accounts for 85% of all steel aerosol-can sales.

The "DOT Specification 2P" can may be used with aerosols having pressures as high as 160 psig (1.20 MPa) at 130°F (54.4°C), but must satisfy several requirements. They must not distort at 160 psig and must not burst below 240 psig. They must have a minimum plate thickness of 0.007-in (0.0178 mm) and be imprinted with both the words "DOT-2P" and the registered canmaker's logo. One can in 25,000 must be pressure-tested to destruction to prove strength. Some canmakers reduce plate inventories by offering the same can specification for "nonspecification" and "DOT Specification 2P" containers; they simply omit the imprints and destructive testing in the case of "nonspecification" cans.

The third grouping consists of "DOT Specification 2Q" cans. They can be used with products having pressures of ≤180 psig (1.34 MPa) at 130°F. The cans must tolerate 180 psig without deformation, and 270 psig (1.96 MPa) without bursting. The minimum plate thickness is 0.008 in. (0.203 mm). Finished cans must be imprinted with the word "DOT-2Q" and the registered logo of the can supplier. One can per 25,000 must be destructively tested and results maintained on file. These cans are about 12–15% heavier than "nonspecification" or DOT-2P cans, and their higher prices reflect the added cost of metal.

The final grouping is the "DOT special exemption" can. For the express purpose of approving (*1*) whipped creams, (*2*) HCFC-22, and (*3*) HFC-134a, the DOT has issued exemptions allowing certain cans to be shipped with products whose 130°F (54.4°C) pressures exceed 180 psig (1.34 MPa). The whipped-cream can is a "DOT Specification 2Q" container, but fitted with a pressure-relief mechanism (PRM) on the top of the top double seam, which effectively prevents the can from bursting at excessive pressures by venting the contents in a controllable fashion. This can may hold whipped cream (only) at pressures of ≤210 psig (1.55 MPa) at 130°F (54.4°C). Such aerosols have a number of product advantages over those packed at lower internal pressures. The DOT also considered that these cans are shipped, stored, and used under refrigeration, making the chance of heat-generated deformation de minimus.

HCFC-22 ($CHClF_2$) has a pressure of 301 psig (2.19 MPa) at 130°F (54.4°C). It requires a very heavy can to safely hold it. The Sexton Can Company produces a two-piece can, with a base-centered pressure-relief attribute, that can do so, and very recently the Advanced Monobloc, Inc. division of CCL Industries, Inc. was able to obtain an exemption for an extra-strong aluminum can, using a rubber "blowout" plug in the bottom, for the same commodity. HFC-134a, another refrigerant gas, is available in a similar Sexton Can Company can.

Aluminum Cans

Aluminum cans have been available since at least 1948. Historically, they have almost always been of one-piece (monobloc) construction, and this is the only type currently made in North America—although one firm in Europe still makes a two-piece specialty can by attaching an aluminum or tinplate dome to a suitably flanged aluminum shell section.

United States aluminum canmakers include the Advanced Monobloc, Inc. firm; the Peerless Tube Company; and the recently formed Exal Corporation. Some pharmaceutical marketers make their own very small cans—and some are imported.

While the smallest commercial tinplate can is 35 × 97 mm (86 mL), aluminum cans are made in sizes down to about 3 mL. At the other end of the size spectrum, the largest aluminum aerosols are about 66 × 275 mm (890 mL).

The essentials of the aluminum canmaking process involve taking a round slug (billet, or puck) and extruding it into a "cup" of essentially the same diameter, using a 25–50-ton ram. The top is trimmed to a fixed length, after which the metal is cleaned and enameled—normally inside and outside. The top is then die-formed into a dome and curl (or bead). The curl engages the aerosol valve—all "one-in" (1-in.) valves and most of the smaller, ferrule-type valves. The very small 13- and 15-mm-dia cans may use noncurl finishes.

Aluminum cans have no seams, and thus the linings can be applied more effectively than on tinplates. Many formulas, such as the popular mousse types, may be commercially packed in aluminum without corrosion, whereas they almost always severely attack lined tinplate cans. The permeation of water-based solutions through aluminum-can linings is in the order of 0.1–1.0% as fast as in the case of double-lined and striped tinplate structures.

The smooth-wall construction of aluminum aerosol cans permits the insertion of plastic, floating-type pistons, so that certain products can be kept "nonaerated." This is important for the popular gel-type (postfoaming) shaving creams and the recently test-marketed toothpaste option, as well as dozens of other product types. In time, as one- and two-piece tinplate cans continue their development, some of this market may go to such containers.

Glass and Plastic

Historically, glass and plastic-coated glass aereosols have been restricted for use as containers for colognes, perfumes, and pharmaceutical liquids. Unfortunately, the flammable and bursting hazards, plus cost and a relative lack of aesthetics, have transferred most of their market to plain-glass bottles (nonpressurized) containing pump-action valves. Then, too, some of their pharmaceutical base has moved over to aluminum aerosols. As a result, the glass and plastic-coated glass market is currently very small, and only a few fillers participate.

During the last decade their has been a lot of interest in the potential of certain plastics as aerosol containers. For most of this time DOT regulations greatly limited such developments, since they would not permit the shipment of nonmetallic aerosols having capacities greater than 4 fl oz (118.2 mL). Meanwhile, products such as furniture polish, cleaning preparations, and even hair sprays have been successfully marketed in Europe and parts of South America—using biaxially oriented PET plastic. Aside from some minor concerns about distortion at temperatures over ~122°F (50°C), this clear plastic seems almost ideal. Special bases and shrink-label attributes were needed to hide the often unappealing aerosol formulation, but these were minor concerns. Currently, the DOT has cancelled their regulations to permit plastic aerosols to be shipped, as well as to reduce hot-waterbath testing requirements.

Valves

The aerosol can is normally partly filled with the product concentrate, a valve inserted into the "one-inch" (nominal 25.4-mm) top opening and hermetically sealed to the can by a crimping process, then gassed with liquid or gaseous propellant, in order to produce the final dispenser, one popular type of which is shown in Figure 2.

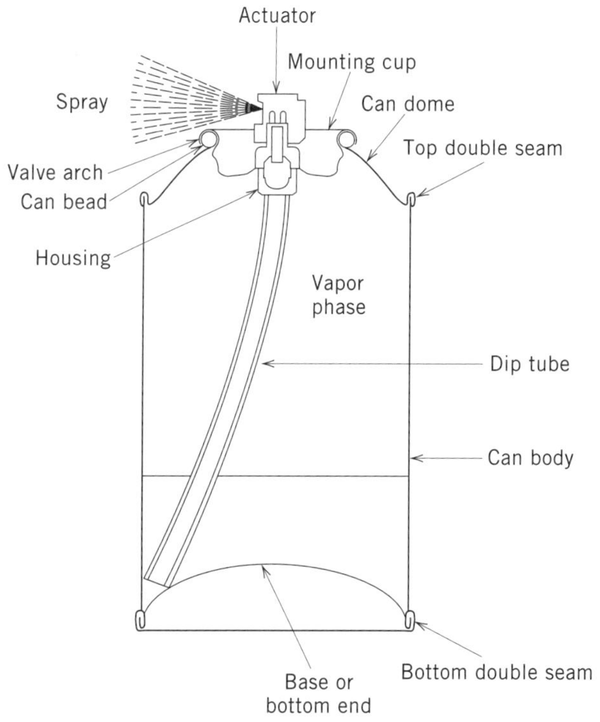

Figure 2. Aerosol dispenser (tinplate can) in operation.

Aerosol valves of the "one-inch" cup varieties are used on perhaps 95% of all aerosol products. They can be characterized as vertical (reciprocating, or up-and-down) valves, or toggle (tilt-action) valves. The vertical type is illustrated in Figure 3.

In the "male" type, vertical-action valve the valve stem protrudes upward from the pedestal and the valve actuator or spout (for foams or gels) is friction-fitted onto it. An alternative is the less-popular "female" valve, where the stem and actuator are one piece, and the stem is pressed into a hole in the pedestal during assembly. Finally, there is the increasingly popular toggle valve, as shown in Figure 4.

Although seven component parts are common to most valves, those designed for inverted use consist of 3–6 pieces, while some meter-spray valves may have as many as 15. Considering the number of components and the amazing diversity of component shapes, designs, orifice sizes, and other configurations, literally billions of valve variables are available from most suppliers. For reasons of confidentiality, valve catalogs only list the more commonly used component variables.

The "one-inch" cup can be obtained in the regular 0.0105-in. (0.267-mm) thickness and high-pressure-resistant 0.0123-in. (0.313-mm) thickness of tinplate, as well as in 0.0160-in. (0.406-mm)-thick aluminum. Several profiles are available, such as regular, conical, cone–conical, and flat. Most valve cups are lined, either with epon–phenolic enamel, organosol enamels, or with 0.008-in. (0.20-mm)-thick polypropylene laminate, bonded to the metal with a special FDA-approved adhesive. The laminate can also serve as a gasket when hermatically sealing the valve cup to the aerosol can.

Other cup gaskets are available. One, known as the "PE-sleeve," has been developed by the Precision Valve Corpora-

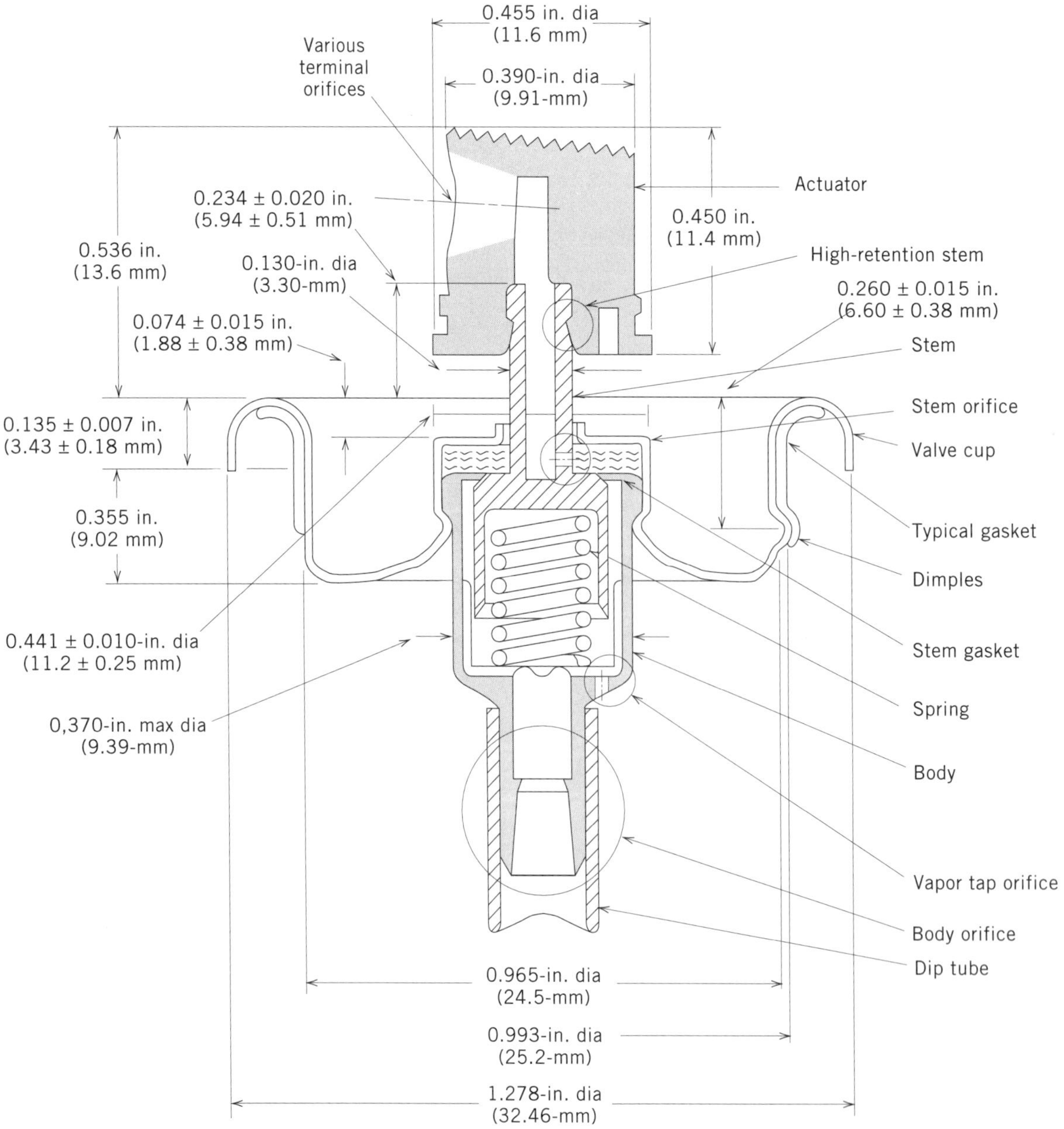

Figure 3. Aerosol valve: "One-inch" cup, vertical action, "male" type.

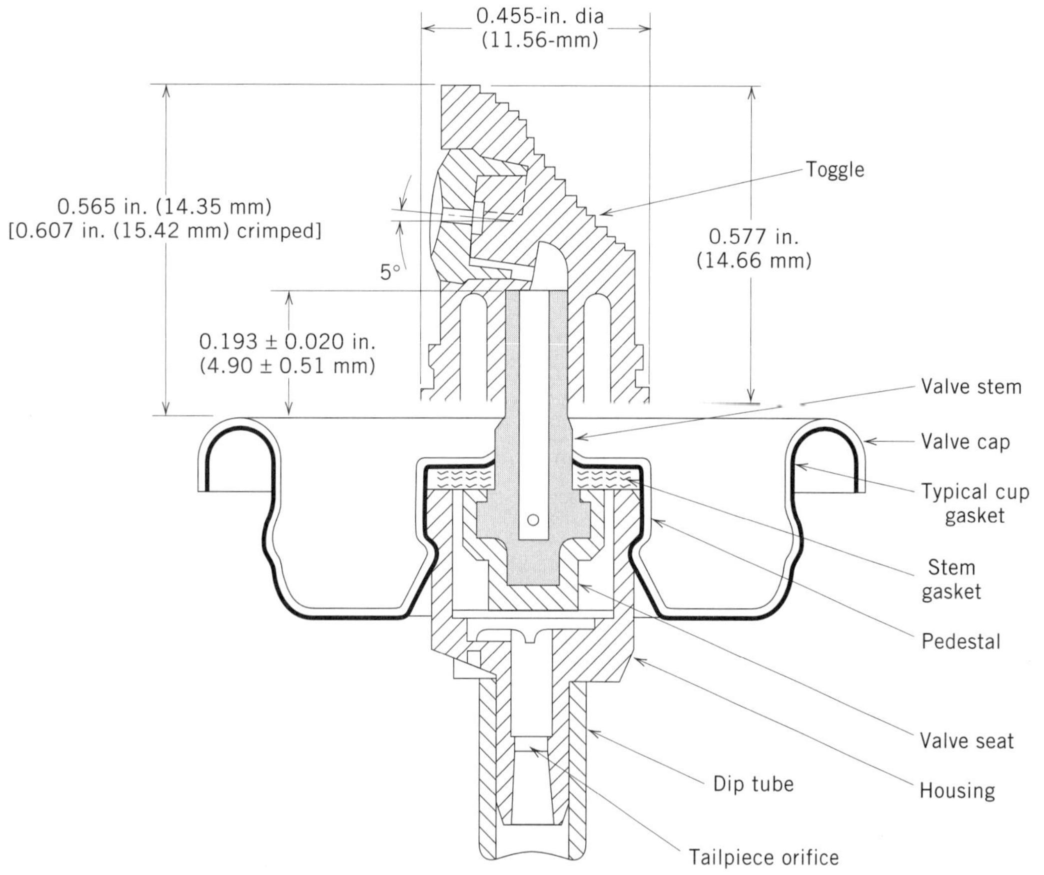

Figure 4. Aerosol valve: "One-inch" cup, toggle-action type.

tion (see Fig. 3). It is a ring of polyethylene, about 0.014-in. (0.356 mm) thick, rammed onto the cup at very high speeds. Another is the "cut gasket": a rubber ring that fits inside the top rim of the valve cup. The older "flowed-in" gasket, used since 1952, has become obsolute for cost and environmental reasons.

The crimping process (called clinching outside the United States) consists of causing a six- or eight-segment collet to descend a fixed distance into the valve cup, after which a mandrel is used to spread the segment toes to a circumscribing diameter of about 1.070 in. (27.18 mm). This spreads the wall of the cup against the can curl profile (throat) of the can, so that the cup gasket can produce a hermetic seal. The crimp depth will vary somewhat, depending on cup and gasket thickness, plus other factors, but will be about 0.178 in. (4.52 mm) from valve cup rim to the bottom of the collet.

Such aspects as valve-stem gaskets, housings, stems, and actuators are not described because they are not technically a part of the aerosol containment system; also, the coverage would necessarily be quite lengthy. In general, the operation of most aerosol valves can be understood from Figures 3 and 4. An increasing number of aerosols now use spray domes, for aesthetic reasons, to better protect the actuator if the unit is dropped, and to provide greater directionality. Otherwise, the valve is protected with a plastic overcap—normally the same diameter as the aerosol can, eg, full-diameter cap. All aerosols are now shipped in strong corrugate cartons, but a recent relaxation in DOT tariff requirements will permit shipment in trays and shrink-wrap packagings. The industry is examining the pros and cons of this regulatory change.

BIBLIOGRAPHY

1. M. A. Johnsen, *The Aerosol Handbook,* 2nd ed., Wayne E. Dorland Co., Mendham, NJ, 1982.
2. *The Aerosol Guide,* 8th ed., The Chemical Specialties Manufacturer's Association, Inc. (CSMA), Washington, DC, 1995.
3. W. Tauscher, *Aerosol Technology* (*Handbook of Aerosol-Packaging*), Melcher Verlag GmbH., Heidelberg/Munich, Germany, 1996 (Engl. ed.).
4. J. J. Sciarra and L. Stoller, *The Science and Technology of Aerosol Packaging,* Wiley, New York, 1974.

Montfort A. Johnsen
Montfort A. Johnsen & Assoc., Ltd.
Danville, Illinois

PRINTING: GRAVURE AND FLEXOGRAPHIC

There are four main printing processes: (*1*) planography or lithography, (*2*) intaglio or gravure, (*3*) porous or screen, and (*4*) relief (flexography or letterpress). In general, the process of printing involves generating two physically different areas: the printing or image area, and the nonprinting or nonimage area. In relief printing, whether flexographic or letterpress, the image or printing area is raised above the nonprinting area. Ink is applied to the raised surface, which is brought into direct contact with the substrate on which the print is to appear. The flexographic relief printing process is used to print on a variety of paper and plastic packaging materials as well as for some magazines and newspapers, labels, and business forms. Water-based or solvent inks are used.

The most typical method of intaglio printing is the gravure process, which uses a nonprinting area that is at a common surface level, while the printing area is recessed, consisting of wells etched or engraved, usually to different depths. Solvent inks are transferred to the printing surface, and a metal doctor blade is used to remove excess ink from the nonprinting surface. Ink is transferred directly to the substrate. Gravure printing is used to print long-run magazines, cartons, bags, labels and gift wraps, as well as plastic laminates, floor coverings, and even textiles. Other types of intaglio printing, such as steel plate or copper-plate printing, use metal plates that are hand- or machine-engraved or chemically etched to produce the lines and characters of the printed piece.

Direct printing refers to the transfer of the image directly from the image carrier to the paper. Most letterpress and gravure printing are done by this method. In indirect or offset printing, the image is transferred from the image carrier to an intermediate rubber-covered blanket cylinder, from which it is transferred to the paper. Letterpress and gravure can also be printed by the offset method.

Images are defined for these printing processes in a number of different ways. The images are produced on a support by chemical, mechanical, or increasingly by electronic imaging means. As of this writing (ca 1996), the greatest number of plates and images are made by photomechanical methods. These systems are characterized by photographic images and light-sensitive coatings that, by using chemical etching or other treatments, lead to the formation of a printing surface. Increasingly, however, this printing surface is produced directly by electronic imaging without the traditional photographic intermediates.

Although there are four types of printing commonly used, gravure and flexographic printing are widely used in packaging printing. They are considered in more detail in the remainder of this article.

Gravure

The gravure printing process, sometimes called *rotogravure,* utilizes a recessed-image plate cylinder to transfer the image to the substrate. The plate cylinder can be either chemically etched or mechanically engraved to generate the image cells. The volume of these cells determines the darkness or lightness of the image. If an area is darker, the cells are larger; if the area is lighter, the cells are smaller.

The gravure market can be considered to comprise three approximately equal segments: publications, packaging, and specialty printing. In publications, gravure retains a significant proportion of the long-run magazine market. In packaging printing, where paperboard and repeat-run cartons are encountered, gravure is the ideal process. The cylinder lasts virtually forever, and color consistency is high. The final third

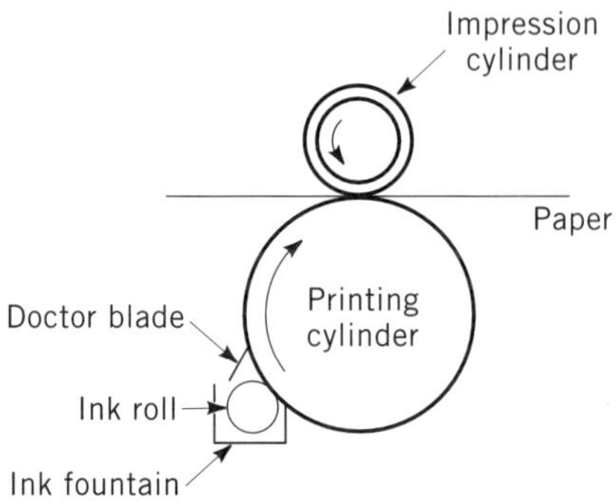

Figure 1. Gravure printing system.

of the gravure market is specialty printing of such items as wallpaper, gift wrap, and floor coverings.

The fundamental strengths of gravure are that the process provides consistent color throughout long print runs, and, because of its ability to apply heavy ink coverage, can be used to print high-quality work or to print on a lower grade of paper than offset lithography and still maintain acceptable print quality.

In contrast, the primary process disadvantages of gravure are long lead time, high cost of manufacture for gravure cylinders, generally long press make-ready times, and environmental hazards associated with the use of solvent-based inks. These disadvantages need to be eliminated if the technique is to remain competitive.

The gravure printing process is based around an inking system that is extremely simple, providing a high degree of consistency, particularly with regard to color printing. This consistency is difficult to match using other printing techniques. The system, shown in Figure 1, utilizes a liquid ink that has traditionally been solvent-based, although environmental pressures have also resulted in the development of aqueous-based inks. The environmental concerns associated with the use of toxic and flammable solvent inks are being addressed by ink manufacturers who are working on the development of water-based products.

The gravure cylinder sits in the ink fountain and is squeegeed off with a doctor blade as it rotates. The impression cylinder is covered with a resilient rubber composition, which presses the paper into contact with the ink in the tiny cells of the printing surface. The image is thus transferred directly from the gravure cylinder to the substrate. Frequently an electrostatic assist is used to help the ink transfer from the gravure cylinder to the substrate (1).

Gravure inks are composed of pigment, resin binder, and, most frequently, a volatile solvent. The ink is quite fluid and dries entirely by evaporation. In multicolor printing, where two or more gravure units operate in tandem, each color dries before the next is printed. This is seen as a particular advantage for gravure over, for example, offset, where the placing of wet ink on wet ink can lead to inferior print quality. The wet ink on dry and the simple ink train, together with a long-run-length capability, have led to the belief that gravure sets the standard for high-quality printing.

Traditionally, a gravure cylinder was prepared chemically by using a chemical-etch process. Since the early 1970s, electromechanical engraving, in which a diamond stylus cuts the cells into cylinders, has become the preferred approach (2).

Newer approaches include laser- and electron-beam engraving.

In gravure, all elements within the image are screened. This is in contrast to flexographic and lithographic plates, which can contain true solids as well as halftones.

In the chemical process for preparing gravure cylinders, a light-sensitive film, typically gelatin containing dichromate salts and carbon black, is exposed first through a gravure screen to establish the cell pattern, then through a continuous-tone photographic negative, hardening the carbon tissue in the light exposed areas in proportion to the amount of light that passed through the negative. The exposed tissue is attached to a fresh copper-plated surface of a gravure cylinder. The unexposed gelatin is washed out, and the cylinder placed in a chemical etching bath that dissolves away the copper layer in proportion to the thickness of the remaining hardened gelatin layer, thus creating the ink-carrying cells. Since the cell pattern is established using a screen pattern, all cells are the same area, but vary in depth. The surface of the etched cylinder is then chrome-plated to impart wear resistance. Gravure cylinders can be recycled many times by removing the copper layer and replating.

In the case of electromechanically engraved cylinders, an electronic scanner reads the density of the photographic negative pixel by pixel, and translates the density into an electromechanical impulse that drives a diamond stylus into the soft copper layer of a cylinder. The completed cylinder is then chrome-plated and is ready for use. Because of the shape of the diamond stylus, the cells produced vary both in depth and area depending on how far into the copper surface the stylus is driven (3).

Electromechanical engraving avoids all the complexity, hazards, and toxic wastes involved with using chemical-etching techniques, at the cost of increased capital investment in the engraving equipment. Electromechanical engraving is a slow process, but this is offset by the ability to use digital data, which better fits a modern electronic prepress workflow. Newer systems drive the engraving stylus (or multiple styli for improved efficiency) directly with digital data from an electronic prepress system, or engrave a cylinder using a high-power laser (4).

Flexography

Flexography is a variation of letterpress printing used mainly for packaging applications. It is characterized by the use of an elastomeric printing plate, fast-drying inks, and an ink-metering (anilox) roll system. The principal advantages of flexography are reflected in the markets where it is most often used. Flexo's ability to print on a wide range of substrates, including plastic films, foils, coated and uncoated paper, paperboard, and corrugated board, make it ideal for many packaging uses as well as for printing continuous patterns such as wallpaper and gift wrap.

Other advantages include low cost and short cycle time, ability to change cylinder diameters to reduce stock waste, precise ink transfer with minimum on-press adjustments, and ability to print one layer and laminate another layer over it in a continuous process.

Limitations of flexography include higher highlight-dot gain (spreading of halftone dots in low-density areas) and lower solid density as compared to gravure and offset lithography. The increased dot gain results from a combination of printing-plate deformation and ink spread on press. Another limitation is the inability to print uniform solids and halftones using the same plate without substantial make-ready. The increased pressure required for uniform solids increases dot gain in highlights.

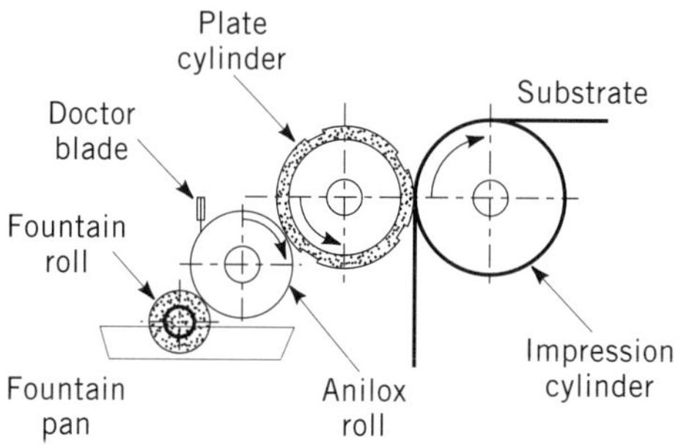

Figure 2. Flexographic printing station, where the fountain pan supplies ink to the rubber fountain roll, which, in turn, supplies ink to the anilox roll. The optional doctor blade removes excess ink from the surface of the anilox roll so that it transfers a uniform layer of ink to the printing plate. The printing plate then transfers this layer of ink to the substrate, which is supported by the impression roll.

A typical flexo print station is shown in Figure 2. The three basic types of flexographic presses are shown in Figure 3. Dryers generally separate individual print stations. In the stack press (Fig. 3**a**), individual print stations are in sequence one on top of the other. This press is used primarily for paper and laminated films. Advantages include accessibility to print stations and the ability to reverse the web, thus allowing printing on both sides in one pass. In the central impression press (Fig. 3**b**) the print stations are distributed around a large central impression cylinder, which is precisely geared to each print station and thus improves the registration. This press is used mainly to print high-quality wide-web films. The in-line press (Fig. 3**c**) is used mainly for printing corrugated and folding carton, as well as for narrow-web tag and label.

The heart of the flexographic printing system is the anilox roll, a steel cylinder optionally coated with ceramic and engraved with a pattern of pits or cells. The function of the anilox roll is to meter a uniform film of ink from the ink fountain to the printing plate without the need for continuous adjustment. There are many types of anilox rolls, distinguished by the mode of engraving, the materials of construction, the pattern of the cells, and the cell geometry. Anilox-roll variables include the screen count and screen angle as well as cell volume. Anilox rolls fall into two main families: mechanically engraved chrome rolls and laser-engraved ceramic rolls.

Mechanically engraved chrome rolls are steel cylinders engraved by a precision tool. Following engraving the rolls are plated with copper, which acts as a bonding layer; and chrome, which hardens the surface. Mechanically engraved chrome rolls have been the mainstay of the industry for many years. However, on the introduction of the reverse-angle doctor blade, which better controls ink metering, and the subsequent wear problems, the industry has been steadily shifting to laser-engraved ceramic rolls.

Ceramic rolls are steel cylinders that have been coated with a ceramic, usually chromium oxide, layer and then engraved with the beam from a high-energy laser. By control-

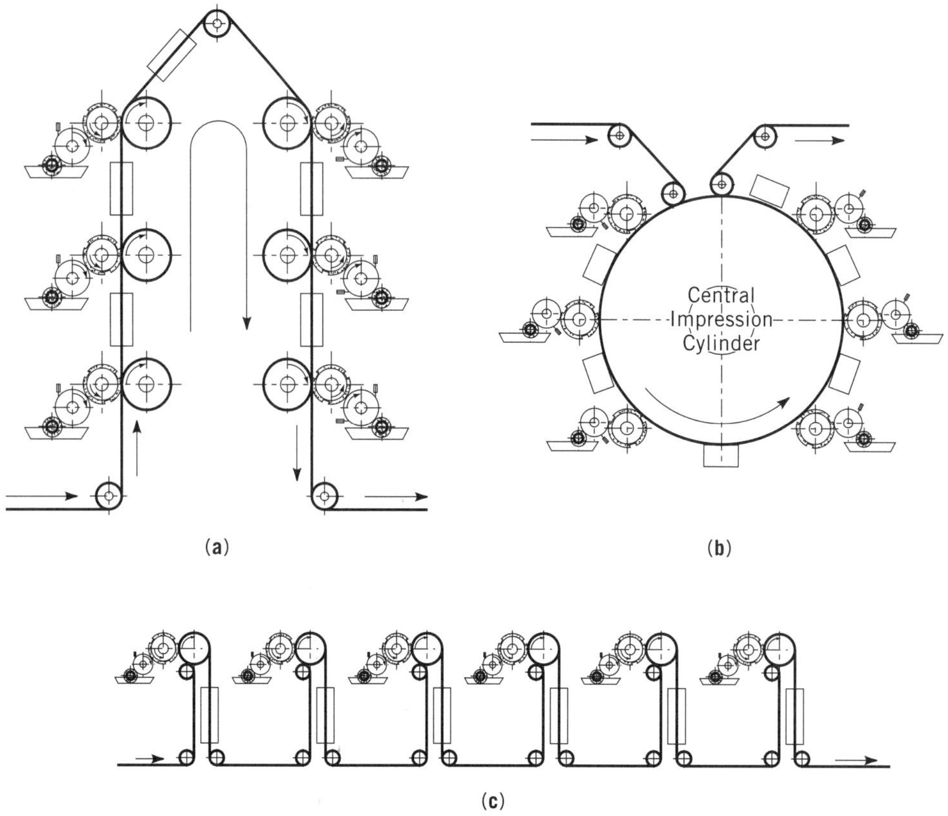

Figure 3. Schematics of flexographic presses, where the arrow designates the direction of the web flow and □ represents an interstation dryer: (**a**) stack press; (**b**) central impression press; (**c**) in-line press.

ling the energy and timing of the laser pulse, the depth and diameter of the anilox cell can be tightly controlled. Unlike mechanically engraved rollers, the volume of laser-engraved rolls is independent of screen count, allowing a variety of rollers having a wider range of volumes and screen counts. The ceramic coating reduces doctor-blade wear by an order of magnitude, thus prolonging the life of the roll.

Flexography typically employs low-viscosity fast-drying inks. These are dispersions having 25–35% solids content, which consist mainly of nearly equal parts of a pigment or pigments dispersed in polymer resins. These resins form the ink film and attach the pigment to the substrate. Inks may also contain additives such as slip agents, surfactants, plasticizers, and antifoams. The solvent portion is usually a blend of alcohols and acetate esters. Drying rate can be varied by changing the ratio of acetate to alcohol in the solvent.

For aqueous inks, the resins are water- or alkali-soluble or dispersible and the solvent is mostly water containing sufficient alcohol (as much as 25%) to help solubilize the resin. To keep the alkali-soluble resin in solution, pH must be maintained at the correct level. Advances include the development of uv inks. These are high viscosity inks that require no drying but are photocurable by uv radiation. In these formulations, the solvent is replaced by monomers and photoinitiators that can be crosslinked by exposure to uv radiation. The advantage of this system is the complete elimination of volatile organic compounds (VOCs) as components of the system and better halftone print quality. Aqueous and uv inks are becoming more popular as environmental pressure to reduce VOC increases.

Three primary types of flexographic printing plates are used: molded rubber, solid-sheet photopolymer, and liquid photopolymer.

Initially, flexographic printing plates were made of handcut or molded rubber. The basic steps of rubber platemaking include preparing an engraving by exposing a photoresist-coated metal plate to uv light through a photographic negative, washing away the unexposed resist with solvent, and acid-etching the unprotected metal surface; molding a phenolic matrix board using the metal engraving; and finally molding the rubber plate using the matrix mold.

Molded-rubber plates transfer ink well, and, when large numbers of identical designs are needed, are inexpensive to make. However, environmental concerns over the use of acid-etching solutions to prepare the metal engraving, poor thickness uniformity, and poor dimensional stability of the molded plates are disadvantages. Photopolymer plates, both solid and liquid, have replaced most engravings.

Solid photopolymer plates consist of an elastomeric, photosensitive layer bonded to a polyester support. The plate formulation usually contains an elastomeric binder that provides the required properties of flexibility and resilience. The plates typically contain acrylic or methacrylic monomers, sensitizers, and photoinitiators, which, in the presence of sufficient ultraviolet (uv) energy, polymerize to reduce the solubility of the material. Photopolymer plates are made by exposure to uv through the back of the plate to consume stabilizers and define relief height, front uv exposure through a photographic negative (which is usually halftone-screened) to form the image, removal of unexposed areas with a solvent to form the relief image, drying to remove solvent, overall exposure to eliminate surface tack, and postexposure to further crosslink and toughen the plate (5). Advantages of solid-sheet photopolymer plates over rubber include improved dimensional stability derived from the polyester support, longer print run length, improved print quality including more predictable rendition of four-color process images, and better thickness uniformity.

Newer technology involves water-processable photopolymer plates. Many platemakers and printers are eager to switch to water processing in order to eliminate volatile organic solvents. The chemistry and process of use are similar to that of the solvent-processable plate except that the plate is formulated with water soluble materials.

The liquid photopolymer plate system is used successfully in corrugated and newspaper markets. The platemaking process is similar to the solid-sheet photopolymer plate except that before exposure to uv, the liquid photopolymer, which consists of liquid rubber oligomers and crosslinkers, is sandwiched between two layers of transparent plastic film to define a nominal and uniform thickness during exposure. Special equipment is used to meter the liquid plate material onto the bottom sheet while covering the top surface with the other plastic sheet. The platemaking process is similar to that of the solid photopolymer plates (6). Lower cost of materials is the main advantage over solid plates. Disadvantages include inferior thickness uniformity and limited run life.

As printers turn more to electronic prepress methods, direct writing of digital data to printing plates will increase in importance. Relief images can be made by laser-engraving a rubber-coated cylinder. The rubber layer is selectively ablated by a high-energy laser beam, leaving a relief printing image. This method eliminates the need for a photographic negative, and is ideal for printing continuous images. Laser imaging of rubber cannot reproduce the fine screen rulings obtainable with solid plates. Other direct digital imaging systems that use photopolymer plates with integral masks have also been shown (4).

BIBLIOGRAPHY

1. J. M. Adams, D. D. Faux, and L. J. Rieber, *Printing Technology,* 3rd ed., Delmar Publishers, Albany, NY, 1988, p. 477.
2. M. H. Bruno, ed., *Pocket Pal, a Graphic Arts Production Handbook,* 15th ed., International Paper, Memphis, TN, 1992, pp. 125–126.
3. *Ibid.,* p. 126.
4. A. J. Taggi and P. Walker, "Printing Processes" in J. Kroschwitz, ed., *Kirk-Othmer Encyclopedia of Chemical Technology,* 4th ed., Vol. 20, Wiley, New York, 1996.
5. *Cyrel® Photopolymer Flexographic Printing Plates Process-of-Use Manual,* DuPont Printing and Publishing, Wilmington, DE, 1995, p. 17.
6. G. Mirolli, "Engravings and Printing Plates" in J. W. Cotton, et al. eds., *Flexography, Principles and Practices,* 3rd ed., Flexographic Technical Association, Inc., Ronkonkoma, New York, 1980, pp. 174–179.

General References

J. W. Cotton, et al. eds., *Flexography, Principles and Practices,* 3rd ed., Flexographic Technical Association, Inc., Ronkonkoma, NY, 1980.

ARTHUR J. TAGGI
DuPont Printing and Publishing
Wilmington, Delaware

PETER A. WALKER
Consultant

PROPELLANTS, AEROSOL

Propellants are an integral part of every aerosol product. They may vary from about 0.5% to 100% of the total composition, and even with specific products there may be rather wide variations in propellant type and percentage. The most basic role of the propellant is to provide pressure inside the dispenser, so that when the valve actuator is pressed, the product emerges. But the propellant fills a number of other rolls as well (see also Aerosol containers).

For products in which the propellant is soluble (usually anhydrous types), the propellant can be used to control particle size. Increasing the percentage of propellant, or the use of a higher-pressure propellant, will serve to decrease the mean aerodynamic particle size of the spray. The importance can be illustrated in the case of insecticides. Sprays designed to kill flying insects must produce a preponderance of particles capable of floating about in the air for significant time periods. In the case of particles larger than ~50 μm, they would fall onto tables and floors so rapidly that the insecticide would lose potency—known as "knockdown" and *cidality* properties. On the other hand, particles that are too small would tend to "airflow" over or under the flying insect. Some insect sprays, more properly identified as bug killers, contain only about 3–20% of propellant in the formula (depending on propellant choice), and these produce coarse sprays designed to coat baseboard areas where many crawling insects travel. The formulas usually contain relatively harsh toxicants that should not be inhaled to any significant degree.

A rather unique insect killer is the wasp-and-hornet type. Using about 0.5% of nitrogen as the sole propellant, and with a high delivery rate, tubular aerosol valve, this dispenser can produce a rather solid stream, sufficient to reach wasp nests and similar targets from distances of 20 ft (6.1 m). As a result, the insecticide can reach nests and hives that are high off the ground, while allowing the operator to remain well out of the path of vengeful biting insects.

The aerosol valve generally works hand-in-hand with the propellant to produce the desired product presentation. Valves are important in controlling delivery rate. They also have terminal orifices that can determine the width of the spray cone. Some of these orifices come with what is known as a mechanical breakup (MBU) insert, that can take what might be a solid stream of product and cause it to be emitted as a coarse spray. The functionality is rather similar to that of a garden-hose nozzle.

In those cases where sufficient dissolved propellant is available to create a spray, as the liquified gas bursts into the gaseous state, the use of an MBU sprayhead can serve to make the spray particles somewhat finer and the spray pattern more uniform and elegant. MBU sprays are generally quieter than other types.

Most propellants are relatively insoluble in water, and thus water-based products, but can be emulsified into them if some surfactant is added. This is the basis of aerosol foams: shaving creams, mousses, foam charcoal starters, and so forth. Typically, 4–8% of hydrocarbon propellant is emulsified in the soap solution by gentle shaking before use. If one uses a spray-type actuator instead of a foam spout, foaming sprays can be generated. Hard-surface cleaners, textile spot removers, over cleaners, and similar products are formulated this way.

About 45–50% of all aerosol valves have vapor-tap orifices: tiny holes pinned or drilled through the body, so that both liquid and gas enter the body chamber when the valve is actuated. The result is a spray with much lower delivery rate, a finer particle size, and a somewhat louder sound. Since much of the emerging propellant is already in the gas phase, the spray is warmer than that of regular sprays. This can be important in the case of underarm deodorants, antiperspirants, body sprays, and medicated foot sprays. The use of vapor-tap valves also allows liquid orifices to be larger. Vapor-taps are always seen on antiperspirants, partly to keep the finely powdered aluminum chlorhydrate astringent from clogging valve orifices. The vapor-tap feature permits the use of large, clog-resistant orifices, but without the stigma of excessive delivery rates.

One important consideration with vapor-tap valves is that sufficient propellant must be used to keep it from running low, due to the continual short-circuiting of the gas phase. For example, depending on the relative size of the liquid-tap and the vapor-tap orifices, a dispenser that starts with 45% propellant could end up with 35% propellant. This might mean a discernably coarser spray and lower delivery rate. A final interesting feature of vapor tap valves is that, if the liquid and vapor orifices are the same diameter, the aerosol will spray just the same way, if it is inverted. In such cases the regular aerosol types would only emit propellant gas.

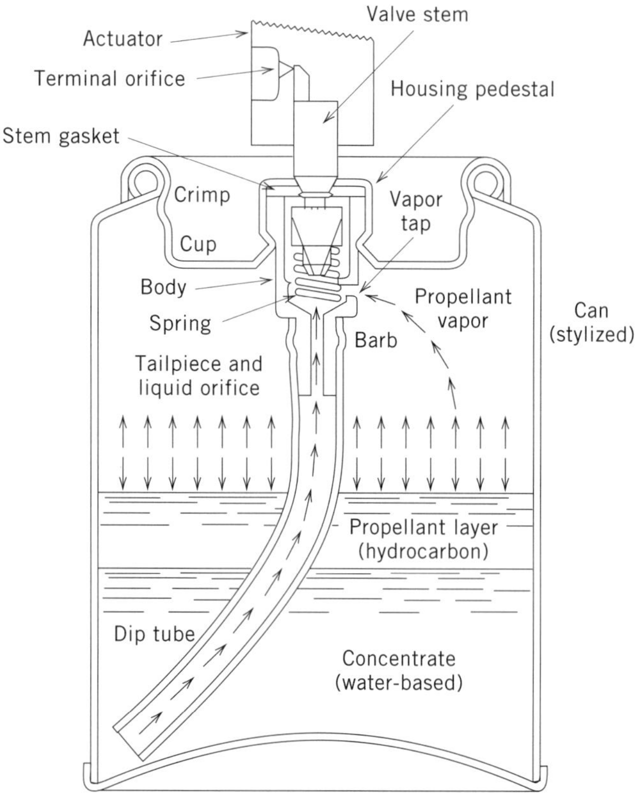

Figure 1. Water-based aerosol spray system.

Propellant Chemistry

The earliest aerosol propellants were carbon dioxide, followed by vinyl chloride, isobutane, dimethyl ether, and a few other gases. In 1938, when the chlorofluorocarbons (CFCs) became commercially available, these very low-odor, nonflammable, and relatively inexpensive gases virtually displaced all the others. Most early aerosols used about 85–90% CFCs—from the heavy "bug bombs" of World War II to the consumer aerosols that first appeared about 1948 in the United States. After 1954 very low-cost, low-odor hydrocarbons became available, and propane, *n*-butane, and isobutane began to be used to propel water-based products, such as window cleaners.

The very light-density hydrocarbon liquefied gases simply sat on top of the product as a discrete liquid layer, conferring a rather constant pressure. Under such conditions, the product itself flowed up the dip tube when the dispenser was actuated, turning into the desired coarse spray as it passed through the MBU valve orifice (see Fig. 1). This type of operation had been impossible with the CFCs, since they are about 1.4 times as dense as water and would have formed a bottom layer of almost pure propellant. Actuating the valve would have swept the propellant up the diptube—and even if the diptube had been shortened, to extend only to the bottom of the product layer, tilting the dispenser during use would still have caused propellant discharge.

The window cleaner, oven cleaner, basin, tub, and tile cleaner and similar products typically ran about 4–6% isobutane propellant (7.3–10.9 vol%). The top layer would diminish in size when the headspace was enlarged during sprayout. But a certain amount of propellant was also added as a contingency against the very real probability of occasional consumer misuse, such as trying to spray the can when it was severely tilted or even upside down.

The aerosol industry first learned about the "ozone/CFC" issue about 1974, at a time when various CFC propellants were used in about 53% of all aerosols. There was disbelief, since the evidence was nonempirical and speculative at first. But in 1978 the EPA and FDA acted to ban the CFCs from all but medical aerosols—while not doing anything about their continuing use in refrigeration, air conditioning, and all other applications. (The United States was also the only country in the world to take such an action.) The resulting press coverage acted to reduce aerosol sales by ~35% in a matter of months. It has taken the industry 16 years to recover this lost volume, and aerosols still retain a certain environmentally related stigma in the minds of about 74% of consumers—down from 87% in 1991, thanks to a massive public-relations program.

Without the ability to continue using nonflammable CFCs for hair sprays and virtually all other aerosols, marketers turned to about the only viable alternatives: the liquefied petroleum gases (LPGs) (eg, propane, *n*-butane, and isobutane).

Elaborate reformulation programs were needed to evolve products of acceptably low flammability, while still maintaining compatibility with active ingredients. One problem involved net weight per can. In the case of antiperspirants, which had used 90% CFC in the formula, the reformulated products now used ~68–85% of very low-density hydrocarbons—generally isobutane. Since the density of the new antiperspirants was about 0.68 g/mL (at room temperature), compared to about 1.38 g/mL for the CFC predecessors, marketers could fit only about half the previous product weight into their cans. Consumers thought the cans were "only half full," and this sparked a degree of discontentment that acted to plummet the sales of this particular product to about 25–30% of what they had been, prior to 1978.

Following the transition away from CFCs, hydrocarbon propellants—especially those deodorized and purified by four companies—were the propellants of choice, and used in about 84% of all aerosols. The annual industry requirement was about 55,000,000 U.S. gallons per year. Carbon dioxide was used in perhaps 8% of all aerosols, nitrous oxide in about 2% (mainly whipped creams), and such propellants as 1,1-difluoromethane (HFC-142a), nitrogen, and the HCFCs were used to fill the remainder. About 1984 DuPont first made commercial quantities of dimethyl ether available, by purifying an intermediate product of their plasticmaking reactors. Finally, about 1993 DuPont and Allied-Signal began to offer 1,1,1,2-tetrafluoroethane (HFC-134a) for certain volume-limited aerosol applications. The global warming potential of HFC-134a (about 550 times that of CO_2) is a matter of concern, and has caused suppliers to scrutinize the health and safety benefits of using this propellant, measured against environmental concerns. As the latter inevitably grow, HFC-134a could well be made less available to the aerosol industry, except, of course, for metered-dose inhalants and related medical products.

In the years following the virtual demise of aerosol CFC products (1978), the industry utilized two additional nonflammable propellants: chlorodifluoromethane (HCFC-22) and 1-chloro-1,1-difluoroethane (HCFC-142b), even though they were 5.0% and 11.2% as active as the major CFCs as stratospheric ozone depleters. Although they were at first viewed as a solution to the problem, later on environmentalists made them a part of the problem. On January 17, 1994 they were banned from aerosol uses, although still permitted for all other applications. They were never widely used for aerosols and are not discussed further here.

Propellants must meet numerous requirements to be ideal for aerosols. Many of these are as follows:

1. Reasonable boiling point and vapor pressure
2. Reasonable cost
3. Acceptable toxicology profile
   a. Acute
   b. Subacute
   c. Chronic
4. Acceptable low odor
5. Acceptable environmental aspects
   a. Minimum effect on stratospheric ozone
   b. Minimum effect on tropospheric ozone effects
6. Flammability consideration
7. Stability consideration
   a. Hydrolysis
   b. Alcoholysis
   c. Polymerization
8. Solvent properties
   a. Useful in some formulas
   b. Sometimes detrimental to elastomers
9. Purity

The CFCs came close to being ideal, except for their great chemical stability, which brought about the free-radical depletion of stratospheric ozone. Current propellants are all considerably less than ideal. The hydrocarbon gases (LPGs) indirectly act to produce tropospheric ozone, or smog, but their major drawback is their extreme flammability. Dimethyl ether is rather uniquely water-soluble, but it is also flammable and has such a strong solvent action that gaskets may be attacked and corrosion possibilities are enhanced. Carbon dioxide is nonflammable, but acidic if added to water. In any event, no more than about ~3–6% can generally be added before aerosol pressures reach the regulatory limits imposed on aerosols by the U.S. Department of Transportation. These and other general properties of the available propellants are given in Tables 1 and 2.

A selection of the physical properties of aerosol propellants is presented in Table 3 (hydrocarbons), Table 4 (DME and HFC-152a) and Table 5 (compressed gases). The propellants may be mixed with other propellants, or with various aerosol concentrates, in order to achieve any reasonable product pressure. Entrapped air, present in all aerosol dispensers to varying degrees, adds about 6–21 psi (0.41–1.45 bar) to the propellant pressure.

The usual aerosol pressures vary within ~30–70 psig at 70°F (2.06–4.83 bar at 21°C). In the special cases of CO_2 and N_2O propellants, pressures are typically 85–95 psig at 70°F (5.86–6.55 bar at 21°C), and where nitrogen or compressed air (CAIR) are used, pressures may be as high as 141 psig at 70°F (9.72 bar at 21°C). All these ranges are designed to keep the aerosol pressure at or below 180 psig at 130°F (12.41 bar at 55.4°C), which is the limit set by the U.S. Department of Transportation (DOT) for the strongest of the three standard aerosol cans. Pressures increase very rapidly with increasing temperature for those formulas that contain large amounts of lower-pressure propellants, such as isobutane. On the other hand, nitrogen, which often has a negligible solubility in the concentrate, increases in pressure more or less in harmony with Charles' law.

Aerosols will buckle (permanently deform) and finally burst if overheated. The label almost invariably warns against heating higher than about 120°F. Some labels list 130°F. Bursting aerosol cans pose a severe hazard, and may be life-threatening. In some cases the contents may also ignite, causing an additional hazard. A limited number of aerosol cans have fitments designed to allow the contents to be released in a controlled fashion if they are overheated. For three-piece ETP cans, about 10 small indentations may be notched into the top of the top double seam. The can is designed so that the dome sections will evert (or "jump up," somewhat) well below the minimum bursting pressure. When this occurs, the work-hardened indentations will re-form into little apertures, releasing the overheated product. This is known as the *pressure-relief mechanism* (PRM). Aluminum

Table 1. General Propellant Comparisons[a]

	Hydrocarbons	DME	HFC	Compressed Gases
Flammability	Flammable	Flammable	Flammable and nonflammable	Nonflammable
Toxicity	Low	Low	Low	Low
Solvency	Poor	Good	Poor	Poor
Density	Low	Low	Intermediate	—
Solubility in water	Low	High	Low	Low
Environmental concerns	VOC	VOC	GWP	None
Cost	Low	Low	High	Low

[a] Abbreviations: DME = dimethyl ether, HFC = hydrofluorocarbon, VOC = volatile organic compound (as defined by regulators), GWP = global warming potential.

cans and others without a top double seam may be relieved by means of various fixtures in the base.

The aerosol industry had to learn, over the years, how to safely handle flammable propellants and gas them into dispensers. They must be manipulated in closed systems. In larger filling plants, flammable propellants are stored in 1 to ~10 bulk tanks, ranging up to 30,000 gal (113,500 L) in nominal capacity. Two-inch (51 mm) double-strength steel pipes lead from the bottom of each tank past valves to a low-pressure pump. From there the line may lead to a blending station, followed by a manifolding unit, to direct the liquified gas (or mixture) to the appropriate gas house. In this enclosure, which is often outside the main establishment because of the potential hazards, the gas supply goes to piston-operated gassers. Two types are used. The earliest, now called a "through-the-valve" (T-t-V) gasser, charges the propellant backward through the valve and into the sealed can. The liquistatic pressure ranges from about 600 to 950 psig (41–66 bars), and the temperature is often controlled to about 104°F (40°C) for a number of technical reasons. Many valves are designed to backflow propellant at very high rates, so that a typical 18-head rotary T-t-V gasser may attain speeds of up to ~350–400 cans per minute.

The other gassing machine performs three consecutive functions: (*1*) after lifting the lossely inserted valve slightly, it draws a partial vacuum on the can; (*2*) then it adds the liquid propellant under high pressure, around the periphery of the lifted valve cup; and (*3*) finally, it presses the cup firmly down on the can bead and performs a crimping (swaging) operation to hermetically seal the can. A minor disadvantage of this machine is that above 3.0 mL of liquid propellant is flashed off when the filling head lifts off the sealed can. This must be exhausted through a floor-sweeping ventilation system. More recently the EPA has become interested in plant emissions of volatile organic compounds (VOCs), especially in the case of facilities located in VOC nonattainment areas, such as larger cities. If one considers the loss of (say) isobutane from an intermediate to large aerosol plant, running a total of 1000 cans per minute; at 3.0-mL loss per can, the total

Table 2. Available Alternatives to CFCs[a]

	Flammable	VP	ODP	GWP
HCFCs				
HCFC 22	No	121	0.05	0.29
HCFC 142b	Yes (marginal)	29	0.06	0.06
HFCs				
HFC 152a	Yes	62	0.0	0.019
HFC 134a	No	71	0.0	0.28
DME	Yes	63	0.0	0.0
Hydrocarbons	Yes		0.0	
Propane		108		
n-Butane		17		
Isobutane		31		
n-Pentane		−6.1		
Blends				
Compressed gases	No			
CO_2			0.0	0.0001
N_2O			Low	0.0017
N_2				

[a] Abbreviations: VP = vapor pressure (psig at 70°F), ODP = ozone-depletion potential (CFC-11 = 1.000), GWP = global warming potential (CFC-11 = 1.000).

Table 3. Properties of the Hydrocarbon Propellants

	Propane	Isobutane	*n*-Butane
Formula	C_3H_8	*i*-C_4H_{10}	*n*-C_4H_{10}
Molecular weight	44.1	58.1	58.1
Boiling point, °F	−43.7	10.9	31.1
Vapor pressure, psig			
70°F	109	31	17
130°F	257	97	67
Density (g/mL), 70°F	0.50	0.56	0.58
Solubility in water (wt%), 70°F	0.01	0.01	0.01
Kauri–butanol value	15	17	20
Flammability limits in air (vol%)	2.2–9.5	1.8–8.4	1.8–8.5
Flash point, °F	−156	−117	−101

Table 4. Properties of DME and HFC-152a

	DME	HFC-152a
Formula	CH_3OCH_3	CH_3CHF_2
Molecular weight	46	66
Boiling point, °F	−12.7	−12.5
Vapor pressure, psig		
70°F	63	62
130°F	174	176
Density (g/mL), 70°F	0.66	0.91
Solubility in water (wt%), autogenous pressure	34	1.7
Kauri–butanol value	60	11
Flammability limits in air (vol%)	3.4–18	3.9–16.9
Flash point, °F	−42	<−58

Table 5. Properties of the Compressed Gases

	Carbon Dioxide	Nitrous Oxide	Nitrogen
Formula	CO_2	N_2O	N_2
Molecular weight	44.0	44.0	28.0
Boiling point, °F	−109	−127	−320
Critical temperature, °F	88	98	−232
Vapor pressure (psig at 70°F)	837	720	477
Solubility in water (vol gas/vol liquid), 70°F at 1 atm	0.82	0.6	0.016
Flammability limits in air (vol%)	Nonflammable	Nonflammable	Nonflammable
Flash point, °F	None	None	None

loss will be ~47.5 gal (180 L) per hour, or 380 gal per 8-h shift. The EPA has required some plants to dramatically reduce their emission rates, either by changing gassers, adding conservation accessories to their U-t-C gassers, or by collecting the bulk of the released gas and compressing it to liquid in a "gasser waste" storage tank. In some cases the "gasser waste" may be used as heating fuel for the plant boiler.

The extreme flammability of hydrocarbon gases has necessitated equipping gas houses with sophisticated sensing devices, halon-powered extinguishing equipment, deluge systems, regular and emergency ventilation attributes, and a complex electroprotective system. Employees must be carefully trained in fire safety. There is a minor but significant trend to depopulate gas houses, so that, if a catastrophe occurs, one or more lives will not be jeopardized.

As little as 0.58 fl oz (9.6 g) of isobutane, poured into an open-top 55-gal (200-L) steel drum and mixed with the air in the drum, will produce about a 500-L "fireball" if ignited. As little as 2.0 g, poured into the drum but unmixed, will produce flames that fill most of the drum if ignited. The following data show that the hydrocarbons cannot be detected by odor, and are substantially heavier than air:

1. Odorless
2. Colorless
3. Heavier than air
   a. Two times
4. Narrow limits of flammability
   a. LEL 2%
   b. HEL 10%
5. Low ignition temperature
   a. 825–900°F
6. Expansion
   a. Liquid to vapor = 1–270

This means they have a tendency to collect in floor-level areas. Mixtures of 1.86–2.04 vol% gas in air are potentially flammable. Fortunately for the industry, practically all hydrocarbon-type aerosols are used for only a few seconds at a time. But for those paints, total-release indoor insect foggers and a few other products, where large amounts are sprayed at a time, good ventilation or dilution is quite important, and such admonitions will be seen on the can labels.

From time to time the industry learns of new propellant possibilities. The Great Lakes Chemical Company offers HFC-227a ($CF_3-CHF-CF_3$) for pharmaceutical uses; 3M, Inc. is now offering $C_4F_9-O-CH_3$ (*n*-perfluorobutylmethyl ether); and Phillips Petroleum has developed an electrochemical process for making CF_3-O-CF_3 (perfluorodimethyl ether) and CF_3-O-CH_3 (trifluoromethylmethyl ether). But all these compounds are very costly and probably pose significant global-warming potentials. CF_3-O-CH_3 is flammable, so its advantages over CH_3-O-CH_3 are hard to imagine.

Several marketers are exhibiting a resurgence of interest in the high-pressure gases, particularly CO_2 and compressed air (CAIR), for certain products. The VOC aspect of the hydrocarbons is sure to become more troublesome in the years ahead, and already one sees quite a number of hair sprays and similar products that are now using non-VOC HFC-152a, despite the fact that it costs about 11 times as much as the hydrocarbons, in bulk quantities. In any event, the aerosol industry is fully confident that they can weather these "environmental storms" in the forseeable future, and extend the U.S. business base to well beyond the current benchmark of 3.3 billion units per year.

MONTFORT A. JOHNSEN
Montfort A. Johnsen and Associates, Ltd.
Danville, Illinois

PULP, MOLDED

Like the paper-making process itself, the origins of pulp molding are pretty well obscured by antiquity. One assumes that artisans in early Oriental, Egyptian, and Greco-Roman civilizations who were involved with screen-felting of vegetable fibers into sheets of paper rapidly discovered benefits that derived from embossing the screening material to obtain contours. It is believed that ornate wall- and ceiling-molding artifacts of those civilizations were fabricated in this manner. With the rapid development of the paper industry (see Paper) around the turn of the century, disposable molded pulp products began to find their way into the mainstream of commercial life as protective vehicles for transport of fragile foods such as pies and eggs. The economics of manufacture of pie transfer plates and egg trays have kept these unglamorous products in the marketplace. Although certainly in the mature part of the viability curve, they remain of considerable commercial interest.

The term "molded pulp" is used to describe three-dimensional packaging and food-service articles that are manufactured by forming from an aqueous slurry of cellulosic fibers into discrete products on screened, formaminated molds in a process analogous to continuous-sheet cylinder-board papermaking. It is not unusual to find the pulp-molding process grouped with other converting processes such as compression

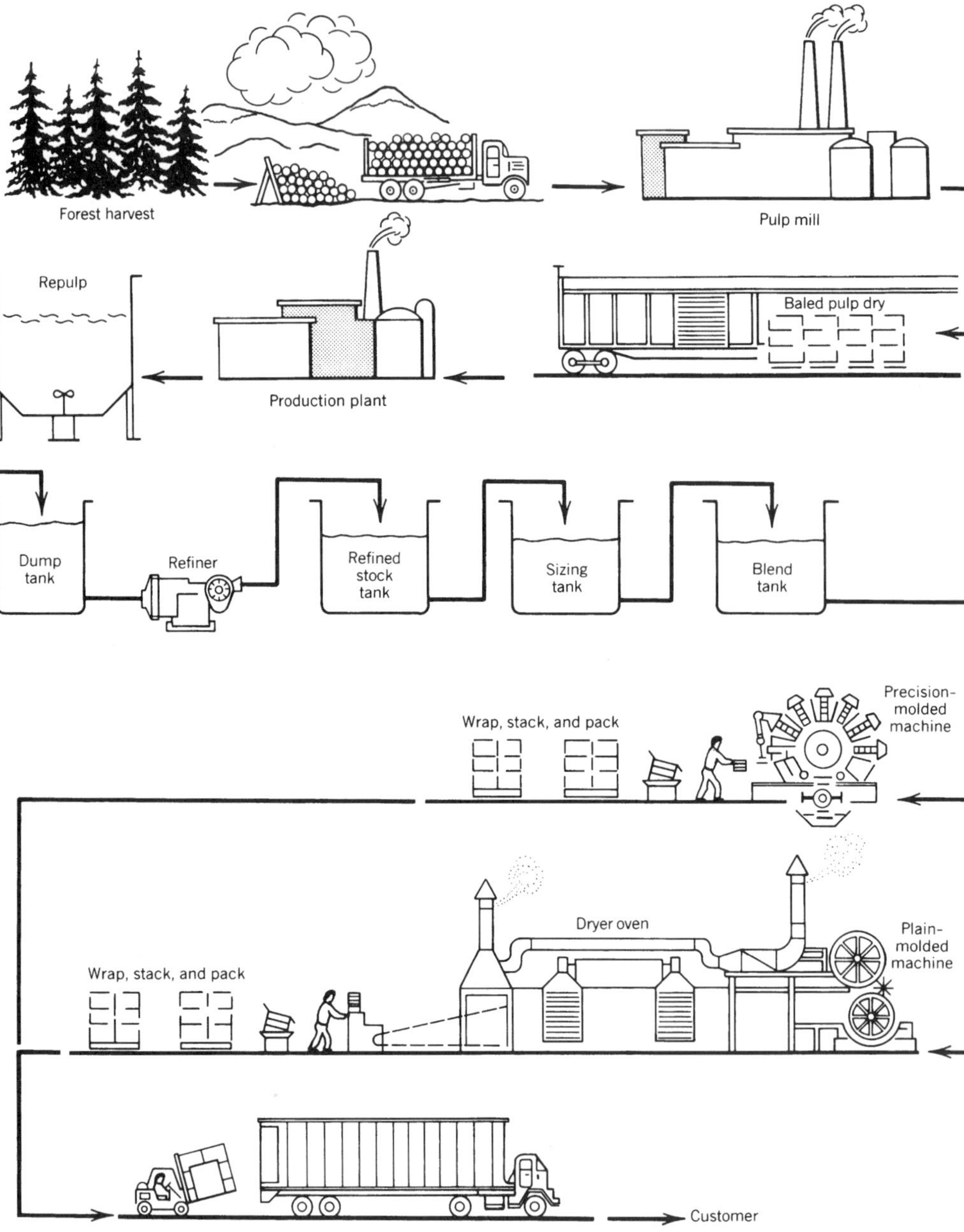

Figure 1. The molded-fiber process.

molding of fiber-reinforced resin parts and pressboard converting because of the similarities in finishing steps and common markets, but the molded-pulp process and its products are fundamentally unique. Two basic methods of fabrication are used: plain molding, and precision molding (see Fig. 1).

Products of plain molding are as fundamental as the triangular corner protector pads used in furniture and appliance packing crates, and as complex hinge-lidded, self-locking, one-dozen egg cartons enhanced by hot afterpressing and graphic-print applications. Between these extremes are products such as berry punnets, peat pots, produce prepackaging trays, and egg and apple-locator trays (see Fig. 2). Such products have commercial relevance because they are produced on highly automated and productive machine modules (>20 tons (>18.1 metric tons) per day), are nestable for efficient transport to users, low in density in comparison to converted paperboard products (see Paperboard), and fabricated from some of the least-costly raw materials available (recycled mechanical cellulosic fibers, groundwood, sphagnum peat fiber, etc).

Precision-molded processing differs from its higher-speed forerunner in drying methodology. Forming of the product and the removal of mechanically bound water by vacuum-assisted compaction is essentially the same; but at that point final drying is accomplished by step- or continuously applied heat between matching mold surfaces (see Fig. 3). Precision-molded products include disposable plates, dishes, bowls, and specialty products such as loudspeaker membranes and cones (see Fig. 4). These products are typically denser, smoother,

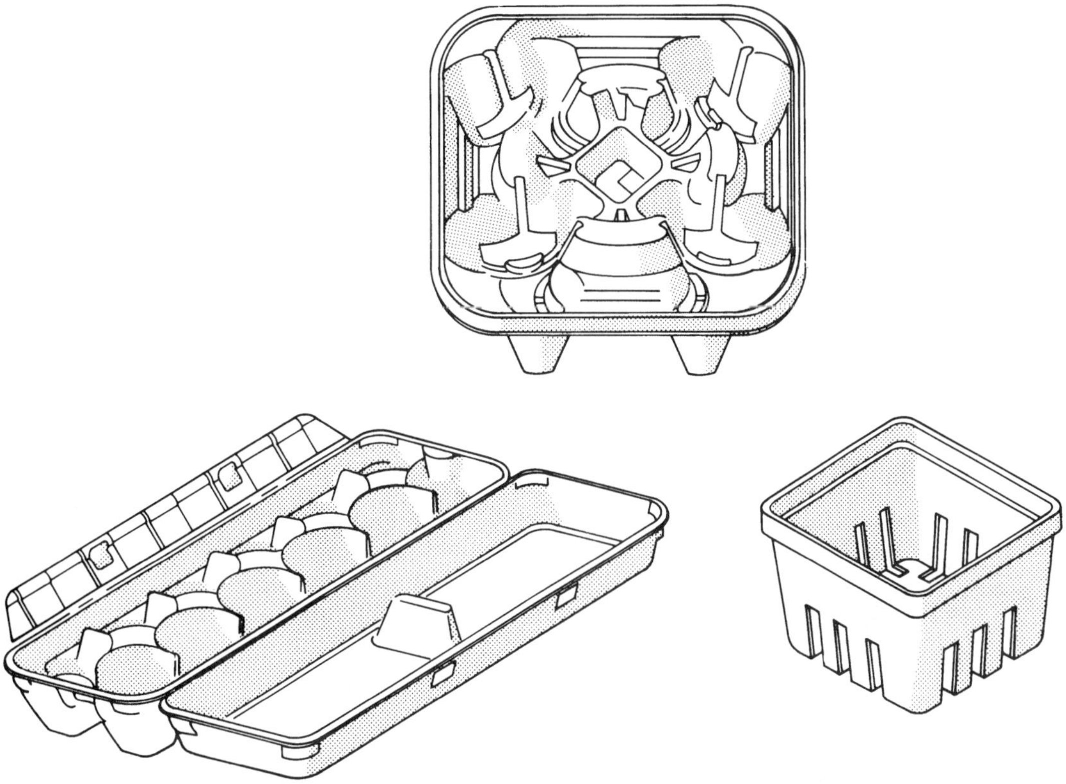

Figure 2. Examples of plain-molded products.

and more exactly dimensioned and contoured than their free-dried counterparts. Virtually any plain-molded product can also be made in this manner, but they would be more costly because of the complexity of tooling and higher electrical energy required.

Many internal treatments are given to the slurries to render them more valuable to their users. By blending fibers that have received different process and refining treatments, "green" matrix strength can be enhanced; drainage, internal bond strength, and shrinkage can be altered; and biodegrad-

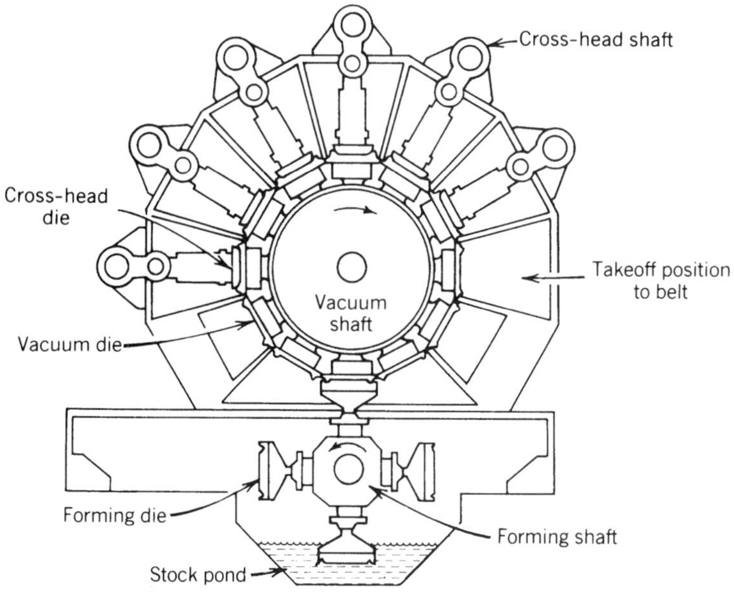

Figure 3. The Chinet precision-molding machine.

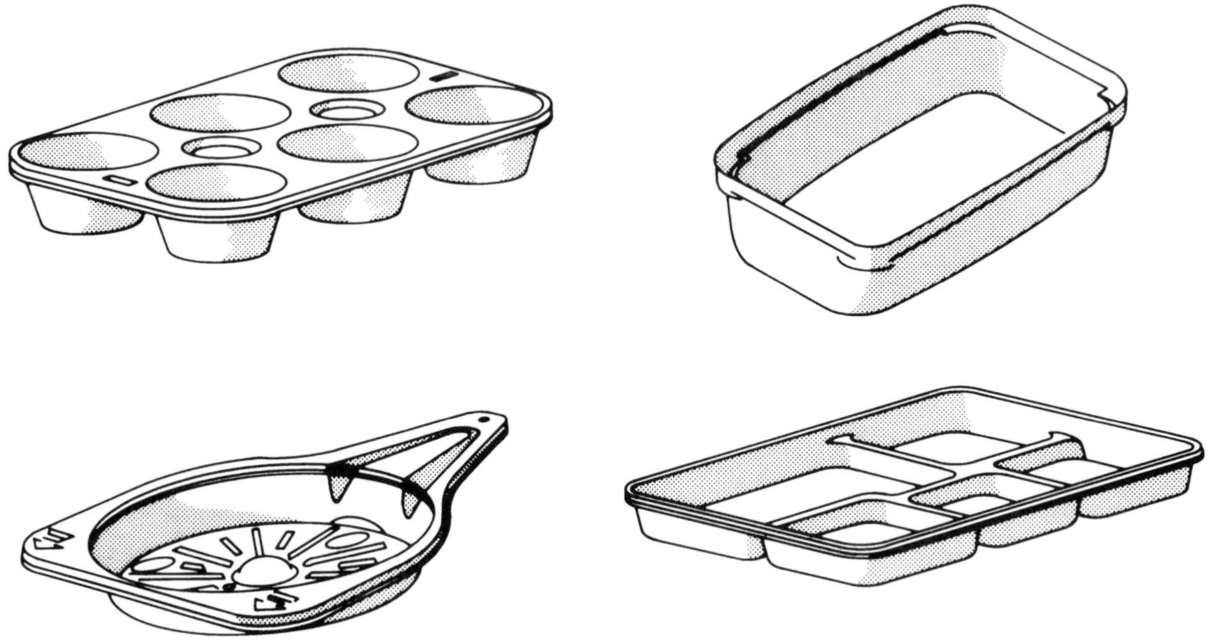

Figure 4. Examples of precision-molded machine.

ability of products after or during use can be controlled. Colloidal rosin or wax emulsions are commonly added together with paper-making alum to convert hydrophilic fiber into the water-repellent products essential to food handling. Fluorocarbon chemicals in concert with cationic retention aids can be added to impart repellency to low-surface-tension oil or greasy liquids. Fertilizers, dyestuffs, flame retardants, and modified starches or wet-strength resins are among other additives that are also commonly added internally where specific end-use effects are desired.

Secondary treatments are sometimes given to nondisposable molded-pulp products. Cafeteria serving trays have been made by after-pressing laminates of molded-fiber preforms loaded with thermosetting polymers. Luggage shells, automobile trunk wells, glove compartments, and door liners are made this way. The hot afterpressing and decorative printing of egg cartons also represents secondary treatment that enhances printability and product automation.

A very interesting recent example of secondary treatment is the lamination of a thin thermoplastic film to one surface of molded-pulp trays by vacuum-thermoforming techniques. Products of this type are being used for frozen dinners because of their "dual ovenability" (ie, their suitability for use in microwave and convection ovens). They could be relevant in gas-flush barrier-packaging applications, and where superior decorator graphics are required. Other significant new product opportunities being actively investigated in nonpackaging applications.

BIBLIOGRAPHY

"Pulp, Molded" in *The Wiley Encyclopedia of Packaging Technology,* 1st ed., by Edwin H. Waldman, Keyes Fibre Company, pp. 559–561.

General Reference

U.S. Pat. 4,337,116 (June 29, 1982), P. D. Foster and C. Stowers.

Q

QUALIFYING

Qualifying is the act of setting up and administering a training program that ensures the people who will be interfacing with the packaging line are given a thorough overview of the packaging line and a detailed program on what they need to know to complete their tasks without hesitation or guessing. All training must address the following questions:

When should training be done?
What training should be done?
How should we do training?
Where should we do training?
How much is enough training?
Manuals and other self-help tools?
Performance reviews and continuing improvement?
Company standards and policy?

QUALITY CONTROL. See SPECIFICATIONS AND QUALITY ASSURANCE.

QUALITY MANAGEMENT. See TOTAL QUALITY MANAGEMENT.

QUALITY PACKAGE. See ISO 9000.

R

RADIATION, EFFECTS ON PACKAGING MATERIALS

The medical, pharmaceutical, and food industries use heat, chemicals, and irradiation to sterilize products (see Canning, food; Health-care packaging; Pharmaceutical packaging) (1). Each of these methods has advantages and disadvantages, and each places special demands on packaging materials. The inherent advantages of irradiation sterilization make it a forerunner. The four important advantages are listed below.

1. Sterilization in a completely sealed, impervious package that excludes any possibility of recontamination.
2. Ability to penetrate into the most inaccessible places, bringing the lethal effect to all vegetative and dormant stages of the living organisms within the package.
3. Great reliability and simple control of the process. There are no sterilization conditions to be monitored, and no residuals that must be removed. The only control is the measurement of the radiation dose to assure that it is above the required minimum.
4. Little or no heat is produced. Gamma irradiation produces no heat. Irradiation with accelerated particles might increase the temperature of plastics by 50°F (10°C) and metals by 104°F (40°C).

Two basic processes are utilized: gamma irradiation (^{60}Co, X rays) and accelerated particles (electron-beam accelerators) (2,3). The highest dose is that which extinguishes all forms of vegetative and dormant life, about 3 Mrad (3×10^4 Gy).

Generally, the effect of irradiation on packaging materials and components is the same for both methods, except for the effect of the small amount of heat generated when accelerated particles are used (4).

The packages consist primarily of the basic packaging materials (glass, metal, paper, and plastics), composite structures of these materials, and packaging components (rubber, adhesives, coatings, and inks).

Irradiation may affect the packages directly or indirectly. It may have direct effects on their physical or chemical properties (5), stability, maintenance of sterility or formation of any toxic, deleterious, or otherwise harmful products. Indirectly, it may affect chemical or biological (6) properties, which leads to interaction with the packaged product, changing its efficacy, potency, palatability, or usability.

The direct effect of irradiation may be predicted to some extent; the indirect effect has to be established for each particular product and package.

Effects on Glass, Metals, and Paper

Glass. Glass is not affected by irradiation, except for discoloration. The color formation is proportional to the irradiation dose and may be utilized as a built-in dosimeter. The color fades with time and temperature; therefore, irradiated glass vials, ampuls, or bottles may vary appreciably from batch to batch depending on their aging history (7).

Metals. Aluminum, tin, and steel are used as containers or as foils in composite materials (see Multilayer flexible packaging). The metals are not affected by irradiation, but may react with the product. Reaction with the product is an indirect effect that must be taken into consideration for each packaging application (8).

Paper and paperboard. These, and all other natural products, are affected by irradiation. Owing to the comparatively small dose used for sterilization, this effect, which manifests itself as discoloration, loss of strength, and embrittlement, is not sufficiently significant to prevent the use of these materials in monolayer or multilayer packaging. The dose effect is cumulative, and repeated sterilization of the same package should be avoided.

Effects on Plastics

Plastics form the bulk of modern packaging materials and must be investigated very carefully as to their usefulness for each application. Although general guidelines may be drawn, each material must be viewed for specific applications. There are two basic effects of irradiation on plastics: degradation or crosslinking. There may be simultaneous degradation and crosslinking of the degradation products, and the net result depends upon the comparative rates of the two reactions.

Crosslinking polymers. Table 1 lists polymers that are predominantly cross-linked by irradiation.

There is virtually no effect on polyethylene at the sterilization dose. The effect of irradiation decreases with increasing density: high-density polyethylene is less sensitive than medium-density polyethylene, which is less sensitive than low-density polyethylene (see Polyethylene, high-density; Polyethylene, low-density). Chemical changes due to irradiation manifest themselves in the changes of infrared spectra, specifically in the type and distribution of unsaturated groups. The presence of oxygen (when irradiated in the air) also has a marked influence on these rearrangements as a result of oxidation. Similarly, spunbonded olefins (see Nonwovens) and ionomers (see Ionomers) are not affected by the sterilization doses.

The effect of irradiation on ethylene copolymers depends

Table 1. Crosslinking Polymers

Polyethylene (PE)
Ethylene-*co*-vinyl acetate (EVA)
Ethylene-*co*-ethyl acrylate (EEA)
Ionomers
Polystyrene (PS)
Styrene-*co*-acrylonitrile (SAN)
Polyesters
Polyurethane
Polyphenylene oxide (PPO)
Polysulfones, aromatic
Poly(vinylidene fluoride) (PVDF)
Ethylene-*co*-tetrafluoroethylene (ETFE)

Table 2. Polymers that Undergo Degradation

Cellulose
Polytetrafluoroethylene (PTFE)
Polychlorotrifluoroethylene (PCTFE)
Poly(vinylidene chloride) (PVDC)
Polyacetals

greatly on the types of monomers used, but radiation improves impact and tensile strength. Both EVA and EEA are suitable for radiation sterilization (9).

Polystyrene undergoes crosslinking on irradiation: an aromatic ring (phenyl group) attached to alternate backbone carbon atoms adds to the resistance of the plastic to degradation. Mechanical properties are changed very little by irradiation, and only by very high doses. High-impact polystyrene, which contains modifiers, is more susceptible to damage, especially to reduction of elongation and impact strength. It can, however, withstand several radiation sterilizing doses without ill effect.

Styrene–acrylonitrile copolymers are not as resistant to radiation as polystyrene itself, but are still fairly stable and able to resist high irradiation doses.

As in polystyrene, the aromatic ring structure in polyesters absorbs a large part of the radiation energy. Thus, the effect of irradiation on strength and elongation of these materials at sterilization dose is negligible. Similarly, aromatic polyurethanes can be sterilized by irradiation without loss of tensile strength.

Other examples of radiation-resistant plastics containing aromatic ring structures are polyphenylene oxide and aromatic polysulfones (10).

Surprisingly, PVDF and ethylene-*co*-tetrafluoroethylene show improved tensile strength and thermal resistance on irradiation, unlike other fluorine-containing plastics that are degraded drastically (11).

Polymers that undergo degradation. Table 2 lists polymers that are degraded by irradiation. Cellulosics and fluorocarbons are examples of materials that are degraded by irradiation. Cellulose, as mentioned in connection with paper, is quite sensitive to irradiation degradation, which manifests itself in a tensile strength loss of about 25% on irradiation at 5 Mrad (5×10^4 Gy). The degradation is owing to main-chain scission and steady reduction of molecular weight. After irradiation with 500 Mrad (5×10^6 Gy), cellulose is almost completely degraded to water-soluble fragments.

Fluorine-containing polymers, namely, PTFE and PCTFE, show rapid deterioration on irradiation. They are the most radiation-sensitive plastics and preferably should not be used for packaging applications with this type of sterilization. PCTFE is less susceptible to degradation than the tetrafluoropolymer. All these changes occur in the presence of air; in the complete absence of oxygen, the damage is less severe.

PVDC is also not recommended for use with irradiation sterilization. Its mechanical properties start to degrade at 4 Mrad (4×10^4 Gy), and the plastic becomes very discolored.

Polyacetals show a high degree of degradation at low irradiation doses. Copolymers are slightly less sensitive than homopolymers, but both lose most of their tensile strength and become brittle when irradiated with doses above 5 Mrad (5×10^4 Gy) (12).

Intermediate polymers. A third group of plastics, listed in Table 3, occupies an intermediate position on the scale between crosslinking and degradation.

These materials are generally affected by higher irradiation doses, but if properly stabilized, they withstand sterilization dose with minimal changes that may permit their use where repeated sterilization is not anticipated.

Polypropylene is a good example of such material. When irradiated in the presence of air, it undergoes marked oxidative degradation (13). With proper stabilization, however, this degradation may be kept to a minimum so as not to impair mechanical properties (14,15).

Polymethylpentene reacts to irradiation in a manner similar to polypropylene. Although it is somewhat more resistant to lower irradiation doses than polypropylene, it is not recommended for repeated exposure.

Indications are that structures with branching alkyl chains are less resistant to irradiation than nonbranched polymer chains. PMMA is an example. Although it can satisfactorily withstand a single radiation sterilization dose, it is not suitable for repeated doses. It degrades in both the main chain and in the ester side chain of the molecule. The plastic becomes brittle and discoloration develops at doses as low as 0.5 Mrad (5×10^3 Gy). Some radiation-grade acrylics are available now for molding medical devices. Polyamides (nylons) react to irradiation in a manner similar to polyethylene insofar as some increase in tensile strength is concerned, but show a much more rapid decrease in impact strength. Therefore, repeated sterilization doses are not recommended.

Esters of cellulose (acetate, propionate, and acetobutyrate), which are of interest in packaging applications, are affected to a lesser degree by irradiation than cellulose itself.

Polycarbonate exhibits some discoloration and becomes brittle at doses considerably above 2.5 Mrad (2.5×10^4 Gy). Although the impact properties are hardly affected, thin irradiated films exhibit some decrease in tensile strength and elongation on aging. Nevertheless, it can be safely exposed to a single sterilization dose (16). Acrylonitrile-*co*-butadiene-*co*-styrene (ABS) is also suitable for a single sterilization, but generally is much less resistant to irradiation than styrene-*co*-acrylonitrile (SAN). PVC does not have a clearly defined position in this scheme. Some investigators report degradation and others crosslinking. Apparently, both statements are true depending on irradiation conditions and the type of additives present in the plastic (17).

Rigid PVC containing 58% chlorine loses HCl on irradia-

Table 3. Intermediate Polymers

Polypropylene
Polymethylpentene
Poly(methyl methacrylate) (PMMA)
Poly(vinyl chloride) (PVC)
Polyamides
Cellulosic esters
Polycarbonate
Acrylonitrile-*co*-butadiene-*co*-styrene (ABS)

Table 4. Radiation Effects on Rubber

Rubber	Stable up to
Polyurethane	500 Mrad (5×10^6 Gy)
Natural	100 Mrad (1×10^6 Gy)
Styrene-*co*-butadiene (SBR)	100 Mrad (1×10^6 Gy)
Nitrile	100 Mrad (1×10^6 Gy)
Silicone (polydimethylsiloxane)	10 Mrad (1×10^5 Gy)
Neoprene	10 Mrad (1×10^5 Gy)
Butyl	1 Mrad (1×10^4 Gy)

tion and also decomposes into unstable fragments that may then undergo crosslinking and recombination. This dehydrochlorination leads to formation of conjugated double-bond structures which manifest themselves in coloration of the polymer (18,19).

This observed change in color may vary from pale yellow to black, depending on irradiation dose and, of course, the stabilizing system (20,21). Also, evolution of HCl gas is noticeable and can cause metal corrosion and changes in pH. Their spectrum also changes. Tensile strength is affected only slightly, but brittleness increases. Other significant changes in mechanical properties proceed at doses that normally are not used in packaging.

The degree of degradation depends therefore on the type and amount of the additives (stabilizers, antioxidants, inhibitors, etc), presence of oxygen during irradiation, aging, and the irradiation dose.

In order to prevent degradation, the plastic must be properly stabilized. When used as packaging material, rigid PVC should retain two important properties: original clarity (be free from discoloration) and low brittleness. There are a multitude of stabilizers to prevent different types of degradation, and combinations of them should be chosen to produce the desired effect. For packaging of foods and drugs, choice of stabilizers is limited owing to toxicity or extractability factors. In fact, a stabilizer should prevent a particular type of degradation, namely, dehydrochlorination, oxidation, destruction of the polymer, or reconstitution (crosslinking) of degraded chains.

As is pointed out in the literature, a combination of several weak stabilizers may provide a very satisfactory synergistic action is used in proper combination. Several firms now offer radiation-resistant PVC compounds that are stabilized to inhibit crosslinking and show low discoloration.

Effects on Rubber

Table 4 shows the effect of irradiation on various types of rubber in an unstressed state. The stability can be influenced by various additives like fillers and antioxidants (22,23).

Effects on Composite Materials

Composite materials may be affected differently than the individual components. For example, the loss of strength of paper or a cellulosic film may not be noticed at all if the film is supported by polyethylene or foil. On the other hand, various types of heat-seal coatings that are not affected by irradiation per se may produce weak seals when used as components of a laminate (see Sealing, heat). This may be caused by the effect of irradiation on the strength of the adhesion between the coating and the substrate rather than on the actual bond between the two sealed surfaces. This effect, if carefully controlled, may be used to produce peelable seals that are so important to ensure sterile delivery of the product.

Aging studies are of utmost importance to determine any progressive degradation that may not have been detected immediately after irradiation. Accelerated aging (high temperature and/or humidity) indicates the trend, but should not be substituted for the actual shelf stability, since it may sometimes cause a change that would not occur under normal conditions (24).

Inks. Commercial printing inks are developed to be light stable; that is, they contain pigments and dyes that do not change color under the influence of visible and ultraviolet light, and they generally resist radiation.

Specially developed inks that are used as radiation dosimeters change color under the influence of radiation due to pH change of the ink system rather than that of the basic colorant (25).

Adhesives and coatings. Most adhesives and coatings are based on polymers and plastics. Their resistance to radiation, therefore, can be predicted from the data developed for the basic components. Generally, thermosetting types are more resistant to radiation than thermoplastic types. Fillers, pigments, plasticizers, and other additives affect radiation stability (26).

Testing. The dose required for sterilization is not usually a satisfactory dose to use for radiation-effects determination. This is because in both gamma and electron-beam sterilization, delivering a specified sterilizing dose to the center of a bulk unit of product involves exposure of the outer regions to a higher dose. The ratio of these two doses is a function of the size and bulk density of the product unit involved. For gamma radiation, this ratio can range from about 1.25 : 1 to 3 : 1; for electron beam it is many times higher. Any exposures for radiation-effects testing should take this into account.

The biological aspects of the irradiation-sterilized package must not be overlooked. Apart from the physical and chemical changes of the packaging materials, their toxicity and any interaction with the product must be determined (27).

Although much work has been done on investigation of the toxicity of various packaging materials, this knowledge is only a starting point for a packaging engineer, as it is obvious that the interaction of the packaging material with the packaged drug or food cannot always be theoretically predicted.

It must first be determined that the product itself as well as the package can withstand irradiation sterilization. The product should be irradiated in an inert medium; a glass container may be used to determine this. The product must then be sterilized in the final package and tested for purity, efficacy, toxicity, palatability, and shelf stability.

BIBLIOGRAPHY

"Radiation, Effects on Packaging Materials" in *The Wiley Encyclopedia of Packaging Technology,* 1st ed., by I. W. Turiansky, Ethicon, Inc., pp. 562–565.

1. R. D. McCormick, *Prepared Foods,* **153**(4), 133 (April 1984).
2. W. J. Maher, "Critical Assessment of Gamma and Electron Radiation Sterilization," presented at *1979 Disposable Medical Packaging–Government Compliance Seminar,* sponsored by TAPPI, Lincolnshire, IL, Oct. 15–16, 1979, pp. 22–26.
3. J. H. Bly, *Radiat. Phys. Chem.* **14,** 403 (1979).
4. D. A. Vroom, *MD & DI* p. 45 (Nov. 1980).
5. J. J. Killoran, *Radiat. Res. Rev.* **3,** 369 (1972).
6. R. K. O'Leary and co-workers, "The Effect of Cobalt-60 Irradiation upon the Biological Properties of Polymeric Materials," presented at *Sterile Disposable Devices: Updated 1973 Technical Symposium,* sponsored by HIA Sterile Disposable Device Committee, Washington, DC, Oct. 17–18, 1973, pp. 160–166.
7. H. J. Zehnder, *Alimenta* **23**(2), 47 (1984).
8. P. S. Elias, *Chem. Ind.* **10,** 336 (1979).
9. A. Barlow, J. Biggs, and M. Maringer, *Radiat. Phys. Chem.* **9,** 685 (1977).
10. J. R. Brown and J. H. O'Donnell, *J. Appl. Polym. Sci.* **19,** 405 (1975).
11. G. G. A. Bohm, W. F. Oliver, and D. M. Pearson, *SPE J.* **27,** 21 (July 1971).
12. M. Možišek, *Plaste Kautsch.* **17**(3), 177 (1970).
13. J. L. Williams and co-workers, *Radiat. Phys. Chem.* **9,** 445 (1977).
14. T. S. Dunn and J. L. Williams, *J. Ind. Irradiat. Technol.* **1**(1), 33 (1983).
15. P. Hornig and P. Klemchuk, *Plast. Eng.* p. 35 (April 1984).
16. B. A. Rohn, "Polycarbonate Coextrusions for Medical Device Thermoform, Fill & Seal (TFFS) Packaging" in *Proceedings of the First International Conference on Medica and Pharmaceutical Packaging, Healthpak '84,* May 8–9, 1984, Düsseldorf, FRG, Schotland Business Research, Inc., Princeton, NJ, pp. 65–84.
17. C. Wippler, *Nucleonics* **18**(8), 68 (1960).
18. A. A. Miller, *J. Phys. Chem.* **63,** 1755 (1959).
19. S. Ohnishi, Y. Nakajima, and I. Nitta, *J. Appl. Polym. Sci.* **6**(24), 629 (1962).
20. R. V. Albarino and E. P. Otocka, *J. Appl. Polym. Sci.* **16,** 61 (1972).
21. M. Foure and P. Rakita, *MD & DI,* 57 (Nov. 1983); 33 (Dec. 1983).
22. D. W. Plester, "The Effects of Radiation Sterilization on Plastics" in G. Briggs Phillips and W. S. Miller, eds., *Industrial Sterilization,* International Symposium, Amsterdam, 1972, Duke University Press, Durham, N.C., pp. 149–150.
23. R. W. King and co-workers, "Polymers" in J. F. Kircher and R. E. Bowman, eds., *Effects of Radiation on Materials and Components,* Reinhold Publishing Corp., New York, 1964, pp. 110–139.
24. J. P. Ferrua, "New Multilayer Materials Specifically Designed for Radiation Sterilization," in Ref. 16, pp. 53–63.
25. D. B. Powell, "Packaging Requirements for Radiation-Sterilized Items," in *Ionizing Radiation and the Sterilization of Medical Products,* Proceedings of the First International Symposium organized by the Panel on Gamma and Electron Irradiation, Dec. 6–9, 1964, Taylor and Francis Ltd., London, pp. 95–103.
26. Ref. 23, pp. 139–144.
27. K. Figge and W. Freytag, *Dtsch. Lebensm. Rundsch.* **73**(7), 205 (1977).

General References

G. O. Payne, Jr., C. J. Spiegl, and F. E. Long, *Study of Extractable Substances and Microbial Penetration of Polymeric Packaging Materials to Develop Flexible Plastic Containers for Radiation Sterilized Foods,* Technical Report 69-57-FL, U.S. Army Natick Laboratories, Food Laboratory FL-65, Natick, MA, Jan. 1969.

A. F. Readdy, Jr., *Applications of Ionizing Radiations in Plastics and Polymer Technology,* Plastic Report R41, Plastics Technical Evaluation Center, 1971.

A. Chapiro, *Radiation Chemistry of Polymeric Systems,* Vol. XV of *High Polymers,* Interscience Publishers, New York, 1962.

R. O. Bolt and J. G. Carroll, *Radiation Effects on Organic Materials,* Academic Press, New York, 1963.

A. Charlesby. *Atomic Radiation and Polymers,* Pergamon Press, New York, 1960.

M. Dole, ed., *The Radiation Chemistry of Macromolecules,* Academic Press, New York, 1972.

J. Silverman and A. R. Van Dyken, eds., *Radiation Processing,* Transactions of the First International Meeting on Radiation Processing, Vols. I and II, Puerto Rico, May 9–13, 1976, Pergamon Press, New York, 1976.

H. K. Mann, "Radiation Sterilization of Plastic Medical Devices," *Radiat. Phys. Chem.* **15**(1) (1980).

O. Sisman and C. D. Bopp, *Physical Properties of Irradiated Plastics,* ORNL-928 Report, TIS, U.S. Atomic Energy Commission, Oak Ridge, TN, June 29, 1951.

T. S. Nikitina, E. V. Zhuravskaya, and A. S. Kuzminsky, *Effect of Ionizing Radiation on High Polymers,* Gordon and Breach Science Publishers, New York, 1963.

J. E. Wilson, *Radiation Chemistry of Monomers, Polymers and Plastics,* Marcel Dekker, New York, 1974.

W. E. Skiens, "Sterilizing Radiation Effects on Selected Polymers," Symposium on Radiation Sterilization of Plastic Medical Products, Cambridge, Mass., March 28, 1979, Report No.: CONF-7903108-1, Dept. of Energy, Washington, DC, PNL-SA-7640.

B. J. Lyons and V. L. Lanza, "Protection Against Ionizing Radiation" in *Polymer Stabilization,* W. L. Hawkins, ed., Wiley-Interscience, New York, 1972, pp. 250–311.

G. de Hollain, *Plast. Rubber: Mater. Applications* pp. 103–108 (Aug. 1980).

RECYCLING

Recycling is steadily increasing in both developed and underdeveloped countries. World population is both growing rapidly and becoming more concentrated in urban areas. By the year 2000, more people will be living in cities than in rural areas. Most of this growth will take place in urban areas (1). Increased urbanization will increase the concentration of packaging wastes, thus improving collection economics. A United Nations study indicates that increased recycling rates are accompanied by job creation and improved living conditions (1) (see also Environmental laws, North America and International Org).

In the United States in 1993, over 34% by weight of the approximately 200 million tons of municipal solid waste was containers and packaging materials (2). Of this waste, 22% was recycled while 62% was placed in landfills and 16% incinerated (3). Packaging and other wastes are collected for recycling by municipalities and private waste-collection firms. Major sources of packaging wastes are grocery stores, restaurants, and office buildings. Residential collection of waste materials is increasing, although sorting of the wastes is often a major expense. The number of U.S. residential collection programs increased from 1000 in 1988 to more than 7000,

Table 1. U.S. Recycling Rates of Various Materials[a]

Material	Recycling Rate, %
Paper and paperboard	40
Ferrous metals (iron and steel)	37
Aluminum	30
Glass	7
Plastics	1
Rubber and leather	3

[a] Ref. 4.

involving more than 100 million people in 1993 (3). Approximate 1994 U.S. recycling rates are given in Table 1.

Both economic factors and governmental regulation are driving recycling. Energy costs associated with recycling are almost always less than in manufacture of products from virgin materials. The energy to recycle aluminum is only 5% of that required to process bauxite into aluminum (5). Steel and glass recycling consume about 50% of the energy needed to make these products from ore and silica. Plastics recycling takes only 10–15% of the energy needed to refine petroleum and manufacture virgin resins. Incineration of plastics is a less efficient means of saving energy. For example, 100 lb (45.4 kg) of high-density polyethylene (HDPE) has a fuel value of 20×10^6 Btu (19 kJ). Recycling saves twice this: 40×10^6 Btu (38 kJ).

Both economic and environmental factors have led to government regulations designed to promote recycling. Economic concerns are related to balance of payment issues when virgin raw materials for packaging products are imported. For example, in Japan, limited forest resources means that many paper products have to be imported. Recycling reduces these imports. In fact, Japan imports used paper products from the United States and elsewhere to recycle them.

In some areas, the number of landfill sites is becoming limited. Although the number of landfills in the United States is declining, the remaining sites are large, modern facilities. Concerns about landfill disposal costs are becoming less of a factor in promoting North American recycling.

Environmental concerns include forest depletion and the effects of wastes on the environment. One concern receiving much publicity is the harmful effects plastics can have on marine animal life such as dolphins when eaten.

Separation of Commingled Materials

Solid wastes, particularly from residential curbside collection programs, arrive at material recovery facilities (MRFs) as a complex mixture. MRFs are typically built to process 100–500 tons of waste per day (2). Unit operations are summarized in Figure 1. The wastes are dumped on a tipping floor, where paper products are separated from metals and plastics. Metals and plastics, mostly containers, are pushed onto a conveyer belt. Two types of magnetic separators remove steel and aluminum from plastics and glass. Density differences or manual sorting are used to separate glass from plastics. The glass containers are hard-sorted by color. The plastics are separated into individual polymer types by the MRF or in separate reclaiming facilities (2). Plastic bottles are classified into: clear polyethylene terephthalate (PET) soft-drink bottles; green PET soft-drink bottles; translucent high density polyethylene milk, water, and juice bottles; pigmented HDPE detergent bottles, poly(vinyl chloride) (PVC) water bottles, and food containers such as polypropylene ketchup bottles (2).

Metals

Ferrous metals. Iron and steel scrap account for more than one-third of the cast iron and steel produced in the United States. Much of this is "home scrap" produced during production of cast iron and steel. Home scrap is always recycled. "Prompt industrial scrap" is produced by industries while manufacturing products from steel or cast iron. Postconsumer scrap comes primarily from automobiles. Other sources include railroad rails and other machinery. Municipal solid waste is a growing source of postconsumer scrap in the United States. In some countries with low labor costs, ship breaking is a major source of postconsumer scrap.

Shredders received whole or baled vehicle bodies from salvage yards. Typical shredders process 200–1500 cars per day. All major automobile producing countries have adequate capacity to shred their available supply of discarded automobiles. In the United States, approximately 1.21 million tons of steel are recovered annually from 9 million scrap vehicles (4). Steel fuel tanks are removed from automobiles before the vehicles are shredded. These are sold for metal recovery.

A hammermill shreds the vehicle into fist-sized fragments. These are separated into a magnetic fraction (ferrous metals) and a nonmagnetic fraction. The nonmagnetic fraction is separated using air or water to form a low-density nonmetallic fraction and a high-density fraction consisting mostly of zinc, copper, and aluminum.

Tin must be removed when recycling tinplate cans. As little as 0.01% tin can form hard spots in steel and causes difficulties in rolling (5). Therefore, only low levels of steel can scrap can be used directly in furnaces. By processing cans with hot caustic and an oxidizing agent, tin can be removed from tinplated steel cans. The use of steel beverage cans is declining. Those used often contain no tin. Therefore, few mills operate detinning operations. Can scrap has been used

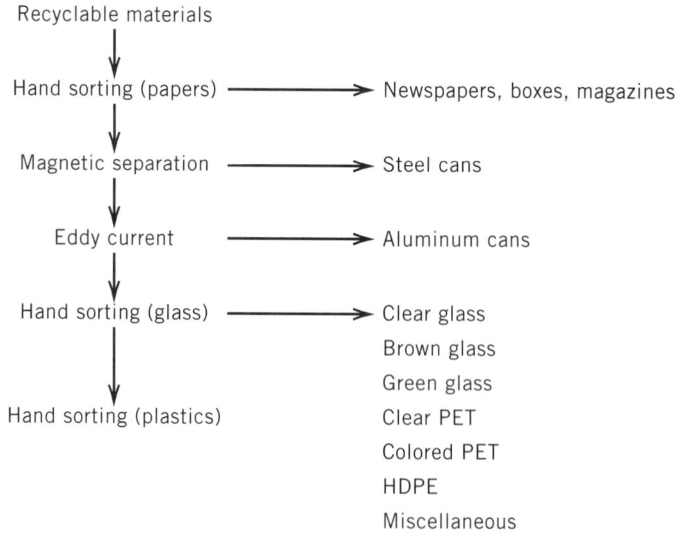

Figure 1. Basic unit operations in sorting waste plastics.

as a precipitant in copper leaching mining. However, this is a low-volume application.

Minimills using electric arc furnaces use scrap iron and steel as their only feedstock. The amount of scrap that may be used in basic oxygen furnishes is limited to about 33% of the charge because the only heat source for melting the scrap is carbon and silicon added with iron ore.

Nonferrous metals. When processing municipal solid wastes, an eddy-current separation unit is often used to separate aluminum and other nonferrous metals from the waste stream. This is done after removal of the ferrous metals (Fig. 1). The eddy-current separator produces an electromagnetic field through which the waste passes. The nonferrous metals produce currents having a magnetic moment which is phased to repel the moment of the applied magnetic field. The repulsion causes the nonferrous metals to be thrown out of the process stream away from nonmetallic objects (6).

Another separation device that may be used is the mineral jig. This unit produces a loose vibrating bed of particles in a liquid medium. The vibrations segregate the solids into layers by density. The dense nonferrous metals—primarily lead, zinc, and copper—are at the bottom while organics are at the top. The middle layer is primarily glass.

Aluminum is the largest-volume nonferrous metal recovered and recycled. More than 95% of beer and carbonated soft-drink containers are two-piece aluminum cans (7). Aluminum is also used for aerosol cans and as foil–plastic–paper laminates to package cereal and frozen foods. Aluminum is also used for food trays and pie plates. All of these are found in and collected from municipal solid wastes. A survey of 13 municipalities indicated that aluminum constituted 0.6–1.1% of their municipal solid waste (8) (see also Cans, aluminum).

The primary economic advantage in recycling aluminum is the energy savings. The energy requirement to manufacture primary aluminum ingot from virgin materials is about 220 MJ/kg (~95,000 Btu/lb). In contrast, 10 MJ/kg (~4300 Btu/lb) is needed to convert aluminum scrap into molten aluminum similar to that produced in primary production facilities. Another advantage is that fluoride emission, long a great concern in aluminum smelting, is reduced to the degree that the plant processes recycled aluminum (9). (The U.S. aluminum industry uses 15 kg of fluoride ion per metric ton of virgin aluminum produced. About 10–15% of this fluoride is lost during the production process.)

Organic coatings must be removed before the recovered aluminum can be used to make new sheet stock (6). This is done in special furnaces. About 90% of recovered aluminum cans is converted into new cans. The remainder are used for other aluminum products.

Removing tin from steel cans is the only U.S. domestic source of this metal. As noted earlier, treatment of steel cans with caustic and an oxidizing agent is used to separate the tin. Recycled lead amounts to about 60% of the U.S. lead industry production. Most recycled lead comes from automobile batteries.

Total U.S. recovery of nonferrous metals from vehicle shredding is 0.8 million tons annually (4). The metals are mostly zinc, copper, and aluminum. About 8 kg of copper can be recovered from each automobile copper radiator. Their recovery and sale contributes significantly to shredder profits. A study of 13 municipalities indicated nonferrous metals other than aluminum constitute 0.1–0.3% of their municipal solid waste (8).

Precious metals are recovered profitably from automobile catalytic converters (4).

Glass

The largest glass packaging market is soft-drink, beer, wine, and hard-liquor bottles (10). Some glass bottles are used in food, drugs, and cosmetics packaging. About 75% of the containers are narrow-neck bottles.

When processing municipal solid wastes, glass is generally removed by hand sorting after separating paper and metals. The mineral jig may be used to separate glass particles from other particles. Particles should be less than 5 cm in their longest dimension. Glass occurs in the middle layer of the slurry formed by the mineral jig. Froth flotation may be used to separate glass particles from denser particles. Froth flotation is most efficient for particles less than 850 μm in size. A cationic fatty-acid amine surfactant has been used to improve the efficiency of froth flotation of glass (11).

The color distribution of glass in municipal solid waste averages 65% colorless, 20% amber, and 15% green. [As manufactured, 85% of container glass is clear (10).] Hand sorting of containers or optical sorting of particles can be used to separate glass of different colors. In optical sorting, interruption of a light beam by a colored or opaque particle will activate an air jet that blows the particle out of the process stream. Mixed colored glass can be used in amber and green glass bottle furnaces in proportions of $\leq$5% and $\leq$20%, respectively (11).

The amount of cullet (recovered glass) used in glass production typically constitutes 46% of the charge used to manufacture amber soda-lime container glass (12). About 25% of this cullet is from postconsumer sources, while the rest is waste from the glassmaking process.

While most glass is used in the manufacture of new glass, glass can be used without purification or other processing by other industries. Crushed glass is used in construction as aggregate for roadways, in building bricks, in synthetic slate products, and for glass-wool insulation and honeycomb structural materials.

Paper

The primary process steps in recycling old corrugated containers (OCCs) and other paper packing materials are pulping, high-density cleaning, coarse and fine screening, centrifugal (reverse) cleaning, fiber fractionation, and refining. Pulping disintegrates the containers into individual fibers dispersed in water. High-density cleaning removes large dense particles such as nails and large staples (13). Coarse screening removes large low-density contaminants such as unpulped tapes and large adhesive particles (14). Coarse screen hole size ranges from 6 to 20 mm. Fine screens are fitted with slots as small as 0.15–0.30 mm. These separate smaller ($\geq$250-μm) contaminant particles such as plastic, wax, and adhesive particles (15). A type of hydrocyclone device called a centrifugal or reverse cleaner is used to remove low-density (adhesive) particles in the 70–250-μm size range (13).

Despite the use of screening and reverse cleaning to remove adhesive particles or "stickies," their removal is often inefficient. Stickies in pulp cause paper-mill operating problems by reducing pulp drainage rates on paper machines (16). They do this by forming deposits on paper-machine wires, dryer fabrics, press rolls, and drying cylinders. Reduced drainage rates force operators to run paper machines at slower speeds or shut them down to clean the fabrics. Both result in lower mill production rates and reduced profitability. Stickies also prevent good fiber–fiber bonding, reducing paper strength, an important consideration for corrugated containers. Both hot-melt adhesives and pressure-sensitive adhesives can form stickies.

Fractionation separates fine particles and short, weak cellulose fibers from longer, stronger fibers (17). Refining is used to develop the desired pulp drainage properties on the paper machine and control paper bulk and density, strength, surface smoothness, and porosity (18,19). Caustic soaking is also used to improve OCC fiber properties (20).

Extensive bleaching of OCC can produce a high-brightness pulp suitable for use in manufacturing office paper (21). Economics compare favorably to those obtained for manufacturing office paper from deinked recovered office paper (22).

A considerable fraction of corrugated containers are coated with wax to increase water resistance. Waxes and hot-melt adhesives can form a thin film on when making linerboard from OCC and result in a slippery surface (16). This can cause "telescoping" when the linerboard is wound on a roll. Removal of this wax when recycling OCC is difficult. Flotation has been found to remove the wax efficiently (23). However, high paper fiber losses make this process uneconomical.

Plastics

Plastics are the next packaging material likely to experience major growth in recycling. In May 1992, the U.S. Food and Drug Administration established the following guidelines to help assure the consumer safety of plastics recycling processes (24). Primary recycling is the recycling of plastics that are plant scrap and have not been sold for consumer use. Secondary recycling is the physical cleaning and processing of postconsumer plastic products. Tertiary recycling is the chemical treatment of polymers. This treatment is usually depolymerization to produce monomers that are purified and then polymerized to produce new polymer. Using tertiary recycling, materials such as fillers and fibers can be physically removed from the monomer. The monomers can also be purified by distillation and other processes prior to polymerization. The leading example of tertiary recycling is polyethylene terephthalate (PET). Tertiary recycling also has been suggested for nylon from discarded carpets (25).

Residential collection programs indicate high collection rates for easily recognized types of containers (Table 2).

Table 2. Residential Recovery Rate by Package Type[a]

Package Type	% Recovery
Beverage bottles	65
Liquid detergent bottles	50
Other rigid containers	10
Packaging film	5
Average of all plastics	30

[a] Ref. 2.

Table 3. Approximate Polymer Recycling Costs[a]

Process Step	Costs: $ per Pound	Costs: %
Collection	0.10	27
Sorting	0.12	32
(Subtotal cost)	(0.22)	(59)
Grinding and cleaning	0.15	41
Total	0.37	100

[a] Ref. 2.

Sorted plastic packaging materials are shipped, usually in bales, to processing plants to be converted to polymer resins. The bales are broken and the bottles sorted to ensure that only one type of polymer is further processed. Processing consists of chopping and grinding the bottles into flakes. These flakes are washed. Processing steps such as flotation are used to remove polymeric contaminants from the flakes. The flakes are melted and converted into pellets. Process costs are summarized in Table 3.

For high-value food-packaging applications, minimal migration of contaminants into food products is critical. Currently the FDA requirement is a maximum 0.5 part per billion (ppb) of noncarcinogenic compounds by dietary exposure (26).

Polyethylene terephthalate. About 1.6 billion lb of PET is used in food-packaging applications annually in the United States (27). Initial polyethylene terephthalate (PET) processing is similar to that outlined in Table 3. Cleaning comprises washing, rinsing, and drying. Poly(vinyl chloride) (PVC) and PET have very similar density values; both will sink to the bottom of the waterbath during rinsing. Therefore it is difficult to separate the two polymers after the bottles have been ground into small particles (28). Melting PET containing PVC will produce black spots due to charring of the PVC during processing to produce new bottles (28,29).

For food applications, improved cleaning of PET produced by secondary recycling is needed. Supercritical fluid extraction using carbon dioxide (24) and solvents such as propylene glycol (30) have been proposed. High temperature and the use of vacuum to remove volatile impurities has also been suggested (31). Stripping of volatile components at temperatures above 160°C for 3 min has been reported (32). Application of a multilayer approach, the manufacture of a bottle with an inner layer of recycled PET sandwiched between surface layers of virgin PET, is used commercially for soft drink applications (33).

Two PET tertiary recycling technologies are used commercially: methanolysis (36) and glycolysis (34). Both cleave the ester linkages in the polymer to form monomers. In methanolysis, methanol is used to cleave the polymer ester linkages, producing stoichiometric amounts of dimethyl terephthalate and ethylene glycol (35,36). The ethylene glycol is separated and purified by distillation. The dimethyl terephthalate is purified by crystallization and distillation. Both bis-hydroxyethylterephthalate and oligomers are formed in glycolysis (34,35). These are recovered and purified by vacuum distillation and then polymerized in the presence of ethylene glycol to form PET. Glycolysis is claimed to be somewhat less costly than methanolysis (36).

Hydrolysis yielding terephthalic acid and ethylene glycol is a third process. High temperatures and pressures are required for this currently noncommercial process. The purification of the terephthalic acid is costly and is the reason the hydrolysis process is no longer commercial.

High-density polyethylene. Process steps in recycling high-density polyethylene (HDPE) are similar to those outlined in Table 3. Cleaning comprises washing, rinsing, and drying (28). Removing labels is the worst problem in washing and drying stages. Detergents are often used to improve the efficiency of label removal. Metal foil labels can introduce metals into the polymer. When metal foil labels are heat-sealed onto plastic, the only way to remove them is using an extrusion melt filter. This leads to plugging of the filter screens causing more frequent changes and increasing production costs.

During the rinse cycle, polyethylene particles float to the surface of the waterbath. The higher density PET and PVC particles sink to the bottom of the bath and can be separated from the polyethylene.

Blends of PET and HDPE have been suggested to exploit the availability of these clean recycled polymers. The blends could combine the inherent chemical resistance of PET with the processing characteristics of HDPE. Since the two polymers are mutually immiscible, about 5% compatibilizer must be added to the molten mixture (37). The properties of polymer blends containing 80–90% PET/20–10% HDPE have been reported (38). Use of 5–15% compatibilizer produces polymers more suitable for extrusion blow molding than pure PET.

Polypropylene. Polypropylene (PP) is used in packaging applications as films and in rigid containers. Battery cases could be considered another packaging application. Dead batteries are often collected at the point of sale of new batteries. In the United States, some states have laws mandating this. Lead, acid, and plastics, particularly PP from battery casings, is recovered and recycled (4). PP is also recovered from bale wrap and other PP fabrics used for wrapping in the textile industry and from other containers (39).

Steps in polypropylene recycling include size reduction by grinding, washing, rinsing, and drying to remove contaminants and produce PP flakes (39). After extrusion, molten polymer is filtered through screen packs. The polymer may be separated into different melt flow ranges to produce more uniform product grades.

Polystyrene. Polystyrene (PS) packaging applications include injection molded products such as beverage containers, dairy product containers, and packaging for personal-care products (40). Extruded solid-sheet PS packaging products include salad boxes, dairy-product containers, baked-goods containers, and vending cups and lids. Extruded foam sheet PS packaging products include poultry and meat trays, produce trays, hinged-lid containers, egg cartons, and foam cups. One formerly large-volume PS use, fast-food clamshell containers, has greatly diminished because of concerns over slow biodegradation of discarded polystyrene containers. Blow- and foam-molded PS packaging products include vitamin bottles, loose-fill packaging, and cushion packaging.

Polystyrene recycling processing steps include densification (for foams), granulation to reduce particle size, washing, drying, extrusion, and pelletizing. High-density baling is used to increase the bulk density of polystyrene, often by a factor of 2. Contaminants are more easily removed before this densification step than after. A demonstration plant chemically decomposes polystyrene to produce monomer (40).

Consumers readily recognize polystyrene foam products and turn these products in for recycling. However, recovery rates for other PS packaging products is significantly less (40).

Table 4. Representative Composition of Automobile Shredder Residue[a]

Material	Percent by Weight
Plastics	27
Rubber	7
Glass	16
Textiles	12
Fluids	17
Other	21

[a] Ref. 4.

Commingled plastic wastes. A relatively small amount of PVC goes into packaging applications and appears in municipal solid waste (29). The greatest concern with PVC is as a contaminant in other polymers being recycled, particularly PET.

The composition of nonmetal residues produced in shredding automobiles is summarized in Table 4. Each vehicle generates 500–800 lb of residue. The annual U.S. total is about 3.5 million tons or about 1.3% of the municipal solid waste generated annually (4). The mixture is too complex to separate and recycle. Depending on the amount of glass, water, metal, and dirt present, the residue has a heating value of 4800–6800 Btu/lb (4,41). Incineration reduces residue weight by 50% and volume by 80%.

Advanced waste recycling is the high-temperature/high-pressure conversion of commingled plastic wastes to form petrochemical process streams (42). Research is in progress to determine the conditions that will favor formation of certain types of chemical feedstocks, including synthesis gas (hydrogen + carbon monoxide), hydrogen, crude pyrolysis oil (containing benzene, toluene, and xylene), olefins, and oxygenates (methanol, esters, methyl formate, etc). This technology has also been evaluated for producing fuels: medium-Btu gas for boilers, and liquid fuels such as diesel oil. None of these processes is currently economic.

Economics and Statistics

Paper. Worldwide use of recycled paper is forecast to increase from nearly 75 million lb in 1988 to 130 million lbs in 2001 (43). In the United States in 1995, recovered paper accounted for 31.5% of the 84.1 million metric tons of paper products produced (44). In 1995, 70% of the corrugated containers produced were recycled (45). The maximum economically feasible recovery rate of old corrugated containers is estimated to be 75%. Alternative recovery paper sources, primarily unsorted office paper, old newspapers, and mixed papers will be increasingly used to supplement old corrugated containers in mills producing paper packaging materials (46).

More than 228 U.S. mills and industrial facilities convert corrugated containers and other paper packaging to new paper products and construction materials (47). More than 60 of these produce linerboard and corrugated medium with a recycle content of 100% (48). In North America, minimills designed to process old corrugated containers are becoming increasingly common. Located in or near major cities, minimills are 25–50% the size of conventional paper mills.

Metals. In 1993, U.S. steel recycling was equivalent to 26% of production (49). In 1991, 61% of all aluminum cans produced in the United States were recycled (50). The reason for this high value is that aluminum cans are easy for consumers to recognize and separate from other waste. Besides curbside collection of segregated aluminum cans, there are thousands of aluminum can collection sites. In addition, many office buildings, restaurants, and airlines separate aluminum cans which are then picked up by waste collectors. The primary source of additional aluminum cans for recycling is municipal solid waste.

Glass. In 1992, U.S. glass container production was more than 41 billion containers containing more than 13 million tons of glass (10). In 1993, glass recycling accounted for 22% of glass production (49). Cullet from recycled glass costs 80–95% of the virgin raw materials used in glass production is 80–95% of the cost of virgin raw materials (11). In addition, the use of cullet saves 15% of the energy costs associated with glass production from virgin raw materials. This energy savings arises from the lower melting temperature of cullet compared to virgin raw materials and lower operating costs for pollution-abatement equipment.

Plastics. U.S. plastics production in 1993 was about 70 billion lb, approximately one-third of which was used for packaging and transportation of goods (2). Plastics recycling continues to increase from its current value of 3.5% in the United States (3,49). However, recycling rates of some packaging products are much higher than this. The recycling rate of PET from soft-drink bottles was 42% in 1993 while that of HDPE from milk and water jugs was 24% (3). This translates to 448 million lb of PET and 450 million lb of HDPE (2). For example, the process cost of recycling PET and HDPE bottles has been given as $100–$150 (all U.S. dollars) per ton (51). The value of the PET and HDPE produced from recycled materials is $470 and $120 per ton, respectively. So these recycling processes can be profitable. The favorable economics of PET recycling have been attributed in part to forward integrated PET recyclers consuming their own product to make bottle resin (28).

However, recycling of many other plastics remains uneconomic (2). The costs of recycling commingled plastics has been estimated at $1700 per ton (52). This is 10 times more expensive that of recycling easily separated homogeneous products such as PET and HDPE. In Germany, federal mandates require recycling of 60% of all plastic packaging. German projects to recycle over one billion pounds a year of plastics have been announced (53,54). These processes will use high-temperature/high-pressure processes to depolymerize commingled plastics to produce petrochemical feedstocks. These processes are not economic (55).

In 1994, U.S. capacity for recycling polypropylene from battery casings was 265 million lb annually (39). About 75% of the recovered polypropylene is used in new battery cases that have a recycled PP content of about 50%. Total polypropylene recycling capacity was 350 million lb annually for an operating rate estimated at 71–90% (39).

BIBLIOGRAPHY

1. R. Knight, *U.S. News & World Report* **120**(22), 12 (June 3, 1996).
2. R. G. Saba and W. E. Pearson, "Curbside Recycling Infrastructure: A Pragmatic Approach" in Ref. 49, Chapter 2, pp. 11–26.
3. Franklin Associates Limited, *Characterization of Municipal Solid Waste in the United States, 1994 Update,* Report No. EPA 530-94-042, Nov. 1994.
4. R. A. Pett, A. Golovny, and S. S. Labana, "Automotive Recycling," in Ref. 49, pp. 47–61.
5. J. Milgrom, *Encyclopedia of Packaging Technology,* 1st ed., Wiley, New York, pp. 965–968.
6. E. Schloeman and D. B. Spencer, *Resource Recovery Conserv.* **1,** 151 (1975).
7. J. Staley and W. Haupin, "Aluminum and its Alloys" in J. I. Kroschwitz and M. Howe-Grant, eds., *Encyclopedia of Chemical Technology,* 4th ed., Vol. 2, Wiley, New York, 1992, pp. 184–251.
8. R. S. DeCesare, F. J. Palumbo, and P. M. Sullivan, *U.S. Bureau of Mines Report of Investigations RI 8429,* Washington, D.C., 1980.
9. K. Grjotheim, H. Kvande, K. Hotzfeldt, and B. J. Welch, *Can. Metall. Quart.,* **11**(4), 585 (1972).
10. Anonymous, *Ceramic Ind.* **141**(6), 14 (1993).
11. P. Marsh, "Recycling, Glass" in M. Grayson and D. Eckroth, eds., *Encyclopedia of Chemical Technology,* 3rd ed., Vol. 19, Wiley, New York, 1982, pp. 961–966.
12. E. D. Spinosa, R. M. Stephan, and J. R. Schorr, *Review of Literature on Control Technology which Abates Air Pollution and Conserves Energy in Glass Melting Furnaces,* to Corning, Inc., EPA-600/2-77-005/2-76-269/2-76-032b, Battelle, Columbus, OH, Nov. 11, 1977.
13. K. Merriman, in R. J. Spandenberg, ed., *Secondary Fiber Recycling,* TAPPI Press, Atlanta, 1993, p. 101.
14. T. Bliss, in Ref. 13, p. 125.
15. C. M. Vitori, *Pulp Pap. Can.* **94**(12), 109 (1993).
16. M. Doshi, *Prog. Paper Recycl.* **4**(2), 103 (1995).
17. C. F. Yu, R. F. DeFoe, and B. R. Crossley in *Proceedings of the TAPPI Pulping Conference,* TAPPI Press, Atlanta, 1994, p. 451.
18. M. Doshi, *Prog. Paper Recycl.* **1**(2), 61 (Feb. 1992).
19. R. DeFoe, *TAPPI J.* **76**(2), 157 (1993).
20. T. Pekkarinen, "Fractionation of OCC Waste Paper with a Pressure Screen" in *Proceedings of the TAPPI Pulping Conference,* TAPPI Press, Atlanta, 1985, 37.
21. T. Nguyen, A. Shariff, P. F. Earl, and R. J. Eamer, *Prog. Paper Recycl.* **2**(3), 25–32 (May 1993).
22. H. M. Bisner, R. Campbell, and W. T. McKean, "Bleached Kraft Pulp from OCC," in *Recycled Paper Technology: An Anthology of Published Papers,* TAPPI Press, Atlanta, 1994, pp. 48–56.
23. G. Galland, Y. Vernac, and J. Brun, "Recycling of Waxed Papers and Boxes," in Ref. 22, pp. 81–89.
24. U.S. Pat. 4,764,323 (Aug. 16, 1988), A. Ghatta (to Cobarr S.p.A.).
25. A. L. Bisio and M. Xanthos, eds., *How to Manage Plastic Waste: Technology and Market Opportunities,* Hanser Publishers, New York, 1995.
26. Anonymous, *Fed. Reg.* **58,** 52719–52729 (Oct. 12, 1993).
27. Anonymous, *Modern Plast.* **71,** 73–94 (Jan. 1994).

28. W. K. Atkins, "A Resin Producer's Perspective and Experiences in Polyethylene Recycling," in Ref. 49, pp. 104–112.
29. R. A. Burnett, "Progress in Poly(vinyl chloride) Recycling," in Ref. 49, pp. 97–104.
30. U.S. Pat. 4,680,060 (July 14, 1987), A. S. Gupta and J. T. Camp (to The Coca-Cola Company).
31. *Points to Consider for the Use of Recycled Plastics in Food Packaging: Chemistry Considerations,* U.S. Food and Drug Administration, Center for Food Safety and Applied Nutrition, Publication HFS-245, Washington, DC, April 1992.
32. D. E. Pierce, D. B. King, and G. D. Sadler, "Analysis of Contaminants in Recycled Poly(ethylene terephthalate) by Thermal-Extraction Gas Chromatography—Mass Spectroscopy," in Ref. 49, pp. 458–471.
33. T. H. Begley and H. C. Hollifield, *Food Technol.* **47,** 109–112 (Nov. 1993).
34. G. Bauer, in K. J. Thome-Kozmensky, ed., *Recycling of Wastes,* EF-Press, Berlin, 1989, p. 293.
35. F. L. Bayer, D. V. Myers, and M. J. Gage, "Consideration of Poly(ethylene terephthalate) Recycling for Food Use," in Ref. 49, pp. 152–160.
36. D. Gisstis, *Makromol. Chem. Makromol. Symp.* **57,** 185 (1992).
37. S. Jabarin and P. Sambaru, *Polym. Eng. Sci.* **33**(13), 827 (July 1993).
38. S. A. Jabarin and V. V. Bhakkad, "Morphology and Properties of Poly(ethylene terephthalate)—High-Density Polyethylene Blends," in Ref. 49, pp. 113–138.
39. R. M. Prioleau, "Recycling of Polypropylene," in Ref. 49, pp. 80–88.
40. D. A. Thomson, "Polystyrene Recycling: An Overview of the Industry in North America," in Ref. 49, pp. 89–96.
41. W. S. Hubble, I. G. Most, and M. R. Wolman, *Investigation of the Energy Value of Automobile Shredder Residue,* Publication DOE/ID/12551-1, U.S. Department of Energy, Washington, D.C., Aug. 1987.
42. S. J. Pearson, G. D. Kryder, R. R. Kopang, and W. R. Seeker, in Ref. 49, pp. 183–193.
43. P. N. Williamson, *TAPPI J.* **79**(1), 55 (Jan. 1996).
44. G. B. Stanley, *TAPPI J.* **79**(1), 37 (Jan. 1996).
45. M. V. Tuomisto, "U.S. Mini-mills: The Activity is in Recycled Containerboard" in *Proceedings of the TAPPI Recycling Symposum,* TAPPI Press, Atlanta, 1996, pp. 51–56.
46. R. P. Hoffman, "The Potential Use of Office Wastepaper in Containerboard Manufacture," in Ref. 48, pp. 103–107.
47. Doshi & Associates, "1996–1997 Directory of Recycled Pulp & Paper Mills," supplement to *Progress in Paper Recycling,* **5**(3) (May 1996).
48. W. E. Franklin, "Old Corrugated Containers Recovery Forecast" in V. Stefan, ed., *Recovery and Use of OCC,"* Miller Freeman, San Francisco, 1993, pp. 12–14.
49. C. P. Rader and R. F. Stockel, "Polymer Recycling: An Overview" in C. P. Rader, S. D. Baldwin, D. D. Cornell, G. B. Sadler, and R. F. Stockel, eds., *Plastics, Rubber, and Paper Recycling: A Pragmatic Approach,* American Chemical Society, Washington, DC, 1995, Chapter 1, p. 3.
50. R. S. Morrow and C. M. Quinn, in J. I. Kroschwitz and M. Howe-Grant, eds., *Encyclopedia of Chemical Technology,* 4th ed., Vol. 5, Wiley, New York, 1993, p. 19–34.
51. J. W. Porter, *Recycling at the Crossroads,* Porter & Associates, Sterling, VA, 1993.
52. J. R. Ellis, "Polymer Recycling: Economic Realities," in Ref. 49, pp. 62–69.
53. P. Mapleston, *Modern Plast.* **70,** 53 (Nov. 1993).
54. Anonymous, *Eur. Chem. News* p. 37 (April 25, 1994).
55. M. W. Meszaros, "Advances in Plastics Recycling: Thermal Depolymerization of Thermoplastic Mixtures," in Ref. 3, pp. 170–182.

JOHN K. BORCHARDT
Shell Chemical Company
Houston, Texas

RECYCLING, EUROPE

Recycling of packaging waste in Europe was given a much higher profile in December 1994 when the European Union institutions approved a Directive on Packaging and Packaging Waste.

The significance of this directive is two fold: first it put down an important market as far as Europe's environmental protection legislation is concerned, and second, while it sets guidelines requesting Member States to develop national legislation for packaging waste adapted to local conditions, it supersedes such laws if member states do not act (see also Environmental laws, international).

The Goals of EU Packaging Waste Directive

This Directive aims to accomplish two objectives: the harmonization of national legislation on packaging and packaging waste in order to progress towards a single European market where goods and services flow freely, and to reduce the impact of packaging waste on the environment. The legislation seeks to reduce the amount of packaging waste from the industrial, business, and private sectors by requiring all Members States to look at minimizing packaging by prevention, reuse, and recovery. While well-defined targets are given for recycling recovery, there are no specific goals for prevention and reuse.

The targets set for recovery recognize the diversity of Europe and their broad range acknowledge that the European community is still at the early stages of understanding what is achievable. After entry into force of the Directive, ie June 30, 1996, there is a five-year initial period during which time the minimum and maximum recovery targets are 50–65% by weight, of which 25–40% must be recycled (recovery includes materials recycling, composting, and energy recovery). A minimum of 15% must be recycled by material.

The so-called "cohesion" countries, Ireland, Greece, and Portugal, are given ten years to reach these targets. Higher targets may be set by other countries but only after demonstrating that the measures taken "do not constitute an arbitrary means of discrimination or a disguised restriction on trade between Member States."

As practical knowledge and experience is gained, the need for a review of these targets and other elements of the text will become apparent. For this reason a Committee comprising representatives of the Member States has been created, chaired by the Commission. This Committee can give its opinion on any Commission proposal to amend or update the law, and is currently studying issues such as marking, material identification and database format. It is named by the Directive article that defines its duties: Article 21 Committee.

National Implementation: The Challenges

As of June 30, 1996, the deadline for notification of national implementation only four countries, Austria, Belgium, Lux-

Table 1. Waste-Stream Quantity and Composition[a]

City	Total Waste Stream (kg/inhab/yr)	Composition (% of Total)					
		Paper	Glass	Metal	Plastic	Organic	Rest
Dublin (Eire)	226	25	5	3	9	46	12
Adur (UK)	244	42	9	8	9	19	13
Sheffield (UK)	300	33	7	11	8	26	15
Dunkirk (F)	344	24	17	3	8	26	22
Barcelona (E)	370	32	6	3	13	37	9
Lemsterland (NL)	380	29	8	3	5	48	7
Prato (I)	386	19	4	3	8	33	33
Saarland (D)	386	21	11	4	7	29	28
Pamplona (E)	387	25	7	2	6	46	14
Queijas (P)	590	15	7	2	14	50	27

[a] Year 1994

embourg, and Sweden, had notified the Commission that they have enacted national legislation to implement the Directive.

Why have only four Members States out of fifteen notified the Commission that they are enacting national legislation? The answer lies partly in the complexity of each national situation, and in the number of players involved in the packaging waste management issue in each country. In order to develop a long-term infrastructure for waste management at a national level, there needs to be a certain level of consensus and commitment among the different players, be they members of the packaging chain, municipal or national decision makers, consumer groups or waste managers.

One could say that the Directive did the easy part. It established political targets, but left it up to national bodies to organize recovery and recycling systems, to establish financial structures, to formalize relationships between all parts of the packaging chain. In short, no system is transferable from one Member State to another. In countries where no national law has been implemented by June 30, one could invoke the EU Directive in the absence of national legislation.

Industry Response

But the legislative situation does not tell the whole story. What is becoming clear is that even without legislation, momentum has been building for collection and separation of household waste. In several Member States, industry schemes have been established to set up and support collection, separation, and recovery systems. This, in turn, generates a review of waste management options by Municipal Solid Waste managers. These industry-led schemes are all different and reflect the enormous diversity of culture and waste generation that characterises Europe.

This trend towards industry organizations shifts the direct responsibility for packaging recovery and recycling from the individual company in the packaging chain towards a collective role within these organizations. This company can demonstrate responsible conduct in its overall packaging strategy by focusing its direct activities on prevention and/or reuse and the packaging recoverability.

Variable Factors

Table 1 demonstrates Europe's diversity in terms of one factor: composition of the waste stream. This analysis is based on several European recovery programs monitored by ERRA, the European Recovery and Recycling Association. Elements impacting such composition include for example, climate, which can trigger different patterns in the Mediterranean than in the Nordic countries; geography, ie, rural versus ur-

Table 2. Schemes in Place vs Status National Legislation

Country	Industry Scheme	Legislative Status	EU Compliance	Impact of Directive
Denmark	No	Existing	No	Issue of can ban
Germany	Yes, DSD	Existing	No	Issue of quotas and targets
Netherlands	Yes, SVM	In progress		Negotiated agreement between industry, government and others
Belgium	Yes, Fost Plus	In progress		Issue of ecotaxes
UK	Yes, VALPAK	In progress		
Ireland	Yes, REPAK	In progress		
France	Yes, Eco-Emballages	Existing	Partial	Complementary legislation in progress
Spain	In progress	In progress		
Portugal	In progress	Existing	No	Refillable/recyclable quotas
Luxembourg	Yes, VALORLUX	In progress		Issue of ecotaxes
Italy	In progress	Existing	No	New legislation in progress
Greece	In progress	In progress		Potential issue regarding refillable/recyclable quotas
Austria	Yes, ARA	Existing	No	Review in progress
Sweden	Yes, REPA	Existing	Partial	No modification expected
Finland	In progress	Existing	No	Review in progress

Table 3. Best Practice Used Packaging Recovery and Recycling Key Principles

1. Responsibility should be shared
2. Waste management should be integrated
3. Process should be environmentally effective
4. Total system should be economically efficient
5. Trade barriers must be avoided
6. Secondary material markets have to be stimulated

ban areas, and availability of materials. Some specific materials may be available in large quantities in one place and virtually nonexistent in another. For instance, PET is present in large quantities in used packaging and justifies material recovery in Italy, but is not present in important volumes in Germany's large cities. This variable state of affairs generates different approaches and different legislation in each Member State.

A Patchwork of Initiatives

The pace of change in approaches to packaging waste management varies also. Implementation of industry recovery schemes and indeed, of national translation of the EU Directive leads ERRA to believe that there will be many modifications in current legislation and approaches to recovery within the next five years. The situation is constantly evolving. In fact as one looks at the various legislative initiatives, the Dutch and French approaches may be the only ones still effective five years from now due to the highly conceptual and nontechnical approaches taken by the French and Dutch legislators.

Table 2 provides an overview of schemes in place versus status of national legislation. This table is a status report, only as the situation is constantly evolving, especially given the June 30 Directive implementation deadline.

Recycling rates differ between Member States. Such activities are either coordinated with national schemes such as DSD or Eco-Emballages, or by material specific organizations at national or EU level. A contact list for the major EU materials organizations who can be consulted to evaluate specific national situations is given later.

Trends

Trends in terms of packaging waste issues are, however, clearly identifiable. Information regarding national legislation, both for EU and neighboring countries, is continuously updated by ERRA. ERRA has also produced a series of Best Practice guidelines which are being used by industry and increasingly by local government as benchmarks for assessing environmental effectiveness and cost efficiency across Europe (Table 3).

It is clear that within the next 18 months all EU Member States will have both legislation implementing the Directive, and industry scheme(s) in response to this legislation.

As an exporter into the European Union, companies must be informed about Member States initiatives, what each national plan consists of, plus the specific requirements linked to the directive. These specifics include such issues as "essential requirements" (in the Directive) ie, marking and materials identification, as well as mandatory information regarding composition of packaging and number of units put on the market.

Once a packaging complies with essential and marking (including material identification) requirements, it can be used without any limitation in all Member States as per article 18 of the Directive.

The Danish can ban and some refillable quotas existing for beverage containers in Germany and Portugal will limit the use of specific packaging in these countries. These limitations are and will be tested in the coming months but represent today limitations to the free circulation of packaging outlined above.

Companies commercializing packaged goods in the EU should integrate the packaging and packaging waste directive packaging requirements into their marketing strategies.

Table 4. List of Material Specific Organization

Europe

1. UNICE
   (Union of Industrial and Employer's Confederations of Europe)
   rue Joseph II, 40 - bte 4
   1000 Brussels, Belgium
2. EUROPEN
   (The European Organization for Packaging and the Environment)
   avenue de Tervueren 113
   1040 Brussels, Belgium
3. ERRA
   (The European Recovery and Recycling Association)
   avenue E. Mounier 83, - bte 4
   1040 Brussels, Belgium

Member States Specific

1. DSD (Duales System Deutschland)
   Frankfurter Straβe 720-726
   D-51145 Köln, Germany
2. Eco-Emballages
   44 avenue Georges Pompidou
   F-92300 Levallois-Perret, France
3. Fost Plus
   rue Martin V, 40
   B-1200 Brussels, Belgium
4. ARA (Altstoff Recycling Austria)
   Schottenfeldgasse 29
   A-1072 Wien, Austria
5. SVM (Stichting Verpakking en Milieu)
   Prinses Beatrixlaan 5/Postbus 95598
   NL-2509 CN Den Haag, Netherlands
6. INCPEN (The Industry Council for Packaging and the Environment)
   Tenterden House
   3 Tenterden Street
   GB-London W1R 9AH
   United Kingdom
7. VALORLUX
   Zone Industrielle AM
   L-3372 Leudelange, Luxembourg
8. REPAK
   c/o IBEC
   84/86 Lower Baggot Street
   IRL-Dublin 2, Ireland
9. REPA
   Klarabergsviadukten 90
   S-Stockholm, Sweden

Table 5. Key Materials European Organizations

1. APEAL ("Association Professionnelle des Producteurs Européens d'Aciers pour Emballage")
   avenue Louise, 89
   B-1050 Brussels, Belgium
2. CEPI (Confederation of European Paper Industries)
   avenue Louise 306, b. 4
   B-1050 Brussels, Belgium
3. EUPC (European Plastic Converters)
   avenue de Cortenbergh 66, b. .4
   B-1000 Brussels, Belgium
4. EAA (European Aluminium Association)
   avenue de Broqueville 12
   B-1150 Brussels, Belgium
5. APME (Association of Plastics Manufacturers in Europe)
   avenue E. Van Nieuwenhuyse 4, B. 3
   B-1160 Brussels, Belgium
6. FEVE (European Container Glass Federation)
   avenue Louise 89
   B-1050 Brussels, Belgium
7. BCME (Beverage Can Makers Europe)
   c/o PRP
   avenue R. Vandendriessche 5
   B-1150 Brussels, Belgium
8. CITPA (International Confederation of Paper and Board Converters in Europe)
   Arndstrasse 47
   D-60325 Frankfurt am Main, Germany
9. ACE (Alliance for Beverage Cartons and the Environment)
   rue Joseph II, 36 - Box 2
   B-1000 Brussels, Belgium

Recovery Capacity

The community is still in the first phase of establishing what is realistic in terms of recovery capacity. The situation is by no means mature. But it is no longer top of the political agenda either. One reason for this is that political pressures have created market conditions whereby the necessary capacity has been generated for recycling and recovery and new technologies. This is particularly true for paper, plastics and laminates. In addition, although prices fluctuate, markets now exist in most Member States for secondary materials, because there is now a constant supply of recycled materials. More information can be obtained from the list of material-specific organizations (see Tables 4 and 5).

Staying Informed

One word of warning is that a Europe-wide approach can be in direct opposition to national interests. The issue of refillable versus recoverable quotas, especially beverage containers, is an example. The existence of minimum refillable quotas (up to 90% in some cases) create trade barriers. This may compromise ease of access to certain markets for certain products.

Although recycling is here to stay in Europe, practices, political will, and industry responses are still in transition. They probably will not stabilize for at least five years. The hope of ERRA is that with industry and government working together, recycling and recovery will become part of a broader initiative to manage waste in an integrated manner, which will allow local waste managers to make environmentally and economically sound decisions based on local infrastructure, resources, and priorities. The goal is surely to optimize all forms of recovery, including materials recycling. How this goal is achieved is not altogether clear at this time: companies need to stay flexible and above all, informed.

J. Fonteyne
ERRA
Brussels, Belgium

RETORTABLE FLEXIBLE AND SEMIRIGID PACKAGES

This article pertains to packages made entirely of flexible materials, eg, pouches, as well as semirigid containers that have at least one flexible body wall or lid. These packages, available in many shapes, sizes, and combinations, are used for food and medical products processed by heat and pressure.

The characteristics of "retortable," as related to packages, has been defined by the American Society for Testing and Materials (ASTM) Committee on Flexible Barrier Materials and its Subcommittee on Retortable Pouches and Related Packages. The standard definition relating to retortable flexible barrier materials is that they are "capable of withstanding specified thermal processing in a closed retort at temperatures above 100°C (1). The legal requirements for retort processing of foods in metal cans, aluminum cans, glass jars, and retort pouches in the United States are clearly defined in the *Code of Federal Regulations* (*CFR*) (2).

Retortable packages must maintain their material integrity as well as their required barrier properties for their designated end-product use during product-to-package handling, thermal processing, and subsequent shipping abuse. For shelf-stable food packages, the materials used must be retortable and still maintain extended barrier characteristics against such effects as light, moisture, oxygen, and microbial penetration.

Steel cans (see Cans, steel) perform very well in a retort environment. Glass jars and lightweight aluminum rigid containers also are well-established containers that can be thermostabilized in this manner, except that they both normally require some special handling in the form of overriding process pressures and critical control of pressure changes. Special handling is required to prevent closure lids on jars from releasing their vacuum seal, easy-open scored lids on aluminum cans from fracturing, and aluminum can bodies from paneling owing to excessive internal or external pressure.

Flexible and semirigid packages also have critical retort process-handling requirements. A variety of high-temperature process and equipment systems other than standard retort pressure vessels (see Canning, food) can be utilized as long as proper pressure control and uniform heat distribution are maintained. Most medical-type autoclaves (see Healthcare packaging) lack the necessary controls, but high-temperature short-time (HTST) systems, vertical or horizontal hydrostatic cookers, microwave chambers, and many variations thereof have been adapted for thermostabilization of flexible and semirigid packages.

Retort processing of these packages may cause degrading effects such as weakened package-seal integrity, flexcrack or flex-stress leading to film delamination or leakage, semirigid container distortion or denting, and even loss of hermetic-seal integrity because of pressure changes during heat processing or between the cooking and chilling processes.

Since flexible and semirigid containers are more sensitive than rigid metal cans to degrading effects during the thermal process, they cannot be retorted in the same way. Rigid metal cans are usually processed in a saturated steam atmosphere, typically at ~250°F (121°C). The external pressure of this process on the container is therefore 15 psig (205 kPa). The internal pressure of the heated product in a rigid metal can can be different from the external pressure of the steam, within limits, and the rigid container supports this pressure difference through its body structure. Flexible and semirigid packages, however, are susceptible to rupture or seal separation if the internal pressure exceeds the external process pressure excessively. Consequently, the thermal processes for these containers must include an overriding pressure beyond that of the saturated steam. This is accomplished by processing in superheated water with overriding air or steam pressure, or by processing in a combination steam–air atmosphere. Application of an overpressure of 10–15 psig (170–205 kPa) is typical, yielding a processing pressure of 25–30 psig (274–308 kPa) for a retort temperature of 250°F (121°C).

The primary reason for specifying a package to be retortable is for thermal sterilization or at least minimal destruction of pathogenic microbial contamination that could cause spoilage of contained products such as foods and prevent chemical reactions and resulting degradation of other packaged products. Numerous studies of these effects of retort processing on retortable packages have been published (3,4). The most positive effects on foods packaged in the retortable flexible and semirigid packages are the resultant shortened process times for commercial sterilization and improved product quality.

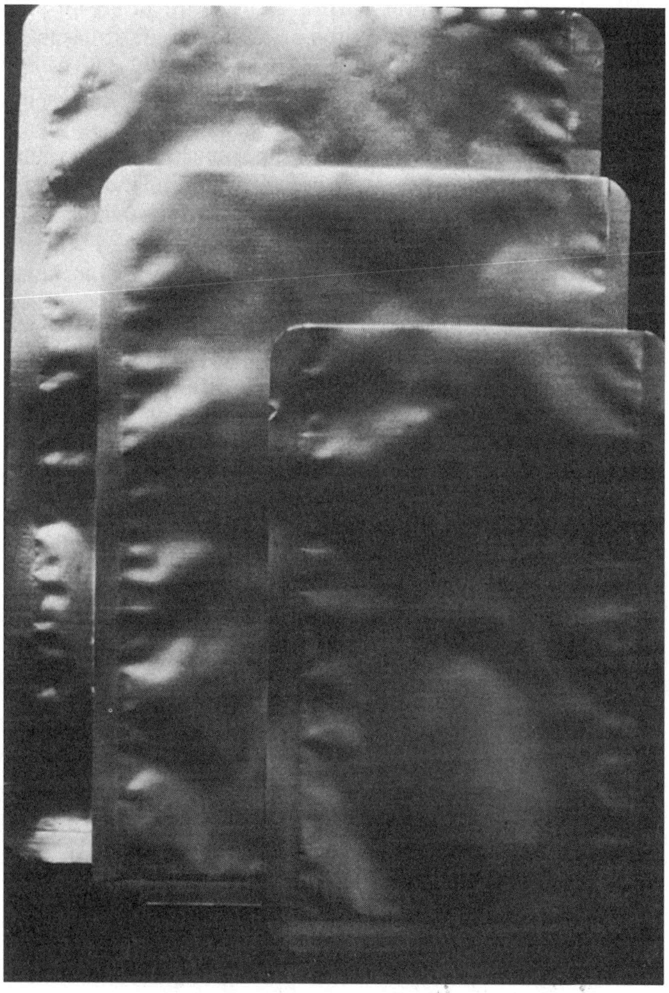

Figure 1. Retortable flexible packages.

Flexible Packages

The descriptive packaging term *flexible* has been defined by the ASTM Committee on Flexible Barrier Materials as "easily hand-folded, flexed, twisted, and bent" (1). The United States Flexible Packaging Association defines *flexible* packages as "packaging utilizing flexible materials (papers, films, foils, and metallized films) which are normally printed or laminated as roll products and which may conform to the shape of their contents."

A wide range of monolayer materials and multilayer structures (see Multilayer flexible packaging) can be considered as flexible-packaging material structures (see Fig. 1). The limitations of each package are determined by their fabrication characteristics, ie, heat sealability, printability, flexibility, and retortability; and by their product-to-package integrity, ie, moisture-, light-, and oxygen-barrier properties, flexcrack resistance, odor, and food-content compatibility.

The term *flexible* in packaging is meaningful both as a package characteristic and as it relates to product application. Although savings can often result from the use of one flexible-packaging material instead of another or rather than rigid containers, flexible packages must meet the specifications of compatibility with the products contained and also offer convenience, aesthetic values, quality benefits, extended shelf life, or other advantages. The application of flexible packages to foods requires development and evaluation of all package and product characteristics, production operations required, and economics. These steps involve the sciences of food and packaging technology (5). Nonfood products in flexible packaging require multiple technical and marketing disciplines as well.

Retortable Pouch

All sides of retortable pouches are flexible and are constructed of one or more layers of plastic and/or foil layers, each having its package functionality. The choice of barrier layers, sealant layers, and food-contact layers depends on the processing and product application. Flexible retortable pouches collapse or compress tightly around the contained products when a vacuum is applied to the package before sealing.

The retortable-pouch concept dates back to the 1940s, when university and container-industry suppliers evaluated various flexible barrier film materials as a means of improving the economics and quality of shelf-stable foods. Researchers in the United States and around the world soon realized the potential advantages of the thin-profile package concept, and development was rapid. An early pioneer in proving the integrity of this package and its product-quality improvement for military field rations was the U.S. Army Natick Research and Development Center (6). Military acceptance worldwide has led to the gradual replacement of metal cans with flexible retortable pouches for individual portions of shelf-stable foods. The pouches offer product quality; carrying, reheating, serving convenience; and weight and cubage savings. The U.S. military terms this packaged-food concept as "Meal, Ready-To-Eat" (MRE).

A chief technical problem resolved to the satisfaction of many producers and users, but unsolved by normal commercial standards, is the state of development of equipment, manufacturing techniques, and quality-control programs for the forming, filling, vacuumization, and sealing of retortable-pouch packaged foods in comparison with the high-volume production-line speeds and economies of other food containers. Achieving package-seal integrity without seal contamination during filling and sealing remains a limitation in the manufacturing process.

The retortable pouch continues to find new applications, new markets, new advantages, and even new equipment and film materials. The impression forming of multiple pouches across a web stock, followed by filling and sealing with another lid layer of film (see Thermoform/fill/seal), produces another package configuration that offers cost-saving potential. In the 1980s developments in multilayer plastic structures (see Coextrusions for flexible packaging) produced packages that could be microwave-processed and reheated for serving convenience without sacrificing the long shelf life previously achieved with foil barrier layers. Improved package integrity also resulted from film laminations offering more pliability and resistance to flexcracking. Food and nonfood uses are being developed in small sizes for individual-consumer servings and in larger-volume institutional sizes.

Retortable pouches are also widely used in the sanitary packaging required by the medical and pharmaceutical industries. Products sterilized within such a package include surgical instruments, sterile water, bandages, intravenous solutions, medicated ointments, and liquid-diet tube-feeding foods. A typical retort pouch for packaged liquid-diet food for enteral feeding (see Fig. 2) features foil and PVDC (see Vinylidene chloride copolymers) barrier layers, spike-and-hang and dispensing features, and extended shelf-life without refrigeration.

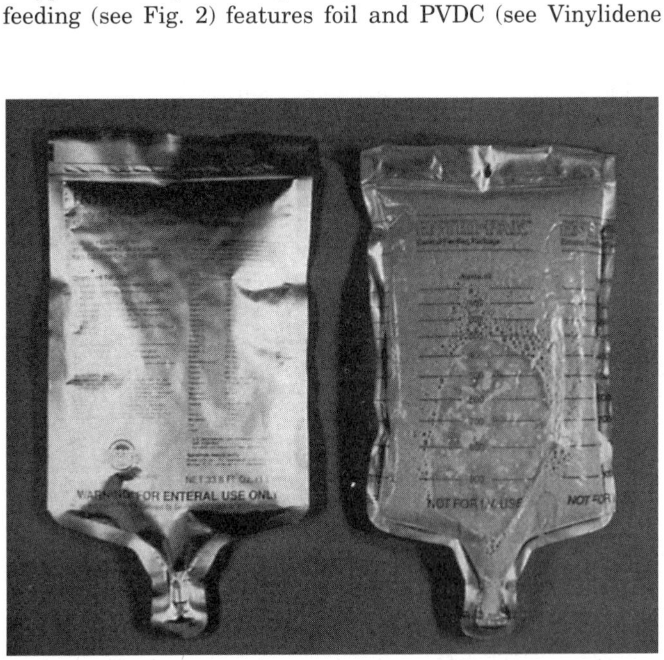

Figure 2. Retortable flexible-pouch package for liquid-diet food.

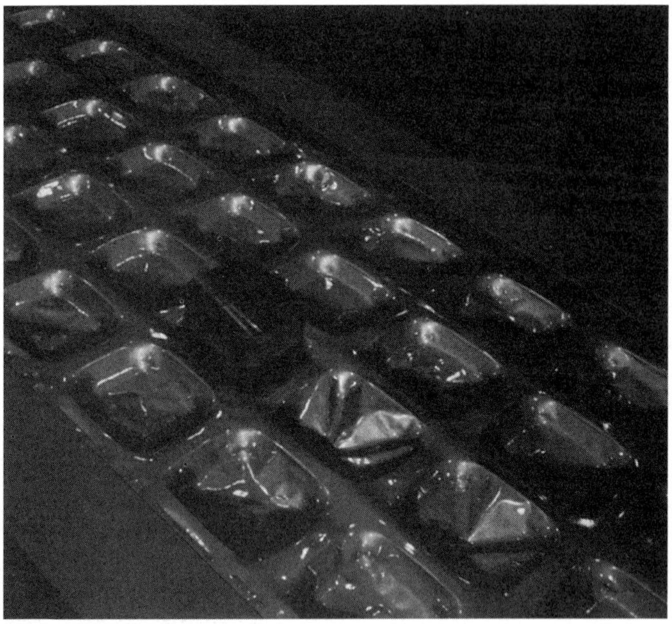

Figure 3. Forming multiple units of flexible packages.

Retortable Trays

The retortable tray usually has a rigid or semirigid structural supporting body and a sealable flexible lid. Composition may be of foil or plastic combinations that are retortable and have barrier properties. The tray package is a thin-profile container that offers all the advantages of the retortable pouch, ie, sterilization and reheating, and can also be used as a serving dish. The advantages of a thin-profile package, ie, reduced heat-sterilization requirements, improved product quality, and usage conveniences, have led to widespread development of semirigid packages. Improved techniques of thermoforming (see Thermoforming) or (cold) impact forming and coextrusion (see Coextrusions for semirigid packaging) have recently advanced the state of the art of this package. Thermoforming is a relatively low-cost technique of heating a thermoplastic sheet to an optimum temperature, then forcing it against contoured molds to achieve desired shapes, and finally cooling to a setting temperature. Forming may utilize either vacuum, pressure, or both. This technique has been applied to medical devices for many years, and more recently to foods.

Advantages of semirigid packages include ease of filling or top-loading of products (most are formed and filled in a horizontal mode) and potential increases in production-line speeds and volume due to multiple pocket forming on a wide-web-style operation (see Fig. 3).

Completely nonmetallic varieties of this package are now commercially available for processing or reheating in microwave units. Economic advantages of these materials will soon become apparent. This lightweight, retortable plastic package is available in a wide range of sizes and shapes: tubs, trays, and cans with double-seamed, heat-sealed, or induction-bonded lids. They can be supplied to users as rollstock for inline form/fill/seal operations or as preformed packages.

The concept of retorting semirigid nonmetallic packages is relatively new. This article has mentioned the use of thermoforming, but retortable plastic containers made by injection blow molding (see Blow molding) have recently been introduced (7). These new developments will necessitate new definitions and standards.

BIBLIOGRAPHY

"Retortable Flexible and Semirigid Packages" in *The Wiley Encyclopedia of Packaging Technology,* 1st ed., by D. D. Duxbury, Pouch Technology, Inc., pp. 568–571.

1. "Paper; Packaging; Flexible Barrier Materials; Business Copy Products"; *1984 Annual Book of ASTM Standards (Sect. 15), General Products, Chemical Specialties, and End Use Products* Vol. 15.09, American Society for Testing of Materials, Philadelphia, 1984, pp. 831–832.
2. *Code of Federal Regulations,* Title 21, Office of the Federal Register, General Services Administration, Washington, DC, 1982, Part 113.
3. R. A. Lampi, *Adv. Food Res.* **23,** 305 (1977).
4. J. P. Adams, W. R. Peterson, and W. S. Ottwell, *Food Technol.* pp. 123–142 (Apr. 1983).
5. F. A. Paine and H. Y. Paine, *A Handbook of Food Packaging.* Leonard Hill, Blackie & Son, London, 1984.
6. *Proceedings of the Symposium on Flexible Packaging for Heat-Processed Foods,* National Research Council, National Academy of Sciences, Washington, DC, Sept. 1973.
7. J. A. Wachtel, B. C. Tsai, and C. J. Farrell, "Retorted EVOH Multilayer Cans with Excellent Oxygen Barrier Properties" in *Proceedings of EUROPAK '85, First Ryder European Conference on Plastics and Packaging,* Sept. 17–18, 1985, Ryder Conferences. Ltd., St. Helier, Jersey, UK, 1985.

ROBOTS

Packaging means "to place compactly in a trunk, box, etc., for storing and carrying" (1). That broard meaning is appropriate to a variety of work tasks common to light industry. The most obvious of these tasks include placing individual product units into packers, placing packers into shipping cases, and palletizing shipping cases for shipment and/or warehouse storage. In high-speed filling operations, where 300 upm (units per minute) is probably typical, unpacking is a common task associated with removing empty containers from their storage medium and placing them on the line in an orderly manner, ready for filling. Another, but not so obvious, set of packaging tasks includes those operations associated with the transfer of work-in-process units through the manufacturing process. Examples here include moving metal parts through a heat-treatment process in a batching, rather than a continuous, mode (2) and transporting glass and mercury clinical thermometers through an air-separation process in basket-sized lots. All of these packaging tasks are suitable for implementation with robots, and actual robot work experience in typical light-industry factories will be used here to illustrate the major installation considerations associated with packag-

Figure 1. Robot transfers four jars at a time from the line into the case.

ing with robots. In these factory examples the emphasis will be on key technical aspects of these installations and those human interface requirements that must be met to achieve a successful robot installation. Although such factors as installation cost and financial justification are important to the overall packaging objective, they are beyond the scope and intent of this article and are excluded from detailed discussion. It should be understood, however, that all the devices and applications selected for illustration fall within reasonable (return-on-investment) (ROI) limits and represent cost-effective applications of robot technology.

Robot Tasks

When considering a robot installation from a job-task definition viewpoint, there are three major factors to define: weight to be moved, speed of movement, and placement accuracy. Placement accuracy includes initial point of pickup and final point of discharge. All three of these factors—weight, speed, and accuracy—are so interrelated that the limitations and capabilities of any one factor can affect the remaining two. For example, increased speed typically is obtained at the expense of reduced weight, reduced accuracy, or both. Thus, in most cases robot installations that approach the operating limits of the selected unit require some degree of application compromise to enable all three factors to share the burden without pushing any one of them over its limit. In Figure 1, the robot arm was not fast enough to move individual lotion jars from the line to the shipper at the required rate of one per second. Moving 10 jars at a time solved the speed and distance problem, but ten 16-oz (0.5-kg) jars exceeded the weight-carrying limits of this particular end effector. Furthermore, the 10-jar-in-a-line discharge did not match the shipper configuration. A compromise of four jars at a time, however, satisfied both weight and speed requirements without exceeding the unit's accuracy capabilities.

In installations where the task requirements are well within the robot design limits, the user can devote more time and resources to the total application environment and less effort to the robot itself. Figure 2 illustrates an application in which all three major task factors are well below the particular robot's design specifications. The same model robot as in Figure 1, a Unimate 500, is shown in Figure 2 executing a liquid makeup pan-filling operation where the weight requirements are measured in grams, placement accuracies are acceptable anywhere within the individual pan dimensions of 1.28 in. × 0.79 in. (3.25 cm × 2 cm), and filling speed is limited by pumping rates rather than end-effector transfer speeds. In this installation particular emphasis was placed on the design and installation of safety guards so that work envelope integrity could be maintained without significantly impeding tray placement and removal by the human operator.

In addition to the major task-factor differences between the applications illustrated in Figures 1 and 2, a significant difference in the degree and level of control interface with the work environment also should be noted. For example, in the pan-filling operation the robot's interface is limited to two binary switch position indicators (3): one switch indicating the safety gate's open or closed position, the other signaling the presence or absence of a pan-ready tray. In contrast to this simple switch arrangement, the lotion-jar transfer from the conveyor to the shipper in the first example required computer-controlled action between six sensory inputs and six action outputs. These requirements were dictated by the added burden of coordinating the robot's actions with those of the screw–feedguide mechanism, the case handler, the case sealer, and their connecting conveyor drive controls The screw-feed unit was used to provide the required spacing between jars and to present the jars in four unit batches for robot acquisition. The case handler presented the empty cases to the robot-loading station for product insertion, and the case sealer glued and sealed the case flaps, and transferred the finished shippers farther down the conveyor to the palletizing station. None of these individual tasks were particularly difficult or complex, but their collective control, necessary for automatic coordination between the work stations, was as time-consuming as the robot installation itself. In this particular installation, the overall system requirements exceeded the robot I/O (input/output) practical limits to such a degree that a separate computer, an Intel 8748 with appropriate hardware, was added to separate the peripherals' program from the robot's program. Interfacing software provided the necessary coordination between these two functions. This separation not only simplified the overall programming task but also, more importantly, enabled the robot to act independently of the peripherals when appropriate. For example, the addition of accumulating conveyors enabled the robot to continue packing even when the primary conveyor was shut down because of palletizing delays.

Figure 2. A Unimate 500 PUMA moving the filling nozzle from pan to pan. Note protective shield in the foreground.

Figure 3. A Pacer robot, manufactured by Production Automation, Livonia, Michigan, in position for vacuum cup acquisition of empty jars.

Unloading and handling. An unloading requirement common to many high-speed filling-line operations is that associated with placing the unfilled containers onto the in-feed conveyor. One hard automation technique for accomplishing this task involves the use of unscrambler units that take randomly fed parts from bins and automatically orient them into position for in-feed line placement. These units, however, have drawbacks, which include excessive changeover times, high initial costs, and exterior product damage caused by forces involved in mechanically driven orientation. In many cases robot line feeders or loaders can circumvent these limitations. For example, Figure 3 shows empty plastic containers being removed from their storage cases and placed onto the in-feed conveyor. This robot application illustrates the variety of end-effector designs available for meeting a broad range of product-handling requirements. In this case the finger extensions equipped with suction cups combine a strong capture force with a positive orientation feature for reliable product placement. A simple end-effector change in conjunction with new-product placement instructions enables the robot to accommodate changes in jar size with a minimum of expended time or effort. This line-loading technique requires known product orientation at the point of acquisition. Newer systems combine machine vision with suitable end effectors in such a

Figure 4. A Unimate 500 PUMA: (**a**) reaching in for glass jars; (**b**) moving eight glass jars at a time from storage container to filling line.

way that bin packing on a randomly oriented basis can be accomplished using robots (4).

Automation

Many factory tasks are selected for automation in order to replace human labor with machine labor and to realize an attendant cost savings. Robots are particularly helpful in this regard because they can be implemented in many situations with a minimal impact on their surrounding environment. Hard automation, on the other hand, often requires major changes in line layouts, space allocation, and worker job assignments. In either case, however, the human workers to be replaced frequently do more than simply transfer materials from one point to another, even when job tasks appear solely manipulative. In addition to transferring materials, the human worker ensures that the correct materials are being moved, that aesthetic values are acceptable, and that the finished goods do indeed contain the type and amount of product promised on the label. When considering job tasks for mechanical automation, therefore, care must be given to the potential impact that the removal of human surveillance will have on that work area. For example, in the robot lotion loading noted earlier in Figure 1, the human case packers also ensured that proper fill heights were maintained, and their replacement with a robot did not satisfy this requirement. As a consequence, it was necessary to add a fill-height inspection device to the line in order to take advantage of the labor savings at the case-packing station.

Loading example. Figure 4 illustrates a glass-jar loading application that, on the surface, seems fairly straightforward. Line speeds of 150 upm were easily met with multiple jar pickups. The light weight of individual empty jars and the packing orientation of eight jars per row were ideally suited for this robot application. Two problems that threatened the viability of this approach, however, soon emerged. The first problem was noted when the end effector's horizontal alignment changed after just a few hours of operation. The cases holding the empty jars were not being consistently aligned with the originally programmed axis, and the end effector was striking the case in these instances, causing the end-effector deflection. The end-effector bar extensions had been deliberately designed to flex during the glass-acquisition sequence to accommodate minor positioning errors without breaking the glass (5). The solution was to add an additional bar guide to the case feeder to provide a positive positioning guide. The second problem, however, was more difficult to solve. Visual inspection of the glass for cross-malfunctions had previously been done by the human operator. This was not a finished-product quality concern, since line inspectors positioned after the filling cycle would detect and remove any such defective glass. It was a concern, however, at the filling station, because poorly formed glass could cause jams and spills, resulting in excessive machine downtime. An intermediate solution to this problem was to add a few seconds per case to the job standard for inspecting for gross glass anomalies at the case-loading station. A better solution, however, would be to equip the robot with machine vision for automatic glass inspection and rejection.

Worker Integration

In these examples of robot applications the human element is the single most important factor, and this factor covers a great deal more than simply the task targeted for the robot installation (7). At least for the present, humans are needed to turn robots on and off, to program them for new tasks, to repair and maintain them, and to work side-by-side with them in the factory environment. If the human worker perceives the robot as an enemy or as a threat to job security, these human-related factors become barriers, rather than aids, to the widespread implementation of robots in industry (8). With worker assistance, however, the task of adapting robots to specific work environment jobs is greatly simplified.

Thermometer example. One example of worker assistance, without which the project would have failed, involved the use of a large robot to carry out the air-separation process in a clinical thermometer manufacturing operation. These thermometers were of the glass and mercury type, and their clinical use required exacting standards for accuracy, reliability, and repeatability. The purpose of this process was to remove any gases that had become entrapped in the mercury during earlier operations by localizing these gases in an attached top chamber destined to be removed in the later stages of the process. This process was primarily mechanical in nature, since the differences in the process gases were used to bring about the desired separation. For example, mercury is ideally suited for centrifuge separation techniques because of its high specific gravity, and a large portion of the gas was expelled in this manner. Cooling and heating the mercury accelerated this separation, and vibrating the thermometers between these steps helped even more. This process is well established in the glass and mercury industry, and the diligence with which the manufacturer adheres to these principles is directly proportional to the desired level of finished product quality. In typical installations these steps are accomplished by a group of "air separators" working in tandem to move the thermometers in hand-sized lots through a series of process steps in a rigidly specific sequence. Missing steps, or changing their order, is fatal to the desired result. Thus, manual operations of this sort are not only unpopular with the workforce but also rife with opportunities for human error. In this particular case the frequent warm, cold, and chilled tank hand-contact requirements were uncomfortable, frequently causing wet clothes and slippery floor conditions. These conditions were unsafe without constant concern for proper floor-mat placement and maintenance. It is not surprising, therefore, that process error frequently occurred when process cycle times were shortened, as might happen with the approach of breaks, lunch, or shift end.

Years of experience with this process indicated that dip/soak-time requirements had to meet their individual minimum time limits, or the desired end results would be placed in jeopardy. For example, if the warm-out tank cycle minimums were set at 5 min per bundle, anything less than this period would lessen the effectiveness of the air-separation process. On the other hand, excessive times in these different process steps were damaging, if only because of the reduced production volume throughput. Unnecessary delays at any one point caused comparable delays at all preceding points, with a subsequent overall negative impact on production rate. This reduced rate not only added unnecessary costs to this particular operation but also created imbalances in the work-in-process inventory, with dire consequences for the remaining operations in the process. It also should be noted that these mandatory step delays are contrary to the natural bent of motivated workers and require constant subjugation of the natural tendency to work faster, not slower. Hence at times, worker zeal, the most valuable of job-related resources, unknowingly contributed to poor process quality.

Another major factor in this air-separation process was associated with thermometer breakage. Each broken thermometer represented a loss of not only its component raw materials but also its accumulated labor costs to that point in the process. Furthermore, the nature of this process was such that the natural flow of product to and from its various steps contributed to the breakage problem. For handling and volume-control reasons, the thermometers were initially sorted into 50 unit bundles, with rubber bands used to provide bundle integrity. This bundle size was appropriate for human-hand manipulation and was light enough not to cause undue worker fatigue. By the same token, however, this low-weight factor contributed to high-speed hand movement; thus, minor errors in positional accuracy, for example, striking the metal surface of a water tank, caused breakage. The point of this somewhat detailed process description is to emphasize another benefit of robot installations, a benefit often ignored, which is the robot's inherent capability of maintaining reliably accurate process cycle times. In those operations where timing is important to the desired result, the robot's advantage over the human operator increases in direct proportion to the number and complexity of these time-critical steps.

In determining whether or not this air-separation process could be accomplished using a robot, four major issues had to be resolved: the placement of the peripheral support equipment, ie, water tanks, vibrating tables, centrifuges, and asso-

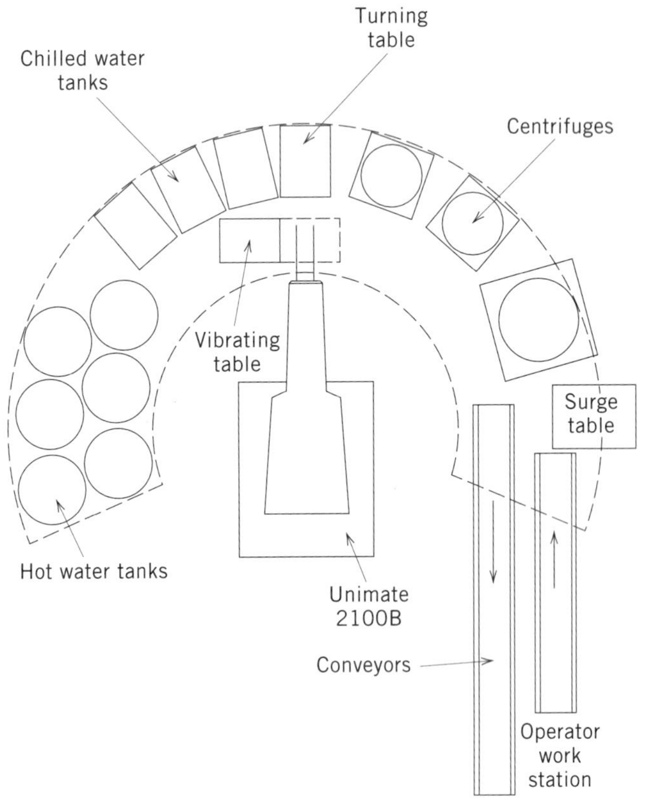

Figure 5. Equipment layout for air-separation process. The extended arm working radius is approximately 6 ft (1.8 m).

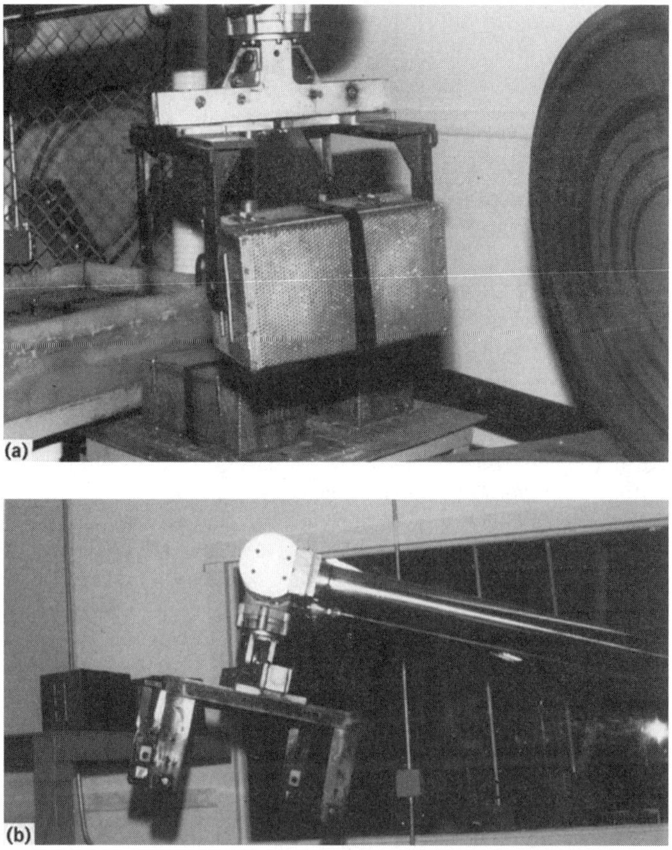

Figure 6. (**a**) Wire basket designed specifically for transporting thermometers in 1100 unit lots; (**b**) closeup of the end effector. Note inserts on finger ends for better control of wire baskets.

ciated holding stations; the unit lot size; total process time as a function of robot speed and individual step cycle times; and end-effector design suitable to lot sizes and required production rates. Figure 5 shows the selected layout, with the peripheral equipment located on an arc of a circle with a radius equal to the operating length of an appropriately sized and commercially available robot. Weight, speed, and accuracy considerations combined to indicate that lot weights could not exceed 25 lb (11 kg). This, combined with end-effector design criteria, indicated that wire baskets measuring 12 in. × 6 in. × 6 in. (30 cm × 15 cm × 15 cm) should be used as the basic product-carrying container (see Fig. 6). The basket size was based on using two baskets per end-effector load. This enabled each basket to carry a 1100 thermometer load; the combined basket and thermometer weight of 12 lb (5.4 kg), yielding a total end-effector load of 24 lb (10.9 kg), was within the 25-lb (11.3-kg) limit. This basket size was also convenient for the centrifuge-loading limits, which will be reviewed later.

The interlocking controls for this application were simple in concept but complex in implementation because of their large number. The process itself involved 14 different load/unload actions, with multiple repeats in a 38-step sequence, and was executed by a robot point-to-point program of 380 steps. For safety purposes, the robot and peripherals were located within a 24-ft × 26-ft (7.3-m × 7.9-m) wire cage (see Fig. 7). Access was limited to one door, equipped with supervisory controlled keys, and monitored by photocells and directed light beams for space intruder warning and automatic robot shutdown (9). Baskets were loaded by human operators outside the cage and delivered to the robot by conveyor. Product was removed from the cage by a similar conveyor, onto which the robot had placed the finished product baskets for delivery to the outside human operators for unloading. Routine daily start-ups consisted of merely turning the robot on a few minutes in advance of shift start for hydraulics warmup, conducting an activation test of area access safety beams and emergency shutdown switches, and switching on power to associated peripherals. Since the robot actions were dictated by product location within the process and by stations' status information conveyed to the robot control panel by means of automatic sensor inputs, no further human control interface was required. The robot simply waited for the product-in-place (PIP) signal from the in-feed conveyor indicating product availability, and the process began.

At any given time and/or process station, robot control throughout the process was based on three general conditions: product in position and ready for next step, next station ready, and next station clear. Product-position signals were generated by means of two position microswitches, photocell beam triggers, and proximity detectors. Control principles were based on positive signals generating action, rather than the lack of signal-allowing action. In practice this meant that the robot had to receive a signal that the target location was clear before moving to the target location, rather than moving there in the absence of any signal to the contrary. This provided additional protection to the system against the more common type of failure of the selected switches. "Station ready" signal inputs included temperature sensors at the water tanks, timer status indicators at the vibrating tables, and rpm indicators at the centrifuges. The centrifuges generated a "basket holder in position and locked" signal in addition to the "product clear" signal. The centrifuges (see Fig. 8) presented their own unique set of problems in that their swing arms had to reach the same alignment after each operating cycle so that the basket load/unload positions would remain the same for the robot. In addition, these arms had to remain steadfast during loading and unloading so that minor end effector contact would not cause holder movement, hence major misalignment problems. These problems were solved by employing solenoid-controlled locking pins for the drive shaft, activated by rpm indications at the desired low-speed set point.

Figure 7. The entire operation is enclosed in a wire cage with controlled access for added safety.

Figure 8. Closeup view of the wire basket being loaded into the centrifuge.

Separate microswitch signal inputs verified swing-arm alignment. A final series of added safety signal inputs were microswitch indicators for each centrifuge cover, arranged so that a "cover closed" signal was required to enable the centrifuge start-up sequence. The net effect of these controls was not only to ensure individual station integrity but also to ensure that station conditions were meeting specifications. For example, temperature sensors controlled the station timers so that temperature dips at the beginning of a warm-out sequence disabled the timers until the desired set point was reached. Varying ambient conditions, therefore, could have no detrimental effects on the quality of the process.

Typically, daily operations started with a conveyor in-feed signal to the robot that product was ready for processing. After receiving "next station clear" and "station status ready" signal inputs as well, the robot moved to acquire the product at the start point and transfer it to the programmed next station. Calling the next station station 2, the robot's next action was to move from station 2 when a "station 2 complete" signal was received and "station 3 clear" and "status ready" signals were received. If new product was ready on the in-feed conveyor, the robot repeated the station 2 cycle while waiting for station 3 completion. If new product was not introduced to the system, the robot would simply advance the existing product through the process in its normal programmed sequence. In the wait state the robot program would continually monitor for action/ready signals at either end of the process for subsequent action. Signal action conflicts, assuming station clear and ready signals for intervening steps, always favored action at the most advanced step in the process. When product was available for processing, action delays beyond programmed time limits generated a flashing yellow light for operator attention. The most common cause of this delay was lower-than-normal ambient temperatures, creating longer than usual tank-temperature recoveries on cold days. This unnecessary annoyance for the operator was eliminated with the use of an ambient-temperature sensor that automatically adjusted the timer warning signal to allow for longer delays on colder days.

Robots with Vision

Advances in robotics that combine precise movement with the ability to "see" are currently enabling manufacturing facilities to assess the quality of packaging components according to fine specifications. Such machine vision-equipped robots can also provide a manufacturing concern with savings in time and money. For example, the robotic workcell installed on a tube-filling production line efficiently and accurately inspects and places acceptable tubes in a vacuum-formed tray and has contributed to a significant increase in throughout per labor-hour, as well as a significant reduction in waste.

The process begins with prelabeled and capped tubes being automatically filled in a tube-filling machine. Once each tube is filled, the machine indexes to a registration mark scanning station where the tube is spun in front of a photodetector that locates an index mark on the label and stops the spinning action. The filling machine then indexes to the tube crimping and sealing location, where the bottom of the tube is sealed. After the next index, the filled and sealed tube is discharged onto a conveyor that carries two tubes side-by-side under the robotic work-cell vision system. The vision system automatically locates each tube and rejects those not conforming to exact label position and placement specifications. Properly labeled tubes are tracked as they move down the conveyor; the robot is then guided by the vision system to pick up and place the finished tubes in the tray.

A second robotic workcell that performs similar label inspection and packaging functions was also installed, although in this application the final package is an individual, clear plastic blister with a decorated hang card rather than a tray containing several filled tubes. For both installations, human workers continued to play a key role in developing the application and in integrating the robots into the workplace. They were also consulted and encouraged to provide information and insights, raise questions, and help find answers to assist in the application. Packaging-line associates and specialist, in fact, formed the core of an acceptance team that tested the new technology before it was shipped from the supplier to the filling-line location. On the basis of their suggestions, several safety and mechanical improvements were made before installation.

Conclusion

The commonality among all of these robotic packaging applications is that each represents some degree of compromise between weight, speed, and accuracy. The greater the difference between robot design limits and application requirements, the greater the chance for long-range application success. As existing factory layouts based on older packaging technology reach maturity, these and other new designs will make the robot user's task even easier. For instance, modern robotic workcells that combine the precise motion control of robots with the sensory precision of machine-vision systems are increasingly utilized to save labor and reduce waste on traditional packaging lines. However, compromises must continue to be made between the current capabilities of robots and the factory environment as it currently exists.

BIBLIOGRAPHY

1. *Webster's New Collegiate Dictionary,* G&C Merriam Co., Springfield, MA, 1976.
2. J. F. Engelberger, *Robotics in Practice: Management and Applications of Industrial Robots,* AMACOM, New York, 1980, p. 249.

3. J. E. Cunningham, *Handbook of Remote Control and Automation Techniques,* Tab Books, Blue Ridge Summit, PA, 1978, pp. 28–29.
4. G. G. Dodd and L. Rossol, eds., *Computer Vision and Sensor-Based Robots,* Plenum Press, New York, 1979.
5. Ref. 2, p. 42.
6. U.S. Pat. 4,135,204 (Jan. 16, 1979), to R. E. Davis, Jr. and co-workers (to Chesebrough-Pond's Inc.); U.S. Pat. 4,344,146 (Aug. 10, 1982), to R. E. Davis, Jr. and co-workers (to Chesebrough-Pond's Inc.); U.S. Pat. 4,445,185 (April 24, 1984), to R. E. Davis, Jr. co-workers (to Chesebrough-Pond's Inc.).
7. R. E. Davis, Jr., "Creating the Proper Environment for the Introduction of Non-Conventional Technology into Typical Manufacturing Operations," in *Proceedings of the Western Plant Engineering Conference,* Anaheim, CA, 1983, Section D-3, pp. 1–14.
8. J. M. Shepard, *Automation and Alienation: A Study of Office and Factory Workers,* MIT Press, Cambridge, MA, 1971, pp. 39–40.
9. Ref. 3, pp. 161–163.

RAY E. DAVIS, JR.
J. KEITH UNGER
Chesebrough-Pond's USA
Greenwich, Connecticut

ROLL HANDLING

In web processes, such as slitting and rewinding, coating and laminating, printing, and other applications, rolls must be loaded from the rewind station after the process. Because of weight and size, handling of rolls is done by general-purpose equipment such as cranes and forklifts. For greater productivity, roll-handling equipment is engineered to suit the characteristics of the web-process equipment. An important fringe benefit of engineered handling systems is the reduction of safety hazards associated with general-purpose equipment.

Figure 1. A duplex slitter/rewinder with roll-handling system. Turret carriages are in full-roll downward position at slitter/rewind station. New mandrels are in an upward position awaiting turret index.

Slitter/rewinder overview. A slitter/rewinder (see Slitting-rewinding machinery) converts master rolls into high-quality, narrow-width, rewound rolls that may or may not be the final product. Variables in the web characteristics and in the final form of the product determine which of the several types of slitting and rewinding techniques is used. The slitter/rewinder machine is typically composed of unwind, slitting, and rewind stations. A turret used for unwinding allows the loading of new master rolls while the machine is slitting and rewinding. Continuous unwinding can be attained by automatic splicing equipment.

The slitting methods used depend on the material characteristics and quality of the slit edge desired. Razor slitting is used primarily on films and score slitting on many types of fabrics, nonwovens, and textiles. Shear slitting is excellent for most materials, but requires a relatively high initial invest-

Figure 2. A slitter/rewinder roll-handling system with a shaft-type unwind stand. This two-motor drive system has an adjustable vacuum winding capability and an overhead mandrel lift/transfer mechanism.

Figure 3. A surface center slitter/rewinder with hydraulically pivoting rewind arms, preset constant surfacing pressing in the center surface rewinding mode, and an automatic minimum gap for center winding.

ment and operator expertise. Locked-core rewinding is successful on good caliper webs and pressure-sensitive tapes with relatively small roll diameters. Differential rewinding is required for off-caliper materials and large-diameter rolls.

An effective method of rewinding is to alternate the slit webs on two rewinding shafts. This method, termed duplex rewinding, is particularly suited for narrow-width slit material. Winding on a single shaft, called simplex winding, is generally effective for slit and unslit paper, nonwovens, and textiles.

Shaft extraction/insertion system. On duplex and simplex slitter/rewinders, webs are slit and rewound on cores supported by rewind mandrels. Rewind mandrels may be extremely heavy, ie, several thousand pounds (kilograms), or cumbersome to handle manually. Each machine cycle requires that a rewind mandrel be loaded up with cores, put into the rewind station, removed from the rewind station, and then extracted from the rewound rolls. This might take place many times a day, and an operator might have to lift and maneuver many tons of rewind mandrels per shift. Apart from the burden on the operator, manual shaft handling hinders production because the cycle time for loading and unloading is relatively long.

The use of a shaft extraction/insertion system minimizes cycle time as well as operator fatigue. With this concept, the finished rolls and rewind mandrels are transferred to a lift table where shaft extrators pull the rewind mandrel from the rolls. The shaft is supported in the extracted position until the rolls are removed from the extractor table. Fixtures called core boxes are sometimes engineered into the extractor table to locate cores in their proper positions. The extractor inserts the shaft back into the cores. The cored shafts are then brought back into the rewind stations.

Shaft extractors are easily interfaced in close proximity to turret winders, unwinders, and other simplex winders, which often require the handling of steel shafts 6–10 in. (15–25 cm) in diameter by several hundred inches (several meters) long.

Handling systems for duplex rewinders. A duplex slitter/rewinder (see Fig. 1), with the rewind stations located above the main machine, interfaces well with an overhead roll-handling system. After the rewind cycle, the shafts with the rewound rolls are lifted out and the new shafts with cores must be inserted into the machine.

A typical system (see Fig. 2) may have an overhead-powered carriage that supports dual vertically adjustable shaft-clamp assemblies. With two clamps for each vertical lifting mechanism, new shafts with cores can be loaded into the machine immediately after the shafts with the finished rolls are removed. In other words, a machine with two rewind stations has four shafts circulating through the system simultaneously. As the new rewind cycle begins, the shafts with rolls are conveyed overhead to the shaft extractors, where the slit and rewound rolls become free of the shafts and may be upended on pallets for conveyor or forklift removal into storage.

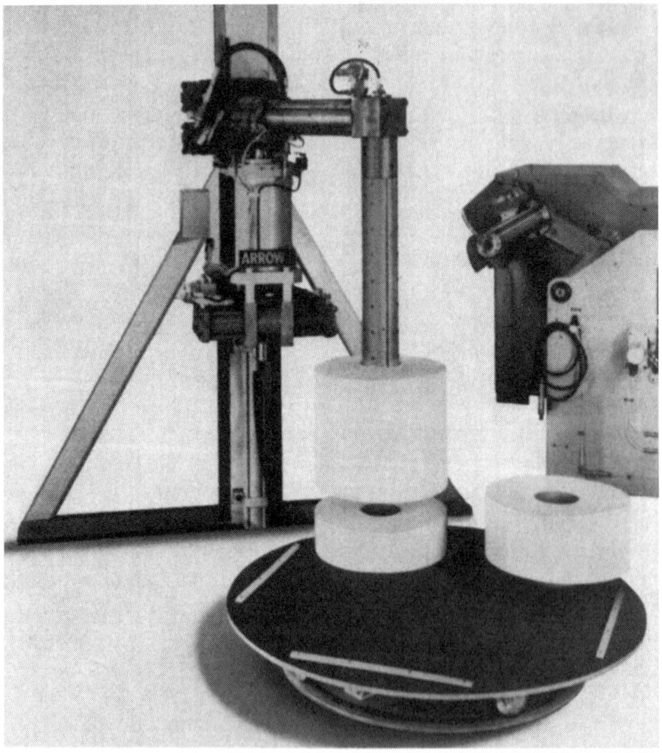

Figure 4. A roll unloading and stacking robot with hydraulically controlled cantilever mounted boom with 90° rotational capability on vertical and horizontal axes for pallet offload.

Figure 5. A fully automatic pressure sensitive tape turret slitter/rewinder with automatic core cutter and loader.

Floor loading and unloading. Floor loading and unloading features are available as an integral part of many unwinds and simplex rewinds. These units use hydraulically pivoting arms to raise and lower rolls to the floor. The need for auxiliary equipment is eliminated. Lift tables, shaft extractors, and upenders are designed to work with simplex (single-position) rewinders that produce large-diameter rolls of paper, laminates, and nonwovens (see Fig. 3). Driven "V"-trough conveyors can then move the rolls from the rewinders to weighing stations, stretch wrapping, and storage conveyors.

Robotic roll handling. An example of robotics in special-purpose roll handling is shown in Figure 4. A simplex slitter/rewinder with an automatic roll pusher pushes the slit and wound rolls onto the robot shaft. The robot shaft then expands to lock the cores to the shaft. The robot is programmed to shift the axes of the rolls 90° and to automatically position them over a pallet. A rotary table supports the pallet and will index every 90° to allow unloading of finished stacks of rolls in the four quadrants of the pallet.

Total automation. The ultimate in roll handling is shown in Figure 5. This slitter/rewinder for duct tape automatically cuts cores, loads cores at their proper location on dual rewind shafts, feeds the loaded shafts into rewind stations, winds the slit tapes, unloads the rewound rolls from the machine, and feeds the empty shafts back in for core loading. This completely "hands-off" machine is microprocessor-controlled. Cycle times are better than 30 s.

BIBLIOGRAPHY

"Roll Handling" in the *Wiley Encyclopedia of Packaging Technology,* 1st ed., by R. E. Mastriani, Arrow Converting Equipment, Inc., pp 571–573.

ROTATIONAL MOLDING

Reusable, large industrial containers can be made by rotational molding. Rotational molding generally is not thought of as a technique for making packaging. However, some manufacturers use this process, which can make complex, hollow, seamless products of all sizes and shapes, to produce large bulk containers for shipment of powders and liquids, 5–55-gal drums, trash and recycling containers, insulated food-product containers, and double-wall shipping trays for equipment components.

Advantages

Compared with blow molding or thermoforming, rotational molding offers significant advantages:

- Mold costs are low and comparable to those for blow molding and thermoforming.
- The process is economical for short production runs.
- Secondary tooling is minimal or not needed.
- There can be little or no scrap.

Numerous design features are possible; for example, double-walled parts, good surface detail, intricate contours and undercuts, stress-free parts, minimal cross-sectional deformation and warping, and molded-in inserts can be used.

Polyethylenes Most Widely Used

Generally, powdered resins are used in rotational molding. However, a few liquid resins are available and some high-flow resins, such as nylons, have been used in small pellet form.

While many thermoplastics can be rotationally molded, only a few actually are used to make commercial products. Polyethylenes—low-density (LDPE), linear low-density (LLDPE), high-density (HDPE), and copolymers—are the most widely used materials (see also Thermosetting polymers).

LLDPE rotational molding grades typically have densities ranging from 0.935 to 0.940 g/mL and melt index (MI) of 3–10 g/10 min. LLDPE rotational molding resins offer good stiffness and low-temperature impact strength, excellent environmental stress-crack resistance (ESCR), and warp resistance.

LDPE rotational molding grades generally have densities of 0.915–0.920 g/mL and MI of 10–25 g/10 min. LDPE rotational molding resins offer good impact strength, low shrinkage, good warp resistance, and flexibility.

HDPE rotational molding grades have densities of 0.942–0.950 g/mL and MI of 2–8 g/10 min. HDPE rotational molding resins offer high stiffness, high impact strength, and excellent chemical resistance.

Ethylene–vinyl acetate (EVA) copolymers for rotational molding have densities of 0.925–0.945 and MI of 10–25 g/10 min. EVAs offer good low-temperature impact strength and flexibility.

Postconsumer polyethylene resins (PCR) are made from HDPE blow-molded containers. HDPE PCRs have much lower melt indices than the MI of typical rotational molding resins. Therefore, for rotational molding, a HDPE PCR is blended with a virgin resin. HDPE PCR content in blends can range within 10–25%. The blends can be made by either melt compounding or dry blending PCR/virgin blends.

How the PCR/virgin resin blend is prepared affects the inner surfaces of rotationally molded parts. Parts rotationally molded from dry blends can have very rough inner surfaces. Parts made from HDPE PCR/virgin resin extrusion blends exhibit smooth inner surfaces.

The color of parts rotationally molded from HDPE PCR/virgin resin blends will be influenced by the type of HDPE PCR in the blend. Since some HDPE PCR are green (eg, PCR based on HDPE copolymers), rotational molding resins made with these resins have a green tint. Other HDPE PCR are gray and have black specks. Thus, parts rotationally molded from blends made with these HDPE PCR have a dirty white color.

Compared to the virgin LLDPE resin, the toughness of HDPE PCR/virgin resin blends is much less, and it decreases as HDPE PCR content increases.

The environmental stress-crack resistance (ESCR) of HDPE PCR/virgin resin blends is considerably lower than that of the virgin polyolefin resins.

Crosslinked polyethylene rotational moldings can be made by using materials that contain a chemical crosslinking agent. Crosslinking improves physical properties, in particular impact strength, dimensional stability, creep resistance, and environmental stress crack resistance (ESCR).

Polyethylene rotational moldings also can be crosslinked using electron-beam treatment. Electron-beam processing allows the degree of crosslinking to be controlled; therefore, the level of property improvements can be controlled. Also, parts can be recycled before they are subjected to the EB treatment. Thus, scrap losses are reduced.

Polypropylene rotational molding resins provide higher heat resistance (containers can be autoclaved) and have higher stiffness than polyethylenes. Reactor-produced polypropylene copolymers (TPOs), ie, copolymers produced in a reactor, have not been subjected to the mechanical and thermal degradation associated with melt blending of rubber compounds into a polypropylene base. Other advantages of the reactor-produced TPOs include their inherent temperature and chemical resistance, low specific gravity, and high flexibility. Also, unlike rubber blends, which can degrade or crosslink, the reactor-produced TPOs are readily rotationally molded.

Nylon rotational molding resins, both liquid and pellet, offer good heat resistance, toughness, good wear and abrasion resistance, high strength and stiffness, and good chemical resistance.

Polycarbonate rotational molding resins offer high heat resistance, good impact strength, and clarity.

Vinyl rotational molding resins, both powder and liquid, are available. Vinyl plastisols are liquid rotational molding resins that offer a wide range of stiffness, from very soft (low-durometer) to very rigid (high durometer).

A Simple, but Versatile Process

There are six basic steps to the rotational molding process:

1. A predetermined amount of resin is placed in the bottom half of a two-piece mold.
2. The mold halves are closed.
3. The mold is placed in a heated oven and rotated biaxially.
4. The resin melts against the inside surface of the mold, fuses, and then densifies into the shape of the mold cavity.
5. The mold is removed from the oven and placed in a cooling chamber. A combination of air and water is used to slowly cool the mold.
6. The mold is transferred from the cooling chamber and opened, and then the finished part is removed.

There are many variations of the basic rotational molding process.

Generally, the walls of rotational moldings are made of one material. However, multilayer rotational moldings can be made. For example, there can be a foam layer, a solid layer, a layer of recycled material, a crosslinked layer, and different-colored layers.

The standard technique for making multilayer rotational moldings is to use a "dump box." After the first layer is molded, the dump box, which is attached to the mold, is opened to allow the material for the second layer to enter. The dump-box cover is closed, and the second layer is molded.

Foam filling, using rigid polyurethane foam, injected into the hollow space between the walls of rotational moldings adds stiffness and structural strength.

Integral skin foam moldings also are possible with resins that contain foaming agents. These one-step foaming resins expand inside the mold, and the material directly against the mold surface forms a dense skin. This approach to multilayer foam molding is faster than other techniques, and tooling costs are less.

Molded-in decoration also is possible with rotational molding. By embedding a graphic into the surface of a rotational molding, the graphic becomes permanent and will not peel, and cannot be scratched or rubbed off. Molded-in graphics eliminate surface pretreating required with other decorating methods. Also, no paints or painting equipment are required.

Precolored polyethylene powders save rotational molders from the hassle of dry-blending pigments with natural-colored resin powder.

Rotolining is a process that chemically bonds a reactive polymer to the inner surface of hollow metal products. The metal product is mounted on the rotational molding arm as opposed to a mold.

Several Types of Equipment

There are many different types of rotational molding equipment.

Open-flame and slush molding machines. These machines (Fig. 1), which are the oldest type of equipment, usually rotate on only one axis. This equipment is used mostly to produce open containers.

Carousel machines. These machines (Fig. 2) are the most common type. Carousels consist of a heating station or oven, a cooling station or enclosed chamber, and a loading and unloading station. The carousel can have three to six spindles or arms for mounting molds. Most carousels have the freedom to rotate in a complete circle.

Clamshell machines. These machines (Fig. 3) have only one arm, and the heating, cooling, and loading/unloading stations are all in the same location.

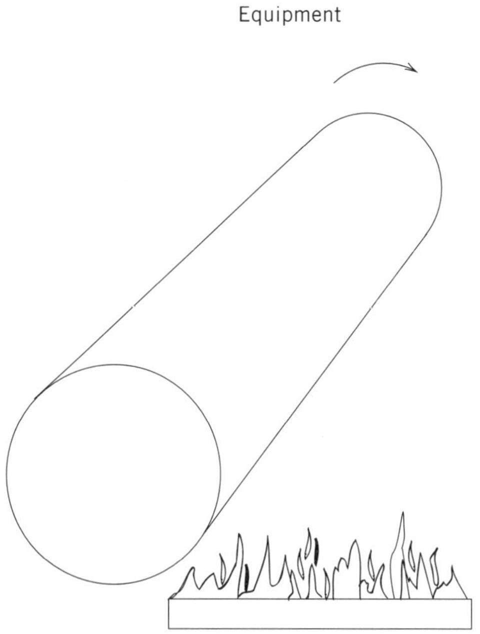

Figure 1. Rotational-molding equipment.

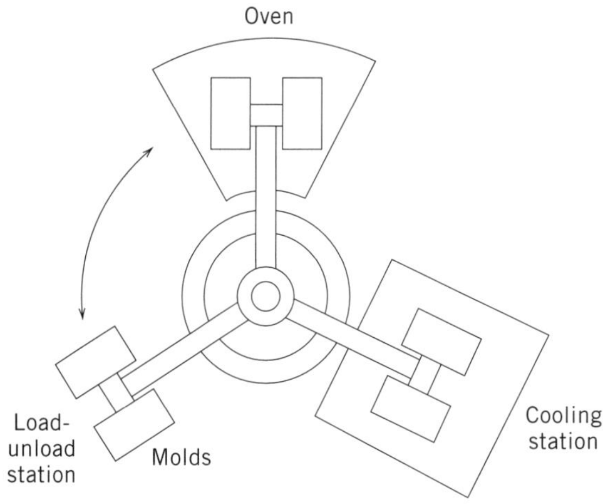

Figure 2. Carousel/turret machines have a center pivot with three, four, or five arms that hold molds. The arms index simultaneously from station to station, ie, load/unload station, to oven and to cooler.

Rock-and-roll machines. These machines (Fig. 4), which rotate on one axis and tilt on another, are used for long items.

Shuttle machines. These machines (Fig. 5) generally are used to make very large parts. A frame for holding one mold is mounted on a movable bed. The bed is on a track that allows the mold and the bed to move into and out of the oven. After the heating cycle is complete, the mold is moved into an open cooling station. A duplicate bed with a mold is then sent into the oven from the opposite end.

Recent Trends

Considerable progress has been made in recent years to improve processing controls. Microprocessors, which indicate processing set points and conditions, greatly improve precision. Mold-cycle data for the oven and cooling chambers can be stored for future production of parts. Cycle time, oven temperature, major and minor axis speeds, and fan and water-spray times are other functions under computer control. New temperature-monitoring systems are also available.

A key benefit of rotational molding is the low relative cost

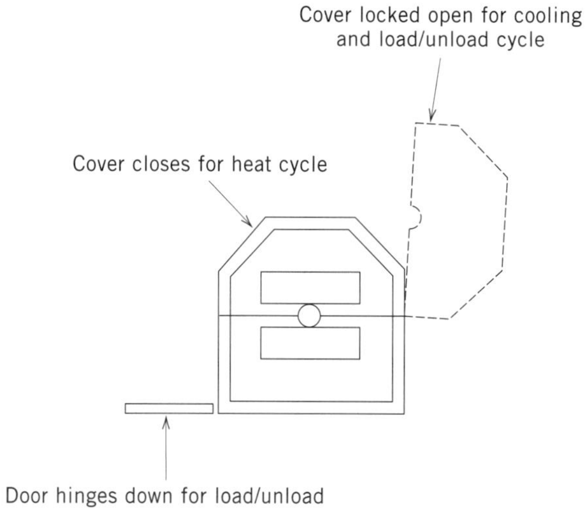

Figure 3. Single-station machines, also called *clamshell machines,* have an oven that has a hinged cover and a hinged front panel. The mold rotation arm can swing into and out of the oven. The cover and front panel are closed during heating and are opened for part cooling, removal, and reloading of the mold.

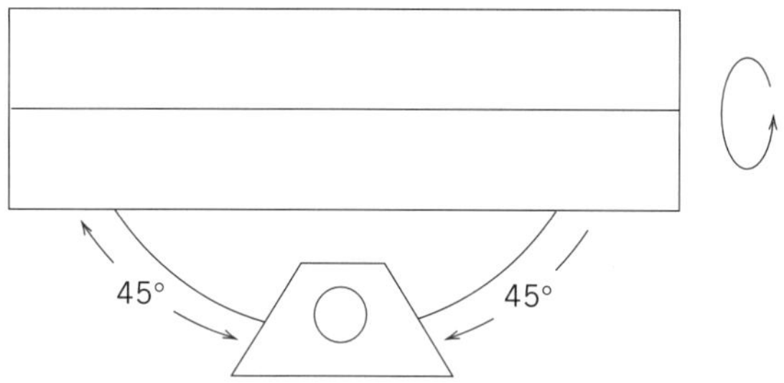

Figure 4. Rotational moldings with very long length:diameter (L/D) ratios can be made on equipment called "rock-and-roll machines." The mold is rocked back and forth on a stationary, horizontal axis while it is rotated about a moving axis that is perpendicular to the rotating axis.

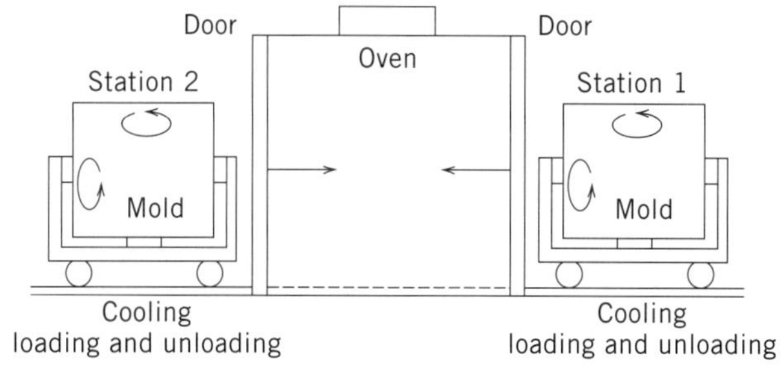

Figure 5. Shuttle machines either move the mold, along an oval or straight track, from the load/unload station, to the oven, and to the cooling station.

of its tooling. Cast aluminum molds are most frequently used, especially for small to medium-sized parts. Cast aluminum has good heat transfer and is cost-effective when several molds for the same part are required.

Sheet-metal molds are normally used for larger parts. These molds are easy to make since sections can be welded together. Other molds, such as electroformed and vapor-formed nickel molds, yield an end product with fine detail. Cost for these is typically higher.

To assist in part removal, most rotational molds require a mold-release agent, which is either baked or wiped on. Many molds now use a fluoropolymer coating to eliminate the use of mold release. Also, waterborne release agents, as a result of the U.S. Clean Air Act of 1993, are being used in place of CFC- and solvent-borne release agents.

Future Trends

The ability of rotational molding to economically make large, complex, hollow, seamless products is becoming more interesting to the industrial packaging industry. Other attractive advantages of this process are its ability to economically make low-volume production runs, the wide range of resins that are available, and its multilayer capabilities.

The rotational molding industry is growing very rapidly. At present, there are well over 200 rotational molding plants in the United States.

Rotational molding technology is constantly improving. Many new developments in rotational molding materials are announced each year.

A great deal of technical information is readily available about rotational molding from the Association of Rotational Molders. The industry organization offers many technical brochures and copies of recent technical papers.

Phil Dodge
Quantum Chemical Company
Cincinnati, Ohio

S

SEALING, HEAT

Most of the materials used in flexible packaging use thermoplastics in their construction. When heat is applied, the thermoplastics melt and act like a glue in effecting a seal. This article describes how heat is applied in both flexible and semirigid packaging in order to achieve acceptable seals.

Flexible-packaging materials fall into two categories: laminated packaging materials (see Laminating; Multilayer flexible packaging) and unsupported films (see Films, plastic). Unsupported films consist of just one or more thermoplastic materials; laminations consist of nonthermoplastic materials with thermoplastic layers for sealing purposes. Even though unsupported films with more than one layer can be laminations, the term *lamination* generally refers to constructions with a portion that does not melt during sealing.

When laminated materials are to be sealed, the outer face of the material is generally nonthermoplastic. This makes it possible to apply a heated bar directly to the outer face in order to get the heat to the sealing interfaces and join the two package members. The outer face may be a thermoplastic if its melting point is sufficiently higher than that of the sealant so that seals can be made with a hot bar.

Heat cannot usually be applied directly to unsupported films because they melt and stick to the surface of the sealing bar. The seal area is destroyed in the process. For this reason, such materials are sealed by impulse sealing (see below). If the package members are too thick and too insulating, other means must be used to get the heat to the sealing interfaces. Since bar sealing uses the least expensive equipment, it is more widely used for sealing than any other method. The other methods described below are used in applications where bar sealing is not suitable.

When sealing thin materials together, it is generally sufficent to introduce heat from one side of the construction. When using thicker materials, or if higher speeds are required with thinner materials, heat may be introduced from both sides. The fundamental principles in heat sealing are to provide heat at the interfaces, pressure to bring them intimately in contact, and complete a weld, all within an acceptable time period. The only exception to this is in radiant sealing, which relies on film orientation and surface tension as a substitute for applied pressure. All of the important sealing methods are listed in Table 1.

Bar sealing. Bar sealing is the most widely used method for sealing. It is used both to make and seal pouches. It is also used in most form/fill/seal equipment as well. When very long seals are to be made, it is essential that the seal bars be designed to avoid any deflection in order to assure uniform pressure throughout the length of the seal. Since it is important to avoid wrinkles in seals, means are frequently provided to stretch out the seal area of the packaging materials in order to assure that they are flat when sealed. Another approach is to have mating serrations incorporated into the seal bar faces, which will stretch the packaging materials, and hopefully, remove any wrinkles. Care must be taken to see that the serrations do not puncture the films during sealing. Serrated bars are used where good mechanical strength is required and some tiny leaks in the seal can be tolerated.

When hermetic and/or liquid-resistant seals are desired, they are best made by a flat-faced heater bar, opposed by a bar that has a resilient surface. Silicone rubber is generally used for this purpose. It is best if the resilient surface is curved when viewed from the end of the seal bar. When the bars come together, they first create a line of pressure throughout the length of the seal bar. As the bars close further, this line broadens into a band. The pressure along the initial line is at a maximum and at a minimum along the edges of this band. This assures that the optimum sealing pressures are present somewhere in between. In the event that drops of liquid are in the seal area, this pressure profile expanding from a line will tend to push these droplets out of the seal area. If the droplets are water and they are not pushed out of the seal area, they become steam and rupture the seal.

As an alternative to curving the section of the resilient bar, the heated metal bar can be curved. This is generally not done since it is more difficult to maintain a curved section than a flat section on a metal seal bar. It is also important that the edges of metal seal bars be gently rounded to avoid puncturing the packaging materials. Figures 1 and 2 show the best constructions used for straight seal bars.

Bar sealing is also used for applying covers to cups and trays. The upper heated bar is shaped to match the shape of the rim of the container being covered. The bottom bar is shaped to fit under the rim and support the container. In order to assure good seals, it is essential that uniform seal pressure be effected around the entire rim of the container. Factors contributing to nonuniform pressure are variations in the thickness of the rim of the container and warping of the opposed sealing bars. Using a resilient sealing surface under the rim of the container generally corrects these deficiencies. Another approach is to design the rim of the container to be curved upward in section. When the seal is made, this curved rim deflects much like a resilient backup member.

For some applications, continuous seals can be made by passing thin materials between heated rollers. Unfortunately,

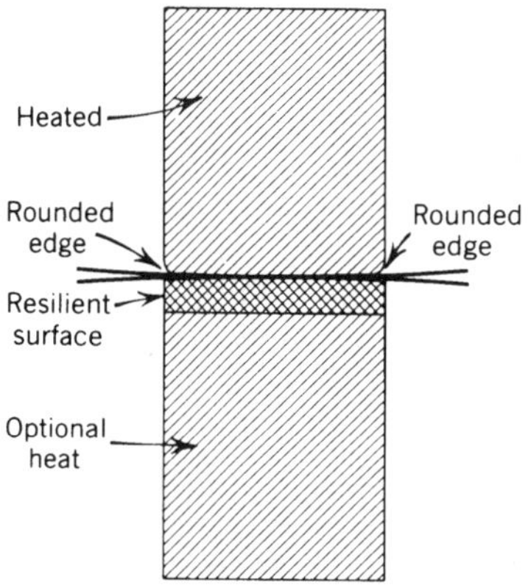

Figure 1. Bar sealer.

Table 1. Sealing Methods

Bar	*Concept:* Jaw sealer with one or two heated opposed bars. *Applications:* Manufacture and closing of laminated pouches, cup lidding, form/fill/seal packaging (eg, potato-chip bags).
Band	*Concept:* Two moving bands backed by heated and cooled metal jaws. *Applications:* Sealing filled pouches, including those made from unsupported materials.
Impulse	*Concept:* Jaw sealer backed with resilient silicone rubber. Electric current flows through Nichrome ribbon stretched over one or both surfaces and covered with high-temperature release film or fabric. *Applications:* Sealing tacky materials, unsupported thermoplastic films (eg, frozen vegetable bags).
Wire or knife	*Concept:* Hot wire or knife seals and cuts film. *Applications:* Bagmaking and bagclosing; overwrap for toys, records.
Ultrasonic	*Concept:* Tooling hammers or rubs materials together at high frequency, generating heat for sealing. *Applications:* Sealing biaxially oriented films, thick webs, aluminum foil, rigid container components.
Friction	*Concept:* Frictional heat generated by rubbing components together generates heat for sealing. *Applications:* Assembly of round containers, sealing ends of strapping.
Gas	*Concept:* Hot air or gas flame applied to both surfaces and removed; molten surfaces are then pressed together. *Applications:* Manufacture and closing of polyethylene-coated paperboard milk containers.
Contact	*Concept:* Plate is placed between surfaces to be sealed, withdrawn, and the molten surfaces pressed together. *Applications:* Sealing ends of strapping, tubing.
Hot-melt	*Concept:* Continuous strip or dots of hot, molten thermoplastic sealant are applied between two surfaces just before pressing them together. The hot melt contains sufficient heat to effect seal, and surfaces absorb heat to cool melt rapidly. *Applications:* Paperboard containers, peelable seals, case packers.
Pneumatic	*Concept:* Heated film is sealed to another surface by application of air pressure. *Applications:* Skin packaging.
Dielectric	*Concept:* High-frequency electric field melts materials held under pressure. *Applications:* Unsupported PVC, PVC-coated paper (not for polyolefins).
Magnetic	*Concept:* Gasket with iron-containing compound pressed between surfaces, assembly placed in magnetic field. *Applications:* Heavy-gauge polyolefins (eg, cap liners or lids).
Induction	*Concept:* Alternating current is induced in metallic foil, usually aluminum, which heats and melts surfaces pressed against the foil. *Applications:* Tamper-evident closure liners; lids.
Radiant	*Concept:* Infrared radiation sealing without pressure. *Applications:* Sealing uncoated highly oriented films and nonwovens, including polyester, nylon, polyolefins.
Solvent	*Concept:* Solvent liquefies surfaces that are then pressed together. *Applications:* Sealing configurations where heat may degrade the thermoplastic or is not practical to apply.

the dwell time for actual sealing under pressure between rotating rollers is extremely short. Good seals can be effected only if the rollers are moving very slowly or if the materials are adequately preheated before passing through the heated rollers.

Band sealing. In band sealing, a pouch mouth is introduced between two moving bands, which are pressed together by heated bars (see Fig. 3). The heat passes through the bands and into the pouch material, softening it for sealing. As the pouch continues along between the bands, the bands are next pressed together by chilled bars that withdraw heat from the pouch seal through the bands. The bands then progress to release the pouch. Band sealing is fast and widely used for closing pouches filled with product. It is important that the pouch mouth be flattened before entering the bands if wrinkles in the seals are to be avoided. Band sealing provides a continuous method for sealing, avoiding the problems found in sealing between rotating hot bars.

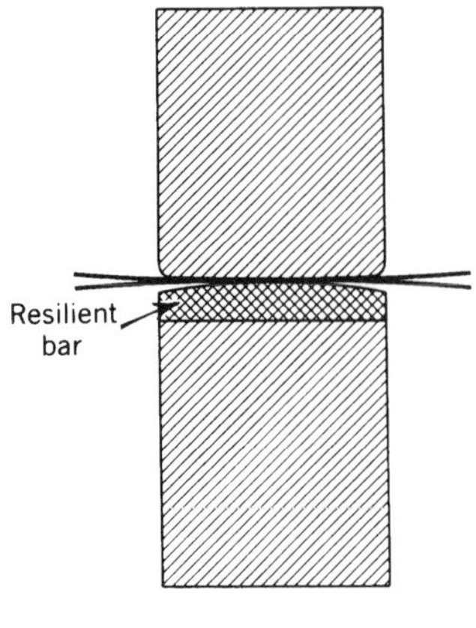

Figure 2. Rounded resilient bar.

Impulse sealing. Impulse sealers have the same general configuration and mechanical construction used for bar sealers; the difference lies in the sealing jaws (see Fig. 4). Each of the opposed jaws is generally covered with a resilient surface, such as silicone rubber. A taut Nichrome ribbon is then laid over the resilient jaw and covered with an electrically insulating layer of thin heat-resistant material, such as silicone-rubber-coated fiberglass, Teflon-coated fiberglass, or Teflon-coated Kapton (DuPont). A pouch mouth is placed between the jaws, and the jaws are closed. An electric current passes through the Nichrome ribbon for a brief period of time and is then turned off. Heat is withdrawn from the pouch mouth and the ribbon through the resilient jaw surfaces, and

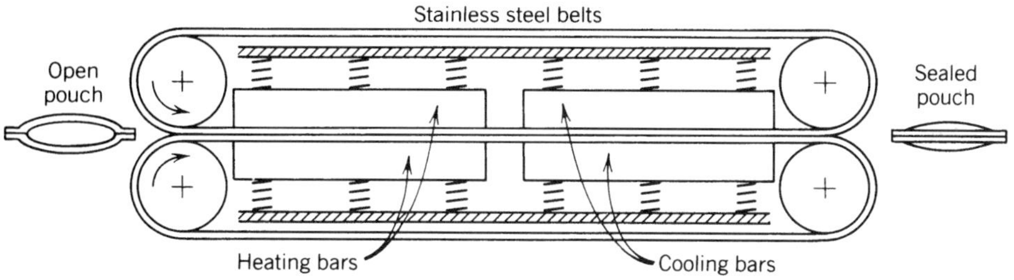

Figure 3. Band sealer.

through the jaws. The jaws are then opened and the sealed pouch removed.

The advantage of impulse sealing over bar sealing is that the seal is cooled to achieve adequate strength before the jaws are opened. This permits the sealing of constructions that are insufficiently tacky when hot to hold together when a bar sealer is opened. It also permits the sealing of unsupported thermoplastic materials, which would stick to heated bars and fall apart when heated bars are opened. The material that covers the Nichrome ribbon has release properties, so that it does not stick to unsupported thermoplastics, or to many of the coatings that are used on the outside of packaging laminations. Such coatings include PVDC, ink, and varnish. Shaped impulse-sealing ribbons are used for securing lids to cups and trays. The dominant drawback of impulse sealing is high maintenance cost. The impulse ribbons slowly deteriorate and the release coverings over the ribbons degrade requiring frequent replacement. On a production basis, impulse sealing should be used only where bar sealing does not do the job properly.

Hot-wire or knife sealing. Hot-wire or knife sealing is adaptable to a very high-speed operation and is used for both sealing and cutting apart polyethylene bags on bagmaking equipment (see Bags, plastic). Unsupported films, when trimmed by this method, tend to form a strong bead in their seal areas due to surface tension and orientation. This method is also used to a limited degree with laminated constructions. Hot-wire cutoff is also used in film-dispensing equipment and with L sealers.

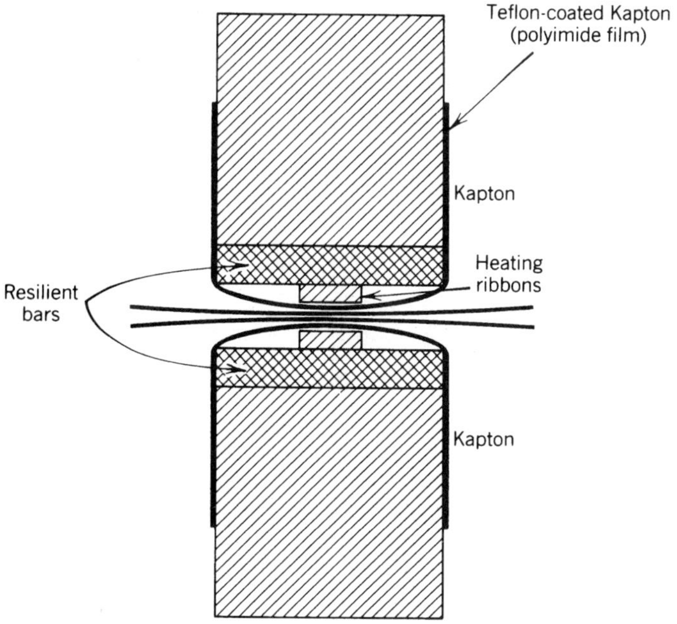

Figure 4. Impulse sealer.

Ultrasonic sealing. In ultrasonic sealing, the sealing heat is produced by mechanically hammering or rubbing the packaging materials together at a high frequency. It is valuable with highly oriented films because sufficient heat is generated to melt the interface, but not enough to heat the rest of the material to the point of degradation. It is also very useful to seal materials that are much too thick to permit heat transfer through them for sealing. It is the only method used for welding aluminum foil in its production.

Friction sealing. In friction sealing, the top and bottom halves of cylindrical containers may be easily joined. The bottom member is held against rotation by a friction brake; the top member is pressed against the bottom member and rotated. The heat generated by friction at the interface between the two halves melts their surfaces and the viscous interface causes the bottom half to rotate in unison with the upper half. This effects the seal. Thermoplastics with surfaces that become slippery when heated are best suited to friction welding; those that become doughlike present problems. Sealing is also effected between the ends of package strapping, which are rubbed together at high speed until the interface melts, causing them to weld. Ultrasonic welding is actually a method of friction welding.

Gas sealing. In gas sealing, where paperboard and thicker thermoplastic materials offer too much resistance to the passage of sealing heat, heat is applied directly to the surfaces to be sealed by a gas. The gas may be either hot-air or a gas flame. The most important application of this method is the manufacture of paper containers for dairy products. A gas flame is played on the surfaces about to be sealed, after which they are clamped together between chilled bars. Sufficient sealant must be used on the surfaces in order to fill all the fissures formed at the joints in the paperboard.

Contact sealing. In contact sealing, an alternative to gas, the sealing interfaces are each contacted by a heated plate before being pressed together. This is also used for strapping as an alternative to friction sealing.

Hot-melt sealing. Hot-melt sealing is effected by depositing either a continuous strip, or dots, of a thermoplastic material on a packaging member, just before the next member is pressed against it. It is used as an alternative to glue, since the packaging members quickly remove heat from the melt, making it tacky enough to hold the members together quickly even though springback may be tending to open them up (in sealing the covers of a shipping container). With glue, the members must be held together until enough of the glue's solvent is absorbed by the packaging material to make it tacky. Hot melts are also used where applied heat might damage a packaging member or where peelability is desired (see Adhesives; Waxes).

Pneumatic sealing. Pneumatic sealing is used for skin packaging (see Carded packaging), where pressure from the atmosphere pushes a hot, tacky film into close conformity with an object and seals the film to a substrate on which the object is placed. It is also used in hermetic skin packaging for fresh red meat.

Dielectric sealing. In dielectric sealing, used principally with PVC materials, a high-frequency electric field generates heat for sealing. The members are pressed together between a cold bar, generally made of brass, and a cold flat metal surface. The field is generated between the bar and the lower surface, and these members are kept cold to withdraw any residual heat from the materials that have been sealed. The polyolefins, unfortunately, do not respond adequately to a dielectric field.

Magnetic sealing. Magnetic sealing is used where heavy polyolefin package members are to be sealed. A gasket, shaped to fit between the sealing faces of the two package members, is made from the same thermoplastic as the package, but has milled into it an iron-containing compound that has a high hysteresis loss. The package portions are pressed together with the gasket between their sealing surfaces and placed in a magnetic field oscillating at a frequency high enough to melt the gasket. This, in turn, melts the sealing surfaces and causes them to weld together. As an alternative to a gasket, an iron-containing material can be preapplied to one or both of the sealing surfaces, producing the same result.

Induction sealing. In induction sealing an alternating electric field heats up a metal, eg, aluminum foil, placed between two members to be sealed. Its most common use is to produce a tamper-evident seal over the top of a bottle. The foil member is generally incorporated under the closure liner (see Closure liners). The assembly is placed in the alternating field, and a current is induced in the foil; this current heats the foil. Foil sticks to polyethylene bottles; with bottles made from other materials, a suitable coating is put on the foil, which will stick the foil to the bottle when the foil is heated.

Radiant sealing. Radiant sealing makes it possible to seal many materials used in packaging that have hitherto been considered either difficult or impossible to seal without coatings (see Fig. 5). Polyester film (see Film, oriented polyester) long used in packaging constructions and many nonwoven (see Nonwovens) materials can now be sealed by this method. These include Tyvek (DuPont), a nonwoven polyolefin used in packaging and for protective garments; Reemay (DuPont), a nonwoven polyester, used in filters; Cerex (Monsanto), a nonwoven nylon, used as a carpet backing and in filters; and nonwovens made from polypropylene and other base materials. This method is used for sealing other oriented films, such as polypropylene and polyethylene. OPET pouches are commercially made for document preservation. OPET is the only material approved by the Library of Congress for use in contact with valuable documents. In addition to making pouches, this sealing method can also be used to produce continuous seals to edge-join rolls of materials at high speed.

Solvent sealing. Solvent sealing is used to join together package configurations that do not lend themselves to heat sealing. It is also used for joining materials that are either not susceptible to heat sealing, or may be damaged by the application of heat (eg, rigid containers made of acrylics or polystyrene).

Selecting a Sealing Method

For laminations, bar sealing is the least expensive of all sealing methods and it should be tried first before considering other methods. If sticking to or contamination of the seal bar is a problem, the following should be tried out before considering impulse sealing:

1. Coating or impregnating the seal bar with Teflon.
2. Periodically wiping the seal bar with a silicone grease.
3. Mounting a release material, such as Kapton (DuPont), between the seal bar and the packaging material.

If all of these fail, impulse sealing should be considered next. For unsupported films, hot-wire, knife, or radiant sealing should be investigated. Although the equipment is more expensive, ultrasonic should also be considered. There are many applications where ultrasonic will be the method of choice when other methods fail. For heavier packaging materials, friction, contact, or magnetic sealing should be considered. As a last resort, solvent sealing should be considered.

When packaging materials are sealed, the package and its function dictate the properties to be achieved. Mechanical strength is important, since the seals should not fail under normal handling. Hermetic integrity is important in food, pharmaceutical, medical, and chemical packaging. Although small wrinkles in a seal area may have slight effect on its mechanical strength, they are completely unacceptable if a hermetic-type package is required. The best approach to avoid seal wrinkles is to hold the seal region of the construction under tension during sealing. If this is impractical, increasing the thickness of the sealant portion of a lamination may assure good seals, even though small wrinkles may be present. Incorporating EVA in polyethylene, or substituting ionomer (see Ionomers) as a sealant, results in improved flow properties, and better tolerance of small wrinkles.

The most reliable seals for hermetic packaging consist of true fusion welds between the opposed sealant members of a lamination. A true weld can be identified by pulling apart the lamination, and noting where seal failure occurs. If it occurs along the interface between the materials, a true weld was not attained. If failure migrates from the interface, through the sealant, and to the outer portion of the lamination, a homogeneous seal has been achieved (see Multilayer flexible packaging).

If a peelable seal is desired, one should not rely on control

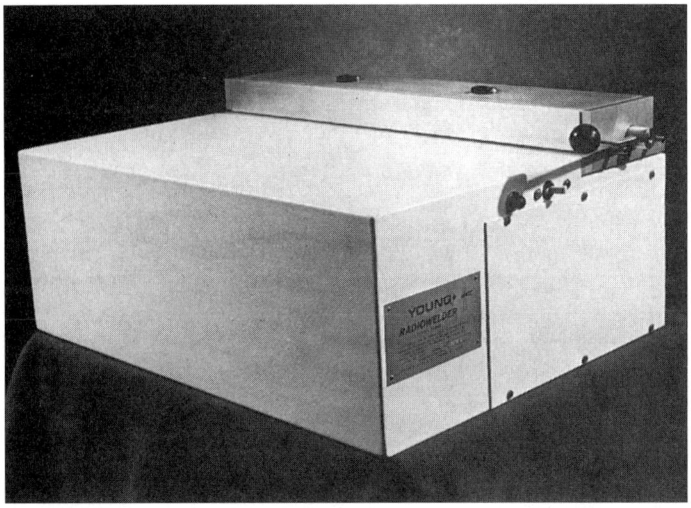

Figure 5. A radiant sealer.

of temperature or pressure to achieve peelability. Seals made in this way may be peelable, but they can come apart by themselves with handling. The only way to achieve dependable peelability is to have an interface formulated to be peelable. An example of true peelability is something that behaves very much like mending tape: the seal should be a tacky seal. Peelable seal laminations have been developed recently that survive steam sterilization and are now being used in sterile food packaging for market testing.

Seal Testing

The American Society of Testing and Materials (ASTM) has set up a standard for the testing of the seal strength of flexible barrier materials. Their designation for the testing procedure is ASTM 88–68. This standard describes how the specimens are to be prepared, along with the equipment and methods to be employed for testing. Tests in accordance with these standards are extremely useful in comparing the seal strength achieved with various materials under consideration for a particular application.

Seal strength, however, is not the only criterion to be evalulated for many applications. Although some seals may test well in the laboratory, they may fail in service. Reasons for this failure include absence of a true fusion weld of the interfaces of the materials being sealed. Occasionally, unwelded seals perform well under mechanical testing, but fail in the actual applications. When there is not a true fusion weld, portions of the packaged product may find their way out of the package through the seal. For seals that must be reliable, particularly where sterility must be assured in a package (see Healthcare packaging), sections of the seals should be examined under the microscope. Unless examination shows that the interface between the sealant portions of the laminations has been completely obliterated, the seals should be suspect; with a true fusion weld, there is no visible interface remaining. The exception to this is peelable seals. The only way to assure that these are acceptable for the long term is to subject them to rough-handling testing for extended times, exceeding the environmental extremes likely to be encountered. If heat processing is involved, good sense dictates that the tests be carried out for at least twice the period of time that the product is heat processed. Tests should also be conducted to simulate environmental pressure fluctuations at something in excess of the maximum temperature expected to be used in processing. For the packaging of chemicals, or other corrosive products, testing should be carried out with filled packages, subjected to rough handling, elevated temperatures, and any other severe conditions that might be experienced.

When dealing with flexible material constructions, even small wrinkles in the seal area produce a statistically significant percentage of leakers. Aside from visual inspection, no satisfactory automatic system has yet been devised to isolate packages with small seal wrinkles, although a great deal of time and money has been spent in this area. When developing a packaging system where seal integrity is paramount, everything should be done in the design of the system and its equipment to eliminate chances for seal wrinkles. The best way is to keep the packaging materials under tension in two directions while they are being sealed. This can do more to achieve success in eliminating seal wrinkles than any other single effort that might be made. Frequently, design of the packaging can be improved for the sole purpose of reducing the possibility of wrinkles in the seals.

BIBLIOGRAPHY

"Sealing, Heat" in *The Wiley Encyclopedia of Packaging Technology* 1st ed., by W. E. Young, William E. Young Company, Inc., pp. 574–578.

General References

R. D. Farkas, *Heat Sealing,* Reinhold, New York, 1964.

W. E. Young, "Sealing" in W. C. Simms, ed., *Packaging Encyclopedia,* Cahners Publishing, Boston, 1984.

SHEET, PETG

PETG copolyester is a clear amorphous polymer with a glass-transition temperature (T_g) of approximately 81°C (178°F) as determined by differential scanning calorimetry (DSC). It is manufactured by addition of a second glycol to polyethylene terephthalate (PET) to eliminate crystallization; therefore, all

PETG-finished products are amorphous. PETG has a number-average molecular weight (M_n) of approximately 26,000. As with all thermoplastic polyesters, PETG is subject to hydrolysis (during processing) in the melt state. While some reduction in molecular weight is a normal consequence of melt extrusion, insufficient drying of PETG can result in an excessive inherent (or intrinsic) viscosity (IV) breakdown that is reflected in lower physical properties—particularly impact strength.

A harmful degree of hydrolytic degradation can be prevented if PETG is dried in a dehumidifying dryer at 65°C (149°F) for 4 h to reduce the moisture level to less than 0.08% before processing.

The molecular weight of either PETG pellets or extruded sheet can be correlated to intrinsic viscosity (IV), which is normally determined with a capillary viscometer; however, since this procedure is rather complex and the required solvents are quite hazardous, a simpler melt-flow-rate test is preferable. The resulting flow-rate number can later be converted (via the calibration curves) to the more standard polyester IV units.

Thermal and electrical properties of PETG copolyester are shown in Tables 1 and 2.

Unstressed sheet extruded from PETG copolyester exhibits good resistance to dilute aqueous solutions of mineral acids, bases, salts, soaps, aliphatic hydrocarbons, alcohols, and a variety of oils.

Halogenated hydrocarbons, low-molecular-weight ketones, and aromatic hydrocarbons will swell or dissolve the plastic.

Plastic sheet extruded from PETG copolyester has good transparency, low haze, and high surface gloss. It has high stiffness and very good impact strength. Physical properties for 3-mm (0.12-in.) PETG sheet are shown in Table 1.

PETG Copolyester Sheet Extrusion

PETG copolyester can be extruded into a clear, tough, amorphous sheet at melt temperatures that usually range within 230–250°C (446–482°F). The sheet thickness is limited only by the cooling capacity of the processing rollstock and/or the capability of the downstream takeoff equipment, but otherwise might typically range from 1 mm (0.04 in.) to 13 mm (0.50 in.) or thicker.

The extrusion of PETG copolyester is straightforward, and good results can be expected when suitable equipment and proper procedures are used, which include pellet drying, melt extrusion into sheet, and nip polishing on a chrome-plated three-roll stack to ensure an excellent surface finish and good transverse-direction gauge control. Machine-direction gauge control is dependent on the uniformity of extruder output. Excellent uniformity can be obtained using a gear pump with PETG, but its use is especially suggested when regrind is included in the PETG feedstock.

At a given speed, a gear pump provides a constant volumetric rate of copolymer melt to the sheet die. Pressure between the extruder discharge and the entrance to the gear pump is controlled by varying the extruder screw rpm. Variation in extruder output caused by changes in percentage of regrind or particle shape and size are isolated from the extrusion die, and the effect on the plastic sheet gauge is minimized. For example, a gear pump set at a suction pressure of 6.9 MPa (1000 psi) would typically control PETG sheet gauge to approximately ±4%.

A barrier screw that is cored for water cooling in the feed zone is typically used to extrude PETG sheet. Industry practice for all polyesters, including PETG copolyester, usually require an extruder screw with a minimum length: diameter (L/D) ratio of 24 : 1. As the L/D ratio increases (eg, 30 : 1), output capacity (lb/h) and output stability (−/+ lb/h) will improve for PETG copolyester, if a suitably designed extrusion screw is used.

Melt filtration is normally used to trap contaminants and/or other types of particulate matter such as "fines," which can be associated with regrind use. A simple screen pack assembly is often used with PETG copolyester, but several different types of automated filtration systems are commercially available for polyesters.

Secondary Fabrication of PETG Sheet

The manufacture of plastic articles from PETG copolyester sheet normally involves secondary operations such as sawing, drilling, bending, decorating, and assembling. PETG sheet can be worked with most tools that are suitable for use with wood or metal; however, the tool speed normally must be reduced to prevent friction melting.

PETG copolyester sheet thickness of ≤2.5 mm (0.10 in.) can be effectively cut with either a power shear (straight cut) or with a steel rule die (irregular cut). The blade-to-bed clearance and alignment for the shear knife edge is an important consideration when cutting PETG sheet. Additionally, a steel rule die press must have a very accurate stroke to prevent damage to the cutting die. Saw cutting is a preferred technique for sheet thicknesses of >2.5 mm (0.10 in.).

Forming

Hot bending. PETG copolyester sheet can be bent around a small radius by preheating an area on both sides with an electric strip heater and then quickly bending the sheet along the heated line. Sheet thicknesses of >3.2 mm (0.13 in.) should be turned periodically during the heating cycle.

Cold bending. Unheated PETG sheet of ≤2.5 mm (0.10 in.) thick can be successfully brake-formed at bending rates somewhat slower than heated sheet.

Thermoforming. Several different thermoforming techniques can be used to force PETG sheet, once heated, into the shape of a mold by mechanical force, vacuum force, or by air pressure. Both male (plug) and female (cavity) molds can be used. Low-cost plaster molds, cast aluminum molds, and/or water-cooled steel molds are commonly used. Typical forming processes include vacuum, drape, and matched mold. Thermoformed items include light fixtures, tote trays, housewares, toys, and a variety of different transparent enclosures.

PETG sheet assembly. A number of solvents, cements, and adhesives can be effectively used with PETG copolyester sheet products. Sheet edges, once aligned, are softened by applying cement and clamping until dry.

PETG copolyester sheet is often fabricated with mechanical fasteners. This technique is preferred if frequent disassembly of a product is required. Common mechanical

Table 1. Typical Properties of 3-mm PETG Copolyester Sheet[a]

Property[b]	Conditions	ASTM Method	Units SI	Units U.S. Customary	Typical Value SI	Typical Value U.S. Customary
General						
Density	23°C (73°F)	D1505	kg/m³	g/cm³	1270	1.27
Water absorption	23°C (73°F), 24-h immersion	D570	%	%	0.2	0.2
Mechanical						
Tensile stress at yield	50 mm/min (2 in./min)	D638	MPa	psi	53	7700
Tensile stress at break	50 mm/min (2 in./min)	D638	MPa	psi	26	3800
Elongation at yield	50 mm/min (2 in./min)	D638	%	%	4.8	4.8
Elongation at break	50 mm/min (2 in./min)	D638	%	%	50	50
Tensile modulus	5.0 mm/min (0.2 in./min)	D638	MPa	10^5 psi	2200	3.2
Flexural modulus	1.27 mm/min (0.05 in./min)	D790	MPa	10^5 psi	2100	3.1
Flexural strength	1.27 mm/min (0.05 in./min)	D790	MPa	psi	77	11,200
Rockwell hardness	—	D785	R scale	R scale	115	115
Izod impact strength, notched	23°C (73°F)	D256	J/m	ft · lbf/in.	88	1.7
	0°C (32°F)	D256	J/m	ft · lbf/in.	66	1.2
	−30°C (−22°F)	D256	J/m	ft · lbf/in.	39	0.7
Impact strength, unnotched	23°C (73°F)	D4812	J/m	ft · lbf/in.	NB[c]	NB[c]
	0°C (32°F)	D4812	J/m	ft · lbf/in.	NB[c]	NB[c]
	−30°C (−22°F)	D4812	J/m	ft · lbf/in.	NB[c]	NB[c]
Impact resistance—puncture, energy at maximum load	23°C (73°F)	D3763	J	ft · lbf	33	24
	0°C (32°F)	D3763	atJ	ft · lbf	40	30
	−10°C (14°F)	D3763	J	ft · lbf	42	31
	−20°C (−4°F)	D3763	J	ft · lbf	43	32
	−30°C (−22°F)	D3763	J	ft · lbf	47	34
Thermal						
Deflection temperature	0.455 MPa (66 psi)	D648	°C	°F	74	164
	1.82 MPa (264 psi)	D648	°C	°F	70	157
Vicat softening temperature	1 kg	D1525	°C	°F	83	181
UL flammability classification	—	UL94	—	—	[d]	[d]
Flammability (France)	—	NFP92501	—	—	[e]	[e]
Flammability (Germany)	—	DIN4102, Part 1	—	—	B2	B2
Flammability (Great Britian)	—	BS476, Part 7	—	—	2	2
Oxygen index[f]	—	D2863	%	%	26	26
Coefficient of linear thermal expansion	—	D696	10^{-5}/°C	10^{-5}/°F	6.8	3.8
Optical						
Haze	—	D1003	%	%	<1	<1
Light transmission	Specular	D1003	%	%	86	86
	Diffuse	D1003	%	%	88	84
Gloss	60° angle	D523	Units	Units	159	159
Color[b*]	CIELAB, Illuminant D6500, 10° observer	E308	Units	Units	<1	<1
Yellowness index	CIELAB, Illuminant D6500, 10° observer	D1925	Units	Units	<1.5	<1.5
Refractive index, N_d	—	D542	—	—	1.57	1.57
Electrical						
Dielectric constant	1 kHz	D150	—	—	2.6	2.6
	1 MHz	D150	—	—	2.4	2.4
Dissipation factor	1 kHz	D150	—	—	0.005	0.005
	1 MHz	D150	—	—	0.02	0.02
Arc resistance	—	D495	s	s	158	158
Volume resistivity	—	D257	ohm · cm	ohm · cm	10^{15}	10^{15}
Surface resistivity	—	D257	ohms/square	ohms/square	10^{16}	10^{16}
Dielectric strength short time	500 V/s rate of rise	D149	kV/mm	V/mil	16.1	410

[a] Properties reported here are typical of average lots. Eastman makes no representation that the material in any particular shipment will conform exactly to the values given.

[b] Unless noted otherwise, all tests are run at 23°C (73°F) and 50% rh, using specimens machined from extruded sheeting with a thickness as indicated.

[c] Nonbreak as defined in ASTM D4812 using specimens with thickness as indicated.

[d] Not tested by UL but would expect to be the same as Spectar copolyester 14471 [94V-2 for thicknesses of 3.13–12.7 mm (0.123–0.5 in.) and 94HB for thicknesses of <3.13 mm (0.123 in.)].

[e] Tests performed on 5-mm sheeting gave a rating of M2.

[f] Dripping and warpage of samples during testing can cause erratic test results.

Table 2. Thermal Properties of 3-mm PETG Copolyester Sheet[a]

Property,[b] Units	Typical Value	Test Method ASTM	Test Method ISO
Deflection temperature			
at 1.83-MPa (264-psi) fiber stress			
°C	64	D648	75
°F	147		
at 0.46-MPa (66-psi) fiber stress			
°C	70		
°F	158		
Vicat softening point			
°C	85	D1525	306
°F	185		
Thermal conductivity			
W/m·K	0.19	C177	—
(Btu·in.)/(h·ft²·°F)	1.3		
Glass-transition temperature			
°C	81	[c]	[c]
°F	178		
Specific heat, kJ/(kg·K) [Btu/(b·°F)]			
at 60°C (140°F)	1.30 (0.31)		
at 100°C (212°F)	1.76 (0.42)		
at 150°C (302°F)	1.88 (0.45)	[c]	[c]
at 200°C (392°F)	1.97 (0.47)		
at 250°C (482°F)	2.05 (0.49)		
Coefficient of linear thermal expansion, mm/(mm·°C) [in./(in.·°F)]	5.1×10^{25} (2.8×10^{25})	D696	—
Flammability [3.2-mm (⅛-in.)-thick specimen]			
cm/min	<2.5	D635	—
in./min	<1		
UL flammability classification			
3.2 mm (0.125 in.)	94V-2	UL94	—
1.6 mm (0.0625 in.)	94HB		
Oxygen index %	24	D2863	—
Melt density			
at 200°C, g/cm³	1.19	—	—
at 250°C, g/cm³	0.98		

[a] Properties reported here are typical of average lots. Eastman makes no representation that the material in any particular shipment will conform exactly to the values given.
[b] Unless noted otherwise, all tests were run at 23°C (73°F) and 50% rh.
[c] Glass-transition temperature and specific heat were determined by DSC.
[d] UL flammability classifications are formula-specific. Spectar copolyester 14471 has the ratings shown.

fasteners include screws, rivets, bolts, clips, hinges, and dowels. This may be a concern, however, if security is a consideration.

PETG sheet finishing. PETG copolyester sheet is best sanded wet to avoid frictional heat buildup. An 80-grit silicon carbide could be initially used, followed by progressively finer abrasives (eg, 280, 400, 600).

PETG sheet decoration. PETG copolyester sheet can be hot-die-stamped through a film carrier that contains either metal or paint. Letters, designs, and trademarks are examples of hot-stamped processes. PETG sheet can also be printed on conventional processes such as letterpress, offset lithography, rotogravure, silk screen, and stenciling. Application of a protective lacquer is often applied over the hot-stamped and printed areas (see the articles on Decorating).

PETG copolyester can be extruded in a variety of thicknesses, coextruded with or without tie layers, and formed or fabricated, to meet the needs of ever-increasing consumer and industrial markets. The price, performance, and availability of PETG copolyester make it a very important material in the plastics market today.

BIBLIOGRAPHY

General References

References include MBS-80J, Eastar *PETG Copolyester 6763.*

DSS-429A, *Physical Properties of Spectar Copolyester 14471 Plastic Sheet.*

TRS-94F, *Secondary Fabrication Techniques for Sheet Extruded from Spectar Copolyester.*

TRS-131, *Extrusion of Plastic Sheet from Spectar Copolymer.*

John Wininger
Eastman Chemical Co.
Kingsport, Tennessee

SHELF LIFE

Definition of Shelf Life

Shelf life is the time after production and packaging that a product remains acceptable under defined environmental conditions. It is a function of the product, the package, and the environment through which the product is transported, stored, and sold. Each of these influencing factors will be discussed below.

The shelf life for any given product can vary greatly dependant upon the formulation, packaging system, and the distribution environment. In general, use of preservatives, increased barrier packaging, and colder distribution environments will extend shelf lives. These options typically also increase costs. The ideal shelf life from the company perspective is one that matches the distribution and use of the product to the company's inventory. The ideal shelf life from a consumer point-of-view allows them to fully use the product. Specification for overly lengthy shelf lifes usually increase costs in terms of materials; overly short shelf lifes usually cost in wasted or discarded product, or in increased liability. It is therefore prudent to define a reasonable shelf life, and assure that the shelf life specification is met.

The principles of shelf life testing hold for virtually any product/package combination. Products which are packaged in impermeable materials, such as glass or metal, degrade primarily through mechanisms that are inherent in the chemistry of the product. Since these mechanisms are essentially product dependent and vary with the product, evaluations of the package would be unproductive. Notable exceptions to this are situations in which the package allows or contributes to deterioration of the product. For example, glass allows light transmission which can promote oxidation reactions, but can be tinted to filter wavelengths that promote these reactions. Metal cans can react with products with either the metal substrate itself (eg, pinholes in tinplate), or with components of the can coating. Improvements in can and coating technology are designed to preclude these interactions. (For further discussion refer to sections of this Encyclopedia on the appropriate container.) Since the reactions of these impermeable containers are specific to the containers, further discussion of packaging influences will be directed toward semipermeable and permeable materials.

Factors that Influence Shelf Life

Product. Products differ greatly in their susceptibility to degradation by various agents. Some products will become unacceptable from a change in moisture content. For example, a moisture gain in ready-to-eat breakfast cereals will destroy the crisp texture; a moisture loss in an intravenous product will alter the declared dosage. Snack items often are sensitive to oxidation of the oils absorbed during frying and can become rancid. Snacks also can be sensitive to moisture change. The mode of failure of the product will have a direct influence on the type of protection which is sought from the packaging material. The match between product susceptibilities and package protection will impact upon the shelf life.

Once the mode of deterioration of the product is determined, acceptance criteria need to be defined. Acceptance criteria is the range of the critical component for which the product is considered acceptable. This is often done by a sensory panel. Ideally sensory scores can be correlated to analytical data, with the result that analysis can provide the index for product quality.

The acceptability criteria for any product has a direct influence on the measurement of shelf life. Some products have a clear point of acceptability. Others have complex deteriorations which make it difficult to identify a critical point. This can be further complicated by products which have multiple or interacting modes of failure. In these cases subjective decisions must be made to set the acceptance criteria. It deserves mention that the ultimate acceptability is that of the consumer. Concordance between a standardized definition of acceptability (as used within the company) and actual consumer acceptance is crucial to the commercial success of the product. The determination of this concordance is more properly within the domain of marketing research, but the degree of correlation determines the validity of the research effort.

The product which is tested for shelf life must be specified. This sounds almost trivial, but reformulations and ingredient cost reductions can alter product characteristics without changing the product name. The result is that assumptions may be made on the basis of past performance which may not hold for the present product. Therefore, the test label should include product name, formulation (or reference to specific formulation in a lab notebook), date of manufacture, date of test, conditions of test, and principal researcher. The packaging system must also be identified in terms of key factors as described below.

Package. The package offers protection for the product against an agent which degrades the product. For moisture sensitive products, this protection is in terms of the water vapor transmission rate (WVTR); for oxygen sensitive it is the oxygen transmission rate (OTR), etc. The mode of protection can change for the same product with different packaging. For example, a snack product which is sensitive to moisture gain (loses crispness) and oxygen (becomes rancid) could be called "moisture sensitive" if the texture degrades before the rancidity becomes objectionable. This same product, if packaged in a sufficient moisture barrier, would be identified as "oxygen sensitive" in terms of package criteria.

Permeation rates for chosen packaging materials need to be determined for the agent in question. WVTR and OTR are measurable by a variety of standard procedures (see Testing, permeation and leakage). Permeation rates for flavors and other vapors can be determined by newer instruments (see Aroma barrier testing) as well as by gas chromatographic, mass spectral, infra red, and other techniques.

The size of the package also influences the shelf life. As the package size increases, the surface to volume ratio decreases. The result is that the amount of permeant that comes through the package increases as a square function, but the volume of product that absorbs that permeant increases as a cube function. The barrier requirements, therefore, decrease with the package size, all other factors remaining equal.

In summary, therefore, the packaging parameters which must be specified for shelf life testing are the material designation and source, the appropriate transmission rate, the surface area, and the net weight of the enclosed product.

Environment. Product distribution causes stress on the product as a function of product sensitivities and all conditions experienced by the product as it is carried through various distribution networks. Actual conditions vary with where it is shipped, how it is shipped, the seasons, warehouse conditions, etc. It is impossible to account for this variety of conditions by using a single storage condition. The manufacturer of sensitive products (food, pharmaceuticals, cosmetics, tobacco, chemicals), therefore, must define the criteria that will be used for the shelf-life determinations. Pharmaceutical manufacturers must package for the worst case situation, so packaging will be designed to survive under the most severe conditions in the distribution chain. Food products are often packaged for less than worst case scenarios. Products which are not rotated on the supermarket shelf (such as if a product is replenished in front of existing stock) may fail, but this loss is considered more desirable than increasing packaging costs for the entire line to accommodate this situation. Conditions which allow 80% of the product to survive distribution is not uncommon. Once the representative distribution mode is chosen, useful statistical indices can be derived which represent this mode.

The shelf life impacts greatly on the distribution system. Products with short shelf lives require a more rapid distribution system. For example, Frito Lay supplies the nation's grocery stores with snack items via hundreds of distribution sites which in turn are supplied from 44 production facilities (1,2). This allows potato chips and other snacks, which are susceptible to both moisture and oxidative degradation, to reach the consumer in a fresh and crisp form. In contrast, Proctor and Gamble Company distributes Pringles from one production facility (1). A casual examination of Pringles vs Frito's potato chip packaging will describe for the reader the consequences of the product/distribution strategies. Pringles uses a multilayered composite can structure with foil barrier as compared to the simpler flexible pouch in the Frito system.

Refrigerated distribution is costly, but provides additional protection for the products. Additional processing/packaging such as hot fill, aseptic, or retort packaging extend the shelf life and eliminate the need for refrigeration. In Europe and Asia, where typical home refrigerator and freezer space is more limited than in typical U.S. households, retort and aseptic packaging are far more important. The Brik-Pack, for example, was developed in Europe to fill the gap that the United States fills with refrigeration.

The measurement of shelf life is becoming increasingly more important. The competitive nature of contemporary business and increased energy costs have resulted in renewed interest in reducing packaging costs. An abundance of new packaging materials facilitates the choices, but also adds to the complexity of the choice. In addition to the above, governmental requirements for open dating necessitate shelf life testing. Open dating is the addition of a date on the package which indicates by what date the product should be consumed. This can take different forms as shown below.

A brief synopsis of different types of open dating is in order. Dating of products on line has been the practice of companies producing food, pharmaceutical, and other sensitive products. Each company developed its own proprietary coding system to enable identification of production dates and production facility when necessary. These codes, which were designed to be unreadable to consumers and competing producers, identify production dates when necessary. Open dating changes this.

Various types of dating are possible. A direct conversion of production codes into consumer readable form gives the date of manufacture. A freshness date ("use by") indicates the date by which the product should be consumed to assure maximum freshness. A pull date is the date after which the product should no longer be sold. Milk and dairy products use pull dates.

It must be noted that all of the above dates assume storage under specific conditions. For example, dairy products are assumed to be refrigerated. If conditions are not held as anticipated, product quality will not be reflected as implied in the dates.

Dating of products will benefit the consumer, but also has the potential of benefiting both manufacturer and retailer. It would improve stock rotation (as mentioned above) which could ultimately reduce the packaging requirements necessitated by existence of old product.

Traditional Approaches to Shelf Life Testing

Shipping tests. One common procedure for testing shelf life is a shipping test. Shipping tests are often used as an adjunct to other shelf life testing, but sometimes are considered a replacement. This is a mistake, because shipping is only a component of the testing environment. This type of testing is characterized by uncontrolled and often unknowable storage conditions. A product is shipped to a warehouse, stored for a specified time, and then returned to headquarters (or R&D), and examined for acceptability. Shipping tests are also used to check for product deterioration from shock and vibration during shipment. At least one company uses shipment through the Chicago post office as the ultimate test for product endurance. (Shock and vibration effects are important for assessing product quality. They are abuse dependent rather than time dependent, however, and are discussed separately (see Testing, product fragility).

Shipping tests show product quality changes with time under real world situations. The specific experience of a specific test may be representative of part of the distribution chain but rarely, if ever, represents all conditions which the product will experience. The test carries the illusion of being a representative real world test. However, the changing conditions, and factors such as the difference in product experience between shipment in carton vs pallet quantities, will necessitate careful evaluation of the results.

Storage tests. Storage tests consist of maintaining product in a static facility, and evaluating product quality with time. Storage conditions can be uncontrolled or controlled. Warehouse storage is often characterized by uncontrolled conditions. Storage conditions can be controlled through the use of storage cabinets that maintain temperature (and often humidity) at specified values. This results in a test which can offer more reproducible data. Various conditions can be tested including accelerated conditions which allow for more rapid results. A typical ambient condition (for U.S. as defined by TAPPI) is 73°F(23°C)/50% rh.

Accelerated conditions vary, but a typical condition may be 39°C (95°F)/80% R.H. Accelerated storage speeds product degradation. However, the factor that defines this increase is related to the kinetics of the chemical reactions which define the degradation. This varies with product and with mode of degradation. Therefore, a standard multiplication factor is meaningless.

Accelerated studies are useful if an initial evaluation is performed at both ambient and accelerated conditions, and compared. This comparison can be through actual kinetic studies (such as Q_{10} analysis described below) or by shelf life comparison. It is imperative to assure that discontinuities do not exist between ambient conditions and accelerated conditions. If either the product or packaging passes through a transition, the results of the accelerated test will have little bearing on comparisons with ambient. Transitions which effect packaging include the glass-transition temperature (T_g). For example, permeation values for polypropylene cannot be extrapolated to frozen conditions due to a T_g at -10°C (14°F). Those for products include melting/freezing of any ingredient (usually water or fats), recrystallizations, etc. In addition, activation energies must be considered, because certain reactions will not occur below certain temperatures. Extrapolation below these temperatures is therefore not recommended.

Q_{10} analysis involves testing products at varying temperatures, and defining the difference in reaction rate for a 10°C (18°F) temperature increase. A Q_{10} of 2 means that the reaction rate will double for each 10°C increase in temperature. This is based on the Arrhenius relationship which is discussed in any text on kinetics or physical chemistry. The Arrhenius relationship is perhaps an oversimplification for the complex chemistries of most products. A straight line relationship with a minimum of three points (over a limited temperature range) indicates that Q_{10} can provide a useful indicator for storage duration. Caution relating to transitions remains, and it also deserves mention that the Q_{10} itself is temperature dependent. However, products are usually studied under a limited range of temperatures so that a constant Q_{10} can be assumed.

Once the Q_{10} is determined (and shown valid through a linear Arrhenius relationship), the value can be used to predict the time necessary to test a product to be equivalent to the shelf life at ambient temperature. For example, a product with a Q_{10} equal to 2.4 can be stored for 10 weeks at 33°C (91°F) to obtain results similar to 24 weeks storage at 23°C (73°F). (One can view these studies as a test in which the conditions extended to the package are exaggerated in an attempt to simulate the accelerated passage of time. The underlying assumption is not always valid, and the technique is only as good as that underlying assumption.)

An alternative type of "accelerated" test is a study in which the permeation rate of the packaging material is "accelerated". A product which is being tested for sensitivity to moisture is packaged in a low moisture/high oxygen barrier material (such as polyacrylonitrile or polyacrylonitrile copolymers). With a known permeability, area, and net weight, the amount of moisture permeating the package can be determined. Product can be sampled for acceptability until failure is reached. This defines the amount of moisture transfer that can be tolerated. Calculations wtih higher moisture barrier materials can provide performance criteria for the additional materials. A similar study can be performed for oxygen sensitive products by using a low oxygen/high moisture barrier material (such as polyolefins). In essence, this test purposefully introduces a known amount of the critical agent to accelerate the time to degradation without changing the conditions experienced by the product. A cautionary note is important. The accelerated test described above includes the inherent assumption that the changes in the product occur faster than the transmission across the package. If an accelerated test allows sufficient permeation such that product changes are slower than the influx of permeant, the results will indicate a shelf life which is longer than the extrapolation to a reasonable barrier material. This is due to a critical permeant level being obtained faster than the evidence of product degradation.

Computer Models for Shelf Life

The shipping and storage tests require that the choice of packaging be made before the study. Chosen materials are then submitted for evaluation with time as the test proceeds. Successful completion of the test verifies that the chosen package will last at least as long as the chosen test duration under the conditions of the test. Failure to survive the test conditions demonstrates insufficient protection. The deficiencies of this procedure are that failure is not known until demonstrated, and overprotection is not shown at all because testing usually terminates after desired time is reached.

The use of a computer to simulate a storage test provides a means to pretest materials before actual storage testing. By performing these calculations rapidly, as only a computer can do, many iterations are possible within a reasonable time, providing a powerful support tool for packaging development. The simulations ae accomplished by separating the product and packaging characteristics. The product is analyzed for changes that occur on exposure to the shelf-life-limiting parameter, which is identified as the mode of failure for the product. The packaging material is analyzed for its barrier properties against that parameter. The computer is then used to combine the protective aspects of the package with the sensitive properties of the product. Since packaging, product, and storage parameters are entered independently, it is possible to test the effect of different packages or conditions simply by entering the new variables.

Computer simulation yields the most cost-effective packaging design. Later real-time storage studies serve the purposes of legal requirements and confirmation of predicted results. Differences between predicted and actual results can be used as indices to check initial assumptions. For example, poor seals will shorten actual shelf life achieved with an otherwise-adequate barrier material. In this example, the false assumption is that the material barrier value is equivalent to the package permeation (see Testing, permeation and leakage).

Simulated studies extend beyond the one-package/one-condition study. Parameters can be changed to effect another study. A survey of potential new packaging materials can be accomplished without need to retest the product. Only the changes in material parameters need be entered to test performance. This is especially important when limited amounts of product material are available, as in the product-development stage. Changing storage conditions follow a similar procedure, but the caution for passing transitions remains.

Product evaluation. The mode of failure for the product must be known prior to using any simulated approach. This could be moisture gain or loss, oxidation, CO_2 loss, flavor loss, chemical degradation, etc. Once the critical factor is known, testing is conducted to show product changes with change in the critical factor. For example, a moisture sensitive product is evaluated for change in moisture content as the relative humidity in the storage environment is varied. This test is the moisture isotherm. The isotherm can be completed in usually one to four weeks, depending on the rapidity at which the samples equilibrate. For oxygen-sensitive products, the testing consists of evaluating product changes with oxygen uptake, and can be performed by inverse gas chromatography, respirometry, Warburg, or a variety of tests for degree of oxidation. For carbonated beverages packaged in permeable bottles, the test would be to measure carbonation loss (soda), or oxidation (beer) or both.

The product testing should be performed at various temperatures if temperature effects are to be modeled. The evaluation of differences in reaction rates vs temperature supplies important information concerning the kinetics of the reaction, and can be used to investigate performance for different distribution environments. A minimum of three tests are recommended. The caution concerning transitions remains valid.

Package evaluation. For computer modeling purposes, the chief packaging characteristics are barrier properties. Materials that do not change significantly with machining and flexing can be tested in flat sheet form for water vapor transmission rate (WVTR), Oxygen transmission rate (OTR), or any other permeant which is critical for the product. For materials which change with handling (waxed glassine, foil, etc) full package testing is recommended. The computer model measures the change of permeant level in the package as a function of the transmission rate. If the flat sheet value is used in a system which is poorly sealed, or has developed fractures in handling, the predicted shelf life will be erroneously long.

The package material characteristics of sealability, printability, compatability, cost, mechanical strength, etc remain the province of the packaging engineer and are not usually incorporated into computer models. Nonbarrier parameters are not included in this discussion of shelf life. Those parameters that do affect shelf life, but are not in the model (eg, light transmission for oxygen-sensitive products), must be considered in evaluating any computer printouts.

Model fundamentals. The internal environment of a package changes as permeants enter or leave the package. In terms of the mathematics, components which enter the package from the outside environment, and those which permeate from the inside to the outside, are identical. Transmission will always occur from the high concentration to the lower concentration side of the material. Therefore, any discussion or examples of permeation rates are valid for either circumstance.

The transmission rate is standardized for package area, film thickness, time of measurement, and difference of partial pressure of permeant across the package. In general, the terms are amount of permeant multiplied by film thickness per package area per time per unit of partial pressure differential. The measurement is also made at a temperature which is implied but not usually stated in the units. In use, the standardized transmission rate is combined with area, thickness, and time in the model to define how much permeant can enter or leave the package. The effect on the product is a function of the net weight of the product upon which the permeant acts. The partial pressure differential across the package becomes difficult to measure, and changes as the internal environment of the package changes.

The partial-pressure-differential change with permeation is the major reason for the use of computers in shelf-life modeling. If this change did not occur, one could simply multiply transmission rate by time and area, and divide by the thickness to obtain amount of permeation for any time. This assumed straight-line relationship overestimates transmission because the permeation process itself reduces the differential of partial pressure across the film and therefore reduces the transmission. In addition to the driving-force change due to the transmission of permeant, some of the permeant is absorbed by the product and further complicates the system. In engineering terms, as the permeation process proceeds, the driving force of that process (the partial pressure differential) decreases. This change can be described by a differential equation. The computer models incorporate these effects and yield a representative manifestation of the process.

Computer models are usually based on a relatively simple mass transfer equation. Many models have been developed that have not been reported in a usable form in the literature. These include models on oxidation-sensitive products, beverages, and products in which product characteristics result in product-dependent models.

The general models are based on a standard differential equation which describes mass transfer across a permeable barrier. This is shown in the following equation (14):

$$\frac{dW}{dt} = \frac{k}{l} \times A\,(P_{\text{out}} - P_{\text{in}})$$

where t = time
W = change in weight of the critical component
k = permeability coefficient for the packaging material
l = thickness of the packaging material
A = area of the package
P_{out} = partial pressure of permeant outside the package
P_{in} = partial pressure of permeant inside the package

This equation can be converted into a usable form by substituting measurable values for the W and P_{in} terms. The P_{out} term is readily obtainable: for oxygen it is 0.21 atm (21.3 kPa); for moisture, it is the saturated vapor pressure (6) times the rh of storage; for flavors and volatile components, it is essentially zero.

The P_{in} term requires knowledge of the reaction of the permeant with the product. In the case of moisture change, this is expressed as the moisture isotherm which describes change in product moisture vs storage relative humidity. An expression which describes the water activity over the product for any moisture value can be used to define the P_{in} term. (Water activity (Aw) is the partial pressure of water above the product divided by the saturated vapor pressure of water at the same temperature.) First the isotherm is determined. Second an expression is derived which describes Aw as a function of moisture content. Finally, P_{in} at any moisture content can be described as the expression for Aw times the saturated vapor pressure. (Note: water activity is traditionally used. However, this is only valid at the temperature used in the isotherm study. The preferred procedure would study moisture content as a function of vapor pressure directly, instead of as a function of Aw. The model converts Aw into pressure terms.) Models for permeants other than moisture require analogous expressions to relate internal pressure to product changes.

With a substituted expression for the P_{in} term in concentration terms rather than weight terms, the W term need be expressed in concentration terms. An expression is therefore needed which describes the change in concentration as a function of weight change.

The above expressions are then substituted into the equation, rearranged to combine variables onto one side of the equation, and the new equation is integrated to form the model. The limits of integration are permeant concentration from initial to critical values, and time from initial to shelf life (time to reach critical amount of permeant).

The various models described in the literature use approaches related to the above procedure. Simplifying assumptions (such as linearity in the isotherm) are used to develop simpler models. These models can be used to advantage as long as the assumptions are kept in mind. The less restrictions used in deriving the expressions, the more universal the model. However, the mathematics become increasingly complex.

The models do not replace the need for actual storage studies. Storage is still necessary to verify product changes, find unusual effects, and in some cases, meet compliance criteria with government regulations. The use of the models serves to reduce the time and effort necessary to identify optimal packaging protection, and eliminate storage studies on materials with insufficient barrier properties to be reasonable candidates for the product.

Other Models

Modeling of permeation through packaging films can be extended to evaluate more complicated scenarios than storage of a product in an ambient atmosphere. As long as permeation of gases and vapors remains the key shelf life determining parameter, modeling of mass transfer is applicable. Modified Atmosphere Packaging (MAP) (see Modified atmosphere packaging) utilizes a gas mixture to extend the life of the product. The gas mixtures are composed of gases with different permeation rates and different concentrations in and out of the pacakge. Computer models can predict how the concentrations will change with time. Shelf life can be directly determined if the shelf life relates directly to a gas concentration. For example, if an oxygen sensitive product would oxidize rapidly above a critical oxygen concentration, the time required for the oxygen concentration within the package to attain that level would determine the shelf life.

Another computer model which relates to shelf life involves Controlled Atmosphere Packaging (CAP). CAP implies that the atmosphere is controlled during the life of the product, and is typically employed in warehouse situations, particularly with fresh produce. CAP can also be used for a consumer package in which a combination of permeation characteristics of a film and respiration rates of a fresh product combine to create the desirable packaging environment. As the produce respires (takes in oxygen and expels carbon dioxide) the concentration of oxygen within the package will be depleted, and that of carbon dioxide will be enhanced. An impermeable package would quickly reach conditions of low oxygen and high carbon dioxide that would asphyxiate most produce. A correctly chosen permeable film, however, can result in a packaging atmosphere that slows senescence without killing the product, thereby enhancing shelf life. Computer models in this case are used to calculate the gas concentrations that result from the dual respiration and permeation effects. The concentrations are then compared with ideal CAP conditions for the particular product and thereby used to extend the shelf life of the product.

Various models have been developed to predict shelf life of products that are limited by microbial growth. Examples of some of these models are referenced in the bibliography (7–16).

BIBLIOGRAPHY

1. *Food Engineering's 1983 Directory of Food and Beverage Plants,* Food Engineering, Chilton Co., Radnor, Pa.
2. C. E. Morris, "Who's Building What & Where", *Food Engineering,* **56**(9), 65–80 (1984).
3. K. S. Marsh, "Computer-Aided Shelf Life Prediction", *1984 Polymers, Laminations and Coatings Conference—Book 1,* TAPPI Press, Atlanta, GA, 1984.
4. S. Gyeszly, "Shelf-life Simulation Predicts Package's 'Total Performance'", *Package Engineering,* 70–73 (June 1980).
5. S. S. Rizvi, "Requirements for Foods Packaged in Polymeric Films", *CRC Critical Reviews,* 111–134 (Feb. 1981).
6. *Handbook of Chemistry and Physics,* CRC Press, Inc., West Palm Beach, FL (any edition will have saturated vapor pressure charts).
7. D. Olds, "Shelf-Life Predictions Using Temperature as an Acceleration Factor", *Food Quality,* 28, (Jan./Feb. 1996).
8. K. S. Marsh, T. J. Ambrosio, and D. K. Morton, "The Determination of Moisture Stability of a Dynamic System under Different Environmental Conditions", in D. Henyon, ed., *Food Packaging Technology,* ASTM STF 1113, American Society for Testing and Materials, Philadelphia, PA, 1191, p. 13.
9. K. R. Anantha-Narayanan; A. Kumar, and G. R. Patil, "Kinetics of various deteriorative changes during storage of UHT—soy beverage and development of a shelf-life prediction model", *Lebensmittel-Wissenschaft-und-Technologie,* **26**(3), 191–197 (1993).
10. M. Ferrer-Gimenez, "[Determination of the commercial [shelf]-life of a dehydrated food.]", *Alimentaria,* No. (169), 43–48 (1986).
11. S. S. Kurade and J. D. Baranowski, "Prediction of shelf-life of frozen minced fish in terms of oxidative rancidity as measured by TBARS number", *Journal of Food Science,* **52**(2), 300–302, 311 (1987).
12. H. Einarsson, "Predicting the shelf the shelf life of cod (Gadus morhua) fillets stored in air and modified atmosphere at temperatures between −4 degree and +16 degree", in *Quality Assurance in the Fish Industry,* ISBN 0-444-89077-7, 1992, pp. 479–488.
13. B. R. Jorgensen, D. M. Gibson, and H. H. Huss, "Microbiological quality and shelf life prediction of chilled fish", *International Journal of Food Microbiology* **6**(4), 295–307 (1988).
14. G. C. Mead, "Predictive aspects of spoilage in fresh poultry meat including the use of biochemical parameters and the application of mathematical models", in *Seventh European symposium on poultry meat quality,* 1985, pp. 171–186.
15. G. J. Newell, "A new shelf-life failure model", *Journal of Food Protection,* **44**(8), 580, 582 (1984).
16. M. G. Geoule, Jr., and J. J. Kubala, "Statistical models for shelf life failures", *Journal of Food Science,* **40**(2), 404–409 (1975).

KENNETH S. MARSH
Kenneth S. Marsh & Associates, Ltd.
Woodstock Institute for Science in Service to Humanity
Woodstock, Illinois

SHOCK

The term *shock* describes a rapid change in something over a very short period of time. In the packaging field, this could relate to mechanical movements, sudden stops, or even rapid changes in environmental conditions such as thermal shock, when temperatures change rapidly with movement of products from inside to outside. We focus our discussions here only on the mechanical shocks related to the physical movement of products and packages.

The most common shocks experienced by packages and products result from handling drops. It is not the fall that hurts, but the sudden stop on impact with the solid ground. The suddenness of the shock is described as the duration of the impact. It is normally expressed in terms of milliseconds (ms) or $\frac{1}{1000}$th of seconds. Typical packaged product impact durations range from 2 to 50 ms from drops onto hard floors. The duration of the shock is dependent primarily on the amount of cushioning provided by the packaging materials used to protect the product and to some extent on the rigidity of the surface impacted.

The term *cushioning* is used to describe the effect of slowing the rate of change of the shock by increasing the time duration of the shock event. This is accomplished by allowing the product to stop over a longer period of time by beginning to stop it at a higher level above the point of final rest. Cushioning requires distance to provide the added time. Other articles in this *Encyclopedia* address the need of how to best design and shape the cushion to effectively use the available stopping distance provided by the cushion thickness to obtain maximum cushioning.

The rate of change of the speed of an item when it impacts is called the *acceleration rate* (or *deceleration rate*) of the impact. The greater the time duration of the impact, the lower the acceleration rate of the shock for impacts of the same speed. The impact acceleration rate is expressed in terms of G values, which are the dimensionless ratio between the acceleration in length per time squared and the acceleration of gravity in the same units—in effect, multiples of the acceleration rate of gravity ($1g$) on earth. The gravitational attraction of the earth is defined as 386 in./(s · s), 32 f^t/(s · s), or 9.8 m/(s · s).

Measured impact acceleration levels as previously discussed depend on the duration of the impact shock resulting from the cushion's effect. With no or very limited cushioning, acceleration levels can easily reach 500–1000G. With effective cushioning, acceleration levels can be reduced to 15–100G. How much cushioning is provided depends on the products' need for protection and the expected height from which the product will be dropped.

In simple terms, if an item is dropped from 30 in., it will fall at a rate of 1G, due to the effect of gravity. To stop the item at a rate of 1G would require an additional stopping distance of 30 in., assuming that you could bring it to a stop at a constant rate of 1G throughout the stopping distance. Fortunately, most products can be stopped at a much higher rate. In our example, if the product could withstand a stopping rate of 30G, it could theoretically be stopped within a minimum stopping distance of 1 in. We will go on to see that by using cushions it will actually be necessary to use much thicker cushions than the 1 in. that is the minimum stopping distance. This is due to the limited compressibility of cushions, 50–80% maximum; and to the nonconstant zero to peak–vs–constant peak deceleration rate.

Shock refers to the maximum rate of acceleration that is experienced during an impact often resulting from a free-fall drop. Obviously a packaged product may experience mechanical shocks resulting from several other events. A common event in shipping may be the impact shock received by a package when another packaged is dropped onto it. Or, it may result from being hit with a diverter on a conveyor belt during an automated sorting process. Whenever there is a sudden rate of change of velocity, an event called a shock occurs.

Measuring Impact Shocks

Measuring shock levels requires the use of instrumentation and the understanding of what the instrumentation is presenting. Although mechanical shock recorders have been used in the past, they have been almost totally replaced with electronic instrumentation whenever accurate measurements of shocks are desired. Most shock-measuring devices rely on the use of accelerometer transducers. The accelerometers change the measured characteristics into electronic signals that can be displayed, analyzed and recorded. Acceleration levels are presented as voltage proportional to the acceleration due to gravity. Usually this is in the range of 1–100 mV/g. The voltages are the result of the transducers converting the forces generated during an impact shock into an electronic signal. Since $F = ma$ (force = mass × acceleration), and the mass in the transducer remains constant, the measured force is directly proportional to the acceleration being experienced.

Accelerometer transducers typically are either piezoelectric or piezoresistive designs. Piezoelectric accelerometers rely on crystalline elements that generate an electronic signal when subjected to an applied force. Both accelerometer designs have their advantages and limitations, but it is sufficient to say that they are the most common means of obtaining records of acceleration levels experienced during impact shocks.

This discussion does not address the signal conditioning, power supplies, filtering, and calibration necessary to obtain records of shock acceleration levels. However, once the signals are obtained, they are often displayed as a function of time and are referred to as *shock signatures, shock pulses,* or *waveforms.*

Shock Waveform Analysis

Figure 1 shows a typical acceleration shock waveform resulting from an impact. We can discuss and characterize components of the recorded shock waveform. The first observation is that of the peak G level or the maximum acceleration experienced during the shock event. Since the entire recording is that of the acceleration levels, we must determine at what point to take a reading, and since it is usually the maximum faired acceleration that is of concern, that is what is most often used to define the shock acceleration level. Care must be taken when recording shock acceleration levels to ensure that any filtering used to clarify the shock signature does not adversely effect the peak G level recorded.

The second observation is that of the time over which the event occurred referred to as the shock pulse duration. By convention, the duration of a shock pulse is measured from points where the acceleration/time record is at a level of 10% of the peak G level. This is done because of the difficulty of precisely determining when the waveform actually crosses the baseline to start and stop the duration-time measurements. Shock pulse durations typically measured in package drop tests range from 2 to 50 ms.

Additional data that can be derived from an analysis of the shock waveform is the velocity change that occurs during the shock event. The velocity change is proportional to the area

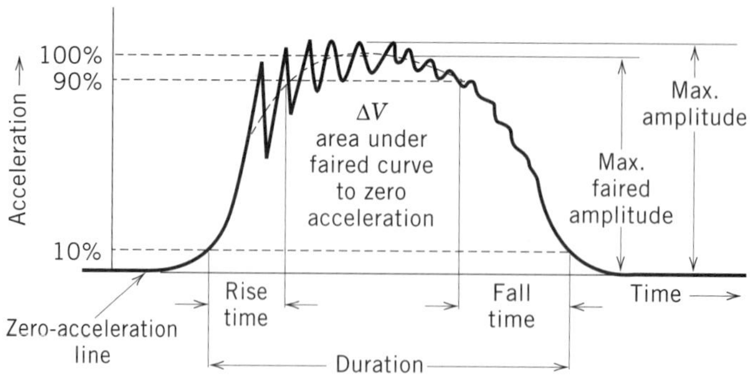

Figure 1. Standard pulse interpretation.

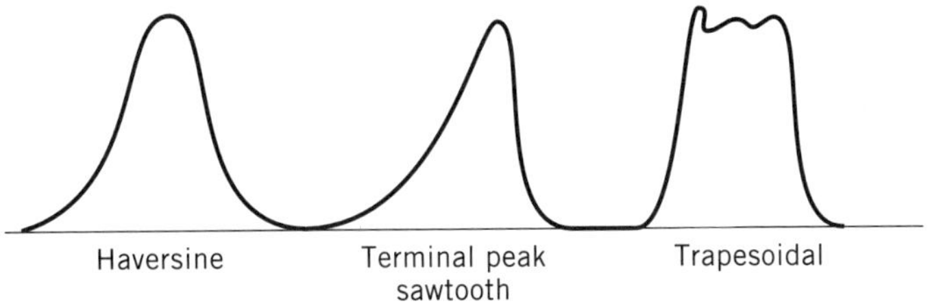

Figure 2. Typical shock waveform shapes.

under the acceleration time curve. It is, in fact, the integral of the acceleration with respect to the time over the total duration of the shock pulse. The velocity change is the sum of the impact velocity from a package's fall plus the rebound velocity resulting from the impact energy stored in the cushion system being returned in the form of rebound motion.

The shape of a shock waveform can be very instructive as to the type of impact that has occurred as shaped by the cushion systems performance (see Fig. 2). The waveshape of an elastic cushion acting as a linear spring will have a haversine, often referred to as a half-sine, waveshape. A nonrebounding cushion such as some corrugated pads may exhibit a waveform that can be described as a terminal peak sawtooth waveform on the basis of its visual shape. The acceleration (deceleration) reaches a peak when the object comes to rest with no rebound energy to create a rebound velocity. Another waveform shape that occurs during package testing is the trapezoidal waveform in which the acceleration reaches a peak *G* level that continues for a period of time as the cushion continues to crush. These trapezoidal waveforms are most often created by honeycomb or cylindrical cushion structures being crushed or in foam-in-place systems where the cushioning is provided as the material is being frangibly torn rather than compressed.

Seasoned package designers can often interpret what may be happening to a product in a cushioned package by analyzing the shock waveform observed during an impact (see Fig. 3). A symmetrical half-sine waveform typically indicates that the static stress loading of the product on the cushion is in the optimum range for maximum cushioning. A waveform approaching a terminal peak sawtooth shock pulse with a sharp rise near that end compared with the beginning of the waveform may show that the product is bottoming out the cushion; the static stress loading may be too high for the cushion or the thickness of the cushion insufficient. A waveform showing a high spike at the beginning of the shock pulse may be indicating that the cushion is too stiff for the static

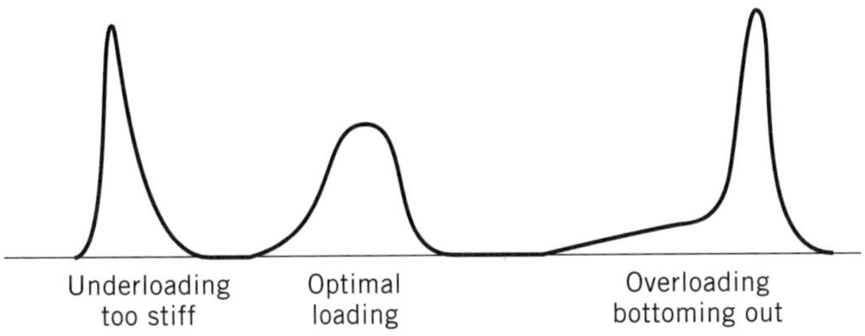

Figure 3. Cushion static stress loading concerns.

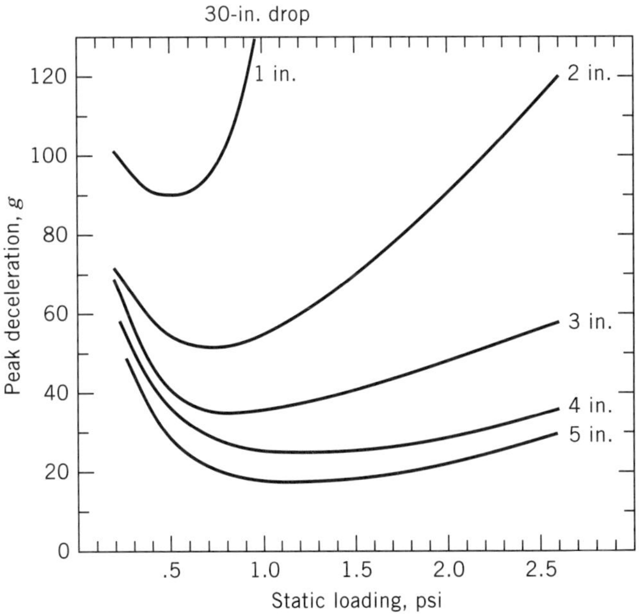

Figure 4. Dynamic cushioning curves.

stress loading being applied by the product. It may be like a diver making a belly flop smack on the water. The thickness of the cushion in this instance is immaterial since none of it is initially deflecting on impact.

Characterizing Protective Cushions

Cushioning materials used to protect products from shocks are normally characterized by their shock-transmitting attributes. "Cushion curve" data present the average level of the peak *G* level experienced on a simulated product dropped from a given height onto the given thickness of cushion (see Fig. 4).

The static stress of the product on the cushion, expressed in psi, is then varied to determine the optimum loading range for the cushion. This data are derived empirically through an extensive series of drop tests during which the variables of drop height, cushion thickness, and loading (in psi) are systematically changed. The cushion-curve data thus presented can then be used as a starting point for predicting the effectiveness of a given cushioning material for a needed application.

From the cushion curve, the peak transmitted shock level is noted for given drop heights. When the maximum acceleration fragility of a product is compared with the cushion curves for the specified environmental drop height, both the minimum cushion thickness and optimum static stress loading can be determined. This provides the initial information for developing protective package systems for products.

The Damaging Effect of Shocks

The use of fragility testing is the most common means of determining a product's resistance to mechanical shocks. The process of fragility testing is described in much greater detail elsewhere in this *Encyclopedia* (see Testing, product fragility). The result of product fragility testing is the characteriza-

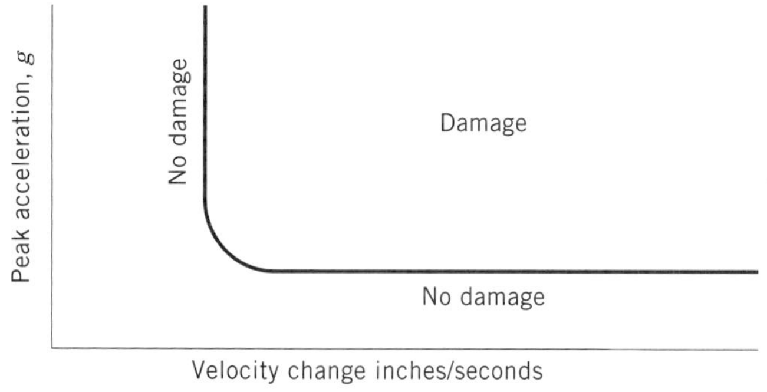

Figure 5. Typical damage boundary curve.

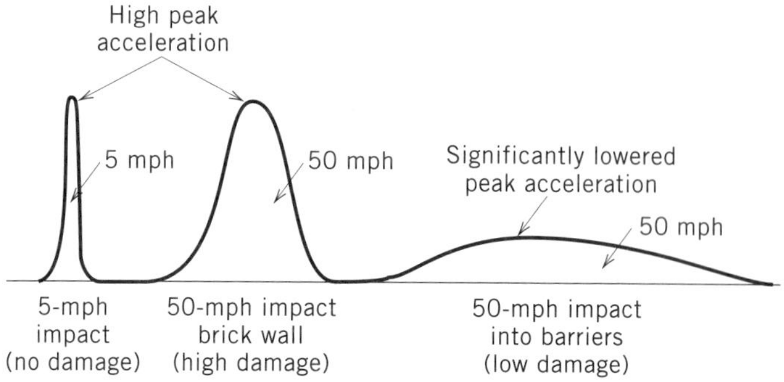

Figure 6. Effects of impact velocity and cushioning on damage in an automobile impact.

tion of a product in terms of its resistance to mechanical shocks.

For a product to be damaged from mechanical shock, two conditions must occur simultaneously. The first is that the mechanical shock must have sufficient energy to cause damage. The second required condition is that the energy be imparted to the product at a rate exceeding the product's ability to absorb the energy without breaking. The first condition is expressed as the velocity-change side of the damage boundary; the second condition, as the acceleration side of the damage boundary (see Fig. 5).

The velocity-change side of the damage boundary curve is an expression of how much energy is required to damage a product. This is directly related to the velocity change or area under an acceleration-time shock-waveform record. Again, the predominant source of velocity in package shocks result from drops. Even if the peak acceleration level were very high, if the pulse duration were very short, the resulting velocity change might provide insufficient energy to damage a product. The product, in effect, does not respond to the high-intensity shock because the duration and energy input are too limited to cause any damage. It is similar to how the human eye can withstand the high intensity of a flash from a camera for a very short duration of time since the light does not have time to cause the eye to respond adversely.

The acceleration side of the damage boundary curve is of concern only with the prior condition, in which sufficient velocity change has been met, and is thus then a question of how fast the velocity change is being imparted to the product. A familiar analogy is that of an automobile in a crash. If the speed is slow, usually less than 5 mph (mi/h), no significant damage will occur because it is below the velocity-change damage boundary. However, if the automobile is moving much faster, say, 50 mph, sufficient energy is present and damage will certainly occur if the automobile is driven into a brick wall, causing the energy to be dissipated at a very high rate (acceleration level) when it comes to a sudden stop. If the same automobile traveling 50 mph is impacted into yellow safety barrels placed in front of the brick wall, the stop rate may be significantly less, resulting from the added stopping time, so that no significant damage occurs (see Fig. 6). It should be apparent that without seat belts and airbags, people in a vehicle will continue moving at the 50-mph speed and not benefiting from the yellow barrier cushions stopping the automobile at a much lower rate.

Although both damage-boundary parameters—velocity change and acceleration—are critical in defining a product's fragility, the velocity-change side of the damage boundary is often considered to be sufficiently low for normally cushioned products, so that it will be exceeded during normal handling conditions. For that reason, often only the acceleration side of the damage boundary is referred to when one discusses the fragility of a product. This is technically insufficient; however, it is a significant improvement over working without any product fragility data.

It is next to impossible to determine the fragility of products through observations or even calculations. It is essential to conduct fragility testing to assess the fragility of any given product. However, there are certain categories of products that because of the severity of their use environments, typically have fragility levels that reach a minimum stated acceleration fragility level. Table 1 lists the typical minimum fragility levels of products found in these categories. Stated again, without testing, one cannot know the fragility of a product. However, it may be sufficient to test only to a minimum-assurance fragility level. In other words, it may be sufficient to subject a product only to a minimum acceptable acceleration shock level to ensure that it at least withstands the minimum assurance level. This then becomes the basis for developing protective packaging.

One critical issue must be addressed when discussing the shock fragility of a product. That is the manner in which the force of a mechanical shock is imparted into a product. As an example, the acceleration portion of a damage boundary for a fresh chicken egg is 35–50*G*, depending on how it is impacted on the side or end of the egg. This is based on testing and

Table 1. Benchmark Fragility Levels of Products

G Factor	Classification	Examples
15–25	Extremely fragile	Precision instruments, first-generation computer hard drives
25–40	Fragile	Benchtop and floor-standing instrumentation and electronics
40–60	Stable	Cash registers, office equipment, desktop computers
60–85	Durable	Television sets, appliances, printers
85–110	Rugged	Machinery, durable appliances, power supplies, video monitors
110	Portable	Laptop computers, optical readers
150	Handheld	Calculators, telephones, microphones, radios

imparting the shock using a flat impact surface. If, on the other hand, the shock is imparted to the fresh egg using a surface conforming to the shape of the egg, the acceleration fragility level exceeds 150*G*. The same holds true for many products. How the cushions transmit shock into the products, on corners, edges, or flat surfaces, can significantly effect the critical fragility of the product.

Product-fragility tests must be conducted with an eye toward the final package cushion configuration that will be used. It almost makes one wonder what comes first, the cushion or the egg.

Summary

In summary, shock occurs usually as the result of stopping a fall. Shocks are characterized by their duration, peak acceleration, velocity change, and waveshape. Waveshapes are the signatures of shock events that can be analyzed to characterize impacts and to analyze protective cushioning. Protective cushions are characterized as cushion curves that depict the cushion's transmitted peak acceleration levels for given drop heights and cushion thicknesses. Product-fragility testing measures the product's resistance to shocks defining in terms of both velocity sensitivity and resistance to peak acceleration shock levels. The use and analysis of shocks provide the measurements essential to engineer products and packages that can withstand their shock-generating environments.

BIBLIOGRAPHY

General References

ASTM D1586, *Method of Test for Dynamic Properties of Package Cushioning Materials,* American Society for Testing and Materials, 1996.

ASTM D3332, *Standard Test Methods for Mechanical-Shock Fragility of Products, Using Shock Machines,* American Society for Testing and Materials, 1996.

ASTM D4003, *Standard Test Methods for Programmable Horizontal Impact Test for Shipping Containers and Systems,* American Society for Testing and Materials, 1996.

ASTM D4168, *Standard Test Methods for Transmitted Shock Characteristics of Foam-in-Place Cushioning Materials,* American Society for Testing and Materials, 1996.

ASTM D5276, *Standard Test Method for Drop Test of Loaded Containers by Free Fall,* American Society for Testing and Materials, 1996.

ASTM D5487, *Standard Test Method for Simulated Drop of Loaded Containers by Shock Machines,* American Society for Testing and Materials, 1996.

ASTM D5277, *Standard Test Method for Performing Programmed Horizontal Impacts Using an Inclined Impact Tester,* American Society for Testing and Materials, 1996.

C. M. Harris and C. E. Crede, *Shock and Vibration Handbook,* Editions 1–4, McGraw-Hill, New York, 1961.

International Electrotechnical Commission, *Basic Environmental Testing Procedures, Part 2: Tests-Test Ea: Shock,* Publication 68-2-27, 2nd ed. 1972.

R. D. Newton, *Fragility Assessment Theory and Test Procedure,* Monterey Research Laboratory, Inc. Monterey, CA, 1968.

ROBERT M. FIEDLER, CPP
Robert Fiedler & Associates
Minneapolis, Minnesota

SHRINK BANDS. See BANDS, SHRINK.

SILK SCREENING. See DECORATING.

SKIN PACKAGING

The innovative, technological developments in skin packaging over the past several years have opened up a whole new arena for this packaging concept. The major developments that have been made in skin packaging are better-quality skin board [solid bleached sulfate (SBS) and recycled] and improved printing techniques. Now, with the new specialized printing equipment that provides higher resolution [133 dots per inch (dpi)], you can obtain beautiful-looking graphics with great sales appeal. Also, the new advances in the separation of the colors allow you to print just about anything on a skin board. Improvements in printing, along with the advancements made in skin equipment and the development of new high-performance skin packaging films, has helped to greatly expand the market opportunities for skin packaging.

As the proliferation of mass merchandising outlets and self-serve retailing continues to expand, the growth of skin packaging will continue because of its merchandising appeal to both the consumer and retailer. Today, manufacturers are striving for a cost-effective eye-appealing package that sets their product apart from competition and enhances the product. They are looking for packaging that will put new life into old products to gain additional market share. A package that offers product visibility and creates impulse sales to help make the buying decision easier. Skin packaging provides all this and in addition has the merchandising advantage of being peggable.

There are two market opportunities for skin packaging: retail and industrial. For the retail visual carded packaging market, skin packages are typically merchandised by hanging them on a peg. This enhances space utilization in that the package requires less shelf space and provides better use of the space allocated by offering "more facings" and "good vertical display." Skin packaging is a cost-effective alternative to other visual carded packaging, ie, blister packaging, clamshells, foldover blisters, and Stretch Pak, because it requires no tooling in that the product itself becomes the mold over which the film is drawn down and around by vacuum.

Industrial skin packaging differs from visual carded packaging in that its primary purpose is to protect products in transit. Products as divergent as computer tapes, lamps, service repair kits, fans, and tabletops may be skin-packaged instead of using die-cut corrugated, foam-in-place, foam peanuts, and other stabilizing or dunnage materials. Industrial skin packaging offers high throughputs and full visibility to check for tampering or missing components and allows quick identification, usually at significant cost reduction. It also considerably reduces the amount of waste that has to be disposed of and is therefore more environmentally friendly.

Skin-Packaging Materials and Equipment

Skin packaging involves placing a product on a substrate material such as paperboard or corrugated board (see Fig. 1). When the film is heated to the proper softening temperature,

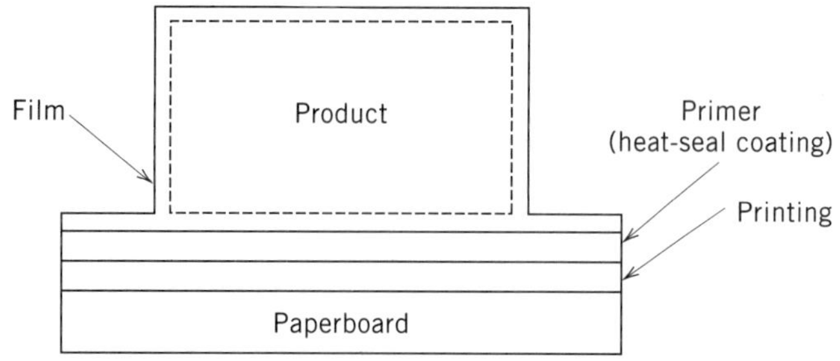

Figure 1. A typical skin package.

it is draped over the product and the substrate material and a vacuum draws the film down and around the product and into the pores of the board to make a secure and attractive-looking package.

A skin package has four components: the plastic film, a heat-seal coating, printing ink, and a paperboard or corrugated substrate card. The product itself becomes the mold over which the heated plastic film is drawn down and around. This is one of the major differences between blister and skin packaging, in that in blister packaging you have to form a blister that requires a mold, for not only the thermoforming machine but also the blister packaging machine if you are buying preformed blisters.

Skin packaging begins with a sturdy card made of high-porosity paperboard or corrugated paperboard. After printing the board, which, with the developments in printing technology and the higher-quality skin board available today, you can now get beautiful four- to six-color graphics, is coated with a heat-seal dispersion coating. Today the majority of printers use an EVA water-based coating. The product is placed on to the board, and then the product and board are moved onto the platen. The plastic film is held tightly in place by a clamping frame, directly below the heater, which is located above the platen.

At the beginning of the packaging cycle, the product is usually loaded onto a printed, coated skin board that is then moved onto the platen of the machine to be skinned. The plastic film that is held in the frame above the product on the skin board is heated to the proper softening temperature and is then lowered over the product and the skin board. As the frame is being lowered, the vacuum system is energized, and as the film is draped over the product and the board, the vacuum pulls the film tightly around the product and into the pores of the skin board, making a tight, attractive-looking package. The hot film instantly activates the coating on the board, and when a milky white coating is used, it becomes crystal clear when coming in contact with the hot film. The paperboard card with the product skinned to it is then moved to the die cutter, which die-cuts it into individual packages (see Fig. 2).

Plastic film. One of the least expensive components of the total package, the film is one of the most important. It is the film that holds the product on the board. The film also reinforces the board and provides a glossy surface for enhanced graphics. Film of DuPont Surlyn ionomer resin is the most widely used film in the retail visual carded packaging market with some LDPE and EVA coextruded films used for the low-end retail market. Film of Surlyn is a very tough clear film and enhances the graphics and appearance of the package to help create impulse sales. It has excellent abrasion resistance and also has very low shrink force, which eliminates any unsightly board curl. Polyethylene (PE) is used primarily for the industrial skin-packaging market where optical properties are not as important as in the retail visual carded packaging market. PE is not as clear as film of Surlyn and requires more heat and dwell time. It shrinks more of the cooling cycle and has a higher shrink force than film of Surlyn, which causes board curl. It is a tough film and adheres well to primed paperboard and corrugated. PE is less expensive but requires a heat-cycle time that is twice that of Surlyn; therefore Surlyn can be more cost-effective because of the increase in productivity.

The film gauge that is needed will depend on the weight of the product, the draw required, and the abuse that the pack-

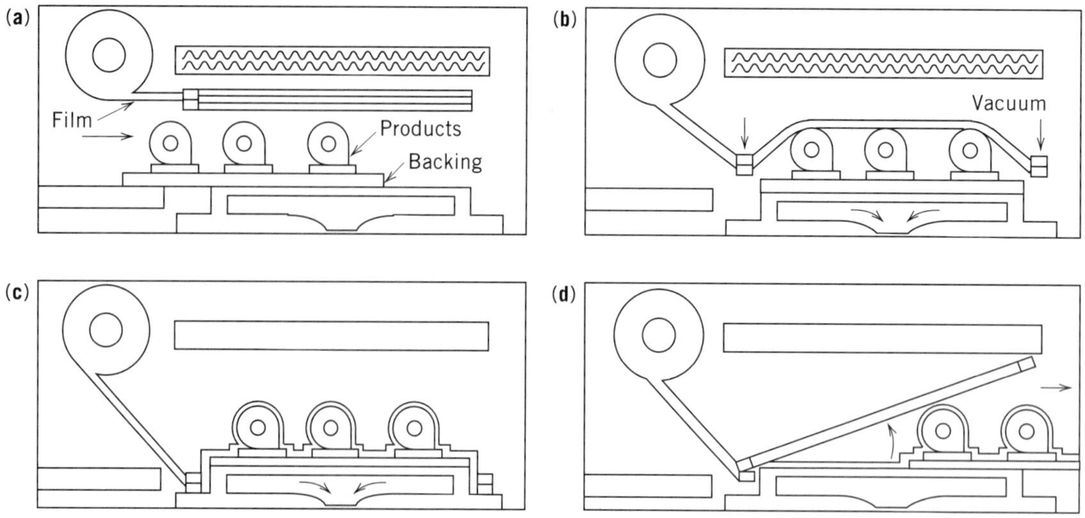

Figure 2. (**a**) Product and backing are moved onto the vacuum platen of skin-packaging machine and film is heated; (**b**) heated film is draped over product and backing; (**c**) vacuum suction draws film around product and backing, creating a skin-tight package; (**d**) finished board is moved out of the machine.

Table 1. Properties of Aqueous Dispersions of Skin Packaging

Solids (weight %)	40–50
Viscosity	500–100 mPa
pH	9.5–10
Odor	None
Appearance unfused	Clear or milky white
Appearance fused	Colorless
Dilution	Water as required
Flammability	Nonflammable
Drying	130–230°F (55–110°C), variable

age is likely to encounter. It is recommended that the film be corona-treated to a level of 42 dyn to further improve the adhesion of the film to the board. If the film is surface-treated on one side, the roll label should designate which side is treated. The treated side should be down to come in contact with the board.

Primer. Most skin-packaging films (Surlyn, LDPE, etc) do not exhibit good natural adhesion to ink or board. Printed and unprinted boards are, therefore, coated with heat-sealable primers to provide strong, long-term adhesion of the film to the ink/board surface.

EVA films and coextruded films containing EVA have good initial board/ink adhesion. However, actual package failures have demonstrated that EVA to ink/board bonds will deteriorate with time. The deterioration is accelerated by high ambient temperatures (like those usually experienced during shipping).

Primers are generally aqueous EVA-based dispersions specifically formulated to provide a good bond between film and board. The primers soak into the board fibers and porous inks. This maximizes mechanical and chemical adhesion between film, ink, and board.

Application of primer. The manufacturers of dispersion coatings can supply specific recommendations for applying their primers to achieve a good fiber tear bond between the film and the board. The general properties of commercially available dispersions are described in Table 1.

Dry coatings are applied commercially with a gravure cylinder. The applied coating should be of the proper weight and one that will not seal the surface of the sheet. Too light a coating will result in poor adhesion. Too heavy a coating will block the pores in the board and ink. This will cause poor evacuation and film forming.

Board priming. Board priming is usually done most economically by a company set up to print and coat the boards in one operation.

Ink. It is important to use the correct compatible heat-resistant ink to ensure good adhesion to the board (see Inks). Using the wrong inks or the wrong board/ink/primer combinations can interfere with the ink's adhesion to the board because of one or more of the following chemical or physical problems:

Chemical Incompatibility. A primer-to-ink bond failure can occur if the ink and primer are chemically incompatible.

Unstable Pigment, Vehicle, or Binder. An ink cohesive failure will occur if the pigment, vehicle, or binder is not stable under heat.

High Board Porosity. The ink may fail to adhere to the board if the ink has a low binder and vehicle : pigment ratio and the board is very porous. The binder and vehicle are absorbed into the board, and the poorly bound pigment remains on the surface. This phenomenon is known as "chalking."

Ink/Board Incompatibility. The ink may not bond to the board if the ink and board are incompatible. This is often caused by contaminated board.

Ink/Primer Incompatibility. A primer-to-ink bond failure will occur when the ink contains lubricants, waxes, or oils incompatible with the primer.

When starting a new job, it is important that preproduction ink evaluation tests be made before any board is printed in large quantities. This is a simple precaution to assure that the ink system is compatible with board and film. This test consists of having proofs prepared by the printer with the ink and board intended for use. Sample skin packages should be prepared with these proofs, checking for proper evacuation of air and adhesion.

Paperboard. Three types of board are used for skin packaging:

- Fourdrinier board (solid bleached sulfate)
- White lined recycled board
- Corrugated board

The board must be compatible with the ink and primer system, be porous enough to permit a vacuum to be drawn through it, and be strong enough to hold and display the product (see Paperboard).

By specifying board density and the weights of any coatings to be used, you can ensure acceptable porosity for your skin board, specify the nature of any coatings or treatments to ensure compatibility of the board with inks and primers, and ensure that the board is strong enough for the application by specifying board type and thickness. The strength of the board is enhanced by the skin-packaging film used. Film of Surlyn, with its superior strength, will produce a stronger package than competitive skin packaging films, so you may sometimes be able to specify a slightly thinner board gauge.

Solid bleached sulfate board (SBS). Solid bleached sulfate board for skin packaging is available in four calipers: 0.020, 0.024, 0.026, and 0.028. The specifications for a SBS board that will be acceptable in skin packaging are as follows:

Density. The density should be low enough to produce a board porosity that will allow a suitable vacuum to be drawn through the skin card. This porosity can be attained with a density that is equal to or less than 1.9 kg/(m^2 · mm) (9.8 lb/1000 ft^2-point).

Finish. A soft finish is recommended for SBS board.

Brightness. The board should have a brightness of 80 or better (General Electric (GE) brightness scale). The printer may have to adjust the inks used to accommodate variations in board tint.

Sizing. Waxes, lubricants, release agents, and surface treatments are not recommended for this board and should be avoided.

Clay coating. Clay coating is not recommended. If a clay coating is used, the coating weight should be less than 0.49 kg/100 m^2 (1.0 lb/1000 ft^2).

Clay binders. Clay binders must be heat-stable. Protein binders are acceptable. Styrene latex binders should be avoided.

White lined recycled board

Density. White lined recycled board for skin packaging is available in five standard calipers: 0.028, 0.032, 0.036, 0.042, and 0.054.

The density of white lined recycled board should be low enough to produce a board porosity that will allow a suitable vacuum to be drawn through the card. This porosity can be attained with a density that is equal to or less than 2.1 kg/(m^2 · mm) (11.1 lb/1000 ft^2-point). A denser board will require perforations.

Finish. A number 2 or number 3 finish is preferred, although number 4 may be acceptable (according to boxboard standards of the National Paperboard Association). Finish refers to density and surface; higher numbers are smoother and more dense.

Brightness. The board should have a brightness of 72 or better (GE brightness). Printers may have to adjust their inks to accommodate variations in board tint.

Gross ink treatment. Gross ink treatment is not recommended for white lined recycled board. If gross ink is used, the treatment should not contain any release agents or wax-based material.

Binders. No wax or styrene binders should be used on white lined recycled board.

Clay coating. Clay coating is not recommended for white lined recycled board. If a clay coating is used, the coating weight should be less than 0.49 kg/100 m^2 (1.0 lb/1000 ft^2).

Clay binders. Clay binders must be heat-stable. Protein binders are acceptable. Styrene latex binders should be avoided.

Corrugated board

Density. Nearly all common corrugated boards are porous enough to permit skin packaging. Normal bleached facing and unbleached facing made from 90-lb liner are satisfactory. The time required to permit 100 mL of air to pass through the board, as measured with a Gurley densometer (TAPPI method T460 OS75), should be less than 100 s. Laminated facings should generally be avoided. All flutings are acceptable; types B, C, and E are the most popular for skin packaging.

Finish. If a printed label is applied to the corrugated board, all the considerations that apply to a solid bleached sulfate board should apply to the label. The adhesive used to bond the label to the board should not form a continuous film as this will destroy board porosity. Pressure-sensitive-type adhesives should be avoided. Converted starch adhesives are acceptable.

Sizing. Waxes, lubricants, release agents, and most surface treatments are not recommended for this board and should be avoided. A normal starch calender coating is acceptable.

Clay coating. Clay coating is not recommended for corrugated board. If a clay coating is used, the coating weight should be less than 0.49 kg/100 m^2 (1.0 lb/1000 ft^2).

Clay binders. Clay binders must be heat-stable. Protein binders are acceptable. Styrene latex binders should be avoided.

Machinery. Skin-packaging equipment is available in a wide range of models and sizes. The most inexpensive manual models will provide only the basic requirements needed to skin-package a product. A manual system will normally run at speeds of 2–3 cycles per minute. The number of packages produced per minute on these systems will depend on the skin-board size and the number of individual cards that can be cut out of the board. The standard skin-board sizes are 18 in. × 24 in., 24 in. × 30 in., 30 in. × 36 in., and 36 in. × 36 in.; however, there are skin-board sizes available up to 30 in. × 96 in. Depending on the size of the package, you can normally get 10–12 individual packages out of an 18-in. × 24-in. skin board. Semiautomatic and fully automatic systems can run at 5–12 or more cycles per minute depending on the sophistication of the machine.

The heating system on a skin machine is mounted above the platen, and a radiant-heat source is used to soften the film. The source of radiant heat is usually a series of electric resistance rods, wire coils, or quartz infrared heat lamps. A diffuse reflecting surface directs all available radiation toward the film and platen and avoids the formation of hot spots.

The vacuum system is connected to the base of the platen and is capable of drawing through the substrate board during the vacuum cycle. There are two types of vacuum systems: turbine and positive displacement. The turbine system is the least expensive and is generally used on the manual and inexpensive skin models. This system has a turbine blower (like the type used in commercial vacuum cleaning equipment) and is capable of drawing a large volume of air at a relatively low vacuum pressure. This system works most effectively with board that has good porosity.

The positive displacement vacuum system consists of a pump and a surge tank that can evacuate the air between the film and the board very rapidly with a strong force. This system is used for the faster, more sophisticated skin machines. In some cases, a dual-vacuum system (turbine and positive displacement) is needed, such as for high-profile and odd-shaped products and for board with poor porosity. It is sometimes worthwhile to purchase both systems if you have a wide range of products to package.

A roller-type or hydraulic die cutter cuts the skin-packaged board into individual packages and die-cuts the hanger hole used to hang the card on a hook. Stainless steel or polypropylene is an acceptable cutting surface.

High-speed fully automatic skin packaging systems can meet the packaging speed requirements of today's packaging applications while solving the concerns traditionally associated with skin packaging. They come equipped with smooth acceleration "no roll" product infeed, adjustable high/low vacuum controls, quick changeover (<5 min), automatic card feeding, ease of loading with templates, and automatic off-load boxing systems to meet industry's high-volume, just-in-time (JIT) production requirements. The new quiet hydraulic die cutters come equipped with in-press die-saver sensor systems to eliminate product damage. Automatic product placement systems are also available today for high-speed automatic packaging.

Applications

A wide variety of applications are being skin packaged today, although not all products are suitable for skin packaging. The product's shape and composition will usually determine whether it can be skin-packaged successfully.

For example, very flimsy products that would be crushed by the force of the vacuum or articles that would stick to the skin-packaging film should not be skin packaged. However, in the past where you could not package round or unstable ferrous metal parts, ie, ball bearings, because of the difficulty of maintaining the product on the card prior to film contact, today you can with moving magnetic plates. You can also package parts that protrude through the back of the card.

The weight and configuration of the product will usually determine the strength of the board and the gauge of the film that is needed. Heavy, high-profile items will require stronger board and heavier-gauge film.

Skin packaging for the retail visual carded display market offers a number of benefits for the packager, retailer, and consumer. The packaging operation is quick and uniform, the film's clarity and sparkle enhance the product, and the display method spurs impulse buying because of product visibility and the attractive colorful graphics.

Burt Spottiswode
DuPont P&IP
Wilmington, Delaware

SLIPSHEETS

Small shipping containers have been combined into unit loads since the 1940s, when the wooden pallet was developed (see Pallets). The weight of wooden pallets, however, and the space they take in transport equipment precluded economical unitized shipping of many lightweight products, such as frozen foods, cereal products, and cotton balls. Also, as palletized shipping programs expanded the difficulties of controlling pallet quality and costs increased substantially. These shortcomings led to the development of the palletless unit-load system that utilizes a slipsheet as the carrying base. The use of slipsheets lowers freight costs by increasing the payload on transport vehicles and eliminates the need to return pallets to the shipper. Their use requires front-end attachments on forklift trucks.

Materials. Slipsheets are manufactured from solid fiber, corrugated board, or plastic. Solid fiber is a lamination of plies of kraft linerboard and cylinder board designed to provide adequate tensile strength and other properties. The thickness of a solid-fiber slipsheet is expressed as caliper of board and is measured in points. One point (mil) is 0.001 in. (25.4 μm). Corrugated board (see Boxes, corrugated) is described in terms of flute size and board test, eg, B flute, 250-lb test (linerboards are 69 and 42 lb/1000 ft^2, ie, 337 and 205 (kg/m^2). Plastic slipsheets are made of polyethylene, polypropylene, or other polymers. Thickness is expressed in terms of mils.

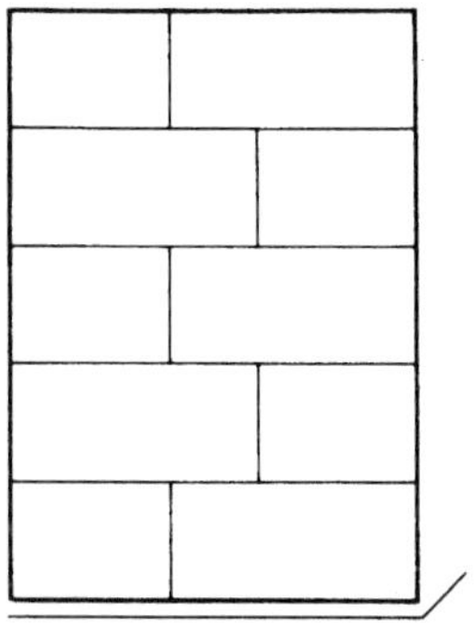

Figure 1. Slipsheet with side tab.

Configurations. A slipsheet is a flat sheet of material that is used as the platform base upon which goods and materials may be assembled, stored, and transported. It may have one

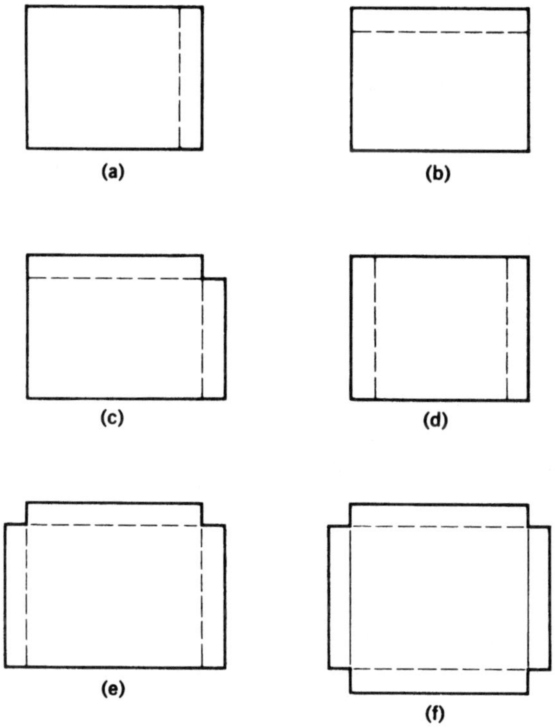

Figure 2. (**a**) One tab on the short side is excellent for unit loads that are positioned in one direction and loaded and unloaded on the short side of the slip sheet. (**b**) One tab on the long side is excellent for unit loads that are positioned in one direction and where it is desirous to load and unload a unitized load on the long side of the slipsheet. (**c**) Two adjoining tabs are perfect for unit loads that are positioned in a pinwheel pattern, that is, one unit load facing in the short-side direction and one unit load facing in the long-side direction, side by side. This results in greater cube utilization of transportation vehicles. (**d**) Two tabs on opposite ends are excellent for those situations where handling requirements demand loading or unloading from opposite ends of the unit load. (**e**) Three tabs, two on the short side and one on the long side, are excellent for those situations where handling requirements demand entry from three sides. It is particularly useful in the doorway area of railcars. (**f**) Four tabs, on all four sides, are excellent for those situations where handling requirements demand entry from all four sides, particularly in the doorway area of railcars.

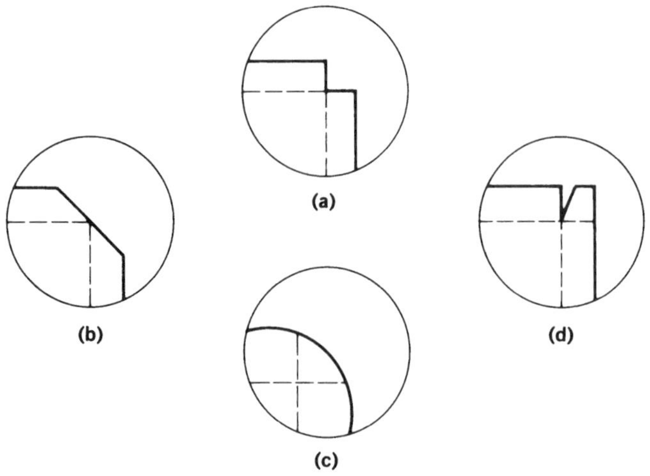

Figure 3. Common tab corner configurations: (**a**) 90° corner cutout; (**b**) diagonal corner cutout; (**c**) rounded corners; (**d**) slit corner.

or more tabs along its side that extend beyond the load to facilitate mechanical handling. The platform base is the area on which product is placed and stacked to form a unit load. The dimensions of the platform base, excluding the tabs, must be at least as large as the base dimensions of the unit load in order to prevent overhang. Any overhang results in dragging of the product or container, which can break up the unit load (see Fig. 1).

The number of tabs, located on one or more sides, is usually determined by the handling and shipping requirements of the user. The recommended width of tabs is 3 in. (76 mm) min to 4 in. (102 mm) max. These dimensions are compatible with the gripper bar of handling trucks. The most common tab configurations are shown in Figure 2.

Overall size. The combination of base and tab dimensions determines the overall slipsheet size. The most common size is 48 in. × 40 in. (1.219 m × 1.016 m), which corresponds to the standard Grocery Manufacturers Association (GMA) pallet. With two adjoining 4-in. (102-mm) tabs, these dimensions would be described as 48 in./4 in. × 40 in./4 in. (1219 mm/102 mm × 1016 mm/102 mm). Sizes in countries using the metric system correspond to standard metric pallets (eg, 1200 mm × 1000 mm).

The base and tabs should be manufactured within a tolerance of $\pm\frac{1}{4}$ in. (±6.4 mm) in the length and width dimensions. Common tab corner configurations are shown in Figure 3.

Properties. *Tensile strength* is the maximum tensile stress developed before failure of the slipsheet. The slipsheet must contain sufficient strength to avoid rupture when the tab is properly pulled by the gripper bar. The gripper bar is that part of the front-end attachment of forklifts and walkie-riders that clamps the tab of the slipsheet uniformly along its length. Tensile strength of paper is tested by either TAPPI D494 or ASTM D828. Tensile strength of plastic is tested by ASTM D638-76.

Scorelines. Slipsheets with tabs are tested through the scorelines after bending scores 90°. A score is an impression or crease that is provided to locate and facilitate folding.

Stiffness. Adequate stiffness is required for handling certain products such as bagged or irregularly shaped materials to avoid excessive sagging of the slipsheet tab, which may prevent proper clamping by the gripper bar of the front-end attachment. Additional stiffness is normally obtained by increasing caliper. Stiffness testing is described in TAPPI D489 and in ASTM D790-71.

BIBLIOGRAPHY

"Slipsheets" in *The Wiley Encyclopedia of Packaging Technology,* 1st ed., by E. M. O'Mara, Union Camp Corporation, pp. 582–583.

General References

J. C. Bouma and P. F. Shaffer, *USDA Marketing Research Report, No. 1075,* United States Department of Agriculture, Washington, DC, 1978.

P. J. Mann, *Food Eng. Internatl.* **2**(10), 32 (1977).

Quick Frozen Foods **38**(7), 52 (1976).

SLITTING AND REWINDING MACHINERY

Flexible plastic packaging materials generally do not have perfectly flat surfaces. The material in a parent roll contains stretch lanes and varies in gauge by about 1–5% across the web width. Slitting and rewinding such rolls requires equipment that can control the rewinding speed of each slit roll independently (1). The rewinding process is further complicated by air entrainment into the coils of the rewinding rolls, which separates the outer convolutions and allows them to slip laterally. Winding difficulty caused by this phenomenon increases when handling high-slip films or films with very smooth surfaces. The primary aim of every slitter design is to provide apparatus to deal selectively or generally with one or more of the diverse products presented to it. All designs are based on center winding (1), surface winding (2), or a combination of both.

Center Winding

Center winder (duplex winder). The most widely used and versatile slitter is a duplex stagger center winder or two-bar winder (see Fig. 1). The web tension at the unwind of the machine is isolated from the rewind section by driven rollers in the slitting section of the machine. These rollers positively control web speed, opposing any overdraw by the rewind. For most applications, the rewinding is done on differential-rewind (slip-core) mandrels in order to maintain suitable rewind tension on each slit roll despite stretch lanes in the web and unequal diameter buildup due to gauge bands commonly found in plastic webs (1). Individual riding rollers forced against each of the rewinding rolls minimize air entrainment into the rewinding roll. For winding paper products, the riding roller also has a surface-winding effect that increases roll hardness without using excessive rewind tension (3).

These machines are suitable for handling web widths up to about 80 in. (2032 mm), rewind diameters up to 24 in. (610 mm), and web speeds of ≤ 1500 ft/min (~460 m/min).

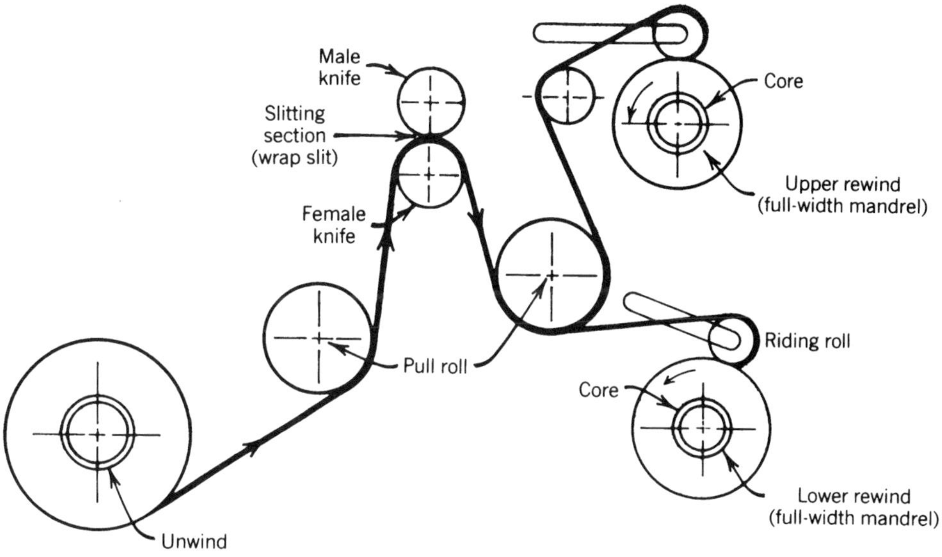

Figure 1. Center winder (duplex winder) illustrating stagger differential winding principle.

They are especially suited for narrow-width slitting. The productivity of machines with removable rewind mandrels can be increased by using two pairs of mandrels so that the slitting operation can continue while the alternative pair is being serviced. Removable mandrels commonly require two operators or one operator and a hoist for handling into and out of the machine. Machines equipped with nonremovable cantilevered mandrels permit one operator to remove the slit rolls, recore, and restart winding quickly without mandrel handling (see Fig. 2). The choice of systems depends on the size and number of slit rolls per setup. Maximum production is achieved by mounting cantilevered mandrels into two-station turrets (see Fig. 3). With automatic cutoff and restart of web around the core, these machines can slit and rewind a web emerging from an extruder on a continuous basis.

Individual-rewind-arm (IRA) surface-center winders. IRA-winder design is based on both center- and surface-winding principles (see Fig. 4). The design stresses the ability to control the surface-contact force and rewind torque of each rewinding roll individually. Each rewind core is supported by chucks (one or both driven by a torque-controlled motor) supported on pivoted arms arranged in stagger-wind fashion on each side of a winding drum. Handling of rewind mandrels is eliminated (1). The contact force between the rewind roll and the winding drum is exerted by air cylinders controlled manually or by microprocessors. For better control of roll hardness, this force can be programmed as a function of rewind diameter and web speed.

These machines are suited for processing webs in excess of 80 in. (2032 mm) in width and for winding in excess of 24-in. (610-mm) diameter. Maximum design web speed is about 2000 ft/min (610 m/min). Actual production operating speed is frequently limited to ≤1200 ft/min (365 m/min) because of web properties, vibration of the rewound roll (4), parent-roll eccentricity, or off-machine roll-handling backup (see Roll handling). Doffing of the rewound rolls can be done in sets

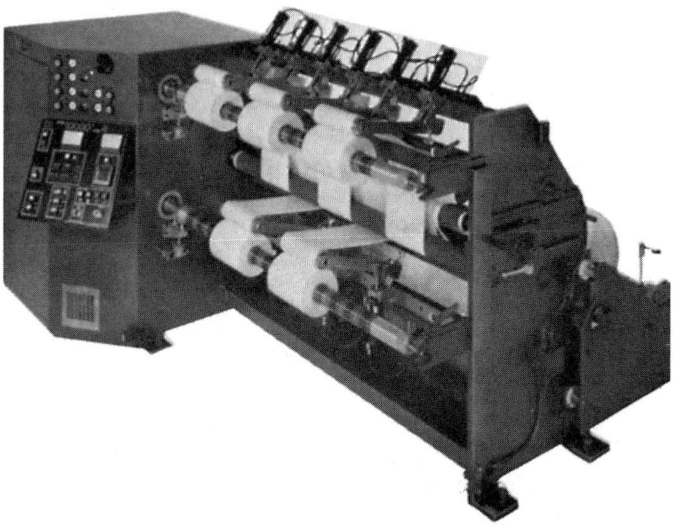

Figure 2. Center winder with cantilevered rewind mandrels and integral unwind station.

Figure 3. Center winder with cantilevered rewind mandrels mounted in turrets for continuous differential winding.

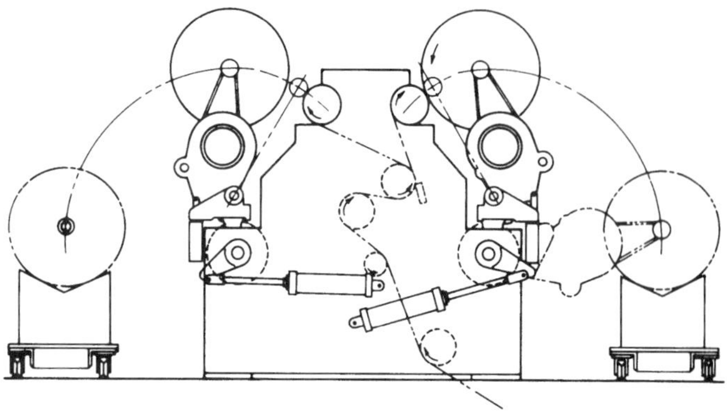

Figure 5. Individual-rewind-arm (IRA) surface-center winder with doffing to floor.

with overhead hoist and special "grabbing" devices. Recent designs feature rotating rewind arm support beams, allowing the rewound rolls to be lowered and released directly onto low-bed, wheeled dollies. This is called doffing to the floor (see Fig. 5).

Individual-rewind-arm (IRA) center winders. This winder is similar to the IRA discussed above except that an individual contact roller, supported by a pair of pivoted arms, contacts the rewound roll in place of a full-width winding drum (see Fig. 6). The contact roller is urged against the rewound roll by air cylinders. The winding tension is developed by a rewind motor driving each core. The primary function of the contact roller is to minimize air entrainment. Sensors detect the buildup of the rewind roll, causing the rewind arms to move outward and maintain a relatively fixed position of the contact roller. This winding method is more responsive to control of stretch lanes in the web and significantly reduces the roll vibration sometimes associated with high-speed surface winders (4). For best results, the contact force and rewind torque must be programmed by microprocessors as a function of web speed and rewind diameter, thereby giving the desired hardness and uniformity in the rewound roll at web speeds upward of 2000 ft/min (610 m/min). These machines also feature doffing to the floor.

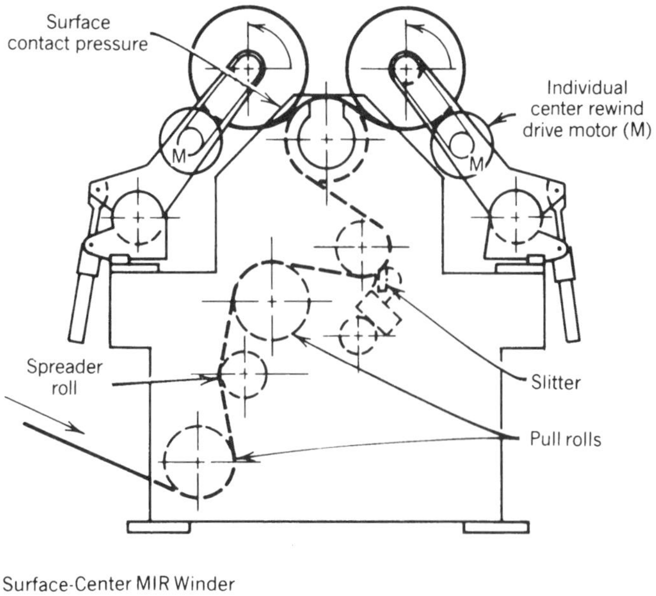

Figure 4. Individual-rewind-arm (IRA) surface-center winder.

Surface Winders

Many versions and sizes of slitting equipment employ the surface-winding principle (2,5). For detailed discussion of larger-size rewinders, see Ref. 6. The machine most widely used for general converting is the two-drum winder with riding roll (see Fig. 7). It is commonly used for slitting plain paper and paperboard, as well as many printed, coated, and laminated paper products.

Good-quality paper has the following physical properties conducive to surface winding: uniformity of gauge, freedom from bags, relatively rough surface with medium friction coefficient, high modulus of elasticity, and high degree of compressibility. The density of the finished roll is determined by

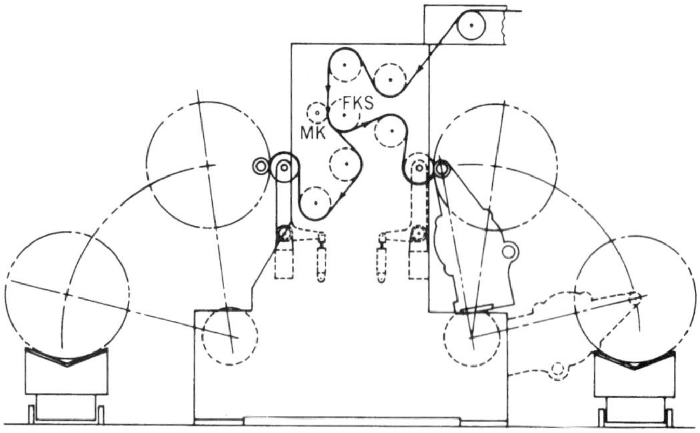

Figure 6. Individual-rewind-arm (IRA) center winder with riding roll and doffing to floor.

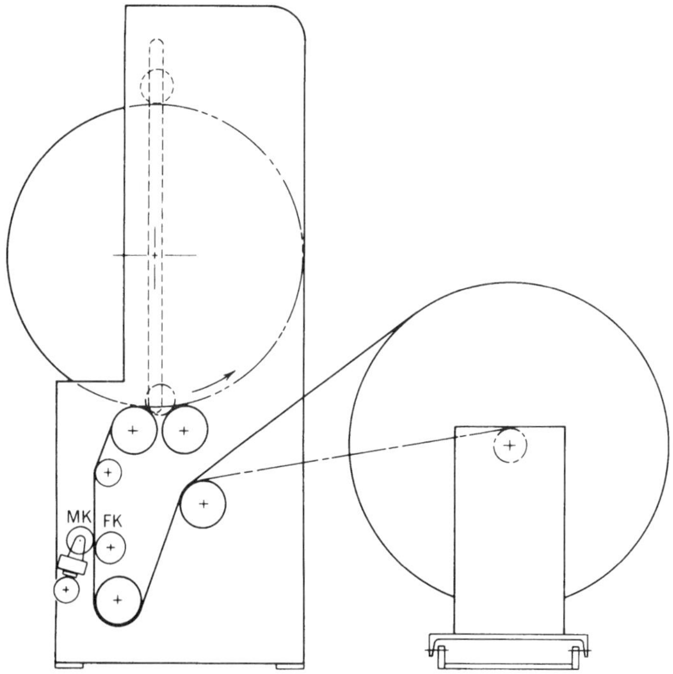

Figure 7. Two-drum surface winder with kiss-shear slitting.

a combination of constant unwind tension and down force on the riding roll (7).

The medium-size, general-purpose converting slitter has the following advantages:

1. The controls and construction are relatively simple.
2. It can wind to large-diameter buildup ratios.
3. It can wind coreless or on almost any diameter core.
4. It winds on one mandrel or without mandrel.
5. The slit-width changeover is accomplished primarily by relocation of knives.
6. It runs at relatively high web speeds, commonly ≥2000 ft/min (610 m/min).

It also has the following disadvantages:

1. The risk of slit rolls interleaving make it difficult or impossible to separate the slit rolls. Bowed spreader rollers can alleviate roll-separation problems (8,9).
2. It lacks versatility for winding baggy or off-caliper webs.
3. It is unsuitable for winding paper with slippery, eg, silicone, coatings.
4. Some grades of paper develop a washboard surface on the periphery of the rewound roll during winding, thereby causing vibrations and restricting operating speed (4).
5. Operator skill is necessary to keep scrap rate low on some products.

Unwinding

Many types of unwind stands are available, from simple shaft-type units with manual side shift to large shaftless turret unwinders (10) with automatic web guiding (11,12). Friction brakes usually provide the holdback torque (13). Regenerative dc drives are occasionally required for parent-roll acceleration, as well as holdback (14). The choice of automatic tension controllers for maintaining constant unwind tension during acceleration and deceleration, as well as compensating for decreasing unwind diameter, includes load-cell sensing rollers (15); dancer roller sensors, recommended for use with eccentric parent rolls (16); and open-loop microprocessor control based on real-time calculations of diameter and weight of parent roll. It has been common practice to use overhead hoists for loading parent rolls into operating position, but hydraulic lifts integral with the unwinder are frequently preferred (17).

Tension Control—Rewind

Tension control for surface winders simply requires constant tension control from the unwind stand and has a secondary effect on rewind roll hardness. Tension control of center winders is much more complex because tension is the primary control factor. Rewind buildup ratios of less than 5:1 may require only constant torque (decreasing tension) windup for many products. However, large buildup ratios and tension-sensitive products require either constant tension or programmed tension (18). Depending on the design of the machine, this may require automatic control of pressures to differential-rewind mandrels or air-operated slip clutches, or control of reference voltages to dc rewind motors (19). When differential winding at high speeds and large buildup ratios, it is necessary to automatically reduce the rpm of the rewind mandrels as a function of rewind diameter. The profile of the tension pattern as a function of diameter must be determined empirically. Microprocessor control is ideal for exploratory profile studies and for maintaining repeatability once the desired results are achieved.

Slitting

Rotary shear knives can be used to slit virtually all packaging-web products. For simplicity of setup and sometimes for cleaner cutting, razor blades are used for plastics thinner than about 0.005 in. (127 μm). For better control, especially when taking narrow trim cuts, the web is wrapped around and is positively supported by a grooved roll at the point of slitting. Similarly, shear slitting is done with the web wrapped around the so-called lower or female knife shaft, known as wrap slitting. If frequent changes of slit width are required when shear cutting, it is desirable to run the web tangentially over the female knife (called "kiss slitting") (1) (see Fig. 7). The prime consideration in the slitting section of the machine is ease of changing slit widths. Equipment is being introduced to accomplish this automatically, but it is not yet in widespread use on the average converter-type slitter primarily because of cost and space requirements.

Other Considerations

In addition to the basic features discussed above, many others require consideration in a slitter specification. The more common items include automatic web guiders, spreader rolls, predetermining rewind footage counters, multiple rewind and main drives, special roller coverings, static eliminators, expanding unwind and rewind mandrels, core-locking differential-rewind mandrels, roll-unloading devices, provision for web inspection, web-splicing devices, safety guards, emergency trip cables, and trim-removal equipment. Last, but not least, increasing effort is being made to automate nearly every operational and setup function associated with slitter operation. About half of all slitters sold requires some design modification or accessory to meet the special needs of the user. New models are constantly being introduced to handle new products, reduce costs, increase productivity, and enhance the quality of slit rolls (20).

The performance of a slitter depends not only on the basic physical properties of the web, but also on the quality consistency of parent rolls. It is difficult to measure some of the more important characteristics, let alone predict how they will affect the slitting operations. Two supposedly identical generic products, each from a different manufacturing source, can exhibit surprising performance differences. Furthermore, problems can be expected when attempting to handle webs that deviate significantly from the norm in physical properties or when web speed and roll size exceed state-of-the-art limits. Consequently, it is always wise to check, if possible, performance of actual parent rolls on existing demonstration equipment to be assured that performance requirements can be achieved.

Nomenclature

Bags	Undulations in a web that otherwise should be a flat surface, usually occurring in lanes of various widths in the machine direction of web. These bands frequently occur where a thickness variation is above average relative to the rest of the web (1).
Buildup ratio	The diameter of a rewound roll divided by the outside diameter of its core.
Doffing	Process of removing or unloading wound rolls from a winder.
Draw	The distance a web must travel unsupported between two web transport rollers.
Gauge	Thickness or caliper of a web. Sometimes expressed in units of points (mil, 0.001 in. or 254 μm); gauge (0.00001 in. or 0.25 μm); or micrometers (0.001 mm or 10^{-6} m).
Gauge band	A machine-direction strip or band of above-average thickness in a web evidenced by a peripheral circumferential bulge on the periphery of a web roll (1).
Guide, edge	Automatic web guide using the edge of a web as reference. Those using an air nozzle as a sensing device are sometimes called air guiders (12).
Guide, line	Automatic web guide using photocells to track a reference line on a printed web (12).
PLI	Pound force per lineal inch (1bf/in. = 175.1 N/m). Unit of measure of web tension or line contact force of riding roll against a rewinding roll.
Roll	A coiled spool of web material usually wound on a paper core.
Roll, parent	A large roll from which smaller rolls are slit and rewound. Also called mill roll, master roll, mill reel, bundle roll, unwind roll, and stock reel.
Roll, rewind	Roll resulting from a slitting or trimming operation. Also called coil, spool, and bobbin.
Roller	General term for any type of rotating cylinder serving as a web-transport device to support and guide a web through a slitter.
Roller, riding	An idler roller that maintains contact with the surface of a rewinding roll. Also called touch roller, top-riding roller, contact roller, layon roller, ironing roller, and squeeze roller. Its purpose is to minimize air entrainment into a rewinding roll.
Slitter	Short term for slitter/rewinder. It is generally understood to include an unwinder.
Winding, differential	A method of stagger winding on two-bar (duplex) center winders whereby the rewinding cores are allowed to slip with controlled torque between keyed spacer sleeves on an overrunning rewind mandrel with the aim of winding each slit strip with equal tension regardless of parent-web defects. Also called slip-core winding (1).
Winding, stagger	Winding alternate slit strips on each of two rewind mandrels so that adjacent slit strips are not wound side by side on the same axis or mandrel (1).
Winding, taper-tension	A reduction of winding tension in a controlled manner from the center of a rewinding roll outward with the aim of giving the desired hardness and uniformity in the roll (18).

BIBLIOGRAPHY

"Slitting and Rewinding Machinery" in *The Wiley Encyclopedia of Packaging,* 1st ed., by R. W. Young, John Dusenbery Company, Inc., pp. 583–587.

1. J. R. Rienau, *Techniques of Slitting and Rewinding,* John Dusenbery Co., Randolph, NJ, 1979.
2. J. D. Pfeiffer, "Mechanics of a Rolling Nip on Paper Webs," *TAPPI* **51**(8), 774 (Aug. 1968).
3. J. D. Pfeiffer, "Nip Forces and Their Effect on Wound-in Tension," *TAPPI* **60**(2), 115 (Feb. 1977).
4. D. A. Daly, "How Paper Rolls on a Winder Generate Vibration and Bouncing," *Pap. Trade. J.* 48 (Dec. 11, 1967).
5. D. Satas, *Web Processing and Converting Technology and Equipment,* Van Nostrand Reinhold, New York, 1984, p. 383.
6. L. Rockstrom, *Control of Residual Strain and Roll Density by Three Winding Methods,* Cameron Machine Co., New Brunswick, NJ 1964.
7. J. Colley, A. J. Kelley, and P. J. Schnackenberg, *Appita* **36**(4), 288 (Jan. 1983).
8. R. G. Lucas, *Pap. Age,* 9 (Sept. 1972, Nov. 1972).
9. Ref. 5, p. 414.
10. H. L. Weiss, *Coating and Laminating Machines,* Converting Technology Co., Milwaukee, WI, 1983, p. 326.

11. Ref. 5, p. 404.
12. H. L. Weiss, *Control Systems for Web-fed Machinery,* Converting Technology Co., Milwaukee, WI, 1983, p. 277.
13. *Ibid.,* p. 105.
14. Ref. 12, p. 207.
15. Ref. 12, p. 61.
16. Ref. 5, p. 400.
17. Ref. 10, p. 336.
18. S. E. Amos, "Winding Webs: A Case of Constant Tension versus Constant Torque," *Pap. Film Foil Converter* p. 56 (Sept. 1970) and p. 62 (Oct. 1970).
19. Ref. 12, p. 239.
20. R. Aylott, *Pap. Film Foil Converter* **58**(10), 128 (Oct. 1984).

General References

D. Satas, *Web Processing and Converting Technology and Equipment,* Van Nostrand Reinhold, New York, 1984. Contains an exhaustive bibliography and comprehensive review of converting machinery.

H. L. Weiss, *Coating and Laminating Machines* (441 pp.) and *Control Systems for Web-fed Machinery*, Converting Technology Co., Milwaukee, WI, 1983. Two exceptionally comprehensive reference sources.

SPECIFICATIONS AND QUALITY ASSURANCE

Specifications and quality-assurance procedures are in the simplest sense the means of communicating information critical to purchasing, production, and distribution functions between the vendor, product manufacturer, and customer. Each of these functions requires the assurance that the correct material is purchased and will be appropriately used and distributed to the end consumer. For this reason, specifications are clear, have extreme detail, and are the focal point for discussions on packaging. This article explains key elements of specifications and quality assurance by covering general packaging specifications, component-specific packaging, manufacturing specifications, and consumer or finished-goods specifications.

General Packaging Specifications

General packaging specifications are developed with the needs of the end user in mind and relate expectations independent of the supplier of the packaging materials. General packaging specifications are employed to outline characteristics common to all packaging components of the same component type. For example, rigid injection-molded components such as jars, tubs, and closures are within one component type. The purpose of the general specification is to specify general and broad standards and expectations for components that may include items mentioned in Figure 1. Included in the general packaging specification is the expectation that the packaging will be within acceptable quality limits (AQLs). Use of AQL quantifies what level of quality is expected and acceptable. As shown in Figure 1, the general specification is very vague and further refinement to a component-specific specification is required before the packaging components can be of known dimensions and consistency.

Closures used by the Charlotte Corporation will conform the following General Specification:

Federal Food, Drug and Cosmetic Act. Tolerances for migration from additives, packaging materials, and processing aids must not exceed regulations established in the Federal Food, Drug, and Cosmetic Act as amended.

General Manufacturing Procedures. This material shall be produced under General Manufacturing Procedures (GMP) to ensure that it is safe for food-contact purposes. The vendor may not make any changes in the materials or equipment used to manufacture the packaging component without prior changes in the Component Specific Specification. Obvious defects in the packaging components will result in an entire lot being rejected.

Inks and Colorants. FDA-approved inks and colorants must be used. One sample from each lot should be sent to Charlotte Graphics Department for approval of the colorant and inks used in each lot. Deviation from the color standard will result in a reject of the packaging components.

Packaging and Distribution. All components must be packed in a polyethylene bag within a corrugated box not to exceed a weight of 50 lb per packed corrugated case. Cases will be labeled on four sides in type size equal to or larger than 14-point. Labels to include production date, lot number, order number, component identification number, vendor name and address, and the number closures per container. Temperatures during storage and distribution of the components should not exceed 90°F and 60% relative humidity.

Acceptable Quality Limits. AQL for dimensions A, 0.5%; AQL for packing/casing C, 6.0%.

Figure 1. Representative general packaging specification for injection-molded components.

Component-Specific Specifications

Component-specific packaging specifications for each package component are developed jointly with the parties involved to ensure both an understanding of compliance and compliance to specifications. As a case study, if the incoming packaging is an injection-molded closure (see Closure), specifications are developed jointly with the resin, tool molding, closure and liner, capping machine manufacturer, and end user of the closure as shown in Figure 2. As shown in Figure 2, all phases of manufacturing the packaging component leading up to the

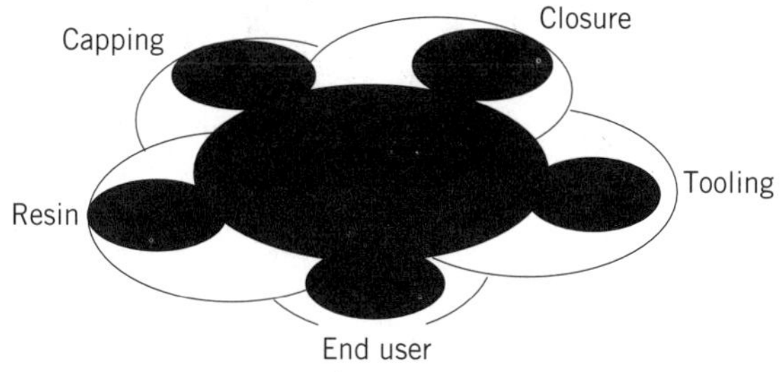

Figure 2. Closure specification development team: manufacturers and quality-assurance functions from each phase of developing a packaging component, a closure.

Charlotte Corp. Division Food, #32	Spec. Code 32-7800-310	Issue Date 12/21/95
Packaging Component 33-mm/400-Polycam dispensing closure		*Application* Products 32-7840

Production Location Approval	*Related Specifications*	
________ Kodiak, Alaska	Shipping case:	32-7800-410
________ Fort Collins, Colorado	Bottle:	32-7800-110
________ Royal Oak, Michigan	Front label:	32-7800-110
________ St. Paul, Minnesota	Back label:	32-7800-510b
Dimensions as shown in print number PS-354-14	Tolerances:	±0.008 inches

Approved Materials

Resins	H-78356jck	*Processing Aids:*	ASBX 198, ASBX 200
Colorants	Charlotte color code blue-#81	*Regulatory Compliance Number:*	15

Acceptable Quality Limits

AQL for dimensions, ovality, and torques	A	0.5%
AQL for packing and casing	C	6.0%

Special Consideration

Removal Torque to be 28–30 inch pounds at 2–6 months shelf life of the finished product. Vendor must comply with General Specifications for all Packaging Components.

Approvals

Specification Originator ________ Supplier ________ Purchasing ________

Figure 3. Representative component-specific packaging specification for a closure.

end user contribute to constructing the specification. In addition, all functions throughout the process as connected to the other manufacturers by a quality-assurance umbrella—which assures the manufacturers and end user that the component produced will be the component that was specified—and all internal quality-control and assurance procedures become part of the specification.

An example of how specifications are derived using a closure specification follows. Let us suppose that the consumers of the end user require that the removal torque on the closure be 28–30 in. · lb between 2 and 6 months of shelf life. This requires the following:

- The resin supplier selects the best resin for the closure supplier to have low variability of closure thread dimensions, which will translate into less torque variability.
- The moldmaker and closure supplier design the tool to accommodate this resin and provide low thread tolerances.
- Quality-assurance specifications detailing defects, AQLs, and actions to be taken regarding defects are established jointly by the closure manufacture and the end user.
- The closure supplier and capping manufacturer ensure that the capping equipment is designed for the closure and establishes the application torque range to attain the required 28–30-in. · lb removal torque.
- The end user verifies that the specification is followed via uniform quality-assurance checks on retained samples.

Specifying the removal torque is accomplished through specifying the optimal resin, tool design, tolerance within the closure and liner material, and ensuring that the capping manufacturer can produce a capping equipment to sufficient tolerances. Figures 3 and 4 show an representative component-specific specification for a closure. Figure 3 expands on the general specification shown in Figure 1 and adds details specific to a closure. Figure 4 provides details in a print form to accompany the written component specification. Each area from the resin selection to the capping is subject to internal quality-assurance specifications to verify that the product specified is the product produced.

Manufacturing Specifications

Manufacturing specifications are developed jointly between the suppliers of the packaging component, often a technology function, and the production facility. The team identifies critical control points throughout the process where a specification is required. Then a test protocol and a specification range are developed to ensure that the production facility operates within the required range. In the example employed, closures, the most critical area is in the removal torque at 2–6 months' shelf life, so controlling application torque at the manufacturing level is a critical control point. A correlation between removal torque at 2–6 months' shelf life and on-line application torque can be made (eg, an application torque range of 45–50 in. · lb results in a removal torque of 28-30 in. · lb at 2–6 months' shelf life). And thus on-line checks to verify machine

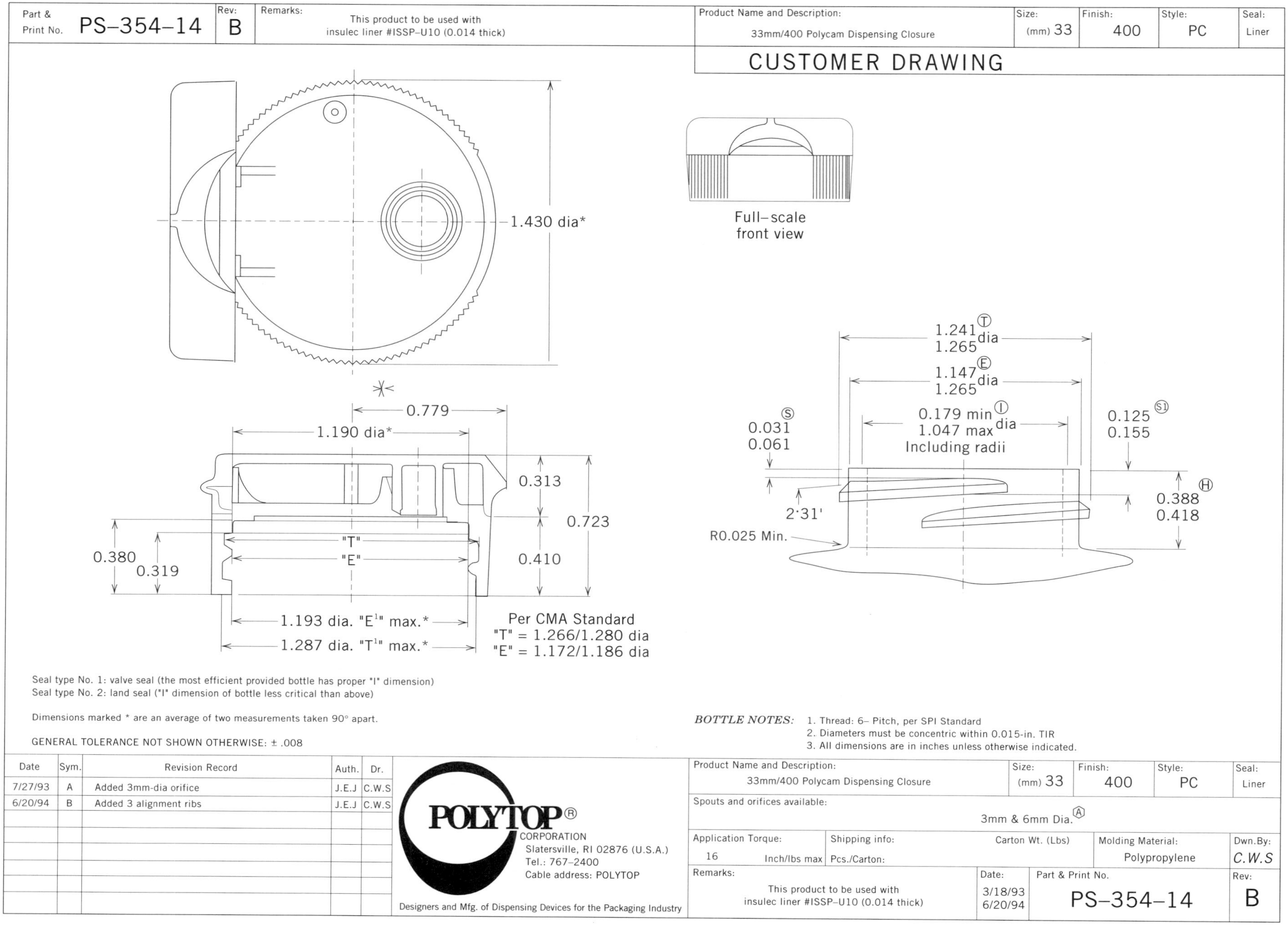

Figure 4. Print for a representative specific packaging specification, a closure.

Inspection Frequency:	*1 full package-per-hour production.*
Action Required:	Submit samples as appropriate and notify line supervisor of deviations immediately after inspection.
Corrective Action:	Required within 1 hour of production time.
Report:	Daily report to include defect and corrective actions taken.

Subject	Specification	Report
Label placement	0.0625 in.	Deviations from horizontal and vertical
Label color	Color standards	Forward suspect deviations to Graphics
Glue coverage	80% of label	Percent covered and location of no or excess coverage
Scuff/tear	No tolerance	Scuffing location and retain label
Removal torque	40 ± 5 in. · lb	Deviations and time of removal torque
Closure color/scuff	Color standards	Forward suspect deviations to Graphics
Bottle scuff	No tolerance	Location
Paneling	0.0625 in.	Deviation and location
Grease/dirt	No tolerance	Location and type

Figure 5. Representative finished-goods specification.

settings of 45–50 in. · lb on each chuck are necessary. Sampling frequency is established.

Consumer or Finished-Goods Specifications

Finished-goods specifications provide the final check in the process to verify that the packaging components were received, were utilized, and will arrive to the consumer packaged as intended. Finished-goods specifications are developed by the manufacturing facility, and marketing, technology, and operations functions. Often visual verifications of package compliance are acceptable, but quantifiable scales of acceptance provide a clear image of shift, manufacturing, and location variability. An example of a finished-goods specification for a package system using the closure specification discussed previously is shown in Figure 5.

Conclusion

Packaging specifications and quality-assurance procedures extend throughout and from an organization to

- Communicate expectations of quality and consistency.
- Translate specific needs and features of packaging components.
- Provide a localized forum for appropriate use of packaging components.
- Convey consumer needs into quantifiable specifications.

Claire Koelsch Sand
Performance Development Systems
Chicago, Illinois

STANDUP FLEXIBLE POUCHES

Conventional standup pouches are those with bottom "horizontal" panel gussets, more commonly known as *bottom gussets.* The gussets are heat-sealed to produce flexible bases on which the pouches may stand without support. This base permits the two sidewalls or facewalls to spread at the bottom when the pouch is filled. In effect, standup flexible pouches are three-panel flexible pouches that are self-standing when full or partially full of product heavy enough to bear down on the bottom panel. In most instances, the bottom panel is a separate sheet of flexible material, but some structures fold a single web sheet into a "W" shape and heat-seal a base from this configuration. Standup flexible pouches with separate bottom gussets have been in existence for more than 30 years.

These pouches are very adaptable for packaging of dry products and liquids. In many applications, they are a suitable replacement for other types of packaging, including plastic or glass bottles, cans, and boxes. The advantages of standup pouches for packaging are numerous. They are environmentally sound, offering source reduction of solid waste ranging from 70% to 90% by both weight and volume. The use of pouches reduces the need for recycling, landfill, and/or incineration. Another advantage of standup pouches is that they offer the use of four- to six-color graphics, for improved shelf appeal and acceptance. Standup pouches offer cost savings, with reduced transportation costs. Unlike the shipping of traditional large plastic empty containers, there is no shipping of "air." There is further savings with reduced inventory space and storage costs for containers. Standup flexible pouches take up only $\frac{1}{80}$th of the cubic feet compared to storing an equal quantity of rigid containers (see Fig. 1).

Other standup flexible pouch structures employ sidewall folds or side gussets and overlapping flat sheet bases. Because they do not have the bottom gusset, these pouches do not stand up quite as well, and have been used more successfully for dry products than for liquids. Both types of standup flexible pouches are currently being used and/or considered for dry products packaging.

Conventional flexible pouches can be fabricated with four

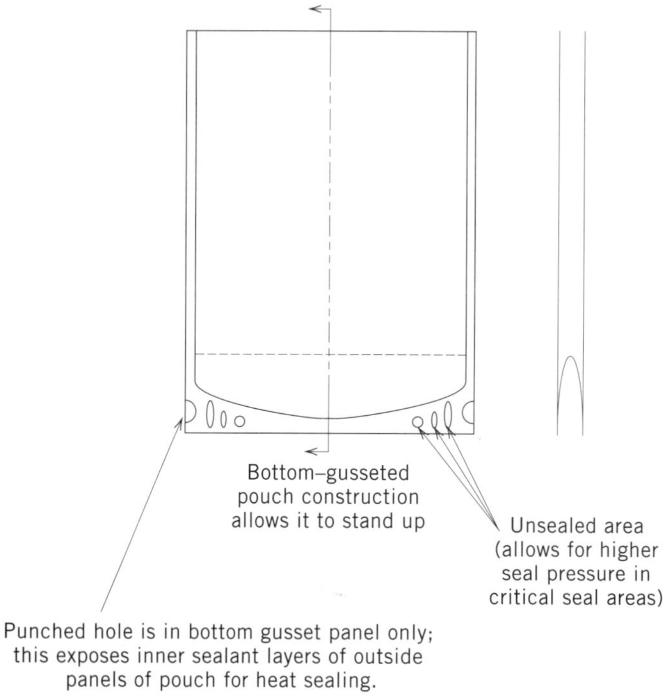

Figure 1. Section view of standup pouch. Unsealed area (Allows for higher seal pressure in critical seal areas)

face-to-face seals, with three face-to-face/one overlap, or with fin seals with a vertical form/fill/seal machine. In any case, they are generally not able to stand independently because they have a single bottom fin seal or fold that brings the face and back panels together. In the standup flexible pouch, rather than a single seal or fold, there are two seals, one of the bottom-to-face panel, and the other of the bottom-to-back panel. In unopened form, the bottom gusset panel is generally folded in and hidden. When opened, the bottom panel (or the bottom fins of the expanded face panels) functions as a base on which the pouch stands (see Fig. 2).

Standup flexible pouches generally employ two face-to-face fusion side seals and a single face-to-face across-the-top seal in addition to the bottom seals. When viewed from the side, it looks like a distorted triangle with the base of the triangle at the bottom. The across-the-top seals may be interrupted by a spout insert, zipper closure, or other structural devices to facilitate opening, dispensing of contents, and reclosing. The application of spouts and closure devices adds versatility and ease of use to the standup flexible pouches (see Figs. 3 and **4a,b**).

Other flexible pouches that have self-standing capabilities include pouches with side gussets that permit the bottom to fold into a flat position, like a paper grocery sack; and those with four-corner side seals to actually form a rectangular solid shape from several webs of flexible materials. In such pouches, the product itself and not the package provides the rigidity to impart the standup feature (see Fig. 5).

History

Historically, standup flexible pouches have been in existence commercially for over 35 years. The first pouches were developed by French machine maker Thimonnier. In fact, standup flexible pouches are often generically (and erroneously) called Doy Packs after their inventor, M. Louis Doyen, majority shareholder of Thimonnier.

By the end of the 1960s, preformed standup flexible pouches were being used in the United States for sugar candies and in Europe for olives in brine and for wine, among the earlier applications of such pouches for liquid contents. During the early to mid-1970s, the standup flexible-pouch concept was applied to the European-initiated Capri Sun, a shelf-stable fruit-flavored beverage. Capri Sun marked the first major application of standup flexible pouches for liquid product contents.

Relatively few new applications appeared until the mid- to late 1980s, with the reemergence of environmentalist pressures for package source reduction. German liquid detergent manufacturers converted polyethylene bottles into standup flexible pouches. By this time, Thimonnier had developed machines that could produce and fill standup pouches directly from rollstock. Other manufacturers had also developed horizontal form/fill/seal machines for standup pouches, although the majority of liquids were still packaged in premade standup pouches (see "Technology" section). The detergent packaging was reasonably successful in Germany, where strict environmental regulations were already in effect. The same

Figure 2. When standup pouch is filled, the bottom gusset opens or expands. Bottom of side panels function as base of pouch.

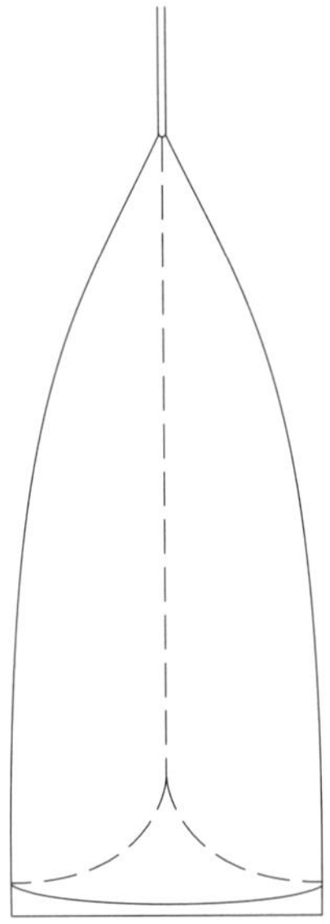

Figure 3. Section view of unopened and filled standup pouch.

concept, although offering environmental and economic benefits, has not been as successful in North American markets.

Actually, standup flexible pouches found greater commercial success in Japan than in Europe during the 1980s, spearheaded by Fujimori.

Technology

All standup flexible pouches are not alike. In general, they may be classified as one of the following types:

1. Preformed Pouches. Pouches are made on a separate converting machine and delivered to a packager in ready-to-open, fill-and-seal form); often with dispensing spouts and zippers already built in.
2. Horizontal Form/Fill/Seal Pouches. Pouches fabricated in-line by the packager from flexible roll stock materials on machines that fold the sidewalls and die-cut openings for heat-sealing into the bottom section (see Form/fill/seal, horizontal).

The most widely used standup flexible pouches for liquids are made from preformed pouches produced by flexible package converters using laminated film structures. Most standup pouch flexible-material constructions are relatively simple webs, such as trap-printed polyester film plus an interior stiffener/sealant of low-density or linear low-density polyethylene (LDPE or LLDPE). Effecting a seal in the pouch requires polyethylene-to-polyethylene or analogous fusion heat sealing. The sealing of the bottom-web material to the two face webs requires that the heat sealants face each other, which dictates die cutting large openings in the bottom material (see Fig. 1).

Positive sealing is essential for liquid containment because liquid is pressing against the base heat seals. Of significance is the hydraulic pressure exerted by the liquid against the bottom seals, particularly when there is an impact or a drop, and the liquid exerts a sudden high force against the heat-seal area. Thus, although the thick gauges of polyethylene sealant serve to enhance the standup feature, an important reason for their presence in such quantities is to provide sufficient strength to resist hydraulic shock. This problem also justifies the preferred use of LLDPE, which has greater stress resistance than low density polyethylene. Many pouches use a nylon cast or biaxially oriented film/polyethylene lamination as the bottom gusset.

In earlier years, the materials employed included nylon film as well as, or in place of, polyester film, plus exterior coatings. The materials for consumer-size pouches now are largely polyester/polyethylene combinations that have proved satisfactory in commercial practice. Output speeds of inter-

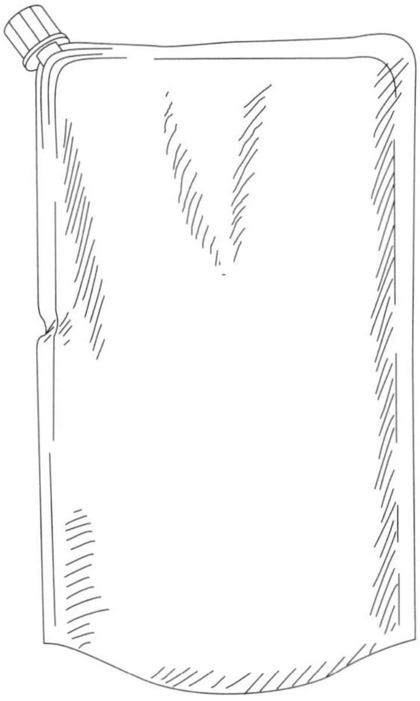

(**a**)

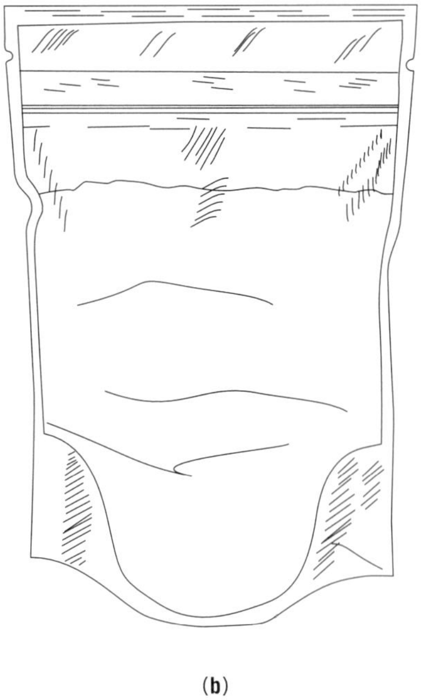

(**b**)

Figure 4. Standup pouches with (**a**) reclosable pour spout and (**b**) reclosable zipper.

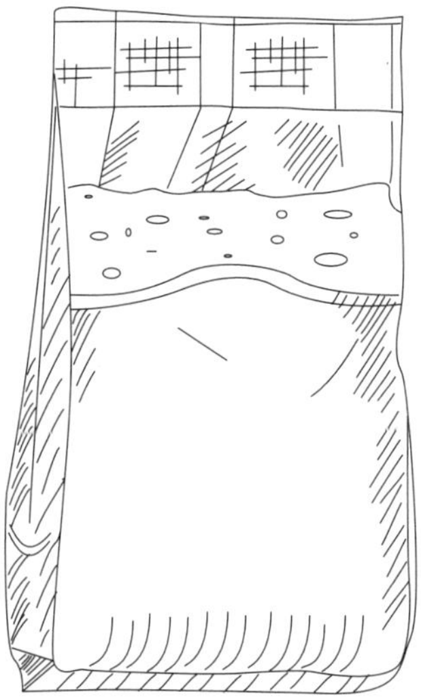

Figure 5. Side-gusseted pouch.

mittent motion preformed pouchmaking machines today are always faster than those on form/fill/seal machines because of multilane possibilities as well as the limiting factors of filling and sealing on the filling machines. Further, it is easier to incorporate dispensing and other fitments on separate converting equipment than on inline form/fill/seal equipment. Preformed pouches, to date, have tended to be more reliable in distribution performance than in-line-made form/fill/seal pouches. Because of this reliability factor, most liquid-containing standup flexible packages to date have used preformed pouches. Since many different fill/seal machines beginning with preformed pouches are commercially available, preformed pouches have commonly been used.

Standup Flexible-Pouch Machinery

In general, there are two types of machines for producing standup pouches: preformed pouchmaking machines and horizontal form/fill/seal machines.

Preformed pouch machines are used by converters to make standup pouches that are supplied to a packaging company that will fill the pouch through the open top. These machines utilize single-wound laminated (nonextensible) rollstock. They make pouches in either two lanes of production, for pouches smaller than approximately 7 in. × 11 in. or one lane for pouches larger than 7 in. × 11 in. The same pouch machines can add zipper reclosures and easy-open features, or can be used to make conventional three-side-seal pouches. Some pouch machines use a separate bottom gusset web and can then easily manufacture in two lanes ("2-up" or two pouches at one time). Other machines use one web and plow the bottom gusset into the center-folded material, but are limited to one lane of production.

The advantages of using the preformed pouch are its superior seal integrity and the ease of adding a reclosure feature, such as the reclosable pour spout. Preformed pouch equipment is used more often for packaging liquids and for small-market-niche products.

Packagers utilize form/fill/seal machinery to fabricate and fill the standup pouch. Again, single-wound laminated sheeting is used on the machine to form the pouch and then fill and seal it in-line. This is known as a *one-step operation,* while the preformed pouches must be shipped from the converter to the packager for filling and sealing in a second operation.

The form/fill/seal (one-step) operation offers economies to the packager. Even though the speeds of the pouch-forming operation are typically much slower than those using preformed pouches, the fact that all operations are done in-line saves on storage and labor costs. Standup pouch form/fill/seal machinery is more typically used for the packaging of large-volume, dry granular products such as nuts and snacks (see Form/fill, seal horizontal and vertical).

Use of Standup Flexible Pouches Worldwide

The worldwide use of flexible standup pouches, exclusive of Capri Sun, was about 700–900 million in 1993, with the largest fraction being in Japan, with the grand total in the 3–3.5 billion range. There has been limited but steady growth during the past few years.

The following is a discussion of the use of standup flexible pouches throughout the world, broken down by area.

North America. Package material source reduction was a major factor behind the development of product packaging in standup flexible pouches in the United States and Canada. Although there is really no practical way to recycle the standup flexible pouch itself, the initial package reduces material use and results in less material going to the landfill. Standup flexible pouches offer the potential for up to 80% package mass reduction for liquid products such as household laundry products, window cleaners, chemicals, detergents, fabric softeners, lotions and creams, and shampoos. Packagers with environmental concerns began to investigate and test standup flexible pouches that had been commercialized in Germany in the mid-1980s. Using structures of polyester film and sometimes nylon film plus LLDPE and a variety of European packaging equipment, packages were introduced as environmentally "friendlier." A popular application was as a refill for high-density polyethylene (HDPE) bottles.

The first use of a large-size standup flexible pouch in the United States was for evaporated milk. It was retorted to achieve ambient temperature shelf stability.

Approximately 10 million standup pouches for liquids were produced in the United States in 1993. A separate figure for Capri Sun stands at ≥500 million pouches for the same year.

Standup flexible pouches appear to have effected a more successful entry into the Canadian market for liquid laundry products under the umbrella of being environmentally more "friendly." Some of these laundry products are packaged in the United States. Other products being marketed in standup flexible pouches in Canada include hair shampoo/conditioner, motor oil, and liquid household cleaner.

Europe. In Europe, materials and/or formed pouches for 300–350 million standup pouches are produced annually by 1 of at least 13 converters. Among the products being packaged in standup flexible pouches in Europe are liquid detergents, fabric softeners, shampoos, hand creams, windshield-washer fluids, and latex paint in ≤5-L sizes. Standup flexible pouches represent a rather large usage for liquid packaging in Europe.

Most European standup flexible pouches for liquids are premade and appear to be three-piece, using 12-μm (0.00048-in.)

PET film/0.004–0.006-in. LLDPE for the two major faces and the softer, stronger nylon film/LLDPE or PET/LLDPE for the bottom panel. Thus, the flexible structures for standup flexible pouches are not especially unique or difficult to fabricate.

Latin America. In Latin America, standup flexible pouches have been well received. One reason is the apparent low cost of flexible materials versus their semirigid plastic package counterparts. The pouches have become established in a number of food markets, including mayonnaise, mustard, ketchup, salad dressings, and hot-filled jams and jellies in the fluid category, and for dry (powdered) gelatin desserts, syrups, toppings, rice, and analogous flowable dry products. In the nonfood category, several companies pack liquid fabric softeners and detergents in standup flexible pouches. Capri Sun is also present in the Latin American market. The number of standup flexible pouches in Latin America is estimated to be in the 200-million-unit range.

Japan. Success in Japan for standup flexible pouches has been led by retort pouches with a volume of over 100 million units annually. Almost all are performed pouches fabricated from polyester film/aluminum foil/cast polypropylene film, with some in glass-coated polyesters as the barrier, packaged on several different types of equipment. A few of the aluminum foil laminated stand-up flexible retort pouches are distributed in the United States.

In Japan, several supply companies developed both equipment and materials that allowed for a relatively healthy market niche for liquid food products. The concept was so well engineered that it could be applied for retort pouches of low-acid fluid foods and beverages such as milk.

India. Standup flexible pouches have been used in India for more than 30 products, in both the food and nonfood areas. About 300 million units have been produced over the past 10 years. Prominent uses include dry rice, motor oil, and related automotive fluids. Environmental factors once again play a role in the choice of pouch packaging. Motor oil represents a significant environmental solid-waste problem, since the conventional extrusion blow-molded HDPE bottles are not recyclable. Large numbers of semirigid bottles become part of the municipal solid-waste stream. The incorporation of motor oil into flexible packages helps reduce this package mass.

Economic Issues

Historically, the standup flexible pouches have not offered lower packaging costs, although this has now begun to change. The actual costs of the standup flexible pouches with liquid dispensing spouts have been more than double the cost of extrusion blow-molded plastic bottles with labels and closures. Another added cost is importing from Europe, Israel, and even Japan. Thus the expected cost savings initially did not materialize, and, in fact, they were negative or at best even. In late 1993, however, a significant change in the costing situation emerged as domestic sources for both preformed and form/fill/seal standup flexible pouches began to develop. In the United States, the costs have come down considerably, and are now competitive with the costs for more traditional packaging, such as an extrusion blow-molded HDPE bottle. On the horizon, however, is the ability of plastic bottle blowers to reduce the wall thicknesses and hence the package mass and the economics even more—and, of course, their increasing use of postconsumer scrap plastic for nonfood/cosmetic contents.

Yet another major problem is that all automatic standup pouch flexible-packaging equipment is much slower than bottle-packaging equipment, some down to 25–30 one-liter units per minute. Maximum output speed for any standup pouch flexible packager is in the range of 120 per minute. Most bottling lines for liquid household products begin at 100 and go up to 400 bottles per minute, with some even higher.

Some pouch equipment manufacturers are developing higher-speed machines, with speeds of $\geq$400 per minute. This new equipment will generate greater attention for standup flexible pouches for liquid packaging.

Standup flexible pouches have almost zero vertical compression strength and so must be totally protected by higher-cost and higher-package-mass secondary and tertiary packaging in distribution. Further, standup flexible pouches are much more difficult to handle than their bag-in-box equivalents.

Another key to the use of standup flexible pouches is a lack of equipment to handle and unitize the pouches into secondary and tertiary packages. Much of the North American output is actually hand-packed into corrugated fiberboard cases. On the other hand, equipment for utilizing flexible pouches is commercially available.

The standup flexible liquid-content pouch market is relatively small, representing only about 1% of the total of all flexible pouches.

BIBLIOGRAPHY

General References

A. L. Brody, Rubbright-Brody, Inc., "The Source Reduction Pros and Cons of Liquid Stand-up Pouches," paper presented at Green Packaging '94, Washington, DC.

A. L. Brody, Rubbright-Brody, Inc., *Worldwide Technology, Markets and Prospects for Stand-up Flexible Pouches,* Packaging Strategies, Inc., 1994.

"Cleaner Refills Spout a Cap," *Packag. Digest* (Sept. 1992).

Liquid Pouch Packaging vs. Rigid Blow Molded Containers: Opportunity or Threat and *Environmental Implications 1991–1996,* Mastio & Company Pouch Study, 1992.

W. J. Noone, "Stand-up Flexible Pouches: Has Their Time Come?" *Packag. Print. Converting* (March 1995).

MICHAEL J. GREELY
Amplas, Inc.
Green Bay, Wisconsin

STAPLES

In packaging, staples are used to assembly wooden pallets, skids, and boxes and crates (see Boxes, wood). In fastening wood-to-wood, the relatively heavy shank of a nail provides greater shear strength, but staples provide greater holding power because their two legs provide more wood-to-metal contact surface area. Staples are used as a replacement for nails in many applications including furniture, housing, fencing,

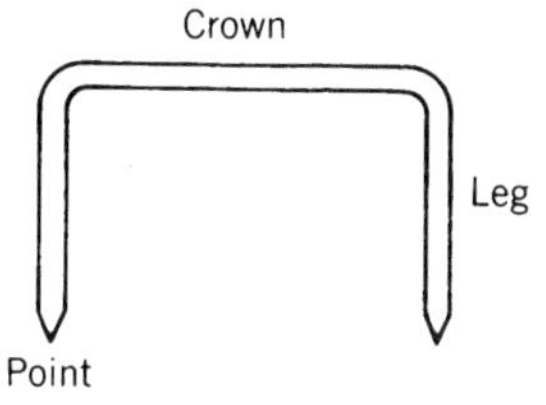

Figure 1. Three basic parts of the staple.

etc, but the following discussion focuses on packaging applications only.

The staple has three basic parts: the crown, two legs, and two points (see Fig. 1). In selecting the proper staple, four factors must be considered: wire size (gauge), crown width, leg length, and type of point.

Wire size. Heavy-duty staples are designated by the gauge (ga) of the wire prior to manufacturing the staples.

$$16 \text{ ga} = 0.062 \text{ in.} = 1.587 \text{ mm}$$
$$15 \text{ ga} = 0.072 \text{ in.} = 1.829 \text{ mm}$$
$$14 \text{ ga} = 0.080 \text{ in.} = 2.032 \text{ mm}$$

In packaging applications, 16-gauge is the most common. Heavier gauge staples are used if a particularly strong combination of holding power and shear strength is required, as in the assembling of some heavy-duty skids. Lightwire staples are designated by the width of the wire after manufacturing; for example, "30" is equivalent to 0.030 in. (0.762 mm) finished width; "50" is equivalent to 0.050 in. (1.270 mm) finished width. A "50" staple is heavier and more power is required to drive it.

Crown width. The widths most common in packaging are shown in Figure 2.

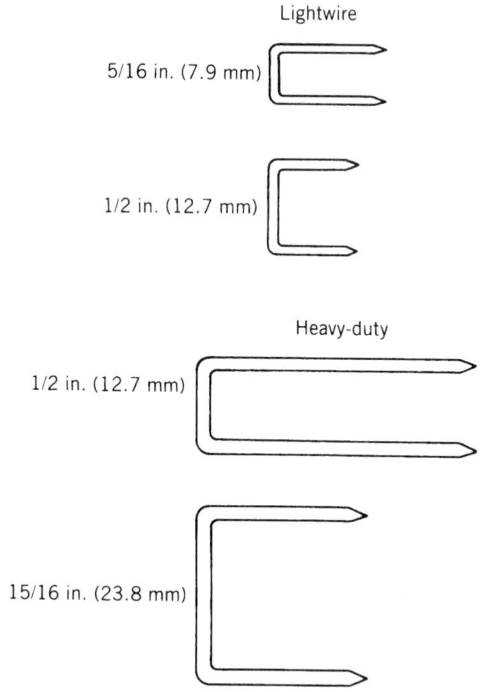

Figure 2. Common crown widths.

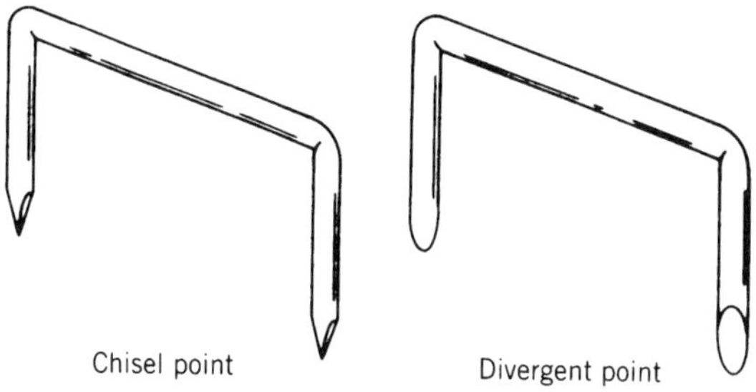

Figure 3. Staple points.

The staple used most frequently for packaging applications is the heavy-duty $\frac{1}{2}$-in. (1.3-cm) crown. The wide-crown $\frac{15}{16}$-in. (2.4-cm) heavy-duty staple is used to prevent pull-through on attachment of porous materials; for example, fastening a corrugated box to a wooden skid. The wide-crown staple is also used over steel strapping to secure it in place. Lightwire staples are used to attach identification tags to the outside of wooden boxes. Plastic liners are attached to boxes with lightwire staples as well. These are just a few of the numerous applications for heavy-duty and lightwire staples in packaging.

Legs. A staple has two legs of equal length. The staple must be long enough to penetrate the receiving member with sufficient length to adequately hold two pieces of material together. The stress of the application must be considered in order to determine the minimum allowable penetration of the staple in order to do the job.

Points. Staple points are cut at different angles to meet various application requirements (see Fig. 3). A divergent point causes the staple legs to divert in opposite directions when it is driven into the wood. This type of point is advantageous when clinching is required. Clinching is the process of bending the staple points over the opposite side of the work piece. The staple legs penetrate the wood members then strike a steel plate diverting the points back into the wood. The chisel point, which is designed to cut through the wood without diverging, is the most common in the industry.

BIBLIOGRAPHY

"Staples" in *The Wiley Encyclopedia of Packaging,* 1st ed., by J. R. Charters, Paslode Corporation, pp. 615–616.

General References

Products 1, Paslode Company, Lincolnshire, IL, 1979.

Paslode Products, Paslode Company, Lincolnshire, IL, 1983.

STATIC CONTROL

Since its inception in the packaging industry approximately seventy years ago, static control has become an integral part of packaging processing. As machinery speeds have increased and materials have changed from paper to plastics and combination laminates and coatings, the need for static control has rapidly expanded.

As a result, the last ten years have seen the greatest increase in the use and development of static control products to enhance manufacturing processes.

Static is the excess or deficit of electrons on a material. A regular nonpolar atom has a nucleus with a positive charge and negatively charged electrons orbiting it. When one atom comes in close proximity to another, electrons from the orbit draw or tear away electrons, leaving one atom with an excess of electrons and the other with a deficit. This process, called ionization, leaves the atom with the excess carrying a negative charge, while the one with the deficit carries a positive charge.

Causes of Static

Static in packaging manufacturing is caused by friction, separation, heat change and improper grounding. Packaging and material handling equipment often incorporate a combination of these to accomplish its function.

Friction. Static is most commonly caused by friction, the rubbing of two materials during packaging processing. When a plastic substrate in web form rubs over a steel roller, or when a pouch is shaped around a forming collar on a vertical form fill seal machine, for example, the plastic picks up the negative charge while the steel roller carries the positive charge. Because metal is such a good conductor, however, the positive charge is bled off. The charge on the plastic is not.

Table 1 shows, the triboelectric series, indicates the likely relative polarity a material will charge if it is rubbed with another. The material higher on the chart will hold a positive charge, while the lower will hold the negative charge.

Table 1. Triboelectric Series

Air	Positive charge
Human skin	
Rabbit fur	
Glass	
Human hair	
Nylon	
Wool	
Silk	
Aluminum	
Paper	
Cotton	
Steel ----------------------------------	
Wood	
Hard rubber	
Nickel, copper	
Brass, silver	
Gold, platinum	
Acetate fiber (rayon)	
Polyester	
Cling film	
Polyethylene	
PVC	
Silicon	
Teflon	Negative charge

Separation. When material separation occurs, such as the unwinding of packaging material on a form–fill–seal machine, static charges are produced at the tangent of the unwind. This can result in very high charges that attract dust and contaminants and can cause operator shock hazards. Anyone who has handled Scotch tape may recognize this phenomena. As a piece of tape is pulled off its roll and the tape wraps around itself, it becomes difficult to handle because static has occurred from the separation.

Heat change. The change in heat during the thermoforming, injection and blow molding of packaging products can cause static charges. As plastic is melted to shape and removed from its appropriate tooling, it undergoes a cooling process which can add more static to existing products. As a part cools, its charge increases and can draw existing dust from the plant atmosphere. In extreme situations, the dust can draw to the product surface and adhere to the hot part, making it nearly impossible to remove later. If the parts undergo additional manual operations, such as bottle printing or container sealing, operators are susceptible to shock (see also Thermoforming).

Improper grounding. One of the common misconceptions in static control is that if a machine is grounded, no static can be present. However, despite grounding efforts from the material to ground, there can be blocked passages in the form of greased bearings, poor metal-to-metal connections and charges too high to be properly grounded. Because many packaging products are insulators, partial static reduction can be attained, but full grounding is impossible.

Results of Untreated Static Electricity

The development of static-related problems are often inevitable in the manufacturing environment and can lead to a variety of different problems.

Production problems and slow downs. A static charge generates an electrical field which acts like a magnet, repelling similar charges and attracting opposite or neutral charges. This accounts for the attraction between charged materials and machine frames or rollers causing machine jams. In many cases, the machine operator will be forced to run at slow speeds to avoid the problems caused by static.

Dust attraction. Many products, such as plastic molding materials and film, develop high static charges on their surfaces. This highly charged surface will attract airborne dust, sometimes from more than three feet away, and hold it tightly to the surface. This can lead to high scrap rates. Subsequent operations (such as molding, printing, or laminating) and their end products can be seriously affected by such difficult-to-remove contamination.

Shocks to personnel. When handling highly charged material, people receive unpleasant shocks either directly from material or indirectly. This can become a safety issue if the operator, afer being shocked, recoils into other moving machinery in the plant.

Fires and explosions. Static is usually found on nonconductive materials where high resistivity prevents the movement of the charge. There are at least two situations where the static charge an move quickly and be dangerous in a combustible atmosphere. The first is where a grounded object intensifies the static field until it overcomes the dielectric strength of the air and allows current to flow in the form of a spark. The second case is where the charge is on a floating conductor such as an isolated metal plate. Here the charge is very mobile and will flash to a proximity ground at the first opportunity.

Damage and interference in electronic components. Strong static fields and static discharges can cause interference with electronic equipment. This can cause metal detectors to malfunction, ink-jet printer boards to burn out and weighing equipment to give false readings.

Types of Static Control Devices

Active/electrical ionizers. Active electrical ionizing equipment, one of the most effective ways of eliminating static charges, is seen in nearly every large packaging plant which processes plastic or paper at high speeds. These units create ionization by forcing high voltage to ground and breaking down air molecules in between into low levels of ozone. This makes available millions of free positively and negatively charged ions and effectively "feeds" away the static charge by "matching up" ions of opposite charges. Active ionizing equipment comes in many shapes and sizes, including alternating current (ac) bars or rods, blowers, nozzles, and air curtains, steady state direct current (dc), and pulsed dc units.

Electrical ionizers are driven by power supplies, which transform 110 V or 220 V into voltage ranging from 4500 V to 14000 V while lowering amperages to a safe level. This equipment is normally custom manufactured and retrofit to the machine it serves. If this equipment is not employed properly, it is useless in packaging applications.

Many bar or rod-type units are used in web and sheet-based applications, while air driven units are used for static control on asymmetrical parts such as closures, jars or thermoformed packages. Most electrical ionizers cannot be used in hazardous areas.

Static bars. Static bars are most often used in web-based applications in the packaging industry. This equipment is necessary on many form–fill–seal, overwrapping, tamper evident sealing, and stretch wrapping machines. Bars are normally placed within close proximity to the web and do not require any air to operate. Most bar-type ionizers must be within three inches of the substrate as it passes beneath the bar(s).

Air ionizers. Air ionizers, driven by small fans, compressed air or large blowers for large parts, are used for static control on asymmetrical products. In conveying, unscrambling, filling, denesting, and sorting operations, air ionizers may be necessary to prevent parts from sticking together, operator shock or dust attraction.

Nozzles, guns, and pinpoint ionizers. *Nozzles* are used in packaging to clean packaging products, and to eliminate static and dust. An example of this is during composite can cleaning. Ionizing *guns* are used in more manual environments where cycle times afford the luxury of cleaning each unit. *Pinpoint ionizers* are used in powder filling, for example, when a small area must be ionized with a small piece of equipment. Powder filling operations often use pinpoint ionizers to discharge materials.

Steady state dc. Steady state dc generators allow the production of a desired polarity, positive or negative or both, which makes them ideal for high speed webs where the polarity of the material is known. This allows control of static with higher voltage where conventional lower voltage ac static control is less effective.

Pulsed dc. Pulsed dc controllers allow adjustment of positive and negative ions and control of the frequency of ions, in hertz, that the unit produces. This type of equipment is used in clean room and benchtop applications where the use of air is not allowed or where large areas need to be ionized. A "cloud" of ions is produced, making it ideal for hand packaging of medical devices, for example.

Passive. Passive static control incorporates grounding and conductive materials. Gold and silver are excellent conductors but cost-prohibitive. Carbon fiber, stainless steel, and phosphor bronze are more commonly used passive static control devices in the packaging industry. However, tinsel remains most commonly seen because of its low cost and availability. It is recognized by its similarity to Christmas garland. Tinsel is usually loosely tied to the frame of a machine's ground. Its effectiveness is considered mixed, but if it reduces static to an acceptable level, it does away with the need for more expensive electrical ionizers.

Nuclear ionizers. Nuclear ionizers, radioactive ionizing devices, incorporate Polonium-210, an alpha emitting isotope. When the alpha particles collide with an air molecule, it creates thousands of positive and negative ions without the use of electricity.

This makes nuclear ionizers ideal for explosion-proof environments where electricity cannot be used and the possibility of an electrically related spark can exist. Coating and chemical packaging are examples of plants where nuclear ionizers must be used for static control. Most nuclear ionizing units have an active life of approximately one year after which they must be returned and refilled.

The Nuclear Regulatory Commission requires yearly registration of these devices, which must be continually leased instead of owned. The cost of leasing nuclear static control devices versus buying electrical ones is much higher and can make them cost prohibitive. If a nuclear unit is damaged, the site can be susceptible to strict clean-up requirements by the NRC.

Static generators. Static generators are used to temporarily bond materials together, generally in web handling processes. Static generators work the same way as static eliminators, except that a ground is not present locally. It is placed on the opposite side of the materials passing between forcing two substrates to accept the ionization, thereby pinning them. Static generators operate on dc, allowing polarity control in packaging operations. Examples of static generator use in the packaging industry include air bubble removal, overwrapping, laminating, in-mold decorating and heatshrink packaging.

Static control devices, when used properly, can make a great deal of difference in production efficiency and overall costs. It is important that you meet face-to-face with your static eliminator supplier or know your application well when you are specifying or purchasing to assure the equipment is necessary and correct for the application intended. When ionizers results are minimal, it is usually because equipment is improperly placed or the wrong equipment is being used.

In the packaging industry, static "problems" are relative. In one plant, such as a plastic bag converting plant, charges of 10,000 volts may be barely noticed. In another plant, such as a medical device packager, charges as low as 1,000 volts may be attracting airborne dust and foreign matter. It is important to know what actions can be taken to remove existing problems relating to static and to what extent they will enhance the packaging manufacturing and material handling operations.

H. Sean Fremon
Meech Static Eliminators
Akron, Ohio

STRAPPING

Strapping is a material used throughout industry to preserve the integrity of a package or load during handling, shipping, and/or storage. Strapping is used to close cartons, bundle or bale items, unitize pallet loads of materials, or brace shipments of goods during transit.

There are four primary types of strapping materials being manufactured today: steel, nylon, polypropylene, and polyester. A choice of the best strapping material for a particular application requires not only knowledge of the characteristics of the types of strapping but also the characteristics of the packages or loads to be strapped.

Strapping Materials

Several characteristics highlight the essential differences among steel, nylon (see Nylon), polypropylene (see Polypropylene), and polyester (see Polyesters, thermoplastic) strapping. These characteristics affect the packaging performance of strapping most directly: (*1*) strength; (*2*) working range; (*3*) retained tension; and (*4*) elongation and recovery.

For the sake of uniformity, the following comparisons are based on strapping samples with single cross sections of $\frac{1}{2}$ in. × 0.020 in. (13 mm × 0.5 mm). The performance characteristics were evaluated under the typical environmental conditions of 72°F (22°C) and 50% rh.

Strength. The strength of strapping materials is measured in two ways: break strength and tensile strength. Table 1 illustrates that most steel strapping has much higher break and tensile strength than the strongest plastic strapping material. If a package or load is heavy enough to require a strapping with great strength, and if it is strong enough to withstand the application of strapping at high tension, steel strapping is often the best selection. On the other hand, if a package is not heavy and strong, one of the plastic strapping materials will probably render effective protection. However, given the relatively small strength difference among the plastic materials, other criteria must be examined before selecting the proper plastic strapping.

Table 1. Strapping Materials, Comparative Strength

Strapping Material	Break Strength, lbf (N)	Tensile Strength, psi (MPa)
Steel	1170 (S204)	117,000 (806)[a]
Polyester	600–800	60,000–80,000
	(2669–3559)[b]	(4 ± 4–551)[b]
Nylon	630 (2802)	63,000 (434)
Polypropylene	500–600	50,000–60,000
	(2224–2669)[b]	(345–414)[b]

[a] There is also a premium-grade steel material with a tensile strength of 145,000 psi (1000 MPa).

[b] Range results from different characteristics of materials from different suppliers.

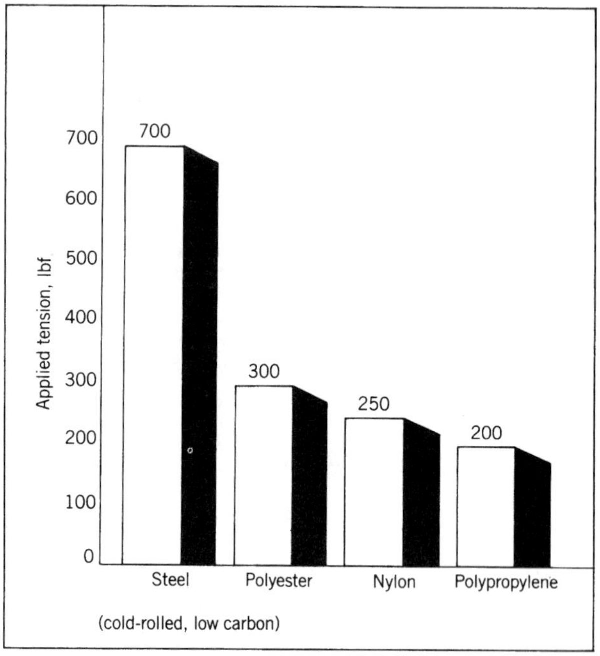

Figure 1. Typical working ranges of $\frac{1}{2}$ in. × 0.020 in. (13 mm × 0.5 mm) strapping. To convert lbf to N, multiply by 4.448.

Working range. Working range can be defined as the range of applied tension and is dependent on the package/product system. The minimum may be virtually zero on a fragile package and the maximum is the highest tension level at which strapping is applied in actual situations. It is within the working range that strapping does nearly all its work securing a package or load. Although impact during handling or shipping may sometimes subject strapping to tensions above the working range, the package itself, not the strapping, usually fails under such conditions.

Figure 1 shows that steel strapping is applied at upper-level tensions of only 700 lbf (3114 N), although its break strength is much higher, 1170 lbf (5204 N). Polyester, nylon, and polypropylene strappings of the same dimension may exhibit break strengths of 600 lbf (2669 N) or more, but they are applied at tensions with typical upper levels of only 300, 250, and 200 lbf (1334, 1112, and 890 N), respectively, Strapping application equipment limitations help explain the wide difference between each material's break strength and its applied tension level. Often, the characteristics of the package or load also restrict applied tension to lower levels.

D3953-91), American Association for Testing and Materials, West Conshohocken, PA.

Standard Guide for Selection and Use of Flat Strapping Materials (ASTM D4675-91), American Association for Testing and Materials, West Conshohocken, PA.

PAUL E. SOWA, CPP
SIGNODE/An ITW Company
Glenview, Illinois

STERILIZATION METHODS. See HEALTH-CARE PACKAGING; CANNING, FOOD.

STRETCH FILMS. See CLOSURES.

STYRENE–BUTADIENE COPOLYMERS

The styrene–butadiene copolymers (SB) that are suitable for packaging applications are those resinous block copolymers that typically contain a greater porportion of styrene than butadiene and that are predominantly polymodal with respect to molecular weight distribution. These copolymers, produced by solution polymerization processes using sequential multiple additions of an organolithium initiator, styrene, and butadiene (1,2), are amorphous in nature and transparent. Compared to other transparent styrenic polymers (eg, general-purpose polystyrene and styrene-acrylonitrile copolymers), the styrene-butadiene copolymers offer outstanding shatter resistance (3). Physical properties are dependent on molecular weight distribution, styrene:butadiene ratio, and the nature of the block structure. Since all three of these parameters are a function of the polymerization recipe, grades having different properties can be produced: Grades tailored for injection molding (see Injection molding) and for extrusion. (see Extrusion) are commercially available (see Table 1).

One of the major applications for SB is single-service cups and lids, but a combination of good mechanical and visual properties, low density compared to other transparent polymers, and ease of fabrication by the typical molding and extrusion techniques make styrene-butadiene copolymers a good choice for many packaging applications. These include bottles and jars, hinge boxes (ball-and-socket or integral-living hinges) (see Boxes, rigid plastic), tubs and lids, blisters, medical devices, packaging, and many film applications. (see Carded packaging), (see Films, plastic).

Table 1. Styrene–Butadiene Copolymer Physical Properties

Process	Injection Molding and Extrusion	Injection Molding
Specific gravity	1.01	1.01
Flow rate, condition G, g/10 min	7.5	8.0
Tensile yield strength, psi (MPa)	3700 (26)	4,400 (30)
Elongation, %	160	20
Flexural yield strength, psi (MPa)	4900 (34)	6,400 (44)
Flexural modulus, psi (MPa)	205,000 (1413)	215,000 (1483)
Hardness, shore D	75	75
Heat-deflection temperature, 264 psi (1.8 MPa), °F(°C)	163 (73)	170 (77)
Vicat softening point, °F(°C)	188 (87)	200 (93)
Light transmission, %	90	90

Table 2. Producers of Styrene–Butadiene Resinous Copolymers

Company	Location	Tradename
Phillips 66 Company	U.S.	K-Resin
BASF	Germany	Styrolux
Denka-Kaguku	Japan	Clearene
Asahi Chemical	Japan	Asaflex
Fina	Belgium	Finaclear

Styrene–butadiene copolymers are often blended with other polymers (4). If transparency is a requirement, the blending polymer must have a refractive index very near that of the styrene–butadiene copolymer. This includes styrenic polymers such as general-purpose polystyrene (GPPS) (see Polystyrene), styrene–acrylonitrile copolymers (SAN) (see Nitrile polymers), and styrene–methyl methacrylate copolymers (SMMA). GPPS is used extensively as a blending polymer for applications requiring sheet extrusion and thermoforming (see Thermoforming) (eg, carded packaging). Blends with SAN or SMMA are mainly used for injection molded applications (eg, hinged boxes, videocassette cases). The impact strength of SB can be either enhanced or reduced when blended with other styrenic polymers. For example, a significant loss of impact strength occurs with the addition of even low levels of GPPS. In contrast, certain blends of SB and SMMA have synergistic impact strength. Therefore, the suitable blend ratio for a given application depends on the required combination of stiffness, toughness, and economics. Blend ratios typically range from 40:60 to 80:20 SB: styrenic polymer. If clarity is not a requirement, blends with polymers such as high impact polystyrene (HIPS), polypropylene (PP), and polycarbonate (PC) can provide synergistic enhancement of certain physical properties, unusual aesthetics such as gloss and pearlescence, or reduced cost. Bottles and jars can be produced from nonblended SB by extrusion or injection blow molding (see Blow molding). Blown and cast films can be produced from nonblended SB or blends of SB/GPPS.

Some U.S.-produced styrene–butadiene resinous copolymers meet FDA specifications (CFR 177.1640). Certain grades are fully documented to qualify for U.S. Pharmacopoeia Class VI-50 medical applications and blood contact uses, and have been shown to be nonmutagenic, nonirritants to sensitive tissues, and nondestructive to living cells. In view of these qualifications and because they can be sterilized by both gamma irradiation and ethylene oxide (ETO) procedures, these copolymers are widely used in medical packaging applications.

There are currently five producers of styrene–butadiene copolymers throughout the world (see Table 2). The 1996 selling price in the United States was $0.86–0.99/lb.

Styrene–butadiene copolymers have no demonstrated toxicity in laboratory studies, and the level of machine-side toxic fumes detected during typical fabrication processes is well below OSHA standards. As with most hydrocarbon based polymers, SBs burn under the right conditions of heat and oxygen supply. Combustion products of any hydrocarbon based material should be considered toxic.

BIBLIOGRAPHY

1. U.S. Pat. 3,639,517 (Feb. 1, 1972), A. G. Kitchen and F. J. Szalla (to Phillips Petroleum Company).
2. U.S. Pat 4,091,053 (May 23, 1978), A. G. Kitchen (to Phillips Petroleum Company).
3. L. M. Fodor, A. G. Kitchen, and C. C. Baird, "K-Resin BDS Polymer: A New Clear Impact-Resistant Polystyrene," in *New Industrial Polymers, ACS Symposium Series 4,* D. Deanin, ed., American Chemical Society, Washington, DC, 1972, pp. 37–48.
4. G. M. Swisher and R. D. Mathis, "A Close-up of Blends Based on Butadiene–Styrene Copolymer," *Plast. Eng.* 40(6), 53–56 (June 1984).

D. L. Hartsock
Phillips 66 Company
Houston, Texas

Table 1. Summary of Methods Used for Improvement of Adhesion and Hydrocarbon-Barrier Properties

Method	Improves	Comments
Corona discharge	Adhesion	Sophisticated, high treat levels possible
Gas plasma	Adhesion	Sophisticated, high treat levels possible
Flaming	Adhesion	Commonly used for plastic bottles
Fluorination	Hydrocarbon barrier	Posttreat or in-mold treat possible
Sulfonation	Hydrocarbon barrier	Posttreat, some surface yellowing
Polymer blends	Hydrocarbon barrier	Polyethylene–nylon, water barrier decreases
Coextrusion	Hydrocarbon barrier	Films, sheets, bottles, generally containing EVOH, PVDC, or nylon
Coating	Hydrocarbon barrier	Films, sheets, bottles, generally coated with PVDC

SURFACE AND HYDROCARBON-BARRIER MODIFICATION

The surfaces of materials used in packaging applications are often modified to improve the properties of the container. The reasons for modification and the methods used vary widely. Black plate to be used in steel cans is electrolytically coated with a thin layer of tin, which aids in soldering the side seam and helps protect the outside of the can from corrosion (see Cans, steel). Tin-free steel has a thin coating of chromates, phosphates, or aluminum to protect the exterior surface from corrosion (see Tin-mill products). Aluminum cans do not need an exterior coating, since a thin layer of aluminum oxide forms automatically on the surface, protecting the bulk of the material (see Cans, aluminum).

Thin coatings are applied to glass containers just before annealing (hot-end coating) and just after annealing (cold-end coating) (see Glass-container manufacturing). The hot-end coating is usually tin chloride or titanium chloride, which improves adhesion of the cold-end coating and hardens the surface to make it more difficult to scratch. The cold-end coating may be silicone, stearates, polyethylene, or other compounds that lubricate the surface to decrease friction, which, in turn, improves scratch resistance. Because of the surface-sensitive nature of glass, any process that will decrease the number of scratches on a container will increase its strength.

Lacquers and varnishes are applied to the surface of folding cartons (see Cartons, folding) and paper labels (see Labels and labeling) to give a glossy appearance and to improve abrasion resistance. The top and bottom surfaces of multiwall bags (see Bags, paper) and corrugated shippers (see Boxes, corrugated) can be coated to increase their coefficient of friction, thereby helping to maintain unity of pallet loads. Fiber surface treatments that cause covalent and ionic crosslinking (fiber-to-fiber primary bonding, instead of the normal weak secondary bonding) can increase the wet strength of materials manufactured from cellulose fibers, such as kraft paper (see Paper) (1). Adhesion of labels and printing inks (see Inks) to plastic containers (especially polyolefins) can be improved by pretreating, and sometimes posttreating, the containers using flame, corona discharge, or gas-plasma techniques.

Barrier properties of materials or containers can be improved with surface coatings or surface treatments that vary in thickness from $4 \times (10^{-3} - 10^{-7})$ in. [$1 \times (10^{-4} - 10^{-8})$m]. Nonselective barriers, such as metallizing (see Metallizing) decrease transmission rates of all gases and vapors. Metallizing also provides a conductive surface that is easily grounded to dissipate static charges. Other barrier materials are selective with respect to which gas- and vapor-transmission rates will be decreased. Vinylidene chloride copolymer (see Vinylidene chloride copolymers), generally called PVDC, is applied as a coating primarily for its oxygen-barrier properties, although it is also a good barrier to water vapor and to many other gases and vapors (see Barrier polymers). Nitrocellulose coatings are often used on cellulosic materials (see Cellophane) to reduce their water-vapor transmission rates. Fluorination or sulfonation of the surface of a polyethylene container improves its hydrocarbon-barrier properties.

Although not a surface treatment, another means by which hydrocarbon transmission rates can be decreased is by molding a container from a blend of polyethylene and nylon. Films, sheets, and bottles can be coextruded to provide a hydrocarbon-barrier layer. It is apparent that the reasons for surface modification are extremely varied. This article deals only with methods for improving adhesion and hydrocarbon-barrier properties. The methods discussed are summarized in Table 1.

Adhesion Improvement

In order to improve the adhesion of printing inks, coatings, and labels to plastic films and containers, the surface can be cleaned by oxidation. For example, corona-discharge techniques improve the adhesion of uv varnishes to inks, which often have a high wax content, on folding cartons, so the varnish can be applied evenly. This prevents separation during folding and gluing; the varnish will adhere to the adhesive used for the carton flaps and side seams. Also, the varnish will not separate when a pressure-sensitive price label is applied in a retail store (2).

In addition to removing dust, oils, greases, processing aids, mold-release agents, etc, the surface is activated. Its energy is increased, so its wettability is enhanced. The formation of carbon–carbon double bonds and carbonyl ($>$C=O) and hydroxyl—OH) groups produces a surface with a higher electron density (polarity) and thus better bondability. In general, two polar surfaces form a stronger bond than one polar and one nonpolar surface or two nonpolar surfaces, since dipole interactions are involved. Although the mechanism of adhesion improvement is not completely understood, experimental evidence points to changes in the chemical nature of the surface (oxidation and chain scission), in addition to morphological changes at the surface (increased amorphous fraction) and in layers below the surface (increased percent crystallinity) (3).

More sophisticated surface-treating techniques are being used because fire and safety hazards are associated with the flame treatment of containers (4) and because the newer water-based inks and coatings require higher treating levels than can be achieved with flaming (5). If too high a treat level is used, problems can be encountered; for example, in heat-sealing films (see Sealing, heat). Optimization of the treat level is required.

Corona-discharge techniques. A plastic film (web) or sheet moves between an electrically grounded roller and an electrode maintained at high voltage. The air between the two surfaces is ionized and a continuous arc discharge (corona) is generated at the surface of the film or sheet, which cleans, oxidizes, and activates the surface. The process is performed in air at atmospheric pressure and elevated temperatures (6,7). Ozone is a byproduct of the corona-discharge method, so its removal from the workplace is required. This can be accomplished with fans that supply fresh air and remove the ozone. Catalysts can be incorporated to destroy the ozone chemically (8). The air in the treating area must be maintained at a low relative humidity, or the oxidation that occurs will result in excessive formation of low-molecular-weight polymer chains, exhibited as a frosted appearance on the surface (9). Multidischarge corona techniques are also being used to remove the rolling oils from aluminum foil (10).

Gas-plasma techniques. This is an automated batch process in which the containers to be treated are placed in a reaction chamber, which is then evacuated. The chamber is subsequently charged with oxygen, argon, helium, or nitrogen while a radiofrequency field ionizes the gas. A glow discharge is produced, which affects the surface of the containers in the way described for corona discharge. In this case, oxygen cleans and oxidizes the surface, and the other gases activate the surface. The process is performed at low pressures and relatively low temperatures (11).

Flame techniques. Containers are passed by a bank of flame jets. The source of energy is usually natural gas. The flame oxidizes the surface and burns off surface contaminants, such as mold-release agents, by a mechanism similar to that of corona discharge.

Hydrocarbon-Barrier Improvement

Polyolefins are widely used as packaging materials (see Polyethylene, high-density; Polyethylene, low-density; Polypropylene). Their low densities result in lightweight packages. Polyolefins are relatively tough materials, able to survive relatively high container drops. Because they are relatively inert chemically, resistant to most acids and bases and many organic solvents, product–package interactions are minimal. Polyolefins are low-cost materials that are easy to blow-mold (see Blow molding), and regrind can be added without significant loss in properties. They also have potential for recycling. Polyolefins are good water-vapor barriers; however, they are poor barriers to gases such as oxygen and to vapors such as nonpolar hydrocarbons. The transmission rates of aromas, flavors, fragrances, and hydrocarbon solvents are important to the shelf life of many products.

By treating the surface of polyethylene containers with fluorine or sulfur trioxide, a very thin polar surface can be created on a nonpolar polymer, decreasing the transmission rates of nonpolar hydrocarbons, eg, gasoline, motor oil, propellants, hexane, carbon tetrachloride, and benzene, through the walls of the container. Generally, nonpolar penetrants have high transmission rates through nonpolar polymers and low transmission rates through polar polymers. The reverse is true for polar penetrants. An added benefit of increasing surface polarity is that it facilitates printing, wetting, and labeling, eg, heat-transfer labeling, (see Decorating). However, heat sealing is made more difficult because of crosslinking of the surface.

Surface fluorination. There are two basic methods of treating the surface of polyolefin containers with fluorine: after fabrication and during fabrication (blow molding) (12–14). The Surface-Modified Plastics (SMP) Fluorination System was developed by the Linde Division of the Union Carbide Corporation, which offers the technology under license. It is a batch process in which the manufactured containers are placed in a heated reaction chamber prior to evacuating it. A mixture containing a low percentage of fluorine in nitrogen is allowed to flow into the chamber. The fluorine reacts with the surface of the container, oxidizing it (fluorine is one of the best oxidants available) and creating a polar, crosslinked surface. This eliminates the need for flame, corona discharge, of gas-plasma techniques to improve printability or label-adhesion properties. The chamber is again evacuated to remove the unreacted fluorine and the containers are removed. Both the inside and the outside surface of the container are fluorine-treated in this method. Since it is a posttreat method, there is no possible corrosion of expensive tooling. The process has been cleared by the FDA for use with containers for foods and pharmaceuticals.

The other fluorination method was developed by Air Products and Chemicals, Inc., which offers the technology under license (15). It involves the use of a mixture containing a low percentage of fluorine in nitrogen in the blow gas during blow molding. The parison is prepurged with nitrogen, the blow gas is introduced to form the container and treat the inside surface with fluorine, and the container is purged to remove the fluorine mixture. Since fluorine is such a reactive molecule, fluorination of the container surface takes place at normal extrusion and blow-molding temperatures. Containers that are manufactured by in-mold fluorination have not yet been cleared by the FDA for use with foods and pharmaceuticals.

In both types of fluorine treatment, a polar fluorocarbon layer is formed that is 0.08–0.16 mil (2–4 μm) thick (15). Even this very thin layer decreases transmission rates of nonpolar hydrocarbons by factors of 10–100 and even higher, depending on the exact chemical nature of the permeating molecule (15,16). This hydrocarbon-barrier improvement is useful for fuel tanks and containers for home, automotive, agricultural, and industrial chemicals. If the vapor pressure of the contained solvent is too high, the container may balloon, and rupture, especially at elevated temperatures. On a smaller scale, container creep may thin out the barrier layer, which is already extremely thin, leading to significantly higher transmission rates than would be expected. If this is the case, a change in the structural design of the container may be necessary in order to minimize expansion and creep.

The barrier properties achieved with fluorine-treated pigmented containers are sometimes inferior to those attainable without pigment. This can be overcome by using a coextruded bottle (see Multilayer plastic bottles) with a clear layer inside and a pigmented layer outside (17,18). The fluorination of the clear plastic on the inside will provide a good barrier, even if the outer surface is not treated at all (in-mold treat) or if its fluorine-treat is not optimal (posttreat).

Surface sulfonation. A posttreat process with a mixture of sulfur trioxide in an inert gas has been developed for HDPE containers by the Dow Chemical Co., which offers the technology under license. In this case, polar sulfonic acid groups are bonded to the surface of the polyethylene. These groups are subsequently neutralized with ammonia or sodium hydroxide. Ammonia is preferred, since it can be used as a gas. The barrier properties, ie, the depth of treat, developed by this process depend on the concentration of sulfur trioxide and the time of exposure. This process has been used commercially for treating the inside surface of fuel tanks. It has potential for organic-solvent containers. Some surface yellowing occurs with this process, but this has not been apparent ion most experimental containers, which contained carbon black. Permeation rates of gasoline through treated containers were less than 1% of those for identical untreated containers (19). Containers that are manufactured by this process have not yet been cleared by the FDA for use with foods and pharmaceuticals.

Polymer blends. Polyethylene is a good water-vapor and polar-hydrocarbon barrier, but a poor barrier to nonpolar hydrocarbons. Nylon is a good barrier to oxygen and nonpolar hydrocarbons, but a poor barrier to water vapor (see Nylon). Selar (DuPont) resin is a blend of a special nylon resin (5–18 wt%) and polyethylene (12–14,20). Bottles can be blow molded from Selar resin with only minor modification of the extrusion screw in a conventional blow-molding machine. Up to 80% regrind can be used. The nylon forms a maze of overlapping, discontinuous barrier plates integral to the end product (21). The blend provides the good properties of both resins without the need for coextrusion or surface treatment with toxic or corrosive chemicals. Compared to polyethylene, the improvement in solvent-, gas-, and odor-barrier properties is comparable to that provided by the fluorination processes (21,22). Strength, toughness, and heat resistance are also improved, but WVTR is higher than that of a polyethylene container of the same configuration. Containers can be pigmented with no adverse effect on their properties.

Potential package forms for this resin are bottles, pails and drums, injection-molded containers, aerosols, and films. All of these package forms except films can be easily fluorine-treated by the posttreat process; only those which are blow molded can be treated by the in-mold process. Potential markets are the same as those targeted for the fluorine and sulfur trioxide treatment processes: agricultural, industrial, and household chemicals; oil-based paints and cosmetics; waxes and polishes; insecticides; fertilizers; etc. These containers have not yet been cleared by the FDA for use with foods and pharmaceuticals.

Other polymer blends are being used in packaging today, and growth in this area is predicted (23). Most of the widely used blends, eg, HDPE–LDPE used for grocery bags, have been selected because of the improvement in mechanical properties. However, Continental Can has been producing an accordion-pleated bag from a blend of nylon and LDPE for use in an aerosol can. This bag must be a good barrier to hydrocarbons in order to keep the propellant separate from the product (24) (see Pressurized containers). Increased use of polymer blends is anticipated in packaging that requires a flavor or fragrance barrier, or both.

Coextrusion. Since ethylene vinyl alcohol (EVOH) (see Ethylene vinyl alcohol) and nylon (see Nylon) have good hydrocarbon-barrier properties (25–27), coextrusion with polyethylene or other plastics can be done to manufacture film, sheet, or bottles (28,29) that have low oxygen- and water-vapor transmission rates (16,30,31) in addition to low transmission of hydrocarbons, eg, solvents such as toluene, xylene, and methyl ethyl ketone, (13) (see Coextrusions for flexible packaging; Multilayer plastic bottles).

Drying agents are being added to some of the layers of plastic cans (see Cans, plastic) that are being developed for retort applications (see Retortable flexible and semirigid containers) in order to maintain a low moisture content in the EVOH layer, since the permeability of EVOH increases as its moisture content is increased (32). The choice of the tie layers between the layers is usually critical to the performance of the package (33).

Coating. Vinylidene chloride copolymers (generally called PVDC) are good barriers to hydrocarbons, as well as to oxygen and water vapor. PVDC has been widely used as a barrier layer in multilayer films and sheets and is now being used as a coating on plastic bottles (24,34,35). It is applied by spray coating, roll coating, or dip coating in layers as thin as 0.08 mil (2 μm) (36,37). Other materials such as epoxies can also be used to coat containers in order to reduce hydrocarbon transmission rates, but their commercial use has been very limited (24).

BIBLIOGRAPHY

"Surface and Hydrocarbon-Barrier Modification" in *The Wiley Encyclopedia of Packaging Technology,* 1st ed., by Mary A. Amini, Center for Packaging Engineering Rutgers, The State University of New Jersey, pp. 620–623.

1. A. N. Neogi and J. R. Jensen, *TAPPI* **63**(8), 86 (1980).
2. J. P. Nixon, *Package Print.* **30**(8), 16 (Aug. 1983).
3. B. Catoire, P. Bouriot, O. Demuth, A. Baszkin, and M. Chevrier, *Polymer* **25,** 771 (1984).
4. R. N. Gidwani, *Am. Lab.* **15**(11), 81 (Nov. 1983).
5. T. W. Sprecher, *Pap. Film Foil Converter,* **57**(11), 114 (Nov. 1983).
6. *Flexo* **10**(4), 54, 62 (1985).
7. R. N. Gidwani, *Am. Lab.* **16**(11), 84 (Nov. 1983).
8. *Pap. Film Foil Converter* **57**(7), 50 (July 1983).
9. D. Briggs, C. R. Kendall, A. R. Blythe, and A. B. Wooton, *Polymer* **24**(1), 51 (1983).
10. B. P. Sherman, *Converting World U.K.* **11,** 6 (1985).
11. R. N. Gidwani, *Am. Lab.* **16**(11), 85, 87 (Nov. 1983).
12. *Chem. Week* **134**(11), 50 (1984).
13. *Chem. Week* **134**(11), 52 (1984).
14. *Chem. Week* **134**(11), 55 (1984).
15. *Packag. Eng.* **26,**(12) 64 (Nov. 1981).
16. M. A. Amini, *Packag. Technol.* **14**(2), 24 (Oct. 1984).
17. R. D. Sackett, *Food Drug Packag.* **47**(10), 34 (Oct. 1983).
18. *Mod. Plast.* **60**(7) 30 (July 1983).
19. *Dow Surface Treatment, Gasoline Barrier Improvement for HDPE Fuel Tanks,* The Dow Chemical Company, Midland, MI, 1974, Form No. 308-184-74.
20. *Modern Plast.* **61**(12), 96 (Dec. 1984).
21. *Modern Plast.* **61**(1), 12 (Jan. 1984).
22. *Packag. Lett.* **28**(25), 1 (1983).
23. P. R. Lantos, *Packaging* **30**(6), 59 (May 1985).
24. R. D. Sackett, *Food Drug Packag.* **47**(10), 30 (Oct. 1983).
25. H. L. Allison, *Packaging* **30**(3), 25 (Mar. 1985).
26. R. D. Sackett, *Food Drug Packag.* **47**(10), 28, 30, 33 (Oct. 1983).
27. *Modern Plast.* 94, **61**(12), 94, 96 (Dec. 1984).
28. G. R. Smoluk, *Modern Plast.,* **60**(12), 44 (Dec. 1983).
29. J. W. Peters and R. Heuer, *Packaging* **29**(1), 31 (Jan. 1984).
30. F. Labell and J. Rice, *Food Process.* **46**(3), 46 (1985).
31. M. A. Amini, *Packag. Technol.* **14**(2), 22 (Oct. 1984).
32. L. H. Doar, *Food Eng.* **57**(2), 59 (Feb. 1985).
33. D. P. Dempster, *Food Can.* **45**(2), 21, 37 (1985).
34. J. R. Newton, *Plast. Eng.,* **41**(5), 83 (May 1985).
35. H. L. Allison, *Packaging* **30**(3), 26 (Mar. 1985).
36. J. W. Peters and R. Heuer, *Packaging* 29(1), 35 (Jan. 1984).
37. G. R. Smoluk, *Modern Plast.* **60**(12), 46 (Dec. 1983).

SURFACE TREATMENT

In almost every industry, the nature of a material's surface can drastically affect a product's success. The reasons can be quite different, varying from purely aesthetic to functional. This is particularly true for packaging applications. To the consumer at the point of purchase a package must appear attractive and clean as well as preserve its contents. Obtaining the appropriate balance of structural, aesthetic, and functional (barrier) properties often requires compounding specific additives into the bulk material or combining several separate materials into a composite structure. The type of specialty additives often used as bulk treatments may include but are not limited to antistats, antiblocks, slip modifiers, plasticizers, fillers, and stabilizers for UV, oxygen, and heat. The operations or procedures used to take various raw materials and fashion them into packaging structures is often referred to as "converting." A surface treatment is frequently employed as part of the conversion process to alter the surface characteristics of the specific material being used. Typical surface treatment processes include altering the wettability of a substrate, improving the bondability of an applied material or the elimination of accumulated static charge. Surface treatment technologies can play a key role in the preparation of surfaces of the most commonly used packaging substrates such as paper, plastic, foil, or metal/inorganic depositions for subsequent processing steps. In many cases, packaging producers are required to select specially formulated and expensive materials (eg, printing inks, adhesives, polymer films, or structures) to ensure satisfactory performance. The alternative to this scenario is to choose a material or combination of materials for their bulk properties and then modify their surfaces to achieve appropriate performance attributes. Surface treatments can allow the necessary modifications to packaging material surfaces without altering their bulk properties so that individual or multilaminate composite packaging structures can meet or exceed end use requirements.

Surface Preparations

Achieving adequate adhesion to polymers is a recurring and difficult problem throughout the packaging industry. Historically, various surface treatments have been used to improve the adhesion of coatings to plastics, including flame and corona, mechanical abrasion, solvent cleaning or swelling followed by wet chemical etching, or the application of specialized coatings in the form of chemical primers. Also, high energy density treatments (1) such as ultraviolet (uv) radiation, electron-beam and cold-gas-plasma methods have gained greater acceptance on a larger scale for substrate surface modification. They provide a medium rich in reactive species, such as energetic photons, electrons, free radicals, and ions, which, in turn, interact with the polymer surface, changing its chemistry and/or morphology. These processes can be readily adopted to modify surface properties of webs, films, and rigid containers, which are commonly incorporated into packaging structures. The available surface treatment technologies are summarized in Table 1.

Abrasive techniques. Among conventional surface treatment techniques, mechanical abrasion serves only to increase the surface area of the material by "roughening" the exposed areas prior to coating or adhesive bonding. Mechanical abrasion can be achieved through dry blasting, wet blasting or hand/machine sanding. These processes can be very operator-sensitive, labor-intensive, dirty, and difficult to perform on the high-production volumes normally associated with packaging applications. To remove particulates or residues, a solvent wash usually follows mechanical abrasion. In many

Table 1. Surface-Treatment Technologies

Technique	Process(es)	Types	Technology Status	Comments
Abrasion	Mechanical	Dry or wet blasting, hand or machine sanding	Obsolete	Labor-intensive, dirty, applicable only for low production volumes, must deal with residuals
Solvent cleaning	Physical and chemical	Wiping, immersion, spraying or vapor degreasing	Obsolete	Safety, disposal and environmental concerns (ie, emissions).
Water-based cleaning	Physical	Multistep power wash systems	Contemporary	Low environmental impact, high volume capacity, and relatively low cost
Chemical etching with acids or bases	Chemical	Immersion, brushing, rinsing, spraying	Obsolete	Safety issues due to the use of corrosive, toxic materials and hazardous-waste-disposal problems
Chemical primers	Chemical	Solution application of polyethyleneamine, polyurethanes, acrylates, chlorinated polymers, nitrocellulose, or shellac	Mature	Requires specific equipment, and different primers are necessary for specific end-use requirements
Flame treatment	Thermal and chemical	Available for flat films or three-dimensional configurations	Mature	Fire hazard, limited to some extent to thermally insensitive materials
Corona discharge	Electrical and chemical	Available for both conductive and dielectric substrates	Contemporary	Applicable primarily to films and webs
Gas plasma	Electrical and chemical	Available for film or three-dimensional applications; can use ac, F, or microwave frequency	Contemporary	Convenient and cost-effective; no toxic materials or disposal issues; can be effective in numerous different configurations
Uv and uv/ozone	Electrical and chemical	For distinct parts in batch systems	Developmental, contemporary	Generally only in batch format and requires longer residence times
Evaporated acrylate coatings	Physical and chemical	Currently for webs and films only	Developmental, contemporary	Still being developed for commercial-scale applications
Fluorination	Chemical	Short exposure to elemental fluorine can be batchwise or continuous	Developmental, contemporary	Specialized equipment required for delivery and monitoring fluorine
Electrostatic-discharge control	Electrical	Can be in the form of charge dissipation or charge neutralization	Contemporary	Equipment can be simple through complex and expensive, depending on the application.

cases, the spent abrasive materials fall under the classification of hazardous substances and must be disposed of accordingly.

Liquid cleaning techniques. Liquid cleaning can be very useful for removing gross contamination. Fluid cleaning techniques for polymer surfaces fall into three main categories: hand wiping, solvent cleaning, and water-based washing. Hand wiping can be done with a variety of solvents, combination of solvents, or an aqueous solution of various chemicals. This process is very labor-intensive and is usually only employed in situations with low production volumes. Hand wiping can result in inconsistencies in quality due to either human error or the redeposition of soils onto the surfaces being cleaned from contaminated rags used in the process. Surface treatment by solvent cleaning is most beneficial in those cases where swelling of the polymer surface, due to solvent absorption, results in a rougher morphology that can improve the adhesion of coatings without adversely affecting the substrate's mechanical properties. The process uses inexpensive equipment and works reasonably well in many cases such as in the surface preparation of molded polymer parts for subsequent paint adhesive, or coating application. Solvent treatment processes can be conducted through wiping, immersion, spraying, or vapor degreasing. Typically, high-vapor-pressure organic solvents (alcohols, ketones, toluene, etc), chlorinated hydrocarbon solvents (eg, Freons, or 1,1,1-trichloroethane) or low-vapor-pressure organic solvents (terpenes, isoparaffins, lactates, esters, etc) are used in these processes (2). The major drawbacks of the technology are the environmental and process hazards associated with the use of large quantities of volatile chemicals, to the extent that any savings in equipment cost are usually offset by the increased cost of obligatory environmental controls. Also, solvent-based surface treatment has limited utility when a distinct change in the chemical nature of the substrate surface is desired. Water-based cleaning processes operate with relatively low costs, have low environmental impact, and are well suited for high production volumes. An industrial power washer usually consists of an overhead or floor conveyor with parts mounted on racks that pass through various spray stages. Most systems are composed of seven separate functions: precleaning, cleaning, rinsing, conditioner or rinse aid, deionized water rinsing, air blowoff, and oven drying (3). However, for most commercial applications this technology requires capital investment for new equipment.

Chemical etching. Generally, chemical or acid etching is more effective in improving adhesion to polymers than liquid cleaning or solvent swelling. These processes cause specific chemical changes to the substrate surface, allowing greater chemical and physical interactions to adhesives or coatings. Some common examples of chemical etching processes for various polymer materials are listed in Table 2 (4). The chemical treatment of polyolefins in many cases incorporates the use of chromic–sulfuric acid mixtures (5). Previous studies have shown that for LDPE and HDPE severe roughening of the surface occurs. The effect of such treatments on polypropylene depends strongly on the prior thermal history of the polymer, and higher etch rates have been observed in areas of low crystallinity. There can also be changes in the polymer surface chemistry after chromic acid oxidation. Reflection ir spectra (6) show that this treatment results in the incorporation of oxygen (hydroxyl, carbonyl, and ester groups) and sulfur (SO_3H)-containing functional groups in the LDPE surface. However, the precise chemical state of a polymer surface after chemical exposure is dependent on the nature and thermal history of the polymer, the composition of the etchant solution, and the time and temperature of the exposure. Often a process will work well for one material but will not be effective for another, necessitating specific treatments for each type of substrate. Also, chemical etching processes must be monitored closely as overexposure can result in overtreated, discolored, or damaged materials. In many instances, the etchant materials used can pose serious safety, hazard, and disposal problems. Although many of these processes can be effective in treating specific polymer materials, numerous users are seeking alternatives because of the concerns for operator safety and the complications of use and disposal.

Chemical priming. Chemical primers can provide improved printing and adhesion characteristics by applying a chemically distinct layer on the substrate. This is usually accomplished by applying a liquid material in the form of a thin film and then drying off the solvents to leave a desired resin coating. Many polymer surfaces in the form supplied by the manufacturer can generate problems with respect to printability or the adhesion of decorative or functional coatings. Many packaging grade polymers are treated for improved adhesion, but chemical priming can also be used to improve productivity of converting processes. When primers are used on low-surface-energy substrates such as polyolefins, printing defects can be greatly reduced and issues such as screening, mottling, and "fisheyes" can be virtually eliminated. In the printing industry, press-speed limitations are seldom a function of solvent retention, but rather of adequate ink adhesion (7). As press speed increases the effectiveness of high-energy-density treatments decreases. The fact that printing primers have the same surface tension characteristics at all press speeds provides a productive advantage as long as there is adequate drying capacity. As a result, maximum press or laminator speeds are attainable as long as the primer and subsequent printing inks can be dried. Unlike corona or flame-treatment methods, primed surfaces tend to remain unaltered and the effect of additive migration to their surfaces appears to be limited. Primers can fall under various chemical classes such as polyethyleneimine, polyurethanes, acrylates, and chlorinated polymers. To prime foil substrates for printing or other subsequent converting operations, solvent-based solutions of nitrocellulose and shellac are still used. However, the trend is toward specific high-performance water-based primers such as ethylene acrylic acid. The main drawback to chemical priming is that there is no universal primer and different materials are needed for specific end-use requirements.

Flame treatment. In flame treatment, the polymer surface

Table 2. Chemical-Etching Processes

Material	Chemical Treatment	Temperature	Immersion Time	Post-Preparation
Polyester (PET & PETG)	Sodium hydroxide, 20 pbw[a] Distilled water, 80 pbw	70–95°C	10 min	Rinse in hot water and hot-air-dry
Fluorocarbon polymers	Naphthalene, 13.0 pbw Sodium, 2.0 pbw Tetrahydrofuran, 85.0 pbw	70–90°C	2 min	Rinse in MEK and toluene, then rinse in distilled water
Low-density polyethylene	Potassium dichromate, 5% Distilled water, 7% Concentrated sulfuric acid, 88%	70°C	30 min	Rinse in distilled water
High-density polyethylene	Potassium dichromate, 0.5% Distilled water, 9.5% Concentrated sulfuric acid, 90%	20°C	5 s	Rinse in distilled water
Polypropylene	Potassium dichromate, 5% Concentrated sulfuric acid, 95%	20°C	1 min	Rinse in distilled water
Acetal homopolymers	1,4-Dioxane, 44.0 pbw Perchloroethylene, 54.0 pbw *p*-Toluene sulfuric acid, 2.0 pbw	80–120°C	5–30 s	Heat-treat at 120C for 1 min; rinse in tap water
Polyamides or polysulfones	Phenol, 80.0 pbw Distilled water, 20.0 pbw	70–90°C	Brushon	Rinse off twice with tap water

[a] Percent body weight.

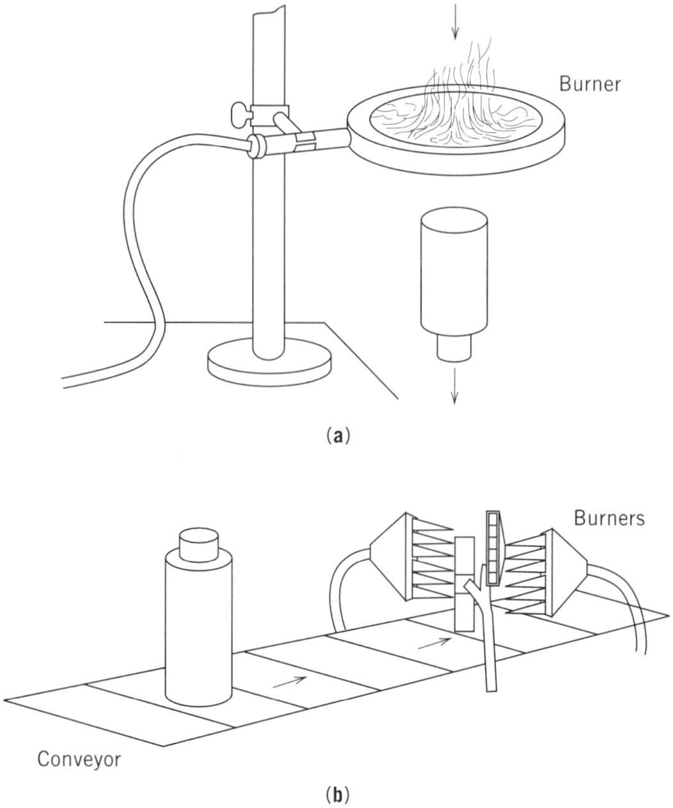

Figure 1. (**a**) Ring burner for round-bottle treatment; (**b**) burner arrangement for treatment of round plastic bottles.

is passed through a flame generated by the combustion of a hydrocarbon (typically natural gas). Flame treatment can be conducted in a variety of configurations (illustrated in Fig. 1). Usually, containers or polymer webs are passed through a bank of flame jets at a given speed to provide the desired properties. In direct flame treatment, the high temperature (adiabatic flame temperature is approximately 33,000°F) is sufficient to dissociate nitrogen and oxygen molecules into free atoms (8). In addition, this high-temperature plasma contains carbon, free electrons, positively charged oxygen, and other ions and excited species. Because of this reaction, polar functional groups such as ether, ester, carbonyl, carboxyl, and hydroxyl are contained in a flame plasma; these are incorporated into the surface and affect the electron density of the polymer material. The result is that the polymer surface is polarized. By changing the polymer surface from nonpolar to polar, the ink adhesion, laminating, and metallizing characteristics are enhanced. Also, exposure to the open flame oxidizes the surface and burns off surface contamination such as material additives, processing aids, or organic contamination such as oils or grease (9). It is probable that some of the polymer chains actually undergo melting, which "locks" their positions on cooling with respect to the three-dimensional configuration of the substrate, restricting rotation of the polymer molecules. Polar functional groups tend to stay in place on the surface, which can explain why the surface change due to flame treatment does not decay like that due to corona treatment. This process is somewhat energy-intensive, and it may be difficult to reach recessed areas and to evenly treat complex shapes. Also, care must be taken to prevent thermal damage to sensitive materials such as thin-walled plastics or film substrates, and higher-energy output is necessary as production speed or throughput are increased.

Corona treatment. In the case of corona treatment, the surface is exposed to a discharge between a grounded and powered electrode at high voltage. A low-frequency (typically 10–20-kHz) generator and stepup transformer usually provide the high voltage to the electrode. In each half-cycle the applied voltage (20-kV peak) increases until it exceeds the threshold value for electrical breakdown of the air gap, causing the atoms and molecules to become ionized and creating an atmospheric plasma discharge. The voltage eventually eventually peaks and falls below the conducting threshold. Each cycle consists of two such events involving current flow in each direction. In continuous operation the discharge appears to be a random series of faint sparks in a blue-purple glow (uv radiation). The point discharge generated across the pair of electrodes ionizes the gas present in the gap, which subsequently induces changes in the chemistry of the surface. Researchers (10) have demonstrated through derivatization reactions that carbonyl, enol, and carboxylic acid groups are formed on polyolefin materials after corona treatment. The most likely mechanism is free-radical in nature. The corona discharge contains ions, electrons, excited neutrals (atoms and molecules, and photons. All of these have sufficient energy to cause bond cleavage in the polymer surface. The resulting polymer chain radicals react extremely rapidly with O_2. Chain scission is involved in the formation of many of these groups, leading to a progressive reduction in the average molecular weight and finally to the production of CO, CO_2, and H_2O. In addition to oxidative degradation, there will also be direct degradation by ion-induced sputtering. These changes can have dramatic effects on the surface energy and functionality of polymer materials. Both dielectric polymer and conductive substrates can be treated with this method as illustrated in Figure 2. With nonconducting polymer films, the grounded roller is covered with a dielectric insulating material and a linear electrode is used. However, with conductive metallic substrates, the process is simply reversed by using a rotating electrode covered with a dielectric insulating material to prevent short-circuiting to ground. In either case, the electrode is always connected to a source of high voltage power, and the roller always remains grounded. However, the corona is a shower of arcs or sparks and each discharge point has the capability of causing localized damage and is difficult to apply consistently on three-dimensional components or structures. With corona treatment the effect on many materials is reported to be short-lived. This can represent a problem in some packaging applications where treatment stability is important.

Cold-gas-plasma treatment. This process consists of exposing a polymer to a low-temperature, low-pressure glow discharge (ie, a plasma). The resulting plasma is a partially ionized gas consisting of large concentrations of excited atomic, molecular, ionic, and free-radical species. Excitation of the gas molecules is accomplished by subjecting the gas, which is enclosed in a vacuum chamber, to an electric field, typically at radiofrequency (rf). Free electrons gain energy from the imposed rf electric field, colliding with neutral gas molecules and transferring energy, dissociating the molecules to form nu-

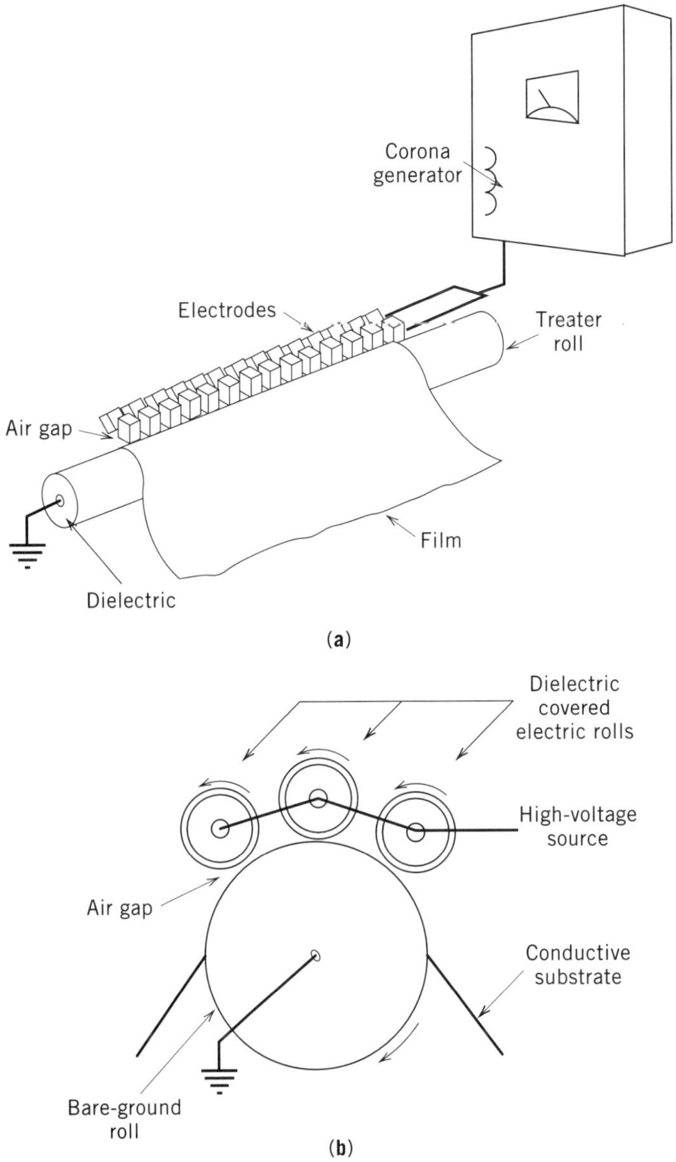

Figure 2. Corona discharge treaters with (**a**) segmented electrodes; (**b**) driven electrode rolls.

merous reactive species. It is the interaction of these excited species with solid surfaces placed in the plasma that results in the chemical and physical modification of the material surface (see Fig. 3).

The effect of a plasma on a given material is determined by the chemistry of the reactions between the surface and the reactive species present in the plasma. At the low exposure energies typically used for surface treatment, the plasma surface interactions only change the surface of the material; the effects are confined to a region only several molecular layers deep and do not change the bulk properties of the substrate. The resulting surface changes depend on the composition of the surface and the gas used. Gases, or mixtures of gases, used for plasma treatment of polymers can include air, nitrogen, argon, oxygen, nitrous oxide, helium, tetrafluoromethane, water vapor, carbon dioxide, methane, or ammonia. Each gas produces a unique plasma composition and results in different surface properties. For example, the surface energy can be increased very quickly and effectively by plasma-induced oxidation, nitration, hydrolyzation, or amination. Depending on the chemistry of the polymer and the source gases, substitution of molecular moieties into the surface can make polymers either wettable or totally nonwettable. The specific type of substituted atoms or groups determines the specific surface potential. For any gas composition, three competing surface processes simultaneously alter the plastic, with the extent of each depending on the chemistry and process variables: ablation, crosslinking, and activation (11). *Ablation* is similar to an evaporation process. In this process, the bombardment of the polymer surface by energetic particles (ie, free radicals, electrons, and ions) and radiation breaks the covalent bonds of the polymer backbone, resulting in lower-molecular-weight polymer chains. As long molecular components become shorter, the volatile oligomer and monomer byproducts boil off (ablate) and are swept away with the vacuum-pump exhaust. *Crosslinking* is done with an inert process gas (argon or helium). The bond breaking occurs on the polymer surface, but since there are no free-radical scavengers, it can form a bond with a nearby free radical on a different chain (crosslink). *Activation* is a process where surface polymer functional groups are replaced with different atoms or chemical groups from the plasma. As with ablation, surface exposure to energetic species abstracts hydrogen or breaks the backbone of the polymer, creating free radicals. In addition, plasma contains very high-energy uv radiation. This uv energy creates additional similar free radicals on the polymer surface. Free radicals, which are thermodynamically unstable, quickly react with the polymer backbone itself or with other free-radical species present at the surface to form stable covalently bonded atoms or more complex groups.

Figure 4 illustrates the components of a typical plasma surface-treatment system. In a conventional plasma process, the chamber is evacuated to a specified pressure using a mechanical vacuum pump and gas is introduced into it through flow controllers. Once the gas flow has stabilized and the de-

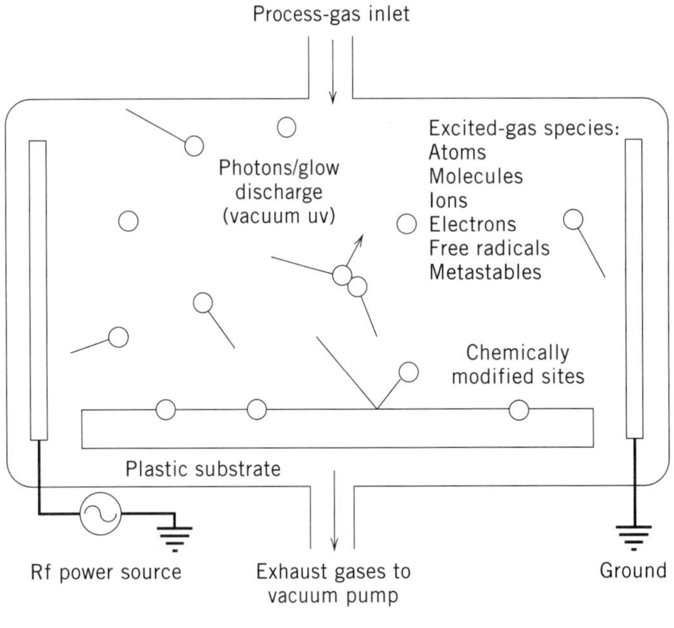

Figure 3. Plasma surface-modification mechanism.

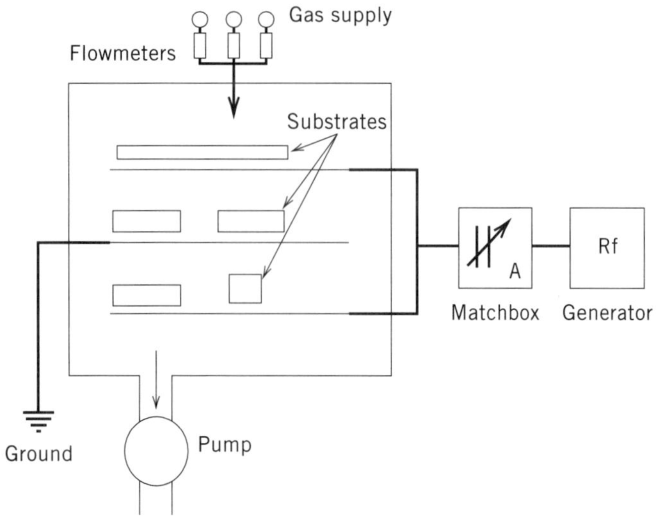

Figure 4. Components of a typical plasma surface-treatment system.

sired operating pressure has been reached, the rf power is applied to the electrodes and the gas is ionized. A capacitance-matching network tunes the chamber impedance to a constant load. During normal operation, gas is being continually introduced into the chamber and the unreacted species and byproducts are continuously evacuated. The chamber thus operates in a steady state. Cold-gas plasma offers the engineer a means of reengineering the polymer surface and introducing the desired functional groups in a controlled and reproducible manner. The nature of plasma surface modification lends itself to precise control and process repeatability. In a majority of applications, plasma surface treatment employs innocuous gases that allow the engineer or scientist to radically modify the surface while maintaining workplace and environmental cleanliness and safety (12). The treatment effect can be very long-lived and can be applied to a variety of configurations such as webs of plastic films or three-dimensional containers.

Cold-gas-plasma processes can also be used to apply transparent thin-film silicon oxide–based gas-barrier depositions for flexible-packaging applications. Common polymer films packaging films such as polyethylene terephthalte (PET) and biaxially oriented polypropylene (OPP) can be used as substrates for plasma-deposition processes. A common process involves the plasma decomposition of 1,1,3,3-tetramethyldisiloxane (TMDSO) or hexamethyldisiloxane (HMDSO) in a helium and oxygen plasma to form a gas and vapor barrier layer of SiO_2 (Fig. 5). The process has been successfully commercialized into production equipment that can accommodate 1.5-m web widths and run up to 100 m/min (13). The resulting barrier films can then be processed through typical converting steps to form transparent high-barrier flexible packaging. This approach can improve the shelf life of packaged foods or products and offer an alternative to conventional packaging gas-barrier technologies.

Ultraviolet/ozone. For this process, the polymer surface is exposed to both uv light and ozone to increase the number of oxygen functional groups incorporated into the material. This approach can be useful in the surface modification of three-dimensional parts. The process has been used on polypropylene and polyester substrates and has shown rapid and reproducible uptake of surface oxygen functional groups (14). The attachment of oxygen groups greatly changes the surface energy and chemistry, which can lead to improved adhesion of functional and decorative coatings. Most of the initial process development has been targeted at the treatment of three-dimensional plastic components. These are suspended or tumbled inside a reaction chamber at room temperature and pressure and exposed simultaneously to uv (mercury-source) light and various concentrations of ozone (O_3) gas. After 5 min of exposure, oxygen functional groups can be attached to as many as 30% carbon atoms on the outermost surface of polypropylene. The stability of this treatment is usually quite good. For example, relatively little change in the receding contact angle occurs on the treated surface of polypropylene after aging in air for a period of $\leq$28 days.

Evaporated acrylate coatings. This is a relatively new high-speed process in which thin highly uniform acrylate coatings are applied to the polymer surface in order to smooth the surface of polymer film substrates so that subsequent desposi-tions are more defect-free. This process involves feeding acrylate monomeric fluids into an ultrasonic atomizer connected to an evaporator in a reaction chamber at reduced pressure. The acrylate material is atomized into a mist that contacts the hot walls of the evaporator, where it is transformed into a gas. The molecular vapor that results exits through a slit in the evaporator nozzle and is condensed on a substrate moving in front of the nozzle. This thin liquid is then irradiated with a low-voltage electron gun and the coating is transformed into a hard, tough thin-film coating. The process is claimed to be compatible with various vacuum coating processes such as sputtering, evaporation, or plasma deposition; and can in principle be conducted in series with other deposition processes within the same reaction vessel. The presence of this coating has been shown (15) to greatly improve the coating uniformity of aluminum metallized barrier depositions on both oriented polypropylene and polyester packaging films. Consequently, the oxygen- and water-vapor-barrier properties are improved. This can have an impact in numerous applications for extending the shelf life of packaged products. The improvement in barrier can be attributed to three factors:

1. The coating forms a smooth layer on the polymer surface, eliminating any surface irregularities.
2. The coating has very good temperature stability which provides a thermally stable platform on which to apply a barrier material.
3. The acrylate surface is more chemically polar than many polymer films, and the density of the resulting film is higher. Barrier layers composed of metals and inorganic oxides tend to grow more readily on a polar substrate than on a nonpolar substrate.

Fluorination treatment. The fluorination process involves exposing polymeric webs continuously to fluorine gas (F_2) diluted with an inert gas (eg, nitrogen) inside a reaction chamber (16). This process can greatly increase the surface energy of polymer materials such that excellent adherence can be at-

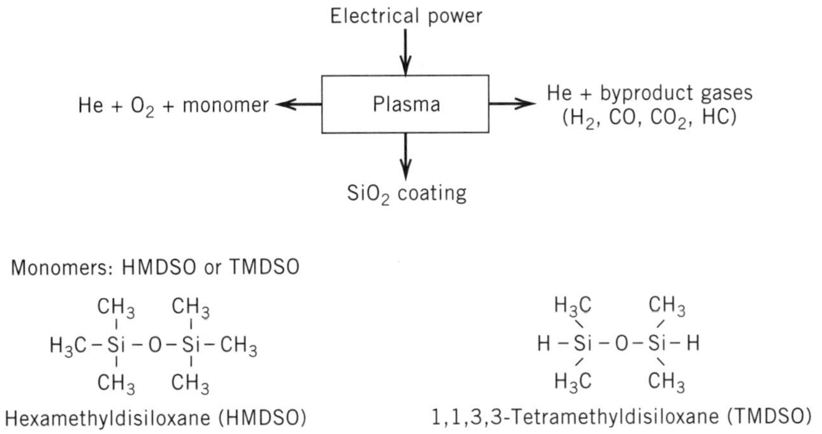

Figure 5. Low-pressure SiO_2 plasma deposition process.

tained to other materials such as lacquers and adhesive agents. Diatomic fluorine, an almost colorless gas, is one of the strongest oxidizing agents; it reacts with almost all organic and inorganic substances (except nitrogen and other inert gases). Fluorine's great reactivity is due to the interaction of the low dissociation energy of the molecule and the very strong bonds it forms with other atoms. Electron spectroscopy for chemical analysis (ESCA) data indicate that the activation of polymer surfaces using this process results from the partial fluorination of the hydrocarbon structure of the polymer molecules. An additional application would be the fluorination of high-density polyethylene gasoline tanks to provide hydrocarbon barrier. Fluorine is routinely transported in its liquid state and is commercially available because of use in the nuclear industry for the refinement of fuels. The safeguards used for this technology are similar to safety measures used and approved for ozone generation. Compared with other pretreatment processes, surface fluorination not only has a wide spectrum of applications but also doesn't require the use of electrical equipment such as corona or plasma treatment. Surfaces treated with fluorine exhibit longlasting, if not irreversible, changes. This can be very important in practical applications in industry, since subsequent converting processes don't have to immediately follow the surface activation.

Electrostatic discharge treatment. Plastic, as opposed to metal, substrates make good electrical insulators because they are electrically nonconductive and possess high electrical resistivity. The higher the surface resistivity, the lower the surface conductivity. However, those plastic insulating materials that have high dielectric constants can generate and store static electricity. Static electricity is generated when two materials in intimate contact are separated by a frictional force causing electrons to be preferentially stripped from one surface and transferred to the other surface. This causes the electron-rich and electron-deficient surfaces to assume positive and negative charges and this surface polarization results in the generation of static electricity. Unless this charge is dissipated, the static buildup can cause the attraction of dust, lint, sparks, materials-handling problems, shocks, and difficulty in wetting or adhering.

Packaging substrates made from polyethylene, polypropylene, polyester, polystyrene, and other dielectric materials at some time during their manufacture are usually subjected to at least one of the many available static control techniques. These fall into two separate categories: charge dissipation and charge neutralization. With electrically conductive materials, dissipation of static charge can be accomplished by simply grounding the charged material. However, this is difficult with nonconducting materials such as polymer films, so one approach is to humidify the work area so that the exposed surface absorbs a thin layer of water that conducts the charge to ground. An alternative method is to shield the surface with antistatic organic compounds. Most antistatic agents fall under the following types: nonionic ethoxylated alkylamine, anionic aliphatic sulfonate/phosphates, and cationic quatenary ammonium compounds (17). Antistats can be applied topically or blended, and their purpose is to retard static buildup and also to rapidly discharge any accumulated charge.

Another approach to static elimination is to neutralize the accumulated charge using devices capable of ionizing the surrounding air. This works by exposing electrically neutral atoms in air to an applied electric field of voltage high enough to create positively and negatively charged ions. Because of the bipolar nature of the ionized air, the static charge on a material can be neutralized by the oppositely charged ions present in the surrounding air. Basically, there are three types of air-ionizing devices available: nonpowered, powered, and self-powered. The nonpowered induction type of static eliminator consists of brass brushes mounted on ground straps that come in light contact with the charged material, causing the surrounding air to ionize. Electrically powered static eliminators are powered with a low-amperage high-voltage power supply for the ionization of the air. Radioactive self-powered units are similar to electrical static eliminators in design and construction except for the source of power. Radioactive devices are self-propagating, usually consisting of a low-energy source of an α-emitting radioisotope such as polonium-210 (^{210}Po). The alpha radiation interacts spontaneously with the air molecules, producing ionization of the surrounding environment.

BIBLIOGRAPHY

1. E. Occhiello and E. Garbassi, "Surface Modifications of Polymers Using High Energy Density Treatments," *Polym. News,* **13,** 365–368 (1988).
2. R. S. Gallagher, "Manual Cleaning Relies on Solvent Alternatives," *Precision Cleaning,* p. 29 (April 1995).

3. T. D. Held, *Rinse Aid Technology for Improved Rinsing of Plastic Surfaces,* Society of Manufacturing Engineers, 1992, EM92 182.
4. L. E. Rantz, "Proper Surface Preparation," *Adhes. Age* p. 10 (May 1987).
5. D. Briggs, "Surface Treatments for Polyolefins" in *Coating Technology Handbook,* Marcel Dekker, New York, 1991, Chapter 9, pp. 216–218.
6. H. A. Willis and V. J. I. Zichy in D. T. Clark and W. J. Feast, eds., *Polymer Surfaces,* Wiley, New York, 1978, p. 287.
7. R. M. Podhany, "Comparing Surface Treatments," *Converting* pp. 48–52 (Nov. 1990).
8. J. DiGiacomo, *Flame Plasma Treatment—a Viable Alternative to Corona Treatment,* Society of Plastic Engineers Regional Technical Conference on Decorating and Joining of Plastics, Sept. 1995, pp. 37–61.
9. G. W. Scott, "Flame Treatment Before Printing Enhances Ink Permanence," *Microelectron. Mfg.* pp. 60–61 (May 1990).
10. D. Briggs and C. R. Kendall, *Polymer* **20,** 1053 (1979).
11. E. Finson, S. L. Kaplan, and L. Wood, "Plasma Treatment of Webs and Films" in *38th Annual Technical Conference Proceedings for the Society of Vacuum Coaters,* Chicago, 1995.
12. S. L. Kaplan, and W. P. Hansen, "Plasma—the Environmentally Safe Method to Prepare Plastics and Composites for Adhesive Bonding and Painting," paper presented at SAMPE Environmental Symposium, San Diego, May 1991.
13. E. Finson, and J. Felts, "Transparent SiO_2 Barrier Coatings: Coinversion and Production Status," *TAPPI J.* **78**(1), 161–165 (Jan. 1995).
14. N. S. Mcintyre and M. J. Walzak, "New UV/Ozone Treatment Improves Adhesiveness of Polymer Surfaces," *Modern Plast.,* 79–81 (March 1995).
15. D. G. Shaw and M. C. Langlois, "Some Performance Characteristics of Evaporated Acrylate Coatings" in *37th Annual Technical Conference Proceedings for the Society of Vacuum Coaters,* Boston, MA, 1994.
16. R. Milker, in D. Satas, ed., *Coating Technology Handbook,* Marcel Dekker, New York, 1991, Chapt. 31, pp. 303–339.
17. R. Gidwani, "Fundamentals of Surface Treatment of Packaging Materials," *Am. Lab.,* 81–89 (Nov. 1983).

ERIC FINSON
BOC Coating Technology
Concord, California

STEPHEN L. KAPLAN
4th State
Belmont, California

T

TAGS

Tags have been used in all areas of business and industry for centuries. The more progressive merchants of the 1800s used tags in the same manner and for the same reasons as merchants today: to inform, instruct, identify, and inventory (see Fig. 1). Traditional cloth and paper materials have been joined by nonwoven and film-based synthetics (see Nonwovens; Plastic paper).

Tags used in retail environments are generally produced from cast-coated paperboard stocks ranging in thickness from 8 to 10 points (0.008–0.010 in., 203–254 μm). Type stands out and colors are brilliant on the smooth surface of cast-coated stocks. Cloth tags are frequently called "law labels." This term describes labels generally sewn onto mattresses and furniture to comply with government regulations regarding the identification of stuffing materials. Tags manufactured of synthetics such as nonwoven Tyvek (DuPont Company) are gradually replacing cloth tags because they exhibit an extremely high tear strength, are insensitive to moisture, and have a smoother, more uniform printing surface than cloth. Plastic tags are generally used for advertising or instructional purposes in the retail industry.

Many fastening methods are available, but the most common in the retail industry for product identification is a knotted string (see Fig. 2**a**). Plastic barbs are generally used for price marking in retail garment applications. Manufactured from an ultrathin plastic, these barbs affix price tags to ready-to-wear merchandise to prevent price-tag switching. Plastic barbs can also be combined with a hook (see Fig. 2**b**) to hang merchandise such as socks.

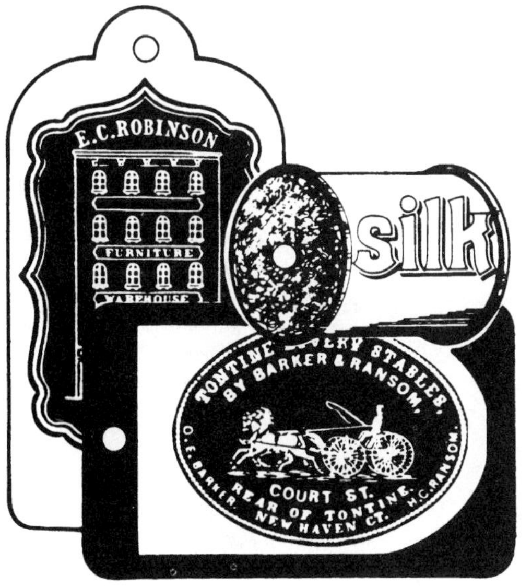

Figure 1. Price marking tags from the 1800s.

In manufacturing, processors, converters, and fabricators have a variety of uses for tags. Tags play an important part in tracking serially numbered goods such as electronic components. The simple tag shown in Figure 3**a** uses transfer tape on two of the three stubs to enable the tag to travel with the component throughout the entire packaging cycle without losing track of the serial numbers. Color coding minimizes the need for processing instructions. The entire tag is affixed to the component at the end of the assembly line, using the transfer tape on the gray stub first. Prior to boxing the product, stubs two and three (white and off-white) are detached and affixed to the outside of the box. On receipt of shipping instructions, the off-white portion is detached and forwarded with a copy of the shipping document to inventory control.

Tags and pressure-sensitive labels are frequently combined to create unusual tag products. The tag in Figure 3**b**

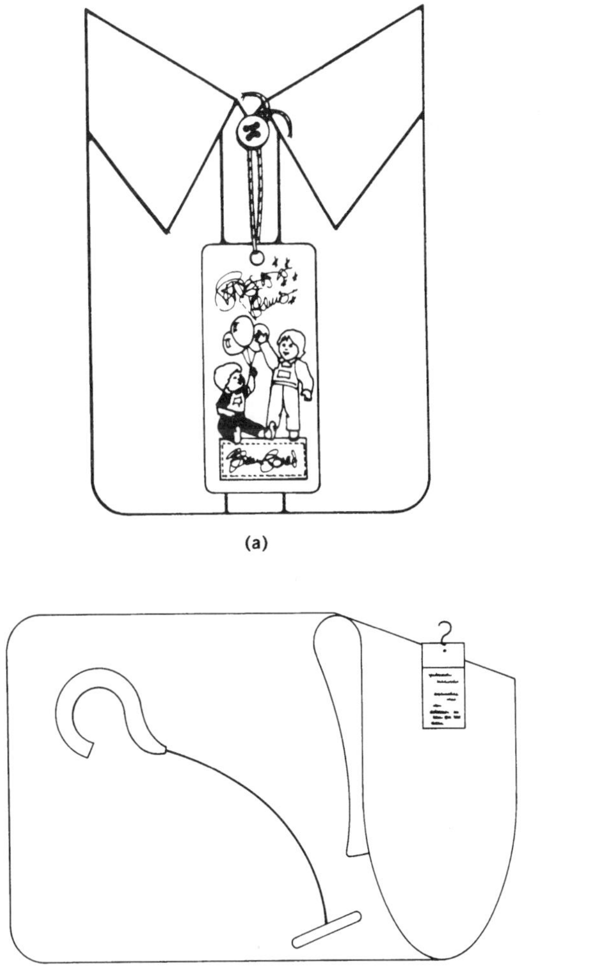

Figure 2. (**a**) Knotted-string fastener. (**b**) Plastic barb with hook.

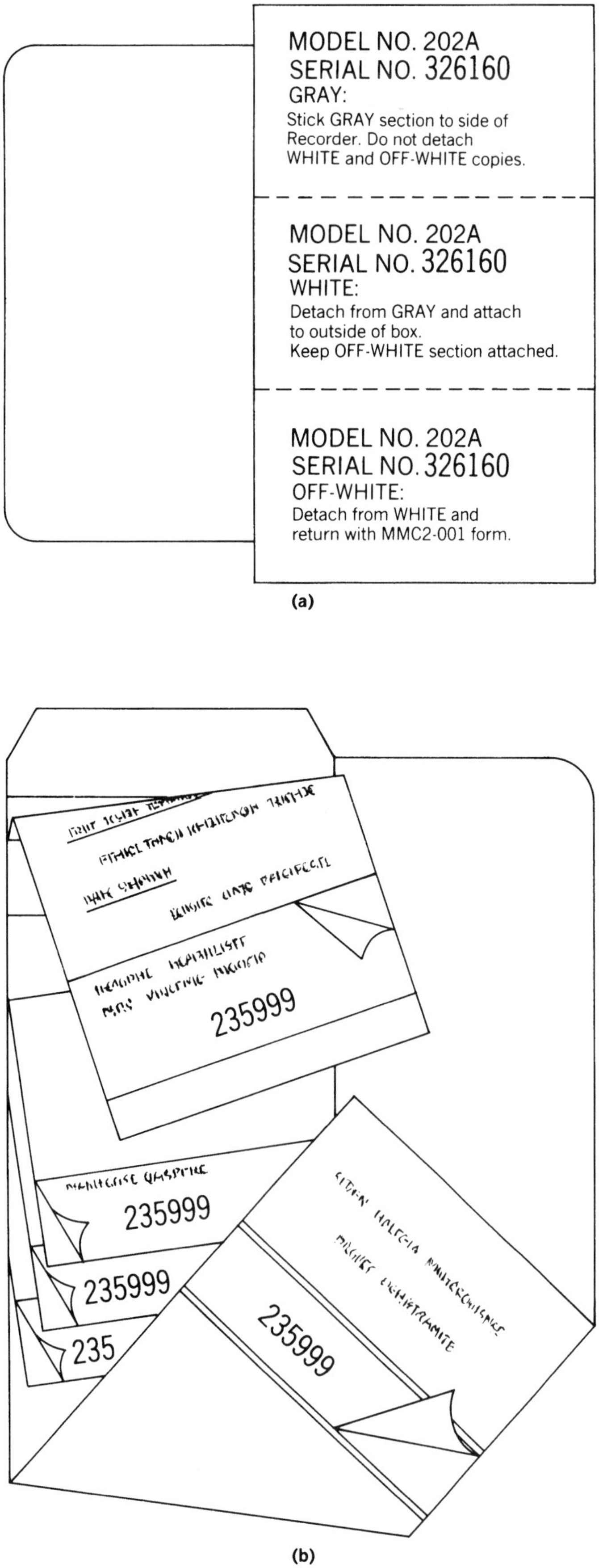

Figure 3. (**a**) A three-stat tag; (**b**) tag-and-label combination.

includes four pressure-sensitive labels, each carrying the serial number of the component, in addition to the nameplate, which in this case is affixed to the product after delivery by the installer. The tag is attached to the component with string or transfer tape, depending on the surface or characteristics of the component. The shipping department lifts up part 1 to reveal part 2 of the set, which consists of a small-face slit label containing the serial number of the component and the instructions “put on shipping document.” Part 2 is torn at the perforation along the top and reveals three additional labels with instructions for distribution (warehouse copy, distributor copy, installer copy). This unusual tag and label combination provides a simple solution to the often difficult task of tracking serially numbered products and components from assembly through shipping to their ultimate destination.

Tag applications in the retail packaging industry include, but are not limited to, product description, content identification (ie, materials used in production), care and use instructions (ie, operating and care directions, identification of potential hazards), guarantee and warranty information, decoration and promotion, inventory control, and price marking (frequently combined with inventory control when using magnetic stripes, OCR, or bar codes) (see Bar code; Code marking and imprinting).

Product marking tags, also called “hang tags,” are printed in up to eight colors on a variety of stocks, and include many features designed to get the attention of the potential consumer. Sizes range from small jewelry marking tags (see Fig. 4**a**) to oversized tags used to identify major appliances.

Hang tags are generally shear-cut, die-cut, or continuous, and offer maximum versatility for product identification, sales promotion, and advertising. The three basic tag styles found in retail tagging applications are shear-cut, die-cut, and continuous tags.

Shear-cut tags. Shear-cut tags, also called “shipping tags,” are the oldest and most common of all tag styles. Shear-cut tags are simply sheared off the end of the web from the bindery section of the press after all manufacturing processes are completed. These tags may be simple, printed on one side with black ink, with a standard $\frac{3}{16}$-in. (4.8-mm) punched hole. They may also be very complex, with such features as multicolor printing on both sides. Eyelets in tags used for product identification are rarely reinforced because the life-expectancy of the tag is short.

Die-cut tags. Die-cut tags are cut out of the web at the end of the bindery section of the printing press using either stock or custom-manufactured dies. Die cutting produces tags that are round, have rounded or scalloped corners, or have unusual shapes as shown (see Fig. 4**b**). Die-cut tags may have all or most of the features of shear-cut tags, but usually do not require features such as large patches or jumbo numbers. They are used primarily as promotional pieces or to convey information regarding a specific product, and are usually not subjected to a great deal of rough handling.

Continuous tags. These are used in applications where variable information is imprinted by a computer printer. Many possibilities exist for the number of perforations and form depths since stop and go presses are used in their manufacture. Continuous tags are generally used for retail price

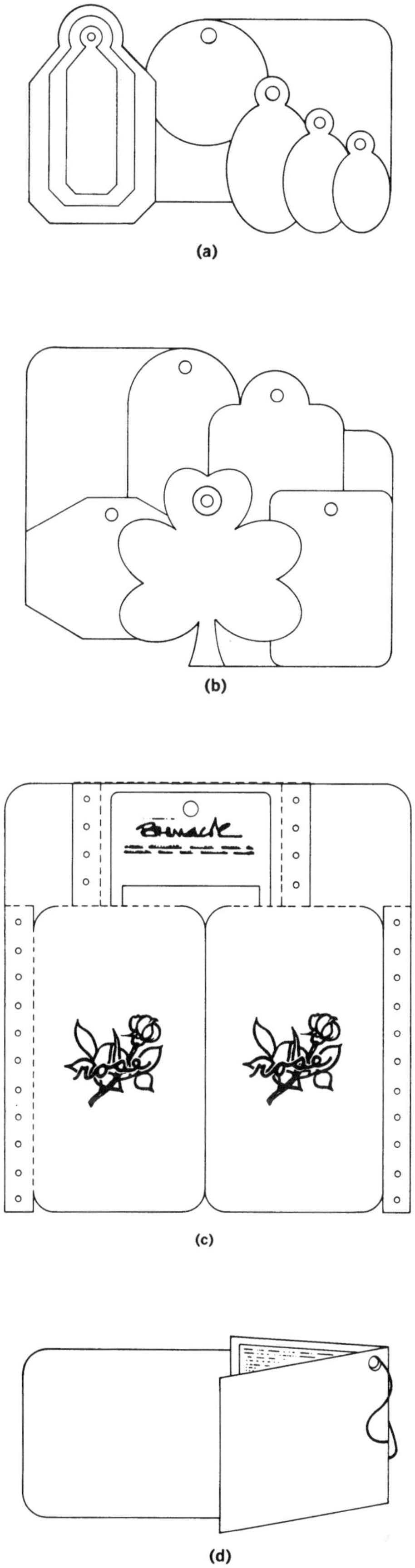

Figure 4. (**a**) Jewelry tags; (**b**) die-cut tags; (**c**) continuous tags; (**d**) booklet tag.

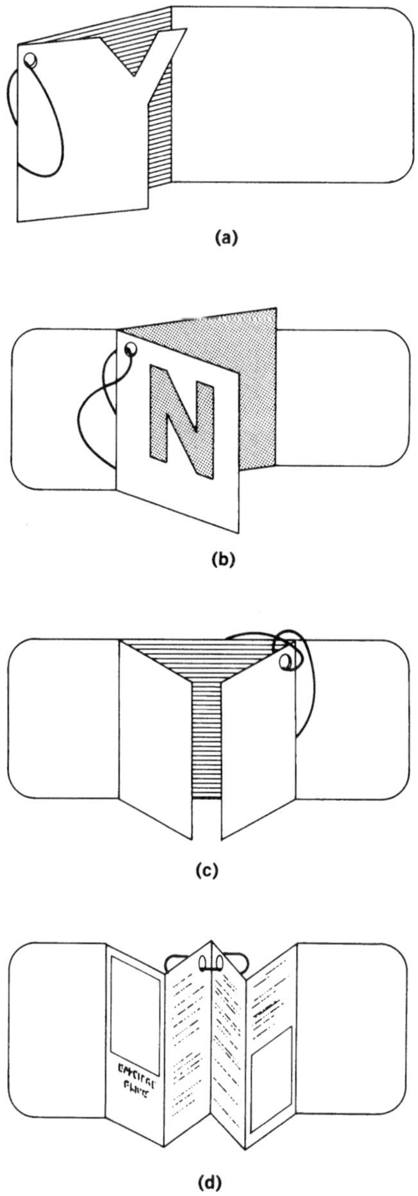

Figure 5. (**a**) Die-cut outline; (**b**) three-dimensional die-cut tag; (**c**) gatefold tag; (**d**) accordion-fold tags.

marking (see Fig. 4**c**). Special features are frequently incorporated to create hang tags that are used effectively as promotional pieces.

Booklet tags are shear-cut tags, scored vertically to form a fold, providing four surfaces for copy instead of two (Fig. 4**d**). Special effects can be added to further enhance the image of these versatile tags. Die cuts on one of the pages, generally the first, can be used to highlight the company logo or trademark or provide a representation of the product (Fig. 5**a**). Die cuts are also used on the front cover to create a window on the cover panel or to create a three-dimensional effect. The die cut is generally backed up with a solid color block or printed design which shows through the die cut (Fig. 5**b**). *Gatefold tags* are a special variety of booklet tag featuring two vertical scores to form three separate double-sided surfaces for a total of six surfaces available for printing (Fig. 5**c**).

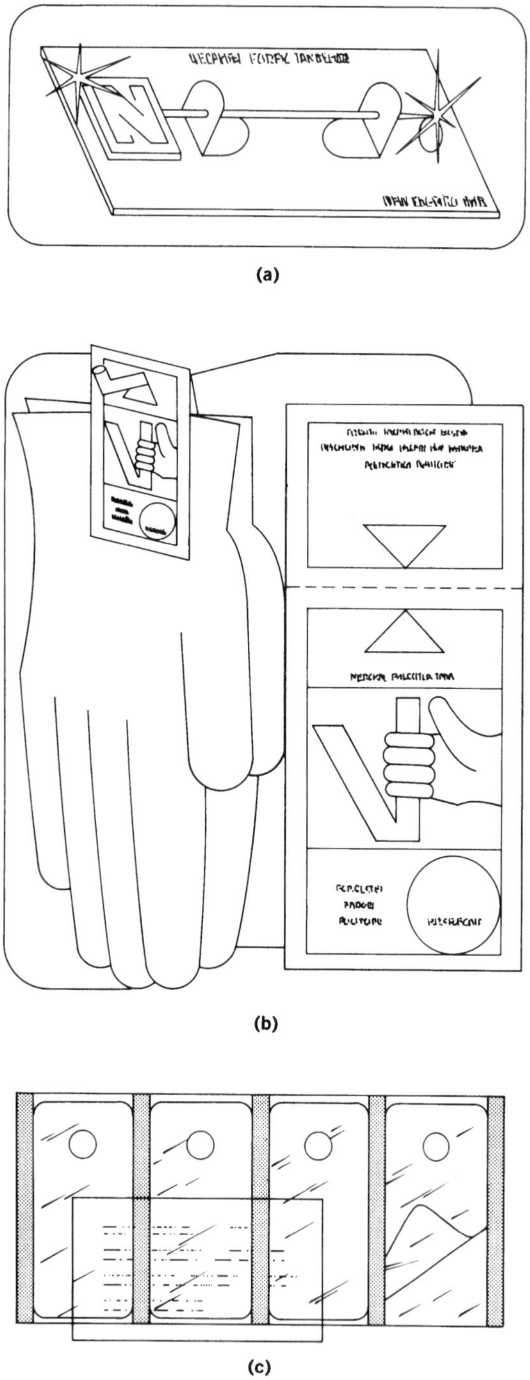

Figure 6. (**a**) Jewelry display tag; (**b**) display-rack tag; (**c**) transparent hang tag.

Accordion-fold tags have three vertical scores that create a total of four "pages" with eight surfaces for printing. Accordion-fold tags also may be produced with additional scores to add pages and sides (Fig. 5**d**).

Many applications require that tags not only carry a message regarding the product but also serve as a carrier for the product itself. In instances where merchandise (eg, jewelry) is hung on racks for display or when merchandise is seasonal and only temporarily displayed, this type of tag is very effective. Figure 6**a** shows a tag with die cuts and additional punches to hold stick pins in a retail jewelry application. The center punch is used to hang the merchandise on a display rack. The tag shown in Figure 6**b** uses two die-cut triangles to hang gloves on display racks in order to conserve space because of the product's seasonal nature. The tag is folded at the score over the cuff and stapled to fasten it to the merchandise. The die-cut triangle is then used to hang the gloves on the display rack. Figure 6**c** shows a hang tag of clear plastic partially coated with a pressure-sensitive adhesive. The coated portion of the tag is applied to the merchandise and the die-cut hole is used to hang the merchandise on the rack.

Tags used for price marking are generally produced as continuous tags. These tags satisfy the special needs of large retail organizations by affording them the opportunity to ship

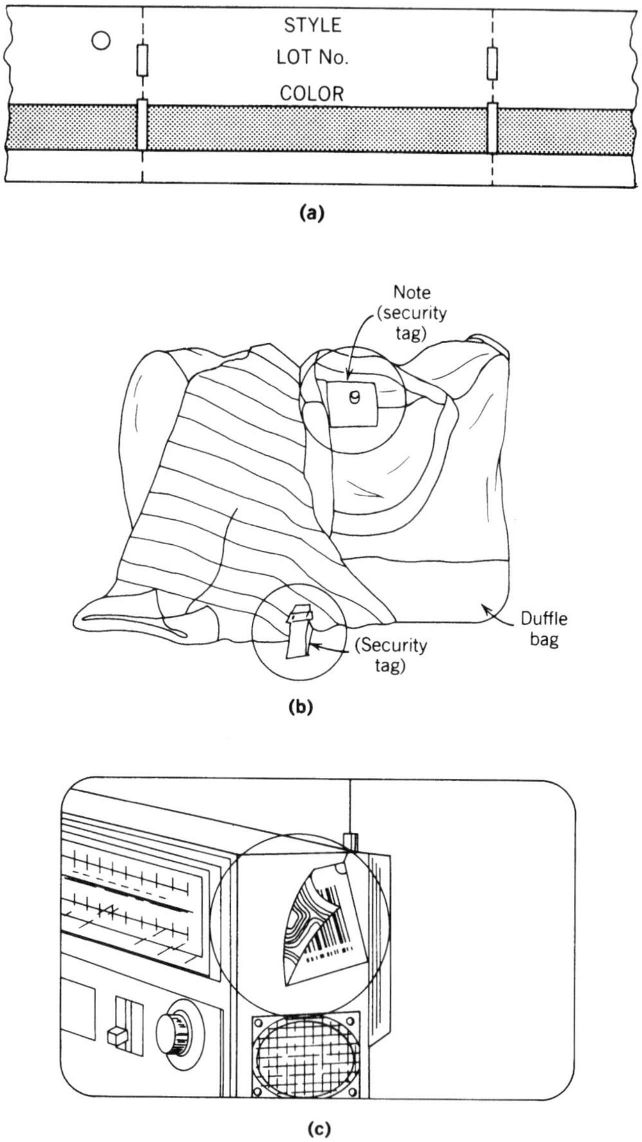

Figure 7. (**a**) Marking tag with magnetic stripe; (**b**) plastic electronic security tags; (**c**) security tag with miniaturized radio circuit.

preprinted price tags along with stock shipments to their various distribution locations. The tags come in a variety of sizes and shapes are are generally affixed with plastic barbs by tagging guns to prevent price switching. The accelerated use of computers in the retail industry has had a dramatic effect on the appearance and use of tags. Price-marking tags are now being produced with magnetic stripes (Fig. 7**a**), which can contain information such as style, lot number, and manufacturer. In addition, bar code and OCR (optical character recognition) technologies have gained acceptance in the retail tag market.

Tags can also aid in theft prevention, which is one of the biggest concerns in the retail industry. Hard plastic tags such as those shown in Figure 7**b** are affixed to articles of clothing in many large retail stores. The tags must be removed at the time of sale to prevent triggering an alarm as the garment passes through a special gate that contains a scanner that is sensitive to the tag. Because of their bulk, these tags are used primarily for clothing. Recent innovations in miniaturization have produced tags with what appears to be a strip of aluminum foil (Fig. 7**c**), but which is actually a built-in security device. If not deactivated by another strip of foil affixed at the time of sale, these circuits interrupt radiowaves being transmitted from the exit gate and set off an alarm. These tagging systems are currently marketed by companies that sell retail-security systems.

Industry indicators predict that the future of the tag industry in retail operations will not be adversely affected by advances in computer technology. As long as there are products to be identified, inventoried, and priced, there will be a market for tags of all kinds in the retail industry.

BIBLIOGRAPHY

"Tags" in *The Wiley Encyclopedia of Technology,* 1st ed., by Linda L. Butler, National Business Forms Association, pp. 624–628.

TAMPER-EVIDENT PACKAGING

History of the Problem

Since 1982 the entire industry, including the consumer has become aware of tamper-evident packaging. This awareness has benefited the consumer by a reduction in loss of life due to consumption to adulterated products from tampering. Never before has an industry reacted so swiftly to resolve a problem that could be the ruination of most manufacturers.

In 1982, seven people in the Chicago area died from consuming cyanide-laced Tylenol capsules. The incident resulted in a total product recall, massive negative publicity for the product, new requirements for safe packaging, and a federal statute making product tampering a crime. Since that time, the packaging industry has become visible to most consumers.

History of Incidents

Although there were incidents prior to 1982, product tampering became highly visible after the Chicago incidents. The exact number of incidents per year is unknown because of various methods of reporting. Beginning with the 1982 deaths, there have been deaths from product tampering about every 4 years: 1982, 1986, and early 1991. A death due to targeted tampering in 1988 is not included, as we are discussing random tampering, not tampering with a specific person in mind. It is probably a coincidence on the timing as the various incidents are unrelated, and in some instances the violator in earlier cases was incarcerated at the time of the ensuing incidents.

According to government figures, the problem peaked in the United States in 1986, when there were almost 1800 claims of possible product tampering. The number has decreased to around 500 per year now. Is the decrease due to better packaging or less interest on the part of potential violators, due to the penalties for violating product-tampering laws? Considering the experience in assisting law-enforcement agencies, the decrease appears to be caused by a change in the way claims are recorded.

If the experience of other developed nations is an indication of the future; it is expected the problem of product tampering will escalate in this country in the future. Every developed nation has experienced product-tampering incidents. The main difference between domestic incidents and what is happening elsewhere is the motive of the tamperers: extortion prior to injury, with money appearing to be the primary motive, versus apparently random tampering without prior threat in this country. Most developed nations are either implementing or modifying their rules on the use of tamper-evident packaging. Some features as they are used in the United States would have to be modified, or the use of a secondary feature would be required to meet the standards of various countries.

The problem of state-sponsored terrorism is so serious that a series of meetings were held in the late 1970s were the problem was addressed in great detail. Participating in the meetings were representatives of the FBI, Commerce Department, Defense Department, State Department, and CIA. One of the chief concerns at the time was the threat of retail-product tampering by a state-sponsored organization. The term "state-sponsored terrorism" refers to any group of terrorists that is supported financially, logistically, or in intelligence by a government body of any country. It has been documented that training in ways to tamper with retail products and how to use such acts to further their agenda is being conducted in certain countries that sponsor terrorist groups. Picture, if you will, the potential for disaster that exists if a potential tamperer has the financial and human resources available to build a complete packaging line and can print duplicate labels. It sounds far-fetched, but a similar incident occurred in South America when a drug organization bought a beverage plant to smuggle cocaine into the United States. In that case at least one bottle in a specially marked case contained the drug in a liquid form. When the contents of the bottle was distilled in the United States, it would yield a powder that could then be cut in strength and distributed to the dealers. Unfortunately, one bottle was overlooked and sold to a consumer who died from a massive cocaine overdose.

While most incidents of retail product tampering in the United States have been perpetrated by individuals, the potential exists for even greater harm from the same acts committed by a state sponsored terrorist group.

The FDA Rule

In 1982 the FDA passed a rule (21 *CFR* 211.132) requiring the use of tamper-evident packaging on all over-the-counter (OTC) drugs and some cosmetics, while ignoring other products they regulate. The fact that your product is not required to use tamper-evident (TE) packaging does not protect you in the event of a claim of product tampering.

According to attorneys specializing in packaging and product liability, product tampering is a foreseeable possibility, and manufacturers have a responsibility to protect consumers against such possible acts. If your product in an adulterated form could harm a consumer, you have a responsibility to protect the product and consumer against such acts. This means that the use of tamper-evident packaging transcends FDA regulations.

The Rule specifies which packaging features are acceptable for use in providing resistance to tampering, and the list was modified in 1988. At the current time there is a proposal to modify the Rule again and to make the acceptability of a feature based on performance.

What if the feature you want to use is not on the approved list of the FDA? The FDA has a procedure by which methods of providing protection that are not on the approved list may obtain approval on a case-specific basis. To obtain approval, samples of the complete package must be submitted to the FDA along with a written request for a waiver. Contact the local FDA compliance officer to find out the current procedure to obtain approval.

The inclusion of a specific form of protection does not warrant that the feature will deter violation, nor does it prevent legal action in the event of a claim of injury related to product tampering. There are variances in designs, and even tooling of the same design that affect the effectiveness of each feature. Any evaluation of a package relates to the exact components used in the tests, and material from a different manufacturer will usually result in a different level of effectiveness, much the same as using a different resin or closure liner will affect a stability study.

Consumer Preferences

Studies into consumer preferences for tamper-evident packaging have consistently revealed that the consumer prefers products that are resistant to tampering and prefer the features to be shelf-visible. The same studies indicated a willingness to pay slightly more than a competing brand that is not TE. The exact amount is inconsistent among the studies, but the consistent replies of concern indicate consumer awareness of packaging. This awareness increases with publicity of tampering incidents, and decreases with time until the next publicized incident.

Cost of Claim of Product Tampering

The cost of responding to a claim of possible product tampering would boggle the mind of the average person. The cost begins when the complaint is received by company personnel and includes.

- *Meetings that Will Be Held to Determine How to Respond to the Claim.* A complaint of possible product tampering requires participation of top management in all phases of meeting the problem.
- *Expenses Related to High-Level Managers Visiting the Locale Affected.* Top management or their representative will usually visit the location involved to get a better feel for the situation.
- *Cost of Picking Up All Stock and Possible Restocking of Retail Displays.* If the brand is to remain viable, fresh supplies must replace all product that may have been vulnerable to violation.
- *Cost of Complete Inspection of Targeted Stocks or Destruction of Those Not Capable of Inspection.* Since it is unlikely that the entire supply of product on the shelves at the time of the incident was affected, inspection of the stock may prevent destruction of the entire inventory.
- *Lost Sales When Consumers Changed to Other Products.* Many consumers will switch to other brands until they regain confidence in the brand affected.
- *Rewards for Information on Incident.* Rewards usually lead to information as to the identity of the violator and result in ending further incidents.
- *Fees for Packaging Consultants.* Since it is impossible for every company to have employees who are experts on every contingency, outside consultants will be required to assist in meeting the challenge.
- *Cost of Legal Counsel to Respond to Litigation.* Self-explanatory.
- *Damage Awards Where Determined in Court.* See above.

The list can go on. Many of these costs will be incurred even in false claims of possible product tampering.

When compared to the potential expense for defending a single claim of tampering, the cost of effective tamper-evident packaging becomes insignificant. Some in the industry will take the moral position that we have an obligation to protect the consumer and they are right. Beyond the moral position is the reality that many firms simply cannot afford the cost of responding to product tampering claims, especially if the firm is a small, with a very limited product line where the reputation of the entire product line can be affected by adverse publicity on one item in the product line. Liability insurance cannot restore lost-customer confidence.

A recent incident, later determined to be suicide, almost caused a household name brand to go out of business because of the high cost of produced recall and package redesign, even though the product and company were linked to the incident only by circumstance.

In considering cost, the cost associated with responding to one claim of product tampering far exceeds the cost of incorporating tamper-evident features in the package design. Would you want to explain to your boss how your company could not afford the penny or two for making each package tamper-evident, and how you can afford the 10 million dollars or more your company is spending on defending a claim of possible product tampering? Who can put a value on human life, that would justify not using effective tamper-evident packaging? The reality is, tamper-evident packaging should be utilized if it will provide any protection for your product, regardless of government regulation.

Selecting Which Feature to Use

Everyone has a preference for one type of package over another. This preference has been developed by either the products we manufacture or our experience with various types of packaging in our everyday living. Selecting which feature to use should not be affected by these preferences, but by objective testing during the package development stage.

During the design stage, the package engineer should consider the function of the product and how the consumer intends to use it. Next, each TE feature that is usable on the package should be tested to determine which feature will offer the greatest protection to the consumer. The test used should be objective, consistent, and replicable. Records of the test results should be retained indefinitely. If a feature selected for use achieves a lower value than others that were rejected, reasons for the selection should be recorded and retained with the test results. Remember—cost cannot be a factor in selecting which feature to use. You would not want to be questioned by attorneys for the other side as to why you were willing to accept less effectiveness to save a few pennies and to compare those few cents to the value of injury to a consumer.

One form of testing the effectiveness of tamper-evident packaging is the *Rosette Protocol,* which measures the degree of difficulty in violating a specific package and restoring it to a near original condition. The Protocol also measures increases in effectiveness through the use of multiple features. The value for a specific combination of features is not equal to the sum of each feature. Some factors cover the combination, rather than each feature separately. For example, the knowledge factor is applied once, regardless of the number of features in the combined package, and only one knowledge level was required. Time is cumulative; if it takes 20 min to violate each feature, the time required is not the value for 20 min multiplied by the number of features used on the package. In this example the time factor is the value for 1 h. Only one category of equipment may be required if all tools or equipment required to violate the different features in the combination are in the same class. The feature visibility values for all used on multiple feature packages are multiplied; even the use of multiple features that are not shelf-visible increases the effectiveness of the package. The feature material is added for each feature replaced or reused to determine the feature material value. The value of the feature, used with the specific package components, on the specific product and form of product tested, is the sum of all the factors.

The FDA and the Non-prescription Drug Manufacturers Association have expressed concern that the industry would gravitate to the feature achieving the highest score in any testing procedure. In reality, the value for a specific feature will vary depending on the exact product and all other packaging components used in a specific package. The same feature from different manufacturers may achieve different values, because there are slight variations in design and manufacture, even though the features may appear to be identical. If a value were to be established at 20 on a scale of 0–50 for a single feature, the value of 20 could be attained through the use of multiple features if necessary. This would preclude any single feature from becoming the industry mandated method to provide protection. The above value of 20 is for illustration purposes only; prior testing has shown the minimum design for FDA acceptance to be around 11, with some minimum values for approved features scoring much higher. A standard requiring 20 as the minimum value would require most packages to be improved before the package would meet the standard. The use of multiple features can result in a value higher than 50. While the Rosette Protocol has been tested and in use for several years, there are others that may be as objective and in use by other companies.

Certain tamper-evident features in use today, and approved for use by the FDA, are at best window dressing. Such features can be violated without the use of tools of any type, the feature removed and replaced after violation of the package, and all without leaving any indication of possible violation. Some very effective alternatives to the features are not being used because of a slightly higher per unit cost. Other features used to provide resistance to violation are being used in a form that does not meet the FDA standards and can be duplicated very easily—all to save a penny per unit. It is time for the packaging industry—the producers and the users—to accept their responsibility and provide effective packaging to their customers. The best way to do this is through the use of an objective approval process that disregards minor costs or inconveniences.

What is the Best TE Feature?

A lot of people ask "What is the best TE feature?" No single TE feature is best for all products! There are variations in effectiveness of similar features from different manufacturers as well as variations in effectiveness where the product contributes to the effectiveness. An example is a metal can that is much more effective for a carbonated product than a noncarbonated product. The product can determine which feature provides the most protection; for instance, a product that can be adulterated effectively by penetration would require a more rigid outer container than one that degrades visibly on violation by penetration. The best feature for your product is the one that provides the greatest resistance to violation for that product in its current form and size. All features can be violated in some manner; effective TE features provide greater difficulty in violating the product than do ineffective features. In some cases the packaging was violated by being opened, just as any consumer would do, and then the tamperer replaced the original product with a toxic substance, not making any attempt to restore the package to its original appearance. The package worked as intended—it showed that it had been opened, but still resulted in injury to the consumer. The violation does not have to be exotic to harm the consumer. Tamperproofness does not exist!

Does Tamper-Evident Packaging Work?

Since the implementation of the FDA rule (21 *CFR* 211.132) in 1982, the consumer has increased awareness of packaging. This awareness has led to an increase in the number of complaints of possible product tampering, although most incidents are later dismissed as unfounded. Tamper-evident packaging prevents tasting in stores, prevents in-store violation, and if the feature is intact assures the consumer, the product is safe. Effective tamper-evident packaging acts as a deterrent to most persons who would commit such acts of violation and makes it difficult for others to violate the package and restore it to its original appearance. Yes—effective tamper-evident packaging works, provided consumers are aware of the feature and pay attention to what they are about to use.

Most experts agree that the consumer should be more aware of what to look for in tamper-evident packaging. Educating the consumer could include pictures of the feature and product on the label, and in media ads. One recent case resulted in loss of life where the package met this test with two indicators—the product was pictured on the label and the actual product consumed did not resemble the picture on the label in two different ways, as well as numerous other indications of violation. The consumer used the product anyway and died from consuming the substituted item. If the consumer had examined the product and package, it should have been obvious that the product was not to be used. Better consumer awareness may have prevented this incident.

The consumer today is different from those in previous generations. You may remember as a child that if a can was dented, it would be passed over by shoppers until it was the last package on the self. Most dented cans were sold at a discount in a special bin. Today consumers are so confident of the quality of the product, and accustomed to manufacturing defects, transporation damage, and other variations in the package, that they accept without question packages their parents and grandparents would have rejected. This demonstrates the confidence the consumers have in our products, which we can be proud of, but insulates them from taking responsibility for their own safety.

Both the products themselves and their manufacturers have taken the entire blame for past incidents of product tampering. Another weak link in the chain of package security is the retailer. In prior cases, the retailer has not been held accountable for the violation of any package, or the sale of packages with obvious indication of prior opening. During research, many stores were visited to observe what products are converting to the use of tamper-evident packaging and to determine which features are used on what types of products. During those visits many packages were observed where the tamper-evident feature indicated possible prior opening. In every case the package was taken to the store manager, with an explanation of what was observed. Their statements have ranged from "I will put it aside for the salesman to pick up" to "there's nothing wrong with it, it happens all the time" to "put it back on the shelf and I will have the company pick it up." What if the package had been violated and a consumer died from using the product? How does the manager know that another consumer will not buy the package before the company picks up the product? Retailers need to participate more in making tamper-evident packaging work as intended, rather than being a weak link in the chain of package security.

Making tamper-evident packaging work as it should requires the efforts of all involved: the manufacturers of packaging components should promote effective packaging for use in providing protection, product manufacturers should use the most effective feature for their product, retailers should be aware of the potential for violated products being in their display, the consumer should maintain an awareness of what to look for in a secure package and refuse to buy packages that look suspicious, law-enforcement agencies should conduct professional investigations and prosecute all violations, government agencies provide for legal relief to companies meeting or exceeding government standards, and the media should report only those incidents that are verified as representing a threat to the health and welfare of the public without causing hysteria through false alarms. It all begins with you as packaging professionals.

Product tampering will not disappear. We as an industry must remain committed to providing the consumer with quality products in packages that provide protection to the consumer. Since 1982, we have come a long way, but we must continue to research and develop improvements in packaging to assure the availability of safe products for the next generation.

BIBLIOGRAPHY

General References

Books

J. L. Rosette, *Development of an Index for Rating the Effectiveness of Tamper-Evident Packaging Features,* master's thesis, California Coast University, Santa Ana, CA, 1985.

J. L. Rosette, *Improving Tamper-Evident Packaging,* Technomic Publishing, Lancaster, PA, 1992.

J. L. Rosette, *Product Tampering Detection,* Forensic Packaging Concepts, Inc., Atlanta, 1993.

Publications

Code of Federal Regulations, Chapter 21, Government Printing Office, Washington, DC.

Federal Register.

Food & Drug Packaging, published monthly.

Packaging, published monthly.

Packaging Digest, published monthly.

Packaging Technology & Engineering, published eight times per year.

Articles and technical papers

D. Lowe, "Crisis Management Marketing, Tampering Strategies for International and Domestic Marketplace Terrorism," *San Francisco State Univ. J.* **1** (1990).

J. L. Rosette, "Defending Against Terroristic Tampering," 1991.

Incident Reports, 1986 to 1994, reports of claims of possible product tampering from various law-enforcement agencies.

J. L. Rosette, "Product Tampering as It Affects Consumers and Law Enforcement," 1992 *FBI Law Enforcement Bull.* (Sept. 1991).

Product Tampering in the Marketplace" Foundation for American Communications, 1986.

Product Tampering Problems, Foundation for American Communications, 1993.

J. Sneden, H. Lockhart, and M. Richmond, *Tamper-Resistant Packaging,* Michigan State University, 1983.

Trends in Product Tampering, 1990, FDA & FBI data on product tampering.

JACK L. ROSETTE
Forensic Packaging Concepts, Inc.
Fort Mill, South Carolina

TAPE, GUMMED

Water-activated gummed paper tapes used to seal the center and/or end seams of corrugated boxes are very different from what they were at the time of their inception, approximately 100 years ago. Technological advancements in paper, laminates, adhesives, reinforcements, and dispensing and application equipment provide today's user with a scientifically produced high-quality product. Depending on the packager's need, gummed tape can be obtained with a wide range of tailor-made special features.

There are two basic forms of gummed tapes; single-ply nonreinforced "paper" sealing tape, and double-ply fiberglass "reinforced" sealing tape. Both varieties begin with what is commercially known as "gumming kraft," which differs from ordinary kraft paper in that it is sized to prevent the adhesive from penetrating too deeply into the paper. Following application of a vegetable-based remoistenable adhesive (see Adhesives), single-ply paper-sealing tape is slit into roll lengths of 375–6000 ft (114–1829 m). Depending on the basis weight of the paper (35#, 60# or 90#) (see Paper), the product is categorized for light-, medium-, and heavy-duty application.

Double-ply reinforced gummed sealing tapes are produced by sandwiching fiberglass yarns between two sheets of kraft paper with either a hot melt of water-based adhesive. Fiberglass yarns generally run in three directions (machine direction and both transverse directions), forming a diamond or "three way" pattern. As in single-ply paper tapes, a water-remoistenable adhesive of vegetable and/or animal glue formulation is applied to the bottom sheet before or after lamination. Depending on the user's needs, paper basis weight, yarn spacing and denier, and laminate- and adhesive-coating weight can be varied to produce products for light-, medium-, and heavy-duty application. Finally, the parent or "jumbo" roll is slit into smaller rolls ranging in size from 360 to 4500 ft (110–1372 m).

Both paper and reinforced sealing tapes are usually wound gummed-side-in on a cardboard core. Most manufacturers also supply coreless rolls. Tape widths generally range from 1 to 3 in. (25–76 mm) in 1/2-in. (13-mm) increments. Gummed tapes are available in a range of colors, widths, and lengths. They can also be custom-printed and can be obtained in strippable grades or with special antitheft, tamper-evident, or inventory-control features.

When properly purchased and applied, both paper and reinforced sealing tape fully meet the requirements and specifications for rail (UCC Rule 41), truck (NCB Rule 222), plane, parcel post, UPS, and other parcel delivery-service shipments. General Services Administration Commercial Descriptions (CIDs) A-A-1492A, A-A-1671A, and A-A-1672A govern the purchase of gummed tapes by the federal government.

The product does not lend itself to differentiation by appearance, but tape users should be aware of the fact that all gummed tape is not the same. Because of diverse manufacturing techniques, adhesive and laminate formulations, and quality-control guidelines, tape-performance differences between producers exist.

Gummed tape dispensing and application equipment fall within several categories: (*1*) hand-operated dispensers; (*2*) electrically operated dispensers; (*3*) automatic taping machinery for fixed-size boxes, and (*4*) automatic taping machinery for random-size boxes.

BIBLIOGRAPHY

"Taped, Gummed" in *The Wiley Encyclopedia of Packaging Technology*, 1st ed., by R. W. McKellar, The Gummed Industries Association, pp. 631–632.

General References

Gumming Industry Voluntary Product Standard for Paper Sealing Tape (single ply), The Gummed Industries Association, Inc., Jericho, NY, 1983.

Gumming Industry Voluntary Product Standard for Fiberglass Reinforced Paper Sealing Tape (Double Ply), The Gummed Industries Association, Inc., Jericho, NY, 1983.

How to Use the New GIA Sealing Tape Standards: A Guide to the Proper Selection and Use of Gummed Sealing Tapes, The Gummed Industries Association, Jericho, NY, 1983.

TAPE, PRESSURE-SENSITIVE

Pressure-sensitive tape is used for closing boxes, combining packages, attaching packaging lists, color coding, pallet unitizing, adding carrying handles, splicing, providing ease of package opening, protecting labels, reinforcing critical package components, holding documents, and a variety of other jobs.

The first pressure-sensitive tape was developed in 1925 for paint masking; it had a paper backing and a glue–glycerol adhesive. Today, there are hundreds of specialty tapes available for specific applications in packaging. The common theme is a backing material coated with an adhesive that adheres with a light touch without a need for activating solvent or heat (see also Adhesives).

Tape Testing

As with all package components, tape testing must be focused on the function of that component on the package. An engineering analysis of the package followed by package testing are excellent beginnings. The critical physical properties of the tape can then be chosen, in conjunction with technical input from a reputable tape supplier.

One of the most important tape properties is often the strength of the backing. ASTM D3759 is the standard laboratory method for measuring the tensile strength and elongation of pressure-sensitive tapes. Some packages require a tape with very little stretch; others require a relatively high elongation to achieve the desired breaking energy. For many narrow tapes, only the machine-direction strength is of importance. For wider tapes, however, the cross-direction strength can be more critical.

Peel adhesion, ASTM D3330, is perhaps the most often quoted adhesive measurement. It measures the force in ounces per inch of width (N/100 mm) required to peel a strip of tape back onto itself (180°) from a stainless-steel surface. Test conditions and rubdown are closely controlled. The relationship of this test to tape functioning correctly on a package is often questionable; the angle of pull for the test is not seen on most packages, and the stainless-steel substrate does not represent package surfaces well. However, for a given adhesive type, the test may be useful as a quality-assurance procedure.

A packaging tape adhesive must have good tack to allow it to adhere to a variety of package surfaces with only a light rubdown. The literature on pressure-sensitive tape includes several methods of measuring the initial tack or "wet grab" of an adhesive. These can be valuable research test methods, but, because most of them use glass or steel as a test surface, it can be difficult to correlate tack test results with the performance of tape on a package.

Tapes are often used on packages where the tape must hold in shear: the force on the tape acts in parallel with the package surface. One test that can relate to this is ASTM D3654-A, holding power of pressure-sensitive tapes to fiberboard. A $\frac{1}{2}$-in. × $\frac{1}{2}$-in. area of tape is applied to a fiberboard surface (NIST Standard Reference Material). A 1-kg mass is attached to an end of the tape. The test measures the time it takes for the mass to pull the tape from the surface.

ASTM test methods and other standards relating to pressure-sensitive tape are listed in Table 1. Other test methods are published by the Pressure Sensitive Tape Council (PSTC). Tape test methods are workable in both the inch-pound system of units and the SI metric system; physical properties are convertible from one system to the other. The

Table 1. ASTM Standards Relating to Pressure-Sensitive Tape

ASTM	Title
D1974	*Practice for Methods of Closing, Sealing and Reinforcing Fiberboard Boxes*
D2860	*Adherence of Pressure Sensitive Tape to Fiberboard at 90° Angle and Constant Stress*
D2979	*Pressure Sensitive Tack of Adhesives Using an Inverted Probe Machine*
D3121	*Tack of Pressure Sensitive Adhesives by Rolling Ball*
D3330	*Peel Adhesion of Pressure Sensitive Tapes a 180° Angle*
D3611	*Accelerated Aging of Pressure Sensitive Tapes*
D3652	*Thickness of Pressure Sensitive Tapes*
D3654	*Holding Power of Pressure Sensitive Tapes*
D3759	*Tensile Strength and Elongation of Pressure Sensitive Tapes*
D3662	*Bursting Strength of Pressure Sensitive Tapes*
D3715	*Quality Assurance of Pressure Sensitive Tapes*
D3811	*Unwind Force of Pressure Sensitive Tapes*
D3813	*Curling and Twisting on Unwinding of Pressure Sensitive Tapes*
D3815	*Accelerated Aging of Pressure Sensitive Tapes by Carbon Arc Exposure Apparatus*
D3816	*Water Penetration of Pressure Sensitive Tapes*
D3833	*Water Vapor Transmission of Pressure Sensitive Tapes*
D3889	*Adherence to Linerboard of Pressure Sensitive Tape at Low Temperature*
D5105	*Performing Accelerated Outdoor Weathering of Pressure Sensitive Tapes Using Natural Sunlight*
D5264	*Abrasion Resistance of Printed Materials by the Sutherland Rub Tester*
D5330	*Specification for Pressure Sensitive Tape for Packaging, Filament Reinforced*
D5375	*Liner Removal at High Speeds from PSA Label Stock*
D5486	*Specification for Pressure Sensitive Tape for Packaging, Box Closure and Sealing*
D5570	*Water Resistance of Tapes and Adhesives Used as a Box Closure*

Table 2. Common Widths of Pressure-Sensitive Tape

Inch-Pound System (in.)	SI Metric System (mm)
$\frac{1}{2}$	12
$\frac{3}{4}$	18
1	24
2	48
3	72
4	96

standard widths of tape are a little different in the two systems. The metric replacement widths are based on uniform slitting increments, resulting in standard widths shown in Table 2; these are based on PSTC-71, *Guide for Width and Length of Pressure Sensitive Tape.* Most packaging uses of pressure-sensitive tape will readily allow these metric sizes.

Box-Sealing Tape

The largest use of pressure-sensitive tape in packaging is the closure of regular slotted containers. Figure 1 depicts a typical construction of a box-sealing tape; a plastic film is coated on one side with a pressure-sensitive adhesive. The film may have a release treatment on one side to allow easy removal of the tape from the roll during dispensing. Some film backings also are treated or coated on the adhesive side to increase the

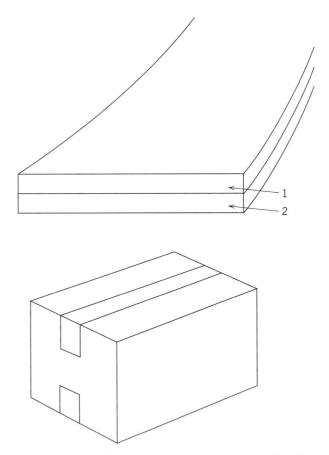

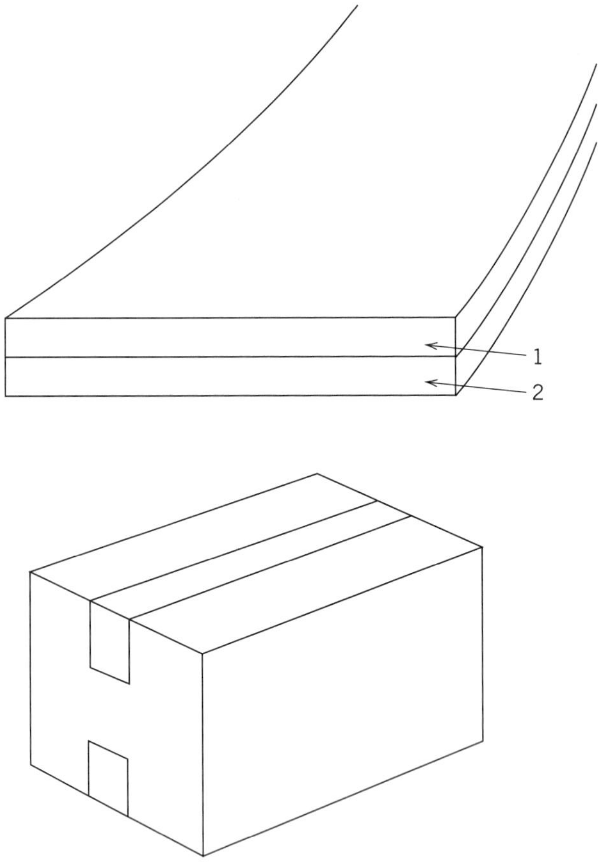

Figure 1. A typical box-sealing tape consists of a backing film (1) and a layer of pressure-sensitive adhesive (2). It is used most often as a closure for regular slotted containers.

bond of the adhesive to the backing, although for a box sealing tape, this is not a critical factor.

The standard application of a box sealing tape is described in ASTM D1974, *Standard Practice for Closing and Sealing Fiber Boxes.* A 2-in. (48-mm)-wide tape is applied over the center seams of a regular slotted container (RSC) and extends about $2\frac{1}{2}$ in. (65 mm) onto the end panels of the box. This seals the center seam and helps keep dust and dirt out of the box. If a total seal is needed, cross-strips of tape can be added at the end edges of the box. This "six-strip seal" or "H-seal" is specified for some military and export packages but is seldom used for domestic shipments.

The choice of tape for box closure is very important and affects the performance of the entire package during storage and distribution. The most common backing material is biaxially oriented polypropylene. Some tape backing is made of a relatively "square" film with the machine-direction and cross-direction properties about the same. Most heavier-duty tapes are made of a tentered film with the cross-direction strength higher than the machine-direction strength; this usually results in better performance on boxes. Polyester, unplasticized PVC, and saturated papers are also used as backings.

A proper backing for a box-sealing tape is a good start, but the tape must have an aggressive adhesive if the backing strength is to be realized. Many box-sealing tapes use an adhesive based on rubber and a tackifying resin, but acrylics and other synthetic adhesive systems are also used. Critical adhesive properties are tack (for production efficiency) and shear-holding power (for package warehousing and shipping).

Tape is applied by handheld dispensers (Fig. 2) or with box-sealing equipment. A range of equipment is available for both semiautomatic (operator feeds the boxes into the box sealer) or fully automatic operation. Box sealers are either adjustable to boxes of a single size or random, able to take a mixture of boxes of varying sizes.

Most pressure-sensitive box-sealing tapes hold well in damp and humid conditions. Water resistance (ASTM D5570) is required by the United Nations (UN) and the U.S. Department of Transportation (DOT) for tapes used on boxes containing hazardous materials.

Standard Specification ASTM D5486, *Pressure Sensitive Tape for Packaging, Box Closure, and Sealing,* replaced Federal Specifications PPP-T-60 and PPP-T-76. Type I is a polyester-backed tape used in H-type closure. Type II is a polyester-backed tape used in single-strip closures. Type III is a polypropylene-backed tape used in single-strip closures. Type IV is a cloth-backed tape, and Type V is a paper-backed tape.

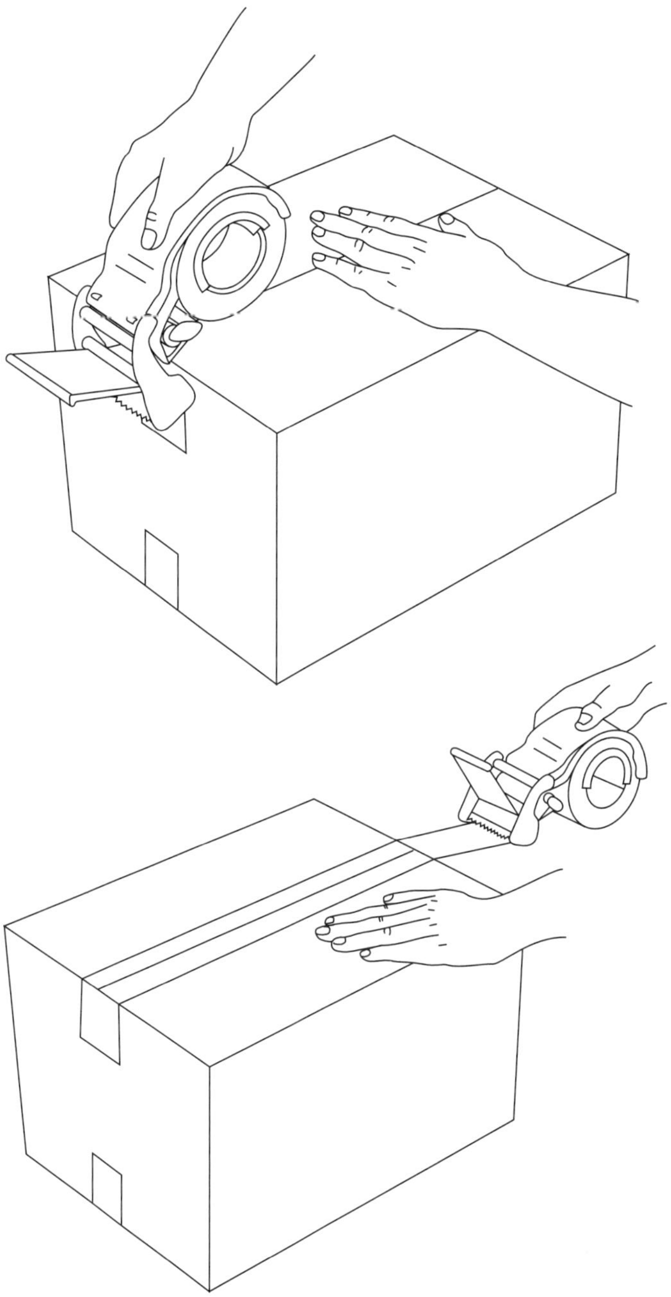

Figure 2. Use of a handheld dispenser for applying box-sealing tape to close RSC.

Filament Tapes

A second broad category of packaging tape is pressure-sensitive filament tape, sometimes known as "strapping tape." Figure 3 shows that this is typically made of a film backing (polyester or polypropylene) with reinforcing filaments embedded in the pressure-sensitive adhesive. The most common filament is fiberglass, which provides a high tensile strength with very little elongation. A few tapes have polyester or rayon filaments for extra impact or cut resistance. Tapes are also available with integral polymeric filaments. The adhesive requirements for filament tapes are at least as critical as those of box-sealing tapes. Care should be taken to choose a tape with a balance of tack (initial adhesion) and shear-holding power.

Box closure can be accomplished with filament tape on a variety of box styles. In ASTM D1974, filament tape is recommended for use with boxes with fully overlapping flaps such as a five-panel folder, full-telescope box, or full-overlap box. A high performance filament tape may be applied in L-shaped strips as small as $\frac{1}{2}$ in (12 mm) wide by $4\frac{1}{4}$ in. (105 mm): two in (50 mm) on each side of the L-clip. A tape with lower tensile and adhesive properties may have to be used in $\frac{3}{4}$-in. × $6\frac{1}{4}$-in. (18-mm × 155-mm) L-clips for equivalent performance on a box.

Filament tape is used in dozens of other packaging applications where high strength is required. These include reinforcing boxes, combining boxes for shipment, bundling, recooperage, and pallet unitizing. Special filament tapes are available for use directly on appliance surfaces to hold doors and drawers in place during shipment.

Standard Specification ASTM D5330, *Pressure Sensitive Tape for Packaging, Filament Reinforced,* replaced Federal Specification PPP-T-97. It includes four classes of filament tape: Type I—cut-resistant, Type II—medium-tensile-strength, Type III—high-tensile-strength, and Type IV—high-tensile-strength, weather-resistant. Other grades of filament tape are also available for industry, institution, and home use.

Specialty Tapes

With a full choice of films, papers, and foils for backings, and a wide range of adhesives available, it has been possible to develop special pressure-sensitive tapes for many packaging uses, including combining and tamper evidence. Tapes are used to attach and protect labels on packages. Tape pouches often enclose packing lists and documents for shipment (see also Tape, gummed).

Several tapes are available not only on traditional rolls but also in precut pads (Fig. 4). These pads provide for the use of presized pieces of tape for covering labels, attaching documents, and closing packages. They do not require a dispenser and offer the advantage of portability.

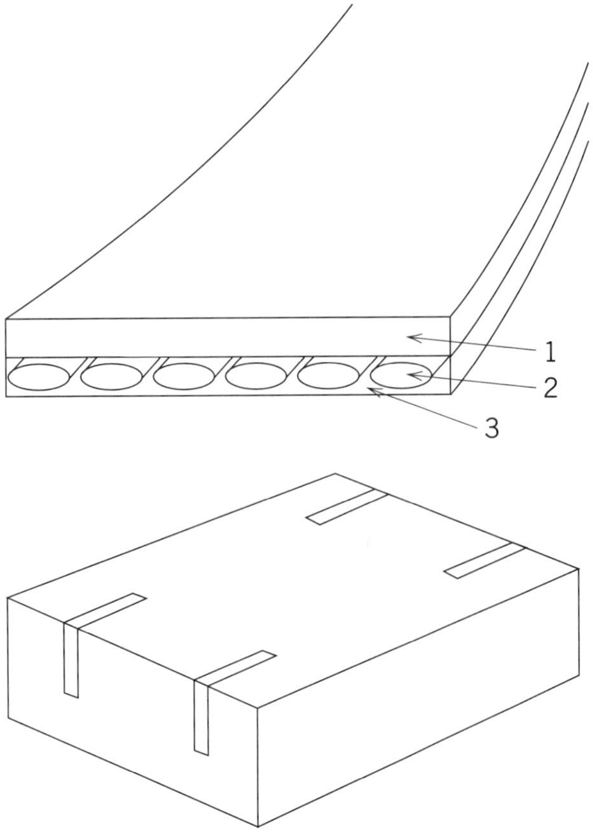

Figure 3. A typical filament tape has a backing film (1) and filaments (2) embedded in a pressure-sensitive adhesive (3). One of the uses is the L-clip closure of full-overlap boxes.

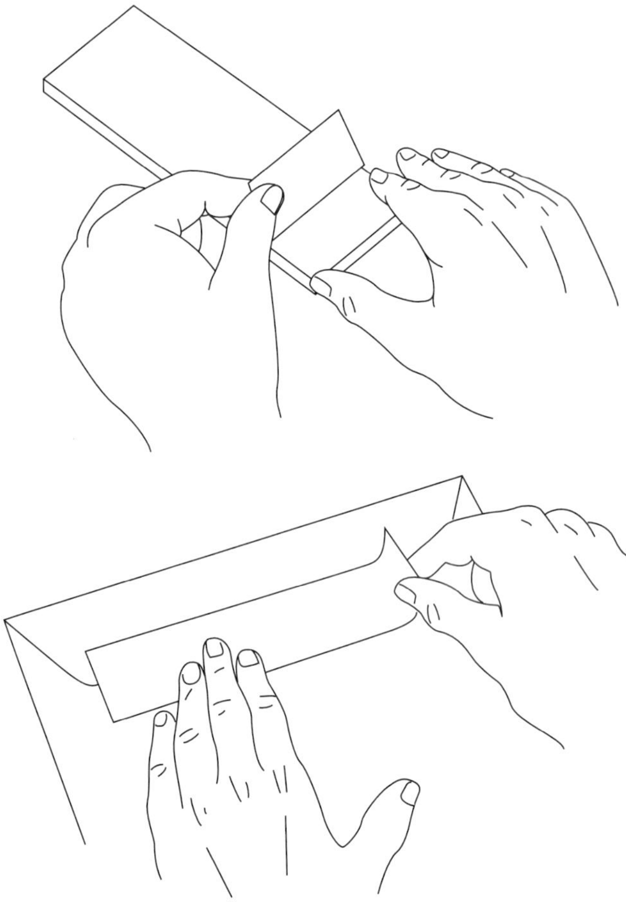

Figure 4. Use of a pressure-sensitive tape from a pad to close a mailing envelope.

Environmental Considerations

The environmental considerations for tapes applied to boxes are given in ASTM D1974. Pressure-sensitive tapes are often a good method of source reduction, considered to be a primary objective in reducing the contribution of packaging to municipal solid waste. Box-closure tapes often use less materials than alternate closures. Small L-clips of tape can close folding cartons, replacing a shrink overwrap and reducing material usage.

Compatibility with recycling operations for larger package components is important. For box recycling, pressure-sensitive tapes do not have to be removed from corrugated boxes prior to recycling operations. When a tape has a water-resistant tape backing, the adhesive stays with the backing in the hydropulper. The tape is easily removed intact by normal pulp-cleaning processes; neither the recycling paper nor recycling water is contaminated by the tape.

Stretchable LLDPE tapes are available for pallet unitizing, offering source reduction from full stretch wraps. When the tape is used in conjunction with stretch wrap, it is compatible with the stretch-wrap recycling operations (see also Recycling; Environmental laws, International and North America).

Pressure-sensitive tape is a versatile packaging material that can offer cost savings, productivity improvements, and source reductions. Innovations beneficial to packagers will continue.

BIBLIOGRAPHY

General References

Anonymous, *Test Methods for Pressure Sensitive Adhesive Tapes,* 11th ed., Pressure Sensitive Tape Council, Chicago, 1994.

K. S. Booth, ed., *Industrial Packaging Adhesives,* Blackie and Son, London, 1990.

D. Eagland, "What Makes Stuff Stick?" *Chemtech* (April 1990).

P. C. Halden, and L. E. Losinski, "Converting from Glue to Pressure Sensitive Tape for Closing Boxes," *J. Packag. Technol.* (June 1989).

T. B. Jensen, "PSA Tapes Offer Environmental Advantages in Packaging," *Adhes. Age* (Sept. 1992).

R. P. Muny, "Basing PSA Tests on End-Use Improves Profits and Quality," *Adhes. Age* (Dec. 1986).

D. Satas, ed., *Handbook of Pressure Sensitive Adhesive Technology,* 2nd ed., Van Nostrand Reinhold, New York, 1989.

R. L. Sheehan, "Box and Closure: Partners in Performance," *J. Packag. Technol.* (Aug. 1988).

R. L. Sheehan and L. E. Gruenewald, "Closure Materials Handling Plays Big Role in Box Recycling," *Packag. Technol. Eng.* (April 1994).

R. L. SHEEHAN
3M Corporation
St Paul, Minnesota

TESTING CONSUMER PACKAGES FOR MARKETING EFFECTIVENESS

Primary and secondary packaging are key to marketing success in bringing brands from production, through distribution, to selection by end users, and in generating the levels of satisfaction that warrant repeat purchase. Survey research is used to help those involved in package development make the informed choices that result in a marketing success.

Many package performance dimensions can be measured in the laboratory. Characteristics related to marketing effectiveness need to be evaluated by users and potential users of a brand—preferably before the package is put into production. A variety of marketing research methods are used for this evaluation.

Marketing Research Options

Most research options fall into three broad categories:

Analysis of existing information.
Original qualitative research.
Original quantitative research.

Analysis of existing information. An analysis of existing information is often undertaken at the start of a package development program. It can be an important first step because it assures that subsequent surveys will add to, rather than duplicate, information.

The information is drawn from public sources such as government, trade groups, and publications; from past studies completed within the organization (especially tracking studies which reveal changes in consumer behavior); from syndicated surveys (sponsored and conducted by a research company that sells the results to any company willing to pay for the information); and from customer- and consumer-contact records. (In large packaged-goods companies, consumer-contact information is available daily, and is often the first clue to product and packaging problems, changing consumer concerns, deficiencies in instructions for use, etc).

The information obtained is summarized and analyzed in a report that contains conclusions about the nature of the market and recommended packaging objectives—criteria or goals that the package must satisfy. The objectives guide the development program and are used as standards for judging packaging effectiveness.

Qualitative research. When people talk about "qualitative research," they are referring to studies among small samples of current and/or prospective users of the brand. These people are interviewed in depth, either individually or in a group. The research is diagnostic, that is exploratory or directional, and its purpose is to obtain early indications of packaging problems and opportunities.

In the individual interviews, one person is questioned at a time using a predetermined question guide (list of subjects to be covered), or a questionnaire that includes mostly open-ended, rather than closed-ended, questions. ("How do you feel about the New York subway system?" is an open-ended question. An example of a closed-ended question is "Compared with other forms of public transportation, do you think the New York subway system is better, about the same or not as good?").

The group interview is called a "focus-group session" and is conducted with eight to ten people. Questioning is in the form of a discussion that is led by a trained leader or moderator who covers subjects listed in a predetermined discussion guide. Each of these qualitative research techniques has certain advantages:

During individual-depth interviews, information is obtained, and feelings are explored in detail. Each person expresses his/her opinions and interpretations without reference to the ideas of others so there are clear indications of individual preferences and understanding.

Participant interaction in focus group interviews provides a rich, multilayered accumulation of information and feelings about the subject being discussed. Comments made by one person may remind others of points they had not thought of. Each idea expressed suggests other ideas. Normally, the discussion results in a group consensus.

Both approaches are helpful in the early stages of an investigation, or as an aid to informed judgment. Individual-depth interviews are recommended for early trial, or where there are questions about consumer understanding of how a package works, packaging copy, or communication of ideas. Group interviews are recommended in order to gain insights about a broad range of feelings and opinions, as well as to identify the language or terminology consumers use to talk about the package and product.

Qualitative research is usually used to develop instructions for use, create hypotheses and questionnaires for larger studies, or get initial indications of opinions and ideas about products, packages, or product/package combinations.

Quantitative Research. "Quantitative studies" are conducted to obtain definitive information for decisionmaking. This type of research can take many forms but is characterized by sample sizes where enough people are interviewed so results can be analyzed using statistical techniques. The interviews may be conducted in person, by mail, on-line, by computer, or via video. Certain kinds of packaging studies are conducted by telephone, for example surveys to determine how and where products are stored and used, or to define the target market (the intended buyer/user).

Questionnaires for quantitative studies usually include more closed-ended, than open-ended, questions. The questions are structured to permit various kinds of statistical analyses which will aid in interpreting the results.

Packaging Evaluation

In considering its marketing effectiveness, management is most likely to question the degree to which a package meets three types of goals, and these are the ones most often subjected to some kind of survey research evaluation: functional effectiveness, visual impact (visibility, shelf attention), and communications effectiveness (brand and product identification and imagery).

Testing functional effectiveness. Functional effectiveness is easier to evaluate than the other marketing dimensions. Some aspects can be handled in the laboratory; others, through product-testing techniques such as one-time or multiple-use tests; others, through individual-depth or focus-group sessions, depending on the nature of the product and the package.

With few exceptions, product users can be questioned directly about whether a package is easy to handle, open, close, store, and use; and whether it might protect the contents over a period of time. The sample of potential brand users interviewed should include people who use the product in different ways, those in a mix of age groups (from children through seniors depending on use), people of various ethnic backgrounds, as well as those with a variety of physical characteristics (left-, or right-handed, have smaller or larger hands, etc).

It is a given that packages must protect the contents. But it is also true that the user must *believe* that the package really does this. The inner, outer and overwrap for many over-the-counter (OTC) remedies, as well as premium food and beauty-aids products are technically unnecessary. However, proposed substitutes intended to reduce packaging costs are often rejected by consumers, primarily because of perceived differences in product protection.

Depending on the nature of the product and package involved, an in-home or at-work trial study may be conducted among people in a number of geographic areas. Where a totally new packaging concept is being tested, such as the early versions of aseptic containers for foods, the study may be restricted to people in areas closest to manufacturing facilities because the samples have a short storage life.

Those people who agree to try the product are interviewed to obtain their initial reactions about its appearance, anticipated benefits and drawbacks, and general expectations. The product may be tried at that time, or left for use later, with an evaluation sheet to be completed immediately after trial. If needed, follow-up interviews can be conducted by telephone the next day.

Testing the effectiveness of graphic design. While the physical package can be evaluated using product-testing techniques such as the ones described above, different approaches are needed to determine the effectiveness of packaging graphics.

There is extensive evidence that overall surface designs including words, are primary contributors to the marketing effectiveness of a package. If you think of packaging as the environment for a product, it is easier to understand its effect. The tray used for a frozen entrée, the baking pan included in a mix, a plastic or glass bottle, a composite or metal can, a see-through or foil pack; words, colors, illustrations and overall design of the container or label; all of these influence consumer expectations of, and satisfaction with a brand.

In all studies involving packaging graphics, the physical appearance of the test and competitive packages must be equal. An apparently unfinished mockup or one that is custom-finished should not be shown with printed competitive packages. The difference flags the package in a way that biases results. In addition, if the test versions are potential alternatives for an existing package, the sample of people interviewed should include enough current users of the brand to analyze their reactions separately.

In general, graphic design can be evaluated in regard to three aspects: visual or shelf impact, communications effectiveness, and aesthetic appeal.

Visual or shelf impact. The most elusive and difficult packaging aspect to measure is visual, or shelf impact. The questions to be answered include the following:

1. When people shop the product category, does the package call attention to the brand?
2. As people shop for other types of product, is the package creating an in-store presence that will enhance familiarity with the brand?
3. When the buying decision has been made *before* the store visit, does the package make the brand easy to find?

To date, there are no research techniques that fully answer these questions, primarily because visual perception involves a complex relationship between the structure of the eye and the functioning of the mind. For example, seeing is selective. Regardless of where or how the eye moves, people "screen out" elements they consider irrelevant and focus on the familiar. In a visibility test, this familiarity bias usually produces a higher score for the current package than for a new package, and for an established brand than for a new one.

In addition, the effects of demonstrated visual "enhancers" such as high contrast, size, and color density are predictable. This makes it possible to design for the test, producing an apparent "winner." The most commonly used methods for testing some aspects of visual impact are a shopping test, a T-scope (tachistoscope) or controlled-exposure test and an eye-movement test of a single, or multiple packages in a competitive environment. To some degree, all of these techniques are deficient, but the shopping test comes closest to indicating what might happen in a store.

In a shopping test, a rough approximation of the in-store environment is reproduced with actual packages on shelves in a central interviewing facility, on a videotape, or on a computer screen. The product arrangement is deliberately structured to give all the test brands an equal opportunity to be noticed, and study participants are free to look at what interests them and to ignore what does not.

Where an actual product display is used, people are able to "shop" the shelves and view the packages in motion at their own pace. They can pick up and examine packages to gain more information about the brand. While shopping tests using videotape and computer simulations are easier to administer, they have some drawbacks: people are seated in a quiet room and concentrating, there may be color and copy distortion due to the limitations of photography, and the product display is "framed" so that some of the normal in-store confusion is reduced.

In the controlled-exposure test, the length of time people can view a picture of a package or shelf display is limited. A device called a tachistoscope or T-scope is attached to the lens of a projector. Operating like the shutter of a camera, it controls the amount of time that a picture is shown (in fractions of a second). Each exposure is shown at an increasing amount of time. After each one, people are asked what they saw.

The controlled-exposure method of testing visual aspects of packaging is a speed test. It measures how quickly something is seen, and has several drawbacks: people are concentrating on a stationary picture that organizes the display, and there may be photographic distortion. For new brands, the interpretation of results is strictly limited to how easily the brand name is read. For established brands the test offers a measure of recognition as well as readability.

T-scope test scores have limited application, but can be helpful in cases where pictures must be used. Some examples are as follows: to develop indications of differentiation within a line of products, to test custom-designed containers where several alternative models are being considered, to provide indications of recognition elements when a redesign is being considered for an established brand.

In an eye-movement test, the respondent controls the length of time the picture will be viewed. A device records where the eye moves and where it stops over the surface of the picture. Results are evaluated on the basis of this record. In addition to drawbacks related to the test environment and possible photographic distortion, this method provides information about where the eye travels but not about what the brain absorbs.

Regardless of the method used, visual-impact scores are rarely used as a basis for accepting or rejecting a package. Packages that score well in a visibility test but do not meet communications objectives are not usually adopted. On the other hand, when communications objectives are met but the visibility score is low, a package is often selected, although graphics may be modified to enhance visual impact.

Communications measurements. Most packaging tests include communications measurements. Designers can predict the visual effectiveness of a package based on their knowledge of the physiology and psychology of seeing. However, they find it more difficult to determine the package's communications effectiveness (the ideas the package suggests about the brand).

Packages have a substantial impact on people's ideas about products and brands. In order to measure this impact, research methods and exhibit materials used need to be appropriate for the product and the particular packaging design problem, as well as economically feasible. There are many options for testing package communications, and no standard techniques or series of techniques.

The majority of package communications studies evaluate the brand in relation to competition, and measure several packaging aspects. Some of the more popular approaches are

1. Experimental design
   Matched samples of potential users are interviewed about one or two competitive brands, as well as the test brand in *one* of the package designs. Questioning and test procedures are identical so that differences in results can be attributed to the different packages. Within this framework, the most often used approaches are as follows:
   a. Brand expectation study, which measures impressions of the brand on the basis of package appearance
   b. Brand trial study, which measures brand and package performance on the basis of a single trial
   c. Location use test which measures brand and package performance in-home or in the workplace on a one-time or multiple-use basis.
2. *Product formulation/structure study,* which treats each package as a separate formulation or structural version of the brand and determines perceived differences.

Aesthetic appeal. If the purpose of the study is to measure aesthetic appeal, consumers are asked to rank proposed packages in regard to how attractive they are.

Conclusion

In addition to standard specifications for conducting useful surveys, the following are the primary requirements for conducting research to evaluate the marketing effectiveness of packaging:

1. The packaging dimensions of interest need to be measurable. Subtle differences in graphic design and some packaging objectives (such as "generate trade excitement") cannot be measured.
2. Study objectives should be clearly stated in writing, related to the packaging objectives, and indicate the kind of information to be obtained and how it will be used.
3. The study plan must be capable of yielding the information specified in the objectives.
4. Action standards should be established in advance. These standards provide a common framework for judging study results, by indicating the order of importance of each measurement used, and the research criteria to be met.
5. The people interviewed need to be capable of providing meaningful information.

6. Alternative test packages must be equally acceptable to management. If, for any reason, a package cannot be adopted, it should not be included in the test.
7. Exhibit materials need to be appropriate and equivalent to avoid biasing results.
8. When developing packaging communications information, the tester is measuring how the package influences ideas about the brand. This means people must be asked about the product and brand before any direct questions about the package.
9. Package designers and developers should be involved in the research to be sure that the study plan is compatible with the packaging objectives, to control the appearance of exhibit materials and to add valuable insights in explaining or interpreting test results.

In conclusion, the package is an integral part of most brands and can provide a competitive marketing edge. Successful package development is the result of careful consideration of alternatives. As each packaging decision is made, it narrows the options to be considered at the next step. Survey research assures that choices will result in a package with maximum marketing effectiveness.

LORNA OPATOW
Opatow Associates
New York, New York

TESTING, CUSHIONING. See CUSHIONING, DESIGN.

TESTING, PACKAGING MATERIALS

The testing of packaging materials covers the evaluation of a wide range of component parts of containers and packages. There are thousands of products and substances shipped via commercial carriers that must be protected in relation to the atmospheric, transportation, handling, and warehouse environments. Many standards and test methods pertinent to particular materials are in existence to meet the necessary requirements specified by the various agencies and classification committees. Sources of U.S. standards and tests are the Technical Association of the Pulp and Paper Industry (TAPPI), the Uniform Freight Classification Committee (UFC), the National Motor Freight Classification Committee (NMFC), the Department of Transportation (DOT), the United Nations (UN), and the International Civil Aviation Organization (ICAO) (see *General References* for further information).

Bags

A bag is a preformed container of tubular construction made of flexible material, generally enclosed on all sides except one that forms an opening that may or may not be sealed after filling (1). It may be made of any flexible material, or multiple plies of the same, or combination of various flexible materials (see Bags, paper; Bags, plastic; Bags, heavy-duty plastic). Since the contents of bags must have considerable resistance to crushing hazards, the outer bag materials must have the flexibility and toughness to withstand the forces induced.

Paper and multiwall bags. Five basic tests are employed to evaluate paper and paper combinations: basis weight, tearing resistance, bursting strength, tensile strength, and water-vapor transmission.

Basis weight. Normally, this term is expressed in lb/1000 ft^2 or g/m^2. After proper conditioning, the material is cut to an appropriate size and weighed on an accurate scale or balance. Using the results of the mass per unit (grammage) area, the basis weight can be calculated (2).

Tearing resistance–internal. This test method is designed to determine the average force in gram-force (gf or N) needed to tear a single sheet of paper after the tear has been originated. An Elmendorf-type tester employs the principle of a pendulum, making a single swing and causing the tearing of one or more paper sheets at one time through a fixed distance. The work required for this tearing is measured in gram-force (actually gf · m or N · m = J) as the loss in potential energy of the pendulum. This test is performed, after proper conditioning, in two directions, one with the tear parallel to the machine direction and the other with the tear perpendicular to the machine direction (3).

Bursting strength. This test method is used to determine the amount of hydrostatic pressure in $lb/in.^2$ (kPa) required to obtain a rupture of the material. The pressure is applied at a controlled increasing rate through a rubber diaphragm. A fixed lower plate and an upper adjustable plate clamp the material in position around the circumference with sufficient pressure to prevent slippage during the test but to allow the center to bulge under pressure during the test (4,5).

Tensile strength. This test method determines the resistance of paper to direct tensile stress. The paper, cut to a specific width, is held in place by two clamps, one fixed and one movable, which are aligned to hold the material in the designated plane without slippage. The load is applied at a rate that induces failure within a specified time period. After proper conditioning, the test is performed on specimens cut from both principal directions of the paper (6,7). Certain bag materials are subjected to stress when wet. A wet-strength device is attached to the lower clamp so that the paper remains saturated for the appropriate period of time. The load is applied again at a rate that involves failure within a specified time period. This test is also performed on specimens cut from both principal directions of the material (8,9).

Water-vapor transmission. There are test methods applicable to the sheet material, such as paper, multiwall sacks, and plastic films. The purpose of these methods is to determine a rate of water-vapor transmission, which is designated as the time rate of water-vapor flow normal to two specified parallel surfaces, under steady conditions, through unit area, under the conditions of test. The accepted unit is one gram per 24 h per square meter ($g/(m^2 \cdot day)$. Open-mouth test dishes are used, with the specimens attached to the top of the dishes by means of a sealant. The dishes contain either a desiccant or water, depending on the method. The complete assemblies are inserted in the appropriate atmosphere, and after specific periods of time, are weighed accurately to determine gain or loss. The test is conducted until a nominal steady state exists (10,11) (see Testing, permeation and leakage).

Plastic film bags. In addition to being evaluated for weight, tearing resistance, bursting strength, tensile strength, and water-vapor transmission, plastic bag and sack materials are also tested for heat-seam strength and impact resistance when applicable (see Films, plastic; Sealing, heat).

Heat-seam strength. 1-in. (25.4-mm)-wide bands are cut perpendicular to the seam, with the seam located in the center of the test length. After proper conditioning, the individual specimen is placed in the grips of the testing machine, which are then tightened evenly to prevent slippage during the test. The movable grip pulls the specimen apart at the seam at a specified rate, and the maximum load is recorded. Specimens are tested from seams, both parallel and perpendicular to the machine direction of the material (12).

Impact resistance, free-falling dart. This method determines the energy that causes plastic film to fail under specified conditions of impact of a free-fall dart having a hemispheric head. The dart mass (g), dropped from a specific height, is adjusted to give a 50% failure rate of the specimen. After proper conditioning, a large enough perfect specimen is clamped in place by an air-operated horizontal upper-gasketed steel ring. Sufficient pressure must be employed to prevent specimen slippage. An adjustable bracket holds the dart head, which is positioned vertically above the center of the film and released from a predetermined height to impact the specimen. The specimen is then inspected to determine if failure has occurred. The dart mass is adjusted accordingly to arrive at a point where 50% of the film specimens fail (13).

Textile bags. Two principal textiles, burlap and cotton, are fabricated into bags. These textiles can be laminated with different combinations of papers or films to provide the necessary protection for the application desired. Tensile strength, tearing resistance, and waterproof characteristics are important factors in determining a suitable material.

Bottles

A bottle is designated as a hollow vessel of glass, earthenware, plastic, or similar substance, with a narrow neck or mouth and without handles. It must have a suitable closure to prevent leakage.

Glass. The primary ingredients of glass are soda ash, limestone, and sand. Coloring agents are added to obtain a wide variety of color combinations. Various tests are performed on the completed container in relation to size, thickness, weight, capacity, impact resistance, internal pressure resistance, thermal shock resistance, and resistance to chemical attack.

Plastic. Many thermoplastic materials are employed to produce plastic bottles. The type of product being packaged determines which thermoplastic is most applicable. Clarity and oxygen and moisture barriers are important considerations. In addition to tests performed on completed containers, such as those specified for glass bottles, evaluations are conducted in relation to heat, cold, sunlight, stiffness, resistance to acids, alkalis, oils, solvents, environmental stress-crack resistance and susceptibility to soot accumulation (14).

Boxes

A box is a rigid container having faces and completely enclosing the contents (15). When this term is used in the classifications, its signifies that if fiber boxes (corrugated or solid fiber) are used, such fiber boxes must comply with all the requirements of NMFC Item 222 and UFC Rule 41. NMFC Item 220 also states that boxes are containers with solid or closely fitted sides, ends, bottoms, and tops. They must be made of wood, metal, plastic, fiberboard, or paperboard and foamed or cellular plastic combined, and completely enclose their contents (15).

Corrugated and solid fiberboard boxes. Corrugated boxes are made of corrugated board, a structure formed from two or more paperboard facings and one or more corrugated members.

Solid fiberboard is a solid board made by laminating two or more plies of container board (15) (see Boxes, corrugated; Boxes, solid fiber). The following tests are required for evaluating corrugated or solid fiberboard: basis weight; bursting and short column- or edgewise-compression strength of corrugated fiberboard; flat crush of corrugating medium; internal tearing resistance of paper; moisture content of paperboard, ply separation (wet); puncture and stiffness; ring crush of paperboard; static bending and thickness (see Testing, shipping containers).

Basis weight. See Ref. 2 for further information.

Bursting strength. See Refs. 4 and 5 for further information.

Short column- or edgewise-compression strength of corrugated fiberboard. This test method determines the edgewise-compressive strength, parallel to the flutes, of combined corrugated fiberboard. It is used to compare different material combinations or different apportions of similar corrugated fiberboard. The specimens are accurately cut by a circular saw or other device to cut clean parallel and perpendicular edges to a specific width and height so that the flutes are in the vertical plane during the test. Each loading or long edge is dipped to an approximately $\frac{1}{4}$-in. (6.4-mm) depth and allowed to dry. After proper conditioning, the individual specimen, held vertically by guide blocks, is placed on a bottom platen of a compression tester. The upper parallel platen applies a load to the top of the specimen, the guide blocks are removed, and the pressure is increased until failure occurs. The force is recorded and the average maximum load per unit width is calculated as lb/in. or kg/cm (16).

Flat crush of corrugated fiberboard. This test method, primarily used for single-face or single-wall corrugated fiberboard, determines the resistance of the flutes to crushing when pressure is applied perpendicularly to the board surface. Its purpose is to give a general level of quality in relation to the corrugated-board fabrication and whether the flute have been damaged during the printing process. The specimens are usually cut very carefully into a 10-in.^{2} (65-cm^2) circular size. After proper conditioning, the specimen is placed in a flat position on the bottom of the platen of a compression tester. The upper parallel platen applies a load to the top of the specimen until the sidewalls of the corrugations collapse and fail. The force is recorded and the load is calculated in lbf/in. or kPa (17).

Internal tearing resistance. See Ref. 3 for further information.

Moisture content of paper paperboard (oven drying). This test method determines the moisture in all papers, paperboard, and paperboard and fiberboard containers, except those containing matter other than water that is volatile at 105°C. The individual specimen is removed from unsealed and unprinted sections of a container, set in a holder with a known tare weight, and weighed on a precision balance. The specimen and holder are then placed in an oven set at 105°C and held at that temperature for a minimum of 2 h. The specimen and holder are set into a desiccator, allowed to cool for 1 h, removed from the desiccator, and reweighed. The difference between the original weight and oven-dried weight is calculated as the percentage of moisture in the specimen (18,19).

Moisture content of corrugated fiberboard (Cobb test modified). This test method had been modified, to determine the moisture absorption characteristics of corrugated fiberboard samples in relation to meeting the requirements call for by the DOT 49 *CFR,* ICAO, and UN. A test specimen, measuring 12.5 cm × 12.5 cm, is removed from the unprinted surface of the corrugated fiberboard box, conditioned, and weighed on a precision balance to the nearest 0.01 g. The sample is then clamped in place with the outer liner in the top position, by means of a steel ring, having an inside diameter of 11.28 cm and a metal cross-bar across the ring with two wing nuts. Then 100 mL of water is poured into the ring, starting the of 30-min test period immediately. At the end of this period, the excess water in the ring is poured out quickly, the wing nuts release the steel ring and the test sample is removed. Blotting paper and a hand roller are used to remove surplus water, and the specimen is reweighed. The difference or gain in weight must not exceed 155 g/m^2 (20).

Ply separation (wet). This test method determines if solid or corrugated fiberboard fabricated with weather-resistive adhesive separates after being exposed directly in water. The individual specimens are placed in a freshly aerated water tank for 24 h (the corrugated fiberboard is set with the flutes in the vertical plane), then removed and allowed to drain. The specimens are then examined at the edges for adhesion between components. If there is delamination, the amount of the separation is specified in in. or cm from the edge (21).

Puncture and stiffness. This test method determines the resistance of paperboard, corrugated fiberboard, and solid fiberboard to puncture. The specimen to be tested is placed between clamping jaws. A triangular pyramid, attached to a pendulum arm, travels through an arc to strike the underside of the specimen. The resistance to puncture or energy required is measured in units, where one unit equals 0.265 lbf-in. or 2.99 N · cm (0.03 J). In determining stiffness, a modified test procedure is used wherein the material is slit prior to releasing the puncture head (22).

Ring crush. This test method determines the resistance of paperboard between 0.28 mm (0.011 in.) and 0.51 mm (0.020 in.) thick to edgewise compression. The specimen, 0.5 in. (12.7 mm) wide and 6 in. (153 mm) long, is placed in a holding block having a circular groove. The protruding specimen and holder are positioned on a bottom platen. An upper parallel platen, moving at a uniform rate, applies pressure to the edge of the specimen until it fails. Specimens are tested from samples in both the machine direction and cross-machine direction of the paper. The maximum load is registered in lbf or N (23).

Flexural stiffness of corrugated board. This test method measures the bending resistance of corrugated fiberboard in the machine direction or cross-direction. Flexural stiffness measurements of the combined board when used with the edge-crush test results can accurately predict the top-to-bottom compression strength of a box. After proper conditioning, the specimen, 2.54–5.08 cm (1–2 in.) wide in each of the two principal directions and 5.08 cm (2 in.) longer than the length of the span on the upper assembly, is cut accurately with a sharp knife or saw. This is a four-point beam test, with two lower round-edge support anvils, adjustable from 5.08 to 30.48 cm (2 to 12 in.). The upper loading assembly has two vertical round-edge loading anvils, adjustable from 10.16 to 45.72 cm (4 to 18 in.). These top anvils are successively loaded by a weight mechanism so that the deflection can be measured (24).

Thickness. This test method determines the thickness of paper, paperboard, and combined board by measuring the perpendicular distance between two principal surfaces. After proper conditioning, the specimen is placed between the contact surface of a dial type micrometer and a pressure foot is lowered to record the thickness. The value is listed in in. or mm (25).

Wood boxes and crates. Crates are containers constructed of members made of sawn wood or structural panels or metal combined with fiberboard, securely nailed, bolted, screwed, riveted, welded, glued, dovetailed, or wired and stapled together, having sufficient strength to hold the article packed therein so as to protect it from damage when handled or transported with ordinary care (26). Crates must be constructed so as to protect contents on the sides, ends, tops, and bottoms, and in such a manner that the crate containing its contents may be taken into or out of the vehicle. Contents must be securely held within crates and no part shall protrude, unless otherwise provided in individual items. Surfaces liable to be damaged must be fully covered and protected. A crate is usually a framework or open container. Boxes have closed faces (see Boxes, wirebound; Boxes, wood). Wood boxes and crates are still in demand when specific requirements are called for, such as high compressive or stacking loads, stiffness, resistance to puncture (as a box), and the ability to maintain their strength characteristics under high humidity or very wet conditions. Certain military specifications (see Military packaging) still call for the use of wooden containers to meet the rigors of transportation, handling, and atmospheric environments under the most adverse conditions. Wood combinations are being used more frequently. This category would include plywood, structural-sandwich construction, woods, and paper-base laminates. Component materials are evaluated for the properties given below.

Physical properties. Appearance, moisture content, shrinkage, density, working qualities, weathering, thermal and electrical properties, and chemical resistance.

Mechanical properties. Elastic strength and vibration characteristics, influence of growth, and effect of manufacturing and service environment.

Plastic boxes. Plastic boxes constructed of high-density polyethylene, self-supporting, rigid construction, not extruded or expanded, must be molded by either an injection-molding, blow-molding, rotational molding, or thermal molding process. Tops or covers must be securely affixed. The basic material-properties tests for plastic boxes are specified as melt index, density, tensile strength, and elongation (26).

Melt index or flow rates. This test is performed on the resin material, as specified in ASTM D1238. This method determines the rate of extrusion of a molten resin through a die under specific temperature, load, and piston position conditions. A deadweight piston plastometer is used, consisting of a thermostatically controlled heated steel cylinder, a die at the lower end, and a piston inside the cylinder. After the test specimen, in the form of powder, granules, or strips, has been inserted in the cylinder base, a preheat cycle is originated. The weighted piston, as designated by the material, is inserted, and a purge sequence takes place. At the specified test temperature, as designated by the material, additional material is inserted after the preheat cycle, and the rate of flow is measured for a specific time period. The extrudate is cooled and weighed accurately in milligrams. The flow rate is designated in terms of grams per 10 min (27).

Density. This test method determines the density of solid plastics for identification purposes, to verify uniformity, and to learn if any physical changes have taken place. A temperature-controlled density-gradient tube is used, with different density liquids inserted. Calibrated glass floats are also inserted to cover the various density ranges. Three test specimens are carefully placed inside the tube and allowed to reach equilibrium. The test specimens' densities are determined by their relation to the position of the calibrated glass floats. The reading is specified in g/cm^3 (28).

Tensile strength and elongation. This test method determines the maximum tensile stress and the increase in length produced in the gauge length of the specimen by a tensile load. Tensile and elongation values are helpful for quality-control purposes and for research and development studies. The specimens are prepared in the form of "dogbones" by machining, die cutting, or molding. The center section is narrower to assure breakage in that area. Gauge marks or extension indicators are placed in the middle of the narrower center section to measure elongation. After proper conditioning, the specimen is placed in the grips of the testing machine, which are then tightened evenly to prevent slippage during the test, but not too tight to cause crushing of the material. The movable grip pulls the specimen apart at a specified rate, and the load and elongation are recorded accordingly until ultimate failure occurs. The load value divided by the unit area determines the tensile strength in psi or MPa. The elongation, measured as an increase in length of inches or millimeters, is calculated as a percentage increase in relation to the original gauge length (29).

Bulk Containers

To reduce shipping and handling costs, dry or solid materials such as plastic pellets, sand, flour, and chemicals are transported in bulk containers. When the materials being shipped are hazardous, the containers must conform to the DOT, 49 *CFR* Regulations, Subpart N, Paragraphs 178–819. Provisions are made in certain corrugated fiberboard containers for rapid product removal. Steel and rigid plastic bins are used to handle automotive parts in bulk. Large plastic bags, capable of transporting loads up to 6000 lb (2722 kg) are employed for both domestic and export shipping (see Intermediate bulk containers). These bulk units are designed with special straps for ease of handling to accommodate commercial land carriers and ship operations. The test methods for the component materials of the various bulk containers have been described above (see Testing, shipping containers).

Cans

A can is a receptacle generally of ≤10-gal (38-L) capacity, normally not used as a shipping container. The body is made of lightweight metal, plastic, or is a composite of paperboard and other materials having the ends made of paperboard, metal, plastic, or a combination thereof. There is intense competition today among can suppliers to furnish the optimum container. Different can designs, innovative composite combinations, and special attractive features help make this type of container an ever-changing, progressive part of the packaging industry. The test methods for the component materials of the various types of cans, such as tensile strength and moisture-vapor transmissions, have been described earlier.

Cartons

A folding box is generally made from boxboard for merchandising consumer quantities of products (see Cartons, folding). A carton serves a variety of purposes. It must protect a specific amount of contents against warehouse and transportation environments, the store shelf, and home storage hazards; be appealing and send a powerful selling message; it must be small and compact, handle easily, and be capable of dispensing the contents readily; preserve freshness, taste, odor, appearance, and original form; and prevent contamination. It must be produced efficiently and economically and be capable of traveling through packaging systems at high speeds. The board stock must be material that can be folded and creased without breaking.

Paperboard cartons. Many paperboards (see Paperboard) are now being treated with coatings or laminated with other materials to provide the necessary protection for the product. The basic materials are tested for basis weight, thickness, water vapor transmission, and folding endurance.

Basis weight. See Ref. 2 for further information.

Thickness. See Ref. 25 for further information.

Water vapor transmission. See Refs. 10 and 11 for further information.

Folding endurance. This test determines the resistance of paper to folding. After proper conditioning, the specimen is placed between jaws of a tester. A specified test is applied to the test strip, and the machine is activated to give the necessary number of double folds per minute until it breaks. Specimen strips are cut from each principal direction of the paper (30–32).

Additional protection tests can be performed in relation to corrosion, mold resistance, possible migration, antitarnishing, heat-sealing characteristics, and abrasion resistance.

Plastic-sheet packages. Although plastic-sheet packages are not designated as cartons, their application and usage are quite similar. They must meet the same wide variety of purposes that were described previously for cartons. The majority of units produced are blister packs (see Carded packaging), trays, cups, and boxes (see Boxes, rigid plastic). Chipboard material is often combined with the formed plastic sheet to provide printing surfaces for visual merchandising. Test methods for the plastic sheets have been described in previous paragraphs (10,28,29).

Drums and Pails

A drum is a cylindrical shipping container having straight sides, and a flat, convex, or embossed ends, designed for storage and shipment as an unsupported outer package that may be shipped without boxing or crating. It may be made of metal, or plywood, or of fiber with wooden, metal, or finer ends. Drums are also made of rubber or plastics (compare barrel.); drums range in capacity from (3 to 165 gal) (1.4–625 L), without bilge, with or without bails or handles (33).

Pails are containers (1–12 gal or 3–50 L capacity) with heads or covers, with or without bails or handles, with bilge, and must be made of plastic or metal.

Fiberboard drums. Fiberboard drums can be used to hold dry or solid materials. They are fabricated with either fiberboard tops and bottoms, steel tops and bottoms or plastic tops and fiberboard bottoms. The fiberboard sidewalls, tops, and bottoms are evaluated for bursting strength and thickness using two test methods described previously. The steel tops and bottoms are checked for thickness. The plastic tops are checked for density, melt index, and tensile strength using three test methods described above. In addition, the material is checked for its vicat softening point and stress cracking.

Vicat softening point. This test method determines the temperature at which a needle penetrates the specimen under a specific condition. The specimen is placed on a support so that it is under the needle. The entire assembly is placed in an oil bath with a temperature-measuring device as close as possible to the specimen. A dial indicator measures the amount of needle penetration. A specified load is set on top of the needle assembly, and the oil, under constant stirring, is heated at a specified rate. The temperature is recorded when the indicator shows 1-mm (0.04-in.) penetration (34).

Stress cracking. This test method determines the susceptibility of polyethylene plastics to failure by cracking when exposed to a surface-active agent such as Igepal CO-630. Individual specimens are cut to size, then notched to a specific length and depth. They are bent into position and placed into a brass channel specimen holder. The entire assembly is set inside a test tube, the agent is induced, and the test tube is sealed. The sealed test tube is inserted into a controlled-temperature hotwater bath for a specified period, after which time the specimens are examined for failure (35).

Plastic drums and pails. Plastic drums (see Drums, plastic) and plastic pails (see Pails, plastic) are either of removable-head or tight-head construction. They are suitable for liquid, dry, or solid materials. The plastic components are subjected to melt index, density, tensile strength, and percent elongation tests, using methods that have been described above. When steel covers are installed on plastic pails, the steel gauge is verified by a precision micrometer.

Aluminum or steel drums. Aluminum or steel drums and pails (see Drums and pails, steel) have their sidewalls and top and bottom heads checked for proper thickness by means of a precision micrometer. Tensile strength and percent elongation tests described above can also be conducted to verify that quality standards have been maintained.

Because of space limitations, certain tests for packaging materials required for other containers and packages, such as aerosols, ampuls, carboys, cups, envelopes, kegs, pouches, tubes, and vials, have not been included in this article. For information on testing of cushioning materials, integral components of many packages. ASTM and TAPPI both publish detailed indexes with cross-references as guides for locating the right tests for the right materials.

BIBLIOGRAPHY

1. ASTM D996-95, *Standard Definition of Terms Relating to Packaging and Distribution Environments,* American Society for Testing and Materials, Conshohocken, PA, 1995.
2. TAPPI T410-93, *Grammage of Paper and Paperboard (Weight per Unit Area),* Technical Association of the Pulp and Paper Industry, Atlanta, 1994/95.
3. TAPPI T414-88, *Internal Tearing Resistance of Paper,* Technical Association of the Pulp and Paper Industry, Atlanta, 1994/95.
4. ASTM D774-92, *Bursting Strength of Paper,* Vol. 15.09, American Society for Testing and Materials, Philadelphia, 1995.
5. TAPPI T403-91, *Bursting Strength of Paper,* Technical Association of the Pulp and Paper Industry, Atlanta, 1994/95.
6. ASTM D828-93, *Tensile Strength of Paper and Paperboard,* Vol. 15.09, American Society for Testing and Materials, Philadelphia, 1995.
7. TAPPI T404-92, *Tensile Breaking Strength and Elongation of Paper and Paperboard,* Technical Association of the Pulp and Paper Industry, Atlanta, 1994/95.
8. ASTM D829-93, *Tensile Strength of Paper and Paperboard (Wet),* Vol. 15.09), American Society for Testing and Materials, Philadelphia, 1995.
9. TAPPI T456-87, *Wet Tensile Breaking Strength of Paper and Paperboard,* Technical Association of the Pulp and Paper Industry, Atlanta, 1994/95.
10. ASTM E96-94, *Water Vapor Transmission of Sheet Material,* Vol. 15.09, American Society for Testing and Materials, Philadelphia, 1995.
11. TAPPI T464-90, *Gravimetric Determination of Water Vapor Transmission Rate of Sheet Materials at High Temperature and Humidity,* Technical Association of the Pulp and Paper Industry, Atlanta, 1994/95.
12. L-P-378D, *Para. 4.3.8, Heat Seal Strength, Plastic Sheet and Strip,* Federal Specification, U.S. Government Printing Office, Washington, DC, 1973.
13. ASTM D1709-91, *Impact Resistance of Polyethylene Film by the Free Falling Dart,* Vol. 08.02, American Society for Testing and Materials, Philadelphia, 1995.
14. ASTM D2741-91, *Susceptibility of Polyethylene Bottles to Soot Accumulation,* Vol. 08.02, American Society for Testing and Materials, Philadelphia, 1995.
15. Item 222-6, *National Motor Freight Classification,* National Motor Freight Classification Committee, Alexandria, VA, 1994; Rule 41, Sect. 14, *Uniform Freight Classification,* Uniform Freight Classification Committee, Atlanta, 1995.
16. TAPPI T811-88, *Edgewise Compressive Strength of Corrugated Fiberboard (Short Column Test),* Technical Association of the Pulp and Paper Industry, Atlanta, 1994/95.
17. TAPPI T808-92, *Flat Crush Test of Corrugated Board,* Technical Association of the Pulp and Paper Industry, Atlanta, 1994/95.
18. ASTM D644-84, *Moisture Content of Paper and Paperboard by Oven Drying,* Vol. 15.09, American Society for Testing and Materials, Philadelphia, 1995.

19. TAPPI T412-94, *Moisture in Paper,* Technical Association of the Pulp and Paper Industry, Altanta, 1994/95.
20. TAPPI T441-90, *Water Absorptiveness of Size (Non-Bibulous) Paper and Paperboard (Cobb Test),* Technical Association of the Pulp and Paper Industry, Atlanta, 1994/95.
21. TAPPI T812-92, *Ply Separation of Solid and Corrugated Fiberboard,* Technical Association of the Pulp and Paper Industry, Atlanta, 1994/95.
22. TAPPI T803-88, *Puncture and Stiffness Test of Container Board,* Technical Association of the Pulp and Paper Industry, Atlanta, 1994/95.
23. TAPPI T818-84, *Compression Resistance of Paperboard (Ring Crush),* Technical Association of the Pulp and Paper Industry, Atlanta, 1994/95.
24. TAPPI T820-85, *Flexural Stiffness of Corrugated Board,* Technical Association of the Pulp and Paper Industry, Atlanta, 1994/95.
25. TAPPI T411-89, *Thickness (Caliper) of Paper, Paperboard and Combined Board,* Technical Association of the Pulp and Paper Industry, Atlanta, 1994/95.
26. Item 245, *National Motor Freight Classification,* National Motor Freight Classification Committee, Alexandria, VA, 1994.
27. ASTM D1238-90b, *Flow Rates of Thermoplastics by Extrusion Plastimeter,* Vol. 08.01, American Society for Testing and Materials, Philadelphia, 1995.
28. ASTM D1505-90; *Density of Plastics by the Density-Gradient Technique,* Vol. 08.01, American Society for Testing and Materials, Philadelphia, 1995.
29. ASTM D638-91, *Tensile Properties of Plastics,* Vol. 08.01, American Society for Testing and Materials, Philadelphia, 1995.
30. ASTM D2176-93, *Folding Endurance of Paper by the M.I.T. Tester,* Vol. 15.09, American Society for Testing and Materials, Phildelphia, 1995.
31. TAPPI T423-89, *Folding Endurance of Paper (Schopper Type Tester),* Technical Association of the Pulp and Paper Industry, Atlanta, 1994/95.
32. TAPPI T511-88, *Folding Endurance of Paper (M.I.T. Tester),* Technical Association of the Pulp and Paper Industry, Atlanta, 1994/95.
33. Item 255, *National Motor Freight Classification,* National Motor Freight Classification Committee, Washington, DC, 1984; Rule 40, Sect. 4; *Uniform Freight Classification,* Uniform Freight Classification Committee, Atlanta, 1995.
34. ASTM D1525-91, *Vicat Softening Temperature of Plastics,* Vol. 08.01, American Society for Testing and Materials, Philadelphia, 1994.
35. ASTM D1693-88, *Condition B, 50°C—Environmental Stress-Cracking of Ethylene Plastics,* Vol. 08.02, American Society for Testing and Materials, Philadelphia, 1995.

General References

Annual Books of ASTM Standards, Vols. 08.01, 08.02, 15.09, ASTM, Conshohocken, PA, 1995

TAPPI Test Methods, T200-T1210, Technical Association of the Pulp and Paper Industry, Atlanta, 1994/95.

ICC-UFC No. 6000K, Uniform Freight Classification Committee, Atlanta, 1994. (Standards pertain to shipment by rail.)

ICC-NMF No. 100-T, National Motor Freight Classification Committee, Alexandria, VA, 1994. (Standards pertain to shipment by truck.)

Code of Federal Regulations, Title 49, Department of Transportation U.S. Government Printing Office, Washington, DC, 1995, PTS. 178-199. (Standards pertain to shipments of hazardous materials.)

Technical Instruction for the Safe Transport of Dangerous Goods by Air, International Civil Aviation Organization, Montreal, Quebec, Canada, 1994–1996.

Chester Gaynes
Gaynes Testing Laboratories, Inc.,
Chicago, Illinois

TESTING, PERMEATION AND LEAKAGE

These two factors, permeation and leakage, can have similar effects on packaged goods, but are totally different processes. When oxygen (O_2), water vapor (H_2O), carbon dioxide (CO_2), odors, and flavors move out of packages or into other packages, quality is lost, and shelf life is shortened.

Great strides have been made in permeation, a science relatively unknown to the average citizen, but leakage, a subject familiar to many, seems to be slower in developing a universally accepted testing technology. Testing methods for both permeation and leakage are discussed here.

Permeation

In general, three separate phenomena occur simultaneously.

Solubility—penetration into polymer
Diffusion—penetration through polymer
Desorption—evaporation from polymer

The total permeation through the barrier is made up of the sum total of all three parameters, and is referred to metrically as

Permeation rate—(mL · mm)/(m^2 · 24 h · atm), dry, at (temperature, 0°)
Transmission rate—(mL/(m^2 · 24 h), dry, at (temperature, 0°)

Note that the transmission rate (TR) has not been normalized for atmospheric pressure or thickness. The laboratory must decide the best units. However, if the instrument is equipped with a pressure transducer, so that results are adjusted to sea level, it is common to state the TR value at sea level and is, therefore, easily compared from one laboratory to another.

Equilibration. A sample of barrier film is mounted in a permeation instrument test cell and the test is started. Note the time data in Figure 1. The results of a test are valid only once the barrier material has reached equilibrium.

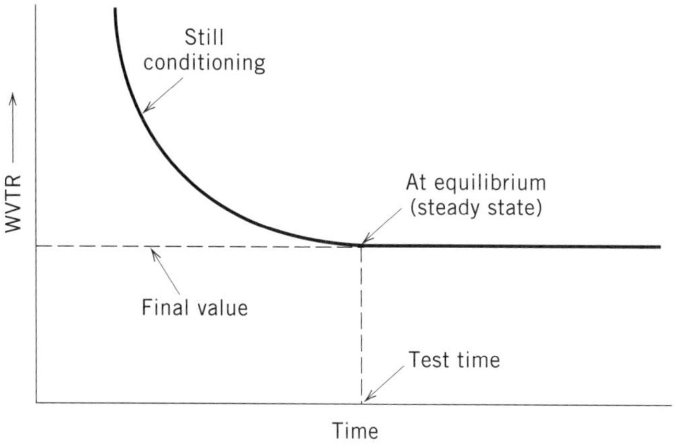

Figure 1. Typical example of time to condition specimen.

- Material thickness
- Relative humidity
- Temperature
- Time
- Barometric pressure

Figure 2. Five major factors affecting permeation rate.

The time for the sample to reach equilibrium cannot be appreciably accelerated or shortened. Time is a physical function of the material. The better the barrier, the longer it takes to get results. For example, 1 mil (25 μm) PET takes 1 h, while 5 mils (125 μm) PET takes 8 h to stabilize. If the result is taken prematurely, it will be wrong, as steady state has not yet been reached.

Five major factors that affect permeation rate through a barrier polymer are listed in Figure 2.

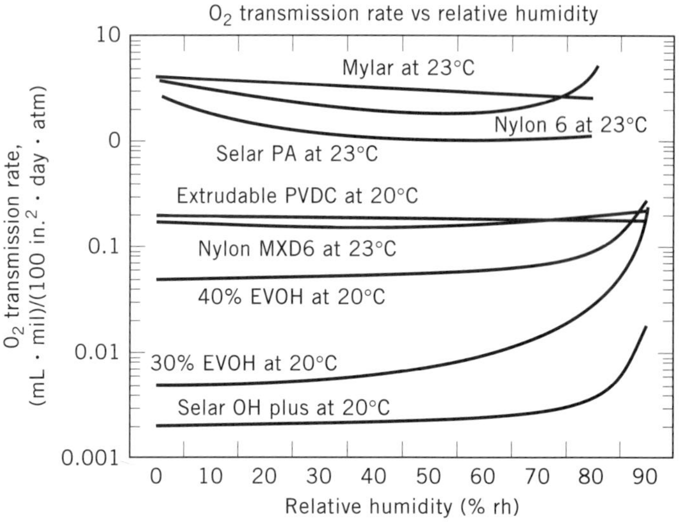

Figure 4. Effect of relative humidity on O_2 transmission rates of several plastic films.

Pioneering Methods

We must acknowledge some of the pioneer work in the field of permeation, in particular the water-vapor permeation cup test (1, 2) (see also Fig. 3) developed in the early 1940s. These test methods determine the loss of water from a cup through a barrier sealed on top of the cup, and are still used in a few locations. Permeation does have disadvantages, such as a long time to obtain results (days vs hours), limited testing range, being labor-intensive, requiring expensive climate-control chambers, and being operator-dependent. However, steady progress has been made in instrumented testing systems that allow shorter test times, greater limits, better repeatability, increased sensitivity, computerized control, and documentation.

Also, recognition is due to other early developments as listed in ASTM D1434 (M) on the manometric cell (also known as the *Dow cell*), and ASTM D1434 (V) on the volumetric cell. These were gallant attempts, but are no longer in widespread use.

Current Methods

Oxygen transmission rate (O_2TR). Oxygen is often referred to as the "thief" of flavor, texture, color, nutrition, and shelf life. The importance of testing O_2TR was recognized early in the development of barrier materials such as coated papers and plastics. The coulometric sensor (also known as the coulox sensor) soon dominated because it could detect O_2 down to the ppb level, was stable, and was easy to maintain and reliable.

Initially only "dry" tests were performed, 0% rh, which was adequate for many of the barriers of that day. However, subsequent barrier materials were developed that had superior O_2 permeation characteristics, but the O_2TR of these new materials was a function of humidity.

Oxygen permeation testing has advanced in capability to allow testing of the most recent developments in barrier materials. The ability to control the percent relative humidity (% rh) independently on each side of the barrier has been one of the most significant advancements. This allows a better simulation of true-to-life shelf conditions and, therefore, more credible results.

The effect of % rh and temperature on O_2TR can be seen in Figures 4 and 5.

Performance expectations are very important. Measurable

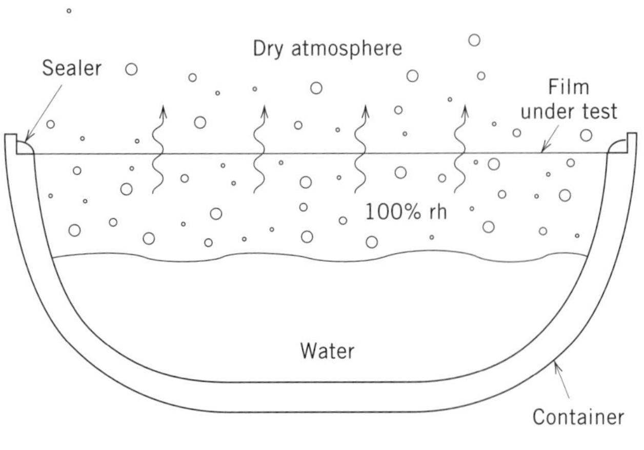

Figure 3. Wet cup test.

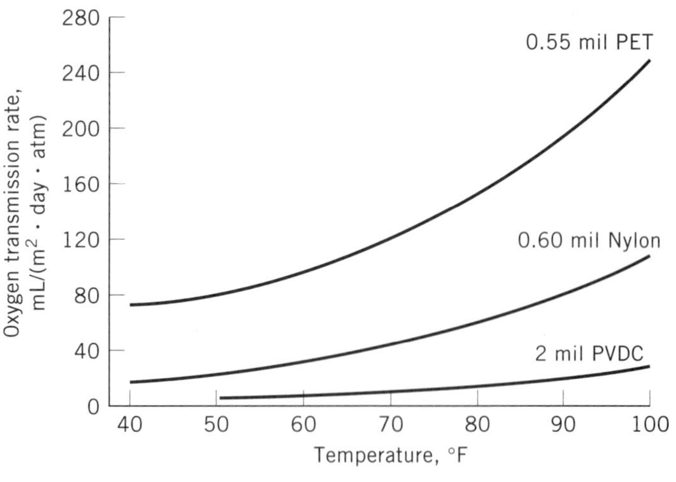

Figure 5. Effect of temperature on O_2 transmission rate of 0% rh for several plastic films.

Oxygen (coulometric method)			
ASTM	D3985	Films	North and South America
DIN	53380	Films	Europe
JIS	K7126	Films	Asia
ASTM	F1307	Packages	North and South America
Water vapor (infrared method)			
ASTM	F1249	Films	North and South America
JIS	K7129	Films	Asia

Figure 6. International standards for permeation and transmission-rate testing.

range of O_2 transmission rates is 0.001 mL/(m^2 · day) to over 1,000,000 mL/(m^2 · day). Following are specifications on Ox-Tran systems,[3] as used in ASTM D3985 and F1307:

Sensitivity—films, 0.01 mL/(m^2 · day) [0.0006 mL/(100 in.^{2} · day)]; packages, 0.00005 mL/package · day)

Repeatability—films, ±0.05 mL/(m^2 · day) [0.003 mL/(100 in.^{2} · day)] or 1% of reading; packages, ±0.00025 mL/(package · day) or 1% of reading

A new high-sensitivity method and apparatus now makes it possible to attain the following specification:

Sensitivity—films, 0.001 mL/(m^2 · day) [0.00006 mL/100 in.^{2} · day)]; packages, 0.000005 mL/(package · day)

Repeatability—films ±0.005 mL/(m^2 · day) [0.0003 mL/(100 in.^{2} · day)] or 1% of reading; packages, ±0.000025 mL/(package · day) or 1% of reading

The importance of temperature control cannot be over emphasized. Relative humidity is a function of temperature, and to control % rh is not an easy task because it can vary by 5% rh/°C in a region of the test.

Measuring high O_2TR values. At first, all efforts were directed toward low permeation readings (high barriers). An emerging need also exists to measure *high* O_2TR values. Initially the use of a mask (to decrease the permeation test-cell area) was used for years. As higher O_2TR were required, new test techniques were developed to yield more repeatable results and to keep the O_2 sensor from swamping. With care, values of ≤ 155,000 mL/(m^2 · day) can be measured. So the measurable range of O_2TR is 0.001 mL/(m^2 · day) to over 155,000 mL/(m^2 · day).

The following international test methods, as shown in Figure 6, were written around the use of the coulox sensor for oxygen and the infrared method for water vapor, making it possible for laboratories in different parts of the world to compare O_2TR and WVTR test results.

Water-vapor transmission rate (WVTR). After the cup test, instrumented systems (ASTM F372, F378, TAPPI 378), and ir (infrared) systems were developed that achieved results in a shorter time (hours vs days). Comparison of WVTR using the ir method and the cup test revealed remarkably similar results. Today, it is estimated that 95% of WVTR testing in the United States is done using ASTM method F1249 and TAPPI method T557 pm-95.

It was standard practice, at one time, to run all WVTR tests at a given rh, (eg, 100% rh). Then, if the WVTR at another % rh was desired, one would simply multiply the 100% rh results by the % rh of interest. For example, the 100% rh result was multiplied by 0.8 if a result at 80% rh was of interest. This was done because there was no convenient, clean method to generate precise % rh values.

The results using this technique, called *factoring,* can be seen in Figure 7.

Notice the excellent correlation between these two methods of creating the driving-force RH. The test results seen on four common packaging films were almost identical, no matter whether the actual rh was created with salts, distilled water was used, and the result then factored.

Considerable advantages have been obtained with factoring, including

1. Elimination of messy, uncontrollable salt solutions
2. Greatly reduced possibility of instrument corrosion
3. Increased ease of equipment operation

This method of testing without salts is widespread today.

By about 1990, however, it became apparent that this technique may not work for all barrier materials, most notably hydrophilic polymers. In these cases, water vapor is so soluble that the "factoring" relationship does not appear to hold true. Nylon 6 is a good example of this (see Fig. 8).

As can be seen, factoring does not work for all barriers. Therefore, a system was developed that is more universally applicable by creating humidity and temperature test conditions close to actual shelf conditions of the package. The advantages are

1. Tests can be conducted close to actual shelf conditions.
2. A range of RH values is available.
3. Relative-humidity values easily changed.
4. No mess.

Water-vapor transmission rate is expressed in g/(m^2 · day), or g/(100 in.^{2} · day). Measurable range of WVTR is 0.0001 g/(m^2 · day) to 1000 g/(m^2 · day). The specifications of some WVTR systems are

Sensitivity—films, 0.0001 g/(m^2 · day) [0.000006 g/(100 in.^{2} · day)]; packages, 0.000001 g/(package · day)

Repeatability—films, ±0.002 g/(m^2 · day) [±0.00015 g/(100 in.^{2} · day) or 1% of reading; packages, ±0.00001 g/(package · day) or 1% of reading

Carbon dioxide transmission rate (CO_2TR). An ir sensor is used to detect CO_2 and when incorporated in an instrument system to detect permeation, yields CO_2 TR. The major applications are for cola drinks (CO_2 escaping from PET containers), cheese packaging, and other CO_2-sensitive products. There is no way to extrapolate the action of one gas to determine the action of another gas. Each gas is independent of the other gases and must be tested as an individual gas to determine its transmission rate (TR).

Measurable range of CO_2 transmission rate is 1.0 mL/(m^2

All values are g/(m2 · 24 h) at 37.8°C (100°F)		WVTR value using zinc sulfate salt to create 84.5% rh	WVTR value using distilled water to create 100% rh, then factored by 0.846	% DEV
Polyethylene	#1	11.2	11.2	0.0
1.9 mils (48.3 μm)	#2	12.1	12.3	1.7
Polypropylene	#1	9.61	9.58	-0.3
1.8 mils (45.7 μm)	#2	9.85	9.67	-1.8
Polyester	#1	21.1	21.8	3.3
0.92 mils (48.3 μm)	#2	21.3	21.6	1.4
Saran Wrap 8	#1	7.06	7.32	3.7
1.0 mils (25.4 μm)	#2	6.80	6.99	2.8

Figure 7. Comparison of four films "factored" with 100% rh vs tests with a saturated salt solution of zinc sulfate. (Saran Wrap is a registered trademark of Dow Chemical USA.)

· day) up to 100,000 mL/(m^2 · day). Specifications for CO_2 systems are

Sensitivity—films, 1 mL/(m^2 · day) [0.07 mL/(100 in.^{2} · day)]; packages, 0.005 mL/(package · day)

Repeatability—films ±5 mL/(m^2 · day) [± 0.35 mL/(100 in.^{2} · day)] or 1% of reading; packages, ±0.025 mL/(package · day) or 1% of reading

CO_2TR permeation systems can detect 100 ppm CO_2. This is an important factor when considering permeation or leakage. Efforts have to be made to keep the CO_2 from being diluted during the testing process. Diluting the CO_2 level can make a bad situation look good. This is just the opposite for O_2. Diluting a low-O_2 signal with air (20.8% O_2) will make it worse, not better.

An outstanding example of *partial pressure* can be seen using CO_2. Fill a PET container with 100% CO_2 and seal securely. The partial-pressure phenomenon happens over a period of time. CO_2 leaves the PET container much faster than O_2 or N_2 air enters the container. As CO_2 leaves, a vacuum is formed, which eventually causes the container to collapse. Even though CO_2 was removed, leaving a vacuum, O_2 and N_2 followed the partial-pressure law. The vacuum seems to play no part in the permeation process in this case.

Aromas and flavors. As stated above, these need to be tested as individual gases, and there are a host of these in organics. This is a developing field with many opportunities, such as citrus flavor, D-limonene from an orange-juice container, citrus flavor into other food products (eg, breakfast cereals), petroleum gases out of packages and into other packaged goods, etc. Fortunately, a technique has been developed to handle all these organic molecules, however, only one at a time.

In brief (referring to the Aromatran schematic in Fig. 9) a carrier gas (helium, N_2, or any number of inert gases) carries the test gas (D-limonene, butane, methane, apple flavor, etc), which has permeated through a barrier, to a *cold trap,* where it is captured and accumulated over a period of time. At the end of the "accumulate" time, the cold trap is heated very quickly, thereby releasing the test gas molecules that had been captured.

The carrier gas, which is not affected by the cold trap, continues to flow and carries the test-gas molecules to a sensor. The sensor is an FID that is set up to detect the amount of test gas captured.

The results are given in common permeation terms:

$$X \text{ TR} = \text{mL/(m}^2 \cdot \text{day) at } 30°\text{C}$$

Permeation becomes more critical as shelf life is extended. Not only is there concern about gases, such as O_2, water vapor, and CO_2 entering and leaving packages, but now the effects of aromas and flavors are playing an important role in the industry.

Leakage

Everyone knows what a leak is—leaks in the roof, a leak in a tire, even a secret can be leaked out. It is something everyone can define, and everyone has an opinion on the subject.

In the previous section we discussed permeation. Both permeation and leakage do the same thing: shorten shelf life.

All values are g/(m2 · 24 h) at 37.8°C (100°F)		WVTR value using zinc sulfate salt to create 84.5% rh	WVTR value using distilled water to create 100% rh, then factored by 0.846	% DEV
Nylon 6	#1	123	213	73.2
	#2	134	207	54.5

Figure 8. Comparison of nylon 6 "factored" with 100% rh vs tested with a saturated salt solution.

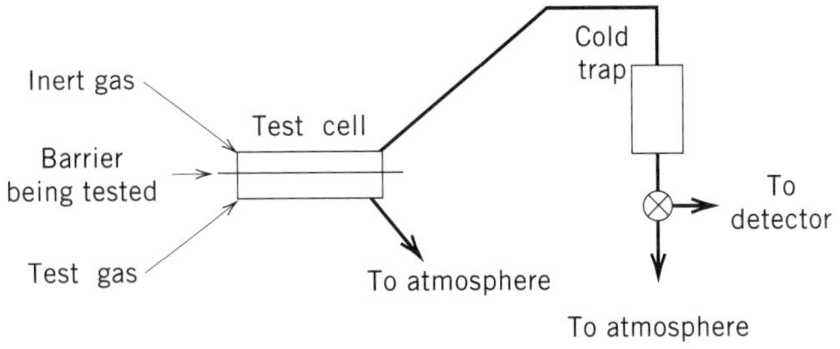

Figure 9. Aromatran schematic.

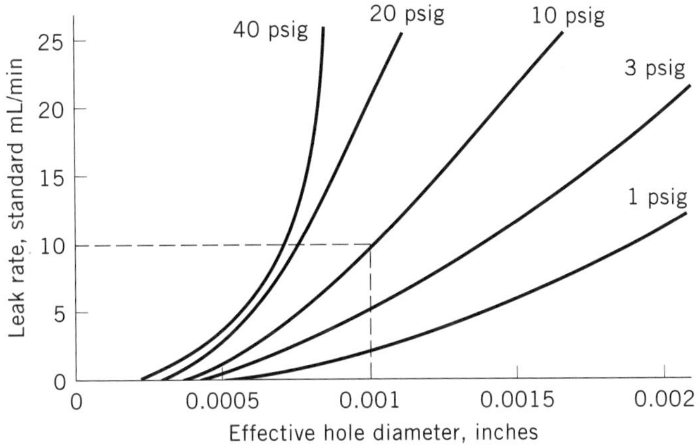

Figure 10. Flow rate of air through a round hole.

Permeation can be determined, as described above; however, leaks are more erratic, evasive, and unpredictable. Where a permeation value might hold true for a production run of plastic film, leaks can vary from one package to the next. The ideal is to have 100% leak testing on a production run. This is ideal but not practical in most applications.

The perpetual issue of "hole size" is debated among manufacturers of food, pharmaceutical, and sterile medical products. Each has different criteria. The hole that will allow bacteria to enter a sterile package is much smaller than a hole that will allow gravy to drip out of a microwave entrée. First, we must agree on a term called "effective hole size." This is the summation of all the holes in the package. It could be the summation of round holes, square holes, holes with long tunnels, slight tears, big tears, hairline cracks, etc. Let one hole represent the combination of all the leaks. By so doing, we provide a way of comparing one package to the next. Figure 10 illustrates the flow rate that one would expect through a round-style hole under various differential pressures across the hole. These curves apply to air, at ambient temperature through a hole in a thin material.

Destructive testing for leaks

Waterbath or dye–leak test. (See Fig. 11). Throughout most of the twentieth century, package integrity has been verified by placing the package under water and looking for bubbles. Variations of the "waterbath" test have generally involved attempts to improve on the problems inherent in looking for bubbles. The "dye test" was an improvement, but the two remaining problems are lack of repeatability and destruction of the package. Advantages and disadvantages of these tests are as follows:

1. The waterbath test
   a. Advantages
      (1) Can provide sensitivity
      (2) Inexpensive equipment requirements
      (3) Low-level technology, easy to train operator
   b. Disadvantages
      (1) Destructive
      (2) Results are very operator-dependent
      (3) Expensive labor requirements
      (4) Messy
      (5) Capillary attraction can give false results
2. The dye test
   a. Advantages
      (1) Ability to test many samples at once
      (2) Inexpensive equipment requirements
      (3) Low-level technology, easy to train operator
   b. Disadvantages
      (1) Destructive

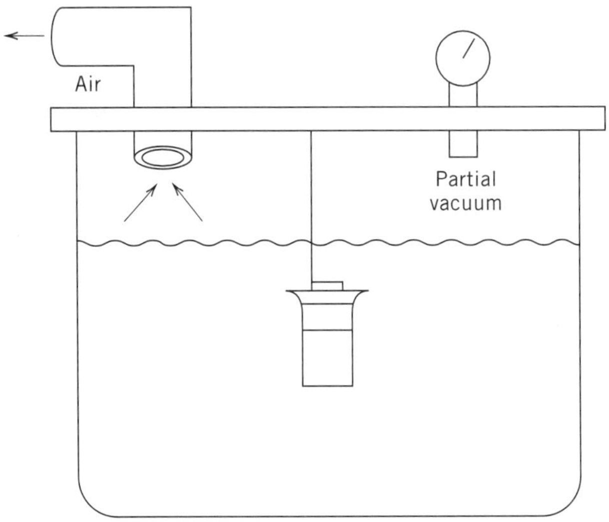

Figure 11. Waterbath or dye leak test.

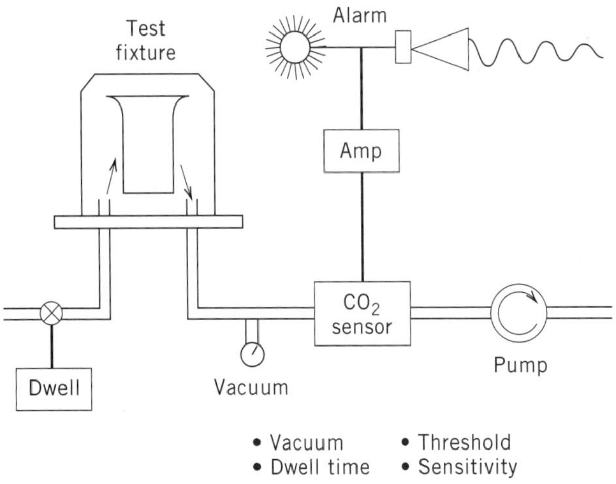

Figure 12. Co_2 trace-gas leak detection; nondestructive.

(2) Unreliable and operator dependent
(3) Slow and labor intensive
(4) Messy
(5) One-way valve effect

Pressure/pressure-decay method. This technique allows two tests to be run sequentially: (*1*) burst/seal strength and (*2*) leak.

Burst/seal strength. This is a practical starting point. It is important to ensure that handling and shipping does not destroy the package, which left the plant without leaks. It must remain sealed to maintain adequate shelf life and/or the desired quality when it reaches the consumer. Air pressure is introduced into the package until it *bursts!* The burst-point pressure is held so that it can be recorded. This burst test should be repeated several times to get an average. Analysis of this data would be required if damaged packages were discovered by a consumer. This also allows R&D to investigate new packaging materials and sealing techniques. Average test time is 15 s.

Leakage. Once the burst/seal strength has been determined, look for leaks. Introduce a pressure into a new package at less than 50% burst pressure and monitor pressure decay. Pressure decay is a function of effective hole size and pressure differential and is inversely proportional to the volume of the package under test. It has been shown that holes as small as 1-mil diameter are detectable in small packages such as pharmaceutical bubbles, but would be almost impossible or impractical to sense a 1-mil hole in a blimp. This method does not indicate the location of leak(s). Average test time is 10 s.

Other techniques that are destructive include: halogen detection where a halogen agent, such as iodine, bromine, or fluorine, is used as a tracer gas. The system is pressurized with the tracer gas, and a probe is used to detect a leakage point. Although this test does detect leakage points, it is highly operator-dependent; thus, a poor operator will get the best results, only on the basis of fewer rejects. The instrument response is slow and lacks quantitative accuracy. Food or drug products are not normally tested by this technique.

Halide torch. Similar to the halogen system described above, in the halide-torch test, the tracer gas is sucked into a burner through a hose attached to the bottom of a burner. A change in the color of the flame indicates a leak.

Thermal-conductivity detector. A tracer gas, such as helium, carbon dioxide, or butane, is pressurized in the package, and a probe senses a change in gas thermoconductivity as the tracer gas enters the probe. This method does detect the leakage point; however, it also is highly operator-dependent because if the probe is moved too fast or too far away from a hole, nothing will be indicated.

Helium leak detector. This has the greatest sensitivity. Inflate the package with helium. Then, using a probe, *sniff* around the package to find a leak. The major disadvantages are, besides high operator dependence, initial costs (for supplies, etc), and unit maintenance, that if the instrument is swamped, it takes hours or days for a skilled technician to get it back on line.

Headspace analysis. A practical way to check for leaks is to measure the headspace of a package, and this can be done at any time, such as immediately following sealing or later, or even years later. Using a syringe, a sample of the headspace is taken from the package and injected into an instrument. If O_2 content is to be determined, a zirconium oxide sensor is used; if CO_2 content is to be determined, an ir sensor is used. The results are given in percentage of contents, such as 0.5% O_2 or 30% CO_2, and within a few seconds after injection of the sample into the instrument. Applications include checking for cap/map effectiveness, shortly after testing to determine whether sealing was effective and met specifications and much later, to determine whether permeation had any effect. Note, in this latter case, that the long-term effect could be either a very slow leak or permeation; this instrument cannot separate the two.

Nondestructive testing for leaks. The CO_2 leak detector is most commonly used in nondestructive testing of food and pharmaceutical packages. It is ideal if the package has been flushed with CO_2 before sealing or if CO_2 is generated inside, as in the case of coffee.

Leak detection utilizing CO_2 tracer gas. Figure 12 shows the flow diagram of a CO_2 leak detector. The pump pictured is always running, sweeping room air in from the left and purging the test fixture and CO_2 sensor. When the test sequence is initiated, the dwell solenoid closes, thus causing the pump to draw a vacuum on the package inside the test fixture, which draws CO_2 out of the leaks. At the end of the test cycle, room air is again allowed to sweep into the test fixture. This flushes any CO_2 gas into the CO_2 sensor. If this CO2 concentration exceeds the alarm threshold, the instrument signals that a leak is present. Throughout this cycle, room CO_2 concentration is viewed as the unit's zero, ie, reference, point.

An important factor is the size of the test fixture compared to the package under test. Sensitivity is inversely proportional to the space outside the sample being tested. As mentioned earlier, care must be taken to avoid diluting the CO_2 signal. Therefore, a specially designed fixture is needed for best results, ie, to sense small leaks, and in most foods and/or pharmaceutical applications, small leaks are important.

In those cases where there is no CO_2 in the package, CO_2 can be forced into the packages through the leaks by use of a pressure chamber filled with CO_2. The vacuum, shown in Figure 14 then draws CO_2 out of the package, which is sent to the CO_2 sensor. Tests using this technique take 15 s on average.

Carbon dioxide trace-gas leak detection has been shown to be a sensitive, repeatable, and nondestructive alternative to traditional waterbath and dye-leak techniques. Justification is easiest when the product is expensive, as this nondestructive test procedure enables product to be exposed to additional laboratory tests, or to be returned to the production line.

Conclusion

The search for better leak-testing methods continues. The challenge is to devise a system that is nondestructive, can do 100% inspection at high speed, and, of course, is low-cost. To date, the diversity of applications has precluded this. Each application must be evaluated to determine the best solution. Testing-systems manufacturers are challenged to write their specification legibly, clearly, and precisely to ensure that they are understood and can be applied.

Permeation is well defined and accepted in ASTM specifications. Leakage requires the same attention.

BIBLIOGRAPHY

1. ASTM E96, American Society for Testing and Materials, Philadelphia, PA
2. TAPPI 464, Technical Association of the Pulp and Paper Industry, Atlanta

BERT JOHNSON
ROBERT DEMOREST
MOCON
Minneapolis, Minnesota

TESTING, PRODUCT FRAGILITY

Most products must undergo a certain amount of handling and transportation from the time they are manufactured until they are ultimately used. Normally, products are enclosed in a protective-package system during the time they are exposed to the distribution environment. This article outlines engineering principles and procedures that optimize the protective function of a package system and help guarantee the product's safe arrival at favorable cost.

The waste of resources caused by improper packaging is enormous. Whether this waste shows up as damaged product or an overly expensive package system, it is still waste that can be avoided. The application of sound engineering principles to optimize the product protection system must receive significant endorsement by product managers, quality assurance (see Specifications and quality assurance), and all those concerned with delivering a quality product to the customer.

To develop a protective-package system, three important pieces of data are necessary: (*1*) information on the likely inputs (shock and vibration) from the distribution environment; (*2*) product fragility characteristics and sensitivities in terms of the environmental inputs; and (*3*) the performance characteristics of commonly available packaging materials.

The engineering process for developing a protective-package system involves a step-by-step problem-solving approach similar to other engineering disciplines. The principal focus is on product characteristics because it is impossible to optimize a package system without first knowing basic engineering data of the product itself.

Once the environment has been defined and product fragility has been determined, the engineer can evaluate the economic feasibility of possible product modifications. For example, if a certain component within the product keeps failing at a relatively low input level, it may be economically more desirable to modify that component and increase its ruggedness rather than design an expensive and elaborate package system for the entire product based on the sensitivity of that one component. If it is economically feasible to modify the product in order to effect a package-cost savings, this option should be studied carefully. Many examples exist where slight product modifications have resulted in substantial cost savings in packaging materials and reduced damage in shipment.

There are many potential hazards in the distribution environment, including shock, vibration, temperature or humidity extremes, electrostatic discharge, magnetic fields, and compression. This article deals primarily with shock and vibration inputs; however, it is important to explore carefully those areas where the product is sensitive or a large environmental input is likely.

The steps involved in designing an optimized package system are as follows:

1. *Define the environment* in terms of shock and vibration inputs likely during the product's manufacturing and distribution cycle.
2. *Define product fragility* in terms of shock and vibration.
3. *Obtain product improvement feedback,* ie, examine product improvements in light of the economic tradeoffs between packaging costs and product modification costs.
4. *Evaluate cushion material performance,* ie, evaluate the shock and vibration characteristics of available cusion materials.
5. *Design the package system,* ie, select cushion materials and design for optimum shock and vibration performance.
6. *Test the final package,* ie, verify that the package system performs as designed and properly protects the product.

The primary purpose of this article is to examine product fragility testing in detail, and therefore the other steps are not covered in a comprehensive fashion. Refer to the bibliography for excellent background material on the other steps in this process (1–6) (see also Cushioning, design; Distribution hazards).

Defining the Environment

The end result of defining the environment should be the establishment of a design drop height and a resonant-frequency spectrum for a particular product in a given distribution environment. That is, based on the size and weight of a packaged product, the engineer selects a drop height that represents a given probability of input. Also, vibration profiles for likely modes of transportation are selected. Both of these pieces of information are used in the last step, testing of the package system (7–10).

Defining Product Fragility

The term "product fragility" is misunderstood by many people and often conjures up images of totally destroyed products, broken bottles, and the like. In reality, product fragility is simply another product characteristic such as size, weight, and color. Just as other product characteristics are determined by measurement, product fragility (or product ruggedness) can be measured with shock inputs. This measurement takes the form of a damage-boundary curve for shock and resonant frequency plots for vibration. In both cases, the importance of determining these characteristics cannot be overemphasized. Most people would not think of buying a pair of shoes based on guessing their foot size. It is just as shortsighted to design a package system by guessing at product fragility.

Shock fragility assessment. The damage boundary is the principal tool used to determine the ruggedness of a product; it takes the general shapes shown in Figure 1. This plot defines an area on a graph bounded by peak acceleration on the vertical axis and velocity change on the horizontal axis. Any shock pulse that can be plotted inside this boundary causes damage to the product regardless of whether it is packaged. (This is a *product* test).

Acceleration is a vector quantity describing the time rate of change of velocity of a body in relation to a fixed reference point, ie, it describes the rate at which velocity is increasing or decreasing (deceleration). The terms *acceleration* and *deceleration* are often used interchangeably because most products respond similarly to a rapid start or a rapid stop. Both terms are usually expressed in G values, multiples of earth's gravitational constant, g.

Velocity change is the difference in a system's velocity magnitude and direction from the start to the end of shock pulse. Velocity change is the integral of the acceleration vs time pulse and is directly related to drop height.

To run a damage boundary, mount the product to be tested on the table of a suitable shock-test machine (see Fig. 2). Secure the product with a rigid fixture that lends even support to the product over its entire surface. The fixture must be as rigid as possible so that it does not distort the shock pulse transmitted to the product.

Set the shock machine to produce a low velocity-change shock pulse with a duration of approximately 2 ms (a half-sine waveform is generally used for this test). After the shock pulse is delivered to the product, examine it to determine if damage has occurred. If not, set the shock machine to produce a slightly higher velocity change and repeat the test. Continue this process with small incremental increases in velocity change until damage occurs. The last nonfailure shock input defines critical velocity change ΔV_c for the product in that orientation.

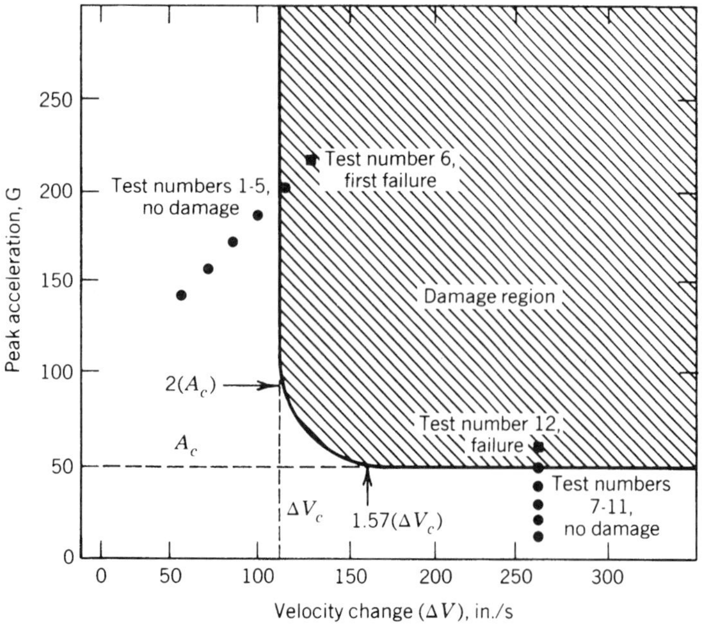

Figure 1. Damage boundary (single orientation). A_c critical acceleration; V_c critical velocity. To convert in. to cm., multiply by 2.54.

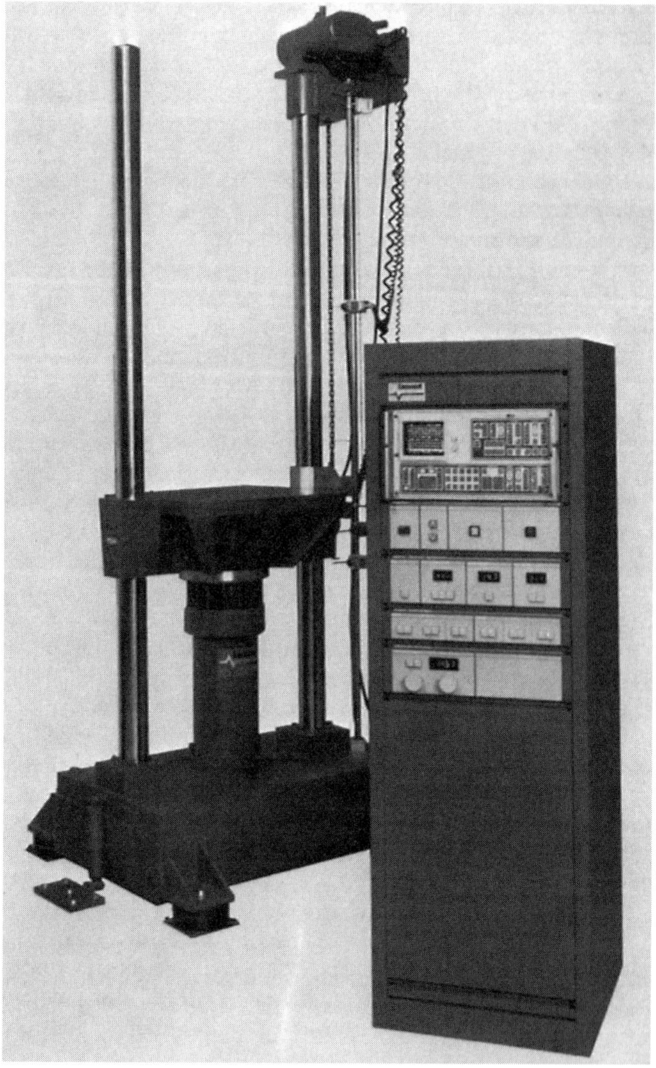

Figure 2. Shock-test machine.

Next, fasten a new test specimen to the shock table and set the machine to produce a trapezoidal pulse with a low acceleration level and a velocity change at least two times that of the critical velocity determined in the previous test. Program the shock pulse into the product and as before, examine the product to determine if failure occurs. If no failure has occurred, set the shock machine to produce a higher acceleration level at approximately the same velocity change. Repeat this procedure with small increments in acceleration until the failure level is reached. The last nonfailure shock input defines the critical acceleration. A_c for the product in that orientation (11).

The damage boundary curve may now be plotted by drawing a vertical line through the critical velocity change point and a horizontal line through the critical acceleration point. The intersection of these two lines (the knee) is a smooth curve as shown in Figure 1. A rectangular corner may be used as a conservative approximation of the damage region (12).

Note that the critical acceleration as determined by a trap-

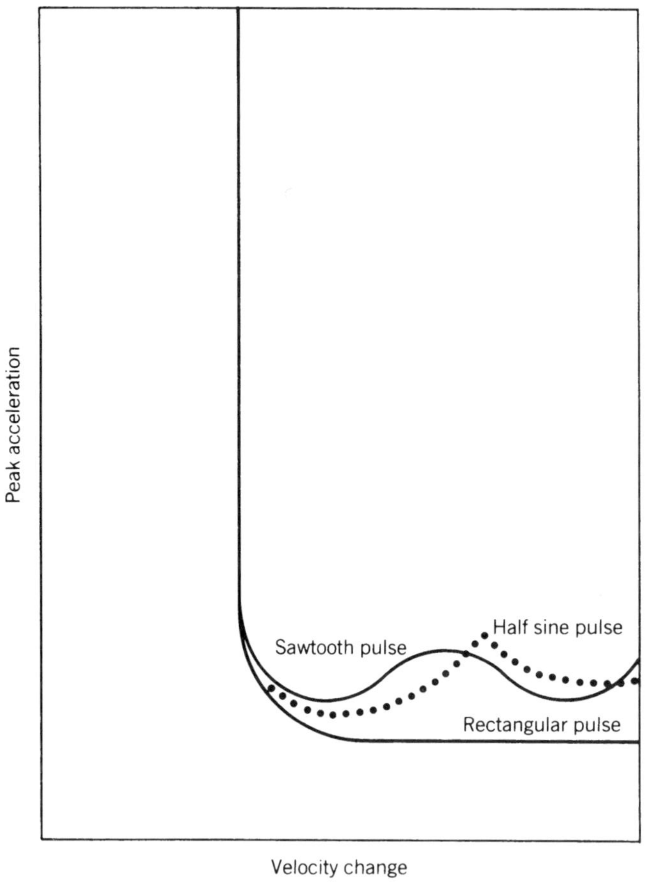

Figure 3. Damage boundary for various pulses.

ezoidal pulse is conservative when compared to that generated through the use of other waveforms; ie, a trapezoidal pulse is more damaging than other waveforms of the same peak acceleration and duration. This is shown graphically in Figure 3.

Since the shape of the waveform transmitted through various packaging materials during impact is generally not known, the use of a trapezoidal wave during damage boundary testing results in a higher confidence level in a packaging system and is recommended for this purpose. It should also be pointed out that the use of the trapezoidal waveform results in a nearly linear abscissa on the damage boundary. This means that it is necessary to determine only one point on that axis in order to determine the critical acceleration for the product in that orientation. Other waveforms result in critical accelerations that are a complex function of the natural frequency of components within the product. The procedure described is based on ASTM D3332 (13).

The damage boundary is a valuable and powerful tool. Critical velocity change is equated to equivalent free drop height from the formula $\Delta V = (1 + e)\sqrt{2gh}$, where e is the coefficient of restitution of the impact surfaces, g is acceleration of gravity [386 in./s^2 at sea level (9.8 m/s^2)], and h is equivalent free-fall drop height in in. (or m). The critical velocity change tells the designer how high the unpackaged product can fall onto a rigid surface before damage occurs in that axis. If this equivalent drop height is likely to be exceeded in the distribution environment, then the product must be cushioned. The performance requirements of the cushion are that no more than the critical acceleration be transmitted to the product.

In theory, the value of e (coefficient of restitution) can vary from zero to one. A value of zero would imply a totally plastic impact with no rebound whatsoever. A value of one implies a perfectly elastic impact here the rebounded velocity is exactly equal to the impact velocity. As a practical matter, a range of $e = 0.25–0.75$ produces good accuracy in calculating the range of equivalent freefall drop height. The chart in Figure 4 shows the effect of e.

The damage boundary also tells the engineer that at low velocity changes infinite accelerations are possible, and, at low acceleration levels infinite velocity changes are possible without product damage. That means that it is necessary to define both critical acceleration and critical velocity change to properly characterize the fragility of a product.

Before running the damage-boundary test, the engineer must define what constitutes damage to the product. On one extreme, damage may be catastrophic failure. However, there are many less severe damage modes that can make a product unacceptable to the customer. In some cases, damage can be determined by observation of the product; at other times it involves running sophisticated functional checks. Once the determination of damage is made, the definition must remain constant throughout the testing and must be consistent with what is deemed unacceptable to the customer.

In general, damage-boundary tests must be run for each axis in each orientation of the product. In the case of a rectangular product such as a television set, this means that a total of 12 specimens are necessary for a rigorous test. However, since this testing should be done in the product prototype stage, this quantity is rarely available for a potentially destructive test. As a practical matter, much information can be gained from a limited number of samples, and multiple dam-

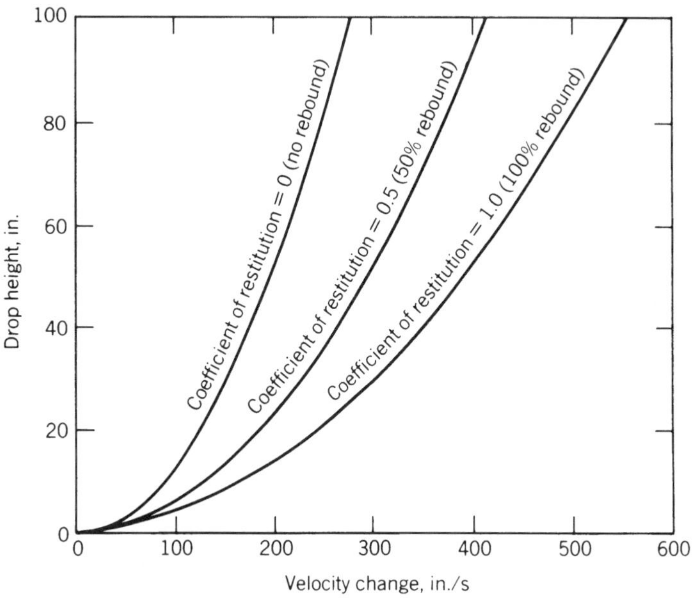

Figure 4. Coefficients of restitution. To convert in. to cm., multiply by 2.54.

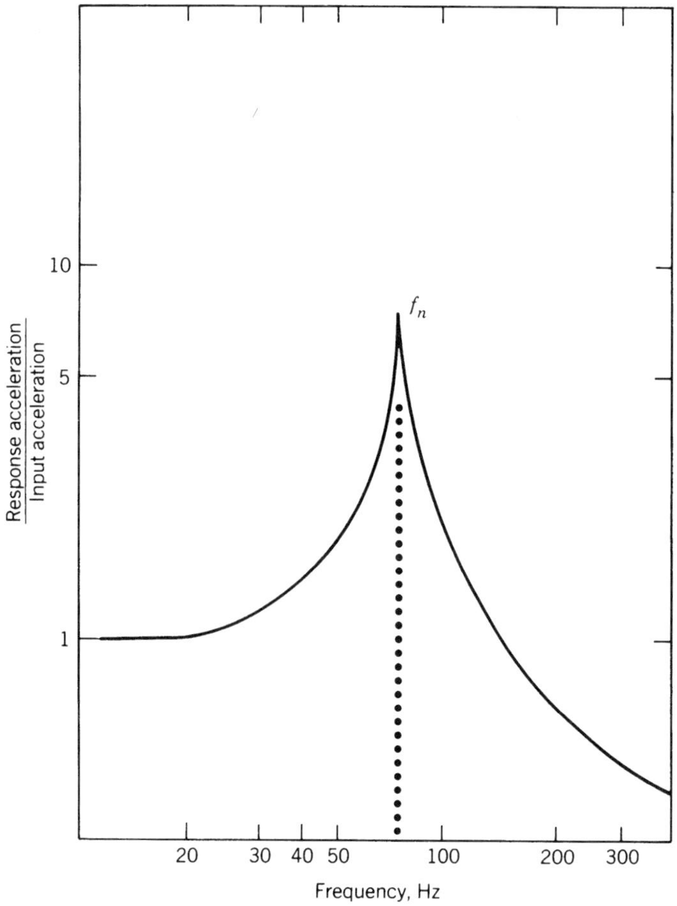

Figure 5. Resonant frequency plot.

age boundaries are often run on the same unit in different orientations.

Vibration-fragility assessment. Determining product-vibration sensitivities involves identifying resonant frequencies of components in the product in each of the principal axes. Resonance is that characteristic displayed by all spring/mass systems wherein at a given frequency, the response acceleration of a component is greater than the input acceleration. This characteristic is shown graphically in Figure 5.

As a general rule, the product will not be damaged, due to nonresonant inertial loading caused by vibration in the *distribution environment.* This is because the acceleration levels of most vehicles are relatively low when compared to the critical acceleration of most products. It is only when a component within a product is excited or forced by vibration at its natural frequency that damage is likely to occur.

At frequencies below the resonant frequency, the response of a critical component is roughly equal to the input (the response : input ratio is 1). At frequencies higher than the resonant or natural frequency, the response acceleration is lower than the input. In this region a component acts as its own isolator and results in a condition known as *attenuation.*

At (and near) the product resonant frequency, however, the response acceleration can be very much greater than the input, causing product fatigue and ultimate failure in a relatively short time. The purpose of vibration sensitivity assessment is to identify those critical frequencies likely to cause damage to the product.

A resonant frequency search test is run by attaching a product to the table of a suitable vibration test machine (Fig. 6) and subjecting it to a constant acceleration input at a low level (typically 0.25–0.5G) over a suitable frequency range, typically 2–300 Hz (cycles per second). An accelerometer is fastened to a critical component within the product in order to determine the component's response to the input acceleration. The response : input ratio is plotted as a function of frequency. This ratio reaches a maximum at the component resonant or natural frequency. The test usually involves monitoring many components in each axis of the product in order to characterize its overall vibration sensitivities (14).

The importance of vibration teting cannot be overemphasized. Any product shipped from point A to point B is subject to vibration because of the transit vehicle it is riding. The probability of this input is 100%. In contrast, the probability of a shock input because of a drop is exactly that, a probability function. In some cases, the drop height experienced by a product may be severe, in other cases, it is hardly measurable. However, any product that is shipped in a vehicle is subject to vibrational input and it should be tested for sensitivity to that input.

Conclusion

At this point, the engineer has sufficient data to make intelligent decisions about tradeoffs between the product modifications and package costs. If a fragile component can be ruggedized at minimal cost, resulting in substantial package savings, then it makes sense to pursue the product modification.

The package-design process uses environmental data product-fragility information, cushion-performance data (see Foam cushioning; Testing, cushion system), and a healthy dose of designer creativity. Knowledge of package-fabrication techniques is essential, as well as other vital information on flammability restrictions, maximum weight and cube for storage and transportation, recyclability of the package components future cost trends of various key materials, etc. (1,3,5,6,15).

Once the design has been finalized and a prototype fabricated, it must be tested to verify compliance with product requirements. It is important to specify the correct inputs, both their magnitude (to duration) and sequence, in order to closely duplicate the potentially damaging effects of the distribution environment. Recently developed test procedures such as ASTM D4169 (7) hold great promise for improving the correlation between laboratory tests and field experience (2,4,16).

A properly engineered packaged system is now within reach of all manufacturers, distributors, and other package-system users. The wasteful practice of overpackaging can virtually disappear along with damage-in-shipment reports. The tools are available and the technology straightforward. Optimized packaging is indeed an attainable goal.

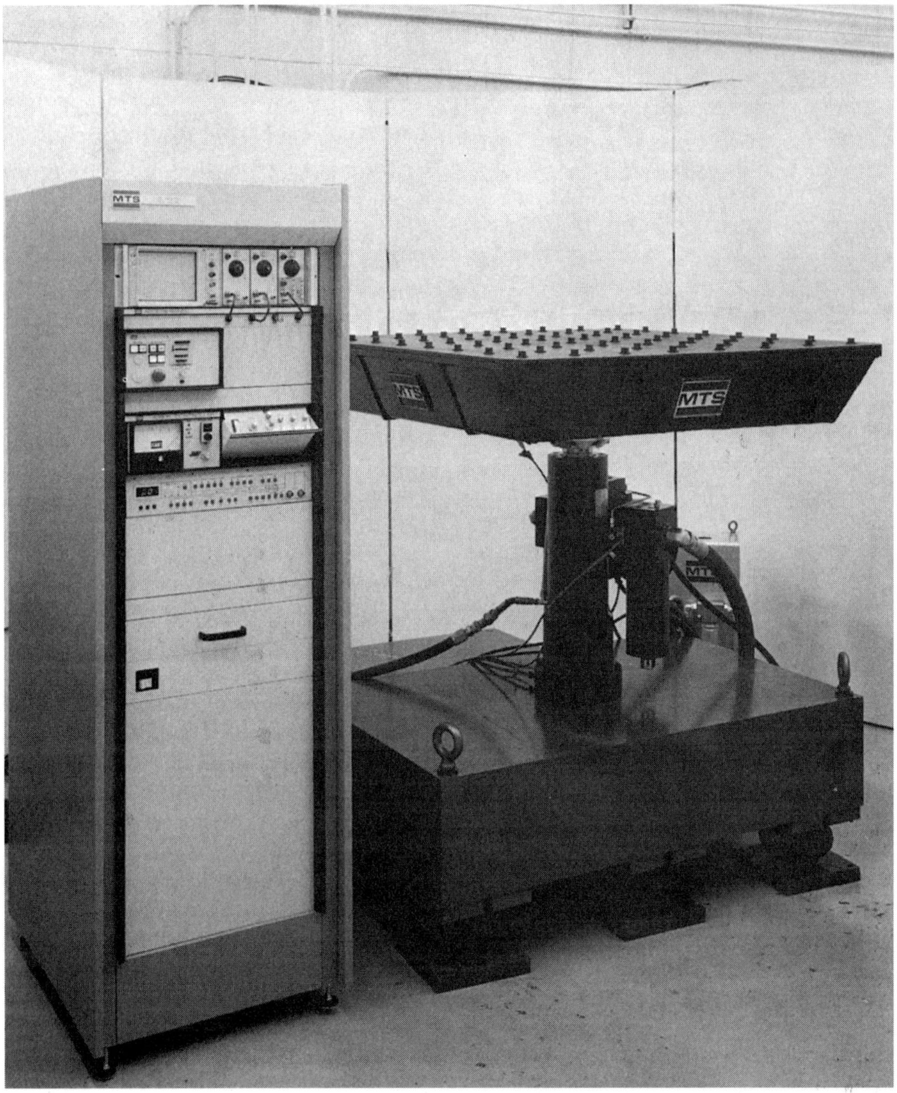

Figure 6. Vibration-test machine.

BIBLIOGRAPHY

1. M. E. Gigliotti, *Design Criteria for Plastic Package Cushioning Materials,* Plastic Technical Evaluation Center, Picatinny Arsenal, Dover, NJ, 1962.
2. T. J. Grabowski, *Design and Evaluation of Packages Containing Cushioned Items Using Peak Acceleration Versus Static Stress Data, Shock, Vibration and Associated Environments Bulletin No. 39,* part II, Office of the Secretary of Defense, Research and Engineering, Washington, DC, 1962.
3. C. Henny and F. Leslie, *An Approach to the Solution of Shock and Vibration Isolation Problems as Applied to Package Cushioning Materials, Shock, Vibration and Associated Environments Bulletin No. 30,* part II, Office of the Secretary of Defense, Research and Engineering, Washington, DC, 1962.
4. M. T. Kerr, "The Importance of package Testing in Today's Data Processing Industry" in *Proceedings of Western Regional Forum Packaging Institute,* USA, 1981.
5. R. D. Mindlin, *Bell System Tech. J.* **24**(304), 352 (1945).
6. S. Mustin, *Theory and Practice of Cushion Design,* Shock and Vibration Information Center, U.S. Department of Defense Washington, DC 1968.
7. ASTM D4169-94, *Standard Practice for Performance Testing of Shipping Containers and Systems,* American Society for Testing and Materials, Philadelphia, 1994.
8. S. G. Guins, in *Shock and Vibration Handbook,* Part II, McGraw-Hill, New York, 1961, Chapter 45.
9. F. E. Ostrem and B. Libovicz, *A Survey of Environmental Conditions Incident to the Transportation of Materials, Report PB-204 442,* General American Transportation Corp., Niles, IL, 1971.
10. R. J. Winne, "What Really Happens to Your Package in Trucks and Trailers" in *Proceedings of Western Regional Forum of the Packaging Institute,* USA, 1977.
11. R. D. Newton, *Fragility Assessment Theory and Test Procedure,* Monterey Research Laboratory, Inc., Monterey, Calif., 1968.
12. D. E. Young and S. R. Pierce, *Development of a Product Protection System, Shock and Vibration Bulletin No. 42,* Office of the Secretary of Defense, Research and Engineering, Washington, DC, 1972.
13. ASTM D3332-93, *Standard Methods for Mechanical-Shock Fragility of Products Using Shock Machines,* American Society for Testing and Materials, 1993.
14. ASTM D3580-80, *Standard Method of Vibration (Vertical Sinusoi-*

dal Motion) Test of Products, American Society for Testing and Materials, Philadelphia, Pa., 1982.

15. J. P. Phillips, "Package Design Consideration for the Distribution Environment," *Proceedings of the 1979 International Packaging Week Assembly, Packaging Institute, Oct. 1979.*
16. J. F. Perry, *A Brief Summary of Dynamic Test Methods for Shipping Containers,* Del Monte Corporation, Walnut Creek, Calif., March, 1982.

General References

C. M. Harris and C. E. Crede, *Shock and Vibration Handbook,* Vols, 1, 2, and 3, McGraw-Hill Book Co., New York, 1961.

C. E. Crede, *Vibration and Shock Isolation,* John Wiley & Sons, Inc., New York, 1951.

J. F. Perry, "Cost Reduction Through Dynamic Testing of Shipping Containers," *Packaging Technology* (June/July 1982).

W. Silver and E. Szymkowiak, "Recommended Shock and Vibration Test for Loose Cargo Transported by Trucks and Railroads," *Proceedings of the Institute of Environmental Sciences,* (1979).

ASTM D 5276-94, *Drop Test for Shipping Containers,* American Society for Testing and Materials, Philadelphia, Pa., 1994.

ASTM D 1596-91, *Method of Testing for Dynamic Properties of Package Cushioning Materials,* American Society for Testing and Materials, Philadelphia, Pa., 1994.

ASTM D 999-96, *Standard Methods for Vibration Testing of Shipping Containers,* American Society for Testing and Materials, Philadelphia, Pa., 1996.

H. H. SCHUENEMAN
Westpak
San Jose, California

TESTING, SHIPPING CONTAINERS

Shipping containers are tested in the laboratory to verify their ability to survive hazards in the distribution environment. Tests may be conducted as a packaging development tool for quantitative purposes, or as a qualitative measure of performance. Objectives may be reduction of damage, cost improvement, assurance of safe product shipment, or other package-design reasons. Laboratory preshipment tests, utilizing proven industry standards, will provide a high level of assurance that a package design will function properly and safely deliver its contents, without cumbersome trial shipments. Such trials require a large number of test packages and significant time delay before an indication of success or failure. They provide no fundamental data, are expensive to conduct, and may severely delay introduction of new products to the marketplace. While some trial shipments are a good idea as a final check, prior laboratory testing is a much quicker indicator of success or failure and can precisely identify causes of inadequacy.

Sources of Test Methods

The oldest and largest developer of packaging test methods, including those for shipping containers, is ASTM (American Society of Testing and Materials). ASTM test methods, which are based on a balanced consensus approval process, have worldwide acceptance. Updated at least every 5 years, their standards for packaging began development in 1914 and now include over 120 test methods, specifications, or practices, which are included in the ASTM *Annual Book of Standards,* Volume 15.09 (1).

The International Safe Transit Association (ISTA) publishes a widely used pre-shipment testing protocol. Their Projects 1 and 1A for domestic shipments and Projects 2 and 2A for overseas (2) utilize ASTM test methods and define levels of test intensities based on weight of the packaged product.

Internationally, the International Organization for Standardization (ISO) develops and maintains test methods for shipping containers that are similar to, but not exactly the same as, those of ASTM. ISO standards may be obtained from the American National Standards Institute (ANSI) in New York City.

Environmental Hazards to Shipping Containers

Shipping containers and interior protective packaging serve the primary purpose of protecting products against environmental hazards occurring in distribution. The best definition of these hazards was set forth in a 1979 report by the Forest Products Laboratory of the USDA (3), and still remains the source of many test methods and levels of intensities. It identifies the following hazards as the major causes of damage in distribution: rough handling, shock and vibration in transit, compressive forces, and temperature and humidity extremes. Experience of many shippers supports the contention that these hazards are still the major culprits, with rough handling clearly the most severe and frequent cause of damage.

Test Methods for Reproducing Shipping Hazards

Each of the major distribution hazards of handling, shock and vibration in transit, and compression can be reproduced by one of several test methods, depending on the type of container, test equipment availability, or purpose of test. The following discussion covers the most widely used ASTM methods.

Drop test. Accidental or purposeful dropping of containers during handling, loading, and unloading is the prime source of damage to most packages under 200 lb. ASTM test method D5276, *Drop Test of Loaded Containers by Free Fall,* describes the equipment and methodology for conducting a laboratory simulation of this hazard. All types of shipping containers may be tested, including boxes, cylindrical containers, bags, and sacks. Accuracy of the drop-test equipment is specified, including a requirement for impact surface mass of at least 50 times that of the heaviest container to be tested. The method may be used to check the ability of the container to survive free fall, evaluate the protective capability of container and inner packing, to compare performance of different package designs, or permit observation of progressive failure of a container and damage to its contents.

A popular free-fall drop-test sequence and level of intensity for packages under 100 lb is the ISTA Project 1A specification. Drop heights begin at 30 in. for packages up to 20 lb, and become progressively lower as weight increases. The test sequence calls for impacts against a bottom corner, with three edges radiating from that corner, and all six flat surfaces of the shipping container.

An alternative to free-fall drop is ASTM test method D5487, *Simulated Drop of Loaded Containers by Shock Ma-*

chine. This method provides more accurate package positioning for impacts, particularly on flat impacts where the shock transmitted to package contents is usually greatest. While the shock machine method has been shown to be quite accurate for most applications, rigid package systems are not recommended for testing by this method (where "rigid" means a natural frequency above 83 Hz for the system).

Incline impact test. For large shipping containers, an alternate method for obtaining an impact is ASTM test method D880, *Impact Testing for Shipping Containers & Systems.* A guided test carriage, tilted at 10°, is loaded with the test specimen on the leading edge and pulled up the incline to a distance sufficient to produce the desired impact velocity. The carriage is then released and travels down the incline until the container strikes the impact surface, also tilted at 10°. When the incline test equipment, dubbed the "Conbur test," is constructed with impact surface at least 50 times the mass of the heaviest test specimen, it is considered equivalent to a free-fall drop test for intensity of shock. ASTM D880 also includes the use of pendulum equipment as an alternative to the incline.

A popular test sequence and level of intensity for the Conbur is the ISTA Project 1 for containers weighing over 100 lb. It specifies an impact velocity of 5.75 ft/s with impacts on all six flat surfaces. For skidded items, the top impact is omitted, but two bottom drops from 8 in. are substituted, either flat or on opposite bottom edges.

The incline impact equipment may also be modified such that it can reproduce shocks similar to railcar switching impacts. ASTM test method D5277, *Performing Programmed Horizontal Impacts Using an Inclined Impact Tester,* describes the required carriage bulkhead, programming material or device, and other modifications that are necessary for simulating standard railcar draft-gear impacts. The method is not appropriate for long-travel draft gears or cushioned railcar underframes.

Horizontal impact test. A more accurate method of reproducing all types of railcar impacts is described in ASTM test method D4003, *Programmable Horizontal Impact Test for Shipping Containers and Systems.* The test equipment specified is a horizontal shock machine with precise control over impact and rebound velocities. Dynamic compression of containers during railcar switching is great, due to backload of other freight pressing against it; therefore, the test method describes how such backload should be simulated during testing. Proper instrumentation and calibration, critical to the method's accuracy, are specified in some detail. ASTM D4003 may also be used for other types of horizontal impact simulation, such as pallet marshalling.

Bridge impact test. Some handling systems place heavy stress on long packages, particularly in small-parcel shipping. ASTM test method D5265, *Bridge Impact Testing,* specifies equipment and methodology for applying impacts to the center of long packages. Two options are included; A, for using a free-fall drop tester; or B, for using a SMITE tester. In both methods, the package is supported on its ends and an impacting device drops onto its center at a specified impact velocity.

Other handling tests. ASTM Standard D1083, *Mechanical Handling of Unitized Loads and Large Shipping Cases & Crates,* contains a dozen test methods particularly suited to reproducing handling hazards for either unitized loads or large containers or both. Included are the following:

Free-fall drop test—utilizes a quick-release hook and sling for equipment to drop test on any container surface.

Rotational edge-drop test—one end of specimen is supported on a block while opposite end is raised to specified height and then dropped to floor.

Corner drop test—one bottom corner is supported on a block while the diagonally opposite bottom corner is raised to specified height and then dropped to floor.

Raised-edge drop test—one end of specimen is raised to specified height while opposite end rests on floor, then raised end is dropped to floor.

Grabhook test—uses grabhooks to determine ability of containers or unit loads to withstand horizontal pressures when grabhooks are applied.

Sling test—uses wire rope, cable, or woven-fiber slings to determine resistance to compression applied by slings.

Rolling test—determines the ability of a container to withstand the effects of rolling.

Tip test—determines the ability of tall or top-heavy containers to resist tipping over.

Tipover test—determines the protective ability of loaded shipping containers when subjected to tipover impacts.

Lift-truck handling tests for unitized loads—a test course and methodology is described for handling tests by four types of lift truck: fork, spade, clamp, and push–pull.

Vibration tests.

Although vibration-induced damage is not a frequent problem for most packaged products, vibration does occur with certainty on every shipment by any mode of transport. It is therefore important to test all types of shipping containers for vibration. Four methods of vibration testing are available, depending on equipment and test objective.

ASTM test method D4728, *Random Vibration Testing of Shipping Containers,* provides the closest simulation to actual vibration received by containers in transit. Input for the test may be generated from field measurements made with accurate recording instruments or from use of the sample vibration spectra for commercial transport that is included in the standard. Test durations frequently run as long as 3 h or more, depending on the packaged product and objective of the test. Resonance buildups during random vibration are less intense than by sinusoidal vibration methods, and unrealistic fatigue damage is therefore minimized.

ASTM Standard D999, *Vibration Testing of Shipping Containers,* actually contains four distinct test methods and utilizes two entirely different types of vibration equipment. Method A is a repetitive shock test, not really vibration in the pure sense, and is divided into A1 for rotary motion and A2 for vertical motion. Both tests are conducted with the package free to bounce on the table, which is running at a frequency causing the package to just lift off during the upstroke of each vibration cycle (known as $1G+$). The operating frequency and table displacement are constant throughout the test, which is usually performed on a mechanical-type vibrator. Typical test duration is about one hour with smaller, nonskidded containers tested one-third of the duration on each axis. The two methods do not create the same type of motion; therefore, they also do not cause the same types of damage.

Methods B and C of D 999 are resonance search and dwell tests, typically performed on electrohydraulic equipment that easily performs at 3–200 Hz as required for most packaging tests. The table displacement varies inversely with the frequency so as to maintain a constant level of force, usually 0.25–1*G*. The purpose of the test is to locate natural frequencies of the packaged product and then dwell for a specified time at each natural frequency while the package resonates. Procedures are described for sweeping the vibration frequency up and down to locate the natural frequency points. Method B is conducted on single containers that are fastened to the table during testing. Method C is conducted on stacks of containers or unitized loads, with no fastening to the table. Typical dwell times at resonance for both methods is 5–15 min for each point, although some users dwell for up to one hour.

Compression tests. The effects of stacking one container on another in storage or transportation can be studied by means of the compression test. Two ASTM test methods are available: D642, *Determining Compressive Resistance of Shipping Containers* and D4577, *Compression Resistance of a Container Under Constant Load.* The former requires a testing machine that applies a steadily increasing force to the container until it fails. A recording device notes applied load and container deflection during testing that generally requires ≤2 min for completion. ASTM test method D4577 may use the same machine, but more typically utilizes a simple stationary apparatus that positions the test specimen under a fixed constant load for the desired time, anywhere from one hour to one year. Both methods may be used on filled shipping containers or on empty ones, depending on test objective.

Water-vapor transmission test. To measure the resistance of shipping containers to water vapor, ASTM D4279, *Water Vapor Transmission of Shipping containers,* provides two test methods. The constant-atmosphere method is conducted at 90% rh and 100°F. The cycle method goes up and down, from 0°F to 100°F., 90% rh and back, for a number of specified cycles. Different procedures are defined for reclosable and nonreclosable containers.

Table 1. Special Atmospheres

Environment	Temperature, °F (°C)	Relative Humidity (rh), %
Cryogenic	−67 ± 6 (−55 ± 3)	
Frozen-food storage	0 ± 4 (−18 ± 2)	
Refrigerated storage	41 ± 4 (5 ± 2)	85 ± 5
Temperate, high-humidity	68 ± 4 (20 ± 2)	85 ± 5
Tropical	104 ± 4 (40 ± 2)	85 ± 5
Desert	140 ± 6 (60 ± 3)	15 ± 2

Water-resistance test. ASTM test method D951, *Water Resistance of Shipping Containers by Spray Method,* determines how well a filled container will resist water spray such as produced by heavy rains or waves over the deck of a ship. The test is frequently conducted in a series with other tests such as drop or incline impact.

Conditioning for Testing

A requirement in all ASTM shipping container test methods is for testing at a specified condition of temperature and humidity. A variety of atmospheric conditions are described in D4332, *Conditioning Containers, Packages, or Packaging Components for Testing.* This standard practice provides special conditions that may be used to simulate particular field conditions a container may encounter during distribution. The standard also describes procedures for preconditioning the containers to ensure that they reach equilibrium in the atmosphere before they are exposed to the final test conditions. Following preconditioning, containers are placed in the conditioning chamber for a specified time, ordinarily at least 72 h, to reach equilibrium at test conditions. Most testing is conducted at the U.S. standard conditioning atmosphere of 73°F and 50% rh, but other test atmospheres may also be used, as listed in Table 1.

Performance Testing

The test methods described to this point may be used in several ways by package designers. Most frequent use is as an

Table 2. Hazard Elements and Corresponding Tests

Code	Hazard Element	Test Simulation of Hazard	ASTM Designation
A	Manual handling	Drop	D5276
B	Mechanical handling	Drop, stability	D1083
C	Warehouse stacking	Compression	D642
D	Vehicle stacking	Compression	D642
E	Truck and rail stacked or unitized vibration	Vibration	D4728, D999
F	Loose-load vibration	Vibration	D999
G	Vehicle vibration	Vibration	D4728, D999
H	Rail switching	Longitudinal shock	D4003, D5277
I	Climate, atmospheric conditions	Temperature, moisture humidity	D4332
J	Military environment	Cyclical exposure	D4169

Table 3. Performance Test Sequence by Distribution Cycle

Distribution Cycle (DC)	DC Hazard Element Number	DC Hazard Element Sequence
General schedule—undefined distribution system	1	A/B, D, E, F, H, A/B
Special—controlled environment, user-specified	2	User-specified
Single-package environment, ≤100 lb	3	A, D, F, G, A
Motor freight		
Single package >100 lb	4	B, D, F, G, B
Truckload, not unitized	5	A/B, D, E, G, A/B
Truckload or LTL, unitized	6	B, D, E, B, C
Rail, carload		
Bulk loaded	7	A, D, E, H, A
Unitized	8	B, D, E, H, B, C
Rail and motor freight		
Not unitized	9	A/B, D, G, H, F, A/B
Unitized	10	B, D, E, H, B, C
Trailer-on-flatcar, container-on-flatcar	11	A/B, D, H, E, F, A/B
Air and motor freight		
>100 lb	12	A/B, D, E, G, A/B
≤100 lb	13	A, D, F, G, A
Warehousing (partial cycle)	14	A/B, C
Export/import shipment		
By intermodal container or roll on/roll off equipment (partial cycle)	15	I, B, D, B
For palletized cargo ship (partial cycle)	16	I, A/B, D, A/B
For breakbulk cargo ship (partial cycle)	17	I, A, D, A
Government shipments	18	A/B, C/D, A/B, J, F, H, A/B

engineering development tool to ascertain how well design and materials are doing their job of protecting the product. But the same methods may also be utilized to indicate whether the package will *perform* as expected for the total distribution cycle. Whereas engineering development testing often overstresses the package to determine its limits when subjected to a particular hazard, the performance use of the same methods is a pass/fail, go/no-go situation with the same container going unopened through the full set of tests. Development testing is a quantitative measure, while performance testing is qualitative.

ASTM D4169, *Standard Practice for Performance Testing of Shipping Containers and Systems,* has received acceptance as the national standard for measuring expected shipping-container performance. It is essentially a matrix of test methods, distribution modes, and levels of test intensities. Putting them together in a sequence that simulates a particular distribution environment produces a "distribution cycle" (DC) containing certain hazards (tests). If the shipping container and contents survive the DC, then it can be assumed that they will also survive in actual handling and shipping by the same method of distribution (DC). ASTM D4169 also requires that the criteria of acceptance (definition of success or failure) be predetermined and documented prior to testing.

Each hazard in distribution is identified in D4169 with a test method to reproduce it. Table 2 lists the nine hazard elements recognized in the performance standard, along with the corresponding test methods.

Presently there are 18 distribution cycles in D4169, each containing a sequence of hazards expected by that particular distribution method. Table 3 is a listing of the cycles with hazards noted by their code letters from Table 2.

For those packaging engineers with a good knowledge of their distribution environment(s), DC 2 has become the performance test plan of choice. It permits them to develop a test plan that fits the often unique sequence of hazards that their packages encounter in distribution.

Regulatory Agency Use of Container Testing

Shipping-container performance requirements, as a substitute for material or design specifications in regulatory documents, has been slowly making progress. The U.S. Department of Transportation (DOT) now requires that many hazardous materials be packaged in conformance to a brief set of performance-test specifications rather than to cumbersome and complex material and design specifications previously required (4). The National Motor Freight Classification in 1995 authorized an alternate rule, Item 180, performance requirements for LTL shipments, which can be utilized at the shipper's discretion instead of other material- and design-based rules or special package numbers (5).

BIBLIOGRAPHY

1. *Annual Book of Standards, Volume 15.09,* American Society of Testing and Materials, Philadelphia, 1995.
2. *Pre-shipment Test Procedures,* International Safe Transit Association, Chicago, 1990.
3. F. E. Ostrem and W. D. Godshall, *An Assessment of the Common Carrier Shipping Environment,* General Technical Report FPL 22, U.S. Forest Products Laboratory, USDA, Madison, WI, 1979.
4. *Docket HM-181,* U.S. Department of Transportation, Office of Hazardous Materials Safety, Washington, DC, 1993.
5. *National Motor Freight Classification,* American Trucking Association, Alexandria, VA, 1995.

Alfred H. McKinlay
Packaging Consultant
Pattersonville, New York

THERMOFORM/FILL/SEAL

Thermoform/fill/seal equipment is used for a variety of food and nonfood packaging applications. Thermoform/fill/seal (tffs) machines use tow continuous webs or rolls of film (see Fig. 1). Typically the lower web is formed into a cup, which is then filled with product. The upper web becomes the lid. Although over 90% of all tffs machines are run with formed bottom webs and unformed lidstock, it should be mentioned that this is not always the case. Sometimes both webs are formed; sometimes the top web is formed, and the bottom one is not.

Packaging Applications

Tffs machines were originally designed as vacuum-packaging machines to package ham, bacon, or sausage. From cured meats, their use spread to other sectors of the meat industry (frozen steaks, certain fresh meats, cook-in-package delicatessen meats), and then to cheese products, baked goods, fresh pasta, and a variety of perishable food products requiring extended shelf life.

Nonfood applications of tffs evolved from the original vacuum-packaging machines for sausage and meat products. These are primarily sterile-medical applications (see Healthcare packaging). Typical medical items packed on tffs equipment are syringes, needles, catheters, kidney-dialysis filters, scrub sponges, surgical drapes and clothing, operating-room kits, and many other items. Although medical tffs machines are structurally and conceptually the same as tffs machines used in the food industry, the materials tend to be more difficult to run and the quality standards for cutting and thermoforming are more difficult to meet. Among the tffs applications that are not food or medical, are hardware items, cigarette lighters, cosmetics, and similar applications where packages made on continuous-web machinery are replacing preformed plastic "blisters" (see Carded packaging).

Food packaging. In food applications, the packages are almost always hermetically sealed, and the films have varying degrees of barrier properties. The primary application for vacuum packaging is to improve shelf life of foods that are sensitive to oxygen or susceptible to dehydration. Almost all cured-meat products (ham, bacon, sausage, corn beef, beef jerky, etc) are vacuum-packaged, as are many cheeses. An off-shoot of vacuum packaging is controlled gas-atmosphere packaging (see Controlled-atmosphere packaging). A controlled-atmosphere package produced on tffs equipment is a step beyond vacuum packaging. First, the unsealed package is indexed automatically into a vacuum chamber. Then all the air is removed from the package. Once the air is removed, the desired gas mixture is injected into the package, usually until the pressure of gas inside the package reaches about one atmosphere. Table 1 lists some commonly used gas mixtures and their application.

Vacuum and controlled-atmosphere packages are generally used to protect refrigerated perishable foods that would otherwise spoil in less than a week. In a controlled atmosphere, they may stay fresh for 4–6 weeks or longer. Many frozen foods are also packed on tffs machinery, and a controlled atmosphere can sometimes replace freezing (see Fig. 2). Some nonperishable foods with shelf life of over one year can be packaged on tffs equipment. In Europe, for example, shelf-stable heat-processed vegetables are marketed in retorted thermoformed packages. Thermoforming machines can be modified to mechanically cold-form aluminum foil for retort processing of various meat and nonmeat dishes. The taste, texture, and nutritional quality of food in retort-sterilized aluminum foil pouches is generally higher than that of conventionally canned foods (see Retortable flexible and semirigid packages).

Table 1. Gases Used in Packages Produced on Tffs Equipment

Gas	Application
N_2	As a pressure-relief agent to prevent external atmosphere from crushing the product. N_2 is an inert gas. It does not react with either the food substances or bacteria. Common applications: bulk-pack bacon and sausage, shredded and sliced cheese, and beef jerky.
CO_2	Depending on the application, 25–100% CO_2 may be used. CO_2 lowers the pH of the food product and can exert a powerful slowing effect on the growth of bacteria and molds. Primary application is for baked goods, cookies, cakes, breads, as well as dough and pasta products. CO_2 tends to be absorbed into the actual body of the food product itself. Accordingly, CO_2 is frequently mixed with N_2 to prevent the package from clinging too tightly to the product. On the other hand, some products that are not sensitive to strong pressure or tight cling, but that are susceptible to spoilage by mold growth, are packed in an atmosphere of 100% CO_2. An example of this is chunk cheese. Many CO_2 gas packages have the appearance of a vacuum package, with much of the CO_2 absorbed into the product itself.
O_2	The applications using O_2 are primarily for red meat. Still experimental in the United States, the concept is widely used in Europe for centrally packed retail cuts. The O_2 is used as an oxygenating agent, at levels in the range of 40–80%, to form the bright-red, fresh-meat color called oxymyoglobin. When O_2 is used, it is usually mixed with CO_2 and N_2: CO_2 for its preservation effect; N_2 to provide a bulking agent. The O_2 tends to disappear inside the package. It can be metabolized by the meat to CO_2 that is absorbed in the water phase of the meat as carbonic acid.

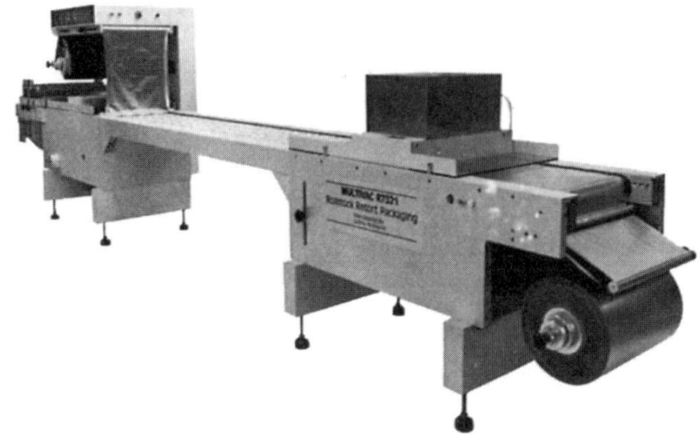

Figure 1. Thermoform/fill/seal machine. Shown with lower web forming station in foreground. Note extensive product-loading area.

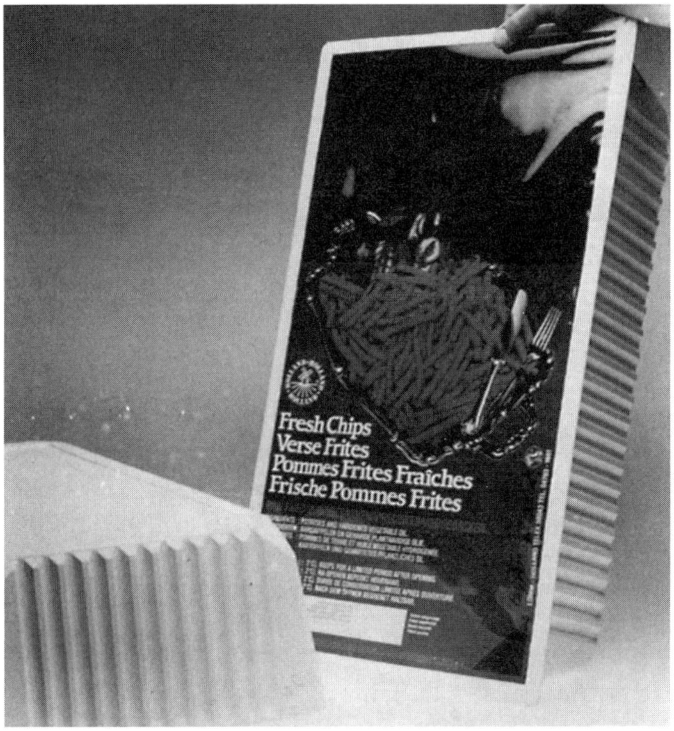

Figure 2. Controlled-gas atmosphere package for perishable foods. Fresh (unfrozen), precooked french fries (10 kg) packed in N_2 atmosphere. Positive (male) formed PVC/PE bottom, polyester/PE cover. Print registration of flexible upper web by means of stretching with a brake.

Tffs Machines

The essence of tffs machines is their modularity. Within any given machine designation an almost infinite number of configurations is possible. There are six basic operations that make up the production of a package on tffs equipment: film advance, thermoforming, loading, sealing, cutting, and labeling/printing (see Fig. 3).

Film advance. Tffs machines are built like miniature assembly lines with separate stations for different operations. There are several film-advance mechanisms for moving the webs through these stations. The drive mechanisms (ie, AC Digital, DC Digital, Geneva, Elliptical Gear, and Planetary Gear) differ as to accuracy of advance, acceleration curve, or suitability for print registration of stretchable or nonstretchable materials.

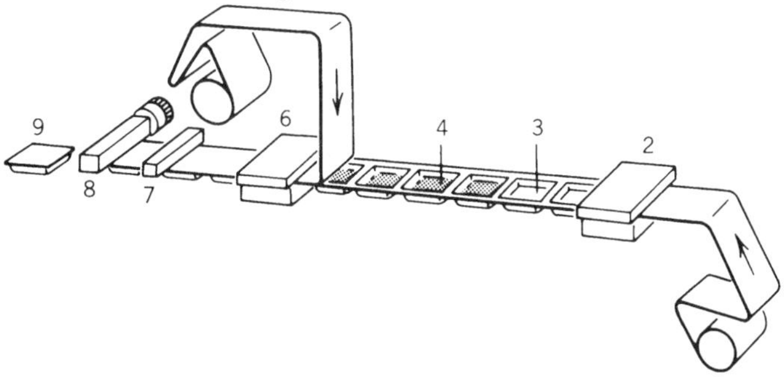

Figure 3. A schematic of the simplest and most basic configuration of a tffs machine as would be used for flexible, not semirigid, upper and lower webs: 1, Lower or forming web; 2, thermoforming die; 3, formed, unfilled pocket; 4, filled pocket; 5, upper or lidding web; 6, vacuum/sealing die; 7, flying knife for across-the-machine direction package cutoff; 8, high speed rotary knives for package cutoff in the machine direction; and 9, finished package (rectangular).

Thermoforming. This is the first step in the formation of a package. Thermoforming (see Thermoforming) may involve heating in the forming die or preheating in a separate station. There are two main ways to thermoform: positive (male) and negative (female). Female forming involves using compressed air or vacuum to pull the heated, softened film into a mold. It is this mold or concave surface that is the die that produces package shape and surface detail. A mechanical plug may help push the film down into the cavity. In this case the designation "plug-assist forming" applies.

Positive (male) forming is the reverse. Positive forming usually permits the production of a package with greater surface detail and more-uniform wall thickness. A male die part operated by a piston provides package shape and surface detail. Air outside the film or vacuum on the inside, or both air and vacuum, force the preheated film to drape itself onto the exterior of the plug. Positive forming requires a separate station or two separate stations to preheat the film. A typical package requires thermoforming of only the lower web, but the machines can also form both lower and upper. Usually each product to be packaged requires its own separate forming die. Major machine manufacturers have designed and built thousands of different forming dies.

Loading. The lower web emerges from the forming station with the pocket formed and ready to accept product. The loading area must be long enough to accommodate the people required for loading and inspecting, or automatic loading equipment.

Sealing. Once the pockets on the web are loaded, the upper web is indexed over them and the two webs are sealed (see Sealing, heat). In food packaging, evacuation and, if needed,

Table 2. Cutting Methods for Rigid Materials

In-the-Machine Direction	Across-the-Machine Direction
Squeezing knives	Steel rule dies with rounded corners
Shear cut knives with or without strip removal	Strip removal for rounded corners by means of matched male and female dies
Complete 360° cut by means of matched male/female dies	

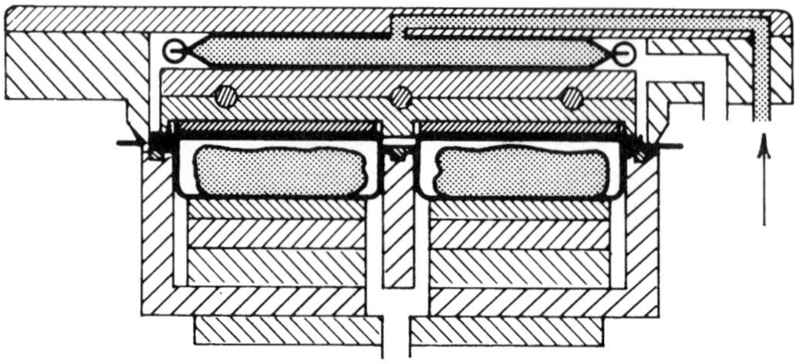

Figure 5. Heat sealing. After the package is evacuated, an air bladder lowers the sealing bar and the package perimeter is clamped shut and heat-sealed. In the next step, air is readmitted into the die, the die opens, and the package is indexed out.

gas injection are performed at the sealing station. In order to provide a more attractive package, sometimes all surfaces of the top and bottom webs not in contact with the product are sealed to each other. Sometimes, using a patented process, steam is injected into the sealing die to shrink the lower web (forming web) around the product to form an attractive skin-tight package.

Cutting. Cutting systems range from the simple and inexpensive to the complex and costly: paper-cutting-type knives (ie, flying knives and high-speed rotary knives); shear cutting (machine direction) with cross-direction strip removal by means of matched male and female dies; shear cutting (machine direction) and cross-direction cutting by means of steel rule dies; and complete package cutting with matched male and female dies (Table 2).

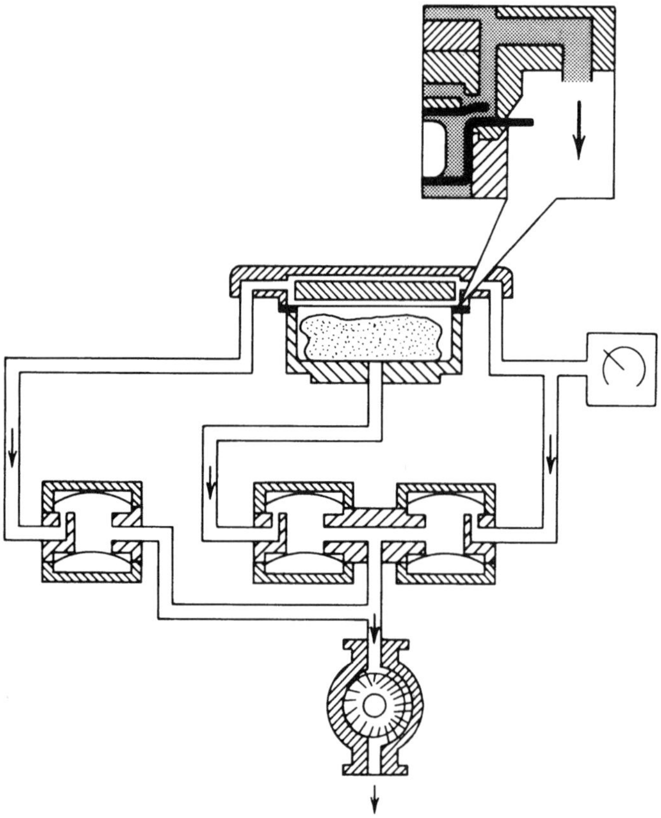

Figure 4. Evacuation. This shows the die after the first or outer box has closed. Vacuum is drawn inside the die chamber. Because vacuum is drawn on the outside of the package from both above and below, air pressure inside the package forces the two webs to separate, providing a large opening for air to flow out of the package. During evacuation, the travel of air from the inside of the package is through the gap between the narrow upper web and the wide lower web. Package air exits along with air from above the upper web. In this figure, the seal bar is in its upper, or retracted, position so that air can leave the package.

Labeling and printing. On-line printing may consist of simply printing, or embossing, a code date or lot number (see Code marking and imprinting), or may consist of printing an entire index, actually "bleeding" finished package edges. Labels may be applied to either surface of either web.

Vacuum and Sealing-Die Operation

At this station the package is evacuated and the two webs are sealed together. In order to understand how this die operates, it helps to think of the die as consisting of two nesting, concentric boxes that close in sequence. The outer box closes first (see Fig. 4). It mechanically seals off that portion of the package represented by the lower web. The upper web is narrower than the lower web and is not sealed at this stage. Therefore, until the inner box closes, atmosphere is free to enter or leave the package under the lip of the upper web. The inner box is made up of the seal mechanism. After the entire die chamber, including the interior of the package, is evacuated, the inner box (the seal mechanism) closes on both webs, first sealing them mechanically, so no air can enter or leave the package, then applying a permanent heat-seal (see Fig. 5). The key operation, package evacuation, takes place when air flows out of the package through the gap between the narrow upper web and the outer wall of the die chamber. This occurs during the interval after the outer box has sealed off the lower web, but before the inner box has closed off the package itself.

This is one of the most common methods of package evacuation. If gas injection is required for controlled-atmosphere packaging, the operation takes place in the sealing die.

Materials

The detailed design of a tffs machine depends on the packaging materials to be run. Therefore, any discussion of how tffs machines operate must refer to materials as well. The two webs consist of a bottom web and a top web on the type of equipment generally known as horizontal thermoform/fill/seal machines. In the usual case, the bottom web is thermoformable and is formed by heat into a cup or cavity, which forms a receptacle for the articles to be packaged. However, all variations are possible:

Top Web	Bottom Web
Nonformable	Formable
Nonformable	Nonformable
Formable	Formable
Formable	Nonformable

When food products are packaged, the webs are usually made of multilayer materials selected for their barrier and strength properties. Typically, a barrier material protects against transmission of water and oxygen into or out of the package. In the medical industry, however, products are typically sterilized by ETO gas. In this case, high gas transmission is desired. As a result, the tffs machines use an upper web (also called lidding material) (see Lidding) selected for high-gas transmission properties as well as strength. Typical high-gas porosity materials include paper and Tyvek (DuPont Company), a tough, spunbonded polyolefin. Tables 3 and 4 list some commonly used materials in tffs applications.

Flexible materials such as nylon (see Fig. 6) can usually be formed by heating and pressure-forming into a female die part. Cutting is usually by means of a flying knife in the cross-direction and high-speed rotary knives in the machine direction. More severe draw ratios call for a preheat station and plug assist. Nonrectangular shapes, such as circles and ovals, can be cut with a flying knife traveling on a cam.

Rigid and semirigid materials (Fig. 7) usually require cost-

Table 3. Common Food-Packaging Webs[a]

Structure	Comments
Nonforming web	
Polyester/polyethylene or polyester/ionomer	Polyester is used for printability, strength, and general resistance to moisture and abrasion. Normally it is reverse (capture) printed (ie, the polyester is printed on its inside surface for protection of the printing). Printed polyester is normally stretch-registered.
These structures may contain an intermediate layer of PVDC or EVOH as oxygen barrier.	
Forming web	
Flexible	
Usually a nylon-base web is used for strength and formability, combined with PE or ionomer as heat sealant and moisture barrier, and, if needed, PVDC or EVOH as oxygen barrier.	Forming of flexible web by compressed air, vacuum, or compressed air and vacuum possibly has a heated plug for severe draw ratios. Cutting by means of flying knife in the cross-direction (XD), and high-speed rotary knives in the machine direction (MD) for rectangular packages. Circular, oval, or shape cutting by means of flying knife following a pattern or cam.
Rigid	
Semirigid PVC, acrylonitrile, polyester, or HIPS, combined with PE or ionomer for moisture barrier and if needed, PVDC or EVOH as oxygen barrier.	Forming of rigid web by either negative (female) forming with plug assist or positive (male) forming for more uniform forming and surface definition. Cutting by steel-rule die, or matched male and female die and strip removal across-the-machine direction with shear, cut strip removal in-machine direction. Complete cut by means of matched male and female dies shaped to the final package size.

[a] See Ionomers; Film, oriented polyester; Nylon; Nitrile polymers; Polyesters, thermoplastic; Polystyrene; Vinylidene chloride copolymers; Ethylene–vinyl alcohol.

Table 4. Common Medical-Packaging Webs

Structure	Comments
Nonforming web	
Tyvek (spunbonded polyolefin) or paper used for wet strength and resistance to puncture, also exhibits minimal fiber generation when cut. Paper is used for superior printability and lower cost.	Tyvek can be stretched registered (by means of photocell and brake). Paper cannot be stretch-registered, so it needs to be registered by controlling the advance with either an ac or dc servo drive. Stretch or advance registration.
Both Tyvek and paper are usually heat-seal coated. Used for ETO gas sterilization because of their porosity. Polyester or polyester/aluminum foil for gamma sterilization or gas or moisture barrier. These materials are used in conjunction with appropriate heat seal coatings and laminating adhesives.	
Forming web	
Flexible forming webs: polyolefin laminates or blends.	Forming methods are similar to those used in flexible food packaging. Shear cutting is normally used in both directions for superior cleanliness of cut with paper or Tyvek upper webs.
Rigid forming webs: PVC or acrylic multipolymer or high impact polystyrene (HIPS) or copolyester.	Forming is typically with temperature controlled plug or positive forming depending on draw ratio and degree of surface definition required. Cutting, typically, matched male-female in across-the-machine direction. Shear cut in-the-machine direction. Or 100% matched male-female cutting. These cutting systems provide relatively particulate-free packages with radiused corners.

Figure 6. Sliced luncheon-meat package. Nylon-ionomer flexible forming web, OPET/ionomer top web. Circle cutting by means of flying knife. Print registration by means of stretching film with a brake.

lier and more-elaborate forming and cutting methods. Rigid films usually call for one or more preheat stations and plug-assist forming. If good surface definition and uniform wall thickness is required, positive (male-plug) forming is usually required.

In order to avoid sharp corners, rigid materials must be cut by one of the methods shown in Table 2.

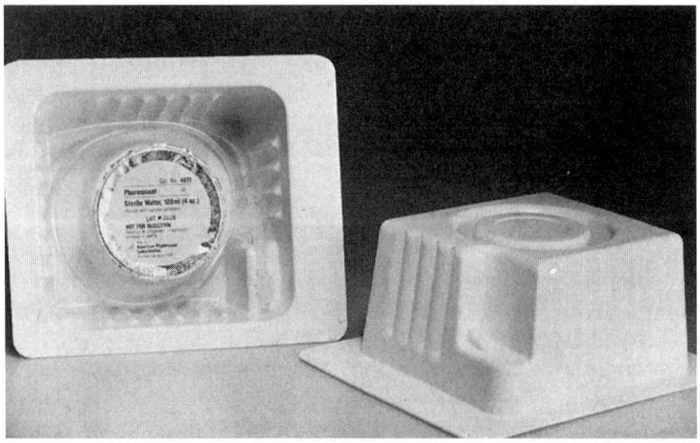

Figure 7. Thermoformed medical-kit package. Typically this type of package would have a semirigid HIPS (high-impact polystyrene) thermoformed bottom and a Tyvek or paper lid. Forming would be positive (male-plug). Note both severe draw ratio and enhanced surface detail. Print registration would be by means of a dc drive since paper cannot be registered by stretching. Cutting could be a complete 360° cut with matched male/female dies, or shear cut (machine direction) with strip removal by means of matched male/female dies (cross-direction). Package is shown with lid removed.

BIBLIOGRAPHY

"Thermoform/Fill/Seal" in *The Wiley Encyclopedia of Packaging Technology* 1st ed., by L. D. Starr, Koch Supplies Inc., pp. 664–668.

General References

Modern Plastics Encyclopedia, McGraw Hill, New York, published annually.

The Packaging Encyclopedia, Cahners Publishing, Denver, published annually.

J. M. Ramsbotton, "Packaging" in J. F. Price and B. S. Schweigert, eds., *The Science of Meat and Meat Products,* 2nd ed., Freeman, San Francisco, 1960, pp. 513–537.

THERMOFORMING

Thermoforming is the means of shaping thermoplastic sheet into a product through the application of heat and pressure. In most cases, the heat-softened plastic assumes the shape by being forced against the mold until it cools and sets up. Forming pressure may be developed by vacuum (atmospheric pressure), positive air pressure, or by mating matched molds.

The word *sheet* as it relates to thermoforming is used to describe flat extruded plastic material that is generally relatively heavy, in contrast to comparatively thin plastic films. It is ambiguous, however, in two ways: thermoforming is also used to shape relatively thin films, and the word sheet is also used to distinguish between thermoforming machines that form cut pieces (sheet-fed) rather than as extruded webs (web-fed). The distinction between sheet-fed and web-fed machines is similar to that between sheet-fed and coil-fed equipment in the metal-can industry, except that in plastics forming, the wide web can comes from either coiled stock or directly from the extruder.

Thermoforming is an important thermoplastic fabrication process that began in the early 1960s when containers and lids formed from high-impact polystyrene (HIPS) were first used by the dairy industry to package cottage cheese, sour cream, yogurt, and other dairy foods. Roll-fed thermoforming machinery is available that can produce beverage cups at rates of 75,000–100,000 pieces/h while consuming plastic sheet in excess of 1 ton/h (910 kg/h). Sheet consumption for larger and heavier containers can exceed 2 tons/h (1.8 metric ton/h), but unit production rates may be lower because fewer mold cavities can be mounted in the machines and more time is required to cool the thicker walls.

The ability to produce extruded HIPS foam sheet (see Foam, extruded polystyrene) has provided additional packaging markets for thermoforming. The first of these was meat and produce trays, and later egg cartons. In the United States, approximately 90% of the meat and produce tray market is served by the foam trays and about 60% of consumer-market eggs are packaged in thermoforming foam cartons. Other applications include fast-food carryout cartons, institutional dinnerware medical trays and blisters, and inserts for rigid boxes.

Process Steps

The thermoforming process is used to make products from thermoplastic sheet by a sequence of heating, shaping, cooling, and trimming. Trimming is not necessarily an integral part the forming cycle, but few applications can use the formed web without some kind of trimming.

Heating. Thermoplastic sheet is typically heated to a temperature range adequate for forming, usually 285–325°F (141–163°C), depending on the material used. Temperature control is critical because of plastic's poor thermoconductivity and because temperature affects the forming characteristics, ie, ductility, of the materials: too much heat and the sheet flows without drawing; too little and it ruptures early in the forming process.

Most thermoplastics absorb infrared energy emitted in the 3.0–3.5-μm wavelength range, which makes them ideally suited to heating by radiation. Heating time is governed by the heating process used, surface conditions of the material, and the combination of low thermoconductivity and relatively high specific-heat capacity. Any of the common transfer processes such as convection, conduction, or radiation can be used. Convection heating would be ideal because the sheet could be soaked in hot air at just the right temperature with the assurance that the entire sheet would be uniformly heated. However, straight convection systems are not practical because they present materials-handling problems relative to transferring the sheet from the oven to the forming station. Therefore, most thermoforming machines are equipped with systems that take advantage of both radiation and convection. Most of these ovens are electrically heated, but there are also a large number of cut-sheet machines using gas-fired ovens. Most are capable of heating the sheet from two sides, which is especially advantageous if the sheet is ≥40 mil (≥1 mm). The ovens should also be appropriately zoned and temperature-regulated by instrumentation enabling the operator to maintain good control over the heating process.

Today any number of electrically powered heating systems can be specified, including the conventional tubular steel rods, glass or ceramic panels, quartz lamps, emitter-strip panels, and small ceramic modules. The most common continues to be the calrod type, but use of rectangular ceramic elements has been steadily increasing. Their relatively small size permits them to be incorporated into elaborate microprocessor-control systems to provide preferential heating in localized areas. Emitter strip panels have been introduced as an alternative to provide temperature uniformity with a loss-complex installation. They are gaining popularity because they provide a full-area heat source and can be mounted closer to the sheet and operated at lower temperatures, thereby conserving energy.

Conduction or contact heating, sometimes called trapped-sheet heating, has been used successfully with materials such as oriented polystyrene (OPS) and PVC. Here, the plastic sheet is trapped between the mold and the temperature-controlled hot plate, which is perforated with extremely small holes, ie, ~0.020-in. (0.5-mm) dia. The holes are drilled in a grid pattern on ≤1 in. (2.5-cm) centers. After the sheet has been in contact with the heated plate for a predetermined time, compressed air is injected through the plate, forcing plastic off the heated platen and into the mold. This system is generally reserved for materials ≤0.015 in. (0.38 mm) thick.

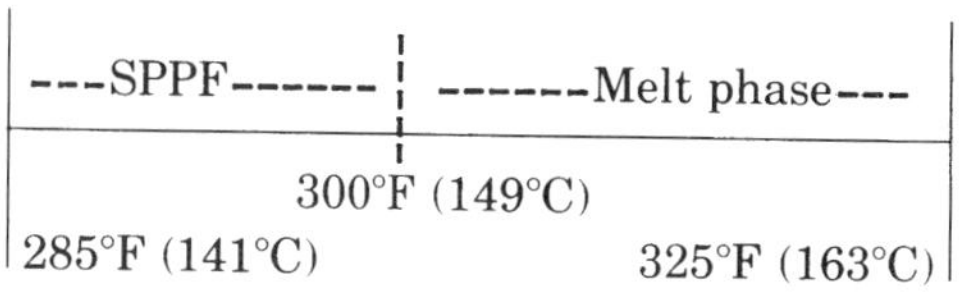

Figure 1. Thermoforming window.

Solid-phase and conventional thermoforming. Solid-phase pressure forming (SPPF) means forming below the crystalline melting point, ie, 5–8% lower than melt-phase forming, depending on the material. For example, polypropylene (PP) is formed in the melt phase at 310–315°F (154–157°C) but at 285–295°F (141–146°C) in SPPF. SPPF does not require special thermoforming equipment. A thermoforming machine forms products within the thermoforming window, and product-configuration requirements, ie, stress, strength, rigidity, flexibility, determine the proper temperature range.

Figure 1 indicates that SPPF is done in a temperature range within the window, as in melt-phase forming. Compared with products formed in the melt-phase, SPPF can pro-

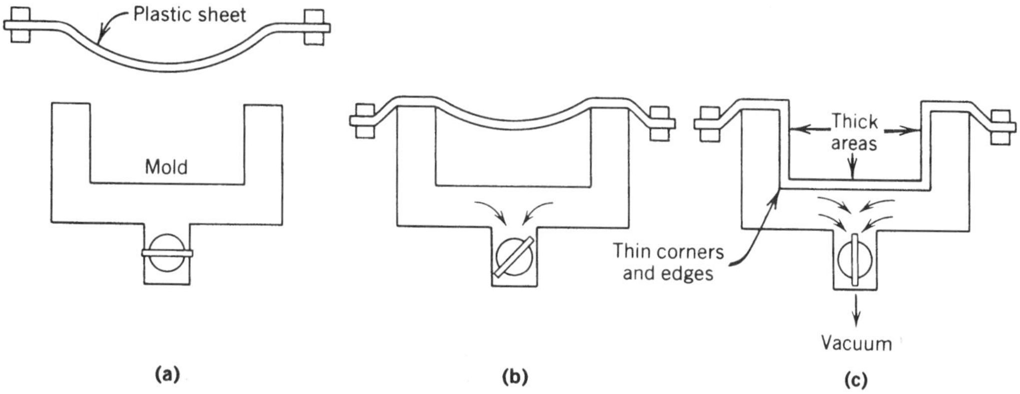

Figure 2. Shallow-draw forming, female-mold bottom; (**a**) the plastic sheet is clamped and heated to th required forming temperature; (**b**) a vacuum is applied through the mold, causing the plastic sheet to be pushed down by atmospheric pressure—contact with the mold cools the newly formed plastic part; (**c**) areas of the sheet reaching the mold first are the thickest. Applications: ice-chest lids, tub enclosures, etc.

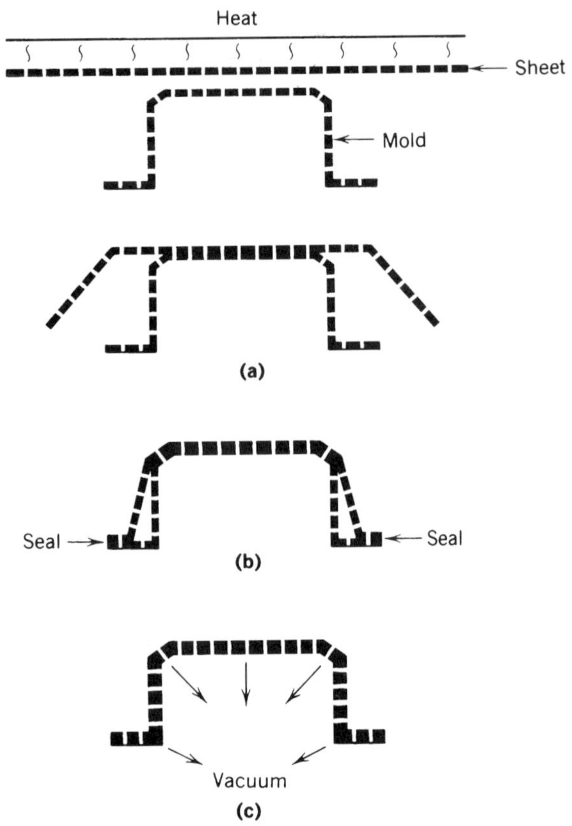

Figure 3. Vacuum (drape) forming; (**a**) the plastic sheet is clamped and heated to the required forming temperature; (**b**) the sheet is sealed over the male mold; (**c**) vacuum from beneath the mold is applied, forcing the sheet over the male mold and forming the sheet. Material distribution is not uniform throughout the part, as the portion of the sheet touching the mold remains nearly the original thickness. The walls are formed from the plastic sheet between the top edges of the mold and the bottom-seal area at the base. Applications: trays, tubs, etc.

duce stiffer parts with less materials. Solid-phase forming improves the sidewall strength of a container and increases the stress factor. When forming cups, trimming in the mold is recommended. Plug assist and high forming pressure, ie, 100 psi (689 kPa), are required, and the plug design is more critical than in conventional melt-phase forming.

Forming. There are two basic forming methods from which all others are derived: drape forming over a positive (male) mold, and forming into a cavity (female) mold. Product configuration, stress and strength requirements, and material specifications all play a part in determining the process technique. Generally, forming into a female mold is used if the draw is relatively deep, eg, cups. Female molds generally provide better material distribution and faster cooling than male molds. Male-mold forming is preferred for certain product configurations, however, particularly if product tolerances on the inside of a part are critical. Male-mold forming produces heavier bottom strength; female-mold forming produces heavier lip or perimeter strength. An advantage of straight forming into a female mold is that parts with vertical sidewalls can be formed and extracted, stress-free, from the molds because of the shrinkage that occurs as the part cools.

In drape forming, when the hot plastic sheet touches the mold as it is being drawn, it chills and start to set up. To successfully drape-form, several variables must be considered. One of the most significant is shrinkage. Since all plastic materials have a high coefficient of thermal expansion and contraction, ie, about 7–10 times that of steel, care must be taken when designing the mold to provide sufficient draft on the sidewalls so the part can be extracted from the mold. It is not unusual for parts to rupture on cooling on an improperly designed drape mold. Another potential problem is that the part may become so highly stressed during forming and cooling that it loses most of the physical properties that the sheet would otherwise provide.

Natural-process evolution has combined the two systems to take advantage of the better parts of each method. The plug-assist process, similar to matched-die forming, involves a male mold (of plug) that ranges in size from ~60 to 90% of the volume of the cavity. By controlling the geometry and size of the plug and its rate and depth of penetration, material distribution can be improved for a broad range of products. The plus-assist technique is used to manufacture cups, containers, and other deep-draw products.

Many thermoforming techniques have been developed to obtain better material distribution and broaden the applicability of the process. Some of the more popular methods are illustrated and described in Figures 2 through 10. Most of these techniques can employ vacuum, pressure, or a combination to apply the force necessary to shape the heat-softened plastic sheet.

Cooling. The time required to cool the heat-softened plastic below its heat-deflection temperature while it is in contact with the mold is often the key to determine the overall forming cycles. Cooling is accomplished by conductive heat loss to the mold and convective heat loss to the surrounding air. Cooling rate depends on the tooling, because in all methods except matched mold, the plastic is in contact with the mold on one side only. The opposite side is cooled convectively by forced air or ambient air. Water sprays are sometimes used but often pose as many problems, eg, water spotting, as they solve. Pressure forming helps minimize cooling time because

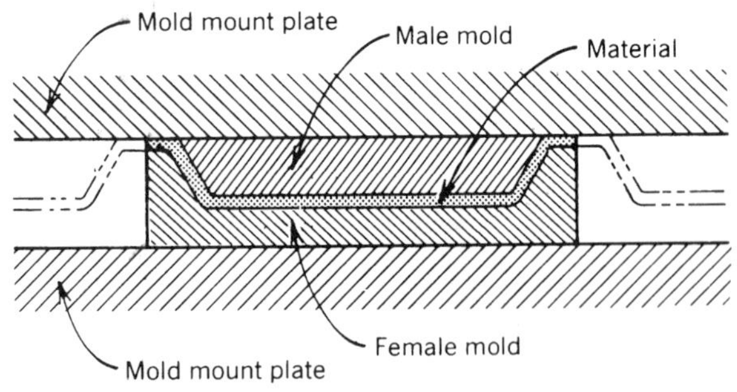

Figure 4. Matched-mold forming. The plastic sheet is clamped and heated to the required forming temperature. The heated sheet can be positioned over the female die, or draped over the male mold. The male and female halves of the mold are closed, forming the sheet, and trapped air is evacuated by vents located inside the mold. Material distribution varies with mold shape. Detailed reproduction and dimensional accuracy are excellent. Applications: foam products, egg cartons, meat trays, etc.

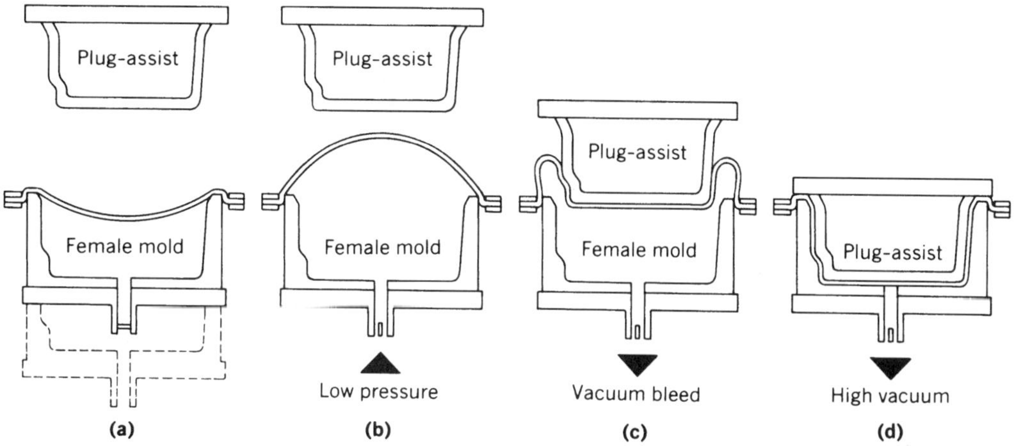

Figure 5. Pressure-bubble/plug-assist vacuum forming. The plastic sheet is clamped and heated to the required forming temperature. (**a**) The heated sheet is positioned over the female mold and sealed. (**b**) Prestretching is accomplished by applying controlled pressure through the female mold, creating a bubble. (**c**) When the sheet is prestretched to the desired degree, the male plug is forced into the sheet. (**d**) When the male plug is fully engaged, a vacuum is applied through the female mold to form the sheet. Pressure may be applied through the plug to assist forming, depending on the material and forming requirements. Applications: refrigerator liners, bathtubs, etc.

the higher air pressure keeps the sheet in more intimate contact with the mold surface.

Trimming. A number of trimming methods are available, including hand cutting, rough shearing of the web, punching parts out from the web, or trimming around a periphery with a steel rule or sharp-edged die used to penetrate the web. The punch-and-die trim provides the greatest accuracy and longest life, but at the highest cost. These dies are usually designed with a hardened-steel punch and will pass through a slightly softer steel die that can be peened when it dulls. These trim dies are designed so the parts are punched through the die successively and exit from the trim station in a nested fashion. This simplifies the material handling required to pack or prepare the parts for the next postforming operation, eg, rim rolling, or printing.

Steel-rule dies are better suited for the lower volume, thin-sheet operations. Steel-rule trim dies are used predominantly in custom thermoforming operations, where the run size does not justify the cost of matched shearing dies. They are used

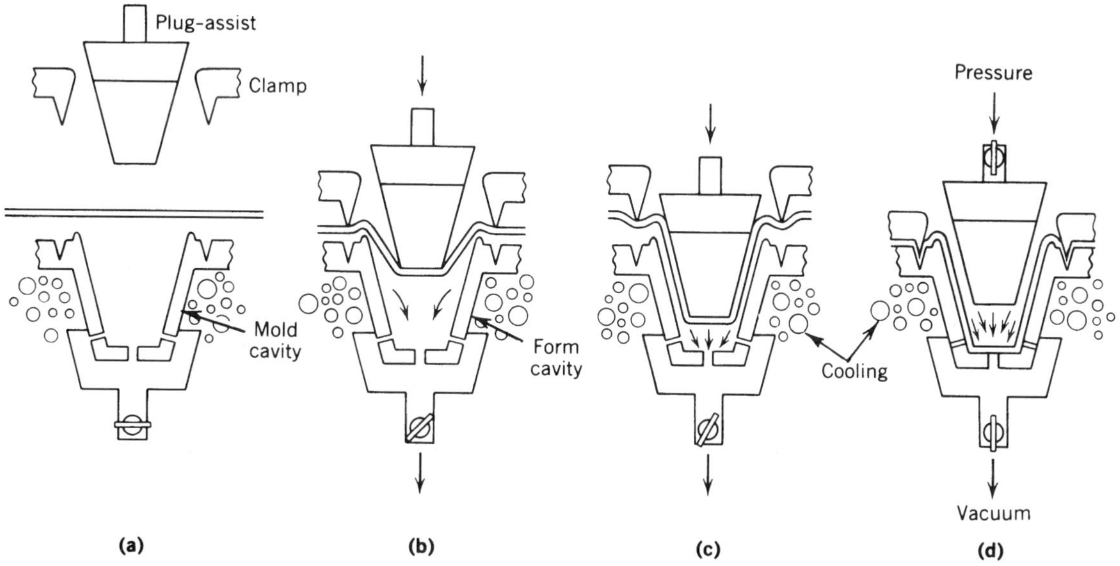

Figure 6. Plug-assist forming sequence: (**a**) the plastic sheet is clamped and heated to the required forming temperature—the sheet is then sealed across the female-mold cavity; (**b, c**) the male plug is forced into the sheet—the depth and speed of penetration, as well as plug size, are the primary factors in material distribution; (**d**) when the plug is fully engaged, vacuum, pressure, or a combination are applied to form the sheet. Applications: cups, food containers, etc.

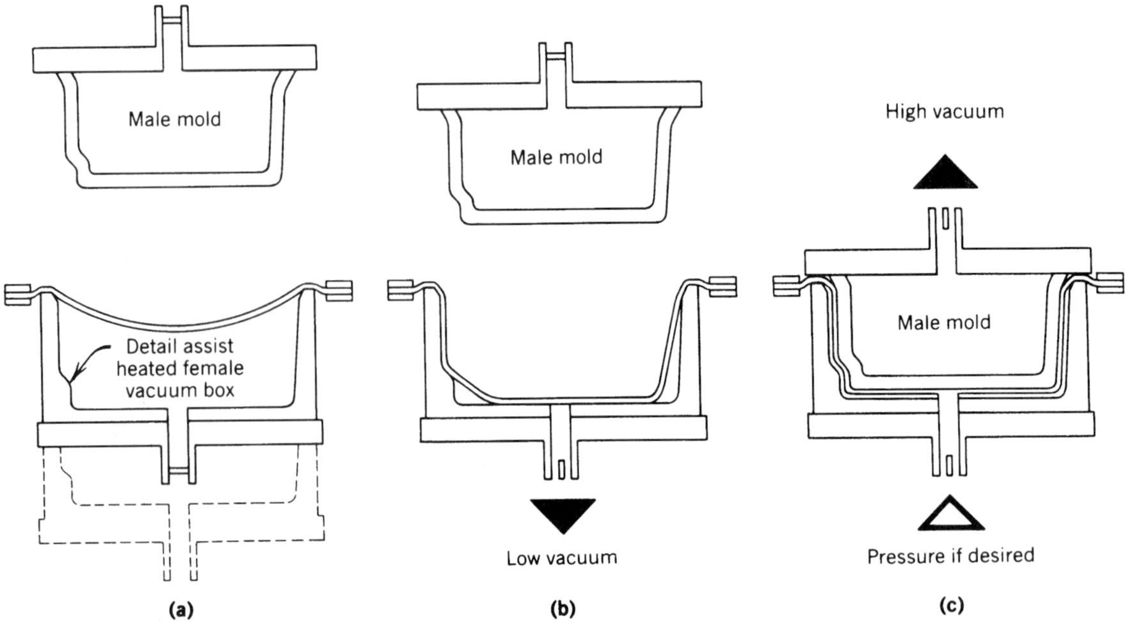

Figure 7. Vacuum snapback: (**a**) the plastic sheet is heated and sealed over the female vacuum box; (**b**) when vacuum is applied to the bottom of the vacuum box, the plastic sheet is prestretched into a concave shape; (**c**) after the plastic sheet is prestretched to the desired degree, the male plug enters the sheet. A vacuum is applied through the male plug while the vacuum box is vented, creating the snapback. Light air pressure may be optionally applied to the vacuum box depending on the materials used and forming requirements. Applications: deep-draw parts (eg, luggage, auto parts), gun cases, cooler liners, etc.

almost exclusively in the manufacture of blister packages and decorative packaging for the cosmetics industry.

The trimming operation can be done inside or outside the mold. The remote-trim configuration utilizes a separate trim press, which punches out 1–3 rows of parts at a time. The parts are discharged in nested horizontal stacks ready for subsequent operations or for packing. Trimming within the mold provides the most accurate alignment of the trim cut to the formed part, but the close proximity of the trimming to the mold body is a complex arrangement that may limit cooling capacity near the cutting surface, and after trimming, precise control of the part ceases. Complex mechanisms are required to sort and nest these parts downstream. Trimming is done by sawing or routing in the production of low volume odd-gemoetry industrial parts, eg, boat hulls, recreational vehicle bodies, bathtubs, but not in packaging.

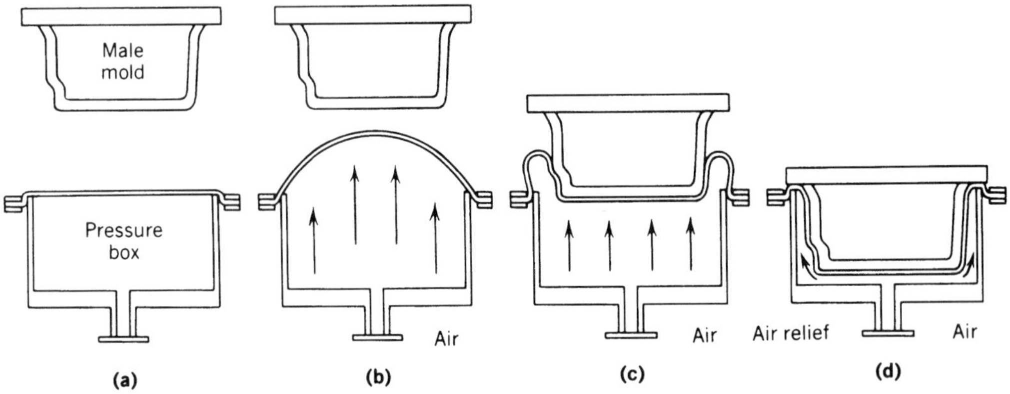

Figure 8. Pressure-bubble vacuum snapback: (**a**) the plastic sheet is clamped and sealed across the pressure box; (**b**) prestretching is accomplished by applying controlled air pressure under the sheet, through the pressure box, creating a bubble; (**c**) when the sheet is prestretched to the desired degree, ie, 35–40%, the male plug is forced into the sheet while the air pressure beneath the sheet remains constant; (**d**) after the male plug is fully engaged, the lower air pressure is increased. A vacuum is simultaneously applied through the male plug, creating the snapback. Applications: intricate parts where material distribution is critical.

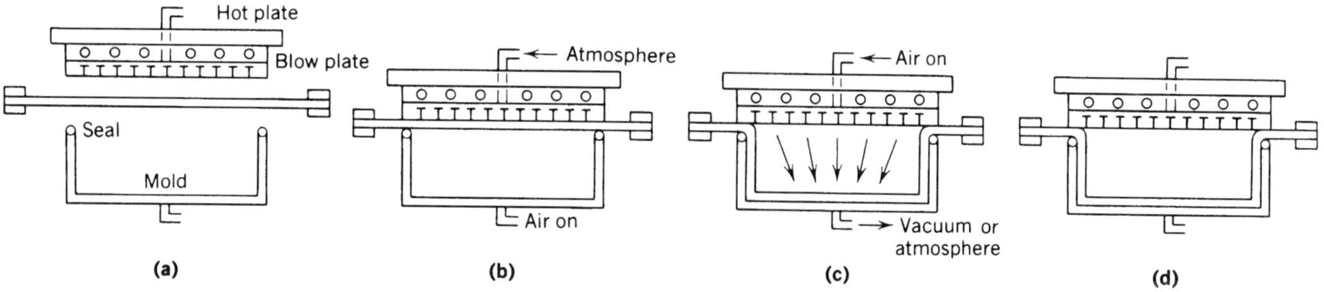

Figure 9. Trapped-sheet, contact-heat, pressure forming; the plastic sheet is clamped between the female-mold cavity and a hot blow plate: (**a**) the hot, porous blow plate allows air or vacuum to be applied through its face; (**b**) the mold cavity seals the sheet against the hot plate; (**c**) controlled air pressure, applied from the mold, blows the plastic sheet in contact with the hot plate to ensure complete contact with the heating surface; (**d**) when the desired heating has been accomplished, air pressure is applied through the hot plate, forcing the plastic sheet into the mold. Venting of the mold can be simultaneous, depending on materials and forming requirements. Steel knives may be inserted in the mold for sealing and in-place trimming if additional closing pressure is available. Applications: OPS containers, covers, etc.

Forming machines. There are basically two types of thermoforming machines: sheet-fed and web-fed. Sheet-fed machines operate from sheet cut into definite lengths and widths for specific applications. The sheet is generally heavy, ie, 0.060–0.5 in. (1.5–13 mm), and the products are industrial. Packaging-related applications include dunnage trays and pallets, large produce bins, shipping-case dividers, box liners, crates, and carrying cases formed to the shape of the product.

Packaging applications generally employ web-fed machines, which use either coil stock or a web that comes directly, ie, in-line, from a sheet extruder. The need to form a coil limits the thickness of the web to ca 0.125 in. (3.2 mm).

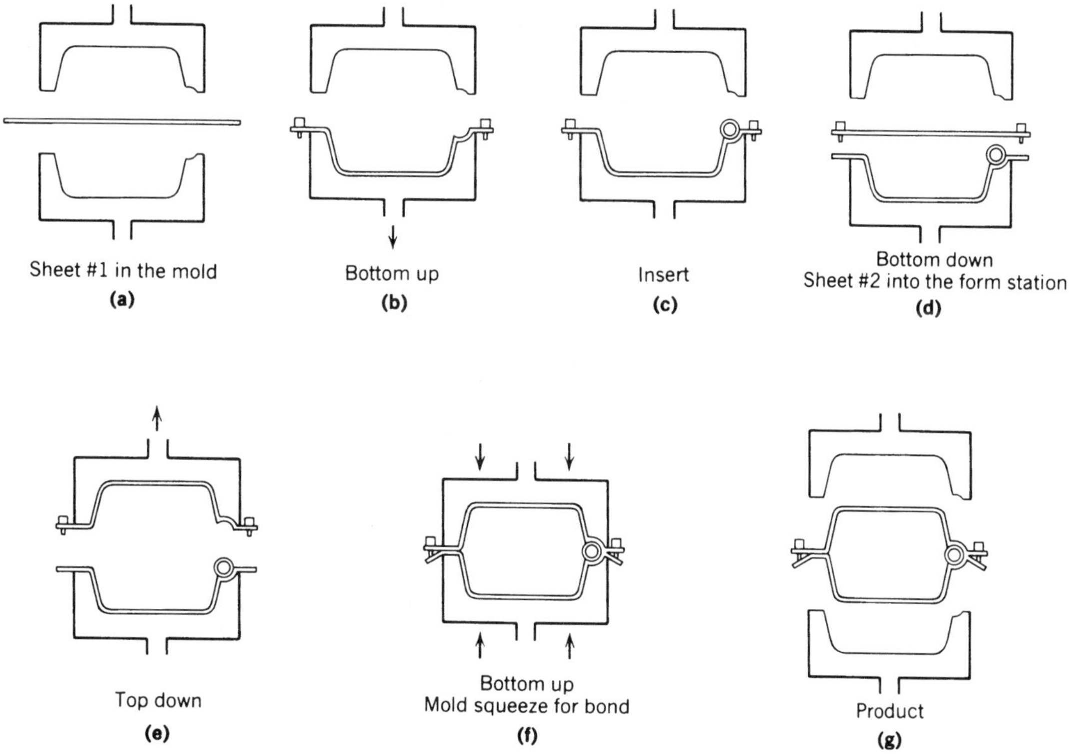

Figure 10. Twin-sheet forming. This technique utilizes a rotary-type machine with two heating stations and bottom-load clamp frames. The first sheet (**a**) is heated and formed into the lower half of the mold (**b, c**) while the second sheet is being heated. After forming the first sheet, the second sheet is indexed into the forming station (**d**) and vacuum is applied to the upper mold half to form the sheet (**e**). A bond is simultaneously made between the parts by pressure applied to the mating surfaces (**f**). After cooling, the molds are removed to give the product (**g**).

Web thickness can be increased by operating directly in line with an extruder.

Tooling

Among the plastic-fabricating processes, thermoforming permits the use of the broadest range of mold-tooling materials. Mold materials such as wood, epoxy, polyesters, or combinations thereof can be used if the volume is not sufficient to warrant the cost of advanced, temperature-controlled metal molds. For high-volume applications, eg, food packaging, molds are normally manufactured from cast or machined aluminum, sometimes with hard coats applied to the wearing surfaces for longer service life. Molds are occasionally made from beryllium copper or brass that is chrome-plated for extended service life.

Vacuum and compressed air are the primary means used to form the heated, thermoplastic sheet to the mold configuration with rigid-type materials. For foam materials, matched metal molds are frequently used to compress the material to the desired thickness without compressed air. Vacuum is transmitted through small holes that are located to allow the sheet to come in intimate contact with the mold surface. The size of these holes ranges from ca 0.013 to 0.030 in. (0.33–0.76 mm) dia, depending on the type of thermoplastic to be formed. Other means of applying vacuum in the mold include the incorporation of narrow grooves or slots, which often are easier to incorporate than vacuum holes and allow greater evacuation of the air in the tool. Polyolefin materials require small vacuum holes and slots; HIPS materials, which cool faster, can use larger holes. Final product design and material selection determine the exact size of the holes required.

In addition to vacuum holes and slots, molds often require a complete sandblasting of the molding surface to minimize the possibility of trapped air between the sheet and the mold. Eliminating trapped air is important because it prevents imperfections on the part surface and aids in rapid transfer of heat out of the material by providing close contact between the plastic and the cavity. Undercut areas of molds are used. HIPS are normally not sandblasted because this material is relatively rigid and sandblasting would make part removal difficult. Molds used for polyolefins are normally sandblasted completely over the forming surface. Vacuum is usually pulled through the forming mold, male or female, and the compressed air is introduced from the opposite side of the sheet material. In foam-forming with matched metal molds, vacuum is normally applied only through the female cavity.

Tooling for high-production applicants is most often aluminum because of its thermal conductivity, light weight, and cost. Most crucial is thermal conductivity, because in any thermoforming process, the residual heat of the plastic must be removed as rapidly as possible for rapid cycling. The temperature control of the mold is accomplished by water channels in the mold and designing them to provide maximum heat removal from the sheet. Ideally, all molds should provide a temperature differential between inlet and exit water temperature of no more than 5°F (3°C).

Many techniques are used to incorporate cooling in the molds for improved cycle times. The most common methods are contact cooling and direct water cooling. Contact cooling consists of coring a mold base and carefully machining cavity inserts so that contact occurs when the cavity insert expands from the heat of the plastic sheet while the mold base contracts from the coolant circulating through it. The most common method today for rigid HIPS is to circulate water through hollow-shelled cavities. This is more expensive but the most effective way to cool the plastic sheet rapidly.

Tooling for low volume work, eg, box inserts, blisters, and other nonfood packaging, and prototyping can often be simple and effective without any of the special cooling features described above. This is because it normally takes longer to heat the sheet than to cool it, and the work is generally done on slow, low production equipment that does not require the rapid cycle rates. With these molds, cooling is very often augmented by the use of fans to increase the air circulation over the material after it is formed. In such applications, low-cost, simple molds of plastic, plaster, wood, or other materials are used effectively. Mold dimensions must allow for the different shrinkage rates of different types of thermoplastics. In the manufacture of a family of products, it is sometimes possible to utilize the same cooling base and interchange only the molding inserts. Depending on the trim dimensions of the products, the same cutting tool might be suitable for more than one product; if not, they can be designed to be easily interchanged from the trim die shoe.

Cut-Sheet Thermoforming

Cut sheet is processed on two types of thermoforming machines: the shuttle style and the rotary, or carousel, style. In single-station shuttle machines, the plastic sheet is placed in a clamping frame that shuttles into the oven for heating and back again for forming. Another very popular shuttle machine is the double-ender, which has a common oven with a form station at each end. This arrangement provides 100% utilization of the oven because heating and forming can be done simultaneously.

Rotary or carousel thermoformers are most frequently used for heavier-gauge materials, ie, starting at ~0.050 in. (1.3 mm). Most rotary machines have three working stations load/unload, heat, and form. In a typical machine, the carousel frame rotates 120° on a time-controlled basis. Generally, the time cycle is dictated by the period required for the plastic sheet to cool below its heat-deflection temperature. When cooling time is faster than the heating time, four-station rotary machines are often used; these have two heating stations. Another type of four-station machine has a separate load and unload for automation and parts handling.

Most cut-sheet thermoformers are designed to operate as vacuum formers, but rotary and single-station (shuttle-type) machines are available with pressure-forming capabilities and sustain forming pressures up to 50 psi (345 kPa), depending on mold area. Most sheet-fed machines can be adapted for any of the standard thermoforming techniques. Cut-sheet machines have been built to accept sheet 10 ft (3.05 m) wide and 20 ft (6.1 m) long with the capability of making parts up to 4 ft (1.22 m) deep.

Twin-Sheet Thermoforming

Twin-sheet thermoforming can produce hollow parts eg, pallets, from cut sheet. These parts can be produced from both cut-sheet and roll-fed machinery. In sheet-fed machinery, the material to be formed can be 0.050–0.500 in. (1.3–13 mm) thick. Roll-fed machines are limited to materials 0.005–0.125

in. (1.3–3.2 mm) thick. With a typical web-fed, twin-sheet system, two rolls of plastic materials are simultaneously fed, one above the other. The webs are transported through the oven on separate sheet-conveyor chains and heated to a formable temperature. At the forming station, a specially designed blow pin enters the space between the two sheets before the mold closes. Air pressure is introduced between the sheets through the blow pin, and, at the same time, vacuum is applied to each mold half. Twin-sheet forming is done by a slightly different method on specially designed rotary thermoformers.

Equipment Improvements

Rapidly expanding markets for thermoformed products are attributable to technological advancements in equipment design. Better process control has enabled converters to increase productivity and improve product quality. Use of microprocessor-control systems allows the operator to enter data, examine existing values in the program, or troubleshoot problems by viewing a television monitor. The conventional control panel includes a cathode-ray tube (CRT) and keyboard, form-on and emergency stop push buttons, and indicator lights to signal form or automatic modes. Machine functions may be displayed on the CRT upon request and function control is obtained by means of the simplified keyboard.

The microprocessor-control systems feature control of oven-temperature orientation in multiple zoned ovens, interface speed between former and trim press, sheet-index length, machinery functions and timing, and control and graphic display of start end points of forming functions, storage of production information, and diagnostic capabilities. Use of the microprocessor results in tremendous savings in labor, energy, and downtime. Many of these systems are available as a retrofit to existing equipment. Equipment has also been designed to monitor several lines of similarly equipped machines from one central location. Such data-management systems provide comparative data, setup parameters, production rates, maintenance schedules, alarms and safety signals, and storage of reference information, resulting in a true online management-information reporting system.

Design improvements that have contributed to improved productivity include redesigned forming presses that provide front and rear access to the forming station to facilitate tooling installation and periodic maintenance; access doors to oven heaters for quick cleaning; ovens that eject automatically if the sheet overheats; and automatic lubrication systems that increase component life substantially and allow faster mechanical operation due to reduced friction.

Microprocessor heating control allows the operator to control the temperature profile of the sheet from edge to edge or to provide localized heating. Solid-state instruments used in conjunction with mercury or semiconductor-controlled rectifier (SCR) relays have made it possible to maintain extremely close heater temperature control.

Forming presses are now available with precision guidance systems on the platens to permit critical operations to take place in the forming molds that formerly had to be done on auxiliary equipment. An example is the method of incorporating the latching slots that are cut in the sidewalls of egg cartons, fast-food cartons, and trays. Previously, these slots were punched in on a special press located downstream from the forming station. Registration problems relative to the slot location were often encountered because of variations in the index length by the thermoformer web-conveyor system. These problems were overcome by expensive drive systems. On machines with precision-guidance systems, mating punches can be incorporated in the forming molds that engage during the last 5–10% of press closure and form a slot in the heat-softened plastic that has reinforced edges similar to button holes.

High-speed trim presses are available with supported moving platens that carry up to 85% of the gross-platen and trim-tool weight. The advantage of this system is that it permits the press tie rods to function primarily as guidance members rather than as both platen transporters and guides, in addition to serving as tie rods. The system reduces trim-tool maintenance by up to 50%.

Significant improvements have been made in product-handling systems, which can now automatically count, stack, and package products at the trim-press discharge. The design of these automated systems varies from relatively simple systems using guide chutes or magazines and transport conveyors to those employing robotics and advanced packaging machinery.

Automation has been extended to the materials handling of large plastic sheets into sheet-fed machines. The development of automatic sheet loaders provides automatic loading systems that enable use of the entire machine area. This results in a 20–30% increase in productivity, achieved largely from savings in loading time, since a manually loaded machine normally used only one quarter of the machine effectively.

BIBLIOGRAPHY

"Thermoforming" in *The Wiley Encyclopedia of Packaging Technology*, 1st ed., by Lynn McKinney, William Kent, and Richard Roe, Brown Machine, pp 668–675.

General References

"Thermoforming Techniques" in J. Agranoff, ed., *Modern Plastics Encyclopedia* 1990, McGraw-Hill, New York, 1900.

THERMOSETTING PLASTICS, PROCESSING SYSTEMS FOR

Thermosetting plastics are synthetic chemical formulations of reactive polymers which undergo an irreversible exothermic molecular cross-linking reaction when the resin component is mixed in correct stoichiometric proportions with the catalyst component. The resin and catalyst may be liquid at room temperature or may be solid at room temperature. Heat may be needed to initiate the reaction of the room temperature liquid resin and catalyst. Heat and pressure are generally applied to initiate, and in some instances to accelerate, the reaction of the resin and catalyst systems which are solids at room temperature. Various fillers such as silica, calcium carbonate, short or long glass fibers, etc, are often added prior to mixing to modify and improve the properties of the molded parts and to minimize shrinkage and possible warpage. Other additives and modifiers are often part of the formulation to provide improved characteristics (see also Thermosetting polymers.)

Physical Characteristics of Thermoset Parts

The resultant cross-linked or polymerized plastics may be rigid or may be elastomeric, depending on the chemistry. They are generally highly resistant to moisture and to many chemicals. The rigid thermosets have excellent dimensional stability, are generally good electrical and thermal insulators, and offer excellent resistance to harsh environments.

Although thermosets do not melt when subjected to elevated temperatures, they will char or burn when the heat is extreme, in ranges from 400°F to 800°F, depending on resin type. Heat distortion temperatures (HDT) are usually between 300°F and 600°F. At the HDT, some minimal softening occurs and physical and electrical properties are reduced.

The elastomeric thermosets remain flexible even at temperatures of 0°F or lower.

Types and General Characteristics

The common thermosetting plastics are formaldehydes (phenol, urea, and melamine). These three are generally solid at room temperature before cure. They require temperatures of 200–325°F to react. They are rigid after cross-linking and epoxies, polyesters, and silicones and they may be solid or liquid at room temperature.

The solid types will react at temperatures of 200°F or higher, whereas the liquid sytems may react at room temperature. Following cross-linking, the epoxies and polyesters are generally rigid, whereas the silicones may be solid or flexible.

Urethanes are generally liquid at room temperatures. They will react when mixed at room temperatures, but reactions are faster at slightly elevated temperatures. When polymerized, they can be very resilient or may be hard, tough, and only slightly resilient, depending on the specific chemistry.

Processing Systems for Producing Thermoset Parts

Room temperature liquid systems

Casting. After mixing, but before polymerization has completed, these systems may be poured into open molds, where they will polymerize, retaining the shape of the mold cavity. Casting may be done at room or elevated temperatures, and may be done atmospherically or under vacuum (to minimize voids or bubbles).

Dipping. After mixing and before curing, liquid systems may be used to coat objects by a dipping process. Following dipping, the coated objects are heated to cure the coating. The cured coating provides physical, environmental, or electrical protection.

Room temperature solid systems

Compression molding. To produce a molding compound, thermosetting resins and catalysts are mixed and heated to initiate the polymerization or cross-linking reaction. The reaction is prematurely halted by cooling the mix to room temperature or below before polymerization has completed. This partially reacted thermoset is said to be in the B-stage. Such B-stage thermosets are subsequently crushed into granules for ease in handling and measuring. They will be stable for several months in the B-stage, especially if stored at 40°F or lower temperatures.

When compression molding is to be performed, a measured quantity of B-stage granules is placed into the lower cavity or cavities of a mold heated 300°F or higher. The mold halves are generally mounted in a compression molding press. Then the upper cavity or "force," at the same temperature, is brought down over the plastics with sufficient force to cause the plastic to melt and flow into the configurations of the cavity or cavities. The mold is held tightly closed until the material has "cured," perhaps for several minutes, depending on thickness, temperature, and other factors. Following cure, the mold is opened and part or parts are removed (see also Compression molding).

Transfer molding. The transfer molding process is similar to compression molding. The heated mold halves are brought together in a press and firmly clamped before B-stage compound is introduced. This compound is introduced into a heated cylindrical "pot" in the upper half of the mold, and a piston or plunger is made to descend into this pot, moving downward to put pressure on the melting compound, from 500 to 5,000 psi, depending on the specific thermoset. Under the heat (300–400°F) and the pressure, the plastic liquefies and flows from the bottom of the pot, at the parting surface of the two halves of the mold, through channels or "runners" to one or more cavities of the closed mold. Entering each cavity through a restricted "gate" or port, the plastic fills and packs each cavity well before final polymerization. Within a minute or more, depending on maximum cross sectional thickness of the part, the plastic hardens. The mold is then opened, the part or parts are ejected from the cavities and removed from the mold area, along with the runners and "cull" (excess material remaining at the bottom of the transfer pot). Thermosetting plastics runners and culls are not reusable because the molding compound has gone through an irreversible chemical reaction.

Transfer molding is often used to mold plastics around an "insert," making the insert integral with the molded part. The insert is then said to be encapsulated. Often the insert is an electrical or electronic component such as a small coil or resistor or transistor or hybrid circuit, etc. Such components are designed such that their current-carrying wires or "leads" are able to support the component in the closed mold cavity before molding, in such a way that liquid plastic flowing through the gate into the cavity may completely surround the critical part of the component while allowing the leads to protrude for subsequent connecting to the electric circuit of the final product, appliance, automobile, etc. Following cure of the plastic, the molded part is removed from the mold. The plastic serves to protect the inner component from physical damage, from moisture or other environments which might otherwise damage the device, and from possible short circuits which could occur if the unencapsulated component touched other metals.

Injection molding. Injection molding of thermosets is similar to transfer molding and to thermoplastics injection molding. The granular B-shaped material is placed into a "hopper" which allows the material to fall through an opening or 'throat" of a portion of the injection press termed the "barrel." The barrel is heated along its length, and an auger-type "screw" inside the barrel moves the plastic from the throat under the hopper to a nozzle at the mold end of the barrel. The mechanical shearing action of the rotating screw, plus the heated barrel, cause the material to become plasticized. As it reaches the end of the barrel, it is temporarily trapped. As the screw continues to rotate, more plastic is pushed in front of the screw, causing the screw to move backward ("reciprocate"), allowing a charge of the plasticized material to build up in front of the screw. When the screw has platicized the desired amount of molding compound for the next cycle, its rotation is stopped. Shortly thereafter, when the press has closed the two mold halves under considerable clamping pressure, the screw advances toward the nozzle end of the barrel, forcing the plasticized charge of material to flow through the nozzle into the sprue part of the mold half (a passage in the mold through which platicized material flows from the barrel) to the runners and cavities of the mold. There it is held under pressure for a minute or more, for curing, following which the mold opens, parts are removed, and the cycle repeats itself, often with no operator intervention.

Applications of Thermosetting Plastics in Packaging

Consumer products. Whereas thermoplastics are often used for "throw away packaging," thermosets are rarely, if ever, so used. In consumer goods packaging, thermosets may be molded into attractive rigid "presentation boxes" for expensive jewelry or other small valuable gifts such as pen and pencil sets, fine board games, etc. These containers are often designed to resemble highly polished black onyx (generally phenol or melamine formaldehydes), or white or veined marble or granite (generally melamine or urea formaldehydes). Such containers offter physical protection, long life, and attractive appearance, weight, and feel.

Historically, 35 to 60 years ago, such usage was widespread. But with growth of a wide variety of thermoplastics, with generally lower costs of production, thermosetting plastics in this application have been displaced by thermoplastics. Today, such consumer packaging with thermosets serves a very limited market.

Transparent thermosetting casting compounds, such as epoxies or polyesters, are sometimes used for embedding exotic items such as coins, insects, butterflies, and other keepsakes. The polymerized material preserves the embedded article while allowing it to be admired through the transparent package.

Commercial and industrial products. Because of the strength, rigidity, durability, and practical moldability of thermosets, they are extensively selected for protective packaging of commercial and industrial products.

One of the principal applications is to enclose and protect electrical articles such as fan motor housings, electric shavers, clippers, hair dryers, etc, and a host of household appliances, where heat resistance, electrical insulation, and attractive three-dimensional shapes are highly desirable.

Literally billions of electrical and electronic components are produced each year, globally, using plastics to insure long life with physical, environmental, electrical, and thermal integrity. Such products include most "active" semiconductor devices such as integrated circuits, transistors, and diodes, as well as "passive" components such as resistors, capacitors, inductors, small coils and transformers, delay lines, etc. This type of packaging of electrical and electrical components, often termed "encapsulation," may use liquid casting epoxies and polyesters for low volume production or B-stage molding epoxies, polyester, and alkyds with transfer molding for high volume production. These soft flowing materials, with reasonably fast cures, prove ideal for protecting such components, especially the more delicate and fragile semiconductor devices. Under-the-hood automotive electrical devices are often packaged in this manner for protection from the harsh environment, including extensive oil and water vapors, temperatures, and shock and vibration.

In some of the above-mentioned applications, newer high temperature-resistant and relatively rigid thermoplastics have replaced the thermosets because of lower cost manufacturing.

Radio and television sets were originally packaged with thermoset housings, but thermoplastics have displaced most thermosets in that market. On the other hand, thermosetting plastics are displacing metals for automotive exterior panels. Such plastic panels are generally compression molded with fiber-reinforced polyesters which enable not only a very attractive noncorrosive surface finish and shape, but also provide considerable structural strength, toughness, and dimensional stability. Their moldability allows great latitude in producing three-dimensional curved shapes to meet demands of styling. Such uses of thermosetting plastics for functional packaging have extended to the appliance industry, even to the machine tool industry.

Future Outlook

Currently, the use of thermosetting plastics in the electrical and electronics industry in the U.S. accounts for 3.2%, or 237 million pounds annually. It is estimated that approximately one-third of those figures represents packaging applications as described above. Growth rate for such usage over the past several years has been flat. For the foreseeable future, the growth rate is expected to remain essentially flat. Principal factors affecting the growth rate are (*1*) increasing number of electrical and electronics components packaged annually; (*2*) decreasing physical size of many such components and circuits; and (*3*) transition from thermosetting plastics to thermoplastics for packaging some types of such items.

Today's use of thermosetting plastics for decorative consumer packaging in nonelectrical applications is estimated to be a miniscule percentage, by weight, of all thermosetting plastics consumption. That low usage is not expected to increase or decrease significantly over the coming years.

In more functional packaging applications, such as automotive panels, and shrouds for industrial and commercial products, reinforced thermosets are expected to experience steadily increasing usage due to their attractive appearance, structural strength, weatherability, and suitability for almost unlimited configurations.

JOHN L. HULL
Hull Corporation
Hatboro, Pennsylvania

THERMOSETTING POLYMERS

Thermoset (TS) plastics provide the packaging industry with applications that have performance and/or cost advantages when compared to the much more popularly used thermoplastic (TP) applications. During the first part of the 20th century practically all packaging products were made of TSs; very few and limited amount of TPs existed at that time. TS products included containers, tanks, bottle caps, trays, displays, and pallets, for packaging foods, cosmetics, paints, chemicals, etc.; also electrical appliances (switches, insulators, etc), mechanical (gears, bearings, etc) and other products. Such products continue to be used, meeting speciality requirements such as aesthetics or decorations, chemical or heat resistance, healthcare sterilization or biodegradablility, strength or light weight, insulation or environment, production or cost, and so on.

Structures

TSs are plastics that undergo chemical change during processing to become permanently insoluble and infusible. Such natural and synthetic rubbers (plastic elastomers) as latex, nitrile, millable polyurethanes, silicone butyl, and neoprene, which attain their properties through the process of vulcanization, are also in the TS family. The best analogy with TSs is that of a hard-boiled egg whose yolk has turned from a liquid to a solid and can not be converted back to a liquid as is done with TPs. Figure 1 provides a simplified example of the heat–time processing files for a TS and TP. Both cases involved plastics that required higher processing temperatures. In general, with their tightly crosslinked structure, TSs resist higher product-performance temperatures and provide greater dimensional stability than do most TPs (1–9).

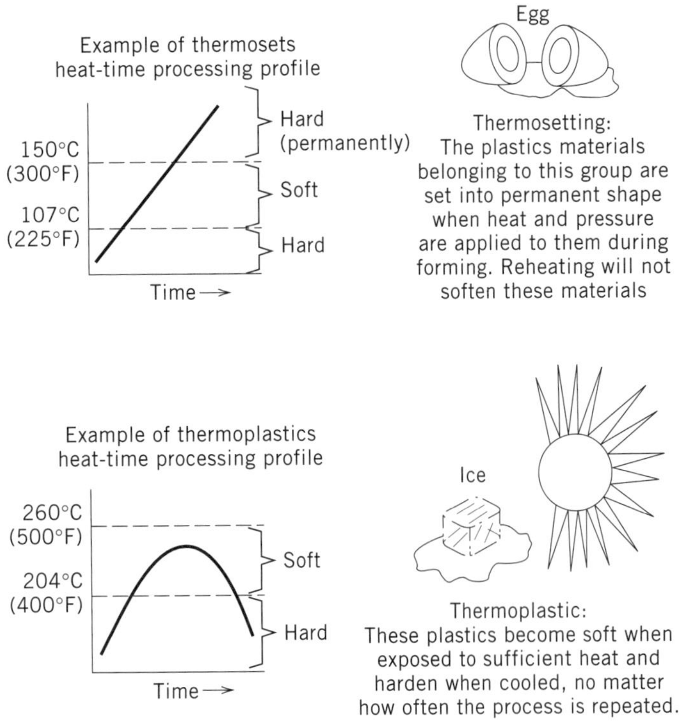

Figure 1. Examples of thermoset and thermoplastic processing heat–time profile cycle.

The structure of TSs, as of TPs, is also chainlike. Prior to molding, TSs are similar to TPs. Cross-linking is the principal difference between TSs and TPs. In TSs, during curing or hardening the crosslinks are formed between adjacent molecules, resulting in a complex, interconnected network that can be related to its viscosity and performance. These cross-bonds prevent the slippage of individual chains, thus preventing plastic flow under the addition of heat. If excessive heat is added after crosslinking has been completed, degradation rather than melting will occur.

TSs generally cannot be used alone in high-performance structural applications such as trays, containers, and tanks (which require the higher performances). They must be filled or reinforced with materials such as calcium carbonate, talc, and/or glass fibers. The most common reinforcement used is glass fiber, but others are used.

TSs pass through one soft "plastic" stage and then harden irreversibly. Like TPs, they develop a time–temperature profile during polymerization. In the soft stage, they can be shaped. During polymerization, monomer molecules are crosslinked to form large molecules. Softening by heating releases these linkages. TSs are usually purchased as liquid monomer–polymer mixtures or as partially polymerized compounds. In this uncured or B-stage condition, they can be formed under pressure or without pressure and polymerized by chemical curing agents and/or heat. Because the first observed irreversible reactions required heat, such plastics were called *thermosetting,* although the reaction may occur at room temperature. Two similar mechanisms account for the reaction crosslinking and interlinking.

With crosslinking the chain molecules may become linked by chemical bonds, just as in mechanical chains may be linked by crosslinks. The mass stiffness and may become hard and brittle. It is not softened by heat, as occurs with true TPs. TS polyesters used in RPs are in this class; so are epoxies and polyurethanes. Nevertheless, these resins can also be fabricated as TPs.

The mechanism of interlinking involves monomeric units linking to each other in three-dimensional arrays, starting with small clumps that eventually interlink completely to form a giant molecule constituting the whole article. As interlinking proceeds, the mass becomes rigid, and cannot be softened, because the links are not broken by heat. Water (and/or other gases depending on type material) may be given off as a byproduct in the form of steam; phenolics and the melamines are examples of this type. No water is produced by TS polyesters (unsaturated polyesters), a characteristic that permits them to be processed with no pressure.

Molding

Compression and transfer molding (CM and TM) are the two main methods used to produce molded parts from TS resins. A limited amount is processed by injection molding (IM), reaction injection molding, rotational molding, and casting. CM was the major method of processing plastics during the first part of this century because of the development of phenolic resin (TS) in 1909 and its rather extensive use at that time. By the 1940s this situation began to change with the development and use of TPs in extrusion and IM. CM originally processed $\sim$ 70 wt% of all plastics, but by the 1950s its share of total production was $<$ 25 wt%, and now that figure is $\sim$ 3 wt%.

Table 1. Properties and Manufacturing Processes of TSs

Thermosets	Properties	Processes
Polyesters	Simplest, most versatile, economical, and most widely used family of resins; good electrical properties, good chemical resistance, especially to acids	Compression molding, filament winding, hand layup, mat molding, pressure bag molding, continuous pultrusion, injection molding, sprayup, centrifugal casting, cold molding, encapsulation
Epoxies	Excellent mechanical properties, dimensional stability, chemical resistance (especially to alkalies), low water absorption, self-extinguishing (when halogenated), low shrinkage, good abrasion resistance, excellent adhesion properties	Compression molding, filament winding, hand layup, continuous pultrusion, encapsulation, centrifugal casting
Phenolic resins	Good acid resistance, good electrical properties (except arc resistance), high heat resistance	Compression molding, continuous lamination
Silicones	Highest heat resistance, low water absorption, excellent dielectric properties, high arc resistance	Compression molding, injection molding, encapsulation
Melamines	Good heat resistance, high impact strength	Compression molding
Diallyl *o*-phthalate	Good electrical insulation, low water absorption	Compression molding

During the 20th century, TSs experienced an extremely low growth rate, whereas TPs expanded at an unbelievably high rate. Regardless of the present situation, CM and TM are still important, particularly in the production of certain low-cost parts as well heat-resistant and dimensionally precise parts. CM and TM are classified as high-pressure processes, requiring 2000–10,000-psi molding pressures Some TSs, however, require only lower pressures of down to 50 psi or even just contact (zero pressure). CM is the most common method of molding TSs. In this process, material is compressed into the desired shape using a press containing a two-part closed mold, and is cured with heat and pressure. The CM and TM processes are not generally used with TPs.

Curing

Different methods are used to monitor TS curing cycles. The basic method involves relating temperature-to-time based on part thickness and temperature profile across the surface of the parts. With dielectric monitoring, the cure of TSs is done by tracking the changes in their electrical properties during the processing, particularly with reinforced plastics (RPs). In a typical application, the sensor is placed at the location where the cure is to be monitored. During molding the capacitance and conductance of the material are recorded. Measurements are taken at several frequencies over several orders of magnitude (0.1 Hz to 100 kHz). From these data, the value of the ionic conductivity is deduced. Research has provided information about the relationship between the ionic conductivity and the viscosity and the degree of cure of reinforced TSs (RTSs) (since 1940s). Results show correlation with fluidity before gellation and with flexibility after gellation. Rugged computer-controlled dielectric monitoring systems are used.

As reviewed, properties of TSs change during curing where temperature–time–pressure interrelate. A time–temperature–transformation (TTT) cure diagram may be used for understanding and comparing the cure and physical properties of TSs. It relates times with different factors that relates to isothermal cure such as the onset of phase separation, gellation, vitrification, full cure, and devitrification (7).

Types

Most of the TS materials are unreinforced TSs; however, reinforced TSs (RTSs) are also used in packaging applications. By far the biggest RP market share is in reinforced TPs (RTPs). Both RTSs and RTPs can be characterized as engineering plastics, competing with engineering unreinforced TPs. When comparing processibility of RTSs and RTPs, the RTPs are usually easier to process and permit faster-molding cycles with efficient processing such as during injection molding (8, 9).

Certain phenolic materials and easy-flowing, granular TSs based on urea, melamine, polyester, and epoxy resins provide the basis of numerous technical applications (Table 1), because of their high thermal and chemical resistance, stiffness, surface hardness, dimensional stability, and low flammability. There are also the popular RTS sheet molding compounds (SMCs) and bulk molding compounds (BMCs); the BMCs are also known as dough molding compounds (DMCs). These are predominently glass fiber-reinforced TS polyester resins used to meet a broad range of properties and applications.

In general, unreinforced TSs exhibit fair to good strength properties, but low stiffness as measured by elastic moduli. In structural components, they are frequently combined with other materials, often as the matrix or binder constituents of plastic composites. The reinforcements usually are fibrous. Fillers are usually included to provide strength, lower costs, and improve other performances. The principal reinforcments are glass fibers (95 wt% of consumption). Higher performancing fibers that are used are high performance glass, aramid, carbon, and graphite. These RTSs are usually called advanced RPs or advance (plastic) composites. There are also whiskers that have a place in very high performance RTSs (8, 9).

BIBLIOGRAPHY

1. D. V. Rosato, *Plastics Processing Data Handbook,* C&H, 1990.
2. D. V. Rosato, *Rosato's Plastics Encyclopedia and Dictionary,* Hanser, New York, 1993.
3. D. V. Rosato, "Thermosets" in J. Kroschwitz, ed., *Encyclopedia of Polymer Science and Engineering,* Vol. 14, Wiley, New York, 1988, pp. 350–391.

4. E. Baer and A. Moet, *High Performance Polymers,* Hanser, New York, 1990.
5. G. Lubin, *Handbook of Composites,* C&H, 1982.
6. A. D. Roberts, *Natural Rubber Science and Technology,* Hanser, New York, 1988.
7. J. K. Gillham, "Curing" in J. I. Kroschwitz, ed., *Concise Encyclopedia of Polymer Science and Engineering,* Wiley, New York, 1990, p. 236.
8. D. V. Rosato, *Injection Molding Higher Performance RP Composites,* SPE-ANTEC, May 1996.
9. D. V. Rosato, *Designing with Reinforced Plastic Composites,* Hanser, New York, 1997.

DONALD V. ROSATO
Plastics FALLO
Waban, Massachusetts

TIME–TEMPERATURE INDICATORS

Temperature is one of the key environmental factors influencing the storage stability of a food product. For most foods, an increase in storage temperature causes a reduction in their storage life. To determine the deleterious effects of temperature, it is often necessary to know for how long a food is exposed to that temperature. The cumulative effects of temperature and time cause irreversible changes in foods. In the commercial food-distribution chain, specific guidelines for temperature management have been developed. However, when these recommended temperature guidelines are not met, food quality is seriously impaired. Time–temperature indicators (TTIs) are devices that provide useful information regarding the temperature history experienced by a food during storage and distribution.

A food material may undergo physical, chemical, or microbiological change during storage. When these changes accumulate to such an extent that they exceed some predetermined criterion of acceptable quality, the food is considered to have reached the end of its shelf life. A variety of changes such as color, flavor, and texture may lead to the end of a food's shelf life. Similarly, the microbiological changes may render a product unsafe for human consumption. In such cases, a monitoring system that allows an early warning of the approching undesirable increase in the microbial level would be necessary. The common practice in the modern food-distribution system is to use some type of open dating that indicates the expiration of the shelf life of a product. The open-dating policy assumes that a product has been kept at the recommended temperature. However, if the temperature during storage and distribution exceeds the recommended conditions, the open-dating policy fails. In these situations, TTIs are more suitable predictors of product's shelf life (see also Shelf life).

A TTI affixed to a food package undergoes a temperature history similar to that of food. The indicator response is measured either through some physical means, or a certain property of an indicator may change causing a visually discernible response. The indicator response may be a change in its color such as green to yellow or change in light reflectance measured with special instruments, or there may be an advance of a boundary between two different colors that is observed visually. When the change in the property of an indicator is irreversible and cumulative with time and temperature exposure, the indicators are referred to as time–temperature integrators, full-history TTIs, or all-temperature TTIs. For certain applications, the indicators may be designed to respond only when the food is exposed to temperatures above some preselected threshold level, thus giving a partial history of the storage period. These types of indicators are called partial history TTIs or threshold–temperature TTIs. The TTIs respond at a rapid rate when the temperature goes up, and their response slows down as the temperature decreases.

For a TTI to correctly predict the end of the shelf life of a food, it must be able to accurately mimic the kinetics of the product quality. Therefore, the first step in the selection of a TTI is to identify one or more quality attributes of a food that are the key determinants of the shelf life of a food. The next step is to quantitatively determine their kinetics. For example, if a food material undergoes an undesirable change in color due to storage at temperatures higher than ideal, and the off-color makes the food unacceptable, then the kinetics of color change in the food, as influenced by temperature, must be known.

The kinetics of quality change in a food material is usually expressed by the rate constants. For example, if the change in the attribute is described by a first-order reaction, then the magnitude of the first-order rate constant will indicate how rapidly the attribute may be undergoing change at that temperature. The rate constants are usually determined for at least three different temperatures and then combined into Arrhenius equation to relate rate constants to temperature. Another common way to express shelf-life kinetics is to use the Arrhenius expression as follows:

$$t_{\text{shelf life}} = Ae^{E_a \neq RT}$$

where $t_{\text{shelf life}}$ is the shelf life (days or months); A is a constant with same units as $t_{\text{shelf life}}$; T is temperature (K or °C + 273), R is a constant (0.001987 kcal/(mole · degrees), and E_a is the activation energy (kcal/mol).

From the Arrhenius equation one obtains the activation energy of a reaction or shelf life. The Arrhenius equation is also used to predict the shelf life at any unknown temperature. The shelf-life data and the activation energy are the key design parameters for a TTI. It is absolutely essential that the activation energy of a TTI match the activation energy of the quality attribute of concern. This equality in kinetic parameters allows the TTI to reach the same level of change as experienced by a food material when exposed to identical storage time and temperature.

The commercially available TTIs use either a biochemical reaction between an enzyme and a substrate, polymerization reaction in some material resulting in a highly colored polymer, or diffusion of a dye along a wick. In each case the reactions are temperature-dependent and the rates of reactions are carefully controlled by controlling the concentration of the indicator's constituents. Some of the indicators are activated at the time of manufacture, requiring their storage at ultralow temperature until they are affixed to a product package. Other indicators are activable at any desired time, such as when the TTI is fixed on a package at the start of the distribution chain.

TTIs differ in how they display their response to time and temperature. For example, a TTI involving polymerization reaction involves the use of a bar-coded form. In this case, a bar of an indicator polymer is combined with bars of other information about the product. This type of TTI requires an electronic reader that is used to obtain information on the shelf life as well as additional related information about the product. A TTI based on biochemical reactions involves a pouch containing an enzyme and a substrate that are initially kept separate by a thin seal. At the time of activation, the seal separating the pouch contents is broken and the contents are mixed. The pouch is made of a transparent film and is placed inside a plastic sleeve with a window. As the enzymatic reaction in the pouch proceeds as a result of the combined effect of temperature and time, the pH of the mix changes, resulting in a change in color. The user observes the color of the pouch's contents to determine the time–temperature exposure. In another type of TTI, the movement of a dye along a wick is indicative of time and temperature exposure because the diffusion process is temperature-dependent. A rectangular wick is typically sandwiched inside an outer sleeve that contains viewing holes and an appropriate scale. The advance of the dye along the wick is visually observed and related to the extent of the reaction. Another method used to display the indicator response is the use of a bull's-eye concept. In this case an indicator is located at the inner circle, while a surrounding ring is painted with a color that is eventually reached by the indicator when the shelf life of the product is deemed to have expired. In this type of a TTI, a user compares the color of the indicator seen in the center with the color of outside ring(s).

TTIs are useful in implementation inventory management policies other than the commonly used first-in–first-out (FIFO) policy. The time-based FIFO policy fails when the food-chain temperature exceeds certain critical limits. In such cases, time–temperature-based policies such as shortest remaining shelf life (SRSL), using the TTIs, offers several advantages. For example, SRSL, policy assures more consistent quality at the final consumption of food items that may have been exposed to different abusive temperature conditions during storage and distribution. In the SRSL policy, using TTIs, those foods that receive an elevated temperature exposure are moved more rapidly through the system.

The TTIs offer a new and improved method to manage the food storage and distribution system, with the ultimate goal of providing a higher-quality food for the consumer.

BIBLIOGRAPHY

1. P. S. Toukis, B. Fu, and T. P. Labuza, "Time–Temperature Indicators," *Food Technol.* **45**(10), 70–82 (1991).
2. J. H. Wells, and R. P. Singh, "The Application of Time–Temperature Indicator Technology to Food Quality Monitoring and Perishable Inventory Management" in S. Thorne, ed., *Mathematical Modeling of Food Processing Operations,* Elsevier Applied Sciences, London, 1992, Chapt 7.

R. Paul Singh
University of California
Davis, California

TOTAL QUALITY MANAGEMENT

Overview

Since the early 1980s the packaging industry, as well as other industries, has been working with an organizational concept called by various names but generally referred to as *continuous quality improvement* or *total quality management.* And, as with other industries, the degree of success among packaging companies using the process ranged from very successful to not so successful.

A significant problem for the packaging community has been its traditional view of quality as an internal process focused on compliance to regulations and/or specifications. This is particularly true of segments that deal with the food, beverage, and pharmaceutical markets. As a result, the industry has maintained much of its historic dependence on quality control, quality assurance, and technical audit concepts. Statistical process control is now used broadly; however, this has been seen as a technical tool rather than as a management process.

Also, generations of managers had been taught to believe that quality was for specialists and certainly not for executives, and that the results of quality techniques were not the appropriate concern of the total organization.

In the last 3 years, certain elements of the general news media, industry press, and the consulting profession have proclaimed that "TQM is dead." These same pundits would also say that if a ship misses a harbor and runs aground, it is the harbor's fault. What we have seen as an industry is the need to refine the concept of quality and apply it as a management system aimed at managing change and optimizing business performance.

It is the purpose of this article to present a summary of lessons learned from past experiences of companies using quality management and the key success factors that will enable organizations to satisfy customers, involve employees, improve work processes, and increase profits.

Managing Change

A strong negative perception of quality management is that it takes too much time to effect a change in the way companies do business and obtain positive results. Certainly one of the primary lessons learned from the experiences of businesses large and small is that the quality process must be managed with short- and long-range objectives in mind.

This is especially true of the packaging industry. With the tremendous changes occurring in industry consolidation, materials substitution, processing and distribution technology, global competition, intensified government regulation, and environmental concerns, no organization can afford to spend years waiting for major improvements to take place in operating results. One industry response to this situation has been reengineering, which has had mixed results to say the least.

The fact is that, while it does take time to complete a culture change in all except the smallest companies, it is not true that *results* need to be a long time in coming.

Identification and quantification of improvement opportunities within critical processes can be accomplished within weeks of the start of a quality initiative. Common analysis techniques employed include process mapping, value-chain analysis, activity-based cost of quality benchmarking, customer service measures, and variability studies. Cross-functional teams of internal subject matter experts can be educated quickly in advanced problem-solving methodologies and supported by specialized software to facilitate planning, decisionmaking, and project tracking, reducing or completely eliminating the need for time-consuming status review meetings. Many companies have shown dramatic results in 6 months' time.

A Rose by Any Other Name

One important principle is so straightforward as to be considered trivial by some. *Focus on organizational and business results and don't worry about what the process is called.* Many organizations fall victim to the name game and fail to endorse quality management because of the negative stigma associated with "TQM."

The U.S. General Accounting Office report, released in May 1991, was one of the first studies that linked certain organizational competences to improved business results (1). These positive characteristics are further defined in the criteria established by the Malcom Baldrige National Quality Award. If customer focus, cost reduction, employee participation, and competitive advantage are among an organization's desired results, then the name of the process used to achieve these should not be a major concern.

A Strategic Imperative

Quality is not a function but a strategic plan for change. Corporate leadership must ensure that quality is defined and integrated into the strategic plan at all levels of the organization. Several areas are critical:

Leadership—executives must provide vision, direction, resources, and ongoing personal support and must remove cultural obstacles.

Commitment—executives must understand first the positive business results to be expected from quality management, and must establish and deploy policies that guide the organization toward the realization of these results.

Evaluation—all levels of management must evaluate the quality process systematically using measurements appropriate to each level.

Alignment

For quality-management processes to produce superior business results, all functions, areas, systems, processes, groups, and individuals must operate congruently. Internal capabilities must match customer needs. Marketing and sales plans must be based on customers' expectations for product and service excellence rather than on quotas or commission schedules. Financial plans must reflect an investment in customer satisfaction and employee involvement, not just an abstract reference to bottom-line results. Employees must be educated and dedicated to performing their roles in the business, and they must understand the rewards that they and their workgroups will receive for minimizing waste and improving their work processes.

Customer Focus

Virtually all companies have learned to say that they are customer-driven. And it is generally agreed that the packaging industry has gotten closer to its customers. However, there is abundant evidence that the voice of the customer may not be heard as clearly as it should be throughout many companies.

Communication with customers is one of the major areas where important lessons have been learned over the past decade. Customers were at one time considered to be the responsibility of a few groups or individuals within a company. Typically, this was a sales/marketing function with some top managements joining in for key accounts. Today, everyone in an organization must be able to listen for and respond to customer messages.

Successful organizations use a variety of tools to monitor present and future customer needs and expectations—surveys, focus groups, formal and informal audits, quality-function-deployment models, and decision matrices, to name a few The old comment, "We know what the customer wants," without actual use of actionable customer data, is being heard less and less.

The key principles are to identify and translate customer requirements from broad categories, eg, "customer satisfaction," "superior quality," and "competitive price," into specific work-process improvements all levels of the organization can implement.

Financial Impact

One of the major reasons given by senior managers for their disaffection with quality-management process is that there is no financial payoff. This attitude is very disturbing to quality professionals because quality management can have a major impact on bottom-line results and stakeholder value.

In their widely quoted book, *The PIMS Principles,* Gale and Buzzell present a compelling case that quality has a major positive effect on market share and profitability (2). Financial impact tools—cost of quality, customer value analysis, and others—have been available for many years but are little used for several reasons:

- General fear or distrust of measurement in general.
- Traditional accounting systems do not lend themselves to process costing.
- Data-collection systems are considered too complex to maintain.
- Cost-of-quality systems are considered to be an extension of the finance function.
- Employees are not properly educated on the use of the financial impact system.
- No constructive action is taken once the data are gathered.
- Interdepartmental turf issues prevent cooperation.
- Financial impact systems are used for punishment or are not linked to compensation.

Financial impact systems are vital to the continued success of quality management processes. They provide a primary means of measuring business success, secure management's involvement and support, and help to identify and prioritize opportunities for improvement.

Employee Involvement

The word "employees" in the subheading above refers to all individuals in the organization, from senior management through all levels of the workforce. All individuals require an understanding of the quality process and their role in the new culture. They should be aware of several factors: the common language of quality, how to apply the concepts to their jobs on a day-to-day basis, and the use of selected skills such as problem solving or statistical process control in order to cause improvements to happen in their workgroups.

Quality tools and skills are organizational, managerial, individual, and technical. Most organizations concentrate on implementing individual and technical tools before they do anything about the organization and its management. This occurs when the senior management does not understand the philosophy behind the quality process. Executives and managers want to satisfy customers in an effective manner, but they must also learn to see quality as a system, and they must learn their role in the system.

When employees have learned their roles, tools, and techniques, they must then be formally (and sometimes forcefully) able to use this knowledge to carry out the business objectives of the company. Many companies are frustrated when recently educated employees fail to participate actively in the quality process. The reasons are clear: employee involvement was not a part of the traditional corporate culture; they did not understand that they had permission to get involved; they did not see this new behavior as consistent with what their supervisors were doing; they did not know how they fit into the new quality plan; and they had not been convinced of what benefits would be theirs if they participated.

Next to the lack of sustained management commitment for the reasons cited in the "Financial Impact" section, the biggest obstacle to most companies in achieving success in their quality processes has been the lack of employee involvement. The keys are education, recognition, and compensation.

Customer–Supplier Relationships

A large number of packaging companies have established successful partnerships with their key customers and suppliers. Several elements are vital to establishing such productive relationships.

As a part of their strategic planning process, successful companies involve customers and suppliers extensively in planning for new and improved products and higher levels of product and service performance. Systems such as quality-function deployment aid customers and suppliers in matching customer needs with supplier capabilities and communicating these factors throughout both organizations.

Measurement systems are developed jointly between partners and used as a means of improvement. Measurement extends to the development and sharing of testing methods, equipment, and research techniques.

Vendor certification and rating systems are being used more and more as companies seek to develop a common language of quality improvement. The exact form of these systems is specific to each customer. Many formats, however, follow the general structure of the Malcom Baldrige or ISO 9000 criteria.

Summary and Conclusions

In the remainder of the 1900s and beyond, the packaging industry will be faced with an increasing challenge to manage within narrower time limits and to be even more focused on customers and customers' customers. At the same time, companies will be required to spend more time and effort developing the "people side" of the business at all levels.

Quality-management processes are an effective system for managing 21st-century organizations and maximizing total business performance in the age of intense competition and dramatic change.

The successful organizations will employ the following key success factors:

- Strong management commitment to leading and supporting a quality-improvement process
- Aggressive customer focus using quantitative techniques for requirement identification
- A planning system that involves broad participation and integrates quality objectives into strategic and operational planning
- Immediate identification of critical work processes and major business success drivers
- Effective, open communication system
- Quality-process education for all employees, suppliers, and customers
- Measurement systems that support continuous improvement
- Collaborative relationships with customers and suppliers
- Compensation system for all employees tied to quality improvement
- Recognition of all who participate in the quality process

BIBLIOGRAPHY

1. A. Medelowitz, *U.S. Companies Improve Performance through Quality Efforts,* GAO/NSIAD-91-190, 1991
2. R. Buzzel and B. Gale, *The Pims Principles,* Macmillan, New York, 1987.

Mark B. Eubanks
MBE Associates
Alpharetta, Georgia

TRANSPORTATION CODES

Two types of transportation codes are in use in the United States with regard to packaging: federal regulations and carrier rules. Carrier rules such as Item 222 and Rule 41 are the type that is more familiar to the packaging community. They have been in effect longest and have changed least over the past several years.

Carrier Rules

Carrier packaging rules are found in the carriers' tariffs and must be followed if the shipper wants coverage by the carrier for damage claims. Carriers reserve the right to refuse articles they consider inadequately packaged, and conformance to packaging rules is implicit in collection of shipping rates as listed in the tariffs. The Airline Tariff Publishing Company (air cargo) and Air Transport Association of America (airline) publish North American tariffs, but these do not include detailed packaging instructions except for special articles such as live animals, human remains, and seafood. Individual carriers, United Parcel Service, Federal Express, and the U.S. Postal Service, publish their own tariffs, which also typically lack detailed packaging instructions.

The American Trucking Association and the National Railroad Freight Committee publish the National Motor Freight Classification (NMFC) and the Uniform Freight Classification (UFC), respectively. These contain detailed packaging rules as well as rates, which apply to the participating individual carriers named in the tariffs. The packaging rules mandate minimum quality values for material of construction based on the total gross weight and united dimensions (length + width + depth) of the package and its contents.

Item 222. Item 222 is the rule for corrugated and solid-fiber packaging of articles transported in the less-than-truckload (LTL) common-carrier mode. The majority of the goods transported in the United States today are transported by this mode at some time in their distribution cycle. If the NMFC specifies "in boxes" in the entry for the article to be transported, the outside of the box must carry a circular box manufacturers certificate that precisely coincides with the instructions in the rule and certifies compliance with the manufacturing instructions in the rule, or damage claims and rates may not be honored. If the entry for the article specifies "in package number . . . ," detailed instructions for that (those) numbered package(s) are listed in the section "Specifications for Numbered Packages," which must be followed, and a rectangular certificate is applied indicating the package number and grade of board used. The numbered package certificate is the only rectangular certificate that is recognized by carriers for tariff compliance. All others are simply advertisements.

Rule 41. Rule 41 is the corrugated and solid-fiber packaging rule for articles shipped by rail. However, piggyback shipments in trucks on railroad flatcars are subject to truck rules. Rule 41 and Item 222 are very similar, but not identical. The numbered packages can be quite different. Rule 41 is the older of the two and is the model on which Item 222 is based, but articles in boxes are shipped most often by truck making Item 222 the rule to be consulted.

Edge-crush test (ECT). In 1990 the corrugated industry trade associations sponsored proposals to revise Item 222 and Rule 41. The primary thrust of the proposals was to have the rules recognize edge crush test criteria as an indicator of box-compression strength, and allow its use as an option to the traditional linerboard basis weight and combined board-burst requirements. ECT is a characteristic of the combined corrugated board that directly predicts compression strength of the corrugated packaging. By providing alternative requirements in the carrier rules, box manufacturers would have more latitude to design and supply boxes that better meet the customer's performance requirements. This concept is supported by organizations representing carriers, users, and box and containerboard producers with the growing recognition that box compression strength may be a more important factor as unitized carrier loading and warehousing has proliferated. Enhanced containerboard materials achieving a guaranteed crush or compression performance at lower basis weights had become available, providing potential environmental advantages and improved quality through source reduced packaging systems and increased recycled content.

Use as standards. Over a long period, the terminology and minimum requirements for corrugated packaging used in the carrier rules became an assumed standard for boxes. Using the carrier rules as standards is often inappropriate, as the carrier rules were not developed to address the box users' warehousing and other distribution needs. They were developed to delineate the carrier's liability. As the box-manufacturing industry and its customers emphasize quality and efficiency, corrugated-box standards should evolve to support the new emphasis.

Federal Regulations

Packaging itself is not typically of federal interest; however certain specific issues are regulated by U.S. government agencies. Consequently, packaging is sometimes regulated by those agencies. Federal regulations that affect packaging are found in the *Code of Federal Regulations* (*CFR*) of the agency under whose jurisdiction the issue resides. For example, the U.S. Department of Transportation (DOT) regulates the transportation of hazardous materials. Regulations for packaging hazardous materials for transport are in DOT's *CFR*, Title 49. Packaging of these materials is based on the capability of the packaging to pass a battery of tests that are dictated by the degree of hazard of the material, or its "packing group." These regulations were adopted by DOT from the UN (United Nations) Recommended Rules, replacing the former DOT specification packages such as the DOT 12B.

DOT shares the regulation of packaging infectious substances and regulated medical wastes with the Department of Labor (29CFR) due to OSHA concerns over bloodborne pathogens, and of packaging hazardous wastes and pesticides with the Environmental Protection Agency (40 *CFR*) due to potential environmental impact. The Food and Drug Administration (21 *CFR*) regulates direct contact packaging of fatty and aqueous foods as indirect food additives. The Department of Agriculture (9 *CFR*) regulates the packaging of meats and poultry under meat inspection and poultry inspection.

Labeling. EPA assisted the Federal Trade Commission (16 *CFR*) in defining the voluntary ecolabel statements that are allowed, such as recycled, recyclable, and biodegradable. Aside from environmental claims, package labeling issued are generally specific to the article inside the package. With transport packaging often doubling as consumer packaging, this has become an important concern. FDA regulates the nutritional labeling of packages with respect to food contents. Both FDA and FTC require metric as well as standard measurement units displayed on the packages of many consumer products. DOT gives very specific instructions for labeling hazardous-materials packages, even specifying the colors to be used. EPA requires products (including packages) that are manufactured using or containing ozone-depleting substances to be specifically labeled. Package labeling of regulated articles should be a joint effort between package manufacturer and user.

BIBLIOGRAPHY

General References

Airline Tariff Publishing Company, Fairfax, VA.

Air Transport Association of America, Annapolis Junction, MD.

National Motor Freight Classification, American Trucking Associations, Alexandria, VA.

Uniform Freight Classification, National Railroad Freight Committee, Atlanta.

Code of Federal Regulations, Superintendent of Documents/U.S. Government Printing Office, Washington, DC.

Fibre Box Handbook, Fibre Box Association, Rolling Meadows, IL.

Mary Alice Opfer
Fibre Box Association
Rolling Meadows, Illinois

TRAYS, BARRIER-FOAM

These composite trays are made of expanded polystyrene with a high-gas-barrier layer and a heat-activated sealant. The tray is designed to be hermetically sealed on its flange for vacuum or gas-flushed packaging of refrigerated fresh, processed, and cooked meat, poultry, and other foods. The refrigerated shelf life of these foods is thereby extended so that they may be packed at a processing plant and shipped to retail outlets. The unique features of this tray are its similar appearance to widely used foam trays and low cost relative to solid plastic barrier trays. The primary process used for making trays is extrusion of foamed polystyrene sheet and thermoforming the trays (see Trays, foam). A high-oxygen-barrier sealant film is laminated to the sheet in-line as it passes through a set of rollers (S-wraps) right after extrusion. This film not only provides the oxygen barrier but also has a top layer for heat sealing a lidding film to provide a hermetic seal. Trays may be formed either by conventional means or with a specially equipped horizontal form/fill/seal machine in the food-processor plant.

History

Amoco Foam Products Company of Atlanta, Georgia (USA) developed the first commercial application of this product in 1986 for Perdue Foods in Bridgewater, Virginia. There were samples and prototypes made earlier in Japan, but no commercial application. For this application, Amoco made foam rollstock, which was formed on specially equipped Multivac R5200 horizontal form/fill/seal machines at Perdue's plant. Perdue packaged precooked chicken products in these trays, which were then evacuated and gas-flushed. Perdue doubled the refrigerated shelf life of the product, guaranteeing the retailer 17 days to sell it. The package received three major industry awards:

- Food & Drug Packaging's 1987 Packaging Achievement of the Year
- 1987 DuPont Diamond Award
- 1988 Package of the Year from Packaging Institute International, now part of the Institute of Packaging Professionals (IOPP).

Individual formed trays became commercially available in early 1991. Because the film layer comprises multiple plastics, reusing scrap was problematic, requiring considerable effort to develop an economically and technically viable system. In addition to Amoco, W. R. Grace fielded an entry to package fresh turkey at Rocco, Dayton, Virginia. Another major application was packaging fresh beef and pork for Tesco supermarkets in the United Kingdom. Like the Perdue package, these trays were gas-flushed to extend shelf life.

Manufacturing Process

The process is basically the same as for manufacturing foam rollstock and trays. The major difference is the film lamination. The film must be able to adhere to polystyrene under heat and pressure. Corona bonding may also improve the adhesion. The film must also meet the oxygen-barrier requirement similar to most processed-meat films of less than 2.0 mL/100 in.^{2} (15.5 mL/m^2) per day at 73°F (23°C) at 0% relative humidity. The film must also have a heat-seal layer, typically low-density or linear low-density polyethylene (LDPE or LLDPE) sometimes blended with ethylene vinyl acetate (EVA), for sealing to a lid material.

There are somewhat different techniques for manufacturing rollstock for form/fill/seal and manufacturing trays. The film for rollstock can generally be made of any material that meets the preceding requirements. There is very little scrap, so reclaiming it is not critical to the economics. The main requirement for making rollstock is precision in width. The form/fill/seal machines have a tolerance of 0.2 in. (5 mm) +5, −0 millimeters (+.2, −0 inches) in width. Since foam has orientation, it usually shrinks after manufacturing. The sheet is 65–75 mils (1.7–1.9 mm) thick and is typically slit to widths of 15.7–25.6 in. (400–650 mm). Rolls are 60–84 in. (1.5–2.1 m) in diameter and 3000–6000 ft (900–1800 m) long to minimize roll changes and maximize loading for shipment. (See Fig. 1.)

Making trays is a much more complex process. The need to reclaim and reuse scrap affects each component of the process back to the manufacture of the film. First, the film must be made of materials that can be reclaimed with polystyrene. While a variety of materials are claimed in patents, the typical film structure is a 2–3-mil (50–75-μm) structure with a

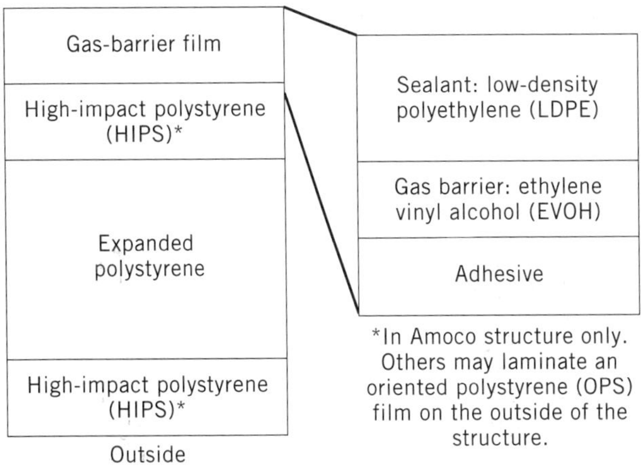

Figure 1. Barrier foam tray: material cross section.

sealant such as LLDPE or LLDPE/EVA, EVOH (ethylene vinyl alcohol) barrier resin, and one or more tie layers of EVA or EMA (ethylene methacrylic acid). After the trays are formed and trimmed, the scrap is blended back into the polystyrene resin for foaming at a rate of up to 50% by weight. A blend containing up to 30% of barrier film by weight can be reprocessed. In addition, a blend of up to 99% scrap and the remainder impact polystyrene can be extruded as 2–3 mils (50–75 μm) films and simultaneously laminated to the foam. This reclaim process may be aided by compatibilizers such as Kraton (registered trademark) styrene–butadiene block copolymers. Although all producers can reclaim scrap internally, the tray is not being widely recycled by consumers because it is classified as "number 7 other," and the final package has remnants of incompatible lidstock.

Key Producers

There are three producers of barrier foam trays: Amoco Foam Products Company (Chippewa Falls, WI), W. R. Grace Formpac (Reading, PA), and Linpac Plastics International (Knottingley, West Yorkshire, England). Grace's trays are comarketed by its parent company, Cryovac (Duncan, SC), the largest producer of films for fresh food. Linpac to date markets only in Europe. Amoco also sells barrier rollstock from its plant in Winchester, Virginia.

Because barrier foam trays can cost two to three times more than polystyrene foam trays of a similar size, storage and transportation is a much smaller component of cost. Trays can be shipped over 1000 m (1600 km) from the manufacturing plant and keep these costs less than 5% of total costs.

Products, Uses, and Applications

Generally, barrier foam trays can be used in any application where a gas-barrier tray is needed and the use temperature is below 212°F (100°C). Typically, this would include modified-atmosphere packaging (MAP) or vacuum packaging of fresh refrigerated foods. Although manufacturers do not warrant the product for microwave re-heating, some food processors have included package instuctions. To date, this package is used for the Perdue precooked poultry mentioned above, fresh turkey and fresh sausage in the United States. Beef and pork are also packed in them, mostly in the United Kingdom. Producers are promoting its use for the latter application in the United States.

Although pricing can vary widely by barrier properties and package size, barrier foam is generally 30–70% less costly than other equivalent rigid plastic barrier sheet materials in tray or sheet form. This would include polyvinylidene chloride–coated polyvinylidene chloride laminated with polyethylene sealant (PVC/PVDC/PE), HIPS/EVOH/PE coextrusions, Barex (registered trademark) acrylonytrile copolymer, and others. However, foam is not clear, is limited in its draw ratio to about 50% of the width of the article, and sidewalls must be at least 25° from vertical (see also Barrier polymers).

Packaging a product in MAP or vacuum with barrier foam is essentially the same as any other barrier tray or rollstock material. The equipment draws a vacuum on a tray loaded with product, and, if desired, inserts a gas mix. For fresh meats where color is important, processors use 20–30% CO_2 and 40–80% O_2, with the remainder nitrogen (if needed).

Table 1. Common Barrier-Foam Tray Sizes

Tray	in. Width × Length × Height	mm Width × Length × Height
	Amoco	
3P	8.81 × 6.75 × 1.03	224 × 171 × 26
3H	8.81 × 6.75 × 1.49	224 × 171 × 38
3D	8.81 × 6.75 × 1.86	224 × 171 × 47
3DD	8.81 × 6.75 × 2.36	224 × 171 × 60
4D	10.00 × 7.63 × 1.86	254 × 194 × 32
10B	10.75 × 6.75 × 1.75	273 × 171 × 44
10P	10.75 × 6.75 × 1.03	273 × 171 × 26
	Linpac (includes UK sizes)	
2–37	7.70 × 6.00 × 1.60	196 × 152 × 41
2–45	7.70 × 6.00 × 1.90	196 × 152 × 48
31	7.96 × 5.96 × 2.25	202 × 151 × 57
3D	8.75 × 6.50 × 1.75	222 × 165 × 44
3–37	8.80 × 6.75 × 1.60	224 × 171 × 41
3–45	8.80 × 6.75 × 1.90	224 × 171 × 48
31–37	9.30 × 6.50 × 1.60	236 × 165 × 41
31–45	9.30 × 6.50 × 1.90	236 × 165 × 48
14–37	9.30 × 7.75 × 1.60	236 × 197 × 41
14–55	9.30 × 7.75 × 2.25	236 × 197 × 57
18–37	10.70 × 9.30 × 1.80	272 × 236 × 46
33	10.75 × 6.50 × 1.80	273 × 165 × 46
21–37	13.40 × 10.00 × 1.80	340 × 254 × 46

Source: Amoco Foam Products Company and Linpac Packaging Systems. Sizes and designations may differ somewhat by producer and location.

Shelf life ranges from 8–12 days for ground beef up to 18 days for fresh pork and ground turkey. For precooked or processed meats, typical mixes are 20–30% CO_2 with the remainder nitrogen, resulting in shelf lives of 30 days or more. Sometimes all nitrogen is used if bacterial growth is slowed by other means. The tray is sealed with a barrier lidstock using a polyethylene sealant. Tray-sealing equipment made by Ross, Mahaffey & Harder, Multivac, Raque, Linfresh, and others is virtually identical to equipment used for other barrier trays. Rollstock equipment is modified to handle the forming, trimming, and handling of foam; thus it is about 40% more expensive. While a rollstock application was the first use of barrier foam, most subsequent uses have been for preformed trays.

The primary target market for barrier foam is the modified-atmosphere packaging of fresh meat and poultry. The tray is very similar to the overwrapped foam tray widely used to package these items. Consumers perceive this readily and also recognize other benefits such as a leakproof seal and abuse-resistant lidding. Between 1991 and 1995 the British supermarket chain Tesco converted all the packaging of beef, pork, lamb, and veal to barrier-foam trays using modified atmosphere on Ross tray sealers. All meat cutting has been removed from the store. Consumer acceptance of this package has been good, and Tesco recently overtook Sainsbury as the leading UK chain in market share.

Since MAP requires considerable headspace around the meat to circulate gases, most barrier-foam trays are an inch deep or more. While nearly all shape or size is possible, most trays are rectangular, partially to minimize scrap. Colors are currently limited to yellow and white, although any color is technically possible. Tray flanges are at least 0.375 in. (10 cm), to guarantee at least a 0.25-in. (6-mm)-wide seal area required by most tray sealing machines. The combination of the flange width and headspace means that the finished package can be as much as 40% larger than the package it replaces. To partly offset this "overpackaged" appearance, Linpac recently developed a tray-sealing machine that will seal a tray with about half the flange width (see Table 1).

While sales are still a tiny fraction of the foam-tray market, or even the barrier-tray market, growth has been quite rapid. The key producers are devoting significant resources to it to realize the potential offered by case-ready meat.

BIBLIOGRAPHY

General References

S. Burns, Tesco Stores, Ltd., interview, May 17, 1996.

H. C. Keith, "Options For Meat Packaging," paper presented at Packaging Challenges for the Meat Industry Expo '89 Pre-Convention Conference, American Meat Institute Center for Continuing Education, Sept. 1989, Chicago, Illinois.

U.S. Pat. 4,847,148 (July 2, 1989), H. G. Schirmer (to W. R. Grace & Co., Connecticut).

U.S. Pat. 5,118,561 (June 2, 1992), G. G. Gusavage, T. A. Hessen, T. R. Hardy, S. R. Flye, and H. G. Schirmer (to W. R. Grace & Co., Connecticut).

U.S. Pat. 5,128,196 (July 7, 1992), M. L. Luetkens, Jr., R. D. Pischke, and J. C. Schubert (to Amoco Corporation).

U.S. Pat. 5,221,395 (June 22, 1993), M. L. Luetkens, Jr., R. D. Pischke, and J. C. Schubert (to Amoco Corporation).

For further information, contact the companies mentioned in the article.

HUSTON KEITH
Keymark Associates
Marietta, Georgia

TRAYS, FOAM

These lightweight, inexpensive trays are made of expanded polystyrene. The major packaging applications are for packaging fresh meat, poultry, eggs, produce, and other foods for retail sale or packaging prepared foods for imminent consumption. Other uses include industrial protective packaging and disposable tableware (plates, bowls, cups, trays, etc). The primary process used for making trays is extrusion of foamed polystyrene sheet (Fig. 1) and thermoforming the trays. A minor quantity of trays are made from steam molding of precharged beads—this process is used mostly for cups, industrial packaging, and insulation. This article focuses on the thermoformed trays as being of primary interest to packaging, but discusses the other applications briefly in context of packaging use.

History

The process of making polystyrene foam sheet for thermoforming was developed in the late 1950s, based on technology first commercialized by Dow Chemical Company in 1943 for extruding foam board for insulation. It was called the "direct-injection process," because gas was injected into the extruder and mixed with melted polymer, as compared to the precharged bead process, where resin beads are impregnated by the gas. Dow then started a plant in Carteret, New Jersey to produce sheet and trays. Because of Dow's prominence in commercializing these processes and continued dominance of the polystyrene resin market, its trade name Styrofoam for foam insulation board became associated with nearly any form of expanded polystyrene. In fact, many people improperly use the trade name Styrofoam for all polystyrene foam products.

The initial application for this process for packaging was for forming meat trays. By the mid-1960s, equipment and techniques were developed for forming egg cartons. Dow formed a joint venture with a major egg producer, Olson Farms, to produce egg cartons. The venture, Dolco Packaging, later became an independent company. In the late 1960s, food-service containers and packaging were developed. Foam trays began replacing pulp trays for meat and egg packaging at an accelerated rate because of lower cost and better durability. In addition, the growth of the supermarket and especially the food-service industry propelled rapid increases in the use of this package.

The rapid growth of extruded-polystyrene foam packaging continued as more users were found. Foam disposable plates and bowls found acceptance, as did carry out containers in an affluent society driven more and more by convenience. The fast growth of polystyrene attracted petrochemical giants Amoco and Mobil into the business. In 1973, polystyrene foam containers were adopted by McDonald's as its primary hamburger package. Better heat retention, enhanced produce protection and concerns for forest depletion and landfill stability were cited as benefits. This led to over two decades of unprecedented industry growth, from 15 million lb (6,800 metric tons) in 1966 to 651 million lb in 1988 (295,900 metric tons) in the United States alone. Costs were driven lower by more efficient production, making the replacement of substitutes (mostly paper) more and more compelling.

Under pressure from environmental groups, led by the Environmental Defense Fund (EDF), McDonald's first placed a moratorium on chlorofluorocarbon (CFC) use in its containers, then announced in November 1989 that it would discontinue the use of polystyrene foam trays for hamburger packaging, replacing them with a paper-based wrap. Other fast-food suppliers followed suit. This caused a loss of 90 million lb of demand in the face of industry overcapacity. One supplier entered Chapter 11 bankruptcy (since recovered), and at least two plants were closed.

After a period of retrenchment, the industry began to slowly recover. While the fast-food packaging market is moribund, the low cost and stiffness of foam relative to alternatives has provided continued growth in disposable tableware. The growth of the poultry market has also pulled along foam tray consumption. While the foam tray may never see the usage in fast food as before, it has proved to be an almost irreplaceable packaging container for fresh meat and poultry.

Manufacturing Process

As mentioned earlier, most foam trays are made by extruding foam sheet and thermoforming it into trays. While this pro-

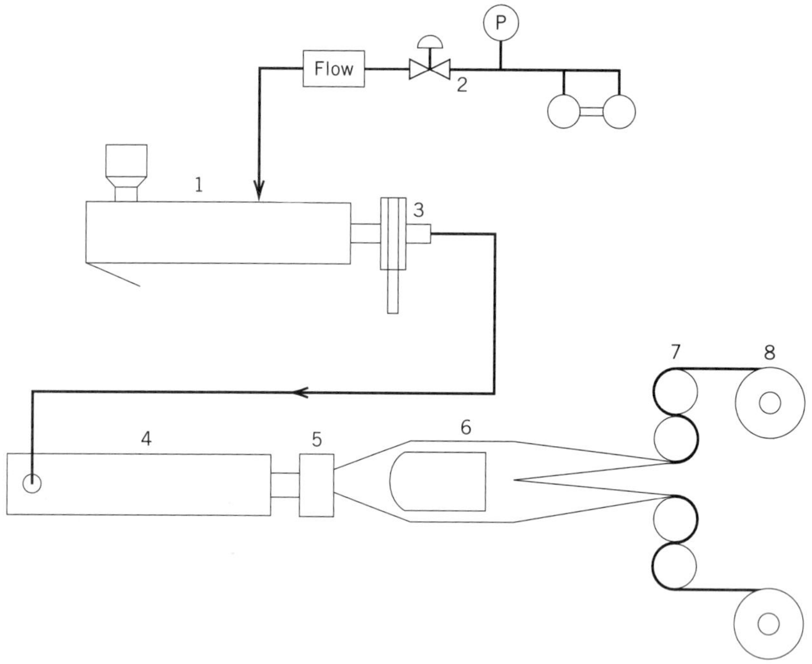

Figure 1. A tandem extrusion process for the manufacture of polystyrene foam sheet: 1 primary extruder, 2 blowing-agent addition system, 3 screen changer, 4 secondary extruder, 5 annular die, 6 cooling mandrel, 7 S-wrap, 8 winders.

cess is similar to the process used to make solid plastic trays, it also differs in several significant aspects. Two extruders are used instead of just one, an annular die is used rather than a flat one, rolls must be aged before forming, and matched metal dies are used for forming. As a result, capital costs are much higher than for solid plastic. On the other hand, raw-material costs are much lower as a percentage.

The typical tandem extrusion system includes a primary extruder for melting and mixing the polymer and a secondary extruder connected in series for allowing the extrudate to cool and expand. While a single extruder may be used, large commercial operations use tandem systems for better-quality sheet. The blowing agent is injected into the primary extruder. The melt then passes into the secondary extruder, which is usually a third larger in diameter (78% larger in volume), primarily for cooling. The extrudate then exits through an annular die, expands, and flows over a mandrel. The resulting tube is slit to make a flat sheet. Sometimes it is slit twice to form two sheets. The diameter of the mandrel determines sheet width and provides biaxial orientation. The sheet is then pulled through a series of two to three rollers (S-wraps) to remove the curl, make the thickness uniform, and cool it for winding.

A typical commercial extrusion setup is 4.5 in. (115 mm in nominal size) in diameter for the primary extruder and 6.0 in. (150 mm) for the secondary extruder for 500–1000 lb (275–450 kg) per hour. A common configuration for output of 800–1500 lb (350–700 kg) involves a 6.0 in. (150-mm) primary and 8.0 in. (200-mm) secondary. The resulting foam has fine, closed cells with a density of 2–10 lb/ft^3 (32–160 g/L). For packaging applications, the resulting sheet typically has a basis weight of 3–25 g/100 $in.^2$ (50–400 g/m^2 with a thickness of 30–200 mils (0.75–5.00 mm). Sheet width typically ranges from 20 to 52 in. (500 to 1325 mm).

A variety of blowing agents may be used. The majority of producers used chlorofluorocarbons until 1988 (see below). Others used butane, pentane, or other hydrocarbon gases. While hydrocarbons are much less expensive, they are highly flammable. Extreme precautions had to be taken to prevent fires and explosions. As noted below in the "Environmental Issues" section, all these gases are being reduced or eliminated for environmental concerns. A leading replacement candidate is carbon dioxide (CO_2). Dow has developed technology for using 100% CO_2 that it is licensing. Other gases are being investigated.

During the extrusion process, special finishes and laminates may be applied for graphics, gloss, thickness reduction without loss in strength, gas barrier, or other purposes. As the foam exits the extruder, it may be hit with a stream of air from completely around the tube. Called an *air-ring coating,* the foam cells near the surface are collapsed, forming a very thin glossy skin on one side of the sheet. Some producers have also developed a technique of coextruding a thin (< 1 mil or 25 μm) layer of polymer (usually polystyrene) through the same annular die that produces the foam. At the S-wraps, a premade and sometimes preprinted film (oriented polystyrene or barrier) may be laminated in the nip. One producer accomplishes the same feat by extruding a thin-impact polystyrene film onto the foam sheet in the nip.

After the foam has been extruded and wound onto rolls 60–84 in. (1.5–2.1 m) in diameter, it is placed in a storage area for 1–3 days to allow the blowing agent to partially escape, for optimal expansion in thermoforming. After aging, the foam is taken and placed on an unwind stand for thermoforming.

Thermoforming foam sheet into trays is somewhat more complex than forming plastic sheet. Because foam is an insulator, heating the sheet to forming temperatures requires more time. Distribution of heat is also critical. During the heating process, the foam reexpands to nearly twice its stating thickness. The expansion needs to be as uniform as possible to produce a good part. Thus on production machines, ovens are much longer than typical thermoformers. When the sheet enters the mold platen, the male and female molds close to a thickness slightly less than the starting sheet thickness. This provides very uniform wall thickness throughout the part. This helps provide the same strength as a much heavier comparable plastic part. Although foam trays can be trimmed in place, most large commercial equipment uses a separate trim station. Punch-and-die trimming is normally used for large commercial operations for higher speed and more durability, although steel-rule dies may be used for lower volume situations.

After molded parts are trimmed, the remaining sheet "skeleton" is ground up immediately to be stored in a silo for reuse. Since the unused scrap sheet is as much as 50% of sheet area, reusing scrap is critical to producing parts economically. While the ground-up foam or "fluff" may be reused directly, it is difficult to blend and feed properly with virgin resin pellets. Thus most producers have a reclaim operation that remelts the fluff and extrudes it as pellets.

Key builders for foam-extrusion equipment are Battenfeld Gloucester (Gloucester, MA) and Berstorff of Germany. Davis Standard (Pawtucket, RI) and Egan (Somerville, NJ), also produce foam lines. Most foam thermoformers are built by Irwin (Yakima, WA) and Brown (Beaverton, MI). The typical resin used is a high-heat grade of crystal polystyrene. Dow Chemical (Midland, MI) is the largest supplier worldwide. Others include BASF (Frankfurt, Germany), Huntsman (Salt Lake City, UT), and Chevron (Houston, TX).

The cost of an extruded-polystyrene foam tray plant is more than double the cost of a thermoformed solid-plastic-tray plant for comparable output in trays. A typical setup of an extrusion line, reclaim line, silos, two thermoformers, and tooling as of 1994 would cost about $3–$5 million. In addition, a foam plant requires more land and building for equipment, roll storage, and finished-goods storage. As a tradeoff, however, raw materials are a much lower proportion of manufacturing costs (30–50% vs 50–80%).

Key Producers

The major producer of foam trays by far is Tenneco Packaging (Evanston, IL) with five plants around the country, which purchased Mobil Chemical Company's foam business in October 1995. Amoco Foam Products Company (Atlanta, GA), with nine plants, is next, and is expected to be sold before this article is published. Both of these companies offer a broad line of both disposables and packaging products. Four other manufacturers have also recently changed ownership. Tekni-Plex (Somerville, NJ) was bought in 1994 by MST, a financial group, which then also purchased Dolco Packaging (Studio City, CA) in February 1996. Sweetheart Cup (Owings Mills, MD), primarily a disposable manufacturer, was acquired in 1993 by American Industrial Partners Capital Fund (San Francisco) from Morgan Stanley (New York). Elm Packaging, (Memphis, TN) management-led financial group acquired Fripp Fibre's operations in 1994. Other foam producers Genpak (Glens Falls, NY), Linpac (Wilson, NC), Solo Cup (Highland Park, IL), Dart Container (Mason, MI), and W. R. Grace (Reading, PA) have not changed ownership since 1993. All producers except Amoco and Tenneco specialize in one or two market segments and often compete within a limited geographic area.

Because foam is very lightweight and parts are generally three or more times thicker than parts made from competing materials, storage and transportation costs represent a much higher proportion of costs. Shipping products within one day's drive (400–500 m or 600–800 km) from the manufacturing facility can hold transportation costs under 5% of the selling cost. Some makers have been able to offset this somewhat by the use of laminates. In some applications, such as fast-food or carryout packaging, where retail storage is minimal, the value of this thickness reduction more than compensates for its increased cost. For the most part, however, foam producers have found it generally more cost-effective to locate near key markets. This easily offsets economies of scale from larger production facilities. Only in the case of specialized products are manufacturers able to ship farther than a day's drive.

Products, Uses, and Applications

The primary packaging applications for foam trays are fresh meat, poultry, and eggs. There is also considerable usage for packaging carryout foods in restaurants, although that market segment has dropped considerably since McDonald's discontinued using clamshells for hamburgers. McDonald's still uses large quantities for packaging breakfasts. There are minor quanitites used for a variety of industrial packaging applications, primarily as cushioning. Since disposable tableware is used for immediate food consumption and is not meant to protect food for transit to another location, it will not be discussed here.

The major application is for packaging fresh meat, with about 8 billion trays. These trays are sold to supermarkets and grocery stores, which cut and package meat in the backroom to display in self-service cases. Since a red, watery fluid called "purge" oozes out of the meat, an absorbent pad is placed in the tray just prior to placement of the meat. The entire package is wrapped with a clear film, usually poly(vinyl chloride) (PVC). Wrapping is now done on manual or semiautomatic equipment. The package is placed on display in a self-service case for a maximum of 2–5 days depending on the freshness life of the meat cut.

The industry has developed fairly standard sizes and colors, although there is some variation by producer. The vast majority of supermarket foam trays are white, although there is growing interest in other colors. Rose has been used for pork more frequently, blue for fish or seafood, green for produce, and yellow for poultry. Within the past 5 years, the use of black has grown dramatically, especially for upscale or niche products such as natural beef, low-fat meats, or marinates. While nearly any shape or size is possible, most of the industry has standardized on a set of shallow, square trays, partly for merchandising and partly for cost savings (see Table 1). Most meat and seafood trays are about $\frac{1}{2}$ in. (13 mm) deep, allowing most cuts to be seen from three sides. Poultry trays are typically $\geq$1 in. deep (25 mm), because of greater purge and typical practice of packing more product in each tray. Product trays are often as much as 3 in. (75 mm) deep because of the greater height of the product.

In the mid-1960s, many supermarkets began to discontinue cutting chicken for sanitation and cost reasons in favor of purchasing it already cut and packed in foam trays. Holly Farms (now Tyson) led the industry with its "deep chill" technique of crust-freezing the chicken with a quick blast at -40°F (-40°C), usually after packing it in the tray. This technique gave chicken up to 18 days' shelf life for travel to the

Table 1. Common Meat and Poultry Tray Sizes

	in.					mm				
Tray	Width	×	Length	×	Height	Width	×	Length	×	Height
Supermarket										
1	$5\frac{1}{4}$	×	$5\frac{1}{4}$	×	1	133	×	133	×	25
1S	$5\frac{1}{4}$	×	$5\frac{1}{4}$	×	$\frac{1}{2}$	133	×	133	×	13
2	$5\frac{3}{4}$	×	8	×	1	146	×	203	×	25
2S	$5\frac{3}{4}$	×	8	×	$\frac{1}{2}$	146	×	203	×	13
4	$7\frac{1}{8}$	×	$9\frac{1}{4}$	×	1	181	×	235	×	25
4S	$7\frac{1}{8}$	×	$9\frac{1}{4}$	×	$\frac{1}{2}$	181	×	235	×	13
7S	$5\frac{3}{4}$	×	$14\frac{3}{4}$	×	$\frac{1}{2}$	146	×	375	×	13
8S	$8\frac{1}{4}$	×	10	×	$\frac{1}{2}$	210	×	254	×	13
10S	$5\frac{1}{2}$	×	$10\frac{3}{4}$	×	$\frac{1}{2}$	140	×	273	×	13
12S	$9\frac{1}{4}$	×	$11\frac{1}{4}$	×	$\frac{1}{2}$	235	×	286	×	13
16S	7	×	12	×	$\frac{1}{2}$	178	×	305	×	13
17S	$4\frac{3}{4}$	×	8	×	$\frac{1}{2}$	121	×	203	×	13
20S	$6\frac{1}{2}$	×	9	×	$\frac{1}{2}$	165	×	229	×	13
23S	8	×	18	×	1	203	×	457	×	25
24S	8	×	16	×	1	203	×	406	×	25
25S	8	×	$14\frac{3}{4}$	×	1	203	×	375	×	25
1014	10	×	$13\frac{3}{4}$	×	1	254	×	349	×	25
1216	12	×	16	×	1	305	×	406	×	25
Processor										
2P	$5\frac{3}{4}$	×	8	×	$1\frac{1}{8}$	146	×	203	×	29
3P	$6\frac{5}{8}$	×	$8\frac{5}{8}$	×	$1\frac{1}{4}$	167	×	219	×	32
4P	$7\frac{1}{8}$	×	$9\frac{1}{4}$	×	$1\frac{1}{4}$	181	×	235	×	32
8P	$8\frac{1}{4}$	×	10	×	$1\frac{1}{4}$	210	×	254	×	32
10P	$5\frac{1}{2}$	×	$10\frac{3}{4}$	×	$1\frac{1}{4}$	140	×	273	×	32
11	$7\frac{5}{8}$	×	$11\frac{5}{8}$	×	$1\frac{1}{4}$	194	×	295	×	32
24P	8	×	$15\frac{5}{8}$	×	$1\frac{3}{8}$	203	×	397	×	35
25P	8	×	$14\frac{3}{4}$	×	$1\frac{1}{3}$	203	×	375	×	33

Source: Amoco Foam Products Company. Sizes and designations may differ somewhat by producer and location.

retail store and sale. The poultry industry also began using high-speed wrapping machines to gain efficiency. The deep-chilling, longer-life, and high-speed wrappers were too demanding for lightweight supermarket trays, so Amoco, Grace, and Tekniplex developed heavier-gauge trays (150–200 mils or 3–5 mm) to address this market. Increasing demands from wrapping machines have led to not only increased tray weights but also design changes such as gussets and bowed sidewalls. The processor segment now accounts for about 3 billion trays and has grown much faster than the total market as a result of increasing poultry consumption.

Because of environmental pressures and easy substitution, pulp has regained considerable market share in the 3-billion-unit egg-carton market, where the wet strength of foam is less of an advantage. The primary size are for one dozen large eggs, although half-dozen and 18-count packages are available. Colors are primarily white or yellow. Most packages are printed off-line with the packer's name. Dolco is the major producer in this market.

Although produce uses almost 2 billion trays, less than a quarter are foam, with the remainder being pulp, oriented polystyrene, and injection-molded. Trays are sold to both supermarkets and packers for packing tomatoes, mushrooms, corn, and a variety of other items. A small number goes to various berry packers. Cut fruits and vegetables are sometimes packed in supermarket trays. Because of consumer preferences for bulk produce, industry fragmentation, environmental pressures, and questionable need for rigid packages for many items, foam trays have made very little progress gaining usage in produce. Seafood is similarly fragmented. Tray usage is confined largely to frozen items, with fresh products sold over the counter wrapped in paper. Miscellaneous other foods such as pasta, deli meats and cheeses, and prepared foods, are also wrapped foam trays using typical processor or supermarket sizes.

A new development in foam trays is the *barrier-foam trays* (see Trays, barrier-foam). This product is a composite of a barrier film laminated to foam sheet, then formed into trays. This tray can be evacuated and/or gas-flushed and sealed with a barrier-film lid to make an extended-shelf-life modified-atmosphere package. It is used for precooked chicken, fresh turkey, and fresh sausage in the United States. Beef and pork are also packed in them, mostly in the United Kingdom.

Other packaging. Hinged-lid or clamshells for carryout packaging address a market of 4–5 billion trays. These trays are only designed to hold food for 1–2 h, or long enough to protect it from the kitchen to the point of consumption. They are popular for this application because of low cost, wet strength, and ability to hold food hot or cold for up to 30 min. However, they are not leakproof, and the tab holding the lid closed does not stand up well to transport abuse. The primary users are fast-food chains, restaurants, and institutional feeders. These customers also use large quantities of foam trays, plates, and bowls.

Protective packaging has been a minor market for foam trays. Volumes per part are much lower; thus mold costs become a significant expense. In addition, thermoforming generally cannot do the vertical sidewalls and intricate details required to hold a product in plate for transit. For these reasons, bead molding of expandable polystyrene beads has been the preferred material for these applications.

Environmental Issues

Polystyrene foam containers came under aggressive attack by environmental groups in the late 1980s and early 1990s. They were fingered for a variety of enviornmental ills, from creating holes in the ozone to filling landfills to over capacity to litter problems to choking water animals. While the industry began a vigorous lobbying and public relations campaign to show that these notions were greatly exaggerated or erroneous, much damage could not be prevented. Several local governments, notably Suffolk County, New York on Long Island; Berkeley, California; and Portland, Oregon, placed outright bans on the product's use. While Suffolk's law was later rescinded, others are still in place. Other locales such as Minneapolis and the State of California have passed sweeping recycling mandates that foam and other plastic containers will have difficulty meeting.

Most producers used a blowing agent called *chlorofluorocarbon* (CFC), also known by the DuPont trade name Freon 12. In 1987, the Montreal protocol established a timetable for halving CFC use by 1999, because these compounds react with ozone in the upper atmosphere, reducing the filtering of ultraviolet rays provided by the ozone layer. Then, McDonald's placed a moratorium on using packages with CFCs in 1988, forcing users to seek alternatives. While foam was actually a very small consumer of CFCs relative to air conditioning, refrigeration, and electronics manufacturing, its visibility with McDonald's caused it to receive an inordinate share of the blame for ozone depletion. Many foam producers switched to a hydrochlorofluorocarbon (HCFC), which had less than 5% of the ozone-depletion potential of a CFC, but at a much higher cost. But in 1990, the Montreal protocol was updated to eliminate all CFCs. Now most producers have been forced to switch to hydrocarbon blowing agents, although Dow's CO_2 process is used by a few.

One of the main issues with foam trays is its supposed contribution to the solid-waste problem. While the industry convincingly proved that polystyrene foam accounted for less than 0.5% of landfill weight, again visibility caused it to be a target. The package was also faulted for lack of biodegradability, although it was clearly shown that nothing degrades in a landfill. The industry, particularly Amoco, Mobil, and Dart, countered this with lobbying, public-relations efforts, and developing three recycling plants, now operated by the National Polystyrene Recycling Company, a joint venture. McDonald's even participated briefly in a recycling program, prior to its decision to discontinue using foam.

While the industry gradually recovered from the CFC and solid-waste issues, it slowly began to face a new one. The hydrocarbon blowing agents are volatile organic compounds (VOCs), which most states are restricting to protect air quality. In many states it is virtually impossible to expand, others require an environmental impact study before issuing a permit, and California is demanding outright reductions. Thus most producers are vigorously pursuing new blowing-agent technologies that will allow for expansion.

BIBLIOGRAPHY

General References

I. E. Cavallo, Amoco Foam Products Company (retired), interview, Feb. 1995.

R. Ramirez, Dolco Packaging Corporation, interview, Feb. 1995.

R. E. Skochdopole and G. C. Welsh "Polystyrene Foams" in *Styrene Polymers,* E. R. Moore, ed., Dow Chemical Company, Midland, MI, 1989, pp. 193–205. Reprinted from the *Encyclopedia of Polymer Science and Engineering,* Vol. 16, Wiley, New York, 1989, pp. 1–246.

P. A. Wagner, "Foam, Extruded Polystyrene" in M. Bakker, ed., *Encyclopedia of Packaging Technology,* Wiley, New York, 1986, pp. 345–346.

For further information, contact the companies mentioned in the article.

HUSTON KEITH
Keymark Associates
Marietta, Georgia

TRAYS, STEAM-TABLE

Reusable steam-table trays are common in institutional food service. These shallow rectangular serving pans are usually made of heavy-gauge stainless steel. Foods prepared in the institutional or commercial kitchen may be transferred to a steam-table tray for holding on a serving line equipped to maintain foods at appropriate serving temperatures. Full-size steam-table trays are dimensionally standardized to fit most serving-line equipment. Such trays are also available in one-half and one-third sizes to serve multiple items in the same space required by full-size trays. Identical trays are also used on cold serving lines. In Europe, the standardized food equipment system is called "Gastronorm."

Disposable steam-table trays are now available to food processors for packaging prepared foods such as entrees, vegetables, fruits, desserts, and some baked goods. Foods packaged in steam-table trays are available in the food-service market in frozen, chilled, and shelf-stable (heat-processed) forms. The advantage of all these packaged forms is that the foods are shipped, stored, reheated (or otherwise prepared for serving), and served from the same disposable containers. These types of fully prepared foods are of particular interest in food-service systems where skilled chefs and cooks are not available or affordable. They also eliminate the cost of cleaning reusable trays.

Frozen prepared foods are commonly offered in full- and half-size steam-table trays (see Fig. 1). The trays are usually made of heavy-gauge press-formed aluminum foil. Lids for shipping, made of fiberboard or aluminum, are removed or loosened for reheating in conventional or convection ovens. Individual pans are usually contained in disposable outer cartons for protection during shipment and storage. In some cases, foods that are meant to be served cold may be packaged in formed plastic trays with low heat resistance.

In the early 1970s, two companies in the United States

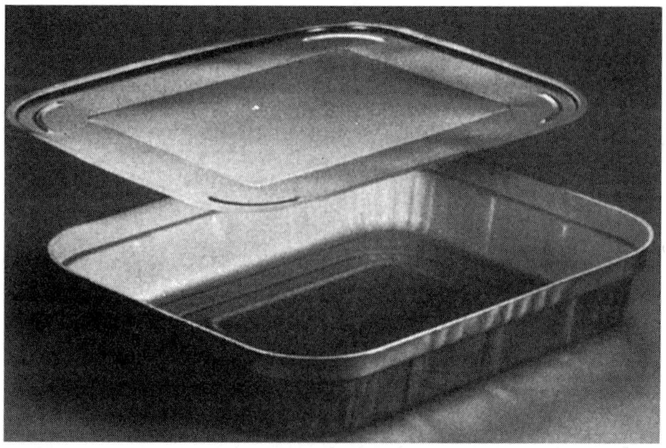

Figure 1. A half-size steam-table tray. Courtesy of Central States Can Company.

(Kraft and Central States Can) began production of steel half-size steam-table tray containers designed for production of thermally stabilized or "canned" food items (see Canning, food) (1,2). These containers are formed (drawn) from tin-free steel (see Tin Mill Products), and shipped to the food processor nested (a space-saving transportation advantage over round cans) with separate steel lids. The processor fills, seals, and retorts trayed food products as in the conventional canning process (see Cans, steel). After processing, these foods are shelf-stable without costly refrigerated transport or storage. Inherent in the retortable steam-table tray design is the potential for shorter heat-processing time to achieve commercial sterility as compared to round (cylindrical) cans. This shorter process time is said to allow processors to produce foods with substantially improved product favor, texture, and color (3). Central States Can Co., now the sole tray maker in the United States, has licensed half-size steam-table trays for manufacture in Europe (Groupe Carnaud) and Japan (Kojima Press) (1,4).

An aluminum one-third-size steam-table tray package was reported in 1982 for a line of high-quality entrées (5). Heat-sterilization times were significantly reduced in this case through use of a rotating retort cooker to produce superior flavor and avoid "canned" flavor and texture. This tray and its package assembly, developed by Food America Corp. and called Entree-Pak, won a 1982 Food and Drug Award (6) for innovative design, but it has yet to become a commercial reality.

Since the early 1970s the United States Department of Defense has been actively interested in the advantages to be gained from shelf-stable foods packaged in half-size steam-table trays (7) (see Shelf life). The objective has been to develop new, simplified combat food-service systems that will achieve substantial savings in transportation, storage, and skilled-manpower requirements. A tray-pack-based ("T-Ration") food-service system of special interest to the U.S. Army (8) has been tested by U.S. Army Natick Research and Development Center for use by the other U.S. military services (9) (see Military packaging.)

Frozen foods are associated with the aluminum steam-table trays in the minds of many food-service operators (10), and will likely continue that way because of long-time usage in the industry. On the other hand, shelf-stable foods in trays have not yet attracted a substantial food-service market (1). A factor that may affect future markets for trayed foods is the successful development of a retortable plastic tray. Plastic composite structures with high oxygen barriers have reportedly already been used successfully in the United States (11) and Europe (12) to form containers that will withstand elevated food-sterilization temperatures. Although size constraints on a plastic steam-table tray have not yet been adequately explored, the appearance on the market of a practical steam-table tray that can be heated in a microwave oven could have a profound effect on future use of shelf stable, tray-packed foods by food-service operators.

The future use of the retortable steam-table tray as a packaging system may depend on the success of the U.S. military in developing a total food system based on shelf-stable foods in half steam-table trays (13). The range of products packaged in trays for a successful military system must include a far greater variety of items and higher quality than is currently available commercially. This U.S. military effort, if successful, could make available the overall benefits of a simplified food-service system to civilian food-service operators.

BIBLIOGRAPHY

"Trays, Steam-Table" in *The Wiley Encyclopedia of Packaging Technology,* 1st ed., by Avalon L. Dungan, L. J. Minor Corporation, pp. 679–680.

1. Central States Can Co., Massillon, Ohio, private communication.
2. *Canner / Packer* **146**(8), 40 (1977).
3. *Food Eng. Internatl.* **4**(5), 38 (1979).
4. G. Booth, *Food Flavour. Ingredients Pack. Process.* **1**(8), 36 (1980).
5. *Nat. Provis* **27**(2), 33 (1982).
6. *Food Drug Packag.* **46**(6), 1 and **46**(12), 31 (1982).
7. J. Szczeblowski, *Activities Report of the Research and Development Associates for Military Food and Packaging Systems,* **25**(1), 77–84 (1973).
8. R. S. Maize, II, *Activities Report of the Research and Development Associates for Military Food and Packaging Systems,* **35**(2), 46–50 (1983).
9. H. J. Kirejczyk, *The Cold Weather 83 Evaluation of Mobile Food Service Unit and Tray Packs,* Technical Report 85/009. U.S. Army R & D Center, Natick, Mass., Aug. 1984; *Field Feeding Systems to Support U.S. Marine Corps Forces in the 1990s,* Technical Report 85/011, July 1984.
10. K. S. Ferguson, *Activities Report of the Research and Development Associates for Military Food and Packaging Systems,* **32**(2), 20–22 (1980).
11. J. Szczeblowski, *Activities Report of the Research and Development Associates for Military Food and Packaging Systems,* **32**(2), 23–31 (1980).
12. *Int. Z. Lebensmittal Technologie Verjahrenstechnik* **34**(6), 541, 544 (1983).
13. A. Dungan, *Activities Report of the Research and Development Associates for Military Food and Packaging Systems,* **35**(2), 41–45 (1983).

TUBE FILLING

Industrial production of collapsible tubes in two processes:

1. The manufacture of an empty tube
2. Filling and closing of the tube

Most tube-filling and tube-closing machines can handle different tube shapes and materials and a large variety of filling products by means of versatile accessories. Different tube shapes such as round, oval, conical, and cylindrical can be handled with the use of size parts. Cylindrical tubes range within 10–52 mm in diameter and 50–250 mm in skirt length. Tubes are made from metal (aluminum, lead, tin) or laminates (foil laminates, plastic laminates). Filling materials range from thin products (shampoo), or cremes (toothpaste, ointment) to extremely viscous materials such as caulking compounds.

Depending on the different materials, production speeds range from within 30–200 tubes per minute. A speed of ≤ 120 min^{-1} single-lane machines are used, while two-lane machines are common for higher speeds. All tube fillers are using an intermittent production process in which tubes, positioned in tube holders or pucks, are indexed at consistent speed and spacing through the work stations.

Tube Functions

The working process of a tube-filling/closing machine can be divided into four main functions.

Tube handling. For lower speed tubes are fed manually by infeed conveyors or chutes into the machine. Medium-speed machines are typically fed by cassettes or magazines. High-speed machines are normally equipped with their own tube loader that picks up tubes row by row from the case and feeds them on to conveyors to the first workstation (Fig. 1). The tube is put into vertical position and pushed into the tube holder. For safe operation, the tube should not be dropped by gravity. The preferred method is tipping of the tube by suction blocks into the tube holders and vacuum cups for tube discharge. A rotary turret, driven by a Ferguson-type indexing gear, running in a oil bath ensures precise positioning of the tube through the workstations. This method guarantees safe and maintenance-free operation for many years.

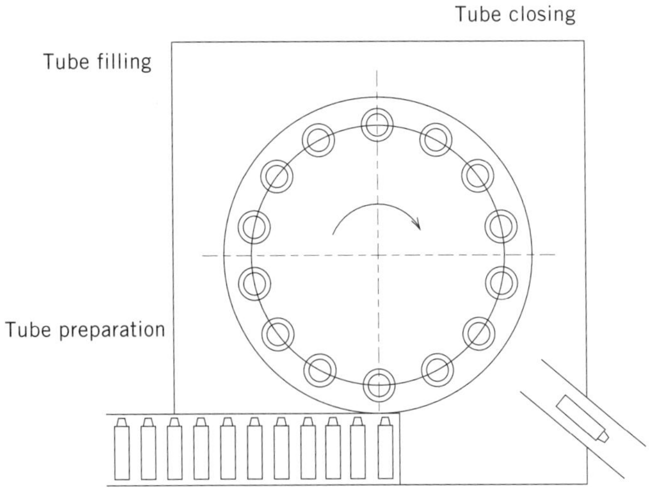

Figure 1. Tube handling.

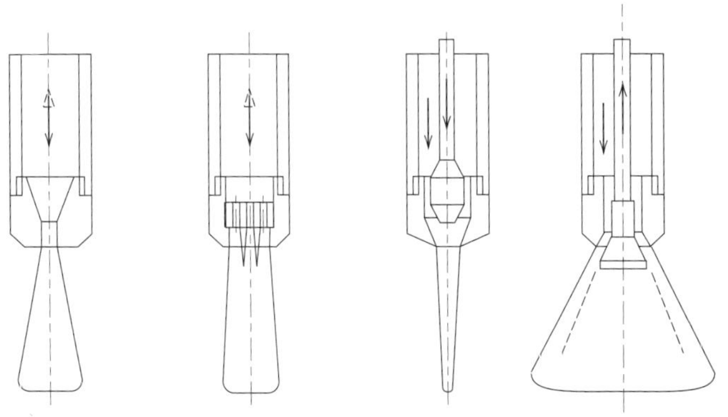

Figure 2. Different types of nozzle systems.

Tube preparation. Several machine stations perform different functions prior to filling:

- Checking for presence of cap and cap tightening by means of sensors and adjustable clutches or torque-controlled motors.
- Checking for damage to the open end of the tube by mechanical or photoelectric sensors.
- Cleaning of the tube by clean-air blast and vacuum; also purging of the tube with inert gas to reduce the amount of residual oxygen in the filled tube.
- Printing registration and code verification by means of photoeyes and scanners. Tubes are rotated and adjusted by means of clutch brakes or servomotors. Ink-jet coding of the tube can be done while the tube is rotating.

Tube filling. The tubes are filled by a filling nozzle that dives into the tube before the filling process. As the material is discharged from the nozzle, the nozzle is retracted out of the tube, keeping a constant distance between the nozzle outlet and the rising level. This relative movement between the tube and the filing system ensures continuous filling without air entrapment. After the predetermined amount of filling material has been discharged, the filling material, which in most cases tends to string, has to be cut off and—depending on the behavior of the material—several different nozzle-cutoff systems are available (Fig. 2).

Most tubes are filled with a single product. But there are more and more applications, mostly in the toothpaste industry, where two or three different components arranged in stripes over the whole tube length have to be filled. This so-called deep-stripe filling system is becoming more and more popular than the conventional surface stripes (Fig. 3).

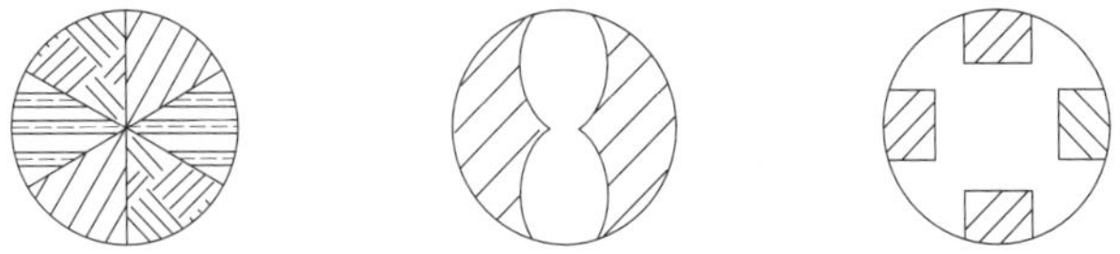

Figure 3. Cross sections of deep-striping.

While filling nozzle systems depend mostly on the filling product, a basic common design for the dosing system is used, comprising a volumetric pump dosing system consisting of cylinder and piston with two valves or a three-way rotary valve. One of the most critical parts of this system is the seal between the piston and the cylinder. For abrasive products such as toothpaste, the necessary sealing materials must be carefully selected. The typical cup seals made from polyurethane, Viton, PTFE, or polyethylene are subject to wear and have to be replaced at regular intervals. As an alternative, ceramic components can be used. Unlike ceramic coatings, solid ceramic parts (eg, aluminum oxide) have absolutely no pores and combine very high hardness and resistance to wear with excellent chemical resistance and food-grade characteristics.

Parts in contact with the filling material in tube fillers can be easily dismantled, cleaned, and sterilized. To avoid any risk of microbiological contamination during assembly of a clean system, the assembled unit can be sterilized in place. These so-called CIP (clean-in-place) systems are used more and more in the pharmaceutical industry.

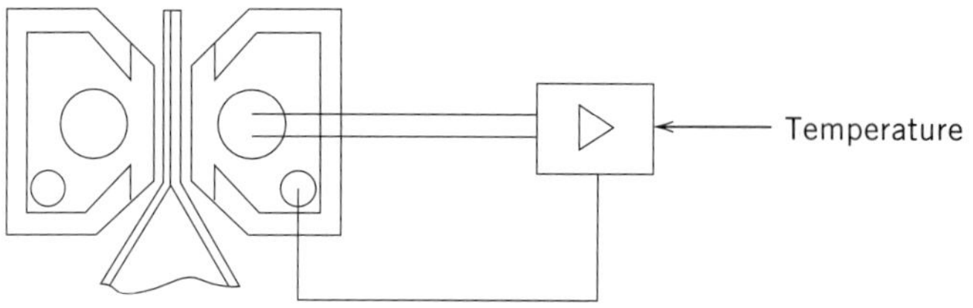

Figure 6. Hot-jaw sealing system.

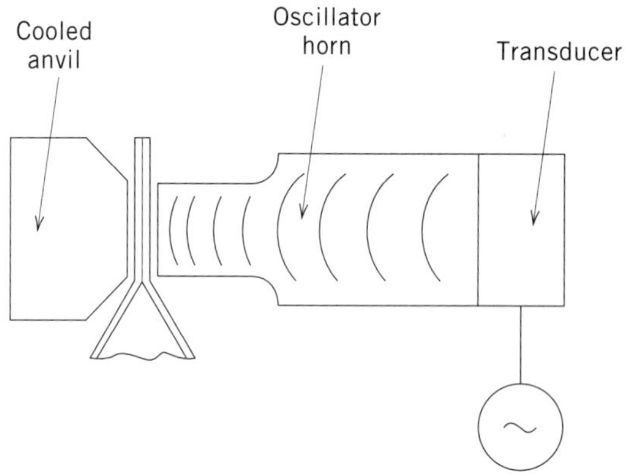

Figure 7. Ultrasonic sealing system.

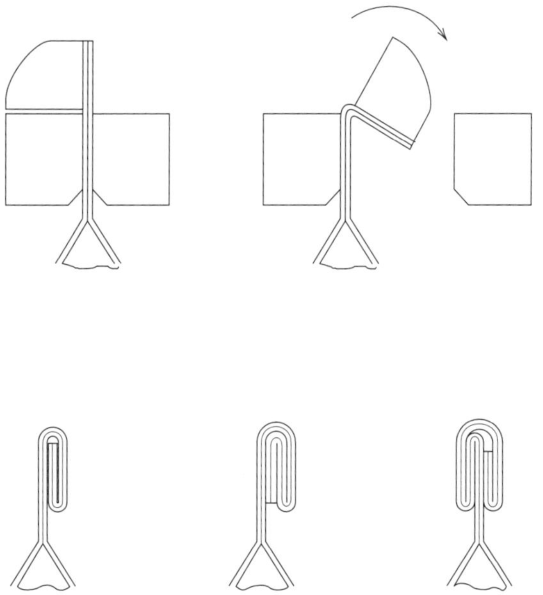

Figure 4. Metal tube folds.

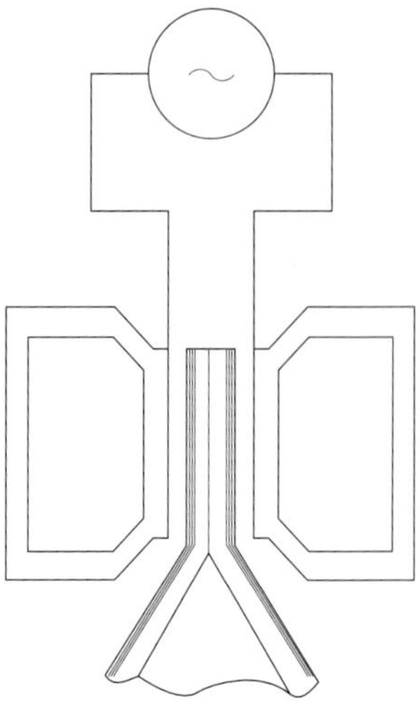

Figure 5. High-frequency sealing system.

Tube closing. Different materials require different types of closure. Normally, most tube fillers are equipped for one type of tube closure.

Tube Types

Metal Tubes. Metal tubes are closed by folding, which means that the tube end is flattened and folded over. Various types of fold are possible: double fold, four fold, and saddle fold (Fig. 4).

The actual folding process is carried out by two folding jaws and a hinged folding tool. For the required precision and speed, the system is cam-driven. In many cases a latex or heat-sealable lacquer is applied inside the fold area for a hermetic seal.

Foil laminate tubes. The safest method of sealing plastic tubes with a thin aluminum foil layer is high-frequency sealing. High-frequency generates an eddy current in the aluminum layer, which heats up the aluminum and the neighboring plastic layers (Fig. 5).

Plastic tubes. Today the most common tube material is plastic. Originally polyethylene was used in a single layer and the tubes were sealed between heated jaws as shown in Figure 6. The system applies heat on the outside of the tube to

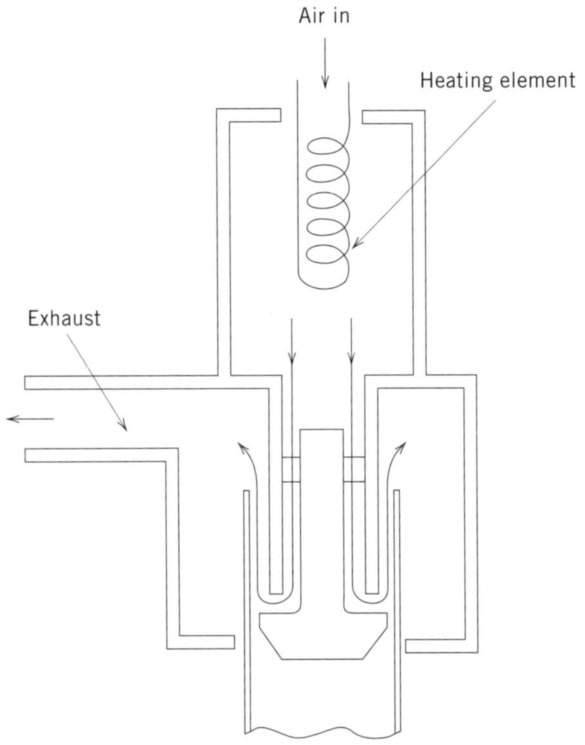

Figure 8. Hot-air sealing system.

heat to the inside for sealing. This is less efficient than high-frequency sealing, where the heat is generated directly at the inside layer to be sealed. Consequently, the hot-jaw system requires a longer heating time and therefore a reduced cycling speed. This is the reason why this system is hardly used today in new machines.

Another method for closing plastic tubes is ultrasonic sealing (Fig. 7). This system applies mechanical vibration to the outside of the tube, which induces friction, which—in turn—generates heat to seal the tube. This sealing method is suitable only for moderate speeds.

Because of the above-mentioned restrictions, the preferred method for sealing laminated and plastic tubes today is hot-air sealing. This system heats the seal area inside the tube by hot air. In a subsequent station, the tube is then pressed and chilled (Fig. 8). The system is suitable for speeds well in excess of 100 min^{-1}.

The pressing station for the hot air system is also necessary for the other types of tube closure. It provides a tight seal and improves the aesthetics of the tubes and is also used to apply a code (ie, batch number or expiration date).

All modern packaging machines are equipped with programmable logic controllers (PLCs) to control and monitor all machine functions. Clear text displays are used to guide the operator and to provide information about the production process such as machine speeds and number of packs produced. Interface for data link for central line controls are available.

Thomas Miller
T. J. Boedekker
IWKA
Fairfield, New Jersey

TUBES, COLLAPSIBLE

The collapsible tube, be they aluminum, plastic or laminate, is a most unique package allowing the user ease of dispensing and, where necessary, application of product directly to the area of use. The tube is a perfect package for viscous products such as dentifrice, pharmaceutical creams and ointments, cosmetic creams and gels, household and industrial products.

The tube definitely is a package for food products where the contents can be conveniently used in precise amounts when required. Examples of food products contained in tube abound in the European and Asian markets. Examples such as tomato paste, fish and meat paté, and dessert toppings can be found in gourmet shops around the world. But other food products in tubes may include preserves, mustards, chocolate sauces, mayonnaise, mustard and cheese spreads. In short, the tube is ideal for delivering the product to the consumer's toothbrush or dinner table given creative application and packaging technology.

The Tube Council of North America reported that Americans purchased nearly 2.7 billion tubes in 1995. The market for products packaged in tubes continues to grow fueled by new materials and the consumers hunger for convenience in their life.

The first tubes, manufactured back in the mid 1800s were for artist colors. Shortly thereafter, a dentist saw the convenience his patients could enjoy if the tube could contain dental cremes, replacing powders and the use of a rather unsanitary common jar. Hence, the toothpaste tube with all the benefits. The original tubes were produced from soft metals, mostly tin or lead or an alloy of these. The aluminum tube was introduced in the mid-1950s and has been consistently improved in the years that followed. Today, the metal tubes are predominantly aluminum. The only tin tubes are for special pharmacal applications, mostly ophthalmic. The 1170 Aluminum Alloy used by tube manufacturers in North America is the highest purity available. The 99.7% pure aluminum is virtually free of trace substances and extrudes a seamless package. New, high barrier internal liners and specially sealed necks made the aluminum tube the package of choice for products requiring an absolute oxygen barrier. Additionally, aluminum tube decorations utilizing the metallic surface add unsurpassed luster to the appearance of the package (see also Decorating).

In the 1950s metal tubes were joined by the plastic tube. The immediate appeal of plastic was obvious to every manufacturer of cosmetics and hair care products. The plastic tube kept its "just home from the store" look through its life in the bathroom or vanity. It could be extruded in clear or colored resins to enhance the package aesthetics. The plastic tube perfectly complemented the aluminum tube and allowed the launch of a whole new range of products not previously found in tubes.

In the early 1970s laminate tubes joined their metal and plastic cousins. The barrier properties of laminate, coupled with the limited suck back accomplished by the inclusion of an aluminum foil layer, gave laminate greater product appeal for products such as toothpaste. In the ensuing years the laminate tube effectively captured the dentifrice market. It has now been selected to package products as diverse as mayonnaise and vaginal cream. The laminate tube surface may be

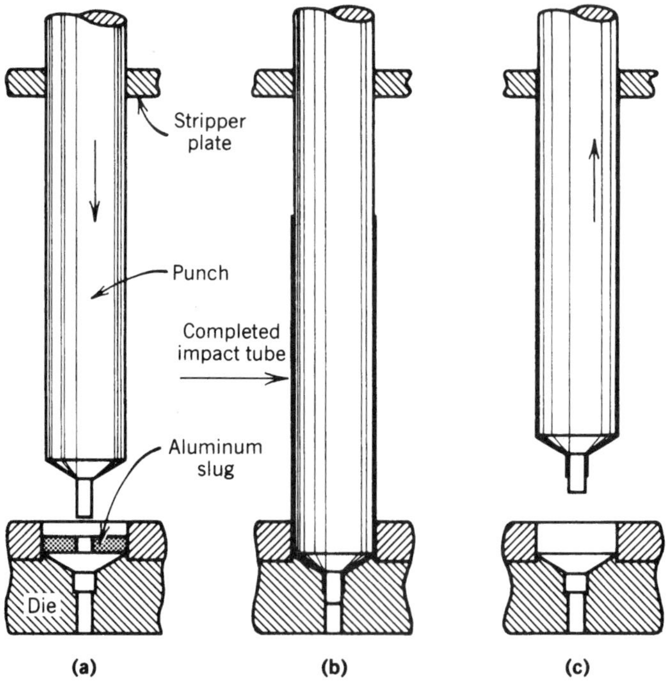

Figure 1. Impact extrusion of a metal tube: (**a**) beginning impact stroke; (**b**) bottom of stroke, (**c**) beginning to strip. Courtesy of Herlan & Co.

printed on the web allowing an eye catching graphic presentation, even including scenic images or the human likeness.

The Metal Tube

The aluminum tube production process starts with an aluminum slug roughly varying in size from a dime to a silver dollar. The slug will be the approximate diameter of the tube and the slug thickness predetermined to minimize trimming to the length prescribed by the customer. The slug will either be a donut, having a center hole, or a solid. The solid slug is most often used to extrude a one piece tube with an integral diaphragm forming the tamper-evident sealed neck feature. The tube manufacturer may elect to use flat or dished slugs, depending upon the design of their production equipment and tooling.

State of the art aluminum tube production line commonly run at speeds varying from 150 to 200 tubes per minute. The slug is impact extruded one up. That is to say that one tube is made one at a time, 3 times per second. Aluminum tube lines are readily available from machinery suppliers in Europe including Polytype and Hinterkopf.

The aluminum tube is formed by impact extrusion accomplished horizontally. The tube slug is first tumbled with a lubricant, generally zinc stearate, USP. It is then feed into position in from of the shaft and the press moves forward with an impact forces of 80 to 100 tons, depending upon the diameter and, to a lesser extent, the thickness of the slug which prescribes the length of the tube. A section of the typical tooling setup is shown in Figure 1. The male tool head matches the inside profile of the final tube. The largest diameter is only $\frac{1}{32}$–$\frac{1}{16}$ in. (0.8–1.6 mm) long, after which it tapers back to a diameter that runs around 0.008 in. (0.2 mm) smaller than the tip. The clearance between the male tool and the female cavity establishes the wall thickness of the tube sleeve.

The slug is automatically fed into the cavity and is compressed by the male tool as it moves down. The first portion of the stroke seats the slug and takes up the bearing slack of the press. The balance of the stroke cold-works the slug, forming the head of the tube. As the plunger continues moving down, the excess metal is extruded out around the sides of the plunger, forming the body. Since the metal never reaches its melting point, the body is cold enough to maintain its shape and additional support from the tooling beyond the head area is not required. As the plunger moves back at the end of the stroke, the completed tube is stripped off, falling into a transfer conveyor, which takes it to the trimming operation. Here, the body is trimmed to exact length, cap threads rolled onto the neck, and the mouth of the tube trimmed.

By now the tube has been severely work-hardened and has the feel of an aluminum can. To relieve this hardness, it is annealed in an 1100°F (593°C) oven and cooled, after which a spray application of a product-resistant lining is applied. A number of tough epoxy–phenolic or acrylic lacquers are available that not only resist the attack of the product but can be creased without chipping or peeling from the base metal. "After the internal lining has been cured, the tube is coated with either a white or colored base coat, usually a polyester enamel or, optionally, an acrylic, polyurathane. This coating gives an opaque printing base to the tube side walls. The shoulder of the tube may also be coated to complete the look desired. Another base sidewall coating involves the use of clear or translucent lacquers that utilize the aluminum surface to achieve a metallic appearance."

Metal tubes are printed by a dry offset process (see Decorating; Printing) using either thermally or uv-cured inks (see Curing; Inks). The tube is pushed on a mandrel and rolled past a curved printing blanket, which applies up to six colors simultaneously. The tube now goes through a final drying cycle to cure the ink, after which the cap is applied and the tube automatically packed in boxes for shipment, usually nested in a dividerless tray, for shipment to customer.

Plastic Tubes

The first practical plastic-tube patent was issued in 1954 (1). The patent covered the process of making a thin plastic sleeve by an extrusion method (see Extrusion) and then injection-molding (see Injection molding) a head on one end to produce a tube. Modern equipment has improved but not changed the basic Strahm concept.

Many plastics can be used to make plastic tubes, but LDPE is the primary material used today (see Polyethylene, low-density). It has high-moisture-barrier properties, low cost, and good appearance. Its lack of oxygen and flavor barrier have been improved with barrier coatings. It was the development of a barrier coating by American Can Co. that made the plastic tube a practical container for general packaging.

HDPE (see Polyethylene, high-density) is used for packaging some hydrocarbon-based products such as grease, and PP (see Polypropylene) is used for applications requiring non-straining, better perfume barrier, or higher temperature resistance. Both HDPE and PP are much stiffer than LDPE for tube sidewalls, and are not as popular.

COEX Tubes, a coextrusion of EVOH and polyethylene,

has been demonstrated to have excellent barrier properties enabling manufacturers of products that are adversely effected by air or moisture transmission the option of using the plastic tube. It was quickly adopted by toothpaste manufacturers where flavor loss in conventional plastic tubes was unacceptable. The COEX tube is often specified with a stand-up hinged cap. The COEX tube, otherwise has all the function of plastic.

The selection of a suitable plastic for producing a tube is critical to its performance. DuPont Alathon 2020T, a 0.92-g/cm^3-density resin with a melt index of 1.0, is the primary LDPE resin used by all tube manufacturers in the United States. It has good processibility, excellent stress-crack resistance to product attack, and an extremely low gel content, which reduces surface irregularities that would affect printing quality.

Plastic tubes are produced by two principal methods in the United States: the Strahm method and the Downs method. Both processes make excellent tubes.

Production of the tube begins with the extrusion of continuous thin-walled tubing (2). This has a wall thickness of 0.014–0.018 in. (0.35–0.46 mm), depending on the diameter. A standard extruder is used with a thin-walled tubing die. As the hot plastic emerges, it is corona-treated (2) (see Surface modification) for later printing-ink adhesion. At the same time, it is drawn over a chilled internal forming mandrel and cooled on the outside with cold water. The tube cools and shrinks to an accurately controlled diameter as it is drawn off. After it has passed through the drive rolls of the haul-off unit, it is cut to exact length with a rotary knife cutter. This piece is called a sleeve to differentiate it from the completed container, which is called a tube.

Printing can be done before or after heading. The location is based on the layout and the relative scrap generated in printing and heading. All tubes are printed by dry offset printing. The same type of printer used for metal tubes is used for plastic sleeves. Thermally dried and UV-dried inks are available in a full range of colors. Good-quality process printing is possible, and for cosmetic applications, postdecoration with hot-stamp foils or silk screening is popular (see Decorating).

After the ink has cured, the sleeves are roller-coated with a high gloss, oxygen- and flavor-barrier coating. High-barrier, two-component amine-cured epoxy coatings are available that reduce the overall tube permeability by a factor of 10. Special coatings more resistant to product staining or having lower coefficients of friction are also used. Most coatings are thermally cured, but UV-curable coatings are beginning to be used. These provide equally high barriers to oxygen and essential oils and cure more quickly. After the coating is cured, the completed sleeves are headed.

The head of a plastic tube must be compatible with the sleeve in order to produce a good bond. Sleeves made of LDPE can be headed with LDPE or HDPE, but PP sleeves must have a PP head. Head thickness is 0.030–0.065 in. (0.76–1.65 mm), depending on tube diameter and application.

The Strahm heading process traps the top end of the sleeve in an injection mold and injects plastic into the cavity to form the head and bond it to the sleeve. The process is done with multiple tools at each station. Slower machines have a female tool that remains in the injection station, and the tube is held there until it has cooled sufficiently before moving to the next

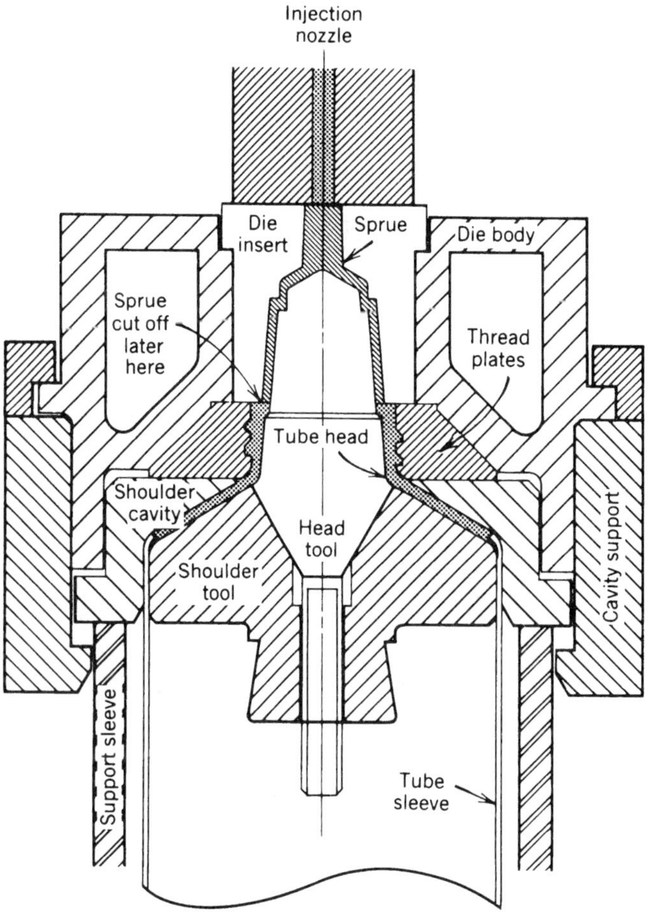

Figure 2. Strahm heading method.

position. The higher speed machines have locked-die tooling, as shown in Figure 2. The male tool containing the sleeve is pushed up into the shoulder cavity. The end of the sleeve enters the shoulder cavity, rolling in slightly as it touches the radius of the shoulder. At the same time, the support sleeve, which moves with the male tool, contacts the shoulder cavity, lifting it up against the thread plates. This forces the plates together in the position shown in Figure 2. The injection nozzle now closes on the assembly, and molten plastic from a small reciprocating-screw extruder is injected to form the head and bond it to the sleeve. Injection-cavity pressures are low since excessive pressure causes "blowby," forcing plastic past the head area and down the sleeve wall. Melt temperature of the plastic is generally over 500°F (260°C), high enough to ensure a good fusion seal to the sleeve wall. After the injection cycle, the nozzle retracts. The tooling remains locked together and continues to the next station. After the head cools, the male tool drops, the thread plates open, and the completed tube is released, remaining on the male tool. The die body and cavity support move off to one side to allow the completed tube to be extracted. A new sleeve is then placed on the tool, the female tooling moves over the top of the male tools, and the cycle is repeated.

After cooling, the completed tube is transferred to a snipper-capper. Here, the sprue is removed and a cap applied, after which the tube is packed in unpartitioned trays for shipment to the customer.

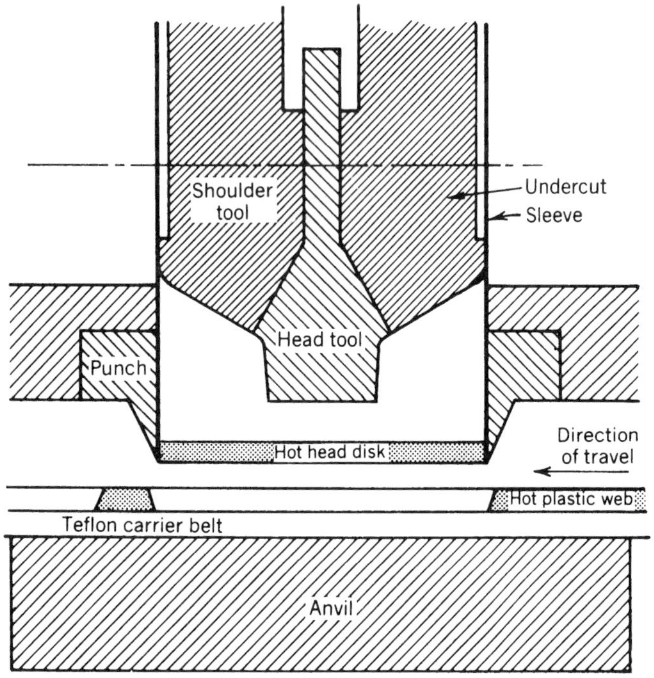

Figure 3. Downs process; head disk just after blanking out.

In the Downs process (3,4), the sleeve production is the same but the heading process considerably different. As shown in Figure 3, the sleeve on its male tool enters a punch, which is placed just above a continuous strip of LDPE heated to well above its softening point. This strip is ~2 in. (51 mm) wide and $\frac{1}{8}$ in. (3.2 mm) thick. When the leading edge of the sleeve is about flush with the cutting edge of the punch, both parts move down into the semimolten mass below. The punch forms a disk of plastic the diameter of the sleeve, and the sleeve inside the punch immediately adheres to it. The tool moves up carrying the sleeve bonded to the hot plastic disk and indexes to the next station. Here, the female tool closes on the sleeve and head disk, compression-molding the head into its final shape. The compression-molding method eliminates the sprue and the snipping operation necessary with the Strahm process.

More recent commercial Downs machines have multiple heads at each station to make a number of tubes simultaneously.

A third process, developed by KMK (Karl Magerle AG Hinwil, Switzerland) (5), injects a "donut" of molten plastic in the female cavity before it closes on the male cavity. The head is formed by compression molding. The Magerle machines are small, with only one tool per station, but they also incorporate capping on the same turret.

Although these are the most common methods, other processes have been developed. In Europe, the Valer Flax process (6) uses a premolded head which is spin-welded to the sleeve. Several companies have developed blow-molding methods (7) where the tube is blown as a bottle and the bottom trimmed off. This approach generates a high amount of scrap that must be put back into the process, and care must be taken to maintain uniform wall thickness for the later printing step.

Laminated Tubes

In 1971 Procter & Gamble switched their Crest dentifrice to a laminate tube. The laminated material (see Laminating Machinery) consisted of 7 layers with paper, plastic and foil acting as the barrier. The foil was far superior to the plastic tubes then available and minimized flavor loss to acceptable levels. The laminate replaced the lead tube then being used by P & G for Crest.

The chief difference between this package and the plastic tube is the sleeve, made from preprinted laminate web (8,9). It contains up to ten layers, each contributing to the function of the structure. Figure 4 shows the material in each layer and describes its use. The complete laminate, 0.013 in. (0.33 mm) thick is produced on an extrusion laminating line (10). This web is made in large rolls that are slit into the proper width for the tube being made. The slit rolls are shipped to the tube plant, where they are formed into sleeves. The sleeve has an impermeable aluminum layer, so the only permeation is along the seam overlap.

The seaming process takes the flat web, folds it into cylindrical form and seams it to form a continuous tube (11). The seam must protect the raw edge of the foil; this is achieved by overlapping the edges of the sleeve, then heating and compressing the overlapped portion to squeeze some of the plastic out around the raw edge of the foil and paper. After the sleeve has been seamed, it is cut to length using the print registration marks on the web.

The present of aluminum foil in the sleeve permits use of rf energy for induction heating (12). This is preferred in the seaming process because it permits better control of heat distribution.

To take full advantage of the vastly improved barrier properties of the laminated sleeve, an oxygen and flavor barrier is necessary in the head area. This is accomplished with a premolded insert (13) of polybutylene terephthalate (PBT), which has good oxygen- and flavor-barrier properties, can withstand the injection pressures and temperatures of the

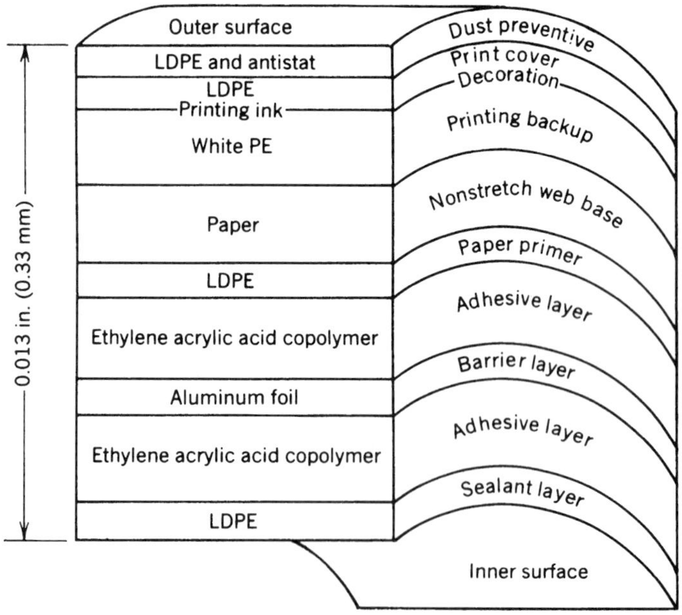

Figure 4. Typical structure of a laminate toothpaste tube.

heading plastic, and does not crack in the head. Urea inserts are also used. The insert is placed on the heading tool before the sleeve is placed in position, and the head plastic locks the insert in position as it bonds to the sleeve. The resultant tube has some permeation windows along the sleeve seam and the area where the head bonds to the sleeve, but the result is still a very low gas and flavor permeability.

Alternative methods of producing laminated stock and tubes have also been developed, as well as alternative materials.

The American Can approach for seaming requires the axis of the tube to be parallel with the length of the laminated web (11). KMK has developed another system (14), where the axis of the tube is perpendicular to the web. In this process, the blank is cut from the web and wrapped around a fixed mandrel and sealed. KMK uses the heading process described under the plastic tube section above.

Another Swiss manufacturer, AISA (Automation Industrielle SA), has produced a machine using premoled heads which are bonded to the sleeve with rf induction (15). To ensure a good bond, a foil laminate disk, called a rondelle (16), is placed over the shoulder of the tube. The sleeve is butted against the rondelle and the rf bond made. This approach still requires a urea or PBT insert to act as a flavor barrier.

New Tube Concepts

The relatively recent appearance of ethylene–vinyl alcohol (EVOH) (see Ethylene–vinyl alcohol) as a clear barrier plastic has made possible the development of an all-plastic barrier tube (see Barrier polymers). Coextrusion methods are available for producing laminates, and the Japanese have produced coextruded sleeves and blow-moled barrier tubes (7). The latter tubes have a continuous barrier, ie, there are no higher permeation areas such as side seams or head bonds. At present, the all-plastic barrier tube is not economically competitive with a laminated structure. EVOH costs about the same as aluminum foil, so there are no material cost savings involved. The advantage of being able to print flat web is an important factor in the cost of the laminated tube which would appear to limit the appeal of this new concept. However, a significant drop in the price of EVOH relative to aluminum foil or the development of a new market where higher-barrier properties are required could quickly change its importance.

BIBLIOGRAPHY

1. U.S. Pat 2,673,374 (Mar. 30, 1954), A. Strahm (to American Can Co.). (Describes basic heading concept.)
2. U.S. Pat. 3,849,286 (Nov. 19, 1974). R. Brandt and J. Piltzecker (to American Can Co.). (Details the sleeve extrusion process.)
3. U.S. 3,047,910 (Aug 7, 1962), M. Downs (to Plastomer Development Corp.). (Offers good description of basic Downs heading machine.)
4. U.S. Pat. 3,591,896 (July 13, 1911), R. Tartaglia (to Peerless Tube Co.). (Contains good drawings of the basic commercial machine using the Downs process.)
5. U.S. Pat. 4,352,775 (Oct. 5, 1982), K. Magerle. (Shows Magerle heading process.)
6. U.S. Pat. 3,824,145 (July 16, 1984), V. Flax (to Continentalplastic AG). (Contains good drawings of spin-welding method of heading.)
7. U.S. Pat. 4,261,482 (Apr. 14, 1981), M. Yamada, T. Sugimoto, and J. Yazaki (to Toyo Seikan Kaisha, Ltd. and Lion Hanigaki Kabushiki Kaisha). (Describes blow-molded tube with EVOH barrier.)
8. U.S. Pat. 3,260,410 (July 12, 1966), R. Brandt and R. Kaercher (to American Can Co.). (Shows basic laminate structure.)
9. U.S. Pat. 3,347,411 (Oct. 17, 1967), R. Brandt and N. Mestanas (to American Can Co.). (Basic article patent of laminated tube.)
10. U.S. Pat. 3,505,143 (Apr. 7, 1970), D. Haas (to American Can Co.). (Shows laminating process.)
11. U.S. Pat. 3,388,017 (June 11, 1968), A Grimsley and C. Scheindel (to American Can Co.). (Describes side seaming of laminated sleeves.)
12. U.S. Pat. 4,210,477 (July 1, 1980), W. Gillespie and H. Inglis (to American Can Co.). (Illustrates rf induction seaming coil design.)
13. U.S. Pat. 3,565,293 (Feb. 23, 1971), A. Grimsley (to American Can Co.). (Describes head inserts.)
14. U.S. Pat. 4,200,482 (Apr. 29, 1980) (to KMK AG). (Shows Magerle side seamer.)
15. U.S. Pat. 4,123,312 (Oct. 31, 1978), G. Schmid and R. Jeker (to Automation Industrielle SA). (Shows AISA seaming and heading process.)
16. U.S. Pat. 4,448,829 (May 15, 1984), A. Kohler (to Automation Industrielle SA). (Shows AISA rondelle.)

General References

N. L. Ward, "Cold (Impact) Extrusion of Aluminum Alloy Parts," in T. Lyman, ed., *Metals Handbook,* 8th ed., Vol. 4, American Society for Metals, Metals Park, OH, 1969, pp. 490–494. (Discusses impact extrusion of aluminum in general, with little specific information on tubes.)

ASTM Committee on Aluminum and Aluminum Alloys, "Introduction to Aluminum," *Metals Handbook,* Vol. 2, 9th ed., American Society for Metals, Metals Park, OH, 1979, pp. 3–23. (Briefly refers to aluminum alloys.)

Packaging in Plastic and Glaminate Tubes, Washington Technical Center, American Can Co., Washington, NJ. (A nontechnical presentation of the manufacture of tubes and a good write-up on the problems of packaging in tubes and what to look for.)

Patents describing the Strahm process of heading, all ultimately assigned to American Can Co.

U.S. Pat. 2,713,369 (July 19, 1955), A. Strahm. (Illustrates sections of tube head and sleeve.)

U.S. Pat. 2,812,548 (Nov. 12, 1957), A. Quinche and E. Lecluyse. (Shows the concept of split-thread plates.)

U.S. Pat. 2,994,107 (Aug. 1, 1961), A. Quinche. (Describes first stationary-die heading machine.)

DOUGLAS A. STEWART
Montabello Packaging
Oak Park, Illinois

ROGER BRANDT
American Can Company

U

UNIT-DOSE PACKAGING.
See PHARMACEUTICAL PACKAGING

UNIVERSAL PRODUCT CODE.
See BAR CODE.

UNSCRAMBLING

In the packaging industry, an "unscrambler" is a machine that orients packaging components from random bulk storage to deliver them in an ordered fashion to the production line. The term *unscrambling* usually refers to the handling of relatively large components such as plastic containers, some metal cans, and glass bottles as opposed to sorting of relatively small components such as closures, caps, spray pumps, actuators, and applicators. With the higher speeds of today's production lines, limited production-line floor space, and the immense range of container shapes and sizes, it is not practical to unscramble containers from bulk manually. Unscrambling offers a means of delivering oriented containers to a packaging line with high speed and reliability, while taking advantage of the lower cost of purchasing the containers in random bulk lots. Containers can be packed in boxes or gaylords directly from the molding, extrusion, or forming machines and shipped directly to the packaging plant. The cost of preorienting into reshippers or onto pallets is eliminated. In some cases, shipping costs are also reduced because of the greater number of components that can be shipped in the same space, with no need for special handling.

The traditional method of shipping glass bottles and metal cans was on pallets or in boxes with the components preoriented and stacked in "tiers" or layers, usually separated by sheets of chipboard or cardboard. The feeding of these containers to the packaging line was accomplished by the use of depalletizers or feed tables for handling boxes of containers (see Palletizers). These feed tables worked basically on the same principle as the depalletizer in that a layer of containers would be pushed off an elevator table onto a feed chain or turntable. Both of these feeders required a full time operator and were limited to simple container shapes.

With the conversion to plastic containers within the packaging industry, more efficient feed systems were needed to accommodate the higher speeds and often exotic shapes and sizes available. The advent of plastic containers also opened the possibility for bulk shipments. Unscramblers, therefore, were created to sort out bulk shipments of the containers and deliver them to a packaging line. Other applications for unscramblers involve integration with printing or decorating equipment, labelers, sleeving or overwrap equipment, cleaning or sterilizing equipment, and pucking or base-cupping equipment.

There are several types of unscramblers on the market today and they employ different means of orientation. There are, however, some similarities:

1. The unscrambler is provided with, or can accommodate, a bulk storage/delivery system. This usually consists of a conventional hopper—elevator combination. Hopper size can vary depending on the type of container being handled and the desired time of unattended operation based on container size and line speed. Actual available floor space is, of course, another consideration when sizing and selecting the hopper-elevator package. The elevator or conveyor discharging the containers from the hopper must be selected to deliver the containers, still in random orientation, at a relatively constant rate to interface with the unscrambler-orienter unit (see Conveying). Most unscramblers provide level sensing through pneumatic sensors, electric photoeyes, or electromechanical level sensors to meter the operation of the hopper-elevator.
2. Another similarity of unscramblers is that the process is carried out in three stages.
   a. Since all containers requiring orientation have an opening for filling or distribution, one end or side of the container is different from the opposite end or side. It is this difference that is used to orient the container. The first stage of unscrambling is to "preorient" the container into one of two conditions: neck leading or neck trailing. If the cross section of the containers is rectangular or oval-shaped, this first stage of orientation can also preorient width with respect to the minor or major container dimension (see Fig. 1). By presizing a channel, pocket, flight track, or lane to just more than the minor dimension, but less than the major dimension, containers can fit into this space only in the minor dimension. This first stage is accomplished by allowing the containers to tumble into the presized channel, pocket, flight track, or lane in which the containers will lie. All other containers which are skewed or standing are cleared away. Air jets, agitators, clearing wheels or other mechanisms are used to move the bulk of containers away from the preoriented neck-leading or neck-trailing containers.
   b. The second stage of orientation is accomplished by turning the containers so that they are all in the same orientation. Turning in accomplished by differentiating the neck from the base and dropping all containers base

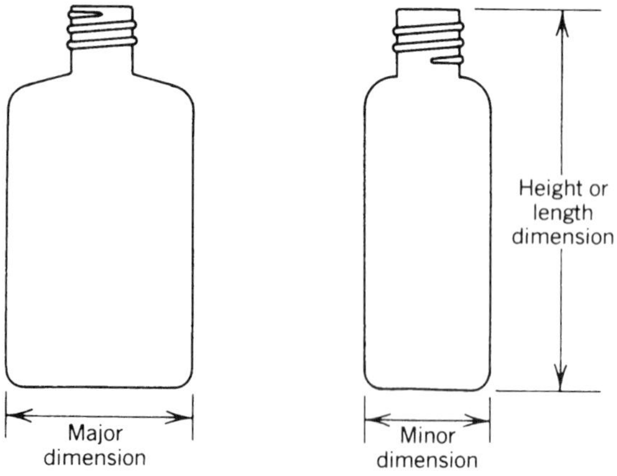

Figure 1. Typical use of terminology for describing container size.

down, or by mechanically catching the neck of the container and rotating those containers which are misoriented. Another method of orienting is to simply reject all containers that are not in the correct orientation and return them back to bulk storage. This method, however, reduces the overall efficiency of the orienter system and presents the possibility of recirculating and handling the container several times, which may be detrimental to scuff-sensitive containers.

c. The third and final common stage of unscrambling is to deliver the container or component to the next successive packaging operation. In most cases, the container is delivered to and stabilized on a flat-top conveyor. This is a very important part of the unscrambling operation because of the relative instability of an empty container on a running conveyor. It is essential that the unscrambler discharge mechanism be synchronized with the discharge conveyor chain. Discharge is usually accomplished through synchronized pocket- or side-belt release onto the conveyor. In many cases the unscrambler discharge is used for other secondary operations such as coding, marking, or container cleaning. It should be noted, however, that the operation performed by most unscramblers is unique with respect to other packaging operations in that the container is empty and is interfacing at a random container-per-minute rate to a constant rate. When installing a new unscrambler or when laying out a new packaging line, allowances have to be made for both "surge" and "backlog" distance between the unscrambler and the next successive packaging machine. "Surge" distance is the amount of conveyor length required between the unscrambler and the next machine on line. Consider, for example, that the next machine is the filler. When the filler starts, containers on the conveyor before the filler will begin to move downline. The line of containers will pass the unscrambler backlog detector (usually a photo eye) and start the unscrambler. As the unscrambler starts and accelerates to full operating speed, containers will be discharged onto the conveyor. The line of containers before the filler must not pass the filler low-level infeed sensor before the containers being discharged from the unscrambler "catch-up." If the filler low-level sensor is allowed to clear, the filler will stop and there will not be a synchronous automatic start-up of the packaging line, resulting in poor line efficiency and lost production. The solution to the "surge" distance problem is more than moving the unscrambler backlog sensor closer to the unscrambler to make it start sooner, because this would lead to a poor "backlog" condition. In this condition, when the filter stops, containers will backlog on the conveyor after the unscrambler and the unscrambler backlog detector will not detect the stoppage soon enough to decelerate and stop the unscrambler and avoid a jam.

All unscrambler manufacturers work with the designer of the packaging line to recommend proper spacing of the equipment for optimum performance. For high speed applications, for example, unscramblers are manufactured with variable speed settings to "ramp up or down" in speed rather than start and stop in automatic operation. Unscramblers are presently used in all of the major markets for plastic bottles, and their use is increasing along with the new capabilities of plastic bottles. Particular areas of growth are in the petroleum and beverage industries.

BIBLIOGRAPHY

"Unscrambling" in *The Wiley Encyclopedia of Packaging Technology,* 1st ed., by L. F. Byron, New England Machinery, Inc., pp. 688–689.

V

VACUUM-BAG COFFEE PACKAGING

The concept of vacuum-bag packaging for ground coffee was introduced in the U.S. market some years ago. It had been developed in Europe years before, but except for nonhermetic paper bags still used in the southeastern states, ground coffee in the United States has remained in the three-piece sanitary can with plastic covercap that replaced the key-opening can about 40 years ago. The new vacuum-bag package has been described as a "flexible can," since it delivers the same vacuum-packed fresh coffee that the consumer buys in the can, but that is where the similarity stops. The "vac-bag" is rectangular rather than round, and it can be shelved standing up, or lying on its side like a "brick," which is one of its less-glamorous names.

Coffee characteristics. Coffee has some unique characteristics. After coffee beans are roasted, they slowly evolve CO_2 amounting to over 1000 cm^3/lb (2200 cm^3/kg) of product depending on the blend and roasting color. During grinding, the volume of CO_2 is reduced to about half and, depending on the grind size, it requires varying holdup times before packing. Grind size varies greatly depending on the type of coffee-brewing equipment used and individual taste preference. Typical fine grind for most European countries is 500–600 μm; North American coarser grinds are 800–1000 μm. Vacuum-ag packaging of coarser grinds present more of a problem due to slower degassing. At 6–8 in. Hg (20–27 kPa) a vac-bag becomes soft and it can even balloon like a football (positive pressure) if it has insufficient degassing. Because rigid metal cans are capable of withstanding full vacuum as well as positive pressure, they require a relatively short holdup time even for coarse grinds. Because ground coffee also varies in density, the height of fixed-cross-section bags varies as well. This means that the equipment and materials must be adaptable to a limited automatic height change during a production run. In addition, coffee must be protected from oxygen to maintain the freshness. Vacuum packing eliminates most of the oxygen from the package by the nature of the process.

The vacuum bag. The first commercial "vac-bag" packages were made in the early 1960s, probably in Sweden. At that time the package was a bag-in-box style (see Bag-in-box, dry product) that later evolved to a lower-cost bag-in-bag style, which is still the most widely used type in Europe and Canada. A still lower-cost style, the single-wall bag, is being used to some extent in Europe and by the major coffee producers in the United States. The bag-in-bag style and the single-wall style function equally well in protecting the product, but they differ in appearance. High-quality graphics on the paper or foil/paper outer wrap of a bag-in-bag package gives a smooth surface appearance. Most single-wall bags have a rough or "orange peel" surface when under full vacuum. As with any package concept with more than one style, there are pros and cons for each; but the current U.S. market is predominately single-wall vac bag and is likely to stay that way.

Equipment. Since Europe spawned the vac-bag's popularity, it is no surprise that all the automatic high-speed equipment is made there. There are three major equipment suppliers who compete in the coffee-packaging area: SIG, Hesser, and Goglio (1–3). All have automatic high-speed (40–120-boxes-per-minute (bpm)] 1-lb (0.45-kg)-size machines capable of forming bags from roll stock material, filling, evacuating, sealing, and folding the top of the bag to deliver a finished vac-bag. Bags can be made in different ways from rollstock. Mandrel- and tubeforming machines form the bag, which is carried horizontally into the filler. After filling it is transferred into a vacuum chamber where it is evacuated and sealed. It is then transferred out of the vacuum chamber to a trimmer and final folding of the top. Tape or hot-melt glue are two methods used to hold the top folds in place. The sealing of the bag is what keeps it hard and under vacuum.

Materials. Each of the major equipment suppliers provides basic information on what packaging material properties are essential for efficient vac-bag production. There are some differences but there are many similarities for achieving a low level of defects from the machine. Each vac-bag must be durable enough to survive the rigors of transportation, warehousing, and retail distribution to remain vacuum-packed at point of purchase. Although many European packers use a three-ply structure made of 48-gauge polyester film (OPET)/35-gauge aluminum foil (AF)/300-gauge sealant (see Film, polyester; Foil, aluminum; Sealing, heat), most U.S. coffee roasters employ a more durable four-ply structure. This can be accomplished by the addition of one of several films such as biaxially oriented polypropylene (BOPP) (see Film, oriented polypropylene) or biaxially oriented nylon (BON) (see Nylon). These structures can be constructed in various ways but most typically as follows:

48-ga OPET/75 ga BOPP/35 ga AF/300 ga sealant
48-ga OPET/35 ga AF/60 ga BON/300 ga sealant.

When these structures are designed for a single-wall bag, the OPET is reverse-printed. Gravure printing (see Printing) on OPET provides excellent graphics, high gloss, and a scuff-resistant package. Vacuum-metallized films (see Metallizing vacuum) have been used successfully in Europe and Canada as lowercost foil replacements. Since their oxygen barrier may be lower than foil, each coffee packer must determine if nonfoil structures provide adequate barrier.

Currently the best method to joint these materials into a finished structure is by adhesive lamination (see Laminating; Multilayer flexible packaging). To produce a finished structure with uniform-gauge control, high internal bond strength, minimum curl, and a controlled coefficient of friction, adhesive lamination has proved to be an excellent converting system. Both solvent- and solventless-type adhesive laminators (see Adhesives; Coating equipment) are now converting materials in the United States and Canada for coffee packers. These new high-speed automatic packaging machines require consistently high quality materials to operate efficiently. All of the material converters acknowledge the sometimes difficult learning experience on vac-bag films and the need to set up a very thorough quality-control system.

Valves and sorbents. Gas-off of CO_2 from ground coffee is a problem for vac-bag packagers that can be handled in a number of ways. The simplest solution is to grind the coffee beans very fine, as in Europe. The fine grind evolves gas more quickly and requires only a short holdup time ($1\frac{1}{2}$–2 h). The North American markets require coarse grinds, however, that need up to 24 h of holdup time and considerable storage capacity. There are two methods to eliminate all holdup time. One is the one-way coffee valve, which allows CO_2 out of the bag without allowing O_2 into the bag (see Valves). The bag, however, is soft. This concept was tested in the United States by two major roaster but it was abandoned in favor of the hard vac bag that consumers seem to prefer. The other method involves a new technology from Canada, whereby a CO_2-sorbent pouch is placed in the coffee bag much like a desiccant pouch is used for moisture-absorbing applications. This new coffee vac-bag technology allows the roaster to package freshly ground coffee without any holdup time and still obtain a hard vacuum bag. This is not possible with any other system.

A growing coffee market has been found in specialty and gourmet stores where whole beans are purchased. The beans are generally not hermetically packaged, so CO_2 gas-off is not an issue. In this case, the use of a one-way valve is an excellent way to achieve an airtight package and keep the beans fresh. The valve can be heat-sealed into the bag film as done by Goglio and SIG. It can also be attached to the outside surface using a pressure-sensitive valve concept developed by Hesser.

Advantages. Vac-bag is seen as low-cost package compared with metal cans. For some coffee packers, this may be true if bag packing line speeds are equal to or greater than those for cans. For others, where the reverse is true, the material-cost advantage is not as large. Current 1-lb-size (0.45-kg-size) metal cans, including plastic overcap, cost about \$0.17–0.19. A high-quality four-ply vac bag costs about \$0.08–0.10, which provides a material-cost advantage of about 9 cents per 1-lb (0.45-kg) package. Additionally, there are advantages in storage, handling, and shelving a vac-bag because of its compact shape. The reduction in cube is quite dramatic (35%). This cube reduction is beneficial to both manufacturer and retailer, since warehousing and shelving space is always a premium. As vac-bags become more widely distributed, consumers will determine the future for this new package.

BIBLIOGRAPHY

"Vacuum-Bag Coffee Packaging" in *The Wiley Encyclopedia of Packaging Technology,* 1st ed., by F. J. Nugent, General Foods Corp., pp. 690–691.

1. Product literature. SIG Swiss Industrial Company, Represented by Raymond Automation Company, Inc., Norwalk, CT.
2. Product literature. Hesser Division of Bosch Package Machinery, S. Plainfield, NJ.
3. Product literature. Goglio, represented by Fres-co System USA, Inc., Telford, PA.

VACUUM PACKAGING

Out of World War II came plastics, the wide commercial application of mechanical refrigeration, self-service supermarkets, the superhighway system, and a growth economy. During the late 1940s, Dewey and Almy Chemical Company purchased the patent and brought a new and unique French packaging development in the United States. The first significant commercial application of this concept was for vacuum packaging of whole turkeys using rubber stretch bags. These were soon replaced with PVDC shrink bags. The concept led to the wide commercial use of vacuum packaging for perishable (refrigerated and frozen) food; the Cryovac (registered trademark) vacuum-packaging process. In the ≥45 years since, many individuals and companies have contributed to the technical advancement and wide successful commerical use of vacuum packaging for the preservation, distribution, and marketing of perishable foods. In future annals of history, vacuum packaging of perishable foods in flexible and semirigid plastic films should be ranked right up there with canning, pasteurizing, and quick-frozen foods as a technological breakthrough that contributed greatly to improving the quality of life for humankind!

Vacuum packaging involves the placing, either manually or automatically, of a perishable food inside a plastic film package and then, by physical or mechanical means, removing air from inside the package so that the packaging material remains in close contact with the product surfaces after sealing.

Packaging in this manner, depending on the product being packaged the barrier properties of the packaging material, the level of air removal, and storage temperature, can substantially retard chemical and/or microbial deterioration of the food product. In many instances, this dramatically extends eating quality life.

Air-Removal Systems and Packaging Equipment

Nozzle vacuuming. Using this method of air removal, a nozzle connected to a vacuum pump is inserted inside the open end of a preformed bag or pouch (see Fig. 1). The vacuum or air removal level using the nozzle system is seldom very high, as the packaging material (depending on its modulus) quickly collapses onto the product surface, blocking significant further air removal. Today nozzle vacuumizing is widely used for packaging whole fresh and frozen poultry, fresh-cut vegetables and bulk packaging of fresh meats, poultry, fish, processed meats, nuts, etc.

For whole poultry, manual and high-speed rotary nozzle machines that provide an aluminum clip closure are used. The twisting of the bag neck and clipping provides for good whole-bird shaping and sealing under high-moisture conditions. Nozzle vacuumizing provides for good bag to product (bird) cling, without collapsing the body cavity as can happen with high-vacuum-chamber systems.

Figure 1. Nozzle vacuuming.

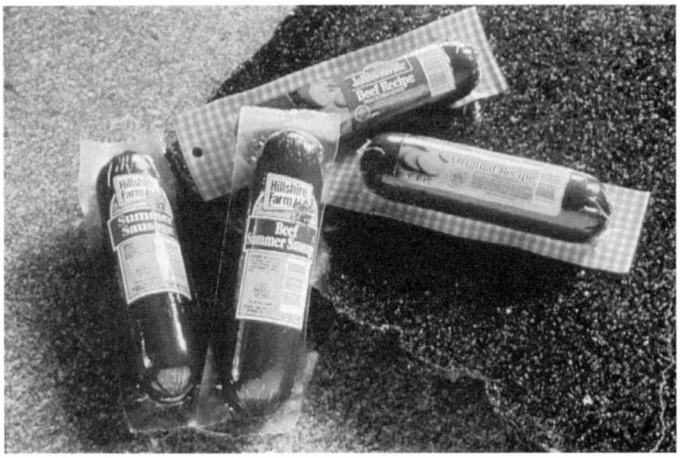

Figure 3. Thermoforming.

For cut vegetables and bulk packaging, manual and semiautomatic nozzle vacuumizing combined with heat sealing predominates.

Chamber vacuumizing. Using this method, the product is placed inside the flexible plastic film package, which is loaded into the bottom section of a vacuum chamber (Fig. 2). Such systems usually consist of a top chamber and a bottom chamber. The chamber is closed and a high vacuum [in both cubic feet per minute (cfm, ie, mL/min) and inches of Hg] mercury drawn on the chamber both inside and outside the package. Some chambers are usually vacuumized to ≥29 in. Hg.

Chamber vacuumizing machines generally employ heat-seal closures and are available from manual through fully automatic rotary high-speed units. Such equipment uses shrink bags and laminate pouches. They are widely used to package fresh primal and subprimal meat cuts, smoked and processed meats, and natural cheeses. Some chamber units are also available that provide an aluminum clip closure inside the chamber.

There are two other widely used chamber vacuumizing systems that utilize rollstock films rather than preformed bags and pouches; thermoforming and vacuum skin packaging.

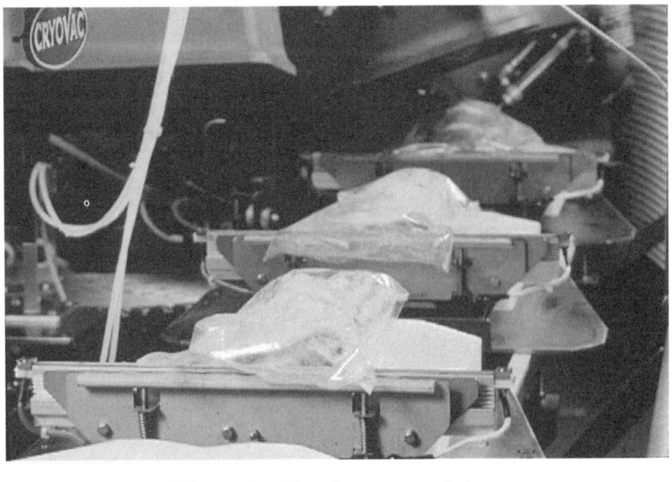

Figure 2. Chamber vacuumizing.

Thermoforming. Thermoforming vacuum-packaging machines dominate the packaging of consumer units of wieners, sliced luncheon meats, sliced bacon, smoked sausages, and some natural cheeses (Fig. 3). Such machines operate using two rolls of multilayer (usually high-O_2-barrier) plastic film (top and bottom webs). The operation of a thermoforming machine involves feeding the bottom web into the machine by clamping the web by its edges and indexing it either (*1*) over a continuously moving train of thermoforming dies (pockets) or (*2*) in a parallel set of clamping chains that first indexes the bottom web over a thermoforming station. In both web-handling modes, the bottom web is heated, then vacuum and/or pressure formed.

Product is manually or automatically loaded into the formed pockets. The top web is then fed down over the bottom web between the two halves of the vacuum-sealing chamber. The two halves of the chamber close and a high vacuum (≥29 in. Hg) is drawn on both sides of both webs and around the product. When the desired vacuum level is reached, the webs are heat-sealed together. The chamber(s) is (are) vented to atmosphere, opened, and the packages indexed out the machine and slit into individual units (see also Thermoforming).

Vacuum skin packaging. There are great similarities between vacuum skin packaging (VSP) (Fig. 4) and thermoforming machines, with the primary differences involving how the forming web (bottom web in thermoforming and top web in VSP) is thermoformed and the plastic film structures used. Like thermoforming in an automatic VSP machine the bottom web is fed through the machine by a parallel set of clamping chains. The bottom web may or may not be thermoformed slightly. It can be an oxygen barrier or nonbarrier flexible or semirigid web. Again, as with thermoforming, the product is loaded onto the bottom web, a top web is fed down over the bottom web, and both are indexed between the two halves of a vacuum (skin) packaging chamber (see also Skin packaging).

As the top web is fed down and into the VSP chamber, it is heated and may be partially thermoformed to fit over the product. When the chamber closes, the bottom web is pulled

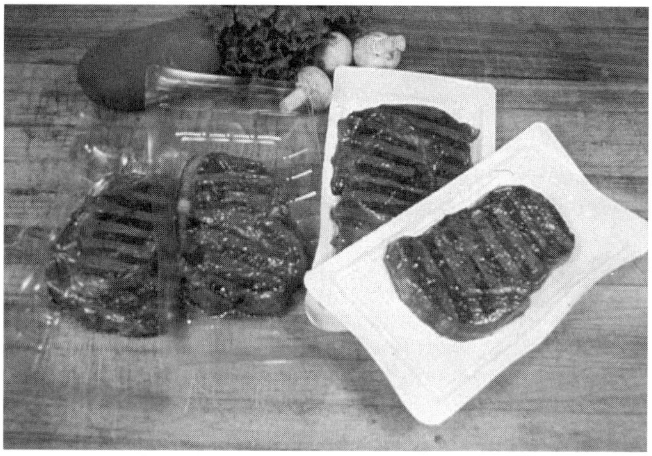

Figure 4. Vacuum skin packaging.

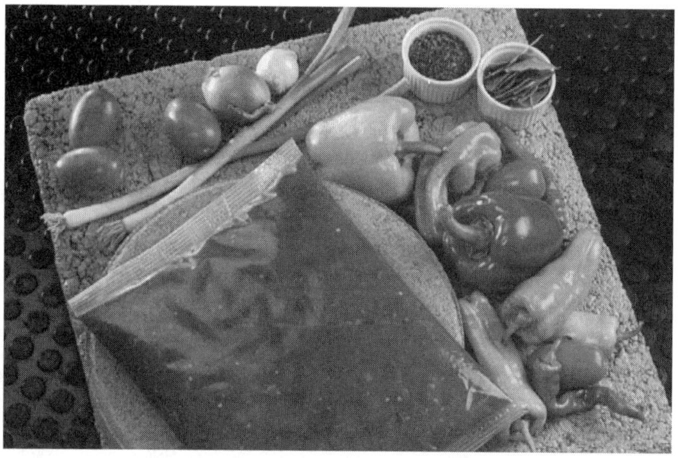

Figure 5. Hot fill.

against the bottom of the VSP chamber with a high vacuum. The top web is either (1) suspended over the product by drawing an equal vacuum over both sides of the web and around the product (2), or is drawn against the heated surfaces of the upper half of the VSP chamber by a high vacuum and the area around the product between the top and bottom webs is vacuumized. When the desired vacuum level is reached, the vacuum to the top of the chamber is vented to atmosphere and the heated top web instantaneously thermoforms to the shape of the product. At the same time, the two webs heat-seal to each other wherever they come in contact. The bottom chamber half is then vented, and the packages index out; being slit into individual units as they do so.

Pressure. Probably the most simple means of producing a vacuum or negative pressure package is through applying pressure to the outside of a bag or pouch containing a product. Two methods have been used commercially. In the early days of fresh-red-meat vacuum packaging, a system of air removal involved holding a bag (with product) by its open end and lowering it into a tank of water so that the product was completely submerged with the open end of the bag still above the water, forcing air out. The open end of the bag was then gathered and closed with an aluminum clip.

Another pressure method used with bags and laminate pouches and today in packaging fresh cut vegetables is to squeeze the air out with sponges and heat seal the package prior to releasing the squeezing action of the sponges.

Hot fill. This process is used to package pumpable prepared foods in flexible, oxygen-barrier plastic film packages. By hot filling, usually above 180°F, steam is present, which helps to exhaust air from the product and headspace of the package just prior to sealing. Such packages are then rapidly chilled, creating a package with a negative pressure inside (Fig. 5).

Following is a discussion of each perishable-food category with emphasis on why specific vacuum-packaging systems are used to preserve, distribute, and market each.

Poultry. Vacuum packaging of fresh poultry is done primarily to retard the growth of aerobic spoilage bacteria, and to provide an attractive leakproof (moisture) package (Fig. 6). When packaging fresh poultry, care must be taken to avoid establishing conditions that support the growth of H_2S-producing bacteria. These bacteria are commonly present on fresh poultry and grow best above 34°F and under low-O_2 conditions. Therefore, fresh whole Rock Cornish, roasters, and turkeys are vacuum packaged using O_2-permeable shrink bags. Some cut-up fresh poultry is now being vacuum-packaged using thermoforming and O_2-permeable structures. Some boneless/skinless parts are being vacuum-skin-packaged.

As for frozen whole and portioned fresh poultry, vacuum packaging in O_2 permeable shrink bags dominates. The bags are coextruded, multilayer polyolefin structures with a modified polypropylene (PP) outer layer to provide high gloss and freezer-case scuff resistance, and an electron-beam cross-linked polyethelene (PE) inner layer. This PE layer provides cold-abuse toughness, high hot-water shrinkability, and both heat and clip sealability. Good wrinkle-free film-to-product cling provided by the vacuumized shrink bag is essential for attractive, frost-free packages. High oxygen barrier is not required as frozen fresh poultry does not develop significant oxidative rancidity during typical storage and distribution.

Figure 6. Vacuum packaging for poultry.

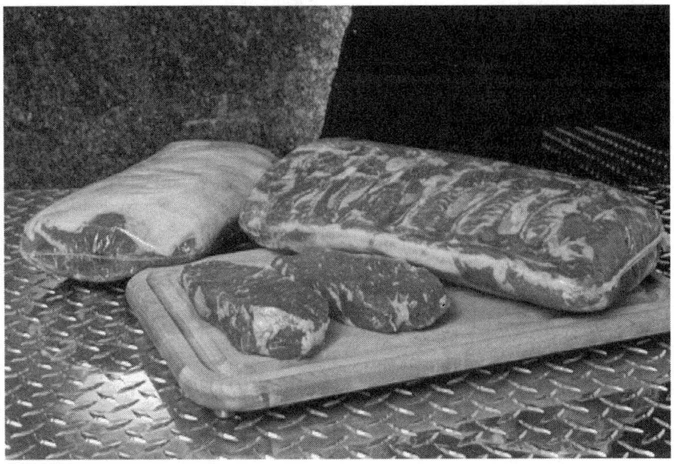

Figure 7. Vacuum packaging for red meat.

Fresh red meats (beef, pork, veal, lamb, venison). Today about 90% of the fresh beef primal and subprimal cuts shipped to food service and retail are vacuum-packaged in high-oxygen-barrier shrink bags (Fig. 7). Such bags are made of coextruded materials consisting of an outer EVA or LLDPE layer, a PVDC barrier layer, and an electron-beam crosslinked EVA or LLDPE heat-sealable/clipable inner layer.

Most fresh-red-meat (FRM) vacuum packages are produced on high-speed, high-vacuum, heat-sealing rotary chamber packaging machines. For bone-in FRM cuts, bags are available with bone-puncture-resistant patches laminated to the outside. Veal and lamb cuts are also vacuum-packaged in this manner. The application of this technology was slow to develop for fresh pork. Several years ago the industry did start using this technology for boneless loins and tenders. With the advent of the "boneguard" shrink bag, bone-in loins were successfully packaged. The industry is now rapidly converting to shrink bag vacuum packaging of bone-in pork loins for retail distribution.

The "aging" of beef is a tenderizing process that takes place as a result of naturally present enzymes. Vacuum packaging was first applied to beef by purveyors to age steaks and roasts for the food-service industry. Beef could be "aged" for up to 30 days without any significant shrink or trim loss. It was then applied to beef (later veal, lamb, and pork) for retail, which dramatically improved the distribution and marketing. In fact, this development ultimately resulted in a total restructuring of the beef industry. The preservation aspects of FRM barrier shrink packaging are threefold: (*1*) retard the growth of aerobic spoilage microorganisms, (*2*) prevent the chemical degradation of myoglobin, and (*3*) minimize the loss of moisture.

As FRM is chemically alive and using some oxygen. Oxygen permeability below 200 ml/m^2 will shift the microbial population from aerobes to facultative anaerobes (lactic acid-producing). To minimize myoglobin degradation and provide adequate red-color (oxymyoglobin) regeneration and stability for retail display, oxygen permeabilities below 50 ml/m^2 are required. Most of the barrier shrink bags used today have O_2 permeabilities below 30 ml/m^2. As shrink bags are excellent moisture barriers, the moisture loss of major concern is "purge" (free liquid that separates from the meat). The cut of FRM packaged, temperature, and other product factors influence moisture separation.

From the package standpoint, the absence of minute negative pressure voids or cavities is critical. The application of high package vacuum to maximize film to product surface cling, and excellent hot-water shrink to minimize film wrinkles, provides for the least purge development.

Fresh ground beef and pork sausage are major FRM product categories. While not truly vacuum-packaged, they are pressure packaged in tubular oxygen-barrier packages. Today, most packages for these products are produced on vertical form/fill/seal machines that provide metal clip closures. The films used are coextruded structures. For ground beef: PE or EVA/nylon/EVOH/ nylon/PE or EVA with appropriate tie layers. For pork sausage: EVA/PVDC/EVA.

For individually cut fresh and frozen steaks distributed primarily to food service, thermoforming vacuum packaging and vacuum skin packaging are widely used.

For fresh steaks, oxygen-barrier structures are required:

Thermoforming: nonforming web-biaxially oriented nylon or PET/PVDC/LLDPE or Surlyn (DuPont)

Forming web: cast nylon/EVOH/LLDPE or Surlyn

Vacuum skin packaging: forming web—PVDC/Surlyn or coextruded, electron-beam crosslinked LLDPE/EVOH/ LLDPE

Nonforming web, similar to the nonforming web used in thermoforming

For frozen steaks, structures are similar to those used for fresh except the PVDC or EVOH layers are left out.

Cured, smoked, and processed meats. This category includes such products a wieners, bacon, hams, loaves, smoked sausages, sliced luncheon meats, and corned beef. Today such products are manufactured using both red meats and poultry. These products are vacuum-packaged in high-oxygen-barrier materials to retard aerobic microbial growth, prevent cured color degradation, and loss or degradation of the cooked and smoked flavors and development of fat rancidity (Fig. 8). As chemical changes are of greatest preservation concern, products in this category are chamber-vacuumized to minimize re-

Figure 8. Vacuum packaging for cured, smoked, and processed meat.

Figure 9. Vacuum packaging for natural cheeses.

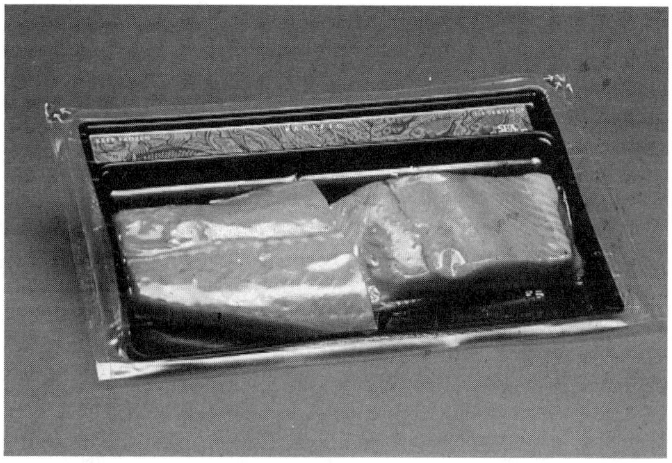

Figure 11. Vacuum packaging for fresh and frozen fish.

sidual O_2 within the packages. Wieners, sliced bacon, and sliced luncheon meats are packaged using materials with O_2 permeabilities below 15 ml/($m^2 \cdot 24$ h). To prevent curing-induced color degradation. They are generally packaged using high-speed laminate rollstock thermoforing machines. The forming web is a coextruded or extrusion structure consisting of nylon/PVOC or EVOH/Surlyn or LLDPE with appropriate tie layers if coextruded. The top layer will be polyester or oriented-nylon PVDC coated and reverse-printed if desired. Adhesive or extrusion laminated to Surlyn or LLDPE sealant.

The larger units such as loaves, hams, and corned beef are usually packaged using high-barrier shrink bags—coextruded EVA or LLDE/PVDC/EVA or LLDPE (electron-beam cross-linked). The packages are vacuumized and heat-sealed using high-speed rotary chamber machines.

Natural cheeses. Vacuum packaging of cheeses for curing and for loaves and consumer units of cured cheeses (Fig. 9).

Curing. Cheddar cheese is formed into 40-lb rectangular blocks and vacuum-packaged (chamber/heat-sealed) using oxygen-barrier shrink bags and laminate pouches (nylon/LLDPE), then placed in strong, tight-fitting corrugated cartons for curing 90 days and longer. The vacuum packaging inhibits mold growth and moisture loss while allowing the non-gas-forming microbial curing to take place. After curing, the 40-lb blocks are cut into consumer units or used as manufacturing cheeses. Other non-gas-forming cheese varieties such as Gouda are also cured in this manner. More traditional shapes such as wheels, horns, and sticks, of non-gas-producing varieties including Italian cheeses are cured using high-O_2-barrier shrink bags (see Fig. 10).

Figure 10. Curing.

For gas-producing cheeses (during curing) such as Swiss, high-CO_2-permeability shrink bags and nylon-based pouches are used. The O_2 permeability has to be kept below 250 ml/($m^2 \cdot 24$ h) to retard mold growth during curing.

A unique application of shrink-bag vacuum packaging is for curing of wheels of Blue and Roquefort cheeses. After vacuum packaging and shrinking the barrier bag tightly around the uncured cheese wheel, which has been inoculated with the desired mold, long needles are forced through the wheel and removed to provide air channels to allow the mold to grow and develop the blue color.

Retail units. Chunks, sticks, blocks, loaves, half moons, etc. of natural cheeses are vacuum packaged using O_2-barrier shrink bags and laminates using material structures and equipment systems very similar to previously described for smoked, cured, and processed meats. For cheese, such packaging is utilized to prevent mold growth, retard light-catalyzed oxidation of the butter fat on the surface, prevent surface color fading under display lights, and minimize moisture loss.

Fresh and frozen fish. With the growth of spoilage, bacteria can be retarded by vacuum packaging fresh fish in oxygen-barrier films. In the United States doing so is looked on with disfavor by the FDA. *Clostridium botulin* type C, which is commonly found on fish, can grow under anaerobic conditions at temperatures as low as 39°F. This is well within the refrigerated distribution temperature ranges especially if abused. A vacuum-packaging system using O_2-permeable plastic films has been approved for improved shelf-life packaging of fresh fish.

Vacuum packaging using O_2-barrier shrink bags is being used successfully for frozen Alaskan salmon to replace ice glazing, to eliminate moisture loss (freezer burn) and rancidity, and for frozen tuna loin storage and transport to tuna-canning plants (Fig. 11).

Fresh produce. A rather new and rapidly growing end use for vacuum packaging is fresh-cut vegetables. Much of the food-service industry is on a fast track to remove the preparation of fresh vegetables from the back of the restaurant to central plants. The processing and extended-shelf-life packaging technology is now being applied to consumer packaging of such vegetables for retail supermarket sales. When the natural skin is removed from a vegetable or is cut in smaller segments (sliced, diced, etc), extended shelf life can be provided through properly designed protective packaging. As fresh-cut vegetables are still alive and respiring, concern must be given to providing adequate O_2 and for release of the CO_2 produced. Fresh vegetables have been classified according to their respiration needs into three general groups:

Figure 12. Vacuum packaging for fresh produce.

Figure 13. Vacuum packaging for fresh produce.

Figure 14. Vacuum packaging for fresh produce.

Respiration group	Vegetables	Film permeabilities [mL/(m² · 24 h)]	
		O_2	CO_2
Light breathers	Carrots, potatoes, turnips	1,000–3,000	10,000–20,000
Medium breathers	Lettuce, celery, pepper	6,000–7,000	30,000–35,000
Heavy breathers	Broccoli, cauliflower, asparagus	15,000–20,000	75,000–100,000

Such cut vegetables are packed in permeability modified polyolefin bags or film or vertical form/fill/seal machines (see Form/fill/seal, vertical). Bags are typically nozzle vacuumized and pressure squeezed to remove residual air from the package prior to heat sealing. For the various vegetables group the polyolefin films are modified through resin variations and additives to achieve the desired O_2/CO_2 permeability.

The polyolefin films are adequate barriers to prevent significant moisture loss. The removal of the air from the package provides for the package quickly becoming a modified atmosphere. Under adequate refrigerated conditions, the O_2 will be maintained below 5% and the CO_2 in the 8–12% range. This slows down respiration and will significantly extend the eating-quality life of the cut vegetable. Large volumes of chopped lettuce, broccoli, and cauliflower florets are now being distributed to food service and retail in this manner. The vacuum or negative-pressure package also provides a good visual-quality measure of the package and product for both the producer and end user (Figs. 12–14).

BIBLIOGRAPHY

General References

"Advances in Fish Science and Technology," in *Proceedings, Torry Research Station Conference,* Aberdeen, Scotland, 1979.

Chilled/Refrigerated Foods Conference, Proceedings, The Packaging Group, Inc. 1988.

A. L. Brody, *Controlled/Modified Atmosphere Packaging of Foods,* Food and Nutrition Press, Turnbull, CT, 1989.

E. Karmas, *Meat, Poultry and Seafood Technology—Recent Developments, Food Technology Review No. 56,* Noyes Data Corporation, Parkridge, NJ, 1982.

F. Kosikowski, *Cheese and Fermented Milk Foods,* published by the author, Ithaca, NY, 1966.

R. Lawrie, *Developments in Meat Science—1,* Applied Science Publishers, London, 1980.

R. Lawrie, *Developments in Meat Science—4,* Elsevier Applied Science, London, 1988.

F. A. Paine and Heather Y. Paine, *A Handbook of Food Packaging*, Blackie Academic and Professional, London, 1992.

M. E. Rhodes, *Food Protection Technology II*, Lewis Publishers, Chelsea, MI, 1990.

S. Sacharow and Roger C. Griffin, Jr., *Principles of Food Packaging*, AVI Publishing Company, Westport, CT, 1980.

F. M. Terlizzi, R. R. Perdue, and L. L. Young, "Processing and Distributing Cooked Meats in Flexible Films," *Foods Technol.* p. 38 (March 1984).

RICHARD PERDUE
Taylor, South Carolina

VIALS. See AMPULS AND VIALS.

VIBRATION

Vibration describes oscillation in mechanical systems. Packages, products, pallet loads, vehicles, machinery, and equipment are mechanical systems. They are subjected to vibration whenever there is movement. In packaging applications, vibration can be either desirable or undesirable, although often it is undesirable, because it can damage products. Examples of packaging applications of desirable vibrations are a filling machine and an ultrasonic heat sealer. Filling machines may have a vibratory tray feeder to dispense the product into a package; an ultrasonic heat sealing uses high-frequency mechanical vibrations to produce localized heat that melts the interface between the sealing materials only, without affecting the product or the rest of the package. However, vibrations encountered in vehicles during distribution are always undesirable. This article concentrates on vibrations in distribution only, or those that are undesirable.

Vibration always occurs in distribution because the operations of transport and handling always move goods. It can influence products to the point of causing physical damage or product spoilage. Studies have shown that vibration in vehicles can cause for example, phase separation to emulsified products, bruising in fruit and vegetables, physical damage in mechanical, and electronic products, damage to labels because of friction (1,2).

Both external and internal sources originate vehicle vibrations. Examples of external sources include road pavements, rail tracks, and ocean waves; internal sources include rotating parts such as unbalanced wheels and the engine itself (3). In a truck, for example, the wheels of the vehicle pick up the pavement up-and-down oscillations as mechanical vibrations; the suspension and structure of the vehicle transmit these, along with other internal vibrations, to the vehicle floor. Packaged products placed on the vehicle floor will then receive the vibrations.

Vibrations that occur in a vehicle floor may be decomposed in three perpendicular directions: vertical, lateral, and longitudinal. Several studies investigated the directions and locations in vehicles where highest vibration levels occur. On the basis of these studies, it is commonly accepted today that in trucks and semitrailers the highest vibrations occur on the rear axle and in the vertical direction (4,5). Product damage due to vibration is more likely to occur if packages are placed over the rear axle in trucks and trailers.

The role of the packaging engineer is to design packages that reduce vibrations. Cushions and other flexible-packaging systems are engineered in such a way that the package attenuates the vibrations that occur in vehicles. No package can completely eliminate vibration. Although still transmitted to products, a package designed for vibration attenuation may virtually eradicate the potential for damage.

Forced versus Free Vibration

Vibrations are generally classified into two categories: *forced* (when there is a continuous input source) and *free* (input source discontinued). The type of vibration that occurs in packaging machinery is a forced vibration since there is a source continuously driving the system to vibrate. The vibration that occurs without external excitation is called *free vibration* (6). Both types, free and forced, occur in all vehicles during transport. As an example, the pavement roughness is a source of vibration that forces the wheels of a vehicle to move up and down, generating vibration. If the source of vibration stops, the system will continually decrease its vibration level (free vibration now) until a full stop, when the energy is totally dissipated. Similarly, when the vehicle hits an obstacle such as a pothole, the vibration will be forced while the obstacle is in contact with the vehicle, and free thereafter until the effect of that particular input ends. An example of free vibration occurs when a mass, placed on top of a spring, is pushed down and released, and then vibrates "freely" after its release. Whenever a system is vibrating freely, the movement will decrease and the vibration will eventually cease, due to *damping*. Damping is always present in mechanical systems (7), and it dissipates the energy that keeps the system vibrating.

Frequency and Acceleration

A state of vibration can be characterized by a frequency and an amplitude. The simplest harmonic motion of a mechanical system, such as of a mass placed on a spring that oscillates up and down, is defined by a *frequency* and a *peak acceleration*. The frequency is the number of times a cycle repeats itself in a unit of time, and the peak acceleration is the maximum value of the acceleration in the entire cycle. If a system is vibrating freely, it vibrates at its natural frequency (8). The natural frequency is mainly a function of the mass and the stiffness of the system. The relationship between peak acceleration, maximum displacement from the position of equilibrium, and frequency of a vibrating system is

$$a_{pk} = \frac{d_{max}(2\pi f)^2}{g} \tag{1}$$

where a_{pk} = peak acceleration in g (where 1 g is equal to the gravitational acceleration), d_{max} = maximum displacement of the oscillating mass from the equilibrium position, f = frequency of oscillation, and g = gravitational acceleration, or 9.81 m/s^2. In applying Equation 1 to ideal, or linear systems, the equation shows that the peak acceleration is directly proportional to the maximum displacement and to the square of the frequency of vibration. In other words, if the maximum displacement is doubled, the peak acceleration will also double (frequency kept constant); if the frequency is doubled (maximum displacement kept constant), the peak accelera-

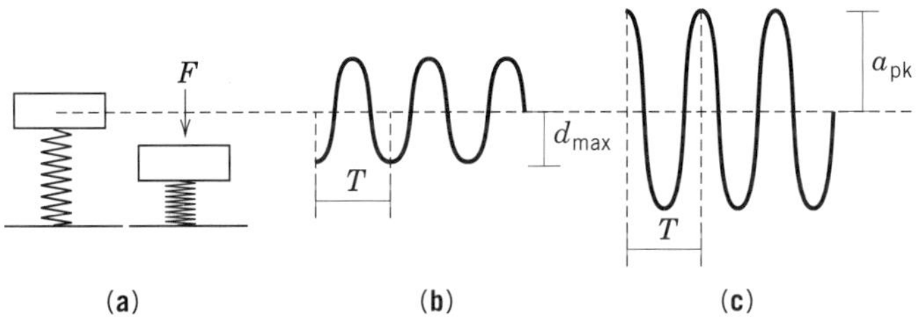

Figure 1. Spring-mass system in free vibration without damping: (**a**) force F causes displacement d_{max} and releases mass to vibrate freely; (**b**) displacement of mass varying with time; and (**c**) acceleration of mass varying with time.

tion will quadruplicate. Figure 1 shows the relationship represented by Equation 1, as applied to a spring-mass system undergoing free vibration. Equation 1 does not take into account the direction of vibration (up or down). In Figure 1, the mass is pushed down with the force F and then released. The situation depicted in Figure 1 does not show the effect of damping. For further information on damping effects on free vibration, the reader may consult Refs. 7–9.

The duration of one cycle of oscillation is shown as the period T. The frequency of oscillation f representing the number of cycles that occur in a unit of time is related to the period by

$$f = \frac{1}{T} \tag{2}$$

For example, if the period T = 0.2 s, and d_{max} = 0.5 cm (0.005 m), the frequency is 1/0.2 s = 5 Hz (using Eq. 2) and a_{pk} = 0.005 m · $(2\pi 5\ \mathrm{Hz})^2/9.81\ \mathrm{m/s^2}$ = 0.5 g (using Eq. 1). This means that the system is vibrating with a peak acceleration of 0.5 g and the mass is oscillating at five cycles per second (5 Hz). If the peak acceleration is 0.5 g, the acceleration at any instant will be somewhere between + 0.5 g and −0.5 g, inclusive.

When vibration levels in vehicles reach values above 1 g (acceleration due to gravity) packages will start to bounce. Since packages are seldom attached to the vehicle floor, if the vertical vibration of the vehicle floor is higher than 1 g, the package would not follow the vehicle floor when it is moving down and will be left suspended in the air momentarily, hitting the floor moments after. This will cause packages to receive impacts rather than only the vibration that occurs at the vehicle floor. Impacts have acceleration levels (G levels) much higher than vibrations (impact accelerations can get up 100 g very easily) and are much more likely to cause damage to products during distribution.

Sinusoidal versus Random Vibration

Regarding the mode of vibration, there are two ways it can occur: *sinusoidal* and *random*. The displacement of the mass shown on the spring in Figure 1 characterizes a sinusoidal motion and it can be expressed mathematically as a function of the time only. For example, when time t = 0, the displacement is at its minimum, when t = T/4, the displacement is *zero*, and when t = T/2, the displacement is at its maximum. This pattern repeats itself every interval of duration T. Vibration encountered in distribution such as in a truck when it travels on the road is not like the periodic or sinusoidal vibration of a spring-mass system as represented in Figure 1; rather, it is complex (10). Vehicles vibrate at several frequencies simultaneously, with the displacement, and therefore the acceleration, varying at each instant and never repeating itself again in the same pattern. Figure 2 shows an example of a typical random signal measured on a truck floor.

Because of this complex characteristic, random vibration cannot be described in a deterministic manner, ie, one cannot write an equation of acceleration versus time to describe this motion. Therefore, a statistical technique, or spectral analysis, is used to study such vibrations and to interpret the motions of the vehicle (9), which are covered in the next paragraphs.

Vibration Measurement and Analysis

Most packaging-dynamics applications use *accelerometers* for measuring vibration. Accelerometers are sensors that produce an electrical output signal proportional to acceleration. These sensors are rigidly mounted on the surface of a truck floor or package item and connected to data-recording devices. Manufacturers of accelerometers supply a calibration certificate with each sensor. One important information in this certificate is the *accelerometer sensitivity* (in output units per g). The sensitivity and the output voltage relate to the acceleration by

$$\text{Acceleration} = \frac{\text{voltage output}}{\text{sensitivity}} \tag{3}$$

For example, consider an accelerometer with a 10 mV/g sensitivity. If it senses a vibration with peak amplitude output of 6 mV, the acceleration is 6 mV/(10 mV/g) = 0.6 g. Accelerometers should be calibrated regularly to ensure accuracy of results acquired (11).

There are two basic ways to measure and store vibration data: *continuous* (or analog) and *intermittent* (or digital). The recording equipment defines the type of measurement and data storage.

In continuous recording, accelerometers are installed in strategic places in the vehicle and vibration data are recorded, without interruption, on a magnetic tape. A voice channel may be used to record a description of the events as they occur, such as the crossing of a rail track, the vehicle speed, or any other occurrence that may be of value for subsequent data analysis and interpretation. This technique re-

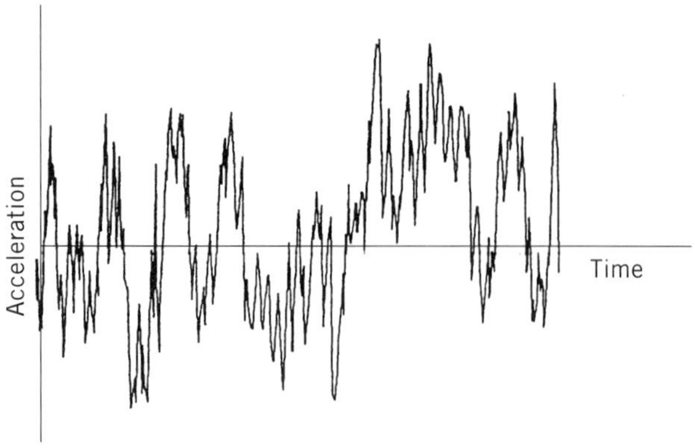

Figure 2. Acceleration versus time in random vibration.

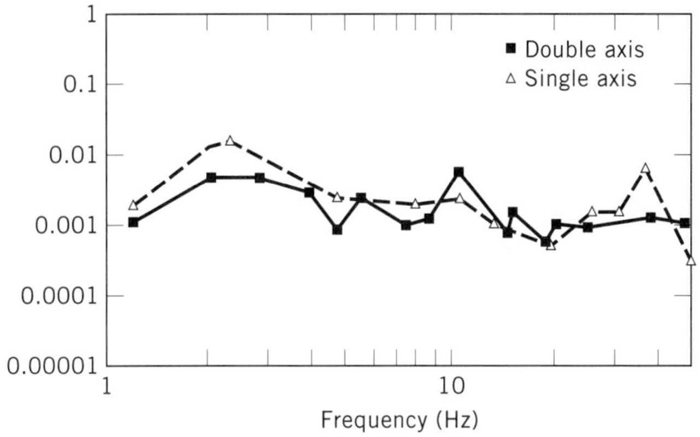

Figure 3. PSDs comparing vibration in single- and double-axle trailers (12). (Originally published in *IPENZ Transactions.*)

quires the continuous presence of an operator and the instrumentation used can be cumbersome, occupying as much space in a vehicle as one whole pallet load. After stored in the magnetic tapes, acceleration recordings are transferred to additional instrumentation for analysis, interpretation and use.

Intermittent monitoring has grown substantially in the last few years with the availability of digital recording and storage instrumentation. Intermittent recording uses self-contained recording units to obtain and store acceleration either at predetermined intervals or above a preset triggering level. Recording units used for intermittent monitoring are usually small and contain a real-time clock and a battery to provide the operating power. After stored, vibration is downloaded to a computer, analyzed, and summarized for subsequent use.

In random-vibration analysis, the complex signal is first decomposed into component sine waves of various frequencies either mathematically using Fourier decomposition (8) or electronically using *bandpass* filters. Whatever the technique used, the complex signal is analyzed in individual component frequency intervals determined by the *bandwidth.* For most packaging applications, the bandwidth has been traditionally set at 1 Hz. The random signal of a vehicle vibration has component sine waves that vary over time but have a zero average, and their peak amplitudes are distributed normally over time so that the Gaussian distribution and all its properties may be used to describe the process.

The acceleration level of any component frequency analyzed is represented by the *power density* (PD), which is calculated as

$$PD(f) = \frac{\sum_{i=1}^{n} (\text{Grms}_i)^2}{n \times BW} \tag{4}$$

where $PD(f)$ = power density at the frequency f, n = number of samples at the frequency f, Grms_i = root mean square acceleration of the ith sample at the frequency f, BW = bandwidth (normalized to 1 Hz).

A *power-spectrum density* (PSD) represents the results of random-vibration analysis. The PSD is a graphic representation of the vibration intensity at each frequency. As an example, Figure 3 shows the results, in PSD format, of vibrations encountered in semi-trailers in urban-area traffic at 60 km/h. This example shows that the single-axle trailer has an overall higher vibration force than the double-axle trailer, except for the frequencies in the range 9–22 Hz (12). Notice that the graphs are in a log–log scale for better clarity at the low-frequency range.

The PSD plots show the component frequencies in the horizontal axis and an indication of their amplitude in the vertical axis. High power-density levels at particular frequencies may indicate problems with the vehicle. This occurs because there are resonant reactions between the vehicle structure and the road surface in this frequency range. For example, for both vehicles represented in Figure 3, there are high power-density levels between 2 and 3 Hz. For the double-axle vehicle, there is another high power-density level at just above 10 Hz; for the single-axle vehicle, a high power density also occurs at 35 Hz.

The PSD plot is interpreted as a "signature" of the random vibration and is used to statistically represent vibrations that occur in vehicles. It can be used to determine the quality of the road or the vehicle, or both. The PSD plots of vehicles are important in the design of protective packages to attenuate how products receive the vibrations that occur in these vehicles. They are also widely used in the execution of laboratory vibration tests to simulate the transportation hazards encountered in these vehicles.

Vibration Tests

In a laboratory, it is typical to perform two types of vibration tests: *sinusoidal* and *random.* Sinusoidal tests are used to identify resonant frequencies of a product, a packaging material or a complete package. Random-vibration tests are usually performed to test finished-packages prototypes and packaging systems (13) but can also be used to test vibration transmissibility of cushioning materials using a frequency analysis technique (14).

Sinusoidal vibration tests should define (*1*) upper and lower frequency of the test, (*2*) level to be maintained at each frequency, (*3*) rate at which frequency will sweep and whether it is logarithmic or linear, and (*4*) the number of sweeps or duration of the test. Random-vibration tests should define (*1*) PSD to be used, (*2*) duration of test at each PSD, and (*3*) the level sequences of each PSD. Both test types are performed using an electrohydraulic vibration tester mounted on a seismic mass (15), although mechanical shakers are still widely used, mainly for sinusoidal vibration tests with a narrow frequency range.

The purpose of a finished-package vibration test is to recreate in the laboratory the damage potential that the packages would encounter in distribution. These tests determine how effective a package system can be in providing product protection (2). ASTM standard D4169 recommends specific PSDs for each transportation mode (truck and rail). It also recommends that random-vibration tests last 3 h (13), although studies have shown that tests with durations as short as 20 min produce valid results (16). There is generally a tradeoff between test severity and test duration (17). Ideally, a test should be only long enough to impose on the packages events that have the same damage potentials of the events the test is supposed to reproduce.

Terminology

The application of mechanical vibrations in packaging involves many terms that may not be familiar to some readers. For definition of common terms used in vibration, refer to Harris and Crede (18). In addition, more specific terminology on distribution packaging may be found in ASTM standard D996-90a (19).

BIBLIOGRAPHY

1. J. P. Mackson and S. P. Singh, "The Effect of Temperature and Vibration on Emulsion Stability of Mayonnaise in Two Different Package Types." *Packag. Technol. Sci.* **4**, 81–90 (1991).
2. S. P. Singh, G. J. Burgess, and M. Xu, "Bruising of Apples in Four Different Packages Using Simulated Truck Vibration," *Packag. Technol. Sci.* **5**, 145–150 (1992).
3. T. D. Gillespie, *Heavy Truck Ride,* Society of Automotive Engineers, Warrendale, PA, 1985.
4. J. A. Marcondes, S. P. Singh, and G. J. Burgess, "Dynamic Analysis of a Less Than Truckload Shipment," ASME Winter Annual Meeting, paper 88-WA/EEP-17, Chicago, 1988.
5. S. P. Singh, J. R. Antle, and G. J. Burgess, "Comparison Between Lateral, Longitudinal, and Vertical Vibration Levels in Commercial Truck Shipments, *Packag. Technol. Sci.* **5**, 71–75 (1992).
6. W. Weaver, Jr., S. P. Timoshenko, and D. H. Young, *Vibration Problems in Engineering,* Wiley, New York, 1990.
7. R. E. Blake, "Basic Vibration Theory" in C. M. Harris, ed., *Shock and Vibration Handbook,* 3rd ed., McGraw-Hill, New York, 1988.
8. W. T. Thomson, *Theory of Vibration with Applications,* Prentice-Hall, Englewood Cliffs, NJ, 1988.
9. D. E. Newland, *An Introduction to Random Vibrations and Spectral Analysis,* 2nd ed., Longman, New York, 1984.
10. S. P. Singh and D. E. Young, "Measurement and Analysis Techniques Used to Simulate the Shipping Environments," *ASME* Winter Annual Meeting, paper 88-WA/EEP-18, Chicago, 1988.
11. H. Schueneman, "Calibrating Issues in Packaging Testing," *Test Eng. Manage.* Vol. **57**, No.(2), 6–7 (1995).
12. J. A. Marcondes and W. H. Feather, "Assessment of Vibration Levels on Truck Shipments" in *Trans. Inst. Prof. Eng. New Zealand* (Wellington, NZ) **19,**(1), 22–27 (1992).
13. ASTM, *Selected ASTM Standards on Packaging,* 4th ed., ASTM, Philadelphia, 1994.
14. J. A. Marcondes, M. Sek, and V. Roulliard, "Testing Foamed Polymeric Materials for Mechanical Protection of Packaged Goods" in *Anais do 2nd Congresso Brasileiro de Polimeros,* Vol. 2, Sao Paulo, Brazil, 1993, pp. 1126–1128.
15. P. J. Vergano and R. F. Testin, "An Inside Look at Seismic Mass Design for Vibration Test System," *Packag. Technol. Eng.* **3,** (4), 38–43 (1994).
16. S. P. Singh and J. A. Marcondes, "Accelerated Lab Simulation of Truck Transport Environment Using Random Vibration," ASME Winter Annual Meeting, paper 89-WA/EEP-17, San Francisco, 1989.
17. D. Charles, "Derivation of Environment Descriptions and Test Severities from Measured Road Transportation Data," *J. Inst. Environ. Sci.* 37–42 (Jan./Feb. 1993).
18. C. M. Harris and C. E. Crede, "Introduction to the Handbook" in C. M. Harris, ed., *Shock and Vibration Handbook,* 3rd ed., McGraw-Hill, New York, 1988.
19. ASTM, "D996-90a Standard Terminology of Packaging and Distribution Environments" in *Selected ASTM Standards on Packaging,* 4th ed., ASTM, Philadelphia, 1994.

General References

R. K. Brandenburg and J. J.-L. Lee, *Fundamentals of Packaging Dynamics,* LAB, Skaneatles, NY, 1991.

R. M. Fiedler, *Distribution Packaging Technology,* Institute of Packaging Professionals, Herndon, VA, 1995.

PCB Piezotronics, *Vibration and Shock—Sensor Selection Guide,* PCB Piezotronics, Depew, NY, 1993.

C. E. Crede and J. E. Ruzicka, "Theory of Vibration Isolation" in C. M. Harris, ed., *Shock and Vibration Handbook,* 3rd ed., McGraw-Hill, New York, 1988.

K. Unholtz, "Vibration Testing Machines" in C. M. Harris, ed., *Shock and Vibration Handbook,* 3rd ed., McGraw-Hill, New York, 1988.

A. J. Curtis, "Concepts in Vibration Data Analysis" in C. M. Harris, ed., *Shock and Vibration Handbook,* 3rd ed., McGraw-Hill, New York, 1988.

JORGE MARCONDES
San Jose State University
San Jose, California

VINYLIDENE CHLORIDE COPOLYMERS

Vinylidene chloride (VDC) copolymers have been an important contributor to the plastic packaging that has increased the fraction of food that is produced to reach the dinner table. VDC copolymers provide exceptionally low permeabilities to gases and liquids. Furthermore, they are resistant to attack by many chemicals and are inert to most foodstuffs. All commercial VDC resins are copolymers with compositions that consist, by definition, of at least 50 wt% vinylidene chloride. These materials are often erroneously called poly(vinylidene chloride) or PVDC. This generally leads to confusion since it formally describes a homopolymer which is not used in commerce.

The overwhelming fraction of VDC copolymers is used for food packaging. Lesser amounts are used for other packaging applications, fibers, specialty binders, and molded parts. Production of VDC copolymers is kept confidential by the various manufacturers; however, one source estimates the worldwide use in 1984 to be 1.1×10^5 mt as barrier copolymers with additional amounts as nonbarrier copolymers (1).

Chemistry

Homopolymer PVDC is not commercially useful because of its crystallinity and thermal instability. The melting temperature of the homopolymer is about 202°C, and the level of crystallinity is greater than 60%. Crystallization from the melt is rapid. As an extrudable resin, the polymer needs to be heated above the melting temperature for a few minutes. The degradation rates above 200°C are too rapid to allow time for extrusion. As a latex, the homopolymer can crystallize in the particles. Then, during drying after coating, the particles will not fuse to give a continuous film. As a resin for solvent coating, the crystals are too stable to be dissolved by convenient methods.

Copolymerization alleviates the problems with the homopolymer. A comonomer decreases the lamellar thickness of the crystals and lowers the melting temperature. A comonomer decreases the rate of the crystallization and allows a latex to remain completely amorphous longer. Hence, the coating will fuse easily and then crystallize. A comonomer gives a semicrystalline material which can dissolve in convenient solvents.

Vinylidene chloride is versatile since it will undergo free-radical polymerization with many other monomers. The commercially important comonomers are vinyl chloride, acrylonitrile, methacrylonitrile, plus various methacrylates and alkyl acrylates. For extrudable resins, vinyl chloride or methyl acrylate is commonly used to depress the melting temperature to the range of 140–175°C.

For latex coatings, an alkylacrylate or alkylmethacrylate is used as comonomer. Terpolymers are common, and in some cases a surface active comonomer is included such as acrylic acid or 2-sulfoethyl methacrylate.

For solvent coatings, acrylonitrile and methacrylonitrile are the most common comonomers.

Most extrudable resins are made in suspension by free-radical polymerization. The monomers are dispersed as droplets in water and stabilized by a protective suspending agent such as poly (vinyl alcohol) or methylcellulose. An oil soluble free-radical initiator is used. The result is a free-flowing collection of nearly spherical, porous particles with diameters in the range 200 to 300 μm.

Latices are made in a free-radical, emulsion polymerization. Often, monomers are added to the reaction vessel during polymerization. The latices must be stabilizes with surface active agents to prevent coagulation.

Resins for solvent coating are made in a free-radical, emulsion polymerization. Often, monomers are added to the reaction vessel during polymerization. A minimum amount of surface active agent is used for the polymerization because coagulation, dewatering, and drying follow the polymerization. Common solvents include tetrahydrofuran (THF) or methyl ethyl ketone with toluene.

In addition to those materials which are required for polymerization, certain additives are commonly used with VDC copolymers. Plasticizers, stabilizers, and extrusion aids are included in extrudable resins at low levels for high barrier applications and moderate levels for less demanding applications. A latex may contain an antimicrobial. A resin for solvent coating will contain some of the salt used to cause coagulation.

VDC copolymers can degrade at elevated temperatures releasing HCl and leaving a double bond in the polymer backbone and, consequently, generating an allylic dichloromethylene unit susceptible to facile thermal radical dehydrohalogenation. Hence, at low levels of degradation a series of conjugated double bonds is formed which may impart color to the polymer. In extreme cases, the degradation can lead to a brittle char and even to carbon. The degradation is also catalyzed by Lewis acids, particularly iron cations formed from the interaction of process equipment with evolved hydrogen chloride. Hence, extrusion of vinylidene chloride resins should be done using the proper materials of construction and with training by experts. Strong bases can also cause degradation of VDC copolymers. For example, contact with organic amines is not recommended. Such agents act to introduce initial double bounds which may serve as initiation sites for rapid thermal degradation.

Table 1. Suppliers of VDC Barrier Copolymers[a]

Producers	Extrudable Resins	Latices	Solvent Coating Resins
Dow Chemical	X		X
Hampshire Chemical		X	
Asahi Kasei	X	X	X
BASF		X	
Kureha	X		
Morton Chemical		X	
Solvay & Cie	X	X	X
Zeneca Resins		X	X

[a] June 1996.

General Principles

All VDC copolymers have certain common characteristics. Commercial copolymers have molecular weight ranges of ca 65,000–150,000. Molecules of lower molecular weights tend to be brittle and have limited commercial value. The physical and chemical properties of copolymers depend largely on the VDC content, which ranges from 72 to 94 wt% in most commercial products. The axial symmetry of the VDC molecule permits tight packing of molecular chains with concomitant formation of crystals constituting 25–45 vol % of the total structure. Specific gravities range from 1.65 to 1.75.

These characteristics of VDC copolymer resins are not significantly altered by the small amounts of processing and stabilizing aids used to facilitate extrusion and molding. A representative extrudable resin is a VDC–VC copolymer with ca 85% VDC and a molecular weight of 120,000, with plasticizer sufficient for ease of extrusion.

Note that the barrier and strength properties of VDC copolymers depend on the chemical compositions of the molecules, the degree of molecular orientation, and the direction of orientation in the finished form. Relatively high VDC content imparts relatively high barrier properties and gas resistance. A comparatively high degree of orientation imparts tensile strength. Higher crystallinity correlates with lower permeability.

Producers

The principal producer/suppliers/ of VDC copolymers in their various forms are listed in Table 1. The Dow Chemical Company produces two forms of Saran resin plus Saran Wrap

Table 2. Suggested Coatings for Various Substrates

Substrates	Coatings	
	Latices	Solutions
Cellophane		X
Nonporous papers[a]	X	X
Porous papers	X	
Polyolefins	X	X
Polyesters	X	X
Polyamides (nylons)	X	X
Styrenics	X	
Vinyls (PVC)[b]	X	

[a] Includes coated and dense forms, such as glassine.
[b] Plasticizers in highly plasticized PVC may migrate to the VDC copolymer coating, thereby damaging the coating's effectiveness as a barrier.

Table 3. Transport Properties of a High Barrier VDC Copolymer Extrudable Resin in Film or Sheet Form

Property	Value	
Gas permeabilities at 23°C (73°F) nmol/m·s·GPa (cc·mil/100·in.^{2}·day·atm)		
Oxygen	0.10	(0.05)
Nitrogen	0.03	(0.015)
Carbon dioxide	0.20	(0.10)
Water vapor transmission rate (WVTR) at 38°C (100°F), 90% rh nmol/ms (g·mil/100·in.^{2}·day)	0.01	(0.03)

films and SARANEX coextruded films. (Saran, Saran Wrap, and SARANEX are trademarks of The Dow Chemical Company).

Table 4. Permeability of Household Films at 25°C to Selected Permeants

Permeant	Plasticized Vinylidene Chloride Copolymer[a]	Film Type Plasticized Poly (Vinyl Chloride)[b]	Polyethylene[c]
Oxygen, nmol/(m·s·GPa)[d]	1.9	220	640
Water vapor[e], nmol/(m·s)[f]	0.055	0.30	0.19
d-Limonene, MZU[g]	130	1.1×10^5	3.3×10^5
Dipropyl disulfide, MZU[g]	1.1×10^4	3.3×10^6	6.8×10^6

[a] Saran Wrap brand plastic film (trademark of The Dow Chemical Company).
[b] Reynolds plastic wrap (trademark of Reynolds Metals Co.).
[c] Handi-Wrap II brand plastic film (trademarkd of DowBrands, Inc.).
[d] Multiply by 0.50 to obtain cc·mil/100in.^{2}·day·atm.
[e] Measured at 37.8°C, 90% in.
[f] Multiply by 3.95 to obtain g·mil/100in^2·day.
[g] MZU = $(10^{-20}$ kg·m)/(m^2·s·Pa).

Applications

Extrudable resens

Flexible films. Monolayer films are used for household wrap, as unit-measure containers for pharmaceuticals and cosmetics, and as drum liners and food bags. Multilayer films, generally coextrusions with polyolefins, are used to package meat, cheese, and other moisture or gas sensitive foods. The structures, which contain 80 to 90% polyolefin with an inner layer of VDC copolymer, are usually shrinkable films that provide a tight barrier seal around the food product.

Semirigid containers. VDC copolymers are used as barrier layers in semirigid thermoformed containers. The sheet can be produced by coextrusion or by laminating monolayer or coextruded VDC copolymer films to semirigid styrenic or olefinic substrates. VDC copolymers can also be used as barrier layers in blow-molded bottles (see also Blow Molding).

Table 5. Properties of Vinylidene Chloride Copolymer Films[a]

Property	ASTM Test Method	Type of Film	
		General-Purpose	High Barrier
Specific gravity	D 1505	1.60–1.71	1.73
Yield, cm^2/(g·m)(in.^{2}/(lb·mil))		6.25–5.85(17,300–16,200)	5.78(16,000)
Haze %	D 1003		2–3
Light transmittance, %		85–88	80–88
Tensile strength, MPa (psi)	D 882		
MD		48.3–100(7,000–14,500)	82.8–86.2(12,000–12,500)
XD		89.7–93.1(13,000–13,500)	138–148(20,000–21,500)
Elongation, %	D 882		
MD		40–100	95–100
XD		40–100	50–60
Tensile modulus, 2% secant, MPa (psi)			
MD		345–759(50,000–110,000)	1103–1138(160,000–165,000)
XD		310–724(45,000–105,000)	931–1034(135,000–150,000)
Tear strength, N/mm(gf/mil) propagating	D 1922	3.9<38.6(10<100)	3.9<38.6(10<100)
Folding endurance, cycles	D 2176	>500,000	>500,000
Change in linear dimensions at 100°C (212°F) for 30 min, %	D 1204		
MD		12–22	6–7
XD		6–18	3–4
Service temperature, °C (°F)		−18 to 135 (0–275)	−18 to 135 (0–275)
Heat-seal temperature, °C (°F)		121–149(250–300)	121–149(250–300)
Oxygen permeability at 23°C (73°F) rh 50% rh[b], nmol/m·s·GPa (cc·mil/(100 in.^{2}·d·atm)	D 1434	1.6–2.2(0.8–1.1)	0.16(0.08)
Water-vapor transmission rate (WVTR) at 100°F (38°C) and 90% rh, nmol/m·s (g·mil/(100 in.^{2}·d))		0.06(0.2)	0.02(0.05)
COF face-to-face, back-to-back, at 23°C (73°F) and 50% rh	D 1894	0.3 to no slip	No slip

[a] Film 1 mil (25.4 μm) thick. Not to be confused with Saran Wrap household brand film.
[b] Humidity has no effect on the permeability of VDC copolymer films.

Table 6. Properties of VDC Copolymer Films[a] Made from Aqueous Latices and from Solutions

Properties	Latices[b] selected for		Solution Resins[d]
	High Barrier	High Seal Strength[c]	
Oxygen permeability at 23°C (73°F) nmol/m·s·GPa (cc·mil/(100 in.^{2}·d·atm)	0.06–0.20(0.03–0.10)	0.20(0.10)	0.04–0.20(0.02–0.10)
Water-vapor transmission rate (WVTR) at 38°C (100°F) and 90% rh, nmol/m·s (g·mil/(100 in.^{2}·day))	0.012–0.015(0.05–0.06)	0.012–0.38(0.05–0.15)	0.005–0.025(0.02–0.10)
Minimum heat-seal temperature, °C (°F)	132(270)	82(180)	104–129(220–265)

[a] Oven-dried films.
[b] Cast on PET film from dispersions.
[c] Heat and dielectric-sealing grades available.
[d] Cast from solutions of up to 20 wt% solids in 65% THF or methyl ethyl ketone/35% toluene (wt/wt) at 23–35°C (73–95°F).

Coatings. Paper and paperboard. Paper and paperboards coated with VDC latex copolymers are used where moisture resistance, grease, resistance, oxygen barrier, and water-vapor barrier are required.

Cellophane. About 90–95% of all cellophane produced in North America is coated with VDC solution coatings to render the films moisture-resistant (thereby retaining the high gas barrier inherent to dry cellophane) and provide the needed moisture barrier.

Plastic films. Latex coatings on plastic films provide barrier to gases, moisture, flavors, and odors, and, in some packaging, heat-seal capability. Heat-seal latices are not especially good gas barriers, and, conversely, the best VDC barriers do not usually provide the best seals. When both heat sealability and barrier are required, it is best to apply two coatings, each designed for one purpose.

Semirigid containers. Latex coatings impart barrier properties to thermoformable plastic sheet used to produce high barrier food containers. Latex coatings on poly(ethylene terephthalate) (PET) bottles impart barrier to oxygen, carbon dioxide, water, and flavors and odors.

Generally preferred forms of VDC copolymers for coating various substrates are shown in Table 2. Pretreatment with primers or electrotreatment may be required on some substrates. The form of the package, ie, film, thermoformed sheet, or blown bottles, influences the choice as well, but the key differentiating elements are the porosity and chemical composition of the substrate.

Properties

The properties of VDC copolymers most pertinent to food packaging include a unique combination of low permeability to atmospheric gases, moisture, and most flavor and aroma bodies and stress-crack resistance to a wide variety of agents. In addition, the ability to withstand the rigors of hot filling and retorting is important in commercial sterilization of foods in multilayer barrier containers.

The range of barrier properties for small molecules in extrudable resins is shown in Table 3.

Furthermore, VDC copolymers are excellent barriers for flavor and aroma compounds. The three major types of household films are compared in Table 4. A high barrier VDC copolymer film would have much lower permeabilities.

The contrast between general purpose and high barrier films is shown in Table 5.

The bulk mechanical properties of VDC-copolymer-coated papers and films and structures such as formed containers and blown bottles depend almost entirely on the properties of the substrate material. The substrate makes essentially no contribution to the specialized attributes provided by the coating, such as barrier to permeation, chemical resistance, or heat or dielectric sealability (see Table 6).

Regulatory Status

VDC–VC copolymers containing ca 10–27 wt% VC are considered to comply with the food-additive provisions of the Federal Food Drug and Cosmetics Act on the basis of "prior sanction." A variety of specific regulations govern VDC copolymers other than those of vinyl chloride. The potential user is encouraged to work with the supplier to compare the current regulations with the application.

BIBLIOGRAPHY

1. Y-C Yen, "Barrier Resins," *SRI International Report #179,* Feb. 1986, p. 19.

General References

R. A. Wessling, D. S. Gibbs, P. T. DeLassus, B. E. Obi, and B. A. Howell, "Vinylidene Chloride Polymers," in J. Kroschwitz and M. Howe-Grant, eds., *Kirk-Othmer Encyclopedia of Chemical Technology,* 4th ed., John Wiley & Sons, Inc., New York, in press.

R. A. Wessling, *Polyvinylidene Chloride,* Gordon and Breach, New York, 1977.

P. T. DeLassus, "Barrier Polymers," in J. Kroschwitz and M. Howe-Grant, eds., *Kirk-Othmer Encyclopedia of Chemical Technology,* 4th ed., Vol. 3, John Wiley & Sons, Inc., New York, 1992, pp. 931–962.

P. T. DeLassus
W. E. Brown
The Dow Chemical Company
Midland, Michigan
B. A. Howell
Central Michigan University

W

WAXES

Introduction

Historically, in the United States, the use of wax-coated folding cartons to provide proper product protection has been a popular concept within the food-packaging industry. A wide assortment of refrigerated and frozen-food entrées have been packaged in this manner since the 1920s.

However, in the late 1950s, a new concept in the manner in which the wax coatings were applied to the folding-carton blank was commercialized. Coupled with the new, commercialized folding carton waxers, a large number of new wax-based folding carton formulations emerged. From the 1960s to present, a large number of new wax-based folding carton formulation concepts have been successfully commercialized and refined in order to fulfill the ever changing packaging requirements of the specialty market.

Over the years, wax-based folding carton formulations have improved on the functional properties that are imparted to the coated carton providing

1. Adequate protection against loss or gain of moisture by [moisture-vapor transmission rate (MVTR)]
2. Effective bond strengths during heat-seal closing operations on packaging equipment
3. The ability to retain good bonds at low-temperature storage conditions
4. High initial gloss and excellent gloss retention
5. Adequate scuff and abrasion resistance
6. Adequate slip properties for high-speed handling on packaging equipment
7. Excellent grease resistance
8. Tamperproof closures before sale, yet open easily at point of use
9. Economical costs commensurate with product protection

Today, a wide variety of specialty foods are now being packaged in wax-coated folding cartons. Wax-blend-coated folding cartons are especially suited for butter, margarine, ice cream, frozen vegetables, frozen pizzas, frozen shrimp, and frozen-dinner entrées, and prepared food specialites, to mention just a few of the limitless applications.

The purpose of this article is to provide examination of the past, present, and future applications of wax-based coated folding cartons.

Hot-Melt Wax Carton Coaters

From the early 1920s to the late 1950s three were two commercially accepted methods of coating folding carton blanks with wax: (*1*) the *hot-wax method,* which imparted a limited amount of protection and release to the carton—the wax coating did not enhance the appearance of the carton itself and, therefore, had very little shelf appeal; and (*2*) the *cold-water waxer,* which gave added surface protection and a reasonably glossy finish that increased the shelf appeal of the carton.

The coating used for both of these methods, wax, at first, nothing more than fully refined paraffin wax. As years went by, it was discovered that additives such as microcrystalline wax and small amounts of low-molecular-weight polyethylene could be added to the fully refined paraffin wax, slightly increasing the appearance and protective functional properties of the carton.

The hot-wax method and the cold-water waxer presented one major problem with the overall wax coating. With the exception of mechanically locked and tuck cartons, and except for those cartons where a dry strip could be left in the waxing operation, the other cartons had to be dewaxed before gluing. The dewaxing operation was time-consuming; very costly, and generated a large volume of wax waste.

In the late 1950s, recognizing the wax coating problem within the folding carton industry, the Oakland Paper Box Company developed the Oakland Hi-Gloss Pattern Coater. The Oakland Hi-Gloss Pattern Coater allowed wax coatings to be applied in any desired pattern, leaving voids in areas to be glued or printed.

The section where the wax is applied has two rubber-covered cylinders, one that coats the underside of a carton blank, while the second waxes the top.

The waxing blanket favored is brass-backed 30–40 durometer neoprene. Like the cutting die or the printing plates, the blanket must be prepared to the needs of a specific carton. Like the cutting die or the printing plate, the blanket will be saved if repeats of the job are anticipated.

A rubber waxing blanket is prepared by stretching an uncut sheet over a cylinder of the same size as the cylinder on which it will be mounted. A template is cut of the specific carton design and this serves as a pattern for cutting the blanket. Wherever wax is not to be applied, the rubber will be cut away. The result is a pad that is identical to a reverse-printing plate.

With the inception of the Oakland Hi-Gloss Pattern Coater, other machine manufacturers began to market similar wax coaters. In the United States, four of the better known wax coaters are the modified Oakland Coater, Tidland "Plastisheen" Carton Coater, the International Paper Box Machinery Company's (IPBM) GM Carton Coater, and the 88-B Carton Coater. Of the five types mentioned, in the United States the majority of the carton coaters in operation are the IPBM GM Carton Coater and the 88-B Carton Center. The two IPBM wax coaters are capable of applying wax coating with viscosities approaching 1000 cP at 300°F (148.9°C).

The IPBM GM Carton Coater has two coating stations. The first station coats the inside (bottom) surface of the carton. The second station coats the outside (top) surface. This arrangement permits the application of different blends on the inside and outside surfaces, if desired. The wax coating weights are controlled by adjustment of the metering roll nip and the reverse or burnishing roll at each coating station. The second coating station is equipped with a weir-type curtain feed to ensure uniform distribution of the wax coating across the applicator roll. This is particularly important in applying high-viscosity wax coatings. Silicone rubber back-up rolls are used to prevent carton blanks sticking to the rolls.

The wax-coated cartons pass through a remelt section consisting of gas-fired infrared heaters. The temperature of the gas-fired infrared heaters is typically 250°F (137.1°C), and are rated at about 500,000 Btu. This exposure to the heat gives a smoother, more evenly dispersed flow to the wax coating and produces a high-gloss finish.

The wax-coated cartons are then passed through a refrigerated, chilled-water curtain and into a water-immersion tank. The temperature of the chilled water must be carefully controlled within a temperature range of 34–40° F (1.1–4.4°C). The purpose of the chilled water is to facilitate the setting of the wax coating without disturbing the surface. This process also enhances the high-gloss finish. If the chilled water is at a temperature above 40°F (4.4°C), the wax coating does not set properly and will produce a carton with a dull, uneven surface. The position of the heaters and the distance to the water-immersion section can be adjusted to obtain the maximum gloss for the particular wax coating being applied.

The waxed cartons are then passed through a set of squeeze rolls to remove the excess water. The waxed cartons are then deposited onto a slow-speed stacker apron.

The ideal wax-coating weight, per carton side, is approximately 3 lb/1000 ft^2 (14.6 g/m^2).

The machines are usually installed with three channels or lanes that automatically feed the folding carton blanks through the wax coater. Each single lane is capable of waxing 15,000 cartons per hour.

Hot-Melt Wax Carton Coatings

In the late 1950s, coupled with the development and commercialization of the "new" folding-carton coaters, new developments and refinements were also taking place in the area of "new" hot-melt wax carton-coating formulations. Specialty wax blends were being researched and developed for exclusive use with the "new" folding carton coaters. The simple formulations of

1. Straight fully refined paraffin-wax blend
2. Fully refined paraffin-wax/microcrystalline wax blend
3. Fully refined paraffin-wax/low-molecular-weight polyethylene blend

These gave way to sophisticated blends of carefully selected petroleum derived waxes with carefully selected wax additives. Because of the significant changes made in the area of folding-carton coaters, a new era in formulating wax-based coatings for this specialty packaging industry began.

In early 1960, a "new" series of products became commercially available that created an entirely new concept in hot-melt wax-based coatings. This "new" series of products were termed *copolymers,* and consisted of various percentages of ethylene and vinyl acetate—or simply EVA. Other wax additives that became commercially available around the time of the EVA copolymers were polyterpene resins, stabilized ester gums, and a wide product diversification of low-molecular-weight and high-molecular-weight polyethylenes. The commercialization of these "new" wax additives gave compounders the tools to manufacture an almost limitless number of combinations or formulations of hot-melt wax-based folding carton coatings.

On the basis of the requirements of the folding carton industry, hot-melt wax-based folding carton coatings are classified into the following two categories:

1. Nonheat-sealable folding-carton waxes
2. Heat-sealable folding-carton waxes

Non-heat-sealable folding-carton waxes are usually applied in a defined pattern onto the folding-carton blank, leaving unwaxed areas where a hot-meal adhesive (HMA) is applied to steal the folding carton closed. The wax coating is applied to both the outside (printed side) and the inside (unprinted side) of the folding carton.

Heat-sealable folding-carton waxes are applied onto the entire surface area of the folding carton blank, and will seal the folding carton *without* the use of a HMA. The wax coating is applied to both the outside (printed side) and the inside (unprinted side of the folding carton). The use of a heat-sealable folding carton coating is advantageous because the gluing operation is eliminated.

As the folding-carton industry developed, it became apparent that only four wax-based folding carton coatings were necessary to fulfill the requirements of the industry. The wax-based coating product line consisted of the following:

1. Non-heat sealable folding-carton waxes
   a. one economical coating
   b. One general-purpose coating
   c. One heavy-duty coating
2. Heat-sealable folding-carton wax
   a. One premium heat-sealable coating

The selection of the proper folding carton coating depends on the physical specifications of the formulation as well as the functional properties that the formulation must impart to the folding carton. Functional properties such as gloss, creased vapor barrier, grease resistance, scuff resistance, freezer-burn resistance, and heat sealability are some of the points to be considered in the section of the proper folding-carton coating. In most instances, the folding-carton coating must be nonflaking, exhibit no ruboff, and exhibit a high antiblock characteristic. These three properties are important when the waxed folding carton is processed through the folder/gluer. The wax coating cannot build up on the runners of the folder/gluer. When the wax coating builds up on the runners of the folder/gluer subsequent waxed folding cartons stick and jam on the runners. The folder/gluer must then be shut down and cleaned (see also Sealing, heat).

National Wax Division's folding-carton coatings are specially formulated to be nonflaking, exhibit no ruboff, and exhibit a high antiblock characteristic.

The economical non-heat-sealable folding-carton coating provides minimal functional properties. The economical formulation is used in applications of generic folding cartons where minimum functional properties are required and the low cost of the wax coating is an important consideration. With a hot-melt wax carton coater containing two separate wax-coating stations, the economical formulation can be used on the inside (unprinted) of the folding carton, while the general-purpose formulation or the heavy-duty formulation can be used on the outside (printed) of the folding carton. This procedure is used for monetary consideration.

The general-purpose nonheat sealable folding-carton coating offers improved functional properties over the economical coating. The general-purpose coating may be used on the outside (printed) area of the folding carton as well as the inside (unprinted) area of the folding carton.

The heavy-duty non-heat-sealable folding-carton coating imparts the optimum appearance, protection, and functional properties of the non-heat-sealable coatings. The heavy-duty coating may be used on the outside (printed) area of the folding carton as well as the inside (unprinted) area of the folding carton.

The premium heat-sealable folding-carton coating imparts the optimum appearance, protection, and functional properties to the folding and will seal the folding carton closed without the use of a HMA. This product will give a fiber-tearing seal over a wide range of sealing temperatures.

Of additional consideration for all four products is that all ingredients and final products must fully qualify with FDA (Federal Drug Administration) regulation as suitable for use in food packaging. The regulation cited most often is Paragraph 21, *CFR* 176.170 for "components of paper and paperboard in contact with aqueous and fatty foods."

With the "new" generation of hot-melt wax-based coatings and the commercialization of the "new" generation of hot-melt carton coaters, the folding-carton industry was well on its way to establishing itself as a strong specialty market within the food-package industry.

New Developments

From the late 1960s until 1980, there were relatively few new developments in the area of hot-melt wax carton coaters or in the area of hot-melt wax-based coatings. The standard wax-based product formulary still consisted of three non-heat-sealable coating formulations and one premium heat-sealable coating formulation.

Later, during this same period, a large number of refineries discontinued the manufacturing and marketing of hot-melt wax blends. This decision reduced the amount of competition in the blended-wax marketplace. With the reduction of competition coupled with the increase of technical background, the National Wax Division became a predominant marketer, manufacturer, and technical center for the folding-carton industry.

From 1980 until present, the folding-carton industry has seen more changes in the past 7 years compared to its "modern conception" in the early 1960s. In an effort to reduce costs, while maintaining a high-quality package, the folding-carton industry now uses a wider variety of boardstock. The new types of boardstock required new folding-carton wax formulations.

In some instances, the boardstock was more porous than previously used. In other instances, the boardstock was more "dry" than previously used. Specialty wax-based coatings had to be developed to maintain the high-quality standards of the folding carton when the newer type of boardstock was used.

Future Outlook

For the National Wax Division, the future of wax-based carton coatings will continue to offer continued opportunities. Future growth is targeted toward the higher-viscosity, premium folding-carton formulations.

Two years ago, the National Wax Division commercialized a low-cost premium heat sealable folding-carton wax. All of our field-test data generated confirms all of our initial laboratory technical test data that this low-cost premium heat-sealable coating will functionally perform just as well, if not better than, the older, higher-cost heat sealable folding-carton formulations. Trials will continue with this new product in order to saturate the heat-sealable folding-carton marketplace.

BIBLIOGRAPHY

General References

R. W. Dannenbrink, *Wax-Polymer Coatings for Folding Cartons,* Shell Oil Company, Houston, TX, May 1969.

E. I. DuPont de Nemours & Company, *Hot-Melt Blends for Frozen Food Carton Coatings,* Wilmington, DE, May 1965.

W. F. Elder, *Sarapoll and Cereplax Coatings for Folding Cartons,* National Wax Company, Skokie, IL, Feb. 1975.

International Paper Box Machine Company (IPMBMCO), Nashua, NH.

J. R. Lauterbach, "Boxmakers Hear How Hot Melt Coatings Aid Carton Users," *Good Packag. Mag.* (Feb. 1966).

E. Sidebotham, (Specialty Automatic Machine Corporation, Burlington, Massachusetts), "What's New in Wax Hot Melt Carton Coaters," paper presented at Packaging Institute Seminar, Chicago, March 1968.

Harry Kannry
Gary Latto
Dussek-Campbell
Skokie, Illinois

WEIGHING.

See Checkweighing; filling machinery, dry.

WELDING, SPIN

Spinwelding is a process that joins plastic container parts together by heat from friction. This is accomplished by rotating either the top or the bottom container part, while holding the other part stationary. The heat produced from friction melts the surface of the two parts where they are in contact with each other, and then the melt solidifies to form the weld. For prototypes or specialty low-volume items, manually operated equipment designed to weld one container at a time can be as simple as a drill press with a tooling drive and a clamp to hold the stationary part. For production runs, fully automated in-line equipment is available that can weld containers at rates exceeding 300 containers per minute.

Several requirements exist for spinwelding that warrant attention. One requirement is that the two container parts to be welded must have round cross sections at the weld band

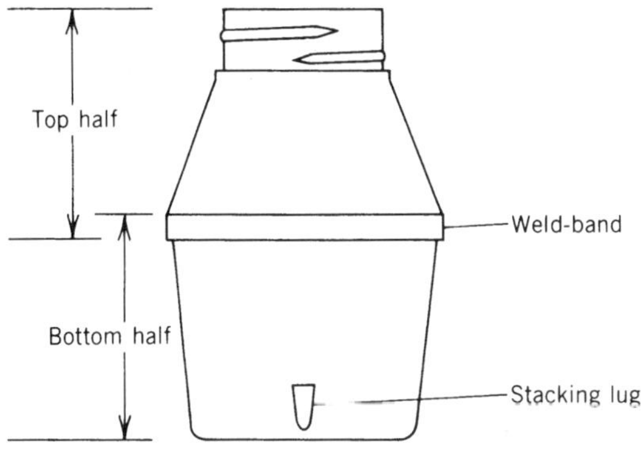

Figure 1. Container parts must have round cross sections at the weld band.

(see Fig. 1) existing in the same plane. Another requirement is that the container parts must be manufactured to extremely close tolerances, by thermoforming, injection molding, or other process. If a container part is out-of-round, too large, or too small, the integrity of the weld is severely jeopardized. A third requirement is that the parts to be welded must be made of compatible materials (ie, polymer type, melt flow, etc). Although dissimiliar materials can be used, they generally result in lower strength welds.

Spinwelding has many important applications in today's markets. The first and probably most obvious is as a product line extension for thermoformers (see Thermoforming). Spinwelding provides a cost-effective entry into plastic-bottle markets traditionally served by blow molding (see Blow molding) and it is of particular interest today as a cost-effective way to produce high barrier containers. Another advantage is the ability to combine container parts of different colors, or a transparent part with an opaque part, providing unique appearance and marketing appeal. A third application involves the ability to apply tamper-evident closure systems (see Tamper-evident packaging). This represents just a few examples of the potential provided by spinwelding.

To automatically spinweld containers at high rates of speed, the following functions are performed sequentially: denest container parts; spinweld container parts together; and transfer finished containers to filling or other process system.

Denesting. Container parts are shipped to the spinwelding location (usually the filling location) in nested stacks. Container tops and bottoms are loaded into their respective magazines in tall vertical stacks. These magazines can be simple as a single stationary stand, or as complex as a multiple-rotating stack fixture. Vacuum-equipped shafts pull the container bottoms down and the container tops are brought down with picker heads. These shafts then feed the container parts into infeed starwheels, which transfer them to the spinwelder. This process is aided by carefully locating the stacking lugs (see Fig. 1) on the container during design. These lugs act to separate the parts a predetermined amount when stacked in order to ease denesting.

Spinwelding. The spinwelding section of the machine accepts parts from the denester system and deposits each half of the container into closely machined tooling. These parts are held in the tooling in many ways, dependent on container design. Existing stacking lugs can be grasped, for example, and/or welding lugs (flutes) can be designed into the container solely for this purpose. The grasping is accomplished either by teeth or by a rubber facing and is usually vacuum assisted. Either part can be held in the tool to be rotated, but usually it is the bottom or the smaller half, again depending on container design.

The rotating tool is brought up to a predetermined speed by a spinner apparatus (see Fig. 2). In some machinery, the container parts are already in contact before the spinning begins. In these systems, the rotation is stopped abruptly by a brake after a very brief period of time, but the pressure is maintained to ensure that the melt solidifies. In other systems, the container parts are brought together only after the rotating tool has reached desired speed. This is accomplished rapidly utilizing a clutch disengages, the parts are brought together, and pressure is maintained as the spinning is stopped by the solidifying melt.

Exit transfer. After welding, the finished containers are removed from the tooling by another starwheel, which transfers them either directly to a filler or onto a conveyor for further processing steps such as weld inspection and rinsing.

Weld integrity. Depending on the desired result, welds can be designed to be no more than a tack, or as much as a full penetration. If the latter is chosen, the weld-band has hoop strength, and is the strongest part of the container. Weld in-

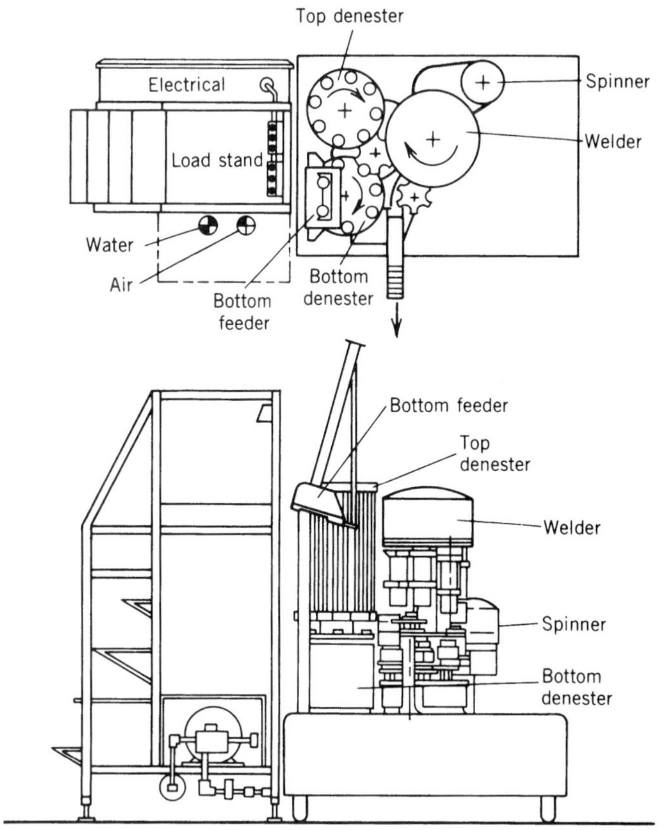

Figure 2. Spinwelding equipment.

tegrity is governed by several variables, including material choice, material thickness, joint pressure, driving force, rotation speed, tolerances of the parts, cycle time, and weld-band design. The number of weld-band design choices, limited only by the imagination, includes butt, tongue and groove, and sleeve. Depending on design, a trimming tool may be required to remove flash created during the weld. One way around this is to design the weld-band incorporating a reservoir for the flash. This is most easily accomplished with a sleeve design as shown in Figure 1. It is sometimes necessary to provide a system to check the weld integrity, but modern commercial units can weld hundreds of millions of containers annually with less than 1% rejects.

BIBLIOGRAPHY

"Welding, Spin" in *The Wiley Encyclopedia of Packaging Technology,* 1st ed., by Scott R. Mitchell, Don MacLaughlin, and Bert Peterson, Vercon, Inc., pp. 700–702.

General References

Modern Plastics Encyclopedia, 1985–1986, McGraw-Hill, New York, 1985, pp. 354, 356.

Packaging Encyclopedia and Yearbook, 1985, Cahners, Boston, 1985, p. 260.

Technical Manual for Vercon Inc., Autowelder-10, Vercon Division of Fina Oil and Chemical Co., Dallas, Texas, 1983.

WRAPPING MACHINERY

The history of packaging automation began in 1869 with the introduction of the first automatic machinery for making paper bags. It was not until the early 1900s that demand for other types of automatic packaging machinery resulted in the formation of a number of small companies devoted to the manufacture of packaging machinery. Early designs of machines for wrapping loaves of bread, cookies, food products, cereal boxes, candy bars, soap tablets, and tobacco products incorporated much of the established bagmaking technology.

The 1936 introduction of cellophane (see Cellophane) as a new transparent wrapping material created new demands, and many new types of wrapping and wrapping-related machines were developed and put into production (see Films, plastic). Progress continued at a fast pace until World War II, when war-related work took precedence over commercial activities. The growth of supermarkets during the war years created a need for new kinds of packaging and new kinds of automated equipment. Firms in the United States and abroad responded to this need with a wide range of new wrapping and wrapping-related machines. These new models featured higher operating speeds, increased efficiency, greater flexibility as to package size, and in many cases, the ability to convert both older and newer models for use with thermoplastic films.

Throughout the history of packaging automation there have been examples of equipment design changes to accommodate marketing trends. Three recent examples are the development and refinement of tamper-resistant packaging, the introduction and growth of generic packaging, and the expansion of promotional packaging programs.

Packing Terms

Many of the trade terms that pertain to wrapping and wrapping-related operations are listed below.

Wrapper/wrap. A cut-to-size sheet of monolayer or multilayer flexible packaging material used to wrap a product or package.

Loose wrap/removable wrap. The forming of a wrap around a product or package in which the folds (overlaps/seams) are sealed to themselves and not to the object being wrapped.

Tight wrap/wet wrap. The forming of a wrapper around a carton or container in which the wrapping material is adhered or sealed to all of the surfaces of the package through the application of wet adhesive or by activating a heat-sealing type of coating preprinted on the underside of the overwrap by the packaging material converter.

Intimate wrap/conforming wrap. These terms describe the type of wrap that is in direct contact with the product, as required in the dairy, candy, and meat industries (see Films. shrink).

Bunched wrap/formed wrap. Describes a style of wrap for irregular-shaped products in which the wrap is either draped over the product or the product is pushed through a die-box so that the wrapping material can be gathered (tucked/folded pleated) and then sealed to itself on the bottom panel of the package.

Bundle wrap/parcel wrap. A low-cost method of wrapping packages together by first combining (collating) a predetermined number of small packages into larger units (collations) preparatory to overwrapping with a heavy-duty packaging material.

Multipack wrap/supermarket multiple-pack wrap. A variation of bundle-wrapping wherein identical products are arranged or collated into multiples of 2-, 6-, 10- or 12-pack units prior to overwrapping.

Multipack band wrap/supermarket banded multiple-pack. Multiple units of identical-size packages arranged or collated together in a pattern that will accommodate a band of stretch or shrink film to be positioned around four sides of the multiple units.

Bag wrap. Prefabricated bags are individually fed into position, opened, and then loaded manually or automatically, prior to being closed in some conventional manner (stapling, crimping, or twist-tying).

Flow pack/tube wrap. Type of style and seal produced on a horizontal form/fill/seal wrapping machine (see Form/fill/seal,

Table 1. Basic Fold Patterns

Figure 1, Fold Patterns		Description	Where Used
(**a**)	Pouch type (pillow style)	Horizontal form/fill/seal style with extended end folds	Irregular shapes
(**b**)	Pouch type	Horizontal form/fill/seal style with end folds tucked under	Irregular shapes
(**c**)	Standard elevator	Double-point end-fold style with long overlap on bottom panel	Cartons, boxes, trays
(**d**)	Special elevator	Standard elevator fold pattern with end folds extended and plowed under and sealed on bottom panel	Extension edge boxes, low-profile items (hosiery boxes)
(**e**)	Cigarette (old style)	Special fold pattern (with cross-feed tucks, folds, seals, made last)	Small packages
(**f**)	Cigarette (new style)	Special fold pattern (with long-side overlap made first plus double-point end seals)	Small packages
(**g**)	Twist wrap	Special fold pattern	Individual candy pieces
(**h**)	Cigar wrap	Special fold pattern	Cigars
(**i**)	Snack wrap	Cigarette-style wrap with end label	Crackers, cookies
(**j**)	Bread wrap	Long overlap on bottom panel plus progressive end-lock folds secured by end labels	Bread, baked goods
(**k**)	Underfold wrap (die-fold wrap) (envelope wrap)	Product placed on bottom board or U-board and then wrapped with all folds tucked and folded under product prior to sealing	Candy bars, gum packs, crackers, tissues
(**l**)	Underfold wrap (diagonal sheet feed)	Irregular-shaped product loaded into a tray (plastic, foil, pulp or board material) and then wrapped using a diagonally positioned sheet, wherein all four corners are folded and sealed against bottom of tray	Trays of meat, poultry, produce, bakery items
(**m**)	Underfold wrap (straight-sheet feed)	Tubes is formed with overlap bottom seam prior to forming end tucks and plowing end folds down and under package before sealing (note: see Shrink packaging)	Soft goods, books, bacon, trays of food products
(**n**)	Underfold wrap (bunched wrap) (formed wrap)	Irregular-shaped product is "pushed" up and through a sheet with extended edges brought together with a series of folds and tucks prior to sealing against the bottom of the package (note: an identifying label is generally applied to the bottom panel to help hold the folds in position)	Paper plates, pot pies, pizzas, heads of lettuce, mints, bearings, tape rolls
(**o**)	Roll wrap (turret wrap)	Product is pushed through sheet into turret where long side overlap is made first followed by making end lock folds progressively	Mints, candies, cookies, biscuits
(**p**)	Roll wrap (fin-seal wrap)	Generally formed on a horizontal form/fill/seal type of wrapping machine	Cookies, biscuits, tray packs

horizontal) generally associated with wrapping irregular-shaped items such as candy bars and bakery items.

Accumulator/collator. Machine attachment designed to collect, assemble, and orient packages in specific patterns or collations, generally in preparation for bundling and multipack operations.

Wrapping, Banding, and Bundling Machines

Packaging machinery designed for wrapping, banding, and bundling operations can be divided into eight classifications:

Wrapping, general purpose. Packagers faced with problems of short production runs, frequent changeovers, and the need for a high degree of flexibility in meeting changing packaging specifications can select from a wide range of new and rebuilt wrapping machines. Examples are supermarket packaging of meats and fruits, cosmetics and perfume packages, pharmaceuticals, and contract packaging (see Contract packaging).

Wrapping, shrink. Standard-model wrapping machines are now available with conversion kits to handle new plastic packaging films, including shrink films (see Films, shrink). The actual shrinking of the film overwrap is accomplished by means of controlled heat supplied by a shrink tunnel or similar attachment (see Wrapping, shrink). Examples are candy multipacks, video and audio cassettes, general range of packages needing protection against tampering, insect infestation, etc.

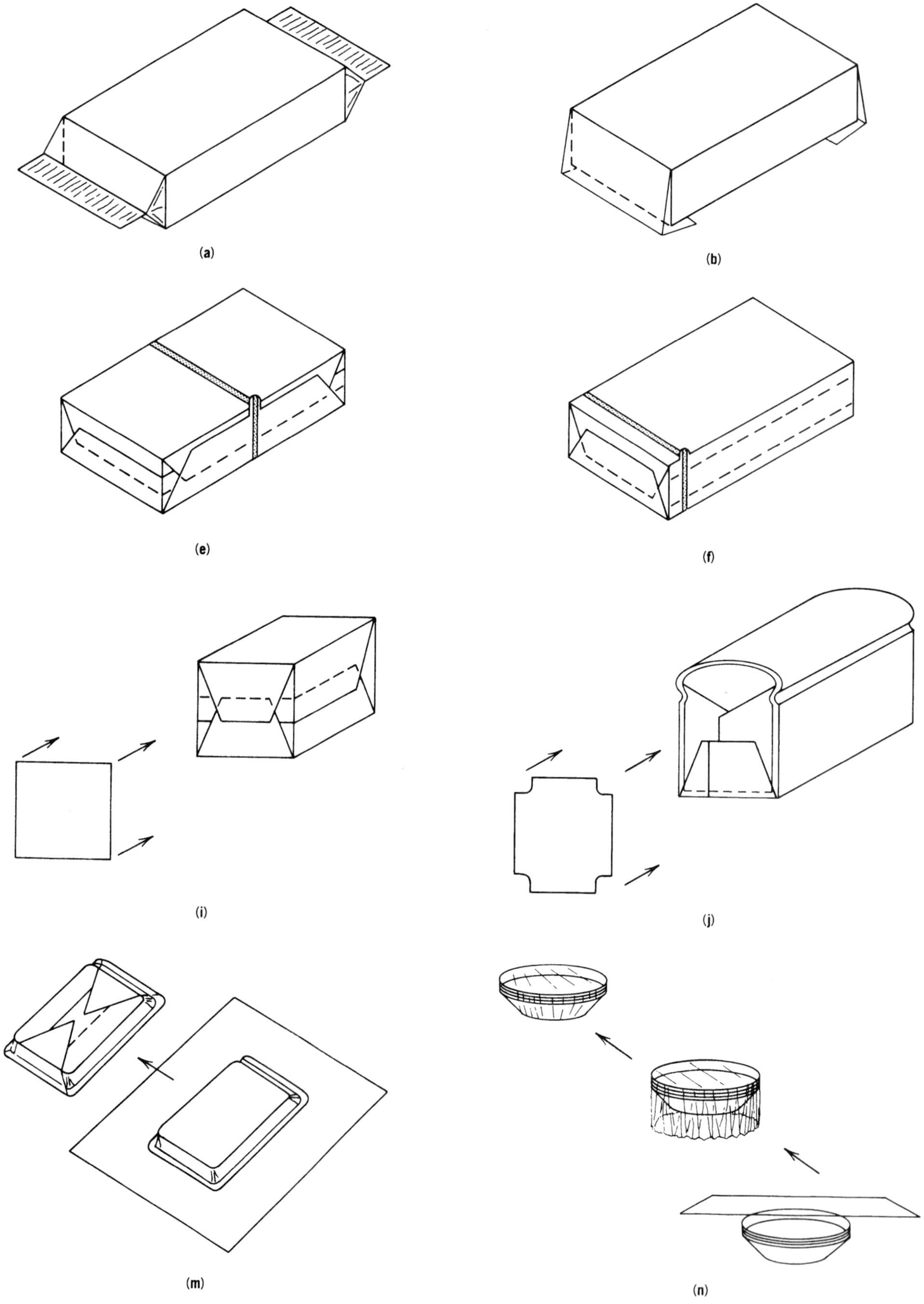

Figure 1. Fold patterns for wrapped packages.

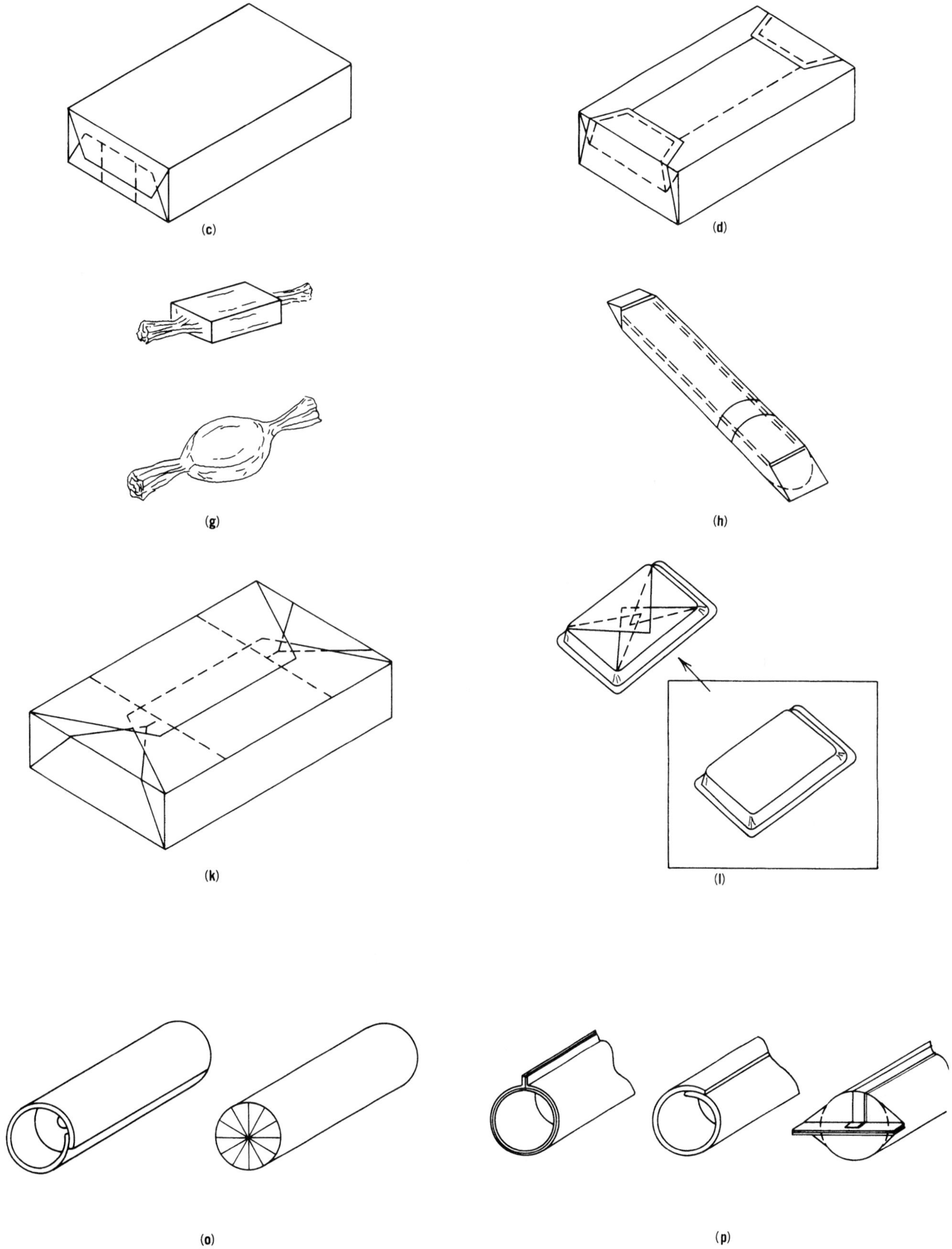

Figure 1. (*Continued*)

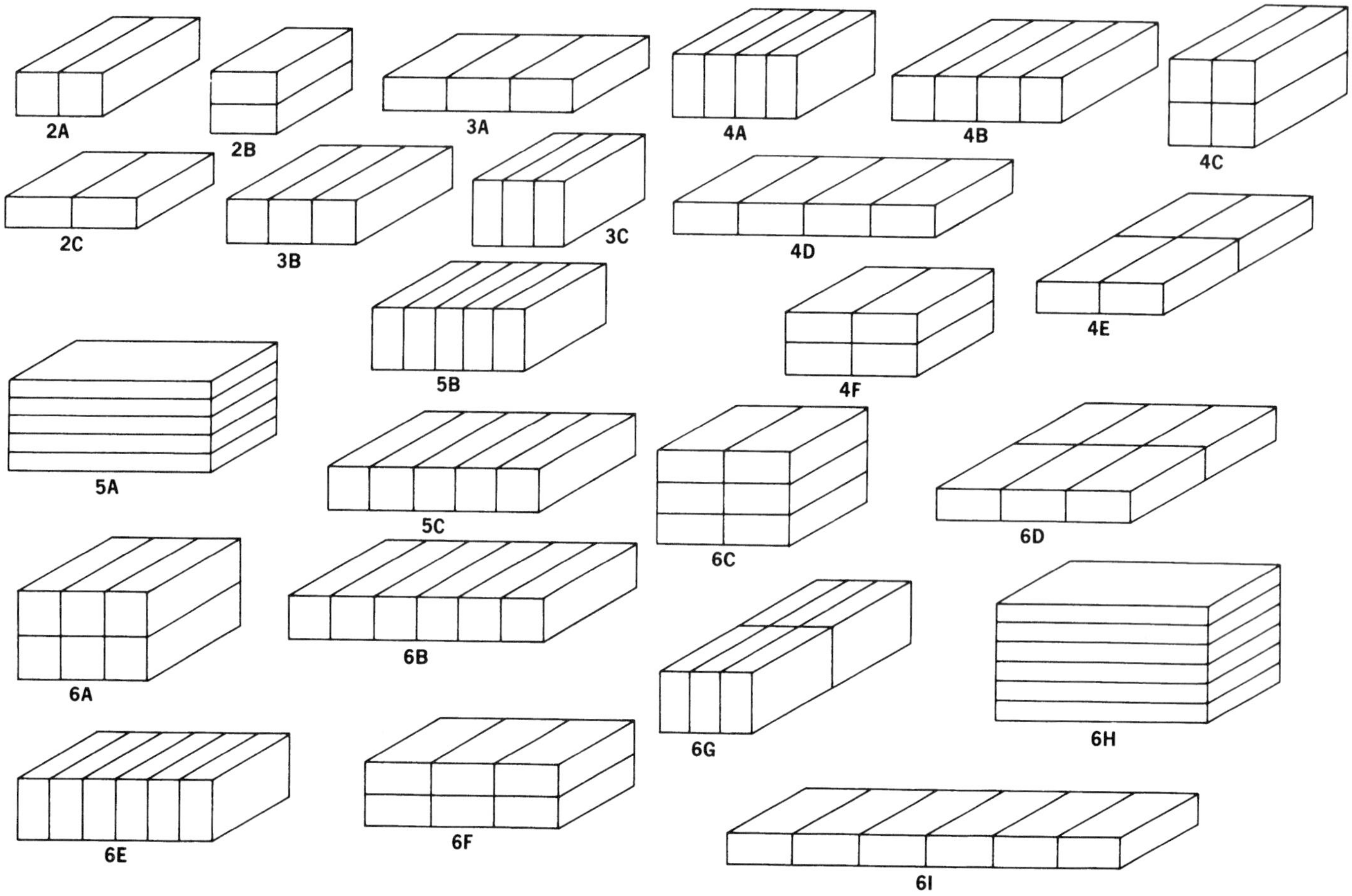

Figure 2. Collation patterns used by bundle-wrapping machines. (*Note:* Collation specifications are based on quantity and arrangement of packages to be bundle-wrapped, eg, 9 patterns for bundles of *6* packages, 6 for *10,* and 12 for *12.*)

Wrapping/banding (multipacks). In "multipack" packaging operations, there is some overlap between wrapping and banding. By combining or collating two or more similar-size packages into a multipack, the packager can choose between wrapping and banding techniques. In some cases the band is adhered directly to the surfaces of the individual packages within the multipack to prevent their sale as spearate items. Examples are multipacks of soap tablets, candies, gum, tobacco products.

Wrapping/bundling. Bundling, parceling, and multipack wrapping are terms applied to packaging machinery designed for collating or accumulating identical-size packages into a larger single unit (dozen count, for example) prior to being overwrapped in a heavy-duty flexible-packaging material such as kraft paper or one of several plastic films. Examples are cookies and crackers, food products, drug packages, candy and confections, and tobacco products.

Wrapping/bunch or die-fold. Special machines designed for wrapping irregular-shaped products wherein the product is "pushed" through a sheet of packaging material into a die-box where the wrapper is gathered and sealed on the underside of the product. Fragile items, such as baked goods, generally require U-cards or bottom boards for protection and sealing of the film wrap. Examples are small baked itmes, soap, typewriter ribbons, friction tape, bearings and sardine cans.

Wrapping/bagging. A simple approach to the packaging of irregular-shaped items is taken in equipment that dispenses prefabricated bags, one at a time, to a station where the bag is opened and loaded (manually or automatically), and then moved to a bag-closing station (twist-tying, stapling with a header-label, crimping, carding, etc). Examples are bagging of bread, baked goods, textile products, and multiples of plastic products such as clothespins.

Wrapping/pouch-style: wrap/tube and wrap/pillow-pack wrap. Widely used for the packaging of unusual and irregular-shaped items, these horizontal form/fill/seal machines call for a continuous "tube" of packaging material to be formed around a mandrel or tube. As the "tubed" product moves forward through the machine, the overwrap material is tucked, sealed, and cut off between packages. Variations include wire seals, foldunder of the ends, and header crimps. Examples are tray packages, baked goods, hardware items, paper products, textiles, foods, and dairy products.

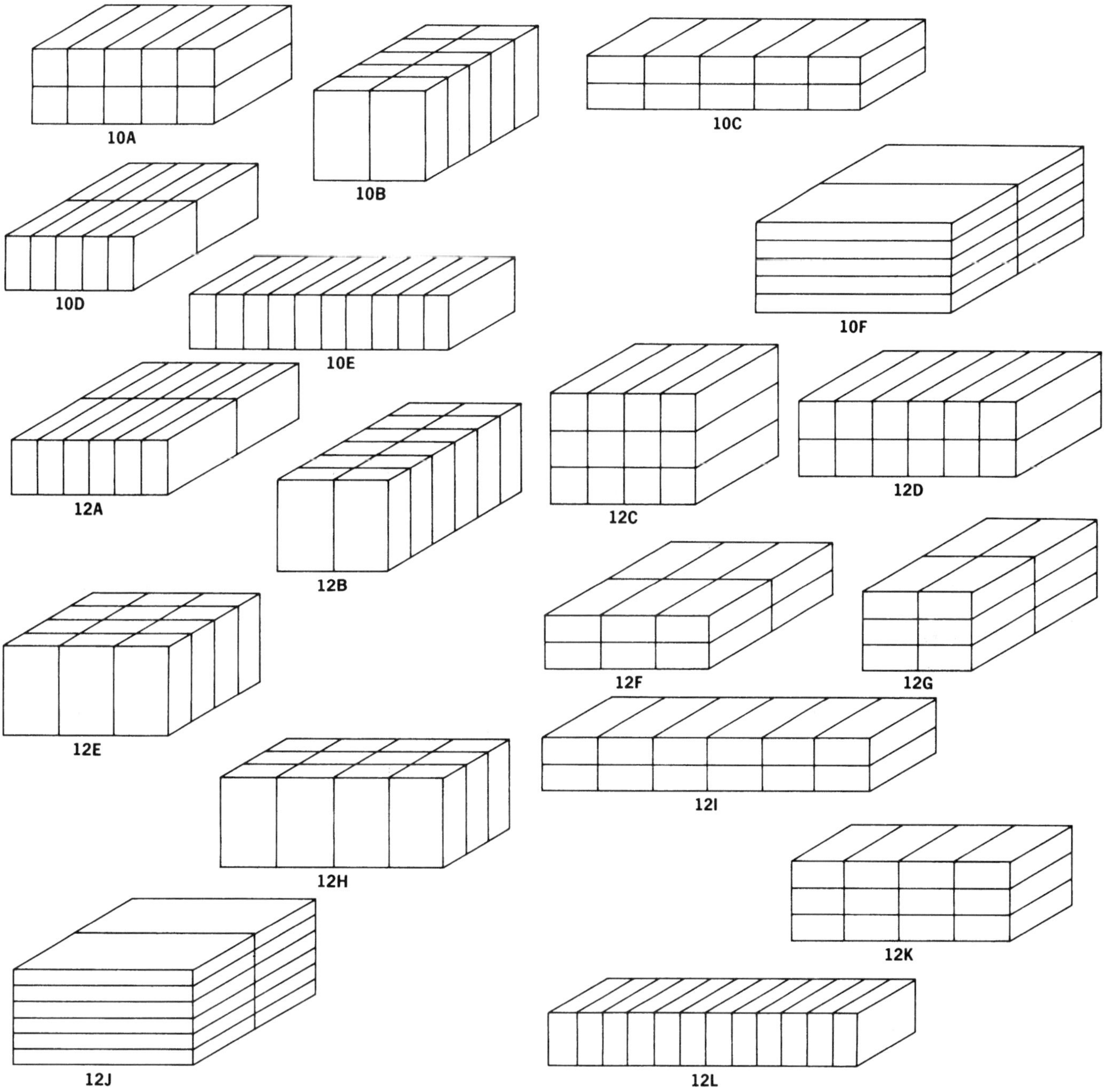

Figure 2. ***(Continued)***

Wrapping-Machine Fold Patterns

Since many wrapping machines are built specifically for a product (eg, bread, cigarettes, candy) the packaging professional must be familiar with the various fold patterns and their use for certain shapes, products, and range of sizes. Some patterns are particularly suitable for higher speeds and improved package protection, as well as better economy in the selection of packaging materials and reduced changeover time where many sizes of product are to be wrapped.

Listed Table 1 are many of the basic fold patterns presently in wide use. Attention is directed to the sketches (Fig. 1) illustrating each of the fold patterns. (It is important to note that although the fold patterns are similar, the tucking-folding-sealing operations can be produced using different mechanical means such as elevators, dies, turrets and the like.)

Wrapping-Machine Attachments

Table 2 lists a number of standard and special attachments available for use on wrapping and wrapping-related packaging equipment.

"Multipack" Wrapping Machines

A very popular variation of overwrapping is the group wrapping of more than one unit into what is called a multipack. Retail multipacks provide an important merchandizing tool for a number of reasons:

Table 2. Standard and Special Attachments

Attachment	Description	Designed For
Package feeds	Magazine single-place/column	Manual feeding of single packages
	Magazine multiple-place/column	Manual feeding of multipacks
	Conveyor lug-type infeed	Manual loading of tray-type packages
	Conveyor automatic infeed	Links previous operation to wrapper
	Conveyor automatic infeed with collator	Receiving packages and then collating into groups of multipacks or collations
Package materials	Conversion kits (necessary in order to wrap with new-generation packaging materials such as polypropylene, polyethylene, polyester–PVC, and numerous laminations)	Special kits designed for converting new- or older-model wrapping machines to handle various types of flexible-packaging materials
	Opening tape (tear-tape, zip tape)	Variety of attachments available for installation on new or older models
	Printing attachments (electric eyes)	Electric-eye registration units used to register the design on printed materials
	Printing attachments (printers)	Code marker and price markers or multipanel printing attachments available
Special feeds	Role-feed attachments, sheet-feed attachments, index feeders (board forms)	Feeding coupons, labels, revenue stamps, price tags, collars, bottom boards, hand-up tabs, header labels, etc
Discharge units	Various designs (rotary, reciprocating, elevator, shingle-style, etc)	Wrapped packages that require collating preparatory for the next operation

1. Most multipacks can be produced on standard-model wrapping machines equipped with standard attachments (opening-tape, electric-eye registration of printed packaging materials, etc) or special attachments if necessary (eg, automatic collating attachments, coupon feeds).
2. Many packaging materials are available to provide different combinations of appearance, tamper resistance, shelf-life, etc.
3. Multipacks offer the packager the opportunity to launch special promotions by simply changing the number of packages within the multipack. Examples are a "bonus" pack added to the regular pack of candy, a special promotion of "Buy one-get one free!," a one-cent sale for an extra tablet of bar soap, and cereal 6- vs 8-packs. The promotion can be set up on short notice and generally with a minimum investment in change-parts for the wrapping machine.

Bundle-Wrapping Machines (Bundlers and Parcelers)

Another variation of overwrapping is bundle-wrapping (called "parcel-wrapping" outside the United States) of a fixed multiple or group of like-size unit packages, using a tough, heavy-duty, flexible kraft paper or plastic film. Most bundling (parceling) operations are replacements for more expensive operations using corrugated cases, chipboard shelf-containers, or display containers. Most food and drug products sold in the United States are bundled using the subdivisions of a gross (144 packages), generally in quarter-dozen, half-dozen, or one dozen counts. For products to be shipped outside of the United States, the most popular metric counts are 10, 20, or 25.

Many of the standard attachments used on standard-model wrapping machines are also available for use on bundling machines; such as electric-eye registration units for printed wrapping materials, easy opening tapes, panel printers, end labelers, perforators, and shrink tunnels for shrink films. There is also a large selection of special attachments, especially in the area of automatic infeed conveyor and collating attachments designed to automatically receive and collate, and then feed the collation (see Fig. 2) into the infeed station of the bundle-wrapping machine.

Regarding collators, the packager should keep in mind that bundle-wrapping machines are subject to two speeds of operation. The first relates to the speed flow of individual packages (as high as 900 packages/min) to the infeed station of the collator attachment. The second speed pertains to the bundler and is determined by dividing the incoming line speed by the number of packages within the bundle. For example, a line speed of 600 packages/min translates into 60 bundles/min for a 10-count bundle, or 50 bundles/min for a 12-count bundle.

Two recent bundling developments warrant special mention, both involving the use of biaxially oriented polypropylene film (see Film, oriented-polypropylene) as the bundling material. The first is a full six-sided wrap that is subjected to controlled shrinking in order to obtain a neat bundle that resists tampering and is strong enough to minimize damage during transit. The second development is a five-sided outgrowth of the first, with one side left unwrapped. The reason for the open side is to permit the price-marking of the individual packages within the bundle at retailer level. Shrinking the five-sided wrapped bundle locked into position despite the open side of the bundle.

BIBLIOGRAPHY

"Wrapping Machinery" in *The Wiley Encyclopedia of Packaging,* 1st ed., by Wilhelm Bronander, Jr., Scandia Packaging Machinery Company, pp. 702–708.

WRAPPING MACHINERY, STRETCH-FILM

Historical Overview

The stretch-wrapper industry is just over 20 years old in 1996. In 1976 there were only four or five manufacturers producing a wide-web model used to wrap standard-height pallet loads with a large 60-in. roll of stretch film. Soon after that the demand to wrap variable pallet load heights produced the *spiral* model, which used a smaller 20-in. roll of stretch film. Unlike the large roll of film, which was stationary when wrapping the entire pallet load, the smaller 20-in. roll traveled up and down on a motorized carriage, wrapping the entire load in a crisscross fashion. The spiral wrapper offered the primary advantage of wrapping a wide variety of load heights at minimal cost and dramatically reduced the use of the wide-web model.

Today the spiral wrapper (Fig. 1), in both semiautomatic and fully automatic formats, represents approximately 90% of all wrapper production. Wrapper manufacturers in the United States now number over 20 ranging from small producers of lower-cost models to major producers offering a complete line of wrapper models to accommodate virtually every stretch-wrap application known.

The Demand for Stretch Wrappers

Prior to the energy crisis of 1974, the use of *shrink* film to unitize pallet loads of product was an increasingly popular packaging medium. The costly and excessive use of heat energy (gas, electric, or propane) to shrink film, however, along with the rising cost/availability of shrink resins placed industry of all types in a cost-prohibitive situation.

Another unitizing method, using strapping (steel, nylon, polyethylene, or polypropylene) to unitize pallet loads, was recognized as labor-intensive, and an average pallet load could take up to 7 min to strap depending on the nature of the product and the acceptable strapping pattern desired.

Still other unitizing mediums such as pressure-sensitive tape, twine, or spot adhesives did not provide inherent protection from the elements and would tend to loosen while in transit.

Stretch film offered a lower cost advantage of being applied by equipment that used only "on/off" electricity as a source of power and could be applied by a semiautomatic stretch wrapper. Significantly less labor time was involved and an average size product load could be wrapped in less than 2 min. As a result, many of the initial products stretch-wrapped were packaged in multiwall paper bags, polyethylene bags or in cartons. The following is a list of very early stretch film product applications:

Product	Shipped in
Kitty (cat) litter	Multiwall paper bags
Dog and cat food	Multiwall paper bags and cartons

Figure 1. A semiautomatic, spiral wrapper (IPM 55UTS30F) equipped with Uni-Tension power prestretch. For unitizing up to 35 pallet loads per hour.

Product	Shipped in
Detergent	Corrugated cartons
Chemicals	Multiwall paper bags and steel drums
Printed forms	Corrugated cartons
Injection-molded bottles	Corrugated cartons
Mortar mix	Multiwall paper bags
Concrete block and brick	Strapped (steel)
Potting soil and mulch	Polyethylene bags
Seed	Polyethylene bags
Canned and jarred food	Corrugated cartons
Liquids in polyethylene/glass bottles	Corrugated cartons or trays

It should be noted that, some 20 years later, these products are still being shipped on pallets and in some cases corrugated slip sheet via the stretch-wrap medium. Thus, the demand for stretch film continues to grow with over 500 million lb being produced annually in the United States. An estimated 15,000 stretch wrappers are sold annually in the United States as this type of machinery continues to replace conventional packaging methods. In addition, virtually every type of product shipped today in a carton, paper/polyethylene bags, drums or polyethylene containers is being stretch-wrapped. Products heretofore having been considered marginal as a stretch-wrap application are now being packaged or unitized with stretch film. Some of the once considered marginal product applications are as follows:

Aluminum/steel coils	Perishable foods (produce and meat)
Tires	Windows
Engine castings	Furniture
Expanded styrofoam sheets	File cabinets
Doors	Frozen fish in ice
Coils of copper tubing	Frozen foods
Batteries	Recycled rubber in polyethylene bags
Signatures	Hazardous chemicals and powders

Prestretching Film

Wrapping a pallet product load with stretch film is generally accomplished by first placing the load on the turntable of the stretch wrapper. The turntable is activated, and rotates and draws film from the supply roll. Originally, film stretch was created by the film roll being gradually retarded from turning while the load pulled the film from the roll. Retarding the film roll was affected by an electromagnetic brake fixed to the film core or through a series of rollers bars. The effective maximum stretch, however, was no more than 120% and in most cases averaged 50–60%. Higher percentages of stretch would generally cause product cartons to crush and could even pull the load from the turntable unless otherwise held in place by an optional top compression platen (electric or air-operated).

Prestretching film, however, became instantly popular in the late 1970s as moderate-to high-volume shippers took advantage of the higher film yields offered by improved stretch films coming into the marketplace. These improved films utilized linear low-density polyethylene (LLDPE) resins either alone or combined with ethyl vinyl acetate (EVA) to enhance puncture resistance, reduce tear propagation and extend film yield up to 400%. The prestretch method of stretching such films was found to reduce stretch film cost per pallet load by 20–45% as sophisticated prestretch machine assemblies became part of an upscale trend in stretch wrappers in the 1980s.

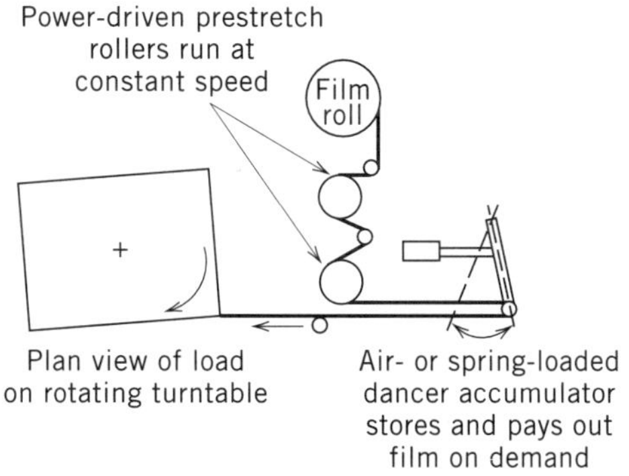

Figure 2. Powered prestretch.

Unlike a stretch wrapper equipped with brake tension, a prestretch assembly stretches the film at the machine by routing the stretch film between two powered rollers prior to its application to the load. Because the two powered rollers are moving mechanically or electrically at different speeds, the film stretches between the rollers.

The percentage of stretch that can be attained through a fixed-ratio or variable-prestretch assembly can reach 300% with the appropriate film quality and gauge. Common or the most desirable level of prestretch is 200–225%, which allows for maximum film recovery (holding force) at an economical level of film usage. Excessive prestretch can cause stretch film to thin, reduce puncture/abrasive resistance, and diminish the "cling" or self-adhering property of the film. Without a reasonable amount of "cling," the trailing segment of the film wrap (pressed against the side of the load at the end of the wrap cycle) can release itself allowing the wrapped load to partially unwrap.

Another advantage of pre-stretch is that most manufacturers have a method to apply tension of the film to the load immediately after the prestretch process occurs. Generally an adjustable spring *or* air operated dancer bar is utilized to regulate film tension to the product load. One such system is the patented IPM Uni-Tension system (Fig. 2), which provides constant and consistent film tension (adjustable) to the product load. As a result, many product loads, which could normally *not* be stretch-wrapped without the use of a top stabilizer platen, *can be wrapped* with power prestretch. Examples of these types of loads are as follows:

- Mixed-product loads
- Partial-pallet loads
- Tall-pallet loads
- Empty drums
- Empty plastic bottles
- Load weights of 150–500 lbs

Prestretch is preferred by many volume shippers because it reduces the amount of stretch film received by their customers. Although stretch film *is* recyclable, prestretch minimizes the film disposal process.

Wrapper Models Available

The stretch–wrapper industry has come a long way since the original *wide-web* machine was introduced. The single-turntable wide-web machine (using a 45–80-in. film roll size) has been expanded to at least several additional models to include wide web pre-stretch. The *spiral* wrapper model has evolved into multiple models designed to accommodate a wide variety of applications and user preferences. The following is a list of the most popular models currently available—also noted is the average production rate for which that model would be applicable and the approximate cost of the model:

Wrapper model	Production rate	Machine cost ($)
Hand wrapper	0–10 pallets/day	50
Power turntable with hand wrapper	0–10 pallets/day	2100
Powered-turntable machine with manually operated film carriage (has mechanical powered carriage)	≤20 pallets/day	2900–4700
Semiautomatic, powered-turntable machine with adjustable programmable controls for number of wraps, speeds of turntable, film carriage, etc and brake tension	≤35 pallets/h	5000–6500
Semiautomatic, powered-turntable machine with adjustable programmable controls for number of wraps, speeds of turntable, film carriage, etc and fixed-ratio powered prestretch	≤35 pallets/h	7500–9000

(*Note:* These powered-turntable machines are available in a high-platform turntable design (for forklift use only) or a low-profile design with a loading ramp for hand/electric pallet jack and forklift use. These machines are also available in wide-web or spiral-type models.).

The machine models listed above continue to withstand the test of time being the most popular and often manufactured with the following most common options to enhance the stretch-wrapping process, increase the production rate, or improve operator efficiency:

- *Photoelectric height sensor* (*for semiautomatic spiral models*)—located on the film carriage to automatically seek the top of the pallet load, halt the film carriage, and permit the top wrap cycle to occur
- *Mast extension* to permit pallet loads in excess of 85 in. in height to be wrapped
- *Top-compression platen* (*electric or air-operated*)—exerts an adjustable amount of stabilizing pressure on the top of a lightweight or unstable product load
- *Additional turntable* (*dual-turntable model*)—increases production rate by 33%
- *Conveyorized turntable* (*gravity roller*)—for in-line capability, ie, machine is installed into existing or new gravity roller production line
- *Pit mount*—machine is installed into the floor with turntable virtually flush with floor; enhances use of pallet jacks
- *Scale*—5000–10,000 lb capacity scale integrated with and under the wrapper turntable eliminates double handling of pallet loads

Fully Automatic Conveyorized Models

Shippers with moderate to high production (15–85 pallets/h) have generated an increased demand for fully automatic stretch wrappers. The most popular fully automatic has a powered conveyorized turntable for in-line powered-conveyor systems or for multiple forklift operations. Unlike a semiautomatic wrapper, the fully automatic wrapper eliminates the operator who necessarily would first attach the stretch film to the load and then press a START button. At the completion of the wrap cycle, the operator would cut the film and adhere the trailing edge of the film to the side of the load. The automatic, however, operates through an electric programmable logic controller (or microprocessor) and a series of electronic photoelectric load sensors. The stretch film is held in a mechanical clamp between load wrap cycles and is cut and brush-wiped onto the pallet load automatically at the completion of the wrap cycle.

A fully automatic wrapper equipped with a power infeed and discharge conveyor can be integrated with an automatic load palletizer (which feeds loads to the wrapper). This same system can be used effectively by having a forklift operator set a load on the power infeed conveyor and then activate that conveyor with an overhead remote pendant control. As that load would feed into the wrapper module, the forklift operator would drive over to the discharge conveyor section to remove a completely wrapped load and place it into storage or a shipping vehicle.

Fully automatic equipment generally costs $25,000–$30,000 for the primary wrapper module. Optional powered-conveyor sections for infeed and discharge are approximately $3000 per section. Automatics are generally equipped with a prestretch assembly for maximum utilization of the stretch film.

Common *optional* equipment to enhance higher-speed, fully automatic systems are as follows:

- Top-compression platen exerts an adjustable amount of stabilizing pressure on the top of a lightweight or unstable product load.
- Automatic top-sheet dispenser that automatically places a 2–3 mil polyethylene top sheet on top of the pallet load prior to the load being conveyed to the wrapping module. A top sheet is usually desirable to prevent the elements from damaging the product.
- The 90° index turntable is used primarily to direct the discharge of the pallet load 90° from which it was ac-

Figure 3. A semiautomatic overhead rotary-arm-type stretch wrapper.

cepted onto the machine. This option is usually desired if there is a floor-space restriction.

- Mast extension to permit pallet loads in excess of 85 in. in height to be wrapped.

Other Wrapper Models

In recent years, *new* stretch-wrapper designs have been created by different manufacturers to accommodate either the end user's specific product application or existing packaging line. One such model is the increasingly popular *overhead rotary-arm*-type stretch wrapper (Fig. 3). With this model, the product load does not rotate on a turntable. The load sits on the floor or on conveyor, and the roll of stretch film travels on the overhead arm *around the load* wrapping it in a spiral fashion. The film stretch delivery system can be either the brake tension type or the powered prestretch type.

The overhead rotary-arm-type machine is advantageous for wrapping pallet loads when

- Loads are extremely heavy
- The product on the pallet is subject to damage if rotated
- Loads are unstable and not easily moved by forklift, pallet jack, or conveyor
- Washdown is required, ie, for use in poultry plants or facilities where the machinery and general work areas must be washed down periodically for sanitation reasons

The overhead-arm-type machine is currently manufactured as a wall- or floor-mounted semiautomatic or can be upgraded and incorporated into a fully automatic conveyorized system for high-production use.

Another very high-production model, although used far less frequently today is the *pass-through* model, in which a product pallet load of constant height is conveyorized through a sheet of 1–3-mil polyethylene film that is heat-sealed on the backside of the load. The film used is stretched no more than 10%. Early pass-through systems were popular with manufacturers of dog and cat food, kitty litter, cereal, and like products.

The *semiautomatic robotic wrapper* is a battery-operated, wheeled vehicle with a vertical mast and film-stretch delivery system. This wrapper is placed next to a product load and when activated, travels around the perimeter of the load while the film carriage travels up and down stretch wrapping the load. A sensing device signals the wrapper to turn as it senses the corner of the load. This type of wrapper has been used to wrap longer-than-average product loads and requires ample floor space in which to operate.

Orbit stretch wrapping has become a popular method to stretch wrap large flat panels such as office dividers, locker-room partitions, etc in which the panel is conveyorized through a large ringlike assembly and a 10–30-in. roll of stretch film travels in a vertical plane with a circular motion around the entire panel wrapping it as it moves through the ring on the conveyor.

The orbital wrapper is also advantageous for wrapping long products either singularly or in bundles. Examples of such products would be rolls of carpet, PVC pipe, fencing, and shovel or rake handles. The orbital wrapper is available in several "ring" sizes to accommodate even small bundling ap-

Figure 4. A downsized fully automatic (IPM Bantam 88) wrapping F-Pack prepared file cabinets.

plications such as a stack of blue jeans, bags of dog food, and like products. Orbital wrappers are largely manufactured as automatic equipment although semiautomatic and manually operated orbital wrappers are available for low-production and individual-product-size applications.

Wrappers have been designed *for* stretch wrapping *doors, windows,* and similar products in the vertical plane. In most cases, the stretch wrapper of choice is a standard semiautomatic, low-profile model with power prestretch for maximum film economy. Wrapping doors on a pallet, however, requires an air-loaded clamp that locks the doors together at the very top of the load. Wrapping windows requires a top-compression stabilizer with an adjustable clamp and a similar adjustable clamp mounted on the turntable. Both clamps hold the window stationary during the wrap process.

Explosion proof wrappers are required for many hazardous materials, flour, and similar powder-type products packaged in cartons, multiwall paper, or polyethylene bags. The manufacturing environment for such products demands that the stretch wrapper be designed as spark-free as possible. Static elminators (for the stretch film) are often used on this type of equipment.

Produce and *perishable* wrappers have been designed to use stretchable and rigid stretch net that allows the wrapped load to "breathe." Produce items such as bananas, melons, peppers, and tomatoes are shipped in ventilated cartons so that damage causing gases generated from certain fruits and vegetables can escape from the carton during the cooling and shipping process.

Stretch net (used in the same way as stretch film on the wrapper) wraps the produce load securely and permits the ventilation to occur. Net, however, cannot be prestretched, and actual stretch is limited to 5–35%.

The product wrapper itself is usually a semiautomatic wrapper and is fitted with a brake-tension delivery system. This same wrap method is being used for perishable products such as cartoned meat, poultry, yogurt, and frozen foods.

A recent innovation to the wrapper is the *roper/bander,* option which allows somewhat less expensive stretch *film* to be reduced in web size (from 20 to 3 in.) by having the film travel through a mechanical fixture as it is being dispensed from the feed roll under tension to the product load. Thus, a 3-in. rope (or band) of film is created that unitizes the load in the standard up/down, crisscross fashion yet does not inhibit the ventilation process. A standard brake tension system or a power prestretch system can be utilized with the roper/bander unitizing method.

Luggage wrappers have recently been designed to stretch-wrap standard luggage at airport departure terminals. The stretch wrap acts as a theft deterrent.

Future Wrappers

The appearance and performance of the future stretch wrapper will like be driven by specific product package designs as many regions of the world further their concept of source reduction:

- *Less* packaging materials per package
- *Less* actual labor involved per package
- *Less* time to prepare the package
- *Less* packaging materials to recycle
- *Less* materials going to landfill
- *Less* factory floor space consumed by packaging material inventory

Packaging engineers are working closely with stretch-wrapper manufacturers to marry the two concepts of source reduction in packaging and the stretch-wrapper design.

A primary example would be the development of the "*clear-view*" package (acronym for *F-Pack*) for products such as file cabinets, bedroom furniture, entertainment centers, and armoires. Driven by consignee resistance to the disposal problem of large corrugated cartons along with the increasing cost of corrugated cartons, product manufacturers have had to develop "abbreviated" package designs.

The F-Pack package design eliminates the use of the corrugated carton and replaces it with a stretch-wrap package utilizing protective corrugated corner posts and corrugated caps and trays. Polyethylene strapping is also used in some cases.

For low-volume producers, the standard semiautomatic stretch wrapper has been modified to accept and wrap F-Pack type packages. For moderate-to high-volume manufacturers, fully automatic stretch-wrapper models have been *downsized* and significantly modified to accept the high-production volume and the specific individual package design. A downsized Bantam 88, conveyorized fully-automatic stretch wrapper (Fig. 4) can be placed in-line to accept F-Pack prepared products like file cabinets. It is likely that additional products such as small appliances, air-conditioning units, and heat pumps, will also shed their corrugated carton in favor of a clear-view package in the future.

Summary

The stretch wrapper continues to be the unitizing medium of choice for palletized product loads, slip-sheeted product loads, and for products having new, lower cost package designs. The use of stretch film (particularly prestretched) and stretch net offers significant savings for the shipper, the carrier, and the consignee. New stretch-film resins such as *metallocene* may enhance prestretch film yield and allow "downgauging" for many applications.

Stretch-wrapper design and performance function should continue to be refined in the areas of manufacturing cost, prestretch assemblies, and ergonomics, resulting in additional benefits to the end user.

BIBLIOGRAPHY

1. NMFC F-Pack numbers: 132-F, 138-F, 141-F, 142-F, 145-F, 146-F, 2418, 2512, 2514 Furniture Packaging; published by Bohman Industrial Traffic Consultants, Inc. P.O. Box 889, Gardner, MA 01440.
2. "A Change to See-through Packaging," *Inst. Packag. Prof. Techn. J.* (Spring issue, 1991). (IPP address: 481 Carlisle Drive, Herndon, VA 22070).
3. National Motor Freight Classification Rule 180, effective Jan. 21, 1995 (National Motor Freight Traffic Association, Inc. 2200 Mill Road, Alexandria, VA 22314).
4. R. F. Nunes (Director of Marketing & Sales, International Packaging Machines, Inc.), New Developments in Stretch Wrapping of Case Goods, Furniture and Small Appliances; Including High Speed Automated Wrapping," International Safe Transit Association (ISTA), 1995 (paper on source reduction).

ROBERT F. NUNES
International Packaging Machines, Inc.
Naples, Florida

Appendix A

CONVERSION FACTORS, ABBREVIATIONS, AND UNIT SYMBOLS

Selected SI Units (Adopted 1960)

Quantity	*Unit*	*Symbol*	*Acceptable equivalent*
BASE UNITS			
length	meter†	m	
mass‡	kilogram	kg	
time	second	s	
electric current	ampere	A	
thermodynamic temperatures§	kelvin	K	
DERIVED UNITS AND OTHER ACCEPTABLE UNITS			
* absorbed dose	gray	Gy	J/kg
acceleration	meters per second squared	m/s^2	
* activity (of ionizing radiation source)	becquerel	Bq	l/s
area	square kilometer	km^2	
	square hectometer	hm^2	ha (hectare)
	square meter	m^2	
density, mass density	kilograms per cubic meter	kg/m^3	g/L; mg/cm^3
* electric potential, potential difference, electromotive force	volt	V	W/A
* electric resistance	ohm	Ω	V/A
* energy, work, quantity of heat	megajoule	MJ	
	kilojoule	kJ	
	joule	J	N · m
	electronvolt^x	eV^x	
	kilowatthour^x	kW · h^x	
* force	kilonewton	KN	
	newton	N	kg · m/s^2
	megahertz	MHz	
* frequency	hertz	Hz	l/s
heat capacity, entropy	joules per kelvin	J/K	
heat capacity (specific), specific entropy	joules per kilogram kelvin	J/(kg · K)	
heat-transfer coefficient	watts per square meter kelvin	W/(m^2 · K)	
linear density	kilograms per meter	kg/m	
magnetic field strength	amperes per meter	A/m	
moment of force, torque	newton-meter	N · m	
momentum	kilogram-meters per second	kg · m/s	
* power, heat flow rate,	kilowatt	kW	
radiant flux	watt	W	J/s
power density, heat-flux density, irradiance	watts per square meter	W/m^2	
* pressure, stress	megapascal	MPa	
	kilopascal	kPa	
	pascal	Pa	
sound level	decibel	dB	
specific energy	joules per kilogram	J/kg	
specific volume	cubic meters per kilogram	m^3/kg	
surface tension	newtons per meter	N/m	
thermal conductivity	watts per meter kelvin	W/(m · K)	
velocity	meters per second	m/s	
	kilometers per hour	km/h	
viscosity, dynamic	pascal-second	Pa · s	
	millipascal-second	mPa · s	
volume	cubic meter	m^3	
	cubic decimeter	dm^3	L (liter)
	cubic centimeter	cm^3	mL

In addition, there are 16 prefixes used to indicate order of magnitude, as follows:

Multiplication factor	*Prefix*	*Symbol*	*Note*
10^{18}	exa	E	
10^{15}	peta	P	
10^{12}	tera	T	
10^{9}	giga	G	
10^{6}	mega	M	
10^{3}	kilo	k	
10^{2}	hecto	h[a]	
10	deka	da[a]	[a] Although hecto, deka, deci, and centi are SI prefixes, their use should be avoided except for SI unit-multiples for area and volume and nontechnical use of centimeter, as for body and clothing measurement.
10^{-1}	deci	d[a]	
10^{-2}	centi	c[a]	
10^{-3}	milli	m	
10^{-6}	micro	μ	
10^{-9}	nano	n	
10^{-12}	pico	p	
10^{-15}	femto	f	
10^{-18}	atto	a	

* The asterisk denotes those units having special names and symbols.
† The spellings "metre" and "litre" are preferred by ASTM; however, "-er" is used in the *Encyclopedia.*
‡ "Weight" is the commonly used term for "mass."
§ Wide use is made of "Celsius temperature" (t) defined by $t = T - T_0$ where T is the thermodynamic temperature, expressed in kelvins, and $T_0 = 273.15$ by definition. A temperature interval may be expressed in degrees Celsius as well as in kelvins.
[x] This non-SI unit is recognized by the CIPM as having to be retained because of practical importance or use in specialized fields.

Factors For Conversion to SI Units

To convert from	*To*	*Multiply by*
acre	square meter (m^2)	4.047×10^{3}
angstrom	meter (m)	1.0×10^{-10}†
atmosphere	pascal (Pa)	1.013×10^{5}
bar	pascal (Pa)	1.0×10^{5}†
barn	square meter (m^2)	1.0×10^{-28}†
barrel (42 U.S. liquid gallons)	cubic meter (m^3)	0.1590
Btu (thermochemical)	joule (J)	1.054×10^{3}
bushel	cubic meter (m^3)	3.524×10^{-2}
calorie (thermochemical)	joule (J)	4.184†
centipoise	pascal-second (Pa · s)	1.0×10^{-3}†
cfm (cubic feet per minute)	cubic meters per second (m^3/s)	4.72×10^{-4}
cubic inch	cubic meter (m^3)	1.639×10^{-5}
cubic foot	cubic meter (m^3)	2.832×10^{-2}
cubic yard	cubic meter (m)	0.7646
dram (apothecaries')	kilogram (kg)	3.888×10^{-3}
dram (avoirdupois)	kilogram (kg)	1.772×10^{-3}
dram (U.S. fluid)	cubic meter (m^3)	3.697×10^{-6}
dyne	newton (N)	1.0×10^{-5}†
dynes per centimeter	newtons per meter (N/m)	1.0×10^{-3}†
fluid ounce (U.S.)	cubic meter (m^3)	2.957×10^{-5}
foot	meter (m)	0.3048†
gallon (U.S. dry)	cubic meter (m^3)	4.405×10^{-3}
gallon (U.S. liquid)	cubic meter (m^3)	3.785×10^{-3}
gallons per minute (gpm)	cubic meter per second (m^3/s)	6.308×10^{-5}
	cubic meter per hour (m^3/h)	0.2271
grain	kilogram (kg)	6.480×10^{-5}
horsepower (550 ft · lbf/s)	watt (W)	7.457×10^{2}
inch	meter (m)	2.54×10^{-2}†
inch of mercury (32°F)	pascal (Pa)	3.386×10^{3}
inch of water (39.2°F)	pascal (Pa)	2.491×10^{2}
kilogram-force	newton (N)	9.807
kilowatt hour	megajoule (MJ)	3.6†
liter (for fluids only)	cubic meter (m^3)	1.0×10^{-3}†
micron	meter (m)	1.0×10^{-6}†
mil	meter (m)	2.54×10^{-5}†
mile (statute)	meter (m)	1.609×10^{3}
mile per hour	meters per second (m/s)	0.4470
millimeter of mercury (0°C)	pascal (Pa)	1.333×10^{2}†

ounce (avoirdupois)	kilogram (kg)	2.835×10^{-2}
ounce (troy)	kilogram (kg)	3.110×10^{-2}
ounce (U.S. fluid)	cubic meter (m^3)	2.957×10^{-5}
ounce-force	newton (N)	0.2780
peck (U.S.)	cubic meter (m^3)	8.810×10^{-3}
pennyweight	kilogram (kg)	1.555×10^{-3}
pint (U.S. dry)	cubic meter (m^3)	4.732×10^{-4}
pint (U.S. liquid)	cubic meter (m^3)	5.506×10^{-4}
poise (absolute viscosity)	pascal-second (Pa · s)	0.10†
pound (avoirdupois)	kilogram (kg)	0.4536
pound (troy)	kilogram (kg)	0.3732
pound-force	newton (N)	4.448
pound-force per square inch (psi)	pascal (Pa)	6.895×10^{3}
quart (U.S. dry)	cubic meter (m^3)	1.101×10^{-3}
quart (U.S. liquid)	cubic meter (m^3)	9.464×10^{-4}
quintal	kilogram (kg)	1.0×10^{-2}†
rad	gray (Gy)	1.0×10^{-2}†
square inch	square meter (m^2)	6.452×10^{-4}
square foot	square meter (m^2)	9.290×10^{-2}
square mile	square meter (m^2)	2.590×10^{6}
square yard	square meter (m^2)	0.8361
ton (long, 2240 lb)	kilogram (kg)	1.016×10^{3}
ton (metric)	kilogram (kg)	1.0×10^{3}†
ton (short, 2000 lb)	kilogram (kg)	9.072×10^{2}
torr	pascal (Pa)	1.333×10^{2}
yard	meter (m)	0.9144†

ABBREVIATIONS AND ACRONYMS

A	ampere
AAMI	Association for the Advancement of Medical Instrumentation
ABS	acrylonitrile–butadiene–styrene
ac	alternating current
adh	adhesive
AF	aluminum foil
AFR	Air Force Regulation
AGV	automated guide vehicle
AM	aluminum metallization
AN	acrylonitrile
AN/MA	acrylonitrile–methacrylate copolymers
ANS	acrylonitrile–styrene copolymers
ANSI	American National Standards Institute
API	American Paper Institute
ASME	American Society of Mechanical Engineers
ASP	asphalt
ASQC	American Society for Quality Control
ASTM	American Society for Testing and Materials
avg	average
BATF	Bureau of Alcohol, Tobacco and Firearms
B&B	blow-and-blow
BBP	butyl benzyl phthalate
BCL	British Cellophane Limited
BHEB	*tert*-butylated hydroxyethylbenzene
BHT	*tert*-butylated hydroxytoluene
BIB	bag-in-box
BK	bleached kraft
BMC	bulk molding compound
BON	biaxially oriented nylon film
BOPP	biaxially oriented polypropylene film
bpm	bottles per minute; bags per minute
BSP	British standard pipe thread
Btu	British thermal unit
BUR	blowup ratio
°C	degree Celsius
CA	controlled atmosphere
ca	approximately (circa)
CAD	computer-aided design
CAE	computer-aided engineering
cal	calorie (4.184 J)
Cal	food calorie (1000 cal)
CAP	cellulose acetate propionate; controlled atmosphere packaging
CELLO	cellophane
CFR	Code of Federal Regulations
CGMP	Current Good Manufacturing Practice
CGPM	Conference Generale des Poids et Mesures (General Conference on Weights and Measures)
CI	central impression
CIPM	Comité International des Poids et Mesures (International Committee on Weights and Measures)
cm	centimeter
CMA	Closure Manufacturers Association
CNC	computer numerical control
coex	coextruded
COF	coefficient of friction
COFC	Container on Flat Car
COREPER	Committee of Permanent Representatives (EEC)
CPE	chlorinated polyethylene
cpm	cans per minute
cPs	centipoise (10^{-3} Pa · s)
CPSC	Consumer Product Safety Commission
CPU	central processing unit
CR	child-resistant
CRC	child-resistant closure
CRT	cathode-ray tube
CT	continuous thread
D & I	drawing and ironing (cans)
dB	decibel

† Exact.

dc	direct current
DEHP	di(2-ethylhexyl)phthalate
DEG	diethylene glycol
DGT	diethylene glycol terephthalate
dia	diameter
DINA	diisononyl adipate
DLAM	Defense Logistics Agency Manual
DME	dimethyl ether
DMF	Drug Master File
DMT	dimethyl terephthalate
DOA	dioctyl adipate
DOD	U.S. Department of Defense
DOP	dioctyl phthalate
DOT	U.S. Department of Transportation
DOZ	dioctyl azelate
DPM	double package maker
DR	double-reduced
DRD	draw–redraw (cans)
DSAM	Defense Supply Agency Manual
DWI	drawn and ironed (cans)
EAA	ethylene–acrylic acid
ECOSOC	Economic and Social Committee (EEC)
ECCS	electrolytic chromium-coated steel
EEA	ethylene–ethyl acrylate
EEC	European Economic Community
eg	for example (*est gratia*)
EG	ethylene glycol
EHMW	extra-high-molecular-weight
EIA	Electronic Industries Association
ELC	end-loading construction
EMA	ethylene methacrylate
EMAA	ethylene–methacrylic acid
EMI/RFI	electromagnetic interference/radio frequency interference
EPA	Environmental Protection Agency
EPC	expanded-polyethylene copolymer
EPE	expanded polyethylene
EPR	ethylene–propylene rubber
EPS	expanded polystyrene
ESC	environmental stress cracking
ESCR	environmental stress-crack resistance
ESD	electrostatic discharge
ESO	epoxidized soybean oil
est	estimated
ETO	ethylene oxide
ETP	electrolytic tinplate
EVA	ethylene–vinyl acetate
EVOH	ethylene–vinyl alcohol
F_{50}	value at which 50% of the specimens have failed
°F	degree Fahrenheit
FDA	U.S. Food and Drug Administration
FD&C	Food, Drug, and Cosmetic
FEP	fluorinated ethylene–polypropylene
ffs	form/fill/seal
FI	flow index
fl oz	fluid ounce (29.57 mL in United States)
FPMR	Federal Property Management Regulations
FRH	full-removable head
FRP	fiberglass-reinforced plastics
ft	foot
ft · lbf	foot-pound force (1.356 J)
G	specific deceleration: gravitational acceleration; specific acceleration: gravitational acceleration
g	gravitational acceleration (9.807 m/s²)
g	gram
gf	gram-force (0.0098 N)
ga	gauge
gal	gallon (3.785 L in United States)
GMA	Grocery Manufacturers Association
GMP	Good Manufacturing Practice
GNP	gross national product
GPPS	general-purpose polystyrene
Gy	gray (10^{-2} rad)
h	hour; height
HBA	health and beauty aids
HCP	healthcare packaging
HD	head diameter
HDPE	high-density polyethylene
HDT	heat-deflection temperature
HF	high-frequency
HFFS	form/fill/seal, horizontal
HIMA	Health Industry Manufacturers' Association
HIPS	high-impact polystyrene
HLMI	high-load melt index
HM	hot melt
HMW	high-molecular-weight
HNR	high-nitrile resin
hp	horsepower (746 W)
HP	high-pressure
HPP	homopolymer polypropylene
HRC	Rockwell hardness (C scale)
HRM	Rockwell hardness (M scale)
HRR	Rockwell hardness (R scale)
HTST	high-temperature–short-time
Hz	hertz (cycles per second)
IBC	intermediate bulk containers
ICAO	International Civil Aviation Organization
ie	that is (*id est*)
IM	intramuscular
IMDG	International Maritime Dangerous Goods
IML	in-mold labeling
IMO	International Maritime Organization
in.	inch (2.54 cm)
I/O	input/output
IPA	isophthalic acid
IS	individual section
ISO	International Standards Organization
ITRI	International Tin-Research Institute
IV	intrinsic (or inherent) viscosity; intravenous
J	joule (energy)
k	kilo (10^3)
K	Kelvin; the molecular weight of a resin
k.d.	knocked-down
kgf	kilogram-force (9.806 N)
kJ	kilojoule
km	kilometer
kPa	kilopascal (0.145 psi)
L	liter (volume)
lb	pound (mass) (453.6 g)
lbf	pound force (4.448 N)
LCB	long-chain branching
LCD	liquid-crystal display
L : D	length : diameter
LDPE	low-density polyethylene
LED	light-emitting diode
LLDPE	linear low-density polyethylene
LMW	low-molecular-weight
LSI	large-scale integration

L:T	length:thickness
LVP	large-volume parenteral
M	mega (10^6)
M_N	number-average molecular weight
M_W	weight-average molecular weight
m	meter; milli (1/1000)
MA	modified atmosphere; methyl acrylate
MAN	methacrylonitrile
max	maximum
MBS	methacrylate–butadiene styrene
m/c	cylinder mold
MD	machine direction
MDPE	medium-density polyethylene
MET	metallized
MF	machine-finished
MFR	melt-flow rate
mg	milligram
MG	mill-glazed; machine-glazed
MGBK	machine-glazed bleached kraft
MGNN	machine grade natural Northern
mi	mile
MI	melt index
MIL	military
min	minute; minimum
MIR	multiple-individual-reward
MMA	methyl methacrylate
MMW	medium molecular weight
mn	millinewton (2.25×10^{-4} lbf)
MN	meganewton (224,909 lbf)
MOE	modulus of elasticity
MOR	modulus of rupture
MP	microprocessor; melting point
MPa	megapascal (145 psi)
ms	millisecond
MSW	municipal solid waste
MTB	Materials Transportation Bureau
MVTR	see WVTR
MW	molecular weight
MWD	molecular-weight distribution
μm	micrometer
N	newton (force)
N/A	not available
NASA	National Aeronautics and Space Administration; National Advertising Sales Association
NBS	National Bureau of Standards
NC	nitrocellulose
NCB	National Classification Board
NDA	New Drug Application
NEC	not elsewhere classified
NF	*National Formulary*
NK	natural kraft
nm	nanometer (10^{-9} meter)
NMFC	National Motor Freight Classification Committee
NODA	*n*-octyl *n*-decyl adipate
NPIRI	National Printing Ink Research Institute
NSTA	National Safe Transit Association
NWPCA	National Wooden Pallet and Container Association
Ω	ohm (resistance)
OD	optical density
OL	overlacquer
ON	oriented nylon
OPET	oriented polyester
OPP	oriented polypropylene
OPS	oriented polystyrene
OSHA	Occupational Safety and Health Administration
OTC	over-the-counter
OTR	oxygen transmission rate
Pa	pascal (pressure)
PA	Proprietary Association
PAN	polyacrylonitrile
P&B	press-and-blow
PBT	poly(butylene terephthalate)
PC	personal computer; polycarbonate
PCTFE	poly(chlorotrifluoroethylene)
PE	polyethylene
PEPS	Packaging of Electronic Products for Shipment
PET	poly(ethylene terephthalate); polyester
phr	parts per hundred of resin
PIB	polyisobutylene
PLI	pound force per lineal inch
PM	Packaging Materials
PMMA	poly(methyl methacrylate)
PMMI	Packaging Machinery Manufacturers Institute
PP	polypropylene
ppb	parts per billion (10^9)
ppm	parts per million (10^6)
PPP	poison-prevention packaging
PPPA	Poison-Prevention Packaging Act
PR	printing
proj	projected
PS	polystyrene
psi	pound (force) per square inch (6.893 kPa)
psig	psi gauge pressure
PSTA	Packaging Science and Technology Abstracts
PSTC	Pressure-Sensitive Tape Council
PTFE	polytetrafluoroethylene
PVAc	poly(vinyl acetate)
PVC	poly(vinyl chloride)
PVDC	poly(vinylidene chloride)
PVF	poly(vinyl fluoride)
PVF_2	poly(vinylidene fluoride)
PVOH	poly(vinyl alcohol)
PX	postexchange
Q_{10}	change in reaction rate for a 10°C temperature increase
QA	quality assurance
QAI	quaternary ammonium inhibitor (nitrite)
qt	quart (946 mL in United States)
R&D	research and development
RCF	regenerated cellulose film
RCPP	random-copolymer polypropylene
RD	root diameter
RDF	refuse-derived fuel
rf	radiofrequency
rh	relative humidity
RIM	reaction injection molding
RM-HNR	rubber-modified high-nitrile resin
rpm	rotations per minute
RSC	regular slotted container
RVR	rim-vent release
s	second
SAN	styrene–acylonitrile
SB	styrene–butadiene
SBS	solid bleached sulfate
SCB	short-chain braching
SCF	Scientific Committee on Food (EEC)
SCR	semiconductor
SIC	Standard Industrial Classification
SMC	sheet molding compound

SMMA	styrene–methyl methacrylate
SP	special packaging
sp gr	specific gravity
SPPS	solid-phase pressure forming
SR	single-reduced
SUS	solid unbleached sulfate
sq	square
SVP	small-volume parenteral
TA	thread angle
TAPPI	Technical Association of the Pulp and Paper Industry
TD	thread-crest diameter
TE	tamper-evident
tffs	thermoform/fill/seal
TFS	tin-free steel
T_g	glass-transition temperature
TH	tight head
T_m	melting temperature
TIS	Technical Information Service
TOFC	trailer on flatcar
TPA	terephthalic acid
TR	tamper-resistant
UCB	Union Chimique Belge
UCC	Uniform Classification Committee
UFC	Uniform Freight Classification Committee
UHMW	ultra-high-molecular-weight
UN	United Nations
UPS	United Parcel Service
USDA	United States Department of Agriculture
USP	*US Pharmacopeia*
uv	ultraviolet
V	volt
VA	vinyl alcohol
VC	vinyl chloride
VCI	volatile corrosion inhibitor
VCM	vinyl chloride monomer
VDC	vinylidene chloride
VFFS	form/fill/seal, vertical
vol	volume
vs	versus
W	watt (J/s)
WD	wire diameter
wk	week
wt	weight
WVTR	water-vapor transmission rate
XD	cross-direction
XKL	extensible kraft linerboard
yr	year
ZCC	zero-crush concept

Appendix B

GLOSSARY OF PACKAGING TERMINOLOGY AND DEFINITIONS

Glossary

The attempt of this Glossary is to communicate, define, and clarify common words used in the packaging industry as they relate to machines and processes. This Glossary has evolved over time and will continue to evolve and refine these definitions and terms and add new definitions and terms as required. Just because someone uses the same word or term as you do does not mean that person understands it the same way as you do. It has been found that miscommunication and misunderstandings are common in the packaging industry even when using the same word. The reader is encouraged to contact the author pertaining to clarifying an existing definition and/or adding more terms.

Wherever possible, definitions from other sources or associations have been maintained, such as terms and definitions from the canning trade, in order to achieve as much alignment as possible with the large diversity in the packaging industry.

Absolute humidity. The actual weight of water vapor contained in a unit volume or weight of air. Referred to also as *relative humidity.*

Absorbent. A substance that has the ability to soak up or retain other substances, such as sugar or salt absorbing water when exposed to high relative humidity atmospheres.

Acid. A substance, usually a liquid, that increases the concentration of hydrogen ions (H^+) in water and reacts with a base to form a salt. On a pH scale, an acid has a pH value of <5; a pH value of 7 indicates a neutral substance. As the pH value declines from pH <5, the substance becomes more acidic and generally more corrosive.

Accumulation. An accumulation apparatus is a device that deliberately amasses inputs or products into one place for the next required operation or process. Taking a single line of product and collating the products into multiple rows to go into an oven, cooler, or similar operation is an example of accumulation. Accumulation should not be confused with buffers. The definition of buffer is to shield or cushion (via conveyor length or off-loading areas such as tables or conveyors to let the inputs gather) an upstream machine from a downstream machine that stops. Generally, a buffer is a technique used as a positive modulating effect on utilization. The two definitions should not be used interchangeably.

Achievable run speed. That speed or rate as tested during commissioning (which also verifies the packaging line design criteria of needs) to be the needed sustainable steady-state speed or rate on a per minute or per hour basis with a wastage factor of <0.01. It is generally 20% less than the packaging line capacity and 50–70% less than the packaging line capacity for a true just-in-time (JIT) packaging line. The proposed run speed, the achievable run speed, and the design run speed can be used interchangeably, although the achievable run speed is the phase used in this book.

Action steps. Specific, tactical activities you take to accomplish goals and objectives.

Activity. An individual step or piece of work that is part of the total job being scheduled.

Activity-based costing (ABC). The use of cost accounting tools that attempt to allocate indirect or overhead costs on the basis of related activities, rather than using surrogate allocation bases such as direct labor or machine hours, floor space, or material costs.

Actual set run speed. That target speed or rate as set or fixed for a given package run cycle on a per minute or per hour basis that is subjectively set by the operator or management. It is generally equal to (ideal) or less than the achievable run speed (usually). It is the dialed run speed set by the operator or management. On dynamically controlled automated packaging lines, it is the instantaneous steady-state target output rate. Note that the effects of wastage rework, and stoppages have not been considered.

Additive. Any substance, the intended use of which results or may reasonably be expected to result, directly or indirectly, in its becoming a part of or otherwise affecting the characteristics of any input.

Adsobent. The material on whose surface adsorption takes place.

Adsorption. The adhesion of a substance to the surface of a solid or liquid.

Adulterant (adulteration). Foreign material in the product, especially substances that are aesthetically objectionable, hazardous to health, or indicate that unsanitary handling or manufacturing practices have been employed.

Aeration. The bringing about of intimate contact between air and a liquid by bubbling air through the liquid or by agitation of the liquid.

Aerator. A device used to or unwittingly causes the promotion of aeration.

Aerosol. The colloidal suspension in which gas is the dispersant. Dispersion or suspension of extremely fine particles of liquid or solid in a gaseous medium.

Affordable capital limits (ACLs). The specified cost or "cap" for a project to achieve maximum value. The ACL is based on the economic facts associated with a specific business need, not a "wish" or arbitrary "grab number."

Agglomerate. To gather, form, or grow into a rounded mass, or to cluster densely.

Aftersales service. Service provided after the sale, such as repair or warranty service, customer problem solving, or follow-up in canvassing customers as to their opinions, needs, quality of products, quality of service (timely delivery), to ensure that the customer is and remains satisfied.

Allocated inventory. Materials that are in inventory or on order but have been assigned to specific production orders in the future. These materials are therefore not available for use in other orders.

Anaerobes. These are microorganisms that grow in the absence of oxygen. Obligate anaerobes cannot survive in the presence of oxygen. Facultative anaerobes normally grow in oxygen but can also grow in its absence.

Anaerobic. Living or active in the absence of free oxygen.

ANSI. American National Standards Institute.

Antimicrobial. A compound that inhibits the growth of a microbe.

Apparent viscosity. See **Viscosity.** Viscosity of a complex (non-Newtonian) fluid under given conditions.

Appropriation. An authorization by the appropriate executive or management committee for funding of capital projects in excess of a specific amount.

Approved-bidders list. A list of prequalified contractors used for the purpose of soliciting competitive bids.

Aseptic processing and packaging. The filling of a commercially sterilized cooled product into presterilized containers, followed by aseptic hermetical sealing, with a presterilized closure, in an atmosphere free of microorganisms.

Asset utilization. Fraction of time of a 24-h day, 7 days a week potential time, that is actually used in producing quality packages until the package run cycle is completed (package run cycle start point to the next package run cycle start point). If the package run cycle (start to next start) takes exactly three days, then the total asset utilization time is $24 \times 3 = 72$ h, even though only one shift per day may be used.

Authorize. To give final approval.

Authorized work. Work that has been approved by higher authority.

Autoclave. A vessel in which high temperatures can be reached by using high steam pressure. Bacteria are destroyed more readily at elevated temperature and autoclaves are used to sterilize food, for example, in cans.

Automation. A technique of making industrial machinery, a process, or a system operate in an independent or self-controlling manner. This is the generic definition of automation. With regard to packaging, this definition should be adjusted to reflect the objectives of packaging, as follows: Automation is controlling the packaging line by using the optimum technique to cause the process to operate at a steady state pace in a self-controlling manner. Note that these definitions say nothing about eliminating labor or guaranteeing profitability, but they both imply that automation will optimize labor and give the potential of profitability.

Availability. The total available time for which a machine or system is in an operable state or between failures divided by the sum of the mean time between failures plus the mean time to repair plus the mean preventive maintenance time. Simply put, it is the fraction of time the machine is in an operable state to the total time period defined. An operable state is the condition that allows the machine or system to function at its achievable run speed in a manner that produces an outcome or an assemblage of inputs within the stated specifications and conformance to customer's needs.

Available inventory. Materials that are in inventory or on order that are not safety stock or allocated to other uses. This available inventory is on immediate call to sustain the needs of production on a timely bases without interruption.

Backcharge. The cost associated with corrective action taken by the purchaser. This cost is chargeable to the supplier under the terms of the contract.

Baffle. A partition or plate that changes the direction or restricts the cross section of a fluid, thus increasing the velocity or turbulence.

Bar chart. A graphic presentation of project activities shown by time-scaled bars with the length of the bar equal to the duration of the activity (Gantt chart).

Base. Alkaline substances (pH $>$7.0) that yield hydroxyl ions (OH^-) in solution. As a general rule, as the pH value increases, the corrosive ability of the solution increases.

Benchmarking. A comparison or scale of product, function, practice, or strategy between identical industry segments. There are four types of benchmarking:

Product benchmarking, which evaluates the current and future strengths and weaknesses of internal and external competitive products or packages.

Functional benchmarking, which can be a comparison of the functional process or manufacturing technique relating to product development, packaging-line design, machinery setup, packaging-line control, and logistics of inputs in and packages out. Utilization when set up correctly can be a form of functional benchmarking.

Best-practices benchmarking, which goes beyond the functional benchmarking by focusing on management practices.

Strategic benchmarking, which compares the goals, direction, and key performance measure of a company within an industry sector against their internal or external competitors.

The *performance index* is a benchmark of functional and best practices.

Bid-clarification management. Meeting with potential suppliers to clarify contract and associated issues prior to a decision on selection of a contractor.

Bid evaluation. Review of a supplier's ability to perform the work requested. The supplier's financial resources, ability to comply with technical criteria and delivery schedules, and past performance are considered.

Bid list. A list of suppliers invited to submit bids for the specified goods or services.

Bid response. A negative or positive response to an invitation to bid from prospective suppliers.

Bill of material (BOM). The list of all end items or products, that is a major element of the material requirements planning (MRP) system. In addition, a bill of material contains a listing of all raw materials, parts, subassemblies, and assemblies that go into each end item or package, as well as how the components go together.

Biodegradability. Susceptibility of a chemical compound to depolymerization by the action of biological agents.

Blanching. Heating by direct contact with hot water or live steam. It softens the tissues, eliminates air from the tissues, destroys enzymes, and washes away raw flavors.

Breakdown maintenance. Breakdown maintenance and emergency maintenance are basically the same, when the predominate method of maintenance is breakdown maintenance. Breakdown maintenance is doing no maintenance other than lubrication between breakdowns. Repairs and adjustments are done as a result of reduced or no production occurring. Emergency maintenance is when a technical person is called on to perform a repair at any time of the day without being forewarned. With most other types of maintenance, emergency maintenance should be the exception, not the rule, as in breakdown maintenance.

British thermal unit (Btu). The British engineering unit of heat quantity. It is approximately the quantity of heat that will raise the temperature of 1 pound of water 1 degree Fahrenheit. 1 Btu = 0.252 calories = 1054 joules.

Buffer. The definition of a buffer is to shield or cushion (via conveyor length or off-load areas to let the inputs gather) an upstream machine from a downstream machine that stops or slows down. Generally, a buffer is a technique used to maintain uptime and effect utilization. It should not be confused with accumulation. The definition of accumulation is a device that deliberately amasses inputs or products into one place for the next required process. The two definitions are similar but different.

Bulk density. Weight per unit volume of a quantity of solid particles that depends on the packing density.

Bursting strength. The strength of material in pounds per square inch, measured by the Cady or Mullen tester.

Business strategy. The organization's reason for being, including purpose, mission, operating principles, objectives, and goals.

CAD. Computer-aided design. This term implies the use of a computer and drafting software such as AutoCAD (trademark) to produce and store prints for layout, installation, machining, assembly, and fabrication. Another definition is the use of computers in interactive engineering drawings and the storage and retrieval of designs.

Caliper. The thickness as related to paperboard, of a sheet measured under specified procedures expressed in thousandths of an inch. Thousands of an inch are sometimes termed "points." The precision instrument used in the paperboard industry to measure thickness. To measure with a caliper.

Calorie. A unit of heat or the amount of heat necessary to raise the temperature of a gram of water 1 degree Celsius. Nutritionists use the large Calorie or kilo-Calorie (spelled with a capital C), which is 1000 calories. One Calorie (kilocalorie) = 4,184 joules or 3.968 Btu.

CAM. Computer-aided manufacturing. This term implies the use of a computer and postprocessing or linking software (such as Smartcam or Mastercam) to manipulate and compile data into a machine language for a machine(s) to execute the desired function. Another definition is the use of computers to program, direct, and control production equipment.

Can, cylinder. A can whose height is relatively large compared to its diameter. Generally called a "tall can."

Can, flat. A can whose height is equal to or smaller than its diameter.

Can, key-opening. A can opened by tearing off a second strip of metal around the body by means of a key, or any can opened by means of a key.

Cap. See also **Closure.** Any form or device used to seal off the opening of the container, so as to prevent loss of its contents.

Cap, lug. A cap closure for glass containers in which impressions in the side of the cap engage appropriately formed mombers on the neck finish to provide a grip when the cap is given a quarter turn, as compared to the full turn necessary with a screw cap.

Cap, screw. A cylindrical closure having a thread on the internal surface of the cylinder capable of engaging a comparable external thread on the finish or neck of a container, such as a glass bottle or collapsible tube.

Cap, snapon. A type of closure for rigid containers. The sealing action of a snapon cap is effected by a gasket in the top of the cap that is held to the neck or spout of the container by means of a friction fit on a circumferential bead. Material of construction is either metal or semirigid plastic.

Cap, two-piece vacuum. Standard CT (continuous-thread) or DS (deep-screw) caps, equipped with a separate disk or lid that is lined with sealing for vacuum-packing processes.

Capability (schedule) (C_p). In simple terms, *capability* is the relative measure of how capable or how effective is the actual packaging process to produce the appropriate volume of needed packages within the planned plant operational time period or schedule. Capability or, more correctly the *schedule capability,* relates to the ratio of the scheduled run cycle time available divided by the actual package run cycle time used. This definition is valid only if the actual time taken is greater than the scheduled time. If the actual time is less than the scheduled time, the capability is the actual package run cycle time divided by the scheduled run cycle time. Capability is always equal to or less than 1.

Capacity. Capacity for a given packaging line and product is the *upper possible sustainable limit* of packages passing a point just before warehousing or shipping in a given amount of time (usually one minute or one hour or one shift). *Sustainable* refers to the ability to maintain consistent production of quality packages at a given speed. One could argue that it is the speed at which the percentage of rejects and/or jams begins to rise in a nonlinear manner. Capacity is usually expressed as a rate (packages/minute, etc). Typically, a line will operate at a somewhat lower speed, called the *actual set run speed.* For just-in-time (JIT) processing, all machines and systems in the packaging process as well as making must have excess capacity.

Capacity requirements planning (CRP). The process of reconciling the master production schedule to the labor and machine capacities of the production departments over the planning horizon. This process is generally used in conjunction with MRP systems.

Case. A non-specific term for a shipping container. In domestic commerce, "case" usually refers to a box made from corrugated or solid fiberboard. In maritime or export usage, "case" refers to wooden or metal box.

Cash flow. The net amount of actual inflows and outlays generated by a project or business unit:

$$\text{Cash flow} = \text{profits} + \text{depreciation} + \text{deferred taxes} - \text{capital} + \text{working capital}$$

Cash-flow analysis. The process of determining monthly and overall total cash flow for projects. The cumulative total is used as a measurement of actual versus budget costs. This activity helps monitor projects and costs.

Catalyst. A substance that alters the rate of chemical change and remains unchanged at the end of a reaction.

CCS charts. For compiled charts that cross-reference conditions (or observations), to causes (or sources) and solutions (or knowledge), a CCS chart is used. This chart lays out all these elements in a methodical and logical pattern. The inputs come from DFA, DA, and/or PRC analysis. After a packaging line has been debugged [and even during commissioning], the CCS chart is an excellent training and troubleshooting tool.

Celsius (°C). The temperature on a scale of 100° between the freezing point (0°) and the boiling point (100°) of water.

Centimeter (cm). One-hundredth of a meter. Equivalent to 0.3937 in. One inch equals 2.54 cm.

Centipoise (cP). Unit of viscosity equal to $\frac{1}{100}$ dyn/($s^2 \cdot cm^2$).

Changeover time (or period). The time to complete the following items:

1. Exchange of changeover parts or tooling for all line elements.
2. Recalibration and/or adjustment of all line elements.
3. Run the first 1000 packages or 15 min of production, whichever comes first

At the end of this procedure, the line and all machine elements are expected to perform the desired functions to produce quality packages at the required output rate.

Clean in place (CIP). A machine's (such as a filler's) ability to be cleaned and sanitized in place on the packaging line without dismantling any components and minimizing time, product loss, cleaning solutions, cleaning procedures, and volume of water required.

Cleanup time or (period). The time required to (*1*) remove change parts from the previous package run cycle and (*2*) clean and/or flush out and inspect areas of the machine and line.

Closure. The joint or seal that is made in attaching the cover to a glass container. Also, the type of closure, such as friction, lug, screw top, etc.

CNC. Computer numerically controlled. This term implies the use of a computer or processor on board the machine that is controlled by compiled program and/or data in APT or ASCII or G codes or other machine languages derived from software (such as Smartcam or Mastercam) to manipulate and instruct the machine in an exact manner.

Commissioning. Commissioning is the act of sequentially and systemically starting up and testing a machine or machines or systems to ensure that they function as specified and that it can meet the needs of production. All components are tested and evaluated as to fit, function, vibration, alignment, integration, control, ergonomics, input flow, installation, and safety.

Commissioning, qualifying, and verifying (CQV). Commissioning, qualifying, and verifying (CQV), or, more simply, "get up and stay up" (GUSU), is basically a management structure to ensure that a new or modified packaging line and all its machinery and systems are tested properly, the operators and maintenance personnel are adequately trained, and the results of the testing and training are verified or observed in the production runs without special assistance of any kind. Some companies restrict CQV to the machinery and systems and keep people and training separate.

Commissioning is the act of sequentially and systematically starting up and testing a machine or machines or systems to ensure that they function as specified. All components are tested and evaluated as to fit, function, vibration, alignment, integration, control, ergonomics, input, flow, installation, and safety.

Qualifying is the act of setting up and administering a training program that ensures the people who will be interfacing with the packaging line are given a thorough overview of the packaging line and a detailed program on what they need to know to complete their tasks without hesitation or guessing. All training must address the following questions:

When should training be done?
What training should be done?
How should we do training?
Where should we do training?
How much is enough training?
Manuals and other self-help tools?
Performance reviews and continuing improvement?
Company standards and policy?

Verifying is the act of being able to determine without hesitation that the machine testing and training of personnel is not only thorough but also effective in the day to day production of the packaging line without senior staff, consultants, and/or machinery service personnel being involved in any manner.

Training of all line and maintenance personnel is critical to the performance of any packaging line. Training must be thorough, to the extent that all personnel working on the line are knowledgeable in the overall operation and be the technical expert in their specific task(s). Management people must be experts in the overall operations of the line, and knowledgeable about the tasks and performance of each person working on the line.

Complete kit (CK). The complete kit (CK) is a managerial method or concept. It suggests that no work should start until all the items required for the completion of the job are available. These items such as components, tools, drawings information, and samples constitute the kit.

Computer-integrated manufacturing (CIM). The integration of the total manufacturing enterprise through the use of integrated systems and data communications coupled with new managerial philosophies that improve organizational and personnel efficiency. In other words, the business enterprise is dependent on human knowledge and information flow in order to operate efficiently. Companies that have implemented CIM successfully tend to say that they get the right information to the right people or devices in the right places at the right times to make the right decisions (SME *Blue Book* series, 1990.) CIM uses such a broad spectrum of technologies that it is best to think of CIM as a goal or strategy or philosophy as outlined by the company's needs and direction.

Computerized maintenance management system (CMMS). A computerized maintenance program is sometimes called a *computerized maintenance management system* (CMMS), which can form the central part of a TMM program. There are many computerized maintenance programs in use today. One or several can be an effective tool for your operation. Always remember that the software has to fit your needs, not the other way around. The main function of the computerized system is to make life easier and faster for maintenance to do their job.

Components. Components can be either input items or product items. Example of components are labels, caps, containers, seals, cartons, cases, and the product such as liquids, pastes, pills, hardware, and powders.

Confidential-disclosure agreement (CDA). A legal agreement between two parties that establishes terms and conditions for protecting the security of specified information.

Consistency. Resistance of a fluid to deformation. For sample (Newtonian) fluids, the consistency is identical with viscosity; for complex (non-Newtonian) fluids, identical with apparent viscosity.

Contingency plan. Alternative strategies to accomplish a project's objectives.

Contract award. The award of the contract to one prospective supplier. Acceptance is usually followed with a purchase order or a signed contract.

Contract manufacturer. An outside company contracted to manufacture or package a product.

Control. The authority or ability to regulate, direct, or dominate a situation or series of events. For any packaging line, whether manual, semiautomatic, or automatic, control is critical.

Cop. Clean-out-of-place.

Criteria. A document that provides objectives, guidelines, procedures, and standards that are to be used to execute the development, design, and/or construction portions of a project.

Critical activity. Any activity on a critical path.

Critical path. The longest path or chain of activities in the project that determines the shortest period of time in which the project can be completed.

Critical-path method (CPM). A scheduling technique using arrow, precedence, or PERT (program review and evaluation technique) diagrams to determine the length of a project and to identify the activities and constraints that are on the critical path.

Culture. Observable work habits and priorities of an organization. It explains how the organization really works.

Customer. The individuals, companies, or their representatives that receive a company's product, use a company's services, or purchase a company's product for resale (trade).

Cycle rate. The cycle rate is the number of machine cycles per minute. We denote the number of cycles as N, so the cycle rate is dN/dt. It is not necessarily equal to the run speed of the machine. Over an 8-h period, the cycle rate of any machine in the packaging line is always higher than any set run speed. Most OEMs specify their machine speeds based on cycle rates not output and therefore the OEM's speeds are the theoretical or design speeds that are possible.

Data-exchange format (DXF). Electronic format used to exchange CAD files from one software brand to another.

Deliverable. A product that satisfies one or more objectives and must be delivered to meet contractual obligations.

Design. The application of process engineering, machine design engineering, or power and controls engineering and the production of detailed engineering documentation, such as drawings and specifications, that explain how the technology defined in the design basis will be embodied in equipment and how the equipment will be constructed, assembled, and installed.

Design basis. A document defining the project scope for designers and communicating how the engineering packages must be structured to fit construction requirements.

Design for manufacture (DFM). A general approach to designing products that can be more effectively manufactured. Often used in conjunction with databases. Includes such concepts as design for assembly, design for serviceability, or design for test.

Design speed. The theoretical capacity obtained in a perfect operating environment. All adverse operating conditions are neglected, and all machinery is assumed to be in optimum operating condition. The design speed usually does not take into consideration the natural handling difficulties or stability of the inputs (K value), environmental concerns, or training. This is usually the machinery builders advertised maximum functioning cycle speed for a given range of optimum inputs under ideal conditions. The concept of line speeds is not well understood in the packaging industry; thus there is a lot of confusion and disappointment over the actual results achieved. The following is a summary of the speed hierarchy from highest to lowest:

1. *Design speed* (usually the OEM's maximum cycle rate)
2. *Capacity* (highest sustainable cycle rate—about 80–90% of design speed)
3. *Achievable run speed* (target set speed or required steady-state condition)
4. *Actual set run speed* (actual packaging line set speed or dial speed)

5. *Output rate* (what goes to the warehouse or shipping per time period)

Note that the actual run speed will be set and its result will equal the output rate plus losses from rework, wastage, and stoppages.

Detergent. Surface-active material or combination of surfactants designed for removal of unwanted contamination from the surface of an article.

Dew Point. The temperature at which air or other gases become saturated with vapor, causing the vapor to deposit as a liquid. The temperature at which 100% RH is reached.

Direct labor. The labor expended in directly adding value to the package. For simplicity, it could be looked at as "touch" labor since direct-labor employees usually physically touch the product or inputs.

Disturbance. Any anomaly in the flow of product, inputs, or packages in the packaging process that may or may not cause production to cease or slow down.

Disturbance frequency. The number of disturbances for a specific output (quantitative amount) as determined by the set or actual run speed of the line.

Disturbance-frequency analysis (DFA). A production process tool that tracks and eliminates disturbances and their source(s). It is an excellent troubleshooting method for the analysis of small or large sections of the packaging process. DFA states that when the causes of disturbances are eliminated the downtime disappears, so why worry about time.

Disturbance-frequency period. The minimum number of outputs allowed for one disturbance. The minimum number of outputs is determined mathematically from the actual set speed of the packaging process. As the actual set speed increases, the minimum number of outputs required increases exponentially.

Downtime. The amount of time a machine or system is not functioning due to stoppages in a given shift or time period. Downtime should not include idle time or time the machine or system is waiting for inputs. Therefore downtime is made up stoppages, and company policies.

Downtime analysis (DA). A production process tool that tracks the amount of time a given machine or system ceases production. Since it is time-based and monitors symptoms, not causes, its effectiveness is limited to specific applications. Historically, downtime analysis has been applied in an ineffective manner.

Drawings and specifications. Engineering details for projects, packages, equipment, systems, and facilities.

Drop test. A test for measuring the properties of a container by subjecting the packaged product to a free fall from predetermined heights onto a surface with prescribed characteristics.

Duration. A unit measure of time for the time interval of a given stoppage.

DXF. See **Data-exchange format.**

Efficiency (*n*). A fundamental engineering term that is broadly defined as the ratio of benefits/penalties or, for packaging, the ratio of output/input. In keeping with this engineering definition, we define it as the average ratio of packaged output multiplied by the number of components over the sum of input components. Unfortunately, in the packaging industry, efficiency has become a generic catch-all term whose definition varies depending on the political needs of the organization, not its real needs.

Elemendorf test. A test for measuring the tearing resistance of paper, paperboard, tape, and other sheet materials.

Element utilization. The fraction of time a given machine or element is actually producing output at a set run speed divided by the total time available for production. Note that changeover, cleanup, and prep work are not included. This definition usually relates to machinery and/or components.

Elements. Any machines, equipment, conveyors, or mechanical components by which the product and/or inputs are manipulated, assembled, transferred, collated, or brought into contact with each other. Examples of elements are fillers, cappers (with sortation), labelers, unscramblers, case packers soap presses, calenders, meat grinders, wrappers, packers, banders, and conveyors. For the purpose of clarification, all elements are considered or designated as machines. For example, we assume that all types of conveyors, collators, or buffers are elements.

Emergency maintenance. Emergency maintenance occurs when a technical person is called on to perform a repair at any time of the day without being forewarned. With most other types of maintenance, emergency maintenance should be the exception not the rule as in breakdown maintenance. Breakdown maintenance and emergency maintenance are basically the same, when the predominate method of maintenance is breakdown maintenance. Breakdown maintenance is doing no maintenance other than lubrication between breakdowns. Repairs and adjustments are done as a result of reduced or no production occurring.

Equipment. A capital-cost-estimating category for hardware used in manufacturing, including manufacturing support.

Equipment acceptance. Criteria for approving equipment for use.

Estimate. A statement of the probable capital appropriation scope and cost associated with changes needed to achieve a specified goal.

Excess capacity (EC). A managerial concept or approach advocating that a system's or process's capacity or limit should exceed the average demand (throughput or output). This position has been developed by Dr. Abraham Grosfeld-Nir from the University of Waterloo. The author agrees strongly with this concept. This concept is in contrast to the "lean and mean" approach where any attempt to have a capacity that exceeds the demand is viewed as wasteful ("fat") and any system or process having a capacity equal to the demand is perceived as ideal. Added to this is the traditional accountants' advise to cut extra capacity, in particular the work force, to force a balance between capacity and demand. In other words, the packaging process capacity is equal to the customer's demand and no more. It is very likely that, at times, EC will recommend to buy more capacity under given conditions, while an accountant would recommend a cut. The need for excess capacity arises from the presence of randomness in demand, processing time, quality, etc. (In a deterministic world, EC would, indeed, be wasteful.) Benefits of operating with EC are small WIP (work in process), quick response time, and increased quality. In observing JIT multistage systems it is clear that each stage is utilized only to a fraction of its capacity as it relates to the speed potential. Each stage is idle some of the time as a result of stoppages. Thus stages of a properly designed JIT system operate with excess capacity. Experience indicates that typically in the automotive industry, stages are utilized 60–80% of their capacity. To use JIT in packaging, it can be argued (because of the speeds involved) that "speed utilization" or the set run speed of 50–70% of the capacity speed are proper for JIT to be effective.

Extrusion. The process of forcing a material in plastic condition through an orifice.

Factory end. Bottom or can manufacturer's end.

Failure analysis. A formal analysis that assumes failure and attempts to define the paths leading to failure and what should be done to prevent it.

Failure rate. The ratio of the total number of failures to the cumulative operating time for a stated period of time. We denote this by the symbol λ and its units are failures per unit of time.

Feasibility studies. Experimental studies on a process or equipment to access what results are achievable and/or what will be required to produce a desired result.

Flow sheet (FS). A schematic presentation of a process.

Flowchart. A graphic representation of a work process in which each step in the process proceeds from start to finish.

Fluidity. Reciprocal of viscosity.

Freight on board (FOB). The term used to signify that the seller is required to bear all costs required to place the goods aboard equipment of the transporting carrier. The stated FOB point is usually the location where title to the goods passes to the buyer. The buyer is liable for all charges and risks after passing of title.

Frequency. The number of occurrences in a given period of time.

Funding. The process that leads to a specific amount of money and resources being set aside for or committed to a specific object.

Gelometer. Instrument used to measure the time required for a fluid to gel. Also, instrument used to determine the firmness of a gel.

General terms and conditions. A package of standard requirements included with all purchase orders.

GMP. See **Good manufacturing processes.**

Good manufacturing practices (GMP). A document that describes agreed-to best or optimal procedures for manufacturing.

Gram (g). Metric unit of weight equal to 0.035 oz. One kilogram is equivalent to 1000 g, and one pound equals 453.6 gs.

Headspace, gross. The vertical distance between the level of the product (generally the liquid surface) and the inside surface of the lid in an upright rigid container (the top of the double seam of a can or the top edge of a glass jar).

Headspace, net. The vertical distance between the level of the product (generally the liquid surface) and the inside surface of the lid in an upright, rigid container having a double seam, such as a can.

Hermetically sealed container. A container designed and intended to be secure against the entry of microorganisms and to maintain the commercial sterility of its contents after processing.

High-performance products. Products with clearly superior attributes such as taste, styling, speed, smell, comfort, and effectiveness with results in the packager receiving a premium price and/or dominate market share more than the perceived normal product.

Hydrometer densimeter. Device used for the measurement of specific gravity or density.

IFT. Institute of Food Technologies. The professional society of food scientists and technologists in the United States.

Impact strength. The ability of a material to withstand mechanical shock.

Indirect labor. Any labor, including supervision and management, that is not direct labor or directly connected to the production or manufacturing process. Overhead activities such as material handling, stockroom, inspection, all engineering functions, maintenance, supervision, cost accounting, and personnel are usually included.

Input (X,x). Input for a given packaging line is the specific item or part required for the package assembly operation or packaging process to form part of the complete package. Examples of input items are bottles, cartons, caps, labels, cases, pallets, tubes, and so on. Inputs are not product entities, but product entities can be classed as inputs. In general, inputs are discard items that are trashed or recycled after product use. Since we use the symbol X for systems input, we will use x_i for the ith machine as the input for a machine. Note that several types of inputs may be required by a machine (and certainly by a line), but in general, only one output is produced. Minimizing input losses reduces cost, increases the performance index, and reduces the waste headed to landfills. From the economic point of view, reducing input item losses, as well as the number of different types of inputs, makes business sense. Minimizing the types of inputs maximizes performance and quality.

Input rate. The input rate for a given packaging line is the amount of a specific item or part (required to form part of the complete package) processed or consumed in a given amount of time.

Integration. *Integration* as it relates to the packaging process is defined as the mechanical, pneumatic, hydraulic, and/or electrical method of physically (mechanically or electronically) connecting machinery and input handling systems to ensure a smooth harmonized throughput operation. It should not be confused with automation and control. Usually control or automation can not solve or correct poor or faulty integration.

Integration principles for input and product handling

1. Once you have it, don't let go. Having it means complete physical or electronic engagement so that the mechanism motion and the input motion are identical or within a tolerance range that is smaller than the functional requirements.
2. Eliminate or minimize input and package manipulations or change of direction, velocity, and/or inertia.
3. Always interface using complete integration or handshaking pass off within a defined boundary, not at a point. The interface position tolerance must be smaller than the operational functional window.
4. If manipulations are required because of overall design constraints, always match motions (intermittent to intermittent, continuous to continuous). Incompatible motions (continuous to intermittent and intermittent to continuous) always yield the highest unreliability.
5. For cycles >60 min^{-1}, continuous is superior to intermittent.

Keep it rugged and simple, meaning the fewer moving parts, the better.

Interfunctional communication. Communications and understandings between manufacturing and other functional areas in a business.

Islands of automation. Pieces of equipment or systems that are not integrated with the packaging line. They can be considered independent and not controlled by the packaging process directly. The output from each island of automation does not affect the other directly.

ISO 9000 Series Protocol. An international set of documents written by members of a worldwide delegation known as the ISO (International Standards Organization)/ Technical Committee 176. Its primary purpose is to harmonize the large number of national and international standards adopted by many countries. This series is also intended to be driven by market and customer needs. The ISO series consists of five main documents:

- Three core quality system documents that are models of quality assurance: ISO 9001, ISO 9002, and ISO 9003.
- Two supporting guideline documents: ISO 9000 and ISO 9004.

Following is a brief outline of each document.

ISO 9000 (ANSI/ASQC Q90-1987): *Quality Management and Quality Assurance Standards: Guidelines for Selection and Use.*

ISO 9001 (ANSI/ASQC Q91-1987): *Quality Systems—Model for Quality Assurance in Design/Development, Production, Installation and Servicing.* Applicable to contractual arrangements, requiring that design effort and the product requirements are stated principally in performance terms or they need to be established. Confidence in product conformance can be attained by adequate demonstration of certain supplier's capabilities in design, development, production, installation, and servicing.

ISO 9002 (ANSI/ASQC Q92-1987): *Quality Systems—Model for Quality Assurance in Production and Installation.* The specific requirements for the product are stated in terms of an established design or specification. Confidence in product conformance can be attained by adequate demonstration of a certain supplier's capabilities in production and installation.

ISO 9003 (ANSI/ASQC Q93-1987): *Quality Systems—Model for Quality Assurance in Final Inspection and Test.* The conformance of the product to specified requirements can be shown with adequate confidence, provided certain supplier's capabilities for inspection and test conducted on the product supplied can be satisfactorily demonstrated on completion.

ISO 9004 (ANSI/ASQC Q94-1987): *Quality Management and Quality System Elements Guidelines.*

Justification. For most industries a justification is a written request for action that consists of

1. Acquisition of valid data
2. Presentation of the problem, need and data
3. A logical course of action with options and alternatives
4. Cost profiles
5. Benefits of undertaking a specific course of action

Just-in-time (JIT). First used by Toyota in Japan as the "Kanban system," it has been successful in reducing inventory while maintaining high throughput and increased quality. Kanbans are cards authorizing production or shipment of material. Mondem in 1981 defined JIT as a production system to produce a kind of units needed, at the time needed and in the quantities needed. Schonberger in 1982 defined JIT as goods produced just in time to be sold and only purchase materials just in time to be transformed into the product. Sohal in 1988 defined JIT as more of a philosophy than a series of techniques, the basic tenet of which is to minimize coste by restricting commitment to expenditure until the last possible moment. Groenvelt in 1991 defined JIT as a management philosophy that fosters change and improvement through inventory reduction. To have the ideal situation of exactly the necessary amount of material available where it is needed and when it is needed. From the book *Benchmarking Global Manufacturing, 1992,* just-in-time is both a philosophy of eliminating waste and a toolset for pacing and controlling production and vendor deliveries on time, with short notice and with little or no inventory. According to JIT, multistage systems as found in packaging lines should be pull (produce only in response to demand) and not push (produce as long as there is raw material to be processed). In simple terms, JIT means that requardless of what disasters happen in my packaging process, you have the excess capacity to guarantee the highest probability of attaining the exact delivery window required by the customer with quality packages.

kg. Kilogram or 1000 g, equivalent to 2.2046 lb.

Kilogram (kg). A unit weight in the metric system equivalent to 1000 g or 2.2046 lb.

Kilopascal (kPa). Unit of pressure. One kilopascal equals

1000 pascals (Pa); 1 atmosphere (atm) equals 1.01325 × 105 pascals. See **Pascal.**

Kraft. A term derived from a German word meaning strength, applied to pulp, paper, or paperboard produced from virgin wood fibers by the sulfate process.

Label. Any display of written, printed, or graphic matter on the container of any consumer commodity, affixed to any package containing a consumer commodity.

Lag. Specified time increment between the start or completion of an activity and the start or completion of a successor activity.

Latent heat. The quantity of heat, measured in Btus or calories, necessary to change the physical state of a substance without changing its temperature, such as in distillation. A definite quantity of heat, the latent heat, must be removed from water at 0°C (32°F) to change it to ice at 0°C.

Lid. Can end applied to open end of can in a cannery. Also known as top, cap, or packer's end.

Liner. Generally, any liner material that separates a product within a container from the basic walls of the container.

Logic. The description of how the activities in a schedule are related to one another.

Logic diagram. Drawings using logic symbols that tell how a piece of equipment or system is to operate.

Lost-time accident frequency. One of the most widely used accident statistics. LTA frequency is a measure of the number of disabling injuries per year, and is calculated as follows:

$$\text{Number of lost days} \times 200{,}000 \\ + \text{ the number of hours worked}$$

Lug. A type of thread configuration, usually thread segments disposed equidistantly around a bottle neck (finish). The matching closure has matching portions that engage each of the thread segments.

Lump sum. A method of paying for equipment fabrication and construction work in which the buyer pays the vendor a single lump sum for all the work involved.

Machine. A *machine* can be defined as a system of components arranged to transmit motion and energy in a predetermined fashion. Another definition is that a machine typically contains mechanisms designed to provide significant forces and transmit significant power.

Machine downtime. Machine (or element) downtime is the total amount of time a given machine in the packaging line stopped or ceased production during the run period.

Maintainability or mean time to repair (MTTR). *Maintainability* is the average time required to repair a device or system. This includes preparation time, active maintenance time, and delay time associated with the repair. It is quantified as the mean time to repair (MTTR).

Maintenance. The physical act of ensuring that all machinery and systems are in operable condition at all times, especially during scheduled periods of demand. Maintenance can also be defined as the ability to keep all mechanisms and machines required for optimum production available to not only produce but also maintain a steady flow of quality packages without interruption. Types of maintenance programs to maximize operational readiness and sustainability: (*1*) breakdown, (*2*) preventive, and (*3*) predictive.

Manufacturing costs. Raw and packing materials, manufacturing expenses, processing costs, and variations.

Manufacturing lead time (MLT). The cumulative time from the beginning of the production cycle until an item is finally finished. Time spent in inventory as work in process, setup times, move times, inspection, and order preparation time are included.

Manufacturing overhead costs (MOC). Those costs that are allocated to unit product costs, including the cost of indirect labor as well as indirect purchased services and supplies but excluding unallocated period costs such as sales and marketing and R&D.

Master production schedule (MPS). The main or overriding schedule of the number and timing of all end items to be produced in a manufacturing plant over a specific planning horizon. An important input to the MRP computer program.

Material requirements planning (MRP). MRP is more of a computerized information system than a managerial philosophy that determines how much of each material, any inventory item with a unique part number, should be purchased or produced in each future time period to support the master production schedule (MPS). It is designed to contain information required for efficient decisionmaking. MRP is mainly used by large industrial organizations but as more improved computer software and hardware emerge, smaller organizations will find MRP advantageous in their operations. Input to MRP is information from the "master production schedule" (MPS) and the "bill of materials" (BOM). The MPS determines starting times for future jobs, while the BOM is a detailed description of all inputs (material and work) required for each job. MRP also monitors all in-stock inventory and outstanding purchase orders. On the basis of information about jobs in progress and waiting jobs, MRP regulates the purchase of material from internal and external suppliers. It also provides information about jobs behind schedule, for potential expediting. Methods such as the "economic order quantity" (EOQ) and "reorder point" (ROP) control material requirements, without taking into account information about job scheduling from the MPS. In contrast, MRP recognizes the handling material is more efficient when the information from the MPS is incorporated. MRP also aggregates requirements from different sources and sets timing for arrivals considering overall system performance, rather than aiming at local satisfaction. In other words, unlike EOQ and ROP, MRP takes advantage of dependent demand. MRP assumes all operations to be deterministic, which is unrealistic. One consequence of this assumption is the frequent need for expediting jobs. MRP requires dealing with an enormous amount of information that needs continuous update. It turns out that when details are required, MRP is unreliable. Maybe this will be taken care of as technology and advanced software are developed.

Mean. The average value of a number of observed data.

Mean time between failures (MTBF). The cumulative operating time divided by the number of failures. This is the reciprocal of the failure rate.

Mean time to failure (MTTF). The average life of nonrepairable items. This does not apply to most machinery in the packaging industry, but it does apply to certain components (eg, seals, bearings).

Mean time to repair (MTTR). *Mean time to repair* (or *maintainability*) is the average time required to repair a device or system. This includes preparation time, active maintenance time, and delay time associated with the repair. It is quantified as the mean time to repair (MTTR).

Mechanism. A system of components arranged to transmit motion in a predetermined fashion. Another definition is that a mechanism is a device that transform motion to some desirable pattern and typically develops very low forces and transmits little power.

Melting. The change from the solid to the liquid state. Also the softening of harder compounds.

Methodology. The sequence or manner in which events are made to occur or a logical thinking process to achieve an end result.

Milestone. A key network event that is of major significance in achieving the program, project, or contract objectives.

Mission. Statement of purpose or reason for the existence of the organization. It answers the question who we are and why we exist. It expresses the distinctive competence or unique contribution of the organization.

mm. Millimeter. Equivalent to 0.001 m, and to 0.0394 in.

Model. A model consists of one or more mathematical equations that describe some idealized form of a system. The idealized form is rendered from the real form by making one or more simplifying assumptions. It is an attempt to explain or predict a portion of reality using mathematics.

Molecular weight. Sum of the atomic weights of all the atoms in a molecule.

Molecule. The smallest theoretical quantity of a material that retains the properties exhibited by the material.

Monitoring. Following the progress of the project. This phase follows the preparation of the CPM plan and schedule.

Mylar. A synthetic polyester fiber or film.

NFPA. National Fire Protection Association.

Neck. The part of a container where the bottle cross section decreases to form the finish.

Occupational Safety and Health Act. Federal organization responsible for health and safety on construction sites and in plants.

Occurrence. Any anomaly or action in the packaging process that actually stops that process. A *stoppage* and an *occurrence* are the same.

Optimized production technology (OPT). A philosophical approach that maximizes throughput through the modification or elimination of bottlenecks. Claiming that (only) throughput or output translates into sales, OPT aims at maximizing throughput. Improvements that do not render increased output are viewed as negligible. OPT concentrates on "bottlenecks" or critical resources. A bottleneck is a resource that is utilized to its full capacity. This aged manufacturing concept is still quite common in many packaging plants. Typically OPT can lead to extremely large inventory levels, which is contrary to JIT. OPT can give substantial gains in short time periods of 3 to 6 months.

OSHA. See **Occupational Safety and Health Administration.**

OSHA No. 200 Log and Summary. The OSHA recordkeeping form used to list injuries and illnesses and to note the extent of each case.

Output (Y,y). Output for a given packaging line and product is the exact quantity of quality packages produced in a package run cycle as required by the customer and shipped to the customer. It is denoted by the symbol Y. Similarly, we define the output for a machine or output rate as the number of items (of acceptable quality) leaving a machine in a given amount of time. We denote this by y_i for the ith machine. The word *throughput* has the same definition as output, but this *Encyclopedia* uses the word output, which is the more correct term.

Output rate. The output rate for a machine or system is the number of items (of acceptable assembled quality) leaving a machine or system in a given amount of time. We denote this by y_i for the ith machine. It is always less than the machine cycle rate and/or the actual set run speed, due to the effects of wastage, rework, and stoppages.

Package. A quality assembly of two or more quality input items and product items. In some cases, the package becomes the product [eg, consider a synthetic fire log—the consumer burns the package (both the product and the input: log and wrapper)]. A good marketing view of a package is one presented by Robert E. Smith, senior vice president of research and development at Nabisco Foods Group. He says that "The package is also a facilitator of new manufacturing systems, adds convenience for consumers, and is beneficial to the environment. The package must play all of these roles to perfection." Therefore, a package is a product that is wrapped in a protective shroud; identified through a label; informs the consumer of its use; and uses color, shape, and name to establish consumer recognition. This *Encyclopedia* is not concerned directly with this marketing definition.

Packaging Changeout (t_{pc}). Packaging change out is the sum of the changeover time, cleanup time, and prep work time. The package change out is the total interval between finishing one run of a given package and starting another run with a different package.

Package-changeout utilization (U_c). The fraction of the package run cycle time the packaging line has available for producing output at a set run speed after changeover, cleanup, and prep time have been completed.

Package run cycle (PRC). The duration of specific events needed to produce a given amount of quality packages. It consists of the total end of the first production run to end of next production run period, which involves change out and the entire run or producing period.

Packaging line. An assemblage of specialty-function machinery or systems and/or manual workstations from depalletizing to palletizing integrated together to carry out a process in which a given product is combined, inspected, and transported with inputs or media. The inputs themselves protect, control, and identify the product: (*1*) bottling lines for liquid products such as beverages, (*2*) canning lines for products such as precooked foods, or (*3*) box-packaging lines for materials such as powered detergents.

Packaging process. The combined execution of specialty-function machinery or systems and/or manual workstations in order to carry out a process in which a given product is combined an/or assembled with inputs or media. Basically the *packaging process* has the same definition as a *packaging line* but is more encompassing since it makes people part of packaging line. Some companies, by nature of their operations, may include the making as part of the packaging process.

Paneling. Distortion (sidewall collapses) of a container caused by development of a reduced pressure (too high to vacuum) inside the container.

Paper, water-resistant. Paper that is treated by the addition of materials to provide a degree of resistance to damage or deterioration by water in liquid form.

Paper, wet-strength. Paper that has been treated with chemical additives to aid in the retention of bursting, tearing, or rupturing resistance when wet.

Parallel elements. Parallel elements (ie, machines) perform the same type of operation in a system. The main line is split into two to feed each machine with consistent product and/or inputs. The elements need not be identical, and the speeds of each can be a factor of the total main-line output. Parallel machines provide system redundancy; the failure of any one of them will not stop the line, but only reduce output by a factor. Usually, the combined capacity, of the two parallel elements should equal over 125% of the total output plus wastage, rework, and stoppage losses.

Pasteurization. A heat treatment of food usually below 212°F, intended to destroy all organisms dangerous to health, or a heat treatment that destroys some but not all microorganisms that cause food spoilage or that interfere with a desirable fermentation.

Percent complete. A comparison of the actual status to the current projection. The percent complete of an activity in a program is determined by inspection of quantities placed as effort hours expended and compared with quantities planned or effort hours planned.

Performance. *Performance* can be defined as the level or effectiveness of carrying out an action or execution of a sequence of events according to a prescribed functional description. A better definition for performance for the packaging industry can be as stated below.

> Performance in the packaging industry is a widely used term such as efficiency that means different things to many people. Therefore the word *performance* is relative and qualitative. The only understanding that may be common to all is that it is a reflection of productivity, output, or effectiveness of time. Unless it is rigorously defined and understood, the word performances has limited value to decisionmakers, other than projecting a sense of being or desire. I would define performance as a measure of profitability based on the ability to produce the needed quantity of quality packages in the time required to fulfill customer needs at the lowest per unit cost over a sustained long period of time (>1 year). To many people, performance is the best bang for the buck based on up-front costs or capital costs only, not on the best value, which is based on capital and ongoing operational costs. If one buys a system based on up-front costs or lowest costs to get in without working out operational costs over a 1-, 3-, or 5-year period, then their anticipated profits (based on marketing targets) will rarely materialize. Too many people are hooked on this false sense of performance that will only contribute to the long-term uncompetitiveness of the company. Short-term or no planning leads to long-term disasters. Anyone can demonstrate excellent performance hour by hour or even day by day, but performance can truly be judged only quarter by quarter and year by year, which translates into consistent steady-state production under complete control (manual or automatic) at all times and under all conditions.

Performance Index (PI). A method of evaluating, benchmarking, tracking, and verifying a company's packaging processes. It gets rid of the old confusing notion of line efficiency and establishes a new framework that establishes a level playing field, especially among multiple corporate plants running the same or similar products. The performance index can also be used to evaluate dissimilar types of packaging lines that are internal or external to the organization. The effectiveness of a packaging operation that relates directly to the profit of the product is the objective of the performance index. PI also utilizes the most effective ideas an/or techniques used in OPT, JIT, TQM, TPM, and MRP and applies them to the packaging process. As such, the PI can be considered an operation's tool, but only a tool in the benchmarking sense, because it is the tip of the pyramid from which everything else expands out to explain the PI value. As a benchmark, it can evaluate functional and best practices of a company against itself or competitors. The performance index is a mathematical model that projects the overall rating of a packaging line. Technically PI is a composite measurement of the productivity of a given packaging line. In mathematical terms, the performance index is the efficiency multiplied by the utilization multiplied by the capability multiplied by the speed factor. Equal weighting is usually given to efficiency and utilization. Under ideal conditions, the PI value would be 1. A typical high-PI packaging line, such as a soft-drink line, should have a value greater than 0.8. Other lines will yield PIs ranging from 0.10 to 0.70. Locating and eliminating the source of low PI will improve productivity. Common symptoms of low PI are a low speed factor or high wastage on the line. Solving the problem is often easier than finding the problem. Note that the index excludes the number of people running the packaging line. When maximizing PI, labor will move to its optimal level, which is minimal labor input. Management must avoid labor replacement and consider instead labor optimization, given the environment and the package. For a given packaging line and package, it is possible to determine the profit to be gained for each point gain in the PI. With this information, one can justify changes very quickly.

Permeability. The passage or diffusion of a gas, vapor, liquid, or solid through a barrier without physically or chemically affecting it.

PERT. Program evaluation and review technique.

pH. The effective acidity or alkalinity of a solution; not to be confused with the total acidity or alkalinity. The pH scale is

Acid solutions	Neutral	Alkaline solutions
0 1 2 3 4 5 6	7	8 9 10 11 12 13 14

where pH 7 is the neutral point (pure water). Decreasing values below 7 indicate increasing activity, while increasing values above 7 indicate increasing acidity, while increasing values above 7 to indicate alkalinity. One pH unit corresponds to a tenfold difference in acidity or alkalinity; hence pH 4 is 10 times as acid as pH 5 and pH 3 is 10 times as acid as pH 4 and so forth. The same relationship holds on the alkaline side of neutrality, where pH 9 is 10 times as alkaline as pH 8, and so on. Most meat and fish products have pH values between 6 and 7, vegetables have pH values between 5 and 7, and fruits have pH values between 3 and 5.

Planned maintenance. A type of maintenance in which a company has a given period or periods of production idle times each month in which maintenance activities can be undertaken. In periods of heavy production demands, this type of maintenance breaks down into simple breakdown maintenance.

Planning. The establishment of the project activities and events, their logical relations and interrelations to each other, and the sequence in which they are to be accomplished.

PO. Acron. The short form for a Purchase Order.

Predictive maintenance. A type of maintenance that utilizes an array of sensors and/or monitoring equipment to determine the status and condition of critical wear components in machinery and systems so that proactive procedures or replacement can be undertaken at a convenient time period before the failure of the component could cause any unscheduled downtime.

Premium time. An allowance to cover the premium portion of overtime pay, that is, the cost differential between straight time and overtime work.

Prep work time or (period). The time required to bring inputs to the line, load and stage ready for production. Some companies may combine change over with clean up and prep work.

Present value. The cash value today of the difference between a project's investment and the related return over the project life.

Preventive maintenance. A type of maintenance determined by historical data and life-cycle testing to determine the optimum time of part life and therefore by extension lead to a program of part changeout prior to the anticipated part failure. Carried to its extreme, it can be a costly venture to a company, and therefore most preventive programs are tempered by budgets and other constraints.

Primary spoilage. See also **Secondary spoilage.** That spoilage due to bacterial or chemical action of product packed within the can.

Process. The succession of actions (filling, capping, labeling, palletizing, etc) undertaken to make a package. Some companies may include the making, if it is essentially part of or integral to the packaging operation.

Procurement. The acquisition (and directly related matters) of equipment, material, and nonpersonnel services (including construction) by such means as purchasing, renting, leasing (including real property), or contracting.

Procurement lead time. The cumulative time from the beginning of the procurement order cycle (order commitment) until the procured item is delivered. It includes vendor lead time, transportation, receiving, and inspection time.

Product. Those items used or consumed by the customer. They are usually items made or modified by human industry. Take shampoo as an example; the consumer uses the shampoo and throws away its container, cap, and labels. The product is the shampoo and the input items are the container, cap, and labels. (The term *product,* as a collective noun, can have a plural sense.)

Product amenities. The extra product features that enhance the basic product and make it easier to use or more enjoyable. For example, a finger pump spray on a bottle versus a pull-and-squeeze cap.

Product run cycle (PRC). The duration of time needed to produce a given amount of quality packages based on the designed or historical output. It is the same as the package run cycle. The word *package* is more applicable than product in most industries and is used in this *Encyclopedia.*

Product specifications. The written description of the products to be manufactured.

Product stability (*K* value). The ability of the product to be handled in a stable and consistent manner. A beer bottle, for example, has all the features necessary to make it a stable and consistent product:

- Cylindrical parallel shape
- Relatively low center of gravity
- Heavy, smooth base for stability and low friction
- Relatively hard to break
- A shape that remains stable throughout the packaging cycle

Generally, a stable product is a product that gives a packaging line the least amount of handling problems. This stability can be defined as a value *K*. A more rigorous definition of *K* value is left for future development. As a packaging line increases its level of automation, the *K* value becomes very critical.

Product support. Those activities that support the customer in the use of a product, such as customer education, information about related products, or services and information hotlines.

Productivity. The level of ability or effectiveness in marketing, manufacturing, distributing, and servicing a package.

Project. The overall work being planned. It will have one specific start point and one finish point in time.

Project manager. The person on a project team responsible for the engineering process and appropriation management.

Psychrometer. An instrument for measuring the humidity (water-vapor) content of air by means of two thermometers, one dry and one wet.

QA. Acronym for quality assurance.

QCP (quick-change process). A new common phase used to describe the techniques used to design and operate a packaging line that is always prepared to change from one product to another within a minute notice. This encompasses quick changeover (QCO), clean-in-place (CIP), eradication of all possible adjustments, matchmark position settings for remaining adjustments, flexible crewing, tooling readiness, and a streamlined results oriented management. A similar phrase is quick changeout.

Qualifying. The act of setting up and administering a training program that ensures that the people who will be interfacing with the packaging line are given a thorough overview of the packaging line and a detailed program on what they need to know to complete their tasks without hesitation or guessing. All training must address the following questions:

Whom should be trained?
When should training be done?
What training should be done?
How should we do training?
Where should we do training?
How much is enough training?
Manuals and other self-help tools?
Performance reviews and continuing improvement?
Company standards and policy?

Quality circles. Teams of employees used to diagnose and solve quality problems relating to fulfilling the needs of the customer. It also includes the use of the work team concept for solving other problems related to productivity improvement, safety, and so on.

Quality control. A system for assuring that commercial products meet certain standards of identity, fill of container, and quality sanitation and adequate plant procedures.

Quality function deployment (QFD). A set of techniques for determining and communicating customer needs and translating them into product and service design specifications and manufacturing methods.

Quality package. A package that meets all design specifications and is manufactured in compliance is ISO 9001 or ISO 9002. Take a bottle of shampoo as an example of a quality package:

- The content must comply with internal and government weight and/or volume regulations.
- The print documentation must be clear, readable, accurate, and acceptable to the end customer and comply with government regulations at point of manufacture to point of end use.
- The container must be clearly coded and identified using industry-standard codes such as UPC and comply with customer and government regulations at point of manufacture to point of end use.
- The seal on the package must be seated correctly to a specified torque and/or form a protective seal against leakage, spoilage, contamination, and/or tampering.

In short, a quality package is the end product envisioned by the marketing group that will fulfill all the design criteria and packaging standards, either internal, governmental, or required by the customer.

Quick-change process (QCP). A new common phase used to describe the techniques used to design and operate a packaging line that is always prepared to change from one product to another within a minute's notice. This encompasses quick changeover (QCO), clean-in-place (CIP), eradication of all possible adjustments, matchmark position settings for remaining adjustments, flexible crewing, tooling readiness, and a streamlined results-oriented management. A similar phrase is quick changeout.

Quick changeout. The total downtime between production runs of different packages or the end of first production run to end of the next production run time. It encompasses the same time-period definition as the quick-change process involving preparation time, cleanup time, and changeover time.

Quick changeover (QCO). A new common phase used to describe the technique of effective tooling used to change a packaging line over from one product to another using no tools and can be done in minutes or less. One such technique is SMED and the other.

Relative humidity. The ratio of actual humidity to the maximum humidity that air can retain without precipitation at a given temperature and pressure. Expressed as percent of saturation at a specified temperature. See also **Absolute humidity.**

Reliability. The probability that a device or system will not fail within a given time frame under given conditions. Quantitatively, this is expressed as a true mathematical probability.

Requisition. A form used to transmit purchasing information to the Buying Department regarding needs.

Reschedule. The process of changing the duration and/or dates of an existing schedule in response to externally imposed conditions or progress.

Resource. Any consumable, except time, required to accomplish an activity.

Return on equity (ROE). An accounting ratio used to evaluate the overall financial performance of an entire company. ROE compares the net earnings generated by a company to the shareholders' equity (net worth) of that business.

Return on investment (ROI). An accounting return measure (similar to RONA) that includes an inflation adjustment to the historical accounting value of the physical assets. ROI approximates the internal rate of return calculated on a computer.

Return on net assets (RONA). An accounting ratio used to evaluate the overall financial performance of an established business. RONA compares the net earning generated by a business to the net book value of the physical assets employed in that business.

Rework (*Q*). Components or packages produced that are of unacceptable quality but are acceptable for reprocessing (denoted by Q and q_i). Furthermore, the product is reclaimable and some input items are reusable. If the product cannot be restored to an acceptable quality, it becomes wastage. In packaging, I would argue that the cost of selling off seconds is greater than the value received. Costs associated with rework are less than those associated with wastage, since rework does not cause large amounts of lost product, only lost time. Some costs incurred in the rework process are input item cost(s), overhead cost, and opportunity cost. Even assuming wastage is negligible, a decrease in rework will, in turn, lower production costs and increase output. In general, rework is an indicator of a process out of control.

RFP. Request for proposal.

RFQ. Request for quotation.

Rheology. Study of the deformation and flow of matter.

Risk. A perceived probability of failure or not achieving the target(s) or goal(s) as *originally* established at the start of the project as well as its consequences. In my opinion, there are two types of risk. Since we do not live in a perfect world, no project or decision is risk-free. There are technical and political risks.

Risk, technical. Risk strictly related to a new process, machine, or component used in packaging a known product, or a new untried product or package design. Since technical risk is a probability of failure and its consequences, it is logical to assume that additional money and time will be expended in a direct relationship to the shortfalls that occurs in the first attempt. Sometimes management will deem a project that is partly successful to be satisfactory since the costs and time to correct the shortfall are not economical. In this case, although the project is technically a failure, politically it was approved. As a guideline, more than 70% of all projects fail to reach the original technical targets or goals. And less than half of all projects that have shortfalls are completely corrected. In some cases where the consequences are death, a low probability is of little comfort and must not be traded off. *Political risk* is a perceived probability of management's ability of not understanding the basic technical and marketing techniques and philosophies in order to make clear timely, and profitable decisions and to accept their consequences in the marketplace. It is also related to management's ability to admit to their mistakes quickly, change direction and to learn from the experience. Finally it relates to ego and power, elements of which make up humanity and the complexity of life. In some cases where the consequences are death, a low probability is of little comfort and must not be traded off.

Run speed. The run speed of a machine is the instantaneous operating rate at some point in time. It is derived in terms of the output rate at that time. For example, if a machine is outputing at a rate of 300 (not necessarily quality) containers per minute (cpm) at a given point in time, then that is the run speed. As the time interval increases, the output rate is always lower than the run speed, due to stoppages, wastage, rework, and so on. When the output approaches the run speed for any given time period, then the line is approaching the steady-state condition. In a perfect world, the output and the run speed would be equal.

Run utilization. The fraction of time the packaging line is producing output at a set run speed divided by the total time available for production. Note that changeover, cleanup, and prep work are not included. This factor can be considered the uptime during the producing time period.

Runup period. The time required after the package changeout has been completed to get the given interval actual run speed (or output rate) to exceed the 80% of achievable run speed. The more correct term is transient period. If the packaging process can not move out of the run up period quick enough, the loss in potential production can be staggering.

Sales and use taxes. Taxes imposed on vendors or contractors and passed on to the purchaser on all applicable expenditures as required by state and local law.

Sanitize. To reduce the microbial flora in or on articles such as food-plant equipment or eating utensils to levels judged safe by public-health authorities.

Sanitizer. A chemical agent that reduces the number of microbial contaminants on food-contact surfaces to safe levels from the standpoint of public-health requirements. Sanitizing can also be done by heating.

Schedule. The plan for completion of a project based on a logical arrangement of activities, resources available, imposed dates, or funding budgets.

Schedule capability (C_p). In simple terms, the schedule capability is the relative measure of how capable or how effective the actual packaging process is in producing the appropriate volume of needed packages within the planned plant operational time period or schedule. The schedule capability relates to the ratio of the scheduled run cycle time available divided by the actual package run cycle time used. This definition is valid only if the actual time taken is greater than the scheduled time. If the actual time is less than the scheduled time, the capability is the actual package run-cycle time divided by the scheduled run-cycle time. The schedule capability is always equal to or less than 1. The term *schedule capability* is sometimes shortened to just *capability*.

Scheduled package run cycle (SPRC). The *scheduled package* (or *product*) run cycle refers to the time that management allots the packaging line to yield a required quantity of quality packages. It can also be defined as the total plant asset available time based on a 24-h clock. If this definition is used, it should be defined as the asset package run cycle (APRC). In an ideal system, the scheduled package run cycle and the package run cycle are equal. Normally, the scheduled product run cycle is far longer than the product run cycle, based on the designed or historical output. For an optimized packaging process, PRC is usually more than 90% of the SPRC.

Scheduling. The assignment of start and finish to all activities belonging to a project that indicate when the activities are expected to be performed.

Secondary spoilage. Consists of those cans rusted or corroded as a result of bursting or leaking cans. May occur during warehousing.

Series elements. Those machines connected in-line that perform unique functions so that the operation of each element is vital to the system. For example, if a line has one filler connected to one capper, these may be represented as series elements, because the failure or jam of either one will stop production almost immediately if no buffers exist.

Setup. This is basically the same as the change over but relates only to a given machine not the packaging line. It is defined as the completion of the following items:

1. The exchange of change over parts or tooling
2. The recalibration and/or adjustment of the machine
3. The preliminary test with samples

At the end of the procedure, the machine is expected to perform the desired function to produce a quality package at the achievable run speed.

Shelf life. The length of time that a container, will maintain market acceptability under specified conditions of storage. Also known as merchantable life.

Simulation. The modeling or effect by computer, scaled models or mathematics to play out a senerio or sequence of events that would give the appearance or outcome that approximates as close as possible the real world. A computer simulation is a means of representing the behavior of real life systems over time, using some combination of models, initial conditions and discrete time steps. Real-time computer simulation is always the optimal condition.

Single minute exchange of die (SMED). As developed by Dr. Shigeo Shingo of Japan, this is a time and motion evaluation using industrial engineering techniques to facilitate the quick exchange of dies as used in the automotive stamping industry. There are some people who are attempting to use this technique and other similar techniques to acquire a quick change over capability for packaging lines. Extensions of these ideas are no tools, no time, no talent concept and Quick Change Over (QCO) as well as the rapid exchange no tool components developed by specialized companies such as Septimatech in Waterloo, Canada and enlightened and progressive OEMs.

SKU. See **Stockkeeping unit.**

Solvent. A substance which dissolves or holds another substance in solution such as common salt in water. Solvents are used in some foods as carriers for flavors, colors, stabilizers, emulsifiers, antioxidants, and other ingredients.

Speed factor (S_f). The actual average output rate or actual run speed of a packaging line divided by the achievable run speed of the packaging line, for a given package run cycle and time period. It does not include changeover, prep time, and cleanup that are already in utilization.

Standard practice. A written description of the minimum necessary to meet the intent of policy and standards. To the extent possible, standards should describe the "what" and not the "how."

Statistical control. The state of predictable stability.

Statistical process control (SPC). A statistical method of using control charts to monitor whether a process is in or out of control. It is a method that requires planning teamwork, methodology of measurement, knowledge of measurement techniques, and acute knowledge of the process. In most causes, the packaging process must be improved first before control charts can be use effectively.

Statistical quality control (SQC). The use of statistical techniques for process control or product and package inspection. It also includes the use of experimental design techniques for process improvement.

Stockkeeping unit (SKU). An industry term that details the assortment or variety of items shipped in "one" physical case.

Stoppage. A short time occurrence that causes the packaging line or any portion of the packaging line to cease production of a quality package. A stoppage is made up of frequency and duration.

Surfactant. Surface-active agent.

System. A collection of one or more elements (eg, machines) that are combined to perform some overall function. A packaging process (line) and a given section of a line having more then two elements or functions are examples of systems.

System downtime. The total amount of time the packaging line stopped or ceased production during the run period. System downtime is generally less than the total sum of all machine and conveyor downtimes, due to buffers and idle machine factors. It is best measured after the last operation in the packaging process, which usually is palletizing. System downtime normally should not include changeover, cleanup, and prep time periods. These times should be documented separately.

System utilization. In general, *system utilization* refers to how effectively time is used during production. With regard to packaging lines, utilization is the fraction of the total package run cycle that is actually used to produce a needed quantity of quality packages. With regard to machines or subsystems, system utilization quantifies the fraction of productive time that remains after factoring in the effects of wastage, rework, changeover, breaks, restraints, stoppages, and so on. *System utilization* refers to how effectively time is used during the production period. With regard to packaging lines, system utilization is the fraction of the total package run cycle that is actually used to produce a needed quantity of quality packages at a set run speed rate. In this *Encyclopedia* the word *utilization* is used to refer to system utilization, except with respect to machine elements. Another way is to say the system utilization is a fraction that quantifies the amount of productive time that remains after factoring in the appropriate effects of downtime over the package run cycle. Some people call this "uptime."

Target cost. The estimated cost to accomplish the project contract as defined by the estimate design basis and the project objectives. The target cost assumes that project execution during design, construction, and initial start-up will meet the normal performance standards of the company. The sum of all direct and indirect costs. By definition, target cost is a tight but realistic figure that should have an equal chance of being overspent or underspent. It excludes "management reserve."

Target date. The date an activity is desired to be started or completed, imposed, or requested by client or program management.

Team approach. Basically, this means using teamwork to accomplish goals. Teamwork is the dedicated work of a team or number of associated persons of different talents and abilities acting together in a joint action or endeavor with reference to coordination of effort and collective efficiency. The team should be made up of internal and external resources, management, operators, and maintenance personnel.

Temper. A measure of the ductility and hardness of steel plate.

Thermocouple. A bimetallic device used to measure temperatures electrically.

Tolerance. A specified allowance for deviations in weighing, measuring, etc from the standard dimensions or weight.

Total quality management (TQM). "Quality" is made up of conformance to specifications as well as conformance to the customer's expectations. Expectations are extremely difficult to measure, since this includes perceived elements. Total quality management is a philosophy of how to bring about quality. Deming and Derant agree that quality is not the problem, it is the solution. Improving quality entails low upfront costs while returns are extremely high. Typically, in the 1980s, 25% of the workforce was busy doing rework. It is estimated that reducing defectives or wastage by 5% increases revenue by 25–85%. The following are TQM milestones:

Quality is achieved by working on the process, striving to eliminate variability, and not by reworking or wasting products or packages.

It is the line operators and maintenance that should control the process, not the QC department. Quality is everyone's job.

Instead of directions, explain intent; take advantage of workers' talents, do not view them as machines or items to be replaced. Workers should be trained for flexibility. Workers should be trained to use "statistical quality control" and other tools such as PRC, DFA, and CCS.

When a problem arises, or a defective product is found, it should be viewed as an opportunity to learn about the process, not as an excuse to blame someone. Learn from your customers by getting their ideas, concerns, and recommendations.

Strive for evolution and innovations, not revolutions and inventions.

Total productive maintenance (TPM). A philosophy or management tool that view maintenance not as a necessary evil, but a vital operation that contributes to the productivity and profitability of the company. For JIT to work, one of the keys is to ensure the process availability to produce is maximized. A coordinated maintenance program with integrates QCP, TQM, preventive maintenance, predictive maintenance, and DFA is what is needed to have a successful TPM program.

Throughput. Throughput for a given packaging line and product is the exact quantity of quality packages produced in a package run cycle as required by the customer and shipped to the customer. It is denoted by the symbol Y. Similarly, we define the throughput for a machine or throughput rate as the number of items (of acceptable quality) leaving a machine in a given amount of time. We denote this by y_i for the ith machine. The word *throughput* has the same definition as *output.*

Training. *Training* is the same as *qualifying.* Training is the act of setting up and administering a training program that ensures the people who will be interfacing with the packaging line are given a thorough overview of the packaging line and a detailed program on what they need to know to complete their tasks without hesitation or guessing. All training should address the following points:

1. Whom should be trained?
2. Why should training be done?
3. When should training be done?
4. What training should be done?
5. How should we do training?
6. Where should we do training?
7. How much is enough training?
8. Manuals and other self-help tools.
9. Performance reviews and continuing improvement.
10. Company standards and policy.
11. Equipment effectiveness analysis.
12. Retrofits, adjustments, and additions to new, modified, or existing packaging lines.
13. The team approach

Training prerequisites. The skills and knowledge that a learner should have before training starts on a new technology.

Training qualification. A test for learners that is both written and hands-on.

Training resources. Training experts such as teachers, photographers, technical writers, and illustrators.

Transient period. The time required after the package changeout has been completed to get the given interval actual run speed (or output rate) to exceed the 80% of achievable run speed. This is sometimes called the "runup period." If the packaging process cannot move out of the transient period quickly enough, the loss in potential production can be staggering.

Translucent. Descriptive of a material or substance capable of transmitting some light, but not clear enough to be seen through.

Transparent. Descriptive of a material or substance capable of a high degree of light transmission (eg, glass).

Uptime. The time a machine or process is available to produce a quality product during the run period of the package run cycle.

USDA. United States Department of Agriculture.

Utilization (U). The fraction of a defined time period that is actually used to produce quality packages. There are many definitions of utilization, but there are five main types of utilization:

1. System utilization
2. Run utilization
3. Package changeout utilization
4. Element or machine utilization
5. Asset utilization

System utilization is the fraction of the total package run cycle that is actually used to produce a needed quantity of quality packages at a set run speed. In this *Encyclopedia* when we use the word utilization, we refer to system utilization. Another way is to say that system utilization is a fraction that quantifies the amount of productive time that remains after factoring in the appropriate effects of downtime over the package run cycle. Some people call this "uptime." *Run utilization* is the fraction of time the packaging line is producing output at a set run speed divided by the total time available for production. Note that changeover, cleanup, and prep work are not included. This factor can be considered the uptime during the producing time period. *Package changeout utilization* is the fraction of the package run cycle time the packaging line has available for producing output at a set run speed after changeover, cleanup, and prep time have been completed. *Element utilization* is the fraction of time a given machine or element is actually producing output at a set run speed divided by the total time available for production. Note that changeover, cleanup, and prep work are not included. This definition usually relates to machinery and/or components. *Asset utilization* is the fraction of a twenty-four hour (24-h)-per-day clock that is actually used in producing quality packages until the package run cycle is completed. If the package run cycle takes exactly 3 days, then the total asset utilization time is $24 \times 3 = 72$ hours, even though only one shift per day is being used. When calculating PI, only the system utilization is used that considers the entire package run cycle with changeover, cleanup, prep work, and run utilization. Run utilization is used to get an overview of the integration, control, and elements that make up the total packaging line. Element utilization is used to get an overview of a given machine, conveyor, or equipment. It can be used as a general guideline for acceptance of a vendor's machine. But for a specific specification for acceptance of a vendor's machine, the disturbance frequency (DF) should be used with the element utilization. In this *Encyclopedia* the word *utilization* refers to system utilization.

Vacuum packaging. Packaging in containers, whether rigid or flexible, from which substantially all gases have been removed prior to final sealing of the container.

Value analysis. A systematic approach to simplification and standardization of products so that they provide needed value at minimum cost. Usually applied to existing products to reduce input counts or amount of packaging.

Value engineering. A corporate engineering tool used in meeting a sharply defined objective that is capable of measurement and control by corporate managers. Most objectives of value engineering are often associated with cost reduction programs. Value engineering uses the systematic job plan consisting of the following steps:

1. Stating the problem
2. Forming the hypothesis
3. Observing and experimenting
4. Interpreting data
5. Drawing conclusions or solutions

The test for value functions are

Reliability	Appearance	Weight
Performance	Initial costs	Packaging
Quality	Lead time	Maintenance and servicing
Human factors		
Productivity		
Capital investment		

Variables. Quantities that describe the current operating state of a machine or system. Output rate, wastage rate, and cycle rates are examples of variables.

Verifying. The act of being able to determine without hesitation that the machine testing and training of personnel is not only thorough but also effective in the day to day production of the packaging line without senior staff, consultants, and/or machinery service personnel being involved in any shape of form.

Viscometer. An instrument to measure viscosity.

Viscosity. Internal friction or resistance to flow of a liquid. The constant ratio of shearing stress to rate of shear. In liquids for which this ratio is a function of stress, the term *apparent viscosity* is defined as this ratio.

Wastage (*W*). Components or packages produced that are of unacceptable quality. The quantity of such items is denoted by the symbol W for a system and w_i for the ith machine. This items of unacceptable quality cannot be reused or recycled back into the system. As a general policy, food and pharmaceutical products that fall onto the production floor are discarded and are therefore wastage. In other industries, wastage may be reworked. Usually, employees use their discretion to decide on wastage items, unless company policy or government regulations clearly define wastage. Wastage increases manufacturing costs. Some basic costs relating to wastage are

1. Initial cost for the product and/or inputs
2. Overhead cost
3. Opportunity cost
4. Disposal cost
5. Recycling cost

A decrease in wastage will, in turn, lower production costs and increase output. In general, wastage is an indicator of a process out of control.

Working capital. The estimated cash investment in inventories and accounts receivable, minus accounts payable.

Conclusions

The attempt of this Glossary is to communicate, define, and clarify common words used in the packaging industry as it relates to machines and processes.

This Glossary has evolved over time and will continue to evolve and refine these definitions and terms and add new definitions and terms as required. The reader is encouraged to contact the author pertaining to clarifying an existing definition and adding more terms.

Just because someone uses the same word or term as you do does not mean that that person understands it the same way as you do.

PAUL ZEPF
Zarpac, Inc.
Oakville, Ontario, Canada

INDEX

Abbreviations, 981–984
Abrasive technique, surface treatment, 867–868
Absorbent paper, described, 716
Acceleration, vibration, 955–956
Acceptable quality limit (AQL), 849
Accordion-fold tags, 878
Acetate, plastic films, 163
Acronyms, 981–984
Acrylate coatings, evaporated, surface treatment, 872
Acrylic adhesives, acrylic plastic polymers, 1
Acrylic-based inks, acrylic plastic polymers, 1
Acrylic multipolymers, nitrile polymers, 670
Acrylic plastic polymers, 1–2
  acrylic adhesives, 1
  acrylic-based inks, 1
  PVC modifiers, 2
Acrylonitrile (AN), hot-fill technology, 495
Acrylonitrile-butadiene-styrene (ABS), nitrile polymers, 670
Active packaging, 2–8
  defined, 2
  film composites, 5–7
  forms of, 3–5
  goals of, 3
  modified atmosphere packaging, 652–653
  problems addressed by, 2–3
  research and development, 7
Additives (plastic), 8–13
  antiblocking agents, 8–9
  antifogging agents, 9
  antimicrobial agents, 9
  antioxidant agents, 9
  antiozonant agents, 9
  antislip agents, 9
  antistatic agents, 9
  barrier polymers, permeability variations, 74
  biodegradable-biocide environment, 9–10
  blowing and foam agents, 10
  catalyst agents, 10
  colorant agents, 10
  coupling agents, 10
  electrically conductive agents, 10–11
  flame retardant agents, 11
  fragrance enhancer agents, 11
  heat stabilizer agents, 11
  impact modifier agents, 11
  lubricant agents, 11
  mold release agents, 11–12
  nucleating agents, 12
  overview, 8
  plasticizer agents, 12
  processing aid agents, 12
  reinforced plastic low-profile agents, 12
  slip agents, 12
  stabilizing agents, 12–13
  ultraviolet stabilizing agents, 13
Adhesive applicators, 13–23
  cold-glue systems, 14
  equipment classification, 14
  hot-melt systems, 15–20
  maximum instantaneous delivery rate calculation, 20–22
  packaging adhesives, 13–14
Adhesive bond strength, polymer properties, 763
Adhesives, 23–25
  corrugated boxes, 101
  extrudable, 25–28
    applications of, 27
    commercial forms, 26–27
    overview, 25–26
    types of, 26
  fiber drums, 311
  hot-melt adhesives, 24–25
  medical packaging, 612–613
  overview, 23
  radiation effects, 798
  solvent-borne adhesives, 25
  surface and hydrocarbon-barrier modification, 864–865
  waterborne systems, 23–24
Advance disposal fees, environmental regulation, 353
Advertising, law and regulation (U.S.), 557
Aerosol containers, 27–31. *See also* Pressure containers
  current technology, 28–29
  future trends, 30–31
  history, 27–28
  tinplate options, 29–30
Aerosol propellants, 787–791
  chemistry, 788–791
  overview, 787–788
Aesthetics
  bottle and jar closures, 208, 214
  consumer packages, 889
Air conveying, 31–35
  benefits of, 34–35
  characteristics of, 33–34
  function of, 31–33
  system design and installation, 274–276
Air-removal system, vacuum packaging, 949–951
Air shipment, export packaging, 370
Alcoholic beverages. *See also* Beer; Beverage
  casks, 71
  food packaging, 703–704
Aliphatic polyesters/thermoplastic starch, biodegradable materials, 79
Aluminum
  dual-ovenable packaging, 645
  radiation effects, 796
  recycling, 345, 801, 804
Aluminum cans, 132–134. *See also* Metal cans; Steel cans
  aerosol propellants, 789–790
  carbonated beverages, 159
  hot-fill technology, 493–494
  pressure containers, 781
Aluminum closures, bottle and jar closures, 216
Aluminum drums, testing, 894
Aluminum foil, 458–463
  applications, 460–461
  aseptic packaging, 462–463
  flexible foil packages, 461–462
  foil lidding, 462
  history, 459
  ionomers, 529
  lidding, 563
  material, 458–459
  microwave ovens, 463
  properties, 459–460
  regulated packages, 462
  rigid packaging, 463
  semirigid packaging, 463
Aluminum pressure containers, 781
American Association of Railroads
  packaging forms, 573
  plastic pails, 706
American Cancer Society, nutrition labeling, 675
American Heart Association, nutrition labeling, 675
American National Standards Institute (ANSI)
  bulk bags (flexible intermediate bulk containers), 53
  filling machinery, still liquid, 396

American National Standards Institute (ANSI) (*Continued*)
ISO standards, 524
steel drums and pails, 321
American Society for Testing and Materials (ASTM)
biodegradable materials, 77
bulk packaging, 53, 122
career development, 167
child-resistant packaging, 203
cushioning design, 289
distribution hazard measurement, 305
distribution packaging, 309
edge-crush concept, 332
electrostatic discharge protective packaging, 342
forensic packaging, 464
heat sealing, 827
permeation testing, 896
plastic drums, 316
polymer properties, 760–764
pressure-sensitive tape, 883–884, 885, 886
product fragility testing, 903, 904
retortable packages, 808
shipping containers, 906–909
slipsheets, 844
solid-fiber boxes, 113
testing, 894
vibration, 957, 958
American Trucking Association
packaging forms, 573
transportation codes, 930
Amorphous-poly-α-olefin (APAO) polymers, hot-melt adhesives, 25
Ampuls and vials, glass, 35–38
Animal glue, waterborne adhesives, 23
ANSI. *See* American National Standards Institute (ANSI)
Antiblocking agents, additives, plastic, 8–9
Antifogging agents, additives, plastic, 9
Antimicrobial agents, additives, plastic, 9
Antimicrobial films, active packaging, 6–7
Antioxidant agents, additives, plastic, 9
Antiozonant agents, additives, plastic, 9
Antislip agents, additives, plastic, 9
Antistatic agents, additives, plastic, 9
Applicators, bottle and jar closures, 211
Apron conveyor, 271
Argon, modified atmosphere packaging, 651
Arm conveyor, 271
Aroma barrier testing, 38–41
apparatus for, 40–41
overview, 38–39
permeation testing, 898
temperature effects, 40
test vapor generation, 41
theory, 39–40
Aromas, permeability (of aromas and solvents), 724–733. *See also* Permeability (of aromas and solvents)
Artificial intelligence (AI), integrated packaging design and development, 514. *See also* Computer applications
Aseptic packaging, 41–45. *See also* Medical packaging; Sterile disposable healthcare products
aluminum foil, 462–463
fiber drums, 312–313
filling systems, 43, 44
food packaging, 702
history, 42
materials, 43–45
multilayer flexible packaging, 661
overview, 41–42
package characteristics, 42
process systems, 42–43
sterile disposable healthcare products, 693–699
thermal process, 42
Aseptic process, blow molding, 91–92
Asia, environmental regulation, 350–351
ASTM. *See* American Society for Testing and Materials (ASTM)
Augers, dry-product filling machinery, 385
Australia, environmental regulation, 350–351
Austria, environmental law and regulation, 550
Automatic wraparound case loading, 193

Backpressure force
air conveying, 31–32
conveying speed, 282
Bag closures, 220
Bag-in-box packaging, 46–51
cartoning machinery (end-load), 584
dry product, 46–48
form/fill/seal pouch, horizontal, 467–468
liquid product, 48–51
oriented polyester film, 412–413
Bagmaking machinery, 54–60
electronic controls, 59–60
multiwall-bag machinery, 54–56
overview, 54
plastic bag machinery, 56–59
Bags
bulk, flexible intermediate bulk containers. *See* Bulk bags (flexible intermediate bulk containers)
bulk packaging, 121
heavy duty, plastic, 60–61
multiwall. *See* Multiwall bag(s)
plastic, 66–69. *See also* Plastic bag(s)
testing, 890–891
Bakery products
bag closures, 220
modified atmosphere packaging, 654
Balanced-pressure fillers, filling machinery, still liquid, 390–391
Bandpass filters, vibration, 957
Bands, shrink, 69–70
Band sealing, heat sealing, 824
Bar chain conveying. *See* Lug or bar chain conveying
Bar code, 225–228
applications, 227
benefits of, 225
computer applications, 228–231
data content, 227
defined, 225
labeling, 537
printing of, 228
reading of, 228
symbology, 225–227
Barges, export packaging, 368
Barrels, 70–71
Barrier films
bag-in-box packaging, liquid product, 49
defined, 177
multilayer flexible packaging, 660–661
Barrier-foam trays, 931–933
Barrier polymers, 71–77
availability, 76–77
modified atmosphere packaging, 652–653
nitrile polymers, 670
overview, 71–72
permeability data, 73
permeability units, 73
permeability variations, 73–76
permeation process, 72–73
polymer composition, 76
polymer properties, 763
Bar sealing, heat sealing, 823–824
"Basic resin doctrine" exemption, law and regulation (U.S.), food packaging, 554
Basis weight
defined, 177
paperboard, 718
Beer. *See also* Alcoholic beverages; Beverage
carbonated beverages compared, 160–161
food packaging, 703–704
glass bottles, 159
metal cans, 159
Belgium, 350, 550
Belt conveying, 266–268
Belt feeders, dry-product filling machinery, 386
Bending moment, defined, marine environment, 592–595
Beverage carriers, 168–170
Beverage industry. *See also* Alcoholic beverages; Beer
air conveying, 31–35
carbonated beverage packaging, 158–161
food packaging, 703–704
Biaxially oriented polypropylene (BOPP), cellophane, 195
Biodegradable-biocide environment, additives, plastic, 9–10
Biodegradable materials, 77–83
cellulose, 79
chitin and chitosan, 79–80
polyamides, 81
polyesters, 80–81
poly(ethylene-*co*-vinyl alcohol), 82
polyethylene oxide, 82
polyurethanes, 81
poly(vinyl alcohol), 81–82
proteins, 80
pullulan, 80
standards, 77
starch-based materials, 78–79
traditional plastics, 77–78
Biological deterioration, active packaging, 2–3
Biplex (duplex), paperboard, 718
Blade coating, coating equipment, 223
Bleached board, defined, 177
Bleached papers, described, 714
Blister packaging. *See also* Carded packaging
components and assembly, 161–162
heat-seal coatings, 163
machinery, 164–165

pharmaceutical packaging, 734
plastic film, 423
plastic films, 163
skin packaging compared, 165
Blocking, polymer properties, 763
Blow and blow (B&B) process, glass container manufacturing, 479–480
Blow and cast film, extrusion, 373, 375–376
Blowing and foam agents, additives, plastic, 10
Blow molding, 83–93
aseptic process, 91–92
defined, 83
extrusion-injection-molded neck process, 87
extrusion process, 83–87
heat-resistant polyester bottles, 90–91
history, 83
injection process, 87
labeling, 297
low-density polyethylene (LDPE), 757–758
multilayer and coextrusion process, 89–90
nylon, 684
plastic bottle design, 93, 94
polypropylene (PP), 767
process basics, 83
product design guidelines, 92
rigid plastic containers, 110
secondary processes, 92
stretch process, 87–89
Blown-film coextrusion, 238. *See also* Coextrusion machinery
Board thickness, defined, 177
Boil-in-bag, oriented polyester film, 412
Booklet tags, 877
Bottle design, plastic. *See* Plastic bottle design
Bottle and jar closures, 206–220
functions, 207–208
future trends, 217–218
history, 206–207
materials, 215–216
methods, 208–210
sealing systems, 214–215
selection of, 216–217
specification, 217
types, 210–214
Bottles, testing, 891
Bottom-discharge bucket conveyor, 271
Bottom seal, plastic bags, 67
Botulism
food canning, 123, 124, 701
oxygen scavengers, 691
Box compression test (BCT), edge-crush concept, 332
Boxes. *See also* Corrugated boxes; Rigid-paperboard boxes; Rigid plastic containers; Wood boxes
solid-fiber, 112–113
testing, 891–893
wirebound, 113–115
wood, 115–117
Box-sealing tape, pressure-sensitive tape, 884–885
Brazil, 117–120
Bread bag closures, 220
Breakage, export packaging, 366
Breakaway caps, bottle and jar closures, 212–213
Bridge impact test, shipping container testing, 907
British Imperial System, 638
British Standards Institute (BSI), 53
Bruce box, 114
Bucket elevator conveyor, 271
Buckling resistance, cushioning design, 289
Budgets, management, 588
Bulk bags (flexible intermediate bulk containers), 51–54, 448–449. *See also* Intermediate bulk containers
disposal and reuse, 53
filling and dispensing, 52
handling, transportation and storage, 52–53
materials, 52
overview, 51
testing, 893
testing and standards, 53
uses for, 51–52
Bulk packaging, 120–122
considerations in, 120–121
materials, 121–122
testing, 122
Bulk palletizer, 709
Bulk-product feed, dry-product filling machinery, 388
Bundle-wrapping machines, wrapping machinery, 972
Burst/seal strength, leakage testing, 900
Butyrate, plastic films, 163

Cable conveying, 273–274
Calcium oxide (lime), active packaging, 3–4
Calender coaters, coating equipment, 223
Calendering, rigid polyvinyl chloride (PVC) film, 428–429
California, 352–353, 354, 556
Caliper (thickness), paperboard, 718
Canada, 344
Candy, food packaging, 703
Can multipacks, beverage carriers, 168
Canning, 123–128, 701–702
defined, 123
future trends, 128
hot-fill technology, 492–496
operations, 124–127
overview, 123
process description, 123–124
regulation, 127–128
Cans
aluminum, 132–134
composite, 134–139
construction, 134–136
defined, 134
future trends, 137
manufacture, 134
self-manufactured, 137–139
corrosion, 139–143. *See also* Corrosion
fabrication of. *See* Metal can fabrication
plastic, 144
steel, 144–155. *See also* Steel cans
testing, 893
Can seamers, 128–132
described, 128–129
features and attachments, 130–131
history, 145
machine types, 129–130
methods, 129
overlap measurement, 131
seam profiles, 131
seam tightness evaluation, 131
setup, 130
steel cans, 151
Capping machinery, 155–158
continuous-thread closures, 155–157
presson closures, 157–158
rollon closures, 157
vacuum closures, 157
Caps. *See* Closures
Carbonated beverage packaging, 158–161
beer market compared, 160–161
deposit laws, 160
food packaging, 703
overview, 158
package types, 158–160
Carbonated liquid filling machinery, 389–390
Carbon dioxide
modified atmosphere packaging, 651
oxygen scavengers, 689
Carbon dioxide absorbers, modified atmosphere packaging, 652
Carbon dioxide tracer gas, leakage testing, 900
Carbon dioxide transmission rate (CO_2TR), permeation testing, 897–898
Carbon monoxide, modified atmosphere packaging, 651
Cardboard boxes. *See* Corrugated boxes
Carded packaging, 161–166
blister packaging, 161–162
heat-seal coatings, 163–164
ionomers, 528
machinery, 164–166
paperboard, 164
plastic films, 163
skin packaging, 162–163
Career development, 166–168
Carrier rules, transportation codes, 929–930
Carriers, beverage, 168–170
Carton finish, defined, 177
Cartoning machinery (end-load), 580–588
carton closing, 587
carton loading, 583
detectors, 587
fully automatic horizontal, 580–582
leaflet feeds, 585–586
microprocessors, 587–588
multipackers, 584–585
overloads, 587
overview, 580
product infeeds, 583–584
semiautomatic horizontal, 580
semiautomatic vertical, 580
side seam gluing, 582–583
Cartoning machinery (top-load), 170–176
carton closing, 174–176
adhesive closure, 175–176
dust-flap-style closure, 174–175
heat-seal closure, 176
lock closure, 175
triple-seal style closure, 175
carton forming, 170–172

Cartoning machinery (top-load) (*Continued*)
conveying, 173
forming capabilities, 173
glue forming, 172–173
heat-seal forming, 173
lock forming, 172
options, 176
product loading
automatic, 174
manual, 173
Cartons. *See also* Folding cartons
folding, 181–187
gabletop, 187–189
testing, 893
Carton terminology, 176–181
carton development, 180, 181
generally, 176–177
guidelines and standards, 178–181
packing, 177
Car-type conveyor, 272
Cascade-filling systems, dry-product filling machinery, 386
Casein, waterborne adhesives, 23–24
Case loading, 189–194
automatic wraparound, 193
fully automated, 190–191
horizontal automatic caser/erector/loader/sealer, 191
horizontal semiautomatic, 190
manual, 189–190
overview, 189
tray former/loader, 193
variations, 193–194
vertical, 191–193
Cask. *See* Barrels
Cast-film process
coextrusion, flexible packaging, 238
stretch film, 437
Catalyst agents, additives, plastic, 10
Cellophane, 194–195
features, 194
history, 194
physical properties, 195
production, 195
types, 194–195
vinylidene chloride copolymer (VDC), 961
Cellular plastic, defined, 451. *See also* Foamed plastics
Cellulose
biodegradable materials, 79
cellophane, 194–195
corrugated boxes, 100
foamed plastics, 451
plastic films, 163
radiation effects, 797
Center winder (duplex winder), slitting and rewinding machinery, 844–845
Certified Packaging Professional (CPP), 589
Checkweighers, 195–199
process control, 196–198
production reporting, 198
regulatory compliance, 195–196
Cheeses, vacuum packaging, 953
Chemical degradation, polymers, 761–763
Chemical deterioration, active packaging, 3
Chemical etching, surface treatment, 869
Chemical priming, surface treatment, 869
Child-resistant packaging, 199–204
bottle and jar closures, 213–214, 216
classification, 203
effectiveness, 203
enforcement, 203
history, 199
law and regulation (Europe), 549
plastic bottle design, 99
regulatory effects, 199–200
testing procedures, 200–202
China, 350
Chitin, biodegradable materials, 79–80
Chitosan, biodegradable materials, 79–80
Chlorinated organics, environment, 344
Chlorine, environment, 344
Chlorofluorocarbons (CFC), aerosol propellants, 788–789, 790
Chub packaging, 204–205
Circulating systems, packaging adhesives, 14
Clay-coated board
defined, 177
skin packaging, 842
Clean Air Act, 458, 556
Clean Water Act, 556
Cloeren system, coextrusion machinery, flat, 234
Clostridium botulinum
food canning, 123, 124, 701
oxygen scavengers, 691
Closure liners, 205–206, 214–215
Closures. *See also* Stretch film
aluminum foil, foil lidding, 462
bottle and jar, 206–220. *See also* Bottle and jar closures
bread bag, 220
continuous-thread, capping machinery, 155–157
presson, capping machinery, 157–158
rollon, capping machinery, 157
vacuum, capping machinery, 157
Coated papers, described, 717
Coated recycled paperboard, folding cartons, 182
Coated solid bleached sulfate (SBS) paperboard, folding cartons, 181
Coated solid unbleached sulfate (SUS) paperboard, folding cartons, 181
Coating equipment, 221–225
coating heads, 221–224
drying, 224–225
metal can fabrication, 627–629
overview, 221
saturators, 224
web handling, 225
Coatings
coextrusion, flexible packaging, 238
evaporated acrylate coatings, surface treatment, 872
extrusion coating, 378–381. *See also* Extrusion coating
metal can fabrication, 626–627
oriented polypropylene film, 422
radiation effects, 798
steel cans, 153–154
surface and hydrocarbon-barrier modification, 866
transparent glass on plastic food-packaging materials, 445–448
vinylidene chloride copolymer (VDC), 961
Cobb test, corrugated box testing, 892
Code, bar, 225–228. *See also* Bar code
Code marking
computer applications, 228–231
labeling, 537
Code of Federal Regulations (CFR)
food additives, FDA, 552–553
steel drums and pails, 322
transportation codes, 930
Coefficient of friction (COF)
coextrusion machinery, tubular, 235
oriented polypropylene film, 418
Coextrusion
blow molding, 89–90
bottles. *See* Blow molding
ethylene-vinyl alcohol (EVOH) copolymers, 359
extrudable adhesives, 27
flexible packaging, 237–240
advantages, 237
manufacturing process, 237–239
raw materials, 239
structures, 239
technology, 237
medical packaging, 612
multilayer flexible packaging, 663–664
nylon, 683–684
oriented polyester film, 411
plastic bags, heavy duty, 61
plastic film, 425
semirigid packaging, 240–242
applications, 241
barrier materials, 240–241
economic factors, 241–242
structural materials, 241
technology, 240
stretch film, 437
surface and hydrocarbon-barrier modification, 866
Coextrusion machinery, 231–237
flat, 231–234
encapsulation and lateral adjustment, 234
equipment, 231–233, 234
methods, 233–234
tubular, 234–237
economic factors, 236
equipment, 235
process design, 235
quality control, 235–236
Coffee
oriented polyester film, 412
vacuum-bag packaging, 948–949
Cohesive bond strength, polymer properties, 763
Cold adhesives, adhesive applicators, 13
Cold-gas-plasma treatment, surface treatment, 870–872
Cold-glue systems, packaging adhesives, 14
Cold-vinyl adhesives, cartoning machinery (top-load), 175
Collagen, biodegradable materials, 80
Collapsible tubes, 941–945
future trends, 945
history, 941–942
laminated, 944–945
metal, 942

plastic, 942–944
Color
bottle and jar closures, 208
folding cartons, 181
Color Additives Amendment of 1960, 255
Colorants, 242–256
additives, plastic, 10
decorating, 294
dyes, 243
glass container design, 472
overview, 242
paper and paperboard, 255
pigments, 242–243, 244–254
plastic bottle design, 97
plastics, 243, 255
regulatory requirements, 255–256
stretch film, 438
supply options, 255
Communication measurement, marketing effectiveness, of consumer packages, 889
Compatibility
food-package compatibility, law and regulation (Europe), 543
product compatibility, steel cans, 149
Component-specific specification, specification and quality assurance, 849–850
Composite cans, 134–139
construction, 134–136
future trends, 137
self-manufactured, 137–139
Composting, environmental concerns, 346
Compounding, extrusion, 372–373
Comprehensive Environmental Response, Compensation, and Liability Act (CERCLA, "Superfund"), 556
Compression molding, 256, 922, 924–925
Compression resistance, cushioning design, 289
Compression strength, edge-crush concept, 332
Compression tests, shipping container testing, 908
Computer applications
cartoning machinery (end-load), 587–588
checkweighers, 198
code marking, 228–231
glass container design, 475
integrated packaging design and development, 514–519
labeling, 296
materials handling, 606–607
pallet patterns, 256–258
plastic bottle design, 97
shelf life, 833–834
thermoforming, 921
Concentrated load, defined, marine environment, 592
Conservation, environment, 343
Constant carton line (CCL), defined, 176
Constant-opening line (COL), defined, 176
Consulting, 260–263
hiring guidelines, 262–263
qualifications, 261–262
rationale for, 260–261
Consumer demand, economics, 328–329
Consumer goods specification, specification and quality assurance, 852
Consumer packages, testing of, for marketing effectiveness, 887–890
Consumer research, 258–260
Contact sealing, heat sealing, 825
Containerized loads, export packaging, 368, 370
Containment closure, bottle and jar closures, 210
Contamination
air conveying, 35
export packaging, 366
Content labeling, 547–548. *See also* Labeling
Content mandates, environmental regulation, 352–353
Continuous-flow heating process, aseptic packaging, 42
Continuous tags, 876–879
Continuous-thread closures
bottle and jar closures, 208–209
capping machinery, 155–157
Contract packaging, 263–264
Convenience closure, bottle and jar closures, 210–212
Conversion factors, 979–981
Conveying, 264–283
air conveying, 31–35
defined, 264–265
guiderails and control components, 281
interconnecting machinery, 281
lightweight containers, 282
power transmission, 276–277
single filing, 282–283
stages, 282
stretch-film wrapping machinery, 975–976
system design and installation, 265–276, 281–282, 283
air design, 274–276
belt design, 266–268
cable design, 273–274
lug or bar chain design, 270, 271–276
mesh-top or open-top design, 268–269
met-top or flat-top chain design, 269–270
roller design, 270, 273, 277, 278, 279
screw design, 273, 280
tabletop chain design, 265–266
vibratory design, 276
technology, 282
transfers, 277–278, 281
Cook/chill food production, 283–285
Copermeant, permeability (of aromas and solvents), 726–728
Copolyester, plastic films, 163
Copolymers
biodegradable materials, 77–83
polyethylene terephthalate (PET), 744
Cork, bottle and jar closures, 206
Corona treatment, 410, 865, 870
Corrosion, 139–143
electrostatic discharge protective packaging, 339
export packaging, 366
hydrogen specificity, 141–142
mechanism, 140–141
overview, 139–140
problems, 142–143
steel cans, 149
vapor-corrosion inhibitor (VCI), marine environment, 590
Corrugated board
edge-crush concept, 331–334
recycling, 345–346
skin packaging, 842
Corrugated boxes, 100–108
board construction, 101–102
bulk packaging, 121
dimensioning, 106
economics, 106–107
equipment, 103–105
future trends, 108
industry organization, 103
joints, 105–106
materials, 100–101
Mullen test versus edge-crush test, 102–103
overview, 100
recycling, 107–108
styles, 107
testing, 891–892
Corrugated pallets, expendable, 710
Corrugated plastic, 285–287
Coupling agents, additives, plastic, 10
Crates
bulk packaging, 122
marine environment, 591–592, 596–597
testing, 892
wood boxes, 117
Creep resistance, cushioning design, 288–289
Crossbar conveyor, 272
Crown bottle and jar closures, 210
Crystallized polyethylene terephthalate (CPET), dual-ovenable packaging, 644
Curing
offset container printing, 300
screen printing, 303
thermosetting plastics, 925
Currently Good Manufacturing Practices (CGMP), 735
Curtain coater, coating equipment, 224
Cushioning
foamed plastics, 455
product fragility testing, 904
shock, 835–836, 837
Cushioning design, 287–293
constraints on, 289–290
cushion characteristics, 288–289
future trends, 293
overview, 287
procedures, 290–293
Customer-supplier relations, total quality management (TQM), 929
Cut-sheet thermoforming, 920

Dairy products, 187–189, 701
Darkening of foods, 143, 149, 153
Databases. *See* Computer applications
Dating. *See also* Shelf life
law, 548
printing, computer applications, 229
shelf life, 832
Dating equipment. *See* Bar code; Code marking
Death rates. *See* Mortality rates
Decals. *See* Decorating; Labeling

Decorating, 294–303. *See also* Labeling
glass container design, 472–473
glass container manufacturing, 484
heat-transfer labeling, 294–296
hot stamping, 296–297
in-mold labeling, 297–298
Japanese packaging, 668–669
labeling, 537
marketing effectiveness, of consumer packages, 888–889
multilayer flexible packaging, 660, 663
offset container printing, 298–300
pad printing, 300–302
PETG copolyester, 830
screen printing, 302–303
surface and hydrocarbon-barrier modification, 864–865
Delaney Clause, Federal Food, Drug and Cosmetic Act, 553
Denmark, 550
Density, polymer properties, 759–760
Depalletizing, 709–710
Deposit laws, 160, 354, 556
Deregulation, logistical/distribution packaging, 573
Design, plastic bottles, 93–100
Design for assembly (DFA), integrated packaging design and development, 514–519
Design for manufacturability (DFM), integrated packaging design and development, 514–519
Dessicants, defined, marine environment, 590
Die-cut tags, 876
Dielectric constant, polymers, 764
Dielectric sealing, heat sealing, 826
Dimension standards, carton terminology, 176
Dioxins, paper, law and regulation (Europe), 546
Direct roll coaters, coating equipment, 221–222
Direct stamping method, hot stamping, 296
Displacement-ram volumetric fillers, filling machinery, still liquid, 393
Distribution hazard measurement, 303–307
data analysis, 305–306
overview, 303–304
process of, 304–305
Distribution packaging, 307–310. *See also* Logistical/distribution packaging
checklist for, 310
components of, 310
design, 309–310
economics, 330
functions of, 307
objective of, 307
protective-package concept, 308–309
system approach, 308
Double-package maker (DPM), bag-in-box packaging, dry products, 46
Dow system, coextrusion machinery, flat, 233
Drag chain conveyor, 272
Drawing and ironing (D&I), metal can fabrication, 615–616, 621, 622–625
Draw and redraw (DRD), metal can fabrication, 615–616, 625–626
Drop test, shipping container testing, 906–907
Drums. *See* Fiber drums; Plastic drums; Plastic pails; Steel drums and pails
Dry bag-in-box packaging, 46–48
Dry foods, food packaging, 702
Dry-product filling machinery, 384–389. *See also* Filling machinery
Dual-ovenable packaging. *See* Microwavable and dual-ovenable packaging
Duplex (biplex), paperboard, 718
Duplex winder (center winder), slitting and rewinding machinery, 844–845
Dust, static control, 858
Dust flap, defined, 177
Dust-flap-style closure, cartoning machinery (top-load), 174–175
Dye-leak test, leakage testing, 899–900
Dyes, colorants, 242, 243. *See also* Colorants

East Asia, 350–351
Eastern Europe. *See also* Europe
environmental regulation, 350
packaging, 360–361, 362
Ecolabeling
environmental regulation, 351, 354
Europe, 549
U.S., 557
Economics, 325–330
consumer demand, 328–329
losses, marine environment, 601–602
macroeconomics, 325
management, budgets, 588
microeconomics, 329–330
recycling, 800, 803–804
standup flexible pouches, 856
supply industries, 325–328
tamper-evident packaging, 880–881
total quality management (TQM), 928
Edge-crush concept, 331–334
background, 331
compression strength, 332
corrugated board, 331–332
models, 332–333
testing, 891
Edge-crush test
corrugated boxes, 102–103, 105
transportation codes, 930
Edible film, 397–401
composition, 397–398
food and drug coating, 398–399
future trends, 400
manufacture, 398
overview, 397
physical properties, 399–400
Education, 334–335
career development, 167
management, 588–589
Electrically conductive agents, additives, plastic, 10–11
Electrical properties, polymers, 764
Electrochemical potential, corrosion, 140–141
Electron-beam sterilization, sterile disposable healthcare products, 697–699
Electronic data processing (EDP), labels, 536, 537
Electronic Industries Association (EIA), 342
Electrostatic charge. *See also* Static control
antistatic agents, additives, plastic, 9
extrusion coating, 380
static control, 857–860
surface treatment, 873
Electrostatic discharge protective packaging, 335–343. *See also* Static control
classification, 336
history, 336
overview, 335–336
standards, 339, 342
technological solutions, 339–342
terms and test methods, 336–339
Embossing, aluminum foil, 461
Employee involvement, total quality management (TQM), 929
Enamel peeling, corrosion, 142
End-load cartoning machinery. *See* Cartoning machinery (end-load)
Environment, 343–348
additives, plastic, 9–10
aerosols, 31, 788–791
biodegradable materials, 77
bulk bags (flexible intermediate bulk containers), 53
colorants, 255
corrugated boxes, 107–108
cushioning design, 289
deposit laws, carbonated beverage packaging, 160
economics, 330
edible film, 397
Europe, 362
foamed plastics, 458
foam trays, 937
green marketing, 347
integrated packaging design and development, 515–516
ISO 14001 environmental management system, 533–535
life-cycle assessment, 347
medical packaging, 614
oriented polyester film, 414
overview, 343
plastic bottle design, 99–100
pollution, 343–344
poly(vinyl chloride) (PVC), 774
pressure-sensitive tape, 886–887
recycling, 799–808
regulation, 347
resource depletion, conservation, and sustainable use, 343
rigid plastic containers, 111
shelf life, 831–832
shipping container testing, 906
solid-waste issues, 344–347
stretch film, 442–443
Environmental Protection Agency (EPA). *See* U.S. Environmental Protection Agency (EPA)
Environmental regulation, 348–355. *See also* Regulation
international, 348–351
ecolabeling, 351, 549
Europe, 348–350, 549–551, 805–808
Pacific Rim and East Asia, 350–351
standards, 351
North America, 351–355, 556–557

advance disposal fees, 353
content mandates, 352–353
deposit laws, 354
green labeling, 354, 557
heavy metals, 354
history, 351–352
landfill bans, 353–354
resin coding, 354–355
EPA. *See* U.S. Environmental Protection Agency (EPA)
Equilibration, permeation testing, 895–896
Ethanol, modified atmosphere packaging, 651
Ethanol emitting sachets, active packaging, 4–5
Ethylene absorbing sachets, active packaging, 4–5
Ethylene-acrylic acid (EAA), coextrusion, flexible packaging, 239
Ethylene-butyl acrylate, hot-melt adhesives, 24
Ethylene-methacrylic acid (EMA), coextrusion
flexible packaging, 239
semirigid packaging, 241
Ethylene oxide sterilization, sterile disposable healthcare products, 696–697
Ethylene-vinyl acetate (EVA)
bag-in-box packaging, liquid product, 48–49
carded packaging, 164
coextruded flexible packaging, 237, 239
heat sealing, 826
hot-melt adhesives, 24–25
hot-melt wax carton, 963
labeling, 298
rotational molding, 819
skin packaging, 840, 841
Ethylene-vinyl alcohol (EVOH) copolymers, 355–360
applications, 360
coextrusion
flexible packaging, 239
semirigid packaging, 240
collapsible tubes, 942, 945
edible film, 399
films, 359–360
hot-fill technology, 495
microwavable packaging, 645
overview, 355–356
packaging structures, 358–359
properties, 356–358
regulation, 358
surface and hydrocarbon-barrier modification, 866
Europe, 360–362. *See also* Law and regulation (Europe)
Eastern Europe, 362
glass, 361
metals, 361
modified atmosphere packaging, 656–659
overview, 360–361
paper, 361
plastic, 361
reduction strategies, 362
regulation, 348–350, 541–552
shelf life, 832
standup flexible pouches, 855–856
technology, 362
European Court of Justice (ECJ), law and regulation, 542
European Standardisation Committee (CEN), biodegradable materials, 77
European Union. *See also* Law and regulation (Europe)
beverage carriers, 170
bulk bags (flexible intermediate bulk containers), 53
child-resistant packaging, 201–202, 203
environmental regulation, 347, 348–350
fiber drums, 315
function and organization of, law and regulation, 541–543
ISO standards, 524–525, 529
plastic drums, 318
recycling, 805–808
steel drums and pails, 322
Evaporated acrylate coatings, surface treatment, 872
Exhibitions, 362–365
Expendable corrugated pallets, 710
Expert witnesses, consulting, 261
Export packaging, 365–370. *See also* Marine environment; Shipping
conditions, 366
guidelines, 370
hazards, 365–366
marine environment, 589–603
marks and symbols, 367, 369
product analysis, 366–367
techniques, 367–368, 370
Extrudable adhesives, 25–28
applications of, 27
commercial forms, 26–27
overview, 25–26
types of, 26
Extruded polystyrene foam, 449–450
Extrusion, 370–378
blow and cast film, 373, 375–376
blow molding, 83–87
compounding, 372–373
foam sheet extrusion, 377
multilayer flexible packaging, 663
nylon, 682–683, 684
oriented polypropylene film, 417
overview, 370
PETG copolyester, 828
plastic film, 425–427
rigid sheet extrusion, 376–377
single-screw extruders, 370–372
stretch film, 437
Extrusion coating, 378–381
applications, 378, 380
folding carton manufacture, 184–185
low-density polyethylene (LDPE), 757
machinery, 379, 380–381
nylon, 684
overview, 378
Extrusion-injection-molded neck process, blow molding, 87
Eye-tracking research, consumer research, 259–260

Fair Packaging and Labeling Act, 676
Fats and oils, food packaging, 702–703
FDA. *See* U.S. Food and Drug Administration (FDA)
Federal Food, Drug and Cosmetic Act of 1938, 255, 553, 676, 758, 961
Federal Hazardous Substances Act, 255
Ferrous metals, recycling, 800–801
Fiber, molded, 382–383
Fiberboard boxes, testing, 891–892
Fiber drums, 310–316. *See also* Plastic drums
applications, 311
construction, 311
defined, 310–311
regulation, 314–315
styles, 312–314
testing, 894
Filament tapes, pressure-sensitive tape, 885–886
Filling machinery
aseptic packaging, 43, 44
bulk bags (flexible intermediate bulk containers), 52
carbonated liquid, 389–390
count measurement, 383–384
dry-product, 384–389
bulk-product feed, 388
equipment, 388–389
overview, 384
product-feed systems, 385–387
weighing systems, 387–388
plastic bottle design, 95
still liquid, 390–397
container positioning, 394–396
design and selection, 396–397
methods, 390–394
overview, 390
tube filling, 939–941
Film(s). *See also* Plastic film; Stretch film
blow and cast film, extrusion, 373, 375–376
edible, 397–401. *See also* Edible film
electrostatic discharge protective packaging, 339–342
ethylene-vinyl alcohol (EVOH) copolymers, 359–360
flexible PVC, 401–403. *See also* Flexible polyvinyl chloride (PVC) film
fluoropolymer, 403–405. *See also* Fluoropolymer film
high-density polyethylene, 405–407. *See also* High-density polyethylene
medical packaging, 612
modified atmosphere packaging, 652–653
multilayer flexible packaging, 660–661
nonoriented polypropylene, 407–408. *See also* Nonoriented polypropylene film
nylon, 683
oriented polyester, 408–415. *See also* Oriented polyester film
oriented polypropylene, 415–422. *See also* Oriented polypropylene film
plastic, 423–427. *See also* Plastic film
polypropylene (PP), 767–768
recycling, 346
rigid polyvinyl chloride (PVC), 427–431. *See also* Rigid polyvinyl chloride (PVC) film
shrink, 431–434

Film(s). *See also* Plastic film; Stretch film (*Continued*)
  stretch, 434–445. *See also* Stretch film
  thermotropic liquid-crystalline polymers (TLCP), 570–572
  transparent glass on plastic food-packaging materials, 445–448
Film composites
  active packaging, 5–7
  composite can construction, 135
Findability testing, consumer research, 259
Finished-goods specification, specification and quality assurance, 852
Finland, 550
Fire
  flame retardant agents, additives, plastic, 11
  foamed plastics, 458
First-in-first-out (FIFO), time-temperature indicators, 927
Fish
  food packaging, 700
  modified atmosphere packaging, 654
  vacuum packaging, 953–954
Fitment closure, bottle and jar closures, 211–212
Fixed-spout closure, bottle and jar closures, 210–211
Flame retardant agents, additives, plastic, 11
Flame technique, 865, 869–870
Flammability, polymers, 762
Flat coextrusion machinery. *See* Coextrusion machinery
Flat-top conveying, 33–34, 269–270
Flavor protection
  ethylene-vinyl alcohol (EVOH) copolymers, 356–357
  oriented polyester film, 409, 410
  permeation testing, 896, 989
Flexible films, ethylene-vinyl alcohol (EVOH) copolymers, 359
Flexible intermediate bulk containers. *See* Bulk bags (flexible intermediate bulk containers)
Flexible packaging
  coextrusion, 237–240
  extrusion coating, 378
  lidding, 561–563
  medical packaging, 611–612
  multilayer flexible packaging, 659–665. *See also* Multilayer flexible packaging
  retortable packages, 809
  standup flexible pouches, 852–856. *See also* Standup flexible pouches
Flexible polyvinyl chloride (PVC) film, 401–403. *See also* Rigid polyvinyl chloride (PVC) film
  composition, 401
  heat stabilizers, 401
  lubricants, 402
  manufacture, 402
  markets, 402–403
  plasticizer, 401
  resin, 401
Flexographic ink, 511–512
Flexography
  described, 785–787
  labeling, 537, 538
Flight conveyor, 272
Florida, 353
Fluorination
  blow molding, 92
  surface treatment, 865–866, 872–873
Fluoropolymer film, 403–405
  applications, 404
  composition, 403
  manufacture, 403–404
  properties, 404
  safe handling, 404
Flutes, corrugated boxes, 101–102
Foam, extruded polystyrene, 449–450
Foamed crystallized polyethylene terephthalate (CPET), dual-ovenable packaging, 644
Foamed plastics, 451–458
  applications, 455–458
  definitions, 451
  environmental concerns, 458
  expansion process, 451
  health and safety factors, 458
  history, 451
  manufacture, 455
  properties, 451–455
Foaming agents, additives, plastic, 10
Foam sheet extrusion, described, 377
Foam trays, 933–937
  applications, 935–937
  environmental concerns, 937
  history, 933
  manufacture, 933–935
  producers, 935
Focus groups
  consumer research, 259
  marketing effectiveness, of consumer packages, 887
Foil, aluminum, 458–463. *See also* Aluminum foil
Foil printing, computer applications, 230
Foils. *See* Film(s)
Foil transfer, holographic packaging, 490
Folding cartons, 181–187
  bag-in-box packaging, dry products, 46
  history, 181
  hot-melt application, 186–187
  manufacture, 184–186
  paperboard selection, 181–182
  styles, 182–184
Food additives, 255, 552–555
Food Additives Amendment of 1958, 255, 552
Food and Drug Administration (FDA). *See* U.S. Food and Drug Administration (FDA)
Food canning. *See* Canning
Food packaging, 699–704. *See also* entries under specific processes and foods
  canned foods, 701–702
  classification, 700
  cook/chill methods, 283–285
  foam trays, 935–937
  fresh foods, 700
  overview, 699–700
  partially processed foods, 700–701
  polycarbonate (PC), 741–742
  steam-table trays, 937–938
  thermoform/fill/seal, 910
Forced vibration, free vibration versus, 955
Forensic packaging, 463–465
Form/fill/seal pouch
  aluminum foil, 462
  horizontal, 465–468
  ionomers, 528
  standup flexible pouches, 855
  vertical, 468–470
Fourdrinier machine, paperboard manufacture, 719
Fragility. *See also* Product fragility testing
  cushioning design, 289
  defined, marine environment, 592
Fragrance enhancer agents, additives, plastic, 11
Fragrance protection, ethylene-vinyl alcohol (EVOH) copolymers, 356–357
France, 550–551
Free vibration, forced vibration versus, 955
Freeze and thaw indicators, described, 501–502
Freight containers, standards, ISO, 525
Frequency, vibration, 955–956
Friction, polymer properties, 763
Friction-fit bottle and jar closures, 209–210
Friction sealing, heat sealing, 825
Frozen foods, food packaging, 702
Fruits and vegetables
  food packaging, 700, 701
  modified atmosphere packaging, 655
  vacuum packaging, 954
FTC. *See* U.S. Federal Trade Commission (FTC)
Furnish, paperboard, 718
F value, food canning, 124

Gabletop cartons, 187–189
Galvanic corrosion, 140–141
Gamma sterilization, sterile disposable healthcare products, 697
Gas-barrier protection
  ethylene-vinyl alcohol (EVOH) copolymers, 356
  modified atmosphere packaging, 652–653
Gas packaging. *See* Modified atmosphere packaging
Gas-plasma technique, surface and hydrocarbon-barrier modification, 865
Gas sealing, heat sealing, 825
Gatefold tags, 877
Gauge randomization, blow and cast film, extrusion, 375–376
Gauging, coextrusion machinery, flat, 234
Gelatin, biodegradable materials, 80
Gel lacquers, carded packaging, 164
Generally recognized as safe (GRAS), food additives, 552, 553
General packaging specification, specification and quality assurance, 849
German Institute for Standardisation (DIN), 77
Germany, 349–350, 551
Glass bottles
  beverage carriers, 169–170
  bottle and jar closures, 207
  carbonated beverages, 159
  testing, 891

Glass closures, bottle and jar closures, 216
Glass container(s)
  ampuls and vials, 35–38. *See also* Ampuls and vials
  economics, 328
  European packaging, 361, 362
  hot-fill technology, 494
  law and regulation (Europe), 546
  microwavable packaging, 646
  pressure containers, 781
  radiation effects, 796
  recycling, 345, 804
  standards, ISO, 524
Glass container design, 471–475
  computer applications, 475
  manufacturing conditions, 471–472
  market factors, 472–473
  shape and dimensions, 471
  strength factors, 473–475
Glass container manufacturing, 475–484
  chemical phase, 476–477
  mechanical/forming phase, 477–478
  overview, 475–476
  process, 478–484
  terminology, 484
Glassine, described, 715
Glass on plastic food-packaging materials, 445–448
Glass-transition temperature, polymer properties, 759
Global warming, 344, 788–791. *See also* Environment
Glue flap, defined, 177
Glue-style carton, defined, 177
Grain direction, defined, 177
Grain products, food packaging, 703
Grammage, defined, 177
Graphics, multiwall bags, 65–66
Gravitational-force indicators, described, 502–503
Gravity-discharge bucket conveyor, 272
Gravity-flow systems, dry-product filling machinery, 385
Gravure coaters, coating equipment, 222–223
Gravure ink, 512–513
Gravure printing
  described, 784–785
  labeling, 538
Greaseproof paper, described, 715
Grease-resistant paper, paper, 715
Greece, 551
Greenhouse gases, environment, 344
Green labeling. *See* Ecolabeling
Green marketing, environment, 347
Gross Domestic Product (GDP), packaging economics, 325
Guidelines, carton terminology, 178–181
Gummed-paper labels, 536
Gummed tape, 883

HACCP system, 485–489
  concept, 485
  origin of, 485
  principles, 485–488
Hand loading. *See* Manual loading
Hazard analysis critical control point (HACCP) system. *See* HACCP system
Hazardous materials
  bulk bags, 449
  export packaging, 366–367
  fiber drums, 310–311, 314–315
  intermediate bulk containers, 519, 521
  law and regulation (Europe), 549
  law and regulation (U.S.), 556
  plastic drum regulation, 317–318
  plastic pails, 705–707
  steel drums and pails, 319, 322
  transportation codes, 930
Hazardous Materials Regulations (HMR), fiber drums, 310–311, 314–315
Headers, defined, marine environment, 592
Health and safety issues
  foamed plastics, 458
  law and regulation (Europe), food packaging, 543–547
  law and regulation (U.S.), food packaging, 552–555
  leak testing, 558–561
  linear and very low-density polyethylene (LLDPE and VLDPE), 752
  logistical/distribution packaging, 575
  low-density polyethylene (LDPE), 758
  nutrition labeling, 674–681
  oxygen scavengers, 690–691
  polypropylene (PP), 768
  static control, 858–859
Healthcare packaging. *See* Aseptic packaging; Medical packaging; Pharmaceuticals; Sterile disposable healthcare products
Heat-resistant polyester bottles, blow molding, 90–91
Heat-seal coatings
  blister packaging, 162, 163
  multilayer flexible packaging, 662–663
  skin packaging, 162, 163–164
Heat-seal forming, cartoning machinery (top-load), 173
Heat sealing, 823–827
  band sealing, 824
  bar sealing, 823–824
  contact sealing, 825
  dielectric sealing, 826
  friction sealing, 825
  gas sealing, 825
  hot-melt sealing, 825–826
  hot-wire or knife sealing, 825
  impulse sealing, 824–825
  induction sealing, 826
  ionomers, 527–529
  magnetic sealing, 826
  method selection, 826–827
  overview, 823
  pneumatic sealing, 826
  polymer properties, 763
  radiant sealing, 826
  secondary conversion, 684
  solvent sealing, 826
  testing, 827
  ultrasonic sealing, 825
Heat-seal labeling machinery, 540
Heat-sensitive labels, 537
Heat-shrink packaging, electrostatic discharge protective packaging, 341
Heat stabilizer agents, additives, plastic, 11
Heat-transfer labeling, described, 294–296
Heavy duty plastic bags, 60–61
Heavy metals, 354, 556
High-density polyethylene (HDPE), 405–407, 745–748
  applications, 405
  blow molding, 83, 86
  bottle and jar closures, 215
  coextruded flexible packaging, 237, 239
  coextruded semirigid packaging, 241
  collapsible tubes, 942
  composite can construction, 136
  defined, 745
  folding cartons, 184
  labeling, 297, 298
  manufacture, 406–407, 745–746
  microwavable packaging, 646
  molecular structure, 746–747
  pallets, 711
  plastic bottle design, 96, 97
  plastic netting, 666
  plastic pails, 705
  properties, 405–406, 747–748
  recycling, 346, 800, 803, 804
  regulation, 748
  rotational molding, 819–820
High-impact polystyrene (HIPS), thermoforming, 914
High-level palletizer, 708
High-nitrile resins, nitrile polymers, 670
High temperature short time (HTST) process, aseptic packaging, 42, 43
Holograms, 414, 489–492
Holt-melt systems, packaging adhesives, 15–20
Homopolymers, polyethylene terephthalate (PET), 744
Hong Kong, 350
Horizontal automatic caser/erector/loader/sealer, 191
Horizontal form/fill/seal pouch, 465–468
Horizontal impact test, shipping container testing, 907
Horizontal semiautomatic case loading, 190
Hot-fill technology, 492–496
  aluminum cans, 493–494
  glass containers, 494
  overview, 492
  plastic packages, 494–495
  processing, 492
  tinplate cans, 493
Hot-melt adhesives
  acrylic plastic polymers, 1
  adhesive applicators, 13–14
  cartoning machinery (top-load), 172–173, 175
  described, 24–25
  folding cartons, 186–187
Hot-melt sealing, heat sealing, 825–826
Hot-melt wax carton, 962–964
Hot-stamping
  computer applications, 230
  described, 296–297
  holographic packaging, 490
  labeling, 538
  oriented polyester film, 414
Hot-wire sealing, heat sealing, 825
Housewares exemption, 554–555

Humidity
electrostatic discharge protective packaging, 338
permeability (of aromas and solvents), 728–729
Humidity indicators, described, 502
Hydrocarbon-barrier modification. *See* Surface and hydrocarbon-barrier modification
Hydrocarbon resistance, ethylene-vinyl alcohol (EVOH) copolymers, 356
Hydrocarbons, aerosol propellants, 789
Hydrogen peroxide, aseptic packaging, 42
Hydrogen specificity, corrosion, 141–142
Hydrogen swell, corrosion, 142–143
Hydroxypropylmethyl cellulose (HPMC), edible film, 397, 398, 399

Impact modifier agents
additives, plastic, 11
poly(vinyl chloride) (PVC), 773
Impact shock measurement, 836
Imprinting, computer applications, 228–231
Impulse sealing, heat sealing, 824–825
Incineration, environmental concerns, 346–347
Incline impact test, shipping container testing, 907
India, 497–498, 856
Indicating devices, 498–503
freeze and thaw indicators, 501–502
gravitational-force indicators, 502–503
humidity indicators, 502
temperature indicators, 500–501
time and temperature indicators, 499–500
Individual-rewind-arm (IRA) winders, slitting and rewinding machinery, 845–846
Induction sealing, heat sealing, 826
Industrial products, extrusion coating, 380
Industrial wraps, extrusion coating, 380
Injection molding, 503–511
blow molding, 87
low-density polyethylene (LDPE), 757
machinery, 505–511
mold design, 503–505
overview, 503
polypropylene (PP), 768
rigid plastic containers, 110
thermosetting plastics, processing systems, 922–923
Ink-jet printing, computer applications, 229–230
Inks, 511–514
acrylic-based, acrylic plastic polymers, 1
blister packaging, 162
colorants, 243, 255–256. *See also* Colorants
corrugated boxes, 101
liquid, 511–513
overview, 511
pad printing, 301–302
paste, 513–514
radiation effects, 798
screen printing, 303
skin packaging, 162, 841
surface and hydrocarbon-barrier modification, 864–865
In-mold labeling, described, 297–298
Innerseals
bottle and jar closures, 214–215
closure liners, 206
Inorganic pigments, colorants, 242. *See also* Colorants
Instantaneous pump delivery rate (IPDR). *See* Maximum instantaneous delivery rate (MIDR) calculation
Institute for Standards Research (ISR), biodegradable materials, 77
Institute of Packaging Professionals (IoPP), 167, 260, 589
Insulation, thermal, foamed plastics, 455–456
Insurance, marine environment, 601–602
Integrated packaging design and development, 514–519
model for, 518
overview, 514–515
rationale for, 515
tools, technologies, and methodologies, 515–517
Intermediate bulk containers, 519–521. *See also* Bulk bags (flexible intermediate bulk containers)
applications, 521
defined, 519
overview, 519
UN code, 519–521
International Air Transport Association (IATA)
fiber drums, 314–315
international standards, 527
International Civil Aviation Organization (ICAO)
fiber drums, 314–315
international standards, 527
steel drums and pails, 322
International environmental regulation. *See* Environmental regulation; International standards
International Maritime Organization (IMO)
fiber drums, 314–315
intermediate bulk containers, 519
international standards, 526
plastic drums, 317
steel drums and pails, 322
International standards, 521–527
defined, 521
International Standards Organization (ISO)
described, 521–523
ISO 9000, 524–526, 529–533
packaging standards, 523–524
organizations, 526–527
International Standards Organization (ISO)
aerosol containers, 30
biodegradable materials, 77
bulk bags (flexible intermediate bulk containers), 53
child-resistant packaging, 199, 201, 202
described, 521–523
ISO 9000, 524–526, 529–533
described, 529–530
European Union, 529
function, 530–532
rationale, 530
registration to, 532–533
ISO 14001 environmental management system, 533–535
membership in, 522
packaging standards of, 523–524
paperboard, 718
shipping containers, 906
steel cans, 145
International System, 638, 639, 642. *See also* Metrication
Interviews, consumer research, 259, 887
Inventory
life-cycle assessment, 565–566
time-temperature indicators, 927
Ionomers, 527–529
applications, 529
development, 527
properties, 527–528
structure, 527
Ireland, 551
Iron, recycling, 800–801
Irradiation, 49. *See also* Radiation effects
ISO 9000. *See* International Standards Organization (ISO)
Isostatic high pressure system, aseptic packaging, 43
Italian Standardization Agency (UNI), biodegradable materials, 77
Italy, 77, 551

Japan, 350, 667–669, 856
Jar closures, 206–220. *See also* Bottle and jar closures
Jones side-seam gluer (SSG), carton terminology, 179
Just-in-time inventory, plastic pallets, 710

Kiss roll coaters, coating equipment, 221–222
Knife and bar coaters, coating equipment, 223
Knife sealing, heat sealing, 825
Korea, 350
Kraft paper
corrugated boxes, 100
described, 714
medical packaging, 611
multiwall bags, 64

Labeling, 536–541. *See also* Decorating; Nutrition labeling; Printing
aluminum foil, 461
application machinery, 538–541
heat-seal labeling, 540
overprinting, 540–541
pressure-sensitive labeling, 539–540
wet-glue labeling, 539
blow molding, 92
bottle and jar closures, 208, 216–217
carbonated beverages, 159
colorants, 255
composite can construction, 136
computer applications, 228–231
corrugated boxes, 101
ecolabeling, environmental regulation, 351, 354
export packaging, marks and symbols, 367, 369
folding cartons, 181, 185

glass containers, 472, 484
hazardous materials, bulk bags, 449
heat-transfer labeling, described, 294–296
holographic packaging, 490
hot stamping, 296–297
in-mold labeling, 297–298
label printing, 537–538
label types, 536–537
law and regulation (Europe), 547–549
law and regulation (U.S.), 556–557
multilayer flexible packaging, 660
nutrition labeling, 674–681
offset container printing, 298–300
oriented polyester film, 414
pad printing, 300–302
printing, computer applications, 228–231
screen printing, 302–303
shrink bands, 69–70
thermoform/fill/seal, 912
Label overprinting machinery, 540–541
Label readership, consumer research, 260
Laminated collapsible tubes, 944–945
Lamination
aluminum foil, 461
coextrusion, flexible packaging, 238
folding carton manufacture, 185
holographic packaging, 490–492
medical packaging, 612
multilayer flexible packaging, 663
secondary conversion, 684–685
Landfill, 347, 353–354
Laser printers, computer applications, 230
Laser scanner, bar codes, 228
Latin America, 856
Law and regulation (Europe), 541–552. *See also* Regulation
child-resistant packaging, 549
environmental, 549–551
European Union, 541–543
food-contact legislation, 543–547
food-package compatibility, 543
labeling, 547–549
overview, 541
trademarks, 549
Law and regulation (U.S.), 552–558. *See also* Regulation
environmental, 556–557
Food and Drug Administration (FDA), 552–555
food additive definition, 552
food packaging, 552–555
generally, 552
Lead, steel cans, 150, 154
Lead pigments, 256
Leaflet feeds, cartoning machinery (end-load), 585–586
Leakage testing, 558–561, 898–901. *See also* Permeation testing
burst/seal strength, 900
CO_2 tracer gas, 900
overview, 898–899
pressure/pressure-decay method, 900
waterbath or dye-leak test, 899–900
Left-hand machine, defined, 176
Legal issues
expert witnesses, consulting, 261
liability, printing, computer applications, 229
Letterpress, labeling, 538
Letterpress ink, 514
Letterset ink, 513–514
Level-sensing fillers, filling machinery, 392–393
Liability, printing, computer applications, 229
Lidding, 561–563
Life-cycle assessment, 563–569
components of, 564–566
defined, 563
environment, 347
impact assessment, 566–568
improvement assessment, 568
limitations, 568
overview, 563
packaging choices, 563–564
uses of, 564
Lighters, export packaging, 368
Linear and very low-density polyethylene (LLDPE and VLDPE), 748–752. *See also* Low-density polyethylene (LDPE)
applications, 750–751, 756–758
defined, 748
history, 748–749
manufacture, 749, 754
markets, 755
pressure-sensitive tape, 886
properties, 751–752
rotational molding, 819–820
safety and health, 752
second-generation, 752
standup flexible pouches, 854, 855
structure and properties, 749–750
Linear low-density resins, plastic bags, heavy duty, 60–61
Liners
closure liners, 205–206, 214–215
paperboard, 718
Liquid
carbonated, filling machinery, 389–390
still, filling machinery, 390–397. *See also* Filling machinery
Liquid bag-in-box packaging, 48–51
Liquid cleaning technique, surface treatment, 868–869
Liquid-crystalline polymers. *See* Thermotropic liquid-crystalline polymers (TLCP)
Liquid-crystal polymers (LCP), dual-ovenable packaging, 645
Liquid inks, 511–513
Liquid packaging, extrusion coating, 378
Lithography. *See* Decorating; Labeling; Printing
Load-bearing floorboards, defined, marine environment, 592
Lock closure, cartoning machinery (top-load), 175
Logistical/distribution packaging, 572–579. *See also* Distribution packaging
innovations, 577
overview, 572–573
packaging forms, 573
packaging performance, 573–576
problems, 576–577
Losses, marine environment, 601–602
Low-density polyethylene (LDPE), 752–758. *See also* Linear and very low-density polyethylene (LLDPE and VLDPE)
applications, 756–758
blow molding, 83
bottle and jar closures, 215
carded packaging, 162, 163, 164
characteristics, 753
coextruded flexible packaging, 237, 239
coextruded semirigid packaging, 241
collapsible tubes, 942, 943
composite can construction, 136
dual-ovenable packaging, 643
edible film, 399
folding cartons, 184
health and safety factors, 758
ionomers, 527, 528
manufacture, 753–754
markets, 755
overview, 752–753
plastic bottle design, 97
properties, 754–755
rotational molding, 819
skin packaging, 840
standup flexible pouches, 854
Low-level palletizer, 708–709
Lubricants
additives, plastic, 11–12
poly(vinyl chloride) (PVC), 773
Lug or bar chain conveying, system design and installation, 270, 271–276
Lug cap bottle and jar closures, 209
Lug closures, capping machinery, 157
Luxembourg, 551

Macroeconomics, packaging, 325
Magnetic sealing, heat sealing, 826
Mall interview, consumer research, 259
Management, 588–589
budgets, 588
career development, 167
future trends, 589
organizational factors, 588
packaging specifications, 589
professional growth, 589
project control, 589
responsibilities, 588
staffing, 588
training, 588
Management information system (MIS), 514
Manual loading
cartoning machinery (top-load), 173
case loading, 189–190
Manufacturer's seam or side seam, defined, 177–178
Manufacturing specifications, specification and quality assurance, 850, 852
Marine environment, 589–603. *See also* Export packaging; Shipping
container problems, 600–601
damage and claims, 602
definitions, 590–595
design considerations, 597–600
insurance and losses, 601–602
marks and numbers, 595
preservation, 595–597
shipping containers, testing of, 906–909
unitization and palletization, 600–601

Marketing effectiveness, of consumer packages, testing for, 887–890
Marks, numbers, and symbols, 367, 369, 590, 595
Massachusetts, 353
Materials handling, 603–610
analytic methods, 604–606
definitions, 603
equipment, 607
objectives, 603–604
overview, 603
packaging and, 607–609
plant layout, 606–607
principles, 604
Maximum instantaneous delivery rate (MIDR) calculation, packaging adhesives, 20–22
McKee formula, edge-crush concept, 332–333
Meat industry
chub packaging, 204–205
edible film, 398
food packaging, 700, 701
modified atmosphere packaging, 653–654
vacuum packaging, 952–953
Meat Inspection Act, 255
Mechanical breakaway caps, bottle and jar closures, 212–213
Medical packaging, 610–615. *See also* Sterile disposable healthcare products
adhesives, 612–613
aluminum foil, foil lidding, 462
environmental concerns, 614
lidding, 562–563
materials selection, 611–614
overview, 610
package function definition, 610–611
polycarbonate (PC), 741
sterile disposable healthcare products, 693–699
sterilization methods, 614
Medications. *See* Pharmaceuticals
Mesh-top conveying, 268–269
Metal can fabrication, 615–629
can types, 615–616
coating equipment, 627–629
coatings, 626–627
overview, 615
three-piece manufacture, 617–621
two-piece manufacture, 621–626
Metal cans. *See also* Aluminum cans; Steel cans
carbonated beverages, 159–160
oriented polyester film, 414
Metal closures, bottle and jar closures, 216
Metal collapsible tubes, 942
Metal containers
bulk packaging, 121–122
standards, ISO, 524
Metallizing. *See* Vacuum metallizing
Metals
economics, 327–328
Europe, 361, 546–547
radiation effects, 796
recycling, 800–801, 804
Methyl cellulose, edible film, 398
Metrication, 638–642
benefits of system, 639
conversion tables, 640–642
history, 638–639
International System, 639, 642
overview, 638
rounding, 642
terms and symbols, 642
Metric Conversion Act of 1975, 639
Met-top conveying, 269–270
Microeconomics, packaging, 329–330
Microprocessors. *See* Computer applications
Microwavable and dual-ovenable packaging, 642–646
active packaging, 7
materials, 643–646
dual-ovenable, 643–645
microwave-only, 645–646
oriented polyester film, 413
overview, 642–643
Microwave ovens, aluminum foil, 463
Microwave pasteurization and sterilization, 646–648
Mildew, export packaging, 366
Military packaging, 648–650
retortable packages, 810
steam-table trays, 938
Modified atmosphere packaging, 650–659
active packaging, 652–653
bakery products, 654
European market, 656–659
fish, 654
fruits and vegetables, 655
gases used in, 651, 657
materials, 652
overview, 650–651
oxygen scavengers, 688, 689
poultry, 654
prepared foods, 654–655
red meat, 653–654
shelf life, 835
Moisture barrier, edible film, 397, 398
Moisture control, active packaging, film composites, 5–6
Moisture vapor transmission rate (MVTR), fiber drums, 311. *See also* Water-vapor transmission rate (WVTR)
Molded fiber, 382–383
Molded pulp
described, 791–794
dual-ovenable packaging, 643–644
Molding
compression molding, 256
injection molding, 503–511. *See also* Injection molding
rigid plastic containers, 110–111
rotational molding, 819–822
thermosetting plastics, processing systems, 921–923, 924–925
Mold release agents, additives, plastic, 11–12
Monsanto v Kennedy, law and regulation (U.S.), food packaging, 554
Mortality rates, child-resistant packaging, 201
Movable-spout closure, bottle and jar closures, 211
Mullen test, corrugated boxes, 102–103, 105
Multilayer films, plastic film, 425
Multilayer flexible packaging, 659–665
appearance, 660
barrier protection, 660–661
future trends, 664
manufacture, 663–664
overview, 659–660
product containment, 661–663
Multilayer plastic bottles. *See* Barrier polymers; Blow molding; Coextrusion
Multilayer process, blow molding, 89–90
Multipackers
cartoning machinery (end-load), 584–585
wrapping machinery, 972
Multiplex, paperboard, 718
Multiwall bag(s), 61–66
constructions, 64
equipment, 65
extrusion coating, 380
graphics, 65–66
history, 61–62
sizing of, 64
specifications, 64–65
testing, 890
transportation, 66
types of, 62–64
Multiwall-bag machinery, 54–56
Municipal solid waste (MSW), 330, 344–347

National Academy of Sciences, nutrition labeling, 675
National Aeronautics and Space Administration (NASA)
electrostatic discharge protective packaging, 336
HACCP system, 485
National Bureau of Standards (NBS), checkweighers, 195
National Classification Committee (NCC), steel drums and pails, 321
National Institute of Standard and Technology (NIST), metrication, 639
National Motor Freight Classification (NMFC)
fiber drums, 314
plastic pails, 706
steel drums and pails, 321
transportation codes, 930
Natural rubber latex, waterborne adhesives, 24
Netherlands, 350, 551
Netting, plastic, 666–667
Nippon packaging. *See* Japan
Nitrile polymers, 669–672
applications, 671–672
copolymers, 670
overview, 669–670
properties, 670–671
Nitrocellulose (NC), 164, 194
Nitrogen, modified atmosphere packaging, 651
Noncirculating systems, packaging adhesives, 14
Nonferrous metals, recycling, 801
Nonoptical systems, filling machinery, count measurement, 383–384
Nonoriented polypropylene film, 407–408
manufacture, 408
properties, 408
Nonreturnable glass bottles, 159, 170

Nonwovens, 672–674
 bonding methods, 673–674
 defined, 672
 fibers, 672
 manufacture, 672–673
 paper, described, 717
North Carolina, 354
Nucleating agents, additives, plastic, 12
Nutrition labeling, 674–681. *See also* Labeling
 claims, 679–680
 computer applications, 228
 Europe, 548–549
 history, 674–675
 Nutrition Facts panel, 677–679
 overview, 674
 regulatory agencies, 675–677
Nutrition Labeling and Education Act of 1990, 674, 675
Nylon, 681–686
 applications, 685
 coextrusion, flexible packaging, 239
 dual-ovenable packaging, 644–645
 overview, 681
 processing methods, 682–684
 properties, 682
 secondary conversion, 684–685
 strapping materials, 862

Occupational injury, logistical/distribution packaging, 575
Occupational Safety and Health Administration (OSHA). *See* U.S. Occupational Safety and Health Administration (OSHA)
Odor protection
 absorbent films, active packaging, 7
 ethylene-vinyl alcohol (EVOH) copolymers, 356–357
 oriented polyester film, 409, 410
Offset container printing, described, 298–300
Offset lithographic ink, 513
Offset printing. *See* Decorating; Printing
Ohmic (electrical) resistance system, aseptic packaging, 43
Oil-resistant paper, 715
Oils. *See* Fats and oils
Open-top conveying, 268–269
Opposed-shelf type vertical chain conveyor, 276
Optical systems, filling machinery, count measurement, 383
Oregon, 352, 556
Organic permeation, aroma barrier testing, 39–40
Organic pigments, colorants, 242. *See also* Colorants
Organic Reclamation and Composting Association (ORCA), 77
Oriented polyester film, 408–415
 applications, 411–414
 environmental concerns, 414
 future trends, 414–415
 manufacture, 409, 411
 overview, 408–409
 properties, 409–410
 surface modifications, 410–411
Oriented polypropylene film, 415–422
 history, 415
 manufacture, 416
 morphology, 416–417
 orientation process, 417–418
 properties, 418–422
 resins, 415
Oriented polystyrene (OPS), plastic films, 163
Outer flat, defined, 178
Overprinting machinery, 540–541
Over-the-counter drugs, pharmaceutical packaging, 735
Overwrap packaging, plastic film, 423
Oxidation, corrosion, 140–141
Oxygen, modified atmosphere packaging, 651
Oxygen scavengers, 687–692
 active packaging, 2, 4, 5–6
 applications, 689
 chemistry, 688–689
 food-safety and regulation, 690–691
 history, 687
 modified atmosphere packaging, 652
 overview, 687
 sizing and selection, 689–690
 spoilage, 687–688
 testing, 690
 troubleshooting, 691
Oxygen transmission rate (OTR)
 permeation testing, 896–897
 shelf life, 831, 833
 surface and hydrocarbon-barrier modification, 866
Ozone, antiozonant agents, additives, plastic, 9
Ozone depletion, 344, 788–791

Pacific Rim, 350–351
Package handling systems. *See* Conveying
Package-integrity issues for sterile disposable healthcare products. *See* Sterile disposable healthcare products
Packaging adhesives. *See* Adhesives
Packaging of food. *See* Food packaging
Pad printing. *See also* Decorating; Printing
 computer applications, 230
 described, 300–302
Pails. *See* Plastic pails; Steel drums and pails
Pallet
 defined, 708
 expendable corrugated, 710
 plastic, 710–714
Palletizing
 described, 708–710
 marine environment, 591, 600–601
 patterns, computer applications, 256–258
Pallet packing
 carton terminology, 179
 standards, ISO, 523
Pallet-type conveyor, 273
Pan conveyor, 273
Paper, 714–717
 absorbent paper, 716
 bleached papers, 714
 coated papers, 717
 colorants, 255
 European packaging, 361
 greaseproof and glassine, 715
 kraft papers, 714
 law and regulation (Europe), 546
 medical packaging, 611
 nonwovens, 717
 overview, 714
 radiation effects, 796
 recycling, 801–802, 803–804
 specialty-treated paper, 716
 synthetic, 724
 tissue papers, 716–717
 vegetable parchment, 715
 water-, grease-, and oil-resistant paper, 715
 waxed papers, 715–716
 wet-strength paper, 716
Paper bags. *See* Multiwall bag(s)
Paperboard, 717–723
 beverage carriers, 168–169
 carded packaging, 162, 164
 colorants, 255
 composite can construction, 134–136, 137–139
 corrugated plastic compared, 286
 dual-ovenable packaging, 643
 economics, 327
 ethylene-vinyl alcohol (EVOH) copolymers, 359
 folding cartons, 181–182
 law and regulation (Europe), 546
 manufacture, 719–721
 overview, 717–718
 physical characteristics, 719
 radiation effects, 796
 recycling, 345–346
 skin packaging, 841–842
 structure and properties, 718–719
 terminology, 718
 types, 721–723
Paperboard boxes, rigid. *See* Rigid-paperboard boxes
Parenteral drugs, pharmaceutical packaging, 733–734
Parison programming, blow molding, 86–87
Pasta products, modified atmosphere packaging, 654
Pasted open mouth (POM) multiwall bag, 62, 63, 64
Pasted valve stepped end (PVSE) multiwall bag, 63, 64
Paste inks, 1, 513–514
Pasteurization, microwave, 646–648
Performance testing, shipping container testing, 908–909
Peristaltic-pump volumetric fillers, filling machinery, 393
Permeability (of aromas and solvents), 724–733
 aroma barrier testing, 39–40
 barrier polymers, 72–76
 chemical composition, 725–726
 copermeant presence, 726–728
 numerical consistency of data, 730–732
 overview, 724–725
 polymer morphology, 726
 relative humidity effects, 728–729

Permeability (of aromas and solvents) (*Continued*)
  temperature effects, 729–730
  transport process, concentration dependence of, 726
Permeation testing, 895–898. *See also* Leakage testing
  carbon dioxide transmission rate (CO_2TR), 897–898
  history, 896
  overview, 895
  oxygen transmission rate (OTR), 896–897
  permeation, 895–896
  water-vapor transmission rate (WVTR), 897
PETG copolyester
  extrusion, 828
  forming, 828
  overview, 827–828
  properties, 829–830
  secondary fabrication, 828
  sheet assembly, 828, 830
pH
  corrosion, 141, 142–143
  food canning, 123
  steel cans, 149
Pharmaceuticals, 733–736
  aluminum foil, foil lidding, 462
  ampuls and vials, glass, 35–38
  bottle and jar closures, 212
  child-resistant packaging, 199, 200
  edible film, 398–399
  ethical, 733–735
  lidding, 562
  manufacture, 735–736
  oriented polyester film, 414
  over-the-counter, 735
  plastic bottle design, 99
Phenolics, bottle and jar closures, 215–216
Photodegradation, polymers, 762
Pigments. *See also* Colorants
  colorants, 242–243
  listing of, table, 244–254
  poly(vinyl chloride) (PVC), 773
  selection of, table, 243
Pilferage, export packaging, 366
Pinch-bottom open-mouth (PBOM) multiwall bag, 63, 64
Pinhole flex test, polymer properties, 761
Pitting corrosion, 142
Pivoted-bucket conveyor, 273
Plain-paper labels, 536
Plastic
  additives, plastic, 8–13
  biodegradable materials, 77–78
  bottle and jar closures, 215–216
  carded packaging, 161–166
  colorants, 243, 255
  corrugated, 285–287
  decorating, 294. *See also* Decorating
  economics, 328
  environmental concerns, 352
  European packaging, 361
  foamed plastics, 451–458. *See also* Foamed plastics
  hot-fill technology, 494–495
  law and regulation (Europe), 543–546
  radiation effects, 796–798
  recycling, 346, 802–803, 804
  thermosetting plastics, processing systems, 921–923
Plastic additives. *See* Additives (plastic)
Plastic bag(s), 66–69
  applications, 68–69
  bulk packaging, 121
  heavy duty, 60–61
  manufacturing methods, 66–68
  testing, 890–891
Plastic bag machinery, 56–59
Plastic bottle(s)
  beverage carriers, 170
  carbonated beverage packaging, 158–159
  testing, 891
Plastic bottle design, 93–100
  environmental concerns, 99–100
  overview, 93
  prototyping and testing, 97–98
  requirements, 93–95
  specialty bottles, 98–99
  specifications, 95–97
  steps in, 93
Plastic boxes, testing, 892–893
Plastic cans, 144
Plastic-clip closure, bag closures, 220
Plastic collapsible tubes, 942–944
Plastic containers. *See* Rigid plastic containers
Plastic drums, 315–318. *See also* Fiber drums
  design, 316–317
  overview, 315
  regulation, 317–318
  resins, 315–316
  testing, 894
Plastic film, 423–427. *See also* Film(s)
  applications, 423–424
  blister packaging, 163
  manufacture, 426–427
  modified atmosphere packaging, 652–653
  multilayer films, 425
  multilayer flexible packaging, 660–661
  overview, 423
  resins, 424–425
  skin packaging, 162, 163, 840–841
  vinylidene chloride copolymer (VDC), 961
Plasticizers
  additives, plastic, 12
  edible film, 397–398
  flexible polyvinyl chloride (PVC) film, 401
  poly(vinyl chloride) (PVC), 773
Plastic netting, 666–667
Plastic pails, 704–708
  design, 707
  manufacture, 707–708
  materials, 705
  overview, 704–705
  performance requirements, 705–706
  specifications, 706–707
  testing, 894
Plastic pallets, 710–714
Plastic ring beverage carriers, 168
Plastic-sheet packages, testing, 893
Platen printers, computer applications, 230
Plug-orifice closure, bottle and jar closures, 211
Ply, paperboard, 718
Ply separation test, corrugated box testing, 892
Pneumatic sealing, heat sealing, 826
Pocket conveyor, 273
Point, defined, 178
Point-of-purchase packaging, 736–740
  future trends, 740
  high visibility versus, 736
  materials, 736
  permanent, 738–740
  temporary, 736–738
Poison-prevention packaging, child-resistant packaging, 199–204. *See also* Tamper-evident packaging
Poison Prevention Packaging Act (PPPA) of 1970, 199, 255, 462
Poisson's ratio, polymer properties, 761
Polarization, corrosion, 141
Politics, environmental regulation, 352
Pollution, environment, 343–344
Polyamides, biodegradable materials, 81
Polycaprolactone, biodegradable materials, 81
Polycarbonate (PC)
  described, 740–742
  dual-ovenable packaging, 645
  electrostatic discharge protective packaging, 339, 340–341
Polyester(s). *See also* Polyethylene terephthalate (PET)
  biodegradable materials, 80–81
  bottles, heat-resistant, blow molding, 90–91
  electrostatic discharge protective packaging, 341
  hot stamping, 296
  medical packaging, 612
  strapping materials, 862
Polyester film. *See* Oriented polyester film
Polyetherimide (PEI), dual-ovenable packaging, 645
Polyethylene naphthalate (PEN), oriented polyester film, 414–415
Polyethylene oxide, biodegradable materials, 82
Polyethylene (PE)
  bag closures, 220
  high-density, 745–748. *See also* High-density polyethylene (HDPE)
  labeling, 295
  linear and very low-density (LLDPE and VLDPE), 748–752. *See also* Linear and very low-density polyethylene (LLDPE and VLDPE)
  low-density, 752–758. *See also* Low-density polyethylene (LDPE)
  microwavable packaging, 646
  modified atmosphere packaging, 652
  plastic bags, heavy duty, 60–61
  plastic drums, 315–316
  plastic pails, 705, 708
  skin packaging, 840
  surface and hydrocarbon-barrier modification, 866
Polyethylene terephthalate glycol (PETG). *See* PETG copolyester
Polyethylene terephthalate (PET), 742–745
  applications, 744–745

beverage carriers, 168
blow molding, 83, 87, 88, 89, 90, 91
carbonated beverages, 158, 160
dual-ovenable packaging, 643–644
folding cartons, 184
history, 742
homopolymers and copolymers, 744
hot-fill technology, 494
labeling, 298
manufacture, 742–744
oriented polyester film, 408–415
plastic bottle design, 97
recycling, 346, 800, 802–803, 804
Polyethylene vinyl alcohol (EVOH)
biodegradable materials, 82
blow molding, 89–90
coextrusion machinery, flat, 231
Polyhydroxyalkanoates, biodegradable materials, 80
Poly(lactic acid)/poly(glycolic acid), biodegradable materials, 80–81
Polymer(s)
acrylic plastic polymers, 1–2
barrier polymers, 71–77. *See also* Barrier polymers
biodegradable materials, 77–83
electrostatic discharge protective packaging, 341–342
permeability (of aromas and solvents), 724–733. *See also* Permeability (of aromas and solvents)
radiation effects, 796–798
Polymer properties, 758–765
barrier properties, 763
density and thermophysical properties, 759–760
electrical properties, 764
mechanical properties, 760–761
optical appearance, 764
overview, 758–759
solubility and chemical degradation, 761–763
surface and adhesion, 763–764
Polyolefins
blow molding, 86
coextruded semirigid packaging, 241
coextrusion, flexible packaging, 239
Polyphenylene oxide/polystyrene (PPO), microwavable packaging, 645–646
Polypropylene, strapping materials, 862
Polypropylene film. *See* Nonoriented polypropylene film; Oriented polypropylene film
Polypropylene (PP), 765–768
applications, 766–768
biaxially oriented, cellophane, 195
blow molding, 83, 87, 88, 89
bottle and jar closures, 215
composite can construction, 136
corrugated plastic, 286
dual-ovenable packaging, 643
folding cartons, 184
health and safety issues, 768
hot-fill technology, 495
labeling, 298
manufacture, 766
microwavable packaging, 645
modified atmosphere packaging, 652
plastic bottle design, 96, 97
properties, 765–766
recycling, 803
Polysaccharide-lipid bilayer film, edible film, 398
Polystyrene foam
extruded, 449–450
foamed plastics, 456–457
Polystyrene (PS), 768–771
applications, 769–770
bag closures, 220
blow molding, 83
bottle and jar closures, 215
coextrusion, semirigid packaging, 241
high-impact polystyrene (HIPS), thermoforming, 914
manufacture, 770–771
overview, 768–769
plastic bottle design, 97
recycling, 346, 351–352, 803
Polysulfone (PSO), dual-ovenable packaging, 645
Polyurethanes, biodegradable materials, 81
Poly(vinyl alcohol), biodegradable materials, 81–82
Poly(vinyl chloride) (PVC), 771–775
additives, plastic, 11, 12
applications, 773–774
beer market, 160
blow molding, 83, 84, 88
carded packaging, 162, 163, 164
compounding, 773
environmental concerns, 344
folding cartons, 182
food packaging, 700
law and regulation (Europe), 544, 547
lidding, 562
manufacture, 772
markets, 772
overview, 771
pharmaceutical packaging, 734
plastic bottle design, 97
polymer properties, 760
recycling, 346, 774, 803
regulation, 774
shrink bands, 69–70
shrink films, 434
structure and properties, 772–773
Poly(vinyl chloride) (PVC) modifiers, acrylic plastic polymers, 2
Polyvinylidene chloride (PVDC) copolymer. *See also* Vinylidene chloride (VDC) copolymer
beer market, 160
cellophane, 194, 195
coextrusion
flexible packaging, 239
semirigid packaging, 240
coextrusion machinery, flat, 231–234
extrusion coating, 380, 381
lidding, 563
oriented polyester film, 410
oriented polypropylene film, 420
plastic films, 424
surface and hydrocarbon-barrier modification, 866
Portable conveyor, 270
Portugal, 551
Positive-displacement volumetric fillers, filling machinery, 393
Pouch packaging. *See also* Standup flexible pouches
cook/chill food production, 285
form/fill/seal pouch, horizontal, 465–468
retortable packages, 809–810
standup flexible pouches, 852–856
Poultry
food packaging, 700
modified atmosphere packaging, 654
vacuum packaging, 951
Poultry Inspection Act, 255
Power-and-free conveyor, 274
Power density, vibration, 957
Power transmission, conveying, 276–277
Prebreak, defined, 178
Preservation, marine environment, 590, 595–597
Press and blow (P&B) process, glass container manufacturing, 479–480
Press-on closures, capping machinery, 157–158
Press-on twist-off closures, capping machinery, 157
Press-on vacuum cap bottle and jar closures, 210
Press-twist bottle and jar closures, 209
Pressure containers, 775–783. *See also* Aerosol containers
aluminum, 781
construction, 779–780
dimensions, 778–779
glass and plastic, 781
history, 775–776
overview, 775
pressure resistance, 780–781
steel, 776–778
tinplate linings, 780
valves, 781–783
Pressure/pressure-decay method, leakage testing, 900
Pressure-sensitive adhesives, acrylic plastic polymers, 1
Pressure-sensitive labeling machinery, 539–540
Pressure-sensitive tape, 883–887
box-sealing tape, 884–885
environmental concerns, 886–887
filament tapes, 885–886
overview, 883
specialty tapes, 886
testing, 883–884
Printing, 783–787. *See also* Colorants; Decorating; Inks; Labeling
aluminum foil, 461
bar codes, 228
colorants, 243, 255–256
computer applications, 228–231
corrugated plastic, 287
flexography, 785–787
gravure, 784–785
holographic packaging, 489–492
inks, 511–514
labeling, 537–538
multilayer flexible packaging, 660, 663
offset container printing, 298–300
oriented polyester film, 414

Printing (*Continued*)
overview, 783–784
pad printing, 301–302
PETG copolyester, 830
screen printing, 302–303
skin packaging, 839, 841
thermoform/fill/seal, 912
Processing aid agents, additives, plastic, 12
Produce, food packaging. *See* Fruits and vegetables
Product fragility testing, 901–906
environment definition, 901
fragility definition, 901–904
overview, 901
Product liability. *See* Liability
Propellants, aerosol. *See* Aerosol propellants
Propionate, plastic films, 163
Protective materials, defined, marine environment, 591
Protective-package concept, distribution packaging, 308–309
Protein
biodegradable materials, 80
waterborne adhesives, 23
Pullulan, biodegradable materials, 80
Pulp. *See* Molded pulp
Pulsed electric field (PEF) system, aseptic packaging, 43
Pump dispenser, bottle and jar closures, 212
Pure Food and Drug Act of 1906, 674
Push bar conveyor, 274

QLF transparent barrier coating, transparent glass on plastic food-packaging materials, 445–448
Qualitative research, marketing effectiveness, of consumer packages, 887–888
Quality assurance. *See* Specification and quality assurance
Quality function deployment (QFD), integrated packaging design and development, 516–517
Quantitative research, marketing effectiveness, of consumer packages, 888
Quenching, blow and cast film, extrusion, 375

Radiant sealing, heat sealing, 826
Radiation effects, 796–799
composites, 798
glass, 796
metals, 796
overview, 796
paper, 796
paperboard, 796
plastics, 796–798
rubber, 798
testing, 798
Radiation sterilization, sterile disposable healthcare products, 697–699
R. A. Jones (RAJ) carton, carton terminology, 176–177
Ramsey proposal, law and regulation (U.S.), food packaging, 553–554
Random vibration, sinusoidal vibration versus, 956
Recall questioning, consumer research, 259
Reciprocating-flight conveyor, 274
Reciprocating printers, computer applications, 230
Recycled materials
board, defined, 178
law and regulation (U.S.), food packaging, 555
paperboard, carded packaging, 164
skin packaging, 842
Recycling, 799–808
corrugated boxes, 107–108
economics, 330, 803–804
edible film, 397
environmental concerns, 343, 345–346, 352–353
Europe, 361, 805–808
history, 351–352
increase in, 799–800
metals, 800–801
paper, 801–802
plastics, 802–803
poly(vinyl chloride) (PVC), 774
separation, 800
Refillable containers
environmental regulation, 350, 354
polycarbonate (PC), 741
Refillable glass bottles
beverage carriers, 169–170
carbonated beverages, 159
Refrigeration, shelf life, 831–832
Regenerated cellulose. *See* Cellophane
Regulation. *See also* Environmental regulation
aerosol propellants, 788–791
aluminum foil, 462
bottle and jar closures, 216
checkweighers, 195–196
child-resistant packaging, 199–203
colorants, 255–256
corrugated boxes, 103, 105
environmental, 347, 348–355
international, 348–351
North America, 351–355
ethylene-vinyl alcohol (EVOH) copolymers, 358
fiber drums, 314–315
food canning, 127–128
high-density polyethylene (HDPE), 748
linear and very low-density polyethylene (LLDPE and VLDPE), 752
low-density polyethylene (LDPE), 758
nutrition labeling, 674–681
oxygen scavengers, 690–691
plastic drums, 317–318
plastic pails, 705–707
polypropylene (PP), 768
poly(vinyl chloride) (PVC), 774
pressure containers, 775, 780, 781
recycling, 800, 805–808
shipping container testing, 909
solid-fiber boxes, 113
steel drums and pails, 321–322
transportation codes, 930
vinylidene chloride copolymer (VDC), 961
Reinforced plastic low-profile agents, additives, plastic, 12
Reinforcing straps, defined, marine environment, 592
Relative humidity. *See* Humidity
Research and development, active packaging, 7
Resin coding, environmental regulation, 354–355
Resin emulsions, waterborne adhesives, 24
Resource Conservation and Recovery Act (RCRA), 556
Resource depletion, environment, 343
Retortable packages, 808–811
flexible, 809
overview, 808–809
pouches, 809–810
trays, 810–811
Retorting
food canning, 124–127
oriented polyester film, 413
Returnable glass bottles. *See* Refillable glass bottles
Reuse, environmental concerns, 345
Reverse roll coaters, coating equipment, 222
Rewind, slitting and rewinding machinery, 847
Rewinding machinery. *See* Slitting and rewinding machinery
Right-hand machine, defined, 176
Rigidity, defined, marine environment, 592
Rigid packaging, aluminum foil, 463
Rigid-paperboard boxes, 108–110
applications, 110
history, 108
manufacture, 108–109
materials, 109–110
testing, 893
Rigid plastic containers, 110–112
bulk packaging, 121
ethylene-vinyl alcohol (EVOH) copolymers, 359
manufacture, 110–111
overview, 110
types of, 111–112
Rigid polyvinyl chloride (PVC) film, 427–431. *See also* Flexible polyvinyl chloride (PVC) film
film and sheet production, 428–429
markets, 431
polymer properties, 760
properties, 428
resins, 427–428
thermoforming, 429–431
Rigid sheet extrusion, described, 376–377
Robotic palletizer, 709
Robots, 811–817
automation, 813
roll handling, 819
task definition, 811–813
vision technology, 816
worker integration, 813–816
Roll coaters, coating equipment, 221–223
Roller coders, computer applications, 230
Roller conveying, 270, 273, 277, 278, 279
Roll handling, 817–819
Rolling chain conveyor, 274
Rolling-diaphragm volumetric fillers, filling machinery, 393
Roll-on, rolloff (Ro-Ro), export packaging, 368
Roll-on bottle and jar closures, 209
Roll-on closures, capping machinery, 157

Roll-on stamping method, hot stamping, 296–297
Rotational molding, 110, 819–822
Rubber, radiation effects, 798
Rubber latex, natural, waterborne adhesives, 24
Rub strips, defined, marine environment, 591–592

Sachets, active packaging, 3–5
Sack industry. *See* Multiwall bag(s)
Sacks, extrusion coating, 380
Safety and health issues. *See* Health and safety issues
Saturators, coating equipment, 224
Scoring, defined, 177, 178
Scraped-surface heat-exchange (SSHE) systems, aseptic packaging, 43
Screen ink, 513
Screen printing, described, 302–303. *See also* Silk screen printing
Screw conveying, 273, 280
Screw feeders, dry-product filling machinery, 386
Sealants. *See* Adhesives
Sealed-container filling system, filling machinery, 390–392
Sealing, bottle and jar closures, 207. *See also* Heat sealing
Self-adhesive labels, 536–537
Self-heating/cooling packages, active packaging, 7
Semirigid packaging
  aluminum foil, 463
  coextrusion, 240–242
  medical packaging, 612–613
  plastic containers, ethylene-vinyl alcohol (EVOH) copolymers, 359
  vinylidene chloride copolymer (VDC), 961
Sewn open mouth (SOM) multiwall bag, 62, 63
Sewn valve (SV) multiwall bag, 63, 64
Shear-cut tags, 876
Sheet polyethylene terephthalate glycol (PETG). *See* PETG copolyester
Shelf impact measurement, consumer research, 259–260
Shelf life, 830–835
  cook/chill food production, 284
  defined, 830–831
  factors influencing, 831–832
  multilayer flexible packaging, 660–661
  testing, 832–835
  time-temperature indicators, 926, 927
Shipping. *See also* Export packaging; Marine environment
  bag-in-box packaging, liquid product, 50–51
  distribution hazard measurement, 303–307
  distribution packaging, 307–310
  export packaging, 365–370
Shipping container testing, 906–909
  compression tests, 908
  conditions, 908
  methods, 906–907
  performance testing, 908–909
  regulation, 909
  sources, 906
  vibration, 907–908
  water-resistance test, 908
  water-vapor transmission test, 908
Shock, 835–839
  cushioning, 288, 837
  damaging effects, 837–839
  impact shock measurement, 836
  overview, 835–836
  product fragility testing, 901–904
  shock waveform analysis, 836–837
Shrinkable oriented polyester film, 414
Shrink bands, 69–70
Shrink films, 431–434
Shrink packaging, plastic film, 423
Sideweld seal, plastic bags, 66–67
Silica gel sachets, active packaging, 3–4
Silk screen printing. *See also* Decorating; Printing
  described, 302–303
  labeling, 538
Singapore, 350
Single-screw extruders, described, 370–372
Sinusoidal vibration, random vibration versus, 956
Skids, defined, marine environment, 591
Skin packaging, 839–843. *See also* Carded packaging
  applications, 843
  blister packaging compared, 165
  components and assembly, 162–163
  equipment, 842–843
  heat-seal coatings, 163–164
  ionomers, 528
  machinery, 165–166
  materials, 839–842
  overview, 839
  plastic films, 163, 423
  sterile disposable healthcare products, 693
Slat conveyor, 274
Sliding-chain conveyor, 275
Slip agents, additives, plastic, 12
Slip depressants, additives, plastic, 9
Slip promoters, additives, plastic, 11
Slipsheets, 843–844
Slit seal, plastic bags, 67–68
Slitting and rewinding machinery, 844–849
  center winders, 844–846
  overview, 844
  slitting, 847
  surface winders, 846–847
  tension control, 847
  terminology, 848
  unwinding, 847
Slot-orifice coater, coating equipment, 224
Snap-fit cap bottle and jar closures, 210
Society for Environmental Toxicology and Chemistry (SETAC), 347
Solid bleached sulfate (SBS) paperboard
  carded packaging, 164
  skin packaging, 841–842
Solid-fiber boxes, 112–113
Solid-phase thermoforming, 915–916
Solid waste, 344–347, 556
Solubility, polymers, 761–763
Solvent-borne adhesives, adhesives, 25
Solvent inks, acrylic-based, acrylic plastic polymers, 1
Solvent resistance, ethylene-vinyl alcohol (EVOH) copolymers, 356
Solvents, permeability (of aromas and solvents), 724–733. *See also* Permeability (of aromas and solvents)
Solvent sealing, heat sealing, 826
Spain, 551
Special construction folding cartons, 184
Speciality labeling, law and regulation (Europe), 549
Specialty tapes, pressure-sensitive tape, 886
Specialty-treated paper, described, 716
Specification and quality assurance, 849–852
  component-specific specification, 849–850
  consumer or finished-goods specification, 852
  general packaging specification, 849
  manufacturing specifications, 850, 852
  overview, 849
  total quality management (TQM), 927–929
Spinwelding, 964–966
Splicing, defined, marine environment, 592
Spray dispenser, bottle and jar closures, 212
Spreads, food packaging, 703
Stabilizing agents
  additives, plastic, 12–13
  poly(vinyl chloride) (PVC), 773
Standards
  biodegradable materials, 77
  bulk bags, 449
  bulk bags (flexible intermediate bulk containers), 53
  carton terminology, 178–181
  child-resistant packaging, 199–203
  corrugated boxes, 102–103, 104, 105
  defined, 521
  electrostatic discharge protective packaging, 339, 342
  environmental regulation, international, 351
  filling machinery, still liquid, 396–397
  international standards, 521–527. *See also* International standards
  military packaging, 649–650
  plastic bottle design, 96–97
  solid-fiber boxes, 113
  steel cans, 145
  steel drums and pails, 321
  styrene-butadiene (SB) copolymers, 864
Standup flexible pouches, 852–856
  applications, 855–856
  economics, 856
  history, 853–854
  machinery, 855
  oriented polyester film, 413–414
  overview, 852–853
  technology, 854–855
Staples, 856–857
Starch, waterborne adhesives, 23
Starch-based materials, biodegradable materials, 78–79
State governments, law and regulation (U.S.), 556, 557
Static. *See* Electrostatic charge
Static control, 857–860. *See also* Electrostatic charge

Static control (*Continued*)
electrostatic discharge treatment, surface treatment, 873
overview, 857–858
rationale, 858–859
static causes, 858
Steam and EtO sterilization, sterile disposable healthcare products, 694–695
Steam injection/infusion, aseptic packaging, 42–43
Steam sterilization, sterile disposable healthcare products, 695–696
Steam-table trays, 937–938
Steel
radiation effects, 796
recycling, 800–801
strapping materials, 862
Steel cans, 144–155. *See also* Aluminum cans; Metal cans
carbonated beverages, 159–160
coatings, 153–154
corrosion, 149
decoration, 154
fabrication, 150–152
history, 144–145
metals, 149–150
performance, 148–149
product compatibility, 149
recycling, 345
shapes and sizes, 145–148
technology, 154
Steel drums and pails, 318–324
construction, 319
history, 318–319
pails, 322–323
protection and lining, 319–321
regulation, 321–322
standards, 321
styles, 319
testing, 894
Steel pressure containers, 776–778
Sterilants, aseptic packaging, 42
Sterile disposable healthcare products, 693–699
ethylene oxide sterilization, 696–697
overview, 693
radiation sterilization, 697–699
requirements, 693–694
steam and EtO sterilization, 694–695
steam sterilization, 695–696
Sterile packaging. *See* Medical packaging
Sterilization. *See also* Aseptic packaging; Canning; Medical packaging; Pharmaceuticals; Sterile disposable healthcare products
ethylene oxide sterilization, sterile disposable healthcare products, 696–697
microwave pasteurization and sterilization, 646–648
radiation effects, 796
radiation sterilization, sterile disposable healthcare products, 697–699
steam and EtO sterilization, sterile disposable healthcare products, 694–695
steam sterilization, sterile disposable healthcare products, 695–696
Still liquid filling machinery, 390–397. *See also* Filling machinery
Storage
bag-in-box packaging, liquid product, 50–51
bulk bags (flexible intermediate bulk containers), 52–53
Strapping, 860–863
applications, 862
materials, 860–862
package and load characteristics, 862
Stress, defined, marine environment, 595
Stress cracking, polymers, 762
Stretch film, 434–445. *See also* Closures
economics, 434–435, 436–437
environmental concerns, 442–443
history, 435–436
inspection and handling, 440–441
manufacture, 437–438
performance measurement, 438–440
problems, 441–442
selection, 437
terminology, 443–445
Stretch-film wrapping machinery, 973–978
applications, 973–974
future trends, 977–978
history, 973
machinery, 975–977
prestretching film, 974–975
Stretch packaging, plastic film, 423–424
Styrene-acrylonitrile (SAN), nitrile polymers, 670
Styrene-butadiene (SB) copolymers, 863–864
Styrenics, plastic films, 163
Sulfide black, corrosion, 142
Sulfonation, surface and hydrocarbon-barrier modification, 866
Sulfur dioxide, modified atmosphere packaging, 651
Sulfur dioxide releasing pads, active packaging, 5
Surface and hydrocarbon-barrier modification, 864–867
adhesion improvement, 864–865
hydrocarbon-barrier improvement, 865–866
overview, 864
Surface resistance, electrostatic discharge protective packaging, 337–338
Surface tension, polymer properties, 763–764
Surface treatment, 867–874
abrasive technique, 867–868
chemical etching, 869
chemical priming, 869
cold-gas-plasma treatment, 870–872
corona treatment, 870
electrostatic discharge treatment, 873
evaporated acrylate coatings, 872
flame treatment, 869–870
fluorination process, 872–873
liquid cleaning technique, 868–869
overview, 867
ultraviolet/ozone process, 872
Surface winders, slitting and rewinding machinery, 846–847
Surgical devices, oriented polyester film, 414
Suspended-tray conveyor, 275
Sustainable use, environment, 343
Sweden, 551
Switzerland, 551
Synthetic paper, 724
Synthetic waterborne adhesives, 24

Tabletop chain conveying, 265–266
Tachistoscopic research, consumer research, 259
Tags, 875–879
Taiwan, 350
Tamper-evident packaging, 879–882
aluminum foil, 462
best feature, 882
bottle and jar closures, 212–213, 216
child-resistant packaging, 199–204
consumer preferences, 880
economics, 880–881
effectiveness, 882–883
FDA rule, 880
history, 879–880
lidding, 562
plastic bottle design, 99
selection, 881
Tape
gummed, 883
pressure-sensitive, 883–887. *See also* Pressure-sensitive tape
Tear bands, bottle and jar closures, 213
Technical Association of the Pulp and Paper Industry (TAPPI), carton terminology, 178
Temperature
aroma barrier testing, 40
aseptic packaging, 42, 43
food canning, 123–124
indicators, described, 500–501
permeability (of aromas and solvents), 729–730
polymer properties, 759–760
time and temperature indicators, 499–500, 926–927
Tension, strapping materials, 861
Tension bands, defined, marine environment, 592
Terrorism, tamper-evident packaging, 879–880
Testing. *See also* Law and regulation
aroma barrier testing, 38–41
bulk bags (flexible intermediate bulk containers), 53
bulk packaging, 122
child-resistant packaging, 200–202
consumer packages, marketing effectiveness, 887–890
corrugated boxes, 102–103, 105
electrostatic discharge protective packaging, 336–339
forensic packaging, 463–465
glass container manufacturing, 483–484
heat sealing, 827
leak testing, 558–561
oxygen scavengers, 690
packaging materials, 890–895
bags, 890–891
bottles, 891
boxes, 891–893
bulk containers, 893
cans, 893

cartons, 893
drums and pails, 894
permeation, 895–898. *See also* Permeation testing
plastic bottle design, 97–98
plastic pails, 706
polymer properties, 758–765
pressure-sensitive tape, 883–884
product fragility, 901–906. *See also* Product fragility testing
radiation effects, 798
shelf life, 832–835
shipping containers, 906–909
slipsheets, 844
vibration, 957
Test marketing, consumer research, 259, 887–890
Textile bags, testing, 891
Themoplastic polyesters. *See* Polyethylene terephthalate (PET)
Thermal insulation, foamed plastics, 455–456
Thermal process, aseptic packaging, 42
Thermal shock, glass container design, 475
Thermal-thermal transfer printers, computer applications, 231
Thermodegradation, polymers, 762–763
Thermoform/fill/seal, 910–914
applications, 910
horizontal form/fill/seal pouch, 467
machinery, 911–912
materials, 912–914
Thermoforming, 914–921
cut-sheet, 920
ionomers, 528–529
machinery, 919–920
overview, 914
polypropylene (PP), 767
process, 914–919
rigid plastic containers, 110–111
rigid polyvinyl chloride (PVC) film, 429–431
technology, 921
tooling, 920
twin-sheet, 920–921
vacuum packaging, 950
Thermophysical properties, polymer properties, 759–760
Thermoplastic closures, bottle and jar closures, 215
Thermoplastic polymer, hot-melt adhesives, 24
Thermoplastic starch, biodegradable materials, 78–79
Thermoset polyester, dual-ovenable packaging, 644
Thermosets, bottle and jar closures, 215
Thermosetting plastics, 921–926
processing systems, 921–923, 924–925
structures, 924
types, 925
Thermotropic liquid-crystalline polymers (TLCP), 569–572
chemistry, 569
films, 570–572
overview, 569
sources, 569–570
Thermx, dual-ovenable packaging, 644
Thickness (caliper), paperboard, 718
Thread-engagement bottle and jar closures, 208–209
Time-fill fillers, filling machinery, still liquid, 394
Time-temperature indicators, described, 499–500, 926–927
Tin
radiation effects, 796
recycling, 800–801
steel cans, 149–150
Tin cans. *See* Steel cans
Tinplate can
aerosol containers, 28–31
hot-fill technology, 493
pressure containers, 780
Tissue papers, described, 716–717
Top-load cartoning machinery. *See* Cartoning machinery (top-load)
Topside, paperboard, 718
Total quality management (TQM), 927–929
alignment, 928
change management, 927–928
customer focus, 928
customer-supplier relations, 929
economics, 928
employee involvement, 929
future trends, 929
integrated packaging design and development, 514
overview, 927
strategy, 928
Tow conveyor, 275
Toxins. *See* Hazardous materials
Trademarks, law and regulation (Europe), 549
Trade shows, exhibitions, 362–365
Transfer molding, thermosetting plastics, processing systems, 922, 924–925
Transfer roll coaters, coating equipment, 222
Transparent glass on plastic food-packaging materials, 445–448
Transportation
bulk bags (flexible intermediate bulk containers), 52–53
distribution hazard measurement, 303–307
distribution packaging, 307–310
economics, 330
export packaging, 365–370. *See also* Export packaging
fiber drums, 314–315
law and regulation (Europe), 549
logistical/distribution packaging, 572–579. *See also* Logistical/distribution packaging
multiwall bags, 66
plastic pails, 705–706
Transportation codes, 929–931
carrier rules, 929–930
federal regulations, 930
Tray former/loader case loading, 193
Trays
barrier-foam, 931–933
foam, 933–937
retortable packages, 810–811
steam-table, 937–938
Tray-style folding cartons, 183
Triboelectricity, 336–337, 338, 764
Trimmer conveyor, 270
Triple-seal style closure, cartoning machinery (top-load), 175
Triplex, paperboard, 718
Trolley conveyor, 276
Tube filling, 939–941
Tubes, collapsible, 941–945
future trends, 945
history, 941–942
laminated, 944–945
metal, 942
plastic, 942–944
Tube-style folding cartons, 182–183
Tubular coextrusion machinery. *See* Coextrusion machinery
Tuck-style carton, defined, 178
Turbine-meter volumetric fillers, filling machinery, still liquid, 393
Twin seal, plastic bags, 67–68
Twin-sheet thermoforming, 920–921
Tyvek, medical packaging, 611–612

Ultrasonic sealing, heat sealing, 825
Ultraviolet-curing technology
offset container printing, 300
screen printing, 303
Ultraviolet light resistance, plastic drums, 316
Ultraviolet/ozone process, surface treatment, 872
Ultraviolet stabilizing agents, additives, plastic, 13
Unbalanced-pressure fillers, filling machinery, still liquid, 391–392
Underliner, paperboard, 718
Uniform Classification Committee (UCC)
gummed tape, 883
plastic pails, 706
steel drums and pails, 321
Uniform Freight Classification (UFC)
fiber drums, 314
steel drums and pails, 321
transportation codes, 930
Uniform load, defined, marine environment, 592
Unit-dose packaging, pharmaceutical packaging, 734–735
United Kingdom, 551–552
United Nations
bulk bags, 449
fiber drums, 315
intermediate bulk containers, 519–521
international standards, 526
plastic drums, 317
steel drums and pails, 321, 322
U.S. Bureau of Alcohol, Tobacco, and Firearms (BATF), filling machinery, still liquid, 397
U.S. Consumer Product Safety Commission (CPSC), child-resistant packaging, 199, 200, 202, 203
U.S. Customary System (USCS), history, 638
U.S. Department of Agriculture (USDA)
colorants, 255–256
edible film, 397

U.S. Department of Agriculture (USDA) (*Continued*)
filling machinery, still liquid, 397
folding cartons, 181
food additives, 552
food canning, 128
nutrition labeling, 675–680
U.S. Department of Commerce, glass container manufacturing, 475
U.S. Department of Defense (DOD)
electrostatic discharge protective packaging, 336
military packaging, 648–650
steam-table trays, 938
U.S. Department of Transportation (DOT)
aerosols, 28, 789
bulk bags, 448–449
corrugated boxes, 103
fiber drums, 310–311, 314–315
hazardous materials labeling, 556
intermediate bulk containers, 519, 521
plastic drums, 317
plastic pails, 705–707
pressure containers, 775, 780, 781
shipping container testing, 909
steel drums and pails, 321, 322
transportation codes, 930
U.S. Environmental Protection Agency (EPA)
aerosol propellants, 788, 790
child-resistant packaging, 199, 202, 203
inks, 511
law and regulation (U.S.), 556
poly(vinyl chloride) (PVC), 344
recycling, 346
solid waste, 344
transportation codes, 930
U.S. Federal Trade Commission (FTC)
environmental regulation, 354
labeling, 556, 557
transportation codes, 930
U.S. Food and Drug Administration (FDA), 552–555
acrylic plastic polymers, 1, 2
additives, plastic, 11
aerosol propellants, 788
aseptic packaging, 42
bottle and jar closures, 212
child-resistant packaging, 199
closure liners, 205
coextruded semirigid packaging, 241
colorants, 255–256
ethylene-vinyl alcohol (EVOH) copolymers, 358
filling machinery, still liquid, 397
folding cartons, 181
food additive definition, 552
food canning, 127–128
food packaging, 552–555
generally, 552
high-density polyethylene (HDPE), 748
hot-melt wax carton, 964
labeling, 556
linear and very low-density polyethylene (LLDPE and VLDPE), 752
low-density polyethylene (LDPE), 758
nutrition labeling, 674, 675–680
pharmaceutical packaging, 733, 734, 735
polypropylene (PP), 768
poly(vinyl chloride) (PVC), 774
printing, computer applications, 229
recycling, 346
steel cans, 150, 154
surface and hydrocarbon-barrier modification, 865–866
tamper-evident packaging, 880, 881
U.S. National Bureau of Standards, metrication, 639
U.S. Occupational Safety and Health Administration (OSHA)
logistical/distribution packaging, 575
styrene-butadiene (SB) copolymers, 864
transportation codes, 930
Unitization, marine environment, 590–591, 600–601
Unitized loads, export packaging, 368
Unit-of-use, pharmaceutical packaging, 735
Universal Product Codes (UPC). *See also* Bar code
beverage carriers, 168
printing, computer applications, 228, 229
Universities, education, 334–335
Unscrambling, 946–947
Unsealed-container filling system, filling machinery, still liquid, 392–394
Unwinding, slitting and rewinding machinery, 847
Urea, bottle and jar closures, 216

Vacuum, food canning, 124
Vacuum-bag coffee packaging, 948–949
Vacuum closures
capping machinery, 157
tamper-evident, bottle and jar closures, 213
Vacuum-filling systems, dry-product filling machinery, 386–387
Vacuum metallizing, 629–638
equipment, 635–637
future trends, 637
nylon, 685
overview, 629–631
thermal sources, 631–635
Vacuum packaging, 949–955
air-removal system, 949–951
cheeses, 953
fish, 953–954
fruits and vegetables, 954
meats, 952–953
modified atmosphere packaging, 651, 653–654, 657
overview, 949
poultry, 951
Valve multiwall bag, 63, 64
Valves, pressure containers, 781–783
Vapor-corrosion inhibitor (VCI), marine environment, 590
Vapor-phase inhibitor (VPI), marine environment, 590
Vegetable parchment paper, described, 715
Vegetables. *See* Fruits and vegetables
Vertical case loading, 191–193
Vertical chain conveyor, opposed-shelf type, 276
Vertical form/fill/seal machinery, bag-in-box packaging, 46, 47, 48
Vertical form/fill/seal pouch, 468–470
Very low-density polyethylene (VLDPE). *See* Linear and very low-density polyethylene (LLDPE and VLDPE)
Vials. *See* Ampuls and vials
Vibrating-bin discharge, dry-product filling machinery, 386
Vibration, 955–958
cushioning design, 288
defined, 955
distribution hazard measurement, 303–307
forced versus free, 955
frequency and acceleration, 955–956
measurement and analysis, 956–957
product fragility testing, 904
shipping container testing, 907–908
sinusoidal versus random, 956
strapping materials, 861
terminology, 958
testing, 957
Vibratory conveying, 276
Vibratory feeders, dry-product filling machinery, 385–386
Vinylidene chloride copolymer (VDC), 958–961. *See also* Polyvinylidene chloride (PVDC) copolymer
applications, 960–961
characteristics, 959
chemistry, 958–959
overview, 958
producers, 959–960
properties, 961
regulation, 961
Vinyls, plastic films, 163
Volume-cup fillers, filling machinery, still liquid, 393

Water-based inks, acrylic-based, acrylic plastic polymers, 1
Waterbath test, leakage testing, 899–900
Waterborne systems, adhesives, 23–24
Water-resistance test, shipping container testing, 908
Water-resistant paper, 715
Water-vapor transmission rate (WVTR). *See also* Moisture vapor transmission rate (MVTR)
cellophane, 195
edible film, 398, 399
high-density polyethylene film, 405
modified atmosphere packaging, 652
permeability (of aromas and solvents), 725. *See also* Permeability (of aromas and solvents)
permeation testing, 897
polymer properties, 763
shelf life, 831, 833
surface and hydrocarbon-barrier modification, 866
testing, 890
transparent glass on plastic food-packaging materials, 445–448
Water-vapor transmission test, shipping container testing, 908
Waxed papers, described, 715–716
Waxes, 962–964
future trends, 964

hot-melt wax carton, 962–964
overview, 962
technology, 964
Web processing, roll handling, 817–819
Weighing systems, dry-product filling machinery, 387–388. *See also* Checkweighers; Filling machinery
Weight fillers, filling machinery, still liquid, 393–394
Welding, spin, 964–966
Welex system, coextrusion machinery, flat, 233–234
Wet-glue labeling machinery, 539
Wet-ink printing, computer applications, 230
Wet-strength paper, described, 716
Wettability, polymer properties, 764
Wirebound boxes, 113–115
Wire ties, bag closures, 220
Wirewound-rod coater, coating equipment, 223
Wisconsin, 353, 354, 556
Wood boxes, 115–117
bulk packaging, 122
marine environment, 591, 596–597, 600
testing, 892
wirebound boxes, 113–115
Wood crate. *See* Crates; Wood boxes
Working range, strapping materials, 860
Workplace injury. *See* Occupational injury
Wrapping machinery, 966–972
bundle-wrapping machines, 972
fold patterns, 967, 968–971
history, 966
machine attachments, 971, 972
multipack machines, 972
stretch-film, 973–978. *See also* Stretch-film wrapping machinery
terminology, 966–967, 970